Organic Chemistry with Biological Topics

Sixth Edition

Janice Gorzynski Smith
University of Hawai'i at Mānoa

ORGANIC CHEMISTRY WITH BIOLOGICAL TOPICS

Published by McGraw-Hill Education, 2 Penn Plaza, New York, NY 10121. Copyright © 2021 by McGraw-Hill Education. All rights reserved. Printed in the United States of America. No part of this publication may be reproduced or distributed in any form or by any means, or stored in a database or retrieval system, without the prior written consent of McGraw-Hill Education, including, but not limited to, in any network or other electronic storage or transmission, or broadcast for distance learning.

Some ancillaries, including electronic and print components, may not be available to customers outside the United States.

This book is printed on acid-free paper.

1 2 3 4 5 6 7 8 9 LWI 24 23 22 21 20

ISBN 978-1-260-57516-3
MHID 1-260-57516-0

Cover Image: *Iñaki Respaldiza/Getty Images*

All credits appearing on page or at the end of the book are considered to be an extension of the copyright page.

The Internet addresses listed in the text were accurate at the time of publication. The inclusion of a website does not indicate an endorsement by the authors or McGraw-Hill Education, and McGraw-Hill Education does not guarantee the accuracy of the information presented at these sites.

mheducation.com/highered

About the Author

©Daniel C. Smith

Janice Gorzynski Smith was born in Schenectady, New York. She received an A.B. degree *summa cum laude* in chemistry from Cornell University, and a Ph.D. in Organic Chemistry from Harvard University under the direction of Nobel Laureate E. J. Corey. During her tenure with the Corey group, she completed the total synthesis of the plant growth hormone gibberellic acid.

Following her postdoctoral work as a National Science Foundation National Needs Postdoctoral Fellow at Harvard, Jan joined the faculty of Mount Holyoke College, where she was employed for 21 years. During this time she was active in teaching organic chemistry lecture and lab courses, conducting a research program in organic synthesis, and serving as department chair. Her organic chemistry class was named one of Mount Holyoke's "Don't-miss courses" in a survey by *Boston* magazine. After spending two sabbaticals amidst the natural beauty and diversity in Hawai'i in the 1990s, Jan and her family moved there permanently in 2000. She has been a faculty member at the University of Hawai'i at Mānoa, where she has taught the two-semester organic chemistry lecture and lab courses. In 2003, she received the Chancellor's Citation for Meritorious Teaching.

Jan resides in Hawai'i with her husband Dan, an emergency medicine physician, pictured with her in Cambodia in 2018. She has four children and six grandchildren. When not teaching, writing, or enjoying her family, Jan bikes, hikes, snorkels, and scuba dives in sunny Hawai'i, and time permitting, enjoys travel and Hawaiian quilting.

For Megan Sarah

Contents in Brief

	Prologue 1
1	Structure and Bonding 7
2	Acids and Bases 58
3	Introduction to Organic Molecules and Functional Groups 91
4	Alkanes 132
5	Stereochemistry 176
6	Understanding Organic Reactions 219
7	Alkyl Halides and Nucleophilic Substitution 254
8	Alkyl Halides and Elimination Reactions 302
9	Alcohols, Ethers, and Related Compounds 337
10	Alkenes and Alkynes 384
11	Oxidation and Reduction 434
12	Conjugation, Resonance, and Dienes 473
Spectroscopy A	Mass Spectrometry 508
Spectroscopy B	Infrared Spectroscopy 528
Spectroscopy C	Nuclear Magnetic Resonance Spectroscopy 553
13	Introduction to Carbonyl Chemistry; Organometallic Reagents; Oxidation and Reduction 598
14	Aldehydes and Ketones—Nucleophilic Addition 646
15	Carboxylic Acids and Nitriles 695
16	Carboxylic Acids and Their Derivatives—Nucleophilic Acyl Substitution 733
17	Substitution Reactions of Carbonyl Compounds at the α Carbon 779
18	Carbonyl Condensation Reactions 812
19	Benzene and Aromatic Compounds 849
20	Reactions of Aromatic Compounds 885
21	Radical Reactions 935
22	Amines 969
23	Amino Acids and Proteins 1021
24	Carbohydrates 1069
25	Lipids 1114
26	Nucleic Acids and Protein Synthesis 1144
27	Metabolism 1169
28	Carbon–Carbon Bond-Forming Reactions in Organic Synthesis (Available Online) 1197
29	Pericyclic Reactions (Available Online) 1226
30	Synthetic Polymers (Available Online) 1255
	Appendices A-1
	Glossary G-1
	Index I-1

Contents

Preface xviii
Acknowledgments xxi

Prologue 1

What Is Organic Chemistry? 1
Some Representative Organic Molecules 2
Marine Natural Products 4

Buttchi 3 Sha Life/ Shutterstock

1 Structure and Bonding 7

- **1.1** The Periodic Table 8
- **1.2** Bonding 10
- **1.3** Lewis Structures 12
- **1.4** Isomers 17
- **1.5** Exceptions to the Octet Rule 17
- **1.6** Resonance 18
- **1.7** Determining Molecular Shape 24
- **1.8** Drawing Organic Structures 28
- **1.9** Hybridization 34
- **1.10** Ethane, Ethylene, and Acetylene 37
- **1.11** Bond Length and Bond Strength 42
- **1.12** Electronegativity and Bond Polarity 43
- **1.13** Polarity of Molecules 45
- **1.14** Oxybenzone—A Representative Organic Molecule 46
  Chapter Review 48
  Problems 51

Comstock/PunchStock

2 Acids and Bases 58

- **2.1** Brønsted–Lowry Acids and Bases 59
- **2.2** Reactions of Brønsted–Lowry Acids and Bases 60
- **2.3** Acid Strength and pK_a 63
- **2.4** Predicting the Outcome of Acid–Base Reactions 66
- **2.5** Factors That Determine Acid Strength 67
- **2.6** Common Acids and Bases 76
- **2.7** Aspirin 77
- **2.8** Lewis Acids and Bases 78
  Chapter Review 81
  Problems 84

vi   Contents

Purestock/SuperStock

3   Introduction to Organic Molecules and Functional Groups   91

- 3.1 Functional Groups   92
- 3.2 An Overview of Functional Groups   93
- 3.3 Intermolecular Forces   101
- 3.4 Physical Properties   104
- 3.5 Application: Vitamins   111
- 3.6 Application of Solubility: Soap   112
- 3.7 Application: The Cell Membrane   113
- 3.8 Functional Groups and Reactivity   116
- 3.9 Biomolecules   117
  - Chapter Review   123
  - Problems   126

Narongsak Nagadhana/ Shutterstock

4   Alkanes   132

- 4.1 Alkanes—An Introduction   133
- 4.2 Cycloalkanes   135
- 4.3 An Introduction to Nomenclature   136
- 4.4 Naming Alkanes   137
- 4.5 Naming Cycloalkanes   142
- 4.6 Common Names   144
- 4.7 Natural Occurrence of Alkanes   145
- 4.8 Properties of Alkanes   145
- 4.9 Conformations of Acyclic Alkanes—Ethane   147
- 4.10 Conformations of Butane   150
- 4.11 An Introduction to Cycloalkanes   153
- 4.12 Cyclohexane   154
- 4.13 Substituted Cycloalkanes   157
- 4.14 Oxidation of Alkanes   163
  - Chapter Review   166
  - Problems   170

George Ostertag/Alamy Stock Photo

5   Stereochemistry   176

- 5.1 Starch and Cellulose   177
- 5.2 The Two Major Classes of Isomers   179
- 5.3 Looking Glass Chemistry—Chiral and Achiral Molecules   179
- 5.4 Stereogenic Centers   182
- 5.5 Stereogenic Centers in Cyclic Compounds   186
- 5.6 Labeling Stereogenic Centers with *R* or *S*   189
- 5.7 Diastereomers   194
- 5.8 Meso Compounds   196
- 5.9 *R* and *S* Assignments in Compounds with Two or More Stereogenic Centers   198
- 5.10 Disubstituted Cycloalkanes   198
- 5.11 Isomers—A Summary   200
- 5.12 Physical Properties of Stereoisomers   201
- 5.13 Chemical Properties of Enantiomers   206
  - Chapter Review   208
  - Problems   212

Contents   vii

Ninikas/Getty Images

6 Understanding Organic Reactions   219

- **6.1** Writing Equations for Organic Reactions   220
- **6.2** Kinds of Organic Reactions   221
- **6.3** Bond Breaking and Bond Making   223
- **6.4** Bond Dissociation Energy   227
- **6.5** Thermodynamics   231
- **6.6** Enthalpy and Entropy   234
- **6.7** Energy Diagrams   236
- **6.8** Energy Diagram for a Two-Step Reaction Mechanism   238
- **6.9** Kinetics   240
- **6.10** Catalysts   243
- **6.11** Enzymes   244

  Chapter Review   245
  Problems   248

Source: Claire Fackler/ CINMS/NOAA

7 Alkyl Halides and Nucleophilic Substitution   254

- **7.1** Introduction to Alkyl Halides   255
- **7.2** Nomenclature   256
- **7.3** Properties of Alkyl Halides   257
- **7.4** Interesting Alkyl Halides   258
- **7.5** The Polar Carbon–Halogen Bond   259
- **7.6** General Features of Nucleophilic Substitution   260
- **7.7** The Leaving Group   261
- **7.8** The Nucleophile   263
- **7.9** Possible Mechanisms for Nucleophilic Substitution   267
- **7.10** Two Mechanisms for Nucleophilic Substitution   267
- **7.11** The S_N2 Mechanism   268
- **7.12** The S_N1 Mechanism   273
- **7.13** Carbocation Stability   277
- **7.14** The Hammond Postulate   279
- **7.15** When Is the Mechanism S_N1 or S_N2?   282
- **7.16** Biological Nucleophilic Substitution   287
- **7.17** Vinyl Halides and Aryl Halides   289
- **7.18** Organic Synthesis   290

  Chapter Review   292
  Problems   295

Source: Forest & Kim Starr

8 Alkyl Halides and Elimination Reactions   302

- **8.1** General Features of Elimination   303
- **8.2** Alkenes—The Products of Elimination Reactions   304
- **8.3** The Mechanisms of Elimination   308
- **8.4** The E2 Mechanism   308
- **8.5** The Zaitsev Rule   312
- **8.6** The E1 Mechanism   314
- **8.7** S_N1 and E1 Reactions   317
- **8.8** Stereochemistry of the E2 Reaction   318
- **8.9** When Is the Mechanism E1 or E2?   321

8.10 E2 Reactions and Alkyne Synthesis 322
8.11 When Is the Reaction S_N1, S_N2, E1, or E2? 323
Chapter Review 327
Problems 331

9 Alcohols, Ethers, and Related Compounds 337

Stephen Orsillo/ Shutterstock

9.1 Introduction 338
9.2 Structure and Bonding 339
9.3 Nomenclature 339
9.4 Properties of Alcohols, Ethers, and Epoxides 343
9.5 Interesting Alcohols, Ethers, and Epoxides 344
9.6 Preparation of Alcohols, Ethers, and Epoxides 346
9.7 General Features—Reactions of Alcohols, Ethers, and Epoxides 348
9.8 Dehydration of Alcohols to Alkenes 350
9.9 Carbocation Rearrangements 352
9.10 Dehydration Using $POCl_3$ and Pyridine 355
9.11 Conversion of Alcohols to Alkyl Halides with HX 356
9.12 Conversion of Alcohols to Alkyl Halides with $SOCl_2$ and PBr_3 359
9.13 Tosylate—Another Good Leaving Group 361
9.14 Reaction of Ethers with Strong Acid 364
9.15 Thiols and Sulfides 366
9.16 Reactions of Epoxides 368
9.17 Application: Epoxides, Leukotrienes, and Asthma 373
Chapter Review 374
Problems 378

10 Alkenes and Alkynes 384

Michael Sewell/ Photolibrary/ Getty Images

10.1 Introduction 385
10.2 Calculating Degrees of Unsaturation 386
10.3 Nomenclature 387
10.4 Properties of Alkenes and Alkynes 391
10.5 Interesting Alkenes and Alkynes 392
10.6 Fatty Acids and Triacylglycerols 393
10.7 Preparation of Alkenes and Alkynes 395
10.8 Introduction to the Reactions of Alkenes and Alkynes 396
10.9 Hydrohalogenation—Electrophilic Addition of HX to Alkenes 399
10.10 Markovnikov's Rule 400
10.11 Stereochemistry of Electrophilic Addition of HX 401
10.12 Hydration—Electrophilic Addition of Water 403
10.13 Halogenation—Addition of Halogen 404
10.14 Stereochemistry of Halogenation 405
10.15 Halohydrin Formation 407
10.16 Hydroboration–Oxidation 408
10.17 Addition of Hydrogen Halides and Halogens to Alkynes 412
10.18 Addition of Water to Alkynes 414
10.19 Hydroboration–Oxidation of Alkynes 416

10.20 Reaction of Acetylide Anions 417
10.21 Synthesis 421
Chapter Review 423
Problems 427

11 Oxidation and Reduction 434

Amarita/Shutterstock

11.1 Introduction 435
11.2 Reducing Agents 436
11.3 Reduction of Alkenes 436
11.4 Application: Hydrogenation of Oils 440
11.5 Reduction of Alkynes 442
11.6 The Reduction of Polar C–X σ Bonds 445
11.7 Oxidizing Agents 446
11.8 Epoxidation 447
11.9 Dihydroxylation 451
11.10 Oxidative Cleavage of Alkenes 453
11.11 Oxidative Cleavage of Alkynes 455
11.12 Oxidation of Alcohols 456
11.13 Biological Oxidation 458
11.14 Sharpless Epoxidation 460
Chapter Review 463
Problems 466

12 Conjugation, Resonance, and Dienes 473

Rene Dulhoste/ Science Source

12.1 Conjugation 474
12.2 Resonance and Allylic Carbocations 476
12.3 Common Examples of Resonance 477
12.4 The Resonance Hybrid 479
12.5 Electron Delocalization, Hybridization, and Geometry 481
12.6 Conjugated Dienes 482
12.7 Interesting Dienes and Polyenes 483
12.8 The Carbon–Carbon σ Bond Length in Buta-1,3-diene 484
12.9 Stability of Conjugated Dienes 485
12.10 Electrophilic Addition: 1,2- Versus 1,4-Addition 486
12.11 Kinetic Versus Thermodynamic Products 488
12.12 The Diels–Alder Reaction 490
12.13 Specific Rules Governing the Diels–Alder Reaction 492
12.14 Other Facts About the Diels–Alder Reaction 496
Chapter Review 498
Problems 501

x   Contents

Adam Bailleaux/
Atomazul/123RF

Spectroscopy A   Mass Spectrometry   508

- **A.1** Mass Spectrometry and the Molecular Ion   509
- **A.2** Alkyl Halides and the M + 2 Peak   514
- **A.3** Fragmentation   516
- **A.4** Fragmentation Patterns of Some Common Functional Groups   518
- **A.5** Other Types of Mass Spectrometry   520
  - *Chapter Review*   523
  - *Problems*   524

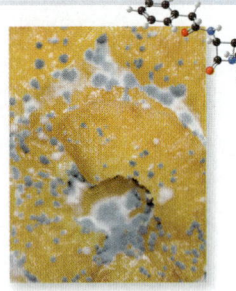
T.Daly/Alamy Stock Photo

Spectroscopy B   Infrared Spectroscopy   528

- **B.1** Electromagnetic Radiation   529
- **B.2** The General Features of Infrared Spectroscopy   531
- **B.3** IR Absorptions   532
- **B.4** Infrared Spectra of Common Functional Groups   539
- **B.5** IR and Structure Determination   544
  - *Chapter Review*   546
  - *Problems*   548

Daniel C. Smith

Spectroscopy C   Nuclear Magnetic Resonance Spectroscopy   553

- **C.1** An Introduction to NMR Spectroscopy   554
- **C.2** ^{1}H NMR: Number of Signals   557
- **C.3** ^{1}H NMR: Position of Signals   562
- **C.4** The Chemical Shift of Protons on sp^2 and sp Hybridized Carbons   565
- **C.5** ^{1}H NMR: Intensity of Signals   567
- **C.6** ^{1}H NMR: Spin–Spin Splitting   568
- **C.7** More-Complex Examples of Splitting   572
- **C.8** Spin–Spin Splitting in Alkenes   575
- **C.9** Other Facts About ^{1}H NMR Spectroscopy   577
- **C.10** Using ^{1}H NMR to Identify an Unknown   579
- **C.11** ^{13}C NMR Spectroscopy   581
- **C.12** Magnetic Resonance Imaging (MRI)   584
  - *Chapter Review*   585
  - *Problems*   589

AS Food studio/
Shutterstock

13   Introduction to Carbonyl Chemistry; Organometallic Reagents; Oxidation and Reduction   598

- **13.1** Introduction   599
- **13.2** General Reactions of Carbonyl Compounds   600
- **13.3** A Preview of Oxidation and Reduction   603
- **13.4** Reduction of Aldehydes and Ketones   604
- **13.5** The Stereochemistry of Carbonyl Reduction   606
- **13.6** Enantioselective Biological Reduction   607
- **13.7** Reduction of Carboxylic Acids and Their Derivatives   608

Contents   xi

- **13.8** Oxidation of Aldehydes   613
- **13.9** Organometallic Reagents   613
- **13.10** Reaction of Organometallic Reagents with Aldehydes and Ketones   617
- **13.11** Retrosynthetic Analysis of Grignard Products   620
- **13.12** Protecting Groups   622
- **13.13** Reaction of Organometallic Reagents with Carboxylic Acid Derivatives   624
- **13.14** Reaction of Organometallic Reagents with Other Compounds   627
- **13.15** α,β-Unsaturated Carbonyl Compounds   629
- **13.16** Summary—The Reactions of Organometallic Reagents   631
- **13.17** Synthesis   632

  Chapter Review   634
  Problems   639

14 Aldehydes and Ketones—Nucleophilic Addition   646

- **14.1** Introduction   647
- **14.2** Nomenclature   648
- **14.3** Properties of Aldehydes and Ketones   652
- **14.4** Interesting Aldehydes and Ketones   654
- **14.5** Preparation of Aldehydes and Ketones   655
- **14.6** Reactions of Aldehydes and Ketones—General Considerations   656
- **14.7** Nucleophilic Addition of H⁻ and R⁻—A Review   659
- **14.8** Nucleophilic Addition of ⁻CN   660
- **14.9** The Wittig Reaction   662
- **14.10** Addition of 1° Amines   666
- **14.11** Addition of 2° Amines   668
- **14.12** Imine and Enamine Hydrolysis   669
- **14.13** Imines in Biological Systems   671
- **14.14** Addition of H_2O—Hydration   673
- **14.15** Addition of Alcohols—Acetal Formation   675
- **14.16** Acetals as Protecting Groups   679
- **14.17** Cyclic Hemiacetals   680
- **14.18** An Introduction to Carbohydrates   682

  Chapter Review   683
  Problems   686

Valentyn Volkov/Shutterstock

15 Carboxylic Acids and Nitriles   695

- **15.1** Structure and Bonding   696
- **15.2** Nomenclature   697
- **15.3** Physical and Spectroscopic Properties   700
- **15.4** Interesting Carboxylic Acids and Nitriles   702
- **15.5** Aspirin, Arachidonic Acid, and Prostaglandins   703
- **15.6** Preparation of Carboxylic Acids   704
- **15.7** Carboxylic Acids—Strong Organic Brønsted–Lowry Acids   705
- **15.8** The Henderson–Hasselbalch Equation   708
- **15.9** Inductive Effects in Aliphatic Carboxylic Acids   710
- **15.10** Extraction   711

Sarka Babicka/Getty Images

15.11 Organic Acids That Contain Phosphorus 714
15.12 Amino Acids 715
15.13 Nitriles 719
Chapter Review 723
Problems 725

Likit Supasai/ Shutterstock

16 Carboxylic Acids and Their Derivatives—Nucleophilic Acyl Substitution 733

16.1 Introduction 734
16.2 Structure and Bonding 736
16.3 Nomenclature 737
16.4 Physical and Spectroscopic Properties 741
16.5 Interesting Esters and Amides 743
16.6 Introduction to Nucleophilic Acyl Substitution 745
16.7 Reactions of Acid Chlorides 749
16.8 Reactions of Anhydrides 750
16.9 Reactions of Carboxylic Acids 751
16.10 Reactions of Esters 756
16.11 Application: Lipid Hydrolysis 758
16.12 Reactions of Amides 760
16.13 Application: The Mechanism of Action of β-Lactam Antibiotics 762
16.14 Summary of Nucleophilic Acyl Substitution Reactions 762
16.15 Acyl Phosphates—Biological Anhydrides 763
16.16 Reactions of Thioesters—Biological Acylation Reactions 766
Chapter Review 768
Problems 771

Surachetkhamsuk/ iStock/Getty Images

17 Substitution Reactions of Carbonyl Compounds at the α Carbon 779

17.1 Introduction 780
17.2 Enols 780
17.3 Enolates 783
17.4 Enolates of Unsymmetrical Carbonyl Compounds 788
17.5 Racemization at the α Carbon 790
17.6 A Preview of Reactions at the α Carbon 790
17.7 Halogenation at the α Carbon 791
17.8 Direct Enolate Alkylation 793
17.9 Malonic Ester Synthesis 796
17.10 Acetoacetic Ester Synthesis 800
17.11 Biological Decarboxylation 802
Chapter Review 803
Problems 806

Contents xiii

Corbis

18 Carbonyl Condensation Reactions 812

- **18.1** The Aldol Reaction 813
- **18.2** Crossed Aldol Reactions 818
- **18.3** Directed Aldol Reactions 821
- **18.4** Intramolecular Aldol Reactions 822
- **18.5** The Claisen Reaction 825
- **18.6** The Crossed Claisen and Related Reactions 827
- **18.7** The Dieckmann Reaction 829
- **18.8** Biological Carbonyl Condensation Reactions 830
- **18.9** The Michael Reaction 833
- **18.10** The Robinson Annulation 835
  - *Chapter Review* 839
  - *Problems* 841

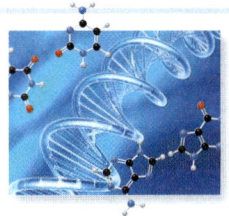
Pasieka/Science Photo Library/Getty Images

19 Benzene and Aromatic Compounds 849

- **19.1** Background 850
- **19.2** The Structure of Benzene 851
- **19.3** Nomenclature of Benzene Derivatives 853
- **19.4** Spectroscopic Properties 855
- **19.5** Interesting Aromatic Compounds 856
- **19.6** Benzene's Unusual Stability 857
- **19.7** The Criteria for Aromaticity—Hückel's Rule 859
- **19.8** Examples of Aromatic Compounds 861
- **19.9** Aromatic Heterocycles 864
- **19.10** What Is the Basis of Hückel's Rule? 869
- **19.11** The Inscribed Polygon Method for Predicting Aromaticity 872
- **19.12** Aromatase Inhibitors for Estrogen-Dependent Cancer Treatment 874
  - *Chapter Review* 876
  - *Problems* 878

Jill Braaten

20 Reactions of Aromatic Compounds 885

- **20.1** Electrophilic Aromatic Substitution 886
- **20.2** The General Mechanism 887
- **20.3** Halogenation 888
- **20.4** Nitration and Sulfonation 890
- **20.5** Friedel–Crafts Alkylation and Friedel–Crafts Acylation 891
- **20.6** Substituted Benzenes 898
- **20.7** Electrophilic Aromatic Substitution of Substituted Benzenes 901
- **20.8** Why Substituents Activate or Deactivate a Benzene Ring 904
- **20.9** Orientation Effects in Substituted Benzenes 905
- **20.10** Limitations on Electrophilic Substitution Reactions with Substituted Benzenes 908
- **20.11** Disubstituted Benzenes 910
- **20.12** Synthesis of Benzene Derivatives 912
- **20.13** Nucleophilic Aromatic Substitution 913

20.14 Reactions of Substituted Benzenes   917
20.15 Multistep Synthesis   921
   Chapter Review   924
   Problems   928

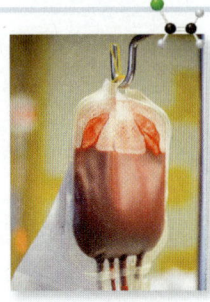
Hin255/Getty Images

21 Radical Reactions   935

21.1 Introduction   936
21.2 General Features of Radical Reactions   937
21.3 Halogenation of Alkanes   939
21.4 The Mechanism of Halogenation   940
21.5 Chlorination of Other Alkanes   943
21.6 Chlorination Versus Bromination   943
21.7 The Stereochemistry of Halogenation Reactions   946
21.8 Application: The Ozone Layer and CFCs   949
21.9 Radical Halogenation at an Allylic Carbon   950
21.10 Application: Oxidation of Unsaturated Lipids   953
21.11 Application: Antioxidants   954
21.12 Radical Addition Reactions to Double Bonds   955
21.13 Polymers and Polymerization   957
   Chapter Review   959
   Problems   962

Werner Arnold

22 Amines   969

22.1 Introduction   970
22.2 Structure and Bonding   970
22.3 Nomenclature   971
22.4 Physical and Spectroscopic Properties   974
22.5 Interesting and Useful Amines   975
22.6 Preparation of Amines   979
22.7 Reactions of Amines—General Features   985
22.8 Amines as Bases   985
22.9 Relative Basicity of Amines and Other Compounds   987
22.10 Amines as Nucleophiles   993
22.11 Hofmann Elimination   995
22.12 Reaction of Amines with Nitrous Acid   997
22.13 Substitution Reactions of Aryl Diazonium Salts   999
22.14 Coupling Reactions of Aryl Diazonium Salts   1004
22.15 Application: Synthetic Dyes and Sulfa Drugs   1005
   Chapter Review   1008
   Problems   1013

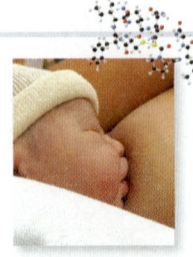
Daniel C. Smith

23 Amino Acids and Proteins   1021

23.1 Amino Acids   1022
23.2 Separation of Amino Acids   1026
23.3 Enantioselective Synthesis of Amino Acids   1029

- 23.4 Peptides 1031
- 23.5 Peptide Sequencing 1035
- 23.6 Peptide Synthesis 1038
- 23.7 Automated Peptide Synthesis 1043
- 23.8 Protein Structure 1045
- 23.9 Important Proteins 1051
- 23.10 Enzymes 1054

    Chapter Review 1058
    Problems 1063

MaraZe/Shutterstock

24 Carbohydrates 1069

- 24.1 Introduction 1070
- 24.2 Monosaccharides 1071
- 24.3 The Family of D-Aldoses 1077
- 24.4 The Family of D-Ketoses 1078
- 24.5 Physical Properties of Monosaccharides 1079
- 24.6 The Cyclic Forms of Monosaccharides 1079
- 24.7 Glycosides 1087
- 24.8 Reactions of Monosaccharides at the OH Groups 1090
- 24.9 Reactions at the Carbonyl Group—Oxidation and Reduction 1091
- 24.10 Reactions at the Carbonyl Group—Adding or Removing One Carbon Atom 1094
- 24.11 Disaccharides 1098
- 24.12 Polysaccharides 1100
- 24.13 Other Important Sugars and Their Derivatives 1103

    Chapter Review 1105
    Problems 1108

Iconotec/Glowimages

25 Lipids 1114

- 25.1 Introduction 1115
- 25.2 Waxes 1116
- 25.3 Triacylglycerols 1116
- 25.4 Phospholipids 1120
- 25.5 Fat-Soluble Vitamins 1123
- 25.6 Eicosanoids 1124
- 25.7 Terpenes 1126
- 25.8 Steroids 1132

    Chapter Review 1137
    Problems 1139

Daniel C. Smith

26 Nucleic Acids and Protein Synthesis 1144

- 26.1 Nucleosides and Nucleotides 1145
- 26.2 Nucleic Acids 1148
- 26.3 The DNA Double Helix 1150
- 26.4 Replication 1153
- 26.5 Ribonucleic Acids and Transcription 1154

26.6 The Genetic Code, Translation, and Protein Synthesis 1156
26.7 DNA Sequencing 1159
26.8 The Polymerase Chain Reaction 1161
26.9 Viruses 1162
Chapter Review 1163
Problems 1165

27 Metabolism 1169

Samuel Borges Photography/ Shutterstock

27.1 Overview of Metabolism 1170
27.2 Key Oxidizing and Reducing Agents in Metabolism 1173
27.3 The Catabolism of Triacylglycerols by β-Oxidation 1176
27.4 The Catabolism of Carbohydrates—Glycolysis 1179
27.5 The Fate of Pyruvate 1184
27.6 The Citric Acid Cycle and ATP Production 1186
Chapter Review 1191
Problems 1193

28 Carbon–Carbon Bond-Forming Reactions in Organic Synthesis (Available Online) 1197

DEA/M. GIOVANOLI/De Agostini Picture Library/ Getty Images

28.1 Coupling Reactions of Organocuprate Reagents 1198
28.2 Suzuki Reaction 1200
28.3 Heck Reaction 1205
28.4 Carbenes and Cyclopropane Synthesis 1207
28.5 Simmons–Smith Reaction 1210
28.6 Metathesis 1211
Chapter Review 1216
Problems 1218

29 Pericyclic Reactions (Available Online) 1226

Premaphotos/Alamy Stock Photo

29.1 Types of Pericyclic Reactions 1227
29.2 Molecular Orbitals 1228
29.3 Electrocyclic Reactions 1231
29.4 Cycloaddition Reactions 1236
29.5 Sigmatropic Rearrangements 1240
29.6 Summary of Rules for Pericyclic Reactions 1246
Chapter Review 1247
Problems 1249

30 Synthetic Polymers (Available Online) 1255

Stuar/Shutterstock

30.1 Introduction 1256
30.2 Chain-Growth Polymers—Addition Polymers 1257
30.3 Anionic Polymerization of Epoxides 1263
30.4 Ziegler–Natta Catalysts and Polymer Stereochemistry 1264
30.5 Natural and Synthetic Rubbers 1265

30.6 Step-Growth Polymers—Condensation Polymers   1267
30.7 Polymer Structure and Properties   1272
30.8 Green Polymer Synthesis   1274
30.9 Polymer Recycling and Disposal   1277
Chapter Review   1279
Problems   1281

Appendix A   Periodic Table of the Elements   A-1
Appendix B   Common Abbreviations, Arrows, and Symbols   A-2
Appendix C   pK_a Values for Selected Compounds   A-4
Appendix D   Nomenclature   A-6
Appendix E   Bond Dissociation Energies for Some Common Bonds [A—B → A• + •B]   A-10
Appendix F   Reactions That Form Carbon–Carbon Bonds   A-11
Appendix G   Characteristic IR Absorption Frequencies   A-12
Appendix H   Characteristic NMR Absorptions   A-13
Appendix I   General Types of Organic Reactions   A-15
Appendix J   How to Synthesize Particular Functional Groups   A-17

Glossary G-1
Index I-1

Preface

Since the publication of *Organic Chemistry* in 2005, chemistry has witnessed a rapid growth in its understanding of the biological world. The molecular basis of many complex biological processes is now known with certainty, and can be explained by applying the basic principles of organic chemistry. Because of the close relationship between chemistry and many biological phenomena, *Organic Chemistry with Biological Topics* presents an approach to traditional organic chemistry that incorporates the discussion of biological applications that are understood using the fundamentals of organic chemistry.

The Basic Features

Organic Chemistry with Biological Topics continues the successful student-oriented approach used in *Organic Chemistry* by Janice Gorzynski Smith. This text uses less prose and more diagrams and bulleted summaries for today's students, who rely more heavily on visual imagery to learn than ever before. Each topic is broken down into small chunks of information that are more manageable and easily learned. Sample Problems illustrate stepwise problem solving, and relevant examples from everyday life are used to illustrate topics. New concepts are introduced one at a time so that the basic themes are kept in focus.

The organization of *Organic Chemistry with Biological Topics* provides the student with a logical and accessible approach to an intense and fascinating subject. The text begins with a healthy dose of review material in Chapters 1 and 2 to ensure that students have a firm grasp of the fundamentals. Stereochemistry, the three-dimensional structure of molecules, is introduced early (Chapter 5) and reinforced often. Certain reaction types with unique characteristics and terminology are grouped together. These include acid–base reactions (Chapter 2), oxidation and reduction (Chapters 11 and 13), reactions of organometallic reagents (Chapter 13), and radical reactions (Chapter 21). Because of its importance in biological molecules, **the chemistry of carbonyl-containing compounds has been moved much earlier** than traditional organic chemistry texts and is now described in Chapters 13–18. Each chapter ends with a Chapter Review, end-of-chapter summaries that succinctly organize the main concepts and reactions.

New to This Edition

Students sometimes ask me if the facts of organic chemistry have significantly changed since the last edition. While the basic principles remain the same—carbon forms four bonds in stable compounds and oppositely charged species attract each other—organic chemistry is a dynamic subject that is continually refined as new facts are determined, and new editions reflect current understanding. Each year, novel compounds are discovered and new drugs are marketed, and these compounds replace older examples to illustrate particular concepts. Also of significance is *how* the material in the text is presented. I continue to endeavor to make this difficult subject as student-friendly as possible, by redesigning sample problems and end-of-chapter material, and rewriting sentences and paragraphs for improved clarity.

General

Expanded Problem-Solving Approach A central component of each chapter of *Organic Chemistry with Biological Topics* is the Sample Problems, which illustrate how to solve key elements of the chapter. In this edition, Sample Problems are always paired with a follow-up Problem to allow students to apply what they have just learned. The Problems are followed by "More Practice," a list of end-of-chapter problems that are similar in concept. Students can find detailed solutions and verify their answers to *all* of the Problems from the book with the Student Study Guide/Solutions Manual for *Organic Chemistry with Biological Topics*.

Chapter Review The end-of-chapter summary sections have been expanded into parts: **Key Concepts, Key Skills, Key Reactions,** and **Key Mechanism Concepts,** with structures and examples to illustrate each part, providing students with a broader and more detailed overview of each chapter's important concepts and skills. Extensive cross-referencing has also been added to connect this material with relevant Sample Problems, Problems, Figures, and Tables within the body of the chapter.

New Chapters

In addition to the six chapters that contained new biological material in the fifth edition—Chapters 3, 6, 15, 16, 18, and 19—two new chapters have been added:

- **Chapter 26** provides an in-depth discussion of the structure and properties of the nucleic acids DNA and RNA. Three key processes are also presented: replication—how DNA makes copies of itself; transcription—how the genetic information in DNA is passed onto RNA; and translation—how the coded genetic information in RNA is used to synthesize proteins. The chapter concludes with discussions of manipulating DNA in the laboratory and how viruses act.
- **Chapter 27** focuses on the biochemical reactions involved in metabolism. The discussion centers on three components: the breakdown of fats, the metabolism of the carbohydrate glucose to the three-carbon unit pyruvate by glycolysis, and the citric acid cycle, a key cyclic metabolic pathway used for amino acids, carbohydrates, and fats.

Spectroscopy

The revisions to the spectroscopy coverage are designed to allow for more flexibility, making these chapters more portable to accommodate various lecture and lab arrangements. Three new spectroscopy chapters have been created for the sixth edition: Spectroscopy A Mass Spectrometry; Spectroscopy B Infrared Spectroscopy; and Spectroscopy C Nuclear Magnetic Resonance Spectroscopy. The coverage and problem sets for these chapters have also been expanded to include material previously covered in other sections of earlier editions. Extensive cross-referencing has been added so that whether spectroscopy is covered early or late in an organic chemistry course, students can readily find the material they need.

Other New Coverage

Examples of biomolecules are sprinkled throughout the chapters to illustrate common organic structural features and reactions, such as Lewis structures (Chapter 1), Lewis acids and bases (Chapter 2), stereochemistry (Chapter 5), and elimination reactions (Chapters 8 and 9). Other changes include the following:

- Section 11.13 on biological oxidation has been expanded to include the treatment of prochirality.
- New material has been added to Sections 13.6 and 13.7, including the biological reduction of acyl phosphates to aldehydes.
- The role of imines in the deamination of amino acids is discussed in Section 14.13B, and a detailed mechanism that illustrates the role of pyridoxal phosphate, vitamin B_6, is presented.
- The coverage of nitriles has been moved to the chapter on carboxylic acids, forming Chapter 15, Carboxylic Acids and Nitriles. This chapter is now placed after Chapter 14, Aldehydes and Ketones, and this move offers two advantages. The chapter places the chemistry of carboxylic acids closer to similar chemistry seen with the acyl derivatives that is covered in Chapter 16. It also places the nucleophilic addition reactions of nitriles in closer proximity to related reactions in Chapter 14.
- A new Section 17.11 on biological decarboxylation has been added to Chapter 17.
- A new Section 23.8D on protein denaturation has been added to Chapter 23.
- Section 23.10 on enzymes illustrates how enzymes work with a specific example, how the serine proteases hydrolyze peptide bonds in proteins. The section concludes with a discussion of how enzymes are used to diagnose and treat diseases.
- The importance of human milk oligosaccharides in breast milk is discussed in Section 24.12D.

Learning Resources for Instructors and Students

The following items may accompany this text. Please consult your McGraw-Hill representative for policies, prices, and availability as some restrictions may apply.

Presentation Tools

Within the Instructor's Resources, instructors have access to editable, accessible PowerPoint lecture outlines, which appear as ready-made presentations that combine art and lecture notes for each chapter of the text. For instructors who prefer to create their lecture notes from scratch, all illustrations, photos, tables, *How To*'s, and Sample Problems are pre-inserted by chapter into a separate set of PowerPoint slides. They are also available as individual .jpg files.

Photos, artwork, and other media types can be used to create customized lectures, visually enhanced tests and quizzes, compelling course websites, or attractive printed support materials. All assets are copyrighted by McGraw-Hill Higher Education, but can be used by instructors for classroom purposes. The visual resources in this collection include:

- **Art** Full-color digital files of all illustrations in the book can be readily incorporated into lecture presentations, exams, or custom-made classroom materials.
- **Photos** The photo collection contains digital files of photographs from the text, which can be reproduced for multiple classroom uses.
- **Tables** Every table that appears in the text has been saved in electronic form for use in classroom presentations and/or quizzes.

Student Study Guide/Solutions Manual

Written by Janice Gorzynski Smith and Erin R. Smith, the Student Study Guide/Solutions Manual provides step-by-step solutions to all in-chapter and end-of-chapter problems. Each chapter begins with an overview of key concepts and includes a short-answer practice test on the fundamental principles and new reactions.

Acknowledgments

Although I have been an author for many years, this edition of *Organic Chemistry with Biological Topics* reflects recent advances in our understanding of organic chemistry, as well as new advances in digital media that allow this work to be better understood by a larger student audience. To produce a high quality text and ancillary materials requires not only my insights as an author, but also the expertise of a group of individuals with whom I work, beginning with the generation of a manuscript, progressing through the publication of the finished product both in print and digital form, and bringing the text to the larger chemistry community by the sales and marketing team.

My special thanks in this edition go out to two individuals who are integral to success of the project. Mary Hurley, Senior Developmental Editor, with whom I have worked for several years, is a master at supervising all the details of this large project and heading off problems before they become crises. I feel that Mary has been key in keeping my projects on a smooth trajectory even when many of the other personnel involved have changed. Amy Gehl, Production Manager, although new to the team, has skillfully and seamlessly managed the conversion of this text from paper manuscript to printed edition. Thanks so much to both of you and my sincere appreciation goes out to the entire chemistry group.

I especially thank my husband Dan and the other members of my immediate family, who have experienced the day-to-day demands of living with a busy author. The joys and responsibilities of the family have always kept me grounded during the rewarding but sometimes all-consuming process of writing a textbook. This book, like the prior edition of *Organic Chemistry with Biological Topics*, is dedicated to my wonderful daughter Megan, who passed away after a nine-year battle with cystic fibrosis.

Among the many others that go unnamed but who have profoundly affected this work are the thousands of students I have been lucky to teach over many years. I have learned so much from my daily interactions with them, and I hope that the wider chemistry community can benefit from this experience.

This edition has evolved based on the helpful feedback of many people who reviewed past editions and digital products, class-tested the book, and attended focus groups or symposiums. These many individuals have collectively provided constructive improvements to the project.

Listed below are the reviewers of *Organic Chemistry with Biological Topics,* fifth edition:

Steven Castle, *Brigham Young University*
Manashi Chatterjee, *Hunter College*
Emma Chow, *Palm Beach State College*
Jeff Corkill, *Eastern Washington University*
Andrew Frazer, *University of Central Florida*
Bob Kane, *Baylor University*
Donna J. Nelson, *University of Oklahoma*
Joshua L. Price, *Brigham Young University*
Elizabeth Walters, *University of North Carolina at Wilmington*
Lisa Whalen, *University of New Mexico*
Alexander Wurthmann, *University of Vermont*

The following individuals helped write and review learning goal-oriented content for **SmartBook for Organic Chemistry with Biological Topics:** David Jones, St. David's School in Raleigh, NC; Adam Keller, Columbus State Community College; and Angela Perkins, University of Minnesota. Andrea Leonard of the University of Louisiana, Lafayette, revised the PowerPoint Lectures, and Ryan Simon also of the University of Louisiana, Lafayette, revised the Test Bank for *Organic Chemistry with Biological Topics*, sixth edition.

Although every effort has been made to make this text and its accompanying Student Study Guide/Solutions Manual as error-free as possible, some errors undoubtedly remain. Please feel free to email me about any inaccuracies, so that subsequent editions may be further improved.

With much aloha,

Janice Gorzynski Smith
jgsmith@hawaii.edu

FOR INSTRUCTORS

You're in the driver's seat.

Want to build your own course? No problem. Prefer to use our turnkey, prebuilt course? Easy. Want to make changes throughout the semester? Sure. And you'll save time with Connect's auto-grading too.

65%
Less Time Grading

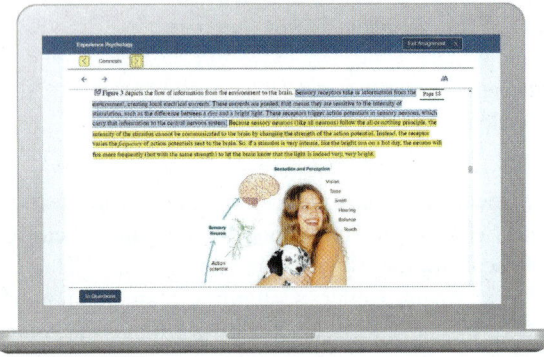

Laptop: McGraw-Hill; Woman/dog: George Doyle/Getty Images

They'll thank you for it.

Adaptive study resources like SmartBook® 2.0 help your students be better prepared in less time. You can transform your class time from dull definitions to dynamic debates. Find out more about the powerful personalized learning experience available in SmartBook 2.0 at **www.mheducation.com/highered/connect/smartbook**

Make it simple, make it affordable.

Connect makes it easy with seamless integration using any of the major Learning Management Systems—Blackboard®, Canvas, and D2L, among others—to let you organize your course in one convenient location. Give your students access to digital materials at a discount with our inclusive access program. Ask your McGraw-Hill representative for more information.

Padlock: Jobalou/Getty Images

Solutions for your challenges.

A product isn't a solution. Real solutions are affordable, reliable, and come with training and ongoing support when you need it and how you want it. Our Customer Experience Group can also help you troubleshoot tech problems—although Connect's 99% uptime means you might not need to call them. See for yourself at **status.mheducation.com**

Checkmark: Jobalou/Getty Images

FOR STUDENTS

Effective, efficient studying.

Connect helps you be more productive with your study time and get better grades using tools like SmartBook 2.0, which highlights key concepts and creates a personalized study plan. Connect sets you up for success, so you walk into class with confidence and walk out with better grades.

Study anytime, anywhere.

Download the free ReadAnywhere app and access your online eBook or SmartBook 2.0 assignments when it's convenient, even if you're offline. And since the app automatically syncs with your eBook and SmartBook 2.0 assignments in Connect, all of your work is available every time you open it. Find out more at **www.mheducation.com/readanywhere**

> *"I really liked this app—it made it easy to study when you don't have your textbook in front of you."*
>
> - Jordan Cunningham, Eastern Washington University

No surprises.

The Connect Calendar and Reports tools keep you on track with the work you need to get done and your assignment scores. Life gets busy; Connect tools help you keep learning through it all.

Calendar: owattaphotos/Getty Images

Learning for everyone.

McGraw-Hill works directly with Accessibility Services Departments and faculty to meet the learning needs of all students. Please contact your Accessibility Services office and ask them to email accessibility@mheducation.com, or visit **www.mheducation.com/about/accessibility** for more information.

Top: Jenner Images/Getty Images, Left: Hero Images/Getty Images, Right: Hero Images/Getty Images

Prologue

What is organic chemistry?
Some representative organic molecules
Marine natural products

Organic chemistry. You might wonder how a discipline that conjures up images of eccentric old scientists working in basement laboratories is relevant to you, a student in the twenty-first century.

Consider for a moment the activities that occupied your past 24 hours. You likely showered with soap, drank a caffeinated beverage, ate at least one form of starch, took some medication, and traveled in a vehicle that had rubber tires and was powered at least partly by fossil fuels. If you did any *one* of these, your life was touched by organic chemistry.

What Is Organic Chemistry?

- **Organic chemistry is the chemistry of compounds that contain the element carbon.**

It is one branch in the entire field of chemistry, which encompasses many classical subdisciplines including inorganic, physical, and analytical chemistry, and newer fields such as bioinorganic chemistry, physical biochemistry, polymer chemistry, and materials science.

Some compounds that contain the element carbon are *not* organic compounds. Examples include carbon dioxide (CO_2), sodium carbonate (Na_2CO_3), and sodium bicarbonate ($NaHCO_3$).

Organic chemistry was singled out as a separate discipline for historical reasons. Originally, it was thought that compounds in living things, termed **organic compounds,** were fundamentally different from those in nonliving things, called **inorganic compounds.** Although we have known for more than 150 years that this distinction is artificial, the name *organic* persists. Today the term refers to the study of the compounds that contain carbon, many of which, incidentally, are found in living organisms.

It may seem odd that a whole discipline is devoted to the study of a single element in the periodic table, when more than 100 elements exist. It turns out, though, that there are far more organic compounds than any other type. **Organic chemicals affect virtually every facet of our lives, and for this reason, it is important and useful to know something about them.**

Clothes, foods, medicines, gasoline, refrigerants, and soaps are composed almost solely of organic compounds. Some, like cotton, wool, and silk, are *naturally occurring;* that is, they can be isolated directly from natural sources. Others, such as nylon and polyester, are *synthetic,* meaning they are produced by chemists in the laboratory. By studying the principles and concepts of organic chemistry, you can learn more about compounds such as these and how they affect the world around you.

Realize, too, what organic chemistry has done for us. Organic chemistry has made available both comforts and necessities that were previously nonexistent, or reserved for only the wealthy. We have seen an enormous increase in life span, from 47 years in 1900 to over 70 years currently. To a large extent this is due to the isolation and synthesis of new drugs to fight infections and the availability of vaccines for childhood diseases. Chemistry has also given us the tools to control insect populations that spread disease, and there is more food for all because

of fertilizers, pesticides, and herbicides. Our lives would be vastly different today without the many products that result from organic chemistry (Figure 1).

Figure 1
Products of organic chemistry used in medicine

a. Oral contraceptives

Comstock Images/PictureQuest

b. Plastic syringes

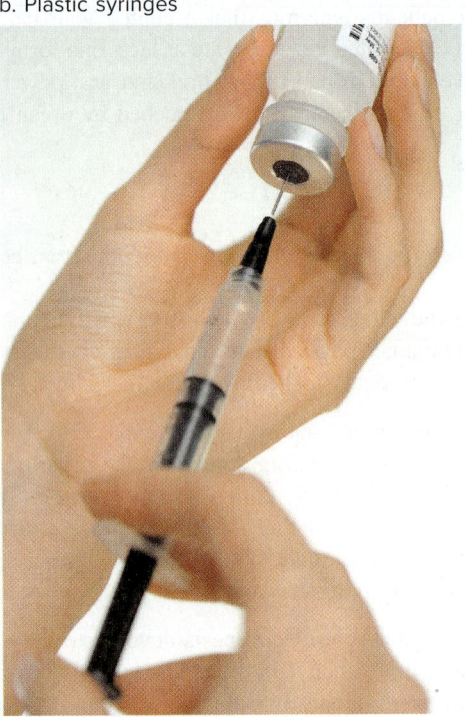

Tom Grill/Corbis Premium/Alamy Stock Photo

c. Antibiotics

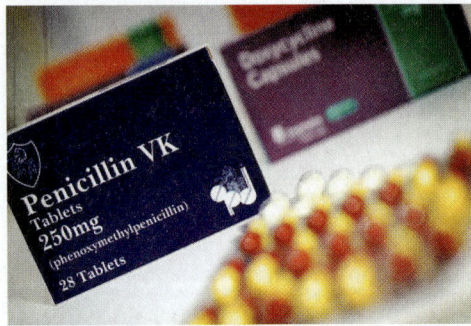

Julian Claxton/Alamy Stock Photo

d. Synthetic heart valves

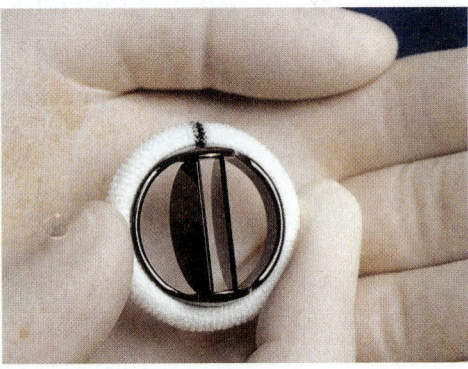

Layne Kennedy/Corbis NX/Getty Images

- Organic chemistry has given us contraceptives, plastics, antibiotics, and the knitted material used in synthetic heart valves.

Some Representative Organic Molecules

Perhaps the best way to appreciate the variety of organic molecules is to look at a few. Three simple organic compounds are **methane, ethanol,** and **trichlorofluoromethane.**

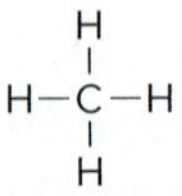

methane

- **Methane,** the simplest of all organic compounds, contains one carbon atom. Methane—the main component of natural gas—occurs widely in nature. Like other **hydrocarbons**—organic compounds that contain only carbon and hydrogen—methane is combustible; that is, it burns in the presence of oxygen. Methane is the product of the anaerobic (without air) decomposition of organic matter by bacteria. The natural gas we use today was formed by the decomposition of organic material millions of years ago. Hydrocarbons such as methane are discussed in Chapter 4.

- **Ethanol,** the alcohol present in beer, wine, and other alcoholic beverages, is formed by the fermentation of sugar, possibly the oldest example of organic synthesis. Ethanol can also be made in the lab by a totally different process, but **the ethanol produced in the lab is *identical* to the ethanol produced by fermentation.** Alcohols including ethanol are discussed in Chapter 9.

- **Trichlorofluoromethane** is a member of a class of molecules called **chlorofluorocarbons,** or **CFCs,** which contain one or two carbon atoms and several halogens. Trichlorofluoromethane is an unusual organic molecule in that **it contains no hydrogen atoms.** Because it has a low molecular weight and is easily vaporized, trichlorofluoromethane has been used as an aerosol propellant and refrigerant. It and other CFCs have been implicated in the destruction of the stratospheric ozone layer, a topic discussed in Chapter 21.

Three complex organic molecules that are important medications are **amoxicillin, fluoxetine,** and **AZT.**

- **Amoxicillin** is one of the most widely used antibiotics in the penicillin family. The discovery and synthesis of such antibiotics in the twentieth century made routine the treatment of infections that were formerly fatal. You were likely given some amoxicillin to treat an ear infection when you were a child. The penicillin antibiotics are discussed in Chapter 16.

Complex organic structures are drawn with shorthand conventions described in Chapter 1.

amoxicillin

- **Fluoxetine** is the generic name for the antidepressant **Prozac.** Prozac was designed and synthesized by chemists in the laboratory, and is now produced on a large scale in chemical factories. Because it is safe and highly effective in treating depression, Prozac is widely prescribed. Over 40 million individuals worldwide have used Prozac since 1986.

fluoxetine

- **AZT, az**idodeoxythymidine, is a drug that treats human immunodeficiency virus (HIV), the virus that causes acquired immune deficiency syndrome (AIDS). Also known by its generic name **zidovudine,** AZT represents a chemical success to a different challenge: synthesizing agents that combat viral infections.

AZT

Other complex organic compounds with interesting properties are **capsaicin** and **DDT.**

- **Capsaicin,** one member of a group of compounds called *vanilloids,* is responsible for the characteristic spiciness of hot peppers. It is the active ingredient in pepper sprays used for personal defense and topical creams used for pain relief.

capsaicin

- **DDT, d**ichloro**d**iphenyl**t**richloroethane, is a pesticide once called "miraculous" by Winston Churchill because of the many lives it saved by killing disease-carrying mosquitoes. DDT use is now banned in the United States and many developed countries because it is a nonspecific insecticide that persists in the environment.

DDT

What are the common features of these organic compounds?

- All organic compounds contain carbon atoms and most contain hydrogen atoms.
- All the carbon atoms have four bonds. A stable carbon atom is said to be *tetravalent*.
- Other elements may also be present. Any atom that is not carbon or hydrogen is called a *heteroatom*. Common heteroatoms include N, O, S, P, and the halogens.
- Some compounds have chains of atoms and some compounds have rings.

These features explain why there are so many organic compounds: **Carbon forms four strong bonds with itself and other elements. Carbon atoms combine together to form rings and chains.**

Marine Natural Products

Nature has generously supplied the organic chemist with a wide variety of complex compounds that have promising therapeutic potential. In the last 40 years, the largely unexplored marine environment has been recognized as a vast resource of unique compounds with novel chemical properties, but the challenges in discovering drug leads among such expansive biodiversity are many. Organisms are often found in waters offshore remote islands, and structure determination must be carried out on minute quantities of material. Even when potential targets are identified, supplying enough compound for preclinical and clinical trials often means that the compound must then be synthesized in the laboratory. Nonetheless, new compounds with useful bioactivity are routinely discovered and synthesized. Among the first available anticancer drugs with origins in the world of marine natural products are **eribulin mesylate** and **trabectedin.**

Eribulin mesylate is a synthetic analogue of the more complex natural product halichondrin B, which is isolated from the black sponge *Halichondria okadai.* Sold under the trade name Halaven, it was approved in the United States in 2010 for the treatment of metastatic breast cancer.

eribulin mesylate

halichondrin B

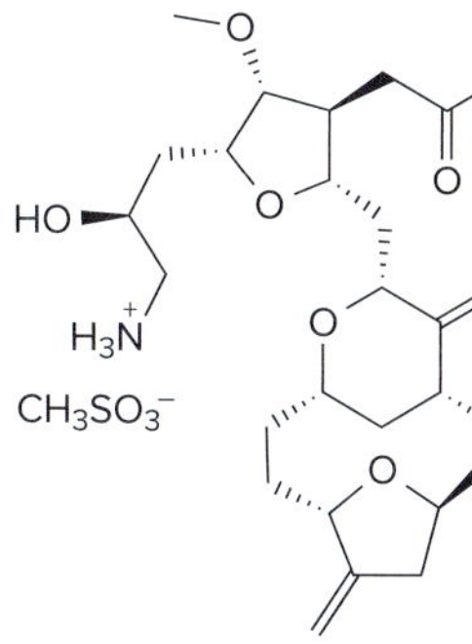

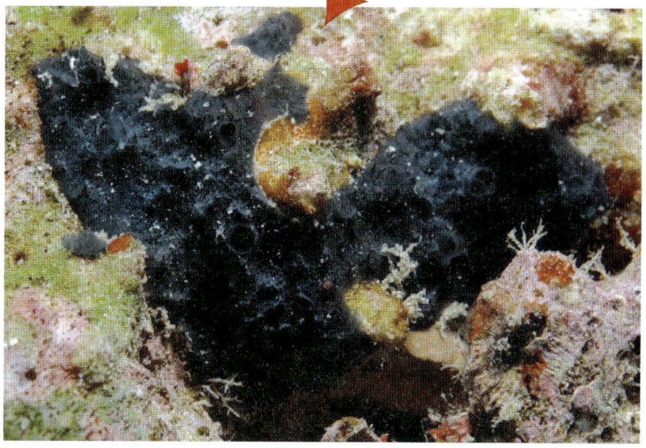

Halichondria okadai

Guido & Philippe Poppe

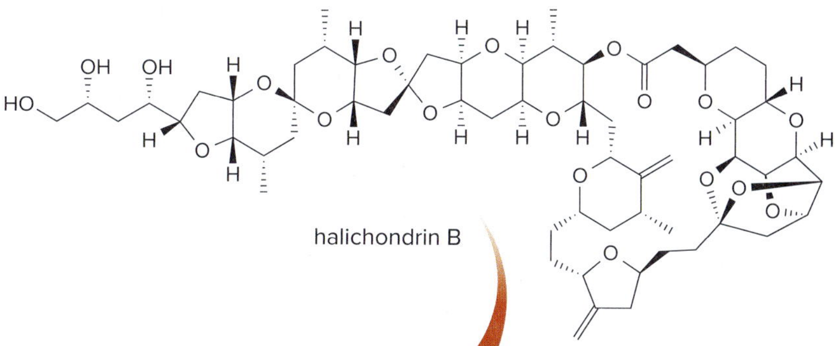

Trabectedin, also known as ecteinascidin 743 or ET-743, is obtained from the sea squirt *Ecteinascidia turbinata.* Sold under the trade name Yondelis, it was approved in the European Union in 2007 for the treatment of advanced soft tissue sarcoma. In 2015, the U.S. Food and Drug Administration approved trabectedin for the treatment of specific soft tissue cancers that cannot be removed by surgery.

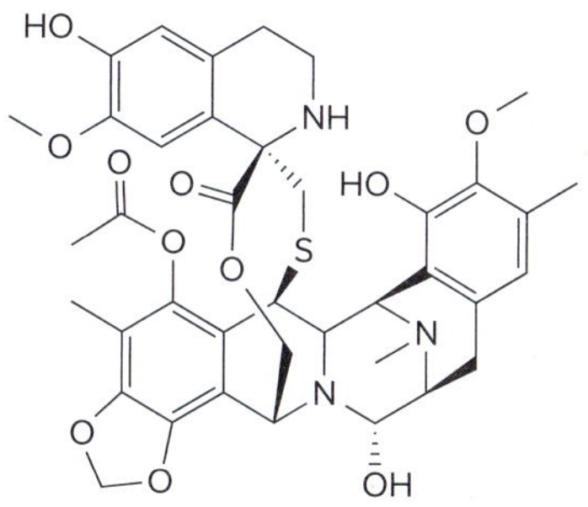

trabectedin, ET-743

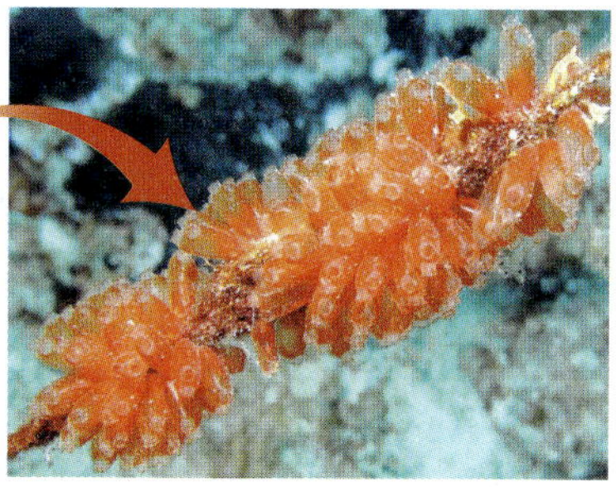

Ecteinascidia turbinata

Florent Charpin 2004-2016. All Rights Reserved

Because isolation of enough trabectedin for clinical trials was not feasible—one ton of organisms yielded one gram of compound—trabectedin was synthesized in the laboratory of Nobel Laureate E. J. Corey in 1996. Now it is readily available by a shorter synthesis from a starting material obtained by a fermentation process.

Hundreds of new biologically active marine natural products are now isolated each year, so the number of compounds in the marine drug pipeline should continue to increase in the near future.

In this introduction, we have seen a variety of molecules that have diverse structures. They represent a miniscule fraction of the organic compounds currently known and the many thousands that are newly discovered or synthesized each year. The principles you learn in organic chemistry will apply to all of these molecules, from simple ones like methane and ethanol, to complex ones like eribulin mesylate and trabectedin. It is these beautiful molecules, their properties, and their reactions that we will study in organic chemistry.

WELCOME TO THE WORLD OF ORGANIC CHEMISTRY!

Structure and Bonding

1

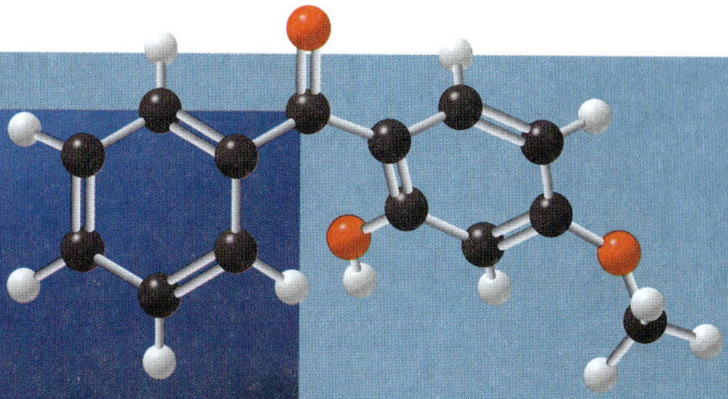

Buttchi 3 Sha Life/Shutterstock

- 1.1 The periodic table
- 1.2 Bonding
- 1.3 Lewis structures
- 1.4 Isomers
- 1.5 Exceptions to the octet rule
- 1.6 Resonance
- 1.7 Determining molecular shape
- 1.8 Drawing organic structures
- 1.9 Hybridization
- 1.10 Ethane, ethylene, and acetylene
- 1.11 Bond length and bond strength
- 1.12 Electronegativity and bond polarity
- 1.13 Polarity of molecules
- 1.14 Oxybenzone—A representative organic molecule

Bleaching is a phenomenon that occurs when corals expel symbiotic algae from their tissues in response to an external stress, causing the coral to turn white. Although coral bleaching is most often associated with an increase in water temperature, recent research at the University of Hawaiʻi suggests that minute amounts of compounds such as **oxybenzone** also contribute to bleaching. Oxybenzone effectively filters a broad spectrum of harmful ultraviolet light, so it is a common sunscreen component, but it can be washed off while swimming, leading to a low but potentially harmful concentration in the water. For this reason, the state of Hawaiʻi now prohibits the sale of sunscreens that contain oxybenzone. In Chapter 1, we learn about the structure, bonding, and properties of organic compounds like oxybenzone.

Why Study . . . Structure and Bonding?

Before examining organic molecules in detail, we must review topics about structure and bonding learned in previous chemistry courses. We will discuss these concepts primarily from an organic chemist's perspective, and spend time on only the particulars needed to understand organic compounds.

Important topics in Chapter 1 include drawing Lewis structures, predicting the shape of molecules, determining what orbitals are used to form bonds, and how electronegativity affects bond polarity. Equally important is Section 1.8 on drawing organic molecules, both shorthand methods routinely used for simple and complex compounds, and three-dimensional representations that allow us to more clearly visualize them.

1.1 The Periodic Table

All matter is composed of the same building blocks called **atoms.** There are two main components of an atom.

- The **nucleus** contains positively charged **protons** and uncharged **neutrons.** Most of the mass of the atom is contained in the nucleus.
- The **electron cloud** is composed of negatively charged **electrons.** The electron cloud comprises most of the volume of the atom.

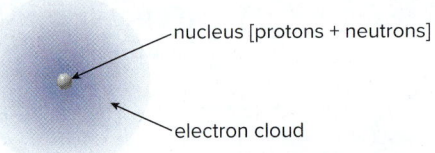

The charge on a proton is equal in magnitude but opposite in sign to the charge on an electron. In a neutral atom, the **number of protons in the nucleus equals the number of electrons.** This quantity, called the **atomic number,** is unique to a particular element. For example, every neutral carbon atom has an atomic number of six, meaning it has six protons in its nucleus and six electrons surrounding the nucleus.

In addition to neutral atoms, we will encounter **charged ions.**

- A *cation* is positively charged and has fewer electrons than protons.
- An *anion* is negatively charged and has more electrons than protons.

The number of neutrons in the nucleus of a particular element can vary. **Isotopes** are two atoms of the same element having a different number of neutrons. The **mass number** of an atom is the total number of protons and neutrons in the nucleus. **Isotopes have *different* mass numbers.** The **atomic weight** of a particular element is the weighted average of the mass of all its isotopes, reported in atomic mass units (amu).

Isotopes of carbon and hydrogen are sometimes used in organic chemistry. The most common isotope of hydrogen has one proton and no neutrons in the nucleus, but 0.02% of hydrogen atoms have one proton and one neutron. This isotope of hydrogen is called **deuterium** and is sometimes symbolized by the letter **D.**

Each atom is identified by a one- or two-letter abbreviation that is the characteristic symbol for that element. Carbon is identified by the single letter **C.** Sometimes the atomic number is indicated as a subscript to the left of the element symbol, and the mass number is indicated as a superscript. Using this convention, the most common isotope of carbon, which contains six protons and six neutrons, is designated as $^{12}_{6}\text{C}.$

1.1 The Periodic Table

The **periodic table** is a schematic arrangement of the more than 100 known elements, arranged in order of increasing atomic number. The periodic table is composed of rows and columns. Each column in the periodic table is identified by a **group number,** an Arabic (1 to 8) or Roman (I to VIII) numeral followed by the letter A or B. Carbon is located in group **4A** in the periodic table in this text.

> A **row** in the periodic table is also called a **period,** and a **column** is also called a **group.** A periodic table is located in Appendix A for your reference.

- Elements in the same row are similar in *size*.
- Elements in the same column have similar *electronic and chemical properties*.

Although more than 100 elements exist, most are not common in organic compounds. Figure 1.1 contains a truncated periodic table, indicating the handful of elements that are routinely seen in this text. **Most elements in organic compounds are located in the first and second rows of the periodic table.**

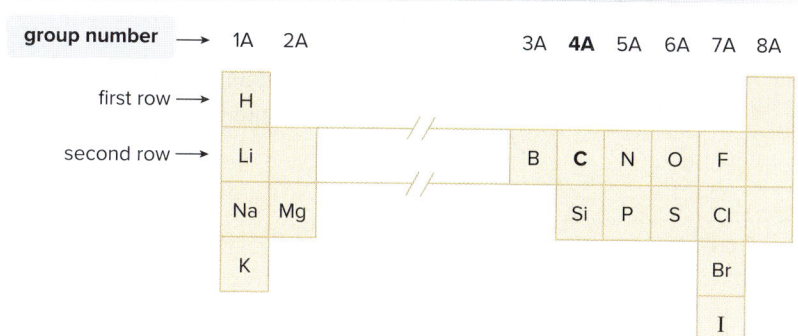

Figure 1.1 A periodic table of the common elements seen in organic chemistry

- Carbon is located in the second row, group **4A**.

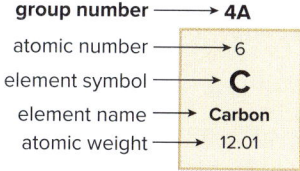

Carbon's entry in the periodic table:
- group number → 4A
- atomic number → 6
- element symbol → C
- element name → Carbon
- atomic weight → 12.01

Across each row of the periodic table, electrons are added to a particular shell of orbitals around the nucleus. Adding electrons to the first shell forms the first row. Adding electrons to the second shell forms the second row. **Electrons are first added to the shells closest to the nucleus.**

Each shell contains a certain number of **orbitals.** An orbital is a region of space that is high in electron density. There are four different kinds of orbitals, called *s, p, d,* and *f.* The first shell has only one orbital, an *s* orbital. The second shell has two kinds of orbitals, *s* and *p,* and so on. Each type of orbital has a particular shape.

For the first- and second-row elements, we must consider only *s* **orbitals** and *p* **orbitals.**

- An *s* **orbital** has a **sphere of electron density.** It is *lower in energy* than other orbitals of the same shell, because electrons are kept closer to the positively charged nucleus.
- A *p* **orbital** has a **dumbbell shape.** It contains a **node of electron density** at the nucleus. A node means there is *no* electron density in this region. A *p* orbital is *higher in energy* than an *s* orbital (in the same shell) because its electron density is farther away from the nucleus.

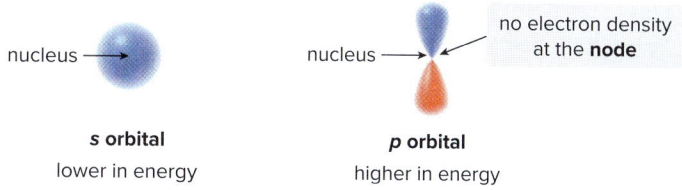

An *s* orbital is filled with electrons before a *p* orbital in the same shell.

1.1A The First Row

The first row of the periodic table is formed by adding electrons to the only orbital in the first shell, called the **1s orbital.**

> • Each orbital can have a maximum of two electrons.

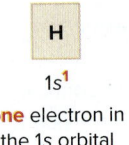

one electron in
the 1s orbital

As a result, there are **two elements in the first row,** one having one electron added to the 1s orbital and one having two. The element **hydrogen (H)** has what is called a $1s^1$ configuration with one electron in the 1s orbital, and **helium (He)** has a $1s^2$ configuration with two electrons in the 1s orbital.

1.1B The Second Row

Every element in the second row has a filled first shell of electrons. Thus, all second-row elements have a $1s^2$ configuration. Each element in the second row of the periodic table also has four orbitals available to accept additional electrons:

- **one 2s orbital,** the s orbital in the second shell
- **three 2p orbitals,** all dumbbell-shaped and perpendicular to each other along the x, y, and z axes

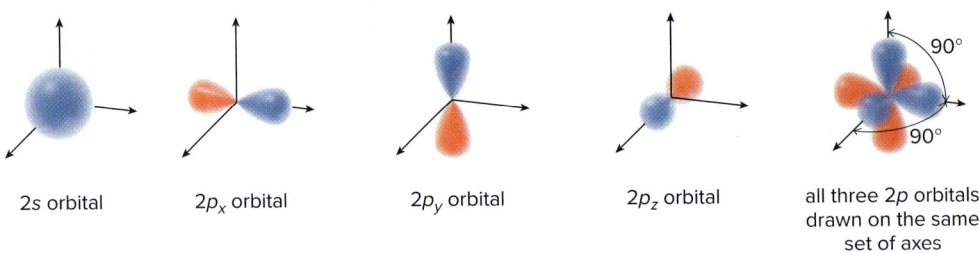

Because each of the four orbitals in the second shell can hold two electrons, there is a **maximum capacity of *eight* electrons** for elements in the second row. The second row of the periodic table consists of eight elements, obtained by adding electrons to the 2s and three 2p orbitals.

The outermost electrons are called **valence electrons.** The valence electrons are more loosely held than the electrons closer to the nucleus, and as such, they participate in chemical reactions. **The group number of a second-row element reveals its number of valence electrons.** For example, carbon in group **4**A has **four** valence electrons, and oxygen in group **6**A has **six.**

Problem 1.1 While the most common isotope of nitrogen has a mass number of 14 (nitrogen-14), a radioactive isotope of nitrogen has a mass number of 13 (nitrogen-13). Nitrogen-13 is used in PET (positron emission tomography) scans by physicians to monitor brain activity and diagnose dementia. For each isotope, give the following information: (a) the number of protons; (b) the number of neutrons; (c) the number of electrons in the neutral atom; (d) the group number; and (e) the number of valence electrons.

1.2 Bonding

Until now our discussion has centered on individual atoms, but it is more common in nature to find two or more atoms joined together.

> • *Bonding* is the joining of two atoms in a stable arrangement.

1.2 Bonding

Joining two or more elements forms **compounds.** Examples of compounds include hydrogen gas (H_2), formed by joining two hydrogen atoms, and methane (CH_4), the simplest organic compound, formed by joining a carbon atom with four hydrogen atoms.

One general rule governs the bonding process.

- Through bonding, atoms attain a complete outer shell of valence electrons.

Because the noble gases in group 8A of the periodic table are especially stable as atoms having a filled shell of valence electrons, the general rule can be restated.

- Through bonding, atoms gain, lose, or share electrons to attain the electronic configuration of the noble gas closest to them in the periodic table.

What does this mean for first- and second-row elements? **A first-row element like hydrogen can accommodate** *two electrons* **around it.** This would make it like the noble gas helium at the end of the same row. **A second-row element is generally most stable with** *eight valence electrons* **around it** like neon. Elements that behave in this manner are said to follow the **octet rule.**

There are two different kinds of bonding: **ionic bonding** and **covalent bonding.**

- *Ionic bonds* result from the *transfer* of electrons from one element to another.
- *Covalent bonds* result from the *sharing* of electrons between two nuclei.

Atoms readily form ionic bonds when they can attain a noble gas configuration by gaining or losing just one or two electrons. NaCl and KI are ionic compounds. *Jill Braaten*

The type of bonding is determined by the location of an element in the periodic table. An **ionic bond** generally occurs when elements on the **far left** side of the periodic table combine with elements on the **far right** side, ignoring the noble gases, which form bonds only rarely. **The resulting ions are held together by extremely strong electrostatic interactions.** A positively charged **cation** formed from the element on the left side attracts a negatively charged **anion** formed from the element on the right side. Examples of ionic inorganic compounds include sodium chloride (NaCl), common table salt, and potassium iodide (KI), an essential nutrient added to make iodized salt.

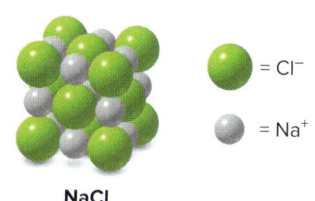

NaCl

Ionic compounds form extended crystal lattices that maximize the positive and negative electrostatic interactions. In NaCl, each positively charged Na^+ ion is surrounded by six negatively charged Cl^- ions, and each Cl^- ion is surrounded by six Na^+ ions.

- The transfer of electrons forms stable salts composed of cations and anions.

A **compound** may have either ionic or covalent bonds. A **molecule** has only covalent bonds.

The second type of bonding, **covalent bonding,** occurs with elements like carbon in the middle of the periodic table, which would otherwise have to gain or lose several electrons to form an ion with a complete valence shell. **A covalent bond is a two-electron bond,** and a compound with covalent bonds is called a **molecule.** Covalent bonds also form between two elements from the same side of the table, such as two hydrogen atoms or two chlorine atoms. H_2, Cl_2, and CH_4 are all examples of covalent molecules.

Problem 1.2 Label each bond in the following compounds as ionic or covalent.
a. F_2 b. LiBr   c. CH_3CH_3 d. $NaNH_2$ e. $NaOCH_3$

How many covalent bonds will a particular atom typically form? As you might expect, it depends on the location of the atom in the periodic table. In the first row, **hydrogen forms one covalent bond** using its one valence electron. When two hydrogen atoms are joined in a bond, each has a filled valence shell of two electrons. **A** *solid line* **indicates a two-electron bond.**

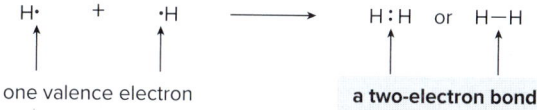

Second-row elements can have no more than eight valence electrons around them. For neutral molecules, two consequences result.

> - **Atoms with one, two, three, or four valence electrons form one, two, three, or four bonds,** respectively, in neutral molecules.
> - **Atoms with five or more valence electrons form enough bonds to give an octet.** In this case, **the predicted number of bonds = 8 − the number of valence electrons.**

For example, B has three valence electrons, so it forms three bonds, as in BF_3. N has five valence electrons, so it also forms three bonds (8 − 5 = 3 bonds), as in NH_3.

These guidelines are used in Figure 1.2 to summarize the usual number of bonds formed by the common atoms in organic compounds. When second-row elements form fewer than four bonds, their octets consist of both **bonding (shared) electrons** and **nonbonding (unshared) electrons.** Unshared electrons are also called **lone pairs.**

> Nonbonded pair of electrons = unshared pair of electrons = lone pair

---

Problem 1.3 How many covalent bonds are predicted for each atom?
a. O   b. Al   c. Br   d. Si

Figure 1.2
The usual number of bonds of common neutral atoms

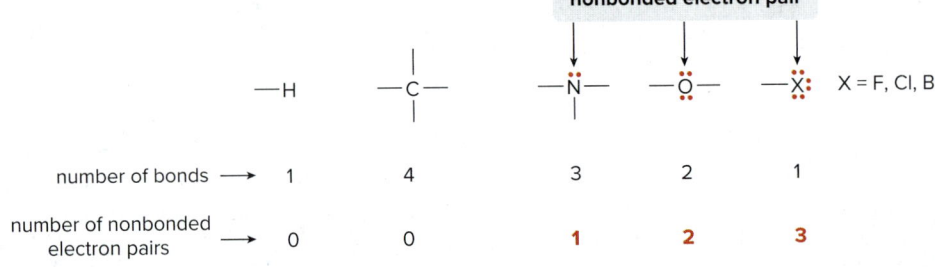

1.3 Lewis Structures

Lewis structures **are electron dot representations for molecules.** Three rules are used for drawing Lewis structures.

> 1. Draw only the valence electrons.
> 2. Give every second-row element no more than *eight* electrons.
> 3. Give each hydrogen *two* electrons.

1.3A A Procedure for Drawing Lewis Structures

Follow a stepwise procedure to draw a Lewis structure.

How To Draw a Lewis Structure

Step [1] Arrange atoms next to each other that you think are bonded together.
- Always place hydrogen atoms and halogen atoms on the periphery because H and X (X = F, Cl, Br, and I) form only one bond each.

> The letter **X** is often used to represent one of the halogens in group 7A: F, Cl, Br, or I.

For CH_3Cl: H C H   *not*   H C H H
                  Cl   Cl
                  H

This H cannot form two bonds.

—Continued

- As a first approximation, use the common bonding patterns in Figure 1.2 to arrange the atoms.

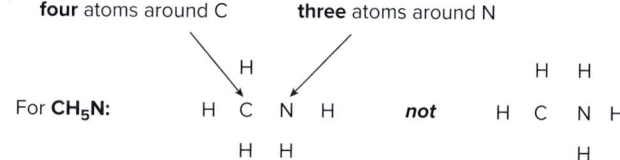

- In truth, the proper arrangement of atoms may not be obvious, or more than one arrangement may be possible (Section 1.4). Even in many simple molecules, the connectivity between atoms must be determined experimentally.

Step [2] **Count the electrons.**
- Count the number of valence electrons from all atoms.
- **Add one electron** for each *negative* charge.
- **Subtract one electron** for each *positive* charge.
- This sum gives the total number of electrons that must be used in drawing the Lewis structure.

Step [3] **Arrange the electrons around the atoms.**
- Place a bond between every two atoms, giving **two electrons to each H** and **no more than eight to any second-row atom.**
- Use all remaining electrons to **fill octets with lone pairs.**
- If all valence electrons are used and an atom does not have an octet, form multiple bonds, as shown in Sample Problem 1.2.

Step [4] **Assign formal charges to all atoms.**
- Formal charges are discussed in Section 1.3C.

Sample Problem 1.1 illustrates how to draw the Lewis structure of a simple organic molecule.

Sample Problem 1.1 Drawing a Lewis Structure for a Simple Molecule

Draw a Lewis structure for methanol, a compound with molecular formula CH_4O.

Solution

Step [1] Arrange the atoms.

```
    H
H   C   O   H
    H
```
- Place the second-row elements, C and O, in the middle.
- Place three H's around C to surround C by four atoms.
- Place one H next to O to surround O by two atoms.

Step [2] Count the electrons.

$$1\,C \times 4\,e^- = 4\,e^-$$
$$1\,O \times 6\,e^- = 6\,e^-$$
$$4\,H \times 1\,e^- = 4\,e^-$$
$$\overline{14\,e^-\text{ total}}$$

Step [3] Add the bonds and lone pairs.

- Add five two-electron bonds to form the C–H, C–O, and O–H bonds, using 10 of the 14 electrons.
- Place two lone pairs on the O atom to use the remaining four electrons and give the O atom an octet.

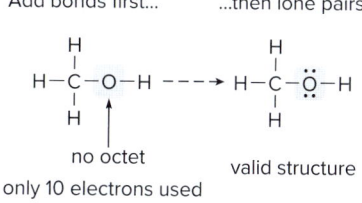

Add bonds first... ...then lone pairs.

no octet
only 10 electrons used

valid structure

This Lewis structure is valid because it uses all 14 electrons, each H is surrounded by two electrons, and each second-row element is surrounded by no more than eight electrons.

Problem 1.4 Draw a valid Lewis structure for each species.

a. CH_3CH_3 b. CH_5N c. C_2H_5Br

More Practice: Try Problem 1.45a.

1.3B Multiple Bonds

Sample Problem 1.2 illustrates an example of a Lewis structure with a double bond.

Sample Problem 1.2 Drawing a Lewis Structure with a Multiple Bond

Draw a Lewis structure for ethylene, a compound of molecular formula C_2H_4, in which each carbon is bonded to two hydrogens.

Solution

Follow Steps [1] to [3] to draw a Lewis structure.

Step [1] Arrange the atoms.

H C C H   • Each C gets 2 H's.
 H H

Step [2] Count the electrons.

$$2\,C \times 4\,e^- = 8\,e^-$$
$$4\,H \times 1\,e^- = 4\,e^-$$
$$\overline{12\,e^-\text{ total}}$$

Step [3] Add the bonds and lone pairs.

Add bonds first...   ...then lone pairs.

H–C–C–H ---→ H–C–C̈–H
 | | | |
 H H   H H
                     (no octet)

After placing five bonds between the atoms and adding the two remaining electrons as a lone pair, one C still has no octet.

To give both C's an octet, **change *one* lone pair into *one* bonding pair of electrons between the two C's, forming a double bond.**

           Move a lone pair.
H–C̈–C̈–H - - - - - → H–C=C–H   ← Each C now has four bonds.
 | | | |
 H H   H H

ethylene
a valid Lewis structure

This uses all 12 electrons, each C has an octet, and each H has two electrons. The Lewis structure is valid. **Ethylene contains a carbon–carbon double bond.**

Problem 1.5 Draw an acceptable Lewis structure for each compound, assuming the atoms are connected as arranged. Formaldehyde (H_2CO) is a preservative, and glycolic acid ($HOCH_2CO_2H$) is used to make dissolving sutures.

a. H_2CO H C O
              H

b. $HOCH_2CO_2H$ H O
                   H O C C O H
                       H

More Practice: Try Problems 1.44, 1.45b–d.

- After placing all electrons in bonds and lone pairs, use a lone pair to form a multiple bond if an atom does not have an octet.

Carbon always forms four bonds in stable organic molecules. Carbon forms single, double, and triple bonds to itself and other elements.

You must change *one* lone pair into *one* new bond for each *two* electrons needed to complete an octet. In acetylene, a compound with molecular formula C_2H_2, placing the 10 valence electrons gives a Lewis structure in which one or both of the C's lack an octet.

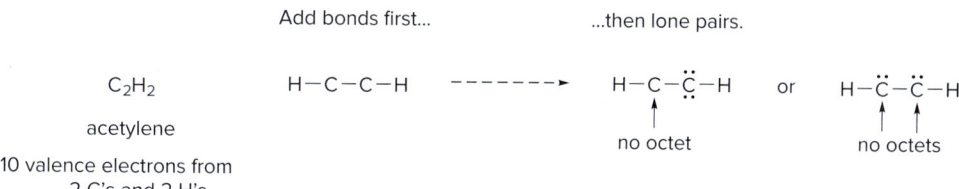

In this case, **change *two* lone pairs into *two* bonding pairs of electrons, forming a triple bond.**

Problem 1.6

Draw an acceptable Lewis structure for each compound, assuming the atoms are connected as arranged.

a. HCN   H C N

b. C_3H_4 H C C C H (with H's on the first C)

1.3C Formal Charge

To manage electron bookkeeping in a Lewis structure, chemists use **formal charge.**

- *Formal charge* is the charge assigned to individual atoms in a Lewis structure.

By calculating formal charge, we determine how the number of electrons around a particular atom compares to its number of valence electrons. Formal charge is calculated as follows:

$$\text{formal charge} = \text{number of valence electrons} - \text{number of electrons an atom "owns"}$$

The number of electrons "owned" by an atom is determined by its number of bonds and lone pairs.

- An atom "owns" *all* of its unshared electrons and *half* of its shared electrons.

$$\text{number of electrons owned} = \text{number of unshared electrons} + \frac{1}{2}\left[\text{number of shared electrons}\right]$$

The number of electrons "owned" by different carbon atoms is indicated in the following examples:

- C shares eight electrons.
- C "owns" **four** electrons.

- Each C shares eight electrons.
- Each C "owns" **four** electrons.

- C shares six electrons.
- C has two unshared electrons.
- C "owns" **five** electrons.

Sample Problem 1.3 illustrates how formal charge is calculated on the atoms of a polyatomic ion. **The sum of the formal charges on the individual atoms equals the net charge on the molecule or ion.**

Sample Problem 1.3 Determining the Formal Charge on an Atom

Determine the formal charge on each atom in the ion H_3O^+.

$$\left[\begin{array}{c} H-\ddot{O}-H \\ | \\ H \end{array} \right]^+$$

Solution

To calculate the formal charge on each atom:

- **Determine the number of valence electrons from the group number.**
- **Determine the number of electrons an atom "owns"** from the number of bonding and nonbonding electrons it has.
- **Subtract** the second quantity from the first to give the formal charge.

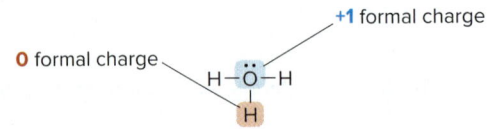

0 formal charge
+1 formal charge

For the O atom (group **6**A):
- number of valence electrons = **6**
- number of bonding electrons = 6
- number of nonbonding electrons = 2

$$\text{formal charge} = 6 - \left[2 + \frac{1}{2}(6) \right]$$
$$= +1$$

For each H atom (group **1**A):
- number of valence electrons = **1**
- number of bonding electrons = 2
- number of nonbonding electrons = 0

$$\text{formal charge} = 1 - \left[0 + \frac{1}{2}(2) \right]$$
$$= 0$$

The formal charge on the O atom is **+1** and the formal charge on each H is **0**. The overall charge on the ion H_3O^+ is the sum of all of the formal charges on the atoms: $1 + 0 + 0 + 0 = +1$.

Problem 1.7 Calculate the formal charge on each second-row atom.

a. $\left[\begin{array}{c} H \\ | \\ H-N-H \\ | \\ H \end{array} \right]^+$
b. $CH_3-N\equiv C:$
c. $:\ddot{O}=\ddot{O}-\ddot{O}:$
d. $CH_3-N=\ddot{O}:$ with $:\ddot{O}:$ below N

More Practice: Try Problems 1.42, 1.43.

Problem 1.8 Draw a Lewis structure for each ion.

a. CH_3O^- b. HC_2^- c. $(CH_3NH_3)^+$ d. $(CH_3NH)^-$

When you first add formal charges to Lewis structures, use the procedure in Sample Problem 1.3. With practice, you will notice that certain bonding patterns always result in the same formal charge. For example, any N atom with four bonds (and thus no lone pairs) has a +1 formal charge. Table 1.1 lists the bonding patterns and resulting formal charges for carbon, nitrogen, and oxygen.

Table 1.1 Formal Charge Observed with Common Bonding Patterns for C, N, and O

Atom	Number of valence electrons	Formal charge +1	Formal charge 0	Formal charge −1
C	4	$-\overset{+}{\underset{|}{\overset{|}{C}}}-$	$-\underset{|}{\overset{|}{C}}-$	$-\underset{|}{\overset{..}{C}}-$
N	5	$-\overset{+}{\underset{|}{\overset{|}{N}}}-$	$-\underset{|}{\overset{..}{N}}-$	$-\underset{..}{\overset{..}{N}}-$
O	6	$-\overset{+}{\underset{..}{\overset{..}{O}}}-$	$-\underset{..}{\overset{..}{O}}-$	$-\underset{..}{\overset{..}{O}}:^-$

Problem 1.9 What is the formal charge on the O atom in each of the following species that contains a multiple bond to O?

a. ≡O: b. =Ö—   c. =Ö:

1.4 Isomers

Sometimes in drawing a Lewis structure, more than one arrangement of atoms is possible for a given molecular formula. For example, there are two acceptable arrangements of atoms for the molecular formula C_2H_6O.

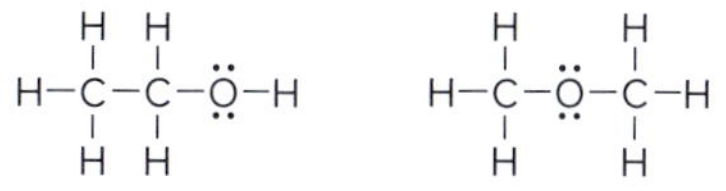

```
    H  H                    H       H
    |  |                    |       |
H — C— C — Ö — H       H — C — Ö — C — H
    |  |                    |       |
    H  H                    H       H

    ethanol              dimethyl ether

           same molecular formula
                  C₂H₆O
                  isomers
```

Both are valid Lewis structures, and both molecules exist. One is called ethanol, and the other, dimethyl ether. These two compounds are called **isomers.**

- *Isomers* are different molecules having the same molecular formula.

Ethanol and dimethyl ether are **constitutional isomers** because they have the same molecular formula, but the *connectivity of their atoms is different.* Ethanol has one C—C bond and one O—H bond, whereas dimethyl ether has two C—O bonds. A second class of isomers, called **stereoisomers,** is introduced in Section 4.13B.

Problem 1.10 Draw Lewis structures for each molecular formula.

a. $C_2H_4Cl_2$ (two isomers)   b. C_3H_8O (three isomers)   c. C_3H_6 (two isomers)

1.5 Exceptions to the Octet Rule

Most of the common elements in organic compounds—**C, N, O, and the halogens**—follow the octet rule. **Hydrogen** is a notable exception, because it accommodates only two electrons in bonding. Additional exceptions include **boron** and **beryllium** (second-row elements in groups 3A and 2A, respectively), and elements in the third row (particularly **phosphorus** and **sulfur**).

Elements in groups 2A and 3A of the periodic table, such as beryllium and boron, do not have enough valence electrons to form an octet in a neutral molecule. Lewis structures for BeH_2 and BF_3 show that these atoms have only four and six electrons, respectively, around the central atom. There simply aren't enough electrons to form an octet. Because the Be and B atoms each have less than an octet of electrons, these molecules are highly reactive.

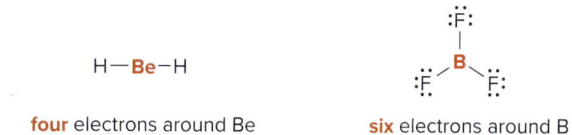

H—Be—H

four electrons around Be   **six** electrons around B

A second exception to the octet rule occurs with some elements located in the third row and later in the periodic table. These elements have empty *d* orbitals available to accept electrons, and thus they may have **more than eight** electrons around them. For organic chemists, the two most common elements in this category are **phosphorus** and **sulfur,** which can have 10 or even 12 electrons around them, as shown in dimethyl sulfoxide (a common solvent), sulfuric acid (a strong inorganic acid), and glyceraldehyde 3-phosphate (an intermediate formed during carbohydrate metabolism).

10 electrons around S   **12** electrons around S   **10** electrons around P

dimethyl sulfoxide
DMSO

sulfuric acid

glyceraldehyde 3-phosphate

1.6 Resonance

Some molecules can't be adequately represented by a single Lewis structure. For example, two valid Lewis structures can be drawn for the anion (HCONH)⁻. One structure has a negatively charged N atom and a C—O double bond; the other has a negatively charged O atom and a C—N double bond. These structures are called **resonance structures** or **resonance forms.** A **double-headed arrow** is used to separate two resonance structures.

double-headed arrow

- **Resonance structures** are two Lewis structures having the *same* placement of atoms but a *different* arrangement of electrons.

Which resonance structure is an accurate representation for (HCONH)⁻? **The answer is *neither* of them.** The true structure is a composite of both resonance forms, and is called a **resonance hybrid.** The hybrid shows characteristics of *both* resonance structures.

Each resonance structure implies that electron pairs are localized in bonds or on atoms. In actuality, resonance allows certain electron pairs to be *delocalized* over two or more atoms, and this delocalization of electron density adds stability. **A molecule with two or more resonance structures is said to be *resonance stabilized.***

1.6A An Introduction to Resonance Theory

Keep in mind the following basic principles of resonance theory.

- Resonance structures are *not* real. An individual resonance structure does not accurately represent the structure of a molecule or ion.
- Resonance structures are *not* in equilibrium with each other. There is no movement of electrons from one form to another.
- Resonance structures are *not* isomers. Two isomers differ in the arrangement of *both* atoms and electrons, whereas resonance structures differ *only* in the *arrangement of electrons.*

For example, ions **A** and **B** are resonance structures because the atom position is the same in both compounds, but the location of an electron pair is different. In contrast, compounds **C** and **D** are isomers because the atom placement is different; **C** has an O—H bond, and **D** has an additional C—H bond.

A   **B**   **C**   **D**

- **A** and **B** are **resonance** structures.
- The position of one electron pair (in red) is different.

- **C** and **D** are **isomers.**
- The position of a H atom (in red) is different.

1.6 Resonance

Problem 1.11 Classify each pair of compounds as isomers or resonance structures.

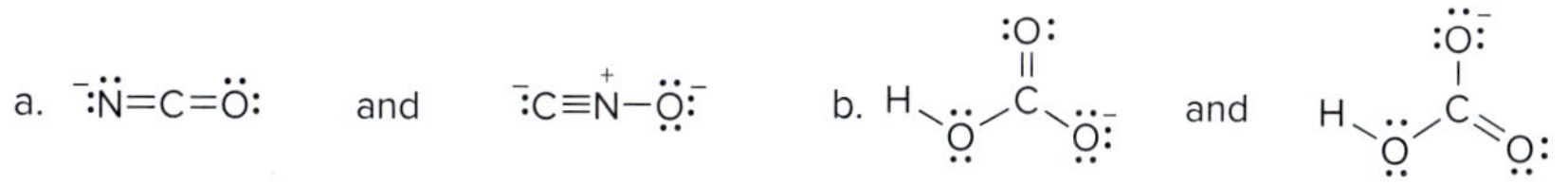

a. :N̈=C=Ö: and :C≡N⁺–Ö:⁻   b. [structure with H, C, O, OH] and [structure with H, C, O, OH]

Problem 1.12 Considering structures **A–D**, classify each pair of compounds as isomers, resonance structures, or neither: (a) **A** and **B**; (b) **A** and **C**; (c) **A** and **D**; (d) **B** and **D**.

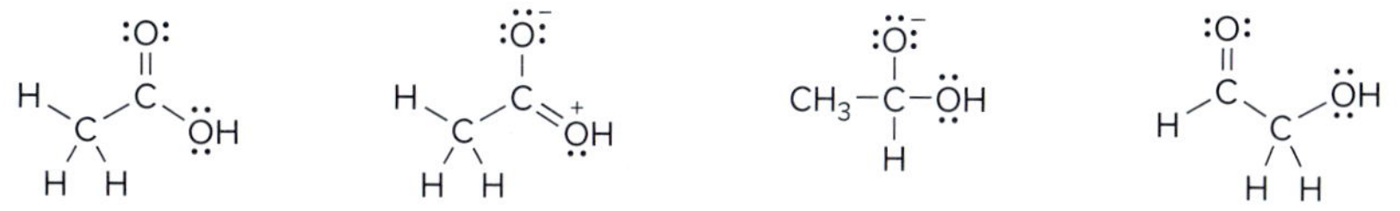

[Structures A, B, C, D shown]

1.6B Drawing Resonance Structures

To draw resonance structures, use three criteria.

Rule [1] Two resonance structures differ in the position of multiple bonds and nonbonded electrons. The placement of atoms and single bonds always stays the same.

[Structures A and B shown]

- The position of a double bond (in blue) is different.
- The position of a lone pair (in red) is different.

Rule [2] Two resonance structures must have the same number of unpaired electrons.

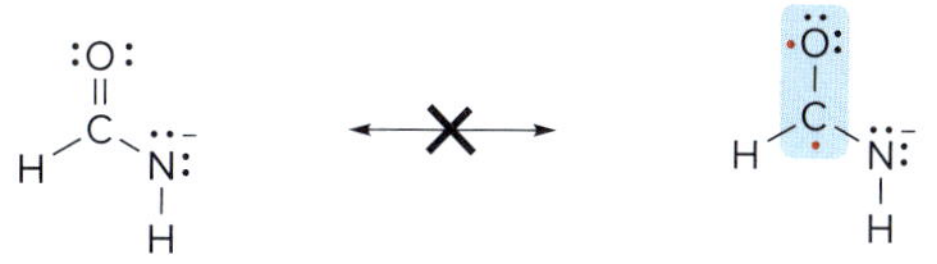

[Structures A and C shown]

A: no unpaired electrons
C: **two** unpaired electrons
not a resonance structure of **A** (or **B**)

Rule [3] Resonance structures must be valid Lewis structures. Hydrogen must have two electrons, and a second-row element can have no more than *eight* electrons.

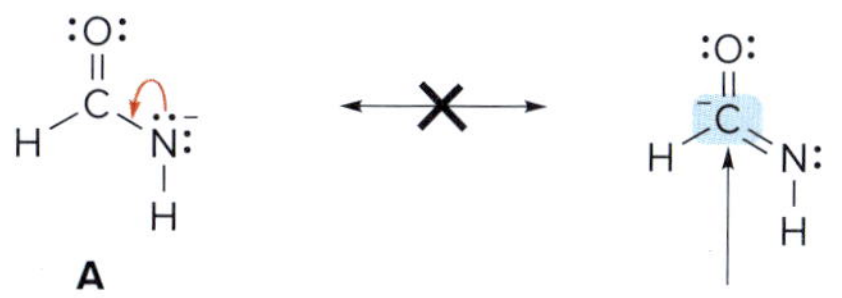

[Structure A shown with arrow, crossed out]

10 electrons around C
not a valid resonance structure of **A**

Curved arrow notation is a convention that shows how electron position differs between the two resonance forms.

- *Curved arrow notation shows the movement of an electron pair.* The tail of the arrow always begins at an electron pair, in either a bond or lone pair. The head points to where the electron pair "moves."

20   Chapter 1   Structure and Bonding

A curved arrow always begins at an electron pair. It ends at an atom or a bond.

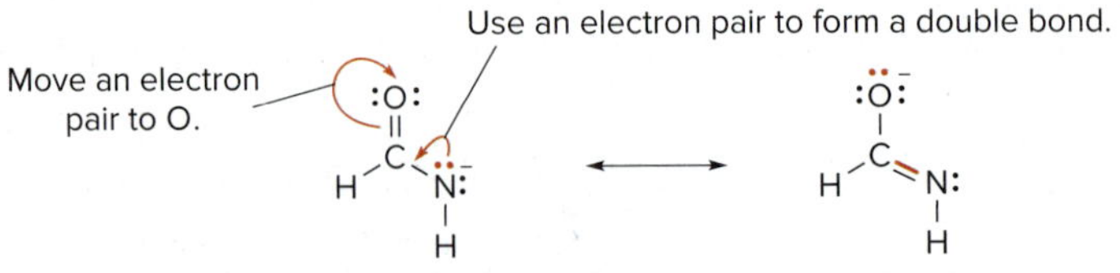

Resonance structures **A** and **B** differ in the location of *two* electron pairs, so *two* curved arrows are needed. To convert **A** to **B**, take the lone pair on N and form a double bond between C and N. Then, move an electron pair in the C–O double bond to form a lone pair on O. Curved arrows thus show how to reposition the electrons in converting one resonance form to another. **The electrons themselves do not actually move.** Sample Problem 1.4 illustrates the use of curved arrows to convert one resonance structure to another.

Sample Problem 1.4   Using Curved Arrows

Follow the curved arrows to draw a second resonance structure for each ion.

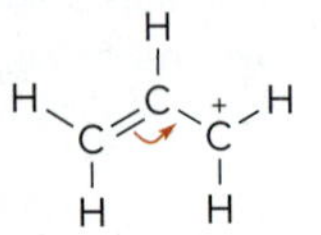

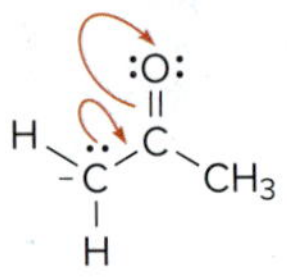

Solution

a. The curved arrow tells us to move **one** electron pair in the double bond to the adjacent C–C bond. Then determine the formal charge on any atom whose bonding is different.

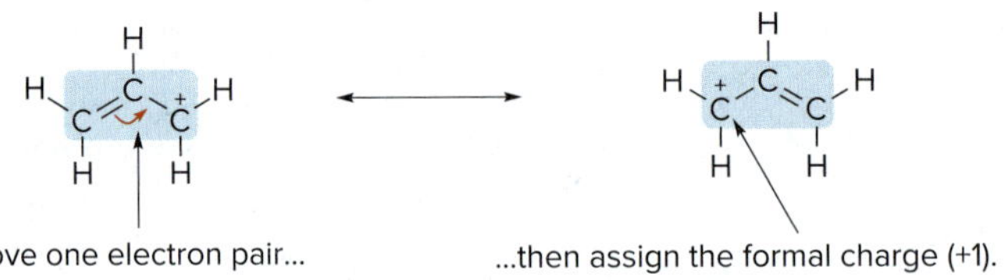

Positively charged carbon atoms are called **carbocations.** Carbocations are unstable intermediates because they contain a carbon atom that is lacking an octet of electrons.

b. **Two** curved arrows tell us to move **two** electron pairs. The second resonance structure has a formal charge of (−1) on O.

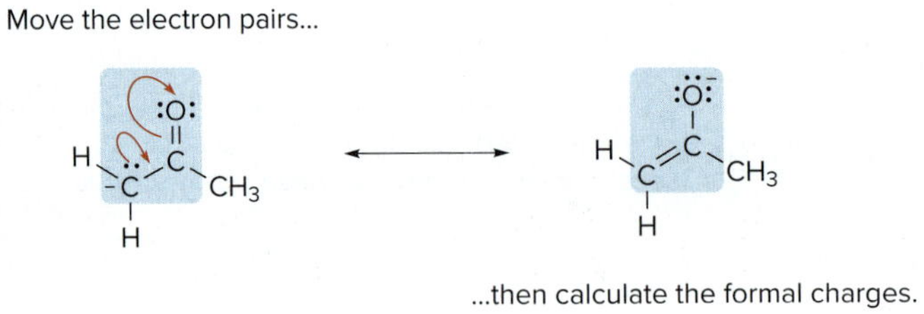

This type of resonance-stabilized anion is called an **enolate anion.** Enolates are important intermediates in many organic reactions, and all of Chapters 17 and 18 is devoted to their preparation and reactions.

Problem 1.13

Follow the curved arrows to draw a second resonance structure for each species.

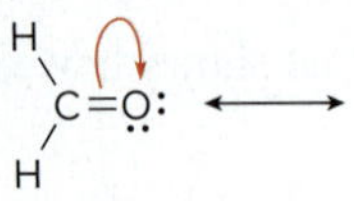

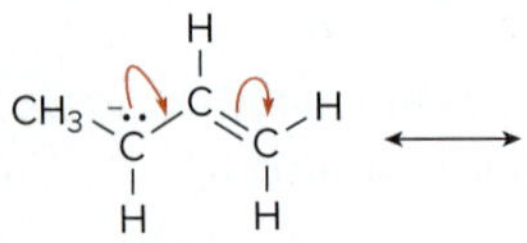

More Practice: Try Problems 1.52, 1.53.

1.6 Resonance

Problem 1.14 Use curved arrow notation to show how the first resonance structure can be converted to the second.

a. [structure: H₂C=C(H)–C⁺(H)(CH₃) with H's ↔ H₂C–C(H)=C(H)(CH₃) carbocation resonance]

b. [structure: carbonate ion resonance structures]

Two resonance structures can have exactly the same kinds of bonds, as they do in the carbocation in Sample Problem 1.4a, or they may have different types of bonds, as they do in the enolate in Sample Problem 1.4b. Either possibility is fine as long as the individual resonance structures are valid Lewis structures.

A resonance structure can have an atom with *fewer* than eight electrons around it. **B** is a resonance structure of **A** even though the carbon atom is surrounded by only six electrons.

[structures A and B: acetone with resonance showing carbocation form]
only six electrons around C

A   **B**
    valid resonance structure

In contrast, a resonance structure can *never* have a second-row element with more than eight electrons. **C** is *not* a resonance structure of **A** because the carbon atom is now surrounded by 10 electrons.

[structures A and C with X between them]
10 electrons around C

A   **C**
    not a valid resonance structure

We will learn much more about resonance in Chapter 12.

The ability to draw and manipulate resonance structures is a necessary skill that will be used throughout your study of organic chemistry. With practice, you will begin to recognize certain common bonding patterns for which more than one Lewis structure can be drawn. For instance, both the carbocation in Sample Problem 1.4a and the enolate anion in Sample Problem 1.4b are specific examples of one general type of resonance observed in certain three-atom systems.

- **In a group of three atoms having a multiple bond X=Y joined to an atom Z having a *p* orbital with zero, one, or two electrons, two resonance structures can be drawn.**

$$X=Y-Z^* \longleftrightarrow {}^*X-Y=Z$$

0, 1, or 2 electrons

The * corresponds to a charge, a lone pair, or a single electron.

* = +, –, ·, or :

Recall from the Prologue that a **heteroatom** is an atom other than carbon or hydrogen.

X, Y, and Z may all be carbon atoms or they may be **heteroatoms** such as nitrogen and oxygen. The atom Z can be charged (positive or negative) or neutral (with a lone pair or a single electron), corresponding to the [*] in the general structure X=Y–Z*. The two resonance structures differ in the location of the multiple bond and the [*].

In the enolate anion in Sample Problem 1.4b, X corresponds to oxygen and [*] is a lone pair, which gives carbon a net negative charge. Moving the double bond and the lone pair and readjusting charges gives the second resonance structure.

[enolate resonance structures labeled with X, Y, Z]

- The position of the double bond changes.
- The location of a lone pair changes.

In Chapter 12, we will learn more about the orbitals involved in this type of resonance.

Sample Problem 1.5 Drawing Resonance Structures

Draw a second resonance structure for acetamide.

acetamide

Solution

Always look for a three-atom system that contains a multiple bond joined to an atom Z with zero, one, or two nonbonded electrons. **Move the double bond (from X=Y to Y=Z) and move the [*] from Z to X.** Recalculate formal charges on X and Z.

acetamide
- C=O
- lone pair on N

second resonance structure
- C=N
- additional lone pair on O

In this example, the three-atom system for resonance (X=Y–Z*) is O=C–N with a lone pair on N. After moving the double bond and the lone pair, the formal charges on O and N are −1 and +1, respectively, calculated using the procedure for determining formal charges.

Problem 1.15 Draw a second resonance structure for each species in parts (a), (b), and (c). Draw two additional resonance structures for the ion in part (d).

More Practice: Try Problems 1.54, 1.55.

1.6C The Resonance Hybrid

The **resonance hybrid** is the composite of all possible resonance structures. In the resonance hybrid, the electron pairs drawn in different locations in individual resonance structures are *delocalized*.

- **The resonance hybrid is more stable than any resonance structure because it delocalizes electron density over a larger volume.**

What does the hybrid look like? When all resonance forms are identical, as they were in the carbocation in Sample Problem 1.4a, each resonance form contributes **equally** to the hybrid.

When two resonance structures are different, the hybrid looks more like the "better" resonance structure. The "better" resonance structure is called the **major contributor** to the hybrid, and all others are **minor contributors.** The hybrid is the weighted average of the contributing resonance structures. What makes one resonance structure "better" than another? There are many factors, but for now, we will learn one fact.

- **A "better" resonance structure is one that has *more bonds* and *fewer charges*.**

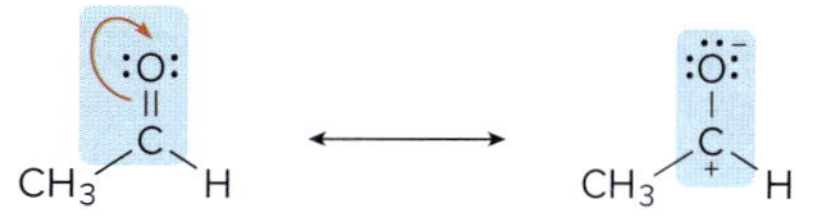

1.6 Resonance 23

X
more bonds
fewer charges
major contributor

Y
minor contributor

Comparing resonance structures **X** and **Y**, **X** is the major contributor because it has more bonds and fewer charges. Thus, the hybrid looks more like **X** than **Y**.

How can we draw a hybrid, which has delocalized electron density? First, we must determine what is different in the resonance structures. Two differences commonly seen are the **position of a multiple bond** and the **site of a charge.** The anion (HCONH)⁻ illustrates two conventions for drawing resonance hybrids.

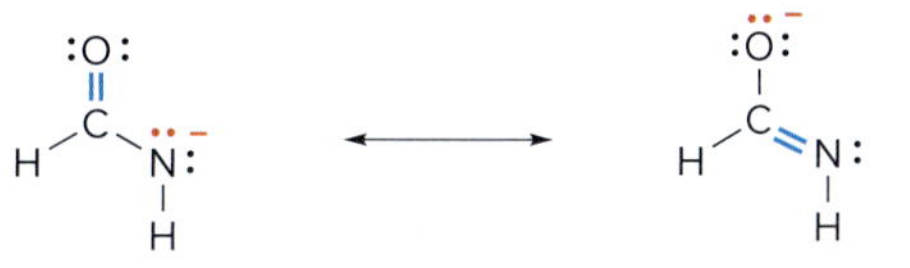

A   B
individual resonance structures

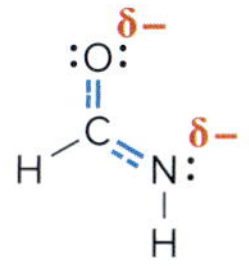

resonance hybrid
• The (–) charge is delocalized on N and O.
• The double bond is delocalized between O, C, and N.

Common symbols and conventions used in organic chemistry are summarized in Appendix B.

- **Double bond position. Use a dashed line for a bond that is single in one resonance structure and double in another.**
- **Location of charge. Use a δ– (partial negative charge) or δ+ (partial positive charge) for an atom that is neutral in one resonance structure and charged in another.**

The hybrid for (HCONH)⁻ shows two dashed bonds, indicating that both the C−O and C−N bonds have partial double bond character. Both the O and N atoms bear a partial negative charge (δ−) because these atoms are neutral in one resonance structure and negatively charged in the other.

This discussion of resonance is meant to serve as an introduction only. You will learn many more facets of resonance theory in later chapters. In Chapter 2, for example, the enormous effect of resonance on acidity is discussed.

Problem 1.16 Label the resonance structures in each pair as major, minor, or equal contributors to the hybrid. Then draw the hybrid.

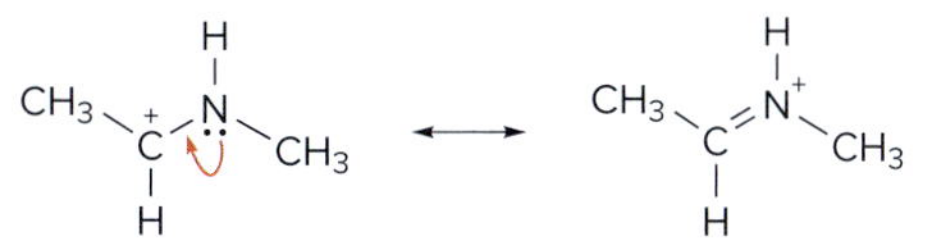

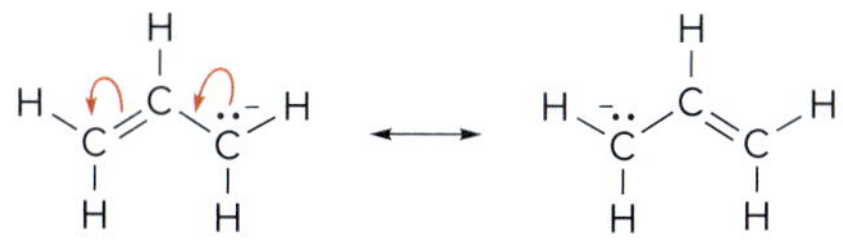

Problem 1.17 (a) Draw a second resonance structure for **A**. (b) Why can't a second resonance structure be drawn for **B**?

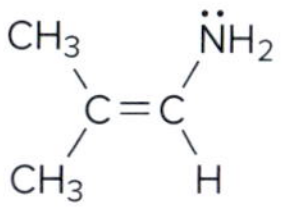

A 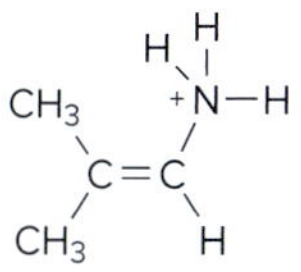   B

1.7 Determining Molecular Shape

Consider the H_2O molecule. The Lewis structure tells us which atoms are connected to each other, but it implies nothing about the geometry. What does the overall molecule look like? Is H_2O a bent or linear molecule? Two variables define a molecule's structure: **bond length** and **bond angle**.

1.7A Bond Length

Although the SI unit for bond length is the picometer (pm), the angstrom (Å) is still widely used in the chemical literature; $1 \text{ Å} = 10^{-10}$ m. As a result, $1 \text{ pm} = 10^{-2}$ Å, and $95.8 \text{ pm} = 0.958$ Å.

Bond length is the average distance between the centers of two bonded nuclei. Bond lengths are typically reported in picometers (pm), where $1 \text{ pm} = 10^{-12}$ m. For example, the O–H bond length in H_2O is 95.8 pm. Average bond lengths for common bonds are listed in Table 1.2.

- Bond length *decreases* across a row of the periodic table as the size of the atom *decreases*.

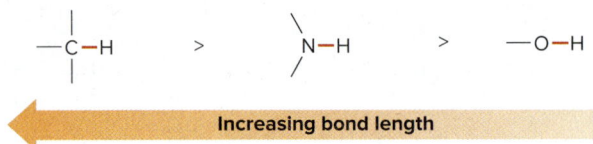

- Bond length *increases* down a column of the periodic table as the size of an atom *increases*.

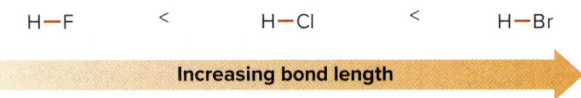

Table 1.2 Average Bond Lengths

Bond	Length (pm)	Bond	Length (pm)	Bond	Length (pm)
H–H	74	H–F	92	C–F	133
C–H	109	H–Cl	127	C–Cl	177
N–H	101	H–Br	141	C–Br	194
O–H	96	H–I	161	C–I	213

1.7B Bond Angle

Bond angle determines the shape around any atom bonded to two other atoms. To determine the bond angle and shape around a given atom, first count how many groups surround the atom. **A group is either an atom or a lone pair of electrons.** Then use the **valence shell electron pair repulsion (VSEPR) theory** to determine the shape. VSEPR is based on the fact that electron pairs repel each other; thus:

- The most stable arrangement keeps the groups around an atom as far away from each other as possible.

A second-row element has only three possible arrangements, defined by the number of groups surrounding it.

To determine geometry:
[1] Draw a valid Lewis structure; [2] count groups around a given atom.

Number of groups	Geometry	Bond angle
• **two** groups	**linear**	180°
• **three** groups	**trigonal planar**	120°
• **four** groups	**tetrahedral**	109.5°

Let's examine several molecules to illustrate this phenomenon. We first need a valid Lewis structure, and then we count groups around a given atom to predict its geometry.

Two Groups Around an Atom

Any atom surrounded by only two groups is linear and has a bond angle of 180°. For example, each carbon atom in **HC≡CH** (acetylene) is surrounded by two atoms and no lone pairs, so each H—C—C bond angle in acetylene is 180°. Therefore all four atoms in HC≡CH are linear.

$$H-C\equiv C-H$$

two atoms around each C

ball-and-stick model

two groups
linear carbons

Acetylene illustrates an important feature: *ignore multiple bonds in predicting geometry.* **Count only atoms and lone pairs.**

We will represent molecules with models having balls for atoms and sticks for bonds, as in the ball-and-stick model of acetylene just shown. These representations are analogous to a set of molecular models. Balls are color-coded using accepted conventions: carbon (black), hydrogen (white or gray), oxygen (red), and so forth, as shown.

> Most students in organic chemistry find that building models helps them visualize the shape of molecules. Invest in a set of models *now*.
>
> Common element colors are also shown in Appendix B.

C H O N F Cl Br I S P

Three Groups Around an Atom

Any atom surrounded by three groups is trigonal planar and has bond angles of 120°. For example, each carbon atom in **CH₂=CH₂** (ethylene) is surrounded by three atoms and no lone pairs, making *each* H—C—C bond angle 120°. All six atoms of CH₂=CH₂ lie in one plane.

three atoms around each C

three groups
trigonal planar carbons

ethylene

Four Groups Around an Atom

Any atom surrounded by four groups is tetrahedral and has bond angles of approximately 109.5°. The simple organic compound methane, **CH₄**, has a central carbon atom with bonds to four hydrogen atoms, each pointing to a corner of a tetrahedron. This arrangement keeps four groups farther apart than a square planar arrangement in which all bond angles would be only 90°.

preferred geometry
larger H—C—H bond angle

square planar arrangement
This geometry does *not* occur.

four groups
tetrahedral molecule

How can we represent the three-dimensional geometry of a tetrahedron on a two-dimensional piece of paper? **Place two of the bonds in the plane of the paper, one bond in front and one bond behind,** using the following conventions:

- A *solid line* is used for a bond *in* the plane.
- A *wedge* is used for a bond *in front* of the plane.
- A *dashed wedge* is used for a bond *behind* the plane.

This is just one way to draw a tetrahedron for CH₄. We can turn the molecule in many different ways, generating many equivalent representations. All of the following are acceptable drawings for CH₄, because each drawing has two solid lines, one wedge, and one dashed wedge.

Finally, **wedges and dashed wedges are used for groups that are really *aligned one behind another*.** It does not matter in the following two drawings whether the wedge or dashed wedge is skewed to the left or right, because the two H atoms are really aligned as shown in the three-dimensional model.

All carbons in stable molecules are **tetravalent,** but the geometry varies with the number of groups around the particular carbon.

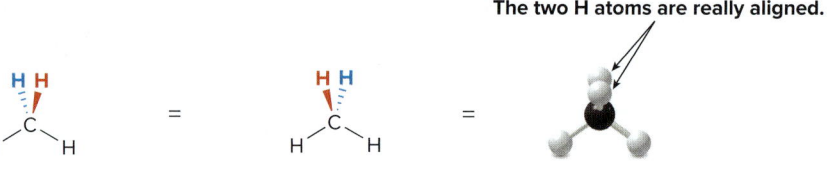

- These representations are equivalent.
- The wedge can be skewed to the left or the right of the dashed wedge.

Ammonia (**NH₃**) and water (**H₂O**) both have atoms surrounded by four groups, some of which are lone pairs. In **NH₃**, the three H atoms and one lone pair around N point to the corners of a tetrahedron. The H–N–H bond angle of 107° is close to the theoretical tetrahedral bond angle of 109.5°. This molecular shape is referred to as **trigonal pyramidal,** because one of the groups around the N is a nonbonded electron pair, not another atom.

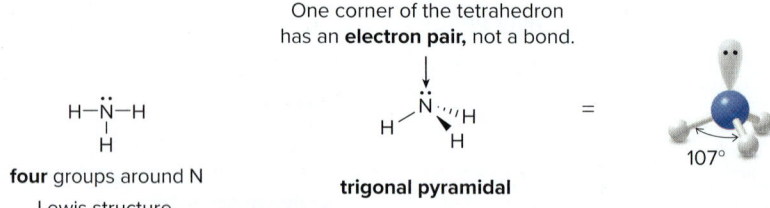

In **H₂O**, the two H atoms and two lone pairs around O point to the corners of a tetrahedron. The H–O–H bond angle of 105° is close to the theoretical tetrahedral bond angle of 109.5°.

Water has a **bent** molecular shape, because two of the groups around oxygen are lone pairs of electrons.

H–Ö–H
four groups around O
Lewis structure

Two corners of the tetrahedron have **electron pairs,** not bonds.

a bent molecule

105°

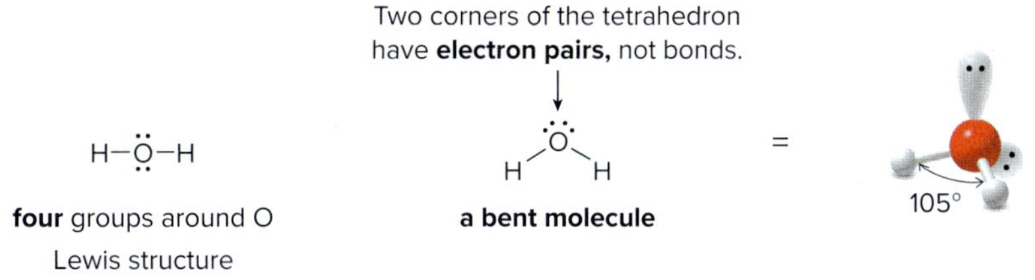

In both NH$_3$ and H$_2$O, the bond angle is somewhat smaller than the theoretical tetrahedral bond angle because of repulsion of the lone pairs of electrons. The bonded atoms are compressed into a smaller space with a smaller bond angle.

Predicting geometry based on counting groups is summarized in Table 1.3.

Table 1.3 Summary: Determining Geometry Based on the Number of Groups

Number of groups around an atom	Geometry	Bond angle	Examples
2	linear	180°	HC≡CH
3	trigonal planar	120°	CH$_2$=CH$_2$
4	tetrahedral	109.5°	CH$_4$, NH$_3$, H$_2$O

Sample Problem 1.6 Determining the Geometry Around a Second-Row Atom

Determine the geometry around the highlighted atom in each species.

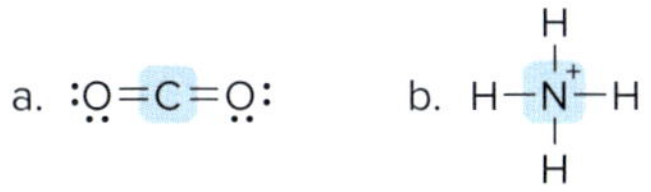

a. :Ö=C=Ö: b. H–N$^+$H$_3$ (H–N–H with H above and below)

Solution

a.
:Ö=C=Ö:
180°
two atoms around C
no lone pairs
two groups
linear

b.
H–N$^+$–H (with H top and bottom) ⟶ tetrahedral structure, 109.5°
four atoms around N
no lone pairs
four groups ⟶ **tetrahedral**

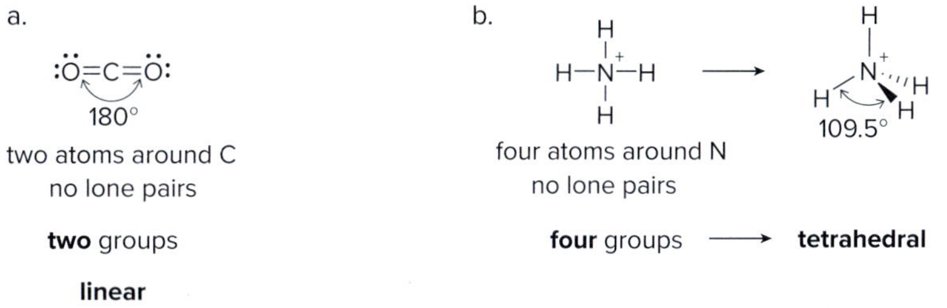

Problem 1.18 Determine the geometry around all second-row elements in each compound drawn as a Lewis structure with no implied geometry.

a. CH$_3$–C(=O:)–CH$_3$ b. CH$_3$–Ö–CH$_3$ c. ⁻:NH$_2$ d. CH$_3$–C≡N:

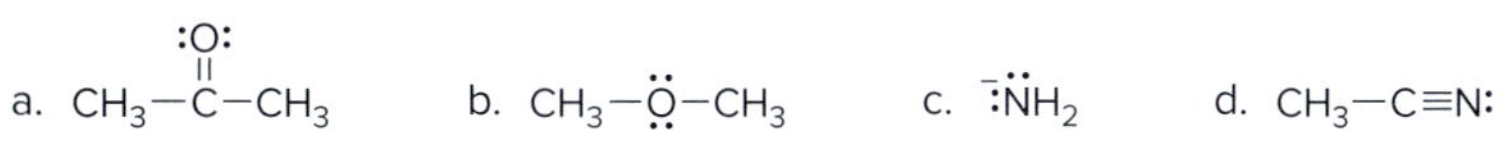

More Practice: Try Problems 1.60, 1.61, 1.77c, 1.79b.

Problem 1.19 Predict the indicated bond angles in each compound drawn as a Lewis structure with no implied geometry.

a. CH$_3$–C≡C–Cl: b. CH$_2$=C(H)–Cl: c. CH$_3$–C(H)(H)–Cl:

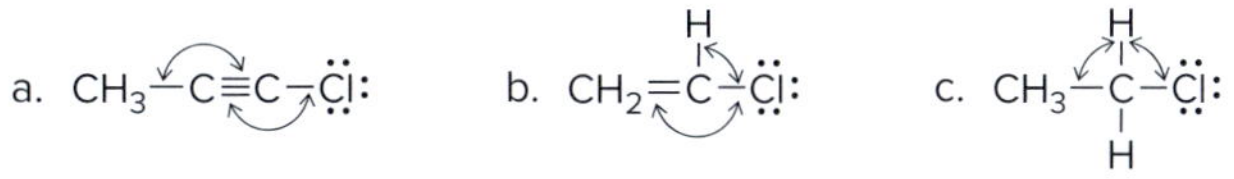

Water hemlock, which grows in wet marshy areas in the western part of North America, is the source of cicutoxin (Problem 1.20), a convulsant toxic to both livestock and humans. *Steven P. Lynch*

Problem 1.20 Using the principles of VSEPR theory, you can predict the geometry around any atom in any molecule, no matter how complex. Cicutoxin is a poisonous compound isolated from water hemlock, a highly toxic plant that grows in temperate regions in North America. Predict the geometry around the highlighted atoms in cicutoxin.

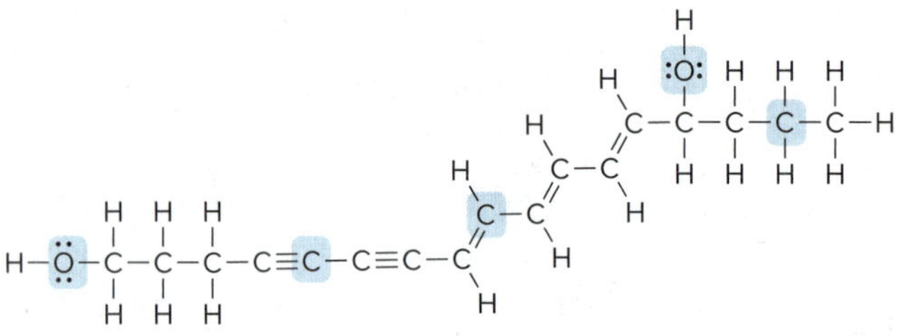

cicutoxin

1.8 Drawing Organic Structures

Drawing organic molecules presents a special challenge. Because they often contain many atoms, we need shorthand methods to simplify their structures. The two main types of shorthand representations used for organic compounds are **condensed structures** and **skeletal structures.**

1.8A Condensed Structures

Condensed structures can be used for compounds having a chain of atoms bonded together. The following conventions are used:

- All of the atoms are drawn in, but the two-electron bond lines are generally omitted.
- Atoms are usually drawn next to the atoms to which they are bonded.
- Parentheses are used around similar groups bonded to the same atom.
- Lone pairs are omitted.

To interpret a condensed formula, it is usually best to start at the *left side* of the molecule and remember that the ***carbon atoms must be tetravalent.*** A carbon bonded to three H atoms becomes **CH₃**; a carbon bonded to two H atoms becomes **CH₂**; and a carbon bonded to one H atom becomes **CH.**

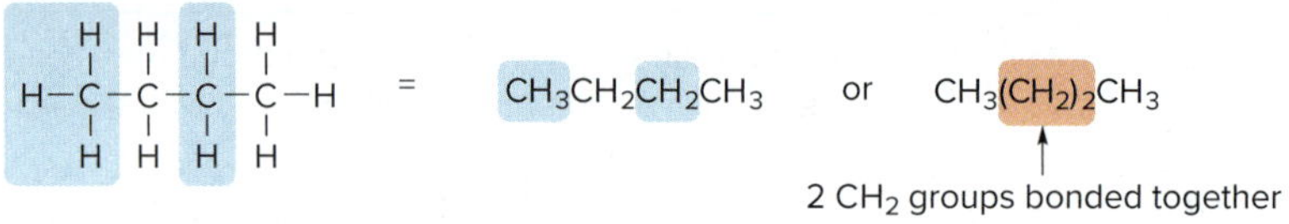

$CH_3CH_2CH_2CH_3$ or $CH_3(CH_2)_2CH_3$

↑
2 CH₂ groups bonded together

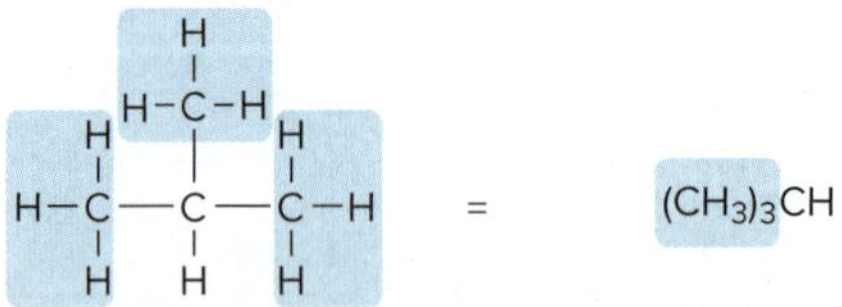

= $(CH_3)_3CH$

Other examples of condensed structures with heteroatoms and carbon–carbon multiple bonds are given in Figure 1.3.

1.8 Drawing Organic Structures

Figure 1.3 Examples of condensed structures

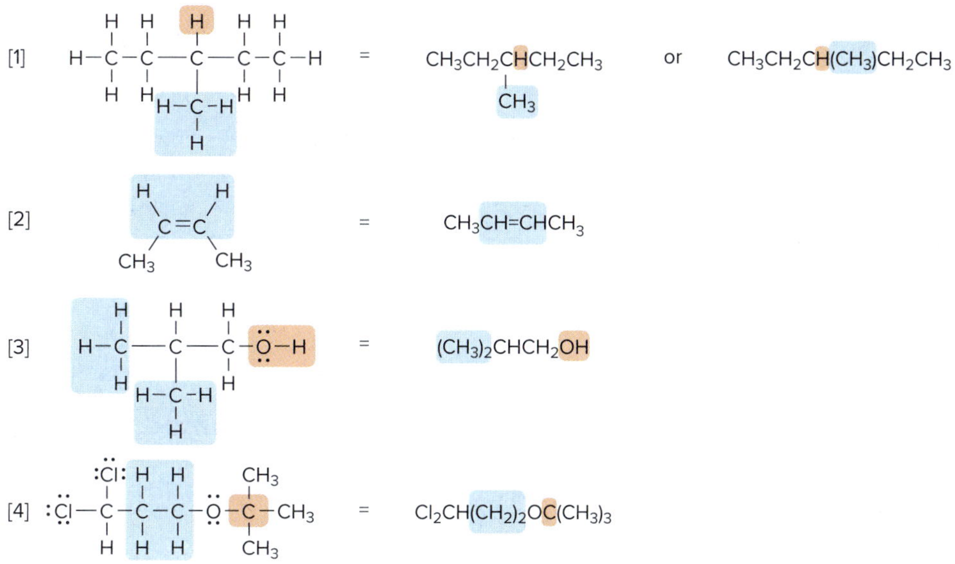

[1] H−C(H)(H)−C(H)(H)−C(H)(CH₃)−C(H)(H)−C(H)(H)−H = CH₃CH₂CHCH₂CH₃ or CH₃CH₂CH(CH₃)CH₂CH₃
                                                              |
                                                             CH₃

[2] (CH₃)(H)C=C(H)(CH₃) = CH₃CH=CHCH₃

[3] H−C(H)(H)−C(H)(CH₃)−C(H)(H)−Ö−H = (CH₃)₂CHCH₂OH

[4] :Cl̈−C(Cl̈)(H)−C(H)(H)−C(H)(H)−Ö−C(CH₃)(CH₃)−CH₃ = Cl₂CH(CH₂)₂OC(CH₃)₃

- Entry [1]: Draw the H atom next to the C to which it is bonded, and use parentheses around CH₃ to show it is bonded to the carbon chain.
- Entry [2]: Keep the carbon–carbon double bond and draw the H atoms after each C to which they are bonded.
- Entry [3]: Omit the lone pairs on the O atom in the condensed structure.
- Entry [4]: Omit the lone pairs on Cl and O and draw the two CH₂ groups as (CH₂)₂.

Translating some condensed formulas is not obvious, and it will come only with practice. This is especially true for compounds containing a carbon–oxygen double bond. Some noteworthy examples in this category are given in Figure 1.4. Whereas carbon–carbon double bonds are generally drawn in condensed structures, carbon–oxygen double bonds are usually omitted.

Figure 1.4 Condensed structures containing a C–O double bond

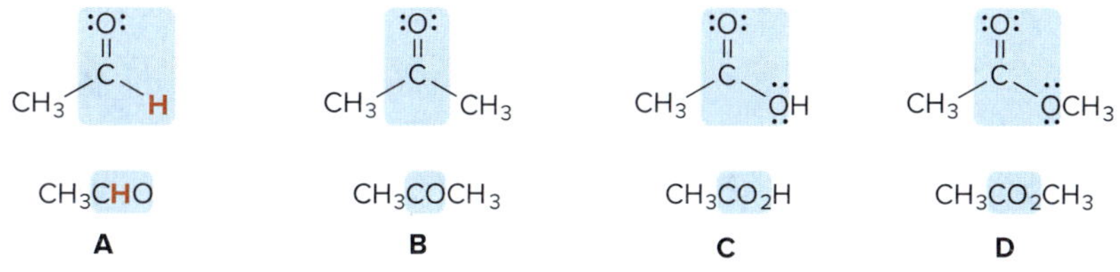

A	B	C	D
CH₃CHO	CH₃COCH₃	CH₃CO₂H	CH₃CO₂CH₃

- In **A**, the **H** atom is bonded to C, *not* O.
- In **B**, each CH₃ group is bonded to C, *not* O.
- In **C** and **D**, the C atom is doubly bonded to one O and singly bonded to the other O.

Sample Problem 1.7 Converting a Condensed Structure to a Lewis Structure

Convert each condensed formula to a Lewis structure.

a. (CH₃)₂CHOCH₂CH₂CH₂OH   b. CH₃(CH₂)₂CO₂C(CH₃)₃

Solution
Start at the left and proceed to the right, making sure that each carbon has four bonds. Give each O atom two lone pairs to have an octet.

a. $(CH_3)_2CHOCH_2CH_2CH_2OH$

b. $CH_3(CH_2)_2CO_2C(CH_3)_3$

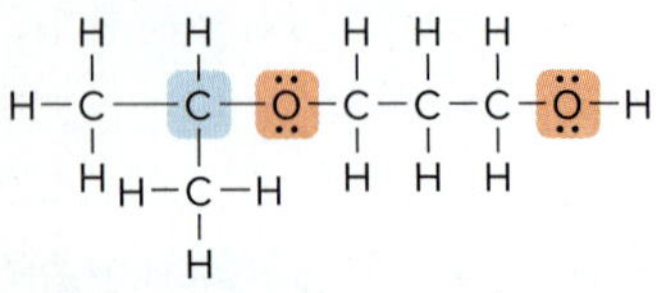

One C atom (labeled in blue) is bonded to 2 CH_3's, 1 H, and 1 O.

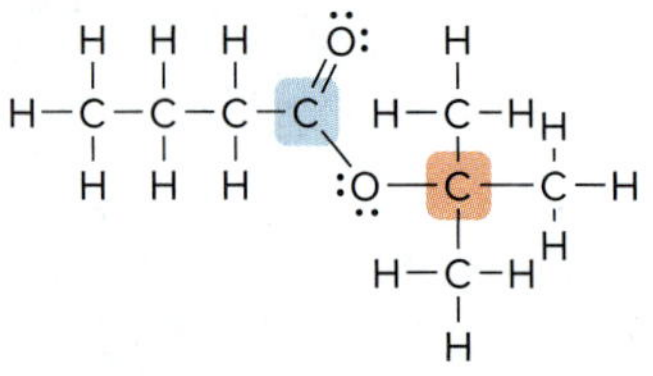

One C atom (labeled in blue) is bonded to both O's.

In part (a), the O atom is singly bonded to two C's, whereas in part (b), a C=O is needed to give each C and O an octet.

Problem 1.21 Convert each condensed formula to a Lewis structure.

a. $CH_3(CH_2)_4CH(CH_3)_2$
b. $(CH_3)_3CCH(OH)CH_2CH_3$
c. $(CH_3)_2CHCHO$
d. $(HOCH_2)_2CH(CH_2)_3C(CH_3)_2CH_2CH_3$

More Practice: Try Problems 1.64a–c, 1.65.

Problem 1.22 During periods of strenuous exercise, the buildup of lactic acid [$CH_3CH(OH)CO_2H$] causes the aching feeling in sore muscles. Convert this condensed structure to a Lewis structure of lactic acid.

1.8B Skeletal Structures

Skeletal structures are used for organic compounds containing both rings and chains of atoms. Three rules are used to draw them.

- Assume a carbon atom is located at the junction of any two lines or at the end of any line.
- Assume each carbon has enough hydrogens to make it tetravalent.
- Draw in all heteroatoms and the hydrogens directly bonded to them.

Carbon chains are drawn in a **zigzag** fashion, and rings are drawn as **polygons,** as shown for hexane and cyclohexane.

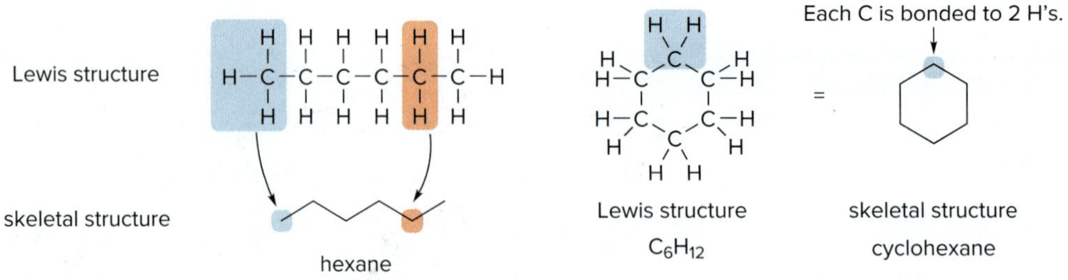

Each C is bonded to 2 H's.

Lewis structure

skeletal structure
hexane

Lewis structure
C_6H_{12}

skeletal structure
cyclohexane

How To Interpret a Skeletal Structure

Example Draw in all C atoms, H atoms, and lone pairs in the following molecule:

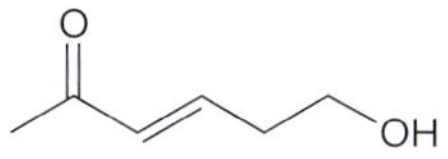

Step [1] Place a C atom at the intersection of any two lines and at the end of any line.

- This molecule has six carbons, including the C labeled in red at the left end of the chain.
- There are two C's (labeled in green) between the C=C and the OH group.

Step [2] Add enough H's to make each C tetravalent.

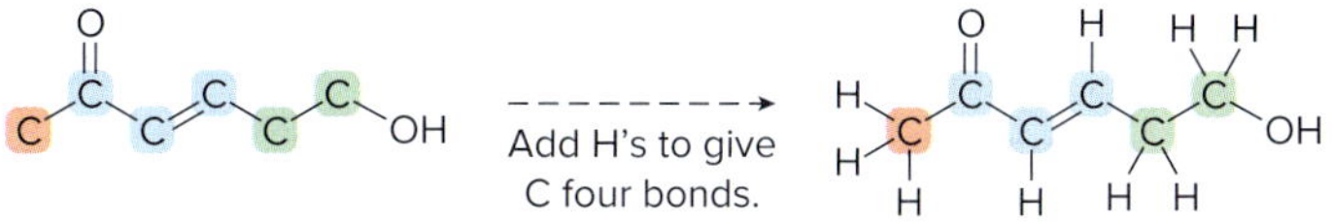

Add H's to give C four bonds.

- The end C labeled in red needs three H's to be tetravalent.
- Each C on the C=C has three bonds already, so only one H must be drawn.
- There are two CH$_2$ groups between the C=C and the OH group.

Step [3] Add lone pairs to give each heteroatom an octet.

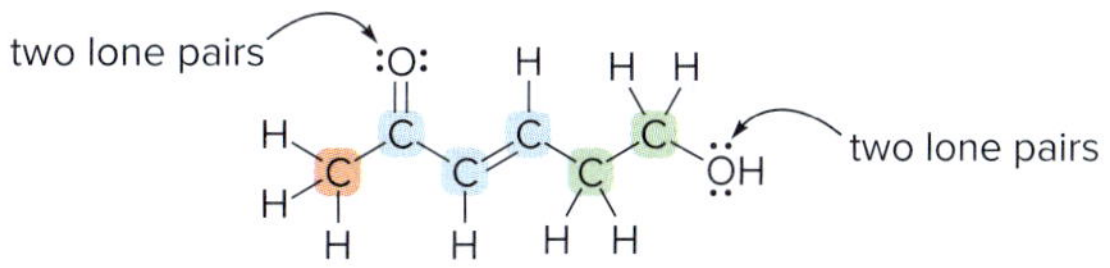

two lone pairs

two lone pairs

Each O needs **two** lone pairs for an octet.

Figure 1.5 shows other examples of skeletal structures, and Sample Problem 1.8 illustrates how to interpret the skeletal structure for a more complex cyclic compound.

Figure 1.5
Interpreting skeletal structures

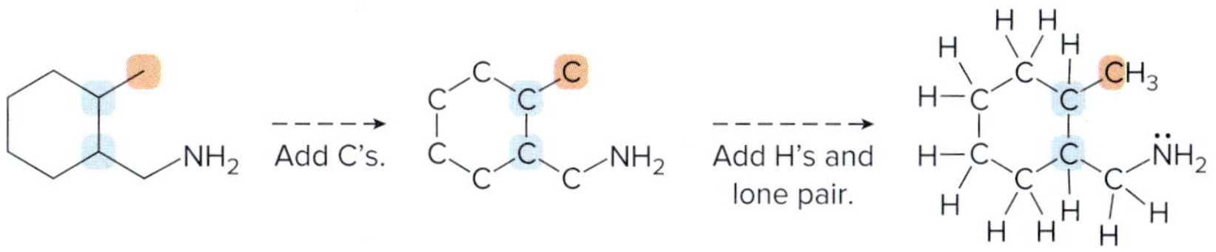

Add C's.

Add H's and lone pair.

- C labeled in red needs 3 H's.
- C's labeled in blue need 1 H.
- All other C's need 2 H's.
- N needs one lone pair.

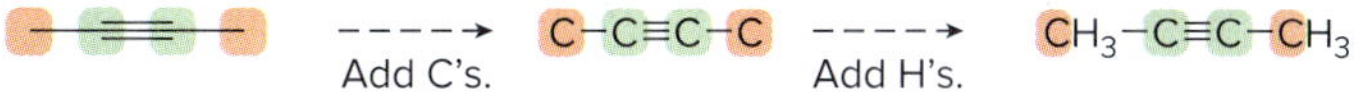

Add C's.   C—C≡C—C   Add H's.   CH$_3$—C≡C—CH$_3$

- C's labeled in red need 3 H's.
- C's labeled in green have four bonds to C.

Chapter 1 Structure and Bonding

Sample Problem 1.8 — Converting a Skeletal Structure to a Lewis Structure

Draw a complete structure for vanillin showing all C atoms, H atoms, and lone pairs, and give the molecular formula. Vanillin is the principal component of the extract of the vanilla bean.

vanilla bean
Vast natalia/Alamy Stock Photo

Solution
- Skeletal structures have a C atom at the junction of any two lines and at the end of any line.
- Each C must have enough H's to make it tetravalent.
- Each O atom needs two lone pairs to have a complete octet.

- C in blue has three H's.
- C in green has four bonds to other C's.
- C in red is doubly bonded to O.

$C_8H_8O_3$

Problem 1.23
How many hydrogen atoms are present around each highlighted carbon atom in the following molecules? What is the molecular formula for each molecule? 2'-Deoxyguanosine is a component of DNA, and riboflavin (vitamin B_2) is a yellow, water-soluble vitamin obtained in the diet from leafy greens, soybeans, almonds, and liver.

a. 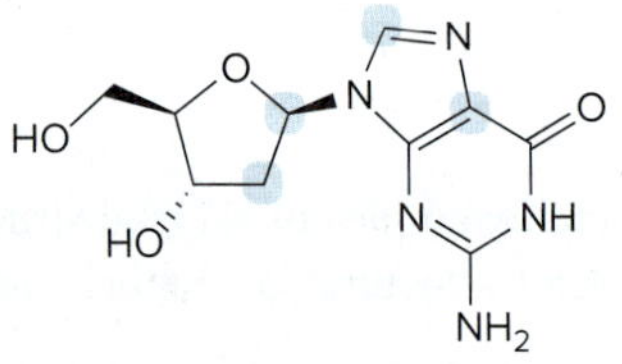 2'-deoxyguanosine
DNA component

b. 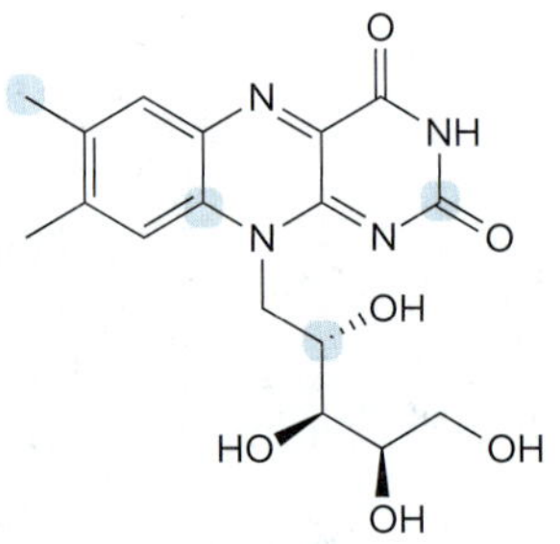 riboflavin
vitamin B_2

More Practice: Try Problems 1.62, 1.63, 1.80a.

Problem 1.24
Convert each skeletal structure to a complete structure with all C's, H's, and lone pairs drawn in.

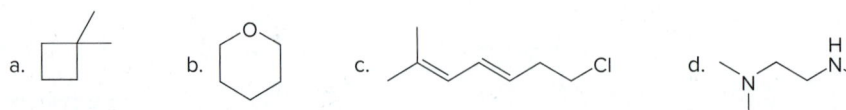

1.8 Drawing Organic Structures

Problem 1.25 What is the molecular formula of each drug?

a. 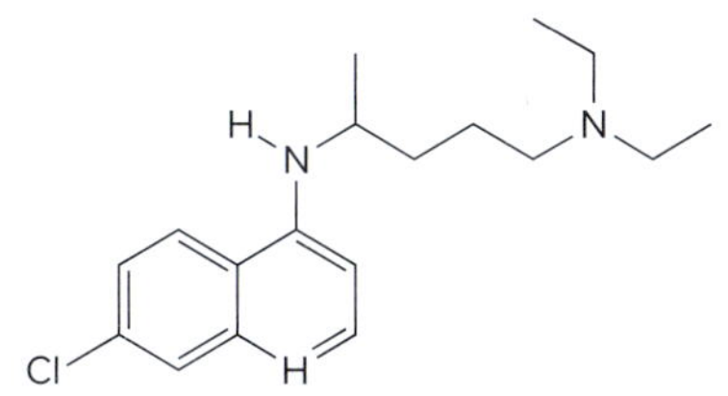 chloroquine (antimalarial medication)

b. 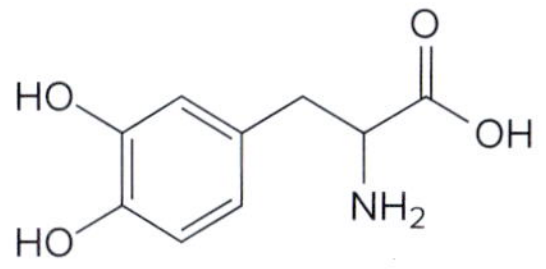 L-dopa (drug used for Parkinson's disease)

When heteroatoms are bonded to a carbon skeleton, the **heteroatom is joined *directly* to the carbon to which it is bonded,** with no H atoms in between. Thus, an OH group is drawn as OH or HO depending on where the OH is located. In contrast, when carbon appendages are bonded to a carbon skeleton, the **H atoms will be drawn to the *right* of the carbon to which they are bonded regardless of the location.**

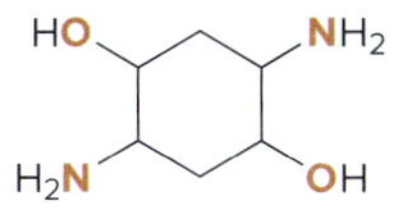

Place the O and N atoms *directly* joined to the ring.

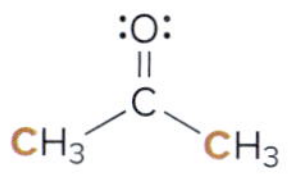

Two C atoms in red are bonded to the middle C.

Two C atoms in red are bonded to the ring.

1.8C Skeletal Structures with Charged Atoms

Take care in interpreting skeletal structures for positively and negatively charged carbon atoms, because *both* the hydrogen atoms *and* the lone pairs are omitted. Keep in mind the following:

- A charge on a carbon atom takes the place of one hydrogen atom.
- The charge determines the number of lone pairs. Negatively charged carbon atoms have one lone pair and positively charged carbon atoms have none.

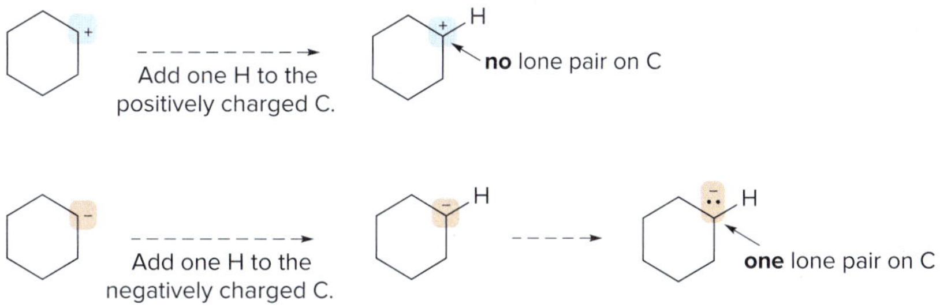

Add one H to the positively charged C. — **no** lone pair on C

Add one H to the negatively charged C. — **one** lone pair on C

Skeletal structures often leave out lone pairs on heteroatoms, but *don't forget about them*. Use the formal charge on an atom to determine the number of lone pairs. For example, a neutral O atom with two bonds needs two additional lone pairs, and a positively charged O atom with three bonds needs only one lone pair.

neutral O atom
two lone pairs

positively charged O atom
one lone pair

Problem 1.26 Draw in all hydrogens and lone pairs on the charged carbons in each ion.

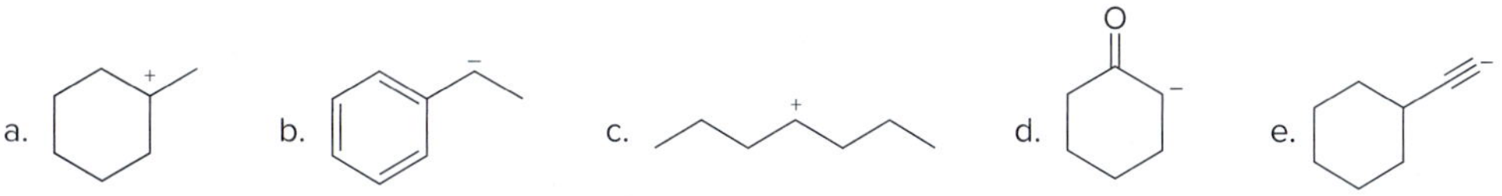

a. b. c. d. e.

Problem 1.27 Use the formal charge to draw in the lone pairs on each N or O atom in the following compounds.

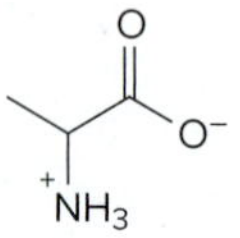

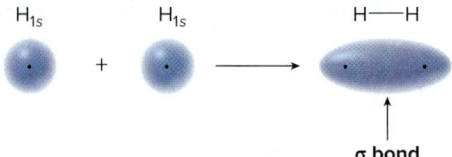

Problem 1.28 Draw a skeletal structure for the molecules in parts (a) and (b), and a condensed structure for the molecules in parts (c) and (d).

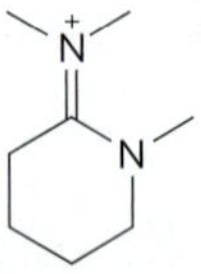

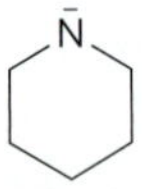

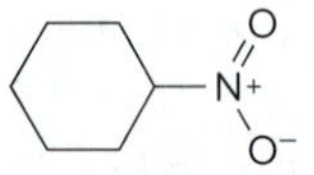

a. $CH_3O(CH_2)_2COCH=C(CH_3)_2$

b. 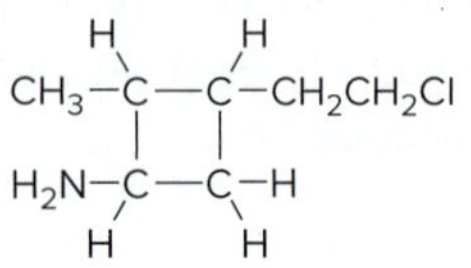 $CH_3-\underset{\underset{H_2N-C-C-H}{|\;\;\;\;\;|}}{\overset{H\;\;\;\;\;H}{C-C}}-CH_2CH_2Cl$

   with bottom carbons bearing H, H

c. (skeletal structure shown)

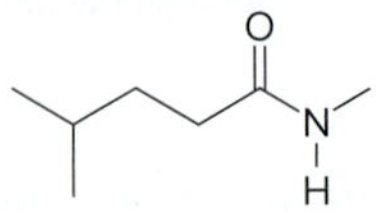

d. 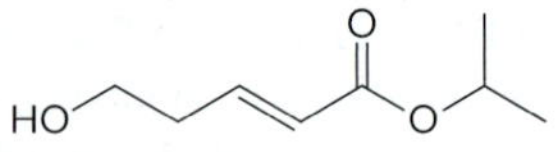 (skeletal structure shown)

1.9 Hybridization

What orbitals do the first- and second-row atoms use to form bonds?

1.9A Hydrogen

Recall from Section 1.2 that two hydrogen atoms share each of their electrons to form H_2. Thus, the 1s orbital on one H overlaps with the 1s orbital on the other H to form a bond that concentrates electron density between the two nuclei. This type of bond, called a **σ (sigma) bond,** is cylindrically symmetrical because the electrons forming the bond are distributed symmetrically about an imaginary line connecting the two nuclei.

H_{1s} + H_{1s} ⟶ H—H
                              ↑
                            σ bond

- A σ bond concentrates electron density on the axis that joins two nuclei. All single bonds are σ bonds.

1.9B Bonding in Methane

To account for the bonding patterns observed in more complex molecules, we must take a closer look at the 2s and 2p orbitals of atoms of the second row. Let's illustrate this with methane, CH_4.

Carbon has **four valence electrons.** To fill atomic orbitals in the most stable arrangement, electrons are placed in the orbitals of lowest energy. For carbon, this places two electrons in the 2s orbital and one each in two 2p orbitals.

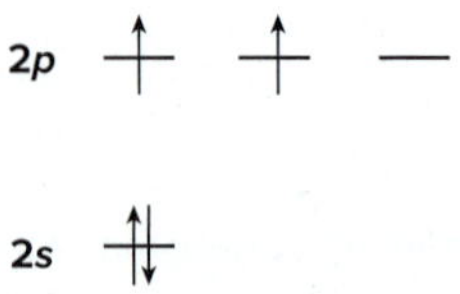

2p ↑ ↑ —

2s ↑↓

ground state for carbon's four valence electrons

- **This lowest-energy arrangement of electrons for an atom is called its ground state.**

In this description, **carbon should form *only two bonds*** because it has only two unpaired valence electrons, and CH₂ should be a stable molecule. In reality, however, CH₂ is a highly reactive species because carbon does not have an octet of electrons.

Because the carbon atom in CH₄ forms four bonds to hydrogen and **all C—H bonds are *identical*,** chemists have proposed that atoms like carbon do *not* use pure *s* and pure *p* orbitals in forming bonds. Instead, atoms use a set of new orbitals called **hybrid orbitals.** The mathematical process by which these orbitals are formed is called **hybridization.**

- *Hybridization* is the combination of two or more atomic orbitals to form the same number of hybrid orbitals, each having the same shape and energy.

Hybridization of *one 2s* orbital and *three 2p* orbitals for carbon forms *four* hybrid orbitals, each with one electron. These new hybrid orbitals are intermediate in energy between the 2s and 2p orbitals.

- These hybrid orbitals are called *sp³* hybrids because they are formed from *one s* orbital and *three p* orbitals.

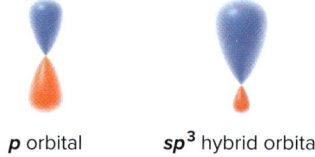

p orbital   *sp³* hybrid orbital

What do these new hybrid orbitals look like? Mixing a spherical 2s orbital and three dumbbell-shaped 2p orbitals together produces four orbitals having one large lobe and one small lobe, oriented toward the corners of a tetrahedron. Each large lobe concentrates electron density in the bonding direction between two nuclei. **Bonds formed from hybrid orbitals are *stronger* than bonds formed from pure *p* orbitals.**

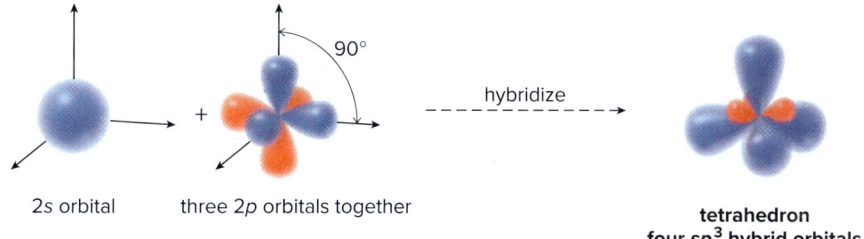

The four hybrid orbitals form four equivalent bonds. We can now explain the observed bonding in CH₄.

- Each bond in CH₄ is formed by overlap of an *sp³* hybrid orbital of carbon with a 1s orbital of hydrogen. These four bonds point to the corners of a tetrahedron.

All four C—H bonds in methane are **σ bonds,** because the electron density is concentrated on the axis joining C and H. An orbital picture of the bonding in CH₄ is given in Figure 1.6.

Figure 1.6
Bonding in CH₄ using *sp³* hybrid orbitals

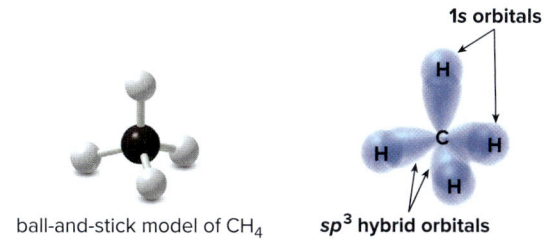

- All four C—H bonds are σ bonds. Each is formed by overlap of an *sp³* hybrid orbital on carbon and a 1s orbital on hydrogen.

Problem 1.29 What orbitals are used to form each of the C—C and C—H bonds in $CH_3CH_2CH_3$ (propane)? How many σ bonds are present in this molecule?

- Any atom surrounded by four groups (atoms and lone pairs) is sp^3 hybridized.

The N atom in **NH₃** and the O atom in **H₂O** are both surrounded by four groups, making them sp^3 hybridized. Each N—H and O—H bond in these molecules is formed by overlap of an sp^3 hybrid orbital with a $1s$ orbital from H. The lone pairs of electrons on N and O also occupy sp^3 hybrid orbitals, as shown in Figure 1.7.

Figure 1.7
Hybrid orbitals of NH₃ and H₂O

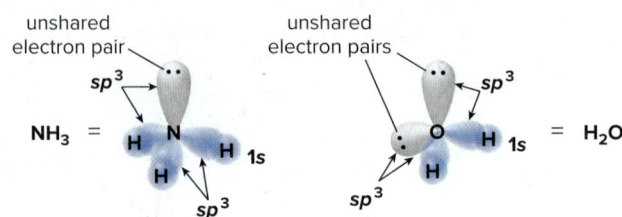

1.9C Other Hybridization Patterns—sp and sp² Hybrid Orbitals

Forming sp^3 hybrid orbitals is just one way that $2s$ and $2p$ orbitals can hybridize. Three common modes of hybridization are seen in organic molecules. The number of orbitals is always conserved in hybridization; that is, **a given number of atomic orbitals hybridizes to form an *equivalent* number of hybrid orbitals.**

- One $2s$ orbital and *three* $2p$ orbitals form *four* sp^3 hybrid orbitals.
- One $2s$ orbital and *two* $2p$ orbitals form *three* sp^2 hybrid orbitals.
- One $2s$ orbital and *one* $2p$ orbital form *two* sp hybrid orbitals.

We have already seen pictorially how four sp^3 hybrid orbitals are formed from one $2s$ and three $2p$ orbitals. Figures 1.8 and 1.9 illustrate the same process for sp and sp^2 hybrids. Each sp and sp^2 hybrid orbital has one large and one small lobe, much like an sp^3 hybrid orbital. Note, however, that both sp^2 and sp hybridization **leave one and two $2p$ orbitals *unhybridized*,** respectively, on each atom.

Figure 1.8
Forming two sp hybrid orbitals

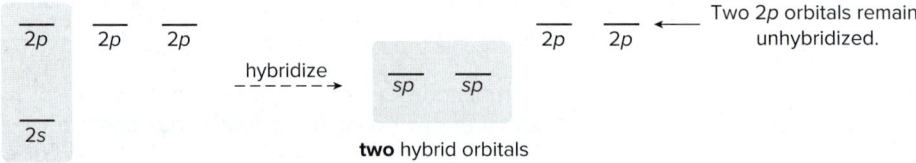

- Forming **two sp hybrid orbitals** uses **one $2s$** and **one $2p$ orbital**, leaving **two $2p$ orbitals unhybridized.**

Figure 1.9
Forming three sp^2 hybrid orbitals

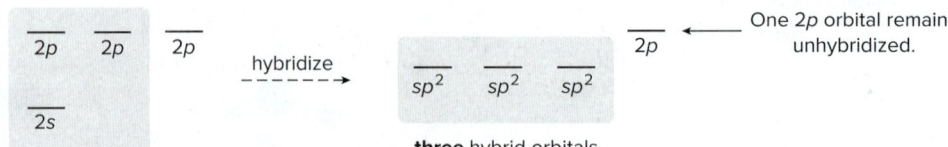

- Forming **three sp^2 hybrid orbitals** uses **one $2s$** and **two $2p$ orbitals**, leaving **one $2p$ orbital unhybridized.**

1.10 Ethane, Ethylene, and Acetylene

To determine the hybridization of an atom in a molecule, we count groups (atoms and lone pairs) around the atom, just as we did in determining geometry.

> The **superscripts** for hybrid orbitals correspond to the **number of atomic orbitals** used to form them. The number "1" is understood.
>
> For example: $sp^3 = s^1p^3$
>
> one 2s + three 2p orbitals used to make each hybrid orbital

- The number of groups around an atom *equals* the number of atomic orbitals that are hybridized to form hybrid orbitals (Table 1.4).

Table 1.4 Three Types of Hybrid Orbitals

Number of groups	Number of orbitals used	Type of hybrid orbital
2	2	two sp hybrid orbitals
3	3	three sp^2 hybrid orbitals
4	4	four sp^3 hybrid orbitals

Hybridization in various carbon compounds is presented in Section 1.10.

Sample Problem 1.9 — Determining the Hybridization of an Atom

What orbitals are used to form each bond in methanol, CH_3OH?

Solution

To solve this problem, **draw a valid Lewis structure** and **count groups around each atom**. Then, use the rule to determine hybridization: **two groups = sp, three groups = sp^2, and four groups = sp^3**.

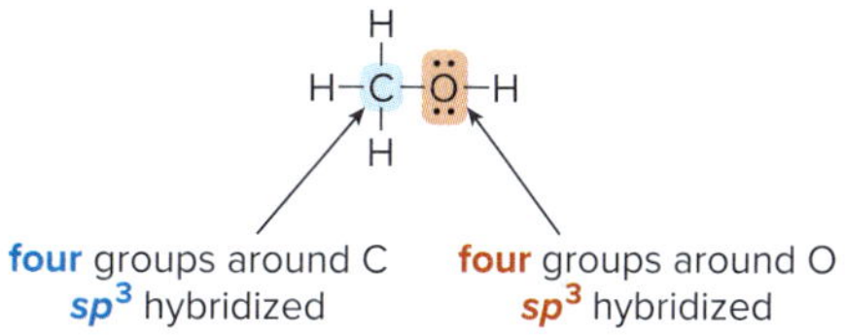

$$H-\underset{H}{\overset{H}{C}}-\ddot{O}-H$$

- **four** groups around C — sp^3 hybridized
- **four** groups around O — sp^3 hybridized

- All C–H bonds are formed from C_{sp^3}–H_{1s}.
- The C–O bond is formed from C_{sp^3}–O_{sp^3}.
- The O–H bond is formed from O_{sp^3}–H_{1s}.

Problem 1.30 What orbitals are used to form each highlighted bond in the following molecule? In what type of orbital do the lone pairs on each O and N reside?

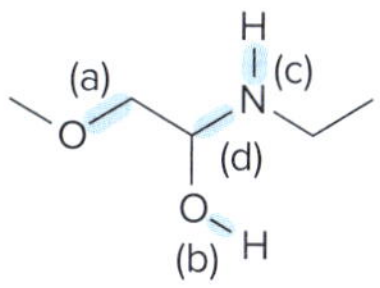

More Practice: Try Problems 1.67a–c, 1.68.

1.10 Ethane, Ethylene, and Acetylene

The principles of hybridization determine the type of bonds in **ethane**, **ethylene**, and **acetylene**.

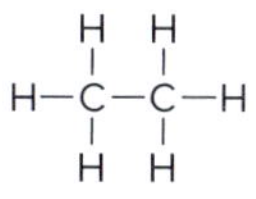

ethane 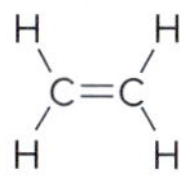   ethylene 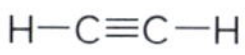   acetylene

1.10A Ethane—CH₃CH₃

According to the Lewis structure for **ethane, CH₃CH₃,** each carbon atom is singly bonded to four other atoms. As a result:

- Each carbon is tetrahedral.
- Each carbon is sp^3 hybridized.

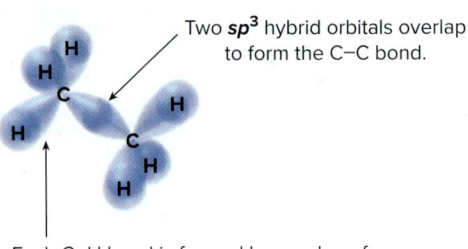

All of the bonds in ethane are σ bonds. The C–H bonds are formed from the overlap of one of the three sp^3 hybrid orbitals on each carbon atom with the 1s orbital on hydrogen. The C–C bond is formed from the overlap of an sp^3 hybrid orbital on each carbon atom.

Ethane is a constituent of natural gas. *Steve Allen/Brand X Pictures*

A model of ethane shows that **rotation can occur around the central C–C σ bond.** The relative position of the H atoms on the adjacent CH₃ groups changes with bond rotation, as seen in the location of the labeled red H atom before and after rotation. This process is discussed in greater detail in Chapter 4.

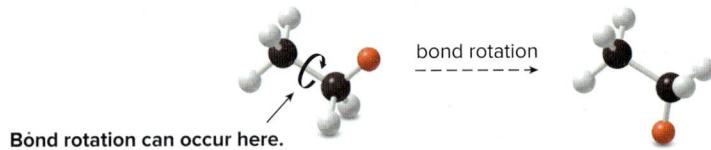

1.10B Ethylene—C₂H₄

Based on the Lewis structure of **ethylene, CH₂=CH₂,** each carbon atom is singly bonded to two H atoms and doubly bonded to the other C atom, so each C is surrounded by three groups. As a result:

- Each carbon is trigonal planar (Section 1.7B).
- Each carbon is sp^2 hybridized.

$$CH_2=CH_2 \quad = \quad \text{ethylene} \quad 120°$$

three groups around each C

Ethylene is an important starting material in the preparation of the plastic polyethylene. *Nextdoor Images/Creatas/PunchStock*

What orbitals are used to form the two bonds of the C–C double bond? Recall from Section 1.9 that sp^2 **hybrid orbitals** are formed from **one 2s and two 2p orbitals,** leaving one **2p orbital unhybridized.** Because carbon has four valence electrons, **each of these orbitals has one electron** that can be used to form a bond.

Each C–H bond results from the end-on overlap of an sp^2 hybrid orbital on carbon and the 1s orbital on hydrogen. Similarly, one of the C–C bonds results from the end-on overlap of

An sp^2 hybridized C in $CH_2=CH_2$ has three sp^2 hybrid orbitals and one higher-energy, unhybridized p orbital:

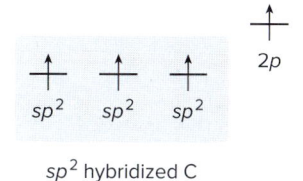

sp^2 hybridized C

an sp^2 hybrid orbital on each carbon atom. Each of these bonds is a **σ bond.** All five σ bonds lie in the same plane, viewed from above in the following representation, and from the side in Figure 1.10a.

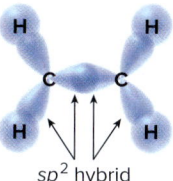

sp^2 hybrid

- Each C has three sp^2 hybrid orbitals.
- The C–H bonds and the C–C bond are σ bonds.

The second C–C bond results from the side-by-side overlap of the $2p$ orbitals on each carbon. Because the unhybridized $2p$ orbitals are located perpendicular to the plane of the molecule, side-by-side overlap creates an area of electron density above and below the plane containing the sp^2 hybrid orbitals (that is, the plane containing the six atoms in the σ bonding system), as shown in Figure 1.10b.

Figure 1.10 The σ and π bonds in ethylene

a. 2p orbitals

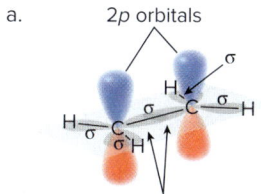

- Overlap of the two sp^2 hybrid orbitals forms the C–C σ bond.

b. π bond

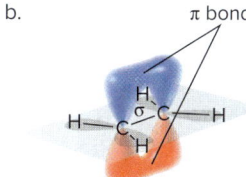

- Overlap of the two $2p$ orbitals forms the C–C π bond.
- The π bond extends **above and below the plane** of the molecule.

In this second bond, the electron density is *not* concentrated on the axis joining the two nuclei. This new type of bond is called a **π bond.** Because the electron density in a π bond is farther from the two nuclei, **π bonds are usually weaker and therefore more easily broken than σ bonds.**

Thus, a carbon–carbon double bond has two components:

- a σ bond, formed by end-on overlap of two sp^2 hybrid orbitals;
- a π bond, formed by side-by-side overlap of two $2p$ orbitals.

Unlike the C–C single bond in ethane, rotation about the C–C double bond in ethylene is **restricted.** It can occur only if the π bond first breaks and then re-forms, a process that requires considerable energy.

All double bonds are composed of one σ and one π bond.

Rotation around a C=C bond does *not* occur.

1.10C Acetylene—C₂H₂

Based on the Lewis structure of **acetylene, HC≡CH,** each carbon atom is singly bonded to one hydrogen atom and triply bonded to the other carbon atom, so each carbon atom is surrounded by two groups. As a result:

- Each carbon is linear (Section 1.7B).
- Each carbon is *sp* hybridized.

Because acetylene produces a very hot flame on burning, it is often used in welding torches. The fire is very bright, too, so it was once used in the lamps worn by spelunkers—people who study and explore caves.
Phillip Spears/Getty Images

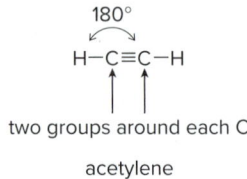

two groups around each C

acetylene

What orbitals are used to form the bonds of the C—C triple bond? Recall from Section 1.9 that *sp* **hybrid orbitals** are formed from **one 2s and one 2p orbital**, leaving **two 2p orbitals unhybridized.** Because carbon has four valence electrons, **each of these orbitals has one electron** that can be used to form a bond.

Each C—H bond results from the end-on overlap of an *sp* hybrid orbital on carbon and the 1*s* orbital on hydrogen. Similarly, one of the C—C bonds results from the end-on overlap of an *sp* hybrid orbital on each carbon atom. Each of these bonds is a **σ bond.**

An *sp* hybridized C in HC≡CH has two *sp* hybrid orbitals and two higher-energy, unhybridized *p* orbitals:

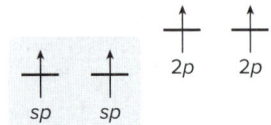

sp hybridized C

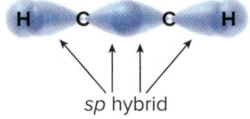

- Each C has two *sp* hybrid orbitals.
- The C—H bonds and C—C bond are σ bonds.

Each carbon atom also has two **unhybridized 2p orbitals** that are perpendicular to each other and to the *sp* hybrid orbitals (Figure 1.11a). Side-by-side overlap between the two 2*p* orbitals on one carbon with the two 2*p* orbitals on the other carbon creates the second and third bonds of the C—C triple bond (Figure 1.11b). The electron density from one of these two bonds is above and below the axis joining the two nuclei, and the electron density from the second of these two bonds is in front of and behind the axis, so both of these bonds are **π bonds.**

Figure 1.11
The σ and π bonds in acetylene

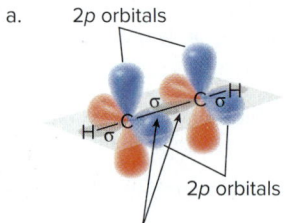

a. 2*p* orbitals

2*p* orbitals

- Overlap of the two *sp* hybrid orbitals forms the C—C σ bond.

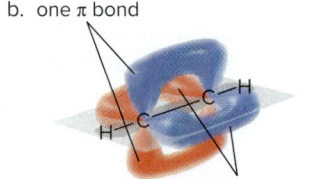

b. one π bond

second π bond

- Overlap of two sets of two 2*p* orbitals forms two C—C π bonds.
- Two π bonds extend out from the axis of the linear molecule.

The side-by-side overlap of two *p* orbitals always forms a π bond.

Thus, a carbon–carbon triple bond has three components:

All triple bonds are composed of one σ and two π bonds.

- a σ bond, formed by end-on overlap of two *sp* hybrid orbitals;
- two π bonds, formed by side-by-side overlap of two sets of 2*p* orbitals.

1.10 Ethane, Ethylene, and Acetylene

Table 1.5 summarizes the three possible types of bonding in carbon compounds.

Table 1.5 A Summary of Covalent Bonding in Carbon Compounds

Number of groups bonded to C	Hybridization	Bond angle	Example	Observed bonding
4	sp^3	109.5°	CH_3CH_3 ethane	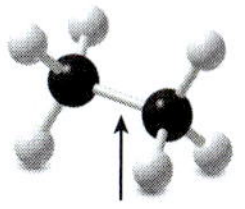 one σ bond $C_{sp^3}-C_{sp^3}$
3	sp^2	120°	$CH_2=CH_2$ ethylene	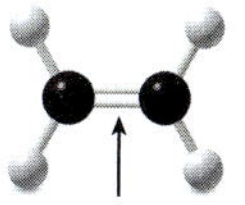 one σ bond $C_{sp^2}-C_{sp^2}$ + one π bond $C_{2p}-C_{2p}$
2	sp	180°	$HC\equiv CH$ acetylene	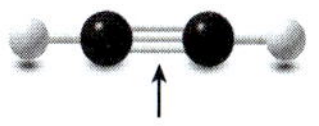 one σ bond $C_{sp}-C_{sp}$ + two π bonds $C_{2p}-C_{2p}$, $C_{2p}-C_{2p}$

Sample Problem 1.10 Determining Hybridization

Answer each question for cyclohexanone.

a. Determine the hybridization of the highlighted atoms.
b. What orbitals are used to form the C–O double bond?
c. In what type of orbital does each lone pair reside?

cyclohexanone

Solution

a.
- three groups around C — sp^2 hybridized
- three groups around O — sp^2 hybridized
- four groups around C — sp^3 hybridized

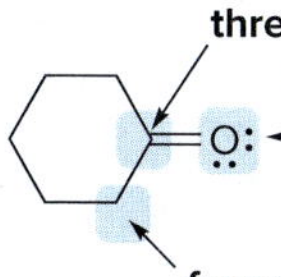

b.
- The σ bond is formed from the end-on overlap of $C_{sp^2}-O_{sp^2}$.
- The π bond is formed from the side-by-side overlap of $C_{2p}-O_{2p}$.

c. The O atom has three sp^2 hybrid orbitals.
- One is used for the σ bond of the double bond.
- The remaining two sp^2 hybrids are occupied by the lone pairs.

Problem 1.31

Determine the hybridization around the highlighted atoms in each molecule.

a. $CH_3-C\equiv CH$ b. 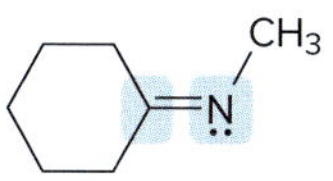 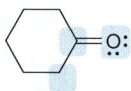 c. $CH_2=C=CH_2$

More Practice: Try Problems 1.40d, e; 1.41d, e; 1.67d, e; 1.69; 1.76a–c.

Problem 1.32 An anion of isocitric acid is formed during the metabolism of many types of organic compounds in cells. (a) How many sp^2 hybridized carbon atoms does isocitric acid contain? (b) What is the hybridization of each O atom shown in red? (c) What orbitals are used to form the highlighted carbon–carbon bond? (d) How many σ bonds does isocitric acid contain? (e) How many π bonds does it contain?

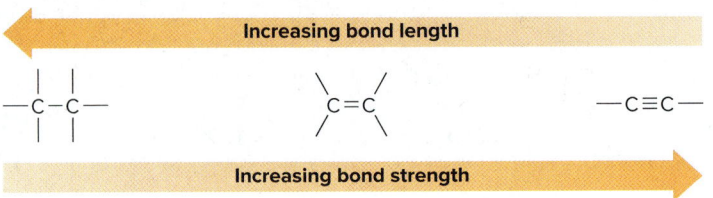

isocitric acid

1.11 Bond Length and Bond Strength

Let's now examine the relative bond length and bond strength of the C–C and C–H bonds in ethane, ethylene, and acetylene.

1.11A A Comparison of Carbon–Carbon Bonds

While the SI unit of energy is the **joule** (J), organic chemists often report energy values in **calories** (cal). For this reason, energy values in the tables in this text are reported in joules, followed by the number of calories in parentheses.
1 cal = 4.18 J

An inverse relationship exists between bond length and bond strength. The shorter the bond, the closer the electron density is kept to the nucleus, and the harder the bond is to break. **Shorter** bonds are *stronger* bonds.

- As the number of electrons between two nuclei *increases*, bonds become shorter and stronger.
- Triple bonds are shorter and stronger than double bonds, which are shorter and stronger than single bonds.

Values for bond lengths and bond strengths for CH_3CH_3, $CH_2\!=\!CH_2$, and $HC\!\equiv\!CH$ are listed in Table 1.6. Be careful not to confuse two related but different principles regarding multiple bonds such as C–C double bonds. **Double bonds, consisting of both a σ and a π bond, are *strong*.** The **π component** of the double bond, however, is usually much *weaker* than the σ component. This is a particularly important consideration when studying alkenes in Chapter 10.

Table 1.6 Bond Lengths and Bond Strengths for Ethane, Ethylene, and Acetylene

Compound	C–C bond length (pm)	Bond strength kJ/mol (kcal/mol)
$CH_3\!-\!CH_3$	153	368 (88)
$CH_2\!=\!CH_2$	134	635 (152)
$HC\!\equiv\!CH$	121	837 (200)

Compound	C–H bond length (pm)	Bond strength kJ/mol (kcal/mol)
$CH_3CH_2\!-\!H$	111	410 (98)
$CH_2\!=\!CH\!-\!H$	110	435 (104)
$HC\!\equiv\!C\!-\!H$	109	523 (125)

1.11B A Comparison of Carbon–Hydrogen Bonds

The length and strength of a C—H bond vary slightly depending on the hybridization of the carbon atom.

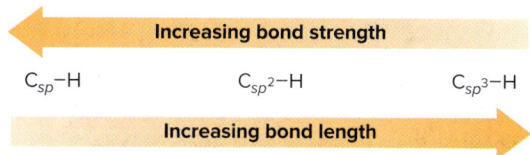

To understand why this is so, we must look at the atomic orbitals used to form each type of hybrid orbital. A single 2s orbital is always used, but the number of 2p orbitals varies with the type of hybridization. The **percent s-character** indicates the fraction of a hybrid orbital due to the 2s orbital used to form it.

sp hybrid	$\dfrac{\text{one 2s orbital}}{\text{two hybrid orbitals}}$	= 50% s-character
sp² hybrid	$\dfrac{\text{one 2s orbital}}{\text{three hybrid orbitals}}$	= 33% s-character
sp³ hybrid	$\dfrac{\text{one 2s orbital}}{\text{four hybrid orbitals}}$	= 25% s-character

Why should the percent s-character of a hybrid orbital affect the length of a C—H bond? A 2s orbital keeps electron density closer to a nucleus compared to a 2p orbital. As the **percent s-character** *increases*, a hybrid orbital holds its electrons closer to the nucleus, and the **bond becomes *shorter* and *stronger*.**

- Increased percent s-character ⇢ Increased bond strength ⇢ Decreased bond length

---

Problem 1.33 Which of the bonds shown in red in each compound or pair of compounds is shorter?

The seeds of some types of sandalwood are rich in santalbic acid (Problem 1.34), an unusual fatty acid that contains a carbon–carbon triple bond.
Bijayakumar/Shutterstock

Problem 1.34 Rank the labeled bonds in santalbic acid, a fatty acid obtained from the seeds of the sandalwood tree used in cosmetics, in order of increasing bond length.

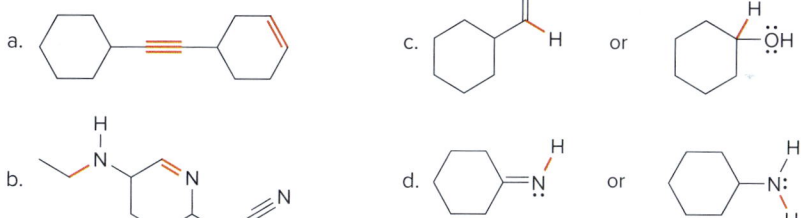

1.12 Electronegativity and Bond Polarity

Electronegativity is a measure of an atom's attraction for electrons in a bond. Electronegativity indicates how much a particular atom "*wants*" electrons.

- Electronegativity *increases* across a row of the periodic table as the nuclear charge increases (excluding the noble gases).
- Electronegativity *decreases* down a column of the periodic table as the atomic radius increases, pushing the valence electrons farther from the nucleus.

As a result, the *most* electronegative elements are located at the **upper right-hand corner** of the periodic table, and the *least* electronegative elements in the **lower left-hand corner.** A scale has been established to represent electronegativity values arbitrarily, from 0 to 4, as shown in Figure 1.12.

Figure 1.12
Electronegativity values for some common elements

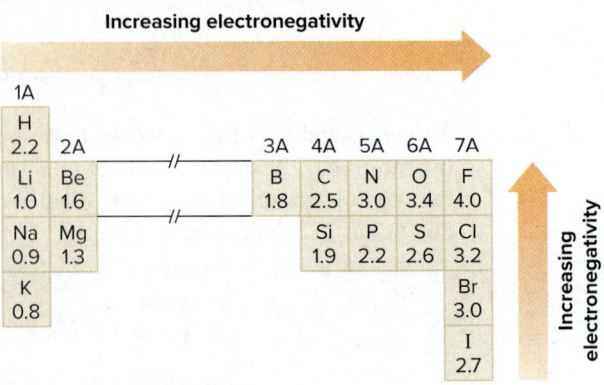

Electronegativity values are relative, so they can be used for comparison purposes only. When comparing two different elements, one is **more electronegative** than the other if it attracts electron density toward itself. One is less electronegative—**more electropositive**—if it gives up electron density to the other element.

Problem 1.35 Rank the following atoms in order of increasing electronegativity. Label the most electronegative and most electropositive atom in each group.

a. Se, O, S   b. P, Na, Cl   c. Cl, S, F   d. O, P, N

Electronegativity values are used as a guideline to indicate whether the electrons in a bond are **equally shared** or **unequally shared** between two atoms. Whenever two identical atoms are bonded together, each atom attracts the electrons in the bond to the same extent. The electrons are equally shared, and the **bond is** *nonpolar*. Thus, a **carbon–carbon bond is nonpolar.** Whenever two different atoms having similar electronegativities are bonded together, the bond is also **nonpolar. C–H bonds are considered to be nonpolar,** because the electronegativity difference between C (2.5) and H (2.2) is small.

The small electronegativity difference between C and H is ignored.

Bonding between atoms of different electronegativity values results in the **unequal sharing** of electrons. In a C–O bond, the electrons are pulled away from C (2.5) toward O (3.4), the element of higher electronegativity. **The bond is** *polar,* or *polar covalent.* The bond is said to have a **dipole**—that is, **a partial separation of charge.**

A C–O bond is a **polar** bond.

C is electron deficient.   O is electron rich.

$$\overset{\delta+\delta-}{\text{—C—O—}}$$

a bond dipole

The direction of polarity in a bond is often indicated by an arrow, with the head of the arrow pointing toward the more electronegative element. The tail of the arrow, with a perpendicular line drawn through it, is positioned at the less electronegative element. Alternatively, the symbols δ+ and δ− indicate this unequal sharing of electron density.

- δ+ means an atom is electron deficient (has a partial positive charge).
- δ− means an atom is electron rich (has a partial negative charge).

Problem 1.36 Show the direction of the dipole in each bond. Label the atoms with δ+ and δ−.

a. H—F   b. \B—C—/| c. —C—Li   d. —C—Cl

Students often wonder how large an electronegativity difference must be to consider a bond polar. That's hard to say. We will set an arbitrary value for this difference and use it as an *approximation*. **Usually, a polar bond will be one in which the electronegativity difference between two atoms is ≥ 0.5 unit.**

The distribution of electron density in a molecule can be shown using an **electrostatic potential map.** These maps are color coded to illustrate areas of high and low electron density. Electron-rich regions are indicated in red, and electron-deficient sites are indicated in blue. Regions of intermediate electron density are shown in orange, yellow, and green.

An electrostatic potential map of CH_3Cl indicates the polar nature of the C—Cl bond (Figure 1.13). The more electronegative Cl atom pulls electron density toward it, making it electron rich. This is indicated by the red around the Cl in the plot. The carbon is electron deficient, and this is shown with blue. When comparing two maps, the comparison is useful only if they are plotted *using the same scale* of color gradation. For this reason, whenever we compare two plots in this text, they will be drawn side by side using the same scale.

Figure 1.13
Electrostatic potential plot of CH_3Cl

a. Color scheme used for electron density

b. Electrostatic potential plot

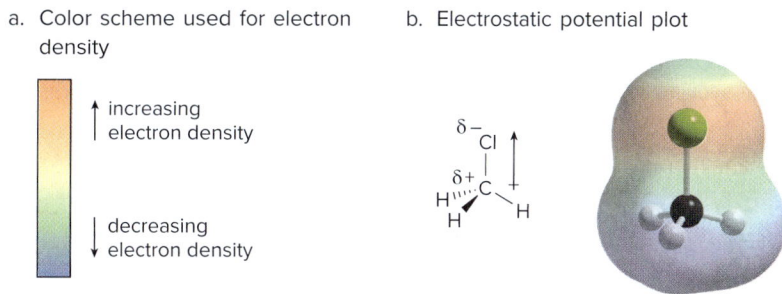

1.13 Polarity of Molecules

A **polar molecule** has either one polar bond, or two or more bond dipoles that reinforce. A **nonpolar molecule** has either no polar bonds, or two or more bond dipoles that cancel.

Thus far, we have been concerned with the polarity of one bond. To determine whether a molecule has a net dipole, use the following two-step procedure:

[1] Use electronegativity differences to **identify all of the polar bonds and the directions of the bond dipoles.**

[2] **Determine the geometry** around individual atoms by counting groups, and decide if individual dipoles **cancel** or **reinforce each other in space.**

The two molecules **H_2O** and **CO_2** illustrate different outcomes of this process. In H_2O, each O—H bond is polar because the electronegativity difference between O (3.4) and H (2.2) is large. Because H_2O is a **bent** molecule, the two dipoles reinforce (both point *up*). Thus, **H_2O has a net dipole, making it a polar molecule.** CO_2 also has polar C—O bonds because the electronegativity difference between O (3.4) and C (2.5) is large. However, CO_2 is a **linear**

46 Chapter 1 Structure and Bonding

Whenever C or H is bonded to N, O, and all halogens, the bond is *polar*. Thus, the C–I bond is considered polar even though the electronegativity difference between C and I is small. Remember, electronegativity is just an approximation.

molecule, so the two dipoles, which are equal and opposite in direction, **cancel**. Thus, CO_2 is a **nonpolar molecule** with **no net dipole**.

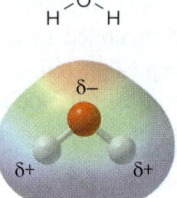

net dipole

no net dipole

Electrostatic potential plots for H_2O and CO_2 appear in Figure 1.14. Additional examples of polar and nonpolar molecules are given in Figure 1.15.

Problem 1.37 Indicate which of the following molecules is polar because it possesses a net dipole. Show the direction of the net dipole if one exists.

a. CH_3Br b. CH_2Br_2 c. CF_4 d. Cl₂C=CH₂ e. cis-CHCl=CHCl

Figure 1.14
Electrostatic potential plots for H_2O and CO_2

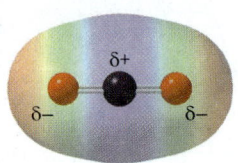

- The electron-rich (red) region is concentrated on the more electronegative O atom. Both H atoms are electron deficient (blue-green).

- Both electronegative O atoms are electron rich (red), and the central C atom is electron deficient (blue).

Figure 1.15
Examples of polar and nonpolar molecules

one polar bond	three polar bonds All dipoles cancel. no net dipole	three polar bonds All dipoles reinforce.	two polar bonds Both dipoles reinforce.	four polar bonds All dipoles cancel. no net dipole
a **polar** molecule	a **nonpolar** molecule	a **polar** molecule	a **polar** molecule	a **nonpolar** molecule

1.14 Oxybenzone—A Representative Organic Molecule

The principles learned in this chapter apply to all organic molecules regardless of size or complexity. We now know a great deal about the structure of the chapter-opening molecule, oxybenzone.

Sample Problem 1.11 Applying the Principles of Bonding, Geometry, and Polarity to a Representative Organic Molecule

Answer each question about oxybenzone, the popular sunscreen component described in the chapter opener.

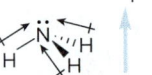

oxybenzone

a. How many lone pairs does oxybenzone contain?
b. What is the molecular formula of oxybenzone?
c. What is the hybridization and geometry around each atom labeled in blue?
d. In what type of orbital(s) are any lone pairs on the O atom in red located?
e. Label all polar bonds.

Solution

a, b. Each O atom needs two lone pairs for an octet, so oxybenzone has six lone pairs. In determining the molecular formula from the skeletal structure, assume there is a C atom at the end of any line and at the intersection of two lines, and that each C has enough H's to make it tetravalent; molecular formula = $C_{14}H_{12}O_3$.

c, d. Count groups to determine hybridization and geometry; with four groups an atom is sp^3 hybridized and tetrahedral; with three groups an atom is sp^2 hybridized and trigonal planar. The O atom in red is surrounded by three groups—one atom and two lone pairs—so it is sp^2 hybridized and its lone pairs occupy sp^2 hybrid orbitals.

e. All C—O and O—H bonds are polar because of the large electronegativity difference between the atoms.

Problem 1.38 Answer each question about L-dopa, a drug used since 1967 to treat Parkinson's disease.

a. Convert the skeletal structure to a Lewis structure.
b. What is the hybridization and geometry around each labeled atom?
c. Label three polar bonds.

More Practice: Try Problems 1.75, 1.77, 1.79, 1.80.

Sample Problem 1.12 illustrates how to derive structural information from a ball-and-stick model.

Sample Problem 1.12

Use the ball-and-stick model of vitamin B_6 to answer each question.

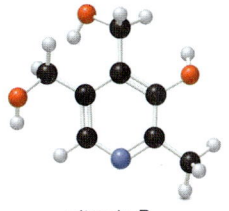

vitamin B_6

a. Draw a skeletal structure of vitamin B_6.
b. How many sp^2 hybridized carbons are present?
c. What is the hybridization of the N atom in the ring?

48   Chapter 1   Structure and Bonding

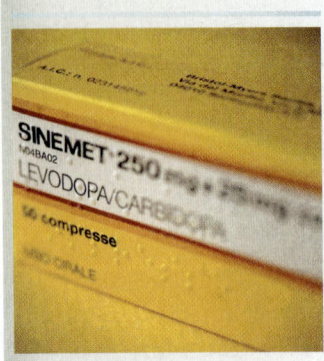

Sinemet, the trade name of a drug used to treat Parkinson's disease, contains a combination of L-dopa (Problem 1.38) and carbidopa (Problem 1.39). Carbidopa increases the effectiveness of L-dopa by inhibiting its metabolism prior to crossing the blood–brain barrier and entering the brain. *Cristina Pedrazzini/Science Source*

Solution

Use the element colors shown in Section 1.7B to convert the 3-D model to a skeletal structure [black (C), gray (H), red (O), blue (N)]. H atoms on carbon are omitted, but H atoms on heteroatoms are drawn. Count groups to determine hybridization. Each O atom needs two lone pairs and the N needs one to give an octet of electrons.

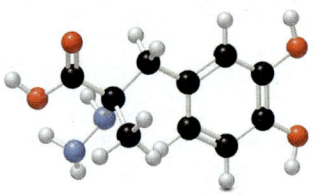

a. skeletal structure of vitamin B_6

b, c.
- Each C labeled in blue is sp^2 hybridized.
- The N atom is surrounded by three groups (two atoms and one lone pair), making it sp^2 hybridized.

Problem 1.39 Use the ball-and-stick model to answer each question about carbidopa, a drug used in combination with L-dopa to treat Parkinson's disease.

carbidopa

a. Draw a skeletal structure of carbidopa.
b. Determine the hybridization around each carbon atom.
c. What is the hybridization and geometry around each N atom?
d. How many polar bonds are present?

More Practice: Try Problems 1.40, 1.41.

Chapter 1 REVIEW

KEY CONCEPTS

Resonance (1.6)

1 Drawing resonance structures

- Look for lone pairs and multiple bonds.
- Atoms and σ bonds do not change location.

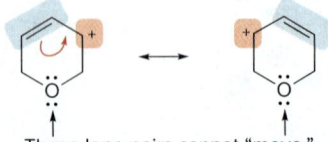

These lone pairs cannot "move."

See Sample Problem 1.5. Try Problems 1.46b, 1.54, 1.55, 1.75d, 1.78b, 1.79e.

2 Drawing the resonance hybrid

- Draw σ bonds and lone pairs that do not move.
- Use a dashed line for a bond that is single in one resonance structure and multiple in another.
- Use a δ+ (or δ−) for an atom that is neutral in one structure and charged in another.

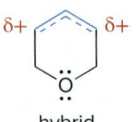

hybrid

Try Problems 1.54, 1.57, 1.78c.

Periodic Trends

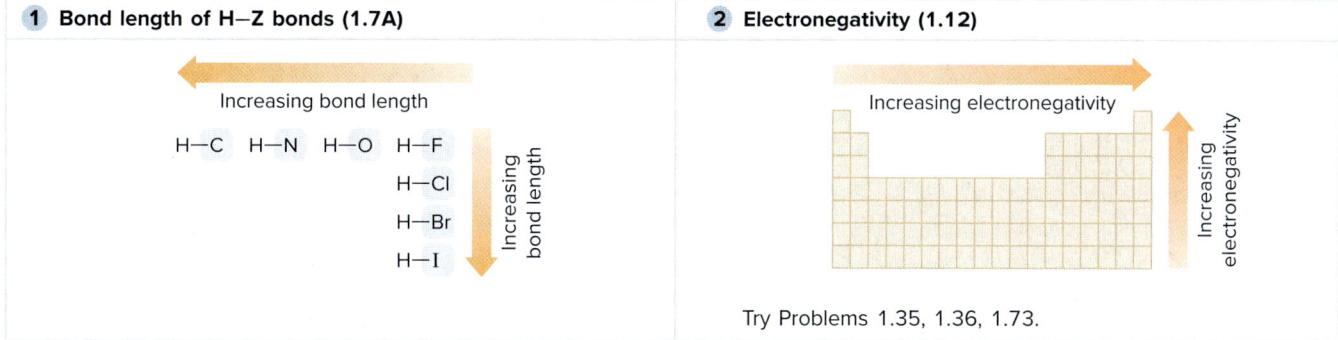

Try Problems 1.35, 1.36, 1.73.

Bond Length and Bond Strength (1.11)

1) Bond length and bond strength of C–C bonds (1.11A)

- Bonds become shorter and stronger as the number of electrons between two nuclei increases.

← Increasing bond length

$CH_3-CH_3 \quad CH_2=CH_2 \quad HC\equiv CH$

Increasing bond strength →

Try Problem 1.71.

2) Bond length and bond strength of C–H bonds (1.11B)

- Bonds become shorter and stronger as the percent s-character increases.

← Increasing bond length

$C_{sp^3}-H \quad C_{sp^2}-H \quad C_{sp}-H$

Increasing bond strength →

See Table 1.6. Try Problem 1.70.

Drawing Organic Structures (1.8)

Abbreviate the structure of complex molecules with skeletal structures or condensed structures.

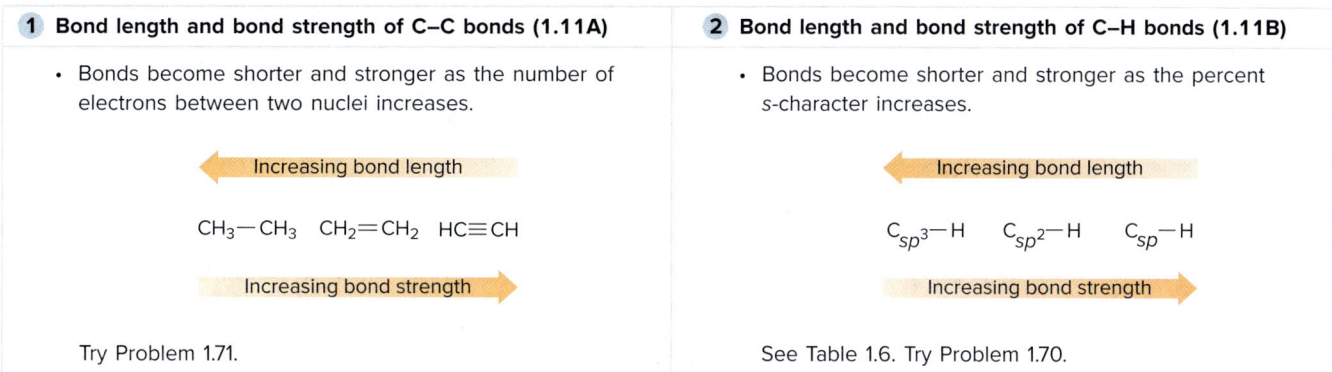

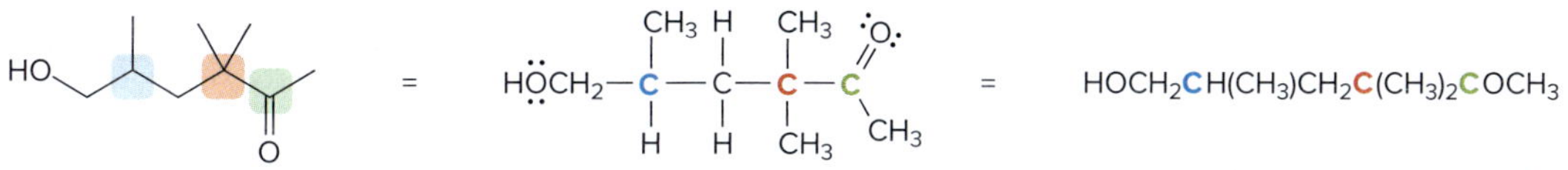

See Figures 1.3, 1.4, 1.5, Sample Problems 1.7, 1.8. Try Problems 1.62–1.65.

KEY SKILLS

[1] Drawing a valid Lewis structure (1.3); example: CH_3CHO

1 Arrange the atoms with H's on the periphery.	2 Count valence electrons.	3 Add single bonds.	4 Complete octets with multiple bonds and lone pairs.
H H H C C O H	2 C's × 4 e⁻ = 8 4 H's × 1 e⁻ = 4 1 O × 6 e⁻ = 6 total e⁻ = 18	H H H–C–C–O H 12 e⁻ used.	H H H–C–C=Ö: H Add one double bond and two lone pairs to complete O and C octets.

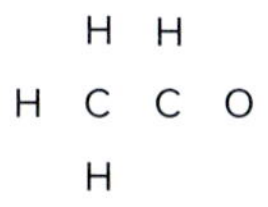

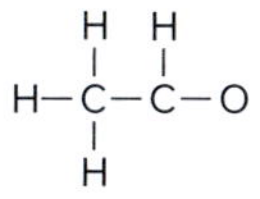

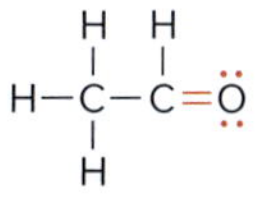

See Sample Problems 1.1, 1.2. Try Problems 1.44, 1.45.

[2] Calculating formal charge (1.3C)

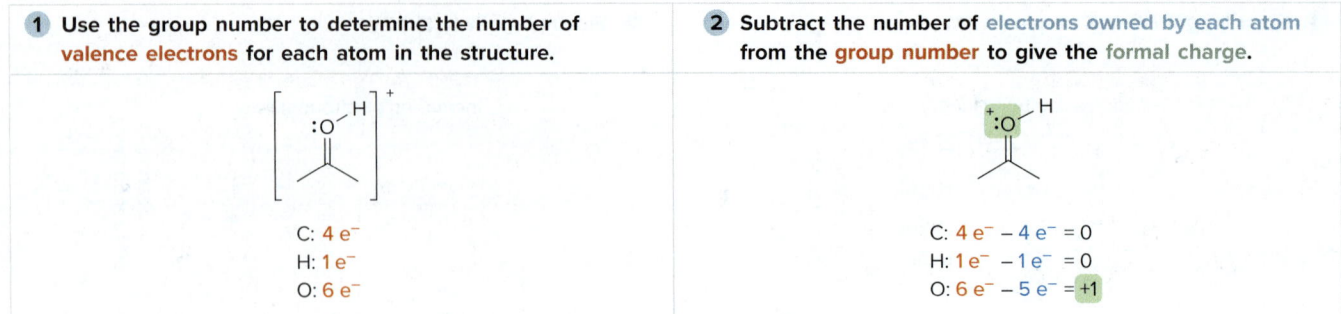

See Sample Problem 1.3. Try Problems 1.42, 1.43.

[3] Predicting geometry from a valid Lewis structure (1.7)

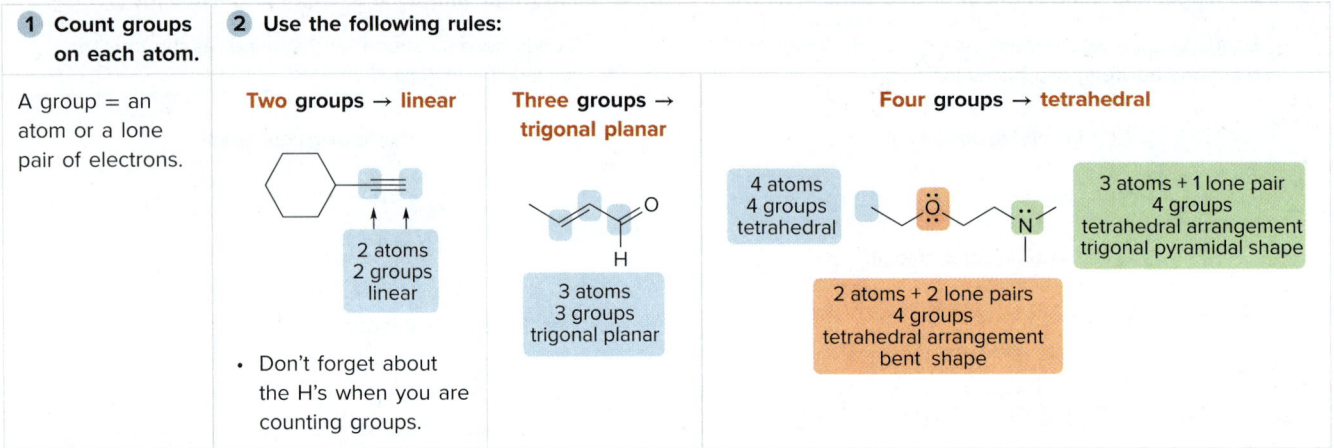

See Sample Problem 1.6. Try Problems 1.60, 1.61, 1.77c, 1.79b.

[4] Identifying isomers and resonance structures (1.4, 1.6)

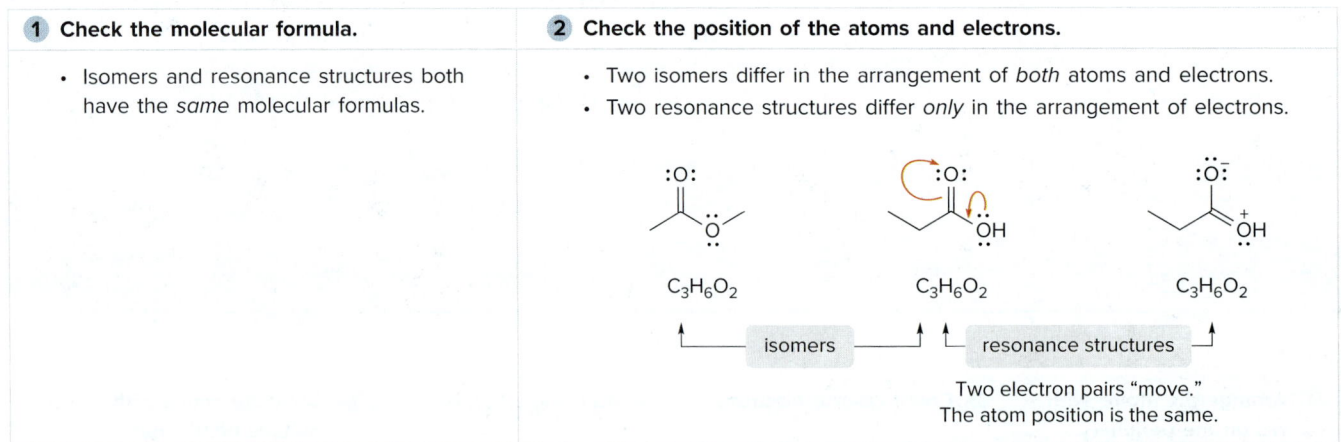

Try Problems 1.49–1.51.

[5] Using curved arrows (1.6B)

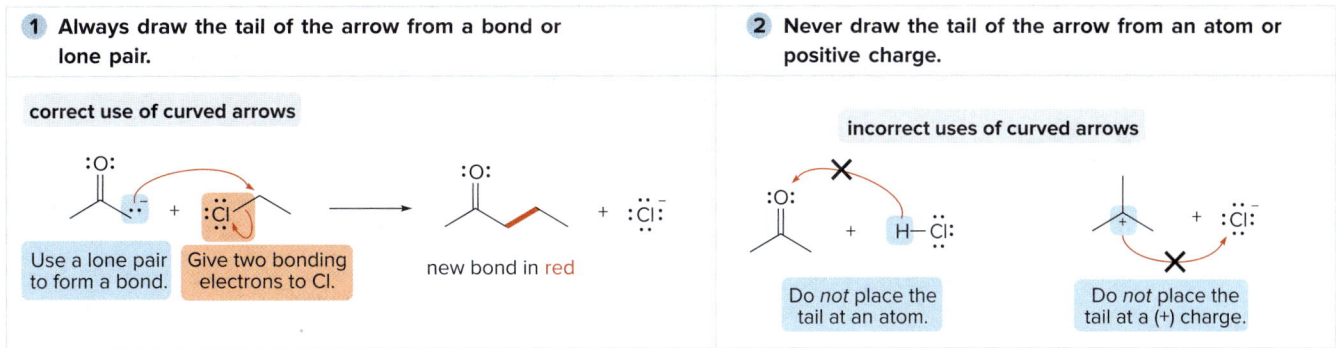

See Sample Problem 1.4. Try Problems 1.52, 1.53.

[6] Predicting hybridization from a valid Lewis structure (1.9)

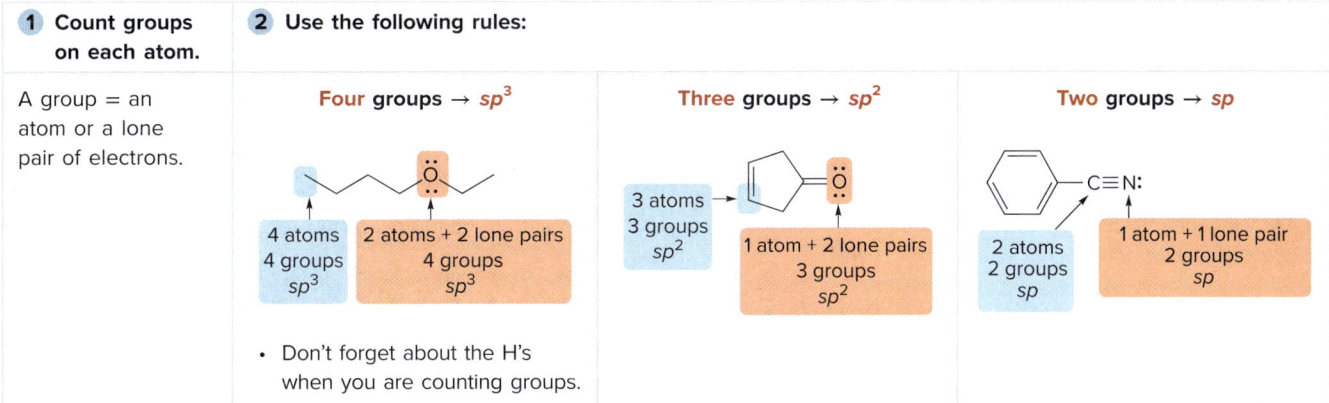

See Sample Problem 1.10. Try Problems 1.67–1.69, 1.75c, 1.77a, 1.79a.

[7] Determining if a molecule has a net dipole from a valid Lewis structure (1.13); example: CH_3OH

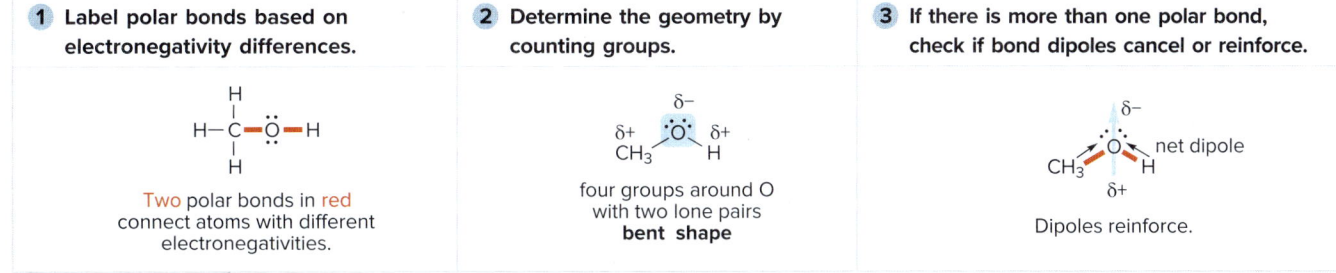

Try Problem 1.74.

PROBLEMS

Problems Using Three-Dimensional Models

1.40 Citric acid is responsible for the tartness of citrus fruits, especially lemons and limes.

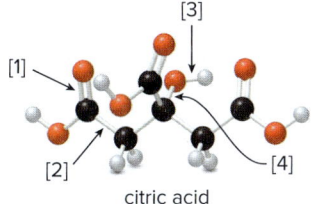

citric acid

a. What is the molecular formula for citric acid?
b. How many lone pairs are present?
c. Draw a skeletal structure.
d. How many sp^2 hybridized carbons are present?
e. What orbitals are used to form each indicated bond ([1]–[4])?

1.41 2'-Deoxyadenosine is a component of DNA.

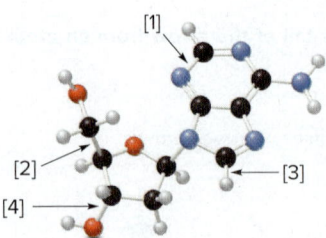

2'-deoxyadenosine

a. What is the molecular formula for 2'-deoxyadenosine?
b. How many lone pairs are present?
c. Draw a skeletal structure.
d. How many sp^2 hybridized carbons are present?
e. What orbitals are used to form each indicated bond ([1]–[4])?

Lewis Structures and Formal Charge

1.42 Give the formal charge on the highlighted carbon in each species. All H's and electrons on the highlighted carbon are drawn in.

a. 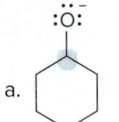 b. 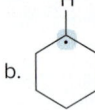 c. 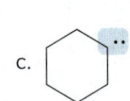 d. 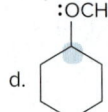 e. 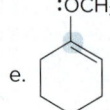 f.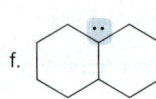

1.43 Assign formal charges to each N and O atom in the given molecules. All lone pairs have been drawn in.

a. 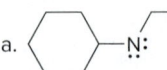 b. :N=N=N: c. 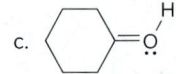 d. cyclopentyl-N=O

1.44 Draw one valid Lewis structure for each compound. Assume the atoms are arranged as drawn.

a. CH_2N_2 H C N N
                H

b. CH_3NO_2 H C N O
                H O

c. CH_3CNO H C C N O
                H

d. $(CH_2CN)^-$ H C C N
                 H

1.45 Draw an acceptable Lewis structure from each condensed structure, such that all atoms have zero formal charge.
a. diethyl ether, $(CH_3CH_2)_2O$, the first general anesthetic used in medical procedures
b. acrylonitrile, CH_2CHCN, starting material used to manufacture synthetic Orlon fibers
c. dihydroxyacetone, $(HOCH_2)_2CO$, an ingredient in sunless tanning products
d. acetic anhydride, $(CH_3CO)_2O$, a reagent used to synthesize aspirin

Isomers and Resonance Structures

1.46 Creatine is a dietary supplement used by some athletes to boost their athletic performance. (a) Draw in all lone pairs in creatine. (b) Draw two additional resonance structures showing all lone pairs and formal charges.

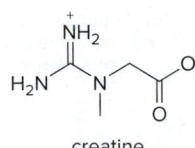

creatine

1.47 Draw all possible isomers for each molecular formula.
a. C_3H_7Cl (two isomers)   b. C_2H_4O (three isomers)   c. C_3H_9N (four isomers)

1.48 Draw Lewis structures for the nine isomers having molecular formula C_3H_6O, with all atoms having a zero formal charge.

1.49 With reference to anion **A**, label compounds **B–E** as an isomer or resonance structure of **A**. For each isomer, indicate what bonds differ from **A**.

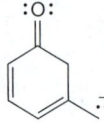

A   B   C   D   E

1.50 Which of the following species is a valid resonance structure of **A**? Use curved arrows to show how **A** is converted to any valid resonance structure. When a compound is not a valid resonance structure of **A**, explain why not.

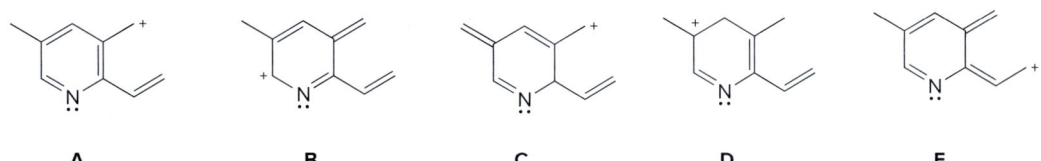

1.51 How are the molecules or ions in each pair related? Classify them as resonance structures, isomers, or neither.

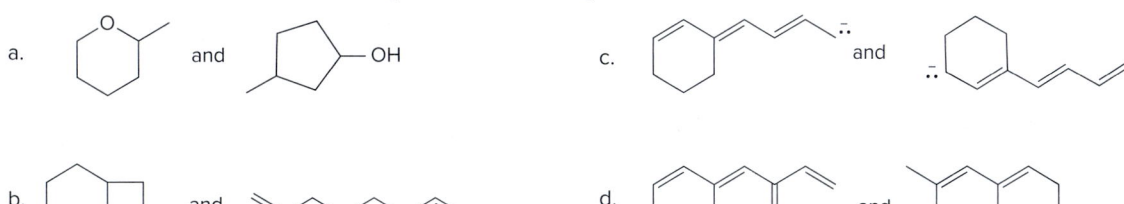

1.52 Add curved arrows to show how the first resonance structure can be converted to the second.

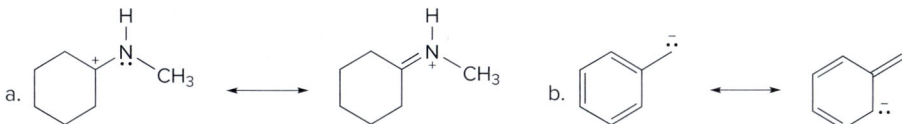

1.53 Follow the curved arrows to draw a second resonance structure for each species.

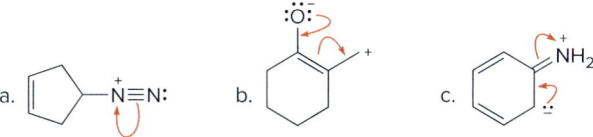

1.54 Draw a second resonance structure for each ion. Then, draw the resonance hybrid.

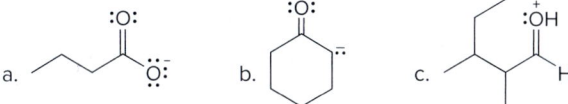

1.55 Draw all reasonable resonance structures for each species.

a. O_3 b. NO_3^- (a central N atom)   c. N_3^- d.   e.

1.56 Consider compounds **A–D**, which contain both a heteroatom and a double bond. (a) For which compounds are no additional Lewis structures possible? (b) When two or more Lewis structures can be drawn, draw all additional resonance structures.

1.57 Draw all reasonable resonance structures for the following cation. Then draw the resonance hybrid.

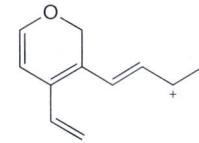

1.58 Which of the given resonance structures (**A**, **B**, or **C**) contributes most to the resonance hybrid? Which contributes least?

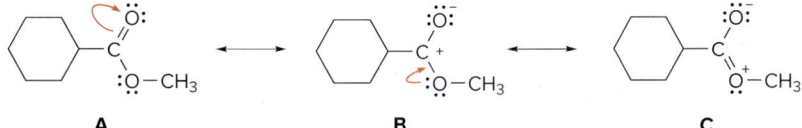

54 Chapter 1 Structure and Bonding

1.59 Consider the compounds and ions with curved arrows drawn below. When the curved arrows give a second valid resonance structure, draw the resonance structure. When the curved arrows generate an invalid Lewis structure, explain why the structure is unacceptable.

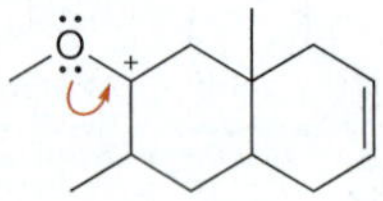

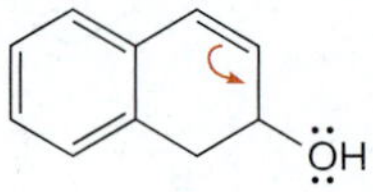

a. 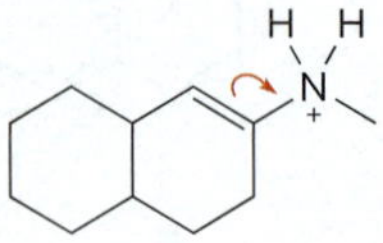 b. 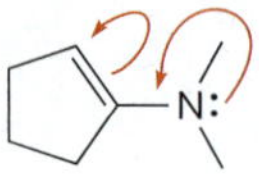 c. d. 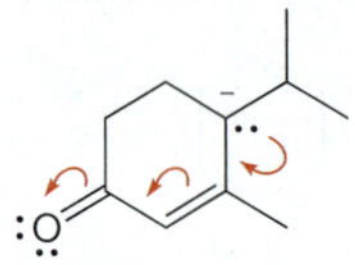 e. f.

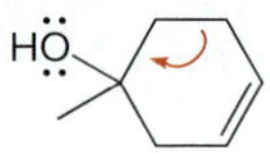

Geometry

1.60 Predict all bond angles in each compound.

a. CH_3Cl b. NH_2OH c. $CH_2=NCH_3$ d. $HC\equiv CCH_2OH$ e.

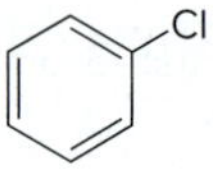

1.61 Predict the geometry around each highlighted atom.

a.    b. $(CH_3)_2N^-$ c. 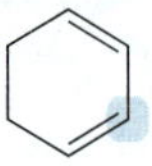   d. 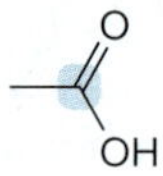   e. $(CH_3)_3N$

Drawing Organic Molecules

1.62 How many hydrogens are present around each carbon atom in the following molecules?

a. 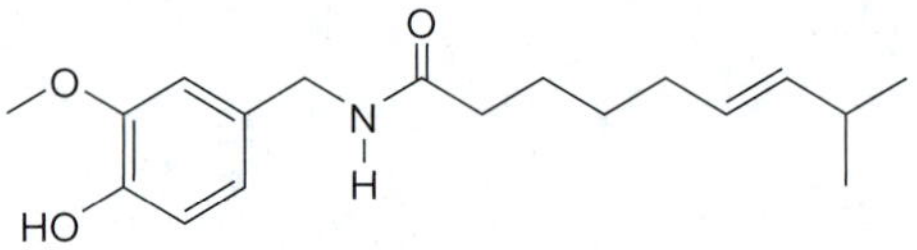 capsaicin (spicy component of hot peppers)

b. 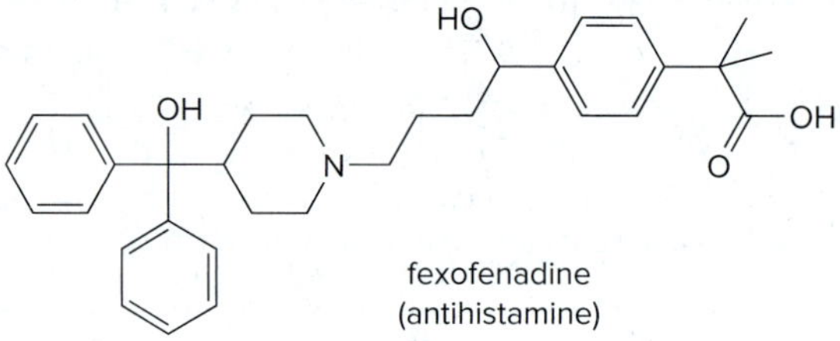 fexofenadine (antihistamine)

1.63 Draw in all the carbon and hydrogen atoms in each molecule.

a. 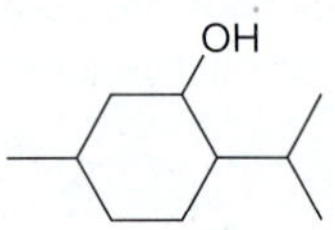 menthol (isolated from peppermint oil)

b. 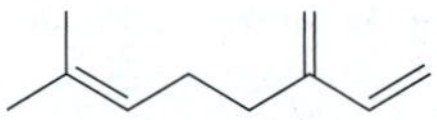 myrcene (isolated from bayberry)

c. 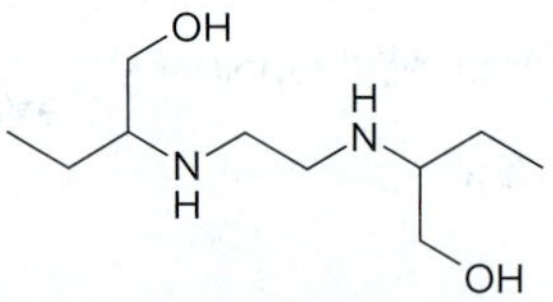 ethambutol (drug used to treat tuberculosis)

d. 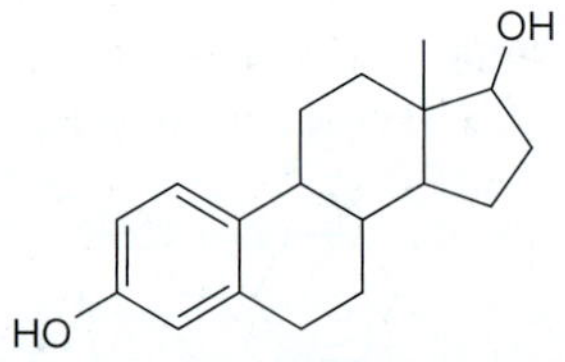 estradiol (a female sex hormone)

1.64 Convert each molecule to a skeletal structure.

a. $(CH_3)_2CHCH_2CH_2CH(CH_3)_2$

b. $CH_3CH(Cl)CH(OH)CH_3$

c. $CH_3(CH_2)_2C(CH_3)_2CH(CH_3)CH(CH_3)CH(Br)CH_3$

d. 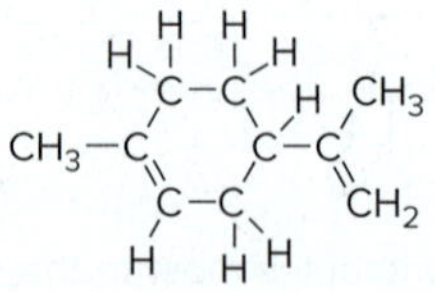 limonene (oil of lemon)

1.65 Convert the following condensed formulas into skeletal structures.

a. $CH_3CONHCH_3$ b. CH_3COCH_2Br c. $(CH_3)_3COH$ d. CH_3COCl e. $CH_3COCH_2CO_2H$ f. $HO_2CCH(OH)CO_2H$

1.66 Draw in all the hydrogen atoms and nonbonded electron pairs in each ion.

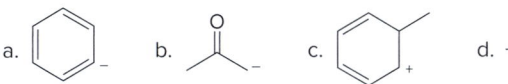

Hybridization

1.67 Predict the hybridization and geometry around each highlighted atom.

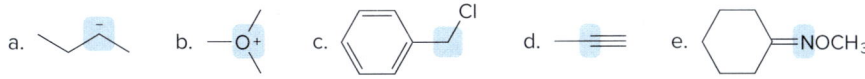

1.68 What orbitals are used to form each highlighted bond? For multiple bonds, indicate the orbitals used in individual bonds.

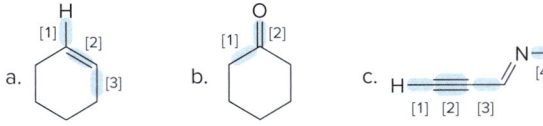

1.69 Ketene, $CH_2=C=O$, is an unusual organic molecule that has a single carbon atom doubly bonded to two different atoms. Determine the hybridization of both C atoms and the O in ketene. Then, draw a diagram showing what orbitals are used to form each bond (similar to Figures 1.10 and 1.11).

Bond Length and Strength

1.70 Rank the following bonds in order of *increasing* bond length.

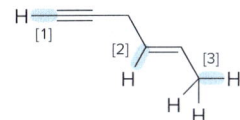

1.71 Answer the following questions about compound **A**.

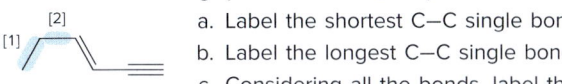

a. Label the shortest C–C single bond.
b. Label the longest C–C single bond.
c. Considering all the bonds, label the shortest C–C bond.
d. Label the weakest C–C bond.
e. Label the strongest C–H bond.
f. Explain why bond [1] and bond [2] are different in length, even though they are both C–C single bonds.

1.72 Two useful organic compounds that contain Cl atoms are vinyl chloride ($CH_2=CHCl$) and chloroethane (CH_3CH_2Cl). Vinyl chloride is the starting material used to prepare poly(vinyl chloride), a plastic in insulation, pipes, and bottles. Chloroethane (ethyl chloride) is a local anesthetic. Why is the C–Cl bond in vinyl chloride stronger than the C–Cl bond in chloroethane?

Bond Polarity

1.73 Use the symbols δ+ and δ– to indicate the polarity of the highlighted bonds.

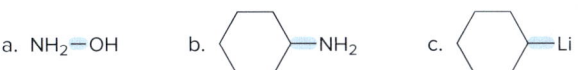

1.74 Label the polar bonds in each molecule. Indicate the direction of the net dipole (if there is one).

a. $CHBr_3$
b. $CH_3CH_2OCH_2CH_3$
c. [structure with two Br on benzene ring]
d. [structure with two Cl on benzene ring]

General Problems

1.75 Anacin is an over-the-counter pain reliever that contains aspirin and caffeine. Answer the following questions about each compound.

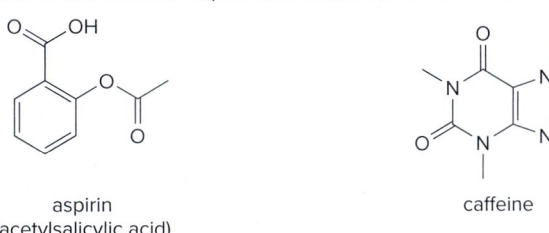

aspirin
(acetylsalicylic acid)

caffeine

a. What is the molecular formula?
b. How many lone pairs are present on heteroatoms?
c. Label the hybridization state of each carbon.
d. Draw three additional resonance structures.

1.76 Answer the following questions about acetonitrile (CH$_3$C≡N:).
a. Determine the hybridization of both C atoms and the N atom.
b. Label all bonds as σ or π.
c. In what type of orbital does the lone pair on N reside?
d. Label all bonds as polar or nonpolar.

1.77 As we will learn in Chapters 16 and 27, acetyl coenzyme A (acetyl CoA) is a key organic reactant in many biochemical transformations in cells.
a. How many lone pairs does acetyl CoA contain?
b. Which atoms in the structure of acetyl CoA do not follow the octet rule?
c. Give the hybridization of each atom highlighted in blue.
d. In what type of orbital do the lone pairs on each atom shown in red reside?
e. Draw three additional resonance structures.

acetyl coenzyme A

1.78 (a) Add curved arrows to show how the starting material **A** is converted to the product **B**. (b) Draw all reasonable resonance structures for **B**. (c) Draw the resonance hybrid for **B**.

A   B

1.79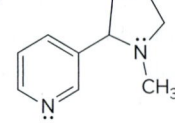

a. What is the hybridization of each N atom in nicotine?
b. What is the geometry around each N atom?
c. In what type of orbital does the lone pair on each N atom reside?
d. Draw a constitutional isomer of nicotine.
e. Draw a resonance structure of nicotine.

nicotine

1.80 Stalevo is the trade name for a medication used for Parkinson's disease, which contains L-dopa, carbidopa, and entacapone.

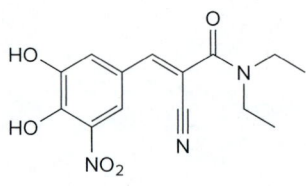

entacapone

a. Draw a Lewis structure for entacapone.
b. Which C–C bond in entacapone is the longest?
c. Which C–C single bond is the shortest?
d. Which C–N bond is the longest?
e. Which C–N bond is the shortest?
f. Use curved arrows to draw a resonance structure that is an equal contributor to the resonance hybrid.
g. Use curved arrows to draw a resonance structure that is a minor contributor to the resonance hybrid.

1.81 CH$_3^+$ and CH$_3^-$ are two highly reactive carbon species.
a. What is the predicted hybridization and geometry around each carbon atom?
b. Two electrostatic potential plots are drawn for these species. Which ion corresponds to which diagram and why?

A

B

Challenge Problems

1.82 The N atom in CH$_3$CONH$_2$ (acetamide) is *sp*^{2} hybridized, even though it is surrounded by four groups. Using this information, draw a diagram that shows the orbitals used by the atoms in the —CONH$_2$ portion of acetamide, and offer an explanation as to the observed hybridization.

1.83 Use the observed bond lengths to answer each question. (a) Why is bond [1] longer than bond [2] (143 pm versus 136 pm)? (b) Why are bonds [3] and [4] equal in length (127 pm), and shorter than bond [2]?

1.84 Draw at least 10 more resonance structures for acetaminophen, the active pain reliever in Tylenol.

acetaminophen

1.85 When two carbons having different hybridization are bonded together, the C—C bond contains a slight dipole. In a C$_{sp^2}$—C$_{sp^3}$ bond, what is the direction of the dipole? Which carbon is considered more electronegative?

1.86 Draw all possible isomers having molecular formula C$_4$H$_8$ that contain one π bond.

1.87 Use the principles of resonance theory to explain why carbocation **A** is more stable than carbocation **B**.

A   **B**

1.88 The curved arrow notation introduced in Section 1.6B is a powerful method used by organic chemists to show the movement of electrons not only in resonance structures, but also in chemical reactions. Because each curved arrow shows the movement of two electrons, following the curved arrows illustrates what bonds are broken and formed in a reaction. Consider the following three-step process. (a) Add curved arrows in Step [1] to show the movement of electrons. (b) Use the curved arrows drawn in Step [2] to identify the structure of **X**. **X** is converted in Step [3] to phenol and HCl.

phenol

2 Acids and Bases

Comstock/PunchStock

- 2.1 Brønsted–Lowry acids and bases
- 2.2 Reactions of Brønsted–Lowry acids and bases
- 2.3 Acid strength and pK_a
- 2.4 Predicting the outcome of acid–base reactions
- 2.5 Factors that determine acid strength
- 2.6 Common acids and bases
- 2.7 Aspirin
- 2.8 Lewis acids and bases

The rich flavor and aroma of a freshly brewed cup of coffee results from a myriad of organic compounds. The mild acidity of coffee made from the beans of plants grown at higher altitudes or in volcanic soil is in part due to **quinic acid,** an organic acid present in low concentration in green coffee beans. Quinic acid concentration increases during processing, as more-complex compounds are degraded by the heat of roasting, and it contributes to the increase in the perceived acidity of coffee that has been warmed for a long time on a hot surface. In Chapter 2, we learn about acidity and acid–base reactions.

Why Study... Acids and Bases?

Chemical terms such as *anion* and *cation* may be unfamiliar to most nonscientists, but *acid* has found a place in everyday language. Commercials advertise the latest remedy for the heartburn caused by excess stomach *acid*. The nightly news may report the latest environmental impact of *acid* rain. Wine lovers know that wine sours because its alcohol has turned to *acid*. Acid comes from the Latin word *acidus,* meaning "sour," because when tasting compounds was a routine method of identification, these compounds were sour.

In Chapter 2, we concentrate on two definitions of acids and bases: the **Brønsted–Lowry** definition, which describes acids as **proton donors** and bases as **proton acceptors;** and the **Lewis** definition, which describes acids as **electron pair acceptors** and bases as **electron pair donors.**

2.1 Brønsted–Lowry Acids and Bases

The Brønsted–Lowry definition describes acidity in terms of protons: positively charged **hydrogen ions, H^+.**

> The general words "acid" and "base" usually mean a *Brønsted–Lowry* acid and *Brønsted–Lowry* base.
>
> H^+ = proton.
>
> **HA** = Brønsted–Lowry acid.
> **B:** = Brønsted–Lowry base.

- A Brønsted–Lowry acid is a *proton donor.*
- A Brønsted–Lowry base is a *proton acceptor.*

A Brønsted–Lowry acid must contain a *hydrogen* atom. This definition of an acid is often familiar to students, because many inorganic acids in general chemistry are Brønsted–Lowry acids. The symbol **HA** is used for a general Brønsted–Lowry acid.

A Brønsted–Lowry base must be able to form a bond to a proton. Because a proton has no electrons, **a base must contain an "available" electron pair** that can be easily donated to form a new bond. These include **lone pairs** or electron pairs in **π bonds.** The symbol **B:** is used for a general Brønsted–Lowry base. Examples of Brønsted–Lowry acids and bases are given in Figure 2.1.

Charged species such as ^-OH and $^-NH_2$ are used as **salts,** with cations such as Li^+, Na^+, or K^+ to balance the negative charge. These cations are called **counterions** or **spectator ions,** and their **identity is usually inconsequential.** For this reason, the counterion is often omitted.

Na**OH**	=	Na^+	$^-$**OH**
K**OH**	=	K^+	$^-$**OH**
salt		counterion	base

Compounds like H_2O and CH_3OH that contain both hydrogen atoms and lone pairs may be either an acid or a base, depending on the particular reaction. These fundamental principles

Figure 2.1 Examples of Brønsted–Lowry acids and bases

a. **Brønsted–Lowry acids (HA)**

HCl
H_2SO_4
HSO_4^-
H_2O
H_3O^+

acetic acid

- All Brønsted–Lowry acids contain a proton.
- The net charge may be zero, (+), or (−).

b. **Brønsted–Lowry bases (B:)**

$H_2\ddot{O}:$ $:\ddot{O}H$ $CH_3\ddot{O}:$

$:NH_3$ $:\ddot{N}H_2$ $CH_3\ddot{N}H_2$

- All Brønsted–Lowry bases contain a lone pair of electrons or a π bond.
- The net charge may be zero or (−).

are true no matter how complex the compound. For example, the addictive pain reliever **morphine** is a Brønsted–Lowry acid because it contains many hydrogen atoms. It is also a Brønsted–Lowry base because it has lone pairs on O and N, and four π bonds.

morphine

- H atoms on O make morphine an acid.
- Lone pairs and π bonds (in blue) make morphine a base.

Morphine is obtained from the opium poppy. *Mafoto/Getty Images*

Problem 2.1

a. Which compounds are Brønsted–Lowry acids: HBr, NH₃, CCl₄?
b. Which compounds are Brønsted–Lowry bases: CH₃CH₃, (CH₃)₃CO⁻, HC≡CH?
c. Classify each compound as an acid, a base, or both: CH₃CH₂OH, CH₃CH₂CH₂CH₃, CH₃CO₂CH₃.

2.2 Reactions of Brønsted–Lowry Acids and Bases

A Brønsted–Lowry acid–base reaction results in transfer of a proton from an acid to a base. These acid–base reactions, also called *proton transfer reactions,* are fundamental to the study of organic chemistry.

Consider, for example, the reaction of the acid HA with the base :B. **In an acid–base reaction, one bond is broken and one is formed.**

- The electron pair of the base B: forms a new bond to the proton of the acid.
- The acid HA loses a proton, leaving the electron pair in the HA bond on A.

Recall from Section 1.6 that a curved arrow shows the movement of an **electron pair**. **The tail of the arrow always begins at an electron pair,** and the head points to where that electron pair "moves."

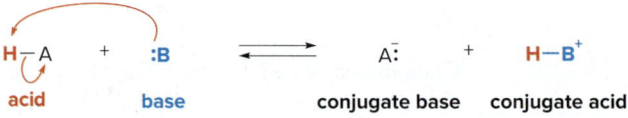

This "movement" of electrons in reactions can be illustrated using curved arrow notation. Because **two electron pairs** are involved in this reaction, **two curved arrows** are needed. Two products are formed.

- Loss of a proton from an acid forms its *conjugate base.*
- Gain of a proton by a base forms its *conjugate acid.*

The **net charge must be the same** on both sides of any equation. In this example, the net charge on each side is zero. Individual charges can be calculated using formal charges. **A double reaction arrow** is used between starting materials and products to indicate that the reaction can proceed in the forward and reverse directions. These are **equilibrium arrows.**

2.2 Reactions of Brønsted–Lowry Acids and Bases

Two examples of proton transfer reactions are drawn here with curved arrow notation.

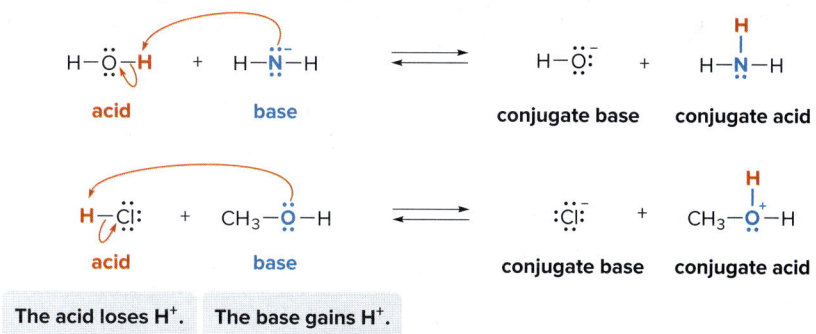

A double reaction arrow indicates equilibrium.

⇌

equilibrium arrows

Remove H⁺ from an acid to form its conjugate base.
Add H⁺ to a base to form its conjugate acid.

The acid loses H⁺. The base gains H⁺.

- Brønsted–Lowry acid–base reactions always result in the transfer of a proton from an acid to a base.

The ability to identify and draw a conjugate acid or base from a given starting material is illustrated in Sample Problems 2.1 and 2.2.

Sample Problem 2.1 | Drawing a Conjugate Acid and a Conjugate Base

a. What is the conjugate acid of CH_3O^-?
b. What is the conjugate base of NH_3?

Solution

a. Add H⁺ to CH_3O^- to form its conjugate acid.

$$CH_3-\ddot{O}:^- \xrightarrow{\text{Add H}^+ \text{ to a lone pair.}} CH_3-\ddot{O}-H$$

base   conjugate acid

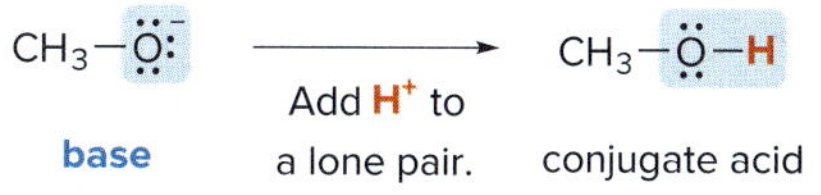

b. Remove H⁺ from NH_3 to form its conjugate base.

acid → conjugate base (The electron pair stays on N.)

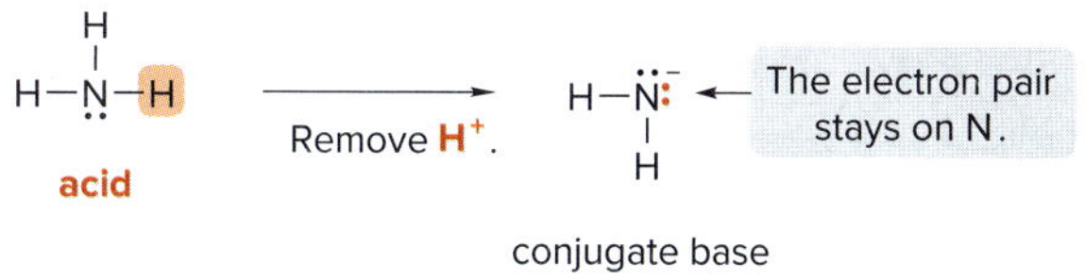

Problem 2.2
a. Draw the conjugate acid of each base: NH_3, Cl^-, $(CH_3)_2C=O$.
b. Draw the conjugate base of each acid: HBr, HSO_4^-, CH_3OH.

More Practice: Try Problems 2.38, 2.39.

Problem 2.3 Label each statement as True or False.

a. $CH_3CH_2^+$ is the conjugate acid of $CH_2=CH_2$.
b. $CH_3CH_2^-$ is the conjugate base of $CH_3CH_2^+$.
c. $CH_2=CH_2$ is the conjugate base of $CH_3CH_2^-$.
d. $CH_2=CH^-$ is the conjugate base of $CH_2=CH_2$.
e. CH_3CH_3 is the conjugate acid of $CH_3CH_2^-$.

Chapter 2 Acids and Bases

Sample Problem 2.2 — Determining the Acid, Base, Conjugate Acid, and Conjugate Base in a Reaction

Label the acid and base, and the conjugate acid and base, in the following reaction. Use curved arrow notation to show the movement of electron pairs.

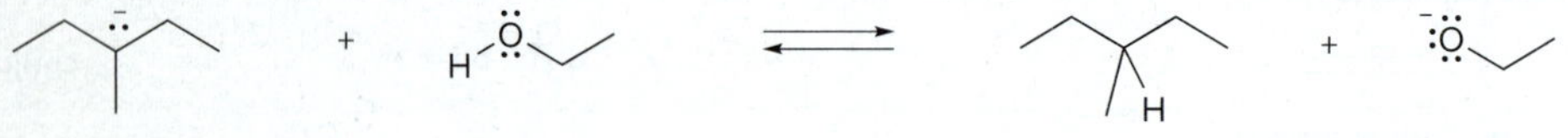

| A | B | C | D |

Solution

A is the base because it accepts a proton, forming its conjugate acid, **C**. **B** is the acid because it donates a proton, forming its conjugate base, **D**. Two curved arrows are needed because two electron pairs are involved. One shows that the lone pair on **A** bonds to a proton of **B**, and the second shows that the electron pair in the O–H bond remains on O.

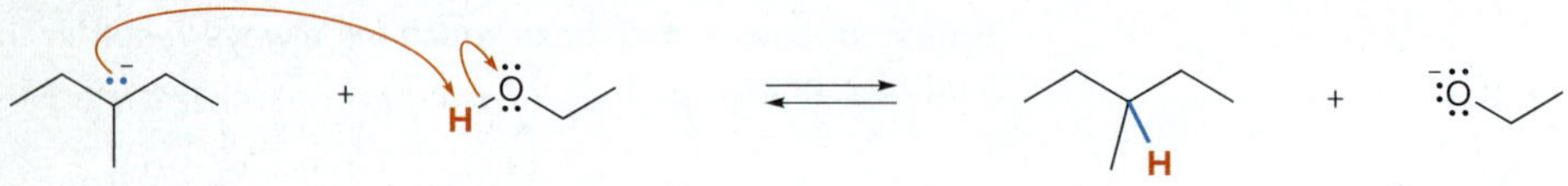

| A | B | C | D |
| base | acid | conjugate acid | conjugate base |

The base gains H$^+$.   The acid loses H$^+$.

Problem 2.4

Label the acid and base, and the conjugate acid and base, in the following reactions. Use curved arrows to show the movement of electron pairs.

a.  H—Cl: + ... ⇌ :Cl:⁻ + ...

b.  ... + ... ⇌ ... + ...

c. 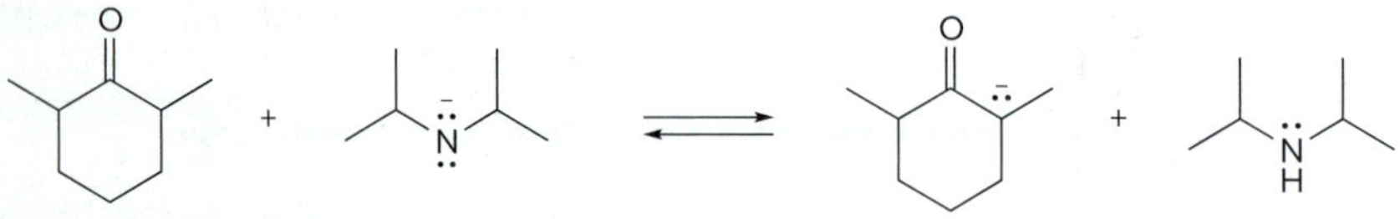 ... + ... ⇌ ... + ...

More Practice: Try Problem 2.40.

In all proton transfer reactions, the **electron-rich base** donates an electron pair to the acid, which usually has a polar HA bond. The H of the acid bears a partial positive charge, making it **electron deficient**. This is the first example of a general pattern of reactivity.

- Electron-rich species react with electron-deficient ones.

Given two starting materials, how do you know which is the acid and which is the base in a proton transfer reaction? Use the following generalizations:

[1] Common acids and bases introduced in general chemistry are used in the same way in organic reactions. HCl and H_2SO_4 are strong acids, and ⁻OH is a strong base.

[2] When only one starting material contains a hydrogen, it must be the acid. If only one starting material has a lone pair or a π bond, it must be the base.

[3] A starting material with a net positive charge is usually the acid. A starting material with a negative charge is usually the base.

2.3 Acid Strength and pK_a

Figure 2.2 shows how to use these generalizations to identify the acid and base with pairs of compounds.

Figure 2.2 Identifying the acid and the base in a proton transfer reaction

a. When one reactant has a net (+) or (−) charge:

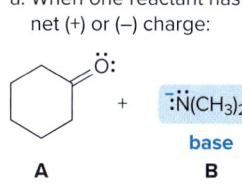

B has a (−) charge, so **B** is the base.

b. When one reactant is an inorganic acid or base:

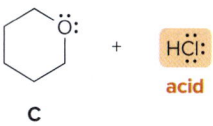

HCl is a strong inorganic acid, so HCl is the acid.

c. When only one reactant has a H or lone pair:

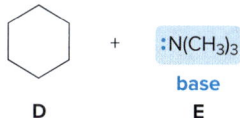

E has a lone pair and **D** does not, so **E** is the base.

Problem 2.5 Decide which compound is the acid and which is the base, and draw the products of each proton transfer reaction.

a. Cl₃C-COOH + ⁻:OCH₃ ⇌

b. Cl₃C-C≡C-H + H:⁻ ⇌

c. cyclopentyl-NH₂ + HCl ⇌

d. CH₃CH₂OH + H₂SO₄ ⇌

Problem 2.6 Draw the products formed from the acid–base reaction of HCl with each compound.

a. CH₃CH₂CH₂CH(OH)CH₃   b. CH₃-O-CH₂CH₃   c. (CH₃)₂N-CH₂CH₃   d. pyrrolidine (NH)

2.3 Acid Strength and pK_a

Acid strength is the tendency of an acid to donate a proton.

- The more readily a compound donates a proton, the *stronger* the acid.

Acidity is measured by an equilibrium constant. When a Brønsted–Lowry acid HA is dissolved in water, an acid–base reaction occurs, and an equilibrium constant K_{eq} can be written for the reaction.

$$H-A \; + \; H-\overset{..}{\underset{..}{O}}-H \;\rightleftharpoons\; A:^- \; + \; H-\overset{H}{\underset{..}{\overset{|}{O}}}-H^+$$

acid   base (solvent)

$$K_{eq} = \frac{[\text{products}]}{[\text{starting materials}]} = \frac{[H_3O^+][A:^-]}{[HA][H_2O]}$$

Because the concentration of the solvent H_2O is essentially constant, the equation can be rearranged and a new equilibrium constant, called the **acidity constant, K_a,** can be defined.

$$K_a = [H_2O]K_{eq} = \frac{[H_3O^+][A:^-]}{[HA]}$$

How is the magnitude of K_a related to acid strength?

- The *stronger the acid*, the farther the equilibrium lies to the right and the *larger the K_a*.

Chapter 2 Acids and Bases

For most organic compounds, K_a is small, typically 10^{-5} to 10^{-50}. This contrasts with the K_a values for many inorganic acids, which range from 10^0 to 10^{10}. Because using exponents can be cumbersome, it is often more convenient to use pK_a values instead of K_a values.

$$pK_a = -\log K_a$$

How does pK_a relate to acid strength?

> Recall that a **log** is an **exponent**; for example, $\log 10^{-5} = -5$.

K_a values of typical organic acids

10^{-5} → 10^{-50}

larger number — stronger acid
smaller number — weaker acid

pK_a values of typical organic acids

$+5$ → $+50$

smaller number — stronger acid
larger number — weaker acid

- The *smaller* the pK_a, the *stronger* the acid.

Problem 2.7 Which compound in each pair is the stronger acid?

a. $CH_3CH_2CH_3$ or CH_3CH_2OH
   $pK_a = 50$ $pK_a = 16$

b. C₆H₅–OH ($K_a = 10^{-10}$) or C₆H₅–CH₃ ($K_a = 10^{-41}$)

Problem 2.8 Use a calculator when necessary to answer the following questions.
a. What is the pK_a for each K_a: 10^{-10}, 10^{-21}, and 5.2×10^{-5}?
b. What is the K_a for each pK_a: 7, 11, and 3.2?

An inverse relationship exists between acidity and basicity.

- A *strong acid* readily donates a proton, forming a *weak conjugate base*.
- A *strong base* readily accepts a proton, forming a *weak conjugate acid*.

Table 2.1 is a brief list of pK_a values for some common compounds, ranked in order of *increasing* pK_a and therefore *decreasing* acidity. Because strong acids form weak conjugate bases, this list also ranks their conjugate bases, in order of *increasing* basicity. CH_4 is the weakest acid in the list, because it has the highest pK_a (50). Its conjugate base, CH_3^-, is therefore the strongest conjugate base. An extensive pK_a table is located in Appendix C.

Table 2.1 Selected pK_a Values

Acid	pK_a	Conjugate base
H–Cl	–7	Cl⁻
CH_3CO_2–H	4.8	$CH_3CO_2^-$
HO–H	15.7	HO⁻
CH_3CH_2O–H	16	$CH_3CH_2O^-$
HC≡CH	25	HC≡C⁻
H–H	35	H⁻
H_2N–H	38	H_2N^-
$CH_2=CH_2$	44	$CH_2=\bar{C}H$
CH_3–H	50	CH_3^-

(Increasing acidity ↑ on left; Increasing basicity ↓ on right)

Comparing pK_a values tells us the **relative acidity of two acids,** and the **relative basicity of their conjugate bases,** as shown in Sample Problem 2.3.

Sample Problem 2.3   Using pK_a Values to Determine Relative Acidity and Basicity

Rank the following compounds in order of increasing acidity, and then rank their conjugate bases in order of increasing basicity.

Solution

Use the pK_a values in Table 2.1 and the rule: **the *lower* the pK_a, the *stronger* the acid.**

$CH_2=CH_2$ pK_a = 44
CH_3CO_2H pK_a = 4.8
$H-Cl$ pK_a = –7

⟶ **Increasing acidity**

Remove a proton to draw the conjugate bases. Because strong acids form weak conjugate bases, the **basicity of conjugate bases *increases* with *increasing* pK_a** of their acids.

⟵ **Increasing basicity**

Problem 2.9 Rank the conjugate bases of each group of acids in order of increasing basicity.
a. NH_3, H_2O, CH_4 b. $CH_2=CH_2$, $HC\equiv CH$, CH_4

More Practice: Try Problem 2.54a, b.

Problem 2.10 Consider two acids: HCO_2H (formic acid, pK_a = 3.8) and pivalic acid [$(CH_3)_3CCO_2H$, pK_a = 5.0]. (a) Which acid has the larger K_a? (b) Which acid is stronger? (c) Which acid forms the stronger conjugate base? (d) When each acid is dissolved in water, for which acid does the equilibrium lie farther to the right?

The pK_a values in Table 2.1 span a large range (–7 to 50). The pK_a scale is logarithmic, so a small difference in pK_a translates into a large numerical difference. The difference between the pK_a values of NH_3 (38) and $CH_2=CH_2$ (44) is six pK_a units, so NH_3 is 10^6 or *one million times more acidic* than $CH_2=CH_2$.

Although Table 2.1 is abbreviated, it is a useful tool for *estimating* the pK_a of a compound similar though not identical to one in the table. Suppose you are asked to estimate the pK_a of the N–H bond of CH_3NH_2. Although CH_3NH_2 is not listed in the table, we have enough information to *approximate* its pK_a. Because the pK_a of the N–H bond of NH_3 is 38, we can estimate the pK_a of the N–H bond of CH_3NH_2 to be 38. Its actual pK_a is 40, so this is a good first approximation.

Problem 2.11 Estimate the pK_a of each of the indicated bonds.

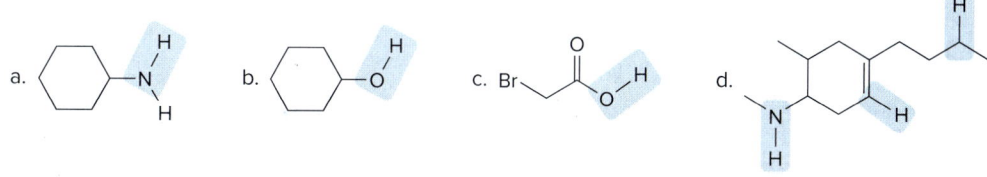

2.4 Predicting the Outcome of Acid–Base Reactions

In a proton transfer reaction, the **stronger** acid reacts with the **stronger** base to form the weaker acid and the weaker base.

A proton transfer reaction represents an equilibrium. Because an acid donates a proton to a base, forming a conjugate acid and conjugate base, there are always two acids and two bases in the reaction mixture. Which pair of acids and bases is favored at equilibrium? **The position of the equilibrium depends on the relative strengths of the acids and bases.**

- Equilibrium always favors formation of the *weaker* acid and base.

Because a strong acid readily donates a proton and a strong base readily accepts one, these two species react to form a weaker conjugate acid and base that do not donate or accept a proton as readily. Comparing pK_a values allows us to determine the position of equilibrium, as illustrated in Sample Problem 2.4.

Sample Problem 2.4 Using pK_a Values to Predict the Direction of Equilibrium

Determine the direction of equilibrium when acetylene (HC≡CH) reacts with $^-NH_2$ in a proton transfer reaction.

Solution

Follow three steps to determine the position of equilibrium:

Step [1] Identify the acid and base in the starting materials.

- $^-NH_2$ is the base because it bears a net negative charge, so HC≡CH is the acid.

Step [2] Draw the products of proton transfer and identify the conjugate acid and base in the products.

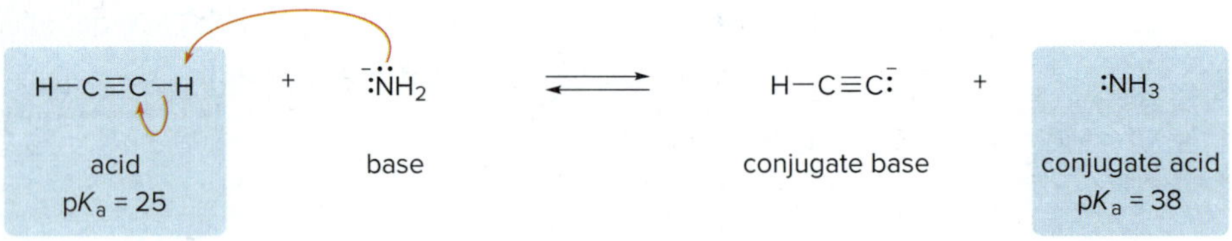

$$H-C\equiv C-H \ (\text{acid}, pK_a = 25) \ + \ {:}NH_2^- \ (\text{base}) \ \rightleftharpoons \ H-C\equiv C{:}^- \ (\text{conjugate base}) \ + \ {:}NH_3 \ (\text{conjugate acid}, pK_a = 38)$$

Step [3] Compare the pK_a values of the acid and the conjugate acid. Equilibrium favors formation of the *weaker* acid with the *higher* pK_a.

$$H-C\equiv C-H \ (pK_a = 25, \text{stronger acid}) \ + \ {:}NH_2^- \ (\text{base}) \ \rightleftharpoons \ H-C\equiv C{:}^- \ (\text{conjugate base}) \ + \ {:}NH_3 \ (pK_a = 38, \text{weaker acid})$$

Use unequal equilibrium arrows (⇀↽ or ⇌) for a reversible reaction in which products or reactants are favored at equilibrium.

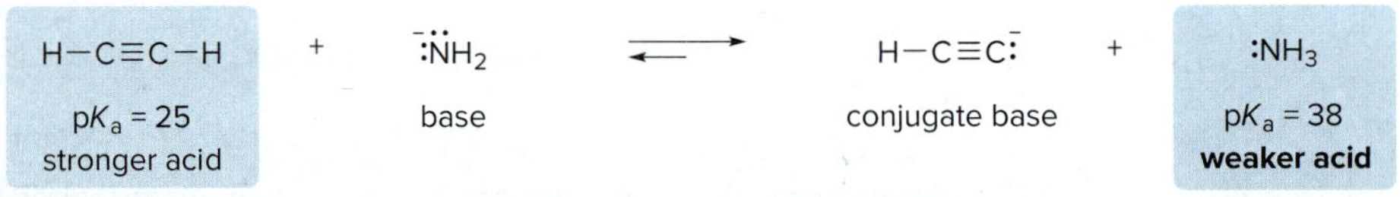

Equilibrium favors the products, forming the weaker acid.

- Because the pK_a of the starting acid (25) is *lower* than the pK_a of the conjugate acid (38), HC≡CH is a *stronger* acid and equilibrium favors the products.

Problem 2.12 Draw the products of each reaction and determine the direction of equilibrium.

a. $H_2C=CH_2 \ + \ H{:}^- \ \rightleftharpoons$

b. $CH_4 \ + \ {:}\ddot{O}H^- \ \rightleftharpoons$

c. $CH_3COOH \ + \ {:}\ddot{O}CH_2CH_3^- \ \rightleftharpoons$

d. $:\ddot{C}l{:}^- \ + \ H-\ddot{O}CH_2CH_3 \ \rightleftharpoons$

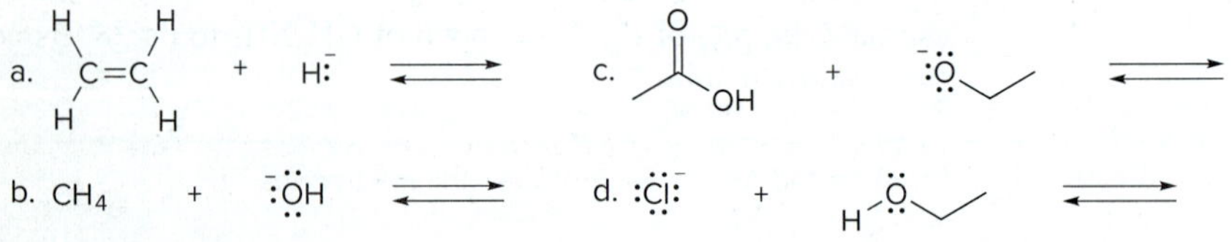

More Practice: Try Problems 2.48, 2.49.

How can we know if a particular base is strong enough to deprotonate a given acid, so that the equilibrium lies to the right? The pK_a table readily gives us this information, as shown in Sample Problem 2.5.

Sample Problem 2.5 Determining if a Base Is Strong Enough to Deprotonate an Acid

Which of the following bases is strong enough to deprotonate N,N-dimethylacetamide [CH$_3$CON(CH$_3$)$_2$, pK_a = 30], so that equilibrium favors the products: (a) NaNH$_2$; (b) NaOH?

Solution

- **Draw the structure of the conjugate acid of each base,** and determine its pK_a from Table 2.1 or Appendix C. **Equilibrium favors the side with the *weaker* acid that has the *higher* pK_a.**
- **Compare the pK_a values of the starting acid and the conjugate acid.** If the conjugate acid has a *higher* pK_a than the starting acid, the conjugate acid is the *weaker* acid and equilibrium favors the *products*. **The base is strong enough to deprotonate the acid.**
- If the conjugate acid has a *lower* pK_a than the starting acid, the conjugate acid is the *stronger* acid and equilibrium favors the *starting materials*. **The base is *not* strong enough to deprotonate the acid.**

a. Na$^+$ is a counterion and $^-$NH$_2$ is the base in NaNH$_2$.

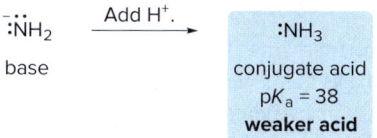

b. Na$^+$ is a counterion and $^-$OH is the base in NaOH.

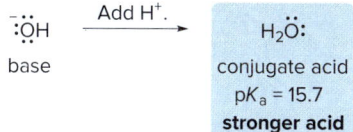

The conjugate acid (NH$_3$) of the base is a *weaker* acid than CH$_3$CON(CH$_3$)$_2$ (pK_a = 30), so the base *is* strong enough to deprotonate the acid, and **equilibrium favors the products.**

The conjugate acid (H$_2$O) of the base is a *stronger* acid than CH$_3$CON(CH$_3$)$_2$ (pK_a = 30), so the base is *not* strong enough to deprotonate the acid, and **equilibrium favors the starting materials.**

Problem 2.13 Using the data in Appendix C, determine which of the following bases is strong enough to deprotonate acetonitrile (CH$_3$CN), so that equilibrium favors the products: (a) NaH; (b) Na$_2$CO$_3$; (c) NaOH; (d) NaNH$_2$; (e) NaHCO$_3$.

More Practice: Try Problems 2.47, 2.59c.

Because Table 2.1 is arranged from low to high pK_a, **an acid can be deprotonated by the conjugate base of any acid below it in the table.**

Sample Problem 2.5 illustrates a fundamental principle in acid–base reactions.

- **An acid can be deprotonated by the conjugate base of any acid having a *higher* pK_a.**

2.5 Factors That Determine Acid Strength

The wide range of pK_a values in Table 2.1 illustrates that a tremendous difference in acidity exists among compounds. HCl (pK_a < 0) is an extremely strong acid, water (pK_a = 15.7) is moderate in acidity, and CH$_4$ (pK_a = 50) is an extremely weak acid. How are these differences explained? One general rule governs acid strength.

- **Anything that stabilizes a conjugate base A:$^-$ makes the starting acid HA more acidic.**

Four factors affect the acidity of HA:

[1] **Element effects**
[2] **Inductive effects**
[3] **Resonance effects**
[4] **Hybridization effects**

No matter which factor is discussed, follow the same procedure. To compare the acidity of any two acids:

- Draw the conjugate bases.
- Determine which conjugate base is more stable.
- The *more stable* the conjugate base, the *more acidic* the acid.

2.5A Element Effects—Trends in the Periodic Table

The most important factor determining the acidity of HA is the location of A in the periodic table.

Comparing Elements in the Same Row of the Periodic Table

To examine acidity trends **across a row** of the periodic table, we compare CH_4 and H_2O, two compounds having H atoms bonded to a second-row element. We know from Table 2.1 that **H_2O has a much *lower* pK_a and therefore is much *more acidic* than CH_4**, but why is this the case?

To answer this question, first draw both conjugate bases and then determine which is more stable. Each conjugate base has a net negative charge, but the negative charge in ^-OH is on oxygen and in CH_3^- it is on carbon.

$$H-\ddot{O}H \xrightarrow{\text{loss of } H^+} {:}\ddot{O}H^-$$
conjugate base
negative charge on **O**
more stable

$$H-\underset{H}{\overset{H}{\underset{|}{\overset{|}{C}}}}-H \xrightarrow{\text{loss of } H^+} H-\underset{H}{\overset{H}{\underset{|}{\overset{|}{C}}}}{:}^-$$
conjugate base
negative charge on **C**
less stable

Because the oxygen atom is much **more electronegative** than carbon, oxygen more readily accepts a negative charge, making ^-OH much more stable than CH_3^-. **H_2O is a stronger acid than CH_4 because ^-OH is a more stable conjugate base than CH_3^-.** This is a specific example of a general trend.

- Across a row of the periodic table, the acidity of HA *increases* as the electronegativity of A increases.

$pK_a \approx 50$	$pK_a \approx 38$	$pK_a \approx 16$	$pK_a = 3.2$
$-\text{C}-\text{H}$	$-\text{N}-\text{H}$	$-\text{O}-\text{H}$	$\text{H}-\text{F}$

Increasing electronegativity
Increasing acidity

The enormity of this effect is evident by comparing the pK_a values for these bonds. **A C–H bond is approximately 10^{47} times *less* acidic than H–F.**

Comparing Elements Down a Column of the Periodic Table

To examine acidity trends down a column of the periodic table, we compare H–F and H–Br. Draw both conjugate bases and then determine which is more stable. In this case, removal of a proton forms F^- and Br^-.

$$H-\ddot{\underset{..}{F}}{:} \xrightarrow{\text{loss of } H^+} {:}\ddot{\underset{..}{F}}{:}^-$$
conjugate base

$$H-\ddot{\underset{..}{Br}}{:} \xrightarrow{\text{loss of } H^+} {:}\ddot{\underset{..}{Br}}{:}^-$$
conjugate base

2.5 Factors That Determine Acid Strength

There are two important differences between F⁻ and Br⁻—electronegativity and size. In this case, **size is more important than electronegativity.** The size of an atom or ion *increases* down a column of the periodic table, so Br⁻ is much *larger* than F⁻, and this stabilizes the negative charge.

- Positive or negative charge is stabilized when it is spread over a larger volume.

smaller anion
less stable
conjugate base

Because Br⁻ is larger than F⁻, Br⁻ is more stable than F⁻, and H–Br is a stronger acid than H–F.

H—F   H—Br
pK_a = 3.2   pK_a = −9
less acidic   more acidic

larger anion
more stable
conjugate base

This again is a specific example of a general trend.

- Down a column of the periodic table, the acidity of HA *increases* as the size of A increases.

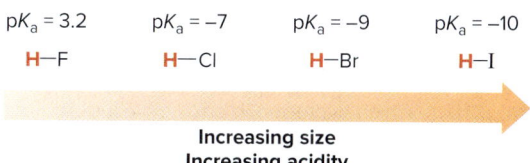

pK_a = 3.2   pK_a = −7   pK_a = −9   pK_a = −10
H—F   H—Cl   H—Br   H—I

Increasing size
Increasing acidity

Because of carbon's position in the periodic table (in the second row and to the left of O, N, and the halogens), **C–H bonds are usually the *least* acidic bonds in a molecule.**

This is *opposite* to what would be expected on the basis of electronegativity differences between F and Br, because F is more electronegative than Br. **Size, *not* electronegativity, determines acidity down a column.** Combining both trends:

- The acidity of HA *increases* both left-to-right across a row and down a column of the periodic table.

Sample Problem 2.6 Using the Identity of X in HX to Determine Relative Acidity

Without reference to a pK_a table, decide which compound in each pair is the stronger acid:

a. H₂O or HF   b. H₂S or H₂O

Solution

a. H₂O and HF both have H atoms bonded to a second-row element. Because the acidity of HA *increases across a row* of the periodic table, the H–F bond is more acidic than the H–O bond. **HF is a stronger acid than H₂O.**

b. H₂O and H₂S both have H atoms bonded to elements in the same column. Because the acidity of HA *increases down a column* of the periodic table, the H–S bond is more acidic than the H–O bond. **H₂S is a stronger acid than H₂O.**

Problem 2.14 Without reference to a pK_a table, decide which compound in each pair is the stronger acid.

a. 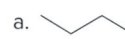 or H₂O   b. ⬠ or H₂S

More Practice: Try Problems 2.51a, b; 2.54a; 2.57.

Chapter 2 Acids and Bases

Problem 2.15 Rank the labeled H atoms in the following compound in order of increasing acidity.

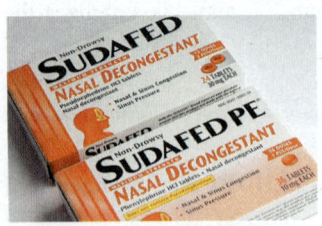

Because the pseudoephedrine (Problem 2.17) in Sudafed can be readily converted to the illegal, addictive drug methamphetamine, products that contain pseudoephedrine are now stocked behind the pharmacy counter so that their sale can be more closely monitored. Sudafed PE is a related product that contains a decongestant less easily converted to methamphetamine.
Jill Braaten/McGraw-Hill Education

When discussing acidity, the most acidic proton in a compound is the one removed first by a base. Although four factors determine the overall acidity of a particular hydrogen atom, **the element effect—the identity of A—is the single most important factor in determining the acidity of the HA bond.**

To decide which hydrogen is most acidic, **first determine what element each hydrogen is bonded to and then decide its acidity based on periodic trends.** For example, $CH_3NHCH_2CH_2CH_2CH_3$ contains only C–H and N–H bonds. Because the acidity of HA increases across a row of the periodic table, the single H on N is the most acidic H in this compound.

most acidic H shown in red

Problem 2.16 Which hydrogen in each molecule is most acidic?

Problem 2.17 Which hydrogen in pseudoephedrine, the nasal decongestant in the commercial medication Sudafed, is most acidic?

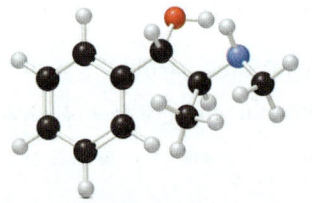

pseudoephedrine

Problem 2.18 Rank the compounds in each group in order of increasing acidity.

2.5B Inductive Effects

A second factor affecting the acidity of HA is the presence of atoms more electronegative than carbon. To illustrate this phenomenon, compare ethanol (CH_3CH_2OH) and 2,2,2-trifluoroethanol (CF_3CH_2OH), two compounds containing O–H bonds. The pK_a table

2.5 Factors That Determine Acid Strength

in Appendix C indicates that CF_3CH_2OH is a stronger acid than CH_3CH_2OH. We are comparing the acidity of the O—H bond in both compounds, so what causes the difference?

ethanol
$pK_a = 16$

2,2,2-trifluoroethanol
$pK_a = 12.4$
stronger acid

Draw both conjugate bases and then determine which is more stable. Both bases have a negative charge on an electronegative oxygen, but the second anion has three very electronegative fluorine atoms. These fluorine atoms withdraw electron density from the carbon to which they are bonded, making it electron deficient. Furthermore, this electron-deficient carbon pulls electron density through σ bonds from the negatively charged oxygen atom, stabilizing the negative charge. This is called an **inductive effect**.

No additional electronegative atoms stabilize the conjugate base.

CF_3 withdraws electron density, stabilizing the conjugate base.

- An *inductive effect* is the pull of electron density through σ bonds caused by electronegativity differences of atoms.

In this case, the electron density is pulled away from the negative charge through σ bonds by the very electronegative fluorine atoms, so it is called an **electron-*withdrawing* inductive effect**. Thus, the three very electronegative fluorine atoms stabilize the negatively charged conjugate base $CF_3CH_2O^-$, making CF_3CH_2OH a stronger acid than CH_3CH_2OH. We have learned two important principles from this discussion:

- More electronegative atoms stabilize regions of high electron density by an *electron-withdrawing* inductive effect.
- The acidity of HA *increases* with the presence of electron-withdrawing groups in A.

Inductive effects result because an electronegative atom stabilizes the negative charge of the conjugate base. **The *more electronegative* the atom and the *closer* it is to the site of the negative charge, the greater the effect.** This effect is discussed in greater detail in Chapter 15.

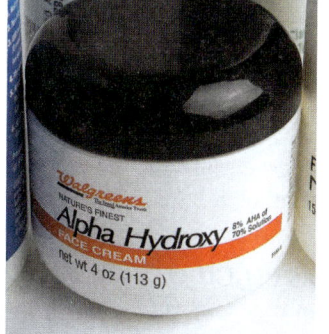

α-Hydroxy acids (Problem 2.20) are used in skin care products that purportedly smooth fine lines and improve skin texture by reacting with the outer layer of skin cells, causing them to loosen and flake off.
Jill Braaten/McGraw-Hill Education

Problem 2.19 Which compound in each pair is the stronger acid?

a. [cyclopentane-COOH with Cl] or [cyclopentane-COOH with F]

b. Cl_2CHCH_2OH or $ClCH_2CHClCH_2OH$ (approximate)

c. [isobutyric acid-like] or [with NO_2 substituent]

Problem 2.20 Glycolic acid, $HOCH_2CO_2H$, is the simplest member of a group of compounds called α-hydroxy acids, ingredients in skin care products that have an OH group on the carbon adjacent to a CO_2H group. Would you expect $HOCH_2CO_2H$ to be a stronger or weaker acid than acetic acid, CH_3CO_2H?

Problem 2.21 Explain the apparent paradox: HBr is a stronger acid than HCl, but HOCl is a stronger acid than HOBr.

2.5C Resonance Effects

> Resonance structures are two Lewis structures having the same placement of atoms but a different arrangement of electrons.

A third factor that determines acidity is resonance. Recall from Section 1.6 that resonance occurs whenever two or more different Lewis structures can be drawn for the same arrangement of atoms. To illustrate this phenomenon, compare ethanol (CH_3CH_2OH) and acetic acid (CH_3CO_2H), two compounds containing O—H bonds. Based on Table 2.1, CH_3CO_2H is a stronger acid than CH_3CH_2OH.

ethanol
pK_a = 16

acetic acid
pK_a = 4.8
stronger acid

Draw the conjugate bases of these acids to illustrate the importance of resonance. For ethoxide ($CH_3CH_2O^-$), the conjugate base of ethanol, only one Lewis structure can be drawn. The negative charge of this conjugate base is *localized* on the O atom.

ethanol
acid

ethoxide
conjugate base
only **one** Lewis structure

With acetate ($CH_3CO_2^-$), however, two resonance structures can be drawn.

acetic acid

acetate
conjugate base
two resonance structures

hybrid
resonance-stabilized
conjugate base

These two resonance structures differ in the **position of a π bond** and a **lone pair.** Although each resonance structure of acetate implies that the negative charge is localized on an O atom, in actuality, charge is *delocalized* over both O atoms. **Delocalization of electron density stabilizes acetate, making it a weaker base.**

Remember that neither resonance form adequately represents acetate. The true structure is a **hybrid** of both structures. In the hybrid, the electron pairs drawn in different locations in individual resonance structures are *delocalized*. With acetate, a dashed line is used to show that each C—O bond has partial double bond character. The symbol δ− (partial negative) indicates that the charge is delocalized on both O atoms in the hybrid.

Thus, **resonance delocalization makes $CH_3CO_2^-$ more stable than $CH_3CH_2O^-$, so CH_3CO_2H is a stronger acid than CH_3CH_2OH.** This is another example of a general rule.

- **The acidity of HA *increases* when the conjugate base A:⁻ is resonance stabilized.**

> Resonance delocalization often produces a larger effect on pK_a than the inductive effects discussed in Section 2.5B. Resonance makes CH_3CO_2H (pK_a = 4.8) a much stronger acid than CH_3CH_2OH (pK_a = 16), whereas the inductive effects due to three electronegative F atoms make CF_3CH_2OH (pK_a = 12.4) a somewhat stronger acid than CH_3CH_2OH.

Electrostatic potential plots of $CH_3CH_2O^-$ and $CH_3CO_2^-$ in Figure 2.3 indicate that the negative charge is concentrated on a single O in $CH_3CH_2O^-$, but delocalized over the O atoms in $CH_3CO_2^-$.

Figure 2.3
Electrostatic potential plots of $CH_3CH_2O^-$ and $CH_3CO_2^-$

$CH_3CH_2O^-$
The negative charge is concentrated on the single oxygen atom, making this anion *less stable*.

$CH_3CO_2^-$
The negative charge is delocalized over both oxygen atoms, making this anion *more stable*.

Problem 2.22 The C–H bond in acetone, $(CH_3)_2C=O$, has a pK_a of 19.2. Draw two resonance structures for its conjugate base. Then, explain why acetone is much more acidic than propane, $CH_3CH_2CH_3$ ($pK_a = 50$).

Problem 2.23 Rank the labeled protons in the following molecule in order of increasing pK_a.

H_aO — CH$_2$CH$_2$ — C(=O) — CH$_2$ — CH(H_b) — CH$_2$ — C(=O) — OH_c

2.5D Hybridization Effects

The final factor affecting the acidity of HA is the hybridization of A. To illustrate this phenomenon, compare ethane (CH_3CH_3), ethylene ($CH_2=CH_2$), and acetylene ($HC\equiv CH$). Appendix C indicates that there is a considerable difference in the pK_a values of these compounds.

ethane	ethylene	acetylene
H_3C-CH_3	$H_2C=CH_2$	$H-C\equiv C-H$
$pK_a = 50$	$pK_a = 44$	$pK_a = 25$

Increasing acidity →

The conjugate bases formed by removing a proton from ethane, ethylene, and acetylene are **carbanions—species with a negative charge on carbon.**

$H_3C-CH_2{:}^-$	$H_2C=CH^-$	$H-C\equiv C{:}^-$
sp^3 hybridized C	sp^2 hybridized C	sp hybridized C
25% s-character	**33% s-character**	**50% s-character**

Increasing percent s-character
Increasing stability →

The hybridization of the carbon bearing the negative charge is different in each anion, so the lone pair of electrons occupies an orbital with a different percent s-character in each case. A higher percent s-character means a hybrid orbital has a larger fraction of the lower-energy s orbital.

- The *higher* the percent s-character of the hybrid orbital, the **more stable** the conjugate base.

Thus, **acidity increases from CH_3CH_3 to $CH_2=CH_2$ to $HC\equiv CH$ as the negative charge of the conjugate base is stabilized by increasing percent s-character.** Once again this is a specific example of a general trend.

- The acidity of HA *increases* as the percent s-character of $A{:}^-$ increases.

Problem 2.24 For each pair of compounds: [1] Which indicated H is more acidic? [2] Draw the conjugate base of each acid. [3] Which conjugate base is stronger?

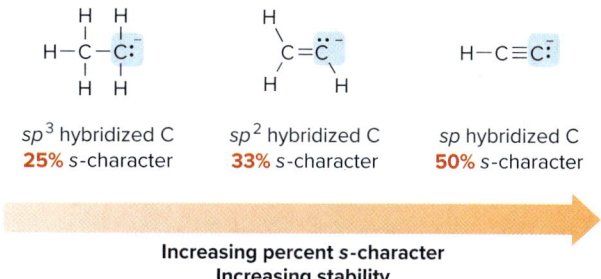

2.5E Summary of Factors Determining Acid Strength

The ability to recognize the most acidic site in a molecule will be important throughout the study of organic chemistry. All the factors that determine acidity are therefore summarized in Figure 2.4. The following two-step procedure shows how these four factors can be used to determine the relative acidity of protons.

Figure 2.4 Summary of the factors that determine acidity

Factor	Example		
1. **Element effects:** The acidity of HA increases both left-to-right across a row and down a column of the periodic table.	CH_4	and	H_2O more acidic
2. **Inductive effects:** The acidity of HA increases with the presence of electron-withdrawing groups in A.	CH_3CH_2-H	and	CF_3CH_2-H more acidic
3. **Resonance effects:** The acidity of HA increases when the conjugate base $A{:}^-$ is resonance stabilized.	CH_3CH_2-H	and	CH_3CO_2-H more acidic
4. **Hybridization effects:** The acidity of HA increases as the percent s-character of $A{:}^-$ increases.	$CH_2{=}CH_2$	and	$H-C{\equiv}C-H$ more acidic

How To Determine the Relative Acidity of Protons

Step [1] Identify the atoms bonded to hydrogen, and use periodic trends to assign relative acidity.
- The most common HA bonds in organic compounds are C–H, N–H, and O–H. Because acidity increases left-to-right across a row, the relative acidity of these bonds is **C–H < N–H < O–H.** Therefore, H atoms bonded to C atoms are usually *less acidic* than H atoms bonded to any heteroatom.

Step [2] If the two H atoms in question are bonded to the same element, draw the conjugate bases and look for other points of difference. Ask three questions:
- Do electron-withdrawing groups stabilize the conjugate base?
- Is the conjugate base resonance stabilized?
- How is the conjugate base hybridized?

Sample Problem 2.7 shows how to apply this procedure to actual compounds.

Sample Problem 2.7 Determining the Relative Acidity of Compounds

Rank the following compounds in order of increasing acidity of their most acidic hydrogen atom.

Solution

[1] Compounds **A**, **B**, and **C** contain C–H, N–H, and O–H bonds. Because acidity increases left-to-right across a row of the periodic table, the **O–H bonds are most acidic.** Compound **C** is thus the least acidic because it has *no* O–H bonds.

[2] The only difference between compounds **A** and **B** is the presence of an electronegative Cl in **A**. The Cl atom stabilizes the conjugate base of **A**, making it more acidic than **B**. Thus,

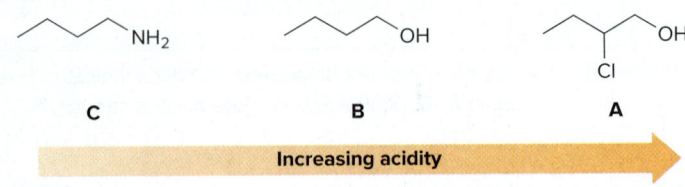

Increasing acidity

2.5 Factors That Determine Acid Strength

Problem 2.25 Rank the compounds in each group in order of increasing acidity.

a. 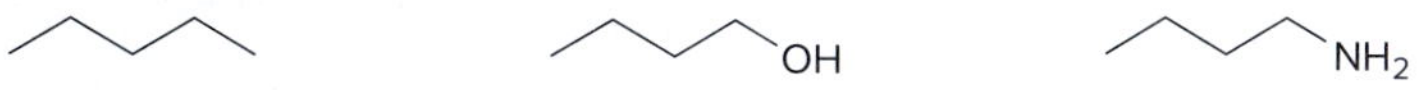 CH₃CH₂CH₂CH₂CH₃   CH₃CH₂CH₂CH₂OH   CH₃CH₂CH₂CH₂NH₂

b. 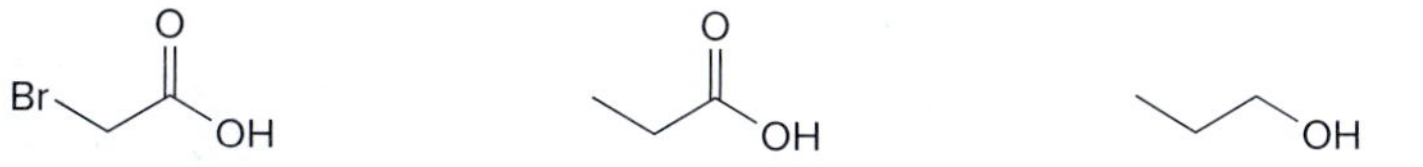 Br-CH₂-COOH   CH₃CH₂-COOH   CH₃CH₂CH₂-OH

c. 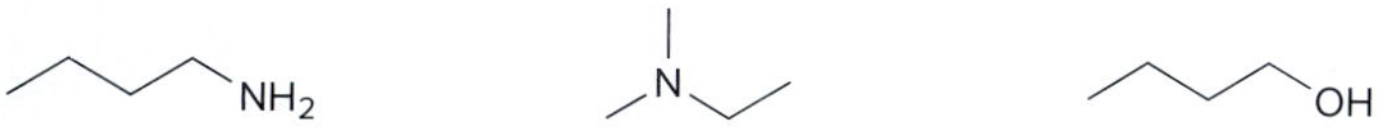 CH₃CH₂CH₂-NH₂   (CH₃)₂N-CH₂CH₃   CH₃CH₂CH₂CH₂-OH

More Practice: Try Problems 2.51; 2.53; 2.54c, d.

Problem 2.26 Which anion (**A** or **B**) is the stronger base?

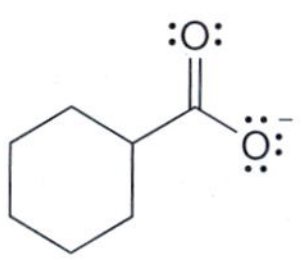

A : cyclohexyl-C(=O)-O:⁻

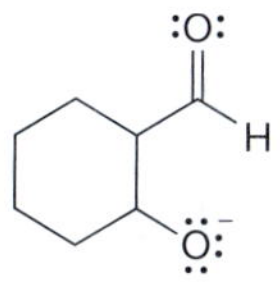

B : cyclohexyl with CHO and O:⁻ substituents

Sample Problem 2.8 — Determining the Most Acidic Proton in a More Complex Molecule

Which proton in quinic acid, the chapter-opening compound present in a brewed cup of coffee, is most acidic?

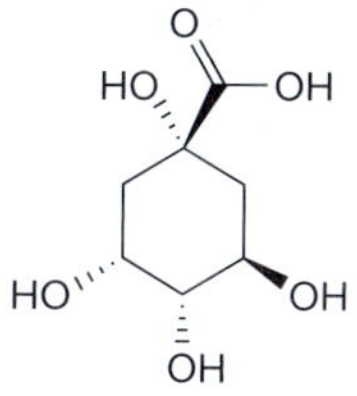

quinic acid

Solution

Because acidity increases left-to-right across a row of the periodic table, **concentrate on the O–H bonds**, which are more acidic than the C–H bonds. By drawing the conjugate bases formed by removal of the O–H protons, we can see that the OH groups bonded to the six-membered ring are different from the OH group bonded to the C=O.

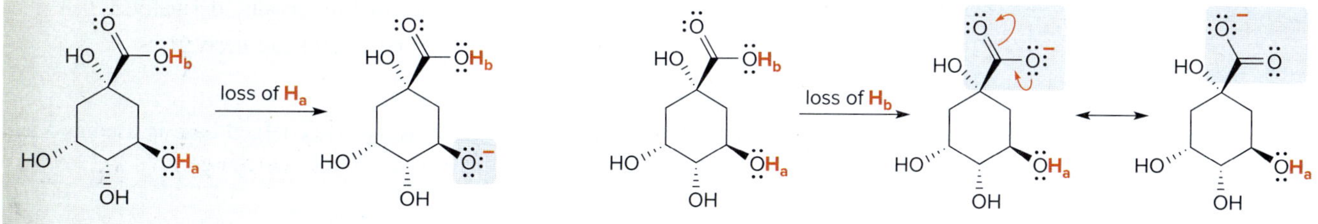

[quinic acid structure with Hₐ and H_b labeled] → loss of Hₐ → conjugate base, only **one** Lewis structure

[structure] → loss of H_b → **two** resonance structures, resonance-stabilized conjugate base

Removal of Hₐ (or any of the protons on the OH groups bonded to the six-membered ring) forms a conjugate base for which only *one* Lewis structure can be drawn, so the negative charge is *localized* on one O atom. Removal of H_b forms a conjugate base for which *two* resonance structures can be drawn, so the negative charge is *delocalized* on two O atoms. Delocalization makes this conjugate base more stable, so **H_b is the most acidic proton in quinic acid.**

Problem 2.27 Which proton in each of the following drugs is most acidic? THC is the active component in marijuana, and ketoprofen is an anti-inflammatory agent.

a. THC
tetrahydrocannabinol

b. ketoprofen

More Practice: Try Problems 2.36a, 2.37a, 2.43, 2.44, 2.60a, 2.62, 2.63.

2.6 Common Acids and Bases

Many strong or moderately strong acids and bases are used as reagents in organic reactions.

Sulfuric acid is the most widely produced industrial chemical. It is also formed when sulfur oxides, emitted into the atmosphere by burning fossil fuels high in sulfur content, dissolve in water. This makes rainwater acidic, forming acid rain, which has destroyed acres of forests worldwide.
Wilmer Stratton

2.6A Common Acids

Several organic reactions are carried out in the presence of strong inorganic acids, most commonly **HCl** and **H$_2$SO$_4$**. These strong acids, with pK_a values $\leq$ **0**, should be familiar from previous chemistry courses.

Two organic acids are also commonly used, namely **acetic acid** and **p-toluenesulfonic acid** (usually abbreviated as **TsOH**). Although acetic acid has a higher pK_a than the inorganic acids, making it a weaker acid, it is more acidic than most organic compounds. p-Toluenesulfonic acid is similar in acidity to the strong inorganic acids. Because it is a solid, small quantities can be easily weighed on a balance and then added to a reaction mixture.

acetic acid
pK_a = 4.8

p-toluenesulfonic acid = TsOH
pK_a = −7

2.6B Common Bases

Three common kinds of strong bases include:

[1] Negatively charged oxygen bases: ⁻OH (hydroxide) and its organic derivatives
[2] Negatively charged nitrogen bases: ⁻NH$_2$ (amide) and its organic derivatives
[3] Hydride (**H⁻**)

Figure 2.5 gives examples of these strong bases. Each negatively charged base is used as a salt with a spectator ion (usually Li⁺, Na⁺, or K⁺) that serves to balance charge.

Figure 2.5 Some common negatively charged bases

Oxygen bases

Na⁺ ⁻OH   sodium hydroxide
Na⁺ ⁻OCH$_3$ sodium methoxide
Na⁺ ⁻OCH$_2$CH$_3$ sodium ethoxide
K⁺ ⁻OC(CH$_3$)$_3$ potassium *tert*-butoxide

Nitrogen bases

Na⁺ ⁻NH$_2$ sodium amide
Li⁺ ⁻N[CH(CH$_3$)$_2$]$_2$ lithium diisopropylamide

Hydride

Na⁺ H⁻   sodium hydride

- Strong bases have weak conjugate acids with high pK_a values, usually > 12.

Strong bases have a net negative charge, but not all negatively charged species are strong bases. For example, none of the halides, F⁻, Cl⁻, Br⁻, or I⁻, is a strong base. These anions have very strong conjugate acids and have little affinity for donating their electron pairs to a proton.

Carbanions, negatively charged carbon atoms discussed in Section 2.5D, are especially strong bases. Perhaps the most common example is **butyllithium.** Butyllithium and related compounds are discussed in greater detail in Chapter 13.

butyllithium

Two other weaker organic bases are **triethylamine** and **pyridine.** These compounds have a lone pair on nitrogen, making them basic, but they are considerably weaker than the amide bases because they are neutral, not negatively charged.

triethylamine   pyridine

Problem 2.28 Draw the products formed when propan-2-ol [$(CH_3)_2CHOH$], the main ingredient in rubbing alcohol, is treated with each acid or base: (a) NaH; (b) H_2SO_4; (c) Li⁺⁻N[CH(CH_3)$_2$]$_2$; (d) CH_3CO_2H.

2.7 Aspirin

Aspirin is one of the most widely used over-the-counter drugs. Whether you purchase Anacin, Bufferin, Bayer, or a generic, the active ingredient is the same—**acetylsalicylic acid.**

Aspirin is the most well known member of a group of compounds called **salicylates.** Although aspirin was first used in medicine for its analgesic (pain-relieving), antipyretic (fever-reducing), and anti-inflammatory properties, today it is commonly used as an antiplatelet agent in the treatment and prevention of heart attacks and strokes. **Aspirin is a synthetic compound;** it does not occur in nature, although some related salicylates are found in willow bark and meadowsweet blossoms (Figure 2.6).

Figure 2.6

Salicin, an analgesic in willow bark

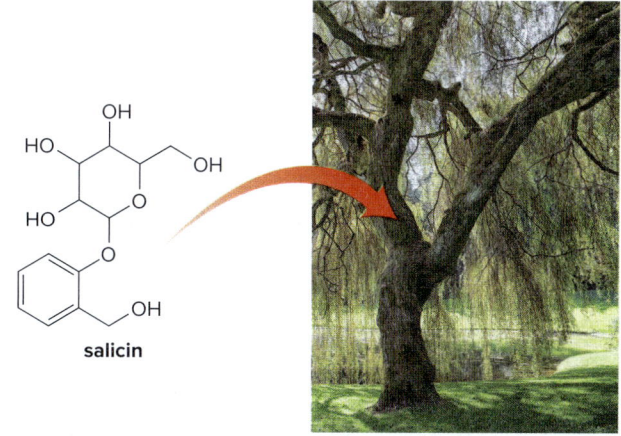

willow tree

Skip Brown/National Geographic/Getty Images

- The modern history of aspirin dates back to 1763 when Reverend Edmund Stone reported on the analgesic effect of chewing on the bark of the willow tree. Willow bark is now known to contain *salicin,* which is structurally related to aspirin.

Like many drugs, aspirin undergoes a proton transfer reaction. Its most acidic proton is the H bonded to O, and in the presence of base, this H is readily removed.

aspirin
acetylsalicylic acid
neutral form
This form exists in the **stomach.**

conjugate base
ionic form
This form exists in the **intestines.**

Why is this acid–base reaction important? After ingestion, aspirin first travels into the stomach and then the intestines. In the acidic environment of the stomach, aspirin remains in its neutral form, but in the basic environment of the small intestine, aspirin is deprotonated to form its conjugate base, an ion. Likewise, in the slightly basic environment of the blood, aspirin exists primarily as its ionic conjugate base.

Whether aspirin is a neutral acid or an ionic conjugate base affects its transport throughout the body and its ability to pass through a cell membrane. In its ionic form, aspirin is readily soluble in the aqueous environment of the blood, so it is transported in the bloodstream to tissues. Once aspirin has reached its target location, however, its conjugate base must be re-protonated to form the neutral acid that can pass through the nonpolar interior of a cell membrane where it inhibits prostaglandin synthesis, as we will learn in Chapter 15. Thus, in the body, aspirin undergoes acid–base reactions and these reactions are crucial in determining its properties and action.

We will learn more about solubility and the cell membrane in Section 3.7.

Problem 2.29 amphetamine

Compounds like amphetamine that contain nitrogen atoms are protonated by the HCl in the gastric juices of the stomach, and the resulting salt is then deprotonated in the basic environment of the intestines to regenerate the neutral form. Write proton transfer reactions for both of these processes. In which form will amphetamine pass through a cell membrane?

2.8 Lewis Acids and Bases

The Lewis definition of acids and bases is more general than the Brønsted–Lowry definition.

All Brønsted–Lowry bases are Lewis bases.

- A Lewis acid is an *electron pair acceptor.*
- A Lewis base is an *electron pair donor.*

Lewis bases are structurally the same as Brønsted–Lowry bases. Both have an **available electron pair**—a lone pair or an electron pair in a π bond. A Brønsted–Lowry base always donates this electron pair to a proton, but a Lewis base donates this electron pair to anything that is electron deficient. Simple Lewis bases are shown in Figure 2.7.

A Lewis acid must be able to accept an electron pair, but there are many ways for this to occur. **All Brønsted–Lowry acids are also Lewis acids, but the reverse is not necessarily true.** Any species that is electron deficient and capable of accepting an electron pair is also a Lewis acid, as shown in Figure 2.7.

2.8 Lewis Acids and Bases

Figure 2.7
Simple Lewis acids and bases

Lewis **bases**

$:\!\ddot{\underset{..}{O}}\!H^-$

$CH_3\ddot{\underset{..}{O}}H$

$\begin{matrix} H \\ \diagdown \\ C=C \\ \diagup \\ H \end{matrix} \begin{matrix} H \\ \diagup \\ \\ \diagdown \\ H \end{matrix}$

Lewis **acids** that are also Brønsted–Lowry acids

$H_2\ddot{\underset{..}{O}}\!:$ $CH_3\ddot{\underset{..}{O}}\mathbf{H}$

Lewis **acids** that are *not* Brønsted–Lowry acids

BF_3 $AlCl_3$

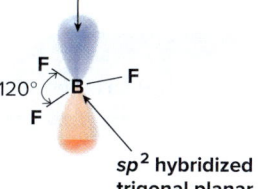

The vacant unhybridized *p* orbital extends above and below the plane of BF$_3$.

120°

sp^2 hybridized trigonal planar

Common examples of Lewis acids (which are not Brønsted–Lowry acids) include **BF$_3$** and **AlCl$_3$**. These compounds contain elements in group 3A of the periodic table that can accept an electron pair because they do not have filled valence shells of electrons. For example, BF$_3$ contains an *sp*^{2} hybridized, trigonal planar B atom with a vacant unhybridized *p* orbital that can accept two electrons.

Problem 2.30   Which species are Lewis bases?

a. NH$_3$ b. CH$_3$CH$_2$CH$_3$ c. H$^-$ d. H—C≡C—H

Problem 2.31   Which species are Lewis acids?

a. BBr$_3$ b. CH$_3$CH$_2$OH   c. (CH$_3$)$_3$C$^+$ d. Br$^-$

Any reaction in which one species donates an electron pair to another species is a Lewis acid–base reaction.

In a Lewis acid–base reaction, a Lewis base donates an electron pair to a Lewis acid. Most reactions in organic chemistry involving movement of electron pairs can be classified as Lewis acid–base reactions. Lewis acid–base reactions illustrate a general pattern of reactivity.

- Electron-rich species react with electron-poor species.

In the simplest Lewis acid–base reaction, one bond is formed and no bonds are broken. This is illustrated with the reaction of BF$_3$ with H$_2$O. BF$_3$ has only six electrons around B, so it is the electron-deficient Lewis acid. H$_2$O has two lone pairs on O, so it is the electron-rich Lewis base.

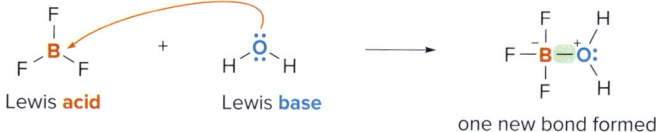

H$_2$O donates an electron pair to BF$_3$ to form one new bond. The electron pair in the new B—O bond comes from the oxygen atom, and a single product, a **Lewis acid–base complex,** is formed. Both B and O bear formal charges in the product, but the overall product is neutral.

Nucleophile = nucleus loving.
Electrophile = electron loving.

- A Lewis acid is called an *electrophile*.
- When a Lewis base reacts with an electrophile other than a proton, the Lewis base is called a *nucleophile*.

In this Lewis acid–base reaction, **BF$_3$** is the electrophile and **H$_2$O** is the nucleophile.

In a Lewis acid–base reaction, the **electron pair is not removed from the Lewis base**; instead, the electron pair is donated to an atom of the Lewis acid, and one new covalent bond is formed.

electrophile   nucleophile   Lewis acid–base complex

Lewis acid   Lewis base   new bond in green

Problem 2.32 For each reaction, label the Lewis acid and base. Use curved arrow notation to show the movement of electron pairs.

a. BF$_3$ + (CH$_3$)$_2$O: ⟶ F$_3$B–O$^+$(CH$_3$)$_2$

b. (CH$_3$)$_2$CH$^+$ + $^-$:ÖH ⟶ (CH$_3$)$_2$CH–ÖH

Problem 2.33 Draw the products of each reaction, and label the nucleophile and electrophile.

a. CH$_3$CH$_2$–Ö–CH$_2$CH$_3$ + BBr$_3$ ⟶

b. cyclohexanone =Ö: + AlCl$_3$ ⟶

Problem 2.34 Draw the product formed when (CH$_3$CH$_2$)$_3$N:, a Lewis base, reacts with each Lewis acid: (a) B(CH$_3$)$_3$; (b) (CH$_3$)$_3$C$^+$; (c) AlCl$_3$.

In some Lewis acid–base reactions, one bond is formed and one bond is broken. To draw the products of these reactions, keep the following steps in mind.

[1] Always identify the Lewis acid and base first.
[2] Draw a curved arrow from the electron pair of the base to the electron-deficient atom of the acid.
[3] Count electron pairs and break a bond when needed to keep the correct number of valence electrons.

For example, draw the Lewis acid–base reaction between cyclohexene and H—Cl. The Brønsted–Lowry acid HCl is also a Lewis acid, and cyclohexene, having a π bond, is the Lewis base.

cyclohexene   H—Cl
Lewis base   δ+ δ−
             Lewis acid

Recall from Section 1.6B that a positively charged carbon atom is called a **carbocation**.

To draw the product of this reaction, the electron pair in the π bond of the Lewis base forms a new bond to the proton of the Lewis acid, forming a carbocation. The H—Cl bond must break, giving its two electrons to Cl, forming Cl$^-$. Because two electron pairs are involved, two curved arrows are needed.

Lewis base   +   H—Cl: ⟶   carbocation   +   :Cl:$^-$
                   Lewis acid   new bond in green

In the reaction of cyclohexene with HCl, the new bond to H could form at **either carbon of the double bond**, because the same carbocation results.

The Lewis acid–base reaction of cyclohexene with HCl is a specific example of a fundamental reaction of compounds containing C—C double bonds, as discussed in Chapter 10.

Lewis acid–base complexes are commonly encountered in biological systems. Metal cations such as Mg^{2+} serve as Lewis acids that complex with the negatively charged oxygens of triphosphates such as **ATP**. The resulting Lewis acid–base complex **B** is a key intermediate in biological processes, as discussed in Section 16.15.

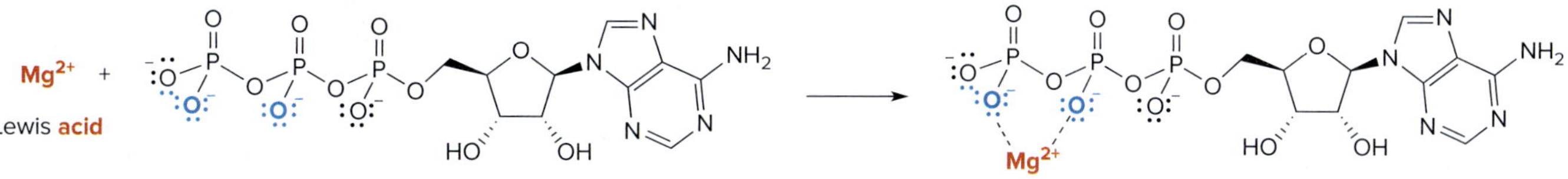

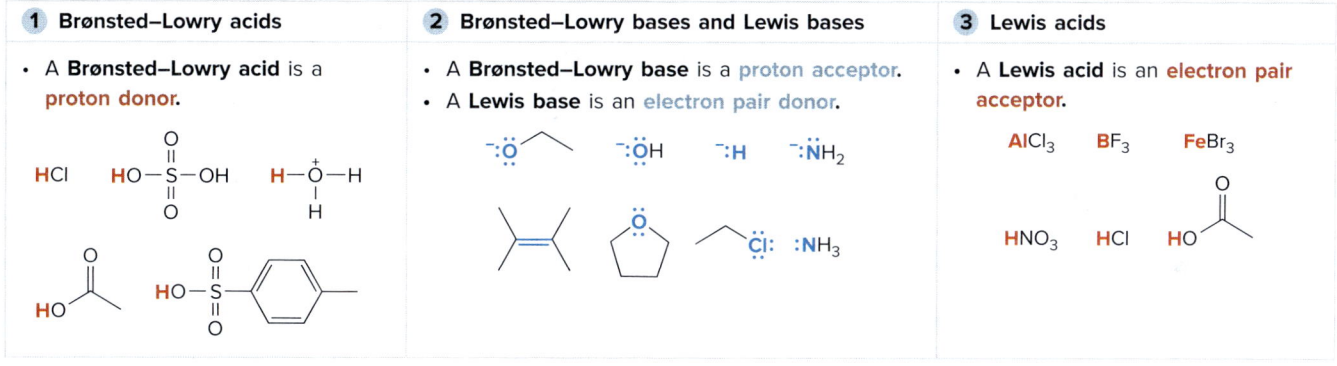

Problem 2.35 Label the Lewis acid and base. Use curved arrow notation to show the movement of electron pairs.

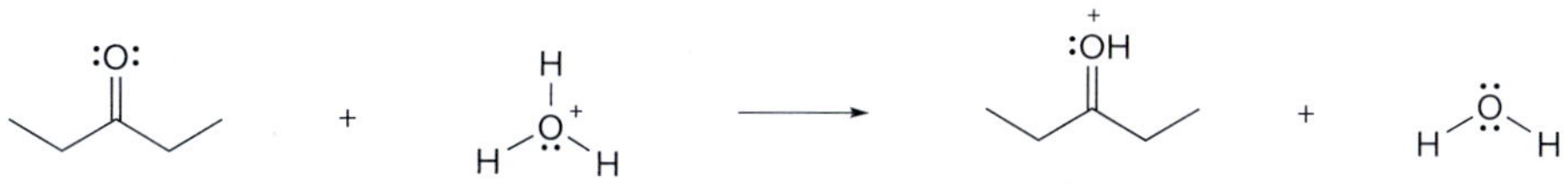

Chapter 2 REVIEW

KEY CONCEPTS

Brønsted–Lowry and Lewis Acids and Bases (2.1, 2.8)

1 Brønsted–Lowry acids

- A **Brønsted–Lowry acid** is a **proton donor**.

  HCl   HO–S(=O)(=O)–OH   H–O(+)(H)–H

  (acetic acid)   (p-toluenesulfonic acid)

2 Brønsted–Lowry bases and Lewis bases

- A **Brønsted–Lowry base** is a **proton acceptor**.
- A **Lewis base** is an **electron pair donor**.

3 Lewis acids

- A **Lewis acid** is an **electron pair acceptor**.

  $AlCl_3$ BF_3 $FeBr_3$

  HNO_3 HCl   (acetate)

See Figures 2.1, 2.7. Try Problems 2.65, 2.66.

Acid–Base Reactions

1 Drawing the products of a Brønsted–Lowry acid–base reaction (2.2)

- A **Brønsted–Lowry acid donates a proton** to a **Brønsted–Lowry base**.

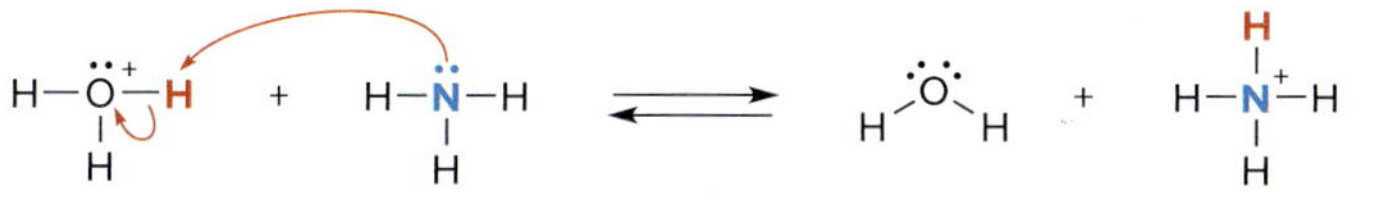

  acid / proton donor   base / proton acceptor   conjugate base   conjugate acid

  See Sample Problems 2.1, 2.2, Figure 2.2. Try Problems 2.40–2.42, 2.48.

2 Drawing the products of a Lewis acid–base reaction (2.8)

- A **Lewis base donates an electron pair** to a **Lewis acid**.

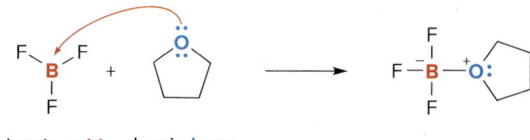

Lewis acid   Lewis base
electrophile   nucleophile

- Electron-rich species react with electron-poor ones.
- Nucleophiles react with electrophiles.

Try Problems 2.67, 2.68.

Periodic Trends (2.5A)

Try Problem 2.51a, b.

Try Problem 2.54a, b.

Acid and Base Strength and pK_a (2.3)

1 Acidity and pK_a

- pK_a = $-\log K_a$
- The lower the pK_a, the stronger the acid.

2 Basicity and pK_a

- As the pK_a of an acid increases, the basicity of its conjugate base increases.
- A strong acid has a weak conjugate base.

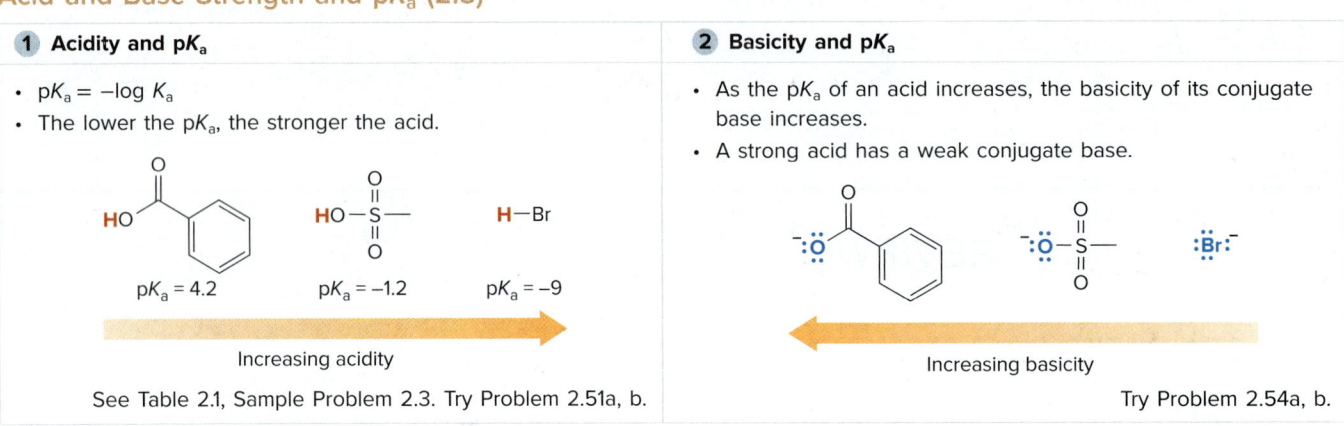

See Table 2.1, Sample Problem 2.3. Try Problem 2.51a, b.

Try Problem 2.54a, b.

KEY SKILLS

[1] Drawing the products of a Brønsted–Lowry acid–base reaction (2.2)

| 1 Identify the acid and the base. | 2 Draw curved arrows to move a proton from the acid to the base. | 3 Draw the products of proton transfer. |

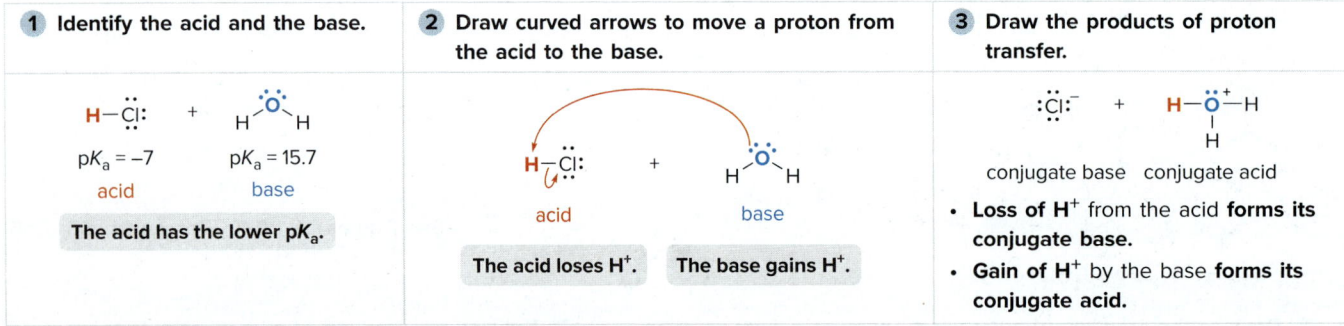

- Loss of H⁺ from the acid **forms its conjugate base.**
- Gain of H⁺ by the base **forms its conjugate acid.**

Try Problems 2.41–2.43.

[2] Determining the direction of equilibrium using pK_a (2.4)

1 Identify the acids on each side of the arrows.

2 Compare the pK_a values of the acids.

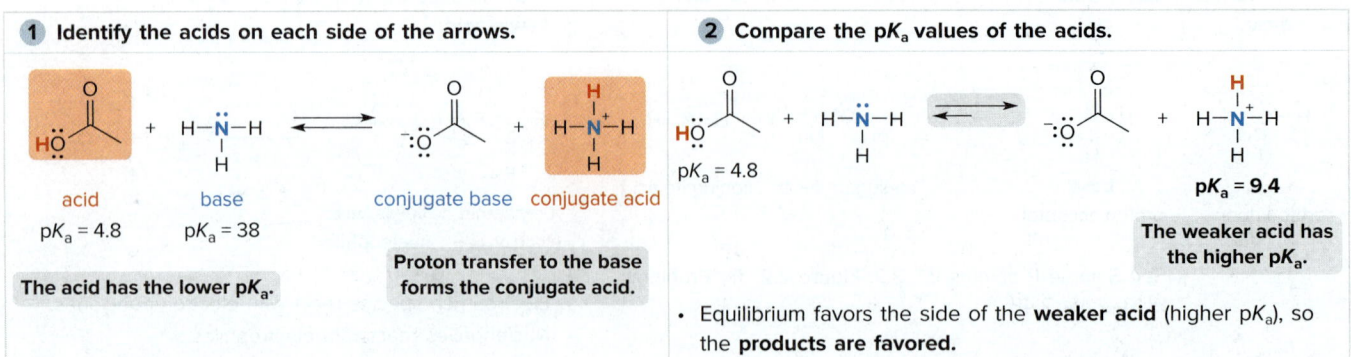

- Equilibrium favors the side of the **weaker acid** (higher pK_a), so the **products are favored.**

See Sample Problems 2.4, 2.5. Try Problem 2.48.

Key Skills

[3] Determining acidity using trends in the periodic table (2.5A)

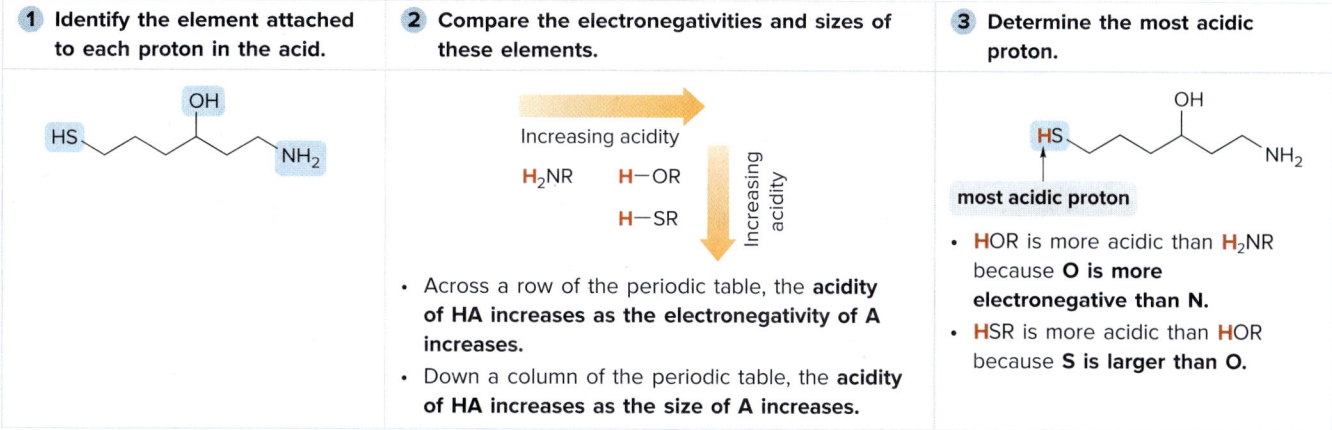

- Across a row of the periodic table, the **acidity of HA increases as the electronegativity of A increases.**
- Down a column of the periodic table, the **acidity of HA increases as the size of A increases.**

- HOR is more acidic than H₂NR because **O is more electronegative than N.**
- HSR is more acidic than HOR because **S is larger than O.**

See Sample Problem 2.6. Try Problems 2.51a, b; 2.54a, b.

[4] Determining acidity using inductive effects (2.5B)

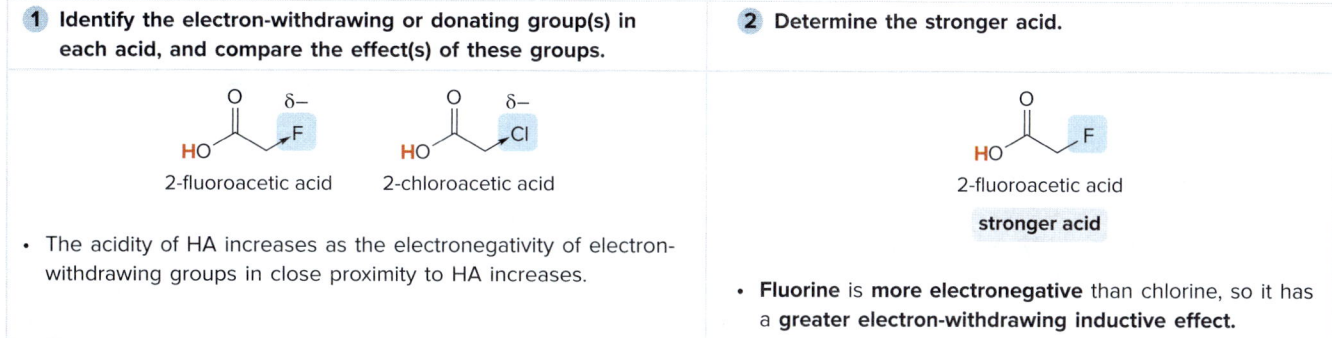

- The acidity of HA increases as the electronegativity of electron-withdrawing groups in close proximity to HA increases.

- **Fluorine is more electronegative** than chlorine, so it has a **greater electron-withdrawing inductive effect.**

Try Problems 2.51c, 2.53.

[5] Determining acidity using resonance effects (2.5C)

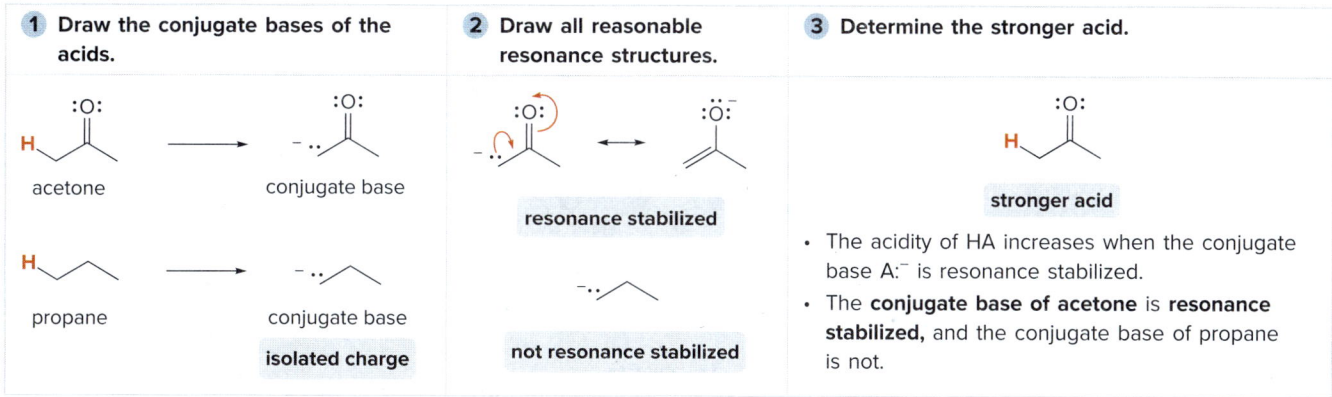

- The acidity of HA increases when the conjugate base A:⁻ is resonance stabilized.
- The **conjugate base of acetone is resonance stabilized,** and the conjugate base of propane is not.

See Figure 2.3. Try Problems 2.54c, 2.55.

[6] Determining the most acidic proton (2.5E)

1 Identify the atoms bonded to hydrogen, and use periodic trends to assign acidity.	2 Draw conjugate bases and resonance structures.	3 Determine the most acidic proton.
 tyramine • HOR is more acidic than H_2NR based on electronegativity. Increasing acidity H_2NR H–OR	 + three additional resonance structures **The oxygen anion is resonance stabilized.** **The nitrogen anion is not resonance stabilized.**	**most acidic** tyramine • Based on both **electronegativity** and **resonance,** HOR is more acidic than H_2NR.

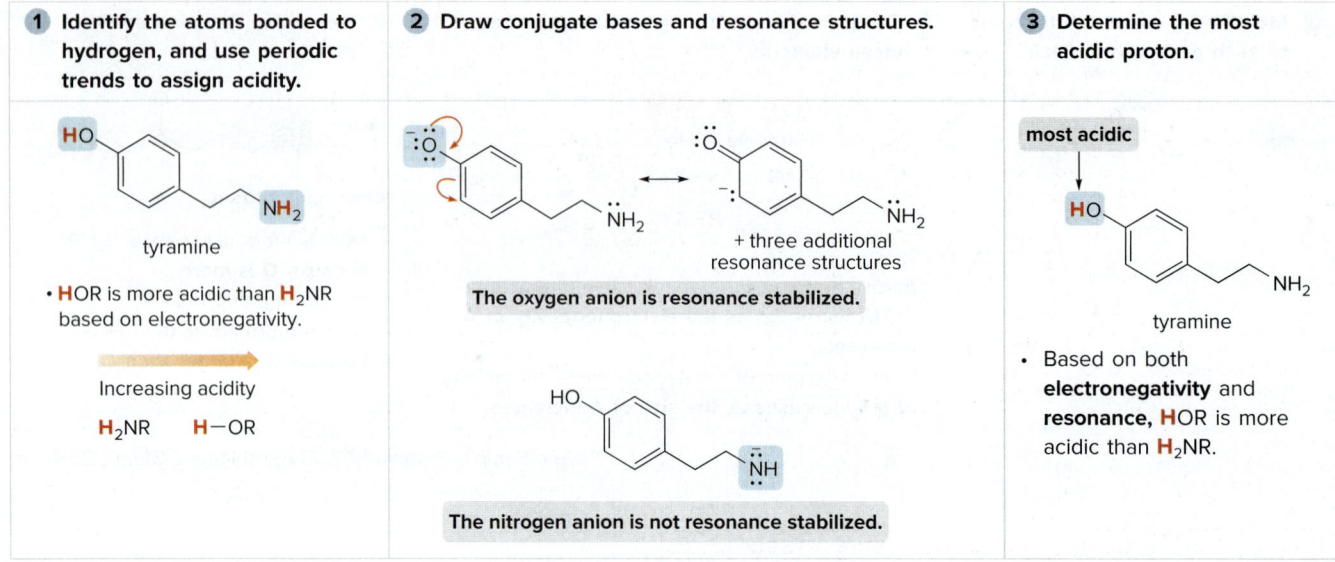

See *How To* (p. 74), Figure 2.4, Sample Problems 2.7, 2.8. Try Problems 2.43, 2.44, 2.52, 2.62.

PROBLEMS

Problems Using Three-Dimensional Models

2.36 Propranolol is an antihypertensive agent—that is, it lowers blood pressure. (a) Which proton in propranolol is most acidic? (b) What products are formed when propranolol is treated with NaH? (c) Which atom is most basic? (d) What products are formed when propranolol is treated with HCl?

propranolol

2.37 Amphetamine is a powerful stimulant of the central nervous system. (a) Which proton in amphetamine is most acidic? (b) What products are formed when amphetamine is treated with NaH? (c) What products are formed when amphetamine is treated with HCl?

amphetamine

Brønsted–Lowry Acids and Bases

2.38 What is the conjugate acid of each base?

a. HCO_3^- b. c. d.

2.39 What is the conjugate base of each acid?

a. HCO_3^- b. c. d.

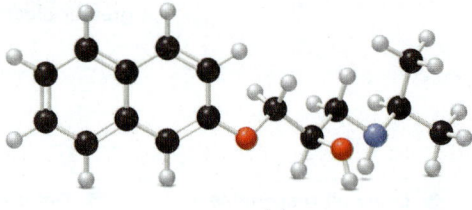

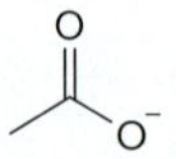

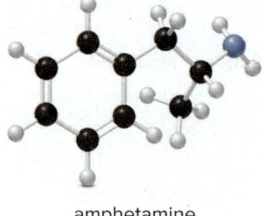

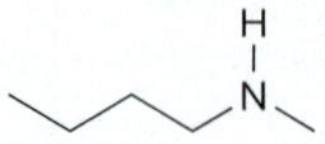

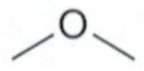

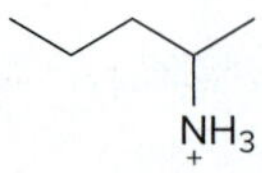

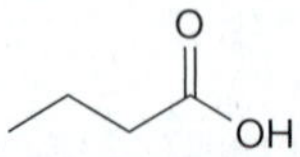

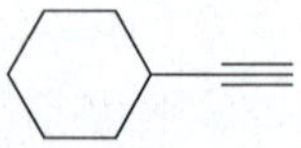

Reactions of Brønsted–Lowry Acids and Bases

2.40 As we will see in later chapters, many steps in key reaction sequences involve acid–base reactions. (a) Draw curved arrows to illustrate the flow of electrons in steps [1]–[3]. (b) Identify the base and its conjugate acid in step [1]. (c) Identify the acid and its conjugate base in step [3].

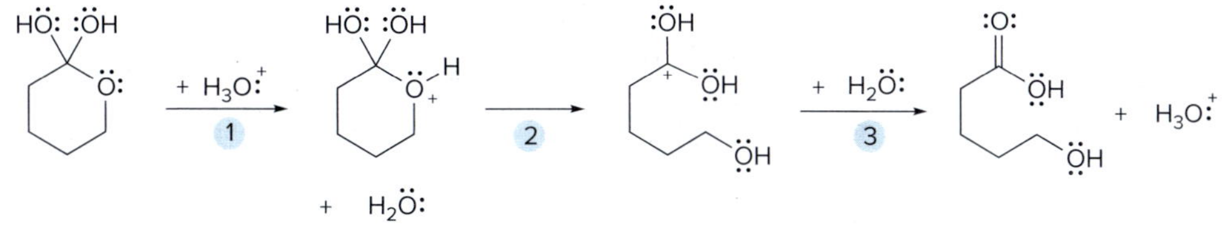

2.41 Draw the products formed from the acid–base reaction of H_2SO_4 with each compound.

a. 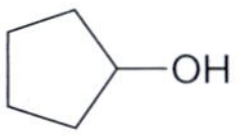 cyclopentyl–OH b. 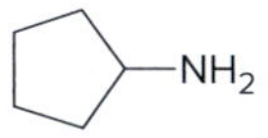 cyclopentyl–NH_2 c. 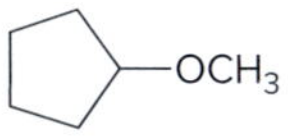 cyclopentyl–OCH_3 d. 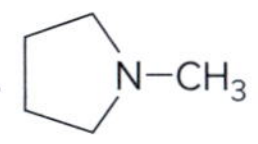 pyrrolidine N–CH_3

2.42 Draw the products of each proton transfer reaction. Label the acid and base in the starting materials, and the conjugate acid and base in the products.

a.  pentanoic acid + $CH_3\ddot{O}{:}^-$

b.  2-pentanol (OH) + HBr

c. 3-methyl-1-pentyne + $NaNH_2$

d. 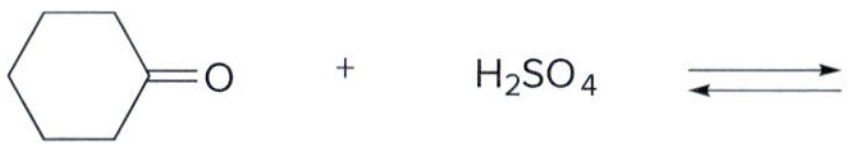 cyclohexanone + H_2SO_4

2.43 Draw the products of each acid–base reaction.

a. 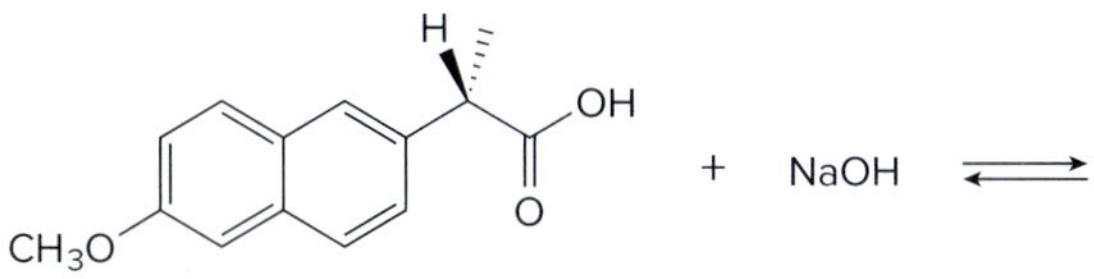 naproxen (anti-inflammatory agent) + NaOH

b. 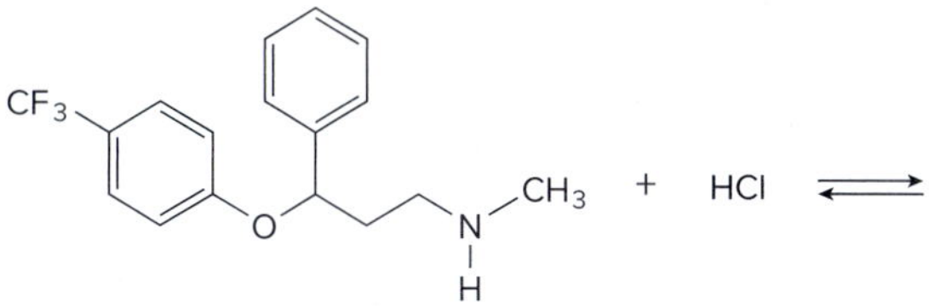 fluoxetine (antidepressant) + HCl

2.44 What product is formed when each compound is treated with NaH? Each of these acid–base reactions was a step in a synthesis of a commercially available drug.

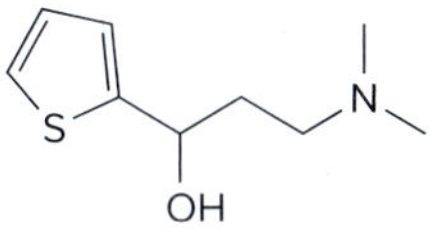

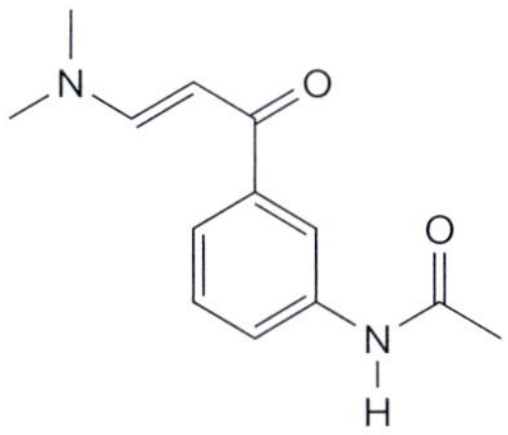

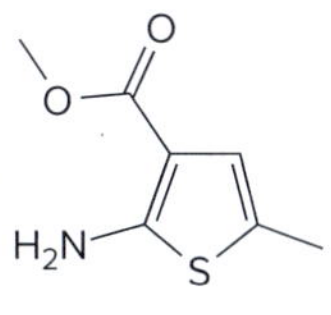

pK_a, K_a, and the Direction of Equilibrium

2.45 What is the K_a for each compound? Use a calculator when necessary.

a. H_2S $pK_a = 7.0$

b. $ClCH_2COOH$ $pK_a = 2.8$

c. HCN $pK_a = 9.1$

2.46 What is the pK_a for each compound?

a. 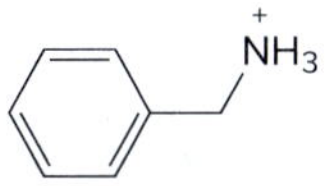 $C_6H_5CH_2\overset{+}{N}H_3$ $K_a = 4.7 \times 10^{-10}$

b. 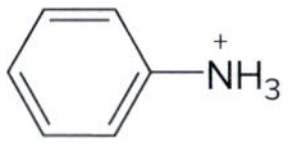 $C_6H_5\overset{+}{N}H_3$ $K_a = 2.3 \times 10^{-5}$

c. 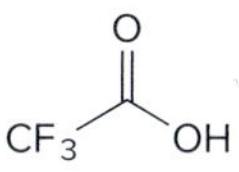 CF_3COOH $K_a = 5.9 \times 10^{-1}$

2.47 Which of the following bases are strong enough to deprotonate C_6H_5OH ($pK_a = 10$) so that equilibrium favors the products: (a) H_2O; (b) NaOH; (c) $NaNH_2$; (d) CH_3NH_2; (e) $NaHCO_3$; (f) NaSH; (g) NaH?

Chapter 2 Acids and Bases

2.48 Draw the products of each reaction. Use the pK_a table in Appendix C to decide if the equilibrium favors the starting materials or products.

a. CH_3NH_2 + H_2SO_4 ⇌

b. CH_3COOH + NaCl ⇌

c. (phenol) + $NaHCO_3$ ⇌

d. $H-C\equiv C-H$ + $CH_3CH_2^- Li^+$ ⇌

2.49 Draw the products of each reaction and decide if equilibrium favors the starting materials or the products.

a. cyclohexyl-NH_2 + benzoate anion ⇌

b. cyclohexyl-NH^- + benzoic acid ⇌

c. cyclohexyl-NH_3^+ + benzoate anion ⇌

d. cyclohexyl-NH_2 + benzoic acid ⇌

2.50 Several biological compounds are derivatives of phosphoric acid (H_3PO_4) and related compounds. H_3PO_4 has three OH groups that can be deprotonated, with pK_a values of 2.1, 6.9, and 12.4. (a) Use these values to estimate the pK_a values of the two most acidic protons in **A**, an intermediate formed during carbohydrate metabolism. (b) What species is present if **A** is treated with excess NaOH? (c) What species is present if **A** is treated with excess pyridine?

A

Relative Acid Strength

2.51 Rank the compounds in each group in order of increasing acidity.

a. propanol (OH), propylamine (NH_2)

b. 2-chloropropane-CH(Cl)-, 2-propanol, 2-propanethiol (SH)

c. propane, 2-chloroethanol (Cl-CH$_2$-CH$_2$-OH), propanol (OH)

d. propyne, propene, propene (alternate)

2.52 Rank the labeled protons in the following molecule in order of increasing pK_a.

(molecule with labeled protons H_a, H_b, H_c, H_d, OH_e)

2.53 Rank the following Brønsted–Lowry acids in order of increasing acidity. Which compound forms the strongest conjugate base?

A (Br-substituted alcohol), **B** (tetrahydropyran with F), **C** (piperidine with NH), **D** (tert-butyl alcohol)

2.54 Rank the ions in each group in order of increasing basicity.

a. $CH_3\bar{C}H_2$, CH_3O^-, $CH_3\bar{N}H$

b. CH_3^-, HO^-, Br^-

c. acetate, alkoxide (RO^-), chloroacetate

d. cyclohexyl-$C\equiv C^-$, cyclohexyl anion, cyclohexyl vinyl anion

2.55 RNA is a biological molecule that translates the genetic information in DNA into protein synthesis. (a) Identify the most acidic proton in uracil, one of the components of RNA, and explain your choice. (b) If uracil is treated with two equivalents of very strong base, what dianion is formed?

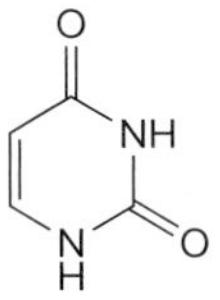

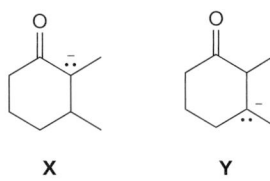
uracil

2.56 Which of the following anions is the stronger base? Explain your choice.

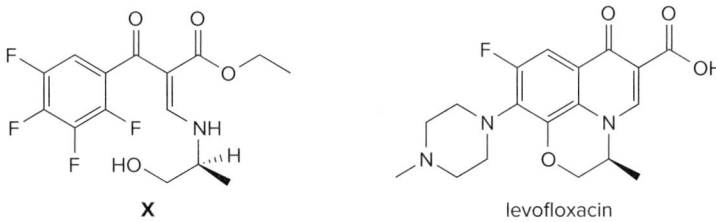

2.57 a. What is the conjugate acid of **A**?
b. What is the conjugate base of **A**?

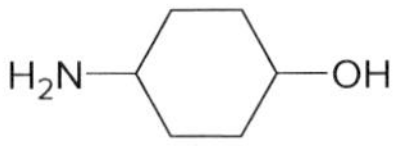

H$_2$N—⬡—OH

A

2.58 Explain why the N—H proton in **X** is more acidic than the O—H proton. **X** was a key intermediate in the synthesis of the antibiotic levofloxacin.

X   levofloxacin

2.59 The pK_a values of the two most acidic protons of the amino acid leucine are shown. (a) Draw the product formed, including all reasonable resonance structures, when leucine is treated with one equivalent of base. (b) Draw the product formed, including all reasonable resonance structures, when leucine is treated with two equivalents of base. (c) Is NaOH a strong enough base to remove both protons? Why or why not?

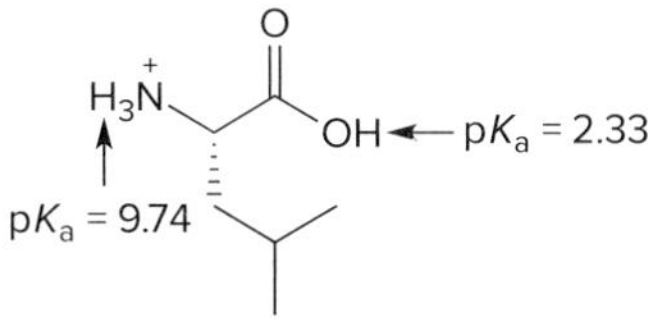

pK_a = 9.74   pK_a = 2.33

leucine

2.60 Many drugs are Brønsted–Lowry acids or bases.
a. What is the most acidic proton in the analgesic ibuprofen? Draw the conjugate base.
b. What is the most basic electron pair in cocaine? Draw the conjugate acid.

ibuprofen   cocaine

2.61 Dimethyl ether (CH$_3$OCH$_3$) and ethanol (CH$_3$CH$_2$OH) are isomers, but CH$_3$OCH$_3$ has a pK_a of 40 and CH$_3$CH$_2$OH has a pK_a of 16. Why are these pK_a values so different?

2.62 Use the principles in Section 2.5 to label the most acidic hydrogen in each drug. Explain your choice.

a. 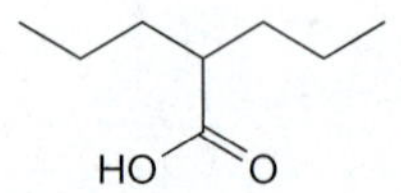 valproic acid (used to treat epilepsy)

b. 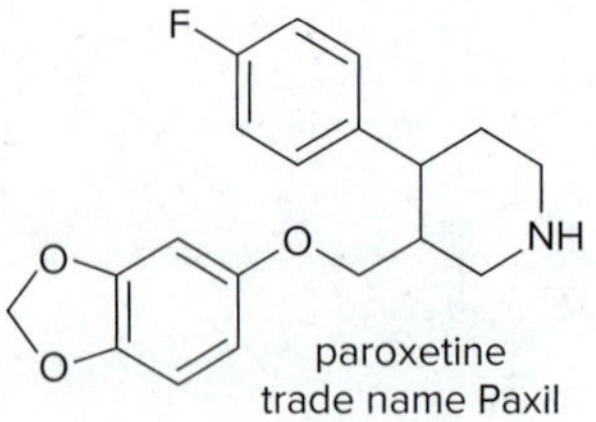 paroxetine trade name Paxil (used to treat depression)

c. 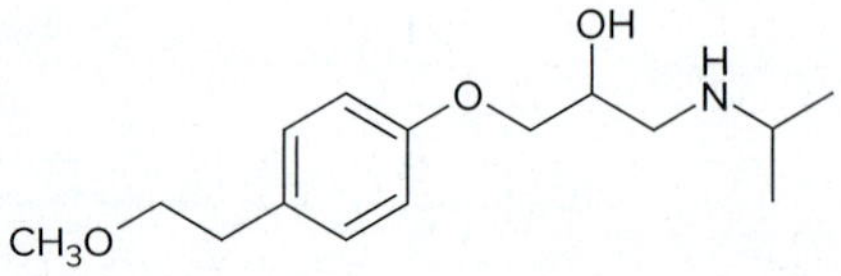 metoprolol (used to treat high blood pressure)

2.63 Label the three most acidic hydrogen atoms in lactic acid, $CH_3CH(OH)CO_2H$, and rank them in order of decreasing acidity. Explain your reasoning.

2.64 Bupivacaine (trade name Marcaine) is a quick-acting anesthetic often used during labor and delivery. Which nitrogen atom in bupivacaine is more basic? Explain your reasoning.

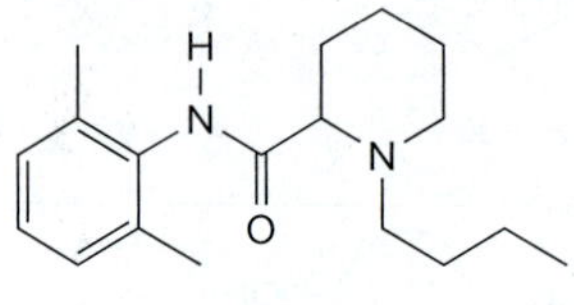

bupivacaine

Lewis Acids and Bases

2.65 Classify each compound as a Lewis base, a Brønsted–Lowry base, both, or neither.

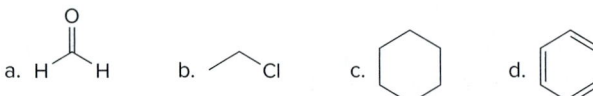

2.66 Classify each species as a Lewis acid, a Brønsted–Lowry acid, both, or neither.
   a. H_3O^+ b. Cl_3C^+ c. BCl_3 d. BF_4^-

Lewis Acid–Base Reactions

2.67 Label the Lewis acid and Lewis base in each reaction. Use curved arrows to show the movement of electron pairs.

a. Cl^- + BCl_3 ⟶

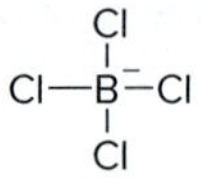

b.

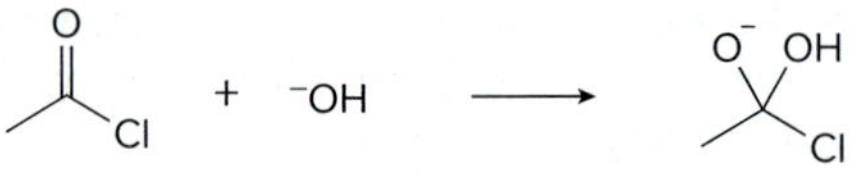

2.68 Draw the products of each Lewis acid–base reaction. Label the electrophile and nucleophile.

a.

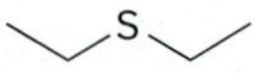

b.

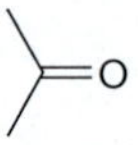

c.

General Problems

2.69 Answer the following questions about the four species **A–D**.

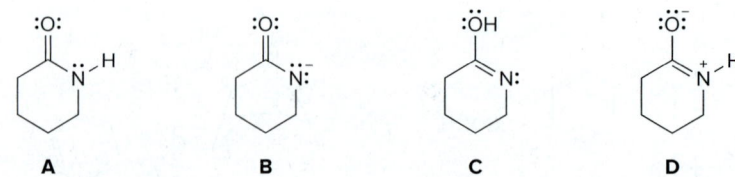

a. Which two species represent a conjugate acid–base pair?
b. Which two species represent resonance structures?
c. Which two species represent constitutional isomers?

2.70 Classify each reaction as either a proton transfer reaction, or a reaction of a nucleophile with an electrophile. Use curved arrows to show how the electron pairs move.

a. (CH₃)₂CHOH + HBr ⟶ (CH₃)₂CHOH₂⁺ + Br⁻ ⟶ (CH₃)₂CHBr + H₂O

b. benzene + Br⁺ ⟶ arenium ion intermediate —Br⁻→ bromobenzene + HBr

2.71 Hydroxide (⁻OH) can react as a Brønsted–Lowry base (and remove a proton) or as a Lewis base (and attack a carbon atom). (a) What organic product is formed when ⁻OH reacts with the carbocation $(CH_3)_3C^+$ as a Brønsted–Lowry base? (b) What organic product is formed when ⁻OH reacts with $(CH_3)_3C^+$ as a Lewis base?

2.72 Answer the following questions about esmolol, a drug used to treat high blood pressure sold under the trade name Brevibloc.

esmolol

a. Label the most acidic hydrogen atom in esmolol.
b. What products are formed when esmolol is treated with NaH?
c. What products are formed when esmolol is treated with HCl?
d. Label all sp^2 hybridized C atoms.
e. Label the only trigonal pyramidal atom.
f. Label all C's that bear a δ+ charge.

Challenge Problems

2.73 Caffeic acid is an organic acid isolated from coffee beans. Predict which labeled hydrogen (H_a or H_b) is more acidic and explain your choice.

caffeic acid

2.74 Molecules like acetamide (CH_3CONH_2) can be protonated on either their O or N atoms when treated with a strong acid like HCl. Which site is more readily protonated and why?

2.75 Two pK_a values are reported for malonic acid, a compound with two COOH groups. Explain why one pK_a is lower and one pK_a is higher than the pK_a of acetic acid (CH_3COOH, pK_a = 4.8).

malonic acid
pK_a = 2.86

pK_a = 5.70

2.76 Amino acids such as glycine (Section 3.9A) are the building blocks of large molecules called proteins that give structure to muscle, tendon, hair, and nails.

glycine   zwitterion form

a. Explain why glycine does not actually exist in the form with all atoms uncharged, but actually exists as a salt called a zwitterion.
b. What product is formed when glycine is treated with concentrated HCl?
c. What product is formed when glycine is treated with NaOH?

Chapter 2 Acids and Bases

2.77 Write a stepwise reaction sequence using proton transfer reactions to show how the following reaction occurs. (Hint: As a first step, use ⁻OH to remove a proton from the CH₂ group between the C=O and C=C.)

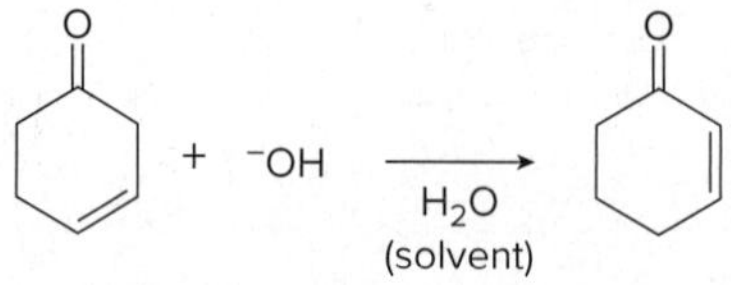

[cyclohex-3-enone] + ⁻OH →(H₂O, solvent)→ [cyclohex-2-enone]

2.78 Which H atom in vitamin C (ascorbic acid) is most acidic?

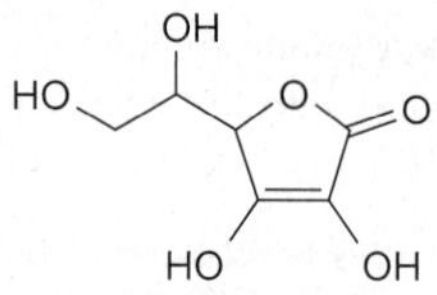

vitamin C
ascorbic acid

Introduction to Organic Molecules and Functional Groups

3

- **3.1** Functional groups
- **3.2** An overview of functional groups
- **3.3** Intermolecular forces
- **3.4** Physical properties
- **3.5** Application: Vitamins
- **3.6** Application of solubility: Soap
- **3.7** Application: The cell membrane
- **3.8** Functional groups and reactivity
- **3.9** Biomolecules

Vitamin C, or **ascorbic acid,** is important in the formation of collagen, a protein that holds together the connective tissues of skin, muscle, and blood vessels. In addition to oranges and grapefruit, kiwi is an excellent source of vitamin C. Grown commercially in Italy, New Zealand, and several other countries, kiwi fruit has a unique sweet flavor, sometimes reminiscent of strawberries. A deficiency of vitamin C causes scurvy, a common disease of sailors in the 1600s when they had no access to fresh fruits on long voyages. In Chapter 3, we learn why some vitamins like vitamin A can be stored in the fat cells in the body, whereas others like vitamin C are excreted in urine.

Why Study...

Functional Groups?

Having learned some basic concepts about structure, bonding, and acid–base chemistry in Chapters 1 and 2, we will now concentrate on organic molecules.

- What are the characteristic features of an organic compound?
- What determines the properties of an organic compound?

After these questions are answered, we can understand some common phenomena. Why do we store some vitamins in the body and readily excrete others? How does soap clean away dirt? Moreover, learning about the structure and properties of organic molecules will give us an understanding of the organic compounds found in biological systems.

3.1 Functional Groups

What are the characteristic features of an organic compound? Most organic molecules have C—C and C—H σ bonds. These bonds are strong, nonpolar, and not readily broken. Organic molecules may have these structural features as well:

- **Heteroatoms—atoms other than carbon or hydrogen.** Common heteroatoms are nitrogen, oxygen, sulfur, phosphorus, and the halogens.
- **π Bonds.** The most common π bonds occur in C—C and C—O double bonds.

These structural features distinguish one organic molecule from another. They determine a molecule's geometry, physical properties, and reactivity, and comprise what is called a **functional group.**

- A *functional group* is an atom or a group of atoms with characteristic chemical and physical properties. It is the *reactive part* of the molecule.

Why do heteroatoms and π bonds confer reactivity on a particular molecule?

- Heteroatoms have lone pairs and create electron-deficient sites on carbon.
- π Bonds are easily broken in chemical reactions. A π bond makes a molecule a base and a nucleophile.

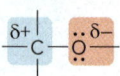

- Lone pairs make O a base and a nucleophile.
- The C atom is electron deficient, making it an electrophile.

- The π bond is easily broken.
- The π bond makes a compound a base and a nucleophile.

Don't think, though, that the C—C and C—H σ bonds are unimportant. They form the **carbon backbone** or **skeleton** to which the functional groups are bonded. A functional group usually behaves the same whether it is bonded to a carbon skeleton having as few as two or as many as 20 carbons. For this reason, we often abbreviate the carbon and hydrogen portion of the molecule by a capital letter **R**, and draw the **R** bonded to a particular functional group.

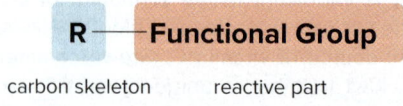

carbon skeleton reactive part

Ethane, for example, has only C—C and C—H σ bonds, so it has *no* functional group. Ethane has no polar bonds, no lone pairs, and no π bonds, so it has **no reactive sites.** Because of this, ethane and molecules like it are very unreactive.

Ethanol, on the other hand, has two carbons and five hydrogens in its carbon backbone, as well as an OH group, a functional group called a **hydroxy** group. Ethanol has lone pairs and polar bonds that make it reactive with a variety of reagents, including the acids and bases discussed in Chapter 2. The hydroxy group makes the properties of ethanol very different from the properties of ethane. Moreover, any organic molecule containing a hydroxy group has properties similar to those of ethanol.

ethane
- all C–C and C–H σ bonds
- no functional group

ethanol
- polar C–O and O–H bonds
- two lone pairs

Most organic compounds can be grouped into a relatively small number of categories, based on the structure of their functional group. Ethane, for example, is an **alkane,** whereas ethanol is a simple **alcohol.**

Problem 3.1 What reaction occurs when CH_3CH_2OH is treated with (a) H_2SO_4? (b) NaH? What happens when CH_3CH_3 is treated with these same reagents?

3.2 An Overview of Functional Groups

The most common functional groups in organic compounds can be subdivided into three types: hydrocarbons, compounds containing a C–Z σ bond (where Z = an electronegative element), and compounds containing a C=O group (Tables 3.1–3.3). In addition, Table 3.4 lists functional groups with phosphorus–oxygen bonds that are found in several biological molecules.

3.2A Hydrocarbons

To review the structure and bonding of the simple aliphatic hydrocarbons, return to Section 1.10.

The word *aliphatic* is derived from the Greek word *aleiphas* meaning "fat." Aliphatic compounds have physical properties similar to those of fats.

*Hydrocarbons are compounds made up of only the elements carbon and hydrogen. They may be **aliphatic** or **aromatic.***

[1] **Aliphatic hydrocarbons.** Aliphatic hydrocarbons can be divided into three subgroups.
- **Alkanes** have only C–C σ bonds and no functional group. Ethane, CH_3CH_3, is a simple alkane.
- **Alkenes** have a C–C double bond as a functional group. Ethylene, $CH_2=CH_2$, is a simple alkene.
- **Alkynes** have a C–C triple bond as a functional group. Acetylene, HC≡CH, is a simple alkyne.

[2] **Aromatic hydrocarbons.** This class of hydrocarbons was so named because many of the earliest known aromatic compounds had strong, characteristic odors.

The simplest aromatic hydrocarbon is **benzene.** The six-membered ring and three π bonds of benzene comprise a *single* functional group.

benzene
molecular formula C_6H_6

phenyl group
C_6H_5–
phenylcyclohexane

When a benzene ring is bonded to another group, it is called a **phenyl group.** In phenylcyclohexane, for example, a phenyl group is bonded to the six-membered cyclohexane ring. Table 3.1 summarizes the four different types of hydrocarbons.

Table 3.1 Hydrocarbons

Type of compound	General structure	Example	Functional group
Alkane	R—H	CH₃CH₃	—
Alkene	\C=C/	H₂C=CH₂	double bond
Alkyne	—C≡C—	H—C≡C—H	triple bond
Aromatic compound	(benzene ring)	(benzene ring)	phenyl group

Polyethylene is a synthetic plastic first produced in the 1930s, and initially used as insulating material for radar during World War II. It is now a plastic used in milk containers, sandwich bags, and plastic wrapping. Over 100 billion pounds of polyethylene are manufactured each year.

Alkanes, which have no functional groups, are notoriously unreactive except under very drastic conditions. For example, **polyethylene** is a synthetic plastic and high-molecular-weight alkane, consisting of chains of $-CH_2-$ groups bonded together, hundreds or even thousands of atoms long. Because it is an alkane with no reactive sites, it is a very stable compound that does not readily degrade and thus persists for years in landfills.

polyethylene
The chain continues in both directions.

Carbon atoms in alkanes and other organic compounds are classified by the number of other carbons directly bonded to them.

- A *primary carbon* (1° carbon) is bonded to *one* other C atom.
- A *secondary carbon* (2° carbon) is bonded to *two* other C atoms.
- A *tertiary carbon* (3° carbon) is bonded to *three* other C atoms.
- A *quaternary carbon* (4° carbon) is bonded to *four* other C atoms.

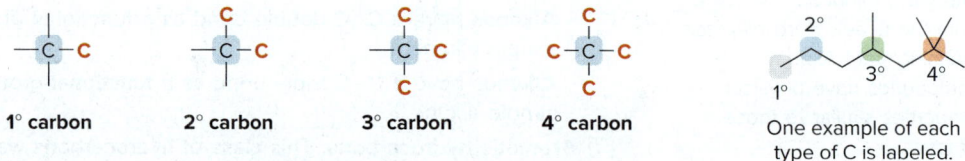

1° carbon 2° carbon 3° carbon 4° carbon

One example of each type of C is labeled.

Hydrogen atoms are classified as **primary (1°), secondary (2°),** or **tertiary (3°)** depending on the **type of carbon atom** to which they are bonded.

- A *primary hydrogen* (1° H) is on a C bonded to one other C atom.
- A *secondary hydrogen* (2° H) is on a C bonded to two other C atoms.
- A *tertiary hydrogen* (3° H) is on a C bonded to three other C atoms.

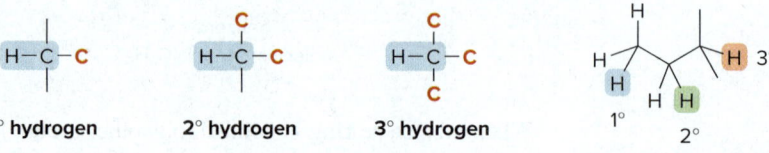

1° hydrogen 2° hydrogen 3° hydrogen

One example of each type of H is labeled.

Sample Problem 3.1 Classifying the Carbons and Hydrogens in a Molecule

Classify the designated carbon atoms in **A** as 1°, 2°, 3°, or 4°. Classify the designated hydrogen atoms in **B** as 1°, 2°, or 3°.

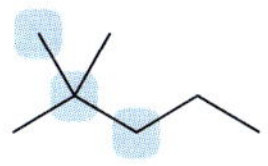

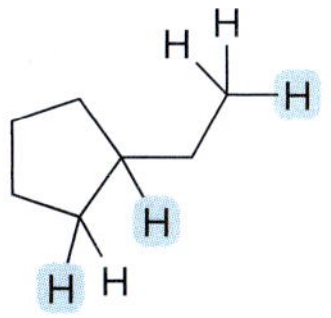

Solution
- Classify C's by the number of other C's bonded to them.
- Classify H's by the type of C to which they are bonded; a 1° H is bonded to a 1° C, etc.

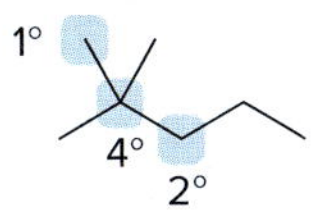

Problem 3.2 (a) Classify the carbon atoms in each compound as 1°, 2°, 3°, or 4°. (b) Classify the hydrogen atoms in each compound as 1°, 2°, or 3°.

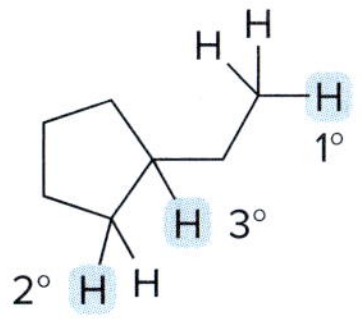

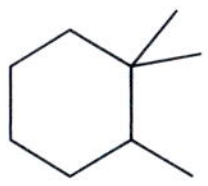

More Practice: Try Problem 3.36.

Problem 3.3 Classifying a carbon atom by the number of carbons to which it is bonded can also be done in more complex molecules that contain heteroatoms. Classify each sp^3 hybridized carbon atom in the carbohydrate sucrose (table sugar) and the antibiotic amoxicillin as 1°, 2°, 3°, or 4°.

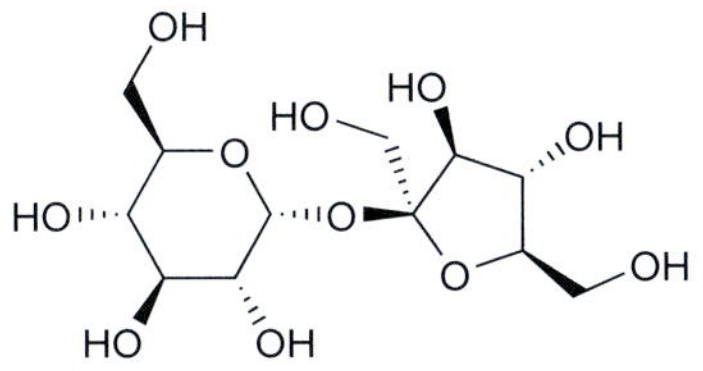

sucrose

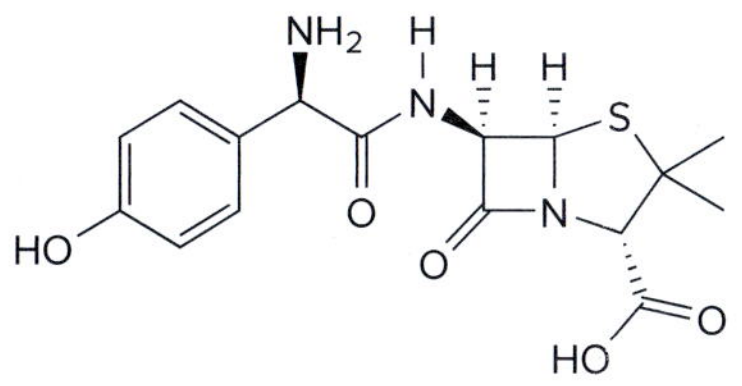

amoxicillin

3.2B Compounds Containing C–Z σ Bonds

Functional groups that contain C–Z σ bonds include **alkyl halides, alcohols, ethers, amines, thiols, sulfides, and disulfides** (Table 3.2). The electronegative heteroatom Z creates a polar bond, making carbon electron deficient. The lone pairs on Z are available for reaction with protons and other electrophiles, especially when Z = N or O.

Molecules containing these functional groups may be simple or very complex. Diethyl ether, the first common general anesthetic, is a simple ether because it contains a single O atom,

Table 3.2 Compounds Containing C–Z σ Bonds

Type of compound	General structure	Example	3-D structure	Functional group
Alkyl halide	R—X: (X = F, Cl, Br, I)	CH$_3$—Br:	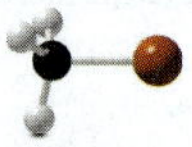	—X halo group
Alcohol	R—ÖH	CH$_3$—Ö—H		—OH hydroxy group
Ether	R—Ö—R	CH$_3$—Ö—CH$_3$		—OR alkoxy group
Amine	R—NH$_2$ or R$_2$NH or R$_3$N	CH$_3$—N(H)—H		—NH$_2$ amino group
Thiol	R—SH	CH$_3$—S—H	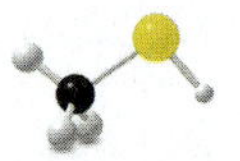	—SH mercapto group
Sulfide	R—S—R	CH$_3$—S—CH$_3$		—SR alkylthio group
Disulfide	R—S—S—R	CH$_3$—S—S—CH$_3$	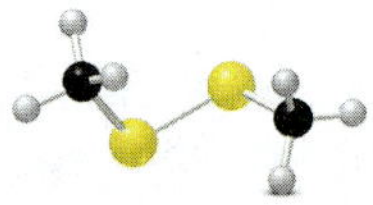	—SS—

Hemibrevetoxin B is a neurotoxin produced by algal blooms referred to as "red tides," because of the color often seen in shallow ocean waters when these algae proliferate.

depicted in red, bonded to two C atoms. Hemibrevetoxin B, on the other hand, contains four ether groups, in addition to other functional groups.

diethyl ether

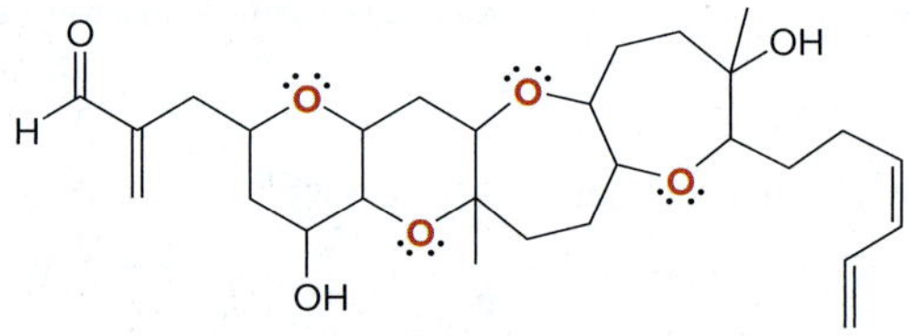

hemibrevetoxin B

Alkyl halides and alcohols are classified as **primary (1°), secondary (2°),** or **tertiary (3°)** based on the number of carbon atoms bonded to the carbon bearing the halogen or OH group. The classification of four functional groups in sucralose, the synthetic sweetener sold as Splenda, is shown.

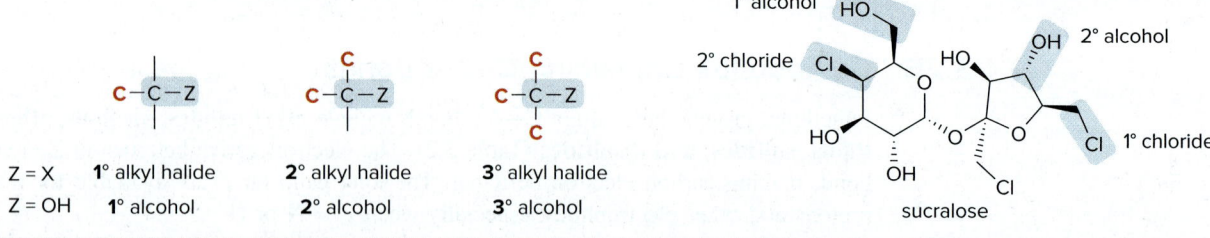

	C—C—Z	C—C(C)—Z	C—C(C)(C)—Z
Z = X	1° alkyl halide	2° alkyl halide	3° alkyl halide
Z = OH	1° alcohol	2° alcohol	3° alcohol

Problem 3.4 Classify each alkyl halide and alcohol as 1°, 2°, or 3°.

a. 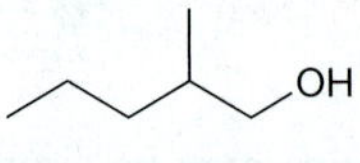 [structure with OH] b. 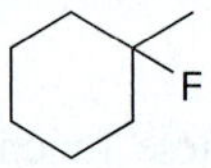 [cyclohexane with F] c. 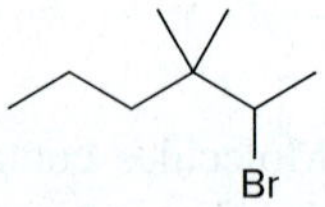 [structure with Br] d. 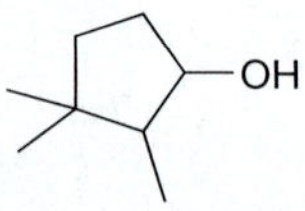 [structure with OH]

Problem 3.5 Classify each OH group and halogen in dexamethasone, a synthetic steroid, as 1°, 2°, or 3°.

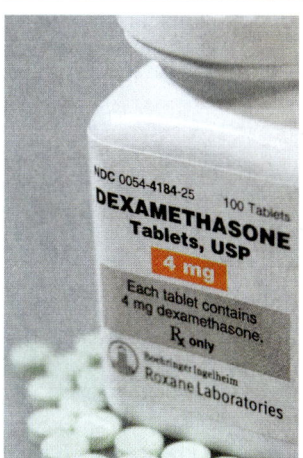

Dexamethasone (Problem 3.5) relieves inflammation and is used to treat some forms of arthritis, skin conditions, and asthma. *Jill Braaten*

dexamethasone

Amines are classified as **primary (1°), secondary (2°),** or **tertiary (3°)** based on the number of carbon atoms bonded to the *nitrogen* atom.

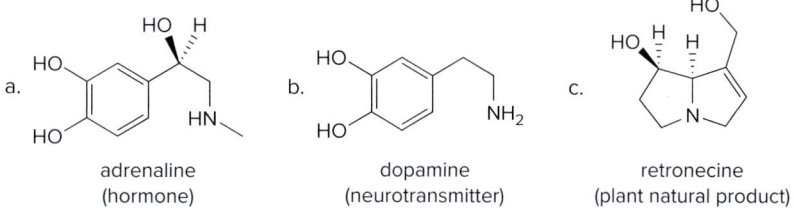

Classifying amines is different from classifying alcohols and alkyl halides as primary (1°), secondary (2°), or tertiary (3°). Amines are classified by the number of carbon–*nitrogen* bonds, whereas alkyl halides and alcohols are classified by the type of *carbon* bonded to the halogen or hydroxy group.

Problem 3.6 Classify each amine in the following compounds as 1°, 2°, or 3°.

a. adrenaline (hormone)   b. dopamine (neurotransmitter)   c. retronecine (plant natural product)

Problem 3.7 Draw the structure of a compound of molecular formula $C_4H_{11}NO$ that fits each description: (a) a compound that contains a 1° amine and a 3° alcohol; (b) a compound that contains a 3° amine and a 1° alcohol.

Sulfur-containing functional groups are especially prevalent in the chemistry of proteins, because cysteine and methionine, two common amino acids, contain a thiol and sulfide, respectively. We will learn more about amino acids and proteins in Section 3.9.

cysteine   methionine

3.2C Compounds Containing a C=O Group

Many different types of functional groups possess a C—O double bond (a **carbonyl group**), including **aldehydes, ketones, carboxylic acids, esters, amides,** and **acid chlorides** (Table 3.3). The polar C—O bond makes the carbonyl carbon an **electrophile,** while the lone pairs on O allow it to react as a **nucleophile** and **base.** The carbonyl group also contains a π bond that is more easily broken than a C—O σ bond.

carbonyl group

98 Chapter 3 Introduction to Organic Molecules and Functional Groups

Table 3.3 Compounds Containing a C=O Group

Type of compound	General structure	Example	Condensed structure	3-D structure	Functional group
Aldehyde	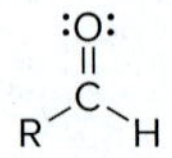 R–C(=O)–H	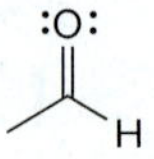 CH₃–C(=O)–H	CH_3CHO	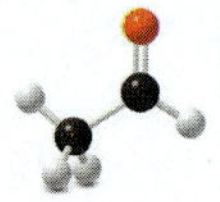	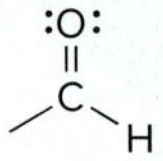 –C(=O)–H
Ketone	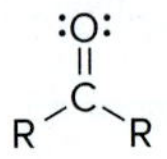 R–C(=O)–R	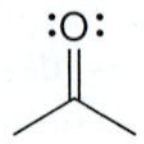 CH₃–C(=O)–CH₃	$(CH_3)_2CO$	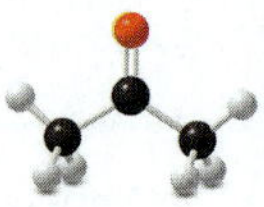	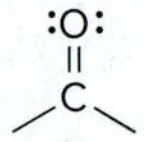 –C(=O)– carbonyl group
Carboxylic acid	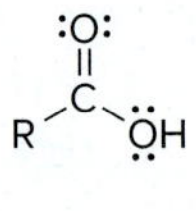 R–C(=O)–OH	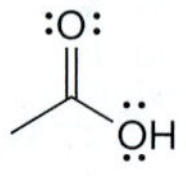 CH₃–C(=O)–OH	CH_3CO_2H	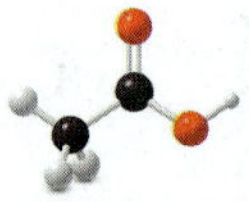	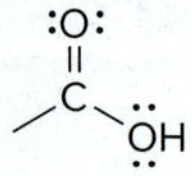 –C(=O)–OH carboxy group
Ester	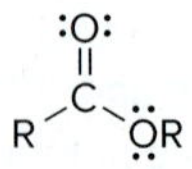 R–C(=O)–OR	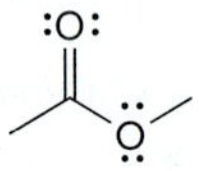 CH₃–C(=O)–OCH₃	$CH_3CO_2CH_3$	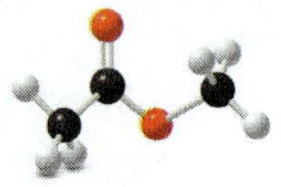	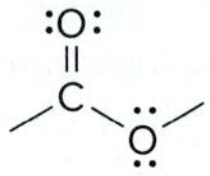 –C(=O)–O–
Amide	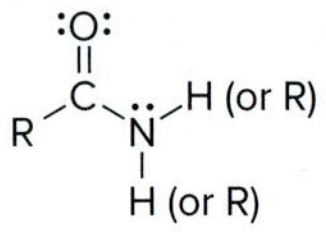 R–C(=O)–N(H or R)(H or R)	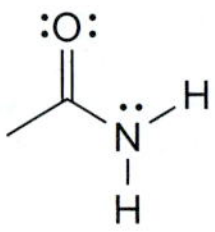 CH₃–C(=O)–NH₂	CH_3CONH_2	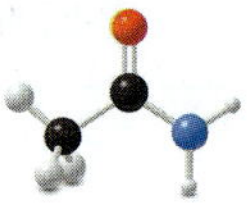	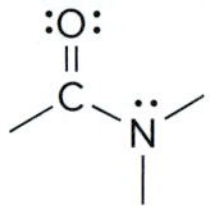 –C(=O)–N<
Acid chloride	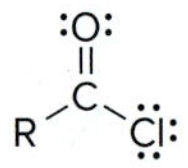 R–C(=O)–Cl	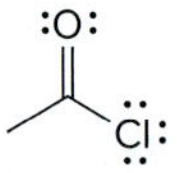 CH₃–C(=O)–Cl	CH_3COCl	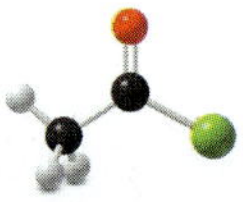	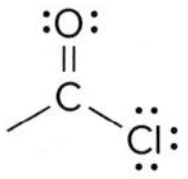 –C(=O)–Cl

Amides, compounds that contain a nitrogen atom bonded directly to the carbonyl carbon, are classified as **primary (1°), secondary (2°),** or **tertiary (3°)** based on the number of carbon atoms bonded to the nitrogen atom.

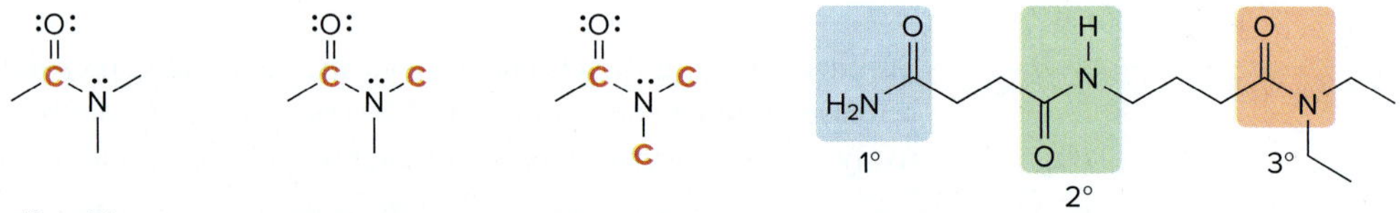

1° amide   2° amide   3° amide

Problem 3.8 Classify the amides in thyrotropin-releasing hormone (TRH), a hormone produced by the hypothalamus, as 1°, 2°, or 3°.

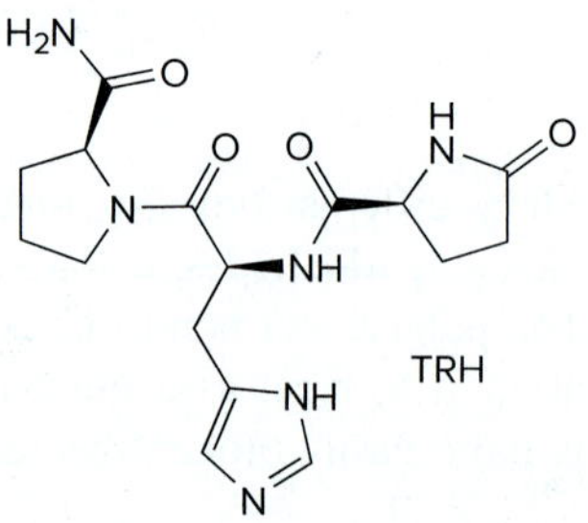

TRH

The importance of a functional group cannot be overstated. A functional group determines a molecule's bonding and shape, type and strength of intermolecular forces, physical properties, nomenclature, and chemical reactivity.

Sample Problem 3.2 — Identifying Functional Groups in a Complex Molecule

Identify the functional groups in two drugs, atenolol and donepezil. Atenolol is a β (beta) blocker, a drug used to treat hypertension (high blood pressure), and donepezil (trade name Aricept) is used to treat mild to moderate dementia associated with Alzheimer's disease.

atenolol
(used to treat high blood pressure)

donepezil
(used to treat Alzheimer's disease)

Solution

Concentrate on the heteroatoms and π bonds. With carbonyl groups, pay attention to what is bonded to the carbonyl carbon—hydrogen, carbon, or a heteroatom.

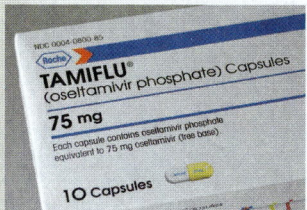

Tamiflu (Problem 3.9) is the trade name for oseltamivir, an antiviral drug used to treat influenza. *Jill Braaten*

Problem 3.9
Oseltamivir can be prepared in 10 steps from shikimic acid. Identify the functional groups in oseltamivir and shikimic acid.

shikimic acid → 10 steps → oseltamivir

More Practice: Try Problems 3.35b; 3.37; 3.38; 3.62a, b; 3.63a, b; 3.64a.

Problem 3.10
Draw the structure of a compound fitting each description:
a. an aldehyde with molecular formula C_4H_8O
b. a ketone with molecular formula C_4H_8O
c. a carboxylic acid with molecular formula $C_4H_8O_2$
d. an ester with molecular formula $C_4H_8O_2$

Problem 3.11
Identify the functional groups in leukotriene C_4, a major contributor to the inflammation associated with asthma.

leukotriene C_4

3.2D Compounds Containing P–O Bonds

Many biological molecules contain functional groups with one or more phosphorus atoms bonded to four oxygens, including monophosphates, diphosphates, triphosphates, and acyl phosphates (Table 3.4). These functional groups bear net negative charges at the pH of cells, a topic discussed in more detail in Chapter 15.

The role of phosphates in biological substitution reactions is discussed in Section 7.16.

The bond between one oxygen of phosphate to an alkyl group results in a phosphorus analogue of an ester—a **phosphate ester.** When two oxygens of phosphate are bonded to alkyl groups, the compound becomes a **phosphate diester** or **phosphodiester.** Phosphodiesters are key elements of nucleic acids, as we will see in Section 3.9.

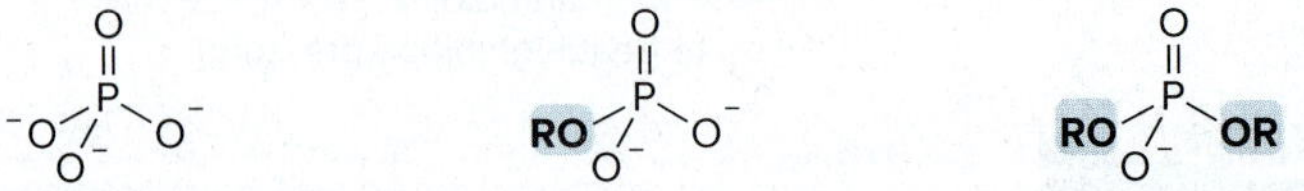

phosphate   phosphate ester   phosphate diester / phosphodiester

Table 3.4 Compounds Containing P–O Bonds

Type of compound	General structure	Example	Condensed structure
Monophosphate	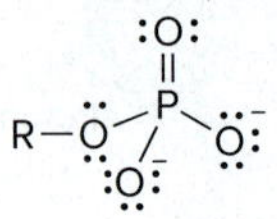	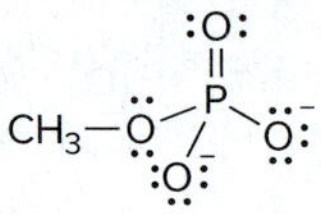	$CH_3OPO_3^{2-}$
Diphosphate	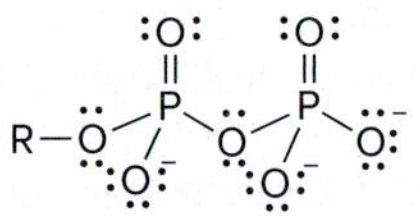	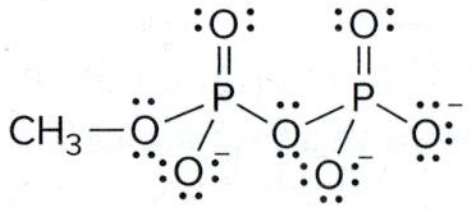	$CH_3OP_2O_6^{3-}$
Triphosphate	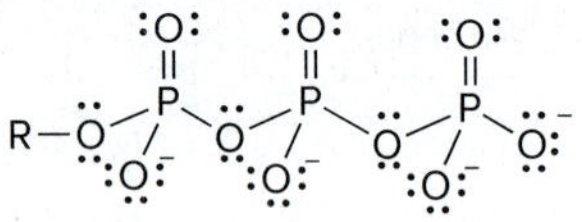	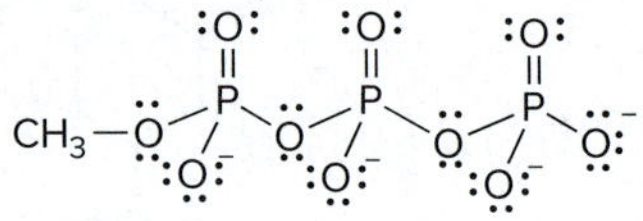	$CH_3OP_3O_9^{4-}$
Acyl phosphate	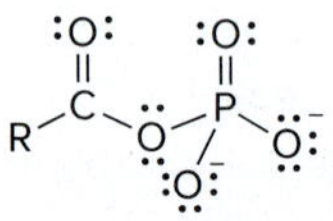	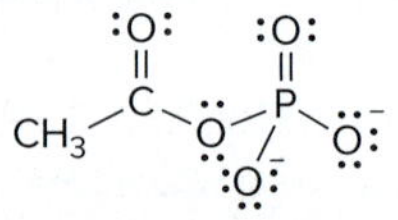	$CH_3CO_2PO_3^{2-}$

Adenosine triphosphate (ATP), the key compound used in energy transfer in metabolism in cells, contains a triphosphate.

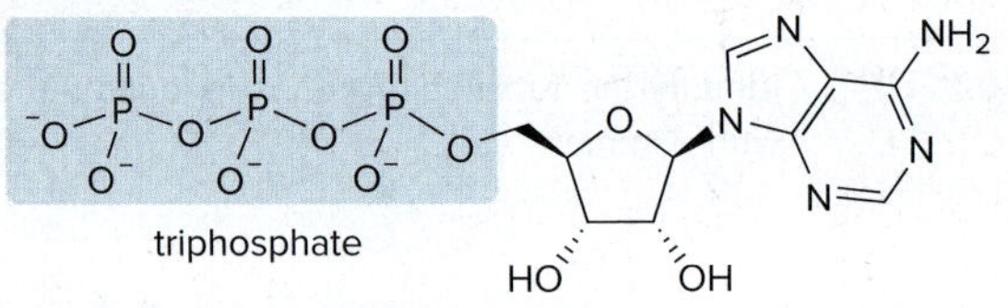

adenosine triphosphate
ATP

Problem 3.12 Label all the functional groups in **A** and **B**, intermediates formed during the metabolism of the simple sugar glucose.

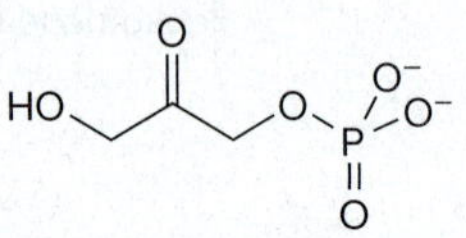

A 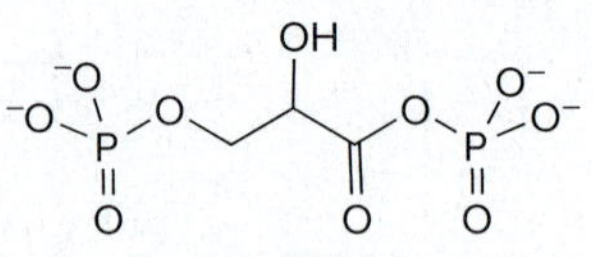   **B**

3.3 Intermolecular Forces

Intermolecular forces are also referred to as **noncovalent interactions** or **nonbonded interactions**.

Intermolecular forces are the interactions that exist *between* molecules. A functional group determines the type and strength of these interactions.

3.3A Ionic Compounds

Ionic compounds, such as NaCl, contain oppositely charged particles held together by **extremely strong electrostatic interactions.** These ionic interactions are much *stronger* than the intermolecular forces present between covalent molecules, so it takes a great deal of energy to separate oppositely charged ions from each other.

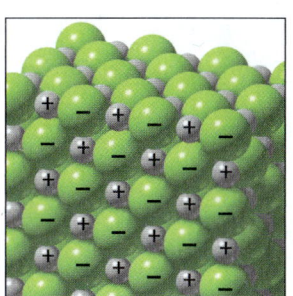

strong electrostatic interaction between Na^+ and Cl^-

3.3B Covalent Compounds

Covalent compounds are composed of discrete molecules. The nature of the forces between the molecules depends on the functional group present. There are three different types of interactions, presented here in order of *increasing strength:* **van der Waals forces, dipole–dipole interactions,** and **hydrogen bonding.**

Van der Waals Forces

Van der Waals forces, also called **London forces,** are very weak interactions caused by the **momentary changes in electron density in a molecule.** Van der Waals forces are the only attractive forces present in nonpolar compounds.

For example, although a nonpolar CH_4 molecule has no net dipole, at any one instant its electron density may not be completely symmetrical, creating a *temporary* dipole. This can induce a temporary dipole in another CH_4 molecule, with the partial positive and negative charges arranged close to each other. **The weak interaction of these temporary dipoles constitutes van der Waals forces.** All compounds exhibit van der Waals forces.

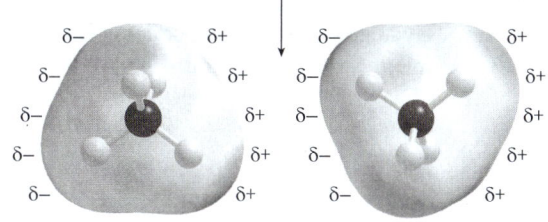

Van der Waals interactions occur between temporary dipoles.

The surface area of a molecule determines the strength of the van der Waals interactions. Long, sausage-shaped molecules such as $CH_3CH_2CH_2CH_2CH_3$ (pentane) have stronger van der Waals interactions than compact, spherical ones like $C(CH_3)_4$ (2,2-dimethylpropane).

- The *larger* the surface area, the *larger* the attractive force between two molecules, and the *stronger* the intermolecular forces.

Another factor affecting the strength of van der Waals forces is **polarizability.**

- *Polarizability* is a measure of how the electron cloud around an atom responds to changes in its electronic environment.

Although any single van der Waals interaction is weak, a large number of van der Waals interactions creates a strong force. For example, geckos stick to walls and ceilings by van der Waals interactions of the surfaces with the 500,000 tiny hairs on each foot.

(top) Don Mennig/Alamy Stock Photo; (bottom) Wrangel/iStockphoto/Getty Images

Larger atoms like iodine, which have more loosely held valence electrons, are more polarizable than smaller atoms like fluorine, which have more tightly held electrons. Because larger atoms have more easily induced dipoles, compounds containing them possess stronger intermolecular interactions.

- Compounds with large, polarizable atoms have *stronger* intermolecular forces than compounds with small, less polarizable atoms.

Dipole–Dipole Interactions

***Dipole–dipole interactions* are the attractive forces between the permanent dipoles of two polar molecules.** In acetone, $(CH_3)_2C=O$, for example, the dipoles in adjacent molecules align so that the partial positive and partial negative charges are in close proximity. These attractive forces caused by permanent dipoles are much *stronger* than weak van der Waals forces.

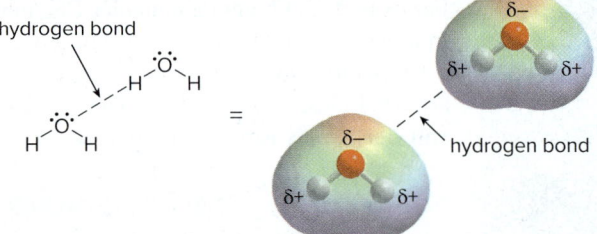

net attraction of permanent dipoles

Hydrogen Bonding

Hydrogen bonding **typically occurs when a hydrogen atom bonded to O, N, or F is electrostatically attracted to a lone pair of electrons on an O, N, or F atom in another molecule.** Thus, H_2O molecules can hydrogen bond to each other. When they do, a H atom covalently bonded to O in one water molecule is attracted to a lone pair of electrons on the O in another water molecule. Hydrogen bonds are the *strongest* of the three types of intermolecular forces, though they are still much weaker than any covalent bond.

> Hydrogen bonding helps determine the three-dimensional shape of large biomolecules such as proteins and carbohydrates. See Chapters 23 and 24 for details.

Hydrogen bonding is important in DNA, the high-molecular-weight compound that stores the genetic information of an organism. As we will learn in Section 3.9, DNA is composed of two long strands of atoms that are held together by an extensive network of hydrogen bonds, as shown in Figure 3.1.

Figure 3.1
Hydrogen bonding and DNA

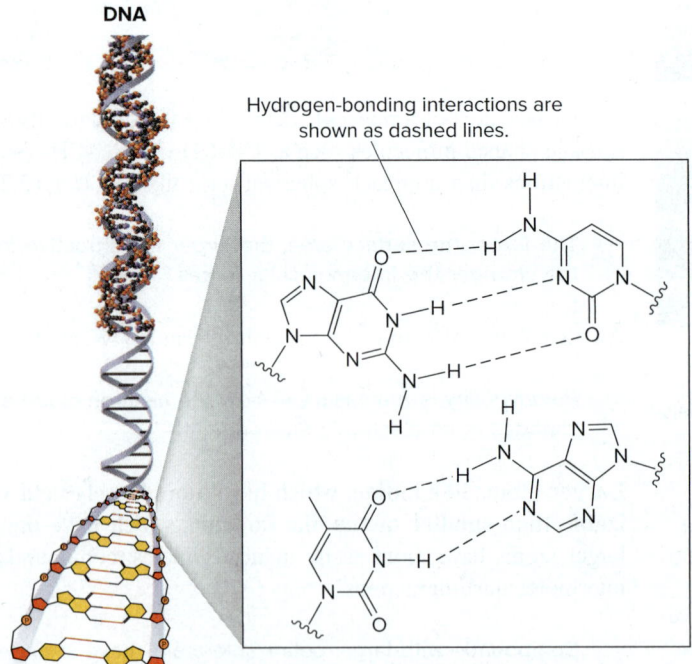

- DNA, which is contained in the chromosomes of the nucleus of the cell, is composed of two long strands of atoms held together by hydrogen bonding.

Sample Problem 3.3 illustrates how to determine the relative strength of intermolecular forces for a group of compounds. Table 3.5 summarizes the four types of interactions that affect the properties of all compounds.

Table 3.5 Summary of Types of Intermolecular Forces

Type of force	Relative strength	Exhibited by	Example
van der Waals	weak	all molecules	$CH_3CH_2CH_2CH_2CH_3$ $CH_3CH_2CH_2CHO$ $CH_3CH_2CH_2CH_2OH$
dipole–dipole	moderate	molecules with a net dipole	$CH_3CH_2CH_2CHO$ $CH_3CH_2CH_2CH_2OH$
hydrogen bonding	strong	molecules with an O–H, N–H, or H–F bond	$CH_3CH_2CH_2CH_2OH$
ion–ion	very strong	ionic compounds	NaCl, LiF

Sample Problem 3.3 Determining Intermolecular Forces in Organic Compounds

Rank the following compounds in order of increasing strength of intermolecular forces: $CH_3CH_2CH_2CH_2CH_3$ (pentane), $CH_3CH_2CH_2CH_2OH$ (butan-1-ol), and $CH_3CH_2CH_2CHO$ (butanal).

Solution

pentane
- nonpolar molecule

butan-1-ol — net dipole
- bent molecule around O
- polar C–O and O–H bonds
- O–H bond for hydrogen bonding

butanal — net dipole
- trigonal planar C
- polar C=O bond

- Pentane has only nonpolar C–C and C–H bonds, so its molecules are held together by only **van der Waals** forces.
- Butan-1-ol is a polar bent molecule, so it can have **dipole–dipole** interactions in addition to **van der Waals** forces. Because it has an O–H bond, butan-1-ol molecules are held together by intermolecular **hydrogen bonds** as well.
- Butanal has a trigonal planar carbon with a polar C=O bond, so it exhibits **dipole–dipole** interactions in addition to **van der Waals** forces. There is *no* H atom bonded to O, so two butanal molecules *cannot* hydrogen bond to each other.

Increasing strength of intermolecular forces →

Problem 3.13
What types of intermolecular forces are present in each compound?

a. (tetrahydrofuran ring with O)

b. (triethylamine)

c. CH₂=CHCl

d. CH₃CH₂CH₂COOH

e. HC≡CH

More Practice: Try Problems 3.39, 3.42.

3.4 Physical Properties

The strength of a compound's intermolecular forces determines many of its physical properties, including its boiling point, melting point, and solubility.

3.4A Boiling Point (bp)

The *boiling point* of a compound is the temperature at which a liquid is converted to a gas. In boiling, energy is needed to overcome the attractive forces in the more ordered liquid state.

- The *stronger* the intermolecular forces, the *higher* the boiling point.

Because **ionic compounds** are held together by extremely strong interactions, they have **very high boiling points.** The boiling point of NaCl, for example, is 1413 °C. **With covalent molecules, the boiling point depends on the identity of the functional group.** For compounds of approximately the same molecular weight:

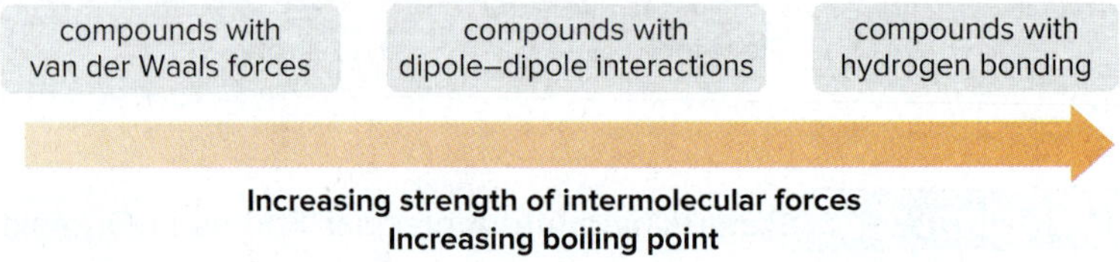

| compounds with van der Waals forces | compounds with dipole–dipole interactions | compounds with hydrogen bonding |

Increasing strength of intermolecular forces
Increasing boiling point

Recall from Sample Problem 3.3, for example, that the relative strength of the intermolecular forces increases from pentane to butanal to butan-1-ol. The boiling points of these compounds increase in the same order.

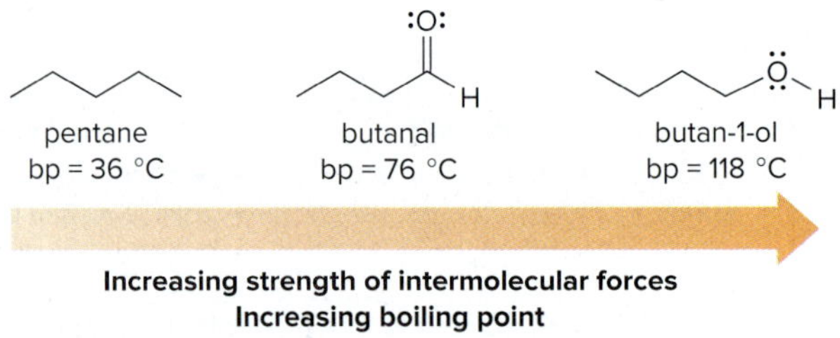

| pentane | butanal | butan-1-ol |
| bp = 36 °C | bp = 76 °C | bp = 118 °C |

Increasing strength of intermolecular forces
Increasing boiling point

Because surface area and polarizability affect the strength of intermolecular forces, they also affect the boiling point. For two compounds with similar functional groups:

- The *larger* the surface area, the *higher* the boiling point.
- The *more polarizable* the atoms, the *higher* the boiling point.

Examples of each phenomenon are illustrated in Figure 3.2. In comparing two ketones that differ in size, pentan-3-one has a higher boiling point than acetone because it has a greater molecular weight and *larger surface area*. In comparing two alkyl halides having the same number of carbon atoms, CH_3I has a higher boiling point than CH_3F because I is *more polarizable* than F.

- In comparing two compounds, the lower-boiling compound is said to be *more volatile* and the higher-boiling compound is said to be *less volatile*.

Figure 3.2
Effect of surface area and polarizability on boiling point

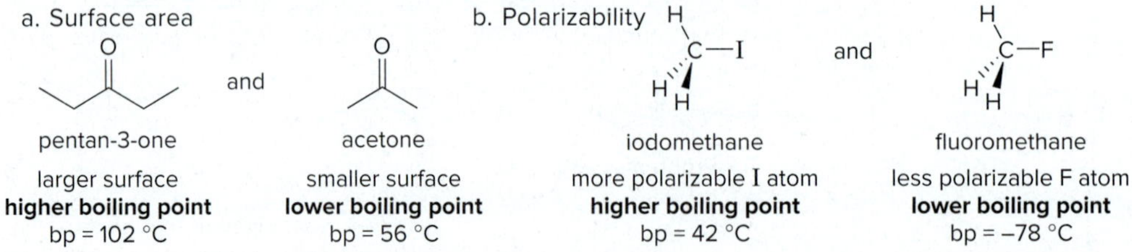

a. Surface area

pentan-3-one
larger surface
higher boiling point
bp = 102 °C

and

acetone
smaller surface
lower boiling point
bp = 56 °C

b. Polarizability

iodomethane
more polarizable I atom
higher boiling point
bp = 42 °C

and

fluoromethane
less polarizable F atom
lower boiling point
bp = –78 °C

3.4 Physical Properties 105

Sample Problem 3.4 Determining Relative Boiling Points

Which compound in each pair has the higher boiling point? Which compound in each pair is more volatile?

a. 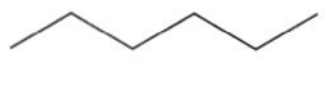 A or B   b.  C or D (OH)

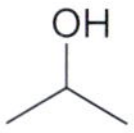

Solution

a. Isomers **A** and **B** have only nonpolar C–C and C–H bonds, so they exhibit only van der Waals forces. Because **B** is more compact, it has less surface area and a lower boiling point. The lower-boiling compound is more volatile, so **B** is more volatile than **A**.
b. Compounds **C** and **D** have approximately the same molecular weight but different functional groups. **C** is a nonpolar alkane, exhibiting only van der Waals forces. **D** is an alcohol with an O–H group available for hydrogen bonding, so it has stronger intermolecular forces and a higher boiling point. **C** is more volatile than **D**.

Problem 3.14 Which compound in each pair has the higher boiling point?

a., 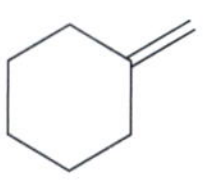 b., c., d. 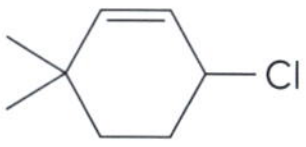 (structures shown)

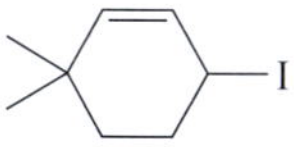

More Practice: Try Problems 3.43, 3.44.

Problem 3.15 Rank the following compounds in order of increasing boiling point.

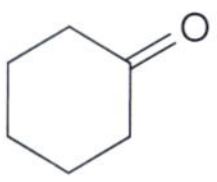

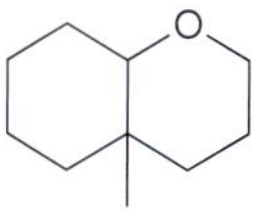

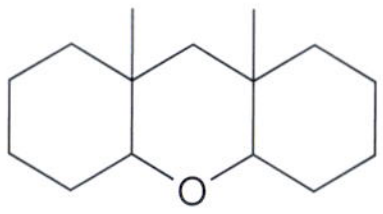

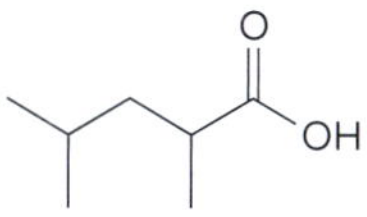

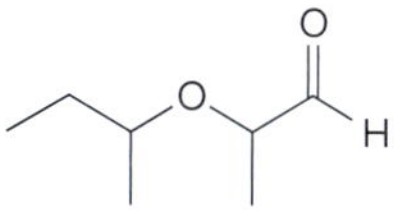

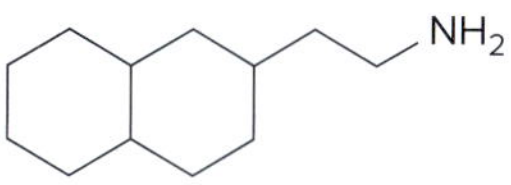

A, 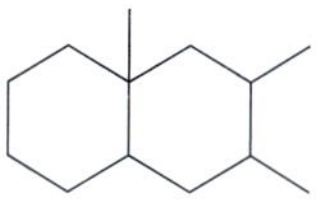 B, 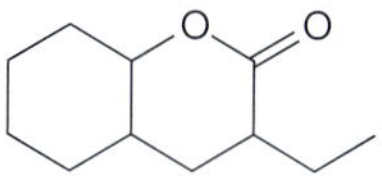 C (structures shown)

3.4B Melting Point (mp)

The *melting point* is the temperature at which a solid is converted to a liquid. In melting, energy is needed to overcome the attractive forces in the more ordered crystalline solid. Two factors determine the melting point of a compound.

- The *stronger* the intermolecular forces, the *higher* the melting point.
- Given the same functional group, the *more symmetrical* the compound, the *higher* the melting point.

Because **ionic compounds** are held together by extremely strong interactions, they have **very high melting points.** For example, the melting point of NaCl is 801 °C. With covalent molecules, the melting point once again depends on the identity of the functional group. For compounds of approximately the same molecular weight:

| compounds with van der Waals forces | compounds with dipole–dipole interactions | compounds with hydrogen bonding |

Increasing strength of intermolecular forces
Increasing melting point

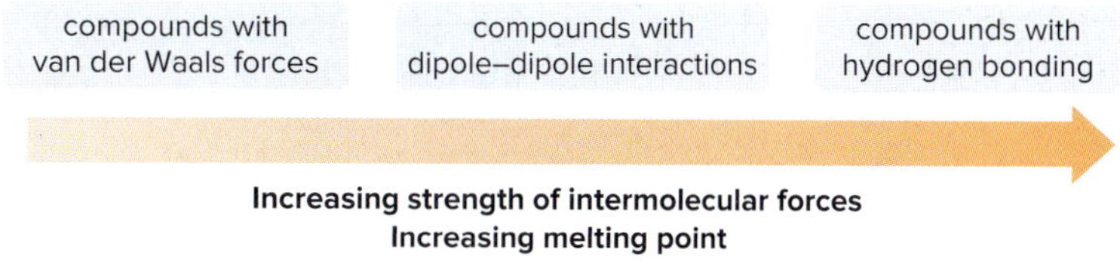

The trend in the melting points of pentane, butanal, and butan-1-ol parallels the trend observed in their boiling points.

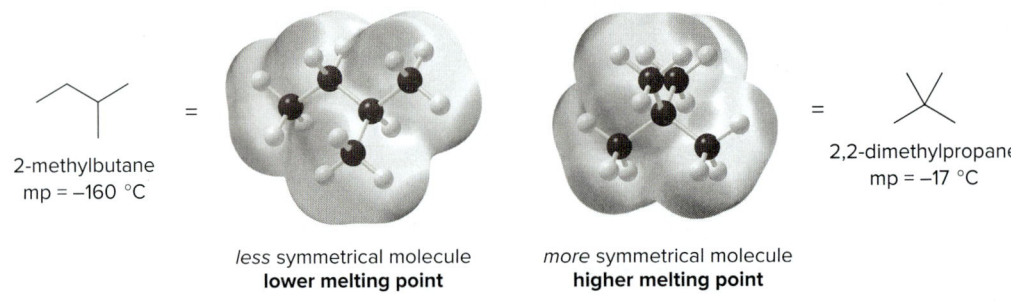

pentane mp = –130 °C

butanal mp = –96 °C

butan-1-ol mp = –90 °C

Increasing strength of intermolecular forces
Increasing melting point

Symmetry also plays a role in determining the melting points of compounds having the same functional group and similar molecular weights, but very different shapes. A compact symmetrical molecule like 2,2-dimethylpropane packs well into a crystalline lattice whereas 2-methylbutane, which has a CH$_3$ group dangling from a four-carbon chain, does not. Thus, 2,2-dimethylpropane has a much higher melting point.

2-methylbutane
mp = –160 °C

less symmetrical molecule
lower melting point

more symmetrical molecule
higher melting point

2,2-dimethylpropane
mp = –17 °C

Problem 3.16 Predict which compound in each pair has the higher melting point.

a. ⌒⌒⌒ or ⌒⌒⌒NH$_2$ b. ⤴⤵ or ⌒⌒⤴

Problem 3.17 Consider acetic acid (CH$_3$CO$_2$H) and its conjugate base, sodium acetate (CH$_3$CO$_2$Na). (a) What intermolecular forces are present in each compound? (b) Explain why the melting point of sodium acetate (324 °C) is considerably higher than the melting point of acetic acid (17 °C).

3.4C Solubility

***Solubility* is the extent to which a compound, called the *solute*, dissolves in a liquid, called the *solvent*.** In dissolving a compound, the energy needed to break up the interactions between the molecules or ions of the solute comes from new interactions between the solute and the solvent.

Quantitatively, a compound may be considered soluble when 3 g of solute dissolves in 100 mL of solvent.

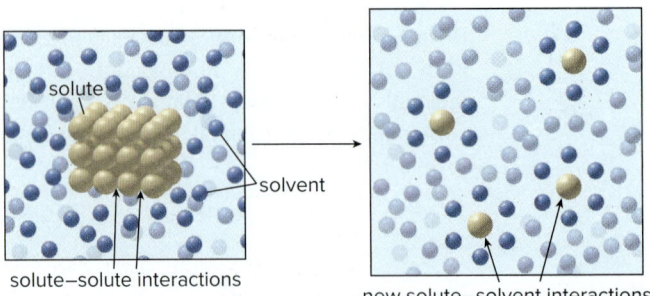

solute

solvent

solute–solute interactions

new solute–solvent interactions

3.4 Physical Properties

Compounds dissolve in solvents having similar kinds of intermolecular forces.

- "Like dissolves like."
- Polar compounds dissolve in polar solvents. Nonpolar or weakly polar compounds dissolve in nonpolar or weakly polar solvents.

Water and **organic liquids** are two different kinds of solvents. Water is very polar because it is capable of hydrogen bonding with a solute. Many organic solvents are either nonpolar, like carbon tetrachloride (CCl_4) and hexane [$CH_3(CH_2)_4CH_3$], or weakly polar like diethyl ether ($CH_3CH_2OCH_2CH_3$).

Ionic compounds are held together by strong electrostatic forces, so they need very polar solvents to dissolve. **Most ionic compounds are soluble in water, but are insoluble in organic solvents.** To dissolve an ionic compound, the strong ion–ion interactions must be replaced by many weaker **ion–dipole interactions,** as illustrated in Figure 3.3.

Figure 3.3
Dissolving an ionic compound in H_2O

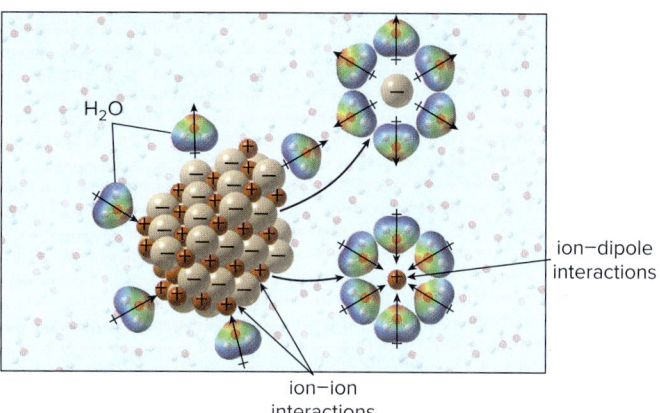

- When an ionic solid is dissolved in H_2O, the ion–ion interactions are replaced by ion–dipole interactions. Though these forces are weaker, there are so many of them that they compensate for the stronger ionic bonds.

Because water is the solvent in cells and many body fluids, many biological organic compounds contain ionic functional groups, so that they are water soluble. Isopentenyl diphosphate, a precursor of cholesterol, and acetylcholine, a neurotransmitter, are examples of ionic, water-soluble biological compounds.

isopentenyl diphosphate   acetylcholine

Most organic compounds are soluble in organic solvents (remember, *like dissolves like*). **An organic compound is water soluble only if it contains one polar functional group capable of hydrogen bonding with the solvent for every five C atoms it contains.** In other words, a water-soluble organic compound has an O- or N-containing functional group that solubilizes its nonpolar carbon backbone.

Compare, for example, the solubility of butane and acetone in H_2O and CCl_4.

butane — CCl_4 → soluble; H_2O → insoluble

acetone — CCl_4 → soluble; H_2O → soluble

Because butane and acetone are both organic compounds having a C–C and C–H backbone, they are soluble in the organic solvent CCl_4. Butane, a nonpolar molecule, is insoluble in the

(CH₃)₂C=O molecules cannot hydrogen bond to each other because they have no OH group. However, (CH₃)₂C=O can hydrogen bond to H₂O because its O atom can hydrogen bond to one of the H atoms of H₂O.

polar solvent H₂O. Acetone, however, is H₂O soluble because it contains only three C atoms and its O atom can hydrogen bond with one H atom of H₂O. In fact, acetone is so soluble in water that acetone and water are **miscible**—they form solutions in all proportions with each other.

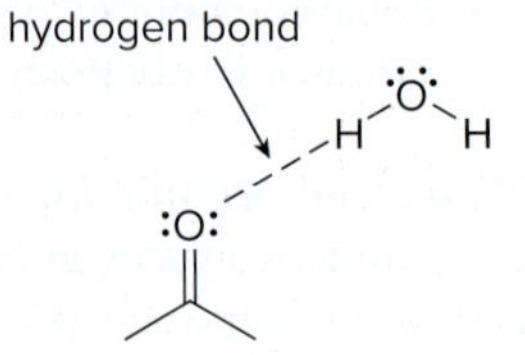

Sample Problem 3.5 Determining Hydrogen Bonding

(a) Which of the following compounds can hydrogen bond to another molecule like itself? (b) Which of the following compounds can hydrogen bond to water?

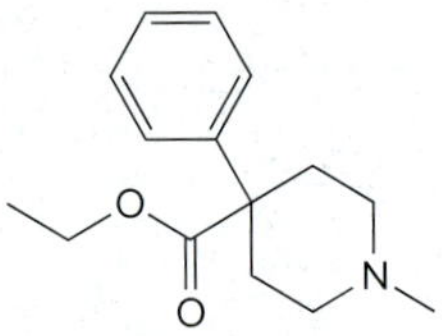

meperidine
(a narcotic)
trade name Demerol

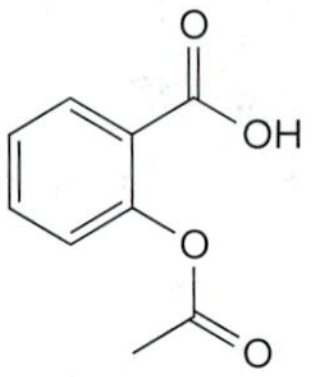

acetylsalicylic acid
(aspirin)

Solution

- To hydrogen bond to another molecule like itself, a compound needs an O—H or N—H bond.
- To hydrogen bond with water, a compound needs an O or N atom.

a. Only acetylsalicylic acid has an O—H bond for intermolecular hydrogen bonding, so two molecules of acetylsalicylic acid can hydrogen bond to each other, but two molecules of meperidine cannot.

b. Both meperidine and acetylsalicylic acid have electronegative O atoms and meperidine has an electronegative N atom, so both compounds can hydrogen bond to water. One possibility for each compound:

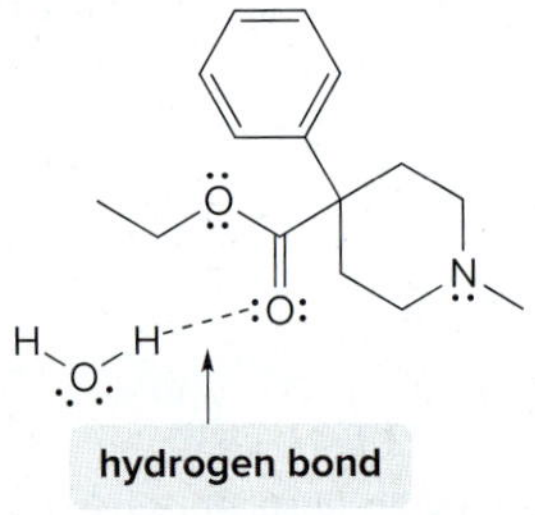

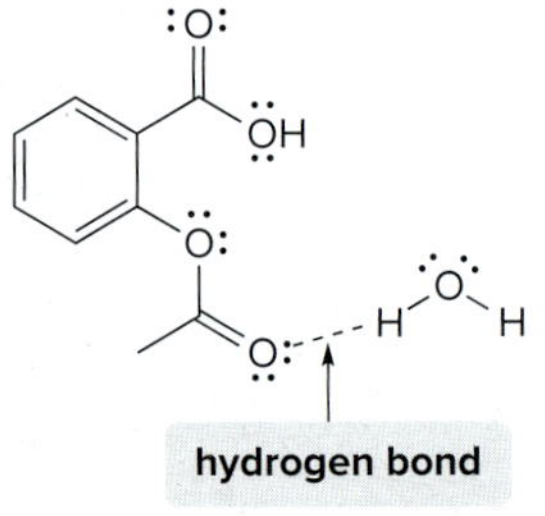

Problem 3.18

(a) At which sites can **C** hydrogen bond to another molecule like itself? (b) At which sites can **D** hydrogen bond to water?

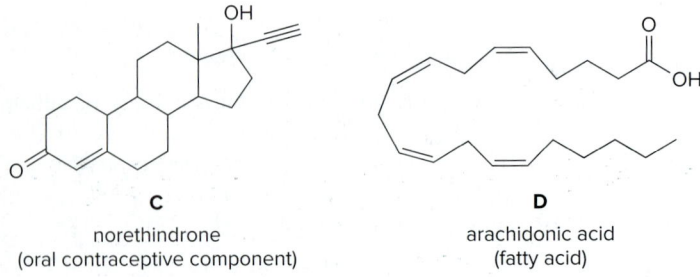

C
norethindrone
(oral contraceptive component)

D
arachidonic acid
(fatty acid)

More Practice: Try Problems 3.40; 3.41; 3.62c; 3.63c, d.

Problem 3.19 Which of the following molecules can hydrogen bond to another molecule like itself? Which can hydrogen bond to water?

a. CBr₃CH₃   b. (CH₃CH₂)₃N   c. CH₃CH₂C(=O)NH₂   d. CH₃CH₂C(=O)OCH₃

The size of an organic molecule with a polar functional group determines its water solubility. A low-molecular-weight alcohol like **ethanol is water soluble** because it has a small carbon skeleton (≤ five C atoms) compared to the size of its polar OH group. Cholesterol, on the other hand, has 27 carbon atoms and only one OH group. Its carbon skeleton is too large for the OH group to solubilize by hydrogen bonding, so **cholesterol is insoluble in water.**

For an organic compound with one functional group, **a compound is water soluble only if it has ≤ five C atoms and contains an O or N atom.**

ethanol
H₂O soluble

cholesterol
H₂O insoluble

- The nonpolar part of a molecule that is not attracted to H₂O is said to be hydrophobic.
- The polar part of a molecule that can hydrogen bond to H₂O is said to be hydrophilic.

Hydrophobic = afraid of H₂O.
Hydrophilic = H₂O loving.

In cholesterol, for example, the **hydroxy group is hydrophilic,** whereas the **carbon skeleton is hydrophobic.**

MTBE (*tert*-butyl methyl ether) and 4,4′-dichlorobiphenyl (a polychlorinated biphenyl, abbreviated as PCB) demonstrate that solubility properties can help determine the fate of organic compounds in the environment.

MTBE
tert-butyl methyl ether

Using **MTBE** as a high-octane additive in unleaded gasoline has had a negative environmental impact. Although MTBE is not toxic or carcinogenic, it has a distinctive, nauseating odor, and **it is water soluble.** Small amounts of MTBE have contaminated the drinking water in several communities, making it unfit for consumption. For this reason, the use of MTBE as a gasoline additive has steadily declined in the United States since 1999.

4,4′-dichlorobiphenyl
(a polychlorinated biphenyl, PCB)

4,4′-Dichlorobiphenyl is a polychlorinated biphenyl (**PCB**), a compound that contains two benzene rings joined by a C–C bond, and substituted by one or more chlorine atoms on each ring. PCBs have been used as plasticizers in polystyrene coffee cups and coolants in transformers. They have been released into the environment during production, use, storage, and disposal, making them one of the most widespread organic pollutants. **PCBs are insoluble in H₂O, but very soluble in organic media,** so they are soluble in fatty tissue, including that found in all types of fish and birds around the world. Although PCBs are not acutely toxic, frequently ingesting large quantities of fish contaminated with PCBs has been shown to retard growth and memory retention in children.

Solubility properties of some representative compounds are summarized in Table 3.6.

Chapter 3 Introduction to Organic Molecules and Functional Groups

Table 3.6 Summary of Solubility

Type of compound	Solubility in H₂O	Solubility in organic solvents (such as CCl₄)
Ionic		
NaCl	soluble	insoluble
Covalent		
CH₃CH₂CH₂CH₃	insoluble (no N or O atom to hydrogen bond to H₂O)	soluble
CH₃CH₂CH₂OH	soluble (≤ 5 C's and an O atom for hydrogen bonding to H₂O)	soluble
CH₃(CH₂)₁₀OH	insoluble (> 5 C's; too large to be soluble even though it has an O atom for hydrogen bonding to H₂O)	soluble

Sample Problem 3.6 Predicting the Water Solubility of a Compound

Which compounds are water soluble?

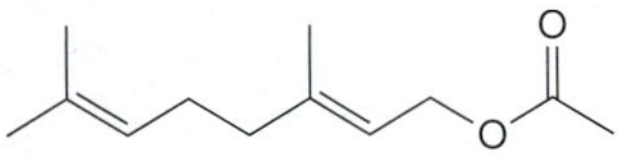

geranyl acetate
(from lily of the valley)
A

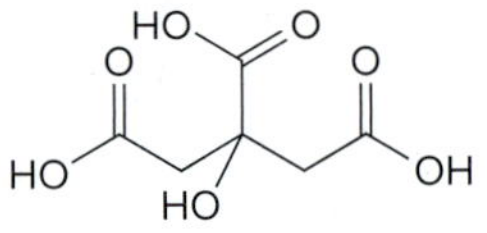

citric acid
(tart acid from citrus fruits)
B

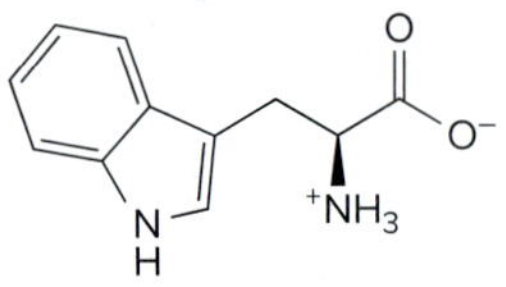

tryptophan
(an amino acid)
C

Solution

Water-soluble compounds are ionic or contain a functional group with an O or N atom that can hydrogen bond to water for every 5 C's.
- **A** has 12 C's and only one oxygen-containing functional group (an ester), so **A** is water insoluble.
- **B** has 6 C's and many OH and C=O's that can hydrogen bond to water, so **B** is water soluble.
- **C** is ionic, so **C** is water soluble.

Problem 3.20 Which compounds are water soluble?

a.

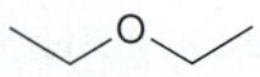

b.

c.

d. cinnamaldehyde (odor of cinnamon)

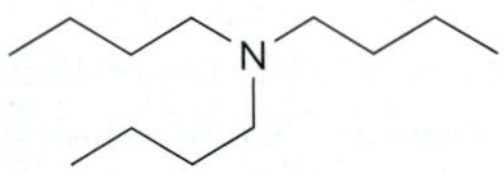

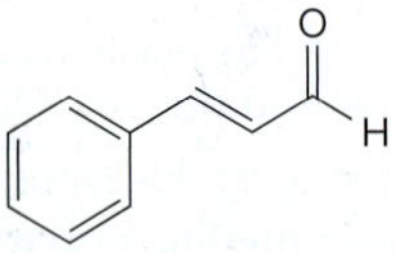

e. 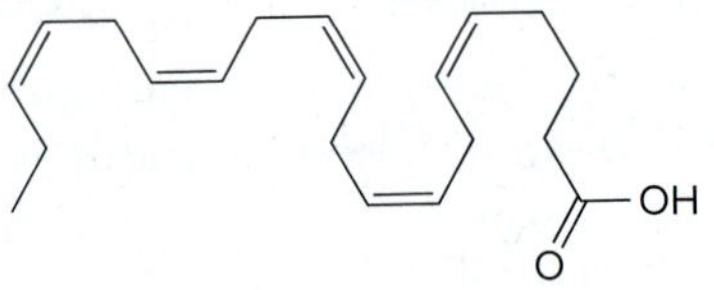 eicosapentaenoic acid (fatty acid from fish oil)

f. 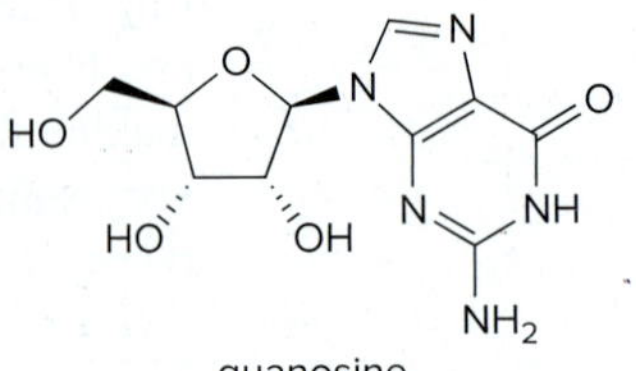 guanosine (RNA component)

More Practice: Try Problems 3.47, 3.49, 3.50, 3.51b.

3.5 Application: Vitamins

***Vitamins* are organic compounds needed in small amounts for normal cell function.** Our bodies cannot synthesize these compounds, so they must be obtained in the diet. Most vitamins are identified by a letter, such as A, C, D, E, and K. There are several different B vitamins, though, so a subscript is added to distinguish them: for example, B_1, B_2, and B_{12}.

Whether a vitamin is **fat soluble** (it dissolves in organic media) or **water soluble** can be determined by applying the solubility principles discussed in Section 3.4C. Vitamins A and C illustrate the differences between fat-soluble and water-soluble vitamins.

3.5A Vitamin A

Vitamin A, or **retinol,** is an essential component of the vision receptors in the eyes. It also helps to maintain the health of mucous membranes and the skin, so many anti-aging creams contain vitamin A. A deficiency of this vitamin leads to a loss of night vision.

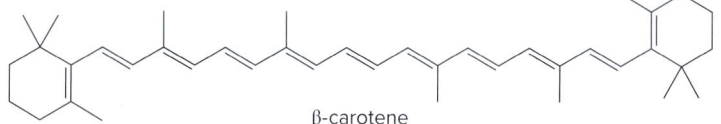

vitamin A

Vitamin A is synthesized from β-carotene, the orange pigment in carrots. *Purestock/SuperStock*

Vitamin A contains 20 carbons and a single OH group, making it **water insoluble.** As a result, vitamin A is insoluble in bodily fluids such as blood, gastric juices in the stomach, and urine, which are largely water with dissolved ions such as Na^+ and K^+. There are also fat cells composed of organic compounds having C—C and C—H bonds. Vitamin A is soluble in this organic environment because it is an uncharged organic compound, and thus it is readily stored in these fat cells, particularly in the liver.

Vitamin A may be obtained directly from the diet. In addition, β-carotene, the orange pigment found in many plants including carrots, is readily converted to vitamin A in our bodies.

β-carotene

Eating too many carrots does not result in an excess of stored vitamin A. If you consume more β-carotene than you need, your body stores this precursor until it needs more vitamin A. Some β-carotene reaches the surface tissues of the skin and eyes, giving them an orange color. This phenomenon may look odd, but it is harmless and reversible. When stored β-carotene is converted to vitamin A and is no longer in excess, these tissues will return to their normal hue.

3.5B Vitamin C

Although most animal species can synthesize vitamin C, humans, guinea pigs, the Indian fruit bat, and the bulbul bird must obtain this vitamin from dietary sources. Citrus fruits, strawberries, kiwi, tomatoes, and sweet potatoes are all excellent sources of vitamin C.

Vitamin C is obtained by eating citrus fruits and a wide variety of other fruits and vegetables. Individuals can also obtain the recommended daily dose of vitamin C by taking tablets that contain vitamin C prepared in the laboratory. Both the "natural" vitamin C in oranges and the "synthetic" vitamin C in vitamin supplements are identical. *Mary Reeg/McGraw-Hill Education*

vitamin C
(ascorbic acid)

Vitamin C has six carbon atoms, each bonded to an oxygen atom that is capable of hydrogen bonding, making it **water soluble.** Vitamin C thus dissolves in urine. Although it has been acclaimed as a deterrent for all kinds of diseases, from the common cold to cancer, the consequences of taking large amounts of vitamin C are not really known, because any excess of the minimum daily requirement is excreted in the urine.

Problem 3.21 Predict the water solubility of each vitamin.

a. vitamin B$_3$ (niacin)

b. vitamin K$_1$ (phylloquinone)

Problem 3.22 (a) Identify the functional groups in the ball-and-stick model of pantothenic acid, vitamin B$_5$. (b) At which sites can pantothenic acid hydrogen bond to water? (c) Predict the water solubility of pantothenic acid.

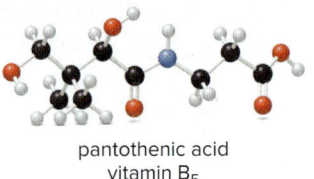

pantothenic acid
vitamin B$_5$

Avocados are an excellent dietary source of pantothenic acid, vitamin B$_5$ (Problem 3.22).
Pixtal/age fotostock

3.6 Application of Solubility: Soap

Soap has been used by humankind for some 2000 years. Historical records describe its manufacture in the first century and document the presence of a soap factory in Pompeii. Before this time clothes were cleaned by rubbing them on rocks in water, or by forming soapy lathers from the roots, bark, and leaves of certain plants. These plants produced natural materials called *saponins,* which act in much the same way as modern soaps.

On a molecular level, soap has two distinct parts:

- **a hydrophilic portion composed of ions, called the *polar head***
- **a hydrophobic carbon chain of nonpolar C–C and C–H bonds, called the *nonpolar tail***

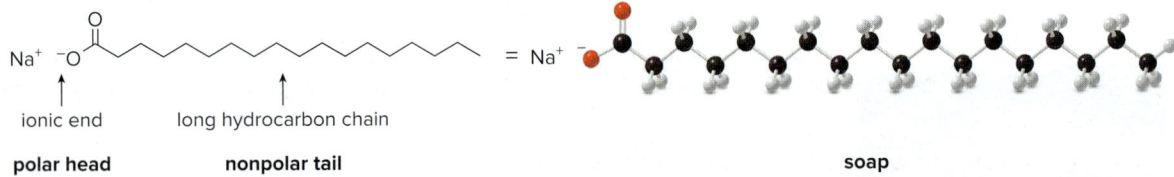

polar head — ionic end | nonpolar tail — long hydrocarbon chain | soap

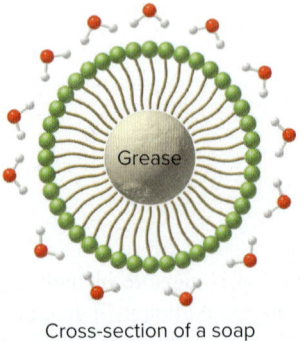

Cross-section of a soap micelle with a grease particle dissolved in the interior

Dissolving soap in water forms **micelles, spherical droplets having the ionic heads on the surface and the nonpolar tails packed together in the interior,** as shown in Figure 3.4. In this arrangement, the ionic heads are solvated by the polar solvent water, thus solubilizing the nonpolar, "greasy" hydrocarbon portion of the soap.

How does soap dissolve grease and oil? Water alone cannot dissolve dirt, which is composed largely of nonpolar hydrocarbons. When soap is mixed with water, however, the nonpolar hydrocarbon tails dissolve the dirt in the interior of the micelle. The polar head of the soap remains on the surface of the micelle to interact with water. The nonpolar tails of the soap are so well sealed off from the water by the polar head groups that the micelles are water soluble, allowing them to separate from the fibers of our clothes and be washed down the drain with water. In this way, soaps do a seemingly impossible task: they remove nonpolar hydrocarbon material from skin and clothes, by solubilizing it in the polar solvent water.

Figure 3.4
Dissolving soap in water

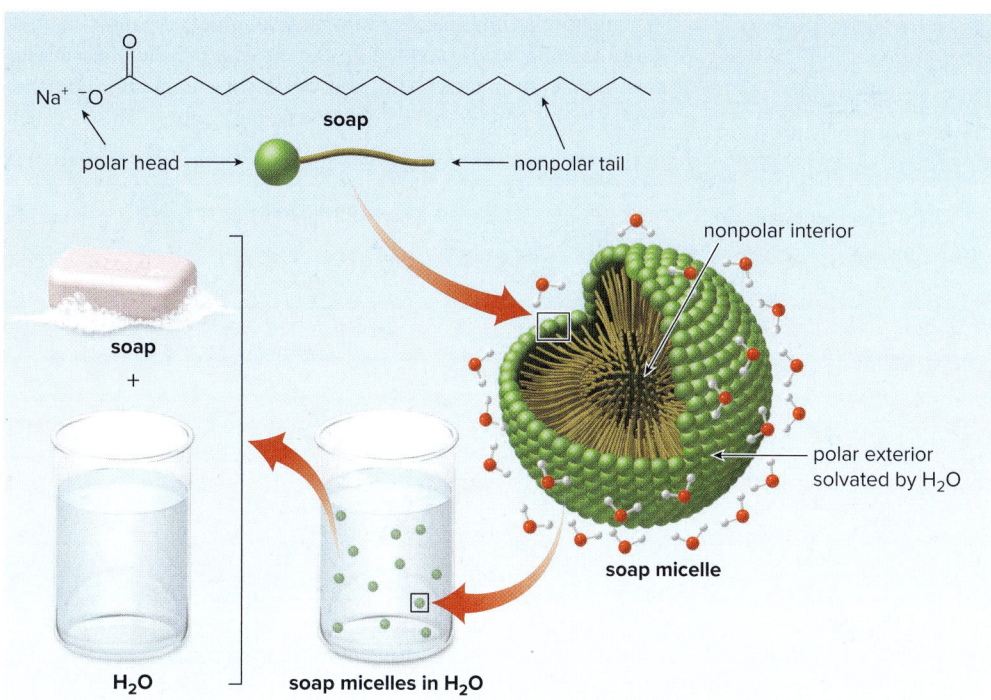

- When soap is dissolved in H₂O, it forms micelles with the nonpolar tails in the interior and the polar heads on the surface. The polar heads are solvated by ion–dipole interactions with H₂O molecules.

Problem 3.23 Today, synthetic detergents like the compound drawn here, not soaps, are used to clean clothes. Explain how this detergent cleans away dirt.

a detergent

3.7 Application: The Cell Membrane

The cell membrane is a beautifully complex example of how the principles of organic chemistry come into play in a biological system.

3.7A Structure of the Cell Membrane

The basic unit of living organisms is the **cell.** The cytoplasm is the aqueous medium inside the cell, separated from water outside the cell by the **cell membrane.** The cell membrane acts as a barrier to the passage of ions, water, and other molecules into and out of the cell, and it is also selectively permeable, letting nutrients in and waste out.

A major component of the cell membrane is a group of organic compounds called **phospholipids.** Like soap, they contain a hydrophilic ionic portion and a hydrophobic hydrocarbon portion, in this case two long carbon chains composed of C–C and C–H bonds. **Phospholipids thus contain a polar head and *two* nonpolar tails.**

ionic end
polar head

phospholipid

two long hydrocarbon chains
nonpolar tails

When phospholipids are mixed with water, they assemble in an arrangement called a **lipid bilayer,** with the ionic heads oriented on the outside and the nonpolar tails on the inside. The polar heads electrostatically interact with the polar solvent H_2O, while the nonpolar tails are held in close proximity by numerous van der Waals interactions. This is schematically illustrated in Figure 3.5.

Figure 3.5 The cell membrane

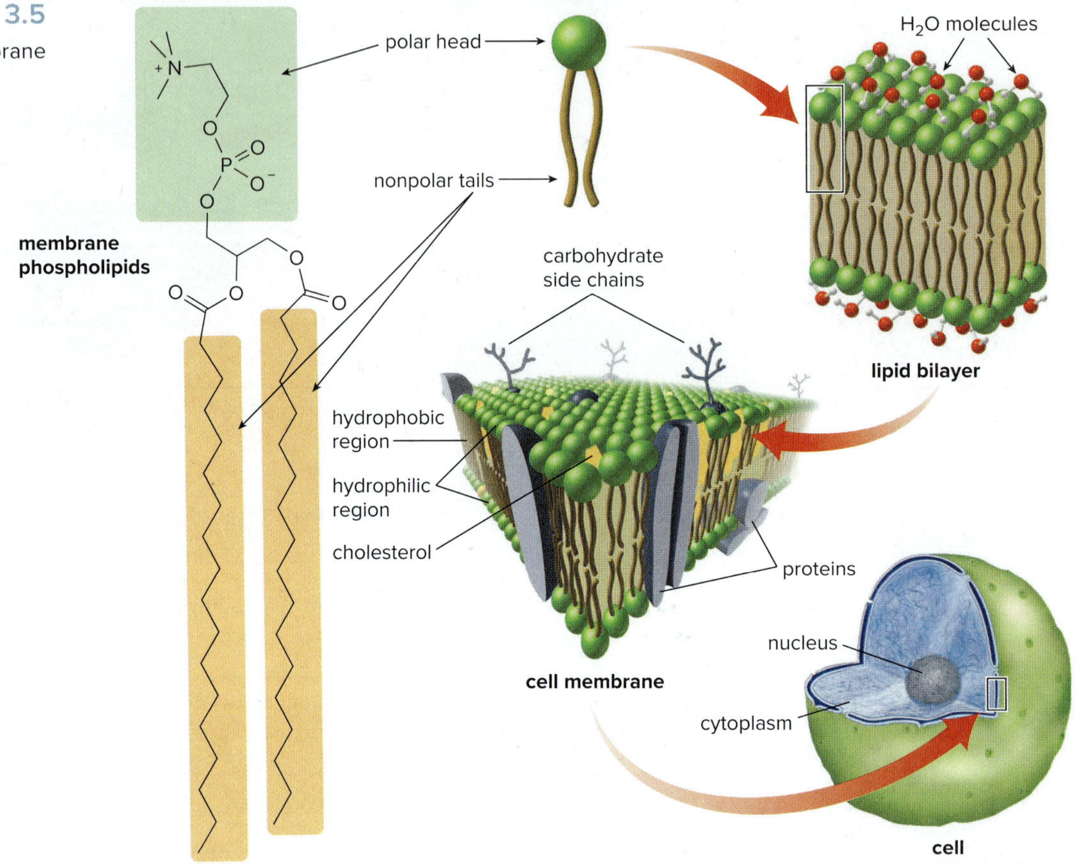

- Phospholipids contain an ionic or polar head, and two long nonpolar hydrocarbon tails. In an aqueous environment, phospholipids form a lipid bilayer, with the polar heads oriented toward the aqueous exterior and the nonpolar tails forming a hydrophobic interior. Cell membranes are composed largely of this lipid bilayer.

Cell membranes are composed of these lipid bilayers. The charged heads of the phospholipids are oriented toward the aqueous interior and exterior of the cell. The nonpolar tails form the hydrophobic interior of the membrane, thus serving as an insoluble barrier that protects the cell from the outside.

The nonpolar interior of the cell membrane is especially important in protecting the human brain from fluctuation in the concentration of compounds in the blood, as well as the passage of unwanted substances into the brain. The blood–brain barrier consists of a tight layer of cells in the blood capillaries of the brain, and all substances must pass through the cell membrane of these capillaries to enter the brain. Because ions are not soluble in the nonpolar interior of the cell membrane, the blood–brain barrier is only slightly permeable to ions. On the other hand, uncharged organic molecules like nicotine, caffeine, and heroin are very soluble in the interior of the cell membrane, so they readily pass into the brain.

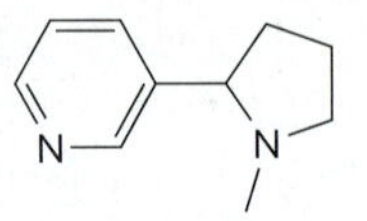

sevoflurane   nicotine 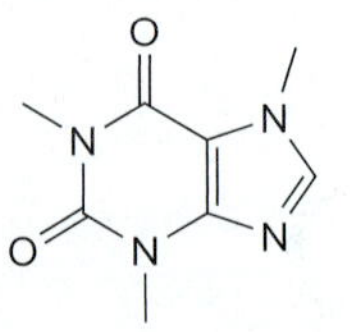   caffeine 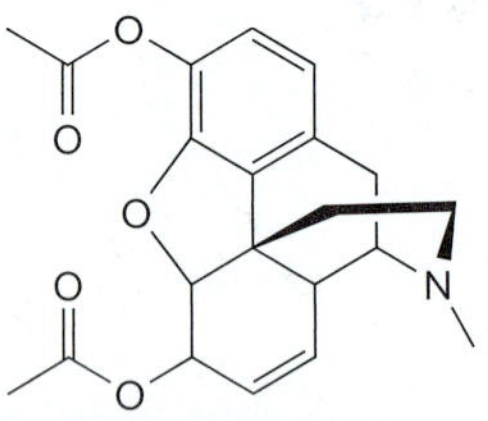   heroin

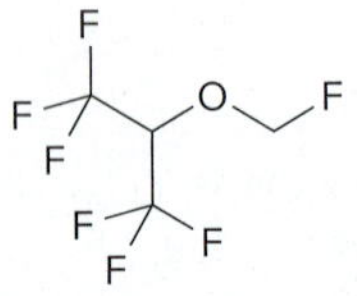

General anesthetics such as sevoflurane are also weakly polar compounds that can penetrate the blood–brain barrier because they are soluble in the lipid bilayer of the blood capillaries.

Problem 3.24 (a) What types of intermolecular forces do morphine and heroin each possess? (b) Which compound can cross the blood–brain barrier more readily, and therefore serve as the more potent pain reliever?

morphine

heroin

Problem 3.25 Explain why the noble gas xenon is a general anesthetic.

3.7B Transport Across a Cell Membrane

How does a polar molecule or ion in the water outside a cell pass through the nonpolar interior of the cell membrane and enter the cell? Some nonpolar molecules like O_2 are small enough to enter and exit the cell by diffusion. Polar molecules and ions, on the other hand, may be too large or too polar to diffuse efficiently. Some ions are transported across the membrane with the help of molecules called **ionophores.**

***Ionophores* are organic molecules that complex cations.** They have a hydrophobic exterior that makes them soluble in the nonpolar interior of the cell membrane, and a central cavity with several oxygen atoms whose lone pairs complex with a given ion. The size of the cavity determines the identity of the cation with which the ionophore complexes. **Nonactin** is a naturally occurring antibiotic that acts as an ionophore.

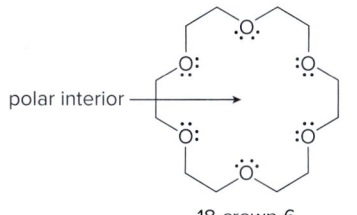
nonactin

Several synthetic ionophores have also been prepared, including one group called **crown ethers.** *Crown ethers* **are cyclic ethers containing several oxygen atoms that bind specific cations depending on the size of their cavity.** Crown ethers are named according to the general format *x*-**crown**-*y,* where *x* is the total number of atoms in the ring and *y* is the number of oxygen atoms. For example, 18-crown-6 contains 18 atoms in the ring, including 6 O atoms. This crown ether binds potassium ions. Sodium ions are too small to form a tight complex with the O atoms, and larger cations do not fit in the cavity.

polar interior →

18-crown-6

How does an ionophore transfer an ion across a membrane? The ionophore binds the ion on one side of the membrane in its polar interior. It can then move across the membrane because its hydrophobic exterior interacts with the hydrophobic tails of the phospholipid. The ionophore then releases the ion on the other side of the membrane. This ion-transfer role is essential for normal cell function. This process is illustrated in Figure 3.6.

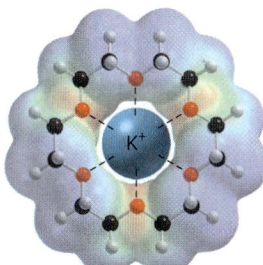
complex with K^+

Figure 3.6 Transport of ions across a cell membrane

- By binding an ion on one side of a lipid bilayer (where the concentration of the ion is high) and releasing it on the other side of the bilayer (where the concentration of the ion is low), an ionophore transports an ion across a cell membrane.

116 Chapter 3 Introduction to Organic Molecules and Functional Groups

In this manner, antibiotic ionophores like nonactin transport ions across a cell membrane of bacteria. This disrupts the normal ionic balance in the cell, thus interfering with cell function and causing the bacteria to die.

Problem 3.26 Now that you have learned about solubility, explain why aspirin (Section 2.7) crosses a cell membrane as a neutral carboxylic acid rather than an ionic conjugate base.

3.8 Functional Groups and Reactivity

Much of Chapter 3 has been devoted to how a functional group determines the strength of intermolecular forces and, consequently, the physical properties of molecules. A functional group also determines reactivity. What type of reaction does a particular kind of organic compound undergo? Begin by recalling two fundamental concepts:

- Functional groups create reactive sites in molecules.
- Electron-rich sites react with electron-poor sites.

All functional groups contain a heteroatom, a π bond, or both, and these features make electron-deficient (or electrophilic) sites and electron-rich (or nucleophilic) sites in a molecule. To predict reactivity, first locate the functional group and then determine the resulting electron-rich or electron-deficient sites it creates. Keep three guidelines in mind:

- An electronegative heteroatom like N, O, or X makes a carbon atom *electrophilic*.

- A lone pair on a heteroatom makes it *basic* and *nucleophilic*.

    base
    nucleophile

- π Bonds create *nucleophilic* sites and are more easily broken than σ bonds.

    one easily broken two easily broken
    π bond   π bonds

Problem 3.27 Label the electrophilic and nucleophilic sites in each molecule.

a. ⬡—Br   b. (ring with S)   c. (bicyclic structure with O, N-H₂, and C=O)

By identifying the nucleophilic and electrophilic sites in a compound you can begin to understand how it will react. In general, electron-rich sites react with electron-deficient sites:

:Nu⁻ = a nucleophile;
E⁺ = an electrophile.

- An electron-deficient carbon atom reacts with a nucleophile, symbolized as :Nu⁻.
- An electron-rich carbon reacts with an electrophile, symbolized as E⁺.

At this point we don't know enough organic chemistry to draw the products of many reactions with confidence. We do know enough, however, to begin to predict if two compounds might react together based solely on electron density arguments, and at what atoms that reaction is most likely to occur.

For example, the carbon–carbon double bond of isopentenyl diphosphate is electron rich, so it reacts with electrophiles, E^+. This reaction is a key step in the early stages of the biosynthesis of cholesterol. On the other hand, the carbonyl carbon of pyruvic acid is electrophilic, so it reacts with electron-rich nucleophiles. This type of reaction is a key step in glucose metabolism.

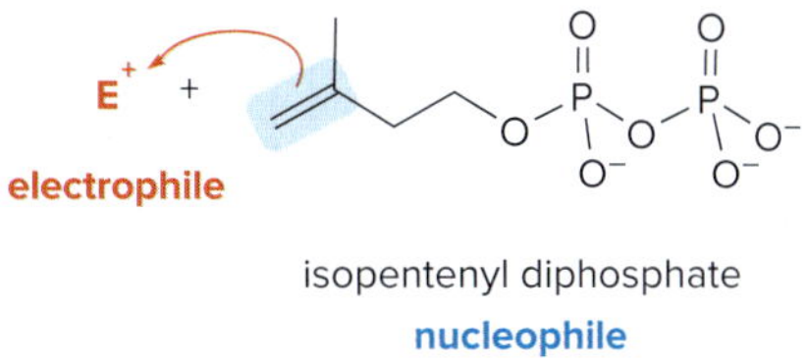

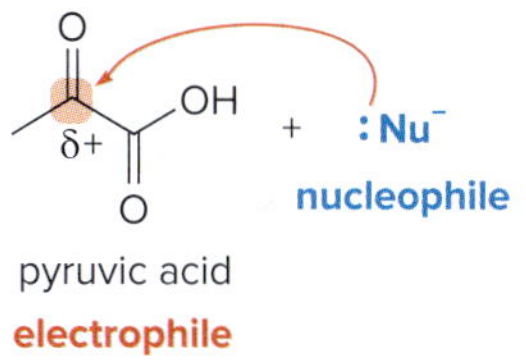

Complex molecules in biological systems may contain many functional groups, but often one of those functional groups determines many of the reactions of the molecule. For example, most reactions of **coenzyme A,** a key molecule in many biological pathways, occur when its electron-rich thiol reacts with electrophiles. As a result, we often abbreviate the structures of these compounds to emphasize the reacting functional group. Coenzyme A is written as **HS–CoA** to emphasize its nucleophilic thiol.

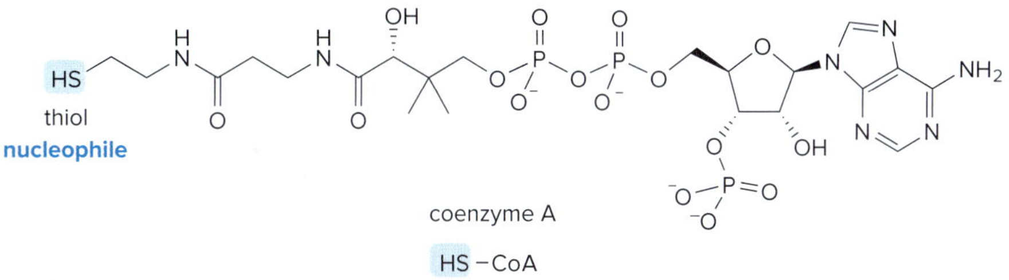

Problem 3.28 Considering only electron density, state whether the following reactions will occur.

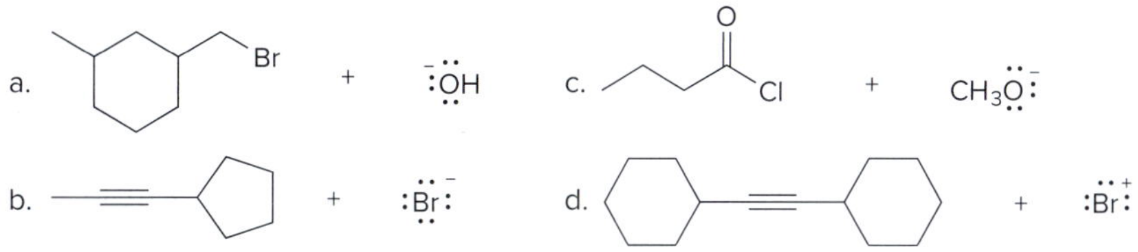

Problem 3.29 Which atom acts as the nucleophile and which atom acts as the electrophile in the starting materials of the following reaction, a key step in the synthesis of *S*-adenosylmethionine from ATP? We will learn more about the preparation and reactions of this biological molecule in Sections 7.16 and 9.15.

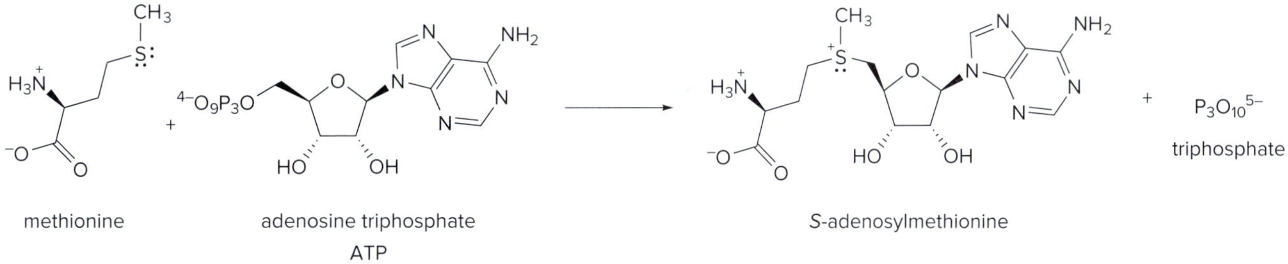

3.9 Biomolecules

***Biomolecules* are organic compounds found in biological systems.** Many are relatively small, with molecular weights of less than 1000 g/mol. There are four main families of these small molecules—**simple sugars, amino acids, lipids, and nucleotides.** Many simple biomolecules are used to synthesize larger compounds that have important cellular functions.

3.9A Amino Acids and Proteins

Of the four major types of biomolecules, **proteins** have the widest range of functions. Proteins like collagen form long insoluble fibers in connective tissue, cartilage, and blood vessels (Figure 3.7), whereas enzymes are proteins that regulate all aspects of cellular function. Hemoglobin, which transports oxygen from the lungs to the tissues, and insulin, a hormone that regulates blood glucose levels, are both proteins.

Figure 3.7 Proteins, large molecules formed from amino acids

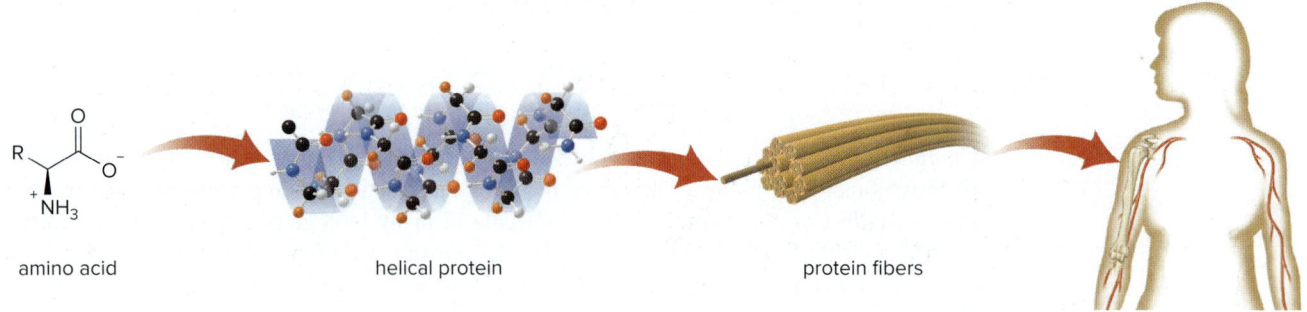

- The fibrous tissue in muscle, cartilage, and tendons is composed of proteins, which are formed from amino acids joined together by amide bonds.

- Proteins are polyamides, formed by joining amino acids together.

A listing of the 20 naturally occurring amino acids is given in Figure 23.2.

Amino acids contain two functional groups—an amino group (NH_2) and a carboxy group (CO_2H)—bonded to the same carbon, called the **α carbon.** The 20 naturally occurring amino acids that form proteins differ in the identity of the R group bonded to the α carbon. The R group is called the **side chain** of the amino acid. An R group can be hydrogen, carbon groups with only C–C and C–H bonds, or phenyl, or it can have additional functional groups such as SH, NH_2, OH, or CO_2H.

An amino acid is both an acid and a base.

- The NH_2 group has a lone pair of electrons, making it a base.
- The CO_2H has an acidic proton, making it an acid.

Even though we sometimes draw amino acids with all neutral atoms, in reality, amino acids exist as salts.

- Proton transfer from the acid to the base forms an ionic salt called a zwitterion, which contains both a positive and negative charge.

We will learn more about the acid–base chemistry of amino acids in Chapter 15.

Problem 3.30 Draw the zwitterionic form of each amino acid.

a. valine b. tyrosine c. proline

3.9B Monosaccharides and Carbohydrates

Carbohydrates, which constitute the largest group of biomolecules in nature, may have as few as three or as many as thousands of carbons. The cellulose in plant stems and tree trunks, the chitin in lobster and crab shells, and the starch in corn are examples of complex carbohydrates (Figure 3.8). One component of the nucleic acid DNA is a simple carbohydrate, as we will see in Section 3.9C.

- Carbohydrates, commonly called sugars and starches, are polyhydroxy aldehydes and ketones, or compounds that can be converted to them by reaction with water.

Figure 3.8 Glucose and starch, a complex carbohydrate

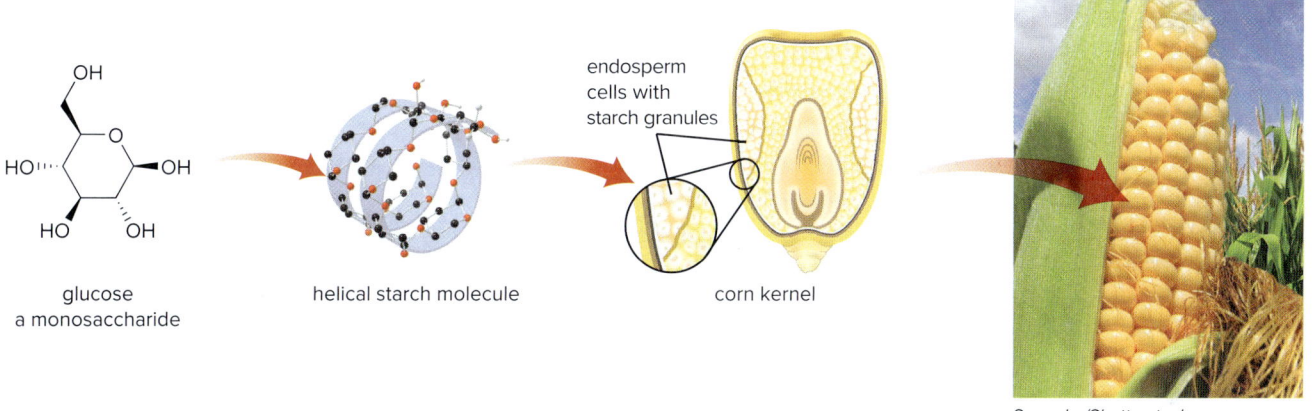

- Kernels of corn contain starch, a complex carbohydrate formed from the simple sugar glucose.

The simplest carbohydrates are called **monosaccharides** or **simple sugars.** Monosaccharides have three to six carbon atoms in a chain, with a carbonyl group at either the terminal carbon or the carbon adjacent to it. In most monosaccharides, each remaining carbon has a hydroxy group. Glyceraldehyde and glucose are monosaccharides. As we will learn in Chapter 14, glucose, the most prevalent monosaccharide, can be drawn as an acyclic compound with an aldehyde or as a cyclic compound with a six-membered ring that contains an oxygen.

glyceraldehyde   glucose acyclic form   glucose cyclic form

Joining two monosaccharides together forms **disaccharides.** Lactose is a disaccharide found in milk. Joining three or more monosaccharides together forms **polysaccharides.** Starch is a common polysaccharide found in corn, rice, and wheat. Disaccharides and polysaccharides are discussed in Chapter 24.

lactose
a disaccharide

amylose
(one form of starch)
a polysaccharide

Problem 3.31 Why are glucose and lactose water soluble, even though they each contain more than five carbon atoms?

3.9C Nucleotides and Nucleic Acids

Nucleic acids are large molecules composed of repeating units called **nucleotides. DNA, deoxyribonucleic acid,** stores the genetic information of an organism and transmits that information from one generation to another (Figure 3.9), whereas **RNA, ribonucleic acid,** translates that information into proteins needed for all cellular functions.

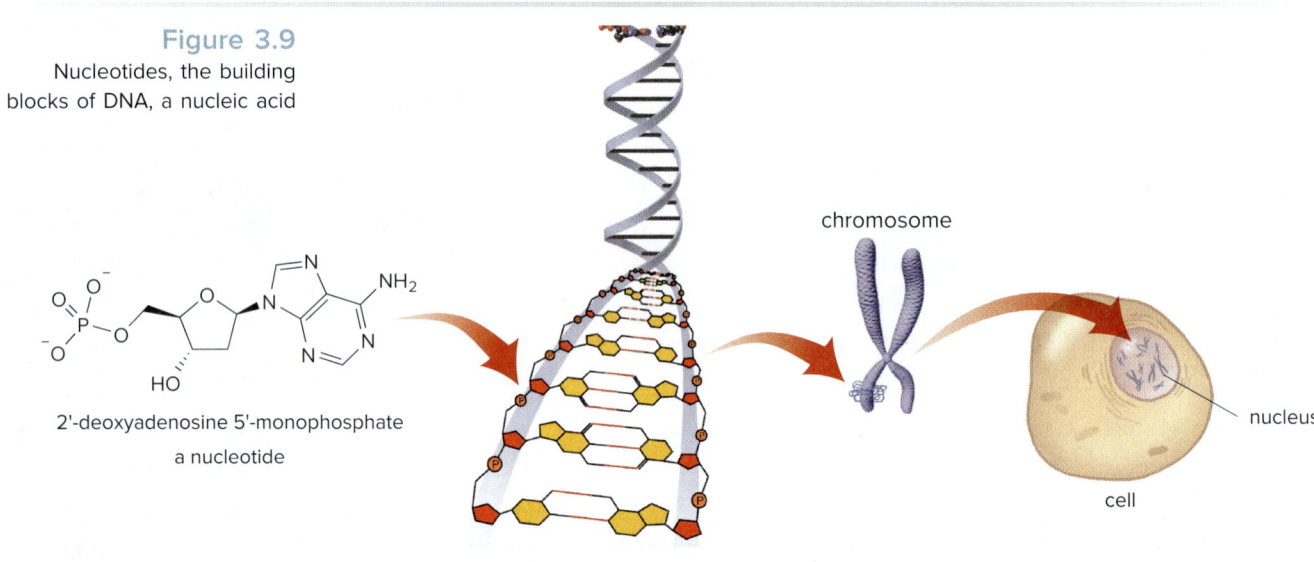

Figure 3.9
Nucleotides, the building blocks of DNA, a nucleic acid

2'-deoxyadenosine 5'-monophosphate
a nucleotide

double helix of DNA

chromosome

nucleus
cell

- DNA, which is formed from smaller units called nucleotides, is contained in the chromosomes of the nucleus. Humans have 46 chromosomes (23 pairs). An individual chromosome is composed of many genes. A **gene** is a portion of the DNA molecule responsible for the synthesis of a single protein.

- A nucleotide consists of three components—a monosaccharide, a nitrogen-containing base, and a phosphate group.

phosphate   monosaccharide   nitrogen-containing base

nucleotide

3.9 Biomolecules

Primes (') are used in numbering the monosaccharides in nucleotides.

In RNA, the monosaccharide is ribose, whereas in DNA, the monosaccharide is 2'-deoxyribose, a compound that lacks a hydroxy group at C2'.

ribose   2'-deoxyribose

Nucleic acids are polynucleotides, formed by joining an OH group of one nucleotide with the phosphate of another in a **phosphodiester linkage.** A nucleic acid contains a backbone consisting of alternating sugar and phosphate groups. The identity and order of the bases distinguish one nucleic acid from another. Figure 3.10 shows a portion of a polynucleotide.

Figure 3.10
A polynucleotide

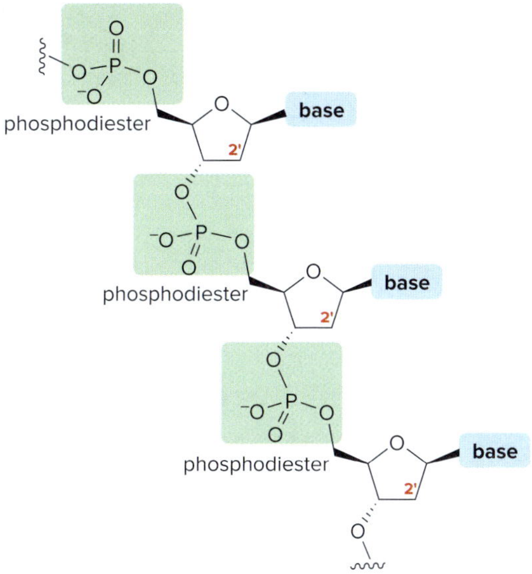

- Phosphodiester bonds that hold the polynucleotide together are highlighted in green. All atoms except the nitrogen-containing bases form the **sugar–phosphate backbone.** Because no OH group is present on C2' in the monosaccharide rings, the polynucleotide is formed from 2'-deoxyribose and represents a segment of DNA.

DNA consists of two polynucleotide chains that wind into a right-handed double helix held together by hydrogen bonding interactions of the nitrogen-containing bases (Figure 3.1). RNA consists of a single chain of nucleotides that can fold back on itself and form loops.

Nucleotides and nucleic acids are discussed in Chapter 26.

Problem 3.32 (a) Label each of the following in dinucleotide **A,** formed by joining two nucleotides together: the monosaccharide units, the phosphodiester, and the nitrogen-containing bases. (b) Can **A** be part of DNA or RNA?

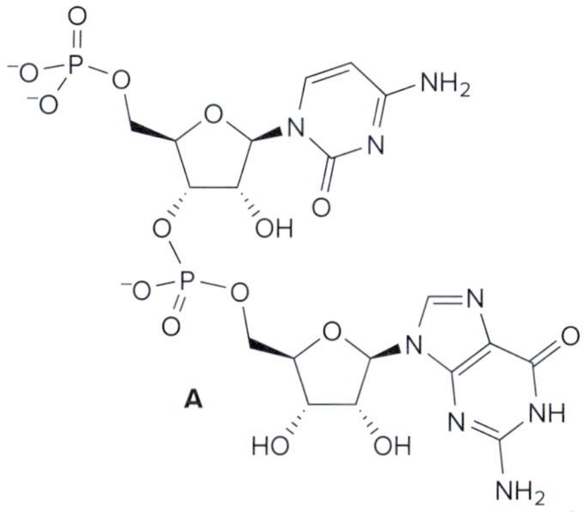

3.9D Lipids

- **Lipids are biomolecules that are soluble in organic solvents and insoluble in water.**

Because lipids are composed mainly of carbon–carbon and carbon–hydrogen bonds, their properties resemble those of the alkanes and other hydrocarbons. Lipids have a wide variety of shapes and sizes. The fat-soluble vitamins like vitamin A (Section 3.5) and the phospholipids in cell membranes (Section 3.7) are two groups of lipids. Triacylglycerols, the most common lipids, compose animal fat and vegetable oil (Figure 3.11).

Figure 3.11
Fatty acids and triacylglycerols, the most common lipids

palmitic acid
a **fatty acid**

triacylglycerol

adipose tissue
Dennis Strete/McGraw-Hill Education

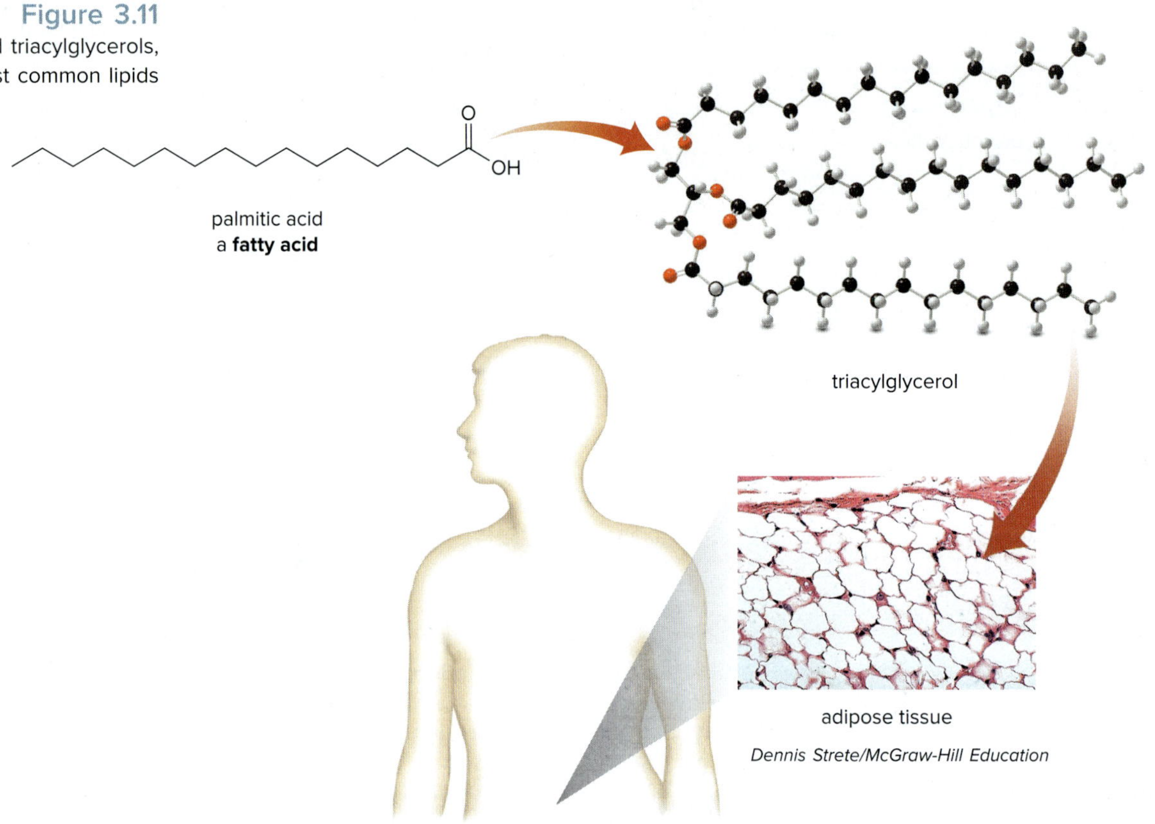

- Triacylglycerols are stored in adipose cells below the skin and are concentrated in some regions of the body. Triacylglycerols are formed from fatty acids like palmitic acid.

- **Triacylglycerols are triesters, formed from glycerol and three fatty acids (water-insoluble carboxylic acids).**

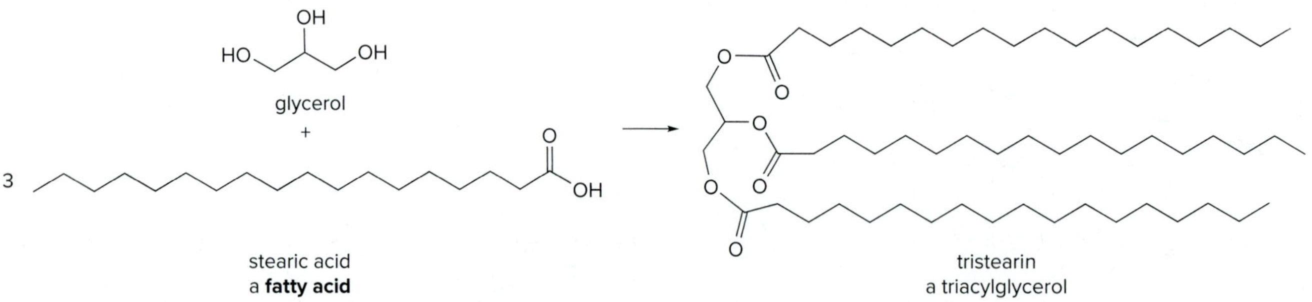

glycerol
+
stearic acid
a **fatty acid**

tristearin
a **triacylglycerol**

We will learn more about lipids in Section 10.6 and Chapter 25.

Problem 3.33 (a) Label the hydrophobic and hydrophilic regions of tristearin. (b) Can two molecules of tristearin hydrogen bond to each other? (c) Can tristearin hydrogen bond to water?

Chapter 3 REVIEW

KEY CONCEPTS

[1] Classifying atoms and functional groups (3.2)

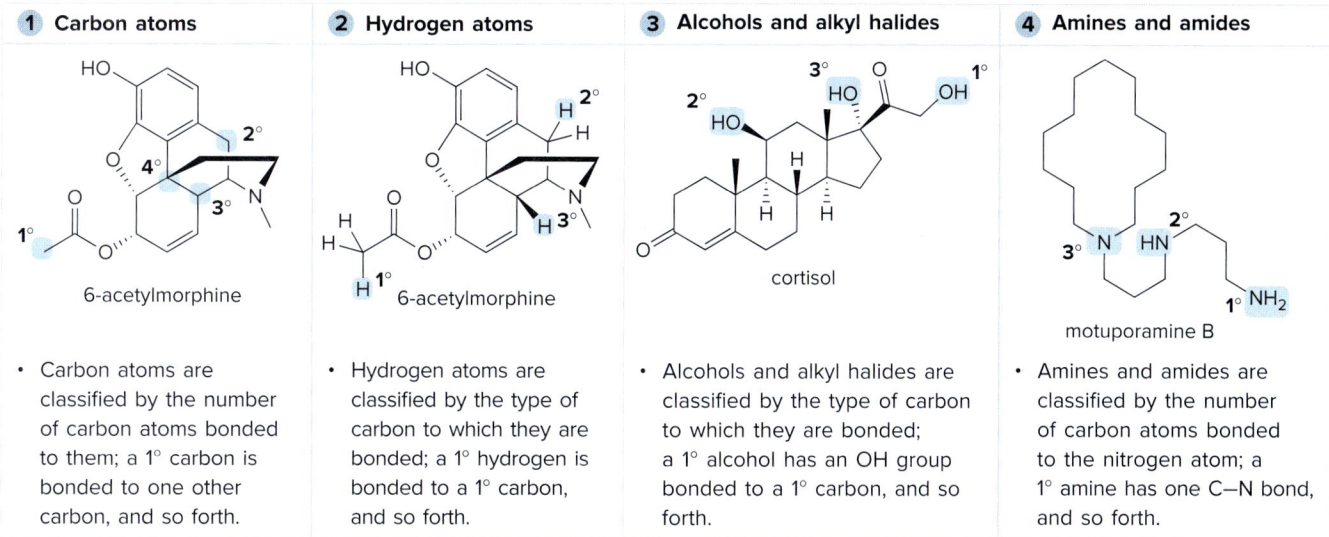

- Carbon atoms are classified by the number of carbon atoms bonded to them; a 1° carbon is bonded to one other carbon, and so forth.

- Hydrogen atoms are classified by the type of carbon to which they are bonded; a 1° hydrogen is bonded to a 1° carbon, and so forth.

- Alcohols and alkyl halides are classified by the type of carbon to which they are bonded; a 1° alcohol has an OH group bonded to a 1° carbon, and so forth.

- Amines and amides are classified by the number of carbon atoms bonded to the nitrogen atom; a 1° amine has one C–N bond, and so forth.

See Sample Problem 3.1. Try Problems 3.36, 3.37, 3.62b, 3.63b.

[2] Types of intermolecular forces (3.3)

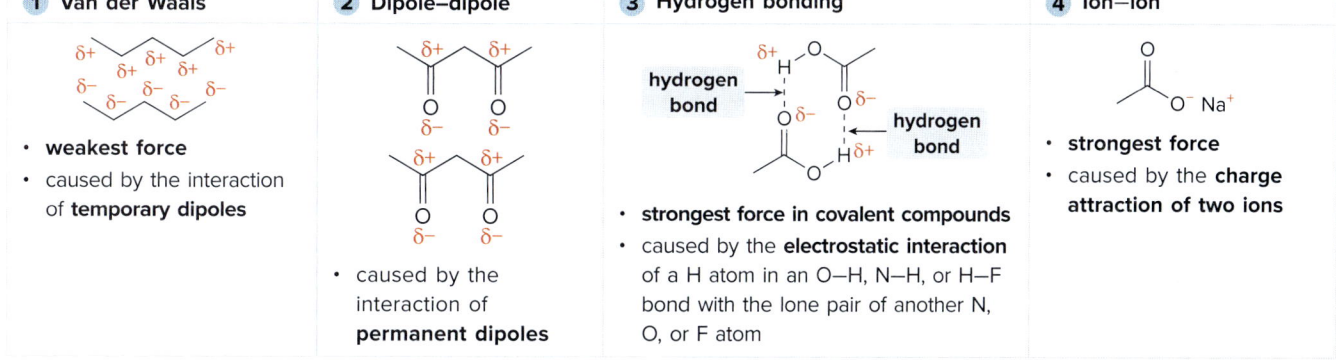

- **Van der Waals**
  - weakest force
  - caused by the interaction of **temporary dipoles**

- **Dipole–dipole**
  - caused by the interaction of **permanent dipoles**

- **Hydrogen bonding**
  - strongest force in covalent compounds
  - caused by the **electrostatic interaction** of a H atom in an O–H, N–H, or H–F bond with the lone pair of another N, O, or F atom

- **Ion–ion**
  - strongest force
  - caused by the **charge attraction of two ions**

See Table 3.5, Sample Problem 3.3. Try Problems 3.39, 3.42, 3.64d.

[3] Factors that determine boiling point (3.4A)

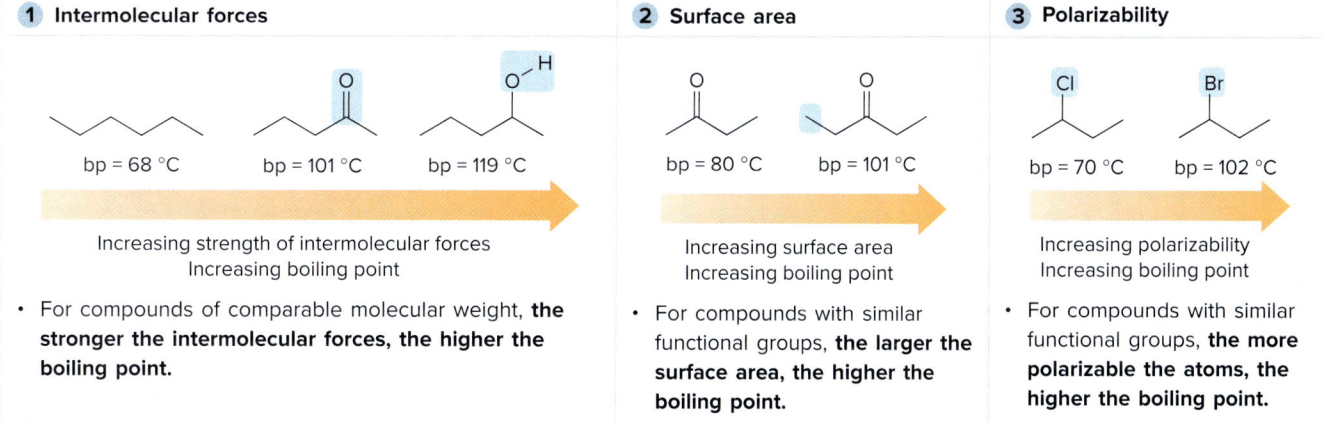

- **Intermolecular forces**: For compounds of comparable molecular weight, **the stronger the intermolecular forces, the higher the boiling point.**

- **Surface area**: For compounds with similar functional groups, **the larger the surface area, the higher the boiling point.**

- **Polarizability**: For compounds with similar functional groups, **the more polarizable the atoms, the higher the boiling point.**

See Figure 3.2, Sample Problem 3.4. Try Problems 3.43, 3.44.

[4] Factors that determine melting point (3.4B)

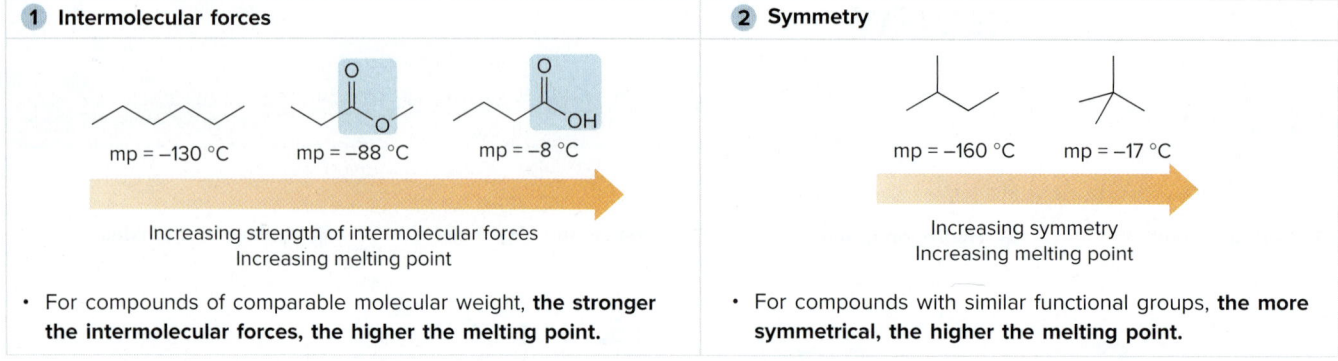

1 Intermolecular forces

- For compounds of comparable molecular weight, **the stronger the intermolecular forces, the higher the melting point.**

2 Symmetry

- For compounds with similar functional groups, **the more symmetrical, the higher the melting point.**

Try Problem 3.45.

[5] Factors that determine solubility (3.4C, 3.5)

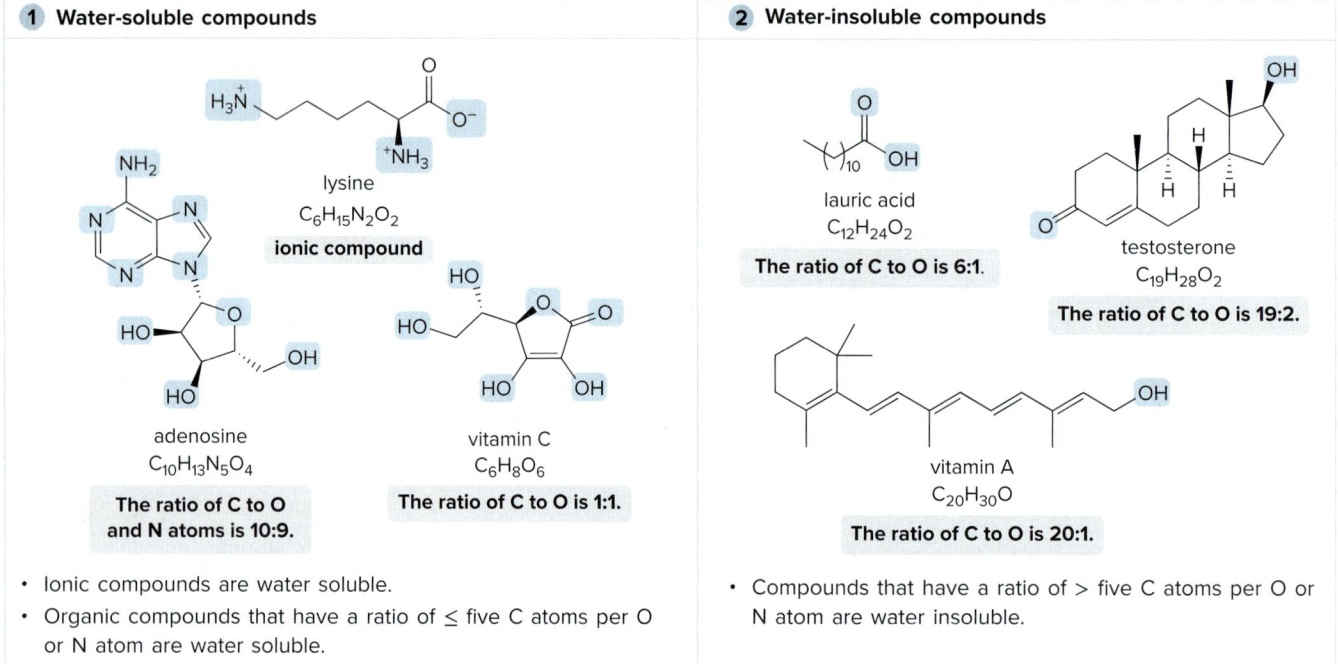

1 Water-soluble compounds

- Ionic compounds are water soluble.
- Organic compounds that have a ratio of ≤ five C atoms per O or N atom are water soluble.

2 Water-insoluble compounds

- Compounds that have a ratio of > five C atoms per O or N atom are water insoluble.

See Table 3.6. Try Problems 3.47, 3.49, 3.50, 3.51b.

[6] Reactivity of functional groups (3.8)

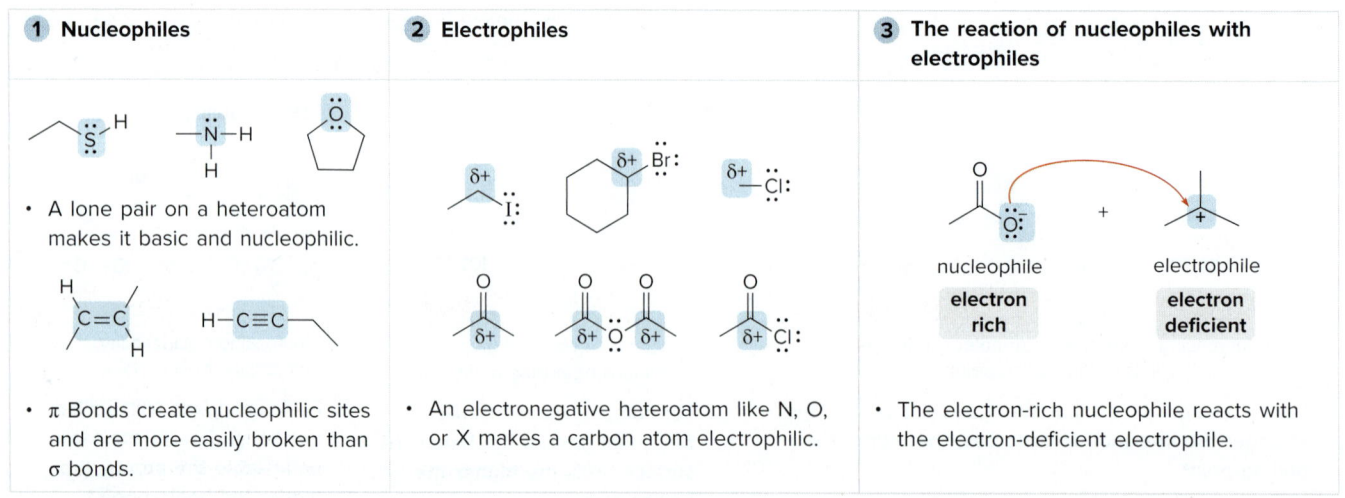

1 Nucleophiles

- A lone pair on a heteroatom makes it basic and nucleophilic.
- π Bonds create nucleophilic sites and are more easily broken than σ bonds.

2 Electrophiles

- An electronegative heteroatom like N, O, or X makes a carbon atom electrophilic.

3 The reaction of nucleophiles with electrophiles

- The electron-rich nucleophile reacts with the electron-deficient electrophile.

Try Problems 3.56b; 3.57; 3.58; 3.62d, e; 3.64g.

KEY SKILLS

[1] Predicting boiling points (3.4A)

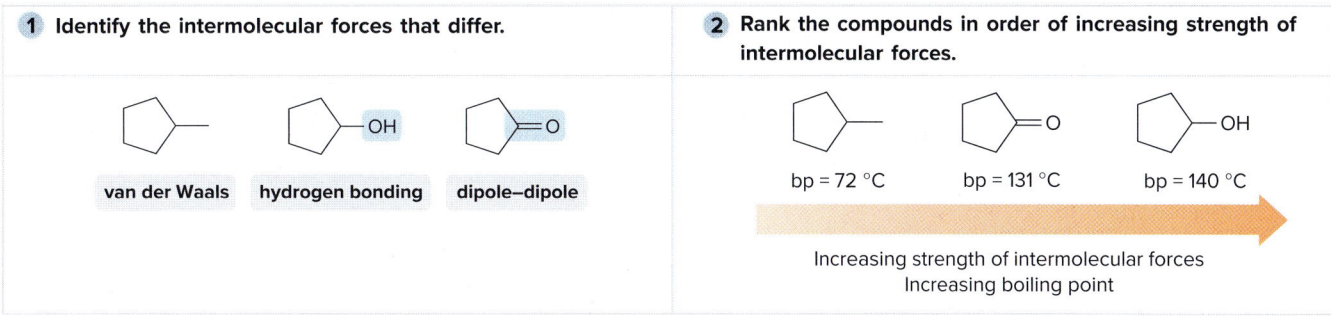

See Sample Problem 3.3. Try Problems 3.43, 3.44.

[2] Determining sites of hydrogen bonding between two identical molecules (3.4C)

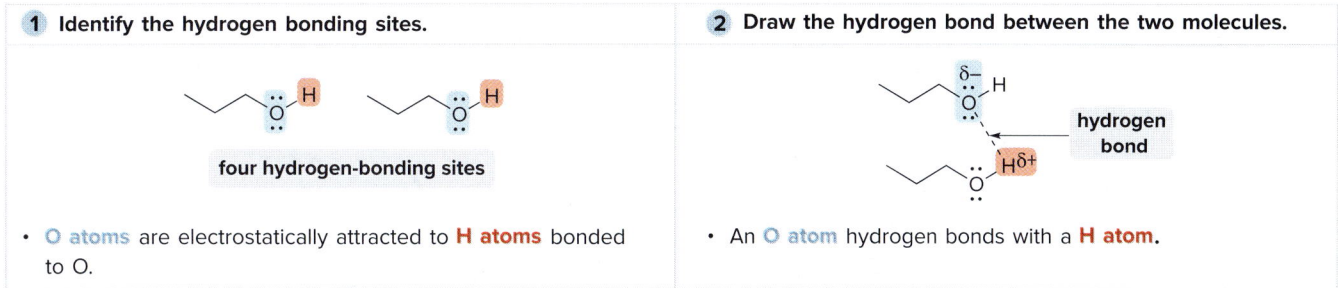

See Sample Problem 3.5. Try Problems 3.40a, 3.41a, 3.62c.

[3] Determining sites of hydrogen bonding between an organic molecule and H₂O (3.4C)

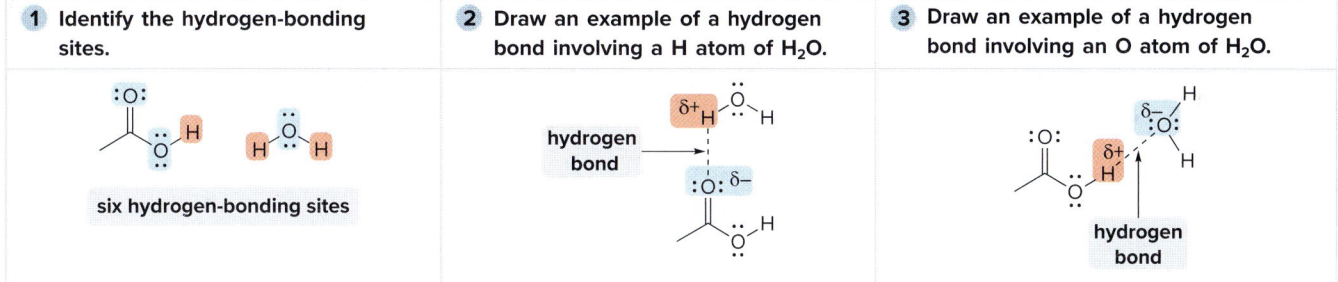

Try Problems 3.40b, 3.41b, 3.63c, 3.64f.

[4] Drawing curved arrows to show the reaction between a nucleophile and an electrophile (3.8)

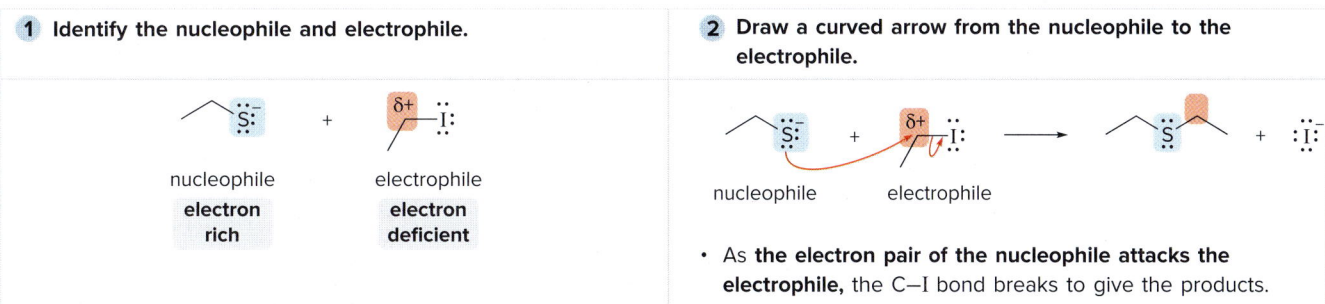

Try Problem 3.58.

PROBLEMS

Problems with Three-Dimensional Models

3.34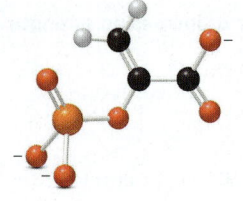
phosphoenolpyruvate

a. Convert phosphoenolpyruvate, an intermediate in glucose metabolism, to a skeletal structure.
b. Draw all reasonable resonance structures that have a phosphorus atom surrounded by 10 electrons.
c. Draw all reasonable resonance structures in which all second- and third-row atoms have an octet.

3.35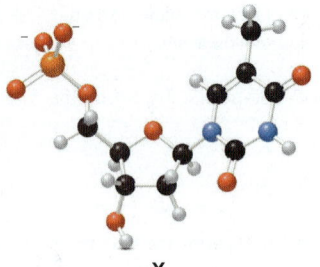
X

a. Convert the ball-and-stick model of nucleotide **X** into a skeletal structure.
b. Identify the phosphorus-containing functional group and classify any hydroxy group as 1°, 2°, or 3°.
c. Can this nucleotide be a component of DNA or RNA?

Functional Groups

3.36 For each alkane: (a) classify each carbon atom as 1°, 2°, 3°, or 4°; (b) classify each hydrogen atom as 1°, 2°, or 3°.

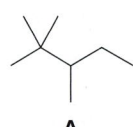 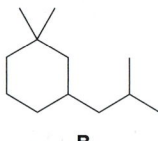

A   B

3.37 Identify the functional groups in each molecule. Classify each alcohol, alkyl halide, amide, and amine as 1°, 2°, or 3°.

a. Darvon (analgesic)

b.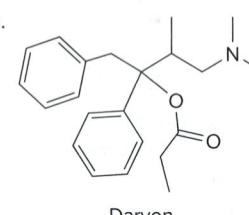
pregabalin
trade name Lyrica
(used in treating chronic pain)

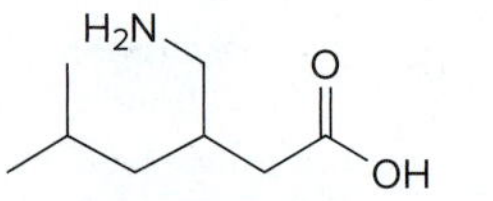

c. salinosporamide A (anticancer agent)

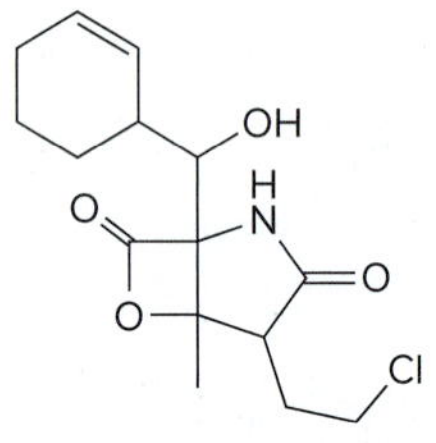

d. histrionicotoxin (poison secreted by a South American frog)

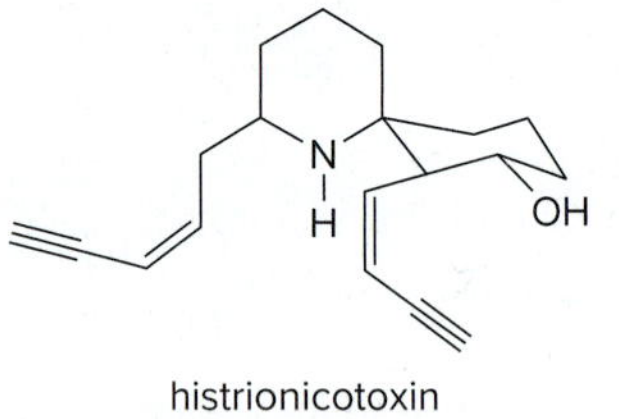

e. penicillin G (an antibiotic)

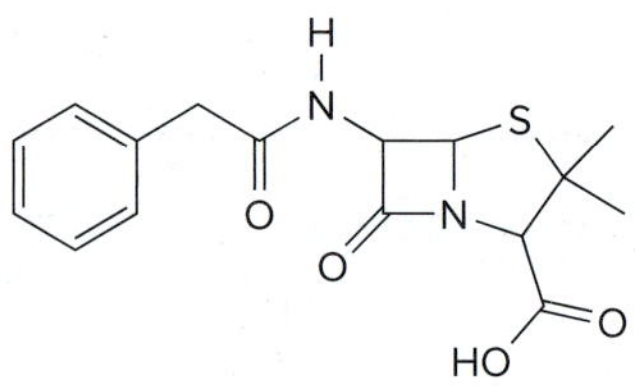

f. pyrethrin I (potent insecticide from chrysanthemums)

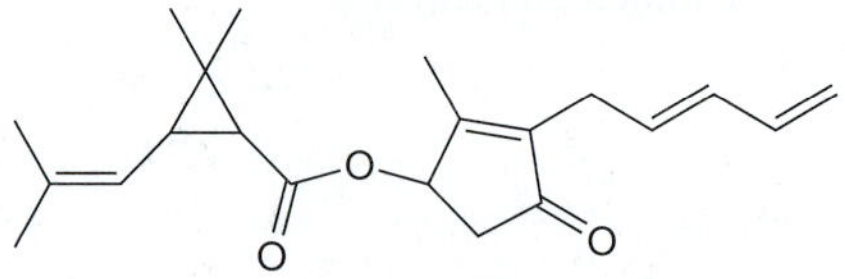

3.38 Draw seven constitutional isomers with molecular formula $C_3H_6O_2$ that contain a carbonyl group. Identify the functional group(s) in each isomer.

Intermolecular Forces

3.39 What types of intermolecular forces are exhibited by each compound?

a.   b. 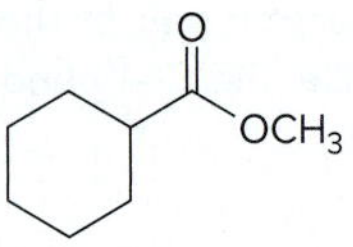   c. 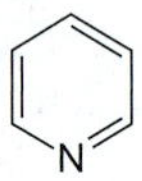   d.

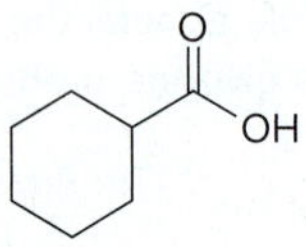

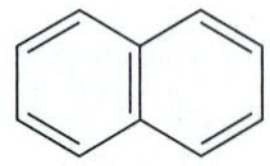

3.40 (a) Which of the following molecules can hydrogen bond to another molecule like itself? (b) Which of the following molecules can hydrogen bond to water?

A

B

C

D

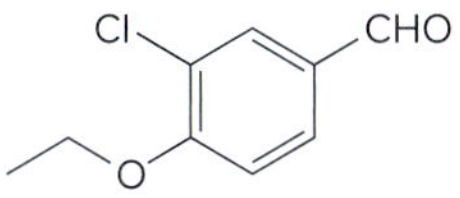

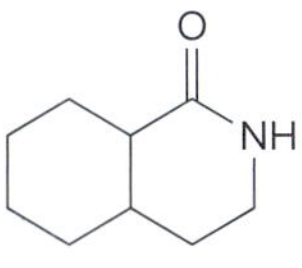

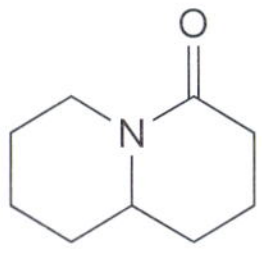

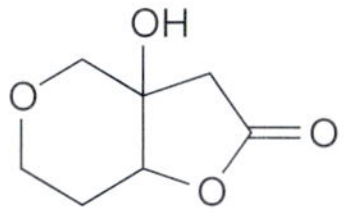

3.41 Indinavir (trade name Crixivan) is a drug used to treat HIV. (a) At which sites can indinavir hydrogen bond to another molecule like itself? (b) At which sites can indinavir hydrogen bond to water?

indinavir

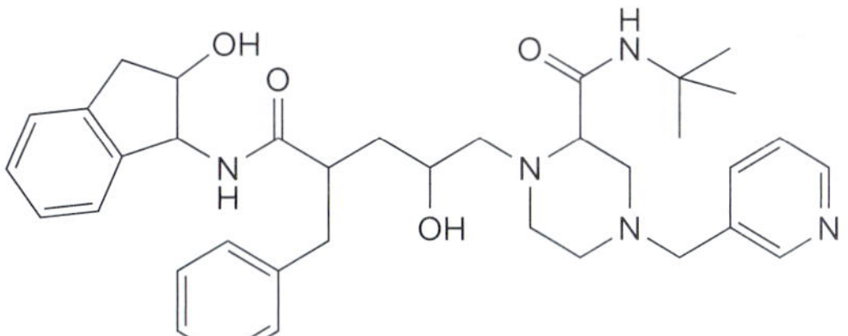

3.42 Intramolecular forces of attraction are often important in holding large molecules together. For example, some proteins fold into compact shapes, held together by attractive forces between nearby functional groups. A schematic of a folded protein is drawn here, with the protein backbone indicated by a blue-green ribbon, and various appendages drawn dangling from the chain. What types of intramolecular forces occur at each labeled site (**A–F**)?

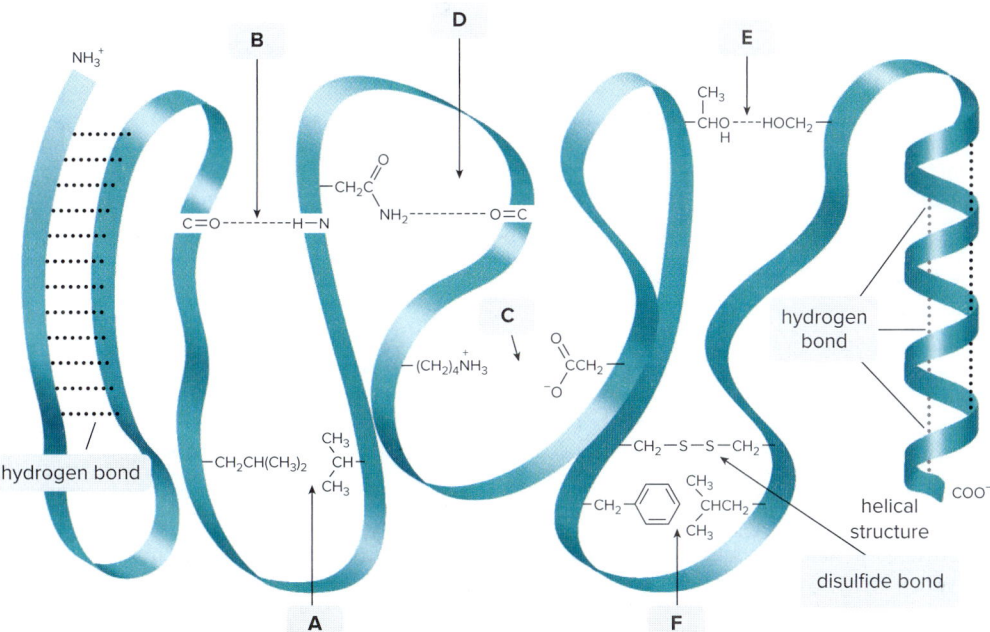

Physical Properties

3.43 Rank the compounds in each group in order of increasing boiling point.

a.

A

C

B

D

b.

E

G

F

H

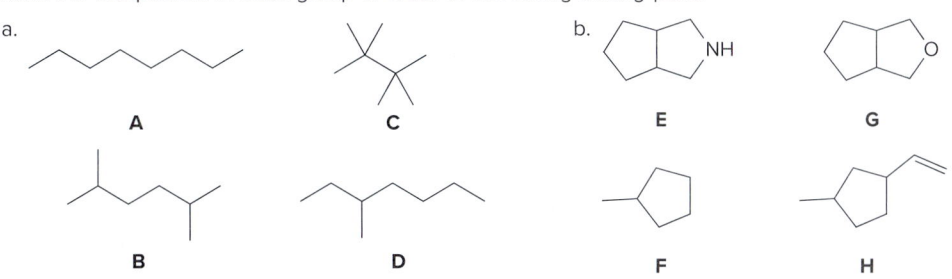

3.44 Explain why $CH_3CH_2NHCH_3$ has a higher boiling point than $(CH_3)_3N$, even though they have the same molecular weight.

3.45 Rank **A–C** in order of increasing melting point.

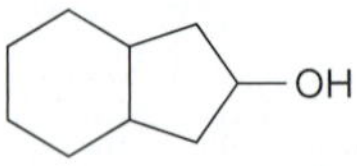

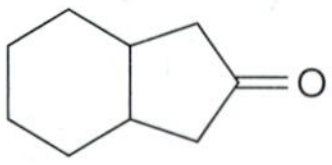

3.46 Explain why benzene has a lower boiling point but much higher melting point than toluene.

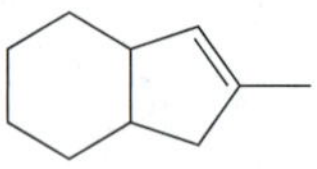

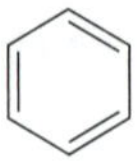

benzene
bp = 80 °C
mp = 5 °C

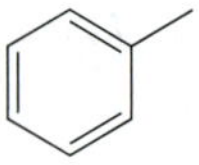

toluene
bp = 111 °C
mp = –93 °C

3.47 Rank the following compounds in order of increasing water solubility.

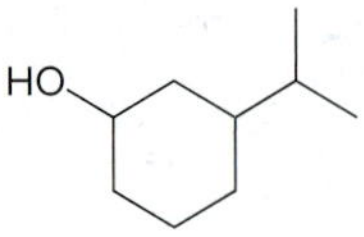

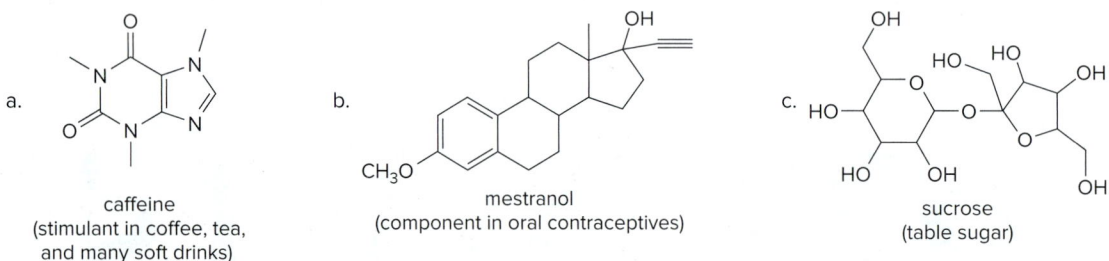

3.48 Explain why diethyl ether ($CH_3CH_2OCH_2CH_3$) and butan-1-ol ($CH_3CH_2CH_2CH_2OH$) have similar solubility properties in water, but butan-1-ol has a much higher boiling point.

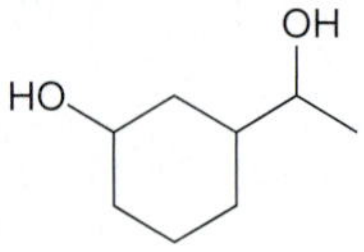

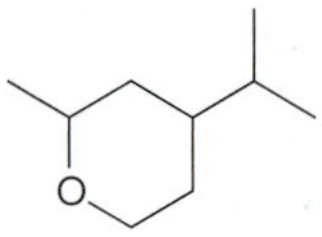

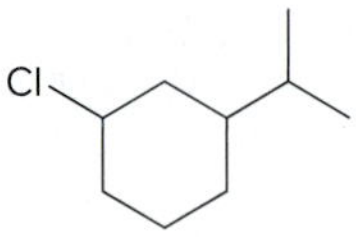

3.49 Predict the water solubility of each of the following organic molecules.

a. caffeine (stimulant in coffee, tea, and many soft drinks)

b. mestranol (component in oral contraceptives)

c. sucrose (table sugar)

Applications and Biomolecules

3.50 Predict the solubility of each of the following vitamins in water and in organic solvents.

a. vitamin E

b. pyridoxine vitamin B_6

3.51 Use the structures of keto acid **A** and amino acid **B** to answer each question. (a) Which compound has the higher melting point? (b) Which compound is more soluble in water? (c) Which compound is more soluble in diethyl ether [$(CH_3CH_2)_2O$]?

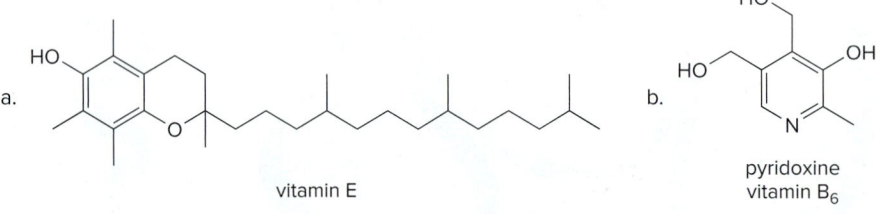

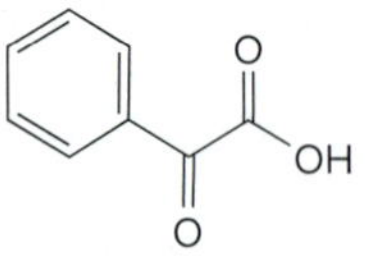

3.52 Avobenzone and dioxybenzone are two commercial sunscreens. Using the principles of solubility, predict which sunscreen is more readily washed off when an individual goes swimming. Explain your choice.

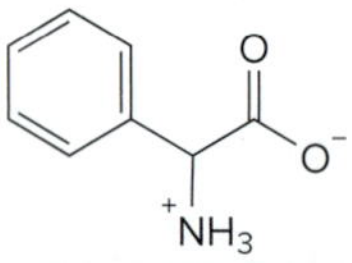

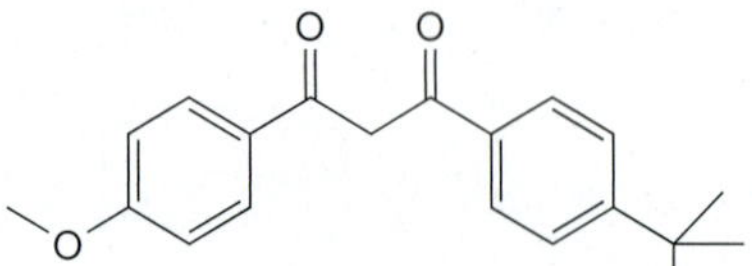

avobenzone

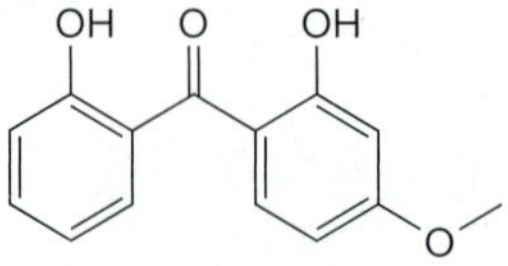

dioxybenzone

3.53 THC is the active component in marijuana, and ethanol is the alcohol in alcoholic beverages. Explain why drug screenings are able to detect the presence of THC but not ethanol weeks after these substances have been introduced into the body.

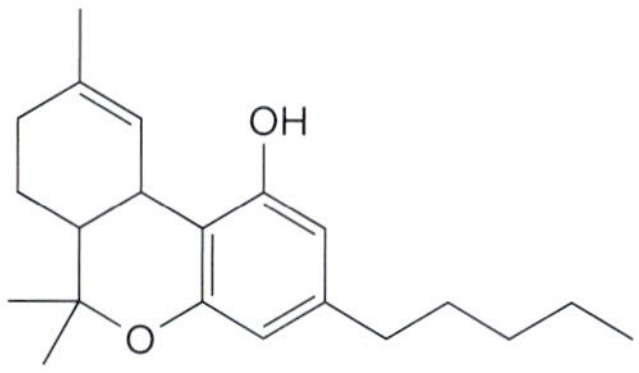

tetrahydrocannabinol
THC

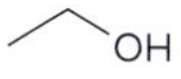

ethanol

3.54 Cocaine is a widely abused, addicting drug. Cocaine is usually obtained as its hydrochloride salt (cocaine hydrochloride) but can be converted to crack (the neutral organic molecule) by treatment with base. Which of the two compounds here has a higher boiling point? Which is more soluble in water? How does the relative solubility explain why crack is usually smoked but cocaine hydrochloride is injected directly into the bloodstream?

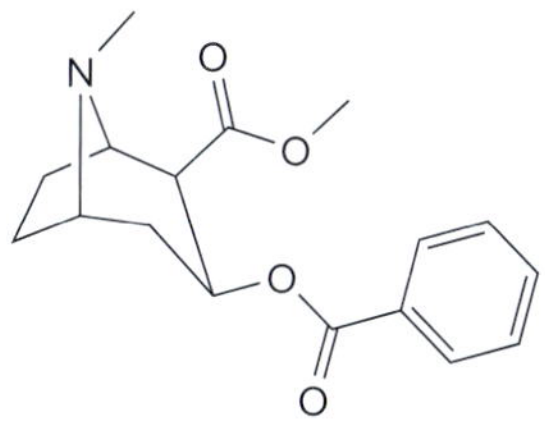

cocaine (crack)
neutral organic molecule

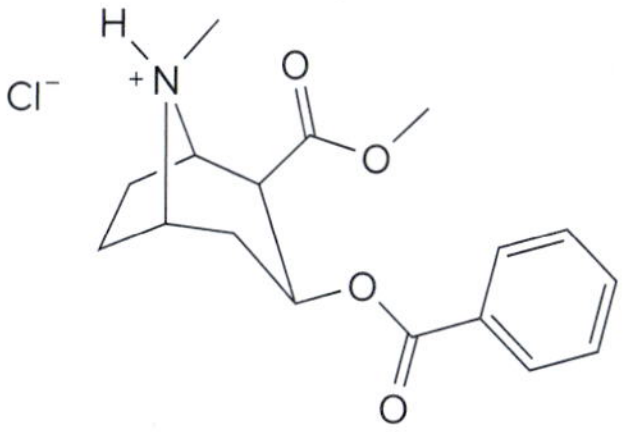

cocaine hydrochloride
a salt

3.55 Many drugs are sold as their hydrochloride salts ($R_2NH_2^+$ Cl^-), formed by reaction of an amine (R_2NH) with HCl.

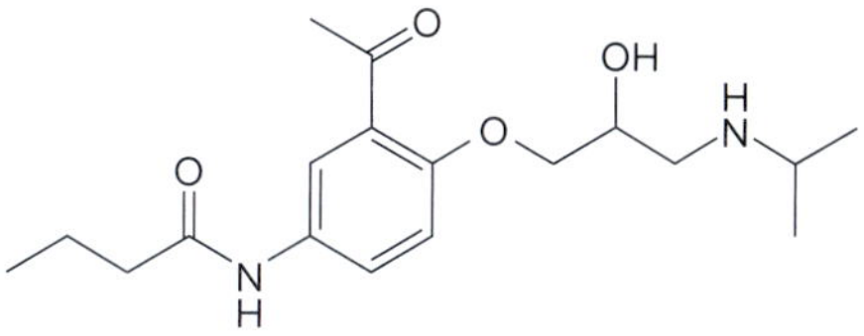

acebutolol

a. Draw the product (a hydrochloride salt) formed by reaction of acebutolol with HCl. Acebutolol is a β blocker used to treat high blood pressure.
b. Discuss the solubility of acebutolol and its hydrochloride salt in water.
c. Offer a reason as to why the drug is marketed as a hydrochloride salt rather than a neutral amine.

3.56 As we will learn in Chapter 16, acetyl coenzyme A (acetyl CoA) is a biological compound involved in reactions in which an acyl group (CH_3CO-) is transferred from one species to another. (a) Identify the functional groups in the highlighted portions of acetyl CoA. (b) Which functional groups serve as the nucleophile and electrophile in the given reaction, which synthesizes the neurotransmitter acetylcholine?

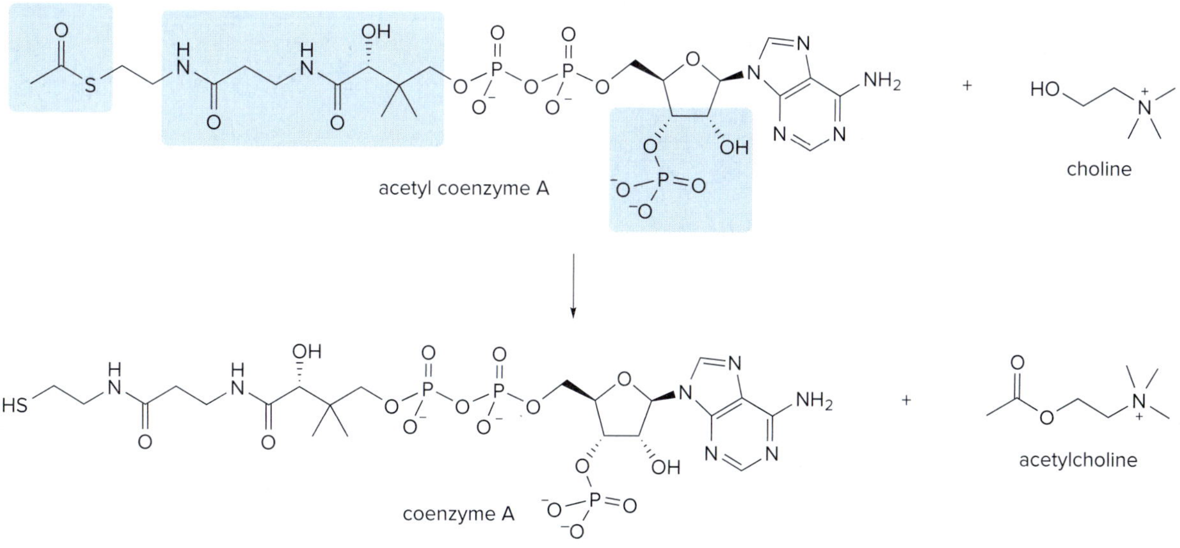

acetyl coenzyme A

choline

coenzyme A

acetylcholine

Reactivity of Organic Molecules

3.57 Label the electrophilic and nucleophilic sites in each molecule.

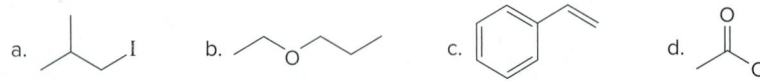

3.58 By using only electron density arguments, determine whether the following reactions will occur.

a. 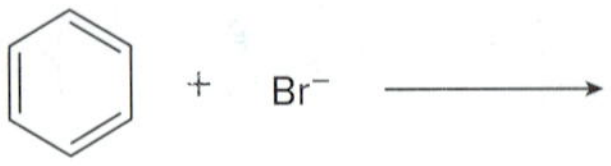 benzene + Br⁻ ⟶

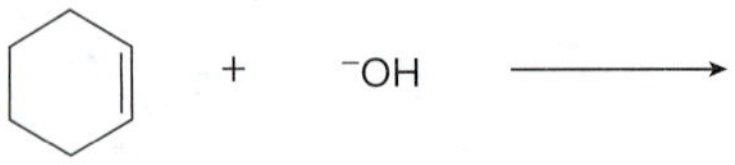

b.  (cyclohexylmethyl chloride) + ⁻CN ⟶

c. cyclohexene + ⁻OH ⟶

d. 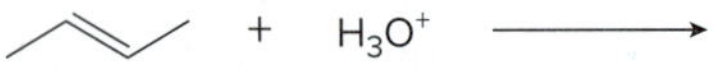 (alkene) + H_3O^+ ⟶

Cell Membrane

3.59 The composition of a cell membrane is not uniform for all types of cells. Some cell membranes are more rigid than others. Rigidity is determined by a variety of factors, one of which is the structure of the carbon chains in the phospholipids that comprise the membrane. One example of a phospholipid was drawn in Section 3.7A, and another, having C—C double bonds in its carbon chains, is drawn here. Which phospholipid would be present in the more rigid cell membrane and why?

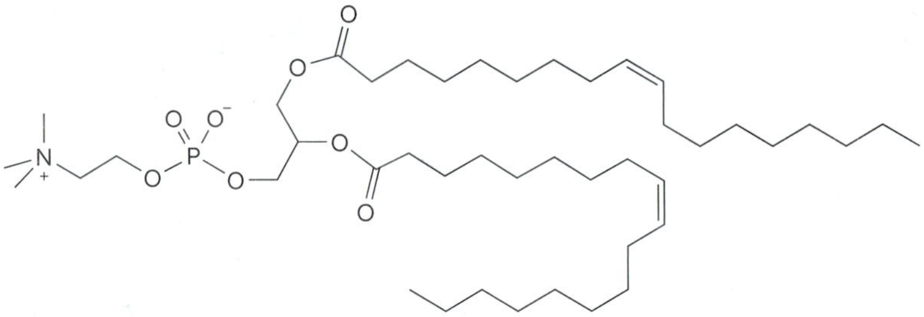

phospholipid

3.60 Sphingomyelins, a group of lipids that resemble the membrane phospholipids discussed in Section 3.7, are a major component of the myelin sheath, the insulating layer that surrounds a nerve fiber. (a) What functional groups are present in sphingomyelin **X**? (b) Classify any alcohol, amine, and amide as 1°, 2°, or 3°. (c) Label the polar head and nonpolar tails of **X**.

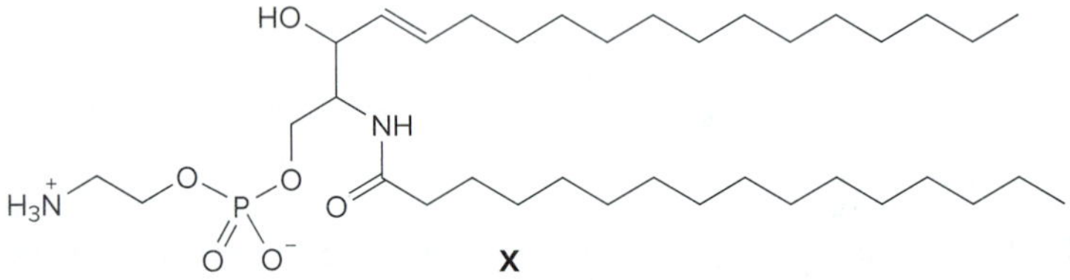

X
a sphingomyelin

3.61 Which compound is more likely to be a general anesthetic? Explain your choice.

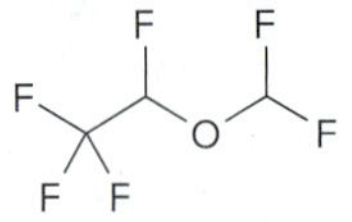

A 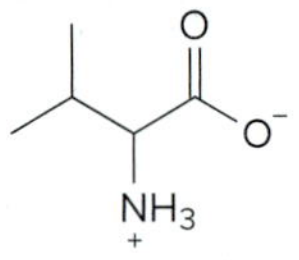   B

General Problems

3.62 Synthadotin is a promising anticancer drug in clinical trials.

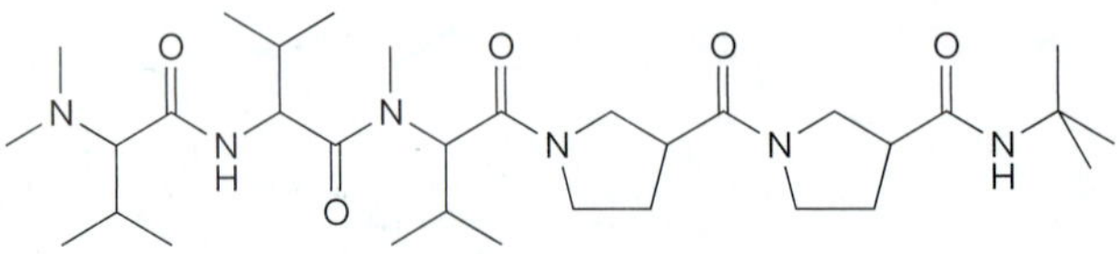

synthadotin

a. Identify the functional groups.
b. Classify any amine or amide as 1°, 2°, or 3°.
c. At which sites can synthadotin hydrogen bond to another molecule like itself?
d. Label two nucleophilic sites.
e. Label two electrophilic sites.
f. What product is formed when synthadotin is treated with HCl?

3.63 Quinapril (trade name Accupril) is a drug used to treat hypertension and congestive heart failure.

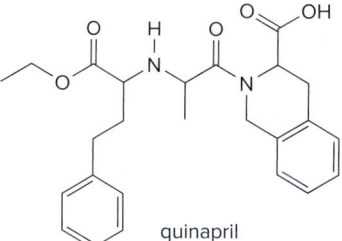
quinapril

a. Identify the functional groups in quinapril.
b. Classify any alcohol, amide, or amine as 1°, 2°, or 3°.
c. At which sites can quinapril hydrogen bond to water?
d. At which sites can quinapril hydrogen bond to acetone [$(CH_3)_2CO$]?
e. Label the most acidic hydrogen atom.
f. Which site is most basic?

Challenge Problems

3.64 Answer the following questions by referring to the ball-and-stick model of fentanyl, a potent narcotic analgesic used in surgical procedures.

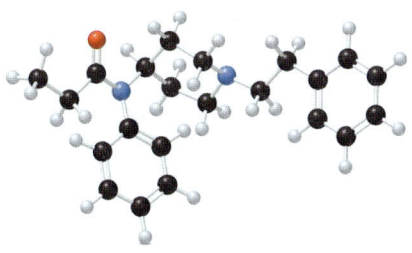
fentanyl

a. Identify the functional groups.
b. Label the most acidic proton.
c. Label the most basic atom.
d. What types of intermolecular forces are present between two molecules of fentanyl?
e. Draw an isomer predicted to have a higher boiling point.
f. Which sites in the molecule can hydrogen bond to water?
g. Label all electrophilic carbons.

3.65 Explain why **A** is less water soluble than **B**, even though both compounds have the same functional groups.

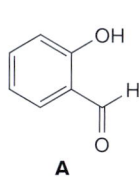

3.66 Recall from Section 1.10B that there is restricted rotation around carbon–carbon double bonds. Maleic acid and fumaric acid are two isomers with vastly different physical properties and pK_a values for loss of both protons. Explain why each of these differences occurs.

	maleic acid	fumaric acid
mp (°C)	130	286
solubility (g/L) in H$_2$O at 25 °C	788	7
pK_{a1}	1.9	3.0
pK_{a2}	6.5	4.5

4 Alkanes

Narongsak Nagadhana/Shutterstock

- 4.1 Alkanes—An introduction
- 4.2 Cycloalkanes
- 4.3 An introduction to nomenclature
- 4.4 Naming alkanes
- 4.5 Naming cycloalkanes
- 4.6 Common names
- 4.7 Natural occurrence of alkanes
- 4.8 Properties of alkanes
- 4.9 Conformations of acyclic alkanes—Ethane
- 4.10 Conformations of butane
- 4.11 An introduction to cycloalkanes
- 4.12 Cyclohexane
- 4.13 Substituted cycloalkanes
- 4.14 Oxidation of alkanes

Alkanes, the simplest hydrocarbons, are found in all shapes and sizes and occur widely in nature. They are the major constituents of petroleum, a complex mixture of compounds that includes hydrocarbons such as **hexane** and **decane**. Crude petroleum spilled into the sea from a ruptured oil tanker or offshore oil well creates an insoluble oil slick on the surface. Petroleum is refined to produce gasoline, diesel fuel, home heating oil, and a myriad of other useful compounds. In Chapter 4, we learn about the properties of alkanes, how to name them (nomenclature), and oxidation—one of their important reactions.

Why Study... Alkanes?

Secretion of **undecane** by a cockroach causes other members of the species to aggregate. Undecane is a *pheromone,* a chemical substance used for communication in an animal species, most commonly an insect population. *God of Insects*

Cyclohexane is one component of the mango, the most widely consumed fruit in the world. *Pixtal/age fotostock*

In Chapter 4, we apply the principles of bonding, shape, and reactivity discussed in Chapters 1–3 to our first family of organic compounds, the **alkanes.** Because alkanes have no functional group, they are much less reactive than other organic compounds, and for this reason, much of Chapter 4 is devoted to learning how to name and draw them, as well as to determining what happens when rotation occurs about their carbon–carbon single bonds. These principles are essential to the understanding of other types of organic compounds that we will discuss in later chapters.

4.1 Alkanes—An Introduction

Recall from Section 3.2 that **alkanes are aliphatic hydrocarbons having only C–C and C–H σ bonds.** Because their carbon atoms can be joined together in chains or rings, they can be categorized as acyclic or cyclic.

- **Acyclic alkanes** have the molecular formula C_nH_{2n+2} (where n = an integer) and contain only linear and branched chains of carbon atoms. Acyclic alkanes are also called *saturated hydrocarbons* because they have the maximum number of hydrogen atoms per carbon.
- **Cycloalkanes** contain carbons joined in one or more rings. Because their general formula is C_nH_{2n}, they have two fewer H atoms than an acyclic alkane with the same number of carbons.

Undecane, an acyclic alkane, and cyclohexane, a cycloalkane, are two naturally occurring alkanes.

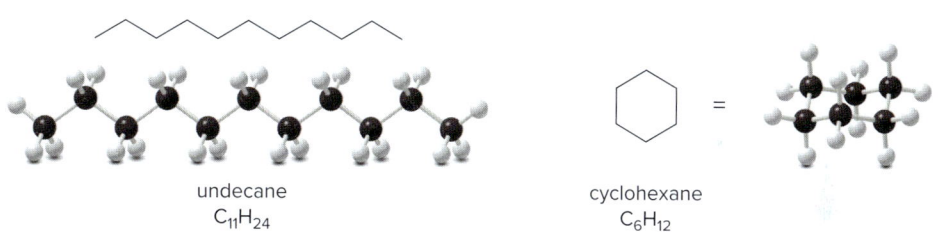

undecane
$C_{11}H_{24}$

cyclohexane
C_6H_{12}

4.1A Acyclic Alkanes Having One to Five C Atoms

Structures for the two simplest acyclic alkanes were given in Chapter 1. **Methane, CH₄,** has a single carbon atom, and **ethane, CH₃CH₃,** has two. All C atoms in an alkane are surrounded by four groups, making them sp^3 **hybridized** and **tetrahedral,** and all bond angles are 109.5°.

> To draw the structure of an alkane, join the carbon atoms together with single bonds, and add enough H atoms to make each C tetravalent.

The three-carbon alkane **CH₃CH₂CH₃, propane,** has molecular formula C_3H_8. Each carbon in the three-dimensional drawing has two bonds in the plane (solid lines), one bond in front (on a wedge), and one bond behind the plane (on a dashed wedge).

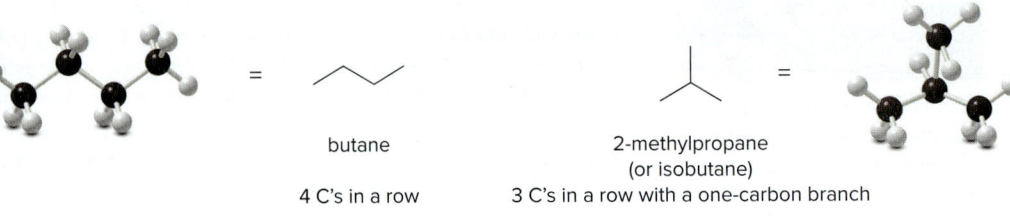

Problem 4.1

Both olives and the leaves of olive trees contain alkanes with long carbon chains. A predominant alkane in olives has 27 carbons, whereas a major alkane component in olive leaves has 31 carbons. What is the molecular formula of each of these alkanes?

There are two different ways to arrange four carbons, giving two compounds with molecular formula C_4H_{10}, named **butane** and **2-methylpropane** (or isobutane).

The alkane content of olives and olive leaves is somewhat different (Problem 4.1), so it is possible to use alkane identity to determine the presence of leaf material in olive oil.
Flickr Open/Getty Images

butane
4 C's in a row
straight-chain alkane

2-methylpropane
(or isobutane)
3 C's in a row with a one-carbon branch
branched-chain alkane

Butane and 2-methylpropane are *isomers,* **two different compounds with the same molecular formula** (Section 1.4). They belong to one of the two major classes of isomers called **constitutional** or **structural isomers.** We will learn about the second major class of isomers, called **stereoisomers,** in Section 4.13B.

• *Constitutional isomers* differ in the way the atoms are connected to each other.

Butane, which has four carbons in a row, is a **straight-chain** or **normal alkane** (an *n*-alkane). 2-Methylpropane, on the other hand, is a **branched-chain alkane.**

> The molecular formulas for methane, ethane, and propane fit into the general molecular formula for an alkane, C_nH_{2n+2}.
> • Methane = CH_4 = $C_1H_{2(1)+2}$
> • Ethane = C_2H_6 = $C_2H_{2(2)+2}$
> • Propane = C_3H_8 = $C_3H_{2(3)+2}$

With alkanes having more than four carbons, the names of the straight-chain isomers are systematic and derive from Greek roots: **pent**ane for five C atoms, **hex**ane for six, and so on. There are three constitutional isomers for the five-carbon alkane, each having molecular formula C_5H_{12}: **pentane, 2-methylbutane** (or isopentane), and **2,2-dimethylpropane** (or neopentane).

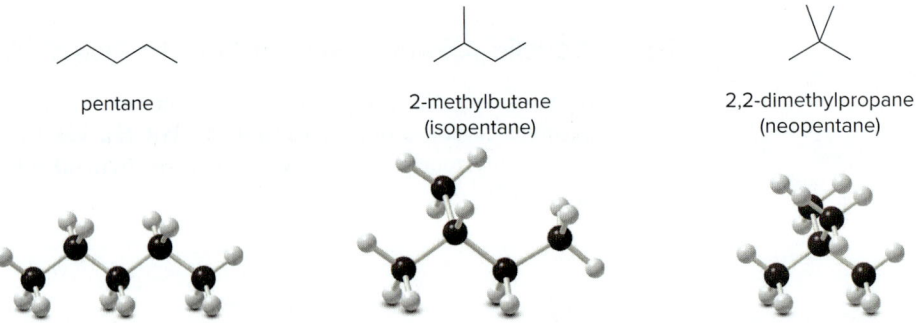

pentane

2-methylbutane
(isopentane)

2,2-dimethylpropane
(neopentane)

Take care in interpreting skeletal structures. Although pentane is typically drawn using a zig-zag structure, the carbon skeleton can be drawn in a variety of ways, and still represent the same compound. Each of the following representations has five carbon atoms in a row, so each represents pentane, not an isomer of pentane.

pentane
C_5H_{12}

5 C's in a row

5 C's in a row

5 C's in a row

Problem 4.2 Which of the following is *not* another representation for 2-methylbutane?

4.1B Acyclic Alkanes Having More Than Five C Atoms

The maximum number of possible constitutional isomers increases dramatically as the number of carbon atoms in the alkane increases, as shown in Table 4.1. For example, there are 75 possible isomers for an alkane having 10 carbon atoms, and 366,319 possible isomers for one having 20 carbons.

Each entry in Table 4.1 is formed from the preceding entry by adding a CH_2 group. **A CH_2 group is called a *methylene group*. A group of compounds that differ by only a CH_2 group is called a *homologous series*.** The names of all alkanes end in the suffix *-ane*, and the syllables preceding the suffix identify the number of carbon atoms in the chain.

Table 4.1 Summary: Straight-Chain Alkanes

Number of C atoms	Molecular formula	Name (n-alkane)	Number of constitutional isomers	Number of C atoms	Molecular formula	Name (n-alkane)	Number of constitutional isomers
1	CH_4	methane	—	9	C_9H_{20}	nonane	35
2	C_2H_6	ethane	—	10	$C_{10}H_{22}$	decane	75
3	C_3H_8	propane	—	11	$C_{11}H_{24}$	undecane	159
4	C_4H_{10}	butane	2	12	$C_{12}H_{26}$	dodecane	355
5	C_5H_{12}	pentane	3	13	$C_{13}H_{28}$	tridecane	802
6	C_6H_{14}	hexane	5	14	$C_{14}H_{30}$	tetradecane	1858
7	C_7H_{16}	heptane	9	15	$C_{15}H_{32}$	pentadecane	4347
8	C_8H_{18}	octane	18	20	$C_{20}H_{42}$	icosane	366,319

Problem 4.3 Draw the five constitutional isomers having molecular formula C_6H_{14}.

Problem 4.4 Review classifying carbons and hydrogens in Section 3.2, and draw the structure of an alkane with molecular formula C_7H_{16} that contains (a) one 4° carbon; (b) only 1° and 2° carbons; (c) 1°, 2°, and 3° hydrogens.

Problem 4.5 (a) Which compounds (**B–F**) are identical to **A**? (b) Which compounds (**B–F**) represent an isomer of **A**?

4.2 Cycloalkanes

Cycloalkanes have molecular formula C_nH_{2n} and contain carbon atoms arranged in a ring. Think of a cycloalkane as being formed by removing two H atoms from the end carbons of a chain, and then bonding the two carbons together. Simple cycloalkanes are named by adding the prefix *cyclo-* to the name of the acyclic alkane having the same number of carbons.

Cycloalkanes with three to six carbon atoms are shown.

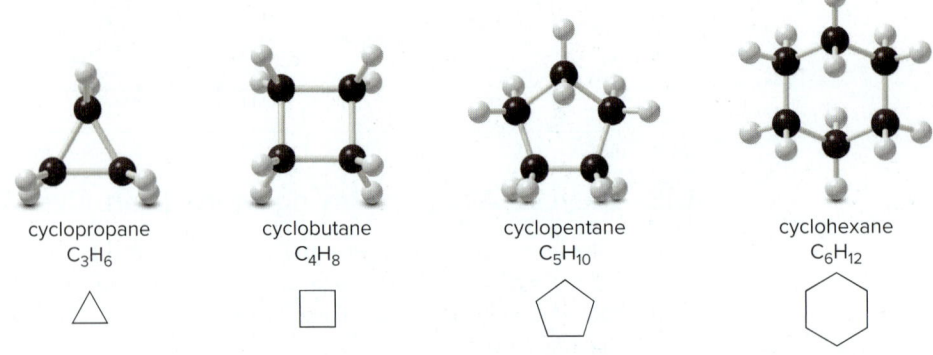

cyclopropane
C$_3$H$_6$

cyclobutane
C$_4$H$_8$

cyclopentane
C$_5$H$_{10}$

cyclohexane
C$_6$H$_{12}$

Problem 4.6 Draw the five constitutional isomers that have molecular formula C$_5$H$_{10}$ and contain one ring.

4.3 An Introduction to Nomenclature

Garlic has been used in Chinese herbal medicine for more than 4000 years, as a form of currency in Siberia, and as a repellent for witches by the Saxons. Today it is used as a dietary supplement because of its reported health benefits. **Allicin,** the molecule largely responsible for garlic's odor, is not stored in the garlic bulb, but instead is produced by the action of enzymes when the bulb is crushed or bruised.
Pixtal/age fotostock

How are organic compounds named? Long ago, the name of a compound was often based on the plant or animal source from which it was obtained. For example, the name for **formic acid,** a caustic compound isolated from certain ants, comes from the Latin word *formica,* meaning "ant"; and **allicin,** the pungent principle of garlic, is derived from the botanical name for garlic, *Allium sativum.*

formic acid
(obtained from certain ants)

allicin
(odor of garlic)

With the isolation and preparation of thousands of new organic compounds it became clear that each organic compound must have an unambiguous name, derived from a set of easily remembered rules. A systematic method of naming compounds was developed by the *I*nternational *U*nion of *P*ure and *A*pplied *C*hemistry. It is referred to as the **IUPAC system of nomenclature;** how it can be used to name alkanes and cycloalkanes is explained in Sections 4.4 and 4.5.

The IUPAC system of nomenclature has been regularly revised since it was first adopted in 1892. Revisions in 1979 and 1993 and extensive recommendations in 2004 have given chemists a variety of acceptable names for compounds. Many changes are minor. For example, the 1979 nomenclature rules assign the name 1-butene to CH$_2$=CHCH$_2$CH$_3$, whereas the 1993 rules assign the name but-1-ene; that is, only the position of the number differs. In this text, more recent IUPAC conventions will be used, and often a margin note will be added to mention the differences between past and recent recommendations.

Naming organic compounds has become big business for drug companies. The IUPAC name of an organic compound can be long and complex, and may be comprehensible only to a chemist. As a result, most drugs have three names:

- **Systematic:** The systematic name follows the accepted rules of nomenclature and indicates the compound's chemical structure; this is the IUPAC name.
- **Generic:** The generic name is the official, internationally approved name for the drug.
- **Trade:** The trade name for a drug is assigned by the company that manufactures it. Trade names are often "catchy" and easy to remember. Companies hope that the public will continue to purchase a drug with an easily recalled trade name long after a cheaper generic version becomes available.

In the world of over-the-counter anti-inflammatory agents, the compound a chemist calls 2-[4-(2-methylpropyl)phenyl]propanoic acid has the generic name ibuprofen. It is marketed under a variety of trade names including Motrin and Advil.

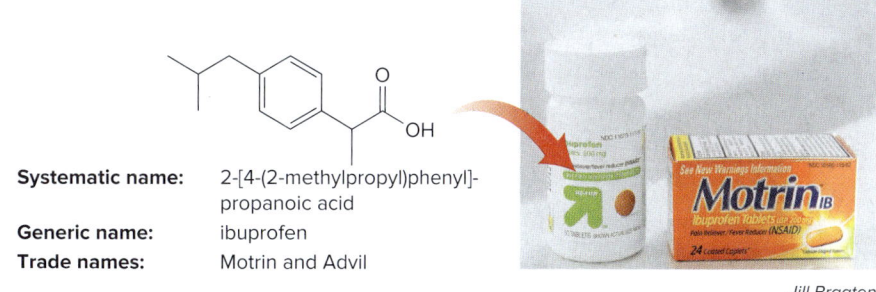

Systematic name: 2-[4-(2-methylpropyl)phenyl]-propanoic acid
Generic name: ibuprofen
Trade names: Motrin and Advil

Jill Braaten

4.4 Naming Alkanes

The name of every organic molecule has three parts:

- The **parent name** indicates the number of carbons in the longest continuous carbon chain in the molecule.
- The **suffix** indicates what functional group is present.
- The **prefix** reveals the identity, location, and number of substituents attached to the carbon chain.

prefix + parent + suffix

What and where are the substituents? | What is the longest carbon chain? | What is the functional group?

The names listed in Table 4.1 of Section 4.1B for the simple *n*-alkanes consist of the parent name, which indicates the number of carbon atoms in the longest carbon chain, and the suffix *-ane,* which indicates that the compounds are alkanes. The parent name for **one carbon is meth-,** for **two carbons is eth-,** and so on. Thus, we are already familiar with two parts of the name of an organic compound.

To determine the third part of a name, the prefix, we must learn how to name the carbon groups or *substituents* that are bonded to the longest carbon chain.

4.4A Naming Substituents

Carbon substituents bonded to a long carbon chain are called **alkyl groups.**

- An *alkyl group* is formed by removing one hydrogen from an alkane.

An alkyl group is a part of a molecule that is now able to bond to another atom or a functional group. **To name an alkyl group, change the *-ane* ending of the parent alkane to *-yl.*** Thus, **methane** (CH_4) becomes **methyl** (CH_3-) and **ethane** (CH_3CH_3) becomes **ethyl** (CH_3CH_2-). As we learned in Section 3.1, **R** denotes a general carbon group bonded to a functional group. **R** thus denotes any alkyl group.

Naming three- and four-carbon alkyl groups is more complicated because the parent hydrocarbons have more than one type of hydrogen atom. Propane has both 1° and 2° H atoms, and removal of each of these H atoms forms a different alkyl group, **propyl** or **isopropyl.**

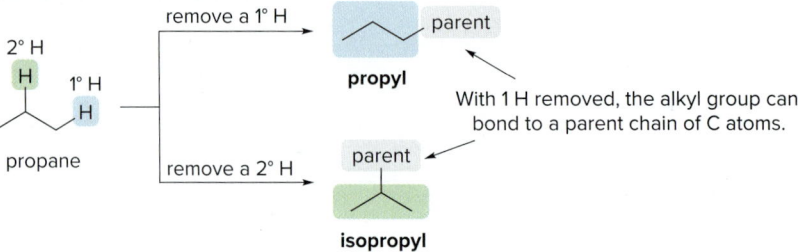

> The prefix *iso-* is part of the words *propyl* and *butyl*, forming a single word: **isopropyl** and **isobutyl.** The prefixes *sec-* and *tert-* are separated from the word *butyl* by a hyphen: ***sec*-butyl** and ***tert*-butyl.**

Because there are two different butane isomers to begin with, each with two different kinds of H atoms, there are *four* possible alkyl groups containing four carbon atoms: **butyl,** *sec*-**butyl, isobutyl,** and *tert*-**butyl.**

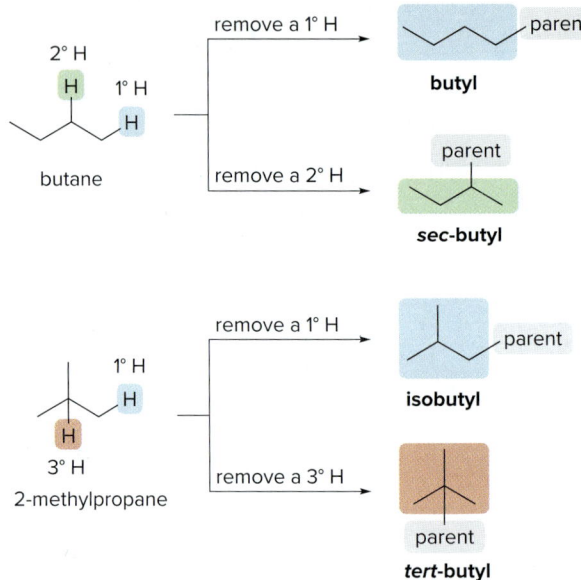

> The prefix *sec-* is short for *secondary.* A *sec*-butyl group is formed by removal of a **2° H.** The prefix *tert-* is short for *tertiary.* A *tert*-butyl group is formed by removal of a **3° H.**

> Abbreviations are sometimes used for certain common alkyl groups.
> - methyl **(Me)**
> - ethyl **(Et)**
> - butyl **(Bu)**
> - *tert*-butyl **(*t*-Bu)**

The names isopropyl, *sec*-butyl, isobutyl, and *tert*-butyl are recognized as acceptable substituent names in both the 1979 and 1993 revisions of IUPAC nomenclature. A general method to name these substituents, as well as alkyl groups that contain five or more carbon atoms, is described in Appendix D.

4.4B Naming an Acyclic Alkane

Four steps are needed to name an alkane.

How To Name an Alkane Using the IUPAC System

Step [1] Find the parent carbon chain and add the suffix.

- Find the *longest continuous* carbon chain, and name the molecule by using the parent name for that number of carbons, given in Table 4.1. To the name of the parent, add the suffix *-ane* for an alkane. Each functional group has its own characteristic suffix.

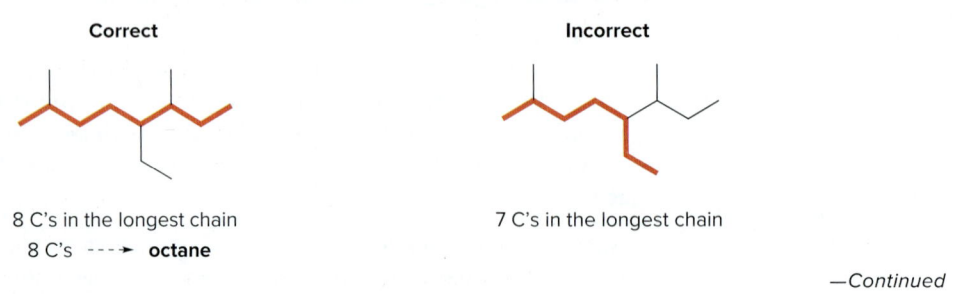

—*Continued*

- Finding the longest chain is a matter of trial and error. Place your pencil on one end of the chain, go to the other end without picking it up, and count carbons. Repeat this procedure until you have found the chain with the largest number of carbons.
- **It does not matter if the chain is *straight* or has *bends*.** All of the following representations are equivalent, and each longest chain has eight carbons.

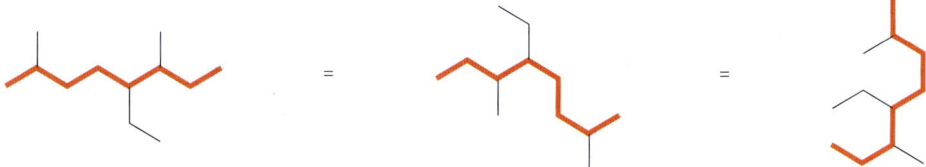

- **If there are two chains of equal length, pick the chain with *more* substituents**. In the following example, two different chains in the same alkane contain 7 C's, but the compound on the left has two alkyl groups attached to its long chain, whereas the compound to the right has only one.

Correct	Incorrect
7 atoms in the longest chain **2** substituents	7 atoms in the longest chain only **1** substituent
more substituents	*fewer* substituents

Step [2]   **Number the atoms in the carbon chain.**
- Number the longest chain to give the *first* substituent the lower number.

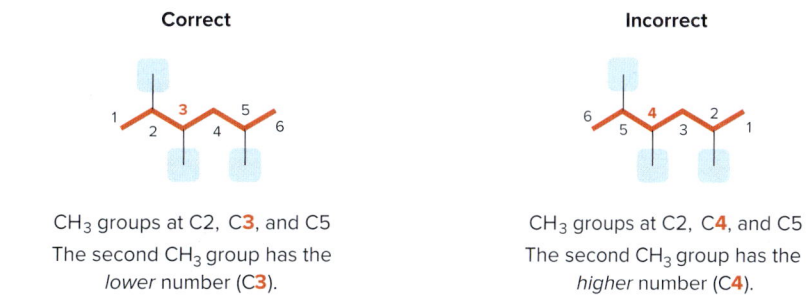

Correct — first substituent at C**2**
Incorrect — first substituent at C**3**

- If the first substituent is the same distance from both ends, number the chain to give the *second* substituent the lower number. **Always look for the first point of difference** in numbering from each end of the longest chain.

Correct: CH₃ groups at C2, C**3**, and C5. The second CH₃ group has the *lower* number (C**3**).

Incorrect: CH₃ groups at C2, C**4**, and C5. The second CH₃ group has the *higher* number (C**4**).

- When numbering a carbon chain results in the *same* numbers from either end of the chain, **assign the lower number alphabetically** to the first substituent.

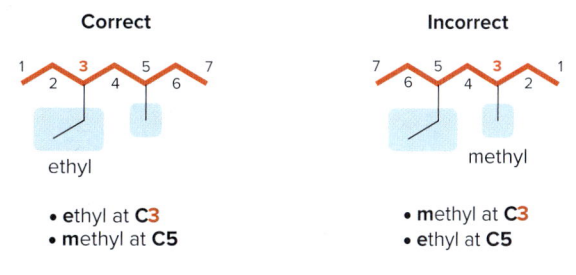

Correct — ethyl
- **ethyl** at C**3**
- **methyl** at C**5**

Incorrect — methyl
- **methyl** at C**3**
- **ethyl** at C**5**

Earlier letter ⟶ *lower* number

—*Continued*

How To, continued...

Step [3] Name and number the substituents.

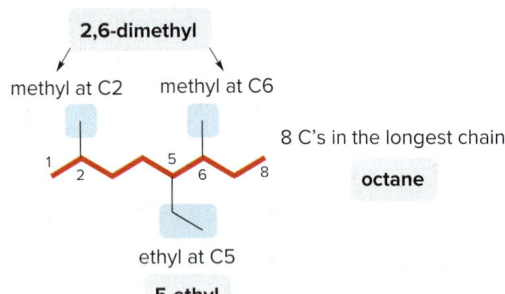

- Name the substituents as alkyl groups, and use the numbers from Step [2] to designate their location.
- Every carbon belongs to *either* the longest chain or a substituent, but *not both*.
- **Each substituent needs its *own* number.**
- If two or more identical substituents are bonded to the longest chain, use prefixes to indicate how many: **di-** for two groups, **tri-** for three groups, **tetra-** for four groups, and so forth. This molecule has two methyl substituents, so its name contains the prefix *di-* before the word methyl → *di*methyl.

Step [4] Combine substituent names and numbers + parent + suffix.

- Precede the name of the parent by the names of the substituents.
- Alphabetize the names of the substituents, **ignoring all prefixes except *iso-*,** as in isopropyl and isobutyl.
- Precede the name of each substituent by the number that indicates its location. There must be **one number for each substituent.**
- Separate numbers by commas and separate numbers from letters by hyphens. The name of an alkane is a single word, with no spaces after hyphens or commas.

[1] Identify all the pieces of a compound, using Steps [1]–[3].

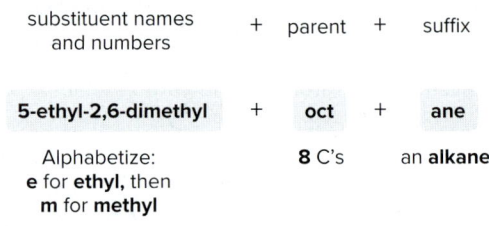

[2] Then, put the pieces of the name together.

substituent names and numbers	+	parent	+	suffix
5-ethyl-2,6-dimethyl	+	oct	+	ane
Alphabetize: e for **e**thyl, then m for **m**ethyl		8 C's		an **alk**ane

Answer: 5-ethyl-2,6-dimethyloctane

Several additional examples of alkane nomenclature are given in Figure 4.1.

Figure 4.1 Examples of alkane nomenclature

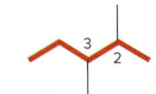

2,3-dimethylpentane

Number to give the 1st methyl group the lower number.

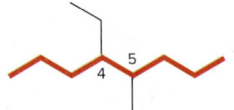

4-ethyl-5-methyloctane

Assign the lower number to the 1st substituent alphabetically: the **e** of **e**thyl before the **m** of **m**ethyl.

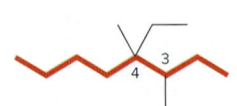

4-ethyl-3,4-dimethyloctane

Alphabetize the **e** of **e**thyl before the **m** of **m**ethyl.

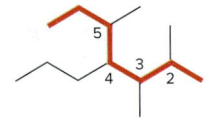

2,3,5-trimethyl-4-propylheptane

Pick the long chain with more substituents.

- The carbon atoms of each long chain are drawn in **red.**

4.4 Naming Alkanes

Sample Problem 4.1 Naming an Alkane

Give the IUPAC name for the following compound.

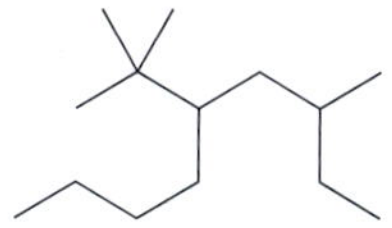

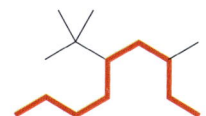

Solution

To help identify which carbons belong to the longest chain and which are substituents, **box in or highlight the atoms of the long chain.** Every other carbon atom then becomes a substituent that needs its own name as an alkyl group.

Step 1: Name the parent.

9 C's in the longest chain
nonane

Step 3: Name and number the substituents.

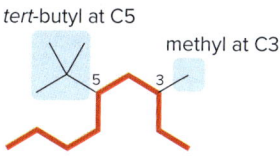

tert-butyl at C5

methyl at C3

Step 2: Number the chain.

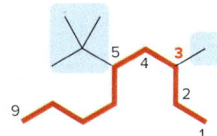

first substituent at C**3**

Step 4: Combine the parts.

- Alphabetize: the **b** of **butyl** before the **m** of **methyl**

Answer: 5-*tert*-butyl-3-methylnonane

Problem 4.7 Give the IUPAC name for each compound.

a.   b. 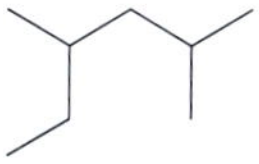   c. 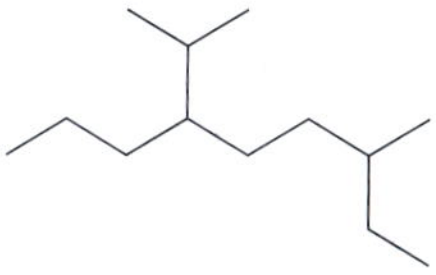   d.

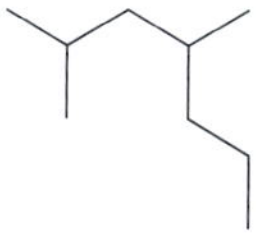

More Practice: Try Problem 4.36a, b, c, d, h, j.

Problem 4.8 Give the IUPAC name for each compound.

a. $(CH_3)_3CCH_2CH(CH_2CH_3)_2$

c. $CH_3(CH_2)_3CH(CH_2CH_2CH_3)CH(CH_3)_2$

b.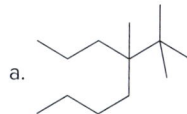

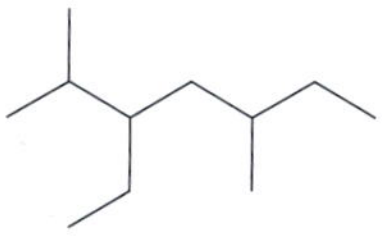

d.

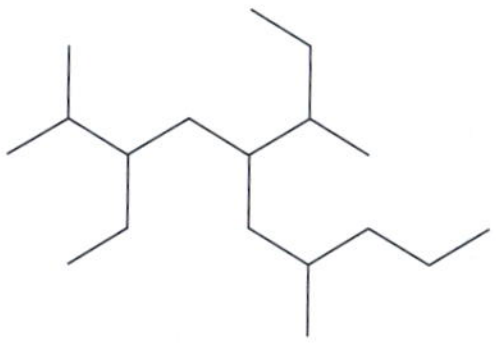

You must also know how to derive a structure from a given name. Sample Problem 4.2 illustrates a stepwise method.

Sample Problem 4.2 Deriving a Structure from a Name

Give the structure corresponding to the following IUPAC name: 6-isopropyl-3,3,7-trimethyldecane.

Solution
Follow three steps to derive a structure from a name.

Step [1] Identify the parent name and functional group found at the *end* of the name.

decane ---→ 10 C's ---→ [structure]

Step [2] Number the carbon skeleton in *either* direction.

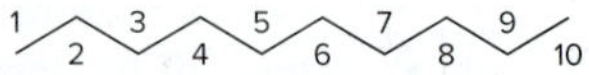

Step [3] Add the substituents at the appropriate carbons.

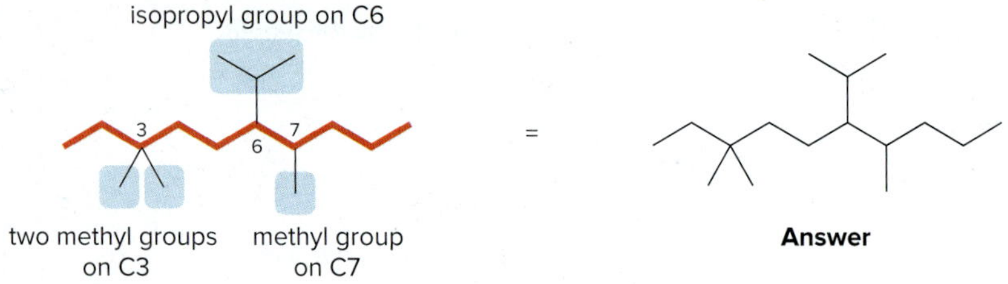

isopropyl group on C6

two methyl groups on C3 | methyl group on C7

Answer

Problem 4.9 Give the structure corresponding to each IUPAC name.
a. 3-methylhexane
b. 3,3-dimethylpentane
c. 3,5,5-trimethyloctane
d. 3-ethyl-4-methylhexane
e. 3-ethyl-5-isobutylnonane

More Practice: Try Problem 4.37a, c, g, h.

Problem 4.10
Give the IUPAC name for each of the five constitutional isomers of molecular formula C_6H_{14} in Problem 4.3.

4.5 Naming Cycloalkanes

Cycloalkanes are named by using similar rules, but the prefix *cyclo-* immediately precedes the name of the parent.

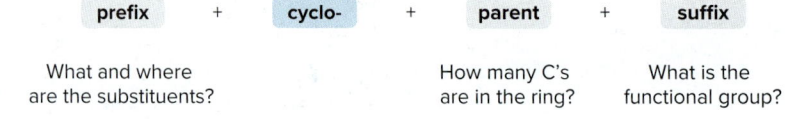

prefix	cyclo-	parent	suffix
What and where are the substituents?		How many C's are in the ring?	What is the functional group?

How To Name a Cycloalkane Using the IUPAC System

Step [1] Find the parent cycloalkane.
- Count the number of carbon atoms in the ring and use the parent name for that number of carbons. Add the prefix *cyclo-* and the suffix *-ane* to the parent name.

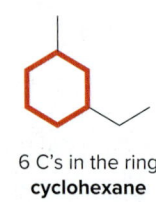

6 C's in the ring
cyclohexane

—Continued

Step [2] **Name and number the substituents.**
- No number is needed to indicate the location of a single substituent.

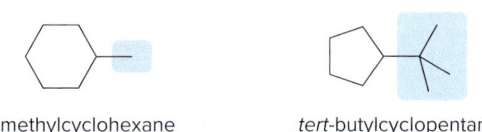

methylcyclohexane   *tert*-butylcyclopentane

- For rings with more than one substituent, **begin numbering at one substituent** and proceed around the ring clockwise or counterclockwise to **give the second substituent the *lower* number.**

CH₃ groups at C1 and C**3**
The 2ⁿᵈ substituent has a lower number.
Correct: 1,3-dimethylcyclohexane

CH₃ groups at C1 and C**5**

Incorrect: 1,5-dimethylcyclohexane

- **With two different substituents,** number the ring to **assign the *lower* number to the substituents *alphabetically*.**

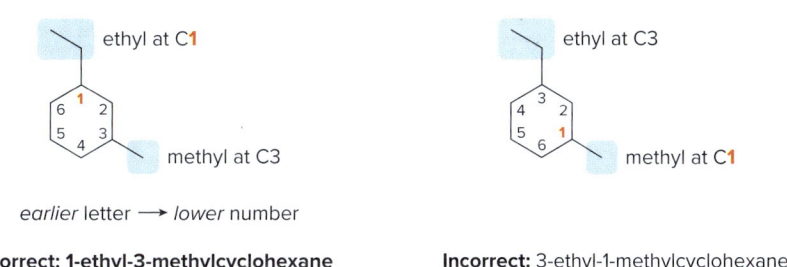

ethyl at C**1**   ethyl at C3
methyl at C3   methyl at C**1**

earlier letter ⟶ *lower* number

Correct: 1-ethyl-3-methylcyclohexane   **Incorrect: 3-ethyl-1-methylcyclohexane**

Several examples of cycloalkane nomenclature are given in Figure 4.2.

Figure 4.2
Examples of cycloalkane nomenclature

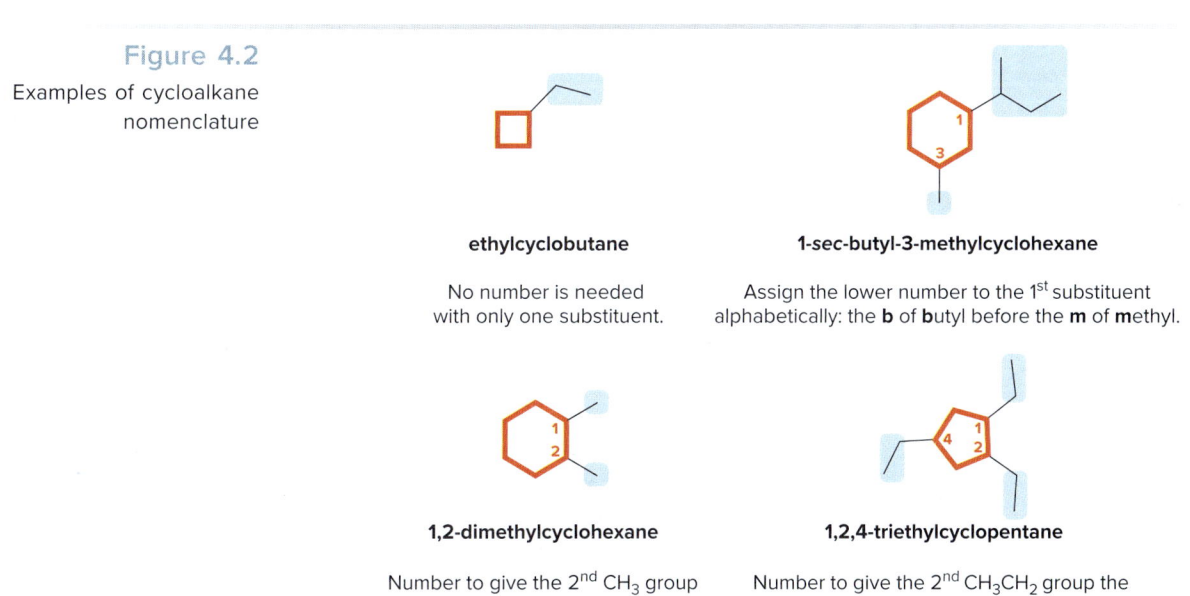

ethylcyclobutane

No number is needed with only one substituent.

1-*sec*-butyl-3-methylcyclohexane

Assign the lower number to the 1ˢᵗ substituent alphabetically: the **b** of **b**utyl before the **m** of **m**ethyl.

1,2-dimethylcyclohexane

Number to give the 2ⁿᵈ CH₃ group the lower number: 1,2- not 1,6-.

1,2,4-triethylcyclopentane

Number to give the 2ⁿᵈ CH₃CH₂ group the lower number: 1,2,4- not 1,3,4- or 1,3,5-.

Problem 4.11
Give the IUPAC name for each compound.

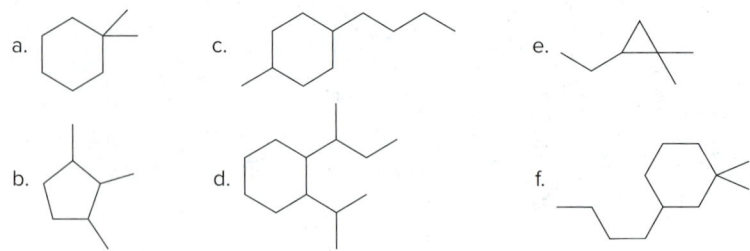

Problem 4.12
Give the structure corresponding to each IUPAC name.

a. 1,2-dimethylcyclobutane
b. 1,1,2-trimethylcyclopropane
c. 4-ethyl-1,2-dimethylcyclohexane
d. 1-sec-butyl-3-isopropylcyclopentane
e. 1,1,2,3,4-pentamethylcycloheptane

4.6 Common Names

Some organic compounds are identified using **common names** that do not follow the IUPAC system of nomenclature. Many of these names were given to molecules long ago, before the IUPAC system was adopted. These names are still widely used. For example, isopentane, an older name for 2-methylbutane, is still allowed by IUPAC rules. We will follow the IUPAC system except in cases in which a common name is widely accepted.

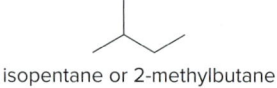
isopentane or 2-methylbutane

dodecahedrane

In the past several years, organic chemists have attempted to synthesize some unusual cycloalkanes not found in nature. **Dodecahedrane,** a beautifully symmetrical compound composed of 12 five-membered rings, is one such molecule. It was first prepared at The Ohio State University in 1982. The IUPAC name for dodecahedrane is undecacyclo-[9.9.0.0^{2,9}.0^{3,7}.0^{4,20}.0^{5,18}.0^{6,16}.0^{8,15}.0^{10,14}.0^{12,19}.0^{13,17}]icosane, a name so complex that few trained organic chemists would be able to identify its structure.

Because these systematic names are so unwieldy, organic chemists often assign a name to a polycyclic compound that is more descriptive of its shape and structure. Dodecahedrane is named because its 12 five-membered rings resemble a dodecahedron. Figure 4.3 shows the names and structures of several other cycloalkanes whose names were inspired by the shape of their carbon skeletons. All the names end in the suffix *-ane,* indicating that they refer to alkanes.

Figure 4.3
Common names for some polycyclic alkanes

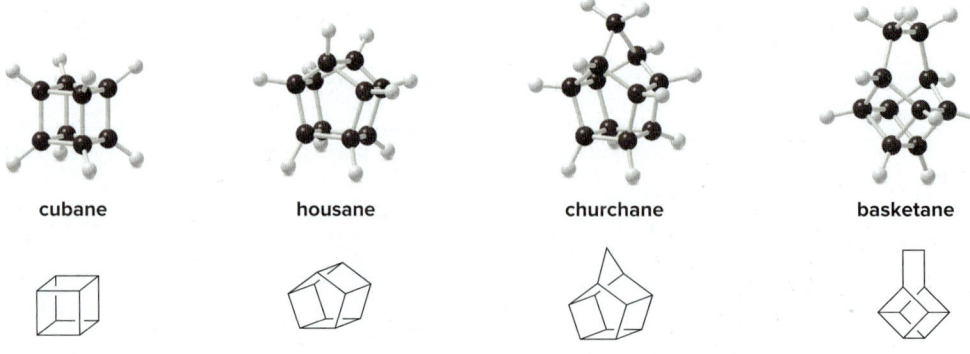

cubane   housane   churchane   basketane

- For a comprehensive list of unusual polycyclic alkanes (including windowpane, davidane, catenane, propellane, and many others), see *Organic Chemistry: The Name Game* by Alex Nickon and Ernest Silversmith, Pergamon Press, 1987.

4.7 Natural Occurrence of Alkanes

Many alkanes occur in nature, primarily in natural gas and petroleum. Both of these fossil fuels serve as energy sources, formed from the degradation of organic material long ago.

Natural gas is composed largely of **methane** (60% to 80% depending on its source), with lesser amounts of ethane, propane, and butane. These organic compounds burn in the presence of oxygen, releasing energy for cooking and heating.

Methane in the atmosphere comes from natural and man-made sources. As global temperatures increase, methane trapped in permafrost and glaciers is released with melting. Microorganisms in the gut of ruminant animals produce methane that is released during defecation and belching. The microorganisms in wetlands and flooded rice fields decompose organic material to form methane when no oxygen is present. Although methane does not persist in the atmosphere as long as carbon dioxide (Section 4.14), methane is a greenhouse gas with significant global warming potential, and its concentration has increased significantly in the last 200 years.

Petroleum is a complex mixture of compounds, most of which are hydrocarbons containing 1–40 carbon atoms. Distilling crude petroleum, a process called **refining,** separates it into usable fractions that differ in boiling point. Most products of petroleum refining provide fuel for home heating, automobiles, diesel engines, and airplanes. Each fuel type has a different composition of hydrocarbons: gasoline (C_5H_{12}–$C_{12}H_{26}$), kerosene ($C_{12}H_{26}$–$C_{16}H_{34}$), and diesel fuel ($C_{15}H_{32}$–$C_{18}H_{38}$).

Petroleum provides more than fuel. About 3% of crude oil is used to make plastics and other synthetic compounds including drugs, fabrics, dyes, and pesticides. These products are responsible for many of the comforts we now take for granted in industrialized countries. Imagine what life would be like without air conditioning, refrigeration, anesthetics, and pain relievers, all products of the petroleum industry.

A significant source of atmospheric methane comes from flooded rice fields. Methane, a greenhouse gas like CO_2 (Section 4.14), is produced by the decomposition of organic matter under anaerobic conditions by soil bacteria. *Daniel C. Smith*

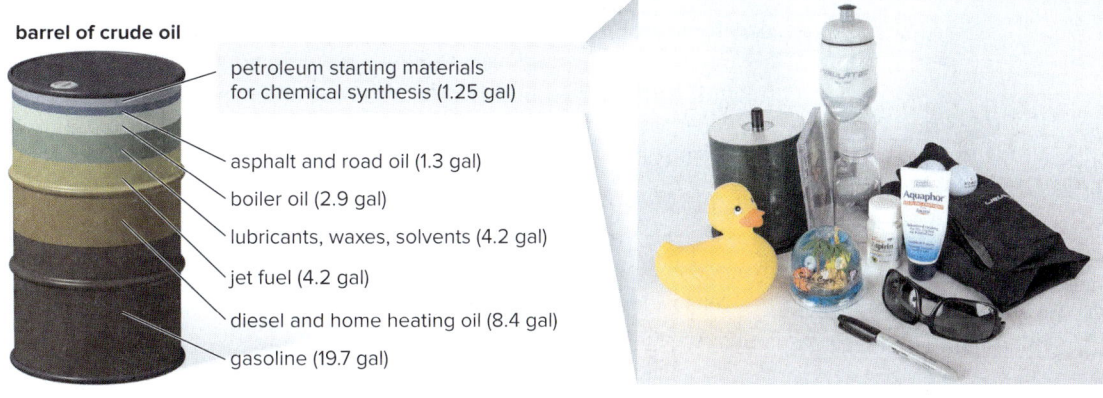

products made from petroleum
Jill Braaten/McGraw-Hill Education

Energy from petroleum is ***nonrenewable,*** and the remaining known oil reserves are limited. Given our dependence on petroleum, not only for fuel, but also for the many necessities of modern society, it becomes clear that we must both conserve what we have and find alternate energy sources.

4.8 Properties of Alkanes

4.8A Physical Properties

Alkanes contain only nonpolar C–C and C–H bonds, and as a result they exhibit only **weak van der Waals forces.** Table 4.2 summarizes how these intermolecular forces affect the physical properties of alkanes.

Table 4.2 Physical Properties of Alkanes

Property	Observation
Boiling point and melting point	• **Alkanes have low bp's and mp's** compared to more polar compounds of comparable size. • Bp and mp increase as the number of carbons increases because of increased surface area. bp = 0 °C, mp = –138 °C bp = 69 °C, mp = –95 °C bp = 139 °C, mp = –78 °C (with OH) Increasing strength of intermolecular forces Increasing boiling point and melting point • The bp of isomers decreases with branching because of decreased surface area. • Mp increases with increased symmetry. bp = 10 °C, mp = –17 °C bp = 30 °C, mp = –160 °C more branching—lower boiling point more symmetry—higher melting point
Solubility	• Alkanes are soluble in organic solvents. • Alkanes are insoluble in water.

Key: bp = boiling point; mp = melting point

The gasoline industry exploits the dependence of boiling point and melting point on alkane size by seasonally changing the composition of gasoline in locations where it gets very hot in the summer and very cold in the winter. Gasoline is refined to contain a larger fraction of higher-boiling hydrocarbons in warmer weather, so it evaporates less readily. In colder weather, it is refined to contain more lower-boiling hydrocarbons, so it freezes less readily.

The mutual insolubility of nonpolar oil and very polar water leads to the common expression "Oil and water don't mix."

Because nonpolar alkanes are not water soluble, crude petroleum that leaks into the sea from an oil tanker or offshore oil well creates an insoluble oil slick on the surface. The insoluble hydrocarbon oil poses a special threat to birds whose feathers are coated with natural nonpolar oils for insulation. Because these hydrophobic oils dissolve in the crude petroleum, birds lose their layer of natural protection and many die.

Problem 4.13   Arrange the following compounds in order of increasing boiling point.

4.8B Spectroscopic Properties

Students who would like to learn about the spectroscopic properties of alkanes are referred to the following sections in later chapters:

- **Mass spectrometry:** Sections A.1A and A.3, especially Figure A.5 and Sample Problem A.6
- **Infrared spectroscopy:** Section B.4A and Table B.2

4.9 Conformations of Acyclic Alkanes—Ethane

Let's now take a closer look at the three-dimensional structure of alkanes. The three-dimensional structure of molecules is called **stereochemistry.** In Chapter 4, we examine the effect of rotation around single bonds. In Chapter 5, we will learn about other aspects of stereochemistry.

Recall from Section 1.10A that **rotation occurs around carbon–carbon σ bonds.** Thus, the two CH₃ groups of ethane rotate, allowing the hydrogens on one carbon to adopt different orientations relative to the hydrogens on the other carbon. These arrangements are called **conformations.**

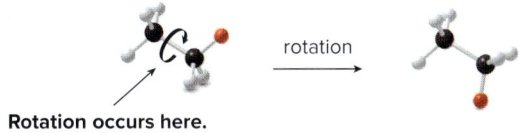

Rotation occurs here.

two different conformations

- *Conformations* are different arrangements of atoms that are interconverted by rotation around single bonds.

Two different arrangements are the **eclipsed conformation** and the **staggered conformation.**

- In the *eclipsed conformation,* the C–H bonds on one carbon are directly *aligned* with the C–H bonds on the adjacent carbon.
- In the *staggered conformation,* the C–H bonds on one carbon *bisect* the H–C–H bond angle on the adjacent carbon.

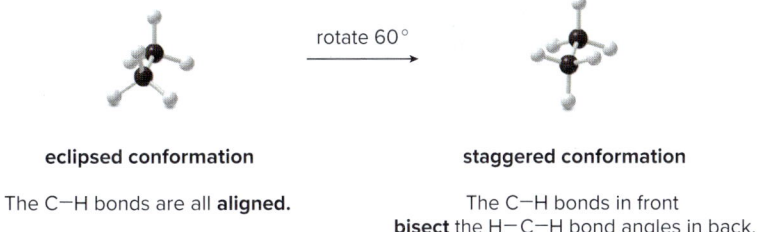

eclipsed conformation

The C–H bonds are all **aligned.**

staggered conformation

The C–H bonds in front **bisect** the H–C–H bond angles in back.

Rotating the atoms on one carbon by 60° converts an eclipsed conformation into a staggered conformation, and vice versa. These conformations are often viewed end-on—that is, looking directly down the carbon–carbon bond. The angle that separates a bond on one atom from a bond on an adjacent atom is called a **dihedral angle.** For ethane in the staggered conformation, the dihedral angle for the C–H bonds is **60°.** For eclipsed ethane, it is **0°.**

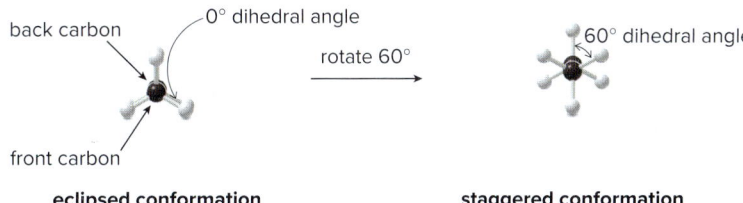

eclipsed conformation

staggered conformation

End-on representations for conformations are commonly drawn using a convention called a **Newman projection.** A Newman projection is a graphic that shows the three groups bonded to each carbon atom in a particular C–C bond, as well as the dihedral angle that separates them.

How To Draw a Newman Projection

Step [1] Look directly down the C–C bond (end-on), and draw a circle with a dot in the center to represent the carbons of the C–C bond.

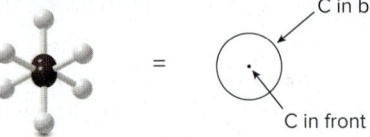

- The circle represents the back carbon and the dot represents the front carbon.

Step [2] Draw in the bonds.

- Draw the bonds on the **front** C as three lines **meeting at the center** of the circle.
- Draw the bonds on the **back** C as three lines coming **out of the edge** of the circle.

Step [3] Add the atoms on each bond.

- Each C has 3 H's in ethane.

Figure 4.4 illustrates the Newman projections for both the staggered and eclipsed conformations for ethane.

Figure 4.4
Newman projections for the staggered and eclipsed conformations of ethane

staggered conformation **eclipsed conformation**

Follow this procedure for any C–C bond. With a Newman projection, **always consider *one* C–C bond only and draw the atoms bonded to the carbon atoms, *not* the carbon atoms in the bond itself.** Newman projections for the staggered and eclipsed conformations of propane are drawn in Figure 4.5.

Figure 4.5
Newman projections for the staggered and eclipsed conformations of propane

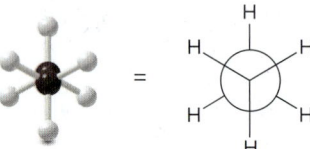

propane

- Arbitrarily pick one C to be in front and one C to be in back.
- 3 H's on one C
- 2 H's and 1 CH₃ on the other C

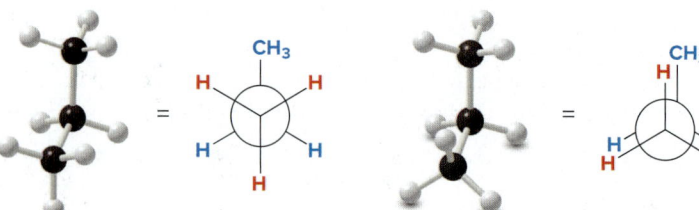

staggered conformation **eclipsed conformation**

Problem 4.14 Convert each representation to a Newman projection around the indicated bond.

Problem 4.15 Which of the following is (are) possible Newman projections for 2-methylpentane?

The staggered and eclipsed conformations of ethane interconvert at room temperature, but **each conformation is *not* equally stable.**

- **The staggered conformations are more stable (lower in energy) than the eclipsed conformations.**

The cause of this stability difference is the subject of some debate in the chemical literature. A contributing factor may be increased electron–electron repulsion between the bonds in the eclipsed conformation compared to the staggered conformation, where the bonding electrons are farther apart.

The difference in energy between the staggered and eclipsed conformations is 12 kJ/mol (2.9 kcal/mol), a small enough difference that the rotation is still very rapid at room temperature, and the conformations cannot be separated. Because three eclipsed C–H bonds increase the energy of a conformation by 12 kJ/mol, **each eclipsed C–H bond results in an increase in energy of 4.0 kJ/mol (1.0 kcal/mol).** The energy difference between the staggered and eclipsed conformations is called **torsional energy.** Thus, eclipsing introduces **torsional strain** into a molecule.

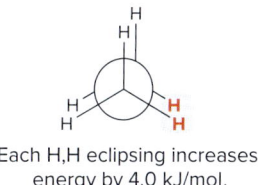
Each H,H eclipsing increases energy by 4.0 kJ/mol.

- *Torsional strain* is an increase in energy caused by eclipsing interactions.

The graph in Figure 4.6 shows how the potential energy of ethane changes with dihedral angle as one CH$_3$ group rotates relative to the other. **The staggered conformation is the most stable arrangement, so it is at an *energy minimum*.** As the C–H bonds on one carbon are rotated relative to the C–H bonds on the other carbon, the energy increases as the C–H bonds get closer until a **maximum is reached after 60° rotation to the eclipsed conformation.** As rotation continues, the energy decreases until after 60° rotation, when the staggered conformation is reached once again.

Strain results in an **increase in energy.** Torsional strain is the first of three types of strain discussed in this text. The other two are **steric strain** (Section 4.10) and **angle strain** (Section 4.11).

- **An energy minimum and maximum occur every 60°** as the conformation changes from staggered to eclipsed. Conformations that are neither staggered nor eclipsed are intermediate in energy.

Problem 4.16 The torsional energy in propane is 14 kJ/mol (3.4 kcal/mol). Because each H,H eclipsing interaction is worth 4.0 kJ/mol (1.0 kcal/mol) of destabilization, how much is one H,CH$_3$ eclipsing interaction worth in destabilization? (See Section 4.10 for an alternate way to arrive at this value.)

Figure 4.6
Graph: Energy versus dihedral angle for ethane

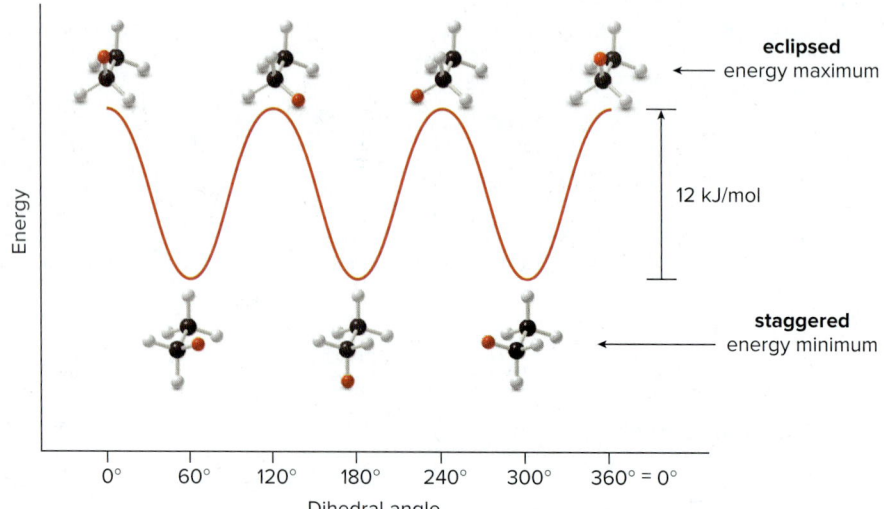

- Note the position of the labeled H atom after each 60° rotation. All three staggered conformations are identical (except for the position of the label), and the same is true for all three eclipsed conformations.

4.10 Conformations of Butane

Butane and higher-molecular-weight alkanes have several carbon–carbon bonds, all capable of rotation.

butane
Consider rotation at C2–C3.
Each C is bonded to 2 H's and 1 CH_3 group.

To analyze the different conformations that result from rotation around the C2–C3 bond, begin arbitrarily with one—for example, the staggered conformation that places two CH_3 groups 180° from each other—then,

It takes six 60° rotations to return to the original conformation.

- **Rotate one carbon atom in 60° increments either clockwise or counterclockwise, while keeping the other carbon fixed. Continue until you return to the original conformation.**

Figure 4.7 illustrates the six possible conformations that result from this process.

Although each 60° bond rotation converts a staggered conformation to an eclipsed conformation (or vice versa), neither all the staggered conformations nor all the eclipsed conformations are the same. For example, the dihedral angle between the methyl groups in staggered conformations **3** and **5** are both 60°, whereas it is 180° in staggered conformation **1**.

- A staggered conformation with two larger groups 180° from each other is called *anti*.
- A staggered conformation with two larger groups 60° from each other is called *gauche*.

Similarly, the methyl groups in conformations **2** and **6** both eclipse hydrogen atoms, whereas they eclipse each other in conformation **4**.

The staggered conformations (**1, 3**, and **5**) are lower in energy than the eclipsed conformations (**2, 4**, and **6**), but how do the energies of the individual staggered and eclipsed conformations

4.10 Conformations of Butane

Figure 4.7
Six different conformations of butane

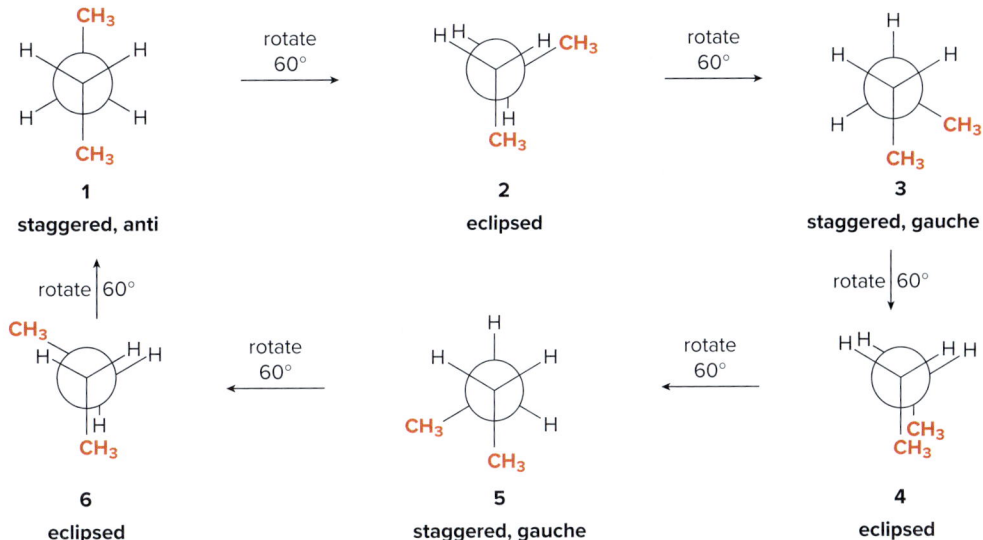

compare to each other? The relative energies of the individual staggered conformations (or the individual eclipsed conformations) depend on their **steric strain.**

- *Steric* strain is an increase in energy resulting when atoms are forced too close to one another.

The methyl groups are farther apart in the anti conformation (**1**) than in the gauche conformations (**3** and **5**), so among the staggered conformations, **1** is lower in energy (more stable) than **3** and **5**. In fact, the anti conformation is 3.8 kJ/mol (0.9 kcal/mol) lower in energy than either gauche conformation because of the steric strain that results from the proximity of the methyl groups in **3** and **5**.

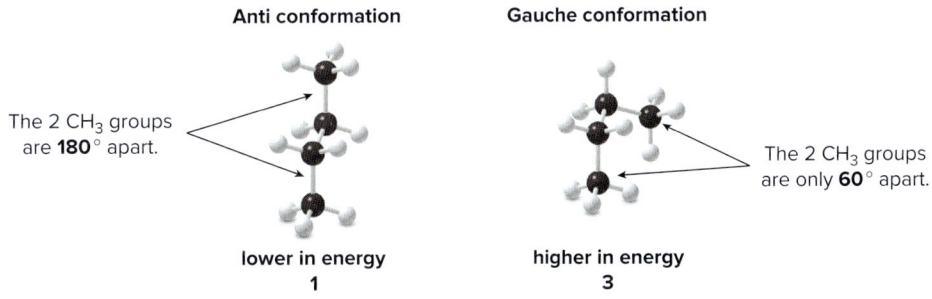

- Gauche conformations are generally *higher* in energy than anti conformations because of steric strain.

Steric strain caused by two eclipsed CH$_3$ groups

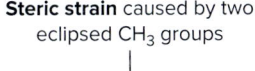
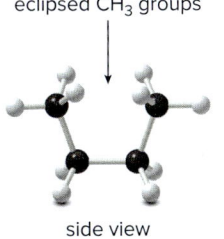

side view
4

Steric strain also affects the relative energies of eclipsed conformations. Conformation **4** is higher in energy than **2** or **6**, because the two larger CH$_3$ groups are forced close to each other, introducing considerable steric strain.

To graph energy versus dihedral angle, keep in mind two considerations:

- Staggered conformations are at energy minima and eclipsed conformations are at energy maxima.
- Unfavorable steric interactions increase energy.

For butane, this means that anti conformation **1** is lowest in energy, and conformation **4** with two eclipsed CH$_3$ groups is the highest in energy. The relative energy of other conformations is depicted in the energy versus rotation diagram for butane in Figure 4.8.

Figure 4.8 Graph: Energy versus dihedral angle for butane

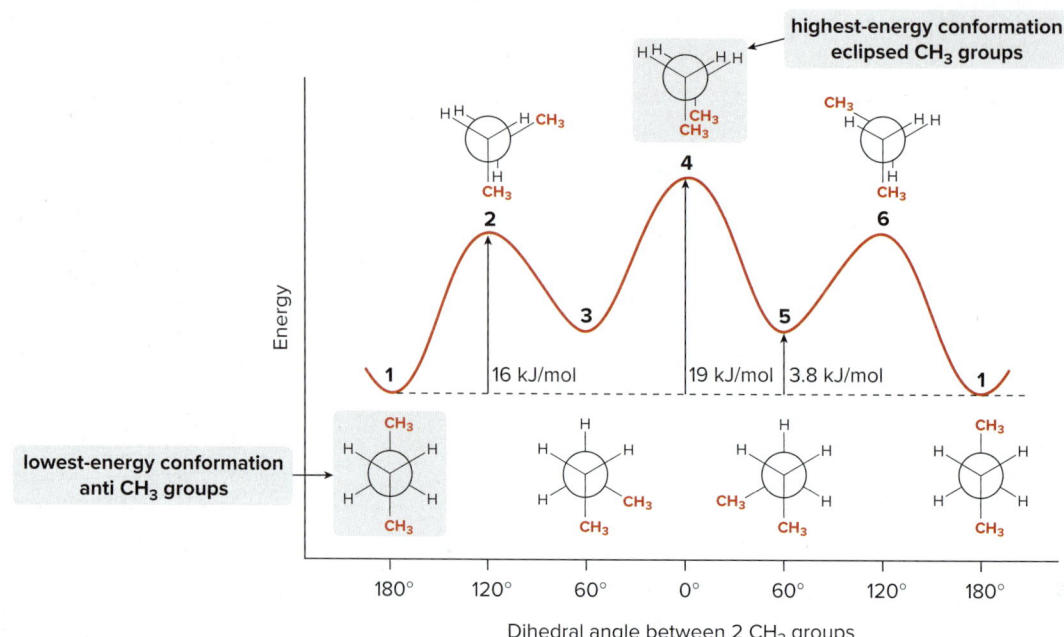

- Staggered conformations **1**, **3**, and **5** are at energy minima.
- Anti conformation **1** is lower in energy than gauche conformations **3** and **5**, which possess steric strain.
- Eclipsed conformations **2**, **4**, and **6** are at energy maxima.
- Eclipsed conformation **4**, which has additional steric strain due to two eclipsed CH₃ groups, is highest in energy.

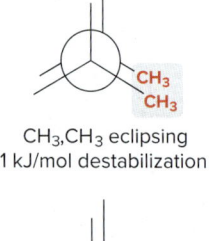

CH₃,CH₃ eclipsing
11 kJ/mol destabilization

H,CH₃ eclipsing
6.0 kJ/mol destabilization

We can now use the values in Figure 4.8 to estimate the destabilization caused by other eclipsed groups. For example, conformation **4** is 19 kJ/mol less stable than the anti conformation **1**. Conformation **4** possesses two H,H eclipsing interactions, worth 4.0 kJ/mol each in destabilization (Section 4.9), and one CH₃,CH₃ eclipsing interaction. Thus, the **CH₃,CH₃ interaction** is worth 19 − 2(4.0) = **11 kJ/mol** of destabilization.

Similarly, conformation **2** is 16 kJ/mol less stable than the anti conformation **1**, and possesses one H,H eclipsing interaction (worth 4.0 kJ/mol of destabilization) and two H,CH₃ interactions. Thus, **each H,CH₃ interaction** is worth 1/2(16 − 4.0) = **6.0 kJ/mol** of destabilization. These values are summarized in Table 4.3.

- The energy difference between the lowest- and highest-energy conformations is called the *barrier to rotation*.

Table 4.3 Summary: Torsional and Steric Strain Energies in Acyclic Alkanes

Type of interaction	Energy increase	
	kJ/mol	kcal/mol
H,H eclipsing	4.0	1.0
H,CH₃ eclipsing	6.0	1.4
CH₃,CH₃ eclipsing	11	2.6
gauche CH₃ groups	3.8	0.9

We can use these same principles to determine conformations and relative energies for any acyclic alkane. Because the **lowest-energy conformation has all bonds staggered and all large groups anti,** alkanes are often drawn in zigzag skeletal structures to indicate this.

Problem 4.17

a. Draw the three staggered and three eclipsed conformations that result from rotation around the bond labeled in **red** using Newman projections.
b. Label the most stable and least stable conformation.

Problem 4.18 Rank the following conformations in order of increasing energy.

A   B   C   D

Problem 4.19 Consider rotation around the carbon–carbon bond in 1,2-dichloroethane ($ClCH_2CH_2Cl$).

a. Using Newman projections, draw all of the staggered and eclipsed conformations that result from rotation around this bond.
b. Graph energy versus dihedral angle for rotation around this bond.

Problem 4.20 Calculate the destabilization present in each eclipsed conformation.

a.   b.

4.11 An Introduction to Cycloalkanes

Besides torsional strain and steric strain, the conformations of cycloalkanes are affected by **angle strain.**

- **Angle** strain is an increase in energy when tetrahedral bond angles deviate from the optimum angle of 109.5°.

Originally cycloalkanes were thought to be flat rings, with the bond angles between carbon atoms determined by the size of the ring. For example, a flat cyclopropane ring would have 60° internal bond angles, a flat cyclobutane ring would have 90° angles, and large flat rings would have very large angles. It was assumed that rings with bond angles so different from the tetrahedral bond angle would be very strained and highly reactive. This is called the **Baeyer strain theory.**

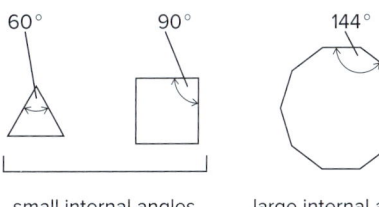

small internal angles
angle strain

large internal angles
angle strain

It turns out, though, that **cycloalkanes with more than three C atoms in the ring are not flat molecules.** They are puckered to **reduce strain,** both angle strain and torsional strain. The three-dimensional structures of some simple cycloalkanes are shown in Figure 4.9. Three- and four-membered rings still possess considerable angle strain, but puckering reduces the internal bond angles in larger rings, thus reducing angle strain.

Figure 4.9
Three-dimensional structure of some cycloalkanes

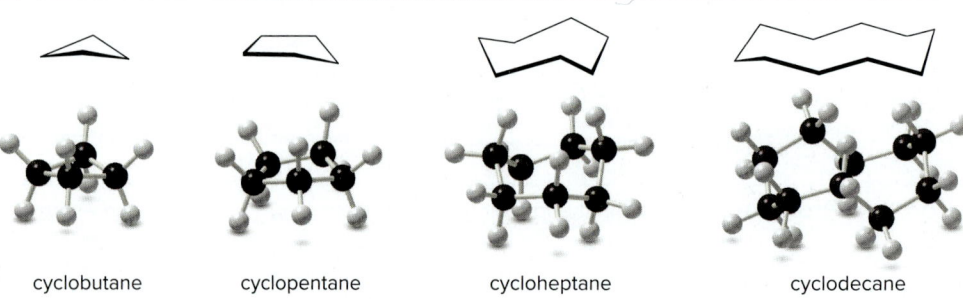

cyclobutane   cyclopentane   cycloheptane   cyclodecane

4.12 Cyclohexane

Let's now examine the conformation of **cyclohexane,** the most common ring size in naturally occurring compounds.

4.12A The Chair Conformation

A planar cyclohexane ring would experience angle strain, because the internal bond angle between the carbon atoms would be 120°, and torsional strain, because all of the hydrogens on adjacent carbon atoms would be eclipsed.

If a cyclohexane ring were flat...

120°
The internal bond angle is > 109.5°.
angle strain

All H's are aligned.
torsional strain

In reality, cyclohexane adopts a puckered conformation, called the **chair** form, which is more stable than any other possible conformation.

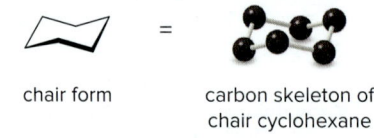

chair form   carbon skeleton of chair cyclohexane

The chair conformation is so stable because it eliminates angle strain (**all C–C–C bond angles are 109.5°**) and torsional strain (all hydrogens on adjacent carbon atoms are **staggered,** not eclipsed).

109.5°   All H's are **staggered.**

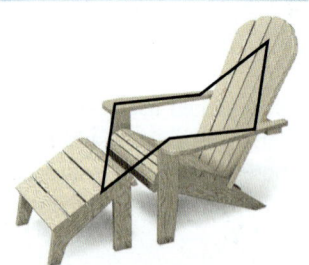

Visualizing the chair. If the cyclohexane chair conformation is tipped downward, we can more easily view it as a chair with a back, seat, and foot support.

- In cyclohexane, three C atoms pucker up and three C atoms pucker down, alternating around the ring. These C atoms are called *up* C's and *down* C's.

4.12 Cyclohexane

Each carbon in cyclohexane has two different kinds of hydrogens.

Each cyclohexane carbon atom has one axial and one equatorial hydrogen.

- **Axial** hydrogens are located above and below the ring (along a perpendicular axis).
- **Equatorial** hydrogens are located in the plane of the ring (around the equator).

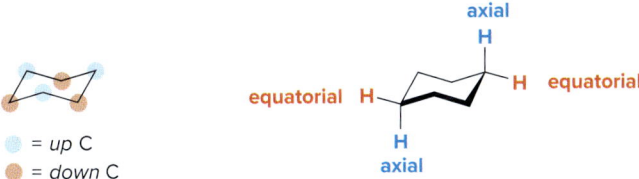

A three-dimensional representation of the chair form is shown in Figure 4.10.

Figure 4.10
A three-dimensional model of the chair form of cyclohexane with all H atoms drawn

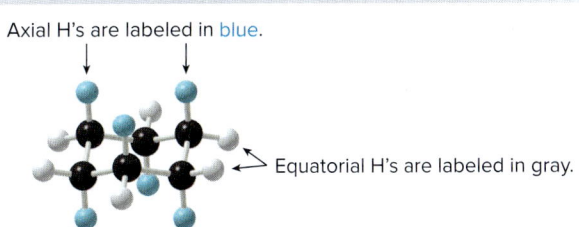

- Cyclohexane has **six axial H's** and **six equatorial H's**.

How To Draw the Chair Form of Cyclohexane

Step [1] Draw the carbon skeleton.

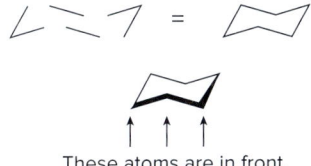

- Draw three parts of the chair: **a wedge, a set of parallel lines**, and **another wedge**.
- Then, join them together.
- The bottom 3 C's come out of the page, and for this reason, bonds to them are sometimes highlighted in bold.

Step [2] Label the *up* C's and *down* C's on the ring.

- There are 3 *up* and 3 *down* C's, and they *alternate* around the ring.

Step [3] Draw in the axial H atoms.

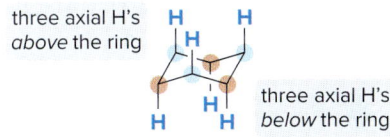

- On an *up* **C** the axial H is *up.*
- On a *down* **C** the axial H is *down.*

Step [4] Draw in the equatorial H atoms.
- **The axial H is down on a down C**, so the equatorial H must be *up.*
- **The axial H is up on an up C**, so the equatorial H must be *down.*

Problem 4.21 Classify the ring carbons as *up* C's or *down* C's. Identify the bonds highlighted in bold as axial or equatorial.

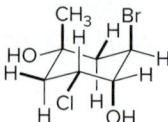

Problem 4.22 Using the cyclohexane with the C's numbered as shown, draw a chair form that fits each description.

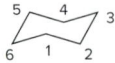

a. The ring has an axial CH₃ group at C1 and an equatorial OH on C2.
b. The ring has an equatorial CH₃ group on C6 and an axial OH group on C4.
c. The ring has equatorial OH groups on C1, C2, and C5.

4.12B Ring-Flipping

Like acyclic alkanes, **cyclohexane does not remain in a single conformation.** The bonds twist and bend, resulting in new arrangements, but the movement is more restricted. One conformational change involves **ring-flipping,** which can be viewed as a two-step process.

- **A *down* carbon flips up.** This forms a new conformation of cyclohexane called a **boat.** The boat form has two carbons oriented above a plane containing the other four carbons.
- The boat form can flip in two possible ways. The carbon labeled with a red circle can flip down, re-forming the initial conformation; or the **second *up* carbon,** labeled with a blue circle, **can flip down. This forms a second chair conformation.**

Because of ring-flipping, the *up* carbons become *down* carbons and the *down* carbons become *up* carbons. Thus, cyclohexane exists as two different chair conformations of equal stability, which rapidly interconvert at room temperature.

The process of ring-flipping also affects the orientation of cyclohexane's hydrogen atoms.

- Axial and equatorial H atoms are interconverted during a ring flip. Axial H atoms become equatorial H atoms, and equatorial H atoms become axial H atoms (Figure 4.11).

Figure 4.11
Ring-flipping interconverts axial and equatorial hydrogens in cyclohexane.

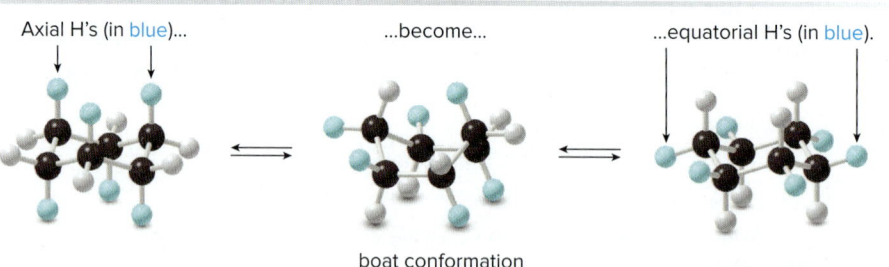

The chair forms of cyclohexane are 30 kJ/mol more stable than the boat forms. The boat conformation is destabilized by torsional strain because the hydrogens on the four carbon atoms in the plane are eclipsed. Additionally, there is steric strain because two hydrogens at either end of the boat—the **flagpole hydrogens**—are forced close to each other, as shown in Figure 4.12.

Figure 4.12 Two views of the boat conformation of cyclohexane

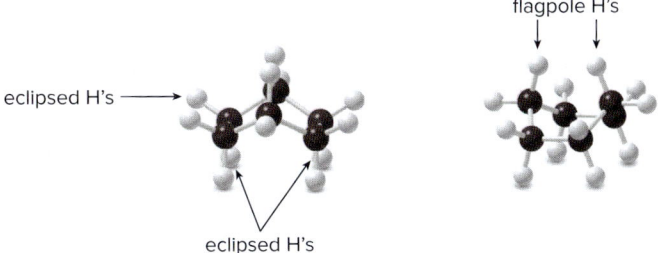

The boat form of cyclohexane is less stable than the chair forms for two reasons:
- Eclipsing interactions between H's cause **torsional strain.**
- The proximity of the flagpole H's causes **steric strain.**

4.13 Substituted Cycloalkanes

What happens when one hydrogen on cyclohexane is replaced by a larger substituent? Is there a difference in the stability of the two cyclohexane conformations? To answer these questions, remember one rule:

- The equatorial position has more room than the axial position, so *larger* substituents are more stable in the *equatorial* position.

4.13A Cyclohexane with One Substituent

There are two possible chair conformations of a monosubstituted cyclohexane, such as methylcyclohexane, as shown in the following *How To*.

How To Draw the Two Conformations for a Substituted Cyclohexane

Step [1] Draw one chair form and add the substituents.
- Arbitrarily pick a ring carbon, classify it as an *up* or *down* carbon, and draw the bonds. **Each C has one axial and one equatorial bond.**
- Add the substituents, in this case H and CH$_3$, arbitrarily placing one axial and one equatorial. In this example, the CH$_3$ group is drawn equatorial.
- This forms one of the two possible chair conformations, labeled **A**.

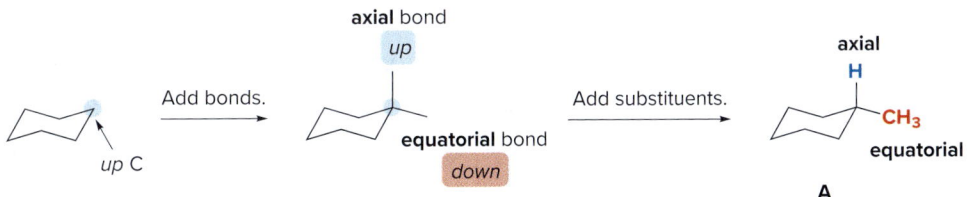

Step [2] Ring-flip the cyclohexane ring.
- Convert *up* C's to *down* C's and vice versa. The chosen *up* C now puckers down.

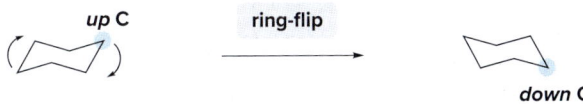

—*Continued*

How To, continued...

Step [3] Add the substituents to the second conformation.
- Draw axial and equatorial bonds. **On a *down* C the axial bond is *down*.**
- Ring-flipping converts axial bonds to equatorial bonds, and vice versa. The equatorial methyl becomes axial.
- This forms the other possible chair conformation, labeled **B**.

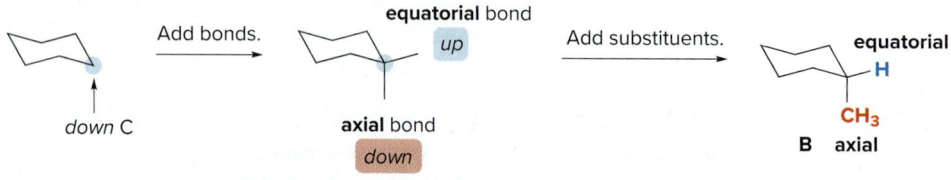

Although the CH$_3$ group flips from equatorial to axial, it starts on a down bond and stays on a down bond. **It *never* flips from below the ring to above the ring.**

- A substituent always stays on the *same side* of the ring—either below or above—during the process of ring-flipping.

Each carbon atom has one *up* and one *down* bond. An *up* bond can be either axial or equatorial, depending on the carbon to which it is attached. **On an *up* C, the axial bond is *up*,** but on a *down* C, the equatorial bond is *up*.

The two conformations of methylcyclohexane are different, so they are not equally stable. In fact, **A,** which places the larger methyl group in the roomier equatorial position, is considerably more stable than **B,** which places it axial.

Why is a substituted cyclohexane ring more stable with a larger group in the equatorial position? Figure 4.13 shows that with an equatorial CH$_3$ group, steric interactions with nearby groups are minimized. An axial CH$_3$ group, however, is close to two other axial H atoms, creating two destabilizing steric interactions called **1,3-diaxial interactions.** Each unfavorable H,CH$_3$ interaction destabilizes the conformation by 3.8 kJ/mol, so **B** is 7.6 kJ/mol less stable than **A**.

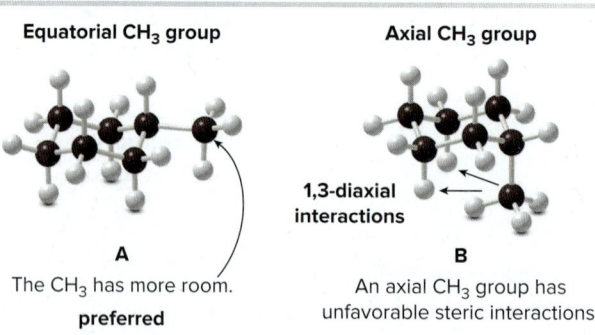

Figure 4.13 Three-dimensional representations for the two conformations of methylcyclohexane

- Larger axial substituents create unfavorable 1,3-diaxial interactions, destabilizing a cyclohexane conformation.

4.13 Substituted Cycloalkanes

The *larger* the substituent on the six-membered ring, the *higher* the percentage of the conformation containing the equatorial substituent at equilibrium. With a very large substituent like *tert*-butyl [$(CH_3)_3C-$], essentially none of the conformation containing an axial *tert*-butyl group is present at room temperature, so **the ring is essentially anchored in a single conformation having an equatorial *tert*-butyl group.**

very crowded
axial *tert*-butyl group
C
highly destabilized

equatorial *tert*-butyl group
D
100%

Problem 4.23 Draw a second chair conformation for each cyclohexane. Then decide which conformation is present in higher concentration at equilibrium.

a. [cyclohexane-Br]   b. [cyclohexane-Cl]   c. [cyclohexane-CH₂CH₃]

Problem 4.24 Draw both conformations for 1-ethyl-1-methylcyclohexane and decide which conformation (if any) is more stable.

4.13B A Disubstituted Cycloalkane

Rotation around the C–C bonds in the ring of a cycloalkane is restricted, so **a group on one side of the ring can *never* rotate to the other side of the ring.** As a result, there are two different 1,2-dimethylcyclopentanes—one having two CH_3 groups on the **same side** of the ring and one having them on **opposite sides** of the ring.

2 CH₃'s above the ring
A
cis isomer

1 CH₃ above and 1 CH₃ below
B
trans isomer

Wedges indicate bonds in front of the plane of the ring, and dashed wedges indicate bonds behind. For a review of this convention, see Section 1.7B. If a ring carbon is bonded to a CH_3 group in **front** of the ring (on a wedge), it is *assumed* that the other atom bonded to this carbon is hydrogen, located **behind** the ring (on a dashed wedge).

Cis and **trans** isomers are named by adding the prefixes *cis* and *trans* to the name of the cycloalkane. Thus, **A** is *cis*-1,2-dimethylcyclopentane, and **B** is *trans*-1,2-dimethylcyclopentane.

A and **B** are **isomers,** because they are different compounds with the same molecular formula, but they represent the second major class of isomers called **stereoisomers.**

- *Stereoisomers* are isomers that differ *only* in the way the atoms are oriented in space.

The prefixes **cis** and **trans** are used to distinguish these stereoisomers.

- The cis isomer has two groups on the *same side* of the ring.
- The trans isomer has two groups on *opposite sides* of the ring.

160   Chapter 4   Alkanes

> **Problem 4.25**   Draw the structure for each compound using wedges and dashed wedges.
> a. *cis*-1,2-dimethylcyclopropane   b. *trans*-1-ethyl-2-methylcyclopentane

> **Problem 4.26**   For *cis*-1,3-diethylcyclobutane, draw (a) a stereoisomer; (b) a constitutional isomer.

4.13C A Disubstituted Cyclohexane

A disubstituted cyclohexane like 1,4-dimethylcyclohexane also has cis and trans stereoisomers. In addition, each of these stereoisomers has two possible chair conformations.

trans-1,4-dimethylcyclohexane   *cis*-1,4-dimethylcyclohexane

All disubstituted cycloalkanes with two groups bonded to *different* atoms have cis and trans isomers.

To draw both conformations for each stereoisomer, follow the procedure in Section 4.13A for a monosubstituted cyclohexane, keeping in mind that two substituents must now be added to the ring.

How To Draw Two Conformations for a Disubstituted Cyclohexane

Step [1] Draw one chair form and add the substituents.
- For *trans*-1,4-dimethylcyclohexane, arbitrarily pick two C's located 1,4- to each other, classify them as *up* or *down* C's, and draw in the substituents.
- The trans isomer must have one group *above* the ring (on an *up* bond) and one group *below* the ring (on a *down* bond). The substituents can be either axial or equatorial, as long as one is up and one is down. The easiest trans isomer to visualize has two axial CH₃ groups. This arrangement is said to be **diaxial**.
- This forms one of the two possible chair conformations, labeled **A**.

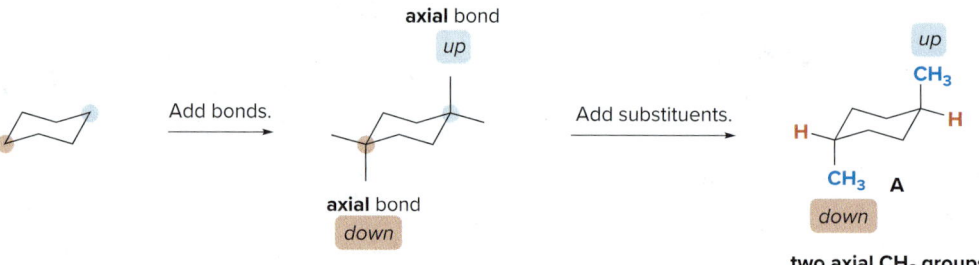

Step [2] Ring-flip the cyclohexane ring.

- The *up* C flips down, and the *down* C flips up.

Step [3] Add the substituents to the second conformation.

two equatorial CH₃ groups
B

- **Ring-flipping converts axial bonds to equatorial bonds, and vice versa.** The diaxial CH₃ groups become **diequatorial**. This trans conformation is less obvious to visualize. It is still trans, because one CH₃ group is above the ring (on an *up* bond), and one is below (on a *down* bond).

4.13 Substituted Cycloalkanes

Conformations **A** and **B** are not equally stable. **Because B has both larger CH₃ groups in the roomier equatorial position, B is *lower* in energy.**

2 CH₃ groups in the more crowded **axial** position

2 CH₃ groups in the more roomy **equatorial** position

A
diaxial conformation

B
diequatorial conformation
more stable

The cis isomer of 1,4-dimethylcyclohexane also has two conformations, as shown in Figure 4.14. Because each conformation has one CH₃ group axial and one equatorial, they are **identical in energy.** At room temperature, therefore, the two conformations exist in a 50:50 mixture at equilibrium.

Figure 4.14
The two conformations of *cis*-1,4-dimethylcyclohexane

C
50%

D
50%

- **A cis isomer has two groups on the same side of the ring,** either both *up* or both *down*. In this example, Conformations **C** and **D** have two CH₃ groups drawn *up*.
- Both conformations have one CH₃ group axial and one equatorial, making them equally stable.

The relative stability of the two conformations of any disubstituted cyclohexane can be analyzed using this procedure.

- A *cis* isomer has two substituents on the *same side,* either both on *up* bonds or both on *down* bonds.
- A *trans* isomer has two substituents on *opposite sides,* one *up* and one *down*.
- Whether substituents are axial or equatorial depends on the relative location of the two substituents (on carbons 1,2-, 1,3-, or 1,4-).

Sample Problem 4.3 Drawing Two Conformations for a Disubstituted Cycloalkane

Draw both chair conformations for *trans*-1,3-dimethylcyclohexane.

Solution

Step [1] Draw one chair form and add substituents.

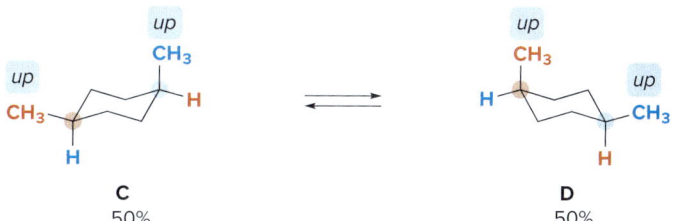

A

- Pick two C's 1,3- to each other.
- **The trans isomer has two groups on opposite sides.** In Conformation **A**, one CH₃ is axial (on an *up* bond), and one group is equatorial (on a *down* bond).

Steps [2–3] Ring-flip and add substituents.

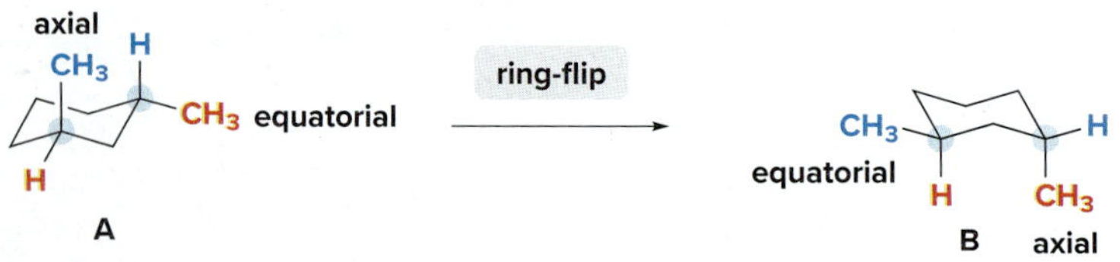

- The two *up* C's flip down.
- The axial CH₃ flips equatorial (still an *up* bond) and the equatorial CH₃ flips axial (still a *down* bond). Conformation **B** is *trans* because the two CH₃'s are still on *opposite* sides.
- **Conformations A and B are equally stable** because each has one CH₃ equatorial and one axial.

Problem 4.27 Consider 1,2-dimethylcyclohexane.
a. Draw structures for the cis and trans isomers using a hexagon for the six-membered ring.
b. Draw the two possible chair conformations for the cis isomer. Which conformation, if either, is more stable?
c. Draw the two possible chair conformations for the trans isomer. Which conformation, if either, is more stable?
d. Which isomer, cis or trans, is more stable and why?

More Practice: Try Problems 4.51a, b, d; 4.52–4.54; 4.56a.

Problem 4.28 Label each compound as cis or trans. Then draw the second chair conformation.

a. 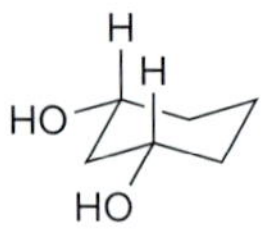 HO, HO substituents   b. 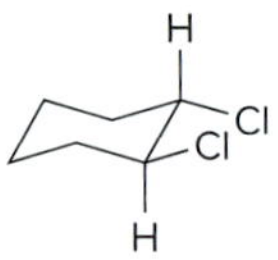 Cl, Cl substituents   c. 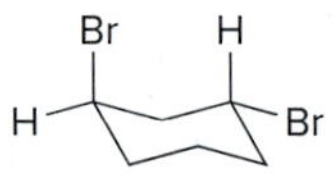 Br, Br substituents

---

Problem 4.29 Draw a chair conformation of cyclohexane with one CH₃CH₂ group and one CH₃ group that fits each description.

a. a 1,1-disubstituted cyclohexane with an axial CH₃CH₂ group
b. a cis-1,2-disubstituted cyclohexane with an axial CH₃ group
c. a trans-1,3-disubstituted cyclohexane with an equatorial CH₃ group
d. a trans-1,4-disubstituted cyclohexane with an equatorial CH₃CH₂ group

---

Sample Problem 4.4 Converting a Hexagon with Substituents to a Chair Form

Draw the two chair forms for the following trisubstituted cyclohexane, and label the more stable conformation.

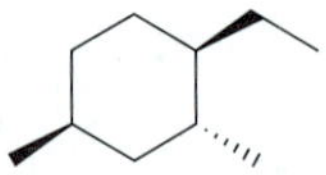

Solution

Use the wedges and dashed wedges to determine what groups are above and below the ring, respectively. Start at a substituent and proceed in the *same* direction around the ring—clockwise or counterclockwise—to convert the hexagon to both chair forms.

4.14 Oxidation of Alkanes

Step [1] Draw one chair form and add substituents.

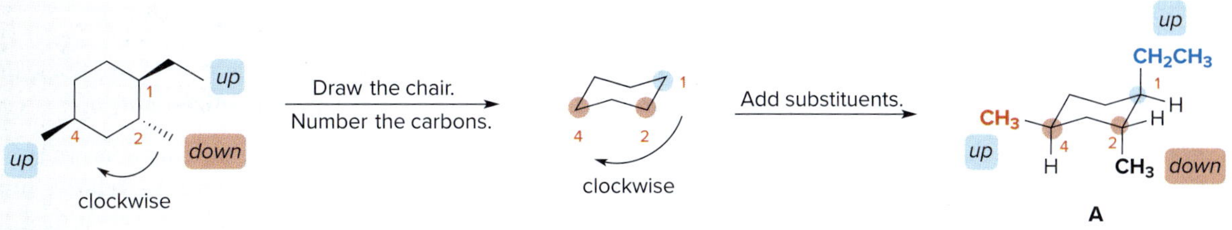

- If the ring C's are numbered 1 → 2 → 4 in the *clockwise* direction in the flat hexagon, number the C's in the chair in the *clockwise* direction. Any C in the chair can be assigned C1.
- The ethyl group on an *up* bond at C1 is axial, the methyl group on a *down* bond at C2 is axial, and the methyl on an *up* bond at C4 is equatorial.
- Conformation **A** has two axial and one equatorial substituents.

Step [2] Ring-flip and add substituents.

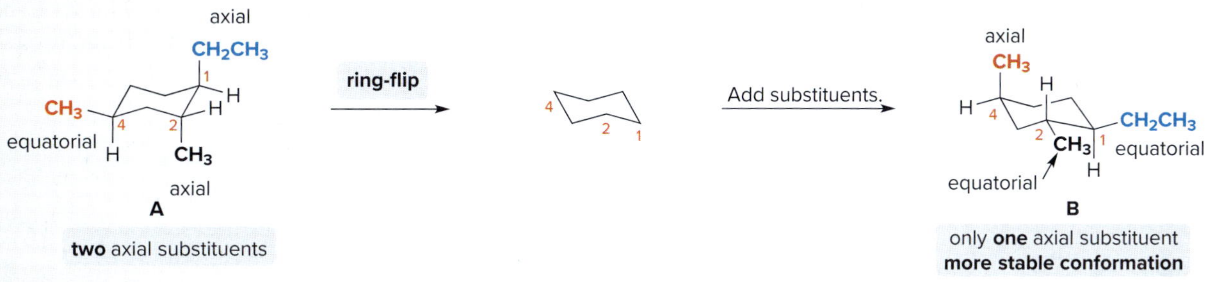

- Ring-flipping the cyclohexane generates the second chair form **B**, which now has one axial and two equatorial substituents, making it the **more stable** conformation.

Problem 4.30 (a) Draw **C** in its more stable chair conformation. (b) Convert **D** to a hexagon with substituents on wedges and dashed wedges.

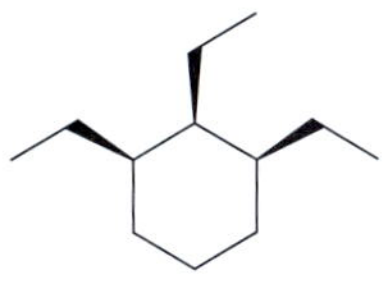

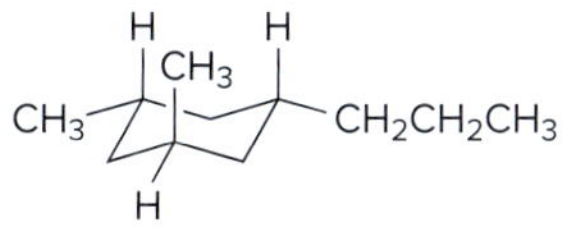

More Practice: Try Problems 4.51c, 4.53a, 4.56b, 4.60c.

4.14 Oxidation of Alkanes

In Chapter 3, we learned that a functional group contains a heteroatom or π bond and constitutes **the reactive part of a molecule.** Alkanes are the only family of organic molecules that has no functional group, and therefore, **alkanes undergo few reactions.** In fact, alkanes are inert to reaction unless forcing conditions are used.

In Chapter 4, we consider only one reaction of alkanes—**combustion.** Combustion is an **oxidation–reduction** reaction.

4.14A Oxidation and Reduction Reactions

Compounds that contain many C–H bonds and few C–Z bonds are said to be in a **reduced state,** whereas those that contain few C–H bonds and more C–Z bonds are in a **more oxidized state.** CH_4 is highly reduced, whereas CO_2 is highly oxidized.

- **Oxidation** is the *loss* of electrons.
- **Reduction** is the *gain* of electrons.

Oxidation and reduction are opposite processes. As in acid–base reactions, there are always two components in these reactions. **One component is oxidized and one is reduced.**

To determine if an organic compound undergoes oxidation or reduction, we concentrate on the carbon atoms of the starting material and product, and **compare the relative number of C–H and C–Z bonds,** where Z = an element *more electronegative* than carbon (usually O, N, or X). Oxidation and reduction are then defined in two complementary ways.

- *Oxidation* results in an *increase* in the number of C–Z bonds; or
- *Oxidation* results in a *decrease* in the number of C–H bonds.

Because Z is more electronegative than C, replacing C–H bonds with C–Z bonds decreases the electron density around C. Loss of electron density = oxidation.

- *Reduction* results in a *decrease* in the number of C–Z bonds; or
- *Reduction* results in an *increase* in the number of C–H bonds.

Figure 4.15 illustrates the oxidation of CH_4 by replacing C–H bonds with C–O bonds (from left to right). The symbol **[O]** indicates oxidation. Because reduction is the reverse of oxidation, the molecules in Figure 4.15 are progressively reduced moving from right to left, from CO_2 to CH_4. The symbol **[H]** indicates reduction.

Figure 4.15 The oxidation and reduction of a carbon compound

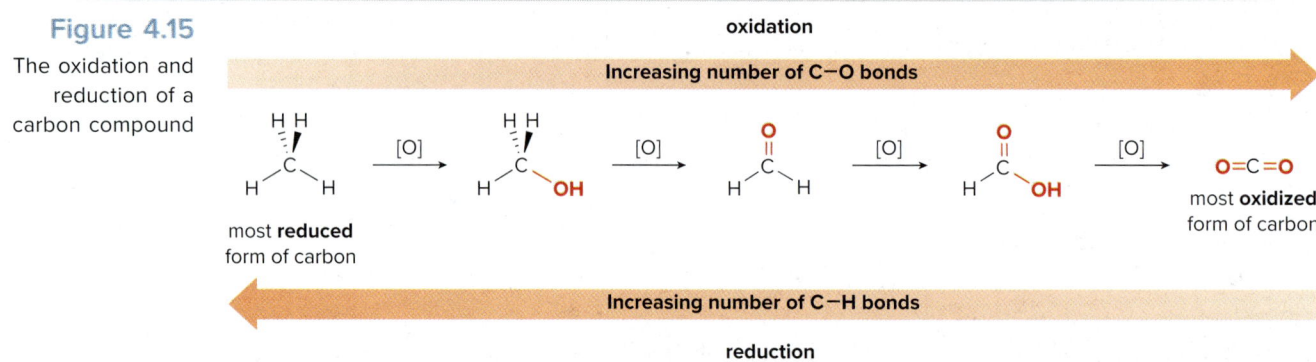

Sample Problem 4.5 Determining Whether a Compound Is Oxidized or Reduced

Determine whether the organic compound is oxidized or reduced in each transformation.

a. ethanol → acetic acid

b. cyclohexene → cyclohexane

Solution

a. The conversion of ethanol to acetic acid is an **oxidation** because the number of C–O bonds increases: CH_3CH_2OH has one C–O bond and CH_3COOH has three C–O bonds.

b. The conversion of cyclohexene (C_6H_{10}) to cyclohexane (C_6H_{12}) is a **reduction** because the number of C–H bonds increases: cyclohexane has two more C–H bonds than cyclohexene.

Problem 4.31 Classify each transformation as an oxidation, reduction, or neither.

a. benzaldehyde → benzoic acid

b. bicyclic ketone → bicyclic alkane (C=O → CH₂)

c. acetone → 2-hydroxypropan-2-ol (gem-diol, (CH₃)₂C(OH)₂)

d. cyclohexanone → cyclohexanol

More Practice: Try Problems 4.62, 4.64a.

4.14B Combustion of Alkanes

When an organic compound is *oxidized* by a reagent, the reagent itself is *reduced*. Similarly, when an organic compound is *reduced* by a reagent, the reagent is *oxidized*. **Organic chemists identify a reaction as an oxidation or reduction by what happens to the *organic* component of the reaction.**

Alkanes undergo **combustion**—that is, **they burn in the presence of oxygen to form carbon dioxide and water.** This is a practical example of oxidation. Every C–H and C–C bond in the starting material is converted to a C–O bond in the product. The products, $CO_2 + H_2O$, are the same, regardless of the identity of the starting material. Combustion of alkanes in the form of natural gas, gasoline, or heating oil releases energy for heating homes, powering vehicles, and cooking food.

$$CH_4 \text{ (methane)} + 2\,O_2 \xrightarrow{\text{flame}} CO_2 + 2\,H_2O + \text{(heat) energy}$$

$$2 \text{ 2,2,4-trimethylpentane (isooctane)} + 25\,O_2 \xrightarrow{\text{flame}} 16\,CO_2 + 18\,H_2O + \text{(heat) energy}$$

reduced starting material → **oxidized** product

Combustion requires a spark or a flame to initiate the reaction. Gasoline, therefore, which is composed largely of alkanes, can be safely handled and stored in the air, but the presence of a spark or match causes immediate and violent combustion.

Driving an automobile 10,000 miles at 25 miles per gallon releases ~10,000 lb of CO_2 into the atmosphere.

The combustion of alkanes and other hydrocarbons obtained from fossil fuels adds a tremendous amount of CO_2 to the atmosphere each year. Quantitatively, data show over a 25% increase in the atmospheric concentration of CO_2 in the last 60 years (from 315 parts per million in 1958 to 406 parts per million in 2018; Figure 4.16). Although the composition of the atmosphere has changed over the lifetime of the earth, this may be the first time that the actions of humankind have altered that composition significantly and so quickly.

An increased CO_2 concentration in the atmosphere may have long-range and far-reaching effects. CO_2 absorbs thermal energy that normally radiates from the earth's surface, and redirects it back to the surface. Higher levels of CO_2 may therefore contribute to an increase in the average temperature of the earth's atmosphere. The global climate change resulting from these effects may lead to melting of the polar ice caps, a rise in sea level, and many more unforeseen consequences.

166 Chapter 4 Alkanes

Figure 4.16
The changing concentration of CO_2 in the atmosphere since 1958

Source: US Department of Commerce, National Oceanic & Atmospheric Administration, "Welcome to Mauna Loa Observatory!"

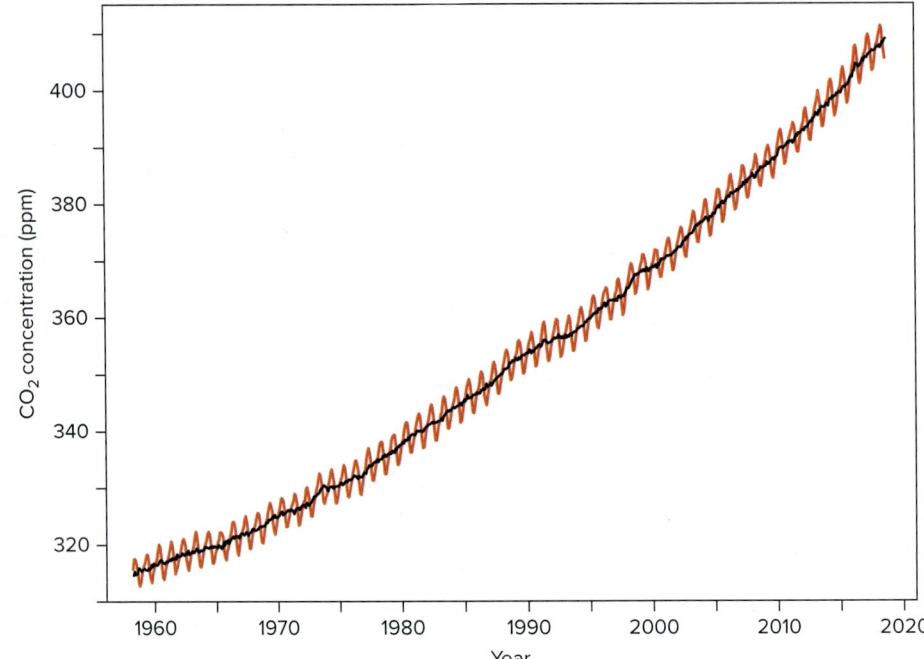

- The increasing level of atmospheric CO_2 is clearly evident on the graph. Two data points are recorded each year. The sawtooth nature of the graph is due to seasonal variation of CO_2 level with the seasonal variation in photosynthesis. (Data recorded at Mauna Loa, Hawai'i)

Problem 4.32 Draw the products of each combustion reaction.

a. (isobutane) + O_2 $\xrightarrow{flame}$

b. (cyclohexane) + O_2 $\xrightarrow{flame}$

Chapter 4 REVIEW

KEY CONCEPTS

[1] General facts about alkanes

① Molecular formula (4.1, 4.2)		② Geometry and hybridization (4.1)	③ Intermolecular forces (4.8)
Acyclic alkanes	**Cyclic alkanes**	• tetrahedral • sp^3 hybridized • all 109.5° bond angles neopentane (109.5°)	• weak **van der Waals forces** • **low boiling point** and **melting point**, increasing as the number of carbons increases • decreasing boiling point with branching • increasing melting point with symmetry
• C_nH_{2n+2} • saturated hydrocarbons pentane — C_5H_{12} 2,4,5-trimethylheptane — $C_{10}H_{22}$	• C_nH_{2n} • two fewer H atoms than acyclic alkanes cyclopentane — C_5H_{10} ethylcyclooctane — $C_{10}H_{20}$		

See Table 4.2. Try Problems 4.40, 4.41.

Key Concepts

[2] Names of alkyl groups (4.4A)

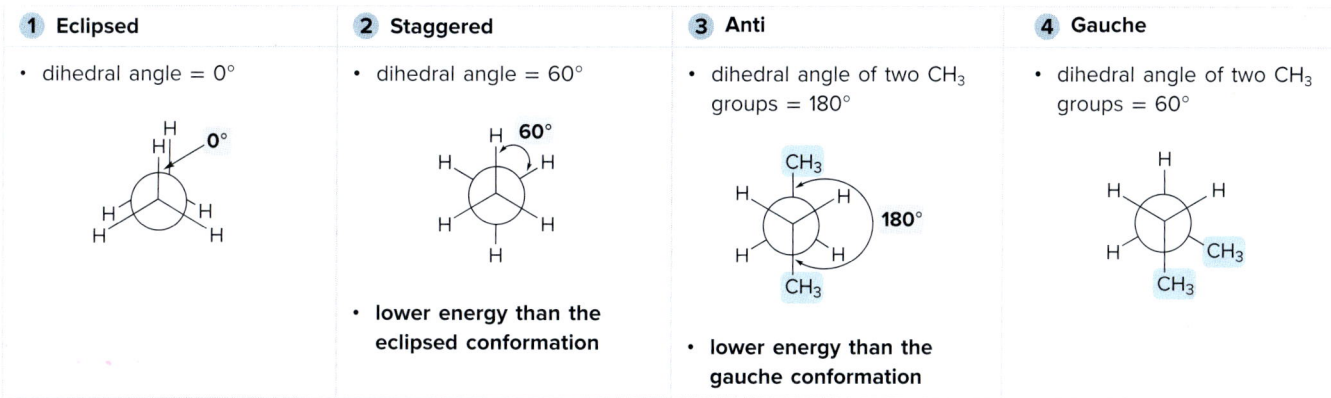

1	Methyl CH₃–	3	Propyl CH₃CH₂CH₂–	5	Butyl CH₃CH₂CH₂CH₂–	7	Isobutyl (CH₃)₂CHCH₂–
2	Ethyl CH₃CH₂–	4	Isopropyl (CH₃)₂CH–	6	sec-Butyl CH₃CH₂CHCH₃	8	tert-Butyl (CH₃)₃C–

[3] Conformations of acyclic alkanes (4.9, 4.10)

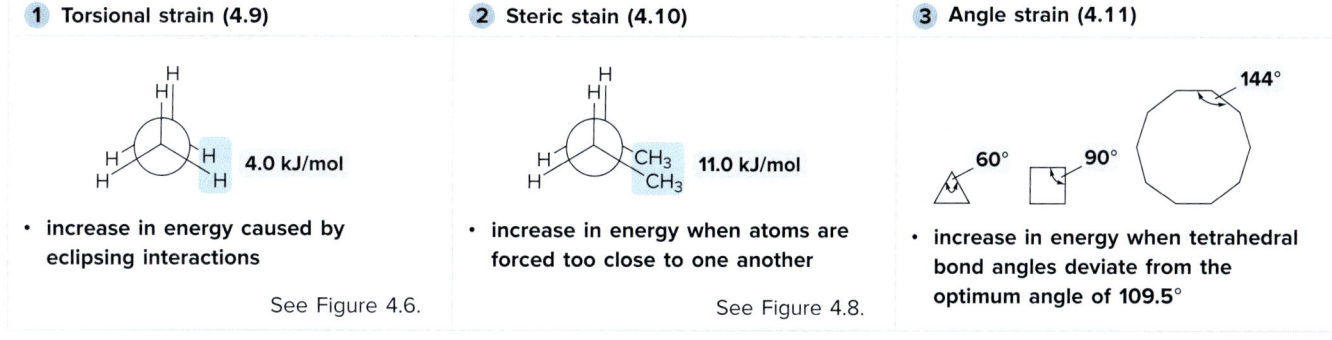

1 Eclipsed	2 Staggered	3 Anti	4 Gauche
• dihedral angle = 0°	• dihedral angle = 60°	• dihedral angle of two CH₃ groups = 180°	• dihedral angle of two CH₃ groups = 60°
	• lower energy than the eclipsed conformation	• lower energy than the gauche conformation	

See Figure 4.7. Try Problems 4.34, 4.43, 4.44, 4.46.

[4] Types of strain

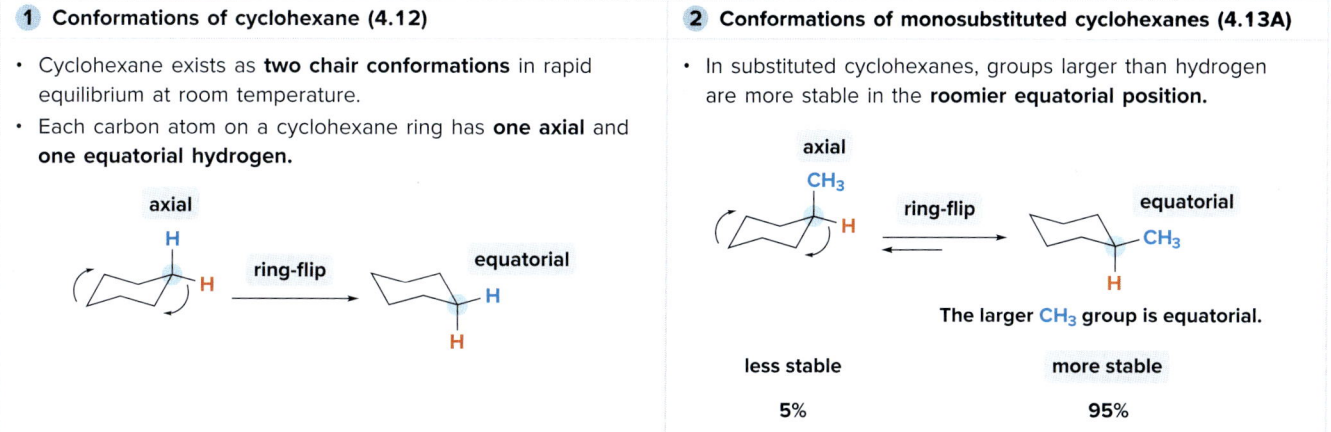

1 Torsional strain (4.9)	2 Steric stain (4.10)	3 Angle strain (4.11)
4.0 kJ/mol	11.0 kJ/mol	
• increase in energy caused by eclipsing interactions	• increase in energy when atoms are forced too close to one another	• increase in energy when tetrahedral bond angles deviate from the optimum angle of 109.5°
See Figure 4.6.	See Figure 4.8.	

Try Problem 4.48.

[5] Chair cyclohexane and monosubstituted cyclohexanes

1 Conformations of cyclohexane (4.12)

- Cyclohexane exists as **two chair conformations** in rapid equilibrium at room temperature.
- Each carbon atom on a cyclohexane ring has **one axial** and **one equatorial hydrogen**.

2 Conformations of monosubstituted cyclohexanes (4.13A)

- In substituted cyclohexanes, groups larger than hydrogen are more stable in the **roomier equatorial position**.

The larger CH₃ group is equatorial.

less stable 5%   more stable 95%

See *How To*'s p. 155, p. 157, Figures 4.11, 4.13. Try Problem 4.51a.

168 Chapter 4 Alkanes

[6] Disubstituted cyclohexanes (4.13C)

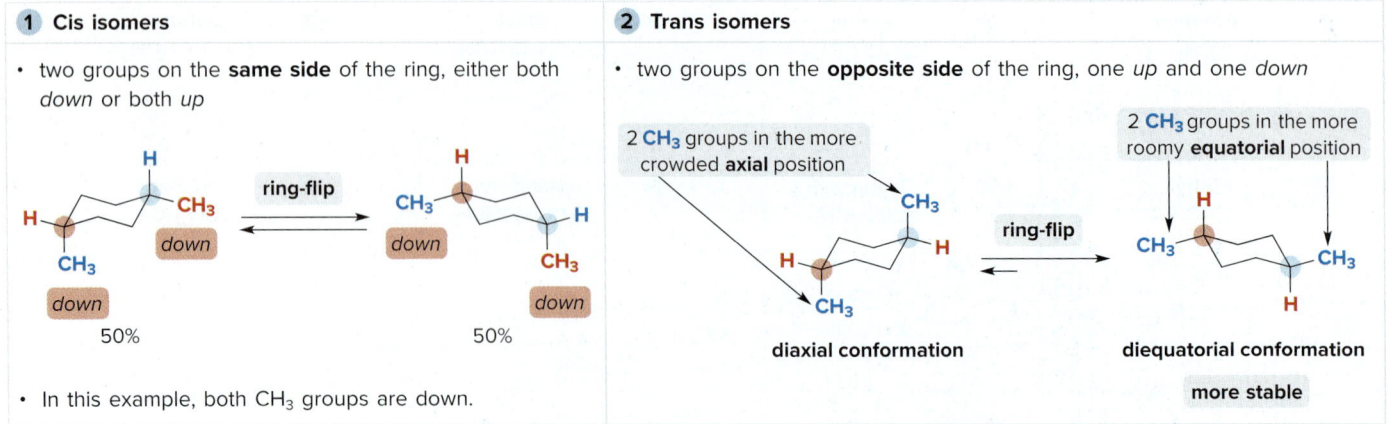

① Cis isomers	**② Trans isomers**
• two groups on the **same side** of the ring, either both *down* or both *up*	• two groups on the **opposite side** of the ring, one *up* and one *down*
• In this example, both CH₃ groups are down.	

See *How To* p. 160, Figure 4.14, Sample Problem 4.3. Try Problems 4.51–4.53, 4.60b.

[7] Two types of isomers

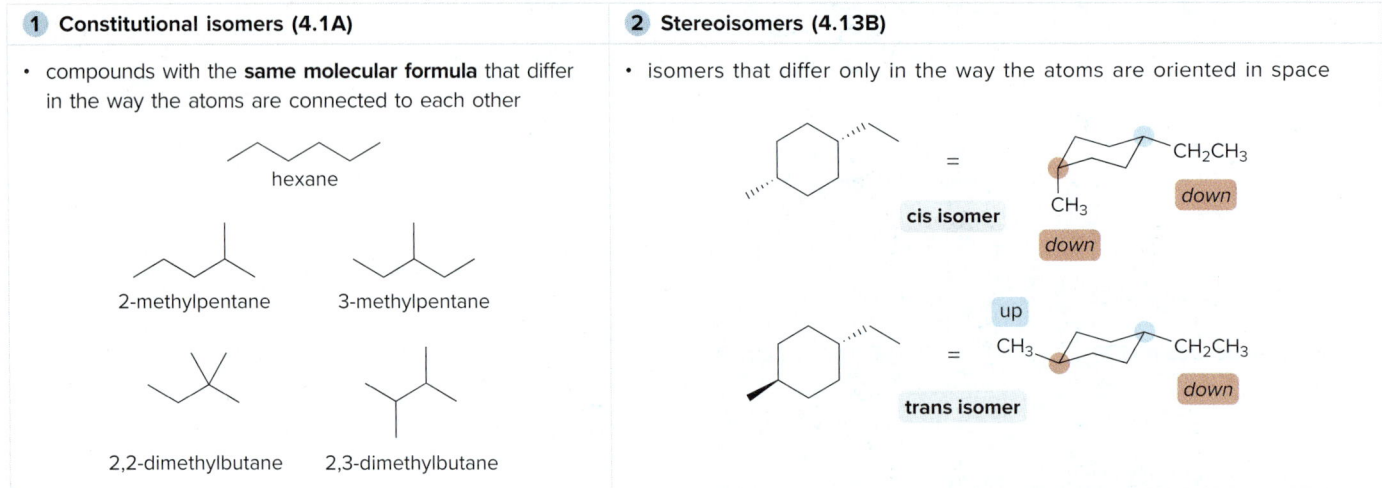

① Constitutional isomers (4.1A)	**② Stereoisomers (4.13B)**
• compounds with the **same molecular formula** that differ in the way the atoms are connected to each other	• isomers that differ only in the way the atoms are oriented in space

Try Problems 4.35, 4.53, 4.57, 4.59, 4.60, 4.61.

KEY SKILLS

[1] Naming an alkane using the IUPAC system (4.4)

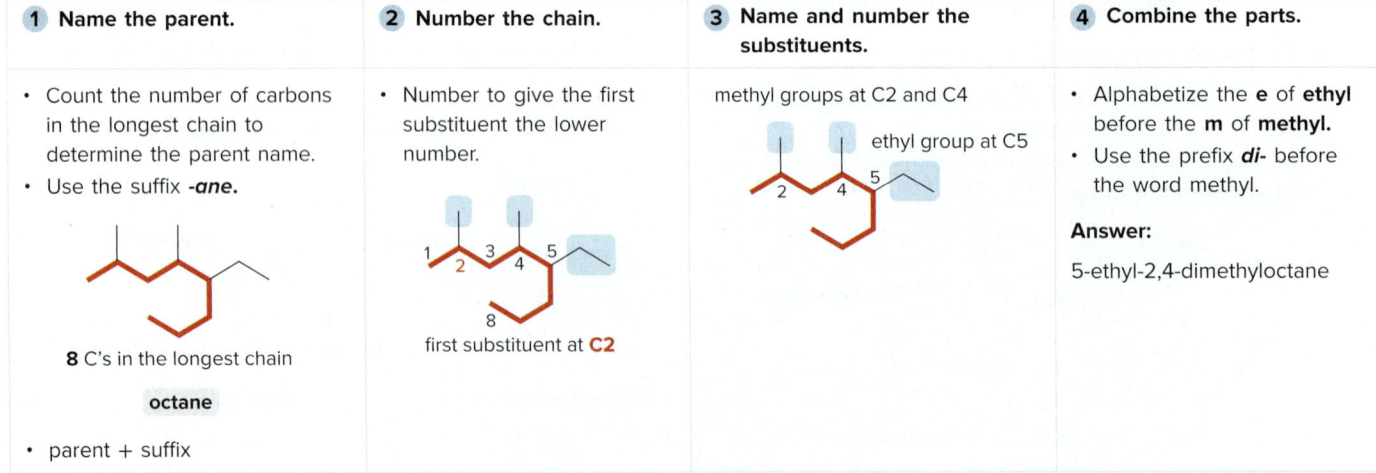

① Name the parent.	**② Number the chain.**	**③ Name and number the substituents.**	**④ Combine the parts.**
• Count the number of carbons in the longest chain to determine the parent name. • Use the suffix **-ane**. 8 C's in the longest chain octane • parent + suffix	• Number to give the first substituent the lower number. first substituent at **C2**	methyl groups at C2 and C4 ethyl group at C5	• Alphabetize the **e** of **ethyl** before the **m** of **methyl**. • Use the prefix **di-** before the word methyl. **Answer:** 5-ethyl-2,4-dimethyloctane

See Table 4.1, *How To* p. 138, Sample Problem 4.1, Figure 4.1. Try Problems 4.36a–d, h, j; 4.39.

[2] Naming a cycloalkane using the IUPAC system (4.5)

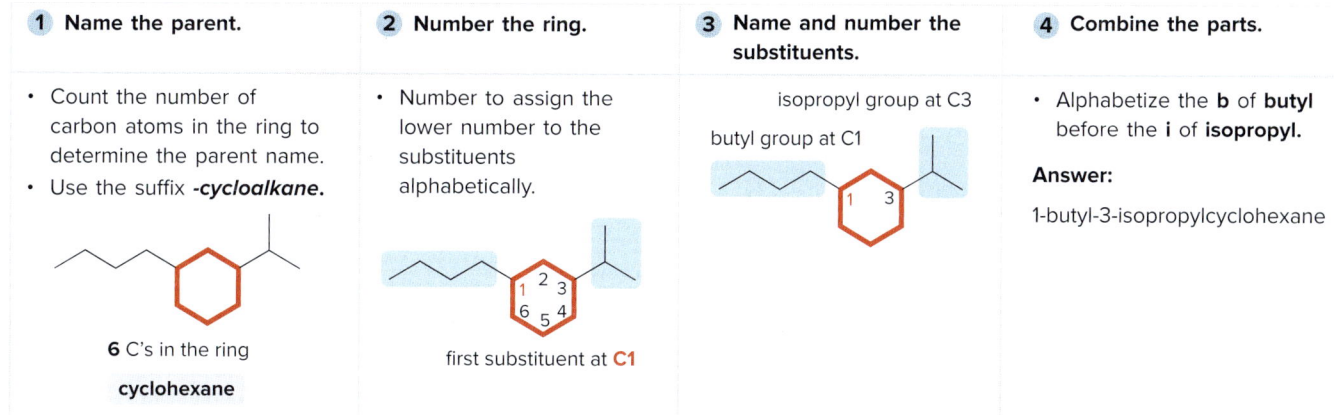

See *How To* p. 142, Figure 4.3. Try Problem 4.36e–g, i.

[3] Determining the highest- and lowest-energy conformations using Newman projections (4.10)

1 Identify the groups around the C–C bond.

- In ClCH$_2$CH$_2$Cl, each **C** is bonded to one **Cl** atom and two **H** atoms.

2 Draw the three eclipsed conformations.

- Begin with an eclipsed conformation, and rotate the groups on the front C atom 120° in the clockwise direction.

- The conformation with the **largest groups eclipsed** has the **highest energy**.

3 Draw the three staggered conformations.

- Begin with a staggered conformation, and rotate the groups on the front C atom 120° in the clockwise direction.

- The conformation with the largest groups anti and the **fewest gauche interactions** has the **lowest energy**.

See *How To* p. 148, Figures 4.4, 4.5, 4.7, 4.8. Try Problems 4.43–4.46.

[4] Drawing two conformations for a disubstituted cyclohexane (4.13C); example: cis-1-tert-butyl-2-methylcyclohexane

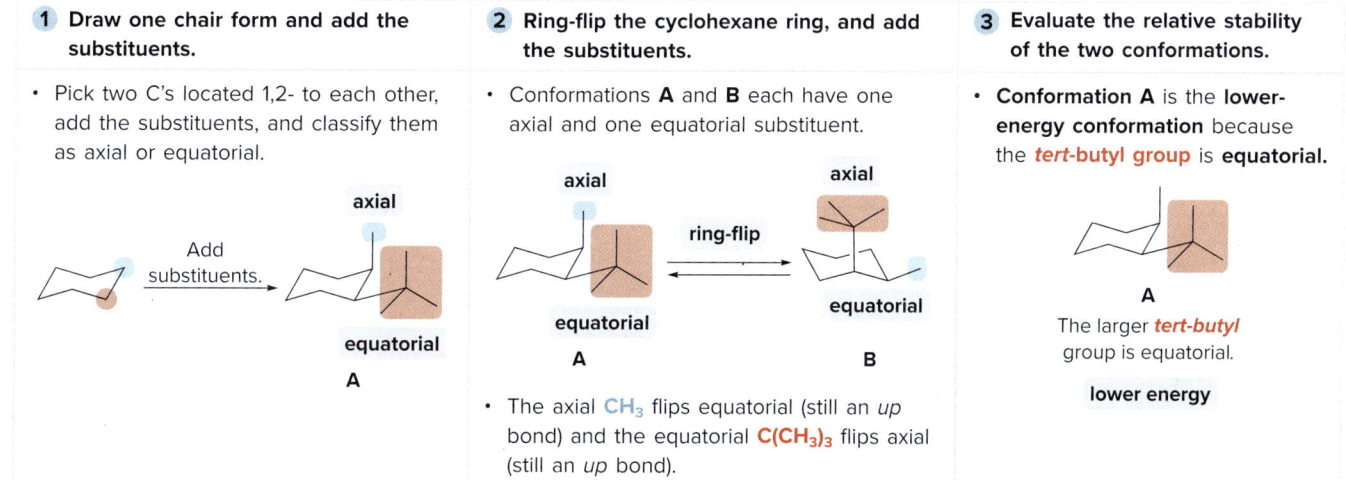

- The axial CH$_3$ flips equatorial (still an *up* bond) and the equatorial C(CH$_3$)$_3$ flips axial (still an *up* bond).

See *How To* p. 160, Figure 4.14, Sample Problem 4.3. Try Problems 4.51, 4.53, 4.60.

170 Chapter 4 Alkanes

[5] Determining whether a compound is oxidized or reduced (4.14A)

① Count the number of C—O bonds in the starting material and compare to the product.

② Count the number of C—H bonds in the starting material and compare to the product.

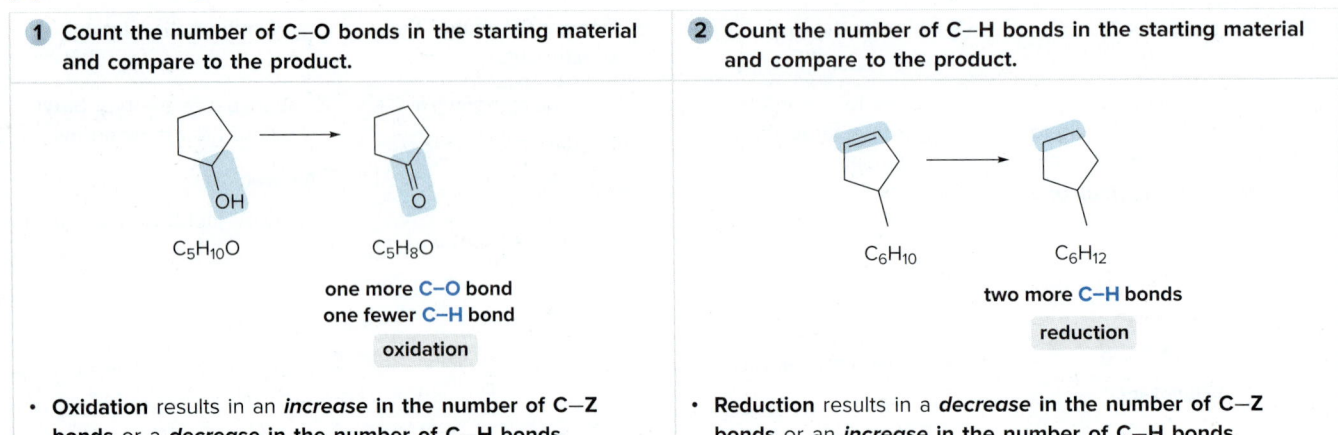

- Oxidation results in an *increase* in the number of C—Z bonds or a *decrease* in the number of C—H bonds.
- Reduction results in a *decrease* in the number of C—Z bonds or an *increase* in the number of C—H bonds.

See Sample Problem 4.5. Try Problems 4.62, 4.64a.

PROBLEMS

Problems Using Three-Dimensional Models

4.33 Consider the substituted cyclohexane shown in the ball-and-stick model.

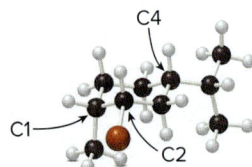

  a. Label the substituents on C1, C2, and C4 as axial or equatorial.
  b. Are the substituents on C1 and C2 cis or trans to each other?
  c. Are the substituents on C2 and C4 cis or trans to each other?
  d. Draw the second possible conformation in the chair form, and classify it as more stable or less stable than the conformation shown in the three-dimensional model.

4.34 Convert each three-dimensional model to a Newman projection around the indicated bond.

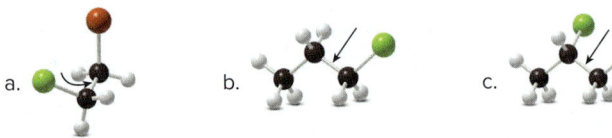

Constitutional Isomers

4.35 Draw the structure of all compounds that fit the following descriptions.
  a. five constitutional isomers having the molecular formula C_4H_8
  b. nine constitutional isomers having the molecular formula C_7H_{16}
  c. twelve constitutional isomers having the molecular formula C_6H_{12} and containing one ring

IUPAC Nomenclature

4.36 Give the IUPAC name for each compound.

4.37 Draw the structure corresponding to each IUPAC name.
   a. 3-ethyl-2-methylhexane
   b. *sec*-butylcyclopentane
   c. 4-isopropyl-2,4,5-trimethylundecane
   d. cyclobutylcycloheptane
   e. 3-ethyl-1,1-dimethylcyclohexane
   f. 4-butyl-1,1-diethylcyclooctane
   g. 6-isopropyl-2,3-dimethyldodecane
   h. 2,2,6,6,7-pentamethyloctane
   i. *cis*-1-ethyl-3-methylcyclopentane
   j. *trans*-1-*tert*-butyl-4-ethylcyclohexane

4.38 Draw the structure of each alkane and cycloalkane from the given incorrect name. Then, give the IUPAC name for each compound.
   a. 7-ethyl-3,6-dimethylnonane
   b. 4-ethyl-3-isopropylheptane
   c. 3-ethyl-1,4-dimethylcycloheptane
   d. 1-ethyl-3-methyl-5-isopropylcyclohexane

4.39 Give the IUPAC name for each compound.

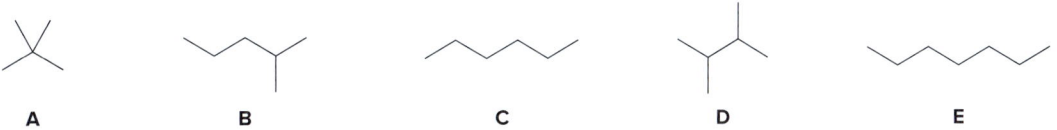

Properties of Alkanes

Students who have already learned about mass spectrometry can try Problems A.1, A.7, A.8, and A.14. Students who have already learned about infrared spectroscopy can try Problem B.12a.

4.40 Rank the following alkanes in order of increasing boiling point.

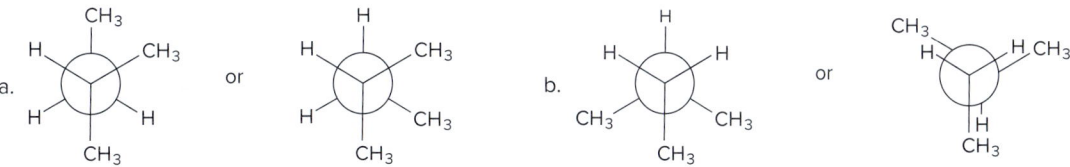

4.41 The melting points and boiling points of two isomeric alkanes are as follows: $CH_3(CH_2)_6CH_3$, mp = −57 °C and bp = 126 °C; $(CH_3)_3CC(CH_3)_3$, mp = 102 °C and bp = 106 °C. (a) Explain why one isomer has a lower melting point but higher boiling point. (b) Explain why there is a small difference in the boiling points of the two compounds, but a huge difference in their melting points.

4.42 Mineral oil, a mixture of high-molecular-weight alkanes, is sometimes used as a laxative. Why are individuals who use mineral oil for this purpose advised to avoid taking it at the same time they consume foods rich in fat-soluble vitamins such as vitamin A?

Conformation of Acyclic Alkanes

4.43 Which conformation in each pair is *higher* in energy? Calculate the energy difference between the two conformations using the values given in Table 4.3.

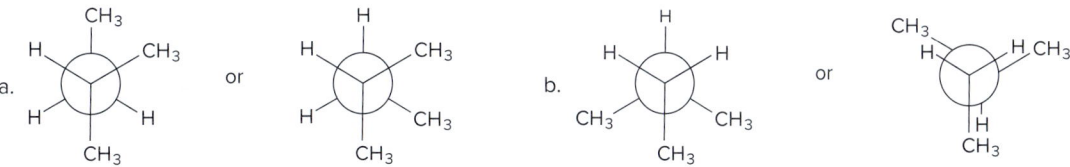

4.44 Considering rotation around the bond highlighted in red in each compound, draw Newman projections for the most stable and least stable conformations.

4.45 Rank the following Newman projections in order of increasing energy.

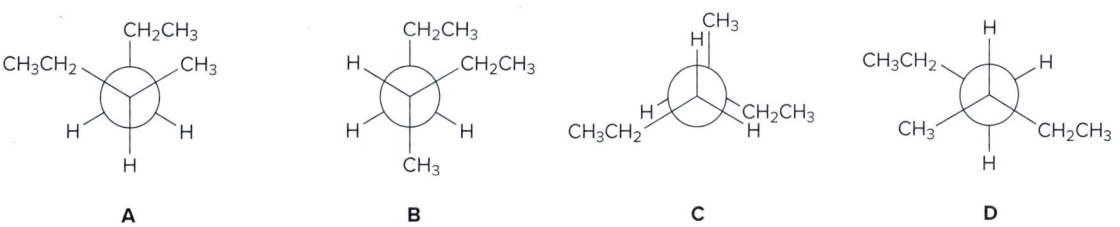

4.46 Classify each conformation as staggered or eclipsed around the indicated bond, and rank the conformations in order of increasing stability.

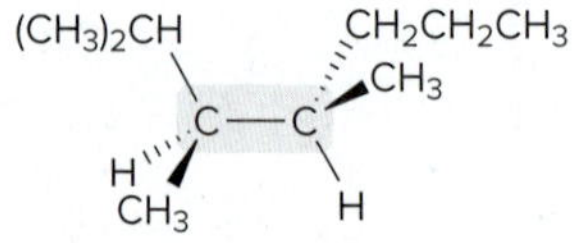

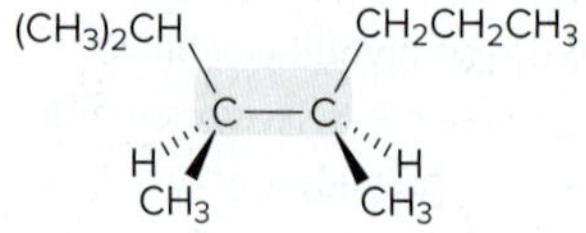

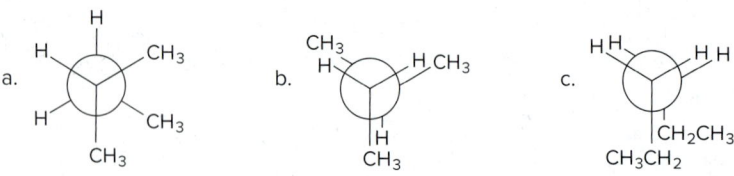

4.47 (a) Using Newman projections, draw all staggered and eclipsed conformations that result from rotation around the bond highlighted in red in each molecule; (b) draw a graph of energy versus dihedral angle for rotation around this bond.

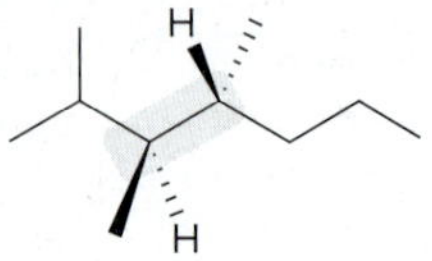

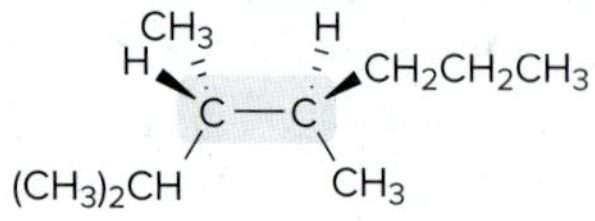

4.48 Label the sites of torsional and steric strain in each conformation.

4.49 Calculate the barrier to rotation for each bond highlighted in red.

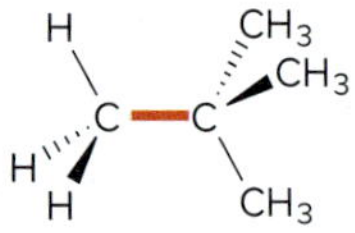

4.50 The eclipsed conformation of CH_3CH_2Cl is 15 kJ/mol less stable than the staggered conformation. How much is the H,Cl eclipsing interaction worth in destabilization?

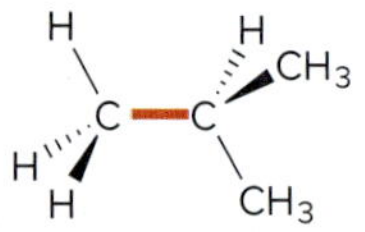

Conformations and Stereoisomers in Cycloalkanes

4.51 For each compound drawn below:
  a. Label each OH, Br, and CH_3 group as axial or equatorial.
  b. Classify each conformation as cis or trans.
  c. Translate each structure into a representation with a hexagon for the six-membered ring, and wedges and dashed wedges for groups above and below the ring.
  d. Draw the second possible chair conformation for each compound.

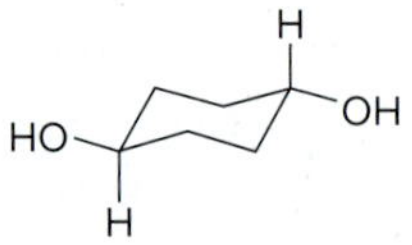

4.52 Draw the more stable chair conformation for each compound.
  a. *trans*-1-isopropyl-3-methylcyclohexane
  b. *cis*-1-*sec*-butyl-4-ethylcyclohexane
  c. *cis*-1-ethyl-2-isobutylcyclohexane
  d. *trans*-1,2-dibutylcyclohexane

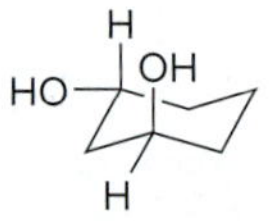

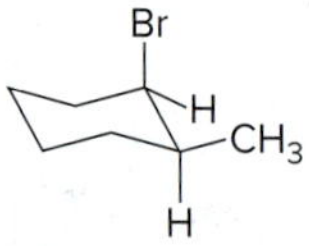

4.53 For each compound drawn below:
  a. Draw representations for the cis and trans isomers using a hexagon for the six-membered ring, and wedges and dashed wedges for substituents.
  b. Draw the two possible chair conformations for the cis isomer. Which conformation, if either, is more stable?
  c. Draw the two possible chair conformations for the trans isomer. Which conformation, if either, is more stable?
  d. Which isomer, cis or trans, is more stable and why?

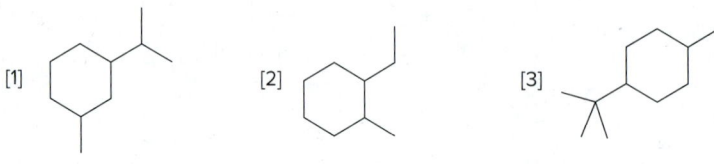

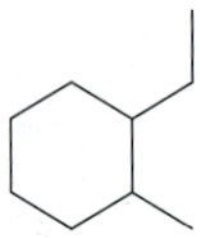

4.54 Draw the more stable chair conformation for each trisubstituted cyclohexane.

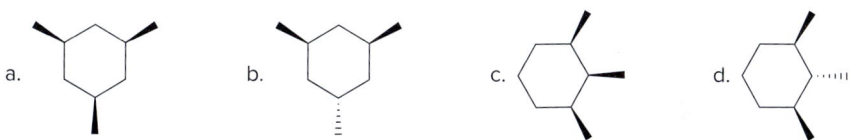

4.55 Answer the following questions about compound **A**, which contains a CH_3 group and OH group bonded to the carbon skeleton that consists of three six-membered rings in the conformation shown.

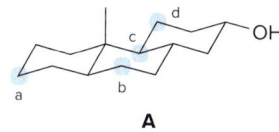

a. Are the CH_3 and OH groups oriented cis or trans to each other?
b. Is a substituent on C_a that is cis to the CH_3 group located in the axial or equatorial position?
c. Is an equatorial Br at C_b oriented cis or trans to the OH group?
d. Is the H atom on C_c located cis or trans to the OH group?
e. Is a substituent on C_d that is trans to the OH group located in the axial or equatorial position?

4.56 Glucose is a simple sugar with five substituents bonded to a six-membered ring.

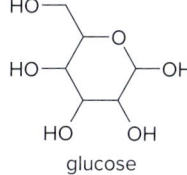

a. Using a chair representation, draw the most stable arrangement of these substituents on the six-membered ring.
b. Convert this representation to one that uses a hexagon with wedges and dashed wedges.
c. Draw a constitutional isomer of glucose.
d. Draw a stereoisomer that has an axial OH group on one carbon.

Constitutional Isomers and Stereoisomers

4.57 Classify each pair of compounds as constitutional isomers, stereoisomers, identical molecules, or not isomers of each other.

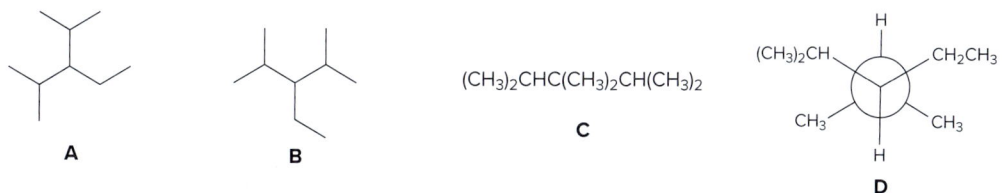

4.58 (a) Are compounds **B–D** identical to or an isomer of **A**? (b) Give the IUPAC name for **A**.

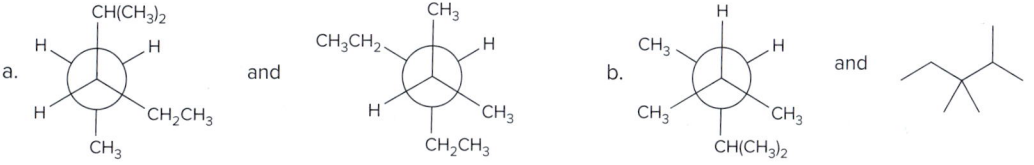

4.59 Classify each pair of compounds as constitutional isomers or identical molecules.

174 Chapter 4 Alkanes

4.60 Answer the following questions about compounds **A–D**.

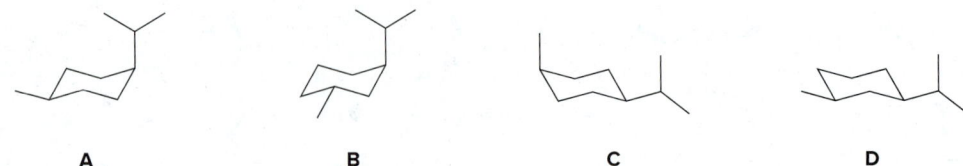

      A   B   C   D

    a. How are the compounds in each pair related? Choose from constitutional isomers, stereoisomers, or identical molecules: **A** and **B**; **A** and **C**; **B** and **D**.
    b. Label each compound as a cis or trans isomer.
    c. Draw **B** as a hexagon with wedges and dashed wedges to show the stereochemistry of substituents.
    d. Draw a stereoisomer of **A** as a hexagon using wedges and dashed wedges to show the orientation of substituents.

4.61 Draw the three constitutional isomers having molecular formula C_7H_{14} that contain a five-membered ring and two methyl groups as substituents. For each constitutional isomer that can have cis and trans isomers, draw the two stereoisomers.

Oxidation and Reduction

4.62 Classify each reaction as oxidation, reduction, or neither.

4.63 Draw the products of combustion of each alkane.

4.64 Hydrocarbons like benzene are metabolized in the body to arene oxides, which rearrange to form phenols. This is an example of a general process in the body, in which an unwanted compound (benzene) is converted to a more water-soluble derivative called a *metabolite,* so that it can be excreted more readily from the body.

    benzene   arene oxide   phenol

    a. Classify each of these reactions as oxidation, reduction, or neither.
    b. Explain why phenol is more water soluble than benzene. This means that phenol dissolves in urine, which is largely water, to a greater extent than benzene.

Challenge Problems

4.65 Cyclopropane and cyclobutane have similar strain energy despite the fact that the C–C–C bond angles of cyclopropane are much smaller than those of cyclobutane. Suggest an explanation for this observation, considering all sources of strain discussed in Chapter 4.

4.66 Although penicillin G has two amide functional groups, one is much more reactive than the other. Which amide is more reactive and why?

penicillin G

4.67 Haloethanes (CH$_3$CH$_2$X, X = Cl, Br, I) have similar barriers to rotation (13.4–15.5 kJ/mol) despite the fact that the size of the halogen increases, Cl → Br → I. Offer an explanation.

4.68 When two six-membered rings share a C—C bond, this bicyclic system is called a **decalin.** There are two possible arrangements: *trans*-decalin having two hydrogen atoms at the ring fusion on opposite sides of the rings, and *cis*-decalin having the two hydrogens at the ring fusion on the same side.

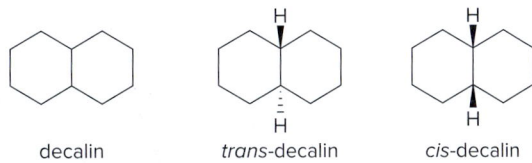

a. Draw *trans*- and *cis*-decalin using the chair form for the cyclohexane rings.
b. The trans isomer is more stable. Explain why.

4.69 Consider the tricyclic structure **B**. (a) Label each substituent on the rings as axial or equatorial. (b) Draw **B** using chair conformations for each six-membered ring. (c) Label the atoms on the ring fusions (the carbons that join each set of two rings together) as cis or trans to each other.

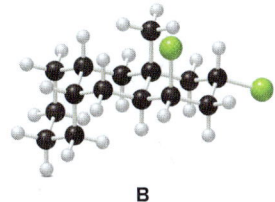

4.70 Read Appendix D on naming branched alkyl substituents, and draw all possible alkyl groups having the formula C$_5$H$_{11}$–. Give the IUPAC names for the eight compounds of molecular formula C$_{10}$H$_{20}$ that contain a cyclopentane ring with each of these alkyl groups as a substituent.

4.71 Read Appendix D on naming bicyclic compounds. Then give the IUPAC name for each of the following compounds.

a.  b. 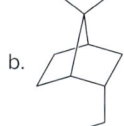 c.  d.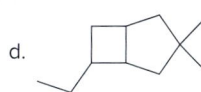

5 Stereochemistry

5.1 Starch and cellulose
5.2 The two major classes of isomers
5.3 Looking glass chemistry—Chiral and achiral molecules
5.4 Stereogenic centers
5.5 Stereogenic centers in cyclic compounds
5.6 Labeling stereogenic centers with *R* or *S*
5.7 Diastereomers
5.8 Meso compounds
5.9 *R* and *S* assignments in compounds with two or more stereogenic centers
5.10 Disubstituted cycloalkanes
5.11 Isomers—A summary
5.12 Physical properties of stereoisomers
5.13 Chemical properties of enantiomers

George Ostertag/Alamy Stock Photo

Paclitaxel (trade name Taxol), a potent anticancer agent active against ovarian, breast, and several other cancers, was discovered in 1962 and approved for use by the Food and Drug Administration in 1992. Initial studies with paclitaxel were carried out with material isolated from the bark of the Pacific yew tree, but stripping the bark killed these magnificent trees. Paclitaxel was synthesized in the laboratory in 1994, and is now produced by a plant cell fermentation process. Like other widely used drugs, paclitaxel is biologically active because of its complex structure and the particular three-dimensional arrangement of its functional groups. In Chapter 5, we learn about the stereochemistry of molecules like paclitaxel.

Why Study... Stereochemistry?

Are you left-handed or right-handed? If you're right-handed, you've probably spent little time thinking about your hand preference. If you're left-handed, though, you probably learned at an early age that many objects—like scissors and baseball gloves—"fit" for righties, but are "backwards" for lefties. **Hands, like many objects in the world around us, are mirror images that are *not* identical.**

In Chapter 5, we examine the "handedness" of molecules, and learn about the importance of the three-dimensional shape of a molecule.

5.1 Starch and Cellulose

Recall from Chapter 4 that *stereochemistry* is the three-dimensional structure of a molecule. How important is stereochemistry? Two biomolecules—starch and cellulose—illustrate how apparently minute differences in structure can result in vastly different properties.

Carbohydrates were introduced in Section 3.9B.

Starch and **cellulose** are carbohydrate polymers (Figure 5.1). **A *polymer* is a large molecule composed of repeating smaller units—called monomers—that are covalently bonded together.**

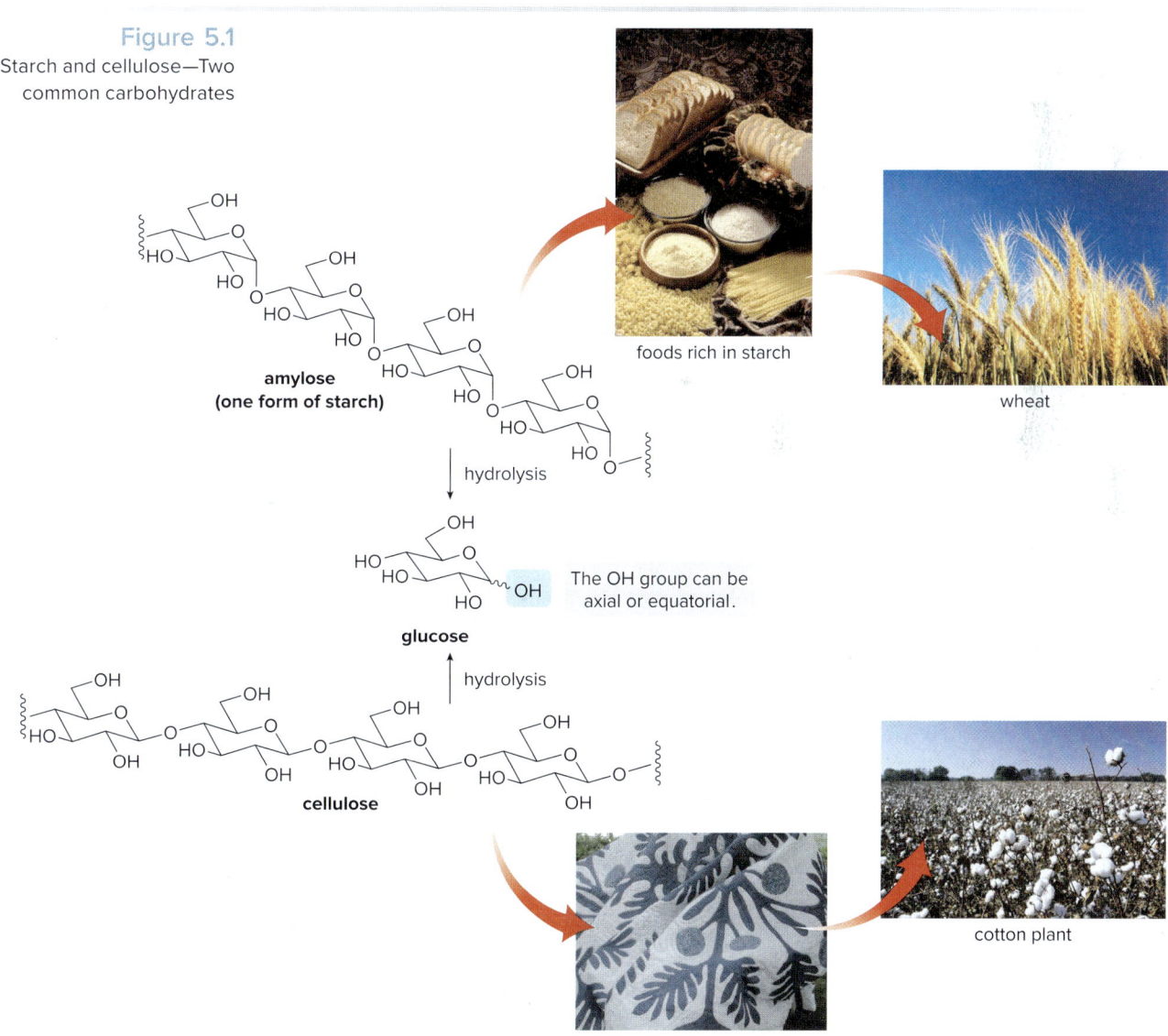

Figure 5.1
Starch and cellulose—Two common carbohydrates

(Top left): Source: Keith Weller/USDA; (top right): Bryan Mullennix/Pixtal/age fotostock; (bottom left): Daniel C. Smith; (bottom right): David Frazier/Corbis

Starch is the main carbohydrate in the seeds and roots of plants. When we humans ingest wheat, rice, or potatoes, we consume starch, which is then hydrolyzed to the simple sugar **glucose,** one of the compounds our bodies use for energy. **Cellulose,** nature's most abundant organic material, gives rigidity to tree trunks and plant stems. Wood, cotton, and flax are composed largely of cellulose. Complete hydrolysis of cellulose also forms glucose, but unlike starch, humans cannot metabolize cellulose to glucose. In other words, we can digest starch but not cellulose.

Cellulose and starch are both composed of the same repeating unit—a six-membered ring containing an oxygen atom and three OH groups—joined by an oxygen atom. They differ in the position of the O atom joining the rings together.

- In cellulose, the O atom joins two rings using two equatorial bonds.
- In starch, the O atom joins two rings using one equatorial and one axial bond.

Starch and cellulose are **isomers** because they are different compounds with the same molecular formula $(C_6H_{10}O_5)_n$. They are **stereoisomers** because only the three-dimensional arrangement of atoms is different.

How the six-membered rings are joined together has an enormous effect on the shape and properties of these carbohydrate molecules. Cellulose is composed of long chains held together by intermolecular hydrogen bonds, forming sheets that stack in an extensive three-dimensional network. The axial–equatorial ring junction in starch creates chains that fold into a helix (Figure 5.2). Moreover, the human digestive system contains the enzyme necessary to hydrolyze starch by cleaving its axial C−O bond, but not an enzyme to hydrolyze the equatorial C−O bond in cellulose.

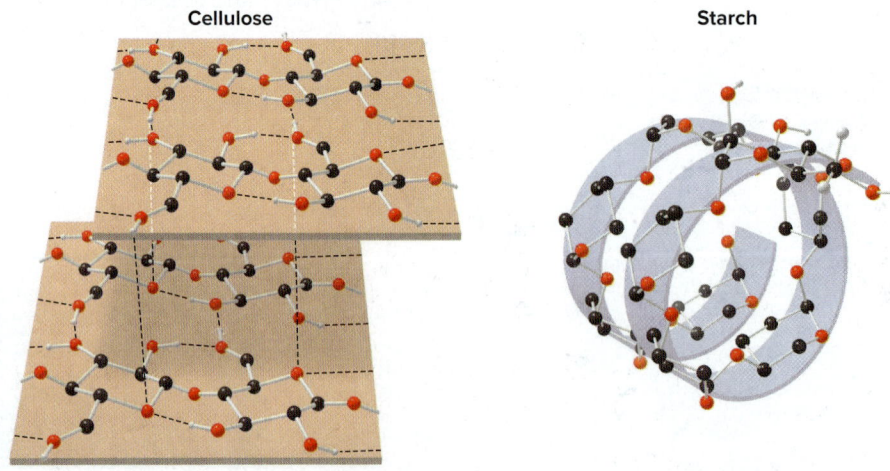

Figure 5.2
Three-dimensional structure of cellulose and starch

- Cellulose consists of an extensive three-dimensional network held together by hydrogen bonds.
- The starch polymer is composed of chains that wind into a helix.

5.2 The Two Major Classes of Isomers

Because an understanding of isomers is integral to the discussion of stereochemistry, let's begin with an overview of isomers.

- Isomers are different compounds with the same molecular formula.

There are two major classes of isomers: **constitutional isomers** and **stereoisomers**. **Constitutional (or structural) isomers differ in the way the atoms are connected to each other.** Constitutional isomers have

- different IUPAC names;
- the same or different functional groups;
- different physical properties, so they are separable by physical techniques such as distillation; and
- different chemical properties. They behave differently or give different products in chemical reactions.

Stereoisomers differ *only* in the way atoms are oriented in space. Stereoisomers have identical IUPAC names (except for a prefix like cis or trans). Because they differ only in the three-dimensional arrangement of atoms, stereoisomers always have the same functional group(s).

A particular three-dimensional arrangement is called a *configuration*. Thus, stereoisomers differ in configuration. The cis and trans isomers in Section 4.13B and the biomolecules starch and cellulose in Section 5.1 are two examples of stereoisomers.

Figure 5.3 illustrates examples of both types of isomers. Chapter 5 concentrates on the types and properties of stereoisomers.

Problem 5.1 Classify each pair of compounds as constitutional isomers or stereoisomers.

a. 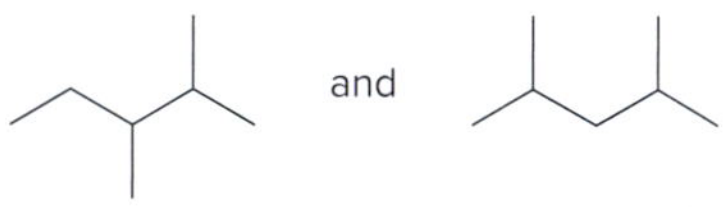 [structure] and [structure]   c. 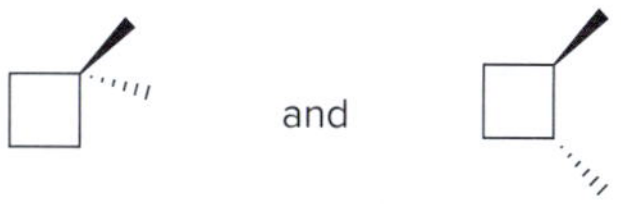 [structure] and [structure]

b. [structure] and [structure]   d.  [structure] and [structure]

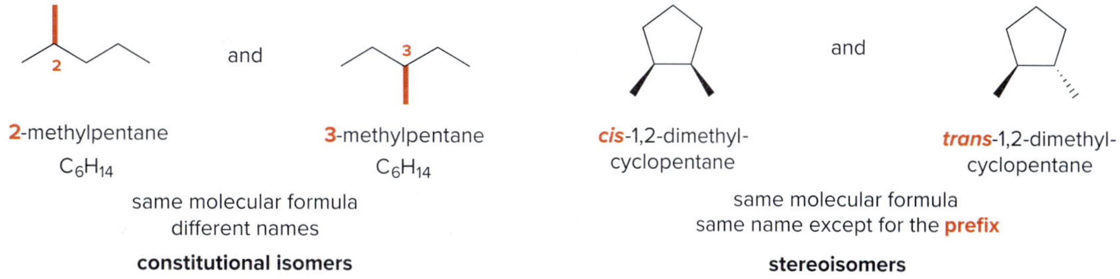

Figure 5.3 A comparison of constitutional isomers and stereoisomers

2-methylpentane C_6H_{14} and   3-methylpentane C_6H_{14}

same molecular formula
different names
constitutional isomers

cis-1,2-dimethyl-cyclopentane   and   *trans*-1,2-dimethyl-cyclopentane

same molecular formula
same name except for the **prefix**
stereoisomers

5.3 Looking Glass Chemistry—Chiral and Achiral Molecules

Everything has a mirror image. What's significant is **whether a molecule is *identical* to or *different* from its mirror image.**

Some molecules are like hands. **Left and right hands are mirror images of each other, but they are *not* identical.** If you try to mentally place one hand inside the other hand, you can never superimpose either all the fingers, or the tops and palms. To *superimpose* an object on its mirror image means to align *all* parts of the object with its mirror image. With molecules, this means aligning all atoms and all bonds.

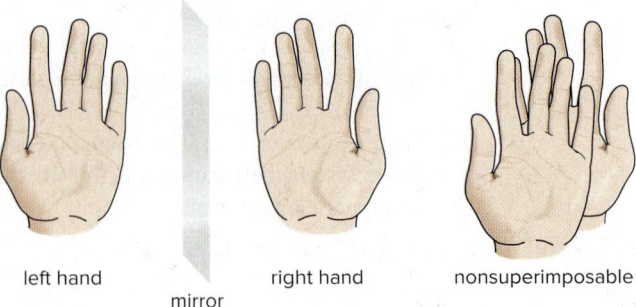

left hand — mirror — right hand — nonsuperimposable

- **A molecule (or object) that is *not* superimposable on its mirror image is said to be *chiral*.**

Other molecules are like socks. **Two socks from a pair are mirror images that *are* superimposable.** One sock can fit inside another, aligning toes and heels, and tops and bottoms. A sock and its mirror image are *identical*.

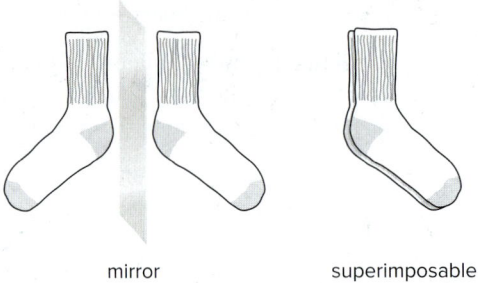

mirror — superimposable

- **A molecule (or object) that *is* superimposable on its mirror image is said to be *achiral*.**

The adjective **chiral** comes from the Greek word *cheir*, meaning "hand." Left and right hands are **chiral:** they are mirror images that do *not* superimpose on each other.

Few beginning students of organic chemistry can readily visualize whether a compound and its mirror image are superimposable by looking at drawings on a two-dimensional page. Molecular models can help a great deal in this process.

Let's determine whether three molecules—H_2O, CH_2BrCl, and $CHBrClF$—are superimposable on their mirror images; that is, **are H_2O, CH_2BrCl, and $CHBrClF$ chiral or achiral?**

To test chirality:

- Draw the molecule in three dimensions.
- Draw its mirror image.
- Try to align all bonds and atoms. To superimpose a molecule and its mirror image, you can perform any rotation but **you cannot break bonds.**

Following this procedure, H_2O and CH_2BrCl are both **achiral** molecules because each molecule is superimposable on its mirror image.

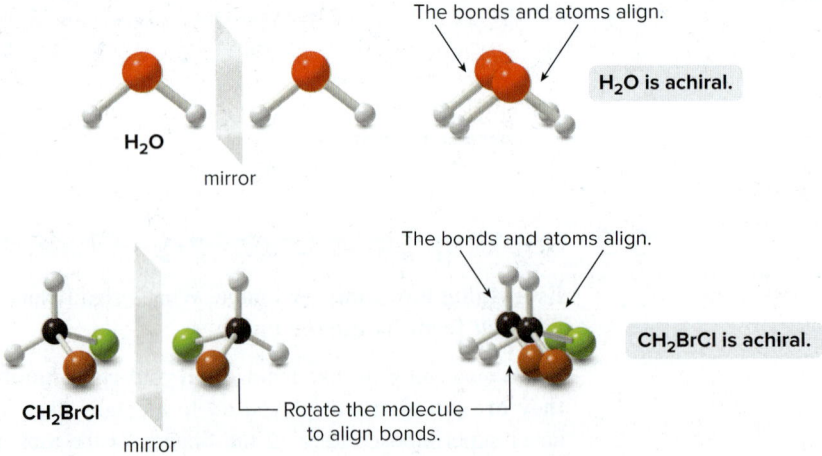

5.3 Looking Glass Chemistry—Chiral and Achiral Molecules

With CHBrClF, the result is different. The molecule (labeled **A**) and its mirror image (labeled **B**) are *not* superimposable. No matter how you rotate **A** and **B**, all the atoms never align. **CHBrClF is thus a chiral molecule,** and **A** and **B** are different compounds.

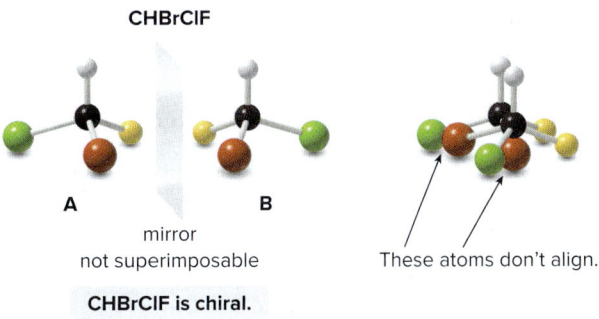

A and **B** are **stereoisomers** because they are isomers differing only in the three-dimensional arrangement of substituents. These stereoisomers are called **enantiomers.**

- *Enantiomers* are mirror images that are not superimposable.

CHBrClF contains a carbon atom bonded to four different groups. **A carbon atom bonded to four different groups is called a tetrahedral *stereogenic center*.** Most chiral molecules contain one or more stereogenic centers.

The general term *stereogenic center* refers to any site in a molecule at which the interchange of two groups forms a stereoisomer. **A carbon atom with four different groups is a *tetrahedral* stereogenic center,** because the interchange of two groups converts one enantiomer into another. We will learn about another type of stereogenic center in Section 8.2B.

We have now learned two related but different concepts, and it is necessary to distinguish between them.

- A molecule that is not superimposable on its mirror image is a *chiral molecule*.
- A carbon atom bonded to four different groups is a *stereogenic center*.

Molecules can contain zero, one, or more stereogenic centers.

- **With no stereogenic centers, a molecule generally is not chiral.** H_2O and CH_2BrCl have *no* stereogenic centers and are *achiral* molecules. (There are a few exceptions to this generalization, as we will learn in Section 19.5.)
- **With one tetrahedral stereogenic center, a molecule is *always* chiral.** CHBrClF is a *chiral* molecule containing *one* stereogenic center.
- **With two or more stereogenic centers, a molecule *may* or *may not* be chiral,** as we will learn in Section 5.8.

> Naming a carbon atom with four different groups is a topic that currently has no firm agreement among organic chemists. The IUPAC recommends the term *chirality center,* but the term has not gained wide acceptance among organic chemists since it was first suggested in 1996. Other terms in common use are chiral center, chiral carbon, asymmetric carbon, stereocenter, and stereogenic center, the term used in this text.

Problem 5.2 Draw the mirror image of each compound. Label each molecule as chiral or achiral.

When trying to distinguish between chiral and achiral compounds, keep in mind:

> - A *plane of symmetry* is a mirror plane that cuts a molecule in half, so that one half of the molecule is a reflection of the other half.
> - Achiral molecules usually contain a plane of symmetry, but chiral molecules do not.

The achiral molecule CH$_2$BrCl has a plane of symmetry, but the chiral molecule CHBrClF does not.

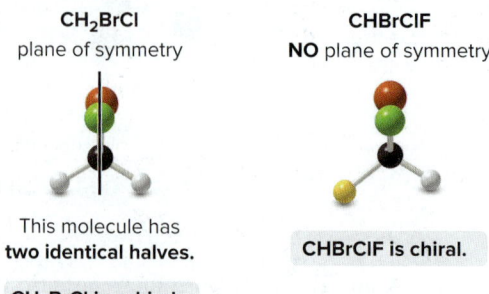

Figure 5.4 summarizes the main facts about chirality we have learned thus far.

Figure 5.4
The basic principles of chirality

- Everything has a mirror image. The fundamental question is whether a molecule and its mirror image are superimposable.
- If a molecule and its mirror image are *not* superimposable, the molecule and its mirror image are **chiral**.
- The terms **stereogenic center** and **chiral molecule** are related but distinct. In general, a chiral molecule must have one or more stereogenic centers.
- The presence of a **plane of symmetry** makes a molecule achiral.

Problem 5.3 Draw in a plane of symmetry for each molecule.

a. b. c. d.

Problem 5.4 A molecule is achiral if it has a plane of symmetry in *any* conformation. Each of the following conformations does not have a plane of symmetry, but rotation around a carbon–carbon bond forms a conformation that does have a plane of symmetry. Draw this conformation for each molecule.

a. b.

When a right-handed shell is held in the right hand with the thumb pointing toward the wider end, the opening is on the right side. *Jill Braaten/McGraw-Hill Education*

Stereochemistry may seem esoteric, but chirality pervades our very existence. On a molecular level, many biomolecules fundamental to life are chiral. On a macroscopic level, many naturally occurring objects possess handedness. Examples include chiral helical seashells shaped like right-handed screws, and plants such as honeysuckle that wind in a chiral left-handed helix. The human body is chiral, and hands, feet, and ears are not superimposable.

5.4 Stereogenic Centers

A necessary skill in the study of stereochemistry is the ability to locate and draw tetrahedral stereogenic centers.

5.4A Stereogenic Centers on Carbon Atoms That Are Not Part of a Ring

Recall from Section 5.3 that any carbon atom bonded to four different groups is a tetrahedral stereogenic center. To locate a stereogenic center, examine each *tetrahedral* carbon atom in a molecule, and look at the four **groups**—not the four *atoms*—bonded to it. CBrClFI has one stereogenic center because its central carbon atom is bonded to four different elements. 3-Bromohexane also has one stereogenic center because one carbon is bonded to H, Br, CH_2CH_3, and $CH_2CH_2CH_3$. We consider all atoms in a group as a *whole unit*, not just the atom bonded directly to the carbon in question. Although C3 of 3-bromohexane is bonded to two carbon atoms, one is part of an ethyl group and one is part of a propyl group.

Always omit from consideration all C atoms that can't be tetrahedral stereogenic centers. These include

- CH_2 and CH_3 groups (more than one H bonded to C); and
- any *sp* or *sp²* hybridized C (less than four groups around C).

Larger organic molecules can have two, three, or even hundreds of stereogenic centers. **Propoxyphene** and **ephedrine** each contain two stereogenic centers, and **fructose**, a simple carbohydrate, has three.

Ephedrine is isolated from ma huang, an herb used to treat respiratory ailments in traditional Chinese medicine. Once a popular drug to promote weight loss and enhance athletic performance, ephedrine has now been linked to episodes of sudden death, heart attack, and stroke. *Mark W. Skinner*

Sample Problem 5.1 Locating Stereogenic Centers

Locate the stereogenic centers in each drug. Albuterol is a bronchodilator—that is, it widens airways—so it is used to treat asthma. Chloramphenicol is an antibiotic used extensively in developing countries because of its low cost.

Heteroatoms surrounded by four different groups are also stereogenic centers. Stereogenic N atoms are discussed in Chapter 22.

Solution

Omit all CH$_2$ and CH$_3$ groups and all doubly bonded (*sp^2* hybridized) C's. In albuterol, one C has three CH$_3$ groups bonded to it, so it can be eliminated as well. Draw in H atoms on tetrahedral C's in skeletal structures to more clearly see the groups. This leaves one C in albuterol and two C's in chloramphenicol surrounded by four different groups, making them stereogenic centers.

a. 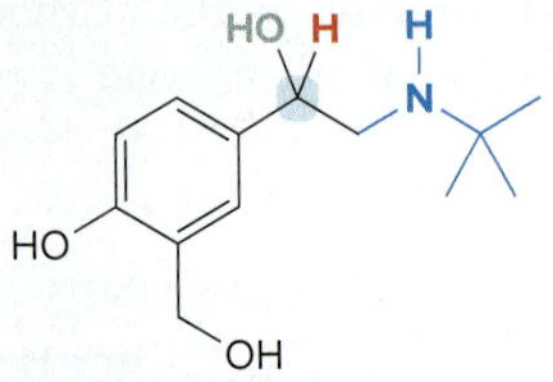 one stereogenic center

b. 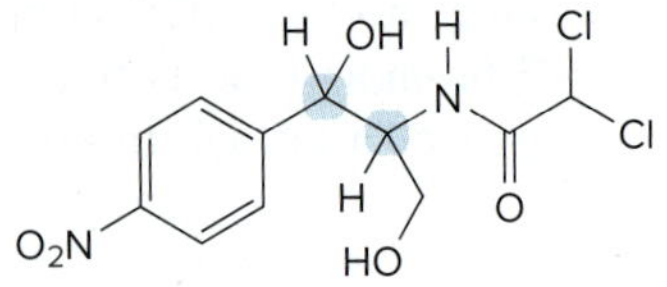 two stereogenic centers

---

Problem 5.5 Locate the stereogenic centers in each molecule. Compounds may have one or more stereogenic centers.

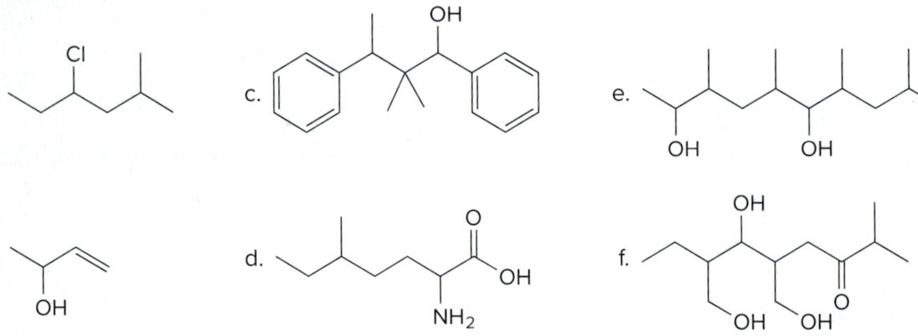

More Practice: Try Problem 5.41b, c, d.

Problem 5.6 The principles in Section 5.4A can be used to locate stereogenic centers in any molecule, no matter how complicated. Always look for carbons surrounded by four different groups. With this in mind, locate the four stereogenic centers in aliskiren, a drug introduced in 2007 for the treatment of hypertension.

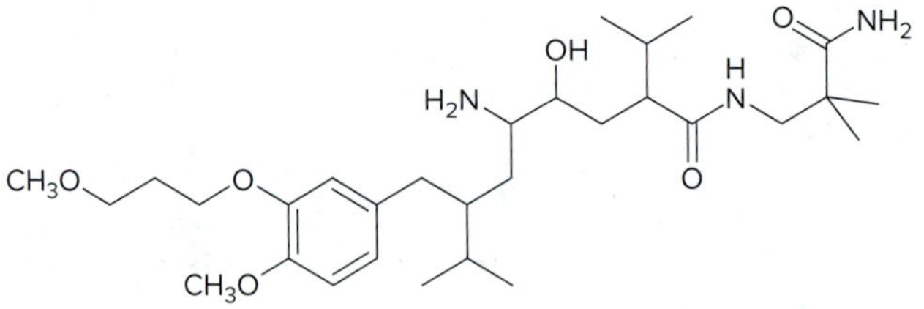

aliskiren

5.4B Drawing a Pair of Enantiomers

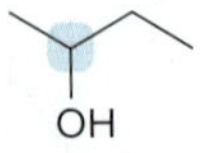

butan-2-ol
one stereogenic center

- Any molecule with one tetrahedral stereogenic center is a chiral compound and exists as a pair of enantiomers.

Butan-2-ol, for example, has one stereogenic center. To draw both enantiomers, use the typical convention for depicting a tetrahedron: **place two bonds in the plane, one in front of the**

plane on a wedge, and one behind the plane on a dashed wedge. Then, to form the first enantiomer **A,** arbitrarily place the four groups—H, OH, CH$_3$, and CH$_2$CH$_3$—on any bond to the stereogenic center.

> In Section 24.2, we will learn about Fischer projection formulas, an older convention used for drawing stereogenic centers utilized mainly in carbohydrate chemistry.

Draw the molecule...then the mirror image.

A | **B**

mirror
not superimposable
enantiomers

Then, draw a mirror plane and arrange the substituents in the mirror image so that they are a reflection of the groups in the first molecule, forming **B.** No matter how **A** and **B** are rotated, it is impossible to align all of their atoms. Because **A** and **B** are mirror images and not superimposable, **A** and **B** are a pair of **enantiomers.**

This is one way to draw an enantiomer, as shown in Figure 5.5a for 3-bromohexane. Another way to draw an enantiomer (Figure 5.5b), especially for compounds with more than one stereogenic center, is to keep the carbon skeleton in the *same* position, but *invert* the configuration at all stereogenic centers by converting bonds in front (on wedges) to bonds in back (on dashed wedges), and vice versa.

Figure 5.5 Different ways of drawing an enantiomer

a. Drawing an enantiomer as a reflection.

b. Drawing an enantiomer by inverting the configuration of a stereogenic center

3-bromohexane   enantiomer

- Groups on wedges and dashed wedges stay the *same*.
- The position of the C's in the long chain is *different*.
- Remember that H and Br are directly aligned.

3-bromohexane   enantiomer

- Groups on wedges and dashed wedges *interchange*.
- The position of the C's in the long chain stays the *same*.

The two representations labeled "enantiomer" are *identical,* just drawn in different ways.

Sample Problem 5.2 Different Ways of Drawing an Enantiomer

Locate the stereogenic center in the amino acid alanine, and draw the enantiomer using the two methods shown in Figure 5.5.

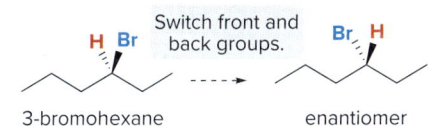

alanine

Solution

The stereogenic center is the carbon with four different groups, labeled in blue.

[1] Project a mirror plane and draw all groups on the stereogenic center as a reflection of the groups in alanine.

Draw all groups as a reflection:

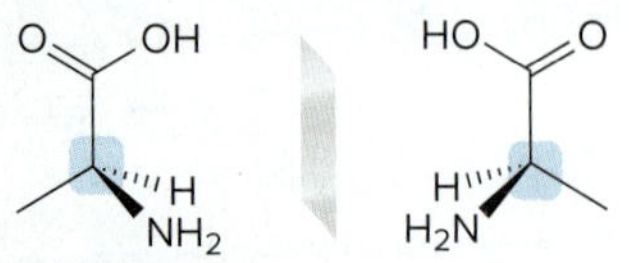

alanine enantiomer

- Groups in front and behind stay in the same position.

[2] Keep the carbon skeleton the same, and switch the position of groups that lie in front of and behind the plane.

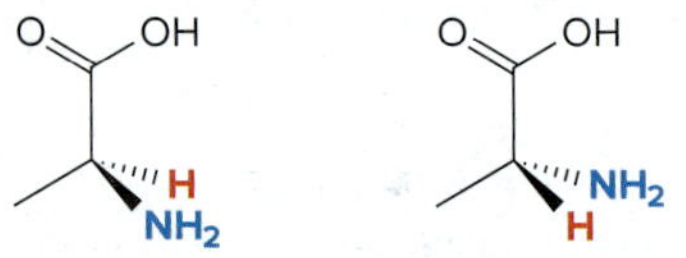

alanine enantiomer

- The H *behind* the plane becomes an H in *front* on a wedge.
- The NH_2 in *front* of the plane becomes an NH_2 in *back* on a dashed wedge.

Problem 5.7 Draw the enantiomer of each compound.

a. 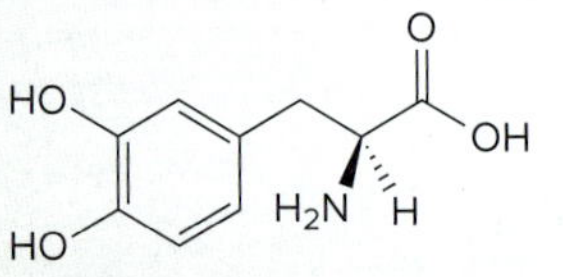

L-dopa
(drug to treat Parkinson's disease)

b. 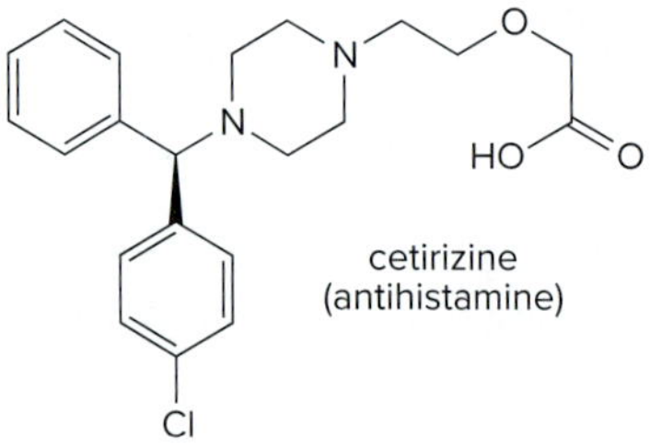

cetirizine
(antihistamine)

c. 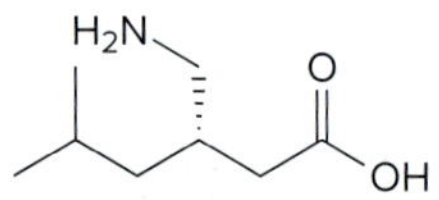

pregabalin
(drug used to treat chronic pain)

More Practice: Try Problems 5.43, 5.64b, 5.65c.

Problem 5.8 Locate the stereogenic center in each compound and draw both enantiomers.

a. 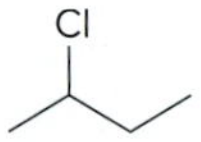

b. 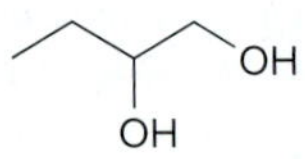

c. 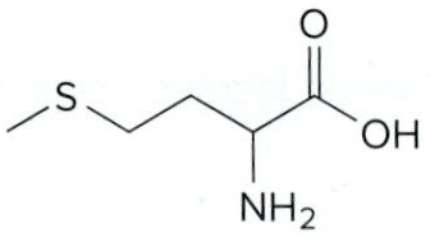

5.5 Stereogenic Centers in Cyclic Compounds

Stereogenic centers may also occur at carbon atoms that are part of a ring. To find stereogenic centers on ring carbons, always **draw the rings as flat polygons,** and look for tetrahedral carbons that are bonded to four different groups, as usual. Each ring carbon is bonded to two other atoms in the ring, as well as two substituents attached to the ring. When the two substituents on the ring are *different,* we must compare the ring atoms equidistant from the atom in question.

5.5 Stereogenic Centers in Cyclic Compounds

In drawing a tetrahedron using solid lines, wedges, and dashed wedges, always **draw the two solid lines first**; then draw the wedge and the dashed wedge on the **opposite side** of the solid lines.

If you draw the two solid lines *down*...

...then add the wedge and dashed wedge *above*.

If you draw the two solid lines to the *left*...

...then add the wedge and dashed wedge to the *right*.

Two enantiomers are *different* compounds. To convert one enantiomer to another, you must **switch the position of two atoms**. This amounts to breaking bonds.

Does methylcyclopentane have a stereogenic center? All of the carbon atoms are bonded to two or three hydrogen atoms except for C1, the ring carbon bonded to the methyl group. Next, compare the ring atoms and bonds on both sides equidistant from C1, and **continue until a point of difference is reached, or until both sides meet,** either at an atom or in the middle of a bond. In this case, there is no point of difference on either side, so C1 is bonded to identical alkyl groups that happen to be part of a ring. **C1, therefore, is *not* a stereogenic center.**

methylcyclopentane — two identical groups, equidistant from C1 — **C1** is *not* a stereogenic center.

With 3-methylcyclohexene, the result is different. All carbon atoms are bonded to two or three hydrogen atoms or are sp^2 hybridized except for C3, the ring carbon bonded to the methyl group. In this case, the atoms equidistant from C3 are different, so C3 is bonded to *different* alkyl groups in the ring. **C3 is therefore bonded to four different groups, making it a stereogenic center.**

3-methylcyclohexene — two different groups, equidistant from C3 — **C3** is a stereogenic center.

Because 3-methylcyclohexene has one tetrahedral stereogenic center, it is a chiral compound and exists as a pair of enantiomers.

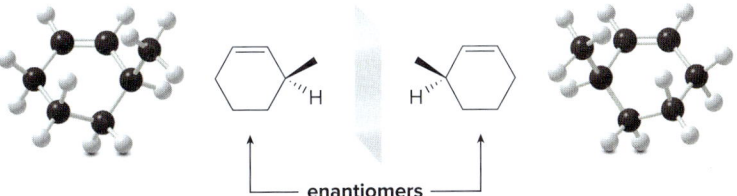

enantiomers

Many biologically active compounds contain one or more stereogenic centers on ring carbons. For example, **thalidomide,** a drug once prescribed as a sedative and anti-nausea agent for pregnant women in Great Britain and Europe, contains one stereogenic center, so it exists as a pair of enantiomers.

anti-nausea drug — teratogen
thalidomide
enantiomers

Today, thalidomide is prescribed under strict controls for the treatment of Hansen's disease (leprosy). Because it was once thought to be highly contagious, individuals in Hawaiʻi with Hansen's disease were sent to Kalaupapa, a remote and inaccessible peninsula on the north shore of the Hawaiian island of Molokaʻi. Hansen's disease is now known to be a curable bacterial infection, which is treated by the sulfa drugs discussed in Section 22.15.

Shallenberger Photography

Unfortunately thalidomide was sold as a mixture of its two enantiomers, and each of these stereoisomers has a different biological activity. This is a property not uncommon in chiral drugs, as we will see in Section 5.13A. Although one enantiomer was an effective sedative and anti-nausea drug, the other enantiomer was responsible for thousands of catastrophic birth defects in children born to women who took the drug during pregnancy. Thalidomide was never approved for use in the United States due to the diligence of Frances Oldham Kelsey, a medical reviewing officer for the Food and Drug Administration, who insisted that the safety data on thalidomide were inadequate.

NAD⁺ and **paclitaxel** (the chapter-opening molecule) are two useful compounds with several stereogenic centers at ring carbons. Identify the stereogenic centers in these more complicated compounds in exactly the same way, **looking at one carbon at a time.** NAD⁺ (nicotinamide adenine dinucleotide), with eight stereogenic centers, is a biological oxidizing agent that we will learn about in Sections 11.13 and 27.2A. Paclitaxel, with 11 stereogenic centers, is an anticancer agent active against ovarian, breast, and some lung tumors.

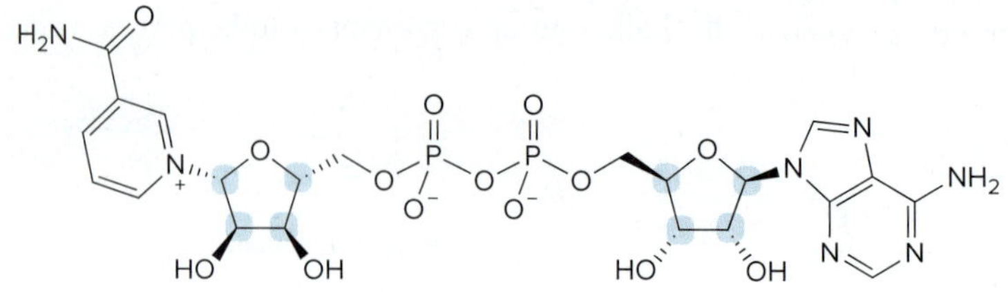

NAD⁺
biological oxidizing agent

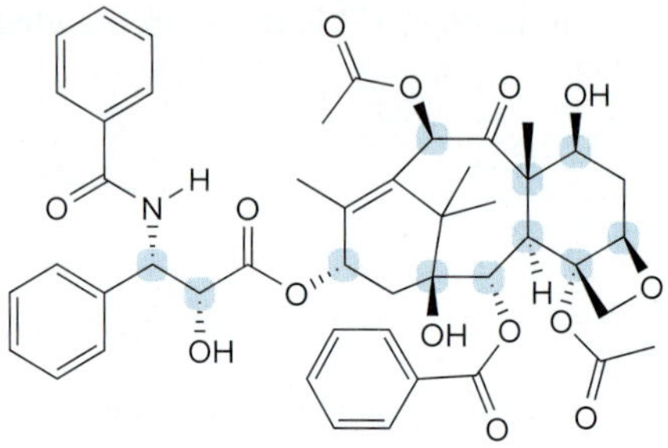

paclitaxel
Trade name Taxol
(anticancer agent)

To draw the enantiomer of a complex compound with many stereogenic centers, change all groups above the plane on wedges to dashed wedges, and all groups behind the plane on dashed wedges to wedges. For example, the inversion of configuration of all eight stereogenic centers in NAD⁺ forms its enantiomer, as shown.

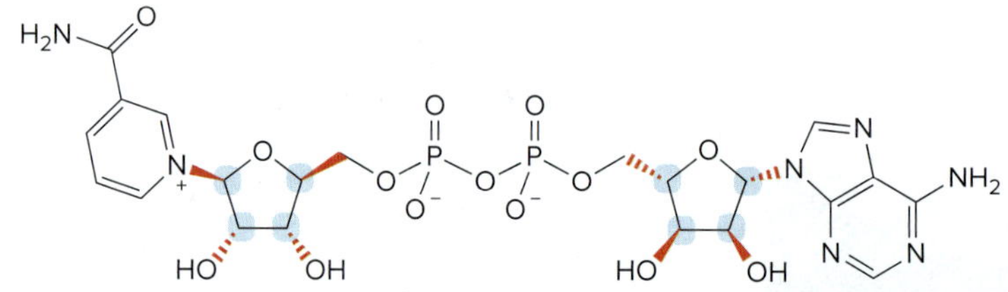

enantiomer of NAD⁺

Problem 5.9 Locate the stereogenic centers in each compound. A molecule may have one or more stereogenic centers. Gabapentin enacarbil [part (d)] is used to treat seizures and certain types of chronic pain.

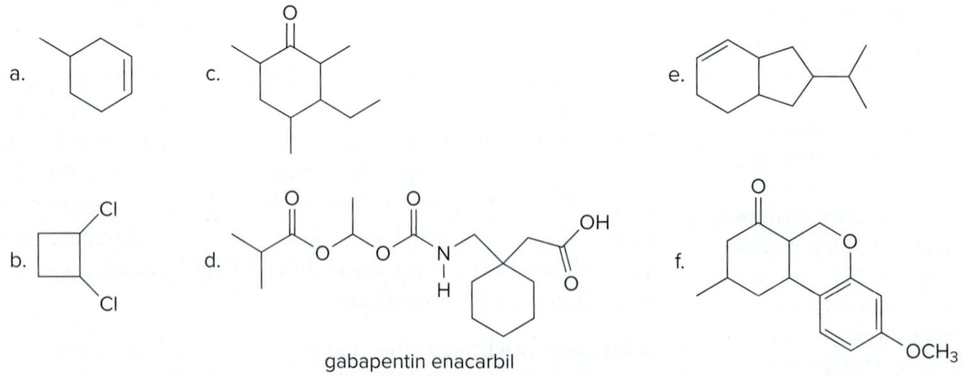

a.

b.

c.

d.
gabapentin enacarbil

e.

f.

Problem 5.10 Locate the stereogenic centers in each compound. Draw the enantiomer of exemestane in part (b).

a. simvastatin
Trade name Zocor
(cholesterol-lowering drug)

b. exemestane
Trade name Aromacin
(anticancer drug)

5.6 Labeling Stereogenic Centers with *R* or *S*

Naming enantiomers with the prefix *R* or *S* is called the Cahn–Ingold–Prelog system after the three chemists who devised it.

Because enantiomers are two different compounds, we need a method to distinguish them by name. This is done by adding the prefix ***R*** or ***S*** to the IUPAC name of the enantiomer. To designate an enantiomer as *R* or *S*, first **assign a priority** (1, 2, 3, or 4) to each group bonded to the stereogenic center, and then use these priorities to label one enantiomer *R* and one *S*.

Rules Needed to Assign Priority

Rule 1 Assign priorities (1, 2, 3, or 4) to the atoms directly bonded to the stereogenic center in order of *decreasing* atomic number. The atom of *highest* atomic number gets the *highest* priority (**1**).

- In CHBrClF, priorities are assigned as follows: Br (1, highest) → Cl (2) → F (3) → H (4, lowest). In many molecules the lowest-priority group will be H.

Rule 2 If two atoms on a stereogenic center are the *same*, assign priority based on the atomic number of the atoms bonded to these atoms. *One* atom of higher atomic number determines a higher priority.

- With butan-2-ol, the O atom gets highest priority (1) and H gets lowest priority (4) using Rule 1. Butan-2-ol also has two carbon atoms bonded to the stereogenic center, one that is part of a CH$_3$ group and one that is part of a CH$_2$CH$_3$ group. To assign priority (either 2 or 3) to the two C atoms, look at what atoms (other than the stereogenic center) are bonded to each C.

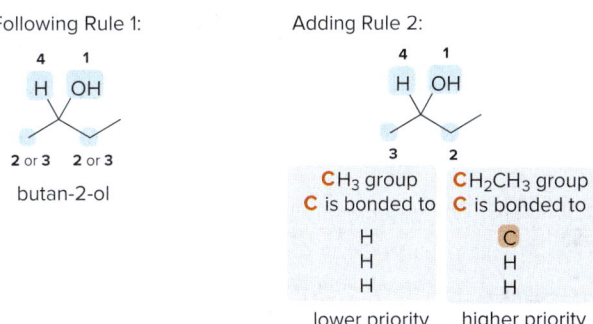

- The CH$_2$CH$_3$ gets higher priority (2) than the CH$_3$ group (priority 3) because the carbon of the ethyl group is bonded to another carbon.
- The order of priority of groups in butan-2-ol is —OH (**1**), —CH$_2$CH$_3$ (**2**), —CH$_3$ (**3**), and —H (**4**).
- If priority still cannot be assigned, continue along a chain until a point of difference is reached.

Rule 3
If two isotopes are bonded to the stereogenic center, assign priorities in order of *decreasing mass number*.

- In comparing two isotopes of the element hydrogen, deuterium, which has a mass number of two (one proton and one neutron), has a higher priority than hydrogen, which has a mass number of one (one proton only).

Rule 4
To assign a priority to an atom that is part of a multiple bond, treat a multiply bonded atom as an equivalent number of singly bonded atoms.

- The C of a C=O is considered to be bonded to two O atoms.

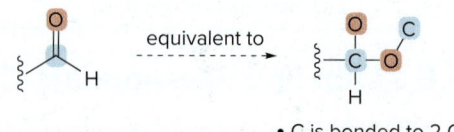

- C is bonded to 2 O's.
- O is bonded to 2 C's.

- Other common multiple bonds are drawn below.

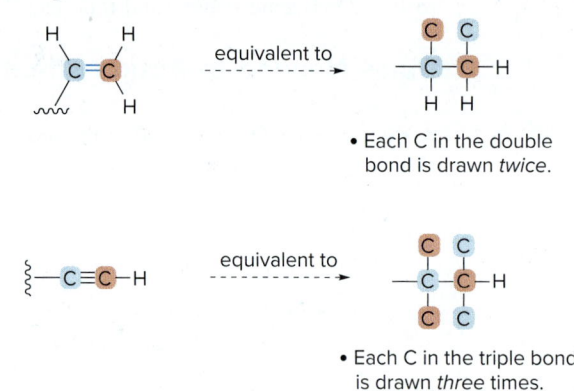

- Each C in the double bond is drawn *twice*.

- Each C in the triple bond is drawn *three* times.

Figure 5.6 gives examples of priorities assigned to stereogenic centers.

Figure 5.6 Examples of assigning priorities to stereogenic centers

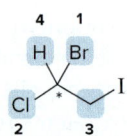

- The stereogenic center is bonded to Br, Cl, C, and H.
- The stereogenic center is *not* bonded directly to I.

- CH(CH$_3$)$_2$ gets the highest priority because the **C** is bonded to 2 other C's.

[* = stereogenic center]

- OH gets the highest priority because O has the highest atomic number.
- CO$_2$H (three bonds to O) gets higher priority than CH$_2$OH (one bond to O).

Problem 5.11 Which group in each pair is assigned the *higher* priority?
- a. –CH$_3$, –CH$_2$CH$_3$
- b. –I, –Br
- c. –H, –D
- d. –CH$_2$Br, –CH$_2$CH$_2$Br
- e. –CH$_2$CH$_2$Cl, –CH$_2$CH(CH$_3$)$_2$
- f. –CH$_2$OH, –CHO

Problem 5.12 Rank the following groups in order of *decreasing* priority.
- a. –COOH, –H, –NH$_2$, –OH
- b. –H, –CH$_3$, –Cl, –CH$_2$Cl
- c. –CH$_2$CH$_3$, –CH$_3$, –H, –CH(CH$_3$)$_2$
- d. –CH=CH$_2$, –CH$_3$, –C≡CH, –H

R is derived from the Latin word *rectus* meaning "right," and *S* is from the Latin word *sinister* meaning "left."

Once priorities are assigned to the four groups around a stereogenic center, we can use three steps to designate the center as either *R* or *S*.

How To Assign *R* or *S* to a Stereogenic Center

Example Label each enantiomer as *R* or *S*.

A **B**

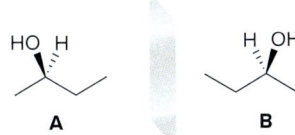

two enantiomers of butan-2-ol

Step [1] **Assign priorities from 1 to 4 to each group bonded to the stereogenic center.**

- The priorities for the four groups around the stereogenic center in butan-2-ol were given in Rule 2, on page 189.

$-OH$	$-CH_2CH_3$	$-CH_3$	$-H$
1	2	3	4
highest			lowest

Decreasing priority →

Step [2] **Orient the molecule with the lowest-priority group (4) *back* (on a *dashed wedge*), and visualize the relative positions of the remaining three groups (priorities 1, 2, and 3).**

- For each enantiomer of butan-2-ol, **look toward the lowest-priority group,** drawn behind the plane, down the C–H bond.

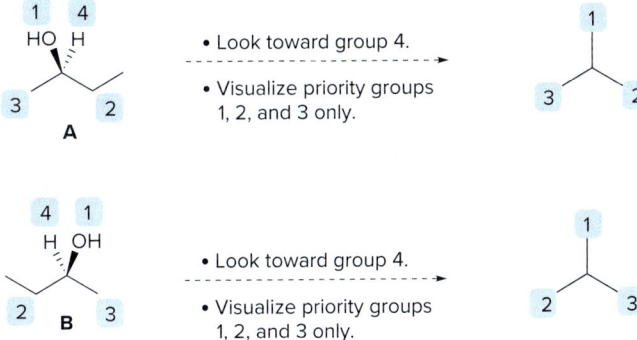

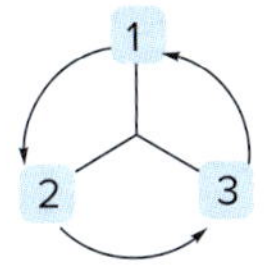

Step [3] **Trace a circle from priority group 1 → 2 → 3.**

clockwise
***R* isomer**

counterclockwise
***S* isomer**

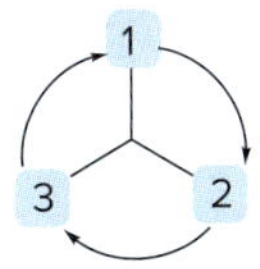

- If tracing the circle goes in the **clockwise** direction—to the right from the noon position—the isomer is named ***R*.**
- If tracing the circle goes in the **counterclockwise** direction—to the left from the noon position—the isomer is named ***S*.**

- The letter *R* or *S* precedes the IUPAC name of the molecule. For the enantiomers of butan-2-ol:

A

clockwise
***R* isomer**
(*R*)-butan-2-ol

B

counterclockwise
***S* isomer**
(*S*)-butan-2-ol

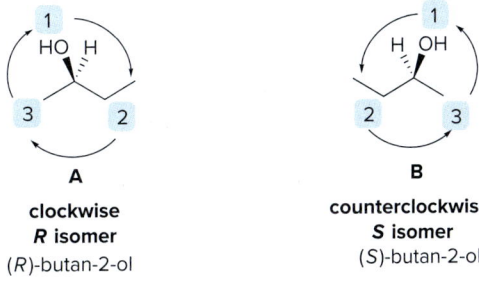

Sample Problem 5.3 — Labeling a Stereogenic Center as R or S

Label the stereogenic center in each compound as R or S.

a. 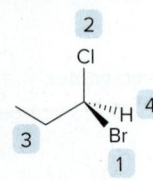   b.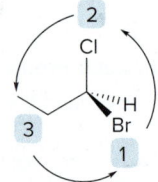

Solution

a. Assign priorities... ...then look toward the lowest-priority group (H), and trace a circle, 1 → 2 → 3.

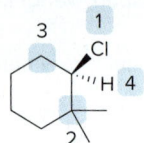

counterclockwise

Answer: **S** isomer

b. Assign priorities... ...then look toward the lowest-priority group (H), and trace a circle, 1 → 2 → 3.

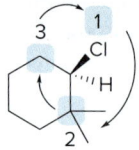

clockwise

Answer: **R** isomer

Problem 5.13 Label each stereogenic center in the following compounds as R or S.

More Practice: Try Problem 5.46a, e.

How do you assign R or S to a molecule when the lowest-priority group is not oriented toward the back, on a dashed wedge? You could rotate and flip the molecule until the lowest-priority group is in the back, as shown in Figure 5.7; then follow the stepwise procedure for assigning the configuration. Or, if manipulating and visualizing molecules in three dimensions is difficult for you, try the procedure suggested in Sample Problem 5.4.

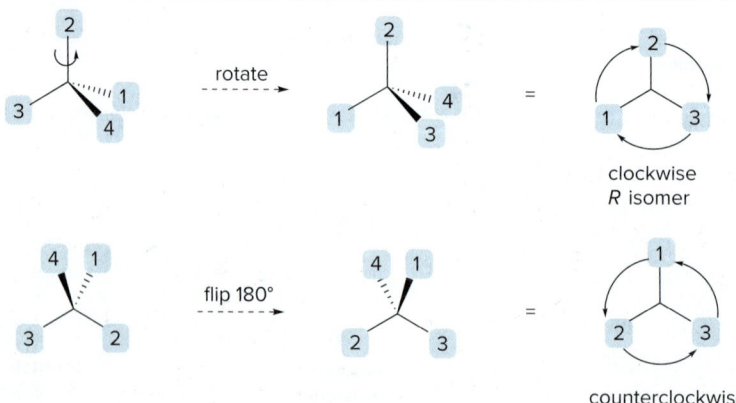

Figure 5.7 Orienting the lowest-priority group in back

- In rotating a molecule about a single bond, the position of *three* groups changes.
- In flipping a molecule 180°, the position of *all four* groups changes.

Sample Problem 5.4 — Designating a Stereogenic Center as *R* or *S* When the Lowest-Priority Group Is Not Drawn Back

Label each stereogenic center as *R* or *S*.

a. 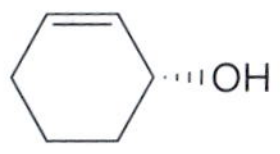 [cyclohexene with ⋯OH] b. 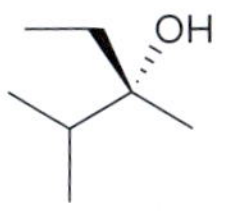 [structure with OH]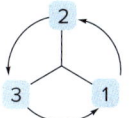

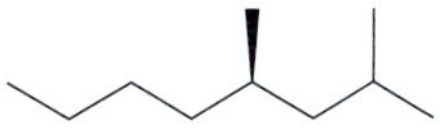

Solution

In both parts, the lowest-priority group is not oriented behind the page. To assign *R* or *S* in this case:

- **Switch** the position of the lowest-priority group with the group located **behind** the page.
- **Determine** *R* or *S* in the usual manner.
- **Reverse the answer.** Because we switched the position of *two* groups on the stereogenic center to begin with, and there are only *two* possibilities, the answer is **opposite** to the correct answer.

a. [1] Assign priorities.   [2] Switch groups 4 and 1. 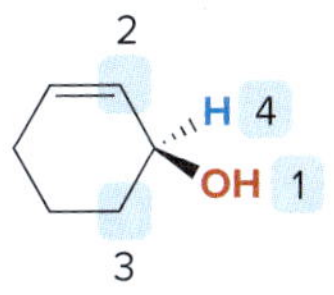   [3] Trace a circle, 1 → 2 → 3, and reverse the answer.

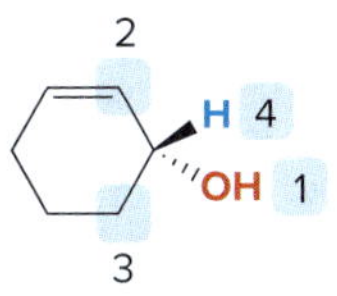

[structures with priorities labeled 2, 3, H 4, OH 1]

counterclockwise
It looks like an *S* isomer, but we must reverse the answer, because we switched groups 1 and 4, *S* → *R*.

Answer: ***R*** isomer

b. [1] Assign priorities.   [2] Switch groups 4 and 1.   [3] Trace a circle, 1 → 2 → 3, and reverse the answer.

[structures with priorities labeled 3, 1 OH, 2, CH₃ 4]

clockwise
It looks like an *R* isomer, but we must reverse the answer, because we switched groups 1 and 4, *R* → *S*.

Answer: ***S*** isomer

Problem 5.14 Label each stereogenic center as *R* or *S*.

a.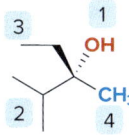

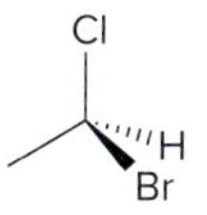

b. 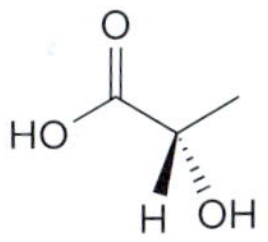 [HO-C(=O)-C(H)(OH)-]

c.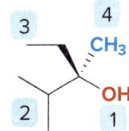

d. 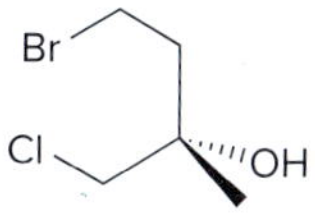 [BrCH₂-CH(Cl)-⋯OH]

e.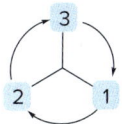

f.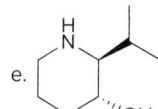

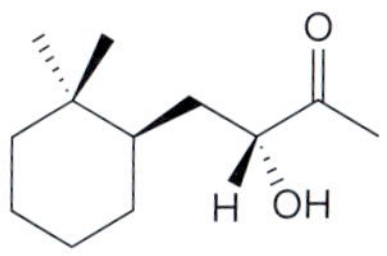

More Practice: Try Problems 5.35b, 5.46, 5.47, 5.64a.

Problem 5.15 FADH$_2$, the reduced form of flavin adenine dinucleotide, is a key biological reducing agent in several metabolic pathways (Section 27.2B). Locate the stereogenic centers in FADH$_2$ and label each stereogenic center as *R* or *S*.

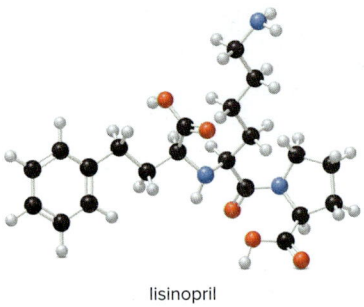

FADH$_2$
biological reducing agent

Lisinopril (trade name Zestril, Problem 5.16) is an ACE inhibitor, a drug that lowers blood pressure by decreasing the amount of angiotensin in the blood. Angiotensin is a polyamide that narrows blood vessels, thus increasing blood pressure. *Alon harel/Alamy Stock Photo*

Problem 5.16 (a) Locate the stereogenic centers in the ball-and-stick model of lisinopril, a drug used to treat high blood pressure. (b) Label each stereogenic center as *R* or *S*.

lisinopril

5.7 Diastereomers

We have now seen many examples of compounds containing one tetrahedral stereogenic center. The situation is more complex for compounds with two stereogenic centers, because more stereoisomers are possible. Moreover, a molecule with two or more stereogenic centers **may** or **may not** be chiral.

- For *n* stereogenic centers, the maximum number of stereoisomers is 2^n.

  - When $n = 1$, $2^1 = 2$. With one stereogenic center, there are always two stereoisomers and they are **enantiomers**.
  - When $n = 2$, $2^2 = 4$. With two stereogenic centers, the maximum number of stereoisomers is four, although sometimes there are *fewer* than four.

Problem 5.17 What is the maximum number of stereoisomers possible for a compound with: (a) three stereogenic centers; (b) eight stereogenic centers?

Let's illustrate a stepwise procedure for finding all possible stereoisomers using 2,3-dibromopentane. Because 2,3-dibromopentane has two stereogenic centers, the maximum number of stereoisomers is four.

In testing to see if one compound is superimposable on another, rotate atoms and flip the entire molecule, but **do not break any bonds.**

2,3-dibromopentane

■ = stereogenic center

To find stereoisomers, use the *eclipsed* conformation.

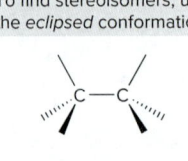

rapid interconversion

eclipsed conformation
easier conformation to visualize

staggered conformation
more stable

5.7 Diastereomers

How To Find and Draw All Possible Stereoisomers for a Compound with Two Stereogenic Centers

Step [1] Draw one stereoisomer by arbitrarily arranging substituents around the stereogenic centers. Then draw its mirror image.

- Arbitrarily add the H, Br, CH_3, and CH_2CH_3 groups to the stereogenic centers, forming **A**. Then draw the mirror image **B** so that substituents in **B** are a reflection of the substituents in **A**.
- Determine whether **A** and **B** are superimposable by flipping or rotating one molecule to see if all the atoms align.
- If you have drawn the compound and the mirror image in the described manner, you have to do only two operations to see if the atoms align. Place **B** directly on top of **A** (either in your mind or use models); and rotate **B** 180° and place it on top of **A** to see if the atoms align.

- H and Br do *not* align.
- **A** and **B** are *different* compounds.

- In this case, the atoms of **A** and **B** do not align, making **A** and **B** nonsuperimposable mirror images—**enantiomers**. **A** and **B** are two of the four possible stereoisomers for 2,3-dibromopentane.

Step [2] Draw a third possible stereoisomer by switching the positions of any two groups on only *one* stereogenic center. Then draw its mirror image.

- Switching the positions of H and Br (or any two groups) on one stereogenic center of either **A** or **B** forms a new stereoisomer (labeled **C** in this example), which is different from both **A** and **B**. Then draw the mirror image of **C**, labeled **D**. **C** and **D** are nonsuperimposable mirror images—**enantiomers**. We have now drawn four stereoisomers for 2,3-dibromopentane, the maximum number possible.

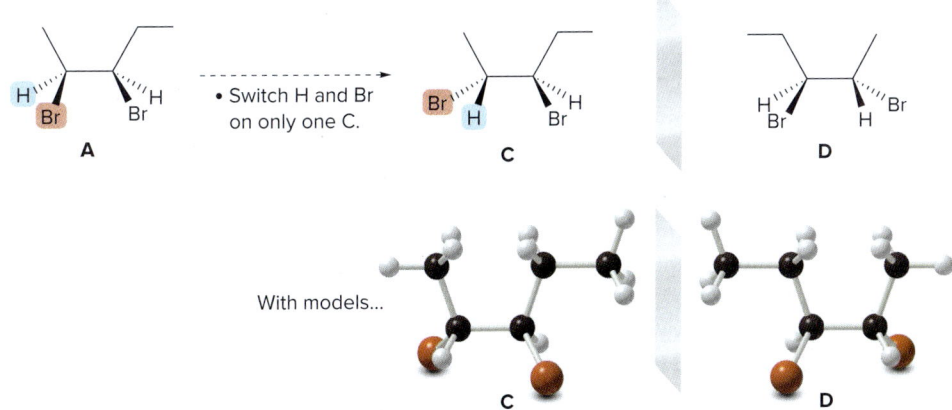

196 Chapter 5 Stereochemistry

> There are only two types of stereoisomers: **Enantiomers** are stereoisomers that *are* mirror images. **Diastereomers** are stereoisomers that are *not* mirror images.

There are four stereoisomers for 2,3-dibromopentane: enantiomers **A** and **B,** and enantiomers **C** and **D.** What is the relationship between two stereoisomers like **A** and **C? A** and **C** represent the second class of stereoisomers, called **diastereomers.** *Diastereomers are stereoisomers that are* not *mirror images of each other.* **A** and **B** are diastereomers of **C** and **D,** and vice versa. Figure 5.8 summarizes the relationships between the stereoisomers of 2,3-dibromopentane.

Problem 5.18 Label the two stereogenic centers in each compound and draw all possible stereoisomers.

a. [structure with Cl and OH on pentane]   b. [structure with Br and Cl on pentane]

Problem 5.19 Compounds **E** and **F** are two isomers of 2,3-dibromopentane drawn in staggered conformations. Which compounds (**A–D**) in Figure 5.8 are identical to **E** and **F?**

[Structures E and F]

Figure 5.8
The four stereoisomers of 2,3-dibromopentane

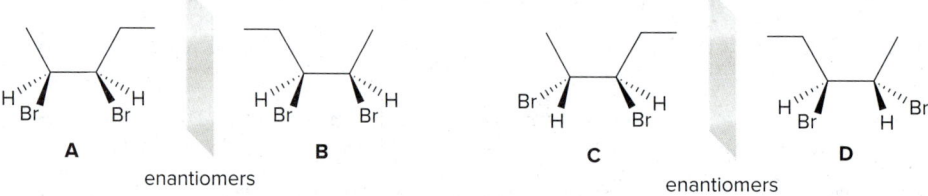

    A   B   C   D
    ←— enantiomers —→   ←— enantiomers —→
           A and **B** are diastereomers of **C** and **D.**

- Pairs of enantiomers: **A** and **B; C** and **D.**
- Pairs of diastereomers: **A** and **C; A** and **D; B** and **C; B** and **D.**

5.8 Meso Compounds

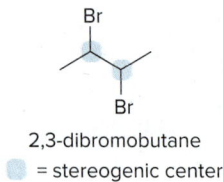

2,3-dibromobutane
▇ = stereogenic center

Whereas 2,3-dibromopentane has two stereogenic centers and the maximum of four stereoisomers, **2,3-dibromobutane** has two stereogenic centers but fewer than the maximum number of stereoisomers.

To find and draw all the stereoisomers of 2,3-dibromobutane, follow the same stepwise procedure outlined in Section 5.7. Arbitrarily add the H, Br, and CH₃ groups to the stereogenic centers, forming one stereoisomer **A,** and then draw its mirror image **B. A** and **B** are nonsuperimposable mirror images—**enantiomers.**

[Ball-and-stick and line structures of A and B, labeled "enantiomers"]

To find the other two stereoisomers (if they exist), switch the position of two groups on *one* stereogenic center of only *one* enantiomer. In this case, switching the positions of H and Br

5.8 Meso Compounds

on one stereogenic center of **A** forms **C**, which is different from both **A** and **B** and is thus a new stereoisomer.

- Switch H and Br on only one C.

identical
C = D

With models...

However, the mirror image of **C**, labeled **D**, is superimposable on **C**, so **C** and **D** are *identical*. Thus, **C** is **achiral**, even though it has two stereogenic centers. **C** is a **meso compound**.

plane of symmetry

C
two identical halves

- A *meso compound* is an achiral compound that contains tetrahedral stereogenic centers.

C contains a **plane of symmetry. Meso compounds generally have a plane of symmetry,** so they possess two identical halves.

Because one stereoisomer of 2,3-dibromobutane is superimposable on its mirror image, there are only three stereoisomers and not four, as summarized in Figure 5.9.

Figure 5.9
The three stereoisomers of 2,3-dibromobutane

A **B** **C**
enantiomers meso compound

A and **B** are diastereomers of **C**.

- Pair of enantiomers: **A** and **B**.
- Pairs of diastereomers: **A** and **C**; **B** and **C**.

Problem 5.20 Draw all the possible stereoisomers for each compound, and label pairs of enantiomers and diastereomers.

a. (2,3-dihydroxybutane structure)   b. (2-chloro-3-hydroxybutane structure)

Problem 5.21 Which compounds are meso compounds?

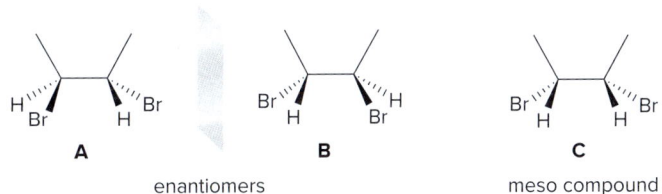

5.9 R and S Assignments in Compounds with Two or More Stereogenic Centers

When a compound has more than one stereogenic center, the *R* or *S* configuration must be assigned to each of them. In the stereoisomer of 2,3-dibromopentane drawn here, C2 has the *S* configuration and C3 has the *R,* so the complete name of the compound is (2*S*,3*R*)-2,3-dibromopentane.

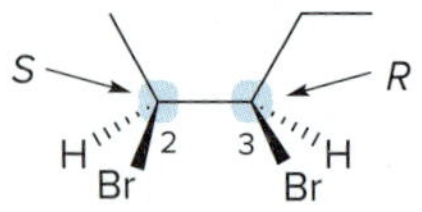

(2*S*,3*R*)-2,3-dibromopentane

R,S configurations can be used to determine whether two compounds are identical, enantiomers, or diastereomers.

- Identical compounds have the *same R,S* designations at every tetrahedral stereogenic center.
- Enantiomers have exactly *opposite R,S* designations.
- Diastereomers have the *same R,S* designation for at least one stereogenic center and the *opposite* for at least one of the other stereogenic centers.

For example, if a compound has two stereogenic centers, both with the *R* configuration, then its enantiomer is *S,S* and the diastereomers are either *R,S* or *S,R*.

Sorbitol (Problem 5.23) occurs naturally in some berries and fruits. It is used as a substitute sweetener in sugar-free—that is, sucrose-free—candy and gum.
Jill Braaten/McGraw-Hill Education

Problem 5.22 Without drawing out the structures, label each pair of compounds as enantiomers or diastereomers.
a. (2*R*,3*S*)-hexane-2,3-diol and (2*R*,3*R*)-hexane-2,3-diol
b. (2*R*,3*R*)-hexane-2,3-diol and (2*S*,3*S*)-hexane-2,3-diol
c. (2*R*,3*S*,4*R*)-hexane-2,3,4-triol and (2*S*,3*R*,4*R*)-hexane-2,3,4-triol

Problem 5.23 (a) Label the four stereogenic centers in sorbitol as *R* or *S*. (b) How are sorbitol and **A** related? (c) How are sorbitol and **B** related?

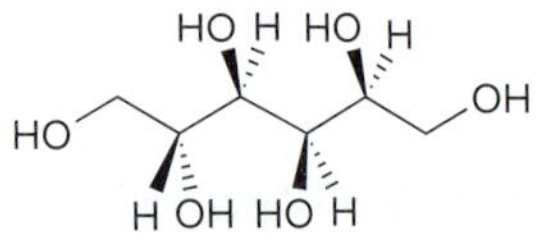

sorbitol 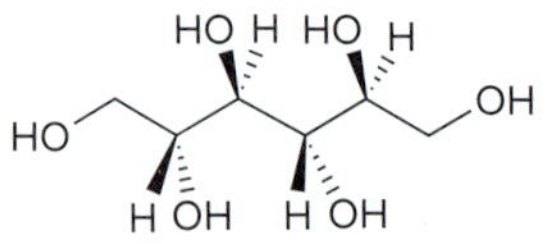   **A** 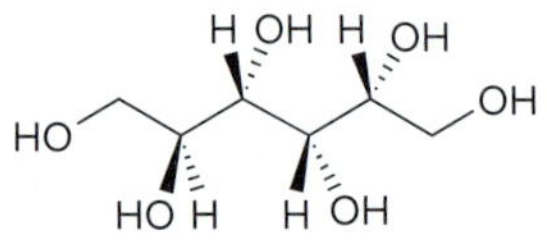   **B**

5.10 Disubstituted Cycloalkanes

Let us now turn our attention to disubstituted cycloalkanes, and draw all possible stereoisomers for **1,3-dibromocyclopentane**. Because 1,3-dibromocyclopentane has two stereogenic centers (labeled in blue), it has a maximum of four stereoisomers.

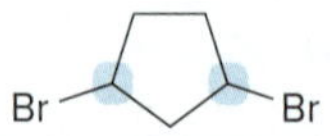

1,3-dibromocyclopentane

To draw all possible stereoisomers, remember that a disubstituted cycloalkane can have two substituents on the *same* side of the ring (**cis isomer,** labeled **A**) or on *opposite* sides of the ring (**trans isomer,** labeled **B**). These compounds are **stereoisomers but not mirror**

images of each other, making them **diastereomers. A** and **B** are two of the four possible stereoisomers.

A
cis isomer

B
trans isomer

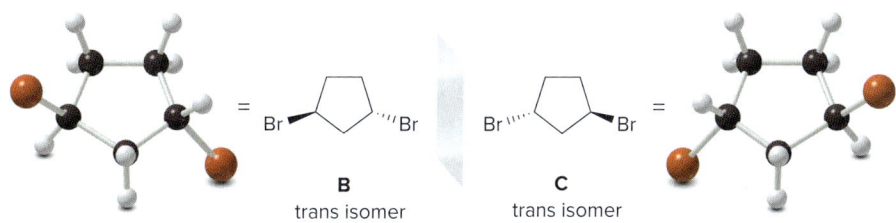

A and **B** are diastereomers.

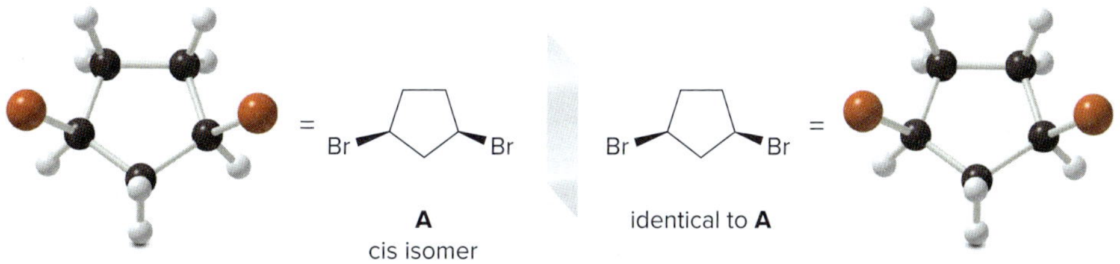

In determining chirality in substituted cycloalkanes, always draw the rings as **flat polygons.** This is especially true for cyclohexane derivatives, where having two chair forms that interconvert can make analysis especially difficult.

cis-1,3-Dibromocyclopentane contains a plane of symmetry.

plane of symmetry

two identical halves

To find the other two stereoisomers (if they exist), draw the mirror image of each compound and determine whether the compound and its mirror image are superimposable.

A
cis isomer

identical to **A**

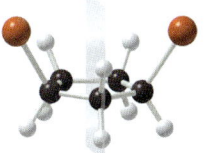

- The cis isomer is superimposable on its mirror image, making them *identical.* Thus, **A** is an **achiral meso compound.**

B
trans isomer

C
trans isomer

B and **C** are enantiomers.

- The trans isomer **B** is *not* superimposable on its mirror image, labeled **C,** making **B** and **C** different compounds. Thus, **B** and **C** are **enantiomers.**

Because one stereoisomer of 1,3-dibromocyclopentane is superimposable on its mirror image, there are only three stereoisomers, not four. **A** is an achiral meso compound, and **B** and **C** are a pair of chiral enantiomers. **A** and **B** are diastereomers, as are **A** and **C.**

Problem 5.24 Which of the following cyclic molecules are meso compounds?

a.

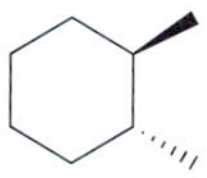

b.

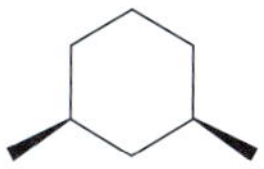

c.

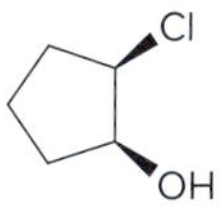

Problem 5.25 Draw all possible stereoisomers for each compound. Label pairs of enantiomers and diastereomers.

a.

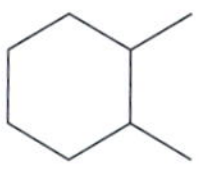

b.

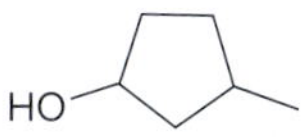

c.

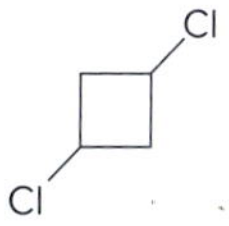

5.11 Isomers—A Summary

Before moving on to other aspects of stereochemistry, take the time to review Figures 5.10 and 5.11. Keep in mind the following facts, and use Figure 5.10 to summarize the types of isomers.

Figure 5.10
Summary—Types of isomers

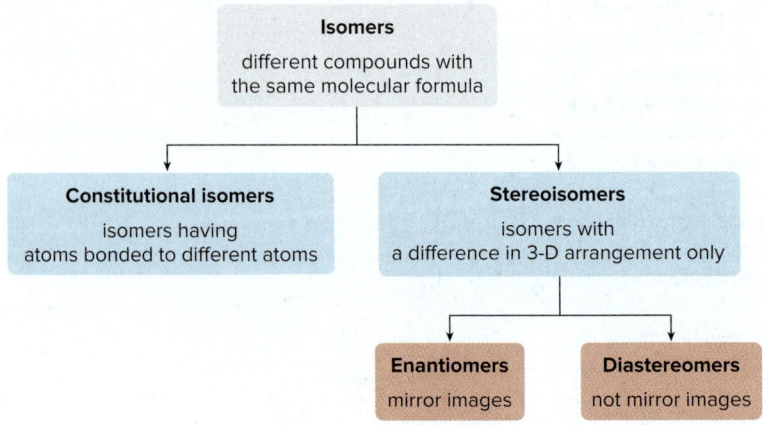

- There are two major classes of isomers: constitutional isomers and stereoisomers.
- There are only two kinds of stereoisomers: enantiomers and diastereomers.

Then, to determine the relationship between two nonidentical molecules, refer to the flowchart in Figure 5.11.

Figure 5.11
Determining the relationship between two nonidentical molecules

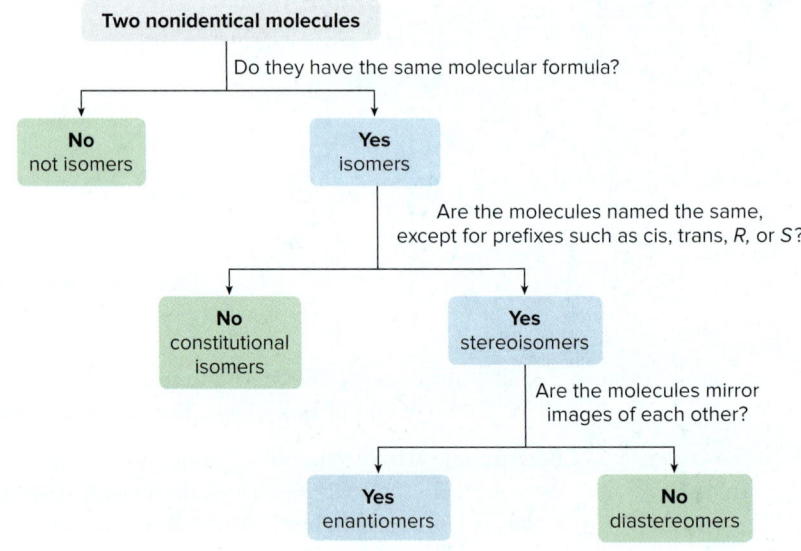

Problem 5.26 State how each pair of compounds is related. Are they enantiomers, diastereomers, constitutional isomers, or identical?

5.12 Physical Properties of Stereoisomers

Recall from Section 5.2 that constitutional isomers have different physical and chemical properties. How, then, do the physical and chemical properties of enantiomers compare?

- The chemical and physical properties of two enantiomers are *identical* except in their interaction with *chiral* substances.

5.12A Optical Activity

Two enantiomers have identical physical properties—melting point, boiling point, solubility—except for how they interact with plane-polarized light.

What is plane-polarized light? Ordinary light consists of electromagnetic waves that oscillate in all planes perpendicular to the direction in which the light travels. Passing light through a polarizer allows light in only one plane to come through, resulting in **plane-polarized light** (or simply **polarized light**). Plane-polarized light has an electric vector that oscillates in a single plane.

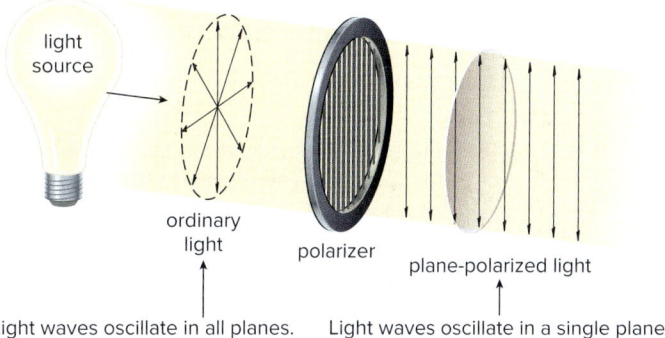

A **polarimeter** is an instrument that allows plane-polarized light to travel through a sample tube containing an organic compound. After the light exits the sample tube, an analyzer slit is rotated to determine the direction of the plane of the exiting polarized light. With **achiral compounds**, the light exits the sample tube *unchanged*, and the plane of the polarized light is in the same position it was before entering the sample tube.

- A compound that does not change the plane of polarized light is said to be *optically inactive*.

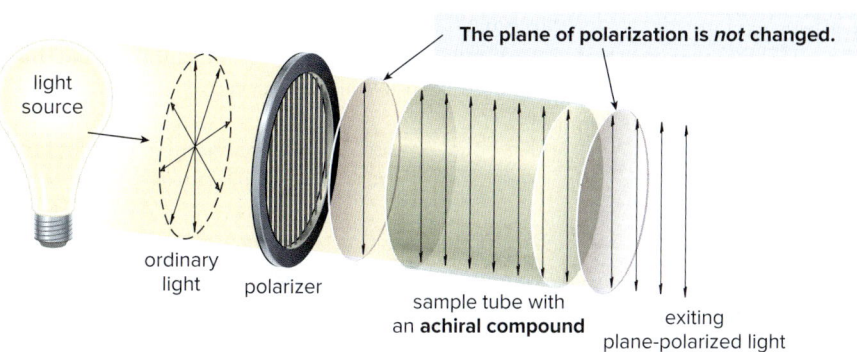

With **chiral compounds,** the plane of the polarized light is rotated through an angle α. The angle α, measured in degrees (°), is called the **observed rotation.**

- A compound that rotates the plane of polarized light is said to be *optically active*.

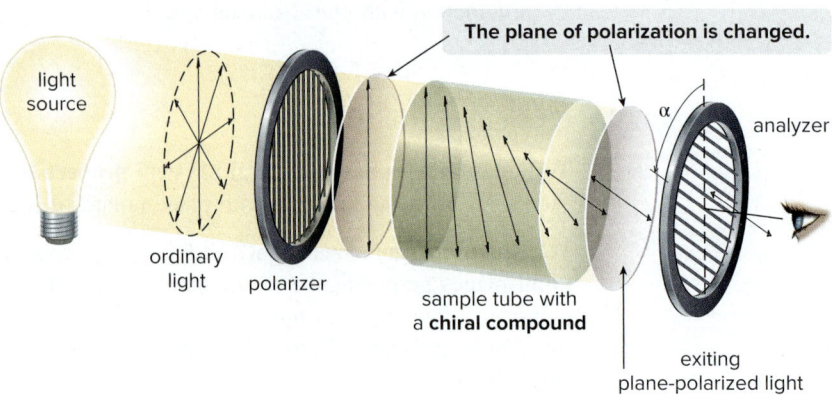

The achiral compound CH_2BrCl is optically *inactive,* whereas a single enantiomer of $CHBrClF$, a chiral compound, is optically *active*.

The rotation of polarized light can be in the **clockwise** or **counterclockwise** direction.

- If the rotation is *clockwise* (to the right from the noon position), the compound is called *dextrorotatory*. The rotation is labeled *d* or (+).
- If the rotation is *counterclockwise* (to the left from noon), the compound is called *levorotatory*. The rotation is labeled *l* or (−).

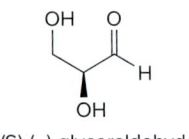

(S)-(−)-glyceraldehyde

(S)-(+)-lactic acid

No relationship exists between the *R* and *S* prefixes that designate configuration and the (+) and (−) designations indicating optical rotation. For example, the *S* enantiomer of lactic acid is dextrorotatory (+), whereas the *S* enantiomer of glyceraldehyde is levorotatory (−).

How does the rotation of two enantiomers compare?

- Two enantiomers rotate plane-polarized light *to an equal extent* but in the *opposite* direction.

Thus, if enantiomer **A** rotates polarized light +5°, then the same concentration of enantiomer **B** rotates it −5°.

5.12B Racemic Mixtures

What is the observed rotation of an equal amount of two enantiomers? Because **two enantiomers rotate plane-polarized light to an equal extent but in opposite directions, the rotations cancel,** and no rotation is observed.

- An equal amount of two enantiomers is called a *racemic mixture* or a *racemate*. A racemic mixture is optically *inactive*.

Besides optical rotation, other physical properties of a racemate are not readily predicted. The melting point and boiling point of a racemic mixture are not necessarily the same as either pure enantiomer, and this fact is not easily explained. The physical properties of two enantiomers and their racemic mixture are summarized in Table 5.1.

Table 5.1 The Physical Properties of Enantiomers **A** and **B** Compared

Property	A alone	B alone	Racemic A + B
Melting point	identical to **B**	identical to **A**	may be different from **A** and **B**
Boiling point	identical to **B**	identical to **A**	may be different from **A** and **B**
Optical rotation	equal in magnitude but opposite in sign to **B**	equal in magnitude but opposite in sign to **A**	0°

5.12C Specific Rotation

The observed rotation depends on the number of chiral molecules that interact with polarized light. This in turn depends on the concentration of the sample and the length of the sample tube. To standardize optical rotation data, the quantity **specific rotation** ($[\alpha]$) is defined using a specific sample tube length (usually 1 dm), concentration, temperature (25 °C), and wavelength (589 nm, the D line emitted by a sodium lamp).

$$\text{specific rotation} = [\alpha] = \frac{\alpha}{l \times c}$$

α = observed rotation (°)
l = length of sample tube (dm)
c = concentration (g/mL)

$$\begin{bmatrix} \text{dm = decimeter} \\ \text{1 dm = 10 cm} \end{bmatrix}$$

Specific rotations are physical constants just like melting points or boiling points, and are reported in chemical reference books for a wide variety of compounds.

Problem 5.27 The amino acid (S)-alanine has the physical characteristics listed under the structure.

(S)-alanine
$[\alpha]$ = +8.5
mp = 297 °C

a. What is the melting point of (R)-alanine?
b. How does the melting point of a racemic mixture of (R)- and (S)-alanine compare to the melting point of (S)-alanine?
c. What is the specific rotation of (R)-alanine, recorded under the same conditions as the reported rotation of (S)-alanine?
d. What is the optical rotation of a racemic mixture of (R)- and (S)-alanine?
e. Label each of the following as optically active or inactive: a solution of pure (S)-alanine; an equal mixture of (R)- and (S)-alanine; a solution that contains 75% (S)- and 25% (R)-alanine.

Problem 5.28 A natural product was isolated in the laboratory, and its observed rotation was +10° when measured in a 1 dm sample tube containing 1.0 g of compound in 10 mL of water. What is the specific rotation of this compound?

5.12D Enantiomeric Excess

Sometimes in the laboratory we have neither a pure enantiomer nor a racemic mixture, but rather a mixture of two enantiomers in which one enantiomer is present in excess of the other. The **enantiomeric excess** (*ee*), also called the **optical purity,** tells how much more there is of one enantiomer.

- Enantiomeric excess = ee = % of one enantiomer − % of the other enantiomer.

Enantiomeric excess tells how much one enantiomer is present in excess of the racemic mixture. For example, if a mixture contains 75% of one enantiomer and 25% of the other, the enantiomeric excess is 75% − 25% = 50%. There is a 50% excess of one enantiomer over the racemic mixture.

Chapter 5 Stereochemistry

Problem 5.29 What is the *ee* for each of the following mixtures of enantiomers **A** and **B**?
a. 95% **A** and 5% **B**   b. 85% **A** and 15% **B**

Knowing the *ee* of a mixture makes it possible to calculate the amount of each enantiomer present, as shown in Sample Problem 5.5.

Sample Problem 5.5 Using Enantiomeric Excess to Calculate the Amount of Each Enantiomer

If the enantiomeric excess is 95%, how much of each enantiomer is present?

Solution

Label the two enantiomers **A** and **B** and assume that **A** is in excess. A 95% *ee* means that the solution contains an excess of 95% of **A**, and 5% of the racemic mixture of **A** and **B**. Because a racemic mixture is an equal amount of both enantiomers, it has 2.5% of **A** and 2.5% of **B**.

- Total amount of **A** = 95% + 2.5% = 97.5%
- Total amount of **B** = 2.5% (or 100% − 97.5%)

Problem 5.30 For the given *ee* values, calculate the percentage of each enantiomer present.
a. 90% *ee*   b. 99% *ee*   c. 60% *ee*

More Practice: Try Problem 5.63b.

The enantiomeric excess can also be calculated if two quantities are known—the specific rotation [α] of a mixture and the specific rotation [α] of a pure enantiomer.

$$ee = \frac{[\alpha] \text{ mixture}}{[\alpha] \text{ pure enantiomer}} \times 100\%$$

Sample Problem 5.6 Calculating Enantiomeric Excess

Pure cholesterol has a specific rotation of −32. A sample of cholesterol prepared in the lab had a specific rotation of −16. What is the enantiomeric excess of this sample of cholesterol?

Solution

Calculate the *ee* of the mixture using the given formula.

$$ee = \frac{[\alpha] \text{ mixture}}{[\alpha] \text{ pure enantiomer}} \times 100\% = \frac{-16}{-32} \times 100\% = 50\% \ ee$$

Answer

Problem 5.31 Pure MSG, a common flavor enhancer, exhibits a specific rotation of +24. (a) Calculate the *ee* of a solution whose [α] is +10. (b) If the *ee* of a solution of MSG is 80%, what is [α] for this solution?

MSG
monosodium glutamate

More Practice: Try Problems 5.63a, d; 5.65f.

Problem 5.32 (S)-Lactic acid has a specific rotation of +3.8. (a) If the ee of a solution of lactic acid is 60%, what is [α] for this solution? (b) How much of the dextrorotatory and levorotatory isomers does the solution contain?

5.12E The Physical Properties of Diastereomers

Diastereomers are not mirror images of each other, and as such, **their physical properties are different, including optical rotation.** Figure 5.12 compares the physical properties of the three stereoisomers of tartaric acid, consisting of a meso compound that is a diastereomer of a pair of enantiomers.

Figure 5.12 The physical properties of the three stereoisomers of tartaric acid

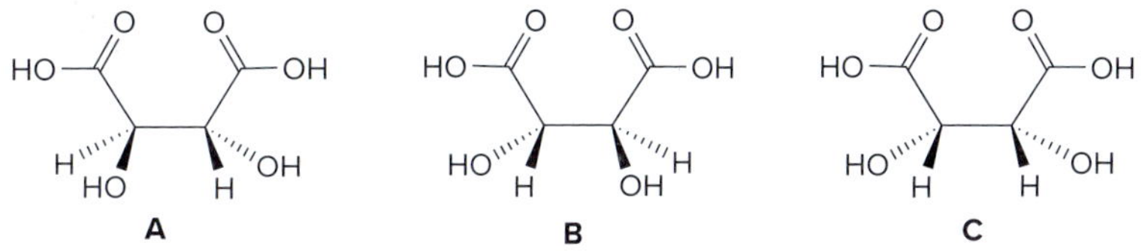

- A and B are enantiomers.
- A and B are diastereomers of C.

Property	A	B	C	A + B (1:1)
melting point (°C)	171	171	146	206
solubility (g/100 mL H_2O)	139	139	125	139
[α]	+13	−13	0	0
R,S designation	R,R	S,S	R,S	—
d,l designation	d	l	none	d,l

- The physical properties of **A** and **B** differ from their diastereomer **C**.
- The physical properties of a racemic mixture of **A** and **B** (last column) can also differ from either enantiomer and diastereomer **C**.
- **C** is an achiral meso compound, so it is optically inactive; [α] = 0.

Whether the physical properties of a set of compounds are the same or different has practical applications in the lab. Physical properties characterize a compound's physical state, and two compounds can usually be separated only if their physical properties are different.

Two enantiomers can be separated by the process of **resolution,** as described in Section 23.2.

- Because two enantiomers have identical physical properties, they cannot be separated by common physical techniques like distillation.
- Diastereomers and constitutional isomers have different physical properties, and therefore they can be separated by common physical techniques.

Problem 5.33 Compare the physical properties of the three stereoisomers of 1,3-dimethylcyclopentane.

three stereoisomers of 1,3-dimethylcyclopentane

a. How do the boiling points of **A** and **B** compare? What about those of **A** and **C**?
b. Characterize a solution of each of the following as optically active or optically inactive: pure **A**; pure **B**; pure **C**; an equal mixture of **A** and **B**; an equal mixture of **A** and **C**.
c. A reaction forms a 1:1:1 mixture of **A**, **B**, and **C**. If this mixture is distilled, how many fractions would be obtained? Which fractions would be optically active and which would be optically inactive?

5.13 Chemical Properties of Enantiomers

When two enantiomers react with an achiral reagent, they react at the same rate, but when they react with a chiral, non-racemic reagent, they react at different rates.

- Two enantiomers have exactly the same chemical properties except for their reaction with chiral, non-racemic reagents.

For an everyday analogy, consider what happens when you are handed an achiral object like a pen and a chiral object like a right-handed glove. Your left and right hands are enantiomers, but they can both hold the achiral pen in the same way. With the glove, however, only your right hand can fit inside it, not your left.

We will examine specific reactions of chiral molecules with both chiral and achiral reagents later in this text. Here, we examine two more general applications.

5.13A Chiral Drugs

A living organism is a sea of chiral molecules. Many drugs are chiral, and often they must interact with a chiral receptor or a chiral enzyme to be effective. One enantiomer of a drug may treat a disease whereas its mirror image may be ineffective. Alternatively, one enantiomer may trigger one biochemical response and its mirror image may elicit a totally different response.

The drugs ibuprofen and fluoxetine each contain one stereogenic center, and thus exist as a pair of enantiomers, only one of which exhibits biological activity. **(S)-Ibuprofen** is the active component of the anti-inflammatory agents Motrin and Advil, and **(R)-fluoxetine** is the active component in the antidepressant Prozac.

Although (R)-ibuprofen shows no anti-inflammatory activity itself, it is slowly converted to the S enantiomer in vivo.

(S)-ibuprofen
anti-inflammatory agent

(R)-fluoxetine
antidepressant

Changing the orientation of two substituents to form a mirror image can also alter biological activity to produce an undesirable side effect in the other enantiomer. The *S* enantiomer of **naproxen** is an active anti-inflammatory agent, but the R enantiomer is a harmful liver toxin.

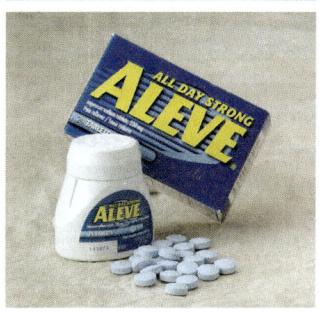

(S)-Naproxen is the active drug in the widely used pain relievers Naprosyn and Aleve. *Elite Images/McGraw-Hill Education*

For more examples of two enantiomers that exhibit very different biochemical properties, see *Journal of Chemical Education,* **1996,** *73,* 481–484.

(S)-naproxen
anti-inflammatory agent

(R)-naproxen
liver toxin

If a chiral drug could be sold as a single active enantiomer, it should be possible to use smaller doses with fewer side effects. Many chiral drugs continue to be sold as racemic mixtures, however, because it is more difficult and therefore more costly to obtain a single enantiomer. An enantiomer is not easily separated from a racemic mixture because the two enantiomers have the same physical properties. In Chapter 11, we will study a reaction that can form a single active enantiomer, an important development in making chiral drugs more readily available.

5.13B Enantiomers and the Sense of Smell

Research suggests that the odor of a particular molecule is determined more by its shape than by the presence of a particular functional group. For example, hexachloroethane (Cl_3CCCl_3) and cyclooctane have no obvious structural similarities, but they both have a camphor-like odor, a fact attributed to their similar spherical shape. Each molecule binds to spherically shaped olfactory receptors present on the nerve endings in the nasal passage, resulting in similar odors (Figure 5.13).

Figure 5.13
The shape of molecules and the sense of smell

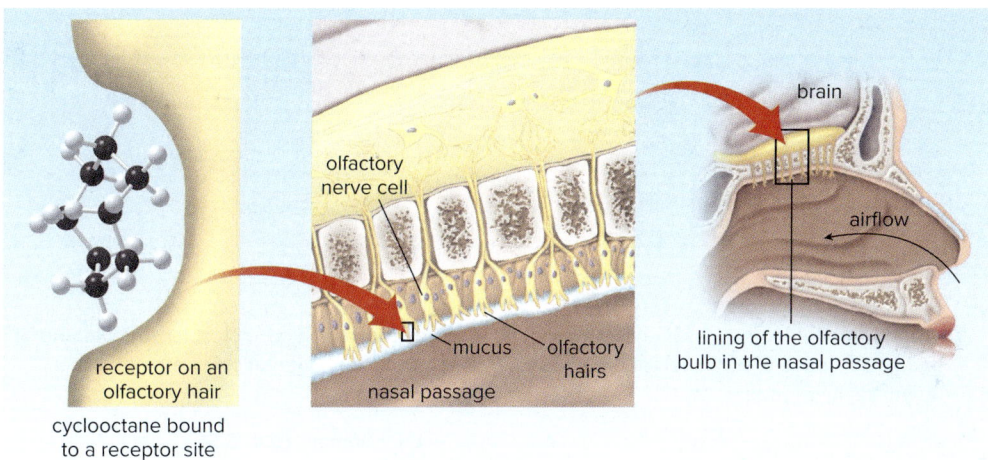

- Cyclooctane and other molecules similar in shape bind to a particular olfactory receptor on the nerve cells that lie at the top of the nasal passage. Binding results in a nerve impulse that travels to the brain, which interprets impulses from particular receptors as specific odors.

Because enantiomers interact with chiral smell receptors, some enantiomers have different odors. There are a few well-characterized examples of this phenomenon in nature. For example,

(S)-carvone is responsible for the odor of caraway, whereas (R)-carvone is responsible for the odor of spearmint.

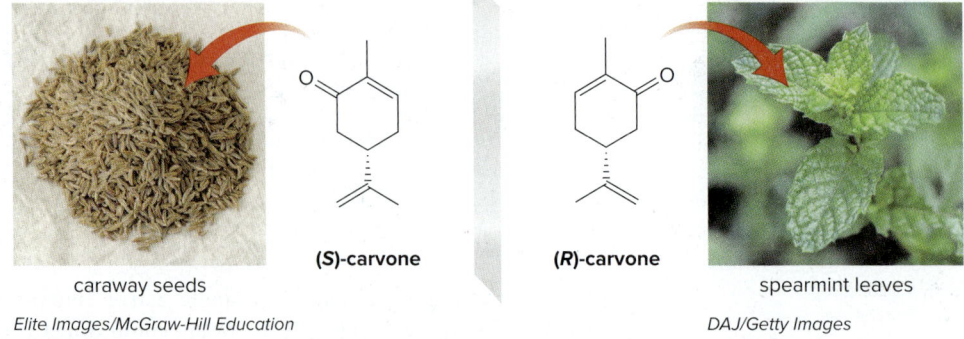

caraway seeds
Elite Images/McGraw-Hill Education

(S)-carvone   (R)-carvone

spearmint leaves
DAJ/Getty Images

These examples demonstrate that understanding the three-dimensional structure of a molecule is very important in organic chemistry.

Problem 5.34 Like carvone, the two enantiomers of celery ketone smell different. The *R* enantiomer smells like celery leaves, whereas the *S* enantiomer smells like licorice. Draw each enantiomer and assign its odor.

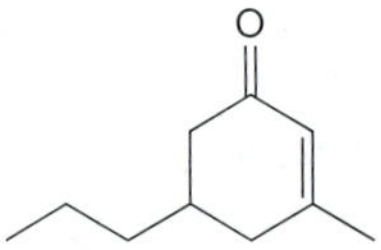

celery ketone

(R)-Celery ketone (Problem 5.34) has an odor reminiscent of celery leaves. *Aaron Roeth Photography*

Chapter 5 REVIEW

KEY CONCEPTS

[1] Two types of isomers (5.2, 5.11); example: $C_5H_{10}BrCl$

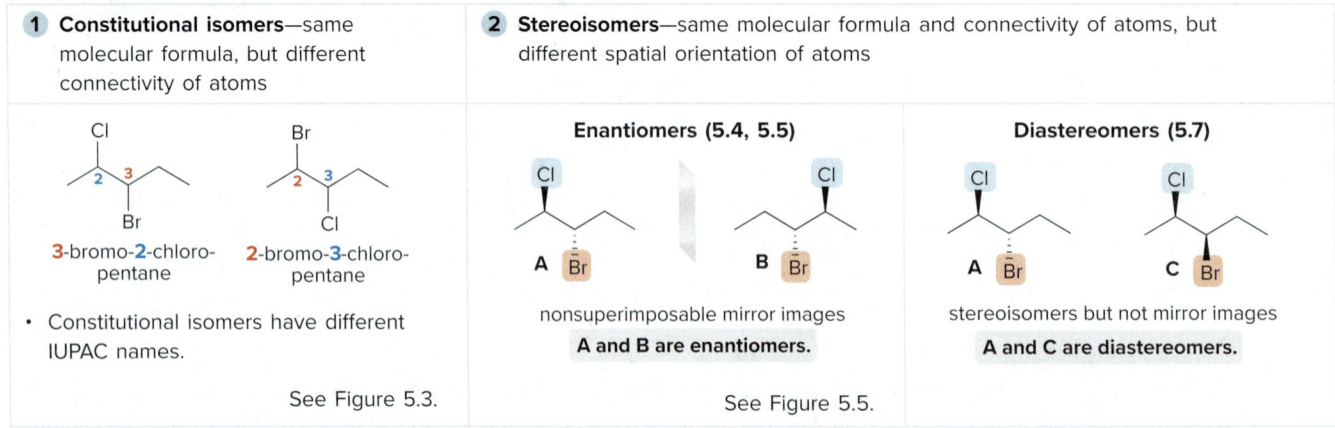

① **Constitutional isomers**—same molecular formula, but different connectivity of atoms

3-bromo-2-chloro-pentane   2-bromo-3-chloro-pentane

- Constitutional isomers have different IUPAC names.

See Figure 5.3.

② **Stereoisomers**—same molecular formula and connectivity of atoms, but different spatial orientation of atoms

Enantiomers (5.4, 5.5)

A   B

nonsuperimposable mirror images
A and B are enantiomers.

See Figure 5.5.

Diastereomers (5.7)

A   C

stereoisomers but not mirror images
A and C are diastereomers.

See Figures 5.8, 5.10, 5.11. Try Problems 5.36, 5.37, 5.39, 5.56–5.60, 5.62a.

Key Concepts

[2] Stereochemical terms

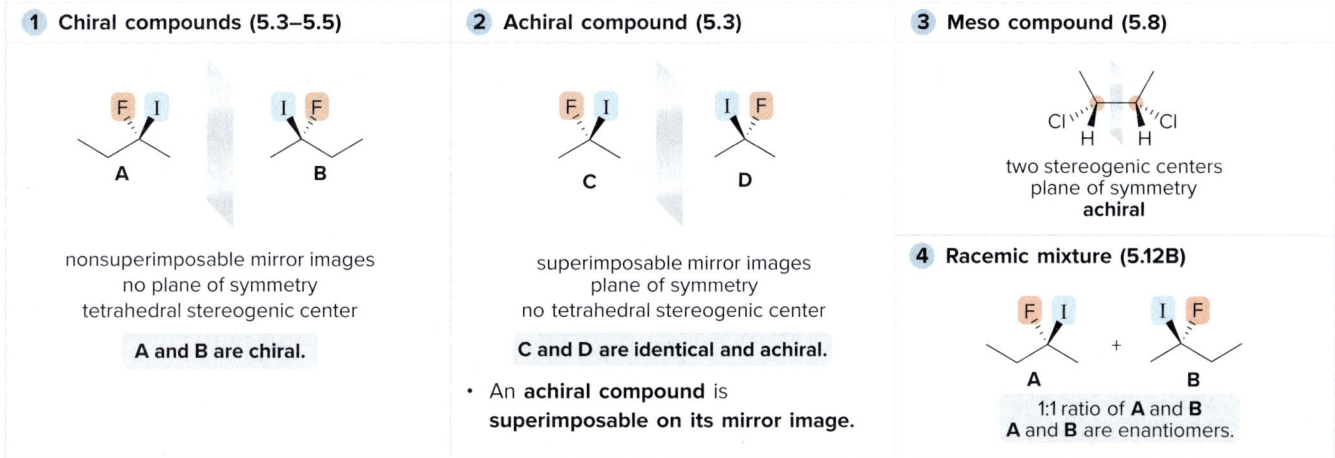

- **1** Chiral compounds (5.3–5.5): **A** and **B** — nonsuperimposable mirror images, no plane of symmetry, tetrahedral stereogenic center. **A and B are chiral.**
- **2** Achiral compound (5.3): **C** and **D** — superimposable mirror images, plane of symmetry, no tetrahedral stereogenic center. **C and D are identical and achiral.** An **achiral compound** is **superimposable on its mirror image**.
- **3** Meso compound (5.8): two stereogenic centers, plane of symmetry, **achiral**.
- **4** Racemic mixture (5.12B): 1:1 ratio of **A** and **B**, **A** and **B** are enantiomers.

Try Problems 5.38, 5.40, 5.62b.

[3] Optical activity (5.12)

1 An optically active solution contains:	2 An optically inactive solution contains one of the following:		
• a chiral compound	• an achiral compound with no stereogenic centers	• a meso compound	• a racemic mixture of two enantiomers

Try Problems 5.61, 5.62c, g.

[4] The prefixes R and S compared with d (+) and l (−) (5.6, 5.12)

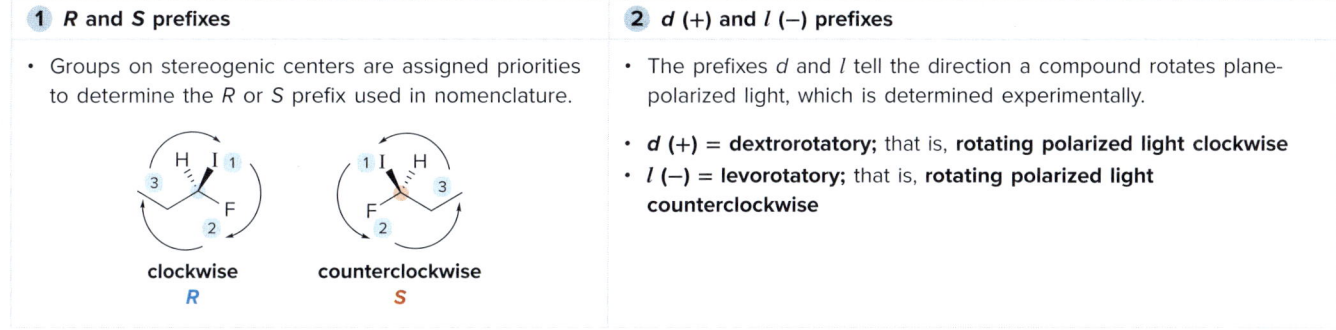

- **1 R and S prefixes**: Groups on stereogenic centers are assigned priorities to determine the R or S prefix used in nomenclature. clockwise = **R**; counterclockwise = **S**.
- **2 d (+) and l (−) prefixes**: The prefixes d and l tell the direction a compound rotates plane-polarized light, which is determined experimentally.
  - d (+) = dextrorotatory; that is, **rotating polarized light clockwise**
  - l (−) = levorotatory; that is, **rotating polarized light counterclockwise**

[5] R and S assignments in compounds with two or more stereogenic centers (5.9); example: 3-bromo-2-chloropentane

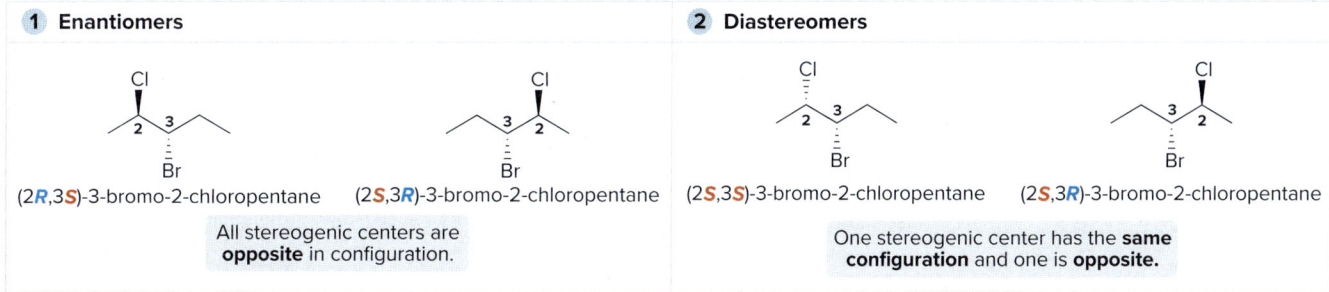

- **1 Enantiomers**: (2R,3S)-3-bromo-2-chloropentane and (2S,3R)-3-bromo-2-chloropentane. All stereogenic centers are **opposite** in configuration.
- **2 Diastereomers**: (2S,3S)-3-bromo-2-chloropentane and (2S,3R)-3-bromo-2-chloropentane. One stereogenic center has the **same configuration** and one is **opposite**.

Try Problem 5.53.

[6] Physical and chemical properties of isomers (5.12, 5.13)

1 Constitutional isomers (5.2)	**2** Enantiomers	**3** Diastereomers
• *different* physical and chemical properties	• *identical* physical properties except for the direction polarized light is rotated • *identical* chemical properties except for their reaction with chiral, non-racemic reagents	• *different* physical and chemical properties

See Figure 5.12. Try Problems 5.61b, 5.62e.

KEY SKILLS

[1] Locating stereogenic centers (5.4, 5.5); example: adenosine

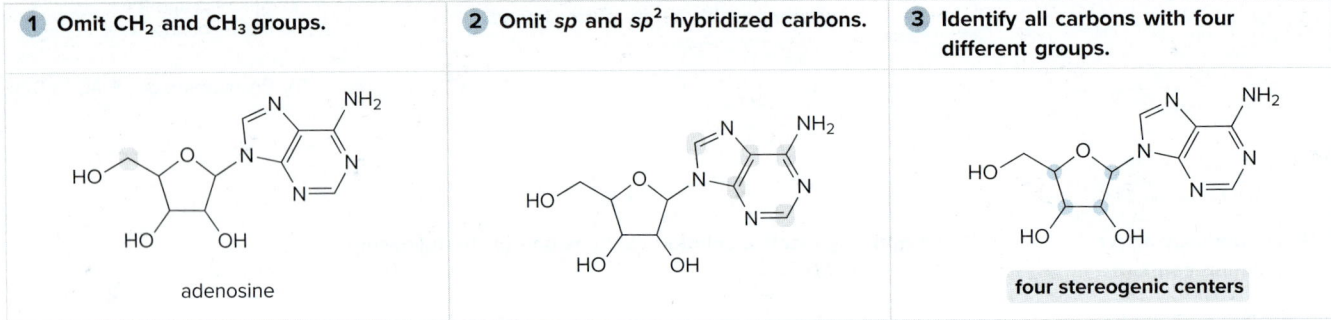

See Sample Problem 5.1. Try Problems 5.35a, 5.41, 5.42, 5.65a.

[2] Labeling stereogenic centers with *R* or *S* (5.6); example: 1-aminoethan-1-ol

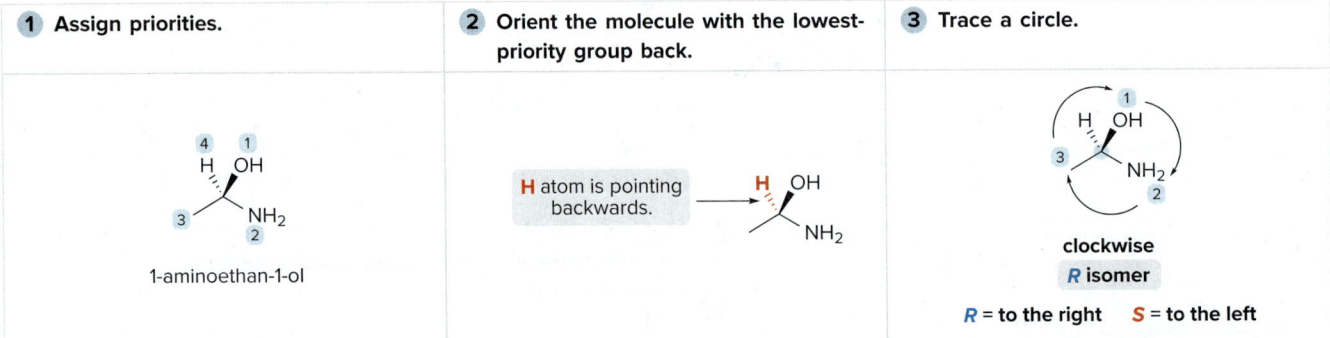

See *How To* p. 191, Sample Problem 5.3, Figure 5.6. Try Problem 5.46a, e.

[3] Assigning *R* or *S* when the lowest-priority group is not oriented toward the back (5.6); example: propranolol

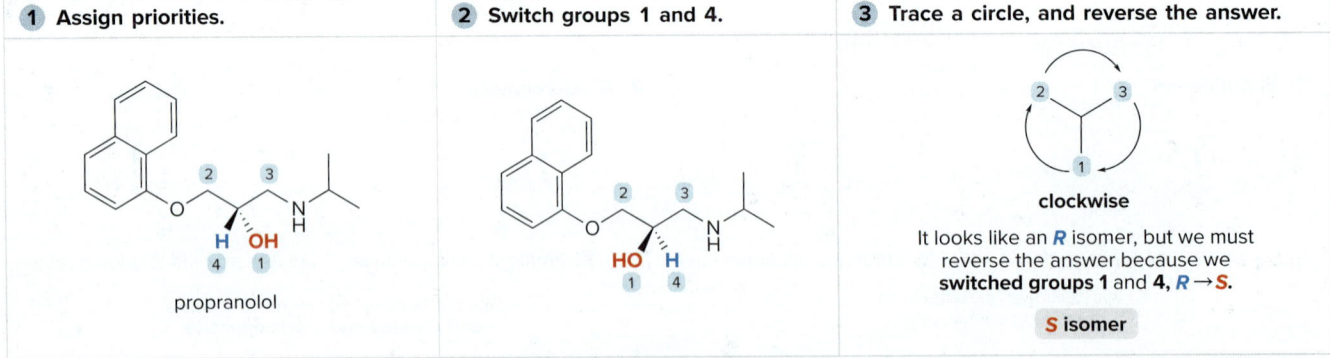

See Figure 5.7, Sample Problem 5.4. Try Problems 5.35b, 5.46, 5.47, 5.51, 5.64a.

[4] Finding and drawing all stereoisomers for a compound with two stereogenic centers (5.7, 5.8)

① Determine how many stereoisomers are possible.

② Draw one stereoisomer and its mirror image.

③ Draw a third stereoisomer and its mirror image.

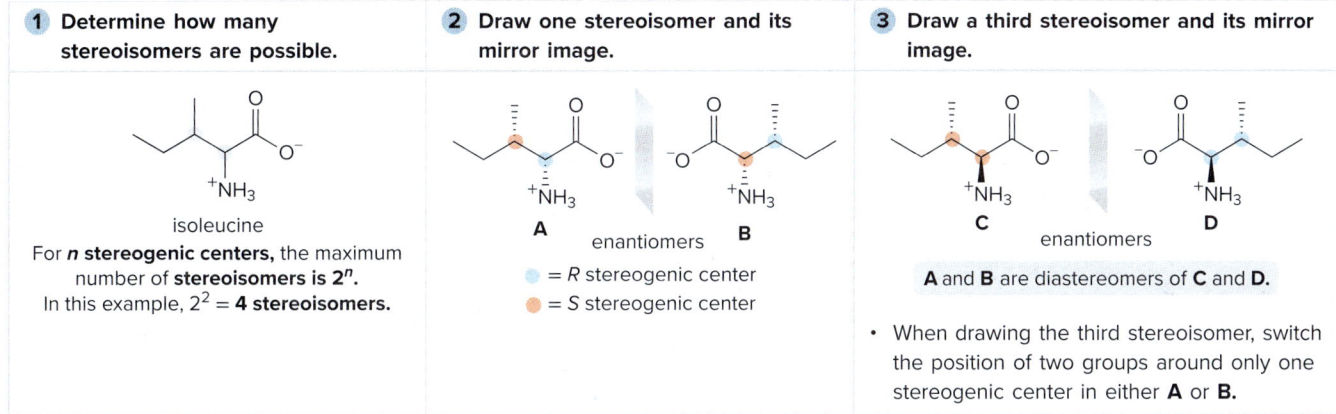

isoleucine

For *n* **stereogenic centers**, the maximum number of **stereoisomers is 2^n**. In this example, $2^2 = $ **4 stereoisomers**.

A   enantiomers   B

● = *R* stereogenic center
● = *S* stereogenic center

C   enantiomers   D

A and **B** are diastereomers of **C** and **D**.

- When drawing the third stereoisomer, switch the position of two groups around only one stereogenic center in either **A** or **B**.

See *How To* p. 195, Figures 5.8, 5.9. Try Problems 5.54, 5.55.

[5] Determining if two nonidentical compounds are constitutional isomers, enantiomers, or diastereomers (5.11); example: menthol and isomers

① Assess the connectivity of atoms, and assign the *R* or *S* configuration to each stereogenic center.

- Menthol and isomers **A** and **B** are **stereoisomers** because they have the same connectivity of atoms, but differ only in the spatial orientation of groups.

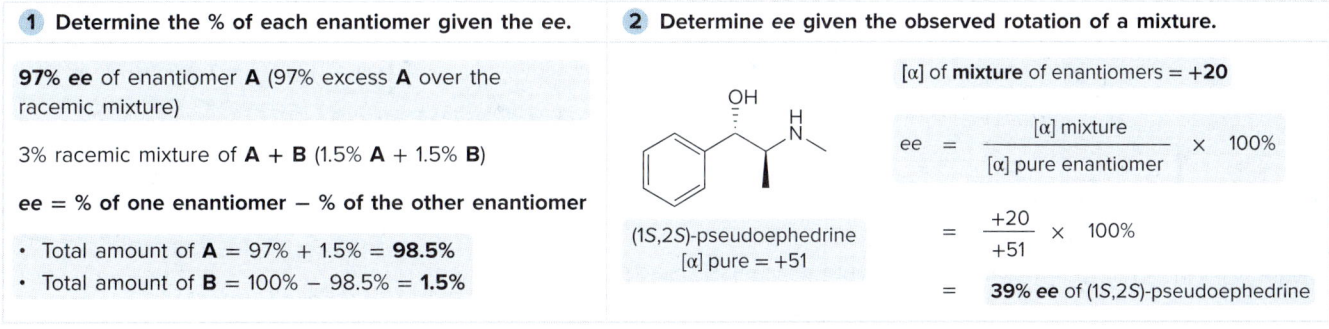

menthol   A   B

● = *R* stereogenic center
● = *S* stereogenic center

② Use configurations to determine whether compounds are enantiomers or diastereomers.

- Menthol and isomer **A** are **enantiomers** because they have exactly *opposite R,S* designations at all stereogenic centers.
- Menthol and isomer **B** are **diastereomers** because they have the *same R,S* designations for two stereogenic centers and the *opposite R,S* designation for one stereogenic center.

See Figure 5.10. Try Problems 5.60, 5.62a.

[6] Calculations involving enantiomeric excess (ee) (5.12D)

① Determine the % of each enantiomer given the ee.

97% **ee** of enantiomer **A** (97% excess **A** over the racemic mixture)

3% racemic mixture of **A** + **B** (1.5% **A** + 1.5% **B**)

ee = % of one enantiomer − % of the other enantiomer

- Total amount of **A** = 97% + 1.5% = **98.5%**
- Total amount of **B** = 100% − 98.5% = **1.5%**

② Determine ee given the observed rotation of a mixture.

[α] of **mixture** of enantiomers = +20

(1*S*,2*S*)-pseudoephedrine
[α] pure = +51

$$ee = \frac{[\alpha] \text{ mixture}}{[\alpha] \text{ pure enantiomer}} \times 100\%$$

$$= \frac{+20}{+51} \times 100\%$$

= **39% ee** of (1*S*,2*S*)-pseudoephedrine

See Sample Problems 5.5, 5.6. Try Problems 5.63, 5.65e, f.

PROBLEMS

Problems Using Three-Dimensional Models

5.35 (a) Locate the stereogenic centers in the ball-and-stick model of ezetimibe (trade name Zetia), a cholesterol-lowering drug.
(b) Label each stereogenic center as *R* or *S*.

ezetimibe

5.36 Consider the ball-and-stick models **A–D**. How is each pair of compounds related: (a) **A** and **B**; (b) **A** and **C**; (c) **A** and **D**; (d) **C** and **D**? Choose from identical molecules, enantiomers, or diastereomers.

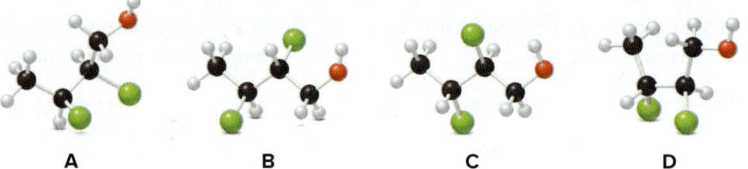

Constitutional Isomers Versus Stereoisomers

5.37 Label each pair of compounds as constitutional isomers, stereoisomers, or not isomers of each other.

Mirror Images and Chirality

5.38 Label each compound as chiral or achiral.

5.39 Determine if each compound is identical to or an enantiomer of **A**.

5.40 Indicate a plane of symmetry for each molecule that contains one. A molecule may require rotation around a carbon–carbon bond to see the plane of symmetry.

Finding and Drawing Stereogenic Centers

5.41 Locate the tetrahedral stereogenic center(s) in each compound. A molecule may have one or more stereogenic centers.

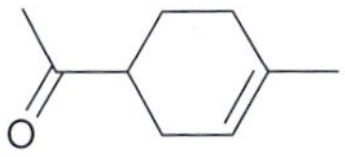

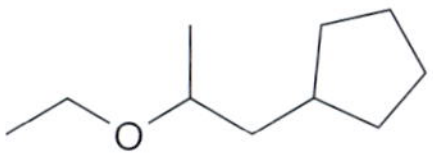

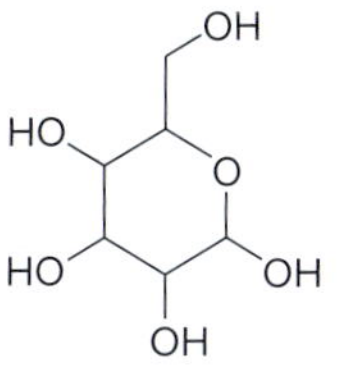

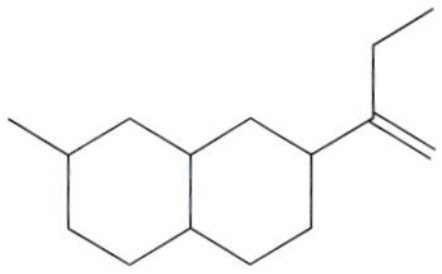

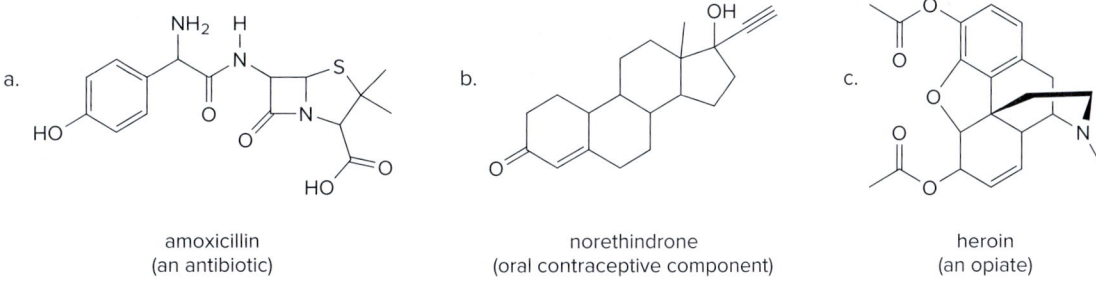

5.42 Locate the stereogenic centers in each drug.

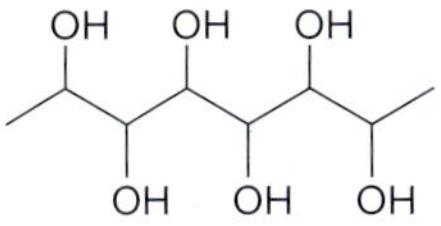

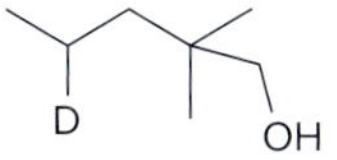

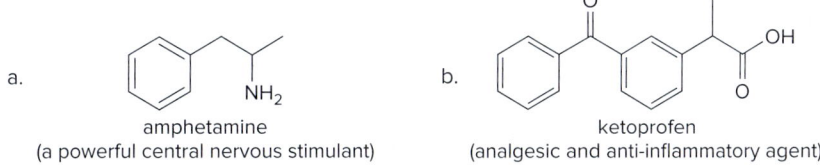

amoxicillin
(an antibiotic)

norethindrone
(oral contraceptive component)

heroin
(an opiate)

5.43 Draw both enantiomers for each biologically active compound.

a. amphetamine
(a powerful central nervous stimulant)

b. ketoprofen
(analgesic and anti-inflammatory agent)

Nomenclature

5.44 Which group in each pair is assigned the higher priority in R,S nomenclature?
- a. $-CD_3$, $-CH_3$
- b. $-CH(CH_3)_2$, $-CH_2OH$
- c. $-CH_2Cl$, $-CH_2CH_2CH_2Br$
- d. $-CH_2NH_2$, $-NHCH_3$

5.45 Rank the following groups in order of decreasing priority.
- a. $-F$, $-NH_2$, $-CH_3$, $-OH$
- b. $-CH_3$, $-CH_2CH_3$, $-CH_2CH_2CH_3$, $-(CH_2)_3CH_3$
- c. $-NH_2$, $-CH_2NH_2$, $-CH_3$, $-CH_2NHCH_3$
- d. $-COOH$, $-CH_2OH$, $-H$, $-CHO$
- e. $-Cl$, $-CH_3$, $-SH$, $-OH$
- f. $-C{\equiv}CH$, $-CH(CH_3)_2$, $-CH_2CH_3$, $-CH=CH_2$

5.46 Label each stereogenic center as R or S.

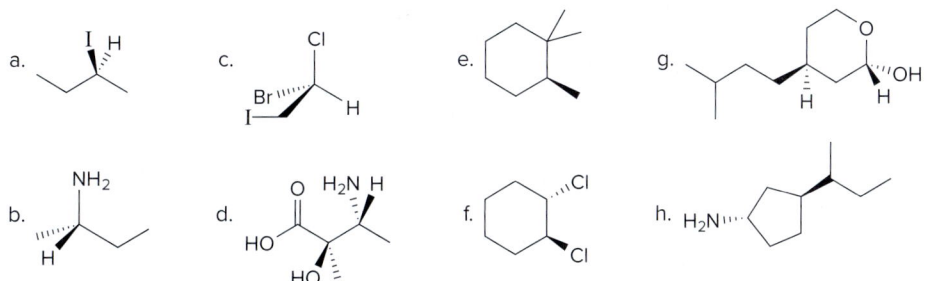

5.47 Locate the stereogenic centers in each Newman projection and label each center as R or S.

a.

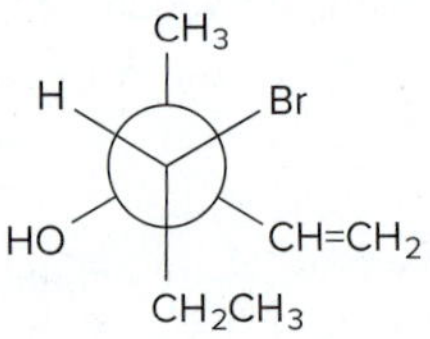

b.

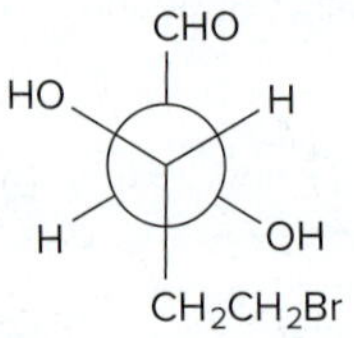

5.48 Draw the structure of (S,S)-ethambutol, a drug used to treat tuberculosis that is 10 times more potent than any of its other stereoisomers.

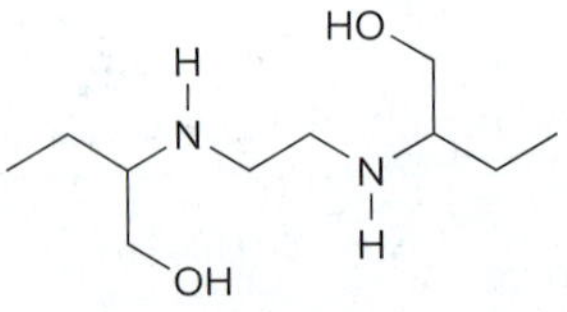

ethambutol

5.49 Draw the structure for each compound.
 a. (R)-3-methylhexane
 b. (4R,5S)-4,5-diethyloctane
 c. (3R,5S,6R)-5-ethyl-3,6-dimethylnonane
 d. (3S,6S)-6-isopropyl-3-methyldecane

5.50 Give the IUPAC name for each compound, including the R,S designation for each stereogenic center.

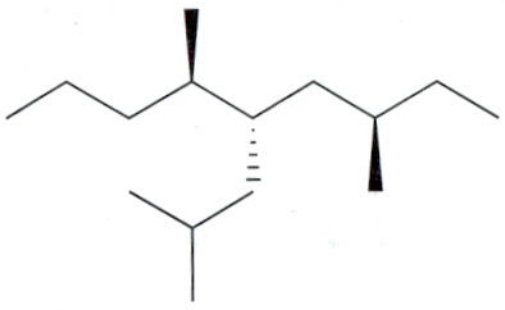

5.51 Locate the stereogenic centers in the following biomolecules and label each stereogenic center as R or S.

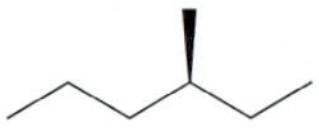

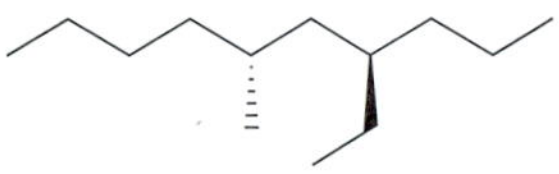

a. 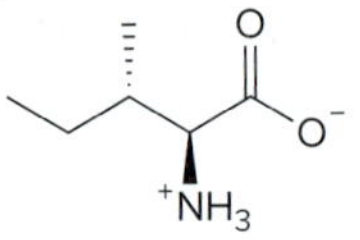 isoleucine (an amino acid)

b. 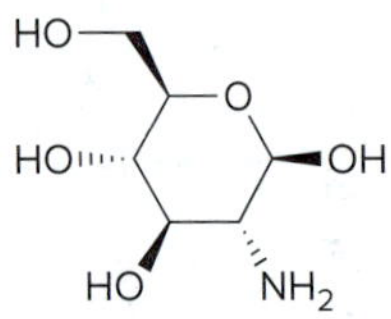 glucosamine (an amino sugar)

c. 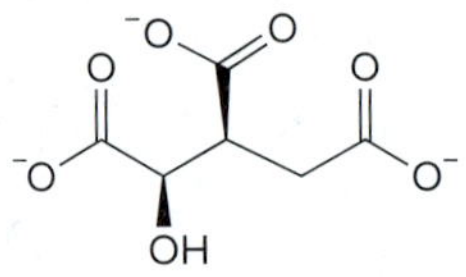 isocitrate (an intermediate in glucose metabolism)

Compounds with More Than One Stereogenic Center

5.52 What is the maximum number of stereoisomers possible for each compound?

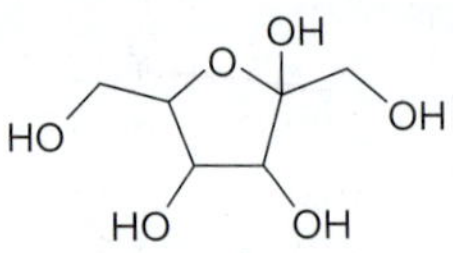

5.53 The shrub ma huang (Section 5.4A) contains two biologically active stereoisomers—ephedrine and pseudoephedrine—with two stereogenic centers as shown in the given structure. Ephedrine is one component of a once-popular combination drug used by body builders to increase energy and alertness, whereas pseudoephedrine is a nasal decongestant.

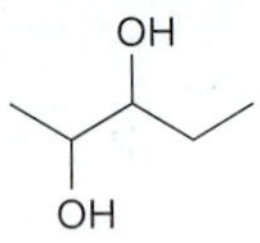

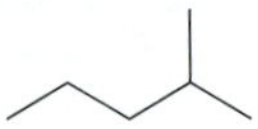

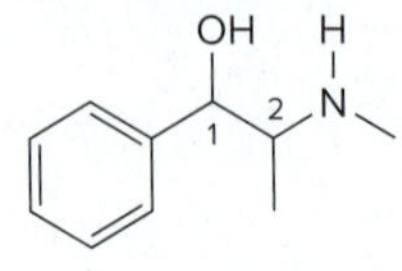

isolated from ma huang

 a. Draw the structure of naturally occurring (−)-ephedrine, which has the 1R,2S configuration.
 b. Draw the structure of naturally occurring (+)-pseudoephedrine, which has the 1S,2S configuration.
 c. How are ephedrine and pseudoephedrine related?
 d. Draw all other stereoisomers of (−)-ephedrine and (+)-pseudoephedrine, and give the R,S designation for all stereogenic centers.
 e. How is each compound drawn in part (d) related to (−)-ephedrine?

5.54 Draw all possible stereoisomers for each compound. Label pairs of enantiomers and diastereomers. Label any meso compound.

a. 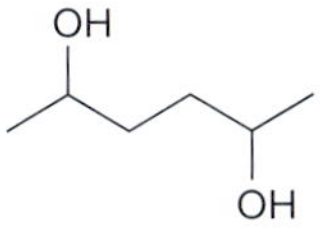 [structure: hexane-2,5-diol] b. 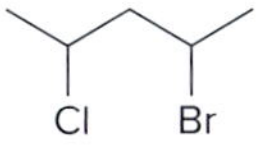 [structure: 2-chloro-3-bromopentane] c. 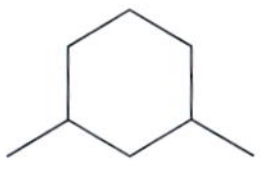 [structure: 1,3-dimethylcyclohexane] d. 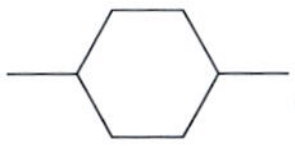 [structure: 1,4-dimethylcyclohexane]

5.55 Hypoglycin A, an amino acid derivative found in unripened lychee, is a compound that is acutely toxic and can lead to death when ingested in large amounts by undernourished children. Draw all possible stereoisomers for hypoglycin A, and give the *R,S* designation for each stereogenic center.

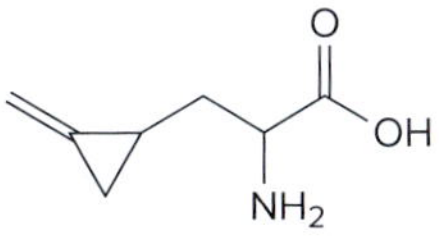

hypoglycin A

Comparing Compounds: Enantiomers, Diastereomers, and Constitutional Isomers

5.56 How is each compound related to the simple sugar D-erythrose? Is it an enantiomer, a diastereomer, or an identical molecule?

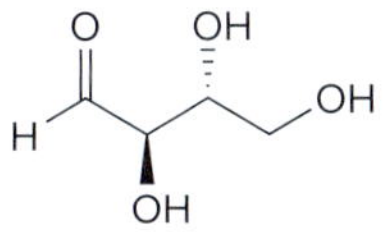

D-erythrose   a. 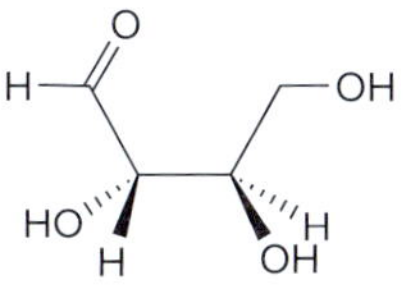   b. 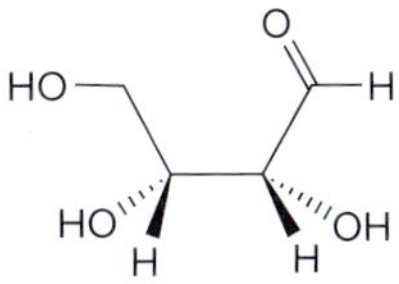   c.

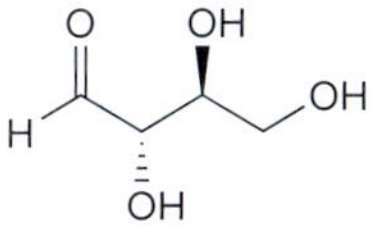

5.57 Consider Newman projections (**A–D**) for four-carbon carbohydrates. How is each pair of compounds related: (a) **A** and **B**; (b) **A** and **C**; (c) **A** and **D**; (d) **C** and **D**? Choose from identical molecules, enantiomers, or diastereomers.

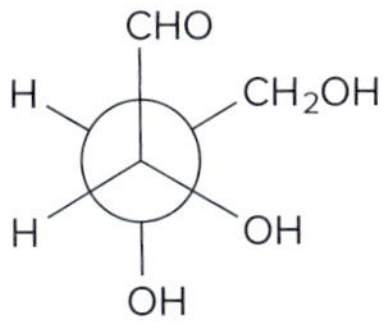

A 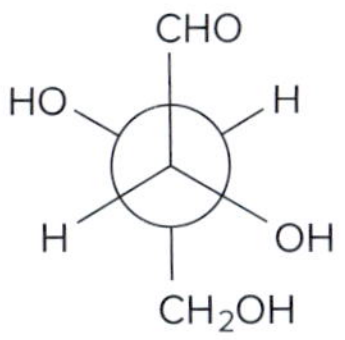   B 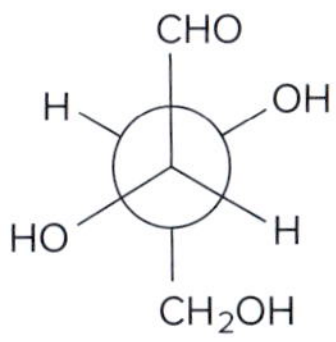   C 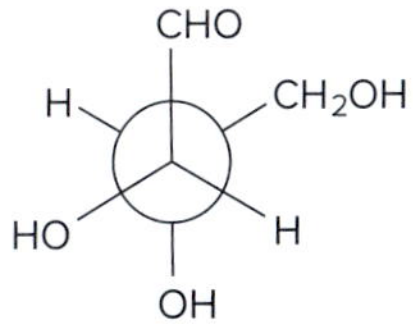   D

5.58 How is compound **A** related to compounds **B–E**? Choose from enantiomers, diastereomers, constitutional isomers, or identical molecules.

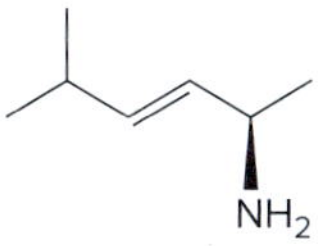

A 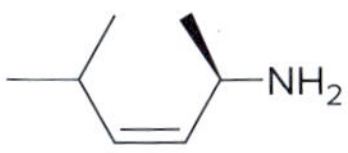   B 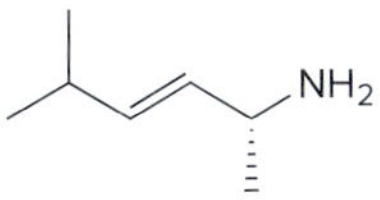   C 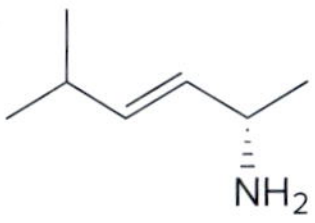   D 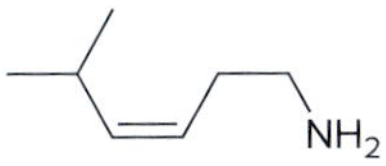   E

5.59 How is each compound (**B–D**) related to **A**? Choose from enantiomers, diastereomers, identical molecules, constitutional isomers, or not isomers of each other.

5.60 How are the compounds in each pair related to each other? Are they identical, enantiomers, diastereomers, constitutional isomers, or not isomers of each other?

Physical Properties of Isomers

5.61 A mixture contains equal amounts of compounds **A–D**.

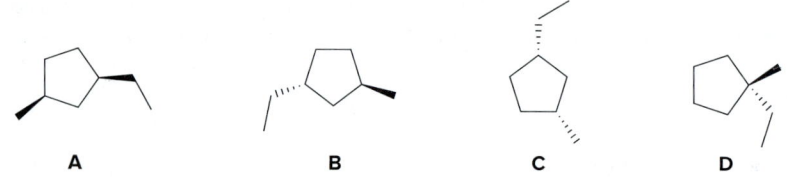

a. Which compounds alone are optically active?
b. If the mixture was subjected to fractional distillation, how many fractions would be obtained?
c. How many of these fractions would be optically active?

5.62 Drawn are four isomeric dimethylcyclopropanes.

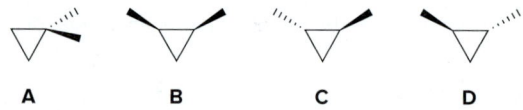

a. How are the compounds in each pair related (enantiomers, diastereomers, constitutional isomers): **A** and **B**; **A** and **C**; **B** and **C**; **C** and **D**?
b. Label each compound as chiral or achiral.
c. Which compounds alone would be optically active?
d. Which compounds have a plane of symmetry?
e. How do the boiling points of the compounds in each pair compare: **A** and **B**; **B** and **C**; **C** and **D**?
f. Which of the compounds are meso compounds?
g. Would an equal mixture of compounds **C** and **D** be optically active? What about an equal mixture of **B** and **C**?

5.63 The [α] of pure quinine, an antimalarial drug, is −165.

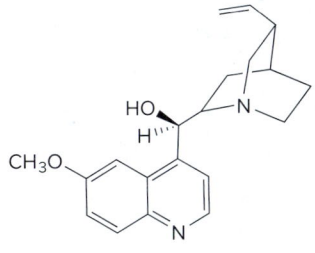

quinine
(antimalarial drug)

a. Calculate the *ee* of a solution with the following [α] values: −50, −83, and −120.
b. For each *ee*, calculate the percent of each enantiomer present.
c. What is [α] for the enantiomer of quinine?
d. If a solution contains 80% quinine and 20% of its enantiomer, what is the *ee* of the solution?
e. What is [α] for the solution described in part (d)?

General Problems

5.64 Captopril is a drug used to treat high blood pressure and congestive heart failure.

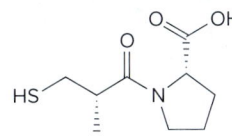

captopril

a. Designate each stereogenic center as *R* or *S*.
b. Draw the enantiomer of captopril.
c. What product is formed when captopril is treated with one equivalent of NaH?
d. What product is formed when captopril is treated with two equivalents of NaH?

5.65 Trabectedin, shown in a ball-and-stick model on the cover of this text, is an anticancer drug sold under the trade name Yondelis.

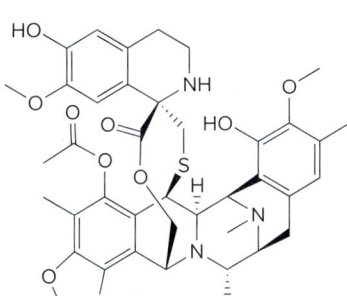

trabectedin

a. Locate the stereogenic centers in trabectedin.
b. What is the maximum number of stereoisomers possible for trabectedin?
c. Draw the enantiomer.
d. Draw a diastereomer.
e. If the specific rotation of trabectedin is +41.5, what is the [α] of a solution that contains 75% trabectedin and 25% of its enantiomer?
f. What is the *ee* of a solution with [α] = +10.5?

Challenge Problems

5.66 A limited number of chiral compounds having no stereogenic centers exist. For example, although **A** is achiral, constitutional isomer **B** is chiral. Make models and explain this observation. Compounds containing two double bonds that share a single carbon atom are called *allenes*. Locate the allene in the antibiotic mycomycin and decide whether mycomycin is chiral or achiral.

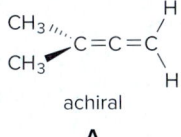

achiral
A

chiral
B

HC≡C−C≡C−CH=C=CH−CH=CH−CH=CH−CH₂CO₂H

mycomycin

5.67 a. Locate all the tetrahedral stereogenic centers in discodermolide, a tumor inhibitor isolated from the Caribbean marine sponge *Discodermia dissoluta*.
b. Certain carbon–carbon double bonds can also be stereogenic centers. With reference to the definition in Section 5.3, explain how this can occur, and then locate the three additional stereogenic centers in discodermolide.
c. Considering all stereogenic centers, what is the maximum number of stereoisomers possible for discodermolide?

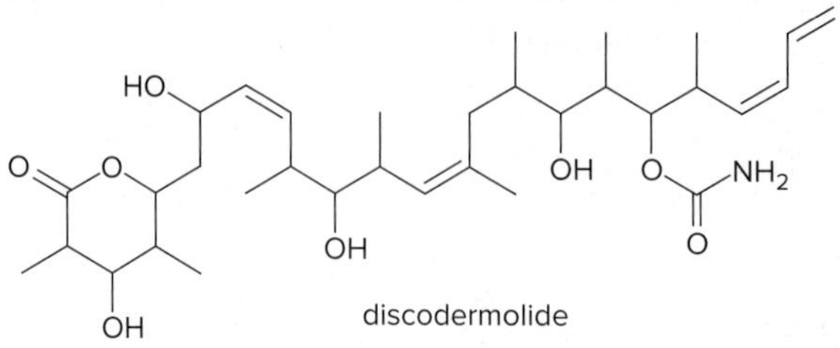

discodermolide

5.68 An acid–base reaction of (*R*)-*sec*-butylamine with a racemic mixture of 2-phenylpropanoic acid forms two products having different melting points and somewhat different solubilities. Draw the structure of these two products. Assign *R* and *S* to any stereogenic centers in the products. How are the two products related? Choose from enantiomers, diastereomers, constitutional isomers, or not isomers of each other.

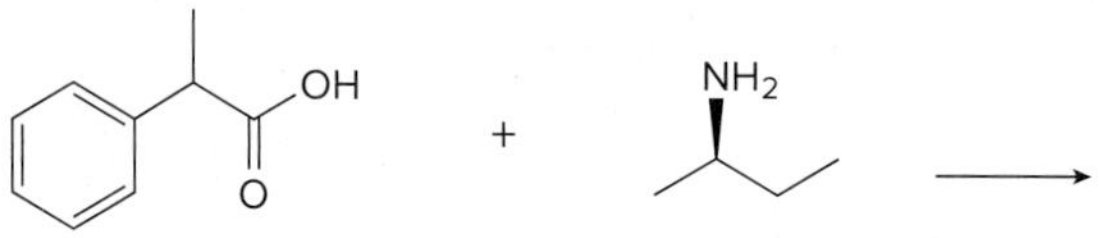

2-phenylpropanoic acid
(racemic mixture)

(*R*)-*sec*-butylamine

Understanding Organic Reactions

6.1 Writing equations for organic reactions
6.2 Kinds of organic reactions
6.3 Bond breaking and bond making
6.4 Bond dissociation energy
6.5 Thermodynamics
6.6 Enthalpy and entropy
6.7 Energy diagrams
6.8 Energy diagram for a two-step reaction mechanism
6.9 Kinetics
6.10 Catalysts
6.11 Enzymes

Ninikas/Getty Images

Glucose, the most abundant simple carbohydrate, is the building block for starch and cellulose and a major sweet-tasting component of honey. Glucose is used as an energy source by most organisms. In humans, when glucose levels are high after a meal is digested, the body stores glucose as glycogen, which is then hydrolyzed when glucose levels fall and energy demands increase. Glucose is transported in the bloodstream and metabolized aerobically to carbon dioxide and water and a great deal of energy. In Chapter 6, we learn about energy changes that accompany chemical reactions.

Why Study... Organic Reactions?

Why do certain reactions occur when two compounds are mixed together, whereas others do not? To answer this question we must learn how and why organic compounds react.

Reactions are at the heart of organic chemistry. The mastery of chemical transformations is essential to our understanding of living organisms as well as laboratory reactions. The most fundamental biological processes, including vision and metabolism, occur because of enzyme-catalyzed organic reactions. Furthermore, our knowledge of these reactions has made possible the conversion of natural substances into new compounds with different, and sometimes superior, properties. Aspirin, ibuprofen, nylon, and polyethylene are all products of chemical reactions between substances derived from petroleum.

Reactions are difficult to learn when each reaction is considered a unique and isolated event. *Avoid this tendency.* **Virtually all chemical reactions are woven together by a few basic themes.** After we learn the general principles, specific reactions then fit neatly into a general pattern.

In our study of organic reactions we will begin with the functional groups, looking for electron-rich and electron-deficient sites, and bonds that might be broken easily. These reactive sites give us a clue as to the general type of reaction a particular class of compound undergoes. Finally, we will learn about how a reaction occurs. Does it occur in one step or in a series of steps? Understanding the details of an organic reaction allows us to determine when it might be used in preparing interesting and useful organic compounds.

6.1 Writing Equations for Organic Reactions

Like other reactions, equations for organic reactions are usually drawn with a single reaction arrow (→) between the starting material and product, but other conventions make these equations look different from those encountered in general chemistry.

The **reagent,** the chemical substance with which an organic compound reacts, is sometimes drawn on the left side of the equation with the other reactants. At other times, the reagent is drawn above or below the reaction arrow itself, to focus attention on the organic starting material by itself on the left side. The solvent and temperature of a reaction may be added above or below the arrow. **The symbols "$h\nu$" and "Δ" are used for reactions that require** *light* **or** *heat,* **respectively.** Figure 6.1 presents an organic reaction in different ways.

When two sequential reactions are carried out without drawing any intermediate compound, the steps are usually numbered above or below the reaction arrow. This convention signifies that the first step occurs *before* the second, and the reagents are added *in sequence,* not at the same time.

> Although chemical reactions are equilibria, which are designated by double reaction arrows (⇌), single reaction arrows (→) are often used instead.

> Often the solvent and temperature of a reaction are omitted from chemical equations, to further focus attention on the main substances involved in the reaction.

> Most organic reactions take place in a **liquid solvent.** Solvents solubilize key reaction components and serve as heat reservoirs to maintain a given temperature. Chapter 7 presents the two major types of reaction solvents and how they affect substitution reactions.

The first reaction...
[1] CH_3MgBr
[2] H_2O
...then the second
(HOMgBr)
inorganic by-product (often omitted)

In this equation only the organic product is drawn on the right side of the arrow. Although the reagent CH_3MgBr contains both Mg and Br, these elements do not appear in the organic product, and they are often omitted on the product side of the equation. These elements have not disappeared. They are part of an inorganic by-product (HOMgBr in this case), and are often of little interest to an organic chemist.

Figure 6.1 Different ways of writing organic reactions

a. cyclohexene + Br_2 → 1,2-dibromocyclohexane

cyclohexene + Br_2 → 1,2-dibromocyclohexane

b. cyclohexane + Br_2, $h\nu$ or Δ, CCl_4 → bromocyclohexane

CCl_4 is the solvent.
$h\nu$—Indicates **light** is needed.
Δ—Indicates **heat** is added.

- The reagent (Br_2) can be on the left side or above the arrow.
- Other reaction parameters can be indicated.

6.2 Kinds of Organic Reactions

Like other compounds, organic molecules undergo acid–base and oxidation–reduction reactions, as discussed in Chapters 2 and 4. Organic molecules also undergo **substitution, elimination,** and **addition** reactions.

6.2A Substitution Reactions

- *Substitution* is a reaction in which an atom or a group of atoms is *replaced* by another atom or group of atoms.

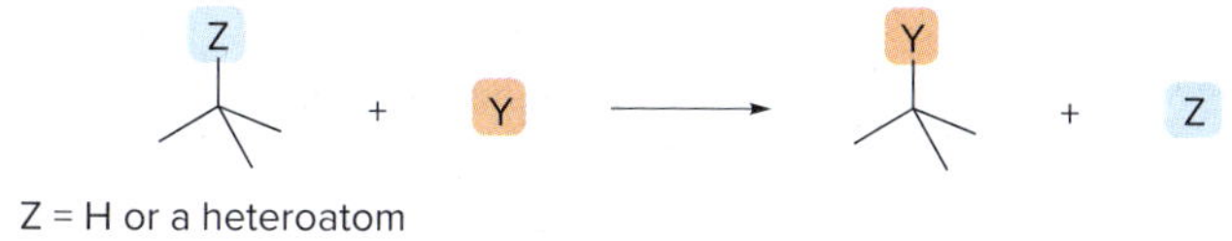

Z = H or a heteroatom

In a general substitution reaction, Y *replaces* Z on a carbon atom. **Substitution reactions involve σ bonds: one σ bond breaks and another forms at the same carbon atom.** The most common examples of substitution occur when Z is hydrogen or a heteroatom that is more electronegative than carbon.

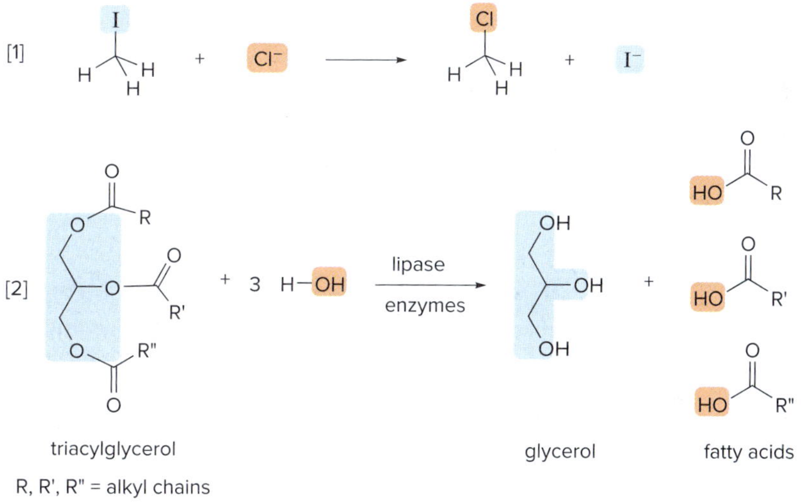

[1]

[2]

triacylglycerol

R, R′, R″ = alkyl chains

glycerol   fatty acids

With a complex starting material, concentrate on the functional groups that *change*. The conversion of the esters in the triacylglycerol to glycerol and three fatty acids is a substitution reaction, because the OH in H_2O replaces the glycerol portion of the ester.

6.2B Elimination Reactions

- *Elimination* is a reaction in which elements of the starting material are "lost" and a π bond is formed.

Two σ bonds are broken.   A π bond is formed.

In an elimination reaction, two groups X and Y are removed from a starting material. **Two σ bonds are broken, and a π bond is formed between adjacent atoms.** The most common examples of elimination occur when X = H and Y is a heteroatom more electronegative than carbon.

> Citrate loses H_2O during the citric acid cycle, an enzyme-catalyzed pathway that occurs during metabolism.

6.2C Addition Reactions

- **Addition is a reaction in which elements are added to a starting material.**

In an addition reaction, new groups X and Y are added to a starting material. **A π bond is broken and two σ bonds are formed.**

> The addition of H_2O to thioesters derived from coenzyme A (HS–CoA) is a key step in fatty acid metabolism (Section 27.3).

> A summary of the general types of organic reactions is given in Appendix I.

Addition and elimination reactions are exactly opposite. A π bond is *formed* in elimination reactions, whereas a π bond is *broken* in addition reactions.

Problem 6.1 Classify each transformation as substitution, elimination, or addition.

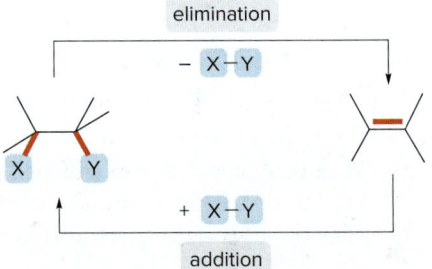

Problem 6.2 To determine the reaction type with complex molecules, concentrate on the functional groups that change. Classify the reaction of **X** and **Y** to form capsaicin as a substitution, elimination, or addition.

Problem 6.3 The following enzyme-catalyzed reactions illustrate the last four steps in the citric acid cycle, a critical part of metabolism discussed in Section 27.6. Classify each reaction as a substitution, elimination, or addition.

succinyl CoA → [1] → succinate → [2] → fumarate → [3] → malate → [4] → oxaloacetate

6.3 Bond Breaking and Bond Making

Having now learned how to write and identify some common kinds of organic reactions, we can turn to a discussion of **reaction mechanism.**

- A *reaction mechanism* is a detailed description of how bonds are broken and formed as a starting material is converted to a product.

A reaction mechanism describes the relative order and rate of bond cleavage and formation. It explains all the known facts about a reaction and accounts for all products formed, and it is subject to modification or refinement as new details are discovered.

A reaction can occur either in one step or in a series of steps.

- **A one-step reaction is called a *concerted reaction*.** No matter how many bonds are broken or formed, a starting material is converted *directly* to a product.

A ⟶ B

- **A stepwise reaction** involves more than one step. A starting material is first converted to an unstable intermediate, called a **reactive intermediate,** which then goes on to form the product.

A ⟶ reactive intermediate ⟶ B

Capsaicin (Problem 6.2) is responsible for the characteristic spicy flavor of jalapeño and habañero peppers. *DNY59/Getty Images*

6.3A Bond Cleavage

Bonds are broken and formed in all chemical reactions. When a bond is broken, the electrons in the bond can be divided **equally** or **unequally** between the two atoms of the bond.

- Breaking a bond by *equally dividing* the electrons between the two atoms in the bond is called **homolysis** or **homolytic cleavage.**

A—B ⟶(homolysis) A• + •B

Each atom gets **one** electron.

- Breaking a bond by *unequally dividing* the electrons between the two atoms in the bond is called **heterolysis** or **heterolytic cleavage**.

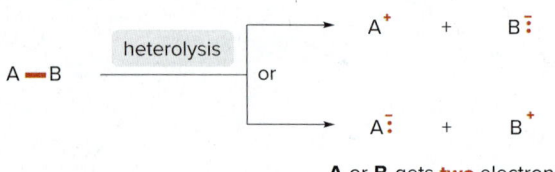

Heterolysis of a bond between **A** and **B** can give either **A** or **B** the two electrons in the bond. When **A** and **B** have different electronegativities, the *electrons normally end up on the more electronegative atom*.

Homolysis and heterolysis require energy. Both processes generate reactive intermediates, but the products are different in each case.

- Homolysis generates uncharged reactive intermediates with *unpaired* electrons.
- Heterolysis generates *charged* intermediates.

Each of these reactive intermediates has a very short lifetime and reacts quickly to form a stable organic product.

6.3B Radicals, Carbocations, and Carbanions

The curved arrow notation first discussed in Section 1.6B works fine for heterolytic bond cleavage because it illustrates the movement of an **electron pair.** For homolytic cleavage, however, one electron moves to one atom in the bond and one electron moves to the other, so a different kind of curved arrow is needed.

- To illustrate the movement of a single electron, use a half-headed curved arrow, sometimes called a *fishhook*.

A full-headed curved arrow (⌒) shows the movement of an electron *pair*. **A half-headed curved arrow** (⌒) shows the movement of a *single* electron.

- Two **half-headed** curved arrows are needed for two **single** electrons.
- One **full-headed** curved arrow is needed for one electron **pair**.

Figure 6.2 illustrates homolysis and two different heterolysis reactions for a carbon compound using curved arrows. Three different reactive intermediates are formed.

Figure 6.2
Three reactive intermediates resulting from homolysis and heterolysis of a C–Z bond

- Radicals are intermediates in **radical** reactions.
- **Ionic intermediates** are seen in **polar** reactions.

Homolysis of the C–Z bond generates two uncharged products with unpaired electrons.

• A reactive intermediate with a single unpaired electron is called a *radical*.

Most radicals are highly unstable because they contain an atom that does not have an octet of electrons. Radicals typically have **no charge. They are intermediates in a group of reactions called** *radical reactions,* which are discussed in detail in Chapter 21.

Heterolysis of the C–Z bond can generate a **carbocation** or a **carbanion.**

• Giving two electrons to Z and none to carbon generates a positively charged carbon intermediate called a *carbocation.*
• Giving two electrons to C and none to Z generates a negatively charged carbon species called a *carbanion.*

Both carbocations and carbanions are unstable reactive intermediates: A carbocation contains a carbon atom surrounded by only six electrons. A carbanion has a negative charge on carbon, which is not a very electronegative atom. **Carbocations (electrophiles)** and **carbanions (nucleophiles)** can be intermediates in *polar reactions—***reactions in which a nucleophile reacts with an electrophile.**

Problem 6.4 By taking into account electronegativity differences, draw the products formed by heterolysis of the carbon–heteroatom bond in each molecule. Classify the organic reactive intermediate as a carbocation or a carbanion.

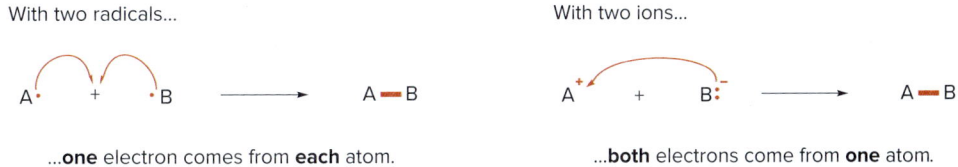

6.3C Bond Formation

Like bond cleavage, bond formation occurs in two different ways. Two radicals can each donate **one electron** to form a two-electron bond. Alternatively, two ions with unlike charges can come together, with the negatively charged ion donating **both electrons** to form the resulting two-electron bond. **Bond formation always releases energy.**

With two radicals...

A• + •B ⟶ A—B

...**one** electron comes from **each** atom.

With two ions...

A⁺ + B:⁻ ⟶ A—B

...**both** electrons come from **one** atom.

6.3D All Kinds of Arrows

Table 6.1 summarizes the many kinds of arrows used in describing organic reactions. Curved arrows are especially important because they explicitly show what electrons are involved in a reaction, how these electrons move in forming and breaking bonds, and if a reaction proceeds via a radical or polar pathway.

226 Chapter 6 Understanding Organic Reactions

A more complete summary of the arrows used in organic chemistry is given in Appendix B, Common Abbreviations, Arrows, and Symbols.

Table 6.1 A Summary of Arrow Types in Chemical Reactions

Arrow	Name	Use
⟶	Reaction arrow	Drawn between the starting materials and products in an equation (6.1)
⇌	Double reaction arrows (equilibrium arrows)	Drawn between the starting materials and products in an equilibrium equation (2.2)
⟷	Double-headed arrow	Drawn between resonance structures (1.6B)
 ↷	Full-headed curved arrow	Shows movement of an electron pair (1.6B, 2.2)
 ↼	Half-headed curved arrow (fishhook)	Shows movement of a single electron (6.3B)

Sample Problem 6.1 Using Curved Arrows in an Equation

Use curved arrows to show the movement of electron pairs in each reaction.

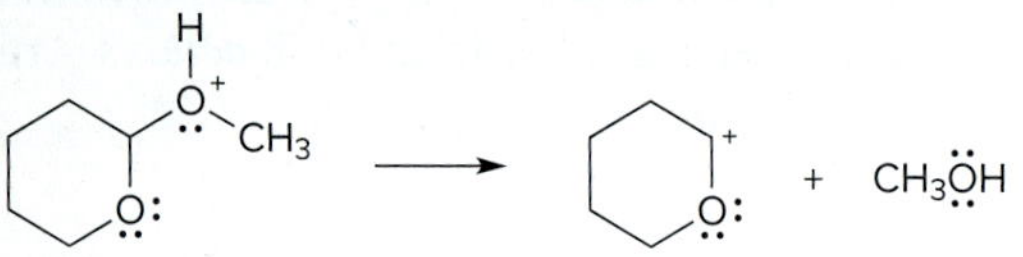

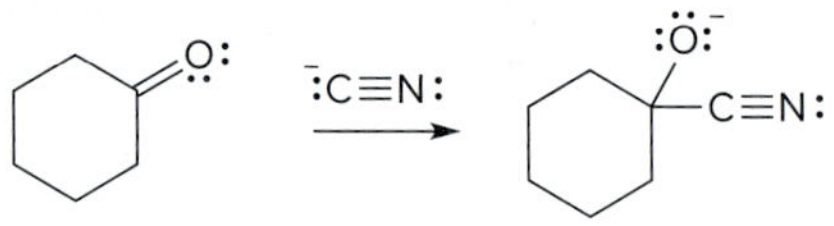

Solution

Concentrate on bonds that are broken or formed, and pay attention to atoms that have different charges in the reactants and products.

a. Only *one* C—O bond is broken, so only *one* curved arrow is needed. The electron pair in the C—O bond (in red) ends up on O.

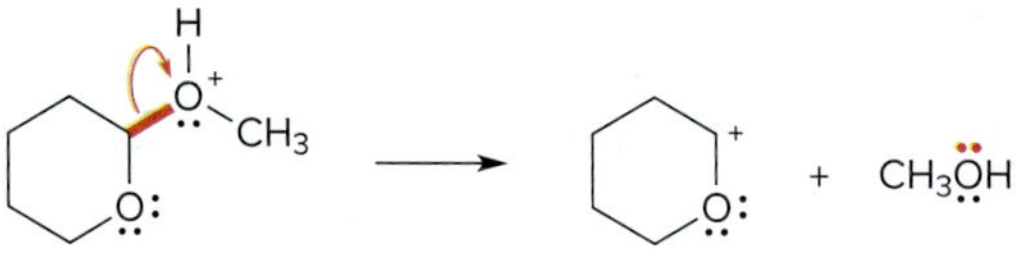

b. *Two* curved arrows are needed because *two* electron pairs are involved. The lone pair on C in ⁻CN forms a new bond to the carbonyl carbon, and an electron pair in the C=O moves onto O.

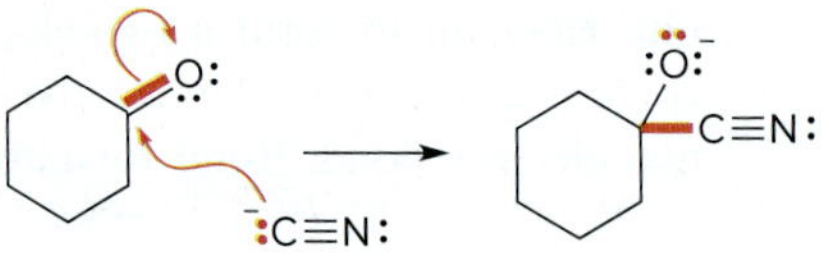

Problem 6.5
Use curved arrows to show the movement of electrons in each equation.

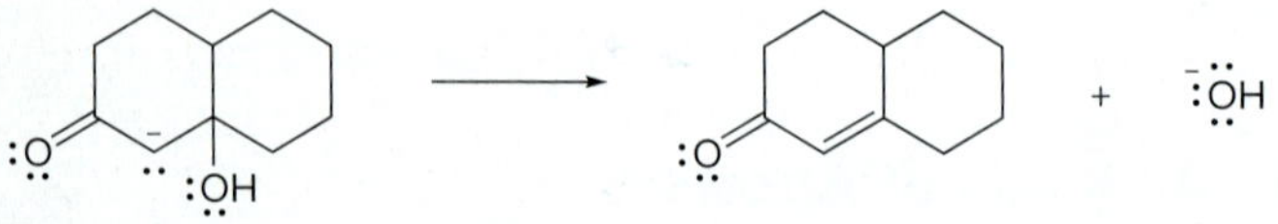

More Practice: Try Problems 6.29, 6.31a, 6.33, 6.34a, 6.44a, 6.49a, 6.51a, 6.52a.

Sample Problem 6.2 — Following Curved Arrows to Draw a Reaction Product

Follow the curved arrows and draw the products of the following reaction.

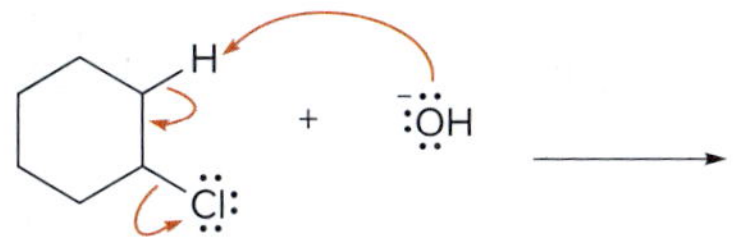

Solution

Three full-headed curved arrows are drawn, so three electron *pairs* take part in the reaction. Arrow **1** shows that a lone pair on ⁻OH forms a new bond to H, forming H₂O. Arrow **2** indicates that the electron pair in the C—H bond forms a carbon–carbon double bond. Arrow **3** shows that the electron pair in the C—Cl bond ends up on Cl, forming Cl⁻. After breaking and making bonds, formal charges on the atoms involved in the reaction are adjusted when necessary.

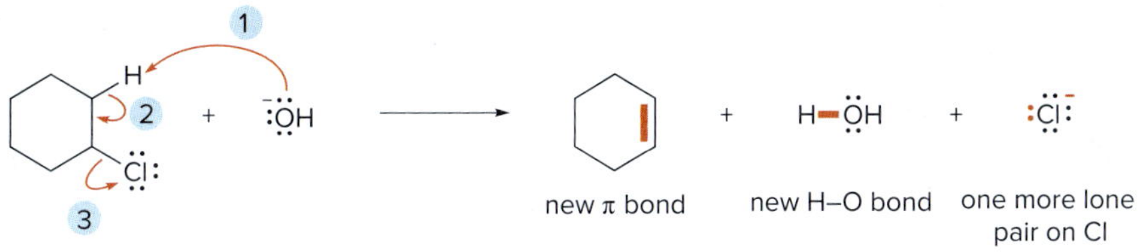

Problem 6.6

Follow the curved arrows and draw the products of each reaction.

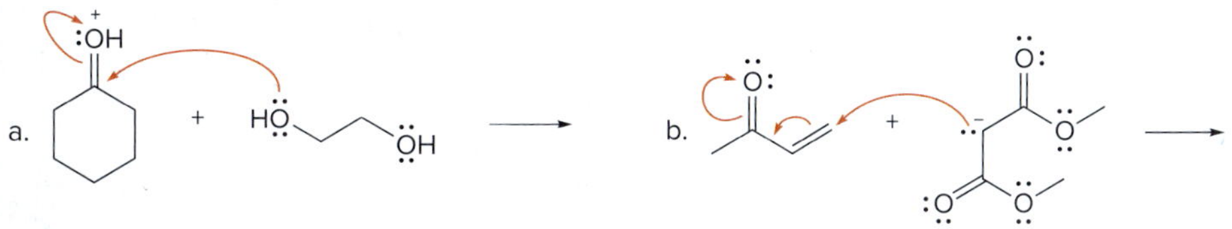

More Practice: Try Problems 6.30, 6.32, 6.34b.

6.4 Bond Dissociation Energy

Bond breaking can be quantified using the bond dissociation energy.

*Bond dissociation **energy** is also called bond dissociation **enthalpy** because it refers to the heat absorbed when bonds are cleaved.*

- The *bond dissociation energy* is the energy needed to homolytically cleave a covalent bond.

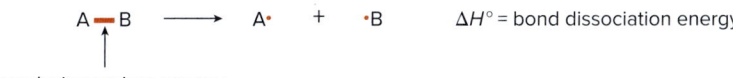

$$A\!-\!B \longrightarrow A\cdot \;+\; \cdot B \qquad \Delta H° = \text{bond dissociation energy}$$

Homolysis requires energy.

The superscript (°) means that values are determined under standard conditions (pure compounds in their most stable state at 25 °C and 1 atm pressure).

The energy absorbed or released in any reaction, symbolized by $\Delta H°$, is called the **enthalpy change** or **heat of reaction.**

- When $\Delta H°$ is positive (+), energy is absorbed and the reaction is *endothermic*.
- When $\Delta H°$ is negative (−), energy is released and the reaction is *exothermic*.

> Additional bond dissociation energies for C—C multiple bonds are given in Table 1.6.
>
> A more extensive table of bond dissociation energies appears in Appendix E.

A bond dissociation energy is the $\Delta H°$ for a specific kind of reaction—the homolysis of a covalent bond to form two radicals. Because bond breaking requires energy, **bond dissociation energies are always *positive* numbers,** and homolysis is always **endothermic.** Conversely, **bond formation always *releases* energy,** so this reaction is always **exothermic.** The H—H bond requires +435 kJ/mol to cleave and releases −435 kJ/mol when formed. Table 6.2 contains a representative list of bond dissociation energies for many common bonds.

$$H-H \longrightarrow H\cdot + \cdot H$$
$$\Delta H = +435 \text{ kJ/mol}$$
endothermic reaction

$$H\cdot + \cdot H \longrightarrow H-H$$
$$\Delta H = -435 \text{ kJ/mol}$$
exothermic reaction

Comparing bond dissociation energies is equivalent to comparing **bond strength.**

- The *stronger* the bond, the *higher* its bond dissociation energy.

For example, the H—H bond is stronger than the Cl—Cl bond because its bond dissociation energy is higher [Table 6.2: 435 kJ/mol (H_2) versus 242 kJ/mol (Cl_2)]. The data in Table 6.2 demonstrate that **bond dissociation energies *decrease* down a column of the periodic table**

Table 6.2 Bond Dissociation Energies for Some Common Bonds [A—B → A• + •B]

Bond	$\Delta H°$ kJ/mol	(kcal/mol)	Bond	$\Delta H°$ kJ/mol	(kcal/mol)
H—Z bonds			**R—X bonds**		
H—F	569	(136)	CH_3—F	456	(109)
H—Cl	431	(103)	CH_3—Cl	351	(84)
H—Br	368	(88)	CH_3—Br	293	(70)
H—I	297	(71)	CH_3—I	234	(56)
H—OH	498	(119)	CH_3CH_2—F	448	(107)
			CH_3CH_2—Cl	339	(81)
Z—Z bonds			CH_3CH_2—Br	285	(68)
H—H	435	(104)	CH_3CH_2—I	222	(53)
F—F	159	(38)	$(CH_3)_2CH$—F	444	(106)
Cl—Cl	242	(58)	$(CH_3)_2CH$—Cl	335	(80)
Br—Br	192	(46)	$(CH_3)_2CH$—Br	285	(68)
I—I	151	(36)	$(CH_3)_2CH$—I	222	(53)
HO—OH	213	(51)	$(CH_3)_3C$—F	444	(106)
			$(CH_3)_3C$—Cl	331	(79)
R—H bonds			$(CH_3)_3C$—Br	272	(65)
CH_3—H	435	(104)	$(CH_3)_3C$—I	209	(50)
CH_3CH_2—H	410	(98)			
$CH_3CH_2CH_2$—H	410	(98)	**R—Z bonds**		
$(CH_3)_2CH$—H	397	(95)	CH_3—OH	389	(93)
$(CH_3)_3C$—H	381	(91)	CH_3CH_2—OH	393	(94)
$CH_2=CH$—H	435	(104)	$CH_3CH_2CH_2$—OH	385	(92)
$HC\equiv C$—H	523	(125)	$(CH_3)_2CH$—OH	401	(96)
$CH_2=CHCH_2$—H	364	(87)	$(CH_3)_3C$—OH	401	(96)
C_6H_5—H	460	(110)	CH_3—NH_2	331	(79)
$C_6H_5CH_2$—H	356	(85)	CH_3—SH	305	(73)

6.4 Bond Dissociation Energy

as the valence electrons used in bonding are farther from the nucleus. Bond dissociation energies for a group of methyl–halogen bonds exemplify this trend.

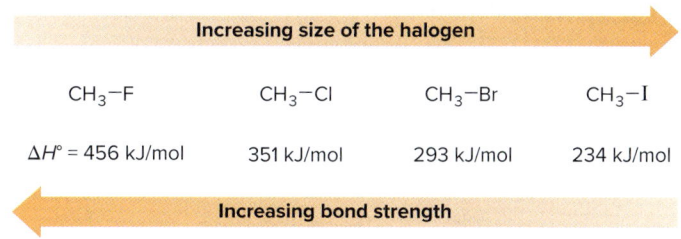

Because bond length increases down a column of the periodic table, bond dissociation energies are a quantitative measure of the general phenomenon noted in Chapter 1—*shorter* **bonds are** *stronger* **bonds.**

Problem 6.7 Which bond in each pair has the higher bond dissociation energy?

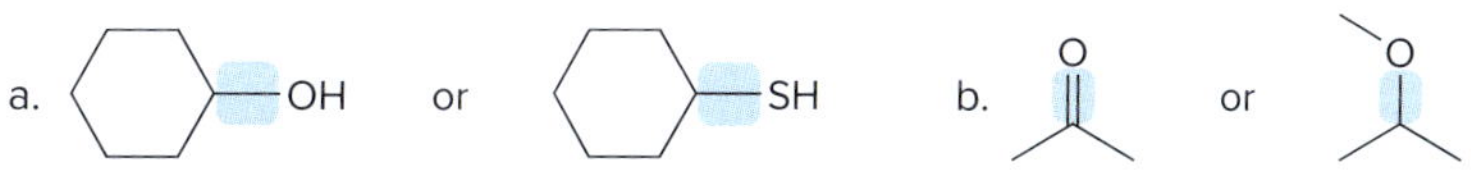

a. cyclohexyl—OH or cyclohexyl—SH   b. (ketone) or (ketone)

Bond dissociation energies are also used to calculate the enthalpy change ($\Delta H°$) in a reaction in which several bonds are broken and formed. **$\Delta H°$ indicates the relative strength of bonds broken and formed in a reaction.**

- When $\Delta H°$ is *positive*, more energy is needed to break bonds than is released in forming bonds. The bonds broken in the starting material are *stronger* than the bonds formed in the product.

- When $\Delta H°$ is *negative*, more energy is released in forming bonds than is needed to break bonds. The bonds formed in the product are *stronger* than the bonds broken in the starting material.

To determine the overall $\Delta H°$ for a reaction:

[1] Beginning with a *balanced* equation, add the bond dissociation energies for all bonds broken in the starting materials. This (+) value represents the **energy needed** to break bonds.

[2] Add the bond dissociation energies for all bonds formed in the products. This (−) value represents the **energy released** in forming bonds.

[3] **The overall $\Delta H°$ is the sum in Step [1] *plus* the sum in Step [2].**

$$\Delta H° \text{ overall enthalpy change} = \text{sum of } \Delta H° \text{ of bonds broken} + (-) \text{ sum of } \Delta H° \text{ of bonds formed}$$

Sample Problem 6.3 Using Bond Dissociation Energies to Calculate $\Delta H°$

Use the values in Table 6.2 to determine $\Delta H°$ for the following reaction.

t-BuCl + H—O—H ⟶ t-BuOH + H—Cl

Solution

[1] Bonds broken		[2] Bonds formed		[3] Overall $\Delta H° =$
	$\Delta H°$ (kJ/mol)		$\Delta H°$ (kJ/mol)	sum in Step [1] + sum in Step [2]
$(CH_3)_3C-Cl$	+331	$(CH_3)_3C-OH$	−401	
$H-OH$	+498	$H-Cl$	−431	
Total	+829 kJ/mol	Total	−832 kJ/mol	+829 kJ/mol −832 kJ/mol
Energy needed to break bonds.		**Energy released in forming bonds.**		Answer: −3 kJ/mol

Because $\Delta H°$ is a negative value, this reaction is **exothermic** and energy is released. **The bonds broken in the starting material are *weaker* than the bonds formed in the product.**

Problem 6.8 Use the values in Table 6.2 to calculate $\Delta H°$ for each reaction. Classify each reaction as endothermic or exothermic.

a. $\diagup\!\!\diagdown\!\!\text{Br}$ + H_2O → $\diagup\!\!\diagdown\!\!\text{OH}$ + HBr

b. CH_4 + Cl_2 → CH_3Cl + HCl

More Practice: Try Problems 6.36, 6.44b.

Certain metabolic compounds, such as adenosine triphosphate (**ATP,** Section 3.2D), are called "high-energy" molecules, because they undergo highly exothermic reactions. Processes such as walking, running, swallowing, and breathing are fueled by the energy released from ATP hydrolysis.

ATP + H−OH $\xrightarrow{\text{Mg}^{2+}\text{, ATPase enzyme}}$ ADP + HPO$_4^{2-}$

ATP hydrolysis

adenosine triphosphate **ATP** → adenosine diphosphate **ADP**
$\Delta H°' = -22$ kJ/mol

Multiple enzymes exist in humans and other organisms to carry out phosphate hydrolysis reactions in the presence of Mg^{2+}.

In the hydrolysis of ATP with H_2O to form ADP and HPO_4^{2-}, a P−O bond and H−O bond are broken, and a P−O bond and H−O bond are formed. Because more energy is released in bond formation than is absorbed in bond cleavage, the reaction is **exothermic** (−22 kJ/mol).

- **Because a substantial amount of heat is given off during ATP hydrolysis, ATP is an excellent energy source in biological systems.**

The superscript (') means the reaction was run at a specified concentration of a species. In biochemical reactions H^+ is often either consumed or produced, so (') is added to designate that the H^+ concentration (pH) remains constant.

In a series of chemical reactions in metabolic biochemical pathways, glucose, the chapter-opening molecule, is oxidized to form CO_2 and H_2O. Several bonds break and form in this process.

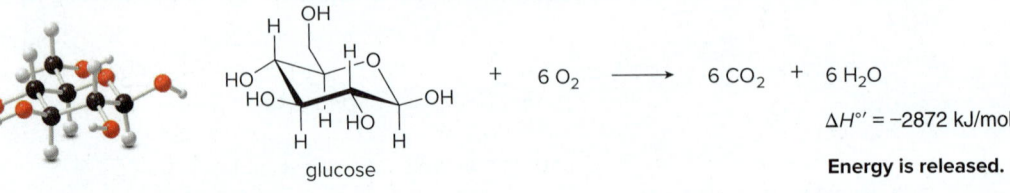

glucose + 6 O_2 → 6 CO_2 + 6 H_2O

$\Delta H°' = -2872$ kJ/mol

Energy is released.

$\Delta H^{\circ\prime}$ is negative for this oxidation, so the overall reaction is exothermic. **Glucose releases heat on oxidation because the reaction series results in a net gain in bond strength in the products versus the reactants.**

Bond dissociation energies have two important limitations. They present only *overall* energy changes. They reveal nothing about the reaction mechanism or how fast a reaction proceeds. Moreover, bond dissociation energies are determined for reactions in the gas phase, whereas most organic reactions are carried out in a liquid solvent where solvation energy contributes to the overall enthalpy of a reaction. As such, bond dissociation energies are imperfect indicators of energy changes in a reaction. Despite these limitations, using bond dissociation energies to calculate ΔH° gives a useful approximation of the energy changes that occur when bonds are broken and formed in a reaction.

Problem 6.9 Calculate ΔH° for each oxidation reaction. Each equation is balanced as written; remember to take into account the coefficients in determining the number of bonds broken or formed. [ΔH° for O_2 = 497 kJ/mol; ΔH° for one C=O in CO_2 = 535 kJ/mol]

a. $CH_4 + 2\ O_2 \longrightarrow CO_2 + 2\ H_2O$
b. $2\ CH_3CH_3 + 7\ O_2 \longrightarrow 4\ CO_2 + 6\ H_2O$

6.5 Thermodynamics

For a reaction to be practical, the equilibrium must favor the products, *and* the reaction rate must be fast enough to form them in a reasonable time. These two conditions depend on the **thermodynamics** and the **kinetics** of a reaction, respectively.

- *Thermodynamics* describes energy and equilibrium. How do the *energies* of the reactants and the products compare? What are the relative *amounts* of reactants and products at equilibrium?

Reaction kinetics are discussed in Section 6.9.

- *Kinetics* describes reaction rates. How *fast* are reactants converted to products?

6.5A Equilibrium Constant and Free Energy Changes

K_{eq} was first defined in Section 2.3 for acid–base reactions.

The **equilibrium constant, K_{eq},** is a mathematical expression that relates the amount of starting material and product at equilibrium. For example, when starting materials **A** and **B** react to form products **C** and **D**, the equilibrium constant is given by the following expression:

$$A + B \rightleftharpoons C + D$$

$$K_{eq} = \frac{[\text{products}]}{[\text{starting materials}]} = \frac{[C][D]}{[A][B]}$$

The size of K_{eq} tells about the position of equilibrium; that is, it expresses whether the starting materials or products predominate once equilibrium has been reached.

- When $K_{eq} > 1$, equilibrium favors the *products* (**C** and **D**) and the equilibrium lies to the *right* as the equation is written.
- When $K_{eq} < 1$, equilibrium favors the *starting materials* (**A** and **B**) and the equilibrium lies to the *left* as the equation is written.

- For a reaction to be useful, the equilibrium must favor the products, and $K_{eq} > 1$.

What determines whether equilibrium favors the products in a given reaction? **The position of equilibrium is determined by the relative energies of the reactants and products.** The free energy of a molecule, also called its **Gibbs free energy,** is symbolized by G°. The **change in free energy** between reactants and products, symbolized by ΔG°, determines whether the starting materials or products are favored at equilibrium.

- $\Delta G°$ is the overall energy difference between reactants and products.

$$\Delta G° = G°_{products} - G°_{reactants}$$

(free energy of the products) (free energy of the reactants)

$\Delta G°$ is related to the equilibrium constant K_{eq} by the following equation:

$$\Delta G° = -2.303RT \log K_{eq}$$

[R = 8.314 J/(K·mol), the gas constant
T = Kelvin temperature (K)]

Using this expression, we can determine the relationship between the equilibrium constant and the free energy change between reactants and products.

At 25 °C, $2.303RT = 5.7$ kJ/mol; thus, $\Delta G° = -5.7 \log K_{eq}$.

$K_{eq} > 1$ when $\Delta G° < 0$, and equilibrium favors the *products*.
$K_{eq} < 1$ when $\Delta G° > 0$, and equilibrium favors the *starting materials*.

- When $K_{eq} > 1$, log K_{eq} is positive, making $\Delta G°$ negative, and energy is *released*. Thus, equilibrium favors the products when the energy of the products is *lower* than the energy of the reactants.
- When $K_{eq} < 1$, log K_{eq} is negative, making $\Delta G°$ positive, and energy is *absorbed*. Thus, equilibrium favors the reactants when the energy of the products is *higher* than the energy of the reactants.

Compounds that are lower in energy have increased stability. Thus, **equilibrium favors the products when they are *more stable* (lower in energy) than the starting materials of a reaction.** This is summarized in Figure 6.3.

Figure 6.3
Summary of the relationship between $\Delta G°$ and K_{eq}

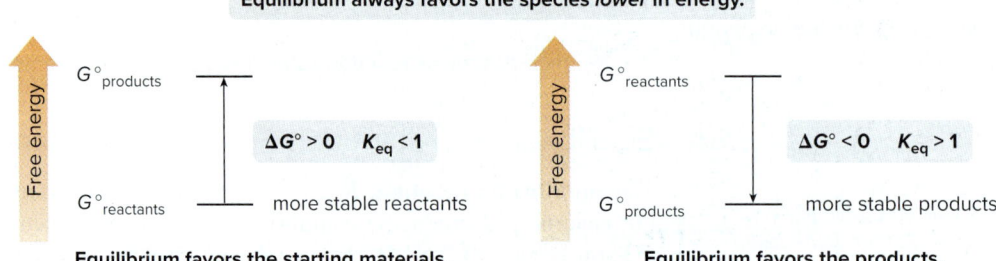

Table 6.3 Representative Values for $\Delta G°$ and K_{eq} at 25 °C, for a Reaction A → B

$\Delta G°$ (kJ/mol)	K_{eq}	Relative amount of A and B at equilibrium
+18	10^{-3}	Essentially all A (99.9%)
+12	10^{-2}	100 times as much **A** as **B**
+6	10^{-1}	10 times as much **A** as **B**
0	1	Equal amounts of **A** and **B**
−6	10^{1}	10 times as much **B** as **A**
−12	10^{2}	100 times as much **B** as **A**
−18	10^{3}	Essentially all B (99.9%)

Increasing [product]

Because $\Delta G°$ depends on the logarithm of K_{eq}, a **small change in energy corresponds to a large difference in the relative amount of starting material and product at equilibrium.** Several values of $\Delta G°$ and K_{eq} are given in Table 6.3. For example, a difference in energy of only ~6 kJ/mol means that there is 10 times as much of the more stable species at equilibrium. A difference in energy of ~18 kJ/mol means that there is essentially only one compound, either starting material or product, at equilibrium.

Problem 6.10 (a) Which K_{eq} corresponds to a negative value of $\Delta G°$, $K_{eq} = 1000$ or $K_{eq} = .001$? (b) Which K_{eq} corresponds to a lower value of $\Delta G°$, $K_{eq} = 10^{-2}$ or $K_{eq} = 10^{-5}$?

The symbol ~ means "approximately."

Problem 6.11 Given each of the following values, is the starting material or product favored at equilibrium?
a. $K_{eq} = 5.5$ b. $\Delta G° = 40$ kJ/mol

Problem 6.12 Given each of the following values, is the starting material or product lower in energy?
a. $\Delta G° = 8.0$ kJ/mol   b. $K_{eq} = 10$ c. $\Delta G° = -12$ kJ/mol   d. $K_{eq} = 10^{-3}$

6.5B Coupled Reactions in Metabolism

As we learned in Section 6.4, the hydrolysis of ATP to ADP is highly exothermic. Because the overall energy difference between the reactants and products ($\Delta G°'$) is also negative (–), **the energy released in ATP hydrolysis can be used to drive a reaction that has an unfavorable energy change by *coupling* the two reactions.**

- Coupled reactions are reactions that are paired together to drive an unfavorable process. The energy released by one reaction provides the energy to drive the other reaction.

Consider the oxidation of glucose to CO_2 and H_2O, resulting from several biochemical reactions. Although the overall conversion releases a great deal of energy, the energy change associated with each individual step is much smaller. In some reactions energy is released, and in others energy is absorbed. The hydrolysis of ATP provides energy to drive reactions that require energy.

In the first step of glucose metabolism, glucose is phosphorylated to form glucose 6-phosphate.

Coupled reactions that involve ATP or other reagents are often drawn using a **combination of horizontal and curved arrows**. The principal reactants and products for a given pathway are drawn from left to right with a reaction arrow as usual, and the other reagents are drawn using a curved arrow.

glucose → (ATP, ADP, Mg^{2+}, hexokinase enzyme) → glucose 6-phosphate $\Delta G°' = -16.7$ kJ/mol

When glucose reacts directly with hydrogen phosphate (HPO_4^{2-}) to form glucose 6-phosphate, 13.8 kJ/mol of energy is required, so this process is energetically *unfavorable*. By coupling glucose phosphorylation with energetically *favorable* ATP hydrolysis (–30.5 kJ/mol of energy released), the coupled reaction becomes an energetically *favorable* process (–16.7 kJ/mol of energy released).

				$\Delta G°'$	
Phosphorylation	glucose + HPO_4^{2-}	→	glucose 6-phosphate + H_2O	+13.8 kJ/mol	← energy required
Hydrolysis	ATP + H_2O	→	ADP + HPO_4^{2-}	–30.5 kJ/mol	← energy released
Coupled reaction	glucose + ATP	→	glucose 6-phosphate + ADP	–16.7 kJ/mol	

Energy is released.

- The coupled reaction is the **net reaction**, written by summing the substances in both equations and eliminating those compounds that appear on both sides of the reaction arrows.
- The overall energy change is found by summing the energies for the individual steps.

Thus, in this example:

- **The hydrolysis of ATP provides the energy for the phosphorylation of glucose.**

Although the coupled reactions are written as two separate equations for emphasis in the preceding example, in reality a single reaction takes place. ATP transfers a phosphate to glucose, forming glucose 6-phosphate and ADP, while both molecules are held in close proximity at the active site of hexokinase, as shown in Figure 6.4.

Figure 6.4
A coupled reaction—
Phosphorylation of
glucose by ATP

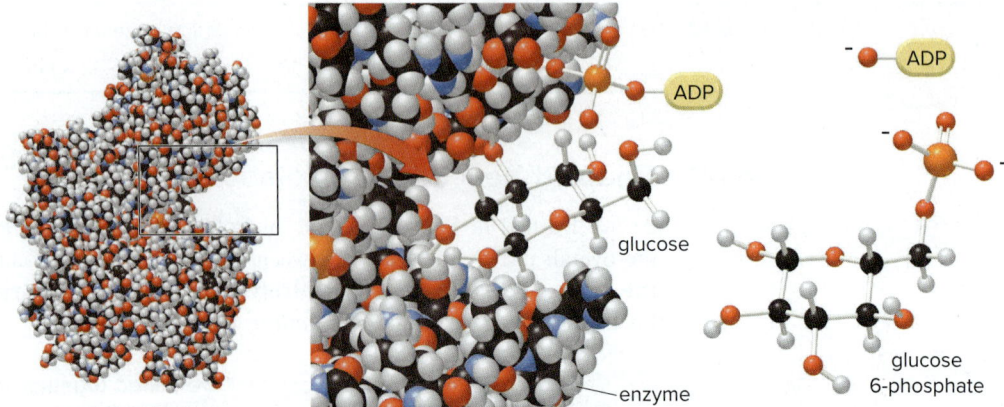

- The enzyme hexokinase binds glucose at its active site and ATP is also bound in close proximity. ATP transfers a phosphate directly to the glucose molecule to form ADP and glucose 6-phosphate. This coupled reaction is energetically favorable.

Problem 6.13 The phosphorylation of fructose to fructose 6-phosphate requires 15.9 kJ/mol of energy. In fructose metabolism, this unfavorable reaction is driven by the hydrolysis of ATP to ADP in the presence of Mg^{2+} and the fructokinase enzyme.

fructose + HPO_4^{2-} ⟶ fructose 6-phosphate + H_2O $\Delta G°' = +15.9$ kJ/mol

a. Write the net (coupled) reaction by summing the substances in both the phosphorylation and hydrolysis equations.
b. How much energy is released in the coupled reaction?
c. Write the equation for the coupled reaction using coupled reaction arrows.

6.6 Enthalpy and Entropy

The **free energy change** ($\Delta G°$) depends on the **enthalpy change** ($\Delta H°$) and the **entropy change** ($\Delta S°$). $\Delta H°$ indicates relative bond strength, but what does $\Delta S°$ measure?

Entropy ($S°$) **is a measure of the randomness in a system.** The more freedom of motion or the more disorder present, the higher the entropy. Gas molecules move more freely than liquid molecules and are higher in entropy. Cyclic molecules have more restricted bond rotation than similar acyclic molecules and are lower in entropy.

The *entropy change* ($\Delta S°$) **is the change in the amount of disorder between reactants and products.** $\Delta S°$ is positive (+) when the products are more disordered than the reactants. $\Delta S°$ is negative (−) when the products are less disordered (more ordered) than the reactants.

- Reactions resulting in an *increase in entropy* are favored.

$\Delta G°$ is related to $\Delta H°$ and $\Delta S°$ by the following equation:

$$\underset{\text{total energy change}}{\Delta G°} = \underset{\text{change in bonding energy}}{\Delta H°} - \underset{\text{change in disorder}}{T\Delta S°}$$

[T = Kelvin temperature]

Entropy is a rather intangible concept that comes up again and again in chemistry courses. One way to remember the relation between entropy and disorder is to consider a handful of chopsticks. Dropped on the floor, they are arranged randomly (a state of high entropy). Placed end-to-end in a straight line, they are arranged intentionally (a state of low entropy). The more disordered, random arrangement is favored and easier to achieve.

This equation tells us that the total energy change in a reaction is due to two factors: the change in the **bonding energy** and the change in **disorder.** The change in bonding energy can be calculated from bond dissociation energies (Section 6.4). Entropy changes, on the other hand, are more difficult to assess, but they are important when the number of molecules of starting material *differs* from the number of molecules of product in the balanced chemical equation. The entropy of a system also changes when an acyclic molecule is *cyclized* to a cyclic one, or a cyclic molecule is converted to an acyclic one.

For example, **when a single starting material forms two products,** as in the homolytic cleavage of a bond to form two radicals, **entropy increases** and favors formation of the products. In contrast, **entropy decreases when an acyclic compound forms a ring,** because a ring has fewer degrees of freedom. In this case, therefore, entropy does *not* favor formation of the product.

single reactant → two products
Entropy *increases.*
Entropy favors the **products.**

acyclic reactant → cyclic product
Entropy favors the **reactants.**
Entropy *decreases.*

The metabolism of glucose (Section 6.4) is favored by entropy because the number of molecules of products formed (6 CO_2 and 6 H_2O) is greater than the number of molecules of reactants ($C_6H_{12}O_6$ and 6 O_2). Moreover, a cyclic reactant is cleaved to form 12 acyclic product molecules.

Problem 6.14 For which reactions does entropy favor the products?

a.
b.

In most reactions that are not carried out at high temperature, the entropy term ($T\Delta S°$) is small compared to the enthalpy term ($\Delta H°$) and it can be neglected. Thus, **we will often approximate the overall free energy change of a reaction by the change in the bonding energy only.** Keep in mind that this is an approximation, but it gives us a starting point from which to decide if the reaction is energetically favorable.

$$\Delta G° \approx \Delta H°$$

According to this approximation:

- The product is favored when $\Delta H°$ is a *negative* value; that is, the bonds in the product are *stronger* than the bonds in the starting material.

- The starting material is favored when $\Delta H°$ is a *positive* value; that is, the bonds in the starting material are *stronger* than the bonds in the product.

Recall from Section 6.4 that a reaction is endothermic when $\Delta H°$ is positive and exothermic when $\Delta H°$ is negative. A reaction is **endergonic when $\Delta G°$ is positive** and **exergonic when $\Delta G°$ is negative.** $\Delta G°$ is usually approximated by $\Delta H°$ in this text, so the terms endergonic and exergonic are rarely used.

Problem 6.15 For a reaction with $\Delta H° = 40$ kJ/mol, decide which of the following statements is (are) true. Correct any false statement to make it true. (a) The reaction is exothermic; (b) $\Delta G°$ for the reaction is positive; (c) K_{eq} is greater than 1; (d) the bonds in the starting materials are stronger than the bonds in the product; and (e) the product is favored at equilibrium.

6.7 Energy Diagrams

An **energy diagram** is a schematic representation of the energy changes that take place as reactants are converted to products. An energy diagram indicates how readily a reaction proceeds, how many steps are involved, and how the energies of the reactants, products, and intermediates compare.

Consider a concerted reaction between molecule **A—B** with anion **C:$^-$** to form products **A:$^-$** and **B—C**. If the reaction occurs in a single step, the bond between **A** and **B** is broken *as* the bond between **B** and **C** is formed. Let's assume that the products are lower in energy than the reactants in this hypothetical reaction.

$$\underset{\text{This bond is broken.}}{\text{A—B}} + \text{C:}^- \longrightarrow \text{A:}^- + \underset{\text{This bond is formed.}}{\text{B—C}}$$

An energy diagram plots **energy on the y axis** versus the progress of reaction, often labeled the **reaction coordinate, on the x axis.** As the starting materials **A—B** and **C:$^-$** approach one another, their electron clouds feel some repulsion, causing an increase in energy, until a maximum value is reached. This unstable energy maximum is called the **transition state.** In the transition state the bond between **A** and **B** is partially broken, and the bond between **B** and **C** is partially formed. Because it is at the top of an energy "hill," **a transition state can never be isolated.**

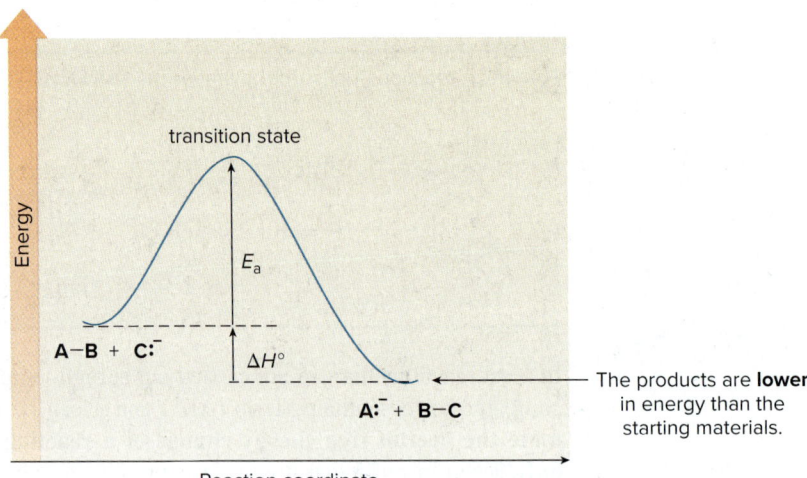

At the transition state, the bond between **A** and **B** can re-form to regenerate starting material, *or* the bond between **B** and **C** can form to generate product. As the bond forms between **B** and **C**, the energy decreases until some stable energy minimum of the products is reached.

- The energy difference between the reactants and products is $\Delta H°$. Because the products are at lower energy than the reactants, this reaction is *exothermic* and energy is *released*.

- The energy difference between the transition state and the starting material is called the **energy of activation,** symbolized by E_a.

The *energy of activation* is the minimum amount of energy needed to break bonds in the reactants. It represents an **energy barrier** that must be overcome for a reaction to occur. The size of E_a tells us about the reaction rate.

A slow reaction has a large E_a.
A fast reaction has a low E_a.

- The *larger* the E_a, the *greater* the amount of energy that is needed to break bonds, and the *slower* the reaction rate.

How can we draw the structure of the unstable transition state? The structure of the transition state is somewhere in between the structures of the starting material and product. Any bond that is partially broken or formed is drawn with a *dashed* line. Any atom that gains or loses a charge contains a *partial charge* in the transition state. Transition states are drawn in brackets, with a superscript double dagger ($\ddagger$).

In the hypothetical reaction between **A—B** and **C:⁻** to form **A:⁻** and **B—C**, the bond between **A** and **B** is partially broken, and the bond between **B** and **C** is partially formed. Because **A** gains a negative charge and **C** loses a charge in the course of the reaction, each atom bears a partial negative charge in the transition state.

$$\left[\overset{\delta-}{A} \text{---} B \text{---} \overset{\delta-}{C} \right]^{\ddagger}$$

This bond is partially broken.   This bond is partially formed.

Several energy diagrams are drawn in Figure 6.5. For any energy diagram:

- E_a determines the height of the energy barrier.
- $\Delta H°$ determines the relative position of the reactants and products.

Figure 6.5
Some representative energy diagrams

Example [1]
- Large E_a ⟶ slow reaction
- (+) $\Delta H°$ ⟶ endothermic reaction

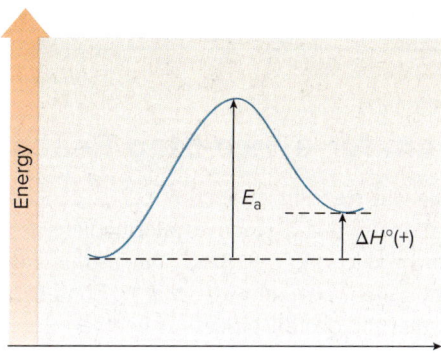

Example [2]
- Large E_a ⟶ slow reaction
- (−) $\Delta H°$ ⟶ exothermic reaction

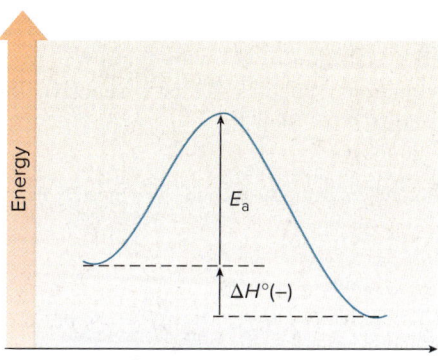

Example [3]
- Low E_a ⟶ fast reaction
- (+) $\Delta H°$ ⟶ endothermic reaction

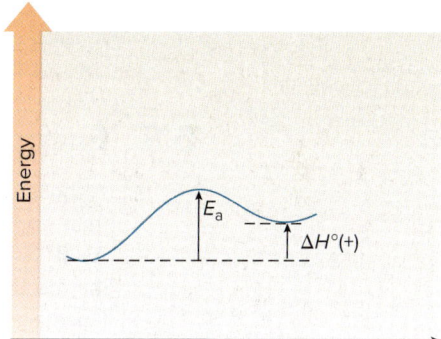

Example [4]
- Low E_a ⟶ fast reaction
- (−) $\Delta H°$ ⟶ exothermic reaction

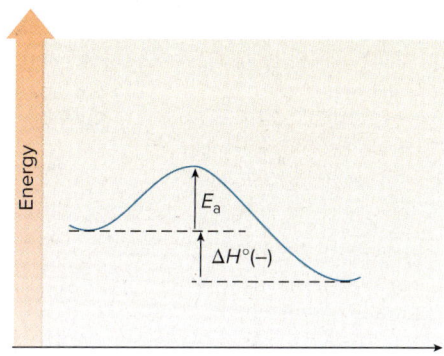

238 Chapter 6 Understanding Organic Reactions

The two variables, E_a **and** $\Delta H°$**, are independent** of each other. Two reactions can have identical values for $\Delta H°$ but very different E_a values. For two exothermic reactions with the same negative value of $\Delta H°$ but different E_a values, the reaction with the lower E_a is faster.

Problem 6.16 Draw an energy diagram for a reaction in which the products are higher in energy than the starting materials and E_a is large. Clearly label all of the following on the diagram: the axes, the starting materials, the products, the transition state, $\Delta H°$, and E_a.

Problem 6.17 Draw the structure for the transition state in each reaction.

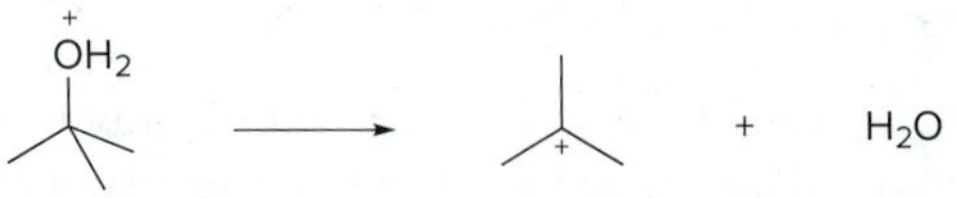

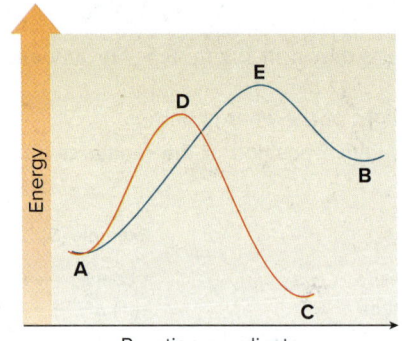

Problem 6.18 Compound **A** can be converted to either **B** or **C**. The energy diagrams for both processes are drawn on the graph below.

a. Label each reaction as endothermic or exothermic.
b. Which reaction is faster?
c. Which reaction generates the product lower in energy?
d. Which points on the graphs correspond to transition states?
e. Label the energy of activation for each reaction.
f. Label the $\Delta H°$ for each reaction.

6.8 Energy Diagram for a Two-Step Reaction Mechanism

Although the hypothetical reaction in Section 6.7 is concerted, many reactions involve more than one step with formation of a reactive intermediate. Consider the same overall reaction, **A–B + C:⁻** to form products **A:⁻ + B–C,** but in this case begin with the assumption that the reaction occurs by a *stepwise* pathway—that is, bond breaking occurs *before* bond making. Once again, assume that the overall process is exothermic.

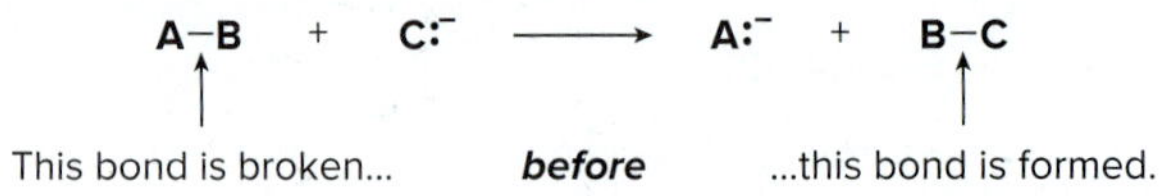

$$\underset{\text{This bond is broken...}}{\text{A–B}} \;+\; \text{C:}^- \;\longrightarrow\; \text{A:}^- \;+\; \underset{\text{...this bond is formed.}}{\text{B–C}}$$

before

One possible stepwise mechanism involves heterolysis of the **A–B** bond to form two ions **A:⁻** and **B⁺,** followed by reaction of **B⁺** with anion **C:⁻** to form product **B–C,** as outlined in the accompanying equations. Species **B⁺** is a **reactive intermediate. B⁺** is a product in Step [1] that reacts with **C:⁻** in Step [2].

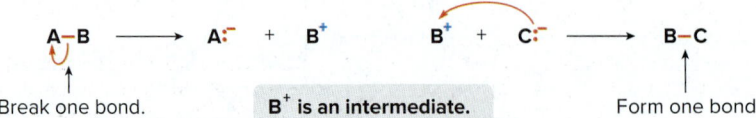

6.8 Energy Diagram for a Two-Step Reaction Mechanism

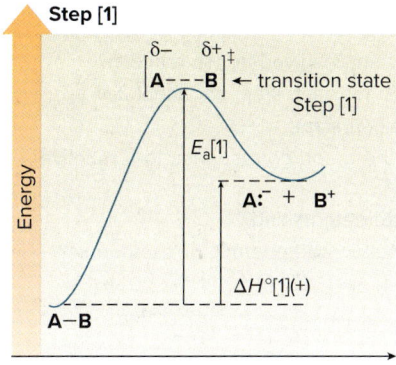

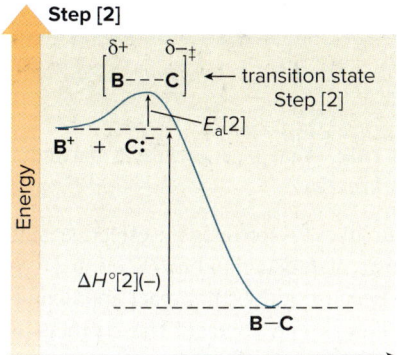

To draw an energy diagram for a two-step mechanism, we must draw an energy diagram for each step, and then combine them. Each step has its own energy barrier, with a transition state at the energy maximum.

Step [1] is endothermic because energy is needed to cleave the **A—B** bond, making $\Delta H°$ a positive value and placing the products of Step [1] at higher energy than the starting materials. In the transition state, the **A—B** bond is partially broken.

Step [2] is exothermic because energy is released in forming the **B—C** bond, making $\Delta H°$ a negative value and placing the products of Step [2] at lower energy than the starting materials of Step [2]. In the transition state, the **B—C** bond is partially formed.

The overall process is shown in Figure 6.6 as a single energy diagram that combines both steps. Because the reaction has two steps, there are two transition states, each corresponding to an energy barrier. The transition states are separated by an energy minimum, at which the reactive intermediate B^+ is located. Because we made the assumption that the overall two-step process is exothermic, the overall energy difference between the reactants and products, labeled $\Delta H°_{overall}$, has a negative value, and the final products are at a lower energy than the starting materials.

The energy barrier for Step [1], labeled $E_a[1]$, is higher than the energy barrier for Step [2], labeled $E_a[2]$, because bond cleavage (Step [1]) is more difficult (requires more energy) than bond formation (Step [2]). A higher-energy transition state for Step [1] makes it the slower step of the mechanism.

- In a multistep mechanism, the step with the highest-energy transition state is called the **rate-determining step**.

In this reaction, the rate-determining step is Step [1].

Figure 6.6 Complete energy diagram for the two-step conversion of **A—B + C:⁻ → A:⁻ + B—C**

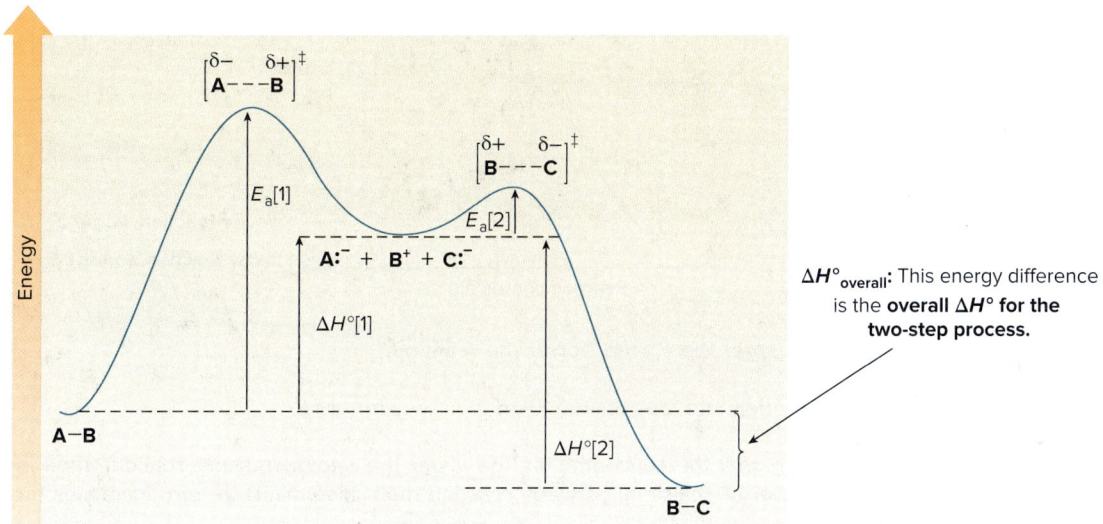

- **The transition states are located at energy maxima, whereas the reactive intermediate B^+ is located at an energy minimum.**
- Each step has its own value of $\Delta H°$ and E_a.
- The overall energy difference between starting material and products is called $\Delta H°_{overall}$. In this example, the products of the two-step sequence are at lower energy than the starting materials.
- Because Step [1] has the higher-energy transition state, it is the **rate-determining step**.

Problem 6.19 Consider the following energy diagram.

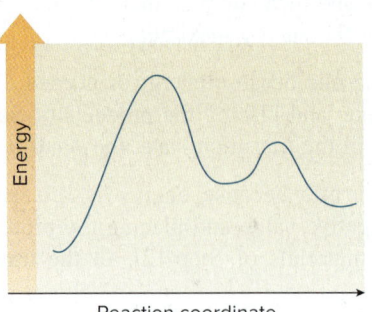

a. How many steps are involved in this reaction?
b. Label $\Delta H°$ and E_a for each step, and label $\Delta H°_{overall}$.
c. Label each transition state.
d. Which point on the graph corresponds to a reactive intermediate?
e. Which step is rate-determining?
f. Is the overall reaction endothermic or exothermic?

Problem 6.20 Draw an energy diagram for a two-step reaction, **A → B → C**, where the relative energy of these compounds is **C < A < B**, and the conversion of **B → C** is rate-determining.

6.9 Kinetics

We now turn to a more detailed discussion of **reaction rate**—that is, how fast a particular reaction proceeds. **The study of reaction rates is called *kinetics*.**

The rate of chemical processes affects many facets of our lives. Aspirin is an effective anti-inflammatory agent because it rapidly inhibits the synthesis of prostaglandins (Section 15.5). DDT (Section 7.4) is a persistent environmental pollutant because it does not react appreciably with water, oxygen, or any other chemical with which it comes into contact. These processes occur at different rates, resulting in beneficial or harmful effects.

Some reactions have a very favorable equilibrium constant ($K_{eq} \gg 1$), but the rate is very slow. Gasoline can be safely handled in the air because its reaction with O_2 is slow unless there is a spark to provide energy to initiate the reaction.
Moodboard/Getty Images

6.9A Energy of Activation

As we learned in Section 6.7, the energy of activation, E_a, is the energy difference between the reactants and the transition state. It is the **energy barrier** that must be exceeded for reactants to be converted to products.

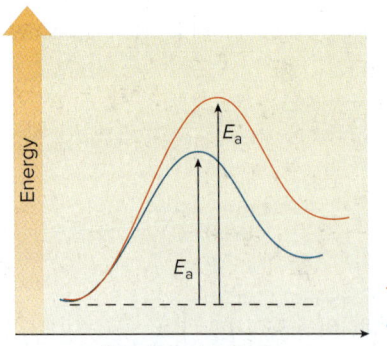

— *slower* reaction, *larger* E_a
— *faster* reaction, *smaller* E_a

Practically, the effect of temperature on reaction rate is used to an advantage in the kitchen. Food is stored in a cold refrigerator to slow the reactions that cause spoilage. *Jill Braaten/McGraw-Hill Education*

- The *larger* the E_a, the *slower* the reaction.

Concentration and temperature also affect reaction rate.

- The *higher* the concentration, the *faster* the rate. Increasing concentration increases the number of collisions between reacting molecules, which in turn increases the rate.

- The *higher* the temperature, the *faster* the rate. Increasing temperature increases the average kinetic energy of the reacting molecules. Because the kinetic energy of colliding molecules is used for bond cleavage, increasing the average kinetic energy increases the rate.

In contrast to laboratory reactions, enzyme-catalyzed reactions must be run at the specific temperature of the organism, and increasing temperatures could deactivate the enzyme.

Keep in mind that certain **reaction quantities** have *no effect* on reaction rate.

- $\Delta G°$, $\Delta H°$, and K_{eq} do *not* determine the rate of a reaction. These quantities indicate the direction of equilibrium and the relative energy of reactants and products.

Problem 6.21 Which value (if any) corresponds to a faster reaction: (a) $E_a = 40$ kJ/mol or $E_a = 4$ kJ/mol; (b) a reaction temperature of 0 °C or a reaction temperature of 25 °C; (c) $K_{eq} = 10$ or $K_{eq} = 100$; (d) $\Delta H° = -10$ kJ/mol or $\Delta H° = 10$ kJ/mol?

Problem 6.22 For a reaction with $K_{eq} = 0.8$ and $E_a = 80$ kJ/mol, decide which of the following statements is (are) true. Correct any false statement to make it true. Ignore entropy considerations. (a) The reaction is faster than a reaction with $K_{eq} = 8$ and $E_a = 80$ kJ/mol. (b) The reaction is faster than a reaction with $K_{eq} = 0.8$ and $E_a = 40$ kJ/mol. (c) $\Delta G°$ for the reaction is a positive value. (d) The starting materials are lower in energy than the products of the reaction. (e) The reaction is exothermic.

6.9B Rate Equations

The rate of a chemical reaction is determined by measuring the decrease in the concentration of the reactants over time, or the increase in the concentration of the products over time. A **rate law** (or **rate equation**) is an equation that shows the relationship between the rate of a reaction and the concentration of the reactants. A rate law is determined *experimentally*, and it depends on the mechanism of the reaction.

A rate law has two important terms: the **rate constant symbolized by k** and the **concentration of the reactants**. Not all reactant concentrations may appear in the rate equation, as we shall soon see.

$$\text{rate} = k[\text{reactants}]$$

k = the rate constant

A rate constant k is a fundamental characteristic of a reaction. It is a complex mathematical term that takes into account the dependence of a reaction rate on temperature and the energy of activation.

> A rate constant k and the energy of activation E_a are inversely related. **A high E_a corresponds to a small k.**

- *Fast* reactions have *large* rate constants.
- *Slow* reactions have *small* rate constants.

What concentration terms appear in the rate equation? That depends on the mechanism. For the organic reactions we will encounter:

- A rate equation contains concentration terms for *all* reactants involved in a *one-step* mechanism.
- A rate equation contains concentration terms for *only* the reactants involved in the *rate-determining step* in a multistep reaction.

In the one-step reaction of **A—B + C:$^-$** to form **A:$^-$ + B—C,** *both* reactants appear in the transition state of the only step of the mechanism. The **concentration of *both* reactants affects the reaction rate,** and *both* terms appear in the rate equation. This type of reaction involving two reactants is said to be **bimolecular**.

$$\text{A—B} + \text{C:}^- \longrightarrow \text{A:}^- + \text{B—C} \qquad \text{rate} = k[\text{AB}][\text{C:}^-]$$

Both reactants are involved in the only step.
Both reactants determine the rate.

sum of the exponents = 2
Second-order rate equation

The *order* of a rate equation equals the sum of the exponents of the concentration terms in the rate equation. In the rate equation for the concerted reaction of **A—B + C:$^-$**, there are two concentration terms, each with an exponent of one. Thus, the sum of the exponents is two and the **rate equation is *second order*** (the reaction follows second-order kinetics).

Because the rate of the reaction depends on the concentration of both reactants, doubling the concentration of *either* **A—B** or **C:⁻** doubles the rate of the reaction. Doubling the concentration of *both* **A—B** and **C:⁻** increases the reaction rate by a factor of *four*.

The situation is different in the stepwise conversion of **A—B + C:⁻** to form **A:⁻ + B—C**. The mechanism shown in Section 6.8 has two steps: a slow step (the **rate-determining** step) in which the **A—B** bond is broken, and a fast step in which the **B—C** bond is formed.

$$A\text{—}B \xrightarrow{\text{Step [1]}} A{:}^- + B^+ \xrightarrow{C{:}^-}_{\text{Step [2]}} B\text{—}C \qquad \text{rate} = k[AB]$$

Step [1] rate-determining

Only **AB** is involved in the rate-determining step. Only [**AB**] determines the rate.

only one concentration term

First-order rate equation

In a multistep mechanism, a reaction can occur no faster than its rate-determining step. **Only the concentrations of the reactants in the rate-determining step appear in the rate equation.** In this example, the rate depends on the concentration of **A—B** *only,* because only **A—B** appears in the rate-determining step. A reaction involving only one reactant is said to be **unimolecular.** Because there is only one concentration term (raised to the first power), the **rate equation is** *first order* (the reaction follows first-order kinetics).

Because the rate of the reaction depends on the concentration of only *one* reactant, doubling the concentration of **A—B** doubles the rate of the reaction, but **doubling the concentration of C:⁻ has** *no effect* **on the reaction rate.**

This might seem like a puzzling result. If **C:⁻** is involved in the reaction, why doesn't it affect the overall rate of the reaction?

The following analogy is useful. Let's say three students must make 20 peanut butter and jelly sandwiches for a class field trip. Student (**1**) spreads the peanut butter on the bread. Student (**2**) spreads on the jelly, and student (**3**) cuts the sandwiches in half. Suppose student (**2**) is very slow in spreading the jelly. It doesn't matter how fast students (**1**) and (**3**) are; they can't finish making sandwiches any faster than student (**2**) can add the jelly. Five more students can spread on the peanut butter, or an entirely different individual can replace student (**3**), and this doesn't speed up the process. How fast the sandwiches are made is determined entirely by the rate-determining step—that is, spreading the jelly.

Rate equations provide very important information about the mechanism of a reaction. Rate laws for new reactions with unknown mechanisms are determined by a set of experiments that measure how a reaction's rate changes with concentration. Then, a mechanism is suggested based on which reactants affect the rate.

Problem 6.23 The rate equation for the reaction of CH_3CH_2Br with ⁻OH is: rate = $k[CH_3CH_2Br][^-OH]$. What effect does the indicated concentration change have on the overall rate of the reaction?

a. tripling the concentration of CH_3CH_2Br only
b. tripling the concentration of ⁻OH only
c. tripling the concentration of both CH_3CH_2Br and ⁻OH

Problem 6.24 Write a rate equation for each reaction, given the indicated mechanism.

a. $\diagup\!\!\diagdown\text{Br}$ + ⁻OH ⟶ $\diagup\!\!=\!\!\diagdown$ + H_2O + Br^-

b. $\underset{\text{Br}}{\diagup\!\!\diagdown\!\!\diagup}$ $\xrightarrow{\text{slow}}$ $\underset{+}{\diagup\!\!\diagdown\!\!\diagup}$ + Br^- $\xrightarrow[\text{fast}]{^-OH}$ $\diagup\!\!=\!\!\diagdown$ + H_2O

6.10 Catalysts

Some reactions do not occur in a reasonable time unless a **catalyst** is added.

- A *catalyst* is a substance that speeds up the rate of a reaction. A catalyst is recovered unchanged in a reaction, and it does not appear in the product.

Common catalysts in organic reactions are **acids** and **metals**. Two examples are shown with the catalyst drawn in red.

$$\text{acetic acid} + \text{ethanol} \xrightarrow{H_2SO_4} \text{ethyl acetate} + H_2O$$

$$\text{cyclohexene} + H_2 \xrightarrow{Pd} \text{cyclohexane}$$

The reaction of acetic acid with ethanol to yield ethyl acetate and water occurs in the presence of an acid catalyst. The acid catalyst is written over or under the arrow to emphasize that it is not part of the starting materials or the products. The details of this reaction are discussed in Chapter 16.

The reaction of cyclohexene with hydrogen to form cyclohexane occurs only in the presence of a metal catalyst such as palladium, platinum, or nickel. The metal provides a surface that binds both the cyclohexene and the hydrogen, and in doing so, facilitates the reaction. We return to this mechanism in Chapter 11.

Catalysts accelerate a reaction by lowering the energy of activation (Figure 6.7). They have no effect on the equilibrium constant, so they do not change the amount of reactant and product at equilibrium. Thus, catalysts affect how *quickly* equilibrium is achieved, but not the relative amounts of reactants and products at equilibrium. If a catalyst is somehow used up in one step of a reaction sequence, it must be regenerated in another step.

Figure 6.7
The effect of a catalyst on a reaction

— uncatalyzed reaction: **larger** E_a—**slower** reaction
— catalyzed reaction: **lower** E_a—**faster** reaction

- The catalyst *lowers* the energy of activation, thus *increasing the rate* of the catalyzed reaction.
- The energy of the reactants and products is the same in both the uncatalyzed and catalyzed reactions, so the **position of equilibrium is unaffected**.

Problem 6.25 Identify the catalyst in each equation.

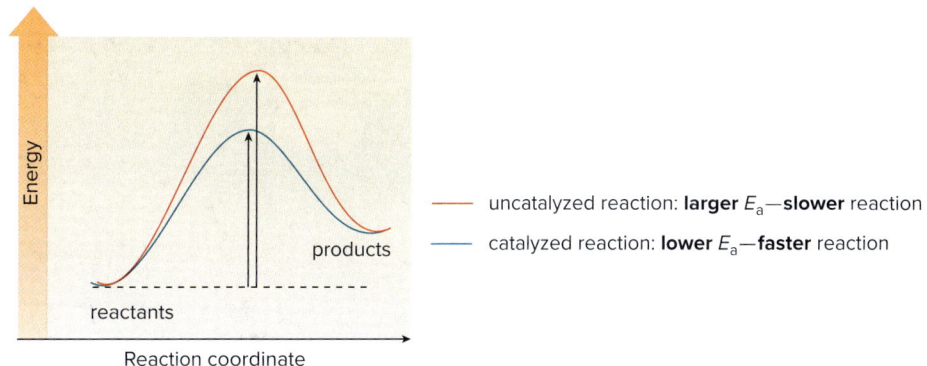

a. $CH_2=CH_2 \xrightarrow[H_2SO_4]{H_2O} CH_3CH_2OH$ b. $CH_3Cl \xrightarrow[^-OH]{I^-} CH_3OH + Cl^-$

6.11 Enzymes

The catalysts that synthesize and break down biomolecules in living organisms are governed by the same principles as the acids and metals in organic reactions. The catalysts in living organisms, however, are usually protein molecules called **enzymes.**

- *Enzymes* are biochemical catalysts composed of amino acids held together in a very specific three-dimensional shape.

An enzyme contains a region called its **active site,** which binds an organic reactant, called a **substrate.** When bound, this unit is called the **enzyme–substrate complex,** as shown schematically in Figure 6.8 for the enzyme lactase, the enzyme that binds lactose, the principal carbohydrate in milk. Once bound, the organic substrate undergoes a very specific reaction at an enhanced rate. In this example, lactose is converted into two simpler sugars, glucose and galactose. When individuals lack adequate amounts of lactase, they are unable to digest lactose, causing abdominal cramping and diarrhea.

An enzyme speeds up a biological reaction in a variety of ways. It may hold reactants in the proper conformation to facilitate reaction, or it may provide an acidic site needed for a particular transformation. Once the reaction is completed, the enzyme releases the substrate and it is then able to catalyze another reaction.

Key distinctions between enzymatic and laboratory reactions are summarized in Table 6.4.

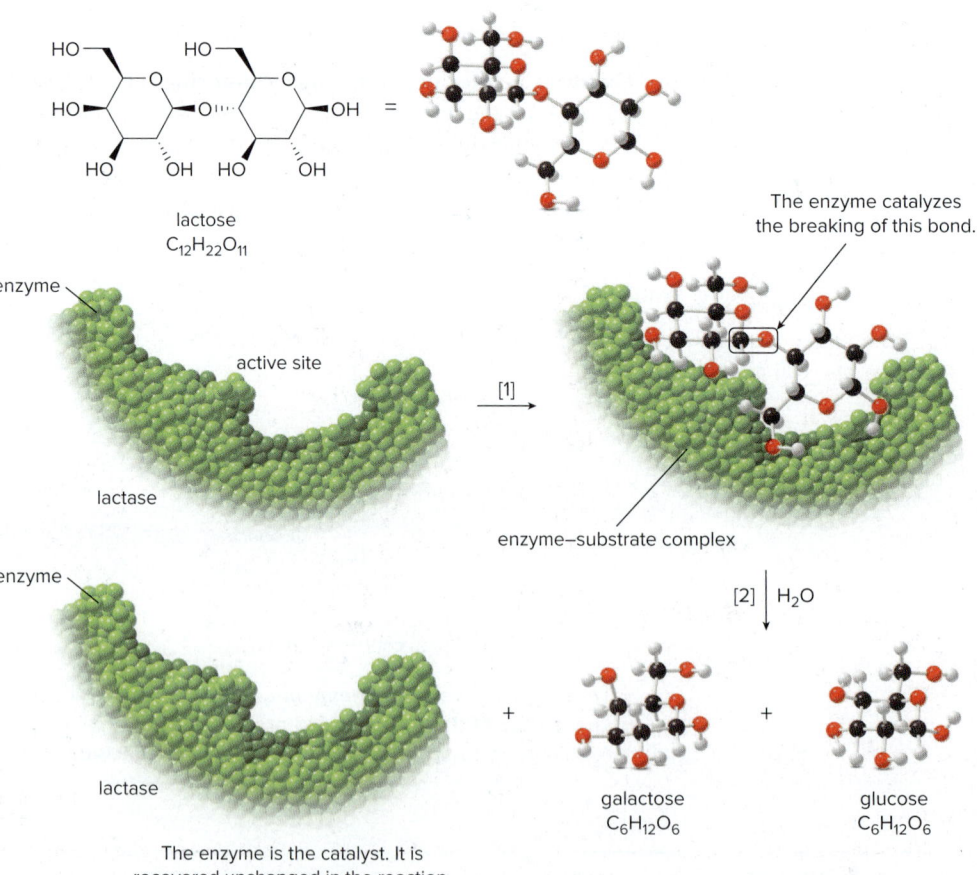

Figure 6.8
Lactase, an example of a biological catalyst

- The enzyme lactase binds the carbohydrate lactose ($C_{12}H_{22}O_{11}$) in its active site in Step [1]. Lactose then reacts with water to break a bond and form two simpler sugars, galactose and glucose, in Step [2]. This process is the first step in digesting lactose, the principal carbohydrate in milk.

Table 6.4 Comparison of Enzymatic and Laboratory Reactions

	Enzymatic reaction	Laboratory reaction
Size	Small part(s) of a large biomolecule	Small molecule
Catalyst	Amino acids and metal ions	Acids, bases, and metal ions
Specificity	Highly specific	Less specific
Solvent	Water	Organic solvent
Temperature and pH	Specific to an organism	Wide range

Chapter 6 REVIEW

KEY CONCEPTS

[1] Types of reactions (6.2)

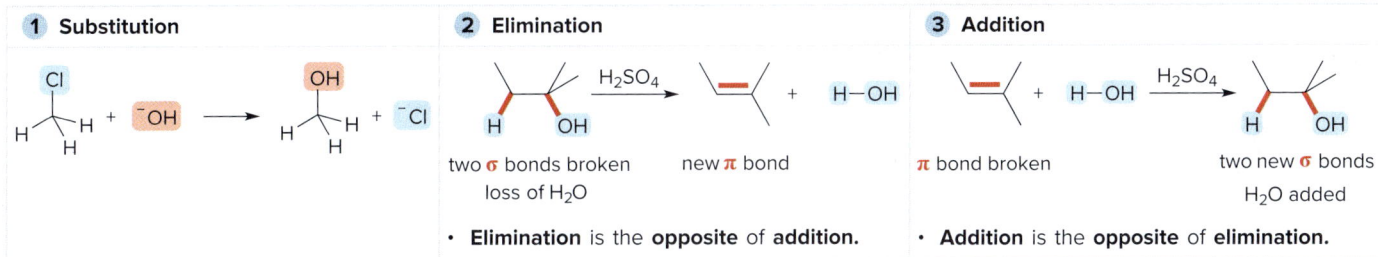

- Elimination is the opposite of addition.
- Addition is the opposite of elimination.

Try Problems 6.28, 6.31b, 6.49e, 6.50a, 6.52e.

[2] Energy trends

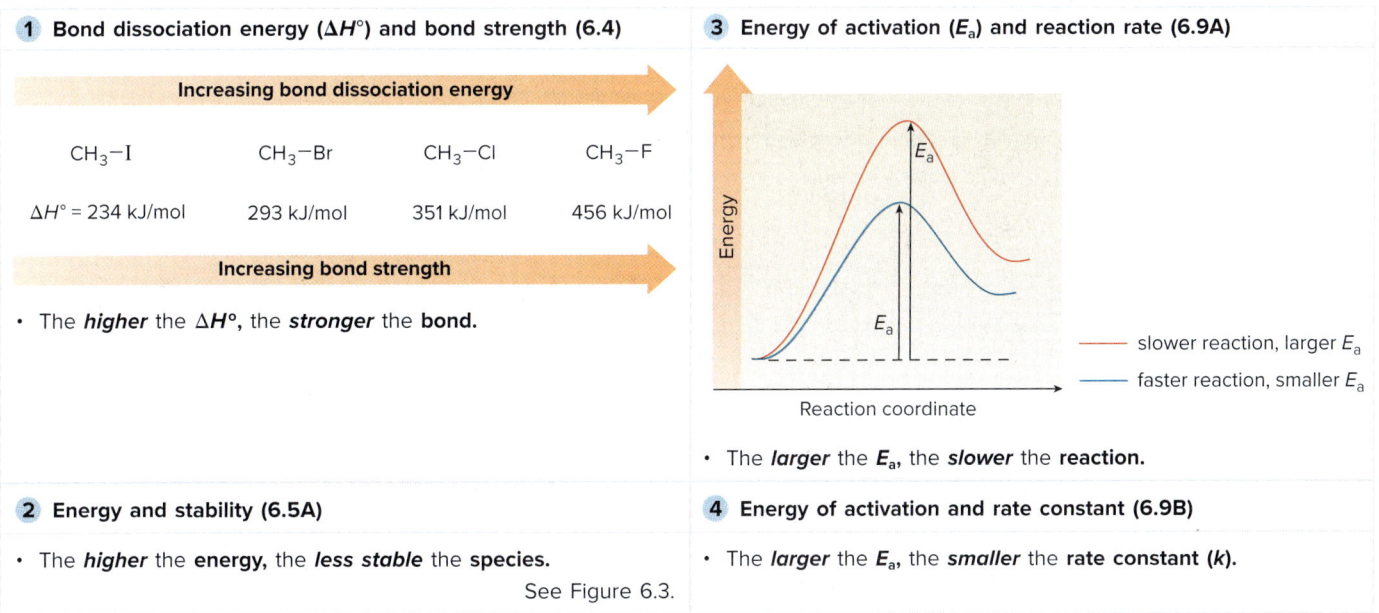

1. **Bond dissociation energy ($\Delta H°$) and bond strength (6.4)**

   Increasing bond dissociation energy →

   CH$_3$–I   CH$_3$–Br   CH$_3$–Cl   CH$_3$–F

   $\Delta H°$ = 234 kJ/mol   293 kJ/mol   351 kJ/mol   456 kJ/mol

   Increasing bond strength →

   - The *higher* the $\Delta H°$, the *stronger* the bond.

2. **Energy and stability (6.5A)**

   - The *higher* the energy, the *less stable* the species.

   See Figure 6.3.

3. **Energy of activation (E_a) and reaction rate (6.9A)**

   — slower reaction, larger E_a
   — faster reaction, smaller E_a

   - The *larger* the E_a, the *slower* the reaction.

4. **Energy of activation and rate constant (6.9B)**

   - The *larger* the E_a, the *smaller* the rate constant (k).

Try Problem 6.35.

[3] Reactive intermediates (6.3)

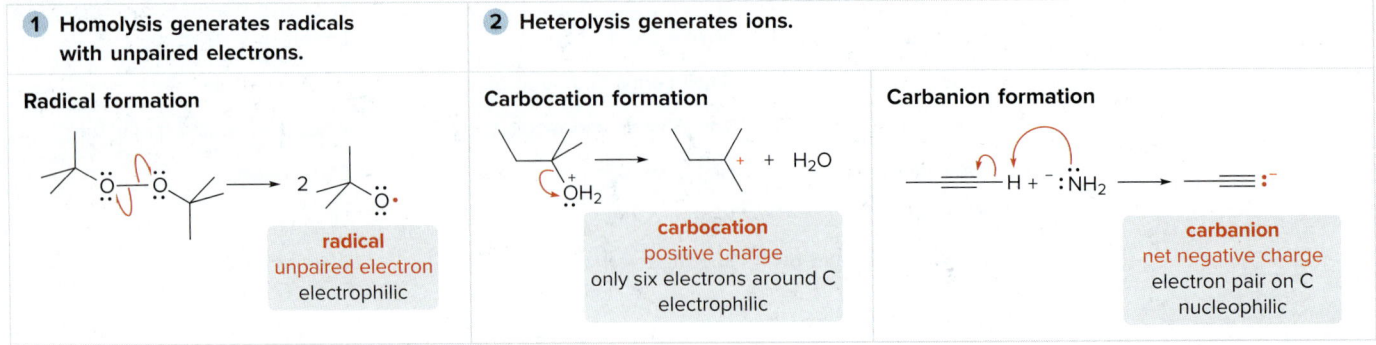

| 1 | Homolysis generates radicals with unpaired electrons. |
| 2 | Heterolysis generates ions. |

Radical formation — radical, unpaired electron, electrophilic

Carbocation formation — carbocation, positive charge, only six electrons around C, electrophilic

Carbanion formation — carbanion, net negative charge, electron pair on C, nucleophilic

Try Problem 6.26.

[4] Energy diagrams

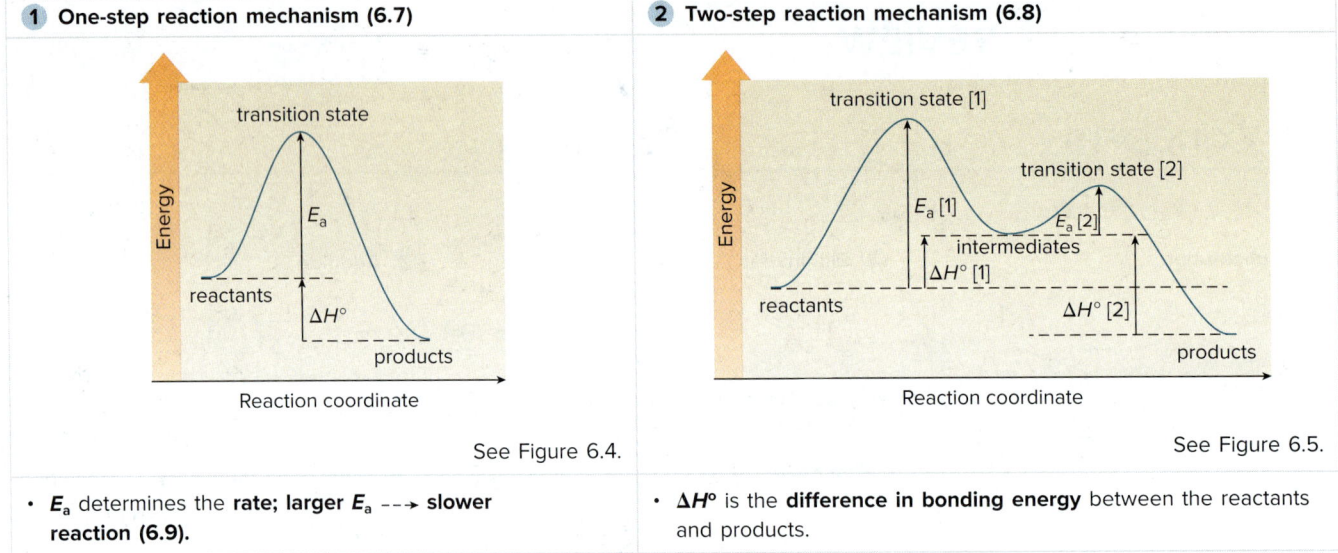

1 One-step reaction mechanism (6.7)

See Figure 6.4.

- E_a determines the **rate**; larger E_a --→ **slower reaction** (6.9).

2 Two-step reaction mechanism (6.8)

See Figure 6.5.

- $\Delta H°$ is the **difference in bonding energy** between the reactants and products.

Try Problems 6.43; 6.44c; 6.45; 6.46e; 6.53e, f.

[5] Conditions favoring product formation (6.5, 6.6)

1	$K_{eq} > 1$	More products than reactants are present at **equilibrium**.
2	$\Delta G° < 0$	The **free energy** of the products is **lower** than the free energy of the reactants.
3	$\Delta H° < 0$	Bonds in the products are **stronger** than bonds in the reactants.
4	$\Delta S° > 0$	The products are **more disordered** than the reactants. When a single starting material forms two products, entropy increases.

Try Problem 6.38–6.40.

KEY EQUATIONS

1 $\Delta G° = -2.303 RT \log K_{eq}$

K_{eq} depends on the energy difference between reactants and products.

[$R = 8.314$ J/(K·mol), the gas constant
T = Kelvin temperature (K)]

2 $\Delta G° = \Delta H° - T\Delta S°$

free energy change — change in bonding energy — change in disorder

[T = Kelvin temperature (K)]

Try Problem 6.39.

KEY SKILLS

[1] Using full-headed curved arrows to show the movement of electron pairs (6.3D)

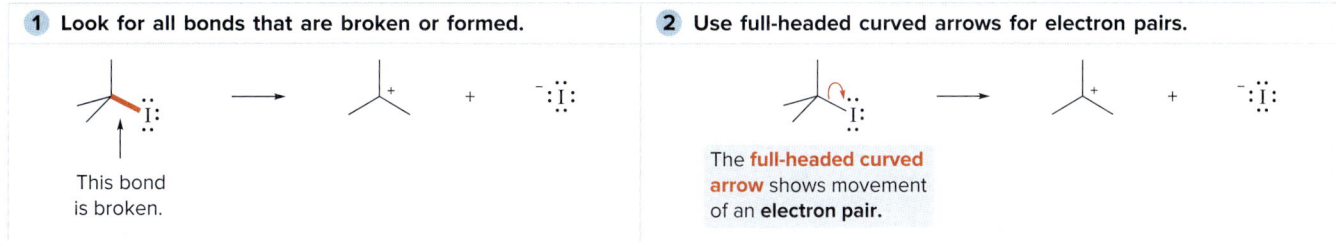

1 Look for all bonds that are broken or formed.

This bond is broken.

2 Use full-headed curved arrows for electron pairs.

The **full-headed curved arrow** shows movement of an **electron pair**.

See Figure 6.2, Sample Problems 6.1, 6.2. Try Problems 6.29a, c, d; 6.30–6.33; 6.49a; 6.51a; 6.52a.

[2] Using half-headed curved arrows to show the movement of single electrons (6.3B)

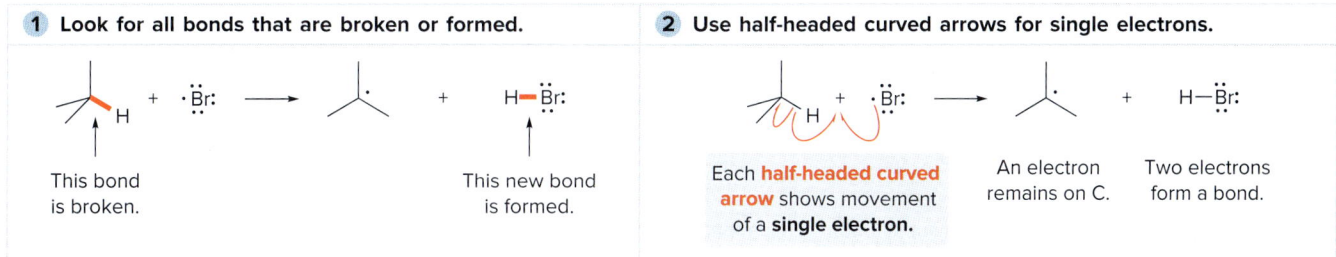

1 Look for all bonds that are broken or formed.

This bond is broken.   This new bond is formed.

2 Use half-headed curved arrows for single electrons.

Each **half-headed curved arrow** shows movement of a **single electron**.

An electron remains on C.   Two electrons form a bond.

See Figure 6.2. Try Problems 6.29b, 6.34, 6.44a.

[3] Calculating $\Delta H°$ of a reaction (6.4)

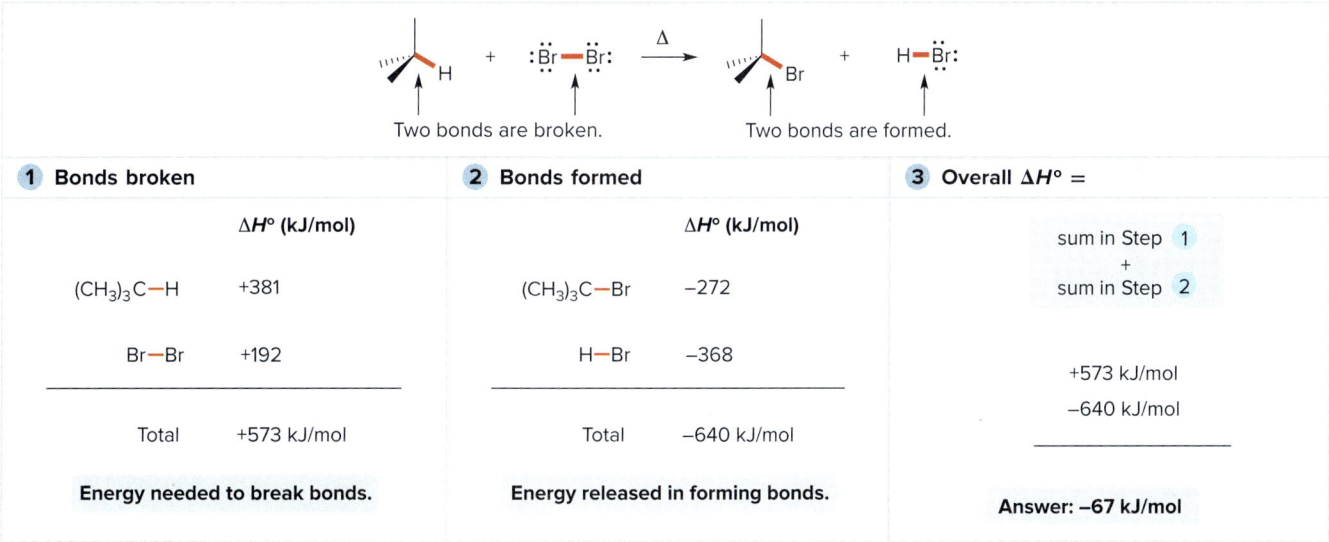

Two bonds are broken.   Two bonds are formed.

1 Bonds broken		**2** Bonds formed		**3** Overall $\Delta H°$ =
	$\Delta H°$ (kJ/mol)		$\Delta H°$ (kJ/mol)	sum in Step **1** + sum in Step **2**
$(CH_3)_3C—H$	+381	$(CH_3)_3C—Br$	−272	
Br—Br	+192	H—Br	−368	+573 kJ/mol
				−640 kJ/mol
Total	+573 kJ/mol	Total	−640 kJ/mol	
Energy needed to break bonds.		**Energy released in forming bonds.**		**Answer: −67 kJ/mol**

See Table 6.2, Sample Problem 6.3. Try Problems 6.36, 6.44b.

PROBLEMS

Problems Using Three-Dimensional Models

6.26 Draw the products of homolysis or heterolysis of each indicated bond. Use electronegativity differences to decide on the location of charges in the heterolysis reaction. Classify each carbon reactive intermediate as a radical, carbocation, or carbanion.

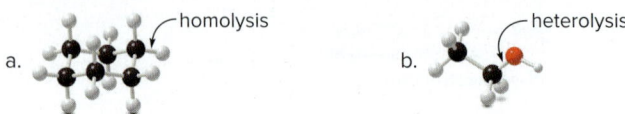

6.27 Explain why the bond dissociation energy for bond (a) is lower than the bond dissociation energy for bond (b).

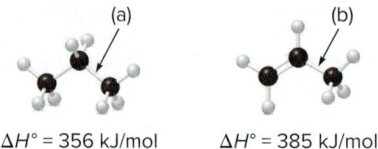

$\Delta H° = 356$ kJ/mol $\Delta H° = 385$ kJ/mol

Types of Reactions

6.28 Classify each transformation as substitution, elimination, or addition.

a. [pyranose with OH groups] $\xrightarrow{\text{CH}_3\text{OH}, \text{HCl}}$ [pyranose with OCH$_3$]

b. ^-O_2C–CH=C(CO$_2^-$)–CO$_2^-$ $\longrightarrow$ ^-O_2C–CH(OH)–C(CO$_2^-$)H–CO$_2^-$

Curved Arrows

6.29 Use full-headed or half-headed curved arrows to show the movement of electrons in each reaction.

a. [cyclohexane with :Ö:⁻ and :Cl: substituents] $\longrightarrow$ [cyclohexanone] + :Cl:⁻

b. [propyl radical] + Br$_2$ $\longrightarrow$ [2-bromopropane] + :Br·

c. [t-butyl cation] + :ÖH⁻ $\longrightarrow$ [isobutylene] + H$_2$Ö:

d. [geranyl diphosphate] $\longrightarrow$ [diene cation] + diphosphate

6.30 Draw the products of each reaction by following the curved arrows.

a.

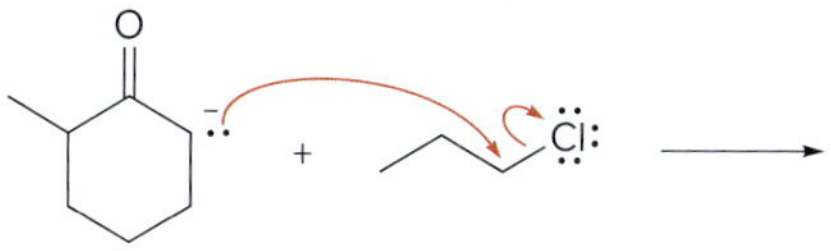

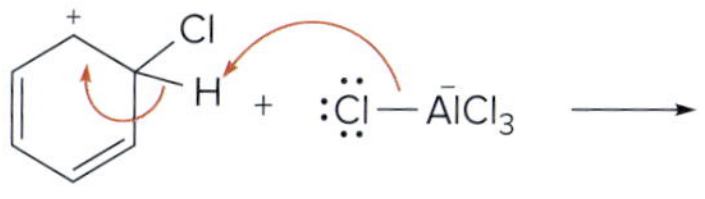

b.

c.

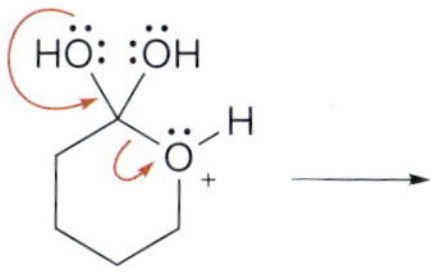

d.

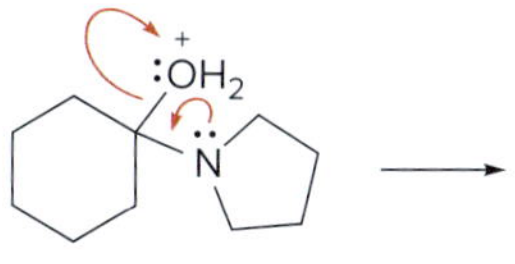

6.31 (a) Add curved arrows for each step to show how **A** is converted to the epoxy ketone **C**. (b) Classify the conversion of **A** to **C** as a substitution, elimination, or addition. (c) Draw one additional resonance structure for **B**.

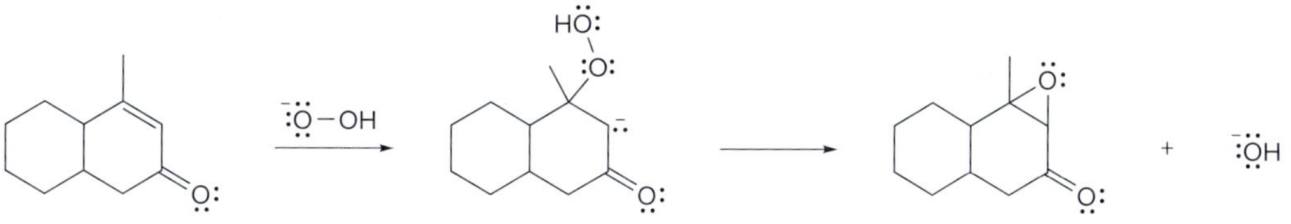

6.32 Biological reactions, which occur in the presence of enzymes, are often shown with two or more bonds broken and formed at the same time. The acid (HA) or base (B:) that may be required in a reaction comes from a functional group located at or near the active site. Follow the curved arrows and draw the products of each reaction involved in metabolism. We will learn about the details of these reactions in Chapter 27.

a. fructose 1,6-bisphosphate

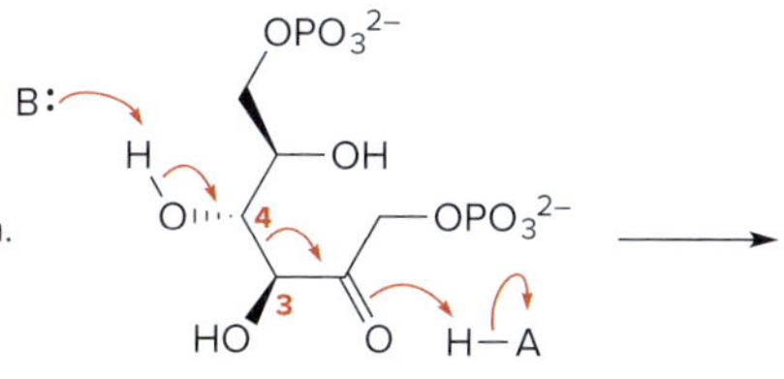

b.

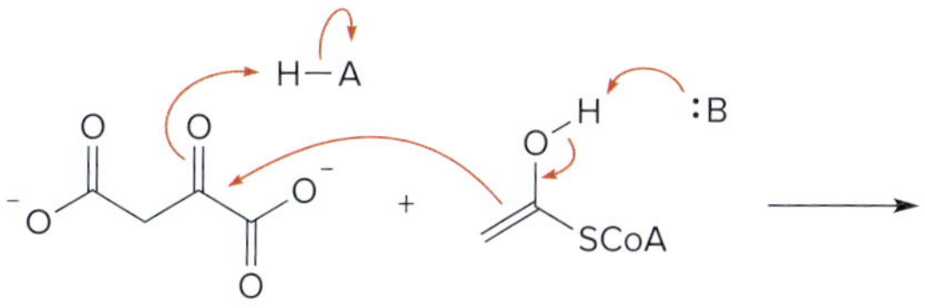

6.33 Add curved arrows to each step in the following reaction sequence.

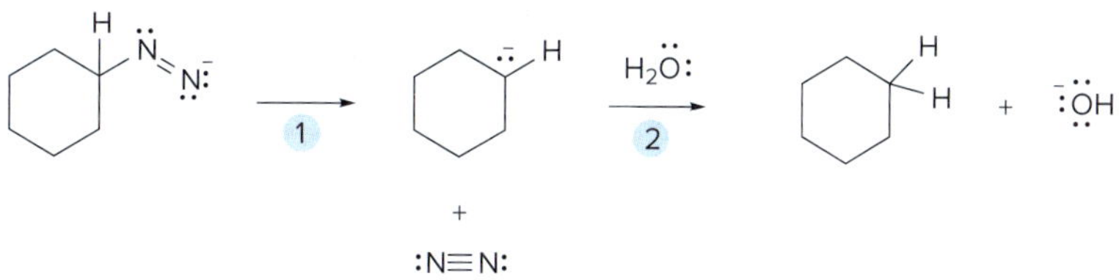

6.34 PGF$_{2\alpha}$, a fatty acid discussed in Section 15.5, is synthesized in cells using a cyclooxygenase enzyme that catalyzes a multistep radical pathway. Two steps in the pathway are depicted in the accompanying equations. (a) Draw in curved arrows to illustrate how **C** is converted to **D** in Step [1]. (b) Identify **Y**, the product of Step [2], using the curved arrows that are drawn on compound **D**.

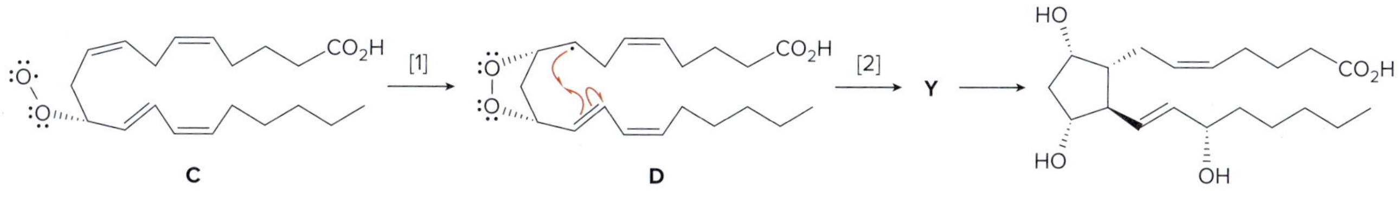

PGF$_{2\alpha}$

Bond Dissociation Energy and Calculating $\Delta H°$

6.35 Rank the indicated bonds in order of increasing bond dissociation energy.

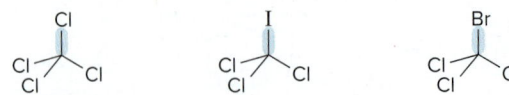

6.36 Calculate $\Delta H°$ for each reaction.
 a. HO• + CH$_4$ ⟶ •CH$_3$ + H$_2$O
 b. CH$_3$OH + HBr ⟶ CH$_3$Br + H$_2$O

6.37 Homolysis of the indicated C–H bond in propene forms a resonance-stabilized radical.

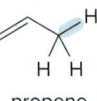
propene

 a. Draw the two possible resonance structures for this radical.
 b. Use half-headed curved arrows to illustrate how one resonance structure can be converted to the other.
 c. Draw a structure for the resonance hybrid.

Thermodynamics, $\Delta G°$, $\Delta H°$, $\Delta S°$, and K_{eq}

6.38 Given each value, determine whether the starting material or product is favored at equilibrium.
 a. $K_{eq} = 0.5$
 b. $\Delta G° = -100$ kJ/mol
 c. $\Delta H° = 8.0$ kJ/mol
 d. $K_{eq} = 16$
 e. $\Delta G° = 2.0$ kJ/mol
 f. $\Delta H° = 200$ kJ/mol
 g. $\Delta S° = 8$ J/(K•mol)
 h. $\Delta S° = -8$ J/(K•mol)

6.39
 a. Which value corresponds to a negative value of $\Delta G°$: $K_{eq} = 10^{-2}$ or $K_{eq} = 10^{2}$?
 b. In a unimolecular reaction with five times as much starting material as product at equilibrium, what is the value of K_{eq}? Is $\Delta G°$ positive or negative?
 c. Which value corresponds to a larger K_{eq}: $\Delta G° = -8$ kJ/mol or $\Delta G° = 20$ kJ/mol?

6.40 For which of the following reactions is $\Delta S°$ a positive value?

 a. 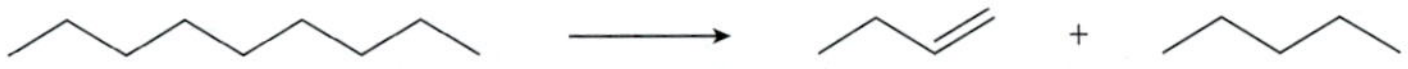 [alkane ⟶ alkene + alkane]

 b. 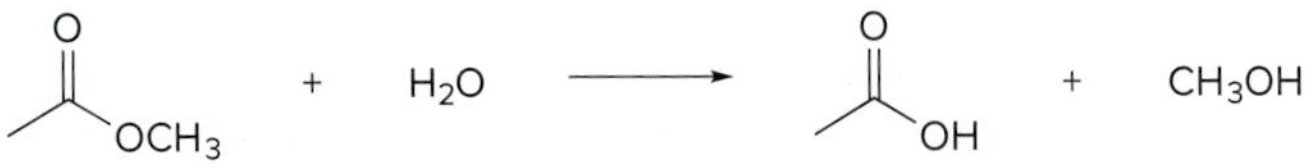 [ester (OCH$_3$) + H$_2$O ⟶ carboxylic acid (OH) + CH$_3$OH]

6.41 The hydrolysis of phosphoenolpyruvate releases 61.9 kJ/mol of energy. In glucose metabolism, this reaction drives the energetically unfavorable phosphorylation of ADP to ATP in the presence of Mg^{2+} and the pyruvate kinase enzyme.

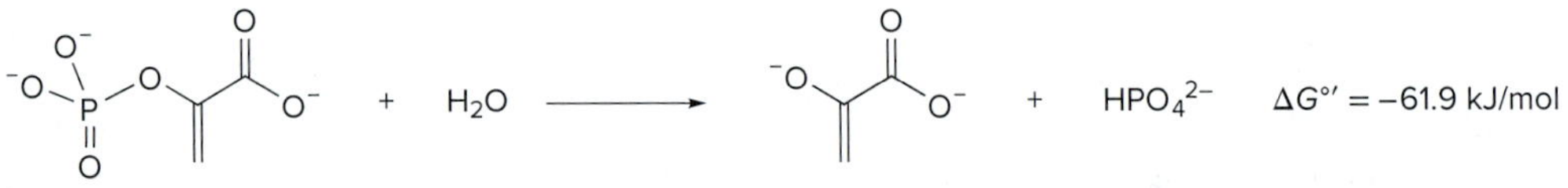

phosphoenolpyruvate + H$_2$O ⟶ pyruvate + HPO$_4^{2-}$ $\Delta G°' = -61.9$ kJ/mol

 a. Write the net (coupled) reaction by summing the substances in both the phosphorylation and hydrolysis equations.
 b. How much energy is released in the coupled reaction?
 c. Write the equation for the coupled reaction using coupled reaction arrows.

Energy Diagrams and Transition States

6.42 Draw the transition state for each reaction.

 a.  cyclohexanol + $^-$NH$_2$ ⟶ cyclohexanolate + NH$_3$

 b.  (t-butyl cation) + H$_2$O ⟶ (isobutylene) + H$_3$O$^+$

6.43 Draw an energy diagram for each reaction. Label the axes, the starting material, product, transition state, $\Delta H°$, and E_a.
 a. a concerted reaction with $\Delta H° = -80$ kJ/mol and $E_a = 16$ kJ/mol
 b. a two-step reaction, **A** → **B** → **C**, in which the relative energy of the compounds is **A** < **C** < **B**, and the step **A** → **B** is rate-determining

6.44 Consider the following reaction: $CH_4 + Cl\cdot \rightarrow \cdot CH_3 + HCl$.
   a. Use curved arrows to show the movement of electrons in this radical reaction.
   b. Calculate $\Delta H°$ using the bond dissociation energies in Table 6.2.
   c. Draw an energy diagram assuming that $E_a = 16$ kJ/mol.
   d. What is E_a for the reverse reaction ($\cdot CH_3 + HCl \rightarrow CH_4 + Cl\cdot$)?

6.45 Consider the following energy diagram for the conversion of **A → G**.

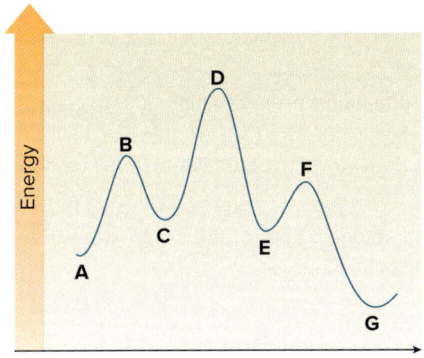

   a. Which points on the graph correspond to transition states?
   b. Which points on the graph correspond to reactive intermediates?
   c. How many steps are present in the reaction mechanism?
   d. Label each step of the mechanism as endothermic or exothermic.
   e. Label the overall reaction as endothermic or exothermic.

6.46 Consider the following two-step reaction:

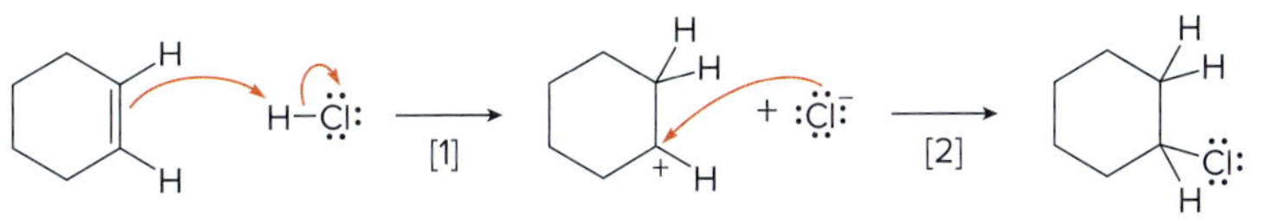

   a. How many bonds are broken and formed in Step [1]? Would you predict the $\Delta H°$ of Step [1] to be positive or negative?
   b. How many bonds are broken and formed in Step [2]? Would you predict the $\Delta H°$ of Step [2] to be positive or negative?
   c. Which step is rate-determining?
   d. Draw the structure for the transition state in both steps of the mechanism.
   e. If $\Delta H°_{overall}$ is negative for this two-step reaction, draw an energy diagram illustrating all of the information in parts (a)–(d).

Kinetics and Rate Laws

6.47 Indicate which factors affect the rate of a reaction.
   a. $\Delta G°$
   b. $\Delta H°$
   c. E_a
   d. temperature
   e. concentration
   f. K_{eq}
   g. k
   h. catalysts

6.48 The following is a concerted, bimolecular reaction: $CH_3Br + NaCN \rightarrow CH_3CN + NaBr$.
   a. What is the rate equation for this reaction?
   b. What happens to the rate of the reaction if [CH_3Br] is doubled?
   c. What happens to the rate of the reaction if [NaCN] is halved?
   d. What happens to the rate of the reaction if [CH_3Br] and [NaCN] are both increased by a factor of five?

6.49 The conversion of acetyl chloride to methyl acetate occurs via the following two-step mechanism:

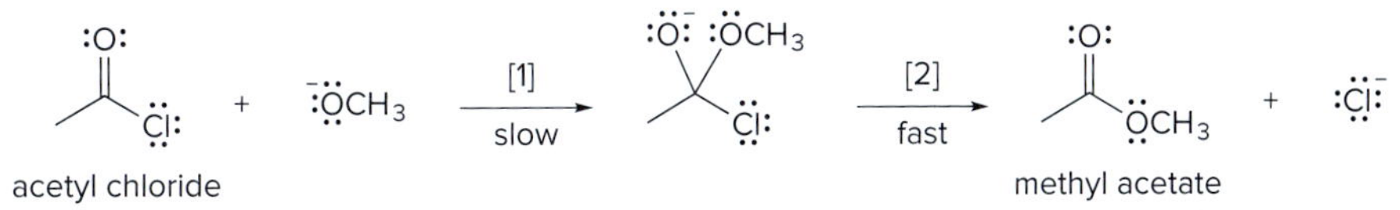

   a. Add curved arrows to show the movement of the electrons in each step.
   b. Write the rate equation for this reaction, assuming the first step is rate-determining.
   c. If the concentration of $^-OCH_3$ were increased 10 times, what would happen to the rate of the reaction?
   d. If the concentrations of both CH_3COCl and $^-OCH_3$ were increased 10 times, what would happen to the rate of the reaction?
   e. Classify the conversion of acetyl chloride to methyl acetate as an addition, elimination, or substitution.

General Problems

6.50 Consider the conversion of alkyl halide **A** to ether **B**.

PhCH$_2$Br + CH$_3$OH ⟶ PhCH$_2$OCH$_3$ + HBr
   A   **B**

a. Classify the conversion of **A** to **B** as a substitution, elimination, or addition.
b. The reaction rate depends on the concentration of **A** only. Write the rate equation for the reaction, and explain why the reaction mechanism must involve more than one step.
c. Heterolysis of the polar bond in **A** forms a resonance-stabilized intermediate. Draw all reasonable resonance structures for this intermediate.

6.51 In Chapter 14, we will learn about the hydrolysis of acetals to aldehydes and ketones. Four of the seven steps in the mechanism for this process are shown in the conversion of acetal **A** to hemiacetal **E**.

A acetal →[H–OSO$_3$H, 1] **B** + HSO$_4^-$ →[2] **C** + CH$_3$OH →[H$_2$O, 3] **D** →[HSO$_4^-$, 4] **E** hemiacetal + H$_2$SO$_4$

a. Add curved arrows for each step.
b. Draw another resonance structure for **C**.
c. Identify the nucleophile and electrophile in Step [3].
d. Which steps are Brønsted–Lowry acid–base reactions?

6.52 The Diels–Alder reaction, a powerful reaction discussed in Chapter 12, occurs when a 1,3-diene such as **A** reacts with an alkene such as **B** to form the six-membered ring in **C**.

A + **B** ⟶ **C**

a. Draw curved arrows to show how **A** and **B** react to form **C**.
b. What bonds are broken and formed in this reaction?
c. Would you expect this reaction to be endothermic or exothermic?
d. Does entropy favor the reactants or products?
e. Is the Diels–Alder reaction a substitution, elimination, or addition?

6.53 The conversion of (CH$_3$)$_3$CI to (CH$_3$)$_2$C=CH$_2$ can occur by either a one-step or a two-step mechanism, as shown in Equations [1] and [2].

[1] (CH$_3$)$_3$CI + $^-$:ÖH ⟶ (CH$_3$)$_2$C=CH$_2$ + I$^-$ + H$_2$Ö:

[2] (CH$_3$)$_3$CI →[slow] (CH$_3$)$_3$C$^+$ + I$^-$ →[$^-$:ÖH] (CH$_3$)$_2$C=CH$_2$ + H$_2$Ö:

a. What rate equation would be observed for the mechanism in Equation [1]?
b. What rate equation would be observed for the mechanism in Equation [2]?

c. What is the order of each rate equation (i.e., first, second, and so forth)?

d. How can these rate equations be used to show which mechanism is the right one for this reaction?

e. Assume Equation [1] represents an endothermic reaction and draw an energy diagram for the reaction. Label the axes, reactants, products, E_a, and $\Delta H°$. Draw the structure for the transition state.

f. Assume Equation [2] represents an endothermic reaction and that the product of the rate-determining step is higher in energy than the reactants or products. Draw an energy diagram for this two-step reaction. Label the axes, reactants and products for each step, and the E_a and $\Delta H°$ for each step. Label $\Delta H°_{overall}$. Draw the structure for both transition states.

Challenge Problems

6.54 Explain why HC≡CH is more acidic than CH_3CH_3, even though the C–H bond in HC≡CH has a higher bond dissociation energy than the C–H bond in CH_3CH_3.

6.55 The use of curved arrows is a powerful tool that illustrates even complex reactions.

a. Add curved arrows to show how carbocation **A** is converted to carbocation **B**. Label each new σ bond formed. Similar reactions have been used in elegant syntheses of steroids.

b. Draw the product by following the curved arrows. This reaction is an example of a [3,3] sigmatropic rearrangement, as we will learn in Chapter 29.

6.56

propylbenzene

a. What carbon radical is formed by homolysis of the C–H_a bond in propylbenzene? Draw all reasonable resonance structures for this radical.

b. What carbon radical is formed by homolysis of the C–H_b bond in propylbenzene? Draw all reasonable resonance structures for this radical.

c. The bond dissociation energy of one of the C–H bonds is considerably less than the bond dissociation energy of the other. Which C–H bond is weaker? Offer an explanation.

6.57 One step in glucose metabolism involves the conversion of dihydroxyacetone phosphate to glyceraldehyde 3-phosphate by way of an intermediate enediol. Each process involves both a protonation and a deprotonation. Draw curved arrows to show the movement of electrons in each step, using HA as an acid for protonation and B: as a base for deprotonation.

dihydroxyacetone phosphate ⇌ enediol ⇌ glyceraldehyde 3-phosphate

7 Alkyl Halides and Nucleophilic Substitution

7.1 Introduction to alkyl halides
7.2 Nomenclature
7.3 Properties of alkyl halides
7.4 Interesting alkyl halides
7.5 The polar carbon–halogen bond
7.6 General features of nucleophilic substitution
7.7 The leaving group
7.8 The nucleophile
7.9 Possible mechanisms for nucleophilic substitution
7.10 Two mechanisms for nucleophilic substitution
7.11 The S_N2 mechanism
7.12 The S_N1 mechanism
7.13 Carbocation stability
7.14 The Hammond postulate
7.15 When is the mechanism S_N1 or S_N2?
7.16 Biological nucleophilic substitution
7.17 Vinyl halides and aryl halides
7.18 Organic synthesis

Source: Claire Fackler/CINMS/NOAA

Giant kelp, a type of marine algae that grows in dense forests in cold ocean waters, is a major source of atmospheric **chloromethane (CH_3Cl),** the simplest alkyl chloride. Chloromethane is also produced by evergreen trees and is released during volcanic eruptions. Although some chloromethane in the atmosphere is man-made, most is natural in origin. In Chapter 7, we learn about alkyl halides like chloromethane and one of their characteristic reactions, nucleophilic substitution.

Why Study... Alkyl Halides?

This is the first of three chapters dealing with an in-depth study of the organic reactions of compounds containing C–Z σ bonds, where Z is an element more electronegative than carbon. In Chapter 7, we learn about **alkyl halides** and one of their characteristic reactions, **nucleophilic substitution,** a key step in the synthesis of several useful drugs and natural products. In Chapter 8, we look at **elimination,** a second general reaction of alkyl halides. We conclude this discussion in Chapter 9 by examining other molecules that also undergo nucleophilic substitution and elimination reactions. In these chapters, we will learn about many specific details that explain how and why key reactions take place.

7.1 Introduction to Alkyl Halides

Alkyl halides are organic molecules containing a halogen atom X bonded to an sp^3 hybridized carbon atom. As we learned in Section 3.2, alkyl halides are classified as **primary (1°), secondary (2°),** or **tertiary (3°)** depending on the number of carbons bonded to the carbon with the halogen. Whether an alkyl halide is 1°, 2°, or 3° is the *most important factor* in determining the course of its chemical reactions.

Alkyl halides have the general molecular formula $C_nH_{2n+1}X$, and are formally derived from an alkane by replacing a hydrogen atom with a halogen.

alkyl halide
X = F, Cl, Br, I
C is sp^3 hybridized.

Four types of organic halides having the halogen atom in close proximity to a π bond are illustrated in Figure 7.1. **Vinyl halides** have a halogen atom bonded to a carbon–carbon double bond, and **aryl halides** have a halogen atom bonded to a benzene ring. These two types of organic halides with X bonded directly to an sp^2 hybridized carbon atom do *not* undergo the reactions presented in Chapter 7, as discussed in Section 7.17.

Figure 7.1
Four types of organic halides (RX) having X near a π bond

vinyl halide   **aryl halide**   **allylic halide**   **benzylic halide**

- C bonded to X is sp^2 hybridized.
- These organic halides are **unreactive** in the reactions discussed in Chapter 7.

- C bonded to X is sp^3 hybridized.
- These organic halides do participate in the reactions discussed in Chapter 7.

Allylic halides and benzylic halides have halogen atoms bonded to sp^3 hybridized carbon atoms and *do* undergo the reactions described in Chapter 7. **Allylic halides** have X bonded to the carbon atom *adjacent* to a carbon–carbon double bond, and **benzylic halides** have X bonded to the carbon atom *adjacent* to a benzene ring.

Problem 7.1 Telfairine, a naturally occurring insecticide, and halomon, an antitumor agent, are two polyhalogenated compounds isolated from red algae. (a) Classify each halide bonded to an sp^3 hybridized carbon as 1°, 2°, or 3°. (b) Label each halide as vinyl, allylic, or neither.

Hundreds of organic halides with diverse structures and biological activities have been isolated from red algae of the genus *Laurencia,* seaweed that grows in shallow water at the edges of reefs. *Michael Guiry*

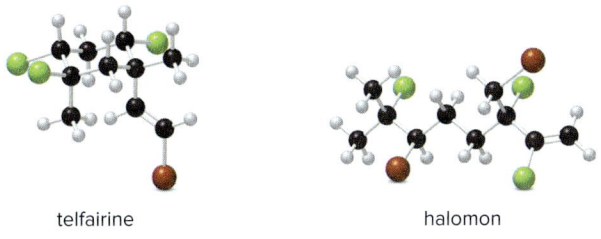

telfairine   halomon

7.2 Nomenclature

The systematic (IUPAC) method for naming alkyl halides follows from the basic rules described in Chapter 4.

7.2A IUPAC System

An alkyl halide is named as an alkane with a halogen substituent—that is, as a **halo alkane.** To name a halogen substituent, change the **-ine** ending of the name of the halogen to the suffix **-o** (chlor*ine* → chlor*o*).

How To Name an Alkyl Halide Using the IUPAC System

Example Give the IUPAC name of the following alkyl halide:

Step [1] Find the parent carbon chain and name it as an alkane.

- Name the parent chain as an *alkane,* with the halogen as a substituent bonded to the longest chain.

7 C's in the longest chain

7 C's ---→ **heptane**

Step [2] Apply all other rules of nomenclature.

a. **Number** the chain.

b. **Name** and **number** the substituents.

- Cl chloro at C2
- methyl at C5

- Begin at the end nearest the first substituent, either alkyl or halogen.

c. **Alphabetize:** **c** for **c**hloro, then **m** for **m**ethyl.

Answer: **2-chloro-5-methylheptane**

7.2B Common Names

Common names for alkyl halides are used only for simple alkyl halides. To assign a common name:

- Name all the carbon atoms of the molecule as a single **alkyl group.**
- Name the halogen bonded to the alkyl group. To name the halogen, change the *-ine* ending of the halogen name to the suffix *-ide;* for example, brom*ine* → brom*ide*.
- Combine the names of the alkyl group and halide, separating the words with a space.

iod*ine* → iod*ide*   chlor*ine* → chlor*ide*

tert-butyl   **tert-butyl iodide**   ethyl   **ethyl chloride**
group   group

Other examples of alkyl halide nomenclature are given in Figure 7.2.

Figure 7.2
Examples: Nomenclature of alkyl halides

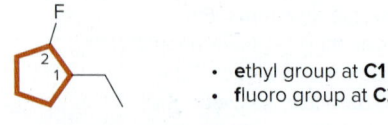

IUPAC: **1-chloro-2-methylpropane**
Common: isobutyl chloride

IUPAC: **1-ethyl-2-fluorocyclopentane**
- **e**thyl group at **C1**
- **f**luoro group at **C2**

earlier letter --→ lower number

[too complex to use a common name]

Problem 7.2 Give the IUPAC name for each compound.

a. [structure with Cl] b. [cyclohexane with Br and methyl] c. [structure with F] d. [structure with Cl]

Problem 7.3 Give the structure corresponding to each name.
a. 3-chloro-2-methylhexane
b. 4-ethyl-5-iodo-2,2-dimethyloctane
c. *cis*-1,3-dichlorocyclopentane
d. 1,1,3-tribromocyclohexane
e. 6-ethyl-3-iodo-3,5-dimethylnonane
f. (*R*)-1-fluoro-2,6,6-trimethylnonane

7.3 Properties of Alkyl Halides

Alkyl halides are weakly polar molecules. They exhibit **dipole–dipole** interactions because of their polar C–X bond, but because the rest of the molecule contains only C–C and C–H bonds they are incapable of intermolecular hydrogen bonding. How this affects their physical properties is summarized in Table 7.1.

The spectroscopic properties of alkyl halides are discussed in Chapters A–C. Of particular note are the characteristic features of the mass spectra of alkyl chlorides and alkyl bromides, which are discussed in Section A.2.

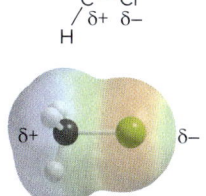

Table 7.1 Physical Properties of Alkyl Halides

Property	Observation
Boiling point and melting point	• Alkyl halides have higher bp's and mp's than alkanes having the same number of carbons. • Bp's and mp's increase as the size of R increases. CH₃CH₃ — mp = –183 °C, bp = –89 °C CH₃CH₂Cl — mp = –136 °C, bp = 12 °C CH₃CH₂CH₂Cl — mp = –123 °C, bp = 47 °C **Increasing boiling point and melting point** • Bp's and mp's increase as the size of X increases. CH₃CH₂Cl — mp = –136 °C, bp = 12 °C and CH₃CH₂Br — mp = –119 °C, bp = 39 °C more polarizable halogen → higher mp and bp
Solubility	• RX is soluble in organic solvents. • RX is insoluble in water.

Problem 7.4 An sp^3 hybridized C–Cl bond is more polar than an sp^2 hybridized C–Cl bond. (a) Explain why this phenomenon arises. (b) Rank the following compounds in order of increasing boiling point.

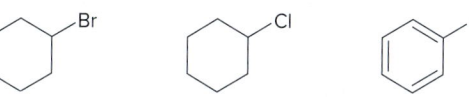

7.4 Interesting Alkyl Halides

Many simple alkyl halides make excellent solvents because they are not flammable and dissolve a wide variety of organic compounds. Compounds in this category include **$CHCl_3$** (chloroform or trichloromethane) and **CCl_4** (carbon tetrachloride or tetrachloromethane). Large quantities of these solvents are produced industrially each year, but like many chlorinated organic compounds, both chloroform and carbon tetrachloride are toxic if inhaled or ingested. Other simple alkyl halides are shown in Figure 7.3.

Figure 7.3
Some simple alkyl halides

CH_2Cl_2

- **Dichloromethane (or methylene chloride, CH_2Cl_2)** is a common solvent, once used to decaffeinate coffee. Coffee is now decaffeinated by using supercritical CO_2 due to concerns over the possible ill effects of trace amounts of residual CH_2Cl_2 in the coffee. Subsequent studies on rats have shown, however, that no cancers occurred when animals ingested the equivalent of over 100,000 cups of decaffeinated coffee per day.

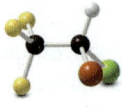
$CF_3CHClBr$

- **Halothane ($CF_3CHClBr$)** is a safe general anesthetic compared to other organic anesthetics such as $CHCl_3$, which causes liver and kidney damage, and $CH_3CH_2OCH_2CH_3$ (diethyl ether), which is very flammable.

Synthetic organic halides are also used in insulating materials, plastic wrap, and coatings. Two such compounds are **Teflon** and **poly(vinyl chloride) (PVC)**.

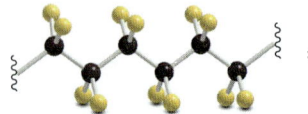
Teflon
(nonstick coating)

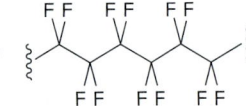

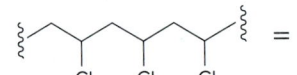

poly(vinyl chloride) (PVC)
(plastic used in films, pipes, and insulation)

Several useful drugs contain one or more fluorine atoms. Examples include fluticasone, an aerosol inhalant used for the treatment of seasonal nasal allergies and asthma, and roflumilast, which was approved by the FDA in 2015 for the treatment of severe cases of chronic obstructive pulmonary disease (COPD).

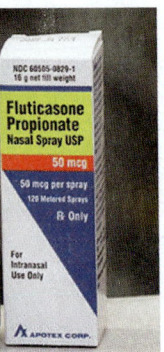

Fluticasone (trade name Flonase) is a synthetic steroid used to treat the chronic inflammation of asthma.
Mark Dierker/McGraw-Hill Education

fluticasone

roflumilast

Although the beneficial effects of many organic halides are undisputed, certain synthetic chlorinated organics such as the **chlorofluorocarbons** and the pesticide **DDT** have caused lasting harm to the environment.

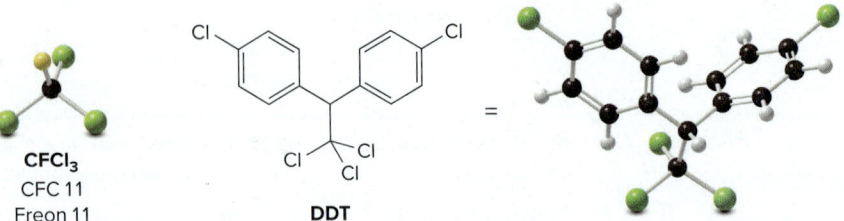

$CFCl_3$
CFC 11
Freon 11

DDT

Chlorofluorocarbons (CFCs) have the general molecular structure CF_xCl_{4-x}. Trichlorofluoromethane [$CFCl_3$, CFC 11, or Freon 11 (trade name)] is an example of these easily vaporized

DDT, a nonbiodegradable pesticide, has been labeled both a "miraculous" discovery by Winston Churchill in 1945 and the "elixir of death" by Rachel Carson in her 1962 book *Silent Spring*. DDT use was banned in the United States in 1973, but because of its effectiveness and low cost, it is still widely used to control inspect populations in developing countries.

compounds, having been extensively used as a refrigerant and an aerosol propellant. CFCs slowly rise to the stratosphere, where sunlight catalyzes their decomposition, a process that contributes to the destruction of the ozone layer, the thin layer of atmosphere that shields the earth's surface from harmful ultraviolet radiation (Section 21.8). Although it is now easy to second-guess the extensive use of CFCs, it is also easy to see why they were used so widely. **CFCs made refrigeration available to the general public.** Would you call your refrigerator a comfort or a necessity?

The story of the insecticide **DDT** (**d**ichloro**d**iphenyl**t**richloroethane) follows the same theme: DDT is an organic molecule with valuable short-term effects that has caused long-term problems. DDT kills insects that spread diseases such as malaria and typhus, and in controlling insect populations, DDT has saved millions of lives worldwide. DDT is a weakly polar organic compound that persists in the environment for years. Because DDT is soluble in organic media, it accumulates in fatty tissues. Most adults in the United States have low concentrations of DDT (or a degradation product of DDT) in their bodies. DDT is acutely toxic to many types of marine life (crayfish, sea shrimp, and some fish), but the long-term effect on humans is not known.

Problem 7.5 Chondrocole A is a marine natural product isolated from red seaweed that grows in regions of heavy surf in the Pacific Ocean. (a) Predict the solubility of chondrocole A in water and CH_2Cl_2. (b) Locate the stereogenic centers and label each as *R* or *S*. (c) Draw a stereoisomer and a constitutional isomer of chondrocole A.

chondrocole A

7.5 The Polar Carbon–Halogen Bond

The properties of alkyl halides dictate their reactivity. The electronegative halogen X of an alkyl halide creates a polar C—X bond, making the carbon atom electron deficient. **The chemistry of alkyl halides is determined by this polar C—X bond.**

What kind of reactions do alkyl halides undergo? **The characteristic reactions of alkyl halides are substitution and elimination.** Because alkyl halides contain an electrophilic carbon, they react with electron-rich reagents—Lewis bases (nucleophiles) and Brønsted–Lowry bases.

- **Alkyl halides undergo substitution reactions with nucleophiles.**

In a substitution reaction of an alkyl halide, **the halogen X is replaced by an electron-rich nucleophile :Nu⁻.** The C—X σ bond is broken and the C—Nu σ bond is formed.

- **Alkyl halides undergo elimination reactions with Brønsted–Lowry bases.**

In an elimination reaction of an alkyl halide, the **elements of HX are removed by a Brønsted–Lowry base :B.**

The remainder of Chapter 7 is devoted to a discussion of the substitution reactions of alkyl halides. Elimination reactions are discussed in Chapter 8.

7.6 General Features of Nucleophilic Substitution

Three components are necessary in any substitution reaction.

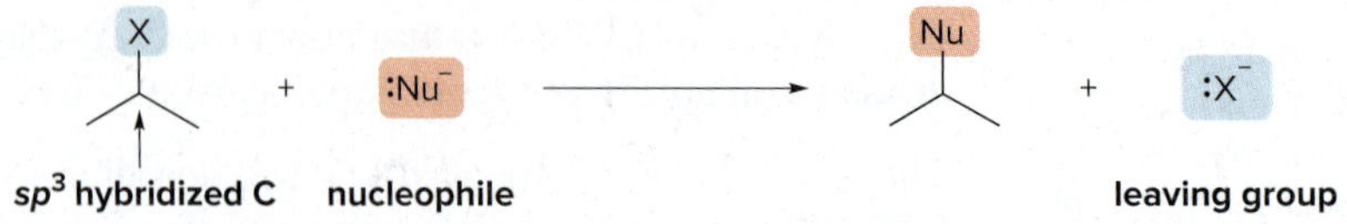

[1] An alkyl group containing an sp^3 hybridized carbon bonded to X.
[2] **X**—An atom X (or a group of atoms) called **a leaving group,** which is able to accept the electron density in the C–X bond. The most common leaving groups are halide anions (X$^-$), but H$_2$O (from ROH$_2^+$) and N$_2$ (from RN$_2^+$) are also encountered.
[3] **:Nu$^-$**—A **nucleophile.** Nucleophiles contain a **lone pair** or a **π bond** but not necessarily a negative charge.

Because these substitution reactions involve electron-rich nucleophiles, they are called *nucleophilic* substitution reactions. Examples are shown in Equations [1] and [2]. **Nucleophilic substitutions are Lewis acid–base reactions.** The nucleophile donates its electron pair, the alkyl halide (Lewis acid) accepts it, and the C–X bond is heterolytically cleaved. Curved arrow notation can be used to show the movement of electron pairs, as shown in Equation [2].

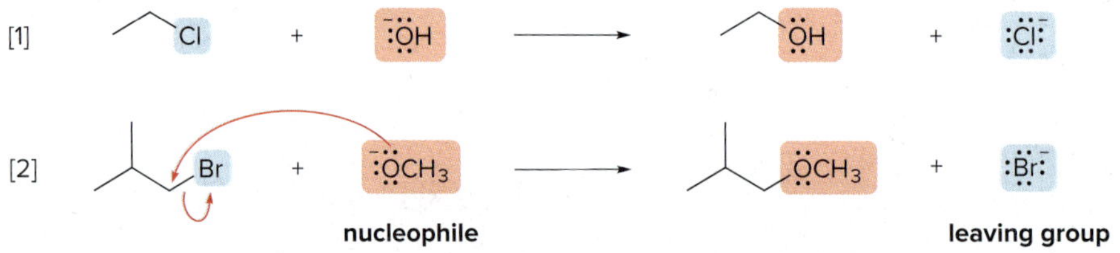

Negatively charged nucleophiles like $^-$OH and $^-$SH are used as **salts** with Li$^+$, Na$^+$, or K$^+$ counterions to balance charge. The identity of the cation is usually inconsequential, and therefore it is often omitted from the chemical equation.

When a neutral nucleophile is used, the substitution product bears a positive charge. **All atoms originally bonded to the nucleophile stay bonded to it after substitution occurs.** All three CH$_3$ groups stay bonded to the N atom in the given example.

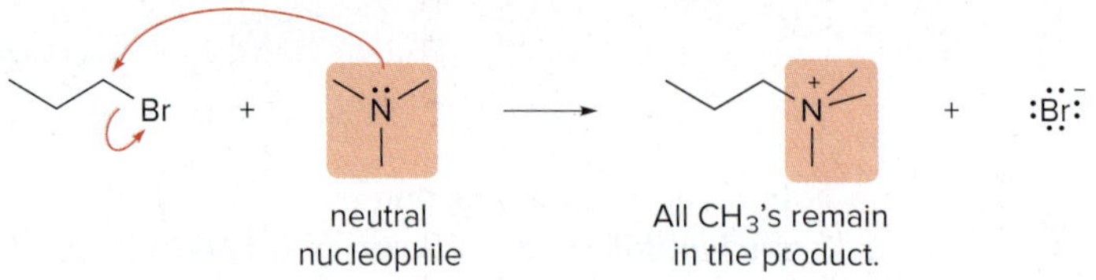

Furthermore, when the substitution product bears a positive charge and also contains a *proton* bonded to O or N, the initial substitution product readily loses a proton in a Brønsted–Lowry acid–base reaction, forming a neutral product.

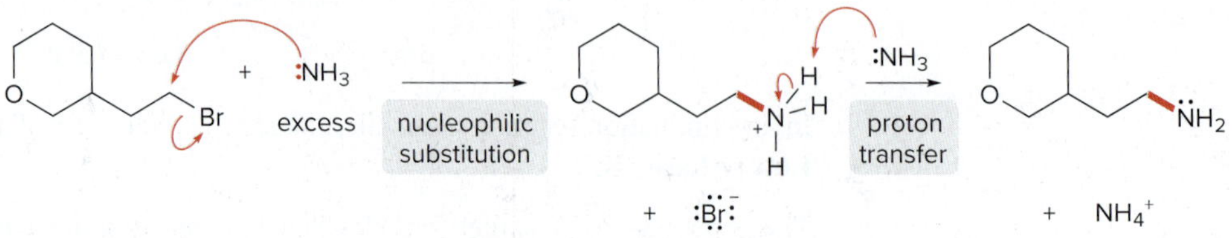

The reaction of alkyl halides with NH$_3$ to form amines (RNH$_2$) is discussed in Chapter 22.

All of these reactions are nucleophilic substitutions and have the same overall result—**replacement of the leaving group by the nucleophile,** regardless of the identity or charge of the nucleophile. To draw any nucleophilic substitution product:

- Find the sp^3 hybridized carbon with the leaving group.
- Identify the nucleophile, the species with a lone pair or π bond.
- Substitute the nucleophile for the leaving group and assign charges (if necessary) to any atom that is involved in bond breaking or bond formation.

Problem 7.6 Identify the nucleophile and leaving group and draw the products of each substitution reaction.

a. 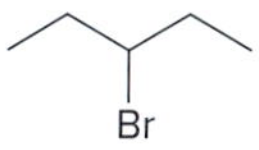 CH₃CH(Br)CH₂CH₃ + ⁻OCH₂CH₃ ⟶

b. 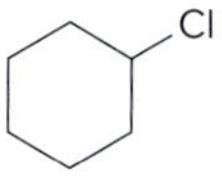 cyclohexyl-Cl + NaOH ⟶

c.  CH₃CH₂CH₂I + N₃⁻ ⟶

d. 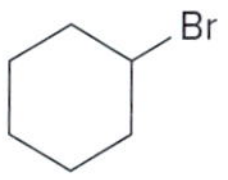 cyclohexyl-Br + NaCN ⟶

Problem 7.7 Draw the product of nucleophilic substitution with each neutral nucleophile. When the initial substitution product can lose a proton to form a neutral product, draw the product after proton transfer.

a. 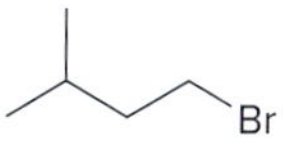 (CH₃)₂CHCH₂CH₂Br + :N(CH₂CH₃)₃ ⟶

b.  (CH₃)₃C-Cl + H₂Ö: ⟶

Problem 7.8 What neutral nucleophile is needed to convert dihalide **A** to ticlopidine, an antiplatelet drug used to reduce the risk of strokes?

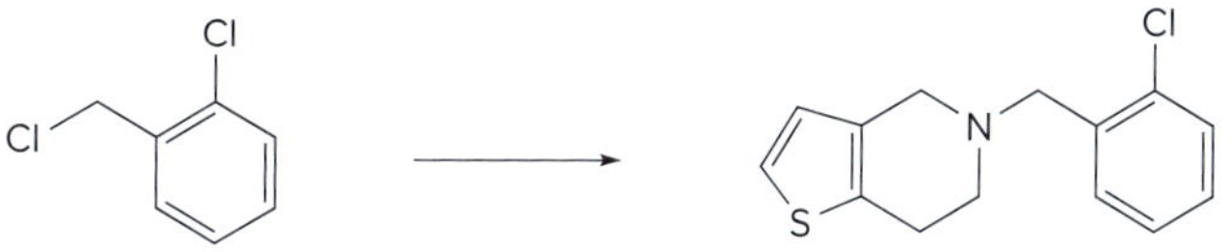

A ⟶ ticlopidine

7.7 The Leaving Group

Nucleophilic substitution is a general reaction of organic compounds. Why, then, are alkyl halides the most common substrates, and halide anions the most common leaving groups? To answer this question, we must understand leaving group ability. **What makes a good leaving group?**

In a nucleophilic substitution reaction of R—X, the C—X bond is heterolytically cleaved, and the leaving group departs with the electron pair in that bond, forming X:⁻. **The more stable the leaving group X:⁻, the better able it is to accept an electron pair,** giving rise to the following generalization:

- In comparing two leaving groups, the *better* leaving group is the *weaker* base.

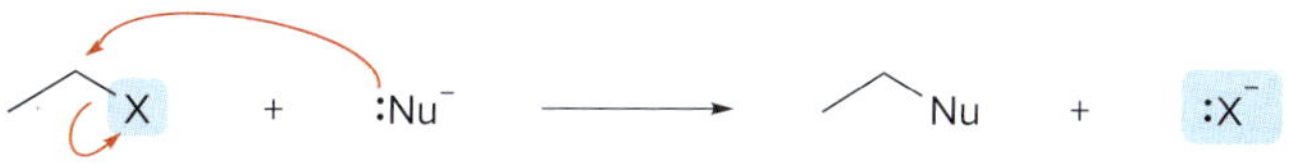

R—X + :Nu⁻ ⟶ R—Nu + :X⁻

Good leaving groups are weak bases.

For example, H₂O is a better leaving group than ⁻OH because H₂O is a weaker base. Moreover, the periodic trends in basicity can now be used to identify **periodic trends in leaving group ability:**

- Left-to-right across a row of the periodic table, basicity *decreases* so leaving group ability *increases*.

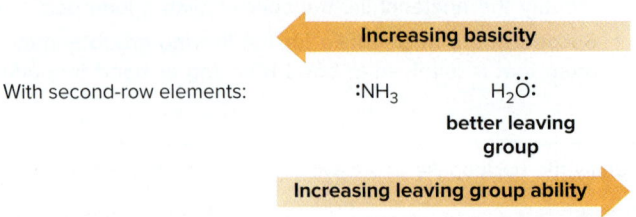

- Down a column of the periodic table, basicity *decreases* so leaving group ability *increases*.

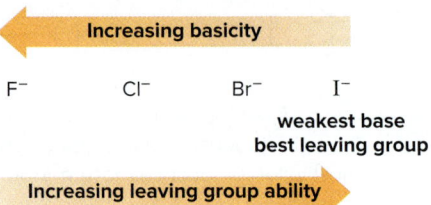

All good leaving groups are weak bases with strong conjugate acids having low pK_a values. Thus, all halide anions except F⁻ are good leaving groups because their conjugate acids (HCl, HBr, and HI) have low pK_a values. Tables 7.2 and 7.3 list good and poor leaving groups for nucleophilic substitution reactions, respectively. Nucleophilic substitution does not occur with any of the leaving groups in Table 7.3 because these leaving groups are strong bases.

Table 7.2 Good Leaving Groups for Nucleophilic Substitution

Starting material	Leaving group	Conjugate acid	pK_a
R−Cl	:Cl:⁻	HCl	−7
R−Br	:Br:⁻	HBr	−9
R−I	:I:⁻	HI	−10
R−OH₂⁺	H₂O:	H₃O⁺	−1.7

Table 7.3 Poor Leaving Groups for Nucleophilic Substitution

Starting material	Leaving group	Conjugate acid	pK_a
R−F	:F:⁻	HF	3.2
R−OH	⁻OH	H₂O	15.7
R−NH₂	⁻NH₂	NH₃	38
R−H	H:⁻	H₂	35
R−R	R:⁻	RH	50

Problem 7.9 Which molecules contain good leaving groups?
a. ∼∼Br   b. ∼∼OH   c. ∼∼OH₂⁺   d. ∼∼

Problem 7.10 (a) Which of the labeled atoms in each molecule is the best leaving group? (b) Which of the labeled atoms in each molecule is the worst leaving group?

Given a particular nucleophile and leaving group, how can we determine whether the equilibrium will favor products in a nucleophilic substitution? We can often correctly predict the direction of equilibrium by comparing the basicity of the nucleophile and the leaving group.

- **Equilibrium favors the products of nucleophilic substitution when the leaving group is a *weaker base* than the nucleophile.**

Sample Problem 7.1 illustrates how to apply this general rule.

Sample Problem 7.1 Using Basicity to Determine If a Substitution Is Likely to Occur

Will the following substitution reaction favor formation of the products?

Solution

Determine the basicity of the nucleophile ($^-$OH) and the leaving group (Cl$^-$) by comparing the pK_a values of their conjugate acids. **The stronger the conjugate acid, the weaker the base, and the better the leaving group.**

			conjugate acids	
nucleophile	$^-$OH	→	H_2O	pK_a = 15.7
leaving group	Cl$^-$	→	HCl	pK_a = −7
	weaker base		stronger acid	

Because Cl$^-$, the leaving group, is a weaker base than $^-$OH, the nucleophile, **the reaction favors the products.**

Problem 7.11 Does the equilibrium favor the reactants or the products in each substitution reaction?

a. ⌒NH$_2$ + Br$^-$ ⟶ ⌒Br + $^-$NH$_2$

b. ⌒⌒I + $^-$CN ⟶ ⌒⌒CN + I$^-$

More Practice: Try Problem 7.46.

7.8 The Nucleophile

Nucleophiles and bases are structurally similar: both have a lone pair or a π bond. They differ in what they attack.

We use the word *base* to mean Brønsted–Lowry base and the word *nucleophile* to mean a Lewis base that reacts with electrophiles *other than protons*.

- **Bases attack protons. Nucleophiles attack other electron-deficient atoms (usually carbons).**

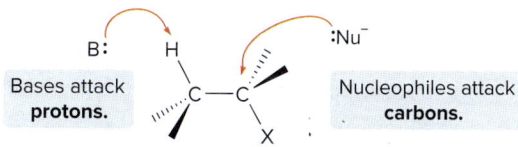

7.8A Nucleophilicity Versus Basicity

How is **nucleophilicity** (nucleophile strength) related to basicity? Although it is generally true that **a strong base is a strong nucleophile**, nucleophile size and steric factors can sometimes change this relationship.

Nucleophilicity parallels basicity in three instances.

[1] For two nucleophiles with the same nucleophilic atom, the *stronger* base is the *stronger* nucleophile.

- The relative nucleophilicity of $^-$OH and $CH_3CO_2^-$, two oxygen nucleophiles, is determined by comparing the pK_a values of their conjugate acids (H_2O and CH_3CO_2H). CH_3CO_2H (pK_a = 4.8) is a stronger acid than H_2O (pK_a = 15.7), so **$^-$OH is a stronger base and stronger nucleophile than $CH_3CO_2^-$.**

[2] A negatively charged nucleophile is always *stronger* than its conjugate acid.

- $^-$OH is a stronger base and stronger nucleophile than H_2O, its conjugate acid.

[3] Right-to-left across a row of the periodic table, nucleophilicity *increases* as basicity *increases*.

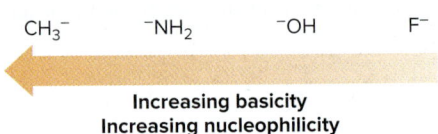

Increasing basicity
Increasing nucleophilicity

Problem 7.12 Identify the stronger nucleophile in each pair.
a. NH_3, $^-NH_2$ b. CH_3NH_2, CH_3OH c. $CH_3CO_2^-$, $CH_3CH_2O^-$

7.8B Steric Effects and Nucleophilicity

Nucleophilicity does not parallel basicity when **steric hindrance** becomes important. *Steric hindrance is a decrease in reactivity resulting from the presence of bulky groups at the site of a reaction.*

All steric effects arise because two atoms cannot occupy the same space. In Chapter 4, for example, we learned that **steric strain** is an increase in energy when big groups (occupying a large volume) are forced close to each other.

For example, although pK_a tables indicate that *tert*-butoxide [$(CH_3)_3CO^-$] is a stronger base than ethoxide ($CH_3CH_2O^-$), **ethoxide is the *stronger* nucleophile.** The three CH_3 groups around the O atom of *tert*-butoxide create steric hindrance, making it more difficult for this big, bulky base to attack a tetravalent carbon atom.

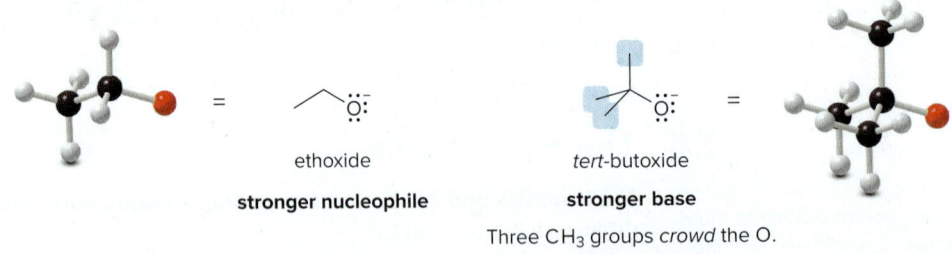

ethoxide
stronger nucleophile

tert-butoxide
stronger base
Three CH_3 groups *crowd* the O.
weaker nucleophile

Steric hindrance decreases nucleophilicity but *not* basicity. Because bases pull off small, easily accessible protons, they are unaffected by steric hindrance. Nucleophiles, on the other hand, must attack a crowded tetrahedral carbon, so bulky groups decrease reactivity.

Sterically hindered bases that are poor nucleophiles are called *nonnucleophilic bases*. Potassium *tert*-butoxide [K^+ $^-OC(CH_3)_3$] is a strong, nonnucleophilic base.

7.8C Comparing Nucleophiles of Different Size—Solvent Effects

Atoms vary greatly in size down a column of the periodic table, and in this case, **nucleophilicity depends on the solvent used in a substitution reaction.** Although solvent has thus far been ignored, most organic reactions take place in a liquid solvent that dissolves all reactants to some extent. Because substitution reactions involve polar starting materials, polar solvents are used to dissolve them. There are two main kinds of polar solvents: **polar *protic* solvents** and **polar *aprotic* solvents.**

Polar Protic Solvents

In addition to dipole–dipole interactions, **polar *protic* solvents are capable of intermolecular hydrogen bonding,** because they contain an O–H or N–H bond. The most common polar protic solvents are water and alcohols (ROH) (Figure 7.4). **Polar protic solvents solvate *both* cations and anions well.**

- Cations are solvated by ion–dipole interactions.
- Anions are solvated by hydrogen bonding.

Figure 7.4 Polar protic solvents	H_2O	CH_3OH methanol	CH_3CH_2OH ethanol	$(CH_3)_3COH$ *tert*-butanol	CH_3CO_2H acetic acid

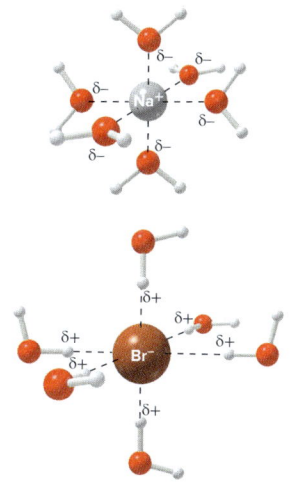

I⁻ is a *weak base* but a *strong nucleophile* in polar protic solvents.

For example, if the salt NaBr is used as a source of the nucleophile Br⁻ in H_2O, the Na⁺ cations are solvated by ion–dipole interactions with H_2O molecules, and the Br⁻ anions are solvated by strong hydrogen bonding interactions.

How do polar protic solvents affect nucleophilicity? **In polar protic solvents, nucleophilicity *increases* down a column of the periodic table as the size of the anion increases. This is *opposite* to basicity.** A small electronegative anion like F⁻ is very well solvated by hydrogen bonding, effectively *shielding* it from reaction. On the other hand, a large, less electronegative anion like I⁻ does not hold onto solvent molecules as tightly. The *solvent does not "hide" a large nucleophile* as well, and the nucleophile is much more able to donate its electron pairs in a reaction. Thus, **nucleophilicity *increases* down a column** even though basicity decreases, giving rise to the following trend in polar protic solvents:

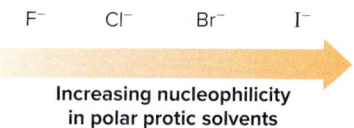

F⁻ Cl⁻ Br⁻ I⁻

Increasing nucleophilicity in polar protic solvents

Polar Aprotic Solvents

Polar *aprotic* solvents also exhibit dipole–dipole interactions, but they have no O–H or N–H bond so they are **incapable of hydrogen bonding.** Examples of polar aprotic solvents are shown in Figure 7.5. **Polar aprotic solvents solvate only cations well.**

- Cations are solvated by ion–dipole interactions.
- Anions are not well solvated because the solvent cannot hydrogen bond to them.

Figure 7.5 Polar aprotic solvents

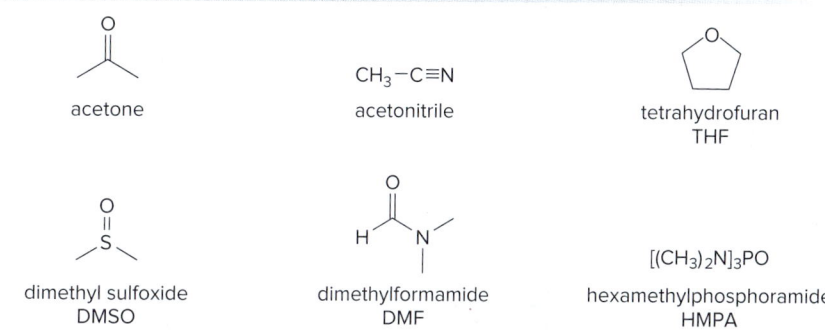

acetone

$CH_3-C\equiv N$ acetonitrile

tetrahydrofuran THF

dimethyl sulfoxide DMSO

dimethylformamide DMF

$[(CH_3)_2N]_3PO$ hexamethylphosphoramide HMPA

Abbreviations are often used in organic chemistry, instead of a compound's complete name. A list of common abbreviations is given in Appendix B.

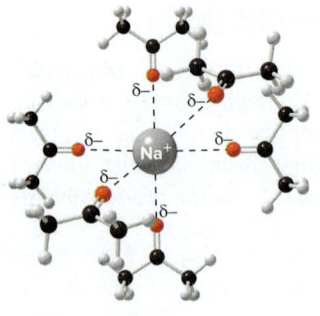

Br⁻ anions are surrounded by solvent but not well solvated by the $(CH_3)_2C=O$ molecules.

When the salt NaBr is dissolved in acetone, $(CH_3)_2C=O$, the Na⁺ cations are solvated by ion–dipole interactions with the acetone molecules, but, with no possibility for hydrogen bonding, the **Br⁻ anions are not well solvated**. Often these anions are called **naked anions** because they are not bound by tight interactions with solvent.

How do polar aprotic solvents affect nucleophilicity? Because anions are not well solvated in polar aprotic solvents, there is no need to consider whether solvent molecules more effectively hide one anion than another. **Nucleophilicity parallels basicity and the *stronger* base is the *stronger* nucleophile.** Because basicity decreases with size down a column, nucleophilicity decreases as well:

F⁻   Cl⁻   Br⁻   I⁻

← Increasing nucleophilicity in polar aprotic solvents

Problem 7.13 Classify each solvent as protic or aprotic.

a. HO~~~OH   b. ~O~   c. CH₃C(=O)O~

Problem 7.14 Identify the stronger nucleophile in each pair of anions.

a. Br⁻ or Cl⁻ in a polar protic solvent   c. HS⁻ or F⁻ in a polar protic solvent
b. HO⁻ or Cl⁻ in a polar aprotic solvent

7.8D Summary

Keep in mind the central relationship between nucleophilicity and basicity in comparing two nucleophiles.

- It is generally true that the *stronger* base is the *stronger* nucleophile.
- In polar *protic* solvents, however, nucleophilicity *increases* with increasing size of an anion (opposite to basicity).
- Steric hindrance *decreases* nucleophilicity without decreasing basicity, making $(CH_3)_3CO^-$ a stronger base but a weaker nucleophile than $CH_3CH_2O^-$.

Table 7.4 lists some common nucleophiles used in nucleophilic substitution reactions.

Problem 7.15 Rank the nucleophiles in each group in order of increasing nucleophilicity.
a. ⁻OH, ⁻NH₂, H₂O   b. ⁻OH, Br⁻, F⁻ (polar aprotic solvent)   c. H₂O, ⁻OH, $CH_3CO_2^-$

Problem 7.16 What nucleophile is needed to convert $(CH_3)_2CHCH_2CH_2Br$ to each product?

a. (CH₃)₂CHCH₂CH₂SH   b. (CH₃)₂CHCH₂CH₂OCH₂CH₃   c. (CH₃)₂CHCH₂CH₂OC(=O)CH₃   d. (CH₃)₂CHCH₂CH₂C≡CH

Table 7.4 Common Nucleophiles in Organic Chemistry

	Negatively charged nucleophiles			Neutral nucleophiles	
Oxygen	⁻OH	⁻OR	$CH_3CO_2^-$	H_2O	ROH
Nitrogen	N_3^-			NH_3	RNH_2
Carbon	⁻CN	HC≡C⁻			
Halogen	Cl⁻	Br⁻	I⁻		
Sulfur	HS⁻	RS⁻		H_2S	RSH

7.9 Possible Mechanisms for Nucleophilic Substitution

Now that you know something about the general features of nucleophilic substitution, you can begin to understand the mechanism.

The σ bond is broken.   The σ bond is formed.

Nucleophilic substitution at an sp^3 hybridized carbon involves two σ bonds: the bond to the leaving group is broken and the bond to the nucleophile is formed. To understand the mechanism of this reaction, though, we must know the timing of these two events; that is, **what is the order of bond breaking and bond making?** Do they happen at the same time, or does one event precede the other? Consider two possibilities:

[1] **The mechanism has one step, and bond breaking and bond making occur at the *same* time.**

The C—X σ bond is broken...   *as*   ...the C—Nu σ bond is formed.

- If the C—X bond is broken *as* the C—Nu bond is formed, the mechanism has **one step.** As we learned in Section 6.9, the rate of such a bimolecular reaction depends on the concentration of *both* reactants; that is, the rate equation is **second order**, and **rate = $k[RX][:Nu^-]$**.

[2] **The mechanism has two steps, and bond breaking occurs *before* bond making.**

The C—X σ bond is broken...   *before*   ...the C—Nu σ bond is formed.

- If the C—X bond is broken *first* and then the C—Nu bond is formed, the mechanism has **two steps** and a **carbocation** is formed as an intermediate. Because the first step is rate-determining, the rate depends on the concentration of RX *only;* that is, the rate equation is **first order**, and **rate = $k[RX]$**.

In Section 7.10, we look at data for two specific nucleophilic substitution reactions and see if those data fit either of these proposed mechanisms.

7.10 Two Mechanisms for Nucleophilic Substitution

Rate equations for two different reactions give us insight into the possible mechanism for nucleophilic substitution.

Reaction of bromomethane (CH_3Br) with acetate ($CH_3CO_2^-$) affords the substitution product methyl acetate with loss of Br^- as the leaving group (Equation [1]). Kinetic data show that the reaction rate depends on the concentration of *both* reactants; that is, the rate equation is **second order**. This suggests a **bimolecular reaction with a one-step mechanism** in which the C—X bond is broken *as* the C—Nu bond is formed.

[1] CH_3-Br + acetate → (second-order kinetics) → CH_3-O methyl acetate + Br^-

Equation [2] illustrates a similar nucleophilic substitution reaction with a different alkyl halide, $(CH_3)_3CBr$, which also leads to substitution of Br^- by $CH_3CO_2^-$. Kinetic data show that this reaction rate depends on the concentration of only *one* reactant, the alkyl halide; that is, the rate

268 Chapter 7 Alkyl Halides and Nucleophilic Substitution

equation is **first order**. This suggests a **two-step mechanism in which the rate-determining step involves the alkyl halide only.**

[2] (CH₃)₃C–Br: + CH₃CO₂⁻ (acetate) →(first-order kinetics) (CH₃)₃C–O–C(=O)CH₃ + :Br:⁻

The numbers **1** and **2** in the names S$_N$1 and S$_N$2 refer to the kinetic order of the reactions. For example, S$_N$2 means that the kinetics are **second** order. The number 2 does *not* refer to the number of steps in the mechanism.

How can these two different results be explained? Although these two reactions have the same nucleophile and leaving group, **there must be two different mechanisms** because there are two different rate equations. These equations are specific examples of two well-known mechanisms for nucleophilic substitution at an sp^3 hybridized carbon:

- S$_N$2 mechanism (substitution nucleophilic bimolecular).
- S$_N$1 mechanism (substitution nucleophilic unimolecular).

The reaction in Equation [1] illustrates an S$_N$2 mechanism, whereas the reaction in Equation [2] illustrates an S$_N$1 mechanism.

7.11 The S$_N$2 Mechanism

The reaction of CH_3Br with $CH_3CO_2^-$ is an example of an **S$_N$2 reaction.** What are the general features of this mechanism?

CH_3–Br: + CH₃CO₂⁻ (acetate) →(S$_N$2 reaction) CH_3–O–C(=O)CH₃ + :Br:⁻

7.11A Kinetics

An S$_N$2 reaction exhibits **second-order kinetics**; that is, the reaction is **bimolecular** and both the alkyl halide and the nucleophile appear in the rate equation.

- rate = $k[CH_3Br][CH_3CO_2^-]$

Changing the concentration of *either* reactant affects the rate. For example, doubling the concentration of *either* the nucleophile or the alkyl halide doubles the rate. Doubling the concentration of *both* reactants increases the rate by a factor of *four*.

Problem 7.17 What happens to the rate of an S$_N$2 reaction under each of the following conditions?
a. [RX] is tripled, and [:Nu⁻] stays the same. c. [RX] is halved, and [:Nu⁻] stays the same.
b. Both [RX] and [:Nu⁻] are tripled. d. [RX] is halved, and [:Nu⁻] is doubled.

7.11B A One-Step Mechanism

The most straightforward explanation for the observed second-order kinetics is a **concerted reaction**—bond breaking and bond making occur at the *same* time, as shown in Mechanism 7.1.

 **Mechanism 7.1** The S$_N$2 Mechanism

One step The C–Br bond breaks as the C–O bond forms.

CH₃CO₂⁻ + CH₃–Br: → CH₃CO₂–CH₃ (new C–O bond) + :Br:⁻

An energy diagram for the reaction of $CH_3Br + CH_3CO_2^-$ is shown in Figure 7.6. Because the equilibrium for this S$_N$2 reaction favors the products, the products are drawn at lower energy than the starting materials.

Figure 7.6

An energy diagram for the S_N2 reaction: $CH_3Br + CH_3CO_2^- \rightarrow CH_3CO_2CH_3 + Br^-$

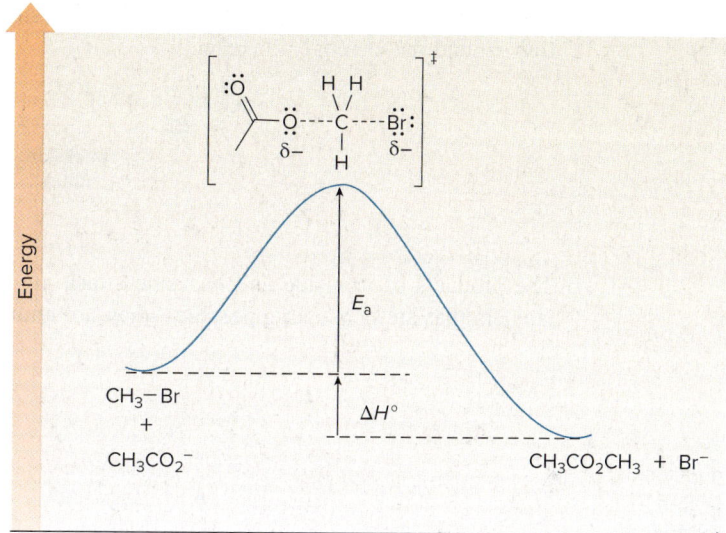

- In the transition state, the C—Br bond is partially broken, the C—O bond is partially formed, and both the attacking nucleophile and the departing leaving group bear a partial negative charge.

Problem 7.18 Draw an energy diagram for the following S_N2 reaction. Label the axes, the starting materials, and the product. Draw the structure of the transition state.

$$\text{CH}_3\text{CH}_2\text{CH}_2\text{Cl} + CH_3\ddot{\text{O}}{:}^- \longrightarrow \text{CH}_3\text{CH}_2\text{CH}_2\ddot{\text{O}}CH_3 + {:}\ddot{\text{C}}\ddot{\text{l}}{:}^-$$

7.11C Stereochemistry of the S_N2 Reaction

From what direction does the nucleophile approach the substrate in an S_N2 reaction? There are two possibilities.

- **Frontside attack:** The nucleophile approaches from the *same* side as the leaving group.
- **Backside attack:** The nucleophile approaches from the side *opposite* the leaving group.

The results of frontside and backside attack of a nucleophile are illustrated with CH₃CH(D)Br as substrate and the general nucleophile :Nu⁻. This substrate has the leaving group bonded to a stereogenic center, thus allowing us to see the structural difference that results when the nucleophile attacks from two different directions.

In frontside attack, the nucleophile approaches from the same side as the leaving group, forming A. In this example, the leaving group was drawn on the *right*, so the nucleophile attacks from the *right*, and all other groups remain in their original positions. Because the nucleophile and leaving group are in the same position relative to the other three groups on carbon, frontside attack results in **retention of configuration** around the stereogenic center.

Recall from Section 1.1 that D stands for the isotope deuterium (²H).

frontside attack

A
Nu replaces Br on the *same* side.

In backside attack, the nucleophile approaches from the opposite side to the leaving group, forming B. In this example, the leaving group was drawn on the *right*, so the nucleophile attacks from the *left*. Because the nucleophile and leaving group are in the opposite position relative to the other three groups on carbon, backside attack results in **inversion of configuration** around the stereogenic center.

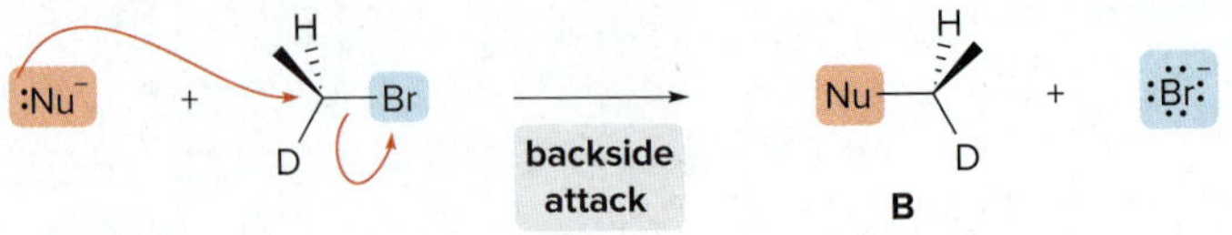

Nu replaces Br on the *opposite* side.

The products of frontside and backside attack are *different* compounds. **A** and **B** are stereoisomers that are nonsuperimposable—they are **enantiomers.**

product with **retention** of configuration

only S$_N$2 product
product with **inversion** of configuration

Inversion of configuration in an S$_N$2 reaction is often called **Walden inversion,** after Latvian chemist Dr. Paul Walden, who first observed this process in 1896.

Backside attack occurs in all S$_N$2 reactions, but we can observe this change only when the leaving group is bonded to a stereogenic center.

Which product is formed in an S$_N$2 reaction? When the stereochemistry of the product is determined, **only B, the product of backside attack, is formed.**

- All S$_N$2 reactions proceed with *backside attack* of the nucleophile, resulting in *inversion* of configuration at a stereogenic center.

One explanation for backside attack is based on an electronic argument. Both the nucleophile and leaving group are electron rich, and these like charges *repel* each other. Backside attack keeps these two groups as far away from each other as possible. In the transition state, the nucleophile and leaving group are 180° away from each other, and the other three groups around carbon occupy a plane, as illustrated in Figure 7.7.

Figure 7.7
Stereochemistry of the S$_N$2 reaction

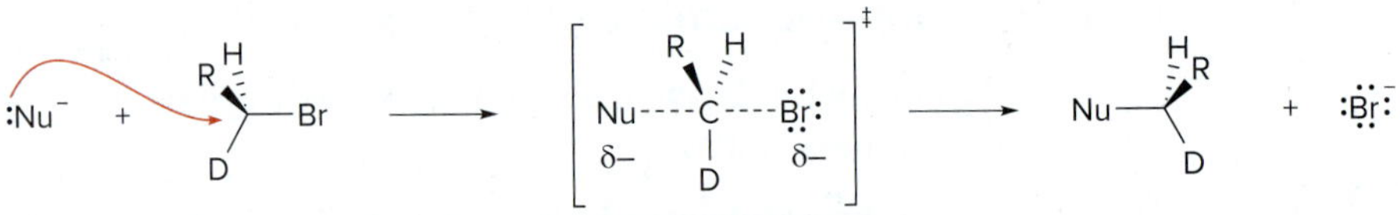

- :Nu$^-$ and Br$^-$ are 180° away from each other, on either side of a plane containing R, H, and D.

Two additional examples of inversion of configuration in S$_N$2 reactions are given in Figure 7.8.

Figure 7.8
Two examples of inversion of configuration in the S$_N$2 reaction

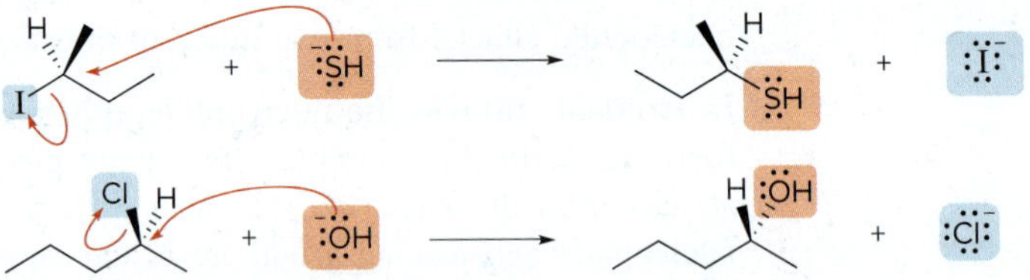

- The bond to the nucleophile in the product is always on the **opposite side** compared to the bond to the leaving group in the starting material. If the leaving group is drawn to the *left*, the nucleophile approaches from the *right*. If the leaving group is drawn in *front* of the plane (on a wedge), the nucleophile approaches from the *back* and ends up on a dashed wedge.

Sample Problem 7.2 Drawing the Product of Inversion in an S$_N$2 Reaction

Label the nucleophile and leaving group, and draw the product (including stereochemistry) of the following S$_N$2 reaction.

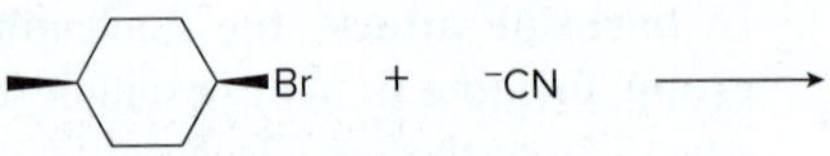

Solution

Br⁻ is the leaving group and ⁻CN is the nucleophile. Because S$_N$2 reactions proceed with **inversion** of configuration and the leaving group is drawn *above* the ring (on a wedge), the nucleophile must come in from *below* (ending up on a dashed wedge).

- **Inversion** of configuration occurs at the C–Br bond.
- **Backside attack** converts the **cis** starting material to a **trans** product because the nucleophile (⁻CN) attacks from *below* the plane of the ring.

Problem 7.19 Draw the product of each S$_N$2 reaction and indicate stereochemistry.

More Problems: Try Problem 7.51.

7.11D The Identity of the R Group

How does the rate of an S$_N$2 reaction change as the alkyl group in the substrate alkyl halide changes from CH$_3$ ---→ 1° ---→ 2° ---→ 3°?

- As the number of R groups on the carbon with the leaving group *increases,* the rate of an S$_N$2 reaction *decreases.*

← **Increasing rate of an S$_N$2 reaction**

- Methyl and 1° alkyl halides undergo S$_N$2 reactions with ease.
- 2° Alkyl halides react more slowly.
- 3° Alkyl halides *do not* undergo S$_N$2 reactions.

This order of reactivity can be explained by steric effects. As small H atoms are replaced by larger alkyl groups, **steric hindrance caused by bulky R groups makes nucleophilic attack from the back side more difficult,** slowing the reaction rate. Figure 7.9 illustrates the effect of increasing steric hindrance in a series of alkyl halides.

Figure 7.9
Steric effects in the S$_N$2 reaction

→ Increasing steric hindrance →

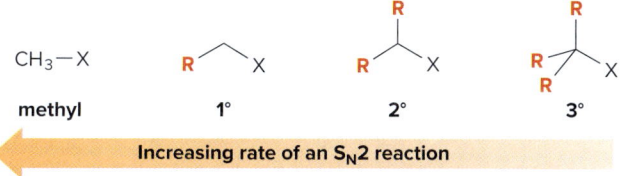

CH$_3$Br   CH$_3$CH$_2$Br   (CH$_3$)$_2$CHBr   (CH$_3$)$_3$CBr

← **Increasing reactivity in an S$_N$2 reaction**

- The S$_N$2 reaction is fastest with unhindered halides.

Chapter 7 Alkyl Halides and Nucleophilic Substitution

Table 7.5 summarizes what we have learned thus far about the S_N2 mechanism.

Table 7.5 Characteristics of the S_N2 Mechanism

Characteristic	Result
Kinetics	• Second-order kinetics; rate = $k[RX][:Nu^-]$
Mechanism	• One step
Stereochemistry	• **Backside attack** of the nucleophile • **Inversion** of configuration at a stereogenic center
Identity of R	• **Unhindered halides react fastest.** • Rate: $CH_3X > RCH_2X > R_2CHX > R_3CX$

Problem 7.20 Which compound in each pair undergoes a faster S_N2 reaction?

a. 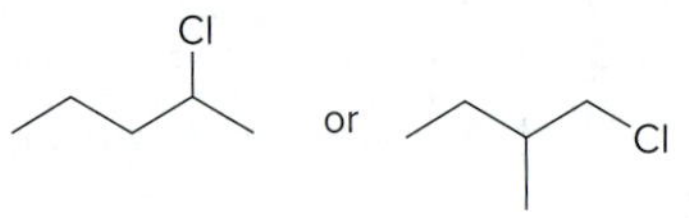 [structure with Cl] or [structure with Cl]   b.  [cyclohexyl-Br] or [tertiary cyclohexyl-Br]

The S_N2 reaction is a key step in the laboratory synthesis of many drugs including **ethambutol** (trade name Myambutol), used in the treatment of tuberculosis. The NH_2 groups in **A** act as neutral nucleophiles to displace halogen. The initial substitution product loses a proton from each N to form ethambutol.

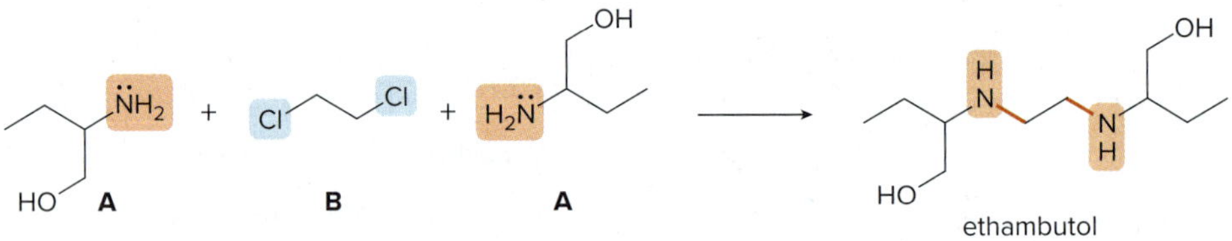

[Reaction scheme: A + B + A → ethambutol (Trade name Myambutol)]

Often an S_N2 reaction is preceded by an acid–base reaction that generates a stronger nucleophile, as shown in Sample Problem 7.3.

Sample Problem 7.3 Drawing an S_N2 product with More Complex Reactants

Identify **C**, the product of an S_N2 reaction in the synthesis of raloxifene, a drug used to reduce the risk of invasive breast cancer in postmenopausal women.

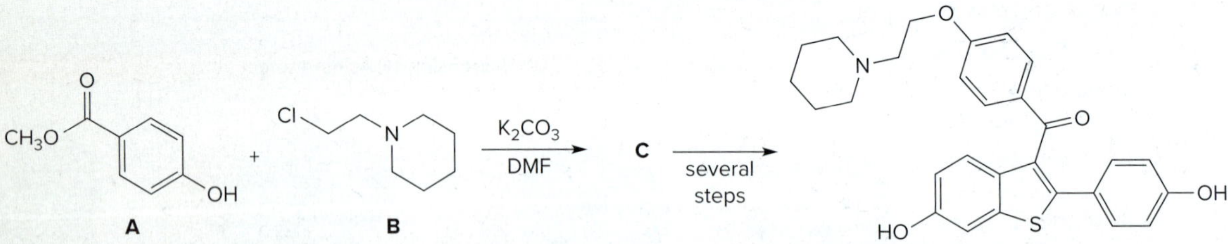

[Reaction scheme: A + B → C (K$_2$CO$_3$, DMF) → several steps → raloxifene]

Solution

Even though both starting materials have two functional groups, follow the same strategy used in simpler reactions.

[1] Identify the nucleophile and the leaving group. There are only a limited number of good leaving groups (Table 7.2). Because **B** contains a Cl bonded to an sp^3 hybridized C, **B** contains the leaving group, so **A** contains the nucleophile.

Because this reaction is carried out in the presence of base (K$_2$CO$_3$) the **most acidic proton in either reactant is removed,** and that is the OH proton in **A**. Removal of the OH proton in **A** forms the negatively charged conjugate base, the **nucleophile.**

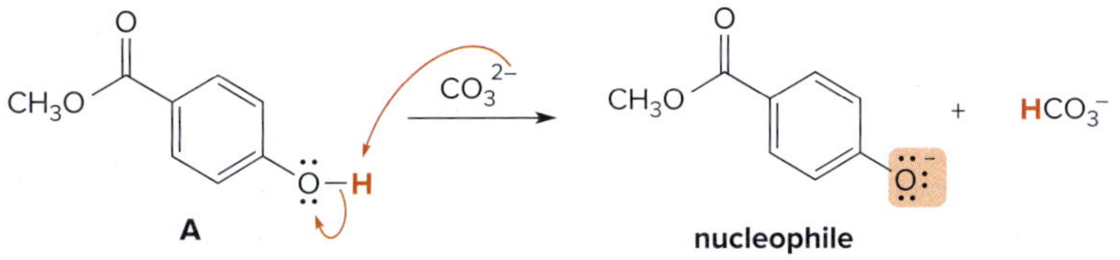

[2] Substitute the nucleophile for the leaving group.

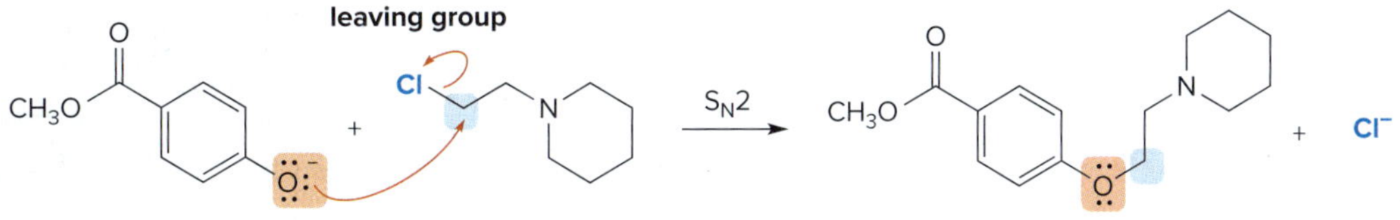

Nucleophilic substitution forms **C** with a new C–O bond.

Problem 7.21 Draw the product **X** of the following S$_N$2 reaction. **X** was a key intermediate in the synthesis of rizatriptan, a drug introduced in 1998 for the treatment of migraines.

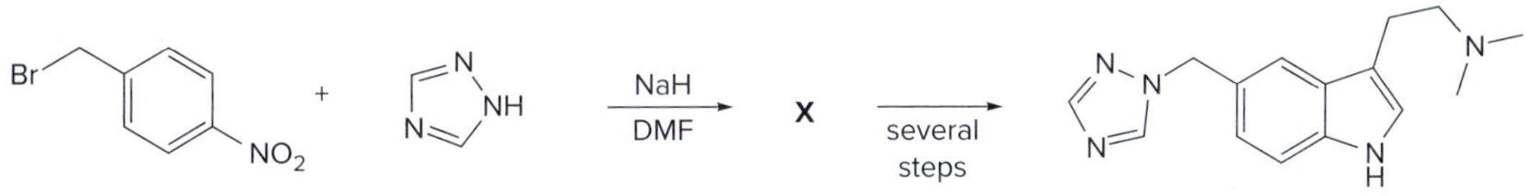

More Practice: Try Problems 7.52–7.54.

7.12 The S$_N$1 Mechanism

The reaction of (CH$_3$)$_3$CBr with CH$_3$CO$_2^-$ is an example of the second mechanism for nucleophilic substitution, the **S$_N$1 mechanism.** What are the general features of this mechanism?

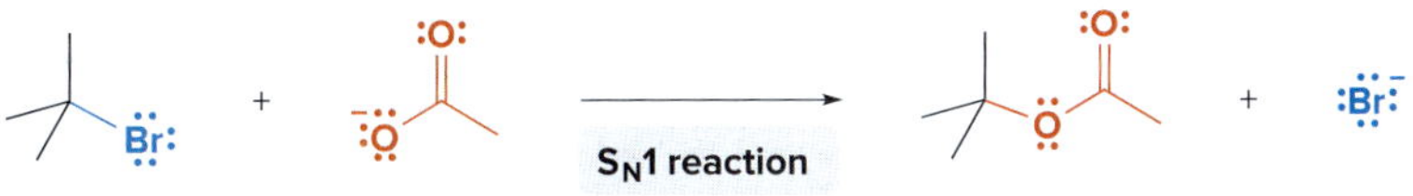

7.12A Kinetics

The S$_N$1 reaction exhibits **first-order kinetics.**

- rate = $k[(CH_3)_3CBr]$

As we learned in Section 7.10, the kinetics suggest that the S$_N$1 mechanism involves **more than one step,** and that the slow step is **unimolecular,** involving *only* the alkyl halide. **The identity and concentration of the nucleophile have *no effect* on the reaction rate.** Doubling the concentration of (CH$_3$)$_3$CBr doubles the rate, but doubling the concentration of the nucleophile has *no effect.*

Problem 7.22 What happens to the rate of an S$_N$1 reaction under each of the following conditions?

a. [RX] is tripled, and [:Nu$^-$] stays the same.
b. Both [RX] and [:Nu$^-$] are tripled.
c. [RX] is halved, and [:Nu$^-$] stays the same.
d. [RX] is halved, and [:Nu$^-$] is doubled.

7.12B A Two-Step Mechanism

The most straightforward explanation for the observed first-order kinetics is a **two-step mechanism** in which **bond breaking occurs *before* bond making,** as shown in Mechanism 7.2.

Mechanism 7.2 The S_N1 Mechanism

[Mechanism diagram showing: step 1 (slow) — heterolysis of C–Br bond from (CH$_3$)$_3$CBr to form a carbocation plus :Br:$^-$; step 2 — nucleophilic attack of acetate on the carbocation to form (CH$_3$)$_3$COCOCH$_3$.]

1. Heterolysis of the C–Br bond forms a **carbocation** in the rate-determining step.
2. **Nucleophilic attack** of acetate (a Lewis base) on the carbocation (a Lewis acid) forms the new C–O bond.

The key features of the S_N1 mechanism are:

- The mechanism has two steps.
- Carbocations are formed as reactive intermediates.

An energy diagram for the reaction of $(CH_3)_3CBr + CH_3CO_2^-$ is shown in Figure 7.10. Each step has its own energy barrier, with a transition state at each energy maximum. Because the transition state for Step [1] is at higher energy, **Step [1] is rate-determining.** $\Delta H°$ for Step [1] has a positive value because only bond breaking occurs, whereas $\Delta H°$ of Step [2] has a negative value because only bond making occurs. The overall reaction is assumed to be exothermic, so the final product is drawn at lower energy than the initial starting material.

Figure 7.10

An energy diagram for the S_N1 reaction:
$(CH_3)_3CBr + CH_3CO_2^- \rightarrow (CH_3)_3COCOCH_3 + Br^-$

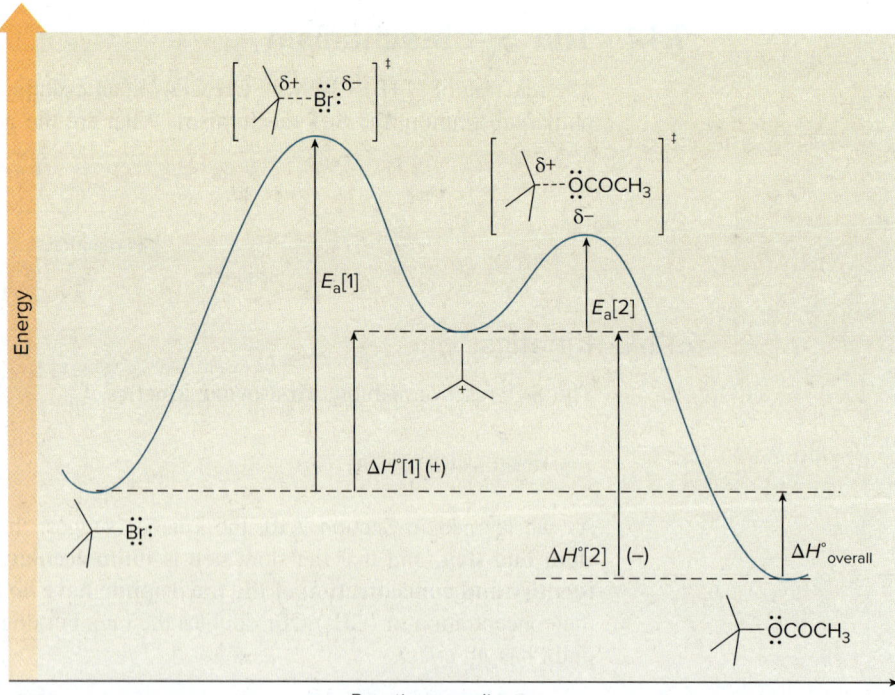

- The S_N1 mechanism has **two steps,** so there are **two energy barriers.**
- $E_a[1] > E_a[2]$ because Step [1] involves bond breaking and Step [2] involves bond formation.
- In each step only one bond is broken or formed, so the transition state for each step has one partial bond.

7.12C Stereochemistry of the S$_N$1 Reaction

To understand the stereochemistry of the S$_N$1 reaction, we must examine the geometry of the carbocation intermediate.

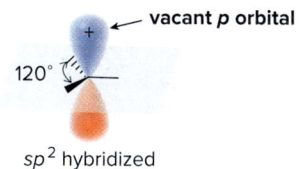

sp^2 hybridized

- A carbocation (with three groups around C) is *sp^2* hybridized and trigonal planar, and contains a vacant *p* orbital extending above and below the plane.

To illustrate the consequences of having a trigonal planar carbocation formed as a reactive intermediate, we examine the S$_N$1 reaction of a 3° alkyl halide **A** having the leaving group bonded to a stereogenic center.

Loss of the leaving group in Step [1] generates a **planar carbocation** that is now *achiral*. Attack of the nucleophile in Step [2] can occur from either the front or the back to afford two products, **B** and **C**. These two products are *different* compounds containing one stereogenic center. **B** and **C** are stereoisomers that are not superimposable—they are **enantiomers**. Because there is no preference for nucleophilic attack from either direction, an equal amount of the two enantiomers is formed—a **racemic mixture**. We say that *racemization* has occurred.

Nucleophilic attack from both sides of a planar carbocation occurs in S$_N$1 reactions, but we see the result of this phenomenon only when the leaving group is bonded to a stereogenic center.

- *Racemization* is the formation of equal amounts of two enantiomeric products from a single starting material.
- S$_N$1 reactions proceed with *racemization* at a single stereogenic center.

Two additional examples of racemization in S$_N$1 reactions are given in Figure 7.11.

Figure 7.11
Two examples of racemization in the S$_N$1 reaction

- Nucleophilic substitution of each starting material by an S$_N$1 mechanism forms a **racemic mixture** of two products.
- With H$_2$O, a neutral nucleophile, the initial product of nucleophilic substitution (ROH$_2^+$) loses a proton to form the final neutral product, ROH (Section 7.6).

Sample Problem 7.4 Drawing the Products of an S$_N$1 Reaction

Label the nucleophile and leaving group, and draw the products (including stereochemistry) of the following S$_N$1 reaction.

Solution

Br⁻ is the leaving group and H_2O is the nucleophile. Loss of the leaving group generates **a trigonal planar carbocation**, which can react with the nucleophile from either direction to form two products.

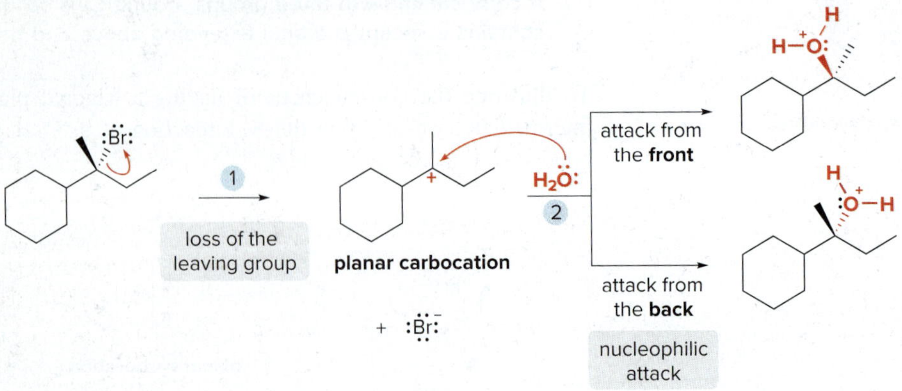

In this example, the initial products of nucleophilic substitution bear a positive charge. They readily lose a proton to form neutral products. The overall process with a neutral nucleophile thus has **three steps**: the first two constitute the **two-step S_N1 mechanism** (loss of the leaving group and attack of the nucleophile), and the third is a **Brønsted–Lowry acid–base reaction** leading to a neutral organic product.

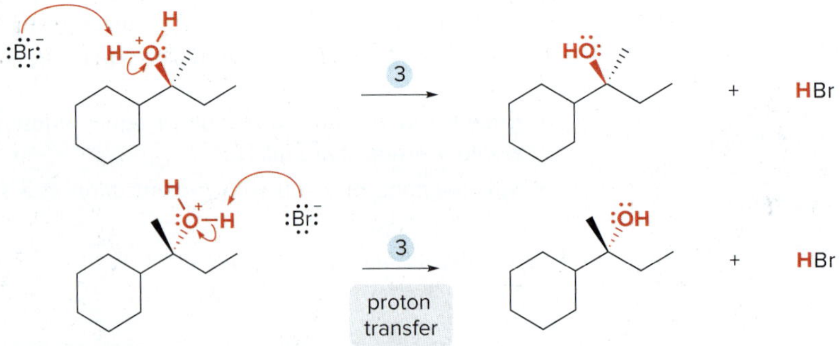

The two products in this reaction are nonsuperimposable mirror images—**enantiomers**. Because nucleophilic attack on the trigonal planar carbocation occurs with equal frequency from both directions, a **racemic mixture is formed**.

Problem 7.23 Draw the products of each S_N1 reaction and indicate the stereochemistry of any stereogenic centers.

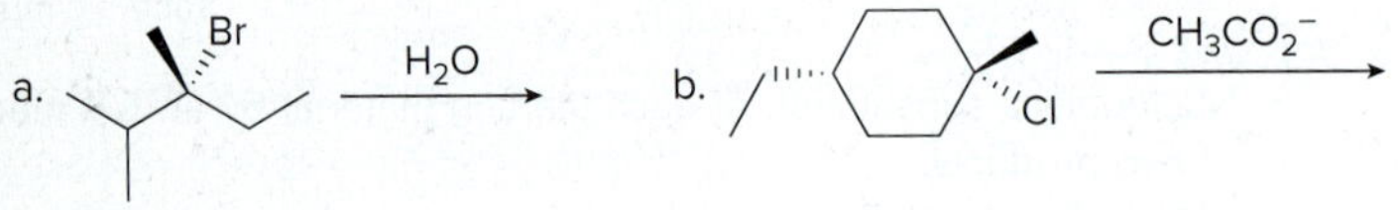

More Practice: Try Problem 7.58.

7.12D The Identity of the R Group

How does the rate of an S_N1 reaction change as the alkyl group in the substrate alkyl halide changes from $CH_3 \dashrightarrow 1° \dashrightarrow 2° \dashrightarrow 3°$?

- As the number of R groups on the carbon with the leaving group *increases,* the rate of an S_N1 reaction *increases.*

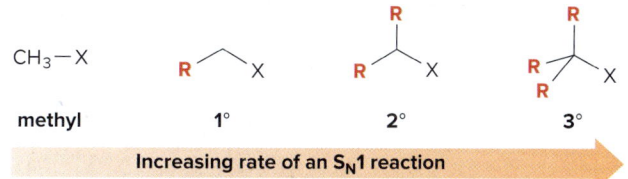

- 3° Alkyl halides undergo S$_N$1 reactions rapidly.
- 2° Alkyl halides react more slowly.
- Methyl and 1° alkyl halides do *not* undergo S$_N$1 reactions.

This trend is exactly opposite to that observed for the S$_N$2 mechanism. To explain this result, we must examine the rate-determining step, the formation of the carbocation, and learn about the effect of alkyl groups on **carbocation stability.** Table 7.6 summarizes the characteristics of the S$_N$1 mechanism.

Table 7.6 Characteristics of the S$_N$1 Mechanism

Characteristic	Result
Kinetics	• First-order kinetics; rate = k[RX]
Mechanism	• Two steps
Stereochemistry	• **Trigonal planar carbocation** intermediate • **Racemization** at a single stereogenic center
Identity of R	• **More-substituted halides react fastest.** • Rate: R$_3$CX > R$_2$CHX > RCH$_2$X > CH$_3$X

7.13 Carbocation Stability

Carbocations are classified as **primary (1°), secondary (2°),** or **tertiary (3°)** by the number of R groups bonded to the charged carbon atom. As the number of R groups on the positively charged carbon atom increases, the stability of the carbocation **increases.**

We will examine the reason for this order of stability by invoking two different principles: **inductive effects** and **hyperconjugation.**

Problem 7.24 Classify each carbocation as 1°, 2°, or 3°.

a. b. c. d.

7.13A Inductive Effects

Inductive effects are electronic effects that occur through σ bonds. In Section 2.5B, for example, we learned that more-electronegative atoms stabilize a negative charge by an **electron-withdrawing inductive effect.**

Electron-donor groups (Z) stabilize a (+) charge; **Z→Y$^+$.** Electron-withdrawing groups (W) stabilize a (−) charge; **W←Y$^-$.**

To stabilize a positive charge, electron-*donating* groups are needed. **Alkyl groups are electron-donor groups that stabilize a positive charge.** An alkyl group with several σ bonds is more polarizable than a hydrogen atom, and more able to donate electron density. Thus, as

R groups successively replace the H atoms in CH_3^+, **the positive charge is more dispersed on the electron-donor R groups, and the carbocation is more stabilized.**

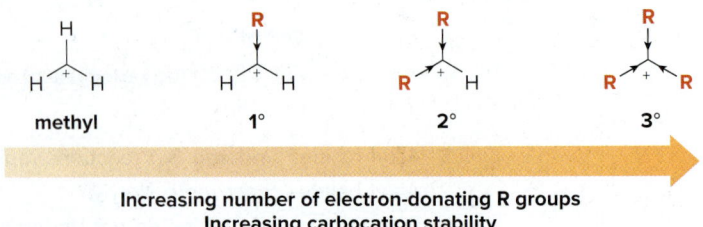

Electrostatic potential maps for four carbocations in Figure 7.12 illustrate the effect of increasing alkyl substitution on the positive charge of the carbocation.

Figure 7.12
Electrostatic potential maps for different carbocations

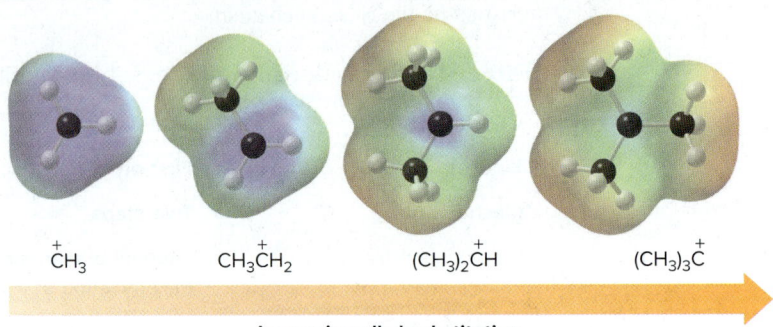

- Dark blue areas in electrostatic potential plots indicate regions low in electron density. As alkyl substitution increases, the region of positive charge is less concentrated on carbon.

Problem 7.25 Rank the following carbocations in order of increasing stability.

7.13B Hyperconjugation

A second explanation for the observed trend in carbocation stability is based on orbital overlap. A 3° carbocation is more stable than a 2°, 1°, or methyl carbocation because the positive charge is *delocalized* over more than one atom.

- **Spreading out charge by the overlap of an empty *p* orbital with an adjacent σ bond is called *hyperconjugation*.**

For example, CH_3^+ cannot be stabilized by hyperconjugation, but $(CH_3)_2CH^+$ can:

no opportunity for hyperconjugation

Hyperconjugation is possible.

Both carbocations contain an sp^2 hybridized carbon, so both are trigonal planar with a vacant *p* orbital extending above and below the plane. There are no adjacent C—H σ bonds with which the *p* orbital can overlap in CH_3^+, but there *are* adjacent C—H σ bonds in $(CH_3)_2CH^+$. This overlap (the **hyperconjugation**) delocalizes the positive charge on the carbocation, spreading it over a larger volume, and this stabilizes the carbocation.

The larger the number of alkyl groups on the adjacent carbons, the greater the possibility for hyperconjugation, and the larger the stabilization. Hyperconjugation thus provides an alternate way of explaining why **carbocations with a larger number of R groups are more stabilized.**

7.14 The Hammond Postulate

The rate of an S_N1 reaction depends on the rate of formation of the carbocation (the product of the rate-determining step) via heterolysis of the C—X bond.

- The rate of an S_N1 reaction *increases* as the number of R groups on the carbon with the leaving group *increases*.
- The stability of a carbocation *increases* as the number of R groups on the positively charged carbon *increases*.

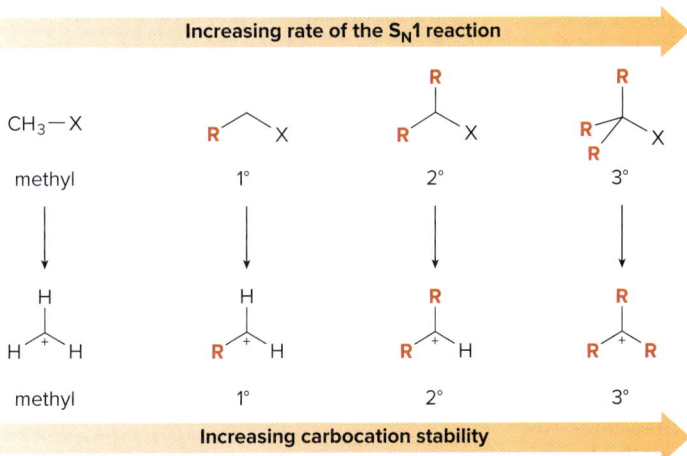

- Thus, the rate of an S_N1 reaction *increases* as the stability of the carbocation *increases*.

The reaction is faster with a more stable carbocation.

The rate of a reaction depends on the magnitude of E_a, and the stability of a product depends on $\Delta G°$. The **Hammond postulate**, first proposed in 1955, **relates rate to stability.**

7.14A The General Features of the Hammond Postulate

The Hammond postulate provides a qualitative estimate of the energy of a transition state. Because the energy of the transition state determines the energy of activation and therefore the reaction rate, predicting the relative energy of two transition states allows us to determine the relative rates of two reactions.

According to the Hammond postulate, the transition state of a reaction resembles the structure of the species (reactant or product) to which it is closer in energy. In endothermic

reactions, the transition state is closer in energy to the **products**. In **exothermic reactions,** the transition state is closer in energy to the **reactants**.

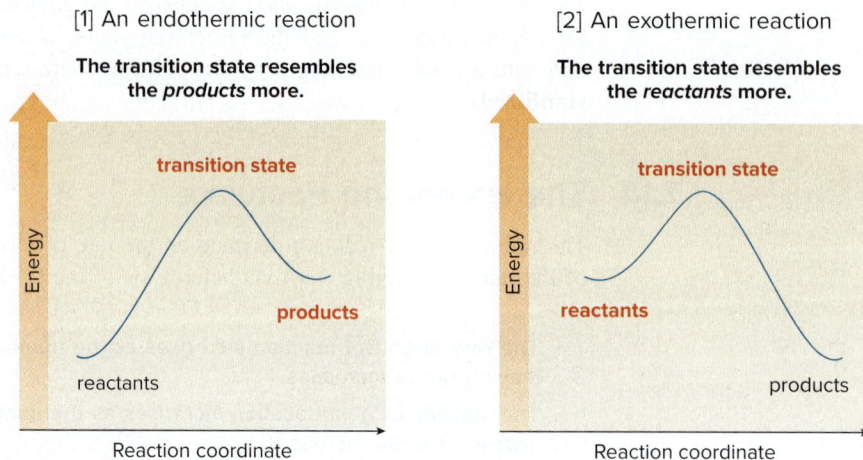

- Transition states in *endothermic* reactions resemble the *products*.
- Transition states in *exothermic* reactions resemble the *reactants*.

What happens to the reaction rate if the energy of the product is lowered? In an **endothermic reaction,** the transition state resembles the products, so anything that stabilizes the product stabilizes the transition state, too. **Lowering the energy of the transition state** *decreases* **the energy of activation (E_a), which** *increases* **the reaction rate.**

Suppose there are two possible products of an endothermic reaction, but one is more stable (lower in energy) than the other (Figure 7.13). According to the Hammond postulate, **the transition state to form the more stable product is lower in energy, so this reaction should occur faster.**

Figure 7.13
An endothermic reaction—how the energies of the transition state and products are related

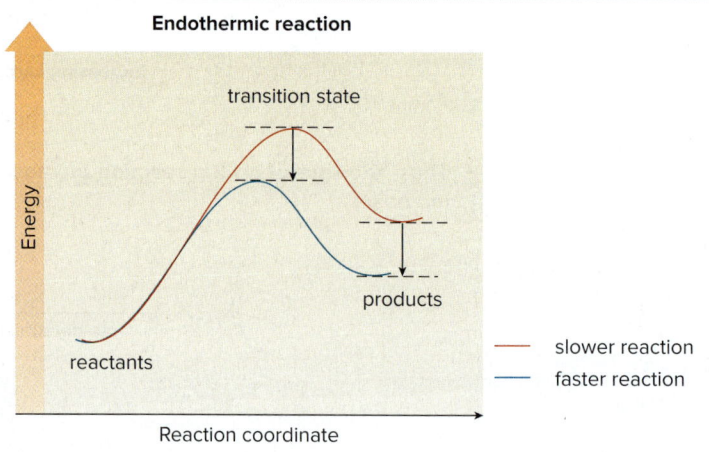

- The *lower* energy transition state leads to the *lower* energy product.

- In an endothermic reaction, the *more stable* product forms *faster*.

What happens to the reaction rate of an **exothermic reaction** if the energy of the product is lowered? The transition state resembles the reactants, so **lowering the energy of the products has little or no effect on the energy of the transition state.** If E_a is unaffected, then the reaction rate is unaffected, too, as shown in Figure 7.14.

- In an exothermic reaction, the more stable product may or may not form faster because E_a is *similar* for both products.

Figure 7.14

An exothermic reaction—how the energies of the transition state and products are related

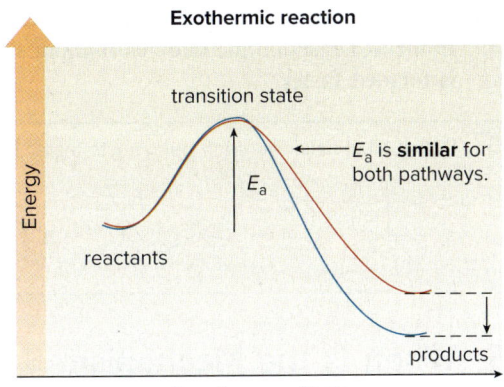

- Decreasing the energy of the product often has *little effect* on the energy of the transition state.

7.14B The Hammond Postulate and the S_N1 Reaction

In the S_N1 reaction, the rate-determining step is the formation of the carbocation, an *endothermic* reaction. According to the Hammond postulate, the **stability of the carbocation determines the rate of its formation.**

For example, heterolysis of the C–Cl bond in $(CH_3)_2CHCl$ affords a less stable 2° carbocation, $(CH_3)_2CH^+$ (Equation [1]), whereas heterolysis of the C–Cl bond in $(CH_3)_3CCl$ affords a more stable 3° carbocation, $(CH_3)_3C^+$ (Equation [2]). The Hammond postulate states that Reaction [2] is faster than Reaction [1], because the transition state to form the more stable 3° carbocation is lower in energy. Figure 7.15 depicts an energy diagram comparing these two endothermic reactions.

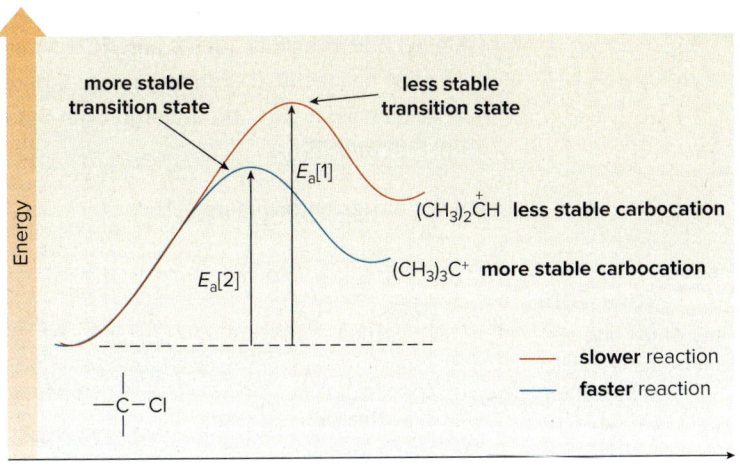

Figure 7.15

Energy diagram for carbocation formation in two different S_N1 reactions

- $(CH_3)_2CH^+$ is less stable than $(CH_3)_3C^+$, so $E_a[1] > E_a[2]$, and Reaction [1] is slower.

In conclusion, the Hammond postulate can be used to predict the relative rates of two reactions. In the S_N1 reaction the rate-determining step is endothermic, so the **more stable carbocation is formed faster.**

Problem 7.26 Which alkyl halide in each pair reacts faster in an S_N1 reaction?

7.15 When Is the Mechanism S_N1 or S_N2?

Given a particular starting material and nucleophile, how do we know whether a reaction occurs by the S_N1 or S_N2 mechanism? Four factors are examined:

- The alkyl halide—CH_3X, RCH_2X, R_2CHX, or R_3CX
- The nucleophile—strong or weak
- The leaving group—good or poor
- The solvent—protic or aprotic

7.15A The Alkyl Halide—The Most Important Factor

The most important factor in determining whether a reaction follows the S_N1 or S_N2 mechanism is the *identity of the alkyl halide.*

- *Increasing* alkyl substitution favors S_N1.
- *Decreasing* alkyl substitution favors S_N2.

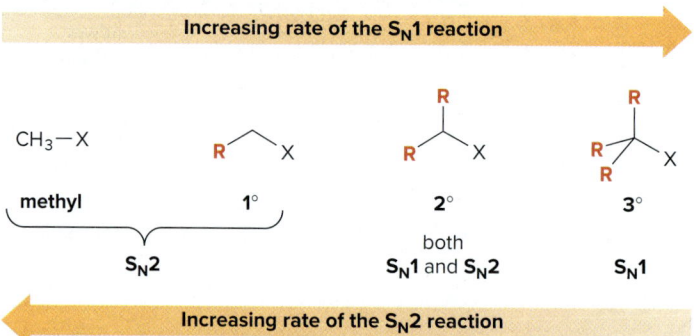

- Methyl and 1° halides (CH_3X and RCH_2X) undergo only S_N2 reactions.
- 3° Alkyl halides (R_3CX) undergo only S_N1 reactions.
- 2° Alkyl halides (R_2CHX) undergo both S_N1 and S_N2 reactions. Other factors determine the mechanism.

Examples are given in Figure 7.16.

Figure 7.16
The identity of RX and the mechanism of nucleophilic substitution

1° halide — S_N2

2° halide — Both S_N2 and S_N1 are possible.

3° halide — S_N1

| Problem 7.27 | What is the likely mechanism of nucleophilic substitution for each alkyl halide? |

a. (2,3-dimethyl-2-bromobutane structure with Br on secondary carbon) b. CH₃CH₂CH₂CH₂CH₂Br c. bromocyclopentane d. 2-bromo-2-methylbutane

7.15B The Nucleophile

How does the strength of the nucleophile affect an S_N1 or S_N2 mechanism? The rate of the S_N1 reaction is unaffected by the identity of the nucleophile because the nucleophile does not appear in the rate equation (rate = k[RX]). The identity of the nucleophile *is* important for the S_N2 reaction, however, because the nucleophile does appear in the rate equation for this mechanism (rate = k[RX][:Nu⁻]).

- **Strong** nucleophiles present in high concentration favor S_N2 reactions.
- **Weak** nucleophiles favor S_N1 reactions by decreasing the rate of any competing S_N2 reaction.

The most common nucleophiles in S_N2 reactions bear a net negative charge. The most common nucleophiles in S_N1 reactions are weak nucleophiles such as H_2O and ROH. The identity of the nucleophile is especially important in determining the mechanism and therefore the stereochemistry of nucleophilic substitution when 2° alkyl halides are starting materials.

Let's compare the substitution products formed when the 2° alkyl halide **A** (*cis*-1-bromo-4-methylcyclohexane) is treated with either the strong nucleophile ⁻OH or the weak nucleophile H_2O. Because a 2° alkyl halide can react by either mechanism, the strength of the nucleophile determines which mechanism takes place.

cis-1-bromo-4-methylcyclohexane
A

⁻OH (**strong** nucleophile)

H_2O (**weak** nucleophile)

The **strong nucleophile ⁻OH favors an S_N2 reaction**, which occurs with **backside** attack of the nucleophile, resulting in **inversion of configuration**. Because the leaving group Br⁻ is *above* the plane of the ring, the nucleophile attacks from *below,* and a single product **B** is formed.

inversion of configuration

:Br: + :ÖH   → S_N2 →   (trans product with ⁻OH) + :Br:⁻

A   strong nucleophile   *trans*-4-methylcyclohexanol   **B**

The **weak nucleophile H_2O favors an S_N1 reaction,** which occurs by way of an intermediate carbocation. Loss of the leaving group in **A** forms the carbocation, which undergoes nucleophilic attack from both above and below the plane of the ring to afford two products, **C** and

D. Loss of a proton by proton transfer forms the final products, **B** and **E**. **B** and **E** are diastereomers of each other (**B** is a trans isomer and **E** is a cis isomer).

Thus, the mechanism of nucleophilic substitution determines the stereochemistry of the products formed.

Problem 7.28 For each alkyl halide and nucleophile: [1] Draw the product of nucleophilic substitution; [2] determine the likely mechanism (S_N1 or S_N2) for each reaction.

a. (structure) + CH_3OH

b. (cyclohexylmethyl bromide) + ^-SH

c. (cyclohexyl iodide) + $CH_3CH_2O^-$

d. (2-bromobutane) + CH_3OH

Problem 7.29 Draw the products (including stereochemistry) for each reaction.

a. (structure with Br) + $H_2O \longrightarrow$

b. (structure with Cl, H, D) + $^-C{\equiv}C{-}H \longrightarrow$

7.15C The Leaving Group

How does the identity of the leaving group affect an S_N1 or S_N2 reaction?

- A *better* leaving group increases the rate of *both* S_N1 and S_N2 reactions.

Because the bond to the leaving group is partially broken in the transition state of the only step of the S_N2 mechanism and the slow step of the S_N1 mechanism, **a better leaving group increases the rate of both reactions.** The better the leaving group, the more willing it is to accept the electron pair in the C–X bond, and the faster the reaction.

For alkyl halides, the following order of reactivity is observed for the S_N1 and the S_N2 mechanisms:

R–F R–Cl R–Br R–I

Increasing leaving group ability
Increasing rate of S_N1 and S_N2 reactions

Problem 7.30 Rank the alkyl halides in the following marine natural product in order of increasing reactivity in the S_N1 reaction.

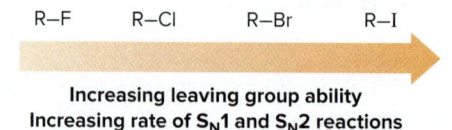

7.15D The Solvent

Polar protic solvents and polar aprotic solvents affect the rates of S_N1 and S_N2 reactions differently.

- Polar *protic* solvents are especially good for S_N1 reactions.
- Polar *aprotic* solvents are especially good for S_N2 reactions.

Polar protic solvents like H_2O and ROH solvate both cations and anions well, and this characteristic is important for the S_N1 mechanism, in which two ions (a carbocation and a leaving group) are formed by heterolysis of the C—X bond. The carbocation is solvated by ion–dipole interactions with the polar solvent, and the leaving group is solvated by hydrogen bonding, in much the same way that Na^+ and Br^- are solvated in Section 7.8C. These interactions stabilize the reactive intermediate.

Polar aprotic solvents exhibit dipole–dipole interactions but not hydrogen bonding, and as a result, they do not solvate anions well. This has a pronounced effect on the nucleophilicity of anionic nucleophiles. Because these nucleophiles are not "hidden" by strong interactions with the solvent, they are **more nucleophilic**. Because stronger nucleophiles favor S_N2 reactions, **polar aprotic solvents are especially good for S_N2 reactions.**

Summary of solvent effects:
- **Polar protic solvents favor S_N1 reactions** because the ionic intermediates are stabilized by solvation.
- **Polar aprotic solvents favor S_N2 reactions** because nucleophiles are not well solvated, and therefore are more nucleophilic.

Problem 7.31 Which solvents favor S_N1 reactions and which favor S_N2 reactions?

a. 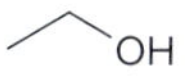 CH₃CH₂CH₂OH   b. CH_3CN c. 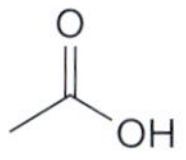 CH₃COOH   d. 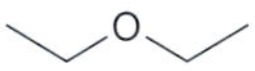 CH₃CH₂OCH₂CH₃

Problem 7.32 Decide on the mechanism for each substitution, and then pick the solvent that affords the faster reaction.

a. $(CH_3CH_2)_2CClCH_3 + CH_3OH$ in CH_3OH or DMSO
b. $CH_3CH_2CH_2Br + {}^-OH$ in H_2O or DMF
c. $(CH_3CH_2)_2CHCl + CH_3O^-$ in CH_3OH or HMPA

7.15E Summary of Factors That Determine Whether the S_N1 or S_N2 Mechanism Occurs

Table 7.7 summarizes the factors that determine whether a reaction occurs by the S_N1 or S_N2 mechanism. Sample Problems 7.5 and 7.6 illustrate how these factors are used to determine the mechanism of a given reaction.

Table 7.7 Summary of Factors That Determine the S_N1 or S_N2 Mechanism

Alkyl halide	Mechanism	Other factors
CH_3X RCH_2X (1°)	S_N2	Favored by • **strong nucleophiles** (usually a net negative charge) • polar **aprotic** solvents
R_3CX (3°)	S_N1	Favored by • **weak nucleophiles** (usually neutral) • polar **protic** solvents
R_2CHX (2°)	S_N1 or S_N2	The mechanism depends on the conditions. • **Strong nucleophiles favor the S_N2 mechanism over the S_N1 mechanism.** RO^- is a stronger nucleophile than ROH, so RO^- favors the S_N2 reaction and ROH favors the S_N1 reaction. • **Protic solvents favor the S_N1 mechanism and aprotic solvents favor the S_N2 mechanism.** H_2O and CH_3OH are polar protic solvents that favor the S_N1 mechanism, whereas acetone [$(CH_3)_2C=O$] and DMSO [$(CH_3)_2S=O$] are polar aprotic solvents that favor the S_N2 mechanism.

Chapter 7 Alkyl Halides and Nucleophilic Substitution

Sample Problem 7.5 Determining the Mechanism of Nucleophilic Substitution

Determine the mechanism of nucleophilic substitution for each reaction and draw the products.

a. 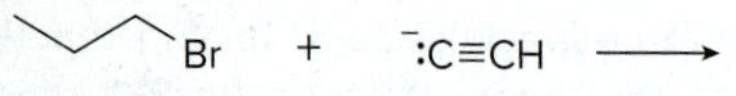 CH₃CH₂CH₂Br + ⁻:C≡CH ⟶

b. 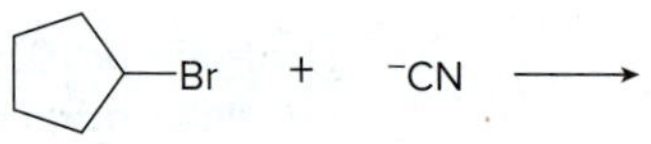 (cyclopentyl)-Br + ⁻CN ⟶

Solution

a. The alkyl halide is 1°, so it must react by an S$_N$2 mechanism with the nucleophile ⁻:C≡CH.

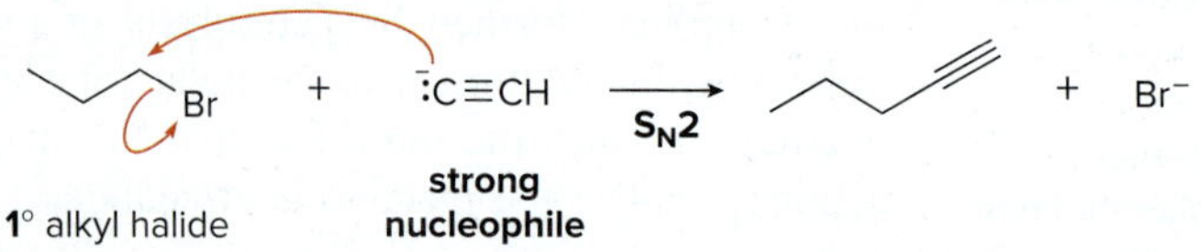

CH₃CH₂CH₂–Br + ⁻:C≡CH $\xrightarrow{S_N2}$ CH₃CH₂CH₂–C≡CH + Br⁻

1° alkyl halide   **strong nucleophile**

b. The alkyl halide is 2°, so it can react by either the S$_N$1 or S$_N$2 mechanism. The strong nucleophile (⁻CN) favors the S$_N$2 mechanism.

(cyclopentyl)–Br + ⁻:CN $\xrightarrow{S_N2}$ (cyclopentyl)–CN + Br⁻

2° alkyl halide   **strong nucleophile**

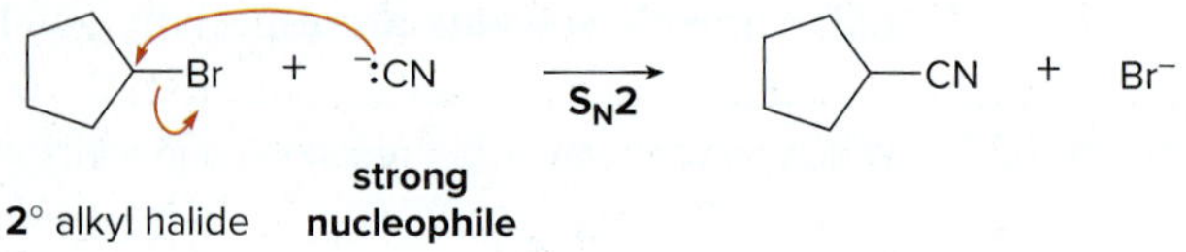

Sample Problem 7.6 Determining the Mechanism and Stereochemistry in Nucleophilic Substitution

Determine the mechanism of nucleophilic substitution for each reaction and draw the products, including stereochemistry.

a. 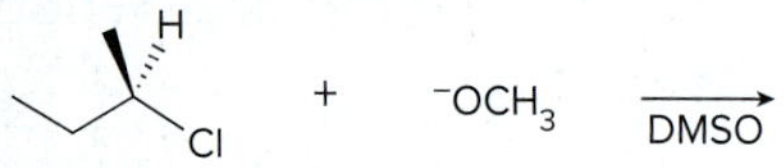 (R)-2-chlorobutane + ⁻OCH₃ $\xrightarrow{DMSO}$

b. 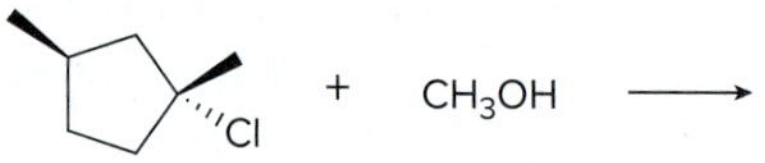 (cis-1-chloro-3-methylcyclopentane) + CH₃OH ⟶

Solution

a. The 2° alkyl halide can react by either the S$_N$1 or S$_N$2 mechanism. **The strong nucleophile (⁻OCH₃) favors the S$_N$2 mechanism,** as does the **polar aprotic solvent** (DMSO). S$_N$2 reactions proceed with **inversion** of configuration.

CH₃O:⁻ + (R-alkyl chloride) $\xrightarrow[S_N2]{DMSO}$ (inverted product with CH₃O) + :Cl:⁻

strong nucleophile   **2° alkyl halide**   **inversion** of configuration

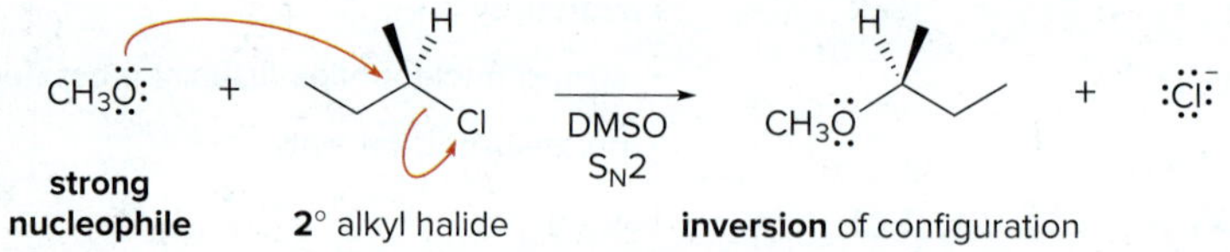

b. The alkyl halide is 3°, so it reacts by an S$_N$1 mechanism with the weak nucleophile CH₃OH. S$_N$1 reactions proceed with **racemization** at a single stereogenic center, so two products are formed.

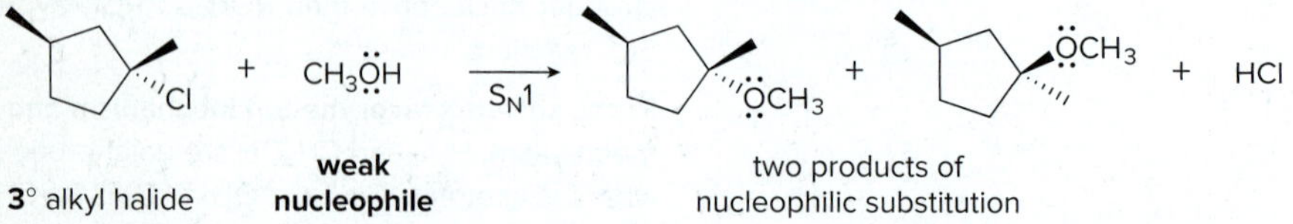

(3° alkyl chloride) + CH₃ÖH $\xrightarrow{S_N1}$ (product with ⁻OCH₃) + (product with OCH₃, opposite) + HCl

3° alkyl halide   **weak nucleophile**   **two products of nucleophilic substitution**

> **Problem 7.33** Determine the mechanism and draw the products of each reaction. Include the stereochemistry at all stereogenic centers.
>
> a. 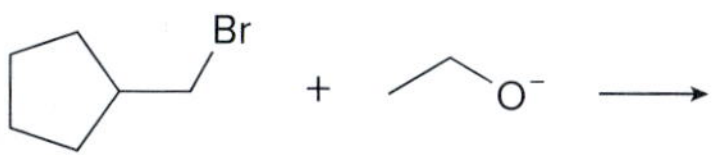 cyclopentyl-CH$_2$-Br + CH$_3$CH$_2$O$^-$ →
>
> b.  trans-1-methyl-3-bromocyclopentane + N$_3^-$ →
>
> c. (CH$_3$)$_2$C(I)CH$_2$CH$_2$CH$_3$ + CH$_3$OH →
>
> d. 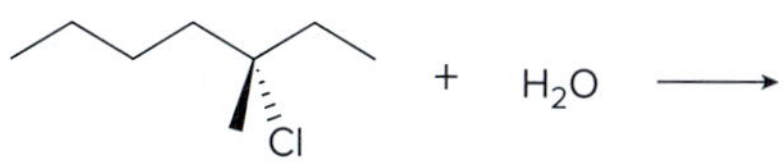 (S)-3-chlorohexane derivative + H$_2$O →
>
> **More Practice:** Try Problem 7.62.

7.16 Biological Nucleophilic Substitution

Nucleophilic substitution occurs in a wide variety of biological reactions.

7.16A Leaving Groups Derived from Phosphorus

In contrast to nucleophilic substitutions run in the laboratory that use alkyl halides as substrates and halide anions as leaving groups, biological substitutions often occur with phosphorus leaving groups, such as phosphate (PO_4^{3-}, abbreviated as P_i for inorganic phosphate), diphosphate ($P_2O_7^{4-}$, abbreviated as PP_i), and triphosphate ($P_3O_{10}^{5-}$, abbreviated as PPP_i). These anions are excellent leaving groups because they are **weak, resonance-stabilized bases.**

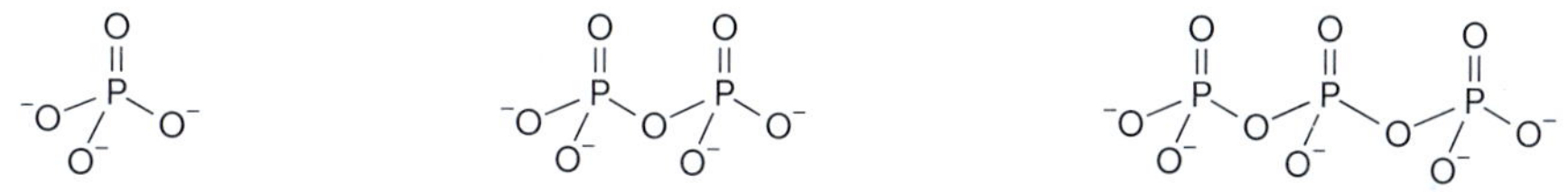

phosphate P_i

diphosphate PP_i

triphosphate PPP_i

As discussed in Section 3.2D, when an organic compound contains a carbon bonded to one of these leaving groups, the compound is called an organic monophosphate, diphosphate, or triphosphate.

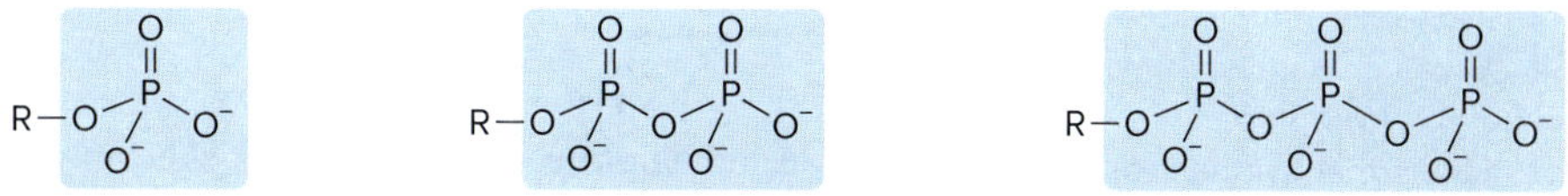

organic monophosphate
R—OP

organic diphosphate
R—OPP

organic triphosphate
R—OPPP

Nucleophilic substitutions with these substrates may proceed by either an S_N2 or S_N1 pathway, as shown with the general diphosphate R_2CHOPP.

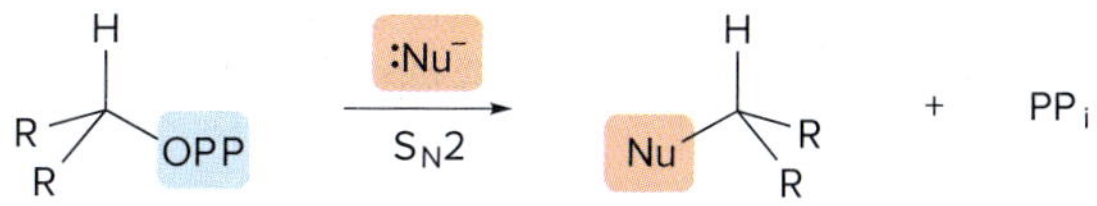

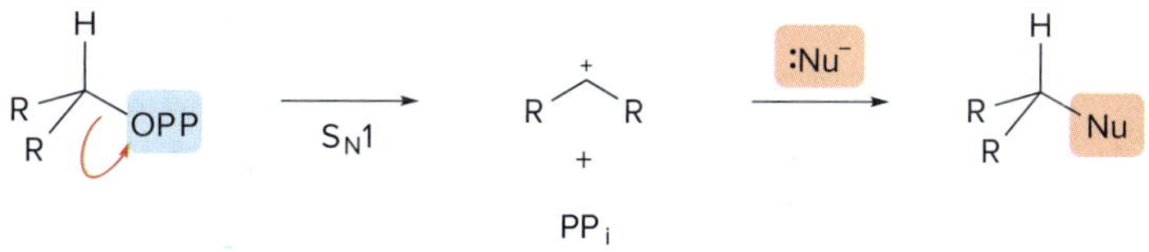

The final step in the biosynthesis of geraniol, a component of rose oil used in perfumery, is an S_N1 reaction of geranyl diphosphate with water. This reaction occurs by way of a resonance-stabilized carbocation. We will learn more about reactions of diphosphates in Chapter 12.

7.16B S-Adenosylmethionine

A common nucleophilic substitution occurs with *S*-adenosylmethionine, or **SAM.** SAM is the cell's equivalent of CH_3I. The many polar functional groups in SAM make it soluble in the aqueous environment in the cell.

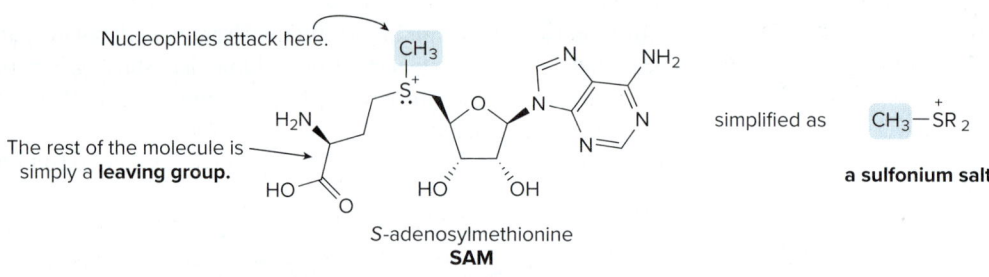

SAM, a nutritional supplement sold under the name SAM-e (pronounced sammy), has been used in Europe to treat depression and arthritis for over 20 years. In cells, SAM is used in nucleophilic substitutions that synthesize key amino acids, hormones, and neurotransmitters.
Jill Braaten

The CH_3 group in SAM [abbreviated as $(CH_3SR_2)^+$] is part of a **sulfonium salt,** a positively charged sulfur species that contains a good leaving group. Nucleophilic attack at the CH_3 group of SAM displaces R_2S, a good neutral leaving group. This reaction is called **methylation,** because a CH_3 group is transferred from one compound (SAM) to another (:Nu⁻).

For example, **adrenaline** (epinephrine) is a hormone synthesized in the adrenal glands from noradrenaline (norepinephrine) by nucleophilic substitution using SAM (Figure 7.17). When an individual senses danger or is confronted by stress, the hypothalamus region of the brain signals the adrenal glands to synthesize and release adrenaline, which enters the bloodstream and then stimulates the formation of glucose, thus providing an energy boost. Heart rate and blood pressure increase, and lung passages are dilated. These physiological changes result from the "rush of adrenaline," and prepare an individual for "fight or flight."

Problem 7.34 Nicotine, a toxic and addictive component of tobacco, is synthesized from **A** using SAM. Write out the reaction that converts **A** into nicotine.

Figure 7.17

Adrenaline synthesis from noradrenaline in response to stress

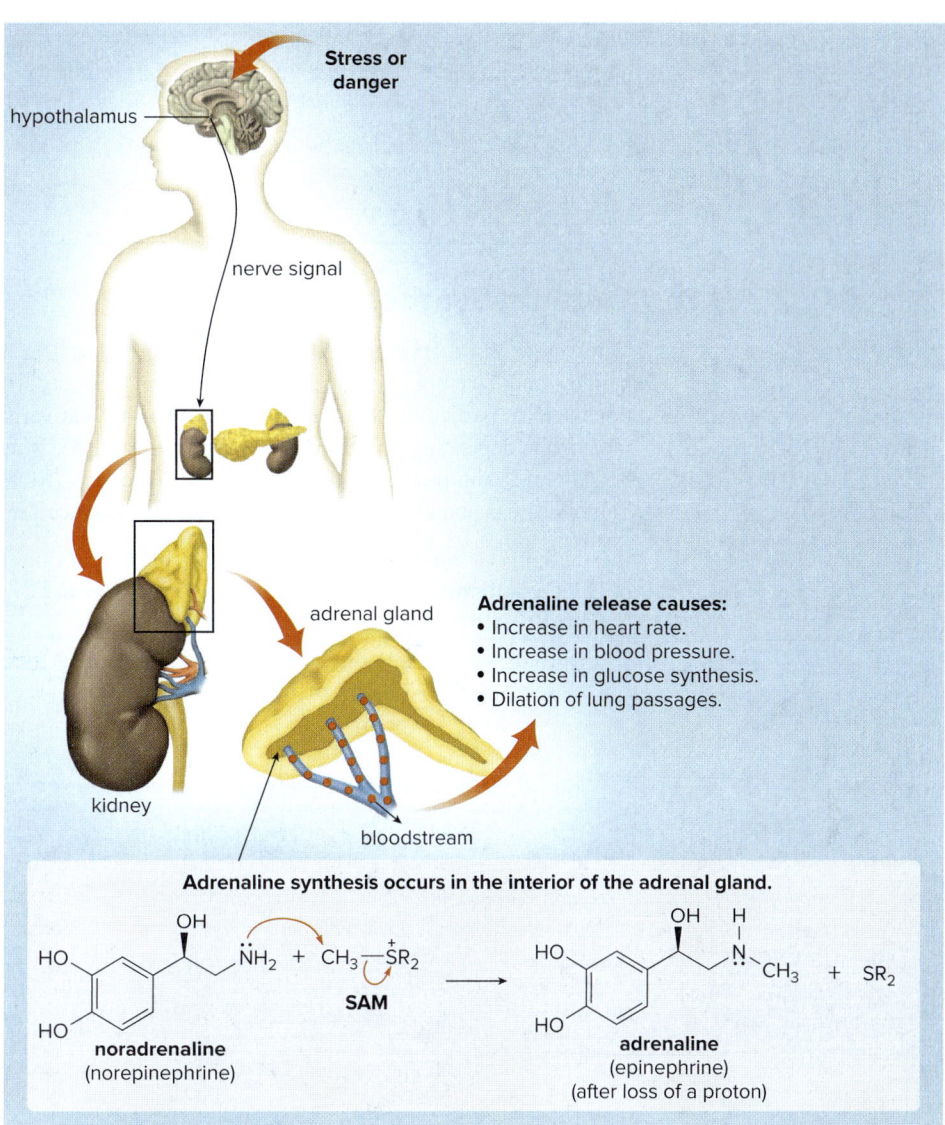

7.17 Vinyl Halides and Aryl Halides

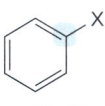

S_N1 and S_N2 reactions occur only at sp^3 hybridized carbon atoms. **Vinyl halides** and **aryl halides,** which have a halogen atom bonded to an sp^2 hybridized C, do *not* undergo nucleophilic substitution by either the S_N1 or S_N2 mechanism. The discussion here centers on vinyl halides, but similar arguments hold for aryl halides as well.

Vinyl halides do not undergo S_N2 reactions in part because of the percent *s*-character in the hybrid orbital of the carbon atom in the C—X bond. The higher percent *s*-character in the sp^2 hybrid orbital of the vinyl halide compared to the sp^3 hybrid orbital of the alkyl halide (33% vs. 25%) makes the bond shorter and stronger.

Vinyl halides do not undergo S_N1 reactions because heterolysis of the C—X bond would form a **highly unstable vinyl carbocation.** Because this carbocation has only two groups around the positively charged carbon, it is sp hybridized. These carbocations are even less stable than 1° carbocations, so the S_N1 reaction does not take place.

Problem 7.35 Rank the alkyl halides in the following marine natural product in order of increasing reactivity in the S_N2 reaction.

7.18 Organic Synthesis

Thus far we have concentrated on the starting material in nucleophilic substitution—the alkyl halide—and have not paid much attention to the product formed. Nucleophilic substitution reactions, and in particular S_N2 reactions, introduce a wide variety of different functional groups in molecules, depending on the nucleophile. For example, when $^-$OH, $^-$OR, and $^-$CN are used as nucleophiles, the products are alcohols (ROH), ethers (ROR), and nitriles (RCN), respectively. Table 7.8 lists some functional groups readily introduced using nucleophilic substitution.

By thinking of **nucleophilic substitution as a reaction that** *makes* **a particular kind of organic compound,** we begin to think about *synthesis*.

- Organic synthesis is the systematic preparation of a compound from a readily available starting material by one or many steps.

Table 7.8 Molecules Synthesized from R–X by the S_N2 Reaction

	Nucleophile (:Nu$^-$)	Product	Name
Oxygen compounds	$^-$OH	R–OH	alcohol
	$^-$OR'	R–OR'	ether
	$^-$O–C(=O)–R'	R–O–C(=O)–R'	ester
Carbon compounds	$^-$CN	R–CN	nitrile
	$:$C≡C–H	R–C≡C–H	alkyne
Nitrogen compounds	N_3^-	R–N_3	azide
	$:$NH$_3$	R–NH$_2$	amine
Sulfur compounds	$^-$SH	R–SH	thiol
	$^-$SR'	R–SR'	sulfide

7.18A Background on Organic Synthesis

Chemists synthesize molecules for many reasons. Sometimes a **natural product,** a compound isolated from natural sources, has useful medicinal properties, but is produced by an organism in only minute quantities. Synthetic chemists then prepare this molecule from simpler starting materials, so that it can be made available to a large number of people.

Sometimes, chemists prepare molecules that do not occur in nature (although they may be similar to those in nature), because these molecules have superior properties to their naturally occurring relatives. **Aspirin, or acetylsalicylic acid** (Section 2.7), is a well-known example.

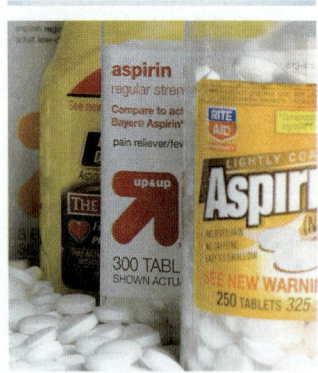

Aspirin is synthesized by a two-step procedure from simple, cheap starting materials.
Jill Braaten

7.18 Organic Synthesis

Figure 7.18
Synthesis of aspirin

phenol →[1] NaOH, [2] CO₂, [3] H₃O⁺ → salicylic acid →(CH₃CO)₂O, acid → aspirin

Acetylsalicylic acid is prepared from phenol, a product of the petroleum industry, by a two-step procedure (Figure 7.18). Aspirin has become one of the most popular and widely used drugs in the world because it has excellent analgesic and anti-inflammatory properties, *and* it is inexpensive and readily available.

Phenol, the starting material for the aspirin synthesis, is a petroleum product, like most of the starting materials used in large quantities in industrial syntheses. A shortage of petroleum reserves thus affects the availability not only of fuels for transportation, but also of raw materials needed for most chemical synthesis.

7.18B Nucleophilic Substitution and Organic Synthesis

To carry out synthesis we must think *backwards*. We examine a compound and ask: **What starting material and reagent are needed to make it?** If we are using nucleophilic substitution, we must determine what alkyl halide and what nucleophile can be used to form a specific product. This is the simplest type of synthesis because it involves only one step. In Chapter 10, we will learn about multistep syntheses.

Suppose, for example, that we are asked to prepare (CH₃)₂CHCH₂OH (2-methylpropan-1-ol) from an alkyl halide and any required reagents. To accomplish this synthesis, we must "fill in the boxes" for the starting material and reagent in the accompanying equation.

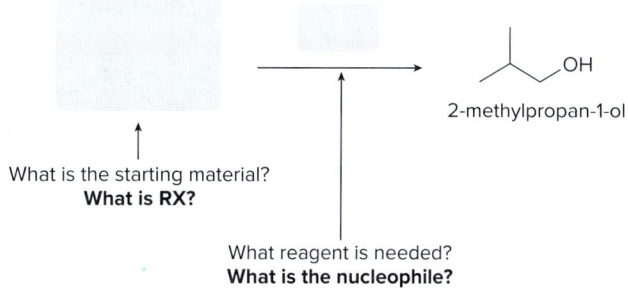

What is the starting material? **What is RX?**

What reagent is needed? **What is the nucleophile?**

To determine the two components needed for the synthesis, remember that the carbon atoms come from the organic starting material, in this case a 1° alkyl halide [(CH₃)₂CHCH₂Br]. The **functional group comes from the nucleophile**, ⁻OH in this case. With these two components, we can "fill in the boxes" to complete the synthesis.

(CH₃)₂CHCH₂Br + ⁻OH → (CH₃)₂CHCH₂OH

The alkyl halide provides the carbon skeleton. The nucleophile provides the functional group.

After any synthesis is proposed, check to see if it is reasonable, given what we know about reactions. Will the reaction written give a high yield of product? The synthesis of (CH₃)₂CHCH₂OH is reasonable, because the starting material is a 1° alkyl halide and the nucleophile (⁻OH) is strong, and both facts contribute to a successful S$_N$2 reaction.

Problem 7.36 What alkyl halide and nucleophile are needed to prepare each compound?

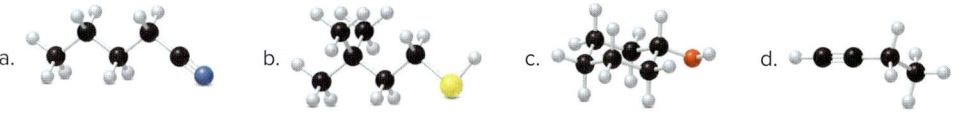

a. b. c. d.

Problem 7.37 The ether, CH₃OCH₂CH₃, can be prepared by two different nucleophilic substitution reactions, one using CH₃O⁻ as nucleophile and the other using CH₃CH₂O⁻ as nucleophile. Draw both routes.

Chapter 7 REVIEW

KEY CONCEPTS

[1] General facts about the reactions of alkyl halides (RX)

① Reactivity with nucleophiles (7.5)

- An alkyl halide is **electrophilic** because of its **polar C—X bond**.

② Reactivity with bases (7.5)

- The proton adjacent to the **polar C—X bond** is removed to form an alkene.

[2] Nucleophilic substitution (7.6)

- A nucleophile replaces a leaving group on an sp^3 hybridized carbon.
- One σ bond is broken and one σ bond is formed. There are two possible mechanisms: S_N1 and S_N2.

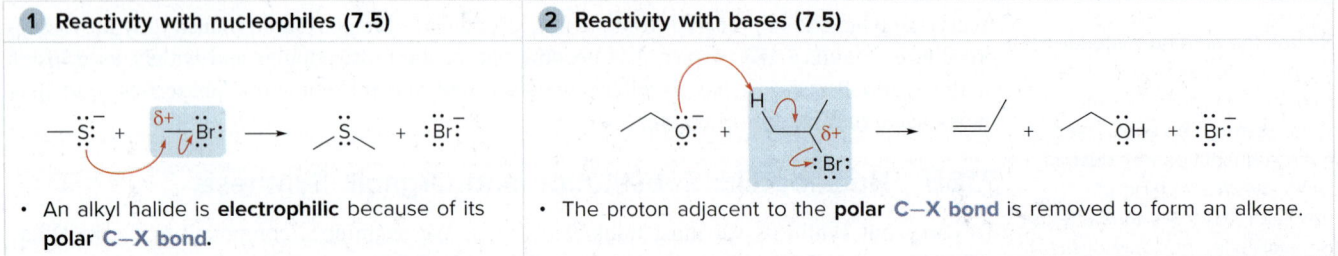

Try Problem 7.43.

[3] Periodic trends

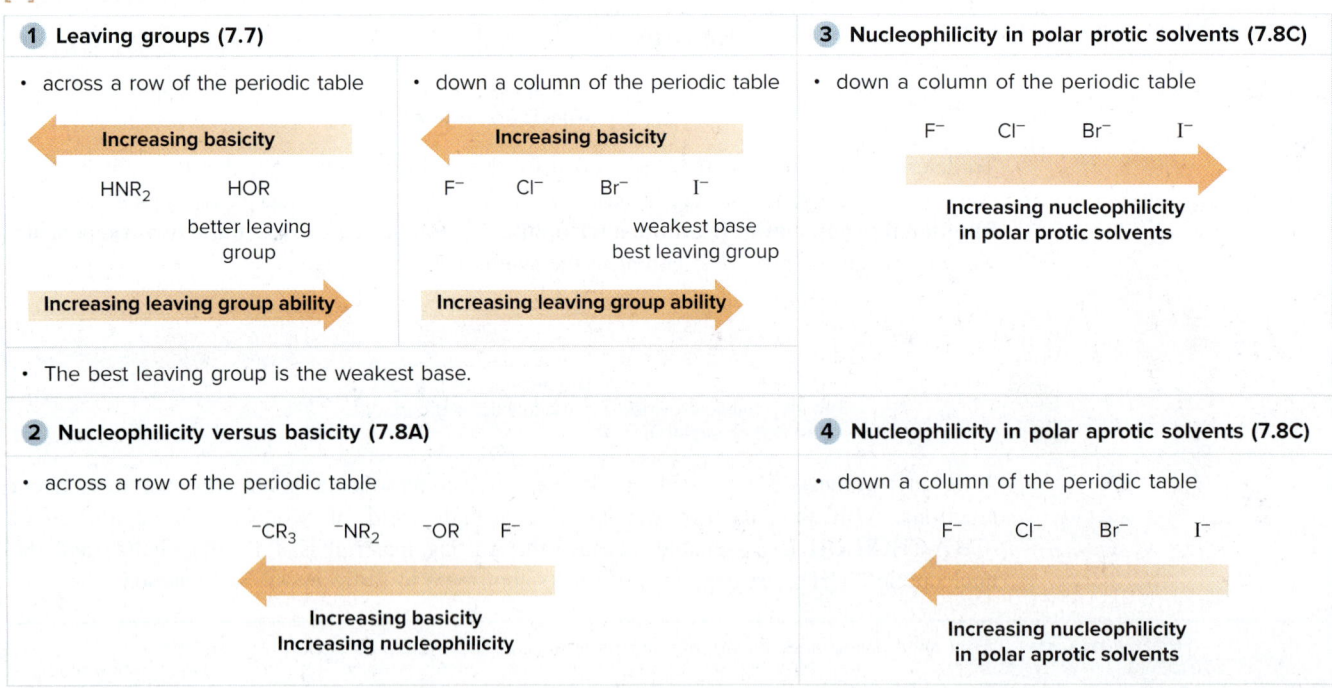

Try Problems 7.45, 7.47, 7.48.

[4] Carbocation stability (7.13)

- The stability of a carbocation increases as the number of electron-donating groups, such as **alkyl** groups, bonded to the **positively charged carbon** increases.

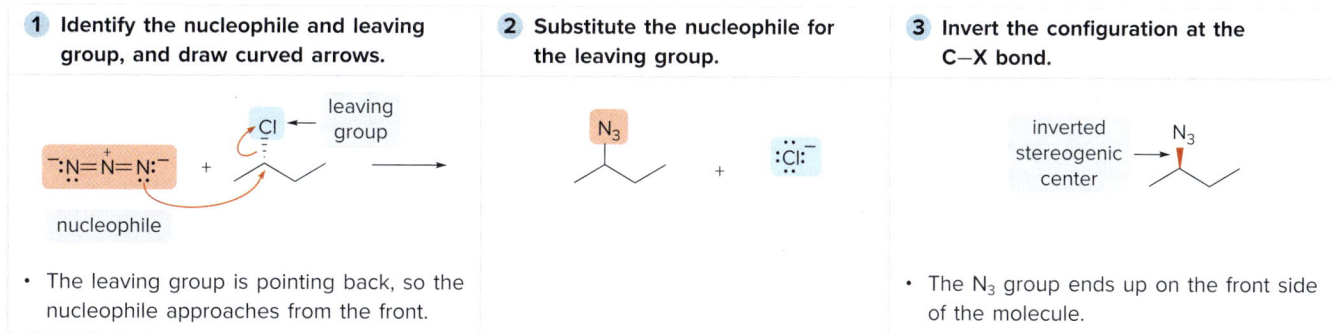

Try Problems 7.55, 7.56.

KEY SKILLS

[1] Drawing the product(s) of an S_N2 reaction (7.11)

1 Identify the nucleophile and leaving group, and draw curved arrows.	**2** Substitute the nucleophile for the leaving group.	**3** Invert the configuration at the C—X bond.

- The leaving group is pointing back, so the nucleophile approaches from the front.
- The N_3 group ends up on the front side of the molecule.

See Sample Problem 7.2, Figures 7.7, 7.8. Try Problems 7.51, 7.52.

[2] Drawing the product(s) of an S_N1 reaction (7.12)

1 Draw the curved arrow for Step [1]—loss of the leaving group.	**2** Draw the curved arrow for Step [2]—nucleophilic attack.	**3** When the initial substitution product bears a positive charge, remove a proton in Step [3].

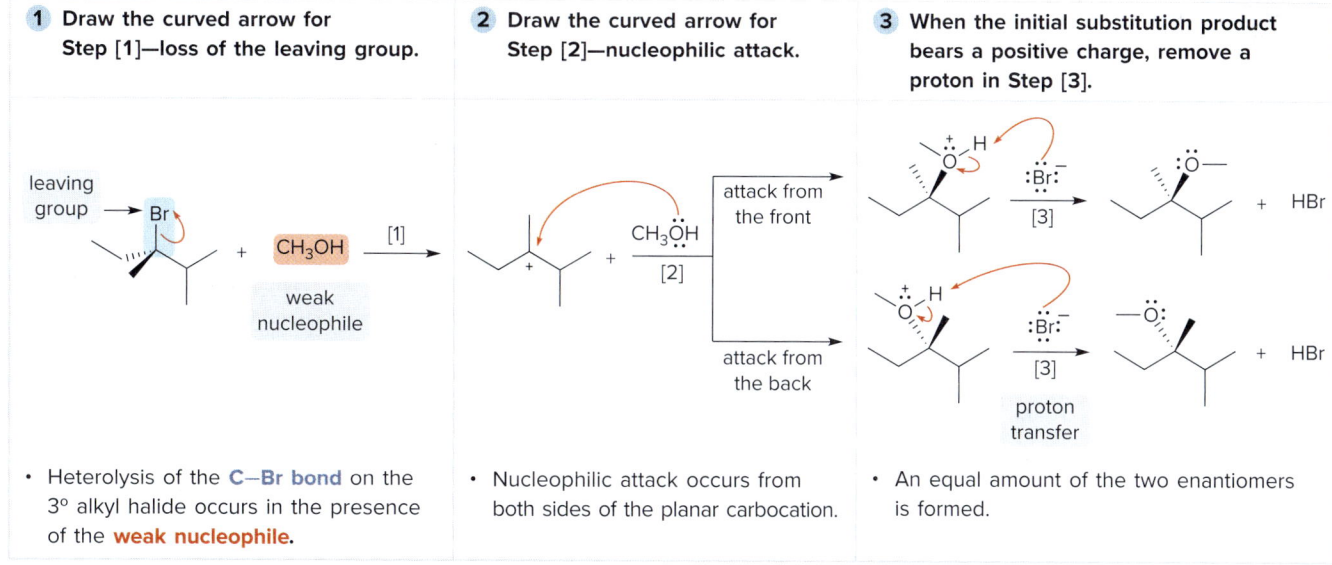

- Heterolysis of the **C—Br bond** on the 3° alkyl halide occurs in the presence of the **weak nucleophile**.
- Nucleophilic attack occurs from both sides of the planar carbocation.
- An equal amount of the two enantiomers is formed.

See Sample Problem 7.4, Figure 7.11. Try Problem 7.58.

[3] Deciding if a reaction proceeds by S_N1 or S_N2 (7.15E)

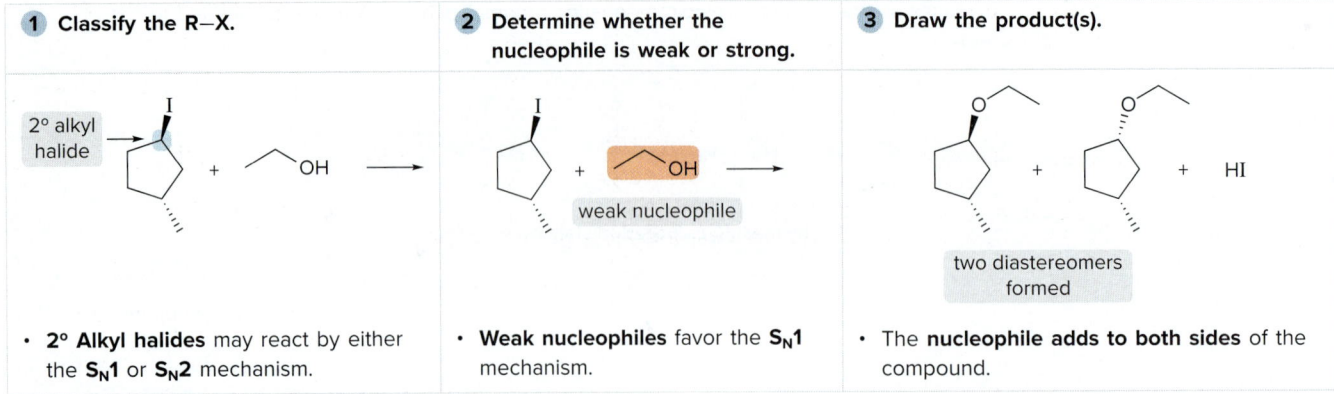

1 Classify the R–X.	2 Determine whether the nucleophile is weak or strong.	3 Draw the product(s).
• 2° Alkyl halides may react by either the S_N1 or S_N2 mechanism.	• Weak nucleophiles favor the S_N1 mechanism.	• The nucleophile adds to both sides of the compound.

See Sample Problems 7.5, 7.6, Table 7.7. Try Problem 7.62.

KEY MECHANISM CONCEPTS

Comparison of S_N1 and S_N2 reactions

	S_N2 mechanism	S_N1 mechanism
1 Mechanism	• one step (7.11B)	• two steps (7.12B)
2 Rate equation	• rate = k[RCl][:Nu$^-$] • second-order kinetics (7.11A)	• rate = k[RCl] • first-order kinetics (7.12A)
3 Alkyl halide	• order of reactivity (7.11D) CH$_3$–Cl, 1°, 2°, 3° methyl ← Increasing rate of an S_N2 reaction • faster with less steric hindrance around the C–X bond	• order of reactivity (7.12D) CH$_3$–Cl, 1°, 2°, 3° methyl Increasing rate of an S_N1 reaction → • faster when more stable (more substituted) carbocations are formed (7.14)
4 Stereochemistry	• backside attack by the nucleophile (7.11C) backside attack → inversion of configuration	• trigonal planar carbocation intermediate (7.12C) attack from the front / attack from the back → two enantiomers formed

5 Nucleophile	• favored by **strong** nucleophiles that usually bear a negative charge (7.15B) • common examples: ^-CN N_3^- ^-OR HS^-	• favored by **weak** nucleophiles that often bear no net charge (7.15B) • common examples: ROH H_2O
6 Solvent	• favored by polar **aprotic** solvents (7.15D) 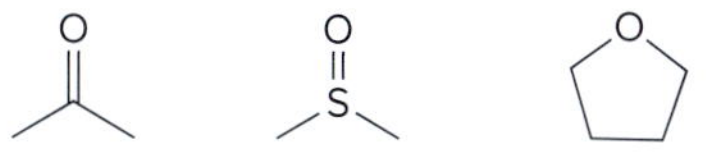 acetone dimethyl sulfoxide tetrahydrofuran See Figure 7.5.	• favored by polar **protic** solvents (7.15D) 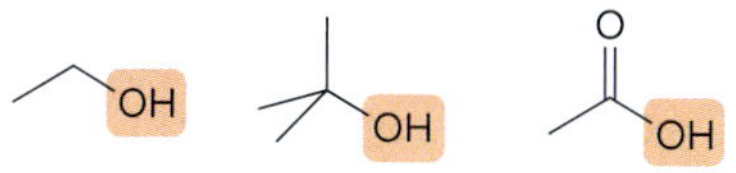 ethanol *tert*-butanol acetic acid See Figure 7.4.
7 Leaving group	• better leaving group ---→ faster reaction for *both* S_N1 and S_N2 (7.15C) R–F R–Cl R–Br R–I **Increasing leaving group ability** **Increasing rate of S_N1 and S_N2 reactions**	

See Tables 7.5, 7.6, 7.7. Try Problems 7.45, 7.47, 7.48, 7.62.

PROBLEMS

Students who have already learned about mass spectrometry can try Problems A.5; A.6a, b; A.15d, e; and A.20(**A**), (**B**).
Students who have learned about nuclear magnetic resonance spectroscopy can try Problem C.50a, b.

Problems Using Three-Dimensional Models

7.38 Give the IUPAC name for each compound, including any *R,S* designation.

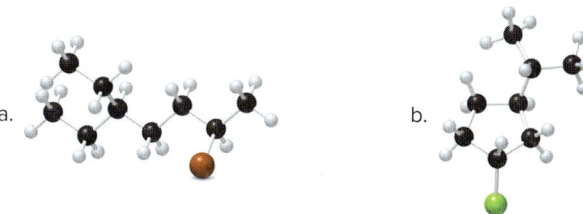

7.39 Draw the products formed when each alkyl halide is treated with NaCN.

Nomenclature

7.40 Give the IUPAC name for each compound.

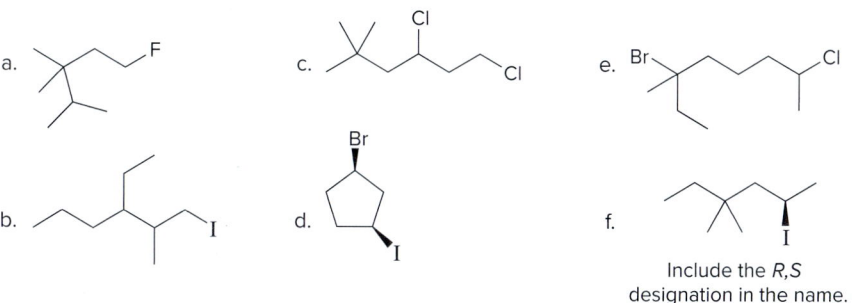

Include the *R,S* designation in the name.

Chapter 7 Alkyl Halides and Nucleophilic Substitution

7.41 Give the structure corresponding to each name.
   a. 3-bromo-4-ethylheptane
   b. 1,1-dichloro-2-methylcyclohexane
   c. 1-bromo-4-ethyl-3-fluorooctane
   d. (S)-3-iodo-2-methylnonane
   e. (1R,2R)-trans-1-bromo-2-chlorocyclohexane
   f. (R)-4,4,5-trichloro-3,3-dimethyldecane

7.42 Draw the eight constitutional isomers having the molecular formula $C_5H_{11}Cl$.
   a. Give the IUPAC name for each compound (ignoring R and S designations).
   b. Classify each alkyl halide as 1°, 2°, or 3°.
   c. Label any stereogenic centers.
   d. For each constitutional isomer that contains a stereogenic center, draw all possible stereoisomers, and label each stereogenic center as R or S.

General Nucleophilic Substitution, Leaving Groups, and Nucleophiles

7.43 Draw the products of each nucleophilic substitution reaction.

7.44 Which of the following molecules contain a good leaving group?

7.45 Rank the following compounds in order of increasing reactivity in a substitution reaction with ^-CN as nucleophile.

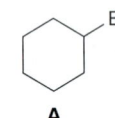
A

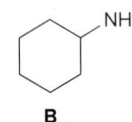
B

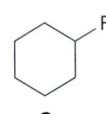
C

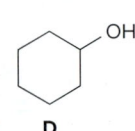
D

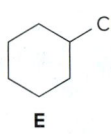
E

7.46 Which of the following nucleophilic substitution reactions will take place?

7.47 Rank the anions in order of increasing nucleophilicity in acetone: CH_3S^-, CH_3NH^-, I^-, Br^-, and CH_3O^-.

7.48 Classify each solvent as protic or aprotic.
   a. $(CH_3)_2CHOH$
   b. CH_3NO_2
   c. CH_2Cl_2
   d. NH_3
   e. $N(CH_3)_3$
   f. $HCONH_2$

7.49 Why is the amine N atom more nucleophilic than the amide N atom in $CH_3CONHCH_2CH_2CH_2NHCH_3$?

The S$_N$2 Reaction

7.50 Consider the following S$_N$2 reaction.

   a. Draw a mechanism using curved arrows.
   b. Draw an energy diagram. Label the axes, the reactants, products, E_a, and $\Delta H°$. Assume that the reaction is exothermic.

c. Draw the structure of the transition state.
d. What is the rate equation?
e. What happens to the reaction rate in each of the following instances? [1] The leaving group is changed from Br⁻ to I⁻; [2] The solvent is changed from acetone to CH₃CH₂OH; [3] The alkyl halide is changed from CH₃(CH₂)₄Br to CH₃CH₂CH₂CH(Br)CH₃; [4] The concentration of ⁻CN is increased by a factor of five; and [5] The concentrations of both the alkyl halide and ⁻CN are increased by a factor of five.

7.51 Draw the products of each S$_N$2 reaction and indicate the stereochemistry where appropriate.

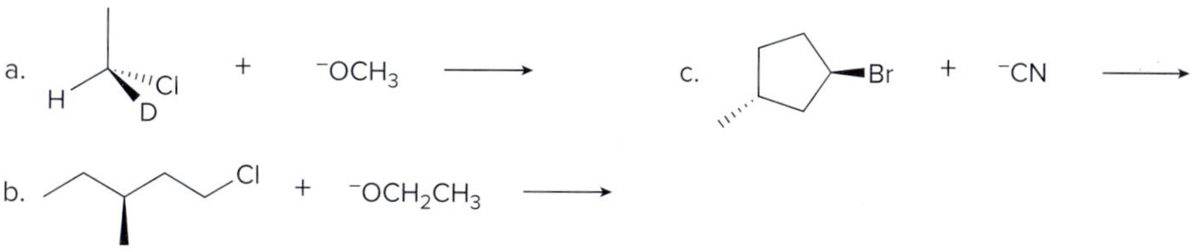

a. [structure] + ⁻OCH₃ →   c. [structure] + ⁻CN →

b. [structure] + ⁻OCH₂CH₃ →

7.52 Draw the product of the following S$_N$2 reaction, including the stereochemistry at all stereogenic centers. The product of this reaction is aprepitant, a drug used to treat nausea and emesis (vomiting) in chemotherapy patients.

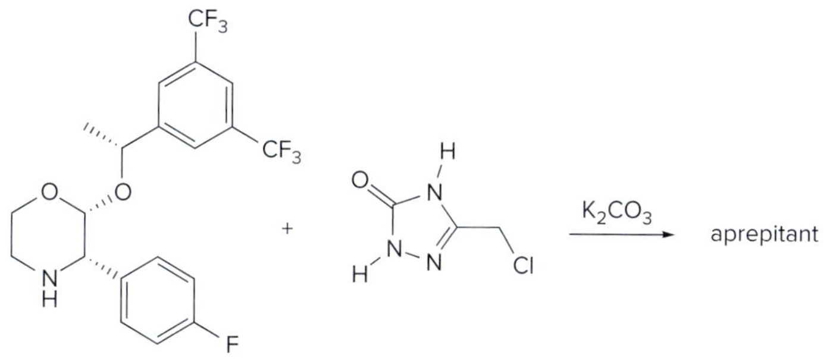

[structures] $\xrightarrow{K_2CO_3}$ aprepitant

7.53 Identify **M** in the following reaction sequence used to prepare the antiulcer drug omeprazole (trade name Prilosec).

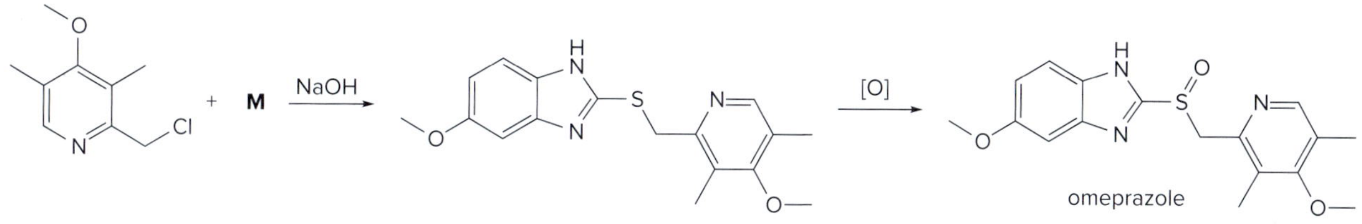

[structure] + **M** $\xrightarrow{NaOH}$ [structure] $\xrightarrow{[O]}$ omeprazole

7.54 The non-sedating antihistamine cetirizine (trade name Zyrtec) is prepared by a reaction sequence that involves two consecutive substitution reactions. Identify **N** and **O** in the following reaction sequence.

[structure] $\xrightarrow{}$ **N** $\xrightarrow[(CH_3)_3COH]{KOC(CH_3)_3}$ **O** $\xrightarrow{HCl}$ cetirizine

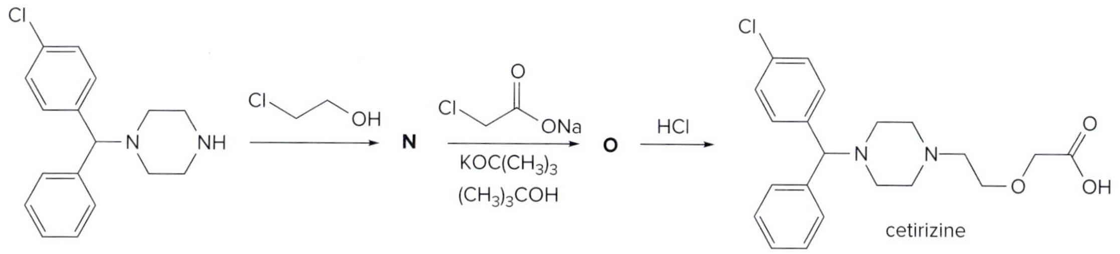

Carbocations

7.55 Classify the carbocations as 1°, 2°, or 3°, and rank the carbocations in each group in order of increasing stability.

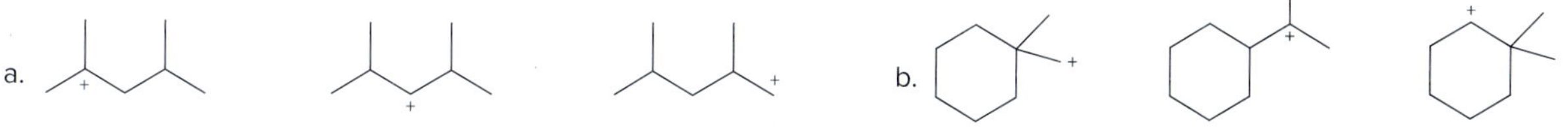

a. [structures]   b. [structures]

Chapter 7 Alkyl Halides and Nucleophilic Substitution

7.56 Which of the following carbocations (**A** or **B**) is more stable? Explain your choice.

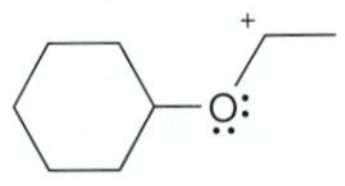

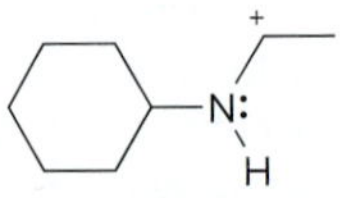

The S$_N$1 Reaction

7.57 Consider the following S$_N$1 reaction.

a. Draw a mechanism for this reaction using curved arrows.
b. Draw an energy diagram. Label the axes, starting material, product, E_a, and $\Delta H°$. Assume that the starting material and product are equal in energy.
c. Draw the structure of any transition states.
d. What is the rate equation for this reaction?
e. What happens to the reaction rate in each of the following instances? [1] The leaving group is changed from I$^-$ to Cl$^-$; [2] The solvent is changed from H$_2$O to DMF; [3] The alkyl halide is changed from (CH$_3$)$_2$C(I)CH$_2$CH$_3$ to (CH$_3$)$_2$CHCH(I)CH$_3$; and [4] The concentrations of both the alkyl halide and H$_2$O are increased by a factor of five.

7.58 Draw the products of each S$_N$1 reaction and indicate the stereochemistry when necessary.

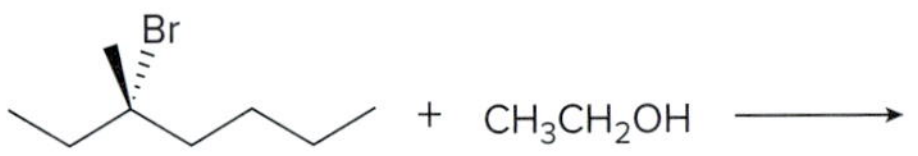

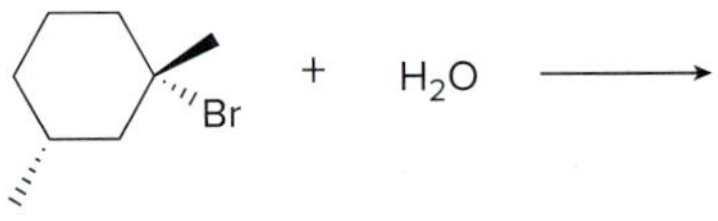

7.59 Draw a stepwise mechanism for the following reaction that illustrates how two substitution products are formed. Explain why 1-bromohex-2-ene reacts rapidly with a weak nucleophile (CH$_3$OH) under S$_N$1 reaction conditions, even though it is a 1° alkyl halide.

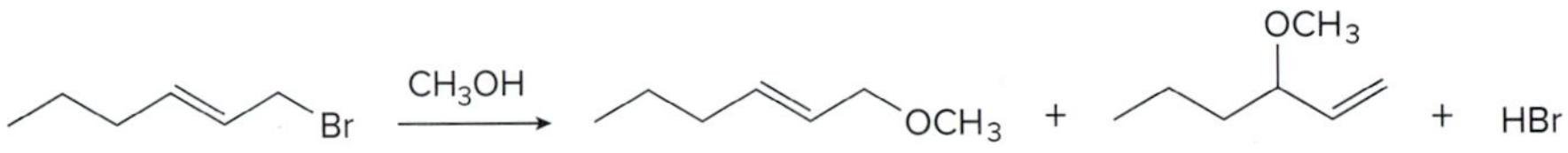

S$_N$1 and S$_N$2 Reactions

7.60 (a) Which halide in the following marine natural product reacts fastest in the S$_N$2 reaction? (b) Which halide in the following marine natural product reacts fastest in the S$_N$1 reaction?

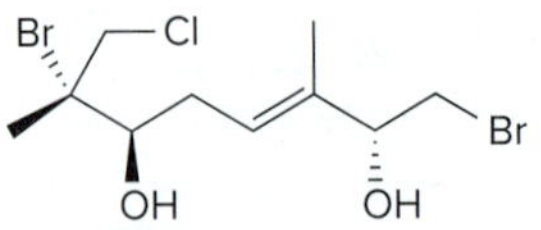

plocamenol A

7.61 (a) Rank **A**, **B**, and **C** in order of increasing S$_N$2 reactivity. (b) Rank **A**, **B**, and **C** in order of increasing S$_N$1 reactivity.

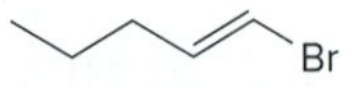

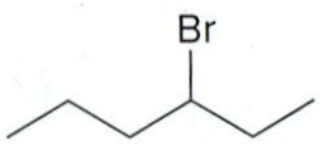

7.62 Determine the mechanism of nucleophilic substitution of each reaction and draw the products, including stereochemistry.

a. 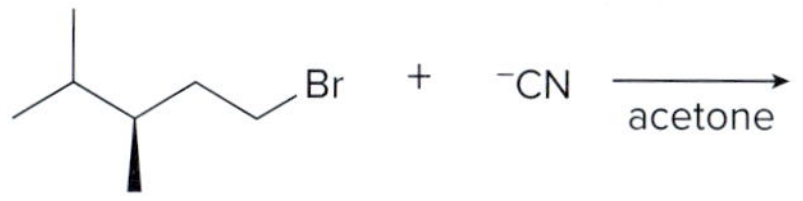 [structure] Br + ⁻CN —acetone→

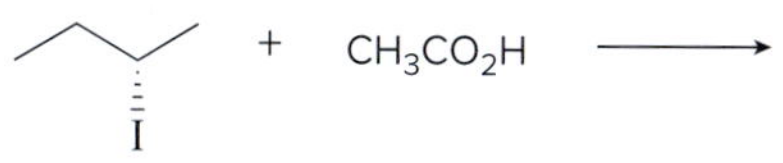

b. 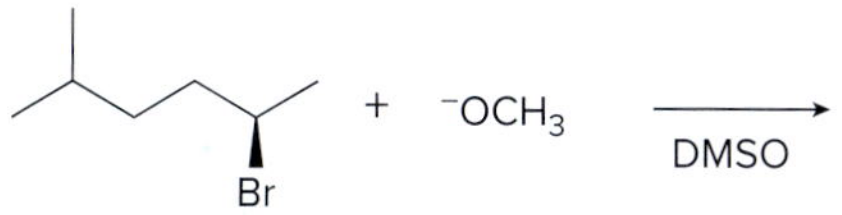 [structure] Br + ⁻OCH₃ —DMSO→

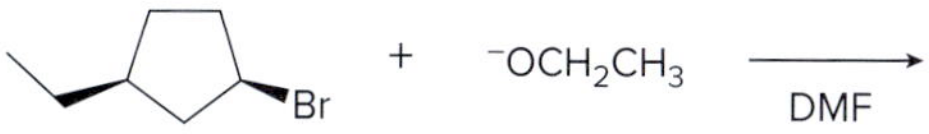

c. 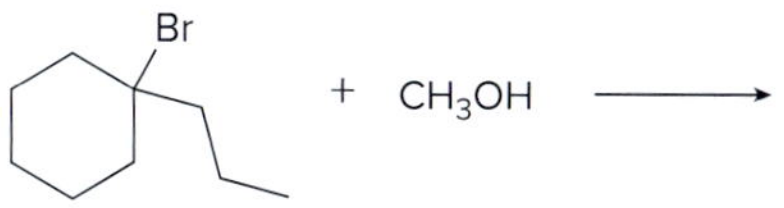 [structure with Br] + CH₃OH →

d. [structure with I] + CH₃CO₂H →

e. [cis-cyclopentyl structure] Br + ⁻OCH₂CH₃ —DMF→

f. 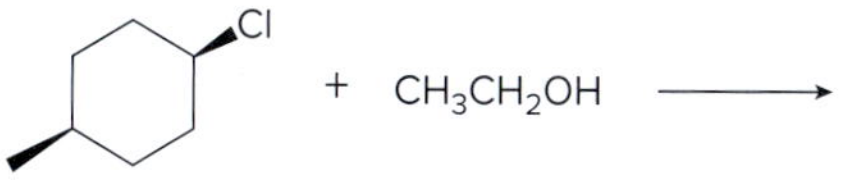 [cyclohexyl Cl structure] + CH₃CH₂OH →

7.63 Uridine monophosphate (UMP) is one of the four nucleotides that compose RNA, the nucleic acid that translates the genetic information of DNA into proteins needed by cells for proper function and development. A key step in the synthesis of UMP is the S$_N$1 reaction of **A** with **B** to form **C**, which is then converted to UMP in one step. Draw a stepwise mechanism for this S$_N$1 reaction.

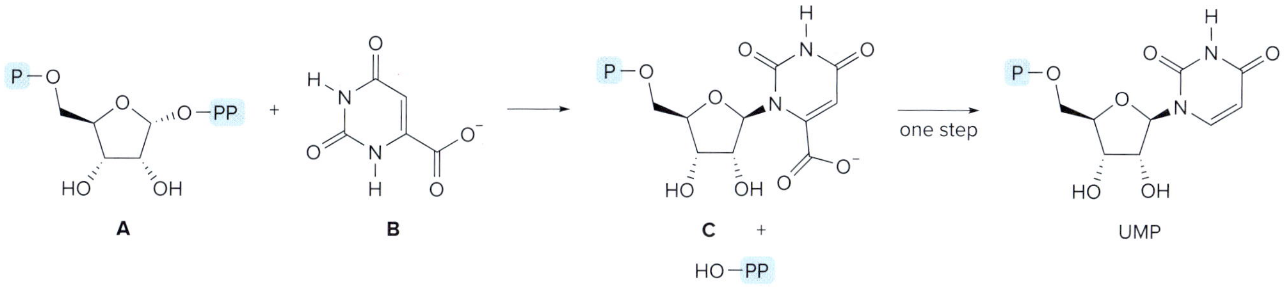

[Structures A, B, C, UMP with "one step" arrow; C + HO–PP]

7.64 Draw a stepwise, detailed mechanism for the following reaction. Use curved arrows to show the movement of electrons.

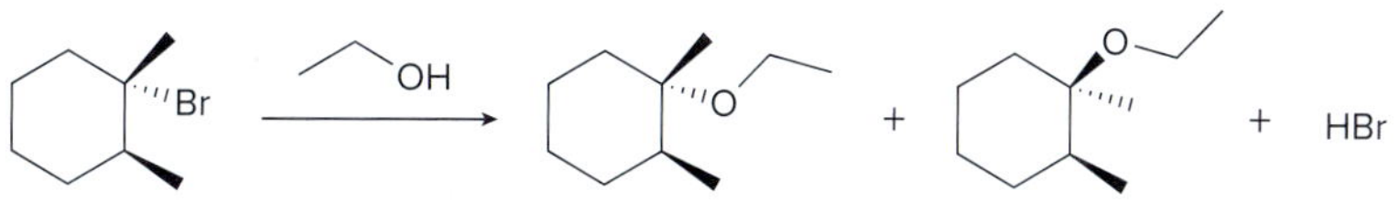

[trans-dimethyl cyclohexyl bromide] + EtOH → [ether products] + HBr

7.65 When a single compound contains both a nucleophile and a leaving group, an **intramolecular** reaction may occur. With this in mind, draw the product of the following reaction.

[cyclohexane carboxylic acid with Br] —⁻OH→ [carboxylate with Br] → C₇H₁₀O₂ + Br⁻

+ H₂O

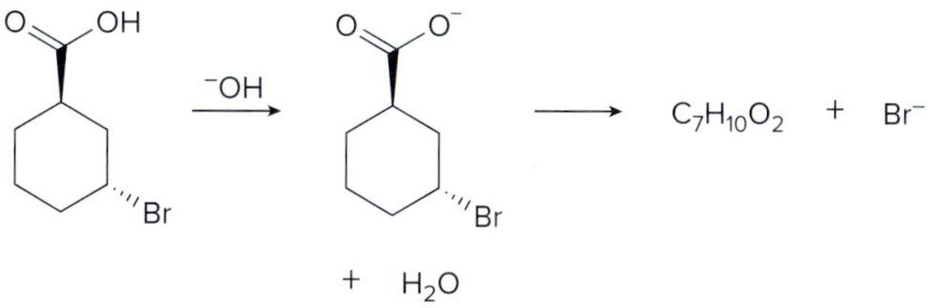

7.66 Nicotine can be made when the following ammonium salt is treated with Na₂CO₃. Draw a stepwise mechanism for this reaction.

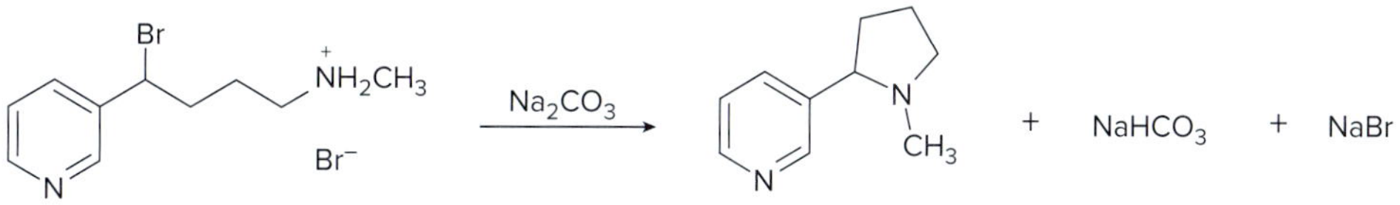

[pyridyl compound with Br and ⁺NH₂CH₃, Br⁻] —Na₂CO₃→ [nicotine structure] + NaHCO₃ + NaBr

nicotine

7.67 A key reaction in the synthesis of some lipids involves the rearrangement of geranyl diphosphate to linalyl diphosphate. Draw a stepwise mechanism for this process.

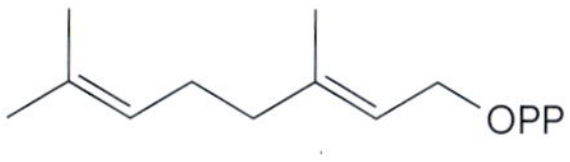

geranyl diphosphate 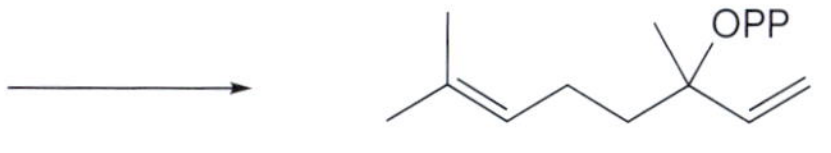 → linalyl diphosphate

Chapter 7 Alkyl Halides and Nucleophilic Substitution

7.68 Draw a stepwise, detailed mechanism for the following reaction.

Cl–CH₂CH₂–N(Et)–CH₂CH₂–Cl + CH₃NH₂ (excess) ⟶ [piperazine with N–Et and N–CH₃] + CH₃N⁺H₃ Cl⁻

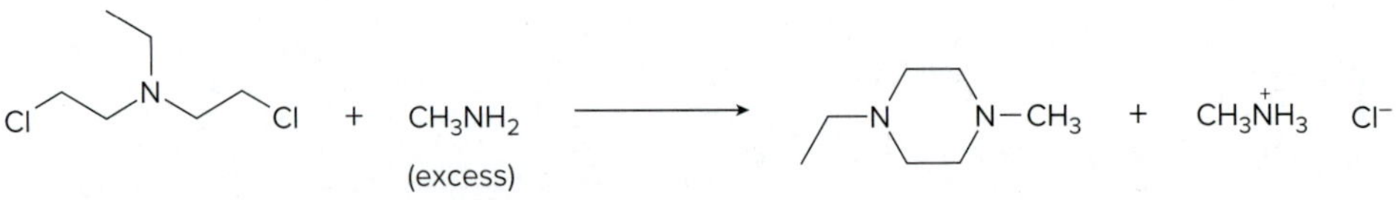

7.69 When (R)-6-bromo-2,6-dimethylnonane is dissolved in CH₃OH, nucleophilic substitution yields an optically inactive solution. When the isomeric halide (R)-2-bromo-2,5-dimethylnonane is dissolved in CH₃OH under the same conditions, nucleophilic substitution forms an optically active solution. Draw the products formed in each reaction, and explain why the difference in optical activity is observed.

Synthesis

7.70 Fill in the appropriate reagent or starting material in each of the following reactions.

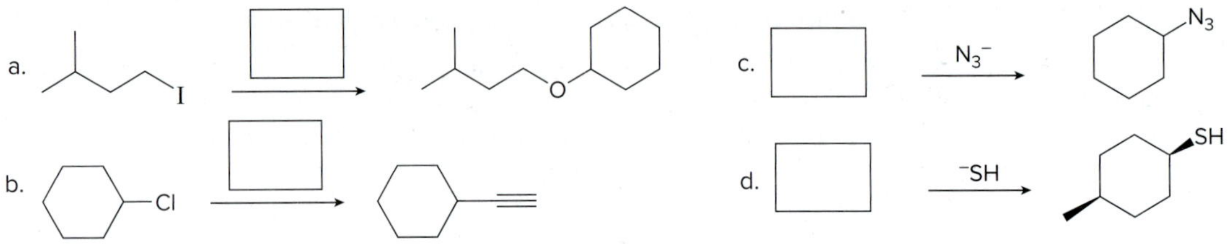

a. (isopentyl iodide) → [ ] → (isopentyl cyclohexyl ether)

b. (cyclohexyl chloride) → [ ] → (cyclohexylacetylene)

c. [ ] —N₃⁻→ (cyclohexyl azide)

d. [ ] —⁻SH→ (trans-4-substituted cyclohexanethiol)

7.71 Devise a synthesis of each compound from an alkyl halide using any other organic or inorganic reagents.

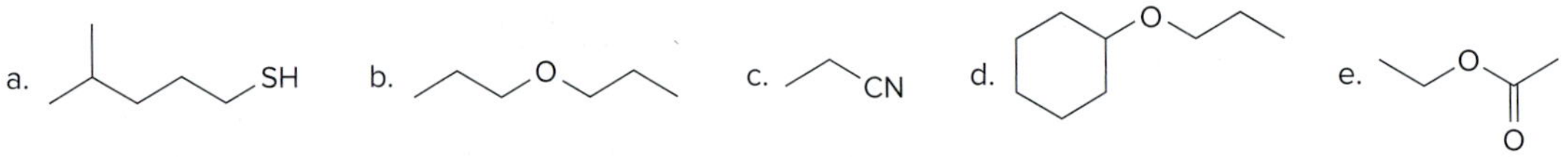

a. (pentyl thiol with branch) — SH
b. CH₃CH₂OCH₂CH₂OCH₃ type ether
c. CH₃CH₂CH₂CN
d. cyclohexyl propyl ether
e. ethyl acetate

7.72 Suppose you have compounds **A–D** at your disposal. Using these compounds, devise two different ways to make **E**. Which one of these methods is preferred, and why?

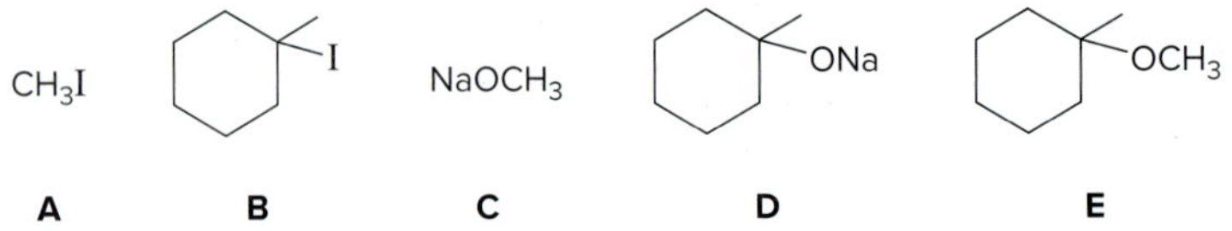

A: CH₃I
B: cyclohexyl iodide (tertiary)
C: NaOCH₃
D: cyclohexyl-ONa (tertiary)
E: cyclohexyl-OCH₃ (tertiary)

7.73 Muscalure, the sex pheromone of the common housefly, can be prepared by a reaction sequence that uses two nucleophilic substitutions. Identify compounds **A–D** in the following synthesis of muscalure.

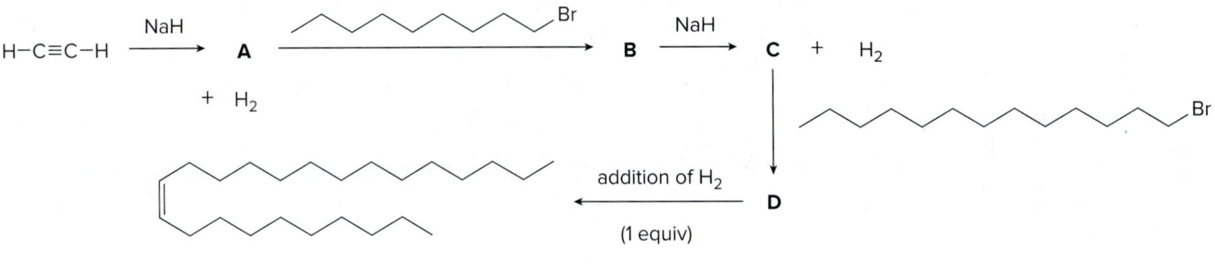

H–C≡C–H —NaH→ A + H₂

A —(CH₃(CH₂)ₙBr)→ B —NaH→ C + H₂

C + (long chain alkyl bromide) → D —addition of H₂ (1 equiv)→ muscalure

Challenge Problems

7.74 Explain why quinuclidine is a much more reactive nucleophile than triethylamine, even though both compounds have N atoms surrounded by three R groups.

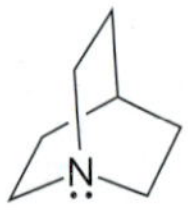

quinuclidine 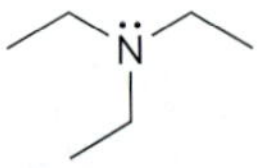   triethylamine

7.75 Draw a stepwise mechanism for the following reaction sequence.

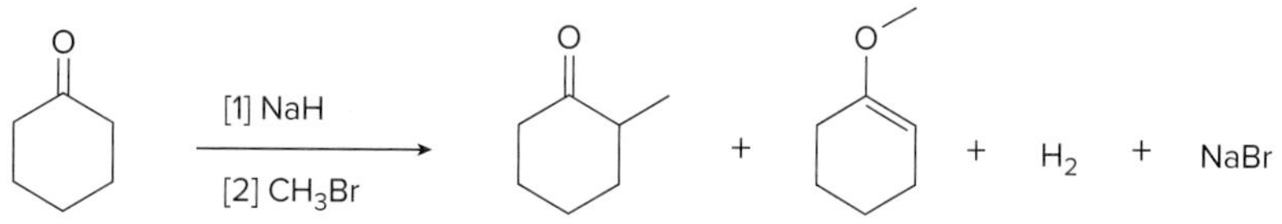

cyclohexanone $\xrightarrow[\text{[2] CH}_3\text{Br}]{\text{[1] NaH}}$ 2-methylcyclohexanone (major product) + 1-methoxycyclohexene (minor product) + H_2 + NaBr

7.76 When trichloride **J** is treated with CH_3OH, nucleophilic substitution forms the dihalide **K**. Draw a mechanism for this reaction and explain why one Cl is much more reactive than the other two Cl's so that a single substitution product is formed.

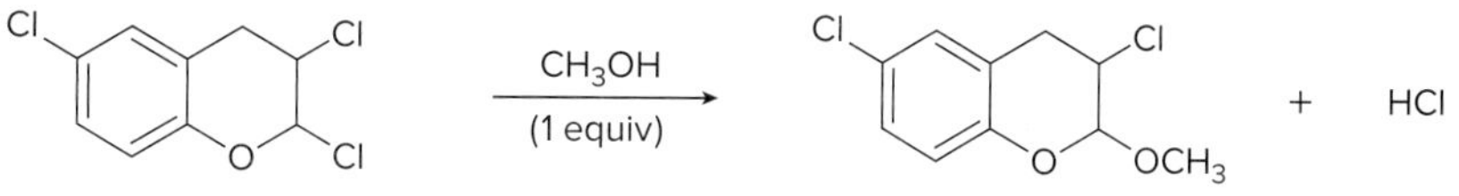

J (6-chloro-3-chloro-2-chlorochroman) $\xrightarrow[\text{(1 equiv)}]{CH_3OH}$ **K** (6-chloro-3-chloro-2-methoxychroman) + HCl

7.77 In some nucleophilic substitutions under S_N1 conditions, complete racemization does not occur and a small excess of one enantiomer is present. For example, treatment of optically pure 1-bromo-1-phenylpropane with water forms 1-phenylpropan-1-ol. (a) Calculate how much of each enantiomer is present using the given optical rotation data. (b) Which product predominates—the product of inversion or the product of retention of configuration? (c) Suggest an explanation for this phenomenon.

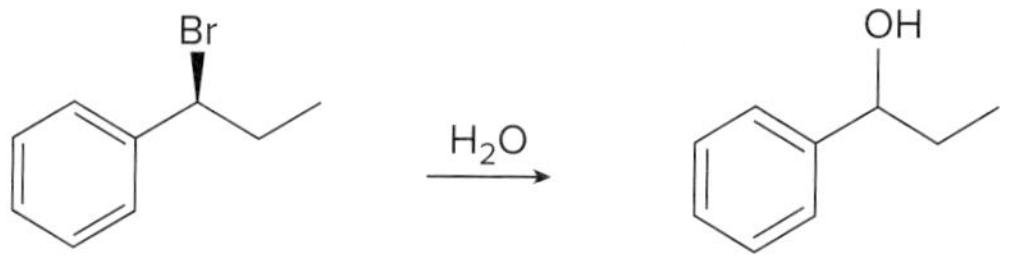

1-bromo-1-phenylpropane $\xrightarrow{H_2O}$ 1-phenylpropan-1-ol
observed $[\alpha] = +5.0$
optically pure S isomer, $[\alpha] = -48$

8 Alkyl Halides and Elimination Reactions

8.1 General features of elimination
8.2 Alkenes—The products of elimination reactions
8.3 The mechanisms of elimination
8.4 The E2 mechanism
8.5 The Zaitsev rule
8.6 The E1 mechanism
8.7 S_N1 and E1 reactions
8.8 Stereochemistry of the E2 reaction
8.9 When is the mechanism E1 or E2?
8.10 E2 reactions and alkyne synthesis
8.11 When is the reaction S_N1, S_N2, E1, or E2?

Source: Forest & Kim Starr

The elegant synthesis of **quinine** in 1944 is considered by many scientists to be the beginning of modern-day organic synthesis. Quinine, a natural product isolated from the bark of the cinchona tree native to the Andes Mountains, is a powerful antipyretic—that is, it reduces fever—and for centuries, it was the only effective treatment for malaria. Its bitter taste gives tonic water its characteristic flavor. One of the steps in a lengthy synthesis of quinine involves elimination, a characteristic reaction of alkyl halides and the subject of Chapter 8.

Why Study... Elimination Reactions?

Elimination reactions introduce π bonds into organic compounds, so they can be used to synthesize **alkenes** and **alkynes**—hydrocarbons that contain one and two π bonds, respectively. Elimination reactions are valuable in organic synthesis because they form functional groups that span two carbons. Like nucleophilic substitution, elimination reactions can occur by two different pathways, depending on the conditions. By the end of Chapter 8, therefore, you will have learned four different reaction mechanisms, two for nucleophilic substitution (S_N1 and S_N2) and two for elimination (E1 and E2).

The biggest challenge with this material is learning how to sort out two different reactions that follow four different mechanisms. **Will a particular alkyl halide undergo substitution or elimination with a given reagent, and by which of the four possible mechanisms?** To answer this question, we conclude Chapter 8 with a summary that allows you to predict which reaction and mechanism are likely for a given substrate.

8.1 General Features of Elimination

All **elimination reactions** involve loss of elements from the starting material to form a new π bond in the product.

- Alkyl halides undergo elimination reactions with Brønsted–Lowry bases. The elements of HX are lost and an alkene is formed.

Equations [1] and [2] illustrate examples of elimination reactions. In both reactions a base removes the elements of an acid, HBr or HCl, from the organic starting material.

Removal of the elements of HX, called **dehydrohalogenation,** is one of the most common methods to introduce a π bond and prepare an alkene. Dehydrohalogenation is an example of **β elimination,** because it involves loss of elements from two adjacent atoms: the **α carbon** bonded to the leaving group X, and the **β carbon** adjacent to it. Three curved arrows illustrate how four bonds are broken or formed in the process.

- The base (B:) removes a proton on the β carbon, thus forming H–B$^+$.
- The electron pair in the β C–H bond forms the new π bond between the α and β carbons.
- The electron pair in the C–X bond ends up on halogen, forming the leaving group :X$^-$.

The most common bases used in elimination reactions are negatively charged oxygen compounds such as $^-$OH and its alkyl derivatives, $^-$OR, called **alkoxides,** listed in Table 8.1. **Potassium *tert*-butoxide, K$^+$ $^-$OC(CH$_3$)$_3$, a bulky nonnucleophilic base, is especially useful** (Section 7.8B).

Table 8.1 Common Bases Used in Dehydrohalogenation

Structure	Name
Na$^+$ $^-$OH	Sodium hydroxide
K$^+$ $^-$OH	Potassium hydroxide
Na$^+$ $^-$OCH$_3$	Sodium methoxide
Na$^+$ $^-$OCH$_2$CH$_3$	Sodium ethoxide
K$^+$ $^-$OC(CH$_3$)$_3$	Potassium *tert*-butoxide

304 Chapter 8 Alkyl Halides and Elimination Reactions

To draw any product of dehydrohalogenation:

- Find the α carbon—the sp^3 hybridized carbon bonded to the leaving group.
- Identify all β carbons with H atoms.
- Remove the elements of H and X from the α and β carbons and form a π bond.

For example, 2-bromo-2-methylpropane has three β carbons (three CH_3 groups), but because all three are *identical*, only *one* alkene is formed upon elimination of HBr. In contrast, 2-bromobutane has two *different* β carbons (labeled $β_1$ and $β_2$), so elimination affords *two* constitutional isomers by loss of HBr across either the α and $β_1$ carbons, or the α and $β_2$ carbons. We learn about which product predominates and why in Section 8.5.

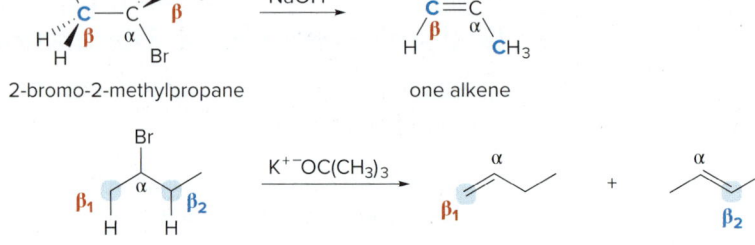

An elimination reaction is the first step in the slow degradation of the **pesticide DDT** (Section 7.4). Elimination of HCl from DDT forms the degradation product **DDE** (dichlorodiphenyldichloroethylene). This stable alkene is found in minute concentration in the fatty tissues of most adults in the United States.

DDE and DDT accumulate in the fatty tissues of predator birds such as osprey. When DDE and DDT concentration is high, female osprey produce eggs with thin shells that are easily crushed, so fewer osprey chicks hatch.
Comstock/PunchStock

Problem 8.1 Label the α and β carbons in each alkyl halide. Draw all possible elimination products formed when each alkyl halide is treated with $K^+{}^-OC(CH_3)_3$.

a. [structure with Cl]   b. [structure with Cl]   c. [cyclohexane with Br]

8.2 Alkenes—The Products of Elimination Reactions

Because elimination reactions of alkyl halides form alkenes, let's review earlier material on alkene structure and learn some additional facts as well.

8.2A Bonding in a Carbon–Carbon Double Bond

Recall from Section 1.10B that alkenes are hydrocarbons containing a carbon–carbon double bond. Each carbon of the double bond is sp^2 hybridized and trigonal planar, and all bond angles are 120°.

ethylene sp^2 hybridized   120°

8.2 Alkenes—The Products of Elimination Reactions

The double bond of an alkene consists of a σ bond and a π bond.

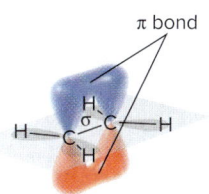

Ethylene, the simplest alkene, is a hormone that regulates plant growth and fruit ripening. A ripe banana placed next to unripe tomatoes speeds up their ripening because the banana gives off ethylene. *Jill Braaten/McGraw-Hill Education*

- The σ bond, formed by end-on overlap of the two sp^2 hybrid orbitals, lies in the plane of the molecule.
- The π bond, formed by side-by-side overlap of two $2p$ orbitals, lies perpendicular to the plane of the molecule. The π bond is formed during elimination.

Alkenes are classified according to the number of carbon atoms bonded to the carbons of the double bond. A **monosubstituted alkene** has *one* carbon atom bonded to the carbons of the double bond. A **disubstituted alkene** has *two* carbon atoms bonded to the carbons of the double bond, and so forth.

R⧵ ⁄H
  C=C
H⧸ ⧹H
	R⧵ ⁄H
	  C=C
	R⧸ ⧹H
		R⧵ ⁄R
		  C=C
		R⧸ ⧹H
			R⧵ ⁄R
			  C=C
			R⧸ ⧹R

monosubstituted (**one** R group)	**di**substituted (**two** R groups)	**tri**substituted (**three** R groups)	**tetra**substituted (**four** R groups)

Figure 8.1 shows several alkenes and how they are classified. You must be able to classify alkenes in this way to determine the major and minor products of elimination reactions, when a mixture of alkenes is formed.

Figure 8.1

Classifying alkenes by the number of R groups bonded to the double bond

monosubstituted	disubstituted	trisubstituted

- Carbon atoms bonded to the double bond are screened in blue.

Problem 8.2 Classify each alkene in the following vitamins by the number of carbon substituents bonded to the double bond.

a. vitamin A

b. vitamin D₃

8.2B Restricted Rotation

Figure 8.2 shows that there is free rotation about the carbon–carbon single bonds of butane, but *not* about the carbon–carbon double bond of but-2-ene. Because of restricted rotation, two stereoisomers of but-2-ene are possible.

Figure 8.2
Rotation around C—C and C=C compared

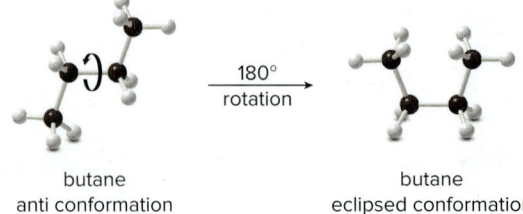

butane
anti conformation

180° rotation

butane
eclipsed conformation

These conformations **interconvert** by rotation.
They represent the **same** molecule.

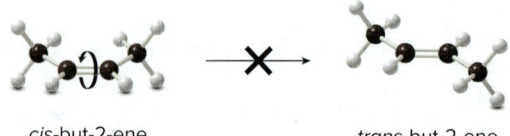

cis-but-2-ene

trans-but-2-ene

These molecules **do *not* interconvert** by rotation.
They are **different** molecules.

The concept of cis and trans isomers was first introduced for disubstituted cycloalkanes in Chapter 4. In both cases, a ring or a double bond restricts motion, preventing the rotation of a group from one side of the ring or double bond to the other.

- The cis isomer has two groups on the *same side* of the double bond.
- The trans isomer has two groups on *opposite sides* of the double bond.

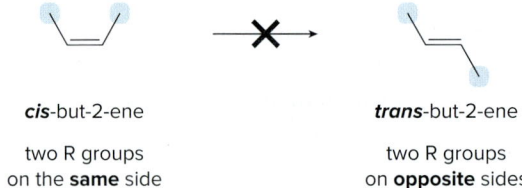

cis-but-2-ene
two R groups on the **same** side

trans-but-2-ene
two R groups on **opposite** sides

cis-But-2-ene and *trans*-but-2-ene are stereoisomers, but not mirror images of each other, so they are **diastereomers.**

The cis and trans isomers of but-2-ene are a specific example of a general type of stereoisomer occurring at carbon–carbon double bonds. **Whenever the two groups on *each* end of a carbon–carbon double bond are *different from each other,* two diastereomers are possible.**

X and X′ must be **different** from *each other*...

...and Y and Y′ must be **different** from *each other*.

Problem 8.3 For which double bonds are stereoisomers possible?

a., b., c. (structures)

Problem 8.4 Label each pair of alkenes as constitutional isomers, stereoisomers, or identical.

a. (structure) and (structure)   c. (structure) and (structure)

b. (structure) and (structure)   d. (structure) and (structure)

8.2C Stability of Alkenes

Some alkenes are more stable than others. For example, **trans alkenes are generally more stable than cis alkenes** because the larger groups bonded to the double bond carbons are farther apart, reducing steric interactions.

The trans isomer has the CH_3 groups *farther* away from each other.

more stable

Steric interactions of the CH_3 groups *destabilize* the cis isomer.

less stable

The stability of an alkene *increases,* moreover, as the **number of R groups bonded to the double bond carbons** *increases*.

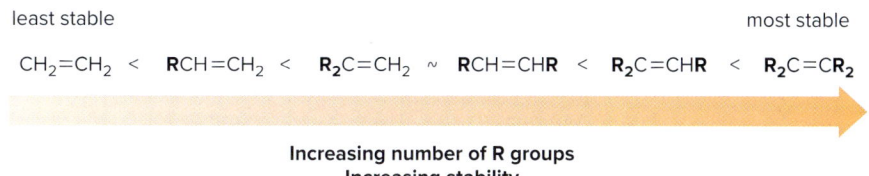

least stable

$CH_2=CH_2$ < $RCH=CH_2$ < $R_2C=CH_2$ ~ $RCH=CHR$ < $R_2C=CHR$ < $R_2C=CR_2$

most stable

Increasing number of R groups
Increasing stability

R groups increase the stability of an alkene because R groups are sp^3 hybridized, whereas the carbon atoms of the double bond are sp^2 hybridized. Recall from Sections 1.11B and 2.5D that the percent *s*-character of a hybrid orbital increases from 25% to 33% in going from sp^3 to sp^2. The higher the percent *s*-character, the more readily an atom accepts electron density. Thus, **sp^2 hybridized carbon atoms are more able to *accept* electron density,** and **sp^3 hybridized carbon atoms are more able to *donate* electron density.**

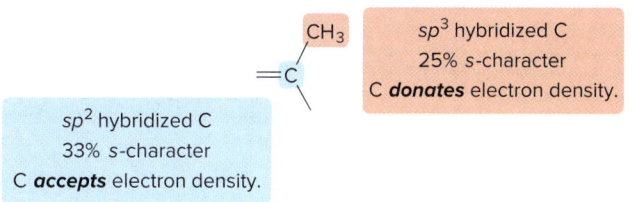

sp^2 hybridized C
33% *s*-character
C **accepts** electron density.

sp^3 hybridized C
25% *s*-character
C **donates** electron density.

- As a result, *increasing* the number of electron-donating R groups on a carbon atom able to accept electron density makes the alkene *more stable*.

Thus, *trans*-but-2-ene (a disubstituted alkene) is more stable than *cis*-but-2-ene (another disubstituted alkene), but both are more stable than but-1-ene (a monosubstituted alkene).

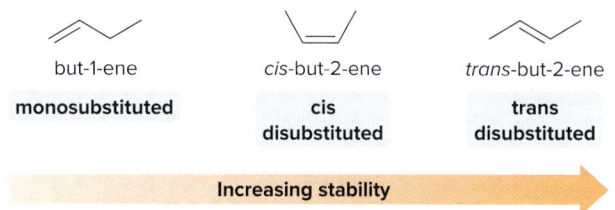

| but-1-ene | *cis*-but-2-ene | *trans*-but-2-ene |
| **monosubstituted** | **cis** **disubstituted** | **trans** **disubstituted** |

Increasing stability

In summary:

- Trans alkenes are *more stable* than cis alkenes because they have fewer steric interactions.
- *Increasing* alkyl substitution *stabilizes* an alkene by an electron-donating inductive effect.

308 Chapter 8 Alkyl Halides and Elimination Reactions

Problem 8.5 Rank the following alkenes in order of increasing stability.

A   B   C   D

8.3 The Mechanisms of Elimination

What is the mechanism for elimination? What is the order of bond breaking and bond making? Is the reaction a one-step process or does it occur in many steps?

There are two mechanisms for elimination—**E2** and **E1**—just as there are two mechanisms for nucleophilic substitution—S_N2 and S_N1.

- The **E2 mechanism (bimolecular elimination)**
- The **E1 mechanism (unimolecular elimination)**

The E2 and E1 mechanisms differ in the timing of bond cleavage and bond formation, analogous to the S_N2 and S_N1 mechanisms. In fact, E2 and S_N2 reactions have some features in common, as do E1 and S_N1 reactions.

8.4 The E2 Mechanism

The most common mechanism for dehydrohalogenation is the E2 mechanism. For example, $(CH_3)_3CBr$ reacts with ^-OH to form $(CH_3)_2C=CH_2$ via an E2 mechanism.

8.4A Kinetics

An E2 reaction exhibits **second-order kinetics;** that is, the reaction is **bimolecular,** and both the alkyl halide and the base appear in the rate equation.

- rate = $k[(CH_3)_3CBr][^-OH]$

8.4B A One-Step Mechanism

The most straightforward explanation for the second-order kinetics is a **concerted reaction: all bonds are broken and formed in a single step,** as shown in Mechanism 8.1.

Mechanism 8.1 The E2 Mechanism

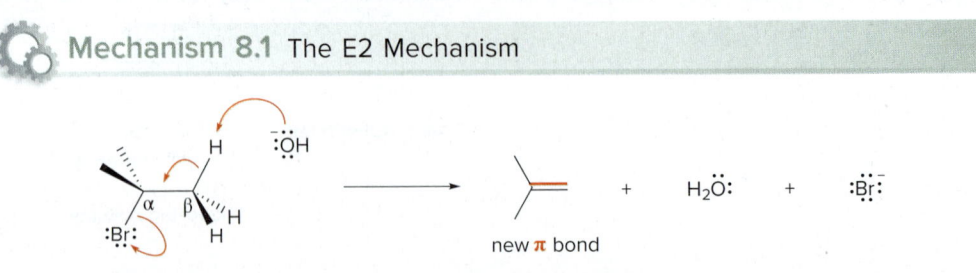

- The base ^-OH **removes a proton** from the β carbon, forming H_2O (a by-product).
- The electron pair in the β C–H bond forms the **new π bond.**
- The **leaving group** Br^- **comes off** with the electron pair in the C–Br bond.

An energy diagram for the reaction of $(CH_3)_3CBr$ with ^-OH is shown in Figure 8.3. Two bonds are broken (C–H and C–Br) and two bonds are formed (H–OH and the π bond) in a single step, so the transition state contains **four partial bonds,** with the negative charge distributed over the base and the leaving group. **Entropy favors the products of an E2 reaction** because two molecules of starting material form three molecules of product.

Figure 8.3

An energy diagram for an E2 reaction: $(CH_3)_3CBr + {}^-OH \rightarrow (CH_3)_2C=CH_2 + H_2O + Br^-$

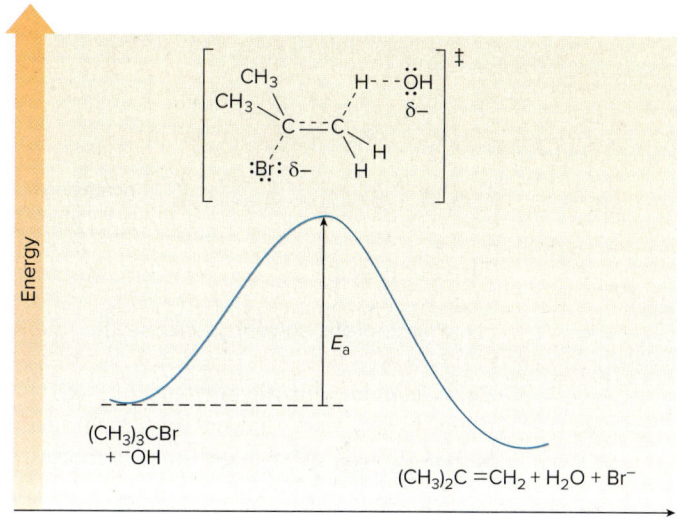

- In the transition state, the C—H and C—Br bonds are partially broken, the O—H and π bonds are partially formed, and both the base and the departing leaving group bear a partial negative charge.

Problem 8.6 Use curved arrows to show the movement of electrons in the following E2 mechanism. Draw the structure of the transition state.

There are close parallels between the E2 and S_N2 mechanisms in how the identity of the base, the leaving group, and the solvent affect the rate.

The Base

- The base appears in the rate equation, so the rate of the E2 reaction *increases* as the strength of the base *increases*.

The IUPAC names for **DBN** and **DBU** are rarely used because the names are complex. **DBN** stands for 1,5-**d**ia**z**a**b**icyclo[4.3.0]-**n**on-5-ene, and **DBU** stands for 1,8-**d**ia**z**a**b**icyclo[5.4.0]-**u**ndec-7-ene.

E2 reactions are generally run with strong, negatively charged bases like $^-$OH and $^-$OR. Two strong, sterically hindered nitrogen bases, called **DBN** and **DBU,** are also sometimes used.

DBN   DBU

For example, reaction of iodide **A** with DBU forms **B,** which contains a new π bond by an E2 elimination.

The Leaving Group

- Because the bond to the leaving group is partially broken in the transition state, the *better* the leaving group the *faster* the E2 reaction.

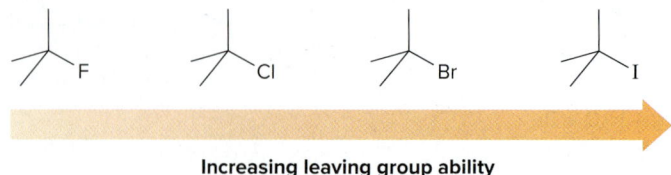

Increasing leaving group ability
Increasing rate of the E2 reaction

The Solvent

- Polar aprotic solvents *increase* the rate of E2 reactions.

Because **polar aprotic solvents** like $(CH_3)_2C=O$ do not solvate anions well, a negatively charged base is not "hidden" by strong interactions with the solvent (Section 7.15D), and the base is stronger. **A stronger base increases the reaction rate.**

Problem 8.7 Consider an E2 reaction between CH_3CH_2Br and $KOC(CH_3)_3$. What effect does each of the following changes have on the rate of elimination? (a) The base is changed to KOH. (b) The alkyl halide is changed to CH_3CH_2Cl.

8.4C The Identity of the Alkyl Halide

The S_N2 and E2 mechanisms differ in how the R group affects the reaction rate.

- As the number of R groups on the carbon with the leaving group *increases*, the rate of the E2 reaction *increases*.

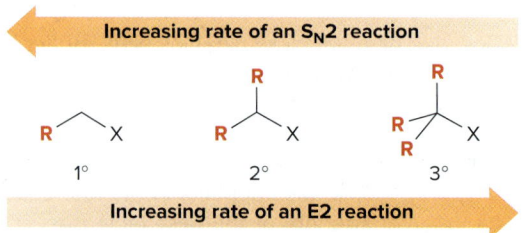

This trend is exactly *opposite* to the reactivity of alkyl halides in S_N2 reactions, where increasing alkyl substitution decreases the rate of reaction (Section 7.11D).

Why does increasing alkyl substitution increase the rate of an E2 reaction? In the transition state, the double bond is partially formed, so *increasing the stability* of the double bond with alkyl substituents *stabilizes* the transition state (i.e., it lowers E_a), which *increases* the rate of the reaction.

$$\left[\begin{array}{c} H\text{----}\ddot{\underset{\delta-}{O}}R \\ \diagdown\diagup \\ C\!=\!=\!=\!C \\ \diagup\diagdown \\ :\!\ddot{X}\!:\delta- \end{array} \right]^{\ddagger}$$

The double bond is partially formed.

- Increasing the number of R groups on the carbon with the leaving group forms more highly substituted, *more stable* alkenes in E2 reactions.

For example, the E2 reaction of a 1° alkyl halide (1-bromobutane) forms a monosubstituted alkene, whereas the E2 reaction of a 3° alkyl halide (2-bromo-2-methylpropane) forms a disubstituted alkene. The disubstituted alkene is more stable, so the 3° alkyl halide reacts faster than the 1° alkyl halide.

1-bromobutane
1° alkyl halide

monosubstituted alkene
less stable

2-bromo-2-methylpropane
3° alkyl halide

disubstituted alkene
more stable

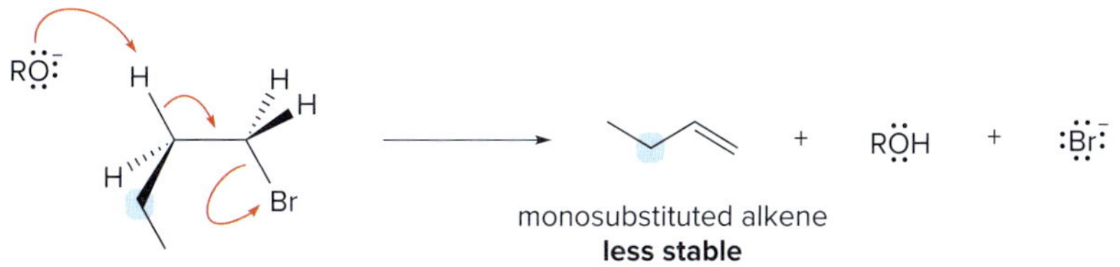

Elimination reactions are often steps in the synthesis of complex natural products. For example, elimination of HCl from compound **A** forms alkene **B,** which was converted to the antimalarial drug quinine, the chapter-opening molecule.

A

B

quinine
antimalarial drug

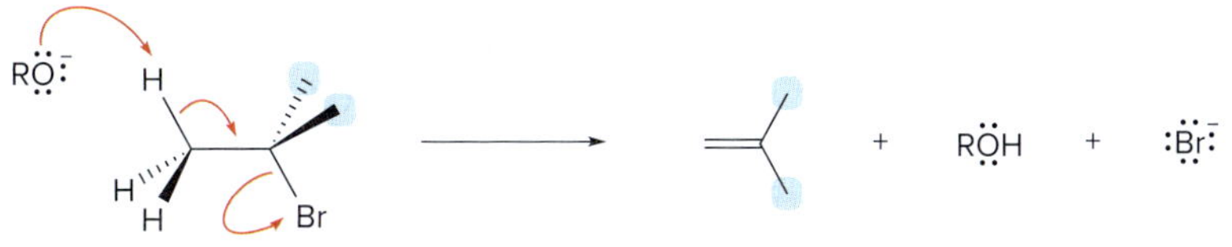

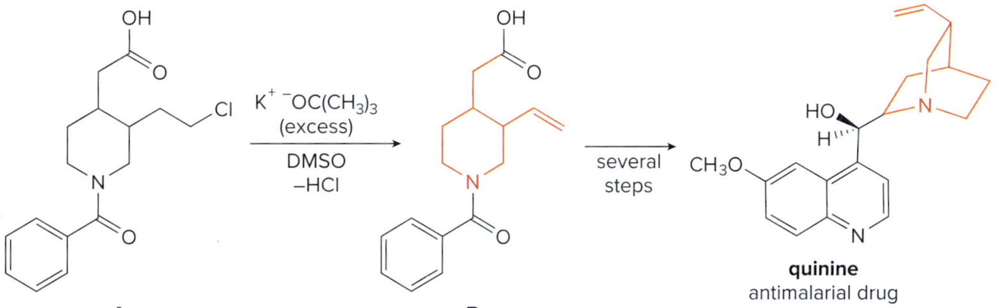

Table 8.2 summarizes the characteristics of the E2 mechanism.

Table 8.2 Characteristics of the E2 Mechanism

Characteristic	Result
Kinetics	• **Second order**
Mechanism	• **One step**
Identity of R	• **More substituted halides react faster.** • Rate: $R_3CX > R_2CHX > RCH_2X$
Base	• Favored by **strong bases**
Leaving group	• **Better leaving group** --→ faster reaction
Solvent	• Favored by **polar aprotic solvents**

Problem 8.8 Rank the alkyl halides in each group in order of increasing reactivity in an E2 reaction.

a.

b.

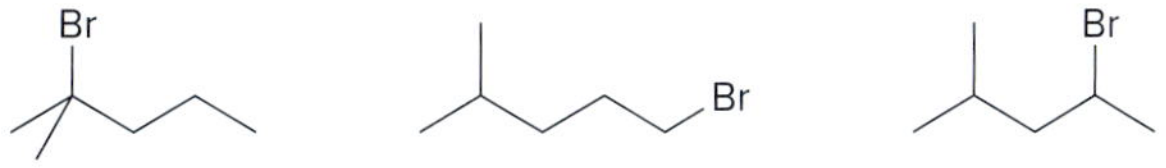

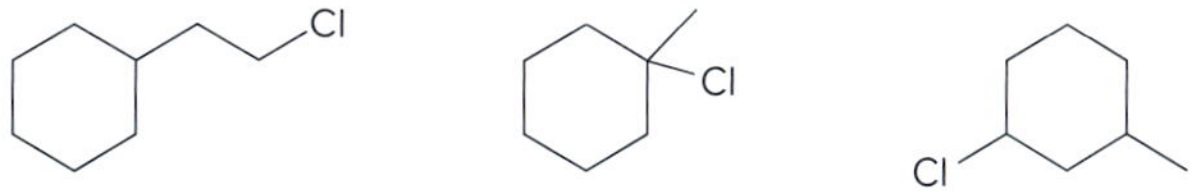

Problem 8.9 How does each of the following changes affect the rate of an E2 reaction?

a. tripling [RX]
b. halving [B:]
c. changing the solvent from CH_3OH to DMSO
d. changing the leaving group from I^- to Br^-
e. changing the base from ^-OH to H_2O
f. changing the alkyl halide from CH_3CH_2Br to $(CH_3)_2CHBr$

8.5 The Zaitsev Rule

Recall from Section 8.1 that a mixture of alkenes can form from the dehydrohalogenation of alkyl halides having two or more different β carbon atoms. When this occurs, one of the products usually predominates. The **major product is the more stable product—the one with the more substituted double bond.** For example, elimination of the elements of H and I from 1-iodo-1-methylcyclohexane yields two constitutional isomers: the trisubstituted alkene **A** (the major product) and the disubstituted alkene **B** (the minor product).

1-iodo-1-methylcyclohexane

A **major product** **tri**substituted alkene

B **minor product** **di**substituted alkene

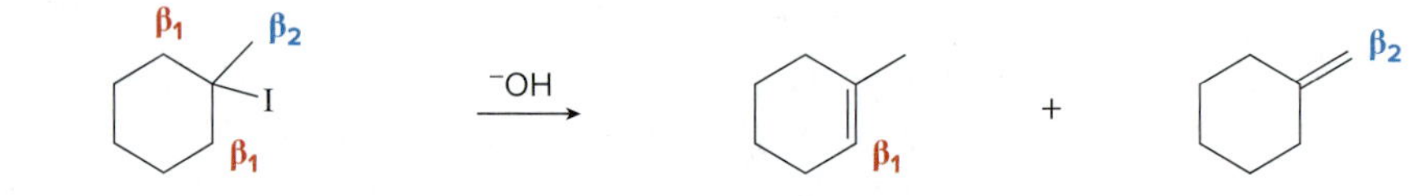

This phenomenon is called the **Zaitsev rule** (also called the **Saytzeff rule,** depending on the translation) for the Russian chemist who first noted this trend.

- **The Zaitsev rule: The major product in β elimination has the more substituted double bond.**

A reaction is *regioselective* when it yields predominantly or exclusively one constitutional isomer when more than one is possible. The E2 reaction is **regioselective** because the more substituted alkene predominates.

The Zaitsev rule results because the double bond is partially formed in the transition state for the E2 reaction. Thus, increasing the stability of the double bond by adding R groups lowers the energy of the transition state, which increases the reaction rate. E2 elimination of HBr from 2-bromo-2-methylbutane yields alkenes **C** and **D. D, having the more substituted double bond, is the major product,** because the transition state leading to its formation is lower in energy.

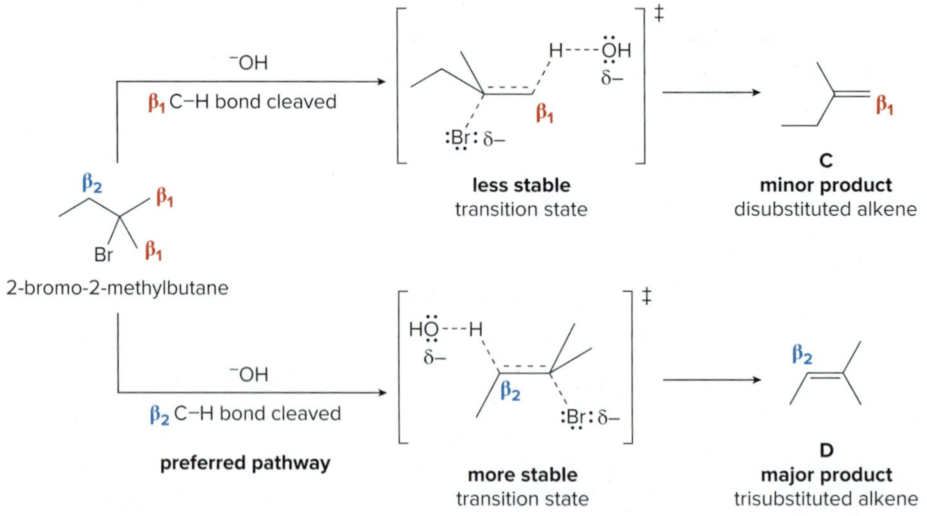

8.5 The Zaitsev Rule

When a mixture of stereoisomers is possible from dehydrohalogenation, the **major product is the more stable stereoisomer.** Dehydrohalogenation of alkyl halide **X** forms a mixture of trans and cis alkenes, **Y** and **Z**. The trans alkene **Y** is the major product because it is more stable.

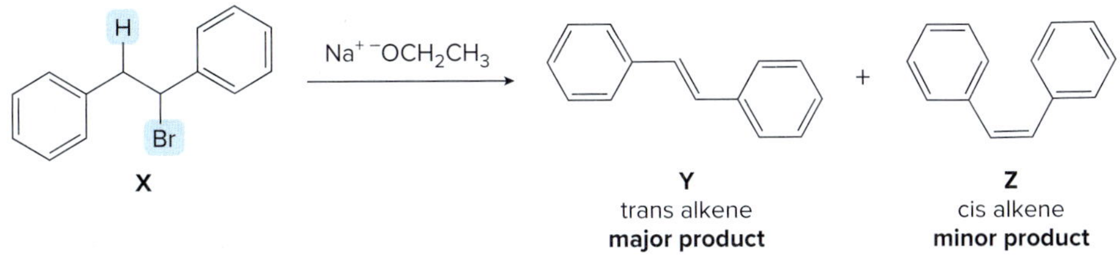

X → Y (trans alkene, major product) + Z (cis alkene, minor product), with Na⁺ ⁻OCH₂CH₃

A reaction is *stereoselective* when it forms predominantly or exclusively one stereoisomer when two or more are possible. The **E2 reaction is stereoselective** because one stereoisomer is formed preferentially.

Sample Problem 8.1 Determining the Major Product of an E2 Reaction

Predict the major product in the following E2 reaction.

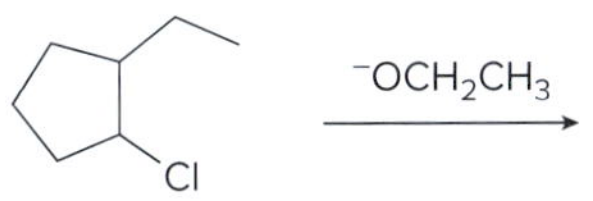

(cyclopentane with ethyl and Cl substituents) + ⁻OCH₂CH₃

Solution

The alkyl halide has two different β C atoms (labeled β₁ and β₂), so two different alkenes are possible: one formed by removal of HCl across the α and β₁ carbons, and one formed by removal of HCl across the α and β₂ carbons. Using the Zaitsev rule, the major product should be **A**, because it has the **more substituted double bond.**

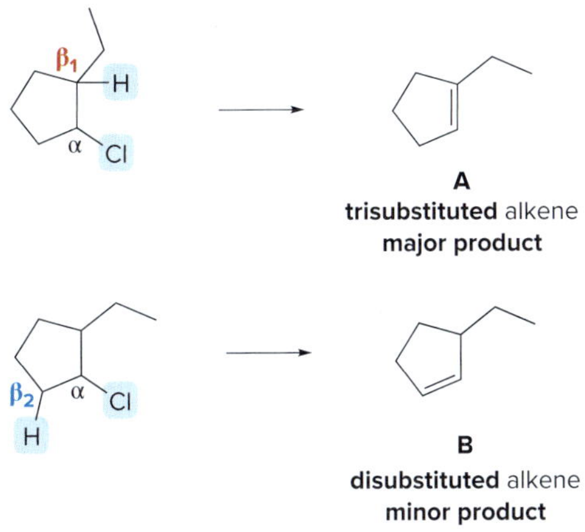

A — trisubstituted alkene, major product

B — disubstituted alkene, minor product

Problem 8.10 What alkenes are formed from each alkyl halide by an E2 reaction? Use the Zaitsev rule to predict the major product.

a. (2-bromo-3-methylpentane) b. (1-bromo-2-methylcyclopentane) c. (2-chloroheptane derivative) d. (1-chloro-2,2-dimethylcyclopentane)

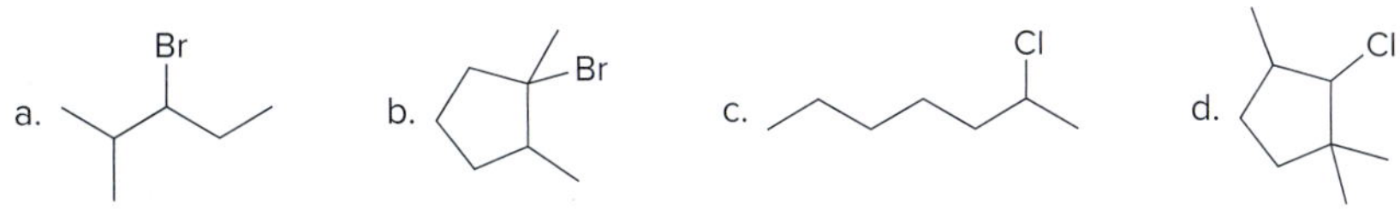

More Practice: Try Problems 8.30, 8.33.

8.6 The E1 Mechanism

The dehydrohalogenation of $(CH_3)_3CI$ with H_2O to form $(CH_3)_2C=CH_2$ can be used to illustrate the second general mechanism of elimination, the **E1 mechanism.**

8.6A Kinetics

An E1 reaction exhibits **first-order kinetics.**

- rate = $k[(CH_3)_3CI]$

Like the S_N1 mechanism, the kinetics suggest that the reaction mechanism involves more than one step, and that the slow step is **unimolecular,** involving *only* the alkyl halide.

8.6B A Two-Step Mechanism

The most straightforward explanation for the observed first-order kinetics is a **two-step reaction: the bond to the leaving group breaks first *before* the π bond is formed,** as shown in Mechanism 8.2.

Mechanism 8.2 The E1 Mechanism

1. Heterolysis of the C—I bond forms a **carbocation** in the rate-determining step.
2. A base (either H_2O or I^-) removes a proton from a carbon adjacent to the carbocation, and the electron pair in the C—H bond forms the π bond.

The E1 and E2 mechanisms both involve the same number of bonds broken and formed. **The only difference is the timing.**

- In an E1 reaction, the leaving group comes off *before* the β proton is removed, and the reaction occurs in *two* steps.
- In an E2 reaction, the leaving group comes off *as* the β proton is removed, and the reaction occurs in *one* step.

An energy diagram for the reaction of $(CH_3)_3CI + H_2O$ is shown in Figure 8.4. Each step has its own energy barrier, with a transition state at each energy maximum. Because its transition state is higher in energy, **Step [1] is rate-determining.** $\Delta H°$ for Step [1] is positive because only bond breaking occurs, whereas $\Delta H°$ of Step [2] is negative because two bonds are formed and only one is broken.

Figure 8.4

Energy diagram for an E1 reaction:
$(CH_3)_3CI + H_2O \rightarrow (CH_3)_2C=CH_2 + H_3O^+ + I^-$

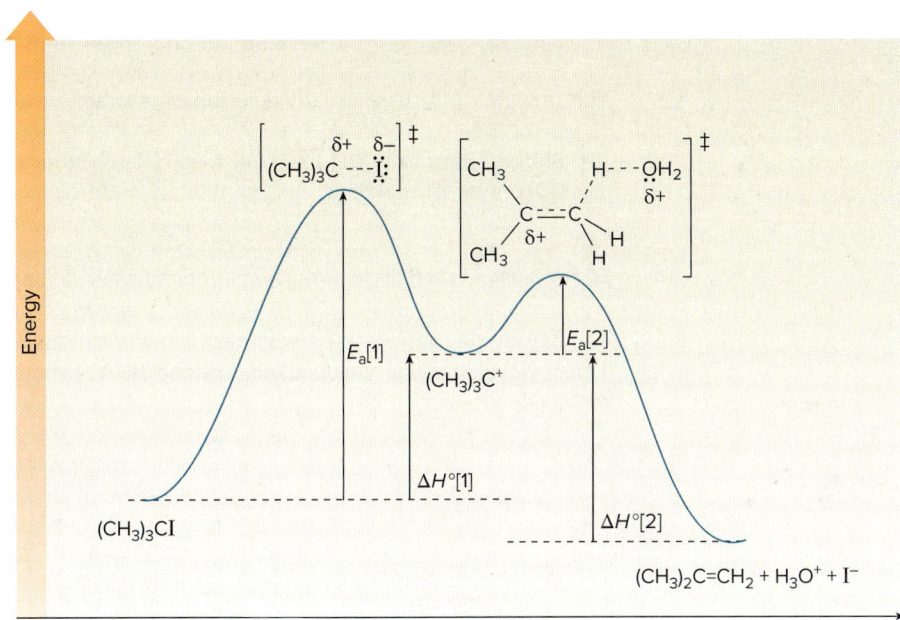

- The E1 mechanism has **two steps,** so there are two energy barriers.
- **Step [1] is rate-determining.**

Problem 8.11 Draw an E1 mechanism for the following reaction. Draw the structure of the transition state for each step.

$$\underset{Cl}{\text{(CH}_3)_3\text{C-C(CH}_3)_2\text{)}} + CH_3OH \longrightarrow \text{(CH}_3)_2C=C(CH_3)_2 + CH_3\overset{+}{O}H_2 + Cl^-$$

8.6C Other Characteristics of E1 Reactions

Three other features of E1 reactions are worthy of note.

[1] The rate of an E1 reaction *increases* as the number of R groups on the carbon with the leaving group *increases*.

Increasing alkyl substitution has the same effect on the rate of *both* an E1 and E2 reaction; increasing rate of the E1 and E2 reactions: RCH_2X (1°) < R_2CHX (2°) < R_3CX (3°).

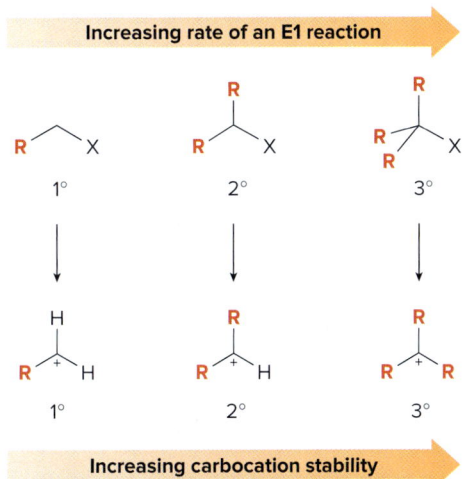

Like an S_N1 reaction, more substituted alkyl halides yield more substituted (and more stable) carbocations in the rate-determining step. **Increasing the stability of a carbocation,** in turn, decreases E_a for the slow step, which **increases the rate of the E1 reaction** according to the Hammond postulate.

[2] Because the base does not appear in the rate equation, *weak bases* favor E1 reactions.

The strength of the base usually determines whether a reaction follows the E1 or E2 mechanism.

- *Strong* bases like ⁻OH and ⁻OR favor E2 reactions, whereas *weaker* bases like H_2O and ROH favor E1 reactions.

[3] E1 reactions are regioselective, favoring formation of the more substituted, more stable alkene.

The Zaitsev rule applies to E1 reactions, too. For example, E1 elimination of HBr from 1-bromo-1-methylcyclopentane yields alkenes **A** and **B**. **A**, having the more substituted double bond, is the major product.

1-bromo-1-methyl-cyclopentane → **A** trisubstituted alkene, major product + **B** disubstituted alkene, minor product

Table 8.3 summarizes the characteristics of E1 reactions.

Table 8.3 Characteristics of the E1 Mechanism

Characteristic	Result
Kinetics	• First order
Mechanism	• Two steps
Identity of R	• More substituted halides react faster. • Rate: $R_3CX > R_2CHX > RCH_2X$
Base	• Favored by **weaker bases** such as H_2O and ROH
Leaving group	• A **better leaving group** makes the reaction faster because the bond to the leaving group is partially broken in the rate-determining step.
Solvent	• **Polar protic solvents** that solvate the ionic intermediates are needed.

Phosphates act as leaving groups not only in the substitution reactions described in Section 7.16, but also in biological elimination reactions. For example, elimination of H and PO_4^{3-} from **X** forms alkene **Y** by an enzyme-catalyzed reaction that resembles an E1 mechanism.

X → (− HPO_4^{2-}) → **Y** ← new π bond

Problem 8.12 What alkenes are formed from each alkyl halide by an E1 reaction? Use the Zaitsev rule to predict the major product.

a. (2-chloro-2-methylbutane) b. (1-iodo-1-methyl-2-ethylcyclopentane)

Problem 8.13 How does each of the following changes affect the rate of an E1 reaction?

a. doubling [RX]
b. doubling [B:]
c. changing the halide from $(CH_3)_3CBr$ to $CH_3CH_2CH_2Br$
d. changing the leaving group from Cl^- to Br^-
e. changing the solvent from DMSO to CH_3OH

8.7 S$_N$1 and E1 Reactions

S$_N$1 and E1 reactions have exactly the same first step—formation of a carbocation. They differ in what happens to the carbocation.

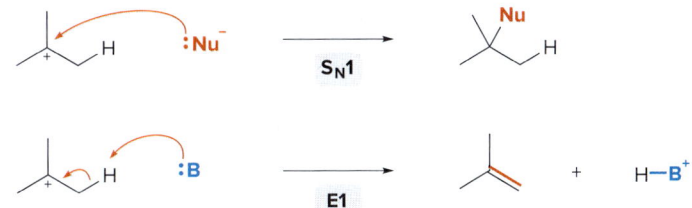

- In an S$_N$1 reaction, a nucleophile attacks the carbocation, forming a substitution product.
- In an E1 reaction, a base removes a proton, forming a new π bond.

The same conditions that favor substitution by an S$_N$1 mechanism also favor elimination by an E1 mechanism: **a 3° alkyl halide as substrate, a weak nucleophile or base as reagent, and a polar protic solvent.** As a result, both reactions usually occur in the same reaction mixture to afford a mixture of products, as illustrated in Sample Problem 8.2.

Sample Problem 8.2 — Drawing the S$_N$1 and E1 Products in a Reaction

Draw the S$_N$1 and E1 products formed in the reaction of (CH$_3$)$_3$CBr with H$_2$O.

Solution

The first step in both reactions is heterolysis of the C–Br bond to form a **carbocation**.

Reaction of the carbocation with H$_2$O as a nucleophile affords the substitution product (Reaction [1]). Alternatively, H$_2$O acts as a base to remove a proton, affording the elimination product (Reaction [2]). **Two products are formed.**

Problem 8.14

Draw both the S$_N$1 and E1 products of each reaction.

More Practice: Try Problems 8.50b, c, h; 8.51a; 8.53a; 8.55.

Because E1 reactions often occur with a competing S$_N$1 reaction, **E1 reactions of alkyl halides are *much less useful* than E2 reactions.**

8.8 Stereochemistry of the E2 Reaction

The transition state of the E2 reaction consists of four atoms that react at the same time, and they react only if they possess a particular stereochemical arrangement.

8.8A General Stereochemical Features

The transition state of an E2 reaction consists of **four atoms** from the alkyl halide—one hydrogen atom, two carbon atoms, and the leaving group (X)—**all aligned in a plane.** There are two ways for the C–H and C–X bonds to be coplanar:

syn periplanar anti periplanar

The dihedral angle for the C–H and C–X bonds equals **0°** for the syn periplanar arrangement and **180°** for the anti periplanar arrangement.

- The H and X atoms can be oriented on the same side of the molecule. This geometry is called *syn periplanar*.
- The H and X atoms can be oriented on opposite sides of the molecule. This geometry is called *anti periplanar*.

All evidence suggests that **E2 elimination occurs most often in the anti periplanar geometry.** This arrangement allows the molecule to react in the lower-energy *staggered* conformation. It also allows two electron-rich species, the incoming base and the departing leaving group, to be farther away from each other, as illustrated in Figure 8.5.

Figure 8.5
Two possible geometries for the E2 reaction

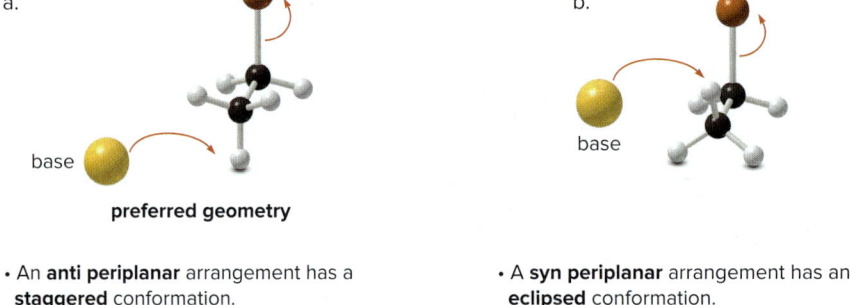

- An **anti periplanar** arrangement has a **staggered** conformation.
- The two electron-rich groups are far apart.

- A **syn periplanar** arrangement has an **eclipsed** conformation.
- The two electron-rich groups are close.

Anti periplanar geometry is the preferred arrangement for any alkyl halide undergoing E2 elimination, regardless of whether it is cyclic or acyclic. This stereochemical requirement has important consequences for compounds containing six-membered rings.

Problem 8.15 Given that an E2 reaction proceeds with anti periplanar stereochemistry, draw the products of each elimination. The alkyl halides in (a) and (b) are diastereomers of each other. How are the products of these two reactions related? Recall from Section 3.2A that C_6H_5- is a phenyl group, a benzene ring bonded to another group.

8.8B Anti Periplanar Geometry and Halocyclohexanes

Recall from Section 4.13 that cyclohexane exists as two chair conformations that rapidly interconvert, and that substituted cyclohexanes are more stable with substituents in the roomier

equatorial position. Chlorocyclohexane exists as two chair conformations, but **X** is preferred because the Cl group is equatorial.

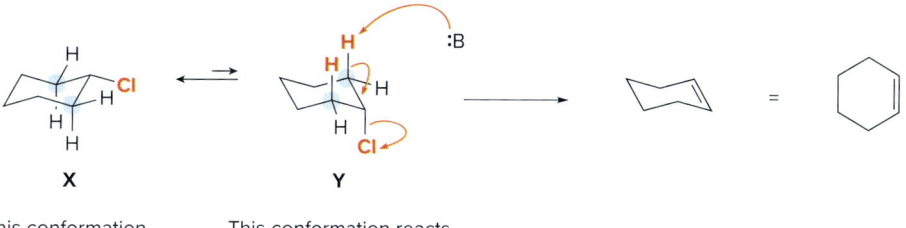

more stable
X

less stable
Y

chlorocyclohexane

For E2 elimination, **the C–Cl bond must be anti periplanar to a C–H bond on a β carbon,** and this occurs only when the H and Cl atoms are both in the **axial** position. This requirement for **trans diaxial geometry** means that E2 elimination must occur from the *less* stable conformation **Y**, as shown in Figure 8.6.

Figure 8.6

The trans diaxial geometry for the E2 elimination in chlorocyclohexane

X — This conformation does *not* react.

Y — This conformation reacts.

- In conformation **X** (**equatorial** Cl group), a β C–H bond and a C–Cl bond are *never* anti periplanar; therefore, **no E2 elimination can occur.** β Carbons are highlighted in blue.
- In conformation **Y** (**axial** Cl group), two β C–H bonds and the C–Cl bond are **trans diaxial**; therefore, **E2 elimination occurs.** Axial H's on β carbons that can react are shown in red.

cis-1-chloro-2-methyl-cyclohexane

trans-1-chloro-2-methyl-cyclohexane

Sometimes this rigid stereochemical requirement affects the regioselectivity of the E2 reaction of substituted cyclohexanes. Dehydrohalogenation of *cis*- and *trans*-1-chloro-2-methylcyclohexane via an E2 mechanism illustrates this phenomenon.

The **cis isomer** exists as two conformations (**A** and **B**), each of which has one group axial and one group equatorial. E2 reaction must occur from conformation **B**, which contains an **axial** Cl atom.

cis isomer
A

cis isomer
B
This conformation reacts.

Because conformation **B** has two different axial β H atoms, labeled H_a and H_b, E2 reaction occurs in two different directions to afford two alkenes. **The major product contains the more stable trisubstituted double bond, as predicted by the Zaitsev rule.**

$[-H_aCl]$ → disubstituted alkene
minor product

$[-H_bCl]$ → trisubstituted alkene
major product

The **trans isomer** exists as two conformations, **C**, having two equatorial substituents, and **D**, having two axial substituents. E2 reaction must occur from conformation **D**, which contains an **axial** Cl atom.

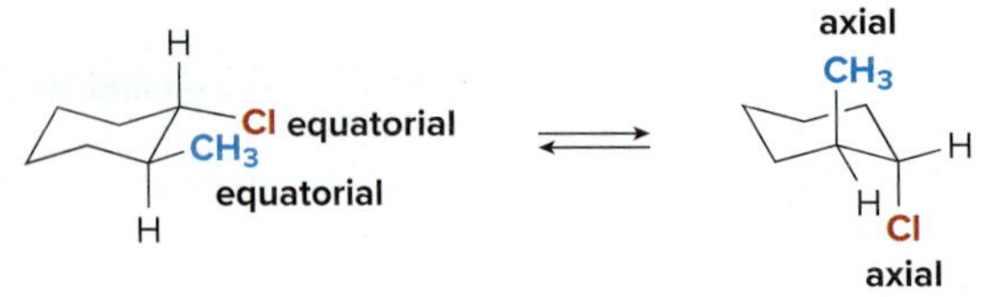

trans isomer
C

trans isomer
D
This conformation reacts.

Because conformation **D** has **only one axial β H,** E2 reaction occurs in only *one* direction to afford a **single product,** having the disubstituted double bond. This is *not* predicted by the Zaitsev rule. **E2 reaction requires H and Cl to be trans and diaxial,** and with the trans isomer, this is possible only when the *less* stable alkene is formed as product.

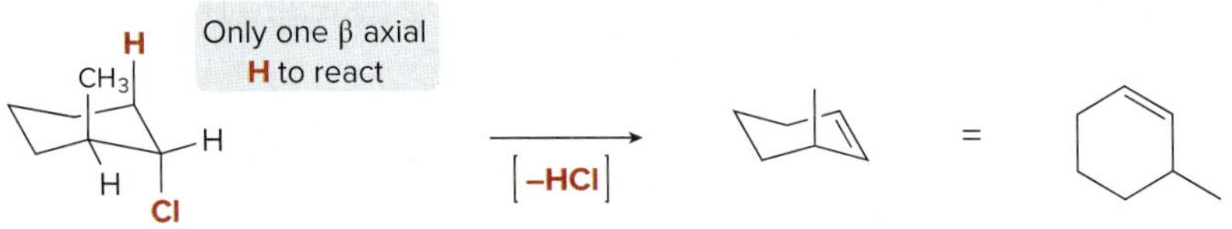

disubstituted alkene
only product

- With substituted cyclohexanes, E2 elimination must occur with a *trans diaxial* arrangement of H and X, and as a result of this requirement, the more substituted alkene is not necessarily the major product.

Sample Problem 8.3 Drawing an E2 Product from a Halocyclohexane

Draw the major E2 elimination product formed from the following alkyl halide.

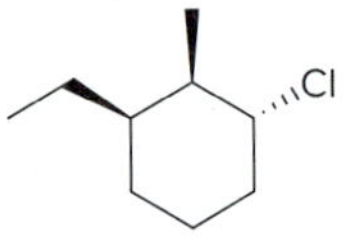

Solution
To draw the elimination products, locate the β carbons and **look for H atoms that are *trans* to the leaving group.** The given alkyl chloride has two different β carbons, labeled β₁ and β₂. Elimination can occur only when the leaving group (Cl) and a H atom on the β carbon are *trans*.

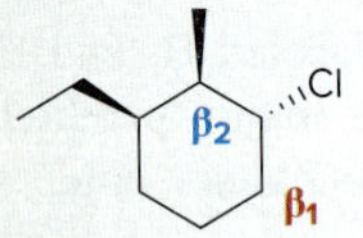

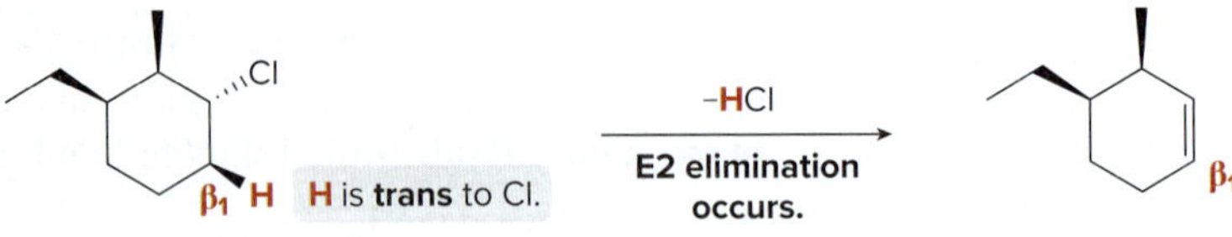

H is **trans** to Cl.

−HCl
E2 elimination
occurs.

disubstituted alkene
only product

H is **cis** to Cl.

E2 elimination
cannot occur.

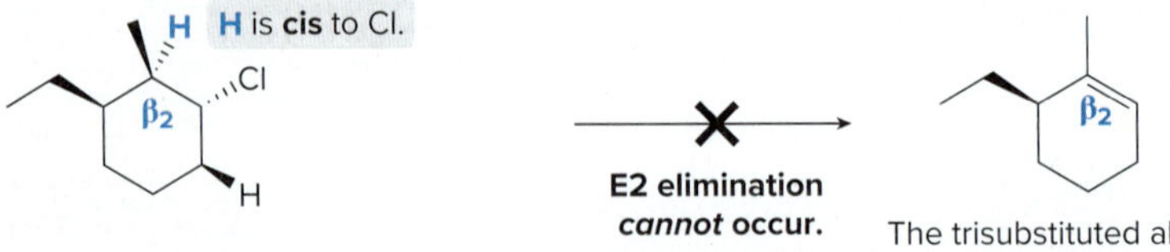

The trisubstituted alkene is **not** formed.

8.9 When Is the Mechanism E1 or E2?

The β₁ C has a H atom **trans** to Cl, so E2 elimination occurs to form a disubstituted alkene. Because there is no trans H on the β₂ C, E2 elimination **cannot** occur in this direction, and the more stable trisubstituted alkene is *not* formed. Although this result is not predicted by the Zaitsev rule, it is consistent with the requirement that the **H and X atoms in an E2 elimination must be located trans to each other.**

Problem 8.16 Draw the major E2 elimination products from each of the following alkyl halides.

a. [cyclohexane with methyl, isopropyl, and Cl substituents] + ⁻OH →

b. [cyclohexane with methyl, isopropyl, and Cl substituents] + ⁻OH →

More Practice: Try Problems 8.33, 8.38, 8.39, 8.41, 8.42, 8.44.

Problem 8.17 Explain why *cis*-1-chloro-2-methylcyclohexane undergoes E2 elimination much faster than its trans isomer.

8.9 When Is the Mechanism E1 or E2?

Given a particular starting material and base, how do we know whether a reaction occurs by the E1 or E2 mechanism?

Because the rate of *both* the E1 and E2 reactions increases as the number of R groups on the carbon with the leaving group increases, **you cannot use the identity of the alkyl halide to decide which elimination mechanism occurs.**

- **The strength of the base is the most important factor in determining the mechanism for elimination. Strong bases favor the E2 mechanism. Weak bases favor the E1 mechanism.**

Table 8.4 compares the E1 and E2 mechanisms.

Table 8.4 A Comparison of the E1 and E2 Mechanisms

Mechanism	Comment
E2 mechanism	• Much more common and useful • Favored by **strong, negatively charged bases,** especially ⁻OH and ⁻OR • The reaction occurs with 1°, 2°, and 3° alkyl halides. Order of reactivity: **R₃CX > R₂CHX > RCH₂X.**
E1 mechanism	• Much less useful because a mixture of S_N1 and E1 products usually results • Favored by **weaker, neutral bases,** such as H₂O and ROH • This mechanism does not occur with 1° RX because they form highly unstable 1° carbocations.

Problem 8.18 Which mechanism, E1 or E2, will occur in each reaction?

a. (CH₃)₃CCl + ⁻OCH₃ →

b. cyclohexyl-CH(I)-isopropyl + H₂O →

c. 1-chloro-1-isopropylcyclohexane + CH₃OH →

d. cyclohexyl-CH(CH₃)-CH₂Br + ⁻OC(CH₃)₃ →

8.10 E2 Reactions and Alkyne Synthesis

Recall from Section 1.10C that the carbon–carbon triple bond of alkynes consists of one σ and two π bonds.

A single elimination reaction produces the π bond of an alkene. **Two consecutive elimination reactions produce the two π bonds of an alkyne.**

alkene
one π bond
One elimination reaction is needed.

alkyne
two π bonds
Two elimination reactions are needed.

- Alkynes are prepared by two successive dehydrohalogenation reactions.

Two elimination reactions are needed to remove two moles of HX from a **dihalide** as substrate. Two different starting materials can be used.

vicinal dihalide   geminal dihalide

The word *geminal* comes from the Latin *geminus*, meaning "twin."

- A vicinal dihalide has two X atoms on *adjacent* carbon atoms.
- A geminal dihalide has two X atoms on the *same* carbon atom.

Equations [1] and [2] illustrate how two moles of HX can be removed from these dihalides with base. Two equivalents of strong base are used and each step follows an **E2 mechanism.**

[1] vicinal dihalide → vinyl halide → R—≡—R

[2] geminal dihalide → vinyl halide → R—≡—R

The relative strength of C–H bonds depends on the hybridization of the carbon atom: $sp > sp^2 > sp^3$. For more information, review Section 1.11B.

Stronger bases are needed to synthesize alkynes by dehydrohalogenation than are needed to synthesize alkenes. The typical base is **amide ($^-NH_2$),** used as the sodium salt **NaNH$_2$** (sodium amide). KOC(CH$_3$)$_3$ can also be used with DMSO as solvent. Because DMSO is a polar aprotic solvent, the anionic base is not well solvated, thus **increasing its basicity** and making it strong enough to remove two equivalents of HX. Examples are given in Figure 8.7.

The strongly basic conditions needed for alkyne synthesis result from the difficulty of removing the second equivalent of HX from the intermediate vinyl halide, RCH=C(R)X. Because H and X are both bonded to sp^2 hybridized carbons, these bonds are shorter and stronger than the sp^3 hybridized C–H and C–X bonds of an alkyl halide, necessitating the use of a stronger base.

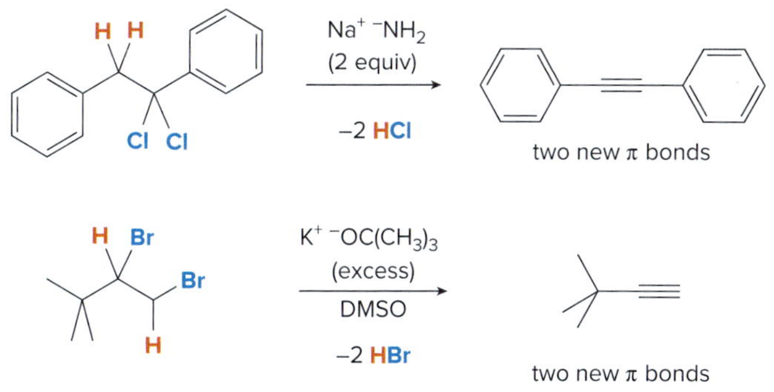

Figure 8.7

Examples of dehydrohalogenation of dihalides to afford alkynes

---

Problem 8.19 Draw the alkynes formed when each dihalide is treated with excess base.

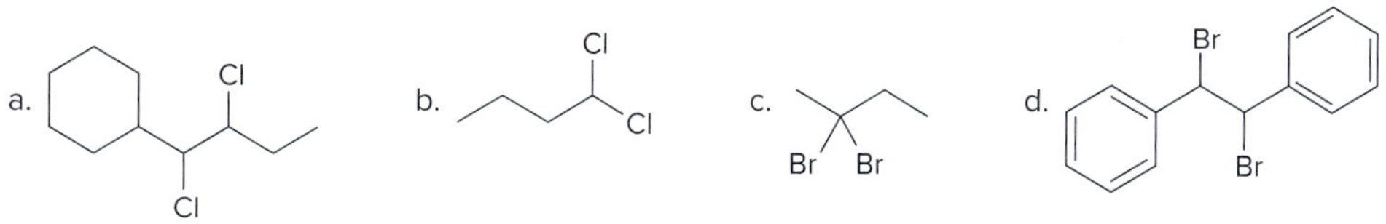

8.11 When Is the Reaction S_N1, S_N2, E1, or E2?

We have now considered two different kinds of reactions (substitution and elimination) and four different mechanisms (S_N1, S_N2, E1, and E2) that begin with one class of compounds (alkyl halides). How do we know if a given alkyl halide will undergo substitution or elimination with a given base or nucleophile, and by what mechanism?

Unfortunately, there is no easy answer, and often mixtures of products result. Two generalizations help to determine whether substitution or elimination occurs.

> **[1]** *Good* nucleophiles that are *weak* bases favor *substitution* over elimination.

Certain anions generally give products of substitution because they are good nucleophiles but weak bases. These include I^-, Br^-, HS^-, ^-CN, and $CH_3CO_2^-$.

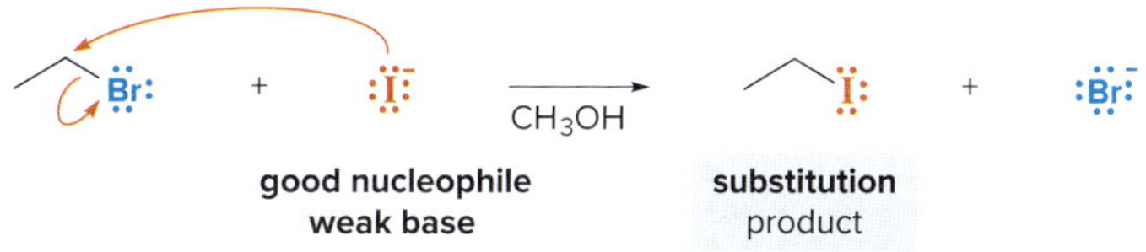

> **[2]** *Bulky*, nonnucleophilic bases favor *elimination* over substitution.

$KOC(CH_3)_3$, DBU, and **DBN** are too sterically hindered to attack a tetravalent carbon, but are able to remove a small proton, favoring elimination over substitution.

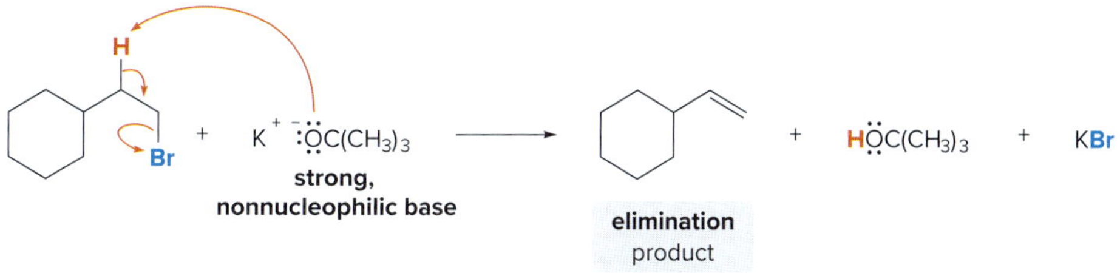

Most often, however, we will have to rely on other criteria to predict the outcome of these reactions. To determine the product of a reaction with an alkyl halide:

[1] Classify the alkyl halide as 1°, 2°, or 3°.
[2] Classify the base or nucleophile as strong, weak, or bulky.

Predicting the substitution and elimination products of a reaction can then be organized by the type of alkyl halide, as summarized in Table 8.5. The explanation that follows the table is organized with 2° alkyl halides last, because their reactions can follow any of the four mechanisms and product mixtures often result.

Table 8.5 Summary of Alkyl Halides and S_N1, S_N2, E1, and E2 Mechanisms

Alkyl halide type	Reaction with	Mechanism
1° RCH_2X	• Strong nucleophile	S_N2
	• Strong bulky base	E2
2° R_2CHX	• Strong base and nucleophile	S_N2 and E2
	• Strong bulky base	E2
	• Weak base and nucleophile	S_N1 and E1
3° R_3CX	• Weak base and nucleophile	S_N1 and E1
	• Strong base	E2

8.11A Tertiary Alkyl Halides

Tertiary alkyl halides react by all mechanisms *except* S_N2.

- With strong bases, elimination occurs by an E2 mechanism.

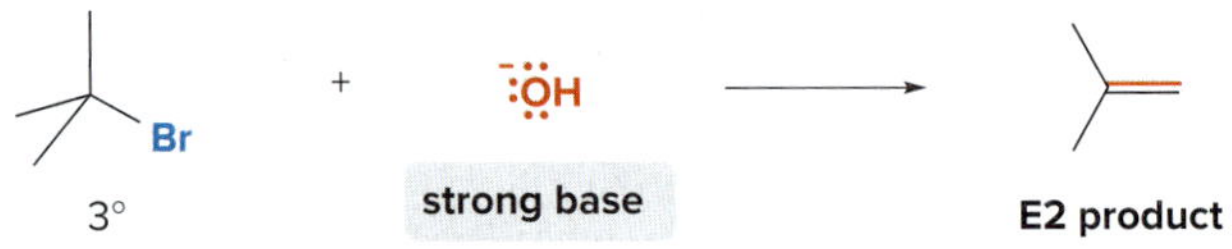

A strong base or nucleophile favors an S_N2 or E2 mechanism, but 3° halides are too sterically hindered to undergo an S_N2 reaction, so only E2 elimination occurs.

- With weak nucleophiles or bases, a mixture of S_N1 and E1 products results.

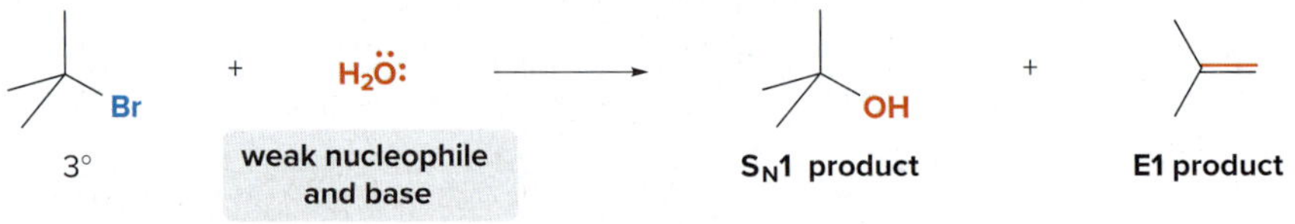

A weak base or nucleophile favors S_N1 and E1 mechanisms and both occur.

8.11B Primary Alkyl Halides

Primary alkyl halides react by S_N2 and E2 mechanisms.

- With strong nucleophiles, substitution occurs by an S_N2 mechanism.

A strong base or nucleophile favors S_N2 or E2, but 1° halides are the *least* reactive halide type in elimination, so only S_N2 reaction occurs.

- With strong, bulky bases, elimination occurs by an E2 mechanism.

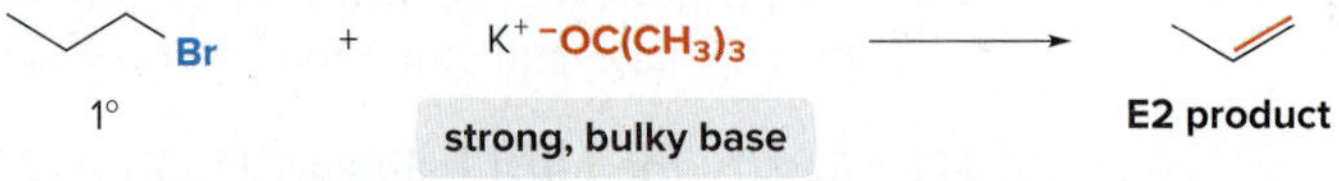

A strong, bulky base cannot act as a nucleophile, so elimination occurs and the mechanism is E2.

8.11C Secondary Alkyl Halides

Secondary alkyl halides react by *all* mechanisms.

- With strong bases and nucleophiles, a mixture of S$_N$2 and E2 products results.

A strong base that is also a strong nucleophile gives a mixture of S$_N$2 and E2 products.

- With strong, bulky bases, elimination occurs by an E2 mechanism.

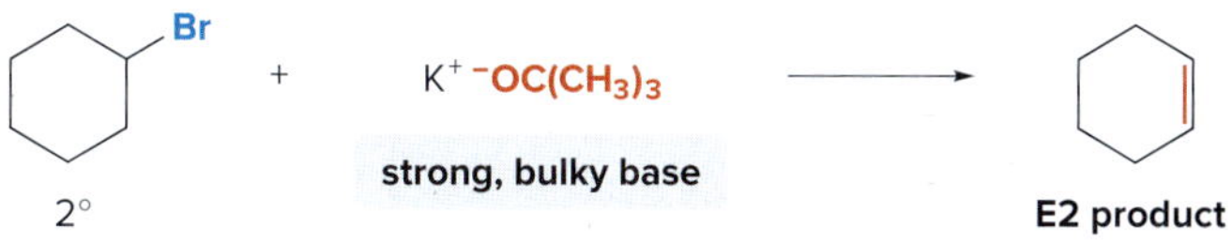

A strong, bulky base cannot act as a nucleophile, so elimination occurs and the mechanism is E2.

- With weak nucleophiles or bases, a mixture of S$_N$1 and E1 products results.

A weak base or nucleophile favors S$_N$1 and E1 mechanisms and both occur.

Sample Problems 8.4–8.6 illustrate how to apply the information in Table 8.5 to specific alkyl halides.

Sample Problem 8.4   Determining the Substitution and Elimination Products from an Alkyl Halide

Draw the products of the following reaction.

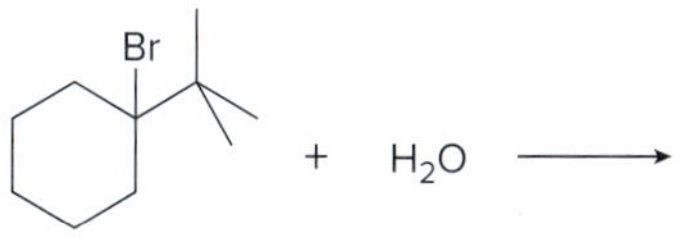

Solution

- **Classify the halide as 1°, 2°, or 3° and the reagent as a strong or weak base (and nucleophile)** to determine the mechanism. In this case, the alkyl halide is 3° and the reagent (H$_2$O) is a weak base and nucleophile, so products of both **S$_N$1** and **E1** mechanisms are formed.
- To draw the S$_N$1 product, **substitute the nucleophile (H$_2$O) for the leaving group (Br⁻)**, and draw the neutral product after loss of a proton.
- To draw the E1 product, **remove the elements of H and Br** from the α and β carbons. There are two identical β C atoms with H atoms, so only one elimination product is possible.

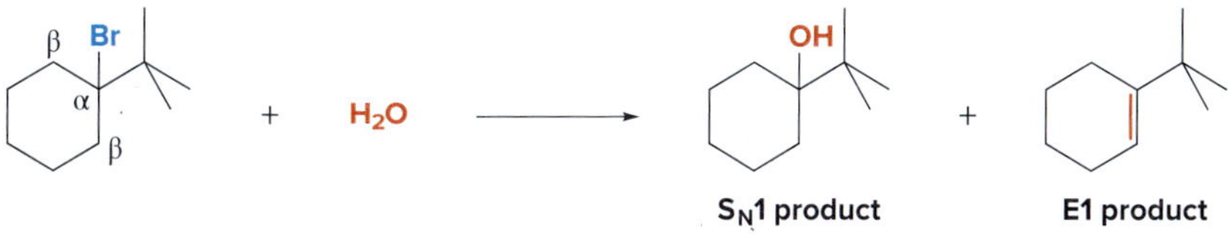

326 Chapter 8 Alkyl Halides and Elimination Reactions

Sample Problem 8.5 Drawing Substitution and Elimination Products from a 2° Alkyl Halide

Draw the products of the following reaction.

cyclopentyl-Br + CH$_3$O$^-$ $\xrightarrow{\text{CH}_3\text{OH}}$

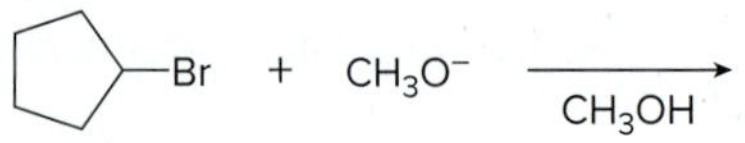

Solution

- **Classify the halide as 1°, 2°, or 3° and the reagent as a strong or weak base (and nucleophile)** to determine the mechanism. In this case, the alkyl halide is 2° and the reagent (CH$_3$O$^-$) is a strong base and nucleophile, so products of both **S$_N$2** and **E2** mechanisms are formed.
- To draw the S$_N$2 product, **substitute the nucleophile (CH$_3$O$^-$) for the leaving group (Br$^-$)**.
- To draw the E2 product, **remove the elements of H and Br** from the α and β carbons. There are two identical β C atoms with H atoms, so only one elimination product is possible.

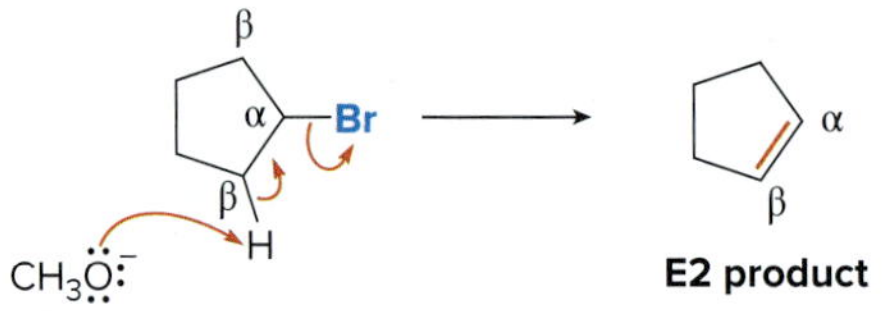

cyclopentyl-Br + CH$_3$Ö:$^-$ ⟶ cyclopentyl-ÖCH$_3$
                nucleophile   **S$_N$2 product**

(cyclopentane with β, α labels, Br, and H being removed by CH$_3$Ö:$^-$) ⟶ cyclopentene (α, β labels)
                                            E2 product

Problem 8.20 Draw the products in each reaction.

a. 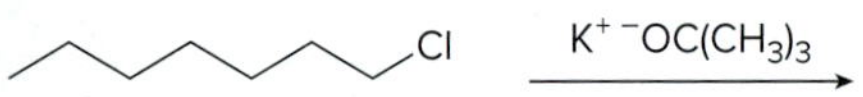 CH$_3$CH$_2$CH$_2$CH$_2$CH$_2$CH$_2$Cl $\xrightarrow{\text{K}^+ \text{ }^-\text{OC(CH}_3)_3}$

b. 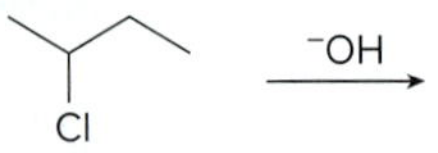 (CH$_3$)$_2$CHCl with Cl labeled $\xrightarrow{^-\text{OH}}$

c. 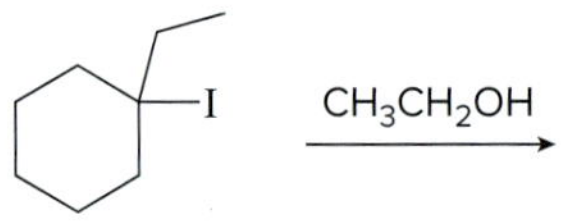 (cyclohexane with ethyl group and I) $\xrightarrow{\text{CH}_3\text{CH}_2\text{OH}}$

d. 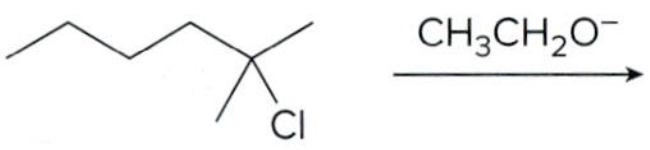 (branched alkyl chloride) $\xrightarrow{\text{CH}_3\text{CH}_2\text{O}^-}$

More Practice: Try Problems 8.48, 8.50, 8.51, 8.54.

Sample Problem 8.6 Drawing the Mechanism When a Reaction Involves Both Substitution and Elimination

Draw the products of the following reaction, and include the mechanism showing how each product is formed.

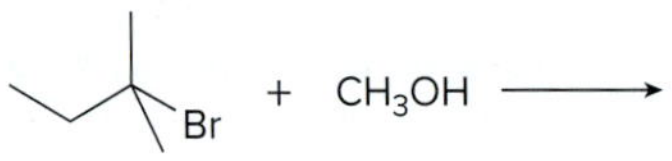

(tertiary alkyl bromide) + CH$_3$OH ⟶

Solution

[1] **Classify the halide as 1°, 2°, or 3° and the reagent as a strong or weak base (and nucleophile)** to determine the mechanism. In this case, the alkyl halide is 3° and the reagent (CH$_3$OH) is a weak base and nucleophile, so products of both **S$_N$1** and **E1** mechanisms are formed.

[2] Draw the steps of the mechanisms to give the products. Both mechanisms begin with the same first step: loss of the leaving group to form a **carbocation**.

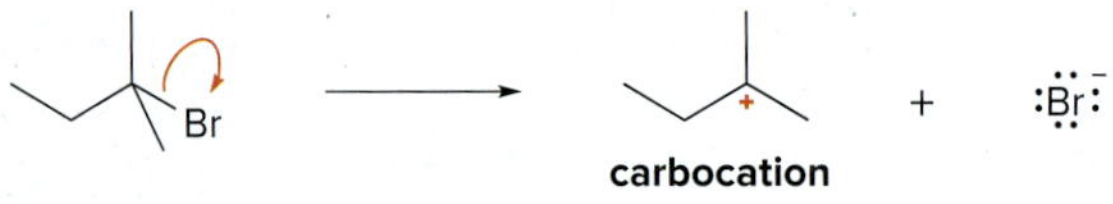

(alkyl bromide) ⟶ (carbocation) + :Br:$^-$
                      carbocation

- **For S$_N$1: The carbocation reacts with a nucleophile.** Nucleophilic attack of CH$_3$OH on the carbocation generates a positively charged intermediate that loses a proton to afford the neutral S$_N$1 product.

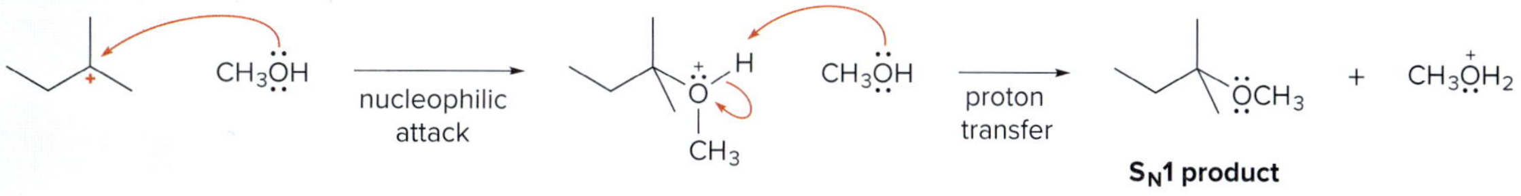

- **For E1: A base (CH$_3$OH or Br$^-$) removes a proton from the carbocation.** Two different products of elimination can form because the carbocation has two different β carbons.

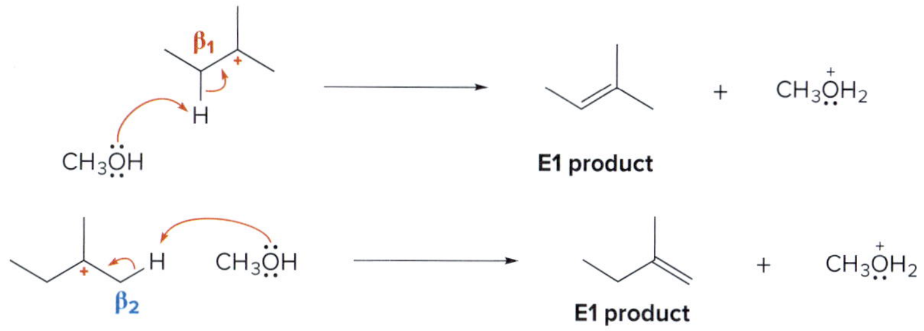

In this problem, three products are formed: one from an S$_N$1 reaction and two from E1 reactions.

Problem 8.21 Draw a stepwise mechanism for the following reaction.

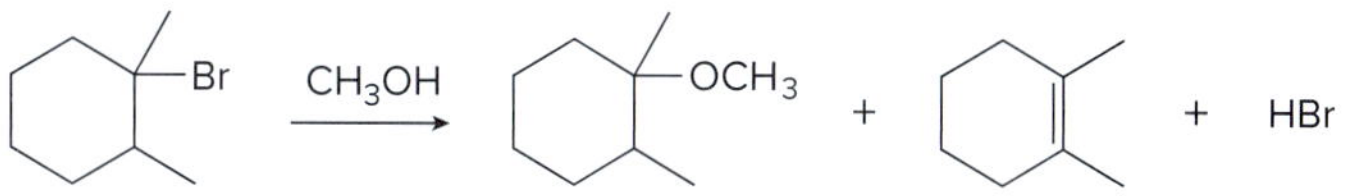

More Practice: Try Problems 8.53, 8.55.

Chapter 8 REVIEW

KEY CONCEPTS

Nucleophiles and bases in S$_N$1, S$_N$2, E1, and E2 reactions (8.11)

1. Nucleophiles that are weak bases	2. Strong, bulky bases	3. Strong nucleophiles and strong bases	4. Weak nucleophiles and weak bases
$^-$SH Br$^-$	$^-$OC(CH$_3$)$_3$	$^-$OH	H$_2$O
$^-$CN I$^-$	DBU	$^-$OR	ROH
CH$_3$CO$_2$$^-$	DBN		
• **Substitution** is favored over elimination.	• **E2** elimination is favored over substitution.	• **S$_N$2** and **E2** mechanisms are favored.	• **S$_N$1** and **E1** mechanisms are favored.

Try Problem 8.54.

KEY SKILLS

[1] Comparing the stability of alkenes (8.2)

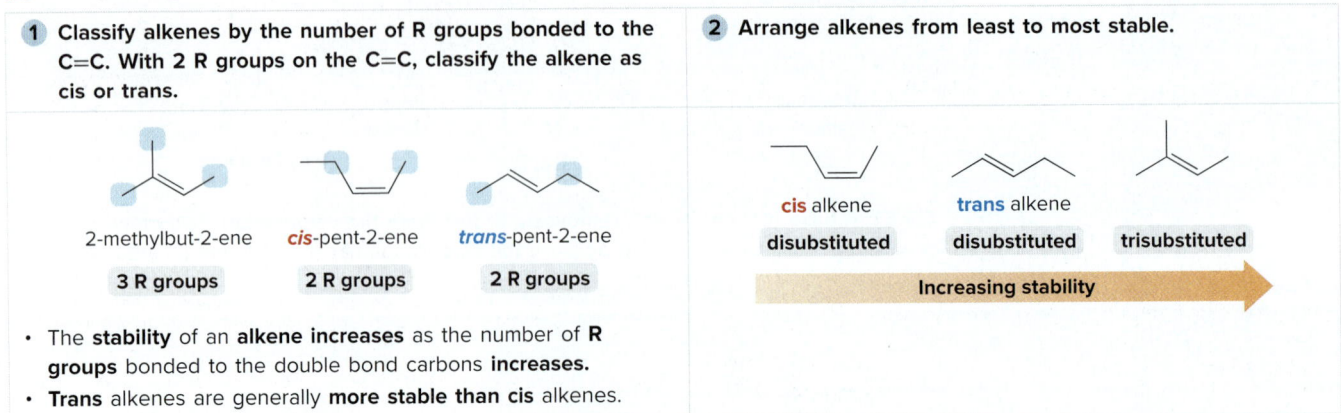

1. Classify alkenes by the number of R groups bonded to the C=C. With 2 R groups on the C=C, classify the alkene as cis or trans.

2. Arrange alkenes from least to most stable.

- The **stability** of an **alkene increases** as the number of **R groups** bonded to the double bond carbons **increases**.
- **Trans** alkenes are generally **more stable than cis** alkenes.

Try Problems 8.22, 8.28.

[2] Drawing all products and predicting the major product of an elimination reaction (8.5)

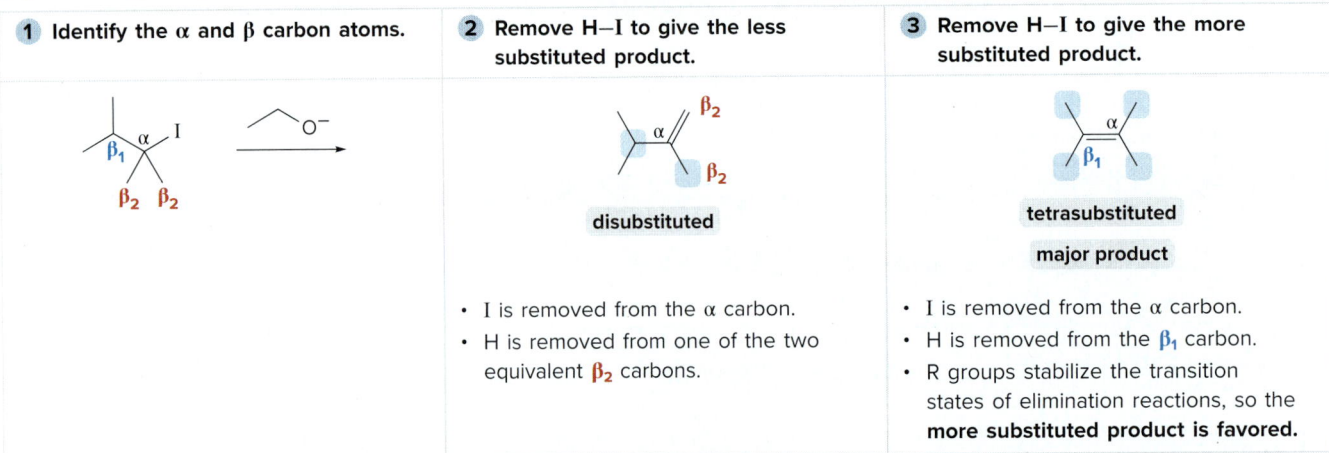

1. Identify the α and β carbon atoms.
2. Remove H–I to give the less substituted product.
3. Remove H–I to give the more substituted product.

- I is removed from the α carbon.
- H is removed from one of the two equivalent β$_2$ carbons.

- I is removed from the α carbon.
- H is removed from the β$_1$ carbon.
- R groups stabilize the transition states of elimination reactions, so the **more substituted product is favored**.

See Sample Problem 8.1. Try Problems 8.30, 8.34.

[3] Drawing the product of an E2 reaction of a halocyclohexane (8.8B)

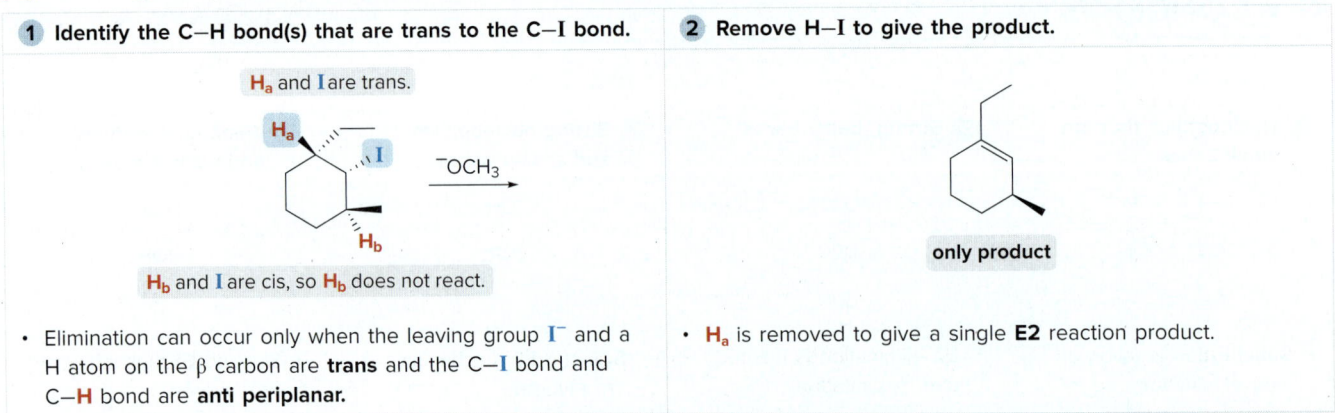

1. Identify the C–H bond(s) that are trans to the C–I bond.
2. Remove H–I to give the product.

H$_b$ and I are cis, so H$_b$ does not react.

- Elimination can occur only when the leaving group I$^-$ and a H atom on the β carbon are **trans** and the C–I bond and C–H bond are **anti periplanar**.

- H$_a$ is removed to give a single **E2** reaction product.

See Figure 8.6, Sample Problem 8.3. Try Problems 8.23, 8.39, 8.40, 8.42.

[4] Deciding if a β elimination reaction proceeds by an E1 or E2 mechanism (8.9)

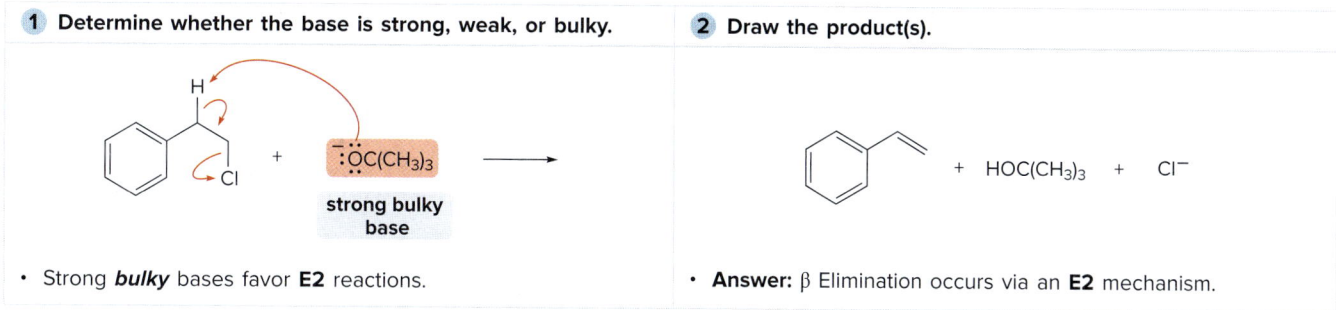

- Strong **bulky** bases favor **E2** reactions.
- **Answer:** β Elimination occurs via an **E2** mechanism.

See Table 8.4. Try Problem 8.37.

[5] Deciding if a reaction proceeds by S_N1, S_N2, E1, or E2 (8.11)

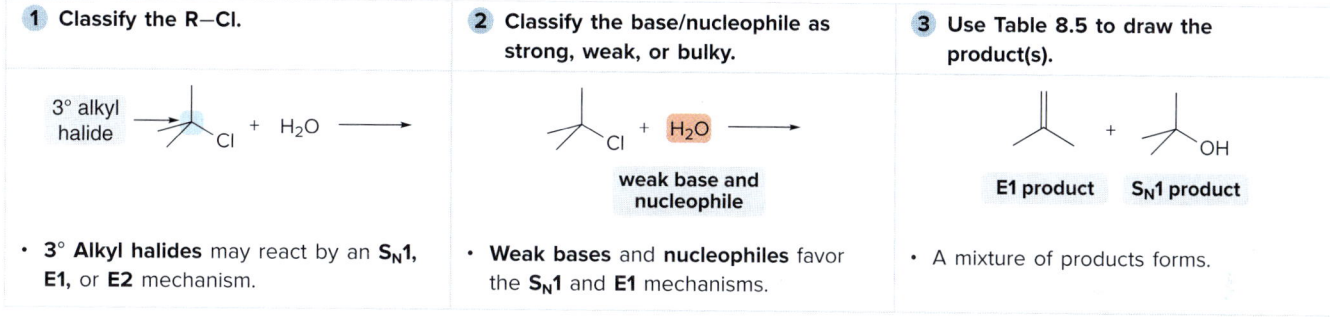

- 3° **Alkyl halides** may react by an S_N1, **E1**, or **E2** mechanism.
- **Weak bases** and **nucleophiles** favor the S_N1 and **E1** mechanisms.
- A mixture of products forms.

Try Problems 8.48, 8.50, 8.51.

[6] Drawing the product(s) of a reaction with a 1° alkyl halide (8.11B)

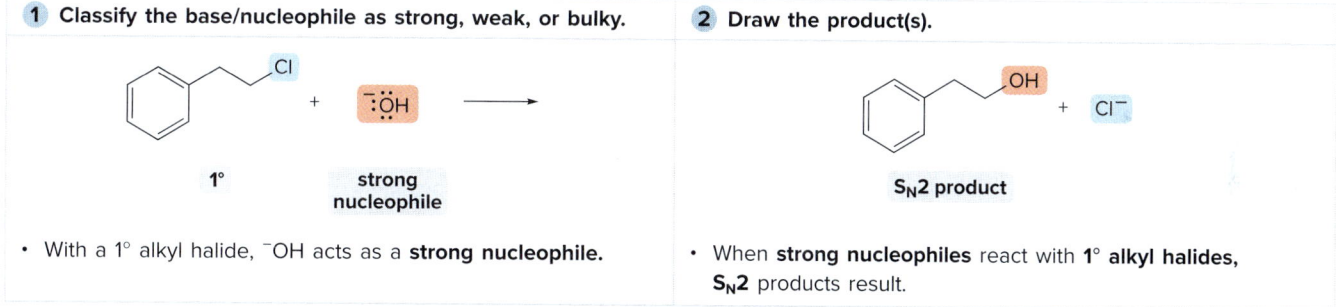

- With a 1° alkyl halide, ⁻OH acts as a **strong nucleophile**.
- When **strong nucleophiles** react with **1° alkyl halides**, S_N2 products result.

Try Problem 8.48a, b, d.

[7] Drawing the product(s) of a reaction with a 2° alkyl halide (8.11C)

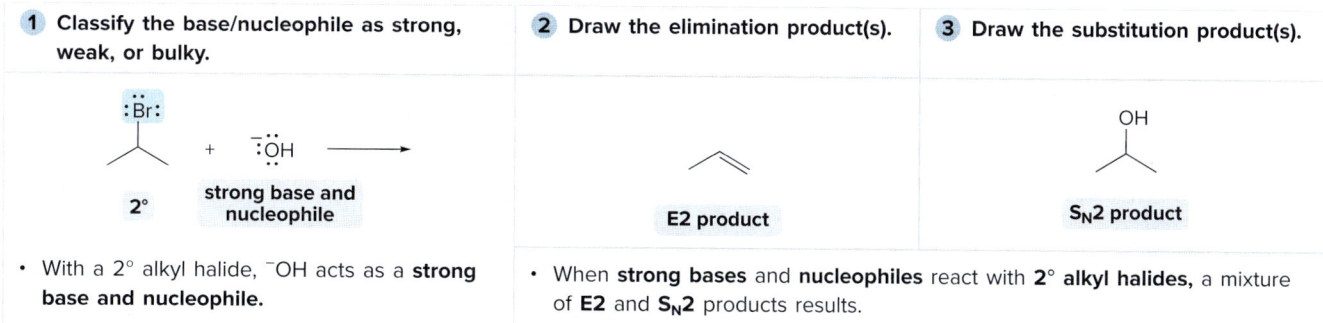

- With a 2° alkyl halide, ⁻OH acts as a **strong base and nucleophile.**
- When **strong bases and nucleophiles** react with **2° alkyl halides,** a mixture of **E2** and S_N2 products results.

See Sample Problem 8.5. Try Problems 8.48e; 8.50a, b, d, e, g, h.

KEY MECHANISM CONCEPTS

[1] Comparison of E1 and E2 reactions

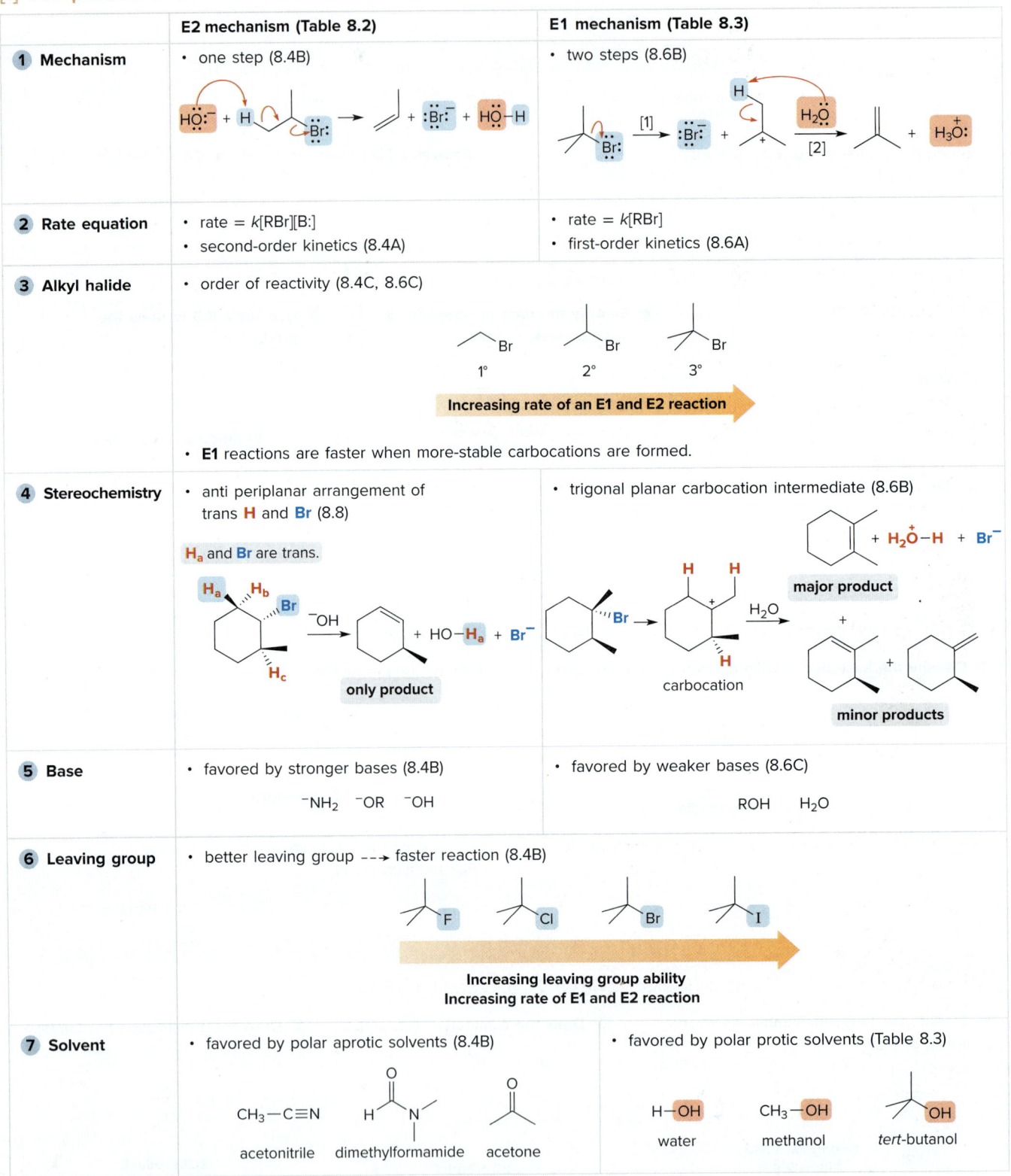

8 Product	• More substituted alkene favored in E2 and E1 reactions (Zaitsev rule, 8.5, 8.6C) 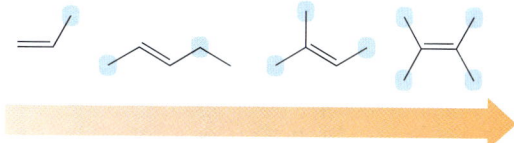

Try Problems 8.32, 8.51–8.53, 8.55.

[2] Summary of S_N1, S_N2, E1, and E2 reactions (8.11)

Alkyl halide type	Reaction with	Mechanism
1 1° RCH_2X	• **strong** nucleophile	S_N2
	• strong **bulky** base	E2
2 2° R_2CHX	• **strong** base and nucleophile	S_N2 + E2
	• strong **bulky** base	E2
	• **weak** base and nucleophile	S_N1 + E1
3 3° R_3CX	• **weak** base and nucleophile	S_N1 + E1
	• **strong** base	E2

PROBLEMS

Problems Using Three-Dimensional Models

8.22 Rank the alkenes shown in the ball-and-stick models (**A–C**) in order of increasing stability.

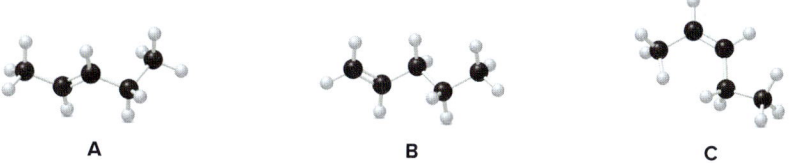

8.23 Name each compound and decide which stereoisomer will react faster in an E2 elimination reaction. Explain your choice.

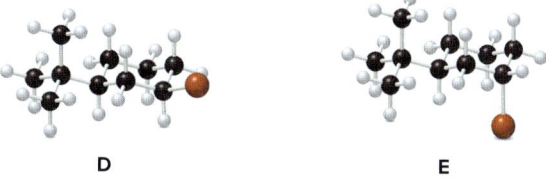

General Elimination

8.24 Draw all possible constitutional isomers formed by dehydrohalogenation of each alkyl halide.

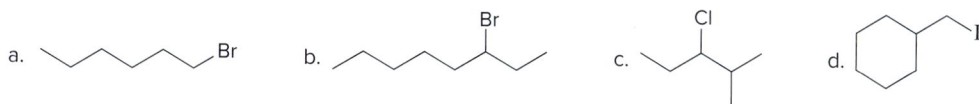

8.25 What alkyl halide forms each of the following alkenes as the *only* product in an elimination reaction?

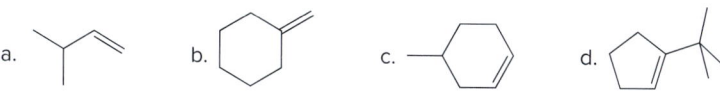

Alkenes

8.26 Which double bonds in the following natural products can exhibit stereoisomerism? Nerolidol is isolated from the angel's trumpet plant, caryophyllene is present in hemp, and humulene comes from hops.

a. 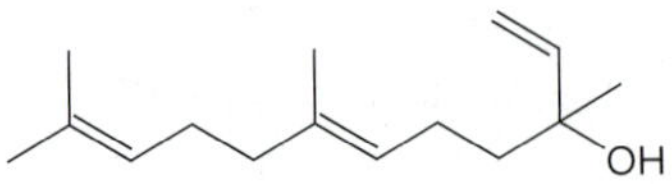 nerolidol   b. 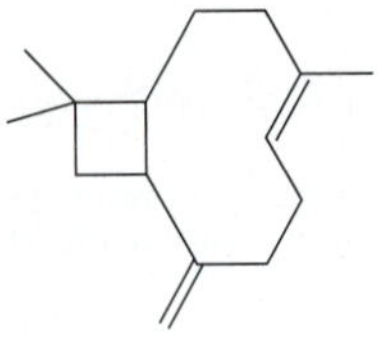 caryophyllene   c. 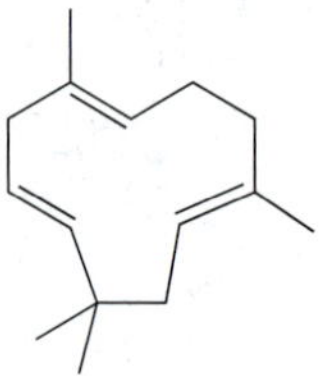 humulene

8.27 Label each pair of alkenes as constitutional isomers, stereoisomers, or identical.

a. 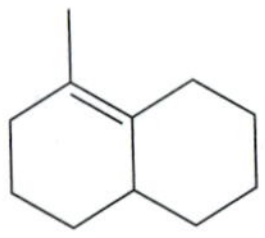   b. 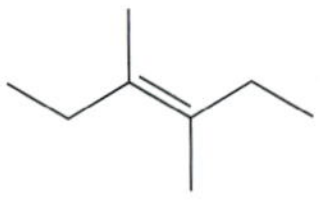   c. 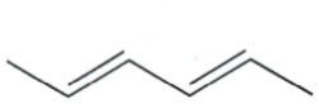   d.

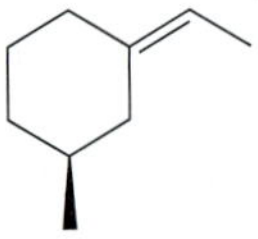

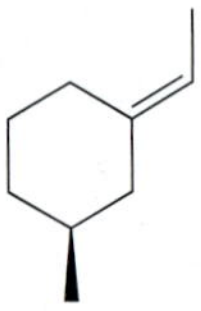

8.28 Rank the following alkenes in order of increasing stability.

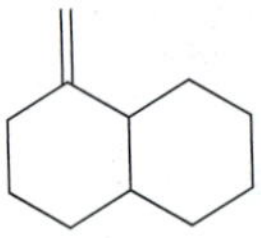

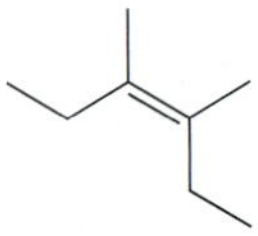

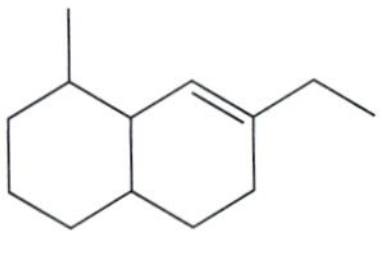

A 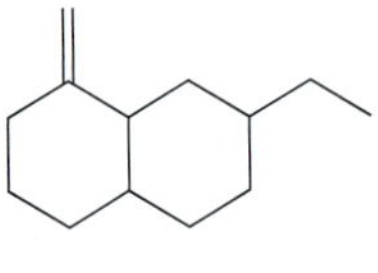   B 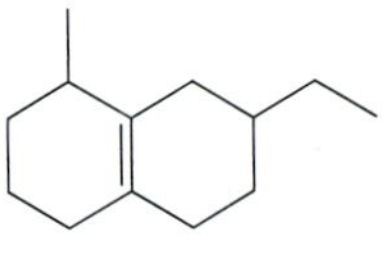   C 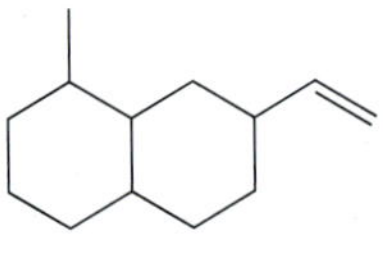   D

8.29 $\Delta H°$ values obtained for a series of similar reactions are one set of experimental data used to determine the relative stability of alkenes. Explain how the following data suggest that cis-but-2-ene is more stable than but-1-ene (Section 11.3A).

but-1-ene + H_2 ⟶ $\Delta H° = -127$ kJ/mol

cis-but-2-ene + H_2 ⟶ $\Delta H° = -120$ kJ/mol

E2 Reaction

8.30 Draw all constitutional isomers formed in each E2 reaction, and predict the major product using the Zaitsev rule.

a.  $(CH_3)_3CO^-$

b. 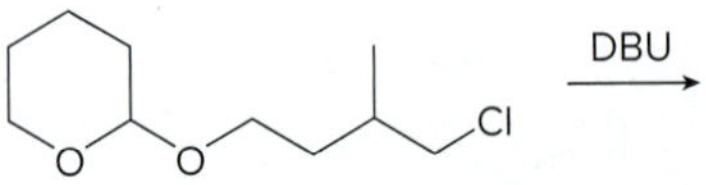 DBU

c. 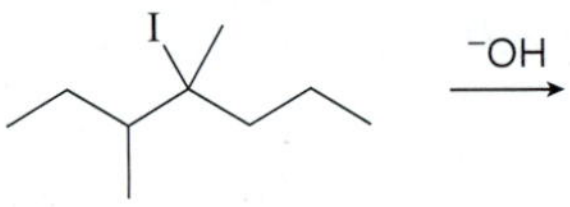 ^-OH

d. 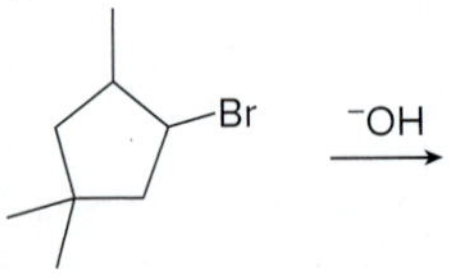 ^-OH

8.31 For each of the following alkenes, draw the structure of two different alkyl halides that yield the given alkene as the only product of dehydrohalogenation.

a. 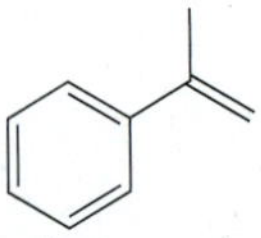   b. 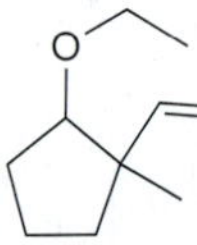   c.

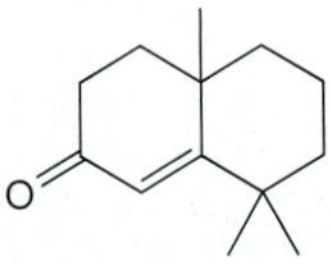

Problems

8.32 Consider the following E2 reaction.

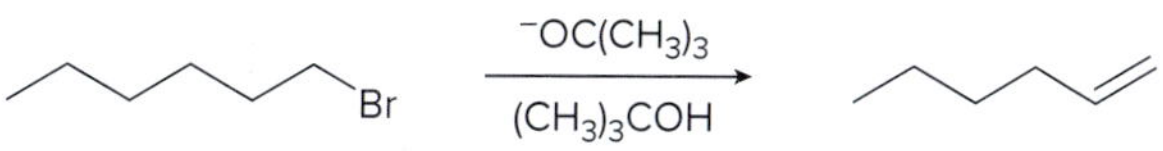

[structure: 1-bromohexane] —OC(CH₃)₃ / (CH₃)₃COH→ [structure: hex-1-ene]

a. Draw the by-products of the reaction and use curved arrows to show the movement of electrons.
b. What happens to the reaction rate with each of the following changes? [1] The solvent is changed to DMF. [2] The concentration of ⁻OC(CH₃)₃ is decreased. [3] The base is changed to ⁻OH. [4] The halide is changed to CH₃CH₂CH₂CH₂CH(Br)CH₃. [5] The leaving group is changed to I⁻.

8.33 What is the major stereoisomer formed when each alkyl halide is treated with KOC(CH₃)₃?

a. 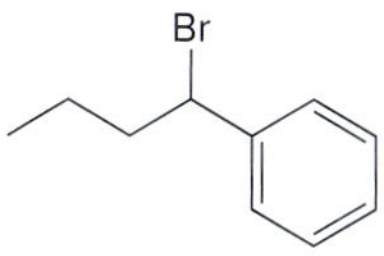 [PhCH(Br)CH₂CH₂CH₃] b. 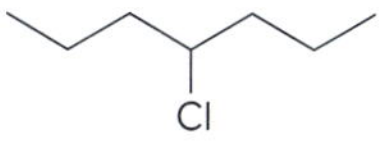 [3-chlorohexane]

E1 Reaction

8.34 What alkene is the major product formed from each alkyl halide in an E1 reaction?

a. 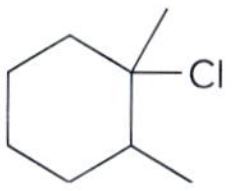 [1-chloro-2-methylcyclohexane] b.  [3-bromo-3-methylhexane] c. 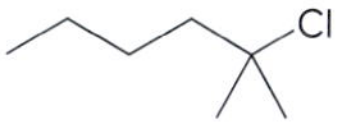 [2-chloro-2-methylpentane]

8.35 Draw a stepwise mechanism for the following reaction, which synthesizes chorismate, an intermediate in the synthesis of aromatic amino acids and folic acid. Assume the reaction follows an E1 mechanism.

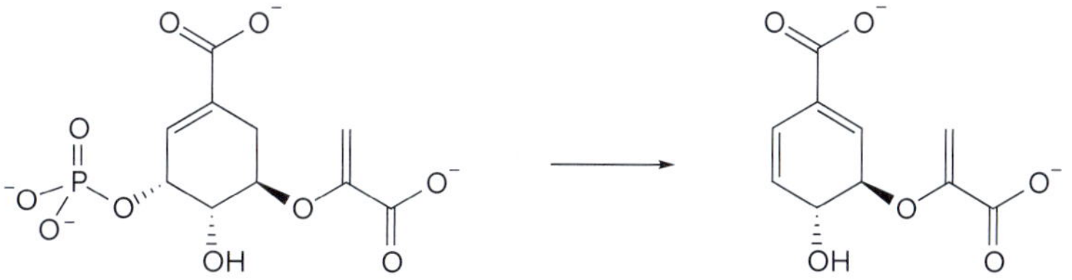

[structure with phosphate and enolpyruvate substituents] → [chorismate structure]

chorismate

E1 and E2

8.36 Rank the following alkyl halides in order of increasing reactivity in E2 elimination. Then do the same for E1 elimination.

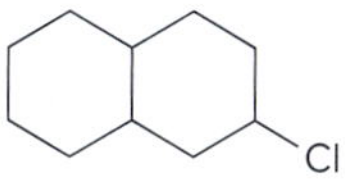

A: [decalin-2-Cl]   B: [decalin-1-Cl (bridgehead-adjacent)]   C: [decalin with Br at ring junction]   D: [decalin-CH₂Cl]

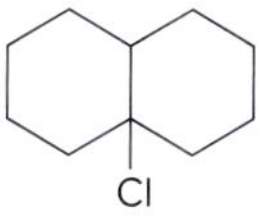

8.37 Draw all constitutional isomers formed in each elimination reaction. Label the mechanism as E2 or E1.

a.  [2-bromopropane derivative] —⁻OCH₃→
b.  [2-bromopropane derivative] —CH₃OH→
c.  [1-iodopentane] —⁻OC(CH₃)₃→
d. 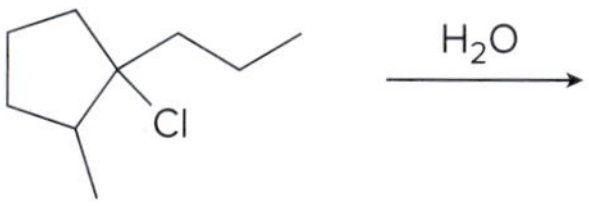 [1-chloro-1-ethyl-2-methylcyclopentane] —H₂O→
e. 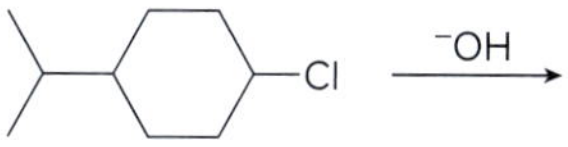 [1-tert-butyl-1-chlorocyclohexane] —⁻OH→
f. 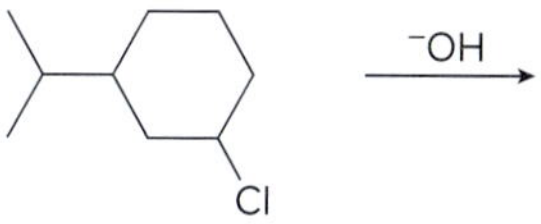 [tert-butyl chlorocyclohexane isomer] —⁻OH→

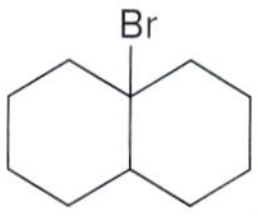

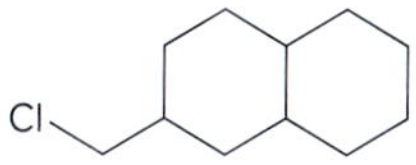

Stereochemistry and the E2 Reaction

8.38 What is the major E2 elimination product formed from each halide?

a. 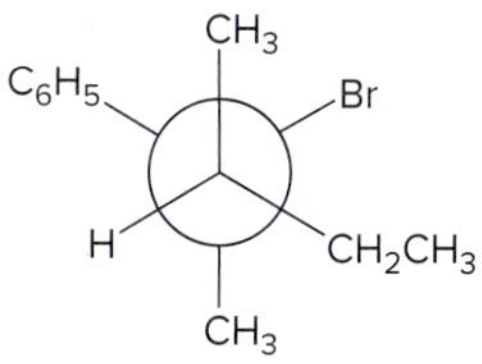 [Newman projection: front C₆H₅, CH₃, Br; back H, CH₃, CH₂CH₃]
b. 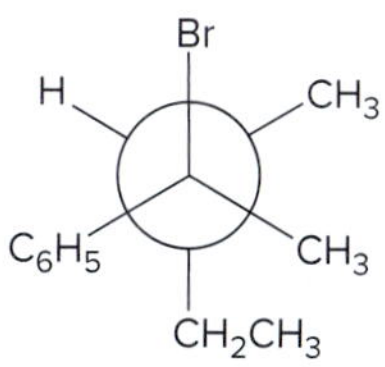 [Newman projection: front H, Br, CH₃; back C₆H₅, CH₃, CH₂CH₃]
c. 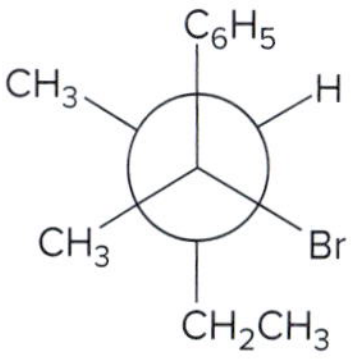 [Newman projection: front CH₃, C₆H₅, H; back CH₃, Br, CH₂CH₃]

Chapter 8 Alkyl Halides and Elimination Reactions

8.39 Taking into account anti periplanar geometry, predict the major E2 product formed from each starting material.

a. b.

8.40 Does *cis*- or *trans*-1-bromo-4-*tert*-butylcylohexane react faster in an E2 reaction?

8.41
  a. Draw three-dimensional representations for all stereoisomers of 2-chloro-3-methylpentane, and label pairs of enantiomers.
  b. Considering dehydrohalogenation across only C2 and C3, draw the E2 product that results from each of these alkyl halides. How many different products have you drawn?
  c. How are these products related to each other?

8.42 Which of the following compounds undergoes E2 elimination with strong base? For compounds that undergo elimination, draw the product. For compounds that do not undergo elimination, explain why they are unreactive.

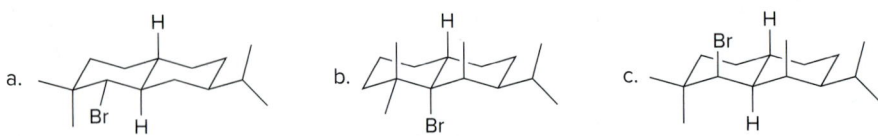

8.43 Draw the structure (including stereochemistry) of an alkyl chloride that forms each alkene as the exclusive E2 elimination product.

a. b. c.

8.44 Draw the major stereoisomer formed when each compound undergoes elimination with strong base (NaOH).

A   B

Alkynes

8.45 Draw the products formed when each dihalide is treated with excess $NaNH_2$.

a. b. c. d.

8.46 Draw the structure of a dihalide that could be used to prepare each alkyne. There may be more than one possible dihalide.

a. b. c.

8.47 Under certain reaction conditions, 2,3-dibromobutane reacts with two equivalents of base to give three products, each of which contains two new π bonds. Product **A** has two *sp* hybridized carbon atoms, product **B** has one *sp* hybridized carbon atom, and product **C** has none. What are the structures of **A**, **B**, and **C**?

S_N1, S_N2, E1, and E2 Mechanisms

8.48 Draw the organic products formed in each reaction.

a.

d.

g.

b.

e.

h.

c.

f.

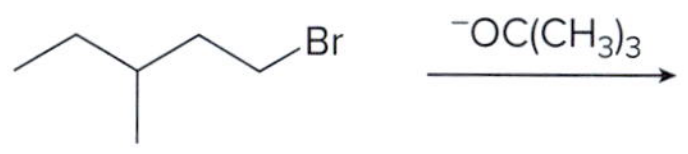

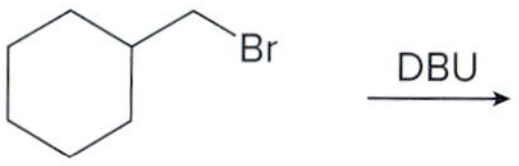

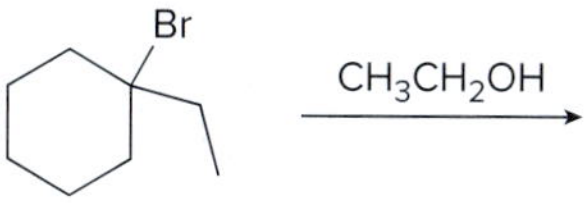

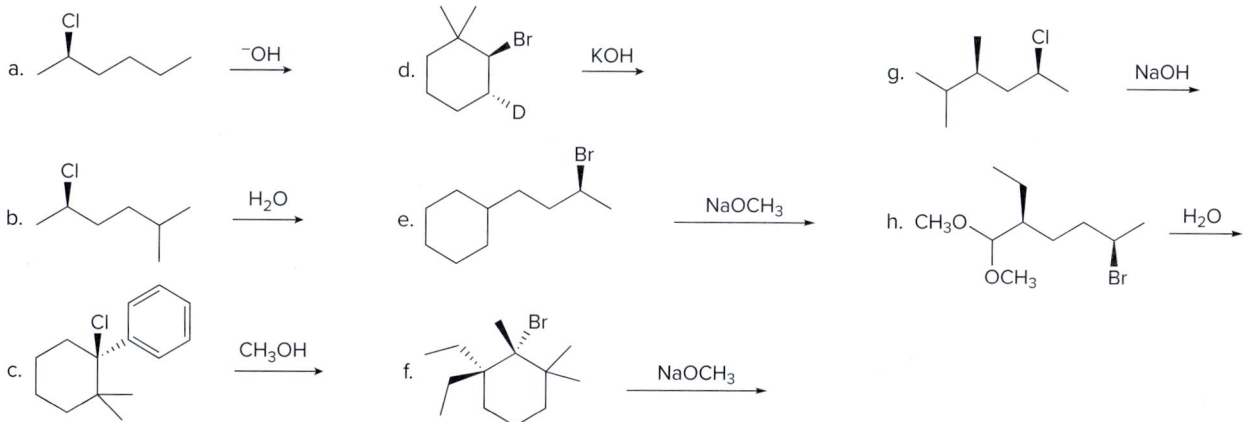

8.49 What reagents and reaction conditions are needed for each of the following conversions?

a.

c.

b.

d.

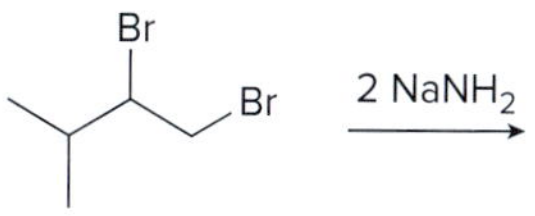

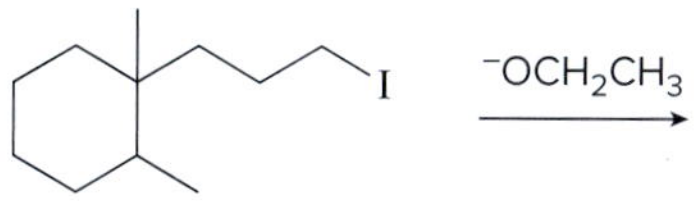

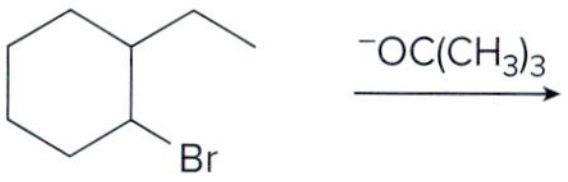

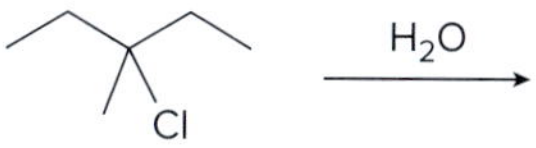

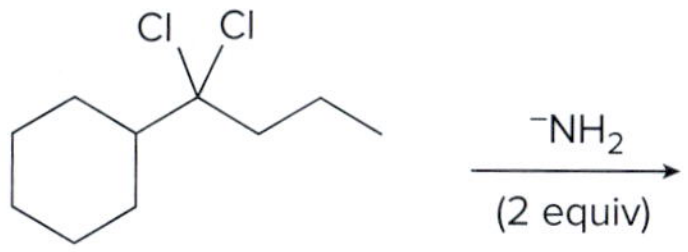

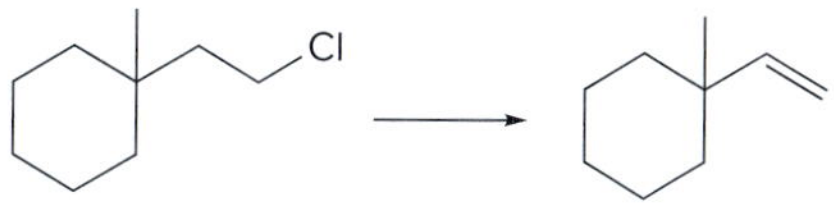

8.50 Draw all products, including stereoisomers, in each reaction.

a.

d.

g.

b.

e.

h.

c.

f.

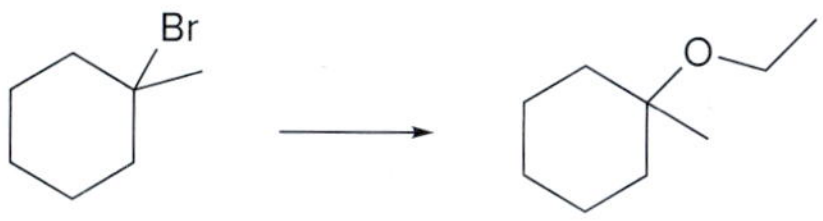

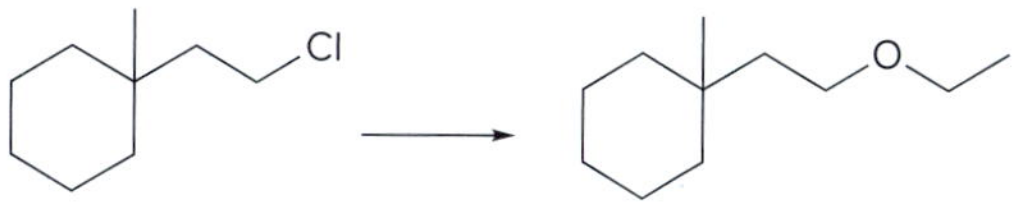

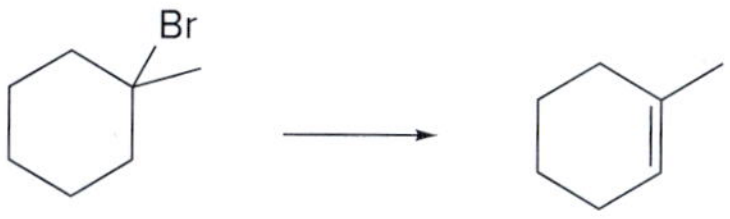

8.51

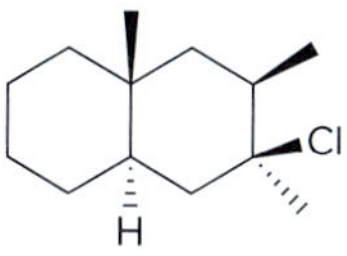

Draw all of the substitution and elimination products formed from the given alkyl halide with each reagent: (a) CH_3OH; (b) KOH. Indicate the stereochemistry around the stereogenic centers present in the products, as well as the mechanism by which each product is formed.

8.52 The following reactions do not afford the major product that is given. Explain why this is so, and draw the structure of the major product actually formed.

a.

c.

b.

d.

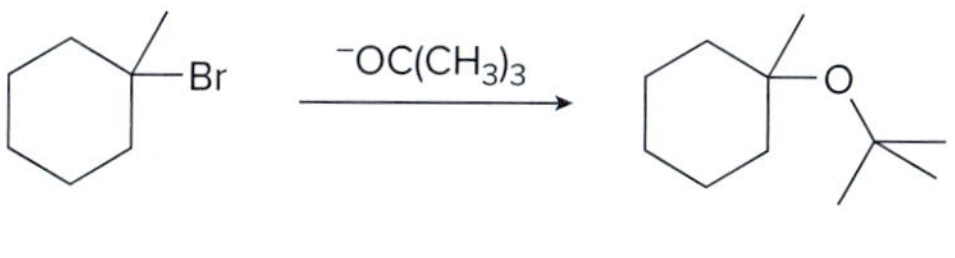

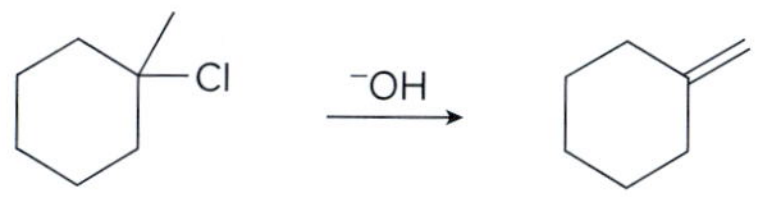

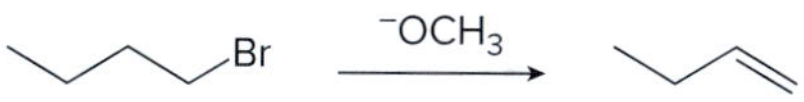

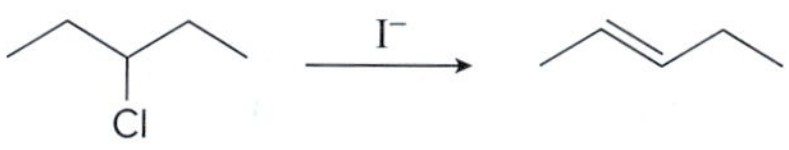

Chapter 8 Alkyl Halides and Elimination Reactions

8.53 Draw a stepwise, detailed mechanism for each reaction.

a. 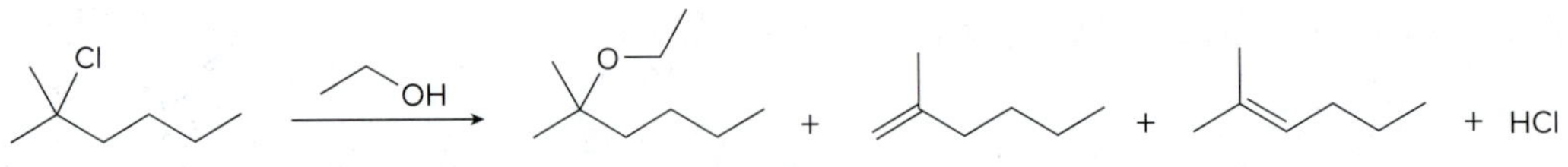 [structure: 2-chloro-2,3-dimethylpentane] $\xrightarrow{\text{CH}_3\text{CH}_2\text{OH}}$ [ether product] + [alkene 1] + [alkene 2] + HCl

b. 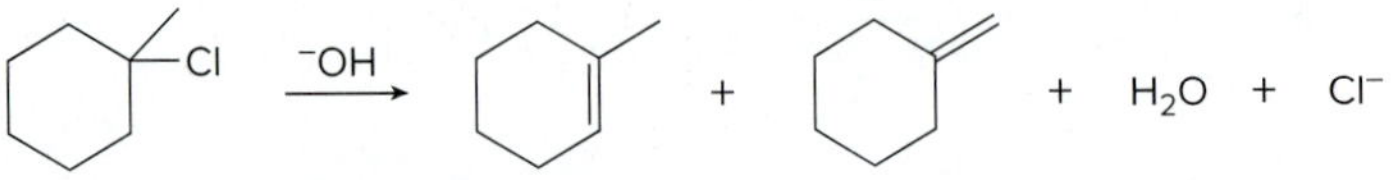 [1-chloro-1-methylcyclohexane] $\xrightarrow{^-\text{OH}}$ [1-methylcyclohexene] + [methylenecyclohexane] + H_2O + Cl^-

8.54 Draw the major product formed when (R)-1-chloro-3-methylpentane is treated with each reagent: (a) $NaOCH_2CH_3$; (b) KCN; (c) DBU.

8.55 Draw a stepwise, detailed mechanism for the following reaction.

[trans-decalin with Br substituent] $\xrightarrow{H_2O}$ [decalin-OH product 1] + [decalin-OH product 2] + [decalin with alkene] + HBr

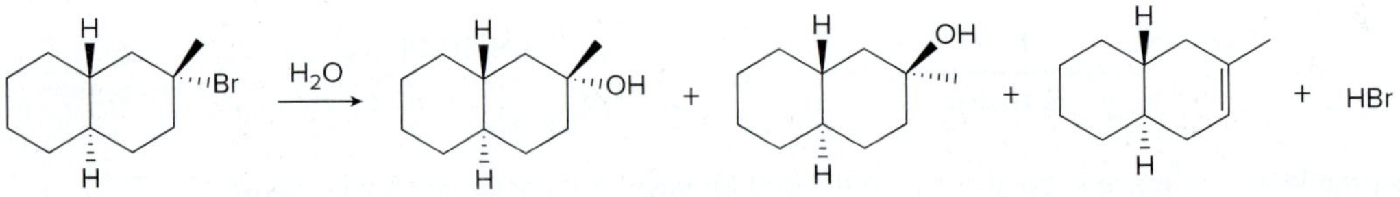

Challenge Problems

8.56 Explain why alkene **A** is more stable than alkene **B**, even though **B** contains more carbon atoms bonded to the double bond. Would you expect **C** to be more or less stable than **A** and **B**?

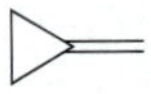

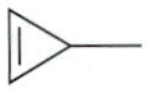

A methylenecyclopropane **B** 1-methylcyclopropene **C** 3-methylcyclopropene

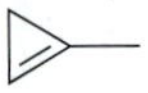

8.57 Draw a stepwise detailed mechanism that illustrates how four organic products are formed in the following reaction.

[1-chloro-1-methylcyclohex-2-ene] $\xrightarrow{CH_3OH}$ [1-methoxy-1-methylcyclohex-2-ene] + [3-methoxy-3-methylcyclohex-1-ene] + [1-methylcyclohexa-1,3-diene] + [methylenecyclohexadiene] + HCl

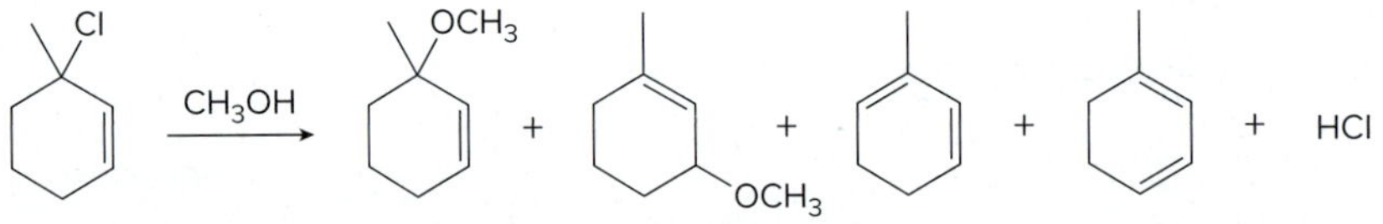

8.58 Although there are nine stereoisomers of 1,2,3,4,5,6-hexachlorocyclohexane, one stereoisomer reacts 7000 times more slowly than any of the others in an E2 elimination. Draw the structure of this isomer and explain why this is so.

8.59 Explain the selectivity observed in the following reactions.

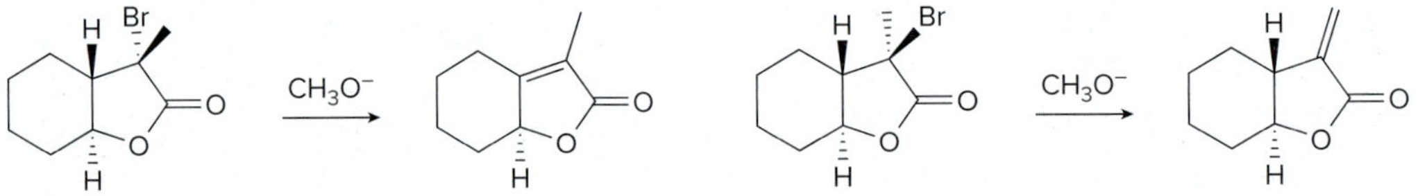

[bicyclic bromo lactone 1] $\xrightarrow{CH_3O^-}$ [bicyclic lactone with endocyclic alkene]

[bicyclic bromo lactone 2 (diastereomer)] $\xrightarrow{CH_3O^-}$ [bicyclic lactone with exocyclic methylene]

8.60 Draw a stepwise mechanism for the following reaction. The four-membered ring in the starting material and product is called a β-lactam. This functional group confers biological activity on penicillin and many related antibiotics, as is discussed in Chapter 16. (Hint: The mechanism begins with β elimination and involves only two steps.)

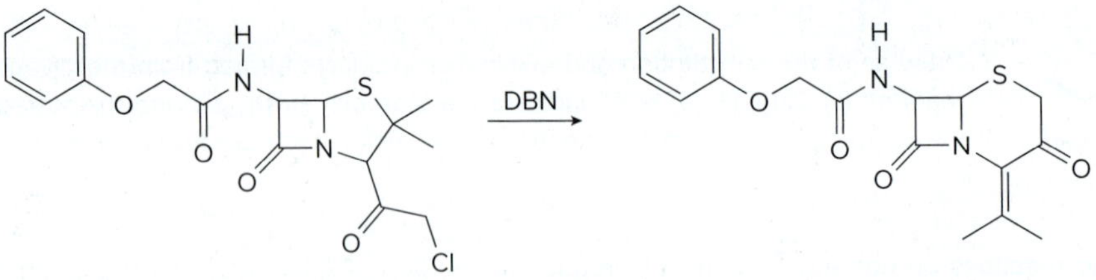

[penicillin-like starting material with CH₂Cl side chain] $\xrightarrow{DBN}$ [product with isopropylidene group]

8.61 (a) Draw all products formed by treatment of each dibromide (**A** and **B**) with one equivalent of $NaNH_2$. (b) Label pairs of diastereomers and constitutional isomers.

a. 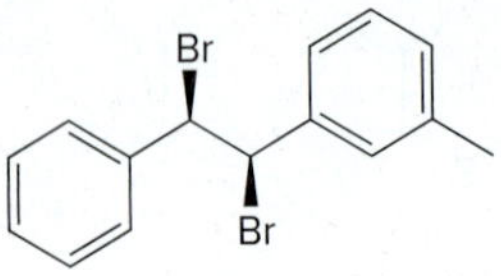 **A** [1,2-dibromo-1-phenyl-2-(3-methylphenyl)ethane, one diastereomer]

b. 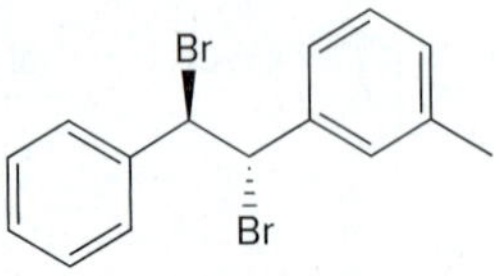 **B** [1,2-dibromo-1-phenyl-2-(3-methylphenyl)ethane, other diastereomer]

Alcohols, Ethers, and Related Compounds

9

- 9.1 Introduction
- 9.2 Structure and bonding
- 9.3 Nomenclature
- 9.4 Properties of alcohols, ethers, and epoxides
- 9.5 Interesting alcohols, ethers, and epoxides
- 9.6 Preparation of alcohols, ethers, and epoxides
- 9.7 General features—Reactions of alcohols, ethers, and epoxides
- 9.8 Dehydration of alcohols to alkenes
- 9.9 Carbocation rearrangements
- 9.10 Dehydration using $POCl_3$ and pyridine
- 9.11 Conversion of alcohols to alkyl halides with HX
- 9.12 Conversion of alcohols to alkyl halides with $SOCl_2$ and PBr_3
- 9.13 Tosylate—Another good leaving group
- 9.14 Reaction of ethers with strong acid
- 9.15 Thiols and sulfides
- 9.16 Reactions of epoxides
- 9.17 Application: Epoxides, leukotrienes, and asthma

Stephen Orsillo/Shutterstock

Patchouli alcohol, a 15-carbon alcohol obtained from the patchouli plant native to Malaysia, has been used in perfumery because of its exotic fragrance. In the 1800s, shawls imported from India were often packed with patchouli leaves to ward off insects, thus permeating the clothing with the distinctive odor. In Chapter 9, we learn about alcohols like patchouli alcohol, as well as related oxygen- and sulfur-containing functional groups.

Why Study...

Alcohols, Ethers, Epoxides, Thiols, and Sulfides?

In Chapter 9, we take the principles learned in Chapters 7 and 8 about leaving groups, nucleophiles, and bases, and apply them to **alcohols, ethers,** and **epoxides,** three new functional groups that contain polar C—O bonds. The hydroxy group (OH) of an alcohol is especially common in many natural products, and the reactions of alcohols are widely used in organic synthesis. In Chapter 9, you will discover that all of the reactions follow one of the four mechanisms introduced in Chapters 7 and 8—S_N1, S_N2, E1, or E2—so there are **no new general mechanisms to learn.**

Later in the chapter, we will also examine **thiols (RSH)** and **sulfides (R_2S),** sulfur analogues of alcohols and ethers, respectively. These functional groups play a key role in the chemistry of biomolecules, especially the proteins discussed in Chapter 23.

9.1 Introduction

Alcohols, ethers, and **epoxides** are three functional groups that contain carbon–oxygen σ bonds.

Alcohols contain a **hydroxy group (OH** group) bonded to an sp^3 hybridized carbon atom. As we learned in Section 3.2, alcohols are classified as **primary (1°), secondary (2°),** or **tertiary (3°)** based on the number of carbon atoms bonded to the carbon with the OH group.

Compounds having a hydroxy group on an sp^2 hybridized carbon atom—**enols** and **phenols**—undergo different reactions than alcohols and are discussed in Chapters 10 and 15, respectively. **Enols** have an OH group on a carbon of a C—C double bond. **Phenols** have an OH group on a benzene ring.

- C bonded to OH is sp^2 hybridized.

Ethers have two alkyl groups bonded to an oxygen atom. An ether is **symmetrical** if the two alkyl groups are the same, and **unsymmetrical** if they are different. *Epoxides* are ethers having the oxygen atom in a three-membered ring. Epoxides are also called **oxiranes.**

Problem 9.1 Label each ether and alcohol in brevenal, a marine natural product. Classify each alcohol as 1°, 2°, or 3°.

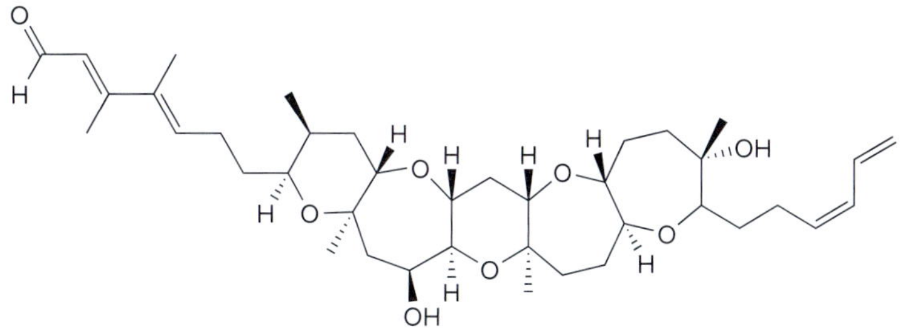

Brevenal (Problem 9.1) is a nontoxic marine polyether produced by *Karenia brevis*, a single-celled organism that proliferates during red tides, vast algal blooms that turn the ocean water red, brown, or green. *Don Paulson Photography/ Purestock/Alamy Stock Photo*

brevenal

9.2 Structure and Bonding

Alcohols, ethers, and epoxides each contain an oxygen atom surrounded by two atoms and two nonbonded electron pairs, making the O atom **tetrahedral** and *sp*³ hybridized. Because only two of the four groups around O are atoms, alcohols and ethers have a **bent** shape like H_2O.

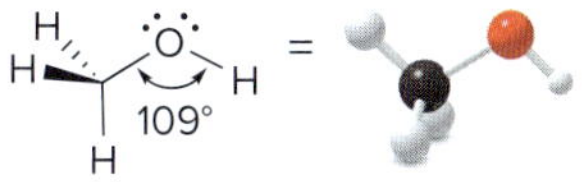

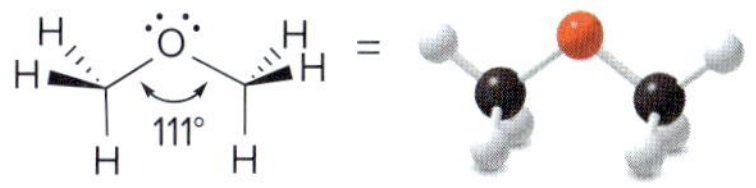

The bond angle around the O atom in an alcohol or ether is similar to the tetrahedral bond angle of **109.5°**. In contrast, the C—O—C bond angle of an epoxide must be **60°**, a considerable deviation from the tetrahedral bond angle. For this reason, **epoxides have angle strain,** making them much more reactive than other ethers.

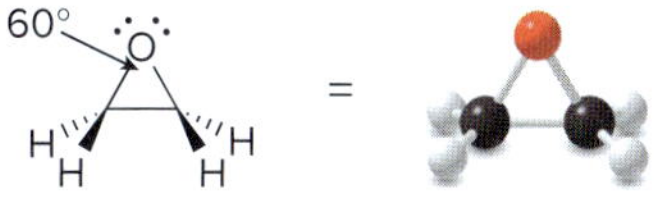

a strained, three-membered ring

Because oxygen is much more electronegative than carbon or hydrogen, the C—O and O—H bonds are all polar, with the O atom electron rich and the C and H atoms electron poor.

9.3 Nomenclature

To name an alcohol, ether, or epoxide using the IUPAC system, we must learn how to name the functional group either as a substituent or by using a suffix added to the parent name.

9.3A Naming Alcohols

- In the IUPAC system, alcohols are identified by the suffix *-ol*.

How To Name an Alcohol Using the IUPAC System

Example Give the IUPAC name of the following alcohol:

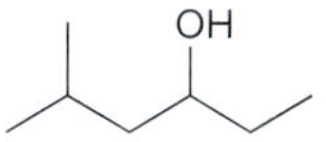

—*Continued*

How To, continued...

Step [1] Find the longest carbon chain containing the carbon bonded to the OH group.

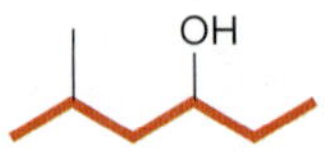

6 C's in the longest chain

6 C's — → hexane — → hexan*ol*

- Change the *-e* ending of the parent alkane to the suffix *-ol.*

Step [2] Number the carbon chain to give the OH group the lower number, and apply all other rules of nomenclature.

a. **Number** the chain.

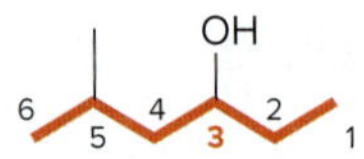

- Number the chain to put the OH group at C**3**, not C4.

hexan-3-ol

b. **Name** and **number** the substituents.

methyl at C5

Answer: 5-methylhexan-3-ol

CH$_3$CH$_2$CH$_2$CH$_2$OH is named as 1-butanol using the 1979 IUPAC recommendations and butan-1-ol using the 1993 IUPAC recommendations.

When an OH group is bonded to a ring, the **ring is numbered beginning with the OH group.** Because the functional group is always at C1, the "1" is usually omitted from the name. The ring is then numbered in a clockwise or counterclockwise fashion to give the next substituent the lower number. Representative examples are given in Figure 9.1.

Figure 9.1
Examples: Naming cyclic alcohols

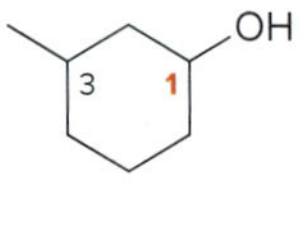

3-methylcyclohexanol

[The OH group is at C**1**; the second substituent (CH$_3$) gets the lower number.]

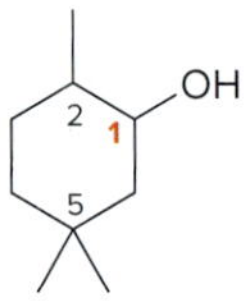

2,5,5-trimethylcyclohexanol

[The OH group is at C**1**; the second substituent (CH$_3$) gets the lower number.]

Common names are often used for simple alcohols. To assign a common name:

- Name all the carbon atoms of the molecule as a single **alkyl group.**
- Add the word *alcohol*, separating the words with a space.

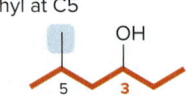

isopropyl group

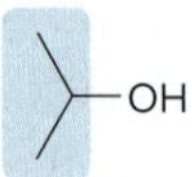

isopropyl alcohol

Compounds with two hydroxy groups are called **diols** (using the IUPAC system) or **glycols.** Compounds with three hydroxy groups are called **triols,** and so forth. To name a diol, for example, the suffix *-diol* is added to the name of the parent alkane, and numbers are used to indicate the location of the two OH groups.

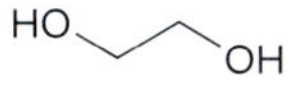

ethylene glycol
(ethane-1,2-diol)

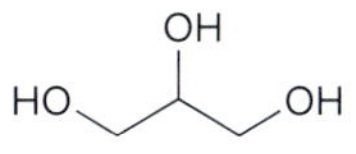

glycerol
(propane-1,2,3-triol)

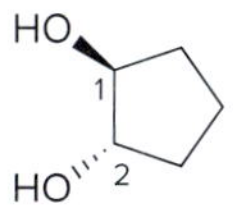

trans-cyclopentane-1,2-diol

Common names are usually used for these simple compounds.

Numbers are needed to show the location of **two** OH groups.

Problem 9.2 Give the IUPAC name for each compound.

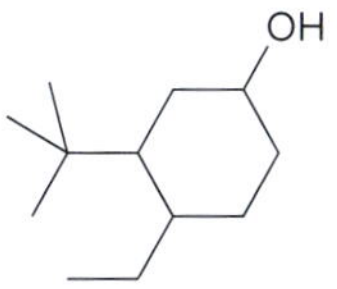

Problem 9.3 Give the structure corresponding to each name.
a. 7,7-dimethyloctan-4-ol
b. 5-methyl-4-propylheptan-3-ol
c. 2-*tert*-butyl-3-methylcyclohexanol
d. *trans*-cyclohexane-1,2-diol

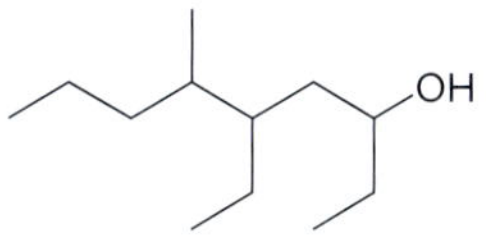

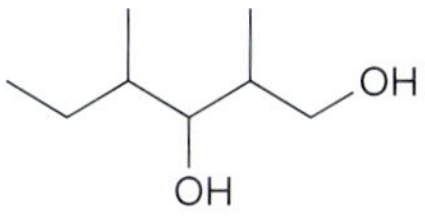

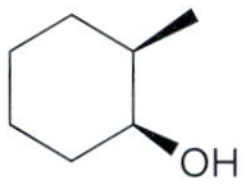

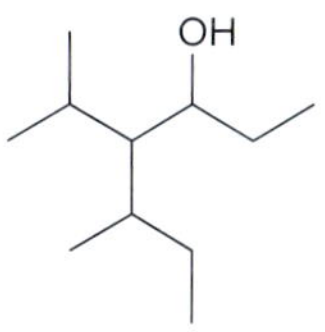

9.3B Naming Ethers

Simple ethers are usually assigned common names. To do so, **name both alkyl groups** bonded to the oxygen, arrange these names alphabetically, and add the word **ether**. For symmetrical ethers, name the alkyl group and add the prefix **di-**.

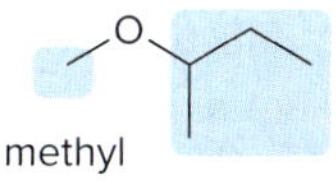

methyl
sec-butyl
sec-butyl methyl ether

[Alphabetize the **b** of **b**utyl before the **m** of **m**ethyl.]

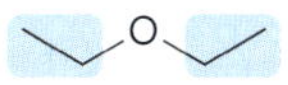

ethyl   ethyl
diethyl ether

More complex ethers are named using the IUPAC system. One alkyl group is named as a hydrocarbon chain, and the other is named as part of a substituent bonded to that chain.

- Name the simpler alkyl group + O atom as an **alkoxy** substituent by changing the *-yl* ending of the alkyl group to *-oxy.*
- Name the remaining alkyl group as an alkane, with the alkoxy group as a substituent bonded to this chain.

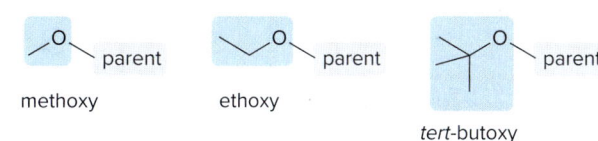

methoxy   ethoxy   *tert*-butoxy

Sample Problem 9.1 — Naming an Ether Using IUPAC Nomenclature

Give the IUPAC name for the following ether.

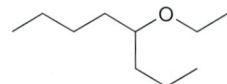

Solution

[1] Name the longer chain as an alkane and the shorter chain as an alkoxy group.

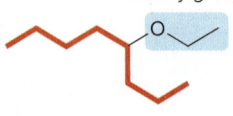

8 C's ----→ octane

[2] Apply the other nomenclature rules to complete the name.

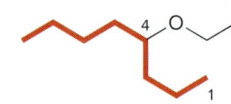

Answer: 4-ethoxyoctane

Problem 9.4
Name each of the following ethers.

a. 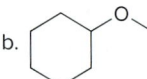   b. 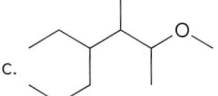   c.    d.

More Practice:
Try Problem 9.37a, b.

tetrahydrofuran
THF

Cyclic ethers have an O atom in a ring. A common cyclic ether is **tetrahydrofuran (THF)**, a polar aprotic solvent used in nucleophilic substitution (Section 7.8C) and many other organic reactions.

9.3C Naming Epoxides

Any cyclic compound containing a heteroatom is called a *heterocycle*.

Epoxides are named in three different ways—**epoxyalkanes, oxiranes,** or **alkene oxides.**

To name an epoxide as an **epoxyalkane,** first name the alkane chain or ring to which the oxygen is attached, and use the prefix *epoxy* to name the epoxide as a substituent. Use two numbers to designate the location of the atoms to which the O's are bonded.

1,2-epoxycyclohexane   1,2-epoxy-2-methylpropane   *cis*-2,3-epoxypentane

Epoxides bonded to a chain of carbon atoms can also be named as derivatives of **oxirane,** the simplest epoxide having two carbons and one oxygen atom in a ring. The oxirane ring is numbered to **put the O atom at position "1" and the first substituent at position "2."** No number is used for a substituent in a monosubstituted oxirane.

   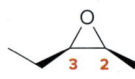

oxirane   2,2-dimethyloxirane

Epoxides are also named as **alkene oxides,** because they are often prepared by adding an O atom to an alkene (Chapter 11). To name an epoxide this way, mentally replace the epoxide oxygen by a double bond, name the alkene (Section 10.3), and then add the word *oxide.* For example, the common name for oxirane is ethylene oxide, because it is an epoxide derived from the alkene ethylene. We will use this method of naming epoxides after the details of alkene nomenclature are presented in Chapter 10.

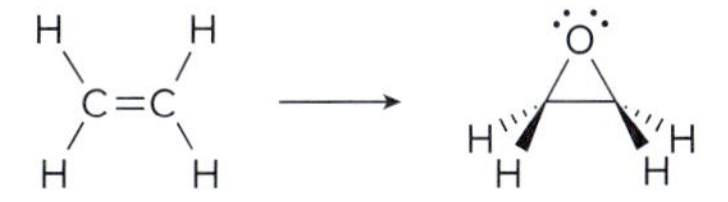

ethylene → ethylene oxide / oxirane

Problem 9.5 Name each epoxide.

a. 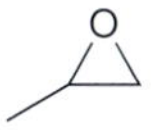 (two ways)   b. 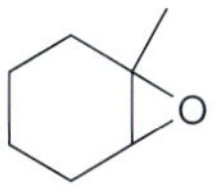   c.  (two ways)

9.4 Properties of Alcohols, Ethers, and Epoxides

Alcohols, ethers, and epoxides exhibit dipole–dipole interactions because they have a bent structure with two polar bonds. **Alcohols are also capable of intermolecular hydrogen bonding** because they possess a hydrogen atom on an oxygen, making alcohols much *more polar* than ethers and epoxides.

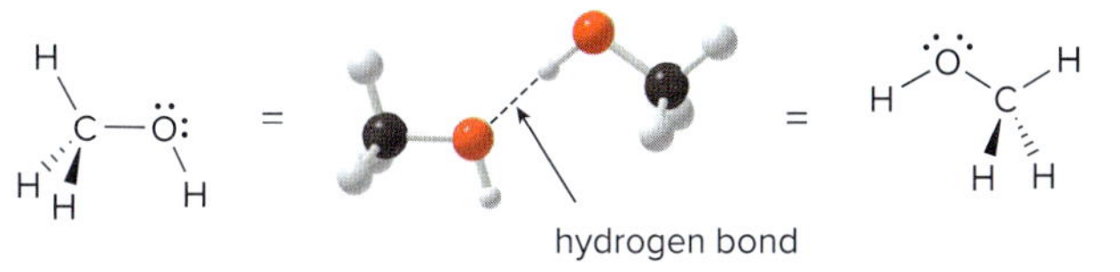

hydrogen bond

Steric factors affect the extent of hydrogen bonding. Although all alcohols can hydrogen bond, *increasing* the number of R groups around the carbon atom bearing the OH group *decreases* the extent of hydrogen bonding. Thus, 3° alcohols are least able to hydrogen bond, whereas 1° alcohols are most able to.

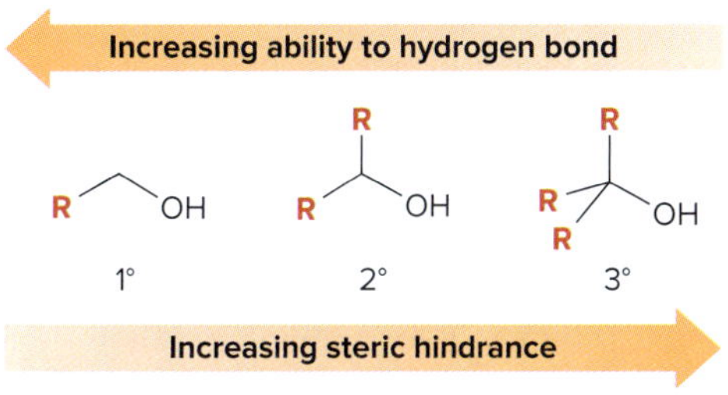

← Increasing ability to hydrogen bond

1°   2°   3°

Increasing steric hindrance →

How these factors affect the physical properties of alcohols, ethers, and epoxides is summarized in Table 9.1.

Problem 9.6 Rank the following compounds in order of increasing boiling point.

a.

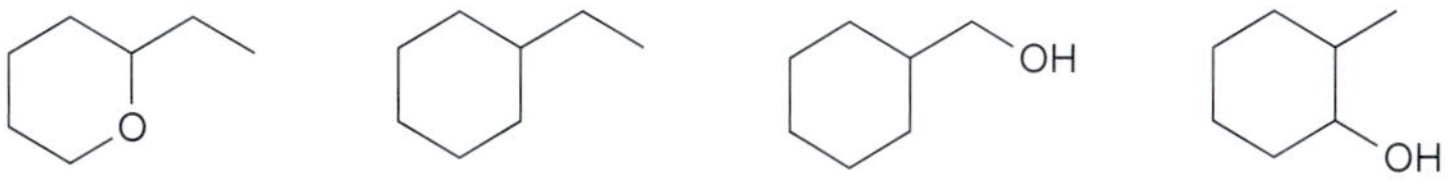

b.

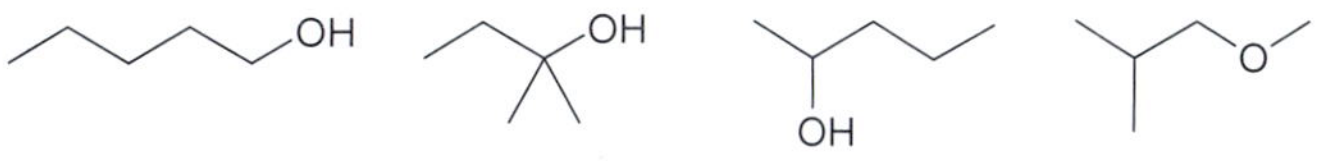

Table 9.1 Physical Properties of Alcohols, Ethers, and Epoxides

Property	Observation
Boiling point and melting point	• For compounds of comparable molecular weight, the **stronger** the intermolecular forces, the **higher** the bp or mp. • Bp's **increase** as the extent of hydrogen bonding **increases.** VDW, DD — bp 35 °C 3° alcohol — bp 83 °C 2° alcohol — bp 98 °C 1° alcohol — bp 118 °C **Increasing boiling point →**
Solubility	• Alcohols, ethers, and epoxides having ≤ 5 C's are **H_2O soluble** because they each have an oxygen atom capable of hydrogen bonding to H_2O (Section 3.4C). • Alcohols, ethers, and epoxides having > 5 C's are H_2O *insoluble* because the nonpolar alkyl portion is too large to dissolve in H_2O. • Alcohols, ethers, and epoxides of any size are soluble in organic solvents.

Key: VDW = van der Waals forces; DD = dipole–dipole

Students who have already been exposed to spectroscopy or who would like to learn about the spectroscopic properties of alcohols and ethers are referred to the following sections of Spectroscopy Chapters A, B, and C:

- Mass spectrometry: Section A.4B and Figure A.6
- Infrared spectroscopy: Section B.4B, Sample Problem B.3
- Nuclear magnetic resonance spectroscopy: Section C.9A, Figures C.12 and C.14a, Sample Problems C.3 and C.5

9.5 Interesting Alcohols, Ethers, and Epoxides

9.5A Ethanol, the Most Common Simple Alcohol

Ethanol (CH_3CH_2OH), formed by the fermentation of the carbohydrates in grains, grapes, and potatoes, is the alcohol present in alcoholic beverages. It is perhaps the first organic compound synthesized by humans, because alcohol production has been known for at least 4000 years. Ethanol depresses the central nervous system, increases the production of stomach acid, and dilates blood vessels, producing a flushed appearance. Ethanol is also a common laboratory solvent, which is sometimes made unfit to ingest by adding small amounts of benzene or methanol (both of which are toxic).

Ethanol is a common gasoline additive, widely touted as an environmentally friendly fuel source. Two common gasoline–ethanol fuels are gasohol, which contains 10% ethanol, and E-85, which contains 85% ethanol. Ethanol is now routinely prepared from the carbohydrates in corn (Figure 9.2). Starch, a complex carbohydrate polymer, can be hydrolyzed to the simple sugar glucose, which forms ethanol by the process of fermentation. Combining ethanol with gasoline forms a usable fuel, which combusts to form CO_2, H_2O, and a great deal of energy.

Because green plants use sunlight to convert CO_2 and H_2O to carbohydrates during photosynthesis, next year's corn crop removes CO_2 from the atmosphere to make new molecules of starch as the corn grows. While in this way ethanol is a *renewable* fuel source, the need for large-scale farm equipment and the heavy reliance on fertilizers and herbicides make ethanol expensive to produce. Moreover, many criticize the use of valuable farmland for an energy-producing crop rather than for food production. As a result, discussion continues on ethanol as an alternative to fossil fuels.

Figure 9.2 Ethanol from corn, a renewable fuel source

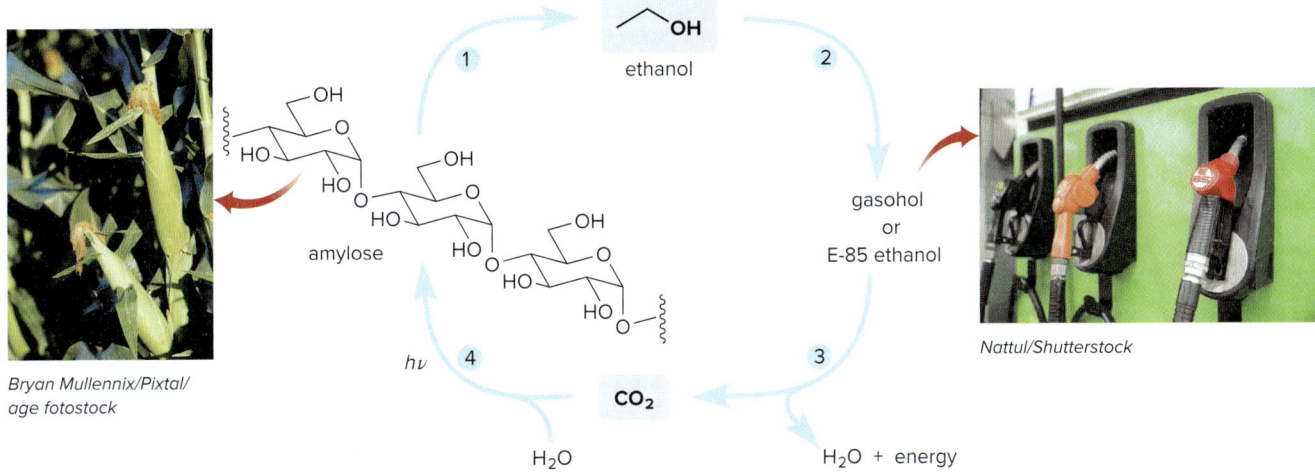

- Hydrolysis of amylose (one form of starch) and **fermentation** of the resulting simple sugars (Step [1]) yield ethanol, which is mixed with hydrocarbons from petroleum refining (Step [2]) to form usable fuels.
- **Combustion** of this ethanol–hydrocarbon fuel forms CO_2 and releases a great deal of energy (Step [3]).
- **Photosynthesis** converts atmospheric CO_2 back to plant carbohydrates in Step [4], and the cycle continues.

9.5B Interesting Ethers

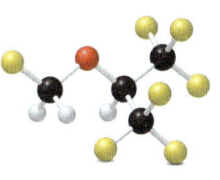

sevoflurane

The discovery that **diethyl ether** ($CH_3CH_2OCH_2CH_3$) is a general anesthetic revolutionized surgery in the nineteenth century. Diethyl ether is an imperfect anesthetic, but given the alternatives in the nineteenth century, it was considered a miracle drug that allowed patients to tolerate the excruciating pain of surgery. It is safe, easy to administer, and causes little patient mortality, but it is highly flammable and causes nausea in many patients. For these reasons, it has largely been replaced by sevoflurane and other halogenated ethers, which are nonflammable and cause little patient discomfort.

Recall from Section 3.7B that some cyclic **polyethers**—compounds with two or more ether linkages—contain cavities that can complex specific-sized cations. For example, 18-crown-6 binds K^+, whereas 12-crown-4 binds Li^+.

Recall from Section 3.7B that crown ethers are named as **x-crown-y**, where **x** is the total number of atoms in the ring and **y** is the number of O atoms.

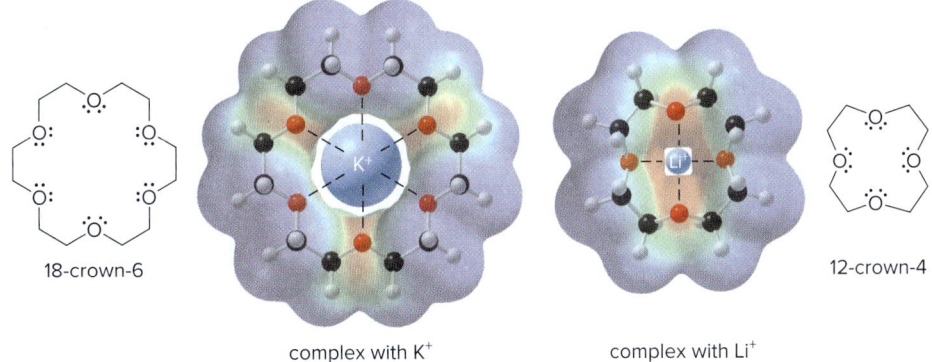

18-crown-6   12-crown-4
complex with K^+ complex with Li^+

- A crown ether–cation complex is called a *host–guest* complex. The crown ether is the *host* and the cation is the *guest*.
- The ability of a host molecule to bind specific guests is called *molecular recognition*.

The ability of crown ethers to complex cations can be exploited in nucleophilic substitution reactions, as shown in Figure 9.3. When 18-crown-6 is added to the reaction of CH_3CH_2Br with KCN, for example, the crown ether forms a tight complex with K^+ that has nonpolar C—H bonds on the outside, making the complex soluble in nonpolar solvents like benzene

Figure 9.3 The use of crown ethers in nucleophilic substitution reactions

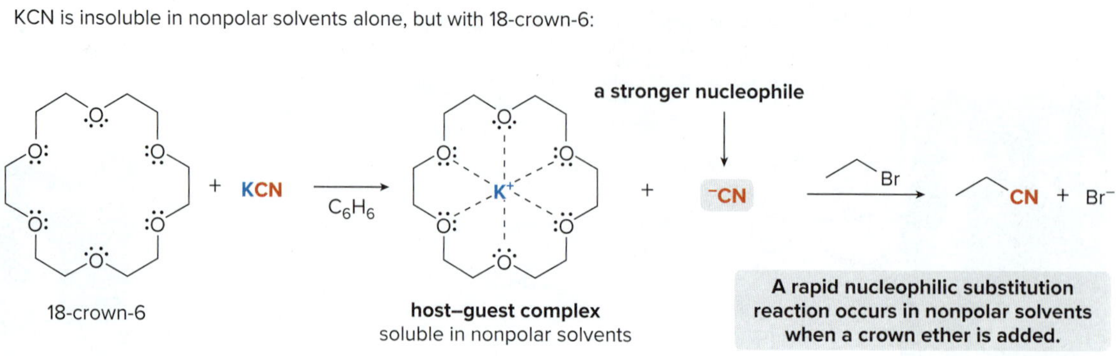

KCN is insoluble in nonpolar solvents alone, but with 18-crown-6:

18-crown-6 + KCN →(C₆H₆) host–guest complex soluble in nonpolar solvents + ⁻CN (a stronger nucleophile) →(Br) CN + Br⁻

A rapid nucleophilic substitution reaction occurs in nonpolar solvents when a crown ether is added.

(C$_6$H$_6$) or hexane. When the crown ether/K$^+$ complex dissolves in the nonpolar solvent, it carries the ⁻CN along with it to maintain electrical neutrality. The result is a solution of tightly complexed cation and relatively unsolvated anion (nucleophile). **The anion, therefore, is extremely nucleophilic because it is not hidden from the substrate by solvent molecules.**

9.5C Interesting Epoxides

Interesting epoxides include two useful drugs, eplerenone and tiotropium bromide. Eplerenone (trade name Inspra) is prescribed to reduce cardiovascular risk in patients who have already had a heart attack. Tiotropium bromide (trade name Spiriva) is a long-acting bronchodilator used to treat the chronic obstructive pulmonary disease (COPD) of smokers and those routinely exposed to secondhand smoke.

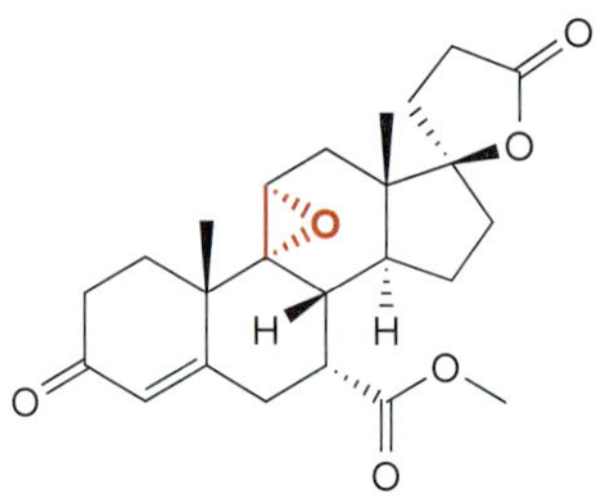

eplerenone

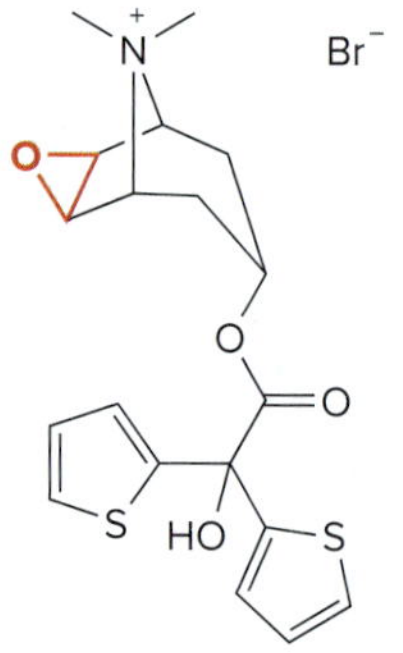

tiotropium bromide

9.6 Preparation of Alcohols, Ethers, and Epoxides

Alcohols and ethers are both common products of nucleophilic substitution. They are synthesized from alkyl halides by S$_N$2 reactions using strong nucleophiles. As in all S$_N$2 reactions, highest yields of products are obtained with unhindered methyl and 1° alkyl halides.

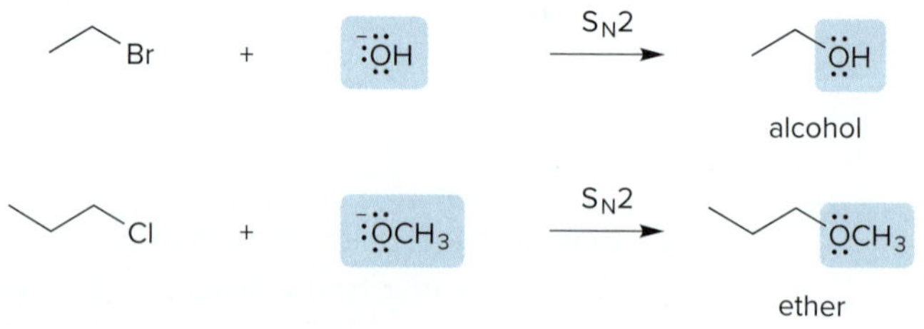

Br + ⁻OH →(S$_N$2) OH (alcohol)

Cl + ⁻OCH$_3$ →(S$_N$2) OCH$_3$ (ether)

The preparation of ethers by this method is called the **Williamson ether synthesis,** and, although it was first reported in the 1800s, it is still the most general method to prepare an ether. Unsymmetrical ethers can be synthesized in two different ways, but often one path is preferred.

9.6 Preparation of Alcohols, Ethers, and Epoxides

For example, ethyl isopropyl ether can be prepared from CH₃CH₂O⁻ and 2-bromopropane (Path [a]), or from (CH₃)₂CHO⁻ and bromoethane (Path [b]). Because the mechanism is S$_N$2, **the preferred path uses the less sterically hindered halide, CH₃CH₂Br—Path [b].**

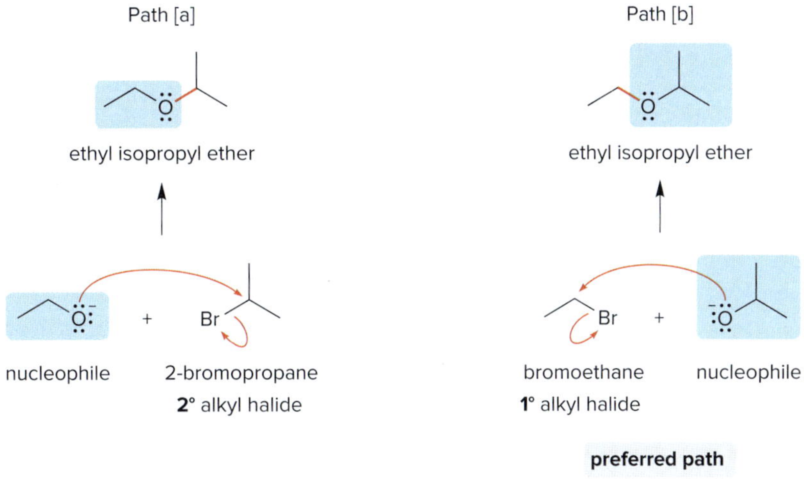

Path [a] — ethyl isopropyl ether formed from nucleophile CH₃CH₂O⁻ + 2-bromopropane (2° alkyl halide).

Path [b] — ethyl isopropyl ether formed from bromoethane (1° alkyl halide) + nucleophile (CH₃)₂CHO⁻. **preferred path**

Problem 9.7 Draw the organic product of each reaction.

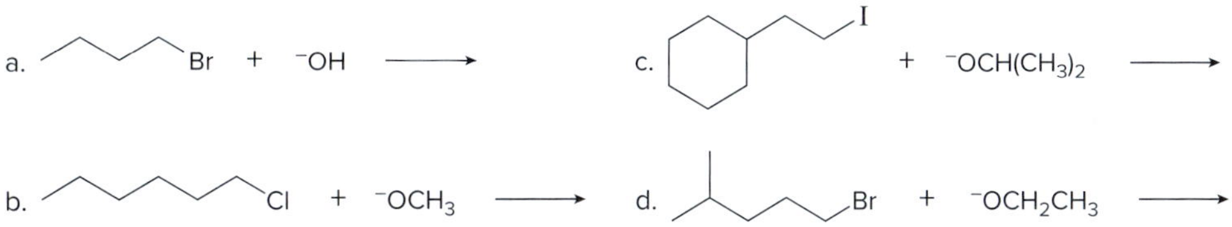

a. CH₃CH₂CH₂CH₂Br + ⁻OH ⟶

b. CH₃(CH₂)₄CH₂Cl + ⁻OCH₃ ⟶

c. cyclohexyl-CH₂CH₂I + ⁻OCH(CH₃)₂ ⟶

d. (CH₃)₂CHCH₂CH₂CH₂Br + ⁻OCH₂CH₃ ⟶

Problem 9.8 A key step in the synthesis of the antidepressant paroxetine (trade name Paxil) involves a Williamson ether synthesis of the acyclic ether in **X**. Draw two different routes to this ether and state which is preferred.

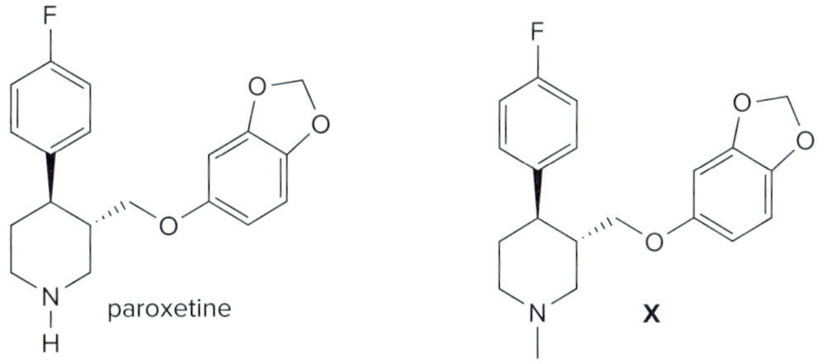

paroxetine   **X**

A **hydroxide** nucleophile is needed to synthesize an alcohol, and salts such as **NaOH** and **KOH** are inexpensive and commercially available. An **alkoxide** salt is needed to make an ether. Simple alkoxides such as sodium methoxide (NaOCH₃) can be purchased, but others are prepared from alcohols by a Brønsted–Lowry acid–base reaction. For example, **sodium ethoxide (NaOCH₂CH₃)** is prepared by treating ethanol with NaH.

CH₃CH₂O–H + Na⁺ H:⁻ ⟶ CH₃CH₂O:⁻ Na⁺ + H₂
                                         alkoxide nucleophile
                                         sodium ethoxide

NaH is an especially good base for forming an alkoxide, because the by-product of the reaction, H₂, is a gas that just bubbles out of the reaction mixture.

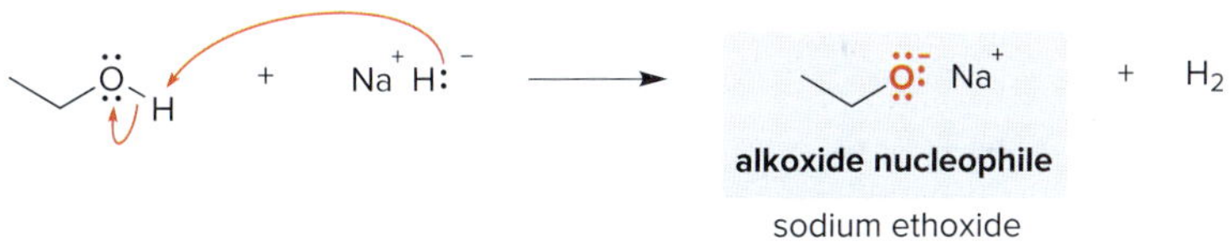

When an organic compound contains both a hydroxy group and a halogen atom on adjacent carbon atoms, an *intramolecular* version of this reaction forms an epoxide. The starting

348   Chapter 9   Alcohols, Ethers, and Related Compounds

material for this two-step sequence, a **halohydrin,** is prepared from an alkene, as we will learn in Chapter 10.

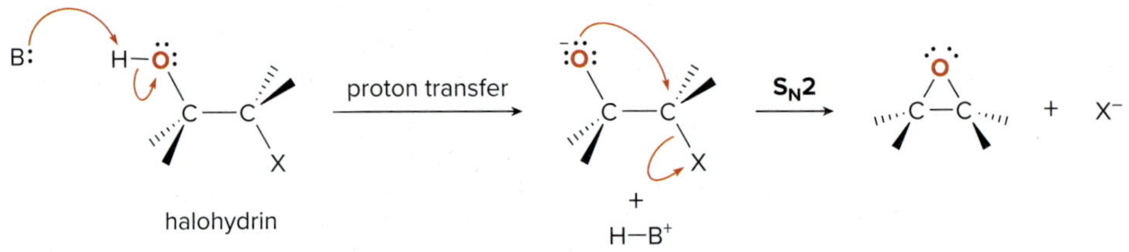

| Sample Problem 9.2 | Synthesizing an Ether by a Two-Step Reaction Sequence |

Draw the product of the following two-step reaction sequence.

cyclohexanol [1] NaH [2] CH₃CH₂Br

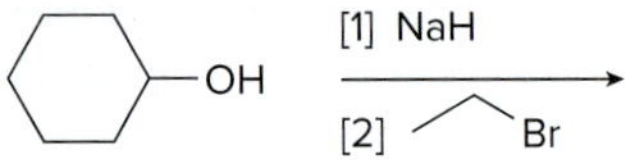

Solution

[1] The base removes a proton from the OH group, forming an alkoxide.

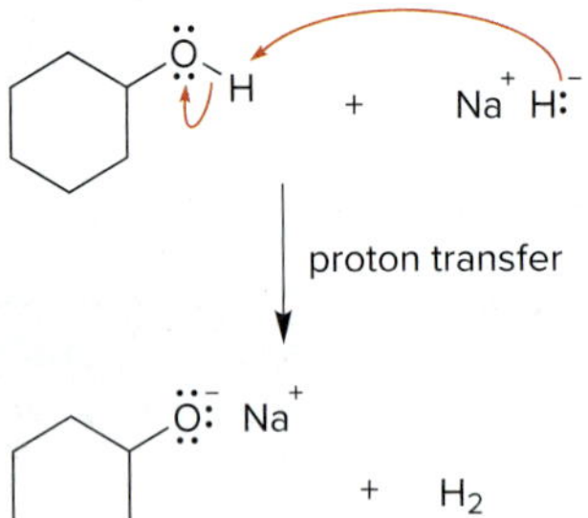

[2] The alkoxide acts as a nucleophile in an S_N2 reaction, forming an ether.

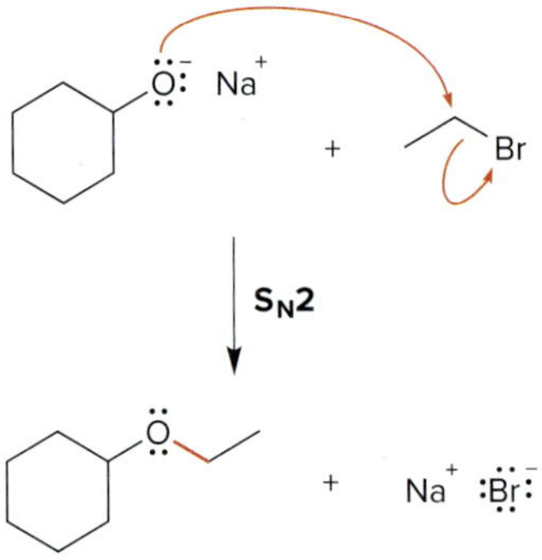

- This two-step sequence converts an alcohol to an ether. The new C–O bond of the ether is shown in red.

Problem 9.9 Draw the products of each reaction.

a.  CH₃CH₂CH₂OH + NaH ⟶

b. 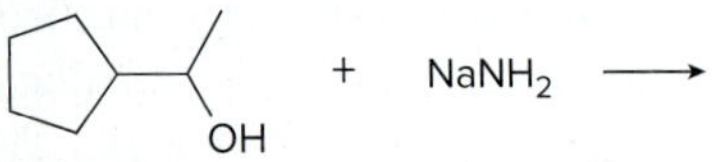 cyclopentyl-CH(OH)CH₃ + NaNH₂ ⟶

c. 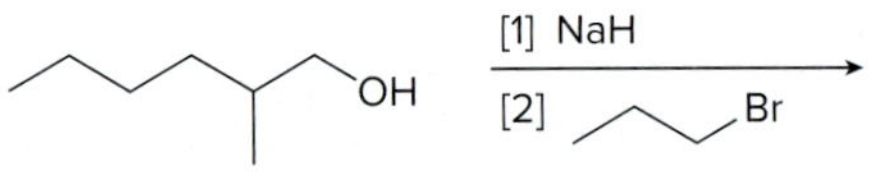 (CH₃)₂CHCH₂CH₂CH₂OH [1] NaH [2] CH₃CH₂CH₂Br

d. 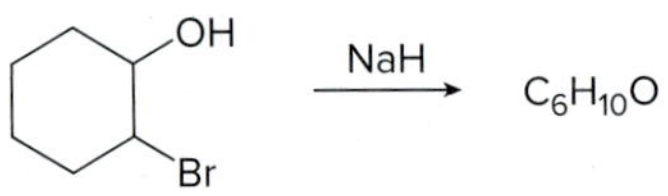 2-bromocyclohexanol NaH ⟶ $C_6H_{10}O$

More Practice: Try Problems 9.40h, 9.43a, 9.60i.

9.7 General Features—Reactions of Alcohols, Ethers, and Epoxides

We begin our discussion of the chemical reactions of alcohols, ethers, and epoxides with a look at the general reactive features of each functional group.

9.7A Alcohols

Unlike many families of molecules, the reactions of alcohols do *not* fit neatly into a single reaction class. In Chapter 9, we discuss only the substitution and β elimination reactions of alcohols. Alcohols are also key starting materials in oxidation reactions (Chapter 11), and their polar O—H bond makes them more acidic than many other organic compounds, a feature we will explore in Chapter 15.

Alcohols are similar to alkyl halides in that both contain an electronegative element bonded to an sp^3 hybridized carbon atom. **Alkyl halides contain a good leaving group (X^-), however, whereas alcohols do *not*.** Nucleophilic substitution with ROH as starting material would displace $^-$OH, a strong base and therefore a poor leaving group.

$$R-X \ + \ :Nu^- \longrightarrow R-Nu \ + \ X^- \quad \textbf{good} \text{ leaving group}$$

$$R-OH \ + \ :Nu^- \ \ \cancel{\longrightarrow} \ \ R-Nu \ + \ ^-OH \quad \textbf{poor} \text{ leaving group}$$

For an alcohol to undergo a nucleophilic substitution or elimination reaction, the **OH group must be converted into a *better* leaving group.** This can be done by reaction with acid. Treatment of an alcohol with a strong acid like HCl or H_2SO_4 protonates the O atom via an acid–base reaction. This transforms the $^-$OH leaving group into H_2O, **a weak base and therefore a *good* leaving group.**

$$R-\ddot{O}H \ + \ H-Cl \ \rightleftharpoons \ R-\overset{+}{\ddot{O}}H_2 \ + \ Cl^-$$
strong acid   weak base
good leaving group

If the OH group of an alcohol is made into a good leaving group, alcohols *can* undergo β elimination and nucleophilic substitution, as described in Sections 9.8–9.12.

> Because the pK_a of $(ROH_2)^+$ is ~−2, protonation of an alcohol occurs only with very strong acids—namely, those having a $pK_a \leq -2$.

β elimination, $-H_2O$ → **Alkenes are formed by β elimination.** (Sections 9.8–9.10)

nucleophilic substitution → **Alkyl halides are formed by nucleophilic substitution.** (Sections 9.11–9.12)

9.7B Ethers and Epoxides

Like alcohols, **ethers do *not* contain a good leaving group,** which means that nucleophilic substitution and β elimination do not occur directly. Ethers undergo fewer useful reactions than alcohols.

$R-\ddot{O}R$ **poor leaving group**

Epoxides don't have a good leaving group either, but they have one characteristic that neither alcohols nor ethers have: **the "leaving group" is contained in a strained three-membered ring.** Nucleophilic attack opens the three-membered ring and relieves angle strain, making nucleophilic attack a favorable process that occurs even with the poor leaving group. Specific examples are presented in Section 9.16.

9.8 Dehydration of Alcohols to Alkenes

The dehydrohalogenation of alkyl halides, discussed in Chapter 8, is one way to introduce a π bond into a molecule. Another way is to eliminate water from an alcohol in a **dehydration** reaction.

- Dehydration is a β elimination reaction in which the elements of OH and H are removed from the α and β carbon atoms, respectively.

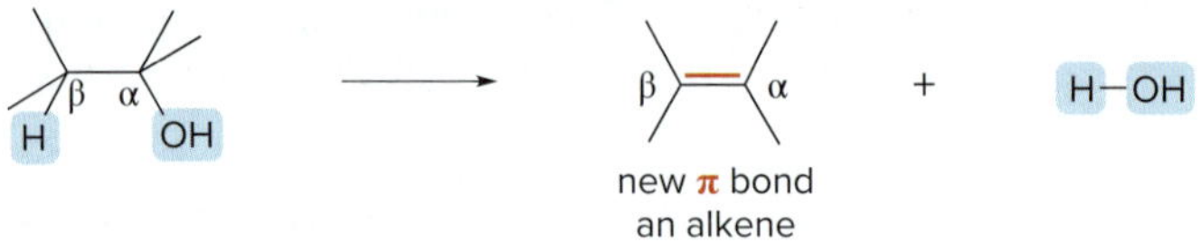

Dehydration is typically carried out using **H₂SO₄** and other strong acids, or phosphorus oxychloride (**POCl₃**) in the presence of an amine base. We consider dehydration in acid first, followed by dehydration with POCl₃ in Section 9.10.

9.8A General Features of Dehydration in Acid

Alcohols undergo dehydration in the presence of strong acid to afford alkenes, as illustrated in Equations [1] and [2]. Typical acids used for this conversion are H₂SO₄ or *p*-toluenesulfonic acid (abbreviated as TsOH).

Recall from Section 2.6 that *p*-toluenesulfonic acid is a strong organic acid (pK_a = −7).

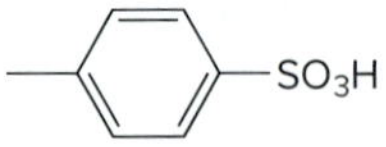

p-toluenesulfonic acid
TsOH

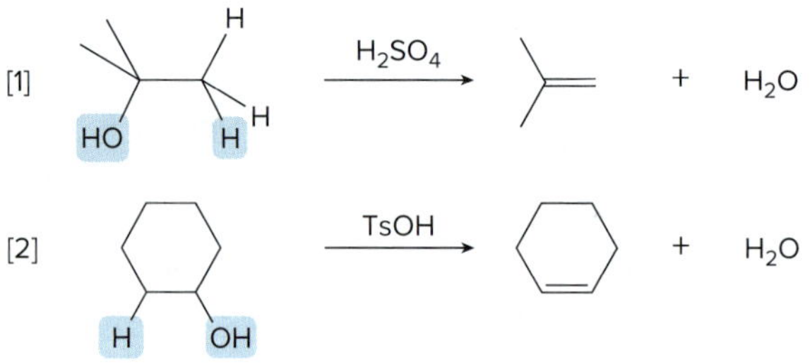

More substituted alcohols dehydrate more readily, giving rise to the following order of reactivity:

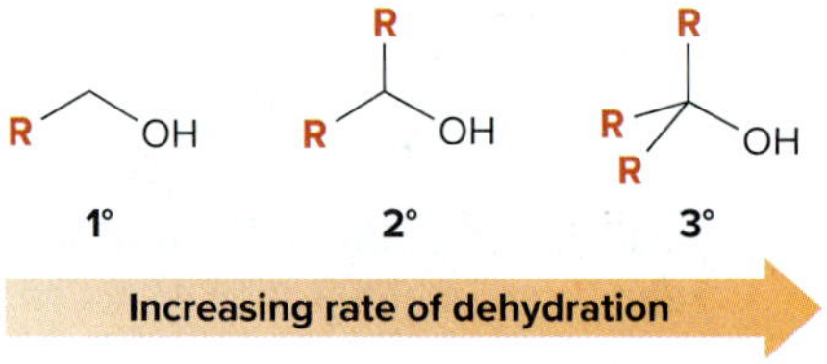

1°   2°   3°

Increasing rate of dehydration

When an alcohol has two or three different β carbons, dehydration is regioselective and follows the Zaitsev rule. **The more substituted alkene is the major product when a mixture of constitutional isomers is possible.** For example, elimination of H and OH from 2-methylbutan-2-ol yields two constitutional isomers: the trisubstituted alkene **A** as *major* product and the disubstituted alkene **B** as *minor* product.

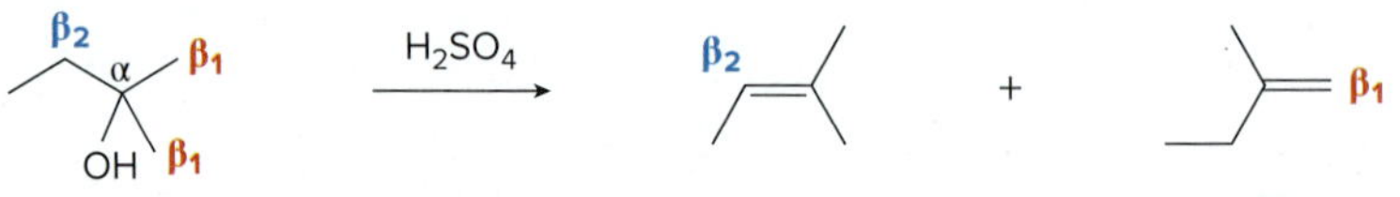

2-methylbutan-2-ol

A
major product
trisubstituted alkene

B
minor product
disubstituted alkene

9.8 Dehydration of Alcohols to Alkenes

Problem 9.10 Draw the products formed when each alcohol undergoes dehydration with TsOH, and label the major product when a mixture results.

a. [structure: 2,5-dimethyl-3-hexanol-like with OH] b. [structure: 3-methyl-3-pentanol-like with OH] c. [structure: cyclohexanol with methyl, OH]

Problem 9.11 Rank the alcohols in order of increasing reactivity when dehydrated with H_2SO_4.

[three alcohol structures: cyclohexylethanol (primary), 1-methylcyclohexanol (tertiary), 3-methylcyclohexanol (secondary)]

9.8B The E1 Mechanism for the Dehydration of 2° and 3° Alcohols

The mechanism of dehydration depends on the structure of the alcohol: **2° and 3° alcohols react by an E1 mechanism, whereas 1° alcohols react by an E2 mechanism.** Regardless of the type of alcohol, however, strong acid is *always* needed to protonate the O atom to form a good leaving group.

The E1 dehydration of 2° and 3° alcohols is illustrated with $(CH_3)_3COH$ (a 3° alcohol) as starting material to form $(CH_3)_2C=CH_2$ as product (Mechanism 9.1). The mechanism consists of **three steps.**

Mechanism 9.1 Dehydration of 2° and 3° ROH—An E1 Mechanism

1. **Protonation** of the oxygen atom converts the poor leaving group (⁻OH) into a **good leaving group** (H_2O).
2. Heterolysis of the C–O bond forms a **carbocation** in the rate-determining step.
3. A base (such as HSO_4^- or H_2O) removes a proton from a carbon adjacent to the carbocation to form the new **π bond.**

Thus, **dehydration of 2° and 3° alcohols occurs via an E1 mechanism with an added first step.** Step [1] protonates the OH group to make a good leaving group. Steps [2] and [3] are the two steps of an E1 mechanism: loss of a leaving group (H_2O in this case) to form a carbocation, followed by removal of a β proton to form a π bond. The acid used to protonate the alcohol in Step [1] is regenerated upon removal of the proton in Step [3], so dehydration is **acid-catalyzed.**

The E1 dehydration of 2° and 3° alcohols with acid gives clean elimination products without by-products formed from an S_N1 reaction. This makes the E1 dehydration of alcohols much more synthetically useful than the E1 dehydrohalogenation of alkyl halides (Section 8.7). Clean elimination takes place because the reaction mixture contains no good nucleophile to react with the intermediate carbocation, so **no competing S_N1 reaction occurs.**

The dehydration of alcohols by an E1 mechanism occurs in biological systems as well. E1 dehydration of diol **A** forms **enol B** with an alkene bonded to an OH group. **B** is an intermediate in the biosynthesis of the amino acid histidine.

9.8C The E2 Mechanism for the Dehydration of 1° Alcohols

Because 1° carbocations are highly unstable, the dehydration of 1° alcohols cannot occur by an E1 mechanism involving a carbocation intermediate. With 1° alcohols, therefore, **dehydration follows an E2 mechanism.** The two-step process for the conversion of $CH_3CH_2CH_2OH$ (a 1° alcohol) to $CH_3CH=CH_2$ with H_2SO_4 as acid catalyst is shown in Mechanism 9.2.

Mechanism 9.2 Dehydration of a 1° ROH—An E2 Mechanism

1. **Protonation** of the oxygen atom converts the poor leaving group ($^-$OH) into a **good leaving group** (H_2O).
2. Two bonds are broken and two bonds are formed. The base (HSO_4^- or H_2O) removes a proton from the β carbon; the electron pair in the β C–H bond forms the new **π bond** and the leaving group (H_2O) departs.

The dehydration of a 1° alcohol begins with the protonation of the OH group to form a good leaving group, just as in the dehydration of a 2° or 3° alcohol. With 1° alcohols, however, loss of the leaving group and removal of a β proton occur at the *same* time, so that **no highly unstable 1° carbocation is generated.**

9.8D Le Châtelier's Principle

Although **entropy favors product formation** in dehydration (one molecule of reactant forms two molecules of products), **enthalpy does *not*,** because the two σ bonds broken in the reactant are stronger than the σ and π bonds formed in the products.

According to **Le Châtelier's principle, a system at equilibrium will react to counteract any disturbance to the equilibrium.** Thus, removing a product from a reaction mixture as it is formed drives the equilibrium to the *right*, forming more product.

Le Châtelier's principle can be used to favor products in dehydration reactions because the alkene product has a lower boiling point than the alcohol reactant. Thus, the alkene can be distilled from the reaction mixture as it is formed, leaving the alcohol and acid to react further, forming more product.

9.9 Carbocation Rearrangements

Sometimes "unexpected" products are formed in dehydration; that is, the carbon skeletons of the starting material and product might be different, or the double bond might be in an unexpected location. For example, the dehydration of 3,3-dimethylbutan-2-ol yields two alkenes, whose carbon skeletons do not match the carbon framework of the starting material.

9.9 Carbocation Rearrangements

[Reaction: 3,3-dimethylbutan-2-ol + H₂SO₄ → two alkene products + H₂O]

This phenomenon sometimes occurs when carbocations are reactive intermediates. **A less stable carbocation can rearrange to a more stable carbocation by shift of a hydrogen atom or an alkyl group.** These **1,2-shifts** involve migration of an alkyl group or hydrogen atom from one carbon to an adjacent carbon atom. The migrating group moves with the two electrons that bonded it to the carbon skeleton.

Because the migrating group in a 1,2-shift moves with two bonding electrons, the carbon it leaves behind now has only three bonds (six electrons), giving it a net positive (+) charge.

- Movement of a hydrogen atom is called a **1,2-hydride shift**.
- Movement of an alkyl group is called a **1,2-alkyl shift**.

Problem 9.12 Show how a 1,2-shift forms a more stable carbocation from each intermediate.

a.   b.   c.

The dehydration of 3,3-dimethylbutan-2-ol illustrates the rearrangement of a 2° to a 3° carbocation by a **1,2-methyl shift,** as shown in Mechanism 9.3. The carbocation rearrangement occurs in Step [3] of the four-step mechanism.

Mechanism 9.3 A 1,2-Methyl Shift—Carbocation Rearrangement During Dehydration

Part [1] Formation of a 2° carbocation and rearrangement

[Mechanism steps: 3,3-dimethylbutan-2-ol → protonated alcohol (+ HSO₄⁻) → 2° carbocation (+ H₂O) → 1,2 shift → 3° carbocation (more stable)]

1 **Protonation** of the oxygen atom converts the poor leaving group (⁻OH) into a **good leaving group** (H₂O).
2 Heterolysis of the C–O bond forms a **2° carbocation.**
3 1,2-Shift of a CH₃ group converts a 2° carbocation to a **more stable 3° carbocation.**

Part [2] Loss of a proton to form the π bond

[Two pathways: loss of H from β₁ or β₂ carbon to form two different alkenes + H₂SO₄]

4 **Loss of a proton** from a β carbon (β₁ or β₂) forms two different alkenes.

Steps [1], [2], and [4] in the mechanism for the dehydration of 3,3-dimethylbutan-2-ol are exactly the same steps previously seen in dehydration: protonation, loss of H₂O, and loss of a proton. Only Step [3], rearrangement of the less stable 2° carbocation to the more stable 3° carbocation, is new.

- **1,2-Shifts convert a less stable carbocation to a more stable carbocation.**

For example, 2° carbocation **A** rearranges to the more stable 3° carbocation by a 1,2-hydride shift, whereas carbocation **B** does not rearrange because it is 3° to begin with.

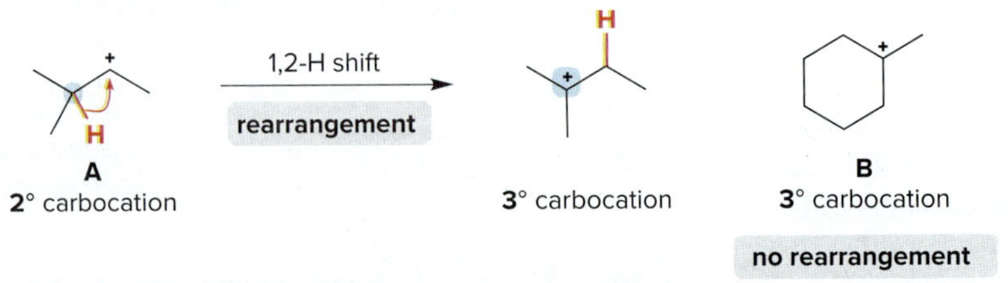

Sample Problem 9.3 illustrates a dehydration reaction that occurs with a **1,2-hydride** shift.

Sample Problem 9.3 — Drawing a Dehydration Reaction with a Rearrangement

Show how the dehydration of alcohol **X** forms alkene **Y** using a 1,2-hydride shift.

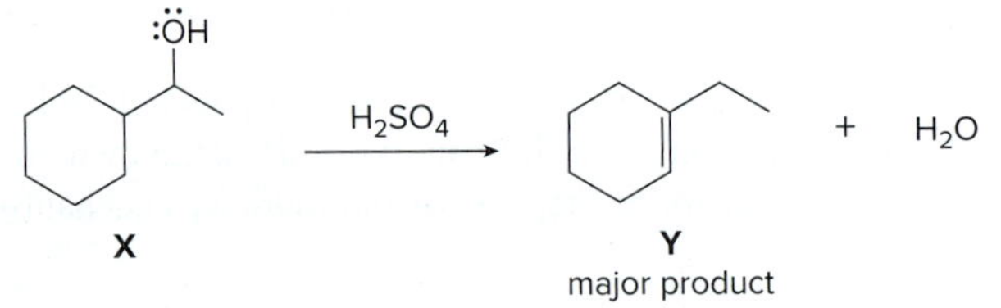

Solution

Steps [1] and [2] Protonation of **X** and loss of H₂O form a 2° carbocation.

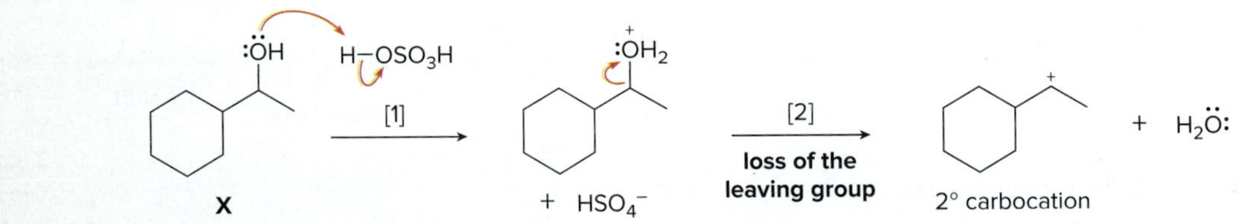

Steps [3] and [4] Rearrangement of the 2° carbocation by a **1,2-hydride shift** forms a more stable 3° carbocation. Loss of a proton from a β carbon forms alkene **Y**.

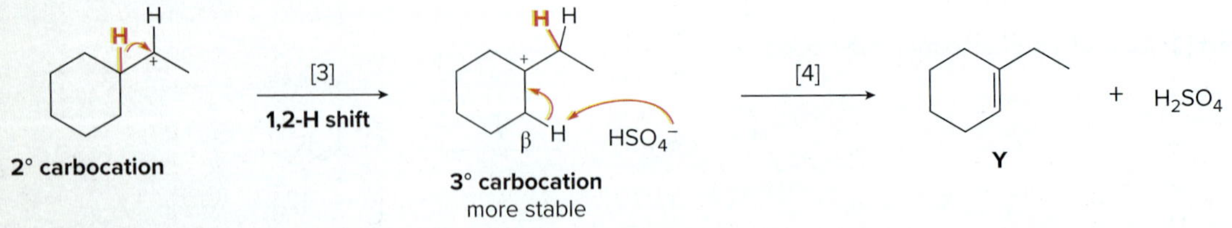

Problem 9.13 What other alkene is also formed along with **Y** in Sample Problem 9.3? What alkenes would form from **X** if no carbocation rearrangement occurred?

More Practice: Try Problems 9.44a, 9.46, 9.47.

Rearrangements are not unique to dehydration reactions. **Rearrangements can occur whenever a carbocation is formed as reactive intermediate,** meaning any S_N1 or E1 reaction. In fact, the formation of rearranged products often indicates the presence of a carbocation intermediate.

Problem 9.14 Explain why two substitution products are formed in the following reaction.

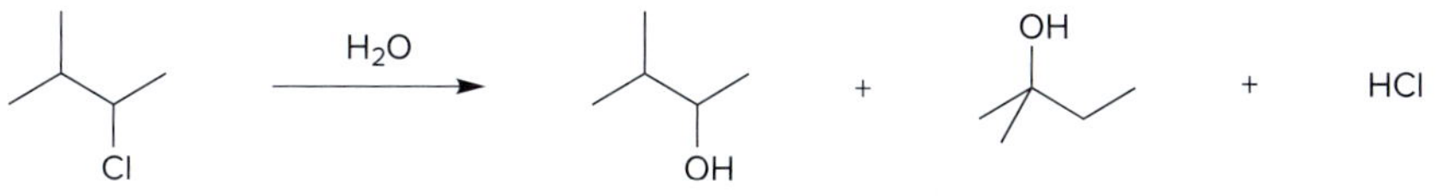

9.10 Dehydration Using POCl₃ and Pyridine

Because some organic compounds decompose in the presence of strong acid, other methods that avoid strong acid have been developed to convert alcohols to alkenes. A common method uses **phosphorus oxychloride ($POCl_3$)** and pyridine (an amine base) in place of H_2SO_4 or TsOH. For example, the treatment of cyclohexanol with $POCl_3$ and pyridine forms cyclohexene in good yield.

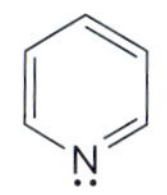

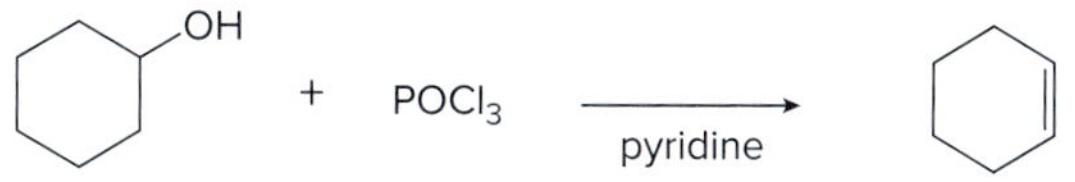

$POCl_3$ serves much the same role as strong acid does in acid-catalyzed dehydration. **It converts a poor leaving group (⁻OH) into a good leaving group.** Dehydration then proceeds by an **E2 mechanism,** as shown in Mechanism 9.4. Pyridine is the base that removes a β proton during elimination.

Mechanism 9.4 Dehydration Using POCl₃ + Pyridine—An E2 Mechanism

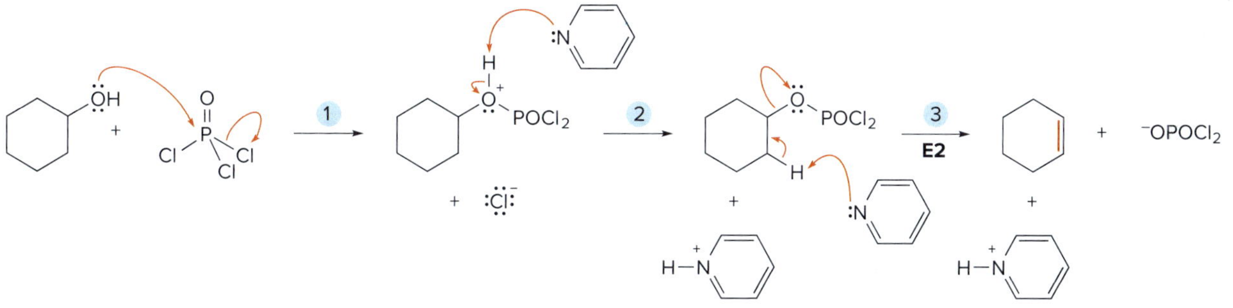

1–2   Reaction of the OH with $POCl_3$ followed by loss of a proton converts a poor leaving group (⁻OH) into a **good leaving group ($^-OPOCl_2$).**

3   Two bonds are broken and two bonds are formed. The base (pyridine) removes a proton; the electron pair in the β C–H bond forms the **π bond,** and the leaving group ($^-OPOCl_2$) departs.

No rearrangements occur during dehydration with $POCl_3$, suggesting that carbocations are *not* formed as intermediates in this reaction. Steps [1] and [2] of the mechanism convert the OH group into a good leaving group. In Step [3], the C–H and C–O bonds are broken and the π bond is formed.

We have now learned about two different reagents for alcohol dehydration—strong acid (H_2SO_4 or TsOH) and $POCl_3$ + pyridine. The best dehydration method for a given alcohol is often hard to know ahead of time, and this is why organic chemists develop more than one method for a given type of transformation.

9.11 Conversion of Alcohols to Alkyl Halides with HX

Alcohols undergo nucleophilic substitution reactions only if the OH group is converted to a better leaving group before nucleophilic attack. Thus, substitution does *not* occur when an alcohol is treated with X⁻ because **⁻OH is a poor leaving group** (Reaction [1]), but substitution *does* occur on treatment of an alcohol with HX because H₂O is now the leaving group (Reaction [2]).

[1] R–ÖH + X⁻ —✗→ R–X + ⁻OH   **poor leaving group** Reaction does *not* occur.

[2] R–ÖH + H–X ⟶ R–X + H₂O   **good leaving group** Reaction occurs.
alkyl halide

- The reaction of alcohols with HX (X = Cl, Br, I) is a general method to prepare 1°, 2°, and 3° alkyl halides.

$$\text{CH}_3\text{CH}_2\text{CH}_2\text{OH} \xrightarrow{\text{HBr}} \text{CH}_3\text{CH}_2\text{CH}_2\text{Br} + \text{H}_2\text{O}$$

cyclohexanol + HCl ⟶ 1-chloro-1-methylcyclohexane + H₂O

More substituted alcohols usually react more rapidly with HX:

RCH₂OH (1°)   R₂CHOH (2°)   R₃COH (3°)

Increasing rate of reaction with HX →

In addition, the reactivity of hydrogen halides *increases* with *increasing* acidity:

H–Cl   H–Br   H–I

Increasing reactivity toward ROH →

Problem 9.15 Draw the products of each reaction.

a. (CH₃)₂C(OH)CH₃ (2-methyl-2-butanol type) $\xrightarrow{\text{HCl}}$

b. cyclopentanol-CH₂OH $\xrightarrow{\text{HI}}$

c. CH₃CH₂CH₂CH(OH)CH₃ $\xrightarrow{\text{HBr}}$

9.11A Two Mechanisms for the Reaction of ROH with HX

How does the reaction of ROH with HX occur? Acid–base reactions are very fast, so the strong acid HX protonates the OH group of the alcohol, forming a **good leaving group** (H₂O) and a **good nucleophile** (the conjugate base, X⁻). Both components are needed for nucleophilic substitution. The mechanism of substitution of X⁻ for H₂O then depends on the structure of the R group.

When there is an oxygen-containing reactant and a strong acid, generally the first step in the mechanism is **protonation of the oxygen atom.**

$$\text{R–ÖH} + \text{H–X} \xrightarrow{\text{proton transfer}} \text{R–ÖH}_2^+ + \text{:X:}^- \xrightarrow[\text{or S}_N2]{\text{S}_N1} \text{R–X:} + \text{H}_2\text{O}$$

good leaving group   good nucleophile

9.11 Conversion of Alcohols to Alkyl Halides with HX

- Methyl and 1° ROH form RX by an S_N2 mechanism.
- Secondary (2°) and 3° ROH form RX by an S_N1 mechanism.

The reaction of CH_3CH_2OH with HBr illustrates the S_N2 mechanism of a 1° alcohol (Mechanism 9.5). Nucleophilic attack on the protonated alcohol occurs in one step: **the bond to the nucleophile X^- is formed *as* the bond to the leaving group (H_2O) is broken.**

Mechanism 9.5 Reaction of a 1° ROH with HX—An S_N2 Mechanism

1. **Protonation** of the OH group forms a **good leaving group** (H_2O).
2. The bond to the nucleophile forms *as* the leaving group departs.

The reaction of $(CH_3)_3COH$ with HBr illustrates the S_N1 mechanism of a 3° alcohol (Mechanism 9.6). Nucleophilic attack on the protonated alcohol occurs in two steps: **the bond to the leaving group (H_2O) is broken *before* the bond to the nucleophile X^- is formed.**

Mechanism 9.6 Reaction of 2° and 3° ROH with HX—An S_N1 Mechanism

1. **Protonation** of the OH group forms a good leaving group (H_2O).
2. Loss of the leaving group forms a **carbocation.**
3. **Nucleophilic attack** of Br^- forms the substitution product.

Both mechanisms begin with the same first step—protonation of the O atom to form a good leaving group—and both mechanisms give an alkyl halide (RX) as product. The mechanisms differ only in the *timing* of bond breaking and bond making.

Knowing the mechanism allows us to predict the stereochemistry of the products when reaction occurs at a stereogenic center.

- Primary (1°) alcohols react by an S_N2 mechanism, so *inversion* occurs at a stereogenic center.
- Secondary (2°) and 3° alcohols react by an S_N1 mechanism, so *racemization* occurs at a stereogenic center.

Sample Problem 9.4 Predicting the Stereochemistry When an Alcohol Reacts with a Hydrogen Halide

Draw the products and stereochemistry for each reaction.

Solution

a. The alcohol is **1°**, so the mechanism of substitution is **S$_N$2**. Because the leaving group OH (which is protonated to form H$_2$O) is drawn on the *right* and S$_N$2 reactions proceed with **inversion of stereochemistry at a stereogenic center,** the nucleophile approaches from the *left* and a single product is formed.

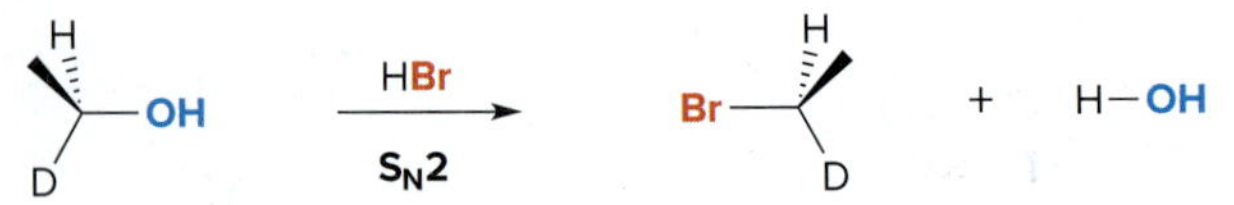

[Structure: 1° alcohol with H, D, OH → HBr / S$_N$2 → product of inversion with H, D, Br + H—OH]

b. The alcohol is **3°**, so the mechanism of substitution is **S$_N$1**. Because S$_N$1 reactions form a **trigonal planar carbocation,** nucleophilic attack of Cl⁻ occurs from in front and behind to afford a **racemic mixture** of two enantiomers.

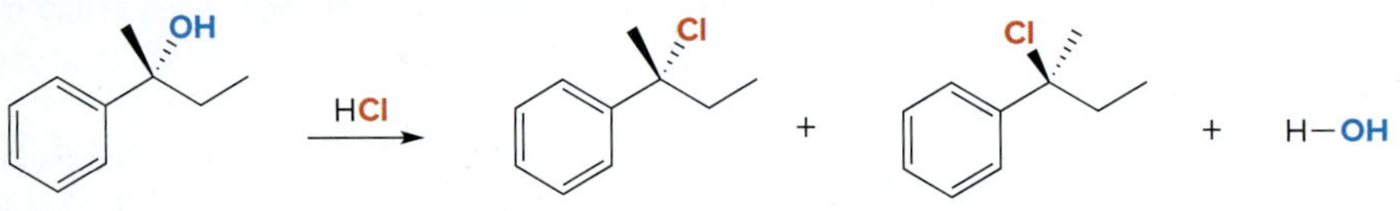

[Structure: 3° alcohol (phenyl, methyl, ethyl, OH) + HCl → two enantiomers (racemic mixture) + H—OH]

Problem 9.16 Draw the products of each reaction, indicating the stereochemistry around any stereogenic centers.

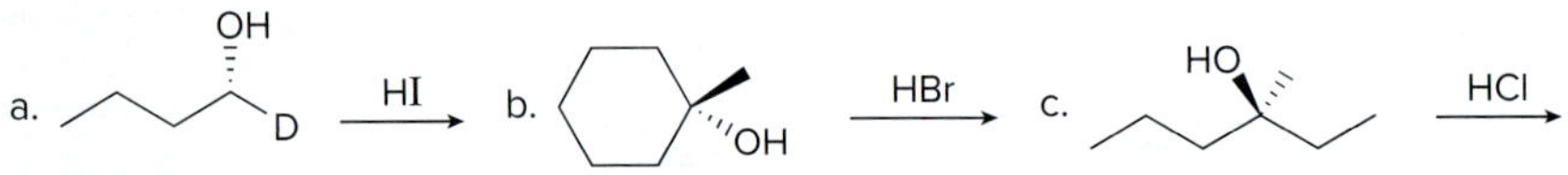

a. [structure with OH and D] —HI→   b. [cyclohexane with methyl and OH] —HBr→   c. [structure with HO] —HCl→

More Practice: Try Problems 9.35a, c; 9.42; 9.60b.

9.11B Carbocation Rearrangement in the S$_N$1 Reaction

Because carbocations are formed in the S$_N$1 reaction of 2° and 3° alcohols with HX, **carbocation rearrangements are possible,** as illustrated in Sample Problem 9.5.

Sample Problem 9.5 Drawing an S$_N$1 Mechanism That Involves a Rearrangement

Draw a stepwise mechanism for the following reaction.

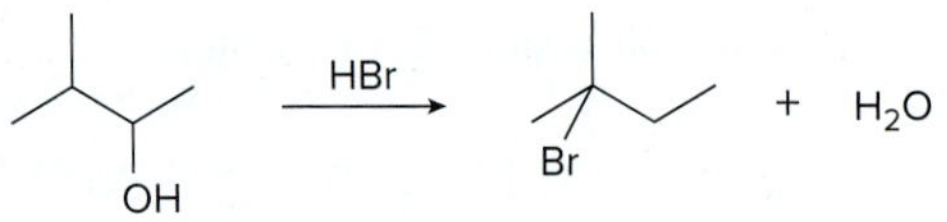

[Structure: 3-methyl-2-butanol] —HBr→ [2-bromo-2-methylbutane] + H$_2$O

Solution

A 2° alcohol reacts with HBr by an S$_N$1 mechanism. Because substitution converts a 2° alcohol to a 3° alkyl halide in this example, a **carbocation rearrangement** must occur.

Steps [1] and [2] Protonation of the O atom and then loss of H$_2$O form a **2° carbocation.**

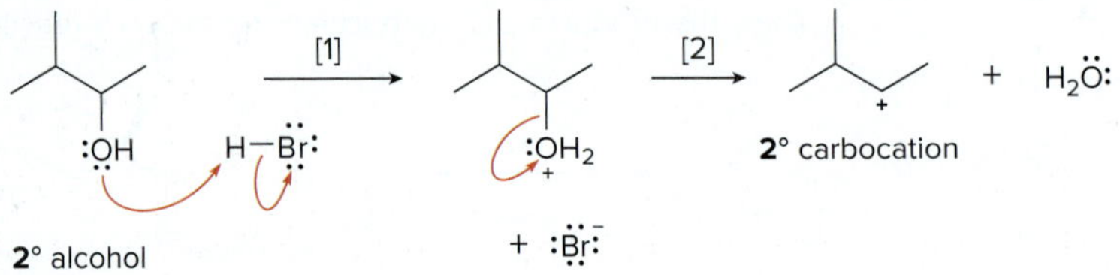

[Mechanism: 2° alcohol with :ÖH, H—Br̈: → [1] → :ÖH$_2$⁺ + :B̈r:⁻ → [2] → 2° carbocation + H$_2$Ö:]

Steps [3] and [4] Rearrangement of the 2° carbocation by a 1,2-hydride shift forms a more stable 3° carbocation. Nucleophilic attack forms the substitution product.

Problem 9.17 What is the major product formed when each alcohol is treated with HCl?

a. (1-methylcyclohexan-1-ol)   b. (3-methylpentan-2-ol)   c. (2-methylcyclohexan-1-ol)

More Practice: Try Problem 9.45.

9.12 Conversion of Alcohols to Alkyl Halides with SOCl₂ and PBr₃

Primary (1°) and 2° alcohols can be converted to alkyl halides using **SOCl₂** and **PBr₃**.

- **SOCl₂** (thionyl chloride) converts alcohols into alkyl chlorides.
- **PBr₃** (phosphorus tribromide) converts alcohols into alkyl bromides.

Both reagents convert ⁻OH into a good leaving group *in situ*—that is, directly in the reaction mixture—as well as provide the **nucleophile**, either Cl⁻ or Br⁻, to displace the leaving group.

9.12A Reaction of ROH with SOCl₂

The treatment of a 1° or 2° alcohol with thionyl chloride, SOCl₂, and pyridine forms an **alkyl chloride**, with SO₂ and HCl as by-products.

$$RCH_2OH + SOCl_2 \xrightarrow{\text{pyridine}} RCH_2Cl \text{ (1° alkyl chloride)} + SO_2 + HCl$$

$$\text{cyclohexanol} + SOCl_2 \xrightarrow{\text{pyridine}} \text{cyclohexyl chloride (2° alkyl chloride)} + SO_2 + HCl$$

The mechanism for this reaction consists of two parts: **conversion of the OH group into a better leaving group, and nucleophilic attack by Cl⁻ via an S_N2 reaction**, as shown in Mechanism 9.7.

Mechanism 9.7 Reaction of ROH with SOCl₂ + Pyridine—An S_N2 Mechanism

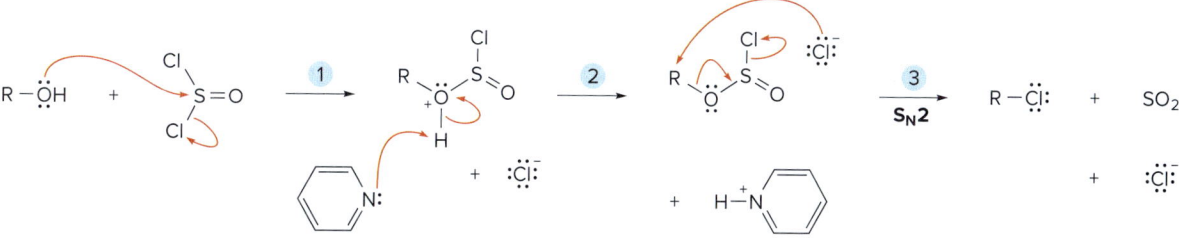

1–2 Reaction of the alcohol with SOCl₂ and loss of a proton convert the OH group to OSOCl, a **good leaving group.**

3 **Nucleophilic attack** of chloride and loss of the leaving group (SO₂ and Cl⁻) form RCl in a single step.

Problem 9.18 If the reaction of an alcohol with SOCl₂ and pyridine follows an S_N2 mechanism, what is the stereochemistry of the alkyl chloride formed from (*R*)-butan-2-ol?

9.12B Reaction of ROH with PBr₃

In a similar fashion, the treatment of a 1° or 2° alcohol with phosphorus tribromide, PBr₃, forms an alkyl bromide.

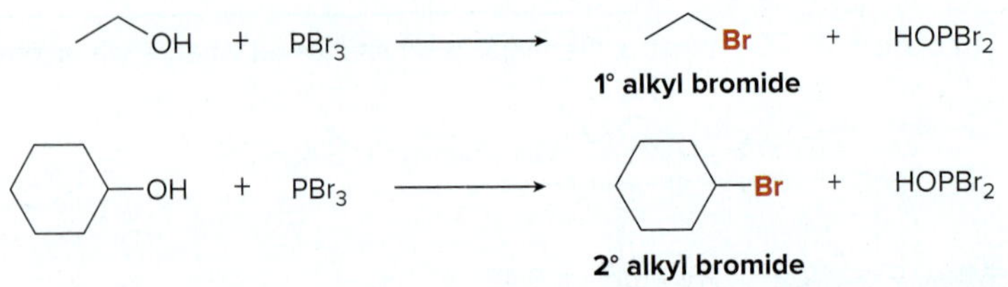

CH₃CH₂CH₂—OH + PBr₃ ⟶ CH₃CH₂CH₂—Br + HOPBr₂
1° alkyl bromide

C₆H₁₁—OH + PBr₃ ⟶ C₆H₁₁—Br + HOPBr₂
2° alkyl bromide

The mechanism for this reaction also consists of two parts: **conversion of the OH group into a better leaving group, and nucleophilic attack by Br⁻ via an S_N2 reaction,** as shown in Mechanism 9.8.

Mechanism 9.8 Reaction of ROH with PBr₃—An S_N2 Mechanism

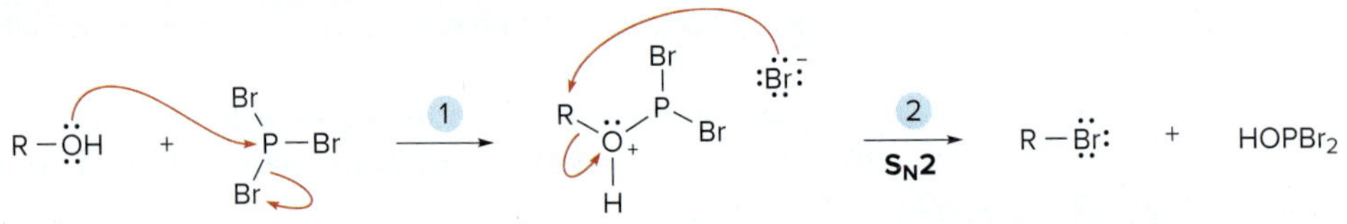

R—ÖH + Br–P–Br (Br) ⟶[1] R–Ö⁺(H)–P(Br)–Br :Br:⁻ ⟶[2, S_N2] R—B̈r: + HOPBr₂

1. Reaction of the alcohol with PBr₃ converts the OH group to OPBr₂, a **good leaving group**, and generates the nucleophile, Br⁻.
2. **Nucleophilic attack** of bromide and loss of the leaving group form RBr in a single step.

Table 9.2 summarizes the methods for converting an alcohol to an alkyl halide presented in Sections 9.11 and 9.12.

Table 9.2 Summary of Methods for ROH → RX

Overall reaction	Reagent	Comment
ROH → RCl	HCl	• Useful for all ROH • An S_N1 mechanism for 2° and 3° ROH; an S_N2 mechanism for CH₃OH and 1° ROH
	SOCl₂	• Best for CH₃OH, and 1° and 2° ROH • An S_N2 mechanism
ROH → RBr	HBr	• Useful for all ROH • An S_N1 mechanism for 2° and 3° ROH; an S_N2 mechanism for CH₃OH and 1° ROH
	PBr₃	• Best for CH₃OH, and 1° and 2° ROH • An S_N2 mechanism
ROH → RI	HI	• Useful for all ROH • An S_N1 mechanism for 2° and 3° ROH; an S_N2 mechanism for CH₃OH and 1° ROH

Problem 9.19 If the reaction of an alcohol with PBr₃ follows an S_N2 mechanism, what is the stereochemistry of the alkyl bromide formed from (R)-butan-2-ol?

Problem 9.20 Draw the organic products formed in each reaction, and indicate the stereochemistry of products that contain stereogenic centers.

a. CH₃CH₂CH₂CH₂OH $\xrightarrow{\text{SOCl}_2, \text{ pyridine}}$

b. (structure with OH on stereocenter) $\xrightarrow{\text{HI}}$

c. (cyclopentane with methyl and OH, stereochemistry shown) $\xrightarrow{\text{PBr}_3}$

9.13 Tosylate—Another Good Leaving Group

We have now learned two methods to convert the OH group of an alcohol to a better leaving group: treatment with strong acids (Section 9.8A), and conversion to an alkyl halide (Sections 9.11–9.12). Alcohols can also be converted to **alkyl tosylates**.

Recall from Section 1.5 that a third-row element like sulfur can have 10 or 12 electrons around it in a valid Lewis structure.

*An alkyl tosylate is often called simply a **tosylate**.*

An alkyl tosylate

R–OH (poor leaving group) → R–O–S(=O)(=O)–C₆H₄–CH₃ (tosylate, good leaving group)

An **alkyl tosylate** is composed of two parts: the **alkyl group R**, derived from an alcohol; and the **tosylate** (short for **p-toluenesulfonate**), which is a good leaving group. A tosyl group, $CH_3C_6H_4SO_2-$, is abbreviated as **Ts**, so an alkyl tosylate becomes **ROTs**.

tosyl group (p-toluenesulfonyl group) = **Ts**

R–O–Ts = **ROTs**

9.13A Conversion of Alcohols to Alkyl Tosylates

A tosylate (TsO⁻) is similar to I⁻ in leaving group ability.

Alcohols are converted to alkyl tosylates by treatment with *p*-toluenesulfonyl chloride (TsCl) in the presence of pyridine. This overall process converts a poor leaving group (⁻OH) into a good one (⁻OTs). A tosylate is a good leaving group because its conjugate acid, *p*-toluenesulfonic acid ($CH_3C_6H_4SO_3H$, TsOH), is a strong acid ($pK_a = -7$, Section 2.6).

R–OH + TsCl (p-toluenesulfonyl chloride, tosyl chloride) $\xrightarrow{\text{pyridine}}$ R–OTs (good leaving group) + pyridinium⁺H + Cl⁻

362 Chapter 9 Alcohols, Ethers, and Related Compounds

(S)-Butan-2-ol is converted to its tosylate with **retention of configuration** at the stereogenic center. Thus, the C–O bond of the alcohol must *not* be broken when the tosylate is formed.

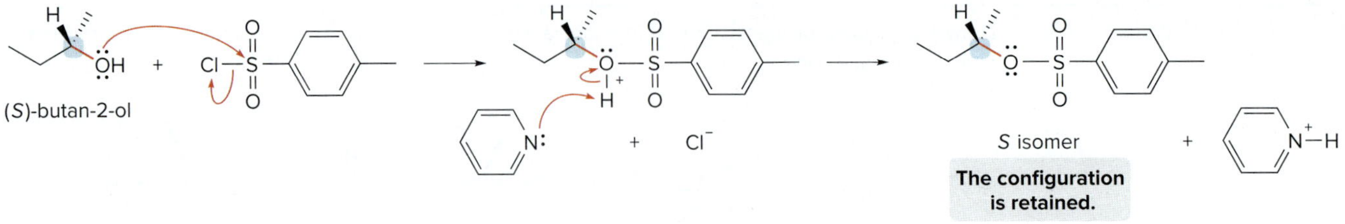

Problem 9.21 Draw the products of each reaction, and indicate the stereochemistry at any stereogenic center.

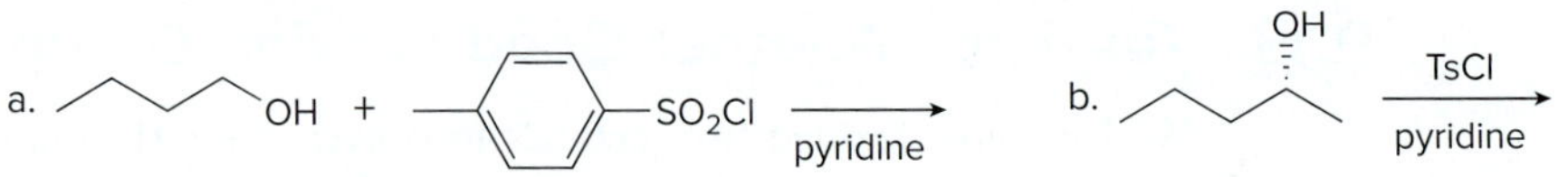

a. [structure: 1-butanol + TsCl/pyridine]

b. [structure: (S)-pentan-2-ol + TsCl/pyridine]

9.13B Reactions of Alkyl Tosylates

Because alkyl tosylates have good leaving groups, **they undergo both nucleophilic substitution and β elimination,** exactly as alkyl halides do. Generally, alkyl tosylates are treated with strong nucleophiles and bases, so that the mechanism of substitution is S_N2 and the mechanism of elimination is **E2**.

For example, propyl tosylate, which has the leaving group on a 1° carbon, reacts with $NaOCH_3$ to yield methyl propyl ether, the product of nucleophilic substitution by an S_N2 mechanism. Propyl tosylate reacts with $KOC(CH_3)_3$, a strong bulky base, to yield propene by an E2 mechanism.

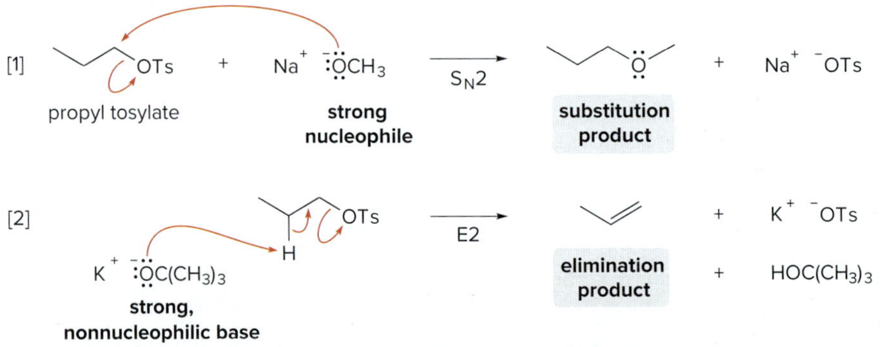

[1] propyl tosylate + Na⁺ ⁻OCH₃ (strong nucleophile) → S_N2 → substitution product + Na⁺ ⁻OTs

[2] propyl tosylate + K⁺ ⁻OC(CH₃)₃ (strong, nonnucleophilic base) → E2 → elimination product + K⁺ ⁻OTs + HOC(CH₃)₃

Because substitution occurs via an S_N2 mechanism, **inversion of configuration** results when the leaving group is bonded to a stereogenic center.

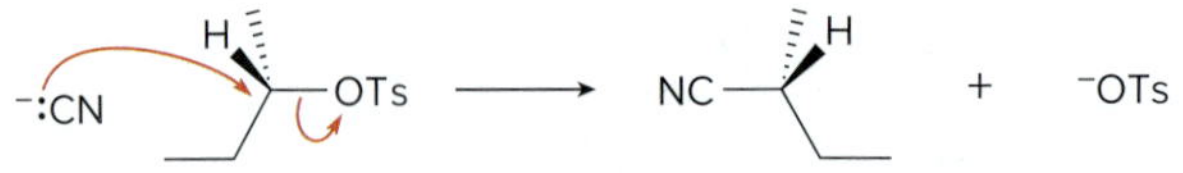

[structure: ⁻:CN attacking R–OTs → NC–R (inverted) + ⁻OTs]

Inversion of configuration

Sample Problem 9.6 Drawing the Substitution Product from an Alkyl Tosylate

Draw the product of the following reaction, including stereochemistry.

[structure: alkyl tosylate with D label + Na⁺ ⁻OCH₂CH₃]

Solution

The 1° alkyl tosylate and the strong nucleophile both favor substitution by an S_N2 mechanism, which proceeds by backside attack, resulting in **inversion** of configuration at the stereogenic center. The leaving group is drawn in front (on a wedge), so the nucleophile approaches from behind, ending up on a dashed wedge.

Problem 9.22 Draw the products of each reaction, and include the stereochemistry at any stereogenic center in the products.

a. CH₃CH₂CH₂CH₂OTs + ⁻CN ⟶

b. CH₃CH₂CH₂OTs + K⁺ ⁻OC(CH₃)₃ ⟶

c. (2-methylbutyl OTs) + ⁻SH ⟶

d. (4-methylcyclohexyl OTs) + NaOCH₂CH₃ ⟶

More Practice: Try Problems 9.42d; 9.43b; 9.60d, f.

9.13C The Two-Step Conversion of an Alcohol to a Substitution Product

We now have another **two-step method to convert an alcohol to a substitution product:** reaction of an alcohol with TsCl and pyridine to form an alkyl tosylate (Step [1]), followed by nucleophilic attack on the tosylate (Step [2]).

$$R-OH \xrightarrow[\text{pyridine}]{\text{TsCl}} R-OTs \xrightarrow[\text{[2]}]{:Nu^-} R-Nu + {}^-OTs$$
                                [1]

Let's look at the stereochemistry of this two-step process.

- Step [1], formation of the tosylate, proceeds with **retention** of configuration at a stereogenic center because the C—O bond remains intact.
- Step [2] is an S_N2 reaction, so it proceeds with **inversion of configuration** because the nucleophile attacks from the back side.
- Overall there is a **net inversion of configuration** at a stereogenic center.

For example, the treatment of *cis*-3-methylcyclohexanol with *p*-toluenesulfonyl chloride and pyridine forms a cis tosylate **A**, which undergoes backside attack by the nucleophile ⁻OCH₃ to yield the trans ether **B**.

cis ⟶ (TsCl, pyridine [1], retention of configuration) ⟶ A (cis) ⟶ (⁻OCH₃ [2], inversion of configuration) ⟶ B (trans)

Problem 9.23 Draw the products formed when (S)-butan-2-ol is treated with TsCl and pyridine, followed by NaOH. Label the stereogenic center in each compound as *R* or *S*. What is the stereochemical relationship between the starting alcohol and the final product?

9.13D A Summary of Substitution and Elimination Reactions of Alcohols

The reactions of alcohols in Sections 9.8–9.13C share two similarities:

- The OH group is converted into a better leaving group by treatment with acid or another reagent.
- The resulting product undergoes either elimination or substitution, depending on the reaction conditions.

Figure 9.4 summarizes these reactions with cyclohexanol as starting material.

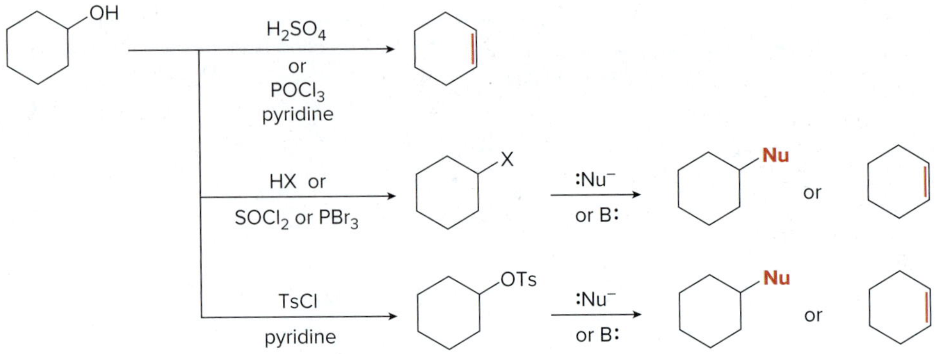

Figure 9.4
Summary: Nucleophilic substitution and β elimination reactions of alcohols

---

Problem 9.24 Draw the product formed when $(CH_3)_2CHOH$ is treated with each reagent.

a. $SOCl_2$, pyridine
b. TsCl, pyridine
c. H_2SO_4
d. HBr
e. PBr_3, then NaCN
f. $POCl_3$, pyridine

9.14 Reaction of Ethers with Strong Acid

Because ethers are so unreactive, diethyl ether and tetrahydrofuran (THF) are often used as solvents for organic reactions.

Recall from Section 9.7B that ethers have a poor leaving group, so they cannot undergo nucleophilic substitution or β elimination reactions directly. Instead, they must first be converted into a good leaving group by reaction with strong acids. Only **HBr** and **HI** can be used, though, because they are strong acids that are also sources of good nucleophiles (Br^- and I^-, respectively). **When ethers react with HBr or HI, both C—O bonds are cleaved and two alkyl halides are formed as products.**

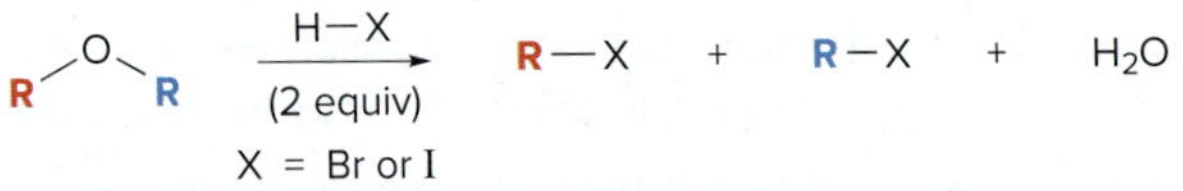

$$R-O-R \xrightarrow[\text{(2 equiv)}]{H-X} R-X + R-X + H_2O$$
$$X = Br \text{ or } I$$

HBr or HI serves as a strong acid that both protonates the O atom of the ether and is the source of a good nucleophile (Br^- or I^-). Because both C—O bonds in the ether are broken, **two successive nucleophilic substitution reactions occur.**

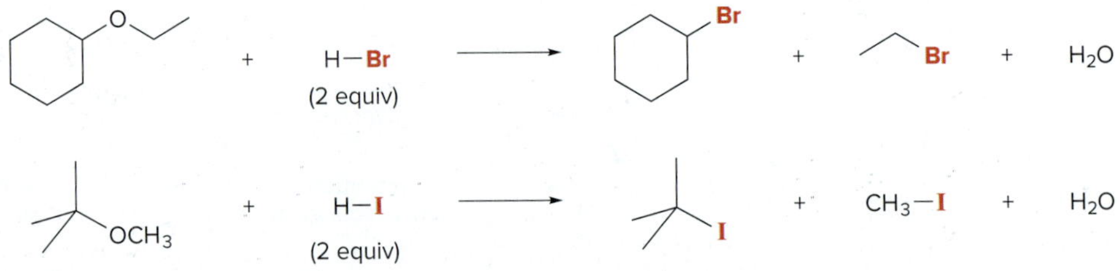

- The mechanism of ether cleavage is S_N1 or S_N2, depending on the identity of R.
- With 2° or 3° alkyl groups bonded to the ether oxygen, the C—O bond is cleaved by an S_N1 mechanism involving a carbocation; with methyl or 1° R groups, the C—O bond is cleaved by an S_N2 mechanism.

9.14 Reaction of Ethers with Strong Acid

For example, cleavage of $(CH_3)_3COCH_3$ with HI occurs at two bonds, as shown in Mechanism 9.9. The 3° alkyl group undergoes nucleophilic substitution by an S_N1 mechanism, resulting in the cleavage of one C—O bond. The methyl group undergoes nucleophilic substitution by an S_N2 mechanism, resulting in the cleavage of the second C—O bond.

- Bond to the 3° C is cleaved by an S_N1 reaction.
- Bond to the methyl C is cleaved by an S_N2 reaction.

Mechanism 9.9 Mechanism of Ether Cleavage in Strong Acid—
$(CH_3)_3COCH_3 + HI \rightarrow (CH_3)_3CI + CH_3I + H_2O$

Part [1] Cleavage of the 3° C—O bond by an S_N1 mechanism

1. **Protonation** of the ether O atom forms a **good leaving group.**
2. **Cleavage** of the C—O bond to the 3° carbon forms a **3° carbocation** and CH_3OH.
3. **Nucleophilic attack of I^-** forms the substitution product.

Part [2] Cleavage of the CH_3—O bond by an S_N2 mechanism

4. **Protonation** of the OH group forms a **good leaving group** (H_2O).
5. **Nucleophilic attack** of iodide forms the second alkyl halide, CH_3I. Because the mechanism is S_N2, the C—O bond is broken as the C—I bond (in red) is formed.

The mechanism illustrates the central role of HX in the reaction:

- HX protonates the ether oxygen, thus making a good leaving group.
- HX provides a source of X^- for nucleophilic attack.

Problem 9.25 What alkyl halides are formed when each ether is treated with HBr?

a., b., c.

Problem 9.26 Explain why the treatment of anisole with HBr yields phenol and CH_3Br, but not bromobenzene.

anisole —OCH_3 →(HBr) phenol —OH + CH_3Br [bromobenzene —Br]

9.15 Thiols and Sulfides

Thiols and **sulfides** are sulfur analogues of alcohols and ethers, respectively.

9.15A Thiols

Thiols, also called mercaptans, contain a mercapto group (SH) bonded to a carbon atom. Because sulfur is below oxygen in the periodic table, the sulfur atom is surrounded by two atoms and two lone pairs, giving thiols a **bent shape.** Unlike alcohols, however, thiols are incapable of intermolecular hydrogen bonds, so thiols have *lower* boiling points and melting points than alcohols with a similar number of carbons.

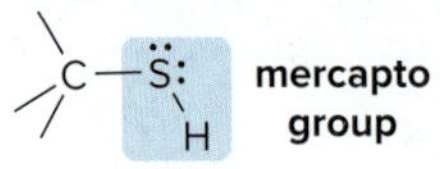

thiol — mercapto group

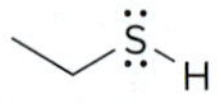

ethanethiol
bp 35 °C

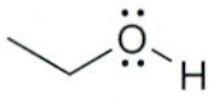

ethanol
bp 78 °C

Many simple thiols have pungent and disagreeable odors. Skunks, onions, and human sweat all contain thiols.

propane-1-thiol
onion odor

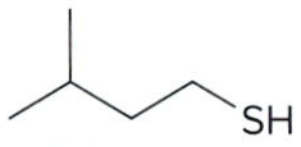

3-methylbutane-1-thiol
skunk odor

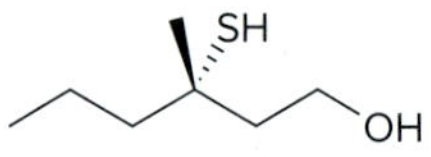

(S)-3-methyl-3-sulfanylhexan-1-ol
onion-like odor in human sweat

Thiols are named in a similar method to alcohols, using the suffix **-thiol** instead of the suffix *-ol.* To name a thiol in the IUPAC system:

- Name the parent carbon chain and add the suffix *-thiol.*
- Number the carbon chain to give the SH group the lower number and apply the other rules of nomenclature.

Examples of thiol nomenclature are given in Figure 9.5.

Figure 9.5 Naming thiols

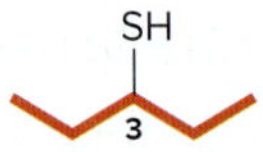

pentane-3-thiol

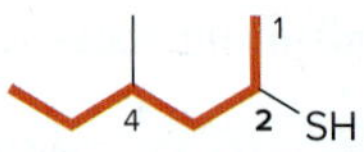

4-methylhexane-2-thiol

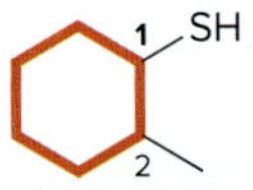

2-methylcyclohexanethiol

Problem 9.27 Name each thiol.

a. 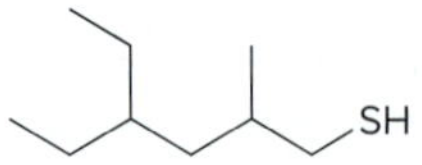   b.

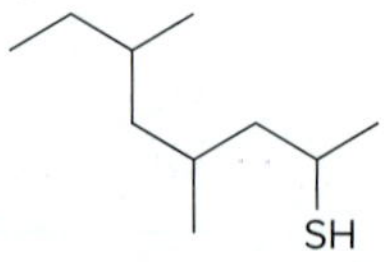

Thiols are prepared by S_N2 reactions of alkyl halides with $^-$SH, a good nucleophile.

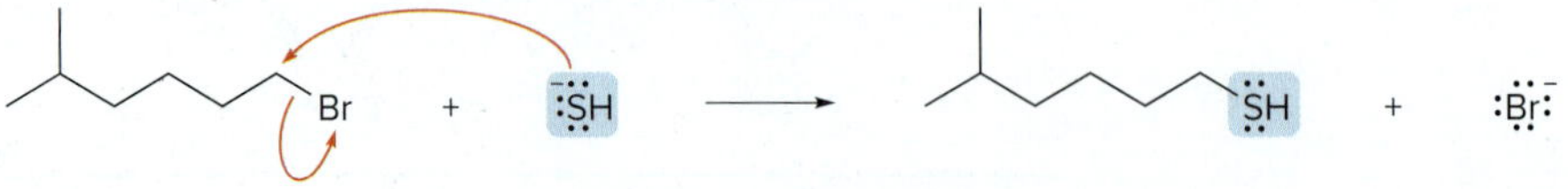

Thiols are easily oxidized with Br_2 or I_2 to **disulfides (RSSR),** compounds that contain a sulfur–sulfur bond. This reaction is an oxidation (Section 4.14) because H atoms are removed from the thiol in forming the disulfide. Disulfides are reduced to thiols with Zn and acid.

Disulfide formation is especially important in determining the shape and properties of some proteins that contain the amino acid cysteine, as we will learn in Chapter 23.

Problem 9.28 Draw the product of each reaction.

9.15B Sulfides

Sulfides contain two alkyl groups bonded to a sulfur atom. Sulfides are named with the same rules used to name ethers. The suffix *sulfide* is used instead of *ether* for simple compounds.

The potent odor of grapefruit mercaptan (Problem 9.28c) contributes to the characteristic aroma of grapefruit.
Purestock/SuperStock

To name more complex sulfides using the IUPAC system, one alkyl group is named as a parent chain and the other is named as part of a substituent bonded to that chain.

- Name the simpler alkyl group + S atom as an *alkylthio* substituent.
- Name the remaining alkyl group as an alkane with an alkylthio substituent using the usual rules of nomenclature.

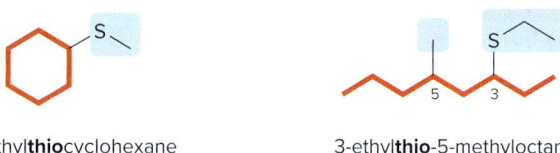

Problem 9.29 Give the IUPAC name for each sulfide.

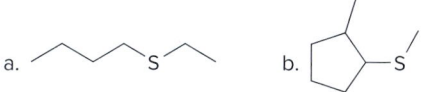

Sulfides are prepared from thiols by an S_N2 reaction that is analogous to the Williamson ether synthesis.

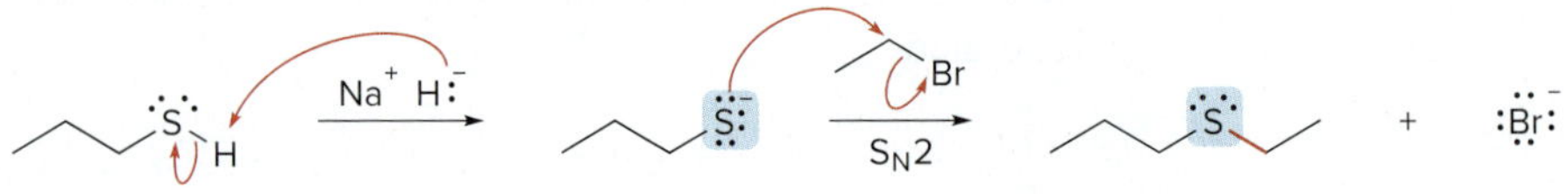

propane-1-thiol → sulfur nucleophile + H_2 → ethyl propyl sulfide + :Br:⁻

Sulfides contain a nucleophilic sulfur atom that reacts readily with unhindered alkyl halides to form **sulfonium ions.**

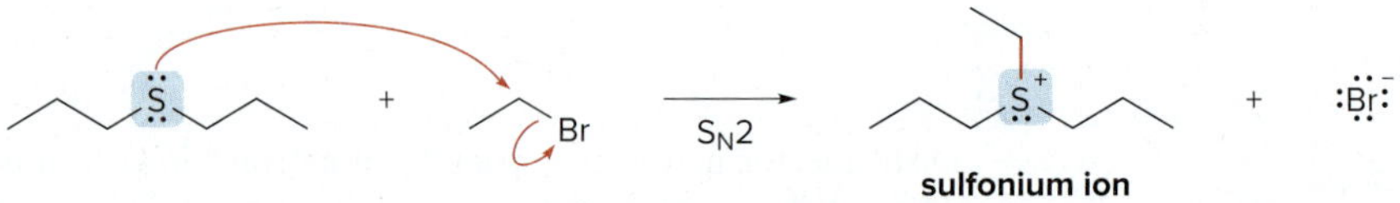

sulfonium ion

S-Adenosylmethionine (SAM), a biological sulfonium ion that was introduced in Section 7.16, is synthesized from the amino acid methionine, which contains a nucleophilic sulfide, and adenosine triphosphate (ATP), which contains a triphosphate leaving group (Section 7.16).

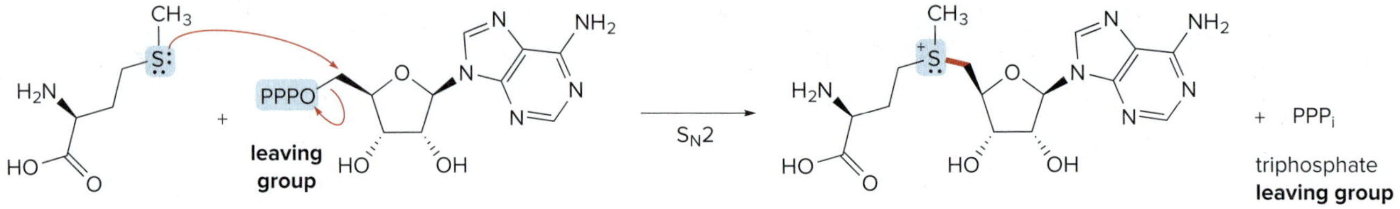

methionine + adenosine triphosphate ATP → *S*-adenosylmethionine + PPP_i triphosphate leaving group

---

Problem 9.30 Draw the product of each reaction.

a. 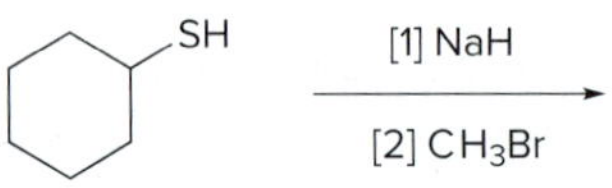 cyclohexyl-SH, [1] NaH, [2] CH_3Br

b. 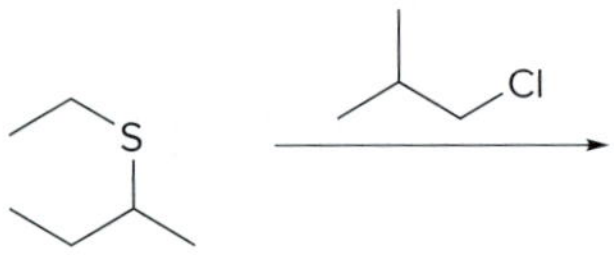 sec-butyl methyl sulfide + isobutyl chloride

---

9.16 Reactions of Epoxides

Although epoxides do not contain a good leaving group, they contain a strained three-membered ring with two polar bonds. **Nucleophilic attack opens the strained three-membered ring, making it a favorable process even with the poor leaving group.**

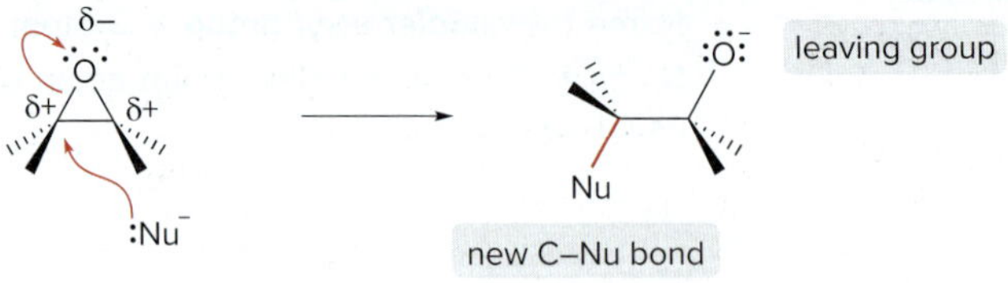

new C–Nu bond   leaving group

This reaction occurs readily with strong nucleophiles like ⁻CN, and with acids like HZ, where Z is a nucleophilic atom.

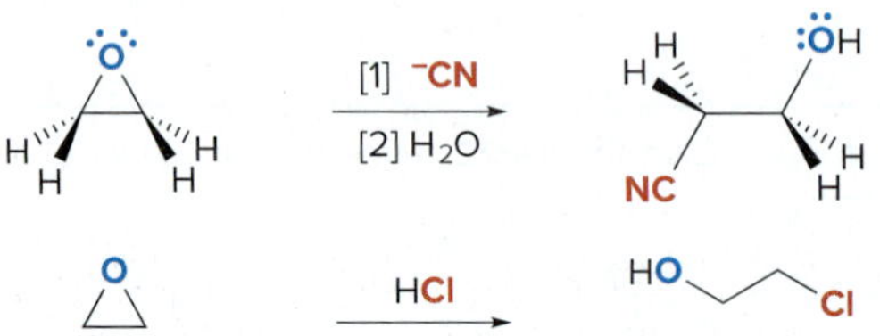

ethylene oxide [1] ⁻CN [2] H_2O → HOCH(H)–CH(H)CN

propylene oxide + HCl → HOCH₂CH₂Cl

9.16A Opening of Epoxide Rings with Strong Nucleophiles

Virtually all strong nucleophiles open an epoxide ring by a two-step reaction sequence.

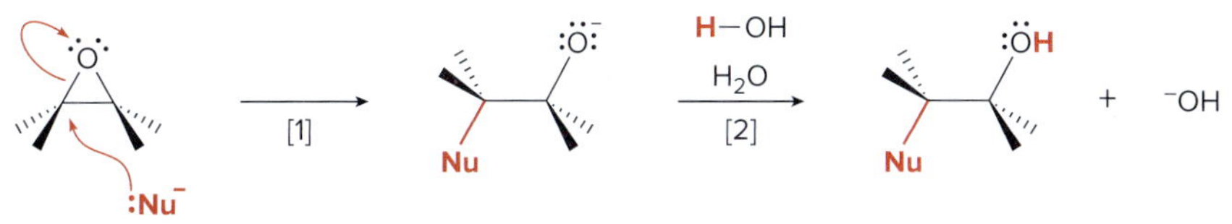

- **Step [1]:** The nucleophile attacks an electron-deficient carbon of the epoxide, cleaving a C–O bond and relieving the strain of the three-membered ring.
- **Step [2]:** Protonation of the alkoxide with water generates a neutral product with two functional groups on adjacent atoms.

Common nucleophiles that open epoxide rings include $^-$OH, $^-$OR, $^-$CN, $^-$SR, and NH$_3$. With these strong nucleophiles, the reaction occurs via an **S$_N$2 mechanism,** resulting in two consequences:

- The nucleophile opens the epoxide ring from the back side.

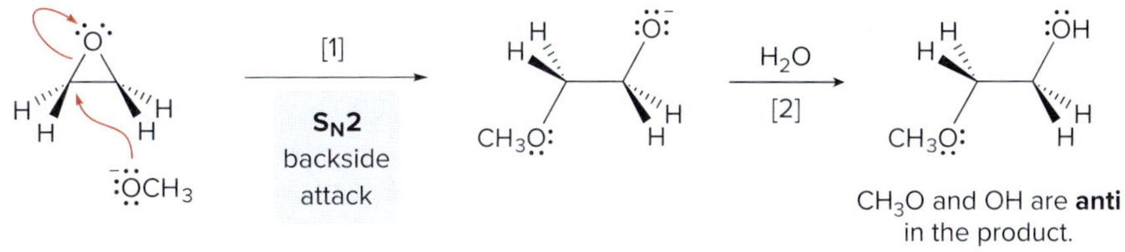

CH$_3$O and OH are **anti** in the product.

Other examples of the nucleophilic opening of epoxide rings are presented in Sections 10.20B and 13.14.

- In an unsymmetrical epoxide, the nucleophile attacks at the *less* substituted carbon atom.

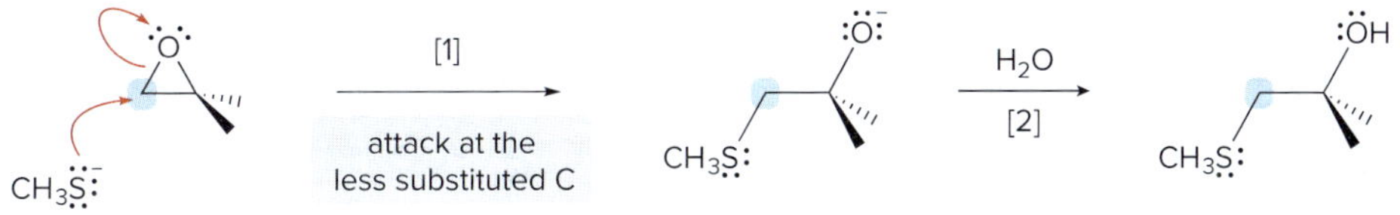

Problem 9.31 Draw the product of each reaction, and indicate the stereochemistry at any stereogenic center.

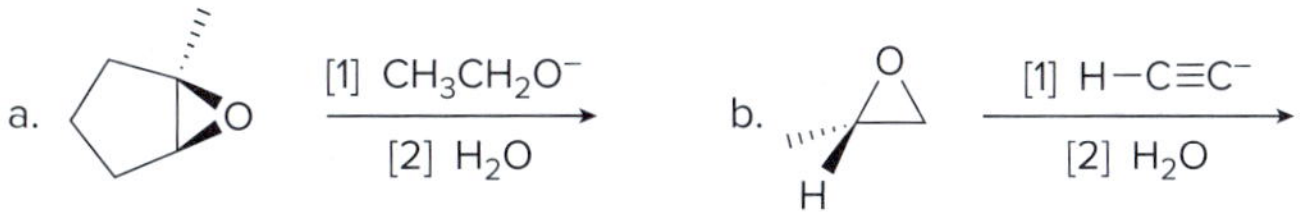

a. [1] CH$_3$CH$_2$O$^-$ [2] H$_2$O

b. [1] H–C≡C$^-$ [2] H$_2$O

1,2-Epoxycyclohexane, an achiral epoxide with a plane of symmetry, reacts with $^-$OCH$_3$ to yield two *trans*-1,2-disubstituted cyclohexanes, **A** and **B**, which are **enantiomers;** each has two stereogenic centers.

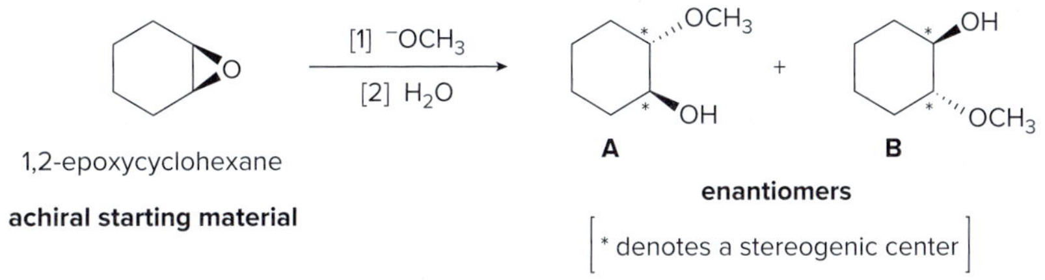

1,2-epoxycyclohexane
achiral starting material

A   **B**
enantiomers

[* denotes a stereogenic center]

Nucleophilic attack of $^-$OCH$_3$ occurs from the back side at *either* C–O bond, because both ends are equally substituted. Because attack at either side occurs with equal probability, an equal amount of the two enantiomers is formed—**a racemic mixture.** This is a specific example of a general rule concerning the stereochemistry of products obtained from an achiral reactant.

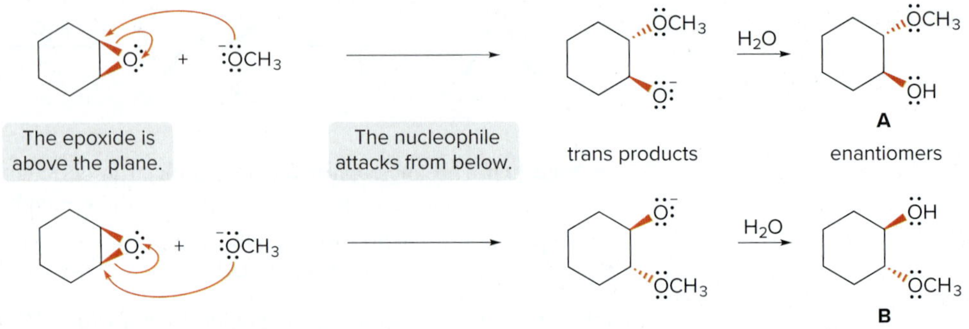

- Whenever an achiral reactant yields a product with stereogenic centers, the product must be achiral (meso) or racemic.

This general rule can be restated in terms of optical activity. Recall from Section 5.12 that achiral compounds and racemic mixtures are optically inactive.

- Optically *inactive* starting materials give optically *inactive* products.

Problem 9.32 The cis and trans isomers of 2,3-dimethyloxirane both react with ⁻OH to give butane-2,3-diol. One stereoisomer gives a single achiral product, and one gives two chiral enantiomers. Which epoxide gives one product and which gives two?

cis-2,3-dimethyloxirane

one enantiomer of *trans*-2,3-dimethyloxirane

9.16B Reaction with Acids HZ

Acids **HZ** that contain a nucleophile **Z** also open epoxide rings by a two-step reaction sequence.

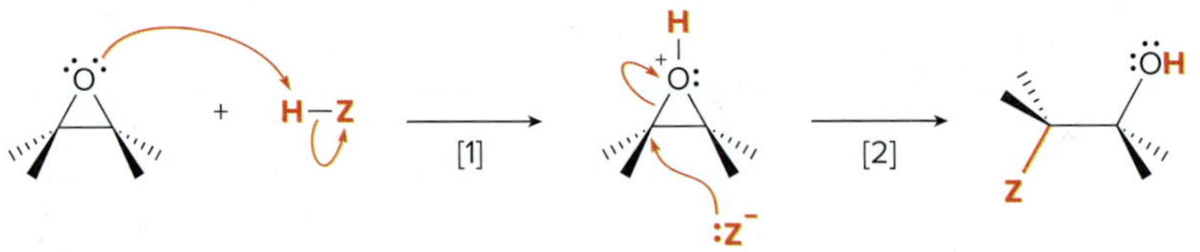

- **Step [1]:** Protonation of the epoxide oxygen with HZ makes the epoxide oxygen into a good leaving group (OH). It also provides a source of a good nucleophile (Z⁻) to open the epoxide ring.
- **Step [2]:** The nucleophile Z⁻ then opens the protonated epoxide ring by **backside** attack.

These two steps—**protonation followed by nucleophilic attack**—are the exact reverse of the opening of epoxide rings with strong nucleophiles, where nucleophilic attack precedes protonation.

HCl, HBr, and **HI** all open an epoxide ring in this manner. H_2O and **ROH** can, too, but acid must also be added. Regardless of the reaction, the product has an OH group from the epoxide on one carbon and a new functional group Z from the nucleophile on the adjacent carbon. With epoxides fused to rings, ***trans*-1,2-disubstituted cycloalkanes** are formed.

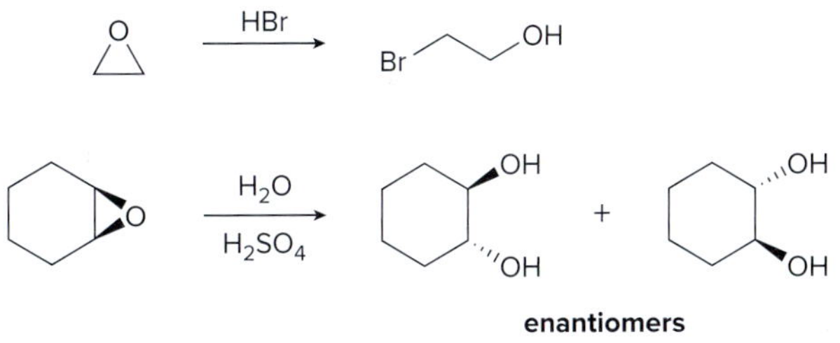

Although backside attack of the nucleophile suggests that this reaction follows an S_N2 mechanism, the regioselectivity of the reaction with unsymmetrical epoxides does not.

- **With unsymmetrical epoxides, nucleophilic attack occurs at the *more* substituted carbon atom.**

For example, the treatment of 2,2-dimethyloxirane with HCl results in nucleophilic attack at the carbon with two methyl groups.

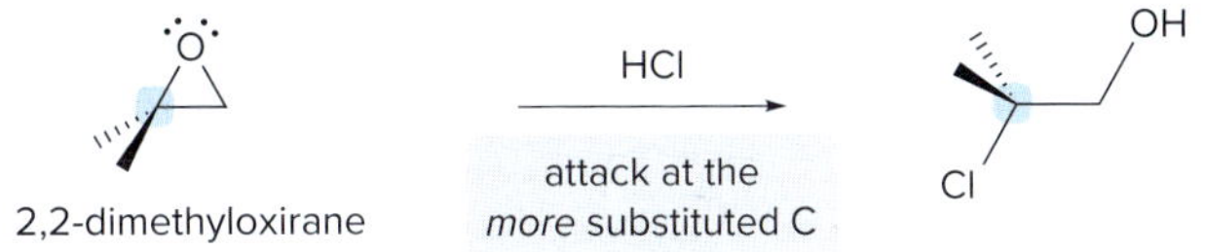

Backside attack of the nucleophile suggests an S_N2 mechanism, but attack at the more substituted carbon suggests an S_N1 mechanism. To explain these results, the **mechanism of nucleophilic attack is thought to be somewhere in between S_N1 and S_N2.**

Figure 9.6 illustrates two possible pathways for the reaction of 2,2-dimethyloxirane with HCl. Backside attack of Cl⁻ at the more substituted carbon proceeds via transition state **A**, whereas backside attack of Cl⁻ at the less substituted carbon proceeds via transition state **B. Transition state A has a partial positive charge on a more substituted carbon, making it more stable.** Thus, the preferred reaction path takes place by way of the lower-energy transition state **A**.

Figure 9.6
Opening of an unsymmetrical epoxide ring with HCl

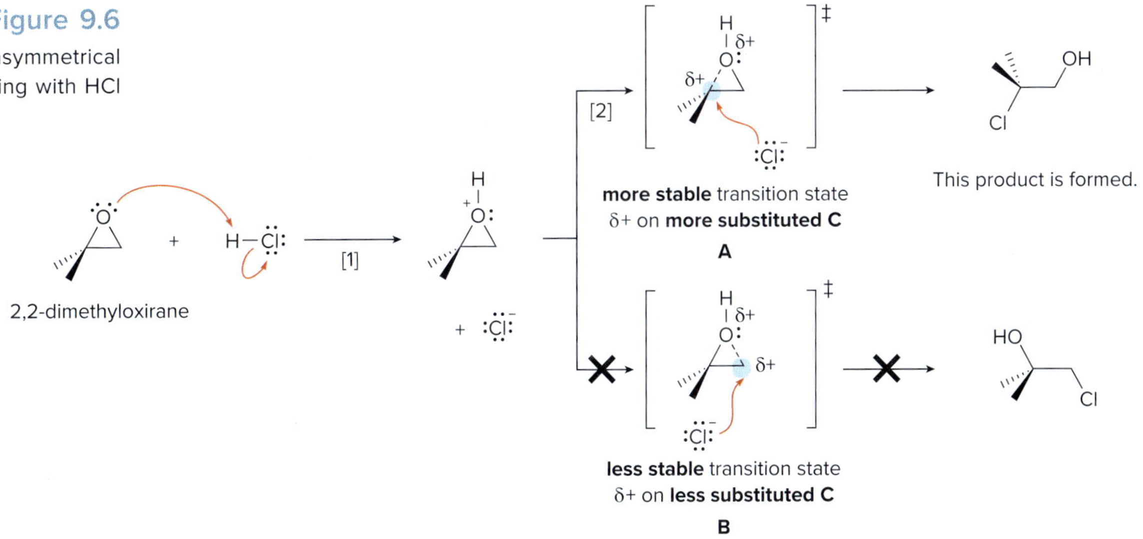

- Transition state **A** is lower in energy because the partial positive charge (δ+) is located on the *more* substituted carbon. In this case, therefore, nucleophilic attack occurs from the back side (an S_N2 characteristic) at the *more* substituted carbon (an S_N1 characteristic).

Chapter 9 Alcohols, Ethers, and Related Compounds

Opening of an epoxide ring with either a strong nucleophile :Nu⁻ or an acid HZ is **regioselective**, because one constitutional isomer is the major or exclusive product. The **site selectivity of these two reactions, however, is *exactly the opposite*.**

- With a strong nucleophile, :Nu⁻ attacks at the *less* substituted carbon.
- With an acid HZ, the nucleophile attacks at the *more* substituted carbon.

Sample Problem 9.7 — Determining the Regioselectivity of Opening an Epoxide Ring

What product is formed when 2,2-dimethyloxirane is treated with each set of reagents: ⁻OCH₃ followed by H₂O, or CH₃OH and H₂SO₄?

Solution
All nucleophiles open an epoxide ring from the back side. Classify the nucleophile to determine if nucleophilic attack occurs at the *more* or *less* substituted carbon. With **strong,** negatively charged nucleophiles, attack occurs at the *less* substituted carbon, whereas with **acids HZ,** nucleophilic attack occurs at the *more* substituted carbon.

2,2-dimethyloxirane —[1] ⁻OCH₃; [2] H₂O→ (attack at the **less** substituted C) → HO–C(CH₃)₂–CH₂–OCH₃
With a strong nucleophile, CH₃O ends up on the **less substituted C.**

2,2-dimethyloxirane —CH₃OH, H₂SO₄→ (attack at the **more** substituted C) → CH₃O–C(CH₃)₂–CH₂–OH
With acid, CH₃O ends up on the **more substituted C.**

Problem 9.33 Draw the product of each reaction.

a. cyclohexene oxide + HBr
b. cyclohexene oxide + [1] ⁻CN; [2] H₂O
c. 2,3-epoxybutane + CH₃CH₂OH, H₂SO₄
d. 2,3-epoxybutane + [1] CH₃O⁻; [2] CH₃OH

More Practice: Try Problems 9.57; 9.60g, h.

The reaction of epoxide rings with nucleophiles is important for the synthesis of many biologically active compounds, including **albuterol**, a bronchodilator used in the treatment of asthma (Figure 9.7).

Figure 9.7 The synthesis of a bronchodilator using the opening of an epoxide ring

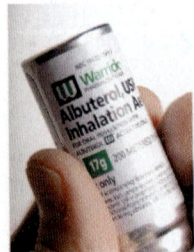

Jill Braaten/McGraw-Hill Education

- A key step in the synthesis is the opening of an epoxide ring with a nitrogen nucleophile to form a new C–N bond, shown in red.

Generic name albuterol
Trade names Proventil, Ventolin

9.17 Application: Epoxides, Leukotrienes, and Asthma

The opening of epoxide rings with nucleophiles is a key step in some important biological processes.

9.17A Asthma and Leukotrienes

Asthma is an obstructive lung disease that affects millions of Americans. Because it involves episodic constriction of small airways, bronchodilators such as albuterol (Figure 9.7) are used to treat symptoms by widening airways. Because asthma is also characterized by chronic inflammation, inhaled steroids that reduce inflammation are also commonly used.

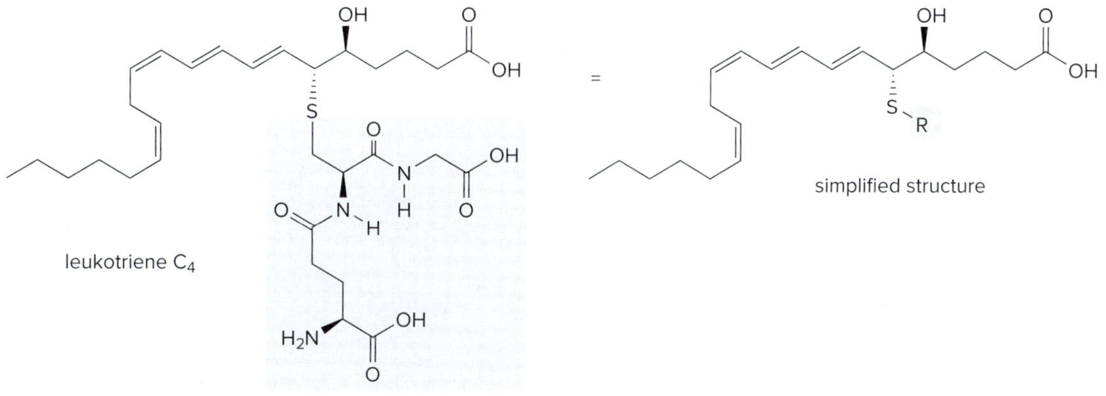

Leukotrienes were first synthesized in 1980 in the laboratory of Professor E. J. Corey, the 1990 recipient of the Nobel Prize in Chemistry.

Leukotrienes are molecules that contribute to the asthmatic response. A typical example, **leukotriene C_4,** is shown. Although its biological activity was first observed in the 1930s, the chemical structure of leukotriene C_4 was not determined until 1979. Structure determination and chemical synthesis were difficult because leukotrienes are highly unstable and extremely potent, and are therefore present in tissues in exceedingly small amounts.

leukotriene C_4

simplified structure

9.17B Leukotriene Synthesis and Asthma Drugs

Leukotrienes are synthesized in cells by the oxidation of **arachidonic acid** to 5-HPETE, which is then converted to an epoxide, **leukotriene A_4.** Opening of the epoxide ring with a sulfur nucleophile **RSH** yields leukotriene C_4.

arachidonic acid

5-HPETE

leukotriene A_4

leukotriene C_4

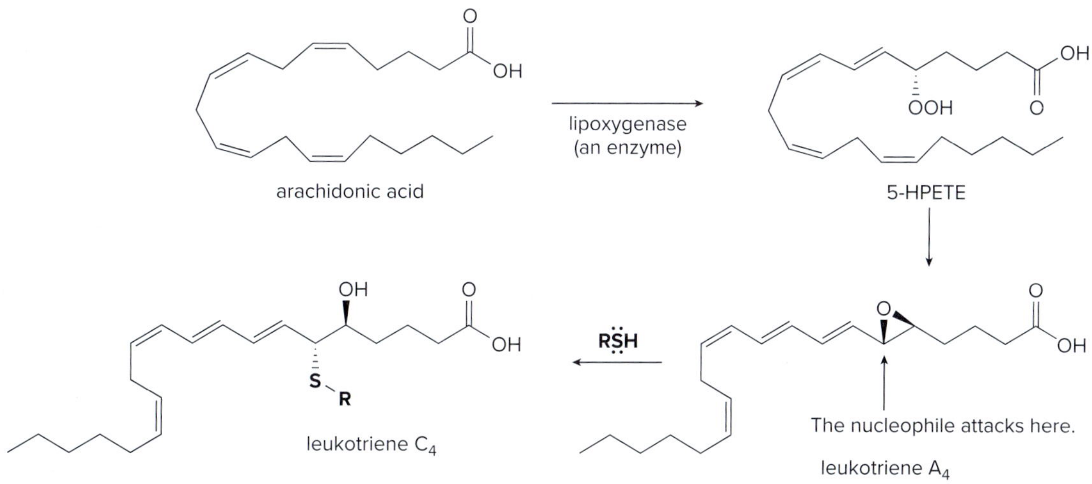

New asthma drugs act by blocking the synthesis of leukotriene C_4 from arachidonic acid. For example, **zileuton** (trade name Zyflo CR) inhibits the enzyme (called a lipoxygenase) needed for the first step of this process. By blocking the synthesis of leukotriene C_4, a compound responsible for the disease, zileuton treats the **cause of asthma,** not just its symptoms.

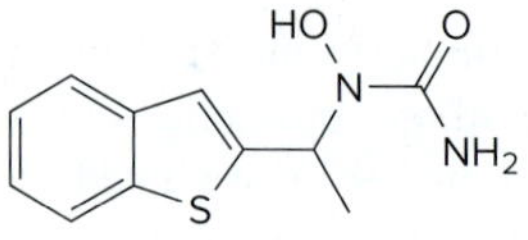

Generic name zileuton
Trade name Zyflo CR
anti-asthma drug

Chapter 9 REVIEW

KEY CONCEPTS

Carbocation rearrangements (9.9)

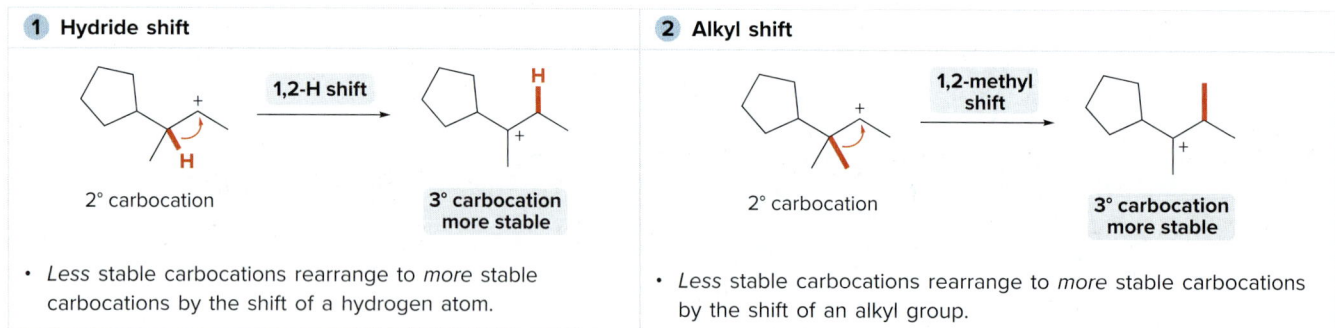

1 Hydride shift

- *Less* stable carbocations rearrange to *more* stable carbocations by the shift of a hydrogen atom.

2 Alkyl shift

- *Less* stable carbocations rearrange to *more* stable carbocations by the shift of an alkyl group.

KEY REACTIONS

[1] Preparation of alcohols, alkoxides, ethers, epoxides, thiols, and sulfides

1. $CH_3{-}X + {}^-OH \xrightarrow[(9.6)]{S_N2} CH_3{-}OH + X^-$ (alcohol); X = Cl, Br, I

2. $CH_3O{-}H + Na^+H^- \xrightarrow[(9.6)]{\text{proton transfer}} CH_3O^-Na^+ + H{-}H$ (alkoxide)

3.

4.

5. $CH_3{-}X + {}^-SH \xrightarrow[(9.15)]{S_N2} CH_3{-}SH + X^-$ (thiol); X = Cl, Br, I

6.

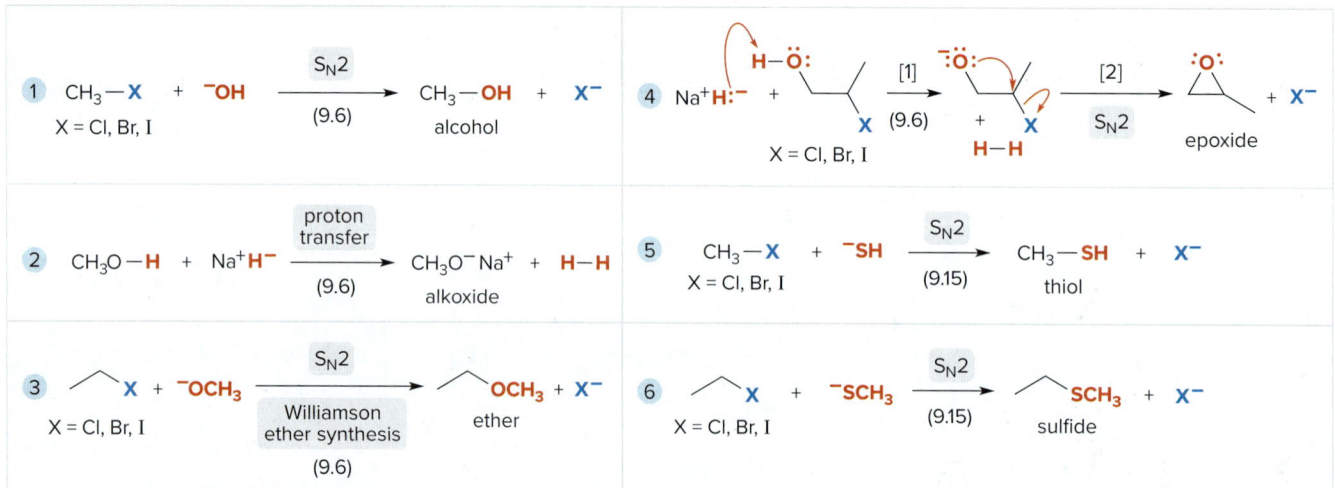

Try Problems 9.40h, i; 9.51; 9.58; 9.60c, d, k.

Key Reactions

[2] Reactions of alcohols

1. Cyclohexanol with H on adjacent C → cyclohexene + H–OH ; H₂SO₄ or TsOH, dehydration (9.8), alkene

2. 2,2-dimethylcyclohexanol with H on adjacent C → alkene + H–OH ; POCl₃, pyridine, dehydration (9.10), alkene

3. 1-methylcyclohexanol + H–X (X = Cl, Br, I) → 1-methyl-1-halocyclohexane + H–OH (9.11), alkyl halide

4. Cyclohexanol + SOCl₂, pyridine, S_N2 → cyclohexyl chloride + pyridinium chloride + SO_2 (9.12), alkyl chloride

5. Cyclohexanol + PBr₃, S_N2 → cyclohexyl bromide + HO–PBr₂ (9.12), alkyl bromide

6. Cyclohexanol + Ts–Cl, pyridine → cyclohexyl tosylate + pyridinium chloride (9.13A), alkyl tosylate

See Table 9.2, Figure 9.4. Try Problems 9.34f, 9.35, 9.40, 9.41.

[3] Reactions of alkyl tosylates

1. Cyclohexyl tosylate (with methyl) + ⁻SCH₃ → cyclohexyl methyl sulfide + ⁻OTs ; S_N2 (9.13B)

2. Cyclohexyl tosylate with β-H + ⁻OCH₃ → cyclohexene + H–OCH₃ + ⁻OTs ; E2 (9.13B)

See Sample Problem 9.6. Try Problems 9.42d, 9.43b, 9.60f.

[4] Reactions of ethers

CH₃CH₂–O–CH₃ + H–X (2 equiv), X = Br, I → CH₃CH₂–X + CH₃–X + H–OH (9.14)

Try Problems 9.53, 9.60j.

[5] Reactions involving thiols and sulfides

1. CH₃–SH + X₂, X = Br, I → CH₃–S–S–CH₃ ; oxidation, disulfide (9.15A)

2. CH₃–S–S–CH₃ + Zn, HCl → 2 CH₃–SH ; reduction (9.15A)

3. CH₃CH₂–S–CH₃ + CH₃CH₂–X, X = Cl, Br, I → (CH₃CH₂)₂S⁺–CH₃ + X⁻ ; S_N2 (9.15B), sulfonium ion

Try Problems 9.60c, k, l; 9.61.

[6] Reactions of epoxides

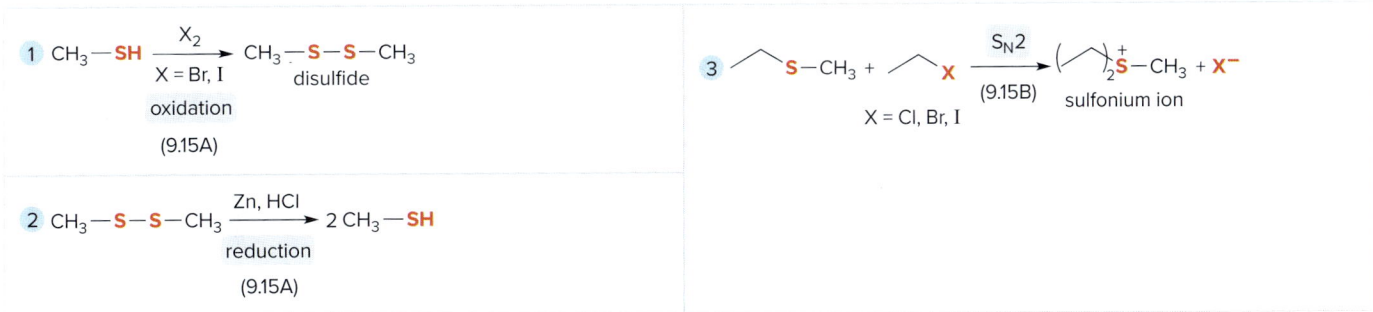

1. Propylene oxide [1] ⁻OCH₃ [2] H–OH → CH₃O–CH₂–CH(OH)–CH₃ + ⁻OH (9.16A)

2. Propylene oxide + H–X, X = Cl, Br, I → HO–CH(CH₃)–CH₂–X (9.16B)

See Sample Problem 9.7, Figure 9.6. Try Problems 9.56; 9.57; 9.60g, h.

KEY SKILLS

[1] Naming an acyclic alcohol using the IUPAC system (9.3A)

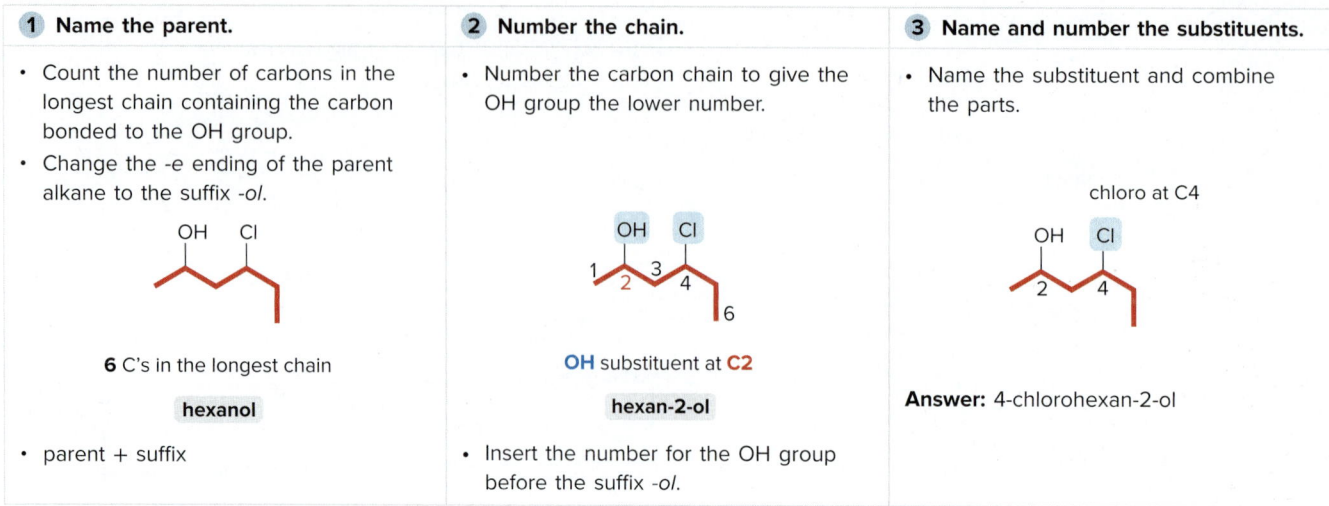

See *How To*, p. 339; Figure 9.1. Try Problems 9.34a, 9.36.

[2] Using the Williamson ether synthesis to convert an alcohol to an ether (9.6)

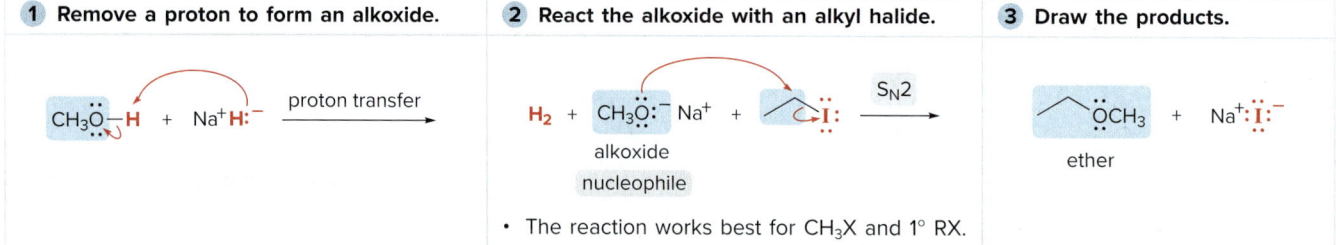

See Sample Problem 9.2. Try Problems 9.51, 9.60i.

[3] Determining when a carbocation rearrangement might occur (9.9)

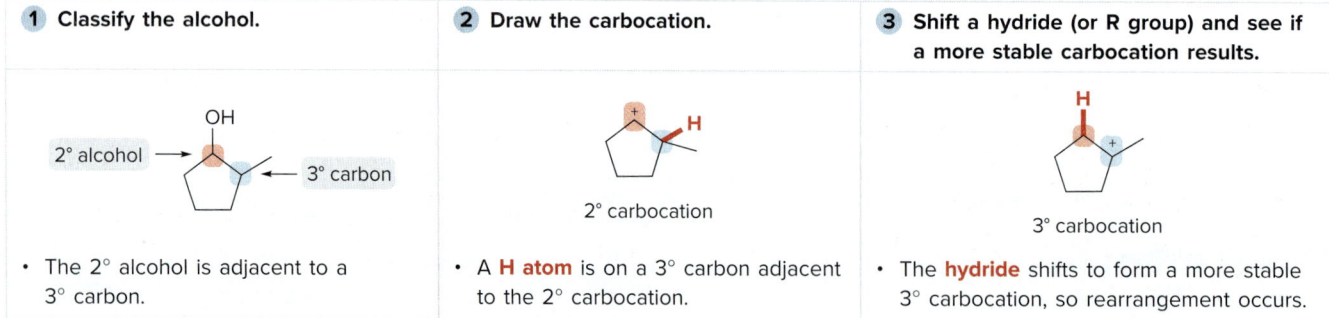

Try Problems 9.44, 9.45, 9.46.

[4] Drawing the products of a dehydration reaction when a 1,2-methyl shift occurs (9.9)

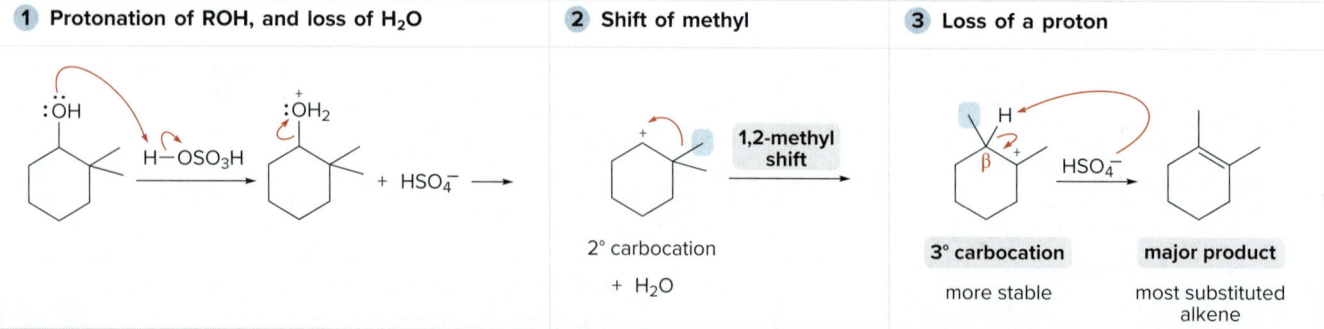

See Sample Problem 9.3. Try Problem 9.41d.

[5] Drawing the products of an S_N1 reaction when a 1,2-hydride shift occurs (9.11B)

1 Protonation of ROH, and loss of H_2O	**2** Shift of hydride	**3** Nucleophilic attack

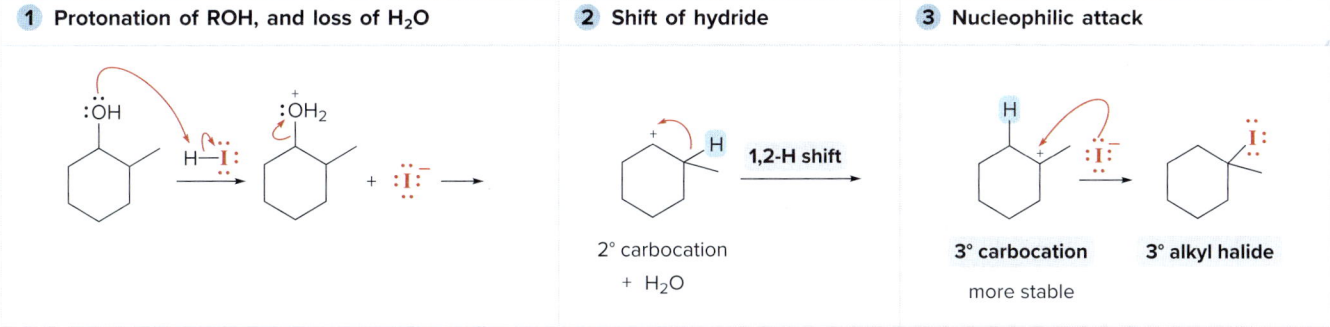

See Sample Problem 9.5. Try Problems 9.44a, 9.45.

[6] Drawing the product of an epoxide ring opening with a strong nucleophile (9.16A)

1 Draw the curved arrows for the nucleophilic attack.	**2** Protonate the alkoxide.

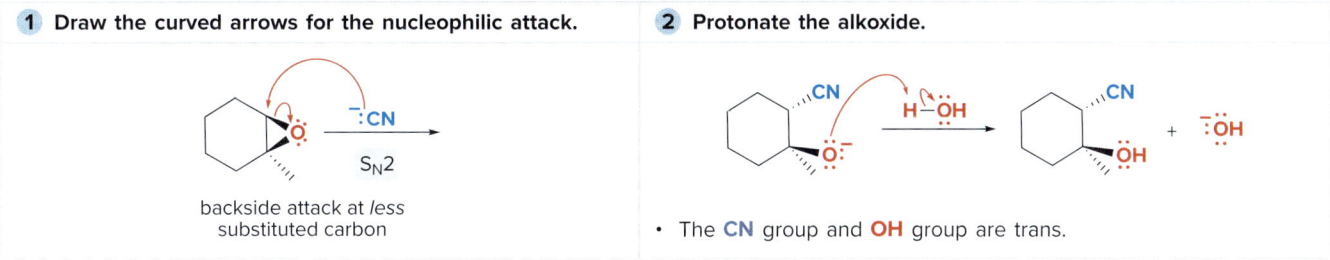

- The **CN** group and **OH** group are trans.

See Sample Problem 9.7. Try Problems 9.57b, d; 9.60h.

[7] Drawing the product of an epoxide ring opening with an acid (9.16B)

1 Protonate the epoxide.	**2** Draw the curved arrows for the nucleophilic attack.

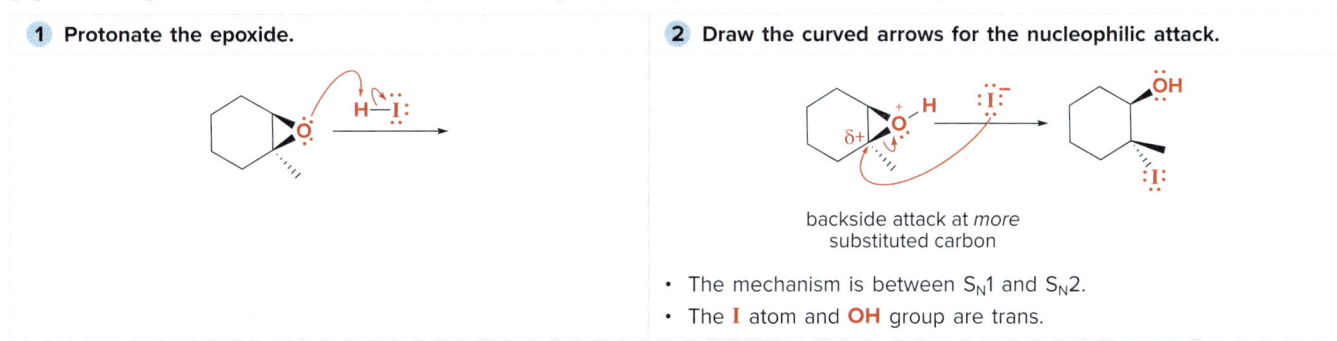

- The mechanism is between S_N1 and S_N2.
- The **I** atom and **OH** group are trans.

See Sample Problem 9.7, Figure 9.6. Try Problems 9.57a, c; 9.60g.

KEY MECHANISM CONCEPTS IN REACTIONS OF ALCOHOLS

1 Reactions of 1° ROH

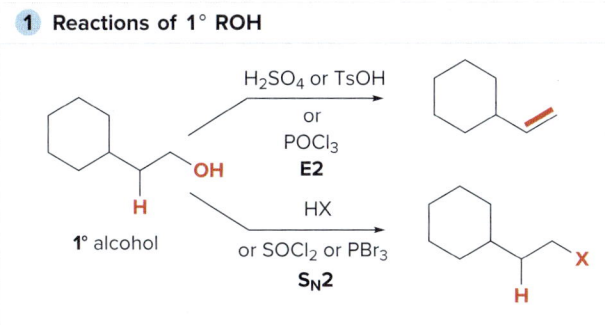

- Dehydration with acid—**E2** mechanism
- Dehydration with $POCl_3$ and pyridine—**E2** mechanism
- Substitution with HX—S_N2 mechanism
- Substitution with $SOCl_2$ and PBr_3—S_N2 mechanism
- No carbocation intermediates
- See Mechanisms 9.2, 9.4, 9.5, 9.7, 9.8.

2 Reactions of 2° and 3° ROH

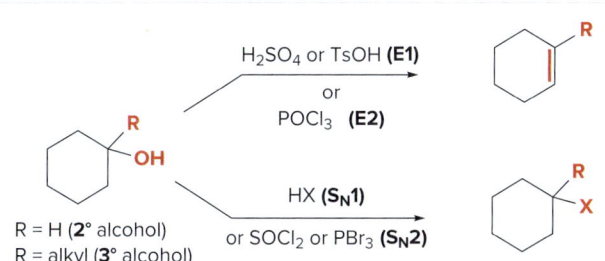

- Dehydration with acid—**E1** mechanism
- Dehydration with $POCl_3$ and pyridine—**E2** mechanism
- Substitution with HX—S_N1 mechanism
- Substitution with $SOCl_2$ and PBr_3 (2° ROH only)—S_N2 mechanism
- Carbocation intermediates in both S_N1 and E1 reactions
- Carbocation rearrangements possible
- See Mechanisms 9.1, 9.3, 9.4, 9.6, 9.7, 9.8.

PROBLEMS

Problems Using Three-Dimensional Models

9.34 Answer each question using the ball-and-stick model of compound **A**.

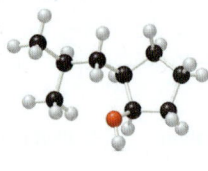

A

a. Give the IUPAC name for **A**, including *R,S* designations for stereogenic centers.
b. Classify **A** as a 1°, 2°, or 3° alcohol.
c. Draw a stereoisomer for **A** and give its IUPAC name.
d. Draw a constitutional isomer that contains an OH group and give its IUPAC name.
e. Draw a constitutional isomer that contains an ether and give its IUPAC name.
f. Draw the products formed (including stereochemistry) when **A** is treated with each reagent:
[1] NaH; [2] H₂SO₄; [3] POCl₃, pyridine; [4] HCl; [5] SOCl₂, pyridine; [6] TsCl, pyridine.

9.35 Draw the product and indicate the stereochemistry when the given alcohol is treated with each reagent: (a) HBr; (b) PBr₃; (c) HCl; (d) SOCl₂ and pyridine.

Nomenclature

9.36 Give the IUPAC name for each alcohol.

9.37 Name each ether, epoxide, thiol, and sulfide.

9.38 Give the structure corresponding to each name.
a. *trans*-2-methylcyclohexanol
b. 2,3,3-trimethylbutan-2-ol
c. 6-*sec*-butyl-7,7-diethyldecan-4-ol
d. 3-chloropropane-1,2-diol
e. 1,2-epoxy-1,3,3-trimethylcyclohexane
f. 1-ethoxy-3-ethylheptane
g. (2*R*,3*S*)-3-isopropylhexan-2-ol
h. (*S*)-2-ethoxy-1,1-dimethylcyclopentane
i. 4-ethylheptane-3-thiol
j. 1-isopropylthio-2-methylcyclohexane

Physical Properties

9.39 Why is the boiling point of propane-1,3-diol (HOCH₂CH₂CH₂OH) higher than the boiling point of propane-1,2-diol [HOCH₂CH(OH)CH₃] (215 °C vs. 187 °C)? Why do both diols have a higher boiling point than butan-1-ol (CH₃CH₂CH₂CH₂OH, 118 °C)?

Alcohols

9.40 Draw the organic product(s) formed when CH₃CH₂CH₂OH is treated with each reagent.
a. H₂SO₄
b. NaH
c. HI
d. HBr
e. SOCl₂, pyridine
f. PBr₃
g. TsCl, pyridine
h. [1] NaH; [2] CH₃CH₂Br
i. [1] TsCl, pyridine; [2] NaSH
j. POCl₃, pyridine

9.41 What alkenes are formed when each alcohol is dehydrated with TsOH? Label the major product when a mixture results.

a. 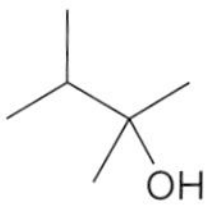 (3-methylbutan-2-ol type structure with OH)   b. 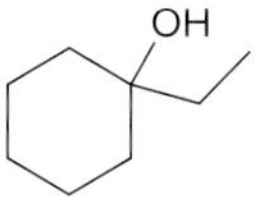 (1-ethylcyclohexan-1-ol)   c. 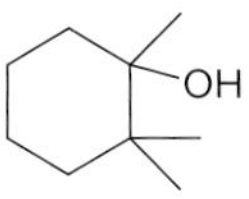 (1,2,2-trimethylcyclohexan-1-ol type)   d. 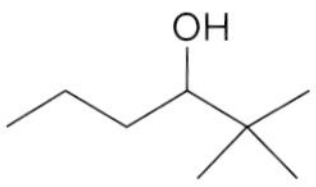 (2-methylhexan-2-ol type with OH)

9.42 Draw the products of each reaction and indicate stereochemistry around stereogenic centers.

a. 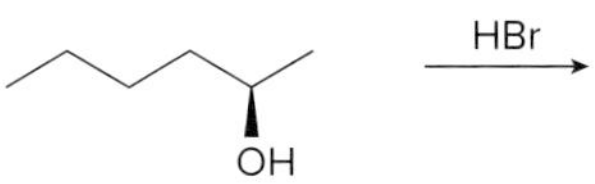 (hexan-2-ol) — HBr →

b. 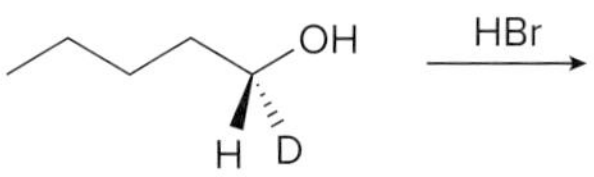 (hexan-2-ol with H, D stereochemistry) — HBr →

c. 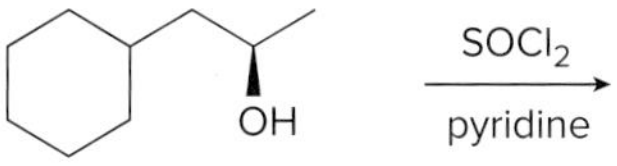 (1-cyclohexyl-propan-2-ol) — SOCl₂, pyridine →

d. (3,3-dimethyl alcohol with stereochem) — TsCl, pyridine; then KI →

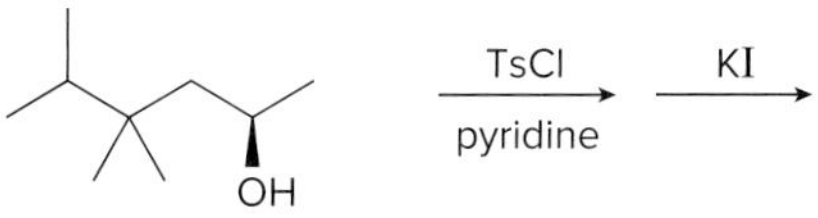

e. 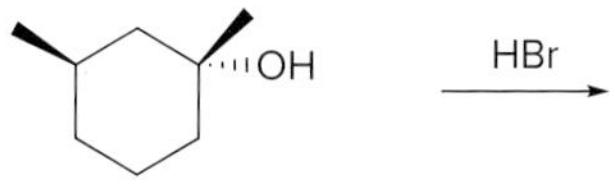 (trans-dimethylcyclohexanol) — HBr →

f. 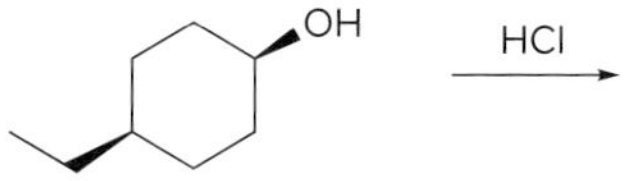 (4-ethylcyclohexanol stereo) — HCl →

9.43 Draw the substitution product formed (including stereochemistry) when (R)-hexan-2-ol is treated with each series of reagents: (a) NaH, followed by CH₃I; (b) TsCl and pyridine, followed by NaOCH₃; (c) PBr₃, followed by NaOCH₃. Which two routes produce identical products?

9.44 (a) What is the major alkene formed when **A** is dehydrated with H₂SO₄? (b) What is the major alkene formed when **A** is treated with POCl₃ and pyridine? Explain why the major product is different in these reactions.

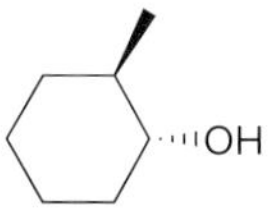

A

9.45 Reaction of 2° alcohol **A** with HCl forms three alkyl chlorides, all of which result from rearrangement of the 2° carbocation initially formed. Draw the structures of these products and a mechanism that illustrates how each is formed.

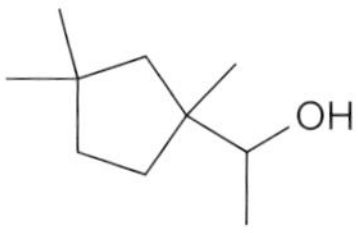

A

9.46 Draw a stepwise mechanism for the following reaction.

(2-methylcyclopentan-1-ol) — H₂SO₄ → (1-methylcyclopentene) + (3-methylcyclopentene) + (methylenecyclopentane) + H₂O

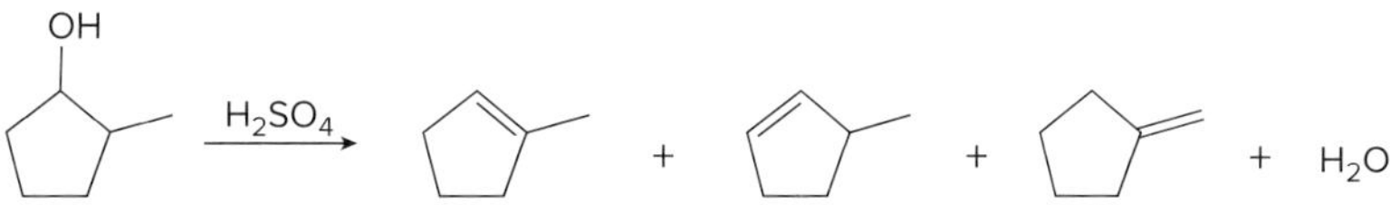

9.47 Sometimes carbocation rearrangements can change the size of a ring. Draw a stepwise, detailed mechanism for the following reaction.

(1-(1-methylcyclopentyl)ethanol) — H₂SO₄ → (1,2-dimethylcyclohexene) + H₂O

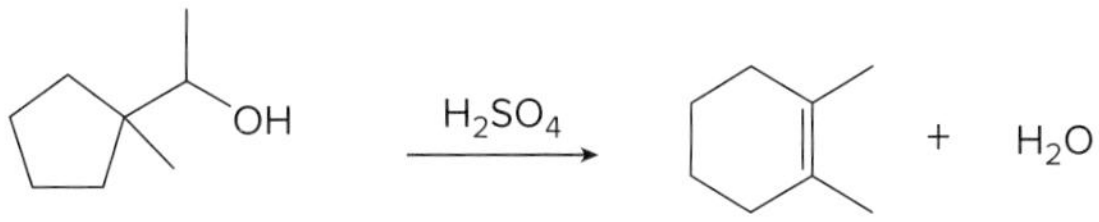

9.48 An allylic alcohol contains an OH group on a carbon atom adjacent to a C–C double bond. Treatment of allylic alcohol **A** with HCl forms a mixture of two allylic chlorides, **B** and **C**. Draw a stepwise mechanism that illustrates how both products are formed.

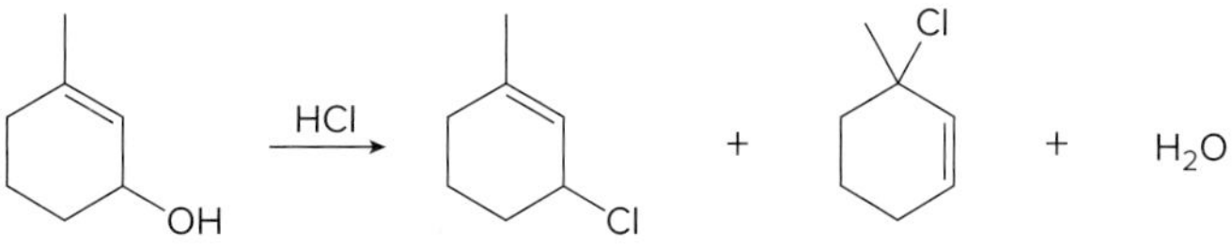

A —HCl→ **B** + **C** + H₂O

9.49 Draw a stepwise, detailed mechanism for the following reaction.

(2-(1-hydroxyethyl)tetrahydrofuran) — H₂SO₄ → (3-methyl-3,4-dihydro-2H-pyran) + H₂O

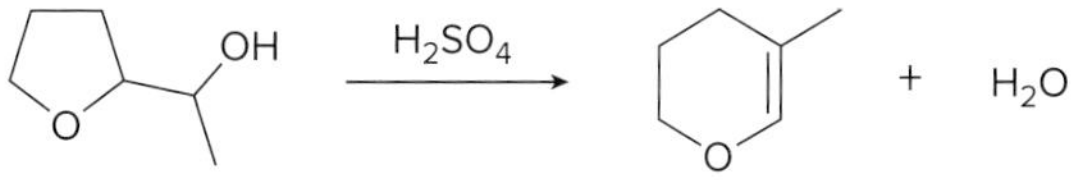

Chapter 9 Alcohols, Ethers, and Related Compounds

9.50 Draw a stepwise mechanism for the dehydration of diol **A** to enol **B** (Section 9.8B). Explain why the 2° OH group (in red) is lost much more readily than the other 2° OH group in **A**.

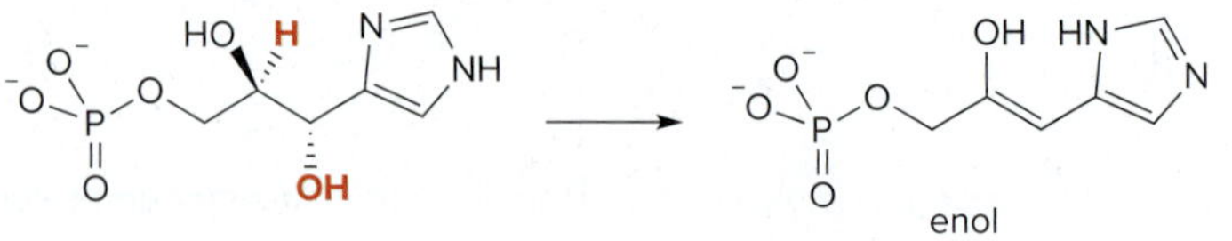

Ethers

9.51 Draw two different routes to each of the following ethers using a Williamson ether synthesis. Indicate the preferred route (if there is one).

a. 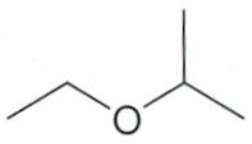 b. 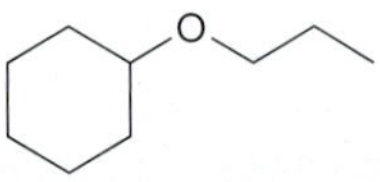 c.

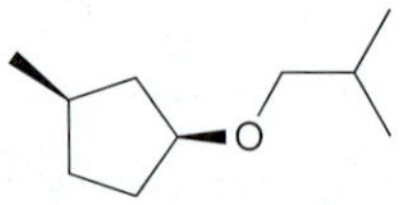

9.52 Explain why it is not possible to prepare *tert*-butyl phenyl ether using a Williamson ether synthesis.

9.53 Draw the products formed when each ether is treated with two equivalents of HBr.

a. 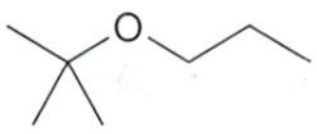 b. 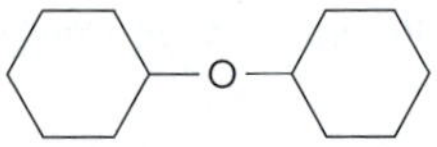 c.

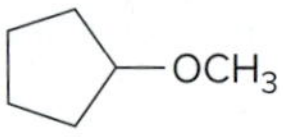

9.54 Draw a stepwise mechanism for the following reaction.

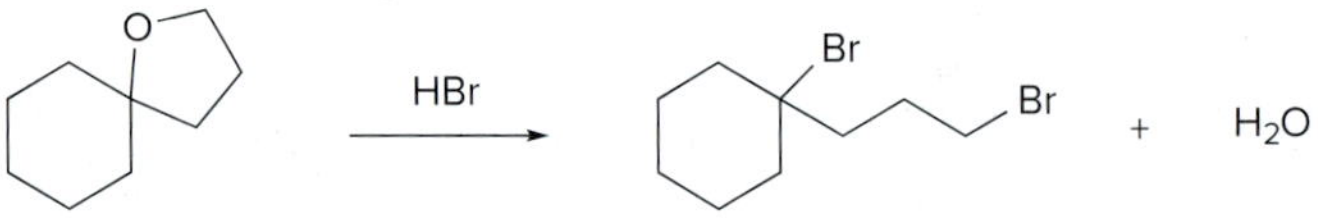

9.55 Draw a stepwise mechanism for the following reaction.

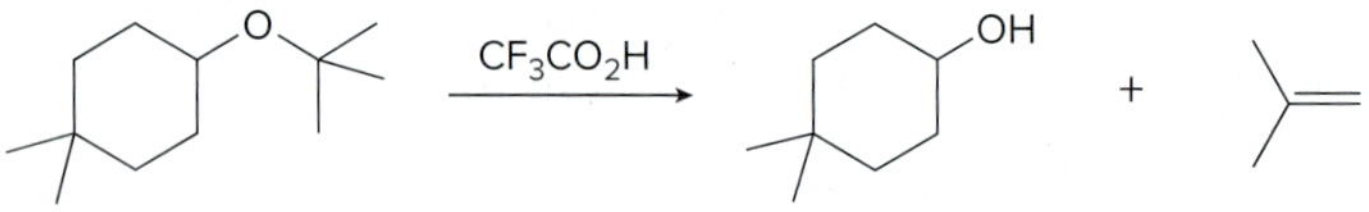

Epoxides

9.56 Draw the products formed when ethylene oxide is treated with each reagent.
- a. HBr
- b. H_2O (H_2SO_4)
- c. [1] $CH_3CH_2O^-$; [2] H_2O
- d. [1] $HC\equiv C^-$; [2] H_2O
- e. [1] ^-OH; [2] H_2O
- f. [1] CH_3S^-; [2] H_2O

9.57 Draw the products of each reaction.

a. 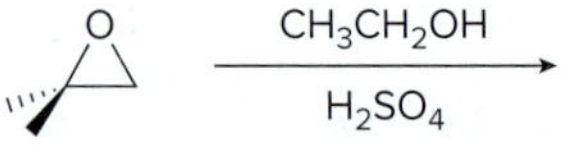 CH_3CH_2OH, H_2SO_4

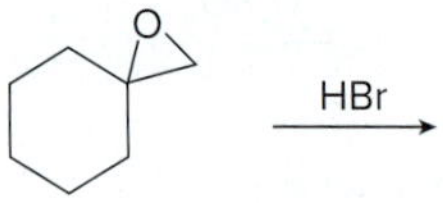

b. 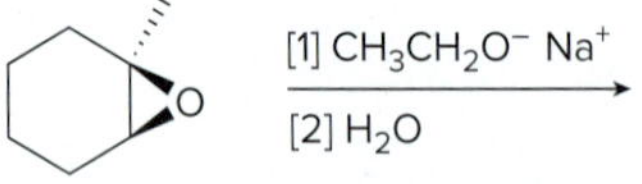 [1] $CH_3CH_2O^-$ Na^+; [2] H_2O

c. HBr

d. 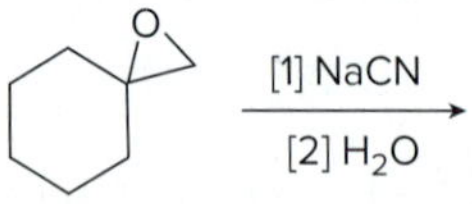 [1] NaCN; [2] H_2O

9.58 When each halohydrin is treated with NaH, a product of molecular formula C_4H_8O is formed. Draw the structure of the product and indicate its stereochemistry.

a. 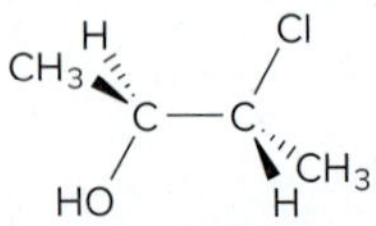 b. 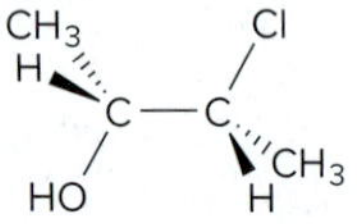 c.

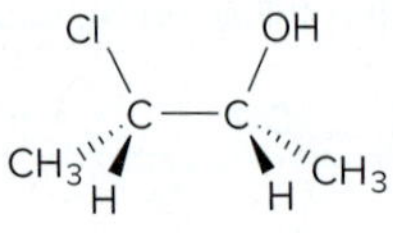

9.59 Draw a stepwise mechanism for the following reaction, which forms the four-membered ring in azelnidipine, a drug used as a calcium channel blocker sold in Japan.

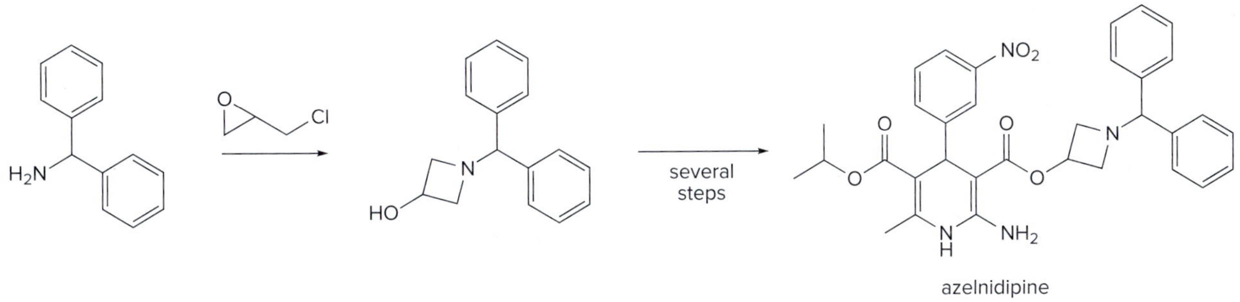

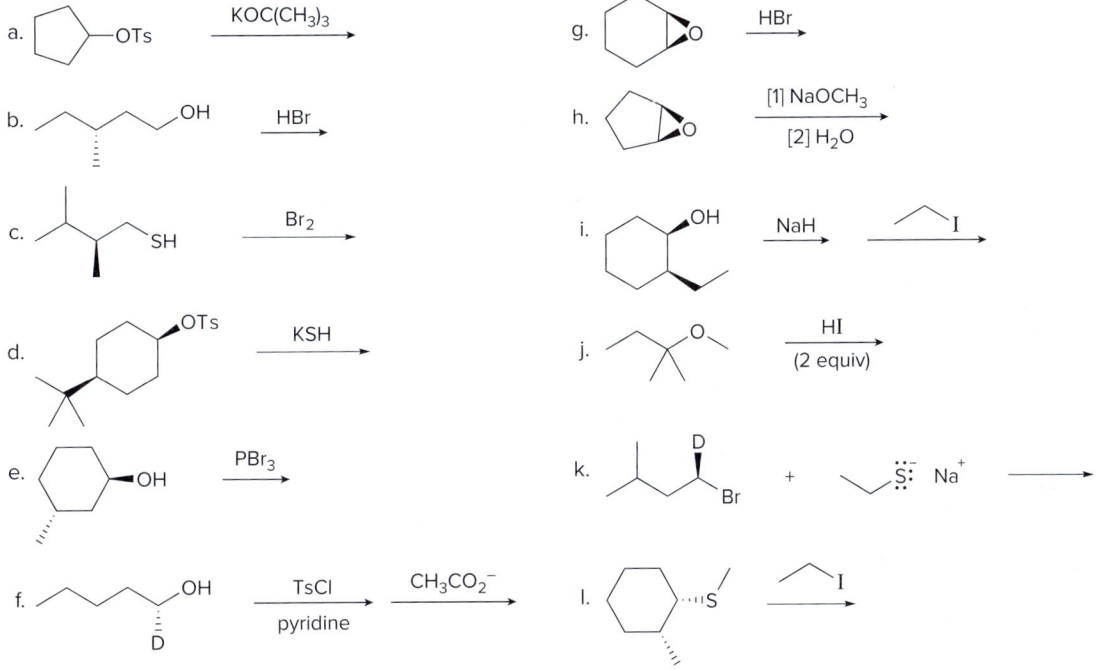

General Problems

9.60 Draw the products of each reaction, and indicate the stereochemistry where appropriate.

a. cyclopentyl-OTs, KOC(CH₃)₃

b. (S)-4-methylpentan-1-ol (with OH), HBr

c. 2-methylbutane-1-thiol (SH), Br₂

d. trans-4-tert-butylcyclohexyl OTs, KSH

e. trans-3-methylcyclohexanol, PBr₃

f. (R)-pentan-2-ol-D (OH, D), TsCl / pyridine, then CH₃CO₂⁻

g. cyclohexene oxide, HBr

h. bicyclic epoxide, [1] NaOCH₃ [2] H₂O

i. trans-2-ethylcyclohexanol, NaH, then CH₃CH₂CH₂I

j. tert-butyl methyl ether, HI (2 equiv)

k. (R)-2-bromo-3-methylbutane-D + CH₃CH₂S⁻ Na⁺

l. trans-2-methylcyclohexyl sulfide + CH₃CH₂CH₂I

9.61 The following two-step procedure was used to prepare a sulfide from a diol. Draw the intermediate formed in Reaction [1] and draw a mechanism for Reaction [2].

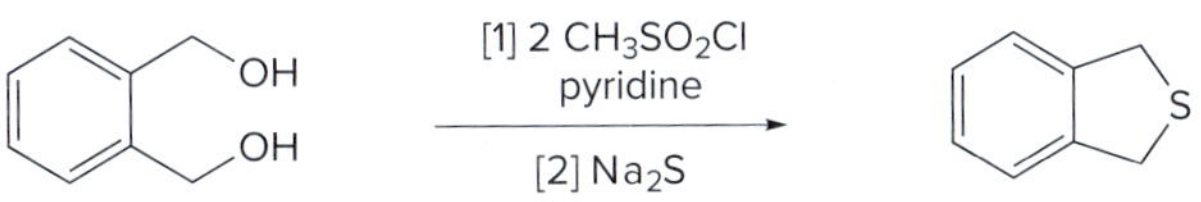

1,2-bis(hydroxymethyl)benzene → [1] 2 CH₃SO₂Cl, pyridine → [2] Na₂S → 1,3-dihydro-2-benzothiophene

9.62 Prepare each compound from cyclopentanol. More than one step may be needed.

a. 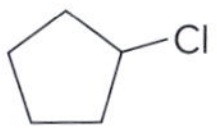 cyclopentyl chloride   b.  cyclopentene   c. 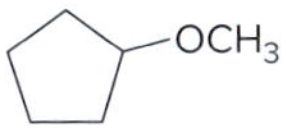 cyclopentyl methyl ether   d. 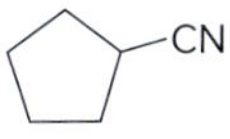 cyclopentanecarbonitrile

9.63 Identify **Y** in the following reaction, one step in the synthesis of methylphenidate, a drug used to treat attention deficit hyperactivity disorder (ADHD).

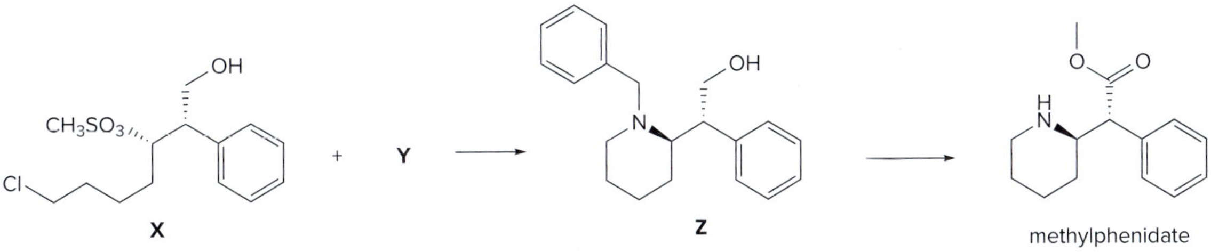

X + Y → Z → methylphenidate

Spectroscopy

Problems 9.64–9.67 are intended for students who have already learned about spectroscopy in Chapters A–C.

9.64 Propose a structure consistent with each set of spectral data:

a. $C_6H_{14}O$: IR peak at 3600–3200 cm^{-1}; NMR (ppm):
0.8 (triplet, 6 H)   1.5 (quartet, 4 H)
1.0 (singlet, 3 H)   1.6 (singlet, 1 H)

b. $C_6H_{14}O$: IR peak at 3000–2850 cm^{-1}; NMR (ppm):
1.10 (doublet, relative area = 6)
3.60 (septet, relative area = 1)

9.65 As we will learn in Chapter 13, reaction of $(CH_3)_2CO$ with LiC≡CH followed by H_2O affords compound **D**, which has a molecular ion in its mass spectrum at 84 and prominent absorptions in its IR spectrum at 3600–3200, 3303, 2938, and 2120 cm^{-1}. **D** shows the following ^1H NMR spectral data: 1.53 (singlet, 6 H), 2.37 (singlet, 1 H), and 2.43 (singlet, 1 H) ppm. What is the structure of **D**?

9.66 Treatment of $(CH_3)_2CHCH(OH)CH_2CH_3$ with TsOH affords two products (**M** and **N**) with molecular formula C_6H_{12}. The ^1H NMR spectra of **M** and **N** are given below. Propose structures for **M** and **N**, and draw a mechanism to explain their formation.

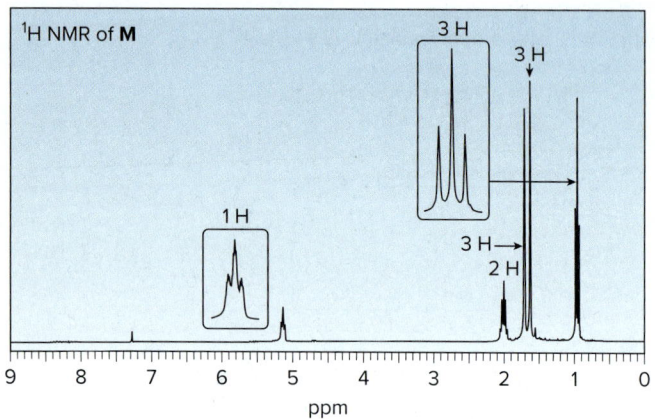
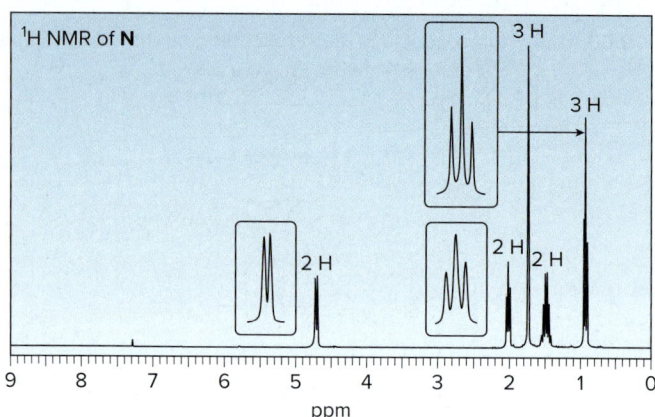

9.67 Use the ^1H NMR and IR spectra given below to identify compound **A**, having molecular formula $C_4H_8O_2$.

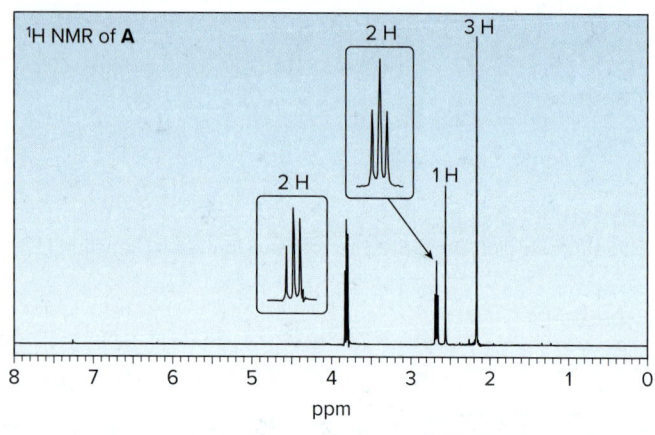
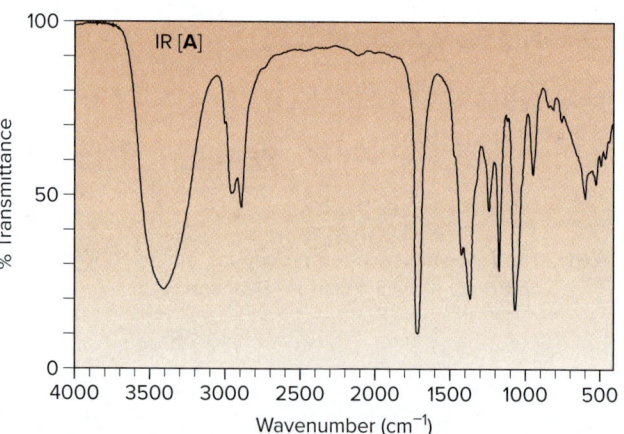

Additional spectroscopy problems involving alcohols, ethers, and epoxides are given in Chapters A–C:

- Mass spectrometry: A.4, A.23, A.24a, A.25, A.27, A.29
- Infrared spectroscopy: B.11; B.12b; B.15b; B.16b; B.19b; B.22; B.25d; B.26(**B**), (**D**); B.27c
- Nuclear magnetic resonance spectroscopy: C.21; C.24; C.25c; C.28a; C.33b, e, g, h; C.37b, c; C.38c, e, f; C.43b, e; C.45a, c; C.59; C.68

Challenge Problems

9.68 Epoxides are converted to allylic alcohols with nonnucleophilic bases such as lithium diethylamide [$LiN(CH_2CH_3)_2$]. Draw a stepwise mechanism for the conversion of 1,2-epoxycyclohexane to cyclohex-2-en-1-ol with this base. Explain why a strong bulky base must be used in this reaction.

9.69 Rearrangements can occur during the dehydration of 1° alcohols even though no 1° carbocation is formed—that is, a 1,2-shift occurs as the C–OH$_2^+$ bond is broken, forming a more stable 2° or 3° carbocation, as shown in Equation [1]. Using this information, draw a stepwise mechanism for the reaction shown in Equation [2]. We will see another example of this type of rearrangement in Section 20.5C.

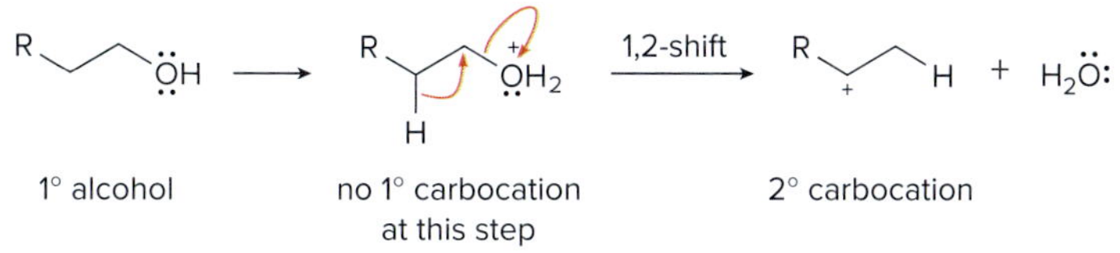

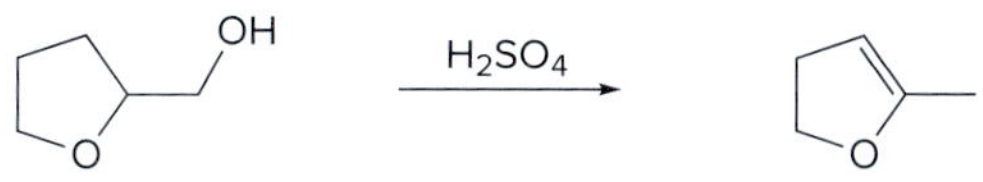

9.70 Dehydration of 1,2,2-trimethylcyclohexanol with H$_2$SO$_4$ affords 1-*tert*-butylcyclopentene as a minor product. (a) Draw a stepwise mechanism that shows how this alkene is formed. (b) Draw other alkenes formed in this dehydration. At least one must contain a five-membered ring.

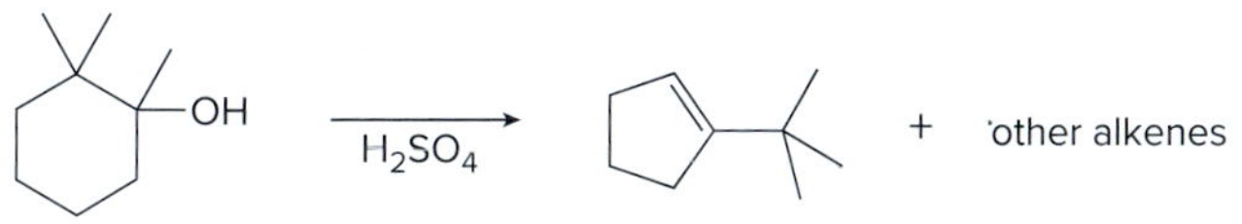

9.71 1,2-Diols are converted to carbonyl compounds when treated with strong acids, in a reaction called the *pinacol rearrangement*. (a) Draw a stepwise mechanism for this reaction. (Hint: The reaction proceeds by way of carbocation intermediates.) (b) Assuming that the pinacol rearrangement occurs via the more stable carbocation, draw the rearrangement product formed from diol **D**.

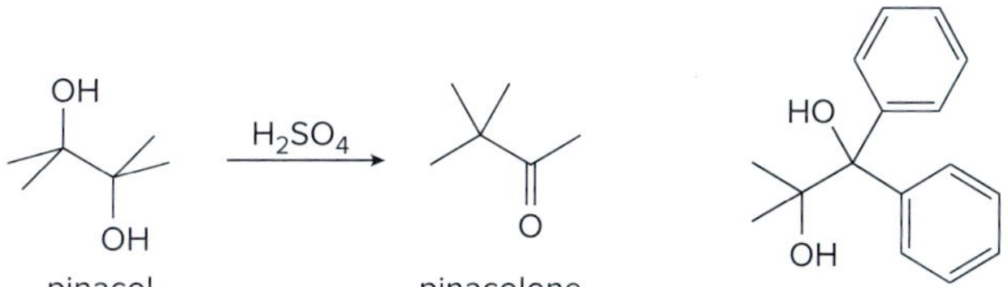

9.72 Draw a stepwise mechanism for the following reaction.

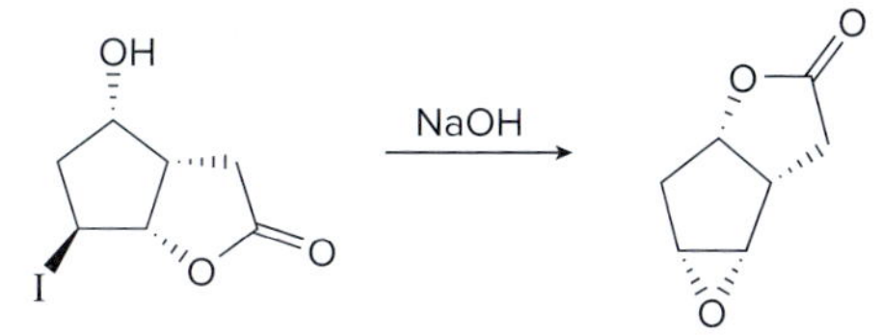

9.73 Draw a stepwise mechanism for the following reaction, a key step in the synthesis of vernakalant, a drug approved in Europe in 2010 for the treatment of atrial fibrillation. Pure **B** was separated from a mixture of diastereomers. Your mechanism must explain the trans stereochemistry of the two substituents on the six-membered ring.

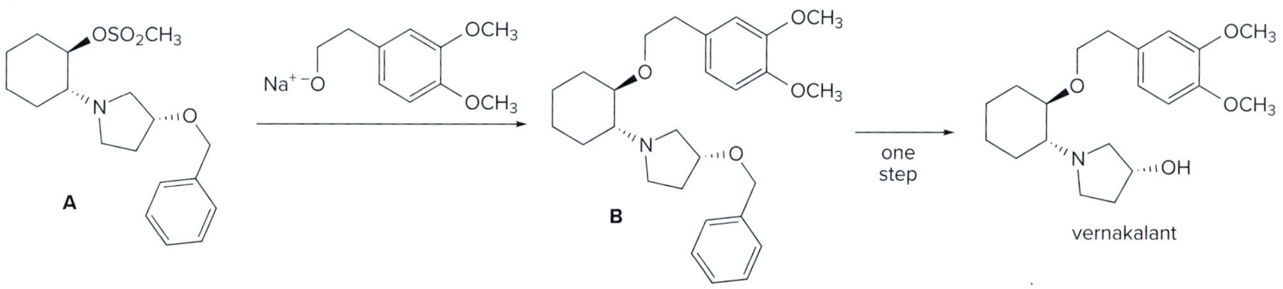

10 Alkenes and Alkynes

10.1 Introduction
10.2 Calculating degrees of unsaturation
10.3 Nomenclature
10.4 Properties of alkenes and alkynes
10.5 Interesting alkenes and alkynes
10.6 Fatty acids and triacylglycerols
10.7 Preparation of alkenes and alkynes
10.8 Introduction to the reactions of alkenes and alkynes
10.9 Hydrohalogenation—Electrophilic addition of HX to alkenes
10.10 Markovnikov's rule
10.11 Stereochemistry of electrophilic addition of HX
10.12 Hydration—Electrophilic addition of water
10.13 Halogenation—Addition of halogen
10.14 Stereochemistry of halogenation
10.15 Halohydrin formation
10.16 Hydroboration–oxidation
10.17 Addition of hydrogen halides and halogens to alkynes
10.18 Addition of water to alkynes
10.19 Hydroboration–oxidation of alkynes
10.20 Reaction of acetylide anions
10.21 Synthesis

Michael Sewell/Photolibrary/Getty Images

Histrionicotoxin is a toxin isolated in small quantities from the skin of *Dendrobates histrionicus,* a colorful South American frog. These small "poison dart" frogs inhabit the moist humid floor of tropical rainforests, and are commonly found in Ecuador and Colombia. Histrionicotoxin is secreted by the frog as a natural defense mechanism, and it acts by interfering with nerve transmission in mammals, resulting in prolonged muscle contraction. The structure of histrionicotoxin contains two carbon–carbon triple bonds and two carbon–carbon double bonds, functional groups that are the subject of Chapter 10.

Why Study... Alkenes and Alkynes?

In Chapter 10, we turn our attention to **alkenes** and **alkynes,** compounds that contain one and two π bonds, respectively. Because π bonds are easily broken, alkenes and alkynes undergo **addition,** the third general type of organic reaction. These multiple bonds make carbon atoms electron rich, so alkenes and alkynes react with a wide variety of electrophilic reagents in addition reactions that are very versatile in organic synthesis.

Alkynes also undergo a reaction that has no analogy in alkene chemistry. Because a C—H bond of an alkyne is more acidic than a C—H bond of an alkene or an alkane, alkynes are readily deprotonated with strong base. The resulting nucleophiles react with electrophiles to form new carbon–carbon σ bonds, so that complex molecules can be prepared from simple starting materials. The study of alkynes thus affords an opportunity to learn more about organic synthesis.

10.1 Introduction

Alkenes are also called **olefins.**

Alkenes are compounds that contain a carbon–carbon double bond. The double bond of an alkene consists of one σ bond and one π bond. Each carbon is sp^2 hybridized and trigonal planar, and all bond angles are approximately 120° (Section 8.2A).

- The π bond is much *weaker* than the σ bond of a C—C double bond, making it much more easily broken. As a result, alkenes undergo many reactions that alkanes do not.

Alkynes contain a carbon–carbon triple bond. A **terminal alkyne** has the triple bond at the end of the carbon chain, so that a hydrogen atom is bonded directly to a carbon atom of the triple bond. An **internal alkyne** has a carbon atom bonded to each carbon atom of the triple bond.

Skeletal structures for alkynes may look somewhat unusual, but they follow the customary convention: a carbon atom is located at the intersection of any two lines and at the end of any line; thus,

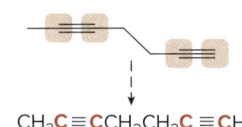

CH$_3$C≡CCH$_2$CH$_2$C≡CH

Each carbon of a triple bond is *sp* hybridized and **linear**, and all bond angles are **180°** (Section 1.10C). The triple bond of an alkyne consists of **one σ bond** and **two π bonds.**

- Both π bonds of a C—C triple bond are weaker than a C—C σ bond, making them much more easily broken. As a result, alkynes undergo many addition reactions.
- Alkynes are more polarizable than alkenes because the electrons in their π bonds are more loosely held.

Problem 10.1 Draw the six alkenes of molecular formula C_5H_{10}. Label one pair of diastereomers.

10.2 Calculating Degrees of Unsaturation

An acyclic alkene has the general molecular formula C_nH_{2n}, giving it *two* fewer hydrogens than an acyclic alkane with the same number of carbons. **An alkyne has the general molecular formula C_nH_{2n-2},** giving it *four* fewer hydrogens than the maximum number possible.

- Alkenes and alkynes are *unsaturated hydrocarbons* because they have fewer than the maximum number of hydrogen atoms per carbon.

In Chapter 11, we will learn how to use the hydrogenation of π bonds to determine how many degrees of unsaturation result from π bonds and how many result from rings.

Cycloalkanes also have the general molecular formula C_nH_{2n}. Thus, **each π bond or ring removes two hydrogen atoms from a molecule, and this introduces one** *degree of unsaturation*. The number of degrees of unsaturation for a given molecular formula can be calculated by comparing the *actual* number of H atoms in a compound and the *maximum* number of H atoms possible. Remember that for *n* carbons, the **maximum number of H atoms is $2n + 2$** (Section 4.1). This procedure gives the total number of rings and π bonds in a molecule.

Sample Problem 10.1   Calculating the Number of Degrees of Unsaturation in a Hydrocarbon

Calculate the number of degrees of unsaturation in a compound of molecular formula C_4H_6, and propose possible structures.

Solution

[1] **Calculate the maximum number of H's possible.**
- For *n* carbons, the maximum number of H's is $2n + 2$; in this example, $2n + 2 = 2(4) + 2 = 10$.

[2] **Subtract the actual number of H's from the maximum number and divide by two.**
- 10 H's (maximum) − 6 H's (actual) = 4 H's fewer than the maximum number.

$$\frac{\text{4 H's fewer than the maximum}}{\text{2 H's removed for each degree of unsaturation}} =$$

Answer: two degrees of unsaturation

A compound with two degrees of unsaturation has:

        two rings   or   **two π bonds**   or   **one ring and one π bond**

Possible structures for C_4H_6:

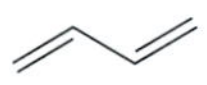

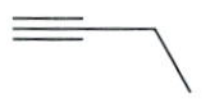

Problem 10.2   Calculate the number of degrees of unsaturation for each molecular formula, and propose two possible structures: (a) C_8H_{12}; (b) $C_{10}H_{10}$.

More Practice: Try Problem 10.42a, b.

This procedure can be extended to compounds that contain heteroatoms such as oxygen, nitrogen, and halogen, as illustrated in Sample Problem 10.2.

Sample Problem 10.2   Calculating the Number of Degrees of Unsaturation in Compounds with O, X, or N

Calculate the number of degrees of unsaturation for each molecular formula: (a) C_5H_8O; (b) $C_6H_{11}Cl$; (c) C_8H_9N.

Solution

a. When a compound contains an oxygen atom, **use the given number of C's and H's and ignore the O atom** in the calculation; that is, C_5H_8O is equivalent to C_5H_8 when calculating degrees of unsaturation.

[1] For 5 C's, the maximum number of H's = 2n + 2 = 2(5) + 2 = 12.
[2] Because the compound contains only 8 H's, it has 12 − 8 = 4 H's fewer than the maximum number.
[3] Each degree of unsaturation removes 2 H's, so the answer in Step [2] must be divided by 2.
Answer: two degrees of unsaturation

b. **A compound with a halogen atom is equivalent to a hydrocarbon having one *more* H;** that is, $C_6H_{11}Cl$ is equivalent to C_6H_{12} when calculating degrees of unsaturation.

[1] For 6 C's, the maximum number of H's = 2n + 2 = 2(6) + 2 = 14.
[2] Because the compound contains only 12 H's, it has 14 − 12 = 2 H's fewer than the maximum number.
[3] Each degree of unsaturation removes 2 H's, so the answer in Step [2] must be divided by 2.
Answer: one degree of unsaturation

c. **A compound with a nitrogen atom is equivalent to a hydrocarbon having one *fewer* H;** that is, C_8H_9N is equivalent to C_8H_8 when calculating degrees of unsaturation.

[1] For 8 C's, the maximum number of H's = 2n + 2 = 2(8) + 2 = 18.
[2] Because the compound contains only 8 H's, it has 18 − 8 = 10 H's fewer than the maximum number.
[3] Each degree of unsaturation removes 2 H's, so the answer in Step [2] must be divided by 2.
Answer: five degrees of unsaturation

Problem 10.3 How many degrees of unsaturation are present in each compound?

a. C_6H_6 b. C_8H_{18} c. C_7H_8O d. $C_7H_{11}Br$ e. C_5H_9N

More Practice: Try Problem 10.42c–h.

Problem 10.4 How many degrees of unsaturation does each of the following drugs contain?

a. zolpidem (sleep aid sold as Ambien), $C_{19}H_{21}N_3O$
b. mefloquine (antimalarial drug), $C_{17}H_{16}F_6N_2O$

10.3 Nomenclature

- An alkene is identified by the suffix *-ene*.
- An alkyne is identified by the suffix *-yne*.

10.3A General IUPAC Rules

How To Name an Alkene or an Alkyne

Example Give the IUPAC name of each compound:

a.    b.

Step [1] Find the longest chain that contains *both* carbon atoms of the multiple bond.

a. Change the *-ane* ending of the parent alkane to *-ene*.

b. Change the *-ane* ending of the parent alkane to *-yne*.

6 C's in the longest chain

hexane ---→ hexene

8 C's in the longest chain

octane ---→ octyne

—Continued

How To, continued...

Step [2] Number the carbon chain to give the multiple bond the lower number, and apply all other rules of nomenclature.

a. **Number** the chain, and name using the *first number* assigned to the C=C.
- Number the chain to put the C=C at C**2**, not C4.
- **Name** and **number** the substituents.

three methyl groups at C2, C3, and C5
Answer: 2,3,5-trimethylhex-2-ene

b. **Number** the long chain; then **name** and **number** the substituents.

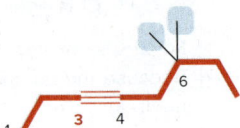

two methyl groups at C6
Answer: 6,6-dimethyloct-3-yne

Compounds with two double bonds are named as **dienes** by changing the *-ane* ending of the parent alkane to the suffix *-adiene*. Compounds with two triple bonds are named as **diynes**. Compounds with both a double and a triple bond are named as **enynes**. The chain is numbered to give the first site of unsaturation (either C=C or C≡C) the lower number.

In naming cycloalkenes, the **double bond is located between C1 and C2,** and the "1" is usually omitted in the name. The ring is numbered clockwise or counterclockwise to give the first substituent the lower number. Representative examples are given in Figure 10.1.

Figure 10.1 Examples of cycloalkene nomenclature

1-methylcyclopentene | 3-methylcycloheptene | 1,6-dimethylcyclohexene

[Number clockwise beginning at the C=C and place the CH₃ at C3.] [Number counterclockwise beginning at the C=C and place the first CH₃ at C1.]

CH₃CH₂CH=CH₂ is named as 1-butene using the 1979 IUPAC recommendations and but-1-ene using the 1993 IUPAC recommendations.

Compounds that contain both a multiple bond and a hydroxy group are named as **alkenols** or **alkynols,** and the chain (or ring) is numbered to **give the OH group the lower number.**

prop-2-yn-1-ol | 6-methylhept-6-en-2-ol

Problem 10.5 Give the IUPAC name for each compound.

Problem 10.6 Give the IUPAC name for each polyfunctional compound.

a. [structure with OH]
b. [structure with OH]
c. [structure with alkyne]
d. [structure with alkynes]

10.3B Common Names

The simplest alkene, $CH_2=CH_2$, named in the IUPAC system as **ethene**, is often called **ethylene**, its common name. The common names for three **alkyl groups** derived from alkenes are also used.

methylene group   **vinyl** group   **allyl** group

The simplest alkyne, HC≡CH, named in the IUPAC system as **ethyne**, is more often called **acetylene**, its common name. The two-carbon alkyl group derived from acetylene is called an **ethynyl group** (HC≡C—). Examples of naming compounds with an alkenyl or alkynyl substituent are shown in Figure 10.2.

Figure 10.2
Naming alkenes with common substituent names

methylenecyclohexane   1-vinylcyclopentene   2-allylcycloheptanol   1-ethynyl-2-isopropylcyclohexane

10.3C Naming Stereoisomers

Whenever the two groups on each end of a C=C are different from each other, two diastereomers are possible (Section 8.2B), and a prefix is needed to distinguish these alkenes by name.

Using Cis and Trans as Prefixes

An alkene having one alkyl group bonded to each carbon atom can be named using the prefixes **cis** and **trans** to designate the relative location of the two alkyl groups. For example, *cis*-hex-3-ene has two ethyl groups on the **same side** of the double bond, whereas *trans*-hex-3-ene has two ethyl groups on **opposite sides** of the double bond.

2 R's on the **same** side
cis-hex-3-ene

2 R's on **opposite** sides
trans-hex-3-ene

E stands for the German word *entgegen* meaning "opposite." *Z* stands for the German word *zusammen,* meaning "together." Using *E,Z* nomenclature, a cis isomer has the *Z* configuration and a trans isomer has the *E* configuration.

Using the Prefixes *E* and *Z*

Although the prefixes cis and trans can be used to distinguish diastereomers when two alkyl groups are bonded to the C=C, they cannot be used when there are three or four alkyl groups bonded to the C=C.

A — 3-methylpent-2-ene

B — 3-methylpent-2-ene

For example, alkenes **A** and **B** are two *different* compounds that are both called 3-methylpent-2-ene. In **A** the two CH₃ groups are cis, whereas in **B** the CH₃ and CH₂CH₃ groups are cis. The **E,Z system of nomenclature** has been devised to unambiguously name these kinds of alkenes.

How To Assign the Prefixes *E* and *Z* to an Alkene

Step [1] Assign priorities to the two substituents on each end of the C=C by using the priority rules for *R,S* nomenclature (Section 5.6).

- Divide the double bond in half, and assign the numbers **1** and **2** to indicate the relative priority of the two groups on each end—the higher-priority group is labeled **1**, and the lower-priority group is labeled **2**.

Assign priorities to each side separately.

Step [2] Assign *E* or *Z* based on the location of the two higher-priority groups (1).

Two higher-priority groups on **opposite sides**

E isomer
(*E*)-3-methylpent-2-ene

Two higher-priority groups on the **same side**

Z isomer
(*Z*)-3-methylpent-2-ene

- The **E** isomer has the two higher-priority groups on the **opposite sides.**
- The **Z** isomer has the two higher-priority groups on the **same side.**

Problem 10.7 Label each C—C double bond as *E* or *Z*.

a. b. c. d.

Problem 10.8 **A** is a toxin produced by the poisonous seaweed *Chlorodesmis fastigiata*. (a) Label each alkene that exhibits stereoisomerism as *E* or *Z*. (b) Draw a stereoisomer of **A** that has all *Z* double bonds.

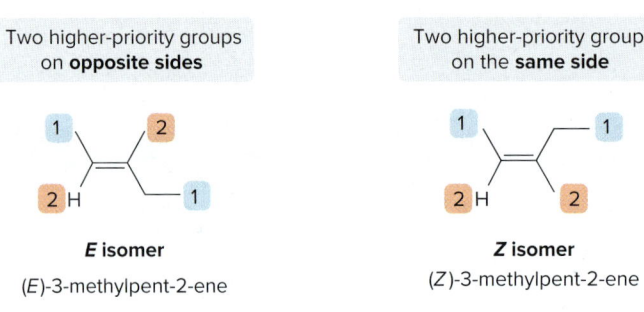

A

In response to a chemical distress signal from the coral *Acropora nasuta*, the goby fish protects the coral by eating the poisonous and invasive seaweed *Chlorodesmis fastigiata* (Problem 10.8).
Danielle Dixson

Problem 10.9 Draw the structure corresponding to each IUPAC name.

a. (*Z*)-4-ethylhept-3-ene   b. (*E*)-3,5,6-trimethyloct-2-ene   c. (*Z*)-2-bromo-1-iodohex-1-ene

10.4 Properties of Alkenes and Alkynes

Most alkenes and alkynes exhibit only weak van der Waals interactions, so their physical properties are similar to those of alkanes of comparable molecular weight.

- Alkenes and alkynes have low melting points and boiling points.
- Melting points and boiling points increase as the number of carbons increases because of increased surface area.
- Alkenes and alkynes are soluble in organic solvents and insoluble in water.

Cis and trans alkenes often have somewhat different physical properties. For example, *cis*-but-2-ene has a higher boiling point (4 °C) than *trans*-but-2-ene (1 °C). This difference arises because the C–C single bond between an alkyl group and one of the double bond carbons of an alkene is slightly polar. **The sp^3 hybridized alkyl carbon donates electron density to the sp^2 hybridized alkenyl carbon.**

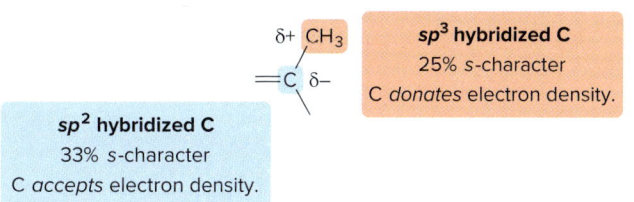

Related arguments involving C_{sp^3}–C_{sp^2} bonds were used in Section 8.2C to explain why the stability of an alkene increases with increasing alkyl substitution.

The bond dipole places a partial negative charge on the alkenyl carbon (sp^2) relative to the alkyl carbon (sp^3) because an sp^2 hybridized orbital has greater percent s-character (33%) than an sp^3 hybridized orbital (25%). **In a cis isomer, the two C_{sp^3}–C_{sp^2} bond dipoles reinforce each other, yielding a small net molecular dipole. In a trans isomer, the two bond dipoles cancel.**

- A cis alkene is more polar than a trans alkene, giving it a slightly higher boiling point and making it more soluble in polar solvents.

Problem 10.10 Rank the following isomers in order of increasing boiling point.

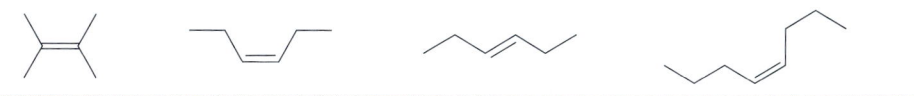

Students who have already been exposed to spectroscopy or who would like to learn about the spectroscopic properties of alkenes and alkynes are referred to the following sections of Spectroscopy Chapters B and C:

- Infrared spectroscopy: Sections B.3A, B.3D, B.4A; Tables B.1, B.2; Sample Problem B.2b
- Nuclear magnetic resonance spectroscopy: Sections C.4, C.8; Tables C.1, C.2, C.4, C.5; Sample Problems C.6c, C.7c

10.5 Interesting Alkenes and Alkynes

Numerous organic compounds containing carbon–carbon double bonds have been isolated from natural sources, including β-carotene, the orange pigment in carrots (Section 3.5A), and zingiberene, a triene in the oil of ginger.

β-carotene
(orange pigment in carrots)

zingiberene
(oil of ginger)

I. Rozenbaum & F. Cirou/PhotoAlto

Pixtal/age fotostock

Ethynylestradiol and **norethindrone** are two components of oral contraceptives that contain a carbon–carbon triple bond (Figure 10.3). Both molecules are synthetic analogues of the naturally occurring female hormones estradiol and progesterone, but are more potent so they can be administered in lower doses. Most oral contraceptives contain two of these synthetic

Figure 10.3 How oral contraceptives work

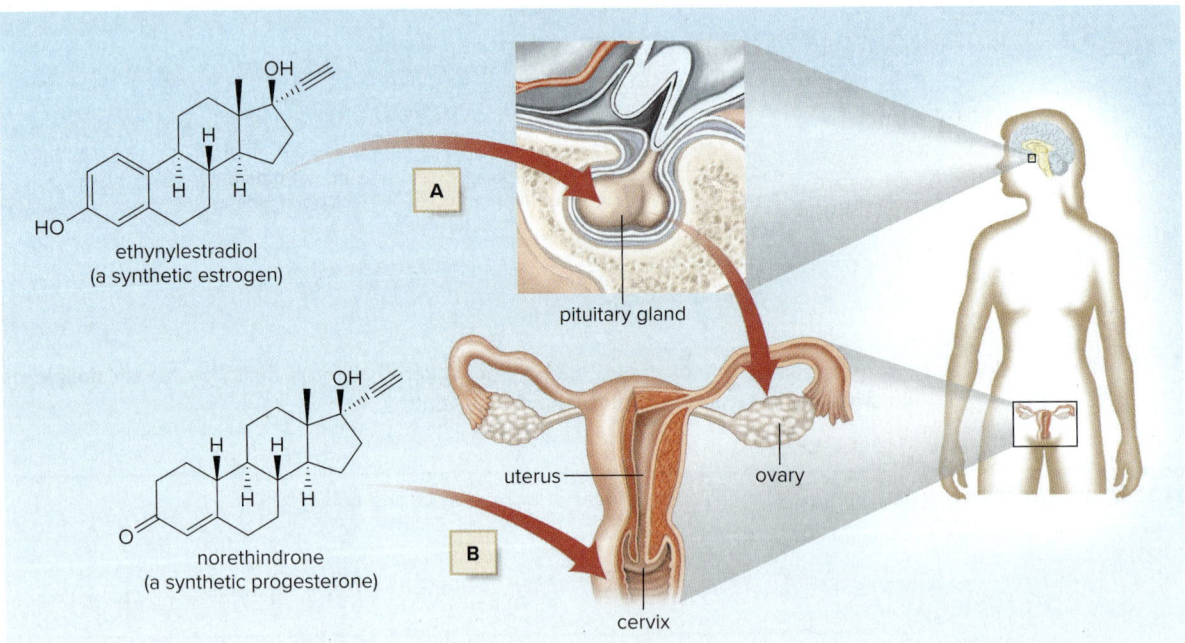

- Monthly cycles of hormones from the pituitary gland cause ovulation, the release of an egg from an ovary. To prevent pregnancy, the two synthetic hormones in many oral contraceptives have different effects on the female reproductive system.
  A: The elevated level of **ethynylestradiol,** a synthetic estrogen, "fools" the pituitary gland into thinking a woman is pregnant, so ovulation does not occur.
  B: The elevated level of **norethindrone,** a synthetic progesterone, stimulates the formation of a thick layer of mucus in the cervix, making it difficult for sperm to reach the uterus.

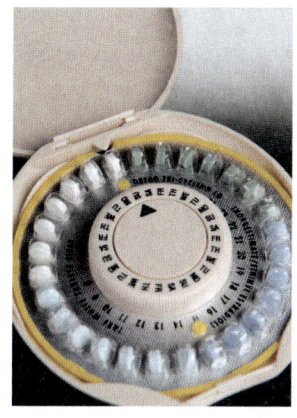

Ethynylestradiol is a synthetic compound whose structure closely resembles the carbon skeleton of female estrogen hormones. *Christopher Kerrigan/McGraw-Hill Education*

hormones. They act by artificially elevating hormone levels in a woman, thereby preventing pregnancy.

estradiol   progesterone

Two other synthetic hormones with alkynyl appendages are **RU 486** and **levonorgestrel.** RU 486 blocks the effects of progesterone and, because of this, prevents implantation of a fertilized egg. RU 486 is used to induce abortions within the first few weeks of pregnancy. Levonorgestrel interferes with ovulation, so it prevents pregnancy if taken within a few days of unprotected sex.

RU 486
(Trade name Mifepristone)

levonorgestrel
(Trade name Plan B)

10.6 Fatty Acids and Triacylglycerols

Understanding the geometry of C—C double bonds provides an insight into the properties of **triacylglycerols,** the most abundant lipids (Section 3.9D). Triacylglycerols contain three ester groups, each having a long carbon chain (abbreviated as R, R', and R'') bonded to a carbonyl group (C=O).

General structure of an ester:

R groups have 11–19 C's.

[Three ester groups are labeled in red.]

triacylglycerol

10.6A Fatty Acids

Triacylglycerols are hydrolyzed to glycerol (a triol) and three **fatty acids** of general structure RCO_2H. Naturally occurring fatty acids contain 12–20 carbon atoms, with a carboxy group (CO_2H) at one end.

Candlenuts, known as kukui nuts in Hawai'i, are rich in linoleic and linolenic acids, two essential fatty acids that cannot be synthesized in the body and must therefore be obtained in the diet. *Inga Spence/Science Source*

triacylglycerol → (H₂O, H⁺ or ⁻OH or enzymes) → glycerol + fatty acids

These fatty acids have **12–20 C's.**

Table 10.1 The Effect of Double Bonds on the Melting Point of Fatty Acids

Name	Structure	Mp (°C)
Stearic acid (**0** C=C)		69
Oleic acid (**1** C=C)		4
Linoleic acid (**2** C=C)		−5
Linolenic acid (**3** C=C)		−11

Increasing number of double bonds →

- Saturated fatty acids have no double bonds in their long hydrocarbon chains, and unsaturated fatty acids have one or more double bonds in their hydrocarbon chains.
- Double bonds in naturally occurring fatty acids have the *Z* configuration.

Table 10.1 lists the structure and melting point of four fatty acids containing 18 carbon atoms. Stearic acid is one of the two most common saturated fatty acids, and oleic and linoleic acids are the most common unsaturated ones. The data show the effect of *Z* double bonds on the melting point of fatty acids.

- As the number of double bonds in the fatty acid *increases,* the melting point *decreases.*

The three-dimensional structures of the fatty acids in Figure 10.4 illustrate how *Z* double bonds introduce kinks in the long hydrocarbon chain, decreasing the ability of the fatty acid to pack well in a crystalline lattice. **The *larger* the number of *Z* double bonds, the more kinks in the hydrocarbon chain, and the *lower* the melting point.**

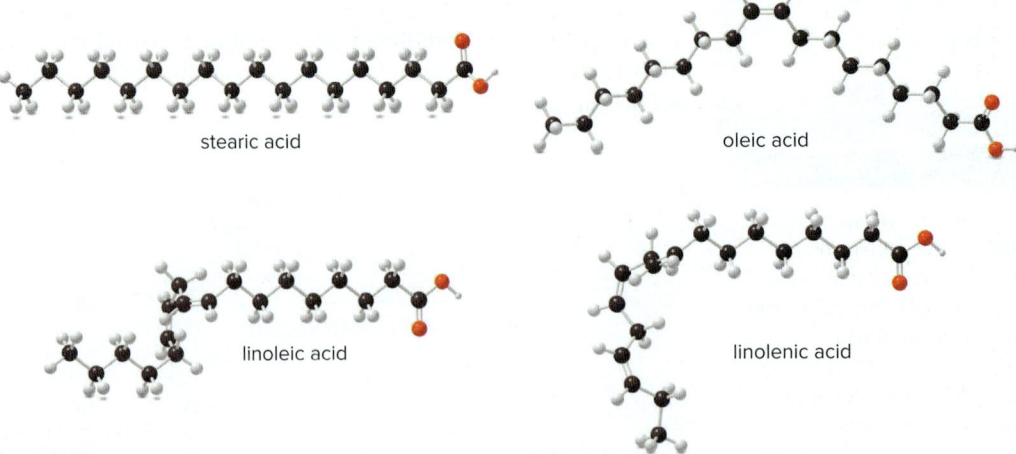

Figure 10.4
Three-dimensional structure of four C$_{18}$ fatty acids

stearic acid

oleic acid

linoleic acid

linolenic acid

Problem 10.11 Linolenic acid (Table 10.1) and stearidonic acid are omega-3 fatty acids, unsaturated fatty acids that contain the first double bond located at C3, when numbering begins at the methyl end of the chain. Predict how the melting point of stearidonic acid compares with the melting points of linolenic and stearic acids. A current avenue of research is examining the use of soybean oil enriched in stearidonic acid as a healthier alternative to vegetable oils that contain fewer degrees of unsaturation.

stearidonic acid

10.6B Fats and Oils

Fats and **oils** are triacylglycerols with different physical properties.

- Fats have higher melting points—they are *solids* at room temperature.
- Oils have lower melting points—they are *liquids* at room temperature.

The identity of the three fatty acids in the triacylglycerol determines whether it is a fat or an oil. **Increasing the number of double bonds in the fatty acid side chains decreases the melting point of the triacylglycerol.**

- Fats are derived from fatty acids having few double bonds.
- Oils are derived from fatty acids having a larger number of double bonds.

Canola, soybeans, and flaxseed are excellent dietary sources of linolenic acid, an essential fatty acid. Oils derived from omega-3 fatty acids (Problem 10.11) are currently thought to be especially beneficial for individuals at risk of developing coronary artery disease. *Jill Braaten/McGraw-Hill Education*

Saturated fats are typically obtained from animal sources, whereas unsaturated oils are common in vegetable sources. Thus, butter and lard are high in saturated triacylglycerols, and olive oil and safflower oil are high in unsaturated triacylglycerols. An exception to this generalization is coconut oil, which is composed largely of saturated alkyl side chains.

Considerable evidence suggests that an elevated cholesterol level is linked to an increased risk of heart disease. Saturated fats stimulate cholesterol synthesis in the liver, thus increasing the cholesterol concentration in the blood.

10.7 Preparation of Alkenes and Alkynes

Recall from Chapters 8 and 9 that alkenes and alkynes can be prepared by elimination reactions.

For example, **dehydrohalogenation of alkyl halides with strong base yields alkenes via an E2 mechanism** (Sections 8.4 and 8.5). **The acid-catalyzed dehydration of alcohols with H_2SO_4 or TsOH yields alkenes,** too (Sections 9.8 and 9.9). These elimination reactions are **stereoselective** and **regioselective,** so the **most stable alkene is usually formed as the major product.**

Alkynes are prepared by the elimination of **two equivalents of HX from a vicinal or geminal dihalide** (Section 8.10).

$$\text{geminal dichloride} \xrightarrow[\text{[-2 HCl]}]{2 \text{ Na}^+ \text{ }^-\text{NH}_2} \text{alkyne}$$

$$\text{vicinal dibromide} \xrightarrow[\substack{\text{DMSO} \\ \text{[-2 HBr]}}]{\substack{\text{K}^+ \text{ }^-\text{OC(CH}_3)_3 \\ \text{(2 equiv)}}} \text{alkyne}$$

Problem 10.12 Draw the products of each elimination reaction.

a. cyclopentanol + H$_2$SO$_4$

b. 2-bromohexane + NaOCH$_2$CH$_3$

c. 2,3-dibromohexane + 2 Na$^+$ $^-$NH$_2$

10.8 Introduction to the Reactions of Alkenes and Alkynes

Most reactions of alkenes and alkynes occur because they have easily broken π bonds. In addition, terminal alkynes contain an acidic *sp* hybridized C—H bond that is readily deprotonated with strong base.

10.8A Addition Reactions

Because a C—C π bond is much *weaker* than a C—C σ bond, the characteristic reaction of alkenes and alkynes is **addition. With an alkene, the π bond is broken and two new σ bonds are formed.**

$$\text{alkene} + \text{X—Y} \longrightarrow \text{product}$$

A π bond is broken.   Two σ bonds are formed.

With an alkyne, two sequential reactions take place: addition of one equivalent of reagent forms an alkene, which then adds a second equivalent of reagent to yield a product having **four new bonds.**

two weak π bonds → weak π bond (*E* or *Z* product) → Four σ bonds are formed.

Alkenes and alkynes are electron rich, as seen in the electrostatic potential plots in Figure 10.5. What kinds of reagents add to the weak, electron-rich π bonds of alkenes and alkynes? There are many of them, and that can make this chemistry challenging. To help you organize this information, keep in mind the following:

> The oxidation and reduction of alkenes and alkynes, reactions that also involve addition, are discussed in Chapter 11.

- Every reaction of the carbon–carbon multiple bonds involves *addition*: π bonds are always broken.
- Because these compounds are electron rich, they do *not* react with nucleophiles or bases, reagents that are themselves electron rich. Alkenes and alkynes react with *electrophiles.*

Figure 10.5
Electrostatic potential plots of ethylene and acetylene

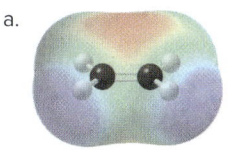

- The red electron-rich region of the π bond is located above and below the plane of the molecule. Because the plane of the alkene depicted in this electrostatic potential plot is tipped, only the red region above the molecule is visible.

- The red electron-rich region is located between the two carbon atoms, forming a cylinder of electron density.

The stereochemistry of addition is often important in delineating a reaction's mechanism. Because the carbon atoms of a double bond are both trigonal planar, the elements of X and Y can be added to them from the **same side** or from **opposite sides.**

- **Syn addition** takes place when both X and Y are added from the *same* side.
- **Anti addition** takes place when X and Y are added from *opposite* sides.

Five reactions of alkenes are discussed in Chapter 10 and each is illustrated in Figure 10.6, using cyclohexene as the starting material. Addition reactions of alkenes are discussed in Sections 10.9–10.16. Four addition reactions of alkynes are discussed and each is illustrated in Figure 10.7 with but-1-yne as the starting material. Additions to alkynes are presented in Sections 10.17–10.19.

Figure 10.6
Five addition reactions of cyclohexene

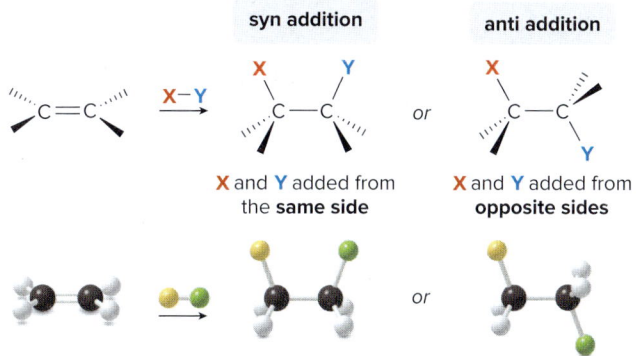

- In each reaction, the π bond is broken and two new σ bonds are formed.

Figure 10.7
Four addition reactions of but-1-yne

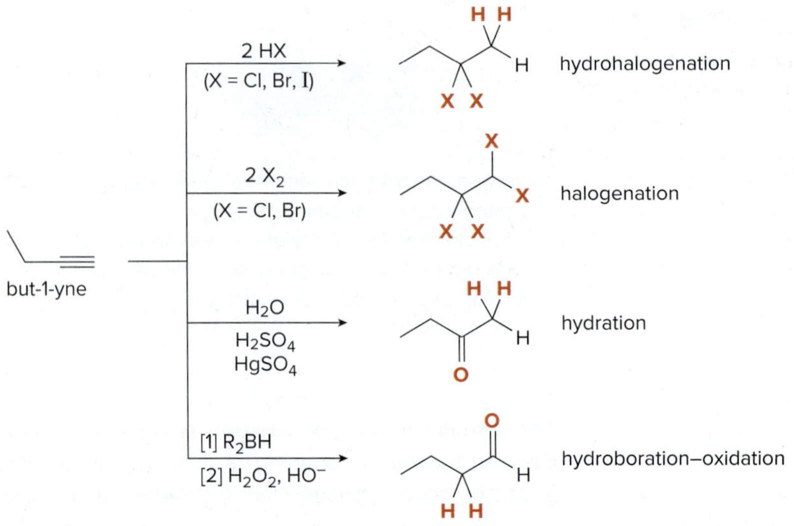

- hydrohalogenation (2 HX, X = Cl, Br, I)
- halogenation (2 X₂, X = Cl, Br)
- hydration (H₂O, H₂SO₄, HgSO₄)
- hydroboration–oxidation ([1] R₂BH; [2] H₂O₂, HO⁻)

- In each addition, both π bonds of the triple bond are broken, and four new bonds are formed.

10.8B Terminal Alkynes—Reaction as an Acid

Because sp hybridized C–H bonds are more acidic than sp^2 and sp^3 hybridized C–H bonds, terminal alkynes are readily deprotonated with strong base in a Brønsted–Lowry acid–base reaction. The resulting anion is called an **acetylide anion**.

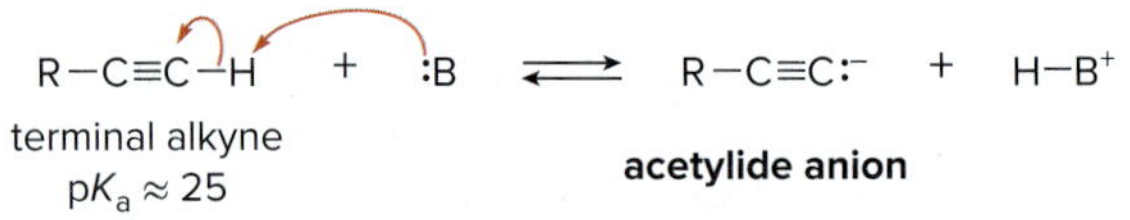

$$R-C\equiv C-H + :B^- \rightleftharpoons R-C\equiv C:^- + H-B$$

terminal alkyne, $pK_a \approx 25$ → acetylide anion

> Recall from Section 2.5D that the acidity of a C–H bond increases as the percent s-character of C increases. Thus, the following order of relative acidity results: $C_{sp^3}-H < C_{sp^2}-H < C_{sp}-H$.

What bases can be used for this reaction? Because an acid–base equilibrium favors the weaker acid and base, only **bases having conjugate acids with pK_a values *higher* than the terminal alkyne—that is, pK_a values > 25—are strong enough** to form a significant concentration of acetylide anion. As shown in Table 10.2, ⁻NH₂ and H⁻ are strong enough to deprotonate a terminal alkyne, but ⁻OH and ⁻OR are not.

Table 10.2 A Comparison of Bases for Alkyne Deprotonation

	Base	pK_a of the conjugate acid
These bases are **strong** enough to deprotonate an alkyne.	⁻NH₂	38
	H⁻	35
These bases are **not** strong enough to deprotonate an alkyne.	⁻OH	15.7
	⁻OR	15.5–18

Why is this reaction useful? The acetylide anions formed by deprotonating terminal alkynes are **strong nucleophiles** that can react with a variety of electrophiles, as shown in Section 10.20.

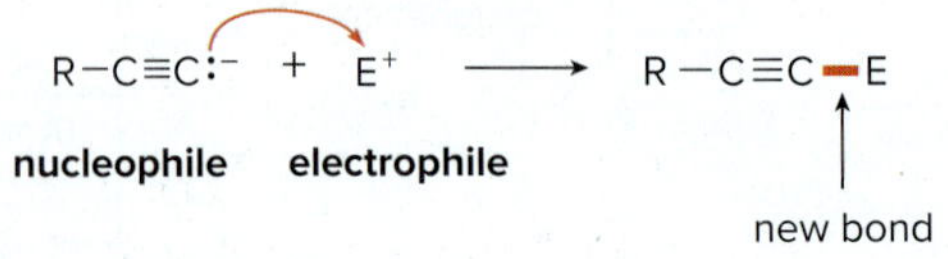

$$R-C\equiv C:^- + E^+ \longrightarrow R-C\equiv C-E$$

nucleophile   electrophile   (new bond)

Problem 10.13 Which bases can deprotonate acetylene? The pK_a values of the conjugate acids are given in parentheses.

a. CH₃NH⁻ (pK_a = 40)   b. CO₃²⁻ (pK_a = 10.2)   c. CH₂=CH⁻ (pK_a = 44)   d. (CH₃)₃CO⁻ (pK_a = 18)

10.9 Hydrohalogenation—Electrophilic Addition of HX to Alkenes

Hydrohalogenation of an alkene to form an alkyl halide is the reverse of the dehydrohalogenation of an alkyl halide to form an alkene, a reaction discussed in detail in Sections 8.4 and 8.5.

Hydrohalogenation results in the addition of hydrogen halides HX (X = Cl, Br, and I) to alkenes to form alkyl halides.

Two bonds are broken in this reaction—the weak π bond of the alkene and the HX bond—and two new σ bonds are formed—one to H and one to X. Because X is more electronegative than H, the H—X bond is polarized, with a partial positive charge on H. Because the electrophilic (H) end of HX is attracted to the electron-rich double bond, these reactions are called **electrophilic additions**. Addition reactions are exothermic because the two σ bonds formed in the product are *stronger* than the σ and π bonds broken in the reactants.

To draw the products of an addition reaction:

- Locate the C—C double bond.
- Identify the σ bond of the reagent that breaks—namely, the H—X bond in hydrohalogenation.
- Break the π bond of the alkene and the σ bond of the reagent, and form two new σ bonds to the C atoms of the double bond.

The mechanism of electrophilic addition of HX consists of **two steps**: addition of H⁺ to form a carbocation, followed by nucleophilic attack of X⁻. The mechanism is illustrated for the reaction of *cis*-but-2-ene with HBr in Mechanism 10.1.

Mechanism 10.1 Electrophilic Addition of HX to an Alkene

1 The π bond of the alkene attacks the H of HBr to form a new C—H bond and a **carbocation** in the rate-determining step.

2 **Nucleophilic attack of Br⁻** on the carbocation forms the new C—Br bond.

The mechanism of electrophilic addition consists of two successive Lewis acid–base reactions. In Step [1], **the alkene is the Lewis base** that donates an electron pair to **H–Br, the Lewis acid**, whereas in Step [2], **Br⁻ is the Lewis base** that donates an electron pair to the **carbocation, the Lewis acid**.

Problem 10.14 What product is formed when each alkene is treated with HCl?

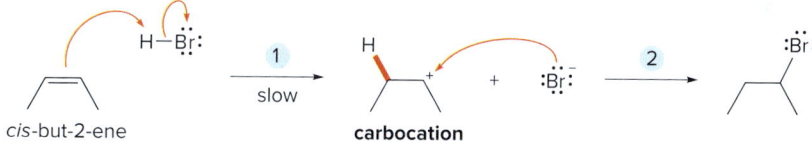

10.10 Markovnikov's Rule

With an unsymmetrical alkene, HX can add to the double bond to give two constitutional isomers.

propene → 1-chloropropane or 2-chloropropane (**only product**)

For example, HCl addition to propene could in theory form 1-chloropropane by addition of H and Cl to C2 and C1, respectively, and 2-chloropropane by addition of H and Cl to C1 and C2, respectively. In fact, **electrophilic addition forms *only* 2-chloropropane.** This is a specific example of a general trend called **Markovnikov's rule,** named for the Russian chemist who first determined the regioselectivity of electrophilic addition of HX.

- Markovnikov's rule: In the addition of HX to an unsymmetrical alkene, the H atom bonds to the *less substituted* carbon atom—that is, the carbon that has more H atoms to begin with.

The basis of Markovnikov's rule is the formation of a carbocation in the rate-determining step of the mechanism. With propene, there are two possible paths for this first step, depending on which carbon atom of the double bond forms the new bond to hydrogen.

[1] → 1° carbocation + Cl⁻

[2] → (preferred path) 2° carbocation + Cl⁻ → 2-chloropropane

The Hammond postulate was first introduced in Section 7.14 to explain the relative rate of S_N1 reactions with 1°, 2°, and 3° RX.

Path [1] forms a highly unstable 1° carbocation, whereas Path [2] forms a **more stable 2° carbocation.** According to the Hammond postulate, Path [2] is faster because formation of the carbocation is an endothermic process, so **the transition state to form the more stable 2° carbocation is lower in energy** (Figure 10.8).

- In the addition of HX to an unsymmetrical alkene, the H atom is added to the *less* substituted carbon to form the *more stable, more substituted* carbocation.

Figure 10.8
Electrophilic addition and the Hammond postulate

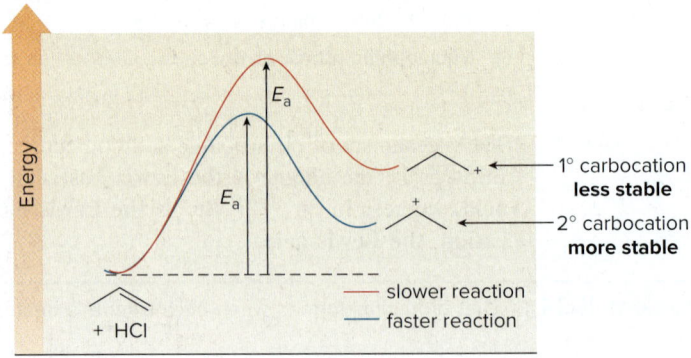

- The E_a for formation of the more stable 2° carbocation is *lower* than the E_a for formation of the 1° carbocation. The 2° carbocation is formed *faster*.

10.11 Stereochemistry of Electrophilic Addition of HX

Similar results are seen in any electrophilic addition involving an intermediate carbocation: **the more stable, *more* substituted carbocation is formed by addition of the electrophile to the *less* substituted carbon.**

Problem 10.15 Draw the products formed when each alkene is treated with HCl.

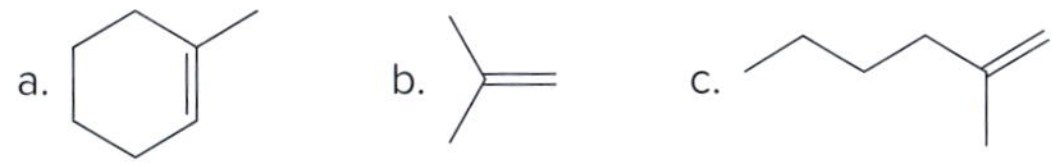

a. [methylcyclohexene] b. [2,3-dimethyl-2-butene] c. [2-methyl-1-pentene]

Because carbocations are formed as intermediates in hydrohalogenation, carbocation rearrangements can occur, as illustrated in Sample Problem 10.3.

Sample Problem 10.3 Drawing Hydrohalogenation with a Carbocation Rearrangement

Draw a stepwise mechanism for the following reaction.

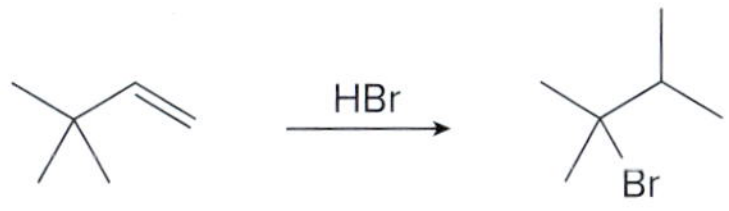

Solution

Because the carbon skeletons of the starting material and product are *different*—the alkene reactant has a 4° carbon and the product alkyl halide does not—a carbocation rearrangement must have occurred.

Markovnikov addition of HBr adds H⁺ to the less substituted end of the double bond, forming a 2° carbocation in Step [1].

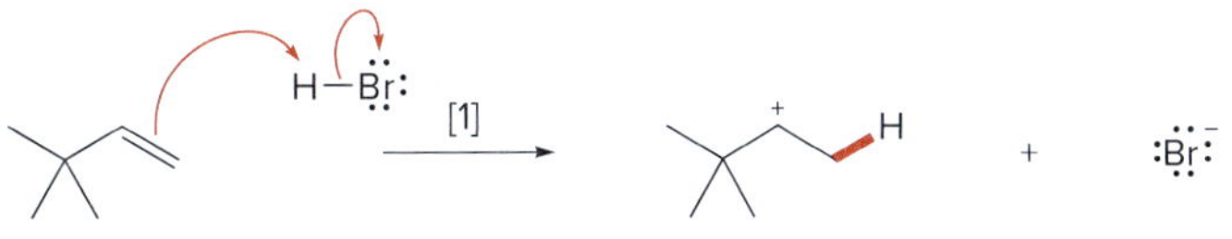

2° carbocation
new bond shown in red

Rearrangement of the 2° carbocation by a 1,2-methyl shift forms a more stable 3° carbocation in Step [2]. Nucleophilic attack of Br⁻ forms the product, a 3° alkyl halide, in Step [3].

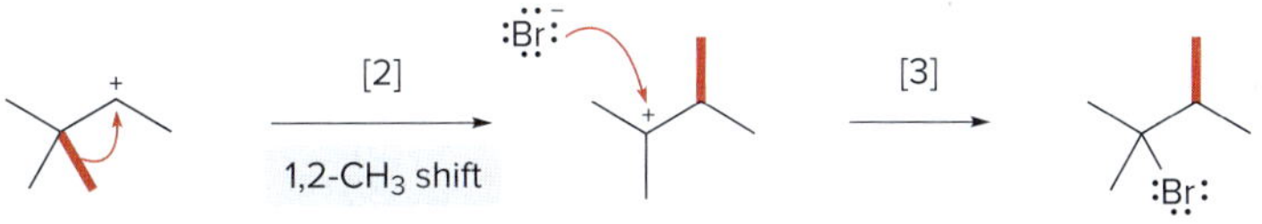

1,2-CH₃ shift

3° carbocation

Problem 10.16 Treatment of 3-methylcyclohexene with HCl yields two products, 1-chloro-3-methylcyclohexane and 1-chloro-1-methylcyclohexane. Draw a mechanism to explain this result.

Problem 10.17 Addition of HBr to which of the following alkenes will lead to a rearrangement?

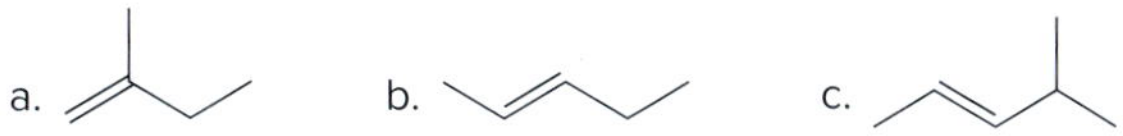

a.   b.   c.

10.11 Stereochemistry of Electrophilic Addition of HX

To understand the stereochemistry of electrophilic addition, recall two stereochemical principles learned in Chapters 7 and 9.

- Trigonal planar atoms react with reagents from two directions with equal probability (Section 7.12C).
- Achiral starting materials yield achiral or racemic products (Section 9.16).

Many hydrohalogenation reactions begin with an **achiral reactant** and form an **achiral product.** For example, the addition of HBr to cyclohexene, an achiral alkene, forms bromocyclohexane, an achiral alkyl halide.

cyclohexene + H—Br ⟶ bromocyclohexane
achiral starting material   **achiral** product

Because addition converts sp^2 hybridized carbons to sp^3 hybridized carbons, sometimes new stereogenic centers are formed from hydrohalogenation. Markovnikov addition of HCl to 1,3,3-trimethylcyclohexene, an achiral alkene, forms one constitutional isomer, 1-chloro-1,3,3-trimethylcyclohexane. Because this product now has a stereogenic center at one of the newly formed sp^3 hybridized carbons (labeled in blue), **an equal amount of two enantiomers—a racemic mixture—must form.**

1,3,3-trimethylcyclohexene
achiral starting material

1-chloro-1,3,3-trimethyl-cyclohexane

new stereogenic center

A   B

The product has one new stereogenic center, so **two enantiomers are formed.**

The mechanism of hydrohalogenation illustrates why two enantiomers are formed. Initial addition of the electrophile H⁺ (from HCl) occurs from **either side of the planar double bond** to form a carbocation. Both modes of addition (from above and below) generate the same **achiral carbocation.** Either representation of this carbocation can then be used to draw the second step of the mechanism.

H⁺ from **above**   H⁺ from **below**

identical carbocations

Nucleophilic attack of Cl⁻ on the trigonal planar carbocation also occurs from two different directions, forming two products, **A** and **B,** having a new stereogenic center. **A** and **B** are not superimposable, so they are **enantiomers.** Because attack from either direction occurs with equal probability, a **racemic mixture** of **A** and **B** is formed.

Cl⁻ from **above**   Cl⁻ from **below**
A   B
syn addition   **anti addition**
H and Cl are added   H and Cl are added
from the **same** side.   from **opposite** sides.

> The terms **cis** and **trans** refer to the arrangement of groups in a particular compound, usually an alkene or a disubstituted cycloalkane. The terms **syn** and **anti** describe the stereochemistry of a process—for example, how two groups are added to a double bond.

Because hydrohalogenation begins with a **planar** double bond and forms a **planar** carbocation, addition of H and Cl occurs in two different ways. The elements of H and Cl can both be added from the same side of the double bond—that is, **syn addition**—or they can be added from opposite sides—that is, **anti addition.** *Both* modes of addition occur in this two-step reaction mechanism.

- **Hydrohalogenation occurs with syn and anti addition of HX.**

Table 10.3 summarizes the characteristics of electrophilic addition of HX to alkenes.

Table 10.3 Summary: Electrophilic Addition of HX to Alkenes

	Observation
Mechanism	• The mechanism involves **two steps.** • The rate-determining step forms a **carbocation.** • **Rearrangements** can occur.
Regioselectivity	• **Markovnikov's rule is followed.** In unsymmetrical alkenes, H bonds to the less substituted C to form the more stable carbocation.
Stereochemistry	• **Syn** and **anti** addition occur.

Problem 10.18 Draw the products, including stereochemistry, of each reaction.

a. [alkene] $\xrightarrow{\text{HBr}}$ b. [alkene] $\xrightarrow{\text{HCl}}$

10.12 Hydration—Electrophilic Addition of Water

Hydration results in the addition of water to an alkene to form an alcohol. H_2O itself is too weak an acid to protonate an alkene, but with added H_2SO_4, H_3O^+ is formed and addition readily occurs.

[alkene] + H—OH ($\delta+$ $\delta-$) $\xrightarrow{H_2SO_4}$ **alcohol**

A π bond is broken.

[cyclohexene] + H_2O $\xrightarrow{H_2SO_4}$ [cyclohexanol with OH and H]

Hydration is simply another example of **electrophilic addition.** The first two steps of the mechanism are similar to those of electrophilic addition of HX—that is, addition of H^+ (from H_3O^+) to generate a carbocation, followed by nucleophilic attack of H_2O. Mechanism 10.2 illustrates the addition of H_2O to cyclohexene to form cyclohexanol.

Mechanism 10.2 Electrophilic Addition of H₂O to an Alkene—Hydration

[Step 1: cyclohexene + H—OH₂⁺ → carbocation, slow]
[Step 2: carbocation + H₂O → protonated alcohol]
[Step 3: protonated alcohol + H₂O → cyclohexanol + H₃O⁺]

carbocation
new bond shown in red

cyclohexanol

1. The π bond of the alkene attacks the H of H_3O^+ to form a new C–H bond and a **carbocation** in the rate-determining step.
2. **Nucleophilic attack of H_2O** on the carbocation forms the new C–O bond.
3. **Removal of a proton** with H_2O forms a neutral alcohol. Because the acid used in Step [1] is regenerated in Step [3], the reaction is acid-catalyzed.

Hydration of an alkene to form an alcohol is the reverse of the dehydration of an alcohol to form an alkene, a reaction discussed in detail in Section 9.8.

There are three consequences to the formation of carbocation intermediates:

- In unsymmetrical alkenes, H adds to the *less* substituted carbon to form the *more* stable carbocation; that is, Markovnikov's rule holds.
- Addition of H and OH occurs in both a syn and anti fashion.
- Carbocation *rearrangements* can occur.

Alcohols add to alkenes, forming ethers, using the same mechanism. Addition of CH_3OH to 2-methylpropene, for example, forms *tert*-butyl methyl ether (**MTBE**), a high octane fuel additive described in Section 3.4C.

tert-butyl methyl ether
MTBE

Problem 10.19 What two alkenes give rise to each alcohol as the major product of acid-catalyzed hydration?

Problem 10.20 What stereoisomers are formed when pent-1-ene is treated with H_2O and H_2SO_4?

10.13 Halogenation—Addition of Halogen

Halogenation results in the addition of halogen X_2 (X = Cl or Br) to an alkene, forming a vicinal dihalide.

A π bond is broken.   vicinal dihalide

Halogenation is synthetically useful only with Cl_2 and Br_2. The dichlorides and dibromides formed in this reaction serve as starting materials for the synthesis of alkynes, as we learned in Section 8.10.

Halogens add to π bonds because halogens are **polarizable.** The electron-rich double bond induces a dipole in an approaching halogen molecule, making one halogen atom electron deficient and the other electron rich ($X^{\delta+}-X^{\delta-}$). **The electrophilic halogen atom is then attracted to the nucleophilic double bond,** making addition possible.

Two facts demonstrate that halogenation follows a different mechanism from that of hydrohalogenation or hydration. First, **no rearrangements** occur, and second, only **anti addition of X_2** is observed. For example, treatment of cyclohexene with Br_2 yields two **trans** enantiomers formed by **anti addition.**

enantiomers

These facts suggest that **carbocations are *not* intermediates in halogenation.** Unstable carbocations rearrange, and both syn and anti addition is possible with carbocation intermediates. The accepted mechanism for halogenation comprises **two steps,** but it does *not* proceed with formation of a carbocation, as shown in Mechanism 10.3.

Mechanism 10.3 Addition of X_2 to an Alkene—Halogenation

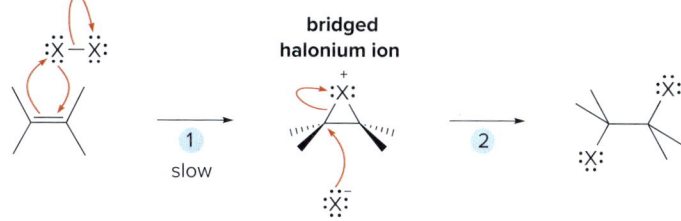

1. Four bonds are broken or formed to generate an unstable **bridged halonium ion** that contains a three-membered ring. The electron pair in the π bond and a lone pair on a halogen are used to form two new C–X bonds, and the X–X bond is cleaved.
2. **Nucleophilic attack of X^-** ring opens the bridged halonim ion and forms a new C–X bond.

Bridged halonium ions resemble carbocations in that they are short-lived intermediates that react readily with nucleophiles. Carbocations are inherently unstable because only six electrons surround carbon, whereas **halonium ions are unstable because they contain a strained three-membered ring** with a positively charged halogen atom.

carbocation

C has no octet.

bridged halonium ion

The ring has angle strain.

Problem 10.21 Draw the products of each reaction, including stereochemistry.

a. ☐ —Br₂→   b. ⬡ —Cl₂→

10.14 Stereochemistry of Halogenation

How does the proposed mechanism invoking a bridged halonium ion intermediate explain the observed **trans products of halogenation?** For example, chlorination of cyclopentene affords both enantiomers of *trans*-1,2-dichlorocyclopentane, with *no* cis products.

trans enantiomers

Initial addition of the electrophile Cl^+ (from Cl_2) occurs from either side of the planar double bond to form the bridged chloronium ion. In this example, both modes of addition (from above and below) generate the same **achiral** intermediate, so either representation can be used to draw the second step.

from **above**   from **below**

identical achiral chloronium ions

406 CHAPTER 10 Alkenes and Alkynes

The opening of bridged halonium ion intermediates resembles the opening of epoxide rings with nucleophiles discussed in Section 9.16.

In the second step, **nucleophilic attack of Cl⁻ must occur from the back side**—that is, from the side of the five-membered ring opposite to the side having the bridged chloronium ion. Because the nucleophile attacks from below in this example and the leaving group departs from above, the two Cl atoms in the product are oriented **trans** to each other. Backside attack occurs with equal probability at either carbon of the three-membered ring to yield an equal amount of two enantiomers—**a racemic mixture.**

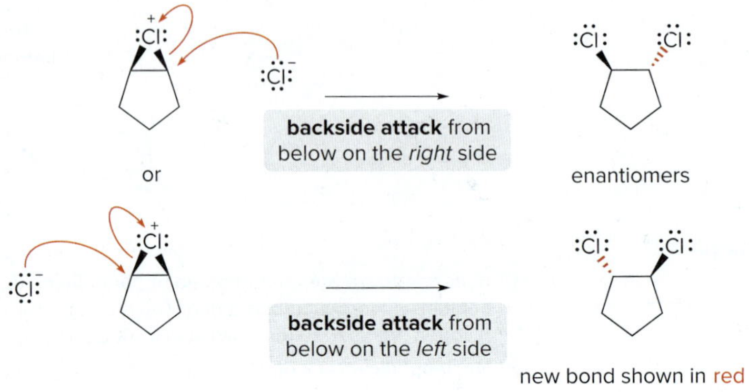

In summary, the mechanism for halogenation of alkenes occurs in two steps:

- Addition of X⁺ forms an unstable bridged halonium ion in the rate-determining step.
- Nucleophilic attack of X⁻ occurs from the *back side* to form trans products. The overall result is *anti addition* of X₂ across the double bond.

Because halogenation occurs exclusively in an anti fashion, cis and trans alkenes yield different stereoisomers. Halogenation of alkenes is a **stereospecific reaction.**

- A reaction is *stereospecific* when each of two specific stereoisomers of a starting material yields a particular stereoisomer of a product.

cis-But-2-ene yields two enantiomers, whereas *trans*-but-2-ene yields a single achiral meso compound, as shown in Figure 10.9.

Figure 10.9
Halogenation of *cis*- and *trans*-but-2-ene

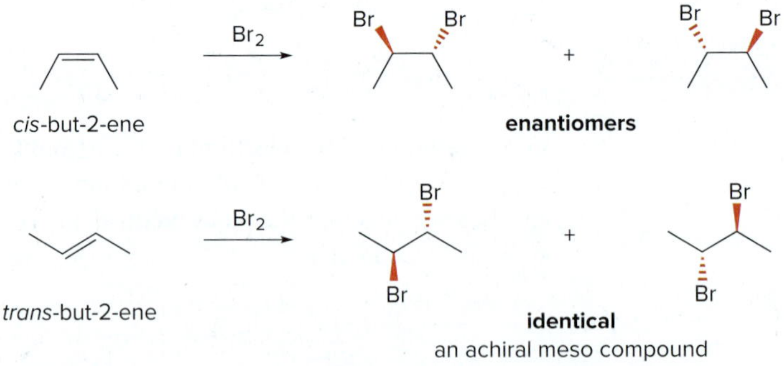

To draw the products of halogenation:
- Add Br₂ in an **anti** fashion across the double bond, leaving all other groups in their original orientations. With the alkene drawn in the plane of the page, **one Br adds from the front (ending up on a wedge), and one Br adds from the back (ending up on a dashed wedge).**
- Sometimes this reaction produces two stereoisomers, as in the case of *cis*-but-2-ene, which forms an equal amount of **two enantiomers.** Sometimes it produces a single compound, as in the case of *trans*-but-2-ene, where a **meso** compound is formed.

Problem 10.22 Draw all stereoisomers formed in each reaction.

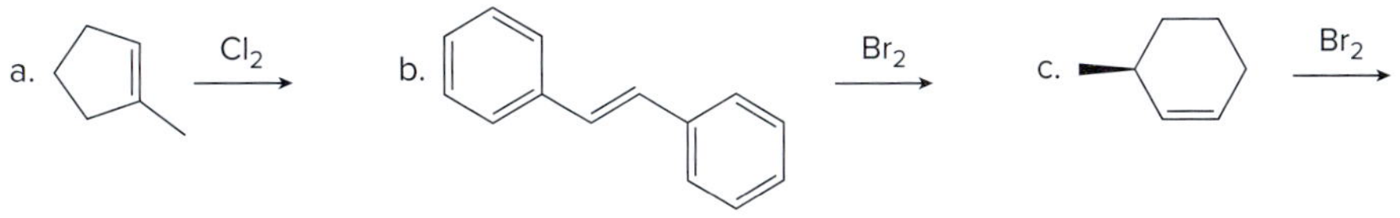

a. [1-methylcyclopentene] $\xrightarrow{Cl_2}$ b. [trans-stilbene] $\xrightarrow{Br_2}$ c. [(R)-4-methylcyclohexene] $\xrightarrow{Br_2}$

10.15 Halohydrin Formation

Treatment of an alkene with a halogen X_2 and H_2O forms a **halohydrin** by addition of the elements of **X** and **OH** to the double bond.

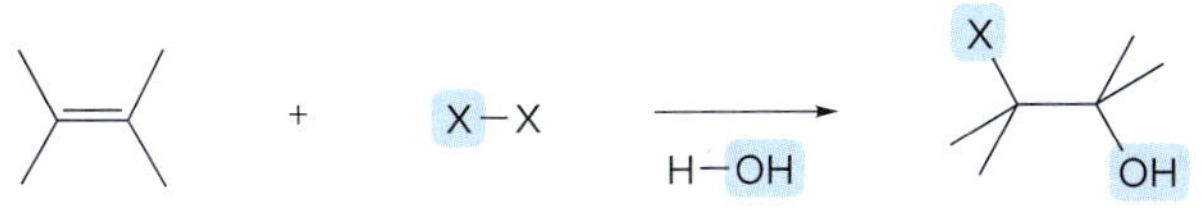

A π bond is broken. → **halohydrin**

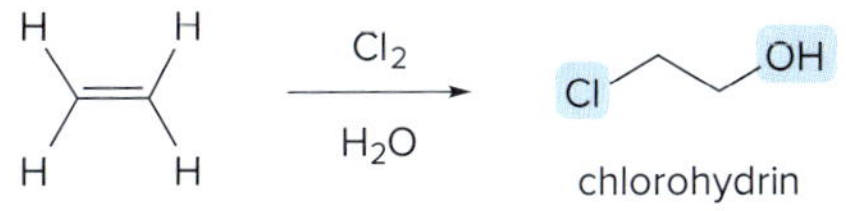

$$CH_2=CH_2 \xrightarrow{Cl_2, H_2O} ClCH_2CH_2OH \text{ (chlorohydrin)}$$

The mechanism for halohydrin formation is similar to the mechanism for halogenation: addition of the electrophile X^+ (from X_2) to form a **bridged halonium ion,** followed by nucleophilic attack by H_2O from the back side on the three-membered ring (Mechanism 10.4). Even though X^- is formed in Step [1] of the mechanism, its concentration is small compared to H_2O (often the solvent), so H_2O and *not* X^- is the nucleophile.

Mechanism 10.4 Addition of X and OH—Halohydrin Formation

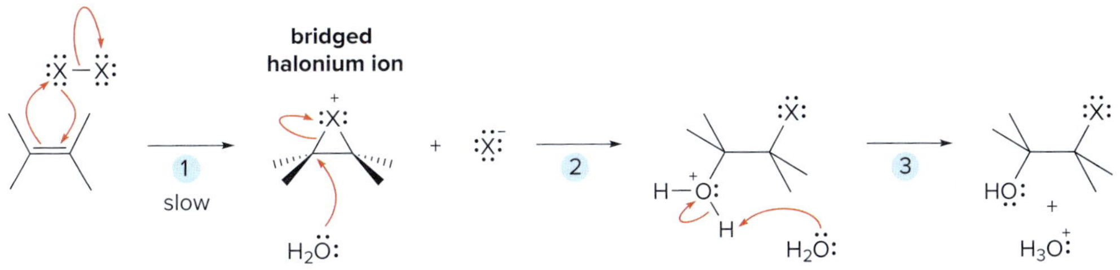

bridged halonium ion

1 slow → 2 → 3

1. Four bonds are broken or formed to generate an unstable **bridged halonium ion** that contains a three-membered ring. The electron pair in the π bond and a lone pair on a halogen are used to form two new C–X bonds, and the X–X bond is cleaved.
2. **Nucleophilic attack of H_2O** ring opens the bridged halonium ion and forms a new C–O bond.
3. Loss of a proton forms the halohydrin.

Recall from Section 7.8C that DMSO (dimethyl sulfoxide) is a polar aprotic solvent.

Although the combination of Br_2 and H_2O effectively forms **bromohydrins** from alkenes, other reagents can also be used. Bromohydrins are also formed with ***N*-bromosuccinimide** (abbreviated as **NBS**) in **aqueous DMSO** [$(CH_3)_2S=O$]. NBS serves as a source of Br_2, which then goes on to form a bromohydrin by the same reaction mechanism.

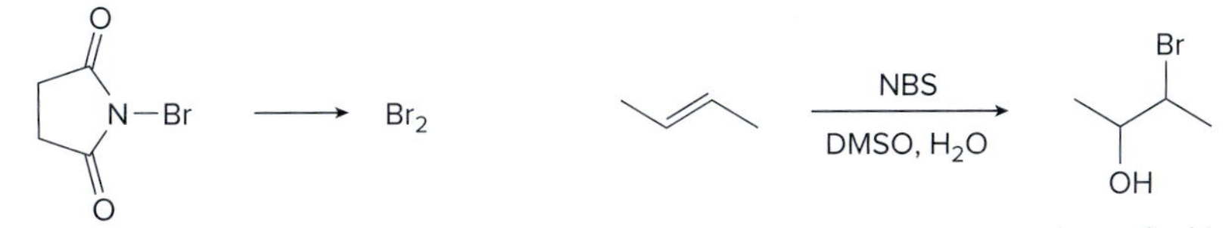

N-bromosuccinimide **NBS** → Br_2

propene $\xrightarrow{\text{NBS, DMSO, }H_2O}$ **bromohydrin**

408   Chapter 10   Alkenes and Alkynes

Because the bridged halonium ion ring is opened by backside attack of H₂O, addition of X and OH occurs in an **anti** fashion and **trans** products are formed.

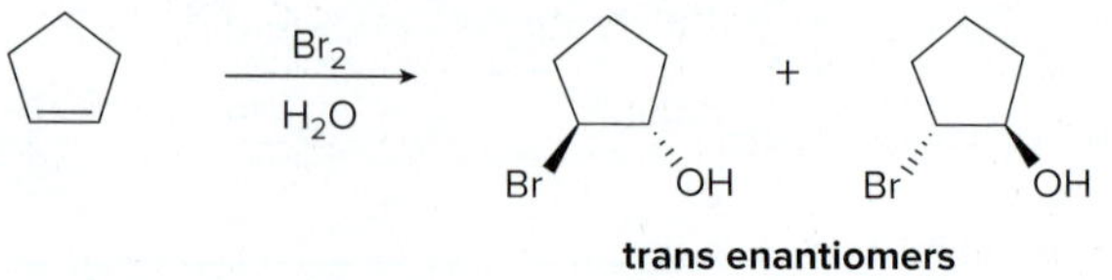

With unsymmetrical alkenes, two constitutional isomers are possible from addition of X and OH, but only one is formed. **The preferred product has the electrophile X⁺ bonded to the *less* substituted carbon atom**—that is, the carbon that has more H atoms to begin with in the reacting alkene. Thus, the **nucleophile (H₂O) bonds to the more substituted carbon.**

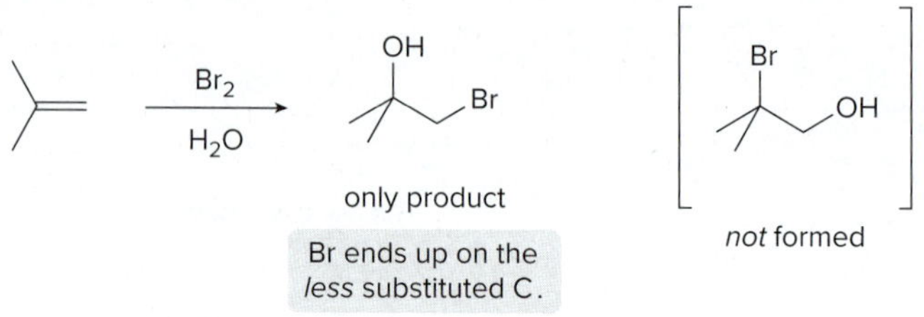

This result is reminiscent of the opening of epoxide rings with acids HZ (Z = a nucleophile), which we encountered in Section 9.16B. As in the opening of an epoxide ring, **nucleophilic attack occurs at the *more* substituted carbon end of the bridged halonium ion** because that carbon is better able to accommodate a partial positive charge in the transition state.

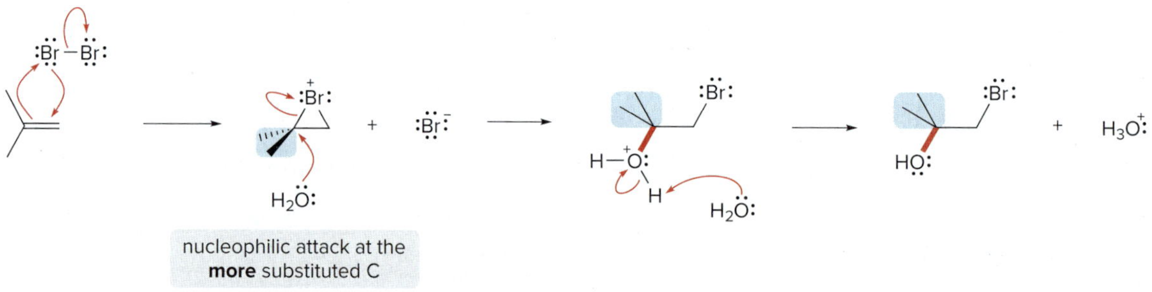

Problem 10.23   Draw the products of each reaction and indicate their stereochemistry.

a. cyclopentene $\xrightarrow{\text{NBS}}_{\text{DMSO, H}_2\text{O}}$

b. cyclohexene $\xrightarrow{\text{Cl}_2}_{\text{H}_2\text{O}}$

10.16 Hydroboration–Oxidation

Hydroboration–oxidation is a two-step reaction sequence that converts an alkene to an alcohol.

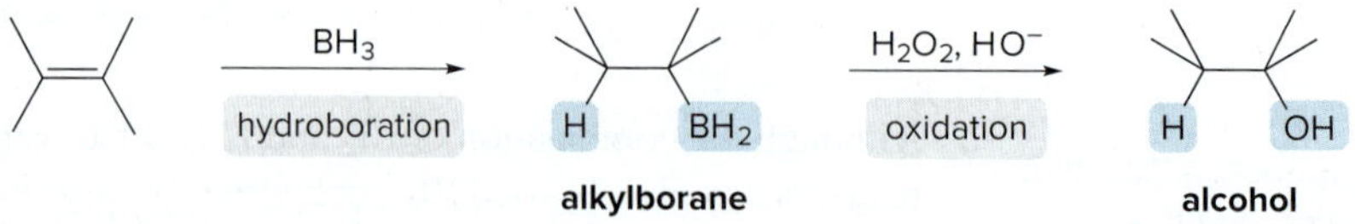

- *Hydroboration* is the addition of borane (BH₃) to an alkene, forming an alkylborane.
- *Oxidation* converts the C–B bond of the alkylborane to a C–O bond.

Hydroboration–oxidation results in **addition of H₂O to an alkene.**

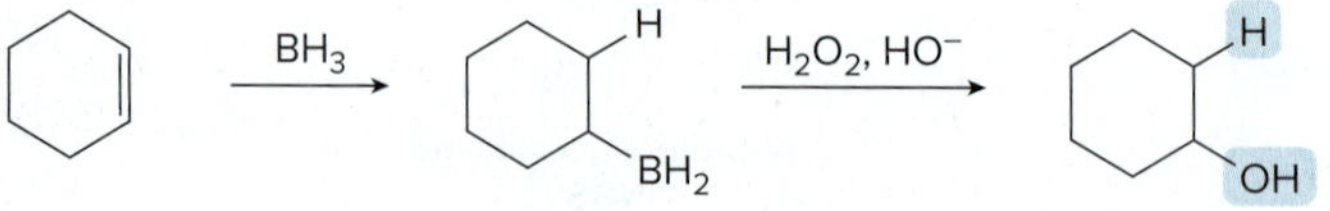

Borane (BH₃) is a reactive gas that exists mostly as the dimer, diborane (B_2H_6). Borane is a strong **Lewis acid** that reacts readily with Lewis bases. For ease in handling in the laboratory, it is commonly used as a complex with tetrahydrofuran (THF).

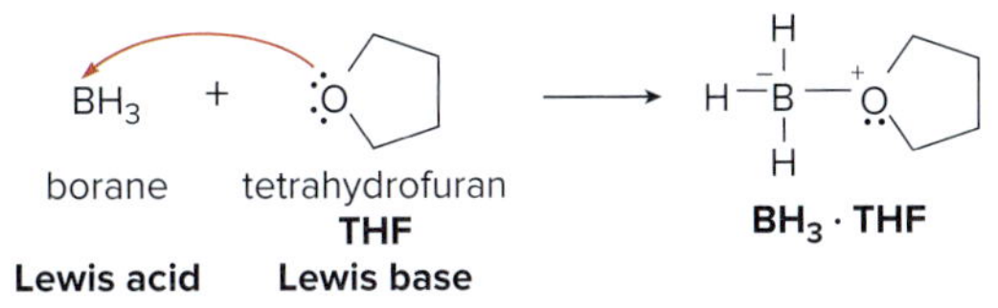

$$BH_3 \text{ (borane, Lewis acid)} + \text{THF (tetrahydrofuran, Lewis base)} \longrightarrow BH_3 \cdot THF$$

10.16A Hydroboration

The first step in hydroboration–oxidation is **addition of the elements of H and BH₂ to the π bond of the alkene**, forming an intermediate alkylborane.

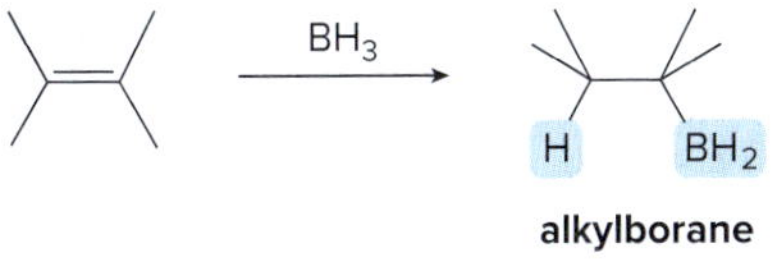

alkene + BH_3 → alkylborane (H and BH_2 added)

Because **syn addition** to the double bond occurs and **no carbocation rearrangements** are observed, carbocations are *not* formed during hydroboration, as shown in Mechanism 10.5. The proposed mechanism involves a **concerted addition of H and BH₂ from the same side of the planar double bond:** the π bond and H–BH₂ bond are broken as two new σ bonds are formed. Because four atoms are involved, the transition state is said to be **four-centered.**

Mechanism 10.5 Addition of H and BH₂—Hydroboration

One step The π bond and H–BH₂ bonds break as the C–H and C–B bonds form.

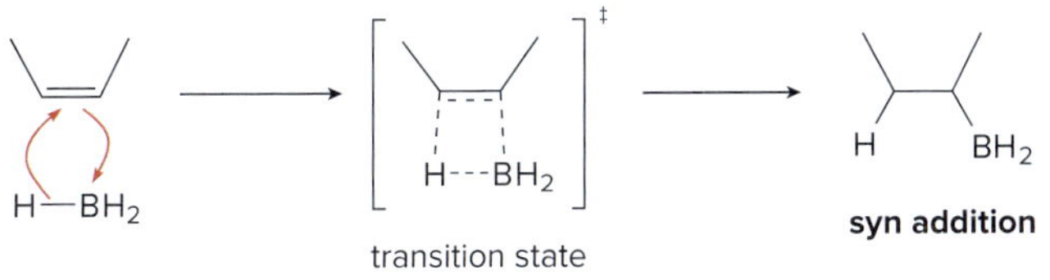

alkene + H–BH₂ → [transition state: H---BH₂] → syn addition product (H and BH₂ on same face)

Because the alkylborane formed by reaction with one equivalent of alkene still has two B–H bonds, it can react with two more equivalents of alkene to form a trialkylborane. This is illustrated in Figure 10.10 for the reaction of $CH_2=CH_2$ with BH_3.

Figure 10.10 Conversion of BH_3 to a trialkylborane with three equivalents of $CH_2=CH_2$

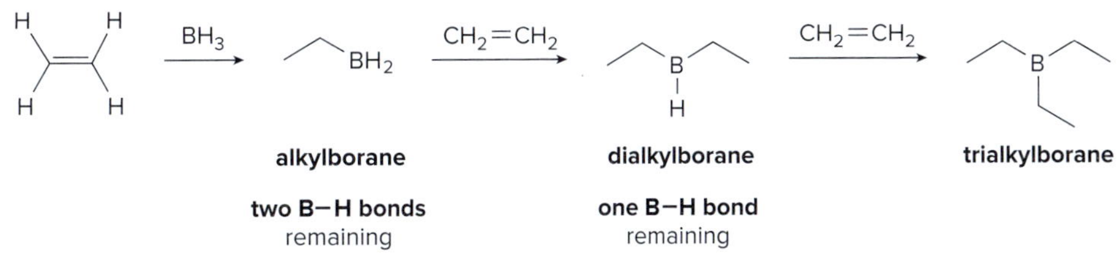

$CH_2=CH_2 \xrightarrow{BH_3}$ alkylborane (two B–H bonds remaining) $\xrightarrow{CH_2=CH_2}$ dialkylborane (one B–H bond remaining) $\xrightarrow{CH_2=CH_2}$ trialkylborane

- We often draw hydroboration as if addition stopped after one equivalent of alkene reacts with BH_3. Instead, all three B–H bonds actually react with three equivalents of an alkene to form a trialkylborane. The term **organoborane** is used for any compound with a carbon–boron bond.

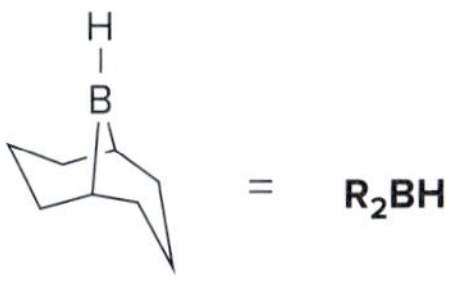

9-borabicyclo[3.3.1]nonane
9-BBN = R_2BH

Because only one B–H bond is needed for hydroboration, commercially available dialkyl-boranes having the general structure R_2BH are sometimes used instead of BH_3. A common example is 9-borabicyclo[3.3.1]nonane (**9-BBN**). 9-BBN undergoes hydroboration in the same manner as BH_3.

Chapter 10 Alkenes and Alkynes

Hydroboration is regioselective. **With unsymmetrical alkenes, the boron atom bonds to the *less* substituted carbon atom.** For example, addition of BH_3 to propene forms an alkylborane with the B bonded to the terminal carbon atom.

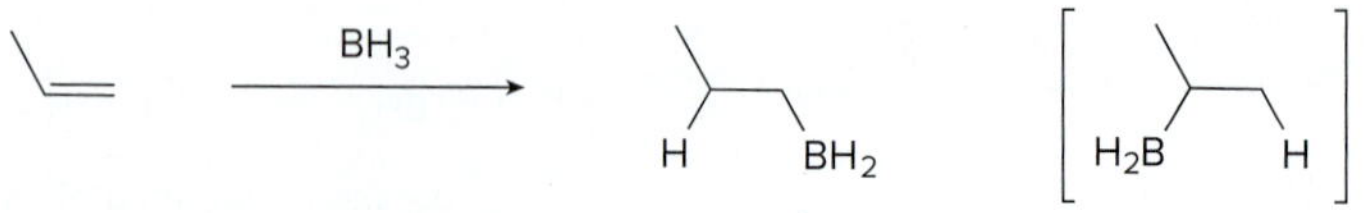

B bonds to the terminal C.

Steric factors can be used to explain this regioselectivity. The larger boron atom bonds to the less sterically hindered, more accessible carbon atom.

- In hydroboration, the boron atom bonds to the *less* substituted carbon.

Problem 10.24
What alkylborane is formed from hydroboration of each alkene?

a.  b.  c.

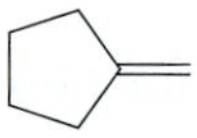

10.16B Oxidation of the Alkylborane

Because alkylboranes react rapidly with water and spontaneously burn when exposed to the air, they are oxidized, without isolation, with basic hydrogen peroxide (H_2O_2, HO^-). **Oxidation replaces the C–B bond with a C–O bond, forming a new OH group with retention of configuration;** that is, the **OH group replaces the BH_2 group in the same position** relative to the other three groups on carbon.

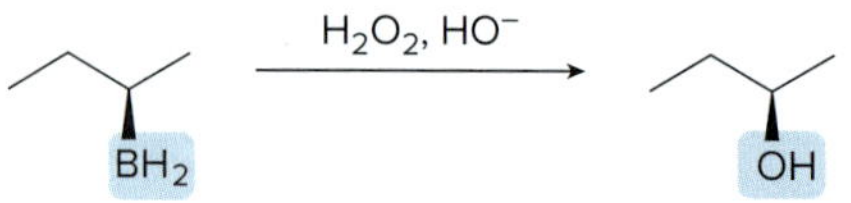

retention of configuration

Thus, to draw the product of a hydroboration–oxidation reaction, keep in mind two stereochemical facts:

- Hydroboration occurs with syn addition.
- Oxidation occurs with retention of configuration.

The overall result of this two-step sequence is **syn addition of the elements of H and OH** to a double bond, as illustrated in Sample Problem 10.4. **The OH group bonds to the *less* substituted carbon.**

Sample Problem 10.4 Drawing the Products of Hydroboration–Oxidation

Draw the product of the following reaction sequence, including stereochemistry.

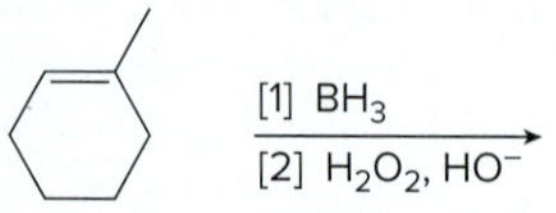

[1] BH_3
[2] H_2O_2, HO^-

Solution
In Step [1], syn addition of BH_3 to the unsymmetrical alkene adds the BH_2 group to the *less* substituted carbon from above and below the planar double bond. Two enantiomeric

10.16 Hydroboration–Oxidation

alkylboranes are formed. In Step [2], oxidation replaces the BH$_2$ group with OH in each enantiomer with **retention of configuration** to yield two alcohols that are also enantiomers.

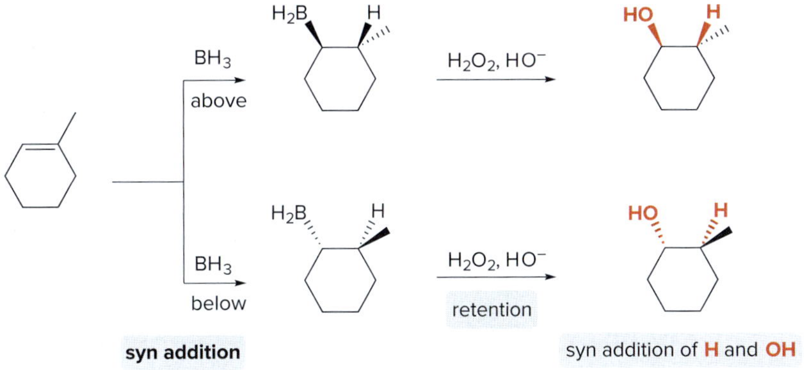

Hydroboration–oxidation results in the **addition of H and OH in a syn fashion** across the double bond. The achiral alkene is converted to an equal mixture of two enantiomers—that is, a **racemic mixture of alcohols.**

Problem 10.25 Draw the products formed when each alkene is treated with BH$_3$ followed by H$_2$O$_2$, HO$^-$. Include the stereochemistry at all stereogenic centers.

a. 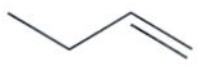 b. 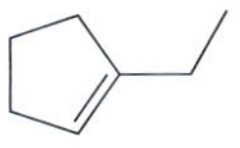 c.

More Practice: Try Problem 10.57d.

Problem 10.26 What alkene can be used to prepare each alcohol as the exclusive product of a two-step hydroboration–oxidation sequence?

a. 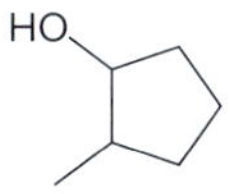 b. 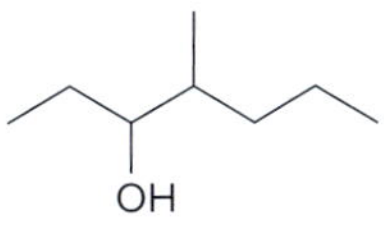 c.

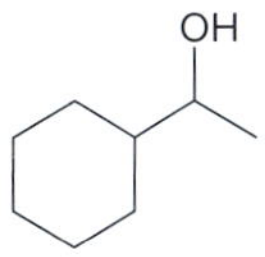

Table 10.4 summarizes the features of hydroboration–oxidation.

Table 10.4 Summary: Hydroboration–Oxidation of Alkenes

	Observation
Mechanism	• The addition of H and BH$_2$ occurs in **one step.** • **No rearrangements** can occur.
Regioselectivity	• The **OH group bonds to the less substituted** carbon atom.
Stereochemistry	• **Syn addition** occurs. • OH replaces BH$_2$ with **retention** of configuration.

Hydroboration–oxidation is a very common method for adding H$_2$O across a double bond. One example is shown in the synthesis of **artemisinin** (or **qinghaosu**), the active component of **qing-hao,** a Chinese herbal remedy used for the treatment of malaria (Figure 10.11).

Figure 10.11
An example of hydroboration–oxidation in synthesis

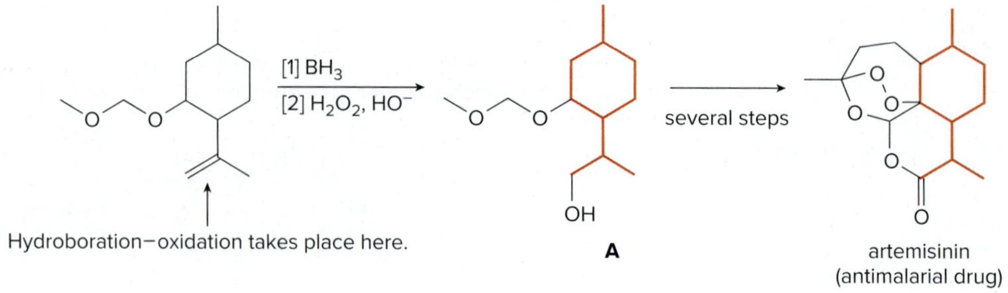

Hydroboration–oxidation takes place here.

A

artemisinin
(antimalarial drug)

- The carbon atoms of artemisinin that come from alcohol **A** are indicated in red.

10.16C A Comparison of Hydration Methods

Hydration (H_2O, H^+) and hydroboration–oxidation (BH_3 followed by H_2O_2, HO^-) both add the elements of H_2O across a double bond. Despite their similarities, these reactions often form different constitutional isomers, as shown in Sample Problem 10.5.

Sample Problem 10.5 | Comparing Two Different Methods of Hydration of an Alkene

Draw the product formed when $CH_3CH_2CH_2CH_2CH=CH_2$ is treated with either (a) H_2O, H_2SO_4; or (b) BH_3 followed by H_2O_2, HO^-.

Solution

With $H_2O + H_2SO_4$, electrophilic addition of H and OH places the **H atom on the *less* substituted carbon** of the alkene to yield a **2° alcohol**. In contrast, addition of BH_3 gives an alkylborane with the **BH_2 group on the *less* substituted terminal carbon** of the alkene. Oxidation replaces BH_2 by OH to yield a **1° alcohol**.

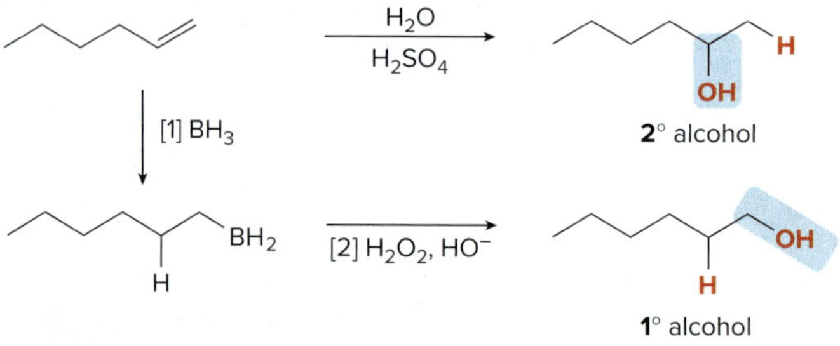

Problem 10.27 Draw the constitutional isomer formed when the following alkenes are treated with each set of reagents: [1] H_2O, H_2SO_4; or [2] BH_3 followed by H_2O_2, ^-OH.

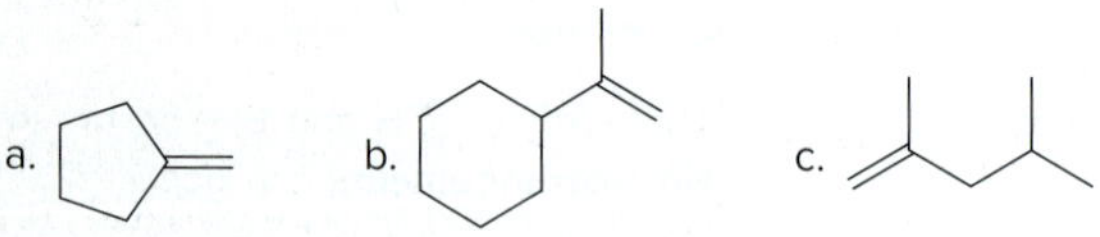

a. b. c.

More Practice: Try Problem 10.55.

10.17 Addition of Hydrogen Halides and Halogens to Alkynes

As discussed in Section 10.8, alkynes contain two weak π bonds, so they undergo addition reactions. The addition of HX and X_2 is described in this section. Hydration and hydroboration–oxidation are discussed in Sections 10.18 and 10.19, respectively.

10.17A Addition of Hydrogen Halides

Alkynes undergo **hydrohalogenation with hydrogen halides, HX** (X = Cl, Br, I). Two equivalents of HX are usually used: addition of one mole forms a **vinyl halide,** which then reacts with a second mole of HX to form a **geminal dihalide.**

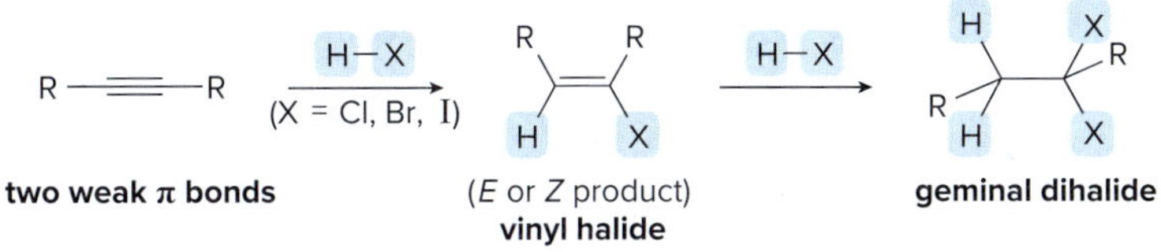

two weak π bonds → (E or Z product) vinyl halide → geminal dihalide

Addition of HX to an alkyne is another example of **electrophilic addition,** because the electrophilic (H) end of the reagent is attracted to the electron-rich triple bond.

- With two equivalents of HX, both H atoms bond to the *same* carbon.
- With a terminal alkyne, both H atoms bond to the *terminal* carbon; that is, the hydrohalogenation of alkynes follows Markovnikov's rule.

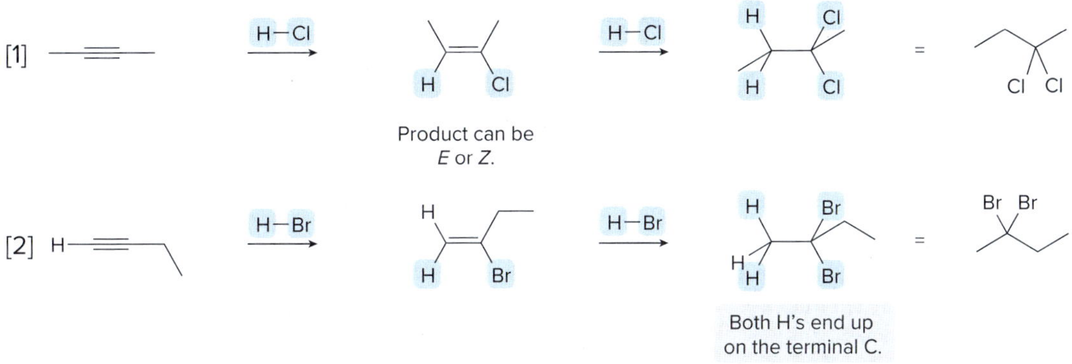

[1] Product can be E or Z.

[2] Both H's end up on the terminal C.

- With only one equivalent of HX, the reaction stops with formation of the vinyl halide.

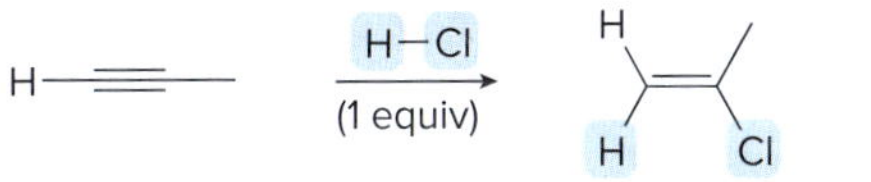

a vinyl chloride
(2-chloropropene)

Problem 10.28 Draw the organic products formed when each alkyne is treated with two equivalents of HBr.

a. 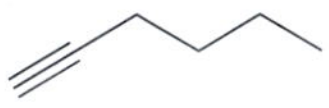   b. 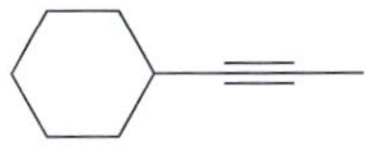   c.

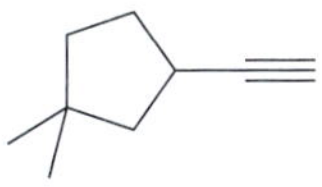

10.17B Addition of Halogen

Halogens, X_2 (X = Cl or Br), add to alkynes in much the same way they add to alkenes (Section 10.13). Addition of one mole of X_2 forms a **trans dihalide,** which can then react with a second mole of X_2 to yield a **tetrahalide.**

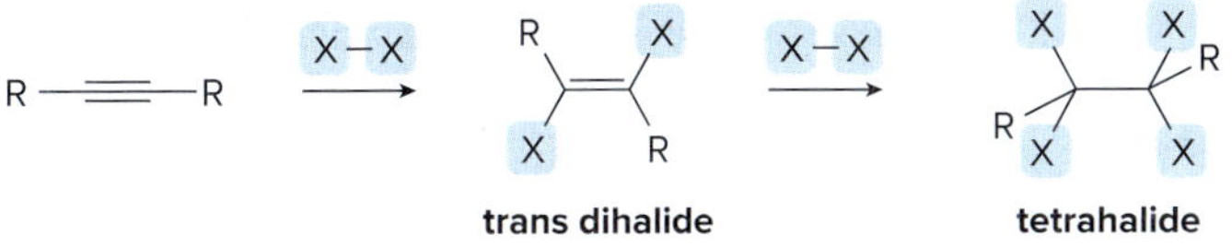

trans dihalide   tetrahalide

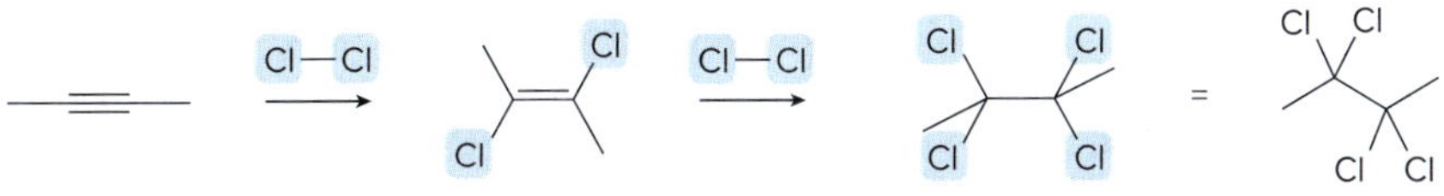

Problem 10.29 Draw the products formed when CH₃CH₂C≡CCH₂CH₃ is treated with each reagent:
(a) Br₂ (2 equiv); (b) Cl₂ (1 equiv).

10.18 Addition of Water to Alkynes

Although the addition of H₂O to an alkyne resembles the acid-catalyzed addition of H₂O to an alkene in some ways, an important difference exists. In the presence of strong acid or Hg²⁺ catalyst, the **elements of H₂O add to the triple bond,** but the initial addition product, an **enol,** is unstable and rearranges to a product containing a **carbonyl group**—that is, a C=O. A carbonyl compound having two alkyl groups bonded to the C=O carbon is called a **ketone.**

R—≡—R $\xrightarrow{\text{H-OH}, \text{H}_2\text{SO}_4, \text{HgSO}_4}$ [less stable **enol**] ⇌ **ketone**

H₂O is added to form a **carbonyl** group.

Internal alkynes undergo hydration with concentrated acid, whereas terminal alkynes require the presence of an additional Hg²⁺ catalyst—usually HgSO₄—to yield methyl ketones by **Markovnikov addition of H₂O.**

Because an enol contains both a C=C and a hydroxy group, the name **enol** comes from alk**ene** + alcoh**ol**.

HgSO₄ is often used in the hydration of internal alkynes as well, because hydration can be carried out under milder reaction conditions.

H—≡— $\xrightarrow{\text{H}_2\text{O}, \text{H}_2\text{SO}_4, \text{HgSO}_4}$ [**enol**] ⇌ **methyl ketone**

Markovnikov addition of H₂O
H adds to the terminal C.

Let's first examine the conversion of a general enol **A** to the carbonyl compound **B**. **A** and **B** are called **tautomers**: **A** is the *enol form* and **B** is the *keto form* of the tautomer.

- *Tautomers* are constitutional isomers that differ in the location of a double bond and a hydrogen atom. Two tautomers are in equilibrium with each other.

enol form **A** ⇌ keto form **B**

Tautomers differ in the position of a double bond and a hydrogen atom. In Chapter 17 an in-depth discussion of keto–enol tautomers is presented.

- An enol tautomer has an O—H group bonded to a C=C.
- A keto tautomer has a C=O and an additional C—H bond.

Equilibrium favors the keto form largely because a C=O is much stronger than a C=C. Tautomerization, the process of converting one tautomer into another, is catalyzed by both acid and base. Under the strongly acidic conditions of hydration, tautomerization of the enol to the keto form occurs rapidly by a two-step process: **protonation,** followed by **deprotonation** as shown in Mechanism 10.6.

10.18 Addition of Water to Alkynes

Mechanism 10.6 Tautomerization in Acid

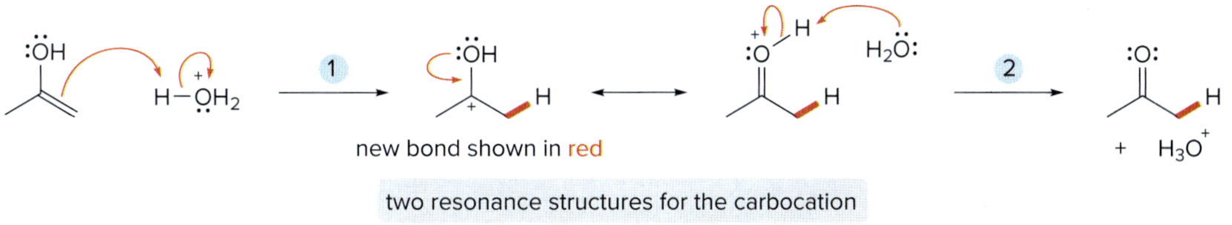

[Mechanism diagram showing protonation of enol and loss of proton to form carbonyl]

new bond shown in red

two resonance structures for the carbocation

1. Protonation of the double bond forms a **resonance-stabilized carbocation**.
2. Loss of a proton, which can be drawn with either resonance structure, forms the **carbonyl group**. Because acid is re-formed in this step, tautomerization is acid-catalyzed.

Hydration of an internal alkyne with strong acid forms an enol by a mechanism similar to that of the acid-catalyzed hydration of an alkene (Section 10.12). Mechanism 10.7 illustrates the hydration of but-2-yne with H_2O and H_2SO_4. Once formed, the enol then tautomerizes to the more stable keto form by protonation followed by deprotonation.

Mechanism 10.7 Hydration of an Alkyne

Part [1] Addition of H_2O to form an enol

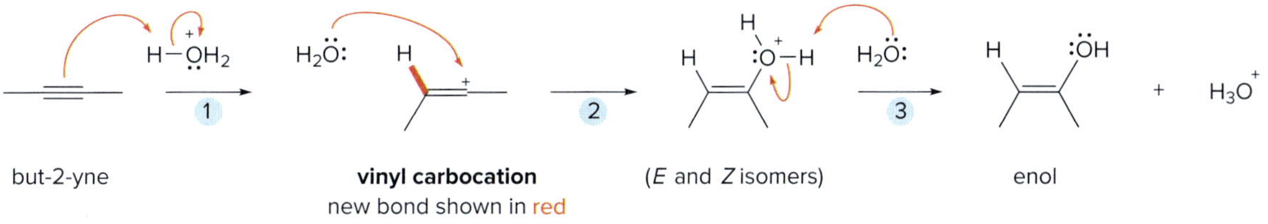

[Mechanism diagram]

but-2-yne | vinyl carbocation (new bond shown in red) | (E and Z isomers) | enol

1. Addition of H^+ forms a **vinyl carbocation**.
2–3. **Nucleophilic attack** followed by loss of a proton forms the enol.

Part [2] Tautomerization

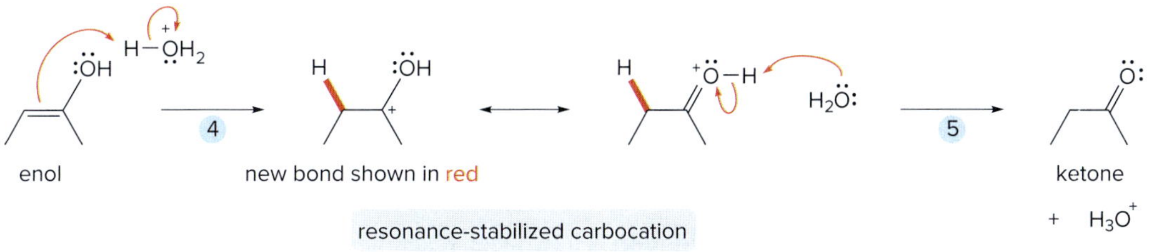

[Mechanism diagram]

enol | new bond shown in red resonance-stabilized carbocation | ketone

4. Tautomerization of the enol to the keto form begins with protonation of the double bond to form a **carbocation**.
5. Loss of a proton, which can be drawn with either resonance structure, forms the **ketone**.

Sample Problem 10.6 Drawing an Enol and a Ketone Formed by Hydration of an Alkyne

Draw the enol intermediate and the ketone product formed in the following reaction.

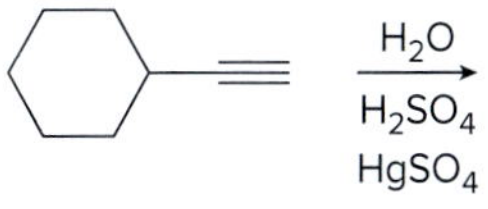

[cyclohexyl acetylene + H_2O, H_2SO_4, $HgSO_4$]

Solution

First, form the enol by adding H₂O to the triple bond with the **H bonded to the less substituted terminal carbon,** according to Markovnikov's rule.

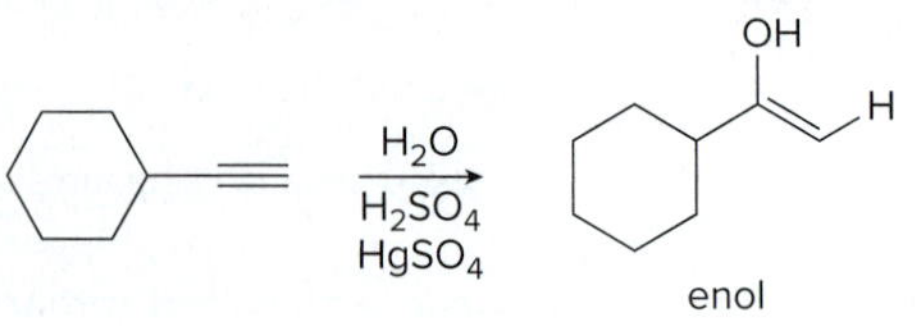

To convert the enol to the keto tautomer, add a proton to the C=C and remove a proton from the OH group. In tautomerization, the C–OH bond is converted to a C=O, and a new C–H bond is formed on the other enol carbon.

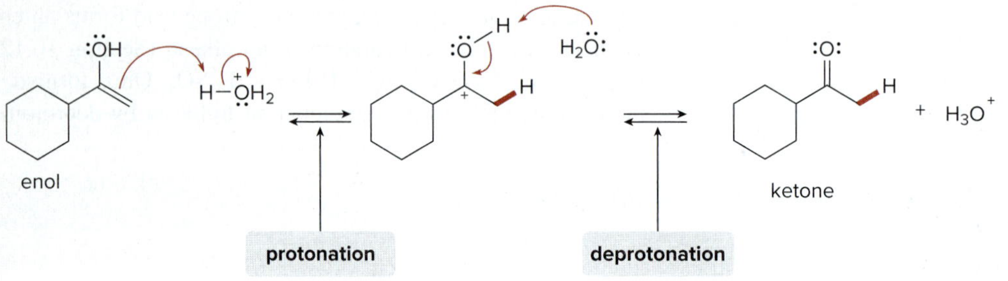

- The overall result is the addition of H₂O to a triple bond to form a ketone.

Problem 10.30 Draw the keto tautomer of each enol.

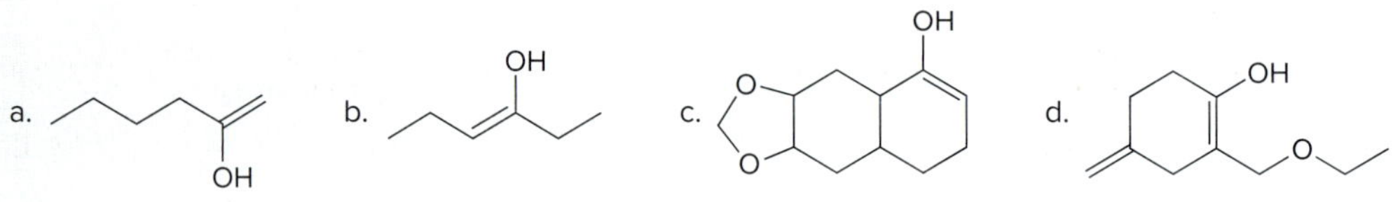

More Practice: Try Problems 10.41, 10.50, 10.51.

Problem 10.31 Ignoring *E* and *Z* isomers, what two enols are formed when pent-2-yne is treated with H₂O, H₂SO₄, and HgSO₄? Draw the ketones formed from these enols after tautomerization.

10.19 Hydroboration–Oxidation of Alkynes

Hydroboration–oxidation is a two-step reaction sequence that converts an alkyne to a carbonyl compound.

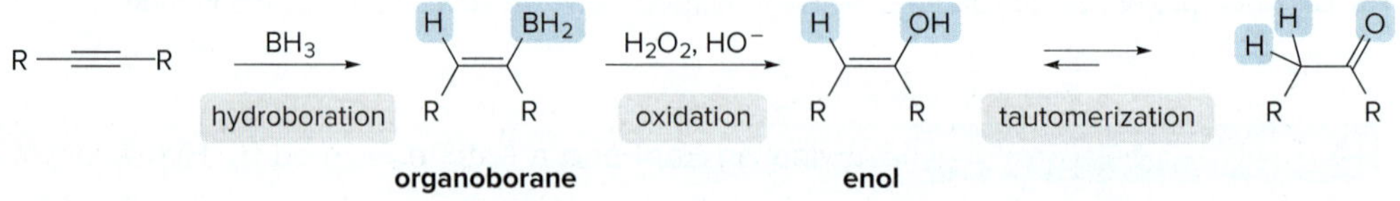

- Addition of borane forms an organoborane.
- Oxidation with basic H₂O₂ forms an enol.
- Tautomerization of the enol forms a carbonyl compound.
- The overall result is addition of H₂O to a triple bond.

Hydroboration–oxidation of an *internal* alkyne forms a **ketone**. Hydroboration of a *terminal* alkyne adds boron to the less substituted, terminal carbon. After oxidation to the enol, tautomerization yields an **aldehyde,** a carbonyl compound having a hydrogen atom bonded to the carbonyl carbon. Hydroboration of a terminal alkyne is generally carried out with a dialkylborane (R_2BH), which has been prepared from BH_3 (Section 10.16).

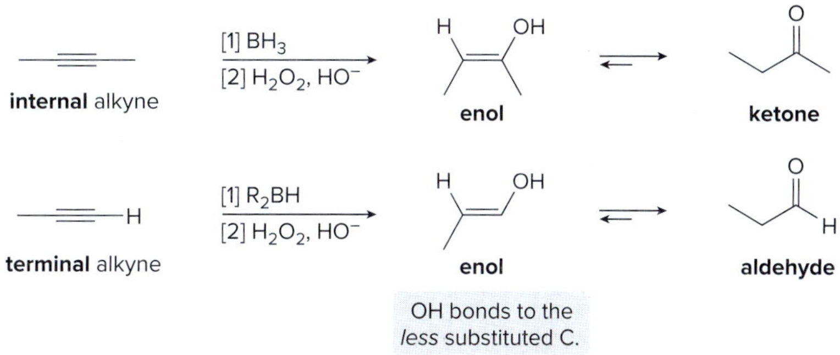

Hydration (H_2O, H_2SO_4, and $HgSO_4$) and **hydroboration–oxidation** (BH_3 or R_2BH followed by H_2O_2, HO^-) both **add the elements of H_2O across a triple bond,** but different constitutional isomers are formed from terminal alkynes in these two reactions.

- Addition of H_2O using H_2O, H_2SO_4, and $HgSO_4$ forms methyl ketones from terminal alkynes.
- Addition of H_2O using an organoborane, then H_2O_2, HO^- forms aldehydes from terminal alkynes.

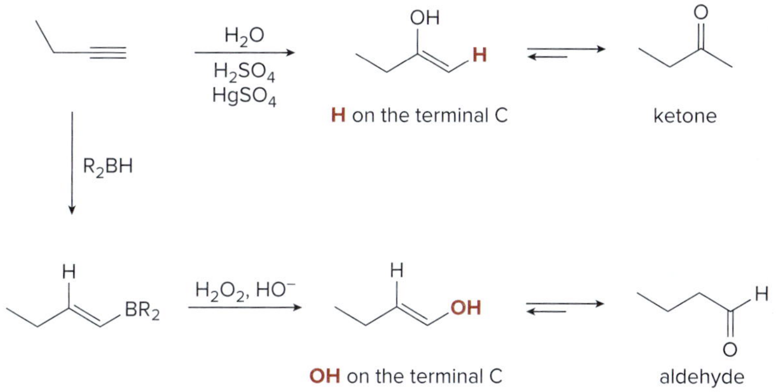

Problem 10.32 Draw the products formed when the following alkynes are treated with each set of reagents: [1] H_2O, H_2SO_4, $HgSO_4$; or [2] R_2BH followed by H_2O_2, ^-OH.

a. 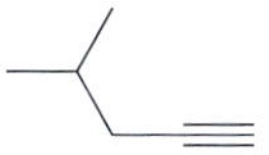 b.

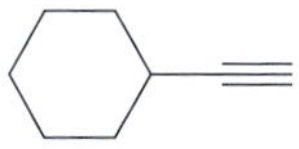

10.20 Reaction of Acetylide Anions

As mentioned in Section 10.8B, terminal alkynes are readily converted to acetylide anions with strong bases such as $NaNH_2$ and NaH. These anions are strong nucleophiles, capable of reacting with electrophiles such as alkyl halides and epoxides.

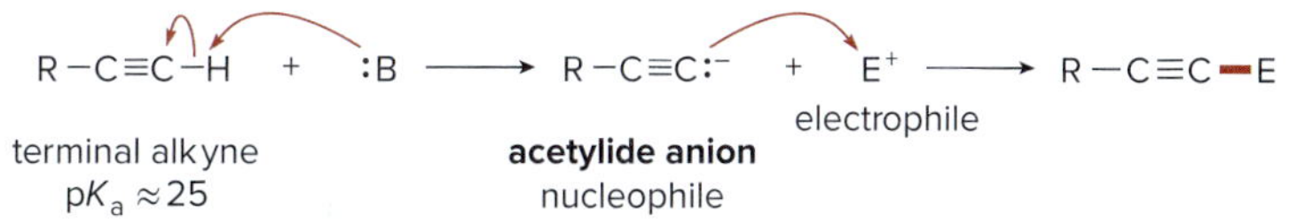

$$R-C\equiv C-H \;+\; :B \longrightarrow R-C\equiv C:^- \;+\; E^+ \longrightarrow R-C\equiv C-E$$

terminal alkyne
$pK_a \approx 25$

acetylide anion
nucleophile

electrophile

418 Chapter 10 Alkenes and Alkynes

10.20A Reaction of Acetylide Anions with Alkyl Halides

Acetylide anions react with unhindered alkyl halides to yield products of nucleophilic substitution.

$$R-X \ + \ {}^-{:}C{\equiv}C-R' \ \xrightarrow{S_N2} \ R-C{\equiv}C-R' \ + \ X{:}^-$$

nucleophile　　　　　　　　　　　　　leaving group

Because acetylide anions are strong nucleophiles, the mechanism of nucleophilic substitution is S_N2, and thus the **reaction is fastest with CH_3X and 1° alkyl halides.** Terminal alkynes (Reaction [1]) or internal alkynes (Reaction [2]) can be prepared depending on the identity of the acetylide anion.

[1] CH₃CH₂Cl + ⁻:C≡C−H → CH₃CH₂−C≡C−H + :Cl:⁻

[2] PhCH₂Br + ⁻:C≡C−(cyclopentyl) → PhCH₂−C≡C−(cyclopentyl) + :Br:⁻

7 C's　　　　　7 C's　　　　　14 C's

new bond drawn in red

- **Nucleophilic substitution with acetylide anions forms new carbon–carbon bonds.**

Because organic compounds consist of a carbon framework, reactions that form carbon–carbon bonds are especially useful. In Reaction [2], for example, nucleophilic attack of a seven-carbon acetylide anion on a seven-carbon alkyl halide yields a 14-carbon alkyne as product.

Although nucleophilic substitution with acetylide anions is a very valuable carbon–carbon bond-forming reaction, it has the same limitations as any S_N2 reaction. **Steric hindrance around the leaving group causes 2° and 3° alkyl halides to undergo elimination by an E2 mechanism,** as shown with 2-bromo-2-methylpropane. Thus, nucleophilic substitution with acetylide anions forms new carbon–carbon bonds in high yield only with unhindered CH_3X and 1° alkyl halides.

(CH₃)₃C−Br + ⁻:C≡C−H (base) → (CH₃)₂C=CH₂ + H−C≡C−H + :Br:⁻

2-bromo-2-methylpropane　　　　　　　　　　E2 product
3° alkyl halide

Steric hindrance prevents an S_N2 reaction.

Problem 10.33 Draw the organic products formed in each reaction.

a. H−C≡C−H $\xrightarrow[\text{[2] (CH}_3\text{)}_2\text{CHCH}_2\text{Cl}]{\text{[1] NaH}}$

b. cyclopentyl−C≡C−H
   - [1] NaH [2] CH₃CH₂CH₂Br
   - [1] NaNH₂ [2] (CH₃)₃CCl

10.20 Reaction of Acetylide Anions

Problem 10.34 What acetylide anion and alkyl halide can be used to prepare each alkyne? Indicate all possibilities when more than one route will work.

a., b., c.

Because acetylene has two *sp* hybridized C—H bonds, two sequential reactions can occur to form **two new carbon–carbon bonds,** as shown in Sample Problem 10.7.

Sample Problem 10.7 — Forming an Internal Alkyne by Two Sequential S$_N$2 Reactions

Identify the terminal alkyne **A** and the internal alkyne **B** in the following reaction sequence.

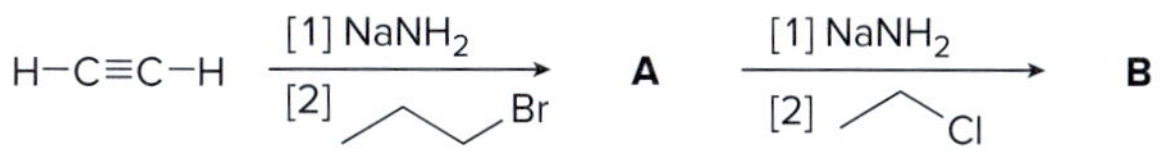

$$H-C\equiv C-H \xrightarrow[\text{[2]} \quad \diagup\diagdown Br]{\text{[1] NaNH}_2} A \xrightarrow[\text{[2]} \quad \diagup Cl]{\text{[1] NaNH}_2} B$$

Solution

In each step, the base $^-NH_2$ removes a proton on an *sp* hybridized carbon, and the resulting acetylide anion reacts as a nucleophile with an alkyl halide to yield an S$_N$2 product. The first two-step reaction sequence forms the **terminal alkyne A** by nucleophilic attack of the acetylide anion on $CH_3CH_2CH_2Br$.

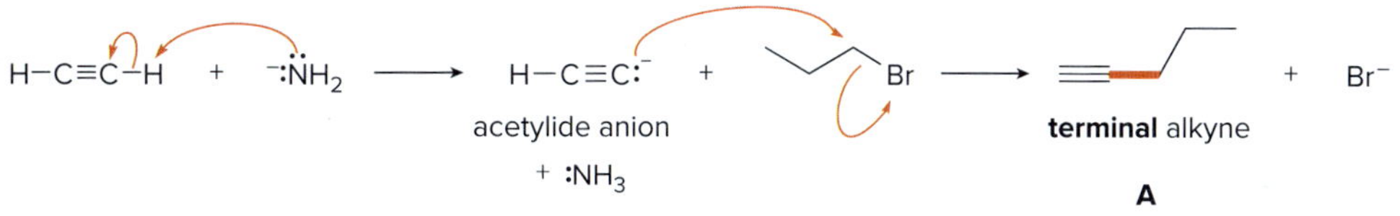

H—C≡C—H + :⁻NH$_2$ ⟶ H—C≡C:⁻ + :NH$_3$
acetylide anion

H—C≡C:⁻ + ⌒⌒Br ⟶ ≡—⌒⌒ + Br⁻
terminal alkyne **A**

The second two-step reaction sequence forms the **internal alkyne B** by nucleophilic attack of the acetylide anion on CH_3CH_2Cl.

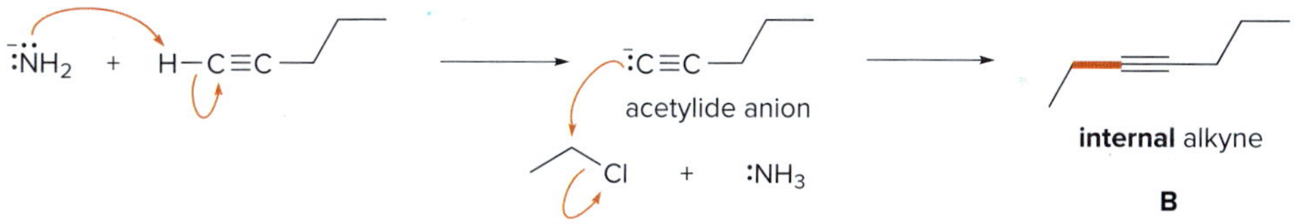

:⁻NH$_2$ + H—C≡C—⌒⌒ ⟶ :⁻C≡C—⌒⌒ + :NH$_3$
acetylide anion

:⁻C≡C—⌒⌒ + ⌒Cl ⟶ ⌒—≡—⌒⌒
internal alkyne **B**

Problem 10.35 Show how HC≡CH, CH_3CH_2Br, and $(CH_3)_2CHCH_2CH_2Br$ can be used to prepare $CH_3CH_2C\equiv CCH_2CH_2CH(CH_3)_2$. Show all reagents, and use curved arrows to show movement of electron pairs.

More Practice: Try Problems 10.53g; 10.59g, i; 10.61a, b; 10.71a.

Sample Problem 10.7 illustrates how a seven-carbon product can be prepared from three smaller molecules by forming two new carbon–carbon bonds.

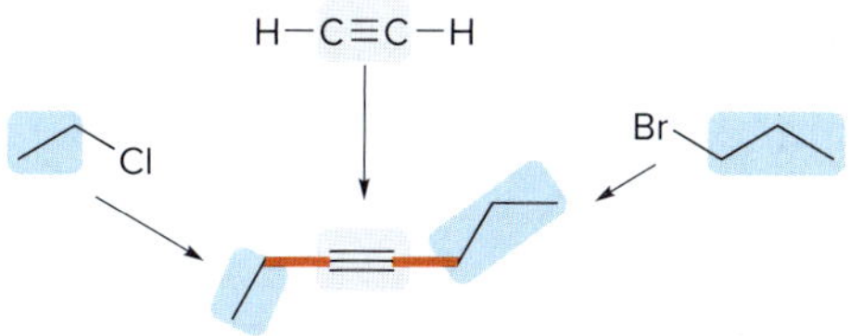

new bonds shown in red

10.20B Reaction of Acetylide Anions with Epoxides

Acetylide anions are strong nucleophiles that open epoxide rings by an S_N2 mechanism. This reaction also results in the formation of a **new carbon–carbon bond.** Backside attack occurs at the **less substituted** end of the epoxide.

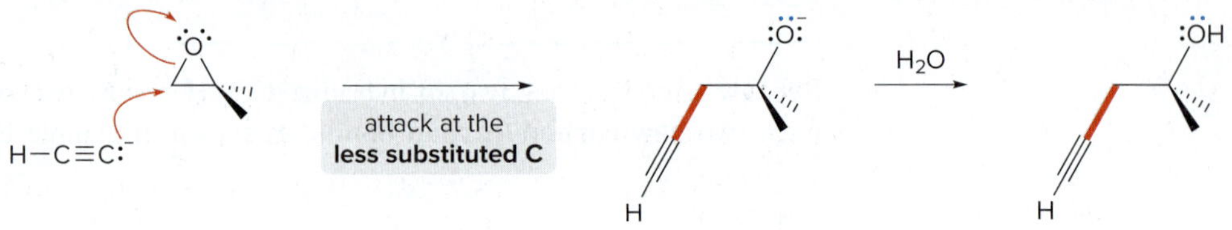

Opening of epoxide rings with strong nucleophiles was first discussed in Section 9.16A.

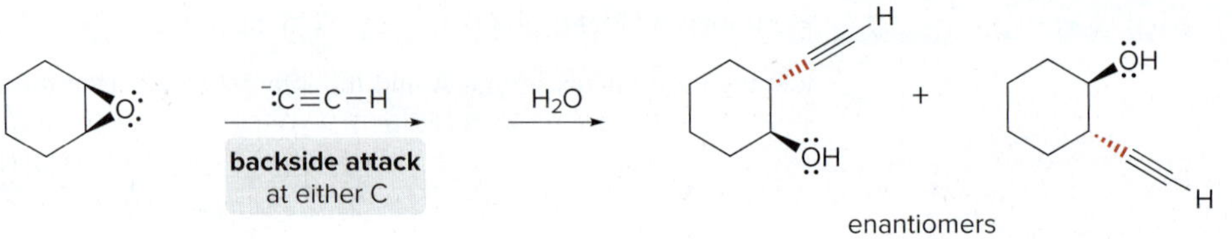

Problem 10.36
Draw the products of each reaction.

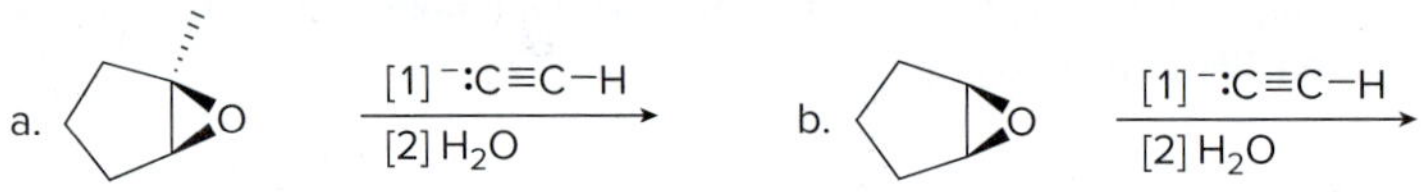

a. [1] $^-:C{\equiv}C{-}H$; [2] H_2O

b. [1] $^-:C{\equiv}C{-}H$; [2] H_2O

Sample Problem 10.8 — Identifying the Acetylide Anion and Epoxide Needed to Synthesize an Alcohol

What acetylide anion and epoxide are needed to synthesize **C**?

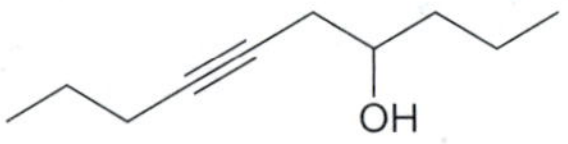

Solution
To identify the bond formed when the acetylide anion opens an epoxide ring, **locate the carbon bonded to the OH and the adjacent carbon bonded to a C≡C.** The new C–C bond results from nucleophilic attack of the acetylide anion with the less substituted end of the epoxide to form an alkoxide that is protonated with water to yield **C**.

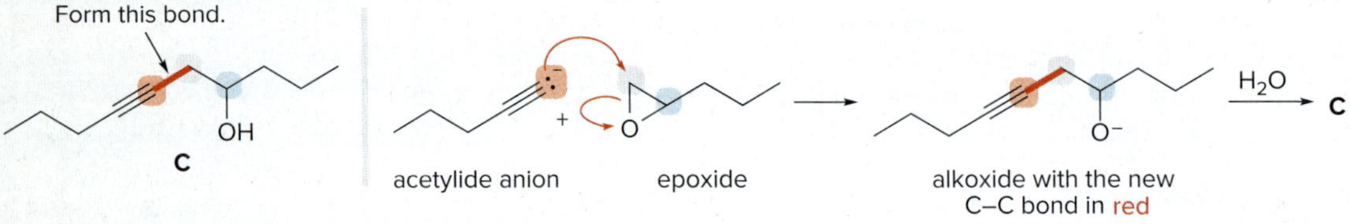

Problem 10.37
What acetylide anion and epoxide are needed to synthesize each compound?

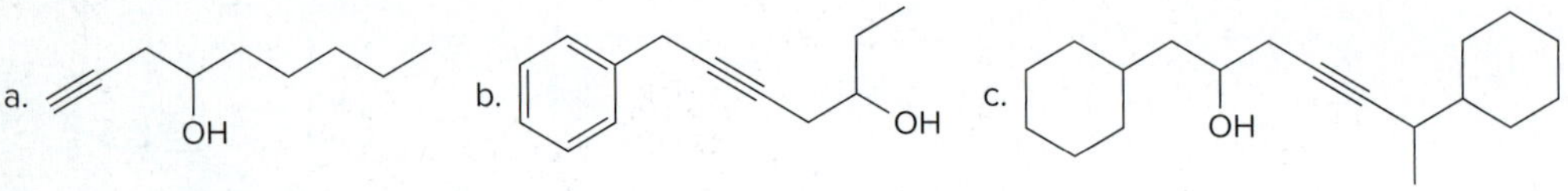

More Practice: Try Problem 10.70.

10.21 Synthesis

The reactions of acetylide anions give us an opportunity to examine organic synthesis more systematically. Performing a multistep synthesis can be difficult. Not only must you know the reactions for a particular functional group, but you must also put these reactions in a logical order, a process that takes much practice to master.

10.21A General Terminology and Conventions

To plan a synthesis of more than one step, we use the process of **retrosynthetic analysis**—that is, **working backwards from the desired product to determine the starting materials from which it is made.** To write a synthesis working backwards from the product to the starting material, an **open arrow** (⇒) is used to indicate that the product is drawn on the left and the starting material on the right.

A **reactive intermediate** is an unstable intermediate like a carbocation, which is formed during the conversion of a stable starting material to a stable product. A **synthetic intermediate** is a stable compound that is the product of one step and the starting material of another in a multistep synthesis.

The product of a synthesis is often called the **target compound.** Using retrosynthetic analysis, we must determine what compound can be converted to the target compound by a single reaction. That is, **what is the immediate precursor of the target compound?** After an appropriate precursor is identified, this process is continued until we reach a specified starting material. Sometimes multiple retrosynthetic pathways are examined before a particular route is decided upon.

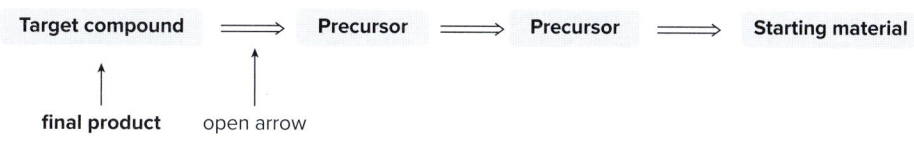

In designing a synthesis, reactions are often divided into two categories:

- Reactions that form new carbon–carbon bonds.
- Reactions that convert one functional group to another—that is, functional group interconversions.

Appendix F lists the carbon–carbon bond-forming reactions encountered in this text.

Carbon–carbon bond-forming reactions are central to organic synthesis because simpler and less valuable starting materials can be converted to more complex products. Keep in mind that whenever the product of a synthesis has more carbon–carbon bonds than the starting material, the synthesis must contain at least one of these reactions.

How To Develop a Retrosynthetic Analysis

Step [1] Compare the carbon skeletons of the starting material and product.
- If the product has more carbon–carbon σ bonds than the starting material, the synthesis must form one or more C–C bonds. If not, only functional group interconversion occurs.
- **Match the carbons in the starting material with those in the product** to see where new C–C bonds must be added or where functional groups must be changed.

Step [2] Concentrate on the functional groups in the starting material and product and ask:
- What methods introduce the functional groups in the product?
- What kind of reactions does the starting material undergo?

Step [3] Work backwards from the product and forwards from the starting material.
- Ask: **What is the immediate precursor of the product?**
- Compare each precursor to the starting material to determine if there is a one-step reaction that converts one to the other. Continue this process until the starting material is reached.
- Always generate *simpler* precursors when working backwards.
- Use *fewer* steps when multiple routes are possible.
- Keep in mind that you may need to evaluate several different precursors for a given compound.

Step [4] Check the synthesis by writing it in the synthetic direction.
- To check a retrosynthetic analysis, write out the steps beginning with the starting material, indicating all necessary reagents.

10.21B Examples of Multistep Synthesis

Retrosynthetic analysis with acetylide anions is illustrated in Sample Problems 10.9 and 10.10.

Sample Problem 10.9 — Devising a Short Synthesis

Devise a synthesis of HC≡CCH$_2$CH$_2$CH$_3$ from HC≡CH and any other organic or inorganic reagents.

Retrosynthetic Analysis

The two C's in the starting material match up with the two sp hybridized C's in the product, so a three-carbon unit must be added.

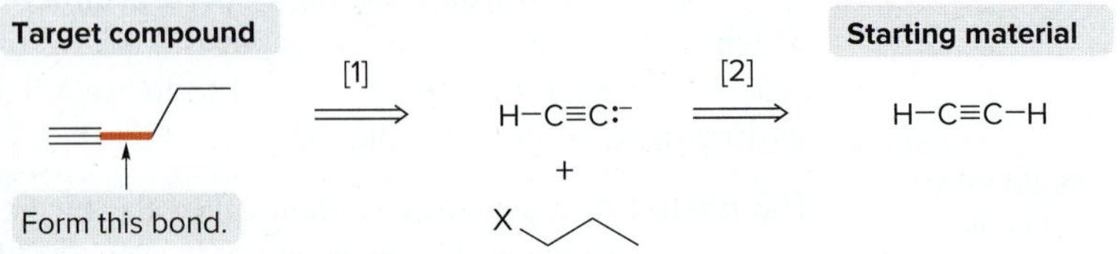

Thinking backwards...

[1] Form a new C–C bond using an acetylide anion and a 1° alkyl halide.
[2] Prepare the acetylide anion from acetylene by treatment with base.

Synthesis

Deprotonation of HC≡CH with NaH forms the acetylide anion, which undergoes S$_N$2 reaction with an alkyl halide to form the target compound, a five-carbon alkyne.

A two-step process:

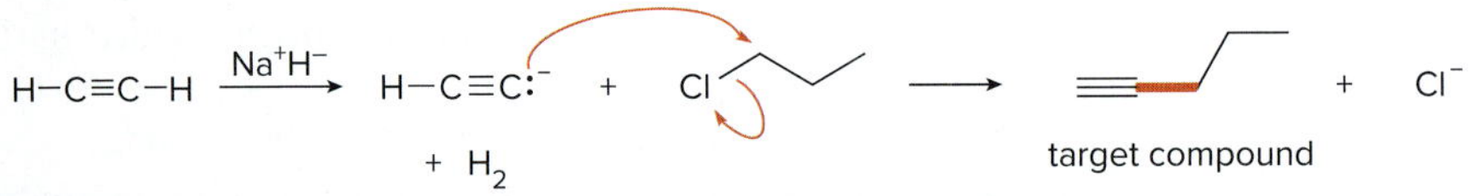

Problem 10.38 Use retrosynthetic analysis to show how hex-3-yne can be prepared from acetylene and any other organic and inorganic compounds. Then draw the synthesis in the synthetic direction, showing all needed reagents.

More Practice: Try Problem 10.69a.

Sample Problem 10.10 — Devising a Synthesis with More Than Two Steps

Devise a synthesis of the following compound from starting materials having two or fewer carbons.

Retrosynthetic Analysis

A carbon–carbon bond-forming reaction must be used to convert the two-carbon starting materials to the four-carbon product.

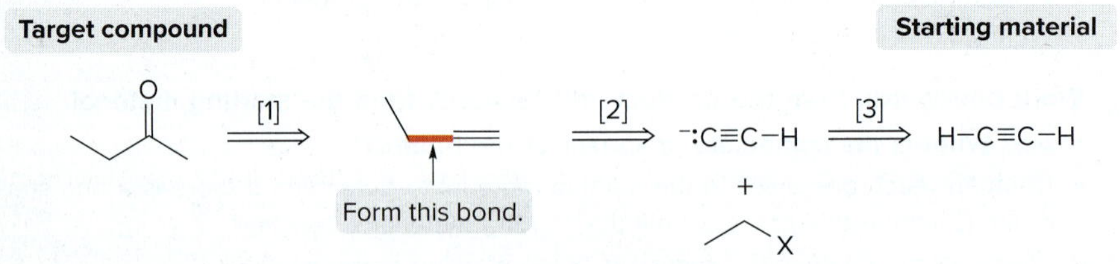

Thinking backwards...

[1] Form the carbonyl group by hydration of a triple bond.
[2] Form a new C–C bond using an acetylide anion and a 1° alkyl halide.
[3] Prepare the acetylide anion from acetylene by treatment with base.

Synthesis

Three steps are needed to complete the synthesis. Treatment of HC≡CH with NaH forms the acetylide anion, which undergoes an S_N2 reaction with an alkyl halide to form a four-carbon terminal alkyne. Hydration of the alkyne with H_2O, H_2SO_4, and $HgSO_4$ yields the target compound.

$$H-C\equiv C-H \xrightarrow{Na^+H^-} H-C\equiv C:^- + Cl\frown \longrightarrow \equiv \xrightarrow[H_2SO_4]{H_2O}{HgSO_4} \text{target compound}$$
$$+ H_2 \qquad\qquad + Cl^-$$

Problem 10.39   Devise a synthesis of $CH_3CH_2CH_2CHO$ from two-carbon starting materials.

More Practice: Try Problems 10.69b, c; 10.70–10.74.

These examples illustrate the synthesis of organic compounds by multistep routes. In Chapter 11, we will learn other useful reactions that expand our capability to do synthesis.

Chapter 10 REVIEW

KEY CONCEPTS

Markovnikov's Rule (10.10, 10.12)

In the addition of HX to an unsymmetrical alkene, the H atom bonds to the *less* substituted carbon.

1 hydrohalogenation → alkyl halide (X = Cl, Br, I) (10.10–10.11)

2 hydration (H_2SO_4) → alcohol (10.12)

Try Problems 10.52a–c; 10.56a, b.

Stereochemistry of Alkene Addition (10.8)

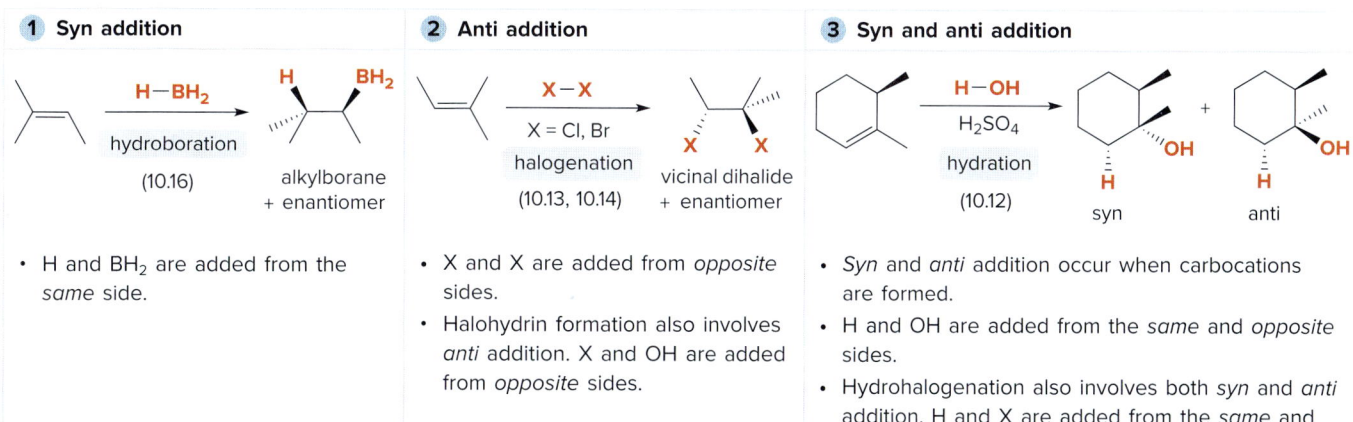

1 Syn addition — hydroboration (10.16) → alkylborane + enantiomer
- H and BH_2 are added from the *same* side.

2 Anti addition — halogenation (X = Cl, Br) (10.13, 10.14) → vicinal dihalide + enantiomer
- X and X are added from *opposite* sides.
- Halohydrin formation also involves *anti* addition. X and OH are added from *opposite* sides.

3 Syn and anti addition — hydration (H_2SO_4) (10.12) → syn and anti
- *Syn* and *anti* addition occur when carbocations are formed.
- H and OH are added from the *same* and *opposite* sides.
- Hydrohalogenation also involves both *syn* and *anti* addition. H and X are added from the *same* and *opposite* sides.

Try Problem 10.57.

KEY REACTIONS

[1] Alkene addition reactions

All reactions of alkenes involve **addition**—the weak π bond is broken and two new σ bonds are formed.

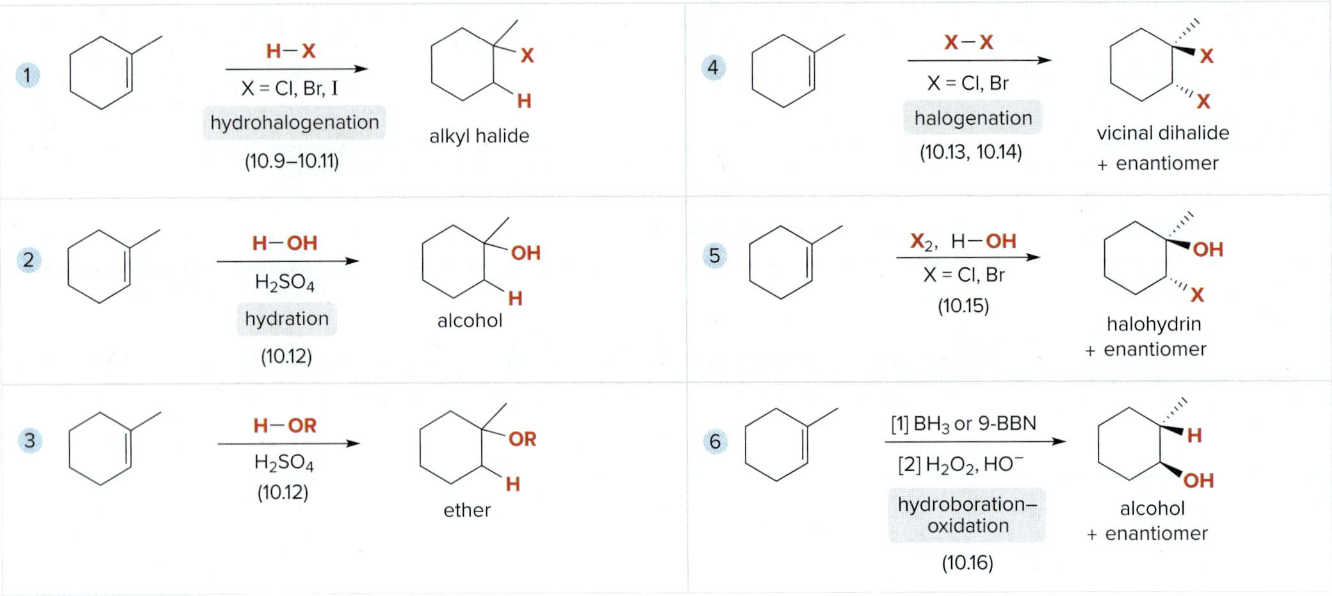

1. Alkene + H–X (X = Cl, Br, I) → alkyl halide
   hydrohalogenation (10.9–10.11)

2. Alkene + H–OH, H$_2$SO$_4$ → alcohol
   hydration (10.12)

3. Alkene + H–OR, H$_2$SO$_4$ → ether
   (10.12)

4. Alkene + X–X (X = Cl, Br) → vicinal dihalide + enantiomer
   halogenation (10.13, 10.14)

5. Alkene + X$_2$, H–OH (X = Cl, Br) → halohydrin + enantiomer
   (10.15)

6. Alkene + [1] BH$_3$ or 9-BBN; [2] H$_2$O$_2$, HO$^-$ → alcohol + enantiomer
   hydroboration–oxidation (10.16)

See Figure 10.6, Sample Problems 10.3, 10.4. Try Problems 10.52, 10.56, 10.57.

[2] Alkyne addition reactions

In each addition, both π bonds of the triple bond are broken, and four new bonds are formed.

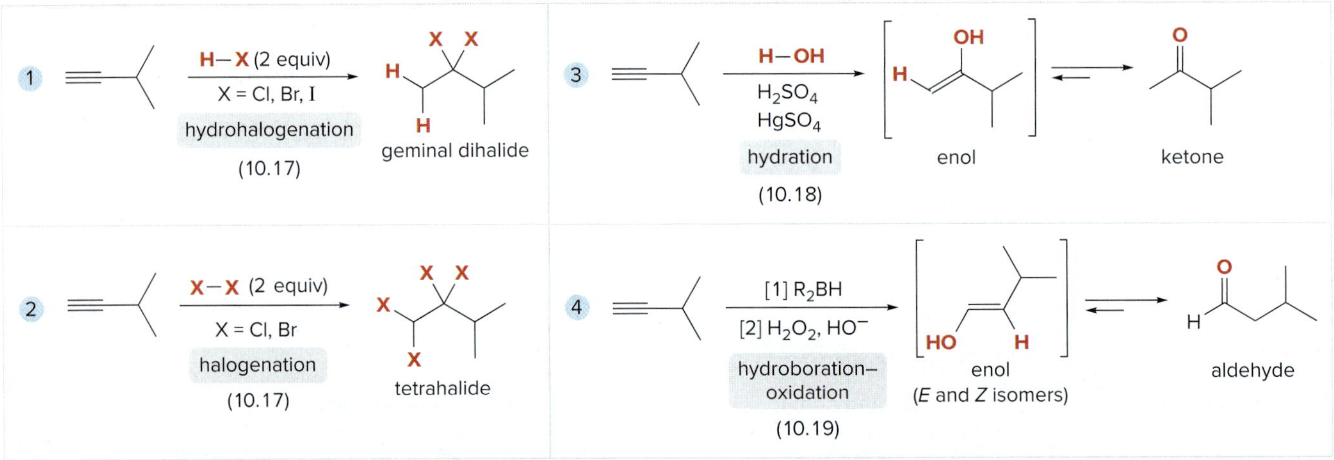

1. Alkyne + H–X (2 equiv) (X = Cl, Br, I) → geminal dihalide
   hydrohalogenation (10.17)

2. Alkyne + X–X (2 equiv) (X = Cl, Br) → tetrahalide
   halogenation (10.17)

3. Alkyne + H–OH, H$_2$SO$_4$, HgSO$_4$ → enol ⇌ ketone
   hydration (10.18)

4. Alkyne + [1] R$_2$BH; [2] H$_2$O$_2$, HO$^-$ → enol (E and Z isomers) ⇌ aldehyde
   hydroboration–oxidation (10.19)

See Figure 10.7. Try Problems 10.53a–e; 10.59a, b, d, f.

[3] Reactions involving acetylide anions

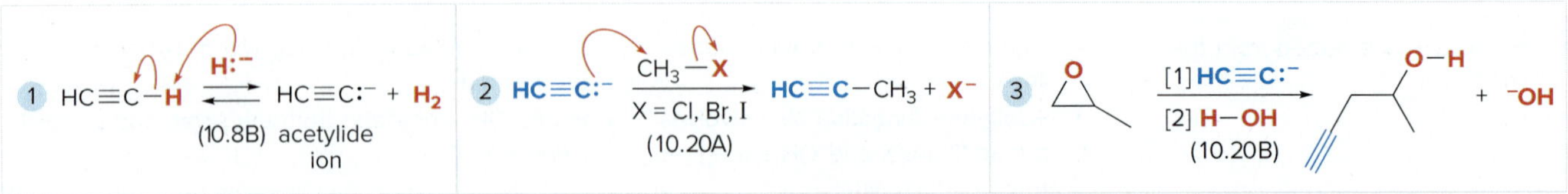

1. HC≡C–H + H:$^-$ ⇌ HC≡C:$^-$ + H$_2$
   (10.8B) acetylide ion

2. HC≡C:$^-$ + CH$_3$–X (X = Cl, Br, I) → HC≡C–CH$_3$ + X$^-$
   (10.20A)

3. Epoxide + [1] HC≡C:$^-$; [2] H–OH → alcohol with alkyne + $^-$OH
   (10.20B)

See Table 10.2. Try Problems 10.53f–h; 10.59e, g–j; 10.61; 10.69; 10.70.

KEY SKILLS

[1] Calculating degrees of unsaturation (10.2)

1 Calculate the maximum number of H's possible.

Example: C_5H_{10}
- For **n** carbons, the maximum number of H's is $2n + 2$; in this example, $2n + 2 = 2(5) + 2 = 12$.

2 Subtract the actual number from the maximum number and divide by two.

- 12 H's (maximum) − 10 H's (actual) = 2 H's fewer than the maximum number.

$$\frac{\text{2 H's fewer than the maximum}}{\text{2 H's removed for each degree unsaturation}} =$$

Answer: one degree of unsaturation

See Sample Problem 10.1. Try Problem 10.42.

[2] Assigning E,Z in naming an alkene (10.3)

1 Assign priorities to the substituents on each end of the C=C.

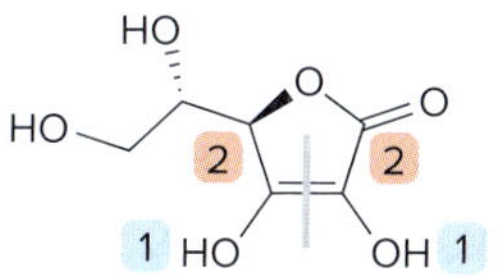

vitamin C
ascorbic acid

2 Assign E or Z based on the location of the two higher-priority groups.

- Two higher-priority groups on *opposite* sides → **E** isomer.
- Two higher-priority groups on the *same* side → **Z** isomer.

Answer: Vitamin C is a **Z** alkene because the two higher-priority groups are on the *same* side of the C=C.

See *How To*, p. 390. Try Problem 10.46a.

[3] Comparing the products of hydration of an alkene (10.12, 10.16)

1 Use the reagents to determine the stereochemistry of H_2O addition.

H_2O, H_2SO_4 (10.12) — A carbocation intermediate gives syn and anti products.

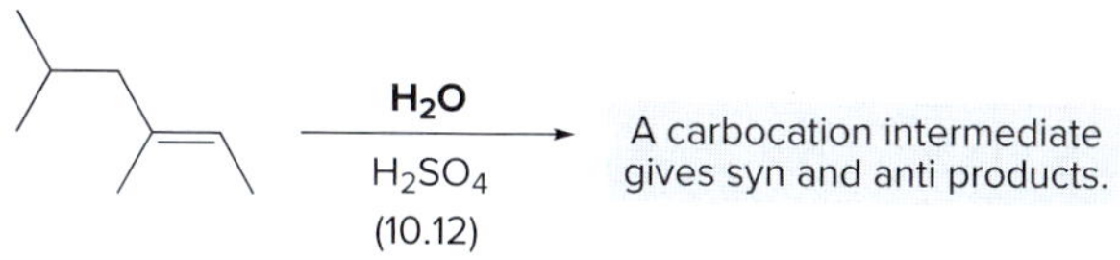

- H and OH are added from the *same* and *opposite* sides.

[1] BH_3 or 9-BBN
[2] H_2O_2, HO^-
(10.16) — The concerted addition gives syn products.

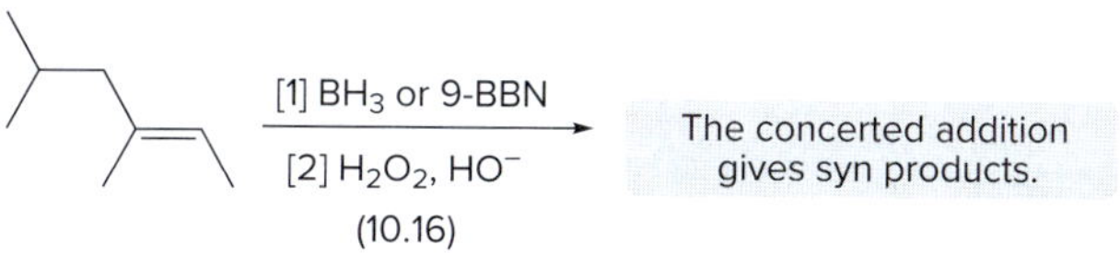

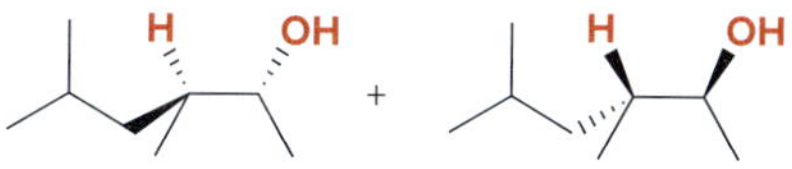

- H and BH_2 are added from the *same* side, and OH replaces BH_2 with retention of configuration.

2 Draw the products.

- H bonds to the *less* substituted C.

- H bonds to the *more* substituted C.

See Sample Problem 10.5. Try Problem 10.55.

[4] Converting an enol to a keto tautomer in acid (10.18)

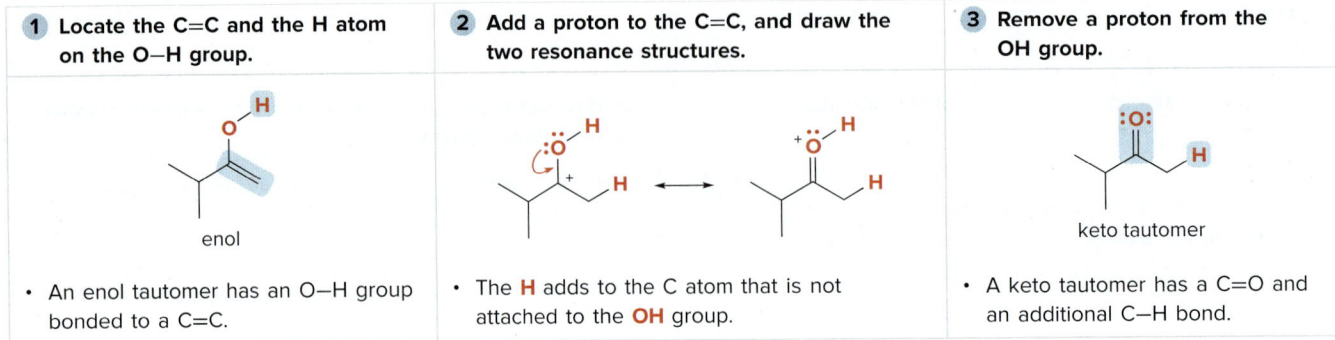

See Mechanism 10.6. Try Problems 10.49–10.51.

[5] Comparing the products of hydration of an alkyne (10.18, 10.19)

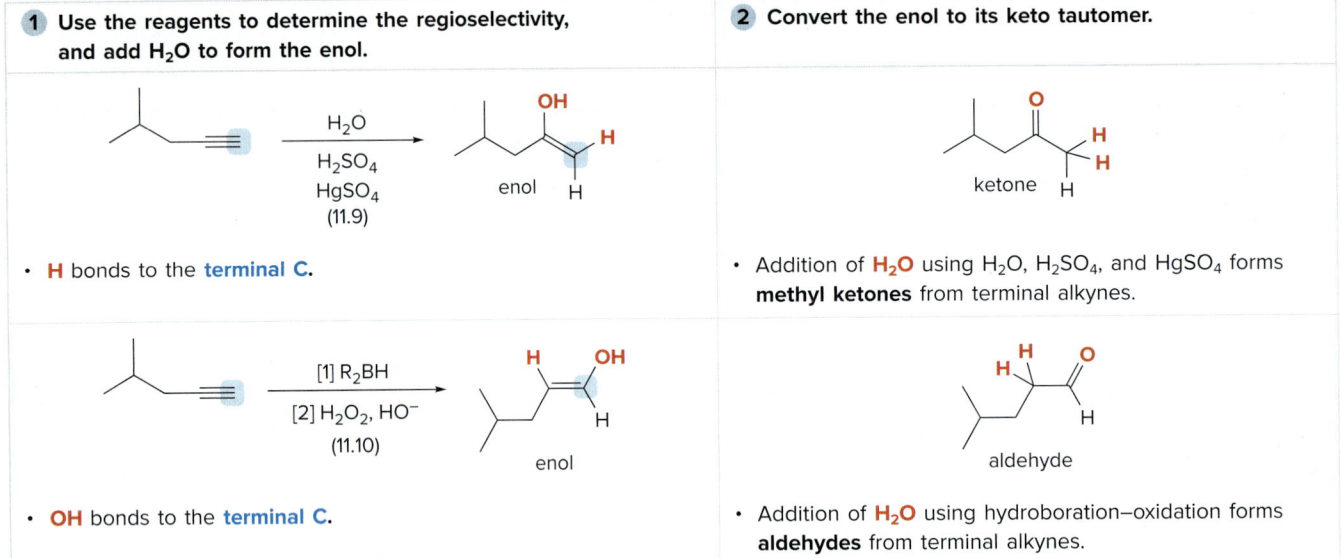

See Sample Problem 10.6. Try Problems 10.53d, e; 10.59d, f.

[6] Devising a synthesis (10.21); example: $(CH_3)_2CHCH_2CHO$ from $(CH_3)_2CHCH=CH_2$

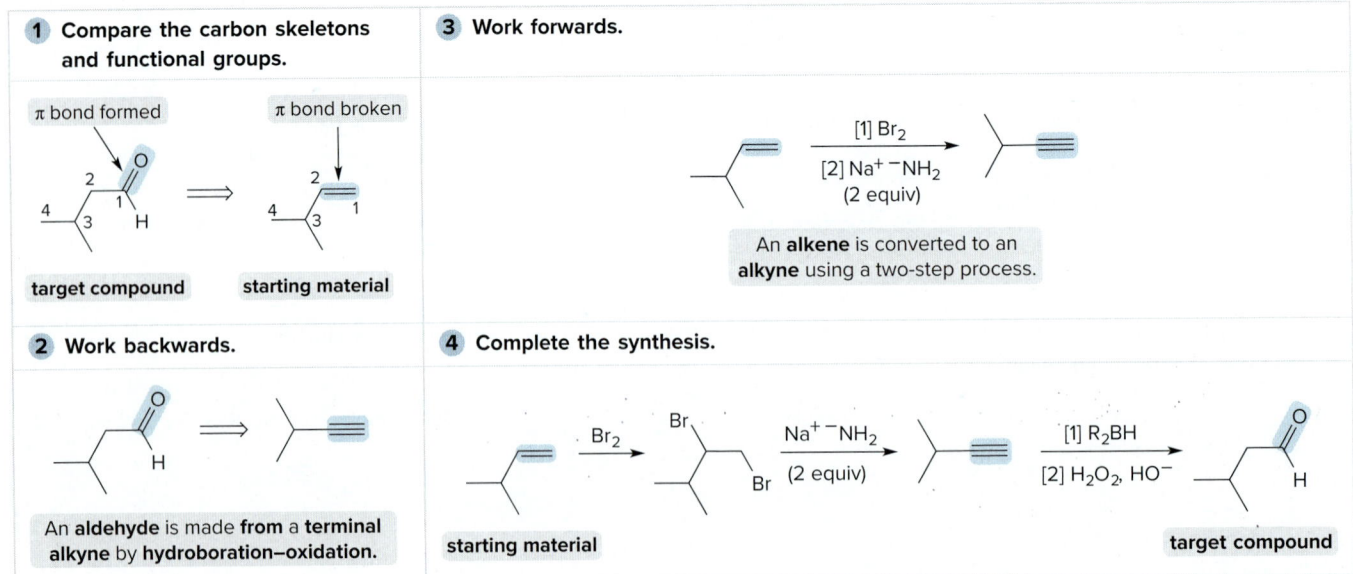

See How To, p. 421; Sample Problems 10.9, 10.10. Try Problems 10.69–10.74.

KEY MECHANISM CONCEPTS

1 Addition of HX and H₂O—Carbocation intermediate

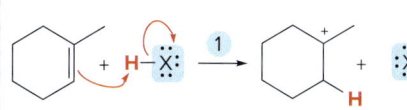

carbocation

- See Mechanisms 10.1 and 10.2.
- **Carbocation** rearrangements are possible.
- Markovnikov's rule is followed.
- **Syn** and **anti addition** occur.

2 Addition of X₂ and X, OH—Halonium ion intermediate

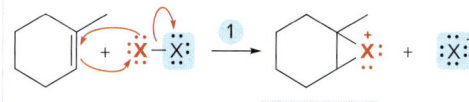

bridged halonium ion

- See Mechanisms 10.3 and 10.4.
- No rearrangements are possible.
- **Anti addition** occurs.
- In the addition of X and OH, X bonds to the *less* substituted C.

3 Addition of BH₃—Concerted reaction

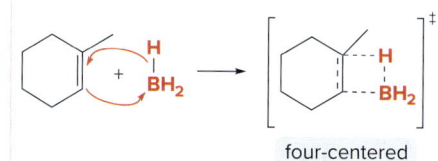

four-centered transition state

- See Mechanism 10.5.
- No rearrangments are possible.
- B bonds to the *less* substituted C.
- **Syn addition** of BH₃ occurs.

See Tables 10.3, 10.4. Try Problems 10.62–10.64.

PROBLEMS

Problems Using Three-Dimensional Models

10.40 Give the IUPAC name for each compound.

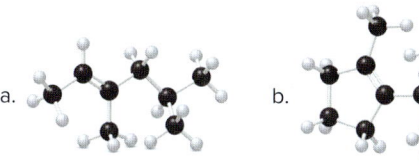

a.   b. 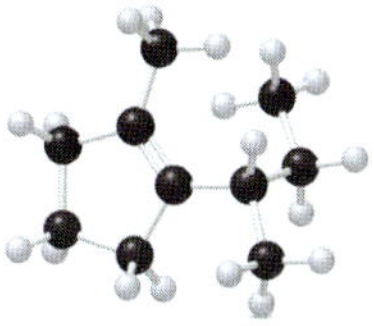   c. 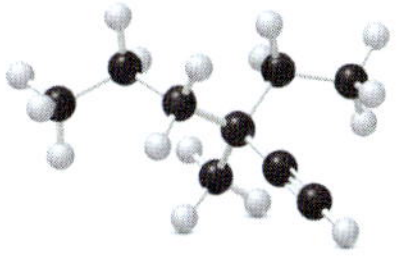   d.

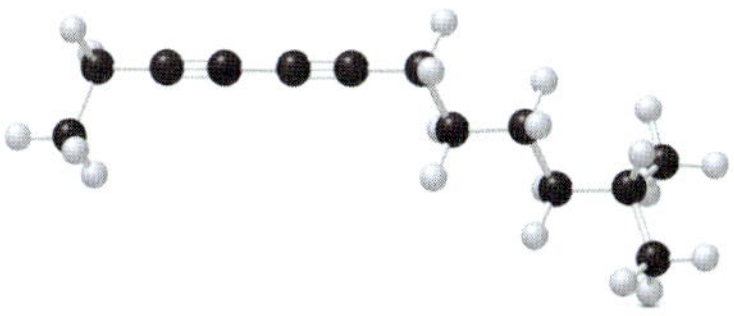

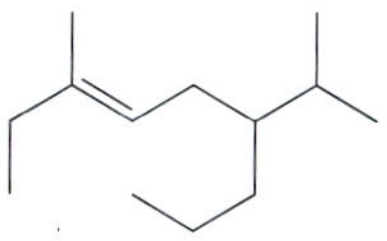

10.41 Draw the enol tautomer of (a) and the keto tautomer of (b).

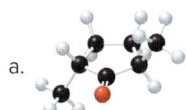 

a.   b.

Degrees of Unsaturation

10.42 Calculate the number of degrees of unsaturation for each molecular formula.
  a. C_6H_8
  b. $C_{40}H_{56}$
  c. $C_{10}H_{16}O_2$
  d. C_8H_9Br
  e. C_8H_9ClO
  f. $C_7H_{11}N$
  g. C_4H_8BrN
  h. $C_{10}H_{18}ClNO$

Structure, Nomenclature, and Stereochemistry

10.43 Give the IUPAC name for each compound.

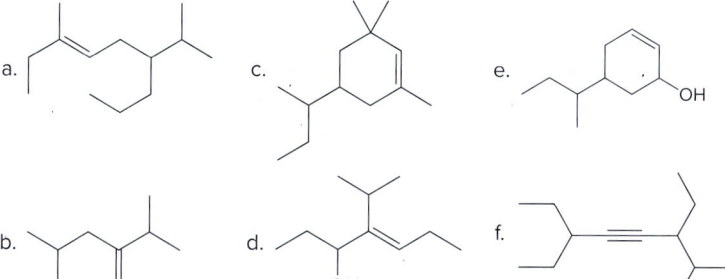

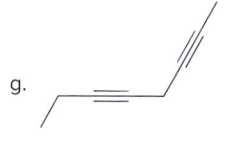

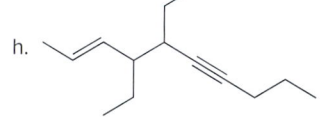

10.44 Give the structure corresponding to each name.
  a. (*E*)-4-ethylhept-3-ene
  b. 3,3-dimethylcyclopentene
  c. 5-*tert*-butyl-6,6-dimethylnon-3-yne
  d. (*Z*)-3-isopropylhept-2-ene
  e. (*Z*)-6-methyloct-6-en-1-yne
  f. 1-isopropyl-4-propylcyclohexene
  g. 3,4-dimethylcyclohex-2-enol
  h. 3,5-diethylhex-5-en-3-ol

10.45 (a) Draw the structure of (1*E*,4*R*)-1,4-dimethylcyclodecene. (b) Draw the enantiomer and name it, including its *E*,*Z* and *R*,*S* prefixes. (c) Draw two diastereomers and name them, including the *E*,*Z* and *R*,*S* prefixes.

10.46 Iejimalide B, an anticancer agent with a 24-membered ring, is isolated from a tunicate found off Ie Island in Okinawa. (a) Label each double bond in iejimalide B as *E* or *Z*. (b) Label each tetrahedral stereogenic center as *R* or *S*. (c) How many stereoisomers are possible for iejimalide B?

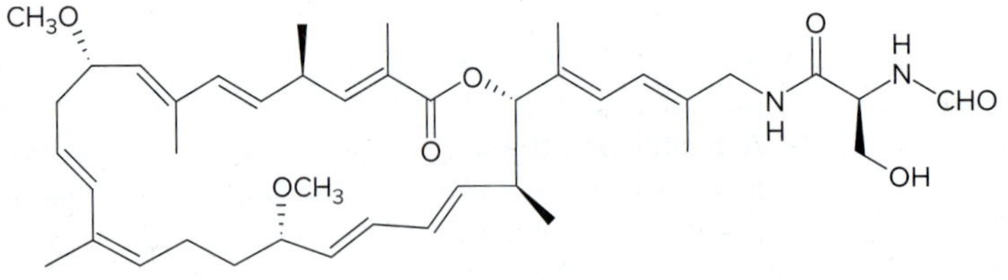

iejimalide B

10.47 Answer the following questions about erlotinib and terbinafine. Erlotinib, sold under the trade name Tarceva, was introduced in 2004 for the treatment of lung cancer. Terbinafine is an antifungal medication used to treat ringworm and fungal nail infections.

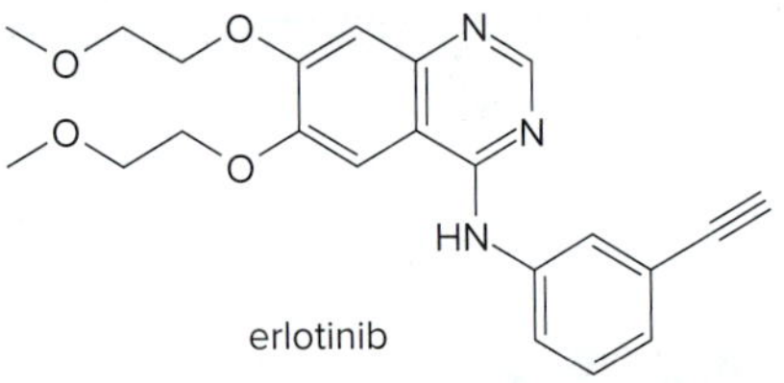

erlotinib

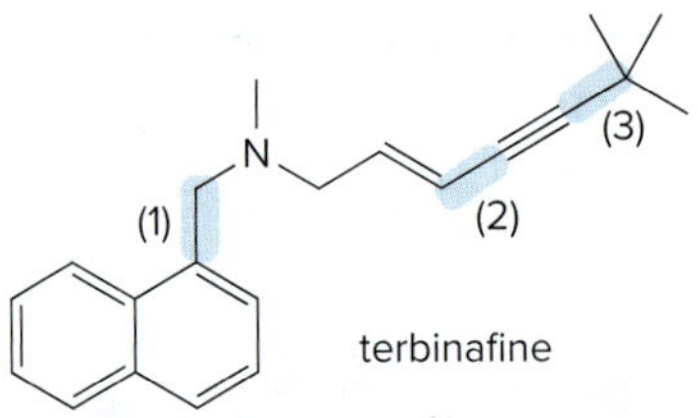

terbinafine

  a. Which C–H bond in erlotinib is most acidic?
  b. What orbitals are used to form the shortest C–C single bond in erlotinib?
  c. Rank the labeled bonds in terbinafine in order of increasing bond strength.
  d. Draw two additional resonance structures for terbinafine that contain all uncharged atoms.

Fatty Acids

10.48 Eleostearic acid is an unsaturated fatty acid obtained from the seeds of the tung oil tree (*Aleurites fordii*), a deciduous tree native to China. (a) Draw the structure of a stereoisomer that has a higher melting point than eleostearic acid. (b) Draw the structure of a stereoisomer that has a lower melting point.

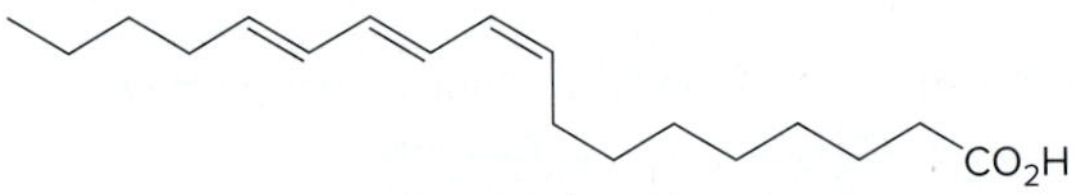

eleostearic acid

Tautomers

10.49 Label each pair of compounds as keto–enol tautomers or constitutional isomers, but not tautomers.

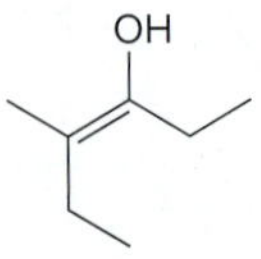

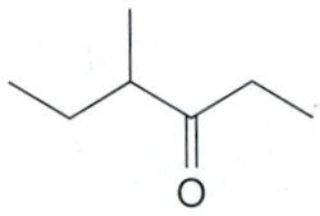

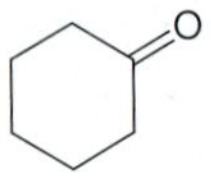

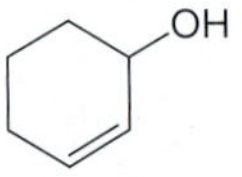

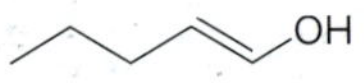

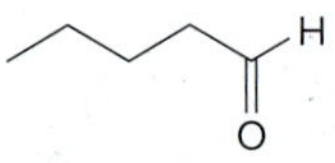

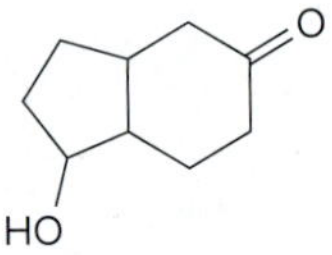

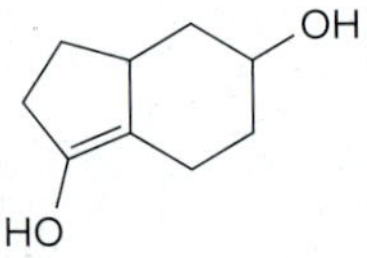

10.50 Draw the enol form of each keto tautomer in parts (a) and (b), and the keto form of each enol tautomer in parts (c) and (d).

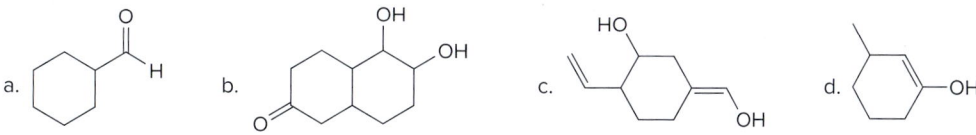

10.51 How is each compound related to **A**? Choose from tautomers, constitutional isomers but not tautomers, or neither.

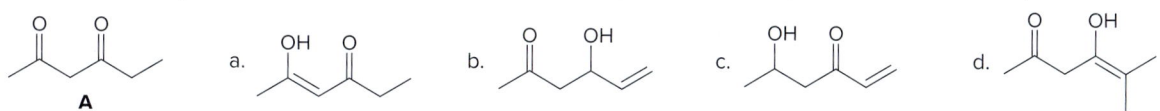

Reactions of Alkenes and Alkynes

10.52 Draw the products formed when $(CH_3)_2C=CH_2$ is treated with each reagent.
 a. HBr
 b. H_2O, H_2SO_4
 c. CH_3CH_2OH, H_2SO_4
 d. Cl_2
 e. Br_2, H_2O
 f. NBS (aqueous DMSO)
 g. [1] BH_3; [2] H_2O_2, HO^-

10.53 Draw the products formed when hex-1-yne is treated with each reagent.
 a. HCl (2 equiv)
 b. HBr (2 equiv)
 c. Cl_2 (2 equiv)
 d. H_2O + H_2SO_4 + $HgSO_4$
 e. [1] R_2BH; [2] H_2O_2, HO^-
 f. NaH
 g. [1] $^-NH_2$; [2] CH_3CH_2Br
 h. [1] $^-NH_2$; [2] △ ; [3] H_2O

10.54 What alkynes give each of the following ketones as the only product after hydration with H_2O, H_2SO_4, and $HgSO_4$?

10.55 Which alcohols can be prepared as a single product by hydroboration–oxidation of an alkene? Which alcohols can be prepared as a single product by the acid-catalyzed addition of H_2O to an alkene?

10.56 Draw the constitutional isomer formed in each reaction.

10.57 Draw the products of each reaction, including stereoisomers.

10.58 (a) What alkene yields **A** and **B** when it is treated with Br₂ in CCl₄? (b) What alkene yields **C** and **D** under the same conditions?

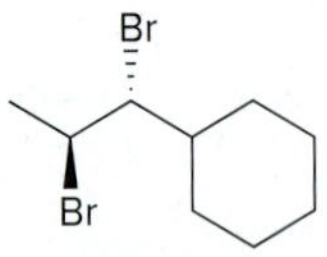

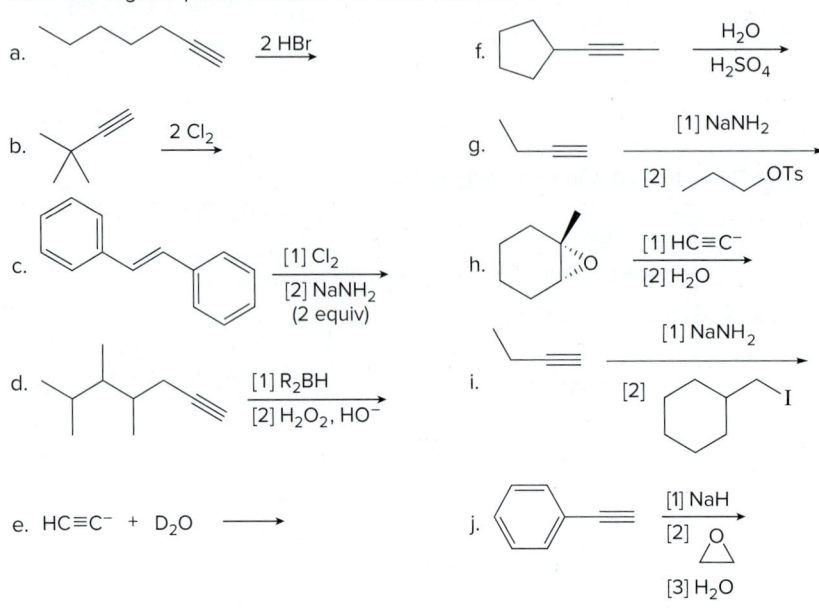

10.59 Draw the organic products formed in each reaction.

a. [hex-1-yne] + 2 HBr

b. [tert-butyl alkyne] + 2 Cl₂

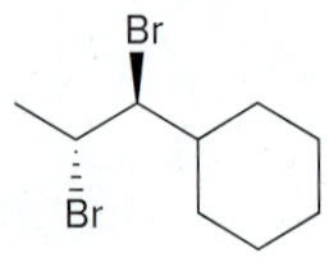

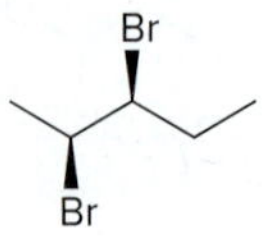

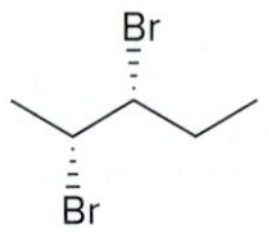

c. [stilbene] [1] Cl₂ [2] NaNH₂ (2 equiv)

d. [branched alkyne] [1] R₂BH [2] H₂O₂, HO⁻

e. HC≡C⁻ + D₂O →

f. [cyclopentyl alkyne] H₂O / H₂SO₄

g. [alkyne] [1] NaNH₂ [2] [CH₃CH₂OTs]

h. [methylcyclohexane epoxide] [1] HC≡C⁻ [2] H₂O

i. [alkyne] [1] NaNH₂ [2] [cyclohexylmethyl I]

j. [phenylacetylene] [1] NaH [2] [epoxide] [3] H₂O

10.60 When alkyne **A** is treated with NaNH₂ followed by CH₃I, a product having molecular formula C₆H₁₀O is formed, but it is *not* compound **B**. What is the structure of the product, and why is it formed?

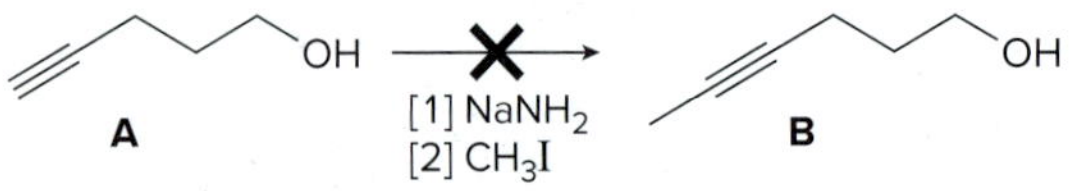

A [HC≡C-CH₂CH₂CH₂-OH] —[1] NaNH₂ [2] CH₃I⟶ ✗ **B** [CH₃-C≡C-CH₂CH₂CH₂-OH]

10.61 Draw the products formed in each reaction and indicate stereochemistry.

a. 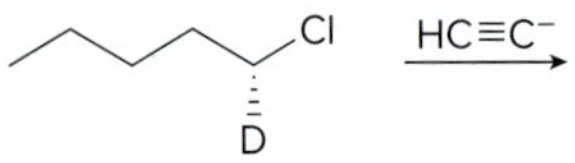 [chloride with D] HC≡C⁻ →

b. 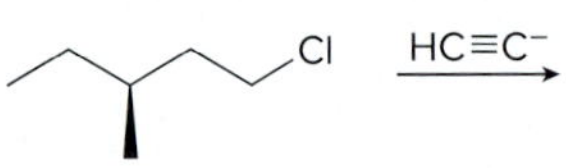 [chloride] HC≡C⁻ →

c. 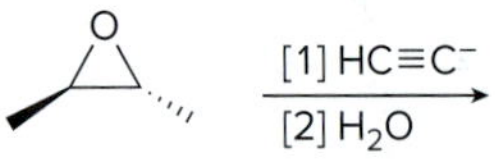 [epoxide] [1] HC≡C⁻ [2] H₂O

d. [epoxide] [1] HC≡C⁻ [2] H₂O

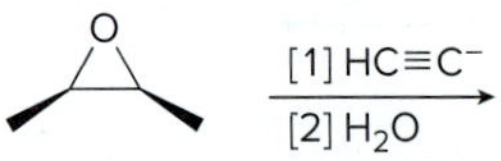

Mechanisms

10.62 Draw a stepwise mechanism for the following reaction, which results in ring expansion of a six-membered ring to a seven-membered ring.

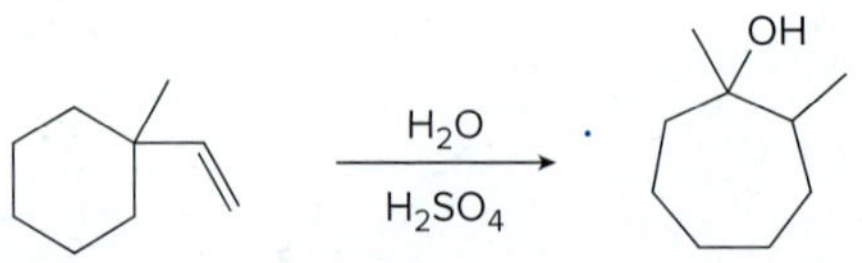

[1-methyl-1-vinylcyclohexane] —H₂O / H₂SO₄⟶ [methylcycloheptanol]

10.63 Draw a stepwise mechanism for the conversion of hex-5-en-1-ol to the cyclic ether **A**.

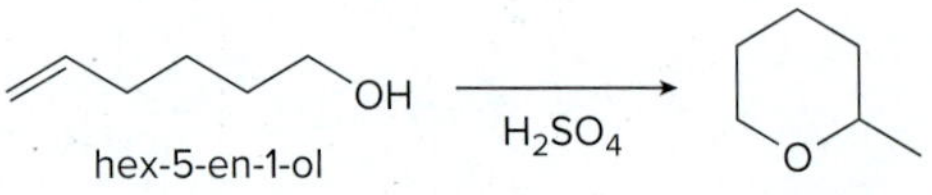

hex-5-en-1-ol —H₂SO₄⟶ **A** [2-methyltetrahydropyran]

10.64 Draw a stepwise mechanism that shows how all three alcohols are formed from the bicyclic alkene.

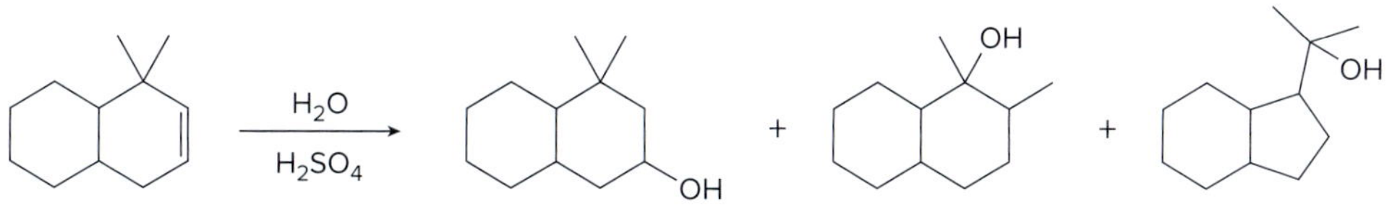

10.65 One step in the synthesis of the antihistamine fexofenadine (Section 22.5) involves acid-catalyzed hydration of the triple bond in **A**. Draw a stepwise mechanism for this reaction and explain why only ketone **B** is formed.

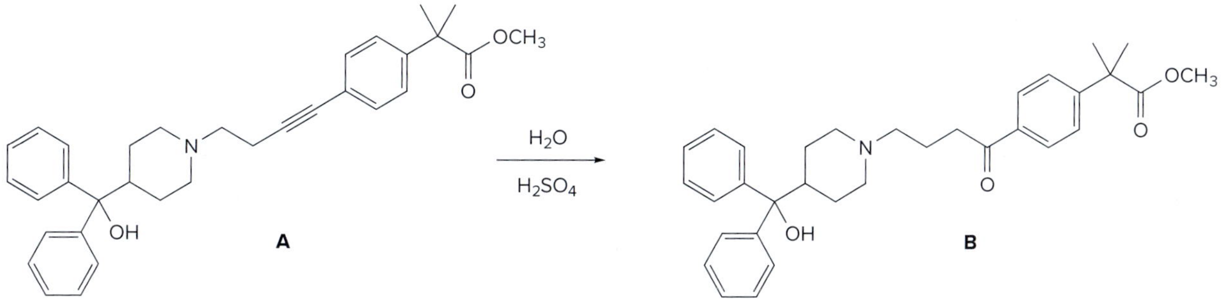

10.66 Tautomerization in base resembles tautomerization in acid, but deprotonation precedes protonation in the two-step mechanism. (a) Draw a stepwise mechanism for the following tautomerization. (b) Then draw a stepwise mechanism for the reverse reaction, the conversion of the keto form to the enol.

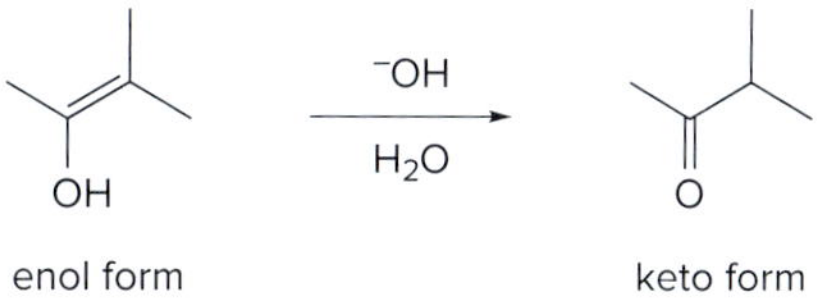

10.67 Draw a stepwise mechanism for the following reaction.

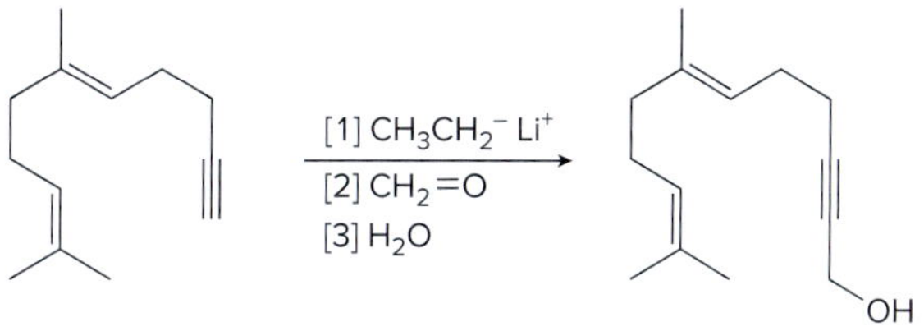

Synthesis

10.68 Devise a synthesis of each product from the given starting material. More than one step is required.

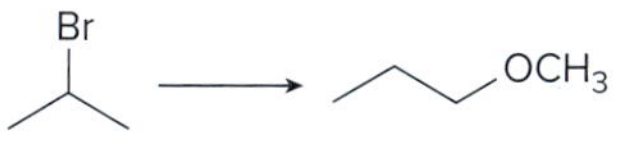

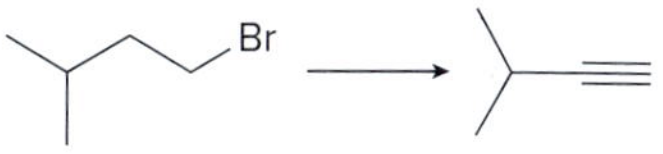

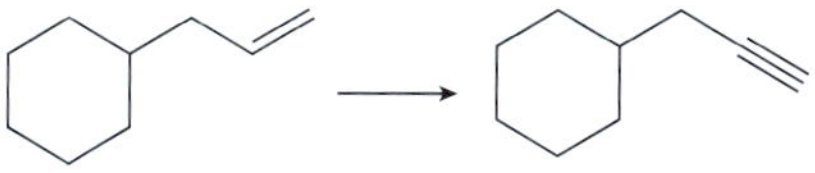

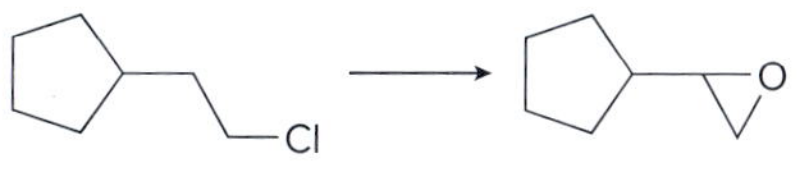

10.69 What acetylide anion and alkyl halide are needed to synthesize each alkyne?

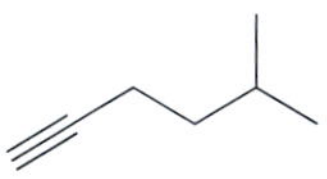

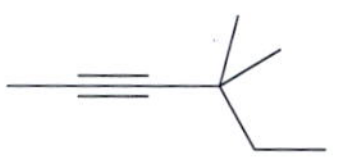

10.70 What acetylide anion and epoxide are needed to synthesize each compound?

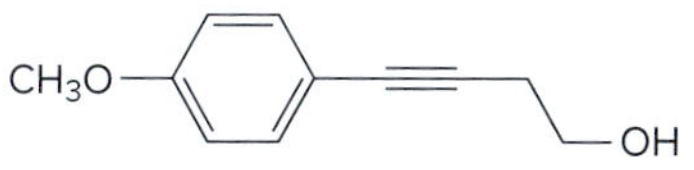

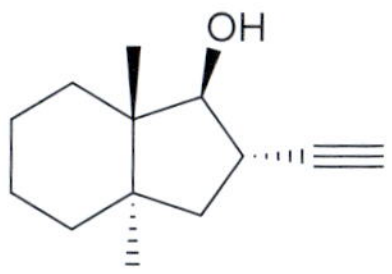

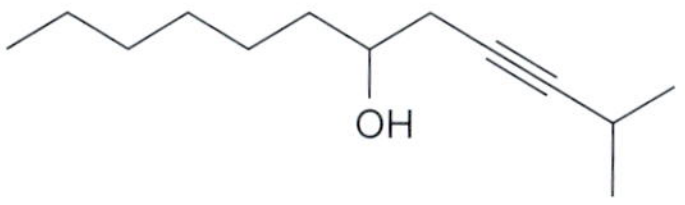

10.71 Synthesize each compound from acetylene. You may use any other organic or inorganic reagents.

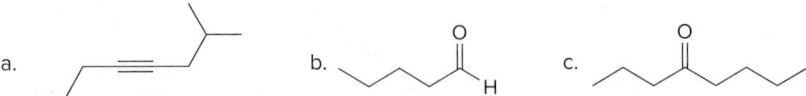

10.72 Devise a synthesis of the ketone hexan-3-one, $CH_3CH_2COCH_2CH_2CH_3$, from CH_3CH_2Br as the only organic starting material; that is, all the carbon atoms in hexan-3-one must come from CH_3CH_2Br. You may use any other needed reagents.

10.73 Devise a synthesis of each compound using $CH_3CH_2CH_2OH$ as the only organic starting material: (a) $CH_3C\equiv CCH_2CH_2CH_3$; (b) $CH_3C\equiv CCH_2CH(OH)CH_3$. You may use any other needed inorganic reagents.

10.74 Devise a synthesis of $CH_3CH_2C\equiv CCH_2CH_2OH$ from CH_3CH_2OH as the only organic starting material. You may use any other needed reagents.

Spectroscopy

Problems 10.75 and 10.76 are intended for students who have already learned about spectroscopy in Chapters A–C.

10.75 When 2-bromo-3,3-dimethylbutane is treated with K^+ $^-OC(CH_3)_3$, a single product **T** having molecular formula C_6H_{12} is formed. When 3,3-dimethylbutan-2-ol is treated with H_2SO_4, the major product **U** has the same molecular formula. Given the following ^1H NMR data, what are the structures of **T** and **U**? Explain in detail the splitting patterns observed for the three split signals in **T**.

^1H NMR of **T**: 1.01 (singlet, 9 H), 4.82 (doublet of doublets, 1 H, $J = 10$, 1.7 Hz), 4.93 (doublet of doublets, 1 H, $J = 18$, 1.7 Hz), and 5.83 (doublet of doublets, 1 H, $J = 18$, 10 Hz) ppm

^1H NMR of **U**: 1.60 (singlet) ppm

10.76 Compound **Y** (molecular formula C_6H_{10}) gives four lines in its ^{13}C NMR spectrum (27, 30, 67, and 93 ppm) and the IR spectrum given here. Propose a structure for **Y**.

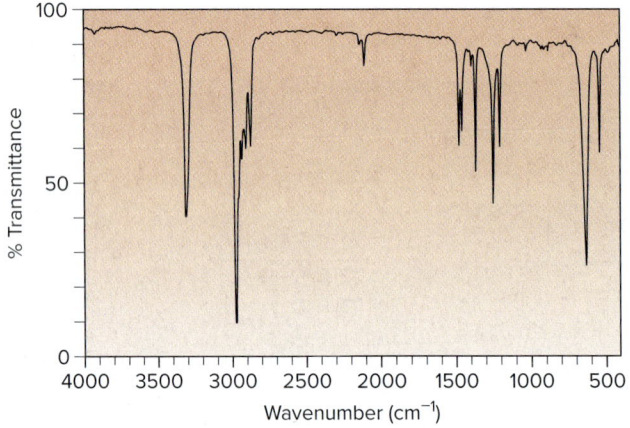

Additional problems on the spectroscopy of alkenes are given in Chapters A–C:
- Mass spectrometry: A.15b, A.18.
- Infrared spectroscopy: B.5, B.7(**A**), B.12c, B16a, B.18c
- Nuclear magnetic resonance spectroscopy: C.11a; C.13d, e; C.25d; C.27d; C.32; C.33d, f; C.38i, j; C.39; C.43d, f; C.44b; C.45c

Additional spectroscopy problems on alkynes are given in Chapters B and C:
- Infrared spectroscopy: B.4a; B.5; B.15a; B.19a; B.21a, d; B.28
- Nuclear magnetic resonance spectroscopy: C.11a

Challenge Problems

10.77 Alkene **A** can be isomerized to isocomene, a natural product isolated from goldenrod, by treatment with TsOH. Draw a stepwise mechanism for this conversion. (Hint: Look for a carbocation rearrangement.)

10.78 Lactones, cyclic esters such as compound **A**, are prepared by **halolactonization,** an addition reaction to an alkene. For example, iodolactonization of **B** forms lactone **C**, a key intermediate in the synthesis of prostaglandin PGF$_{2\alpha}$ (Section 15.5). Draw a stepwise mechanism for this addition reaction.

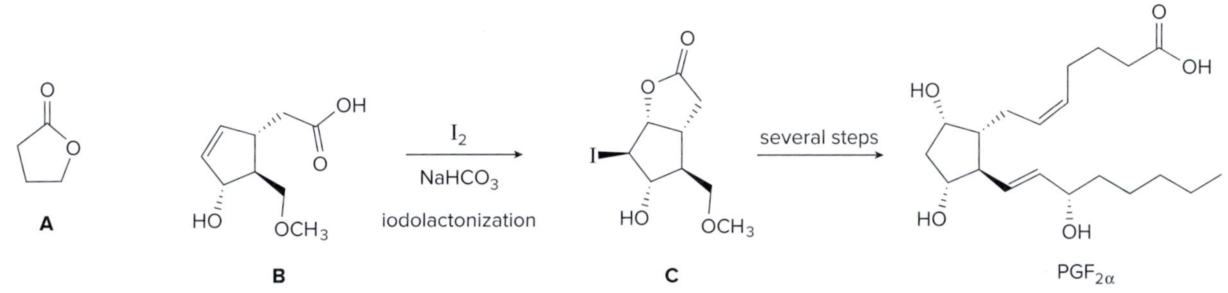

10.79 Explain why the C=C of an enol is more nucleophilic than the C=C of an alkene, despite the fact that the electronegative oxygen atom of the enol inductively withdraws electron density from the carbon–carbon double bond.

10.80 Draw a stepwise mechanism for the following reaction.

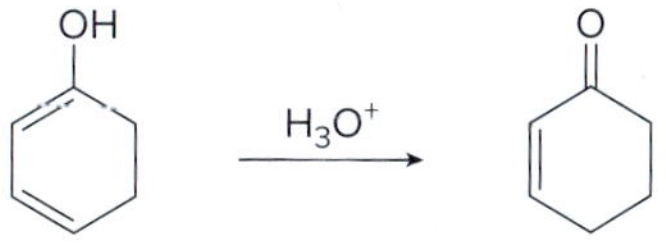

10.81 Draw a stepwise mechanism for the following intramolecular reaction.

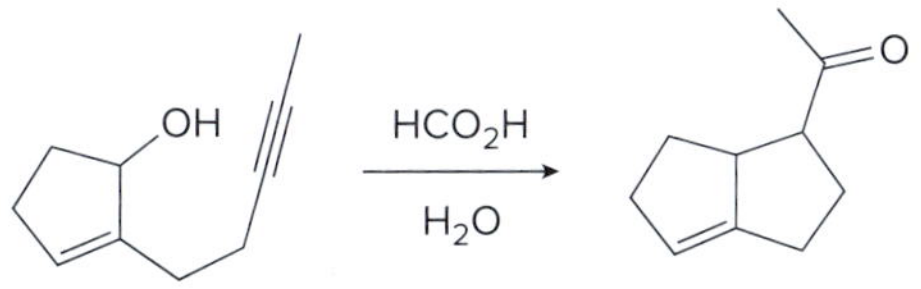

10.82 Explain why an optically active solution of (R)-α-methylbutyrophenone loses its optical activity when dilute acid is added to the solution.

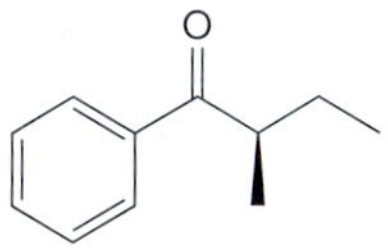

(R)-α-methylbutyrophenone

10.83 Like other electrophiles, carbocations add to alkenes to form new carbocations, which can then undergo substitution or elimination reactions depending on the reaction conditions. With this in mind, draw a stepwise mechanism for the following reaction, which involves the addition of an electrophile—a carbocation—to a double bond.

General reaction

R$^+$ = a carbocation   new carbocation

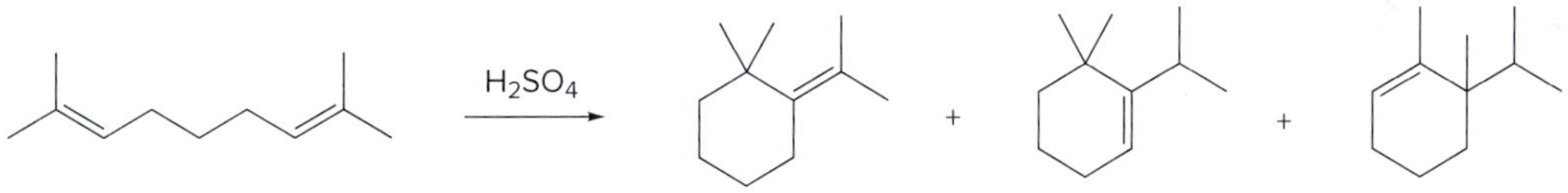

10.84 Draw a stepwise mechanism for the following reaction. This reaction combines two processes: the opening of an epoxide ring with a nucleophile and the addition of an electrophile to a carbon–carbon double bond. (Hint: Begin the mechanism by protonating the epoxide ring.)

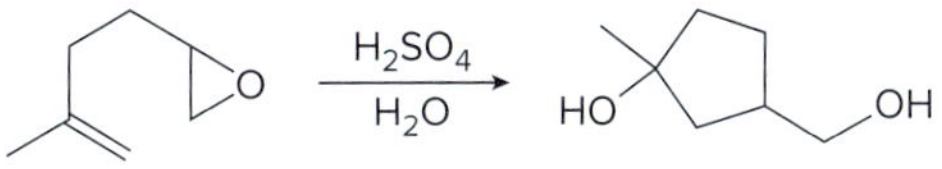

11 Oxidation and Reduction

Amarita/Shutterstock

- **11.1** Introduction
- **11.2** Reducing agents
- **11.3** Reduction of alkenes
- **11.4** Application: Hydrogenation of oils
- **11.5** Reduction of alkynes
- **11.6** The reduction of polar C–X σ bonds
- **11.7** Oxidizing agents
- **11.8** Epoxidation
- **11.9** Dihydroxylation
- **11.10** Oxidative cleavage of alkenes
- **11.11** Oxidative cleavage of alkynes
- **11.12** Oxidation of alcohols
- **11.13** Biological oxidation
- **11.14** Sharpless epoxidation

Soybean oil is rich in **oleic** and **linoleic acids,** two unsaturated fatty acids. When a vegetable oil containing unsaturated fatty acids is treated with hydrogen, some or all of the π bonds add hydrogen, decreasing the number of degrees of unsaturation and increasing the melting point. Adding hydrogen to an alkene is a reduction reaction that increases the number of carbon–hydrogen bonds in the product. In Chapter 11, we learn about oxidation and reduction reactions of alkenes and several other functional groups.

Why Study...

Oxidation and Reduction?

In **Chapter 11,** we discuss the oxidation and reduction of **alkenes** and **alkynes,** as well as compounds with **polar C–X σ bonds**—alcohols, alkyl halides, and epoxides. Although there will be many different reagents and mechanisms, discussing these reactions as a group allows us to more easily compare and contrast them.

The word *mechanism* will often be used loosely here. In contrast to the S_N1 reaction of alkyl halides or the electrophilic addition reactions of alkenes, the details of some of the mechanisms presented in Chapter 11 are known with less certainty. For example, although the identity of a particular intermediate might be confirmed by experiment, other details of the mechanism are suggested by the structure or stereochemistry of the final product.

Oxidation and reduction reactions are very versatile, and knowing them allows us to design many more complex organic syntheses.

11.1 Introduction

Recall from Section 4.14 that the way to determine whether an organic compound has been oxidized or reduced is to compare the **relative number of C–H and C–Z bonds** (Z = an element *more electronegative* than carbon) in the starting material and product.

Two components are always present in an oxidation or reduction reaction—**one component is oxidized and one is reduced.** When an organic compound is *oxidized* by a reagent, the reagent itself must be *reduced*. Similarly, when an organic compound is *reduced* by a reagent, the reagent becomes *oxidized*.

- *Oxidation* results in an *increase* in the number of C–Z bonds (usually C–O bonds) or a *decrease* in the number of C–H bonds.
- *Reduction* results in a *decrease* in the number of C–Z bonds (usually C–O bonds) or an *increase* in the number of C–H bonds.

Thus, an organic compound such as CH_4 can be oxidized by replacing C–H bonds with C–O bonds, as shown in Figure 11.1. Reduction is the opposite of oxidation, so Figure 11.1 also shows how a compound can be reduced by replacing C–O bonds with C–H bonds. The symbols **[O]** and **[H]** indicate oxidation and reduction, respectively.

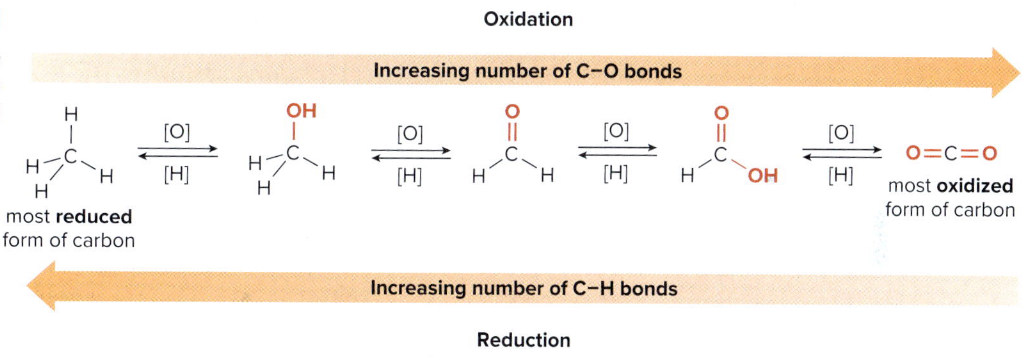

Figure 11.1
A general scheme for the oxidation and reduction of a carbon compound

Sometimes two carbon atoms are involved in a single oxidation or reduction reaction, and the net change in the number of C–H or C–Z bonds at *both* atoms must be taken into account. The conversion of an **alkyne to an alkene** and an **alkene to an alkane** are examples of **reduction,** because each process adds two new C–H bonds to the starting material, as shown in Figure 11.2.

Figure 11.2
Oxidation and reduction of hydrocarbons

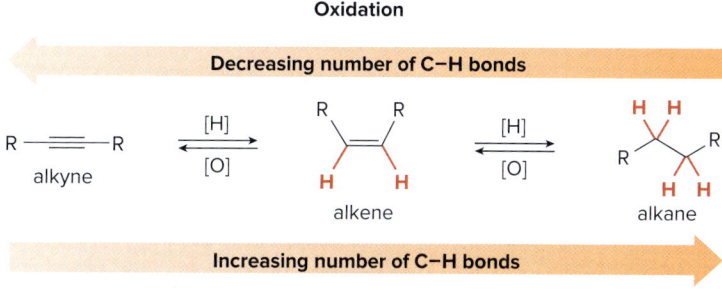

Problem 11.1 Classify each reaction as oxidation, reduction, or neither.

a. [2,3,5-trimethylcyclohexanone] → [1,2,4-trimethylcyclohexane]

b. [cyclopentanone] → [δ-valerolactone]

c. [phenylacetylene] → [phenylacetic acid]

d. [cyclohexene] → [chlorocyclohexane]

11.2 Reducing Agents

Reducing agents provide the equivalent of two hydrogen atoms, but **there are three types of reductions,** differing in how H_2 is added. The simplest reducing agent is molecular H_2. Reductions of this sort are carried out in the presence of a metal catalyst that acts as a surface on which the reaction occurs.

The second way to deliver H_2 in a reduction is to add two protons and two electrons to a substrate—that is, $H_2 = 2\ H^+ + 2\ e^-$. Reducing agents of this sort use alkali metals as a source of electrons and liquid ammonia (NH_3) as a source of protons. Reductions with **Na in NH_3** are called **dissolving metal reductions.**

$$2\ Na \longrightarrow 2\ Na^+ + 2\ e^-$$
$$2\ NH_3 \longrightarrow 2\ ^-NH_2 + 2\ H^+$$

an equivalent of H_2 for reduction

The third way to deliver the equivalent of two hydrogen atoms is to add **hydride (H^-)** and a **proton (H^+)**. The most common hydride reducing agents contain a hydrogen atom bonded to boron or aluminum. Simple examples include **sodium borohydride (NaBH$_4$)** and **lithium aluminum hydride (LiAlH$_4$)**. These reagents deliver H^- to a substrate, and then a proton is added from H_2O or an alcohol.

$Na^+\ H-\overset{H}{\underset{H}{\overset{|}{B}}}-H$ $Li^+\ H-\overset{H}{\underset{H}{\overset{|}{Al}}}-H$

sodium borohydride   lithium aluminum hydride
NaBH$_4$ **LiAlH$_4$**

- Metal hydride reagents act as a source of H^- because they contain polar metal–hydrogen bonds that place a partial negative charge on hydrogen.

$$\underset{\delta+\ \ \delta-}{M-H} = H\!:^-$$

a polar metal–hydrogen bond

M = B or Al

11.3 Reduction of Alkenes

Reduction of an alkene forms an alkane by addition of H_2. Two bonds are broken—the weak **π bond** of the alkene and the H_2 σ bond—and two new C–H σ bonds are formed.

[alkene] + H–H → [alkane with two new C–H bonds]
metal catalyst

A **π bond** is broken.   **Two C–H σ bonds** are formed.

11.3 Reduction of Alkenes

Hydrogenation catalysts are insoluble in common solvents, thus creating a **heterogeneous** reaction mixture. This insolubility has a practical advantage. These catalysts contain expensive metals, but they can be filtered away from the other reactants after the reaction is complete, and then reused.

The addition of H_2 occurs only in the presence of a **metal catalyst,** and thus, the reaction is called **catalytic hydrogenation.** The catalyst consists of a metal—usually Pd, Pt, or Ni—adsorbed onto a finely divided inert solid, such as charcoal. For example, the catalyst 10% Pd on carbon is composed of 10% Pd and 90% carbon, by weight. H_2 adds in a **syn** fashion, as shown in Equation [2].

Problem 11.2 What alkane is formed when each alkene is treated with H_2 and a Pd catalyst?

Problem 11.3 Draw all alkenes that react with one equivalent of H_2 in the presence of a palladium catalyst to form each alkane. Consider constitutional isomers only.

11.3A Hydrogenation and Alkene Stability

Hydrogenation reactions are **exothermic** because the bonds in the product are stronger than the bonds in the starting materials, making them similar to other alkene addition reactions. The $\Delta H°$ for hydrogenation, called the **heat of hydrogenation,** can be used as a measure of the relative stability of two different alkenes that are hydrogenated to the same alkane.

Recall from Chapter 8 that **trans alkenes are generally more stable than cis alkenes.**

For example, both *cis-* and *trans-*but-2-ene are hydrogenated to butane, and the heat of hydrogenation for the trans isomer is less than that for the cis isomer. **Because less energy is released in converting the trans alkene to butane, it must be *lower* in energy (more stable) to begin with.** The relative energies of the butene isomers are illustrated in Figure 11.3.

Figure 11.3
Relative energies of *cis-* and *trans-*but-2-ene

- When hydrogenation of two alkenes gives the same alkane, the more stable alkene has the *smaller* heat of hydrogenation.

Problem 11.4 Which alkene in each pair has the larger heat of hydrogenation?

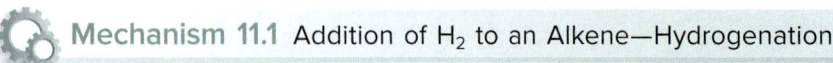

Problem 11.5 Explain why heats of hydrogenation cannot be used to determine the relative stability of 2-methylpent-2-ene and 3-methylpent-1-ene.

11.3B The Mechanism of Catalytic Hydrogenation

In the generally accepted mechanism for catalytic hydrogenation, the surface of the metal catalyst binds both H_2 and the alkene, and H_2 is transferred to the π bond in a rapid but stepwise process (Mechanism 11.1).

Mechanism 11.1 Addition of H_2 to an Alkene—Hydrogenation

① H_2 **adsorbs to the catalyst surface** with partial or complete cleavage of the H–H bond.
② The π bond of the alkene complexes with the metal.

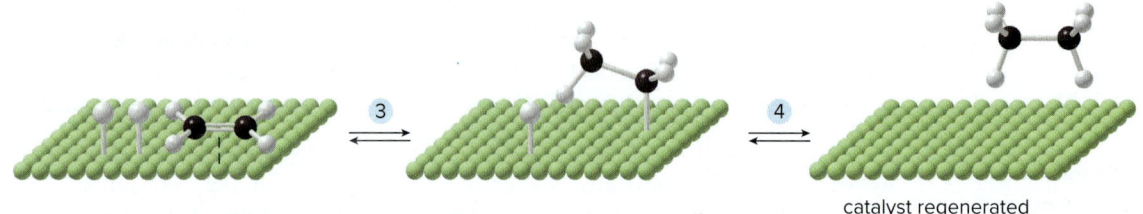

③ – ④ **Two H atoms are transferred sequentially** to the π bond in Steps [3] and [4], forming the alkane. Because the product alkane no longer has a π bond with which to complex to the metal, it is released from the catalyst surface.

The mechanism explains two facts about hydrogenation:

- Rapid, sequential addition of H_2 occurs from the side of the alkene complexed to the metal surface, resulting in syn addition.
- Less crowded double bonds complex more readily to the catalyst surface, resulting in *faster* reaction.

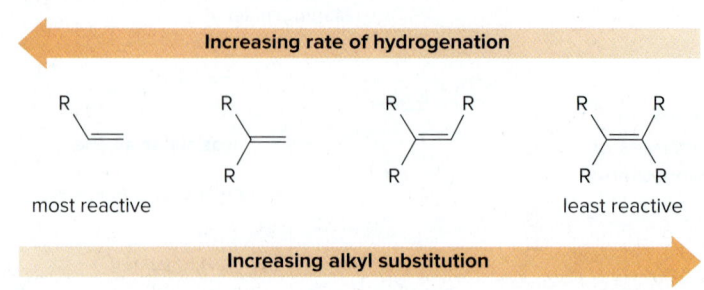

Problem 11.6 Given that syn addition of H₂ occurs from both sides of a trigonal planar double bond, draw all stereoisomers formed when each compound is treated with H₂.

a. b. c. d.

11.3C Hydrogenation Data and Degrees of Unsaturation

Recall from Section 10.2 that the **number of degrees of unsaturation gives the *total* number of rings and π bonds in a molecule.** Because H₂ adds to π bonds but does *not* add to the C–C σ bonds of rings, hydrogenation allows us to determine how many degrees of unsaturation are due to π bonds and how many are due to rings. This is done by comparing the number of degrees of unsaturation before and after a molecule is treated with H₂, as illustrated in Sample Problem 11.1.

Sample Problem 11.1 Using Hydrogenation Data to Determine the Number of Rings and π Bonds in a Molecule

How many rings and π bonds are contained in a compound of molecular formula C_8H_{12} that is hydrogenated to a compound of molecular formula C_8H_{14}?

Solution

[1] Determine the number of degrees of unsaturation in the compounds before and after hydrogenation.

Before H₂ addition—C_8H_{12}
- The maximum number of H's possible for n C's is **$2n + 2$**; in this example, $2n + 2 = 2(8) + 2 = 18$.
- 18 H's (maximum) − 12 H's (actual) = 6 H's fewer than the maximum number.

$$\frac{\text{6 H's fewer than the maximum}}{\text{2 H's removed for each degree of unsaturation}} =$$

three degrees of unsaturation

After H₂ addition—C_8H_{14}
- The maximum number of H's possible for n C's is **$2n + 2$**; in this example, $2n + 2 = 2(8) + 2 = 18$.
- 18 H's (maximum) − 14 H's (actual) = 4 H's fewer than the maximum number.

$$\frac{\text{4 H's fewer than the maximum}}{\text{2 H's removed for each degree of unsaturation}} =$$

two degrees of unsaturation

[2] Assign the number of degrees of unsaturation to rings or π bonds as follows:
- The number of degrees of unsaturation that remain in the product after H₂ addition = the **number of rings** in the starting material.
- The number of degrees of unsaturation that react with H₂ = the **number of π bonds.**

In this example, **two** degrees of unsaturation remain after hydrogenation, so the starting material has **two** rings. Thus:

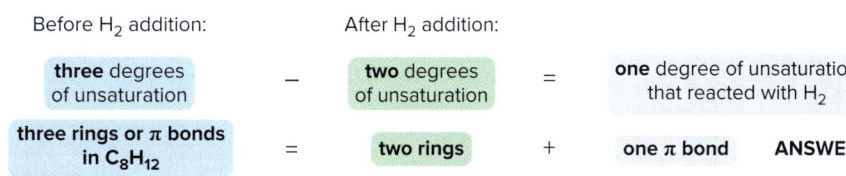

Before H₂ addition: **three** degrees of unsaturation − After H₂ addition: **two** degrees of unsaturation = **one** degree of unsaturation that reacted with H₂

three rings or π bonds in C_8H_{12} = **two rings** + **one π bond** ANSWER

Problem 11.7 Complete the missing information for compounds **A, B,** and **C,** each subjected to hydrogenation. The number of rings and π bonds refers to the reactant (**A, B,** or **C**) prior to hydrogenation.

Compound	Molecular formula before hydrogenation	Molecular formula after hydrogenation	Number of rings	Number of π bonds
A	$C_{10}H_{12}$	$C_{10}H_{16}$	?	?
B	?	C_4H_{10}	0	1
C	C_6H_8	?	1	?

More Practice: Try Problem 11.33.

11.3D Hydrogenation of Other Double Bonds

Compounds that contain a carbonyl group also react with H_2 and a metal catalyst. For example, **aldehydes and ketones are reduced to 1° and 2° alcohols,** respectively. We return to this reaction in Chapter 13.

aldehyde → 1° alcohol ketone → 2° alcohol

11.4 Application: Hydrogenation of Oils

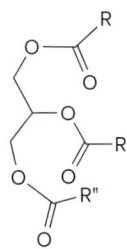

triacylglycerol

The number of double bonds in the R groups of the triacylglycerol determines whether it is a fat or an oil.

Many processed foods, such as peanut butter, margarine, and some brands of crackers, contain *partially hydrogenated* vegetable oils. These oils are produced by hydrogenating the long hydrocarbon chains of triacylglycerols.

In Section 10.6 we learned that **fats and oils are triacylglycerols that differ in the number of degrees of unsaturation** in their long alkyl side chains.

- Fats—usually animal in origin—are solids with triacylglycerols having few degrees of unsaturation.
- Oils—usually vegetable in origin—are liquids with triacylglycerols having a larger number of degrees of unsaturation.

When an unsaturated vegetable oil is treated with hydrogen, some (or all) of the π bonds add H_2, decreasing the number of degrees of unsaturation (Figure 11.4). This increases the melting point of the oil. For example, margarine is prepared by partially hydrogenating vegetable oil to give a product having a semi-solid consistency that more closely resembles butter. This process is sometimes called ***hardening.***

If unsaturated oils are healthier than saturated fats, why does the food industry hydrogenate oils? There are two reasons—aesthetics and shelf life. Consumers prefer the semi-solid consistency of margarine to a liquid oil. Imagine pouring vegetable oil on a piece of toast or pancakes.

Furthermore, unsaturated oils are more susceptible than saturated fats to oxidation at the **allylic carbon atoms**—the carbons adjacent to the double bond carbons—a process discussed in Chapter 21. Oxidation makes the oil rancid and inedible. Hydrogenating the double bonds reduces the number of allylic carbons (also illustrated in Figure 11.4), thus reducing the likelihood of oxidation and increasing the shelf life of the food product. This process reflects a

Figure 11.4   Partial hydrogenation of the double bonds in a vegetable oil

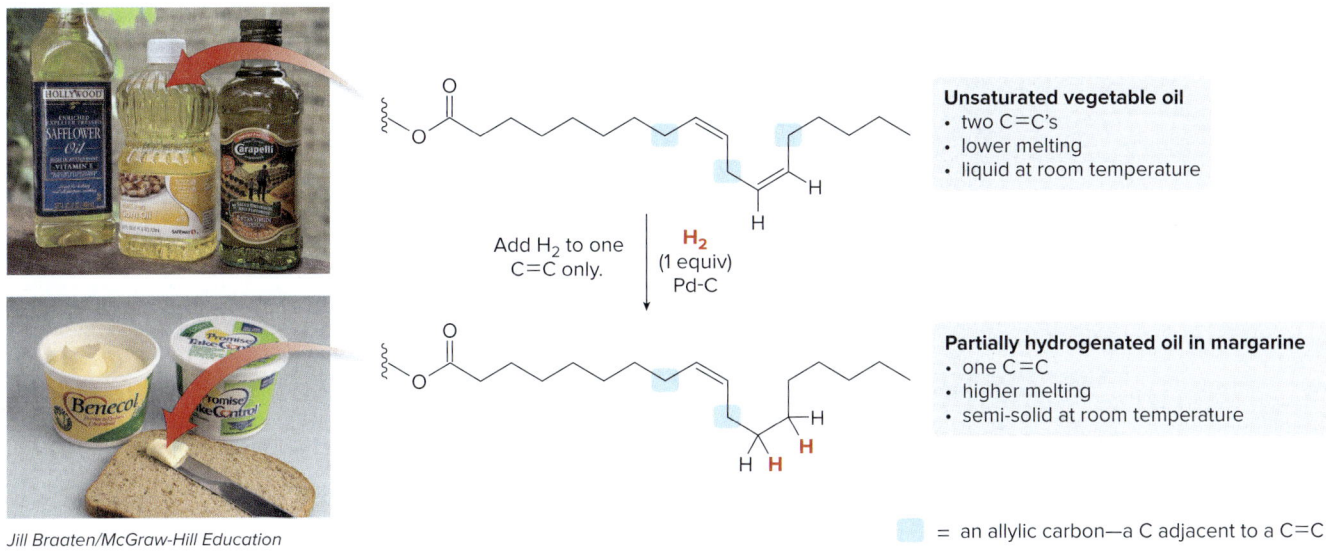

Jill Braaten/McGraw-Hill Education

- **Decreasing** the number of degrees of unsaturation **increases** the melting point. Only one long chain of the triacylglycerol is drawn.
- When an oil is *partially* hydrogenated, some double bonds react with H_2, whereas some double bonds remain in the product.
- Partial hydrogenation **decreases** the number of allylic sites (shown in blue), making a triacylglycerol **less** susceptible to oxidation, thereby increasing its shelf life.

Peanut butter is a common consumer product that contains partially hydrogenated vegetable oil. *Elite Images/McGraw-Hill Education*

delicate balance between providing consumers with healthier food products, while maximizing shelf life to prevent spoilage.

One other fact is worthy of note. Because the steps in hydrogenation are reversible and H atoms are added in a sequential rather than concerted fashion, a **cis double bond can be isomerized to a trans double bond.** After addition of one H atom (Step [3] in Mechanism 11.1), an intermediate can lose a hydrogen atom to re-form a double bond with either the cis or trans configuration.

As a result, some of the cis double bonds in vegetable oils are converted to trans double bonds during hydrogenation, forming so-called **"trans fats."** The shape of the resulting fatty acid chain is very different, closely resembling the shape of a *saturated* fatty acid chain. Consequently, trans fats are thought to have the same negative effects on blood cholesterol levels as saturated fats; that is, trans fats stimulate cholesterol synthesis in the liver, thus increasing blood cholesterol levels, a factor linked to increased risk of heart disease.

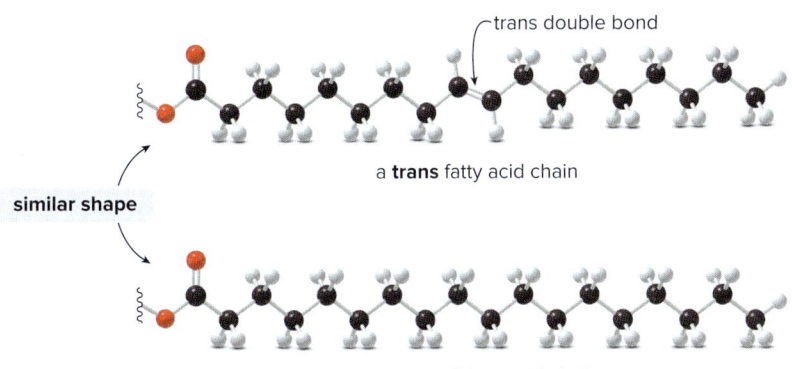

Problem 11.8 Draw the products formed when triacylglycerol **A** is treated with each reagent, forming compounds **B** and **C**. Rank **A**, **B**, and **C** in order of increasing melting point.

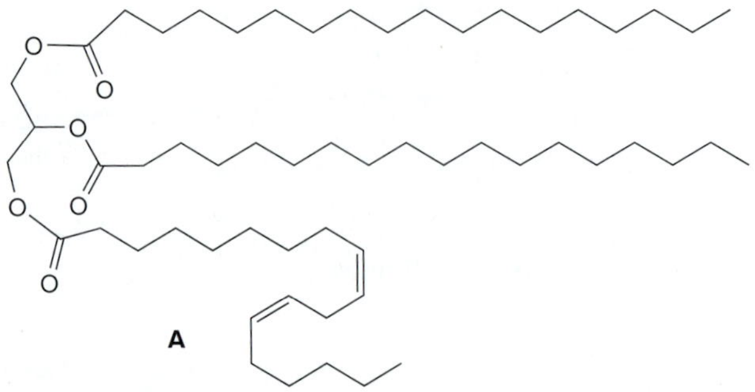

a. H$_2$ (excess), Pd-C (Compound **B**)
b. H$_2$ (1 equiv), Pd-C (Compound **C**)

11.5 Reduction of Alkynes

Reduction of an alkyne adds H$_2$ to one or both of the π bonds. There are three different ways by which the elements of H$_2$ can be added to a triple bond.

- Adding two equivalents of H$_2$ forms an alkane.

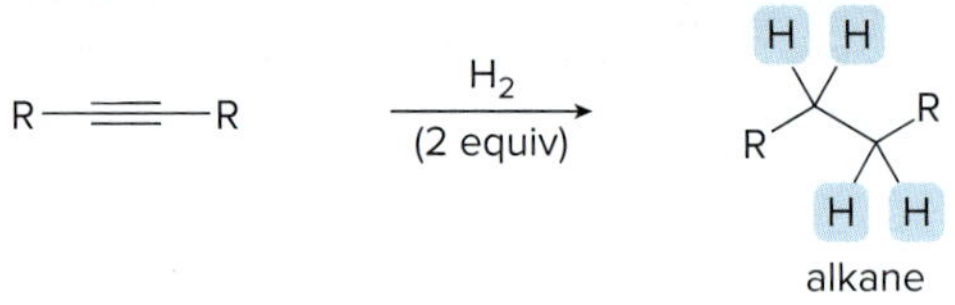

$$R-\!\!\equiv\!\!-R \xrightarrow[\text{(2 equiv)}]{H_2} \text{alkane}$$

- Adding one equivalent of H$_2$ in a syn fashion forms a cis alkene.

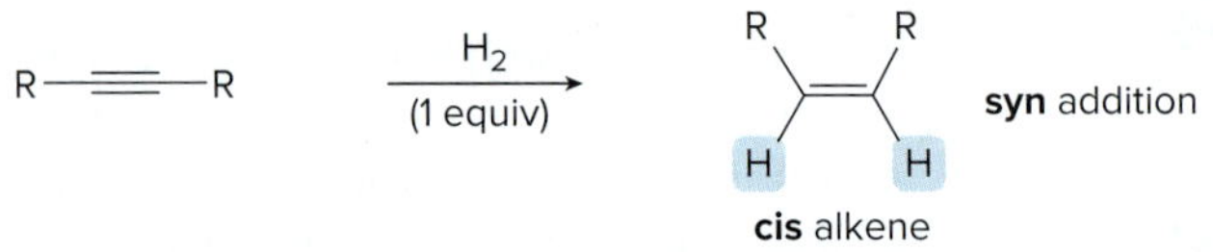

$$R-\!\!\equiv\!\!-R \xrightarrow[\text{(1 equiv)}]{H_2} \text{cis alkene} \quad \text{syn addition}$$

- Adding one equivalent of H$_2$ in an anti fashion forms a trans alkene.

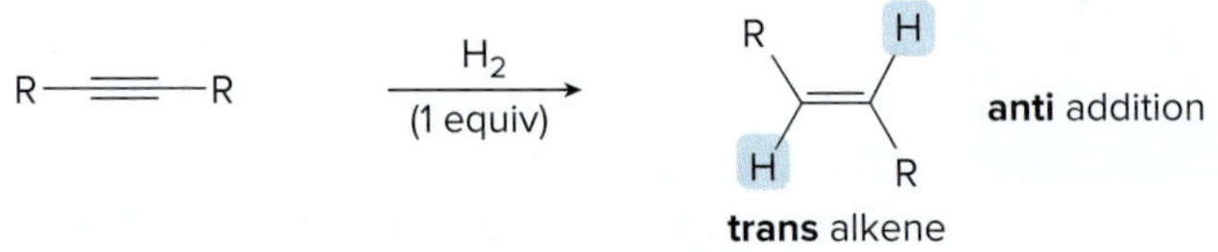

$$R-\!\!\equiv\!\!-R \xrightarrow[\text{(1 equiv)}]{H_2} \text{trans alkene} \quad \text{anti addition}$$

11.5A Reduction of an Alkyne to an Alkane

When an alkyne is treated with two or more equivalents of H$_2$ and a Pd catalyst, reduction of *both* π bonds occurs. **Syn addition** of one equivalent of H$_2$ forms a cis alkene, which adds a second equivalent of H$_2$ to form an **alkane. Four new C–H bonds are formed.** By using a Pd-C catalyst, it is not possible to stop the reaction after addition of only one equivalent of H$_2$.

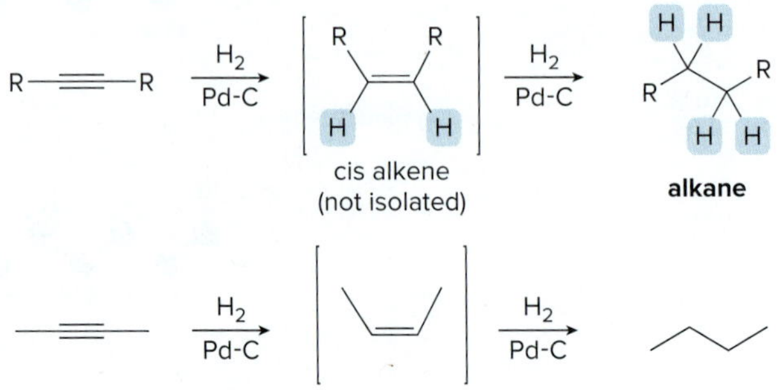

$$R-\!\!\equiv\!\!-R \xrightarrow{H_2, \text{Pd-C}} [\text{cis alkene (not isolated)}] \xrightarrow{H_2, \text{Pd-C}} \text{alkane}$$

Problem 11.9 Which alkyne has the smaller heat of hydrogenation, HC≡CCH$_2$CH$_2$CH$_3$ or CH$_3$C≡CCH$_2$CH$_3$? Explain your choice.

11.5B Reduction of an Alkyne to a Cis Alkene

Palladium metal is too active a catalyst to allow the hydrogenation of an alkyne to stop after one equivalent of H$_2$. To prepare a cis alkene from an alkyne and H$_2$, a less active Pd catalyst is used—Pd adsorbed onto CaCO$_3$ with added lead(II) acetate and quinoline. This catalyst is called the **Lindlar catalyst** after the chemist who first prepared it. Compared to Pd metal, the **Lindlar catalyst is deactivated or "poisoned."**

Pd on CaCO$_3$
+ Pb(OCOCH$_3$)$_2$ + quinoline

Lindlar catalyst   quinoline

Reduction of an alkyne to a cis alkene is a **stereoselective reaction,** because only one stereoisomer is formed.

With the Lindlar catalyst, one equivalent of H$_2$ adds to an alkyne, and the cis alkene product is unreactive to further reduction.

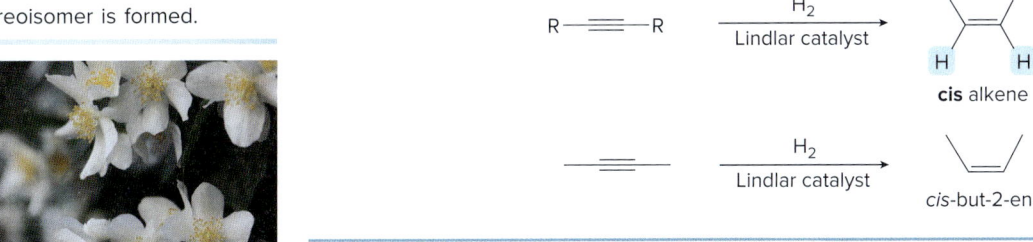

Jasmine flowers are the source of *cis*-jasmone, a perfume component (Problem 11.10).
Charlotte Bjornstrom/EyeEm/Getty Images

Problem 11.10 What is the structure of *cis*-jasmone, a natural product isolated from jasmine flowers, formed by treatment of alkyne **A** with H$_2$ in the presence of the Lindlar catalyst?

Problem 11.11 (a) Draw the structure of a compound of molecular formula C$_6$H$_{10}$ that reacts with H$_2$ in the presence of Pd–C but does not react with H$_2$ in the presence of Lindlar catalyst. (b) Draw the structure of a compound of molecular formula C$_6$H$_{10}$ that reacts with H$_2$ when either catalyst is present.

11.5C Reduction of an Alkyne to a Trans Alkene

Although catalytic hydrogenation is a convenient method for preparing cis alkenes from alkynes, it cannot be used to prepare trans alkenes. With a **dissolving metal reduction** (such as Na in NH$_3$), however, the elements of H$_2$ are added in an **anti** fashion to the triple bond, thus forming a **trans alkene.** For example, but-2-yne reacts with Na in NH$_3$ to form *trans*-but-2-ene.

444 Chapter 11 Oxidation and Reduction

The **mechanism** for the dissolving metal reduction using Na in NH$_3$ features sequential addition of electrons and protons to the triple bond. Half-headed arrows denoting the movement of a single electron must be used in two steps when Na donates *one* electron. The mechanism can be divided conceptually into two parts, each of which consists of two steps: **addition of an electron followed by protonation of the resulting negative charge,** as shown in Mechanism 11.2.

Mechanism 11.2 Dissolving Metal Reduction of an Alkyne to a Trans Alkene

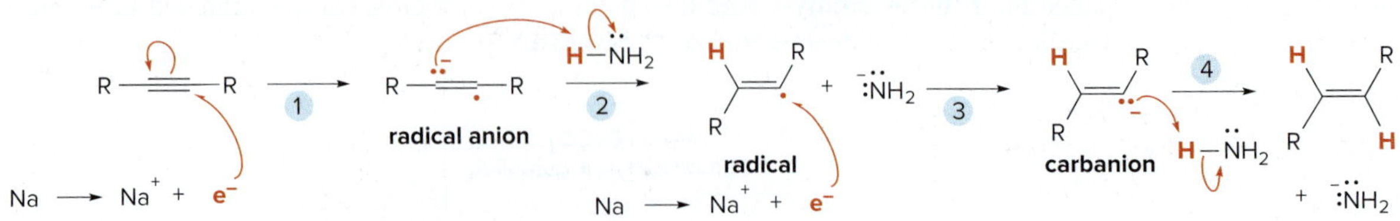

1. Addition of an electron to the triple bond forms a **radical anion**, a species that contains *both* a negative charge *and* an unpaired electron.
2. Protonation of the anion with the solvent NH$_3$ yields a **radical**. The net result of the first two steps is the addition of a H atom.
3. Addition of a second electron forms a **carbanion**.
4. Protonation of the carbanion forms the **trans alkene**. Steps [3] and [4] add the second H atom to the triple bond.

Although the vinyl carbanion formed in Step [3] could have two different arrangements of its R groups, only the trans alkene is formed from the more stable vinyl carbanion; this carbanion has the larger R groups farther away from each other to avoid steric interactions. Protonation of this anion leads to the more stable trans product.

Dissolving metal reduction of a triple bond with Na in NH$_3$ is a **stereoselective reaction** because it forms a trans product exclusively.

The larger R groups are farther away from each other.

This **more stable vinyl carbanion** forms the trans alkene.

Steric interactions between closer R groups **destabilize** this carbanion.

- **Dissolving metal reductions always form the more stable trans product preferentially.**

The three methods to reduce a triple bond are summarized in Figure 11.5 using hex-3-yne as starting material.

Figure 11.5

Summary: Three methods to reduce a triple bond

hex-3-yne

2 H$_2$, Pd-C → hexane

H$_2$, Lindlar catalyst → *cis*-hex-3-ene

Na, NH$_3$ → *trans*-hex-3-ene

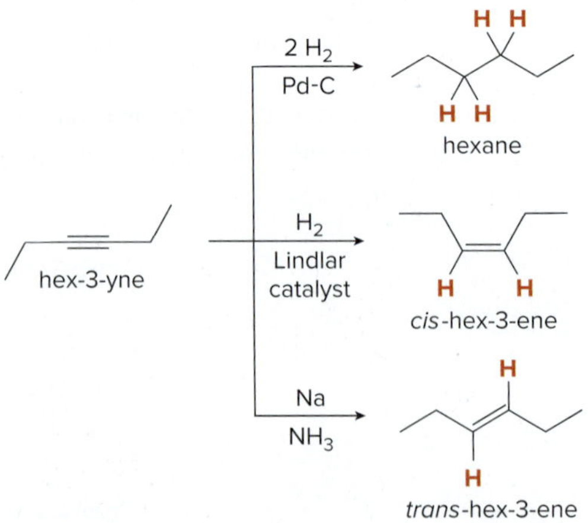

Problem 11.12 What product is formed when $CH_3OCH_2CH_2C\equiv CCH_2CH(CH_3)_2$ is treated with each reagent:
(a) H_2 (excess), Pd-C; (b) H_2 (1 equiv), Lindlar catalyst; (c) H_2 (excess), Lindlar catalyst; (d) Na, NH_3?

Problem 11.13 A chiral alkyne **A** with molecular formula C_6H_{10} is reduced with H_2 and Lindlar catalyst to **B** having the *R* configuration at its stereogenic center. What are the structures of **A** and **B**?

11.6 The Reduction of Polar C–X σ Bonds

Compounds containing polar C–X σ bonds that react with strong nucleophiles are reduced with metal hydride reagents, most commonly lithium aluminum hydride. Two functional groups possessing both of these characteristics are **alkyl halides** and **epoxides**. Alkyl halides are reduced to alkanes with loss of X⁻ as the leaving group. Epoxide rings are opened to form alcohols.

$$R-X \xrightarrow[\text{[2] H}_2\text{O}]{\text{[1] LiAlH}_4} R-H$$
alkyl halide → alkane

epoxide $\xrightarrow[\text{[2] H}_2\text{O}]{\text{[1] LiAlH}_4}$ alcohol

Reduction of these C–X σ bonds is another example of nucleophilic substitution, in which LiAlH$_4$ serves as a source of a hydride nucleophile (H⁻). Because **H⁻ is a strong nucleophile,** the reaction follows an **S$_N$2 mechanism,** illustrated for the one-step reduction of an alkyl halide in Mechanism 11.3.

Mechanism 11.3 Reduction of RX with LiAlH$_4$

$$R-X + H-\bar{A}lH_3 \; Li^+ \longrightarrow R-H + Li^+X^- + AlH_3$$

LiAlH$_4$ donates **H⁻**.

- The **nucleophile H⁻** replaces the **leaving group X⁻** in a single step.

Because the reaction follows an S$_N$2 mechanism:

- Unhindered CH$_3$X and 1° alkyl halides are more easily reduced than more substituted 2° and 3° halides.
- In unsymmetrical epoxides, nucleophilic attack of H⁻ (from LiAlH$_4$) occurs at the *less* substituted carbon atom.

Examples are shown in Figure 11.6.

Figure 11.6
Examples of reduction of C–X σ bonds with LiAlH$_4$

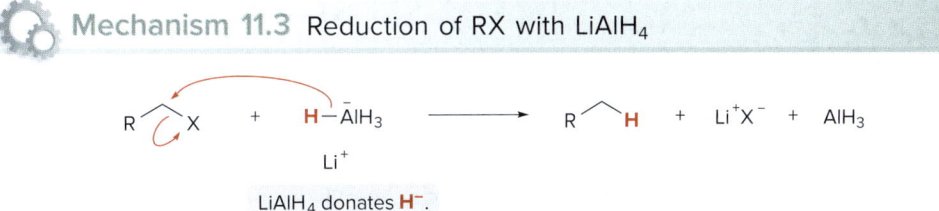

attack at the less substituted C

Problem 11.14 Draw the products of each reaction.

a. (CH₃)₂CHCH₂CH₂CH₂Cl [1] LiAlH₄ / [2] H₂O

b. (bicyclic epoxide) [1] LiAlH₄ / [2] H₂O

11.7 Oxidizing Agents

Oxidizing agents fall into two main categories:

- **Reagents that contain an oxygen–oxygen bond**
- **Reagents that contain metal–oxygen bonds**

Oxidizing agents containing an O–O bond include O_2, O_3 (ozone), H_2O_2 (hydrogen peroxide), $(CH_3)_3COOH$ (*tert*-butyl hydroperoxide), and peroxyacids. **Peroxyacids,** a group of reagents with the general structure **RCO₃H,** have one more O atom than carboxylic acids (RCO_2H). Some peroxyacids are commercially available whereas others are prepared and used without isolation. Two common peroxyacids are peroxyacetic acid and *meta*-chloroperoxybenzoic acid, abbreviated as **mCPBA.**

peroxyacid   peroxyacetic acid   *meta*-chloroperoxybenzoic acid
                                     mCPBA

The most common oxidizing agents with metal–oxygen bonds contain either chromium in the +6 oxidation state (six Cr–O bonds) or manganese in the +7 oxidation state (seven Mn–O bonds). Common Cr^{6+} reagents include chromium(VI) oxide **(CrO₃)** and sodium or potassium dichromate **($Na_2Cr_2O_7$ and $K_2Cr_2O_7$)**. **These reagents are strong oxidants** used in the presence of a strong aqueous acid such as H_2SO_4. **Pyridinium chlorochromate (PCC),** a Cr^{6+} reagent that is soluble in halogenated organic solvents, can be used without strong acid present. This makes it a **more selective Cr^{6+} oxidant,** as described in Section 11.12.

chromium(VI) oxide   pyridinium chlorochromate
CrO₃   **PCC**

The most common Mn^{7+} reagent is **KMnO₄** (potassium permanganate), a strong, water-soluble oxidant. Other oxidizing agents that contain metals include **OsO₄** (osmium tetroxide) and **Ag₂O** [silver(I) oxide].

In the remainder of Chapter 11, the oxidation of alkenes, alkynes, and alcohols—three functional groups already introduced in this text—is presented (Figure 11.7). Addition reactions to alkenes and alkynes that increase the number of C–O bonds are described in Sections 11.8–11.11. Oxidation of alcohols to carbonyl compounds appears in Sections 11.12–11.13.

Figure 11.7
Oxidation reactions of alkenes, alkynes, and alcohols

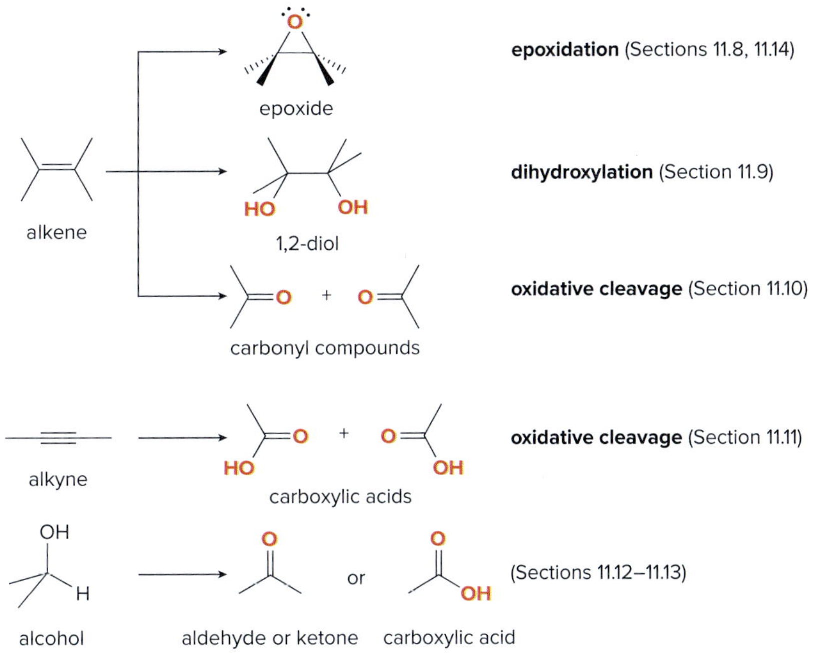

11.8 Epoxidation

Epoxidation is the addition of a single oxygen atom to an alkene to form an **epoxide.**

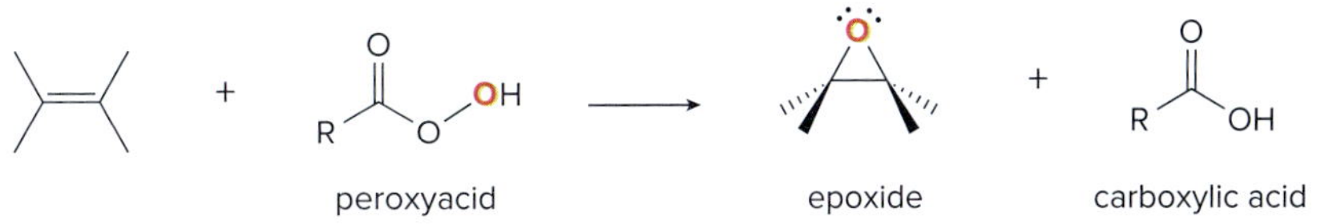

The weak π bond of the alkene is broken and two new C–O σ bonds are formed. Epoxidation is typically carried out with a peroxyacid, resulting in cleavage of the weak O–O bond of the reagent.

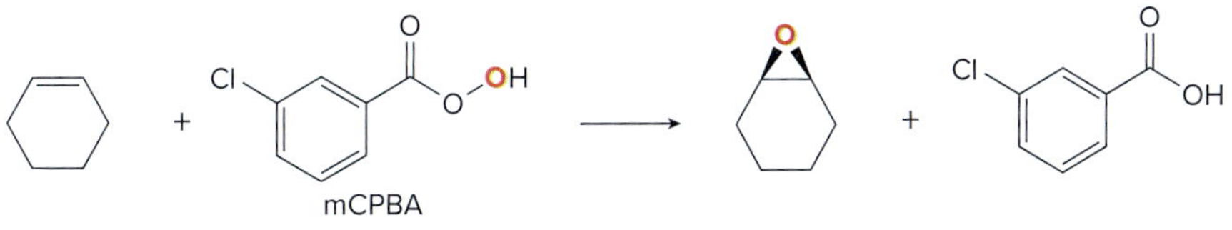

Epoxidation occurs via the **concerted addition** of one oxygen atom of the peroxyacid to the π bond as shown in Mechanism 11.4. Epoxidation resembles the formation of the bridged halonium ion in Section 10.13, in that two bonds in a three-membered ring are formed in one step.

Mechanism 11.4 Epoxidation of an Alkene with a Peroxyacid

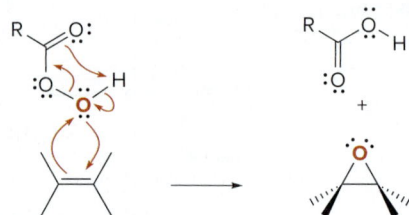

- **All bonds are broken and formed in a single step.** The two epoxide C–O bonds are formed from one electron pair of the π bond and one lone pair of the peroxyacid. The **weak O–O bond is broken.**

Problem 11.15 What epoxide is formed when each alkene is treated with mCPBA?

a. b. c.

11.8A The Stereochemistry of Epoxidation

Epoxidation occurs via **syn addition** of an O atom from either side of the planar double bond, so that both C–O bonds are formed on the same side. The relative position of substituents in the alkene reactant is **retained** in the epoxide product.

- A cis alkene gives an epoxide with cis substituents. A trans alkene gives an epoxide with trans substituents.

Epoxidation is a **stereospecific** reaction because cis and trans alkenes yield different stereoisomers as products, as illustrated in Sample Problem 11.2.

Sample Problem 11.2 Drawing the Stereoisomers Formed in Epoxidation

Draw the stereoisomers formed when cis- and trans-but-2-ene are epoxidized with mCPBA.

Solution

To draw each product of epoxidation, add an O atom from either side of the alkene, and keep all substituents in their *original* orientations. The **cis** methyl groups in *cis*-but-2-ene become **cis** substituents in the epoxide. Addition of an O atom from either side of the trigonal planar alkene leads to the same compound—an **achiral meso compound that contains two stereogenic centers,** labeled in blue.

11.8 Epoxidation

Epoxidation of *cis*- and *trans*-but-2-ene illustrates the general rule about the stereochemistry of reactions: **an achiral starting material gives achiral or racemic products.**

The **trans** methyl groups in *trans*-but-2-ene become **trans** substituents in the epoxide. Addition of an O atom from either side of the trigonal planar alkene yields an equal mixture of two enantiomers—a **racemic mixture**—with two stereogenic centers labeled in blue.

$$\underset{\text{trans-but-2-ene}}{\underset{\text{trans CH}_3 \text{ groups}}{\text{CH}_3\text{CH}=\text{CHCH}_3}} \xrightarrow{\text{mCPBA}} \underset{\substack{\text{trans CH}_3 \text{ groups} \\ \text{O added from above}}}{} + \underset{\substack{\text{trans CH}_3 \text{ groups} \\ \text{O added from below}}}{}$$

Products are **enantiomers.**

Problem 11.16 Draw all stereoisomers formed when each alkene is treated with mCPBA.

a. (propene)   b. (cis-pent-2-ene)   c. (1-methylcyclohexene)

More Practice: Try Problems 11.29b; 11.35d, k; 11.36b.

11.8B The Synthesis of Disparlure

In 1869, the gypsy moth was introduced into New England in an attempt to develop a silk industry. Some moths escaped into the wild and the population flourished. Mature gypsy moth caterpillars eat an average of one square foot of leaf surface per day, defoliating shade trees and entire forests. Many trees die after a single defoliation.
Source: USDA APHIS PPQ, Bugwood.org

Disparlure, the sex pheromone of the female gypsy moth, is synthesized by a stepwise reaction sequence that uses an epoxidation reaction as the final step. Retrosynthetic analysis of disparlure illustrates three key operations:

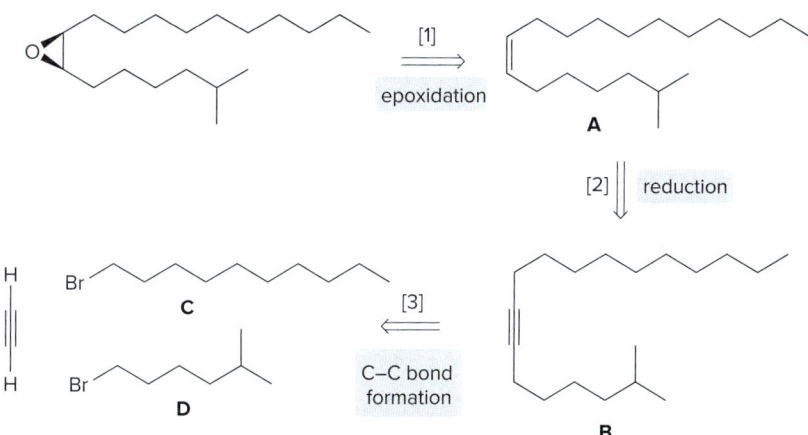

- **Step [1]** The cis epoxide in disparlure is prepared from a cis alkene **A** by epoxidation.
- **Step [2]** **A** is prepared from an internal alkyne **B** by reduction.
- **Step [3]** **B** is prepared from acetylene and two 1° alkyl halides (**C** and **D**) by using S_N2 reactions with acetylide anions.

Figure 11.8 illustrates the synthesis of disparlure beginning with acetylene. The synthesis is conceptually divided into three parts:

- **Part [1]** Acetylene is converted to an internal alkyne **B** by forming two C—C bonds. Each bond is formed by treating an alkyne with base ($NaNH_2$) to form an acetylide anion, which reacts with an alkyl halide (**C** or **D**) in an S_N2 reaction (Section 10.20A).
- **Part [2]** The internal alkyne **B** is reduced to a cis alkene **A** by syn addition of H_2 using the Lindlar catalyst (Section 11.5B).
- **Part [3]** The cis alkene **A** is epoxidized to disparlure using a peroxyacid such as mCPBA.

Figure 11.8 The synthesis of disparlure

Part [1] Formation of two C–C bonds using acetylide anions (Section 10.20A)

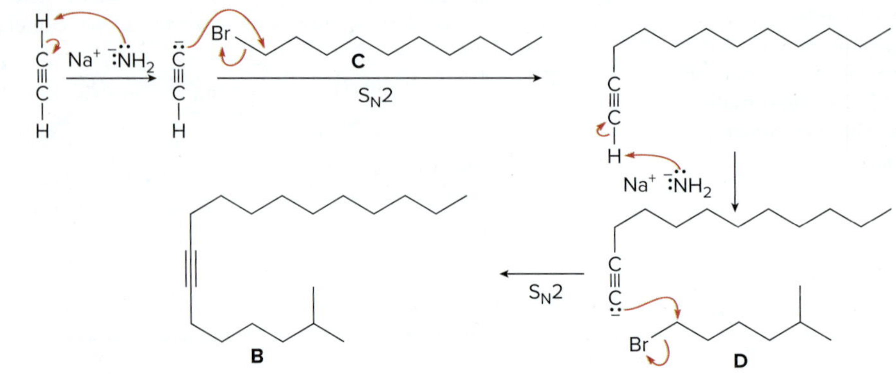

Part [2] Reduction of alkyne **B** to form cis alkene **A** (Section 11.5B)

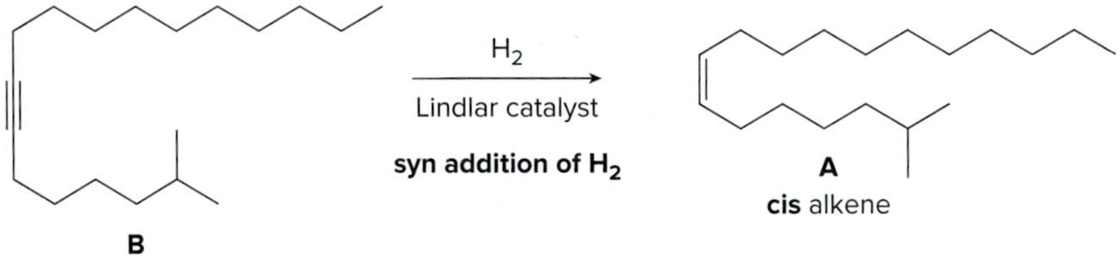

Part [3] Epoxidation of **A** to form disparlure (Section 11.8)

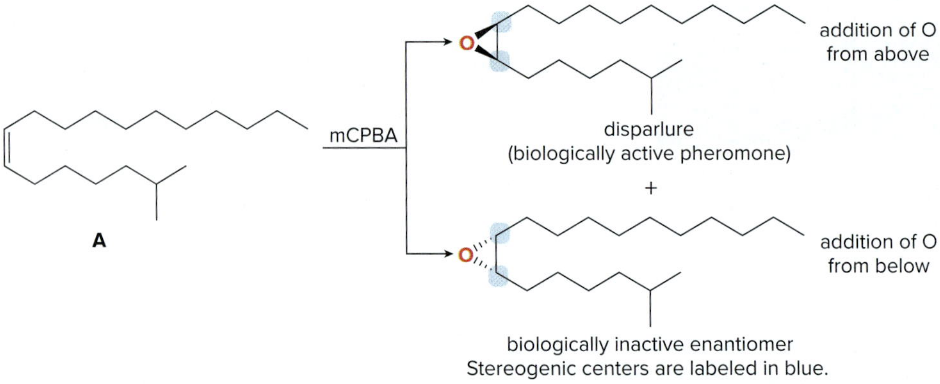

- Disparlure has been used to control the spread of the gypsy moth caterpillar, a pest that has periodically devastated forests in the northeastern United States by defoliating many shade and fruit-bearing trees. The active pheromone is placed in a trap containing a poison or sticky substance, and the male moth is lured to the trap by the pheromone. Alternatively, thousands of disparlure-baited traps are placed along the edges of infestation. When the pheromone permeates the air, males are confused and can't locate individual females, so that mating is disrupted. Such a species-specific method presents a way of controlling an insect population that avoids the widespread use of harmful, nonspecific pesticides.

How to separate a racemic mixture into its component enantiomers is discussed in Section 23.2.

Epoxidation of the cis alkene **A** from two different sides of the double bond affords two cis epoxides in the last step—a racemic mixture of two enantiomers. Thus, half of the product is the desired pheromone disparlure, but the other half is its biologically inactive enantiomer. Separating the desired from the undesired enantiomer is difficult and expensive, because both compounds have identical physical properties. A reaction that affords a chiral epoxide from an achiral precursor without forming a racemic mixture is discussed in Section 11.14.

11.9 Dihydroxylation

Dihydroxylation is the addition of two hydroxy groups to a double bond, forming a **1,2-diol** or **glycol.** Depending on the reagent, the two new OH groups can be added to the opposite sides (**anti** addition) or the same side (**syn** addition) of the double bond.

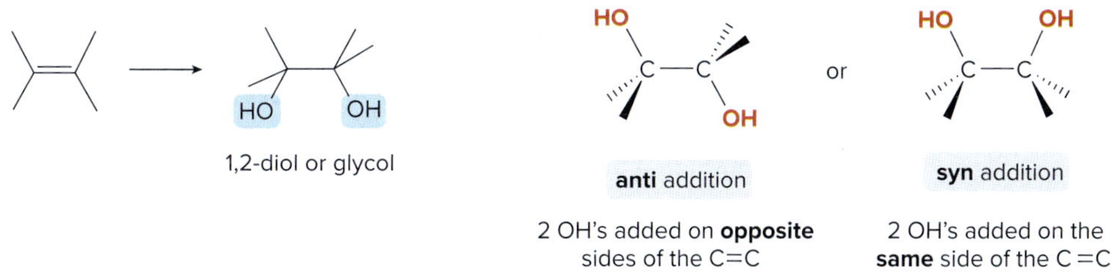

1,2-diol or glycol

anti addition
2 OH's added on **opposite** sides of the C=C

or

syn addition
2 OH's added on the **same** side of the C=C

11.9A Anti Dihydroxylation

Anti dihydroxylation is achieved in two steps—epoxidation followed by opening of the ring with ⁻OH or H_2O. Cyclohexene, for example, is converted to a racemic mixture of two *trans*-cyclohexane-1,2-diols by anti addition of two OH groups.

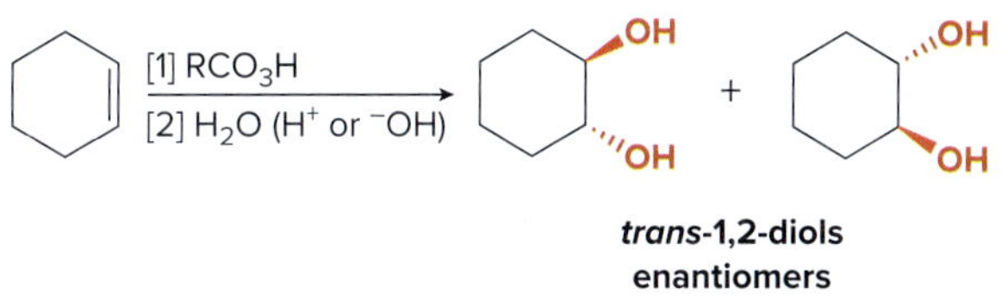

[1] RCO_3H
[2] H_2O (H^+ or ⁻OH)

trans-1,2-diols
enantiomers

The stereochemistry of the products can be understood by examining the stereochemistry of each step.

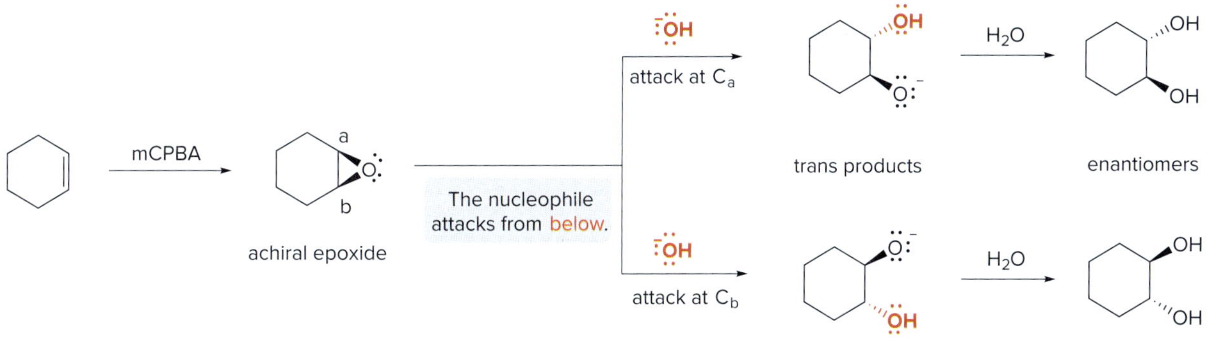

mCPBA

achiral epoxide

The nucleophile attacks from below.

attack at C_a

attack at C_b

trans products

enantiomers

Epoxidation of cyclohexene adds an O atom from either above or below the plane of the double bond to form a single **achiral epoxide,** so only one representation is shown. Opening of the epoxide ring then occurs with **backside attack at either C—O bond.** Because the epoxide is drawn above the plane of the six-membered ring, nucleophilic attack occurs from **below** the plane. This reaction is a specific example of the opening of epoxide rings with strong nucleophiles, first presented in Section 9.16A.

Because one OH group of the 1,2-diol comes from the epoxide and one OH group comes from the nucleophile (⁻OH), the overall result is **anti addition of two OH groups** to an alkene.

Problem 11.17 Draw the products formed when both *cis*- and *trans*-but-2-ene are treated with a peroxyacid followed by ⁻OH (in H_2O). Explain how these reactions illustrate that anti dihydroxylation is stereospecific.

11.9B Syn Dihydroxylation

Syn dihydroxylation results when an alkene is treated with either **KMnO₄** or **OsO₄**.

$$\text{cyclohexene} \xrightarrow[\text{H}_2\text{O, HO}^-]{\text{KMnO}_4} \text{\textit{cis}-cyclohexane-1,2-diol}$$

$$\text{cyclopentene} \xrightarrow[\text{[2] NaHSO}_3, \text{H}_2\text{O}]{\text{[1] OsO}_4} \text{\textit{cis}-cyclopentane-1,2-diol}$$

Each reagent adds two oxygen atoms to the same side of the double bond—that is, in a **syn** fashion—to yield a cyclic intermediate. Hydrolysis of the cyclic intermediate cleaves the metal–oxygen bonds, forming the *cis*-1,2-diol. With OsO₄, sodium bisulfite (NaHSO₃) is also added in the hydrolysis step.

Two O atoms are added to the **same** side of the C=C.

[Mechanism diagrams showing syn addition of MnO₄⁻ and OsO₄ to cyclohexene, followed by hydrolysis with H₂O or NaHSO₃/H₂O, to give *cis*-1,2-diol.]

Although KMnO₄ is inexpensive and readily available, its use is limited by its insolubility in organic solvents. To prevent further oxidation of the product 1,2-diol, the reaction mixture must be kept basic with added ⁻OH.

Although OsO₄ is a more selective oxidant than KMnO₄ and is soluble in organic solvents, it is toxic and expensive. To overcome these limitations, dihydroxylation can be carried out by using a *catalytic* amount of OsO₄, if the oxidant **N-methylmorpholine N-oxide (NMO)** is also added.

> NMO is an **amine oxide.** It is not possible to draw a Lewis structure of an amine oxide having only neutral atoms.
>
> $$R-\overset{R}{\underset{R}{\overset{|}{N^+}}}-\overset{..}{\underset{..}{O}}:^-$$
>
> amine oxide

N-methylmorpholine *N*-oxide
NMO

In the catalytic process, dihydroxylation of the double bond converts the Os^{8+} oxidant into an Os^{6+} product, which is then re-oxidized by NMO to Os^{8+}. This Os^{8+} reagent can then be used for dihydroxylation once again, and the catalytic cycle continues.

$$\text{alkene} + \text{Os}^{8+} \text{ oxidant} \xrightarrow{\text{catalyst}} \text{diol} + \text{Os}^{6+} \text{ product}$$

NMO oxidizes the **Os^{6+} product** back to **Os^{8+}** to begin the cycle again.

Problem 11.18 Draw the products formed when both *cis-* and *trans-*but-2-ene are treated with OsO$_4$, followed by hydrolysis with NaHSO$_3$ + H$_2$O. Explain how these reactions illustrate that syn dihydroxylation is stereospecific.

11.10 Oxidative Cleavage of Alkenes

Oxidative cleavage of an alkene breaks both the σ and π bonds of the double bond to form two carbonyl groups. Depending on the number of R groups bonded to the double bond, oxidative cleavage yields either **ketones** or **aldehydes**.

Lightning produces O$_3$ from O$_2$ during an electrical storm. Moreover, the pungent odor around a heavily used photocopy machine is O$_3$ produced from O$_2$ during the process. O$_3$ at ground level is an unwanted atmospheric pollutant. In the stratosphere, however, it protects us from harmful ultraviolet radiation, as discussed in Chapter 21.
Balazs Kovacs/Getty Images

One method of oxidative cleavage relies on a two-step procedure using **ozone (O$_3$) as the oxidant** in the first step. Cleavage with ozone is called **ozonolysis**.

Addition of ozone to the π bond of the alkene forms an unstable intermediate called a **molozonide**, which then rearranges to an **ozonide** by a stepwise process. The unstable ozonide is then reduced without isolation to afford carbonyl compounds. **Zn (in H$_2$O)** and **dimethyl sulfide (CH$_3$SCH$_3$)** are two common reagents used to convert the ozonide to carbonyl compounds.

To draw the product of any oxidative cleavage:

- Locate all π bonds in the molecule.
- Replace each C=C by two C=O bonds.

Sample Problem 11.3 Drawing the Oxidative Cleavage Products from an Alkene

Draw the products when each alkene is treated with O$_3$ followed by CH$_3$SCH$_3$.

Solution

a. Cleave the double bond and replace it with two carbonyl groups.

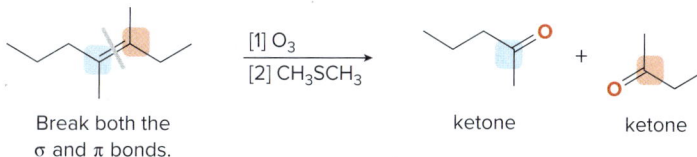

454   Chapter 11   Oxidation and Reduction

b. For a cycloalkene, oxidative cleavage results in a **single molecule with two carbonyl groups—a dicarbonyl compound.**

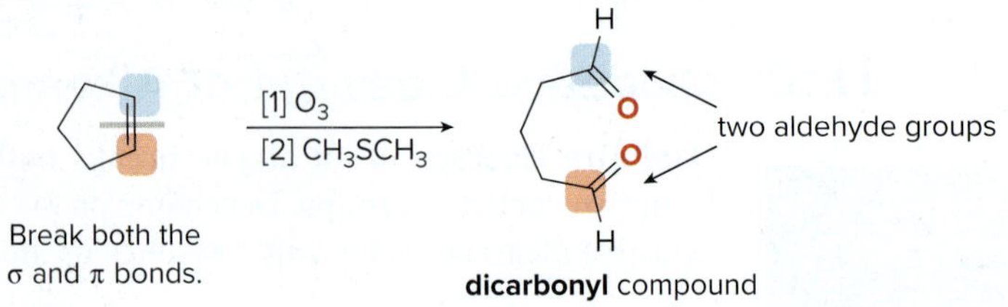

[cyclopentene] $\xrightarrow{\text{[1] O}_3}{\text{[2] CH}_3\text{SCH}_3}$ **dicarbonyl** compound (two aldehyde groups)

Break both the σ and π bonds.

Problem 11.19   Draw the products formed when each alkene is treated with O_3 followed by Zn, H_2O.

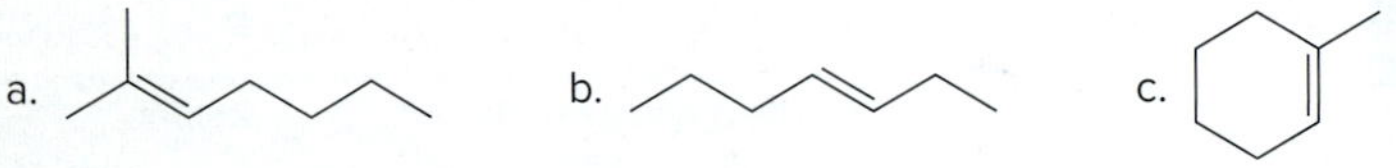

a. [alkene structure]   b. [alkene structure]   c. [cyclohexene with methyl]

More Practice: Try Problems 11.35i; 11.37c; 11.44a, b; 11.47.

Ozonolysis of dienes (and other polyenes) results in oxidative cleavage of all C=C bonds. The number of carbonyl groups formed in the products is *twice* the number of double bonds in the starting material. The *two* double bonds in limonene are converted to products containing *four* carbonyl groups.

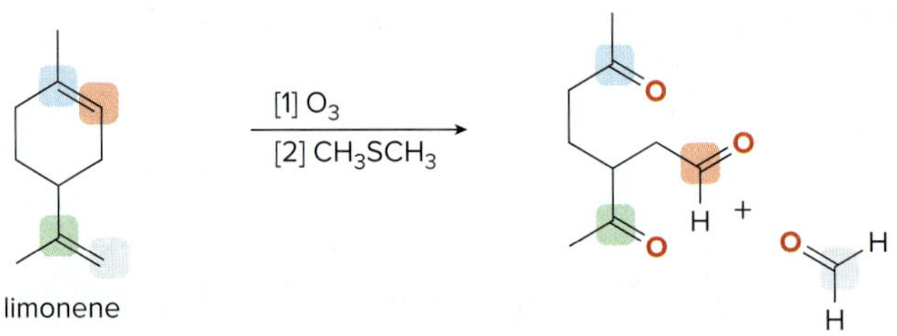

limonene $\xrightarrow{\text{[1] O}_3}{\text{[2] CH}_3\text{SCH}_3}$ [products]

Oxidative cleavage is a valuable tool for structure determination of unknown compounds. The ability to determine what alkene gives rise to a particular set of oxidative cleavage products is thus a useful skill, illustrated in Sample Problem 11.4.

Sample Problem 11.4   Determining the Alkene That Forms a Set of Oxidative Cleavage Products

What alkene forms the following products after reaction with O_3 followed by CH_3SCH_3?

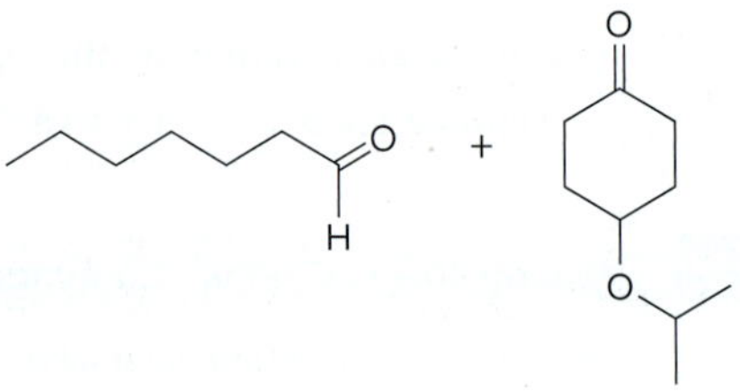

[aldehyde] + [cyclohexanone with isopropoxy]

Solution

To draw the starting material, **ignore the O atoms** in the carbonyl groups and **join the carbonyl carbons together by a C=C.**

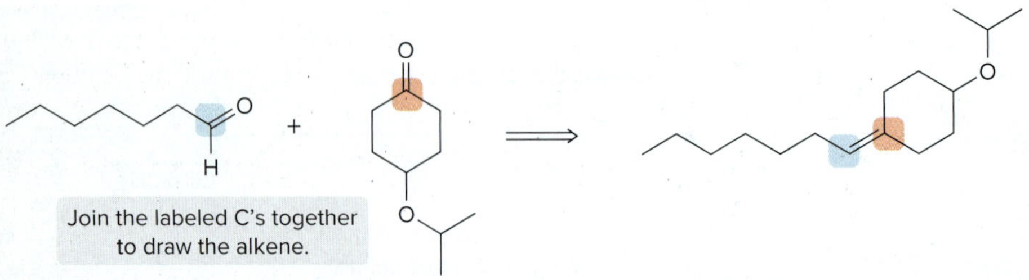

[aldehyde] + [cyclohexanone with isopropoxy] ⟹ [alkene product]

Join the labeled C's together to draw the alkene.

Problem 11.20
What alkene yields each set of oxidative cleavage products?

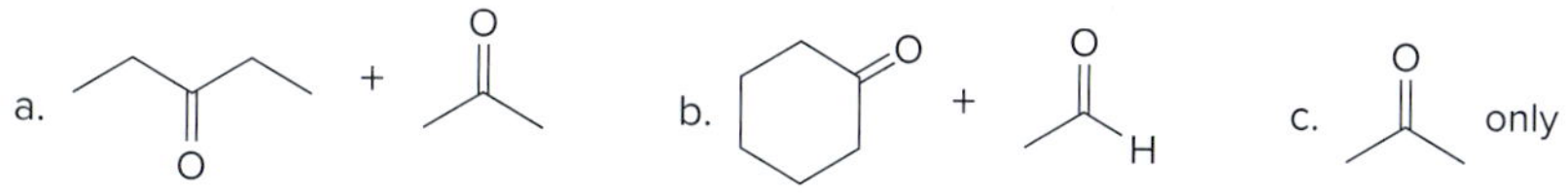

a. [ethyl ketone] + [acetone-like dione] b. [cyclohexanone] + [acetaldehyde] c. [acetone] only

More Practice: Try Problems 11.45a, b; 11.46.

Problem 11.21
Draw the products formed when cembrene A is treated with O_3 followed by CH_3SCH_3. Label each product as chiral or achiral.

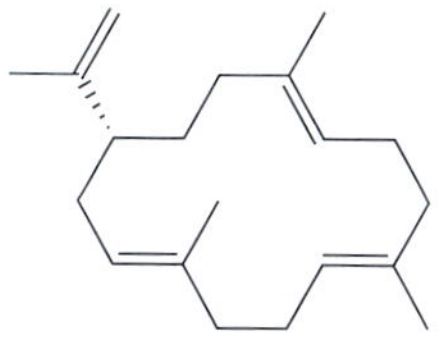

cembrene A

Cembrene A (Problem 11.21) is isolated from soft corals of the genus *Naphthea*.
Magnusdeepbelow/Shutterstock

11.11 Oxidative Cleavage of Alkynes

Alkynes also undergo oxidative cleavage of the σ bond and both π bonds of the triple bond. Internal alkynes are oxidized to **carboxylic acids (RCOOH)**, whereas terminal alkynes afford carboxylic acids and CO_2 from the *sp* hybridized C—H bond.

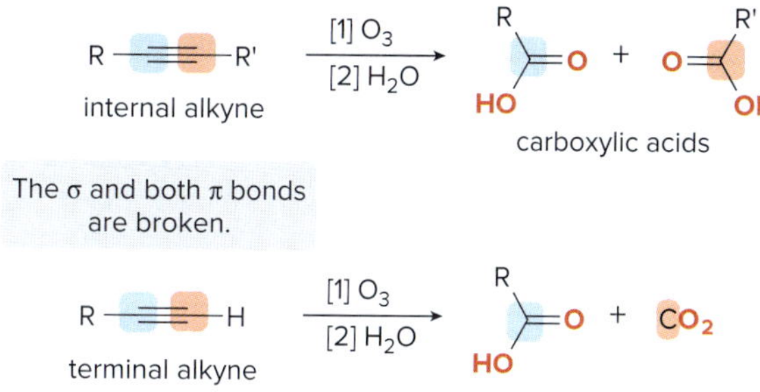

$$R-\equiv-R' \xrightarrow[\text{[2] H}_2\text{O}]{\text{[1] O}_3} \underset{\text{HO}}{R}\!\!\!>\!\!=\!\!O + O\!\!=\!\!<\!\!\!\underset{\text{OH}}{R'}$$

internal alkyne → carboxylic acids

The σ and both π bonds are broken.

$$R-\equiv-H \xrightarrow[\text{[2] H}_2\text{O}]{\text{[1] O}_3} \underset{\text{HO}}{R}\!\!\!>\!\!=\!\!O + CO_2$$

terminal alkyne

Oxidative cleavage is commonly carried out with O_3, followed by cleavage of the intermediate ozonide with H_2O.

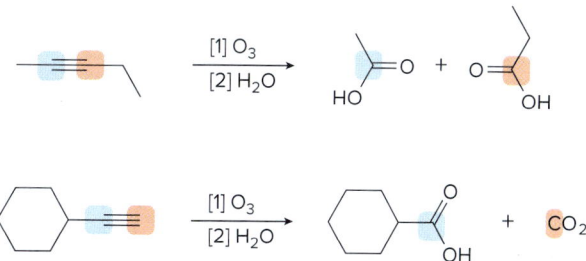

Problem 11.22
Draw the products formed when each alkyne is treated with O_3 followed by H_2O.

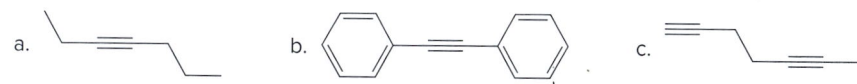

Problem 11.23 What alkyne (or diyne) yields each set of oxidative cleavage products?

a. CO_2 + CH₃(CH₂)₇COOH

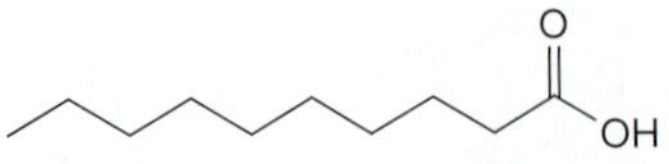

b. 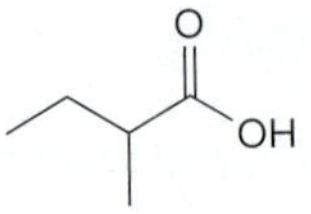 (CH₃CH₂)(CH₃)CHCOOH only

c. 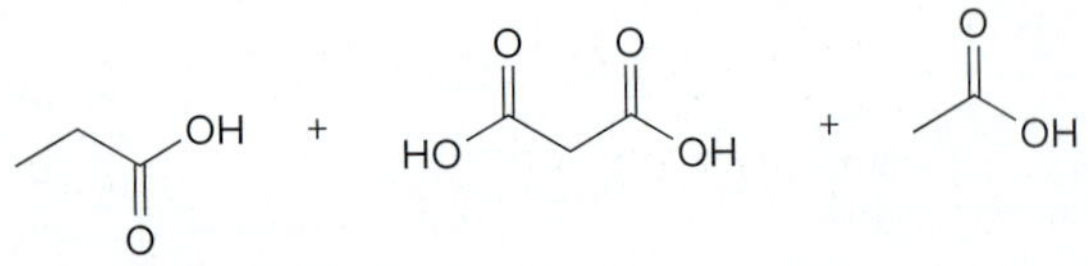 CH₃CH₂COOH + HOOC-CH₂-COOH + CH₃COOH

d. 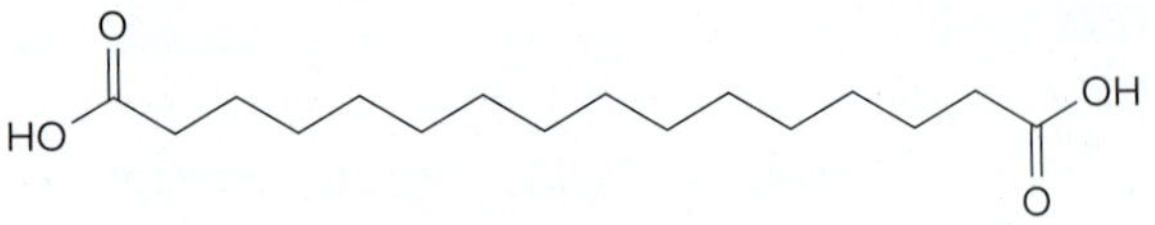 HOOC-(CH₂)₁₂-COOH

11.12 Oxidation of Alcohols

Alcohols are oxidized to a variety of carbonyl compounds, depending on the type of alcohol and reagent. Oxidation occurs by replacing the C–H bonds *on the carbon bearing the OH group* by C–O bonds.

- **1° Alcohols** are oxidized to either **aldehydes** or **carboxylic acids** by replacing either one or two C–H bonds by C–O bonds.

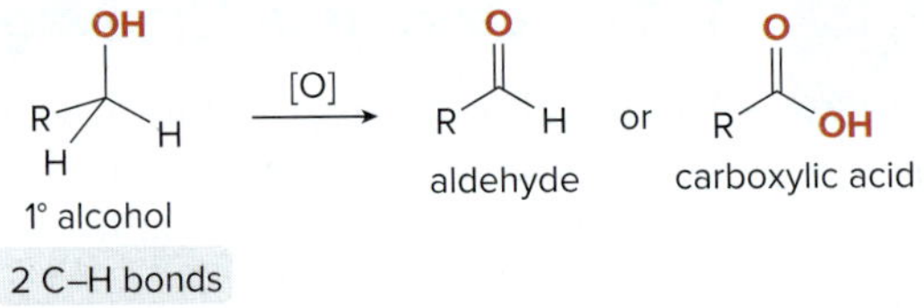

  1° alcohol (2 C–H bonds) → aldehyde or carboxylic acid

- **2° Alcohols** are oxidized to **ketones** by replacing the one C–H bond by a C–O bond.

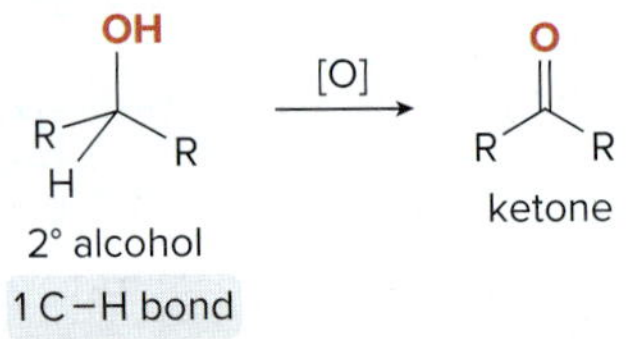

  2° alcohol (1 C–H bond) → ketone

- **3° Alcohols** have no H atoms on the carbon with the OH group, so they are *not* easily oxidized.

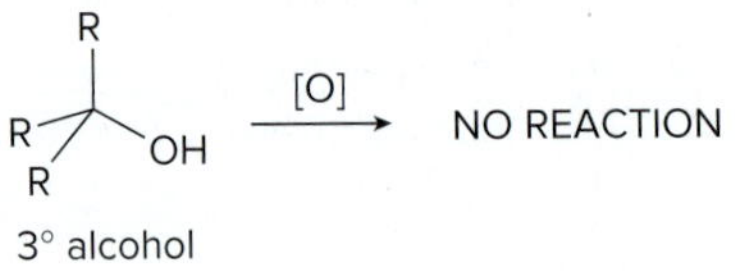

  3° alcohol → NO REACTION

Alcohol oxidations often occur by a pathway that involves bonding a leaving group Z to the oxygen, where Z is typically a metal in a high oxidation state. Elimination with a base then forms a C=O and a metal in a lower oxidation state.

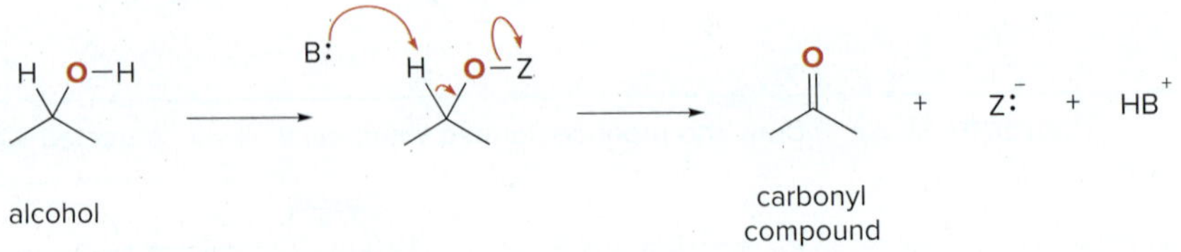

alcohol → carbonyl compound + Z^- + HB^+

The oxidation of alcohols to carbonyl compounds is typically carried out with Cr^{6+} oxidants, which are reduced to Cr^{3+} products.

- **CrO_3, $Na_2Cr_2O_7$, and $K_2Cr_2O_7$** are **strong, nonselective oxidants** used in aqueous acid ($H_2SO_4 + H_2O$).
- **PCC** (Section 11.7) is soluble in CH_2Cl_2 (dichloromethane), and can be used without strong acid present, making it a **more selective, milder oxidant.**

11.12A Oxidation of 2° Alcohols

Any of the Cr^{6+} oxidants effectively oxidizes 2° alcohols to ketones.

2° alcohol → ketone

The mechanism for alcohol oxidation has two key parts: **formation of a chromate ester** and **loss of a proton.** Mechanism 11.5 is drawn for the oxidation of a general 2° alcohol with CrO_3.

Mechanism 11.5 Oxidation of an Alcohol with CrO_3

1 – 2 Nucleophilic attack of the alcohol on the electrophilic metal (Cr^{6+} oxidation state) followed by proton transfer forms a **chromate ester.**

3 A base removes a proton and the electron pair in the C–H bond forms the **new π bond** of the C=O. Carbon is oxidized because the **number of C–O bonds increases,** and Cr^{6+} is reduced to Cr^{4+}.

These three steps convert the Cr^{6+} oxidant to a Cr^{4+} product, which is then further reduced to a Cr^{3+} product by a series of steps.

11.12B Oxidation of 1° Alcohols

1° Alcohols are oxidized to either aldehydes or carboxylic acids, depending on the reagent.

- 1° Alcohols are oxidized to aldehydes (RCHO) under mild reaction conditions—using PCC in CH_2Cl_2.
- 1° Alcohols are oxidized to carboxylic acids (RCOOH) under harsher reaction conditions: $Na_2Cr_2O_7$, $K_2Cr_2O_7$, or CrO_3 in the presence of H_2O and H_2SO_4.

1° alcohol → aldehyde

1° alcohol → carboxylic acid

The mechanism for the oxidation of 1° alcohols to aldehydes parallels the oxidation of 2° alcohols to ketones detailed in Section 11.12A. Oxidation of a 1° alcohol to a carboxylic acid requires three operations: **oxidation first to the aldehyde, reaction with water,** and then further **oxidation to the carboxylic acid,** as shown in Mechanism 11.6.

Mechanism 11.6 Oxidation of a 1° Alcohol to a Carboxylic Acid

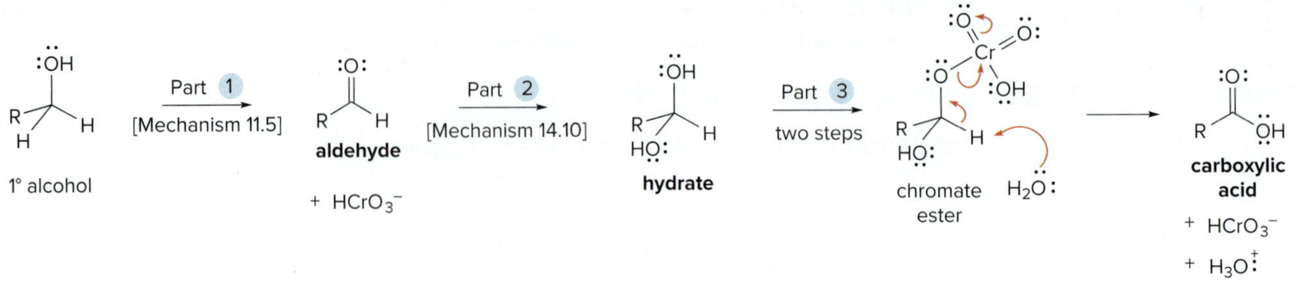

Part 1 The 1° alcohol is oxidized to an aldehyde by the three-step sequence in Mechanism 11.5.

Part 2 Water adds to the C=O to form a **hydrate,** a compound with two OH groups bonded to the same carbon, by a mechanism discussed in Section 14.14.

Part 3 Oxidation of the C–H bond of the hydrate follows Mechanism 11.5—formation of a chromate ester and loss of a proton.

Problem 11.24 Draw the organic products in each of the following reactions.

a. 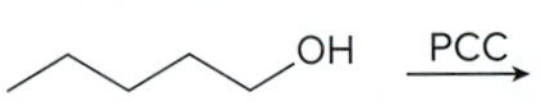 (pentanol) + PCC →

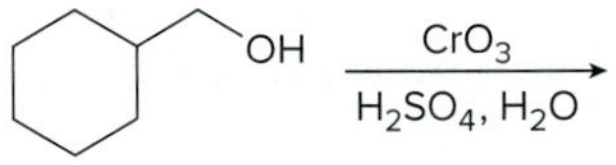

b. 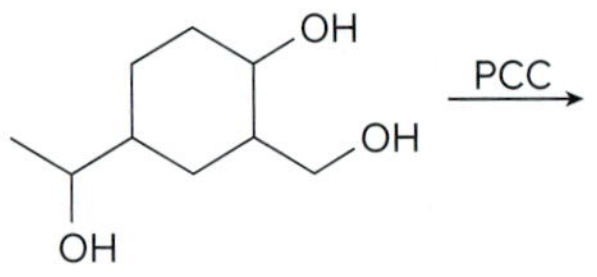 (cyclohexane with OH, CH₂OH, and CH(OH)CH₃ substituents) + PCC →

c. (cyclohexylmethanol) + CrO₃, H₂SO₄, H₂O →

d. 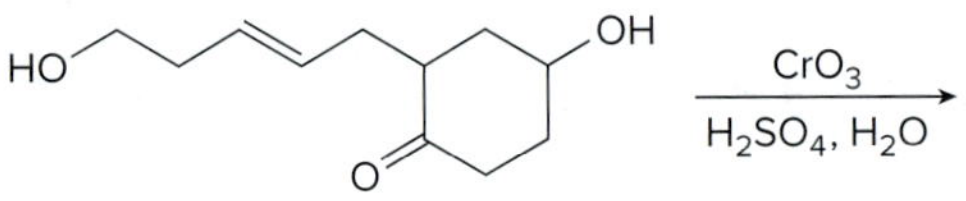 HO–CH₂–CH₂–CH=CH–CH₂–(cyclohexanone with OH substituent) + CrO₃, H₂SO₄, H₂O →

11.13 Biological Oxidation

Many reactions in biological systems involve oxidation or reduction. Instead of using Cr^{6+} reagents for oxidation, cells use two organic compounds—a high-molecular-weight **enzyme** and a simpler **coenzyme** that serves as the oxidizing agent.

Because enzyme-catalyzed reactions generally proceed with complete specificity, two identical groups on a tetrahedral carbon can be distinguished by enzymes. To understand the stereochemistry of biological oxidation and reduction, therefore, we must learn some additional concepts and terminology relating to chirality.

11.13A Prochirality

When an sp^3 hybridized carbon with two identical groups can be converted to a stereogenic center by replacement of one of those groups, the carbon is said to be **prochiral.** For example, ethanol (CH_3CH_2OH) has a **prochiral carbon** because substitution of one hydrogen of the CH_2 group by another group Z forms a stereogenic center. The two hydrogens bonded to the prochiral carbon are called **prochiral hydrogens.**

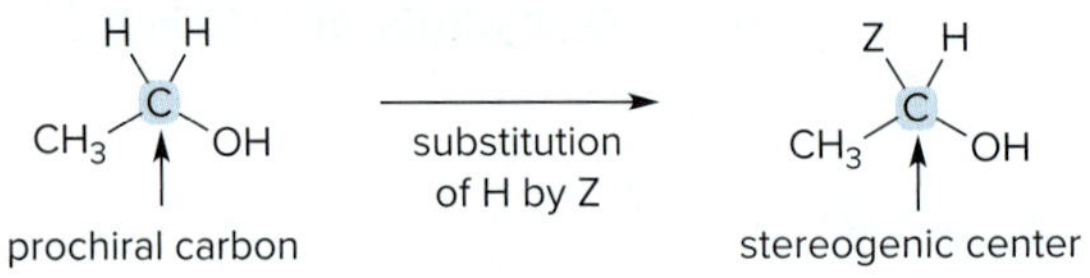

$$CH_3-\overset{H}{\underset{H}{C}}-OH \xrightarrow{\text{substitution of H by Z}} CH_3-\overset{Z}{\underset{H}{C}}-OH$$

prochiral carbon → stereogenic center

> The *R,S* system of nomenclature was first discussed in Section 5.6.

The two prochiral hydrogens are distinguished as **pro-R** and **pro-S** using the *R,S* system of nomenclature. To label the H atoms as *pro-R* or *pro-S:*

- Replace the H atom by another group (such as deuterium) that creates a stereogenic center, but does *not* alter the priority order of the other groups on the prochiral carbon.
- When replacement of H by D forms a stereogenic center with the *R* configuration, the H atom is said to be *pro-R.*
- When replacement of H by D forms a stereogenic center with the *S* configuration, the H atom is said to be *pro-S.*

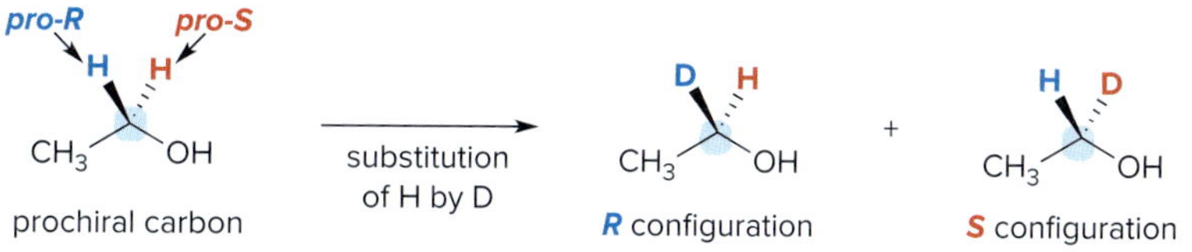

11.13 Biological Oxidation

Problem 11.25 Label the indicated hydrogens as *pro-R* or *pro-S*.

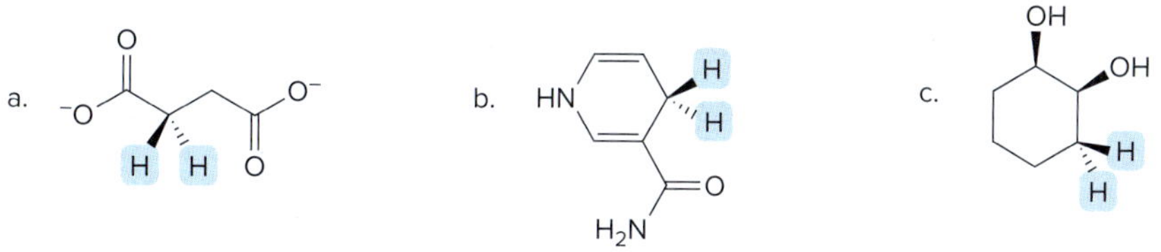

11.13B Biological Oxidation with NAD⁺

The coenzyme often used to oxidize alcohols in biological systems is **nicotinamide adenine dinucleotide,** abbreviated as **NAD⁺**. Although the structure is complex, only a portion of the molecule, drawn in red, participates in redox reactions.

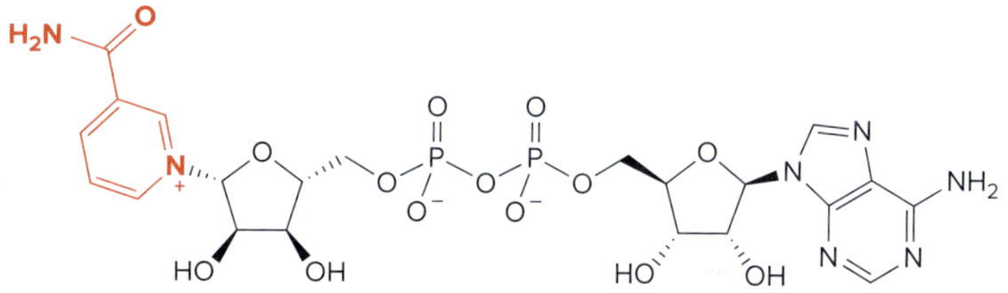

NAD⁺
biological oxidizing agent

Biological oxidation of an alcohol occurs by transferring a **hydride,** a hydrogen atom with two electrons, from the alcohol to NAD⁺ to form a carbonyl group. In the process, NAD⁺ is reduced to nicotinamide adenine dinucleotide (reduced form), abbreviated as **NADH. NADH is a biological reducing agent** that converts carbonyl compounds to alcohols, as discussed in Section 13.6. The reaction is illustrated with the oxidation of ethanol to acetaldehyde.

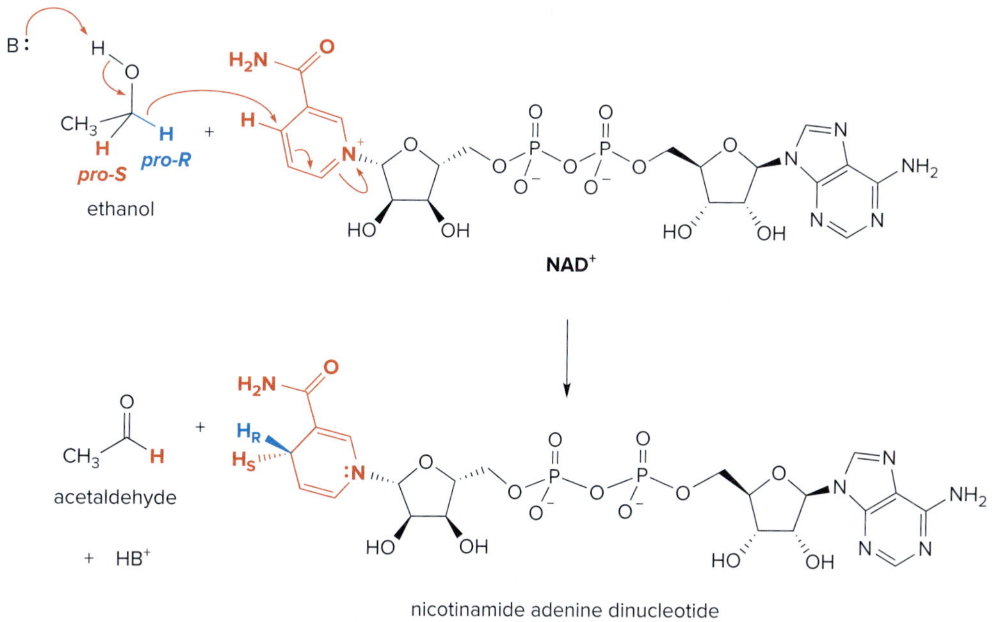

nicotinamide adenine dinucleotide
(reduced form)
NADH

460 Chapter 11 Oxidation and Reduction

Furthermore, it is the *pro-R* hydrogen of ethanol that adds exclusively to one side of the pyridinium ring of NAD⁺, forming a new prochiral carbon. This newly added hydrogen (labeled in blue) has the *pro-R* configuration. Because the oxidation occurs at the active site of a chiral enzyme, the *pro-R* and *pro-S* hydrogens of ethanol are enzymatically distinguishable, and only the *pro-R* hydrogen is removed.

Biological oxidations are the key reactions in the metabolism of ethanol. When CH_3CH_2OH is ingested, it is oxidized in the liver by NAD⁺ to CH_3CHO (acetaldehyde), and then to CH_3COO^- (acetate anion, the conjugate base of acetic acid). Acetate is the starting material for the synthesis of fatty acids and cholesterol. Both oxidations are catalyzed by a dehydrogenase enzyme.

If more ethanol is ingested than can be metabolized in a given time, the concentration of acetaldehyde builds up. This toxic compound is responsible for the feelings associated with a hangover.

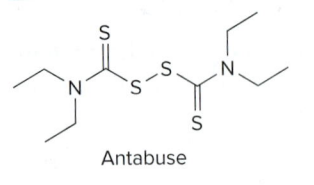
Antabuse

Antabuse, a drug given to alcoholics to prevent them from consuming alcoholic beverages, acts by interfering with the normal oxidation of ethanol. Antabuse inhibits the oxidation of acetaldehyde to the acetate anion. Because the first step in ethanol metabolism occurs but the second does not, the concentration of acetaldehyde rises, causing an individual to become violently ill.

Problem 11.26 Suppose that each of the following alcohols is oxidized with NAD⁺ in the presence of a dehydrogenase enzyme to form an aldehyde. Label the H's of the CH_2OH group as *pro-R* and *pro-S*, and draw the product that results if only the *pro-R* H atom is removed.

a. cinnamyl alcohol   b. geraniol

11.14 Sharpless Epoxidation

In all of the laboratory reactions discussed so far, an **achiral starting material has reacted with an achiral reagent to give either an achiral product or a racemic mixture of two enantiomers.** If you are trying to make a chiral product, this means that only half of the product mixture is the desired enantiomer and the other half is the undesired one. The synthesis of disparlure, outlined in Figure 11.8, exemplifies this dilemma.

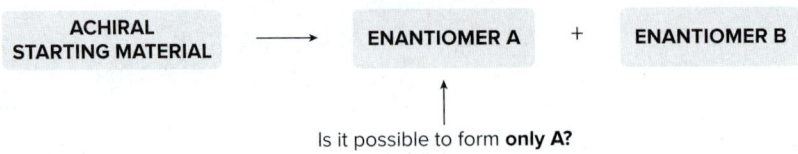

Is it possible to form **only A**?

K. Barry Sharpless shared the 2001 Nobel Prize in Chemistry for his work on chiral oxidation reactions.

K. Barry Sharpless, of The Scripps Research Institute, reasoned that using a chiral reagent might make it possible to favor the formation of one enantiomer over the other.

- An *enantioselective* reaction affords predominantly or exclusively one enantiomer.
- A reaction that converts an achiral starting material into predominantly one enantiomer is also called an *asymmetric reaction*.

The Sharpless asymmetric epoxidation is an enantioselective reaction that oxidizes alkenes to epoxides. Only the double bonds of **allylic alcohols**—that is, alcohols having a hydroxy group on the carbon adjacent to a C=C—are oxidized in this reaction.

11.14 Sharpless Epoxidation

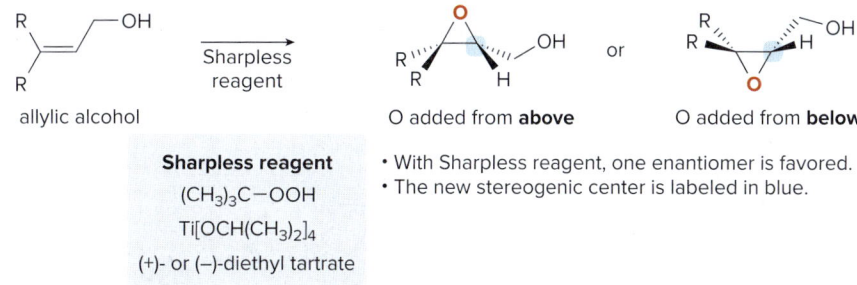

The **Sharpless reagent** consists of three components: *tert*-butyl hydroperoxide, **(CH₃)₃COOH**; a titanium catalyst—usually titanium(IV) isopropoxide, **Ti[OCH(CH₃)₂]₄**; and **diethyl tartrate (DET)**. There are two different chiral diethyl tartrate isomers, labeled as (+)-DET or (−)-DET to indicate the direction in which they rotate polarized light.

The identity of the DET isomer determines which enantiomer is the major product obtained in the epoxidation of an allylic alcohol with the Sharpless reagent.

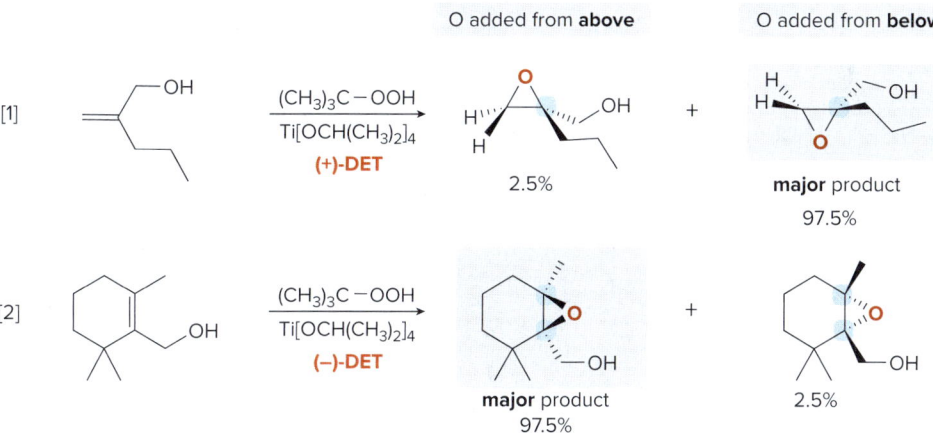

Stereogenic centers are labeled in blue.

The degree of enantioselectivity of a reaction is measured by its enantiomeric excess (**ee**) (Section 5.12D). Reactions [1] and [2] are highly enantioselective because each has an enantiomeric excess of 95% (97.5% of the major enantiomer − 2.5% of the minor enantiomer).

To determine which enantiomer is formed for a given isomer of DET, draw the allylic alcohol in a plane, with the **C=C horizontal and the OH group in the upper right corner**; then:

- Epoxidation with (−)-DET adds an oxygen atom from *above* the plane.
- Epoxidation with (+)-DET adds an oxygen atom from *below* the plane.

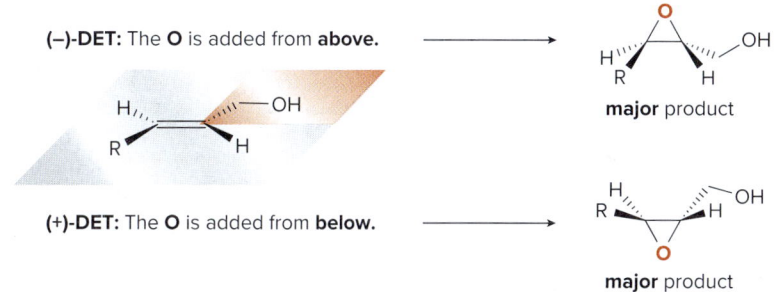

(+)-DET is prepared from (+)-(R,R)-tartaric acid [HO₂CCH(OH)CH(OH)CO₂H], a naturally occurring carboxylic acid found in grapes and sold as a by-product of the wine industry. *Jenny Cundy/Image Source*

Enantiomeric excess = **ee** = % of one enantiomer − % of the other enantiomer.

462 Chapter 11 Oxidation and Reduction

Sample Problem 11.5 Drawing the Product of a Sharpless Epoxidation

Predict the major product in each epoxidation.

a. [allylic alcohol] + (CH₃)₃C—OOH / Ti[OCH(CH₃)₂]₄ / (+)-DET

b. HO—[allylic alcohol] + (CH₃)₃C—OOH / Ti[OCH(CH₃)₂]₄ / (−)-DET

Solution

To draw an epoxidation product:

- Draw the allylic alcohol with the **C=C horizontal and the OH group in the upper right corner** of the alkene. Re-draw the alkene if necessary.
- **(+)-DET** adds the O atom from **below**, and **(−)-DET** adds the O atom from **above**.

a. Because the C=C is drawn horizontal with the OH group in the upper right corner, it is not necessary to re-draw the alkene. With **(+)-DET**, the O atom is added from **below**.

[structure with OH] + (CH₃)₃C—OOH / Ti[OCH(CH₃)₂]₄ / **(+)-DET** → [epoxide product, O below plane]

OH in the upper right corner

The O atom is added from **below** the plane.

b. The allylic alcohol must be re-drawn with the C=C horizontal and the OH group in the **upper right corner**. Because **(−)-DET** is used, the O atom is then added from **above**.

HO—[structure] → **Flip the molecule and re-draw.** → [structure with OH] + (CH₃)₃C—OOH / Ti[OCH(CH₃)₂]₄ / **(−)-DET** → [epoxide product, O above plane]

OH in the upper right corner

The O atom is added from **above** the plane.

Problem 11.27 Draw the products of each Sharpless epoxidation.

a. [structure with OH] + (CH₃)₃C—OOH / Ti[OCH(CH₃)₂]₄ / (+)-DET

b. [cyclohexene with CH₂OH] + (CH₃)₃C—OOH / Ti[OCH(CH₃)₂]₄ / (−)-DET

More Practice: Try Problems 11.29e; 11.35j; 11.36e, f; 11.51; 11.52.

The Sharpless epoxidation has been used to synthesize many chiral natural products, including (−)-frontalin, a pheromone of the western pine beetle.

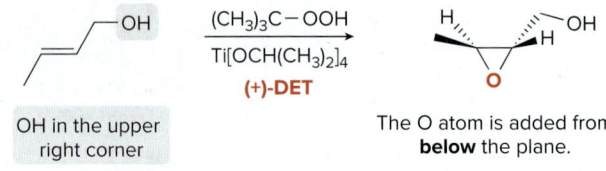

(−)-frontalin
pheromone of the western pine beetle

Problem 11.28 Explain why only one C=C of geraniol is epoxidized with the Sharpless reagent.

geraniol

Chapter 11 REVIEW

KEY CONCEPTS

Reaction Selectivity

1 Regioselective reaction (8.5)

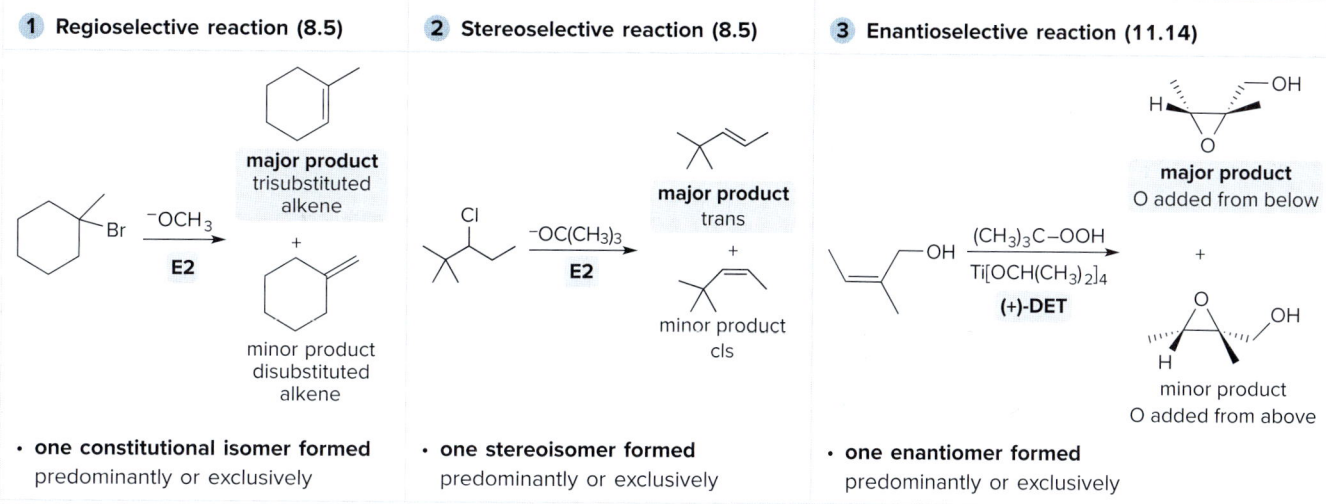

- **one constitutional isomer formed** predominantly or exclusively

2 Stereoselective reaction (8.5)

- **one stereoisomer formed** predominantly or exclusively

3 Enantioselective reaction (11.14)

- **one enantiomer formed** predominantly or exclusively

KEY REACTIONS

Reduction Reactions

[1] Reduction of alkenes

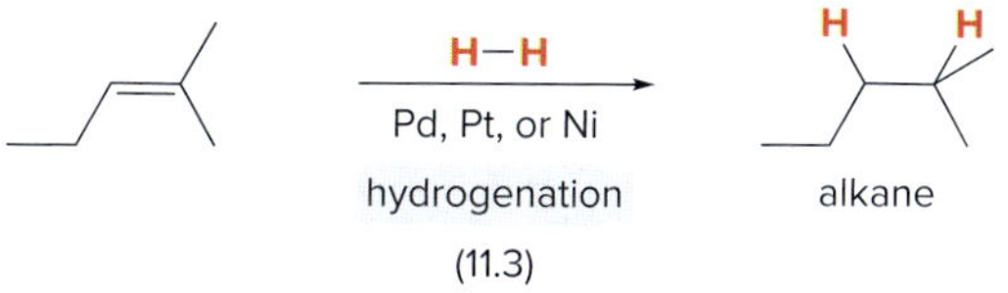

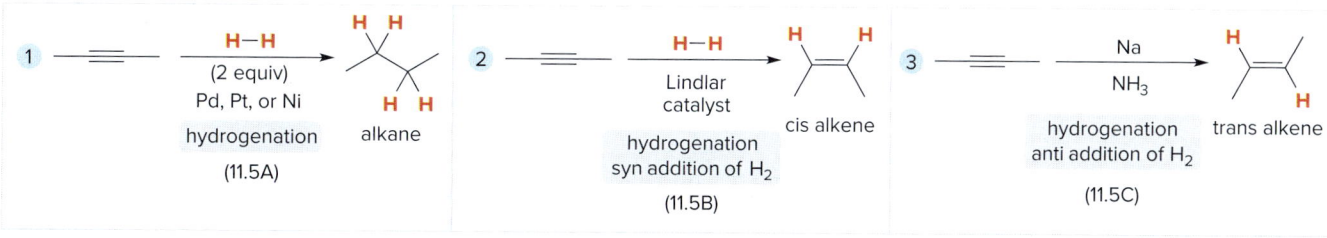

Try Problems 11.29a, 11.32, 11.35a, 11.36a.

[2] Reduction of alkynes

See Figure 11.5. Try Problem 11.38d.

[3] Reduction of alkyl halides and epoxides

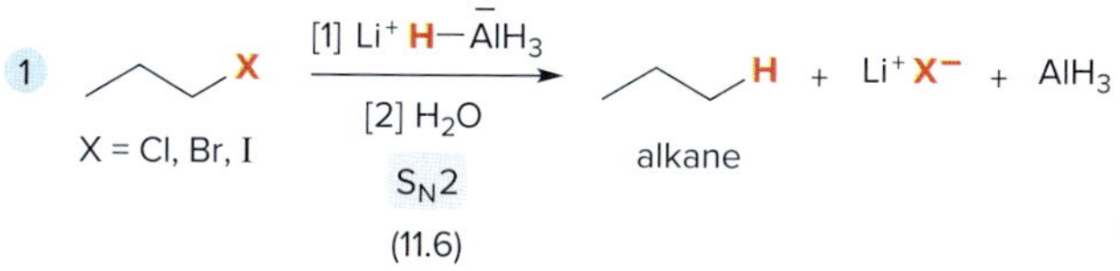

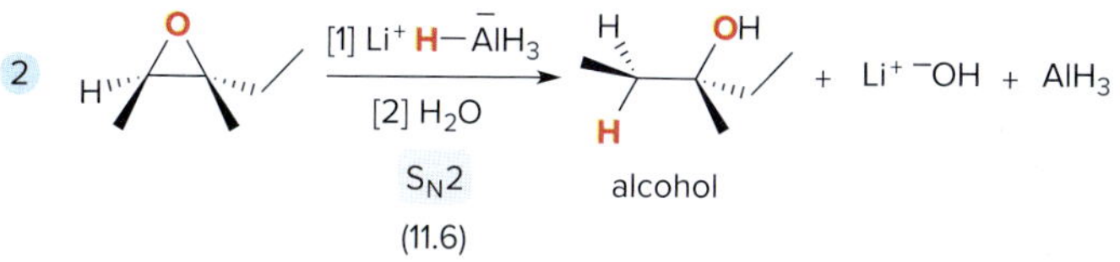

See Figure 11.6. Try Problems 11.35l; 11.36g; 11.38a, c.

Oxidation Reactions

[1] Oxidation of alkenes

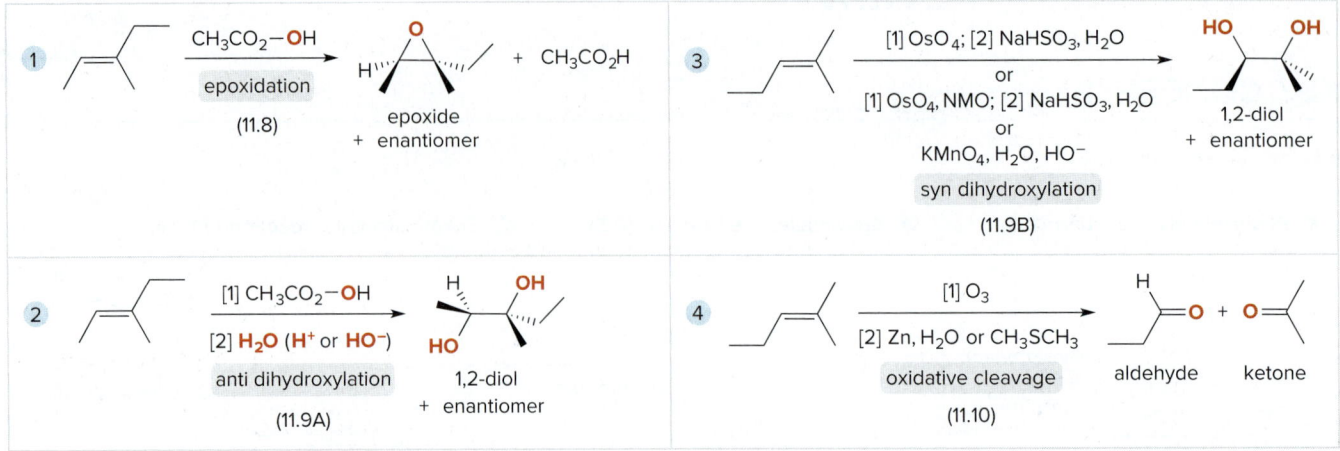

See Sample Problems 11.2, 11.3, Figure 11.7. Try Problems 11.29b, c, d; 11.35d–g, i, k; 11.36b; 11.37c; 11.38b; 11.45a, b.

[2] Oxidative cleavage of alkynes

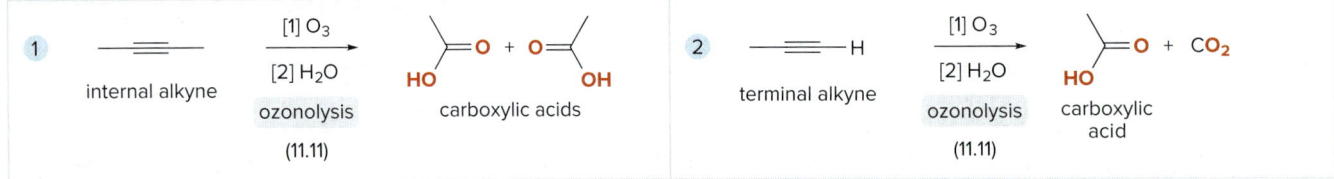

Try Problem 11.44c, d.

[3] Oxidation of alcohols

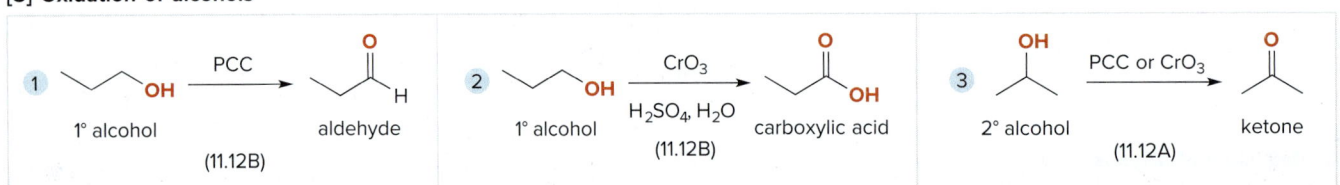

Try Problems 11.29c, d; 11.36c, d.

[4] Asymmetric epoxidation of allylic alcohols (11.14)

See Sample Problem 11.5.
Try Problems 11.29e; 11.35j; 11.36e, f; 11.51; 11.52.

KEY SKILLS

[1] Determining the number of rings and π bonds in a compound (C₁₄H₂₀) hydrogenated to a compound of molecular formula C₁₄H₂₆ (11.3C)

1 Determine the degrees of unsaturation before and after hydrogenation.

Before H₂ addition—$C_{14}H_{20}$
- The maximum number of H's possible for n C's is $2n + 2$; in this example, $2n + 2 = 2(14) + 2 = 30$.
- 30 H's (maximum) − 20 H's (actual) = 10 H's fewer than the maximum number.

$$\frac{10 \text{ H's fewer than the maximum}}{2 \text{ H's removed for each degree of unsaturation}} =$$

five degrees of unsaturation

After H₂ addition—$C_{14}H_{26}$
- The maximum number of H's possible for n C's is $2n + 2$; in this example, $2n + 2 = 2(14) + 2 = 30$.
- 30 H's (maximum) − 26 H's (actual) = 4 H's fewer than the maximum number.

$$\frac{4 \text{ H's fewer than the maximum}}{2 \text{ H's removed for each degree of unsaturation}} =$$

two degrees of unsaturation

2 Assign degrees of unsaturation.

Before H₂ addition: After H₂ addition:

five degrees of unsaturation − **two** degrees of unsaturation = **three** degrees of unsaturation that reacted with H₂

five rings or π bonds in $C_{14}H_{20}$ =   **two** rings   +   **three** π bonds   **Answer**

See Sample Problem 11.1. Try Problem 11.33.

---

[2] Drawing the stereoisomers from alkene epoxidation with mCPBA (11.8A); example: (Z)-3-methylpent-2-ene

1 Draw the starting materials.	**2** Add an O atom from above the alkene.	**3** Add an O atom from below the alkene.	**4** Determine the stereochemistry of the products.
			• Both stereogenic centers are opposite in configuration. • There is no plane of symmetry. • The compounds are enantiomers.

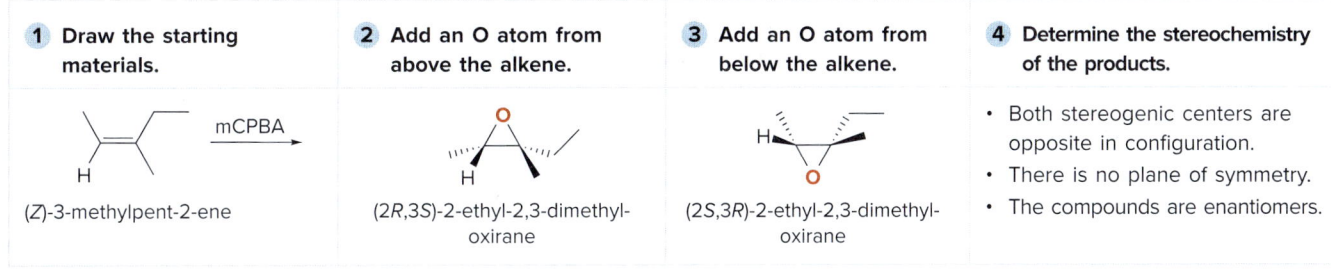

(Z)-3-methylpent-2-ene   (2R,3S)-2-ethyl-2,3-dimethyl-oxirane   (2S,3R)-2-ethyl-2,3-dimethyl-oxirane

See Sample Problem 11.2. Try Problems 11.29b; 11.35d, k; 11.36b.

---

[3] Drawing the products of an ozonolysis reaction (11.10)

1 Cleave the double bond.   **2** Replace the double bond with two carbonyl groups.

Break both the σ and π bonds.   ketone   ketone

See Sample Problem 11.3. Try Problems 11.35i; 11.37c; 11.44a, b; 11.47.

466 Chapter 11 Oxidation and Reduction

[4] Identifying an alkene from ozonolysis products (11.10)

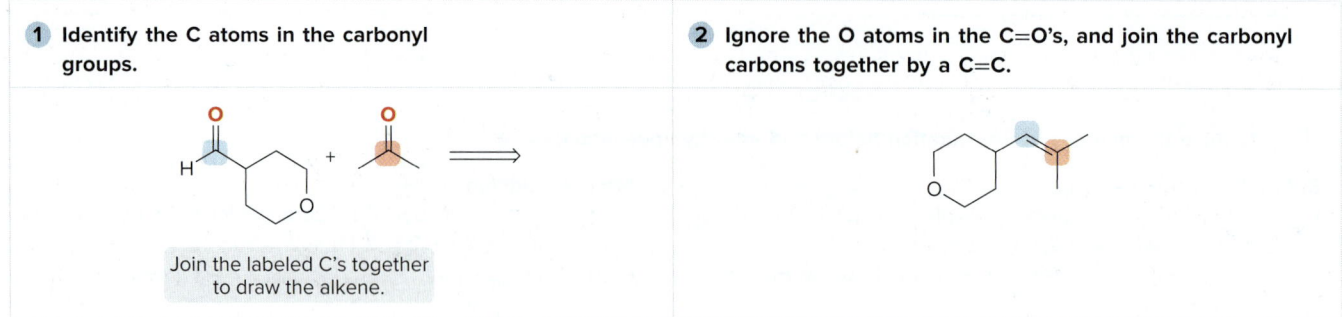

See Sample Problem 11.4. Try Problems 11.45a, b; 11.46.

[5] Predicting the product of a Sharpless reaction (11.14)

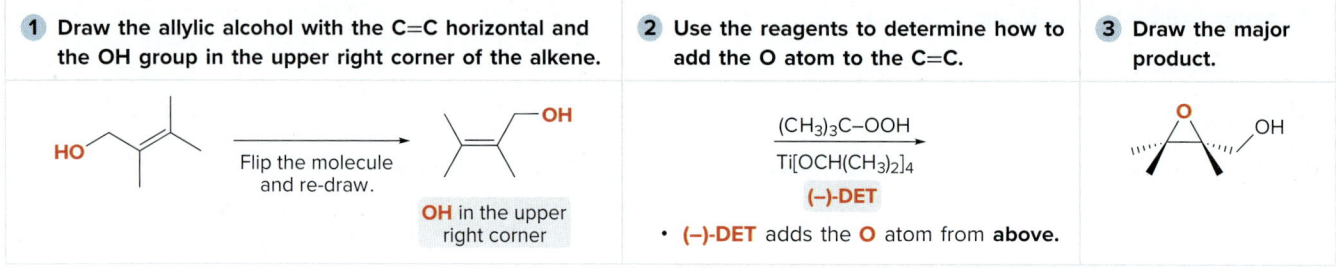

See Sample Problem 11.5. Try Problems 11.29e; 11.35j; 11.36e, f; 11.51; 11.52.

PROBLEMS

Problems Using Three-Dimensional Models

11.29 Draw the products formed when **A** is treated with each reagent: (a) H_2 + Pd-C; (b) mCPBA; (c) PCC; (d) CrO_3, H_2SO_4, H_2O; (e) Sharpless reagent with (+)-DET.

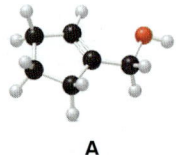

A

11.30 Devise a synthesis of the following compound from acetylene and organic compounds containing two or fewer carbons. You may use any other required reagents.

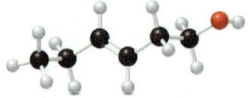

Classifying Reactions as Oxidation or Reduction

11.31 Label each reaction as oxidation, reduction, or neither.

Hydrogenation

11.32 Draw the organic products formed when each compound is treated with H₂, Pd-C. Indicate the three-dimensional structure of all stereoisomers formed.

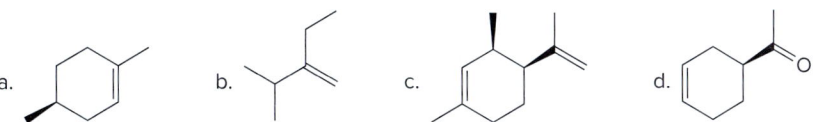

11.33 How many rings and π bonds are contained in compounds **A–C**? Draw one possible structure for each compound.
   a. Compound **A** has molecular formula C_5H_8 and is hydrogenated to a compound having molecular formula C_5H_{10}.
   b. Compound **B** has molecular formula $C_{10}H_{16}$ and is hydrogenated to a compound having molecular formula $C_{10}H_{18}$.
   c. Compound **C** has molecular formula C_8H_8 and is hydrogenated to a compound having molecular formula C_8H_{16}.

11.34 Stearidonic acid ($C_{18}H_{28}O_2$) is an unsaturated fatty acid obtained from oils isolated from hemp and blackcurrant (see also Problem 10.11).

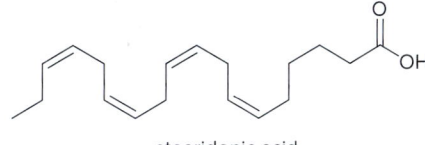

stearidonic acid

   a. What fatty acid is formed when stearidonic acid is hydrogenated with excess H₂ and a Pd catalyst?
   b. What fatty acids are formed when stearidonic acid is hydrogenated with one equivalent of H₂ and a Pd catalyst?
   c. Draw the structure of a possible product formed when stearidonic acid is hydrogenated with one equivalent of H₂ and a Pd catalyst, and one double bond is isomerized to a trans isomer.
   d. How do the melting points of the following fatty acids compare: stearidonic acid; one of the products formed in part (b); the product drawn in part (c)?

Reactions—General

11.35 Draw the organic products formed when cyclopentene is treated with each reagent. With some reagents, no reaction occurs.
   a. H₂ + Pd-C
   b. H₂ + Lindlar catalyst
   c. Na, NH₃
   d. CH₃CO₃H
   e. [1] CH₃CO₃H; [2] H₂O, HO⁻
   f. [1] OsO₄ + NMO; [2] NaHSO₃, H₂O
   g. KMnO₄, H₂O, HO⁻
   h. [1] LiAlH₄; [2] H₂O
   i. [1] O₃; [2] CH₃SCH₃
   j. (CH₃)₃COOH, Ti[OCH(CH₃)₂]₄, (–)-DET
   k. mCPBA
   l. Product in (k); then [1] LiAlH₄; [2] H₂O

11.36 Draw the organic products formed when allylic alcohol **A** is treated with each reagent.

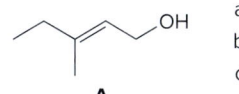

A

   a. H₂ + Pd-C
   b. mCPBA
   c. PCC
   d. CrO₃, H₂SO₄, H₂O
   e. (CH₃)₃COOH, Ti[OCH(CH₃)₂]₄, (+)-DET
   f. (CH₃)₃COOH, Ti[OCH(CH₃)₂]₄, (–)-DET
   g. [1] PBr₃; [2] LiAlH₄; [3] H₂O

11.37 For alkenes **A, B, C,** and **D**: (a) Rank **A–D** in order of increasing heat of hydrogenation; (b) rank **A–D** in order of increasing rate of reaction with H₂, Pd-C; (c) draw the products formed when each alkene is treated with ozone, followed by Zn, H₂O.

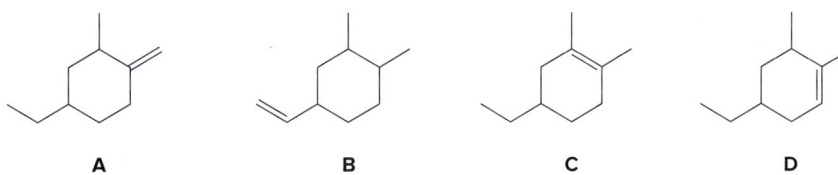

Chapter 11 Oxidation and Reduction

11.38 Draw the organic products formed in each reaction.

a. 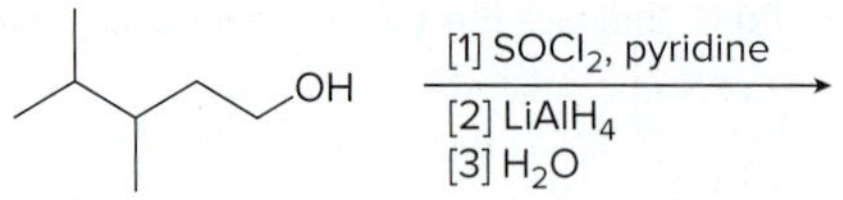 [structure: 3-methylpentan-1-ol type alcohol] → [1] SOCl₂, pyridine; [2] LiAlH₄; [3] H₂O

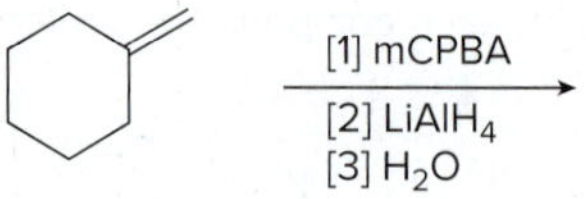

b. 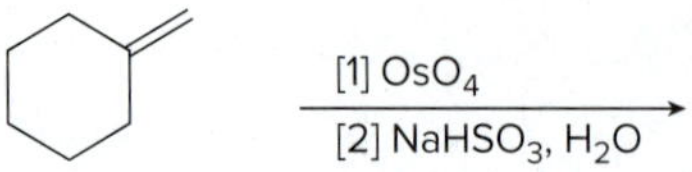 [methylenecyclohexane] → [1] OsO₄; [2] NaHSO₃, H₂O

c. [methylenecyclohexane] → [1] mCPBA; [2] LiAlH₄; [3] H₂O

d. 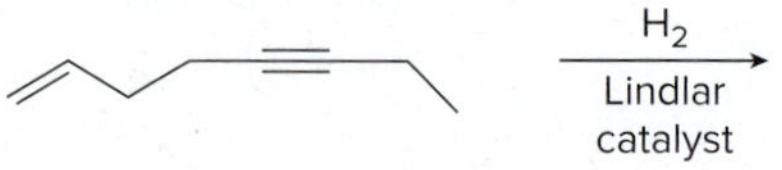 [hex-1-en-4-yne type structure] → H₂, Lindlar catalyst

11.39 One step in the degradation of fats involves the reaction of (R)-glycerol phosphate with NAD⁺ in the presence of the enzyme glycerol phosphate dehydrogenase. What products are formed if reaction occurs at only the 2° alcohol?

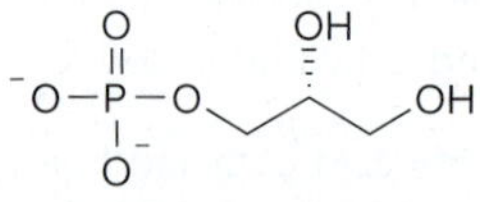

[structure of (R)-glycerol phosphate]

(R)-glycerol phosphate

11.40 Hydrogenation of alkene **A** with D₂ in the presence of Pd-C affords a single product **B**. Keeping this result in mind, what compound is formed when **A** is treated with each reagent: (a) mCPBA; (b) Br₂, H₂O followed by base? Explain these results.

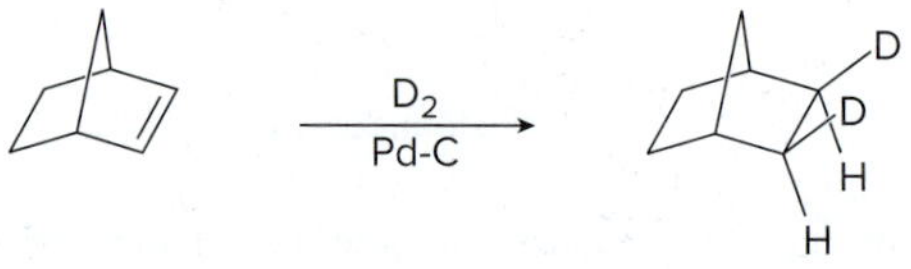

[norbornene A] → D₂, Pd-C → [norbornane with D,D,H,H labels B]

A   **B**

11.41 What alkene is needed to synthesize each 1,2-diol using [1] OsO₄ followed by NaHSO₃ in H₂O; or [2] CH₃CO₃H followed by ⁻OH in H₂O?

a. 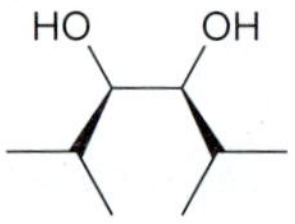 [HO, OH diol structure]

b. 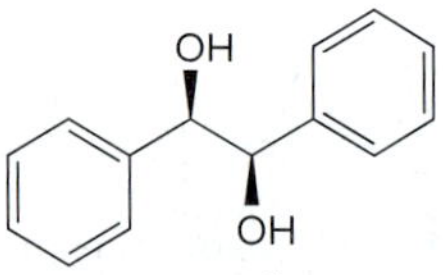 [Ph-CH(OH)-CH(OH)-Ph structure] + enantiomer

c. 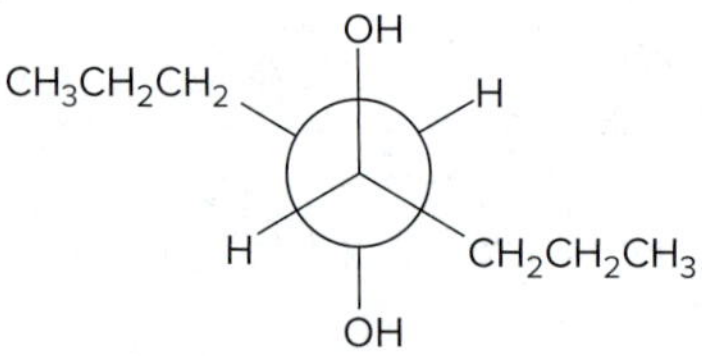 [Newman projection with OH, CH₃CH₂CH₂, H, CH₂CH₂CH₃, OH, H]

11.42 (a) What product is formed in Step [1] of the following reaction sequence? (b) Draw a mechanism for Step [2] that accounts for the observed stereochemistry. (c) What reaction conditions are necessary to form chiral **A** from prop-2-en-1-ol (CH₂=CHCH₂OH)?

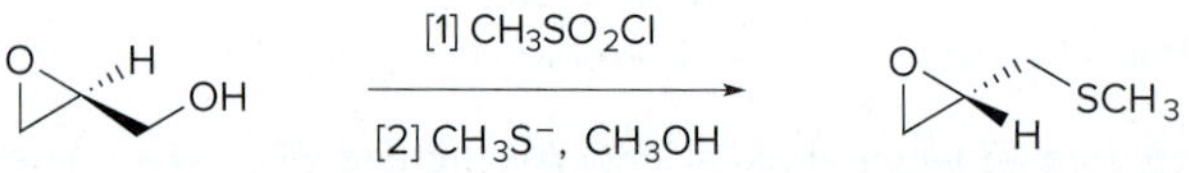

[epoxide with H, CH₂OH, A] → [1] CH₃SO₂Cl; [2] CH₃S⁻, CH₃OH → [epoxide with H, SCH₃, B]

A   **B**

11.43 Draw the products formed after Steps [1] and [2] in the following three-step sequence. Then draw stepwise mechanisms for each step.

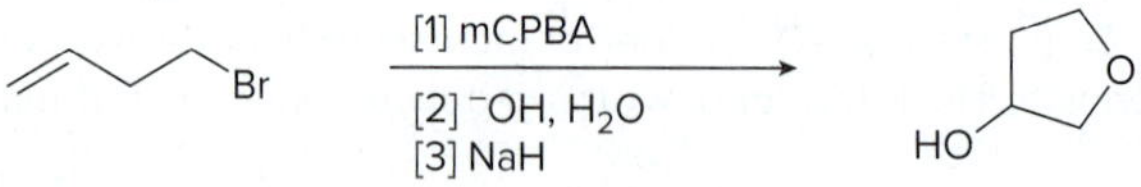

[CH₂=CHCH₂CH₂Br] → [1] mCPBA; [2] ⁻OH, H₂O; [3] NaH → [tetrahydrofuran with HO substituent]

Oxidative Cleavage

11.44 Draw the products formed in each oxidative cleavage.

a. 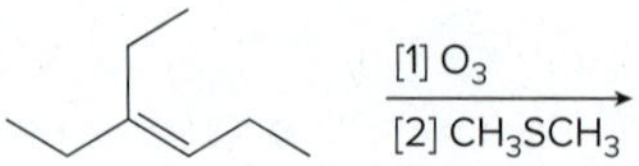 [trisubstituted alkene] → [1] O₃; [2] CH₃SCH₃

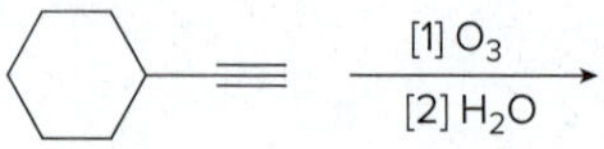

b. 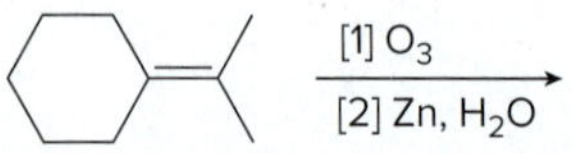 [isopropylidenecyclohexane] → [1] O₃; [2] Zn, H₂O

c. [phenylacetylene / cyclohexyl alkyne] → [1] O₃; [2] H₂O

d.  [diphenylacetylene type / dicyclohexyl alkyne] → [1] O₃; [2] H₂O

11.45 What alkene or alkyne yields each set of products after oxidative cleavage with ozone?

a. [2-methylcyclopentanone with =O] and [cyclopentanone with =O]

b. [CH₃COCH₂COCH₃] and two equivalents of CH₂=O

c. [CH₃CH₂CH₂CH₂COOH] and CO₂

d. [benzoic acid, PhCOOH] and [CH₃COOH]

11.46 Identify the starting material in each reaction.

a. $C_{10}H_{18}$ $\xrightarrow{\text{[1] O}_3}{\text{[2] CH}_3\text{SCH}_3}$ [(CH₃)₂CHCH₂C(O)CH₂CH₂CH₂CHO]

b. $C_{10}H_{16}$ $\xrightarrow{\text{[1] O}_3}{\text{[2] CH}_3\text{SCH}_3}$ [cyclodecane-1,6-dione]

11.47 Draw the products formed when each naturally occurring compound is treated with O_3 followed by Zn, H_2O.

a. squalene

b. linolenic acid

c. zingiberene

Identifying Compounds from Reactions

11.48 Identify compounds **A**, **B**, and **C**.
 a. Compound **A** has molecular formula C_8H_{12} and reacts with two equivalents of H_2. **A** gives $HCOCH_2CH_2CHO$ as the only product of oxidative cleavage with O_3 followed by CH_3SCH_3.
 b. Compound **B** has molecular formula C_6H_{10} and gives $(CH_3)_2CHCH_2CH_2CH_3$ when treated with excess H_2 in the presence of Pd. **B** reacts with $NaNH_2$ and CH_3I to form compound **C** (molecular formula C_7H_{12}).

11.49 Oximene and myrcene, two hydrocarbons isolated from alfalfa that have the molecular formula $C_{10}H_{16}$, both yield 2,6-dimethyloctane when treated with H_2 and a Pd catalyst. Ozonolysis of oximene forms $(CH_3)_2C=O$, $CH_2=O$, $CH_2(CHO)_2$, and CH_3COCHO. Ozonolysis of myrcene yields $(CH_3)_2C=O$, $CH_2=O$ (two equiv), and $HCOCH_2CH_2COCHO$. Identify the structures of oximene and myrcene.

11.50 One compound that contributes to the "seashore smell" at beaches in Hawai'i is dictyopterene D', a component of a brown edible seaweed called limu lipoa. Hydrogenation of dictyopterene D' with excess H_2 in the presence of a Pd catalyst forms butylcycloheptane. Ozonolysis with O_3 followed by $(CH_3)_2S$ forms $CH_2(CHO)_2$, $HCOCH_2CH(CHO)_2$, and CH_3CH_2CHO. What are possible structures of dictyopterene D'?

Sharpless Asymmetric Epoxidation

11.51 Draw the product of each asymmetric epoxidation reaction.

a. [PhCH=CHCH₂OH] $\xrightarrow{\text{(CH}_3)_3\text{COOH}}{\text{Ti[OC(CH}_3)_2]_4,\ (-)\text{-DET}}$

b. [(CH₃)₃CCH=CHCH₂OH] $\xrightarrow{\text{(CH}_3)_3\text{COOH}}{\text{Ti[OC(CH}_3)_2]_4,\ (+)\text{-DET}}$

Chapter 11 Oxidation and Reduction

11.52 Epoxidation of the following allylic alcohol using the Sharpless reagent with (−)-DET gives two epoxy alcohols in a ratio of 87:13.

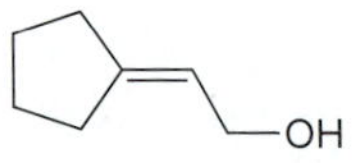

a. Assign structures to the major and minor product.
b. What is the enantiomeric excess in this reaction?

11.53 Identify **A** in the following reaction sequence, and draw a mechanism for the conversion of **A** to **B**. **B** has been converted to (S,S)-reboxetine, an antidepressant marketed outside the United States.

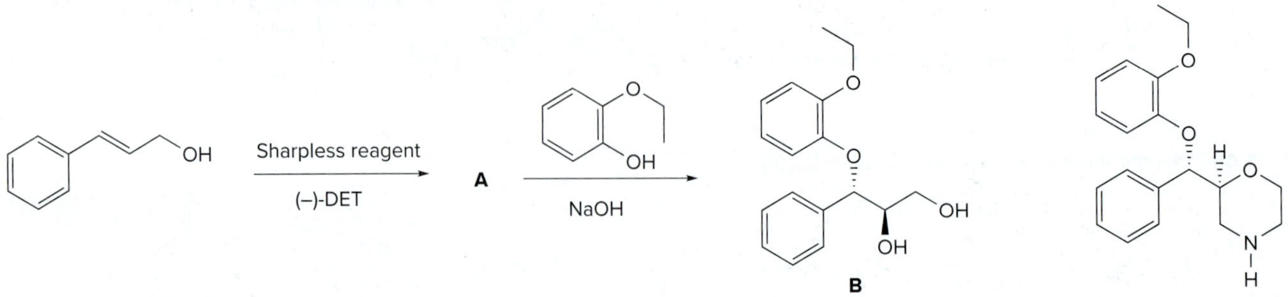

Synthesis

11.54 Devise a synthesis of muscalure, the sex pheromone of the common housefly, from acetylene and any other required reagents.

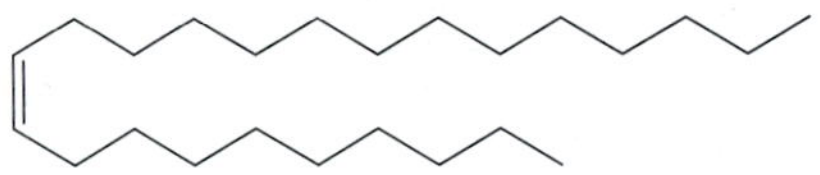

muscalure

11.55 It is sometimes necessary to isomerize a cis alkene to a trans alkene in a synthesis, a process that cannot be accomplished in a single step. Using the reactions you have learned in Chapters 8–11, devise a stepwise method to convert cis-but-2-ene to trans-but-2-ene.

11.56 Devise a synthesis of each compound from acetylene and any other required reagents.

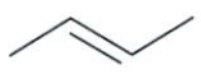

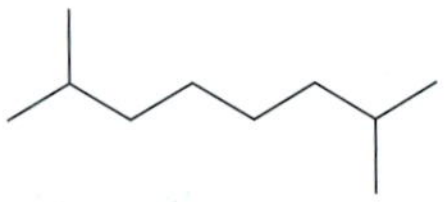

11.57 Devise a synthesis of compound **A** from the given starting materials. You may use any other inorganic reagents or organic alcohols. **A** was used to prepare aliskiren, a drug used to treat hypertension (Problem 5.6).

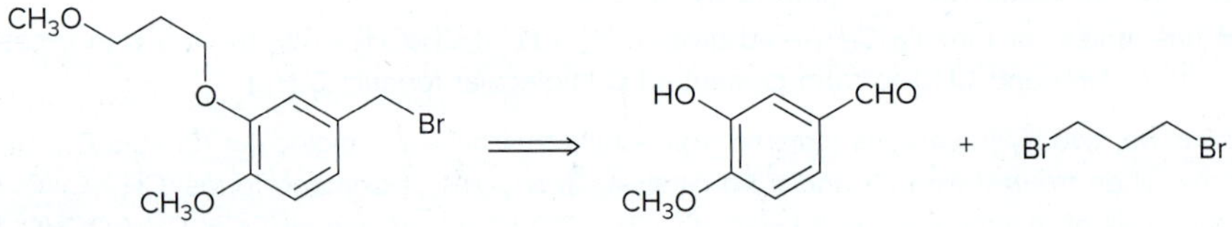

11.58 Devise a synthesis of each compound from the indicated starting material, organic compounds containing one or two carbons, and any other required reagents.

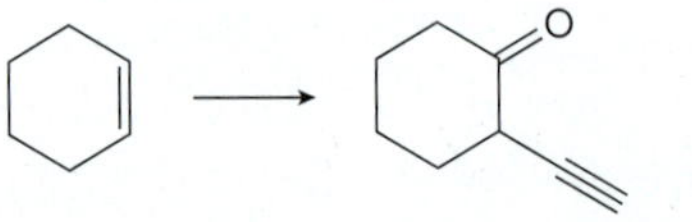

c. H−C≡C−H ⟶

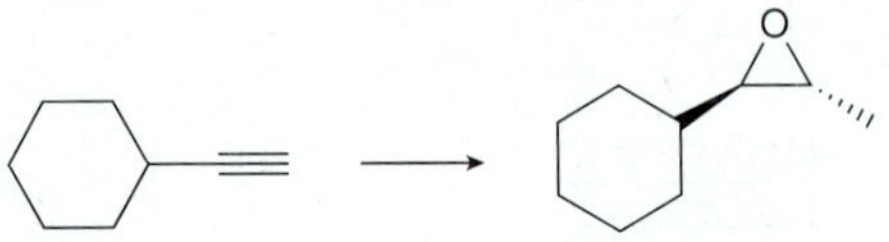

d. H−C≡C−H ⟶

(+ enantiomer)

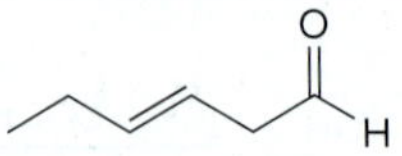

11.59 Devise a synthesis of each compound from the indicated starting material. You may use any other needed organic or inorganic reagents.

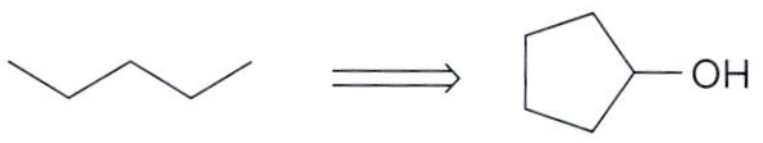

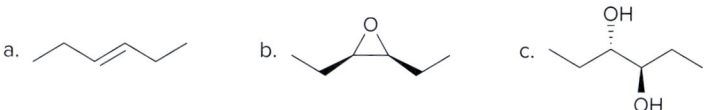

11.60 Devise a synthesis of (3*R*,4*S*)-3,4-dichlorohexane from acetylene and any needed organic compounds or inorganic reagents.

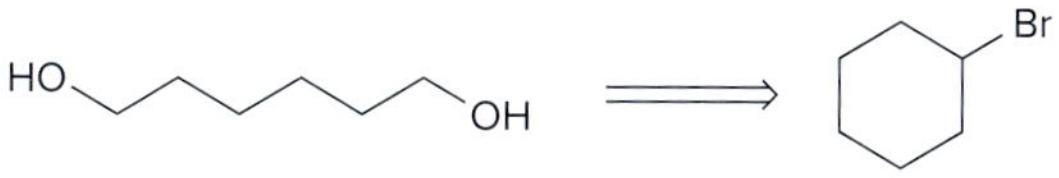

11.61 Devise a synthesis of each compound from CH_3CH_2OH as the only organic starting material; that is, every carbon in the product must come from a molecule of ethanol. You may use any other needed inorganic reagents.

11.62 Devise a synthesis of **A** from the three starting materials given. You may use any other needed organic or inorganic reagents.

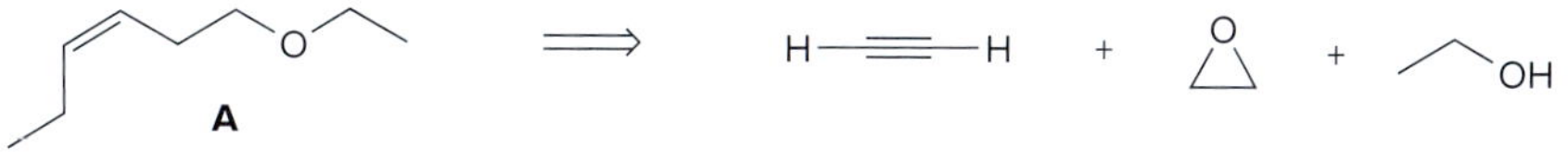

Spectroscopy

Problems 11.63 and 11.64 are intended for students who have already learned about spectroscopy in Chapters A–C.

11.63 Treatment of alcohol **A** (molecular formula $C_5H_{12}O$) with CrO_3, H_2SO_4, and H_2O affords **B** with molecular formula $C_5H_{10}O$, which gives an IR absorption at 1718 cm^{-1}. The ^1H NMR spectrum of **B** contains the following signals: 1.10 (doublet, 6 H), 2.14 (singlet, 3 H), and 2.58 (septet, 1 H) ppm. What are the structures of **A** and **B**?

11.64 Treatment of compound **C** (molecular formula $C_9H_{12}O$) with PCC affords **D** (molecular formula $C_9H_{10}O$). Use the ^1H NMR and IR spectra of **D** to determine the structures of both **C** and **D**.

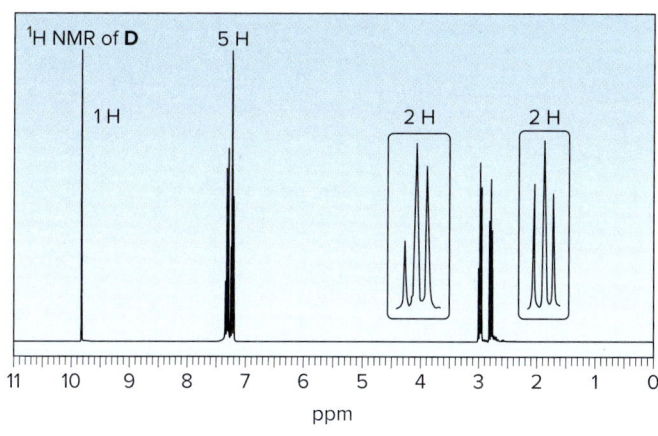
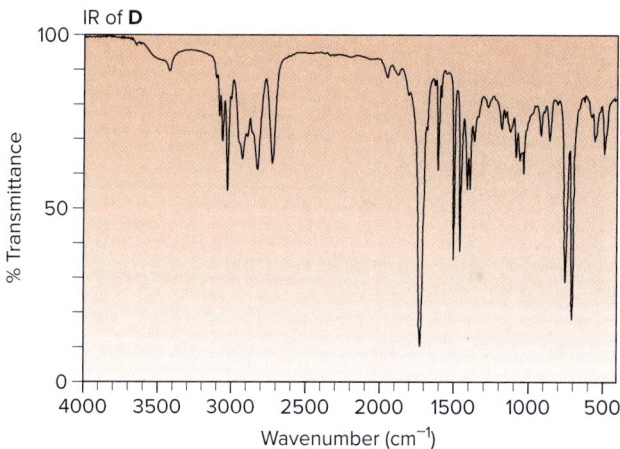

Challenge Problems

11.65 The Birch reduction is a dissolving metal reaction that converts substituted benzenes to cyclohexa-1,4-dienes using Li and liquid ammonia in the presence of an alcohol. Draw a stepwise mechanism for the following Birch reduction.

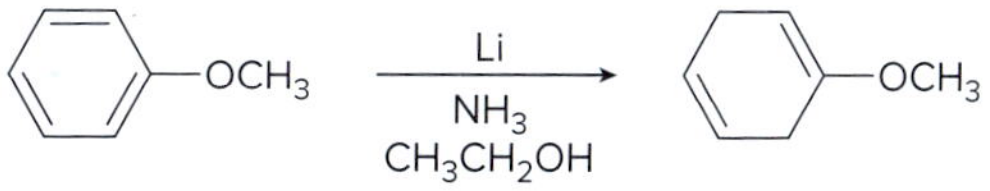

11.66 Identify the starting material in the following reaction sequence.

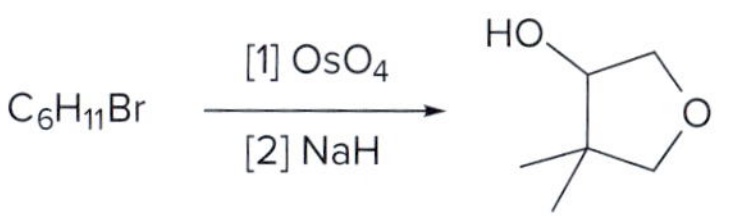

$$C_6H_{11}Br \xrightarrow[\text{[2] NaH}]{\text{[1] OsO}_4} $$

11.67 In the Cr^{6+} oxidation of cyclohexanols, it is generally true that sterically hindered alcohols react faster than unhindered alcohols. Which of the following alcohols should be oxidized more rapidly?

11.68 Dihydroxylation of an alkene can be carried out with H_2O_2 in HCO_2H. In this reaction, *trans*-but-2-ene affords (2R,3S)-butane-2,3-diol, whereas *cis*-but-2-ene affords a mixture of (2R,3R)-butane-2,3-diol and (2S,3S)-butane-2,3-diol. Does dihydroxylation by this method occur with syn or anti addition?

11.69 Draw a stepwise mechanism for the following reaction.

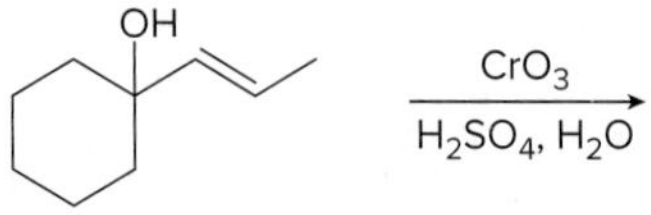

11.70 Sharpless epoxidation of allylic alcohol **X** forms compound **Y**. Treatment of **Y** with NaOH and C_6H_5SH in an alcohol–water mixture forms **Z**. Identify the structure of **Y** and draw a mechanism for the conversion of **Y** to **Z**. Account for the stereochemistry of the stereogenic centers in **Z**. **Z** has been used as an intermediate in the synthesis of chiral carbohydrates.

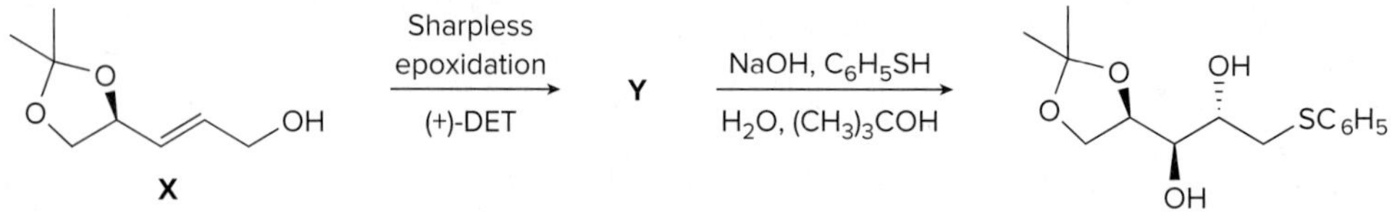

Conjugation, Resonance, and Dienes

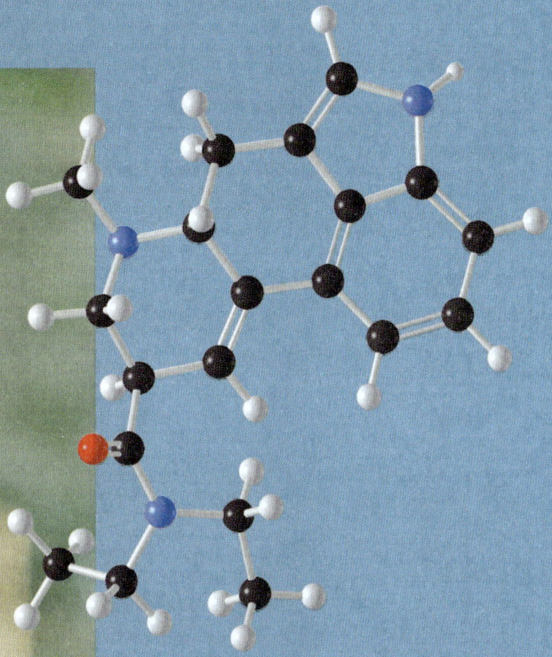

12.1 Conjugation
12.2 Resonance and allylic carbocations
12.3 Common examples of resonance
12.4 The resonance hybrid
12.5 Electron delocalization, hybridization, and geometry
12.6 Conjugated dienes
12.7 Interesting dienes and polyenes
12.8 The carbon–carbon σ bond length in buta-1,3-diene
12.9 Stability of conjugated dienes
12.10 Electrophilic addition: 1,2- versus 1,4-addition
12.11 Kinetic versus thermodynamic products
12.12 The Diels–Alder reaction
12.13 Specific rules governing the Diels–Alder reaction
12.14 Other facts about the Diels–Alder reaction

Lysergic acid diethyl amide (LSD) is a powerful hallucinogen prepared from lysergic acid, a natural product derived from an ergot fungus that attacks rye and other grains. Ergot has a long history as a dreaded poison, affecting individuals who become ill from eating ergot-contaminated bread. A key step in the synthesis of lysergic acid involves a Diels–Alder reaction, a powerful reaction of conjugated dienes discussed in Chapter 12.

Why Study . . .
Conjugated Systems?

Chapter 12 discusses the chemistry of conjugated molecules—molecules with overlapping *p* orbitals on three or more adjacent atoms. Much of Chapter 12 is devoted to the properties and reactions of 1,3-dienes, most notably the Diels–Alder reaction, which is widely used in the synthesis of naturally occurring compounds. To understand 1,3-dienes, however, we must first learn about the consequences of having *p* orbitals on three or more adjacent atoms. Because the ability to draw resonance structures is also central to mastering this material, the key aspects of resonance theory are presented in detail.

12.1 Conjugation

The conjugated systems in benzene and related compounds are discussed in Chapters 19 and 20.

Conjugation occurs whenever *p* orbitals can overlap on three or more adjacent atoms. Two common conjugated systems are 1,3-dienes and allylic carbocations.

1,3-diene   allylic carbocation

12.1A 1,3-Dienes

1,3-Dienes such as buta-1,3-diene contain **two carbon–carbon double bonds joined by a single σ bond.** Each carbon atom of a 1,3-diene is bonded to three other atoms and has no nonbonded electron pairs, so each carbon atom is sp^2 hybridized and has one *p* orbital containing an electron. **The four *p* orbitals on adjacent atoms make a 1,3-diene a conjugated system.**

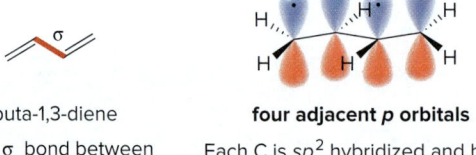

buta-1,3-diene

one σ bond between the double bonds

four adjacent *p* orbitals

Each C is sp^2 hybridized and has a *p* orbital containing one electron.

What is special about conjugation? Having three or more *p* orbitals on adjacent atoms allows *p* orbitals to overlap and **electrons to delocalize.**

The electron density in the two π bonds is **delocalized.**

overlap of adjacent *p* orbitals

- When *p* orbitals overlap, the electron density in each of the π bonds is spread out over a larger volume, thus lowering the energy of the molecule and making it more stable.

Conjugation makes buta-1,3-diene inherently different from penta-1,4-diene, a compound having two double bonds separated by more than one σ bond. The π bonds in penta-1,4-diene are too far apart to be conjugated.

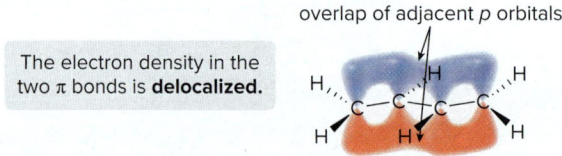

A conjugated diene   **An isolated diene**

buta-1,3-diene   penta-1,4-diene

The electrons in the π bonds are **delocalized.**

The electrons in the π bonds are **localized.**

Penta-1,4-diene is an **isolated diene.** The electron density in each π bond of an isolated diene is *localized* between two carbon atoms. In buta-1,3-diene, however, the electron density of both π bonds is *delocalized* over the four atoms of the diene. Electrostatic potential maps in Figure 12.1 clearly indicate the difference between these localized and delocalized π bonds.

Figure 12.1

Electrostatic potential plots for a conjugated and an isolated diene

A conjugated diene

buta-1,3-diene
The red electron-rich region is spread over four adjacent atoms.

An isolated diene

penta-1,4-diene
The red electron-rich regions are **localized** in the π bonds on the two ends of the molecule.

Problem 12.1 Classify each carbon–carbon double bond as isolated or conjugated.

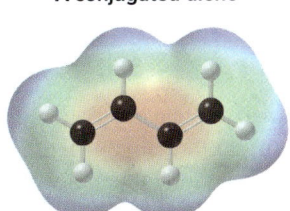

12.1B Allylic Carbocations

The **allyl carbocation** is another example of a conjugated system. The three carbon atoms of the allyl carbocation—the positively charged carbon atom and the two that form the double bond—are sp^2 hybridized and have an unhybridized *p* orbital. The *p* orbitals for the double bond carbons each contain an electron, whereas the *p* orbital for the carbocation is empty.

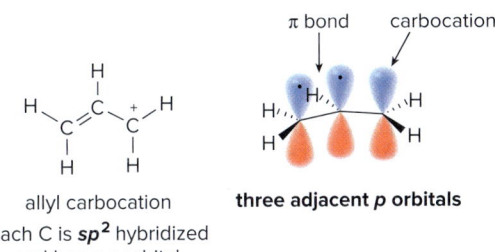

allyl carbocation
Each C is **sp^2** hybridized and has a *p* orbital.

three adjacent *p* orbitals

- Three *p* orbitals on three adjacent atoms, even if one of the *p* orbitals is empty, make the allyl carbocation conjugated.

Conjugation stabilizes the allyl carbocation because overlap of three adjacent *p* orbitals delocalizes the electron density of the π bond over three atoms.

overlap of adjacent *p* orbitals

Problem 12.2 Which of the following species are conjugated?

a. b. c. d. e.

12.2 Resonance and Allylic Carbocations

Recall from Section 1.6 that resonance structures are two or more different Lewis structures for the same arrangement of atoms. Being able to draw correct resonance structures is crucial to understanding conjugation and the reactions of conjugated dienes.

- Two resonance structures differ in the placement of π bonds and nonbonded electrons. The placement of atoms and σ bonds stays the same.

12.2A The Stability of Allylic Carbocations

We have already drawn resonance structures for the acetate anion ($CH_3CO_2^-$, Section 2.5C). The **conjugated allyl carbocation** is another example of a species for which two resonance structures can be drawn. Drawing resonance structures for the allyl carbocation is a way to use Lewis structures to illustrate how conjugation delocalizes electrons.

resonance structures for the allyl carbocation   hybrid

The true structure of the allyl carbocation is a **hybrid** of the two resonance structures. In the hybrid, the π bond is delocalized over all three atoms. As a result, the positive charge is also delocalized over the two terminal carbons. **Delocalizing electron density lowers the energy of the hybrid,** thus stabilizing the allyl carbocation and making it more stable than a normal 1° carbocation. Experimental data show that its stability is comparable to a more substituted 2° carbocation.

least stable   1° < 2° ≈ allyl < 3°   most stable

Increasing stability

Problem 12.3 Draw a second resonance structure for each carbocation. Then draw the hybrid.

a. b. c.

Problem 12.4 Use resonance theory and the Hammond postulate to explain why 3-chloroprop-1-ene ($CH_2=CHCH_2Cl$) is more reactive than 1-chloropropane ($CH_3CH_2CH_2Cl$) in S_N1 reactions.

12.2B Allylic Carbocations in Biological Reactions

Allylic carbocations formed from diphosphates (Section 7.16) are key intermediates in a variety of biological reactions, including the synthesis of geranyl diphosphate from two five-carbon

12.3 Common Examples of Resonance

substrates—dimethylallyl diphosphate and isopentenyl diphosphate. Geranyl diphosphate is the precursor of many lipids that occur in plants and animals.

organic diphosphate

R—OPP

diphosphate leaving group
PP$_i$

dimethylallyl diphosphate + isopentenyl diphosphate ⟶ geranyl diphosphate

This biological process results in the formation of a new carbon–carbon bond and involves two key steps—loss of a good leaving group (diphosphate, $P_2O_7^{4-}$, abbreviated as PP$_i$) to form an allylic carbocation, followed by nucleophilic attack with an electron-rich double bond. The steps of the mechanism are shown in Mechanism 12.1.

Mechanism 12.1 Biological Formation of Geranyl Diphosphate

1. dimethylallyl diphosphate → allylic carbocation + PP$_i$
2. isopentenyl diphosphate attacks, forming new bond shown in red
3. geranyl diphosphate + HB$^+$

1 Loss of the diphosphate leaving group forms an **allylic carbocation.**
2 Nucleophilic attack of isopentenyl diphosphate on the allylic carbocation forms the **new C–C σ bond.**
3 **Loss of a proton** (shown with the general base, B:) forms geranyl diphosphate.

We will learn more about biological reactions involving allylic carbocations derived from diphosphates in Chapter 25.

Problem 12.5 Farnesyl diphosphate is synthesized from isopentenyl diphosphate and **X** by a pathway similar to Mechanism 12.1. Draw the structure of **X**.

isopentenyl diphosphate + **X** ⟶ farnesyl diphosphate

12.3 Common Examples of Resonance

When are resonance structures drawn for a molecule or reactive intermediate? Because resonance involves delocalizing **π bonds** and **nonbonded electrons,** one or both of these structural features must be present to draw additional resonance forms. There are four common bonding patterns for which more than one Lewis structure can be drawn.

Type [1] The Three Atom "Allyl" System, X=Y–Z*

- For any group of three atoms having a double bond X=Y and an atom Z that contains a *p* orbital with zero, one, or two electrons, two resonance structures are possible:

$$X=Y-Z^* \longleftrightarrow {}^*X-Y=Z$$

The asterisk [*] corresponds to a charge, a radical, or a lone pair.

* = +, –, ·, or ··

This is called **allyl** type resonance because it can be drawn for allylic carbocations, allylic carbanions, and allylic radicals.

X, Y, and **Z** may all be carbon atoms, as in the case of an allylic carbocation (resonance structures **A** and **B**), or they may be heteroatoms, as in the case of the acetate anion (resonance structures **C** and **D**). The atom **Z** bonded to the multiple bond can be charged (a net positive or negative charge) or neutral (having zero, one, or two nonbonded electrons). **The two resonance structures differ in the location of the double bond, and in the charge, or the radical, or the lone pair, generalized by [*].**

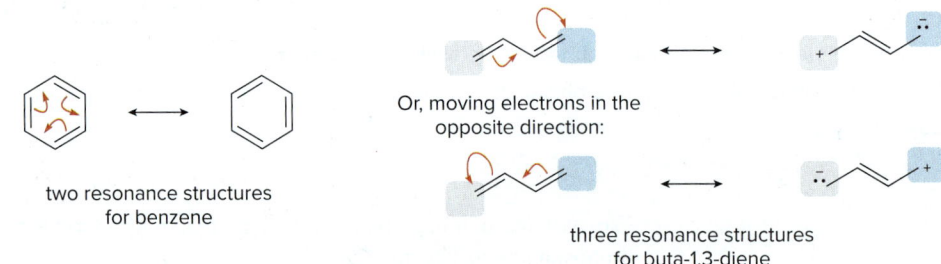

Type [2] Conjugated Double Bonds

Cyclic, completely conjugated rings like benzene have two resonance structures, drawn by moving the electrons in a cyclic manner around the ring. **Three resonance structures can be drawn for conjugated dienes,** two of which involve charge separation.

two resonance structures for benzene

Or, moving electrons in the opposite direction:

three resonance structures for buta-1,3-diene

Type [3] Cations Having a Positive Charge Adjacent to a Lone Pair

- When a lone pair and a positive charge are located on adjacent atoms, two resonance structures can be drawn.

The overall charge is the same in both resonance structures. Based on formal charge, a neutral **X** in one structure must bear a (+) charge in the other.

Type [4] Double Bonds Having One Atom More Electronegative Than the Other

- For a double bond X=Y in which the electronegativity of Y > X, a second resonance structure can be drawn by moving the π electrons onto Y.

Electronegativity of **Y > X**.

Sample Problem 12.1 illustrates how to apply these different types of resonance to actual molecules.

Sample Problem 12.1 Drawing Resonance Structures

Draw two more resonance structures for each species.

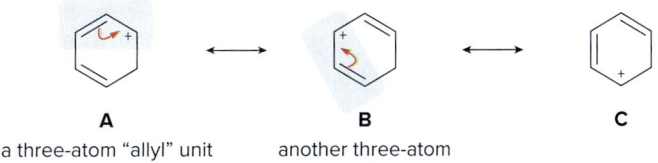

Solution
Mentally breaking a molecule into two- or three-atom units can make it easier to draw additional resonance structures.

a. Think of the top three atoms of the six-membered ring in **A** as an "allyl" unit. Moving the π bond forms a new "allyl" unit in **B**, and moving the π bond in **B** generates a third resonance structure **C**. No new valid resonance structures are generated by moving electrons in **C**.

b. Compound **D** contains a carbonyl group, so moving the electron pair in the double bond to the more electronegative oxygen atom separates the charge and generates structure **E**. **E** now has a three-atom "allyl" unit, so the remaining π bond can be moved to form structure **F**.

Problem 12.6
Draw additional resonance structures for each ion.

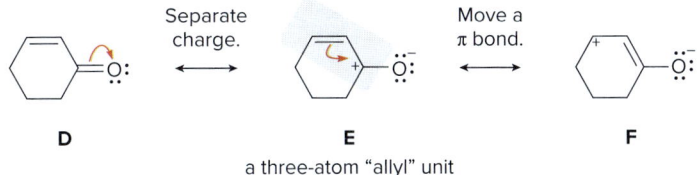

More Practice: Try Problems 12.32, 12.34.

12.4 The Resonance Hybrid

The lower its energy, the more a resonance structure contributes to the overall structure of the hybird.

Although the resonance hybrid is some combination of all of its valid resonance structures, the **hybrid more closely resembles the best resonance structure.** Recall from Section 1.6C that the best resonance structure is called the **major contributor** to the hybrid, and other resonance structures are called the **minor contributors.** Two identical resonance structures are equal contributors to the hybrid.

Use three rules to evaluate the relative energies of two or more valid resonance structures.

Rule [1] Resonance structures with more bonds and fewer charges are better.

all neutral atoms   charge separation
one more bond
better resonance structure

Chapter 12 Conjugation, Resonance, and Dienes

Rule [2] Resonance structures in which every atom has an octet are better.

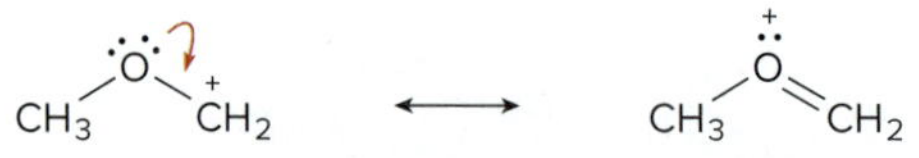

$$CH_3-\overset{..}{\underset{..}{O}}-\overset{+}{C}H_2 \longleftrightarrow CH_3-\overset{+}{\underset{..}{O}}=CH_2$$

All second-row elements have an **octet**.
better resonance structure

In this example, the resonance structure in which all atoms have octets is better, even though it places a (+) charge on a more electronegative O atom.

Rule [3] Resonance structures that place a negative charge on a *more* electronegative atom are better.

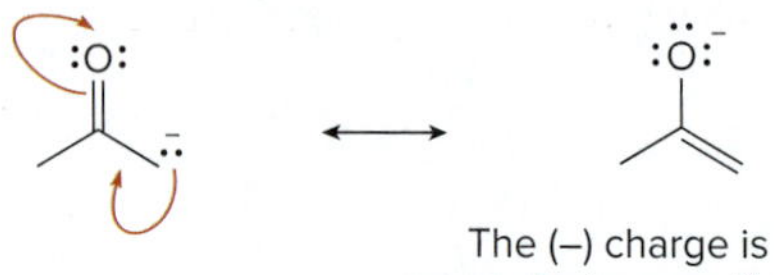

The (−) charge is on the **more electronegative O atom**.
better resonance structure

Sample Problem 12.2 illustrates how to determine the relative energy of contributing resonance structures and the hybrid.

Sample Problem 12.2 Determining the Relative Energy of Resonance Structures and the Hybrid

Draw a second resonance structure for carbocation **A**, as well as the hybrid of both resonance structures. Then use Rules [1]–[3] to rank the relative stability of both resonance structures and the hybrid.

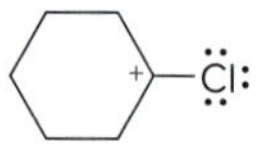

A

Solution

Because **A** contains a positive charge and a lone pair on adjacent atoms, a second resonance structure **B** can be drawn. Because **B** has more bonds and all second-row atoms have octets, **B** is a **better resonance** structure than **A**, making it the **major contributor** to the hybrid **C**. Because the hybrid is more stable than either resonance contributor, the order of stability is:

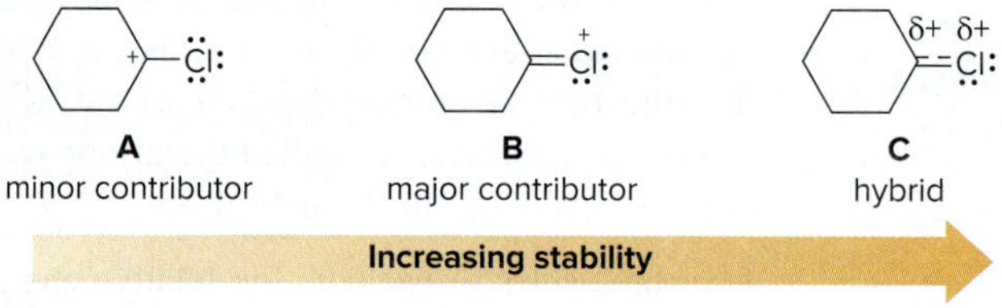

A	**B**	**C**
minor contributor	major contributor	hybrid

Increasing stability →

Problem 12.7 Draw a second resonance structure and the hybrid for each species, and then rank the two resonance structures and the hybrid in order of increasing stability.

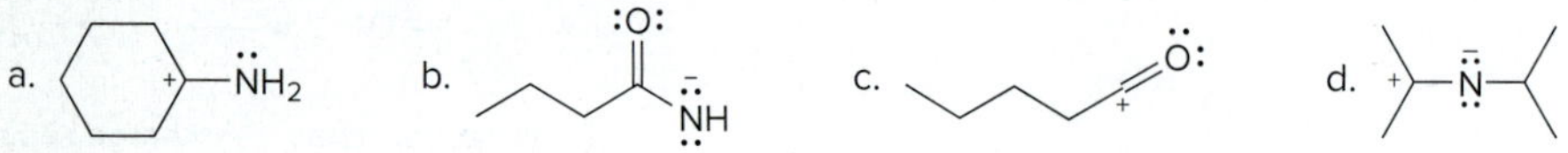

a., b., c., d.

More Practice: Try Problem 12.33.

Problem 12.8 Draw all possible resonance structures for the following cation, and indicate which structure makes the largest contribution to the resonance hybrid.

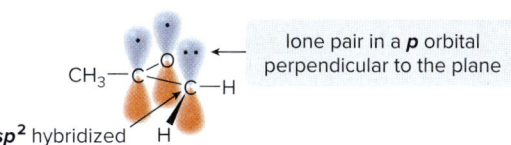

12.5 Electron Delocalization, Hybridization, and Geometry

To delocalize nonbonded electrons or electrons in π bonds, there must be p orbitals that can overlap. This may mean that the hybridization of an atom is *different* than would have been predicted using the rules first outlined in Chapter 1.

For example, there are two Lewis structures (**A** and **B**) for the resonance-stabilized anion $(CH_3COCH_2)^-$.

A
The labeled C is surrounded by four groups—three atoms and one nonbonded electron pair.
Is it sp^3 hybridized?

B
The labeled C is surrounded by three groups—three atoms and no nonbonded electron pairs.
Is it sp^2 hybridized?

Based on structure **A**, the labeled carbon is sp^3 hybridized, with the lone pair of electrons in an sp^3 hybrid orbital. Based on structure **B**, though, it is sp^2 hybridized with the unhybridized p orbital forming the π portion of the double bond.

Delocalizing electrons stabilizes a molecule. The electron pair on the carbon atom adjacent to the C=O can only be delocalized, though, if it has a p orbital that can overlap with two other p orbitals on two adjacent atoms. Thus, the terminal carbon atom is sp^2 hybridized and structure **B** reflects the needed trigonal planar geometry. **Three adjacent p orbitals make the anion conjugated.**

- In a system X=Y−Z:, Z is generally sp^2 hybridized, and the nonbonded electron pair occupies a p orbital to make the system conjugated.

Sample Problem 12.3 Determining Hybridization in a Conjugated System

Determine the hybridization around the labeled carbon atom in the following anion.

Solution

Because this is an example of an allyl-type system (X=Y−Z:), a second resonance structure can be drawn that "moves" the lone pair and the π bond. To delocalize the lone pair and make the system conjugated, the **labeled carbon atom must be sp^2 hybridized with the lone pair occupying a p orbital.**

The labeled C atom must be sp^2 hybridized, with the lone pair in a p orbital.

Problem 12.9 Determine the hybridization of the labeled atom in each species.

a. 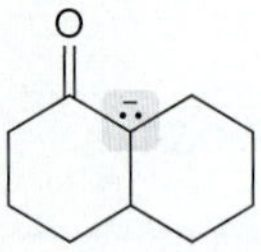 b. 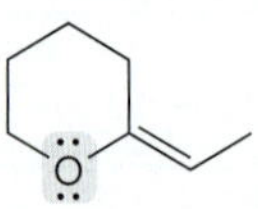 c. 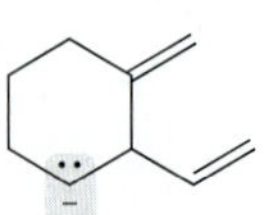 d.

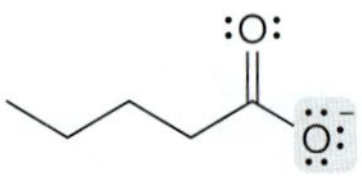

12.6 Conjugated Dienes

Compounds with many π bonds are called **polyenes**.

In the remainder of Chapter 12 we examine **conjugated dienes,** compounds having two double bonds joined by one σ bond. Conjugated dienes are also called **1,3-dienes.** Buta-1,3-diene (CH_2=CH–CH=CH_2) is the simplest conjugated diene.

Three stereoisomers are possible for 1,3-dienes with alkyl groups bonded to each end carbon of the diene (RCH=CH–CH=CHR).

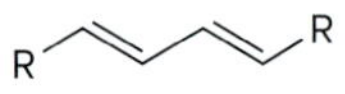

both double bonds **trans**

trans,trans-1,3-diene
or
(E,E)-1,3-diene

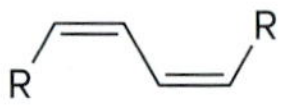

both double bonds **cis**

cis,cis-1,3-diene
or
(Z,Z)-1,3-diene

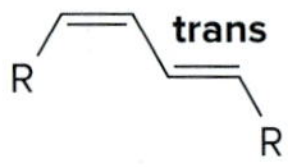

cis,trans-1,3-diene
or
(Z,E)-1,3-diene

Two possible conformations result from rotation about the C–C bond that joins the two double bonds.

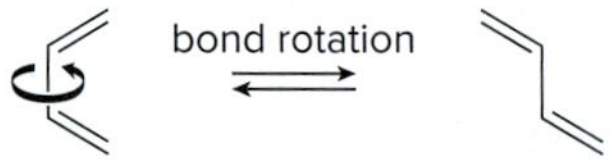

s-cis conformation   s-trans conformation

- The *s*-cis conformation has two double bonds on the *same* side of the single bond.
- The *s*-trans conformation has two double bonds on *opposite* sides of the single bond.

Keep in mind that **stereoisomers are discrete molecules,** whereas **conformations interconvert.** Three structures drawn for hexa-2,4-diene illustrate the differences between stereoisomers and conformations in a 1,3-diene:

cis, cis isomer   **trans, trans** isomer *s-trans* conformation   **trans, trans** isomer *s-cis* conformation

two stereoisomers   two conformations

Sample Problem 12.4   Classifying Compounds as Stereoisomers or Different Conformations

Classify each pair of compounds as stereoisomers or conformations: (a) **X** and **Y**; (b) **X** and **Z**.

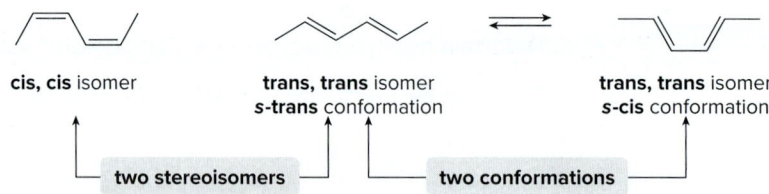

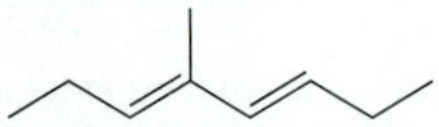

X 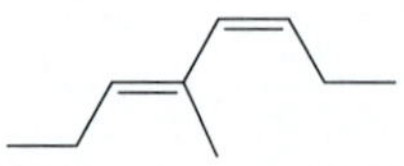   Y 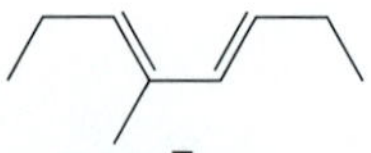   Z

Solution

- **Stereoisomers are *different* compounds.** Groups on each end of a carbon–carbon double bond are arranged differently.
- **Two conformations are the *same* compound,** which interconvert by bond rotation.

a. **X** and **Y** are **stereoisomers** because the groups around the C=C in blue are arranged differently; in **X** two groups are trans, and in **Y** two groups are cis.

b. Each C=C in **X** and **Z** is bonded to the same groups and has the *E* configuration. **X** has the two double bonds on opposite sides of the C–C in blue, whereas **Z** has two double bonds on the same side of the single bond that joins them together. **X** and **Z** are different **conformations.**

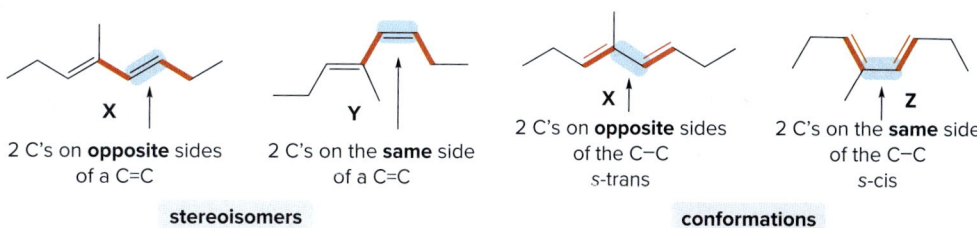

X	Y	X	Z
2 C's on **opposite** sides of a C=C	2 C's on the **same** side of a C=C	2 C's on **opposite** sides of the C–C *s*-trans	2 C's on the **same** side of the C–C *s*-cis
stereoisomers		**conformations**	

Problem 12.10 Label compounds **B–D** as stereoisomers, conformations, or constitutional isomers of **A**.

More Practice: Try Problem 12.38.

Problem 12.11 Draw the structure consistent with each description.

a. (2*E*,4*E*)-octa-2,4-diene in the *s*-trans conformation
b. (3*E*,5*Z*)-nona-3,5-diene in the *s*-cis conformation
c. (3*Z*,5*Z*)-4,5-dimethyldeca-3,5-diene. Draw both the *s*-cis and *s*-trans conformations.

Problem 12.12 Neuroprotectin D1 (NPD1) is synthesized in the body from highly unsaturated essential fatty acids. NPD1 is a potent natural anti-inflammatory agent.

a. Label each carbon–carbon double bond as conjugated or isolated.
b. Label each double bond as *E* or *Z*.
c. For each conjugated system, label the given conformation as *s*-cis or *s*-trans.

12.7 Interesting Dienes and Polyenes

Isoprene and **lycopene** are two naturally occurring compounds with conjugated double bonds.

isoprene
(2-methylbuta-1,3-diene)

11 conjugated double bonds shown in red

lycopene

Figure 12.2
Biologically active organic compounds that contain conjugated double bonds

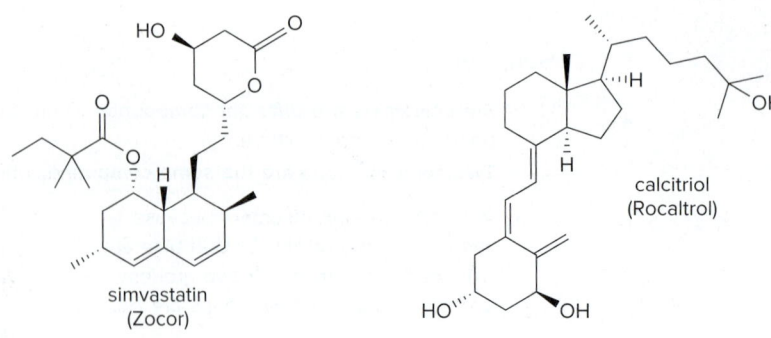

simvastatin (Zocor)

calcitriol (Rocaltrol)

Lycopene is the red pigment found in tomatoes, watermelon, papaya, guava, and pink grapefruit. Lycopene is not destroyed when fruits and vegetables are processed, so tomato juice and ketchup are high in lycopene. *C Squared Studios/Getty Images*

Isoprene, the common name for 2-methylbuta-1,3-diene, is given off by plants as the temperature rises, a process thought to increase a plant's tolerance for heat stress. Isoprene is a component of the blue haze seen above forested hillsides, such as Virginia's Blue Ridge Mountains.

Lycopene is a naturally occurring molecule responsible for the red color of tomatoes and other fruits. The 11 conjugated double bonds of lycopene absorb light in the blue-green region of the visible spectrum. When a compound absorbs visible light, it takes on the color of the light it does *not* absorb. Because lycopene does not absorb red light, it appears red.

Simvastatin and calcitriol are two drugs that contain conjugated double bonds in addition to other functional groups (Figure 12.2). Simvastatin is the generic name of the widely used cholesterol-lowering medicine Zocor. Calcitriol, a biologically active hormone formed from vitamin D_3 obtained in the diet, is responsible for regulating calcium and phosphorus metabolism. Sold under the trade name of Rocaltrol, calcitriol is used to treat patients who are unable to convert vitamin D_3 to the active hormone. Because calcitriol promotes the absorption of calcium ions, it is also used to treat hypocalcemia, the presence of low calcium levels in the blood.

12.8 The Carbon–Carbon σ Bond Length in Buta-1,3-diene

Three features distinguish conjugated dienes from isolated dienes:

[1] The C–C single bond joining the two double bonds is unusually short.
[2] Conjugated dienes are more stable than similar isolated dienes.
[3] Some reactions of conjugated dienes are different than reactions of isolated double bonds.

Hybridization can explain why the central carbon–carbon single bond is shorter than the C–C bond in ethane (148 pm vs. 153 pm).

ethylene 134 pm CH_3—CH_3 153 pm   buta-1,3-diene 134 pm / 148 pm / 134 pm

Each carbon atom in buta-1,3-diene is sp^2 hybridized, so the central C–C single bond is formed by the overlap of **two sp^2 hybridized orbitals,** rather than the sp^3 hybridized orbitals used to form the C–C bond in CH_3CH_3.

CH_3—CH_3
sp^3 C's
25% s-character
lower percent s-character
longer bond

sp^2 C's
33% s-character
higher percent s-character
shorter bond

Recall from Section 1.11B that increasing percent *s*-character decreases bond length.

- Based on hybridization, a C_{sp^2}–C_{sp^2} bond should be shorter than a C_{sp^3}–C_{sp^3} bond because it is formed from orbitals having a *higher* percent s-character.

Problem 12.13 Using hybridization, predict how the bond length of the C–C σ bond in HC≡C–C≡CH should compare with the C–C σ bonds in CH_3CH_3 and $CH_2=CH–CH=CH_2$.

Problem 12.14 Use resonance theory to explain why the labeled C–O bond lengths (in red) are equal in the acetate anion.

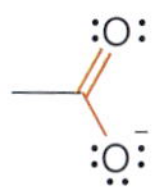

acetate

12.9 Stability of Conjugated Dienes

In Section 11.3, we learned that hydrogen adds to alkenes to form alkanes and that the heat released in this reaction, the **heat of hydrogenation**, can be used as a measure of alkene stability.

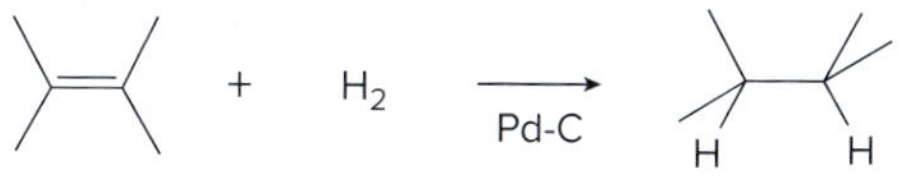

$\Delta H° =$ **heat of hydrogenation**

The relative stability of conjugated and isolated dienes can also be determined by comparing their heats of hydrogenation.

- When hydrogenation gives the same alkane from two dienes, the more stable diene has the *smaller* heat of hydrogenation.

For example, both penta-1,4-diene (an isolated diene) and (*E*)-penta-1,3-diene (a conjugated diene) are hydrogenated to pentane with two equivalents of H_2. Because *less* energy is released in converting the conjugated diene to pentane, it must be *lower in energy* (more stable) to begin with. The relative energies of these isomeric pentadienes are illustrated in Figure 12.3.

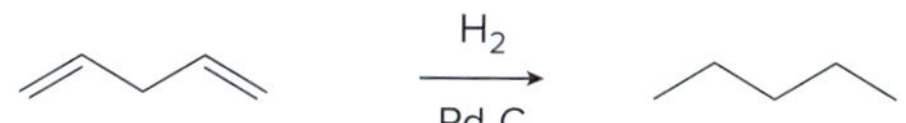

penta-1,4-diene
isolated diene

$\Delta H° = -255$ kJ/mol

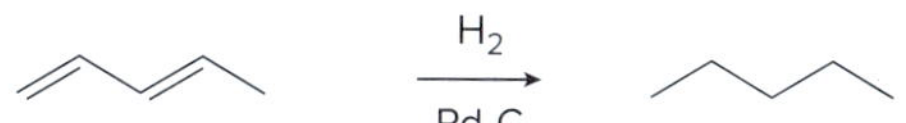

(*E*)-penta-1,3-diene
conjugated diene

$\Delta H° = -226$ kJ/mol

Less energy is released.

more stable
starting material

- A conjugated diene has a *smaller* heat of hydrogenation and is more stable than a similar isolated diene.

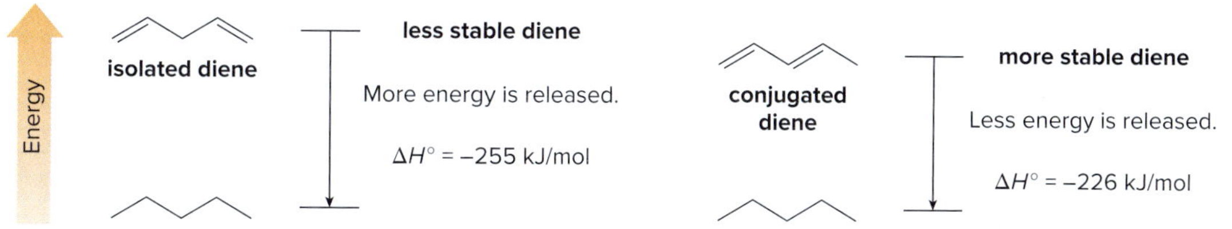

Figure 12.3
Relative energies of an isolated and conjugated diene

isolated diene — less stable diene
More energy is released.
$\Delta H° = -255$ kJ/mol

conjugated diene — more stable diene
Less energy is released.
$\Delta H° = -226$ kJ/mol

In Section 12.1, we learned why a conjugated diene is more stable than an isolated diene. A conjugated diene has overlapping *p* orbitals on four adjacent atoms, so its **π electrons are delocalized over four atoms, thus stabilizing the diene.** This delocalization cannot occur in an isolated diene, so an isolated diene is less stable than a conjugated diene.

Problem 12.15 Which diene in each pair has the larger heat of hydrogenation?

a. [structure] or [structure]   b. [structure] or [structure]

Problem 12.16 Rank the following compounds in order of increasing stability.

[three biphenyl-type structures]

12.10 Electrophilic Addition: 1,2- Versus 1,4-Addition

Recall from Chapter 10 that the characteristic reaction of compounds with π bonds is **addition.** The π bonds in conjugated dienes undergo addition reactions, too, but they differ in two ways from the addition reactions to isolated double bonds.

- Electrophilic addition in conjugated dienes gives a mixture of products.
- Conjugated dienes undergo a unique addition reaction not seen in alkenes or isolated dienes.

We learned in Chapter 10 that HX adds to the π bond of alkenes to form alkyl halides.

[reaction scheme: alkene + H–X (δ+ δ−), a π bond is broken → alkyl halide]

With an **isolated diene**, electrophilic addition of one equivalent of HBr yields *one* product and Markovnikov's rule is followed. The H atom bonds to the less substituted carbon—that is, the carbon atom of the double bond that had more H atoms to begin with.

[reaction: isolated diene + H–Br (1 equiv) → product; H bonds to the *less* substituted carbon.]

With a **conjugated diene**, electrophilic addition of one equivalent of HBr affords *two* products.

[reaction: conjugated diene + H–Br (1 equiv) → 1,2-product + 1,4-product]

The ends of the 1,3-diene are called C1 and C4 arbitrarily, without regard to IUPAC numbering.

- The **1,2-addition product** results from Markovnikov addition of HBr across two adjacent carbon atoms (C1 and C2) of the diene.
- The **1,4-addition product** results from addition of HBr to the two end carbons (C1 and C4) of the diene. 1,4-Addition is also called **conjugate addition.**

The mechanism of electrophilic addition of HX involves **two steps:** addition of H⁺ (from HX) to form a resonance-stabilized carbocation, followed by nucleophilic attack of X⁻ at either electrophilic end of the carbocation to form two products. Mechanism 12.2 illustrates the reaction of buta-1,3-diene with HBr.

12.10 Electrophilic Addition: 1,2- Versus 1,4-Addition

> **Mechanism 12.2** Electrophilic Addition of HBr to a 1,3-Diene—1,2- and 1,4-Addition

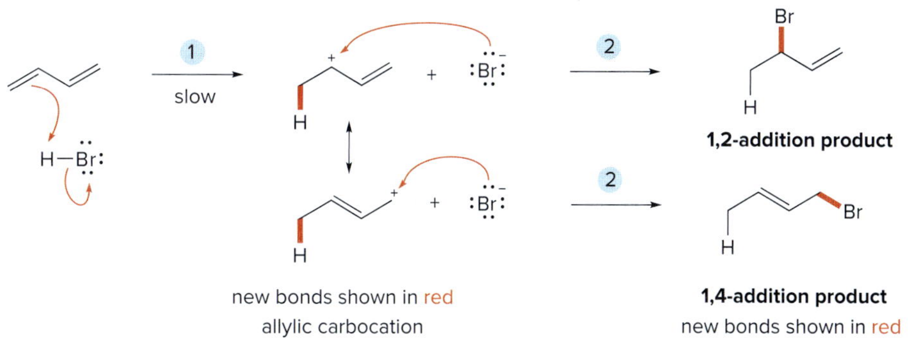

① H⁺ of HBr adds to a terminal carbon of the 1,3-diene to form a **resonance-stabilized allylic carbocation**.

② Nucleophilic attack of Br⁻ occurs at either site of the resonance-stabilized carbocation that bears a (+) charge, forming the 1,2- and 1,4-addition products.

Like the electrophilic addition of HX to an alkene, the addition of HBr to a conjugated diene forms the more stable carbocation in Step [1], the rate-determining step. In this case, however, the carbocation is both 2° and **allylic**, and thus two Lewis structures can be drawn for it. In the second step, nucleophilic attack of Br⁻ can then occur at two different electrophilic sites, forming two different products.

- Addition of HX to a conjugated diene forms 1,2- and 1,4-products because of the resonance-stabilized allylic carbocation intermediate.

Sample Problem 12.5 Drawing the Products of 1,2- and 1,4-Addition

Draw the products of the following reaction.

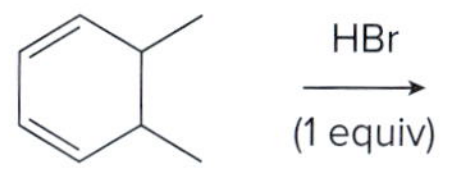

Solution

Write the steps of the mechanism to determine the structure of the products. Addition of H⁺ forms the more stable 2° allylic carbocation, for which two resonance structures can be drawn. **H⁺ always bonds to a terminal carbon of the 1,3-diene,** labeled in blue. Nucleophilic attack of Br⁻ at either end of the allylic carbocation gives two constitutional isomers, formed by 1,2-addition and 1,4-addition to the diene.

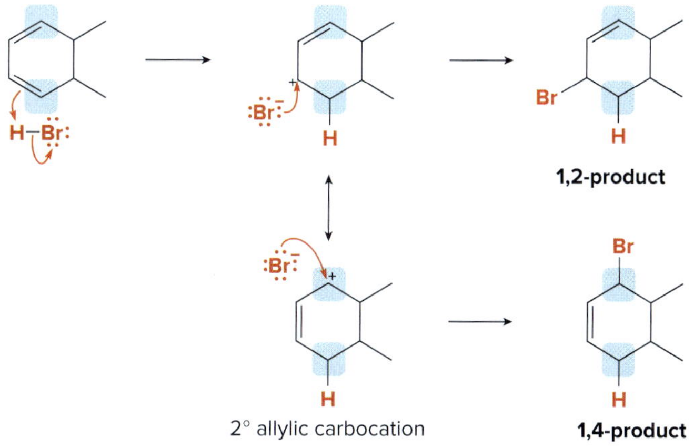

Problem 12.17
Draw the products formed when each diene is treated with one equivalent of HCl.

a. 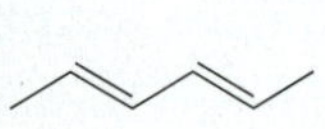 [penta-1,3-diene] b. 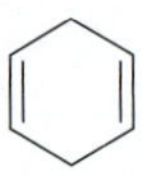 [cyclohexa-1,3-diene] c.  [cyclohexa-1,4-diene] d. 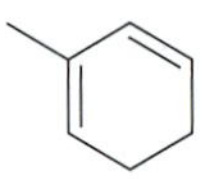 [1-methylcyclohexa-1,3-diene]

More Practice: Try Problems 12.40; 12.55a, d.

Problem 12.18
Draw a stepwise mechanism for the following reaction.

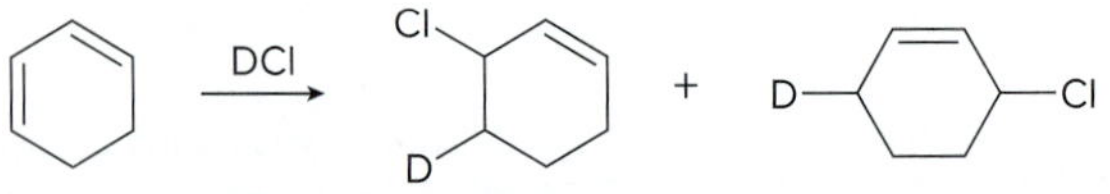

[cyclohexa-1,3-diene] →(DCl) [3-chloro-4-deuterio-cyclohexene] + [3-chloro-6-deuterio-cyclohexene]

12.11 Kinetic Versus Thermodynamic Products

The amount of 1,2- and 1,4-addition products formed in the electrophilic addition reactions of buta-1,3-diene, a conjugated diene, depends greatly on the reaction conditions.

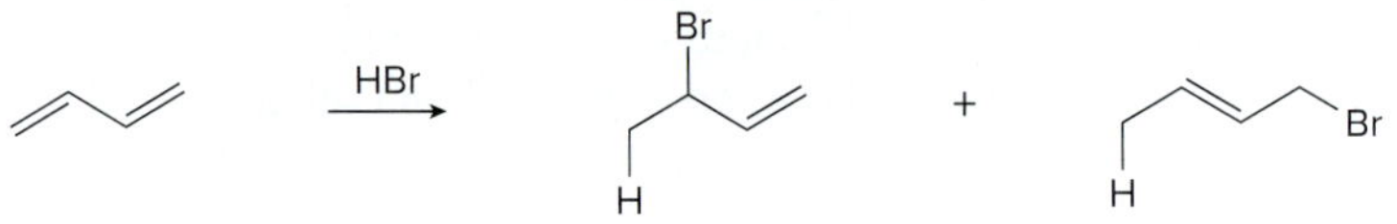

	1,2-product	1,4-product
low temperature (−80 °C)	80%	20%
high temperature (40 °C)	20%	80%

- **At low temperature the major product is formed by 1,2-addition.**
- **At higher temperature the major product is formed by 1,4-addition.**

Moreover, when a mixture containing predominately the 1,2-product is heated, the 1,4-addition product becomes the major product at equilibrium.

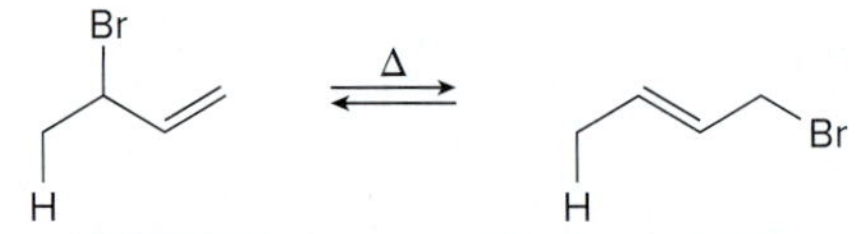

1,2-product ⇌(Δ) 1,4-product

major product at low temperature — major product at equilibrium

kinetic product — **thermodynamic product**

- The 1,2-product is formed *faster* so it predominates at low temperature. The product that is formed faster is called the *kinetic product*.
- The 1,4-product must be *more stable* because it predominates at equilibrium. The product that predominates at equilibrium is called the *thermodynamic product*.

In many of the reactions we have learned thus far, the more stable product is formed faster—that is, the kinetic and thermodynamic products are the same. The electrophilic addition of HBr to buta-1,3-diene is different, in that **the more stable product is formed more slowly**—that is, the kinetic and thermodynamic products are *different*. Why is the more stable product formed more slowly?

To answer this question, recall that the **rate of a reaction is determined by its energy of activation (E_a)**, whereas the **amount of product present at equilibrium is determined by**

12.11 Kinetic Versus Thermodynamic Products

Figure 12.4
How kinetic and thermodynamic products form in a reaction: **A → B + C**

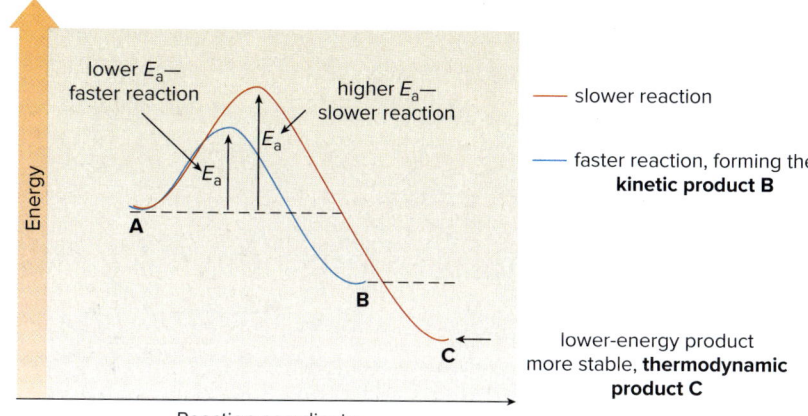

- The conversion of **A → B** is a faster reaction because the energy of activation leading to **B** is *lower*. **B** is the **kinetic product**.
- Because **C** is *lower* in energy, **C** is the **thermodynamic product**.

its stability (Figure 12.4). When a single starting material **A** forms two different products (**B** and **C**) by two exothermic pathways, the relative height of the energy barriers determines how fast **B** and **C** are formed, whereas the relative energies of **B** and **C** determine the amount of each at equilibrium. In an exothermic reaction, the relative energies of **B** and **C** do not determine the relative energies of activation to form **B** and **C**.

Why, in the addition of HBr to buta-1,3-diene, is the 1,4-product the more stable thermodynamic product? The 1,4-product (1-bromobut-2-ene) is more stable because it has two alkyl groups bonded to the carbon–carbon double bond, whereas the 1,2-product (3-bromobut-1-ene) has only one.

3-bromobut-1-ene — **1,2-product** — monosubstituted alkene — **less stable**

1-bromobut-2-ene — **1,4-product** — disubstituted alkene — **more stable** — **thermodynamic product**

- The more substituted alkene—1-bromobut-2-ene in this case—is the thermodynamic product.

A **proximity effect** occurs because one species is close to another.

The 1,2-product is the kinetic product because of a **proximity effect.** When H⁺ (from HBr) adds to the double bond, Br⁻ is *closer* to the adjacent carbon (C2) than it is to C4. Even though the resonance-stabilized carbocation bears a partial positive charge on both C2 and C4, attack at C2 is faster simply because Br⁻ is closer to this carbon.

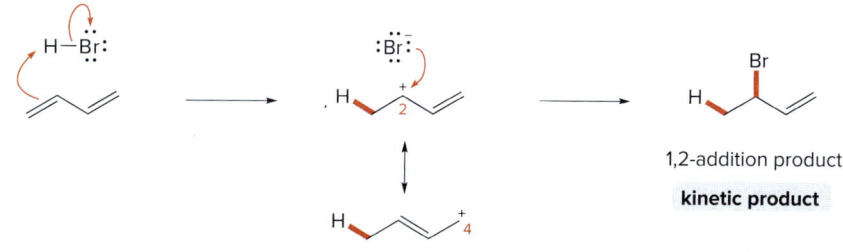

Br⁻ is closer to C2 than C4.

- The 1,2-product forms faster because of the proximity of Br⁻ to C2.

Figure 12.5
Energy diagram for the two-step addition of HBr to $CH_2=CH-CH=CH_2$

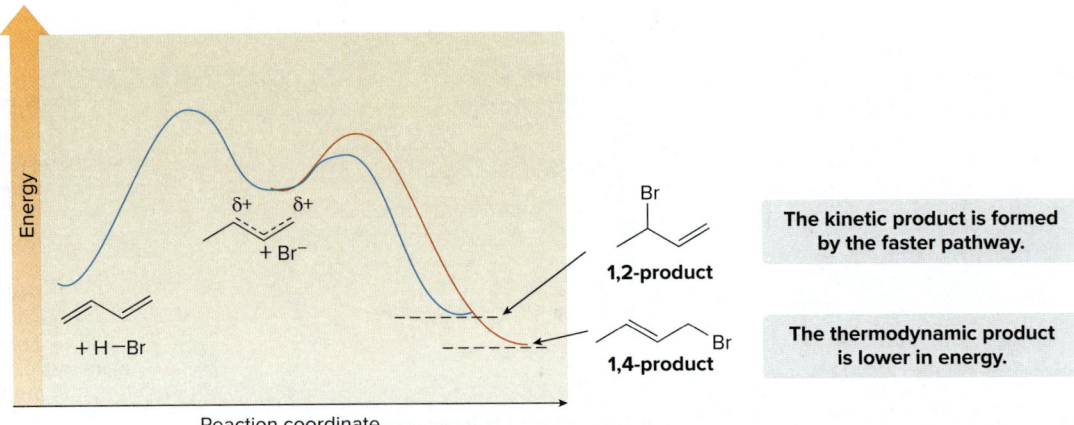

The kinetic product is formed by the faster pathway.

The thermodynamic product is lower in energy.

The overall two-step mechanism for addition of HBr to buta-1,3-diene, forming a 1,2-addition product and 1,4-addition product, is illustrated with the energy diagram in Figure 12.5.

Why is the ratio of products temperature dependent?

- **At low temperature, the energy of activation is the more important factor.** Because most molecules do not have enough kinetic energy to overcome the higher energy barrier at lower temperature, they react by the faster pathway, forming the kinetic product.
- **At higher temperature,** most molecules have enough kinetic energy to reach either transition state. The two products are in equilibrium with each other, and the **more stable compound**—which is lower in energy—**becomes the major product.**

Problem 12.19 Label each product in the following reaction as a 1,2-product or a 1,4-product, and decide which is the kinetic product and which is the thermodynamic product.

12.12 The Diels–Alder Reaction

The **Diels–Alder reaction,** named for German chemists Otto Diels and Kurt Alder, is an addition reaction between a **1,3-diene** and an alkene called a **dienophile,** to form a new six-membered ring.

Diels and Alder shared the 1950 Nobel Prize in Chemistry for unraveling the intricate details of this remarkable reaction.

The arrows may be drawn in a clockwise or counterclockwise direction to show the flow of electrons in a Diels–Alder reaction.

1,3-diene   dienophile

Three curved arrows are needed to show the cyclic movement of electron pairs because three π bonds break and two σ bonds and one π bond form. Because each new σ bond is ~100 kJ/mol stronger than a π bond that is broken, a typical Diels–Alder reaction releases ~200 kJ/mol of energy. The following equations illustrate two examples of the Diels–Alder reaction.

12.12 The Diels–Alder Reaction

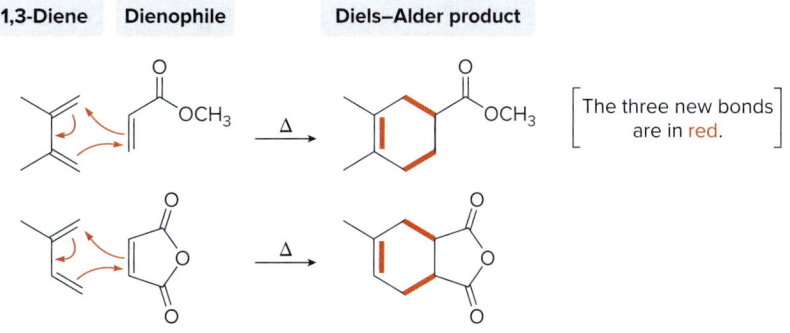

All Diels–Alder reactions have these features in common:

[1] They are initiated by heat; that is, the Diels–Alder reaction is a *thermal* reaction.
[2] They form new six-membered rings.
[3] Three π bonds break, and two new C–C σ bonds and one new C–C π bond form.
[4] They are concerted; that is, all bonds are broken and formed in a single step.

Diels–Alder reactions may seem complicated at first, but they are really less complicated than many of the reactions you have already learned, especially those with multistep mechanisms and carbocation intermediates. **The key is to learn how to arrange the starting materials** to more easily visualize the structure of the product.

How To Draw the Product of a Diels–Alder Reaction

Example Draw the product of the following Diels–Alder reaction:

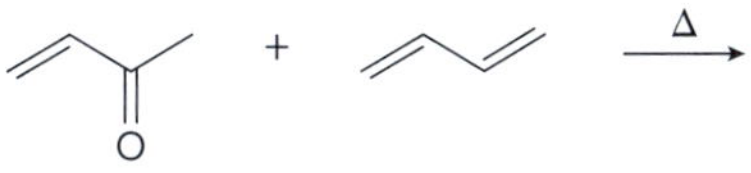

Step [1] Arrange the 1,3-diene and the dienophile next to each other, with the diene drawn in the s-cis conformation.
- This step is key: **Rotate the diene** so that it is drawn in the **s-cis** conformation, and **place the end C's of the diene close to the double bond of the dienophile.**

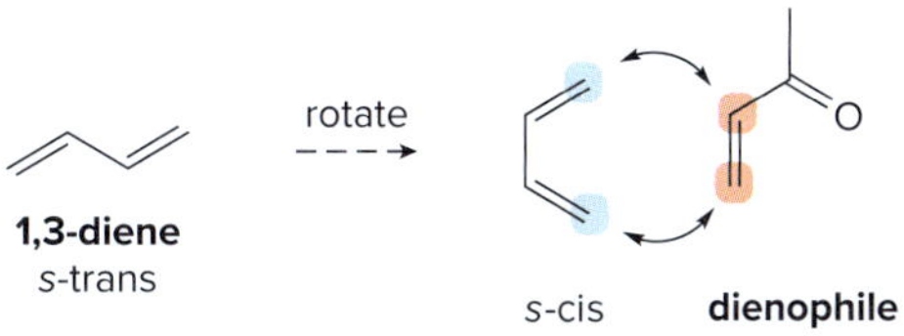

Step [2] Cleave the three π bonds and use arrows to show where the new bonds will be formed.

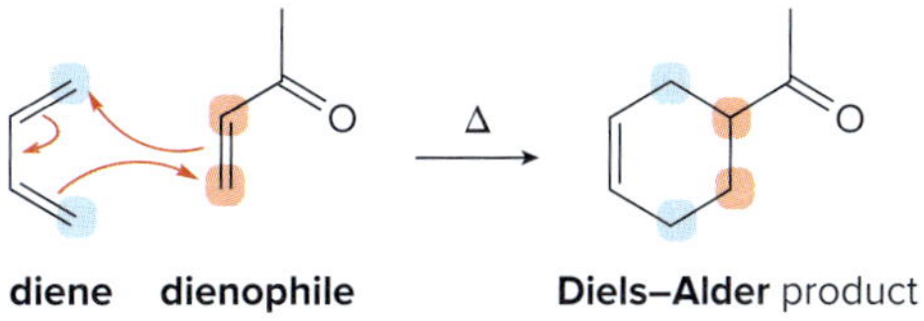

Problem 12.20 Draw the product formed when each diene and dienophile react in a Diels–Alder reaction.

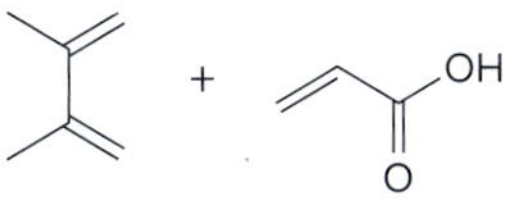

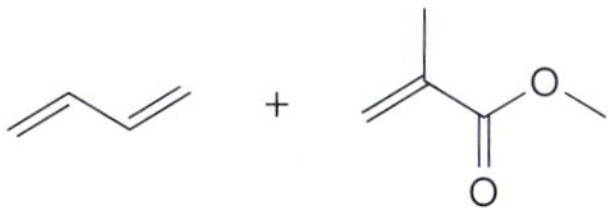

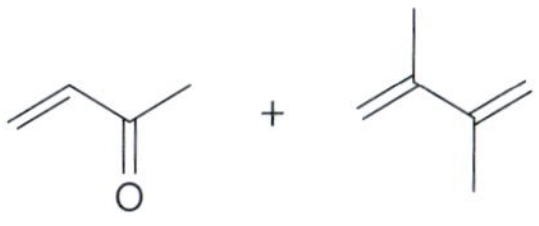

12.13 Specific Rules Governing the Diels–Alder Reaction

Several rules govern the course of the Diels–Alder reaction.

12.13A Diene Reactivity

Rule [1] The diene can react only when it adopts the *s-cis* conformation.

Both ends of the conjugated diene must be close to the π bond of the dienophile for reaction to occur. Thus, an acyclic diene in the *s-trans* conformation must rotate about the central C—C σ bond to form the **s-cis conformation** before reaction can take place.

This rotation is prevented in cyclic dienes. As a result:

- When the two double bonds are constrained in the *s-cis* conformation, the diene is unusually *reactive*.
- When the two double bonds are constrained in the *s-trans* conformation, the diene is *unreactive*.

Problem 12.21 Label each diene as reactive or unreactive in a Diels–Alder reaction.

Zingiberene and β-sesquiphellandrene (Problem 12.22) are trienes obtained from ginger root. Ginger is used as a spice in Indian and Chinese cooking. Ginger candy is sometimes used to treat nausea resulting from seasickness. *Alvis Upitis/Getty Images*

Problem 12.22 Zingiberene and β-sesquiphellandrene, natural products obtained from ginger root, contain conjugated diene units. Which diene reacts faster in the Diels–Alder reaction and why?

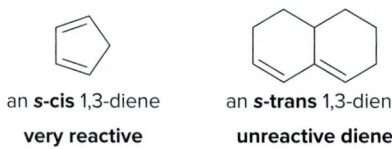

zingiberene   β-sesquiphellandrene

12.13B Dienophile Reactivity

Rule [2] Electron-withdrawing substituents in the dienophile increase the reaction rate.

In a Diels–Alder reaction, **electron-withdrawing groups make the dienophile more electrophilic (and, thus, more reactive)** by withdrawing electron density from the carbon–carbon

12.13 Specific Rules Governing the Diels–Alder Reaction

double bond. If Z is an electron-withdrawing group, then the reactivity of the dienophile increases as follows:

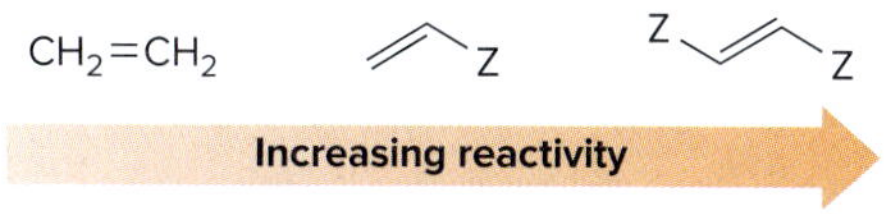

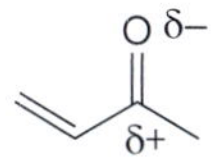

$$CH_2=CH_2 \quad \diagup\!\!\!\diagup Z \quad Z\diagup\!\!\!\diagup Z$$

Increasing reactivity

A carbonyl group is an effective electron-withdrawing group because the carbonyl carbon bears a partial positive charge (δ+), which withdraws electron density from the carbon–carbon double bond of the dienophile. Common dienophiles that contain a carbonyl group are shown in Figure 12.6.

electron-deficient carbonyl carbon

Figure 12.6
Common dienophiles in the Diels–Alder reaction

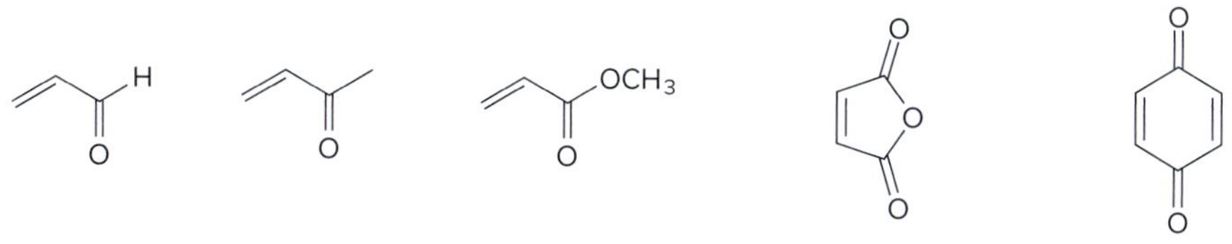

acrolein | methyl vinyl ketone | methyl acrylate | maleic anhydride | benzoquinone

---

Problem 12.23 Rank the following dienophiles in order of increasing reactivity.

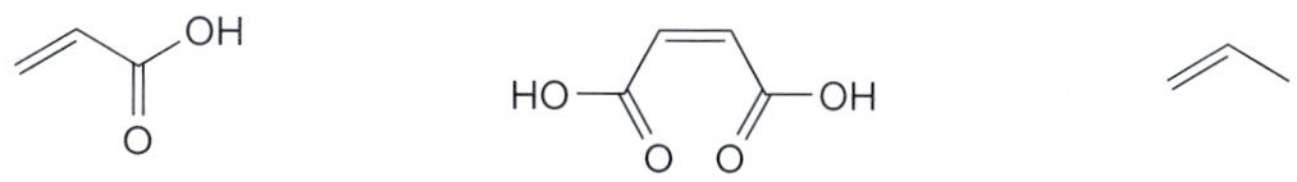

12.13C Stereospecificity

Rule [3] The stereochemistry of the dienophile is retained in the product.

- A cis dienophile forms a cis-substituted cyclohexene.
- A trans dienophile forms a trans-substituted cyclohexene.

The two **cis** CO_2H groups of maleic acid become two **cis** substituents in a Diels–Alder adduct. The CO_2H groups can be drawn both above or both below the plane to afford a single achiral **meso** compound. The **trans dienophile** fumaric acid yields two enantiomers with **trans** CO_2H groups.

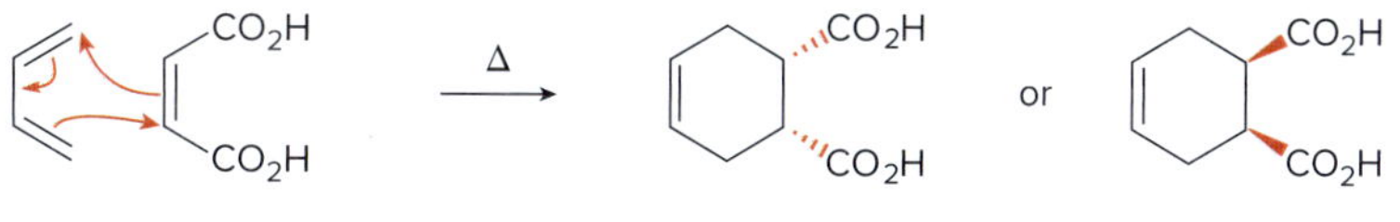

maleic acid
cis dienophile

an achiral meso compound → **cis product**

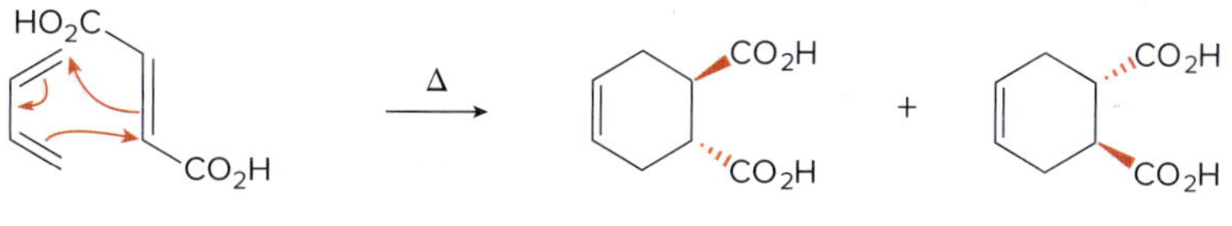

fumaric acid
trans dienophile

enantiomers → **trans product**

494 Chapter 12 Conjugation, Resonance, and Dienes

A **cyclic dienophile** forms a **bicyclic product.** A bicyclic system in which the two rings share a common C–C bond is called a **fused ring system.** The two H atoms at the ring fusion must be cis, because they were cis in the starting dienophile.

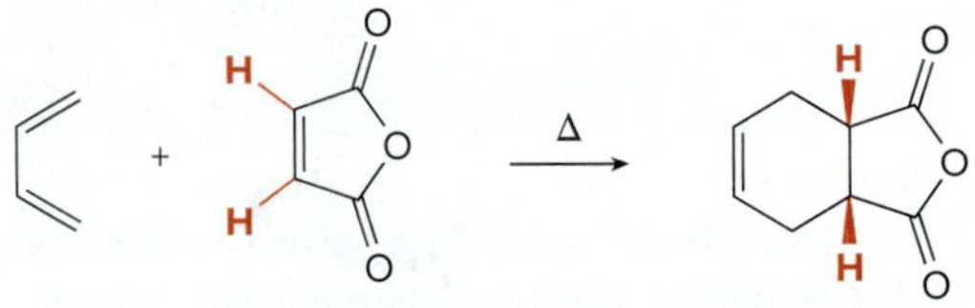

cyclic dienophile   bicyclic product
cis H's in the dienophile   **cis H's** in the product

Problem 12.24 Draw the products of each Diels–Alder reaction, and indicate the stereochemistry.

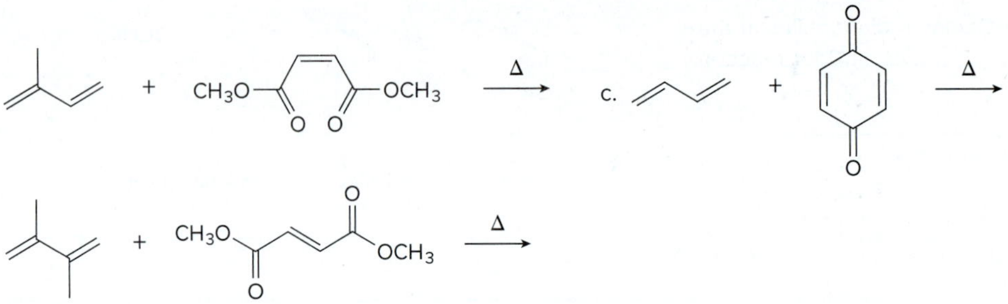

12.13D The Rule of Endo Addition

Rule [4] When endo and exo products are possible, the endo product is preferred.

To understand the rule of endo addition, we must first examine Diels–Alder products that result from cyclic 1,3-dienes. When cyclopentadiene reacts with a dienophile such as ethylene, a new six-membered ring forms, and above the ring there is a **one atom "bridge,"** labeled in green. This carbon atom originated as the sp^3 hybridized carbon of the diene that was not involved in the reaction.

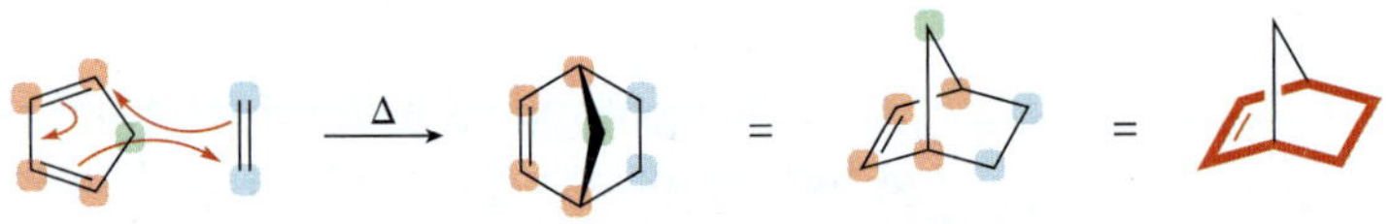

cyclic 1,3-diene   a bridged bicyclic ring system

The product of the Diels–Alder reaction of a cyclic 1,3-diene is bicyclic, but the carbon atoms shared by both rings are *non-adjacent*. Thus, this bicyclic product differs from the fused ring system obtained when the dienophile is cyclic.

- A bicyclic ring system in which the two rings share non-adjacent carbon atoms is called a *bridged* ring system.

Fused and bridged bicyclic ring systems are compared in Figure 12.7.

Figure 12.7
Fused and bridged bicyclic ring systems compared

a. A fused bicyclic system   b. A bridged bicyclic system

- One bond (in red) is shared by two rings.
- The shared C's are adjacent.

- Two non-adjacent atoms (labeled in blue) are shared by both rings.

12.13 Specific Rules Governing the Diels–Alder Reaction

When cyclopentadiene reacts with a substituted alkene as the dienophile (CH$_2$=CHZ), the substituent Z can be oriented in one of two ways in the product. The terms **endo** and **exo** are used to indicate the position of Z.

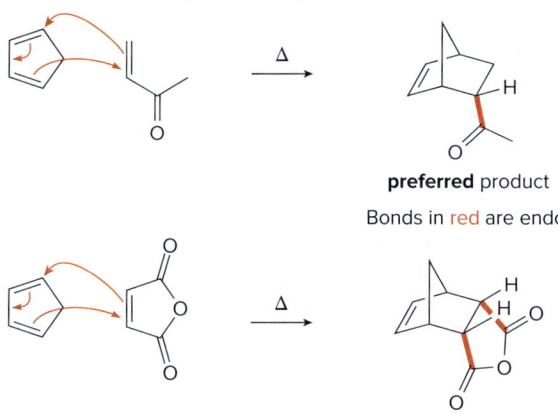

- A substituent on one bridge is *endo* if it is closer to the *longer* bridge that joins the two carbons common to both rings.
- A substituent is *exo* if it is closer to the *shorter* bridge that joins the carbons together.

> To help you distinguish endo and exo, remember that e*n*do is *u*nder the newly formed six-membered ring.
>
> [newly formed ring (in red), Z endo]

In a Diels–Alder reaction, the **endo** product is preferred, as shown in two examples.

[Cyclopentadiene + methyl vinyl ketone → preferred product; Bonds in red are endo.]

[Cyclopentadiene + maleic anhydride → preferred product]

> More details on the Diels–Alder reaction are given in Section 29.4.

The Diels–Alder reaction is **concerted**, and the reaction occurs with the diene and the dienophile arranged one above the other, as shown in Figure 12.8, not side-by-side. In theory,

Figure 12.8 How endo and exo products are formed in the Diels–Alder reaction

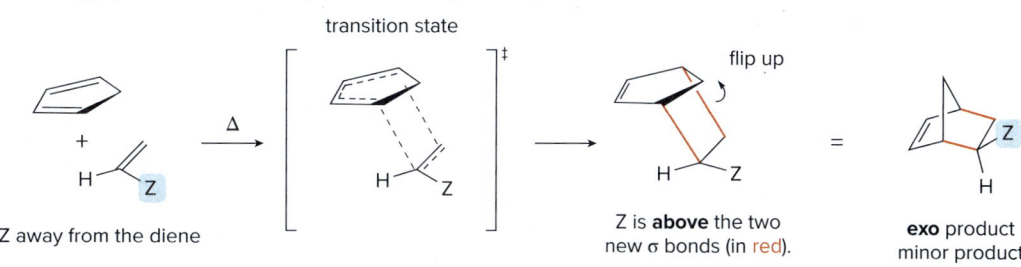

Pathway [1] With Z oriented under the diene, the endo product is formed.

Z under the diene → transition state (The electron-withdrawing Z group is closer to the electron-rich diene.) → flip up; Z is **below** the two new σ bonds (in red). = **endo** product, major product

Pathway [2] With Z oriented away from the diene, the exo product is formed.

Z away from the diene → transition state → flip up; Z is **above** the two new σ bonds (in red). = **exo** product, minor product the substituent Z can be oriented either directly *under* the diene to form the endo product (Pathway [1] in Figure 12.8) or *away* from the diene to form the exo product (Pathway [2] in Figure 12.8). In practice, though, **the endo product is the major product. The transition state leading to the endo product allows more interaction between the electron-rich diene and the electron-withdrawing substituent Z on the dienophile,** an energetically favorable arrangement.

Problem 12.25 Draw the product of each Diels–Alder reaction.

12.14 Other Facts About the Diels–Alder Reaction

12.14A Retrosynthetic Analysis of a Diels–Alder Product

The Diels–Alder reaction is used widely in organic synthesis, so you must be able to look at a compound and determine what conjugated diene and what dienophile were used to make it. To draw the starting materials from a given Diels–Alder adduct:

- Locate the six-membered ring that contains the C=C.
- Draw three arrows around the cyclohexene ring, beginning with the π bond. Each arrow moves two electrons to the adjacent bond, cleaving one π bond and two σ bonds, and forming three π bonds.
- Retain the stereochemistry of substituents on the C=C of the dienophile. Cis substituents on the six-membered ring give a cis dienophile.

This stepwise retrosynthetic analysis gives the 1,3-diene and dienophile needed for any Diels–Alder reaction, as shown in the two examples in Figure 12.9.

Figure 12.9 Finding the diene and dienophile needed for a Diels–Alder reaction

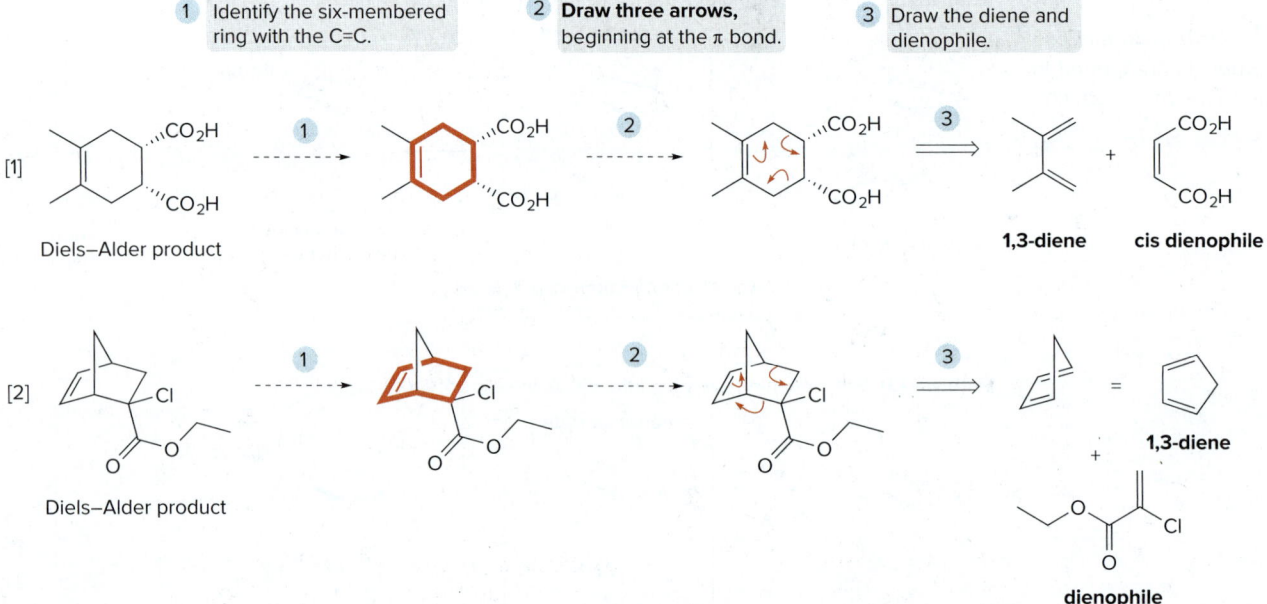

12.14 Other Facts About the Diels–Alder Reaction

Problem 12.26 What diene and dienophile are needed to prepare each product?

12.14B Retro Diels–Alder Reaction

A reactive diene like cyclopenta-1,3-diene readily undergoes a Diels–Alder reaction with *itself;* that is, **cyclopenta-1,3-diene dimerizes because one molecule acts as the diene and another acts as the dienophile.**

The formation of dicyclopentadiene is so rapid that it takes only a few hours at room temperature for cyclopentadiene to completely dimerize. How, then, can cyclopentadiene be used in a Diels–Alder reaction if it really exists as a dimer?

When heated, dicyclopentadiene undergoes a **retro Diels–Alder reaction,** and two molecules of cyclopentadiene are re-formed. If cyclopentadiene is immediately treated with a different dienophile, it reacts to form a new Diels–Alder adduct with this dienophile.

Problem 12.27 Draw the products of each reaction sequence.

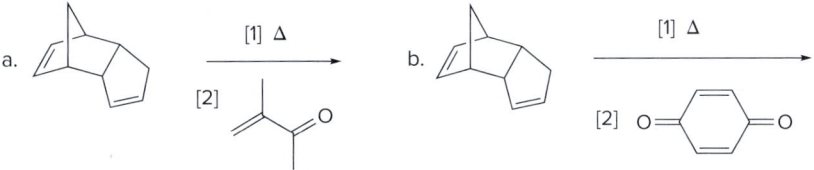

12.14C Application: Diels–Alder Reaction in the Synthesis of Steroids

Steroids **are tetracyclic lipids containing three six-membered rings and one five-membered ring.** The four rings are designated as **A, B, C,** and **D.**

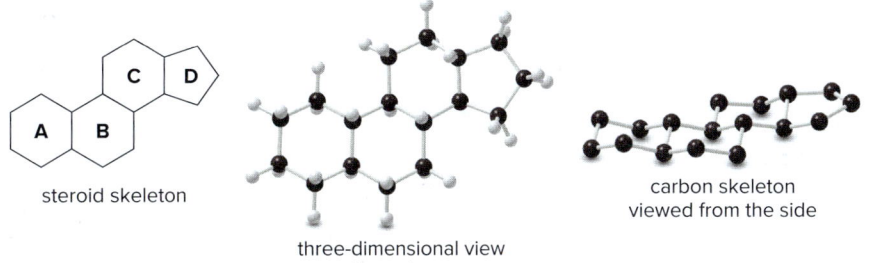

Steroids exhibit a wide range of biological properties, depending on the substitution pattern of functional groups on the rings. They include **cholesterol** (a component of cell membranes that is implicated in cardiovascular disease), **estrone** (a female sex hormone responsible for the regulation of the menstrual cycle), and **cortisone** (a hormone responsible for the control of inflammation and the regulation of carbohydrate metabolism).

cholesterol   estrone   cortisone

Diels–Alder reactions have been used widely in the laboratory syntheses of steroids. The key Diels–Alder reaction used to prepare the C ring of estrone is drawn.

diene   dienophile   Diels–Alder product   estrone

Problem 12.28 Draw the product (**A**) of the following Diels–Alder reaction. **A** was a key intermediate in the synthesis of the addicting pain reliever morphine.

morphine

Chapter 12 REVIEW

KEY CONCEPTS

[1] Four common examples of resonance (12.3)

① The three-atom "allyl" system	② Conjugated double bonds	③ Cations having a positive charge adjacent to a lone pair	④ Double bonds involving one atom more electronegative than the other

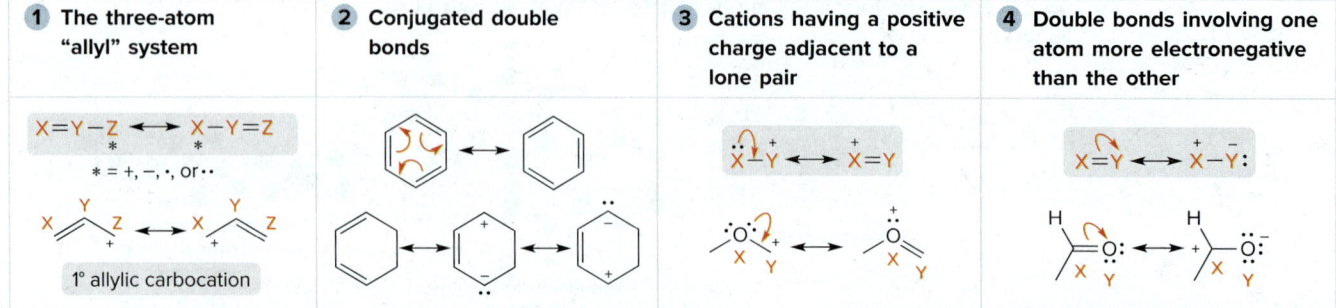

See Sample Problem 12.1. Try Problems 12.32, 12.34.

Key Skills 499

[2] The difference between two conformations and two stereoisomers in 1,3-dienes (12.6)

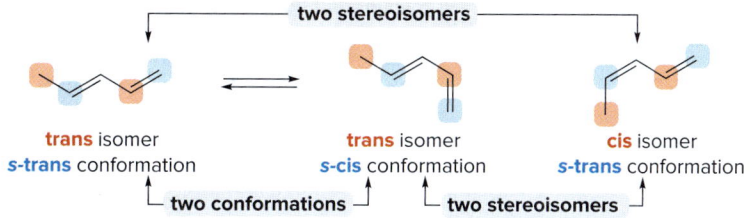

See Sample Problem 12.4. Try Problem 12.38.

[3] Relative reactivity in the Diels–Alder reaction

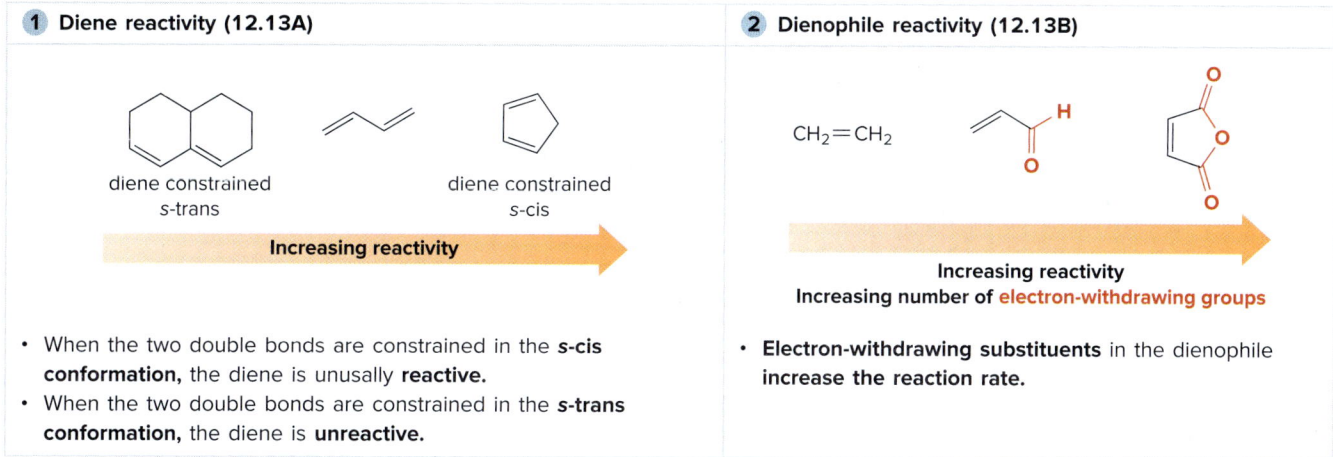

- When the two double bonds are constrained in the **s-cis** conformation, the diene is unusually **reactive**.
- When the two double bonds are constrained in the **s-trans** conformation, the diene is **unreactive**.

- **Electron-withdrawing substituents** in the dienophile increase the reaction rate.

Try Problem 12.53b, c.

KEY REACTIONS

Reactions of conjugated dienes

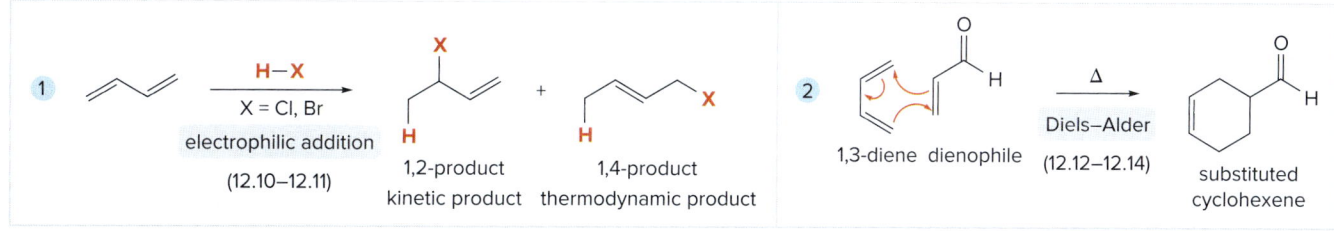

Try Problems 12.40b, c; 12.46; 12.55.

KEY SKILLS

[1] Drawing resonance structures for a conjugated compound (12.3, 12.4)

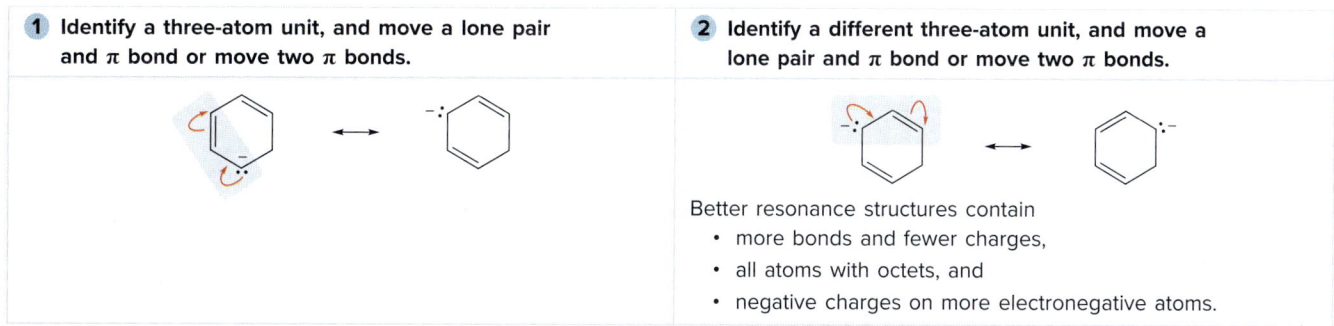

Better resonance structures contain
- more bonds and fewer charges,
- all atoms with octets, and
- negative charges on more electronegative atoms.

See Sample Problem 12.1. Try Problems 12.32, 12.34.

[2] Drawing a resonance hybrid from three resonance structures (12.4); example: acrolein

1 Draw the three resonance structures.	**2** Draw the bonds and partial bonds in the resonance hybrid.	**3** Draw the partial charges.
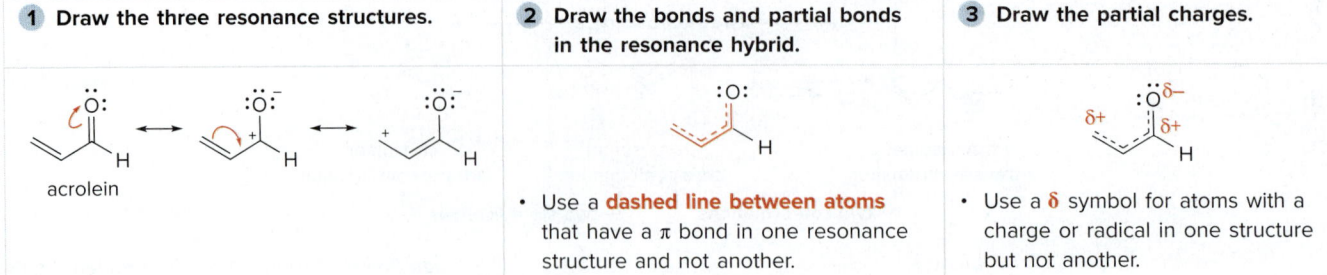	• Use a **dashed line between atoms** that have a π bond in one resonance structure and not another.	• Use a **δ** symbol for atoms with a charge or radical in one structure but not another.

See Sample Problem 12.2. Try Problem 12.33.

[3] Determining the hybridization around an atom when there is resonance (12.5)

1 Draw the two resonance structures.	**2** Determine the hybridization based on orbitals used in both resonance structures.
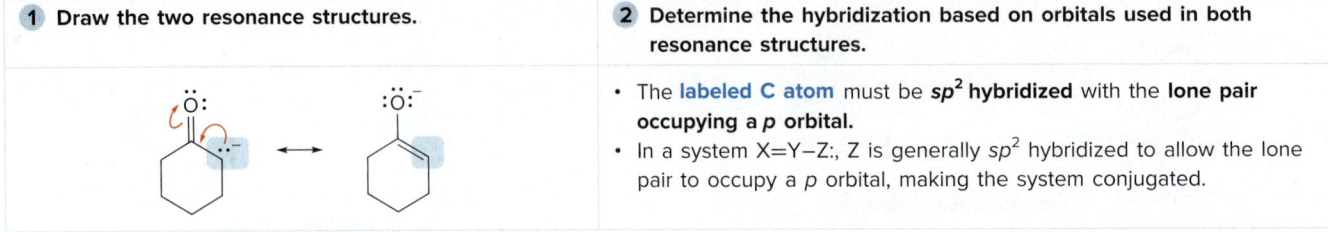	• The **labeled C atom** must be sp^2 hybridized with the **lone pair** occupying a p orbital. • In a system X=Y–Z:, Z is generally sp^2 hybridized to allow the lone pair to occupy a p orbital, making the system conjugated.

See Sample Problem 12.3.

[4] Drawing the products from HX addition to a diene (12.10)

1 Add H⁺ to form an allylic carbocation.	**2** Draw both resonance structures for the carbocation.	**3** Draw the products resulting from nucleophilic attack at the positive charge of both resonance structures.

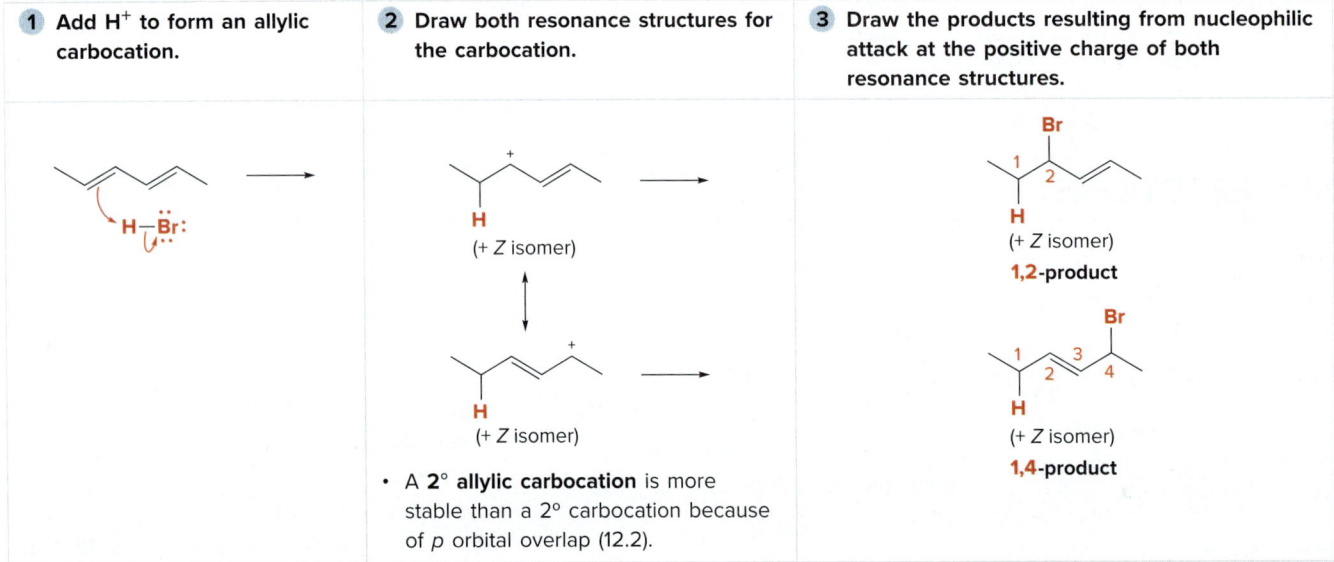

• A 2° allylic carbocation is more stable than a 2° carbocation because of p orbital overlap (12.2).

See Sample Problem 12.5. Try Problems 12.40b, c; 12.55a, d.

[5] Determining the kinetic and thermodynamic products (12.11)

1 Consider HCl addition at the indicated C=C (in red).	**2** Draw both products, and identify the kinetic product and the thermodynamic product.

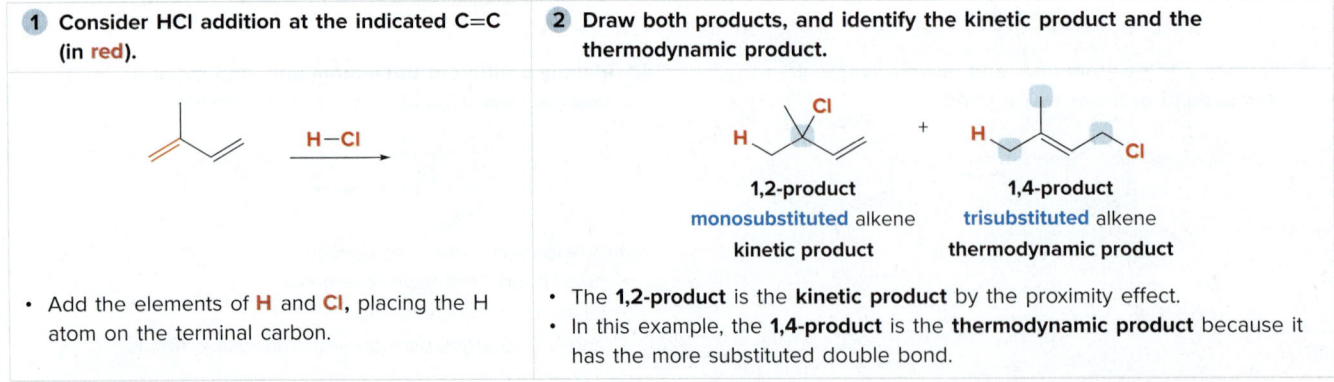

• Add the elements of **H** and **Cl**, placing the H atom on the terminal carbon.
• The **1,2-product** is the **kinetic product** by the proximity effect.
• In this example, the **1,4-product** is the **thermodynamic product** because it has the more substituted double bond.

Try Problems 12.43, 12.44.

Problems

[6] Drawing the product of a Diels–Alder reaction (12.12)

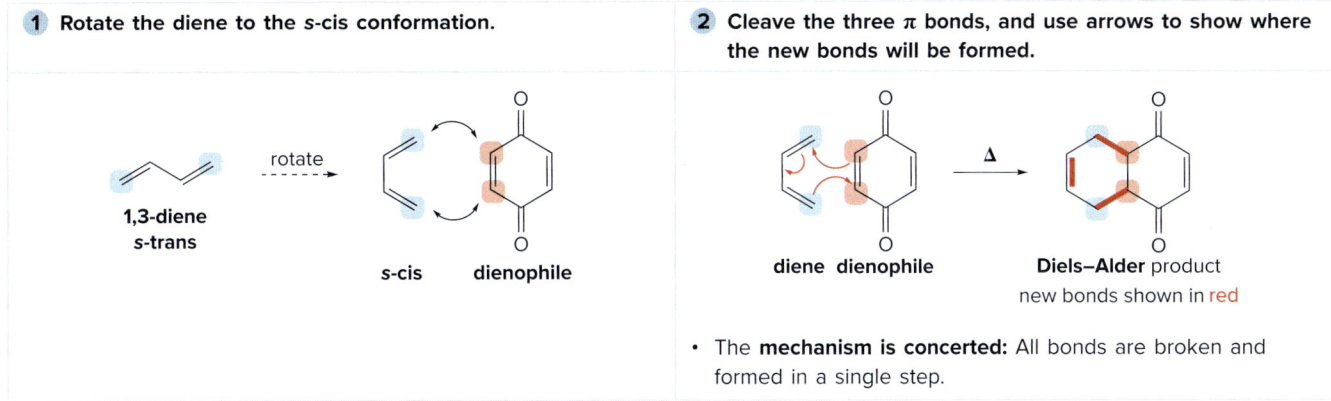

- The **mechanism is concerted:** All bonds are broken and formed in a single step.

See *How To*, p. 491. Try Problem 12.46.

[7] Finding the diene and dienophile needed for a Diels–Alder reaction (12.14)

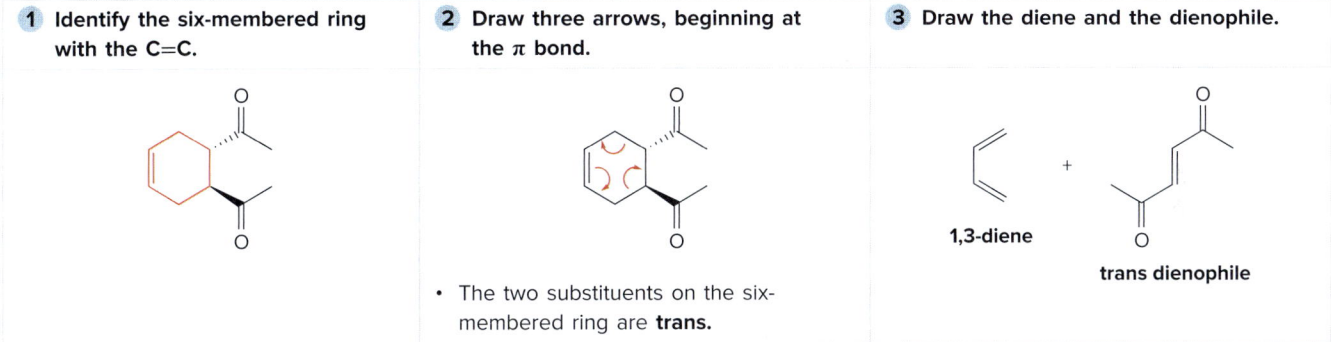

- The two substituents on the six-membered ring are **trans.**

See Figure 12.9. Try Problems 12.30, 12.47, 12.48.

PROBLEMS

Problems Using Three-Dimensional Models

12.29 Name each diene and state whether the ball-and-stick model shows the diene in the *s*-cis or *s*-trans conformation.

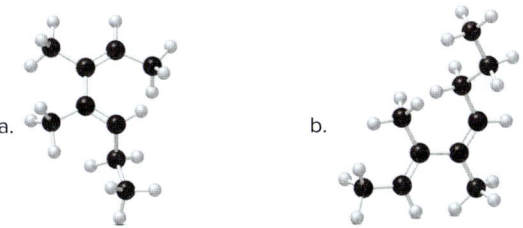

12.30 What diene and dienophile are needed to prepare each compound by a Diels–Alder reaction?

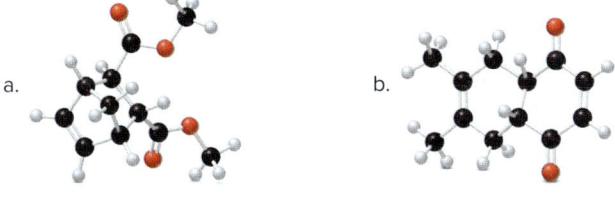

Conjugation

12.31 Which of the following systems are conjugated?

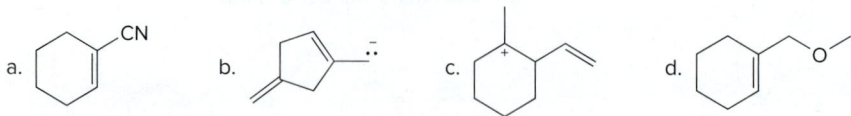

Resonance and Hybridization

12.32 Draw all reasonable resonance structures for each species.

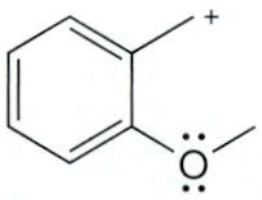

12.33 For which compounds can a second resonance structure be drawn? Draw an additional resonance structure and the hybrid for each resonance-stabilized compound.

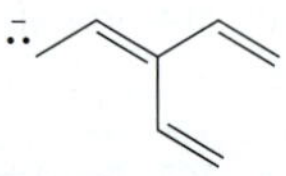

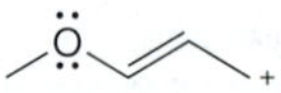

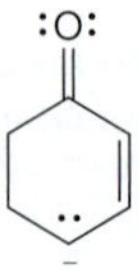

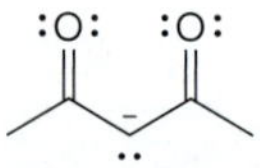

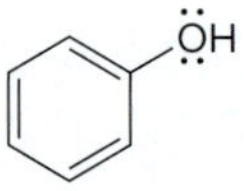

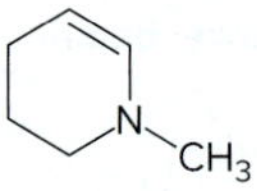

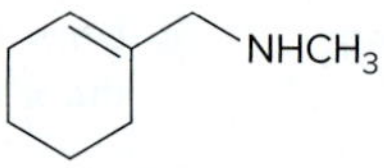

12.34 Draw all reasonable resonance structures for each compound.

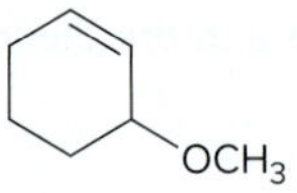

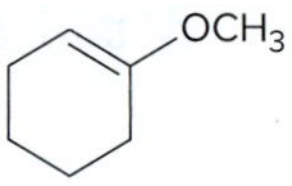

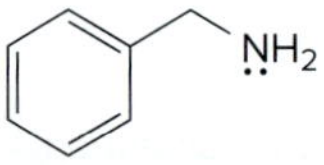

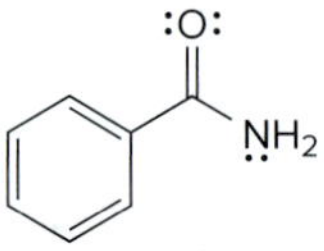

12.35 Why is the bond dissociation energy for the C—C bond in ethane much higher than the bond dissociation energy for the labeled C—C bond in but-1-ene?

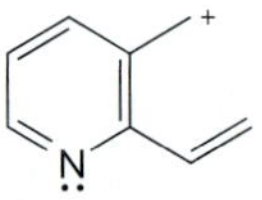

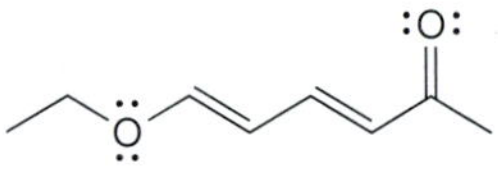

$$CH_3 \text{—} CH_3$$
ethane
+368 kJ/mol

but-1-ene
+301 kJ/mol

Nomenclature and Stereoisomers in Conjugated Dienes

12.36 Draw the structure of each compound.
 a. (Z)-penta-1,3-diene in the s-trans conformation
 b. (2E,4Z)-1-bromo-3-methylhexa-2,4-diene
 c. (2E,4E,6E)-octa-2,4,6-triene
 d. (2E,4E)-3-methylhexa-2,4-diene in the s-cis conformation

12.37 Name each compound and indicate the conformation around the σ bond that joins the two double bonds.

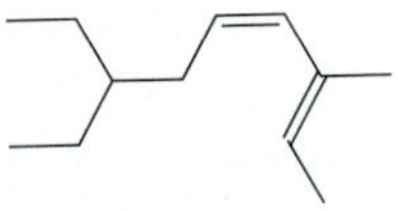

12.38 Label each pair of compounds as stereoisomers, conformations, or constitutional isomers: (a) **A** and **B**; (b) **A** and **C**; (c) **A** and **D**; (d) **C** and **D**.

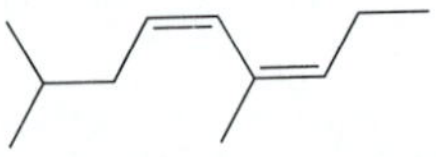

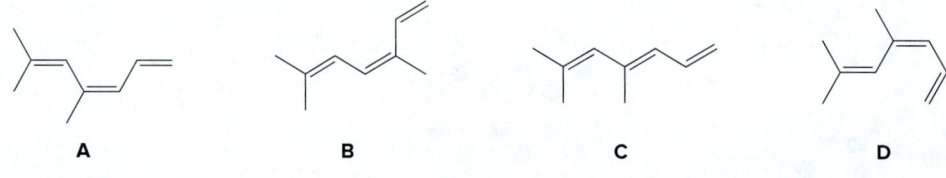

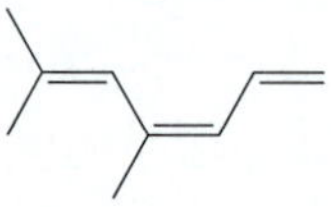

A   B   C   D

12.39 Rank the following dienes in order of increasing heat of hydrogenation.

Electrophilic Addition

12.40 Draw the products formed when each compound is treated with one equivalent of HBr.

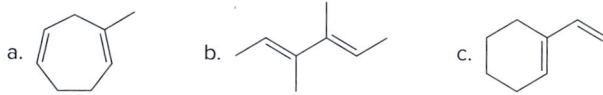

12.41 Ignoring stereoisomers, draw all products that form by addition of HBr to (E)-hexa-1,3,5-triene.

12.42 Treatment of alkenes **A** and **B** with HBr gives the same alkyl halide **C**. Draw a mechanism for each reaction, including all reasonable resonance structures for any intermediate.

12.43 Addition of HCl to alkene **X** forms two alkyl halides **Y** and **Z**.

a. Label **Y** and **Z** as a 1,2-addition product or a 1,4-addition product.
b. Label **Y** and **Z** as the kinetic or thermodynamic product and explain why.
c. Explain why addition of HCl occurs at the indicated C=C (called an exocyclic double bond), rather than the other C=C (called an endocyclic double bond).

12.44 The major product formed by addition of HBr to $(CH_3)_2C=CH-CH=C(CH_3)_2$ is the same at low and high temperature. Draw the structure of the major product, and explain why the kinetic and thermodynamic products are the same in this reaction.

12.45 From what you have learned about the reaction of conjugated dienes in Section 12.10, predict the products of each of the following electrophilic additions.

Diels–Alder Reaction

12.46 Draw the products of the following Diels–Alder reactions. Indicate stereochemistry where appropriate.

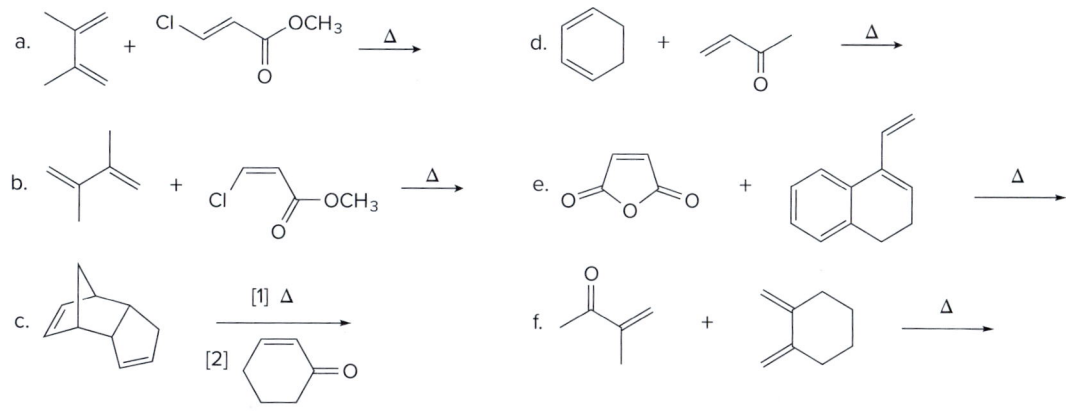

12.47 What diene and dienophile are needed to prepare each Diels–Alder product?

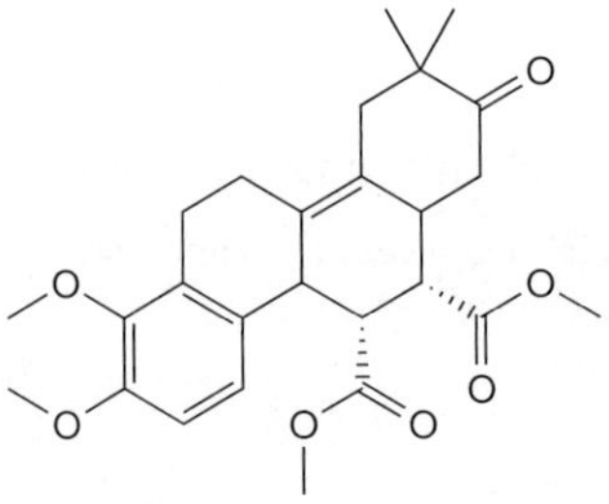

12.48 Give two different ways to prepare the following compound by the Diels–Alder reaction. Explain which method is preferred.

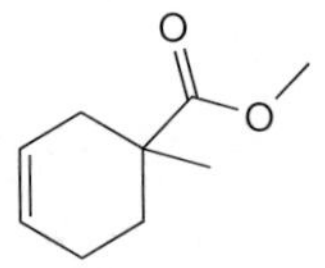

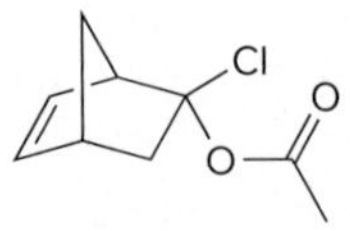

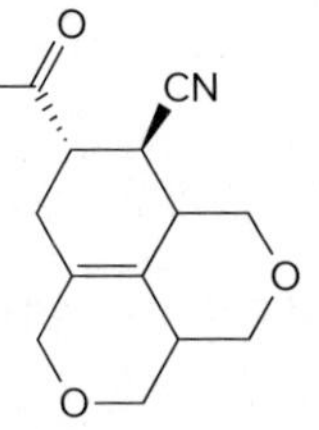

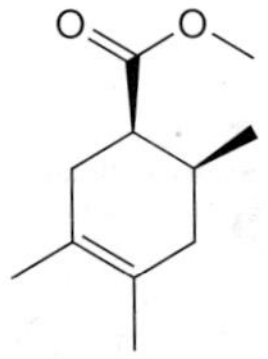

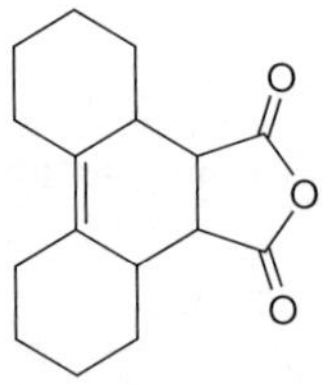

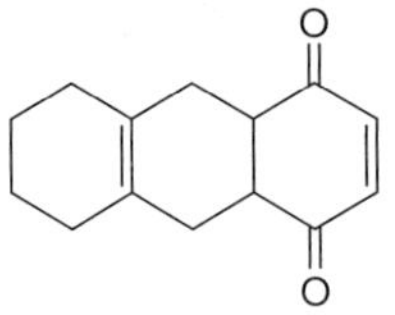

12.49 Compounds containing triple bonds are also Diels–Alder dienophiles. With this in mind, draw the products of each reaction.

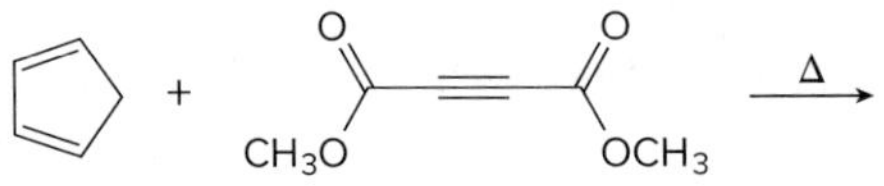

12.50 Diels–Alder reaction of a monosubstituted diene (such as $CH_2=CH-CH=CHOCH_3$) with a monosubstituted dienophile (such as $CH_2=CHCHO$) gives a mixture of products, but the 1,2-disubstituted product often predominates. Draw the resonance hybrid for each reactant, and use the charge distribution of the hybrids to explain why the 1,2-disubstituted product is the major product.

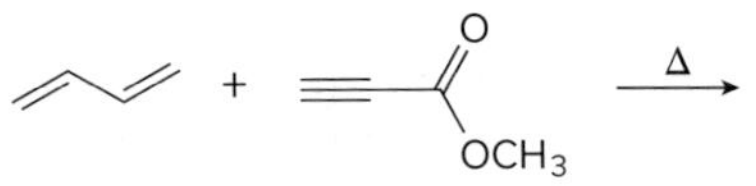

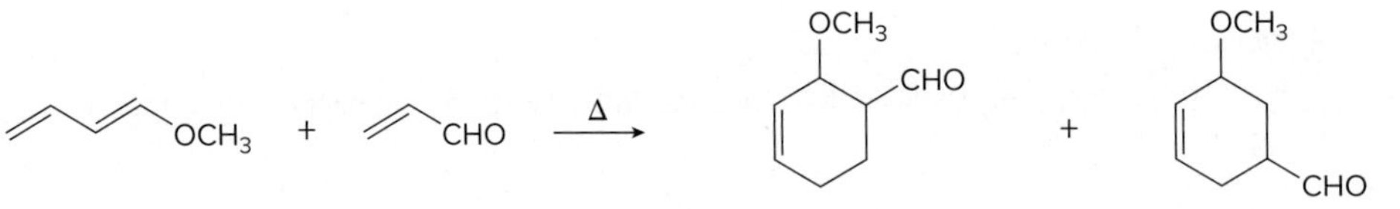

1,2-disubstituted product
major

1,3-disubstituted product
minor

12.51 Devise a stepwise synthesis of each compound from dicyclopentadiene using a Diels–Alder reaction as one step. You may also use organic compounds having ≤ 4 C's, and any required organic or inorganic reagents.

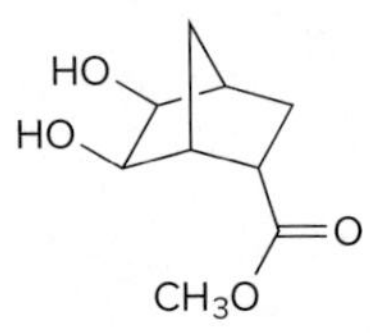

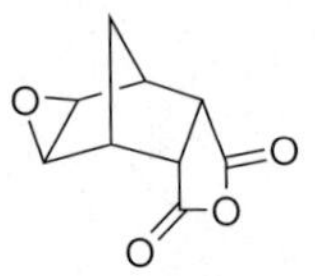

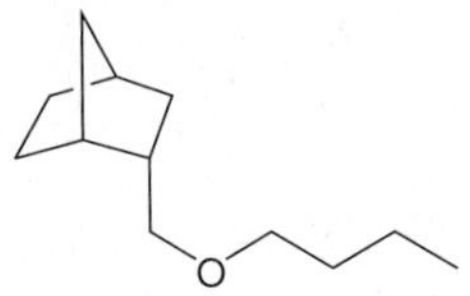

12.52 Intramolecular Diels–Alder reactions are possible when a substrate contains both a 1,3-diene and a dienophile, as shown in the following general reaction.

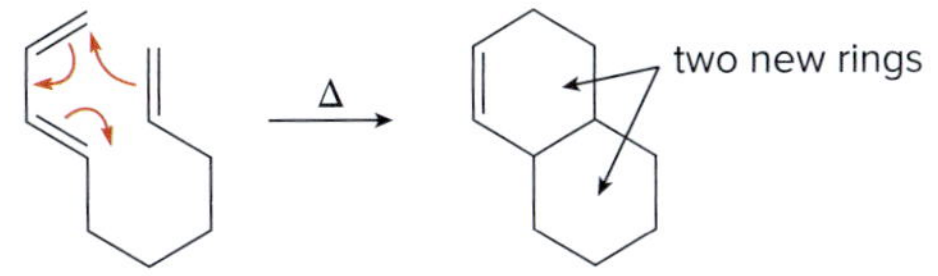

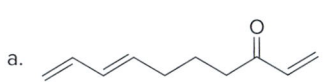

With this in mind, draw the product when each compound undergoes an intramolecular Diels–Alder reaction.

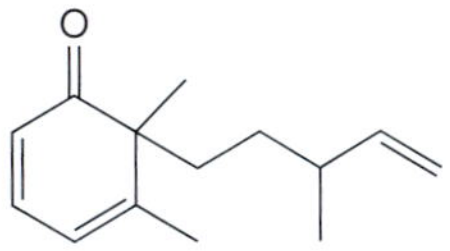

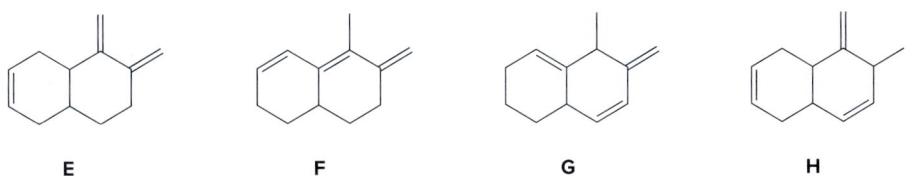

General Problems

12.53 Consider the four trienes **E–H**.

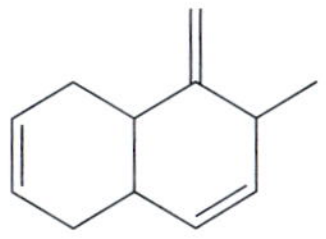

a. Rank compounds **E–H** in order of increasing heat of hydrogenation.
b. Which compound is most reactive in the Diels–Alder reaction?
c. Which compound(s) are unreactive in the Diels–Alder reaction?

12.54 Draw a stepwise mechanism for the following reaction.

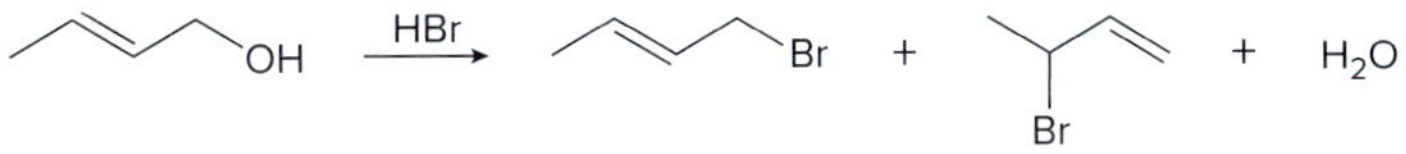

12.55 Draw the products of each reaction. Indicate the stereochemistry of Diels–Alder products.

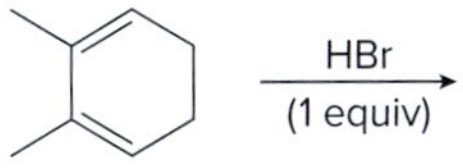

12.56 Draw a stepwise mechanism for the biological conversion of linalyl diphosphate to limonene.

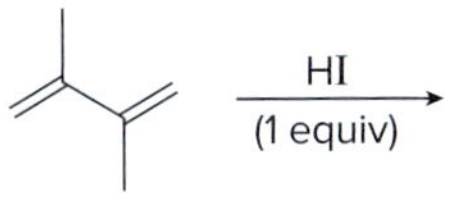

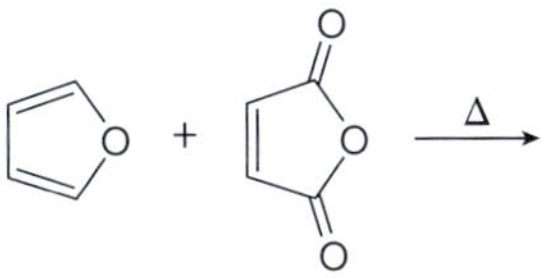

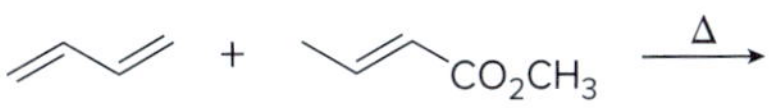

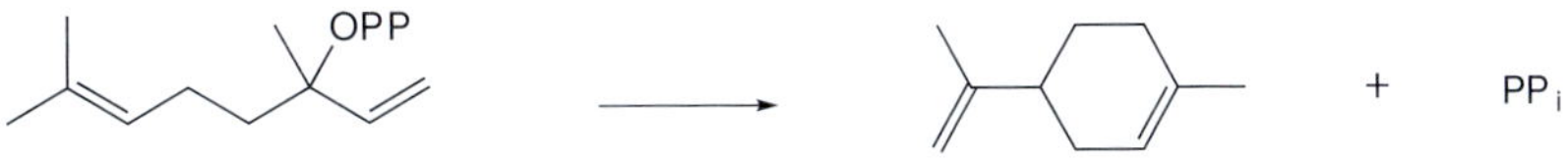

linalyl diphosphate → limonene + PP$_i$

12.57 Which benzylic halide reacts faster in an S_N1 reaction? Explain.

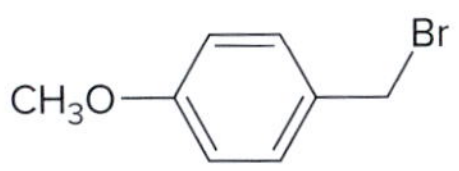

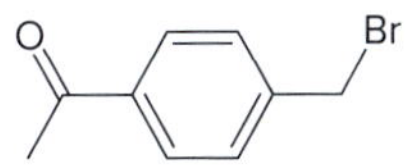

506 Chapter 12 Conjugation, Resonance, and Dienes

12.58 Like alkenes, conjugated dienes can be prepared by elimination reactions. Draw a stepwise mechanism for the acid-catalyzed dehydration of 3-methylbut-2-en-1-ol [$(CH_3)_2C=CHCH_2OH$] to isoprene [$CH_2=C(CH_3)CH=CH_2$].

12.59 (a) Draw the two isomeric dienes formed when $CH_2=CHCH_2CH(Cl)CH(CH_3)_2$ is treated with an alkoxide base. (b) Explain why the major product formed in this reaction does not contain the more highly substituted alkene.

Spectroscopy

Problem 12.60 is intended for students who have already learned about spectroscopy in Chapters A–C.

12.60 The treatment of isoprene [$CH_2=C(CH_3)CH=CH_2$] with one equivalent of mCPBA forms **A** as the major product. **A** gives a molecular ion at 84 in its mass spectrum, and peaks at 2850–3150 cm^{-1} in its IR spectrum. The ^{1}H NMR spectrum of **A** is given below. What is the structure of **A**?

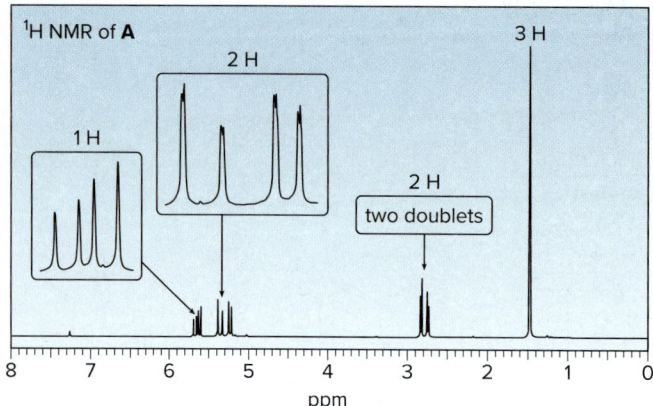

Challenge Problems

12.61 Addition of HBr to allene ($CH_2=C=CH_2$) forms 2-bromoprop-1-ene rather than 3-bromoprop-1-ene, even though 3-bromoprop-1-ene is formed from an allylic carbocation. Considering the arrangement of orbitals in the allene reactant, explain this result.

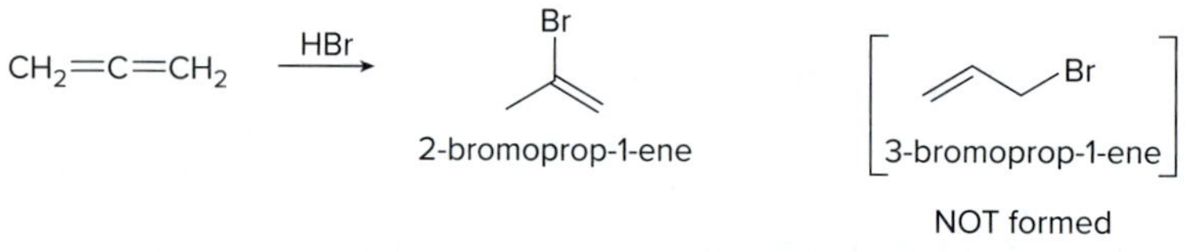

$CH_2=C=CH_2$ $\xrightarrow{HBr}$ 2-bromoprop-1-ene   [3-bromoprop-1-ene] NOT formed

12.62 Determine the hybridization around the N atom in each amine, and explain why cyclohexanamine is 10^6 times more basic than aniline.

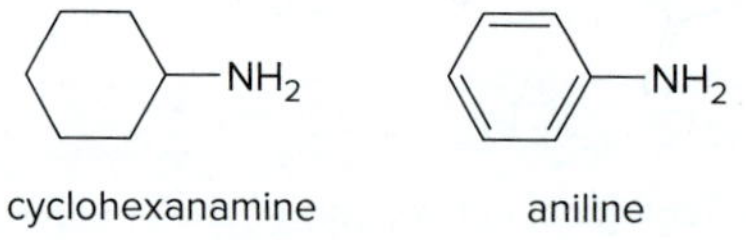

cyclohexanamine   aniline

12.63 Devise a synthesis of **X** from the given starting materials. You may use any organic or inorganic reagents. Account for the stereochemistry observed in **X**.

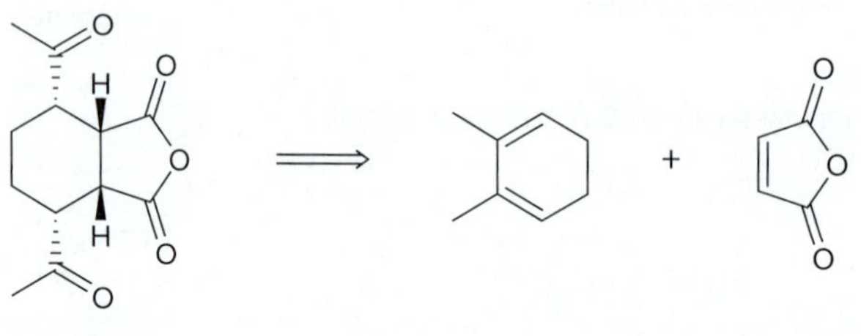

X

12.64 One step in the synthesis of occidentalol, a natural product isolated from the eastern white cedar tree, involves the following reaction. Identify the structure of **A** and show how **A** is converted to **B**.

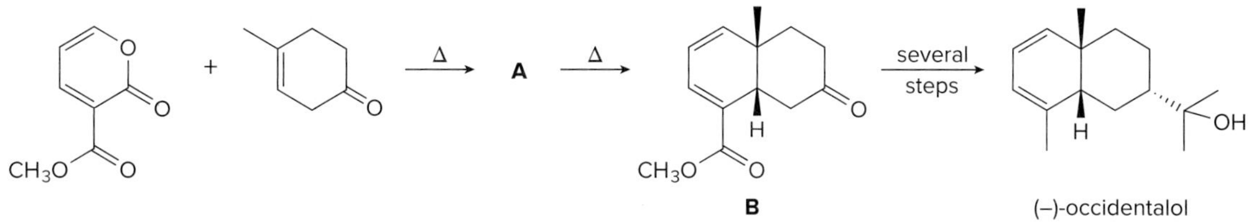

12.65 Dodecahedrane is a polycyclic hydrocarbon that contains 12 five-membered rings joined together to form a sphere. One step in the synthesis of dodecahedrane involves reaction of the tetraene **C** with dimethylacetylene dicarboxylate (**D**) to afford two compounds having molecular formula $C_{16}H_{16}O_4$. This reaction has been called a domino Diels–Alder reaction. Identify the two products formed.

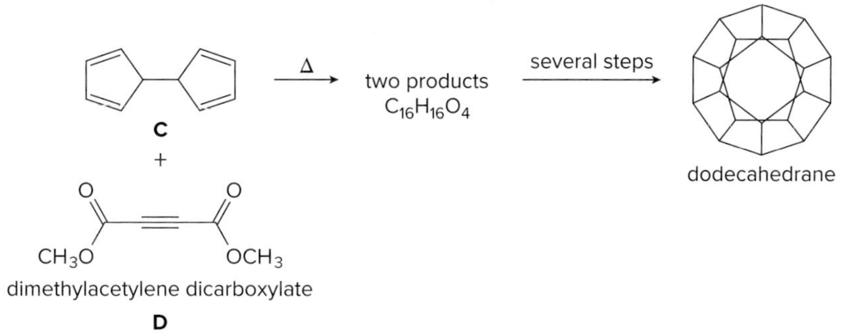

12.66 Devise a stepwise mechanism for the conversion of **M** to **N**. **N** has been converted in several steps to lysergic acid, a naturally occurring precursor of the hallucinogen LSD, the chapter-opening molecule.

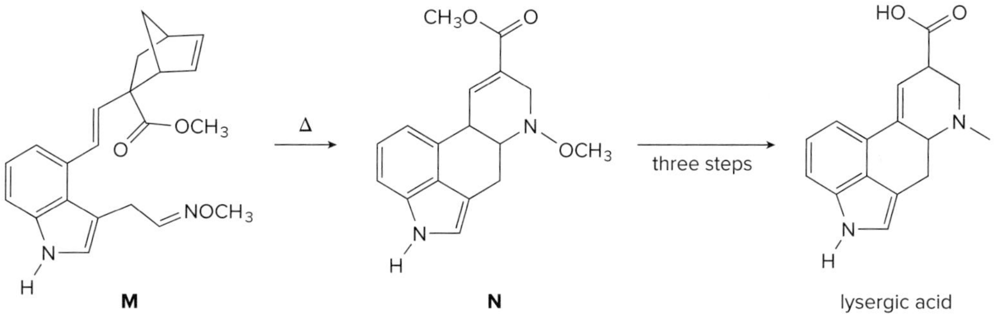

SPECTROSCOPY A

Mass Spectrometry

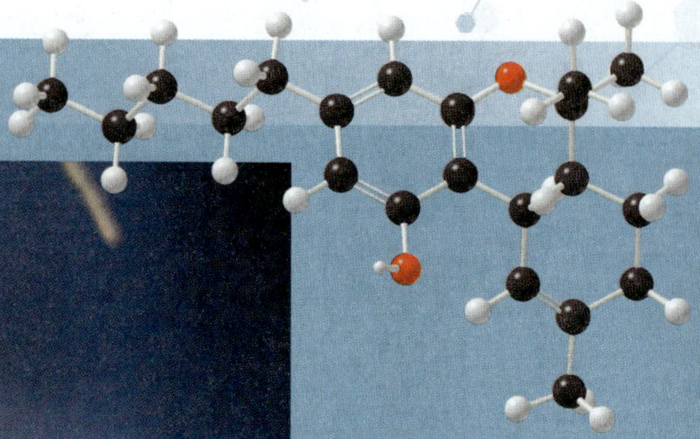

Adam Bailleaux/Atomazul/123RF

- **A.1** Mass spectrometry and the molecular ion
- **A.2** Alkyl halides and the M + 2 peak
- **A.3** Fragmentation
- **A.4** Fragmentation patterns of some common functional groups
- **A.5** Other types of mass spectrometry

Tetrahydrocannabinol (THC), first isolated from Indian hemp, is the primary active constituent of cannabis. The recreational use of cannabis has been legalized in several parts of the United States, and the medical use of THC as an anti-nausea agent for chemotherapy patients and as an appetite stimulant for AIDS-related anorexia is well documented. Like other controlled substances, THC can be detected in minute amounts using modern instrumental methods. In Spectroscopy Part A, we examine mass spectrometry, a method to determine the molecular weight of an organic compound.

508

Why Study... Spectroscopy?

Whether a compound is prepared in the laboratory or isolated from a natural source, a chemist must determine its identity. Seventy years ago, determining the structure of an organic compound involved a series of time-consuming operations: measuring physical properties (melting point, boiling point, solubility, and density), identifying the functional groups using a series of chemical tests, and converting an unknown compound into another compound whose physical and chemical properties were then characterized as well.

Although still a challenging task, structure determination has been greatly simplified by modern instrumental methods. These techniques have both decreased the time needed for compound characterization, and increased the complexity of compounds whose structures can be completely determined.

In Spectroscopy A, we are introduced to **mass spectrometry (MS),** which is used to determine the molecular weight and molecular formula of a compound. In Spectroscopy B, we learn how **infrared (IR) spectroscopy** is used to identify a compound's functional groups. Spectroscopy C is devoted to **nuclear magnetic resonance (NMR) spectroscopy,** which is used to identify the carbon–hydrogen framework in a compound, making it the most powerful spectroscopic tool for organic structure analysis. Each method provides valuable information for determining the structure of an organic compound. These three methods rely on the interaction of an energy source with a molecule to produce a change that is recorded in a spectrum.

A.1 Mass Spectrometry and the Molecular Ion

Mass spectrometry **is a technique used for measuring the molecular weight and determining the molecular formula of an organic molecule.**

A.1A General Features

In the most common type of **mass spectrometer,** a molecule is vaporized and ionized, usually by bombardment with a beam of high-energy electrons, as shown in Figure A.1. The energy

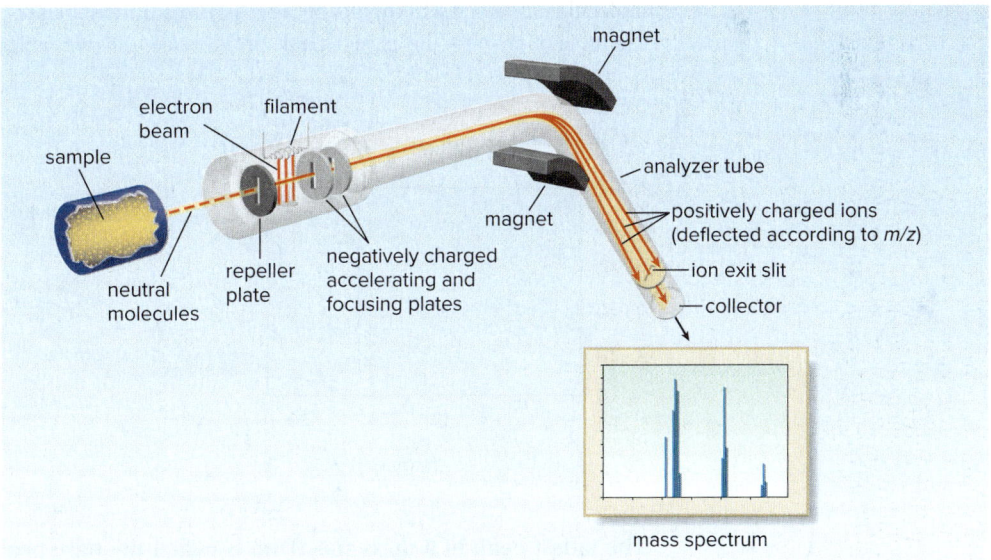

Figure A.1 Schematic of a mass spectrometer

- In a mass spectrometer, a sample is vaporized and bombarded by a beam of electrons to form an unstable radical cation, which then decomposes to smaller fragments. The positively charged ions are accelerated toward a negatively charged plate, and then passed through a curved analyzer tube in a magnetic field, where they are deflected by different amounts depending on their ratio of mass to charge (*m/z*). A mass spectrum plots the intensity of each ion versus its *m/z* ratio.

Spectroscopy A Mass Spectrometry

> The term **spectroscopy** is usually used for techniques that use electromagnetic radiation as an energy source. Because the energy source in MS is a beam of electrons, the term **mass spectrometry** is used instead.

of these electrons is typically about 6400 kJ, or 70 electron volts (eV). This electron beam ionizes a molecule by causing it to eject an electron.

$$\text{M} \xrightarrow{\text{e}^-\text{ beam of electrons}} \text{M}^{+\cdot} + \text{e}^-$$
molecule → radical cation

The species formed is a **radical cation,** symbolized $\text{M}^{+\cdot}$. It is a radical because it has an unpaired electron, and it is a cation because it has one fewer electron than it started with.

- The radical cation $\text{M}^{+\cdot}$ is called the *molecular ion* or the *parent ion.*

A single electron has a negligible mass, so the **mass of $\text{M}^{+\cdot}$ represents the molecular weight** of M. Because the molecular ion $\text{M}^{+\cdot}$ is inherently unstable, it decomposes. Single bonds break to form *fragments,* radicals and cations having a lower molecular weight than the molecular ion. A mass spectrometer analyzes the masses of cations only. The cations are accelerated in an electric field and deflected in a curved path in a magnetic field, thus sorting the molecular ion and its fragments by their **mass-to-charge (*m/z*) ratio.** Because z is almost always +1, *m/z* actually measures the mass (m) of the individual ions.

$$\text{M} \xrightarrow{\text{e}^-} \text{M}^{+\cdot} \longrightarrow \text{radicals} + \text{cations}$$
molecule   unstable radical cation   These fragments are analyzed.

- A *mass spectrum* plots the amount of each cation (its relative abundance) versus its mass.

> The whole-number mass of CH_4 is (1 C × 12 amu) + (4 H × 1 amu) = 16 amu; amu = atomic mass unit.

A mass spectrometer analyzes the masses of *individual* molecules, not the weighted average mass of a group of molecules, so the whole-number masses of the most common individual isotopes must be used to calculate the mass of the molecular ion. Thus, the mass of the molecular ion for CH_4 should be 16. As a result, the mass spectrum of CH_4 shows a line for the molecular ion—the parent peak or **M** peak—at $m/z = 16$.

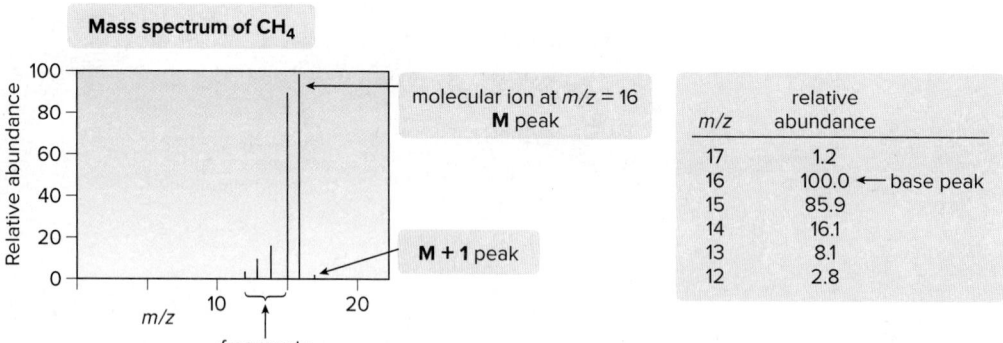

The tallest peak in a mass spectrum is called the base peak. For CH_4, the base peak is also the M peak, although this may *not* always be the case for all organic compounds.

The mass spectrum of CH_4 consists of more peaks than just the M peak. What is responsible for the peaks at $m/z < 16$? Because the molecular ion is unstable, it fragments into other cations and radical cations containing one, two, three, or four fewer hydrogen atoms than methane itself. Thus, the peaks at $m/z = 15, 14, 13,$ and 12, are due to these lower-molecular-weight fragments. The decomposition of a molecular ion into lower-molecular-weight fragments is called **fragmentation**.

$$CH_4 \xrightarrow{e^-} \underset{\text{mass 16}}{(CH_4)^{+\cdot}} \xrightarrow{-H\cdot} \underset{\text{mass 15}}{CH_3^+} \xrightarrow{-H\cdot} \underset{\text{mass 14}}{CH_2^{+\cdot}} \xrightarrow{-H\cdot} \underset{\text{mass 13}}{CH^+} \xrightarrow{-H\cdot} \underset{\text{mass 12}}{C^{+\cdot}}$$

molecular ion | fragments

What is responsible for the small peak at $m/z = 17$ in the mass spectrum of CH_4? Although most carbon atoms have an atomic mass of 12, 1.1% of them have an additional neutron in the nucleus, giving them an atomic mass of 13. When one of these carbon-13 isotopes forms methane, it gives a molecular ion peak at $m/z = 17$ in the mass spectrum. This peak is called the **M + 1** peak.

These key features—the molecular ion, the base peak, and the M + 1 peak—are illustrated in the mass spectrum of hexane in Figure A.2.

Figure A.2

Mass spectrum of hexane ($CH_3CH_2CH_2CH_2CH_2CH_3$)

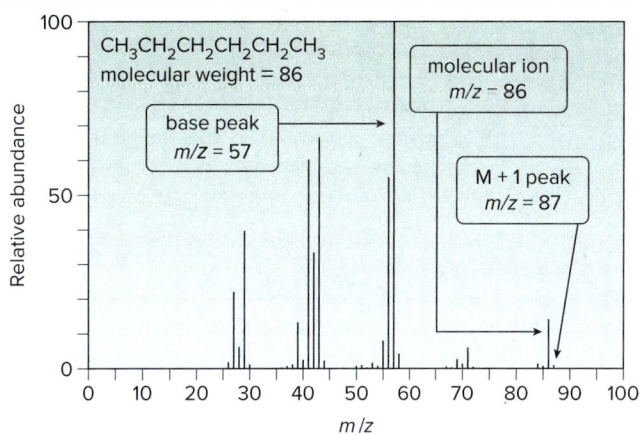

- The molecular ion for hexane (molecular formula C_6H_{14}) is at $m/z = 86$.
- The base peak (relative abundance = 100) occurs at $m/z = 57$.
- A small M + 1 peak occurs at $m/z = 87$.

Problem A.1 Label the molecular ion, the base peak, and the M + 1 peak in the mass spectrum of pentane (C_5H_{12}).

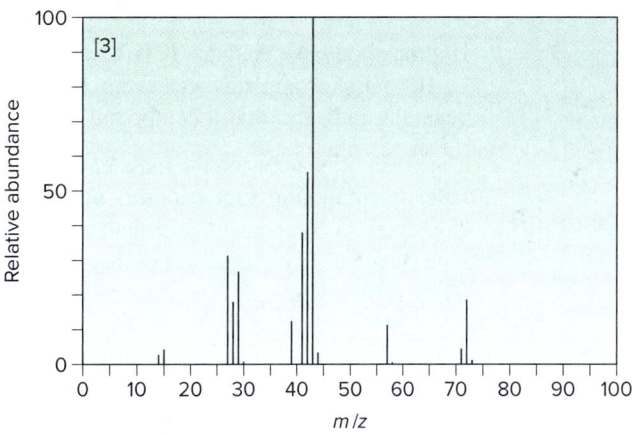

A.1B Analyzing Unknowns Using the Molecular Ion

Because the **mass of the molecular ion equals the molecular weight of a compound**, a mass spectrum can be used to distinguish between compounds that have similar physical properties but different molecular weights, as illustrated in Sample Problem A.1.

Sample Problem A.1 — Using the Molecular Ion to Identify a Compound

Pent-1-ene and pent-1-yne are low-boiling hydrocarbons that have different molecular ions in their mass spectra. Match each hydrocarbon to its mass spectrum.

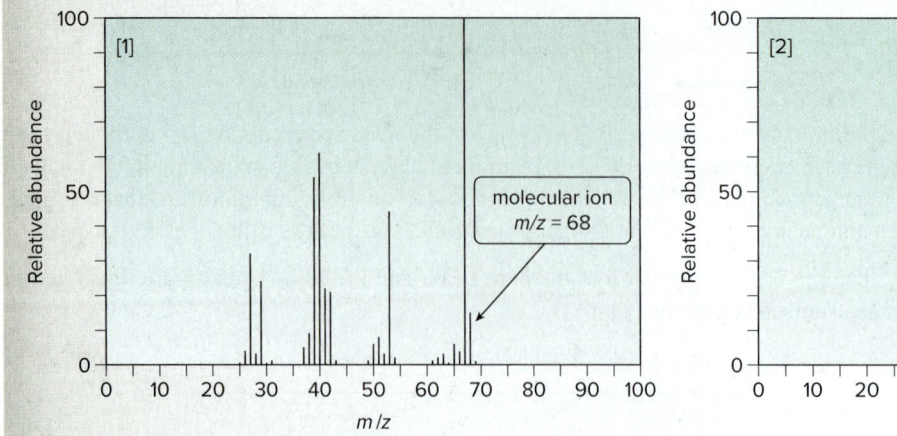
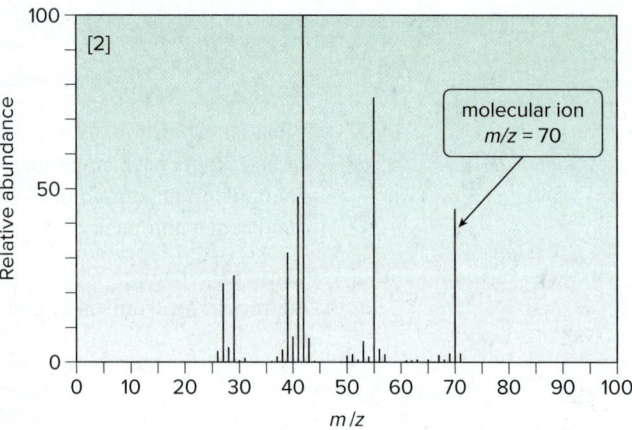

Solution

To solve this problem, first determine the molecular formula and molecular weight of each compound. Then, because the molecular weight of the compound equals the mass of the molecular ion, match the molecular weight to m/z for the molecular ion:

Compound	Molecular formula	Molecular weight = m/z of molecular ion	Spectrum
pent-1-ene	C_5H_{10}	70	[2]
pent-1-yne	C_5H_8	68	[1]

Problem A.2 What is the mass of the molecular ion formed from compounds having each molecular formula: (a) C_3H_6O; (b) $C_{10}H_{20}$; (c) $C_8H_8O_2$; (d) methamphetamine ($C_{10}H_{15}N$)?

More Practice: Try Problem A.15.

Hydrocarbons like methane (CH_4) and hexane (C_6H_{14}), as well as compounds that contain only C, H, and O atoms, always have a molecular ion with an *even* mass. An odd molecular ion generally indicates that a compound contains nitrogen.

The effect of N atoms on the mass of the molecular ion in a mass spectrum is called the **nitrogen rule: A compound that contains an *odd* number of N atoms gives an *odd* molecular ion.** Conversely, a compound that contains an *even* number of N atoms (including *zero*) gives an *even* molecular ion. Two "street" drugs that mimic the effects of heroin illustrate this principle: 3-methylfentanyl (two N atoms, even molecular weight) and MPPP (one N atom, odd molecular weight).

3-methylfentanyl
$C_{23}H_{30}N_2O$
molecular weight = 350

MPPP
(1-methyl-4-phenyl-4-propionoxypiperidine)
$C_{15}H_{21}NO_2$
molecular weight = 247

A.1C Using the Molecular Ion to Propose Molecular Formulas

How to use the molecular ion to propose molecular formulas for an unknown is shown in the stepwise procedure and Sample Problem A.2.

How To Use the Mass of a Molecular Ion to Propose Molecular Formulas for an Unknown

Example Propose possible molecular formulas for a compound with a molecular ion at $m/z = 154$.

Step [1] With an even mass of a molecular ion, the compound likely contains C, H, and possibly O atoms. Use the molecular ion to determine the maximum number of C's possible for a hydrocarbon.

- Divide 154 by 12, the mass of 1 C atom. The remainder gives the number of H's.

$$\frac{154}{12} = \text{12 C's maximum (remainder = 10)} \longrightarrow C_{12}H_{10}$$

Step [2] To determine another possible molecular formula for a hydrocarbon, replace 1 C by 12 H's. Repeat the process until the formula has more than the maximum number of H's possible.

$$C_{12}H_{10} \xrightarrow[+12\text{ H's}]{-1\text{ C}} C_{11}H_{22}$$

$$C_{11}H_{22} \xrightarrow[+12\text{ H's}]{-1\text{ C}} C_{10}H_{34}$$
too many H's
not a possible formula

- Because the maximum number of H's for a compound with 11 C's is 24 ($C_{11}H_{2(11)+2}$), $C_{10}H_{34}$ is not a possible formula.

Step [3] To determine possible molecular formulas for compounds with O atoms, replace CH_4 (mass 16) by O (mass 16) in each formula. Repeat the process to give possible molecular formulas for compounds with two or more O atoms.

- Four possibilities are shown.

$$C_{12}H_{10} \xrightarrow[+1\text{ O}]{-1\text{ CH}_4} C_{11}H_6O$$

$$C_{11}H_{22} \xrightarrow[+1\text{ O}]{-1\text{ CH}_4} C_{10}H_{18}O \xrightarrow[+1\text{ O}]{-1\text{ CH}_4} C_9H_{14}O_2 \xrightarrow[+1\text{ O}]{-1\text{ CH}_4} C_8H_{10}O_3$$

Sample Problem A.2 Using the Molecular Ion to Propose a Molecular Formula

Propose possible molecular formulas for a compound with a molecular ion at $m/z = 86$.

Solution

Because the molecular ion has an **even** mass, the compound likely contains C, H, and possibly O atoms. Begin by determining the molecular formula for a hydrocarbon having a molecular ion at 86. Then, because the mass of an O atom is 16 (the mass of CH_4), replace CH_4 by O to give a molecular formula containing one O atom. Repeat this last step to give possible molecular formulas for compounds with two or more O atoms.

For a molecular ion at $m/z = 86$:

Possible hydrocarbons:

- Divide 86 by 12 (mass of 1 C atom). This gives the maximum number of C's possible.

$$\frac{86}{12} = \text{7 C's maximum (remainder = 2)} \longrightarrow C_7H_2$$

- Replace 1 C by 12 H's for another possible molecular formula.

$$C_7H_2 \xrightarrow[+12\text{ H's}]{-1\text{ C}} C_6H_{14}$$

Possible compounds with C, H, and O:

- Substitute 1 O for CH_4. (This can't be done for C_7H_2.)

$$C_6H_{14} \xrightarrow[+1\text{ O}]{-\text{CH}_4} C_5H_{10}O$$

- Repeat the process.

$$C_5H_{10}O \xrightarrow[+1\text{ O}]{-\text{CH}_4} C_4H_6O_2$$

Problem A.3

Propose two molecular formulas for each of the following molecular ions: (a) 72; (b) 100; (c) 73.

More Practice: Try Problems A.16, A.17.

Sample Problem A.3 — Using the Molecular Ion and Degrees of Unsaturation to Propose a Molecular Formula

Propose a molecular formula for nootkatone, a compound that contains the elements C, H, and O, has five degrees of unsaturation, and has a molecular ion in its mass spectrum at $m/z = 218$.

Solution

Determine possible molecular formulas using the procedure in Sample Problem A.2. Because each degree of unsaturation removes 2 H's, the correct molecular formula has 10 fewer H's than the maximum number.

For a molecular ion at $m/z = 218$:

$$\frac{218}{12} = 18 \text{ C's maximum (remainder} = 2)$$ → $C_{18}H_2$ (not enough H's)

$C_{18}H_2 \xrightarrow[+12\text{ H's}]{-1\text{ C}} C_{17}H_{14} \xrightarrow[+1\text{ O}]{-1\text{ CH}_4} C_{16}H_{10}O$ (not enough H's)

$C_{17}H_{14} \xrightarrow[+12\text{ H's}]{-1\text{ C}} C_{16}H_{26} \xrightarrow[+1\text{ O}]{-1\text{ CH}_4} C_{15}H_{22}O$

five degrees of unsaturation
Answer

Nootkatone (Sample Problem A.3) occurs naturally in grapefruits, and has been used for many years as a flavoring in foods and beverages.
MizC/Getty Images

The maximum number of H's for a compound with 15 C's is $2n + 2 = 2(15) + 2 = 32$. A compound with 22 H's has 10 fewer H's than the maximum number and thus five degrees of unsaturation.

Problem A.4

Propose a molecular formula for cedrol, an alcohol found in cedar oil. Cedrol has three degrees of unsaturation and a molecular ion in its mass spectrum at $m/z = 222$.

More Practice: Try Problems A.18, A.19.

A.2 Alkyl Halides and the M + 2 Peak

Most of the elements found in organic compounds, such as carbon, hydrogen, oxygen, nitrogen, sulfur, phosphorus, fluorine, and iodine, have one major isotope. **Chlorine** and **bromine,** on the other hand, have two, giving characteristic patterns to the mass spectra of their compounds.

Chlorine has two common isotopes, ^{35}Cl and ^{37}Cl, which occur naturally in a 3:1 ratio. Thus, **there are two peaks in a 3:1 ratio for the molecular ion of an alkyl chloride.** The larger peak—the M peak—corresponds to the compound containing ^{35}Cl, and the smaller peak—the M + 2 peak—corresponds to the compound containing ^{37}Cl.

- When the molecular ion consists of two peaks (M and M + 2) in a 3:1 ratio, a Cl atom is present.

A.2 Alkyl Halides and the M + 2 Peak

Sample Problem A.4 Determining the Molecular Ions for an Alkyl Chloride

What molecular ions will be present in a mass spectrum of 2-chloropropane, $(CH_3)_2CHCl$?

Solution

Calculate the molecular weight using each of the common isotopes of Cl.

Molecular formula	Mass of molecular ion (m/z)
$C_3H_7{}^{35}Cl$	78 (M peak)
$C_3H_7{}^{37}Cl$	80 (M + 2 peak)

There should be two peaks in a ratio of 3:1, at $m/z = 78$ and 80, as illustrated in the mass spectrum of 2-chloropropane in Figure A.3.

Problem A.5

What molecular ions will be present in the mass spectrum of the antihistamine chlorpheniramine?

chlorpheniramine

More Practice: Try Problem A.20.

Figure A.3
Mass spectrum of 2-chloropropane [$(CH_3)_2CHCl$]

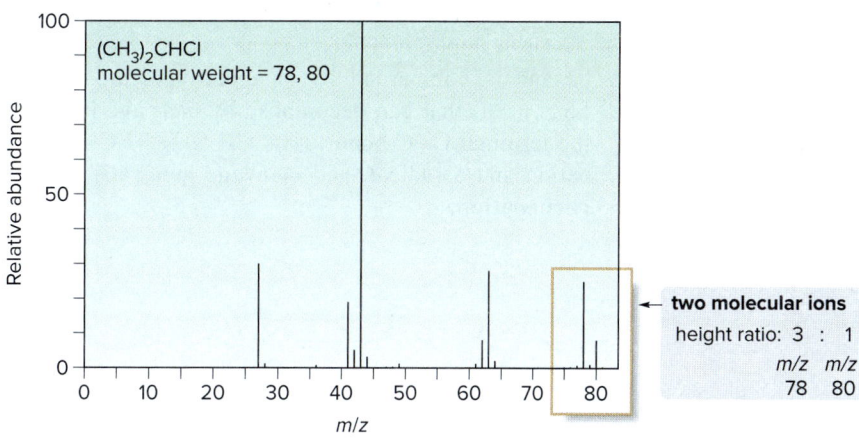

Bromine has two common isotopes, ^{79}Br and ^{81}Br, which occur naturally in a 1:1 ratio. Thus, **there are two peaks in a 1:1 ratio for the molecular ion of an alkyl bromide.** In the mass spectrum of 2-bromopropane (Figure A.4), for example, there is an M peak at $m/z = 122$ and an M + 2 peak at $m/z = 124$.

- When the molecular ion consists of two peaks (M and M + 2) in a 1:1 ratio, a Br atom is present in the molecule.

Figure A.4
Mass spectrum of 2-bromopropane [(CH$_3$)$_2$CHBr]

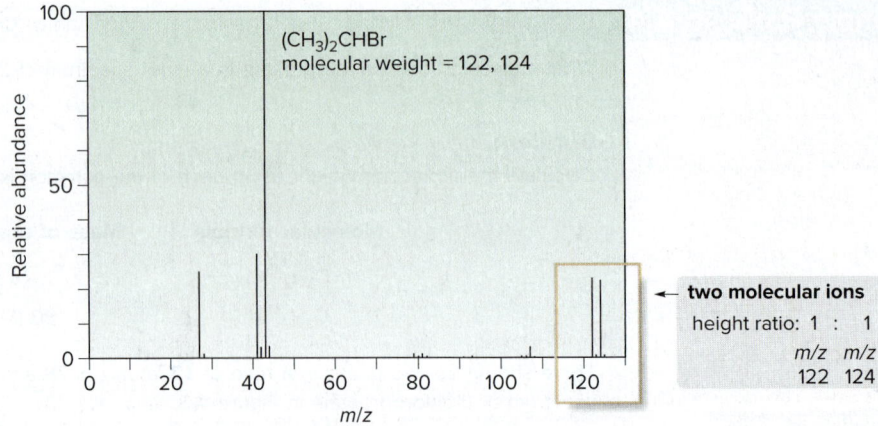

Problem A.6 What molecular ions would you expect for compounds having each of the following molecular formulas: (a) C$_4$H$_9$Cl; (b) C$_3$H$_7$F; (c) C$_4$H$_{11}$N; (d) C$_4$H$_4$N$_2$?

A.3 Fragmentation

While many chemists use a mass spectrum to determine only a compound's molecular weight and molecular formula, additional useful structural information can be obtained from fragmentation patterns. Although each organic compound fragments in a unique way, a particular functional group exhibits common fragmentation patterns.

As an example, consider hexane, whose mass spectrum was shown in Figure A.2. When hexane is bombarded by an electron beam, it forms a highly unstable radical cation ($m/z = 86$) that can decompose by cleavage of any of the C—C bonds. Thus, cleavage of the terminal C—C bond forms CH$_3$CH$_2$CH$_2$CH$_2$CH$_2$$^+$ and CH$_3$·. Fragmentation generates a cation and a radical, and **cleavage generally yields the more stable, more substituted carbocation.**

- Loss of a CH$_3$ group always forms a fragment with a mass 15 units less than that of the molecular ion.

As a result, the mass spectrum of hexane shows a peak at $m/z = 71$ due to CH$_3$CH$_2$CH$_2$CH$_2$CH$_2$$^+$. Figure A.5 illustrates how cleavage of other C—C bonds in hexane gives rise to other fragments that correspond to peaks in its mass spectrum.

Figure A.5
Identifying fragments in the mass spectrum of hexane

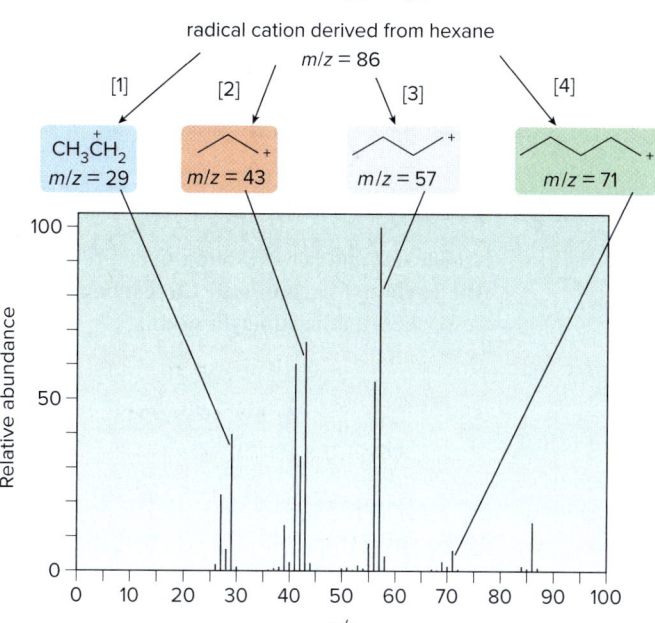

- Cleavage of C—C bonds (labeled [1]–[4]) in hexane forms lower-molecular-weight fragments that correspond to lines in the mass spectrum. Although the mass spectrum is complex, possible structures can be assigned to some of the fragments, as shown.

Sample Problem A.5 — Assigning Possible Structures to Fragments in a Mass Spectrum

The mass spectrum of 2,3-dimethylpentane [(CH$_3$)$_2$CHCH(CH$_3$)CH$_2$CH$_3$] shows fragments at m/z = 85 and 71. Propose possible structures for the ions that give rise to these peaks.

Solution
To solve a problem of this sort, first calculate the mass of the molecular ion. Draw out the structure of the compound, break a C—C bond, and calculate the mass of the resulting fragments. Repeat this process on different C—C bonds until fragments of the desired mass-to-charge ratio are formed.

In this example, 2,3-dimethylpentane has a molecular ion at m/z = 100. Cleavage of bond [1] forms a 2° carbocation with m/z = 85 and CH$_3$·. Cleavage of bond [2] forms another 2° carbocation with m/z = 71 and CH$_3$CH$_2$·. Thus, the fragments at m/z = 85 and 71 are possibly due to the two carbocations drawn.

Problem A.7 The mass spectrum of 2,3-dimethylpentane also shows peaks at m/z = 57 and 43. Propose possible structures for the ions that give rise to these peaks.

More Practice: Try Problem A.14.

Problem A.8 The base peak in the mass spectrum of 2,2,4-trimethylpentane [(CH$_3$)$_3$CCH$_2$CH(CH$_3$)$_2$] occurs at $m/z = 57$. What ion is responsible for this peak and why is this ion the most abundant fragment?

A.4 Fragmentation Patterns of Some Common Functional Groups

Each functional group exhibits characteristic fragmentation patterns that help to analyze a mass spectrum.

A.4A Aldehydes and Ketones

Aldehydes and ketones often undergo the process of **α cleavage, breaking the bond between the carbonyl carbon and the carbon adjacent to it.** Cleavage yields a neutral radical and a resonance-stabilized acylium ion.

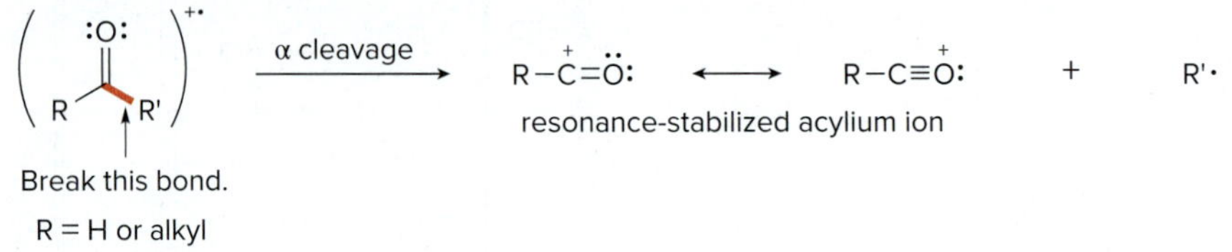

For example, α cleavage of benzophenone forms a fragment at $m/z = 105$ due to a resonance-stabilized acylium ion.

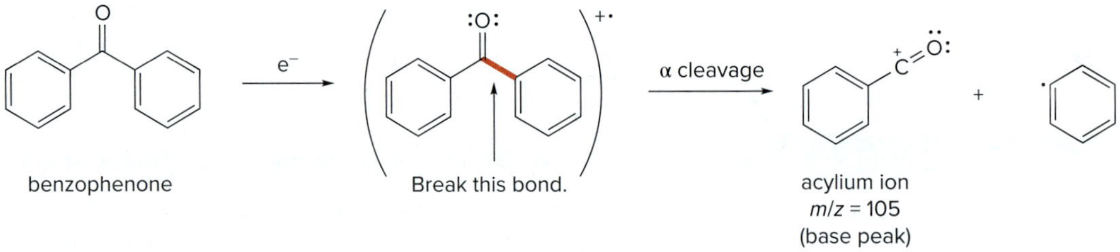

Sample Problem A.6 Drawing the Fragments Formed from α Cleavage

What mass spectral fragments are formed from α cleavage of pentan-2-one, CH$_3$COCH$_2$CH$_2$CH$_3$?

Solution

Alpha (α) cleavage breaks the bond between the carbonyl carbon and the carbon adjacent to it, yielding a neutral radical and a resonance-stabilized acylium ion. A ketone like pentan-2-one with two different alkyl groups bonded to the carbonyl carbon has two different pathways for α cleavage.

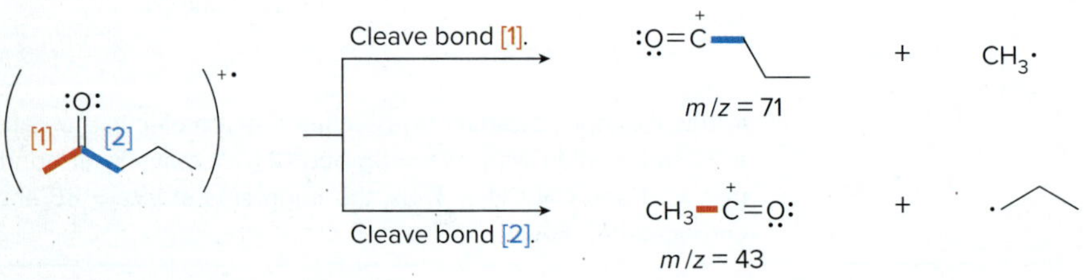

As a result, two fragments are formed by α cleavage of pentan-2-one, giving peaks at $m/z = 71$ and 43.

Problem A.9
What cations are formed in the mass spectrometer by α cleavage of each of the following compounds?

a. [methyl cyclohexyl ketone structure] b. [butanal structure]

More Practice: Try Problems A.24b, c; A.28.

A.4B Alcohols

Alcohols undergo fragmentation in two different ways—**α cleavage** and **dehydration**. Alpha (α) cleavage occurs by breaking a bond between an alkyl group and the carbon that bears the OH group, resulting in an alkyl radical and a resonance-stabilized carbocation.

[α cleavage scheme showing alcohol radical cation breaking into resonance-stabilized carbocation + R·]

Likewise, alcohols undergo dehydration, the elimination of H$_2$O, from two adjacent atoms. Unlike fragmentations discussed thus far, dehydration results in the cleavage of two bonds and forms H$_2$O and the radical cation derived from an alkene. For example, dehydration of cyclohexanol forms the radical cation of cyclohexene, a fragment with a mass 18 units less than that of the molecular ion, as shown in Figure A.6.

[dehydration scheme]

cyclohexanol → cyclohexene + H$_2$O
m/z = 100   m/z = 82

- Loss of H$_2$O from an alcohol always forms a fragment with a mass 18 units less than the molecular ion.

Figure A.6
The mass spectrum of cyclohexanol

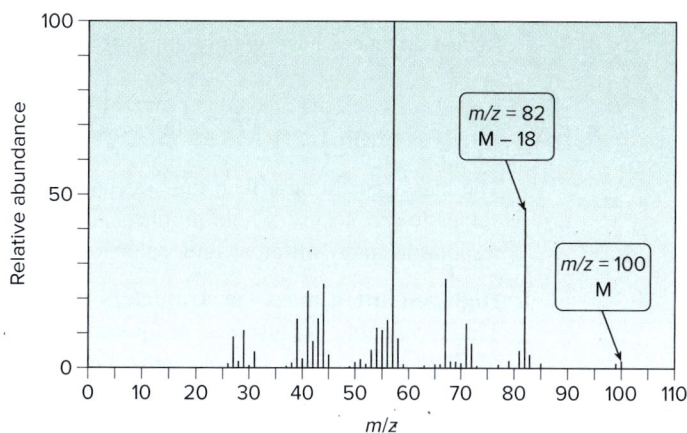

Problem A.10 (a) What mass spectral fragments are formed by α cleavage of butan-2-ol, $CH_3CH(OH)CH_2CH_3$?
(b) What fragments are formed by dehydration of butan-2-ol?

A.4C Amines

Like alcohols, amines undergo fragmentation by **α cleavage**. Alpha (α) cleavage occurs by breaking the bond between an alkyl group and the carbon that bears the amine nitrogen, forming an alkyl radical and a resonance-stabilized carbocation. For example, α cleavage of triethylamine (molecular ion at $m/z = 101$) forms $CH_3\cdot$ and a resonance-stabilized cation at $m/z = 86$.

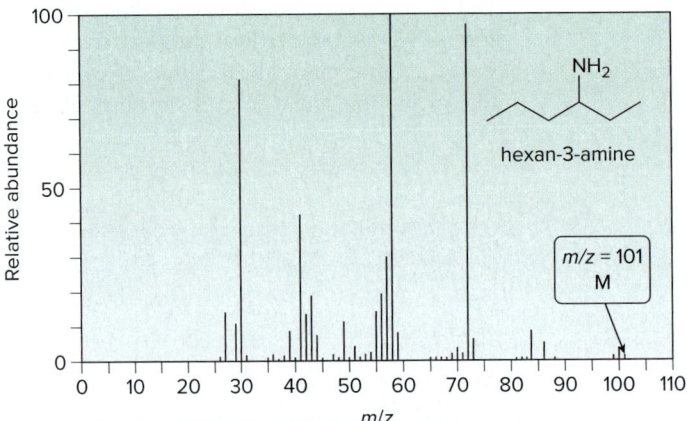

Problem A.11 Propose structures for the two fragments of highest abundance in the mass spectrum of hexan-3-amine.

A.5 Other Types of Mass Spectrometry

Recent advances have greatly expanded the information obtained from mass spectrometry.

A.5A High-Resolution Mass Spectrometry

The mass spectra described thus far have been low-resolution spectra; that is, they report m/z values to the nearest whole number. As a result, the mass of a given molecular ion can correspond to many different molecular formulas, as shown in Sample Problem A.2.

High-resolution mass spectrometers measure m/z ratios to four (or more) decimal places. This is valuable because except for carbon-12, whose mass is defined as 12.0000, the masses of all other nuclei are very close to—but not exactly—whole numbers. Table A.1 lists the exact mass values of a few common nuclei. Using these values, it is possible to determine the single molecular formula that gives rise to a molecular ion.

Table A.1
Exact Masses of Some Common Isotopes

Isotope	Mass
^{12}C	12.0000
^{1}H	1.00783
^{16}O	15.9949
^{14}N	14.0031

For example, a compound having a molecular ion at $m/z = 60$ using a low-resolution mass spectrometer could have the following molecular formulas:

Formula	Exact mass
C_3H_8O	60.0575
$C_2H_4O_2$	60.0211
$C_2H_8N_2$	60.0688

If the molecular ion had an exact mass of 60.0578, the compound's molecular formula is C_3H_8O, because its mass is closest to the observed value.

Problem A.12 The low-resolution mass spectrum of an unknown analgesic **X** had a molecular ion of 151. Possible molecular formulas include $C_7H_5NO_3$, $C_8H_9NO_2$, and $C_{10}H_{17}N$. High-resolution mass spectrometry gave an exact mass of 151.0640. What is the molecular formula of **X**?

A.5B Gas Chromatography–Mass Spectrometry (GC–MS)

Two analytical tools—**gas chromatography (GC)** and **mass spectrometry (MS)**—can be combined into a single instrument **(GC–MS)** to analyze mixtures of compounds (Figure A.7a).

Figure A.7 Compound analysis using GC–MS

a. Schematic of a GC–MS instrument

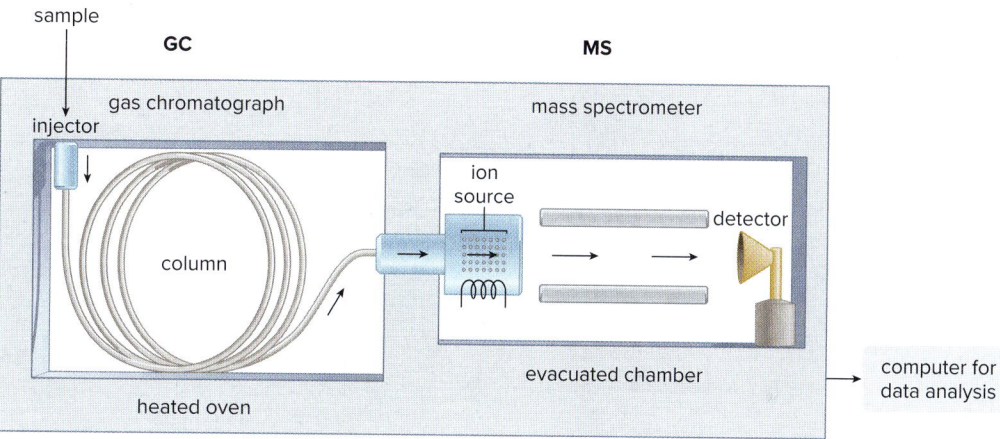

The gas chromatograph separates the mixture into its components.

The mass spectrometer records a spectrum of the individual components.

b. GC trace of a three-component mixture. The mass spectrometer gives a spectrum for each component.

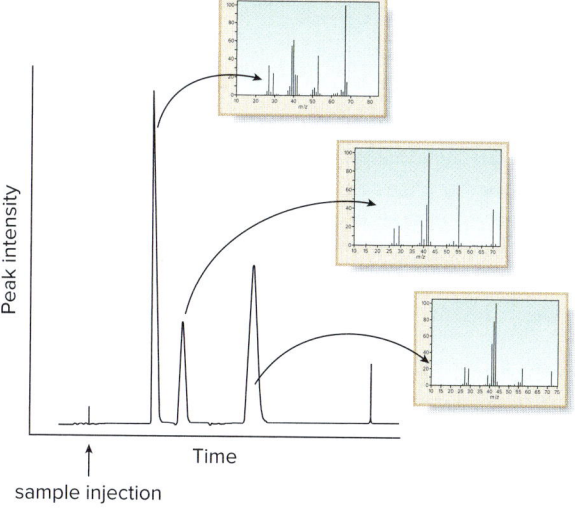

The gas chromatograph separates the mixture, and then the mass spectrometer records a spectrum of the individual components.

A gas chromatograph consists of a thin capillary column containing a viscous, high-boiling liquid, all housed in an oven. When a sample is injected into the GC, it is vaporized and swept by an inert gas through the column. The components of the mixture travel through the column at different rates, often separated by boiling point, with lower-boiling compounds exiting the column before higher-boiling compounds. Each compound then enters the mass spectrometer, where it is ionized to form its molecular ion and lower-molecular-weight fragments. The GC–MS records a gas chromatogram for the mixture, which plots the amount of each component versus its **retention time**—that is, the time required to travel through the column. Each component of a mixture is characterized by its retention time in the gas chromatogram and its molecular ion in the mass spectrum (Figure A.7b).

GC–MS is widely used for characterizing mixtures containing environmental pollutants. It is also used to analyze urine and hair samples for the presence of illegal drugs or banned substances thought to improve athletic performance.

To analyze a urine sample for THC (tetrahydrocannabinol), the principal psychoactive component of marijuana that opened this chapter, the organic compounds are extracted from urine, purified, concentrated, and injected into the GC–MS. THC appears as a GC peak with a characteristic retention time (for a given set of experimental parameters), and gives a molecular ion at 314, its molecular weight, as shown in Figure A.8.

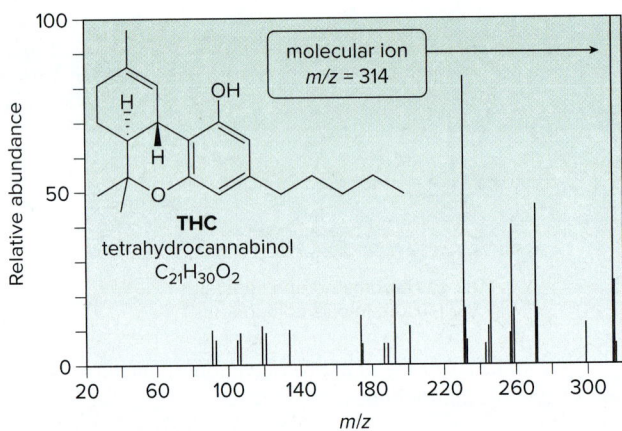

Figure A.8

Mass spectrum of tetrahydrocannabinol (THC)

Problem A.13 Benzene, toluene, and *p*-xylene (BTX) are often added to gasoline to boost octane ratings. What would be observed if a mixture of these three compounds were subjected to GC–MS analysis? How many peaks would be present in the gas chromatogram? What would be the relative order of the peaks? What molecular ions would be observed in the mass spectra?

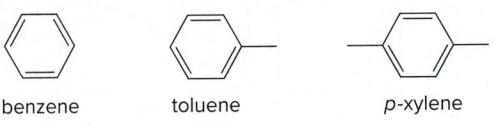

A.5C Mass Spectra of High-Molecular-Weight Biomolecules

Dr. John Fenn shared the 2002 Nobel Prize in Chemistry for his development of ESI mass spectrometry.

Until the 1980s mass spectra were limited to molecules that could be readily vaporized with heat under vacuum, and thus had molecular weights of < 800. In the last 35 years, new methods have been developed to generate gas phase ions of large molecules, allowing mass spectra to be recorded for large biomolecules such as proteins and carbohydrates. **Electrospray ionization (ESI),** for example, forms ions by creating a fine spray of charged droplets in an electric field. Evaporation of the charged droplets forms gaseous ions that are then analyzed by their m/z ratio. ESI and related techniques have extended mass spectrometry into the analysis of nonvolatile compounds with molecular weights greater than 100,000 daltons (atomic mass units).

Spectroscopy A CHAPTER REVIEW

KEY CONCEPTS

Molecular ion (M) in mass spectrometry (A.1, A.2)

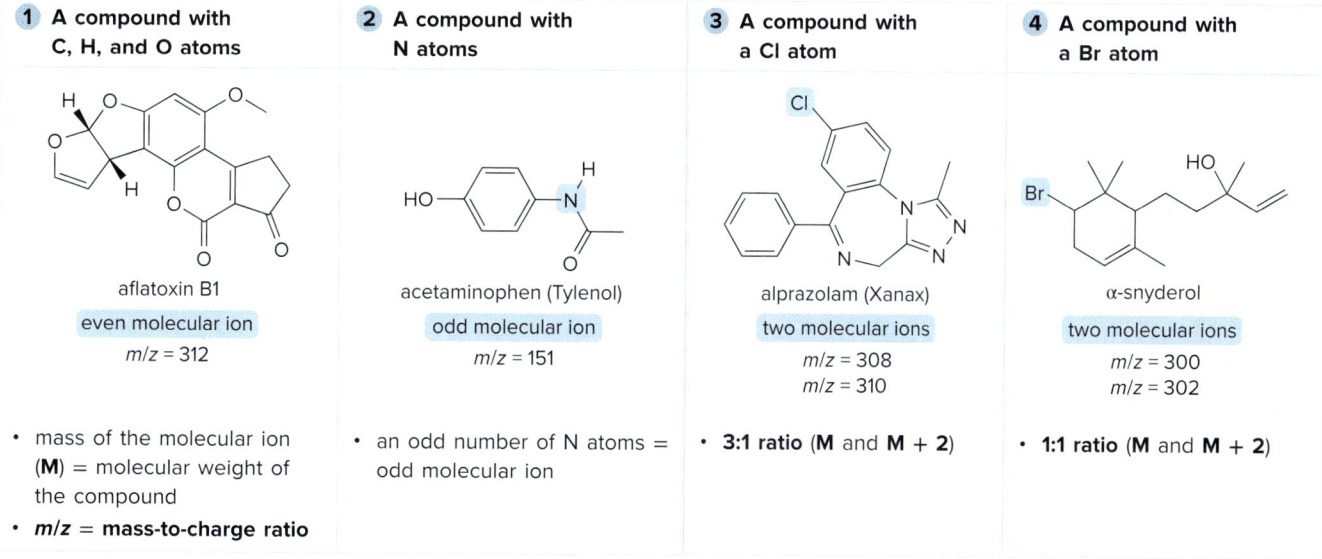

1 A compound with C, H, and O atoms	2 A compound with N atoms	3 A compound with a Cl atom	4 A compound with a Br atom
aflatoxin B1	acetaminophen (Tylenol)	alprazolam (Xanax)	α-snyderol
even molecular ion	odd molecular ion	two molecular ions	two molecular ions
$m/z = 312$	$m/z = 151$	$m/z = 308$ $m/z = 310$	$m/z = 300$ $m/z = 302$

- mass of the molecular ion (**M**) = molecular weight of the compound
- m/z = mass-to-charge ratio
- an odd number of N atoms = odd molecular ion
- 3:1 ratio (**M** and **M + 2**)
- 1:1 ratio (**M** and **M + 2**)

Try Problems A.15, A.20, A.21a–c.

KEY SKILLS

[1] Proposing possible molecular formulas for a compound that contains C, H, and perhaps O with a given molecular ion (A.1); example: $m/z = 100$

Possible hydrocarbons		Possible compounds with C, H, and O	
1 Divide 100 by 12.	**2** Replace one **C** atom by 12 **H** atoms.	**3** Substitute one **O** atom for **CH₄**.	**4** Repeat the process.
$\frac{100}{12}$ = 8 **C**'s maximum (remainder = 4) ⟶ C$_8$H$_4$	C$_8$H$_4$ $\xrightarrow[+12\text{ H's}]{-1\text{ C}}$ C$_7$H$_{16}$	C$_7$H$_{16}$ $\xrightarrow[+1\text{ O}]{-\text{CH}_4}$ C$_6$H$_{12}$O	C$_6$H$_{12}$O $\xrightarrow[+1\text{ O}]{-\text{CH}_4}$ C$_5$H$_8$O$_2$

See *How To* p. 513, Sample Problems A.2, A.3. Try Problems A.16–A.19.

[2] Proposing possible structures for fragmentation by α cleavage (A.3, A.4)

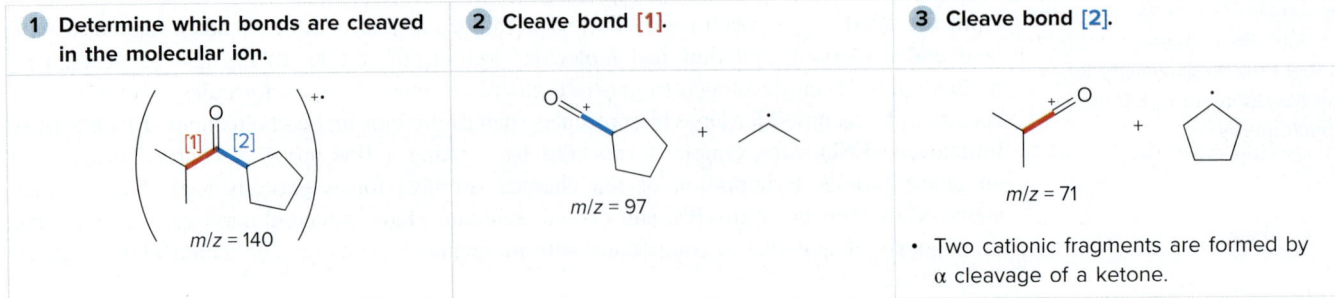

See Sample Problem A.6, Figure A.5. Try Problems A.23–A.29.

PROBLEMS

Problems that combine mass spectrometry and infrared spectroscopy are located at the end of Spectroscopy B. Problems that combine mass spectrometry, infrared spectroscopy, and nuclear magnetic resonance spectroscopy are found at the end of Spectroscopy C.

Problem Using a Three-Dimensional Model

A.14 The mass spectrum of the following compound shows fragments at $m/z = 127$, 113, and 85. Propose structures for the ions that give rise to these peaks.

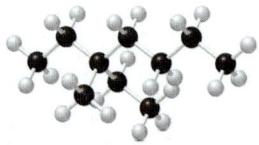

Molecular Ions and Molecular Formulas

A.15 What molecular ion is expected for each compound?

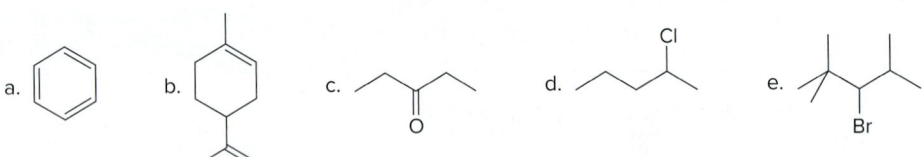

A.16 Propose two molecular formulas for each molecular ion: (a) 102; (b) 98; (c) 119; (d) 74.

A.17 Propose four possible structures for a hydrocarbon with a molecular ion at $m/z = 112$.

A.18 What is the molecular formula for α-himachalene, a hydrocarbon obtained from cedar wood, which has four degrees of unsaturation and has a molecular ion in its mass spectrum at $m/z = 204$?

A.19 Propose a molecular formula for rose oxide, a rose-scented compound isolated from roses and geraniums, which contains the elements of C, H, and O, has two degrees of unsaturation, and has a molecular ion in its mass spectrum at $m/z = 154$.

A.20 Match each structure to its mass spectrum.

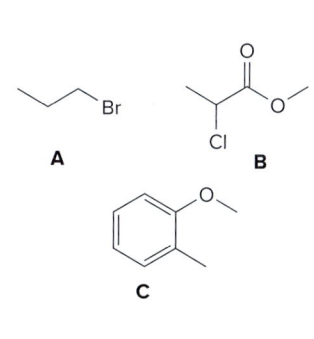

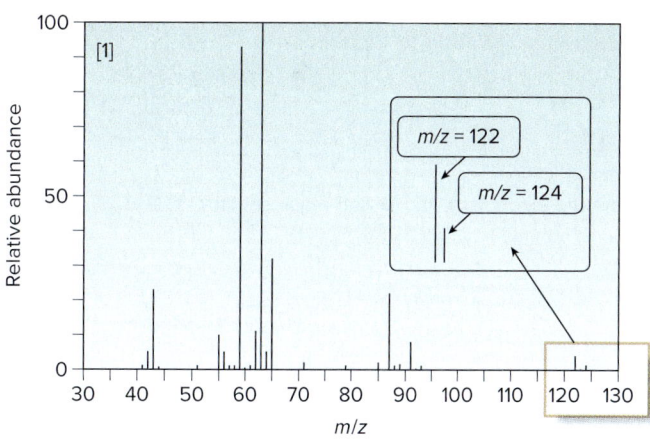

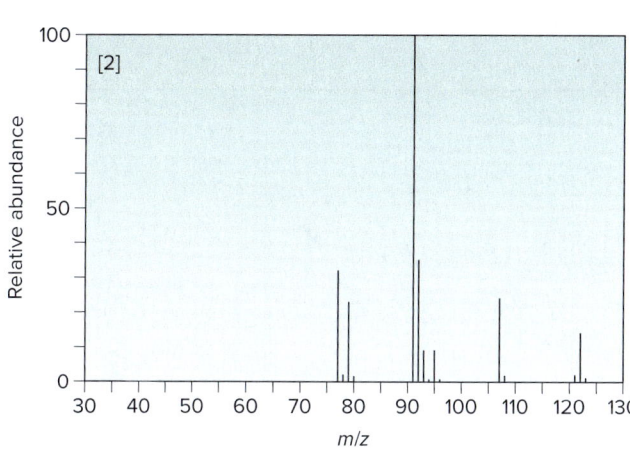

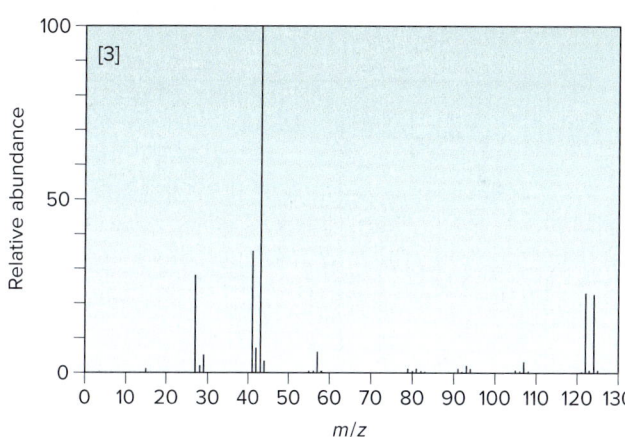

A.21 Propose a structure consistent with each set of data.
   a. a compound that contains a benzene ring and has a molecular ion at $m/z = 107$
   b. a hydrocarbon that contains only sp^3 hybridized carbons and a molecular ion at $m/z = 84$
   c. a compound that contains a carbonyl group and gives a molecular ion at $m/z = 114$
   d. a compound that contains C, H, N, and O and has an exact mass for the molecular ion at 101.0841

A.22 A low-resolution mass spectrum of the neurotransmitter dopamine gave a molecular ion at $m/z = 153$. Two possible molecular formulas for this molecular ion are $C_8H_{11}NO_2$ and $C_7H_{11}N_3O$. A high-resolution mass spectrum provided an exact mass at 153.0680. Which of the possible molecular formulas is the correct one?

Fragmentation

A.23 Label each of the following in the mass spectrum of hexan-2-ol [$CH_3CH(OH)CH_2CH_2CH_2CH_3$]: the molecular ion, the base peak, the fragment resulting from the loss of H_2O, and α cleavage fragments.

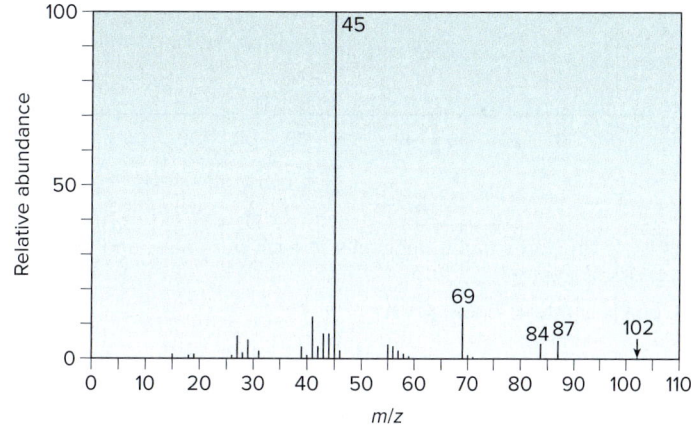

A.24 What cations are formed in the mass spectrometer by α cleavage of each of the following compounds?

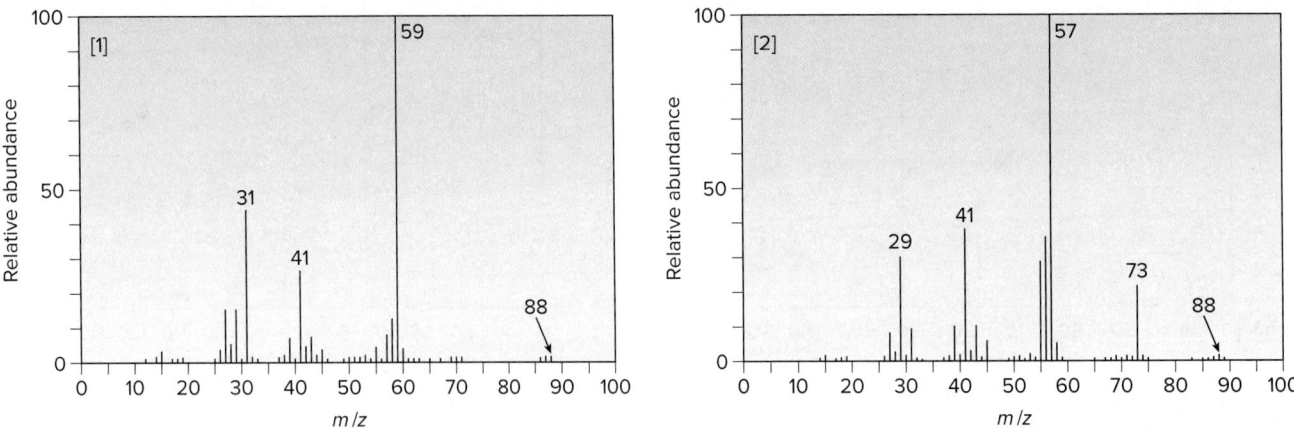

A.25 Consider isomeric alcohols **A** and **B** and mass spectra [1] and [2].

(a) Label the molecular ion and base peak in each spectrum. (b) Use the fragmentation patterns to determine which mass spectrum corresponds to isomer **A** and which corresponds to isomer **B**.

A.26 Consider the mass spectrum of hexan-2-amine. Label the molecular ion and base peak and propose a structure for the fragment that corresponds to the base peak.

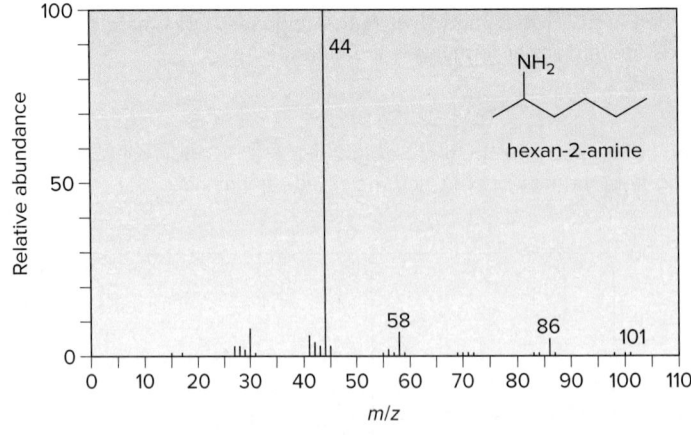

A.27 For each compound, assign likely structures to the fragments at each m/z value, and explain how each fragment is formed.
  a. $C_6H_5CH_2CH_2OH$: peaks at m/z = 104, 91
  b. $CH_2=C(CH_3)CH_2CH_2OH$: peaks at m/z = 71, 68, 41, 31

A.28 Suppose you have two bottles, labeled ketone **A** and ketone **B**. You know that one bottle contains $CH_3CO(CH_2)_5CH_3$ and one contains $CH_3CH_2CO(CH_2)_4CH_3$, but you do not know which ketone is in which bottle. Ketone **A** gives a fragment at $m/z = 99$ and ketone **B** gives a fragment at $m/z = 113$. What are the likely structures of ketones **A** and **B** from these fragmentation data?

A.29 Like alcohols, ethers undergo α cleavage by breaking a carbon–carbon bond between an alkyl group and the carbon bonded to the ether oxygen atom; that is, the red C—C bond in R—CH$_2$OR' is broken. With this in mind, propose structures for the fragments formed by α cleavage of $(CH_3)_2CHCH_2OCH_2CH_3$. Suggest a reason why an ether fragments by α cleavage.

Challenge Problems

A.30 What molecular ions would be present in the mass spectrum of a compound that contains C, H, and (a) 1 Br and 1 Cl; (b) 3 Br's? Give the relative peak intensities of the molecular ions in each case.

A.31 In addition to α cleavage, some aldehydes and ketones undergo the McLafferty rearrangement. In the McLafferty rearrangement, a hydrogen on a carbon three atoms from the C=O is transferred to the carbonyl oxygen and a carbon–carbon bond is broken. This process forms an alkene and the radical cation derived from an enol, which appears as a fragment in the mass spectrum.

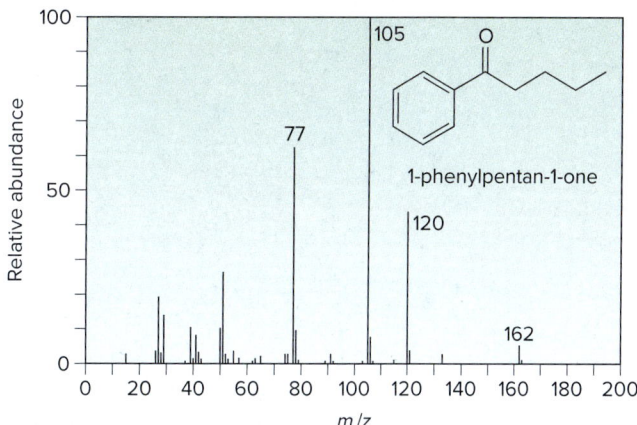

a. Draw the products formed from the McLafferty rearrangement of 1-phenylpentan-1-one, and identify the fragment that results in the given mass spectrum.

b. If a mass spectrum of the ester ethyl pentanoate ($CH_3CH_2CH_2CH_2CO_2CH_2CH_3$) is recorded, what is the mass of the radical cation formed by the McLafferty rearrangement?

c. Which of the following compounds can undergo a McLafferty rearrangement?

SPECTROSCOPY B

Infrared Spectroscopy

T.Daly/Alamy Stock Photo

- **B.1** Electromagnetic radiation
- **B.2** The general features of infrared spectroscopy
- **B.3** IR absorptions
- **B.4** Infrared spectra of common functional groups
- **B.5** IR and structure determination

The serendipitous discovery of **penicillin** from a mold of the genus *Penicillium* by Scottish bacteriologist Sir Alexander Fleming in 1928 is considered one of the single most important events in the history of medicine. Penicillin G and related compounds are members of the β-lactam family of antibiotics, all of which contain a strained four-membered amide ring that is responsible for their biological activity. Penicillin was first used to cure a streptococcal infection in 1942, and by 1944 penicillin production was given high priority by the U.S. government, because it was needed to treat the many injured soldiers in World War II. The unusual structure of penicillin was elucidated by modern instrumental methods in the 1940s. In Spectroscopy Part B, we learn about infrared spectroscopy, which is used to determine the functional groups in organic compounds like penicillin.

Why Study... Infrared Spectroscopy?

Although mass spectrometry tells us the molecular weight and molecular formula for an organic compound, other forms of spectroscopy must be used to completely delineate the structure of a complex compound. **Infrared spectroscopy** is a technique that uses infrared light to interact with compounds, causing bonds to bend and vibrate, and giving a **spectrum with characteristic absorptions for particular functional groups.** Because the properties and reactions of an organic compound are determined in large part by what functional groups it contains, infrared spectroscopy is a valuable method for determining the structure of compounds isolated from natural sources, and for monitoring the progress of reactions that result in the addition or removal of functional groups.

We begin this chapter by learning about infrared light, the energy source used in infrared spectroscopy.

B.1 Electromagnetic Radiation

Infrared (IR) spectroscopy and **nuclear magnetic resonance (NMR)** spectroscopy (Part C) both use a form of electromagnetic radiation as their energy source. To understand IR and NMR, therefore, you need to understand some of the properties of **electromagnetic radiation**—radiant energy having dual properties of both waves and particles.

The particles of electromagnetic radiation are called **photons,** each having a discrete amount of energy called a **quantum.** Because electromagnetic radiation also has wave properties, it can be characterized by its **wavelength** and **frequency.**

- Wavelength (λ) is the distance from one point on a wave (e.g., the peak or trough) to the same point on the adjacent wave. A variety of different length units are used for λ, depending on the type of radiation.
- Frequency (ν) is the number of waves passing a point per unit time. Frequency is reported in cycles per second (s^{-1}), which is also called hertz (Hz).

Length units used to report wavelength include:

Unit	Length
meter (m)	1 m
centimeter (cm)	10^{-2} m
micrometer (μm)	10^{-6} m
nanometer (nm)	10^{-9} m
Angstrom (Å)	10^{-10} m

You come into contact with many different kinds of electromagnetic radiation in your daily life. You use visible light to see the words on this page, you may cook with microwaves, and you should use sunscreen to protect your skin from the harmful effects of ultraviolet radiation.

The different forms of electromagnetic radiation make up the **electromagnetic spectrum.** The spectrum is arbitrarily divided into different regions, as shown in Figure B.1. All electromagnetic radiation travels at the speed of light (c), 3.0×10^8 m/s.

Figure B.1 The electromagnetic spectrum

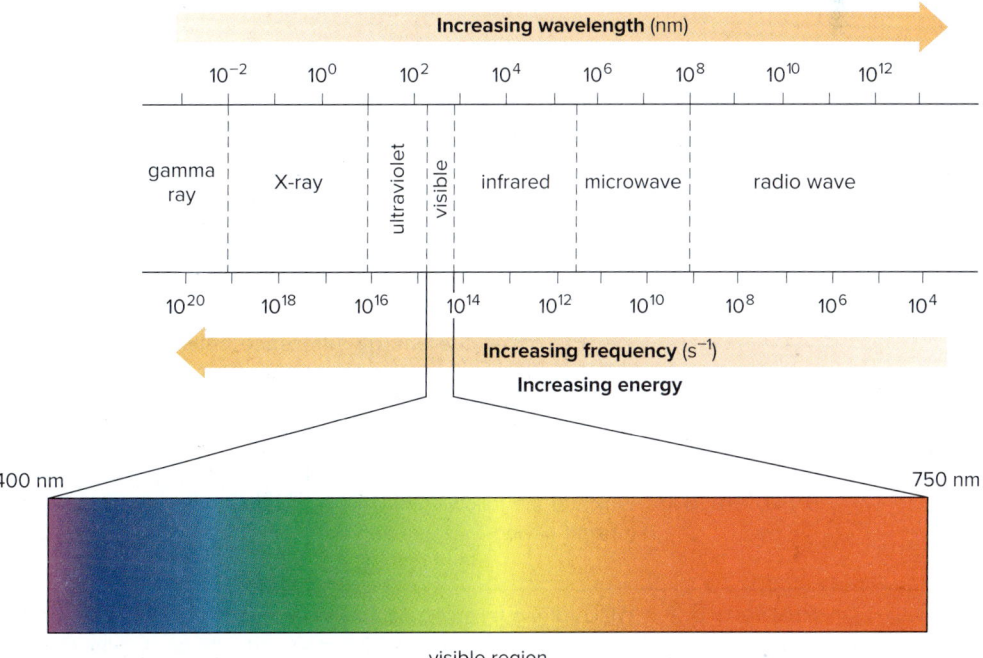

- Visible light occupies only a small region of the electromagnetic spectrum.

The speed of electromagnetic radiation (c) is directly proportional to its wavelength and frequency:

$$c = \lambda \nu$$

The speed of light (c) is a constant, so wavelength and frequency are *inversely* related:

- $\lambda = c/\nu$: Wavelength increases as frequency decreases.
- $\nu = c/\lambda$: Frequency increases as wavelength decreases.

The energy (E) of a photon is directly proportional to its frequency where h = Planck's constant (6.63×10^{-34} J · s).

$$E = h\nu$$

Frequency and wavelength are *inversely* proportional ($\nu = c/\lambda$), however, so energy and wavelength are *inversely* proportional:

$$E = h\nu = \frac{hc}{\lambda}$$

- The energy of electromagnetic radiation increases as frequency increases and wavelength decreases.

When electromagnetic radiation strikes a molecule, some wavelengths—but not all—are absorbed. Only some wavelengths are absorbed because molecules have discrete energy levels. The energies of their electronic, vibrational, and nuclear spin states are *quantized,* not *continuous.*

- For absorption to occur, the energy of the photon must match the difference between two energy states in a molecule.

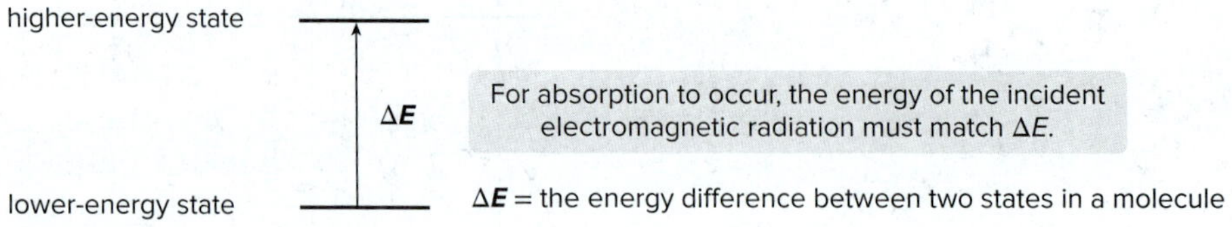

higher-energy state ———

ΔE

For absorption to occur, the energy of the incident electromagnetic radiation must match ΔE.

lower-energy state ——— ΔE = the energy difference between two states in a molecule

- The *larger* the energy difference between two states, the *higher* the energy of radiation needed for absorption, the *higher* the frequency, and the *shorter* the wavelength.

Problem B.1 Which of the following has the higher frequency: (a) light having a wavelength of 10^2 or 10^4 nm; (b) light having a wavelength of 100 nm or 100 µm; (c) red light or blue light?

Problem B.2 Which of the following has the higher energy: (a) light having a ν of 10^4 Hz or 10^8 Hz; (b) light having a λ of 10 nm or 1000 nm; (c) red light or blue light?

B.2 The General Features of Infrared Spectroscopy

Organic chemists use infrared (IR) spectroscopy to identify the functional groups in a compound.

B.2A Background

> Using the wavenumber scale results in IR values in a numerical range that is easier to report than the corresponding frequencies given in hertz (4000–400 cm^{-1} compared to 1.2×10^{14}–1.2×10^{15} Hz).

Infrared radiation ($\lambda = 2.5$–25 µm) is the energy source in infrared spectroscopy. Infrared light has somewhat longer wavelengths than visible light, making infrared light lower in frequency and lower in energy than visible light. Frequencies in IR spectroscopy are reported using a unit called the **wavenumber** ($\tilde{\nu}$):

$$\tilde{\nu} = \frac{1}{\lambda}$$

Wavenumber is *inversely* proportional to wavelength and reported in reciprocal centimeters (**cm^{-1}**). Wavenumber ($\tilde{\nu}$) is *proportional* to frequency (ν). **Frequency (and therefore energy) increases as the wavenumber increases.** Using the wavenumber scale, IR absorptions occur from **4000 cm^{-1} to 400 cm^{-1}**.

- Absorption of IR light causes changes in the vibrational motions of a molecule.

A bond can stretch.

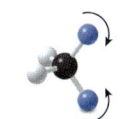

Two bonds can bend.

Covalent bonds are not static. They are more like springs with weights on each end. When two atoms are bonded to each other, the bond stretches back and forth. When three or more atoms are joined together, bonds can also bend. These bond stretching and bending vibrations represent the different vibrational modes available to a molecule.

These vibrations are quantized, so they occur only at specific frequencies, which correspond to the frequency of IR light. **When the frequency of IR light matches the frequency of a particular vibrational mode, the IR light is absorbed,** causing the amplitude of the particular bond stretch or bond bend to increase.

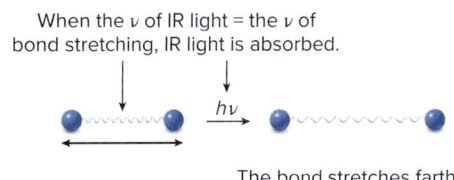

When the ν of IR light = the ν of bond stretching, IR light is absorbed.

The bond stretches farther.
The amplitude increases.

- Different kinds of bonds vibrate at different frequencies, so they absorb different frequencies of IR light.
- IR spectroscopy distinguishes between the different kinds of bonds in a molecule, so it is possible to determine the functional groups present.

Problem B.3 Which of the following has higher energy: (a) IR light of 3000 cm^{-1} or 1500 cm^{-1} in wavenumber; (b) IR light having a wavelength of 10 µm or 20 µm?

B.2B Characteristics of an IR Spectrum

In an IR spectrometer, light passes through a sample. Frequencies that match vibrational frequencies are absorbed, and the remaining light is transmitted to a detector. A spectrum plots the amount of transmitted light versus its wavenumber. The IR spectrum of propan-1-ol, $CH_3CH_2CH_2OH$, illustrates several important features of IR spectroscopy.

- The absorption peaks go *down* on a page. The y axis measures **percent transmittance:** 100% transmittance means that all the light shone on a sample is transmitted and none is

532   Spectroscopy B   Infrared Spectroscopy

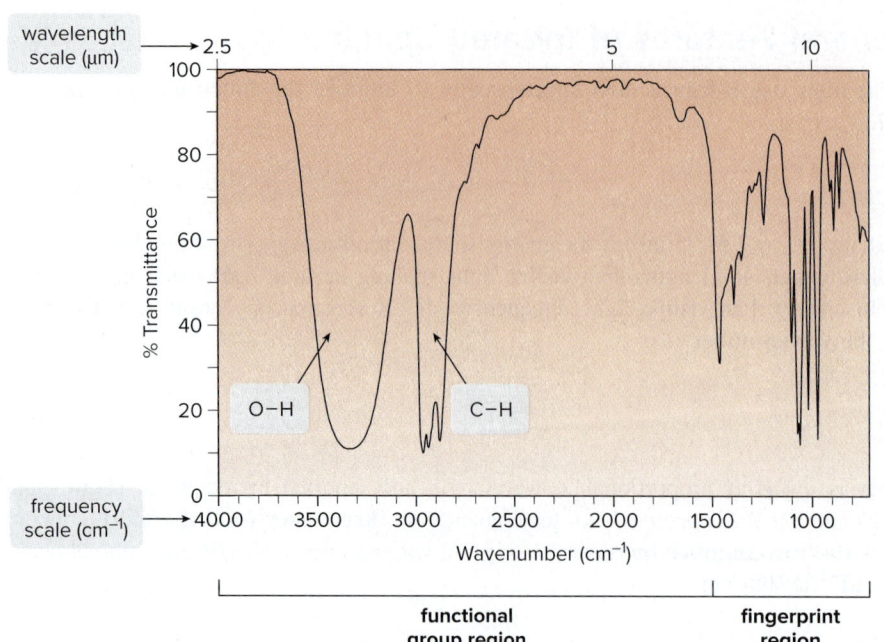

absorbed; 0% transmittance means that none of the light shone on a sample is transmitted and all is absorbed. **A strong absorption has a low % transmittance because much light is absorbed.**

- Each peak corresponds to a particular kind of bond, and each bond type (such as O–H and C–H) occurs at a characteristic frequency.
- IR spectra have both a wavelength and a wavenumber scale on the x axis. Wavelengths are recorded in μm (2.5–25). Wavenumber, frequency, and energy *decrease* from left to right. Where a peak occurs is reported in reciprocal centimeters (cm^{-1}).

Conceptually, the IR spectrum is divided into two regions:

- The functional group region occurs at ≥ 1500 cm^{-1}. Common functional groups give one or two peaks in this region, at a characteristic frequency.
- The fingerprint region occurs at < 1500 cm^{-1}. This region often contains a complex set of peaks and is unique for every compound.

B.3   IR Absorptions

B.3A   Where Particular Bonds Absorb in the IR

Where a particular bond absorbs in the IR depends on **bond strength** and **atom mass.**

- Bond strength: stronger bonds vibrate at higher frequency, so they absorb at higher $\tilde{\nu}$.
- Atom mass: bonds with lighter atoms vibrate at higher frequency, so they absorb at higher $\tilde{\nu}$.

Thinking of bonds as springs with weights on each end illustrates these trends. The strength of the spring is analogous to bond strength, and the mass of the weights is analogous to atomic mass. For two springs with the same weights on each end, the **stronger spring vibrates at a higher frequency.** For two springs of the same strength, **springs with lighter weights vibrate at higher frequency** than those with heavier weights. Hooke's law, as shown in Figure B.2, describes the relationship of frequency to mass and bond strength.

Figure B.2
Hooke's law: How the frequency of bond vibration depends on atom mass and bond strength

The frequency of bond vibration can be derived from Hooke's law, which describes the motion of a vibrating spring:

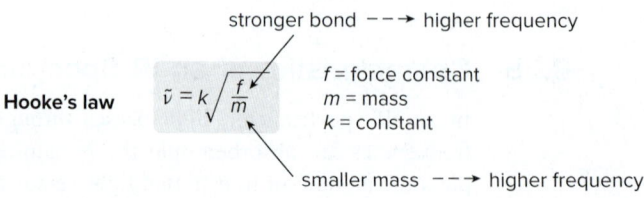

Hooke's law
$$\tilde{\nu} = k\sqrt{\frac{f}{m}}$$

stronger bond --→ higher frequency
f = force constant
m = mass
k = constant
smaller mass --→ higher frequency

- The force constant (f) is the strength of the bond (or spring). The larger the value of f, the stronger the bond, and the higher the $\tilde{\nu}$ of vibration.
- The mass (m) is the mass of atoms (or weights). The smaller the value of m, the higher the $\tilde{\nu}$ of vibration.

As a result, **bonds absorb in four predictable regions in an IR spectrum.** These four regions, and the bonds that absorb there, are summarized in Figure B.3. Remembering the information in this figure will help you analyze the spectra of unknown compounds. To help you remember it, keep in mind these two points:

- Absorptions for bonds to hydrogen always occur on the *left* side of the spectrum (the high wavenumber region). H has so little mass that H—Z bonds (where Z = C, O, and N) vibrate at *high* frequencies.
- Bond strength decreases in going from C≡C → C=C → C—C, so the frequency of vibration *decreases*—that is, the absorptions for these bonds move farther to the *right* side of the spectrum.

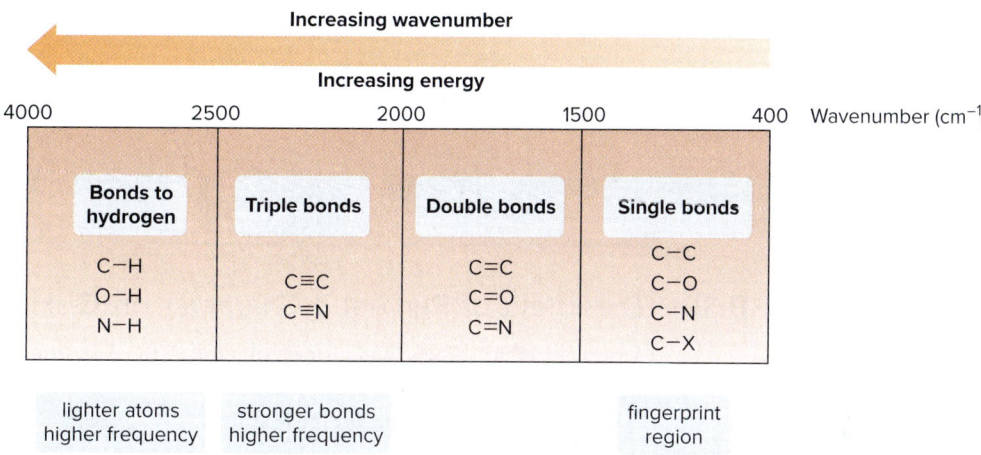

Figure B.3
Summary: The four regions of the IR spectrum

The functional group region consists of absorptions for single bonds to hydrogen (all H—Z bonds), as well as absorptions for all multiple bonds. Most absorptions in the functional group region are due to bond stretching (rather than bond bending). The fingerprint region consists of absorptions due to all other single bonds (except H—Z bonds), often making it a complex region that is very difficult to analyze.

Besides learning the general regions of the IR spectrum, it is useful to learn the specific absorption values for common bonds. Table B.1 lists the most important IR absorptions in the functional group region. Other details of IR absorptions will be presented in later chapters when new functional groups are introduced. Appendix G contains a detailed list of the characteristic IR absorption frequencies for common bonds.

Table B.1 Important IR Absorptions

Bond type	Approximate $\tilde{\nu}$ (cm^{-1})	Intensity
O—H	3600–3200	strong, broad
N—H	3500–3200	medium
C—H	~3000	
• C_{sp^3}—H	3000–2850	strong
• C_{sp^2}—H	3150–3000	medium
• C_{sp}—H	3300	medium
C≡C	2250	medium
C≡N	2250	medium
C=O	1800–1650 (often ~1700)	strong
C=C	1650	medium
(benzene ring)	1600, 1500	medium

Almost all bonds in a molecule give rise to an absorption peak in an IR spectrum, but a few do not. **For a bond to absorb in the IR, there must be a change in dipole moment during the vibration.** Thus, symmetrical, nonpolar bonds do *not* absorb in the IR. The carbon–carbon triple bond of but-2-yne, for example, does not have an IR stretching absorption at 2250 cm^{-1} because the C≡C bond is nonpolar and there is no change in dipole moment when the bond stretches along its axis. This type of vibration is said to be **IR inactive.**

Stretching along the bond axis
does not change the dipole moment.

$$CH_3-C{\equiv}C-CH_3$$

nonpolar bond
IR inactive

Problem B.4 Which highlighted bond in each pair absorbs at higher wavenumber?

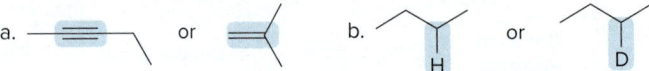

B.3B The Effect of Percent *s*-Character on C–H Absorptions

Any factor that affects bond strength affects the location of an IR absorption. Recall from Section 1.11 that **the strength of a C–H bond *increases* as the percent *s*-character of the hybrid orbital on carbon *increases*;** thus:

C–H	C–H	≡C–H
sp^3 hybridized C	*sp^2* hybridized C	*sp* hybridized C
25% *s*-character	**33%** *s*-character	**50%** *s*-character

Increasing percent *s*-character
Increasing $\tilde{\nu}$

- The *higher* the percent *s*-character, the *stronger* the C–H bond and the *higher* the wavenumber of the absorption.

Problem B.5 Rank the indicated bonds in the following compound in order of increasing (a) strength; (b) bond length; (c) percent *s*-character; (d) wavenumber of absorption.

B.3C The Effect of Resonance on IR Absorptions

When a compound contains a carbonyl group (C=O), often the carbonyl absorption is the most intense peak in the IR spectrum. The exact location of that absorption depends on what groups are bonded directly to the carbonyl carbon.

When a carbonyl group is bonded to a carbon–carbon double bond or a benzene ring, the **two sites of unsaturation are separated by one σ bond and the system is *conjugated*. Conjugation** affects the location of the carbonyl absorption.

[Structures: cyclohexenyl methyl ketone — 1 σ bond — conjugated system; phenyl methyl ketone — 1 σ bond — conjugated system; cyclohexenyl methyl ketone with 2 σ bonds — not conjugated]

- **Conjugation of the carbonyl group with a C=C or a benzene ring shifts the absorption to lower wavenumber by ~30 cm^{-1}.**

The effect of conjugation on the frequency of the C=O absorption is explained by **resonance**. An α,β-unsaturated carbonyl compound can be written as three resonance structures, two of which place a single bond between the carbon and oxygen atoms of the carbonyl group. Thus, the π bond of the carbonyl group is delocalized, giving the conjugated carbonyl group some single bond character, and making it somewhat **weaker** than an unconjugated C=O. **Weaker bonds absorb at lower frequency (lower wavenumber) in an IR spectrum.**

[Resonance structures of α,β-unsaturated carbonyl group. Two resonance contributors have a C–O single bond. Hybrid: The π bonds are delocalized.]

Figure B.4 illustrates the effects of conjugation on the location of the carbonyl absorption in some representative compounds.

Figure B.4
The effect of conjugation on the carbonyl absorption in an IR spectrum

[Structures with IR frequencies: cyclohexyl methyl ketone 1709 cm^{-1}; phenyl methyl ketone 1685 cm^{-1} (conjugated C=O, lower wavenumber); cyclohexanone 1715 cm^{-1}; cyclohexenone 1685 cm^{-1} (conjugated C=O, lower wavenumber)]

Resonance also affects the relative position of the carbonyl absorptions of compounds RCOZ, when Z contains a nonbonded electron pair. Three resonance structures can be drawn for RCOZ.

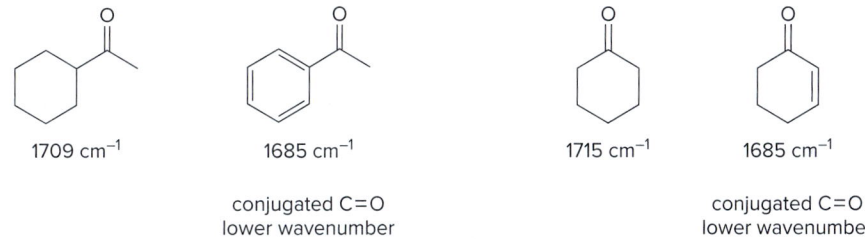

Because resonance structures **2** and **3** contain a carbon–oxygen single bond, the more these structures contribute to the resonance hybrid, the more single bond character the carbonyl group possesses, and the *lower* the frequency of the carbonyl absorption.

- **The more basic Z is, the more it donates its electron pair and the more resonance structure 3 contributes to the hybrid.**
- **As a result, as the basicity of Z *increases*, the frequency of the carbonyl absorption *decreases*.**

536 Spectroscopy B Infrared Spectroscopy

To compare the carbonyl absorptions of an ester (RCO$_2$R') and an amide (RCONR'$_2$), we look at the relative basicity of an OR' group and an NR'$_2$ group. Basicity decreases across a row of the periodic table, so an NR'$_2$ group is more basic than an OR' group. Thus, **an amide carbonyl has more single bond character than an ester carbonyl, and the carbonyl absorption occurs at *lower* wavenumber.**

For example, the carbonyl absorptions of the ester ethyl acetate and the amide N,N-dimethylacetamide occur at 1743 and 1662 cm^{-1}, respectively.

ethyl acetate
1743 cm^{-1}
higher wavenumber

N,N-dimethylacetamide
1662 cm^{-1}
lower wavenumber

O is **less basic** than N.

N is **more basic** than O.

Sample Problem B.1 illustrates how resonance affects the position of the carbonyl absorption in three compounds.

Sample Problem B.1 Predicting the Relative Position of Carbonyl Absorptions

Rank compounds **A**, **B**, and **C** in order of increasing frequency of the carbonyl absorption in their IR spectra.

A   B   C

Solution
Compare pairs of compounds. **A** and **B** are both amides, but **B** contains a carbonyl that is conjugated with a benzene ring. Because conjugation shifts the carbonyl absorption to lower wavenumber, **B** absorbs at lower wavenumber than **A**. **C** is an ester that is not conjugated. In comparing carbonyl absorptions in **A** and **C**, the amide **A** contains a more basic N atom, so the carbonyl absorption of **A** occurs at lower wavenumber than that of ester **C**. Thus, in order of increasing frequency (wavenumber): **B < A < C**.

Problem B.6 (a) Considering compounds **A**, **B**, **C**, and **D**, which compound has a C=O that absorbs at the *highest* wavenumber? (b) Which compound has a C=O that absorbs at the *lowest* wavenumber?

A   B   C   D

More Practice: Try Problems B.23, B.24.

B.3D Analyzing an IR Spectrum

The principles learned in this section can be used to determine what types of bonds are present in a compound, as shown in the stepwise *How To*.

B.3 IR Absorptions

How To Analyze an IR Spectrum

Example Which compound—**A**, **B**, or **C**—gives rise to the given IR spectrum?

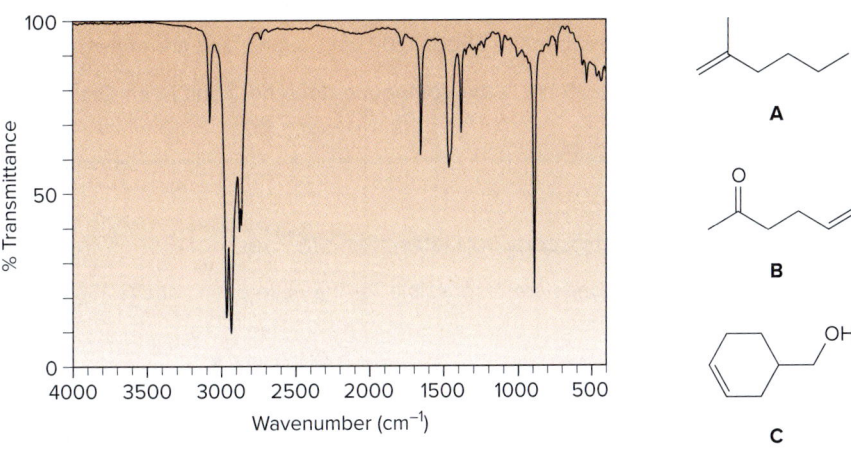

Step [1] Concentrate on the functional group region above 1500 cm^{-1}, and examine the two sections where double and triple bonds absorb, using the values in Table B.1.
- A C=C absorbs at ~1650 cm^{-1}.
- A C=O absorbs between 1650 and 1800 cm^{-1}, often around 1700 cm^{-1}.
- A C≡C or C≡N absorbs at ~2250 cm^{-1}.

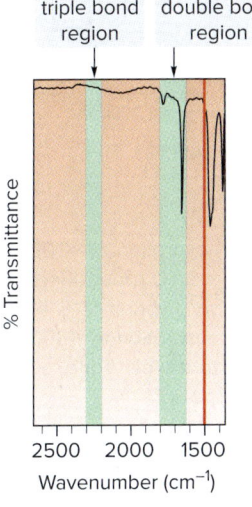

- In this example, there are no absorptions in the triple bond region and an absorption of medium intensity at 1650 cm^{-1}, suggesting that the compound contains a C=C.

- Because there is no strong absorption at ~1700 cm^{-1}, the compound does *not* contain a C=O.

- Using these data, we can eliminate **B** as a possibility because **B** contains a C=O.

Step [2] Examine the C–H region around 3000 cm^{-1}.

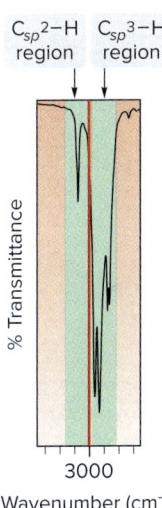

- C_{sp^3}–H bonds absorb at 3000–2850 cm^{-1}.
- C_{sp^2}–H bonds absorb at 3150–3000 cm^{-1}.
- C_{sp}–H bonds absorb at 3300 cm^{-1}.

- In addition to C_{sp^3}–H bonds, which are present in almost all organic compounds, the compound contains an absorption in the C_{sp^2}–H region.

- Both **A** and **C** contain C_{sp^2}–H, so both compounds are still possibilities.

—Continued

538 Spectroscopy B Infrared Spectroscopy

How To, continued...

Step [3] Examine the region above 3000 cm^{-1} for O–H and N–H bonds.
- O–H bonds appear as strong, broad peaks at 3600–3200 cm^{-1}.
- N–H bonds of amines and amides absorb in the 3500–3200 cm^{-1} region, and are of medium intensity.

Because the IR shows no absorption at 3600–3200 cm^{-1}, the compound does not contain an OH group. This eliminates **C** as a possibility, so the IR spectrum is due to **A**.

Sample Problem B.2 — Using IR Spectroscopy to Determine the Types of Bonds in a Compound

What types of bonds are responsible for the absorptions above 1500 cm^{-1} in compounds **A** and **B**?

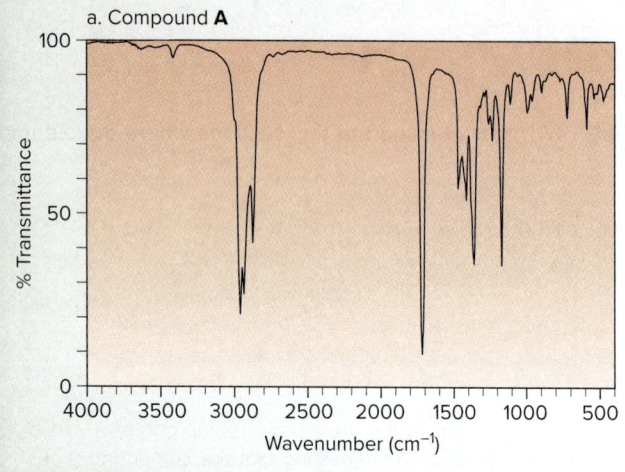
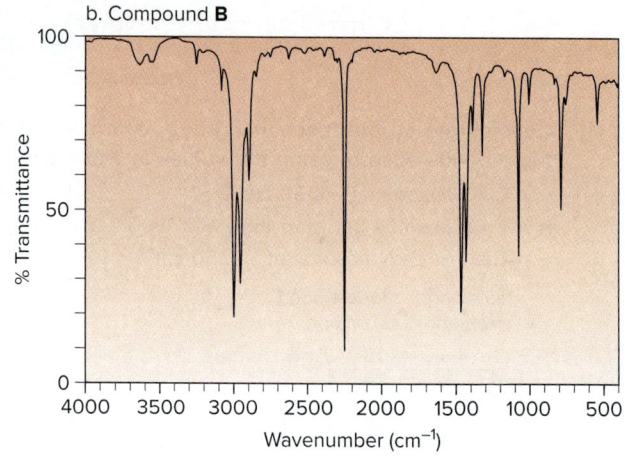

Solution

a. Compound **A** has two major absorptions above 1500 cm^{-1}: The absorption at ~3000 cm^{-1} is due to C–H bonds and the absorption at ~1700 cm^{-1} is due to a C=O group. Because the C–H absorption occurs at < 3000 cm^{-1}, all C–H bonds contain sp^3 hybridized C atoms.

b. Compound **B** has two major absorptions above 1500 cm^{-1}: The absorption at ~3000 cm^{-1} is due to C_{sp^3}–H bonds and the absorption at ~2250 cm^{-1} is due to a triple bond, either a C≡C or a C≡N.

Problem B.7 What types of bonds are responsible for the absorptions above 1500 cm^{-1} in the IR spectra for compounds **A** and **B**?

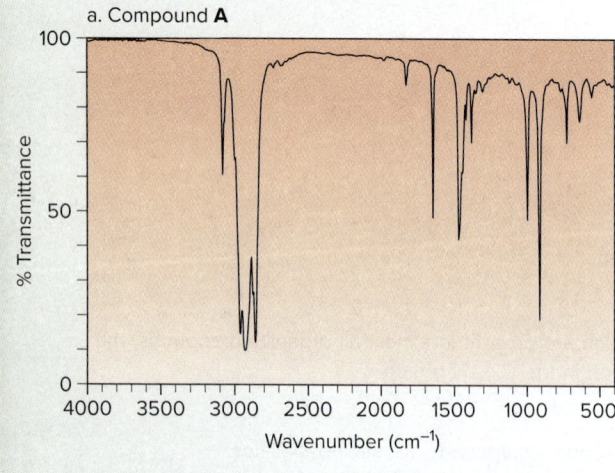
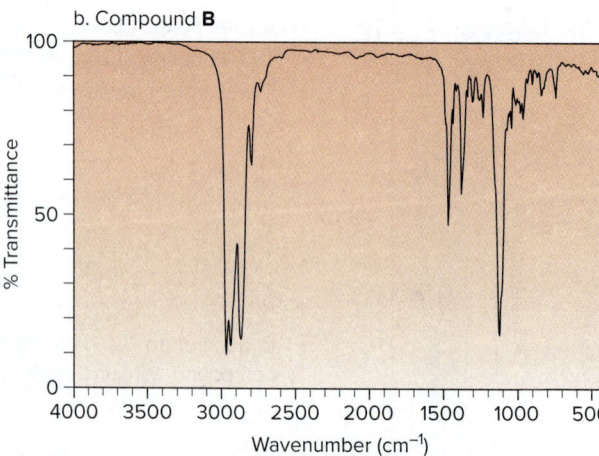

More Practice: Try Problems B.16, B.19, B.26.

B.4 Infrared Spectra of Common Functional Groups

Each class of compounds exhibits characteristic absorptions in the infrared.

B.4A IR Absorptions in Hydrocarbons

The IR spectra of an alkane, an alkene, an alkyne, and an aromatic compound with a benzene ring illustrate characteristic differences.

Alkanes

An **alkane** like hexane has only C—C single bonds and sp^3 hybridized C atoms. Therefore, it has only one major absorption above 1500 cm^{-1}:

- C_{sp^3}–H absorption at 3000–2850 cm^{-1}

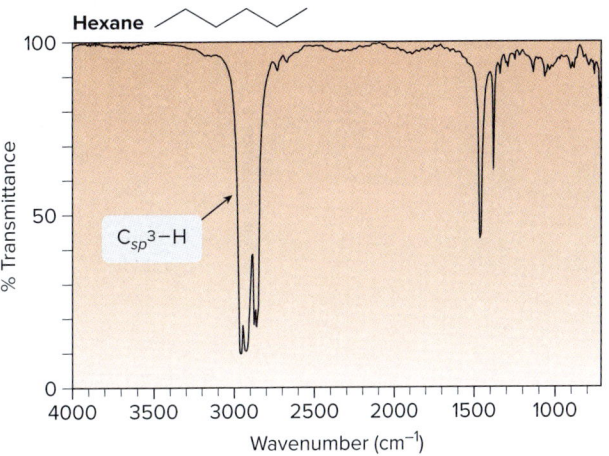

Alkenes and Alkynes

An **alkene** like hex-1-ene has a C=C and C_{sp^2}–H, in addition to its sp^3 hybridized C atoms. Therefore, there are three major absorptions above 1500 cm^{-1}:

- C_{sp^2}–H at 3150–3000 cm^{-1}
- C_{sp^3}–H at 3000–2850 cm^{-1}
- C=C at 1650 cm^{-1}

An **alkyne** like hex-1-yne has a C≡C and C_{sp}–H, in addition to its sp^3 hybridized C atoms. Therefore, there are three major absorptions:

- C_{sp}–H at 3300 cm^{-1}
- C_{sp^3}–H at 3000–2850 cm^{-1}
- C≡C at ~2250 cm^{-1}

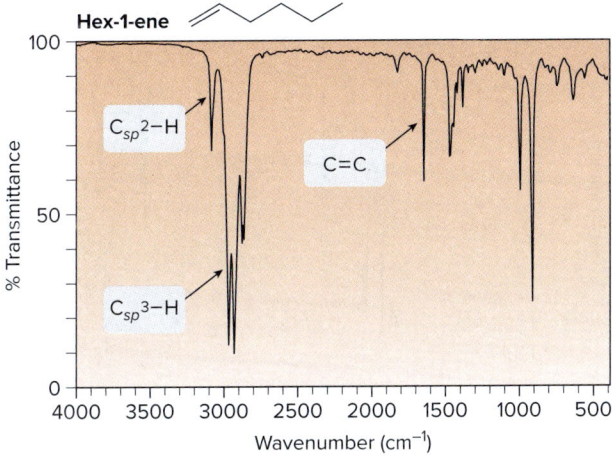
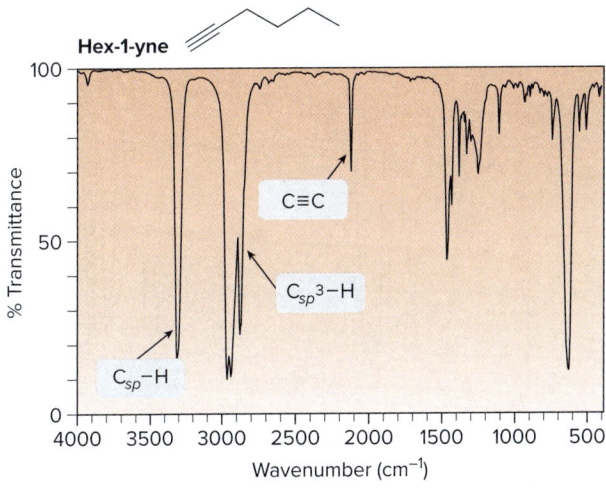

Problem B.8   How do the IR spectra of the isomers cyclopentane and pent-1-ene differ?

Aromatic Compounds with Benzene Rings

An **aromatic compound** like isopropylbenzene contains a benzene ring and C_{sp^2}–H, in addition to its sp^3 hybridized C atoms. Thus, there are three major absorptions:

- C_{sp^2}–H at 3150–3000 cm^{-1}
- C_{sp^3}–H at 3000–2850 cm^{-1}
- Benzene ring at 1600, 1500 cm^{-1}

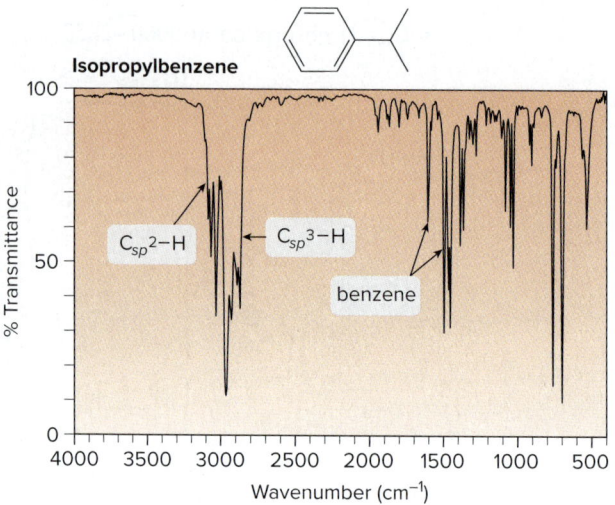

B.4B   IR Absorptions in Oxygen-Containing Compounds

The most important IR absorptions for oxygen-containing compounds occur at **3600–3200 cm^{-1} for an OH group** and at approximately **1700 cm^{-1} for a C=O.**

Alcohols and Ethers

The most prominent absorption for an **alcohol** like butan-2-ol is the broad, strong absorption at **3600–3200 cm^{-1}** due to the **OH** group. An **ether** like diethyl ether has neither an OH group nor a C=O, so its only absorption above 1500 cm^{-1} occurs at ~3000 cm^{-1}, due to sp^3 hybridized C–H bonds. **Compounds that contain an oxygen atom but do not show an OH or C=O absorption are ethers.**

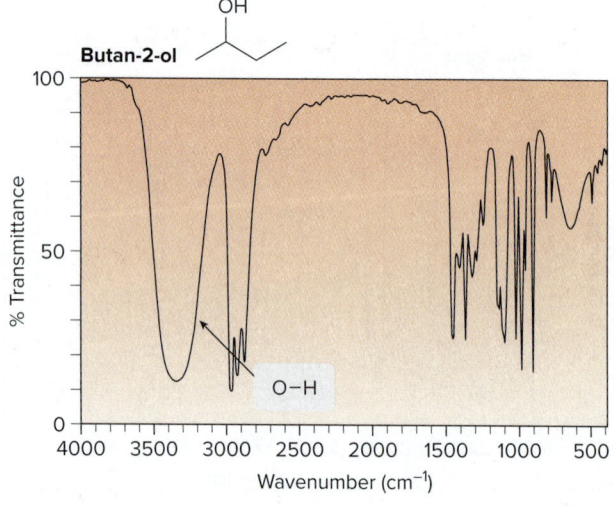

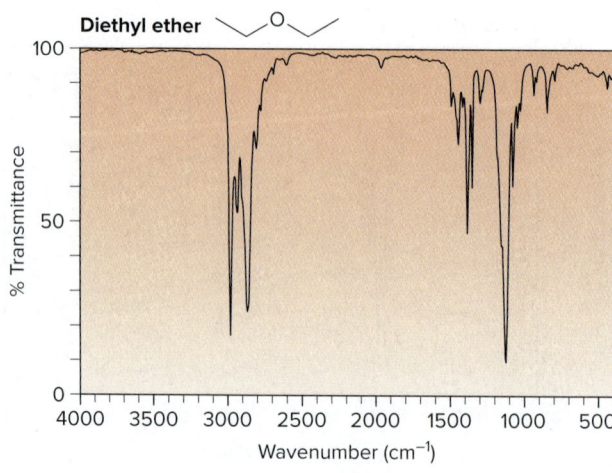

Aldehydes and Ketones

An **aldehyde** like propanal has a C=O and C_{sp^2}–H. In addition to the absorption of its C_{sp^3}–H, there are two major absorptions:

- C_{sp^2}–H of the aldehyde C–H at 2830–2700 cm^{-1} (one or two peaks)
- C=O at ~1700 cm^{-1}

In addition to the absorption of its C_{sp^3}–H, a **ketone** like butan-2-one has one major absorption:

- C=O at ~1700 cm^{-1}

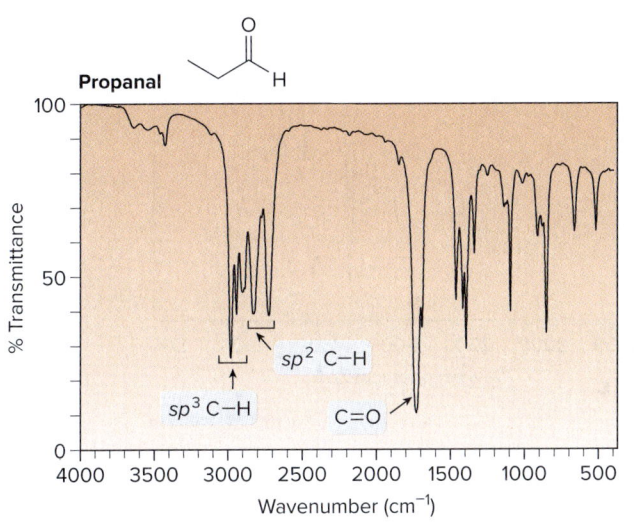
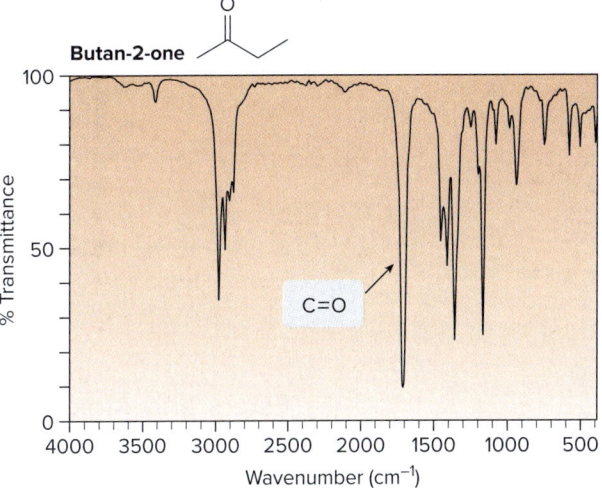

The exact location of the carbonyl absorption provides additional information about a compound. In Section B.3C, we learned that conjugation of the C=O with a C=C or a benzene ring shifts the absorption to lower wavenumber.

When the carbonyl carbon is located in a ring, **ring size** also affects the location of the carbonyl absorption.

- The carbonyl absorption of cyclic ketones shifts to *higher* wavenumber as the size of the ring *decreases* and the ring strain *increases*.

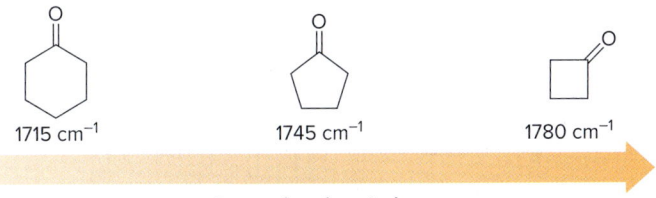

Problem B.9 Rank the following compounds in order of increasing frequency of the carbonyl absorption.

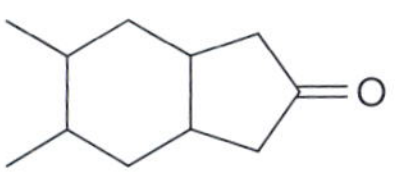

A 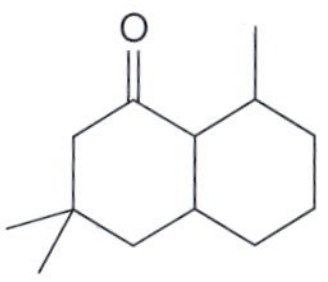   B 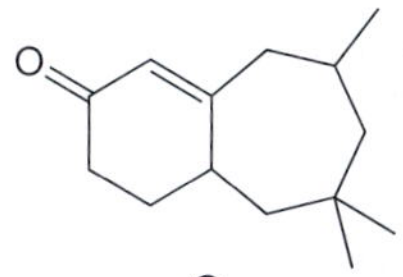   C

Carboxylic Acids

A carboxylic acid like butanoic acid has two characteristic IR absorptions:

- O–H at 3500–2500 cm^{-1}, a broad, strong absorption that almost obscures the C–H peak at ~3000 cm^{-1}.
- C=O at ~1710 cm^{-1}

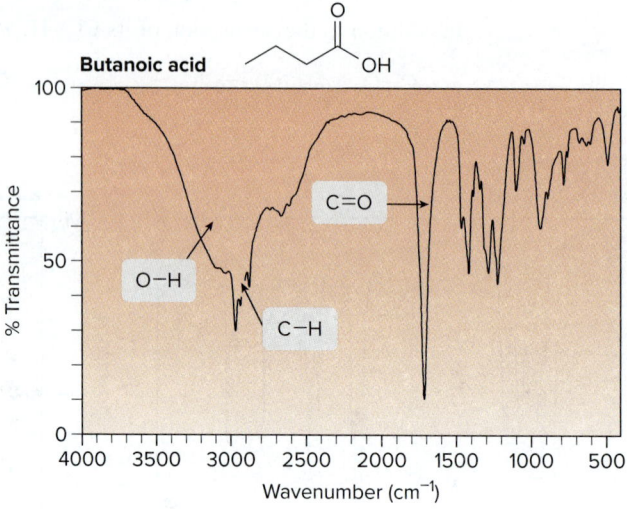

Esters and Amides

In addition to the absorption of its C$_{sp^3}$–H, an **ester** like methyl propanoate has one major absorption:

- C=O at ~1745–1735 cm^{-1}

An **amide** like propanamide has three characteristic absorptions:

- N–H stretching peaks at 3400–3200 cm^{-1} (one or two peaks)
- C=O at 1680–1630 cm^{-1}
- N–H bending absorption at ~1640 cm^{-1}

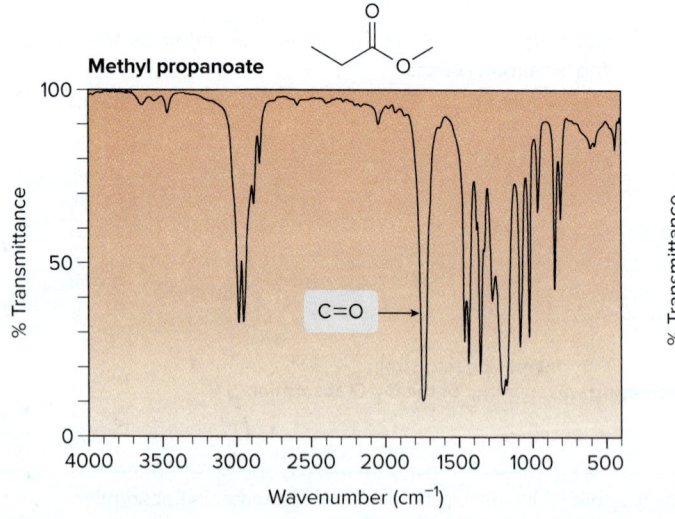
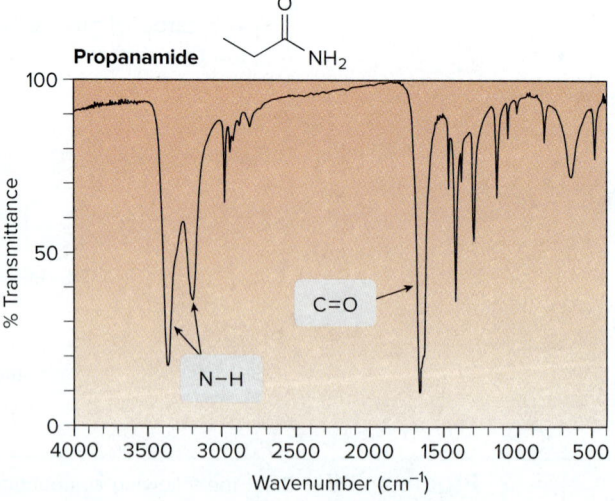

Problem B.10 How would compounds **X** and **Y** differ in their IR spectra?

B.4C IR Absorptions in Amines and Nitriles

Common functional groups that contain nitrogen atoms are also distinguishable by their IR absorptions above 1500 cm^{-1}.

The **N—H** bonds in an **amine** like octylamine give rise to two weak absorptions at 3300 and 3400 cm^{-1}. The **C≡N** group of a **nitrile** like octanenitrile absorbs in the triple bond region at ~2250 cm^{-1}.

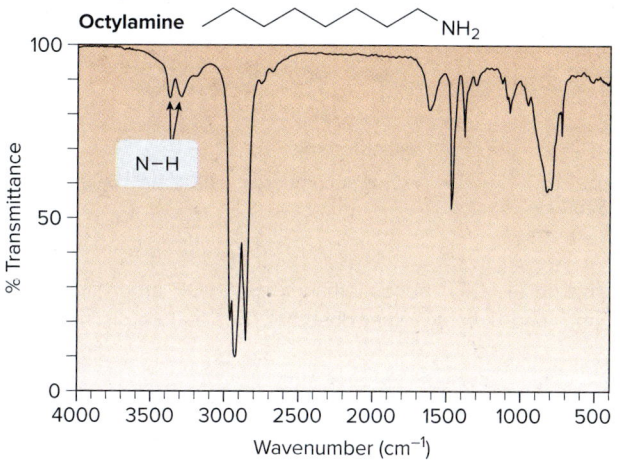
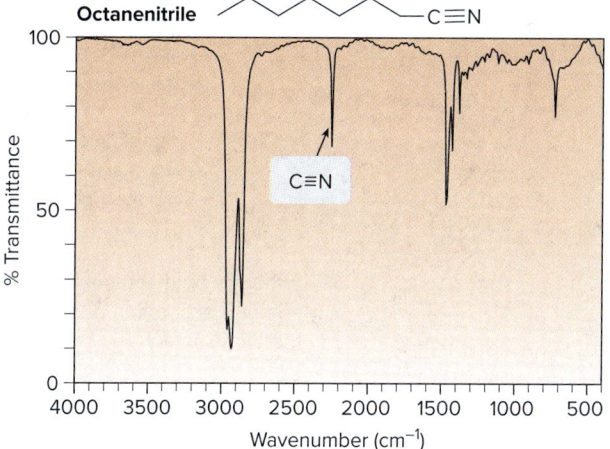

B.4D Summary of IR Absorptions for Common Functional Groups

Table B.2 summarizes the typical IR peaks for common functional groups.

Table B.2 Characteristic IR Absorptions in the Functional Group Region

Compound type	Absorption (cm^{-1})	Intensity
Alkane		
C_{sp^3}—H	3000–2850	strong
Alkene		
C_{sp^2}—H	3150–3000	medium
C=C	1650	medium
Alkyne		
C_{sp}—H	3300	medium
C≡C	2250	medium
Benzene	1600, 1500	medium
Alcohol		
O—H	3600–3200	strong, broad
Amine		
N—H	3500–3300	medium
Carbonyl compounds		
Aldehyde C_{sp^2}—H	2830–2700	medium
Aldehyde C=O	1730	strong
Ketone C=O	1715	strong
Ester C=O	1745–1735	strong
Amide C=O	1680–1630	strong
Amide N—H	3400–3200	medium
Carboxylic acid		
O—H	3500–2500	strong, very broad
C=O	1710	strong
Nitrile		
C≡N	2250	medium

Sample Problem B.3 Using IR Spectroscopy to Distinguish Isomers

How can the two isomers having molecular formula C_2H_6O be distinguished by IR spectroscopy?

Solution

First, draw the structures of the compounds and then locate the functional groups. One compound is an alcohol and one is an ether.

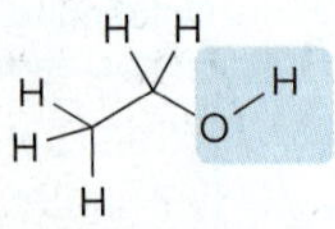

ethanol — hydroxy group
- C–H absorption at ~3000 cm^{-1}
- O–H absorption at 3600–3200 cm^{-1}

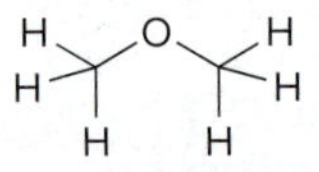

dimethyl ether — no OH group
- C–H absorption at ~3000 cm^{-1} **only**

Although both compounds have sp^3 hybridized C–H bonds, ethanol has an OH group that gives a strong absorption at 3600–3200 cm^{-1}, and dimethyl ether does not. This feature distinguishes the two isomers.

Problem B.11 How do the three isomers of molecular formula C_3H_6O (**A, B,** and **C**) differ in their IR spectra?

A (acetone), B (methyl vinyl ether), C (cyclopropanol)

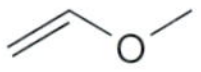

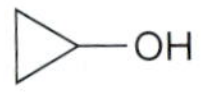

More Practice: Try Problems B.21, B.22, B.25.

Problem B.12 What are the major IR absorptions in the functional group region for each compound?

a.  (heptane)
b. 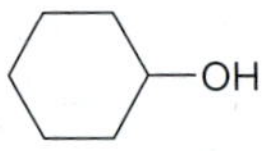 (cyclohexanol)
c. 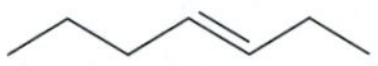 (alkene)
d. 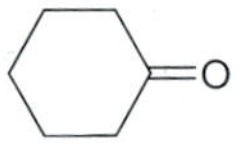 (cyclohexanone)
e. 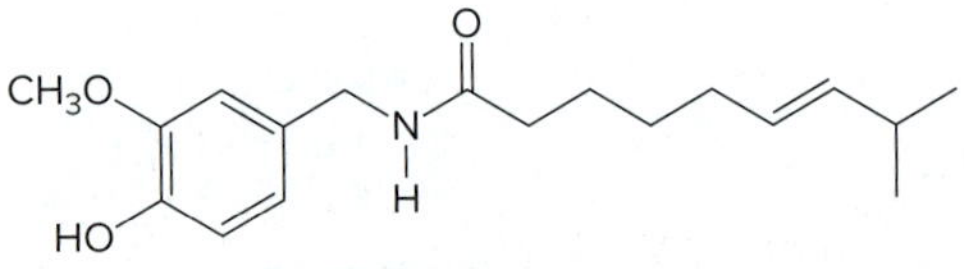 capsaicin (spicy component of hot peppers)

B.5 IR and Structure Determination

Since its introduction, IR spectroscopy has proven to be a valuable tool for determining the functional groups in organic molecules.

In the 1940s, IR spectroscopy played a key role in elucidating the structure of the antibiotic penicillin G, the chapter-opening molecule. **β-Lactams,** four-membered rings that contain an amide, have a carbonyl group that absorbs at ~1760 cm^{-1}, a much higher frequency than that observed for most amides and many other carbonyl groups. Because penicillin G had an IR absorption at this frequency, **A** became the leading candidate for the structure of penicillin rather than **B**, a possibility originally considered more likely. Structure **A** was later confirmed by X-ray analysis.

B.5 IR and Structure Determination

Correct structure

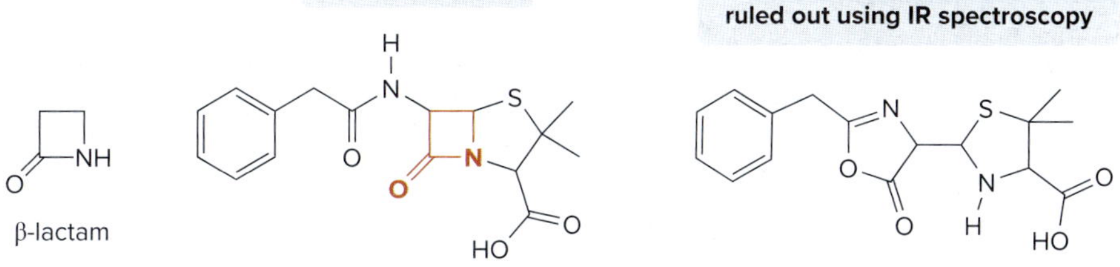

β-lactam

A
penicillin G
(β-lactam in red)

Incorrect structure, ruled out using IR spectroscopy

B

IR spectroscopy is often used to determine the outcome of a chemical reaction. For example, oxidation of the hydroxy group in **C** to form the carbonyl group in periplanone B is accompanied by the disappearance of the OH absorption (3600–3200 cm^{-1}) and the appearance of a carbonyl absorption near 1700 cm^{-1} in the IR spectrum of the product. Periplanone B is the sex pheromone of the female American cockroach.

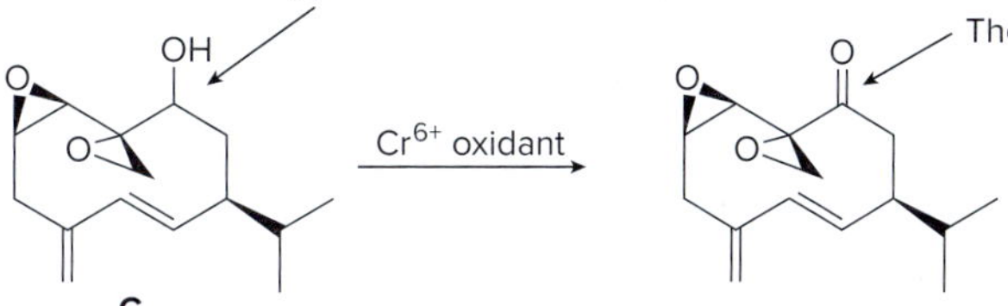

The absorption at 3600–3200 cm^{-1} disappears.

Cr^{6+} oxidant

The absorption at ~1700 cm^{-1} appears.

C

periplanone B

Problem B.13 How can IR spectroscopy be used to determine when the following reaction is complete?

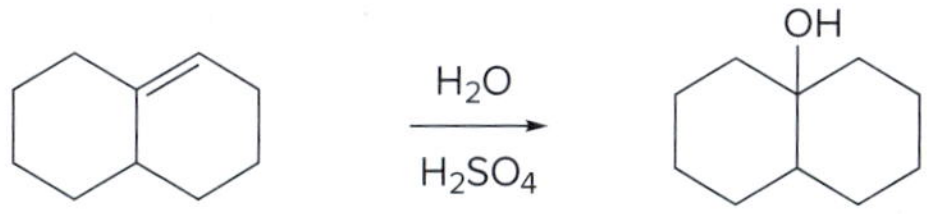

$$\xrightarrow[H_2SO_4]{H_2O}$$

The combination of IR and mass spectral data provides key information on the structure of an unknown compound. The mass spectrum reveals the molecular weight of the unknown (and the molecular formula if an exact mass is available), and the IR spectrum helps to identify the important functional groups.

How To Use MS and IR for Structure Determination

Example What information is obtained from the mass spectrum and IR spectrum of an unknown compound **X**? Assume **X** contains the elements C, H, and O.

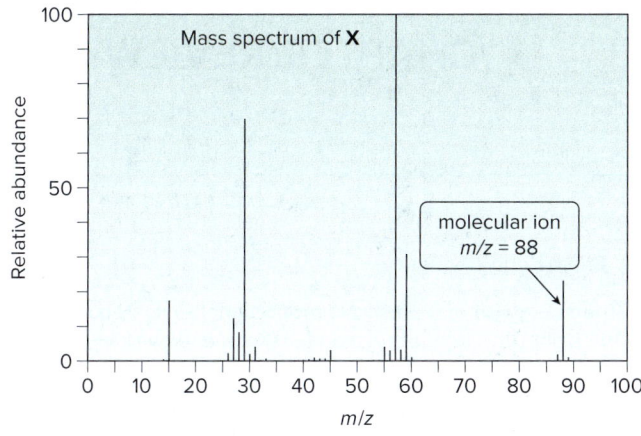

Mass spectrum of **X**

molecular ion
m/z = 88

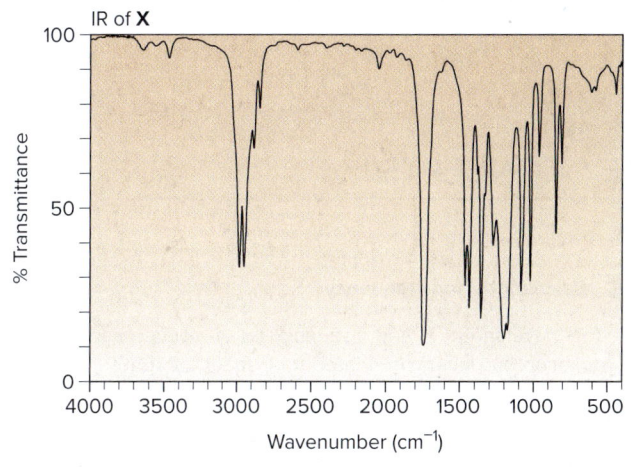

IR of **X**

—Continued

Spectroscopy B Infrared Spectroscopy

How To, continued...

Step [1] Use the molecular ion to determine possible molecular formulas. Use an exact mass (when available) to determine a molecular formula.

- Use the procedure outlined in Sample Problem A.2 to calculate possible molecular formulas. For a molecular ion at $m/z = 88$:

$$\frac{88}{12} = 7 \text{ C's} \text{ maximum (remainder = 4)} \rightarrow C_7H_4 \xrightarrow[+1\text{ O}]{-CH_4} C_6O \xrightarrow[+12\text{ H's}]{-1\text{ C}} C_5H_{12}O \xrightarrow[+1\text{ O}]{-CH_4} C_4H_8O_2 \xrightarrow[+1\text{ O}]{-CH_4} C_3H_4O_3$$

three possible formulas

- Discounting C_7H_4 (a hydrocarbon) and C_6O (because it contains no H's) gives three possible formulas for **X**.
- If high-resolution mass spectral data are available, the molecular formula can be determined directly. If the molecular ion had an exact mass of 88.0580, the molecular formula of **X** is $C_4H_8O_2$ (exact mass = 88.0524) rather than $C_5H_{12}O$ (exact mass = 88.0888) or $C_3H_4O_3$ (exact mass = 88.0160).

Step [2] Calculate the number of degrees of unsaturation (Section 10.2).

- For a compound of molecular formula $C_4H_8O_2$, the maximum number of H's = $2n + 2 = 2(4) + 2 = 10$.
- Because the compound contains only 8 H's, it has $10 - 8 = 2$ H's fewer than the maximum number.
- Because each degree of unsaturation removes 2 H's, **X** has one degree of unsaturation. **X has one ring or one π bond.**

Step [3] Determine what functional group is present from the IR spectrum.

- The two major absorptions in the IR spectrum above 1500 cm^{-1} are due to sp^3 hybridized C–H bonds (~3000–2850 cm^{-1}) and a C=O group (1740 cm^{-1}). Thus, the one degree of unsaturation in **X** is due to the presence of the **C=O**.

Mass spectrometry and IR spectroscopy give valuable but limited information on the identity of an unknown. Although the mass spectral and IR data reveal that **X** has a molecular formula of $C_4H_8O_2$ and contains a carbonyl group, more data are needed to determine its complete structure. In Spectroscopy C, we will learn how other spectroscopic data can be used for that purpose.

Problem B.14 Which of the following possible structures for **X** can be excluded on the basis of its IR spectrum?

a. [structure]   b. HO-CH₂CH₂CH₂-CHO   c. [structure]   d. [structure with OH]

Problem B.15 Propose structures consistent with each set of data: (a) a hydrocarbon with a molecular ion at $m/z = 68$ and IR absorptions at 3310, 3000–2850, and 2120 cm^{-1}; (b) a compound containing C, H, and O with a molecular ion at $m/z = 60$ and IR absorptions at 3600–3200 and 3000–2850 cm^{-1}.

Spectroscopy B CHAPTER REVIEW

KEY CONCEPTS

[1] Electromagnetic radiation (B.1)

1 Wavelength and frequency

- The wavelength (λ) and frequency (ν) of electromagnetic radiation are **inversely** related (c = speed of light):

$$\lambda = c/\nu \quad \text{or} \quad \nu = c/\lambda$$

2 Energy and frequency

- The energy (E) of a photon is **proportional** to its frequency (ν); the higher the frequency, the higher the energy [h = Planck's constant (6.63×10^{-34} J·s)]:

$$E = h\nu$$

Try Problems B.1–B.3.

[2] Bond strength and IR absorption (B.3)

1 The higher the percent s-character, the stronger the bond, and the higher the $\tilde{\nu}$ of absorption.

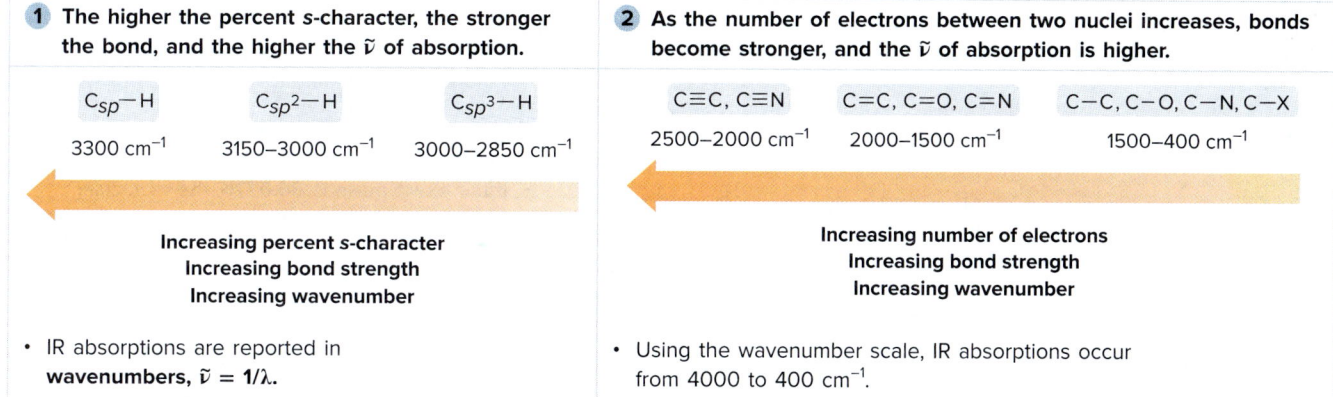

- IR absorptions are reported in wavenumbers, $\tilde{\nu} = 1/\lambda$.

2 As the number of electrons between two nuclei increases, bonds become stronger, and the $\tilde{\nu}$ of absorption is higher.

- Using the wavenumber scale, IR absorptions occur from 4000 to 400 cm^{-1}.

See Table B.1. Try Problem B.18.

[3] Factors affecting the location of a carbonyl absorption (B.3C, B.4B)

1 Conjugation shifts a C=O absorption to lower wavenumber.

2 For cyclic compounds, a smaller ring size shifts a C=O absorption to higher wavenumber.

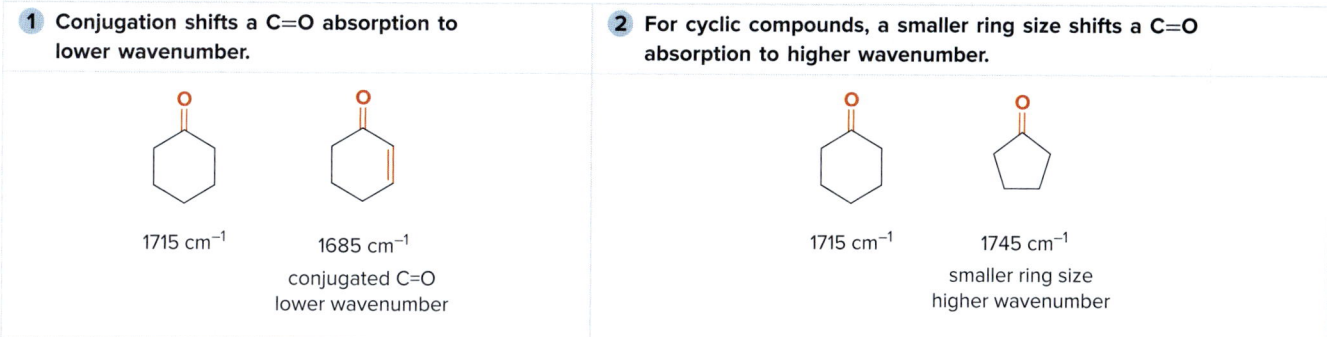

See Sample Problem B.1, Table B.2. Try Problems B.23, B.24.

KEY SKILLS

[1] Using the functional groups to distinguish two compounds by IR spectroscopy (B.4D)

1 Locate the functional groups.

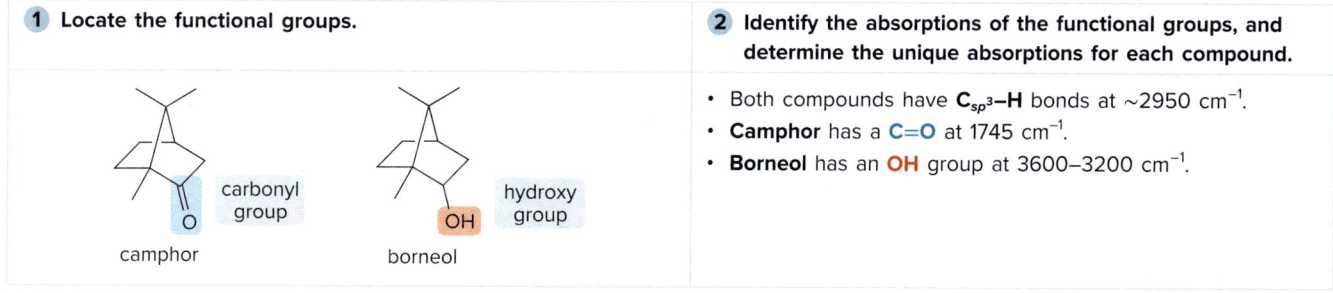

2 Identify the absorptions of the functional groups, and determine the unique absorptions for each compound.

- Both compounds have C_{sp^3}–H bonds at ~2950 cm^{-1}.
- **Camphor** has a **C=O** at 1745 cm^{-1}.
- **Borneol** has an **OH** group at 3600–3200 cm^{-1}.

See Sample Problems B.2, B.3, Table B.2. Try Problems B.21, B.22, B.25.

[2] Using MS and IR to determine possible structures of a compound that contains C, H, and O (B.5); example: m/z = 86

1. Use the molecular ion to determine the possible molecular formulas.

molecular ion at m/z = 86

[1] $\dfrac{86}{12}$ = 7 **C**'s maximum (remainder = 2) → C$_7$H$_2$

[3] C$_6$H$_{14}$ $\xrightarrow[+1\,\text{O}]{-\text{CH}_4}$ C$_5$H$_{10}$O

[2] C$_7$H$_2$ $\xrightarrow[+12\,\text{H's}]{-1\,\text{C}}$ C$_6$H$_{14}$
  ↑ hydrocarbon

[4] C$_5$H$_{10}$O $\xrightarrow[+1\,\text{O}]{-\text{CH}_4}$ C$_4$H$_6$O$_2$

  ↑ hydrocarbon

2. Calculate the number of degrees of unsaturation (10.2).

C$_5$H$_{10}$O

For *n* carbons, the maximum number of H's is 2*n* + 2; in this example, 2*n* + 2 = 2(5) + 2 = 12.

12 H's (maximum) − 10 H's (actual) = 2 H's fewer than the maximum number

2 H's fewer than the maximum
2 H's per degree of unsaturation

Answer: one degree of unsaturation

C$_4$H$_6$O$_2$

For *n* carbons, the maximum number of H's is 2*n* + 2; in this example, 2*n* + 2 = 2(4) + 2 = 10.

10 H's (maximum) − 6 H's (actual) = 4 H's fewer than the maximum number

4 H's fewer than the maximum
2 H's per degree of unsaturation

Answer: two degrees of unsaturation

3. Use an exact mass to determine a molecular formula.

- High-resolution mass spectrometry gives the molecular formula of a compound.
- If the **exact mass** is **86.0775**, the molecular formula of **X** is **C$_5$H$_{10}$O** (exact mass = 86.0732) rather than C$_4$H$_6$O$_2$ (exact mass = 86.0368).

4. Determine the functional groups by IR.

- C$_{sp^3}$–H bonds at 2973–2877 cm^{-1}
- C=O bond (1718 cm^{-1})

pentan-3-one   pentan-2-one   3-methylbutan-2-one

Three structures containing a ketone are consistent with the data.

See *How To* p. 545, Table B.2. Try Problems B.27–B.35.

PROBLEMS

Problems that combine mass spectrometry, infrared spectroscopy, and nuclear magnetic resonance spectroscopy are found at the end of Spectroscopy C.

Problem Using Three-Dimensional Models

B.16 What major IR absorptions are present above 1500 cm^{-1} for each compound?

a.   b.

B.17 What are the major IR absorptions in the functional group region for oleic acid, a common unsaturated fatty acid (Section 10.6A)?

oleic acid

Infrared Spectroscopy

B.18 Which of the highlighted bonds absorbs at higher $\tilde{\nu}$ in an IR spectrum?

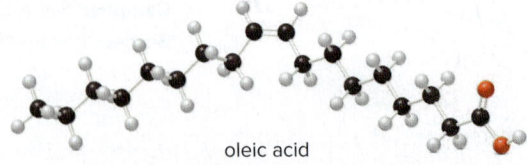

a. =O or —OH   b. =N or —N   c. —H or —H

B.19 What major IR absorptions are present above 1500 cm^{-1} for each compound?

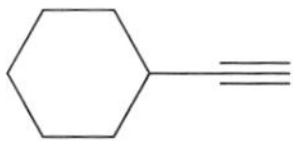

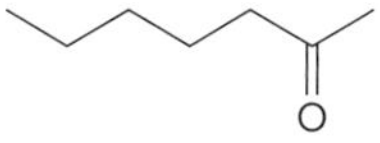

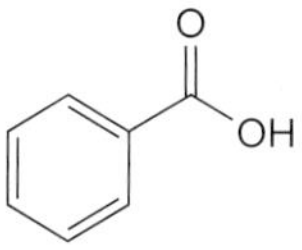

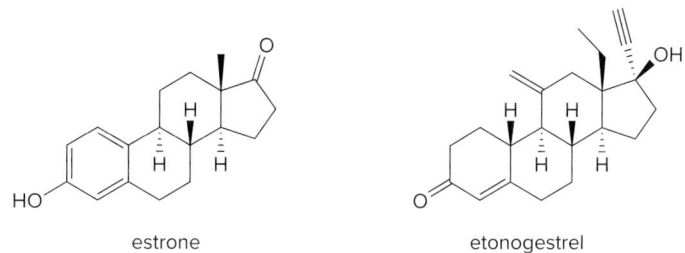

B.20 Estrone is a female sex hormone, and etonogestrel is a synthetic hormone used in contraceptive implants to prevent pregnancy. (a) Identify the prominent IR absorptions resulting from the functional groups in each compound. (b) How do the locations of the carbonyl absorptions in these two compounds compare? Explain your reasoning.

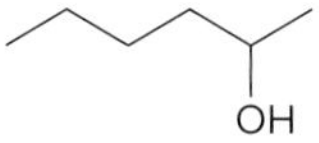

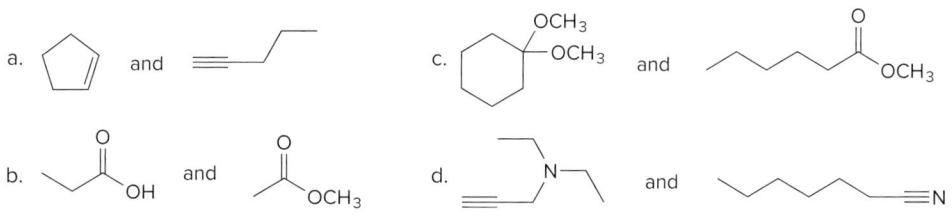

estrone   etonogestrel

B.21 How would each of the following pairs of compounds differ in their IR spectra?

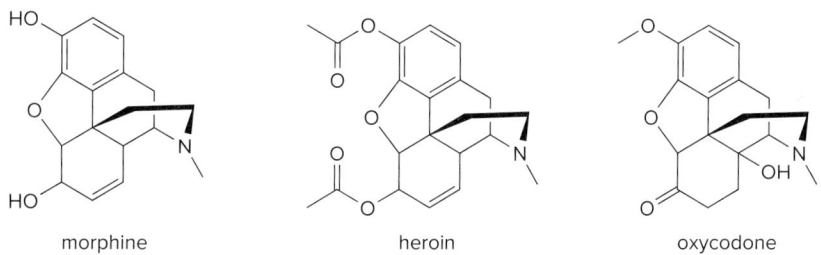

B.22 Morphine, heroin, and oxycodone are three addicting analgesic narcotics. How could IR spectroscopy be used to distinguish these three compounds from each other?

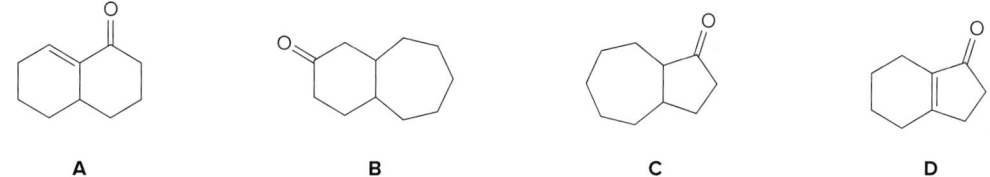

morphine   heroin   oxycodone

B.23 (a) Which of the following compounds has a C=O that absorbs at the *highest* wavenumber? (b) Which of the following compounds has a C=O that absorbs at the *lowest* wavenumber?

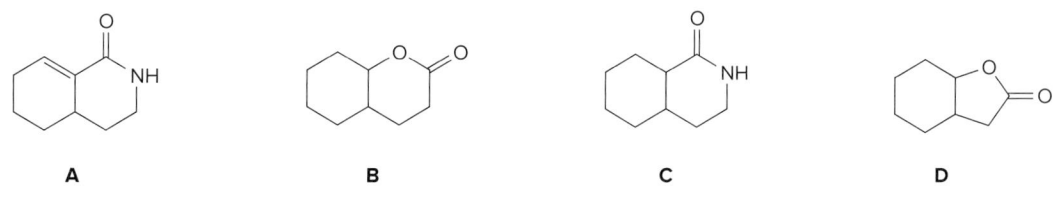

A   B   C   D

B.24 Rank the following compounds in order of increasing wavenumber of the carbonyl absorption in the IR.

A   B   C   D

B.25 Tell how IR spectroscopy could be used to determine when each reaction is complete.

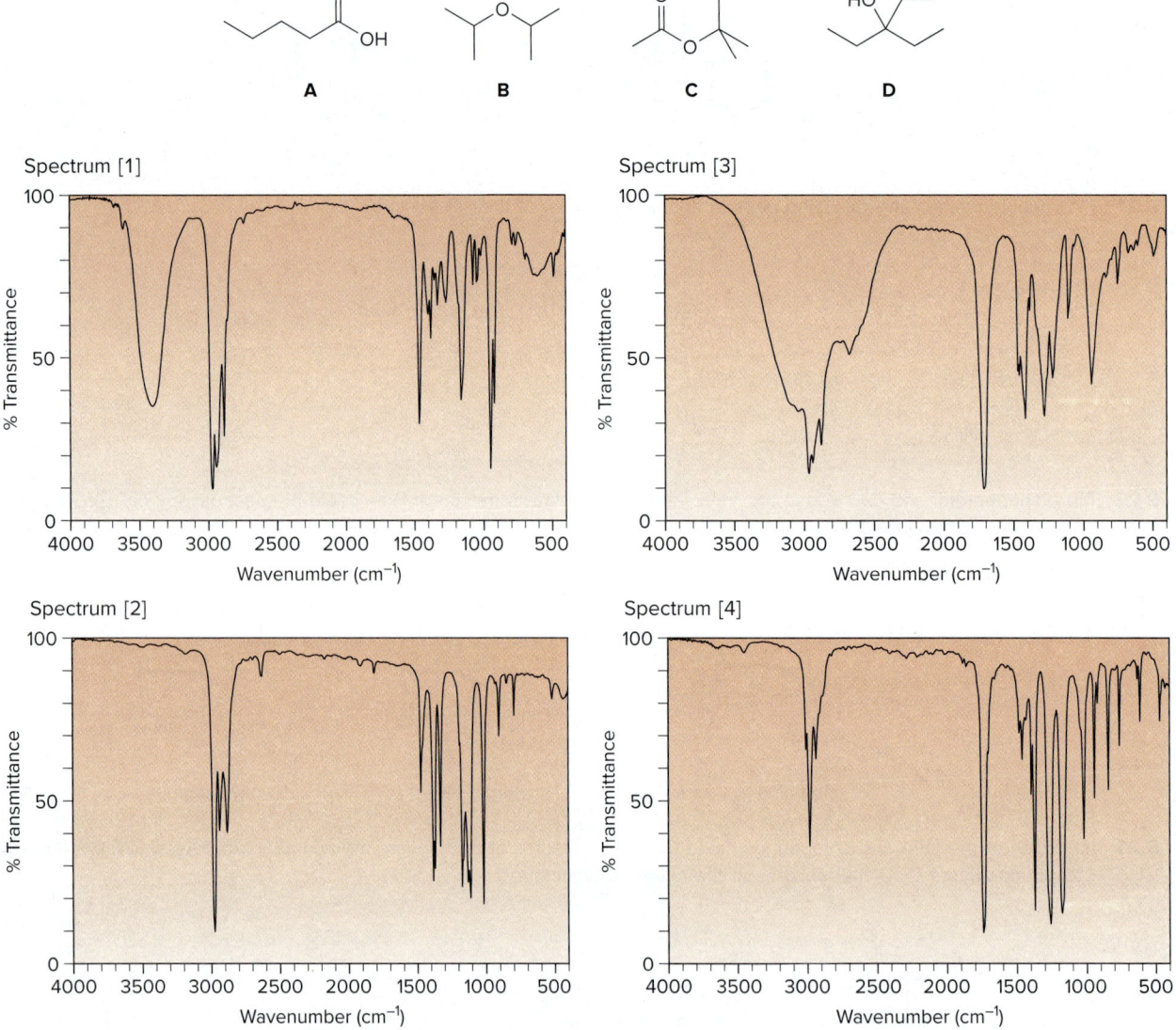

B.26 Match each compound to its IR spectrum.

Spectroscopy Problems That Combine Mass Spectrometry and Infrared Spectroscopy

B.27 Propose possible structures consistent with each set of data. Assume each compound has an sp^3 hybridized C—H absorption in its IR spectrum, and that other major IR absorptions above 1500 cm^{-1} are listed.
   a. a compound having a molecular ion at 72 and an absorption in its IR spectrum at 1725 cm^{-1}
   b. a compound having a molecular ion at 55 and an absorption in its IR spectrum at ~2250 cm^{-1}
   c. a compound having a molecular ion at 74 and an absorption in its IR spectrum at 3600–3200 cm^{-1}

B.28 A chiral hydrocarbon **X** exhibits a molecular ion at 82 in its mass spectrum. The IR spectrum of **X** shows peaks at 3300, 3000–2850, and 2250 cm^{-1}. Propose a structure for **X**.

B.29 A chiral compound **Y** has a strong absorption at 2970–2840 cm^{-1} in its IR spectrum and gives the following mass spectrum. Propose a structure for **Y**.

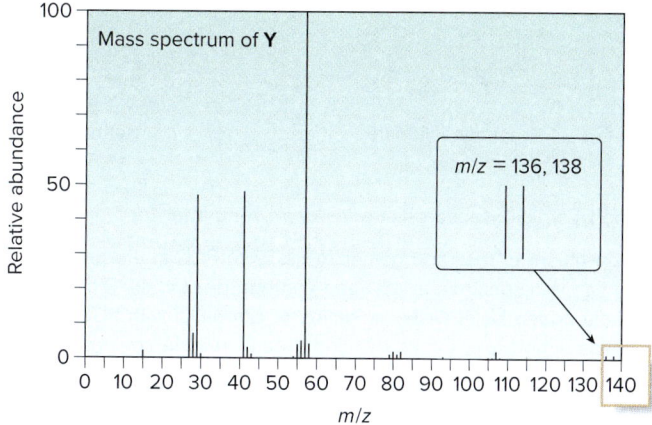

B.30 Treatment of benzoic acid (C$_6$H$_5$CO$_2$H) with NaOH followed by 1-iodo-3-methylbutane forms **H**. **H** has a molecular ion at 192 and IR absorptions at 3064, 3035, 2960–2872, and 1721 cm^{-1}. Propose a structure for **H**.

B.31 Reaction of 2-methylpropanoic acid [(CH$_3$)$_2$CHCO$_2$H] with SOCl$_2$ followed by 2-methylpropan-1-ol forms **X**. **X** has a molecular ion at 144 and IR absorptions at 2965, 2940, and 1739 cm^{-1}. Propose a structure for **X**.

B.32 Reaction of pentanoyl chloride (CH$_3$CH$_2$CH$_2$CH$_2$COCl) with lithium dimethyl cuprate [LiCu(CH$_3$)$_2$] forms a compound **J** that has a molecular ion in its mass spectrum at 100, as well as fragments at m/z = 85, 57, and 43 (base). The IR spectrum of **J** has strong peaks at 2962 and 1718 cm^{-1}. Propose a structure for **J**.

B.33 Benzonitrile (C$_6$H$_5$CN) is reduced to two different products depending on the reducing agent used. Treatment with lithium aluminum hydride followed by water forms **K**, which has a molecular ion in its mass spectrum at 107 and the following IR absorptions: 3373, 3290, 3062, 2920, and 1600 cm^{-1}. Treatment with a milder reducing agent forms **L**, which has a molecular ion in its mass spectrum at 106 and the following IR absorptions: 3086, 2820, 2736, 1703, and 1600 cm^{-1}. **L** shows fragments in its mass spectrum at m/z = 105 and 77. Propose structures for **K** and **L**, and explain how you arrived at your conclusions.

B.34 Treatment of anisole (CH$_3$OC$_6$H$_5$) with Cl$_2$ and FeCl$_3$ forms **P**, which has peaks in its mass spectrum at m/z = 142 (M), 144 (M + 2), 129, and 127. **P** has absorptions in its IR spectrum at 3096–2837 (several peaks), 1582, and 1494 cm^{-1}. Propose possible structures for **P**.

B.35 Reaction of BrCH$_2$CH$_2$CH$_2$CH$_2$NH$_2$ with NaH forms compound **W**, which gives the IR and mass spectra shown here. Propose a structure for **W** and draw a stepwise mechanism that accounts for its formation.

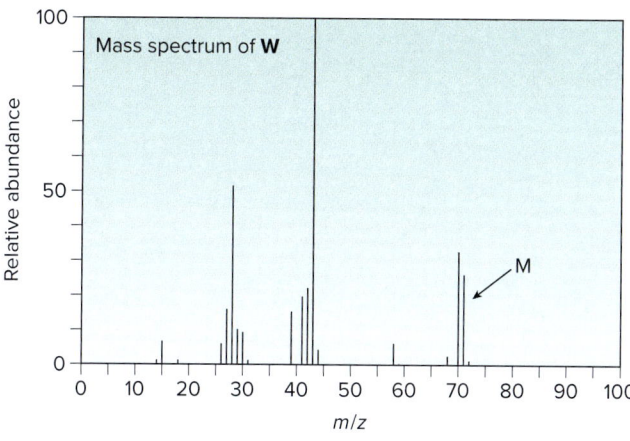
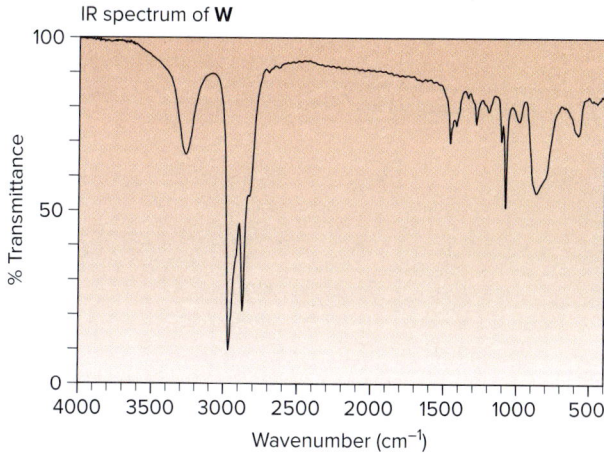

Challenge Problems

B.36 Acid chlorides (RCOCl) constitute another family of compounds that contains a carbonyl group. Would you expect the C=O of an acid chloride to absorb at a higher or lower wavenumber than an ester? Explain your reasoning. We will learn more about acid chlorides in Chapter 16.

B.37 Suggest an explanation for the following observation. The carbonyl group of methyl salicylate absorbs at a significantly lower wavenumber than the carbonyl group of methyl benzoate.

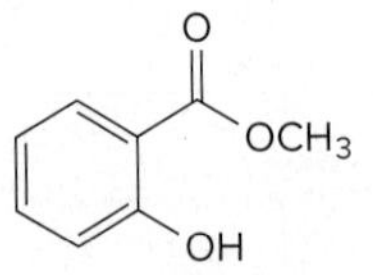

methyl salicylate
$\tilde{\nu} = 1680$ cm^{-1}

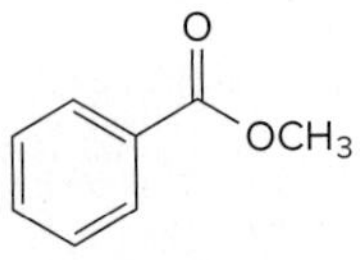

methyl benzoate
$\tilde{\nu} = 1728$ cm^{-1}

B.38 Explain why a ketone carbonyl typically absorbs at a lower wavenumber than an aldehyde carbonyl (1715 vs. 1730 cm^{-1}).

B.39 Oxidation of citronellol, a constituent of rose and geranium oils, with PCC in the presence of added NaOCOCH$_3$ forms compound **A**. **A** has a molecular ion in its mass spectrum at 154 and a strong peak in its IR spectrum at 1730 cm^{-1}, in addition to C–H stretching absorptions. Without added NaOCOCH$_3$, oxidation of citronellol with PCC yields isopulegone, which is then converted to **B** with aqueous base. **B** has a molecular ion at 152 and a peak in its IR spectrum at 1680 cm^{-1}, in addition to C–H stretching absorptions.

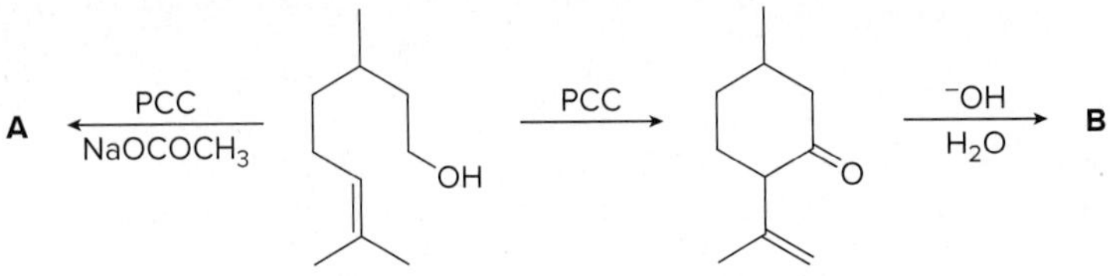

a. Identify the structures of **A** and **B**.
b. Draw a mechanism for the conversion of citronellol to isopulegone.
c. Draw a mechanism for the conversion of isopulegone to **B**.

B.40 The carbonyl absorptions of esters **X** and **Y** differ by 25 cm^{-1}. Which compound absorbs at higher wavenumber and why?

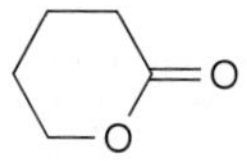

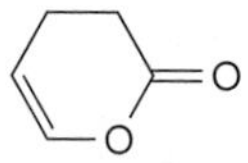

Nuclear Magnetic Resonance Spectroscopy

SPECTROSCOPY C

Daniel C. Smith

- **C.1** An introduction to NMR spectroscopy
- **C.2** ^{1}H NMR: Number of signals
- **C.3** ^{1}H NMR: Position of signals
- **C.4** The chemical shift of protons on sp^2 and sp hybridized carbons
- **C.5** ^{1}H NMR: Intensity of signals
- **C.6** ^{1}H NMR: Spin–spin splitting
- **C.7** More-complex examples of splitting
- **C.8** Spin–spin splitting in alkenes
- **C.9** Other facts about ^{1}H NMR spectroscopy
- **C.10** Using ^{1}H NMR to identify an unknown
- **C.11** ^{13}C NMR spectroscopy
- **C.12** Magnetic resonance imaging (MRI)

Palau'amine is a complex natural product isolated from the sea sponge *Hymeniacidon agminata* (formerly *Stylotella agminata*) collected in the Pacific Ocean near the Republic of Palau. The initial structure proposed for palau'amine in 1993 was revised in 2007 using a variety of modern spectroscopic techniques, including nuclear magnetic resonance spectroscopy. The dense array of functional groups in palau'amine and its antitumor and immunosuppressive properties attracted the attention of dozens of organic chemists, leading to its total synthesis in the laboratory in early 2010. In Spectroscopy Part C, we learn how nuclear magnetic resonance spectroscopy plays a key role in structure determination.

Why Study... Nuclear Magnetic Resonance Spectroscopy?

In **Spectroscopy C, we continue** our study of organic structure determination by learning about **nuclear magnetic resonance (NMR)** spectroscopy. NMR spectroscopy is the most powerful tool for characterizing organic molecules, because it can be used to **identify the carbon–hydrogen framework in a compound.**

C.1 An Introduction to NMR Spectroscopy

Two common types of NMR spectroscopy are used to characterize organic structure:

- ^{1}H NMR (proton NMR) is used to determine the number and type of hydrogen atoms in a molecule; and
- ^{13}C NMR (carbon NMR) is used to determine the type of carbon atoms in a molecule.

NMR stems from the same basic principle as all other forms of spectroscopy: Energy interacts with a molecule, and absorptions occur only when the incident energy matches the energy difference between two states.

C.1A The Basis of NMR Spectroscopy

The source of energy in NMR is radio waves. Radiation in the radiofrequency region of the electromagnetic spectrum (so-called **RF radiation**) has very long wavelengths, so its corresponding frequency and energy are both low. **When these low-energy radio waves interact with a molecule, they can change the nuclear spins of some elements, including ^{1}H and ^{13}C.**

When a charged particle such as a proton spins on its axis, it creates a magnetic field. For the purpose of this discussion, therefore, a nucleus is a tiny bar magnet, symbolized by . Normally these nuclear magnets are randomly oriented in space, but in the presence of an external magnetic field, B_0, they are oriented with or against this applied field. More nuclei are oriented *with* the applied field because this arrangement is lower in energy, but the **energy difference between these two states is very small** (< 0.4 J/mol).

A spinning proton creates a magnetic field.

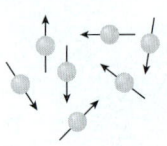

With no external magnetic field...

The nuclear magnets are randomly oriented.

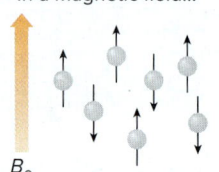

In a magnetic field...

B_0

The nuclear magnets are oriented with or against B_0.

In a magnetic field, there are now two different energy states for a proton:

- In the *lower-energy state* the nucleus is aligned in the *same direction* as B_0.
- In the *higher-energy state* the nucleus is aligned *opposed* to B_0.

When an external energy source ($h\nu$) that matches the energy difference (ΔE) between these two states is applied, energy is absorbed, causing the **nucleus to "spin flip" from one orientation to another.** The energy difference between these two nuclear spin states corresponds to the low-frequency radiation in the RF region of the electromagnetic spectrum.

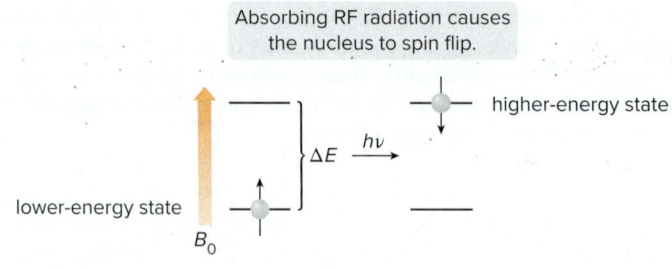

Absorbing RF radiation causes the nucleus to spin flip.

- A nucleus is in *resonance* when it absorbs RF radiation and "spin flips" to a higher-energy state.

Thus, two variables characterize NMR:

- **An applied magnetic field, B_0.** Magnetic field strength is measured in tesla (T).
- **The frequency ν of radiation used for resonance,** measured in hertz (Hz) or megahertz (MHz); (1 MHz = 10^6 Hz).

The frequency needed for resonance and the applied magnetic field strength are proportionally related:

$$\underset{\text{frequency}}{\nu} \quad \propto \quad \underset{\substack{\text{applied magnetic} \\ \text{field strength}}}{B_0}$$

- The *stronger* the magnetic field, the *larger* the energy difference between the two nuclear spin states, and the *higher* the ν needed for resonance.

NMR spectrometers are referred to as 300 MHz instruments, 500 MHz instruments, and so forth, depending on the frequency of RF radiation used for resonance.

Early NMR spectrometers used a magnetic field strength of ~1.4 T, which required RF radiation of 60 MHz for resonance. Modern NMR spectrometers use stronger magnets, thus requiring higher frequencies of RF radiation for resonance. For example, a magnetic field strength of 7.05 T requires a frequency of 300 MHz for a proton to be in resonance. These spectrometers use very powerful magnetic fields to create a small, but measurable energy difference between the two possible spin states. A schematic of an NMR spectrometer is shown in Figure C.1.

If all protons absorbed at the same frequency in a given magnetic field, the spectra of all compounds would consist of a single absorption, rendering NMR useless for structure determination. Fortunately, however, this is not the case.

Figure C.1 Schematic of an NMR spectrometer

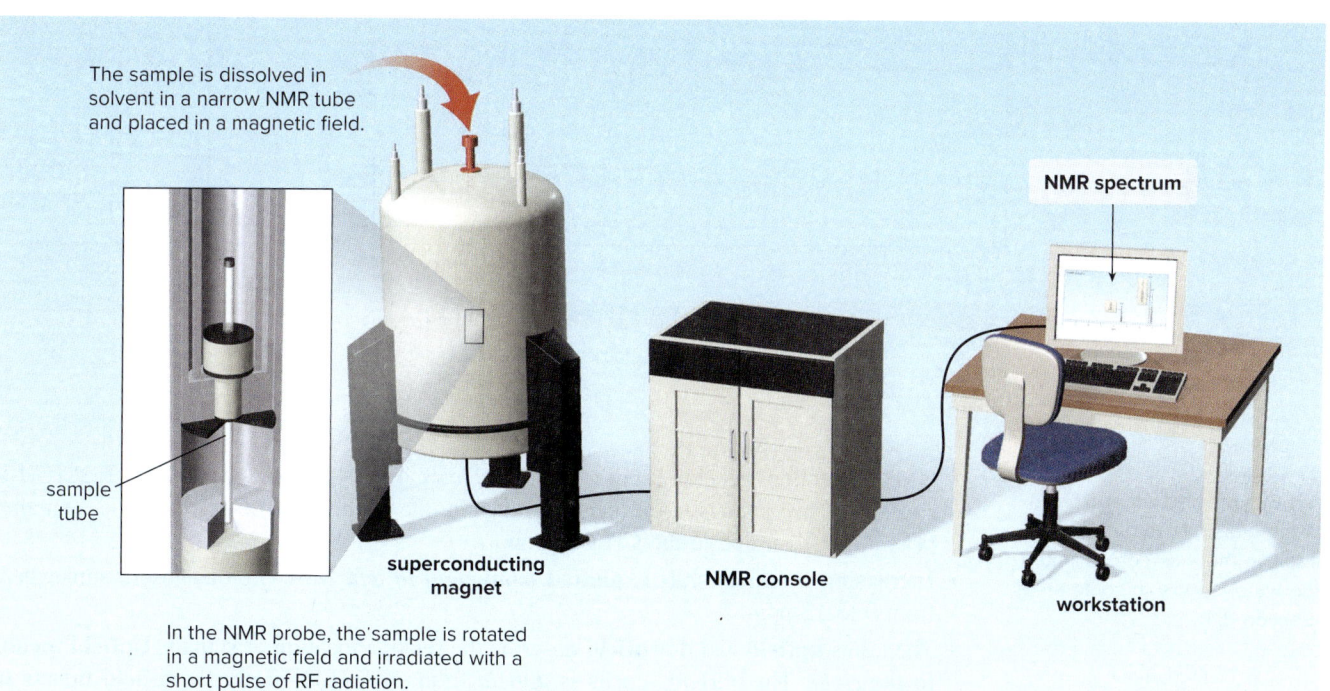

- **An NMR spectrometer.** The sample is dissolved in a solvent, usually $CDCl_3$ (deuterochloroform), and placed in a magnetic field. A radiofrequency generator then irradiates the sample with a short pulse of radiation, causing resonance. When the nuclei fall back to their lower-energy state, the detector measures the energy released, and a spectrum is recorded. The superconducting magnets in modern NMR spectrometers have coils that are cooled in liquid helium and conduct electricity with essentially no resistance.

- All protons do *not* absorb at the same frequency. Protons in different environments absorb at slightly different frequencies, so they are distinguishable by NMR.

The frequency at which a particular proton absorbs is determined by its electronic environment, as discussed in Section C.3. Because electrons are moving charged particles, they create a magnetic field opposed to the applied field B_0, and the size of the magnetic field generated by the electrons around a proton determines where it absorbs. Modern NMR spectrometers use a constant magnetic field strength B_0, and then a narrow range of frequencies is applied to achieve the resonance of all protons.

Only nuclei that contain odd mass numbers (such as ^1H, ^{13}C, ^{19}F, and ^{31}P) or odd atomic numbers (such as ^2H and ^{14}N) give rise to NMR signals. Because both ^1H and ^{13}C, the less abundant isotope of carbon, are NMR active, NMR allows us to map the carbon and hydrogen framework of an organic molecule.

C.1B A ^1H NMR Spectrum

An NMR spectrum plots the **intensity of a signal** against its **chemical shift** measured in **parts per million (ppm)**. The common scale of chemical shifts is called the **δ (delta) scale**. The proton NMR spectrum of *tert*-butyl methyl ether [$CH_3OC(CH_3)_3$] illustrates several important features:

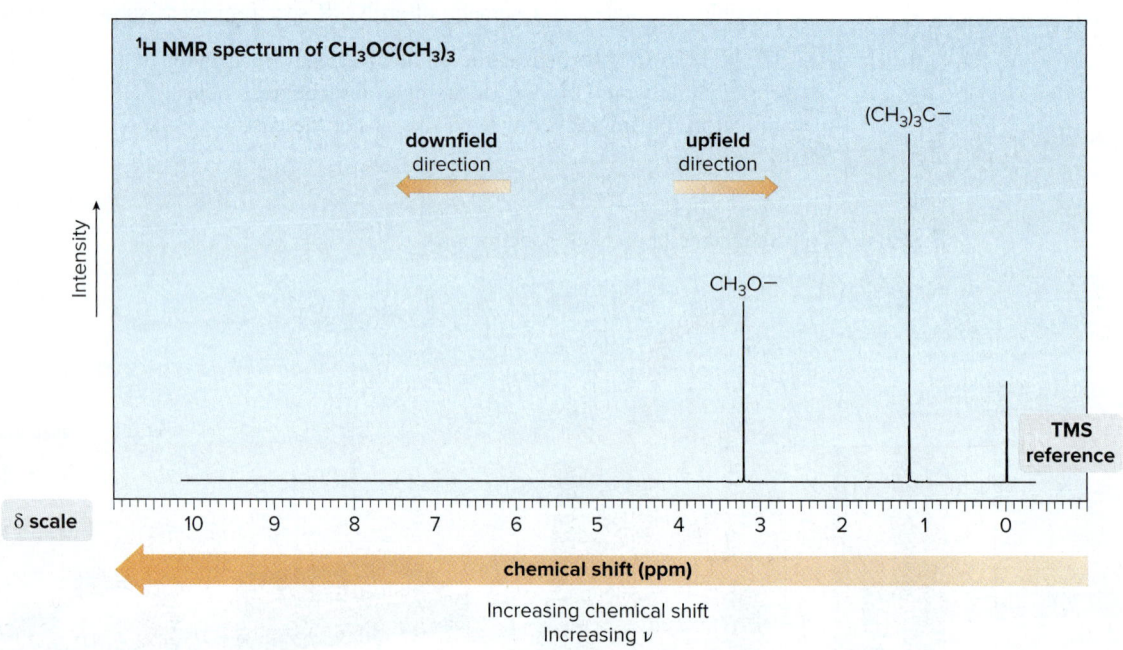

tert-Butyl methyl ether (MTBE) is the high-octane gasoline additive that has contaminated the water supply in some areas (Section 3.4).

$(CH_3)_4Si$
tetramethylsilane
TMS

- NMR absorptions generally appear as sharp signals. The ^1H NMR spectrum of $CH_3OC(CH_3)_3$ consists of two signals: a tall peak at 1.2 ppm due to the $(CH_3)_3C-$ group, and a smaller peak at 3.2 ppm due to the CH_3O- group.
- **Increasing chemical shift is plotted from *right to left*.** Most protons absorb somewhere from 0 to 12 ppm.
- The terms **upfield** and **downfield** describe the relative location of signals. **Upfield means to the *right*.** The $(CH_3)_3C-$ peak is *upfield* from the CH_3O- peak. **Downfield means to the *left*.** The CH_3O- peak is *downfield* from the $(CH_3)_3C-$ peak.

NMR absorptions are measured relative to the position of a reference signal at 0 ppm on the δ scale due to **tetramethylsilane (TMS)**. **TMS** is a volatile and inert compound that gives a single peak upfield from other typical NMR absorptions.

Although chemical shifts are measured relative to the TMS signal at 0 ppm, this reference is often not plotted on a spectrum.

The *positive* direction of the δ scale is *downfield* from TMS. A very small number of absorptions occur upfield from the TMS signal, which is defined as the negative direction of the δ scale. (See Problem C.67.)

The **chemical shift** on the x axis gives the position of an NMR signal, measured in ppm, according to this equation:

$$\text{chemical shift (in ppm on the } \delta \text{ scale)} = \frac{\text{observed chemical shift (in Hz) downfield from TMS}}{\nu \text{ of the NMR spectrometer (in MHz)}}$$

Because the frequency of the radiation required for resonance is proportional to the strength of the applied magnetic field, B_0, reporting NMR absorptions in frequency would be meaningless unless the value of B_0 was also reported. By reporting the absorption as a fraction of the NMR operating frequency, though, we get units—ppm—that are independent of the spectrometer.

Sample Problem C.1 Calculating Chemical Shift

Calculate the chemical shift of an absorption that occurs at 1500 Hz downfield from TMS using a 300 MHz NMR spectrometer.

Solution

Use the equation that defines the chemical shift in ppm:

$$\text{chemical shift} = \frac{1500 \text{ Hz downfield from TMS}}{300 \text{ MHz operating frequency}} = 5 \text{ ppm}$$

Problem C.1 The ^{1}H NMR spectrum of CH_3OH recorded on a 500 MHz NMR spectrometer consists of two signals, one due to the CH_3 protons at 1715 Hz and one due to the OH proton at 1830 Hz, both measured downfield from TMS. (a) Calculate the chemical shift of each absorption. (b) Do the CH_3 protons absorb upfield or downfield from the OH proton?

More Practice: Try Problems C.35, C.36.

Problem C.2 The ^{1}H NMR spectrum of 1,2-dimethoxyethane ($CH_3OCH_2CH_2OCH_3$) recorded on a 300 MHz NMR spectrometer consists of signals at 1017 Hz and 1065 Hz downfield from TMS. (a) Calculate the chemical shift of each absorption. (b) At what frequency would each absorption occur if the spectrum were recorded on a 500 MHz NMR spectrometer?

Four different features of a ^{1}H NMR spectrum provide information about a compound's structure:

[1] **Number of signals** (Section C.2)
[2] **Position of signals** (Sections C.3 and C.4)
[3] **Intensity of signals** (Section C.5)
[4] **Spin–spin splitting of signals** (Sections C.6–C.8)

C.2 ^{1}H NMR: Number of Signals

How many ^{1}H NMR signals does a compound exhibit? The number of NMR signals *equals* the number of different types of protons in a compound.

C.2A General Principles

- Protons in different environments give different NMR signals. Equivalent protons give the same NMR signal.

In many compounds, deciding whether two protons are in identical or different environments is intuitive.

Any CH$_3$ group is different from any CH$_2$ group, which is different from any CH group in a molecule. Two CH$_3$ groups may be identical (as in CH$_3$OCH$_3$) or different (as in CH$_3$OCH$_2$CH$_3$), depending on what each CH$_3$ group is bonded to.

tert-Butyl methyl ether [CH$_3$OC(CH$_3$)$_3$] (Section C.1) exhibits two NMR signals because it contains two different kinds of protons: one CH$_3$ group is bonded to –OC(CH$_3$)$_3$, whereas the other three CH$_3$ groups are each bonded to the same group, [–C(CH$_3$)$_2$]OCH$_3$.

CH$_3$–O–CH$_3$ (H$_a$)
All equivalent H's
1 NMR signal

CH$_3$–CH$_2$–Cl (H$_a$, H$_b$)
2 types of H's
2 NMR signals

CH$_3$–O–CH$_2$CH$_3$ (H$_a$, H$_b$, H$_c$)
3 types of H's
3 NMR signals

- **CH$_3$OCH$_3$:** Each CH$_3$ group is bonded to the same group (–OCH$_3$), making both CH$_3$ groups equivalent.
- **CH$_3$CH$_2$Cl:** The protons of the CH$_3$ group are different from those of the CH$_2$ group.
- **CH$_3$OCH$_2$CH$_3$:** The protons of the CH$_2$ group are different from those in each CH$_3$ group. The two CH$_3$ groups are also different from each other; one CH$_3$ group is bonded to –OCH$_2$CH$_3$ and the other is bonded to –CH$_2$OCH$_3$.

In some cases, it is less obvious by inspection if two protons are equivalent or different. To rigorously determine whether two protons are in identical environments (and therefore give rise to one NMR signal), replace each H atom in question by another atom Z (for example, Z = Cl). **If substitution by Z yields the same compound or enantiomers, the two protons are equivalent,** as shown in Sample Problem C.2.

Sample Problem C.2   Determining the Different Types of H's in a Molecule

How many different kinds of H atoms does CH$_3$CH$_2$CH$_2$CH$_2$CH$_3$ contain?

Solution

In comparing two H atoms, replace each H by Z (for example, Z = Cl), and examine the substitution products that result. The two CH$_3$ groups are identical because substitution of one H by Cl on each carbon gives the same product, 1-chloropentane.

substitution at **C1** or **C5** by Cl → 1-chloropentane

There are two different types of CH$_2$ groups. Substitution of Cl for H on C2 or C4 gives the same product, 2-chloropentane, so these H's are identical. Substitution of Cl for H on C3 gives a different product, 3-chloropentane, so this CH$_2$ group is different from the other two CH$_2$ groups.

2-chloropentane — substitution at **C2** or **C4** by Cl

or

3-chloropentane — substitution at **C3** by Cl

Thus, CH$_3$CH$_2$CH$_2$CH$_2$CH$_3$ has three different types of protons and gives three different NMR signals.

CH$_3$–CH$_2$–CH$_2$–CH$_2$–CH$_3$
H$_a$ H$_b$ H$_c$ H$_b$ H$_a$

Problem C.3   How many ^{1}H NMR signals does each compound show?

a., b., c., d., e., f., g., h.

More Practice: Try Problems C.31a, C.32a, C.33, C.34, C.48a, C.49d.

Figure C.2
The number of ¹H NMR signals of some representative organic compounds

Cl—CH₂—CH₂—Cl
  Hₐ   Hₐ
1 type of H
1 NMR signal

Cl—CH₂—CH₂—CH₂—Br
  Hₐ   H_b   H_c
3 types of H's
3 NMR signals

CH₃—C(=O)—OCH₃
  Hₐ   H_b
2 types of H's
2 NMR signals

CH₃—CH₂—OH
 Hₐ   H_b H_c
3 types of H's
3 NMR signals

Figure C.2 gives the number of NMR signals exhibited by four additional molecules. All protons—not just protons bonded to carbon atoms—give rise to NMR signals. Ethanol (CH₃CH₂OH), for example, gives three NMR signals, one of which is due to its OH proton.

C.2B Determining Equivalent Protons in Alkenes and Cycloalkanes

To determine equivalent protons in cycloalkanes and alkenes that have restricted bond rotation, always **draw in all bonds to hydrogen.**

Draw [cyclopropane with all H's shown and Cl]   NOT [triangle—Cl]   Draw [CH₂=CHCl with all H's shown]   NOT ClCH=CH₂

Then, in comparing two H atoms on a ring or double bond, **two protons are equivalent only if they are cis (or trans) to the same groups,** as illustrated with 1,1-dichloroethylene, 1-bromo-1-chloroethylene, and chloroethylene.

1,1-dichloroethylene: H cis to Cl / H cis to Cl
1 type of H
1 NMR signal

1-bromo-1-chloroethylene: Hₐ cis to Cl / H_b cis to Br
2 types of H's
2 NMR signals

chloroethylene: H_b cis to Cl / H_c cis to Hₐ
3 types of H's
3 NMR signals

- **1,1-Dichloroethylene:** The two H atoms on the C=C are both cis to a Cl atom. Thus, both H atoms are equivalent.
- **1-Bromo-1-chloroethylene:** Hₐ is cis to a Cl atom and H_b is cis to a Br atom. Thus, Hₐ and H_b are different, giving rise to two NMR signals.
- **Chloroethylene:** Hₐ is bonded to the carbon with the Cl atom, making it different from H_b and H_c. Of the remaining two H atoms, H_b is cis to a Cl atom and H_c is cis to a H atom, making them different. All three H atoms in this compound are different.

Proton equivalency in cycloalkanes can be determined similarly.

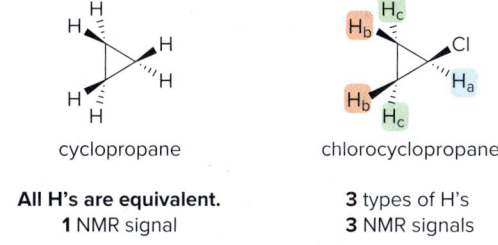

cyclopropane
All H's are equivalent.
1 NMR signal

chlorocyclopropane
3 types of H's
3 NMR signals

- **Cyclopropane:** All H atoms are equivalent, so there is only one NMR signal.
- **Chlorocyclopropane:** There are now three kinds of H atoms: Hₐ is bonded to a carbon bonded to a Cl; both H_b protons are cis to the Cl, whereas both H_c protons are cis to another H.

Sample Problem C.3 — Determining Proton Equivalency in Cyclic Compounds

How many ^{1}H NMR signals does **A** exhibit?

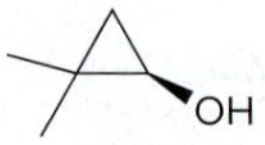

A: 1,1-dimethyl-2-hydroxycyclopropane (methyl groups on C1, OH on C2)

Solution

Use wedges and dashed wedges to emphasize the relative location of groups on a ring. **Two protons are equivalent only if they are cis (or trans) to the same groups.** Start with the protons that can be assigned most easily. In this example, the OH and the CH on the ring look different from all other protons, so they give two NMR signals. The two CH$_3$ groups are different from each other because one CH$_3$ is cis to H$_b$ and one is trans to H$_b$. Likewise, H$_e$ and H$_f$ are different from each other, because H$_e$ is trans to H$_b$, whereas H$_f$ is cis to H$_b$.

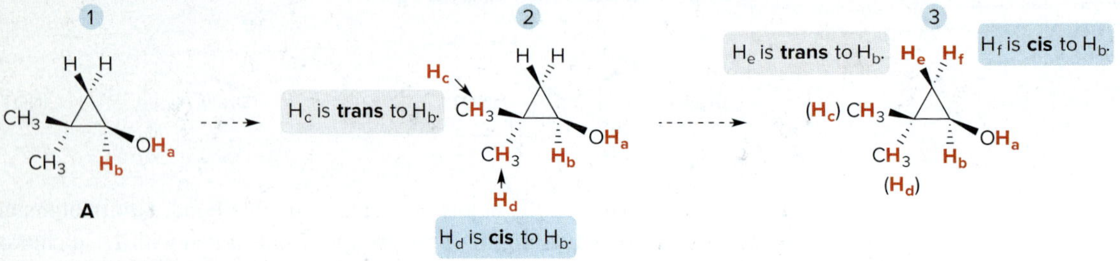

1 — structure A with CH$_3$, CH$_3$, H, H, OH$_a$, H$_b$ labeled

2 — H$_c$ is trans to H$_b$; H$_d$ is cis to H$_b$

3 — H$_e$ is trans to H$_b$; H$_f$ is cis to H$_b$

Thus, **A** contains **six** different types of H's and gives **six** ^{1}H NMR signals.

Problem C.4
How many ^{1}H NMR signals does each dimethylcyclopropane show?

a.  (1,1-dimethylcyclopropane)   b.  (cis-1,2-dimethylcyclopropane)   c.  (trans-1,2-dimethylcyclopropane)

More Practice: Try Problem C.33h, i, j.

Problem C.5
How many ^{1}H NMR signals does each alkene exhibit?

a. 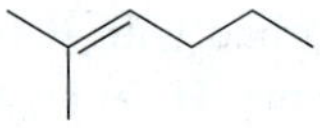   b. 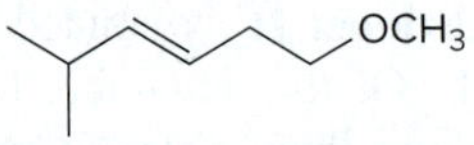 ...OCH$_3$ c. 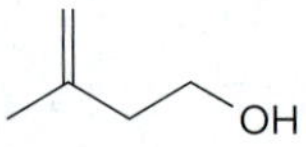 ...OH

Problem C.6
How many ^{1}H NMR signals does each compound give?

a. 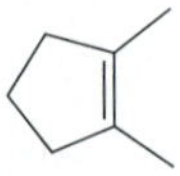   b. 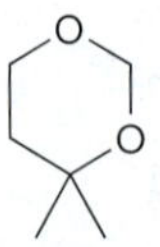   c. 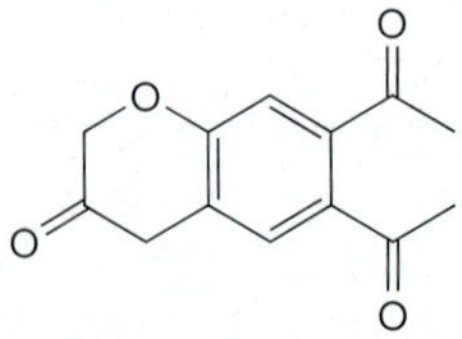   d.

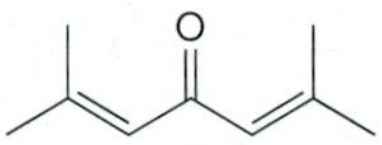

C.2C Homotopic, Enantiotopic, and Diastereotopic Protons

Let's look more closely at the protons of a single sp^3 hybridized CH$_2$ group to determine whether these two protons are always equivalent to *each other*. Three examples illustrate different outcomes.

CH$_3$CH$_2$CH$_3$ has two different types of protons—those of the CH$_3$ groups and those of the CH$_2$ group—meaning that the two H atoms of the CH$_2$ group are *equivalent to each other*. Replacement of each H by Z forms the *same* product, so they give *one* NMR signal.

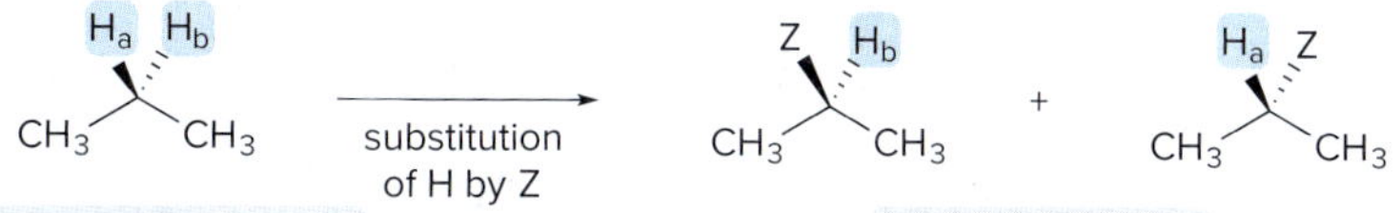

H$_a$ and H$_b$ are **homotopic**.

identical products

- When substitution of two H atoms by Z forms the *same* product, these equivalent hydrogens are called *homotopic* protons.

CH$_3$CH$_2$Br has two different types of protons—those of the CH$_3$ group and those of the CH$_2$ group—meaning that the two H atoms of the CH$_2$ group are *equivalent to each other*. Replacement of each H of the CH$_2$ group by an atom Z creates a new stereogenic center, forming two products that are **enantiomers.**

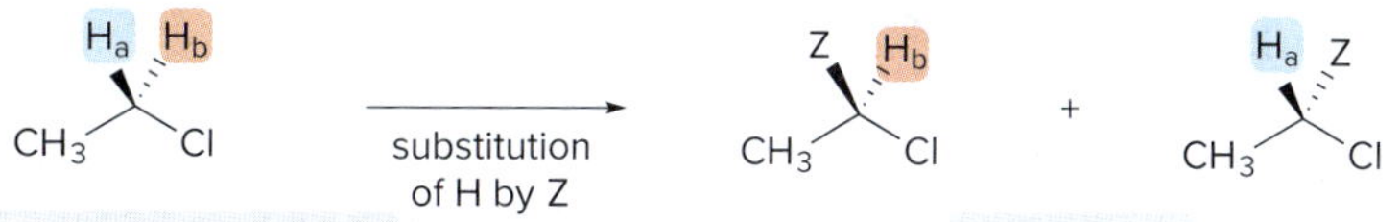

H$_a$ and H$_b$ are **enantiotopic**.

enantiomers

- When substitution of two H atoms by Z forms *enantiomers*, the two H atoms are equivalent and give a single NMR signal. These two H atoms are called *enantiotopic* protons.

In contrast, the two H atoms of the CH$_2$ group in (R)-2-chlorobutane, which contains one stereogenic center, are *not* equivalent to each other. Substitution of each H by Z forms two **diastereomers,** and thus, these two H atoms give *different* NMR signals.

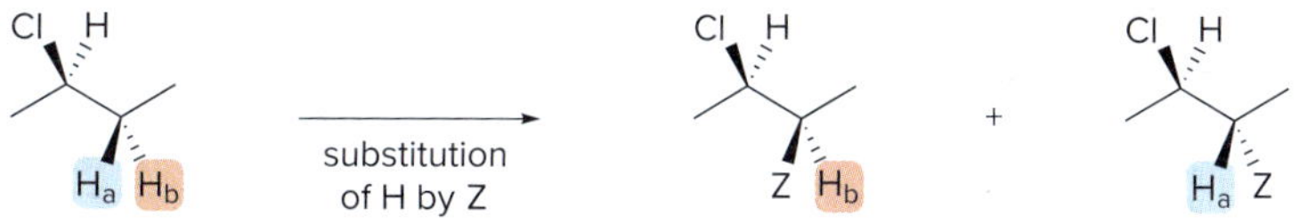

(R)-2-chlorobutane

H$_a$ and H$_b$ are **diastereotopic**.

diastereomers

- When substitution of two H atoms by Z forms *diastereomers*, the two H atoms are *not* equivalent, and give two NMR signals. These two H atoms are called *diastereotopic* protons.

Problem C.7 Label the protons in each highlighted CH$_2$ group as enantiotopic, diastereotopic, or homotopic.

a. 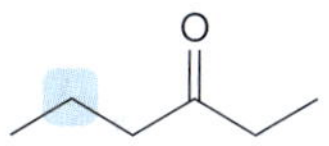 b.  c. 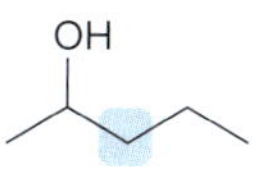 d.

Problem C.8 How many ^{1}H NMR signals would you expect for each compound?

a. 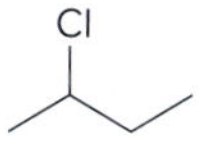 b. 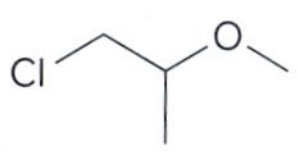 c.

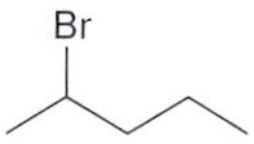

C.3 ^{1}H NMR: Position of Signals

In the NMR spectrum of *tert*-butyl methyl ether in Section C.1B, why does the CH_3O- group absorb downfield from the $-C(CH_3)_3$ group?

- Where a particular proton absorbs depends on its electronic environment.

C.3A Shielding and Deshielding Effects

To understand how the electronic environment around a nucleus affects its chemical shift, recall that in a magnetic field, an electron creates a small magnetic field that opposes the applied magnetic field, B_0. **Electrons are said to *shield* the nucleus from B_0.**

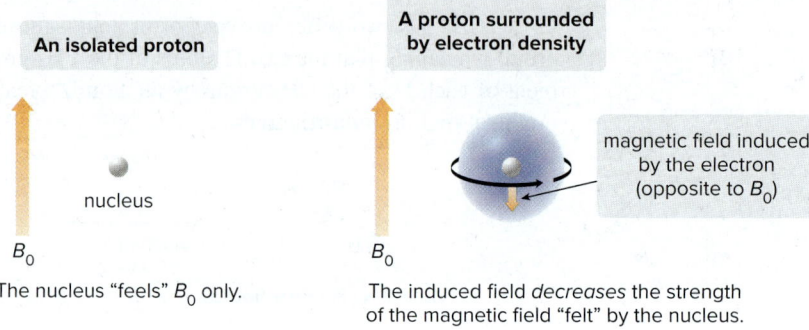

In the vicinity of the nucleus, therefore, the magnetic field generated by the circulating electron *decreases* the external magnetic field that the proton "feels." Because the proton experiences a lower magnetic field strength, it needs a lower frequency to achieve resonance. Lower frequency is to the right in an NMR spectrum, toward lower chemical shift, so **shielding shifts an absorption *upfield*,** as shown in Figure C.3a.

What happens if the electron density around a nucleus is *decreased*, instead? For example, how do the chemical shifts of the protons in CH_4 and CH_3Cl compare?

The less shielded the nucleus becomes, the more of the applied magnetic field (B_0) it feels. This *deshielded* nucleus experiences a higher magnetic field strength, so it needs a higher

Figure C.3   How chemical shift is affected by electron density around a nucleus

a. **Shielding effects**
- An electron shields the nucleus.
- The absorption shifts *upfield*.

b. **Deshielding effects**
- Decreased electron density deshields a nucleus.
- The absorption shifts *downfield*.

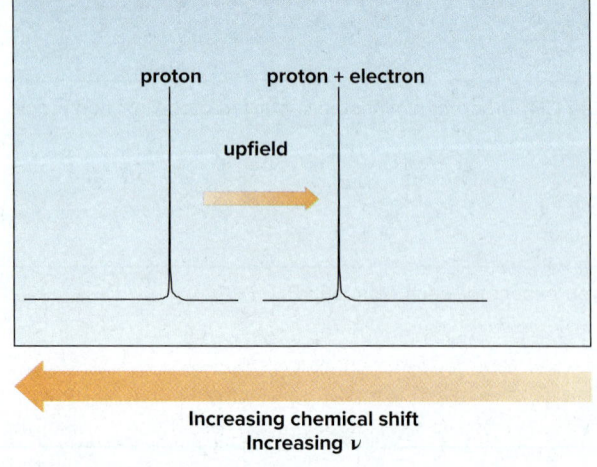
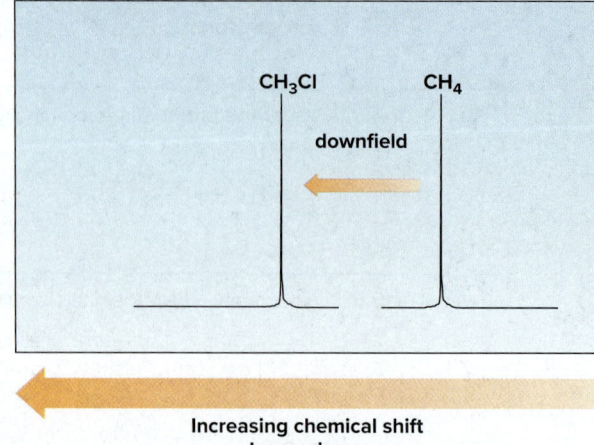

Remember the trend:
Decreased electron density deshields a nucleus and an absorption moves *downfield*.

frequency to achieve resonance. Higher frequency is to the *left* in an NMR spectrum, toward higher chemical shift, so **deshielding shifts an absorption downfield,** as shown in Figure C.3b for CH$_3$Cl versus CH$_4$. The electronegative Cl atom withdraws electron density from the carbon and hydrogen atoms in CH$_3$Cl, thus deshielding them relative to those in CH$_4$.

- **Protons near electronegative atoms are deshielded, so they absorb downfield.**

Figure C.4 summarizes the effects of shielding and deshielding.

These electron density arguments explain the relative position of NMR signals in many compounds.

CH$_3$—CH$_2$—Cl
H$_a$ H$_b$
- The H$_b$ protons are **deshielded** because they are closer to the electronegative Cl atom, so they absorb **downfield** from H$_a$.

Br—CH$_2$—CH$_2$—F
H$_a$ H$_b$
- Because F is more electronegative than Br, the H$_b$ protons are more **deshielded** than the H$_a$ protons and absorb farther **downfield.**

Cl—CH$_2$—CHCl$_2$
H$_a$ H$_b$
- The larger number of electronegative Cl atoms (two vs. one) **deshields** H$_b$ more than H$_a$, so it absorbs **downfield** from H$_a$.

Figure C.4
Shielding and deshielding effects

a. A shielded nucleus

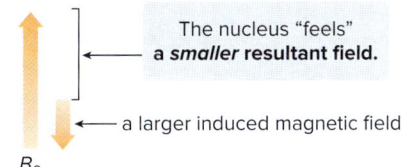

- As the electron density around the nucleus increases, the nucleus feels a *smaller* resultant magnetic field, so a *lower* frequency is needed to achieve resonance.
- **The absorption shifts *upfield*.**

b. A deshielded nucleus

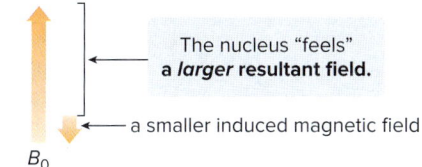

- As the electron density around the nucleus decreases, the nucleus feels a *larger* resultant magnetic field, so a *higher* frequency is needed to achieve resonance.
- **The absorption shifts *downfield*.**

Sample Problem C.4 Determining Shielding and Deshielding Effects

Which of the labeled protons in each pair absorbs farther downfield: (a) CH$_3$CH$_2$C**H**$_3$ or CH$_3$OC**H**$_3$; (b) CH$_3$OC**H**$_3$ or CH$_3$SC**H**$_3$?

Solution

a. The CH$_3$ group in CH$_3$OCH$_3$ is deshielded by the electronegative O atom. **Deshielding shifts the absorption downfield.**
b. Because oxygen is more electronegative than sulfur, the CH$_3$ group in CH$_3$OCH$_3$ is more **deshielded and absorbs downfield.**

Problem C.9 For each compound, which of the protons on the highlighted carbons absorbs farther downfield?

a.    b. 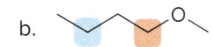   c.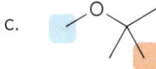

More Practice: Try Problem C.37.

C.3B Chemical Shift Values

Not only is the *relative* position of NMR absorptions predictable, but it is also possible to predict the approximate chemical shift value for a given type of proton.

- **Protons in a given environment absorb in a predictable region in an NMR spectrum.**

564 Spectroscopy C Nuclear Magnetic Resonance Spectroscopy

A more detailed list of characteristic chemical shift values is found in Appendix H.

Table C.1 lists the typical chemical shift values for the most common bonds encountered in organic molecules.

Table C.1 also illustrates that absorptions for a given type of C–H bond occur in a narrow range of chemical shift values, usually 1–2 ppm. For example, all sp^3 hybridized C–H bonds in alkanes and cycloalkanes absorb between 0.9 and 2.0 ppm. By contrast, absorptions due to N–H and O–H protons can occur over a broader range. For example, the OH proton of an alcohol is found anywhere in the 1–5 ppm range. The position of these absorptions is affected by the extent of hydrogen bonding, making it more variable.

Table C.1 Characteristic Chemical Shifts of Common Types of Protons

Type of proton	Chemical shift (ppm)	Type of proton	Chemical shift (ppm)
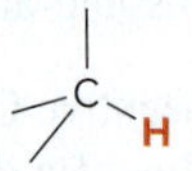 $\overset{\mid}{\underset{\mid}{C}}-H$	0.9–2	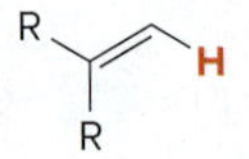 R₂C=CHR	4.5–6
• RC**H**₃	~0.9		
• R₂C**H**₂	~1.3	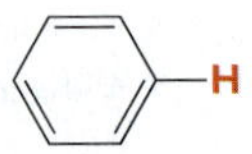 Ar–H (benzene)	6.5–8
• R₃C**H**	~1.7		
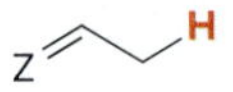 Z–CH₂–H (Z = C, O, N)	1.5–2.5	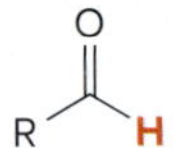 RC(=O)H	9–10
 ≡C–H	~2.5	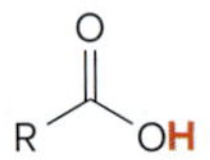 RC(=O)OH	10–12
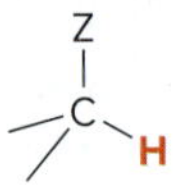 Z–C(–H) (Z = N, O, X)	2.5–4	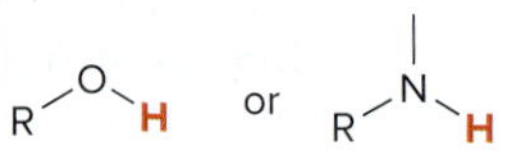 R–O–H or R–N–H	1–5

The chemical shift of a particular type of C–H bond is also affected by the number of R groups bonded to the carbon atom.

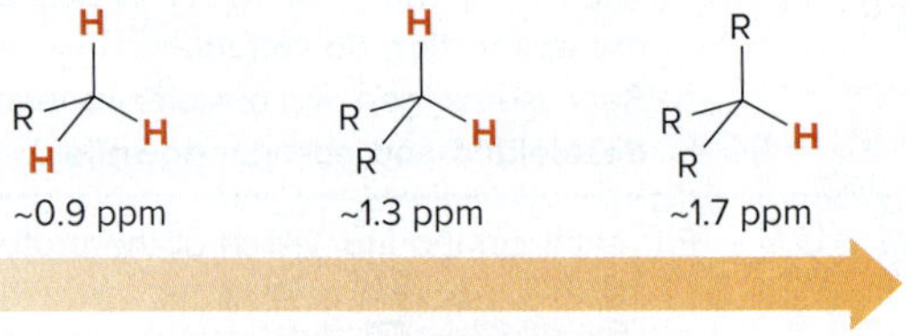

$$\underset{\sim 0.9 \text{ ppm}}{R-CH_3} \quad \underset{\sim 1.3 \text{ ppm}}{R_2CH_2} \quad \underset{\sim 1.7 \text{ ppm}}{R_3CH}$$

Increasing alkyl substitution
Increasing chemical shift

- **The chemical shift of a C–H bond increases with increasing alkyl substitution.**

Problem C.10 For each compound, first label each different type of proton and then rank the protons in order of increasing chemical shift.

a. 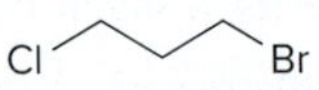 Cl–CH₂CH₂CH₂–Br

b. 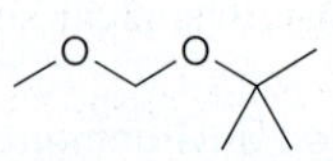 CH₃CH₂–O–CH₂–O–C(CH₃)₃

c. CH₃C(=O)CH₂CH₃

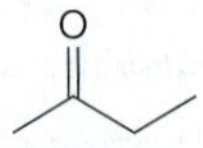

C.4 The Chemical Shift of Protons on sp^2 and sp Hybridized Carbons

The chemical shift of protons bonded to benzene rings, C—C double bonds, and C—C triple bonds merits additional comment.

Ph—H   CH₂=CH—H   HC≡C—H

7.3 ppm   4.5–6 ppm   2.5 ppm

Each of these functional groups contains π bonds with **loosely held π electrons.** When placed in a magnetic field, these π electrons move in a circular path, inducing a new magnetic field. How this induced magnetic field affects the chemical shift of a proton depends on the direction of the induced field *in the vicinity of the absorbing proton.*

Protons on Benzene Rings

In a magnetic field, the six π electrons in **benzene** circulate around the ring, creating a ring current. The magnetic field induced by these moving electrons *reinforces* the applied magnetic field in the vicinity of the protons. The protons thus feel a stronger magnetic field and a higher frequency is needed for resonance, so the **protons are deshielded and the absorption is** *downfield.*

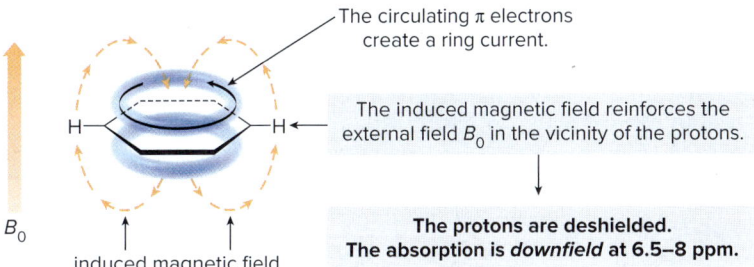

Protons on Carbon–Carbon Double Bonds

A similar phenomenon occurs with protons on carbon–carbon double bonds. In a magnetic field, the loosely held π electrons create a magnetic field that *reinforces* the applied field in the vicinity of the protons. Because the protons now feel a stronger magnetic field, they require a higher frequency for resonance. **The protons are deshielded and the absorption is** *downfield.*

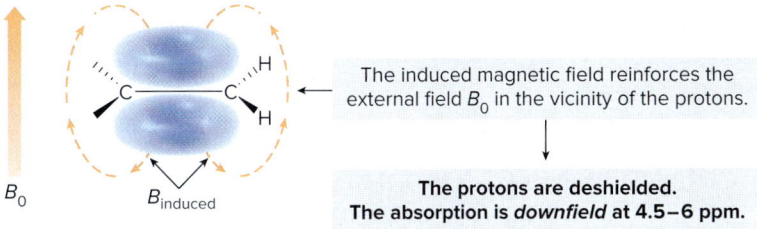

Protons on Carbon–Carbon Triple Bonds

In a magnetic field, the π electrons of a carbon–carbon triple bond induce a magnetic field that *opposes* the applied magnetic field (B_0). The proton thus feels a weaker magnetic field, so a lower frequency is needed for resonance. **The nucleus is shielded and the absorption is** *upfield.*

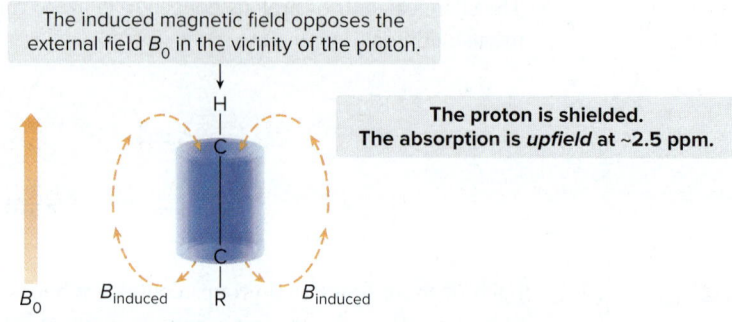

Table C.2 Effect of π Electrons on Chemical Shift Values

Proton type	Chemical shift (ppm)
⌬—H (benzene)	6.5–8 (highly deshielded)
⟩=⟨—H (alkene)	4.5–6 (deshielded)
≡—H (alkyne)	~2.5 (shielded)

Table C.2 summarizes the shielding and deshielding effects due to circulating π electrons.

To remember the chemical shifts of some common bond types, it is helpful to think of a ^{1}H NMR spectrum as being divided into six different regions (Figure C.5).

Figure C.5 Regions in the ^{1}H NMR spectrum

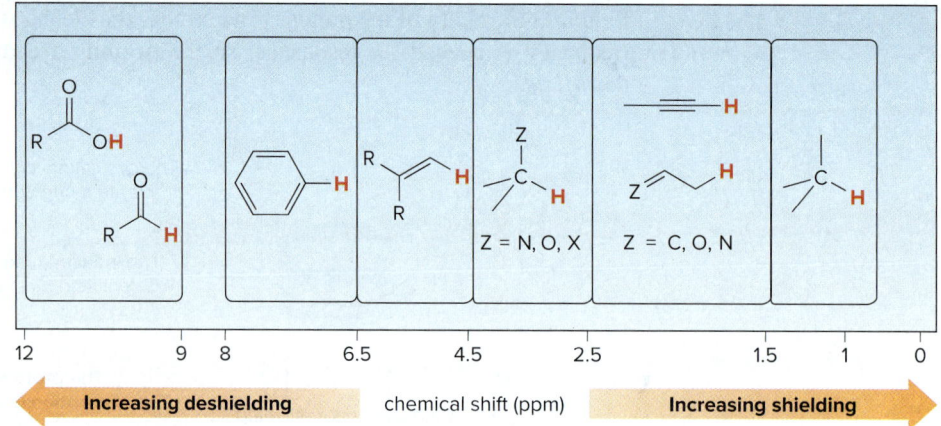

- **Shielded** protons absorb at **lower** chemical shift (to the **right**).
- **Deshielded** protons absorb at **higher** chemical shift (to the **left**).
- Note: The drawn chemical shift scale is not linear.

Sample Problem C.5   Predicting the Relative Chemical Shift of Protons

Rank H_a, H_b, and H_c in order of increasing chemical shift.

$$H_c\!\!\diagdown\!\!C\!\!=\!\!C\!\!-\!\!OCH_2CH_3$$
$$H_b \quad H_a$$

Solution

The H_a protons are bonded to an sp^3 hybridized carbon, so they are shielded and absorb upfield compared to H_b and H_c. Because the H_b protons are deshielded by the electronegative oxygen atom on the C to which they are bonded, they absorb downfield from H_a. The H_c proton is deshielded by two factors. The electronegative O atom withdraws electron density from H_c. Moreover, because H_c is bonded directly to a C=C, the magnetic field induced by the π electrons causes further deshielding. Thus, in order of increasing chemical shift, $H_a < H_b < H_c$.

Problem C.11 Rank each group of protons in order of increasing chemical shift.

a. ≡—H$_a$ CH$_2$=CH—H$_b$ CH$_3$CH$_2$CH$_2$—H$_c$ b. CH$_3$—C(=O)—OCH$_2$CH$_3$
                                                                    H$_a$ H$_b$ H$_c$

More Practice: Try Problem C.37.

C.5 ^{1}H NMR: Intensity of Signals

The relative intensity of ^{1}H NMR signals also provides information about a compound's structure.

- **The area under an NMR signal is proportional to the number of absorbing protons.**

For example, in the ^{1}H NMR spectrum of CH$_3$OC(CH$_3$)$_3$, the ratio of the area under the downfield peak (due to the CH$_3$O– group) to the upfield peak [due to the –C(CH$_3$)$_3$ group] is 1:3. An NMR spectrometer automatically integrates the area under the peaks, and prints out a digital display of the *relative* areas of the NMR signals. Older NMR spectrometers print out a stepped curve (an **integral**) on the spectrum. The height of each step is proportional to the area under the peak, which is in turn proportional to the number of absorbing protons.

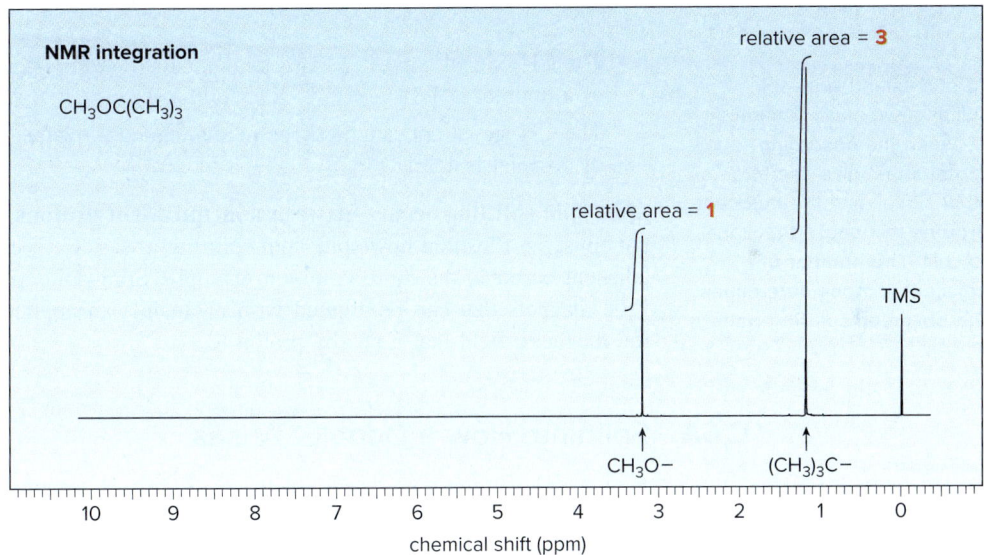

Integrals can be manually measured, but modern NMR spectrometers automatically calculate and plot the value of each integral in arbitrary units. If the heights of two integrals are in a 1:3 ratio, then the ratio of absorbing protons is 1:3, or 2:6, or 3:9, and so forth. This tells the *ratio,* not the absolute number of protons.

Problem C.12 Which compounds give a ^{1}H NMR spectrum with two signals in a ratio of 2:3?

a. CH$_3$CH$_2$Cl   b. CH$_3$CH$_2$CH$_3$ c. CH$_3$CH$_2$OCH$_2$CH$_3$ d. CH$_3$OCH$_2$CH$_2$OCH$_3$

C.6 ¹H NMR: Spin–Spin Splitting

The ¹H NMR spectra you have seen up to this point have been limited to one or more single absorptions called **singlets**. In the ¹H NMR spectrum of $BrCH_2CHBr_2$, however, the two signals for the two different kinds of protons are each split into more than one peak. The splitting patterns, the result of **spin–spin splitting**, can be used to determine how many protons reside on the carbon atoms near the absorbing proton.

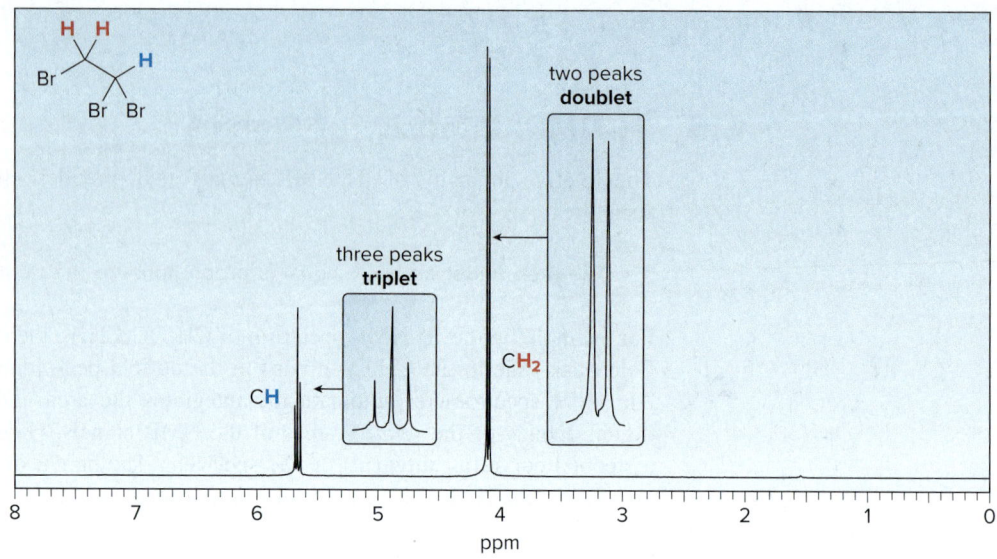

To understand spin–spin splitting, we must distinguish between the *absorbing protons* that give rise to an NMR signal, and the *adjacent protons* that cause the signal to split. **The number of adjacent protons determines the observed splitting pattern.**

- The CH_2 signal appears as **two peaks,** called a *doublet.* The relative area under the peaks of a doublet is 1:1.
- The CH signal appears as **three peaks,** called a *triplet.* The relative area under the peaks of a triplet is 1:2:1.

Spin–spin splitting occurs between nonequivalent protons on the same carbon or adjacent carbons. To illustrate how spin–spin splitting arises, we'll examine nonequivalent protons on adjacent carbons, the more common example. Spin–spin splitting arises because protons are little magnets that can be aligned with or against an applied magnetic field, and this affects the magnetic field that a nearby proton feels.

C.6A Splitting: How a Doublet Arises

First, let's examine how the doublet due to the CH_2 group in $BrCH_2CHBr_2$ arises. The CH_2 group contains the absorbing protons and the CH group contains the adjacent proton that causes the splitting.

When placed in an applied magnetic field (B_0), the adjacent proton ($CHBr_2$) can be aligned with (↑) or against (↓) B_0. As a result, the absorbing protons (CH_2Br) feel two slightly different magnetic fields—one slightly larger than B_0 and one slightly smaller than B_0. Because the absorbing protons feel *two* different magnetic fields, they absorb at *two* different frequencies in the NMR spectrum, thus splitting a single absorption into a *doublet*.

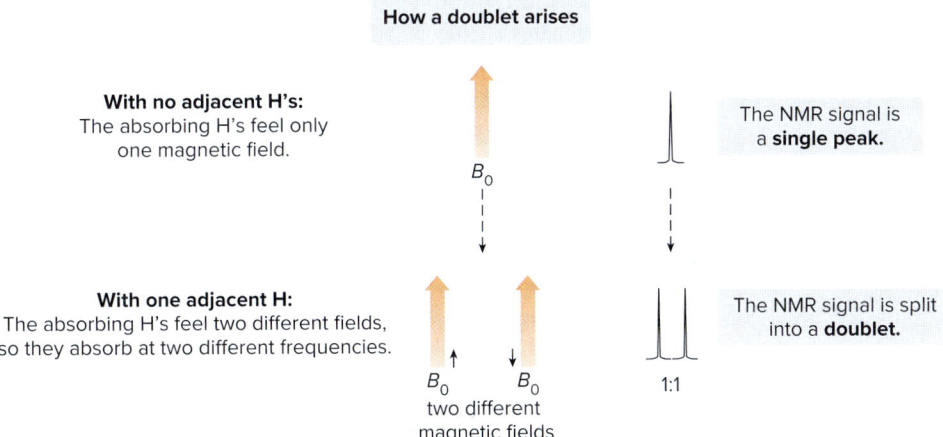

Keep in mind the difference between an **NMR signal** and an **NMR peak**. An NMR signal is the entire absorption due to a particular kind of proton. NMR peaks are contained within a signal. **A doublet constitutes one *signal* that is split into two *peaks*.**

- One adjacent proton splits an NMR signal into a doublet.

coupling constant, *J*, in Hz

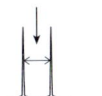

The two peaks of a doublet are approximately equal in area. The area under both peaks—the entire NMR signal—is due to both protons of the CH_2 group of $BrCH_2CHBr_2$.

The frequency difference (measured in Hz) between the two peaks of the doublet is called the **coupling constant,** denoted by ***J***. Coupling constants are usually in the range of 0–18 Hz, and are **independent of the strength of the applied magnetic field, B_0**.

C.6B Splitting: How a Triplet Arises

Now let's examine how the triplet due to the CH group in $BrCH_2CHBr_2$ arises. The CH group contains the absorbing proton and the CH_2 group contains the adjacent protons (H_a and H_b) that cause the splitting.

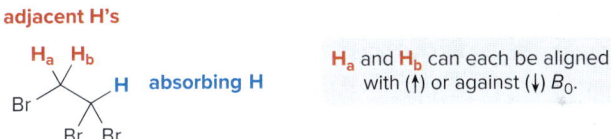

When placed in an applied magnetic field (B_0), the adjacent protons H_a and H_b can each be aligned with (↑) or against (↓) B_0. As a result, the absorbing proton feels three slightly different magnetic fields—one slightly larger than B_0, one slightly smaller than B_0, and one the same strength as B_0.

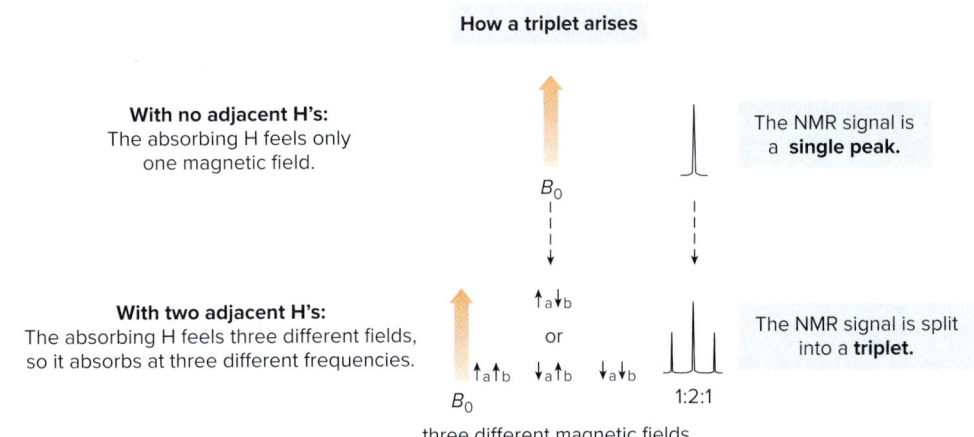

Because the absorbing proton feels *three* different magnetic fields, it absorbs at *three* different frequencies in the NMR spectrum, thus splitting a single absorption into a *triplet*. Because there are two different ways to align one proton with B_0 and one proton against B_0—that is, $\uparrow_a\downarrow_b$ and $\downarrow_a\uparrow_b$—the middle peak of the triplet is twice as intense as the two outer peaks, making the ratio of the areas under the three peaks 1:2:1.

- Two adjacent protons split an NMR signal into a triplet.

When two protons split each other's NMR signals, they are said to be *coupled*. In $BrCH_2CHBr_2$, the CH proton is coupled to the CH_2 protons. **The spacing between peaks in a split NMR signal, measured by the *J* value, is *equal* for coupled protons.**

C.6C Splitting: The Rules and Examples

Three general rules describe the splitting patterns commonly seen in the ^1H NMR spectra of organic compounds.

Rule [1] Equivalent protons don't split each other's signals.

Rule [2] A set of *n* nonequivalent protons splits the signal of a nearby proton into *n* + 1 peaks.

- In $BrCH_2CHBr_2$, *one* adjacent CH proton splits an NMR signal into *two* peaks (a doublet), and *two* adjacent CH_2 protons split an NMR signal into *three* peaks (a triplet). Names for split NMR signals containing two to seven peaks are given in Table C.3. An NMR signal having more than seven peaks is called a **multiplet**.
- The inside peaks of a split NMR signal are always most intense, with the area under the peaks decreasing from the inner to the outer peaks in a given splitting pattern.

Rule [3] Splitting is observed for nonequivalent protons on the *same* carbon or *adjacent* carbons.

If H_a and H_b are not equivalent, splitting is observed in each of the following cases.

Splitting is not generally observed between protons separated by more than three σ bonds. Although H_a and H_b are not equivalent to each other in butan-2-one and ethyl methyl ether, H_a and H_b are separated by *four* σ bonds, so they are too far away to split each other's NMR signals.

butan-2-one
H_a and H_b are separated by **four** σ bonds.
no splitting between H_a and H_b

ethyl methyl ether
H_a and H_b are separated by **four** σ bonds.
no splitting between H_a and H_b

Table C.4 illustrates common splitting patterns observed for adjacent nonequivalent protons.

Predicting splitting is always a two-step process:

- **Determine if two protons are equivalent or different.** Only *nonequivalent* protons split each other.
- **Determine if two nonequivalent protons are close enough to split each other's signals.** Splitting is observed only for nonequivalent protons on the *same* carbon or *adjacent* carbons.

Table C.3
Names for a Given Number of Peaks in an NMR Signal

Number of peaks	Name
1	singlet
2	doublet
3	triplet
4	quartet
5	quintet
6	sextet
7	septet
> 7	multiplet

The splitting of an NMR signal reveals the number of nearby nonequivalent protons. It tells nothing about the absorbing proton itself.

C.6 ¹H NMR: Spin–Spin Splitting

Table C.4 Common Splitting Patterns Observed in ¹H NMR

Example		Pattern	Analysis					
[1]	H–C–C–H (H, H on each C)		• H: one adjacent H proton	--→	two peaks	--→	a **doublet**	
			• H: one adjacent H proton	--→	two peaks	--→	a **doublet**	
[2]	H–C–CH₂–		• H: two adjacent H protons	--→	three peaks	--→	a **triplet**	
			• H: one adjacent H proton	--→	two peaks	--→	a **doublet**	
[3]	–CH₂CH₂–		• H: two adjacent H protons	--→	three peaks	--→	a **triplet**	
			• H: two adjacent H protons	--→	three peaks	--→	a **triplet**	
[4]	–CH₂CH₃		• H: three adjacent H protons	--→	four peaks	--→	a **quartet***	
			• H: two adjacent H protons	--→	three peaks	--→	a **triplet**	
[5]	H–C–CH₃		• H: three adjacent H protons	--→	four peaks	--→	a **quartet***	
			• H: one adjacent H proton	--→	two peaks	--→	a **doublet**	

*The relative area under the peaks of a quartet is 1:3:3:1.

Several examples of spin–spin splitting in specific compounds illustrate the result of this two-step strategy.

ClCH₂CH₂Cl (H H on each C)

- All protons are equivalent, so there is no splitting and the NMR signal is *one singlet*.

ClCH(H$_a$)(H$_a$)–CH(H$_b$)(H$_b$)Br

- There are two NMR signals. H$_a$ and H$_b$ are nonequivalent protons bonded to adjacent C atoms, so they are close enough to split each other's NMR signals. The H$_a$ signal is split into a *triplet* by the two H$_b$ protons. The H$_b$ signal is split into a *triplet* by the two H$_a$ protons.

CH₃–C(=O)–CH₂CH₃
H$_a$ H$_b$ H$_c$

- There are three NMR signals. H$_a$ has no adjacent nonequivalent protons, so its signal is a *singlet*. The H$_b$ signal is split into a *quartet* by the three H$_c$ protons. The H$_c$ signal is split into a *triplet* by the two H$_b$ protons.

Cl\C=C/H$_a$
Br/   \H$_b$

- There are two NMR signals. H$_a$ and H$_b$ are nonequivalent protons on the same carbon, so they are close enough to split each other's NMR signals. The H$_a$ signal is split into a *doublet* by H$_b$. The H$_b$ signal is split into a *doublet* by H$_a$.

Problem C.13 Into how many peaks will each proton shown in red be split?

a. CH₃CH₂C(=O)Cl

b. CH₃CHBr₂ (Br Br H)

c. CH₃C(=O)CH₂CH₂Br

d. (H)(Cl)C=C(Br)(H)

e. (CH₃)(H)C=C(CH₃)(H)

f. CH₃O–CH(OCH₃)–... ClCH₂ H

Problem C.14 For each compound, give the number of ^{1}H NMR signals and then determine how many peaks are present for each NMR signal.

a. b. c. d.

Problem C.15 Sketch the NMR spectrum of CH_3CH_2Cl, giving the approximate location of each NMR signal.

C.7 More-Complex Examples of Splitting

Up to now you have studied examples of spin–spin splitting where the absorbing proton has nearby protons on *one* adjacent carbon only. What happens when the absorbing proton has nonequivalent protons on *two* adjacent carbons? Different outcomes are possible, depending on whether the adjacent nonequivalent protons are *equivalent to* or *different from* each other.

For example, 2-bromopropane [$(CH_3)_2CHBr$] has two types of protons—H_a and H_b—so it exhibits two NMR signals, as shown in Figure C.6.

- The H_a protons have only one adjacent nonequivalent proton (H_b), so they are split into two peaks, a **doublet**.
- H_b has three H_a protons on each side. Because the six H_a protons are *equivalent to each other*, the ***n* + 1** rule can be used to determine splitting: 6 + 1 = 7 peaks, a **septet**.

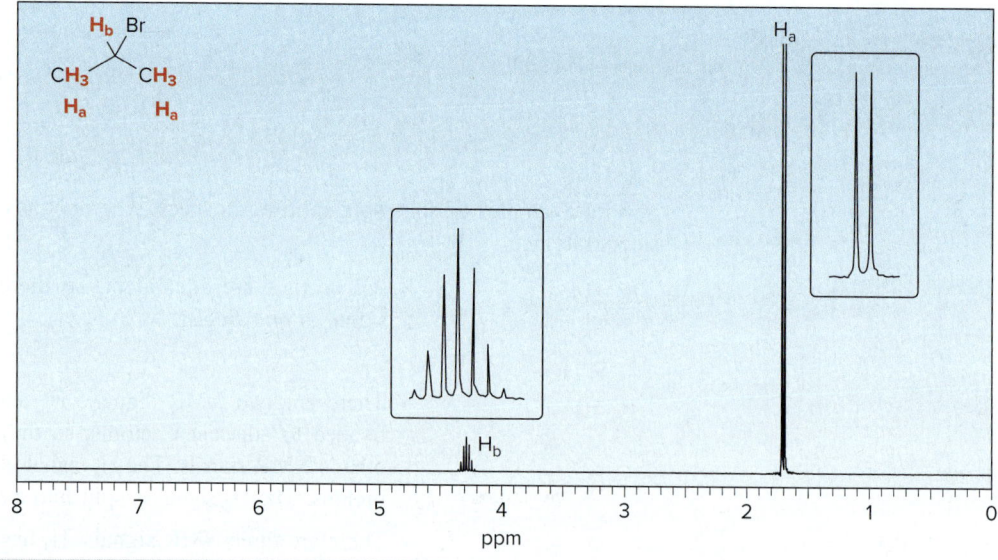

Figure C.6 The ^{1}H NMR spectrum of 2-bromopropane, $(CH_3)_2CHBr$

This is a specific example of a general rule:

- Whenever two (or three) sets of adjacent protons are *equivalent to each other*, use the *n* + 1 rule to determine the splitting pattern.

When an absorbing proton is flanked by two sets of adjacent protons that are *not equivalent to each other*, the outcome depends on the coupling constant (*J*) between the absorbing proton and its neighboring protons.

Let us begin with the result that occurs in **flexible alkyl chains**; that is, **the absorbing and adjacent protons are *not* bonded to a ring or double bond,** as illustrated with 1-bromopropane, $CH_3CH_2CH_2Br$.

$$CH_3CH_2CH_2-Br$$
$$H_a \ \ H_b \ \ H_c$$

$CH_3CH_2CH_2Br$ has three different types of protons—H_a, H_b, and H_c— so it exhibits three NMR signals. The H_a and H_c signals are both triplets because they are adjacent to two H_b protons, as shown in Figure C.7.

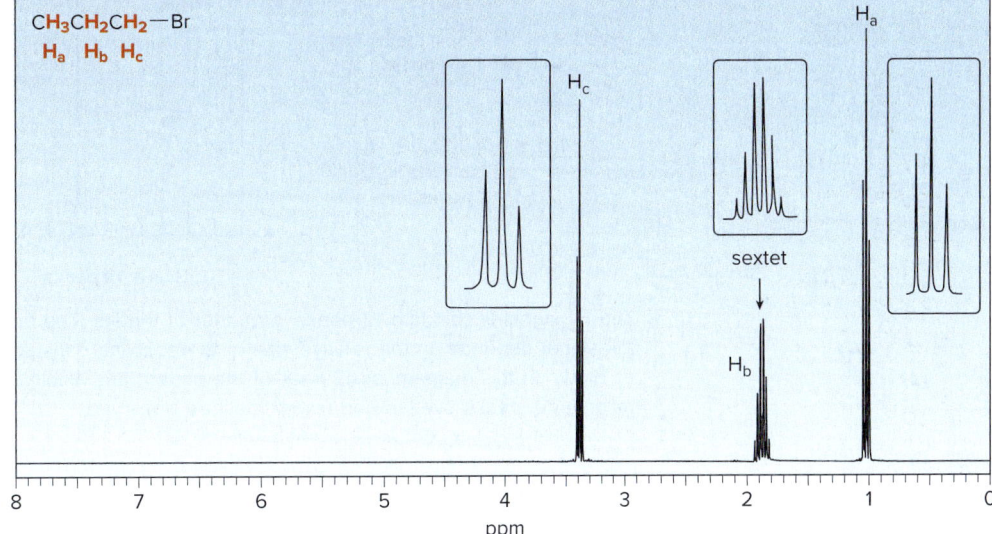

Figure C.7

The ^1H NMR spectrum of 1-bromopropane, $CH_3CH_2CH_2Br$

- H_a and H_c are both triplets.
- The signal for H_b appears as a multiplet of six peaks (a sextet), due to peak overlap; the number of peaks = $n + m + 1 = 3 + 2 + 1 = 6$ peaks.

What splitting is observed for the H_b protons, which have protons on both adjacent carbons, and H_a and H_c are *not equivalent* to each other? In acyclic molecules of this sort, which are not constrained by the geometry of a ring or double bond, the coupling constants between the absorbing proton and both sets of adjacent protons are equal (or close to it); that is, $J_{ab} = J_{bc}$. In this case, even though the H_a and H_c protons are not equivalent to each other, **we can just add the number of protons on both adjacent carbons together.** The 3 H_a protons and the 2 H_c protons split the NMR signal of the H_b protons into $3 + 2 + 1 = 6$ peaks, a **sextet**. This is a specific example of a general phenomenon:

- **In a flexible alkyl chain, the *n* alkyl protons on one adjacent carbon and the *m* protons on the other adjacent carbon split the observed signal into *n + m* + 1 peaks.**

Now let's consider the splitting pattern of the H_b protons in the general compound $CH_3CH_2CH_2Z$ when the coupling constants between the absorbing proton H_b and both sets of adjacent protons (H_a and H_c) are different; that is, $J_{ab} \neq J_{bc}$.

$CH_3CH_2CH_2—Z$
H_a H_b H_c

In this case, to determine the splitting of the H_b signal, we must consider the effect of the H_a protons and the H_c protons *separately*. The three H_a protons split the H_b signal into four peaks and the two H_c protons split each of these four peaks into three peaks—that is, the NMR signal due to H_b consists of $4 \times 3 = \mathbf{12 \text{ peaks}}$. Figure C.8 shows a splitting diagram that illustrates how these 12 peaks arise. This is a specific example of a general phenomenon:

The three possibilities for determining splitting patterns when an absorbing proton has nonequivalent protons on two adjacent carbons are shown with examples in Sample Problem C.6 and in the Key Skills section of the Chapter Review.

- **When two sets of adjacent protons are *different from each other* (*n* protons on one adjacent carbon and *m* protons on the other), the number of peaks in an NMR signal is (*n* + 1)(*m* + 1).**

Complex splitting of this sort is seen with protons on carbon–carbon double bonds in Section C.8. Sample Problem C.6 illustrates how to determine splitting in three different compounds.

The $(n + 1)(m + 1)$ rule in splitting always gives the **maximum** number of peaks that is possible when an absorbing proton has *n* adjacent protons on one side and *m* protons on the other,

Figure C.8
A splitting diagram for the H_b protons in $CH_3CH_2CH_2Z$

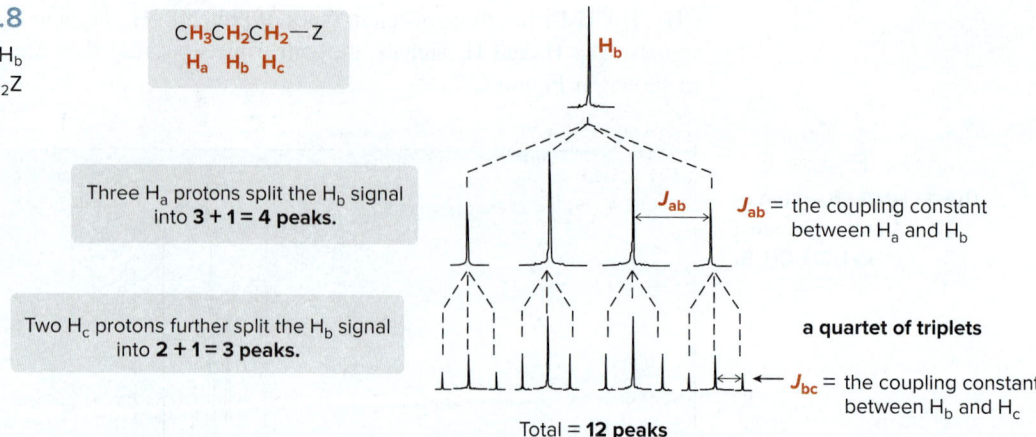

- Three H_a protons split the H_b signal into $3 + 1 = 4$ peaks.
- Two H_c protons further split the H_b signal into $2 + 1 = 3$ peaks.
- Total = **12 peaks**

J_{ab} = the coupling constant between H_a and H_b

a quartet of triplets

J_{bc} = the coupling constant between H_b and H_c

- The H_b signal is split into 12 peaks, a quartet of triplets. The number of peaks actually seen for the signal depends on the relative size of the coupling constants, J_{ab} and J_{bc}. When $J_{ab} \gg J_{bc}$, as drawn in this diagram, all 12 lines of the pattern are visible. When J_{ab} and J_{bc} are similar in magnitude, peaks overlap and fewer lines are observed.

and the coupling constants between nearby protons are different. As the difference between J values decreases, peaks overlap and fewer than the maximum number of peaks is observed.

Sample Problem C.6 — Determining the Number of Peaks in an NMR Signal

How many peaks are present in the NMR signal of the labeled protons of each compound?

a. $ClCH_2CH_2CH_2Cl$ (labeled H H on middle carbon)
b. $ClCH_2CH_2CH_2Br$ (labeled H H on middle carbon)
c. $CH_3-CH=CH-OH$ (with H on each alkene carbon)

Solution

When an absorbing proton is flanked by two sets of adjacent protons, there are three possibilities for determining the splitting pattern, as seen in parts (a), (b), and (c).

a. **5 peaks for H_b**
$(n + 1) = (4\ H_a + 1)$
Structure: $Cl-CH_2-CH_2-CH_2-Cl$ with $H_b\ H_b$ on middle C and $H_a\ H_a\ H_a\ H_a$ on outer Cs.
4 adjacent H_a protons

- H_b has two H_a protons on each adjacent C. Because the four H_a protons are equivalent to each other, the $n + 1$ rule can be used to determine splitting: $4 + 1 = 5$ peaks, a quintet.

b. **5 peaks for H_b**
$(n + m + 1) = (2\ H_a + 2\ H_c + 1)$
Structure: $Cl-CH_2-CH_2-CH_2-Br$ with $H_b\ H_b$ on middle C, $H_a\ H_a$ and $H_c\ H_c$ on outer Cs.
2 H_a and 2 H_c protons bonded to an alkyl chain

- Even though H_a and H_c are not equivalent to each other, they are bonded to a flexible alkyl chain. In this case, we can add the number of protons on both adjacent carbons together, so the number of peaks for $H_b = n + m + 1 = 2 + 2 + 1 = 5$ peaks.

c. **8 peaks for H_b**
$(n + 1)(m + 1) = (3\ H_a + 1)(1\ H_c + 1)$
Structure: $CH_3-CH=CH-OH$ with H_b on one alkene C, H_a (3) on CH_3, H_c on the other alkene C.
3 H_a protons and 1 H_c proton
H_b is bonded to a C=C.

- H_b has three H_a protons on one adjacent C and one H_c proton on the other. Because H_a and H_c are not equivalent to each other, the number of peaks for $H_b = (n + 1)(m + 1) = (3 + 1)(1 + 1) = 8$ peaks.

Problem C.16 How many peaks are present in the NMR signal of each labeled proton?

a. 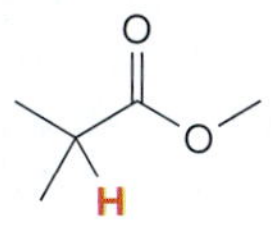 [structure] b. 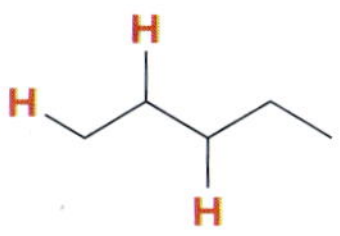 [structure] c. 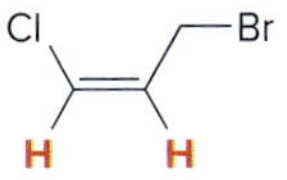 [structure] d. 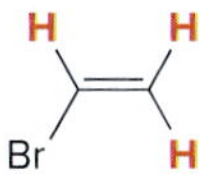 [structure]

More Practice: Try Problems C.32b, C.38, C.48b, C.49e.

Problem C.17 Describe the ^{1}H NMR spectrum of each compound. State how many NMR signals are present, the splitting pattern for each signal, and the approximate chemical shift.

a. 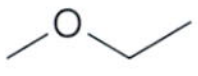 [structure] b. 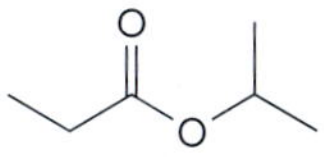 [structure] c.  [structure] d.  [structure]

C.8 Spin–Spin Splitting in Alkenes

Protons on carbon–carbon double bonds often give characteristic splitting patterns. A disubstituted double bond can have two **geminal protons** (on the same carbon atom), two **cis protons**, or two **trans protons**. When these protons are different, each proton splits the NMR signal of the other, so that each proton appears as a doublet. **The magnitude of the coupling constant J for these doublets depends on the arrangement of hydrogen atoms.**

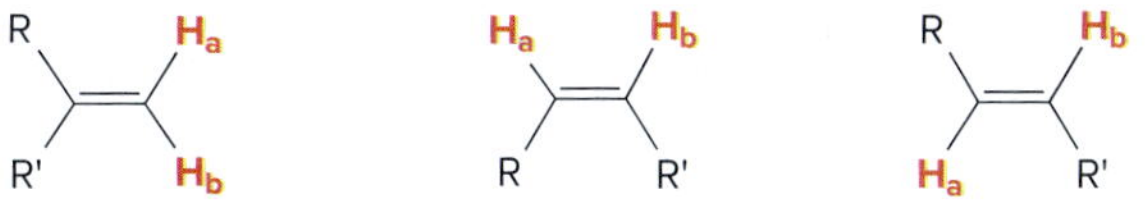

geminal H's	cis H's	trans H's
$J_{geminal}$	J_{cis}	J_{trans}
0–3 Hz	5–10 Hz	11–18 Hz

$J_{geminal}$ < J_{cis} < J_{trans}

Thus, the E and Z isomers of 3-chloropropenoic acid both exhibit two doublets for the two alkenyl protons, but the coupling constant is larger when the protons are trans compared to when the protons are cis, as shown in Figure C.9.

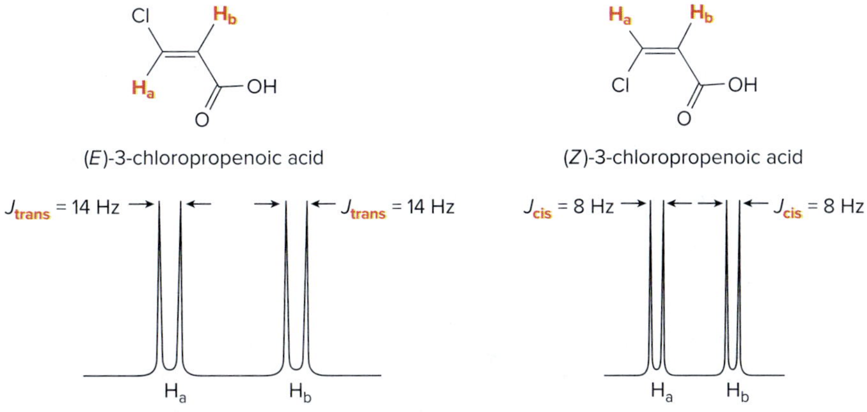

Figure C.9
^{1}H NMR spectra for the alkenyl protons of (E)- and (Z)-3-chloropropenoic acid

(E)-3-chloropropenoic acid: J_{trans} = 14 Hz

(Z)-3-chloropropenoic acid: J_{cis} = 8 Hz

- Although both (E)- and (Z)-3-chloropropenoic acid show two doublets in their ^{1}H NMR spectra for their alkenyl protons, $J_{trans} > J_{cis}$.

When a double bond is monosubstituted, there are three nonequivalent protons, and all three protons are coupled to each other. For example, vinyl acetate (CH_2=CHOCOCH$_3$) has four different types of protons, three of which are bonded to the double bond. Besides the singlet for the CH_3 group, each proton on the double bond is coupled to two other different protons on the double bond, giving the spectrum in Figure C.10. Because the protons are bonded to a double bond, we determine the splitting using the $(n+1)(m+1)$ rule.

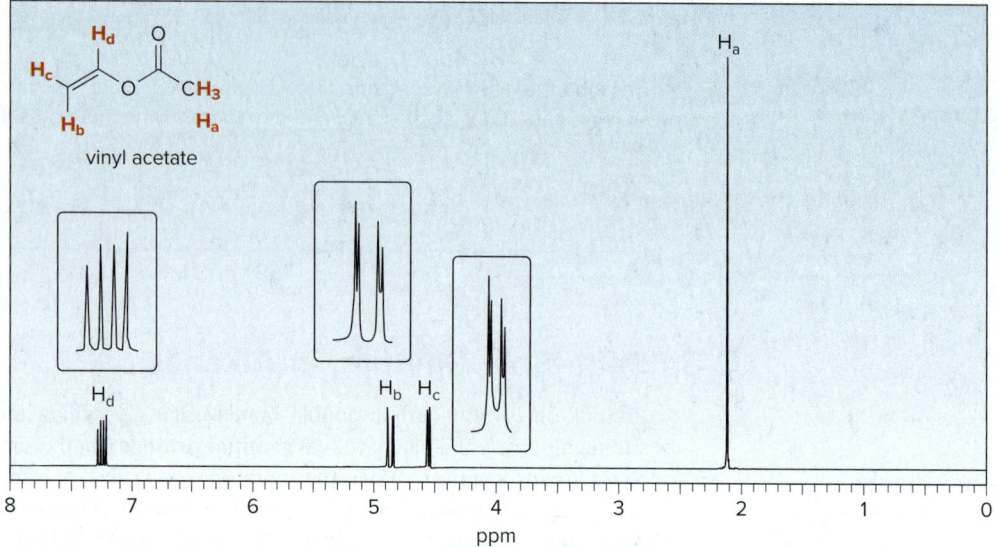

Figure C.10

The ^{1}H NMR spectrum of vinyl acetate (CH_2=CHOCOCH$_3$)

- H_b has two nearby nonequivalent protons that split its signal—the geminal proton H_c and the trans proton H_d. H_d splits the H_b signal into a doublet, and the H_c proton splits the doublet into two doublets. This pattern of four peaks is called a **doublet of doublets**.
- H_c has two nearby nonequivalent protons that split its signal—the geminal proton H_b and the cis proton H_d. H_d splits the H_c signal into a doublet, and the H_b proton splits the doublet into two doublets, forming another **doublet of doublets**.
- H_d has two nearby nonequivalent protons that split its signal—the trans proton H_b and the cis proton H_c. H_b splits the H_d signal into a doublet, and the H_c proton splits the doublet into two doublets, forming another **doublet of doublets**.

Splitting diagrams for the three alkenyl protons in vinyl acetate are drawn in Figure C.11. Note that each pattern is different in appearance because the magnitude of the coupling constants forming them is different.

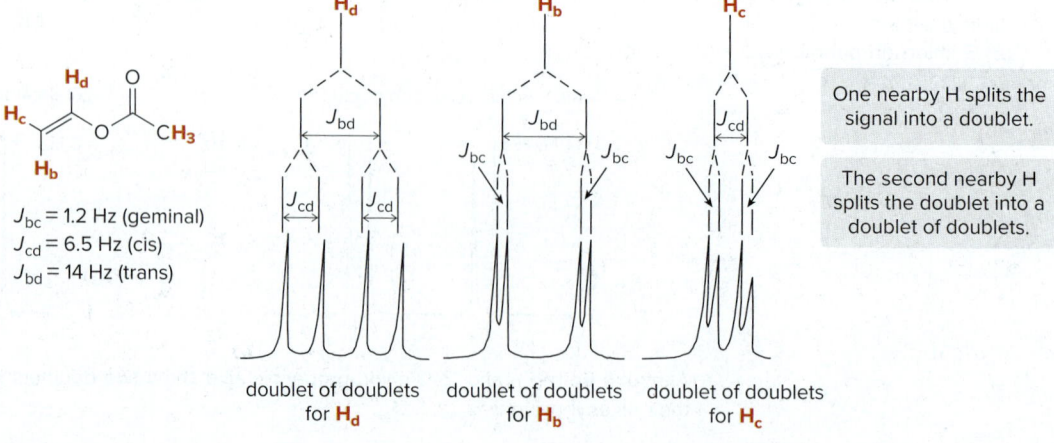

Figure C.11

Splitting diagram for the alkenyl protons in vinyl acetate (CH_2=CHOCOCH$_3$)

Problem C.18 Draw a splitting diagram for H_b in *trans*-1,3-dichloropropene, given that J_{ab} = 13.1 Hz and J_{bc} = 7.2 Hz.

trans-1,3-dichloropropene

Problem C.19 Identify **A** and **B**, isomers of molecular formula $C_3H_4Cl_2$, from the given ^1H NMR data: Compound **A** exhibits signals at 1.75 (doublet, 3 H, J = 6.9 Hz) and 5.89 (quartet, 1 H, J = 6.9 Hz) ppm. Compound **B** exhibits signals at 4.16 (singlet, 2 H), 5.42 (doublet, 1 H, J = 1.9 Hz), and 5.59 (doublet, 1 H, J = 1.9 Hz) ppm.

C.9 Other Facts About ^1H NMR Spectroscopy

C.9A OH Protons

- Under usual conditions, an OH proton does not split the NMR signal of adjacent protons.
- The signal due to an OH proton is not split by adjacent protons.

Ethanol (CH_3CH_2OH), for example, has three different types of protons, so there are three signals in its ^1H NMR spectrum, as shown in Figure C.12.

- The H_a signal is split by the two H_b protons into three peaks, a **triplet.**
- The H_b signal is split by only the three H_a protons into four peaks, a **quartet.** The adjacent OH proton does *not* split the signal due to H_b.
- H_c is a **singlet** because OH protons are *not* split by adjacent protons.

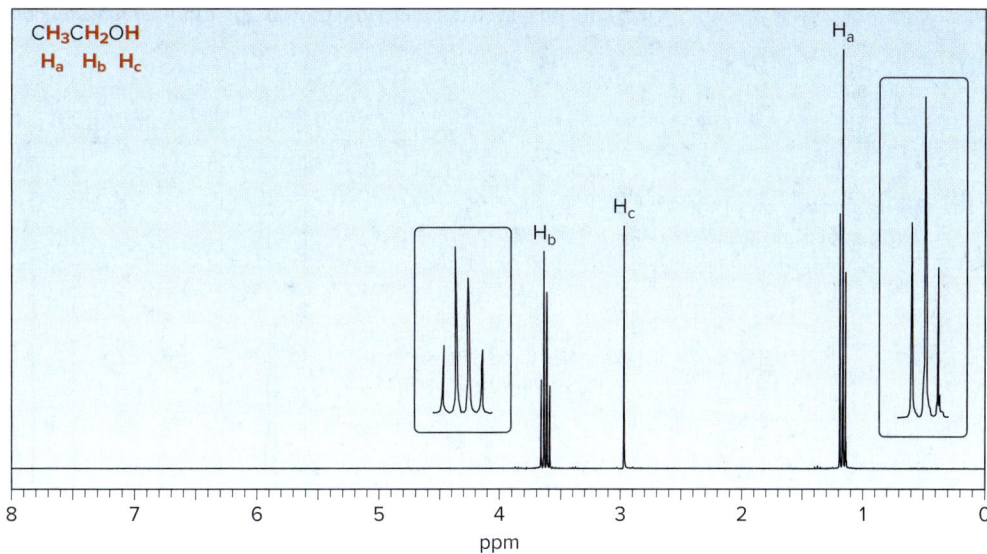

Figure C.12 The ^1H NMR spectrum of ethanol (CH_3CH_2OH)

Why is a proton bonded to an oxygen atom a singlet in a ^1H NMR spectrum? Protons on electronegative elements rapidly **exchange** between molecules in the presence of trace amounts of acid or base. It is as if the CH_2 group in ethanol never "feels" the presence of the OH proton, because the OH proton is rapidly moving from one molecule to another. We therefore see a peak due to the OH proton, but it is a single peak with no splitting. This phenomenon usually occurs with NH and OH protons.

Problem C.20 How many signals are present in the ¹H NMR spectrum for each molecule? What splitting is observed in each signal?

a. (CH₃)₂CHCH₂OH b. CH₃CH₂OH c. (CH₃)₂CHNH₂

C.9B Cyclohexane Conformations

How do the rotation around carbon–carbon σ bonds and the ring flip of cyclohexane rings affect an NMR spectrum? Because these processes are rapid at room temperature, an NMR spectrum records an **average** of all conformations that interconvert.

Thus, even though each cyclohexane carbon has two different types of hydrogens—one axial and one equatorial—the two chair forms of cyclohexane rapidly interconvert them, and an **NMR spectrum shows a single signal for the average environment** that it "sees."

H$_a$ axial ⇌ H$_a$ equatorial
H$_b$ H$_b$

> Axial and equatorial H's rapidly interconvert. NMR sees an average environment and shows one signal.

C.9C Protons on Benzene Rings

We will learn more about the spectroscopic absorptions of benzene derivatives in Chapter 19.

Benzene has six equivalent, deshielded protons and exhibits a single peak in its ¹H NMR spectrum at 7.27 ppm. Monosubstituted benzene derivatives—that is, benzene rings with one H atom replaced by another substituent Z—contain five deshielded protons that are no longer all equivalent to each other. The identity of Z determines the appearance of this region of a ¹H NMR spectrum (6.5–8 ppm), as shown in Figure C.13. We will not analyze the splitting patterns observed for the ring protons of monosubstituted benzenes.

Figure C.13
The 6.5–8 ppm region of the ¹H NMR spectrum of three benzene derivatives

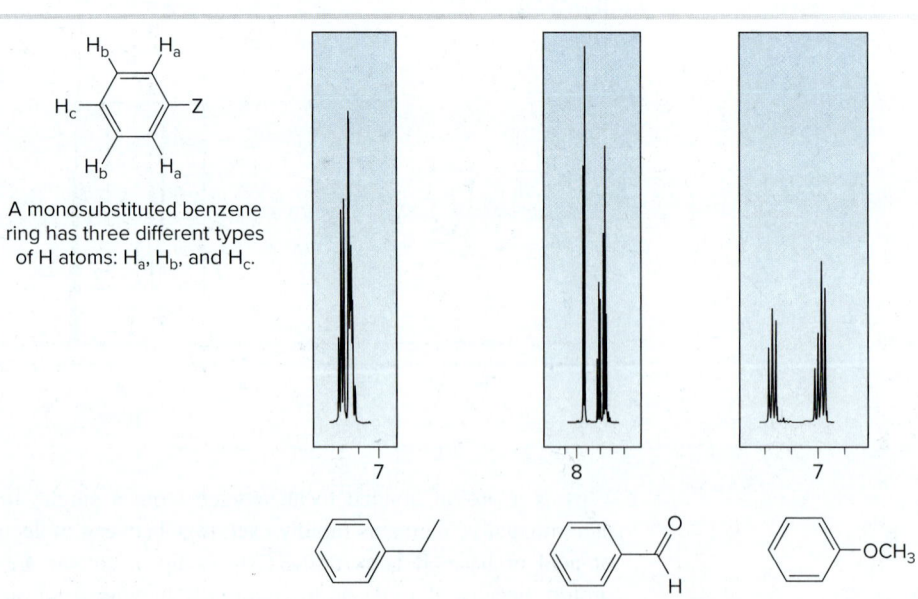

A monosubstituted benzene ring has three different types of H atoms: H$_a$, H$_b$, and H$_c$.

- The appearance of the signals in the 6.5–8 ppm region of the ¹H NMR spectrum depends on the identity of Z in C₆H₅Z.

Problem C.21 What protons in alcohol **A** give rise to each signal in its ¹H NMR spectrum? Explain all splitting patterns observed for absorptions between 0 to 7 ppm.

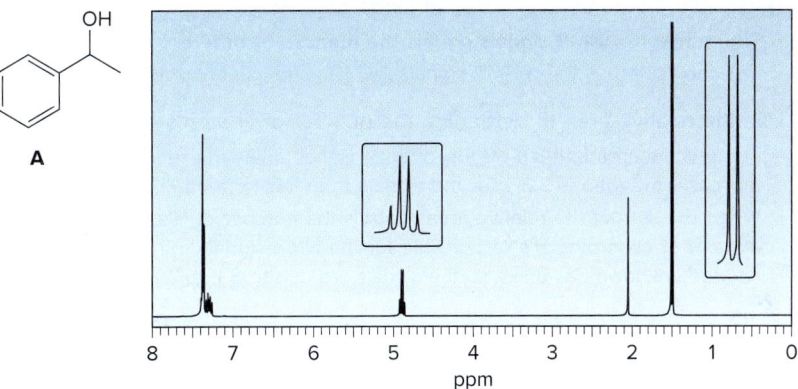

Problem C.22 How many peaks are observed in the ¹H NMR signal for each proton shown in red in palau'amine, the complex chapter-opening molecule?

C.10 Using ¹H NMR to Identify an Unknown

Combined with mass spectrometry (which gives a compound's molecular formula) and infrared spectroscopy (which identifies a compound's functional group), we can then use its ¹H NMR spectrum to determine the structure of an unknown. A suggested procedure is illustrated for compound **X**, whose molecular formula ($C_4H_8O_2$) and functional group (C=O) were determined in Section B.5.

How To Use ¹H NMR Data to Determine a Structure

Example Using its ¹H NMR spectrum, determine the structure of an unknown compound **X** that has molecular formula $C_4H_8O_2$ and contains a C=O absorption in its IR spectrum.

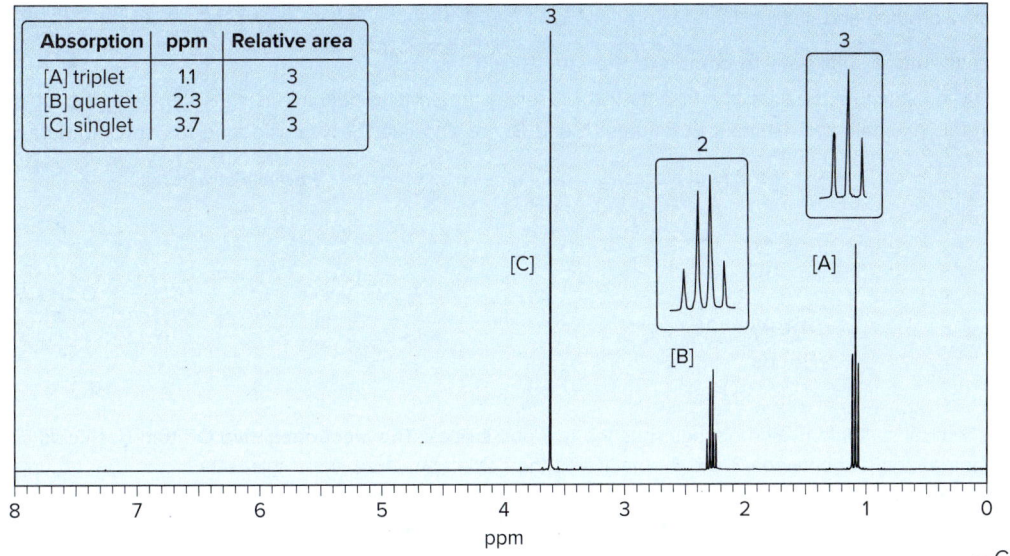

Absorption	ppm	Relative area
[A] triplet	1.1	3
[B] quartet	2.3	2
[C] singlet	3.7	3

—Continued

How To, continued...

Step [1] **Determine the number of different kinds of protons.**
- **The number of NMR signals equals the number of different types of protons.**
- This molecule has three NMR signals ([A], [B], and [C]) and therefore **three** types of protons (H_a, H_b, and H_c).

Step [2] **Use the relative area to determine the number of H atoms giving rise to each signal.**
- The relative area (printed on top of each signal) gives the *ratio* of absorbing protons responsible for each signal. In this case, the ratio is 3:2:3 for the signals from left to right.
- **When the sum of the relative areas *equals* the number of H's in the molecular formula, the relative area gives the number of absorbing H's responsible for the NMR signal.** In this example, the sum of the relative areas is $3 + 2 + 3 = 8$, and the unknown has 8 H's, so the signals are due to 3 H's, 2 H's, and 3 H's from left to right in the spectrum.

3 H_a protons signal [A]	2 H_b protons signal [B]	3 H_c protons signal [C]
Three equivalent H's usually means a CH_3 group.	Two equivalent H's usually means a CH_2 group.	Three equivalent H's usually means a CH_3 group.

Step [3] **Use individual splitting patterns to determine what carbon atoms are bonded to each other.**
- Start with the singlets. Signal [C] is due to a CH_3 group with no adjacent nonequivalent H atoms. Possible structures include:

$$CH_3O- \quad \text{or} \quad CH_3-\overset{\overset{O}{\|}}{C}- \quad \text{or} \quad CH_3-\overset{\diagup}{\underset{\diagdown}{C}}-$$

- Because signal [A] is a **triplet**, there must be **2 H's** (CH_2 group) on the adjacent carbon.
- Because signal [B] is a **quartet**, there must be **3 H's** (CH_3 group) on the adjacent carbon.
- This information suggests that **X** has an **ethyl** group --→ CH_3CH_2-.

An adjacent CH_3 group causes the splitting. — 2 absorbing H's — signal [B]

3 absorbing H's — An adjacent CH_2 group causes the splitting. — signal [A]

→ CH_3CH_2-

To summarize, **X** contains CH_3-, CH_3CH_2-, and C=O (from the IR). Comparing these atoms with the molecular formula shows that one O atom is missing. Because O atoms do not absorb in a ^1H NMR spectrum, their presence can be inferred only by examining the chemical shift of protons near them. O atoms are more electronegative than C, thus deshielding nearby protons and shifting their absorption *downfield*.

Step [4] **Use chemical shift data to complete the structure.**
- Put the structure together in a manner that preserves the splitting data and is consistent with the reported chemical shifts.
- In this example, two isomeric structures (**A** and **B**) are possible for **X** considering the splitting data only:

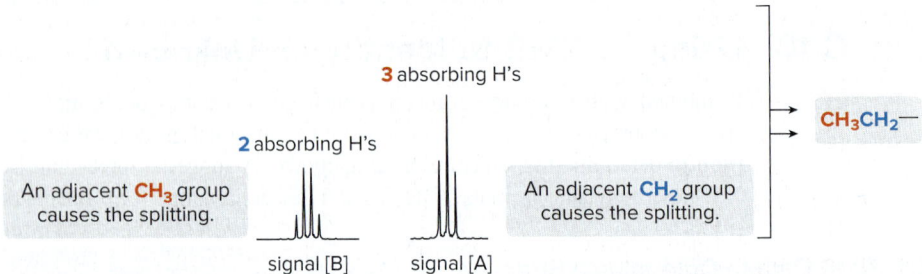

- Chemical shift information distinguishes the two possibilities. **The electronegative O atom deshields adjacent H's, shifting them downfield** between 3 and 4 ppm. If **A** is the correct structure, the singlet due to the CH_3 group (H_c) should occur downfield, whereas if **B** is the correct structure, the quartet due to the CH_2 group (H_b) should occur downfield.
- Because the NMR of **X** has a singlet (not a quartet) at 3.7, **A is the correct structure.**

Problem C.23 Propose a structure for a compound of molecular formula $C_7H_{14}O_2$ with an IR absorption at 1740 cm^{-1} and the following ^1H NMR data:

Absorption	ppm	Relative area
singlet	1.2	9
triplet	1.3	3
quartet	4.1	2

Problem C.24 Propose a structure for a compound of molecular formula C_3H_8O with an IR absorption at 3600–3200 cm^{-1} and the following NMR spectrum:

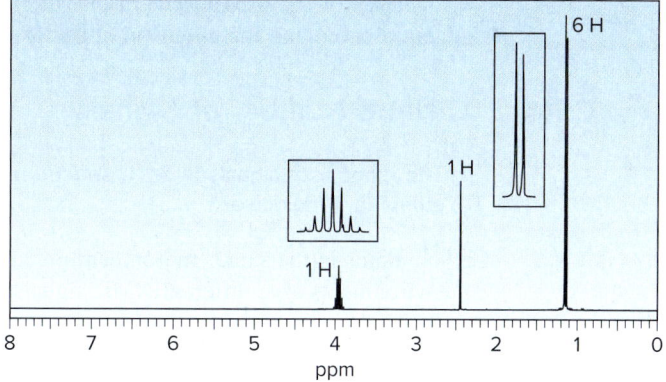

C.11 ^{13}C NMR Spectroscopy

^{13}C NMR spectroscopy is also an important tool for organic structure analysis. The physical basis for ^{13}C NMR is the same as for ^1H NMR. When placed in a magnetic field, B_0, ^{13}C nuclei can align themselves with or against B_0. More nuclei are aligned with B_0 because this arrangement is lower in energy, but these nuclei can be made to spin flip against the applied field by applying RF radiation of the appropriate frequency.

^{13}C NMR spectra, like ^1H NMR spectra, plot peak intensity versus chemical shift, using TMS as the reference signal at 0 ppm. ^{13}C occurs in only 1.1% natural abundance, however, so ^{13}C NMR signals are much weaker than ^1H NMR signals. To overcome this limitation, modern spectrometers irradiate samples with many pulses of RF radiation and use mathematical tools to increase signal sensitivity and decrease background noise. The spectrum of acetic acid (CH_3COOH) illustrates the general features of a ^{13}C NMR spectrum.

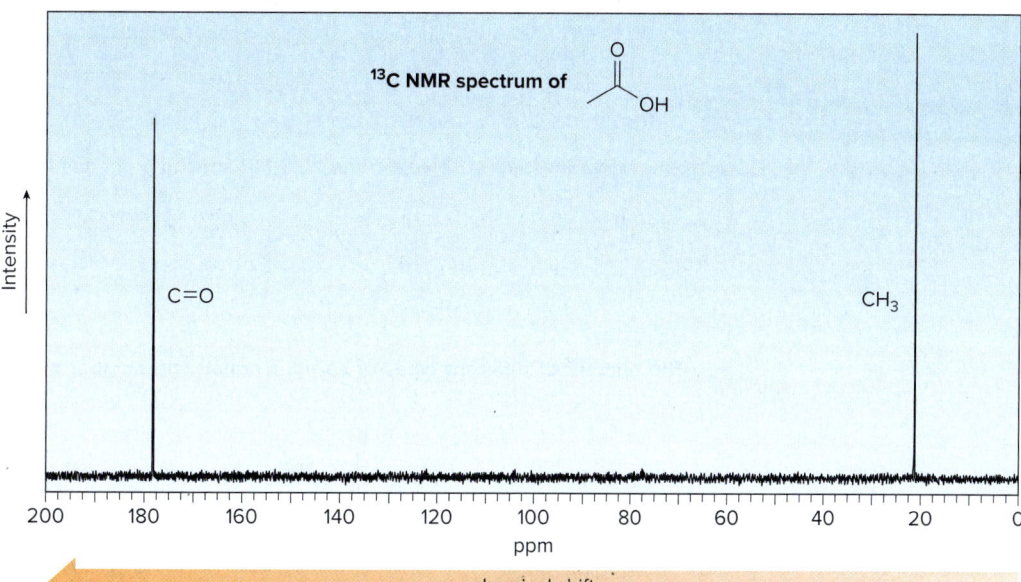

^{13}C NMR spectra are easier to analyze than ^{1}H spectra because signals are not split. **Each type of carbon atom appears as a single peak.**

Why aren't ^{13}C signals split by nearby carbon atoms? Recall from Section C.6 that splitting occurs when two NMR active nuclei—like two protons—are close to each other. Because of the low natural abundance of ^{13}C nuclei (1.1%), the chance of two ^{13}C nuclei being bonded to each other is very small (0.01%), so no carbon–carbon splitting is observed.

A ^{13}C NMR signal can also be split by nearby protons. This ^{1}H–^{13}C splitting is usually eliminated from a spectrum, however, by using an instrumental technique that decouples the proton–carbon interactions, so that every signal in a ^{13}C NMR spectrum is a singlet.

Two features of ^{13}C NMR spectra provide the most structural information: the **number of signals** observed and the **chemical shifts** of those signals.

C.11A ^{13}C NMR: Number of Signals

- The number of signals in a ^{13}C spectrum gives the number of different types of carbon atoms in a molecule.

Carbon atoms in the same environment give the same NMR signal, whereas carbons in different environments give different NMR signals. The ^{13}C NMR spectrum of CH_3COOH has two signals because there are two different types of carbon atoms—the C of the CH_3 group and the C of the carbonyl (C=O).

- Because ^{13}C NMR signals are not split, the number of signals equals the number of lines in the ^{13}C NMR spectrum.

Thus, the ^{13}C NMR spectra of dimethyl ether, chloroethane, and methyl acetate exhibit one, two, and three lines, respectively, because these compounds contain one, two, and three different types of carbon atoms.

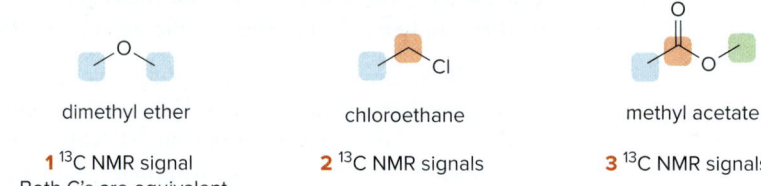

dimethyl ether — 1 ^{13}C NMR signal — Both C's are equivalent.

chloroethane — 2 ^{13}C NMR signals

methyl acetate — 3 ^{13}C NMR signals

In contrast to what occurs in proton NMR, peak intensity is not proportional to the number of absorbing carbons, so ^{13}C NMR signals are not integrated.

Sample Problem C.7 Determining the Number of Lines in a ^{13}C NMR Spectrum

How many lines are observed in the ^{13}C NMR spectrum of each compound?

a. b. c.

Solution
The number of different types of carbons equals the number of lines in a ^{13}C NMR spectrum.

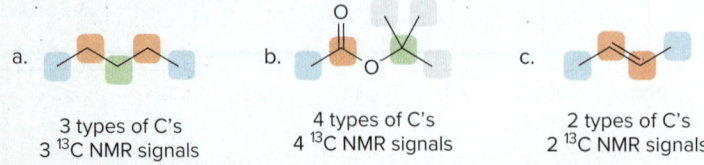

a. 3 types of C's — 3 ^{13}C NMR signals

b. 4 types of C's — 4 ^{13}C NMR signals

c. 2 types of C's — 2 ^{13}C NMR signals

C.11 ^{13}C NMR Spectroscopy

Problem C.25 How many lines are observed in the ^{13}C NMR spectrum of each compound?

a. 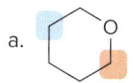 b.  c. 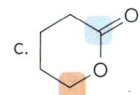 d.

More Practice: Try Problems C.31b, C.41, C.43, C.48c, C.49c.

Problem C.26 Esters of chrysanthemic acid are naturally occurring insecticides. How many lines are present in the ^{13}C NMR spectrum of chrysanthemic acid?

chrysanthemic acid

Esters of chrysanthemic acid are obtained from the flowers of *Chrysanthemum cinerariifolium*. Because they are biodegradable and active against numerous insect species, these esters are widely used insecticides.
Gail Whitfield/Alamy Stock Photo

C.11B ^{13}C NMR: Position of Signals

In contrast to the small range of chemical shifts in ^{1}H NMR (0–12 ppm usually), ^{13}C NMR absorptions occur over a much broader range, 0–220 ppm. The chemical shifts of carbon atoms in ^{13}C NMR depend on the same effects as the chemical shifts of protons in ^{1}H NMR:

- The sp^3 hybridized C atoms of alkyl groups are shielded and absorb upfield.
- Electronegative elements like halogen, nitrogen, and oxygen shift absorptions downfield.
- The sp^2 hybridized C atoms of alkenes and benzene rings absorb downfield.
- Carbonyl carbons are highly deshielded, and absorb farther downfield than other carbon types.

Table C.5 lists common ^{13}C chemical shift values. The ^{13}C NMR spectra of propan-1-ol (CH$_3$CH$_2$CH$_2$OH) and methyl acetate (CH$_3$CO$_2$CH$_3$) in Figure C.14 illustrate these principles.

Table C.5 Common ^{13}C Chemical Shift Values

Type of carbon	Chemical shift (ppm)	Type of carbon	Chemical shift (ppm)
>C<	5–45	C=C	100–140
>C—Z (Z = N, O, X)	30–80	benzene C—	120–150
—C≡C—	65–100	C=O	160–210

Problem C.27 Which of the highlighted carbon atoms in each molecule absorbs farther downfield?

Figure C.14 Representative ^{13}C NMR spectra

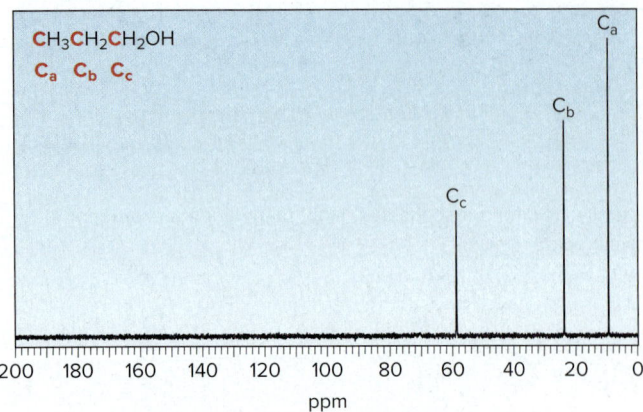
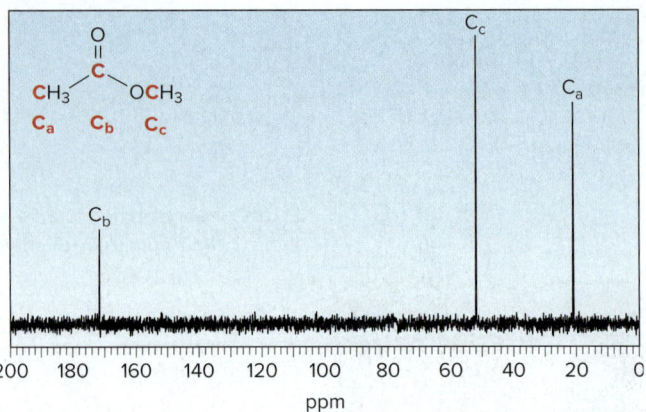

- The three types of C's in propan-1-ol—identified as C_a, C_b, and C_c—give rise to three ^{13}C NMR signals.
- Deshielding increases with increasing proximity to the electronegative O atom, and the absorption shifts downfield; thus, in order of increasing chemical shift: $C_a < C_b < C_c$.

- The three types of C's in methyl acetate—identified as C_a, C_b, and C_c—give rise to three ^{13}C NMR signals.
- **The carbonyl carbon (C_b) is highly deshielded, so it absorbs farthest downfield.**
- C_a, an sp^3 hybridized C that is not bonded to an O atom, is the most shielded, so it absorbs farthest upfield.
- Thus, in order of increasing chemical shift: $C_a < C_c < C_b$.

Problem C.28 Identify the carbon atoms that give rise to each NMR signal.

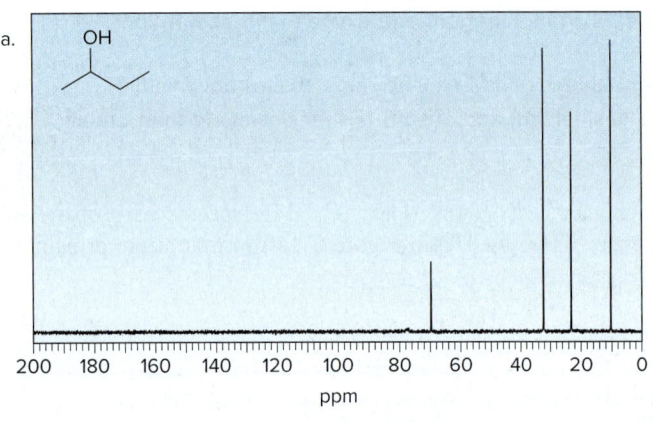
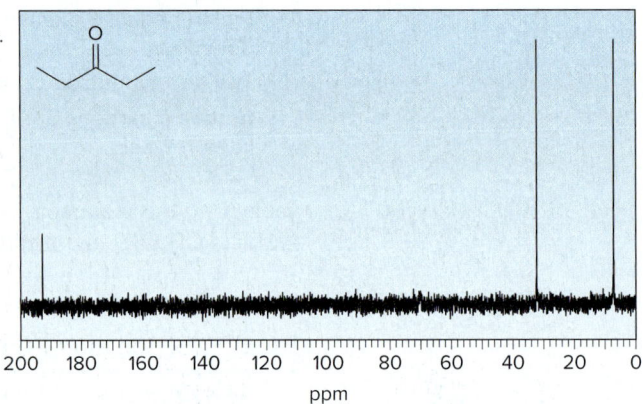

Problem C.29 A compound of molecular formula $C_4H_8O_2$ shows no IR peaks at 3600–3200 or 1700 cm^{-1}. It exhibits one singlet in its ^{1}H NMR spectrum at 3.69 ppm, and one line in its ^{13}C NMR spectrum at 67 ppm. What is the structure of this unknown?

Problem C.30 Draw the structure of a compound of molecular formula C_4H_8O that has a signal in its ^{13}C NMR spectrum at > 160 ppm. Then draw the structure of an isomer of molecular formula C_4H_8O that has all of its ^{13}C NMR signals at < 160 ppm.

C.12 Magnetic Resonance Imaging (MRI)

Magnetic resonance imaging (MRI)—NMR spectroscopy in medicine—is a powerful diagnostic technique (Figure C.15a). The "sample" is the patient, who is placed in a large cavity in a magnetic field, and then irradiated with RF energy. Because RF energy has very low frequency and low energy, the method is safer than X-rays or computed tomography (CT) scans that employ high-frequency, high-energy radiation that is known to damage living cells.

Living tissue contains protons (especially the H atoms in H_2O) in different concentrations and environments. When irradiated with RF energy, these protons are excited to a higher-energy spin state, and then fall back to the lower-energy spin state. These data are analyzed by a computer that generates a plot that delineates tissues of different proton density (Figure C.15b). MRIs can be

Figure C.15
Magnetic resonance imaging

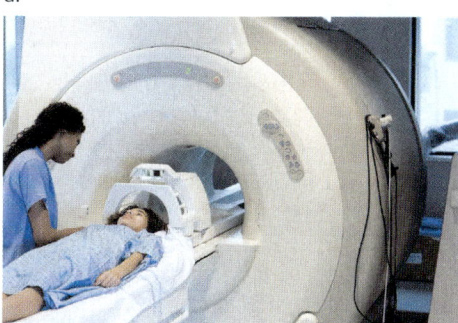
ERproductions Ltd/Blend Images LLC

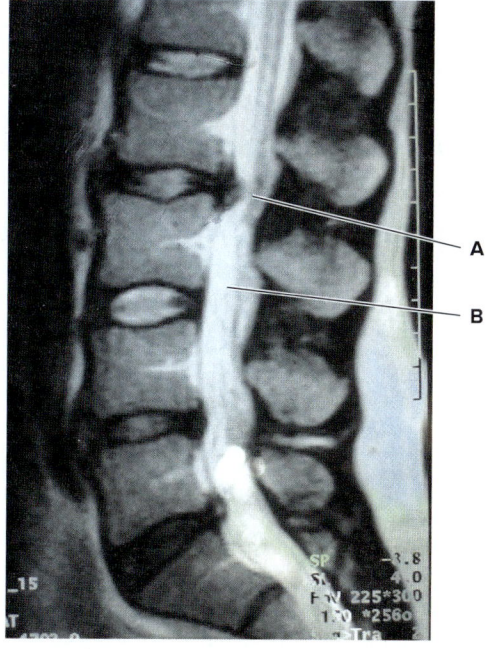
Daniel C. Smith

a. An MRI instrument: An MRI instrument is especially useful for visualizing soft tissue. The 2003 Nobel Prize in Physiology or Medicine was awarded to chemist Paul C. Lauterbur and physicist Sir Peter Mansfield for their contributions in developing magnetic resonance imaging.

b. An MRI image of the lower back: **A** labels spinal cord compression from a herniated disc. **B** labels the spinal cord, which would not be visualized with conventional X-rays.

recorded in any plane. Moreover, because the calcium present in bones is not NMR active, an MRI instrument can "see through" bones such as the skull and visualize the soft tissue underneath.

Spectroscopy C   CHAPTER REVIEW

KEY CONCEPTS

Homotopic, enantiotopic, and diastereotopic protons (C.2C)

1 Two protons are homotopic when replacement of H by **Z** yields the same compound.

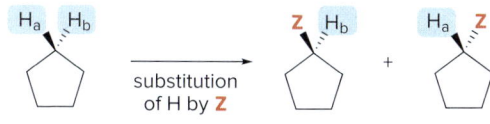

H_a and H_b are **homotopic**.   **identical products**

2 Two protons are enantiotopic when replacement of H by **Z** yields enantiomers.

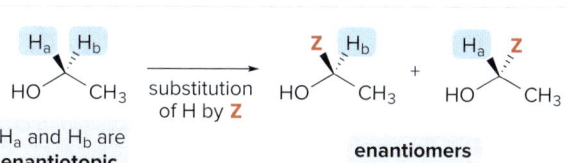

H_a and H_b are **enantiotopic**.   **enantiomers**

3 Two protons are diastereotopic when replacement of H by **Z** yields diastereomers.

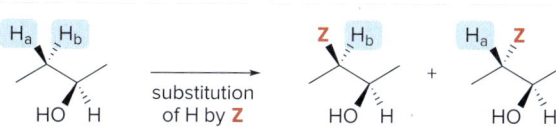

H_a and H_b are **diastereotopic**.   **diastereomers**

Try Problem C.7.

KEY SKILLS

[1] Calculating the chemical shift of an absorption that occurs at 1000 Hz downfield from TMS using a 400 MHz NMR spectrometer (C.1)

1 Use the equation that defines chemical shift in ppm.	**2** Insert the values into the equation.
$$\text{chemical shift (in ppm on the } \delta \text{ scale)} = \frac{\text{observed chemical shift (in Hz) downfield from TMS}}{\nu \text{ of the NMR spectrometer (in MHz)}}$$	$$= \frac{1000 \text{ Hz downfield from TMS}}{400 \text{ MHz operating frequency}} = 2.5 \text{ ppm}$$

See Sample Problem C.1. Try Problems C.35, C.36.

[2] Determining the different types of protons in a compound (C.2A); example: 1,4-dichlorobutane

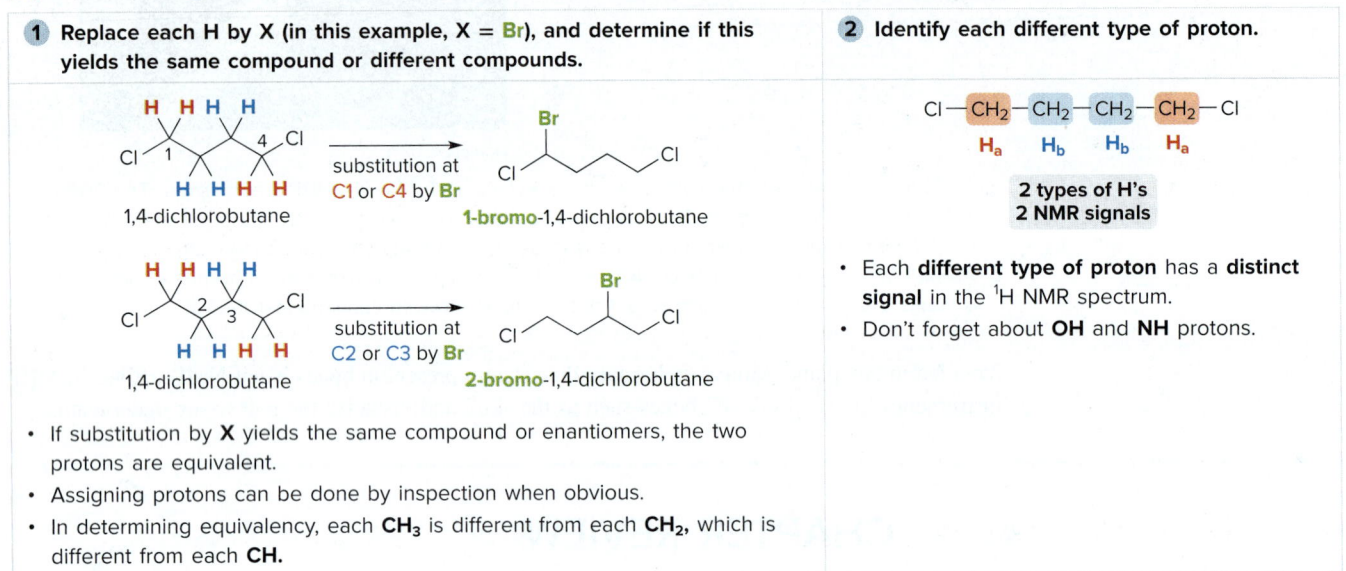

- If substitution by **X** yields the same compound or enantiomers, the two protons are equivalent.
- Assigning protons can be done by inspection when obvious.
- In determining equivalency, each **CH₃** is different from each **CH₂**, which is different from each **CH**.
- Each **different type of proton** has a **distinct signal** in the ¹H NMR spectrum.
- Don't forget about **OH** and **NH** protons.

See Sample Problem C.2, Figure C.2. Try Problems C.31a, C.32a, C33, C.34, C.48a, C.49d.

[3] Determining equivalency in a cycloalkane (C.2B)

1 Draw in all bonds to hydrogen using wedges and dashed wedges for tetrahedral carbons.	**2** Determine if two protons are cis (or trans) to the same groups.

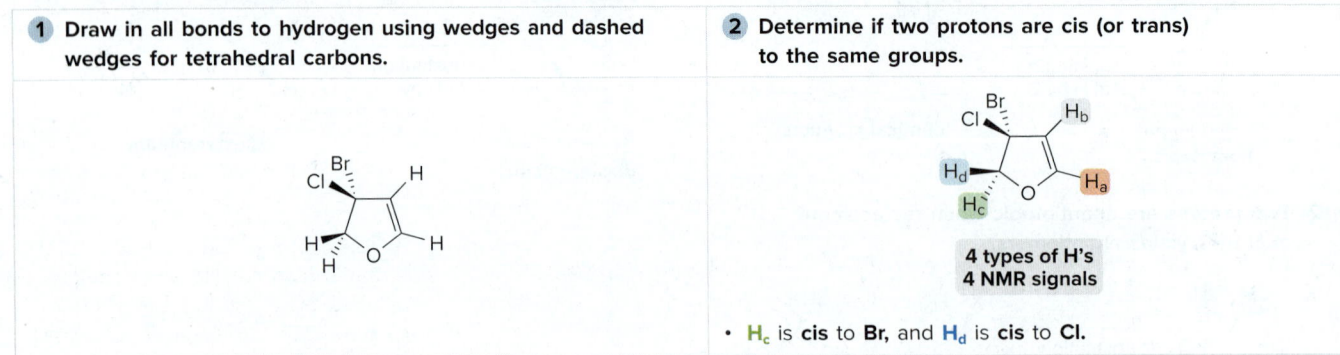

- H_c is **cis** to **Br**, and H_d is **cis** to **Cl**.

See Sample Problem C.3. Try Problems C.33h, i, j; C.48a.

[4] Determining which protons absorb farther downfield; two factors

1 Use the presence of nearby electronegative atoms to determine deshielding effects (C.3A).

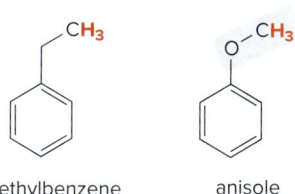

ethylbenzene   anisole

- The CH$_3$ group in anisole is **deshielded** by the **electronegative O atom**.
- Electronegative atoms withdraw electron density, **deshield** a nucleus, and shift an absorption **downfield**.
- **Shielding** shifts an absorption **upfield**.

See Sample Problem C.4.

2 Determine shielding and deshielding effects when protons are bonded to sp^2 and sp hybridized carbons (C.4).

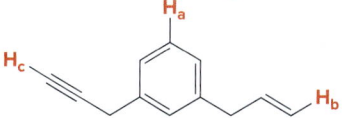

- H_c is **shielded** because it is bonded to an sp hybridized carbon.
- H_b is **deshielded** because it is bonded to an sp^2 hybridized carbon.
- H_a is **highly deshielded** because it is bonded to an sp^2 hybridized carbon on a **benzene ring**.

Answer:
In order of increasing chemical shift, $H_c < H_b < H_a$

See Sample Problem C.5.

Try Problem C.37.

[5] Determining the ^{1}H NMR integration ratio for a compound (C.5); example: CH$_3$CH$_2$OCH$_3$

1 Identify the nonequivalent protons.

CH$_3$—CH$_2$—O—CH$_3$

- three types of protons

2 Count the number of protons in each group.

3 H's, 2 H's, 3 H's

3 Determine the integration ratio.

Answer: 3:2:3

- The area under an NMR signal is proportional to the number of absorbing protons.

Try Problem C.12.

[6] Determining the splitting pattern for a molecule using the $n + 1$ rule (C.6)

1 Identify the nonequivalent protons.

HOCH$_2$CH$_2$—C(=O)—OCH$_3$

4 types of H's

2 Determine if two sets of nonequivalent protons are close enough to split each other's signals.

no splitting with OH and NH protons
singlet → HOCH$_2$CH$_2$—C(=O)—OCH$_3$ ← singlet
↑ ↑
nonequivalent protons on adjacent carbons

- Equivalent protons do not split each other's signals.

3 Apply the $n + 1$ rule.

H_a H_b
HOCH$_2$CH$_2$—C(=O)—OCH$_3$

H_a: two adjacent H's
3 peaks
triplet

H_b: two adjacent H's
3 peaks
triplet

- A set of n **nonequivalent protons** on the same carbon or adjacent carbons **splits** an **NMR signal** into $n + 1$ **peaks**.

See Tables C.3, C.4. Try Problems C.32b, C.38, C.48b, C.49e.

[7] Determining splitting patterns when an absorbing proton has nonequivalent protons on two adjacent carbons (C.7–C.8); three possibilities

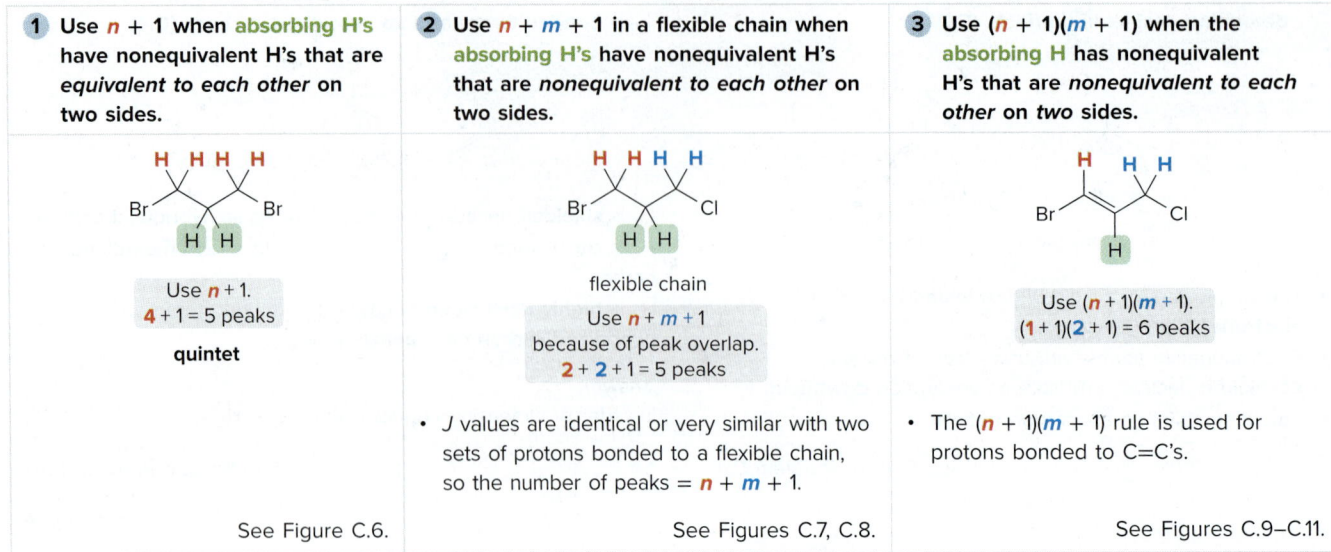

See Sample Problem C.6, Tables C.3, C.4. Try Problem C.38d–j.

[8] Using a molecular formula and ¹H NMR data to determine a structure (C.10); example: $C_4H_{10}O$ with the given ¹H NMR data

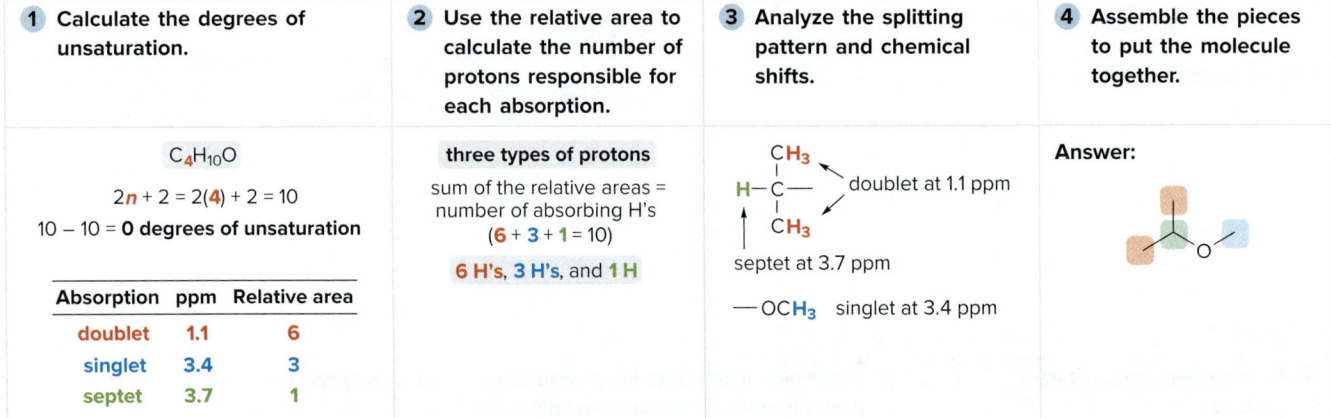

See How To p. 579. Try Problem C.50.

[9] Determining the different types of C atoms in a compound (C.11A)

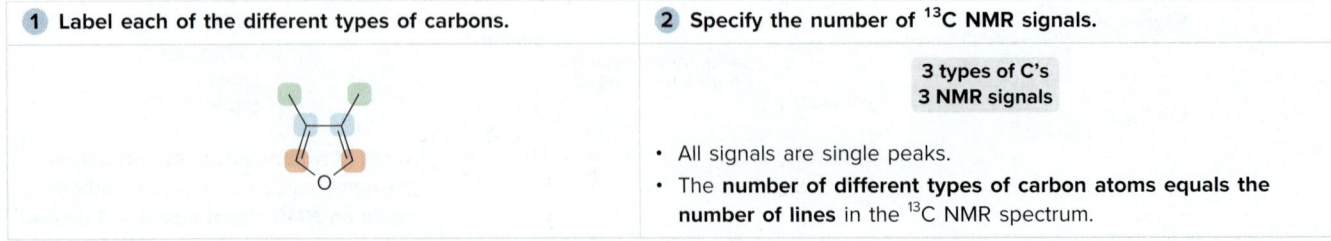

See Sample Problem C.7. Try Problems C.41, C.42.

Problems 589

[10] Using a molecular formula, IR, ¹H NMR, and ¹³C NMR for structure determination (C.10); example: C_3H_5ClO

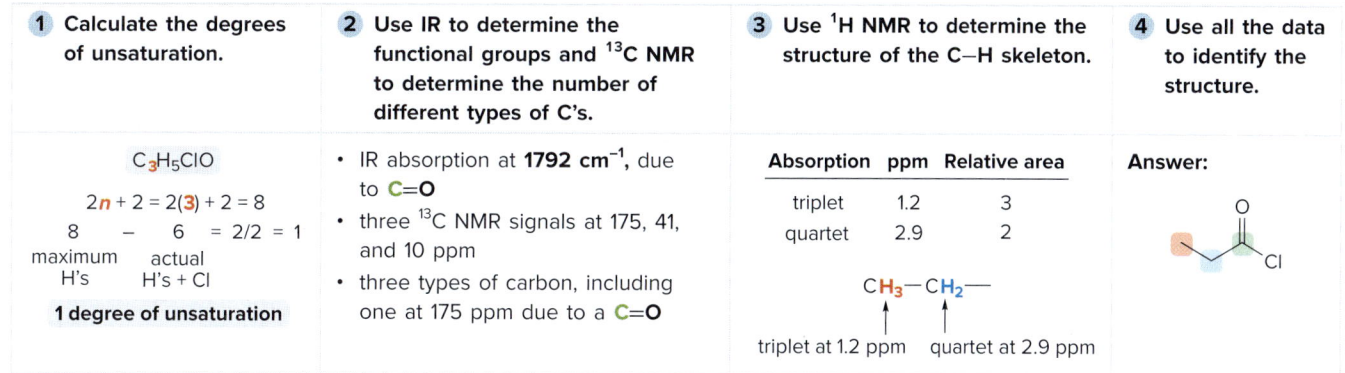

Try Problems C.50–C.65.

PROBLEMS

Problems Using Three-Dimensional Models

C.31 (a) How many ¹H NMR signals does each of the following compounds exhibit? (b) How many ¹³C NMR signals does each compound exhibit?

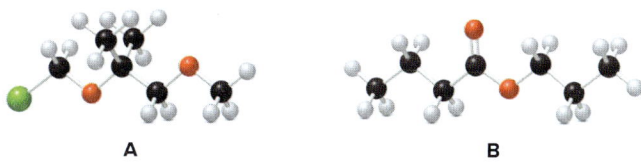

C.32 (a) How many ¹H NMR signals does each compound show? (b) Into how many peaks is each signal split?

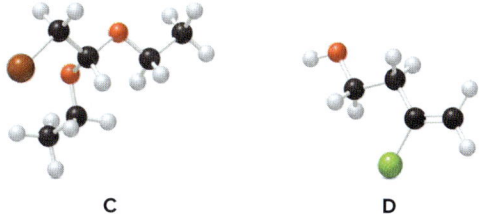

¹H NMR Spectroscopy—Determining Equivalent Protons

C.33 How many different types of protons are present in each compound?

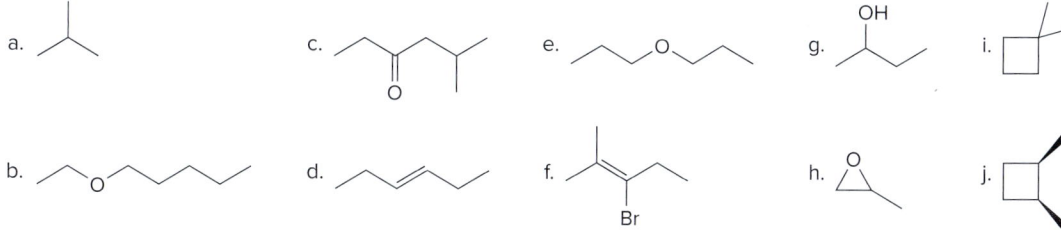

C.34 How many ^{1}H NMR signals does each natural product exhibit?

a.

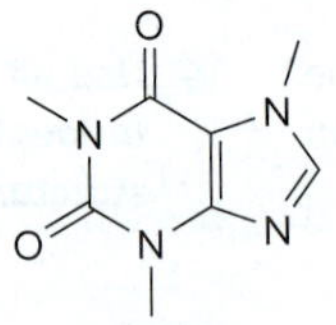

caffeine
(from coffee and tea leaves)

b.

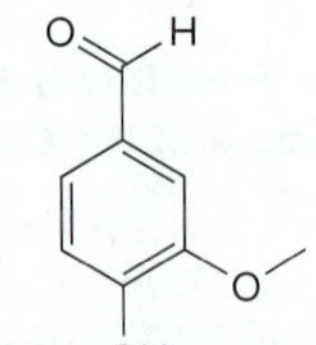

vanillin
(from the vanilla bean)

c.

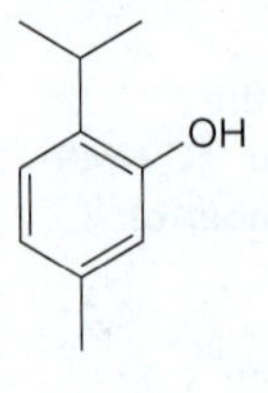

thymol
(from thyme)

d.

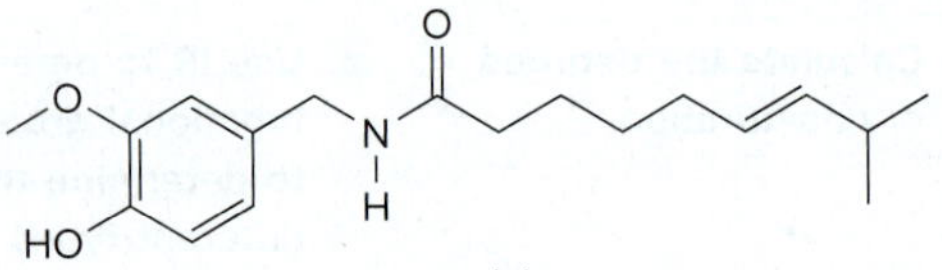

capsaicin
(from hot peppers)

^{1}H NMR—Chemical Shift

C.35 Using a 300 MHz NMR instrument:

a. How many Hz downfield from TMS is a signal at 2.5 ppm?

b. If a signal comes at 1200 Hz downfield from TMS, at what ppm does it occur?

c. If two signals are separated by 2 ppm, how many Hz does this correspond to?

C.36 What effect does increasing the operating frequency of a ^{1}H NMR spectrum have on each value: (a) the chemical shift in δ; (b) the frequency of an absorption in Hz; (c) the magnitude of a coupling constant *J* in Hz?

C.37 Rank the labeled protons in order of increasing chemical shift.

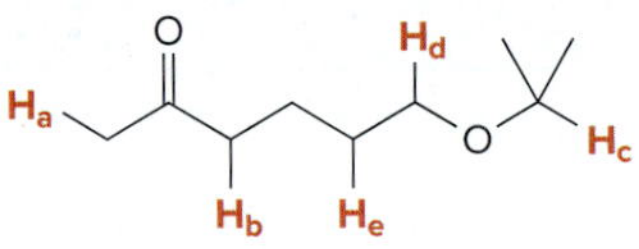

a.

b.

c.

^{1}H NMR—Splitting

C.38 Into how many peaks will the signal for each of the labeled protons be split?

a.

c.

e.

g.

i.

b.

d.

f.

h.

j.

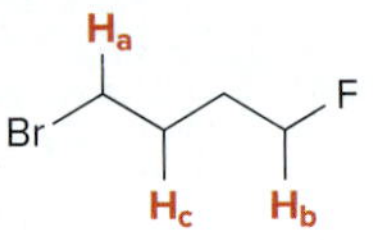

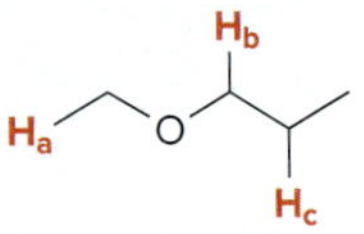

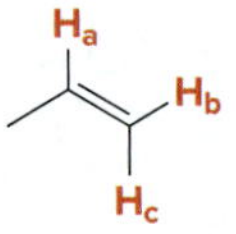

C.39 Label the signals due to H_a, H_b, and H_c in the ^{1}H NMR spectrum of acrylonitrile (CH_2=CHCN). Draw a splitting diagram for the absorption due to the H_a proton.

Hb
Ha
Hc
CN
Jab = 11.8 Hz
Jbc = 0.9 Hz
Jac = 18 Hz
6.6
6.2
5.7
chemical shift (ppm)

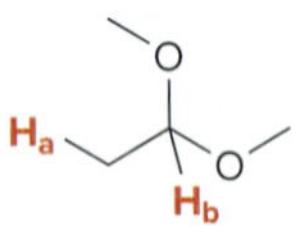

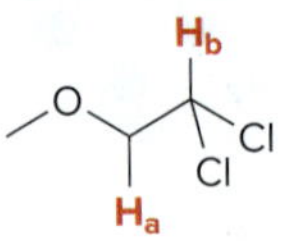

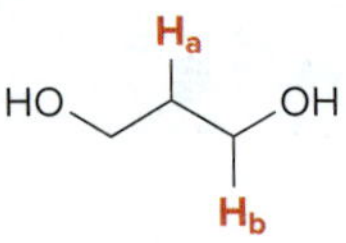

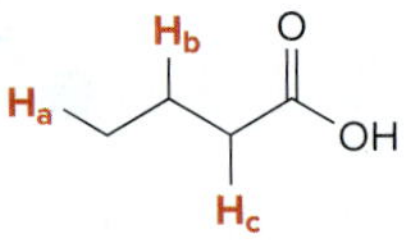

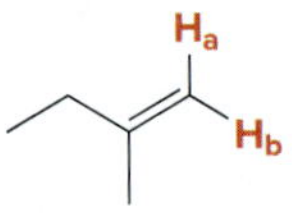

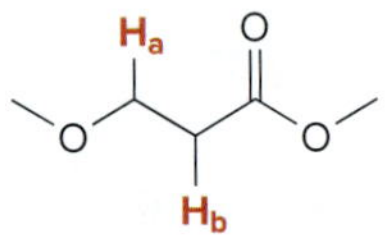

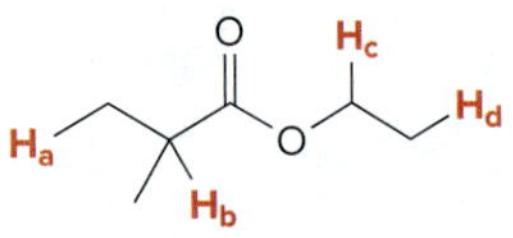

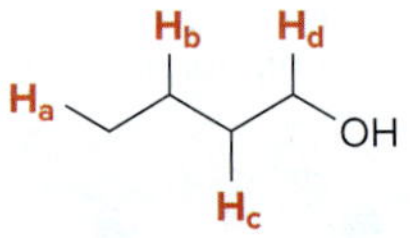

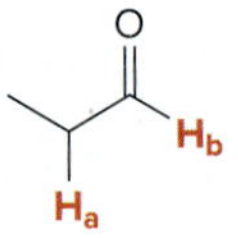

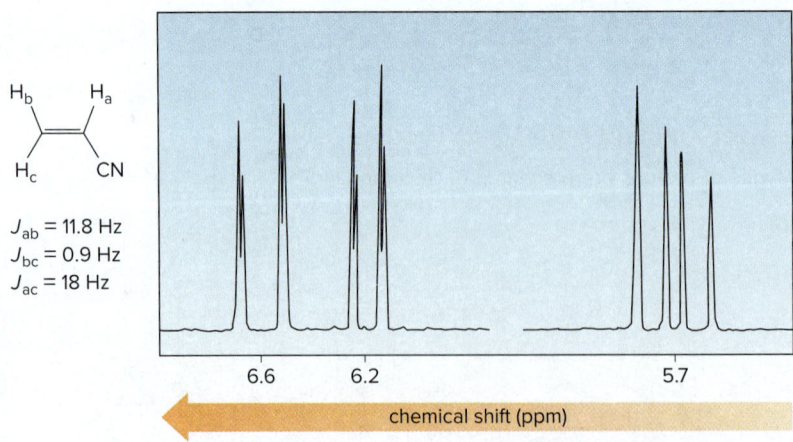

C.40 Draw a splitting diagram for H_b in compound **X** given the following coupling constants: (a) $J_{ab} \gg J_{bc}$; (b) $J_{ab} = J_{bc}$. Clearly indicate how many peaks are visible in the H_b signal in each circumstance.

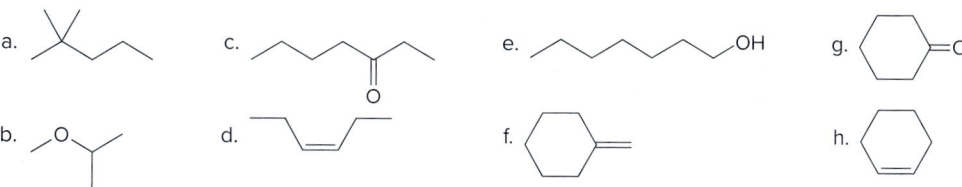

13C NMR

C.41 Draw the four constitutional isomers having molecular formula C_4H_9Br and indicate how many different kinds of carbon atoms each has.

C.42 Explain why the carbonyl carbon of an aldehyde or ketone absorbs farther downfield than the carbonyl carbon of an ester in a ^{13}C NMR spectrum.

C.43 How many ^{13}C NMR signals does each compound exhibit?

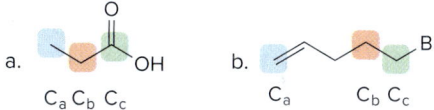

C.44 Rank the highlighted carbon atoms in each compound in order of increasing chemical shift.

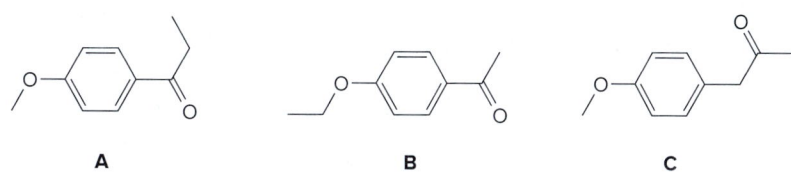

C.45 Identify the carbon atoms that give rise to the signals in the ^{13}C NMR spectrum of each compound.
   a. $CH_3CH_2CH_2CH_2OH$; ^{13}C NMR: 14, 19, 35, and 62 ppm
   b. $(CH_3)_2CHCHO$; ^{13}C NMR: 16, 41, and 205 ppm
   c. $CH_2=CHCH(OH)CH_3$; ^{13}C NMR: 23, 69, 113, and 143 ppm

Identifying Isomers Using NMR Spectroscopy

C.46 How could ^1H NMR spectroscopy be used to distinguish among isomers **A**, **B**, and **C**?

C.47 How could ^1H NMR spectroscopy be used to distinguish between each pair of compounds?

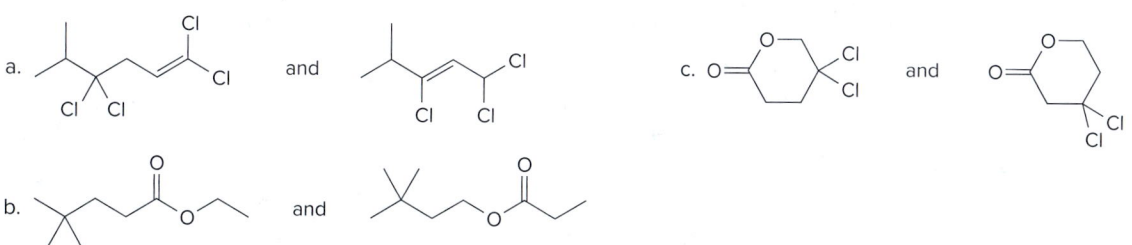

Combined Spectroscopy Problems

Additional spectroscopy problems are located at the end of Chapters 9–17, and 19–22.

C.48 Answer the following questions for compounds **L**, **M**, and **N** drawn below.

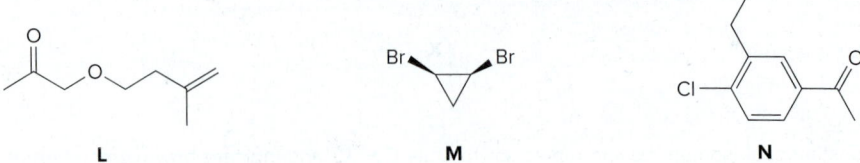

a. How many signals are expected in the ^{1}H NMR spectrum?
b. Into how many peaks is each signal in the ^{1}H NMR spectrum split?
c. How many lines are expected in the ^{13}C NMR spectrum?

C.49 Answer the following questions about each of the hydroxy ketones: 1-hydroxybutan-2-one (**A**) and 4-hydroxybutan-2-one (**B**).

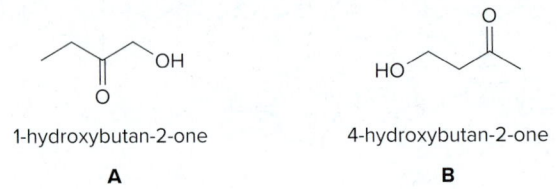

a. What is the molecular ion in the mass spectrum?
b. What IR absorptions are present in the functional group region?
c. How many lines are observed in the ^{13}C NMR spectrum?
d. How many signals are observed in the ^{1}H NMR spectrum?
e. Give the splitting observed for each type of proton as well as its approximate chemical shift.

C.50 Propose a structure consistent with each set of spectral data:

a. $C_4H_8Br_2$: IR peak at 3000–2850 cm^{-1}; NMR (ppm):
   1.87 (singlet, 6 H)
   3.86 (singlet, 2 H)

b. $C_3H_6Br_2$: IR peak at 3000–2850 cm^{-1}; NMR (ppm):
   2.4 (quintet)
   3.5 (triplet)

c. $C_5H_{10}O_2$: IR peak at 1740 cm^{-1}; NMR (ppm):
   1.15 (triplet, 3 H)   2.30 (quartet, 2 H)
   1.25 (triplet, 3 H)   4.72 (quartet, 2 H)

d. C_3H_6O: IR peak at 1730 cm^{-1}; NMR (ppm):
   1.11 (triplet)
   2.46 (multiplet)
   9.79 (triplet)

C.51 Reaction of $C_6H_5CH_2CH_2OH$ with CH_3COCl affords compound **W**, which has molecular formula $C_{10}H_{12}O_2$. **W** shows prominent IR absorptions at 3088–2897, 1740, and 1606 cm^{-1}. **W** exhibits the following signals in its ^{1}H NMR spectrum: 2.02 (singlet), 2.91 (triplet), 4.25 (triplet), and 7.20–7.35 (multiplet) ppm. What is the structure of **W**? We will learn about this reaction in Chapter 16.

C.52 Treatment of 2-methylpropanenitrile [$(CH_3)_2CHCN$] with $CH_3CH_2CH_2MgBr$, followed by aqueous acid, affords compound **V**, which has molecular formula $C_7H_{14}O$. **V** has a strong absorption in its IR spectrum at 1713 cm^{-1}, and gives the following ^{1}H NMR data: 0.91 (triplet, 3 H), 1.09 (doublet, 6 H), 1.6 (multiplet, 2 H), 2.43 (triplet, 2 H), and 2.60 (septet, 1 H) ppm. What is the structure of **V**? We will learn about this reaction in Chapter 15.

C.53 Compound **C** has a molecular ion in its mass spectrum at 146 and a prominent absorption in its IR spectrum at 1762 cm^{-1}. **C** shows the following ^{1}H NMR spectral data: 1.47 (doublet, 3 H), 2.07 (singlet, 6 H), and 6.84 (quartet, 1 H) ppm. What is the structure of **C**?

C.54 Treatment of compound **D** with LiAlH₄ followed by H₂O forms compound **E**. **D** shows a molecular ion in its mass spectrum at $m/z = 71$ and IR absorptions at 3600–3200 and 2263 cm^{-1}. **E** shows a molecular ion in its mass spectrum at $m/z = 75$ and IR absorptions at 3636 and 3600–3200 cm^{-1}. Propose structures for **D** and **E** from these data and the given ¹H NMR spectra.

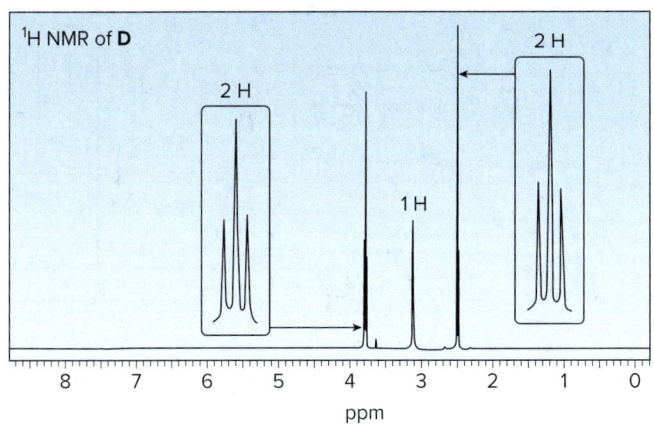
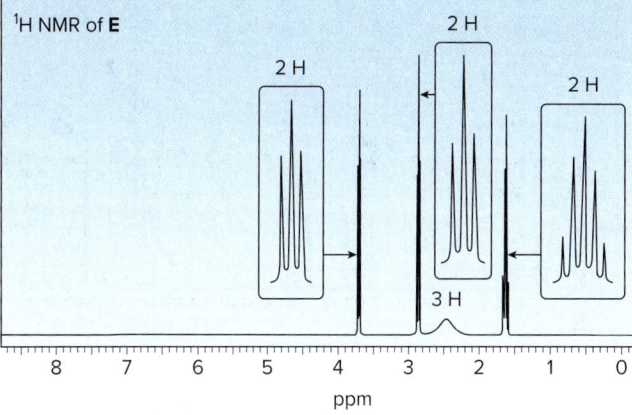

C.55 Identify the structures of isomers **E** and **F** (molecular formula C₄H₈O₂). Relative areas are given above each signal.

a. **Compound E:** IR absorption at 1743 cm^{-1}

b. **Compound F:** IR absorption at 1730 cm^{-1}

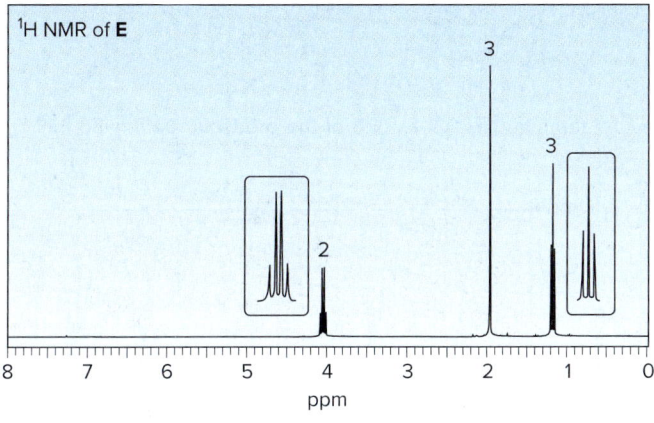
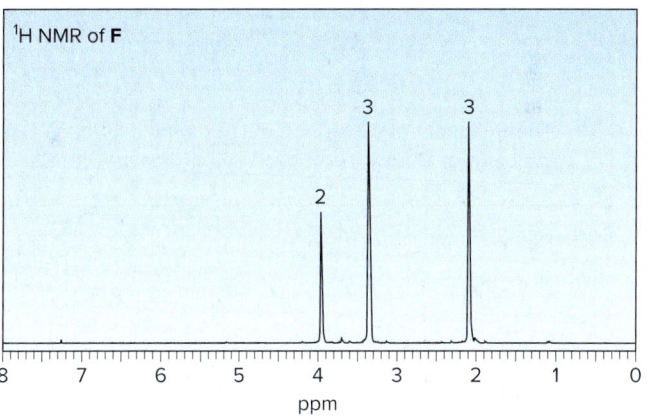

C.56 Identify the structures of isomers **H** and **I** (molecular formula C₈H₁₁N).

a. **Compound H:** IR absorptions at 3365, 3284, 3026, 2932, 1603, and 1497 cm^{-1}

b. **Compound I:** IR absorptions at 3367, 3286, 3027, 2962, 1604, and 1492 cm^{-1}

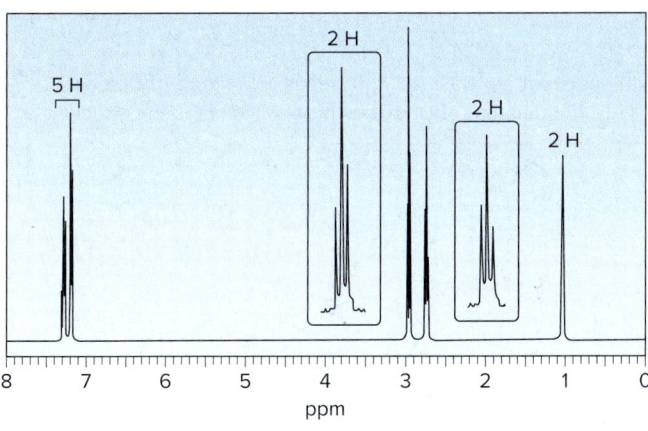
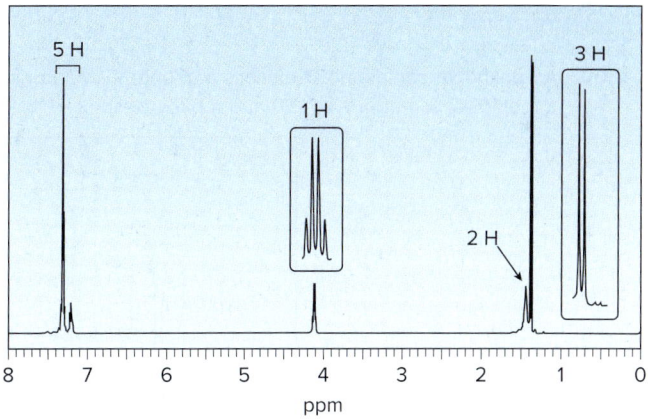

C.57 Propose a structure consistent with each set of data.

a. $C_9H_{10}O_2$: IR absorption at 1718 cm^{-1}

b. C_9H_{12}: IR absorption at 2850–3150 cm^{-1}

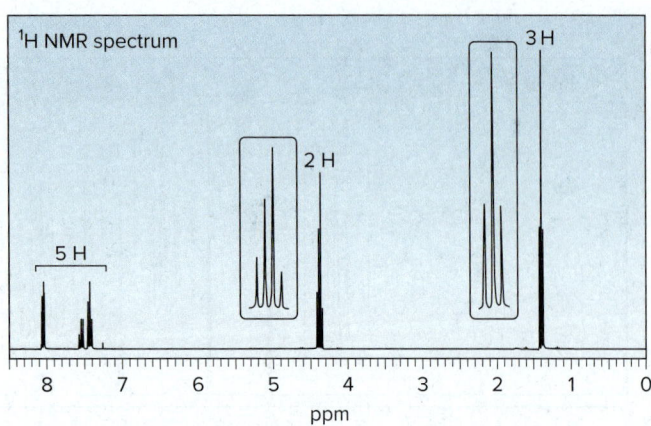
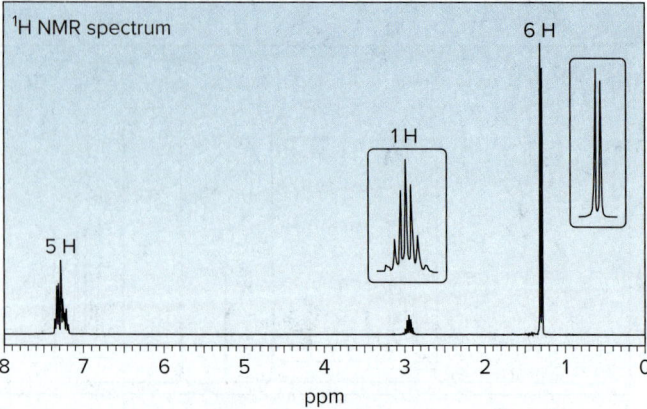

C.58 Reaction of aldehyde **D** with amino alcohol **E** in the presence of NaH forms **F** (molecular formula $C_{11}H_{15}NO_2$). **F** absorbs at 1730 cm^{-1} in its IR spectrum. **F** also shows eight lines in its ^{13}C NMR spectrum, and gives the following ^{1}H NMR spectrum: 2.32 (singlet, 6 H), 3.05 (triplet, 2 H), 4.20 (triplet, 2 H), 6.97 (doublet, 2 H), 7.82 (doublet, 2 H), and 9.97 (singlet, 1 H) ppm. Propose a structure for **F**. We will learn about this reaction in Chapter 20.

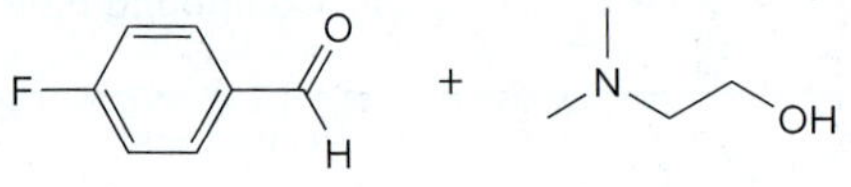

C.59 The treatment of $(CH_3)_2C=CHCH_2Br$ with H_2O forms **B** (molecular formula $C_5H_{10}O$) as one of the products. Determine the structure of **B** from its ^{1}H NMR and IR spectra.

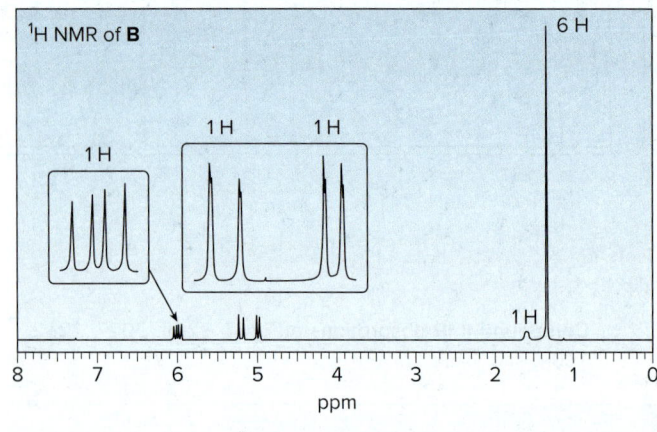
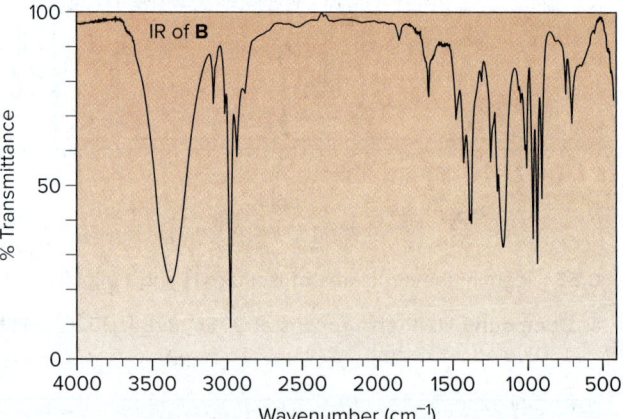

C.60 An unknown compound **D** exhibits a strong absorption in its IR spectrum at 1692 cm^{-1}. The mass spectrum of **D** shows a molecular ion at $m/z = 150$ and a base peak at 121. The ^{1}H NMR spectrum of **D** is shown below. What is the structure of **D**?

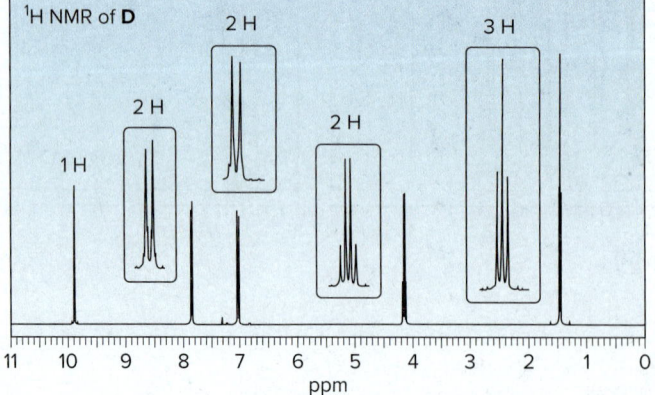

C.61 In the presence of a small amount of acid, a solution of acetaldehyde (CH₃CHO) in methanol (CH₃OH) was allowed to stand and a new compound **L** was formed. **L** has a molecular ion in its mass spectrum at 90 and IR absorptions at 2992 and 2941 cm⁻¹. **L** shows three signals in its ¹³C NMR at 19, 52, and 101 ppm. The ¹H NMR spectrum of **L** is given below. What is the structure of **L**?

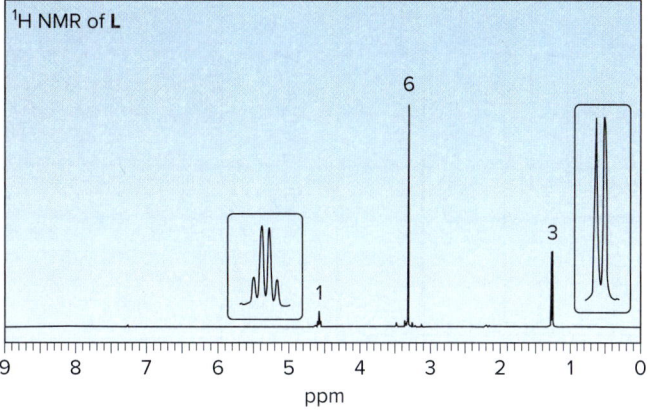

C.62 Compound **O** has molecular formula C₁₀H₁₂O and shows an IR absorption at 1687 cm⁻¹. The ¹H NMR spectrum of **O** is given below. What is the structure of **O**?

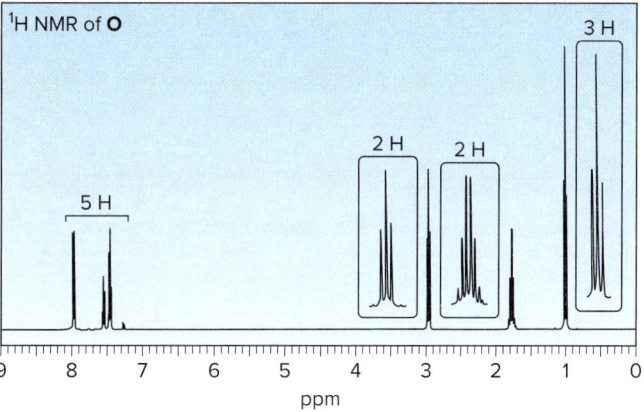

C.63 Compound **P** has molecular formula C₅H₉ClO₂. Deduce the structure of **P** from its ¹H and ¹³C NMR spectra.

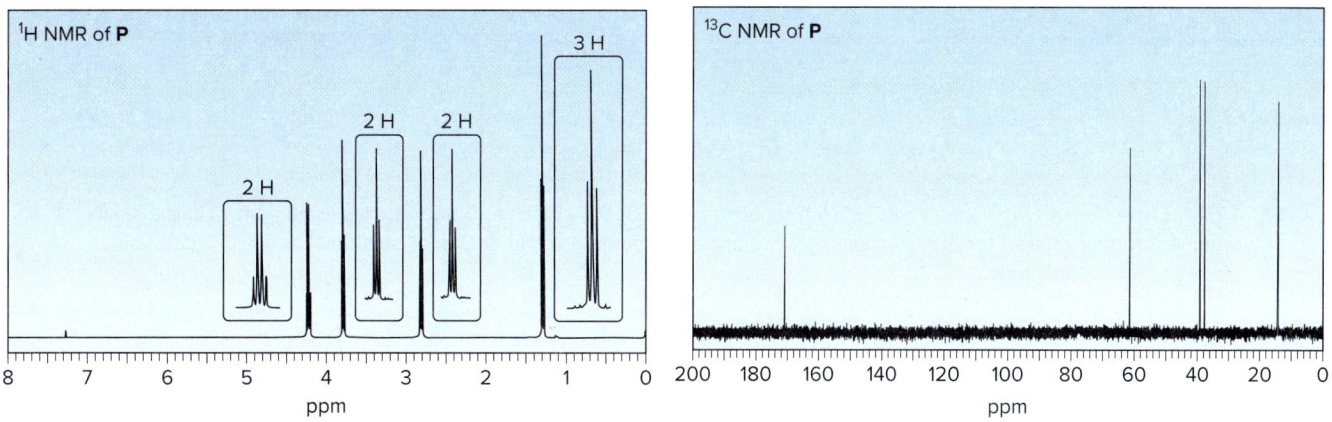

C.64 Treatment of butan-2-one ($CH_3COCH_2CH_3$) with strong base followed by CH_3I forms a compound **Q**, which gives a molecular ion in its mass spectrum at 86. The IR (> 1500 cm^{-1} only) and ^1H NMR spectra of **Q** are given below. What is the structure of **Q**?

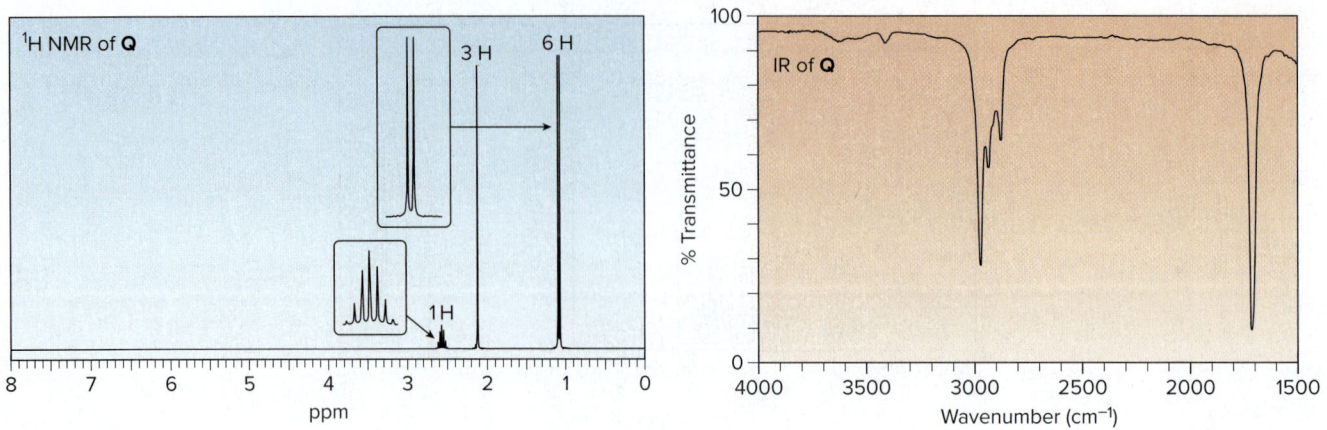

C.65 Compound **X** (molecular formula $C_6H_{12}O_2$) gives a strong peak in its IR spectrum at 1740 cm^{-1}. The ^1H NMR spectrum of **X** shows only two singlets, including one at 3.5 ppm. The ^{13}C NMR spectrum is given below. Propose a structure for **X**.

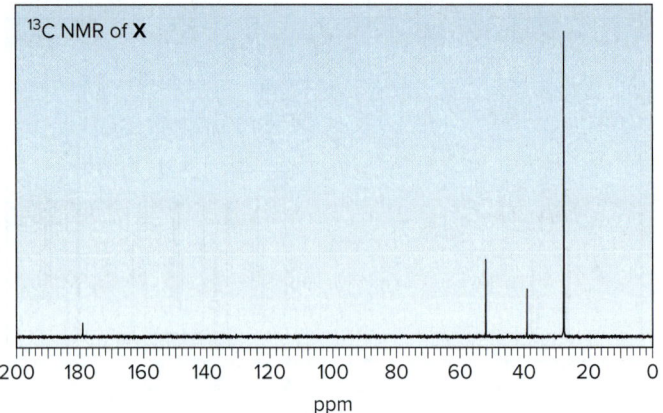

Challenge Problems

C.66 The ^1H NMR spectrum of *N,N*-dimethylformamide shows three singlets at 2.9, 3.0, and 8.0 ppm. Explain why the two CH_3 groups are not equivalent to each other, thus giving rise to two NMR signals.

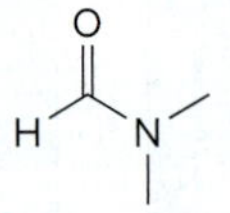

N,N-dimethylformamide

C.67 18-Annulene shows two signals in its ^1H NMR spectrum, one at 8.9 (12 H) and one at −1.8 (6 H) ppm. Using a similar argument to that offered for the chemical shift of benzene protons, explain why both shielded and deshielded values are observed for 18-annulene.

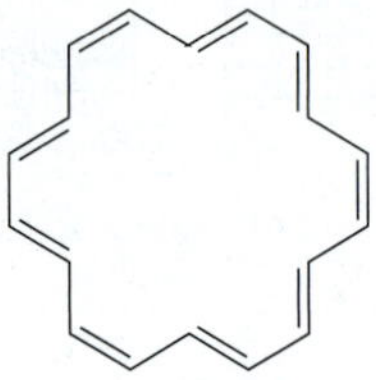

18-annulene

C.68 Explain why the ^{13}C NMR spectrum of 3-methylbutan-2-ol shows five signals.

C.69 Because ^{31}P has an odd mass number, ^{31}P nuclei absorb in the NMR and, in many ways, these nuclei behave similarly to protons in NMR spectroscopy. With this in mind, explain why the ^1H NMR spectrum of methyl dimethylphosphonate, $CH_3PO(OCH_3)_2$, consists of two doublets at 1.5 and 3.7 ppm.

C.70 Cyclohex-2-enone has two protons on its carbon–carbon double bond (labeled H_a and H_b) and two protons on the carbon adjacent to the double bond (labeled H_c). (a) If J_{ab} = 11 Hz and J_{bc} = 4 Hz, sketch the splitting pattern observed for each proton on the sp^2 hybridized carbons. (b) Despite the fact that H_a is located adjacent to an electron-withdrawing C=O, its absorption occurs upfield from the signal due to H_b (6.0 vs. 7.0 ppm). Offer an explanation.

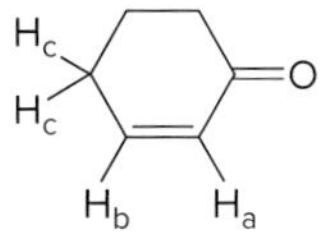

cyclohex-2-enone

13 Introduction to Carbonyl Chemistry; Organometallic Reagents; Oxidation and Reduction

AS Food studio/Shutterstock

- **13.1** Introduction
- **13.2** General reactions of carbonyl compounds
- **13.3** A preview of oxidation and reduction
- **13.4** Reduction of aldehydes and ketones
- **13.5** The stereochemistry of carbonyl reduction
- **13.6** Enantioselective biological reduction
- **13.7** Reduction of carboxylic acids and their derivatives
- **13.8** Oxidation of aldehydes
- **13.9** Organometallic reagents
- **13.10** Reaction of organometallic reagents with aldehydes and ketones
- **13.11** Retrosynthetic analysis of Grignard products
- **13.12** Protecting groups
- **13.13** Reaction of organometallic reagents with carboxylic acid derivatives
- **13.14** Reaction of organometallic reagents with other compounds
- **13.15** α,β-Unsaturated carbonyl compounds
- **13.16** Summary—The reactions of organometallic reagents
- **13.17** Synthesis

Resiniferatoxin, obtained from the flowering cactus *Euphorbia resinifera,* is a compound that produces the same hot, numbing sensation in the mouth that the capsaicin in chili peppers triggers, but it is 1000 times more potent. Like capsaicin, resiniferatoxin desensitizes neurons to pain, so it has potential as an analgesic for treating pain and inflammation. In fact, a thirteenth-century manuscript illustrates that extracts of *Euphorbia resinifera* were used for pain management over 1000 years ago. Although its complex structure was not elucidated until 1975, resiniferatoxin has now been synthesized in the laboratory by a multistep method that utilizes some of the key reactions presented in Chapter 13.

Why Study...
Carbonyl Compounds and Their Reactions?

Chapters 13 through 18 of this text discuss carbonyl compounds—aldehydes, ketones, acid halides, esters, amides, and carboxylic acids. **The carbonyl group is perhaps the most important functional group in organic chemistry,** because its electron-deficient carbon and easily broken π bond make it susceptible to a wide variety of useful reactions.

We begin by examining the similarities and differences between two broad classes of carbonyl compounds. We will then spend the remainder of Chapter 13 on reactions that are especially important in organic synthesis. Chapters 14 and 16 present specific reactions that occur at the carbonyl carbon, and Chapters 17 and 18 concentrate on reactions occurring at the carbon bonded to the carbonyl group. Chapter 15 covers carboxylic acids, which can react at both their OH and C=O groups, and nitriles (RCN), which undergo reactions similar to those of carbonyl compounds.

Although Chapter 13 is "jam-packed" with reactions, most of them follow one of two general pathways, so they can be classified in a well-organized fashion, provided you remember a few basic principles. Keep in mind these fundamental themes about reactions:

- **Nucleophiles attack electrophiles.**
- **π Bonds are easily broken.**
- **Bonds to good leaving groups are easily cleaved.**

13.1 Introduction

Two broad classes of compounds contain a *carbonyl group:*

carbonyl group

[1] Compounds that have only carbon and hydrogen atoms bonded to the carbonyl group

- An **aldehyde** has at least one H atom bonded to the carbonyl group.
- A **ketone** has two alkyl groups bonded to the carbonyl group.

[2] Compounds that contain an electronegative atom bonded to the carbonyl group

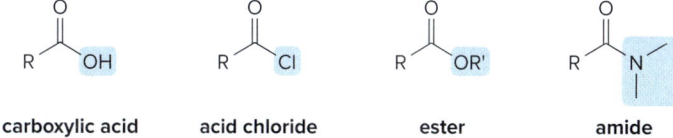

These include **carboxylic acids, acid chlorides, esters,** and **amides,** as well as other similar compounds discussed in Chapter 16. Each of these compounds contains an atom (Cl, O, or N) more electronegative than carbon, capable of acting as a **leaving group.** Acid chlorides, esters, and amides are often called **carboxylic acid derivatives,** because they can be synthesized from carboxylic acids (Chapter 16). Each compound contains an acyl group (RCO—), so they are also called **acyl derivatives.**

- The presence or absence of a leaving group on the carbonyl carbon determines the type of reactions these compounds undergo (Section 13.2).

The carbonyl carbon atom is sp^2 **hybridized** and **trigonal** planar, and all bond angles are ~120°. The double bond of a carbonyl group consists of one σ bond and one π bond. The π bond is formed by the overlap of two *p* orbitals, and extends above and below the plane. In these features the carbonyl group resembles the trigonal planar, sp^2 hybridized carbons of a C—C double bond.

The aldehyde α-sinensal (Problem 13.1) is the major compound responsible for the orange-like odor of mandarin oil, obtained from the mandarin tree in southern China. *Carr Botanical Consultation*

In one important way, though, a C=O and a C=C are very different. **The electronegative oxygen atom in the carbonyl group means that the bond is polarized, making the carbonyl carbon electron deficient.** Using a resonance description, the carbonyl group is represented by two resonance structures, with a charge-separated resonance structure a minor contributor to the hybrid.

the major contributor to the hybrid ⟷ a minor contributor to the hybrid   hybrid polarized carbonyl

---

Problem 13.1   α-sinensal

a. What orbitals are used to form the indicated bonds in α-sinensal?

b. In what type of orbitals do the lone pairs on O reside?

13.2 General Reactions of Carbonyl Compounds

With what types of reagents should a carbonyl group react? The electronegative oxygen makes the carbonyl carbon **electrophilic**, and because it is trigonal planar, a carbonyl carbon is **uncrowded.** Moreover, a carbonyl group has an **easily broken π bond.**

π bond → ← electrophilic carbon
uncrowded
sp^2 hybridized carbon

As a result, **carbonyl compounds react with nucleophiles.** The outcome of nucleophilic attack, however, depends on the identity of the carbonyl starting material.

- Aldehydes and ketones undergo nucleophilic *addition*.

R′ = H or alkyl   [1] :Nu⁻ [2] H—OH   H and Nu are added.

nucleophilic addition

- Carbonyl compounds that contain leaving groups undergo nucleophilic *substitution*.

$$\underset{R}{\overset{O}{\|}}C-Z \xrightarrow{:Nu^-} \underset{R}{\overset{O}{\|}}C-Nu \quad \text{Nu replaces Z.}$$

nucleophilic substitution

Z = OH, Cl, OR, NH$_2$

Let's examine each of these general reactions individually.

13.2A Nucleophilic Addition to Aldehydes and Ketones

Aldehydes and ketones react with nucleophiles to form addition products by the two-step process shown in Mechanism 13.1: **nucleophilic attack** followed by **protonation.**

Mechanism 13.1 Nucleophilic Addition—A Two-Step Process

$$\underset{R}{\overset{:\ddot{O}:}{\|}}C-R' \;+\; :Nu^- \xrightarrow{1} \underset{R'}{\overset{:\ddot{O}:^-}{\underset{|}{R-C-Nu}}} \xrightarrow[H-\ddot{O}H]{2} \underset{R'}{\overset{:\ddot{O}H}{\underset{|}{R-C-Nu}}} \;+\; :\ddot{O}H^-$$

① **The nucleophile attacks the electrophilic carbonyl.** The π bond is broken, moving an electron pair out on oxygen and forming an sp^3 hybridized carbon.

② Protonation of the negatively charged oxygen by H$_2$O forms the **addition product.**

More examples of nucleophilic addition to aldehydes and ketones are discussed in Chapter 14.

The net result is that the π bond is broken, two new σ bonds are formed, and the elements of H and Nu are *added* across the π bond. Nucleophilic addition with two different nucleophiles—**hydride (H:$^-$)** and **carbanions (R:$^-$)**—is discussed in Chapter 13.

Aldehydes are more reactive than ketones toward nucleophilic attack for both steric and electronic reasons.

$$\underset{R \quad H}{\overset{\delta-\;:O:}{\overset{\|}{}\delta+}}C \qquad\qquad \underset{R \quad R'}{\overset{\delta-\;:O:}{\overset{\|}{}\delta+}}C$$

aldehyde   **ketone**

- less crowded
- less stable
- more reactive

- more crowded
- more stable
- less reactive

- The two R groups bonded to the ketone carbonyl group make it *more crowded,* so nucleophilic attack is more difficult.
- The two electron-donor R groups stabilize the partial charge on the carbonyl carbon of a ketone, making it *more stable* and less reactive.

13.2B Nucleophilic Substitution of RCOZ (Z = Leaving Group)

Carbonyl compounds with leaving groups react with nucleophiles to form substitution products by the two-step process shown in Mechanism 13.2: **nucleophilic attack,** followed by **loss of the leaving group.**

Mechanism 13.2 Nucleophilic Substitution—A Two-Step Process

[Mechanism diagram: RCOZ + :Nu⁻ → tetrahedral intermediate → RCONu + :Z⁻]

Z = OH, Cl, OR′, NH₂

① **The nucleophile attacks the electrophilic carbonyl.** The π bond is broken, moving an electron pair out on oxygen and forming an sp^3 hybridized carbon.

② An electron pair on oxygen re-forms the π bond and **Z comes off as a leaving group** with the electron pair in the C—Z bond.

The net result is that Nu replaces Z—a nucleophilic substitution reaction. This reaction is often called **nucleophilic acyl substitution** to distinguish it from the nucleophilic substitution reactions at sp^3 hybridized carbons discussed in Chapter 7. Nucleophilic substitution with two different nucleophiles—**hydride (H:⁻)** and **carbanions (R:⁻)**—is discussed in Chapter 13. Other nucleophiles are examined in Chapter 16.

Carboxylic acid derivatives differ greatly in their reactivity toward nucleophiles. The order in which they react parallels the leaving group ability of the group Z bonded to the carbonyl carbon.

- **The *better* the leaving group Z, the *more reactive* RCOZ is in nucleophilic acyl substitution.**

Recall from Section 7.7 that the *weaker* the base, the *better* the leaving group.

Thus, the following trends result:

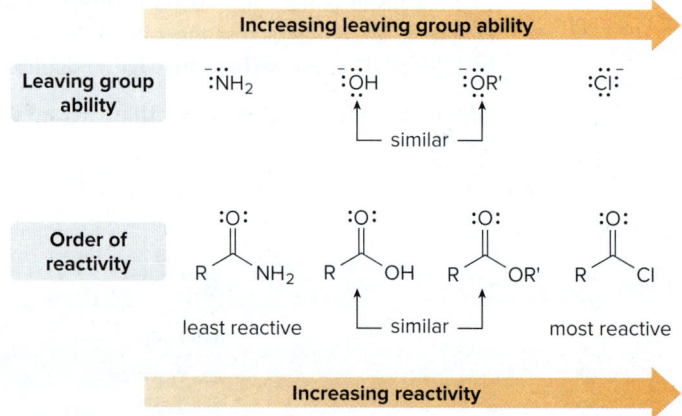

- Acid chlorides (RCOCl), which have the best leaving group (Cl⁻), are the most reactive carboxylic acid derivatives, and amides (RCONH₂), which have the worst leaving group (⁻NH₂), are the least reactive.
- Carboxylic acids (RCOOH) and esters (RCOOR′), which have leaving groups of similar basicity (⁻OH and ⁻OR′), fall in the middle.

Nucleophilic addition and nucleophilic acyl substitution involve the *same* first step—**nucleophilic attack on the electrophilic carbonyl group** to form a tetrahedral intermediate. The difference between them is what then happens to this intermediate. **Aldehydes and ketones cannot undergo substitution because they have no leaving group** bonded to the newly formed sp^3 hybridized carbon. Nucleophilic substitution with an aldehyde, for example, would form H:⁻, an extremely strong base and therefore a very poor (and highly unlikely) leaving group.

An aldehyde does *not* undergo nucleophilic substitution...

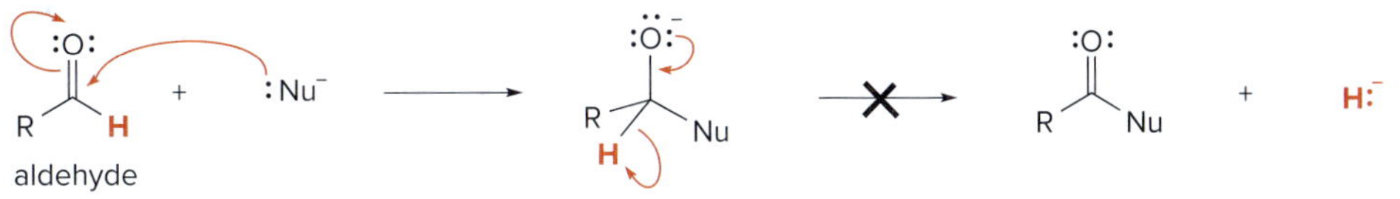

...because H:⁻ is a very *poor* leaving group.

Problem 13.2 Which carbonyl groups in the anticancer drug Taxol (Section 5.5) will undergo nucleophilic addition, and which will undergo nucleophilic substitution?

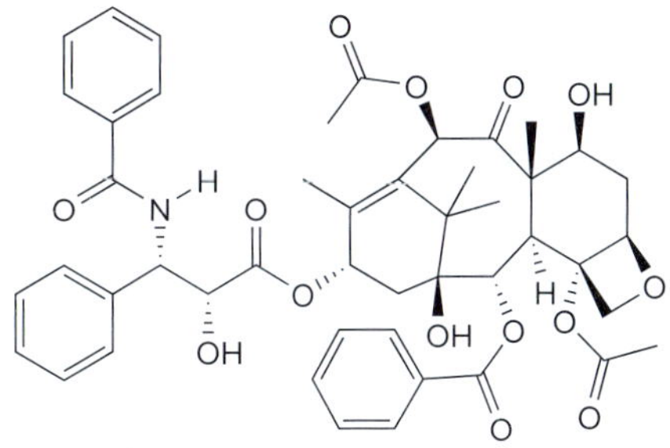

Taxol

Problem 13.3 Rank the compounds in each group in order of increasing reactivity toward nucleophilic attack.

a.

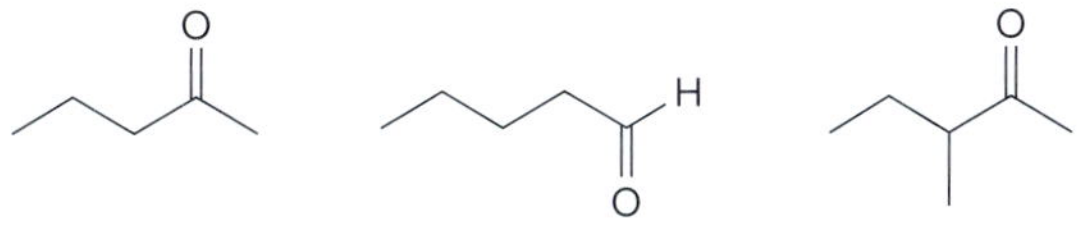

b.

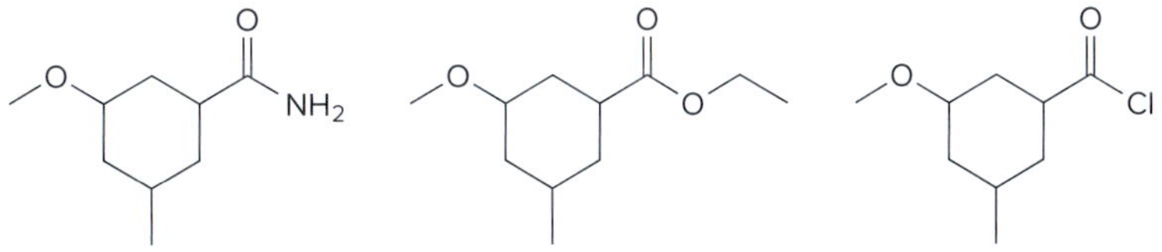

To show how these general principles of nucleophilic substitution and addition apply to carbonyl compounds, we are going to discuss oxidation and reduction reactions, and reactions with organometallic reagents—compounds that contain carbon–metal bonds. We begin with reduction to build on what you learned previously in Chapter 11.

13.3 A Preview of Oxidation and Reduction

Recall the definitions of oxidation and reduction presented in Section 11.1:

- Oxidation results in an *increase* in the number of C–Z bonds (usually C–O bonds) or a *decrease* in the number of C–H bonds.
- Reduction results in a *decrease* in the number of C–Z bonds (usually C–O bonds) or an *increase* in the number of C–H bonds.

Carbonyl compounds are either reactants or products in many of these reactions, as illustrated in the accompanying diagram. For example, because aldehydes fall in the middle of this scheme, they can be both oxidized and reduced. Carboxylic acids and their derivatives

(RCOZ), on the other hand, are already highly oxidized, so their only useful reaction is reduction.

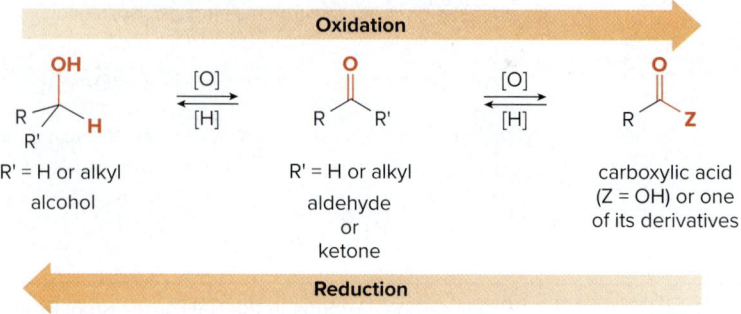

The three most useful oxidation and reduction reactions of carbonyl starting materials can be summarized as follows:

[1] Reduction of aldehydes and ketones to alcohols (Sections 13.4–13.6)

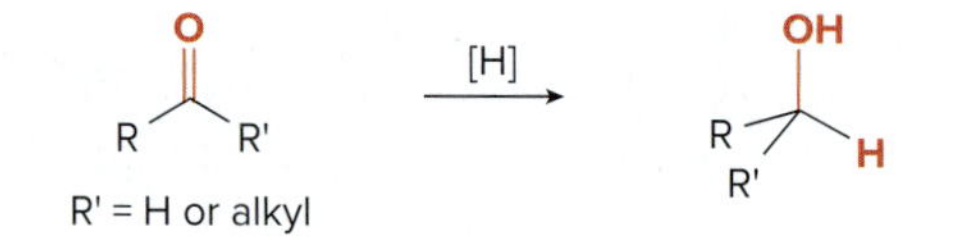

$$\underset{\substack{\text{R' = H or alkyl} \\ \text{aldehyde or ketone}}}{\text{R}-\underset{\text{R'}}{\overset{\text{O}}{\|}}-} \xrightarrow{[H]} \underset{\substack{\text{1° or 2° alcohol}}}{\text{R}-\underset{\underset{\text{R'}}{|}}{\overset{\text{OH}}{|}}-\text{H}}$$

Aldehydes and ketones are reduced to 1° and 2° alcohols, respectively.

[2] Reduction of carboxylic acids and their derivatives (Section 13.7)

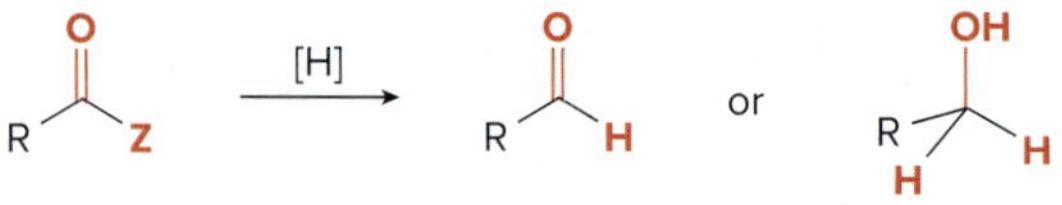

$$\underset{}{\text{R}-\underset{\text{Z}}{\overset{\text{O}}{\|}}-} \xrightarrow{[H]} \underset{\text{aldehyde}}{\text{R}-\underset{\text{H}}{\overset{\text{O}}{\|}}-} \text{ or } \underset{\text{1° alcohol}}{\text{R}-\underset{\underset{\text{H}}{|}}{\overset{\text{OH}}{|}}-\text{H}}$$

The reduction of carboxylic acids and their derivatives gives a variety of products, depending on the identity of Z and the nature of the reducing agent. The usual products are aldehydes or 1° alcohols.

[3] Oxidation of aldehydes to carboxylic acids (Section 13.8)

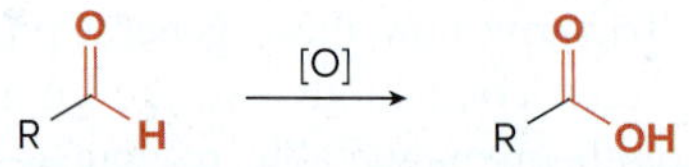

$$\underset{\text{aldehyde}}{\text{R}-\underset{\text{H}}{\overset{\text{O}}{\|}}-} \xrightarrow{[O]} \underset{\text{carboxylic acid}}{\text{R}-\underset{\text{OH}}{\overset{\text{O}}{\|}}-}$$

The most useful oxidation reaction of carbonyl compounds is the oxidation of aldehydes to carboxylic acids.

We begin with reduction, because the mechanisms of reduction reactions follow directly from the general mechanisms for nucleophilic addition and substitution.

13.4 Reduction of Aldehydes and Ketones

The most useful reagents for reducing aldehydes and ketones are the metal hydride reagents (Section 11.2). The two most common metal hydride reagents are **sodium borohydride (NaBH$_4$)** and **lithium aluminum hydride (LiAlH$_4$)**. These reagents contain a polar metal–hydrogen bond that serves as a source of the nucleophile hydride, **H:$^-$. LiAlH$_4$ is a stronger reducing agent than NaBH$_4$,** because the Al—H bond is more polar than the B—H bond.

LiAlH$_4$ and NaBH$_4$ serve as a source of H:$^-$, but there are no free H:$^-$ ions present in reactions with these reagents.

13.4A Reduction with Metal Hydride Reagents

Treating an aldehyde or a ketone with $NaBH_4$ or $LiAlH_4$, followed by water or some other proton source, affords an **alcohol**. This is an addition reaction because **the elements of H_2 are added across the π bond**, but it is also a **reduction** because the product alcohol has fewer C–O bonds than the starting carbonyl compound.

The product of this reduction reaction is a **1° alcohol** when the starting carbonyl compound is an aldehyde, and a **2° alcohol** when it is a ketone.

LiAlH₄ reductions must be carried out under anhydrous conditions, because water reacts violently with the reagent. Water is added to the reaction mixture (to serve as a proton source) *after* the reduction with LiAlH₄ is complete.

$NaBH_4$ selectively reduces aldehydes and ketones in the presence of most other functional groups. Reductions with $NaBH_4$ are typically carried out in CH_3OH as solvent. $LiAlH_4$ reduces aldehydes and ketones and many other functional groups as well (Sections 11.6 and 13.7).

Problem 13.4 What alcohol is formed when each compound is treated with $NaBH_4$ in CH_3OH?

Problem 13.5 What aldehyde or ketone is needed to prepare each alcohol by metal hydride reduction?

13.4B The Mechanism of Hydride Reduction

Hydride reduction of aldehydes and ketones occurs via the general mechanism of nucleophilic addition—that is, **nucleophilic attack** followed by **protonation**. Mechanism 13.3 is shown using $LiAlH_4$, but an analogous mechanism can be written for $NaBH_4$.

Mechanism 13.3 $LiAlH_4$ Reduction of RCHO and $R_2C=O$

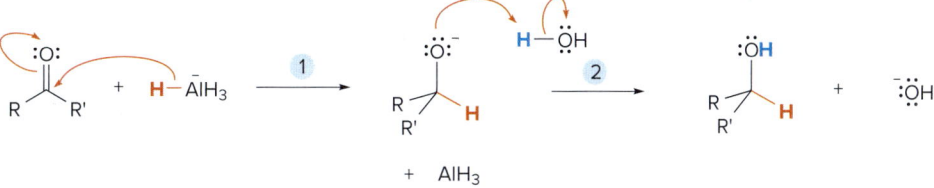

1. **The nucleophile (AlH_4^-) donates H:$^-$ to the carbonyl group,** breaking the π bond and moving an electron pair out on oxygen. This forms a new C–H bond.
2. Protonation of the negatively charged oxygen by H_2O (or CH_3OH) forms the **reduction product** with a new O–H bond.

- The net result of adding H:⁻ (from NaBH₄ or LiAlH₄) and H⁺ (from H₂O) is the addition of the elements of H₂ to the carbonyl π bond.

13.4C Catalytic Hydrogenation of Aldehydes and Ketones

Catalytic hydrogenation also reduces aldehydes and ketones to 1° and 2° alcohols, respectively, using H₂ and Pd-C (or another metal catalyst). H₂ adds to the C=O in much the same way that it adds to the C=C of an alkene (Section 11.3). The metal catalyst (Pd-C) provides a surface that binds the carbonyl starting material and H₂, and two H atoms are sequentially transferred with cleavage of the π bond.

| aldehyde | →(H₂, Pd-C) | 1° alcohol | ketone | →(H₂, Pd-C) | 2° alcohol |

Problem 13.6 Draw the products formed when CH₃COCH₂CH₂CH₃, is treated with each reagent: (a) LiAlH₄, then H₂O; (b) NaBH₄ in CH₃OH; (c) H₂, Pd-C; (d) NaBD₄ in CH₃OH.

13.5 The Stereochemistry of Carbonyl Reduction

> Recall from Section 9.16 that an achiral starting material gives a racemic mixture when a new stereogenic center is formed.

The stereochemistry of carbonyl reduction follows the same principles we have previously learned. Reduction converts a **planar sp^2 hybridized carbonyl carbon to a tetrahedral sp^3 hybridized carbon.** What happens when a new stereogenic center is formed in this process? With NaBH₄ or LiAlH₄, **a racemic product is obtained.** For example, NaBH₄ in CH₃OH solution reduces butan-2-one, an achiral ketone, to butan-2-ol, an alcohol that contains a new stereogenic center. Both enantiomers of butan-2-ol are formed in equal amounts.

butan-2-one →(NaBH₄, CH₃OH) butan-2-ol (new stereogenic center) → (S)-butan-2-ol + (R)-butan-2-ol

achiral starting material — Two enantiomers are formed.

Why is a racemic mixture formed? Because the carbonyl carbon is sp^2 hybridized and planar, hydride can approach the double bond with equal probability from both sides of the plane, forming two alkoxides, which are **enantiomers** of each other. Protonation of the alkoxides gives an equal amount of two alcohols, which are also **enantiomers.**

[NaBH₄ is the source of H:⁻.]

from the front → →(CH₃OH) (S)-butan-2-ol

enantiomers

from behind → →(CH₃OH) (R)-butan-2-ol

- **Conclusion:** Hydride reduction of an achiral ketone with LiAlH₄ or NaBH₄ gives a racemic mixture of two alcohols when a new stereogenic center is formed.

Problem 13.7 Draw the products formed (including stereoisomers) when each compound is reduced with NaBH₄ in CH₃OH.

a. (pentan-3-one structure) b. (butanal structure) c. (4-tert-butylcyclohexanone structure)

13.6 Enantioselective Biological Reduction

Although the laboratory reductions discussed in Section 13.5 give a mixture of enantiomers, biological reductions that occur in cells *always* proceed with complete selectivity, forming a single enantiomer. In cells, the reducing agent is **NADH**, the reduced form of nicotinamide adenine dinucleotide (Section 11.13). In biological reduction, **NADH donates H:⁻,** in much the same way as a metal hydride reagent. Nucleophilic attack of hydride and protonation thus form an alcohol from a carbonyl group, and **NADH is converted to NAD⁺.**

Pyruvate is formed during the metabolism of glucose (Section 27.4). During periods of strenuous exercise, when there is insufficient oxygen to metabolize pyruvate to CO_2, pyruvate is reduced to lactate. The tired feeling of sore muscles is a result of lactate accumulation.

Niacin can be obtained from foods such as soybeans, which contain it naturally, and from breakfast cereals, which are fortified with it to help people consume their recommended daily allowance of this B vitamin.
C Squared Studios/Getty Images

This reaction is completely enantioselective. Addition of the *pro-R* hydrogen (in blue) of NADH to pyruvate catalyzed by lactate dehydrogenase affords a single enantiomer of lactate with the *S* configuration. NADH reduces a variety of different carbonyl compounds in biological systems. The configuration of the product (*R* or *S*) depends on the enzyme used to catalyze the process.

pyruvate →[NADH + H⁺ / NAD⁺][lactate dehydrogenase]→ (S)-lactate

As we learned in Section 11.13, **NAD⁺, the oxidized form of NADH, is a biological oxidizing agent** capable of oxidizing alcohols to carbonyl compounds, forming NADH in the process. NAD⁺ is synthesized from the vitamin niacin, which can be obtained from soybeans among other dietary sources.

niacin
vitamin B₃

Problem 13.8 NADPH, reduced nicotinamide adenine dinucleotide phosphate, resembles NADH in structure and reactivity, but it contains an additional phosphate bonded to one of the carbohydrate rings. Draw the products formed when NADPH reacts with ketone **A** using the *pro-R* hydrogen of the six-membered ring of NADPH to form the *R* enantiomer of the product. This reaction is one step in the biosynthesis of fatty acids.

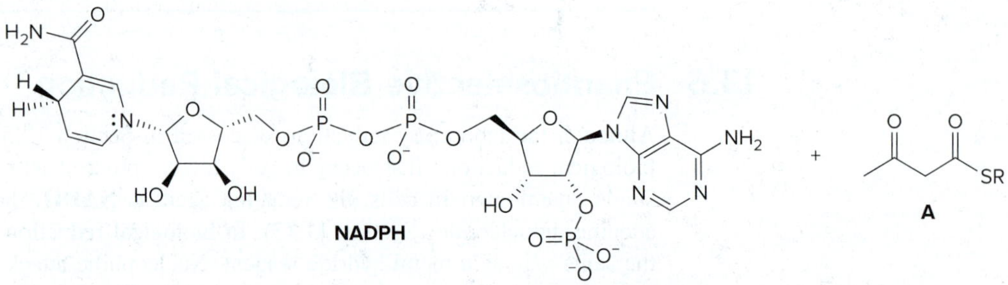

13.7 Reduction of Carboxylic Acids and Their Derivatives

The reduction of carboxylic acids and their derivatives (**RCOZ**) is complicated because the products obtained depend on the identity of both the leaving group (Z) and the reducing agent. Metal hydride reagents are the most useful reducing reagents. **Lithium aluminum hydride is a strong reducing agent that reacts with *all* carboxylic acid derivatives.** Two other related but more selective reducing agents are also used:

[1] **Diisobutylaluminum hydride, [(CH₃)₂CHCH₂]₂AlH,** abbreviated as **DIBAL-H,** has two bulky isobutyl groups, which make this reagent less reactive than LiAlH₄.

[2] **Lithium tri-*tert*-butoxyaluminum hydride, LiAlH[OC(CH₃)₃]₃,** has three electronegative oxygen atoms bonded to aluminum, which make this reagent less nucleophilic than LiAlH₄.

LiAlH₄ is a strong, nonselective reducing agent. DIBAL-H and LiAlH[OC(CH₃)₃]₃ are milder, more selective reducing agents.

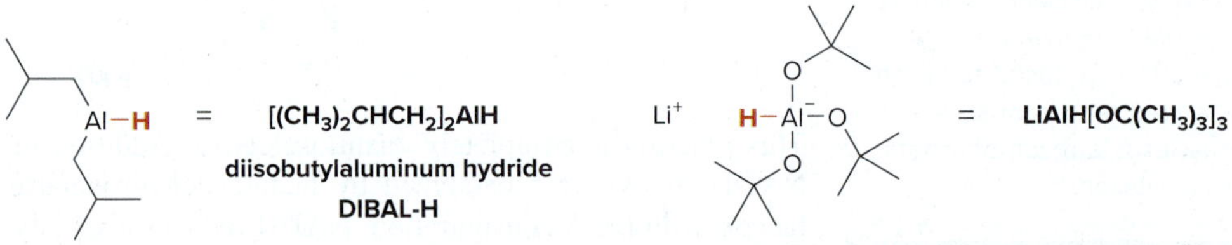

In both reagents, the **single H atom bonded to Al is donated as H:⁻** in hydride reductions.

13.7A Reduction of Acid Chlorides and Esters

Acid chlorides and esters can be reduced to either aldehydes or alcohols, depending on the reagent.

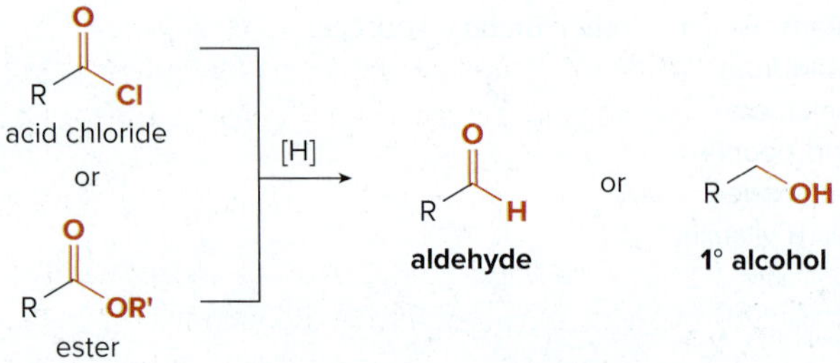

13.7 Reduction of Carboxylic Acids and Their Derivatives

- LiAlH$_4$ converts RCOCl and RCOOR' to alcohols.
- A milder reducing agent (DIBAL-H or LiAlH[OC(CH$_3$)$_3$]$_3$) converts RCOCl or RCOOR' to RCHO at low temperatures.

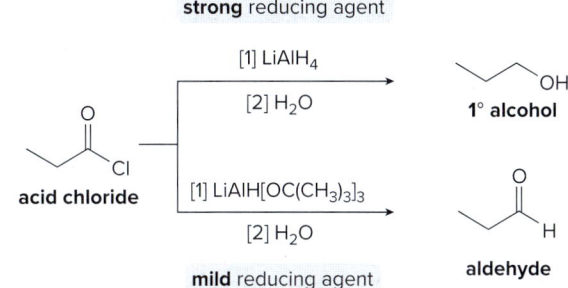

In the reduction of an acid chloride, Cl$^-$ comes off as the leaving group.

In the reduction of the ester, CH$_3$O$^-$ comes off as the leaving group, which is then protonated by H$_2$O to form CH$_3$OH.

Mechanism 13.4 illustrates why two different products are possible. It can be conceptually divided into two parts: **nucleophilic substitution** to form an aldehyde (Steps [1] and [2]), followed by **nucleophilic addition** to the aldehyde to form an alcohol (Steps [3] and [4]). A general mechanism is drawn using LiAlH$_4$ as reducing agent.

Mechanism 13.4 Reduction of RCOCl and RCOOR' with a Metal Hydride Reagent

1. **Nucleophilic attack of H:$^-$** forms a tetrahedral intermediate with a leaving group Z.
2. The π bond is re-formed and **the leaving group Z departs.** The overall result of addition of H:$^-$ and elimination of Z:$^-$ is substitution of H for Z.
3. **Nucleophilic attack of H:$^-$** forms an alkoxide with no leaving group.
4. Protonation of the alkoxide by H$_2$O forms the **alcohol** reduction product. The overall result of Steps [3] and [4] is **addition of H$_2$**.

With less nucleophilic reducing agents such as DIBAL-H and LiAlH[OC(CH$_3$)$_3$]$_3$, the process stops after reaction with one equivalent of H:$^-$ and the aldehyde is formed as product (Steps [1] and [2] of Mechanism 13.4). With a stronger reducing agent like LiAlH$_4$, two equivalents of H:$^-$ are added and an alcohol is formed.

In biological systems, acyl phosphates (RCO$_2$PO$_3^{2-}$, Table 3.4) are carboxylic acid derivatives that undergo similar reductions with NADH. NADH adds one equivalent of H:$^-$ to an acyl phosphate to form an aldehyde by nucleophilic substitution.

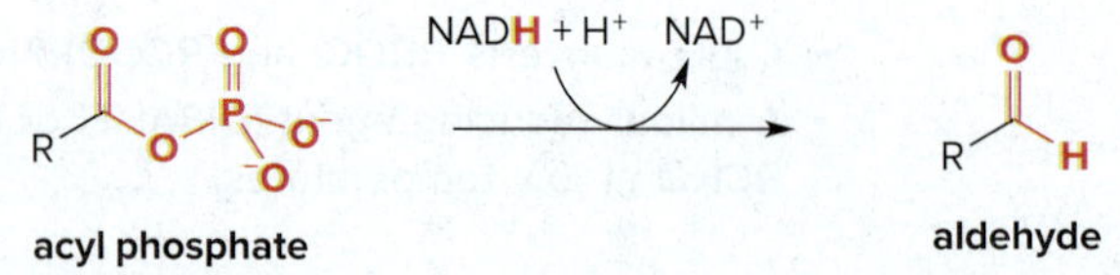

[acyl phosphate] + NADH + H⁺ → [aldehyde] + NAD⁺

For example, one step in the biosynthesis of the amino acid proline involves reduction of the acyl phosphate in glutamate 5-phosphate to form glutamate 5-semialdehyde

glutamate 5-phosphate + NADH + H⁺ → glutamate 5-semialdehyde + NAD⁺

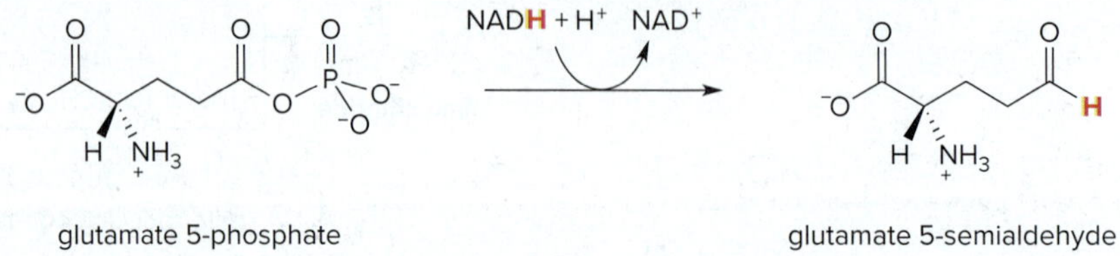

Problem 13.9 Draw a stepwise mechanism for the following reaction.

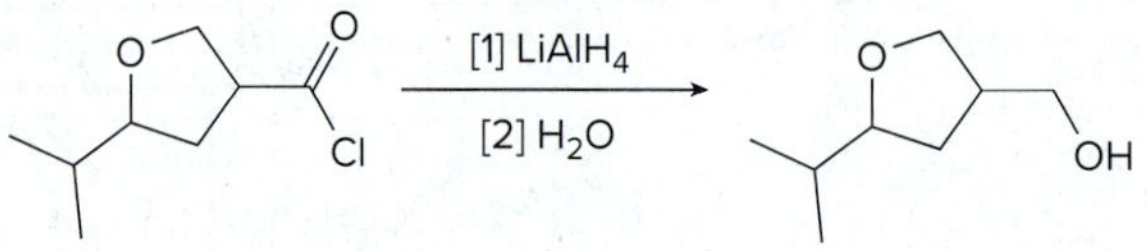

[isopropyl-tetrahydrofuran-acyl chloride] →[1] LiAlH₄ [2] H₂O→ [isopropyl-tetrahydrofuran-CH₂OH]

Problem 13.10 Draw the structure of both an acid chloride and an ester that can be used to prepare each compound by reduction.

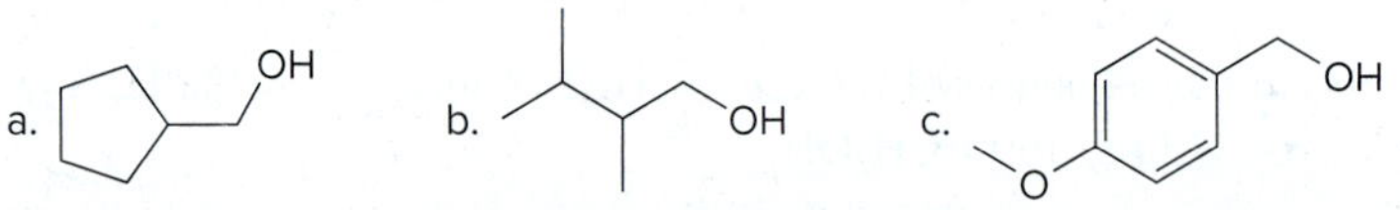

a. cyclopentyl-CH₂OH b. (CH₃)₂CHCH(CH₃)CH₂OH c. 4-methoxybenzyl alcohol (ArCH₂OH where Ar = p-MeO-C₆H₄)

13.7B Reduction of Carboxylic Acids and Amides

Carboxylic acids are reduced to alcohols with LiAlH₄. LiAlH₄ is too strong a reducing agent to stop the reaction at the aldehyde stage, but milder reagents are not strong enough to initiate the reaction in the first place, so this is the only useful reduction reaction of carboxylic acids.

R–COOH —[1] LiAlH₄, [2] H₂O→ R–CH₂OH (1° alcohol)

CH₃CH₂CH₂COOH —[1] LiAlH₄, [2] H₂O→ CH₃CH₂CH₂CH₂OH

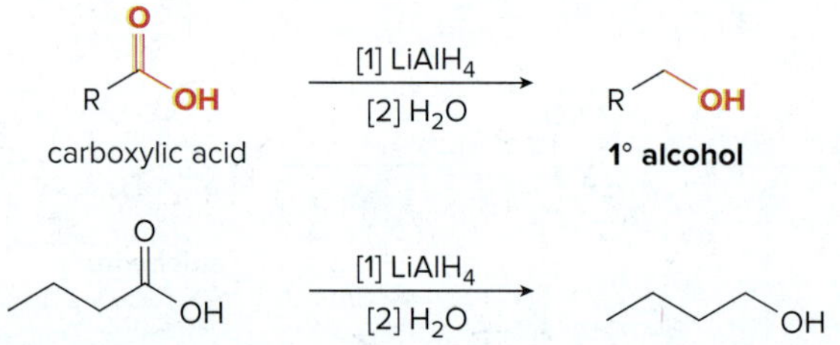

Unlike the LiAlH₄ reduction of all other carboxylic acid derivatives, which affords alcohols, the **LiAlH₄ reduction of amides forms amines.**

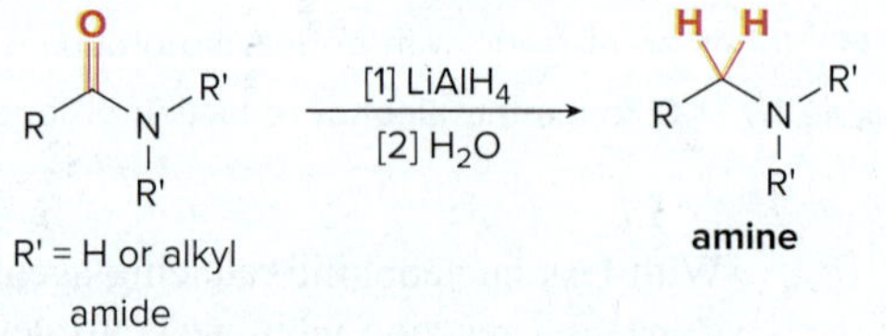

R–C(=O)–NR'₂ —[1] LiAlH₄, [2] H₂O→ R–CH₂–NR'₂ (amine)

R' = H or alkyl
amide

Both C–O bonds are reduced to C–H bonds by LiAlH₄, and any H atom or R group bonded to the amide nitrogen atom remains bonded to it in the product. Because ⁻NH₂ (or ⁻NHR or ⁻NR₂) is a *poorer* leaving group than Cl⁻ or ⁻OR, **⁻NH₂ is never lost during reduction,** and therefore it forms an amine in the final product.

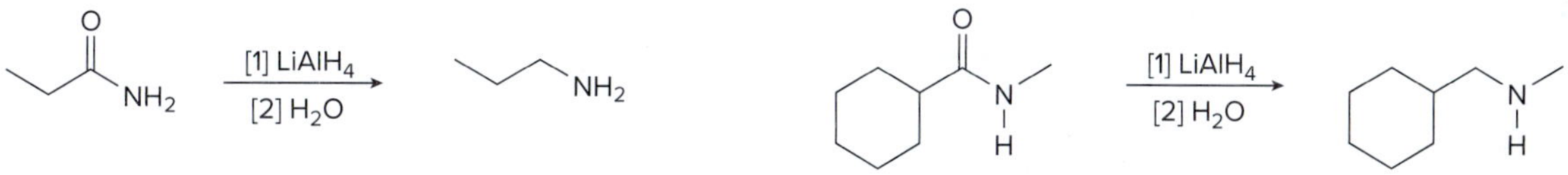

13.7 Reduction of Carboxylic Acids and Their Derivatives

[Reaction schemes showing LiAlH₄ reduction of propanamide to propylamine, and N-methylcyclohexanecarboxamide to N-methyl(cyclohexylmethyl)amine]

Imines and related compounds are discussed in Chapter 14.

The mechanism, illustrated in Mechanism 13.5 with RCONH₂ as starting material, is somewhat different than the previous reductions of carboxylic acid derivatives. Amide reduction proceeds with formation of an intermediate *imine*, **a compound containing a C—N double bond,** which is then further reduced to an amine.

Mechanism 13.5 Reduction of an Amide to an Amine with LiAlH₄

Part [1] Reduction of an amide to an imine

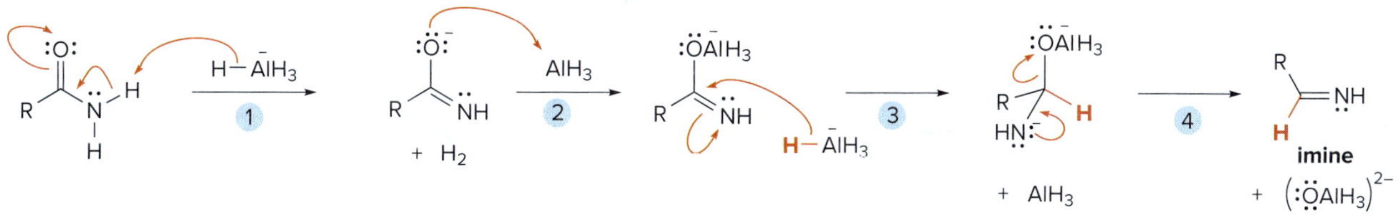

[Mechanism diagram showing steps 1–4]

1 – 2 AlH₄⁻ removes a proton from the amide to form a Lewis base that complexes with AlH₃ in Step [2].

3 – 4 Nucleophilic attack of H:⁻ and loss of a leaving group, (OAlH₃)²⁻, form an imine.

Part [2] Reduction of an imine to an amine

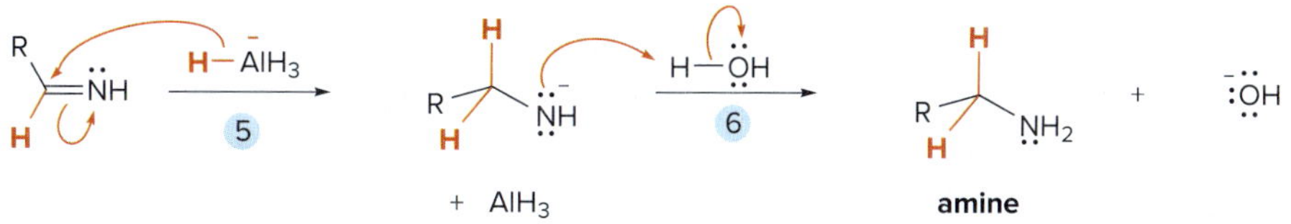

[Mechanism diagram showing steps 5–6]

5 – 6 Nucleophilic addition of H:⁻ and protonation form the **amine**.

Problem 13.11 Draw the products formed from LiAlH₄ reduction of each compound.

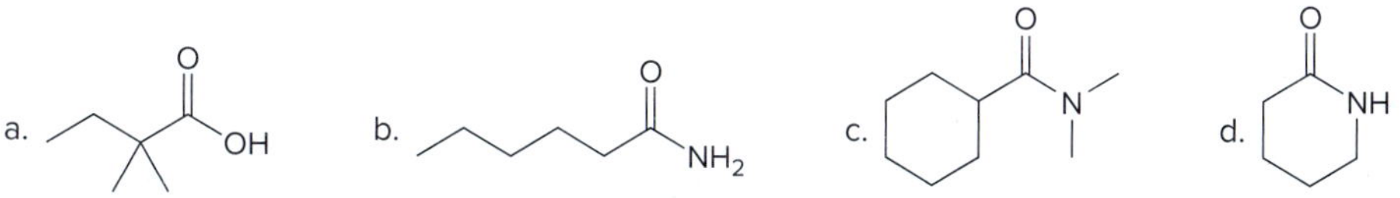

a. [2,2-dimethylpropanoic acid structure]   b. [hexanamide structure]   c. [N,N-dimethylcyclohexanecarboxamide structure]   d. [δ-valerolactam / piperidinone structure]

Sample Problem 13.1 Determining the Amide That Forms an Amine by Reduction

What amide(s) form amine **X** on treatment with LiAlH₄?

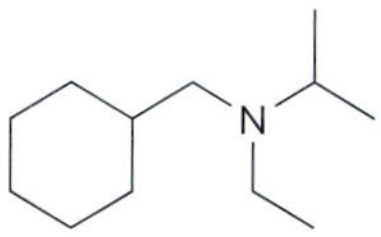

[Structure of amine X: cyclohexylmethyl group, isopropyl group, and ethyl group bonded to N]

X

Solution

LiAlH₄ reduction of an amide converts a **C=O** to a **CH₂** group, so we must examine the alkyl groups bonded to the amine nitrogen. Any alkyl group with a CH₂ group bonded *directly* to the N can be formed by reduction of a C=O. An alkyl group with zero or one H atom *cannot* be formed by reduction of a C=O. For amine **X**, the CH$_2$ groups in blue can be formed from C=O's, but the CH group labeled in red cannot be formed from a C=O.

only 1 H on C

The CH bond in red *cannot* be formed from a C=O.

A **B**

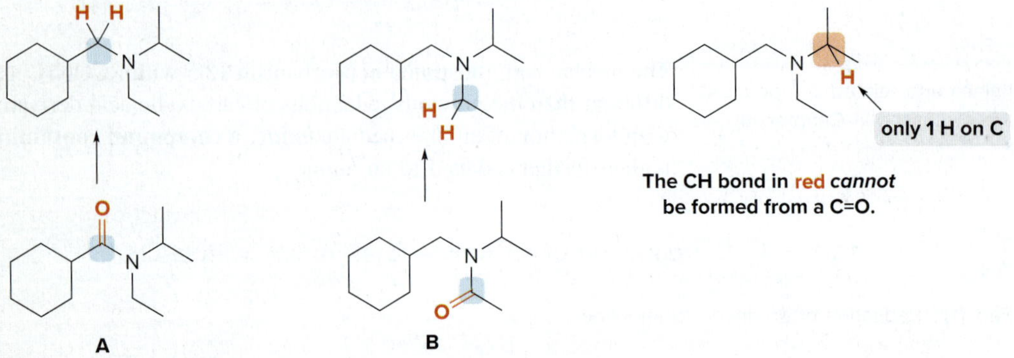

Thus, amides **A** and **B** can be converted to amine **X** by reduction of a C=O with LiAlH$_4$.

Problem 13.12 What amide(s) will form each of the following amines on treatment with LiAlH$_4$?

a. 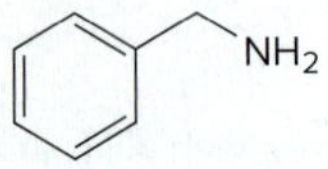 b. 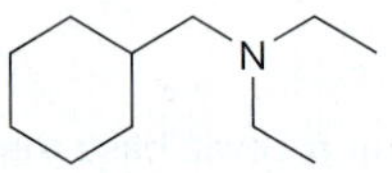 c.

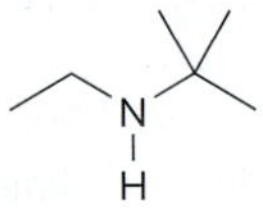

More Practice: Try Problem 13.54.

13.7C A Summary of the Reagents for Reduction

The many available metal hydride reagents reduce a wide variety of functional groups. Keep in mind that **LiAlH$_4$ is such a strong reducing agent that it *nonselectively* reduces most polar functional groups.** All other metal hydride reagents are more selective, and each has its particular reactions that best utilize its reduced reactivity. The reagents and their uses are summarized in Table 13.1.

Problem 13.13 What product is formed when each compound is treated with either LiAlH$_4$ (followed by H$_2$O), or NaBH$_4$ in CH$_3$OH?

a. 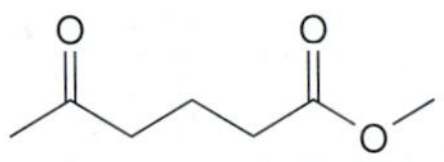 b. 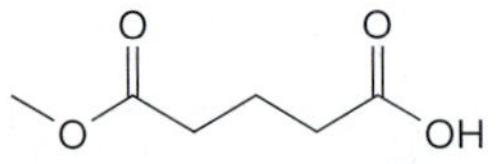 c.

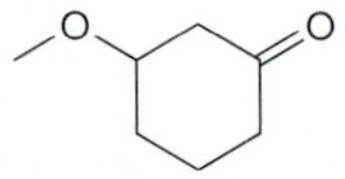

Table 13.1 A Summary of Metal Hydride Reducing Agents

	Reagent	Starting material	→	Product
Strong reagent	LiAlH$_4$	RCHO	→	RCH$_2$OH
		R$_2$CO	→	R$_2$CHOH
		RCOOH	→	RCH$_2$OH
		RCOOR'	→	RCH$_2$OH
		RCOCl	→	RCH$_2$OH
		RCONH$_2$	→	RCH$_2$NH$_2$
Milder reagents	NaBH$_4$	RCHO	→	RCH$_2$OH
		R$_2$CO	→	R$_2$CHOH
	LiAlH[OC(CH$_3$)$_3$]$_3$	RCOCl	→	RCHO
	DIBAL-H	RCOOR'	→	RCHO

13.8 Oxidation of Aldehydes

The most common oxidation reaction of carbonyl compounds is the oxidation of **aldehydes to carboxylic acids.** A variety of oxidizing agents can be used, including CrO_3, $Na_2Cr_2O_7$, $K_2Cr_2O_7$, and $KMnO_4$. Cr^{6+} reagents are also used to oxidize 1° and 2° alcohols, as discussed in Section 11.12. Because ketones have no H on the carbonyl carbon, they do *not* undergo this oxidation reaction.

Aldehydes are oxidized selectively in the presence of other functional groups using **silver(I) oxide in aqueous ammonium hydroxide (Ag_2O in NH_4OH).** This is called **Tollens reagent.** Oxidation with Tollens reagent provides a distinct color change, because the Ag^+ reagent is reduced to silver metal (Ag), which precipitates out of solution.

Aldehydes give a positive Tollens test; that is, they react with Ag^+ to form RCOOH and Ag. When the reaction is carried out in a glass flask, a silver mirror is formed on its walls. Other functional groups give a negative Tollens test, because no silver mirror forms. *Charles D. Winters/McGraw-Hill Education*

Only the aldehyde is oxidized.

Problem 13.14 What product is formed when each compound is treated with either Ag_2O, NH_4OH or $Na_2Cr_2O_7$, H_2SO_4, H_2O?

a. b.

Problem 13.15 Review the oxidation reactions using Cr^{6+} reagents in Section 11.12. Then draw the product formed when compound **B** is treated with each reagent.

B

a. $NaBH_4$, CH_3OH
b. [1] $LiAlH_4$; [2] H_2O
c. PCC
d. Ag_2O, NH_4OH
e. CrO_3, H_2SO_4, H_2O

13.9 Organometallic Reagents

We will now discuss the reactions of carbonyl compounds with organometallic reagents, another class of nucleophiles.

- *Organometallic reagents* contain a carbon atom bonded to a metal.

organometallic reagent = R—M   M = metal

Most common metals:
M = Li, Mg, Cu

Lithium, magnesium, and copper are the most commonly used metals in organometallic reagents, but others (such as Sn, Si, Tl, Al, Ti, and Hg) are known. General structures of the three common organometallic reagents are shown. R can be alkyl, phenyl, allyl, benzyl, sp^2 hybridized, and with M = Li or Mg, sp hybridized. Because metals are *more electropositive* (less electronegative) than carbon, they donate electron density toward carbon, so that **carbon bears a partial negative charge.**

$$R-Li \qquad R-Mg-X \qquad R-\overset{R}{\underset{|}{Cu}}^- \ Li^+$$

organolithium reagents | organomagnesium reagents or Grignard reagents | organocopper reagents or organocuprates

- The *more polar* the carbon–metal bond, the *more reactive* the organometallic reagent.

Because both Li and Mg are very electropositive metals, **organolithium (RLi)** and **organomagnesium reagents (RMgX) contain very polar carbon–metal bonds and are therefore *very reactive* reagents.** Organomagnesium reagents are called **Grignard reagents,** after Victor Grignard, who received the Nobel Prize in Chemistry in 1912 for his work with them.

Organocopper reagents (R_2CuLi), also called organocuprates, have a less polar carbon–metal bond and are therefore *less reactive*. Although organocuprates contain two alkyl groups bonded to copper, only one R group is utilized in a reaction.

Regardless of the metal, organometallic reagents are useful synthetically because they react as if they were free carbanions; that is, carbon bears a partial *negative* charge, so the **reagents react as bases and nucleophiles.**

$$\overset{M^{\delta+}}{\underset{|}{C^{\delta-}}} \quad \text{reacts like} \quad :\overset{|}{C}$$

carbanion
a base and a nucleophile

Electronegativity values for carbon and the common metals in R—M reagents are C (2.5), Li (1.0), Mg (1.3), and Cu (1.8).

13.9A Preparation of Organometallic Reagents

Organolithium and Grignard reagents are typically prepared by reaction of an organic halide with the corresponding metal, as shown in the accompanying equations.

$$R-X + 2\,Li \longrightarrow R-Li + LiX$$
organolithium reagent

$$R-X + Mg \xrightarrow{\text{Et}_2O} R-Mg-X$$
Grignard reagent

$$CH_3-Br + 2\,Li \longrightarrow CH_3-Li + LiBr$$
methyllithium

$$CH_3-Br + Mg \xrightarrow{\text{Et}_2O} CH_3-Mg-Br$$
methylmagnesium bromide

With lithium, the halogen and metal exchange to form the organolithium reagent. With magnesium, the metal inserts in the carbon–halogen bond, forming the Grignard reagent. Grignard reagents are usually prepared in diethyl ether ($CH_3CH_2OCH_2CH_3$) as solvent. It is thought that two ether oxygen atoms complex with the magnesium atom, stabilizing the reagent.

Two molecules of diethyl ether complex with the Mg atom of the Grignard reagent.

Organocuprates are prepared from organolithium reagents by reaction with a Cu^+ salt, often CuI.

$$2\,R-Li + CuI \longrightarrow R-\overset{R}{\underset{|}{Cu}}^-\,Li^+ + LiI$$
organocopper reagent

$$2\,CH_3-Li + CuI \longrightarrow CH_3-\overset{CH_3}{\underset{|}{Cu}}^-\,Li^+ + LiI$$
lithium dimethylcuprate

13.9 Organometallic Reagents

Problem 13.16 Write the step(s) needed to convert CH_3CH_2Br to each reagent: (a) CH_3CH_2Li; (b) CH_3CH_2MgBr; (c) $(CH_3CH_2)_2CuLi$.

13.9B Acetylide Anions

The **acetylide anions** discussed in Chapter 10 are another example of organometallic compounds. These reagents are prepared by an acid–base reaction of an alkyne with a base such as $NaNH_2$ or NaH. We can think of these compounds as **organosodium** reagents. Because sodium is even more electropositive (less electronegative) than lithium, the C—Na bond of these organosodium compounds is best described as **ionic**, rather than polar covalent.

$$R-C\equiv C-H \;+\; Na^+ \;:\!NH_2^- \;\rightleftharpoons\; R-C\equiv C:^- \; Na^+ \;+\; :NH_3$$

acetylide anion
an organosodium compound

An acid–base reaction can also be used to prepare *sp* hybridized organolithium compounds. Treatment of a terminal alkyne with CH_3Li affords a lithium acetylide. Equilibrium favors the products because the *sp* hybridized C—H bond of the terminal alkyne is more acidic than the sp^3 hybridized conjugate acid, CH_4, that is formed.

$$R-C\equiv C-H \;+\; CH_3-Li \;\rightleftharpoons\; R-C\equiv C-Li \;+\; CH_3-H$$

$pK_a \approx 25$ base   **a lithium acetylide** $pK_a = 50$

stronger acid   **weaker acid**

Problem 13.17 Which of the following species represent organometallic compounds: (a) $BrMgC\equiv CCH_2CH_3$; (b) $NaOCH_2CH_3$; (c) $KOC(CH_3)_3$; (d) $PhLi$?

13.9C Reaction as a Base

- Organometallic reagents are strong bases that readily abstract a proton from water to form hydrocarbons.

The electron pair in the carbon–metal bond is used to form a new bond to the proton. Equilibrium favors the products of this acid–base reaction because H_2O is a much stronger acid than the alkane product.

$$CH_3-Li \;+\; H-OH \;\rightleftharpoons\; CH_3-H \;+\; Li^+ \;:\!OH^-$$

base   acid $pK_a = 50$
         $pK_a = 15.7$

        stronger acid   **a very weak acid**

Similar reactions occur for the same reason with the O—H proton in alcohols and carboxylic acids, and the N—H protons of amines.

$$Ph-MgBr \;+\; H-OCH_3 \;\longrightarrow\; Ph-H \;+\; CH_3O:^- \;(MgBr)^+$$

strong base   acid

$$CH_3CH_2CH_2-Li \;+\; CH_3COOH \;\longrightarrow\; CH_3CH_2CH_2-H \;+\; Li^+ \; CH_3COO:^-$$

strong base   acid

Chapter 13 Introduction to Carbonyl Chemistry; Organometallic Reagents; Oxidation and Reduction

Because organolithium and Grignard reagents are themselves prepared from alkyl halides, a two-step method converts an alkyl halide to an alkane (or another hydrocarbon).

$$R-X \xrightarrow{M} R-M \xrightarrow{H_2O} R-H$$

alkyl halide → alkane

Problem 13.18 Draw the product formed when each organometallic reagent is treated with H_2O.

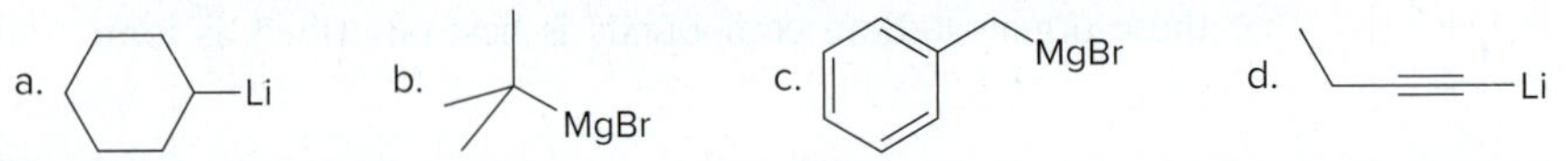

a. cyclohexyl–Li b. (CH_3)_3C–MgBr c. PhCH_2–MgBr d. alkynyl–Li

---

13.9D Reaction as a Nucleophile

Organometallic reagents are also strong nucleophiles that react with electrophilic carbon atoms to form new carbon–carbon bonds. These reactions are very valuable in forming the carbon skeletons of complex organic molecules. The following reactions of organometallic reagents are examined in Sections 13.10, 13.13, and 13.14:

[1] Reaction of R–M with aldehydes and ketones to afford alcohols (Section 13.10)

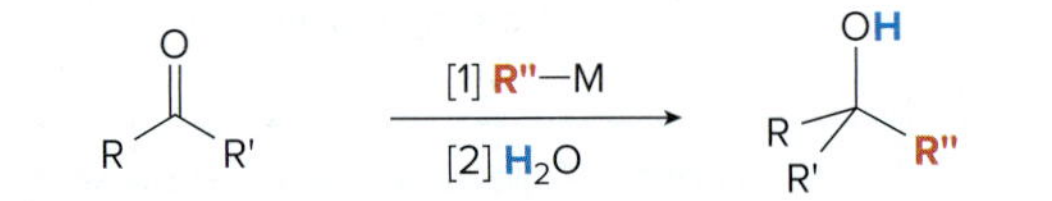

aldehyde or ketone (R' = H or alkyl) → [1] R''–M, [2] H_2O → 1°, 2°, or 3° alcohol

Aldehydes and ketones are converted to 1°, 2°, or 3° alcohols with R''Li or R''MgX.

[2] Reaction of R–M with carboxylic acid derivatives (Section 13.13)

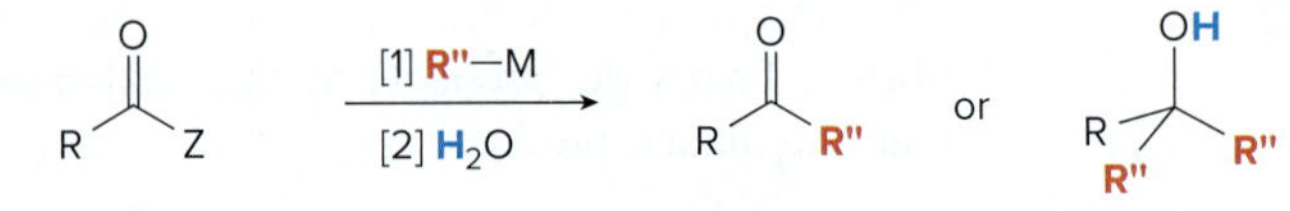

R–CO–Z (Z = Cl or OR') → [1] R''–M, [2] H_2O → ketone or 3° alcohol

Acid chlorides and esters can be converted to ketones or 3° alcohols with organometallic reagents. The identity of the product depends on the identity of R''–M and the leaving group Z.

[3] Reaction of R–M with other electrophilic functional groups (Section 13.14)

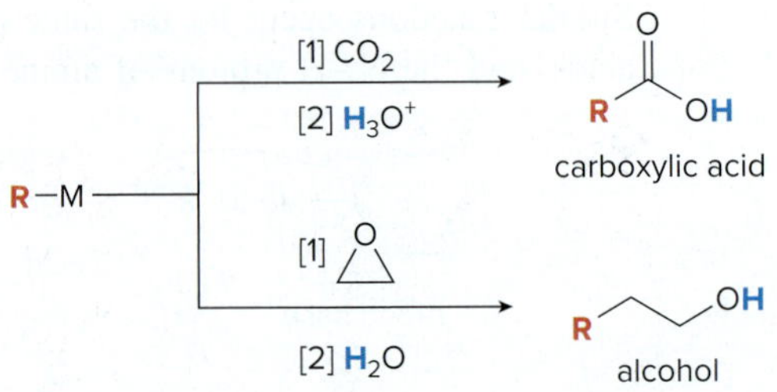

R–M
- [1] CO_2, [2] H_3O^+ → carboxylic acid (R–COOH)
- [1] epoxide, [2] H_2O → alcohol (R–CH_2CH_2–OH)

Organometallic reagents also react with CO_2 to form carboxylic acids and with epoxides to form alcohols.

13.10 Reaction of Organometallic Reagents with Aldehydes and Ketones

Treatment of an aldehyde or ketone with either an organolithium or Grignard reagent followed by water forms an alcohol with a new carbon–carbon bond. This reaction is an **addition reaction** because the elements of R″ and H are added across the π bond.

$$\underset{\text{aldehyde or ketone} \atop R' = H \text{ or alkyl}}{\overset{O}{\underset{R}{\|}}\underset{R'}{C}} \xrightarrow[\text{or } R''Li]{R''MgX} \xrightarrow{H-\ddot{O}H} \underset{1°, 2°, \text{ or } 3° \text{ alcohol}}{\overset{OH}{\underset{R'}{\underset{|}{C}}\underset{R''}{}}} \text{new C–C bond}$$

13.10A General Features

This reaction follows the general mechanism for nucleophilic addition (Section 13.2A)—that is, **nucleophilic attack** by a carbanion followed by **protonation**. Mechanism 13.6 is shown using R″MgX, but the same steps occur with organolithium reagents and acetylide anions.

Mechanism 13.6 Nucleophilic Addition of R″MgX to RCHO and R$_2$C=O

$$\underset{R R'}{\overset{:\ddot{O}:}{\|}} + R''-MgX \xrightarrow{1} \underset{R' R''}{\overset{:\ddot{O}:^-}{\underset{|}{C}}} \overset{H-\ddot{O}H}{\xrightarrow{2}} \underset{R' R''}{\overset{:\ddot{O}H}{\underset{|}{C}}} + :\ddot{O}H^-$$
$$+ \text{MgX}^+$$

1 **The nucleophile (R″)⁻ attacks the carbonyl group,** breaking the π bond and yielding an alkoxide. This forms a new carbon–carbon bond.

2 Protonation of the alkoxide by H$_2$O forms the **addition product** with a new O–H bond. The overall result is addition of R″ and H to the carbonyl group.

This reaction is used to prepare **1°, 2°, and 3° alcohols,** depending on the number of alkyl groups bonded to the carbonyl carbon of the aldehyde or ketone.

[1] formaldehyde + R″—MgX → intermediate → (H$_2$O) → **1° alcohol**

[2] aldehyde (R ≠ H) + R″—MgX → intermediate → (H$_2$O) → **2° alcohol**

[3] ketone + R″—MgX → intermediate → (H$_2$O) → **3° alcohol**

- [1] Addition of R″MgX to formaldehyde (CH$_2$=O) forms a 1° alcohol.
- [2] Addition of R″MgX to all other aldehydes forms a 2° alcohol.
- [3] Addition of R″MgX to ketones forms a 3° alcohol.

Each reaction results in addition of one new alkyl group to the carbonyl carbon, and forms one new carbon–carbon bond. The reaction is general for all organolithium and Grignard reagents, and works for acetylide anions as well, as illustrated in Equations [1]–[3].

[1] 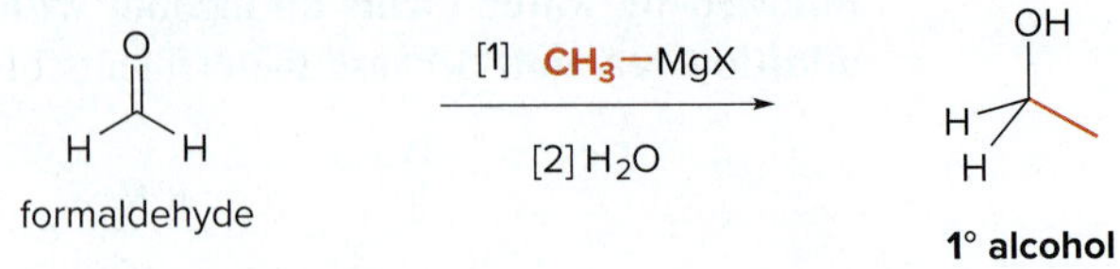 formaldehyde $\xrightarrow[\text{[2] H}_2\text{O}]{\text{[1] CH}_3-\text{MgX}}$ 1° alcohol

[2] 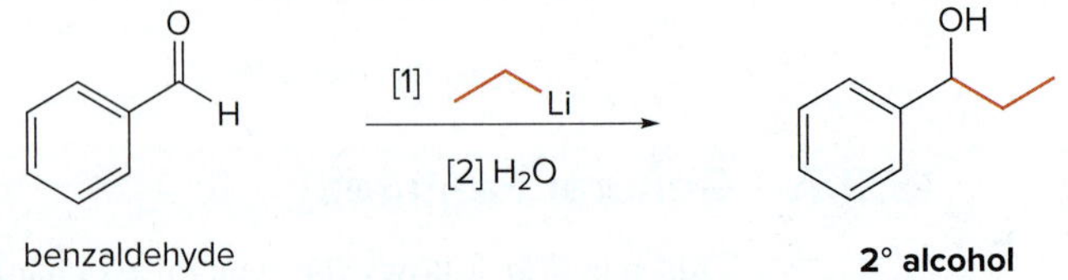 benzaldehyde $\xrightarrow[\text{[2] H}_2\text{O}]{\text{[1] CH}_3\text{CH}_2\text{Li}}$ 2° alcohol

[3] cyclohexanone $\xrightarrow[\text{[2] H}_2\text{O}]{\text{[1] HC}\equiv\text{C-Li}}$ 3° alcohol

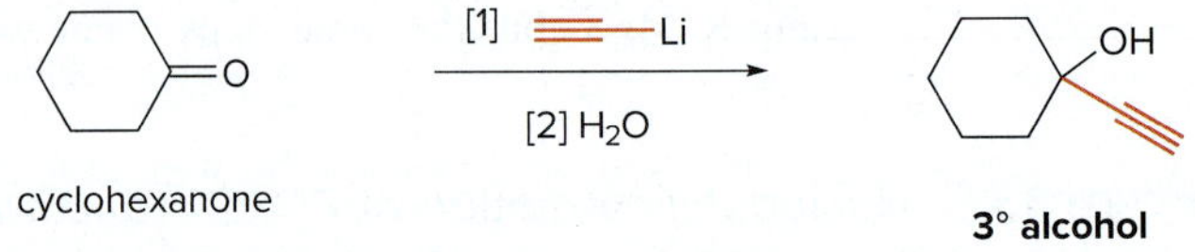

Because organometallic reagents are strong bases that rapidly react with H_2O (Section 13.9C), the addition of the new alkyl group must be carried out under anhydrous conditions to prevent traces of water from reacting with the reagent, thus reducing the yield of the desired alcohol. Water is added *after* the addition to protonate the alkoxide.

Problem 13.19 Draw the product of each reaction.

a. 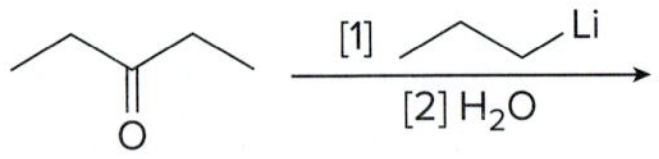 CH₃CH₂COCH₂CH₃ [1] CH₃CH₂CH₂Li; [2] H₂O

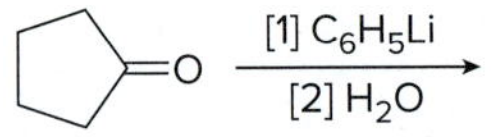

b. 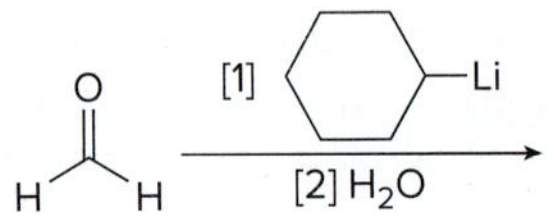 HCHO [1] cyclohexyl-Li; [2] H₂O

c. cyclopentanone [1] C₆H₅Li; [2] H₂O

d. 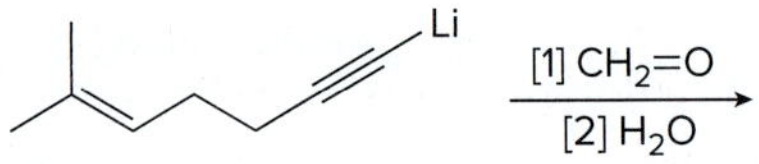 (CH₃)₂C=CHCH₂CH₂C≡C-Li [1] CH₂=O; [2] H₂O

13.10B Stereochemistry

Like reduction, addition of organometallic reagents converts an sp^2 hybridized carbonyl carbon to a tetrahedral sp^3 hybridized carbon. Addition of R—M always occurs from both sides of the trigonal planar carbonyl group. **When a new stereogenic center is formed from an achiral starting material, an equal mixture of enantiomers results,** as shown in Sample Problem 13.2.

Sample Problem 13.2 Drawing the Stereoisomers Formed During Grignard Addition.

Draw all stereoisomers formed in the following reaction.

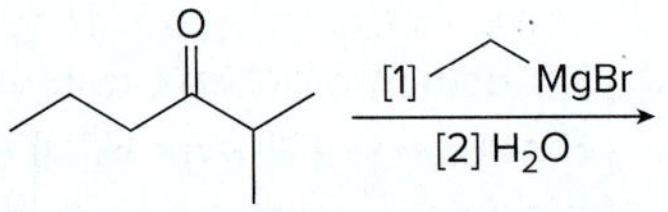

Solution

The Grignard reagent adds from both sides of the trigonal planar carbonyl group, forming two alkoxides, each containing a new stereogenic center labeled in blue. Protonation with water yields **an equal amount of two enantiomers—a racemic mixture.**

Problem 13.20 Draw the products (including stereochemistry) of the following reactions.

More Practice: Try Problems 13.35c, d; 13.46a.

13.10C Applications in Synthesis

Many syntheses of useful compounds utilize the nucleophilic addition of a Grignard or organolithium reagent to form carbon–carbon bonds. For example, one of the last steps in the synthesis of the C_{18} juvenile hormone, a member of a group of structurally related molecules that regulate the complex life cycle of an insect, is the addition of a Grignard reagent to a ketone (Figure 13.1).

Juvenile hormones maintain the juvenile stage of an insect until it is ready for adulthood. This property has been exploited to control mosquitoes and other insects infecting livestock and crops. Although juvenile hormone itself is too unstable in light and too expensive to synthesize for use in controlling insect populations, related compounds, called **juvenile hormone *mimics*,**

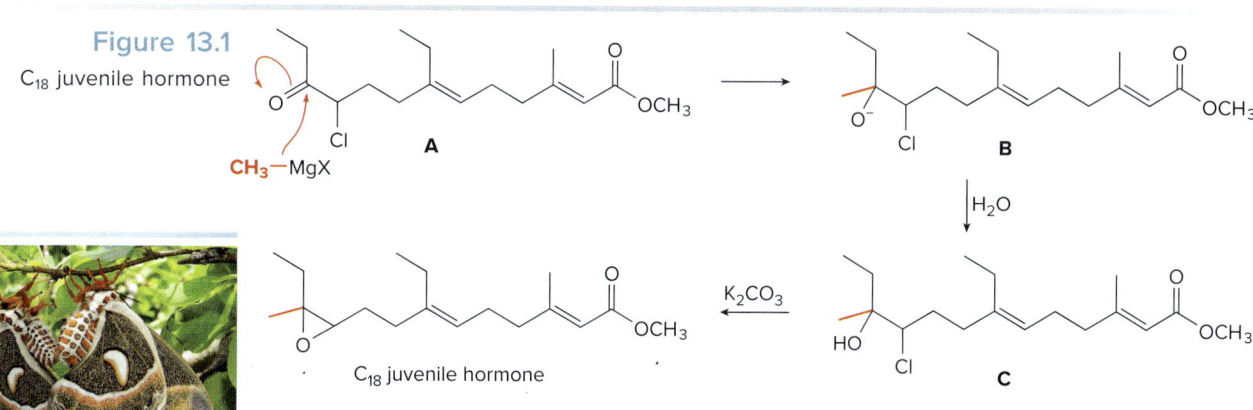

Figure 13.1
C_{18} juvenile hormone

Juvenile hormones regulate the life cycle of the cecropia moth.
Matt Jeppson/Shutterstock

- Addition of CH_3MgX to ketone **A** gives an alkoxide, **B**, which is protonated with H_2O to form 3° alcohol **C**. Although the ester group (–$COOCH_3$) can also react with the Grignard reagent (Section 13.13), it is less reactive than the ketone carbonyl. Thus, with control of reaction conditions, nucleophilic addition occurs selectively at the ketone.
- Treatment of halohydrin **C** with K_2CO_3 forms the C_{18} juvenile hormone in one step. Conversion of a halohydrin to an epoxide was discussed in Section 9.6.

have been used effectively. Application of these synthetic hormones to an egg or larva of an insect prevents maturation. With no sexually mature adults to propagate the next generation, the insect population is reduced. The best-known example of a synthetic juvenile hormone is called **methoprene,** sold under such trade names as Altocid, Precor, and Diacon. Methoprene is used in cattle salt blocks to control hornflies, in stored tobacco to control pests, and on dogs and cats to control fleas.

methoprene
juvenile hormone mimic

13.11 Retrosynthetic Analysis of Grignard Products

To use the Grignard addition in synthesis, you must be able to determine what carbonyl and Grignard components are needed to prepare a given compound—that is, **you must work backwards, in the retrosynthetic direction.** This involves a two-step process:

Step [1] Find the carbon bonded to the OH group in the product.

Step [2] Break the molecule into two components: One alkyl group bonded to the carbon with the OH group comes from the organometallic reagent. The rest of the molecule comes from the carbonyl component.

To synthesize pentan-3-ol [$(CH_3CH_2)_2CHOH$] by a Grignard reaction, locate the carbon bonded to the OH group, and then break the molecule into two components at this carbon. Thus, retrosynthetic analysis shows that one of the ethyl groups on this carbon comes from a Grignard reagent (CH_3CH_2MgX), and the rest of the molecule comes from the carbonyl component, a three-carbon aldehyde.

Retrosynthetic analysis

three-carbon aldehyde ⇐ pentan-3-ol ⇒ two-carbon Grignard reagent

Form this bond by Grignard addition.

Then, writing the reaction in the synthetic direction—that is, from starting material to product—shows whether the analysis is correct. In this example, a three-carbon aldehyde reacts with CH_3CH_2MgBr to form an alkoxide, which can then be protonated by H_2O to form pentan-3-ol, the desired alcohol.

In the synthetic direction:

pentan-3-ol

There is often more than one way to synthesize a 2° alcohol by Grignard addition, as shown in Sample Problem 13.3.

13.11 Retrosynthetic Analysis of Grignard Products

Sample Problem 13.3 — Determining the Starting Materials in a Grignard Synthesis

Show two different methods to synthesize alcohol **A** using a Grignard reaction.

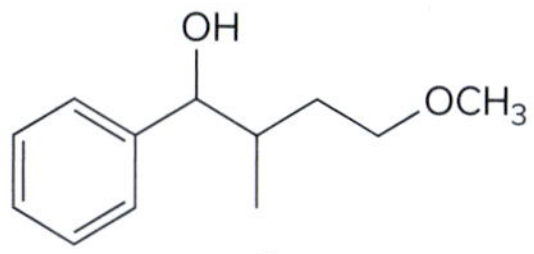

A

Solution

Because **A** has two different R groups bonded to the carbon bearing the OH group, there are two different ways to form a new carbon–carbon bond by Grignard addition.

Possibility [1] Use C_6H_5MgBr and aldehyde **B**. **Possibility [2]** Use Grignard reagent **C** and benzaldehyde.

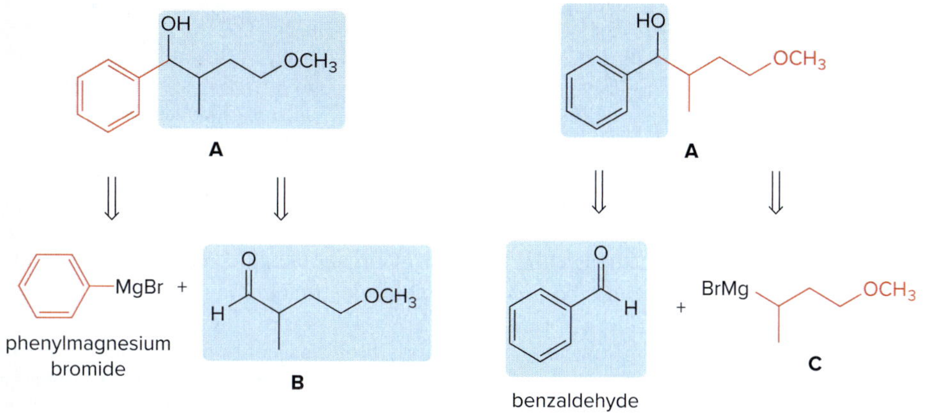

phenylmagnesium bromide

B

benzaldehyde

C

Both methods give the desired product **A**, as can be seen by writing the reactions from starting material to product.

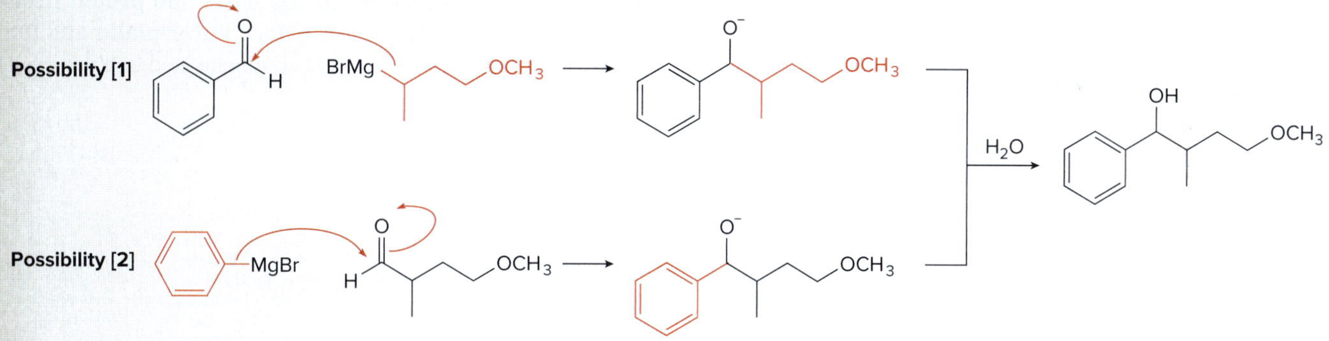

Possibility [1]

Possibility [2]

Problem 13.21

What Grignard reagent and carbonyl compound are needed to prepare each alcohol? As shown in part (d), 3° alcohols with three different R groups on the carbon bonded to the OH group can be prepared by three different Grignard reactions.

a. 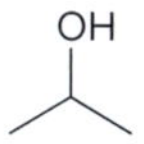 b. 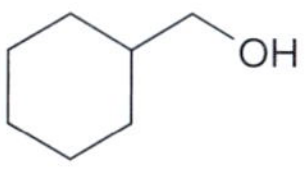 c. 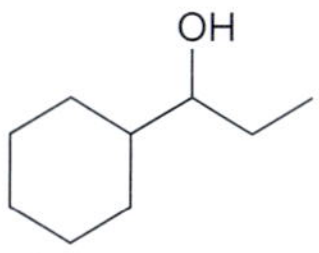 (two methods) d. 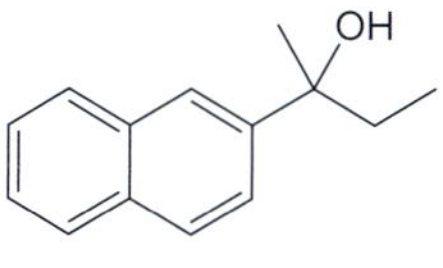 (three methods)

More Practice: Try Problems 13.55, 13.56, 13.58.

The *R* enantiomer of linalool is found in lavender oil, whereas the *S* enantiomer is found in coriander and sweet orange flowers. Linalool is used commercially in scented soaps and lotions. *Daniel C. Smith*

Problem 13.22 Linalool and lavandulol are two of the major components of lavender oil.
(a) What organolithium reagent and carbonyl compound can be used to make each alcohol?
(b) How might lavandulol be formed by reduction of a carbonyl compound? (c) Why can't linalool be prepared by a similar pathway?

linalool
(three methods)

lavandulol

Problem 13.23 What Grignard reagent and carbonyl compound can be used to prepare the antidepressant venlafaxine (trade name Effexor)?

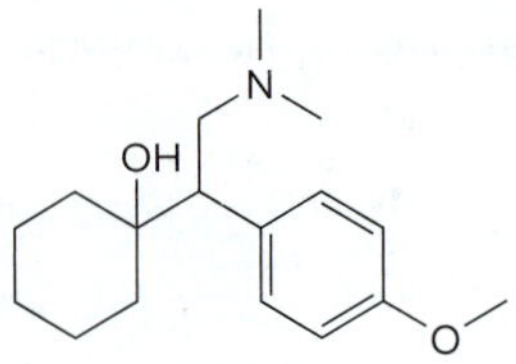

venlafaxine

13.12 Protecting Groups

Although the addition of organometallic reagents to carbonyls is a very versatile reaction, it cannot be used with molecules that contain both a carbonyl group and N—H or O—H bonds.

- Carbonyl compounds that also contain N—H or O—H bonds undergo an acid–base reaction with organometallic reagents, *not* nucleophilic addition.

Rapid acid–base reactions occur between organometallic reagents and all of the following functional groups: ROH, RCOOH, RNH$_2$, R$_2$NH, RCONH$_2$, RCONHR, and RSH.

Suppose, for example, that you wanted to add methylmagnesium chloride (CH$_3$MgCl) to the carbonyl group of 5-hydroxypentan-2-one to form a diol. Nucleophilic addition will *not* occur with this substrate. Instead, **because Grignard reagents are strong bases and proton transfer reactions are fast, CH$_3$MgCl removes the O—H proton before nucleophilic addition takes place.** The stronger acid and base react to form the weaker conjugate acid and conjugate base, as we learned in Section 13.9C.

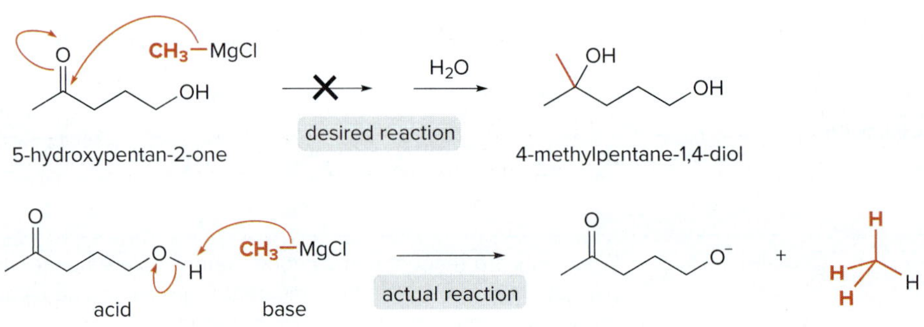

5-hydroxypentan-2-one

desired reaction

4-methylpentane-1,4-diol

acid   base

actual reaction

products of proton transfer

Solving this problem requires a three-step strategy:

Step [1] Convert the OH group to another functional group that does not interfere with the desired reaction. This new blocking group is called a **protecting group**, and the reaction that creates it is called *protection*.

Step [2] Carry out the desired reaction.

Step [3] Remove the protecting group. This reaction is called *deprotection*.

Application of the general strategy to the Grignard addition of CH₃MgCl to 5-hydroxypentan-2-one is illustrated in Figure 13.2.

Figure 13.2
General strategy for using a protecting group

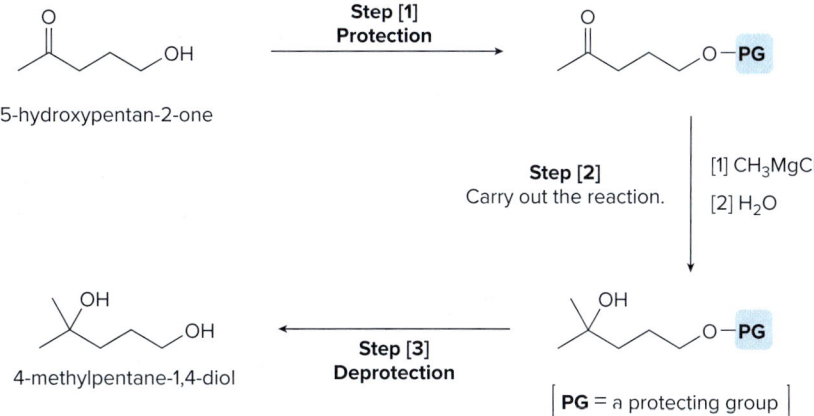

- In Step [1], the OH proton in 5-hydroxypentan-2-one is replaced with a protecting group, written as **PG**. Because the product of Step [1] no longer has an OH proton, it can now undergo nucleophilic addition.
- In Step [2], CH₃MgCl adds to the carbonyl group to yield a 3° alcohol after protonation with water.
- Removal of the protecting group in Step [3] forms the desired product, 4-methylpentane-1,4-diol.

A common OH protecting group is a **silyl ether**. A silyl ether has a new O—Si bond in place of the O—H bond of the alcohol. The most widely used silyl ether protecting group is the **tert-butyldimethylsilyl ether,** abbreviated as **TBDMS**.

tert-Butyldimethylsilyl ethers are prepared from alcohols by reaction with *tert*-butyldimethylsilyl chloride and an amine base, usually imidazole.

The silyl ether is typically removed with a fluoride salt, usually **tetrabutylammonium fluoride** $(CH_3CH_2CH_2CH_2)_4N^+F^-$, drawn as $Bu_4N^+F^-$ (Bu = butyl).

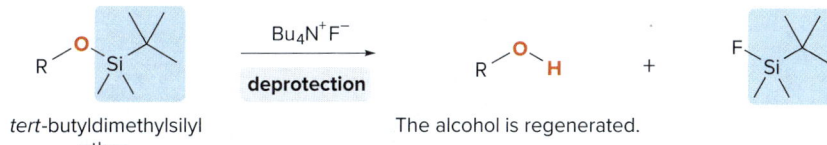

The use of a *tert*-butyldimethylsilyl ether as a protecting group makes possible the synthesis of 4-methylpentane-1,4-diol by a three-step sequence.

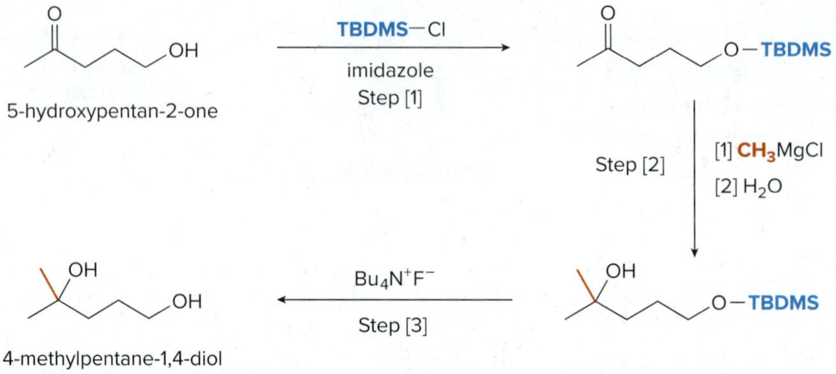

- **Step [1] Protect the OH group** as a *tert*-butyldimethylsilyl ether by reaction with *tert*-butyldimethylsilyl chloride and imidazole.
- **Step [2] Carry out nucleophilic addition** by using CH_3MgCl, followed by protonation.
- **Step [3] Remove the protecting group** with tetrabutylammonium fluoride to form the desired addition product.

Protecting groups block interfering functional groups, and in this way, a wider variety of reactions can take place with a particular substrate. For more on protecting groups, see the discussion of acetals in Section 14.16.

Problem 13.24 Using protecting groups, show how estrone can be converted to ethynylestradiol, a widely used oral contraceptive.

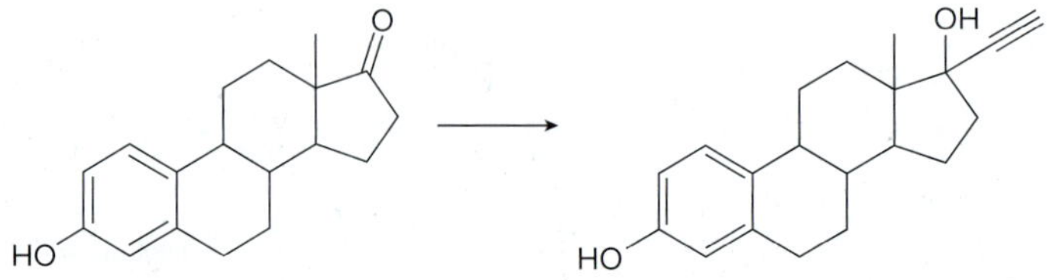

estrone ethynylestradiol

13.13 Reaction of Organometallic Reagents with Carboxylic Acid Derivatives

Organometallic reagents react with carboxylic acid derivatives (RCOZ) to form two different products, depending on the identity of both the leaving group Z and the reagent R—M. The most useful reactions are carried out with esters and acid chlorides, forming either **ketones** or **3° alcohols.**

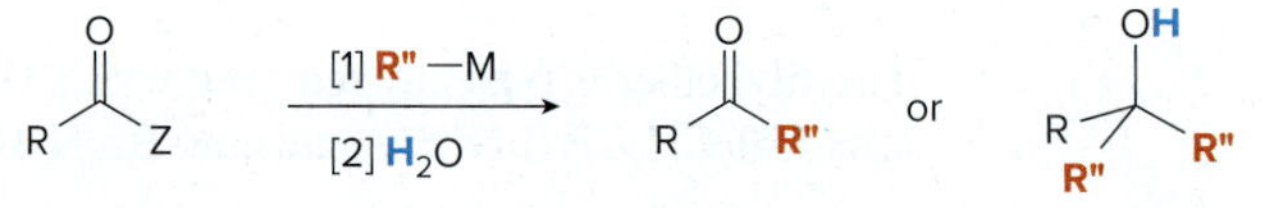

Z = Cl or OR' **ketone** **3° alcohol**

- Keep in mind that RLi and RMgX are very reactive reagents, whereas R_2CuLi is much less reactive. This reactivity difference makes selective reactions possible.

13.13A Reaction of RLi and RMgX with Esters and Acid Chlorides

Both esters and acid chlorides form 3° alcohols when treated with two equivalents of either Grignard or organolithium reagents. Two new carbon–carbon bonds are formed in the product.

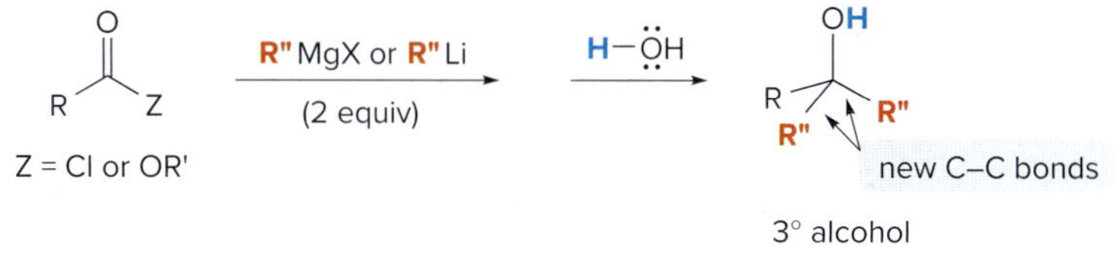
R"MgX or R"Li
(2 equiv)
H–ÖH
Z = Cl or OR'
new C–C bonds
3° alcohol

Two examples using Grignard reagents are shown.

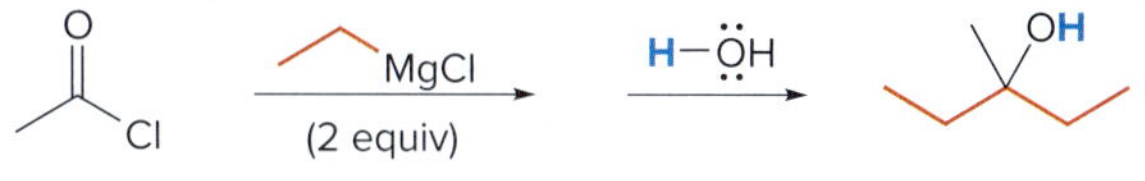
MgCl
(2 equiv)
H–ÖH

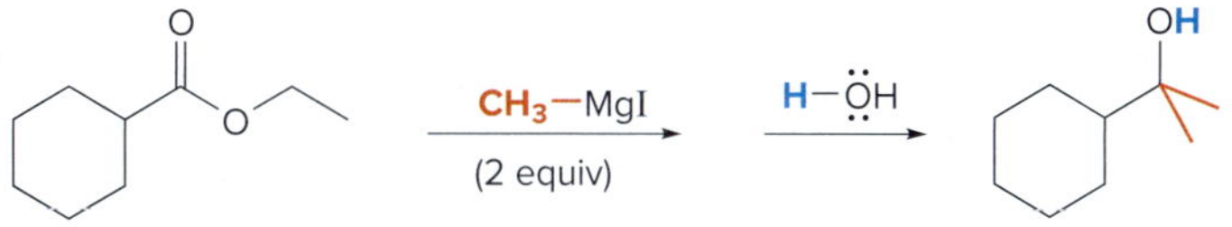
CH_3–MgI
(2 equiv)
H–ÖH

Problem 13.25 Draw the product formed when each compound is treated with two equivalents of $CH_3CH_2CH_2CH_2MgBr$ followed by H_2O.

a. 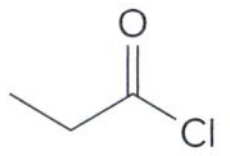

b. 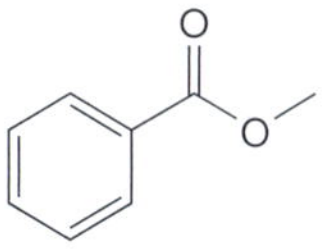

c. 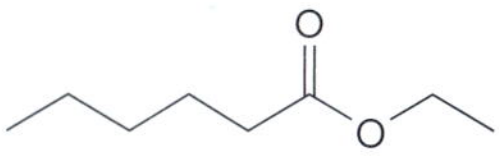

The mechanism for this addition reaction resembles the mechanism for the metal hydride reduction of acid chlorides and esters discussed in Section 13.7A. The mechanism is conceptually divided into two parts: **nucleophilic substitution** to form a ketone (Steps [1] and [2]), followed by **nucleophilic addition** to form a 3° alcohol (Steps [3] and [4]), as shown in Mechanism 13.7.

Mechanism 13.7 Reaction of R"MgX or R"Li with RCOCl and RCOOR'

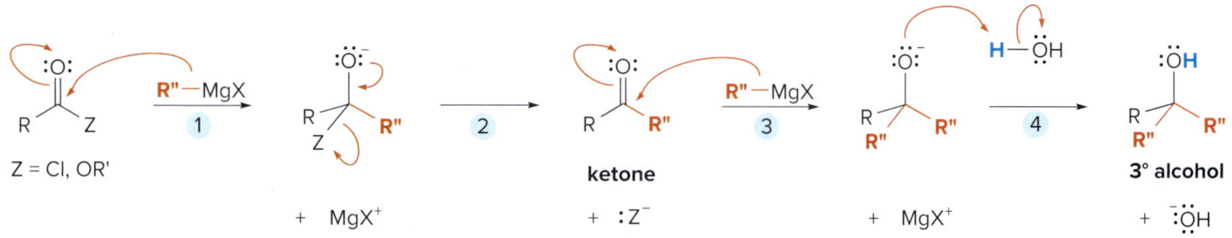
R"–MgX
1
2
3
4
Z = Cl, OR'
ketone
3° alcohol
+ MgX⁺
+ :Z⁻
+ MgX⁺
+ :ÖH⁻

1 **Nucleophilic attack of (R")⁻** forms a tetrahedral intermediate with a leaving group Z.

2 The π bond is re-formed and the **leaving group Z departs** to form a ketone. The overall result of addition of (R")⁻ and elimination of Z:⁻ is **substitution of R" for Z.**

3 **Nucleophilic attack of (R")⁻** forms an alkoxide with no leaving group.

4 Protonation of the alkoxide by H_2O forms a **3° alcohol.**

Organolithium and Grignard reagents afford 3° alcohols when they react with esters and acid chlorides. As soon as the ketone forms by addition of one equivalent of reagent to RCOZ (Steps [1] and [2] of the mechanism), it reacts with a second equivalent of reagent to form the 3° alcohol.

This reaction is more limited than the Grignard addition to aldehydes and ketones, because only 3° alcohols having **two identical alkyl groups** can be prepared. Nonetheless, it is still a valuable reaction because it forms two new carbon–carbon bonds.

626 Chapter 13 Introduction to Carbonyl Chemistry; Organometallic Reagents; Oxidation and Reduction

| Sample Problem 13.4 | Identifying the Ester and Grignard Reagent Needed to Prepare an Alcohol |

What ester and Grignard reagent are needed to prepare the following alcohol?

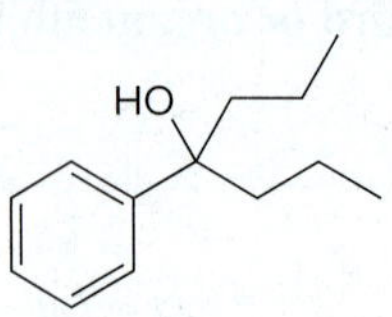

Solution

A 3° alcohol formed from an ester and Grignard reagent must have **two identical R groups**, and these **R groups come from RMgX. The remainder of the molecule comes from the ester.** The carbon (labeled in blue) bonded to the OH group comes from the carbonyl carbon.

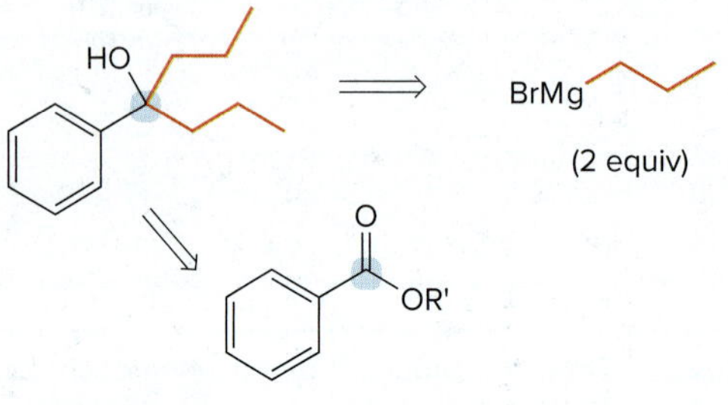

R' = any alkyl group

Checking in the synthetic direction:

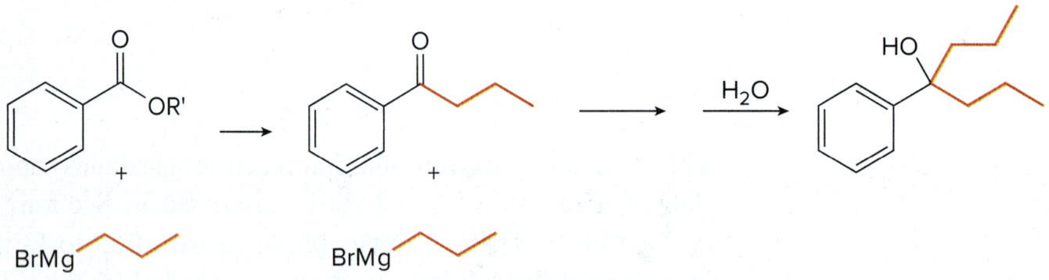

BrMg— first equivalent

BrMg— second equivalent

Problem 13.26 What ester and Grignard reagent are needed to prepare each alcohol?

a. 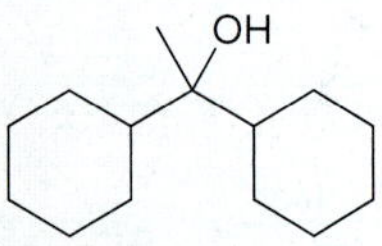 b. 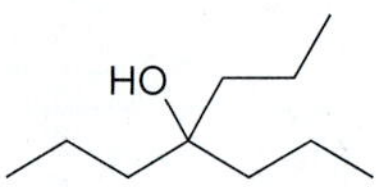 c.

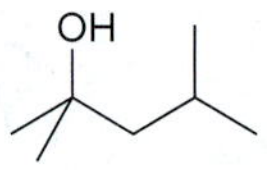

More Practice: Try Problem 13.57.

13.13B Reaction of R₂CuLi with Acid Chlorides

To form a ketone from a carboxylic acid derivative, a less reactive organometallic reagent—namely, an **organocuprate**—is needed. **Acid chlorides, which have the best leaving group (Cl⁻) of the carboxylic acid derivatives, react with R'₂CuLi, to give a ketone as product.** Esters, which contain a poorer leaving group (⁻OR), do *not* react with R'₂CuLi.

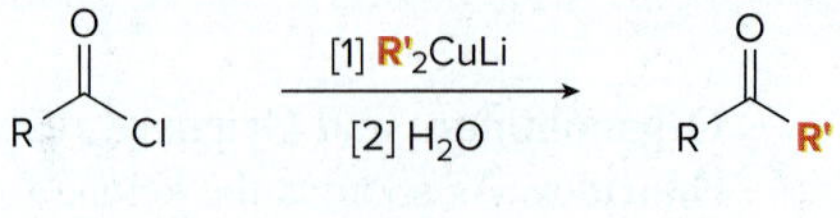

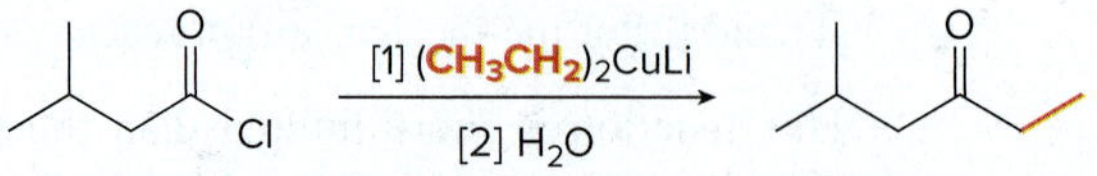

$$R-C(=O)-Cl \xrightarrow[\text{[2] H}_2\text{O}]{\text{[1] R'}_2\text{CuLi}} R-C(=O)-R'$$

$$(CH_3)_2CHCH_2-C(=O)-Cl \xrightarrow[\text{[2] H}_2\text{O}]{\text{[1] (CH}_3\text{CH}_2)_2\text{CuLi}} (CH_3)_2CHCH_2-C(=O)-CH_2CH_3$$

This reaction results in **nucleophilic substitution of an alkyl group R' for the leaving group Cl,** forming one new carbon–carbon bond.

13.14 Reaction of Organometallic Reagents with Other Compounds

Problem 13.27 What organocuprate reagent is needed to convert CH_3CH_2COCl to each ketone?

a. 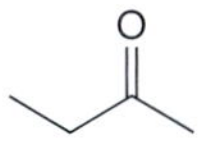 (propan-2-one / acetone structure)   b. 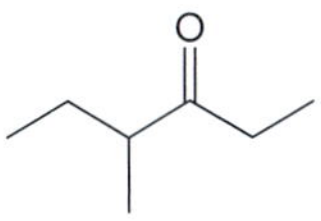 (4-methylhexan-3-one structure)   c. 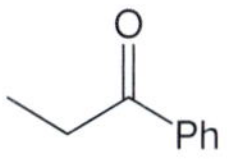 (PhCOCH₃-type: phenyl methyl ketone with extra CH₂)

Problem 13.28 What reagent is needed to convert $(CH_3)_2CHCH_2COCl$ to each compound?

a. 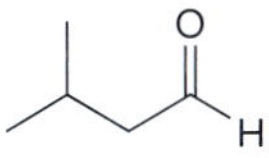 (aldehyde with isobutyl group)   b. 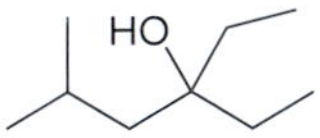 (tertiary alcohol, HO)   c. 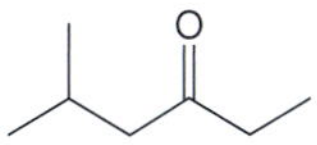 (ketone)   d. 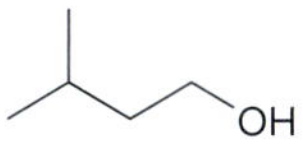 (alcohol OH)

A ketone with two different R groups bonded to the carbonyl carbon can be made by two different methods, as illustrated in Sample Problem 13.5.

Sample Problem 13.5 Determining the Acid Chloride and Organocuprate Needed to Prepare a Ketone

Show two different ways to prepare pentan-2-one from an acid chloride and an organocuprate reagent.

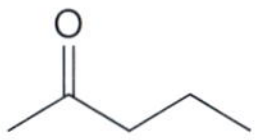

pentan-2-one

Solution

In each case, one alkyl group comes from the organocuprate and one comes from the acid chloride.

Possibility [1] Use $(CH_3)_2CuLi$ and a four-carbon acid chloride.

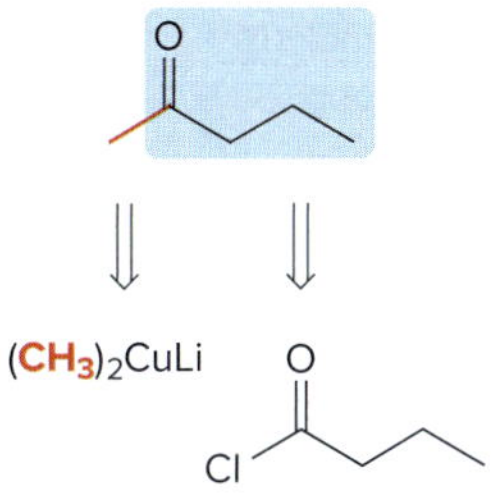

$(CH_3)_2CuLi$ + Cl-CO-CH₂CH₂CH₃

Possibility [2] Use $(CH_3CH_2CH_2)_2CuLi$ and a two-carbon acid chloride.

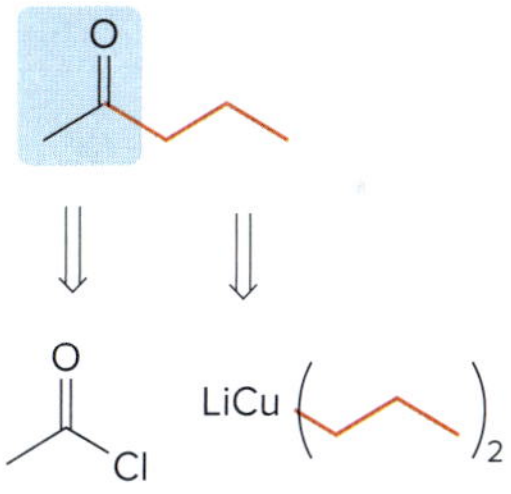

CH₃COCl + $LiCu(CH_2CH_2CH_3)_2$

Problem 13.29 Draw two different ways to prepare each ketone from an acid chloride and an organocuprate reagent.

a. 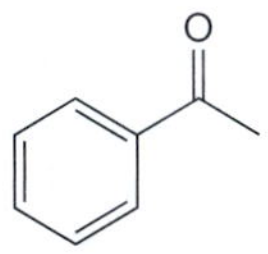 (acetophenone-like: Ph–CO–CH₃)
b. 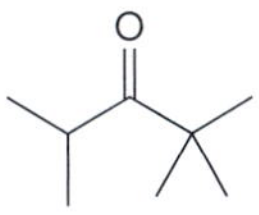 (2-methyl-1-(tert-butyl) ketone structure)

More Practice: Try Problem 13.38a.

13.14 Reaction of Organometallic Reagents with Other Compounds

Because organometallic reagents are strong nucleophiles, they react with many other electrophiles in addition to carbonyl groups. Because these reactions always lead to the formation of new carbon–carbon bonds, they are also valuable in organic synthesis. In Section 13.14, we examine the reactions of organometallic reagents with **carbon dioxide** and **epoxides.**

13.14A Reaction of Grignard Reagents with Carbon Dioxide

Grignard reagents react with CO_2 to give carboxylic acids after protonation with aqueous acid. This reaction, called **carboxylation**, forms a carboxylic acid with one more carbon atom than the Grignard reagent from which it is prepared.

$$R-MgX \xrightarrow[\text{[2] } H_3O^+]{\text{[1] } CO_2} R-\underset{OH}{\overset{O}{\|}}C\text{—carboxylic acid}$$

Because Grignard reagents are made from organic halides, RX can be converted to a **carboxylic acid having one more carbon atom** by a two-step reaction sequence: **formation of a Grignard reagent,** followed by **reaction with CO_2.**

PhCl $\xrightarrow{Mg}$ PhMgCl $\xrightarrow[\text{[2] } H_3O^+]{\text{[1] } CO_2}$ PhCOOH

new C–C bond in red

The mechanism resembles earlier reactions of nucleophilic Grignard reagents with carbonyl groups, as shown in Mechanism 13.8.

Mechanism 13.8 Carboxylation—Reaction of RMgX with CO_2

$$R-MgX + O=C=O \xrightarrow{1} R-CO_2^- + MgX^+ \xrightarrow{2, H-OH_2^+} R-COOH + H_2O$$

1. The nucleophilic Grignard reagent attacks the electrophilic carbon of CO_2, cleaving the π bond and forming a new carbon–carbon bond.
2. Protonation of the carboxylate anion with aqueous acid forms the carboxylic acid.

Problem 13.30 What carboxylic acid is formed from each alkyl halide on treatment with [1] Mg; [2] CO_2; [3] H_3O^+?

a. cyclohexyl bromide   b. 1-chloropropane (CH₃CH₂CH₂Cl — shown as propyl chloride)   c. 4-methoxybenzyl bromide

13.14B Reaction of Organometallic Reagents with Epoxides

Like other strong nucleophiles, **organometallic reagents—RLi, RMgX, and R_2CuLi—open epoxide rings to form alcohols.**

epoxide $\xrightarrow[\text{[2] } H_2O]{\text{[1] RLi, RMgX, or } R_2CuLi}$ alcohol

ethylene oxide $\xrightarrow[\text{[2] } H_2O]{\text{[1] PhMgBr}}$ PhCH₂CH₂OH

13.15 α,β-Unsaturated Carbonyl Compounds

The opening of epoxide rings with negatively charged nucleophiles was discussed in Section 9.16A.

The reaction follows the same two-step process as the opening of epoxide rings with other negatively charged nucleophiles—that is, **nucleophilic attack from the back side of the epoxide ring, followed by protonation of the resulting alkoxide.** In unsymmetrical epoxides, nucleophilic attack occurs at the *less* substituted carbon atom.

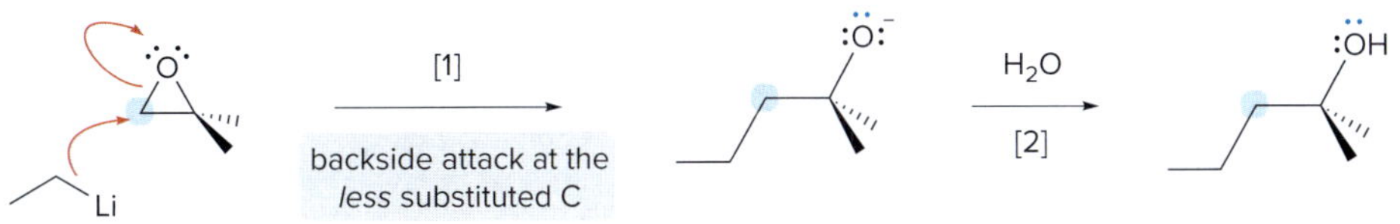

Problem 13.31 What epoxide is needed to convert CH₃CH₂MgBr to each of the following alcohols, after quenching with water?

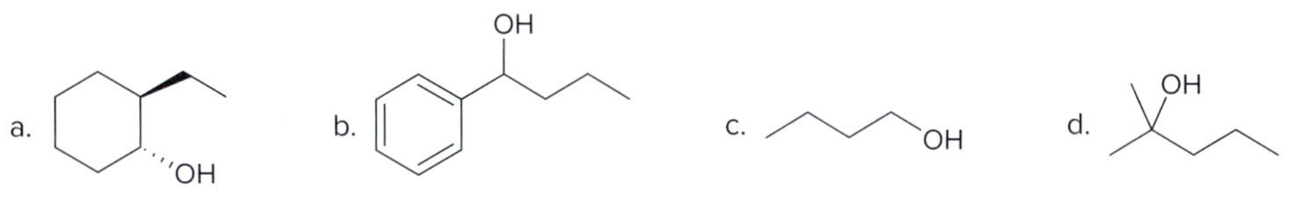

a. (+ enantiomer) b. c. d.

13.15 α,β-Unsaturated Carbonyl Compounds

α,β-Unsaturated carbonyl compounds are conjugated molecules containing a carbonyl group and a carbon–carbon double bond, separated by a single σ bond.

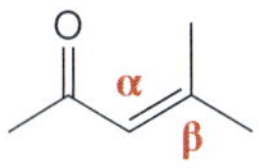

α,β-unsaturated carbonyl compound

Both functional groups of α,β-unsaturated carbonyl compounds have π bonds, but individually, they react with very different kinds of reagents. Carbon–carbon double bonds react with electrophiles (Chapter 10) and carbonyl groups react with nucleophiles (Section 13.2). What happens, then, when these two functional groups having opposite reactivity are in close proximity?

Because the two π bonds are conjugated, the electron density in an α,β-unsaturated carbonyl compound is *delocalized over four atoms*. Three resonance structures show that the carbonyl carbon and the β carbon bear a partial positive charge. This means that **α,β-unsaturated carbonyl compounds can react with nucleophiles at two different sites.**

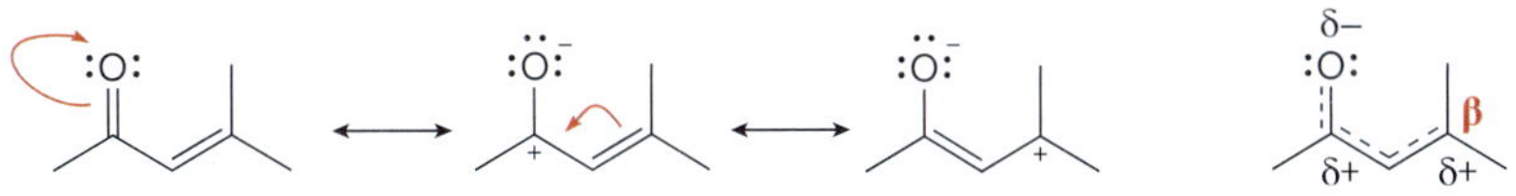

three resonance structures for an α,β-unsaturated carbonyl compound

hybrid
two electrophilic sites

- Addition of a nucleophile to the carbonyl carbon, called **1,2-addition**, adds the elements of H and Nu across the C=O, forming an allylic alcohol.

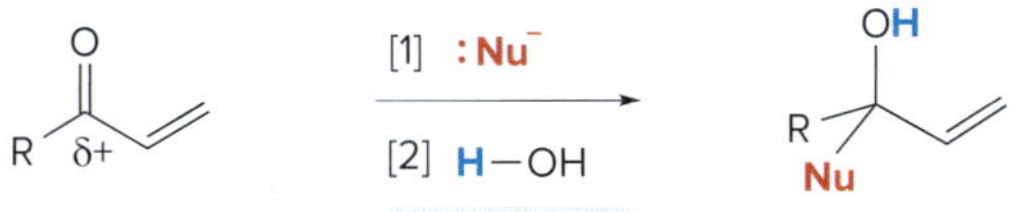

1,2-addition   allylic alcohol

- Addition of a nucleophile to the β carbon, called **1,4-addition or conjugate addition,** forms a carbonyl compound.

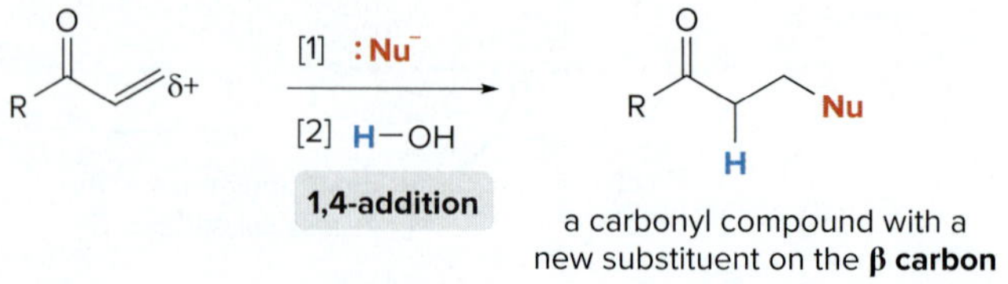

Both 1,2- and 1,4-addition result in nucleophilic **addition of the elements of H and Nu.**

13.15A The Mechanisms for 1,2-Addition and 1,4-Addition

The steps for the mechanism of 1,2-addition are exactly the same as those for the nucleophilic addition to an aldehyde or ketone—that is, **nucleophilic attack,** followed by **protonation,** as shown in Mechanism 13.6 in Section 13.10A.

The mechanism for 1,4-addition also begins with nucleophilic attack, and then protonation and tautomerization add the elements of H and Nu to the α and β carbons of the carbonyl compound, as shown in Mechanism 13.9.

Mechanism 13.9 1,4-Addition to an α,β-Unsaturated Carbonyl Compound

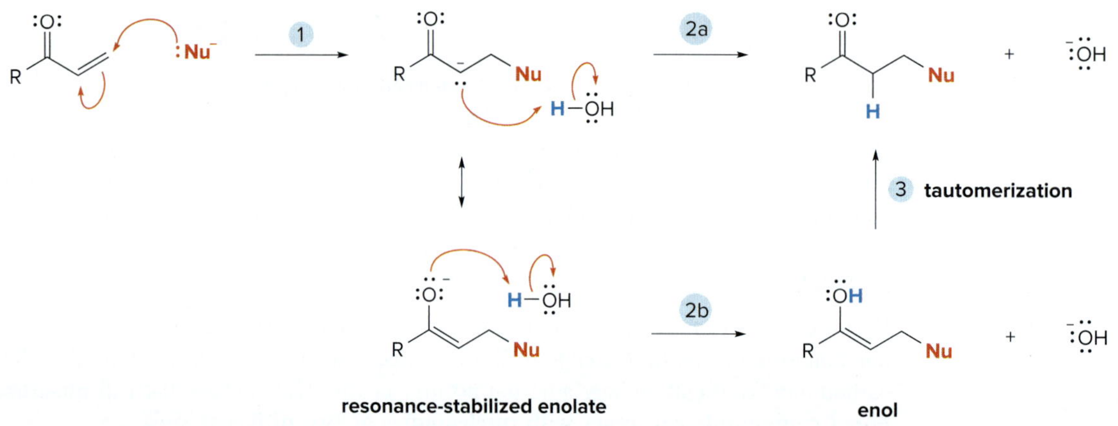

1. Nucleophilic attack at the electrophilic β carbon forms a **resonance-stabilized enolate anion,** which can react on either carbon or oxygen in the second step.
2a. Protonation of the carbon end of the enolate forms the 1,4-addition product directly.
2b – 3. Protonation of the oxygen end of the enolate forms an **enol,** which undergoes **tautomerization** by the two-step process described in Section 10.18. This forms the same 1,4-addition product that results from protonation on carbon.

13.15B Reaction of α,β-Unsaturated Carbonyl Compounds with Organometallic Reagents

The **identity of the metal** in an organometallic reagent determines whether it reacts with an α,β-unsaturated aldehyde or ketone by 1,2-addition or 1,4-addition.

13.16 Summary—The Reactions of Organometallic Reagents

- Organolithium and Grignard reagents form 1,2-addition products.

Why is conjugate addition also called 1,4-addition? If the atoms of the enol are numbered beginning with the O atom, then the elements of H and Nu are bonded to atoms "1" and "4," respectively.

- Organocuprate reagents form 1,4-addition products.

Sample Problem 13.6 Drawing the Products of 1,2- and 1,4-Addition

Draw the products of each reaction.

Solution

The characteristic reaction of α,β-unsaturated carbonyl compounds is nucleophilic addition. The reagent determines the mode of addition (1,2- or 1,4-).

a. **Grignard reagents undergo 1,2-addition.** CH_3CH_2MgBr adds a new CH_3CH_2 group at the carbonyl carbon.

b. **Organocuprate reagents undergo 1,4-addition.** The cuprate reagent adds a new vinyl group ($CH_2=CH$) at the β carbon.

Problem 13.32

Draw the product when each compound is treated with either $(CH_3)_2CuLi$, followed by H_2O, or HC≡CLi, followed by H_2O.

More Practice: Try Problems 13.38c; 13.41c–e; 13.44b, e.

13.16 Summary—The Reactions of Organometallic Reagents

We have now seen many different reactions of organometallic reagents with a variety of functional groups, and you may have some difficulty keeping them all straight. Rather than memorizing them all, keep in mind the following three concepts:

[1] Organometallic reagents (R–M) attack electrophilic carbon atoms, especially the carbonyl carbon.

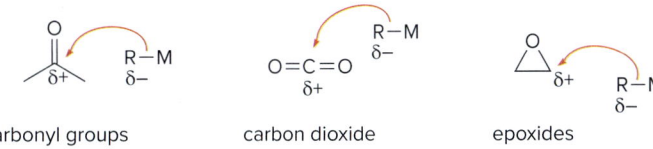

carbonyl groups   carbon dioxide   epoxides

632   Chapter 13   Introduction to Carbonyl Chemistry; Organometallic Reagents; Oxidation and Reduction

[2] After an organometallic reagent adds to a carbonyl group, the fate of the intermediate depends on the presence or absence of a leaving group.

- Without a leaving group, the characteristic reaction is *nucleophilic addition*.
- With a leaving group, the reaction is *nucleophilic substitution*.

[3] The polarity of the R–M bond determines the reactivity of the reagents.

- RLi and RMgX are very reactive reagents.
- R_2CuLi is much less reactive.

13.17 Synthesis

The reactions learned in Chapter 13 have proven extremely useful in organic synthesis. Oxidation and reduction reactions interconvert two functional groups that differ in oxidation state. Organometallic reagents form new carbon–carbon bonds.

Synthesis is perhaps the most difficult aspect of organic chemistry. It requires you to remember both the new reactions you've just learned and the ones you've encountered in previous chapters. In a successful synthesis, you must also put these reactions in a logical order. Don't be discouraged. Learn the basic reactions and then practice them over and over again with synthesis problems.

In Sample Problems 13.7 and 13.8 that follow, keep in mind that the products formed by the reactions of Chapter 13 can themselves be transformed into many other functional groups. For example, hexan-2-ol, the product of Grignard addition of butylmagnesium chloride to acetaldehyde, can be transformed into a variety of other compounds, as shown in Figure 13.3.

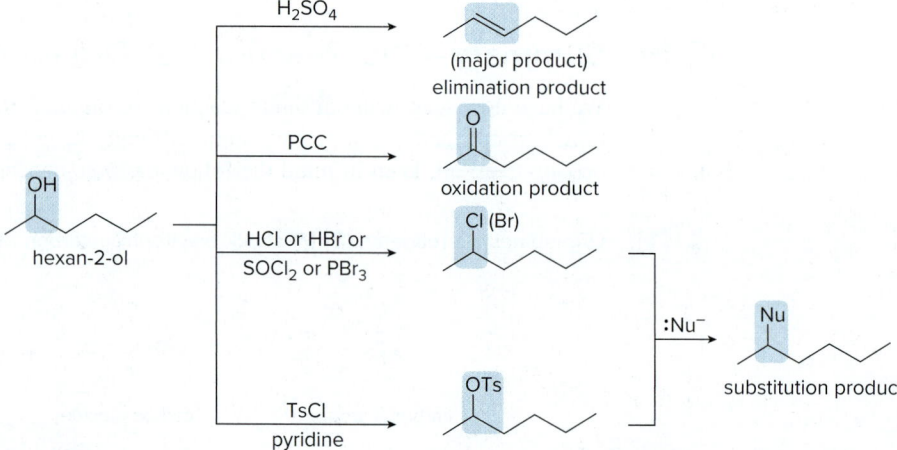

Figure 13.3
Conversion of hexan-2-ol to other compounds

13.17 Synthesis

Before proceeding with Sample Problems 13.7 and 13.8, you should review the stepwise strategy for designing a synthesis found in Section 10.21.

Sample Problem 13.7 — Devising a Synthesis with a Carbon–Carbon Bond-Forming Reaction

Synthesize 2,4-dimethylhexan-3-one from four-carbon alcohols.

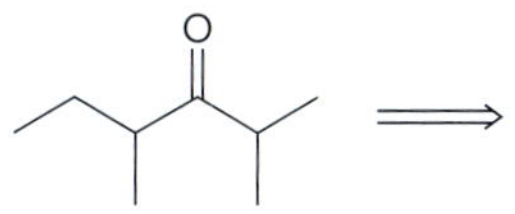

2,4-dimethylhexan-3-one ⟹ alcohols containing 4 C's

Retrosynthetic Analysis

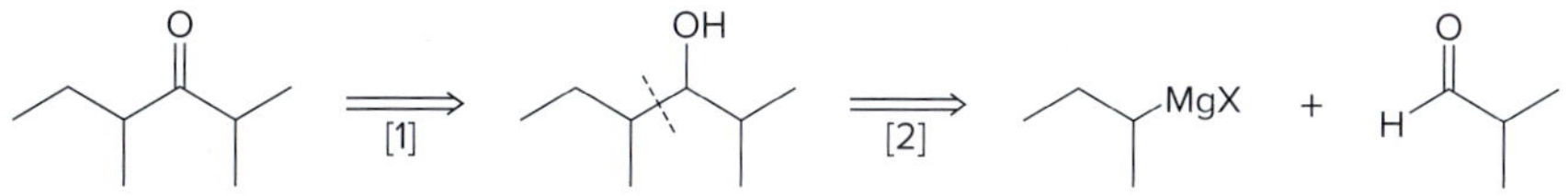

[1] Ketone ⟹ [2] 2° alcohol ⟹ Grignard (MgX) + aldehyde

Synthesize each of these components.

Thinking backwards:
- [1] Form the ketone by oxidation of a 2° alcohol.
- [2] Make the 2° alcohol by Grignard addition to an aldehyde. Both of these compounds have 4 C's, and each must be synthesized from an alcohol.

Synthesis

First, make both components needed for the Grignard reaction.

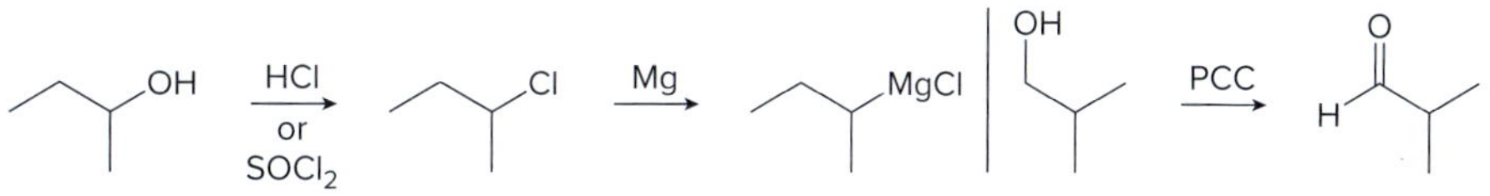

alcohol —HCl or SOCl₂→ alkyl chloride —Mg→ RMgCl

alcohol —PCC→ aldehyde

Then complete the synthesis with Grignard addition, followed by oxidation of the alcohol to the ketone.

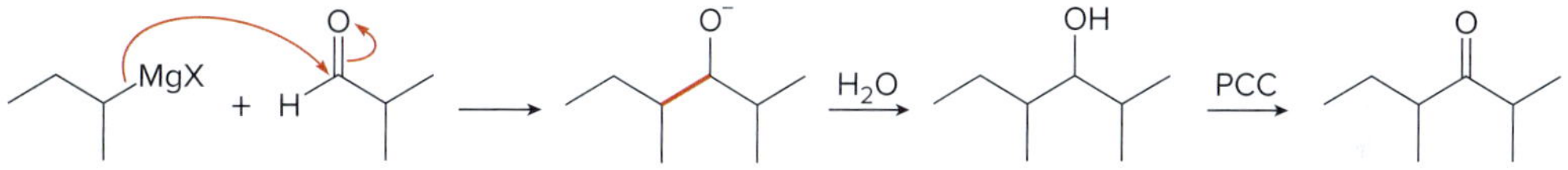

RMgX + aldehyde → alkoxide —H₂O→ 2° alcohol —PCC→ ketone

new C–C bond in red

Problem 13.33 Convert propan-2-ol [(CH₃)₂CHOH] to each compound. You may use any other organic or inorganic compounds.

a. 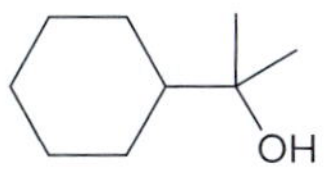 1-isopropylcyclohexan-1-ol   b. 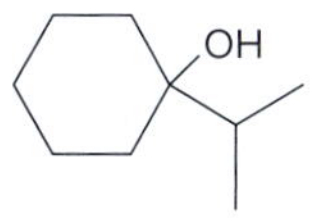 1-cyclohexyl-2-methylpropan-1-ol

More Practice: Try Problems 13.36; 13.61a, d.

Sample Problem 13.8 — Devising a Synthesis with a Grignard Addition

Synthesize isopropylcyclopentane from alcohols having ≤ 5 C's.

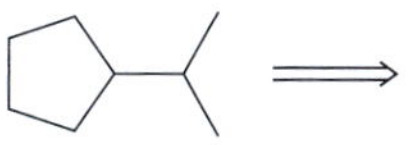

isopropylcyclopentane ⟹ alcohols having ≤ 5 C's

634   Chapter 13   Introduction to Carbonyl Chemistry; Organometallic Reagents; Oxidation and Reduction

Retrosynthetic Analysis

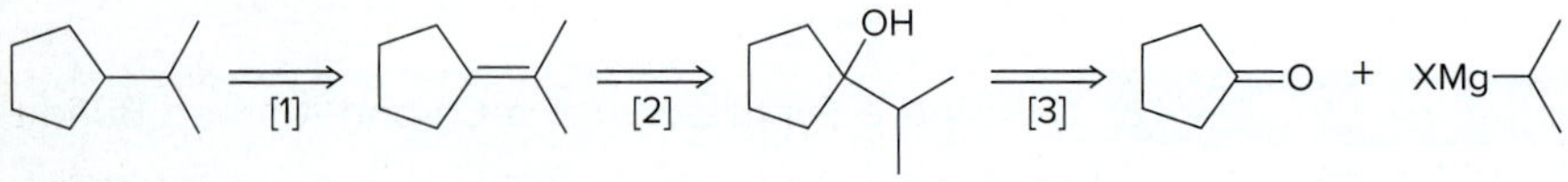

[Retrosynthetic scheme: isopropylcyclopentane ⇒[1] isopropylidenecyclopentane ⇒[2] 1-isopropylcyclopentan-1-ol ⇒[3] cyclopentanone + XMg–CH(CH₃)₂]

Thinking backwards:

- [1] Form the alkane by hydrogenation of an alkene.
- [2] Introduce the double bond by dehydration of an alcohol.
- [3] Form the 3° alcohol by Grignard addition to a ketone. Both components of the Grignard reaction must then be synthesized.

Synthesis

First, make both components needed for the Grignard reaction.

[Scheme: cyclopentanol —PCC→ cyclopentanone ; HO–CH(CH₃)₂ —PBr₃→ Br–CH(CH₃)₂ —Mg→ BrMg–CH(CH₃)₂]

Complete the synthesis with Grignard addition, dehydration, and hydrogenation.

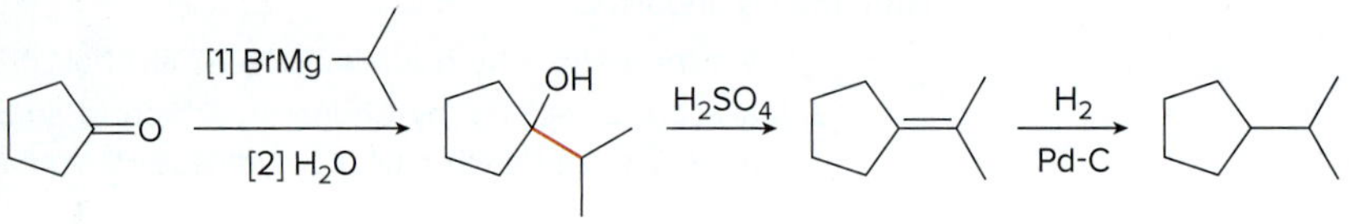

[Scheme: cyclopentanone —[1] BrMg–iPr, [2] H₂O→ 1-isopropylcyclopentan-1-ol (new C–C bond in red) —H₂SO₄→ isopropylidenecyclopentane (major product, tetrasubstituted double bond) —H₂, Pd-C→ isopropylcyclopentane]

Problem 13.34   Synthesize each compound from cyclohexanol, ethanol, and any other needed reagents.

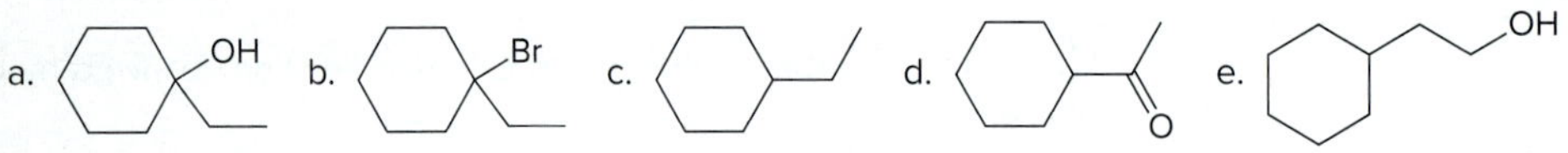

a. 1-methylcyclohexan-1-ol   b. 1-bromo-1-methylcyclohexane   c. ethylcyclohexane   d. 1-cyclohexylethan-1-one   e. 2-cyclohexylethan-1-ol

More Practice: Try Problems 13.60; 13.61b, c; 13.62–13.64.

Chapter 13 REVIEW

KEY REACTIONS

Reduction Reactions

[1] Reduction of aldehydes and ketones

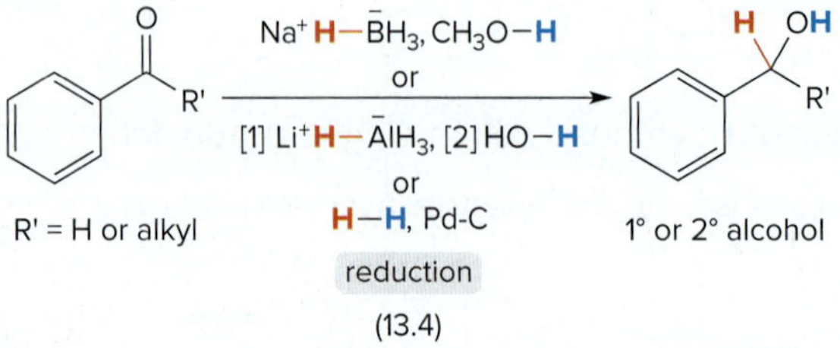

[Scheme: PhC(=O)R' (R' = H or alkyl) —Na⁺ H–BH₃, CH₃O–H or [1] Li⁺ H–AlH₃, [2] HO–H or H–H, Pd-C reduction→ PhCH(OH)R' (1° or 2° alcohol)]

(13.4)

Try Problems 13.35(**A**) a, b; 13.37a, b, c.

Key Reactions

[2] Reduction of acid chlorides

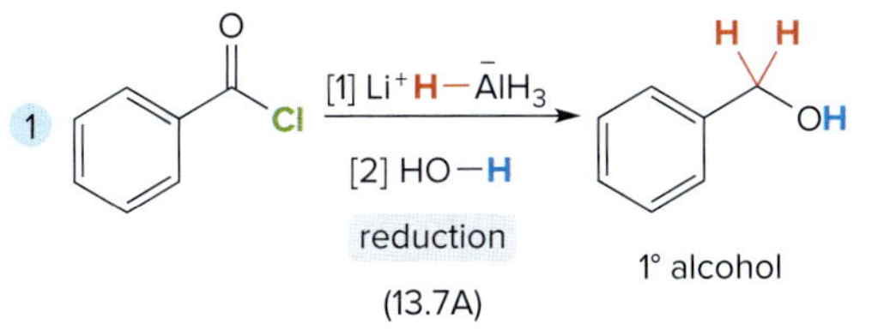

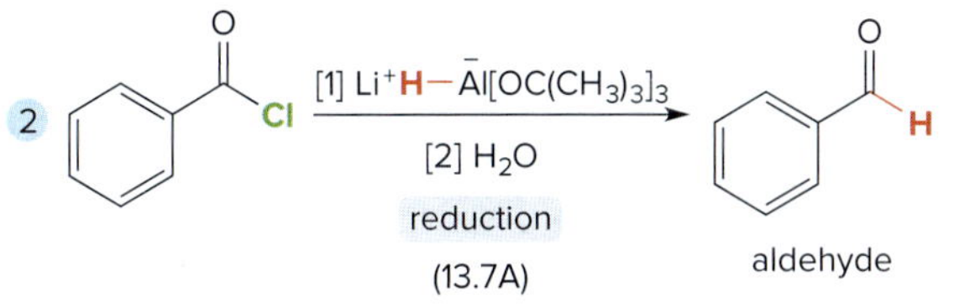

1. PhCOCl → PhCH₂OH (1° alcohol)
   Reagents: [1] Li⁺H–ĀlH₃; [2] HO–H
   reduction (13.7A)

2. PhCOCl → PhCHO (aldehyde)
   Reagents: [1] Li⁺H–Āl[OC(CH₃)₃]₃; [2] H₂O
   reduction (13.7A)

Try Problem 13.42d.

[3] Reduction of esters

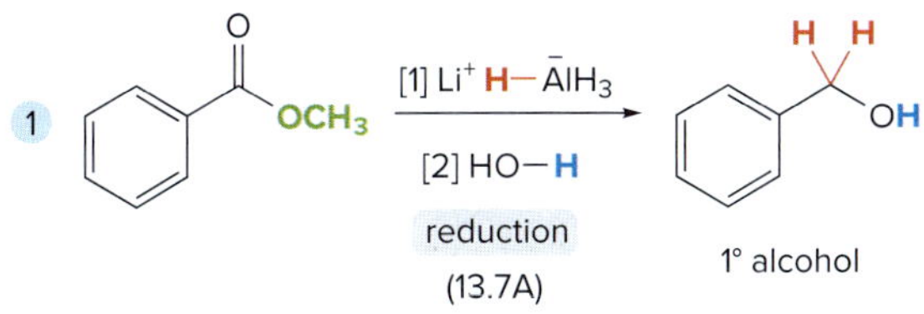

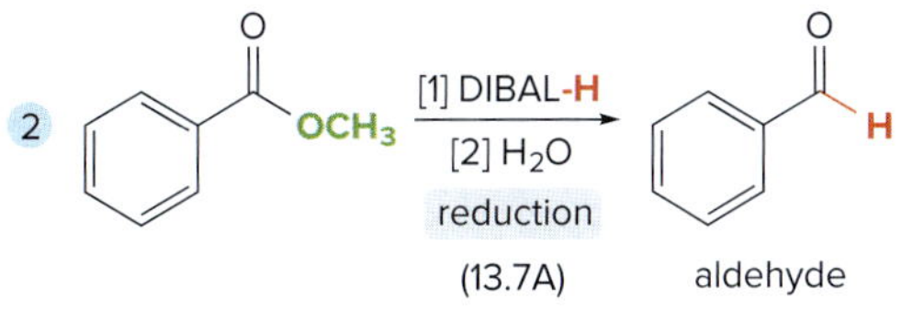

1. PhCO-OCH₃ → PhCH₂OH (1° alcohol)
   Reagents: [1] Li⁺H–ĀlH₃; [2] HO–H
   reduction (13.7A)

2. PhCO-OCH₃ → PhCHO (aldehyde)
   Reagents: [1] DIBAL-H; [2] H₂O
   reduction (13.7A)

Try Problems 13.35(**B**) a, b; 13.42a, b; 13.46c.

[4] Reduction of carboxylic acids and amides

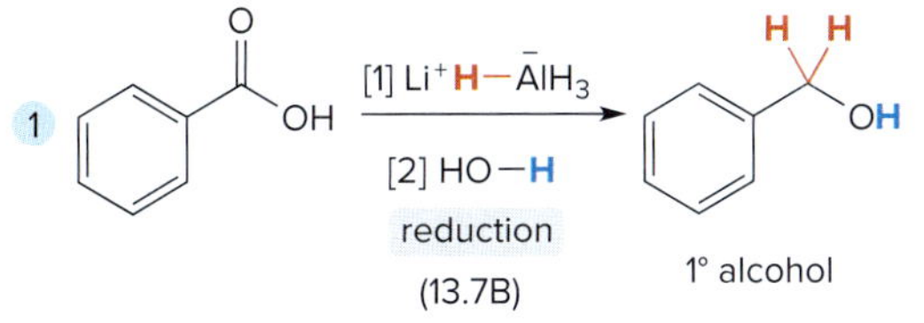

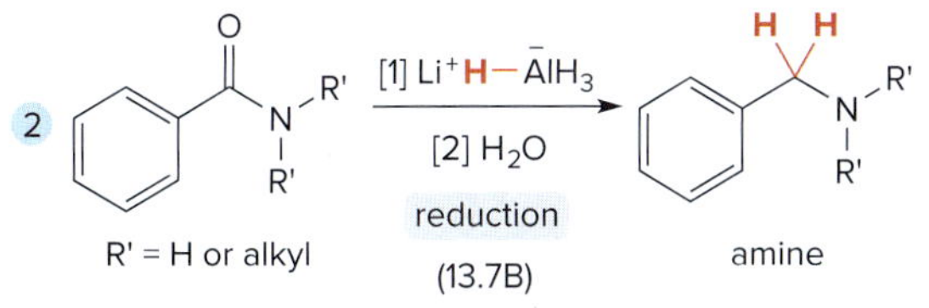

1. PhCOOH → PhCH₂OH (1° alcohol)
   Reagents: [1] Li⁺H–ĀlH₃; [2] HO–H
   reduction (13.7B)

2. PhCON(R')₂ → PhCH₂N(R')₂ (amine)
   R' = H or alkyl
   Reagents: [1] Li⁺H–ĀlH₃; [2] H₂O
   reduction (13.7B)

Try Problems 13.42c, 13.54.

Oxidation of Aldehydes to Carboxylic Acids

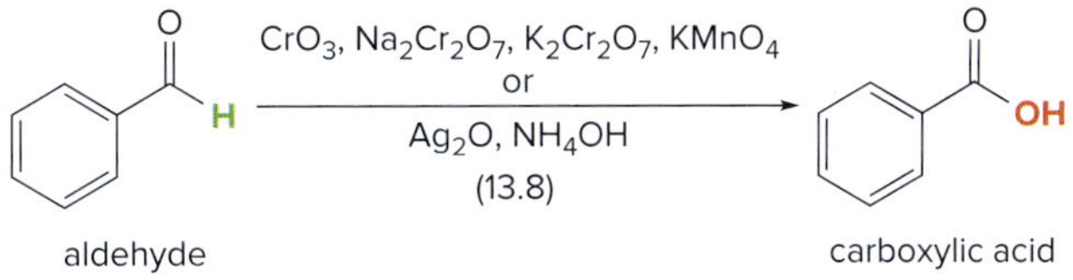

PhCHO (aldehyde) → PhCOOH (carboxylic acid)
Reagents: CrO₃, Na₂Cr₂O₇, K₂Cr₂O₇, KMnO₄ or Ag₂O, NH₄OH
(13.8)

Try Problems 13.37d–f, 13.43c–e.

Preparation of Organometallic Reagents

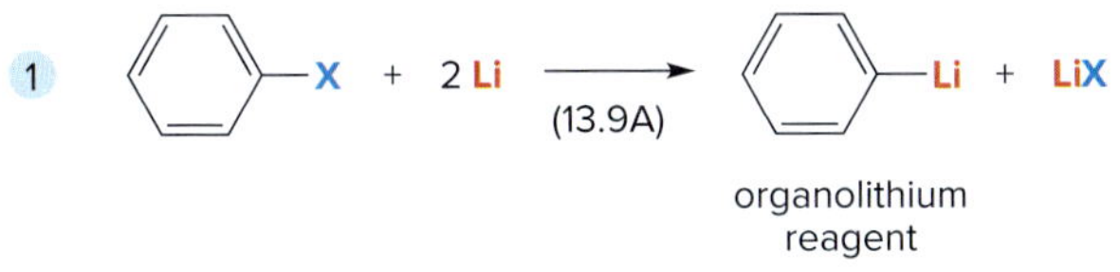

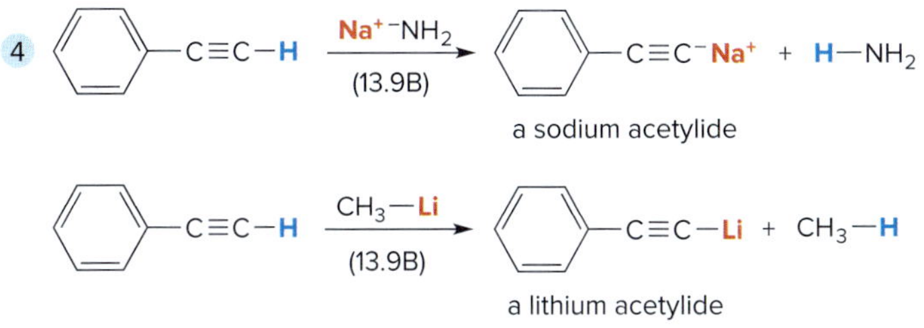

1. Ph–X + 2 Li → Ph–Li + LiX
   (13.9A) organolithium reagent

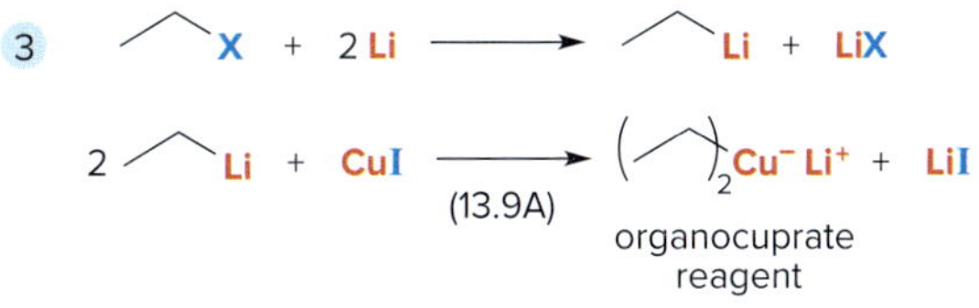

2. Ph–X + Mg → Ph–Mg–X
   (in diethyl ether) (13.9A) Grignard reagent

3. R–X + 2 Li → R–Li + LiX
   2 R–Li + CuI → (R)₂Cu⁻Li⁺ + LiI
   (13.9A) organocuprate reagent

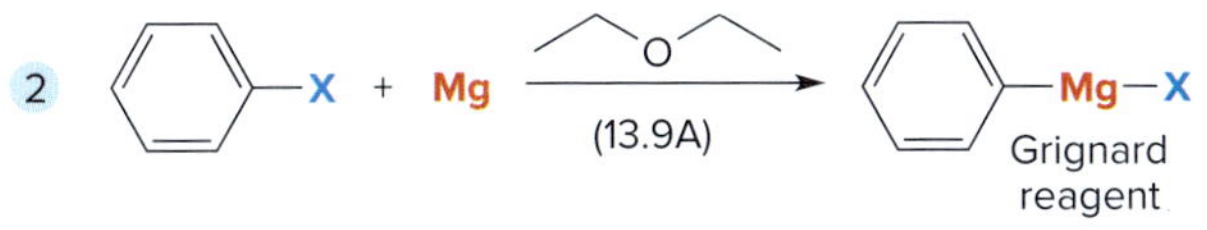

4. Ph–C≡C–H + Na⁺ ⁻NH₂ → Ph–C≡C⁻Na⁺ + H–NH₂
   (13.9B) a sodium acetylide

   Ph–C≡C–H + CH₃–Li → Ph–C≡C–Li + CH₃–H
   (13.9B) a lithium acetylide

Reactions with Organometallic Reagents

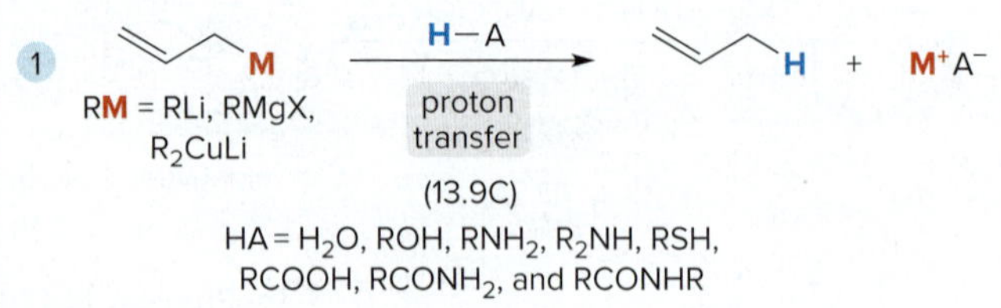

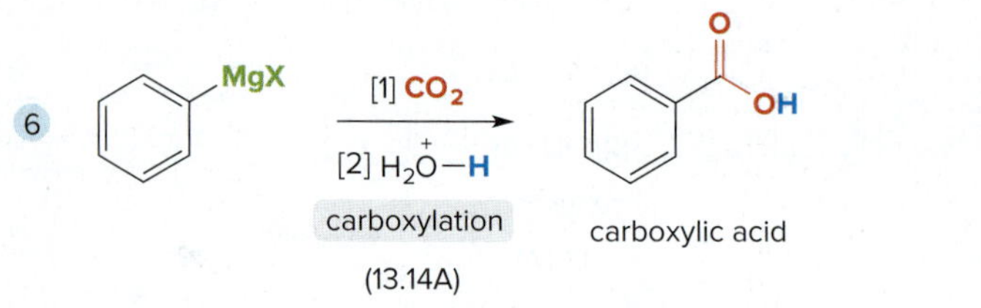

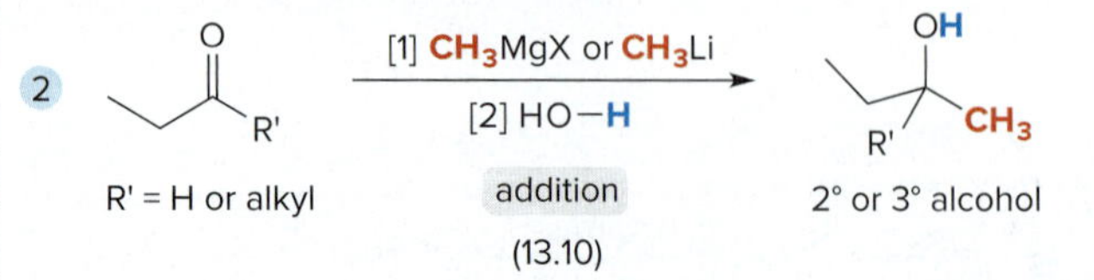

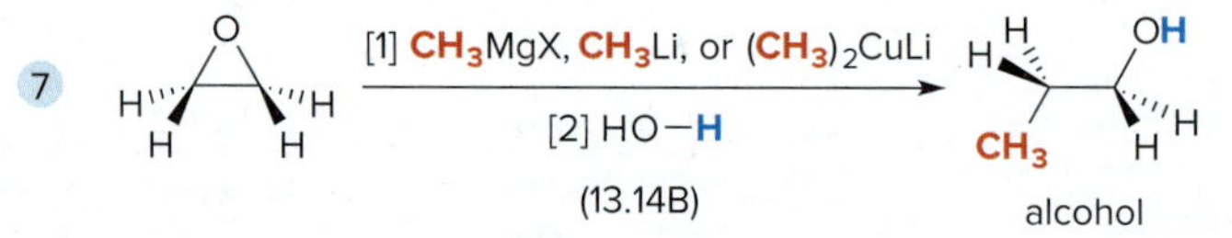

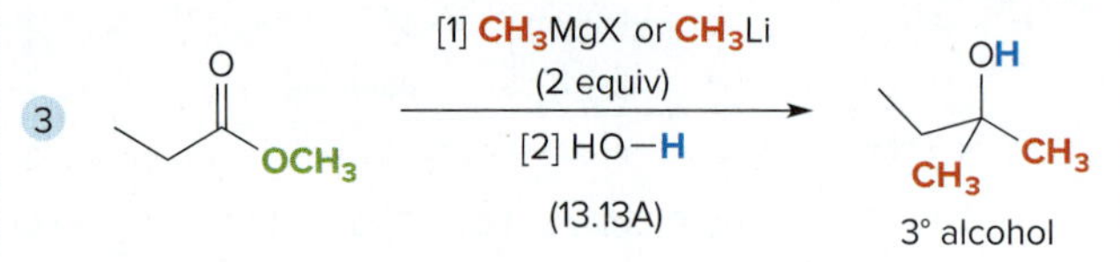

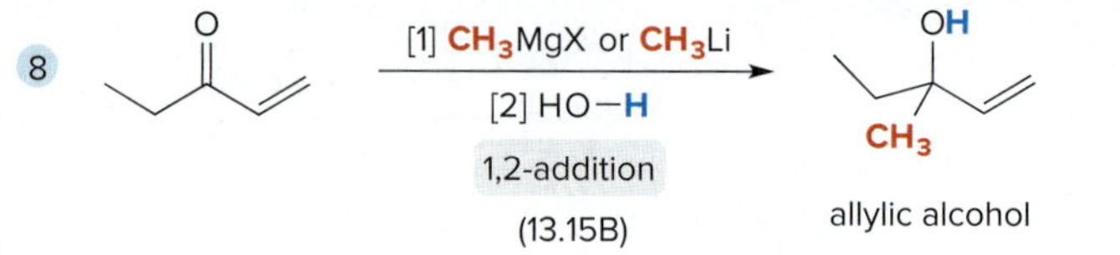

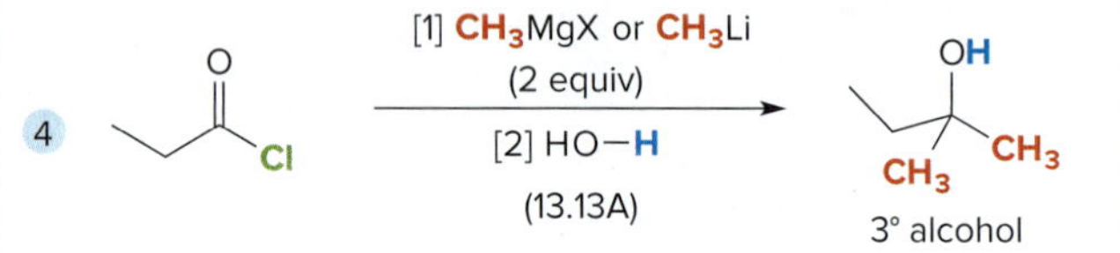

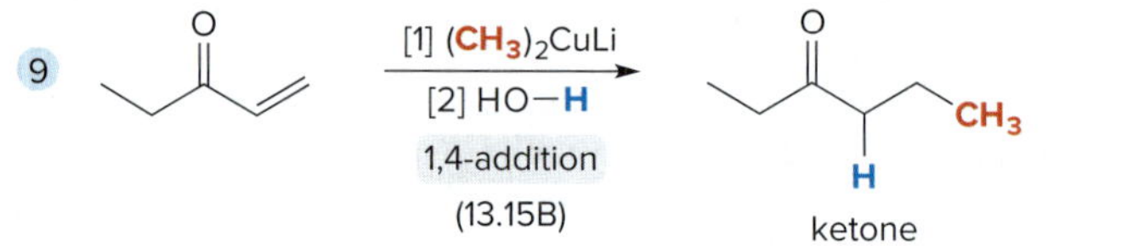

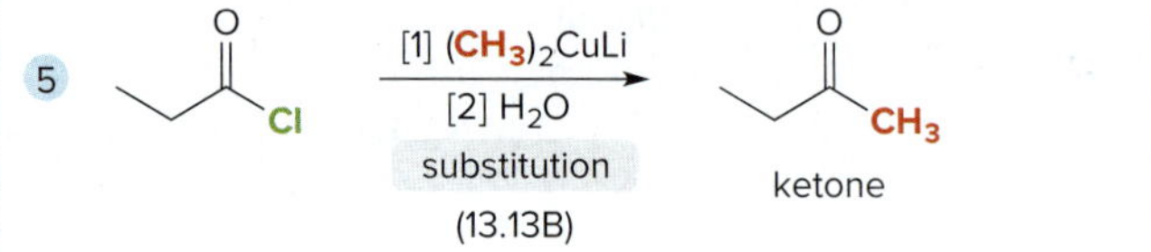

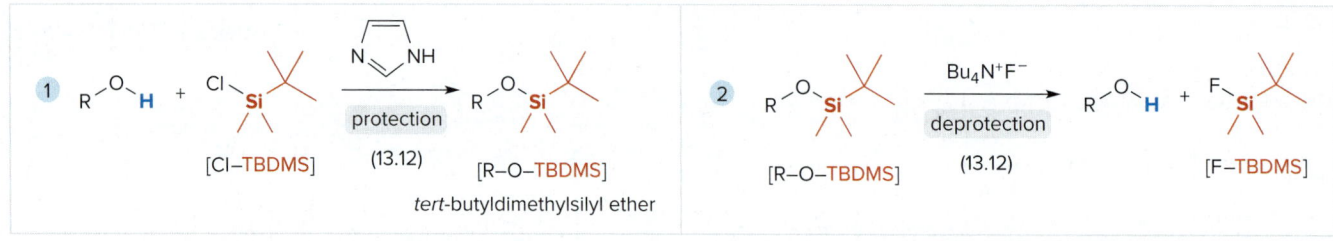

Try Problems 13.35c, d; 13.37g–k; 13.38; 13.41c–e; 13.44; 13.46a, b.

Protecting Groups

Try Problems 13.37l, 13.47.

KEY SKILLS

[1] Drawing all stereoisomers that form in a Grignard reaction (13.10B)

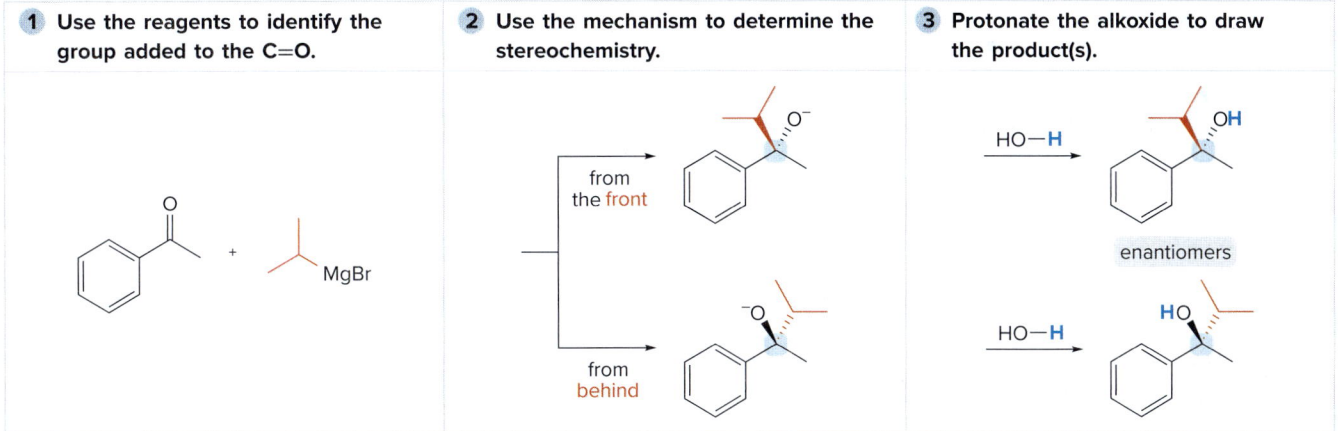

See Sample Problem 13.2. Try Problem 13.46a.

[2] Determining the starting materials for the preparation of an alcohol from an organolithium reagent and an ester (13.13); example: 3-ethyl-2-methylpentan-3-ol

See Sample Problem 13.4. Try Problem 13.57.

[3] Devising a synthesis of a ketone (13.11); example: 4-methylpentan-2-one from acetyl chloride and an alkyl bromide

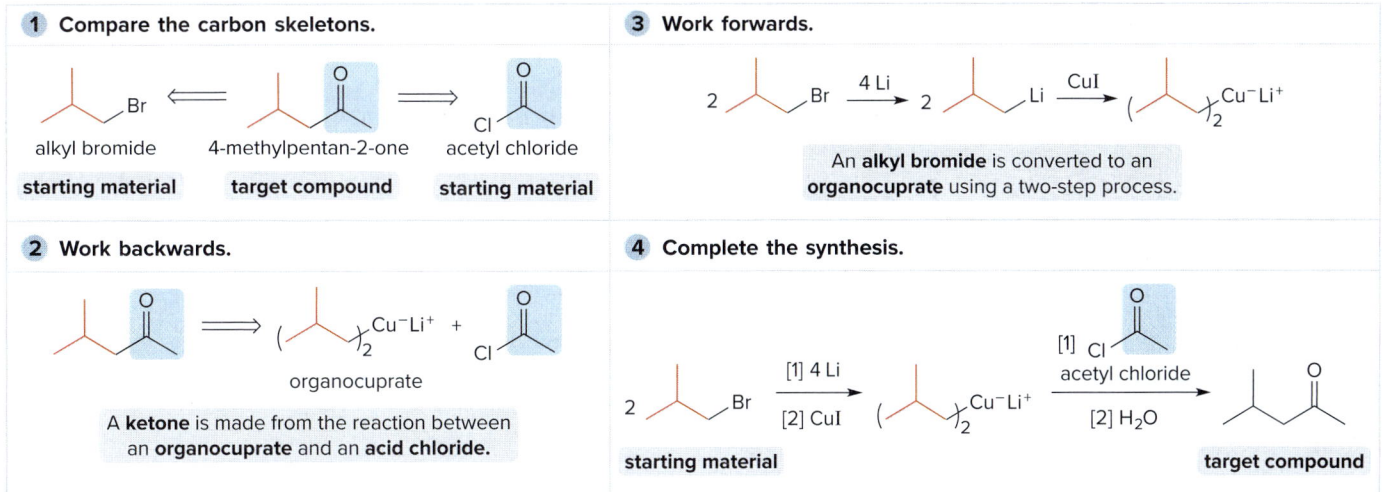

See Sample Problem 13.5. Try Problems 13.61d, 13.62a.

[4] Using a protecting group (13.12)

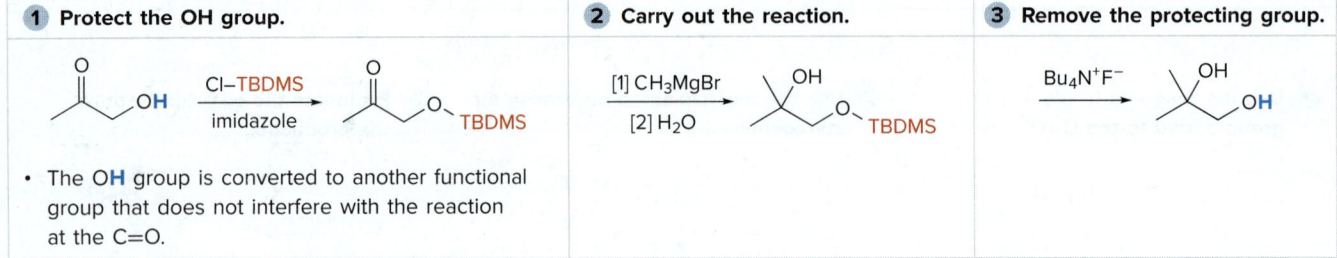

1 Protect the OH group.	**2** Carry out the reaction.	**3** Remove the protecting group.

- The **OH** group is converted to another functional group that does not interfere with the reaction at the C=O.

See Figure 13.2. Try Problem 13.47.

[5] Drawing the product that forms in the reaction of an α,β-unsaturated carbonyl compound with an organometallic reagent (13.10B)

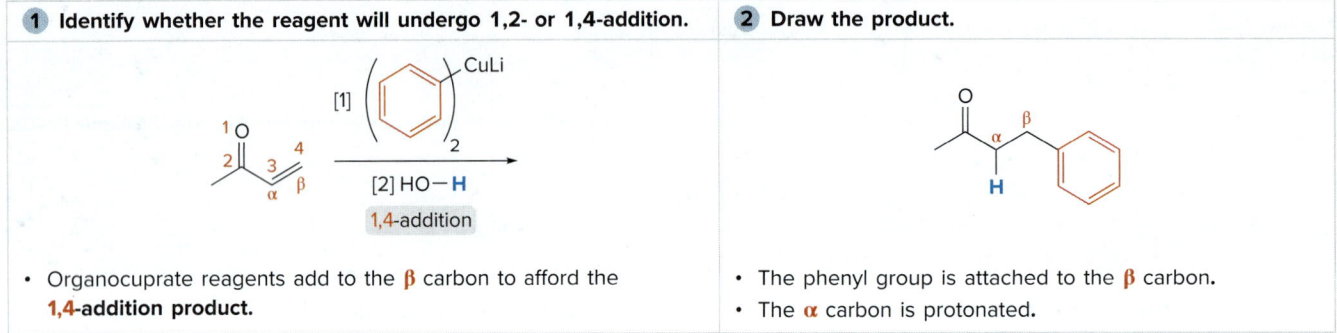

1 Identify whether the reagent will undergo 1,2- or 1,4-addition.

2 Draw the product.

- Organocuprate reagents add to the **β** carbon to afford the **1,4-addition product**.
- The phenyl group is attached to the **β** carbon.
- The **α** carbon is protonated.

See Sample Problem 13.6. Try Problems 13.38c; 13.41c–e; 13.44b, e.

KEY MECHANISM CONCEPTS

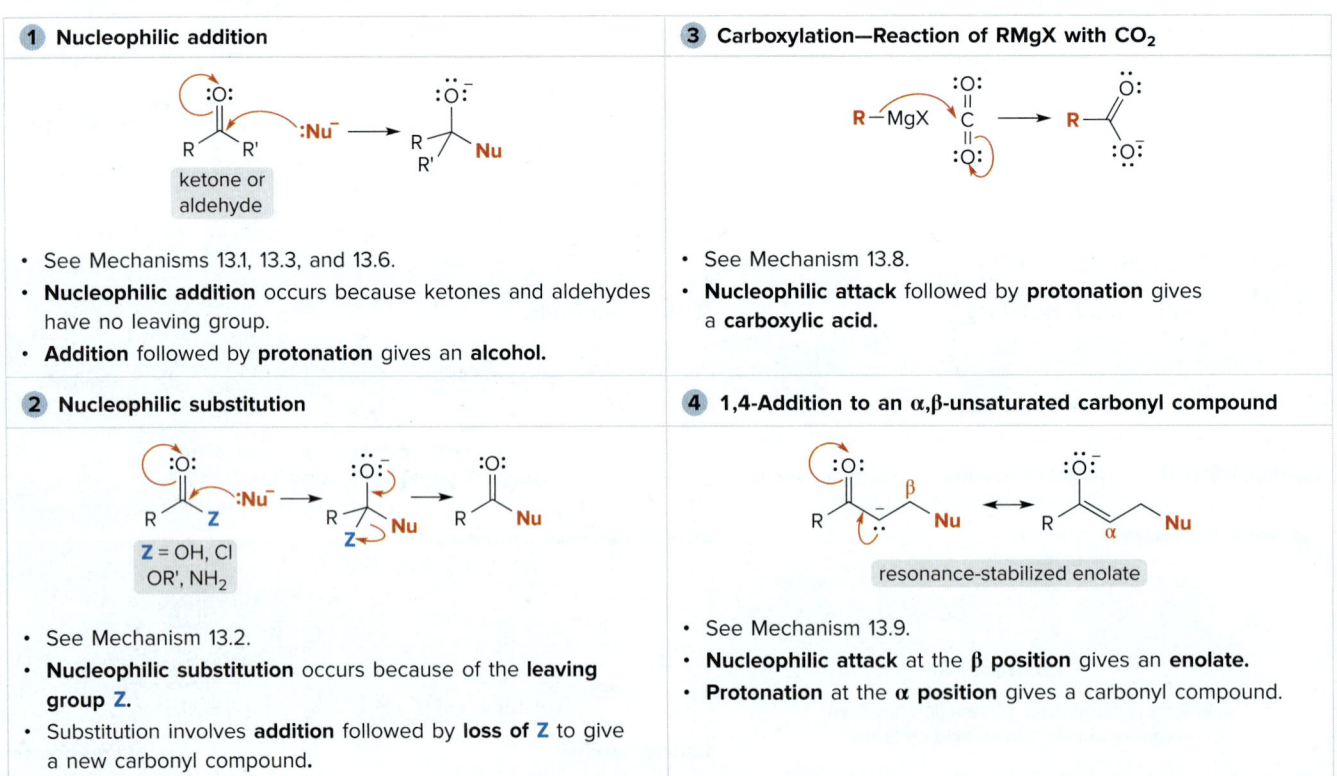

1 Nucleophilic addition

- See Mechanisms 13.1, 13.3, and 13.6.
- **Nucleophilic addition** occurs because ketones and aldehydes have no leaving group.
- **Addition** followed by **protonation** gives an **alcohol**.

2 Nucleophilic substitution

Z = OH, Cl, OR', NH$_2$

- See Mechanism 13.2.
- **Nucleophilic substitution** occurs because of the **leaving group Z**.
- Substitution involves **addition** followed by **loss of Z** to give a new carbonyl compound.

3 Carboxylation—Reaction of RMgX with CO$_2$

- See Mechanism 13.8.
- **Nucleophilic attack** followed by **protonation** gives a **carboxylic acid**.

4 1,4-Addition to an α,β-unsaturated carbonyl compound

resonance-stabilized enolate

- See Mechanism 13.9.
- **Nucleophilic attack** at the **β** position gives an **enolate**.
- **Protonation** at the **α** position gives a carbonyl compound.

Try Problems 13.49–13.51, 13.53.

PROBLEMS

Problems Using Three-Dimensional Models

13.35 Draw the products formed when **A** or **B** is treated with each reagent. In some cases, no reaction occurs.

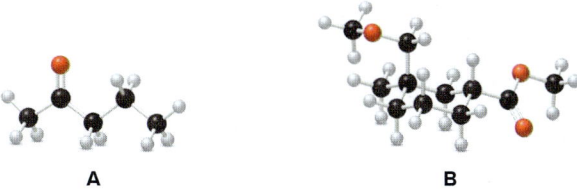

a. NaBH$_4$, CH$_3$OH
b. [1] LiAlH$_4$; [2] H$_2$O
c. [1] CH$_3$MgBr (excess); [2] H$_2$O
d. [1] C$_6$H$_5$Li (excess); [2] H$_2$O
e. Na$_2$Cr$_2$O$_7$, H$_2$SO$_4$, H$_2$O

13.36 Devise a synthesis of each alcohol from organic alcohols having one or two carbons and any required reagents.

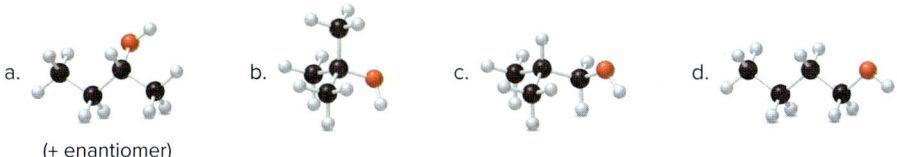

(+ enantiomer)

Reactions and Reagents

13.37 Draw the product formed when pentanal (CH$_3$CH$_2$CH$_2$CH$_2$CHO) is treated with each reagent. With some reagents, no reaction occurs.

a. NaBH$_4$, CH$_3$OH
b. [1] LiAlH$_4$; [2] H$_2$O
c. H$_2$, Pd-C
d. PCC
e. Na$_2$Cr$_2$O$_7$, H$_2$SO$_4$, H$_2$O
f. Ag$_2$O, NH$_4$OH
g. [1] CH$_3$MgBr; [2] H$_2$O
h. [1] C$_6$H$_5$Li; [2] H$_2$O
i. [1] (CH$_3$)$_2$CuLi; [2] H$_2$O
j. [1] HC≡CNa; [2] H$_2$O
k. [1] CH$_3$C≡CLi; [2] H$_2$O
l. The product in (a), then TBDMS–Cl, imidazole

13.38 Draw the product formed when (CH$_3$CH$_2$CH$_2$CH$_2$)$_2$CuLi is treated with each compound. In some cases, no reaction occurs.

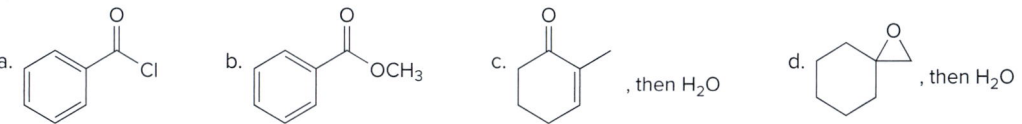

13.39 Draw the product formed when each compound reacts with NADH (or NADPH, Problem 13.8) in the presence of an enzyme. Each reaction occurs during the biosynthesis of an amino acid, a monosaccharide, or a lipid.

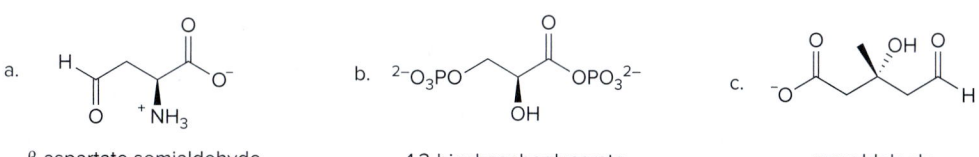

a. β-aspartate semialdehyde
b. 1,3-bisphosphoglycerate
c. mevaldehyde

13.40 What product is formed when dihydroxyacetone phosphate is reacted with NADH in the presence of the enzyme glycerol 3-phosphate dehydrogenase? Assume any stereogenic center has the R configuration. This reaction is one step in the synthesis of glycerol (propane-1,2,3-triol), a starting material needed for the biosynthesis of triacylglycerols.

dihydroxyacetone phosphate

640 Chapter 13 Introduction to Carbonyl Chemistry; Organometallic Reagents; Oxidation and Reduction

13.41 Draw the product formed when the α,β-unsaturated ketone **A** is treated with each reagent.

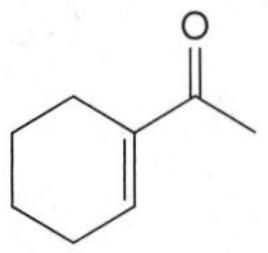

a. NaBH$_4$, CH$_3$OH
b. H$_2$ (excess), Pd-C
c. [1] CH$_3$Li; [2] H$_2$O
d. [1] CH$_3$CH$_2$MgBr; [2] H$_2$O
e. [1] (CH$_2$=CH)$_2$CuLi; [2] H$_2$O

A

13.42 Draw the products of each reduction reaction.

a. 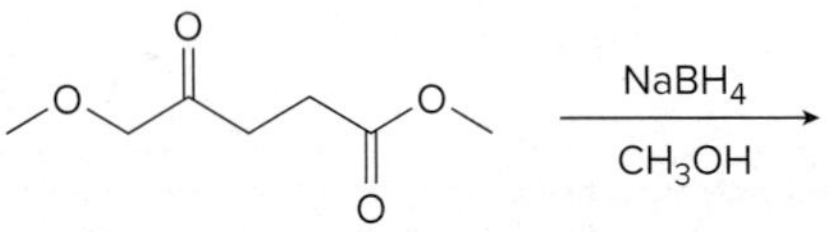 (diketoester) $\xrightarrow{\text{NaBH}_4,\ \text{CH}_3\text{OH}}$

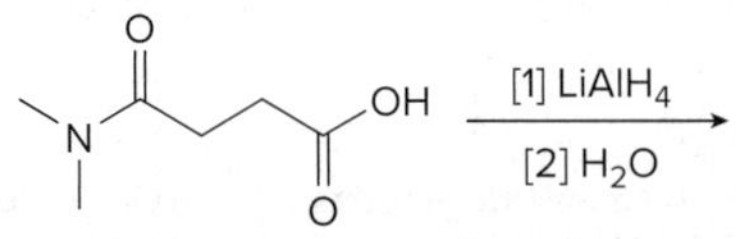

b. 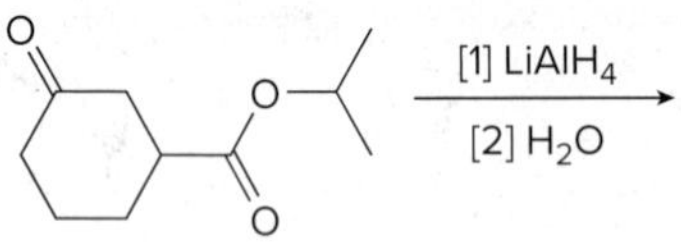 (ketoester) $\xrightarrow{\text{[1] LiAlH}_4,\ \text{[2] H}_2\text{O}}$

c. (amide-acid) $\xrightarrow{\text{[1] LiAlH}_4,\ \text{[2] H}_2\text{O}}$

d. 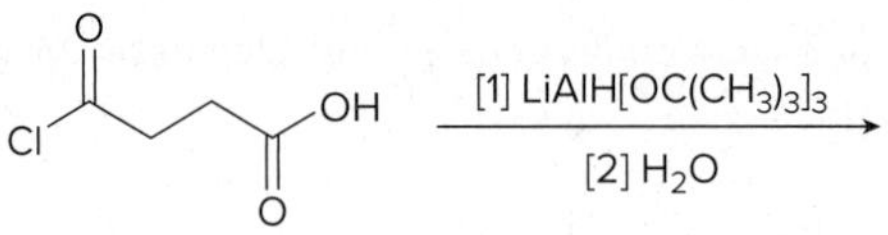 Cl-(ketoacid) $\xrightarrow{\text{[1] LiAlH[OC(CH}_3)_3]_3,\ \text{[2] H}_2\text{O}}$

13.43 Draw the product(s) formed when **A** is treated with each reagent.

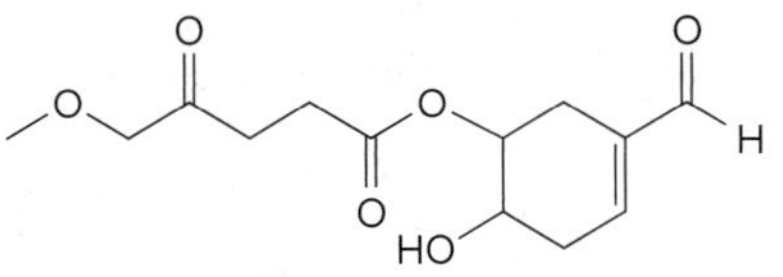

A

a. NaBH$_4$, CH$_3$OH
b. LiAlH$_4$, then H$_2$O
c. Ag$_2$O, NH$_4$OH
d. CrO$_3$, H$_2$SO$_4$, H$_2$O
e. PCC

13.44 Draw the products of the following reactions with organometallic reagents.

a. 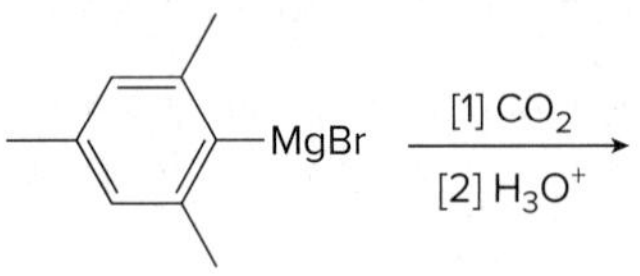 ArMgBr $\xrightarrow{\text{[1] CO}_2,\ \text{[2] H}_3\text{O}^+}$

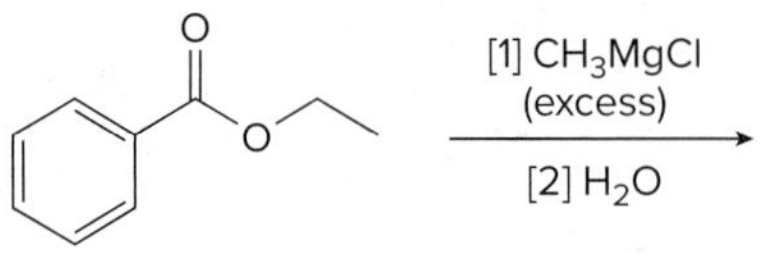

b. 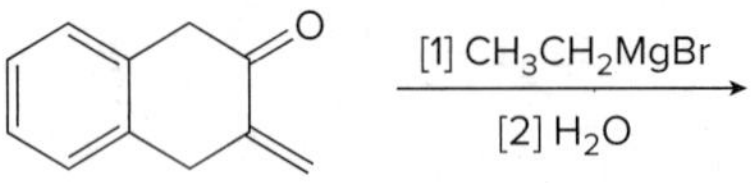 (tetralone with exocyclic =CH$_2$) $\xrightarrow{\text{[1] CH}_3\text{CH}_2\text{MgBr},\ \text{[2] H}_2\text{O}}$

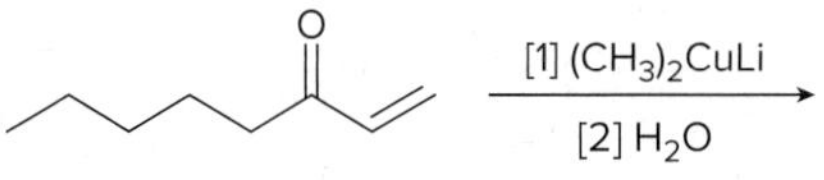

c. 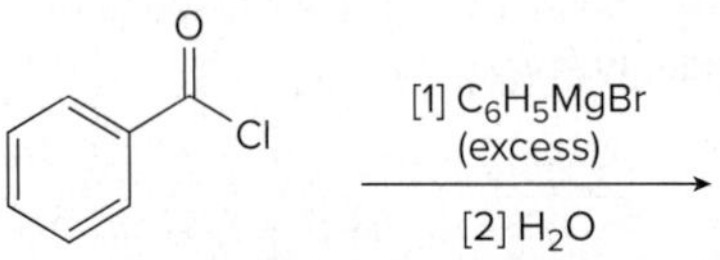 PhC(O)Cl $\xrightarrow{\text{[1] C}_6\text{H}_5\text{MgBr (excess)},\ \text{[2] H}_2\text{O}}$

d. PhCO$_2$Et $\xrightarrow{\text{[1] CH}_3\text{MgCl (excess)},\ \text{[2] H}_2\text{O}}$

e. (α,β-unsaturated ketone) $\xrightarrow{\text{[1] (CH}_3)_2\text{CuLi},\ \text{[2] H}_2\text{O}}$

f. 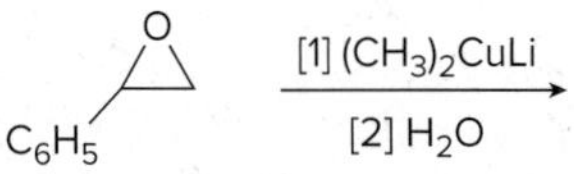 styrene oxide (C$_6$H$_5$) $\xrightarrow{\text{[1] (CH}_3)_2\text{CuLi},\ \text{[2] H}_2\text{O}}$

13.45 Identify the product **X**, formed by the reaction sequence shown. These steps were used in the synthesis of resiniferatoxin, the complex chapter-opening molecule.

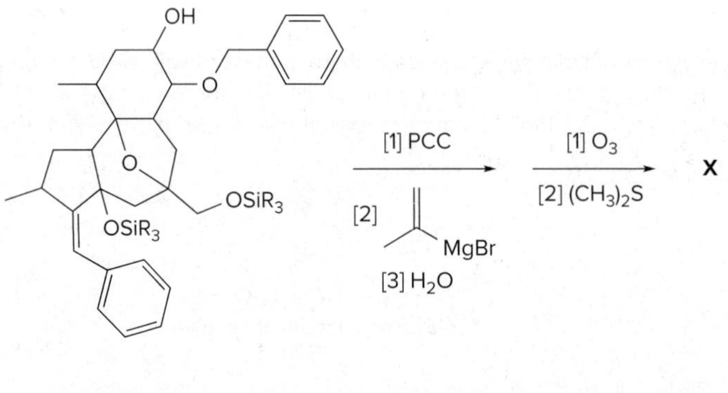

$\xrightarrow{\text{[1] PCC},\ [2]\ \text{CH}_2\text{=C(CH}_3)\text{MgBr},\ [3]\ \text{H}_2\text{O}} \xrightarrow{\text{[1] O}_3,\ \text{[2] (CH}_3)_2\text{S}}$ **X**

Problems   641

13.46 Draw all stereoisomers formed in each reaction.

a. 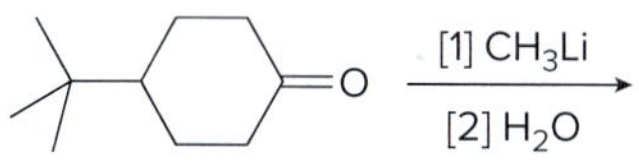 *tert*-butylcyclohexanone + [1] CH₃Li [2] H₂O

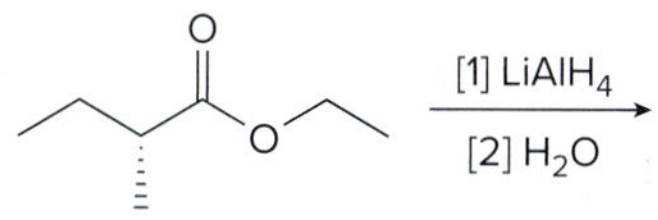

b. 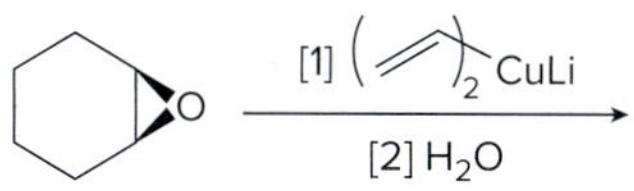 cyclohexene oxide + [1] (CH₂=CH)₂CuLi [2] H₂O

c. ethyl 2-methylbutanoate + [1] LiAlH₄ [2] H₂O

d. 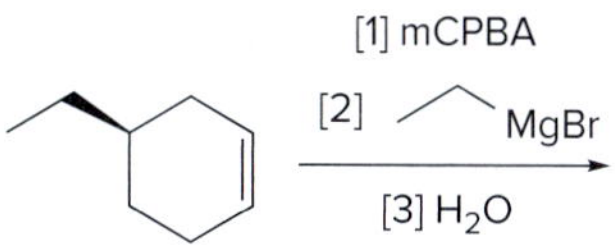 3-ethylcyclohex-1-ene + [1] mCPBA [2] CH₂=CHCH₂MgBr [3] H₂O

13.47 A student tried to carry out the following reaction sequence, but none of diol **A** was formed. Explain what was wrong with this plan, and design a successful stepwise synthesis of **A**.

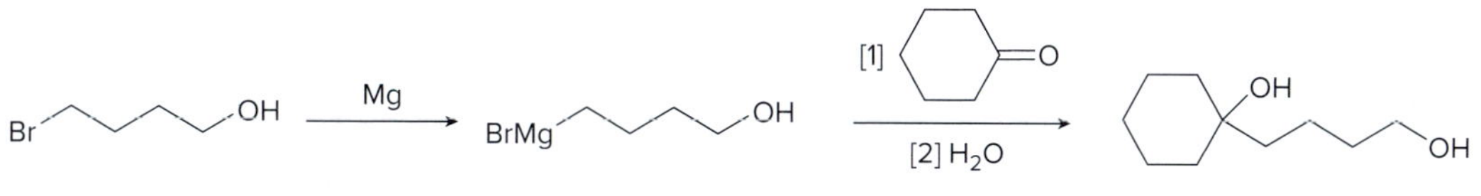

Br(CH₂)₄OH →(Mg)→ BrMg(CH₂)₄OH →[1] cyclohexanone [2] H₂O→ **A** (1-(4-hydroxybutyl)cyclohexan-1-ol)

13.48 Identify the lettered compounds in the following reaction scheme. Compounds **F**, **G**, and **K** are isomers of molecular formula $C_{13}H_{18}O$. How could ^1H NMR spectroscopy distinguish these three compounds from each other?

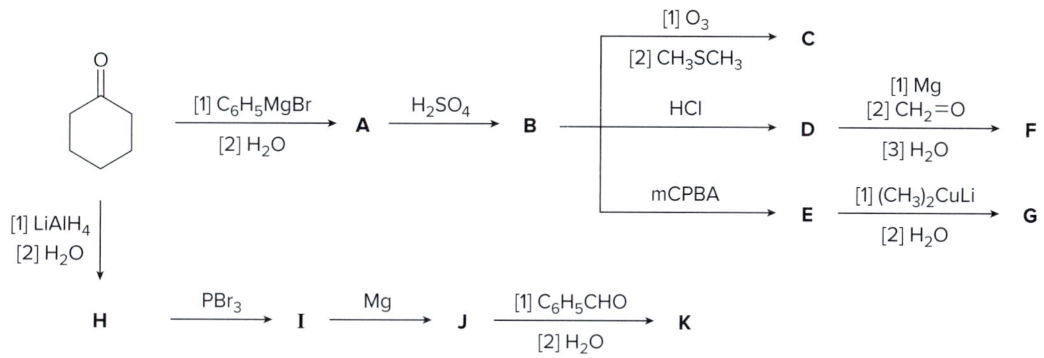

cyclohexanone →[1] C₆H₅MgBr [2] H₂O→ **A** →H₂SO₄→ **B**

B →[1] O₃ [2] CH₃SCH₃→ **C**

B →HCl→ **D** →[1] Mg [2] CH₂=O [3] H₂O→ **F**

B →mCPBA→ **E** →[1] (CH₃)₂CuLi [2] H₂O→ **G**

cyclohexanone →[1] LiAlH₄ [2] H₂O→ **H** →PBr₃→ **I** →Mg→ **J** →[1] C₆H₅CHO [2] H₂O→ **K**

Mechanism

13.49 Draw a stepwise mechanism for the following reaction. Your mechanism must show how both organic products are formed.

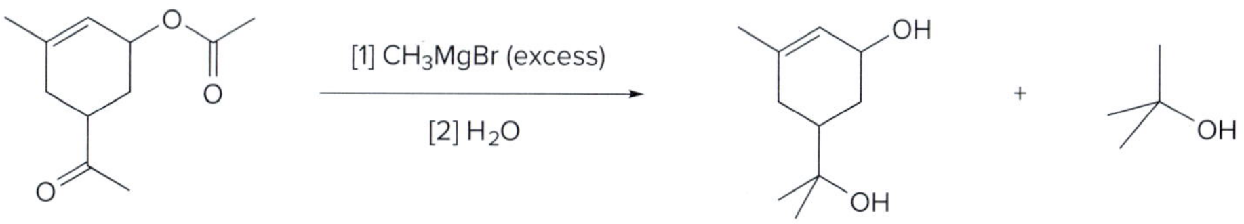

[acetate ester of methylcyclohexenol with acetyl group] →[1] CH₃MgBr (excess) [2] H₂O→ [methylcyclohexenol with C(CH₃)₂OH group] + (CH₃)₃COH

13.50 Draw a stepwise mechanism for the following reaction.

[benzofused lactone with fused cyclohexane] →[1] CH₃CH₂Li (2 equiv) [2] H₂O→ [2-(2-(3-hydroxypentan-3-yl)cyclohexyl)phenol]

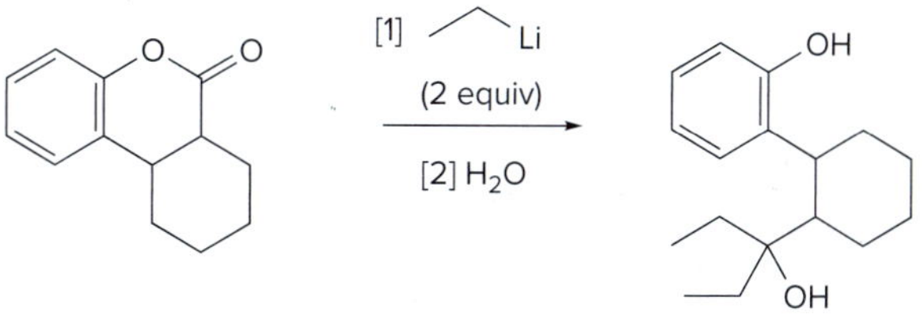

642 Chapter 13 Introduction to Carbonyl Chemistry; Organometallic Reagents; Oxidation and Reduction

13.51 Slow addition of organolithium reagent **A** to **B** afforded **C**, an intermediate in the synthesis of the chapter-opening molecule, resiniferatoxin. Draw a stepwise mechanism for this process.

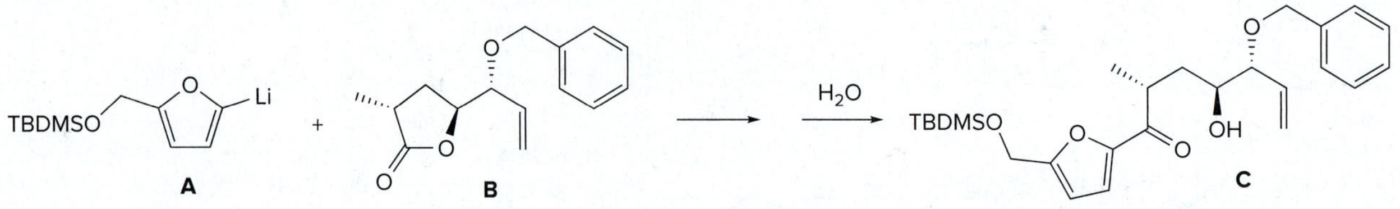

13.52 Draw a stepwise mechanism for the following reaction, the last step in the biosynthesis of the amino acid proline.

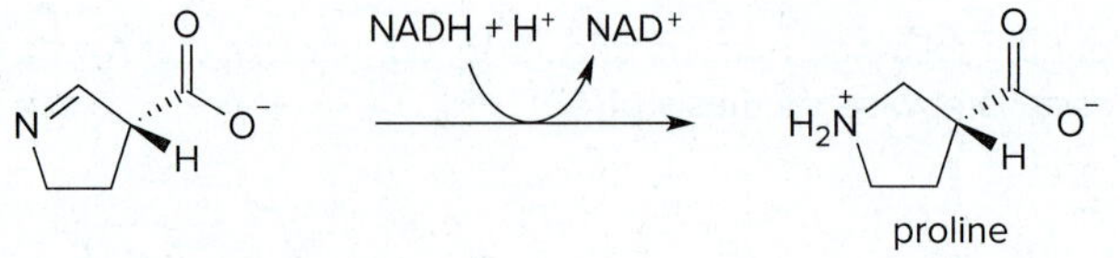

13.53 Draw a stepwise mechanism for the following reaction.

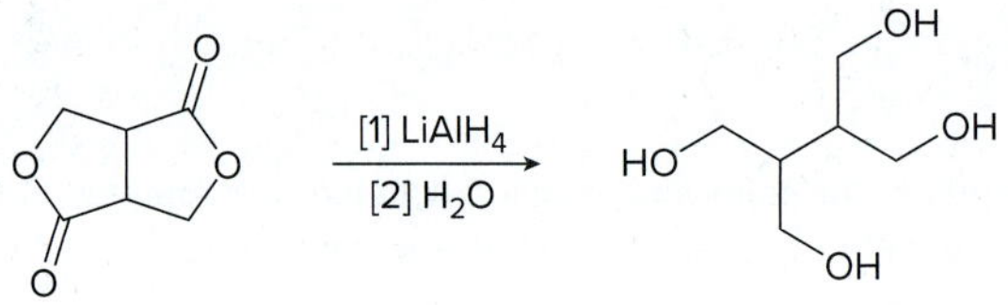

Synthesis

13.54 What amides will form each amine on treatment with LiAlH$_4$?

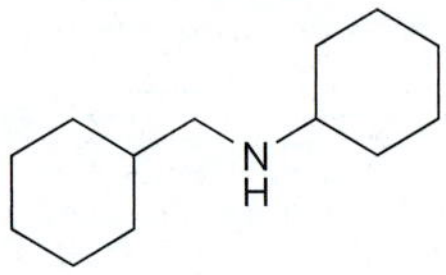

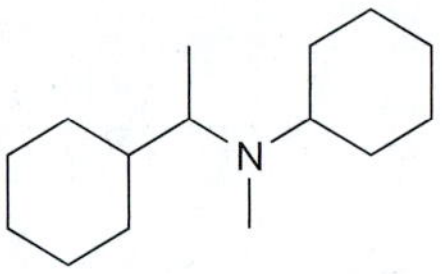

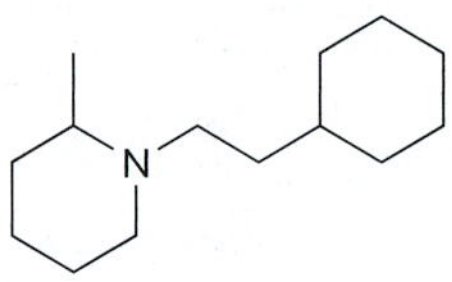

13.55 What Grignard reagent and aldehyde (or ketone) are needed to prepare each alcohol? Show all possible routes.

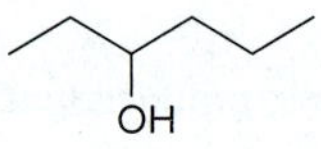

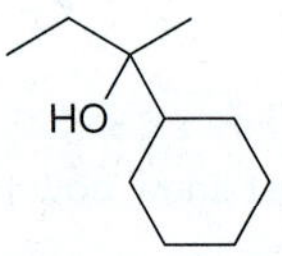

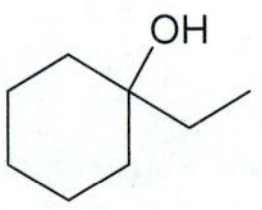

13.56 Procyclidine is a drug that has been used to treat the uncontrolled body movements associated with Parkinson's disease. Draw three different methods to prepare procyclidine using a Grignard reagent.

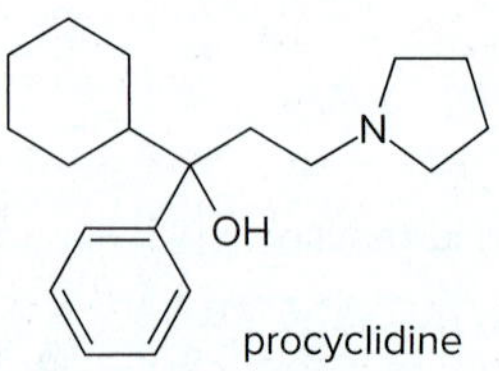

13.57 What ester and Grignard reagent are needed to synthesize each alcohol?

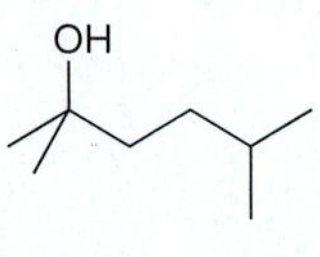

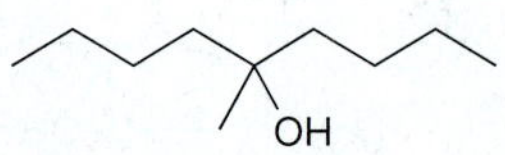

13.58 What organolithium reagent and carbonyl compound can be used to prepare each of the following compounds? You may use aldehydes, ketones, or esters as carbonyl starting materials.

a. 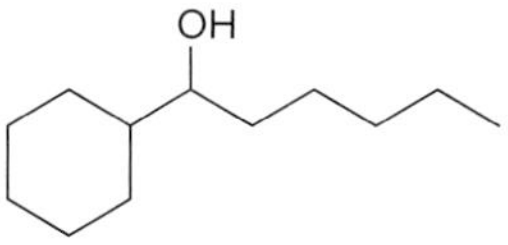 cyclohexyl-CH(OH)-butyl (two ways)

b. 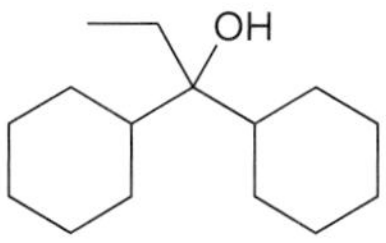 dicyclohexyl-C(OH)(ethyl) (three ways)

13.59 What epoxide and organometallic reagent are needed to synthesize each alcohol?

a. 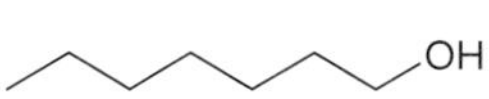 hexan-1-ol (CH₃(CH₂)₅OH)

b. 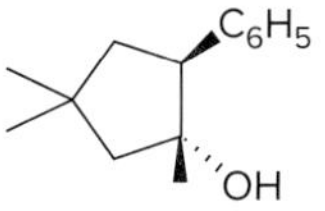 1-phenyl-3,3-dimethylcyclopentan-... with C₆H₅ and OH substituents

c. 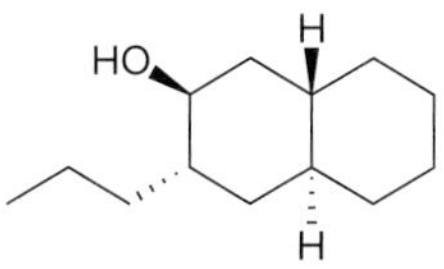 decalin derivative with HO and propyl substituents

13.60 Propose at least three methods to convert $C_6H_5CH_2CH_2Br$ to $C_6H_5CH_2CH_3$.

13.61 Synthesize each compound from cyclohexanol using any other organic or inorganic compounds.

a. 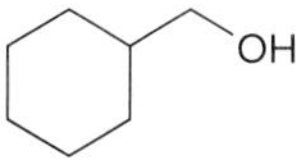 cyclohexylmethanol

b. 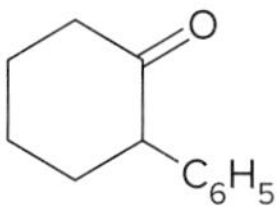 2-phenylcyclohexan-1-one

c. 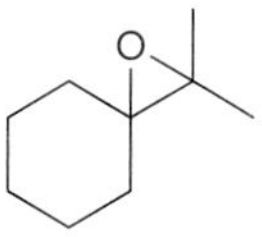 1-methyl-1,2-epoxycyclohexane (spiro epoxide)

d. 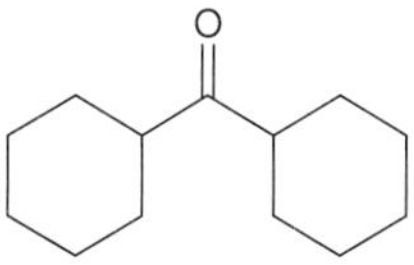 dicyclohexyl ketone (Each cyclohexane ring must come from cyclohexanol.)

13.62 Design a synthesis of each compound from alcohols having four or fewer carbons, acetylene, and ethylene oxide as the only organic starting materials. You may use any other inorganic reagents you choose.

a. 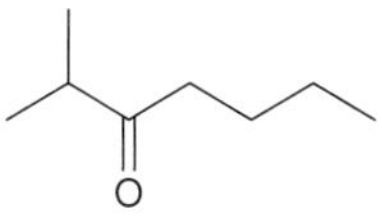 2-methyloctan-3-one (isopropyl pentyl ketone)

b. 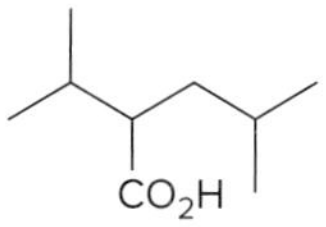 2,5-dimethyl-3-... carboxylic acid

c. 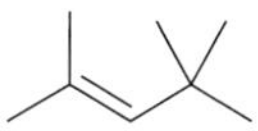 2,4,4-trimethylpent-2-ene

d. 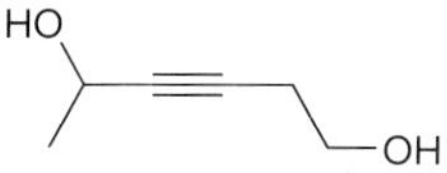 HO-CH(CH₃)-C≡C-CH₂CH₂-OH

13.63 Devise a synthesis of each compound from cyclohex-2-enone and organic halides having one or two carbons. You may use any other required inorganic reagents.

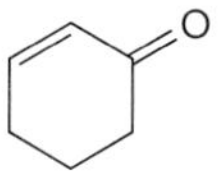

cyclohex-2-enone

a. 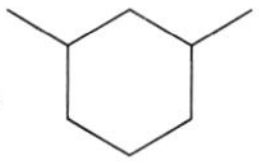 1,3-dimethylcyclohexane

b. 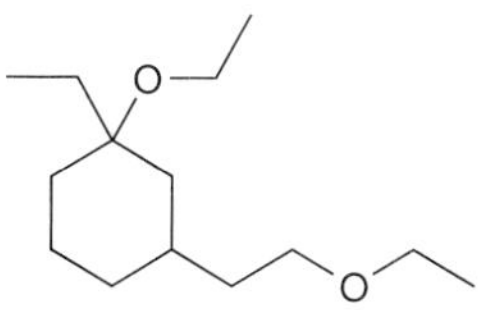 1-ethyl-1-ethoxy-3-(2-ethoxyethyl)cyclohexane

13.64 Devise a synthesis of (E)-tetradec-11-enal, a sex pheromone of the spruce budworm, a pest that destroys fir and spruce forests, from acetylene, $Br(CH_2)_{10}OH$, and any needed organic compounds or inorganic reagents.

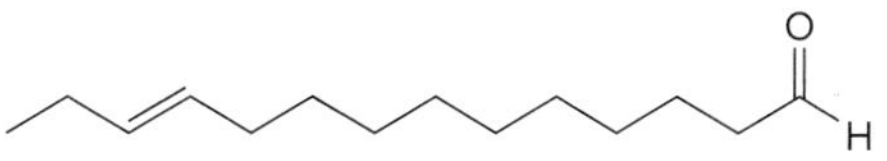

(E)-tetradec-11-enal

Spectroscopy

13.65 An unknown compound **A** (molecular formula $C_7H_{14}O$) was treated with $NaBH_4$ in CH_3OH to form compound **B** (molecular formula $C_7H_{16}O$). Compound **A** has a strong absorption in its IR spectrum at 1716 cm^{-1}. Compound **B** has a strong absorption in its IR spectrum at 3600–3200 cm^{-1}. The ^1H NMR spectra of **A** and **B** are given. What are the structures of **A** and **B**?

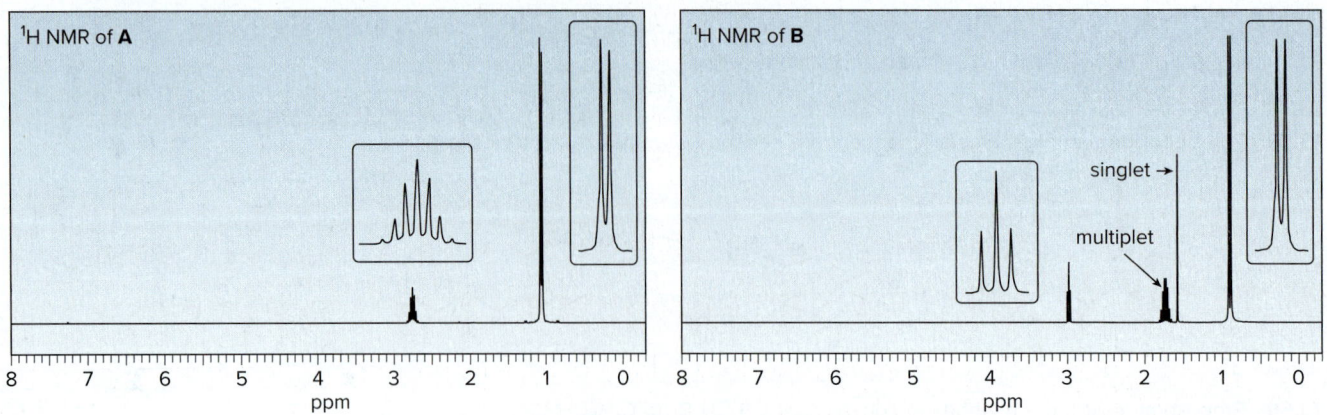

13.66 Reaction of butanenitrile ($CH_3CH_2CH_2CN$) with methylmagnesium bromide (CH_3MgBr), followed by treatment with aqueous acid, forms compound **G**. **G** has a molecular ion in its mass spectrum at $m/z = 86$ and a base peak at $m/z = 43$. **G** exhibits a strong absorption in its IR spectrum at 1721 cm^{-1} and has the ^1H NMR spectrum given below. What is the structure of **G**? We will learn about the details of this reaction in Chapter 15.

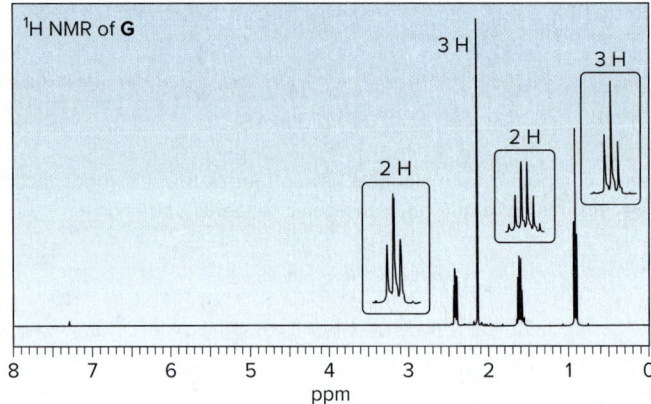

13.67 Treatment of isobutene [$(CH_3)_2C=CH_2$] with $(CH_3)_3CLi$ forms a carbanion that reacts with $CH_2=O$ to form **H** after water is added to the reaction mixture. **H** has a molecular ion in its mass spectrum at $m/z = 86$, and shows fragments at 71 and 68. **H** exhibits absorptions in its IR spectrum at 3600–3200 and 1651 cm^{-1}, and has the ^1H NMR spectrum given below. What is the structure of **H**?

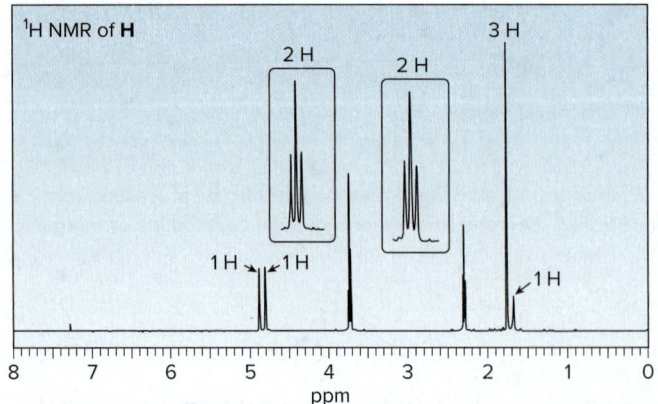

Challenge Problems

13.68 Draw a stepwise mechanism for the following reaction.

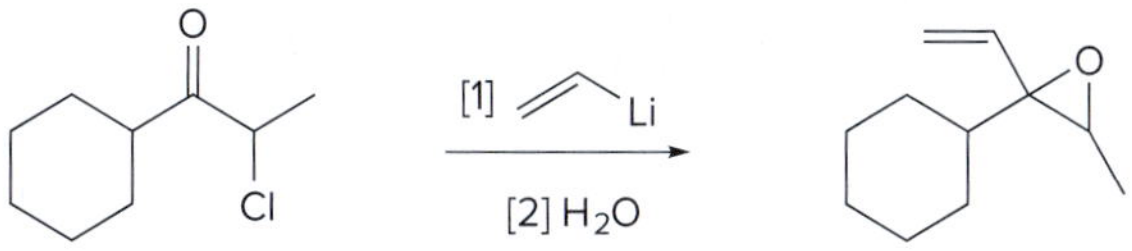

13.69 Explain why the β carbon of an α,β-unsaturated carbonyl compound absorbs farther downfield in the ^{13}C NMR spectrum than the α carbon, even though the α carbon is closer to the electron-withdrawing carbonyl group. For example, the β carbon of mesityl oxide absorbs at 150.5 ppm, whereas the α carbon absorbs at 122.5 ppm.

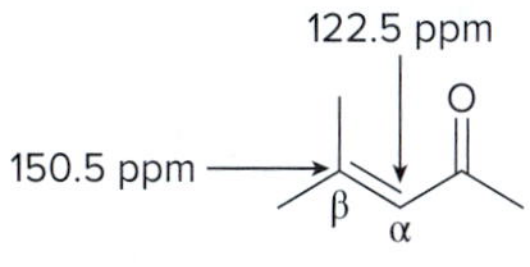

mesityl oxide

13.70 Identify **X** and **Y**, two of the intermediates in a synthesis of the antidepressant venlafaxine (trade name Effexor), in the following reaction scheme. Write a mechanism for the formation of **X** from **W**.

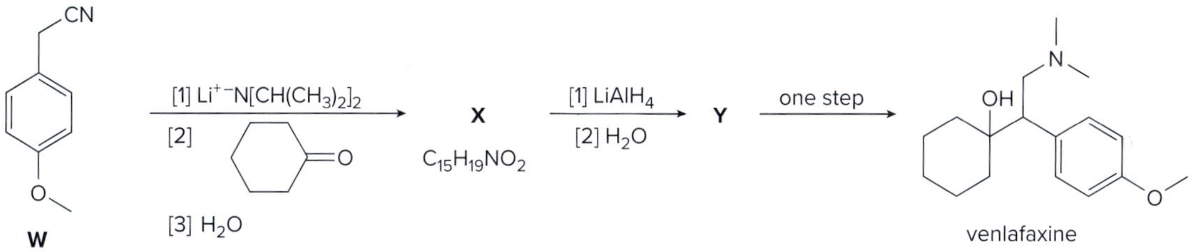

venlafaxine

13.71 Reaction of benzylmagnesium chloride with formaldehyde yields alcohols **N** and **P** after protonation. Draw a stepwise mechanism that shows how both products are formed.

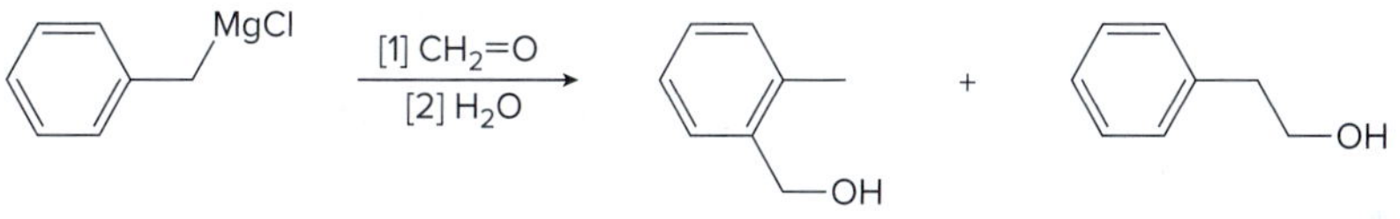

N major product

P minor product

13.72 Draw a stepwise mechanism for the following reaction of a Grignard reagent with a cyclic amide.

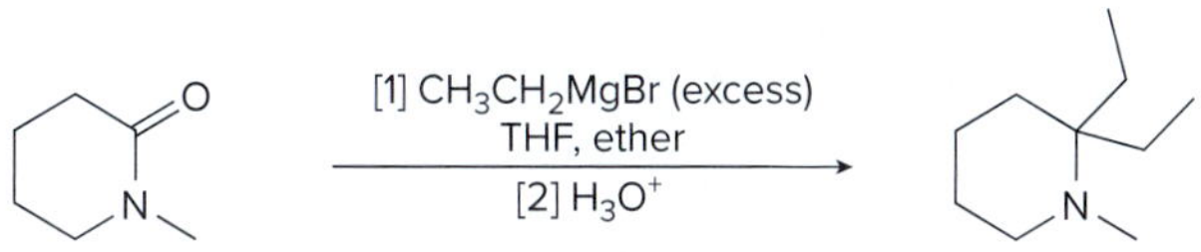

14 Aldehydes and Ketones—Nucleophilic Addition

- 14.1 Introduction
- 14.2 Nomenclature
- 14.3 Properties of aldehydes and ketones
- 14.4 Interesting aldehydes and ketones
- 14.5 Preparation of aldehydes and ketones
- 14.6 Reactions of aldehydes and ketones—General considerations
- 14.7 Nucleophilic addition of H$^-$ and R$^-$—A review
- 14.8 Nucleophilic addition of $^-$CN
- 14.9 The Wittig reaction
- 14.10 Addition of 1° amines
- 14.11 Addition of 2° amines
- 14.12 Imine and Enamine Hydrolysis
- 14.13 Imines in Biological Systems
- 14.14 Addition of H$_2$O—Hydration
- 14.15 Addition of alcohols—Acetal formation
- 14.16 Acetals as protecting groups
- 14.17 Cyclic hemiacetals
- 14.18 An introduction to carbohydrates

Pyridoxal phosphate (PLP) and several structurally similar compounds are collectively called vitamin B$_6$. Fortified breakfast cereals, beef liver, salmon, chickpeas, and pistachios are excellent food sources of vitamin B$_6$, but a significant amount of the vitamin can be lost when foods are heated and processed. Pyridoxal phosphate is a key coenzyme involved in the metabolism of amino acids, using a nucleophilic addition reaction, the characteristic reaction of aldehydes and ketones and the subject of Chapter 14.

Why Study... Aldehydes and Ketones?

In Chapter 14, we continue the study of carbonyl compounds with a detailed look at **aldehydes** and **ketones**. We will first learn about the nomenclature, physical properties, and spectroscopic absorptions that characterize aldehydes and ketones. The remainder of Chapter 14 is devoted to **nucleophilic addition** reactions. Although we have already learned two examples of this reaction in Chapter 13, nucleophilic addition to aldehydes and ketones is a general reaction that occurs with many nucleophiles, forming a wide variety of products, including carbohydrates and molecules central to the process of vision.

Every new reaction in Chapter 14 involves nucleophilic addition, so the challenge lies in learning the specific reagents and mechanisms that characterize each reaction.

14.1 Introduction

As we learned in Chapter 13, **aldehydes and ketones contain a carbonyl group.** An aldehyde contains at least one H atom bonded to the carbonyl carbon, whereas a ketone has two alkyl groups bonded to it.

*An aldehyde is often written as **RCHO**. Remember that the **H atom is bonded to the carbon atom,** not the oxygen. Likewise, a ketone is written as **RCOR** or, if both alkyl groups are the same, **R$_2$CO**. Each structure must contain a C=O for every atom to have an octet.*

Two structural features determine the chemistry and properties of aldehydes and ketones.

- The carbonyl group is sp^2 hybridized and trigonal planar, making it relatively *uncrowded*.
- The electronegative oxygen atom polarizes the carbonyl group, making the carbonyl carbon *electrophilic*.

As a result, **aldehydes and ketones react with nucleophiles.** The relative reactivity of the carbonyl group is determined by the number of R groups bonded to it. **As the number of R groups around the carbonyl carbon *increases*, the reactivity of the carbonyl compound *decreases*,** resulting in the following order of reactivity:

Increasing the number of alkyl groups on the carbonyl carbon decreases reactivity for both steric and electronic reasons, as discussed in Section 13.2B.

Problem 14.1 Rank the following compounds in order of increasing reactivity toward nucleophilic attack.

Problem 14.2 Explain why benzaldehyde is less reactive than cyclohexanecarbaldehyde toward nucleophilic attack.

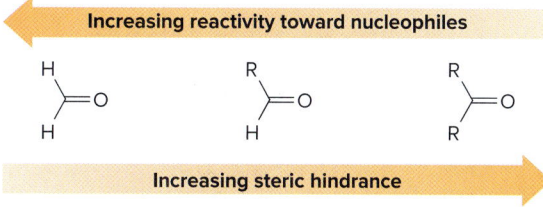

benzaldehyde   cyclohexanecarbaldehyde

14.2 Nomenclature

Both IUPAC and common names are used for aldehydes and ketones.

14.2A Naming Aldehydes in the IUPAC System

In IUPAC nomenclature, aldehydes are identified by a suffix added to the parent name of the longest chain. Two different suffixes are used, depending on whether the CHO group is bonded to a chain or a ring.

To name an aldehyde using the IUPAC system:

[1] If the CHO is bonded to a chain of carbons, find the longest chain containing the CHO group, and change the *-e* ending of the parent alkane to the suffix *-al.* If the CHO group is bonded to a ring, name the ring and add the suffix *-carbaldehyde.*

[2] Number the chain or ring to put the CHO group at C1, but omit this number from the name. Apply all of the other usual rules of nomenclature.

Sample Problem 14.1   Naming an Aldehyde Using the IUPAC System

Give the IUPAC name for each compound.

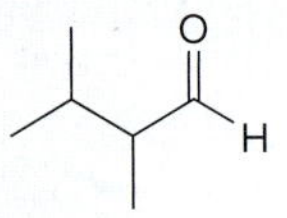

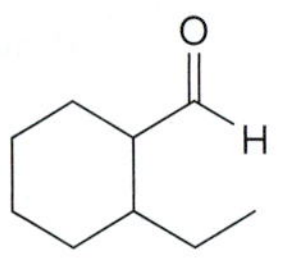

Solution

a. [1] Find and name the longest chain containing the CHO:

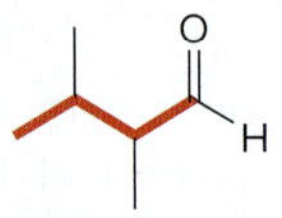

butane → butan*al*
(4 C's)

[2] Number and name substituents:

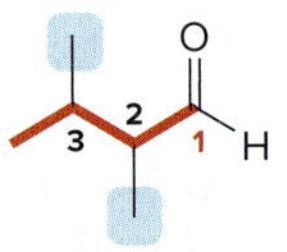

Answer: 2,3-dimethylbutanal

b. [1] Find and name the ring bonded to the CHO group:

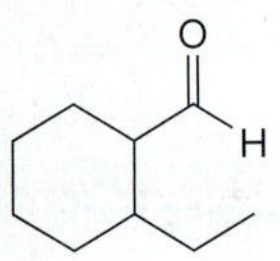

cyclohexane + carbaldehyde
(6 C's)

[2] Number and name substituents:

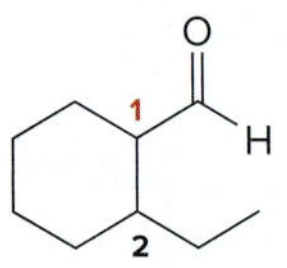

Answer: 2-ethylcyclohexanecarbaldehyde

Problem 14.3   Give the IUPAC name for each aldehyde.

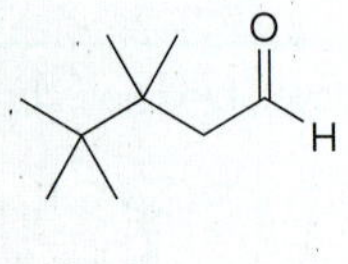

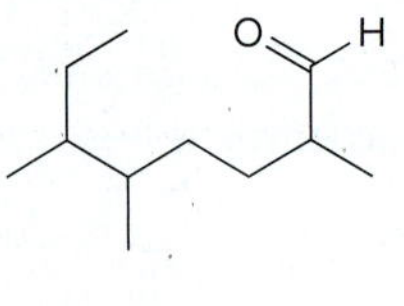

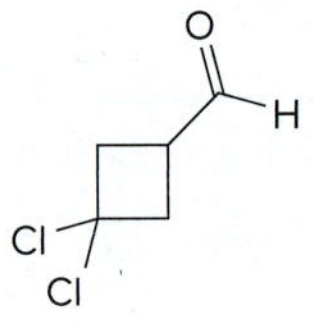

More Practice: Try Problems 14.36a; 14.38b, d.

Problem 14.4 Give the structure corresponding to each IUPAC name.
 a. 2-isobutyl-3-isopropylhexanal
 b. *trans*-3-methylcyclopentanecarbaldehyde
 c. 1-methylcyclopropanecarbaldehyde
 d. 3,6-diethylnonanal

14.2B Common Names for Aldehydes

Many simple aldehydes have common names that are widely used.

> • A common name for an aldehyde is formed by taking the common parent name and adding the suffix *-aldehyde*.

Table 14.1 lists common parent names for some simple aldehydes. These parent names are used in the nomenclature of many other carbonyl compounds (Chapters 15 and 16). The common names **formaldehyde, acetaldehyde,** and **benzaldehyde** are virtually always used instead of their IUPAC names.

Table 14.1 Common Names for Some Simple Aldehydes

Number of C atoms	Structure	Parent name	Common name
1	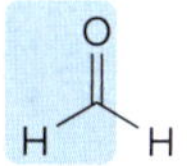 H–CHO	**form-**	formaldehyde
2	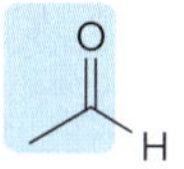 CH₃–CHO	**acet-**	acetaldehyde
3	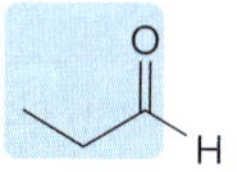 CH₃CH₂–CHO	**propion-**	propionaldehyde
4	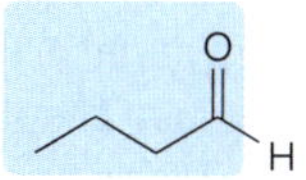 CH₃CH₂CH₂–CHO	**butyr-**	butyraldehyde
5	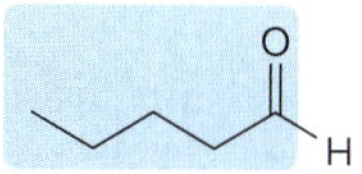 CH₃(CH₂)₃–CHO	**valer-**	valeraldehyde
6	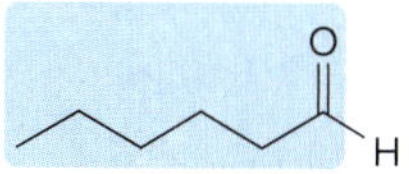 CH₃(CH₂)₄–CHO	**capro-**	caproaldehyde
	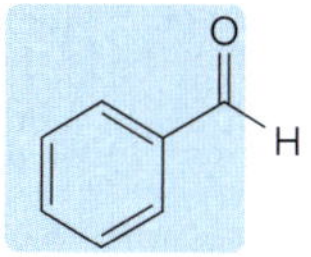 C₆H₅–CHO	**benz-**	benzaldehyde

Greek letters are used to designate the location of substituents in common names.

- The carbon adjacent to the CHO is called the α carbon.
- The carbon bonded to the α carbon is the β carbon, followed by the γ (gamma) carbon, the δ (delta) carbon, and so forth down the chain. The last carbon in the chain is sometimes called the Ω (omega) carbon.

IUPAC numbering begins at the C=O.
Greek lettering begins at the C bonded to the C=O.

Figure 14.1 gives the common and IUPAC names for three aldehydes.

Figure 14.1
Three examples of aldehyde nomenclature

2-chloropropanal
(α-chloropropionaldehyde)

3-methylpentanal
(β-methylvaleraldehyde)

phenylethanal
(phenylacetaldehyde)

(Common names are in parentheses.)

14.2C Naming Ketones in the IUPAC System

- In the IUPAC system, all ketones are identified by the suffix *-one*.

To name an acyclic ketone using IUPAC rules:

[1] Find the longest chain containing the carbonyl group, and change the *-e* ending of the parent alkane to the suffix *-one*.

[2] Number the carbon chain to give the carbonyl carbon the lower number. Apply all of the other usual rules of nomenclature.

With cyclic ketones, numbering always begins at the carbonyl carbon, but the "1" is usually omitted from the name. The ring is then numbered clockwise or counterclockwise to give the *first* substituent the lower number.

Sample Problem 14.2   Naming a Ketone Using the IUPAC System

Give the IUPAC name for each ketone.

a.

b.

Solution

a. [1] Find and name the longest chain containing the carbonyl group:

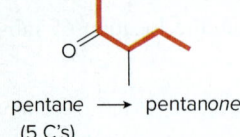

pentane ⟶ pentan*one*
(5 C's)

[2] Number and name substituents:

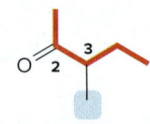

Answer: 3-methylpentan-2-one

b. [1] Name the ring:

cyclohexane → cyclohexan*one*
(6 C's)

[2] Number and name substituents:

Answer:
3-isopropyl-4-methylcyclohexanone

Problem 14.5 Give the IUPAC name for each ketone.

a.   b.   c.

More Practice: Try Problems 14.36a; 14.38a, c.

14.2D Common Names for Ketones

Most common names for ketones are formed by **naming both alkyl groups** on the carbonyl carbon, **arranging them alphabetically,** and adding the word ***ketone.*** Using this method, the common name for butan-2-one becomes ethyl methyl ketone.

IUPAC name: **butan-2-one**   Common name: **ethyl methyl ketone**

Three widely used common names for some simple ketones do not follow this convention:

acetone   acetophenone   benzophenone

Figure 14.2 gives acceptable names for two ketones.

Figure 14.2
Two examples of ketone nomenclature

IUPAC name: 2-methylpentan-3-one
Common name: ethyl isopropyl ketone

2-bromoacetophenone
or
α-bromoacetophenone

14.2E Additional Nomenclature Facts

Do not confuse a **benzyl** group with a **benzoyl** group.

Sometimes **acyl groups (RCO—)** must be named as substituents. To name an acyl group, take either the IUPAC or common parent name and add the suffix *-yl* or *-oyl.* The three most common acyl groups are drawn below.

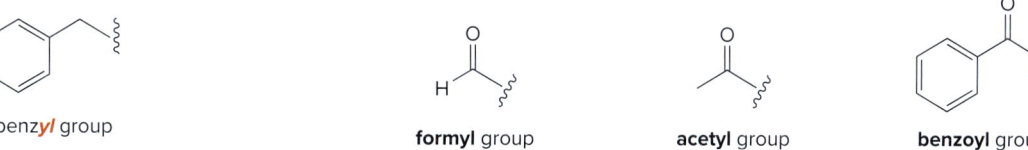

benz**yl** group   **formyl** group   **acetyl** group   **benzoyl** group

Compounds containing both a C—C double bond and an aldehyde are named as **enals,** and compounds that contain both a C—C double bond and a ketone are named as **enones.** The chain is numbered to **give the carbonyl group the *lower* number.**

2,2-dimethylbut-3-enal

4-methylpent-3-en-2-one

Problem 14.6 Give the structure corresponding to each name: (a) *sec*-butyl ethyl ketone; (b) methyl vinyl ketone; (c) 3-benzoyl-2-benzylcyclopentanone; (d) 6,6-dimethylcyclohex-2-enone; (e) 3-ethylhex-5-enal.

Problem 14.7 Give the IUPAC name (including any *E,Z* designation) for each unsaturated aldehyde.

a. neral
found in lemongrass

b. cucumber aldehyde

c. found in stinkbugs and cilantro

14.3 Properties of Aldehydes and Ketones

14.3A Physical Properties

Aldehydes and ketones exhibit dipole–dipole interactions because of their polar carbonyl group. Because they have no O—H bond, two molecules of RCHO or RCOR are incapable of intermolecular hydrogen bonding, making them *less polar* than alcohols. How these intermolecular forces affect the physical properties of aldehydes and ketones is summarized in Table 14.2.

Table 14.2 Physical Properties of Aldehydes and Ketones

Property	Observation
Boiling point and melting point	• For compounds of comparable molecular weight, bp's and mp's follow the usual trend: The stronger the intermolecular forces, the higher the bp or mp. VDW, MW = 72, bp 36 °C VDW, DD, MW = 72, bp 76 °C VDW, DD, HB, MW = 74, bp 118 °C Increasing strength of intermolecular forces Increasing boiling point
Solubility	• RCHO and RCOR are soluble in organic solvents regardless of size. • RCHO and RCOR having ≤ 5 C's are H_2O soluble because they can hydrogen bond with H_2O (Section 3.4C). • RCHO and RCOR having > 5 C's are H_2O insoluble because the nonpolar alkyl portion is too large to dissolve in the polar H_2O solvent.

Key: VDW = van der Waals, DD = dipole–dipole, HB = hydrogen bonding, MW = molecular weight

Problem 14.8 The boiling point of butan-2-one (80 °C) is significantly higher than the boiling point of diethyl ether (35 °C), even though both compounds exhibit dipole–dipole interactions and have comparable molecular weights. Offer an explanation.

14.3B Spectroscopic Properties

Many details of the spectroscopy of aldehydes and ketones have been presented in Spectroscopy Parts A, B, and C:

- Fragmentation patterns in mass spectra: Section A.4A and Sample Problem A.6
- The carbonyl absorption in infrared spectra: Sections B.3C and B.4B
- ^{1}H and ^{13}C NMR absorptions: Section C.11B and Tables C.1 and C.5

Key NMR and IR absorptions for aldehydes and ketones are summarized in Table 14.3, and Figure 14.3 illustrates ^{1}H and ^{13}C NMR spectra for a simple aldehyde.

Table 14.3 Characteristic Spectroscopic Absorptions of Aldehydes and Ketones

Type of spectroscopy	Type of C, H	Absorption
IR absorptions	R—CHO	2700–2830 cm^{-1} (one or two peaks)
	R—CO—R(H)	~1700 cm^{-1} (increasing ν with decreasing ring size)
	R—CO—CH=CH$_2$ (conjugated)	1680 cm^{-1}
^{1}H NMR absorptions	R—CHO	9–10 ppm
	R—CO—CH (α-H)	2–2.5 ppm
^{13}C NMR absorption	R—CO—R(H)	190–215 ppm

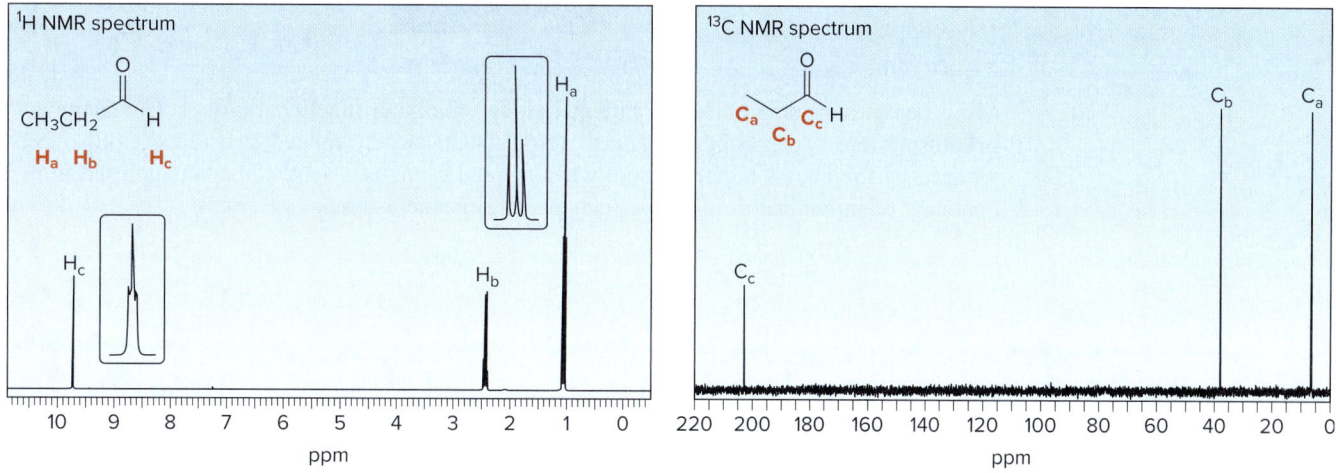

Figure 14.3 The ^{1}H and ^{13}C NMR spectra of propanal, CH$_3$CH$_2$CHO

- **^{1}H NMR:** There are three signals due to the three different kinds of hydrogens, labeled H$_a$, H$_b$, and H$_c$. The **deshielded CHO proton** occurs downfield at 9.8 ppm. The H$_c$ signal is split into a triplet by the adjacent CH$_2$ group, but the coupling constant is small.
- **^{13}C NMR:** There are three signals due to the three different kinds of carbons, labeled C$_a$, C$_b$, and C$_c$. The **deshielded carbonyl carbon** absorbs downfield at 203 ppm.

Problem 14.9 Rank the following compounds in order of increasing frequency of their carbonyl absorption in the infrared.

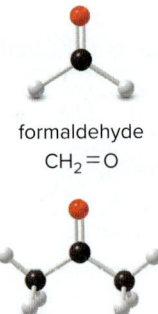

A   B   C

14.4 Interesting Aldehydes and Ketones

Because it is a starting material for the synthesis of many resins and plastics, billions of pounds of **formaldehyde** are produced annually in the United States by the oxidation of methanol (CH_3OH). Formaldehyde is also sold as a 37% aqueous solution called **formalin,** which has been used as a disinfectant, antiseptic, and preservative for biological specimens. Formaldehyde, a product of the incomplete combustion of coal and other fossil fuels, is partly responsible for the irritation caused by smoggy air.

Acetone is an industrial solvent and a starting material in the synthesis of some organic polymers. Acetone is produced in vivo during the breakdown of fatty acids. In diabetes, a common endocrine disease in which normal metabolic processes are altered because of the inadequate secretion of insulin, individuals often have unusually high levels of acetone in their bloodstreams. The characteristic odor of acetone can be detected on the breath of diabetic patients when their disease is poorly controlled.

formaldehyde
$CH_2=O$

acetone
$(CH_3)_2C=O$

Many aldehydes with characteristic odors occur in nature, including vanillin from vanilla beans and cinnamaldehyde from cinnamon.

vanillin
(flavoring agent from vanilla beans)

cinnamaldehyde
(odor of cinnamon)

Jill Braaten/McGraw-Hill Education

Many steroid hormones contain a carbonyl along with other functional groups. **Cortisone** and **prednisone** are two anti-inflammatory steroids with closely related structures. Cortisone is secreted by the body's adrenal gland, whereas prednisone is a synthetic analogue used in the treatment of inflammatory diseases such as arthritis and asthma.

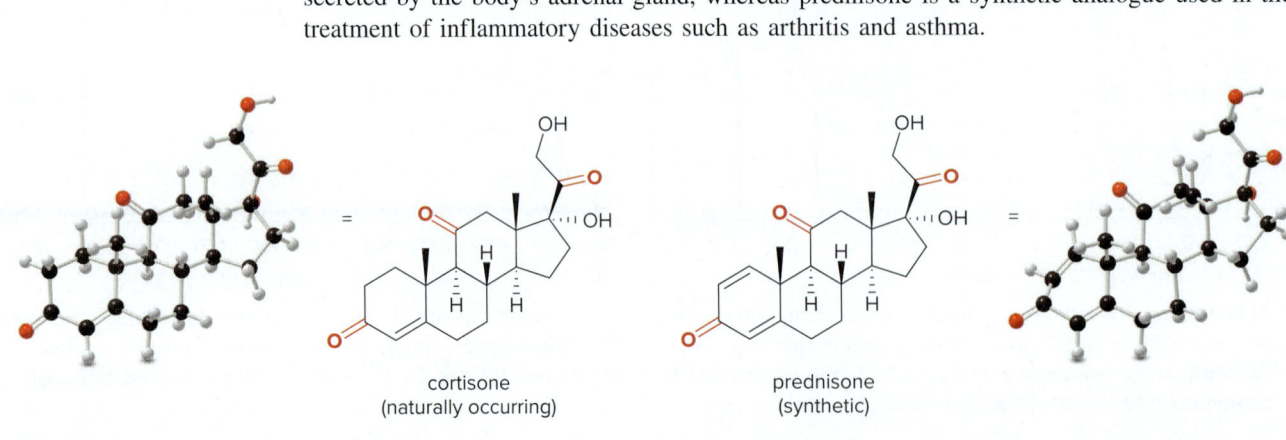

cortisone
(naturally occurring)

prednisone
(synthetic)

14.5 Preparation of Aldehydes and Ketones

Aldehydes and ketones can be prepared by a variety of methods. Because these reactions are needed for many multistep syntheses, Section 14.5 briefly summarizes earlier reactions that synthesize an aldehyde or ketone.

Aldehydes are prepared from 1° alcohols, esters, acid chlorides, and alkynes (Table 14.4).

Table 14.4 Common Methods to Synthesize Aldehydes

Method	Reaction
[1] Oxidation of 1° alcohols with PCC (Section 11.12B)	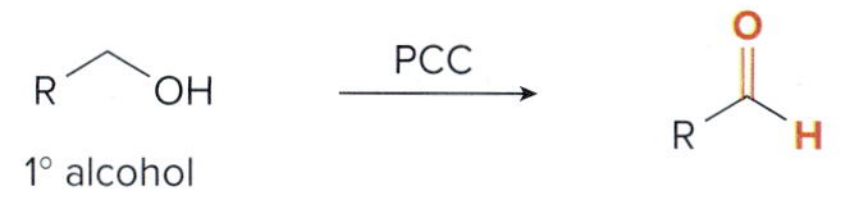 R–OH (1° alcohol) $\xrightarrow{\text{PCC}}$ RCHO
[2] Reduction of esters (Section 13.7A)	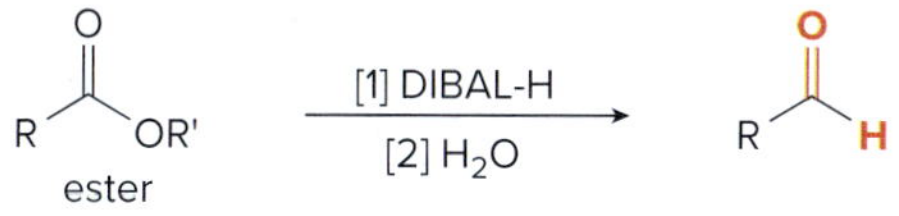 RCO-OR' (ester) $\xrightarrow[\text{[2] H}_2\text{O}]{\text{[1] DIBAL-H}}$ RCHO
[3] Reduction of acid chlorides (Section 13.7A)	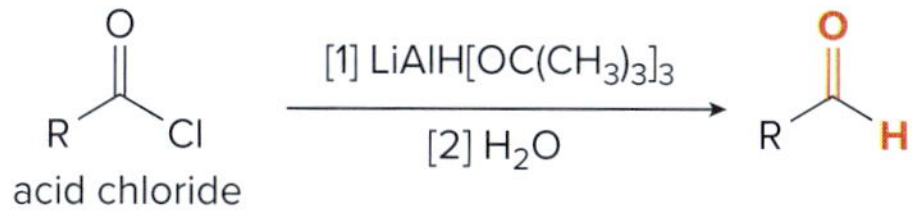 RCO-Cl (acid chloride) $\xrightarrow[\text{[2] H}_2\text{O}]{\text{[1] LiAlH[OC(CH}_3\text{)}_3\text{]}_3}$ RCHO
[4] Hydroboration–oxidation of an alkyne (Section 10.19)	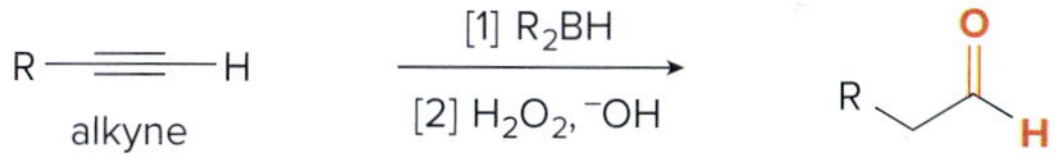 R–≡–H (alkyne) $\xrightarrow[\text{[2] H}_2\text{O}_2, \ ^-\text{OH}]{\text{[1] R}_2\text{BH}}$ RCH$_2$CHO

Ketones are prepared from 2° alcohols, acid chlorides, and alkynes (Table 14.5).

Table 14.5 Common Methods to Synthesize Ketones

Method	Reaction
[1] Oxidation of 2° alcohols with Cr^{6+} reagents (Section 11.12A)	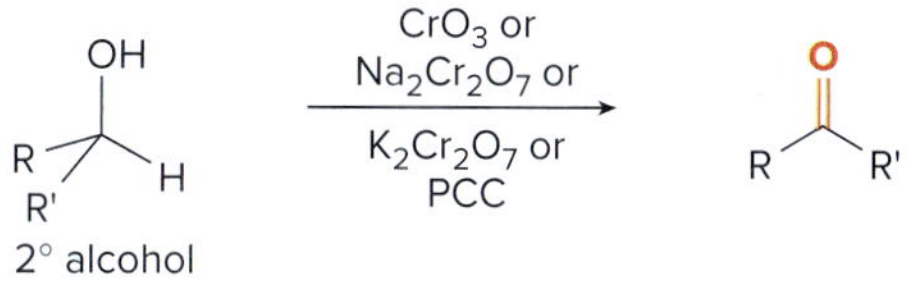 RR'CH–OH (2° alcohol) $\xrightarrow{\text{CrO}_3 \text{ or Na}_2\text{Cr}_2\text{O}_7 \text{ or K}_2\text{Cr}_2\text{O}_7 \text{ or PCC}}$ RCOR'
[2] Reaction of acid chlorides with organocuprates (Section 13.13)	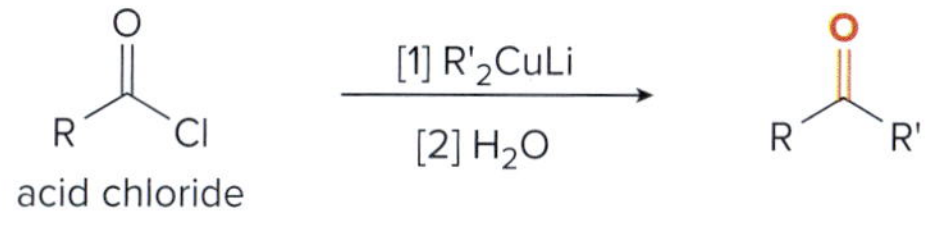 RCO-Cl (acid chloride) $\xrightarrow[\text{[2] H}_2\text{O}]{\text{[1] R'}_2\text{CuLi}}$ RCOR'
[3] Hydration of an alkyne (Section 10.18)	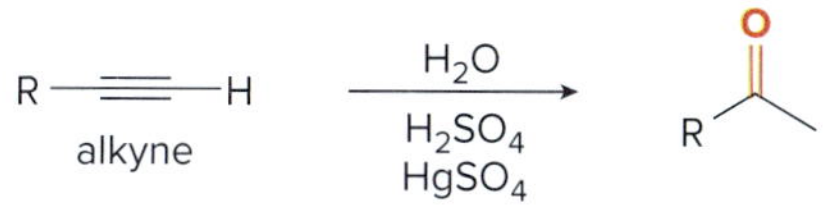 R–≡–H (alkyne) $\xrightarrow[\text{H}_2\text{SO}_4, \text{ HgSO}_4]{\text{H}_2\text{O}}$ RCOCH$_3$

Aldehydes and ketones are also both obtained as products of the oxidative cleavage of alkenes (Section 11.10).

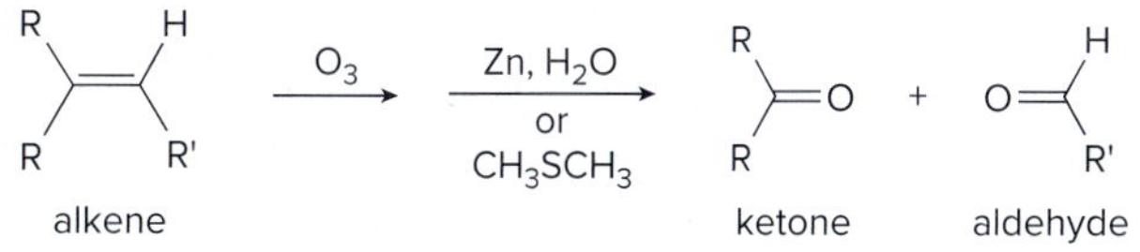

R₂C=CHR' (alkene) $\xrightarrow[\text{or CH}_3\text{SCH}_3]{\text{O}_3, \text{ Zn, H}_2\text{O}}$ R$_2$C=O (ketone) + O=CHR' (aldehyde)

Problem 14.10 What reagents are needed to convert each compound to butanal (CH$_3$CH$_2$CH$_2$CHO)?

a. [structure: methyl butanoate]   b. [structure: 1-butanol]   c. [structure: 1-butyne]   d. [structure: trans-2-heptene]

Problem 14.11 What reagents are needed to convert each compound to acetophenone (C$_6$H$_5$COCH$_3$): (a) C$_6$H$_5$COCl; (b) C$_6$H$_5$C≡CH?

14.6 Reactions of Aldehydes and Ketones—General Considerations

Let's begin our discussion of carbonyl reactions by looking at the two general kinds of reactions that aldehydes and ketones undergo.

[1] Reaction at the carbonyl carbon

Recall from Chapter 13 that the uncrowded, electrophilic carbonyl carbon makes aldehydes and ketones susceptible to **nucleophilic addition** reactions.

[Reaction scheme: R(C=O)R' with R' = H or alkyl, reagents [1] :Nu⁻ [2] H—OH or HNu:, H⁺, giving R(C(OH)(Nu))R'. H and Nu are added. Labeled "nucleophilic addition".]

The elements of H and Nu are added to the carbonyl group. In Chapter 13, you learned about this reaction with hydride (H:⁻) and carbanions (R:⁻) as nucleophiles. In Chapter 14, we will discuss similar reactions with other nucleophiles.

[2] Reaction at the α carbon

A second general reaction of aldehydes and ketones involves reaction at the **α carbon.** A C—H bond on the α carbon to a carbonyl group is more acidic than many other C—H bonds, because reaction with base forms a resonance-stabilized enolate anion.

- Enolates are nucleophiles, so they react with electrophiles (E⁺) to form new bonds on the α carbon.

[Mechanism scheme: R(C=O)CH(α)H + :B → resonance-stabilized enolate anion + H—B⁺, then reaction with E⁺ gives R(C=O)CH(α)E — labeled "reaction at the α carbon".]

Chapters 17 and 18 are devoted to reactions at the α carbon to a carbonyl group.

- Aldehydes and ketones react with nucleophiles at the carbonyl carbon.
- Aldehydes and ketones form enolates that react with electrophiles at the α carbon.

14.6A The General Mechanism of Nucleophilic Addition

Two general mechanisms are usually drawn for nucleophilic addition, depending on the nucleophile (negatively charged versus neutral) and the presence or absence of an acid catalyst. With negatively charged nucleophiles, nucleophilic addition follows the two-step process first discussed in Chapter 13—**nucleophilic attack** followed by **protonation,** as shown in Mechanism 14.1.

14.6 Reactions of Aldehydes and Ketones—General Considerations

Mechanism 14.1 General Mechanism—Nucleophilic Addition

1. The **nucleophile attacks** the electrophilic carbonyl. The π bond is broken, moving an electron pair out on oxygen and forming an sp^3 hybridized carbon.
2. Protonation of the negatively charged oxygen by H_2O forms the **addition product**.

In this mechanism, **nucleophilic attack *precedes* protonation.** This process occurs with strong neutral or negatively charged nucleophiles.

With some neutral nucleophiles, however, nucleophilic addition does not occur unless an **acid catalyst** is added. The general mechanism for this reaction consists of three steps (not two), but the same product results because H and Nu add across the carbonyl π bond. In this mechanism, **protonation *precedes* nucleophilic attack.** Mechanism 14.2 is shown with the neutral nucleophile H—Nu: and a general acid H—A.

Mechanism 14.2 General Mechanism—Acid-Catalyzed Nucleophilic Addition

1. Protonation of the carbonyl oxygen forms a **resonance-stabilized cation**.
2–3. Nucleophilic attack and deprotonation form the neutral addition product. The overall result is **addition of H and Nu** to the carbonyl group.

The effect of protonation is to convert a neutral carbonyl group to one having a net positive charge. **This protonated carbonyl group is much more electrophilic,** and much more susceptible to attack by a nucleophile. This step is unnecessary with strong nucleophiles like hydride (H:⁻) that were used in Chapter 13. With weaker nucleophiles, however, nucleophilic attack does not occur unless the carbonyl group is first protonated.

This step is a specific example of a general phenomenon:

- Any reaction involving a carbonyl group and a strong acid begins with the same first step—protonation of the carbonyl oxygen.

14.6B The Nucleophile

What nucleophiles add to carbonyl groups? This cannot be predicted solely on the trends in nucleophilicity learned in Chapter 7. Only *some* of the nucleophiles that react well in nucleophilic substitution at sp^3 hybridized carbons give reasonable yields of nucleophilic addition products.

658 Chapter 14 Aldehydes and Ketones—Nucleophilic Addition

Cl⁻, Br⁻, and I⁻ are good nucleophiles in substitution reactions at sp^3 hybridized carbons, but they are *ineffective* nucleophiles in addition. Addition of Cl⁻ to a carbonyl group, for example, would cleave the C–O π bond, forming an alkoxide. Because Cl⁻ is a much *weaker* base than the alkoxide formed, equilibrium favors the starting materials (the weaker base, Cl⁻), *not* the addition product.

The situation is further complicated because some of the initial nucleophilic addition adducts are unstable and undergo elimination to form a stable product. For example, amines (RNH₂) add to carbonyl groups in the presence of mild acid to form unstable **carbinolamines**, which readily lose water to form **imines**. This addition–elimination sequence replaces a C=O by a C=N. The details of this process are discussed in Section 14.10.

Figure 14.4 lists nucleophiles that add to a carbonyl group, as well as the products obtained from nucleophilic addition using cyclohexanone as a representative ketone. These reactions are discussed in the remaining sections of Chapter 14. In cases in which the initial addition adduct is unstable, it is enclosed within brackets, followed by the final product.

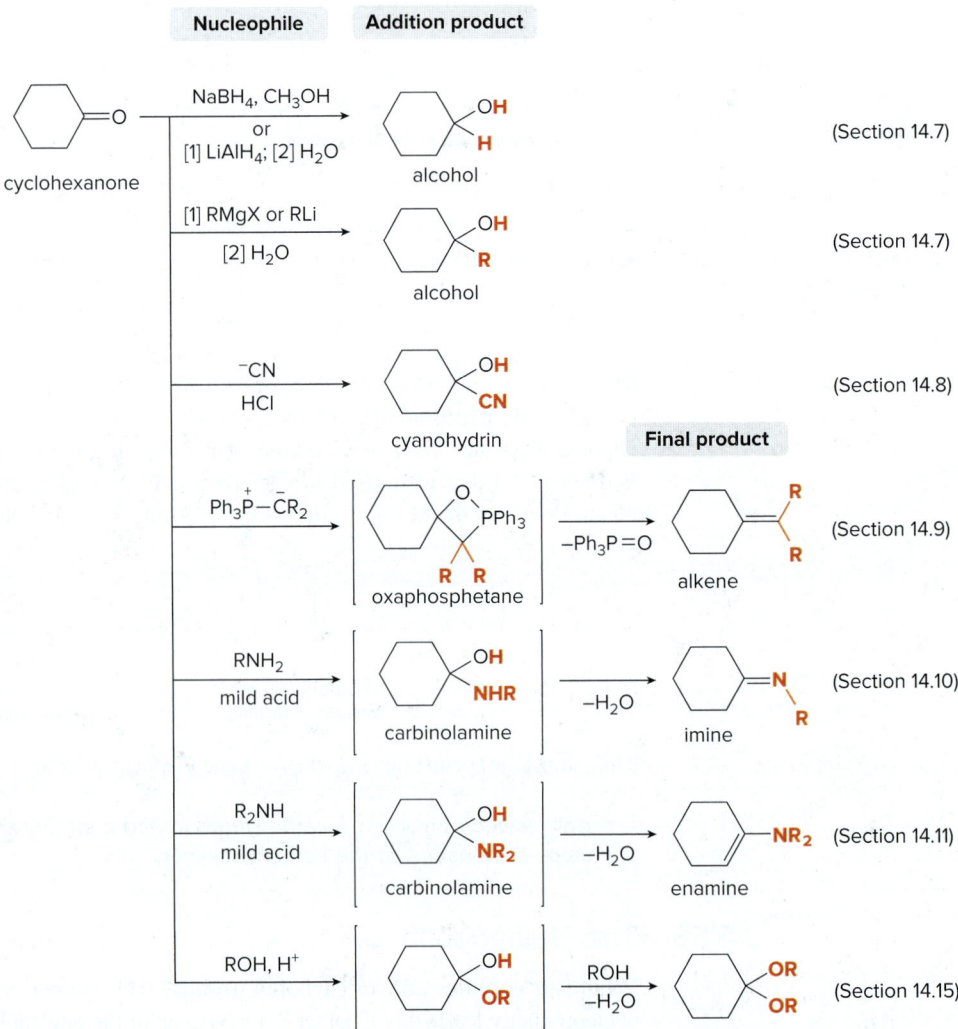

Figure 14.4

Specific examples of nucleophilic addition

14.7 Nucleophilic Addition of H⁻ and R⁻—A Review

We begin our study of nucleophilic additions to aldehydes and ketones by briefly reviewing nucleophilic addition of hydride and carbanions, two reactions examined in Sections 13.4 and 13.10, respectively.

Treatment of an aldehyde or ketone with either NaBH$_4$ or LiAlH$_4$ followed by protonation forms a 1° or 2° alcohol. NaBH$_4$ and LiAlH$_4$ serve as a source of **hydride, H:⁻—the nucleophile**—and the reaction results in addition of the elements of H$_2$ across the C–O π bond. Addition of H$_2$ reduces the carbonyl group to an alcohol.

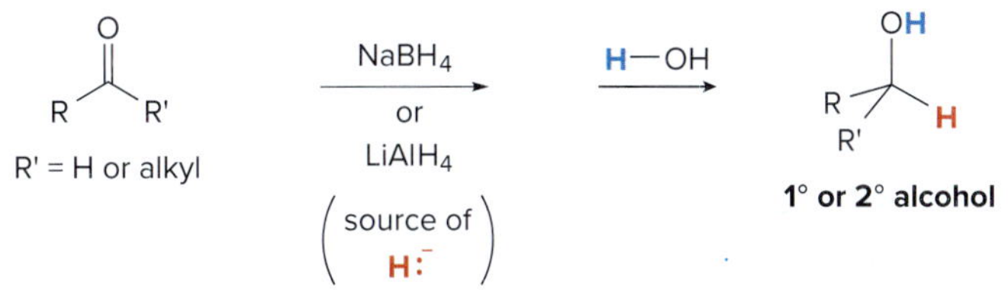

Hydride reduction of aldehydes and ketones occurs via the two-step mechanism of nucleophilic addition—that is, **nucleophilic attack of H:⁻ followed by protonation**—shown in Section 13.4B.

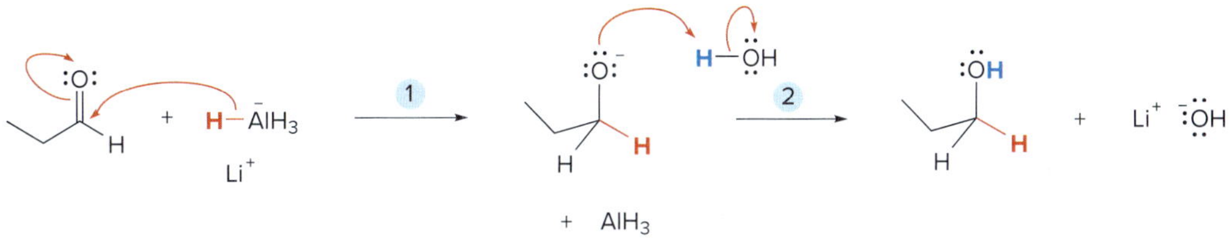

Treatment of an aldehyde or ketone with either an organolithium (R″Li) or Grignard reagent (R″MgX) followed by water forms a 1°, 2°, or 3° alcohol containing a new carbon–carbon bond. R″Li and R″MgX serve as a source of a **carbanion (R″)⁻—the nucleophile**—and the reaction results in addition of the elements of R″ and H across the C–O π bond.

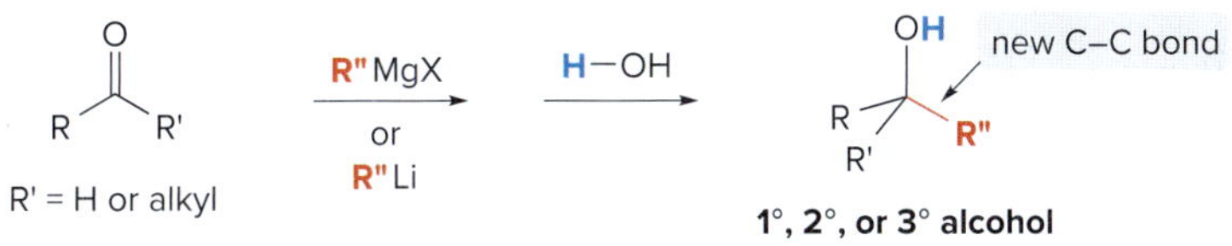

The stereochemistry of hydride reduction and Grignard addition was discussed in Sections 13.5 and 13.10B, respectively.

The nucleophilic addition of carbanions to aldehydes and ketones occurs via the two-step mechanism of nucleophilic addition—that is, **nucleophilic attack of (R″)⁻ followed by protonation**—shown in Section 13.10A. The nucleophile, a carbanion, attacks the trigonal planar sp^2 hybridized carbonyl from both sides, so that when a new stereogenic center is formed, a mixture of stereoisomers results, as shown in Sample Problem 14.3.

Sample Problem 14.3 Drawing the Products with Stereochemistry in Nucleophilic Addition

Draw the products (including the stereochemistry) formed in the following reaction.

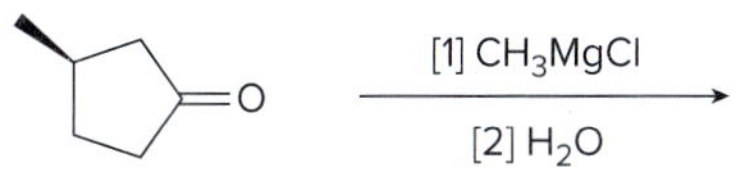

[1] CH$_3$MgCl
[2] H$_2$O

(R)-3-methylcyclopentanone

Solution

The Grignard reagent adds CH₃⁻ from both sides of the trigonal planar carbonyl group, yielding a mixture of 3° alcohols after protonation with water. In this example, the starting ketone and both alcohol products are chiral. The two products, which contain two stereogenic centers, are stereoisomers but not mirror images—that is, they are **diastereomers**.

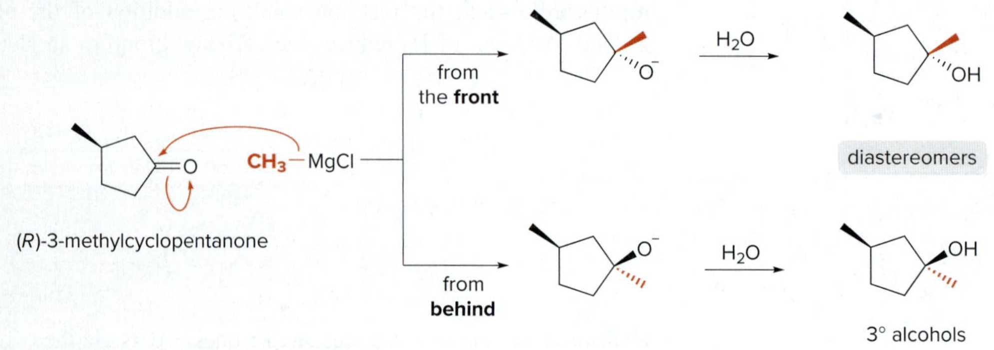

Problem 14.12 Draw the products of each reaction. Include all stereoisomers formed.

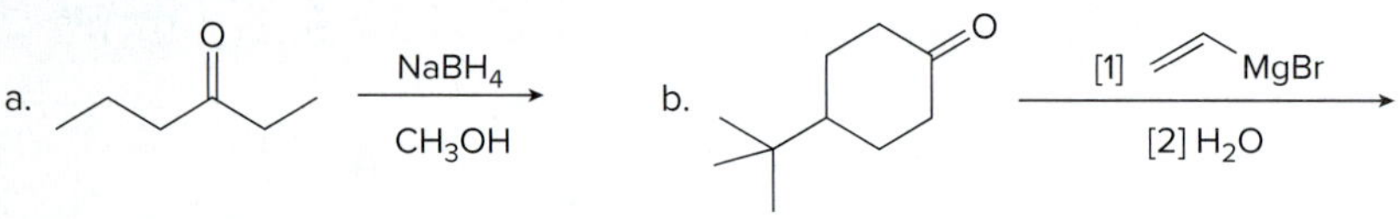

a. [hexan-3-one] $\xrightarrow{\text{NaBH}_4 / \text{CH}_3\text{OH}}$

b. [4-tert-butylcyclohexanone] $\xrightarrow[\text{[2] H}_2\text{O}]{\text{[1] CH}_2=\text{CH-MgBr}}$

More Practice: Try Problems 14.36b [1], [2]; 14.43c.

14.8 Nucleophilic Addition of ⁻CN

Treatment of an aldehyde or ketone with NaCN and a strong acid such as HCl adds the elements of HCN across the carbon–oxygen π bond, forming a **cyanohydrin**.

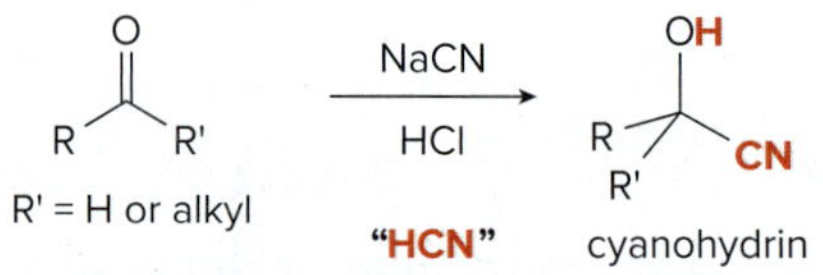

R(C=O)R' $\xrightarrow[\text{HCl}]{\text{NaCN}}$ R(OH)(CN)R'

R' = H or alkyl   "HCN"   cyanohydrin

This reaction adds one carbon to the aldehyde or ketone, forming a **new carbon–carbon bond**.

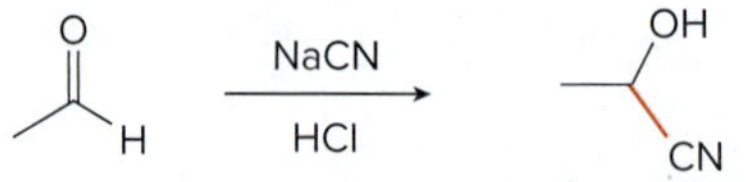

CH₃CHO $\xrightarrow[\text{HCl}]{\text{NaCN}}$ CH₃CH(OH)CN

acetaldehyde cyanohydrin
new C–C bond in red

14.8A The Mechanism

The mechanism of cyanohydrin formation involves the usual two steps of nucleophilic addition: **nucleophilic attack followed by protonation** as shown in Mechanism 14.3.

Mechanism 14.3 Nucleophilic Addition of ⁻CN—Cyanohydrin Formation

1. Nucleophilic attack of ⁻CN forms a **new carbon–carbon bond** with cleavage of the C–O π bond.
2. Protonation of the negatively charged oxygen by HCN forms the **addition product**. The HCN used in this step is formed by the acid–base reaction of ⁻CN with the strong acid, HCl.

This reaction does not occur with HCN alone. The **cyanide anion** makes addition possible because it is a **strong nucleophile** that attacks the carbonyl group.

Cyanohydrins can be reconverted to carbonyl compounds by treatment with base. This process is just the reverse of the addition of HCN: **deprotonation followed by elimination of ⁻CN.**

Note the difference between two similar terms. **Hydration** results in *adding* water to a compound. **Hydrolysis** results in *cleaving bonds* with water.

The cyano group (CN) of a cyanohydrin is readily hydrolyzed to a carboxy group (COOH) by heating with aqueous acid or base. **Hydrolysis replaces the three C–N bonds by three C–O bonds.**

Problem 14.13
Draw the products of each reaction.

a. PhCHO $\xrightarrow{\text{NaCN, HCl}}$

b. cyclopentanone cyanohydrin $\xrightarrow{H_3O^+, \Delta}$

14.8B Application: Naturally Occurring Cyanohydrin Derivatives

Although the cyanohydrin is an uncommon functional group, **linamarin** and **amygdalin** are two naturally occurring cyanohydrin derivatives. Both contain a carbon atom bonded to both an oxygen atom and a cyano group, analogous to a cyanohydrin.

linamarin, amygdalin, laetrile

Peach and apricot pits are a natural source of the cyanohydrin derivative amygdalin. *Jill Braaten/McGraw-Hill Education*

662   Chapter 14   Aldehydes and Ketones—Nucleophilic Addition

Cassava is a widely grown root crop, first introduced to Africa by Portuguese traders from Brazil in the sixteenth century. The peeled root is eaten after boiling or roasting. If the root is eaten without processing, illness and even death can result from high levels of HCN.
Daniel C. Smith

Linamarin is isolated from cassava, a woody shrub grown as a root crop in the humid tropical regions of South America and Africa. **Amygdalin** is present in the seeds and pits of apricots, peaches, and wild cherries. Amygdalin and the related synthetic compound **laetrile** were once touted as anticancer drugs, although their effectiveness is unproven.

Linamarin, amygdalin, and laetrile are toxic compounds because they are metabolized to cyanohydrins, which are hydrolyzed to carbonyl compounds and **toxic HCN gas,** a cellular poison with a characteristic almond odor. This second step is merely the reconversion of a cyanohydrin to a carbonyl compound, a process that occurs with base in reactions run in the laboratory (Section 14.8A). If cassava root is processed with care, linamarin is enzymatically metabolized by this reaction sequence and the toxic HCN is released before the root is ingested, making it safe to eat.

linamarin cyanohydrin derivative → (enzyme) → acetone cyanohydrin → (enzyme) → acetone + **HCN** (toxic by-product)

Problem 14.14   What cyanohydrin and carbonyl compound are formed when amygdalin is metabolized in a similar manner to linamarin?

14.9 The Wittig Reaction

The additions of H^-, R^-, and ^-CN all involve the same two steps—**nucleophilic attack followed by protonation.** Other examples of nucleophilic addition in Chapter 14 are somewhat different. Although they still involve attack of a nucleophile, the initial addition adduct is converted to another product by one or more reactions.

The first reaction in this category is the **Wittig reaction,** named for German chemist Georg Wittig, who was awarded the Nobel Prize in Chemistry in 1979 for its discovery. The Wittig reaction uses a carbon nucleophile, the **Wittig reagent,** to form **alkenes.** When a carbonyl compound is treated with a Wittig reagent, the carbonyl oxygen atom is replaced by the negatively charged alkyl group bonded to the phosphorus—that is, **the C=O is converted to a C=C.**

R₂C=O (R' = H or alkyl) + Ph₃P⁺–C⁻R''₂ (**Wittig reagent**) → R₂C=CR''₂ (**alkene**) + Ph₃P=O (**triphenylphosphine oxide**)

- A Wittig reaction forms two new carbon–carbon bonds—one new σ bond and one new π bond—as well as a phosphorus by-product, Ph₃P=O (triphenylphosphine oxide).

HCHO + Ph₃P⁺–CH₂⁻ → CH₂=CH₂ + Ph₃P=O

cyclohexanone + Ph₃P⁺–CHCH₃⁻ → cyclohexylidene-ethane + Ph₃P=O

14.9A The Wittig Reagent

A **Wittig reagent** is an **organophosphorus reagent**—a reagent that contains a carbon–phosphorus bond. A typical Wittig reagent has a phosphorus atom bonded to three phenyl groups, plus another alkyl group that bears a negative charge.

$$Ph_3\overset{+}{P}-\overset{R''}{\underset{R''}{C:}} \quad \text{an ylide}$$

(+) and (−) charges on adjacent atoms

*Phosphorus ylides are also called **phosphoranes**.*

A Wittig reagent is an **ylide, a species that contains two oppositely charged atoms bonded to each other, and both atoms have octets.** In a Wittig reagent, a negatively charged carbon atom is bonded to a positively charged phosphorus atom.

Because phosphorus is a third-row element, it can be surrounded by more than eight electrons. As a result, a second resonance structure can be drawn that places a double bond between carbon and phosphorus. Regardless of which resonance structure is drawn, a **Wittig reagent has no net charge.** In one resonance structure, though, the **carbon atom bonded to phosphorus (labeled in blue) bears a net negative charge, so it is** *nucleophilic.*

$$Ph_3\overset{+}{P}-\overset{R''}{\underset{R''}{C:}} \longleftrightarrow Ph_3P=\overset{R''}{\underset{R''}{C}}$$

10 electrons around P (five bonds)

Wittig reagents are synthesized by a two-step procedure.

Step [1] S_N2 reaction of triphenylphosphine with an alkyl halide forms a phosphonium salt.

$$Ph_3P: \; + \; R-X: \xrightarrow{S_N2} Ph_3\overset{+}{P}-R \quad :X:^-$$

triphenylphosphine nucleophile → phosphonium salt

Because phosphorus is located below nitrogen in the periodic table, a neutral phosphorus atom with three bonds also has a lone pair of electrons.

Triphenylphosphine (Ph_3P:), which contains a lone pair of electrons on P, is the nucleophile. Because the reaction follows an S_N2 mechanism, it works best with **unhindered CH_3X and 1° alkyl halides (RCH_2X).** Secondary alkyl halides (R_2CHX) can also be used, although yields are often lower.

Step [2] Deprotonation of the phosphonium salt with a strong base (:B) forms the ylide.

Bu—Li
strong base

$$Ph_3\overset{+}{P}-\underset{R}{\overset{H}{C}}-:X:^- \; + \; :B \; \xrightarrow{\text{strong base}} \; Ph_3\overset{+}{P}-\overset{..}{\underset{R}{C:}} \; + \; H-B^+$$

phosphonium salt → ylide

Section 13.9C discussed the reaction of organometallic reagents as strong bases.

Because removal of a proton from a carbon bonded to phosphorus generates a resonance-stabilized carbanion (the ylide), this proton is somewhat more acidic than other protons on an alkyl group in the phosphonium salt. Very strong bases are still needed, though, to favor the products of this acid–base reaction. Common bases used for this reaction

664   Chapter 14   Aldehydes and Ketones—Nucleophilic Addition

are the organolithium reagents such as **butyllithium, CH$_3$CH$_2$CH$_2$CH$_2$Li,** abbreviated as **BuLi.**

To synthesize the Wittig reagent, Ph$_3$P=CH$_2$, use these two steps:

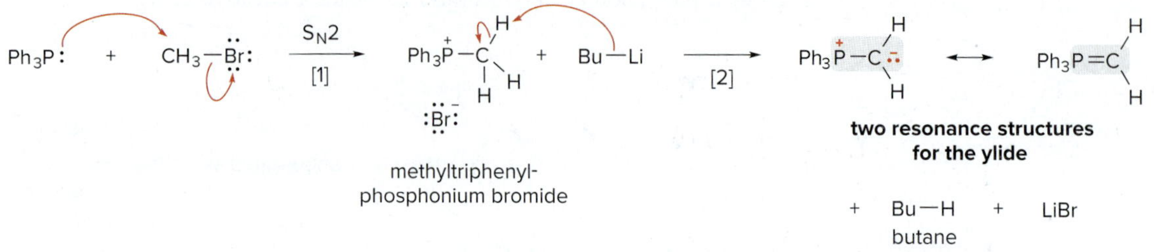

- **Step [1]** Form the **phosphonium salt** by S$_N$2 reaction of Ph$_3$P: and CH$_3$Br.
- **Step [2]** Form the **ylide** by removal of a proton using BuLi as a strong base.

Problem 14.15   Draw the products of the following Wittig reactions.

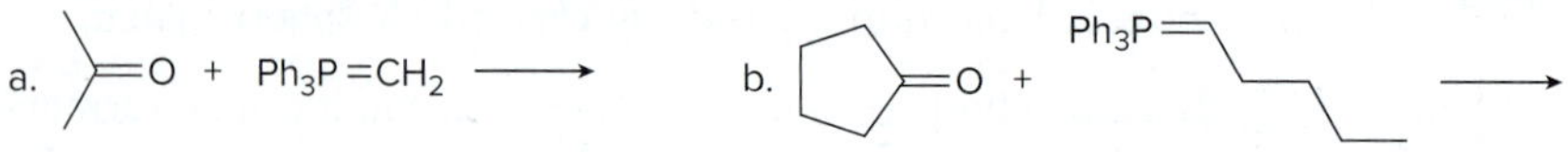

Problem 14.16   Outline a synthesis of each Wittig reagent from Ph$_3$P and an alkyl halide.

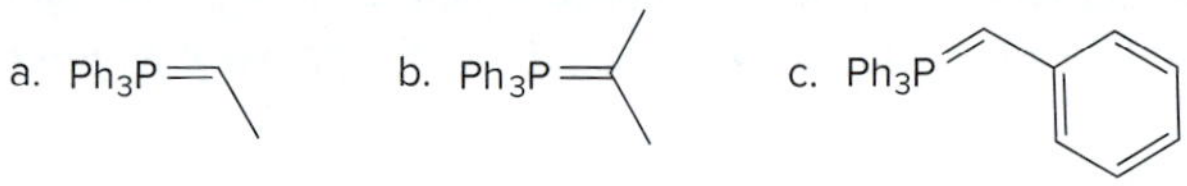

14.9B Mechanism of the Wittig Reaction

The currently accepted mechanism of the Wittig reaction involves two steps. Like other nucleophiles, the Wittig reagent attacks an electrophilic carbonyl carbon, but then the initial addition adduct undergoes elimination to form an alkene. Mechanism 14.4 is drawn using Ph$_3$P=CH$_2$.

Mechanism 14.4   The Wittig Reaction

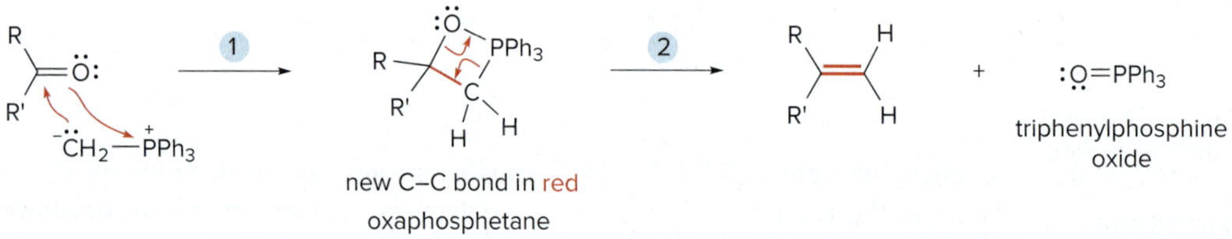

① The negatively charged carbon of the ylide attacks the carbonyl carbon as the carbonyl oxygen attacks the positively charged P atom. This step forms **two bonds** and generates a **four-membered ring** called an **oxaphosphetane.**

② **Elimination of triphenylphosphine oxide forms two new π bonds.** The formation of the strong P=O provides the driving force for the Wittig reaction.

One limitation of the Wittig reaction is that a mixture of alkene stereoisomers sometimes forms. For example, reaction of propanal (CH$_3$CH$_2$CHO) with a Wittig reagent forms the mixture of *E* and *Z* isomers shown.

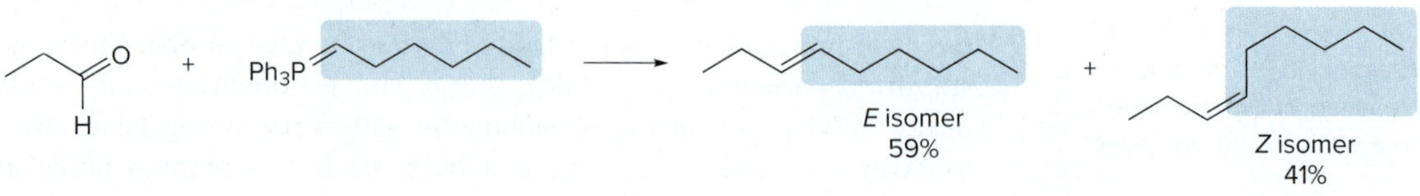

Figure 14.5
A Wittig reaction used to synthesize β-carotene

E alkene

β-carotene
orange pigment found in carrots
(vitamin A precursor)

- The more stable *E* alkene is the major product in this Wittig reaction.

Because the Wittig reaction forms two carbon–carbon bonds in a single reaction, it has been used to synthesize many natural products, including β-carotene, shown in Figure 14.5.

Problem 14.17 Draw the products (including stereoisomers) formed when benzaldehyde (C_6H_5CHO) is treated with each Wittig reagent.

a. Ph_3P⸺ b. Ph_3P⸺ c. Ph_3P⸺

14.9C Retrosynthetic Analysis

To use the Wittig reaction in synthesis, you must be able to determine what carbonyl compound and Wittig reagent are needed to prepare a given compound—that is, **you must work backwards, in the retrosynthetic direction.** There can be two different Wittig routes to a given alkene, but one is often preferred on steric grounds.

How To Determine the Starting Materials for a Wittig Reaction Using Retrosynthetic Analysis

Example What starting materials are needed to synthesize alkene **X** by a Wittig reaction?

X

Step [1] Cleave the carbon–carbon double bond into two components.

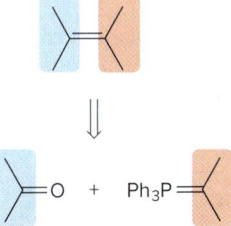

- Part of the molecule becomes the carbonyl component, and the other part becomes the Wittig reagent.

—*Continued*

666 Chapter 14   Aldehydes and Ketones—Nucleophilic Addition

How To, continued . . .

There are usually two routes to a given alkene using a Wittig reaction:

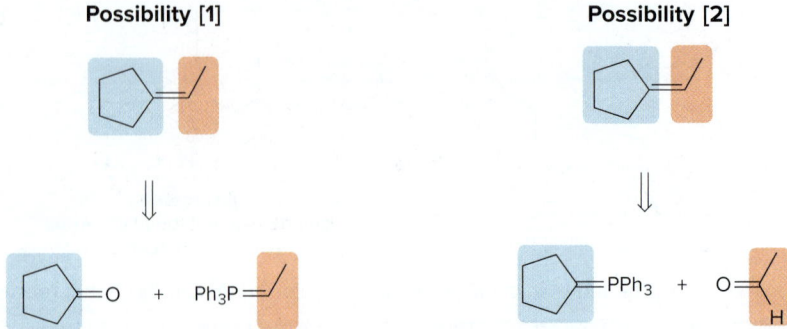

Step [2] Compare the Wittig reagents. The preferred pathway uses a Wittig reagent derived from an unhindered alkyl halide—CH$_3$X or RCH$_2$X.

Determine what alkyl halide is needed to prepare each Wittig reagent:

Possibility [1]   Ph$_3$P=⟶   Ph$_3$P⁺— ⟶   Ph$_3$P: + X—
                                    X⁻   **1° halide**
                                                               preferred pathway

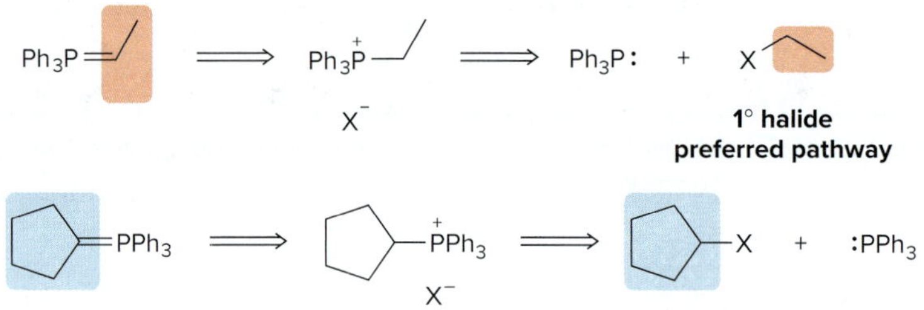

Possibility [2]   ⬡=PPh$_3$ ⟶ ⬡—PPh$_3$⁺ ⟶ ⬡—X + :PPh$_3$
                                    X⁻   **2° halide**

Because the synthesis of the Wittig reagent begins with an **S$_N$2** reaction, **the preferred pathway begins with an unhindered methyl halide or 1° alkyl halide.** In this example, retrosynthetic analysis of both Wittig reagents indicates that only one of them (Ph$_3$P=CHCH$_3$) can be synthesized from a **1° alkyl halide,** making Possibility [1] the preferred pathway.

Problem 14.18   What starting materials are needed to prepare each alkene by a Wittig reaction? When there are two possible routes, indicate which route, if any, is preferred.

a. 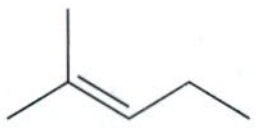 (CH$_3$)$_2$C=CHCH$_2$CH$_3$ b. CH$_3$CH$_2$CH=CHCH$_2$CH$_3$ c. PhCH=CHCH$_3$

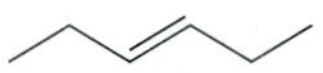

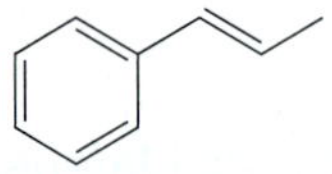

14.10 Addition of 1° Amines

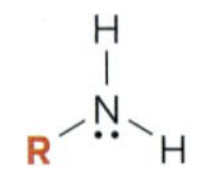

We now move on to the reaction of aldehydes and ketones with nitrogen and oxygen heteroatoms. **Amines are organic nitrogen compounds that contain a nonbonded electron pair on the N atom.** As we learned in Section 3.2, amines are classified as 1°, 2°, or 3° by the number of alkyl groups bonded to the *nitrogen* atom.

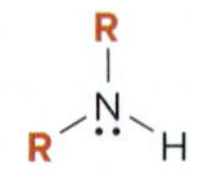

1° amine (1 R group on N): R—NH$_2$ (with H's shown)

2° amine (2 R groups on N): R$_2$N—H

Both 1° and 2° amines react with aldehydes and ketones. We begin by examining the reaction of aldehydes and ketones with 1° amines.

Treatment of an aldehyde or ketone with a 1° amine affords an **imine** (also called a **Schiff base**). Nucleophilic attack of the 1° amine on the carbonyl group forms an unstable **carbinolamine,** which loses water to form an imine. The overall reaction results in **replacement of C=O by C=NR.**

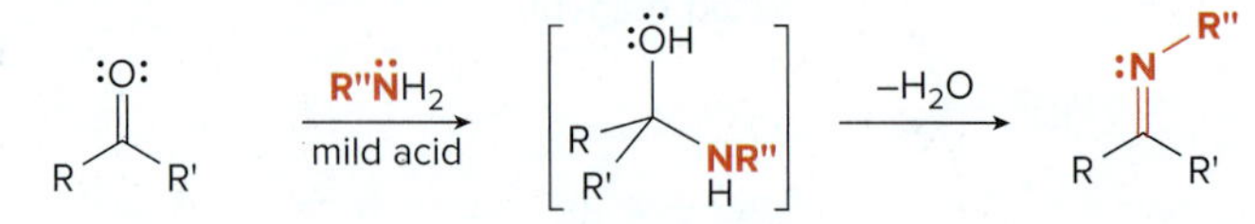

R(R')C=O →[R″NH$_2$, mild acid]→ [R(R')C(OH)(NR″H)] →[−H$_2$O]→ R(R')C=N—R″

R' = H or alkyl   carbinolamine   imine

Because the N atom of an imine is surrounded by three groups (two atoms and a lone pair), it is sp^2 hybridized, making the C—N—R" bond angle ~120° (*not* 180°). Imine formation is fastest when the reaction medium is weakly acidic.

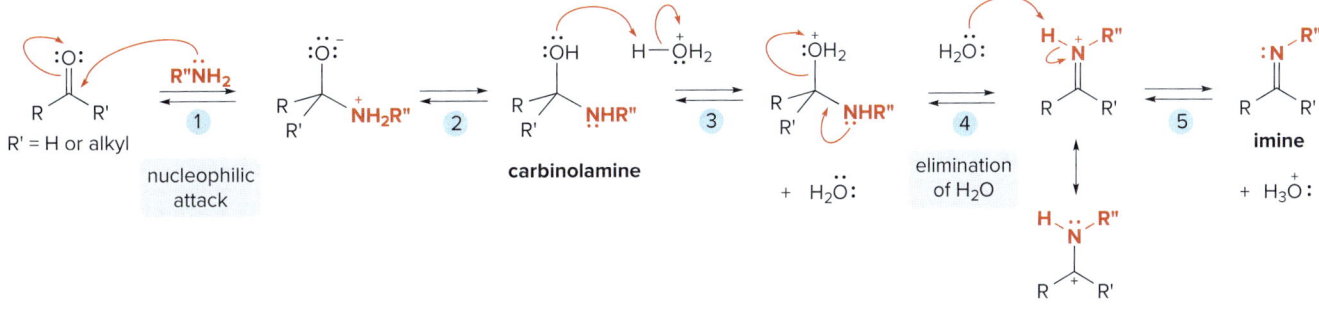

The mechanism of imine formation (Mechanism 14.5) can be divided into two distinct parts: **nucleophilic addition of the 1° amine (Steps [1] and [2]), followed by elimination of H_2O (Steps [3]–[5])**. Each step involves a reversible equilibrium, so that the reaction is driven to completion by removing H_2O.

Mechanism 14.5 Imine Formation from an Aldehyde or a Ketone

1–2 **Nucleophilic attack of the amine** followed by proton transfer forms the **carbinolamine**.
3   Protonation of the OH group forms a **good leaving group**.
4   Loss of H_2O forms a **resonance-stabilized iminium ion**.
5   Loss of a proton forms the **imine**.

Imine formation is most rapid at pH 4–5. Mild acid is needed for protonation of the hydroxy group in Step [3] to form a **good leaving group**. Under strongly acidic conditions, the reaction rate decreases because the amine nucleophile is protonated. With no free electron pair, it is no longer a nucleophile, and so nucleophilic addition cannot occur.

Problem 14.19 Draw the product formed when $CH_3CH_2CH_2CH_2NH_2$ reacts with each carbonyl compound in the presence of mild acid.

a. Ph—CHO   b. (acetone)   c. cyclopentanone

Problem 14.20 What 1° amine and carbonyl compound are needed to prepare each imine?

a.   b.

14.11 Addition of 2° Amines

A 2° amine reacts with an aldehyde or a ketone to give an **enamine**. *Enamines have a nitrogen atom bonded to a double bond* (alk*ene* + *amine* = *enamine*).

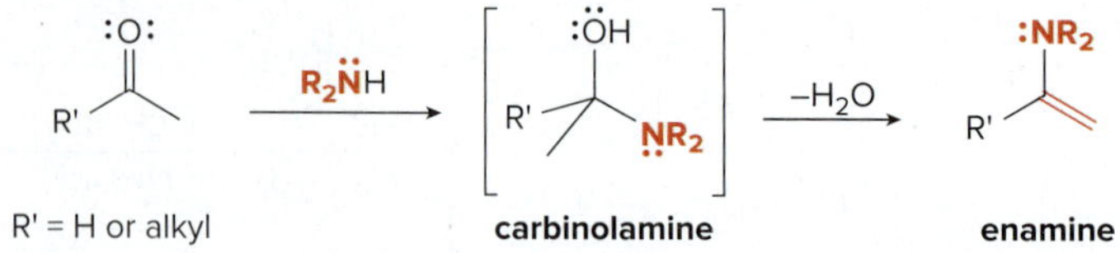

Like imines, enamines are also formed by the addition of a nitrogen nucleophile to a carbonyl group followed by elimination of water. In this case, however, **elimination occurs across two adjacent *carbon* atoms** to form a new carbon–carbon π bond.

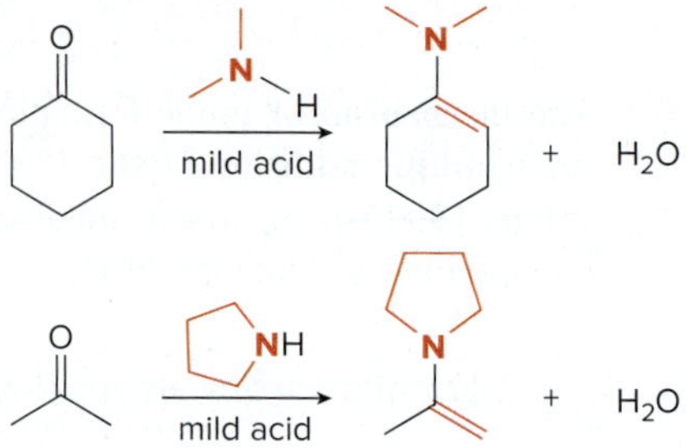

The mechanism for enamine formation (Mechanism 14.6) is identical to the mechanism for imine formation except for the *last step*, involving formation of the π bond. The mechanism can be divided into two distinct parts: **nucleophilic addition of the 2° amine (Steps [1] and [2]), followed by elimination of H₂O (Steps [3]–[5])**. Each step involves a reversible equilibrium once again, so that the reaction is driven to completion by removing H₂O.

Mechanism 14.6 Enamine Formation from an Aldehyde or a Ketone

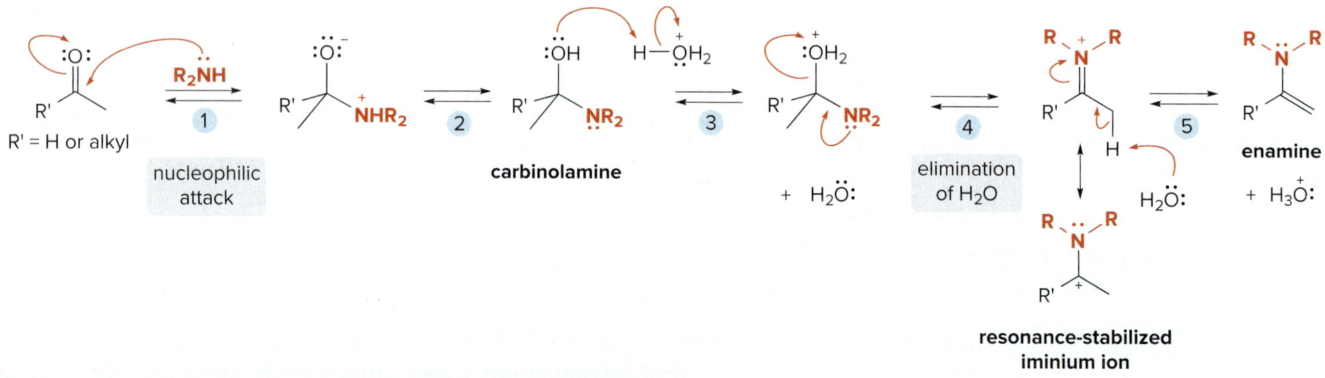

1–2 Nucleophilic attack of the amine followed by proton transfer forms the **carbinolamine**.
3   Protonation of the OH group forms a **good leaving group**.
4   Loss of H₂O forms a **resonance-stabilized iminium ion**.
5   Loss of a proton from the adjacent C–H bond forms the **enamine**.

The mechanisms illustrate why **the reaction of 1° amines with carbonyl compounds forms *imines*, but the reaction with 2° amines forms *enamines***. In Figure 14.6, the last step of both mechanisms is compared using cyclohexanone as starting material. The position of the double bond depends on which proton is removed in the last step. **Removal of an N–H proton forms a C=N, whereas removal of a C–H proton forms a C=C.**

Problem 14.21   What two enamines are formed when 2-methylcyclohexanone is treated with (CH₃)₂NH?

Figure 14.6
The formation of imines and enamines compared

- With a **1° amine**, the intermediate iminium ion still has a proton on the N atom that may be removed to form a **C=N**.
- With a **2° amine**, the intermediate iminium ion has *no* proton on the N atom. A proton must be removed from an adjacent C—H bond, and this forms a **C=C**.

14.12 Imine and Enamine Hydrolysis

Because imines and enamines are formed by a set of reversible reactions, **both can be converted back to carbonyl compounds by hydrolysis with mild acid.**

- Hydrolysis of imines and enamines forms aldehydes and ketones.

The mechanism of these reactions is exactly the *reverse* of the mechanism written for the formation of imines and enamines. In the hydrolysis of enamines shown in Mechanism 14.7, the carbonyl carbon in the product comes from the sp^2 hybridized carbon bonded to the N atom in the starting material.

Mechanism 14.7 Hydrolysis of an Enamine

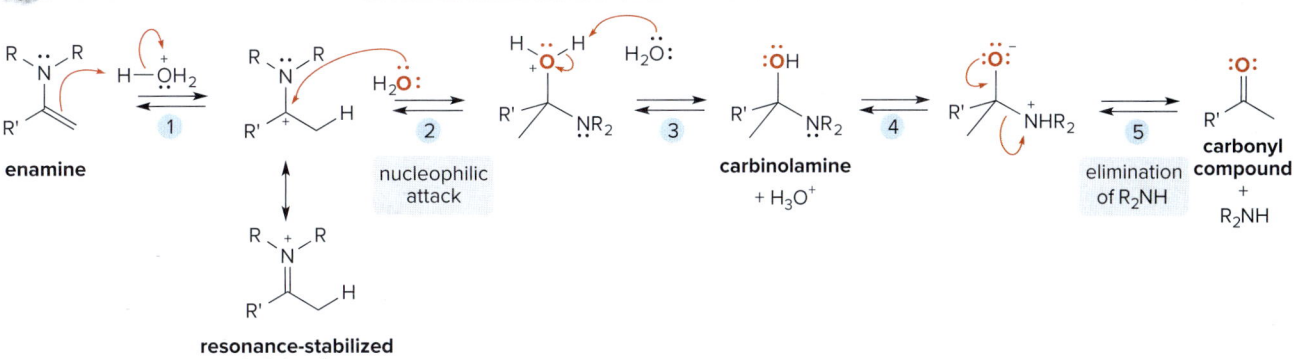

1. Protonation of the enamine forms a **resonance-stabilized iminium ion**.
2. – 3. Nucleophilic attack of H_2O and deprotonation form a **carbinolamine**.
4. – 5. Proton transfer and loss of R_2NH form the **carbonyl group**.

Sample Problem 14.4 Drawing the Products of Imine and Enamine Hydrolysis

Draw the products formed by the hydrolysis of each compound.

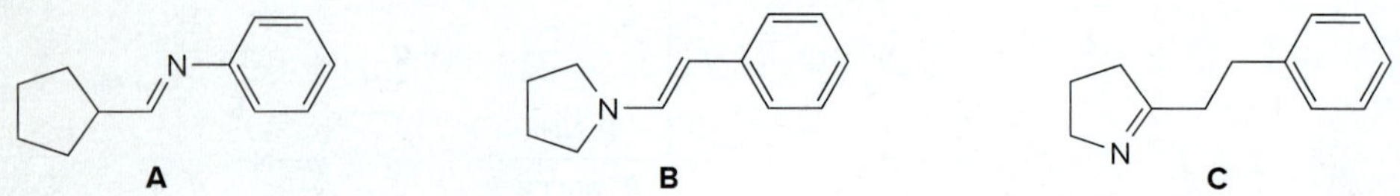

Solution

- An imine contains a C=N, which is converted to a **C=O and a 1° amine** during hydrolysis.
- An enamine, which contains a N atom bonded to a C=C, is hydrolyzed to a **2° amine and a carbonyl compound**.

The carbon in **A, B,** and **C** labeled in blue is converted to the carbonyl carbon.

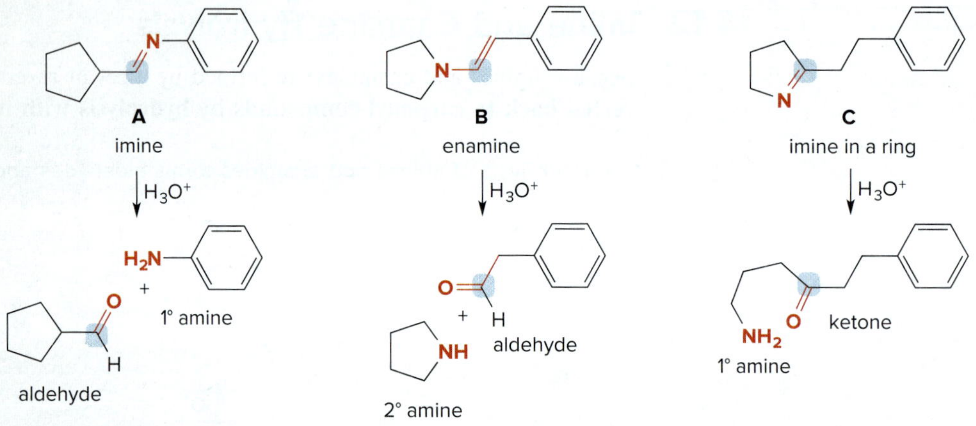

- The imine **A** is converted to a 1° amine and an aldehyde.
- The enamine **B** is converted to an aldehyde and a 2° amine. The alkenyl carbon bonded to N is converted to the carbon of the C=O.
- The imine **C** is converted to a 1° amine and a ketone. Because **C** is cyclic, both functional groups end up in the *same* compound.

Problem 14.22
What carbonyl compound and amine are formed by the hydrolysis of each compound?

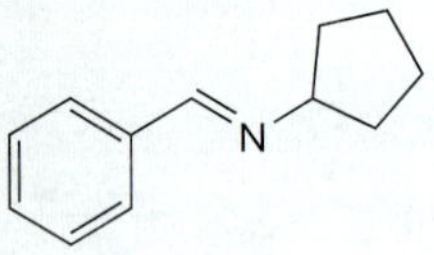

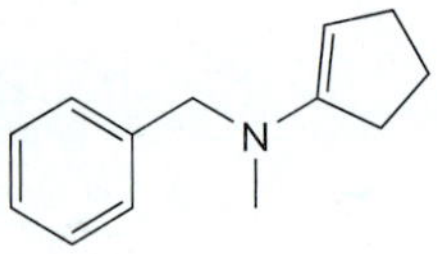

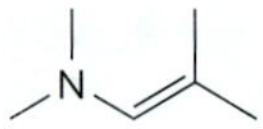

More Practice: Try Problems 14.41c, g; 14.47.

Problem 14.23
Draw a stepwise mechanism for the following imine hydrolysis.

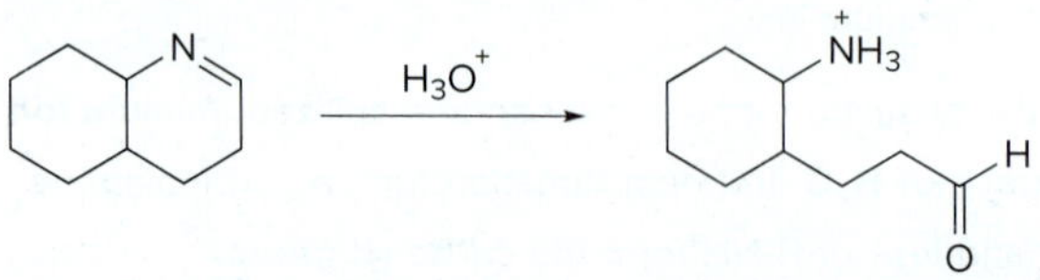

14.13 Imines in Biological Systems

Many imines play vital roles in biological systems.

14.13A Application: Retinal, Rhodopsin, and the Chemistry of Vision

A key molecule in the chemistry of vision is the highly conjugated imine **rhodopsin**, which is synthesized in the rod cells of the eye from **11-*cis*-retinal** and a 1° amine in the protein **opsin**.

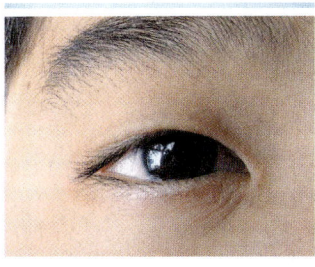

11-*cis*-Retinal is the light-sensitive aldehyde that plays a key role in the chemistry of vision for all vertebrates, arthropods, and mollusks. *Daniel C. Smith*

The central role of rhodopsin in the visual process was delineated by Nobel Laureate George Wald of Harvard University.

The complex process of vision centers around this imine derived from retinal (Figure 14.7). The 11-cis double bond in rhodopsin creates crowding in the rather rigid side chain. When light strikes the rod cells of the retina, it is absorbed by the conjugated double bonds of rhodopsin, and the **11-cis double bond is isomerized to the 11-trans arrangement.** This isomerization is accompanied by a drastic change in shape in the protein, altering the concentration of Ca^{2+} ions moving across the cell membrane, and sending a nerve impulse to the brain, which is then processed into a visual image.

Figure 14.7
The key reaction in the chemistry of vision

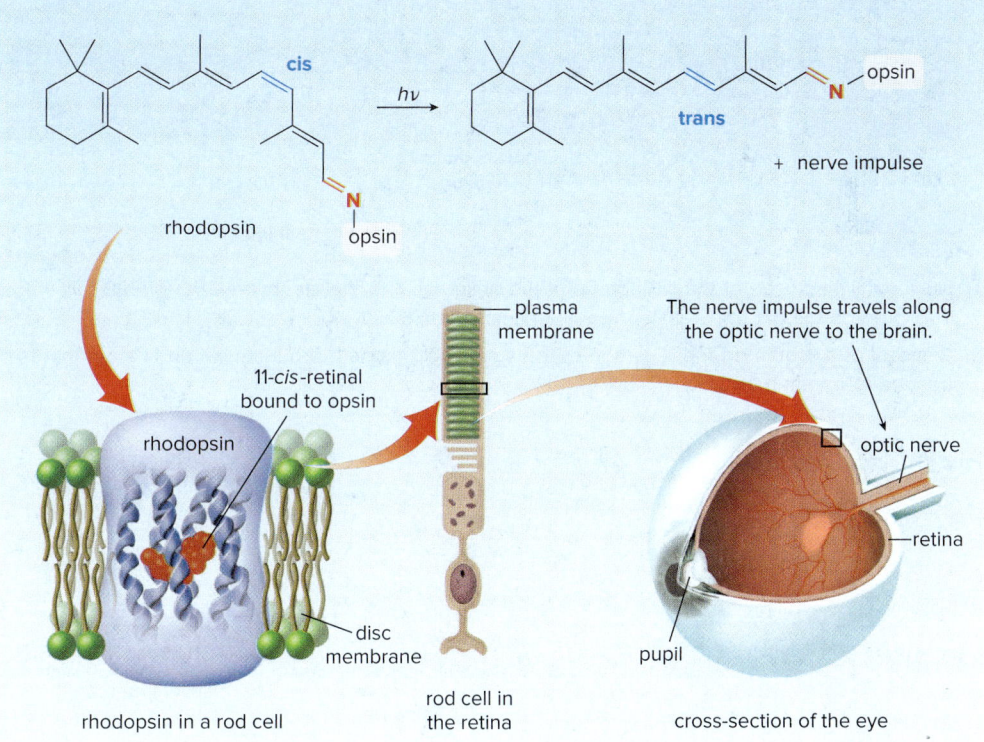

- Rhodopsin is a light-sensitive compound located in the membrane of the rod cells in the retina of the eye. Rhodopsin contains the protein opsin bonded to 11-*cis*-retinal via an imine linkage. When light strikes this molecule, the **crowded 11-cis double bond isomerizes to the 11-trans isomer,** and a nerve impulse is transmitted to the brain by the optic nerve.

14.13B Pyridoxal Phosphate and the Deamination of α-Amino Acids

The metabolism of α-amino acids differs from the metabolic degradation of carbohydrates and lipids, which involves the oxidation of only carbon atoms. With α-amino acids, the amino group (NH_2) must be metabolized as well.

The degradation of α-amino acids begins with **deamination,** the removal of the α amino group, with pyridoxal phosphate (PLP), the coenzyme mentioned in the chapter opener, catalyzed by an aminotransferase enzyme. The α-amino acid is converted to an α-keto acid, and pyridoxal phosphate is converted to pyridoxamine phosphate (PMP).

α-amino acid + pyridoxal phosphate ⟶ α-keto acid + pyridoxamine phosphate

Imine formation and hydrolysis are central in the PLP-dependent deamination of an amino acid, as shown in Mechanism 14.8. While the mechanism shows the overall process that occurs during deamination, the steps numbered [1] and [4] actually consist of a series of operations that have been enumerated in Mechanisms 14.5 and 14.7. Like many other biological mechanisms, the acid (HA) and base (B:) needed for a specific transformation are usually acidic or basic sites at the active site of the enzyme that catalyzes the reaction.

Mechanism 14.8 Deamination of an α-Amino Acid with PLP

Part [1] Imine formation

PLP + α-amino acid →(1, nucleophilic addition, several steps)→ imine →(2)→

① Nucleophilic addition of the NH_2 group of the amino acid to the aldehyde C=O of PLP followed by loss of water forms an **imine.** This process follows the steps shown in Mechanism 14.5.

② Removal of a proton on the α carbon of the amino acid forms a product with an N atom that is part of one double bond and bonded to another C=C.

Part [2] Imine hydrolysis

→(3)→ imine + A:⁻ →(4, hydrolysis, several steps)→ PMP + α-keto acid

③ Protonation of the C=C bonded to N re-forms the positively charged pyridinium ring. The result of Steps [2] and [3] is to move the position of the C=N to the α carbon of the original amino acid.

④ Hydrolysis of the imine occurs by a multistep path similar to Mechanism 14.7.

Problem 14.24 For each amino acid: (a) Draw the structure of the imine formed by reaction with PLP; (b) draw the structure of the α-keto acid formed after imine hydrolysis.

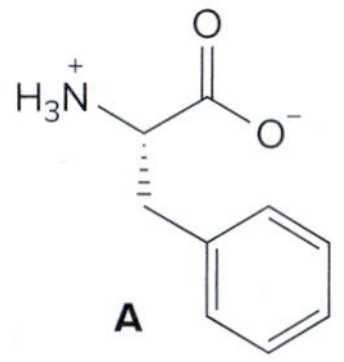

14.14 Addition of H₂O—Hydration

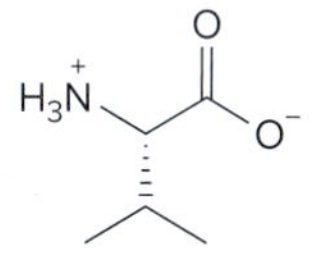

Treatment of a carbonyl compound with H₂O in the presence of an acid or base catalyst **adds the elements of H and OH across the carbon–oxygen π bond**, forming a *gem*-diol or **hydrate.**

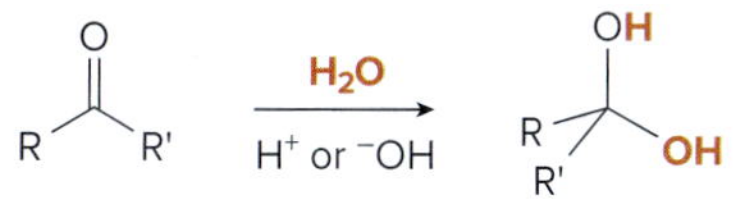

Hydration of a carbonyl group gives a good yield of *gem*-diol only with an **unhindered aldehyde** like formaldehyde, and with aldehydes containing nearby **electron-withdrawing groups.**

14.14A The Thermodynamics of Hydrate Formation

Whether addition of H₂O to a carbonyl group affords a good yield of the *gem*-diol depends on the relative energies of the starting material and the product. With *less stable* carbonyl starting materials, equilibrium favors the *hydrate* product, whereas with *more stable* carbonyl starting materials, equilibrium favors the *carbonyl starting material*. Because **alkyl groups stabilize a carbonyl group** (Section 13.2A):

- *Increasing* the number of alkyl groups on the carbonyl carbon *decreases* the amount of hydrate at equilibrium.

This can be illustrated by comparing the amount of hydrate formed from formaldehyde, acetaldehyde, and acetone.

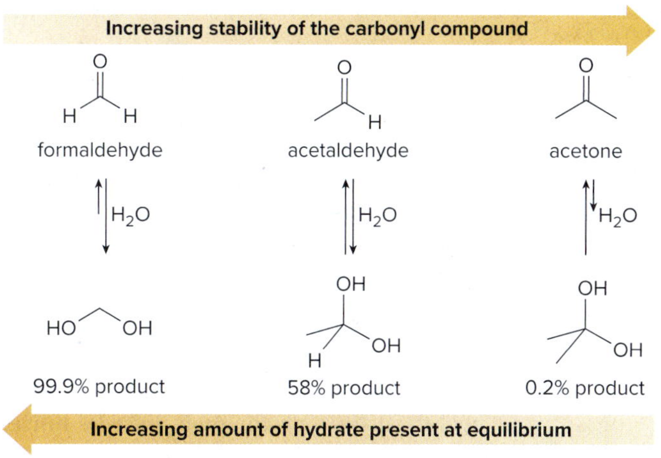

Formaldehyde, the least stable carbonyl compound, forms the largest percentage of hydrate. On the other hand, acetone and other ketones, which have two electron-donor R groups, form < 1% of the hydrate at equilibrium. Other electronic factors come into play as well:

- Electron-*donating* groups near the carbonyl carbon stabilize the carbonyl group, *decreasing* the amount of the hydrate at equilibrium.
- Electron-*withdrawing* groups near the carbonyl carbon destabilize the carbonyl group, *increasing* the amount of the hydrate at equilibrium.

This explains why chloral (trichloroacetaldehyde) forms a large amount of hydrate at equilibrium. Three electron-withdrawing Cl atoms place a partial positive charge on the α carbon to the carbonyl, destabilizing the carbonyl group, and therefore increasing the amount of hydrate at equilibrium.

Chloral hydrate, a sedative sometimes administered to calm a patient prior to a surgical procedure, has also been used for less reputable purposes. Adding it to an alcoholic beverage makes a so-called knock-out drink, causing an individual who drinks it to pass out. Because it is addictive and care must be taken in its administration, chloral hydrate is a controlled substance.

Adjacent like charges (δ+) *destabilize* the carbonyl and *increase* the amount of hydrate.

Problem 14.25 Rank the following carbonyl compounds in order of increasing percentage of hydrate present at equilibrium.

A   B   C

14.14B The Kinetics of Hydrate Formation

Although H_2O itself adds slowly to a carbonyl group, both acid and base catalyze the addition. In base, the nucleophile is ^-OH, and the mechanism follows the usual two steps for nucleophilic addition: **nucleophilic attack followed by protonation,** as shown in Mechanism 14.9.

Mechanism 14.9 Base-Catalyzed Addition of H_2O to a Carbonyl Group

R' = H or alkyl

gem-diol

① **The nucleophile (^-OH) attacks the carbonyl,** breaking the π bond and moving an electron pair out on oxygen.

② Protonation of the negatively charged oxygen by H_2O forms the **hydration product.**

The acid-catalyzed addition follows the general mechanism presented in Section 14.6A. For a poorer nucleophile like H_2O to attack a carbonyl group, the **carbonyl must be protonated by acid first; thus, protonation *precedes* nucleophilic attack.** The overall mechanism has three steps, as shown in Mechanism 14.10.

Mechanism 14.10 Acid-Catalyzed Addition of H₂O to a Carbonyl Group

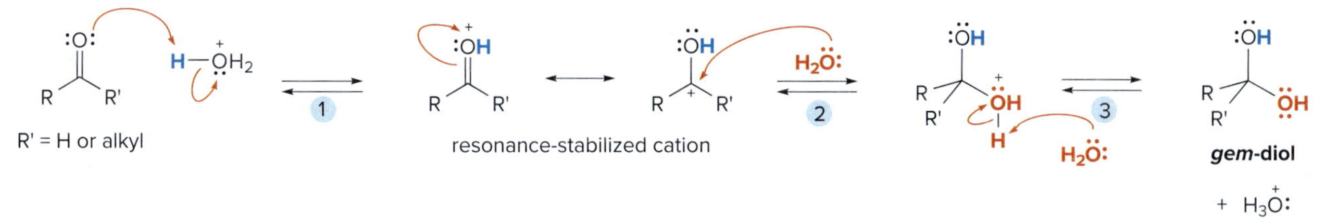

[Mechanism scheme showing: R(R')C=O + H—OH₂ → protonated carbonyl (resonance-stabilized cation) → nucleophilic attack by H₂O → protonated gem-diol → gem-diol + H₃O⁺, with steps labeled 1, 2, 3]

1 Protonation of the carbonyl oxygen forms a **resonance-stabilized cation**.

2–3 Nucleophilic attack and deprotonation form the **gem-diol**. The overall result is addition of H and OH to the carbonyl group.

Acid and base increase the rate of reaction for different reasons:

- Base converts H₂O to ⁻OH, a *stronger nucleophile*.
- Acid protonates the carbonyl group, making it *more electrophilic* toward nucleophilic attack.

These catalysts increase the rate of the reaction, but they do not affect the equilibrium constant. Starting materials that give a low yield of *gem*-diol do so whether or not a catalyst is present.

Problem 14.26 Draw a stepwise mechanism for the following reaction.

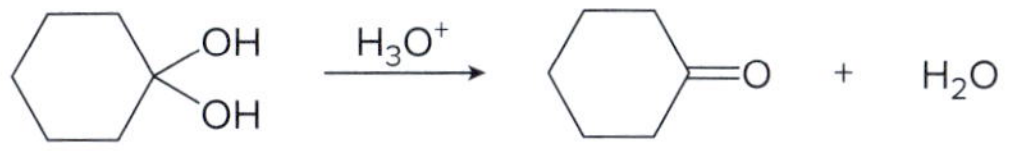

[cyclohexane-1,1-diol →(H₃O⁺) cyclohexanone + H₂O]

14.15 Addition of Alcohols—Acetal Formation

The term *acetal* refers to any compound derived from an aldehyde or ketone, having two OR groups bonded to a single carbon. The term *ketal* is sometimes used when the starting carbonyl compound is a ketone; that is, the carbon bonded to the alkoxy groups is *not* bonded to a H atom and the general structure is R₂C(OR')₂. Because ketals are considered a subclass of acetals in the IUPAC system, we will use the single general term *acetal* for any compound having two OR groups on a carbon atom.

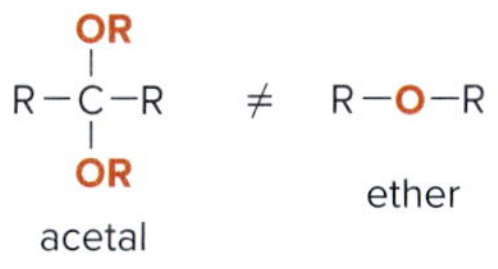

Aldehydes and ketones react with *two* equivalents of alcohol to form acetals. In an acetal, the carbonyl carbon from the aldehyde or ketone is now singly bonded to **two OR" (alkoxy) groups**.

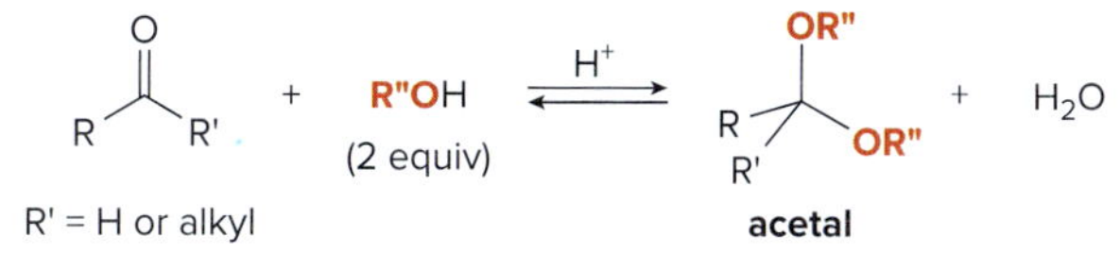

[RCOR' + R"OH (2 equiv) ⇌(H⁺) R(R')C(OR")₂ acetal + H₂O]

This reaction differs from other additions we have seen thus far, because **two equivalents of alcohol are added to the carbonyl group,** and two new C—O σ bonds are formed. Acetal formation is catalyzed by acids, commonly *p*-toluenesulfonic acid (TsOH).

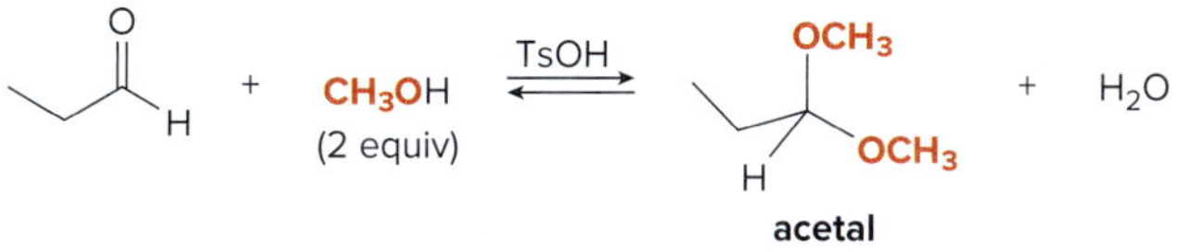

[propanal + CH₃OH (2 equiv) ⇌(TsOH) CH₃CH₂CH(OCH₃)₂ acetal + H₂O]

When a diol such as ethylene glycol is used in place of two equivalents of ROH, a cyclic acetal is formed. Both oxygen atoms in the cyclic acetal come from the diol.

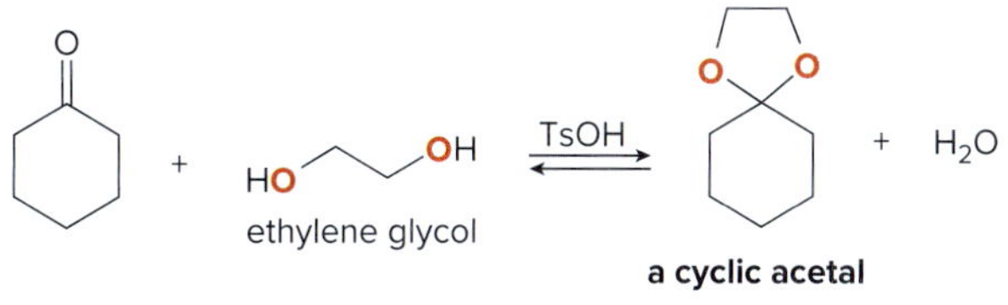

[cyclohexanone + HOCH₂CH₂OH (ethylene glycol) ⇌(TsOH) cyclic acetal + H₂O]

Acetals are *not* ethers, even though both functional groups contain a C—O σ bond. Having two C—O σ bonds on the *same* carbon atom makes an acetal very different from an ether.

R—C(OR)(OR)—R ≠ R—O—R
acetal   ether

Like *gem*-diol formation, the synthesis of acetals is reversible, and often the equilibrium favors reactants, not products. In acetal synthesis, however, water is formed as a by-product, so the equilibrium can be driven to the right by **removing the water as it is formed.** This can be done in a variety of ways in the laboratory. A drying agent can be added that reacts with the water, or more commonly, the water can be distilled from the reaction mixture as it is formed. Driving an equilibrium to the right by removing one of the products is an application of Le Châtelier's principle (see Section 9.8).

676 Chapter 14 Aldehydes and Ketones—Nucleophilic Addition

Problem 14.27 Draw the products of each reaction.

a. cyclopentanone + 2 CH₃OH, TsOH →
b. pentan-3-one + HOCH₂CH₂OH, TsOH →

14.15A The Mechanism

The mechanism for acetal formation can be divided into two parts: **the addition of one equivalent of alcohol** to form a **hemiacetal**, followed by the **conversion of the hemiacetal** to the **acetal**. A hemiacetal has a carbon atom bonded to one OH group and one OR group.

$$R_2C=O \xrightarrow[\text{Part [1]}]{R'OH, H^+} R_2C(OH)(OR') \xrightarrow[\text{Part [2]}]{R'OH, H^+} R_2C(OR')_2 + H_2O$$

hemiacetal → acetal

Like *gem*-diols, hemiacetals are often higher in energy than their carbonyl starting materials, making the direction of equilibrium unfavorable for hemiacetal formation. The elimination of H₂O, which can be removed from the reaction mixture to drive the equilibrium to favor product, occurs during the conversion of the hemiacetal to the acetal. This explains why two equivalents of ROH react with a carbonyl compound, forming the acetal as product.

Mechanism 14.11 is written in two parts with a general acid HA.

Mechanism 14.11 Acetal Formation

Part [1] Formation of a hemiacetal

[Mechanism diagram showing: carbonyl + H—A → resonance-stabilized cation (+ :A⁻) → nucleophilic attack by R'OH → protonated hemiacetal → hemiacetal + H—A]

1. Protonation of the carbonyl oxygen forms a **resonance-stabilized cation**.
2–3. Nucleophilic attack by R'OH and deprotonation form the **hemiacetal**. The overall result is addition of H and OR' to the carbonyl group.

Part [2] Formation of an acetal

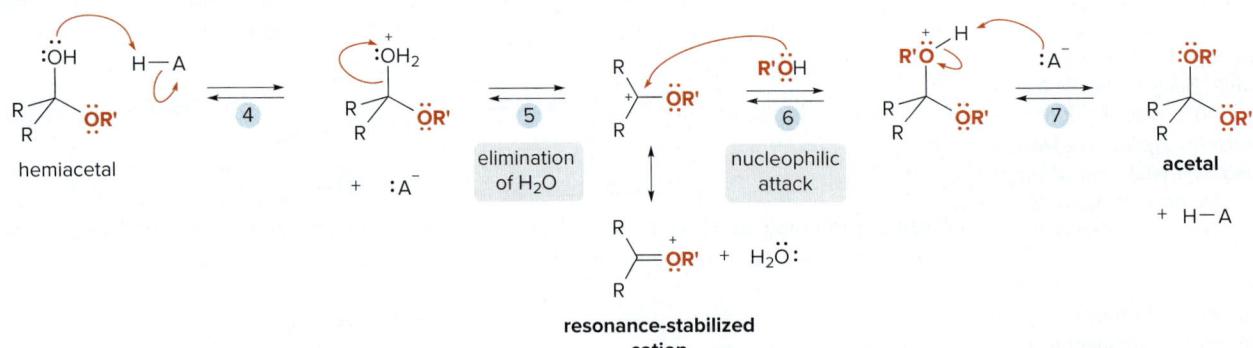

4. Protonation of the OH group of the hemiacetal forms a **good leaving group**.
5. Loss of H₂O forms a **resonance-stabilized cation**.
6–7. Nucleophilic attack by R'OH followed by loss of a proton forms the **acetal**. The overall result of Part [2] is the addition of a second OR' group to the carbonyl.

Although this mechanism is lengthy—there are seven steps—there are only three different kinds of reactions: **addition of a nucleophile, elimination of a leaving group,** and **proton transfer.** Steps [2] and [6] involve nucleophilic attack, and Step [5] eliminates H₂O. The other four steps in the mechanism shuffle protons from one oxygen atom to another, to make a better leaving group or a more electrophilic carbonyl group.

14.15 Addition of Alcohols—Acetal Formation 677

Problem 14.28 Draw a stepwise mechanism for the following reaction.

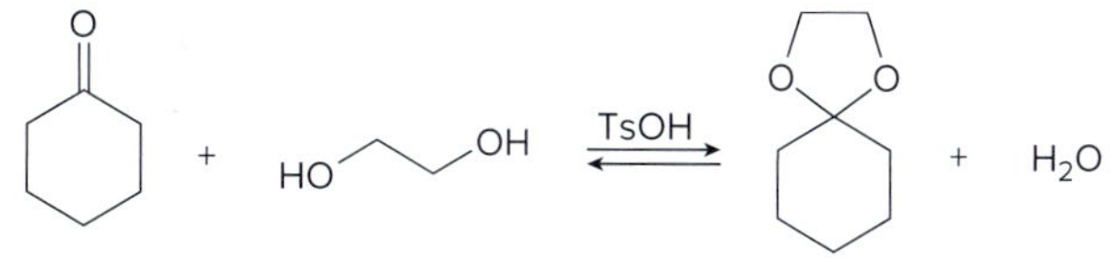

cyclohexanone + HOCH₂CH₂OH ⇌ (TsOH) → cyclohexanone ethylene ketal + H₂O

14.15B Hydrolysis of Acetals

Conversion of an aldehyde or ketone to an acetal is a **reversible reaction, so an acetal can be hydrolyzed to an aldehyde or ketone by treatment with aqueous acid.** Because this reaction is also an equilibrium process, it is driven to the right by using a large excess of water for hydrolysis.

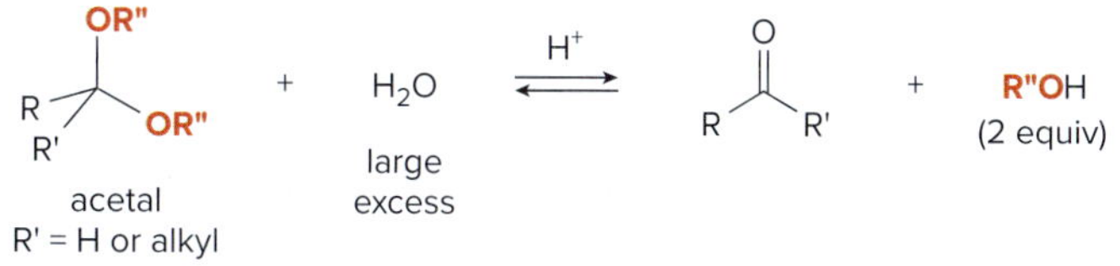

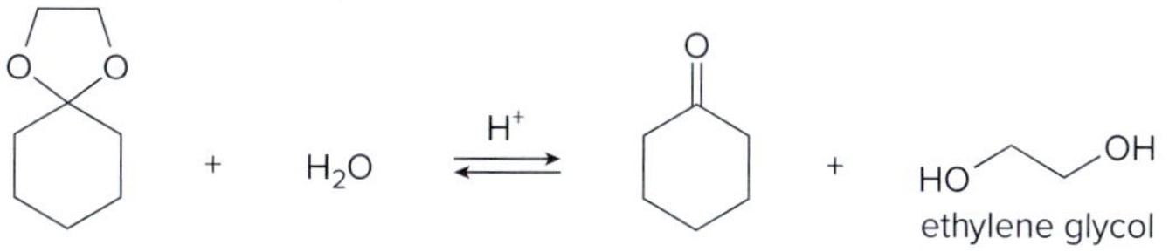

acetal (R' = H or alkyl) + H₂O (large excess) ⇌ (H⁺) carbonyl + R"OH (2 equiv)

cyclohexanone ethylene ketal + H₂O ⇌ (H⁺) cyclohexanone + HOCH₂CH₂OH (ethylene glycol)

The mechanism for this reaction is the reverse of acetal synthesis, as shown in Mechanism 14.12. Acetal hydrolysis requires a strong acid to make a good leaving group (ROH). Acetal hydrolysis does not occur in base.

Mechanism 14.12 Acetal Hydrolysis

Part [1] Conversion of an acetal to a hemiacetal

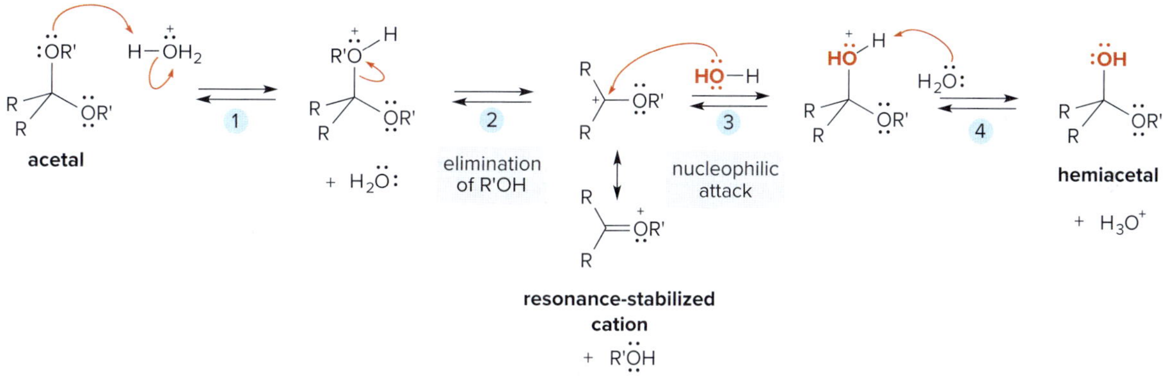

acetal → [1, protonation] → [2, elimination of R'OH] → resonance-stabilized cation + R'OH → [3, nucleophilic attack by HO—H] → [4] → hemiacetal + H₃O⁺

1. Protonation of one OR' group of the acetal forms a **good leaving group.**
2. Loss of R'OH forms a **resonance-stabilized cation.**
3 – 4. Nucleophilic attack by H₂O followed by loss of a proton forms the **hemiacetal.** The overall result of Part [1] is the **substitution** of one OR' group by OH.

Part [2] Conversion of a hemiacetal to a carbonyl group

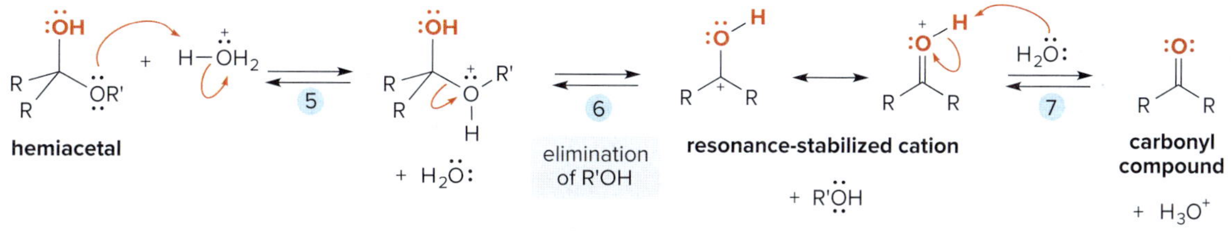

hemiacetal → [5] → [6, elimination of R'OH] → resonance-stabilized cation + R'OH → [7] → carbonyl compound + H₃O⁺

5. Protonation of the OR' group of the hemiacetal forms a **good leaving group.**
6 – 7. Loss of R'OH and deprotonation form a **carbonyl compound.**

Chapter 14 Aldehydes and Ketones—Nucleophilic Addition

Sample Problem 14.5 Drawing the Products of Acetal Hydrolysis

Identify the acetal carbons in **A**, and draw the products formed by hydrolysis of **A** with aqueous acid.

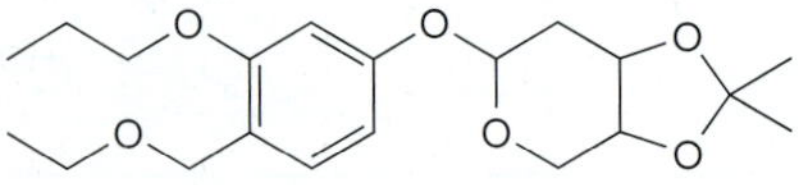

A

Solution

Determine the identity of the functional group that contains each O atom.

- An acetal contains *two* oxygen atoms bonded to the *same* carbon.
- An ether contains *one* oxygen atom bonded to *two* carbons.

A contains two ethers (in gray) and two acetals (with O atoms in red and acetal carbons labeled in blue).

- The **acetal C bonded to two O's is converted to a carbonyl C** during hydrolysis.
- The **acetal O's are converted to OH groups**.

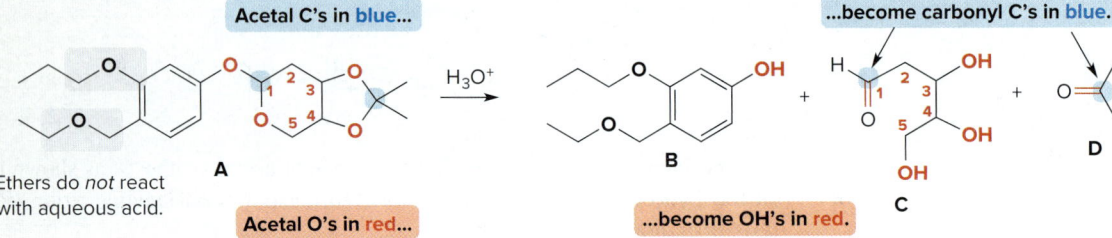

Ethers do *not* react with aqueous acid.

All C—O bonds of the acetals are broken and three products are formed.

Problem 14.29 Draw the products formed when each acetal is treated with aqueous acid.

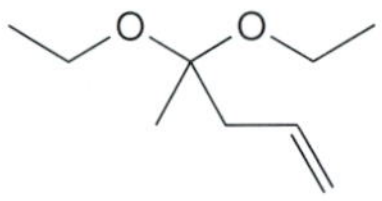

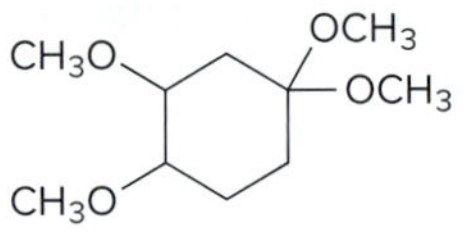

a. 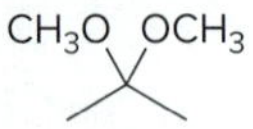 b. 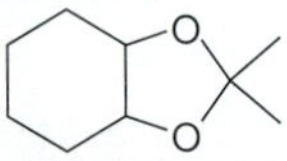 c. d. 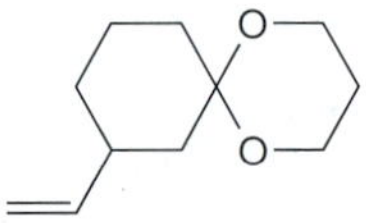 e.

More Practice: Try Problems 14.42, 14.46, 14.49.

Oleandrin (Problem 14.30) and related compounds are responsible for the toxicity of the sap of oleander, a common ornamental shrub that grows in tropical and subtropical regions. *Alessandro0770/Getty Images*

Problem 14.30 Identify the acetal in oleandrin, and draw the products formed by acid-catalyzed hydrolysis of the acetal.

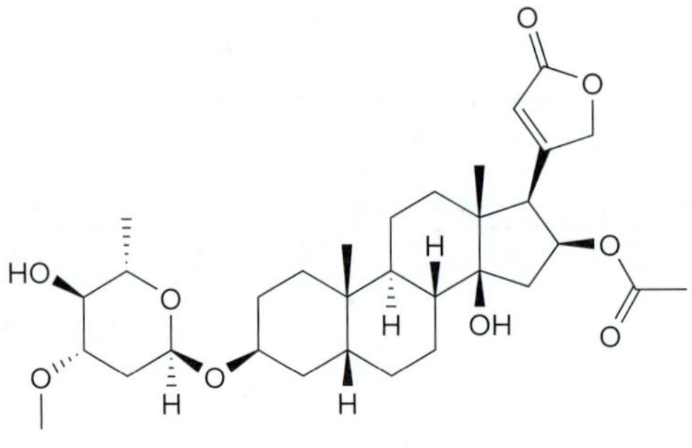

oleandrin

14.16 Acetals as Protecting Groups

Just as the *tert*-butyldimethylsilyl ethers are used as protecting groups for alcohols (Section 13.12), **acetals are valuable protecting groups for aldehydes and ketones.**

Suppose a starting material **A** contains both a ketone and an ester, and it is necessary to selectively reduce the ester to an alcohol (6-hydroxyhexan-2-one), leaving the ketone untouched. Such a selective reduction is *not* possible in one step. Because ketones are more readily reduced, methyl 5-hydroxyhexanoate is formed instead.

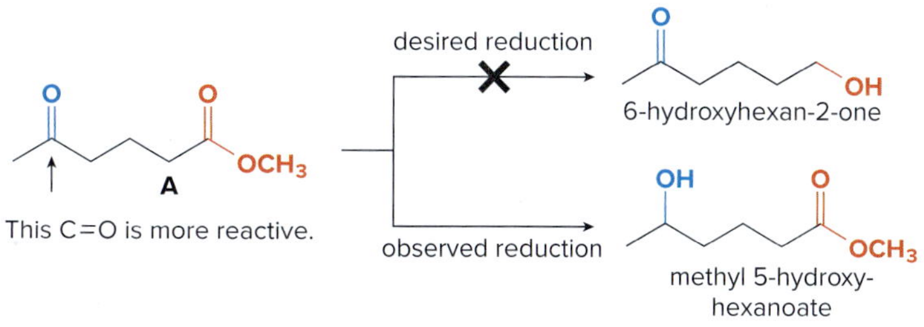

To solve this problem, we can use a protecting group to block the more reactive ketone carbonyl group. The overall process requires three steps.

[1] Protect the interfering functional group—the ketone carbonyl.
[2] Carry out the desired reaction—reduction.
[3] Remove the protecting group.

The following three-step sequence using a cyclic acetal leads to the desired product.

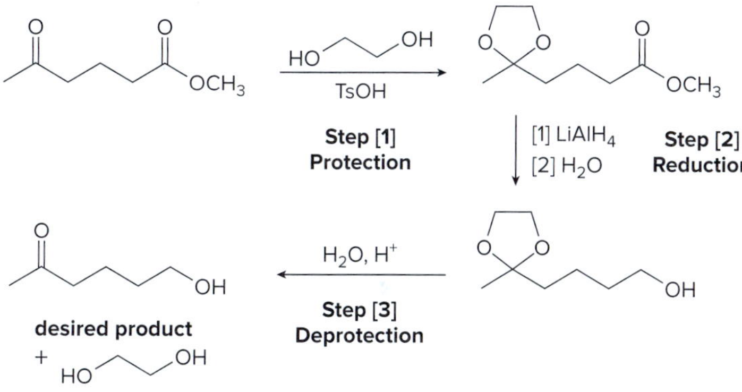

- **Step [1]** The ketone carbonyl is protected as a cyclic acetal by reaction of the starting material with $HOCH_2CH_2OH$ and TsOH.
- **Step [2]** Reduction of the ester is then carried out with $LiAlH_4$, followed by treatment with H_2O.
- **Step [3]** The acetal is then converted back to a ketone carbonyl group with aqueous acid.

Acetals are widely used protecting groups for aldehydes and ketones because they are easy to add and easy to remove, and they are stable to a wide variety of reaction conditions. Acetals do *not* react with base, oxidizing agents, reducing agents, or nucleophiles. Good protecting groups must survive a variety of reaction conditions that take place at other sites in a molecule, but they must also be selectively removed under mild conditions when needed.

Problem 14.31 How would you use a protecting group to carry out the following transformation?

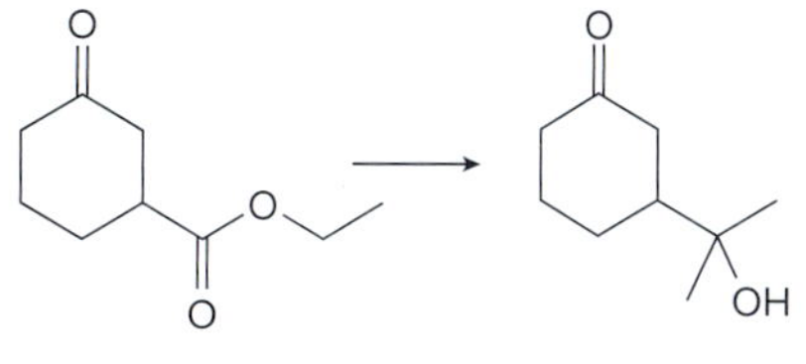

14.17 Cyclic Hemiacetals

Cyclic hemiacetals are also called **lactols**.

Although acyclic hemiacetals are generally unstable and therefore not present in appreciable amounts at equilibrium, **cyclic hemiacetals containing five- and six-membered rings are stable compounds** that are readily isolated.

hemiacetal

One C is bonded to:
- an **OH** group
- an **OR** group

cyclic hemiacetals

Each C is bonded to:
- an **OH** group
- an **OR** group that is part of a ring

14.17A Forming Cyclic Hemiacetals

All hemiacetals are formed by nucleophilic addition of a hydroxy group to a carbonyl group. In the same way, cyclic hemiacetals are formed by **intramolecular cyclization of hydroxy aldehydes.**

5-hydroxypentanal   6%   94%

4-hydroxybutanal   11%   89%

[Equilibrium proportions of each compound are given.]

Such intramolecular reactions to form five- and six-membered rings are faster than the corresponding intermolecular reactions. The two reacting functional groups, in this case OH and C=O, are held in close proximity, increasing the probability of reaction.

Problem 14.32 What lactol (cyclic hemiacetal) is formed from intramolecular cyclization of each hydroxy aldehyde?

Hemiacetal formation is catalyzed by both acid and base. The acid-catalyzed mechanism is identical to Part [1] of Mechanism 14.11, except that the reaction occurs in an **intramolecular** fashion, as shown for the acid-catalyzed cyclization of 5-hydroxypentanal to form a six-membered cyclic hemiacetal in Mechanism 14.13.

Mechanism 14.13 Acid-Catalyzed Cyclic Hemiacetal Formation

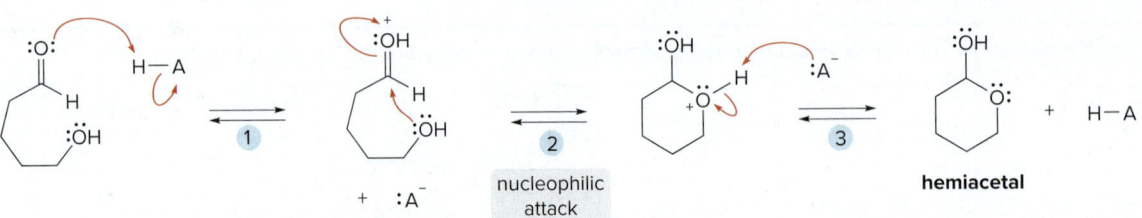

1 – 2 Protonation of the carbonyl oxygen followed by **intramolecular nucleophilic attack** forms the six-membered ring.

3 Deprotonation forms the neutral **cyclic hemiacetal**.

14.17 Cyclic Hemiacetals

Intramolecular cyclization of a hydroxy aldehyde forms a **hemiacetal with a new stereogenic center, so that an equal amount of two enantiomers** results.

14.17B The Conversion of Hemiacetals to Acetals

Cyclic hemiacetals can be converted to acetals by treatment with an alcohol and acid. This reaction converts the OH group that is part of the hemiacetal to an OR group.

Mechanism 14.14, which is similar to Part [2] of Mechanism 14.11, illustrates the conversion of a cyclic hemiacetal to an acetal.

Mechanism 14.14 A Cyclic Acetal from a Cyclic Hemiacetal

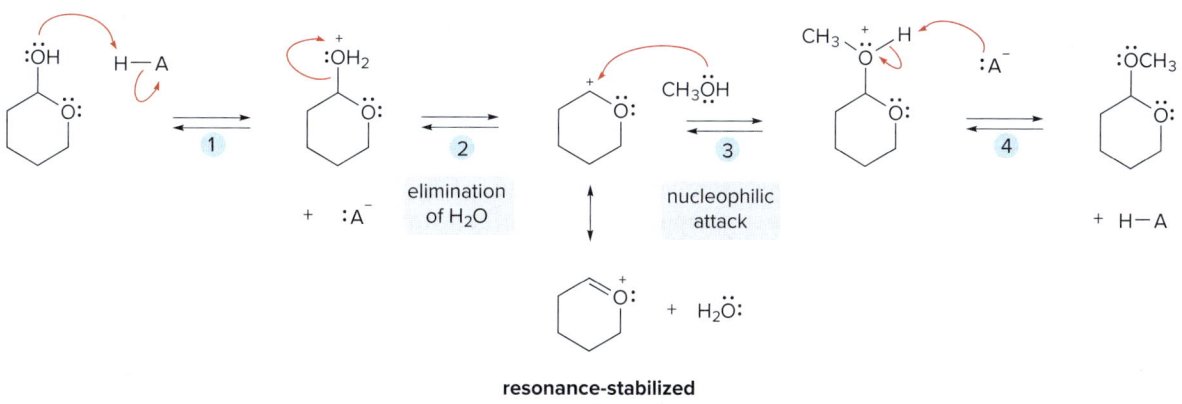

1. Protonation of the OH group of the hemiacetal forms a **good leaving group**.
2. Loss of H_2O forms a **resonance-stabilized cation**.
3–4. Nucleophilic attack by CH_3OH followed by loss of a proton forms the **acetal**.

The overall result of this reaction is the **replacement of the hemiacetal OH group by an OCH_3 group.** This substitution reaction readily occurs because the carbocation formed in Step [2] is stabilized by resonance. This fact makes the OH group of a hemiacetal different from the hydroxy group in other alcohols.

682   Chapter 14   Aldehydes and Ketones—Nucleophilic Addition

Thus, when a compound that contains both an alcohol OH group and a hemiacetal OH group is treated with an alcohol and acid, **only the hemiacetal OH group reacts** to form an acetal. The alcohol OH group does *not* react.

The conversion of cyclic hemiacetals to acetals is an important reaction in carbohydrate chemistry, as discussed in Chapter 24.

Only the hemiacetal OH reacts.

Problem 14.33   Two naturally occurring compounds that contain stable cyclic hemiacetals and acetals are monensin and digoxin. Monensin, a polyether antibiotic produced by *Streptomyces cinnamonensis*, is used as an additive in cattle feed. Digoxin is a widely prescribed cardiac drug used to increase the force of heart contractions. Label each acetal, hemiacetal, and ether in both compounds.

monensin

digoxin

Problem 14.34   Draw the products of each reaction.

Digoxin (Problem 14.33) is obtained by extraction of the leaves of the woolly foxglove plant, which is grown in the Netherlands and shipped to the United States for processing. *Richo Cech/Horizon Herbs*

14.18 An Introduction to Carbohydrates

Carbohydrates, commonly referred to as sugars and starches, are polyhydroxy aldehydes and ketones, or compounds that can be hydrolyzed to them. Along with proteins, lipids, and nucleic acids, they form one of the four main groups of biomolecules responsible for the structure and function of all living cells (Section 3.9).

Many carbohydrates contain cyclic acetals or hemiacetals. Examples include **glucose**, the most common simple sugar, and **lactose**, the principal carbohydrate in milk. Hemiacetal carbons are labeled in blue, whereas the acetal carbon is labeled in red.

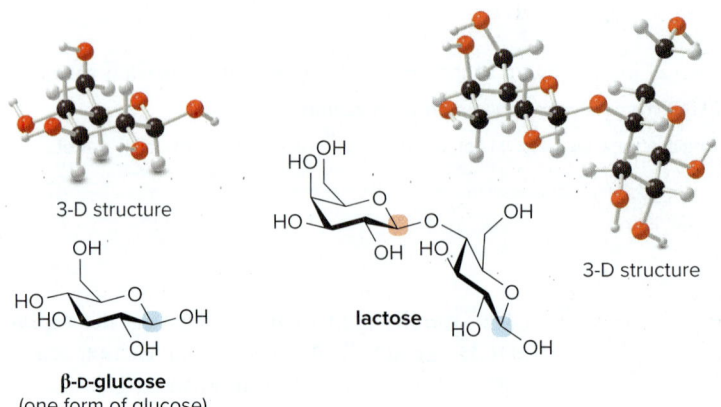

3-D structure

β-D-glucose
(one form of glucose)

lactose

3-D structure

Glucose is the carbohydrate that is transported in the blood to individual cells. The hormone insulin regulates the level of glucose in the blood. Diabetes is a common disease that results from a deficiency of insulin, resulting in increased glucose levels in the blood and other metabolic abnormalities. Insulin injections control glucose levels.

Hemiacetals in sugars are formed in the same way that other hemiacetals are formed—that is, by **cyclization of hydroxy aldehydes.** Thus, the hemiacetal of glucose is formed by cyclization of an acyclic *poly*hydroxy aldehyde **A**, as shown in the accompanying equation. This process illustrates two important features.

- When the OH group on C5 is the nucleophile, **cyclization yields a six-membered ring,** and this ring size is preferred.
- **Cyclization forms a new stereogenic center** (labeled in blue), exactly analogous to the cyclization of the simpler hydroxy aldehyde (5-hydroxypentanal) in Section 14.17A. **The new OH group of the hemiacetal can occupy either the equatorial or axial position.**

For glucose, this results in two cyclic forms, called **β-D-glucose** (having an equatorial OH group) and **α-D-glucose** (having an axial OH group). Because β-D-glucose has the new OH group in the more roomy equatorial position, this cyclic form of glucose is the major product. At equilibrium, only a trace of the acyclic hydroxy aldehyde **A** is present.

Many more details on this process and other aspects of carbohydrate chemistry are presented in Chapter 24.

Problem 14.35

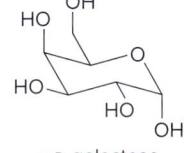

α-D-galactose

a. How many stereogenic centers are present in α-D-galactose?
b. Label the hemiacetal carbon in α-D-galactose.
c. Draw the structure of β-D-galactose.
d. Draw the structure of the polyhydroxy aldehyde that cyclizes to α- and β-D-galactose.
e. From what you learned in Section 14.17B, what product(s) is (are) formed when α-D-galactose is treated with CH₃OH and an acid catalyst?

Chapter 14 REVIEW

KEY CONCEPTS

The relationship between the stability of a carbonyl compound and hydrate formation (14.14)

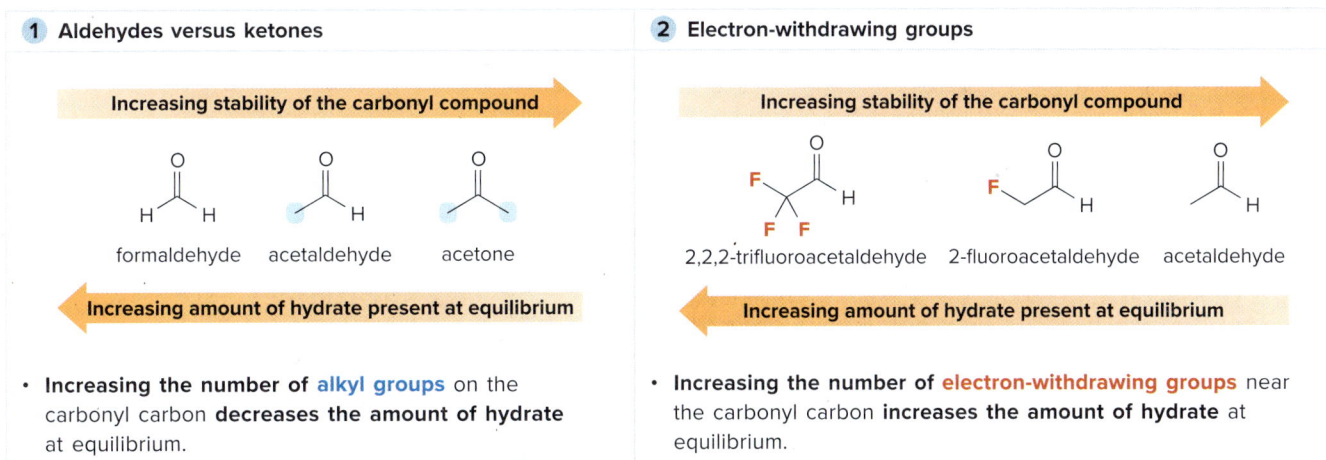

- Increasing the number of **alkyl groups** on the carbonyl carbon **decreases the amount of hydrate** at equilibrium.

- Increasing the number of **electron-withdrawing groups** near the carbonyl carbon **increases the amount of hydrate** at equilibrium.

Try Problem 14.50a, b.

KEY REACTIONS

Nucleophilic Addition Reactions

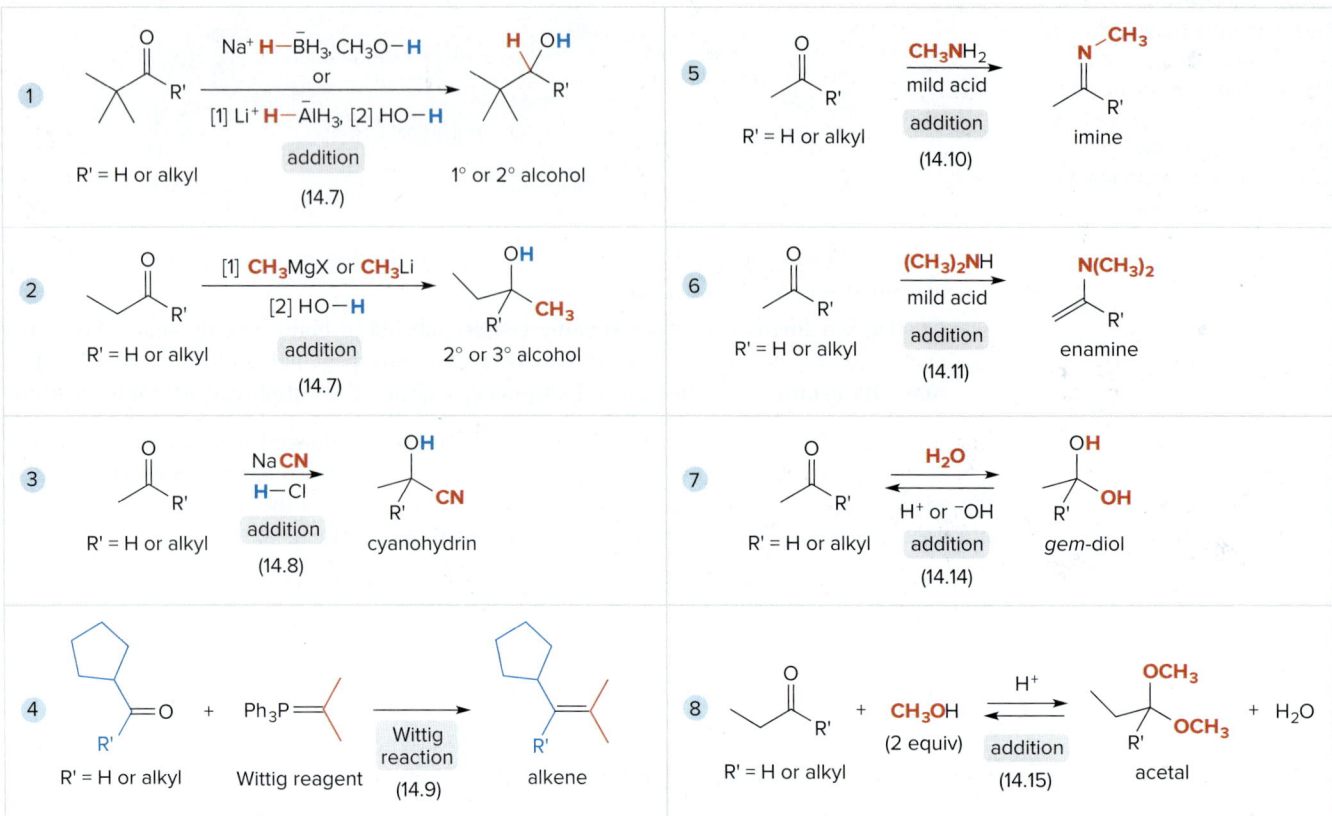

Try Problems 14.36b; 14.40; 14.41a, b, d, f, h; 14.43; 14.44.

Other Reactions

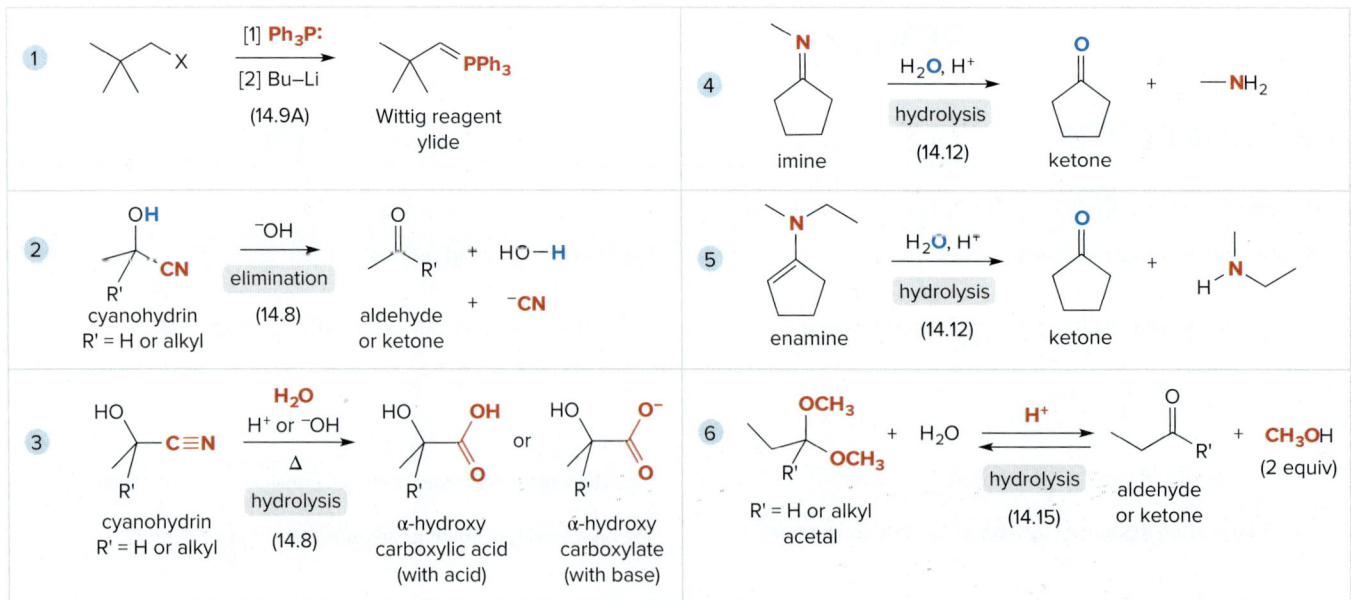

See Sample Problem 14.4. Try Problems 14.40; 14.41c, e, g; 14.42; 14.46b; 14.47; 14.49.

KEY SKILLS

[1] Synthesizing Wittig reagents by a two-step procedure (14.9A)

1 React triphenylphosphine with an alkyl halide.

2 Deprotonate the phosphonium salt with a strong base.

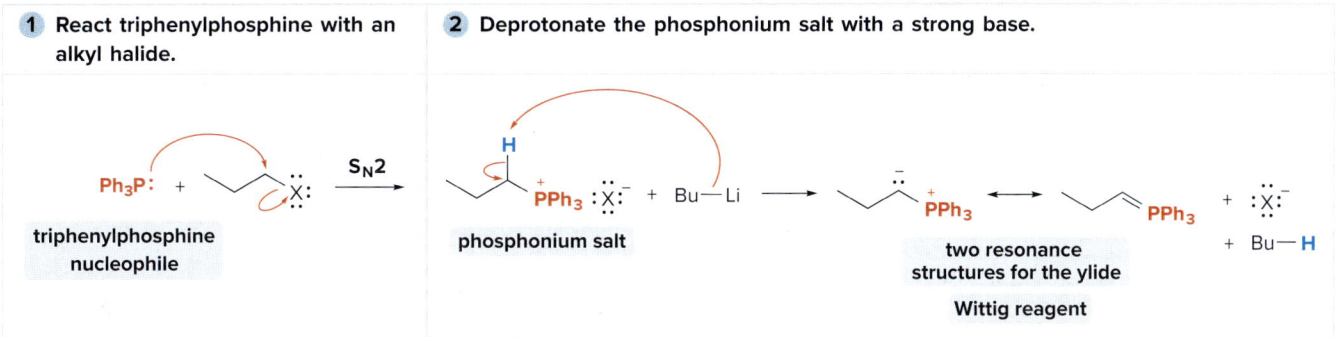

Try Problem 14.40.

[2] Determining the starting materials for a Wittig reaction using retrosynthetic analysis (14.9C)

1 Cleave the carbon–carbon double bond into two components.

2 Compare the Wittig reagents.

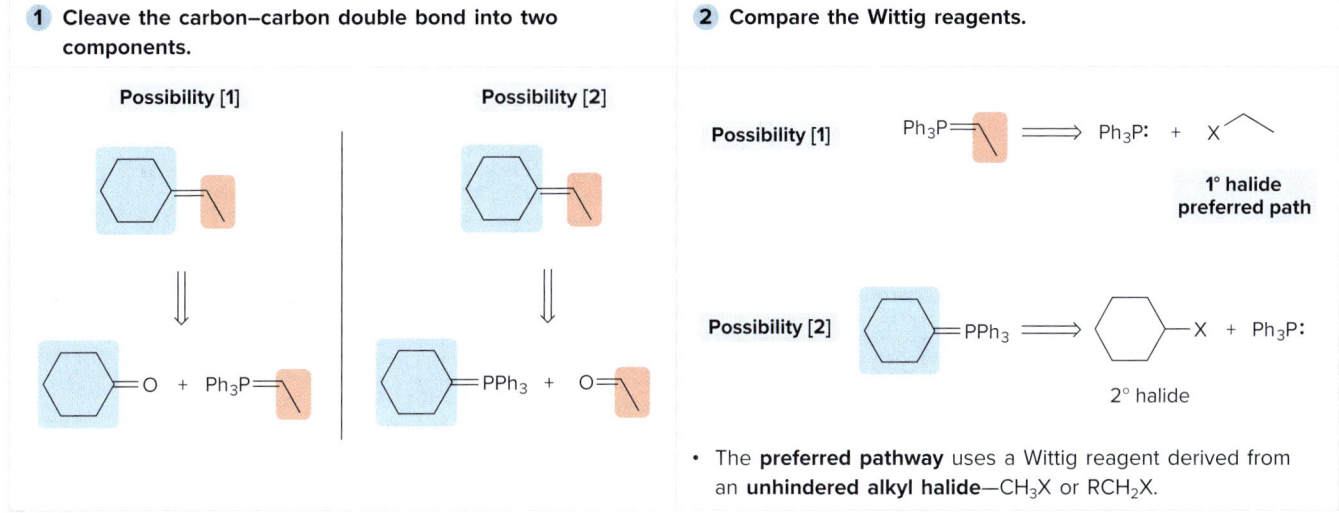

- The **preferred pathway** uses a Wittig reagent derived from an **unhindered alkyl halide**—CH_3X or RCH_2X.

See *How To*, p. 665. Try Problem 14.51.

[3] Using an acetal as a protecting group (14.16)

1 Protect the aldehyde.

2 Carry out the reaction.

3 Remove the protecting group.

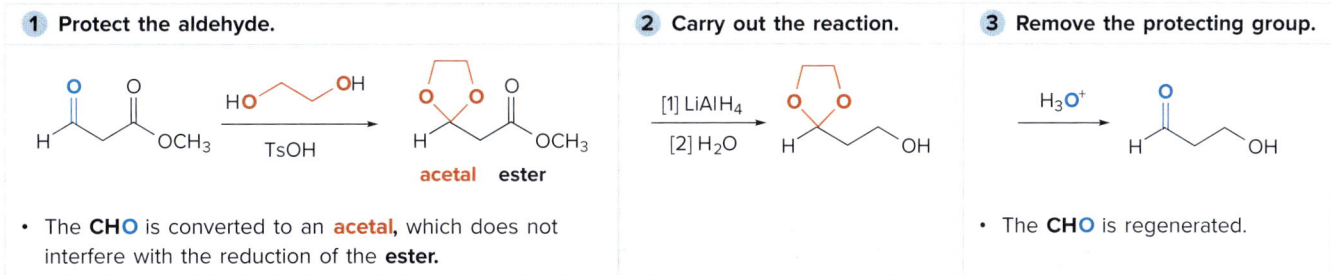

- The **CHO** is converted to an **acetal**, which does not interfere with the reduction of the **ester**.

- The **CHO** is regenerated.

Try Problem 14.57.

686 Chapter 14 Aldehydes and Ketones—Nucleophilic Addition

[4] Drawing the stereoisomers that form in the intramolecular cyclization of a hydroxy aldehyde (14.17A)

① Identify the **OH** group five or six atoms away from the **C=O**.	② Use the mechanism to draw the products and determine the stereochemistry.

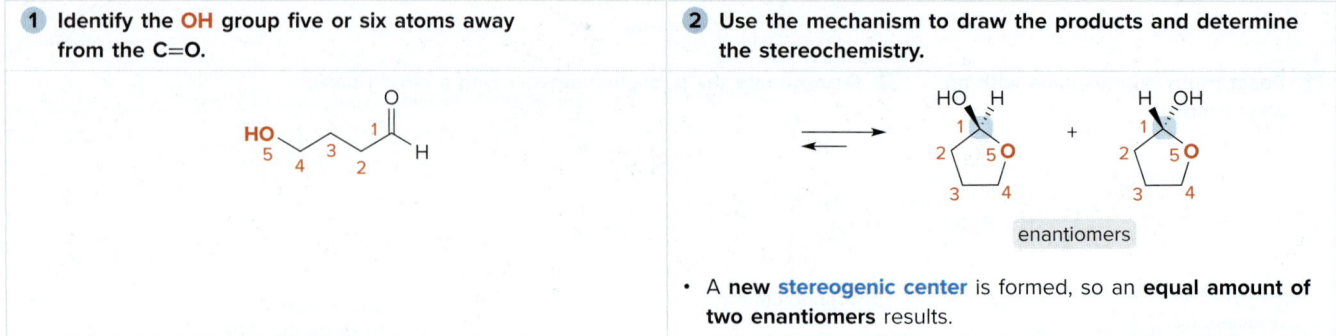

- A **new stereogenic center** is formed, so an **equal amount of two enantiomers** results.

Try Problems 14.45, 14.71.

[5] Determining the reactive OH group that forms an acetal when treated with an alcohol and acid (14.17B)

① Identify the hemiacetal.	② Draw the product(s).

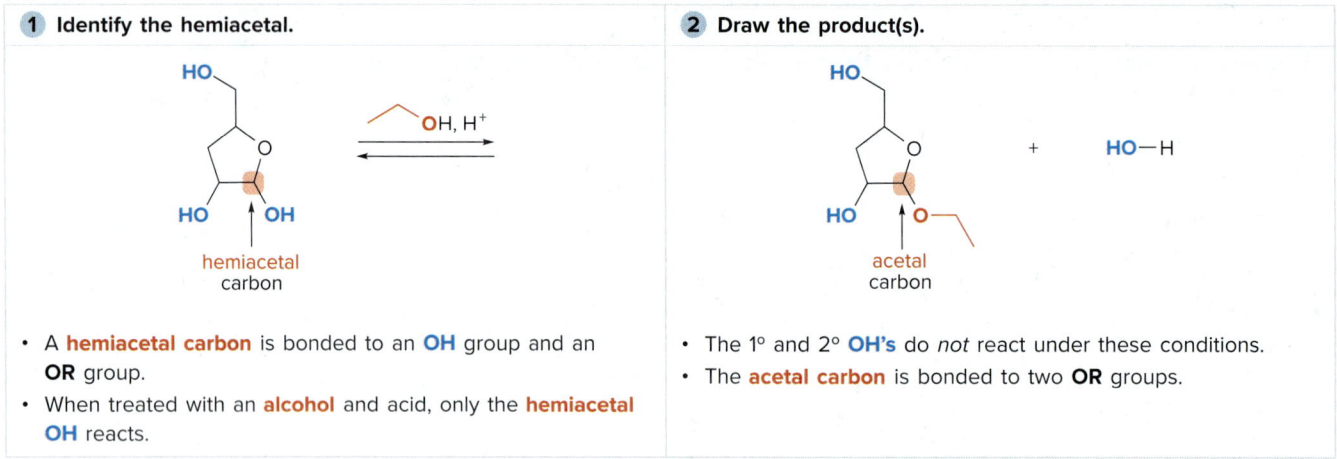

- A **hemiacetal carbon** is bonded to an **OH** group and an **OR** group.
- When treated with an **alcohol** and acid, only the **hemiacetal OH** reacts.

- The 1° and 2° **OH's** do *not* react under these conditions.
- The **acetal carbon** is bonded to two **OR** groups.

Try Problem 14.43d.

PROBLEMS

Problems Using Three-Dimensional Models

14.36 (a) Give the IUPAC name for **A** and **B**. (b) Draw the product formed when **A** or **B** is treated with each reagent: [1] NaBH$_4$, CH$_3$OH; [2] CH$_3$MgBr, then H$_2$O; [3] Ph$_3$P=CHOCH$_3$; [4] CH$_3$CH$_2$CH$_2$NH$_2$, mild acid; [5] HOCH$_2$CH$_2$CH$_2$OH, H$^+$.

14.37 What carbonyl compound and diol are needed to prepare each compound?

a. 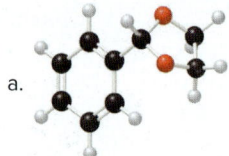 b.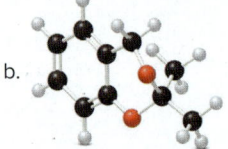

Nomenclature

14.38 Give the IUPAC name for each compound.

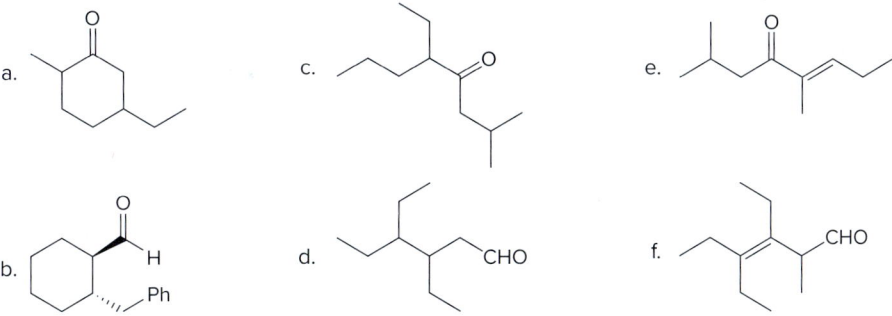

14.39 Give the structure corresponding to each name.
   a. 2-methyl-3-phenylbutanal
   b. 3,3-dimethylcyclohexanecarbaldehyde
   c. 3-benzoylcyclopentanone
   d. 2-formylcyclopentanone
   e. (R)-3-methylheptan-2-one
   f. 2-sec-butylcyclopent-3-enone
   g. 5,6-dimethylcyclohex-1-enecarbaldehyde

Reactions

14.40 Draw the products formed in each reaction sequence.

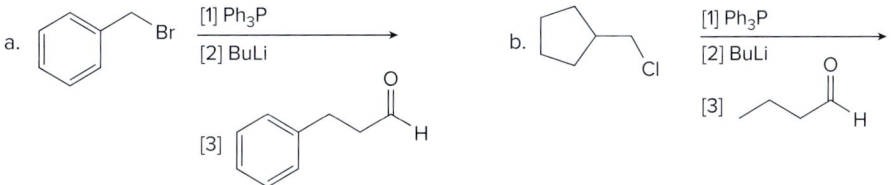

14.41 Draw the products of each reaction.

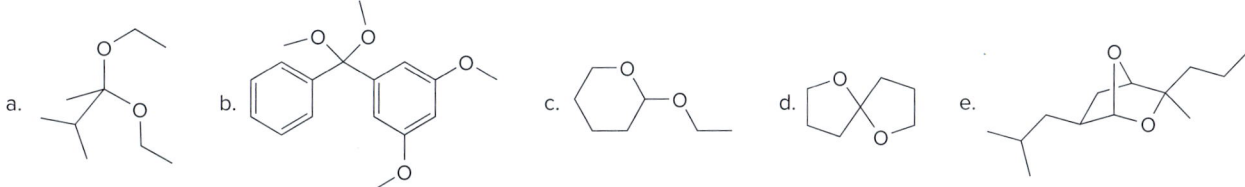

14.42 What products are formed by hydrolysis of each acetal?

14.43 Draw all stereoisomers formed in each reaction.

a. 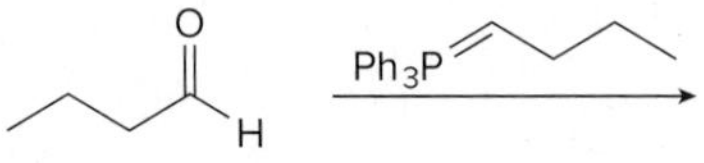 butanal + Ph₃P=CHCH₂CH₃

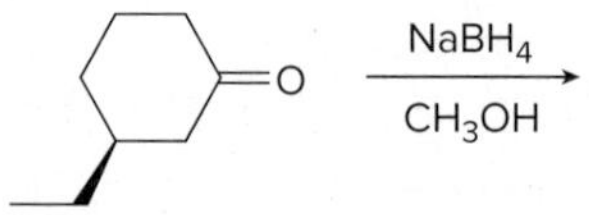

b. 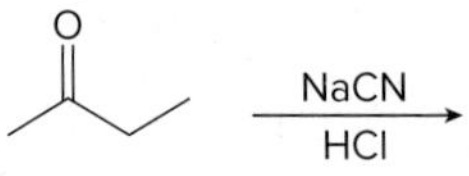 2-butanone + NaCN / HCl

c. 3-ethylcyclohexanone + NaBH₄ / CH₃OH

d. 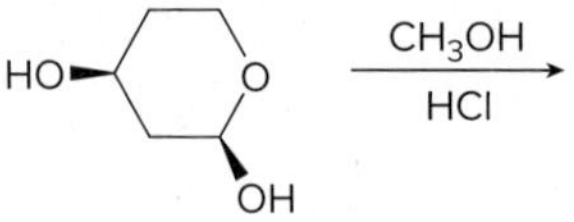 (hydroxy-tetrahydropyran with OH) + CH₃OH / HCl

14.44 What product is formed when each compound undergoes an intramolecular reaction in the presence of acid?

a. 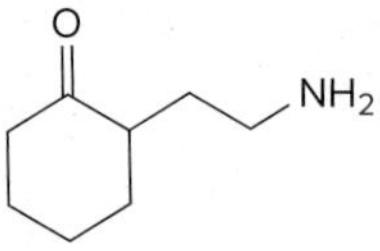 2-(2-aminoethyl)cyclohexanone

b. 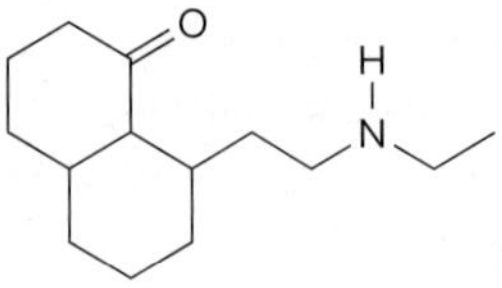 decalone with ethylaminoethyl side chain

c. 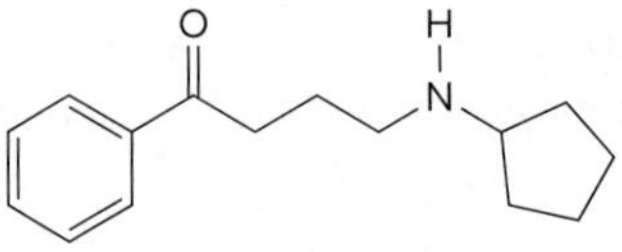 PhC(O)CH₂CH₂CH₂NH-cyclopentyl

14.45 Hydroxy aldehydes **A** and **B** readily cyclize to form hemiacetals. Draw the stereoisomers formed in this reaction from both **A** and **B**. Explain why this process gives an optically inactive product mixture from **A** and an optically active product mixture from **B**.

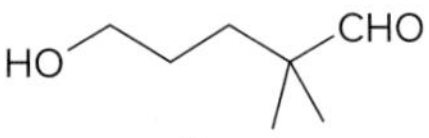

A: HO–CH₂CH₂CH₂–C(CH₃)₂–CHO

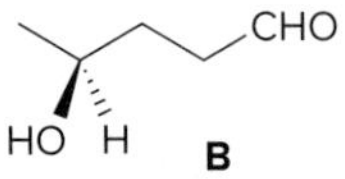

B: CH₃–CH(OH)–CH₂–CH(CH₃)–CHO (with stereochemistry at HO–H carbon)

14.46 Attenol A and pinnatoxin A are natural products isolated from marine sources. (a) Locate the acetals, hemiacetals, imines, and enamines in both compounds. (b) Draw the hydrolysis product formed when attenol A is treated with aqueous acid. Include stereochemistry at all stereogenic centers.

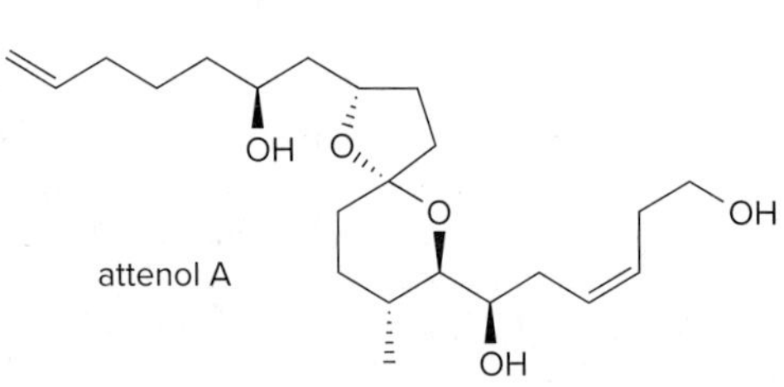

attenol A

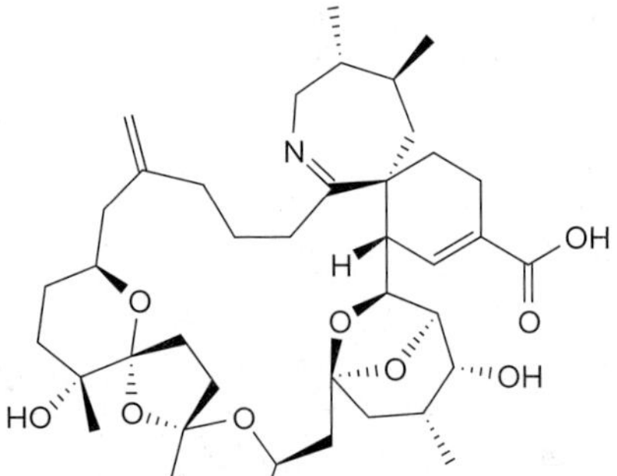

pinnatoxin A

14.47 What products are formed by hydrolysis of each imine or enamine?

a. 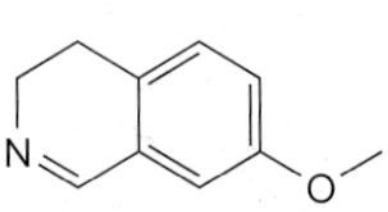 7-methoxy-3,4-dihydroisoquinoline

b. 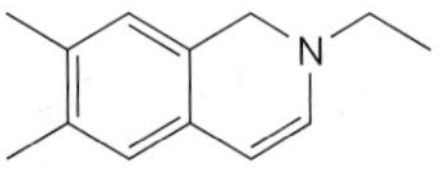 6,7-dimethyl-2-ethyl-1,2-dihydroisoquinoline (enamine)

c. 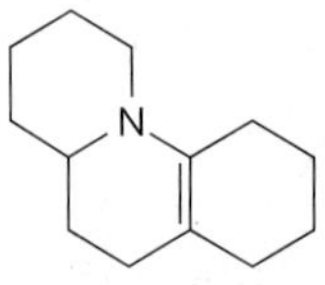 quinolizidine enamine

14.48 Draw the structure of the amino acid that forms each of the following α-keto acids after reaction with PLP.

a. 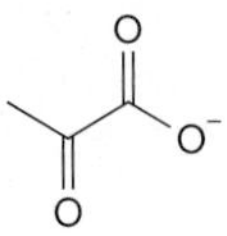 CH₃–C(O)–C(O)O⁻ (pyruvate)

b. CH₃–S–CH₂CH₂–C(O)–C(O)O⁻

c. ⁻O(O)C–CH₂–C(O)–C(O)–NH₂

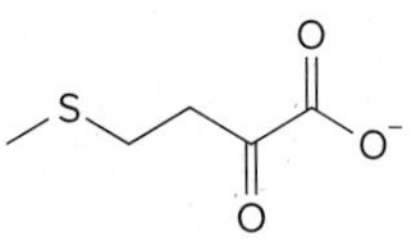

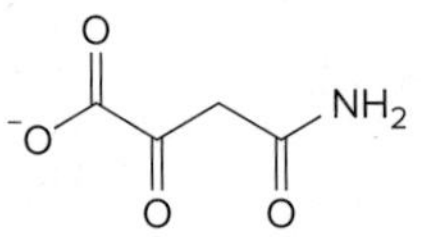

14.49 Etoposide, sold as a phosphate derivative with the trade name of Etopophos, is used for the treatment of lung cancer, testicular cancer, and lymphomas. (a) Locate the acetals in etoposide. (b) What products are formed when all of the acetals are hydrolyzed with aqueous acid?

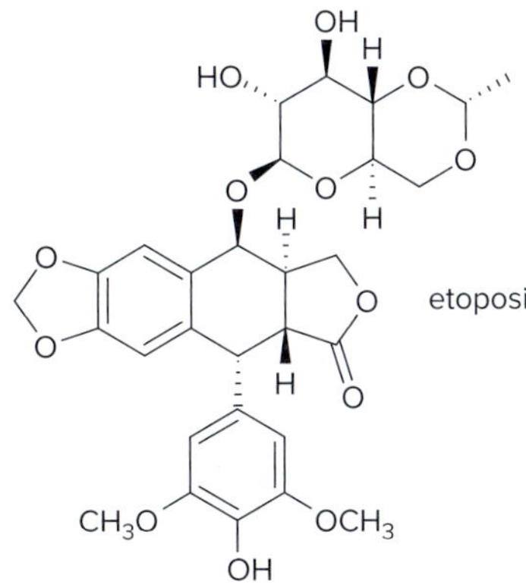

etoposide

Properties of Aldehydes and Ketones

14.50 Consider carbonyl compounds **A–E** drawn below. (a) Rank **A–E** in order of increasing stability. (b) Rank **A–E** in order of increasing amount of hydrate formed when treated with aqueous acid. (c) Which compound is most reactive in nucleophilic addition? (d) From what you learned about the position of the carbonyl absorption in the IR in Sections B.3C and B.4B, which compound has a carbonyl absorption at lowest frequency?

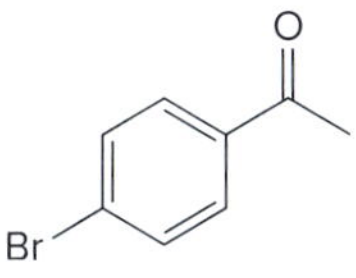

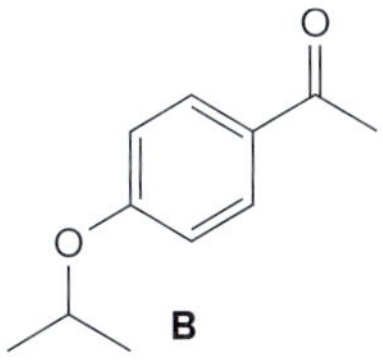

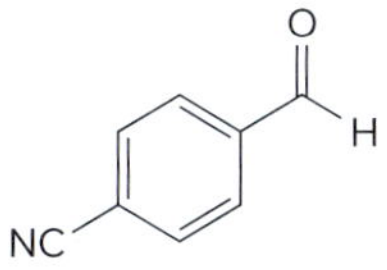

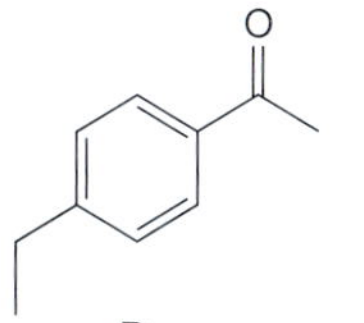

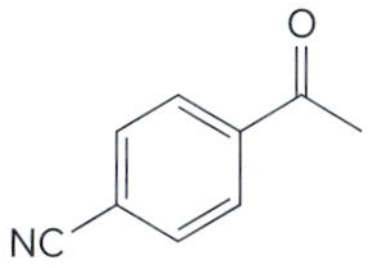

Synthesis

14.51 What Wittig reagent and carbonyl compound are needed to prepare each alkene? When two routes are possible, indicate which route, if any, is preferred.

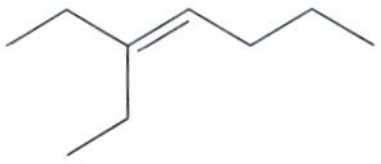

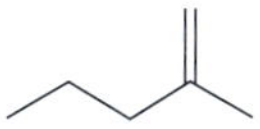

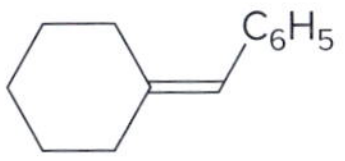

14.52 What carbonyl compound and amine or alcohol are needed to prepare each product?

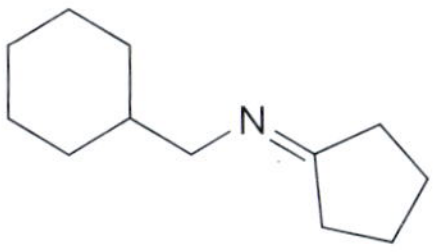

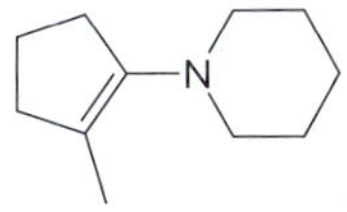

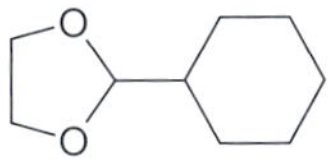

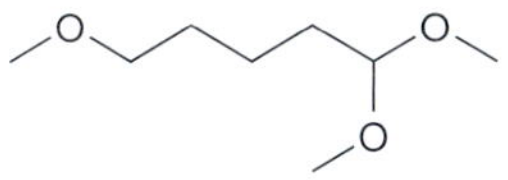

14.53 Devise a synthesis of each alkene using a Wittig reaction to form the double bond. You may use benzyl alcohol ($C_6H_5CH_2OH$) and organic alcohols having four or fewer carbons as starting materials and any required reagents.

a. 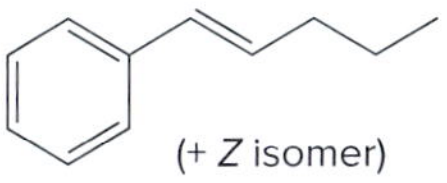 (+ Z isomer)

14.54 Devise a synthesis of each acetal from 1-bromo-2-methylhexane, alcohols (and diols) containing one or two carbons, and any needed inorganic reagents.

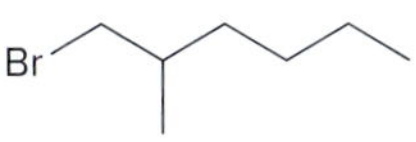

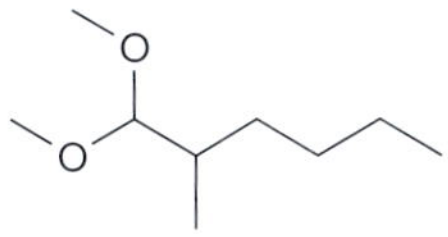

1-bromo-2-methylhexane

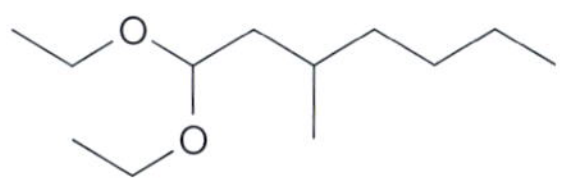

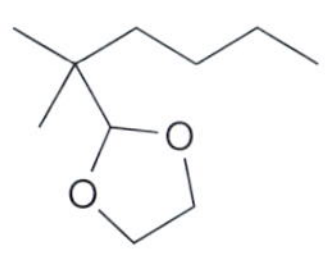

14.55 Devise a synthesis of each compound from cyclohexene and organic alcohols. You may use any other required organic or inorganic reagents.

a. 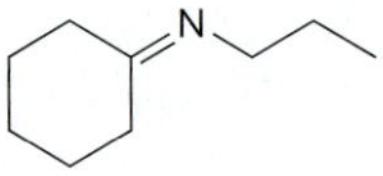 [cyclohexylidene-N-propyl imine]   b. 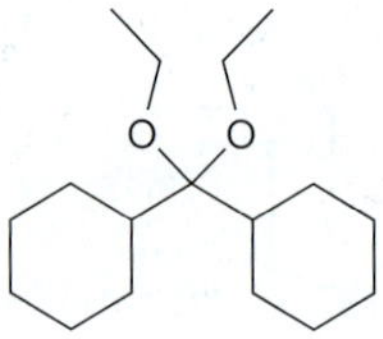 [1,1-diethoxy-dicyclohexylmethane acetal]

14.56 Devise a synthesis of each compound from ethanol (CH₃CH₂OH) as the only source of carbon atoms. You may use any other organic or inorganic reagents you choose.

a. 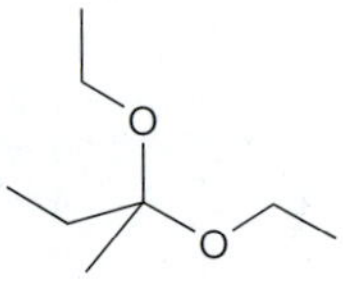 [mixed acetal structure]   b. 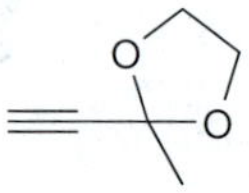 [alkyne with cyclic acetal]

Protecting Groups

14.57 Design a stepwise synthesis to convert cyclopentanone and 4-bromobutanal to hydroxy aldehyde **A**.

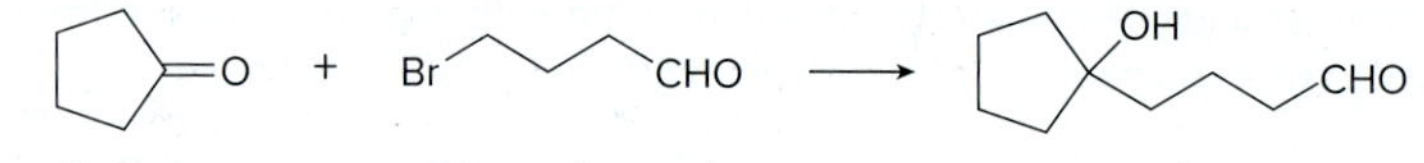

cyclopentanone + Br~~~CHO (4-bromobutanal) → **A** (1-hydroxy-1-(4-oxobutyl)cyclopentane)

Mechanism

14.58 Draw a stepwise mechanism for the following reaction.

[octahydroquinoline] $\xrightarrow{H_3O^+}$ [2-(3-ammoniopropyl)cyclohexanone]

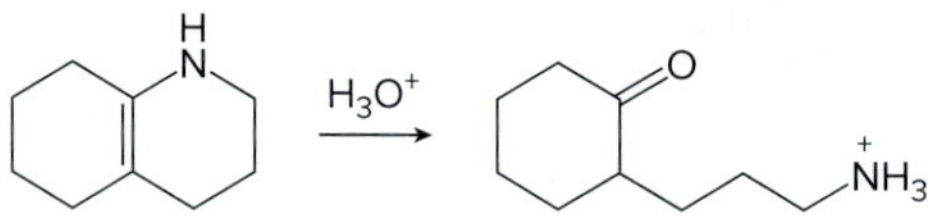

14.59 One acetal can be converted to a different acetal by reaction with a diol in the presence of acid, a process called transacetalization. Draw a stepwise mechanism for the following transacetalization.

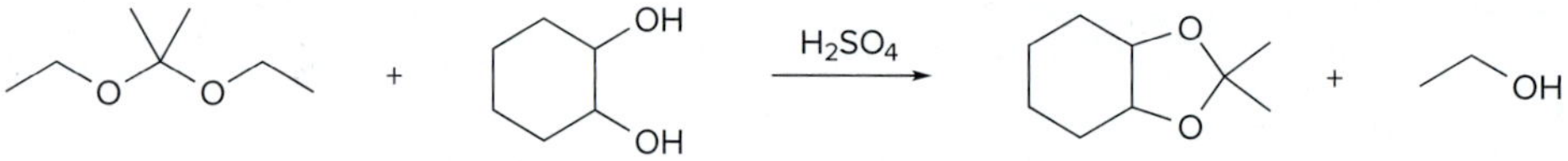

[diethyl acetal] + [trans-1,2-cyclohexanediol] $\xrightarrow{H_2SO_4}$ [cyclic acetal] + EtOH

14.60 Draw a stepwise mechanism for the following reaction, a key step in the synthesis of the anti-inflammatory drug celecoxib (trade name Celebrex).

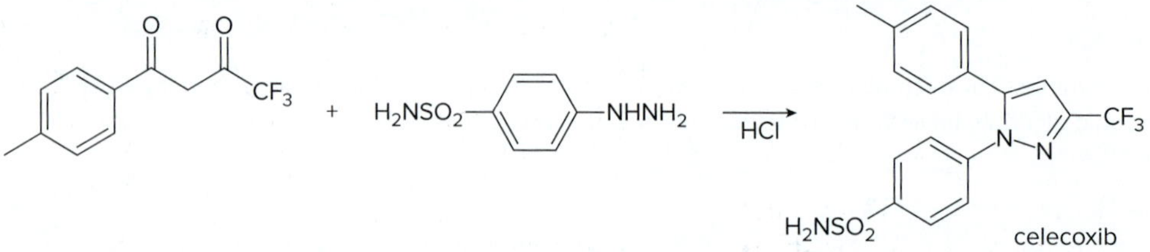

[4-methylphenyl trifluoromethyl 1,3-diketone] + H₂NSO₂–C₆H₄–NHNH₂ $\xrightarrow{HCl}$ celecoxib

14.61 Treatment of (HOCH₂CH₂CH₂CH₂)₂CO with acid forms a product of molecular formula C₉H₁₆O₂ and a molecule of water. Draw the structure of the product and explain how it is formed.

14.62 Draw a stepwise mechanism for the following reaction.

[trans-1,2-cyclohexanediol] + C₆H₅CHO $\xrightarrow{H_2SO_4}$ [2-phenyl-1,3-dioxolane fused to cyclohexane] + H₂O

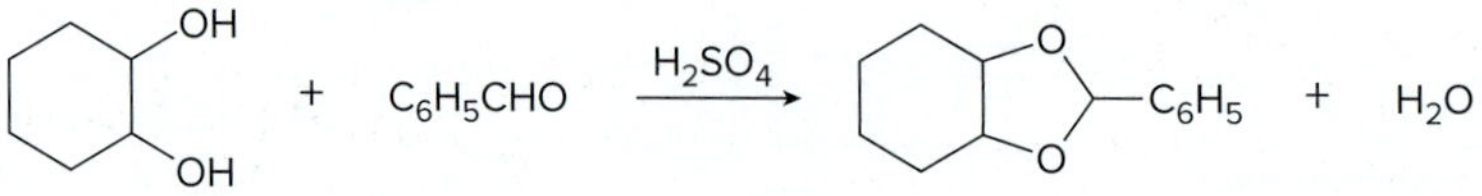

14.63 Draw a stepwise mechanism for the hydrolysis of imine **A,** derived from pyridoxal phosphate (PLP) and the amino acid leucine, to form an α-keto acid and pyridoxamine phosphate (PMP). This reaction is one step in the metabolism of leucine.

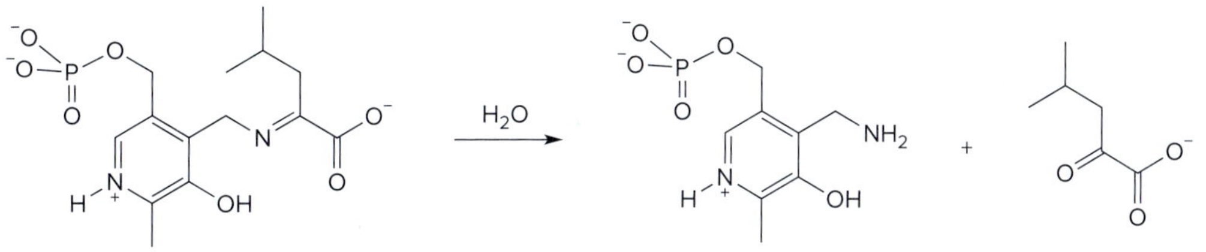

A pyridoxamine phosphate α-keto acid

14.64 Draw a stepwise mechanism for the following reaction, a key step in the synthesis of ticlopidine, a drug that inhibits platelet aggregation. Ticlopidine has been used to reduce the risk of stroke in patients who cannot tolerate aspirin.

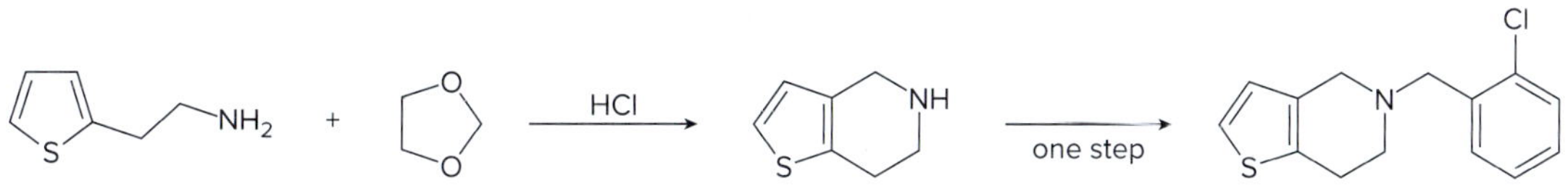

ticlopidine

14.65 Salsolinol is a naturally occurring compound found in bananas, chocolate, and several foods derived from plant sources. Salsolinol is also formed in the body when acetaldehyde, an oxidation product of the ethanol ingested in an alcoholic beverage, reacts with dopamine, a neurotransmitter. Draw a stepwise mechanism for the formation of salsolinol in the following reaction.

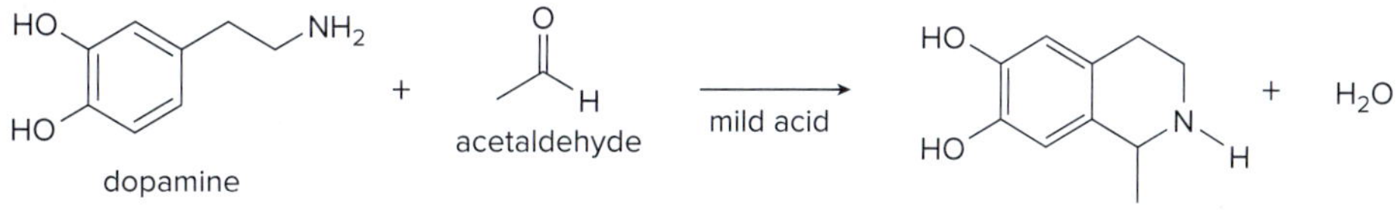

dopamine acetaldehyde salsolinol

14.66 Reaction of 5,5-dimethoxypentan-2-one with methylmagnesium iodide followed by treatment with aqueous acid forms cyclic hemiacetal **Y.** Draw a stepwise mechanism that illustrates how **Y** is formed.

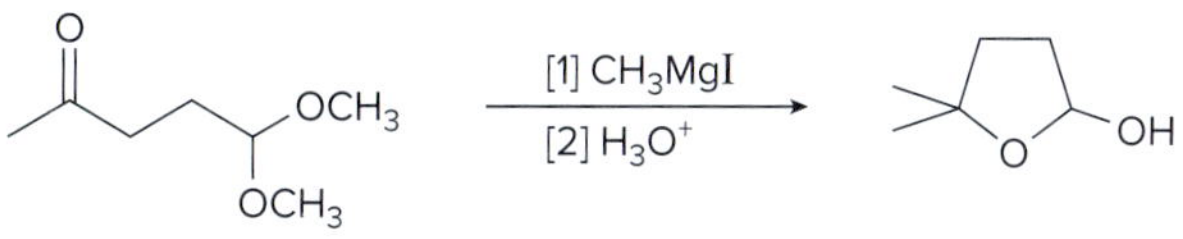

5,5-dimethoxypentan-2-one **Y**

Spectroscopy

14.67 Use the ^{1}H NMR and IR data to determine the structure of each compound.

Compound **A**	Molecular formula:	$C_{10}H_{12}O$
	IR absorption at	1686 cm^{-1}
	^{1}H NMR data:	1.21 (triplet, 3 H), 2.39 (singlet, 3 H), 2.95 (quartet, 2 H), 7.24 (doublet, 2 H), and 7.85 (doublet, 2 H) ppm
Compound **B**	Molecular formula:	$C_{10}H_{12}O$
	IR absorption at	1719 cm^{-1}
	^{1}H NMR data:	1.02 (triplet, 3 H), 2.45 (quartet, 2 H), 3.67 (singlet, 2 H), and 7.06–7.48 (multiplet, 5 H) ppm

14.68 A solution of acetone [(CH₃)₂C=O] in ethanol (CH₃CH₂OH) in the presence of a trace of acid was allowed to stand for several days, and a new compound of molecular formula $C_7H_{16}O_2$ was formed. The IR spectrum showed only one major peak in the functional group region around 3000 cm⁻¹, and the ¹H NMR spectrum is given here. What is the structure of the product?

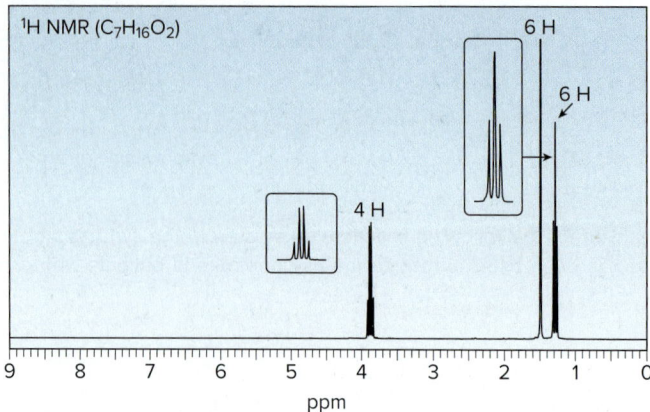

14.69 Identify the structure of compound **A** (molecular formula $C_9H_{10}O$) from the ¹H NMR and IR spectra given.

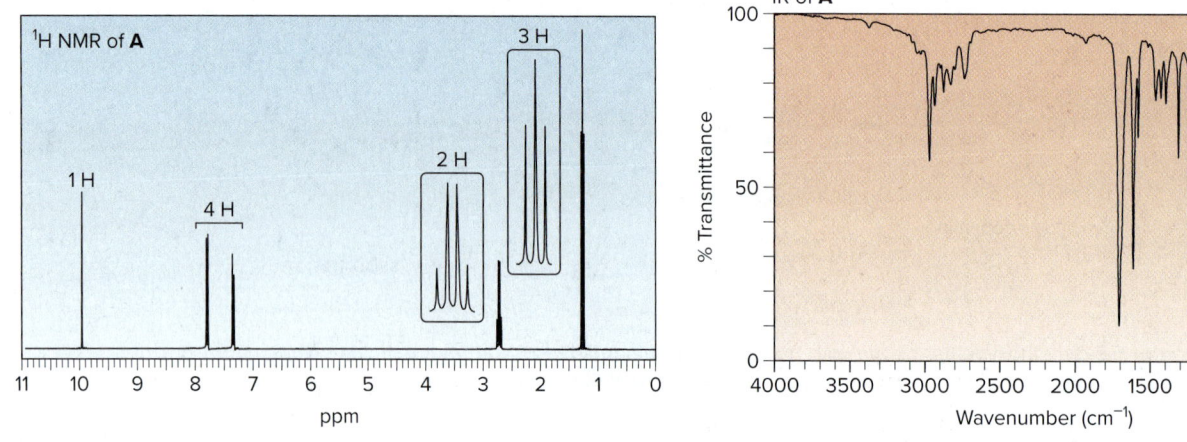

14.70 An unknown compound **C** of molecular formula $C_6H_{12}O_3$ exhibits a strong absorption in its IR spectrum at 1718 cm⁻¹ and the given ¹H NMR spectrum. What is the structure of **C**?

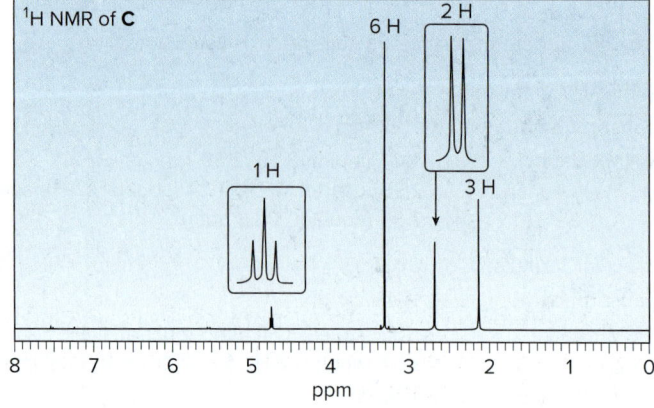

Carbohydrates

14.71 Draw the structure of the acyclic polyhydroxy aldehyde that cyclizes to each hemiacetal.

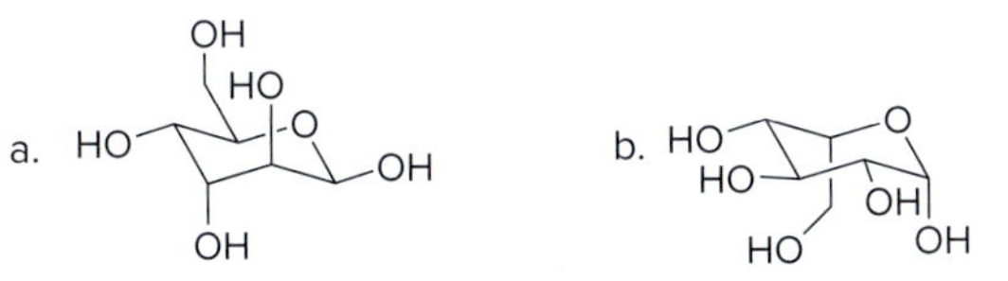

a. [structure]   b. [structure]

14.72 β-D-Glucose, a hemiacetal, can be converted to a mixture of acetals on treatment with CH_3OH in the presence of acid. Draw a stepwise mechanism for this reaction. Explain why two acetals are formed from a single starting material.

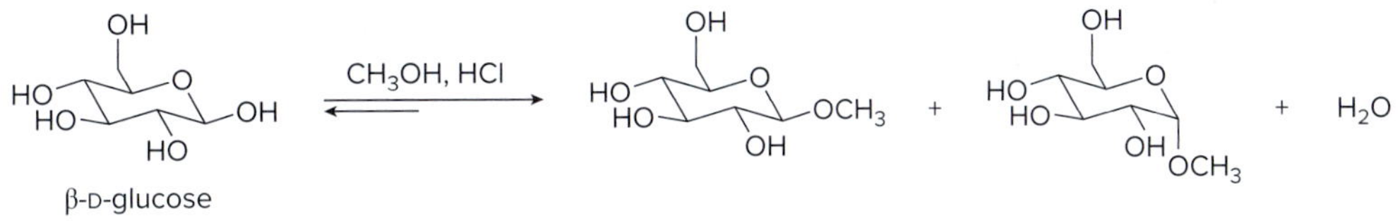

β-D-glucose $\xrightleftharpoons{CH_3OH, HCl}$ [acetal with OCH₃] + [acetal with OCH₃] + H_2O

Challenge Problems

14.73 Brevicomin, the aggregation pheromone of the western pine bark beetle, contains a bicyclic bridged ring system and is prepared by the acid-catalyzed cyclization of 6,7-dihydroxy-nonan-2-one.

a. Suggest a structure for brevicomin.
b. Devise a synthesis of 6,7-dihydroxynonan-2-one from 6-bromohexan-2-one. You may also use three-carbon alcohols and any required organic or inorganic reagents.

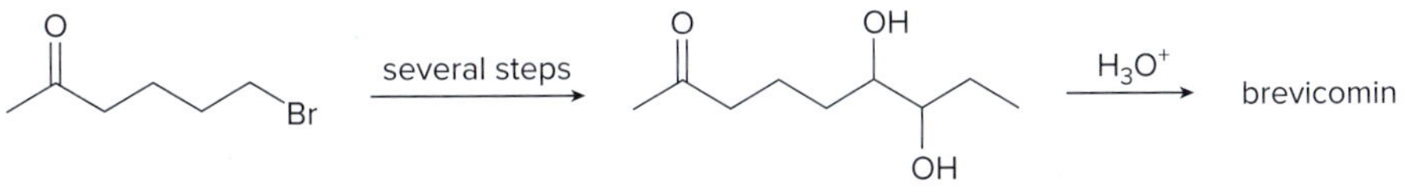

6-bromohexan-2-one $\xrightarrow{\text{several steps}}$ 6,7-dihydroxynonan-2-one $\xrightarrow{H_3O^+}$ brevicomin

14.74 Draw a stepwise mechanism for the following reaction.

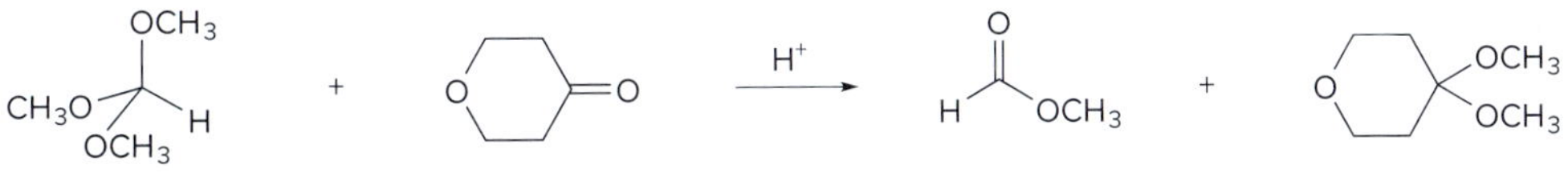

[CH(OCH₃)₃] + [tetrahydropyranone] $\xrightarrow{H^+}$ [HCO-OCH₃] + [tetrahydropyran with two OCH₃]

14.75 Maltose is a carbohydrate present in malt, the liquid obtained from barley and other grains. Although maltose has numerous functional groups, its reactions are explained by the same principles we have already encountered.

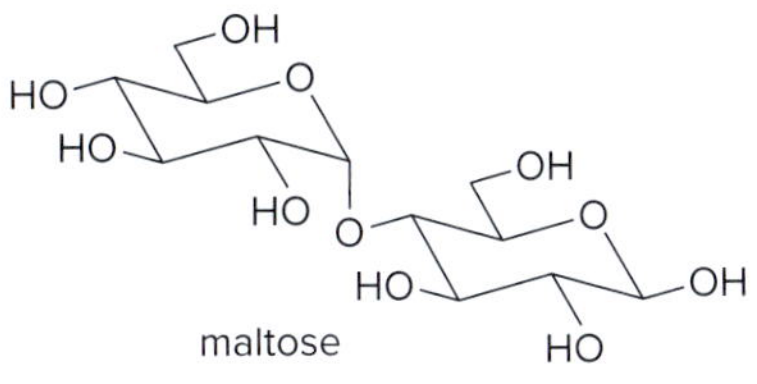

maltose

a. Label the acetal and hemiacetal carbons.
b. What products are formed when maltose is treated with each of these reagents: [1] H_3O^+; [2] CH_3OH and HCl; [3] excess NaH, then excess CH_3I?
c. Draw the products formed when the compound formed in Reaction [3] of part (b) is treated with aqueous acid.

The reactions in parts (b) and (c) are used to determine structural features of carbohydrates like maltose. We will learn much more about maltose and similar carbohydrates in Chapter 24.

694   Chapter 14   Aldehydes and Ketones—Nucleophilic Addition

14.76 Identify **R** and **S** in the following reaction sequence, and draw a mechanism for the conversion of **R** to **S** (molecular formula $C_6H_{10}O_3$). **S** was used in the synthesis of darunavir (trade name Prezista), used to treat HIV.

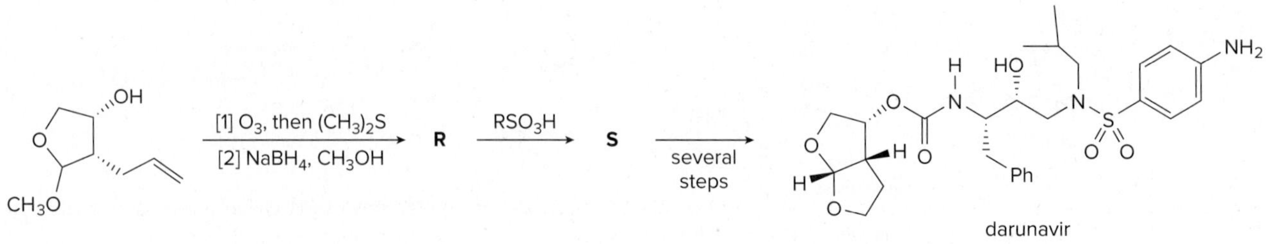

14.77 Draw a stepwise mechanism for the following reaction, a key step in the synthesis of conivaptan (trade name Vaprisol), a drug used in the treatment of low sodium levels.

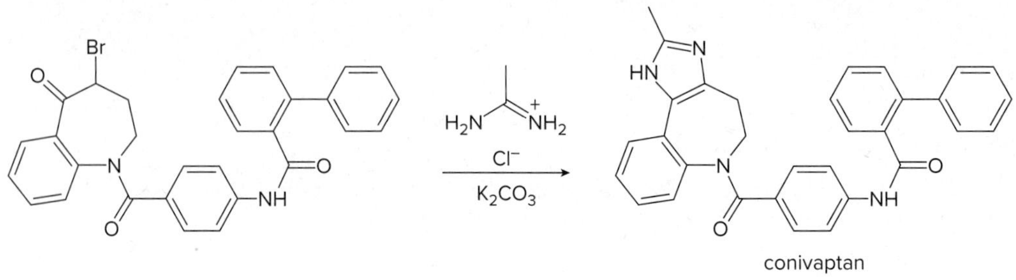

Carboxylic Acids and Nitriles

15

Sarka Babicka/Getty Images

15.1 Structure and bonding
15.2 Nomenclature
15.3 Physical and spectroscopic properties
15.4 Interesting carboxylic acids and nitriles
15.5 Aspirin, arachidonic acid, and prostaglandins
15.6 Preparation of carboxylic acids
15.7 Carboxylic acids—Strong organic Brønsted–Lowry acids
15.8 The Henderson–Hasselbalch equation
15.9 Inductive effects in aliphatic carboxylic acids
15.10 Extraction
15.11 Organic acids that contain phosphorus
15.12 Amino acids
15.13 Nitriles

Lysine is an essential amino acid that is needed for protein synthesis. Because lysine cannot be synthesized by humans and is not stored in the body, it must be ingested on a regular basis. Common food sources of lysine are meat, beans, peas, soy, and peanuts. Although most grains are low in lysine, quinoa is relatively high in lysine content and a good source of essential amino acids for a vegetarian diet. Like other amino acids, lysine contains both a carboxylic acid and an amine base. In Chapter 15, we learn about carboxylic acids and a related family of compounds, nitriles.

Why Study...

Carboxylic Acids and Nitriles?

Chapter 15 concentrates on two classes of compounds, **carboxylic acids (RCO₂H)** and **nitriles (RCN)**. With a polarized C=O and an acidic O—H bond, carboxylic acids undergo a variety of reactions. In this chapter we concentrate on one feature only—the **acidity of carboxylic acids**. Aspirin, a synthetic pain reliever, and naturally occurring fatty acids and prostaglandins are all carboxylic acids.

Nitriles are less common, but this useful functional group can be transformed into many other common functional groups. Moreover, several drugs that contain one or more cyano groups (C≡N) are used in the treatment of breast cancer and depression.

15.1 Structure and Bonding

The word *carboxy* (for a COOH group) is derived from *carb*onyl (C=O) + hydr*oxy* (OH).

Carboxylic acids are organic compounds containing a carboxy group (COOH). Although the structure of a carboxylic acid is often abbreviated as **RCOOH or RCO₂H,** keep in mind that the central carbon atom of the functional group is doubly bonded to one oxygen atom and singly bonded to another.

carboxylic acid   carboxy group

The carbon atom of a carboxy group is surrounded by three groups, making it sp^2 **hybridized** and **trigonal planar,** with bond angles of approximately 120°. The C=O of a carboxylic acid is *shorter* than its C—O. Because oxygen is more electronegative than either carbon or hydrogen, **the C—O and O—H bonds are polar.**

acetic acid

121 pm
136 pm
119°

Nitriles are compounds that contain a cyano group, C≡N, bonded to an alkyl group. Nitriles have no carbonyl group, so they are structurally distinct from carboxylic acids. The carbon atom of the cyano group, however, has the same oxidation state as the carbonyl carbon of a carboxylic acid, so there are certain parallels in their chemistry.

R—C≡N: cyano group
nitrile

R—C≡N:

Each labeled C has three bonds to a more electronegative element.

The structure and bonding in nitriles is very different from that in carboxylic acids, and it resembles the carbon–carbon triple bond of alkynes. Unlike alkynes, however, **nitriles contain an electrophilic carbon atom,** making them susceptible to nucleophilic attack.

CH₃—C≡N:
↑
sp hybridized

180°
δ+ δ−
↑
Nucleophiles attack here.

- The carbon atom of the C≡N group is *sp* hybridized, making it linear with a bond angle of 180°.
- The triple bond consists of one σ and two π bonds.

15.2 Nomenclature

Both IUPAC and common names are used for carboxylic acids and nitriles.

15.2A Naming Carboxylic Acids

In IUPAC nomenclature, carboxylic acids are identified by a suffix added to the parent name of the longest chain, and two different endings are used depending on whether the carboxy group is bonded to a chain or a ring.

To name a carboxylic acid using the IUPAC system:

[1] If the COOH is bonded to a *chain* of carbons, find the longest chain containing the COOH group, and change the *-e* ending of the parent alkane to the suffix *-oic acid.* If the COOH group is bonded to a *ring,* name the ring and add the words *carboxylic acid.*

[2] Number the carbon chain or ring to put the **COOH group at C1,** but omit this number from the name. Apply all the other usual rules of nomenclature.

Sample Problem 15.1 Naming a Carboxylic Acid Using the IUPAC System

Give the IUPAC name of each compound.

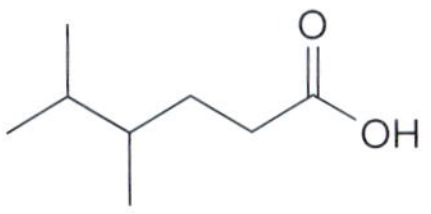

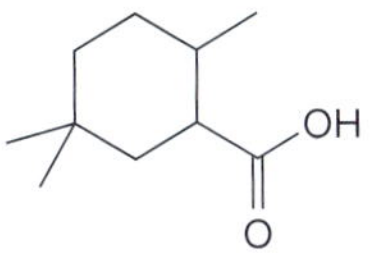

Solution

a. [1] Find and name the longest chain containing COOH:

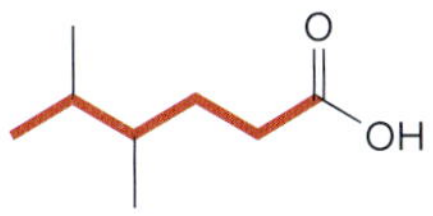

    hexane ⟶ hexan*oic acid*
    (6 C's)

The COOH contributes one C to the longest chain.

[2] Number and name the substituents:

two methyl substituents on C4 and C5

Answer: 4,5-dimethylhexanoic acid

b. [1] Find and name the ring bonded to COOH.

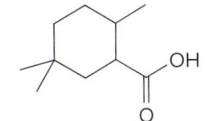

cyclohexane + carboxylic acid
(6 C's)

[2] Number and name the substituents:

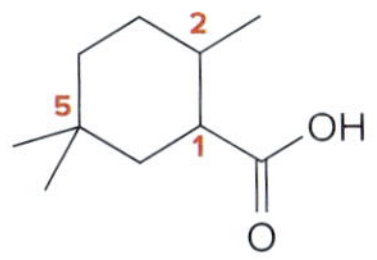

Number to put COOH at C1 and give the second substituent (CH₃) the lower number (C2).

Answer: 2,5,5-trimethylcyclohexanecarboxylic acid

Problem 15.1 Give the IUPAC name for each compound.

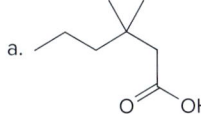 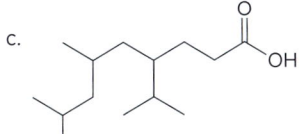

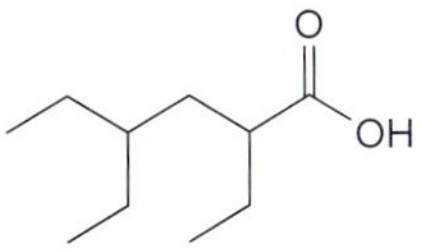

More Practice: Try Problems 15.29a; 15.31a, c, e.

Problem 15.2 Give the structure corresponding to each IUPAC name.

a. 2-bromobutanoic acid
b. 2,3-dimethylpentanoic acid
c. 3,3,4-trimethylheptanoic acid
d. 2-sec-butyl-4,4-diethylnonanoic acid
e. 3,4-diethylcyclohexanecarboxylic acid
f. 1-isopropylcyclobutanecarboxylic acid

Most simple carboxylic acids have common names that are more widely used than their IUPAC names.

- A common name is formed by using a common parent name followed by the suffix *-ic acid*.

The common parent names for simple carboxylic acids are similar to those used for aldehydes (Table 14.1). The common names formic acid, acetic acid, and benzoic acid are virtually always used instead of their IUPAC names.

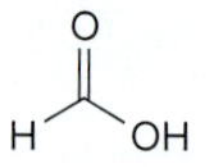

formic acid
(methanoic acid)

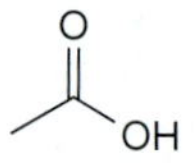

acetic acid
(ethanoic acid)

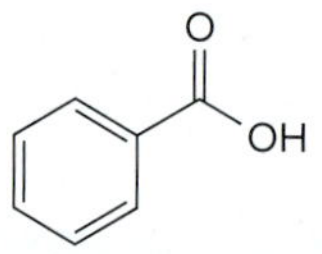

benzoic acid
(benzenecarboxylic acid)

Problem 15.3 Draw the structure corresponding to each common name:

a. α-methoxyvaleric acid
b. β-phenylpropionic acid
c. α,β-dimethylcaproic acid
d. α-chloro-β-methylbutyric acid

15.2B Naming Dicarboxylic Acids and Carboxylates

Many compounds containing two carboxy groups are also known. In the IUPAC system, **diacids** are named by adding the suffix *-dioic acid* to the name of the parent alkane. Many diacids are formed in the citric acid cycle, an enzyme-catalyzed pathway that takes place during the metabolism of carbohydrates, amino acids, and lipids (Section 27.6). Three of these diacids, which are most often identified by their common names, are shown.

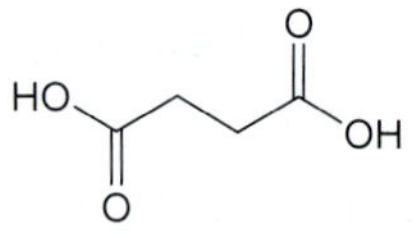

succinic acid
(butanedioic acid)

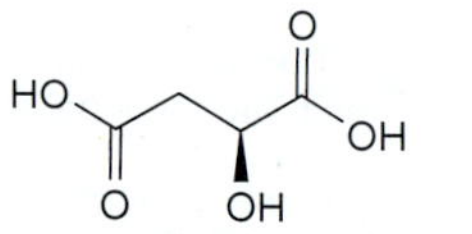

malic acid
[(S)-hydroxybutanedioic acid]

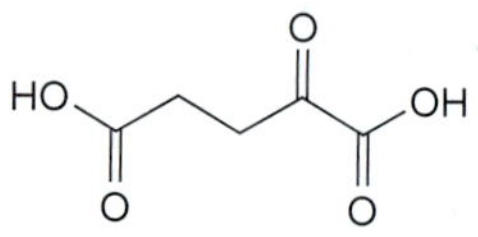

α-ketoglutaric acid
(2-oxopentanedioic acid)

Metal salts of carboxylate anions are formed from carboxylic acids in many reactions in Chapter 15. To name the **metal salt of a carboxylate anion,** change the *-ic acid* ending of the carboxylic acid to the suffix *-ate* and put three parts together:

name of the metal cation + parent + suffix
                            common   -ate
                            or
                            IUPAC

Two examples are shown in Figure 15.1.

Figure 15.1 Naming the metal salts of carboxylate anions

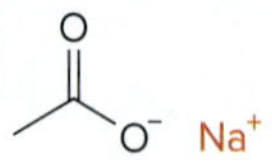

parent + suffix
acet-   -ate
sodium acetate

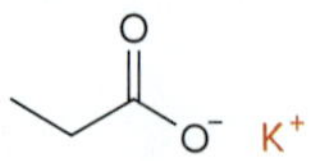

parent + suffix
propano- -ate
potassium propanoate

Problem 15.4 Give the IUPAC name for each metal salt of a carboxylate anion.

a. 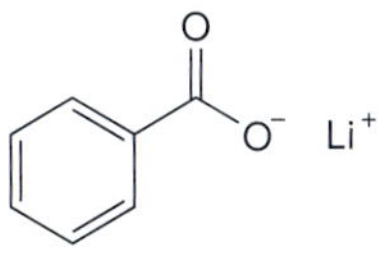 [benzoate] O⁻ Li⁺

b. 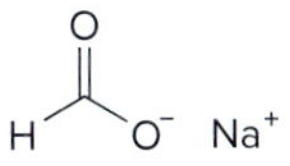 H–C(=O)–O⁻ Na⁺

c. [isobutyrate] O⁻ K⁺

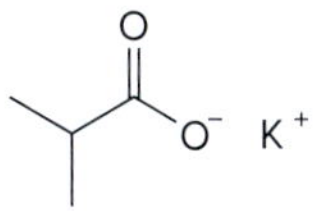

d.  [structure with ethyl branch, Br, and carboxylate] O⁻ Na⁺

Problem 15.5 Depakote, a drug used to treat seizures and bipolar disorder, consists of a mixture of valproic acid [(CH₃CH₂CH₂)₂CHCO₂H] and its sodium salt. Give IUPAC names for each of these compounds.

15.2C Naming Nitriles

In contrast to the carboxylic acids, **nitriles are named as alkane derivatives.** To name a nitrile using IUPAC rules:

In naming a nitrile, the CN carbon is one carbon atom of the longest chain. CH₃CH₂CN is propanenitrile, not ethanenitrile.

- Find the longest chain that contains the CN and add the word *nitrile* to the name of the parent alkane. Number the chain to put CN at C1, but omit this number from the name.

Common names for nitriles are derived from the names of the carboxylic acid having the same number of carbon atoms by replacing the *-ic acid* ending of the carboxylic acid by the suffix *-onitrile*.

When CN is named as a substituent, it is called a *cyano* group. Figure 15.2 illustrates features of nitrile nomenclature.

Figure 15.2 Summary of nitrile nomenclature

a. IUPAC name for a nitrile

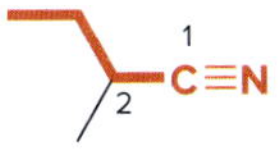

[structure] butane + nitrile (4 C's)
2-methylbutanenitrile

b. Common name for a nitrile

CH₃—C≡N
derived from ace*tic* acid
aceto*nitrile*

c. CN as a substituent

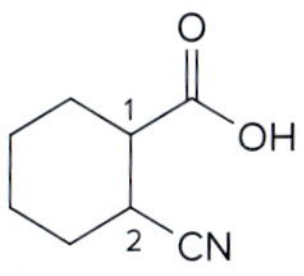

[cyclohexane structure with COOH and CN]
2-cyanocyclohexanecarboxylic acid

Sample Problem 15.2 Naming a Nitrile Using the IUPAC System

Give the IUPAC name for the following nitrile.

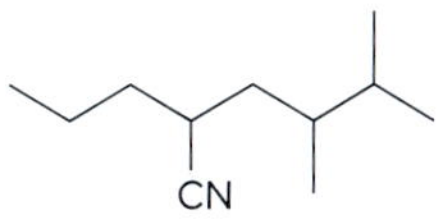

[structure with CN]

Solution

[1] Find and name the longest chain containing the CN.

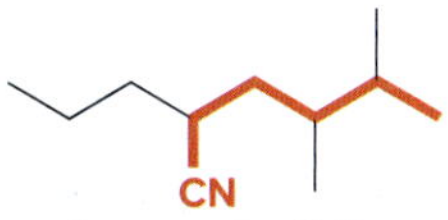

[structure] CN
hexane + nitrile
(6 C's)

The CN contributes one C to the longest chain.

[2] Number and name the substituents.

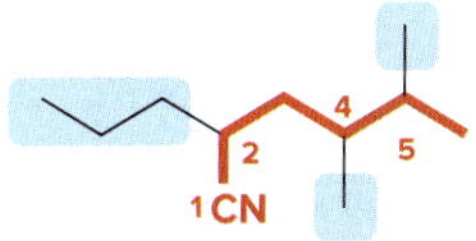

[numbered structure] 1 CN, 2, 4, 5
methyls at C4 and C5
propyl at C2

Answer: 4,5-dimethyl-2-propylhexanenitrile

> **Problem 15.6** Give the IUPAC name for each nitrile.
>
> a. [structure with CN and Cl]   b. [structure with CN]   c. [structure with NC]
>
> **More Practice:** Try Problems 15.30a; 15.31f.

15.3 Physical and Spectroscopic Properties

15.3A Physical Properties

Carboxylic acids and nitriles exhibit **dipole–dipole** interactions because they have polar C—O, C—N, and O—H bonds. Carboxylic acids also exhibit intermolecular **hydrogen bonding** because they possess a hydrogen atom bonded to an electronegative oxygen atom. Carboxylic acids often exist as **dimers,** held together by *two* intermolecular hydrogen bonds between the carbonyl oxygen atom of one molecule and the OH hydrogen atom of another molecule (Figure 15.3). Carboxylic acids are the **most polar** organic compounds we have studied so far.

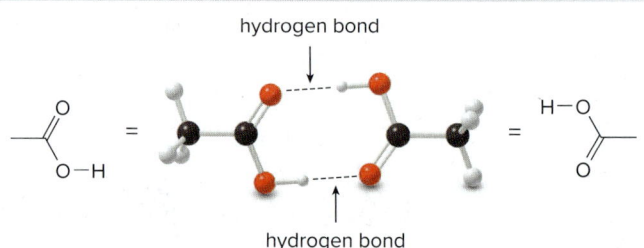

Figure 15.3
Two molecules of acetic acid (CH₃COOH) held together by two hydrogen bonds

How these intermolecular forces affect the physical properties of carboxylic acids is summarized in Table 15.1.

Table 15.1 Physical Properties of Carboxylic Acids

Property	Observation

Boiling point and melting point	• Carboxylic acids have higher boiling points and melting points than other compounds of comparable molecular weight.
	
	     VDW VDW, DD VDW, DD, HB VDW, DD, two HB
	     bp 0 °C bp 48 °C bp 97 °C bp 118 °C
	
	 → Increasing strength of intermolecular forces
	     Increasing boiling point

Solubility	• Carboxylic acids are soluble in organic solvents regardless of size. • Carboxylic acids having ≤ 5 C's are water soluble because they can hydrogen bond with H₂O (Section 3.4C). • Carboxylic acids having > 5 C's are water insoluble because the nonpolar alkyl portion is too large to dissolve in the polar H₂O solvent. These "fatty" acids dissolve in a nonpolar fat-like environment but do not dissolve in water.

Key: VDW = van der Waals, DD = dipole–dipole, HB = hydrogen bonding

15.3 Physical and Spectroscopic Properties

Problem 15.7 Rank the following compounds in order of increasing boiling point. Which compound is the most water soluble? Which compound is the least water soluble?

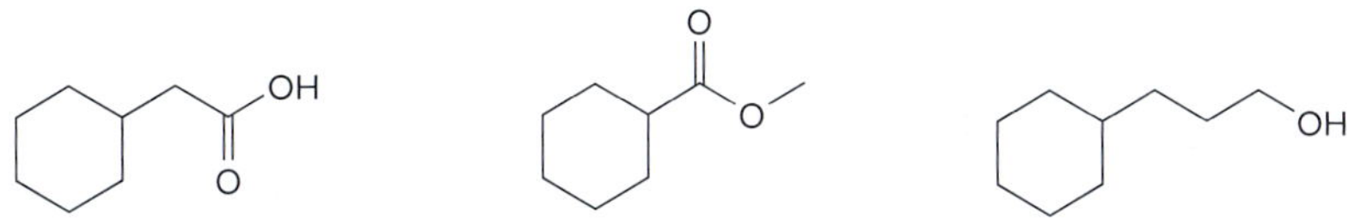

15.3B Spectroscopic Properties

Many details of the spectroscopy of carboxylic acids and nitriles have been presented in Spectroscopy Parts B and C:

- The infrared absorptions of carboxylic acids: Section B.4B and Table B.2
- The infrared absorption of nitriles: Section B.4C and Table B.2
- ^{1}H and ^{13}C NMR absorptions: Tables C.1 and C.5

Key NMR and IR absorptions for carboxylic acids and nitriles are summarized in Table 15.2, and Figure 15.4 illustrates ^{1}H and ^{13}C NMR spectra for a simple carboxylic acid.

Table 15.2 Characteristic Spectroscopic Absorptions of Carboxylic Acids and Nitriles

Compound	Type of spectroscopy	Type of C, H	Absorption
Carboxylic acid	IR absorptions	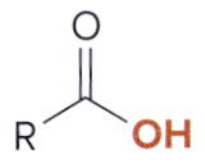 R–C(=O)–OH (O–H)	2500–3500 cm^{-1} (very broad, strong)
		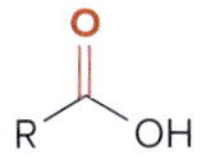 R–C(=O)–OH (C=O)	1710 cm^{-1} (strong)
	^{1}H NMR absorptions	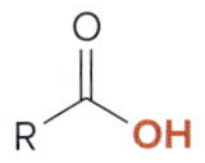 R–C(=O)–OH	10–12 ppm
		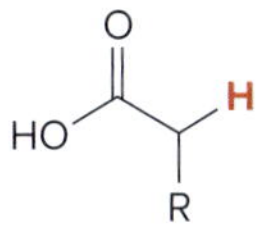 HO–C(=O)–CHR (α-H)	2–2.5 ppm
	^{13}C NMR absorption	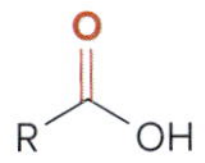 R–C(=O)–OH	170–210 ppm
Nitrile	IR absorption	—C≡N	2250 cm^{-1}
	^{13}C NMR absorption	—C≡N	115–120 ppm

Problem 15.8 Explain how you could use IR spectroscopy to distinguish among the following three compounds.

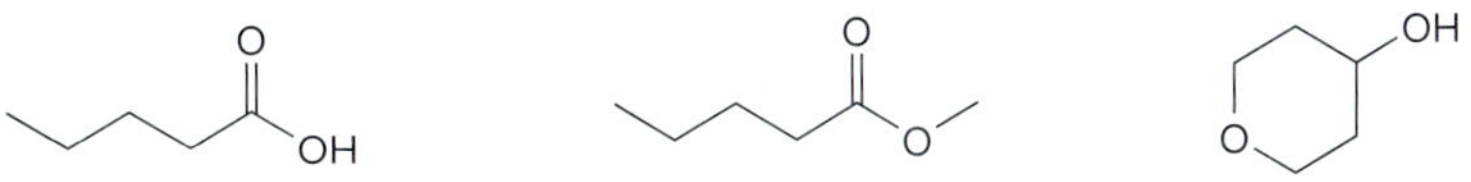

Figure 15.4 The ^{1}H and ^{13}C NMR spectra of propanoic acid

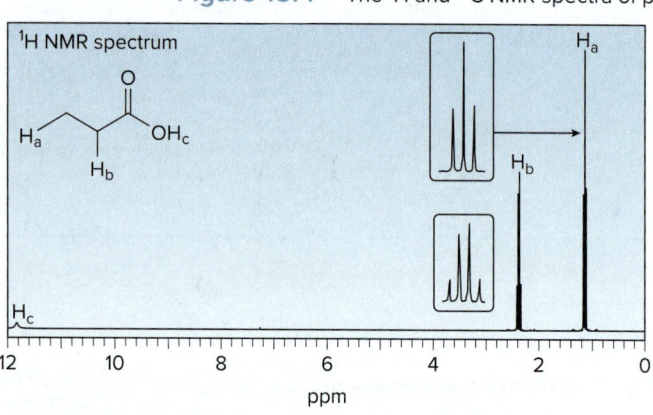
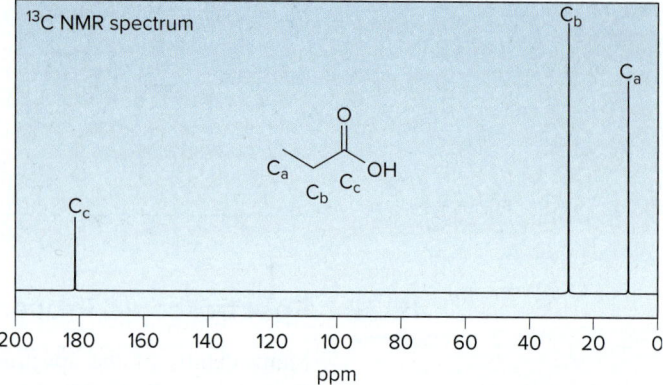

- **^{1}H NMR spectrum:** There are three signals due to three different kinds of H atoms. The H_a and H_b signals are split into a triplet and quartet, respectively. The H_c signal, a singlet, is due to the highly deshielded OH proton.
- **^{13}C NMR spectrum:** There are three signals due to three different kinds of carbon atoms. The carbonyl carbon is highly deshielded.

acetic acid

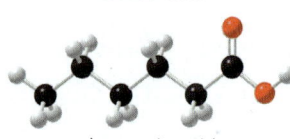

hexanoic acid

Female ginkgo trees produce seeds with an unpleasant odor due to the presence of hexanoic acid. *PicturePartners/Getty Images*

Although oxalic acid is toxic, you would have to eat about nine pounds of spinach at one time to ingest a fatal dose. *Katarzyna Bialasiewicz/123RF*

Soaps, the sodium salts of fatty acids, were discussed in Section 3.6.

15.4 Interesting Carboxylic Acids and Nitriles

Several simple carboxylic acids have characteristic odors and flavors.

Acetic acid (CH_3COOH) is the sour-tasting component of vinegar. The name comes from the Latin word *acetum*, meaning "vinegar." The air oxidation of ethanol to acetic acid is the process that makes "bad" wine taste sour. Pure acetic acid is often called *glacial* acetic acid because it freezes just below room temperature (mp = 17 °C), forming white crystals reminiscent of the ice in a glacier.

Hexanoic acid [$CH_3(CH_2)_4COOH$] is a low-molecular-weight carboxylic acid with the foul odor of dirty socks and locker rooms. Its common name, caproic acid, is derived from the Latin word *caper*, meaning "goat." The fleshy coat of seeds that are produced by female ginkgo trees contains hexanoic acid, giving the seeds an unpleasant and even repulsive odor.

Oxalic acid and **lactic acid** are simple carboxylic acids quite prevalent in nature. Oxalic acid occurs naturally in spinach and rhubarb. Lactic acid gives sour milk its distinctive taste.

Salts of carboxylic acids are commonly used as preservatives. Sodium benzoate, a fungal growth inhibitor, is a preservative used in soft drinks, and potassium sorbate is an additive that prolongs the shelf life of baked goods and other foods.

Although nitriles are much less common than carboxylic acids, the naturally occurring cyanohydrin derivatives discussed in Section 14.8 constitute one group of compounds that contain a nitrile. In addition, several widely used drugs contain one or more cyano groups, including anastrozole, used to reduce the recurrence of breast cancer in women whose tumors are estrogen positive; escitalopram, used to treat depression and anxiety; and verapamil for high blood pressure and chest pain.

Generic name anastrozole
Trade name Arimidex

Generic name escitalopram
Trade names Cipralex, Lexapro

Generic name verapamil
Trade names Calan, Verelan

Anastrozole is called an **aromatase inhibitor** because it blocks the activity of the aromatase enzyme, which is responsible for estrogen synthesis. This inhibits tumor growth in those forms of breast cancer that are stimulated by estrogen.

15.5 Aspirin, Arachidonic Acid, and Prostaglandins

Recall from Chapter 2 that **aspirin (acetylsalicylic acid)** is a synthetic carboxylic acid, similar in structure to **salicin,** a naturally occurring compound isolated from willow bark, and **salicylic acid,** found in meadowsweet.

aspirin
(acetylsalicylic acid)

salicin
(isolated from willow bark)

salicylic acid
(isolated from meadowsweet)

sodium salicylate
(sweet carboxylate salt)

The word *aspirin* is derived from the prefix *a-* for *acetyl* + *spir* from the Latin name *spirea* for the meadowsweet plant.
Biopix.dx http://www.biopix.dk

Aspirin is the most widely used pain reliever and anti-inflammatory agent in the world, yet its mechanism of action remained unknown until the 1970s. John Vane, Bengt Samuelsson, and Sune Bergstrom shared the 1982 Nobel Prize in Physiology or Medicine for unraveling the details of its mechanism.

Both salicylic acid and sodium salicylate (its sodium salt) were widely used analgesics in the nineteenth century, but both had undesirable side effects. Salicylic acid irritated the mucous membranes of the mouth and stomach, and sodium salicylate was too sweet for most patients. Aspirin, a synthetic compound, was first sold in 1899 after Felix Hoffmann, a German chemist at Bayer Company, developed a feasible commercial synthesis. Hoffmann's work was motivated by personal reasons: his father suffered from rheumatoid arthritis and was unable to tolerate the sweet taste of sodium salicylate.

How does aspirin relieve pain and reduce inflammation? Aspirin blocks the synthesis of **prostaglandins,** 20-carbon fatty acids with a five-membered ring that are responsible for pain, inflammation, and a wide variety of other biological functions. **PGF$_{2\alpha}$** contains the typical carbon skeleton of a prostaglandin.

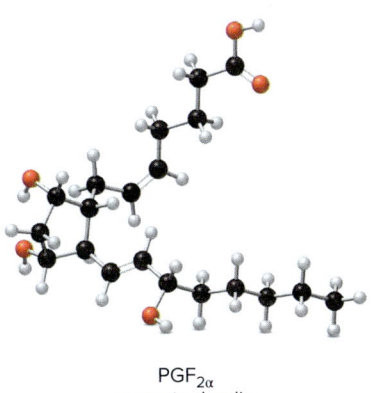

PGF$_{2\alpha}$
a prostaglandin

Prostaglandins are not stored in cells. Rather, they are synthesized from arachidonic acid, a polyunsaturated fatty acid having four cis double bonds. Unlike hormones, which are transported in the bloodstream to their sites of action, prostaglandins act where they are synthesized. Aspirin acts by blocking the synthesis of prostaglandins from arachidonic acid. Aspirin inactivates cyclooxygenase, an enzyme that converts arachidonic acid to PGG_2, an unstable precursor of $PGF_{2\alpha}$ and other prostaglandins. **Aspirin lessens pain and decreases inflammation because it prevents the synthesis of prostaglandins, the compounds responsible for both of these physiological responses.**

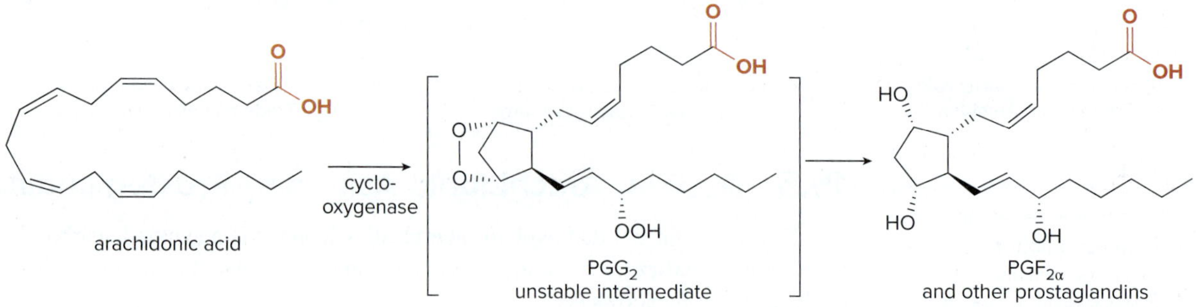

Although prostaglandins have a wide range of biological activity, their inherent instability often limits their usefulness as drugs. Consequently, more-stable analogues with useful medicinal properties have been synthesized. For example, latanoprost (trade name Xalatan) and bimatoprost (trade name Lumigan) are prostaglandin analogues used to reduce eye pressure in individuals with glaucoma.

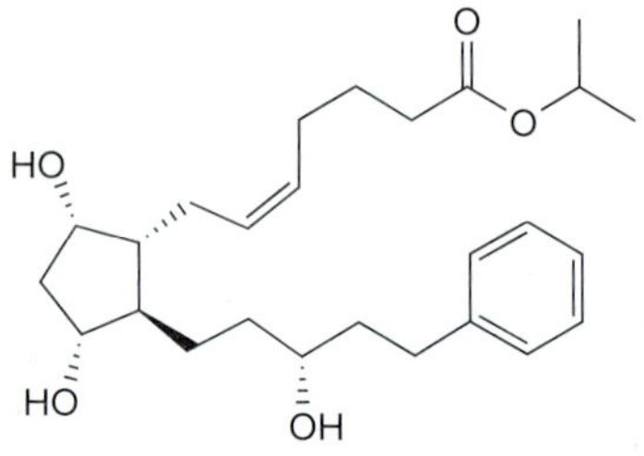

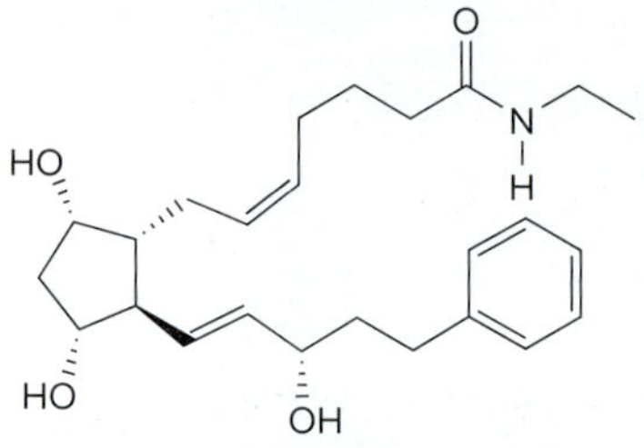

15.6 Preparation of Carboxylic Acids

We begin our study of the reactions involving carboxylic acids and nitriles by summarizing methods that introduce a carboxy group presented in earlier chapters. In Sections 15.7–15.10, we then concentrate on the acidity of carboxylic acids, and in Section 15.13, we examine the preparation and reactions of nitriles.

Where have we encountered carboxylic acids as reaction products before? The carbonyl carbon is highly oxidized, because it has three C–O bonds, so **carboxylic acids are often prepared by oxidation reactions.** Three oxidation methods and one carbon–carbon bond-forming reaction are listed in Table 15.3.

Problem 15.9 What alcohol can be oxidized to each carboxylic acid?

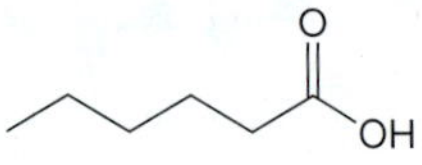

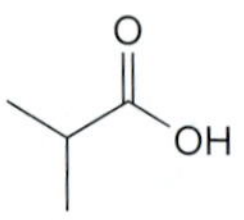

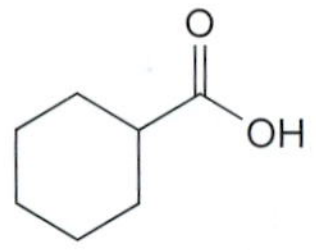

15.7 Carboxylic Acids—Strong Organic Brønsted–Lowry Acids

Table 15.3 Methods That Synthesize Carboxylic Acids

Method	Reaction
[1] **Oxidation of 1° alcohols** (Section 11.12B)	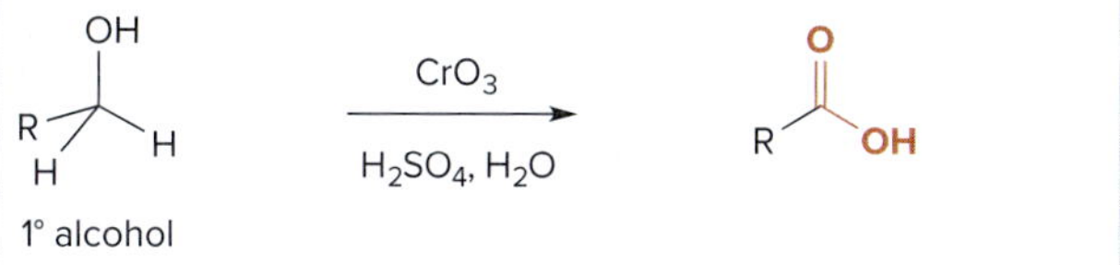 1° alcohol → carboxylic acid with CrO_3, H_2SO_4, H_2O
[2] **Oxidation of aldehydes** (Section 13.8)	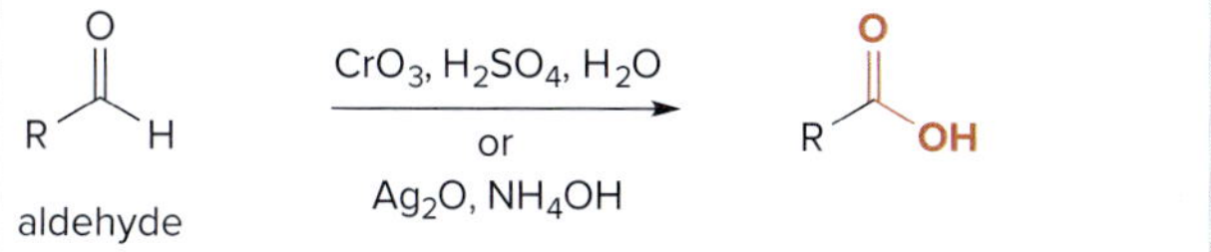 aldehyde → carboxylic acid with CrO_3, H_2SO_4, H_2O or Ag_2O, NH_4OH
[3] **Carboxylation of Grignard reagents** (Section 13.14)	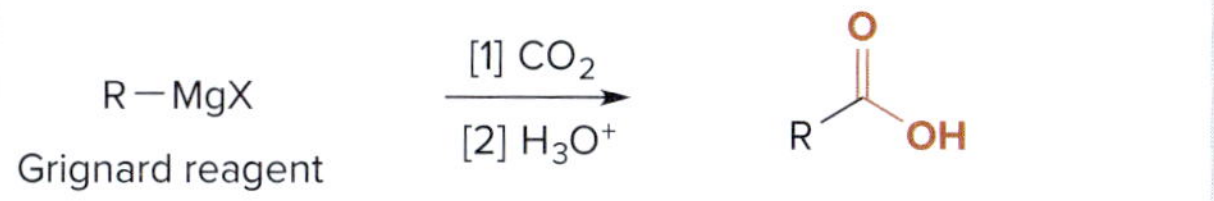 R–MgX (Grignard reagent) → RCOOH with [1] CO_2 [2] H_3O^+
[4] **Oxidative cleavage of alkynes** (Section 11.11)	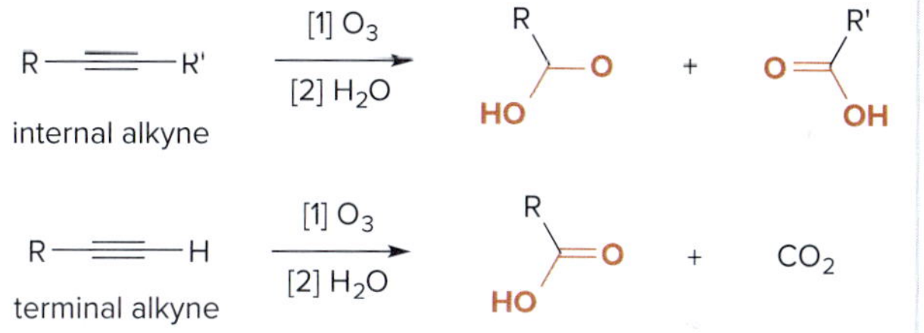 internal alkyne R–≡–R' → RCOOH + R'COOH with [1] O_3 [2] H_2O; terminal alkyne R–≡–H → RCOOH + CO_2 with [1] O_3 [2] H_2O

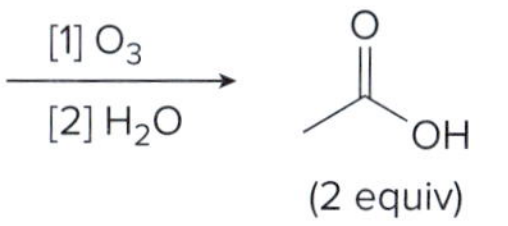

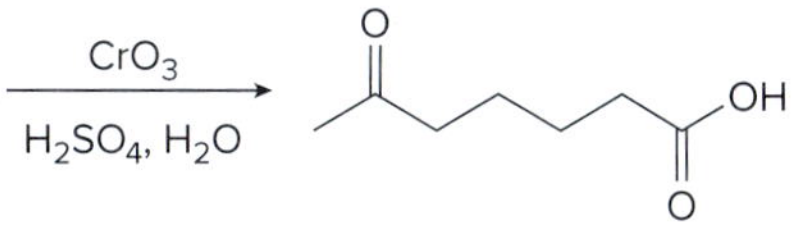

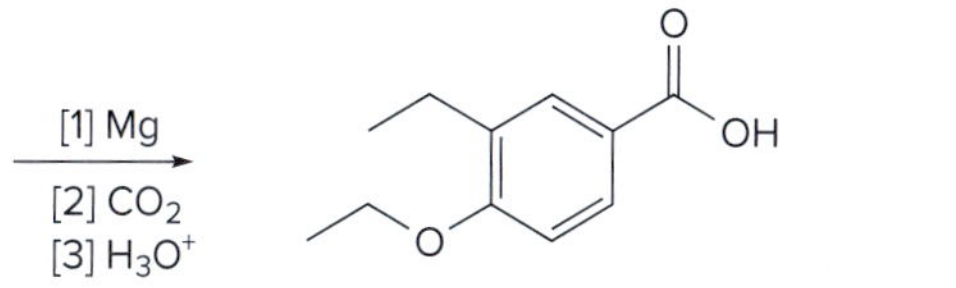

Problem 15.10 Identify **A–E** in the following reactions.

a. **A** $\xrightarrow{CrO_3,\ H_2SO_4,\ H_2O}$ benzoic acid

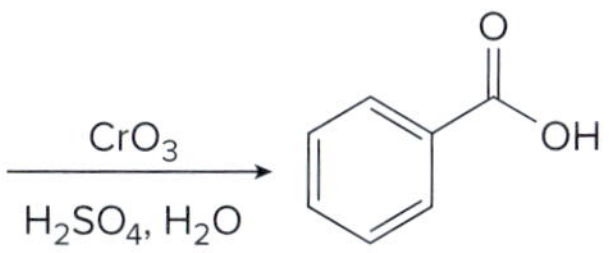

b. **B** $\xrightarrow[{[2]\ H_2O}]{[1]\ O_3}$ acetic acid (2 equiv)

c. **C** $\xrightarrow{CrO_3,\ H_2SO_4,\ H_2O}$ 5-oxohexanoic acid (keto acid)

d. **D** $\xrightarrow[{[3]\ H_3O^+}]{[1]\ Mg,\ [2]\ CO_2}$ substituted benzoic acid (with ethyl and ethoxy groups)

e. **E** $\xrightarrow{Ag_2O,\ NH_4OH}$ keto-hydroxy carboxylic acid

---

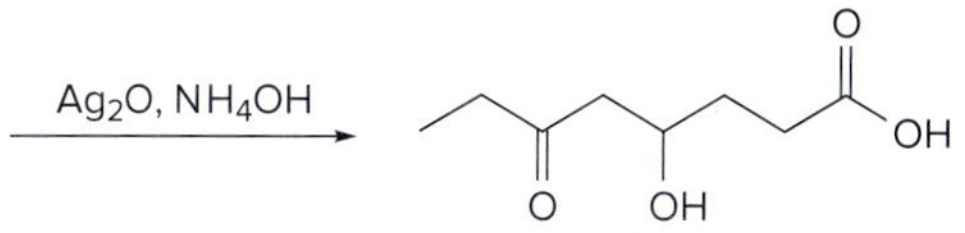

15.7 Carboxylic Acids—Strong Organic Brønsted–Lowry Acids

The polar C—O and O—H bonds, nonbonded electron pairs on oxygen, and the π bond give a carboxylic acid many reactive sites, complicating its chemistry somewhat. By far, **the most important reactive feature of a carboxylic acid is its polar O—H bond, which is readily cleaved with base.**

- **Carboxylic acids are strong organic acids**, and as such, readily react with Brønsted–Lowry bases to form carboxylate anions.

Recall from Section 2.3 that **the lower the pK_a, the stronger the acid.**

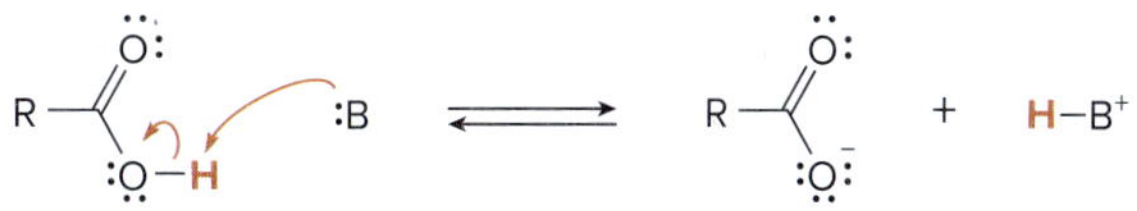

RCOOH + :B ⇌ RCOO⁻ + H–B⁺

carboxylate anion

What bases are used to deprotonate a carboxylic acid? As we learned in Section 2.3, **equilibrium favors the products of an acid–base reaction when the weaker base and acid are formed.** Because a weaker acid has a higher pK_a, this general rule results:

- An acid can be deprotonated by a base that has a conjugate acid with a *higher* pK_a.

Because the pK_a values of many carboxylic acids are ~5, bases that have conjugate acids with pK_a values *higher* than 5 are strong enough to deprotonate them. Thus, acetic acid (pK_a = 4.8) and benzoic acid (pK_a = 4.2) can be deprotonated with NaOH and NaHCO$_3$, as shown in the following equations.

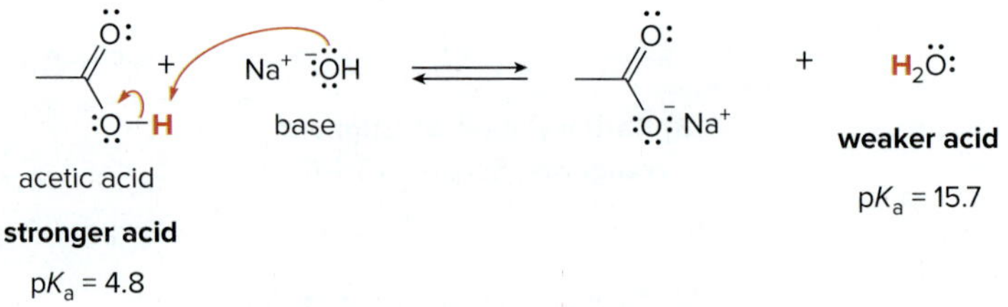

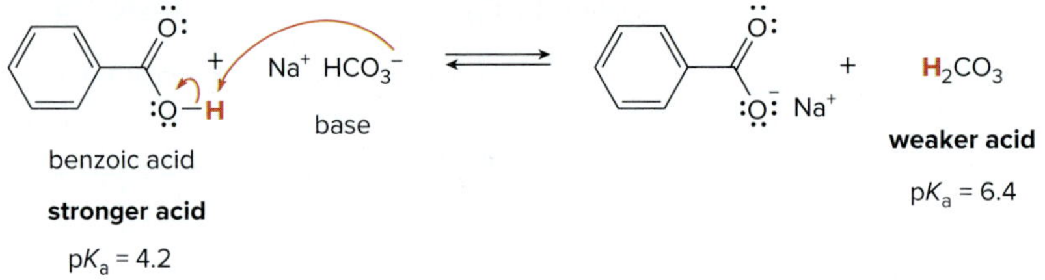

Table 15.4 lists common bases that can be used to deprotonate carboxylic acids. It is noteworthy that even a weak base like NaHCO$_3$ is strong enough to remove a proton from RCOOH.

Table 15.4 Common Bases Used to Deprotonate Carboxylic Acids

Base	Conjugate acid (pK_a)
Na$^+$ HCO$_3^-$	H$_2$CO$_3$ (6.4)
NH$_3$	NH$_4^+$ (9.4)
Na$_2$CO$_3$	HCO$_3^-$ (10.2)
Na$^+$ $^-$OCH$_3$	CH$_3$OH (15.5)
Na$^+$ $^-$OH	H$_2$O (15.7)
Na$^+$ $^-$OCH$_2$CH$_3$	CH$_3$CH$_2$OH (16)
Na$^+$ H$^-$	H$_2$ (35)

Increasing basicity ↓

Why are carboxylic acids such strong organic acids? Remember that a strong acid has a weak, stabilized conjugate base. **Deprotonation of a carboxylic acid forms a resonance-stabilized conjugate base—a carboxylate anion.** Two equivalent resonance structures can be drawn for acetate (the conjugate base of acetic acid), both of which place a negative charge on an electronegative O atom. In the resonance hybrid, therefore, the negative charge is delocalized over two oxygen atoms.

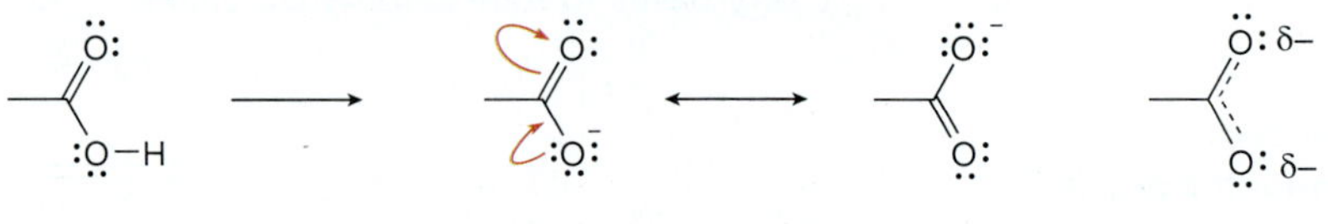

15.7 Carboxylic Acids—Strong Organic Brønsted–Lowry Acids

> How resonance affects acidity was first discussed in Section 2.5C.

Resonance stabilization accounts for why carboxylic acids are more acidic than other compounds with O–H bonds—namely, alcohols and phenols. For example, the pK_a values of ethanol (CH_3CH_2OH) and phenol (C_6H_5OH) are 16 and 10, respectively, both higher than the pK_a of acetic acid (4.8).

ethanol
pK_a = 16

phenol
pK_a = 10

acetic acid
pK_a = 4.8

Increasing acidity →

To understand the relative acidity of ethanol, phenol, and acetic acid, we must compare the stability of their conjugate bases and use this rule:

- **Anything that stabilizes a conjugate base A:⁻ makes the starting acid H–A more acidic.**

Ethoxide, the conjugate base of ethanol, bears a negative charge on an oxygen atom, but there are no additional factors to further stabilize the anion. Because ethoxide is less stable than acetate, **ethanol is a weaker acid than acetic acid.**

ethanol → ethoxide

> The resonance hybrid of phenoxide illustrates that its negative charge is dispersed over four atoms—three C atoms and one O atom.

Like acetate, **phenoxide** ($C_6H_5O^-$, the conjugate base of phenol) is also resonance stabilized. In the case of phenoxide, however, there are *five* resonance structures that disperse the negative charge over a total of *four* different atoms (three different carbons and the oxygen).

hybrid   phenol   1   2   3   4   5

Phenoxide is more stable than ethoxide, but less stable than acetate, because acetate has two electronegative oxygen atoms upon which to delocalize the negative charge, whereas phenoxide has only one. Additionally, phenoxide resonance structures **2–4** have the negative charge on a carbon, a less electronegative element than oxygen. As a result, structures **2–4** are less stable than structures **1** and **5**, which have the negative charge on oxygen.

Moreover, resonance structures **1** and **5** have intact aromatic rings, whereas structures **2–4** do not. This, too, makes structures **2–4** less stable than **1** and **5**. Figure 15.5 summarizes this

Figure 15.5
The relative energies of the five resonance structures for phenoxide and its hybrid

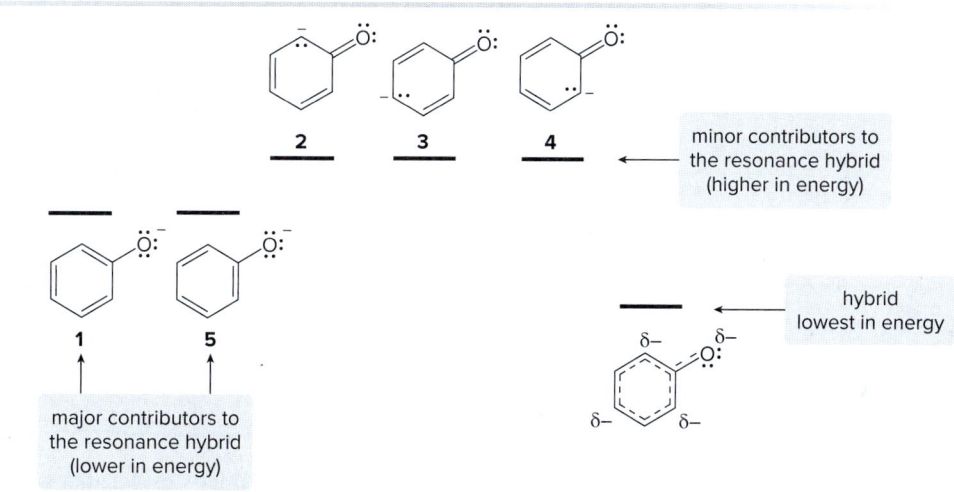

2   3   4   ← minor contributors to the resonance hybrid (higher in energy)

1   5
major contributors to the resonance hybrid (lower in energy)

← hybrid lowest in energy

707

information about phenoxide by displaying the approximate relative energies of its five resonance structures and its hybrid.

As a result, resonance stabilization of the conjugate base is important in determining acidity, but **the absolute number of resonance structures alone is not what's important.** We must evaluate their relative contributions to predict the relative stability of the conjugate bases.

> - Because of their O—H bond, RCOOH, ROH, and C₆H₅OH are *more acidic* than most organic hydrocarbons.
> - A carboxylic acid is a *stronger* acid than an alcohol or a phenol because its conjugate base is more effectively resonance stabilized.

Keep in mind that although carboxylic acids are strong organic acids, they are still *much weaker* than strong inorganic acids like HCl and H_2SO_4, which have pK_a values < 0.

Because alcohols and phenols are weaker acids than carboxylic acids, stronger bases are needed to deprotonate them. To deprotonate C₆H₅OH ($pK_a = 10$), a base whose conjugate acid has a $pK_a > 10$ is needed. Thus, of the bases listed in Table 15.4, NaOCH₃, NaOH, NaOCH₂CH₃, and NaH are strong enough. To deprotonate CH₃CH₂OH ($pK_a = 16$), only NaH is strong enough.

Problem 15.11 Draw the products of each acid–base reaction.

Problem 15.12 Given the pK_a values in Appendix C, which of the following bases are strong enough to deprotonate CH₃COOH: (a) F⁻; (b) (CH₃)₃CO⁻; (c) CH₃⁻; (d) ⁻NH₂; (e) Cl⁻?

Problem 15.13 Rank the labeled protons (H_a–H_c) in mandelic acid, a naturally occurring carboxylic acid in plums and peaches, in order of increasing acidity. Explain in detail why you chose this order.

mandelic acid

15.8 The Henderson–Hasselbalch Equation

What happens when a particular acid is dissolved in an aqueous solution? **Whether or not the acid will lose a proton depends on two factors—its pK_a and the pH of the solution.** The amount of acid (HA) and its conjugate base (A:⁻) can be calculated using the **Henderson–Hasselbalch equation,** which is derived from the expressions for K_a and pK_a (Section 2.3). The derivation of the Henderson–Hasselbalch equation is shown in Figure 15.6.

Henderson–Hasselbalch equation

$$pK_a = pH + \log \frac{[HA]}{[A:^-]}$$

The Henderson–Hasselbalch equation tells us whether a compound will exist in its acidic form (HA) or as its conjugate base (A:⁻) at a particular pH.

15.8 The Henderson–Hasselbalch Equation

Figure 15.6
Derivation of the Henderson–Hasselbalch equation

- Use the definitions of K_a and pK_a to write the equation.

$$pK_a = -\log K_a = -\log \frac{[H_3O^+][A:^-]}{[HA]}$$

- Separate the terms.

$$pK_a = -\log [H_3O^+] - \log \frac{[A:^-]}{[HA]}$$

- Use the definition of pH (pH = $-\log [H_3O^+]$) to simplify.

$$pK_a = pH - \log \frac{[A:^-]}{[HA]}$$

- Invert the terms in the logarithm and change the sign that precedes this term.

$$pK_a = pH + \log \frac{[HA]}{[A:^-]}$$

Henderson–Hasselbalch equation

Recall that a log is an exponent: $\log 10^2 = 2$; $\log 10^{-2} = -2$.

- When the pH of the solution *equals* the pK_a of the acid, the concentration of HA and A:⁻ must be equal, because log ([HA]/[A:⁻]) equals zero.
- When the pH of the solution is *less* than the pK_a of the acid, the concentration of HA is greater than the concentration of A:⁻ because log ([HA]/[A:⁻]) is positive.
- When the pH of the solution is *higher than* the pK_a of the acid, the concentration of A:⁻ is greater than the concentration of HA because log ([HA]/[A:⁻]) is negative.

We can summarize these consequences of the Henderson–Hasselbalch equation as follows:

- An acid exists in its protonated form HA in solutions that are more *acidic* than its pK_a.
- An acid exists as its conjugate base A:⁻ in solutions that are more *basic* than its pK_a.

In what form does a carboxylic acid like pyruvic acid ($pK_a = 2.4$), a product of glucose metabolism, exist in the bloodstream, which is buffered to a pH of 7.4? Because the pH of the solution is *higher* than the pK_a of pyruvic acid, the acid is deprotonated and exists primarily as its conjugate base, pyruvate.

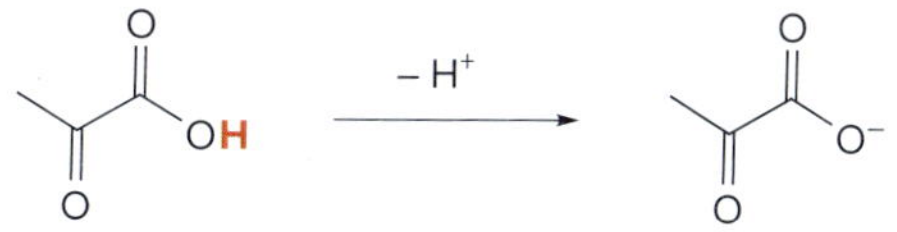

pyruvic acid
$pK_a = 2.4$

pyruvate
predominant form at pH = 7.4

The cellular pH of 7.4 is called **physiological pH**.

Sample Problem 15.3 Using a Compound's pK_a to Determine the Predominant Species at a Given pH

What is the predominant form of each compound in a solution of pH 5.0:
(a) FCH_2CO_2H ($pK_a = 2.7$); (b) CF_3CH_2OH ($pK_a = 12.4$)?

Solution

a. Because the pH of the solution (5.0) is higher *(more basic)* than the pK_a of FCH_2CO_2H (2.7), the compound will exist primarily as its conjugate base, $FCH_2CO_2^-$, formed by removal of its most acidic proton.

b. Because the pH of the solution (5.0) is lower *(more acidic)* than the pK_a of CF_3CH_2OH (12.4), the compound will exist primarily as CF_3CH_2OH, the neutral acid that is *not* deprotonated.

Problem 15.14 What form(s) of each compound predominate at the given pH?

a. 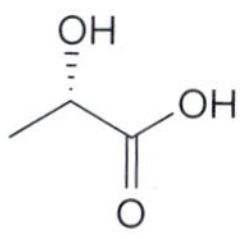 lactic acid, $pK_a = 3.9$, at pH = 7.3
b. acetylsalicylic acid, $pK_a = 3.5$, at pH = 3.5
c. ethanol, $pK_a = 16$, at pH = 2.0

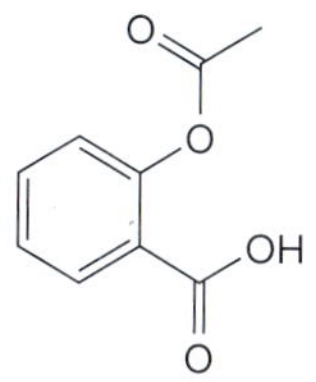

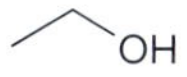

More Practice: Try Problems 15.47–15.49.

Some compounds, such as the dicarboxylic acids mentioned in Section 15.2B and the amino acids discussed in Section 15.12, contain two or more functional groups that can lose protons, so two pK_a values are reported. For succinic acid (pK_{a1} = 4.2 and pK_{a2} = 5.6), both pK_a values are *less than* the physiological pH of 7.4, so both carboxy groups are deprotonated and the predominant species in cells is the dianion succinate.

Problem 15.15 Citric acid, a metabolic intermediate in the citric acid cycle (Section 27.6), contains three carboxy groups with pK_a values of 3.1, 4.8, and 6.4. (a) What is the predominant form of citric acid in the stomach (pH = 2)? (b) What is the predominant form of citric acid in the intestines (pH = 8)?

15.9 Inductive Effects in Aliphatic Carboxylic Acids

The pK_a of a carboxylic acid is affected by nearby groups that inductively donate or withdraw electron density.

- Electron-withdrawing groups *stabilize* a conjugate base, making a carboxylic acid *more* acidic.
- Electron-donating groups *destabilize* the conjugate base, making a carboxylic acid *less* acidic.

The relative acidity of CH_3COOH, $ClCH_2COOH$, and $(CH_3)_3CCOOH$ illustrates these principles in the following equations.

We first learned about inductive effects and acidity in Section 2.5B.

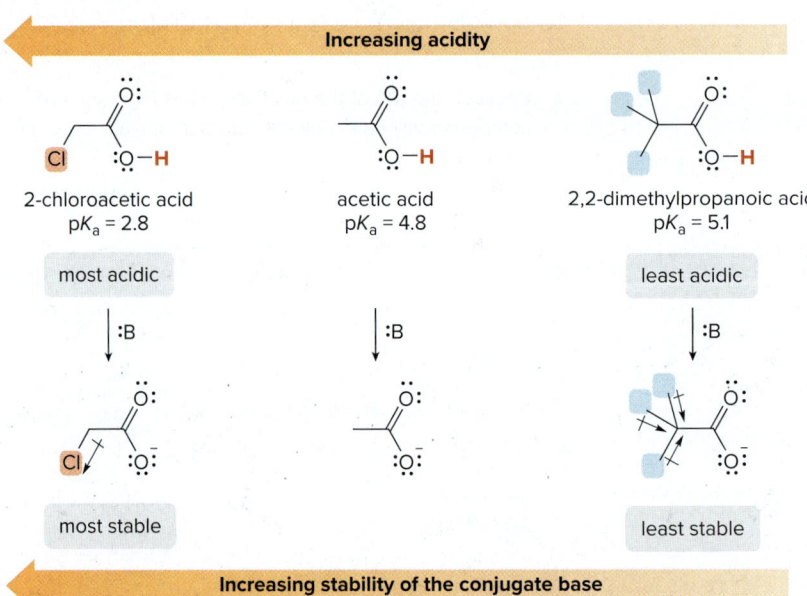

- ClCH$_2$COOH is *more* acidic (pK_a = 2.8) than CH$_3$COOH (pK_a = 4.8) because its conjugate base is stabilized by the **electron-withdrawing inductive effect of the electronegative Cl.**
- (CH$_3$)$_3$CCOOH is *less* acidic (pK_a = 5.1) than CH$_3$COOH because the **three polarizable CH$_3$ groups donate electron density and destabilize the conjugate base.**

The number, electronegativity, and location of substituents also affect acidity.

- **The larger the number of electronegative substituents, the stronger the acid.**

[Structures: ClCH$_2$COOH pK_a = 2.8; Cl$_2$CHCOOH pK_a = 1.3; Cl$_3$CCOOH pK_a = 0.9]

Increasing acidity
Increasing number of electronegative Cl atoms

- **The more electronegative the substituent, the stronger the acid.**

[Structures: ClCH$_2$COOH pK_a = 2.8; FCH$_2$COOH pK_a = 2.6]

F is more electronegative than Cl.
stronger acid

- **The closer the electron-withdrawing group to the COOH, the stronger the acid.**

4-chlorobutanoic acid
pK_a = 4.5

3-chlorobutanoic acid
pK_a = 4.1

2-chlorobutanoic acid
pK_a = 2.9

Increasing acidity
Increasing proximity of Cl to COOH

Problem 15.16 Match each of the following pK_a values (3.2, 4.9, and 0.2) to the appropriate carboxylic acid: (a) CH$_3$CH$_2$COOH; (b) CF$_3$COOH; (c) ICH$_2$COOH.

Problem 15.17 Rank the following compounds in order of increasing acidity.

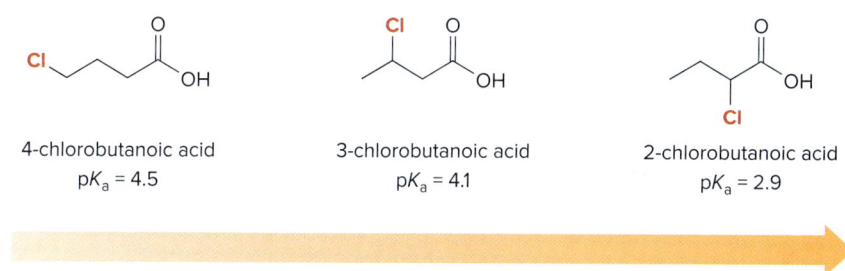

15.10 Extraction

Extraction has long been and remains the first step in isolating a natural product from its source.

An organic chemist in the laboratory must separate and purify mixtures of compounds. One particularly useful technique is **extraction**, which uses solubility differences and acid–base principles to separate and purify compounds.

Two solvents are used in extraction: water or an aqueous solution such as 10% NaHCO$_3$ or 10% NaOH; and an organic solvent such as dichloromethane (CH$_2$Cl$_2$), diethyl ether, or hexane. **Compounds are separated by their solubility differences in an aqueous and organic solvent.**

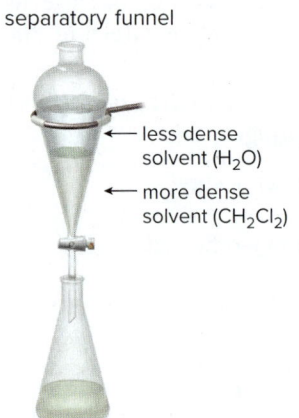

separatory funnel

When two insoluble liquids are added to a separatory funnel, two layers are visible. To separate the layers, the lower layer can be drained from the bottom of the separatory funnel by opening the stopcock. The top layer can then be poured out the top neck of the funnel.

An item of glassware called a **separatory funnel** is used for the extraction. When two insoluble liquids are added to the separatory funnel, two layers form, with the less dense liquid on top and the more dense liquid on the bottom.

Suppose a mixture of benzoic acid (C_6H_5COOH) and NaCl is added to a separatory funnel containing H_2O and CH_2Cl_2. The benzoic acid would dissolve in the organic layer, and the NaCl would dissolve in the water layer. Separating the organic and aqueous layers and placing them in different flasks separates the benzoic acid and NaCl from each other.

How could we separate a mixture of benzoic acid and cyclohexanol? **Both compounds are organic, and as a result, both are soluble in an organic solvent such as CH_2Cl_2 and insoluble in water.** If a mixture of benzoic acid and cyclohexanol were added to a separatory funnel with CH_2Cl_2 and water, both would dissolve in the CH_2Cl_2 layer, and the two compounds would *not* be separated from each other. Is it possible to use extraction to separate two compounds of this sort that have similar solubility properties?

benzoic acid   cyclohexanol

- insoluble in water   *similar* solubility   • insoluble in water
- soluble in CH_2Cl_2 properties   • soluble in CH_2Cl_2

If a carboxylic acid is one of the compounds, the answer is *yes*, because we can use acid–base chemistry to change its solubility properties.

When benzoic acid (a strong organic acid) is treated with aqueous NaOH, benzoic acid is deprotonated, forming sodium benzoate. **Because sodium benzoate is ionic, it is *soluble* in water, but *insoluble* in organic solvents.**

benzoic acid   base   sodium benzoate $pK_a = 15.7$
$pK_a = 4.2$

Recall from Tables 9.1 and 15.1 that alcohols and carboxylic acids having more than five carbons are water insoluble.

- insoluble in water   *different* solubility   • soluble in water
- soluble in CH_2Cl_2 properties   • insoluble in CH_2Cl_2

A similar acid–base reaction does *not* occur when cyclohexanol is treated with NaOH because organic alcohols are much weaker organic acids, so they can be deprotonated only by a *very strong base* such as NaH. **NaOH is not strong enough to form significant amounts of the sodium alkoxide.**

cyclohexanol   base $pK_a = 15.7$
$pK_a \sim 17$ Because equilibrium favors the starting
materials, little alkoxide is formed.

This difference in acid–base chemistry can be used to separate benzoic acid and cyclohexanol by the stepwise extraction procedure illustrated in Figure 15.7. This extraction scheme relies on two principles:

- Extraction can separate only compounds having different solubility properties. One compound must dissolve in the aqueous layer and one must dissolve in the organic layer.
- A carboxylic acid can be separated from other organic compounds by converting it to a water-soluble carboxylate anion by an acid–base reaction.

Figure 15.7 Separation of benzoic acid and cyclohexanol by an extraction procedure

Step [1] Dissolve benzoic acid and cyclohexanol in CH_2Cl_2.

Step [2] Add 10% NaOH solution to form two layers.

Step [3] Separate the layers.

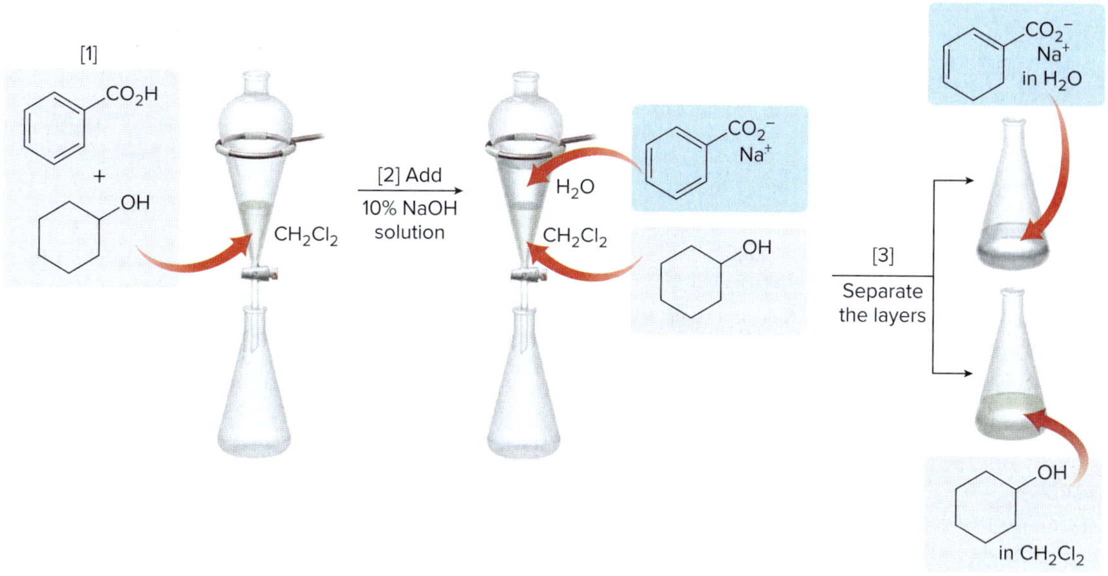

- Both compounds dissolve in the organic solvent CH_2Cl_2.

- Adding 10% aqueous NaOH solution forms two layers. When the two layers are mixed, the **NaOH deprotonates $C_6H_5CO_2H$ to form $C_6H_5CO_2^-Na^+$, which dissolves in the aqueous layer.**
- The cyclohexanol remains in the CH_2Cl_2 layer.

- Draining the lower layer out the bottom stopcock separates the two layers, and the separation process is complete.
- Cyclohexanol (dissolved in CH_2Cl_2) is in one flask. The sodium salt of benzoic acid, $C_6H_5CO_2^-Na^+$ (dissolved in water) is in another flask.

Thus, the water-soluble salt, $C_6H_5CO_2^-Na^+$ (derived from $C_6H_5CO_2H$ by an acid–base reaction), can be separated from water-insoluble cyclohexanol by an extraction procedure.

Sample Problem 15.4 Separating Compounds by Extraction

A mixture of **A**, **B**, and **C** was added to a separatory funnel containing CH_2Cl_2 and 10% aqueous NaOH solution. Which compound(s) are present in the aqueous layer, and which compound(s) are present in the organic layer?

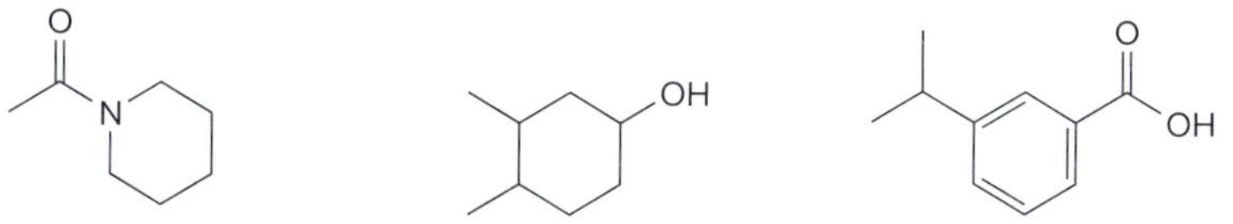

A   **B**   **C**

Solution
Recall the principles of solubility:

- **Organic compounds are soluble in organic solvents.**
- **Organic compounds that can hydrogen bond to H_2O are water soluble if they have ≤ 5 C's.**
- **Uncharged organic compounds with > 5 C's are not water soluble.**
- **Ionic compounds are water soluble.**

A, B, and **C** are uncharged organic compounds, so they are *soluble* in CH_2Cl_2, and because they each have > 5 C's, they are *insoluble* in H_2O. **C**, however, has a CO_2H group with an acidic H

714 Chapter 15 Carboxylic Acids and Nitriles

atom that can be removed with NaOH. Deprotonation forms a **carboxylate anion** that is now *water soluble*.

A
- insoluble in water
- soluble in CH$_2$Cl$_2$

B
- insoluble in water
- soluble in CH$_2$Cl$_2$

C
- insoluble in water
- soluble in CH$_2$Cl$_2$

carboxylate anion
- soluble in water
- insoluble in CH$_2$Cl$_2$

As a result, in a separatory funnel with CH$_2$Cl$_2$ and 10% aqueous NaOH solution, **A** and **B** are soluble in the CH$_2$Cl$_2$ layer, and **C** is deprotonated to form a carboxylate anion that is now soluble in the aqueous layer.

Problem 15.18 Which of the following pairs of compounds can be separated from each other by an extraction procedure?

a. [carboxylic acid] and [alkene]

b. [alkene] and [ether]

c. [carboxylic acid] and NaCl

d. NaCl and KCl

More Practice: Try Problems 15.50–15.52.

15.11 Organic Acids That Contain Phosphorus

As we learned in Sections 3.2 and 7.16, several biological compounds are derivatives of phosphoric acid (H$_3$PO$_4$) and related compounds. Phosphoric acid itself has three OH groups that can be deprotonated, with pK_a values of 2.1, 6.9, and 12.4, forming H$_2$PO$_4^-$, HPO$_4^{2-}$, and PO$_4^{3-}$, respectively.

All of the phosphoric acid esters in Table 3.4 were drawn as negatively charged species. Now that we have learned about the Henderson–Hasselbalch equation, we can understand why these functional groups are ionized at the physiological pH of 7.4.

phosphoric acid ester $\xrightarrow[\text{p}K_{a1} \sim 2.1]{-H^+}$ [intermediate] $\xrightarrow[\text{p}K_{a2} \sim 6.9]{-H^+}$ monophosphate

The dianion predominates at pH = 7.4.

We can use the pK_a values of phosphoric acid to approximate the pK_a values for the two hydroxy protons of a phosphoric acid ester [ROPO(OH)$_2$] as 2.1 and 6.9. Because both pK_a values are *less* than the physiological pH of 7.4, both OH groups are deprotonated and the predominant species at this pH is the dianion of the monophosphate. Thus, glucose 6-phosphate and glyceraldehyde 3-phosphate, two intermediates in glucose metabolism, are monophosphate dianions.

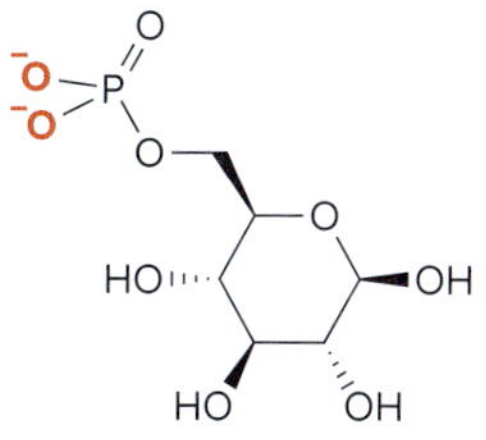

glucose 6-phosphate

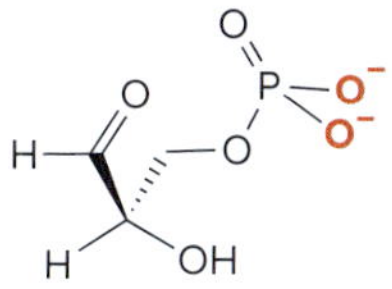

glyceraldehyde 3-phosphate

Problem 15.19 Use the pK_a values of pyrophosphoric acid (0.9, 2.1, 6.7, and 9.3) to approximate the pK_a values for the nucleotide ADP (adenosine 5'-diphosphate), an intermediate formed during metabolism, and draw the predominant form of ADP at physiological pH.

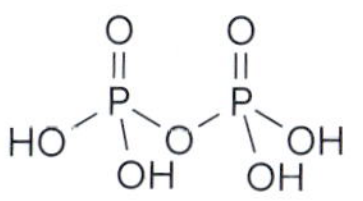

pyrophosphoric acid

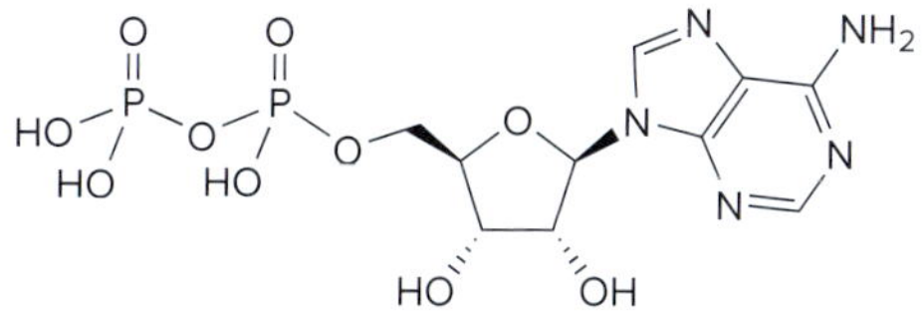

adenosine 5'-diphosphate
ADP

15.12 Amino Acids

Chapter 23 discusses the conversion of amino acids to proteins.

Amino acids, one of four kinds of small biomolecules that have important biological functions in the cell (Section 3.9), also undergo proton transfer reactions.

15.12A Introduction

Amino acids contain two functional groups—an amino group (NH_2) and a carboxy group (COOH). In most naturally occurring amino acids, the amino group is bonded to the α carbon, so they are called **α-amino acids.** Amino acids are the building blocks of proteins, biomolecules that comprise muscle, hair, fingernails, and many other biological tissues.

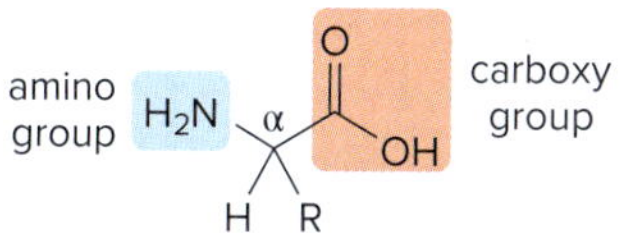

α-amino acid

As we learned in Section 3.9A, an amino acid is both an acid and a base.

- The NH_2 group has a nonbonded electron pair, making it a base.
- The COOH group has an acidic proton, making it an acid.

Amino acids are never uncharged neutral compounds. They exist as salts, so they have very high melting points and are very soluble in water. Proton transfer from the acidic carboxy group to the basic amino group forms a **zwitterion,** which contains both a positive and a negative charge.

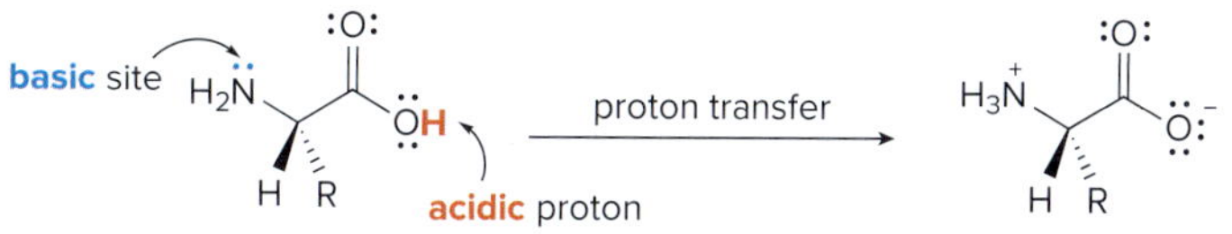

zwitterion

Humans can synthesize only 10 of the 20 amino acids needed for protein synthesis. The remaining 10, called **essential amino acids,** must be obtained from the diet and consumed on a regular, almost daily basis. Vegetarian diets must be carefully balanced to obtain all the essential amino acids. Grains—wheat, rice, and corn—are low in lysine (Figure 23.2), and legumes—beans, peas, and peanuts—are low in methionine, but a combination of these foods provides all the needed amino acids. Thus, a diet of corn tortillas and beans, or rice and tofu, provides all essential amino acids. A peanut butter sandwich on wheat bread does, too. *Brent Hofacker/123RF*

The 20 amino acids that occur naturally in proteins differ in the identity of the R group bonded to the α carbon. **The simplest amino acid, called glycine, has R = H.** When the R group is any other substituent, **the α carbon is a stereogenic center,** and there are two possible enantiomers.

glycine
no stereogenic centers

L amino acid
Only this isomer occurs in proteins.

D amino acid

Amino acids exist in nature as only one of these enantiomers. Except when the R group is CH_2SH, the stereogenic center on the α carbon has the *S* configuration. An older system of nomenclature names the **naturally occurring enantiomer of an amino acid as the L isomer, and its unnatural enantiomer the D isomer.**

The R group of an amino acid can be H, alkyl, or an alkyl chain containing an N, O, or S atom. Representative examples are listed in Table 15.5. All amino acids have common names, which are abbreviated by a three-letter or one-letter designation. For example, glycine is often written as the three-letter abbreviation **Gly,** or the one-letter abbreviation **G.** These abbreviations are also given in Table 15.5. A complete list of the 20 naturally occurring amino acids is found in Figure 23.2. The pK_a values listed in Table 15.5 are discussed further in Section 15.12B.

Table 15.5 Representative Amino Acids

Amino acid	Name	Abbreviations		pK_a (CO_2H)	pK_a (NH_3^+)
	glycine	Gly	G	2.35	9.78
	alanine	Ala	A	2.35	9.87
	serine	Ser	S	2.21	9.15
	valine	Val	V	2.29	9.72
	methionine	Met	M	2.28	9.21
	glutamine	Gln	Q	2.17	9.13

15.12 Amino Acids

Problem 15.20 Draw both enantiomers of the amino acid homocysteine, and label the stereogenic center as R or S. A high blood level of homocysteine, which is formed from methionine by loss of a methyl group, is considered a risk factor for coronary artery disease.

homocysteine

15.12B Acid–Base Properties

Amino acids resemble the dicarboxylic acids discussed in Section 15.8, in that they possess two functional groups—the carboxy group (CO_2H) and the protonated amino group (NH_3^+)—that can lose protons, so two pK_a values are reported. Alanine, [pK_a (CO_2H) = 2.35 and pK_a (NH_3^+) = 9.87] is a representative example from Table 15.5.

- The pK_a of protonated alanine, containing a carboxy group (CO_2H) and a protonated amino group (NH_3^+), is 2.35. Loss of a proton from the CO_2H group yields the zwitterionic form.
- The pK_a of the zwitterion of alanine is 9.87. Loss of a proton from the NH_3^+ group yields an anionic form of alanine, in which both functional groups are now deprotonated.

> Some amino acids contain an additional functional group in the side chain that can be protonated or deprotonated in aqueous solution. The discussion in Section 15.12B is limited to the 13 amino acids that contain only two ionizable functional groups.

protonated form of alanine
pK_a = 2.35

zwitterion

zwitterion
pK_a = 9.87

anionic form of alanine

When alanine is dissolved in an aqueous solution, we can use the Henderson–Hasselbalch equation to determine which functional groups retain their protons and which are deprotonated.

- The ionization of the acidic and basic functional groups of an amino acid depends on the pH of the solution in which it is dissolved.

To determine the state of the amino acid, we use the pK_a values and consider the effect of pH on each functional group separately.

- When the pH = 1, the pH of the solution is *less* than both pK_a values, so both functional groups are protonated, and the amino acid has a net +1 charge.
- When the pH = 12, the pH of the solution is *higher* than both pK_a values, so both functional groups are deprotonated, and the amino acid has a net −1 charge.

alanine in strong acid
form at pH = 1

alanine in strong base
form at pH = 12

What happens between pH 1 and 12?

The carboxy group: As the pH is gradually increased above 1, the more acidic CO_2H group is deprotonated. **When the pH = pK_a (CO_2H) = 2.35, the concentrations of the protonated and zwitterionic forms of alanine are equal.** When the pH is increased further and the solution is more basic than the pK_a of CO_2H, the CO_2H group is largely deprotonated, and the zwitterionic form predominates.

predominant form at **pH < 2.35** ←equal amounts at pH = 2.35→ The carboxylate anion predominates at **pH > 2.35.**

The NH_3^+ group: When the pH is less than the pK_a of NH_3^+ (< 9.87), the NH_3^+ group remains protonated. **When the pH = pK_a (NH_3^+) = 9.87, the concentration of the zwitterionic and anionic forms of alanine are equal.** Above pH 9.87, the solution is more basic than the pK_a of NH_3^+, the NH_3^+ group is largely deprotonated, and the anionic form predominates.

The NH_3^+ group predominates at **pH < 9.87.** ←equal amounts at pH = 9.87→ predominant form at **pH > 9.87**

Thus, **alanine exists in one of three different forms depending on the pH of the solution in which it is dissolved.** If the pH of a solution is gradually increased from 1 to 12, the following process occurs (Figure 15.8).

- At low pH alanine has a net (+) charge (form **B**).
- As the pH is increased to ~6, the carboxy group is deprotonated, and the amino acid exists as a zwitterion with no overall charge (form **A**).
- At high pH, the ammonium cation is deprotonated, and the amino acid has a net (−) charge (form **C**).

Problem 15.21 Draw the positively charged, neutral, and negatively charged forms for the amino acid glycine. Which species predominates at pH 11? Which species predominates at pH 1?

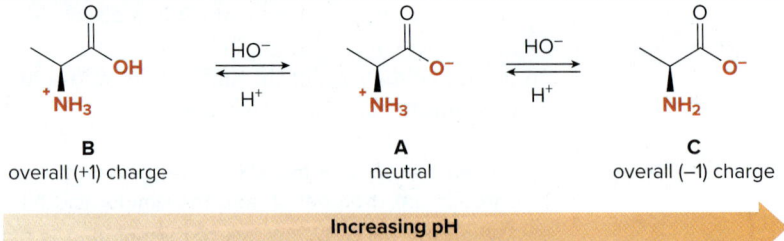

Figure 15.8
Summary of the acid–base reactions of alanine

B
overall (+1) charge

A
neutral

C
overall (−1) charge

Increasing pH

15.12C Isoelectric Point

As we learned in Section 15.12B, a protonated amino acid has at least two different protons that can be removed, so a pK_a value is reported for each of these protons. Table 23.1 lists these values for all 20 amino acids.

- The pH at which the amino acid exists primarily in its neutral form is called its *isoelectric point,* abbreviated as p*I*.

15.13 Nitriles

For the amino acids listed in Table 15.5, the isoelectric point is the average of both pK_a values of an amino acid:

$$\text{Isoelectric point} = pI = \frac{pK_a(\text{COOH}) + pK_a(\text{NH}_3^+)}{2}$$

For alanine: $\quad pI = \dfrac{2.35 + 9.87}{2} = \dfrac{6.12}{pI \text{ (alanine)}}$

Problem 15.22 The pK_a values for the carboxy and ammonium protons of valine are 2.29 and 9.72, respectively. What is the isoelectric point of valine? Draw the structure of valine at its isoelectric point.

15.13 Nitriles

We end Chapter 15 with the chemistry of nitriles. Nitriles are readily prepared by S_N2 substitution reactions of unhindered methyl and 1° alkyl halides with ^-CN. This reaction adds one carbon to the alkyl halide and **forms a new carbon–carbon bond.**

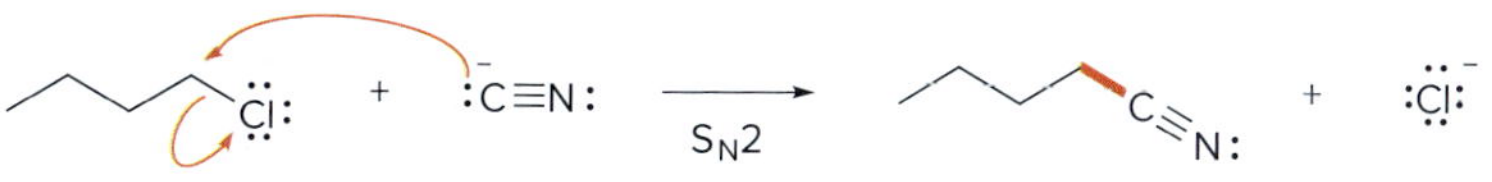

new C–C bond in red

Because a nitrile contains an electrophilic carbon atom that is part of a multiple bond but no leaving group, a nitrile reacts with nucleophiles by a **nucleophilic addition reaction.** The nature of the nucleophile determines the structure of the product.

$$\overset{\delta+}{R}-\overset{\delta-}{C}\equiv N:$$

Nucleophiles attack here.

The reactions of nitriles with water, hydride, and organometallic reagents as nucleophiles are as follows:

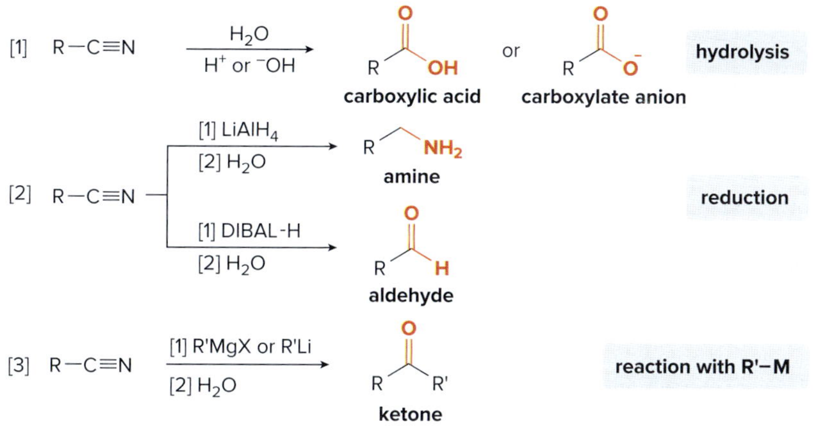

[1] $R-C\equiv N \xrightarrow{H_2O,\ H^+ \text{ or } ^-OH}$ carboxylic acid or carboxylate anion — **hydrolysis**

[2] $R-C\equiv N$
   - $\xrightarrow{[1]\ LiAlH_4,\ [2]\ H_2O}$ amine ($R-CH_2-NH_2$)
   - $\xrightarrow{[1]\ DIBAL-H,\ [2]\ H_2O}$ aldehyde — **reduction**

[3] $R-C\equiv N \xrightarrow{[1]\ R'MgX \text{ or } R'Li,\ [2]\ H_2O}$ ketone — **reaction with R'–M**

15.13A Hydrolysis of Nitriles

Nitriles are hydrolyzed with water in the presence of acid or base to yield **carboxylic acids** or **carboxylate anions**. In this reaction, the three C–N bonds are replaced by three C–O bonds.

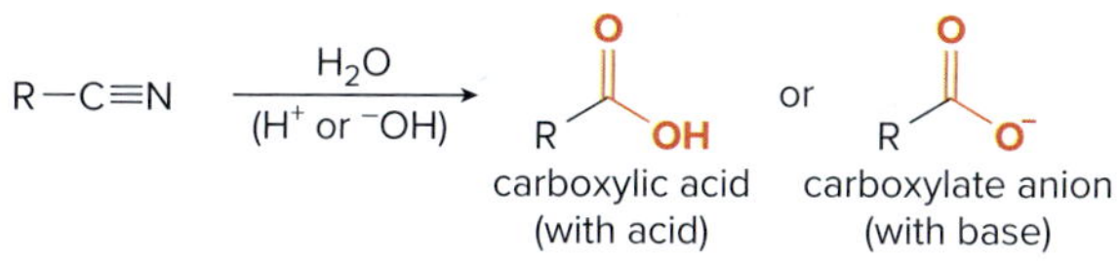

$$R-C\equiv N \xrightarrow{H_2O\ (H^+ \text{ or } ^-OH)} \text{carboxylic acid (with acid) or carboxylate anion (with base)}$$

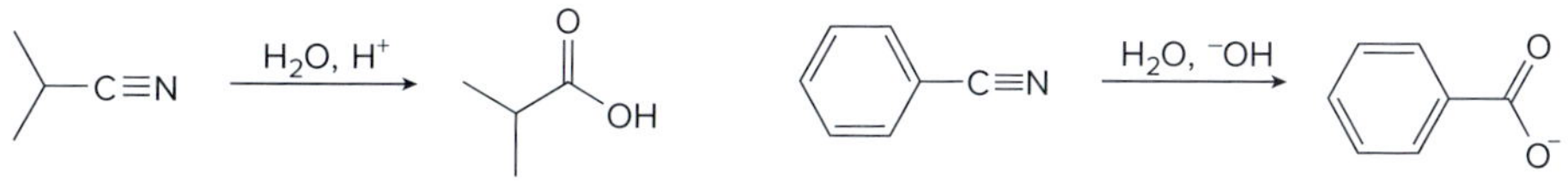

The mechanism of this reaction involves the formation of an **amide tautomer.** Two tautomers can be drawn for any carbonyl compound, and those for a 1° amide are as follows:

- C=N
- O–H bond

imidic acid tautomer

- C=O
- N–H bond

amide tautomer
more stable form

Recall from Chapter 10 that tautomers are constitutional isomers that differ in the location of a double bond and a proton.

- The amide form is the more stable tautomer, having a C=O and an N–H bond.
- The imidic acid tautomer is the less stable form, having a C=N and an O–H bond.

The imidic acid and amide tautomers are interconverted by treating with acid or base, analogous to the keto–enol tautomers of other carbonyl compounds. In fact, the two amide tautomers are exactly the same as keto–enol tautomers except that a nitrogen atom replaces a carbon atom bonded to the carbonyl group.

The mechanism of nitrile hydrolysis in both acid and base consists of two parts: [1] **nucleophilic addition** to form the imidic acid tautomer followed by **tautomerization** to form the amide, and [2] **hydrolysis of the amide** to form RCO_2H or RCO_2^-. The mechanism is shown for the basic hydrolysis of RCN to RCO_2^- (Mechanism 15.1).

Mechanism 15.1 Hydrolysis of a Nitrile in Base

Part [1] Conversion of a nitrile to a 1° amide

1–2 Nucleophilic attack of ⁻OH followed by protonation forms an **imidic acid.**
3–4 **Tautomerization** occurs by a two-step sequence—deprotonation followed by protonation.

Part [2] Hydrolysis of the 1° amide to a carboxylate anion

Conversion of the amide to the carboxylate occurs by a three-step sequence that will be discussed in Chapter 16 (Mechanism 16.9).

Problem 15.23 Draw the products of each reaction.

a. CH₃CH₂Br + NaCN →

b. o-dicyanobenzene + H₂O, H⁺ →

c. (pentyl-CN) + H₂O, ⁻OH →

Problem 15.24 Draw a tautomer of each compound.

a. H₂N-C(=O)-(1-cyclohexenyl)   b. CH₃-CH(OH)-CH₂-C(=O)-NH-CH₃   c. CH₃-C(=NH)-OH

15.13B Reduction of Nitriles

Nitriles are reduced with metal hydride reagents to form either 1° amines or aldehydes, depending on the reducing agent.

- Treatment of a nitrile with LiAlH₄ followed by H₂O adds two equivalents of H₂ across the triple bond, forming a 1° amine.

$$(CH_3)_3C-C{\equiv}N \xrightarrow[\text{[2] H}_2\text{O}]{\text{[1] LiAlH}_4} (CH_3)_3C-CH_2-NH_2$$

- Treatment of a nitrile with a milder reducing agent such as DIBAL-H followed by H₂O forms an aldehyde.

$$Ph-C{\equiv}N \xrightarrow[\text{[2] H}_2\text{O}]{\text{[1] DIBAL-H}} Ph-CHO$$

The mechanism of both reactions involves **nucleophilic addition of hydride (H⁻) to the polarized C–N triple bond.** Mechanism 15.2 illustrates that reduction of a nitrile to an amine requires addition of two equivalents of H:⁻ from LiAlH₄. It is likely that intermediate nitrogen anions complex with AlH₃ (formed in situ) to facilitate the addition. Protonation of the dianion in Step [4] forms the amine.

Mechanism 15.2 Reduction of a Nitrile with LiAlH₄

R−C≡N: →[H–ĀlH₃, 1] R-CH=N:⁻ →[AlH₃] R-CH=N-ĀlH₃ →[H–ĀlH₃, 2] (+ AlH₃) →[3] R-CH₂-N(ĀlH₃)(ĀlH₃) →[2 H₂O, 4] R-CH₂-NH₂ + 2 H₃ĀlOH

1–2 Addition of one equivalent of H:⁻ from LiAlH₄ forms an intermediate with **one new C–H bond**, which complexes with AlH₃.
3–4 Nucleophilic attack of a second equivalent of H:⁻ and complexation with AlH₃ form a dianion, which reacts with water to form **two new N–H bonds,** giving the **1° amine.**

With **DIBAL-H,** nucleophilic addition of one equivalent of hydride forms an anion (Step [1]), which is protonated with water to generate an **imine,** as shown in Mechanism 15.3. As described in Section 14.12, imines are hydrolyzed in water to form aldehydes. Mechanism 15.3 is written without complexation of aluminum with the anion formed in Step [1], to emphasize the identity of intermediates formed during reduction.

Mechanism 15.3 Reduction of a Nitrile with DIBAL-H

R—C≡N: →[H—AlR₂, 1] R-CH=N:⁻ →[H—OH, 2] R-CH=NH (imine) →[H₂O: (Mechanism 14.5)] R-CHO + :OH⁻

1. Addition of H:⁻ from DIBAL-H (drawn as R₂AlH) forms the new C—H bond.
2. Protonation forms an **imine,** which is hydrolyzed to an aldehyde by a stepwise sequence that is the reverse of Mechanism 14.5.

Problem 15.25 Draw the product of each reaction.

a. 3-methoxyphenylethyl bromide [1] NaCN [2] LiAlH₄ [3] H₂O

b. cyclohexyl-C≡N [1] DIBAL-H [2] H₂O

15.13C Addition of Grignard and Organolithium Reagents to Nitriles

Both Grignard and organolithium reagents react with nitriles to form ketones with a new carbon–carbon bond.

C₆H₅—C≡N [1] CH₃CH₂MgBr [2] H₂O → C₆H₅—CO—CH₂CH₃

The reaction occurs by nucleophilic addition of the organometallic reagent to the polarized C—N triple bond to form an anion (Step [1]), which is protonated with water to form an **imine.** Water then hydrolyzes the imine, replacing the C=N by C=O as described in Section 14.12. The final product is a ketone with a new carbon–carbon bond (Mechanism 15.4).

Mechanism 15.4 Addition of Grignard and Organolithium Reagents (R—M) to Nitriles

R—C≡N: →[R'—M, 1] R'C(R)=N:⁻ →[H—OH, 2] R'C(R)=NH (imine) →[H₂O: (Mechanism 14.5)] R'C(R)=O + :OH⁻

1. Addition of R:⁻ from R'M (M = MgX or Li) forms a **new C—C bond.**
2. Protonation forms an **imine,** which is hydrolyzed to a ketone by a stepwise sequence that is the reverse of Mechanism 14.5.

Problem 15.26 Draw the products of each reaction.

a. 2-methoxybenzonitrile [1] CH₃CH₂MgCl [2] H₂O

b. CH₃CH₂CH(CN)CH₂CH₃ [1] C₆H₅Li [2] H₂O

Problem 15.27 What reagents are needed to convert phenylacetonitrile ($C_6H_5CH_2CN$) to each compound: (a) $C_6H_5CH_2COCH_3$; (b) $C_6H_5CH_2COC(CH_3)_3$; (c) $C_6H_5CH_2CHO$; (d) $C_6H_5CH_2COOH$?

Problem 15.28 Outline two different ways that butan-2-one can be prepared from a nitrile and a Grignard reagent.

Chapter 15 REVIEW

KEY CONCEPTS

[1] Factors that affect acidity

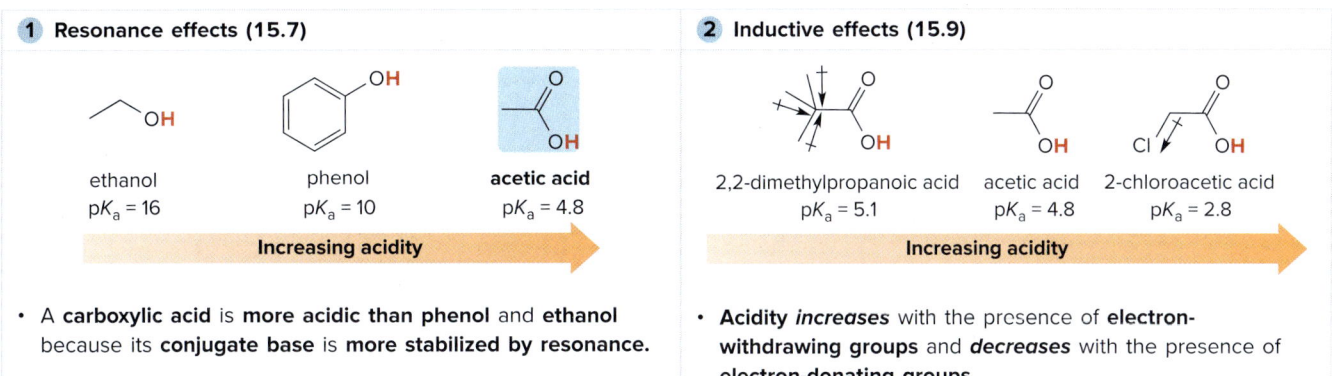

- A **carboxylic acid** is **more acidic than phenol** and **ethanol** because its **conjugate base** is **more stabilized by resonance**.

- **Acidity** *increases* with the presence of **electron-withdrawing groups** and *decreases* with the presence of **electron-donating groups**.

Try Problems 15.35, 15.67.

[2] Using the Henderson–Hasselbalch equation to determine the predominant species at a given pH (15.8); example: propanoic acid ($CH_3CH_2CO_2H$, $pK_a = 4.88$) and its conjugate base $CH_3CH_2CO_2^-$

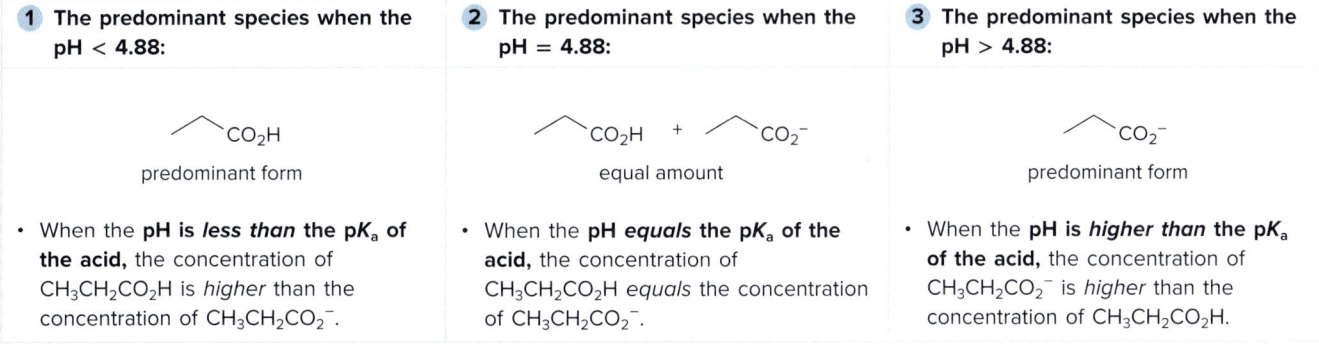

① The predominant species when the pH < 4.88:	② The predominant species when the pH = 4.88:	③ The predominant species when the pH > 4.88:
predominant form	equal amount	predominant form
• When the **pH** is *less than* the **pK_a** of the acid, the concentration of $CH_3CH_2CO_2H$ is *higher* than the concentration of $CH_3CH_2CO_2^-$.	• When the **pH** *equals* the **pK_a** of the acid, the concentration of $CH_3CH_2CO_2H$ *equals* the concentration of $CH_3CH_2CO_2^-$.	• When the **pH** is *higher than* the **pK_a** of the acid, the concentration of $CH_3CH_2CO_2^-$ is *higher* than the concentration of $CH_3CH_2CO_2H$.

See Sample Problem 15.3. Try Problems 15.47–15.49.

[3] Positively charged, neutral, and negatively charged forms of an amino acid (15.12); example: phenylalanine; pK_a (α-COOH) = 2.58, pK_a (α-NH_3^+) = 9.24, pI = 5.91

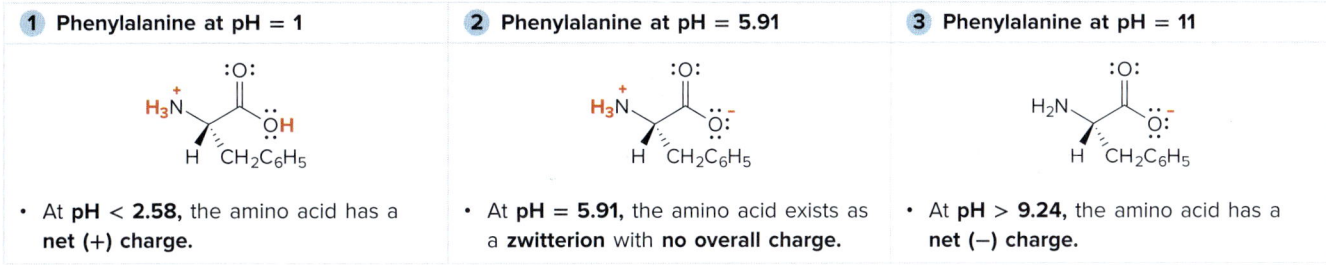

① Phenylalanine at pH = 1	② Phenylalanine at pH = 5.91	③ Phenylalanine at pH = 11
• At **pH < 2.58**, the amino acid has a **net (+) charge**.	• At **pH = 5.91**, the amino acid exists as a **zwitterion** with **no overall charge**.	• At **pH > 9.24**, the amino acid has a **net (−) charge**.

See Figure 15.8. Try Problems 15.54, 15.56.

KEY REACTIONS

[1] Nitrile Synthesis

$$\text{R-CH}_2\text{-X} \xrightarrow[S_N2]{\text{NaCN}} \text{R-CH}_2\text{-CN} + \text{X}^-$$

X = Cl, Br, I   nitrile
(15.13)

Try Problems 15.43d, 15.46a.

[2] Reactions of Nitriles

1. nitrile $\xrightarrow[\Delta]{\text{H}_2\text{O}, \text{H}^+ \text{ or } ^-\text{OH}}$ carboxylic acid (with acid) or carboxylate (with base) — hydrolysis (15.13A)

2. nitrile $\xrightarrow[\text{[2] 2 HO-H}]{\text{[1] Li}^+ \text{H-AlH}_3}$ 1° amine — reduction (15.13B)

3. nitrile $\xrightarrow[\text{[2] H}_2\text{O}]{\text{[1] DIBAL-H}}$ aldehyde — reduction (15.13B)

4. nitrile $\xrightarrow[\text{[2] H}_2\text{O}]{\text{[1] CH}_3\text{MgX or CH}_3\text{Li}}$ ketone (15.13C)

Try Problems 15.30b; 15.43c, d; 15.45; 15.46a, c, d.

KEY SKILLS

[1] Drawing the products of an acid–base reaction involving a carboxylic acid (15.7)

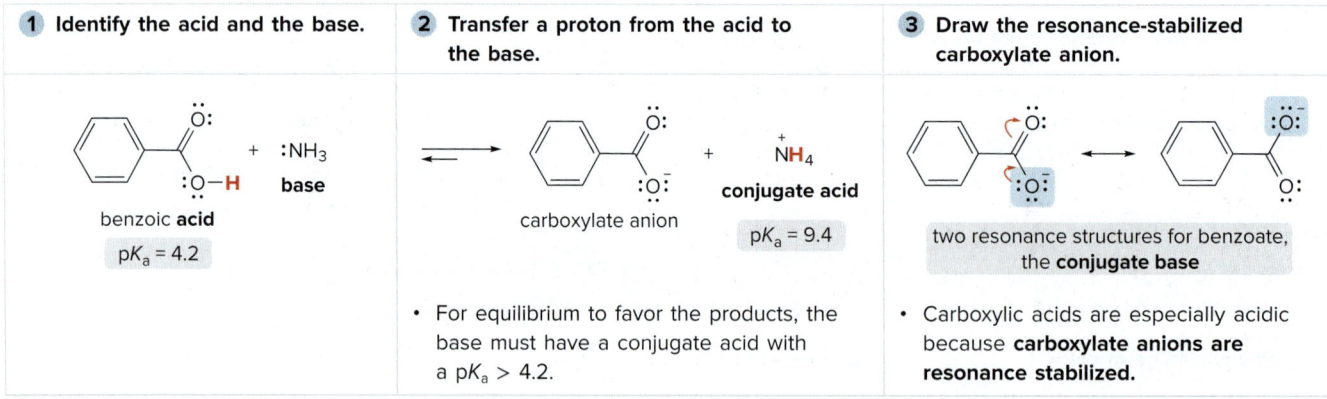

1 Identify the acid and the base.	2 Transfer a proton from the acid to the base.	3 Draw the resonance-stabilized carboxylate anion.
benzoic **acid** $pK_a = 4.2$ + :NH$_3$ **base**	carboxylate anion + $^+$NH$_4$ conjugate acid $pK_a = 9.4$	two resonance structures for benzoate, the **conjugate base**
	• For equilibrium to favor the products, the base must have a conjugate acid with a $pK_a > 4.2$.	• Carboxylic acids are especially acidic because **carboxylate anions are resonance stabilized**.

See Table 15.4. Try Problems 15.29b, 15.34c.

[2] Separating a carboxylic acid from an alcohol by extraction (15.10)

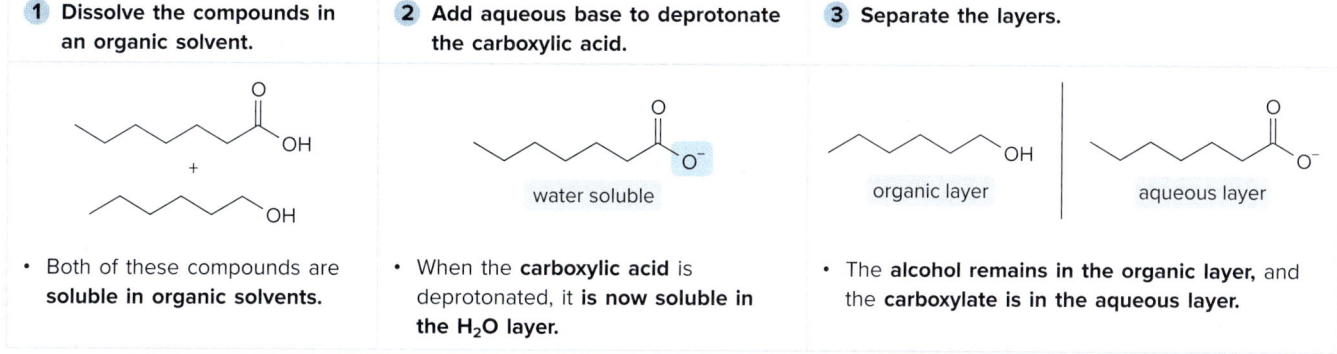

See Sample Problem 15.4, Figure 15.7. Try Problems 15.50–15.52.

[3] Devising a synthesis; example: heptan-4-one from 1-chloropropane (15.13C)

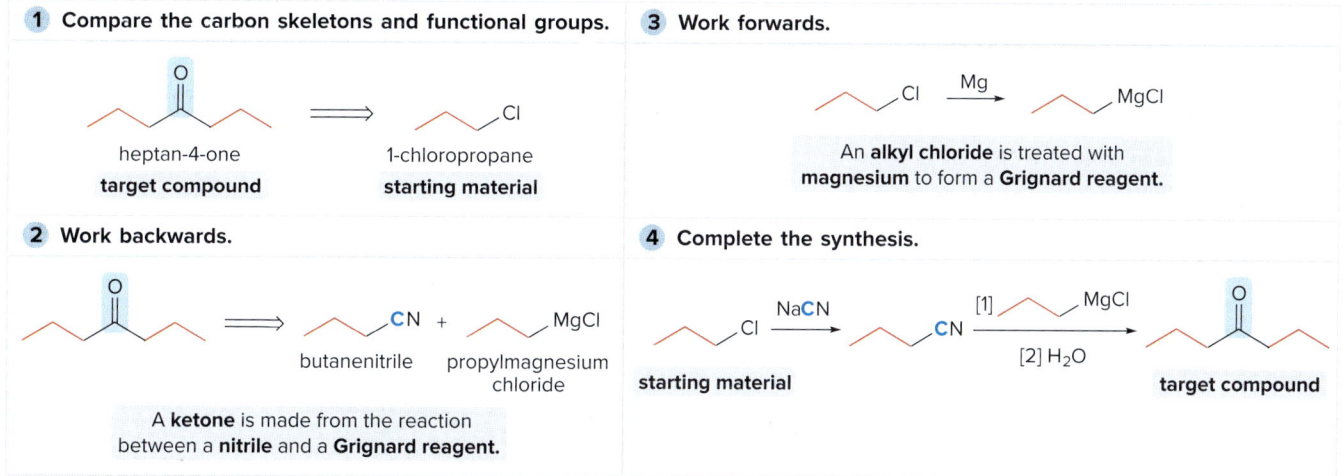

Try Problems 15.60, 15.61.

PROBLEMS

Problems Using Three-Dimensional Models

15.29 Answer each question for **A** and **B** depicted in the ball-and-stick models.

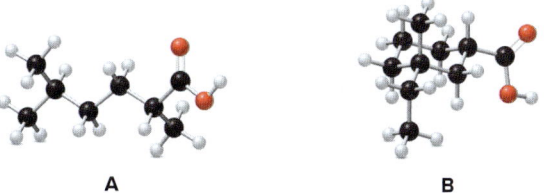

a. What is the IUPAC name for each compound?
b. What product is formed when each compound is treated with NaOH?
c. Name the products formed in part (b).
d. Draw the structure of an isomer that is at least 10^5 times less acidic than each compound.

15.30 (a) Give an acceptable name for compound **C**. (b) Draw the organic products formed when **C** is treated with each reagent: [1] H_3O^+; [2] ⁻OH, H_2O; [3] $CH_3CH_2CH_2MgBr$ (excess), then H_2O; [4] $LiAlH_4$, then H_2O.

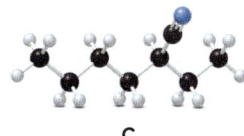

Nomenclature

15.31 Give the IUPAC name for each compound.

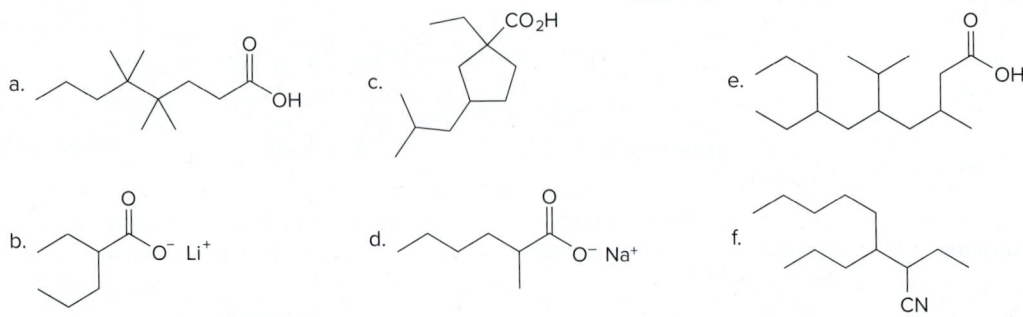

15.32 Draw the structure corresponding to each name.
   a. 3,3-dimethylpentanoic acid
   b. 4-chloro-3-phenylheptanoic acid
   c. (R)-2-chloropropanoic acid
   d. potassium acetate
   e. sodium α-bromobutyrate
   f. 2,2-dichloropentanedioic acid
   g. 4-isopropyl-2-methyloctanedioic acid
   h. 3,3-dimethylpentanenitrile
   i. 4,5-diethyl-2-isopropylnonanenitrile

Acid–Base Reactions; General Questions on Acidity

15.33 Using the pK_a table in Appendix C, determine whether each of the following bases is strong enough to deprotonate the three compounds listed below. Bases: [1] $^-$OH; [2] $CH_3CH_2^-$; [3] $^-NH_2$; [4] NH_3; [5] $HC{\equiv}C^-$.

a. 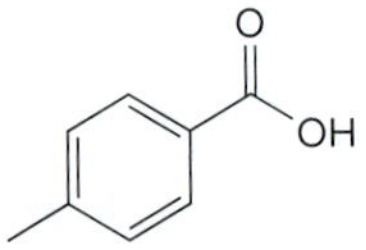 $pK_a = 4.3$ b. 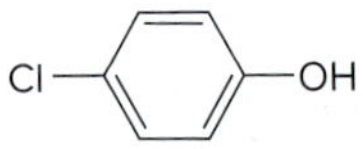 $pK_a = 9.4$ c. 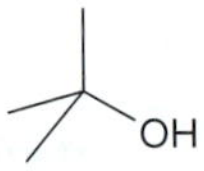 $pK_a = 18$

15.34 Draw the products of each acid–base reaction, and using the pK_a table in Appendix C, determine if equilibrium favors the reactants or products.

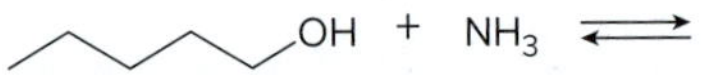

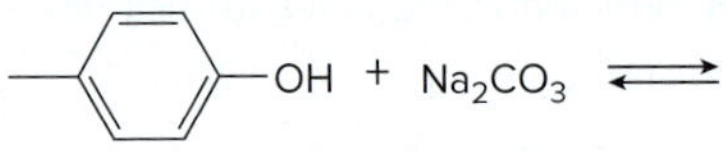

15.35 Caftaric acid is found in grapes, wine, and raisins. Rank the labeled protons in caftaric acid in order of increasing acidity.

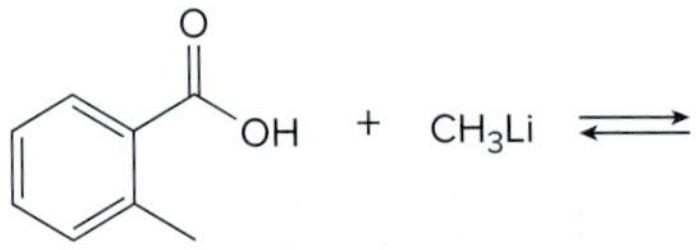

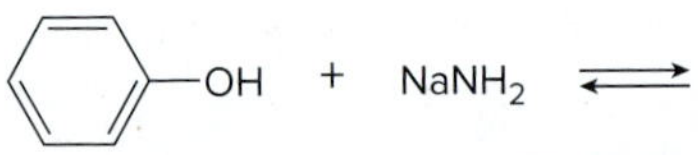

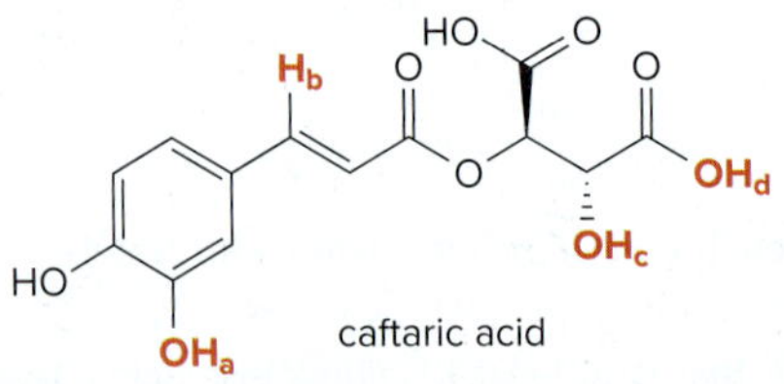

caftaric acid

15.36 Rank the following compounds in order of increasing basicity.

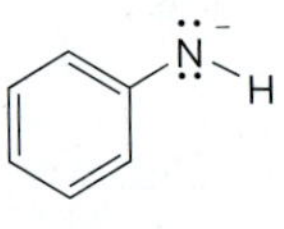

A 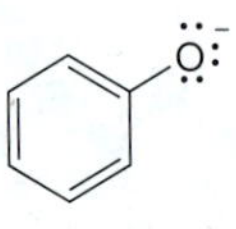   B 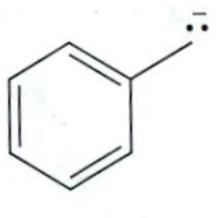   C 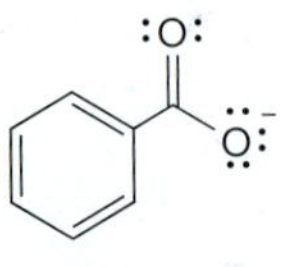   D

15.37 Although codeine occurs in low concentration in the opium poppy, most of the codeine used in medicine is prepared from morphine (the principal component of opium) by the following reaction. Explain why selective methylation occurs at only one OH in morphine to give codeine. Codeine is a less potent and less addictive analgesic than morphine.

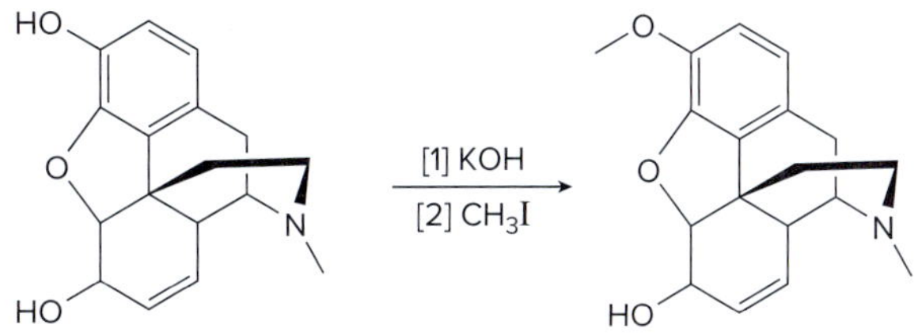

morphine — [1] KOH, [2] CH_3I → codeine

15.38 Explain why the pK_a of compound **A** is lower than the pK_a's of both compounds **B** and **C**.

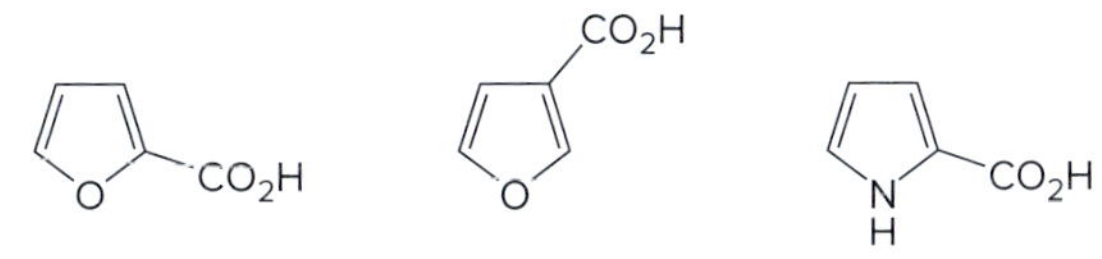

A $pK_a = 3.2$ **B** $pK_a = 3.9$ **C** $pK_a = 4.4$

15.39 Phthalic acid and isophthalic acid have protons on two carboxy groups that can be removed with base. (a) Explain why the pK_a for loss of the first proton (pK_{a1}) is lower for phthalic acid than isophthalic acid. (b) Explain why the pK_a for loss of the second proton (pK_{a2}) is higher for phthalic acid than isophthalic acid.

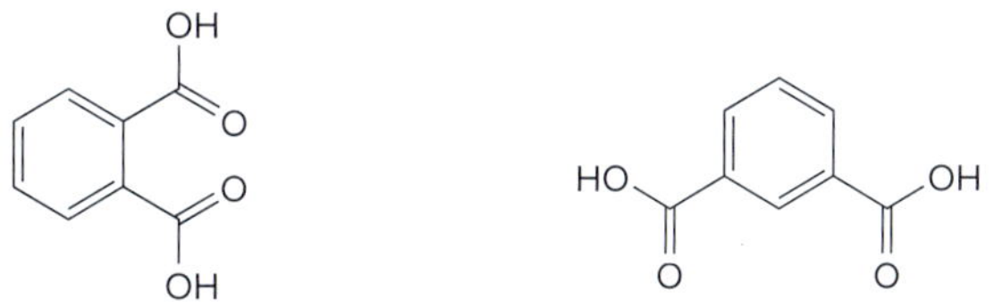

phthalic acid
$pK_{a1} = 2.9$
$pK_{a2} = 5.4$

isophthalic acid
$pK_{a1} = 3.7$
$pK_{a2} = 4.6$

15.40 Explain this result: Acetic acid (CH_3COOH), labeled at its OH oxygen with the uncommon ^{18}O isotope (shown in red), was treated with aqueous base, and then the solution was acidified. Two products having the ^{18}O label at different locations were formed.

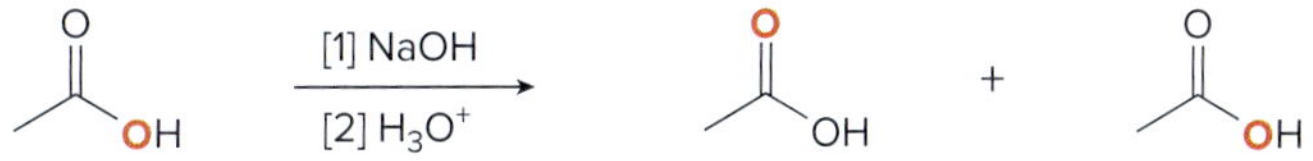

15.41 Draw all resonance structures of the conjugate bases formed by removal of the labeled protons (H_a, H_b, and H_c) in cyclohexane-1,3-dione and acetanilide. For each compound, rank these protons in order of increasing acidity and explain the order you chose.

a. 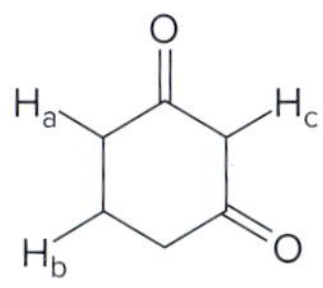 cyclohexane-1,3-dione   b. 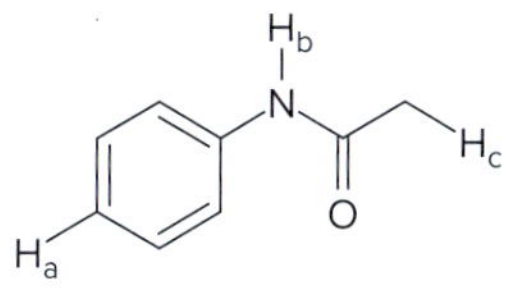 acetanilide

15.42 The pK_a of acetamide (CH_3CONH_2) is 16. Draw the structure for its conjugate base, and explain why acetamide is less acidic than CH_3COOH.

General Reactions

15.43 Draw the organic products formed in each reaction.

a. cyclohexyl-CH$_2$OH $\xrightarrow{\text{CrO}_3,\ \text{H}_2\text{SO}_4,\ \text{H}_2\text{O}}$

b. CH$_3$(CH$_2$)$_6$CH$_2$OH $\xrightarrow{\text{Na}_2\text{Cr}_2\text{O}_7,\ \text{H}_2\text{SO}_4,\ \text{H}_2\text{O}}$

c. C$_6$H$_5$CN $\xrightarrow{\text{[1] CH}_3\text{CH}_2\text{CH}_2\text{MgBr};\ \text{[2] H}_2\text{O}}$

d. CH$_3$CH$_2$CH$_2$CH$_2$Br $\xrightarrow{\text{[1] NaCN};\ \text{[2] H}_2\text{O},\ ^-\text{OH}}$

15.44 Identify the lettered compounds in each reaction sequence.

a. methylenecyclohexane $\xrightarrow{\text{[1] BH}_3;\ \text{[2] H}_2\text{O}_2,\ \text{HO}^-}$ **A** $\xrightarrow{\text{CrO}_3,\ \text{H}_2\text{SO}_4,\ \text{H}_2\text{O}}$ **B**

b. HC≡CH $\xrightarrow{\text{[1] NaNH}_2;\ \text{[2] CH}_3\text{I}}$ **C** $\xrightarrow{\text{[1] NaNH}_2;\ \text{[2] CH}_3\text{CH}_2\text{I}}$ **D** $\xrightarrow{\text{[1] O}_3;\ \text{[2] H}_2\text{O}}$ **E** + **F**

c. 1,2-bis(bromomethyl)benzene (o-dibromomethyl) $\xrightarrow{\text{NaCN}}$ **I** $\xrightarrow{\text{[1] DIBAL-H};\ \text{[2] H}_2\text{O}}$ **J** $\xrightarrow{\text{Ag}_2\text{O},\ \text{NH}_4\text{OH}}$ **K**

15.45 Draw the product formed when phenylacetonitrile (C$_6$H$_5$CH$_2$CN) is treated with each reagent.
a. H$_3$O$^+$
b. H$_2$O, $^-$OH
c. [1] CH$_3$MgBr; [2] H$_2$O
d. [1] CH$_3$CH$_2$Li; [2] H$_2$O
e. [1] DIBAL-H; [2] H$_2$O
f. [1] LiAlH$_4$; [2] H$_2$O

15.46 Draw the products of each reaction, and indicate the stereochemistry at all stereogenic centers.

a. (R)-2-bromopentane $\xrightarrow{\text{[1] NaCN};\ \text{[2] H}_2\text{O, H}^+}$

b. (S)-3-bromo-3-methylpentane-type $\xrightarrow{\text{[1] Mg};\ \text{[2] CO}_2;\ \text{[3] H}_3\text{O}^+}$

c. HO-CO-CH$_2$CH$_2$-CH(CH$_3$)-CN (with stereocenter) $\xrightarrow{\text{[1] LiAlH}_4;\ \text{[2] H}_2\text{O}}$

d. trans-1-cyano-2-methoxycyclohexane $\xrightarrow{\text{[1] DIBAL-H};\ \text{[2] H}_2\text{O};\ \text{[3] CH}_3\text{CH}_2\text{NH}_2,\ \text{mild H}^+}$

Henderson–Hasselbalch Equation

15.47 Use the Henderson–Hasselbalch equation to determine the ratio of phenol (C$_6$H$_5$OH, pK_a = 10.0) to phenoxide (C$_6$H$_5$O$^-$) at each of the following pH values: (a) 8; (b) 13; (c) 10; (d) 6.

15.48 The pK_a values for the ionization of the carboxy groups in glutaric acid (**A**) are 4.3 and 5.4.
(a) At what pH will there be an equal amount of **A** and **B**? (b) At what pH will there be an equal amount of **B** and **C**?

A: HOOC-CH$_2$CH$_2$CH$_2$-COOH
B: $^-$OOC-CH$_2$CH$_2$CH$_2$-COOH
C: $^-$OOC-CH$_2$CH$_2$CH$_2$-COO$^-$

15.49 Consider a carboxylic acid (RCO$_2$H, pK_a = 4.8) and a protonated amine (RNH$_3^+$, pK_a = 9). Indicate the predominant form of each compound at the following pH values: (a) 8; (b) 3; (c) 10.

Extraction

15.50 Write out the steps needed to separate hydrocarbon **A** and carboxylic acid **B** by using an extraction procedure.

A: naphthalene
B: 2-naphthoic acid (naphthalene-COOH)

15.51 Because phenol (C$_6$H$_5$OH) is less acidic than a carboxylic acid, it can be deprotonated by NaOH but not by the weaker base NaHCO$_3$. Using this information, write out an extraction sequence that can be used to separate C$_6$H$_5$OH, benzoic acid, and cyclohexanol. Show what compound is present in each layer at each stage of the process, and if it is present in its neutral or ionic form.

15.52 A mixture of **A, B,** and **C** was added to a separatory funnel containing CH_2Cl_2, and an aqueous layer was added. In which layer is each compound dissolved when the aqueous layer consists of (a) pure water; (b) 10% NaOH solution; (c) 10% $NaHCO_3$ solution?

A   B   C

Amino Acids

15.53 Threonine is a naturally occurring amino acid that has two stereogenic centers.

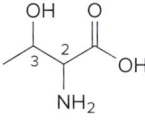
threonine

a. Draw the four possible stereoisomers using wedges and dashed wedges.
b. The naturally occurring amino acid has the 2S,3R configuration at its two stereogenic centers. Which structure does this correspond to?

15.54 Hypoglycin A, an amino acid derivative found in unripened lychee, is an acutely toxic compound that produces seizures, coma, and sometimes death in undernourished children when ingested on an empty stomach (Problem 5.60). (a) Draw the neutral, positively charged, and negatively charged forms of hypoglycin A. (b) Which form predominates at pH = 1, 6, and 11? (c) What is the structure of hypoclycin A at its isoelectric point?

hypoglycin A

15.55 Calculate the isoelectric point for each amino acid.
a. asparagine: pK_a (COOH) = 2.02; pK_a (α-NH_3^+) = 8.80
b. methionine: pK_a (COOH) = 2.28; pK_a (α-NH_3^+) = 9.21

15.56 Glutamic acid is a naturally occurring α-amino acid that contains a carboxy group in its R group side chain. (Glutamic acid is drawn in its neutral form with no charged atoms, a form that does not actually exist at any pH.)

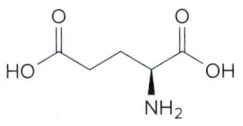
glutamic acid

a. What form of glutamic acid exists at pH = 1?
b. If the pH is gradually increased, what form of glutamic acid exists after one equivalent of base is added? After two equivalents? After three equivalents?
c. Propose a structure of monosodium glutamate, the common flavor enhancer known as MSG.

15.57 Consider γ-aminobutyric acid, a neurotransmitter in the brain that contains an amino group on the γ carbon to the carboxy group. (a) If pK_a (CO_2H) = 4.23 and pK_a (NH_3^+) = 10.43, draw the predominant form(s) of the amino acid at each of the following pH values: [1] 1.21; [2] 12.1; [3] 10.43; [4] 4.23. (b) What is the isoelectric point for this amino acid? (c) Explain why the pK_a of the CO_2H group of γ-aminobutyric acid is considerably higher than the pK_a of the CO_2H group of each amino acid in Table 15.5.

γ-aminobutyric acid

15.58 The amino acid cysteine has three ionizable functional groups with the indicated pK_a values. Draw the predominant form(s) of the amino acid at each of the following pH values: (a) 12; (b) 1; (c) 8; (d) 3.

pK_a = 8.00 → HS   OH ← pK_a = 2.05
             $^+NH_3$ ← pK_a = 10.25
cysteine

Synthesis

15.59 Two methods convert an alkyl halide to a carboxylic acid having one more carbon atom.

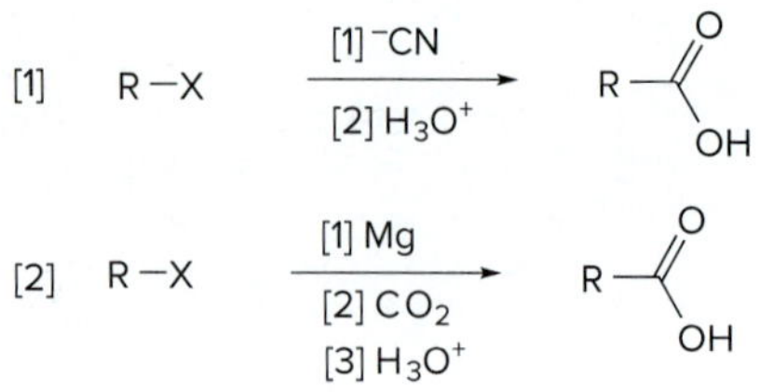

[1] R—X $\xrightarrow{\text{[1] }^-\text{CN}}_{\text{[2] H}_3\text{O}^+}$ R—COOH (Section 15.13)

[2] R—X $\xrightarrow[\text{[3] H}_3\text{O}^+]{\text{[1] Mg, [2] CO}_2}$ R—COOH (Section 13.14A)

Depending on the structure of the alkyl halide, one or both of these methods may be employed. For each alkyl halide, write out a stepwise sequence that converts it to a carboxylic acid with one more carbon atom. If both methods work, draw both routes. If one method cannot be used, state why it can't.

a. CH_3Cl b. 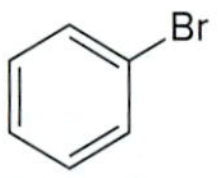 bromobenzene   c. 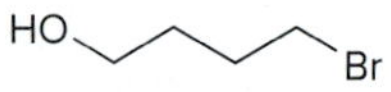 HO–(CH_2)_4–Br

15.60 Synthesize each compound from benzonitrile (C_6H_5CN) as the only organic starting material; that is, every carbon in the product must originate in benzonitrile.

a. 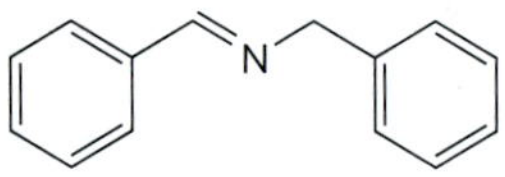 PhCH=N–CH_2Ph   b. 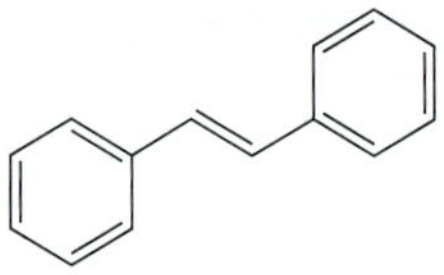 PhCH=CHPh   c. 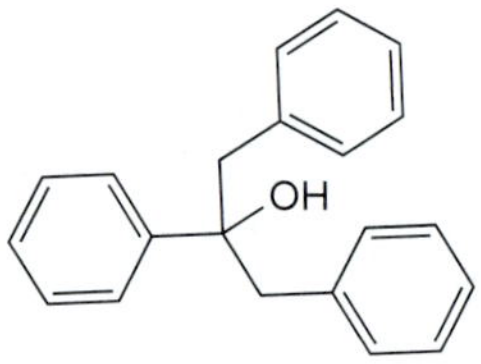 Ph_2C(OH)CH_2Ph (with additional CH_2Ph)

15.61 Devise a synthesis of each compound from the indicated starting material. You may also use any organic compounds with one or two carbons and any needed inorganic reagents.

a. 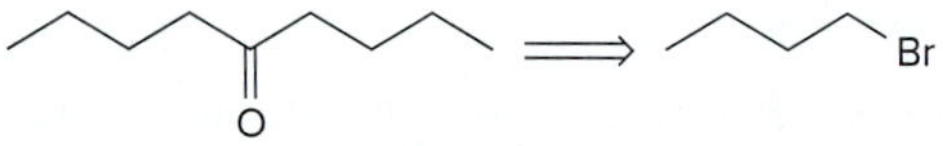 nonan-5-one ⟹ CH_3CH_2CH_2CH_2Br

b. 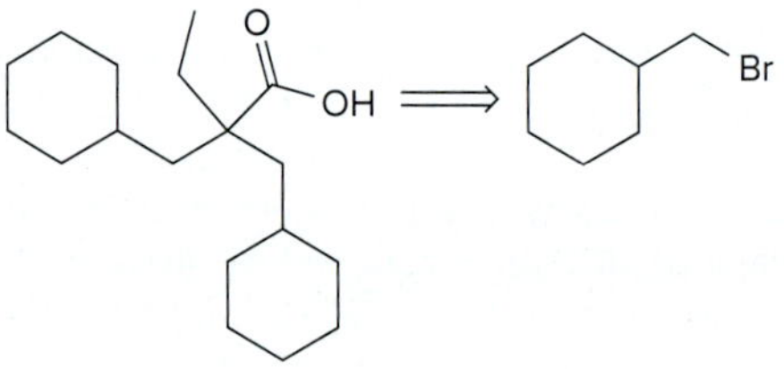 2,2-bis(cyclohexylmethyl)butanoic acid ⟹ cyclohexylmethyl bromide

Spectroscopy

15.62 Identify each compound from its spectral data.

a. Molecular formula: $C_3H_5ClO_2$
   IR: 3500–2500 cm^{-1}, 1714 cm^{-1}
   ^{1}H NMR data: 2.87 (triplet, 2 H), 3.76 (triplet, 2 H), and 11.8 (singlet, 1 H) ppm

b. Molecular formula: $C_8H_8O_3$
   IR: 3500–2500 cm^{-1}, 1710 cm^{-1}
   ^{1}H NMR data: 4.7 (singlet, 2 H), 6.9–7.3 (multiplet, 5 H), and 11.3 (singlet, 1 H) ppm

c. Molecular formula: C_4H_7N
   IR: 2250 cm^{-1}
   ^{1}H NMR data: 1.08 (triplet, 3 H), 1.70 (multiplet, 2 H), and 2.34 (triplet, 2 H) ppm

15.63 Use the ¹H NMR and IR spectra given below to identify the structure of compound **B** (molecular formula $C_4H_8O_2$).

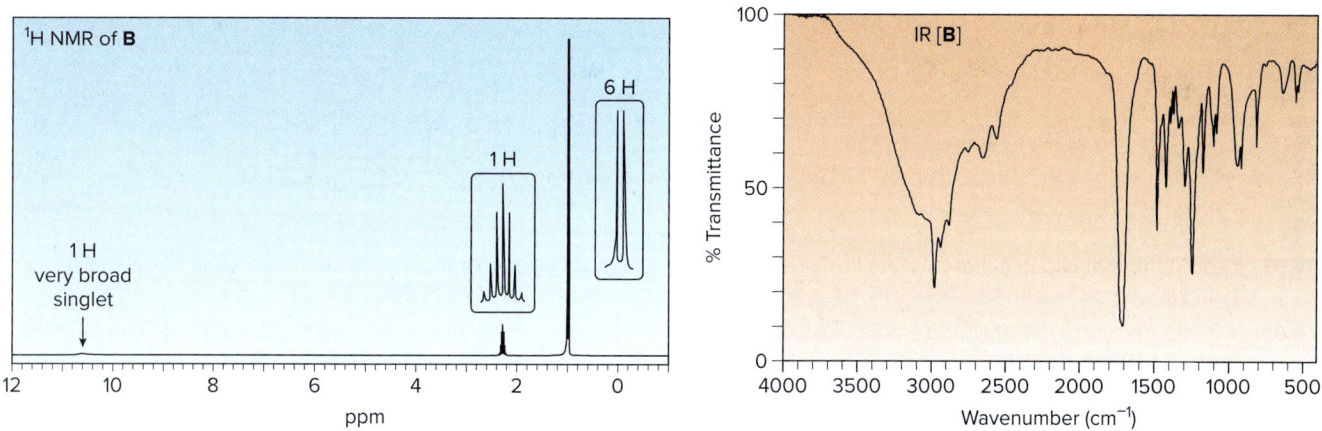

15.64 An unknown compound **C** (molecular formula $C_4H_8O_3$) exhibits IR absorptions at 3600–2500 and 1734 cm^{-1}, as well as the following ¹H NMR spectrum. What is the structure of **C**?

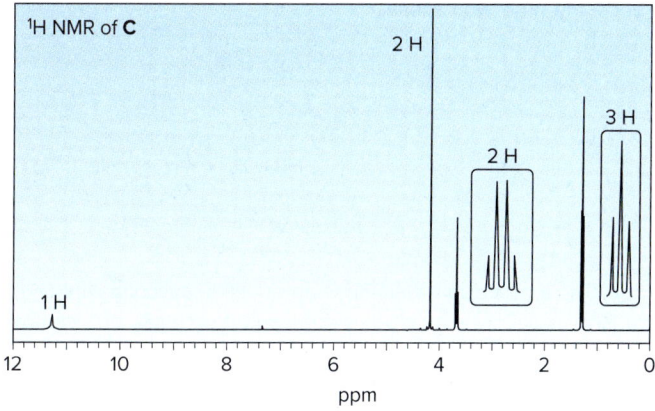

15.65 Propose a structure for **D** (molecular formula $C_9H_9ClO_2$) consistent with the given spectroscopic data.

¹³C NMR signals at 30, 36, 128, 130, 133, 139, and 179 ppm

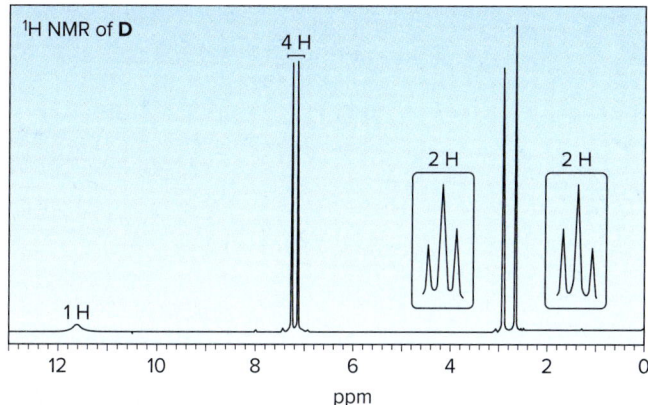

Challenge Problems

15.66 Explain why using one or two equivalents of NaH results in different products in the following reactions.

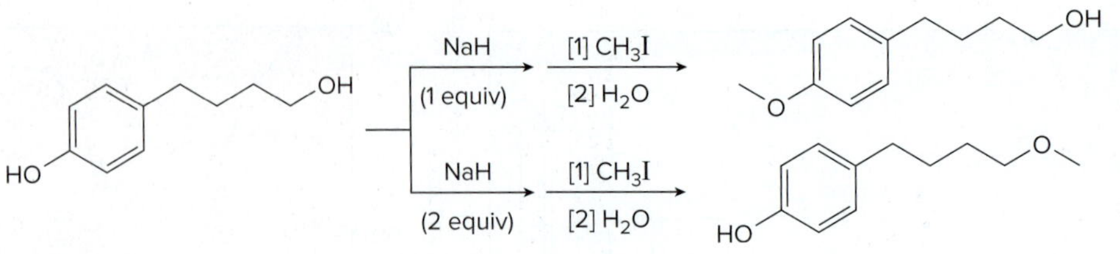

15.67 Explain this statement: Although 2-methoxyacetic acid (CH_3OCH_2COOH) is a stronger acid than acetic acid (CH_3COOH), p-methoxybenzoic acid ($CH_3OC_6H_4COOH$) is a weaker acid than benzoic acid (C_6H_5COOH).

15.68 Although p-hydroxybenzoic acid is less acidic than benzoic acid, o-hydroxybenzoic acid is slightly more acidic than benzoic acid. Explain this result.

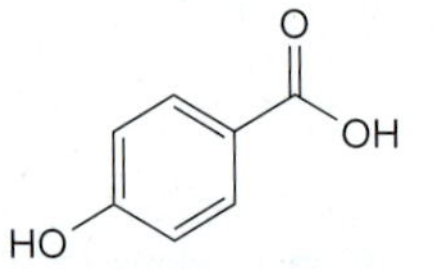

p-hydroxybenzoic acid

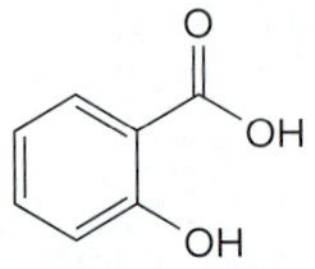

o-hydroxybenzoic acid

15.69 2-Hydroxybutanedioic acid occurs naturally in apples and other fruits. Rank the labeled protons (H_a–H_e) in order of increasing acidity, and explain in detail the order you chose.

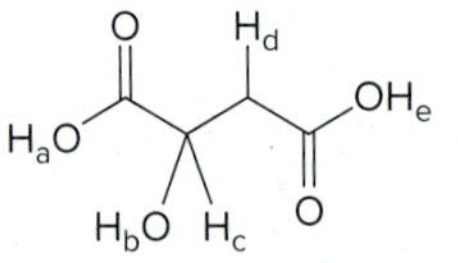

2-hydroxybutanedioic acid

15.70 Although it was initially sold as a rat poison, warfarin is an effective anticoagulant used to prevent blood clots. Label the most acidic proton in warfarin, and explain why its pK_a is comparable to the pK_a of a carboxylic acid.

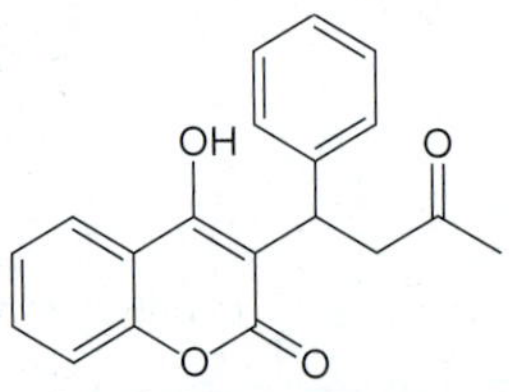

warfarin

Carboxylic Acids and Their Derivatives—Nucleophilic Acyl Substitution

16

Likit Supasai/Shutterstock

- 16.1 Introduction
- 16.2 Structure and bonding
- 16.3 Nomenclature
- 16.4 Physical and spectroscopic properties
- 16.5 Interesting esters and amides
- 16.6 Introduction to nucleophilic acyl substitution
- 16.7 Reactions of acid chlorides
- 16.8 Reactions of anhydrides
- 16.9 Reactions of carboxylic acids
- 16.10 Reactions of esters
- 16.11 Application: Lipid hydrolysis
- 16.12 Reactions of amides
- 16.13 Application: The mechanism of action of β-lactam antibiotics
- 16.14 Summary of nucleophilic acyl substitution reactions
- 16.15 Acyl phosphates—Biological anhydrides
- 16.16 Reactions of thioesters—Biological acylation reactions

Cocaine is an addictive stimulant obtained from the leaves of the coca plant, *Erythroxylon coca*. Chewing coca leaves for pleasure has been practiced by the indigenous peoples of South America for over a thousand years, and coca leaves were a very minor ingredient in Coca-Cola for the first 20 years of its production. Cocaine is a widely abused recreational drug, and the possession and use of cocaine is currently illegal in most countries. Cocaine contains two esters, carboxylic acid derivatives discussed in Chapter 16.

Why Study . . .

Carboxylic Acid Derivatives?

Chapter 16 continues the study of carbonyl compounds with a detailed look at **nucleophilic acyl substitution,** a key reaction of carboxylic acids and their derivatives. Substitution at sp^2 hybridized carbon atoms was introduced in Chapter 13 with reactions involving carbon and hydrogen nucleophiles. In Chapter 16, we learn that nucleophilic acyl substitution is a general reaction that occurs with a variety of heteroatomic nucleophiles. ***Every* reaction in Chapter 16 that begins with a carbonyl compound involves nucleophilic substitution.** Nucleophilic acyl substitutions are useful reactions in both the laboratory and biological systems. Penicillin is an effective antibiotic because it kills bacteria by a nucleophilic substitution mechanism.

16.1 Introduction

Chapter 16 focuses on carbonyl compounds that contain an **acyl group bonded to an electronegative atom.** These include the **carboxylic acids,** as well as carboxylic acid derivatives that can be prepared from them: **acid chlorides, anhydrides, acyl phosphates, esters, thioesters,** and **amides.**

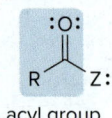

acyl group

Z = an atom more electronegative than carbon

Z = OH

carboxylic acid

R = CH_3
acetic acid

Z = Cl

acid chloride

R = CH_3
acetyl chloride

Z = OCOR

anhydride

R = CH_3
acetic anhydride

Z = OPO_3^{2-}

acyl phosphate

R = CH_3
acetyl phosphate

Z = OR'

ester

R = R' = CH_3
methyl acetate

Z = SR'

thioester

R = R' = CH_3
methyl thioacetate

Z = NR'_2

R' = H or alkyl
amide

R = CH_3, R' = H
acetamide

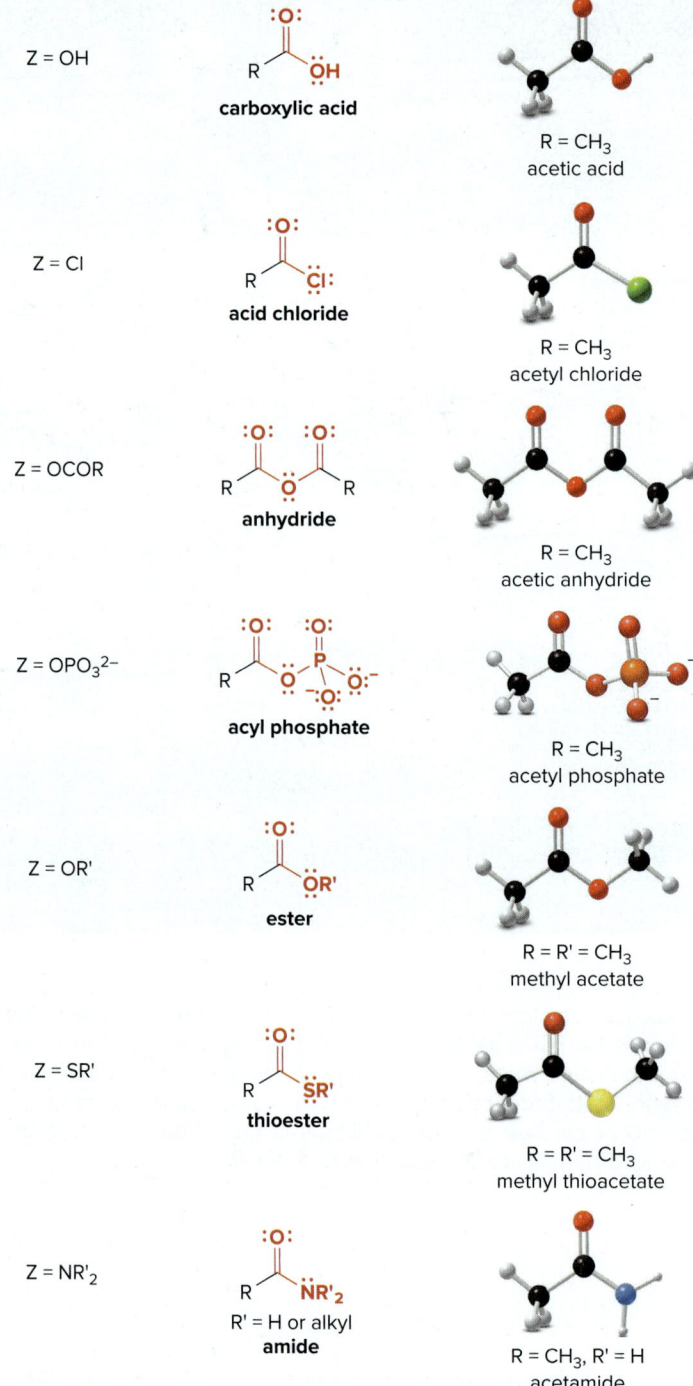

16.1 Introduction

Anhydrides contain two carbonyl groups joined by a single oxygen atom. **Symmetrical anhydrides** have two identical alkyl groups bonded to the carbonyl carbons, and **mixed anhydrides** have two different alkyl groups. **Cyclic anhydrides** are also known. An **acyl phosphate** is a mixed anhydride with a carbonyl group joined to a phosphate by a single oxygen atom. Acyl phosphates, such as glycyl phosphate, are common intermediates in biological pathways.

symmetrical anhydride   mixed anhydride   cyclic anhydride   acyl phosphate / glycyl phosphate

As we learned in Section 3.2, **amides** are classified as **1°, 2°,** or **3°** depending on the number of carbon atoms bonded directly to the *nitrogen* atom.

1° amide
1 C—N bond

2° amide
2 C—N bonds

3° amide
3 C—N bonds

Cyclic esters and amides are called **lactones** and **lactams**, respectively. The ring size of the heterocycle is indicated by a Greek letter. An amide in a four-membered ring is called a **β-lactam**, because the β carbon to the carbonyl is bonded to the heteroatom. An ester in a five-membered ring is called a **γ-lactone**.

γ-lactone   β-lactam

All of these compounds contain an acyl group bonded to an electronegative atom Z that can serve as a **leaving group.** As a result, these compounds undergo **nucleophilic acyl substitution.** Recall from Chapters 13 and 14 that aldehydes and ketones do *not* undergo nucleophilic substitution because they have no leaving group on the carbonyl carbon.

Z = OH, Cl, OCOR, OPO$_3^{2-}$, OR', SR', NR'$_2$

Nu replaces Z.

Nucleophilic acyl substitution was first discussed in Chapter 13 with R⁻ and H⁻ as the nucleophiles. This substitution reaction is general for a variety of nucleophiles, making it possible to form many different substitution products, as discussed in Sections 16.7–16.12, 16.15, and 16.16.

Problem 16.1 Oxytocin, sold under the trade name Pitocin, is a naturally occurring hormone used to stimulate uterine contractions and induce labor. Classify each amide in oxytocin as 1°, 2°, or 3°.

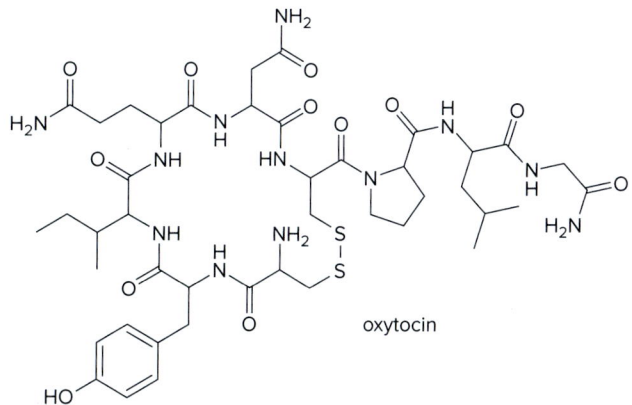

oxytocin

16.2 Structure and Bonding

The two most important features of any carbonyl group, regardless of the other groups bonded to it, are the following:

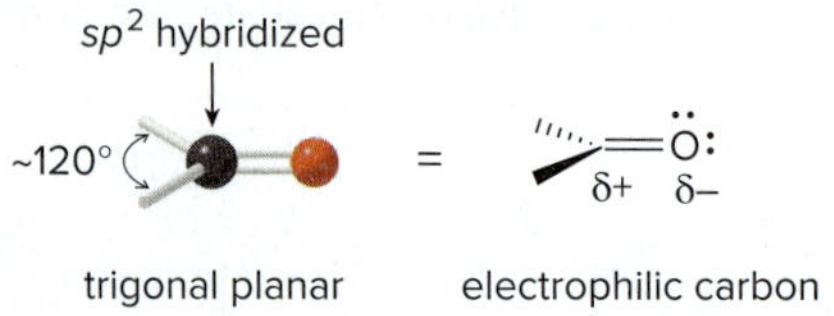

- The carbonyl carbon is sp^2 hybridized and trigonal planar, making it relatively *uncrowded*.
- The electronegative oxygen atom polarizes the carbonyl group, making the carbonyl carbon *electrophilic*.

As we learned in Spectroscopy Section B.3C, three resonance structures can be drawn for RCOZ, compared to just two for aldehydes and ketones (Section 13.1). These three resonance structures stabilize RCOZ by delocalizing electron density. In fact, **the more resonance structures 2 and 3 contribute to the resonance hybrid, the *more* stable RCOZ is.**

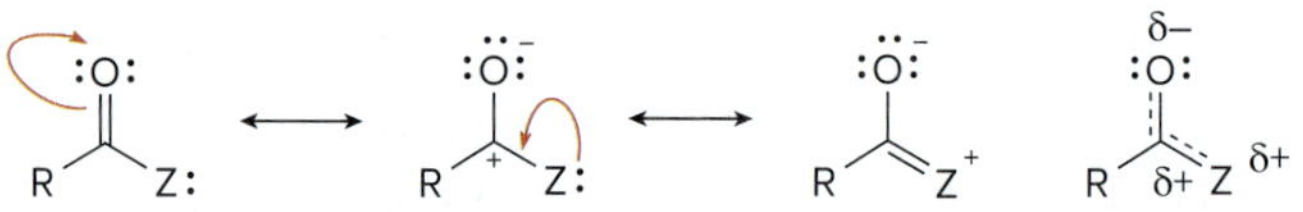

- The *more* basic Z is, the *more* it donates its electron pair, and the *more* resonance structure **3** contributes to the hybrid.

To determine the relative basicity of the leaving group Z, we compare the pK_a values of the conjugate acids HZ, given in Table 16.1. The following order of basicity results:

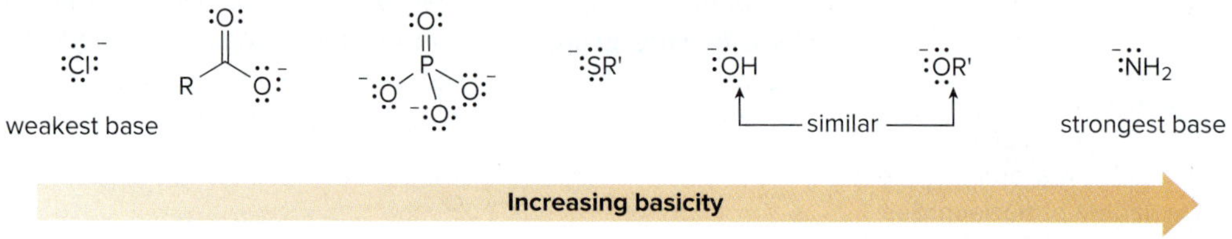

$:\!\ddot{\text{Cl}}\!:^-$ $R\text{CO}_2^-$ PO_4^{3-} $^-\!:\!\ddot{\text{S}}\text{R}'$ $^-\!:\!\ddot{\text{O}}\text{H}$ $^-\!:\!\ddot{\text{O}}\text{R}'$ $:\!\ddot{\text{N}}\text{H}_2^-$

weakest base ———————— similar ———————— strongest base

Increasing basicity →

Table 16.1 pK_a Values of the Conjugate Acids (HZ) for Common Z Groups of Acyl Compounds (RCOZ)

Structure	Leaving group (Z⁻)	Conjugate acid (HZ)	pK_a
RCOCl acid chloride	Cl⁻	HCl	−7
(RCO)₂O anhydride	RCO_2^-	RCO_2H	3–5
RCO₂PO₃²⁻ acyl phosphate	PO_4^{3-}	HPO_4^{2-}	6.9
RCOSR' thioester	⁻SR'	HSR'	9–11
RCO₂H carboxylic acid	⁻OH	H_2O	15.7
RCO₂R' ester	⁻OR'	R'OH	15.5–18
RCONR'₂ amide	⁻NR'₂	R'₂NH	38–40

Increasing basicity of Z ↓   Increasing acidity of HZ ↑

Because the basicity of Z determines the relative stability of the carboxylic acid derivatives, the following **order of stability** results:

| acid chlorides | anhydrides | acyl phosphates | thioesters | carboxylic acids | esters | amides |

Increasing stability →

Thus, **an acid chloride is the *least* stable carboxylic acid derivative** because Cl⁻ is the weakest base. **An amide is the *most* stable carboxylic acid derivative** because ⁻NR'₂ is the strongest base.

- In summary: As the basicity of Z *increases*, the stability of RCOZ *increases* because of added resonance stabilization.

Problem 16.2 Draw the three possible resonance structures for an acid bromide, RCOBr. Then, using the pK_a values in Appendix C, decide if RCOBr is more or less stabilized by resonance than a carboxylic acid (RCOOH).

Problem 16.3 How do the following experimental results support the resonance description of the relative stability of acid chlorides compared to amides? The C—Cl bond lengths in CH₃Cl and CH₃COCl are identical (178 pm), but the C—N bond in HCONH₂ is shorter than the C—N bond in CH₃NH₂ (135 pm versus 147 pm).

16.3 Nomenclature

The names of carboxylic acid derivatives are formed from the names of the parent carboxylic acids discussed in Section 15.2. Keep in mind that the common names **formic acid, acetic acid,** and **benzoic acid** are generally used for the parent acid, so these common parent names are used for their derivatives as well.

16.3A Naming an Acid Chloride (RCOCl) and an Acyl Phosphate (RCO₂PO₃²⁻)

Acid chlorides and acyl phosphates are named by naming the acyl group and adding the word *chloride* (for RCOCl) and *phosphate* (for RCO₂PO₃²⁻). Two different methods are used.

[1] For acyclic compounds: Change the suffix *-ic acid* of the parent carboxylic acid to the suffix *-yl chloride* for acid chlorides and *-yl phosphate* for acyl phosphates; or
[2] When the functional group is bonded to a ring: Change the suffix *-carboxylic acid* to *-carbonyl chloride* for acid chlorides and *-carbonyl phosphate* for acyl phosphates.

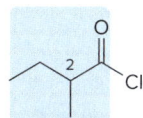

derived from
2-methylbutano*ic acid*

2-methylbutanoyl chloride

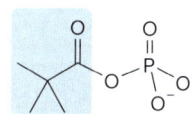

derived from
2,2-dimethylpropano*ic acid*

2,2-dimethylpropanoyl phosphate

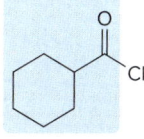

derived from
cyclohexane*carboxylic acid*

cyclohexanecarbonyl chloride

16.3B Naming an Anhydride

The word *anhydride* means "without water." Removing one molecule of water from two molecules of carboxylic acid forms an anhydride.

Symmetrical anhydrides are named by changing the *acid* ending of the parent carboxylic acid to the word *anhydride*. **Mixed anhydrides,** which are derived from two different carboxylic acids, are named by alphabetizing the names for both acids and replacing the word *acid* by the word *anhydride*.

derived from acetic acid
acetic anhydride

derived from acetic acid and benzoic acid
acetic benzoic anhydride

16.3C Naming an Ester (RCOOR') and a Thioester (RCOSR')

Esters are often written as RCOOR', where the alkyl group (R') is written *last*. When an ester is named, however, the R' group appears *first* in the name.

Esters and thioesters have two parts to their structures, each of which must be named: an **acyl group (RCO—)** and an **alkyl group** (designated as **R'**) bonded to an oxygen atom in an ester and bonded to a sulfur atom in a thioester.

- In the IUPAC system, esters are identified by the suffix *-ate*.
- Thioesters are identified by the suffix *-thioate*.

How To Name an Ester (RCO₂R') and a Thioester (RCOSR') Using the IUPAC System

Example Give a systematic name for each compound:

a. b.

Step [1] Name the R' group bonded to the oxygen atom in an ester and bonded to a sulfur atom in a thioester as an alkyl group. For a thioester, use the prefix "**S-**" preceding the name of the alkyl group.
- The name of the alkyl group, ending in the suffix *-yl*, becomes the **first** part of the name.

a. **ethyl** group

b. **isopropyl** group
S-isopropyl

Step [2] Name the acyl group (RCO—) with the suffix *-ate* for an ester and *-thioate* for a thioester.
- The name of the acyl group becomes the **second** part of the name.

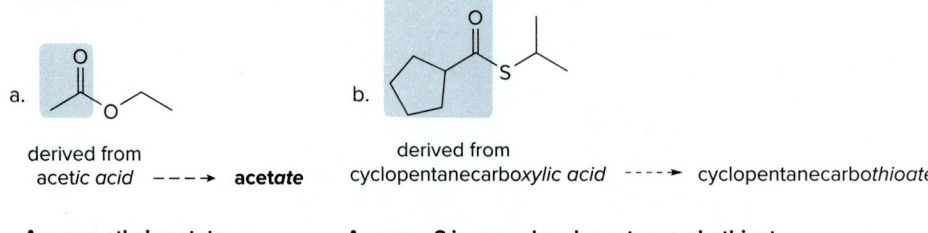

a. derived from acet*ic acid* - - - → acet**ate**

b. derived from cyclopentanecarb**o***xylic acid* - - - → cyclopentanecarb**o***thioate*

Answer: ethyl acetate

Answer: *S*-isopropyl cyclopentanecarbothioate

16.3D Naming an Amide

All 1° amides are named by replacing the *-ic acid, -oic acid,* or *-ylic acid* ending of the parent carboxylic acid with the suffix *amide*.

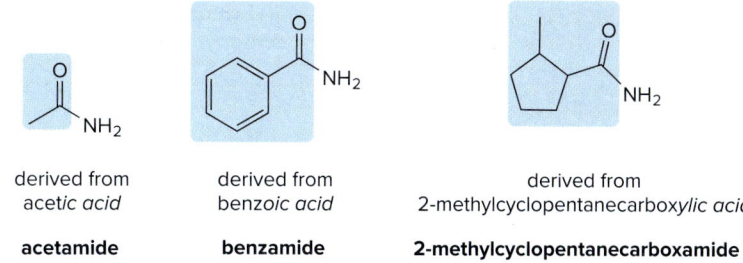

derived from
acet*ic acid*
acetamide

derived from
benz*oic acid*
benzamide

derived from
2-methylcyclopentanecarbox*ylic acid*
2-methylcyclopentanecarboxamide

A 2° or 3° amide has two parts to its structure: an **acyl group** that contains the carbonyl group (**RCO–**) and one or two **alkyl groups** bonded to the nitrogen atom.

How To Name a 2° or 3° Amide

Example Give a systematic name for each amide:

a. [structure] b. [structure]

Step [1] Name the alkyl group (or groups) bonded to the N atom of the amide. Use the prefix "**N-**" preceding the name of each alkyl group.

- The names of the alkyl groups form the **first** part of each amide name.
- For 3° amides, use the prefix **di-** if the two alkyl groups on N are the same. If the two alkyl groups are different, **alphabetize** their names. One "**N-**" is needed for each alkyl group, even if both R groups are identical.

a. [structure with ethyl group]

- The compound is a 2° amide with one ethyl group → **N-ethyl**.

b. [structure with two methyl groups]

- The compound is a 3° amide with two methyl groups.
- Use the prefix *di-* and two "**N-**" to begin the name → **N,N-dimethyl**.

Step [2] Name the acyl group (RCO–) with the suffix *-amide*.

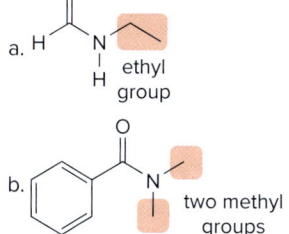

derived from
form*ic acid* – – – → form**amide**

- Change the *-ic acid* or *-oic acid* suffix of the parent carboxylic acid to the suffix **-amide**.
- Put the two parts of the name together.
- **Answer: N-ethylformamide**

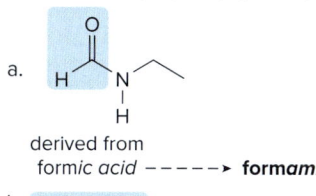

derived from
benz*oic acid* – – – → benz**amide**

- Change benzoic acid to **benzamide**.
- Put the two parts of the name together.
- **Answer: N,N-dimethylbenzamide**

Table 16.2 summarizes the most important points about the nomenclature of carboxylic acid derivatives.

Chapter 16 Carboxylic Acids and Their Derivatives—Nucleophilic Acyl Substitution

Table 16.2 Summary: Nomenclature of Carboxylic Acid Derivatives

Compound	Name ending	Example	Name
acid chloride	-yl chloride or -carbonyl chloride		benzoyl chloride
acyl phosphate	-yl phosphate or -carbonyl phosphate		acetyl phosphate
anhydride	anhydride		benzoic anhydride
ester	-ate		ethyl benzoate
thioester	-thioate		S-methyl cyclohexanecarbothioate
amide	-amide		N-methylbenzamide

Sample Problem 16.1 Naming Carboxylic Acid Derivatives

Give the IUPAC name for each compound.

a. [structure]   b. [structure]

Solution

a. The functional group is an acyl phosphate bonded to a chain of atoms, so the name ends in **-yl phosphate.**

[1] Find and name the longest chain containing the $CO_2PO_3^{2-}$:

hexanoic acid (6 C's) → hexanoyl phosphate

[2] Number and name the substituents:

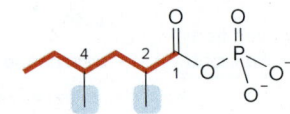

Answer:
2,4-dimethylhexanoyl phosphate

b. The functional group is an ester, so the name ends in -*ate*.

[1] Find and name the longest chain containing the carbonyl group:

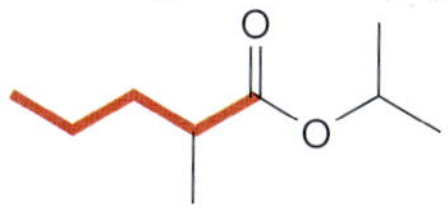

pentano*ic acid* ⟶ pentano*ate*
(5 C's)

[2] Number and name the substituents:

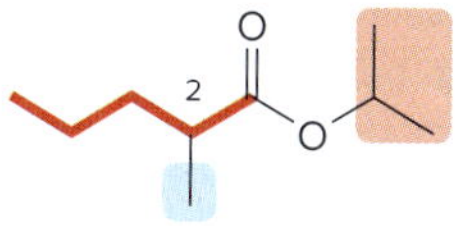

isopropyl group

Answer: isopropyl 2-methylpentanoate
The name of the alkyl group on the O atom goes **first** in the name.

Problem 16.4 Give an IUPAC or common name for each compound.

a., 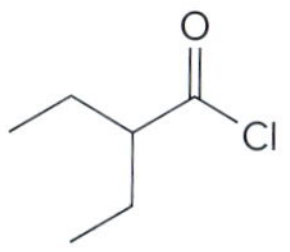 b., c., d., e., f., g., h.

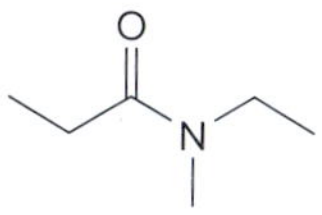

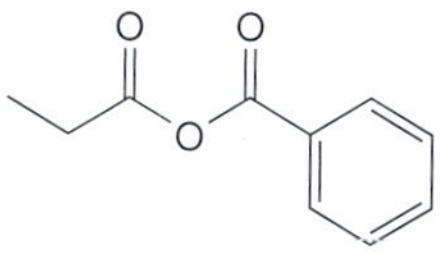

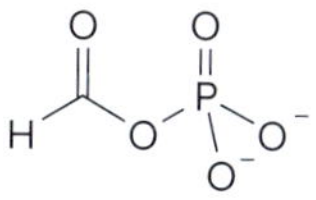

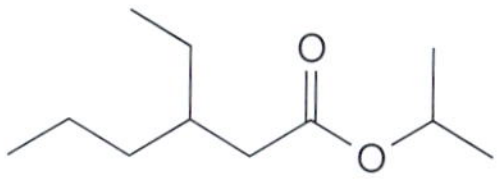

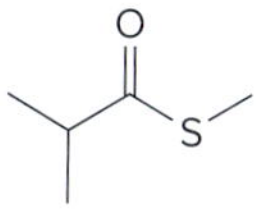

More Practice: Try Problems 16.32a, 16.33.

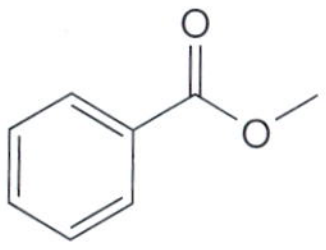

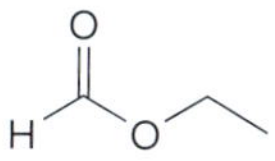

Problem 16.5 Draw the structure corresponding to each name.
a. 5-methylheptanoyl chloride
b. *S*-isopropyl 3-ethylcyclobutanecarbothioate
c. acetic formic anhydride
d. *N*-isobutyl-*N*-methylbutanamide
e. *sec*-butyl 2-methylhexanoate
f. 3-ethyl-2-methylpentanoyl phosphate

16.4 Physical and Spectroscopic Properties

16.4A Physical Properties

Because all carbonyl compounds have a polar carbonyl group, they exhibit **dipole–dipole interactions.** Primary (1°) and 2° amides are capable of intermolecular hydrogen bonding because they contain one or two N–H bonds. The N–H bond of one amide intermolecularly hydrogen bonds to the C=O of another amide, as shown using two acetamide molecules (CH_3CONH_2) in Figure 16.1.

Figure 16.1
Intermolecular hydrogen bonding between two CH_3CONH_2 molecules

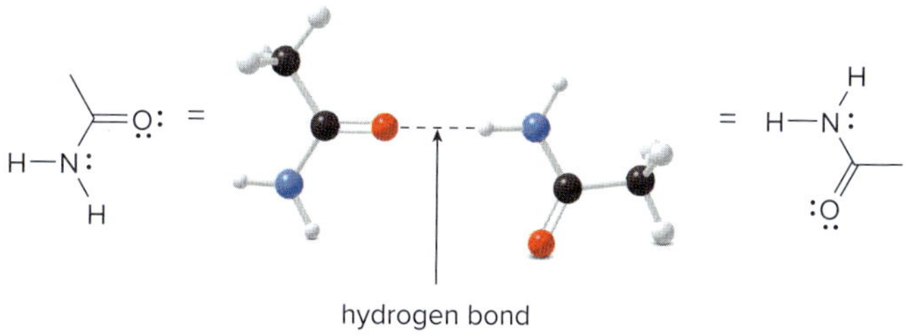

hydrogen bond

Problem 16.6 Explain why the boiling point of CH_3CONH_2 (221 °C) is significantly higher than the boiling point of CH_3CO_2H (118 °C).

How these factors affect the physical properties of carboxylic acid derivatives is summarized in Table 16.3.

Table 16.3 Physical Properties of Carboxylic Acid Derivatives

Property	Observation
Boiling point and melting point	• **Primary (1°) and 2° amides have *higher* boiling points and melting points** than compounds of comparable molecular weight. • The boiling points and melting points of other carboxylic acid derivatives are similar to those of other polar compounds of comparable size and shape. 　　MW = 78.5　　　MW = 74　　　MW = 72　　　MW = 73 　　bp 52 °C　~　bp 58 °C　~　bp 80 °C　<　bp 213 °C 　　└─────── similar boiling points ───────┘　　　higher boiling point 　　　　　　　　　　　　　　　　　　　　　　　　　　　1° amide
Solubility	• Carboxylic acid derivatives are soluble in organic solvents regardless of size. • Most carboxylic acid derivatives having ≤ 5 C's are H_2O soluble because they can hydrogen bond with H_2O (Section 3.4C). • Carboxylic acid derivatives having > 5 C's are H_2O insoluble because the nonpolar alkyl portion is too large to dissolve in the polar H_2O solvent.

Key: MW = molecular weight

16.4B Spectroscopic Properties

Many details of the spectroscopy of carboxylic acid derivatives have been presented in Spectroscopy Parts B and C.

- The infrared absorption of the carbonyl group of carboxylic acid derivatives: Sections B.3C and B.4B, Sample Problem B.1, and Table B.2
- ^1H and ^{13}C NMR absorptions: Tables C.1 and C.5

Key NMR and IR absorptions for carboxylic acid derivatives are summarized in Table 16.4. Recall from Section B.3C that the location of the carbonyl absorption depends on the identity of Z in RCOZ.

> • As the basicity of Z *increases*, resonance stabilization of RCOZ *increases*, and the C=O absorption shifts to *lower* frequency for acid chlorides, anhydrides, esters, and amides.

Basicity trends do *not* correctly predict the location of the carbonyl absorptions of acyl phosphates and thioesters. Due to the significant electron density of the phosphate and the low electronegativity of the sulfur, the carbonyl π electrons on these compounds are more delocalized and thus shifted to lower frequencies than basicity trends suggest.

Problem 16.7 Rank the following compounds in order of increasing frequency of the C=O absorption in their IR spectra.

A　　　　B　　　　C　　　　D

Table 16.4 Characteristic Spectroscopic Absorptions of Carboxylic Acid Derivatives

Type of spectroscopy	Compound	Type of C, H	Absorption
IR absorptions	Acid chloride	R−C(=O)−Cl	1800 cm^{-1}
	Anhydride	R−C(=O)−O−C(=O)−R	1820 and 1760 cm^{-1} (two peaks)
	Ester	R−C(=O)−OR'	1735–1745 cm^{-1}
	Acyl phosphate	R−C(=O)−O−P(=O)(O$^-$)(O$^-$)	1700–1730 cm^{-1}
	Thioester	R−C(=O)−SR'	1690–1720 cm^{-1}
	Amide	R−C(=O)−NR'$_2$ R' = H or alkyl	1630–1680 cm^{-1}
		R−C(=O)−N(H)−	3200–3400 cm^{-1} (one or two N–H stretching peaks) 1640 cm^{-1} (N–H bending)
^{1}H NMR absorptions	All acyl derivatives	Z−C(=O)−C(H)(R)	2–2.5 ppm
	Amide (1° and 2°)	R−C(=O)−N(H)−	7.5–8.5 ppm
^{13}C NMR absorption	All acyl derivatives	R−C(=O)−Z	160–180 ppm

16.5 Interesting Esters and Amides

16.5A Esters

Many low-molecular-weight esters have pleasant and very characteristic odors.

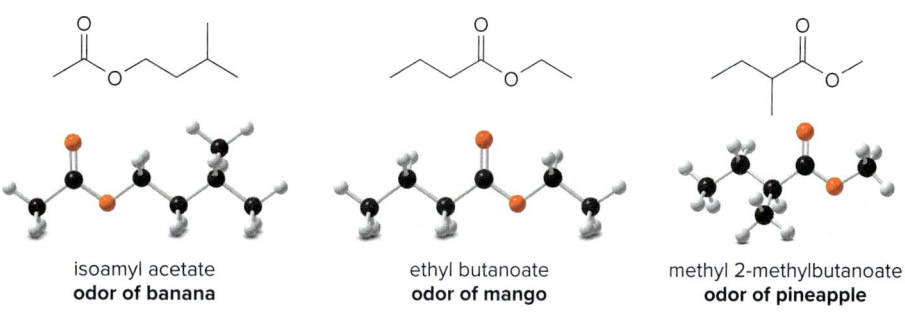

isoamyl acetate
odor of banana

ethyl butanoate
odor of mango

methyl 2-methylbutanoate
odor of pineapple

The characteristic odor of many fruits is due to low-molecular-weight esters. *Jill Braaten*

Several esters, including vitamin C and ginkgolide B, have important biological activities.

Jill Braaten

Vitamin C (or **ascorbic acid**) is a water-soluble vitamin containing a five-membered lactone that we first discussed in Section 3.5B. Although vitamin C is synthesized in plants, humans do not have the necessary enzymes to make it, so they must obtain it from their diet.

Ginkgolide B is a major constituent of the extracts of the ginkgo tree, *Ginkgo biloba*. Ginkgo extracts are widely used herbal supplements, taken to enhance memory and treat dementia. Recent findings of the National Institutes of Health, however, have cast doubt on their efficacy in providing long-term improvement in cognition.

16.5B Amides

An important group of naturally occurring amides consists of **proteins, polymers of amino acids joined together by amide linkages** (Section 3.9A). Proteins differ in the length of the polymer chain, as well as in the identity of the R groups bonded to it. The word *protein* is usually reserved for high-molecular-weight polymers composed of 40 or more amino acid units, whereas the designation *peptide* is given to polymers of lower molecular weight.

portion of a protein molecule
[Amide bonds are shown in red.]

Peptides and proteins are discussed in detail in Chapter 23.

Proteins and peptides have diverse functions in the cell. They form the structural components of muscle, connective tissue, hair, and nails. They catalyze reactions and transport ions and molecules across cell membranes. **Met-enkephalin,** for example, a peptide with four amide bonds found predominately in nerve tissue cells, relieves pain and acts as an opiate by producing morphine-like effects.

met-enkephalin
[The four amide bonds are shown in red.]

Penicillins are a group of structurally related antibiotics, known since the pioneering work of Sir Alexander Fleming led to the discovery of penicillin G in the 1920s. All penicillins contain a strained β-lactam fused to a five-membered ring, as well as a second amide located α to the β-lactam carbonyl group. Particular penicillins differ in the identity of the R group in the amide side chain.

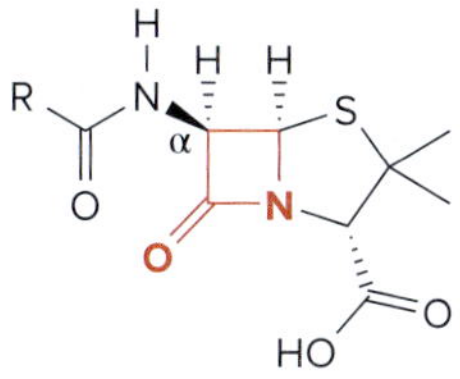

penicillin
β-lactam shown in red

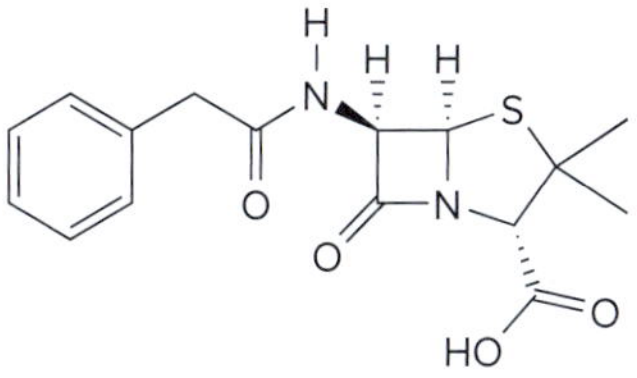

penicillin G

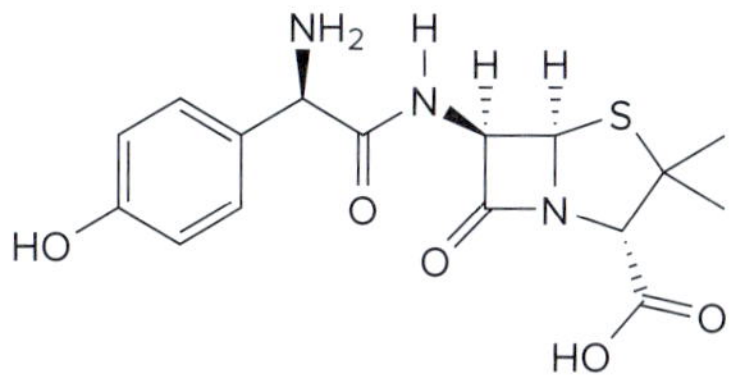

amoxicillin

Cephalosporins represent a second group of β-lactam antibiotics that contain a four-membered ring fused to a six-membered ring. Cephalosporins are generally active against a broader range of bacteria than penicillins.

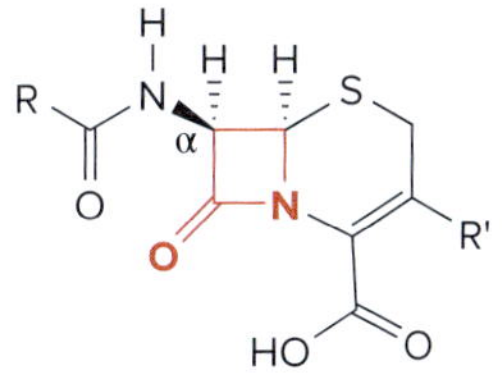

cephalosporin
β-lactam shown in red

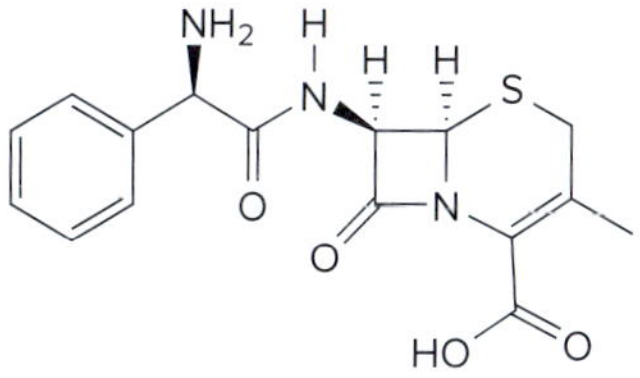

cephalexin
Trade name Keflex

16.6 Introduction to Nucleophilic Acyl Substitution

The characteristic reaction of carboxylic acid derivatives is *nucleophilic acyl substitution*. This is a general reaction that occurs with both negatively charged nucleophiles (Nu:⁻) and neutral nucleophiles (HNu:).

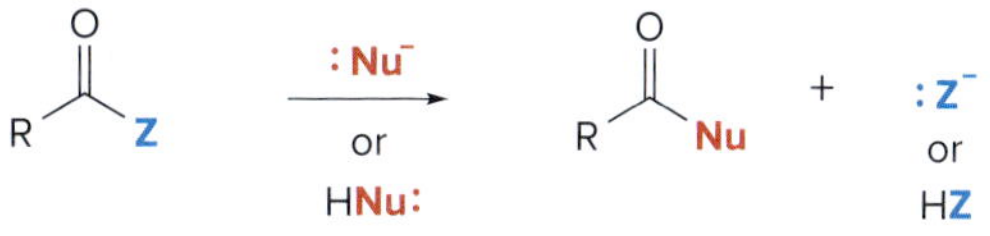

Nu replaces Z.

- Carboxylic acid derivatives (RCOZ) react with nucleophiles because they contain an electrophilic, unhindered carbonyl carbon.
- Substitution, *not* addition, occurs because carboxylic acid derivatives (RCOZ) have a leaving group Z on the carbonyl carbon.

The mechanism for nucleophilic acyl substitution was first presented in Section 13.2.

16.6A The Mechanism

The general mechanism for nucleophilic acyl substitution is a two-step process: **nucleophilic attack** followed by **loss of the leaving group**, as shown in Mechanism 16.1.

Mechanism 16.1 General Mechanism—Nucleophilic Acyl Substitution

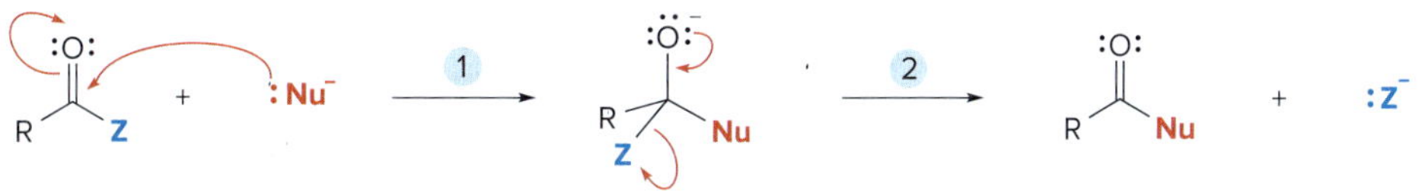

1. **The nucleophile attacks the electrophilic carbonyl group.** The π bond is broken, moving an electron pair out on oxygen and forming an sp^3 hybridized carbon.

2. An electron pair on oxygen re-forms the π bond and **Z comes off as a leaving group** with the electron pair in the C–Z bond.

The overall result of addition of a nucleophile and elimination of a leaving group is *substitution* of the nucleophile for the leaving group. Recall from Chapter 13 that nucleophilic substitution occurs with carbanions (R⁻) and hydride (H⁻) as nucleophiles. A variety of oxygen, nitrogen, and sulfur nucleophiles also participate in this reaction.

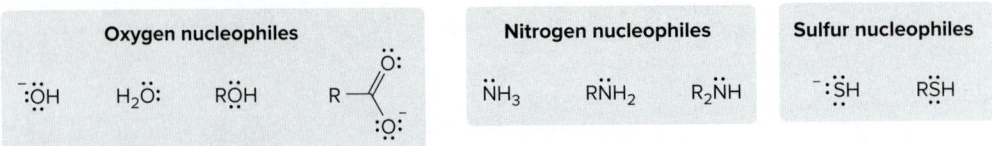

Nucleophilic acyl substitution using heteroatomic nucleophiles results in the conversion of one carboxylic acid derivative to another, as shown in two examples.

Each reaction results in the replacement of the leaving group by the nucleophile, regardless of the identity of or charge on the nucleophile. To draw any nucleophilic acyl substitution product:

- Find the *sp*² hybridized carbon with the leaving group.
- Identify the nucleophile.
- Substitute the nucleophile for the leaving group. With a neutral nucleophile, a proton must be lost to obtain a neutral substitution product.

16.6B Relative Reactivity of Carboxylic Acids and Their Derivatives

As discussed in Section 13.2B, carboxylic acids and their derivatives differ greatly in reactivity toward nucleophiles. The order of reactivity parallels the leaving group ability of the group Z.

- The better the leaving group, the more reactive RCOZ is in nucleophilic acyl substitution.

Recall that the **best leaving group is the weakest base.** The relative basicity of the common leaving groups, Z, is given in Table 16.1.

Thus, the following trends result:

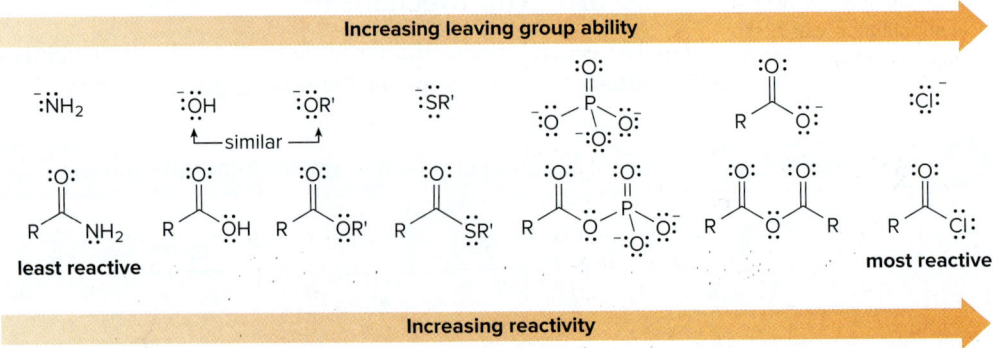

Based on this order of reactivity, *more reactive* acyl compounds (acid chlorides, anhydrides, and acyl phosphates) can be converted to *less reactive* ones (thioesters, carboxylic acids, esters, and amides). The reverse is not usually true.

16.6 Introduction to Nucleophilic Acyl Substitution

To see why this is so, recall that nucleophilic addition to a carbonyl group forms a tetrahedral intermediate with two possible leaving groups, Z⁻ or :Nu⁻. The group that is subsequently eliminated is the *better* of the two leaving groups. For a reaction to form a substitution product, therefore, Z⁻ must be the better leaving group, making the starting material RCOZ a more reactive acyl compound.

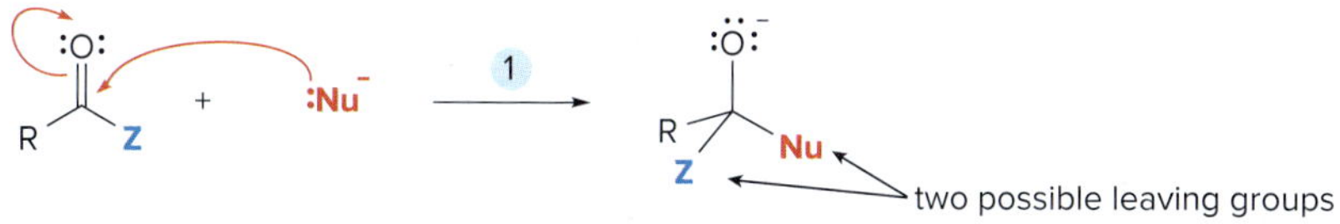

For a reaction to occur, **Z** must be a better leaving group than **Nu**.

To evaluate whether a nucleophilic substitution reaction will occur, **compare the leaving group ability of the incoming nucleophile and the departing leaving group,** as shown in Sample Problem 16.2.

Sample Problem 16.2 — Using Basicity to Determine Whether a Nucleophilic Acyl Substitution Might Occur

Determine whether each nucleophilic acyl substitution is likely to occur.

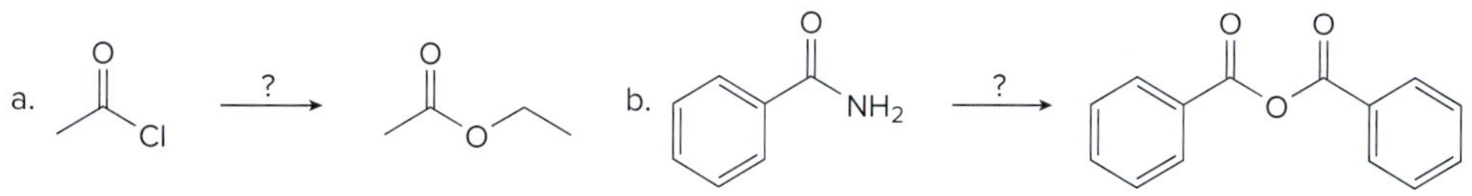

Solution

a. Conversion of CH_3COCl to $CH_3COOCH_2CH_3$ requires the **substitution of Cl⁻ by ⁻OCH₂CH₃**. Because **Cl⁻ is a weaker base** and therefore a better leaving group than ⁻OCH₂CH₃, **this reaction occurs.**

b. Conversion of $C_6H_5CONH_2$ to $(C_6H_5CO)_2O$ requires the **substitution of ⁻NH₂ by ⁻OCOC₆H₅**. Because **⁻NH₂ is a stronger base** and therefore a poorer leaving group than ⁻OCOC₆H₅, **this reaction does *not* occur.**

Problem 16.8 Without reading ahead in Chapter 16, state whether it should be possible to carry out each of the following nucleophilic substitution reactions.

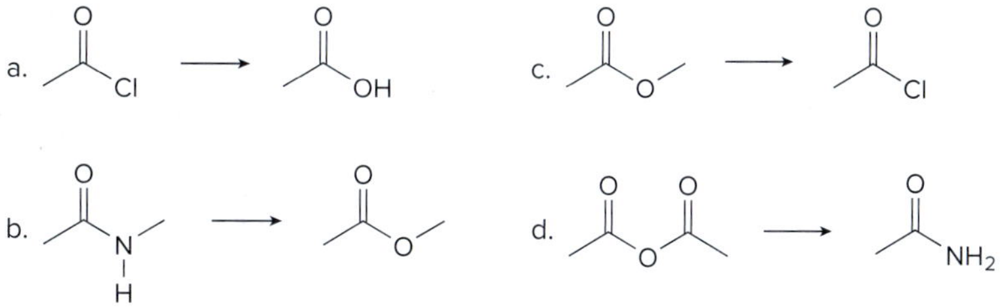

More Practice: Try Problem 16.31.

Learn the order of reactivity of carboxylic acid derivatives. Keeping this in mind allows you to organize a very large number of reactions.

To summarize:

- Nucleophilic substitution occurs when the leaving group Z⁻ is a *weaker* base and therefore *better* leaving group than the attacking nucleophile :Nu⁻.
- *More* reactive acyl compounds can be converted to *less* reactive acyl compounds by nucleophilic substitution.

Problem 16.9 Rank the compounds in each group in order of increasing reactivity in nucleophilic acyl substitution.

a. $C_6H_5CO_2CH_3$, C_6H_5COCl, $C_6H_5CONH_2$
b. $CH_3CH_2CO_2H$, $(CH_3CH_2CO)_2O$, $CH_3CH_2CONHCH_3$

Problem 16.10 Explain why trichloroacetic anhydride [$(Cl_3CCO)_2O$] is more reactive than acetic anhydride [$(CH_3CO)_2O$] in nucleophilic acyl substitution reactions.

16.6C A Preview of Specific Reactions

Sections 16.7–16.12, 16.15, and 16.16 are devoted to specific examples of nucleophilic acyl substitution using heteroatoms as nucleophiles. There are a great many reactions, and it is easy to confuse them unless you learn the general order of reactivity of carboxylic acid derivatives. **Keep in mind that every reaction that begins with an acyl starting material involves nucleophilic substitution.**

We begin with the reactions of acid chlorides, the most reactive acyl compounds, then proceed to less and less reactive carboxylic acid derivatives, ending with amides. Acid chlorides undergo many reactions, because they have the best leaving group of all acyl compounds, whereas amides undergo only one reaction, which must be carried out under harsh reaction conditions, because amides have a poor leaving group.

In general, we will examine nucleophilic acyl substitution with four different nucleophiles, as shown in the following equations.

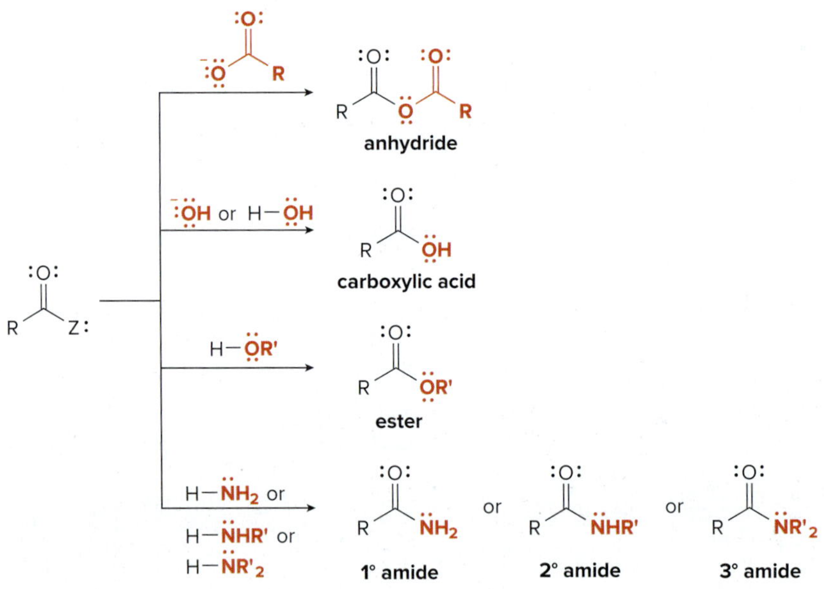

These reactions are used to make anhydrides, carboxylic acids, esters, and amides, but not acid chlorides, from other acyl compounds. Acid chlorides are the most reactive acyl compounds (they have the best leaving group), so they are not easily formed as a product of nucleophilic substitution reactions. **Acid chlorides can only be prepared from carboxylic acids using special reagents,** as discussed in Section 16.9A.

A fifth nucleophile, a thiol (HSR), is introduced in our discussion of biological acyl phosphates and thioesters.

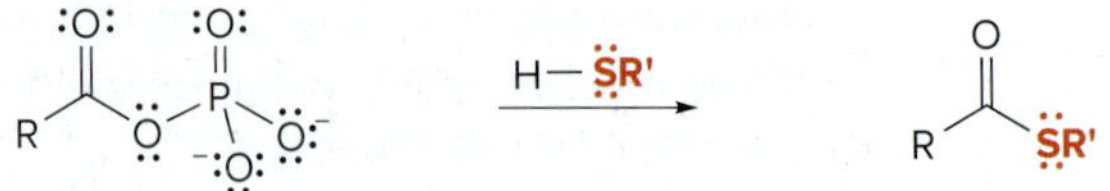

16.7 Reactions of Acid Chlorides

Acid chlorides readily react with nucleophiles to form nucleophilic substitution products, with HCl usually formed as a reaction by-product. A weak base like pyridine is added to the reaction mixture to remove this strong acid, forming an ammonium salt.

Acid chlorides react with oxygen nucleophiles to form anhydrides, carboxylic acids, and esters.

Insect repellents containing DEET have become particularly popular because of the recent spread of many insect-borne diseases such as West Nile virus and Lyme disease. DEET does not kill insects—it repels them. It is thought that DEET somehow confuses insects so that they can no longer sense the warm moist air that surrounds a human body.
Source: Scott Bauer/USDA-ARS

Acid chlorides also react with ammonia and 1° and 2° amines to form 1°, 2°, and 3° amides, respectively. Two equivalents of NH_3 or amine are used. One equivalent acts as a nucleophile to replace Cl and form the substitution product, while the second equivalent reacts as a base with the HCl by-product to form an ammonium salt.

As an example, reaction of an acid chloride with diethylamine forms the 3° amide *N,N*-diethyl-*m*-toluamide, popularly known as **DEET**. DEET, the active ingredient in the most widely used insect repellents, is effective against mosquitoes, fleas, and ticks.

Problem 16.11 Draw the products formed when benzoyl chloride (C_6H_5COCl) is treated with each nucleophile: (a) H_2O, pyridine; (b) CH_3COO^-; (c) NH_3 (excess); (d) $(CH_3)_2NH$ (excess).

With a carboxylate nucleophile, the mechanism follows the general, two-step mechanism discussed in Section 16.6A: **nucleophilic attack followed by loss of the leaving group,** as shown in Mechanism 16.2.

> ### Mechanism 16.2 Conversion of Acid Chlorides to Anhydrides
>
> [mechanism showing acid chloride + carboxylate → tetrahedral intermediate → anhydride + Cl⁻]
>
> ① The nucleophilic carboxylate anion attacks the carbonyl group, forming an sp^3 hybridized carbon.
>
> ② Elimination of the leaving group (Cl⁻) forms the **substitution product,** an anhydride.

Problem 16.12 Draw a stepwise mechanism for the formation of **A** from an alcohol and acid chloride. **A** was converted in one step to blattellaquinone, the sex pheromone of the female German cockroach, *Blattella germanica*.

16.8 Reactions of Anhydrides

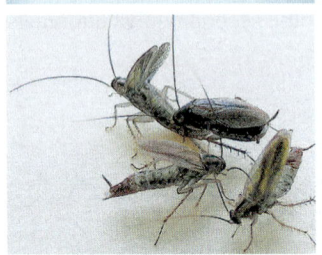

A short laboratory synthesis of blattellaquinone (Problem 16.12), the sex pheromone of the female German cockroach, opens new possibilities for cockroach population control using pheromone-baited traps. *Coby Schal*

Nucleophilic substitution occurs only when the leaving group is a weaker base and therefore a better leaving group than the attacking nucleophile.

Although somewhat less reactive than acid chlorides, anhydrides nonetheless readily react with most nucleophiles to form substitution products. Nucleophilic substitution reactions of anhydrides are no different than the reactions of other carboxylic acid derivatives, even though anhydrides contain two carbonyl groups. **Nucleophilic attack occurs at one carbonyl group, while the second carbonyl becomes part of the leaving group.**

Anhydrides can't be used to make acid chlorides, because RCOO⁻ is a stronger base and therefore a poorer leaving group than Cl⁻. Anhydrides can be used to make other acyl derivatives, however. Reaction with water and alcohols yields **carboxylic acids** and **esters,** respectively. Reaction with two equivalents of NH_3 or amines forms **1°, 2°,** and **3° amides.** A molecule of carboxylic acid (or a carboxylate salt) is always formed as a by-product.

R' = H or alkyl   carboxylic acid, R' = H   by-product
                   ester, R' = alkyl

(2 equiv)   1° amide   by-product

Problem 16.13 Draw the products formed when benzoic anhydride [(C₆H₅CO)₂O] is treated with each nucleophile: (a) H₂O; (b) CH₃OH; (c) NH₃ (excess); (d) (CH₃)₂NH (excess).

The conversion of an anhydride to an amide illustrates the mechanism of nucleophilic acyl substitution with an anhydride as starting material (Mechanism 16.3). Besides the usual steps of **nucleophilic addition** and **elimination of the leaving group,** an additional proton transfer is needed.

Mechanism 16.3 Conversion of an Anhydride to an Amide

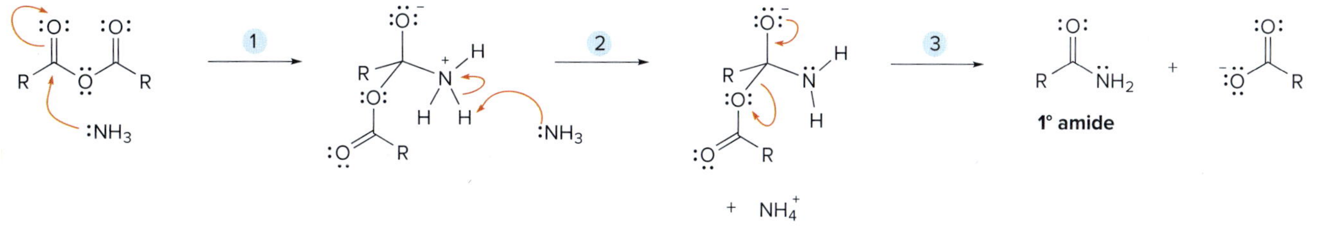

1. **The nucleophile (NH₃) attacks the carbonyl,** forming an sp^3 hybridized carbon.
2–3. Loss of a proton and elimination of the leaving group (RCO_2^-) form the **substitution product,** a 1° amide.

Anhydrides react with alcohols and amines with ease, so they are often used in the laboratory to prepare esters and amides. For example, acetic anhydride is used to prepare **heroin** from morphine, an analgesic compound isolated from the opium poppy. Both OH groups of morphine readily react with acetic anhydride to form the diester present in heroin. This is called an **acetylation** reaction because it results in the transfer of an acetyl group, CH₃CO–, from one heteroatom to another.

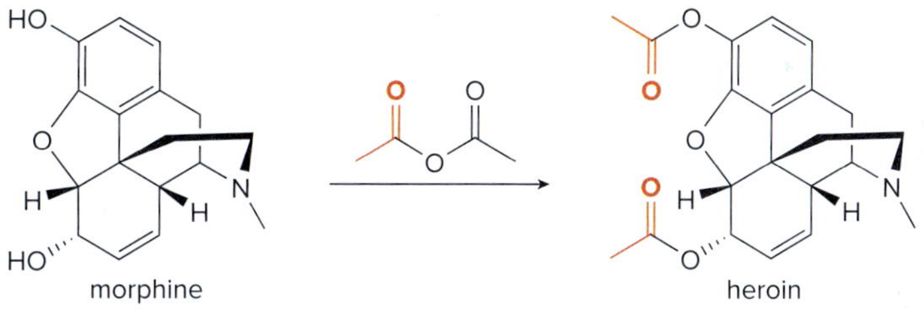

morphine → heroin

Problem 16.14 If anhydrides react like acid chlorides with the nucleophiles described in Chapter 13, draw the products formed when each of the following nucleophiles reacts with benzoic anhydride [(C₆H₅CO)₂O]: (a) CH₃MgBr (2 equiv), then H₂O; (b) LiAlH₄, then H₂O; (c) LiAlH[OC(CH₃)₃]₃, then H₂O.

16.9 Reactions of Carboxylic Acids

Carboxylic acids are strong organic acids. Because acid–base reactions proceed rapidly, any nucleophile that is also a strong base will react with a carboxylic acid by removing a proton *first,* before any nucleophilic substitution reaction can take place.

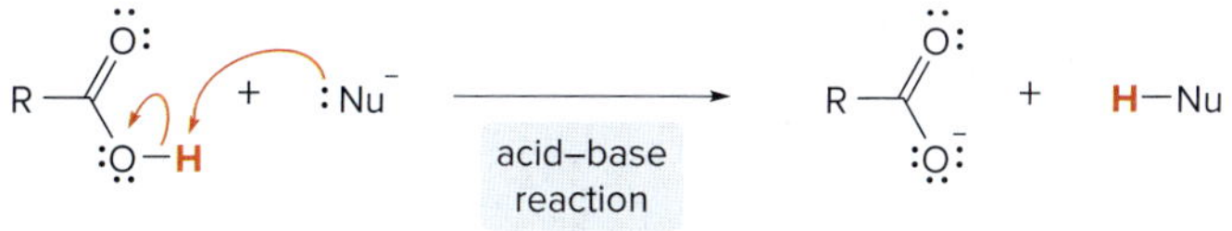

An acid–base reaction occurs with ⁻OH, NH₃, and amines, all common nucleophiles used in nucleophilic acyl substitution reactions. Nonetheless, carboxylic acids do undergo nucleophilic acyl substitution and can be converted to a variety of other acyl derivatives using special reagents, with acid catalysis or, sometimes, by using rather forcing reaction conditions. These reactions are summarized in Figure 16.2 and detailed in Sections 16.9A–16.9D.

Figure 16.2
Nucleophilic acyl substitution reactions of carboxylic acids

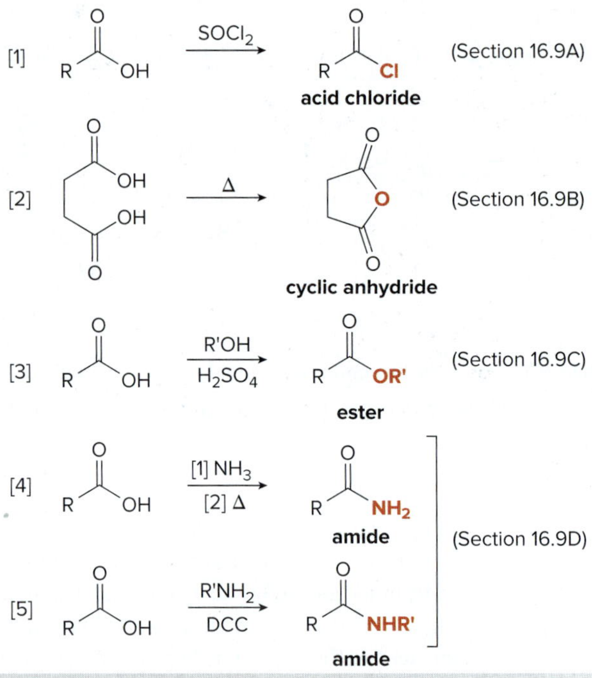

[1] RCOOH $\xrightarrow{SOCl_2}$ RCOCl (acid chloride) (Section 16.9A)

[2] HOOC–CH₂–CH₂–COOH $\xrightarrow{\Delta}$ cyclic anhydride (Section 16.9B)

[3] RCOOH $\xrightarrow[H_2SO_4]{R'OH}$ RCOOR' (ester) (Section 16.9C)

[4] RCOOH $\xrightarrow[\text{[2] }\Delta]{\text{[1] }NH_3}$ RCONH₂ (amide) (Section 16.9D)

[5] RCOOH $\xrightarrow[DCC]{R'NH_2}$ RCONHR' (amide)

16.9A Conversion of RCOOH to RCOCl

Carboxylic acids can't be converted to acid chlorides by using Cl⁻ as a nucleophile, because the attacking nucleophile Cl⁻ is a weaker base than the departing leaving group, ⁻OH. But carboxylic acids *can* be converted to acid chlorides using thionyl chloride, **SOCl₂**, a reagent that was introduced in Section 9.12 to convert alcohols to alkyl chlorides.

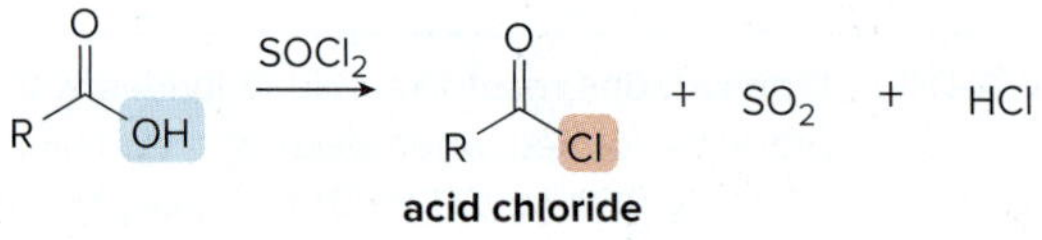

RCOOH $\xrightarrow{SOCl_2}$ RCOCl (acid chloride) + SO₂ + HCl

Treatment of benzoic acid with SOCl₂ forms benzoyl chloride. This reaction converts a less reactive acyl derivative (a carboxylic acid) to a more reactive one (an acid chloride). This is possible because **thionyl chloride converts the OH group of the acid to a better leaving group, and because it provides the nucleophile (Cl⁻) to displace the leaving group.** The steps in the process are illustrated in Mechanism 16.4.

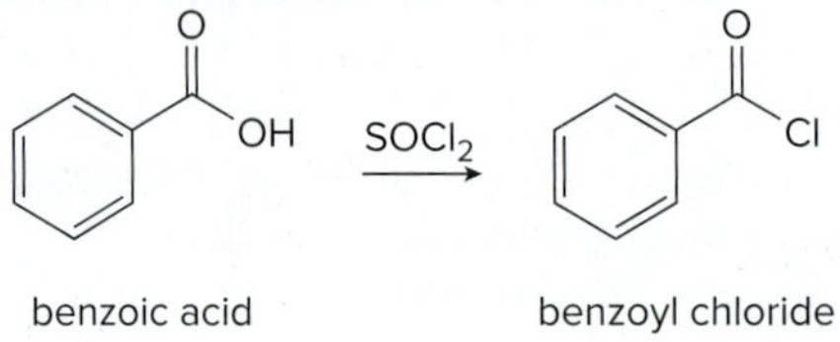

benzoic acid $\xrightarrow{SOCl_2}$ benzoyl chloride

16.9 Reactions of Carboxylic Acids 753

Mechanism 16.4 Conversion of Carboxylic Acids to Acid Chlorides

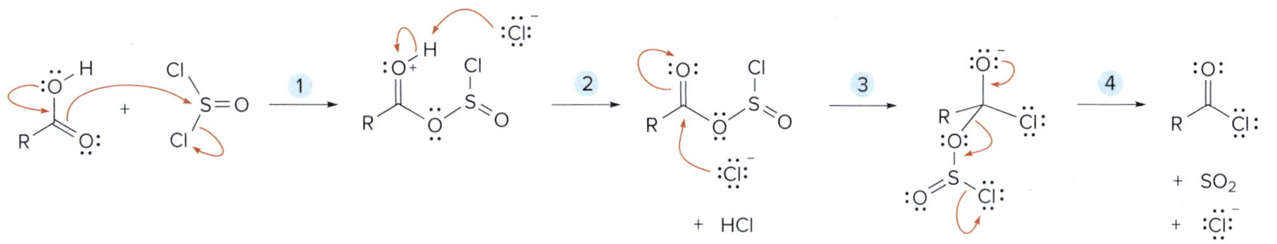

1 – 2 Reaction of the carboxylic acid with SOCl₂ and loss of a proton convert the OH group to OSOCl, a **good leaving group**.

3 – 4 Nucleophilic attack of chloride generates a tetrahedral intermediate, and loss of the leaving group (SO₂ and Cl⁻) forms the **acid chloride**.

Problem 16.15 Draw the products of each reaction.

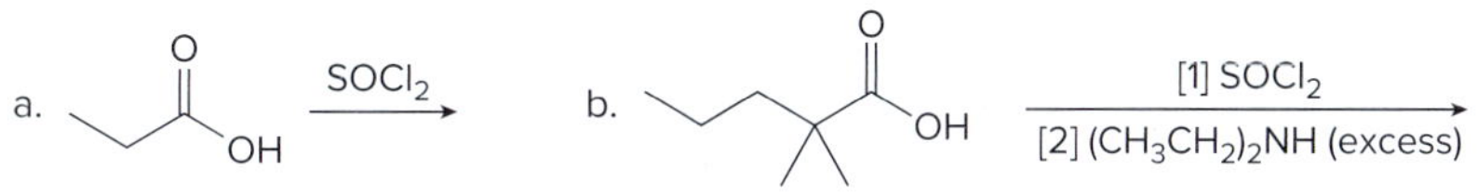

a. [structure] $\xrightarrow{SOCl_2}$

b. [structure] $\xrightarrow[\text{[2] }(CH_3CH_2)_2NH \text{ (excess)}]{\text{[1] SOCl}_2}$

16.9B Conversion of RCOOH to (RCO)₂O

Carboxylic acids cannot be readily converted to anhydrides, but dicarboxylic acids can be converted to cyclic anhydrides by heating to high temperatures. This is a **dehydration** reaction because a water molecule is lost from the diacid.

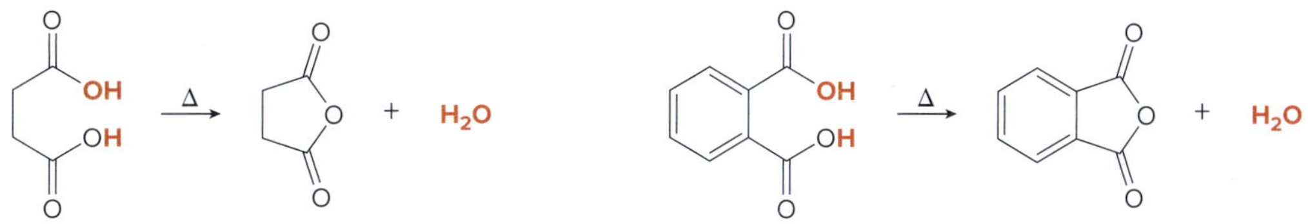

16.9C Conversion of RCOOH to RCOOR'

Treatment of a carboxylic acid with an alcohol in the presence of an acid catalyst forms an ester. This reaction is called a **Fischer esterification.**

$$R\text{-COOH} + H\text{-OR'} \xrightleftharpoons{H_2SO_4} R\text{-COOR'} + H\text{-OH}$$

ester

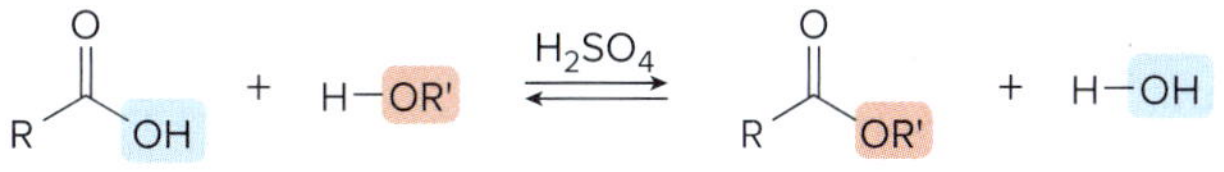

This reaction is an equilibrium. According to Le Châtelier's principle (Section 9.8), it is driven to the right by using excess alcohol or by removing the water as it is formed.

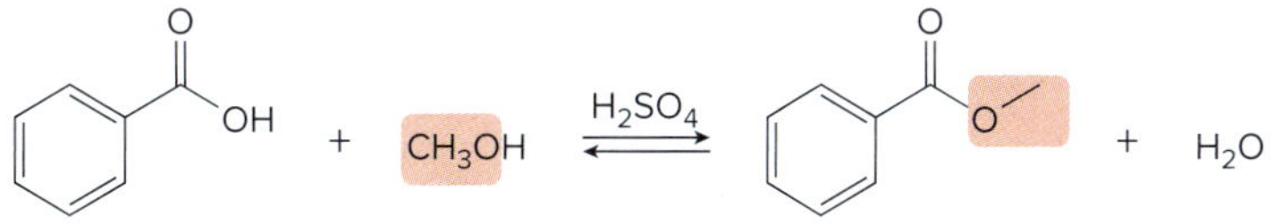

PhCOOH + CH₃OH $\xrightleftharpoons{H_2SO_4}$ methyl benzoate + H₂O

The mechanism for the Fischer esterification involves the usual two steps of nucleophilic acyl substitution—that is, **addition of a nucleophile followed by elimination of a leaving group.** Because the reaction is acid catalyzed, however, there are additional protonation and deprotonation steps. As always, though, the first step of any mechanism with an oxygen-containing starting material and an acid is to **protonate an oxygen atom** as shown with a general acid HA in Mechanism 16.5.

Mechanism 16.5 Fischer Esterification—Acid-Catalyzed Conversion of Carboxylic Acids to Esters

Part [1] Addition of the nucleophile R'OH

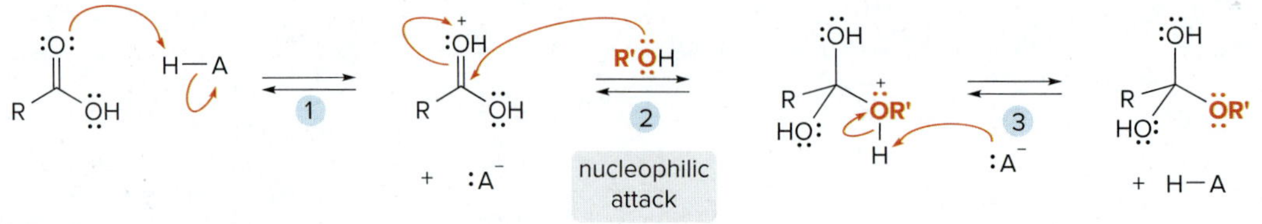

1. Protonation of the carbonyl oxygen makes the carbonyl more electrophilic.
2–3. **Nucleophilic attack** by R'OH forms a tetrahedral intermediate, and deprotonation gives the addition product.

Part [2] Elimination of the leaving group H_2O

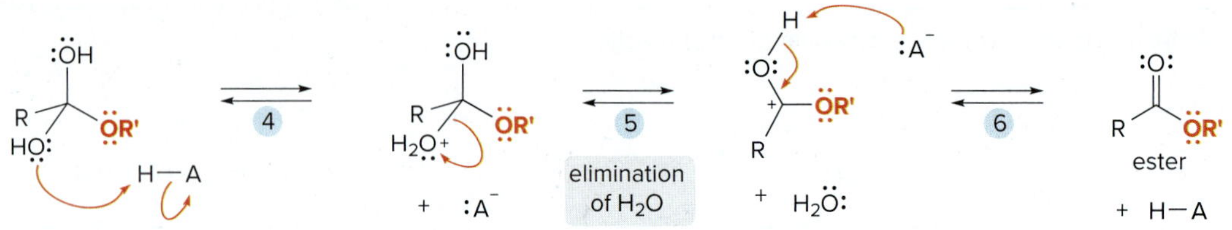

4. Protonation of the OH group forms a **good leaving group.**
5–6. Loss of H_2O and deprotonation give the **ester.**

Esterification of a carboxylic acid occurs in the presence of acid but *not* in the presence of base. Base removes a proton from the carboxylic acid, forming an electron-rich carboxylate anion, which does not react with an electron-rich nucleophile.

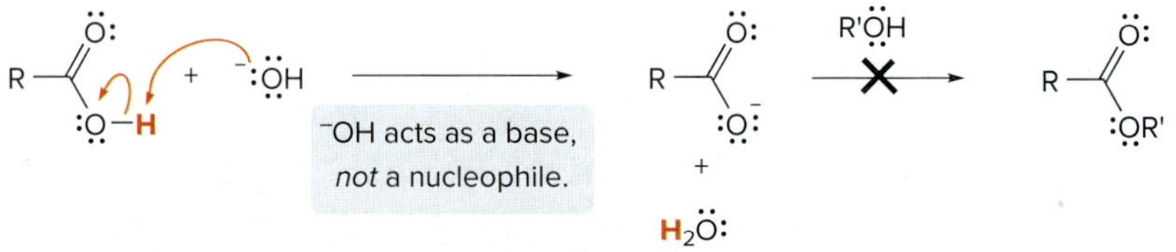

Intramolecular esterification of γ- and δ-hydroxy carboxylic acids forms five- and six-membered lactones.

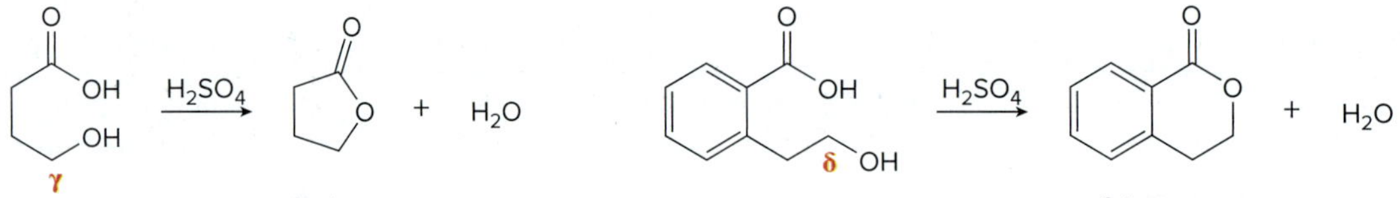

γ-lactone

δ-lactone

Problem 16.16 Draw the products of each reaction.

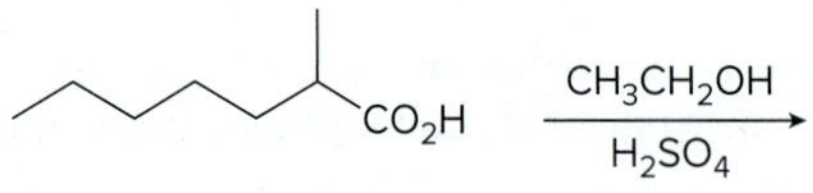

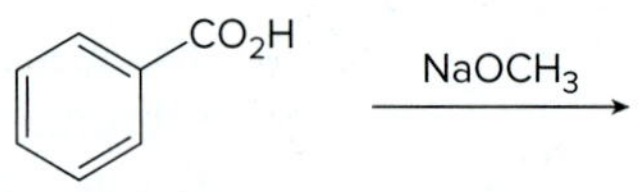

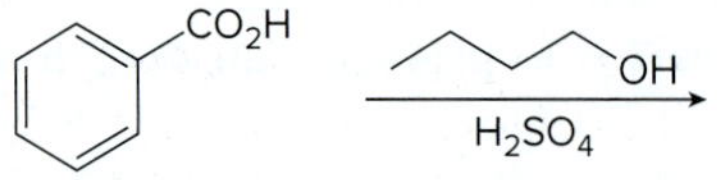

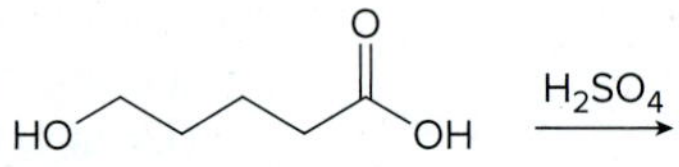

16.9 Reactions of Carboxylic Acids 755

Problem 16.17 Draw the products formed when benzoic acid (C₆H₅CO₂H) is treated with CH₃OH having its O atom labeled with ¹⁸O (CH₃¹⁸OH). Indicate where the labeled oxygen atom resides in the products.

Problem 16.18 Draw a stepwise mechanism for the following reaction.

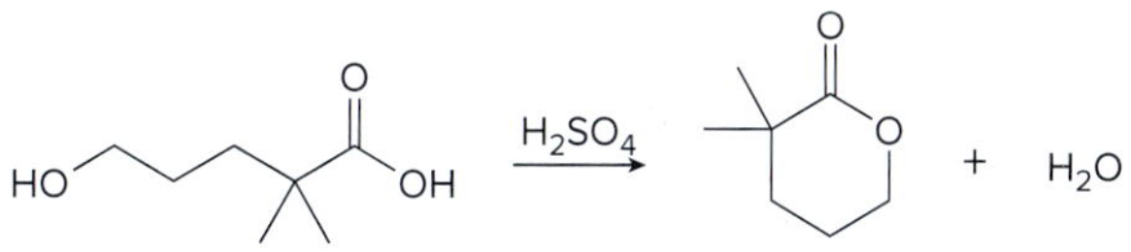

16.9D Conversion of RCOOH to RCONR′₂

The direct conversion of a carboxylic acid to an amide with NH₃ or an amine is very difficult, even though a more reactive acyl compound is being transformed into a less reactive one. The problem is that carboxylic acids are strong organic acids and NH₃ and amines are bases, so they undergo an **acid–base reaction to form an ammonium salt** before any nucleophilic substitution occurs.

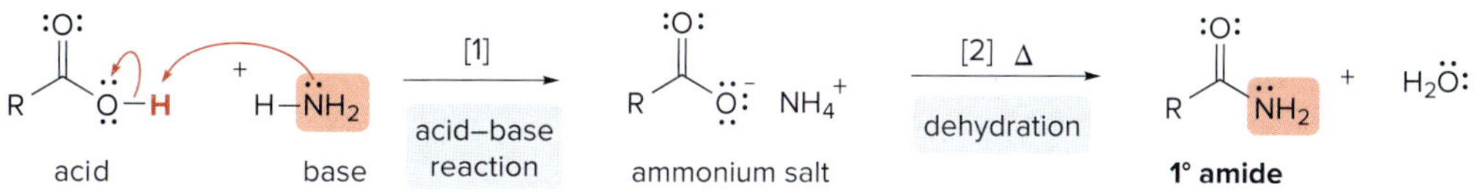

Amides are much more easily prepared from acid chlorides and anhydrides, as discussed in Sections 16.7 and 16.8.

Heating at high temperature (>100 °C) dehydrates the resulting ammonium salt of the carboxylate anion to form an amide, though the yield can be low.

Therefore, the overall conversion of RCOOH to RCONH₂ requires two steps:

[1] **Acid–base reaction of RCOOH with NH₃ to form an ammonium salt**
[2] **Dehydration at high temperature (>100 °C)**

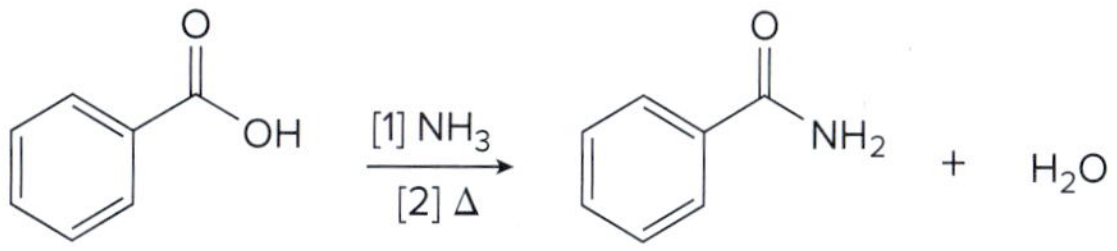

A carboxylic acid and an amine readily react to form an amide in the presence of an additional reagent, **dicyclohexylcarbodiimide (DCC)**, which is converted to the by-product dicyclohexylurea in the course of the reaction.

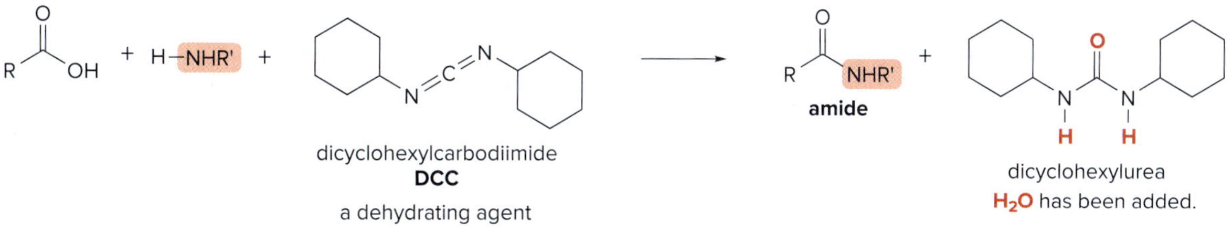

DCC is a dehydrating agent. The dicyclohexylurea by-product is formed by adding the elements of H₂O to DCC. DCC promotes amide formation by converting the carboxy OH group to a better leaving group.

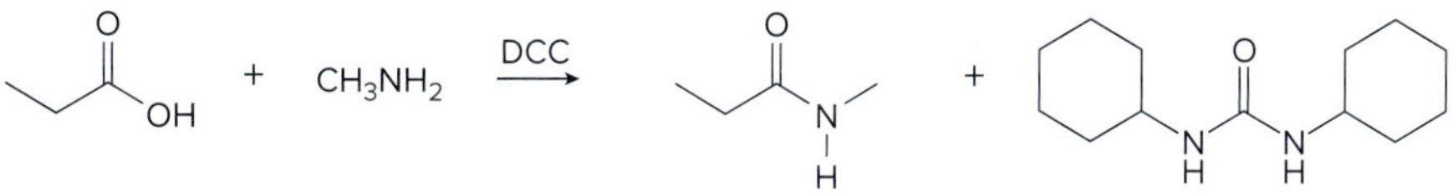

The mechanism consists of two parts: [1] conversion of the OH group to a better leaving group, followed by [2] **addition of the nucleophile and loss of the leaving group** to form the product of nucleophilic acyl substitution (Mechanism 16.6).

Mechanism 16.6 Conversion of Carboxylic Acids to Amides with DCC

Part [1] Conversion of OH to a better leaving group

1. Acid–base reaction results in transfer of a proton from the carboxylic acid to DCC.
2. Nucleophilic attack of RCO₂⁻ on the conjugate acid of DCC forms an addition product. The overall result of Steps [1] and [2] is conversion of OH to a **better leaving group**.

Part [2] Addition of the nucleophile and loss of the leaving group

3. Nucleophilic attack of the amine on the activated carboxy group forms a tetrahedral intermediate.
4–5. Proton transfer and elimination of dicyclohexylurea as the leaving group form the **amide**.

The reaction of an acid and an amine with DCC is often used in the laboratory to form the amide bond in peptides, as is discussed in Chapter 23.

Problem 16.19 What product is formed when acetic acid is treated with each reagent: (a) CH_3NH_2; (b) CH_3NH_2, then heat; (c) CH_3NH_2 + DCC?

16.10 Reactions of Esters

Esters can be converted to carboxylic acids and amides.

- Esters are hydrolyzed with water in the presence of either acid or base to form carboxylic acids or carboxylate anions.

- Esters react with NH_3 and amines to form 1°, 2°, or 3° amides.

16.10A Ester Hydrolysis in Aqueous Acid

The hydrolysis of esters in aqueous acid is a reversible equilibrium reaction that is driven to the right by using a large excess of water.

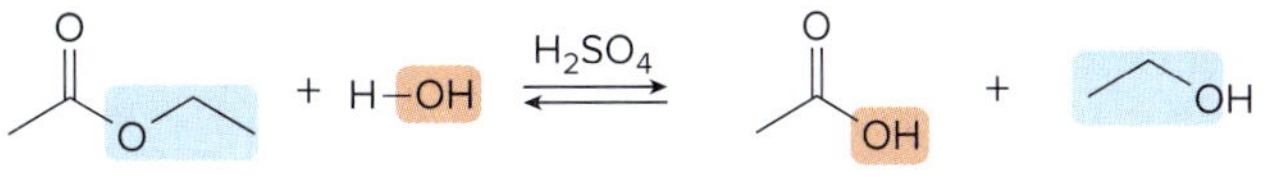

*The first step in acid-catalyzed ester hydrolysis is **protonation on oxygen**, the same first step of any mechanism involving an oxygen-containing starting material and an acid.*

The mechanism of ester hydrolysis in acid (shown in Mechanism 16.7) is the reverse of the mechanism of ester synthesis from carboxylic acids (Mechanism 16.5). Thus, the mechanism consists of the **addition of the nucleophile and the elimination of the leaving group,** the two steps common to all nucleophilic acyl substitutions, as well as several proton transfers, because the reaction is acid-catalyzed.

Mechanism 16.7 Acid-Catalyzed Hydrolysis of an Ester to a Carboxylic Acid

Part [1] Addition of the nucleophile H_2O

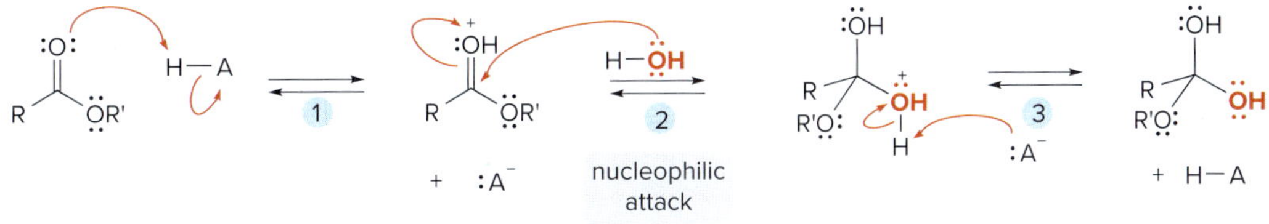

1. Protonation of the carbonyl oxygen makes the carbonyl more electrophilic.
2 – 3. **Nucleophilic attack by H_2O** forms a tetrahedral intermediate, and deprotonation gives the addition product.

Part [2] Elimination of the leaving group R'OH

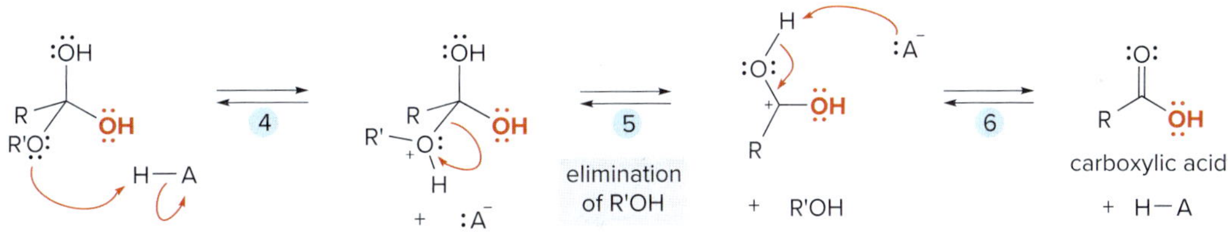

4. Protonation of the OR' group forms a **good leaving group.**
5 – 6. Loss of R'OH and deprotonation give the **carboxylic acid.**

16.10B Ester Hydrolysis in Aqueous Base

Esters are hydrolyzed in aqueous base to form carboxylate anions. Basic hydrolysis of an ester is called **saponification.**

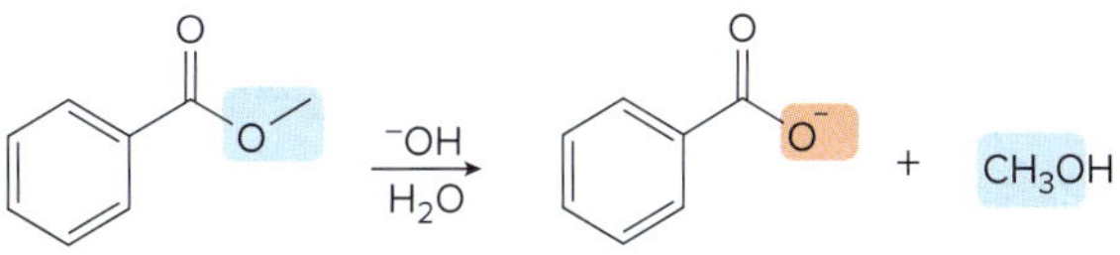

*The word **saponification** comes from the Latin sapo, meaning "soap." Soap is prepared by hydrolyzing esters in fats with aqueous base, as explained in Section 16.11B.*

The mechanism for this reaction has the usual two steps of the general mechanism for nucleophilic acyl substitution presented in Section 16.6A—**addition of the nucleophile** followed by **loss of a leaving group**—plus an additional step involving proton transfer (Mechanism 16.8).

Mechanism 16.8 Base-Promoted Hydrolysis of an Ester to a Carboxylate Anion

1–2 Addition of the nucleophile (⁻OH) followed by **elimination of the leaving group (⁻OR')** form a carboxylic acid. These two steps are reversible.

3 Because the carboxylic acid is a strong organic acid and the leaving group (⁻OR') is a strong base, an acid–base reaction forms the **carboxylate anion.**

The carboxylate anion is resonance stabilized, and this drives the equilibrium in its favor. Once the reaction is complete and the carboxylate anion is formed, it can be protonated with strong acid to form the neutral carboxylic acid.

Hydrolysis is base promoted, *not* base catalyzed, because the base (⁻OH) is the nucleophile that adds to the ester and forms part of the product. It participates in the reaction and is not regenerated later.

Where do the oxygen atoms in the product come from? **The C–OR' bond in the ester is cleaved,** so the OR' group becomes the alcohol by-product (R'OH) and **one of the oxygens in the carboxylate anion product comes from ⁻OH** (the nucleophile).

Problem 16.20 Fenofibrate is a cholesterol-lowering medication that is converted to fenofibric acid, the active drug, by hydrolysis during metabolism. What is the structure of fenofibric acid?

fenofibrate

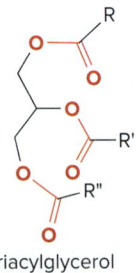

triacylglycerol
R groups have
11–19 C's.
[Three ester groups are labeled in red.]

16.11 Application: Lipid Hydrolysis

16.11A Olestra—A Synthetic Fat

The most prevalent naturally occurring esters are the **triacylglycerols,** which were first discussed in Section 10.6. **Triacylglycerols are the lipids that comprise animal fats and vegetable oils.**

- Each triacylglycerol is a triester, containing three long hydrocarbon side chains.
- Unsaturated triacylglycerols have one or more double bonds in their long hydrocarbon chains, whereas saturated triacylglycerols have none.

Figure 16.3 contains a ball-and-stick model of a saturated fat.

Figure 16.3

The three-dimensional structure of a saturated triacylglycerol

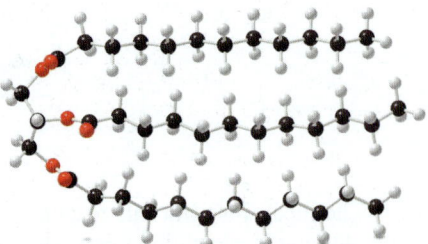

- This triacylglycerol has no double bonds in the three R groups (each with 11 C's) bonded to the ester carbonyls, making it a saturated fat.

16.11 Application: Lipid Hydrolysis

Animals store energy in the form of triacylglycerols, kept in a layer of fat cells below the surface of the skin. This fat serves to insulate the organism, as well as provide energy for its metabolic needs for long periods. The first step in the metabolism of a triacylglycerol is **hydrolysis of the ester bonds to form glycerol and three fatty acids.** In cells, this reaction is carried out with enzymes called **lipases.**

[The three bonds drawn in red are cleaved in hydrolysis.]

The fatty acids produced on hydrolysis are then oxidized in a stepwise fashion, ultimately yielding CO_2 and H_2O, as well as a great deal of energy. Oxidation of fats yields twice as much energy per gram as oxidation of an equivalent weight of carbohydrate.

Diets high in fat content lead to a large amount of stored fat, ultimately causing an individual to be overweight. One attempt to reduce calories in common snack foods has been to substitute "fake fats" such as **olestra** (trade name **Olean**) for triacylglycerols.

Some snack foods contain the "fake fat" olestra, giving them fewer calories than snack foods containing triacylglycerols for the calorie-conscious consumer.
Jill Braaten/McGraw-Hill Education

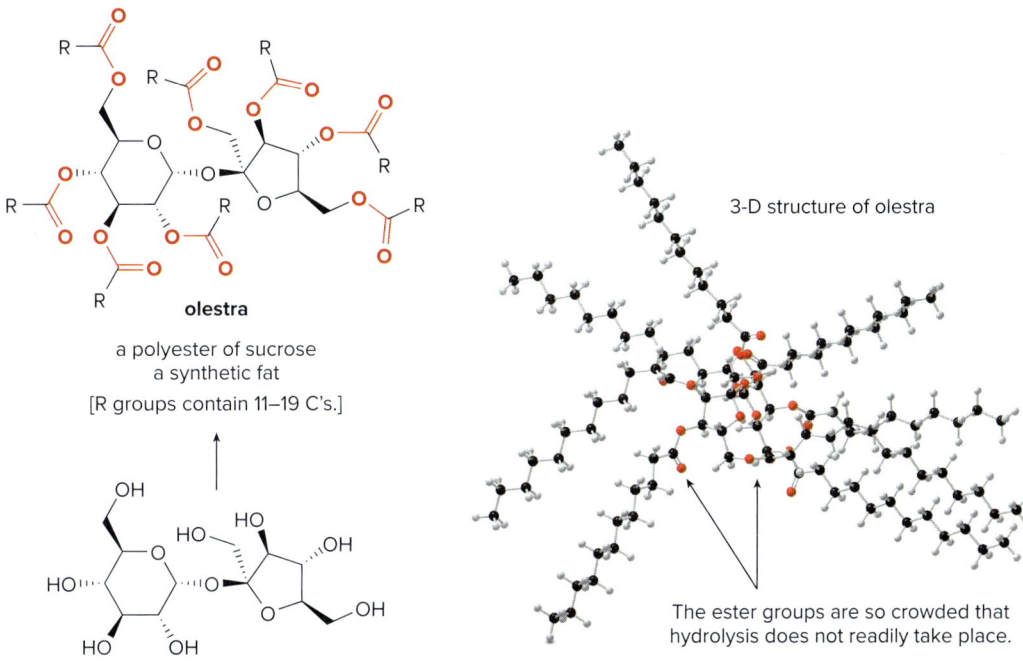

olestra
a polyester of sucrose
a synthetic fat
[R groups contain 11–19 C's.]

sucrose

3-D structure of olestra

The ester groups are so crowded that hydrolysis does not readily take place.

Olestra is a polyester formed from long-chain fatty acids and sucrose, the sweet-tasting carbohydrate in table sugar. Naturally occurring triacylglycerols are also polyesters formed from long-chain fatty acids, but olestra has so many ester units clustered together in close proximity that they are too hindered to be hydrolyzed, so it passes through the body unchanged, providing no calories to the consumer.

Thus, olestra's many C–C and C–H bonds make it similar in solubility to naturally occurring triacylglycerols, but its three-dimensional structure makes it inert to hydrolysis because of steric hindrance.

Problem 16.21 How would you synthesize olestra from sucrose?

16.11B The Synthesis of Soap

Soap was discussed in Section 3.6.

Soap is prepared by the basic hydrolysis or saponification of a triacylglycerol. Heating an animal fat or vegetable oil with aqueous base hydrolyzes the three esters to form glycerol and sodium salts of three fatty acids. These carboxylate salts are **soaps,** which clean away dirt because of their two structurally different regions. The nonpolar tail dissolves grease and oil and the polar head makes it soluble in water (Figure 3.4).

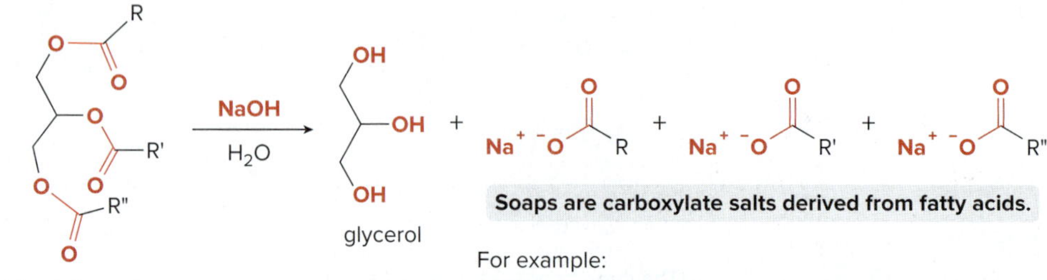

Soaps are carboxylate salts derived from fatty acids.

For example:

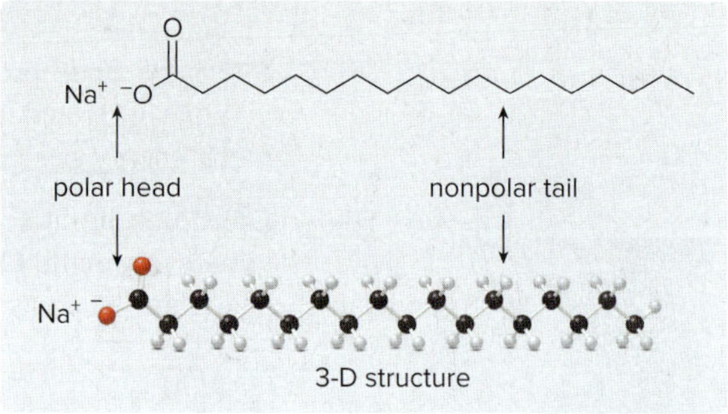

All soaps are salts of fatty acids. The main difference between soaps is the addition of other ingredients that do not alter their cleaning properties: dyes for color, scents for a pleasing odor, and oils for lubrication. Soaps that float are aerated, so that they are less dense than water. *Jill Braaten/McGraw-Hill Education*

Soaps are typically made from lard (from hogs), tallow (from cattle or sheep), coconut oil, or palm oil. Most triacylglycerols have two or three different R groups in their hydrocarbon chains, so soaps are usually mixtures of two or three different carboxylate salts.

Problem 16.22 What is the composition of the soap prepared by hydrolysis of the following triacylglycerol?

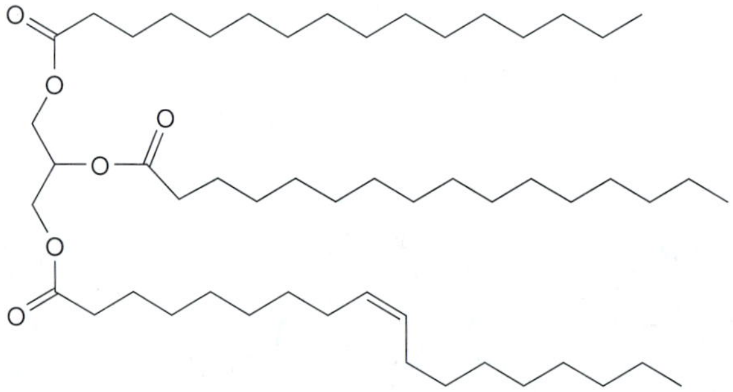

16.12 Reactions of Amides

Because amides have the poorest leaving group of all the carboxylic acid derivatives, they are the least reactive. Under strenuous reaction conditions, **amides are hydrolyzed in acid or base to form carboxylic acids or carboxylate anions.**

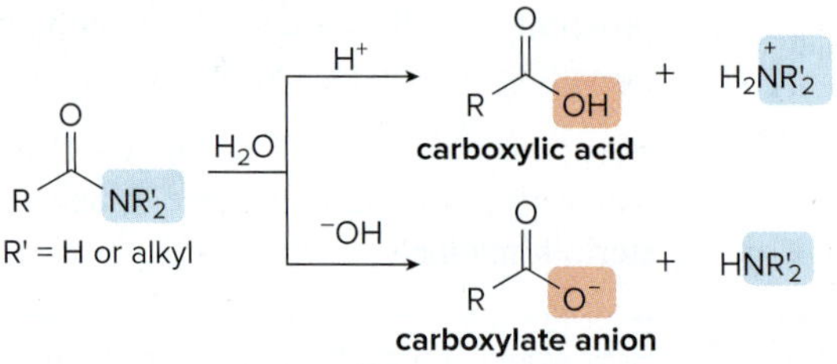

16.12 Reactions of Amides

In acid, the amine by-product is protonated as an ammonium ion, whereas in base, a neutral amine is formed.

CH₃C(O)NH₂ + H₃O⁺ → CH₃C(O)OH + NH₄⁺

PhC(O)N(H)CH₃ + H₂O/⁻OH → PhC(O)O⁻ + CH₃NH₂

The relative lack of reactivity of the amide bond is notable in proteins, which are polymers of amino acids connected by amide linkages (Section 16.5B). Proteins are stable in aqueous solution in the absence of acid or base, so they can perform their various functions in the aqueous cellular environment without breaking down. The hydrolysis of the amide bonds in proteins requires a variety of specific enzymes.

The mechanism of amide hydrolysis in acid is exactly the same as the mechanism of ester hydrolysis in aqueous acid (Section 16.10A) except that the leaving group is different.

The mechanism of amide hydrolysis in base has the usual two steps of the general mechanism for nucleophilic acyl substitution—**addition of the nucleophile** followed by **loss of a leaving group**—plus an additional proton transfer. The initially formed carboxylic acid reacts further under basic conditions to form the resonance-stabilized carboxylate anion, and this drives the reaction to completion. Mechanism 16.9 is written for a 1° amide.

Mechanism 16.9 Amide Hydrolysis in Base

1–2 Addition of the nucleophile (⁻OH) followed by **elimination of the leaving group** (⁻NH₂) form a carboxylic acid. These two steps are reversible.

3 Because the carboxylic acid is a strong organic acid and the leaving group (⁻NH₂) is a strong base, an acid–base reaction forms the **carboxylate anion**.

Step [2] of Mechanism 16.9 deserves additional comment. For amide hydrolysis to occur, the tetrahedral intermediate must lose ⁻NH₂, a *stronger* base and therefore *poorer* leaving group than ⁻OH. This means that loss of ⁻NH₂ does not often happen. Instead, ⁻OH is lost as the leaving group most of the time, and the starting material is regenerated. But, when ⁻NH₂ is occasionally eliminated, the carboxylic acid product is converted to a lower-energy carboxylate anion in Step [3], and this drives the equilibrium to favor its formation.

Problem 16.23 With reference to the structures of acetylsalicylic acid (aspirin) and acetaminophen (the active ingredient in Tylenol), explain why acetaminophen tablets can be stored in the medicine cabinet for years, but aspirin tablets slowly decompose over time.

acetylsalicylic acid   acetaminophen

16.13 Application: The Mechanism of Action of β-Lactam Antibiotics

Penicillin and related β-lactams kill bacteria by a nucleophilic acyl substitution reaction. All penicillins have an unreactive amide side chain and a very reactive amide that is part of a β-lactam. The β-lactam is more reactive than other amides because it is part of a strained, four-membered ring that is readily opened with nucleophiles.

Unlike mammalian cells, bacterial cells are surrounded by a fairly rigid cell wall, which allows the bacterium to live in many different environments. This protective cell wall is composed of carbohydrates linked together by peptide chains containing amide linkages, formed using the enzyme **glycopeptide transpeptidase**.

Penicillin interferes with the synthesis of the bacterial cell wall. A nucleophilic OH group of the glycopeptide transpeptidase enzyme cleaves the β-lactam ring of penicillin by a **nucleophilic acyl substitution reaction**. The opened ring of the penicillin molecule remains covalently bonded to the enzyme, thus deactivating the enzyme, halting cell wall construction, and killing the bacterium. Penicillin has no effect on mammalian cells because they are surrounded by a flexible membrane composed of a lipid bilayer (Chapter 3) and not a cell wall.

Thus, penicillin and other β-lactam antibiotics are biologically active precisely because they undergo a nucleophilic acyl substitution reaction with an important bacterial enzyme.

Problem 16.24 Some penicillins cannot be administered orally because their β-lactam is rapidly hydrolyzed by the acidic environment of the stomach. What product is formed in the following hydrolysis reaction?

16.14 Summary of Nucleophilic Acyl Substitution Reactions

To help you organize and remember all of the nucleophilic acyl substitution reactions that can occur at a carbonyl carbon, keep in mind these two principles:

- The *better* the leaving group, the *more reactive* the carboxylic acid derivative.
- More reactive acyl compounds can always be converted to less reactive ones. The reverse is not usually true.

This results in the following order of reactivity:

$RCONR'_2$ RCO_2H ≈ RCO_2R' $(RCO)_2O$ $RCOCl$

Increasing reactivity →

Table 16.5 summarizes the specific nucleophilic acyl substitution reactions. Use it as a quick reference to remind you which products can be formed from a given starting material.

Table 16.5 Summary of the Nucleophilic Substitution Reactions of Carboxylic Acids and Their Derivatives

Starting material		Product				
		RCOCl	(RCO)$_2$O	RCO$_2$H	RCO$_2$R'	RCONR'$_2$
[1] RCOCl	→	—	✓	✓	✓	✓
[2] (RCO)$_2$O	→	✗	—	✓	✓	✓
[3] RCO$_2$H	→	✓	✓	—	✓	✓
[4] RCO$_2$R'	→	✗	✗	✓	—	✓
[5] RCONR'$_2$	→	✗	✗	✓	✗	—

Table key: ✓ = A reaction occurs.
✗ = No reaction occurs.

16.15 Acyl Phosphates—Biological Anhydrides

As we learned in Section 15.8, carboxylic acids typically exist as carboxylates at the physiological pH of 7.4 in cells. **For a carboxylate anion to undergo nucleophilic acyl substitution, it must first be converted to a more reactive acyl derivative.** Acid chlorides and anhydrides react too rapidly with water to survive in the aqueous environment of a biological system, so carboxylates must be "activated" to nucleophilic attack by conversion to other reactive acyl compounds. Typically, a carboxylate is converted to an **acyl phosphate** (or similar phosphorus derivative) or a **thioester.**

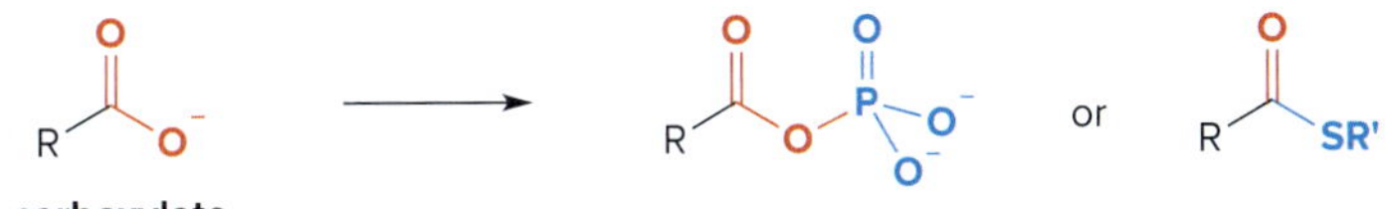

carboxylate → acyl phosphate or thioester

The synthesis and reactions of acyl phosphates are discussed in this section, followed by the reactions of thioesters in Section 16.16.

16.15A The Conversion of RCOO⁻ to Acyl Phosphates

Carboxylates are converted to more reactive phosphorus derivatives by reaction with **adenosine 5'-triphosphate, ATP** (Sections 2.8 and 6.4). ATP activates a carboxylate toward nucleophilic attack by two different mechanisms, involving the attack of the carboxylate at either the γ phosphorus (in red) or the α phosphorus (in green).

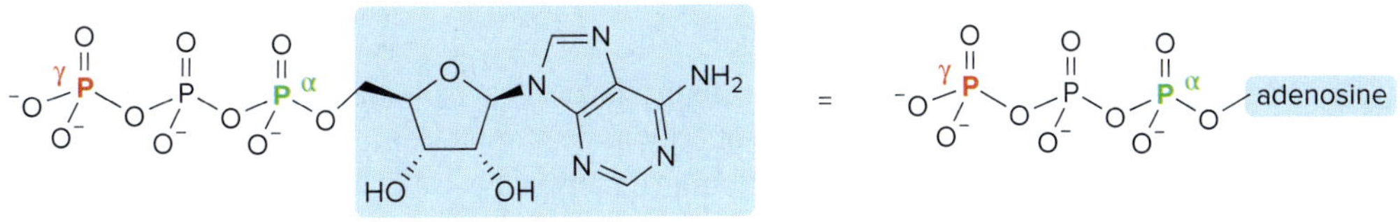

adenosine 5'-triphosphate
ATP

Both reactions form the phosphorus analogue of a mixed anhydride (Section 16.1) and a resonance-stabilized phosphorus leaving group.

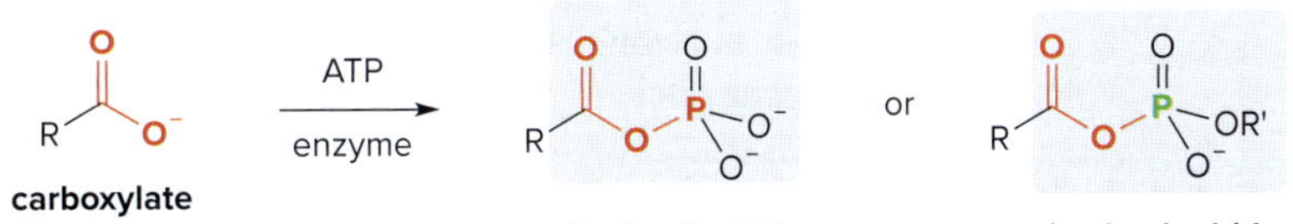

carboxylate —ATP/enzyme→ mixed anhydride from attack the γ P or mixed anhydride from attack the α P

764 Chapter 16 Carboxylic Acids and Their Derivatives—Nucleophilic Acyl Substitution

The enzyme-catalyzed reaction of a carboxylate with ATP at the **γ phosphorus** forms an **acyl phosphate** and adenosine 5′-diphosphate, ADP.

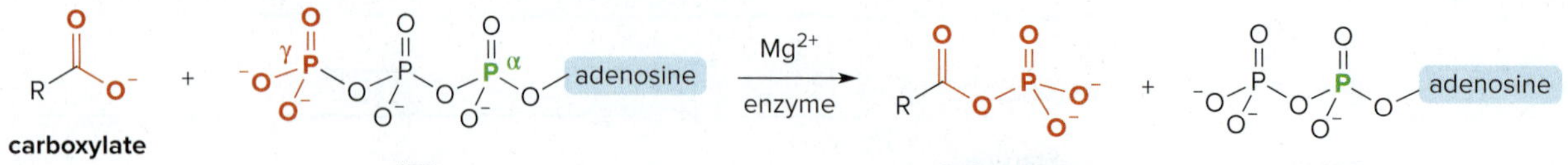

As mentioned in Section 2.8, the reaction occurs in the presence of Mg^{2+}, which acts as a Lewis acid to activate a phosphate and increase the electrophilicity of the phosphorus, as shown in the charge-separated resonance structure, which places a full positive charge on phosphorus.

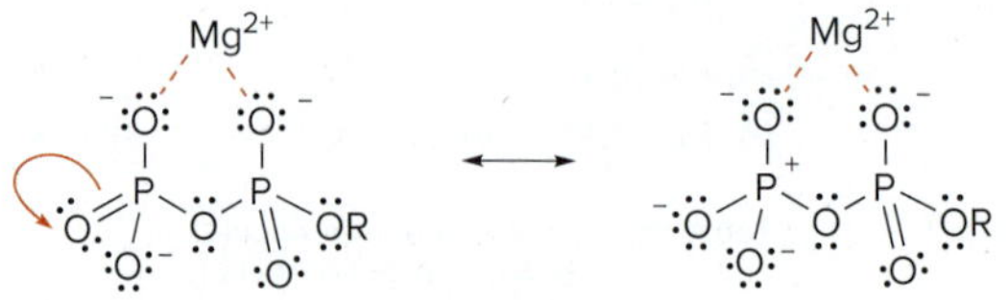

The conversion of a carboxylate into an acyl phosphate is a nucleophilic substitution reaction analogous to nucleophilic acyl substitution, but with a P=O electrophile and a carboxylate nucleophile. Mechanism 16.10 illustrates the steps for this magnesium-activated conversion.

Mechanism 16.10 Biological Conversion of a Carboxylate to an Acyl Phosphate

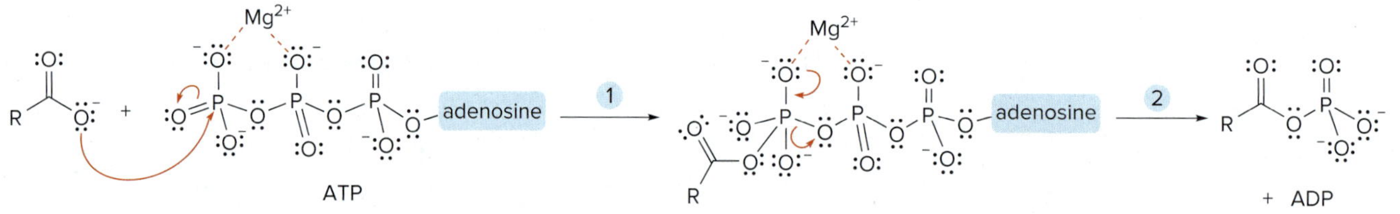

1. The nucleophile (RCO_2^-) attacks the magnesium-activated phosphate.
2. Elimination of the leaving group (ADP) forms the substitution product, an acyl phosphate.

The enzyme-catalyzed reaction of a carboxylate with ATP at the **α phosphorus** forms a mixed anhydride called an **acyl adenosyl phosphate** and a resonance-stabilized diphosphate as leaving group.

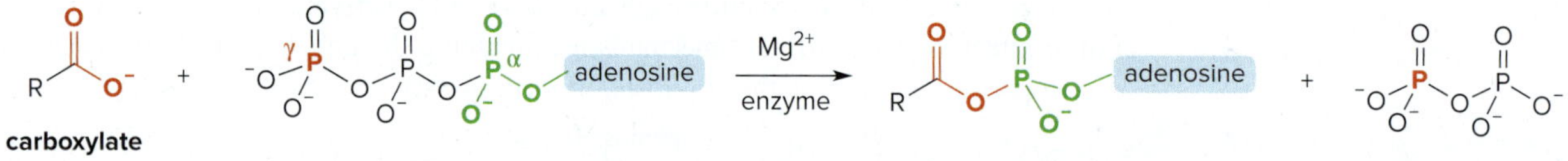

When this reaction involves a carboxylate derived from a fatty acid, it is the first step in the metabolism of fatty acids, a multistep process discussed in Section 27.3. The acyl adenosyl phosphate derived from a fatty acid is called a **fatty acyl AMP** (fatty acyl adenosine 5′-monophosphate).

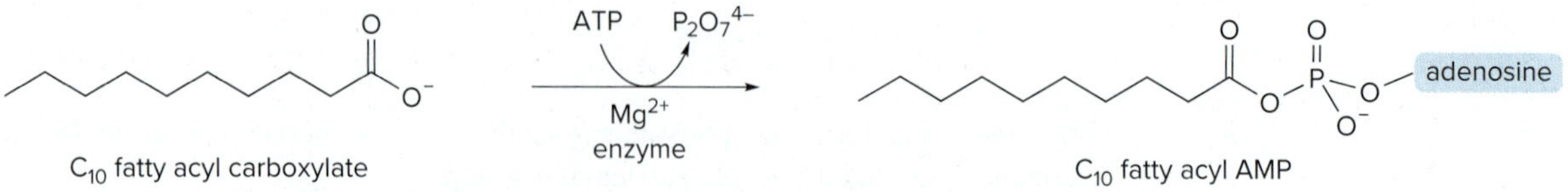

C_{10} fatty acyl carboxylate → C_{10} fatty acyl AMP (ATP, $P_2O_7^{4-}$, Mg^{2+}, enzyme)

Whether nucleophilic attack of a carboxylate occurs at the α or γ phosphorus of ATP depends on the enzyme.

Problem 16.25 Draw a stepwise mechanism for the formation of a fatty acyl AMP from a fatty acyl carboxylate and ATP.

16.15B Reactions of Acyl Phosphates

The acyl phosphate is the most reactive biological carboxylic acid derivative. Nucleophilic acyl substitution reactions occur with these mixed anhydrides just as they do with other carboxylic acid derivatives.

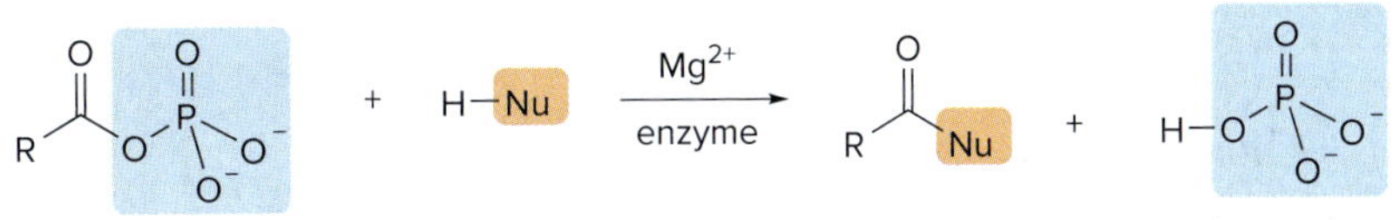

acyl phosphate

by-product

For example, cells synthesize glutamine (an amino acid) from an acyl phosphate derived from glutamate (another amino acid). In this reaction, ammonia (the nucleophile) displaces the phosphate group on γ-glutamyl phosphate to give glutamine.

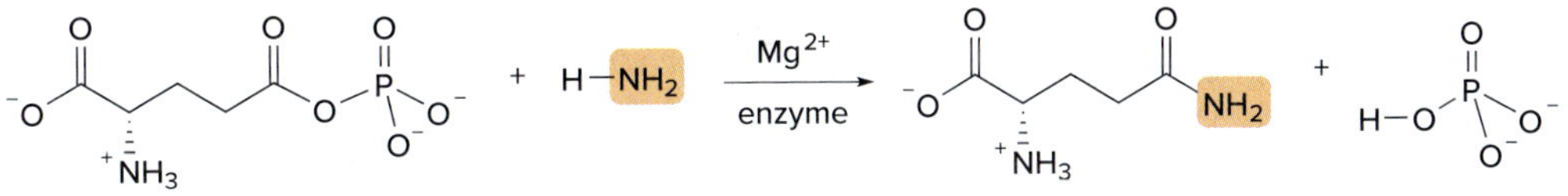

γ-glutamyl phosphate

glutamine

Problem 16.26 a. What is the structure of glutamate, the carboxylate precursor of γ-glutamyl phosphate?

b. Using the information in Section 16.15A, write out the reaction scheme for the formation of γ-glutamyl phosphate from glutamate.

Thiols are also common biological nucleophiles. During fatty acid metabolism, the fatty acyl AMP prepared in Section 16.15A is converted into a thioester, **fatty acyl CoA,** using the thiol in coenzyme A (Section 3.8) as the nucleophile.

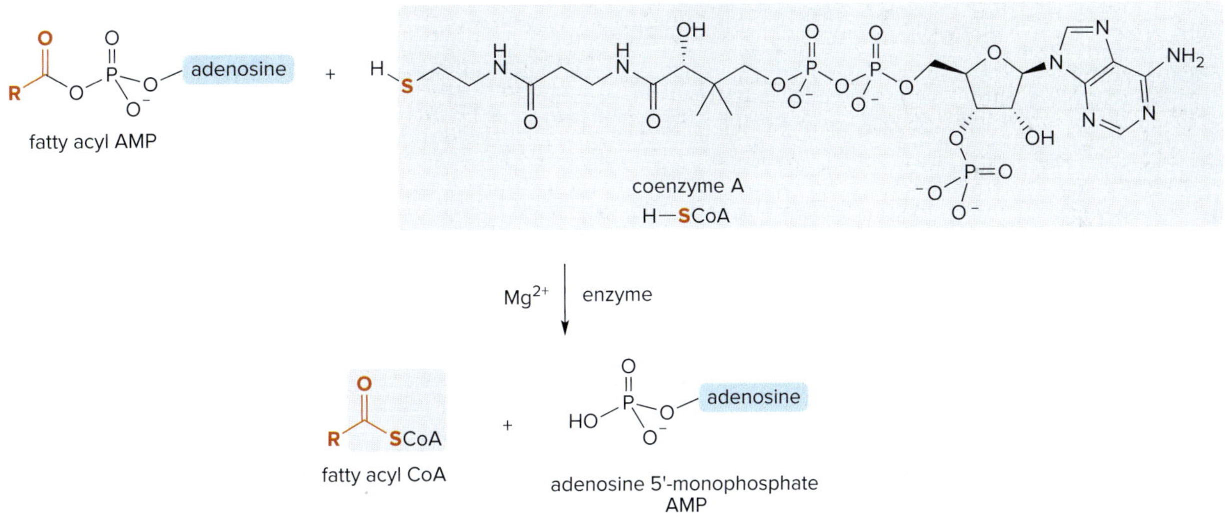

The conversion of an acyl phosphate to a thioester illustrates the mechanism of a biological nucleophilic acyl substitution with an acyl phosphate as starting material (Mechanism 16.11). Mg^{2+} activates the acyl group to increase the electrophilicity of the carbonyl carbon. It is typical in enzyme mechanisms to show the proton transfers as part of the **nucleophilic addition** and **elimination** steps.

Problem 16.27 Draw the products of the following reaction, one step in the synthesis of the amino acid asparagine.

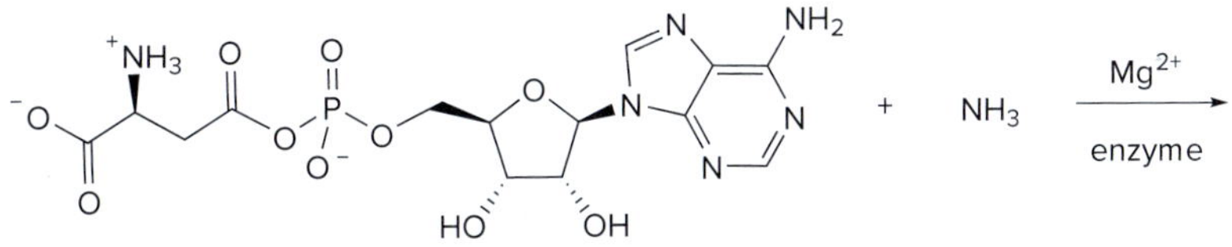

Mechanism 16.11 Biological Conversion of an Acyl Phosphate to a Thioester

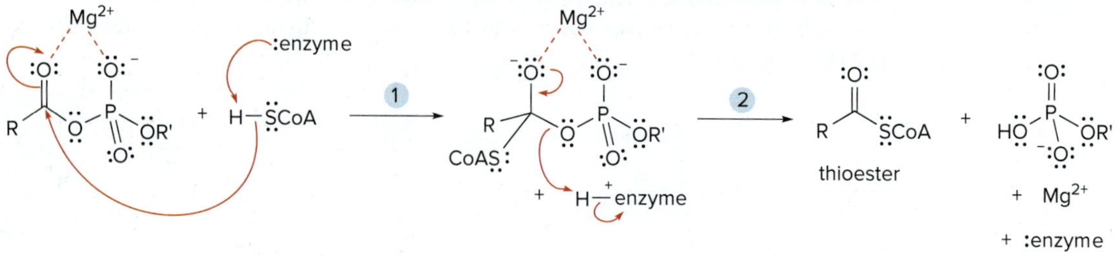

① The nucleophile (HSCoA) attacks the magnesium-activated carbonyl, forming an sp^3 hybridized carbon.
② Elimination of the leaving group and protonation form the substitution product, a thioester.

Problem 16.28 Draw a stepwise mechanism for the following reaction. Coenzyme A (HSCoA) reacts with acetyl AMP to give acetyl CoA, a key biomolecule involved in metabolism, and adenosine 5′-monophosphate (AMP). You may use "R" groups to simplify the mechanism.

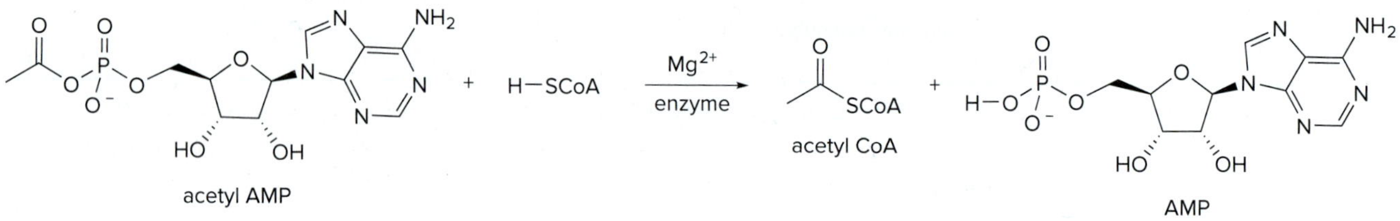

16.16 Reactions of Thioesters—Biological Acylation Reactions

Thioesters are common intermediates in cellular processes and typically undergo nucleophilic acyl substitution to transfer their acyl groups to cellular nucleophiles. These acylation reactions are called **acyl transfer reactions** because they result in the transfer of an acyl group from one atom to another (from Z to Nu in this case).

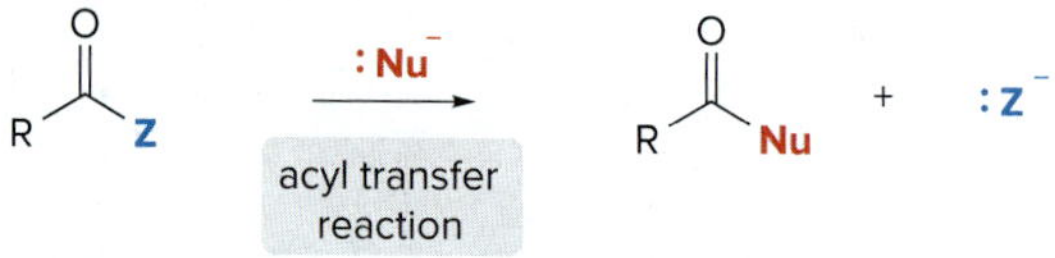

The most common thioester is **acetyl coenzyme A,** usually called **acetyl CoA.**

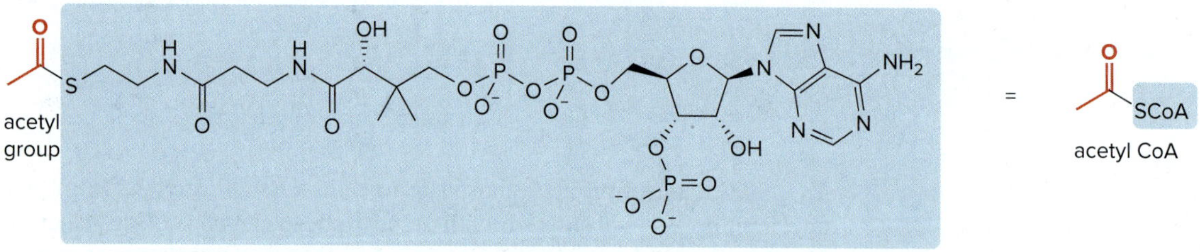

- A thioester (RCOSR′), like other acyl compounds, undergoes substitution reactions with nucleophiles. With acetyl CoA, an acetyl group is transferred from SCoA to a nucleophile, Nu.

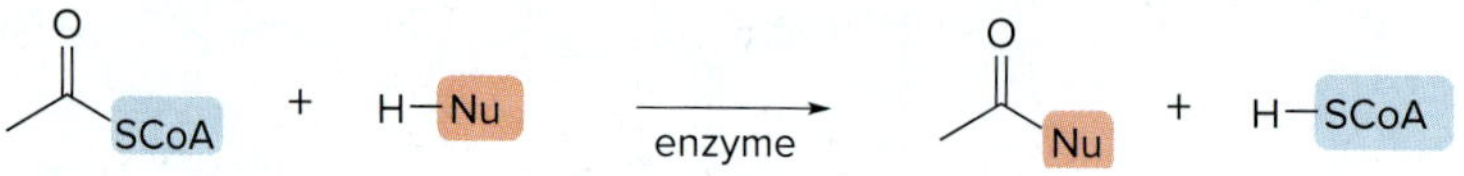

16.16 Reactions of Thioesters—Biological Acylation Reactions

For example, acetyl CoA undergoes enzyme-catalyzed nucleophilic acyl substitution with choline, forming acetylcholine, a charged compound that transmits nerve impulses between nerve cells.

[Reaction: acetyl CoA + choline → acetylcholine (neurotransmitter) + H–SCoA]

In another example, the acyl group of S-citryl CoA is converted to citrate, an intermediate in the citric acid cycle, an energy-generating metabolic cycle (Section 27.6). In this reaction, water is the nucleophile that displaces SCoA to ultimately form a carboxylate from a thioester.

[Reaction: S-citryl CoA + H–O–H → citrate + H–SCoA + H–enzyme⁺]

> **Problem 16.29** Using a protonated enzyme (H–enzyme$^+$) to activate the carbonyl, draw a stepwise mechanism for the formation of citrate from S-citryl CoA and H$_2$O.

Many other acyl transfer reactions are important cellular processes. Thioesters of fatty acids react with cholesterol, forming **cholesteryl esters** in an enzyme-catalyzed reaction (Figure 16.4). These esters are the principal form in which cholesterol is stored and transported in the body. Because cholesterol is a lipid, insoluble in the aqueous environment of the blood, it travels

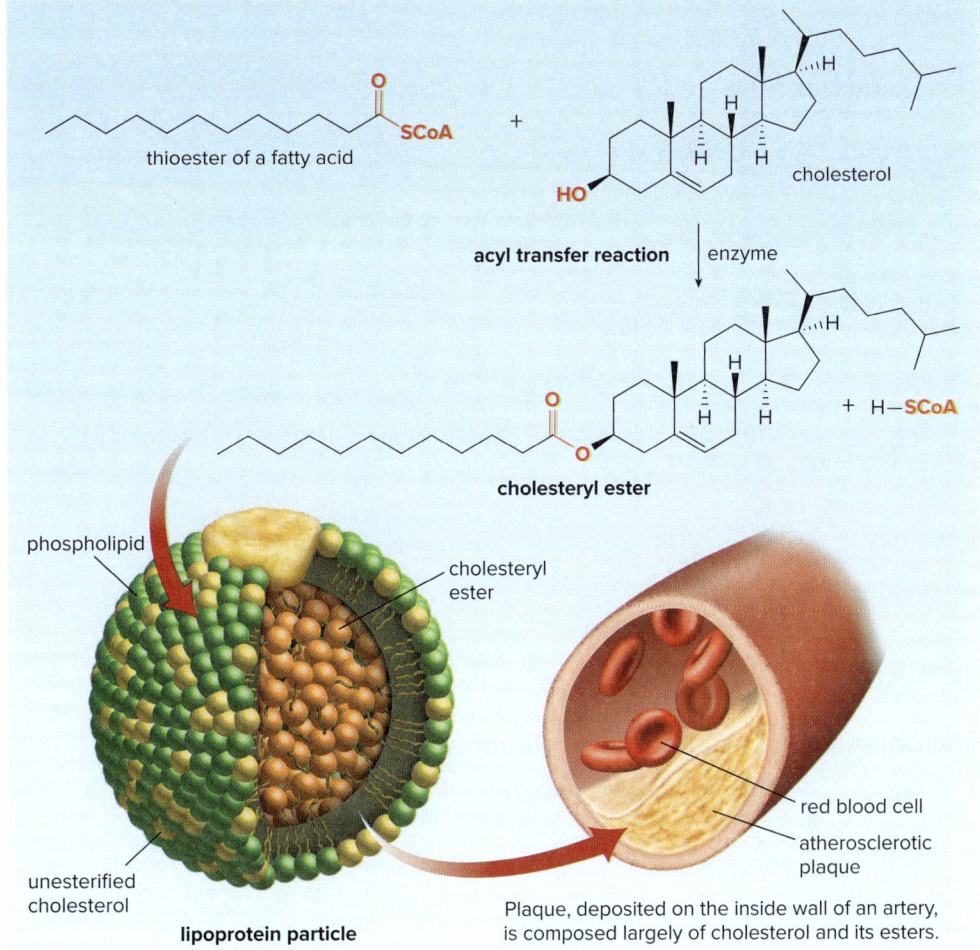

Figure 16.4
Cholesteryl esters and lipoprotein particles

Plaque, deposited on the inside wall of an artery, is composed largely of cholesterol and its esters.

through the bloodstream in particles that also contain proteins and phospholipids. These particles are classified by their density.

- **LDL particles** (low-density lipoproteins) transport cholesterol from the liver to the tissues.
- **HDL particles** (high-density lipoproteins) transport cholesterol from the tissues back to the liver, where it is metabolized or converted to other steroids.

Atherosclerosis is a disease that results from the buildup of fatty deposits on the walls of arteries, forming deposits called **plaque.** Plaque is composed largely of the cholesterol (esterified as an ester) of LDL particles. LDL is often referred to as "bad cholesterol" for this reason. In contrast, HDL particles are called "good cholesterol" because they reduce the amount of cholesterol in the bloodstream by transporting it back to the liver.

Problem 16.30 Glucosamine is a dietary supplement available in many over-the-counter treatments for osteoarthritis. Reaction of acetyl CoA with glucosamine forms NAG, *N*-acetylglucosamine, the monomer used to form chitin, the carbohydrate that forms the rigid shells of lobsters and crabs. What is the structure of NAG?

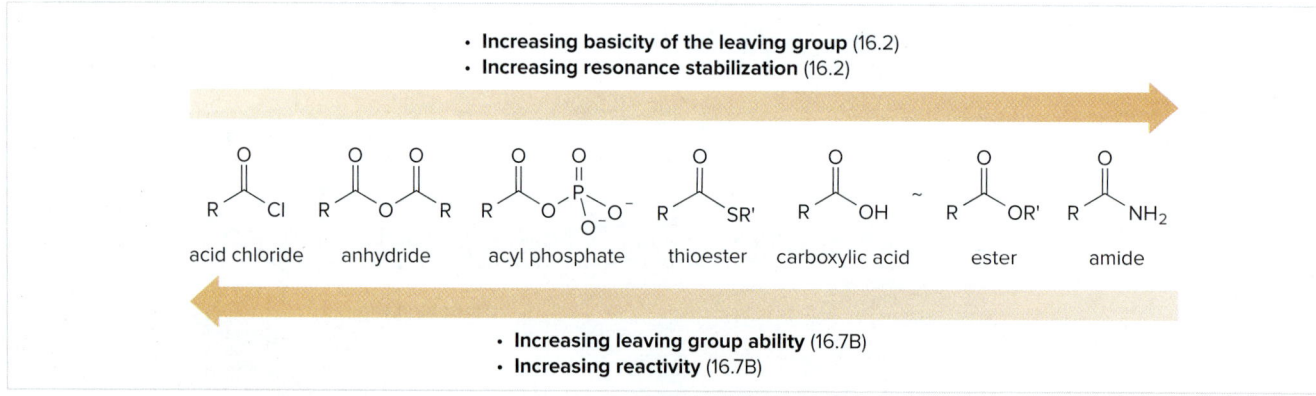

Chapter 16 REVIEW

KEY CONCEPTS

The relationship between the basicity of Z⁻ and the properties of RCOZ

- Increasing basicity of the leaving group (16.2)
- Increasing resonance stabilization (16.2)

acid chloride → anhydride → acyl phosphate → thioester → carboxylic acid → ester → amide

- Increasing leaving group ability (16.7B)
- Increasing reactivity (16.7B)

Try Problems 16.31, 16.58.

KEY REACTIONS

Nucleophilic Acyl Substitution Reactions

[1] Reactions that produce acid chlorides (RCOCl)

carboxylic acid + SOCl₂ (16.9A) → acid chloride + SO₂ + HCl

Try Problem 16.37c.

[2] Reactions that produce anhydrides [(RCO)$_2$O]

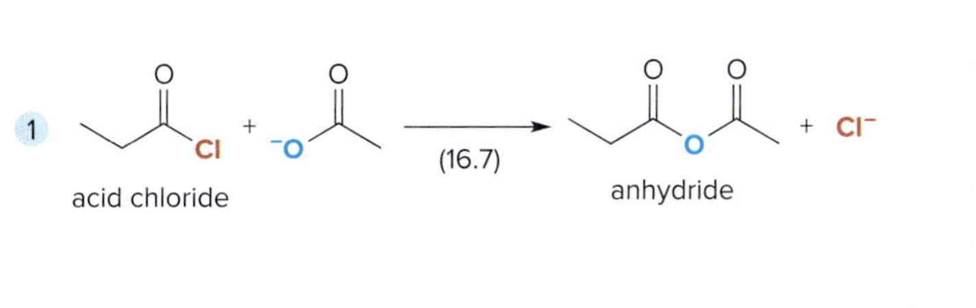

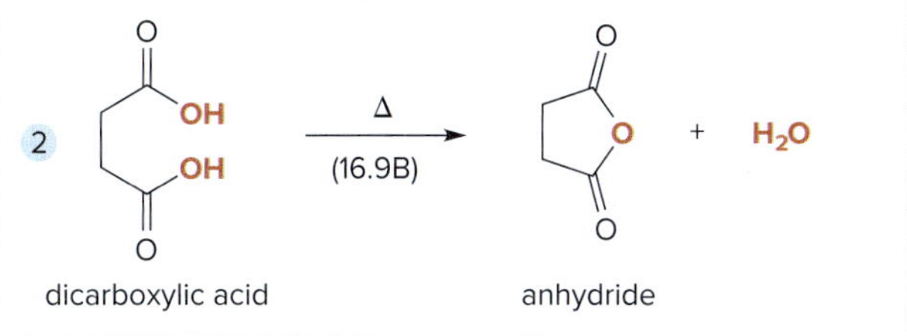

1. acid chloride + carboxylate → anhydride + Cl$^-$ (16.7)

2. dicarboxylic acid $\xrightarrow{\Delta}$ anhydride + H$_2$O (16.9B)

Try Problems 16.37i, 16.38g.

[3] Reactions that produce acyl phosphates (RCO$_2$PO$_3^{2-}$)

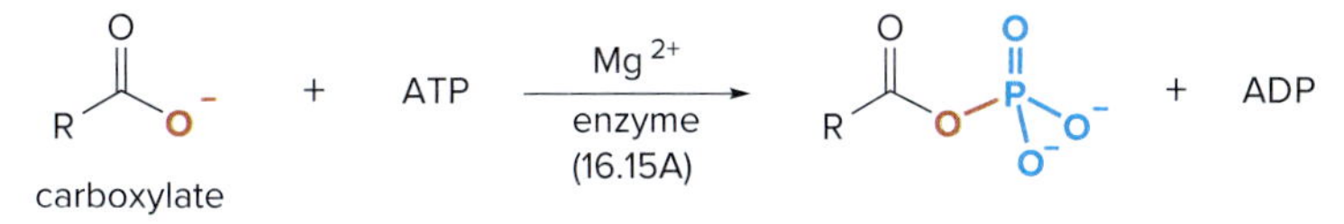

carboxylate + ATP $\xrightarrow[\text{enzyme}]{\text{Mg}^{2+}}$ acyl phosphate + ADP (16.15A)

[4] Reactions that produce thioesters (RCOSR')

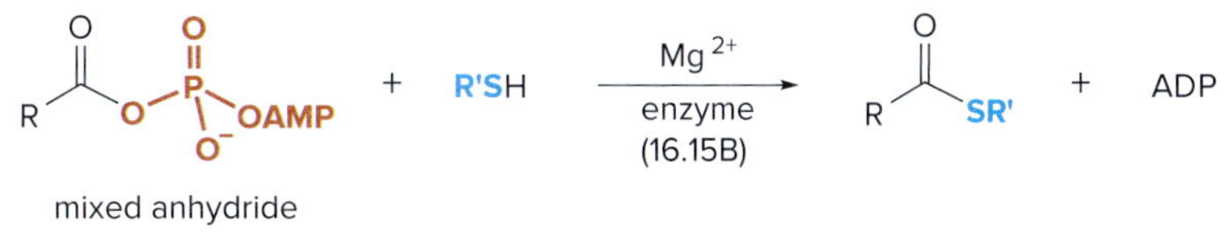

mixed anhydride + R'SH $\xrightarrow[\text{enzyme}]{\text{Mg}^{2+}}$ thioester (RCOSR') + ADP (16.15B)

[5] Reactions that produce carboxylic acids (RCOOH) and carboxylates (RCOO$^-$)

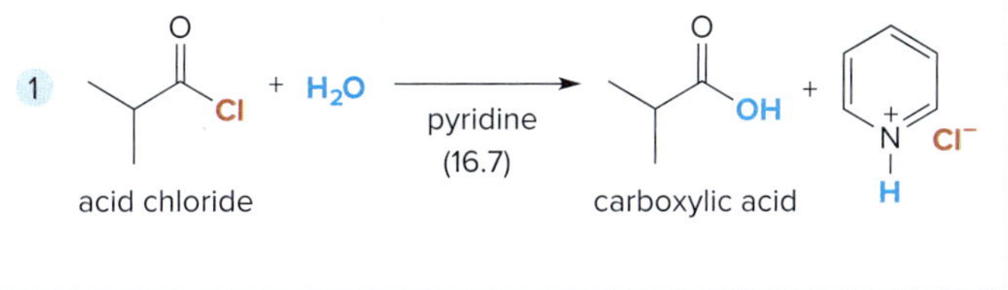

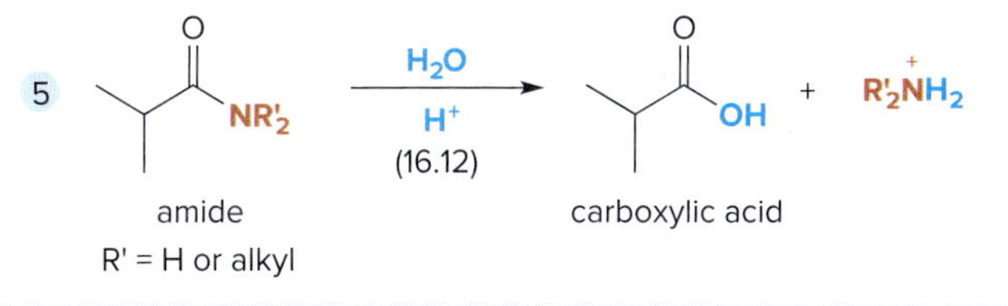

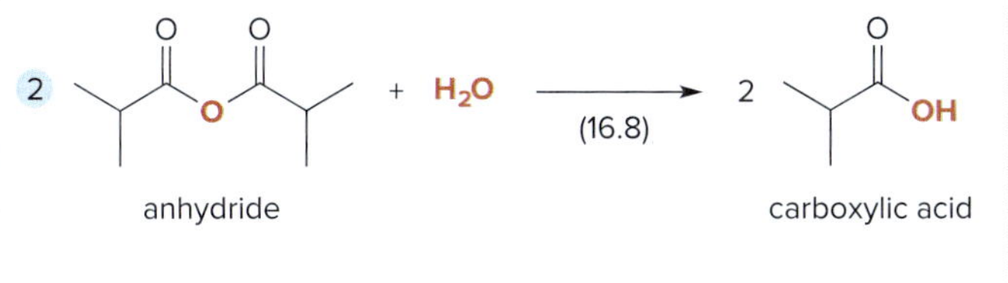

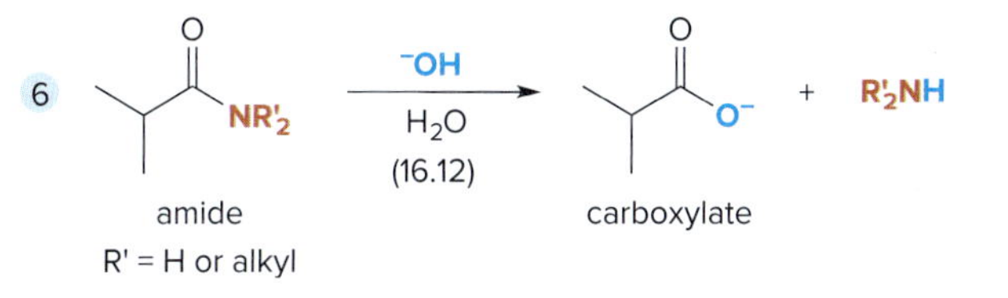

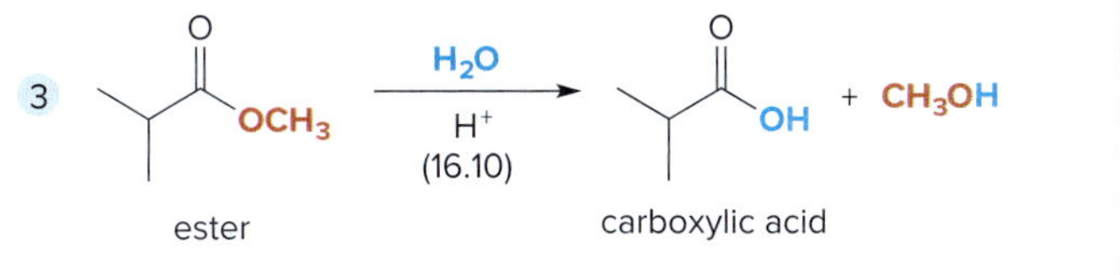

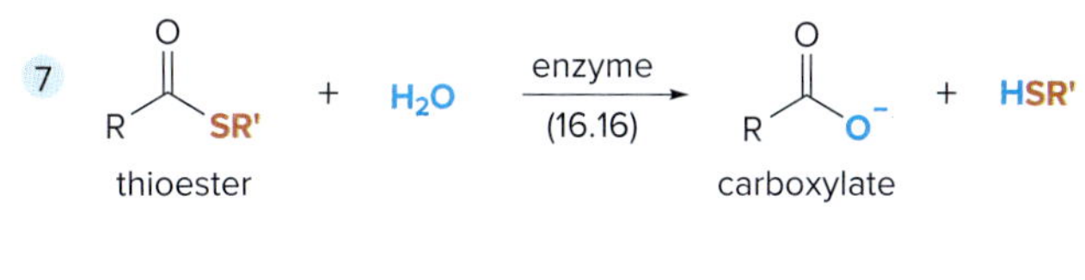

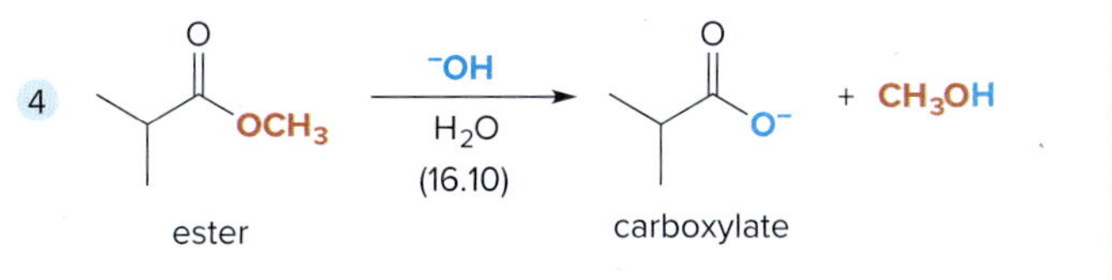

1. acid chloride + H$_2$O $\xrightarrow{\text{pyridine}}$ carboxylic acid + pyridinium Cl$^-$ (16.7)

2. anhydride + H$_2$O → 2 carboxylic acid (16.8)

3. ester + H$_2$O $\xrightarrow{\text{H}^+}$ carboxylic acid + CH$_3$OH (16.10)

4. ester + $^-$OH / H$_2$O → carboxylate + CH$_3$OH (16.10)

5. amide (R' = H or alkyl) + H$_2$O $\xrightarrow{\text{H}^+}$ carboxylic acid + R'$_2$NH$_2^+$ (16.12)

6. amide (R' = H or alkyl) + $^-$OH / H$_2$O → carboxylate + R'$_2$NH (16.12)

7. thioester + H$_2$O $\xrightarrow{\text{enzyme}}$ carboxylate + HSR' (16.16)

Try Problems 16.32b [1], [2]; 16.38c, d; 16.40; 16.41b; 16.42.

[6] Reactions that produce esters (RCOOR')

[7] Reactions that produce amides (RCONR'₂)

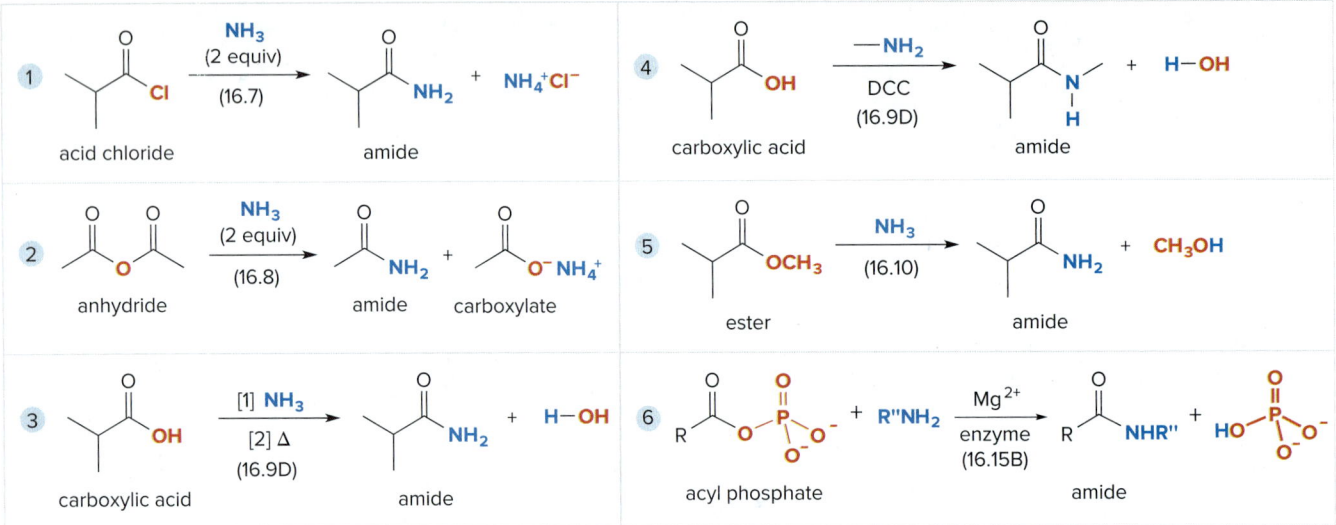

Try Problems 16.37g, l; 16.38b; 16.41a, c.

Try Problems 16.37f, j, k; 16.38a, h; 16.41d.

KEY SKILLS

[1] Determining whether a nucleophilic acyl substitution will occur (16.6B)

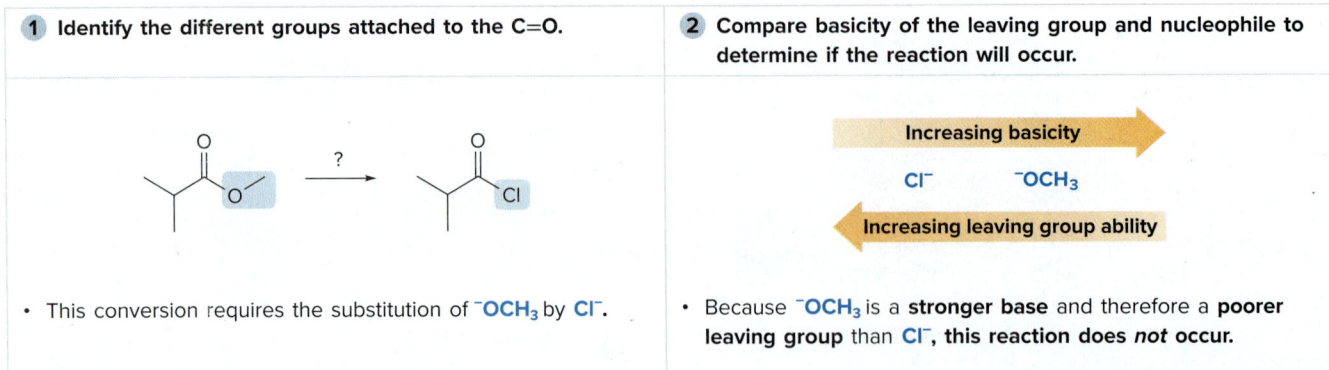

See Sample Problem 16.2. Try Problem 16.37d.

[2] Determining the carboxylic acid and alcohol needed for a Fischer esterification (16.9C)

① Cleave the carbon–oxygen single bond attached to the carbonyl.	② Draw the RCOOH, which becomes the RC=O.	③ Draw the HOR, which becomes the OR group.
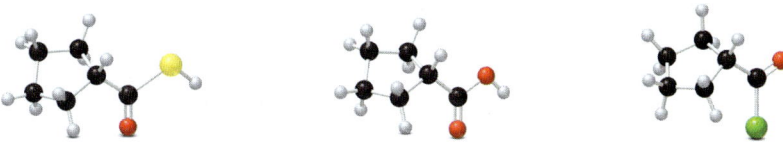		

Try Problem 16.54.

PROBLEMS

Problems Using Three-Dimensional Models

16.31 Rank the following compounds in order of increasing reactivity in nucleophilic acyl substitution.

16.32 (a) Give an acceptable name for compound **A**. (b) Draw the organic products formed when **A** is treated with each reagent: [1] H_3O^+; [2] ⁻OH, H_2O; [3] $CH_3CH_2CH_2MgBr$ (excess), then H_2O; [4] $LiAlH_4$, then H_2O.

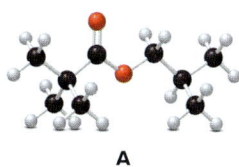

A

Nomenclature

16.33 Give the IUPAC or common name for each compound.

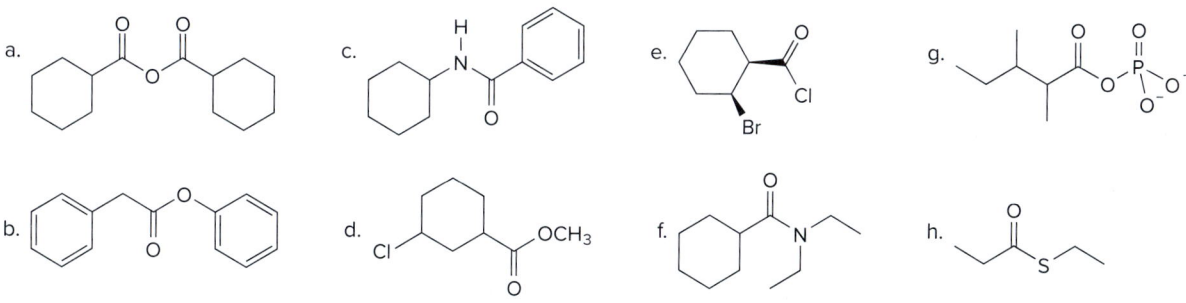

16.34 Give the structure corresponding to each name.
  a. cyclohexyl propanoate
  b. cyclohexanecarboxamide
  c. benzoic propanoic anhydride
  d. 3-methylhexanoyl chloride
  e. 3-ethylcyclobutanecarbonyl phosphate
  f. N,N-dibenzylformamide

Properties of Carboxylic Acid Derivatives

16.35 Explain why CH_3CONH_2 is a stronger acid and a weaker base than $CH_3CH_2NH_2$.

16.36 (a) Propose an explanation for the difference in the frequency of the carbonyl absorptions of phenyl acetate (1765 cm^{-1}) and cyclohexyl acetate (1738 cm^{-1}). (b) Which carbonyl group is more effectively stabilized by resonance? (c) Which ester reacts faster when treated with aqueous base?

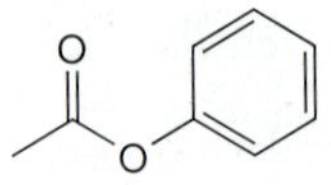

phenyl acetate 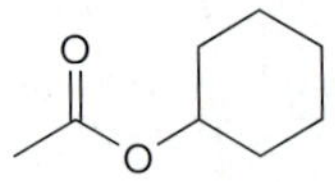   cyclohexyl acetate

Reactions

16.37 Draw the product formed when phenylacetic acid (C$_6$H$_5$CH$_2$COOH) is treated with each reagent. With some reagents, no reaction occurs.

a. NaHCO$_3$
b. NaOH
c. SOCl$_2$
d. NaCl
e. NH$_3$ (1 equiv)
f. NH$_3$, Δ
g. CH$_3$OH, H$_2$SO$_4$
h. CH$_3$OH, $^-$OH
i. [1] NaOH; [2] CH$_3$COCl
j. CH$_3$NH$_2$, DCC
k. [1] SOCl$_2$; [2] CH$_3$CH$_2$CH$_2$NH$_2$ (excess)
l. [1] SOCl$_2$; [2] (CH$_3$)$_2$CHOH

16.38 Draw the organic products formed in each reaction.

a. 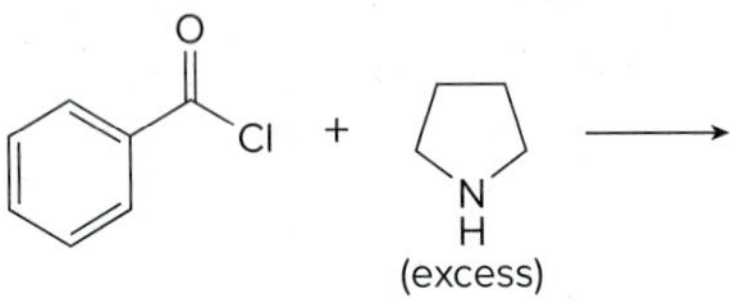 PhCOCl + pyrrolidine (excess)

b. 3 R-SCoA + glycerol →(enzyme) (fatty acyl CoA)

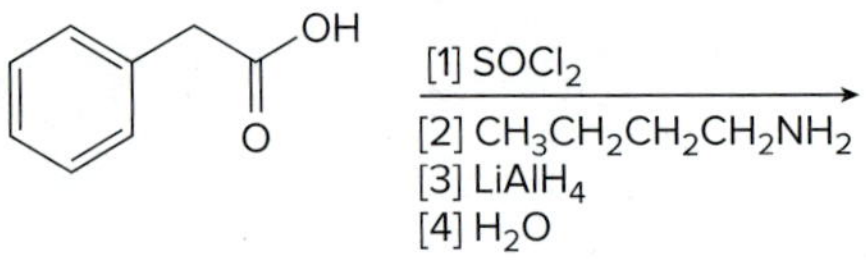

c. 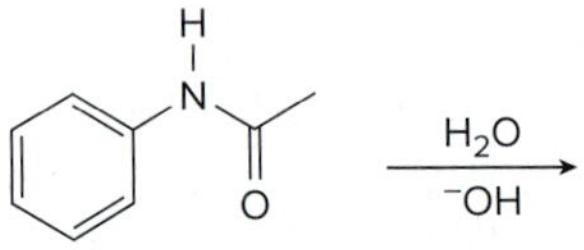 PhNHC(O)CH$_3$ →(H$_2$O, $^-$OH)

d. bicyclic lactone →(H$_3$O$^+$)

e. CH$_3$O-CH(CH$_3$)-CH$_2$-C(O)OH →[1] CH$_3$OH, H$^+$ [2] CH$_3$CH$_2$MgBr (2 equiv) [3] H$_2$O

f. PhCH$_2$C(O)OH →[1] SOCl$_2$ [2] CH$_3$CH$_2$CH$_2$CH$_2$NH$_2$ [3] LiAlH$_4$ [4] H$_2$O

g. HOOC-CH=CH-COOH →Δ

h. acetic anhydride + cyclohexyl-NH$_2$ (excess) →

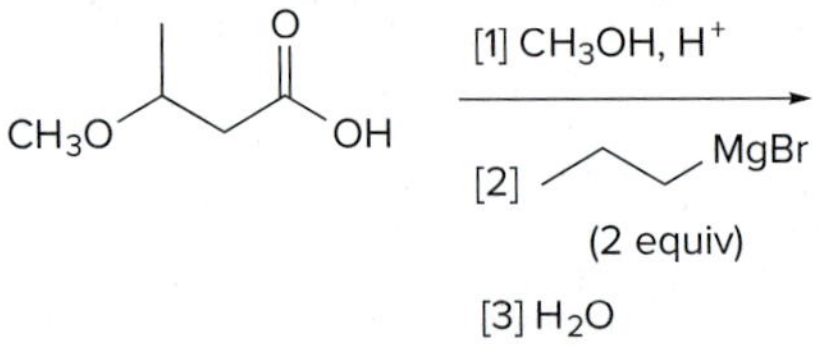

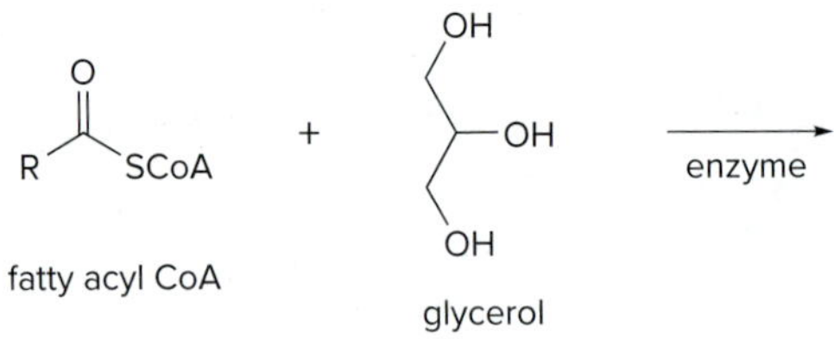

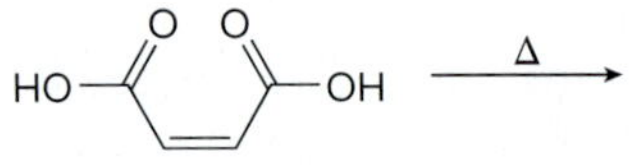

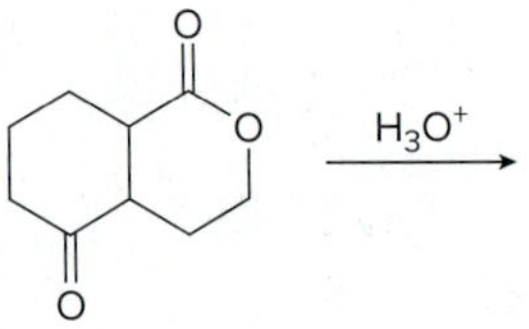

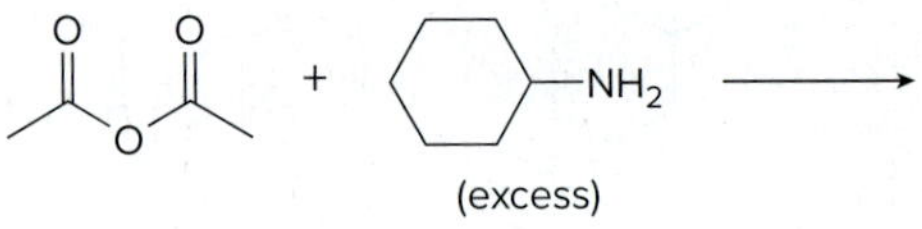

16.39 Cinnamoylcocaine, a natural product that occurs in coca leaves, can be converted to cocaine, the chapter-opening molecule, by the following reaction sequence. Identify the structure of cinnamoylcocaine, as well as intermediates **X** and **Y**.

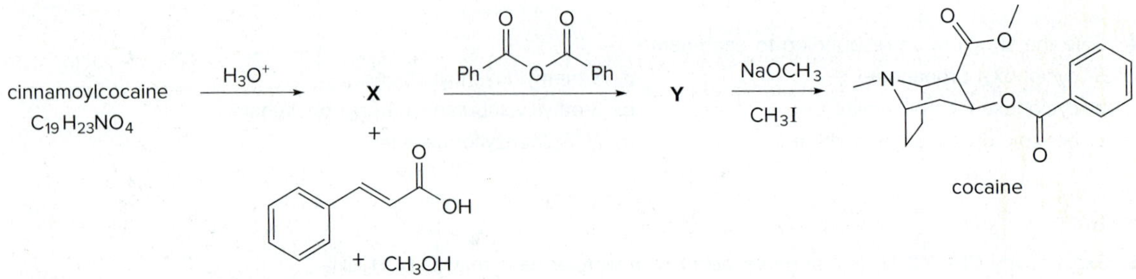

cinnamoylcocaine C$_{19}$H$_{23}$NO$_4$ →(H$_3$O$^+$) **X** + PhCH=CHCOOH → Ph-C(O)-O-C(O)-Ph → **Y** →(NaOCH$_3$, CH$_3$I) cocaine

+ CH$_3$OH

16.40 What products are formed by hydrolysis of each lactone or lactam with acid?

a. 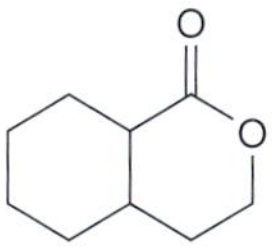 b. 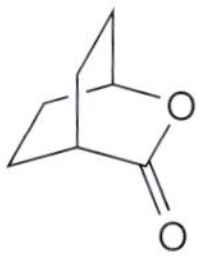 c. 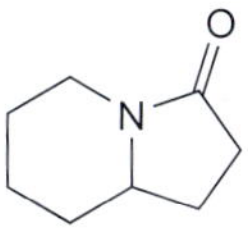 d.

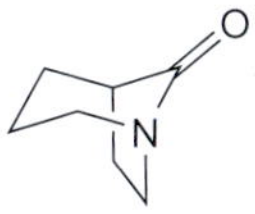

16.41 Draw the products of each reaction and indicate the stereochemistry at any stereogenic centers.

a. 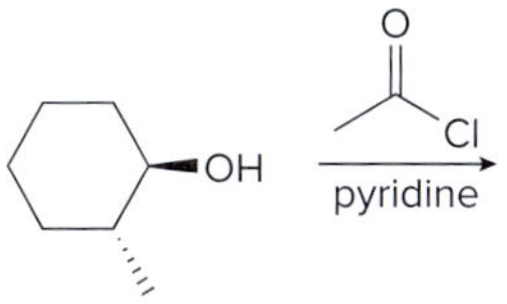 [trans-2-methylcyclohexanol] + CH₃COCl / pyridine

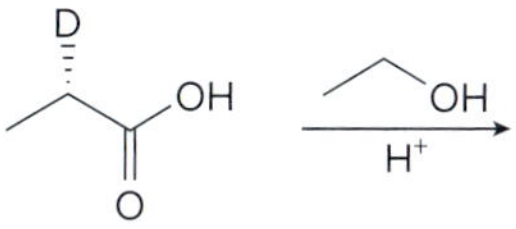

b. 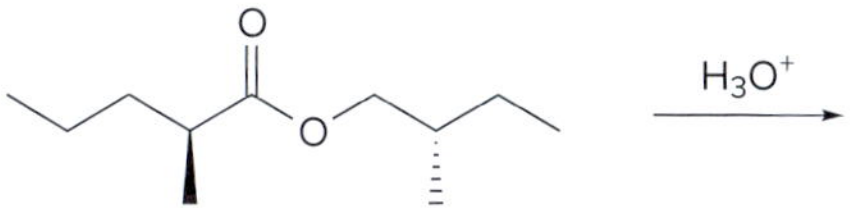 [ester] + H₃O⁺

c. [(D-labeled) lactic acid] + EtOH / H⁺

d. 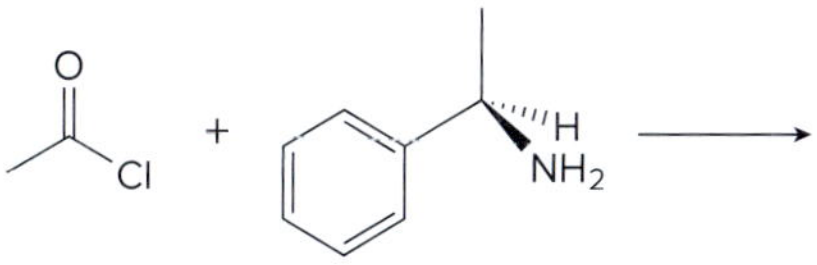 CH₃COCl + PhCH(CH₃)NH₂ (2 equiv)

16.42 What products are formed when all of the amide and ester bonds are hydrolyzed in each of the following compounds? **Tamiflu** [part (a)] is the trade name of the antiviral agent oseltamivir, thought to be the most effective agent in treating influenza. **Aspartame** [part (b)] is the artificial sweetener used in Equal and many diet beverages. One of the products of this hydrolysis reaction is the amino acid phenylalanine. Infants afflicted with phenylketonuria cannot metabolize this amino acid, so it accumulates, causing mental retardation. When the affliction is identified early, a diet limiting the consumption of phenylalanine (and compounds like aspartame that are converted to it) can make a normal life possible.

a. 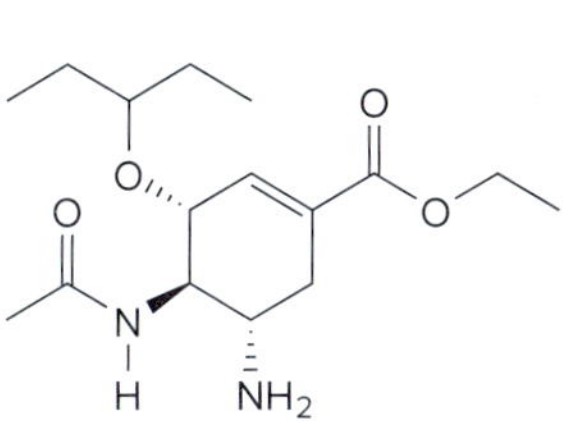 oseltamivir

b. 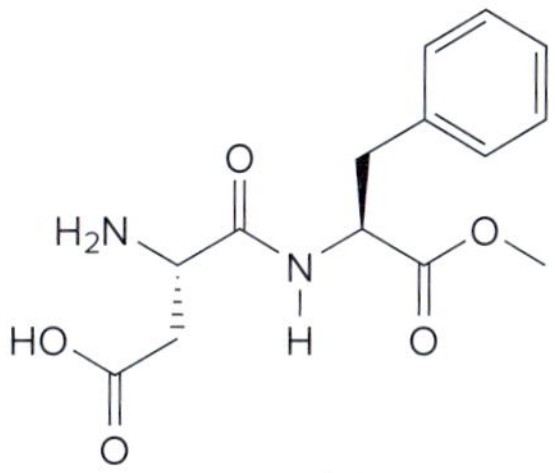 aspartame

16.43 Identify **F** in the following reaction sequence. **F** was converted in several steps to the antidepressant paroxetine (trade name Paxil; see also Problem 9.8).

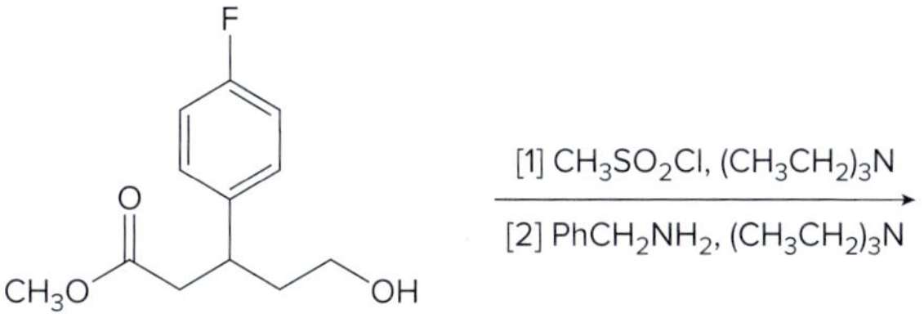

[starting material with 4-fluorophenyl, CH₃O ester, and OH]

[1] CH₃SO₂Cl, (CH₃CH₂)₃N
[2] PhCH₂NH₂, (CH₃CH₂)₃N

F
$C_{18}H_{18}FNO$

Mechanism

16.44 Although γ-butyrolactone is a biologically inactive compound, it is converted in the body to 4-hydroxybutanoic acid (GHB), an addictive and intoxicating recreational drug. Draw a stepwise mechanism for this conversion in the presence of acid.

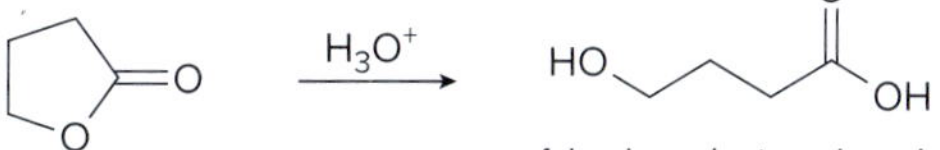

γ-butyrolactone — H_3O^+ → 4-hydroxybutanoic acid (GHB)

16.45 Aspirin is an anti-inflammatory agent because it inhibits the conversion of arachidonic acid to prostaglandins by the transfer of its acetyl group (CH$_3$CO–) to an OH group at the active site of an enzyme (Section 15.5). This reaction, called transesterification, results in the conversion of one ester to another by a nucleophilic acyl substitution reaction. Draw a stepwise mechanism for the given transesterification.

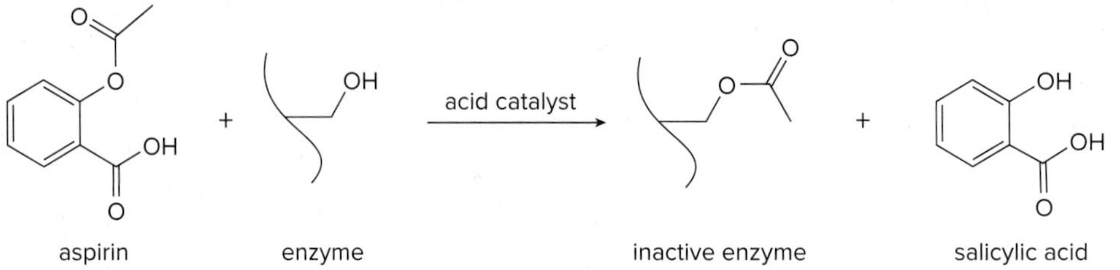

aspirin + enzyme —acid catalyst→ inactive enzyme + salicylic acid

16.46 Glycinamide ribonucleotide is an intermediate in the biosynthesis of the nitrogen bases in DNA. Draw a stepwise mechanism for the formation of glycinamide ribonucleotide from glycyl phosphate and 5-phosphoribosylamine.

glycyl phosphate + 5-phosphoribosylamine —Mg^{2+}, enzyme→ glycinamide ribonucleotide + HO–P + Mg^{2+} + enzyme

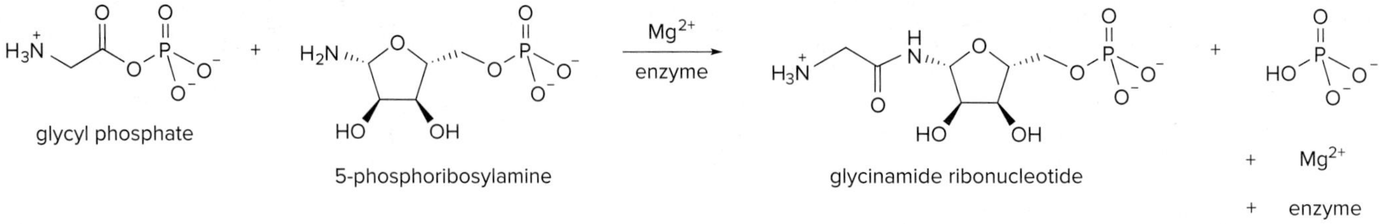

16.47 Using a protonated enzyme (H–enzyme$^+$) to activate the carbonyl, draw a stepwise mechanism for the formation of succinate from succinyl CoA, a step in the citric acid cycle.

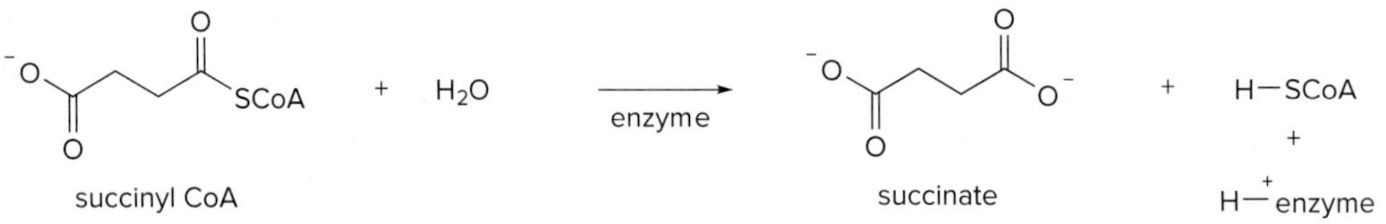

succinyl CoA + H$_2$O —enzyme→ succinate + H–SCoA + H–enzyme$^+$

16.48 Draw a stepwise mechanism for the following reaction, one step in the synthesis of the cholesterol-lowering drug ezetimibe.

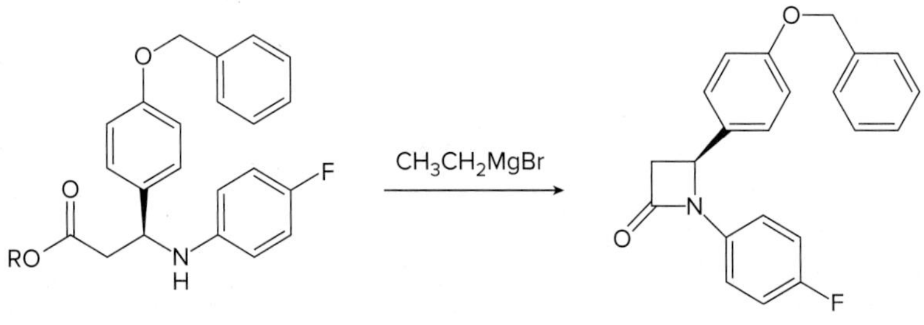

(structure) —CH$_3$CH$_2$MgBr→ (β-lactam structure)

16.49 Draw a stepwise mechanism for the following reaction, which involves both a Diels–Alder reaction and a nucleophilic acyl substitution.

(diene-alcohol) + (maleic anhydride) —Δ→ (bicyclic lactone-CO$_2$H)

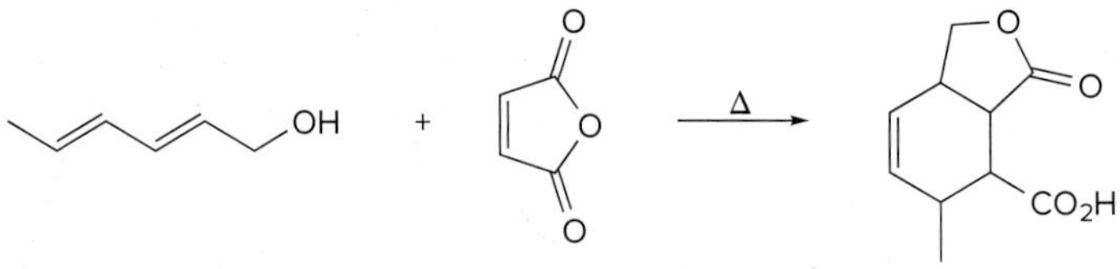

16.50 Draw a stepwise mechanism for the conversion of lactone **C** to carboxylic acid **D**. **C** is a key intermediate in the synthesis of prostaglandins (Section 15.5) by Nobel Laureate E. J. Corey and co-workers at Harvard University.

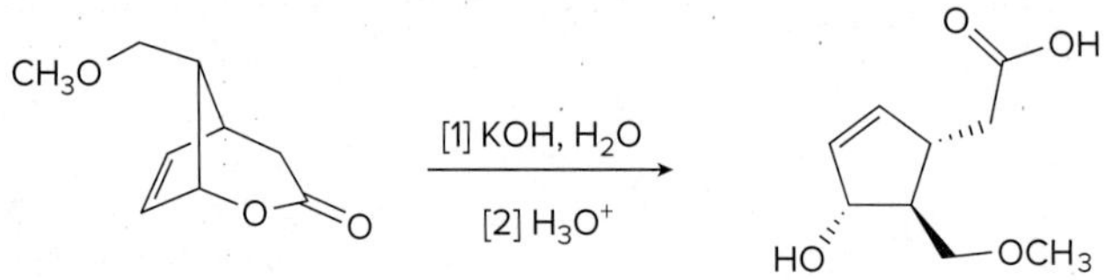

C —[1] KOH, H$_2$O; [2] H$_3$O$^+$→ **D**

16.51 Two steps in the synthesis of tadalafil, a drug sold under the trade name Cialis for the treatment of erectile dysfunction, are shown. Identify intermediate **A**, and draw a mechanism for the conversion of **A** to tadalafil.

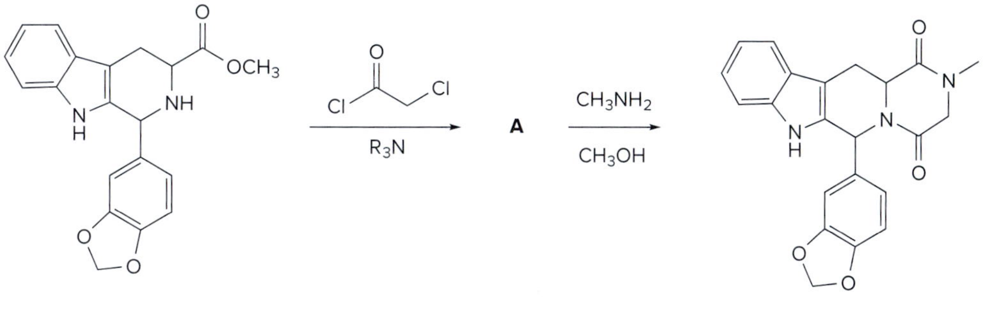

16.52 Draw a stepwise mechanism for the following reaction.

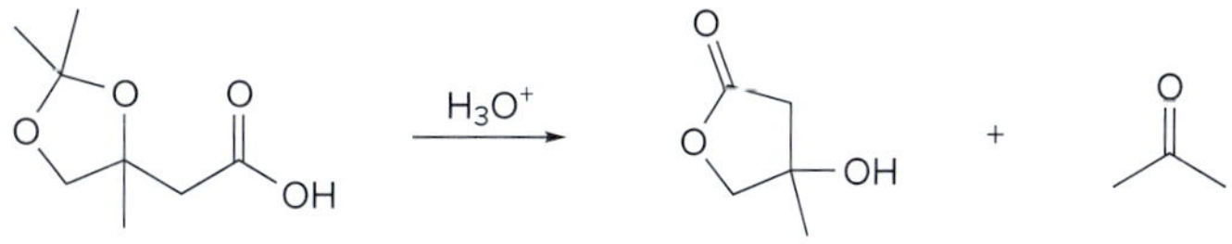

16.53 Three steps in the synthesis of the anticancer drug Taxol (paclitaxel, Chapter 5 opening molecule) involve the conversion of **A** to **B**. Draw stepwise mechanisms for Steps [2] and [3] in this reaction scheme.

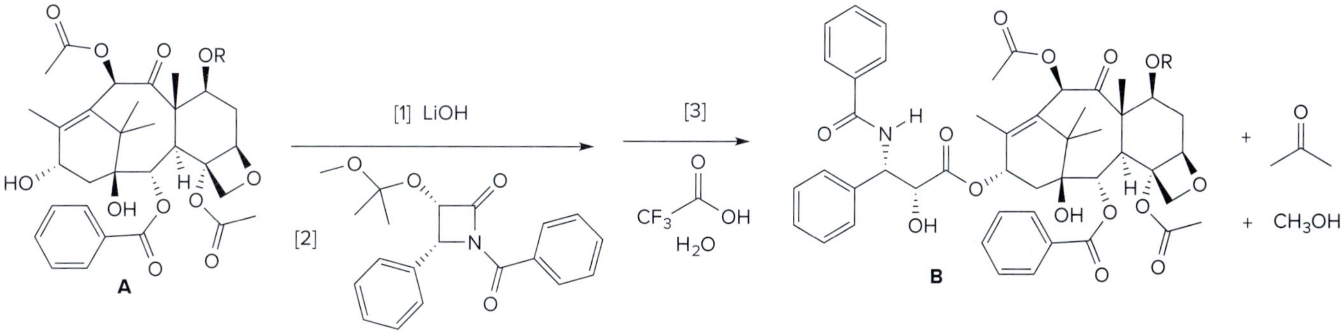

Synthesis

16.54 What carboxylic acid and alcohol are needed to prepare each ester by Fischer esterification?

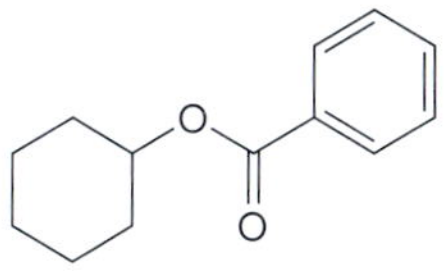

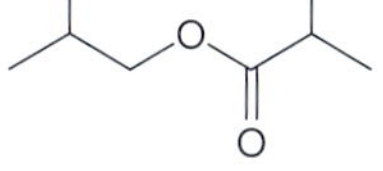

16.55 Write out the steps in the biological conversion of aspartate to asparagine.

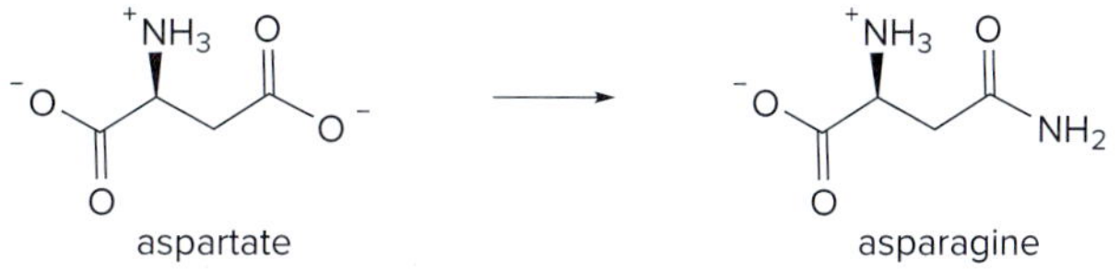

16.56 Devise a synthesis of each compound using 1-bromobutane ($CH_3CH_2CH_2CH_2Br$) as the only organic starting material. You may use any other inorganic reagents.

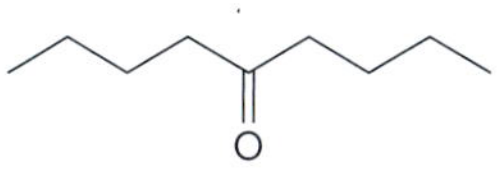

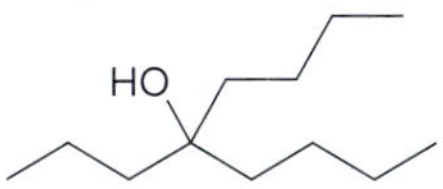

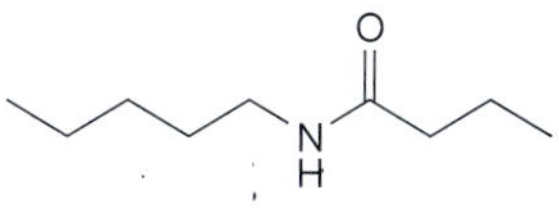

16.57 How would you convert benzoic acid ($C_6H_5CO_2H$) to each compound?

a. phenyl isopropyl ketone   b. 2-phenyl-1,3-dioxolane   c. N-benzylacetamide   d. 4-phenyl-4-heptanol

Spectroscopy

16.58 Rank compounds **A–D** in order of increasing frequency of the C=O absorption in their IR spectra.

A. methyl cyclopentanecarboxylate   B. α-isopropenyl-γ-butyrolactone   C. 1-benzoylpiperidine   D. methyl benzoate

16.59 Identify the structures of each compound from the given data.
  a. Molecular formula $C_6H_{12}O_2$
     IR absorption: 1738 cm^{-1}
     ^{1}H NMR: 1.12 (triplet, 3 H), 1.23 (doublet, 6 H), 2.28 (quartet, 2 H), and 5.00 (septet, 1 H) ppm
  b. Molecular formula C_8H_9NO
     IR absorptions: 3328 and 1639 cm^{-1}
     ^{1}H NMR: 2.95 (singlet, 3 H), 6.95 (singlet, 1 H), and 7.3–7.7 (multiplet, 5 H) ppm

16.60 Identify the structures of **A** and **B**, isomers of molecular formula $C_{10}H_{12}O_2$, from their IR data and ^{1}H NMR spectra.

  a. IR absorption for **A** at 1718 cm^{-1}
  b. IR absorption for **B** at 1740 cm^{-1}

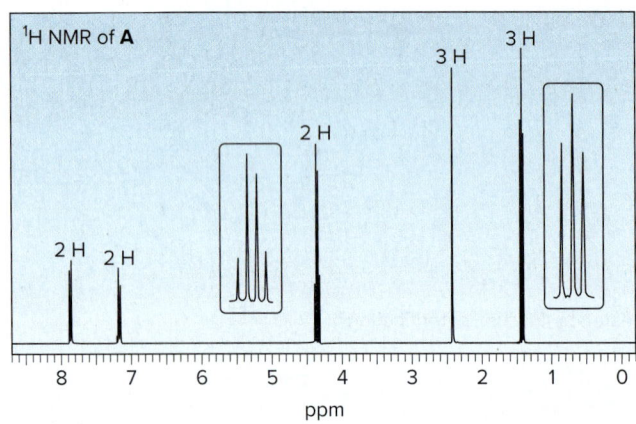

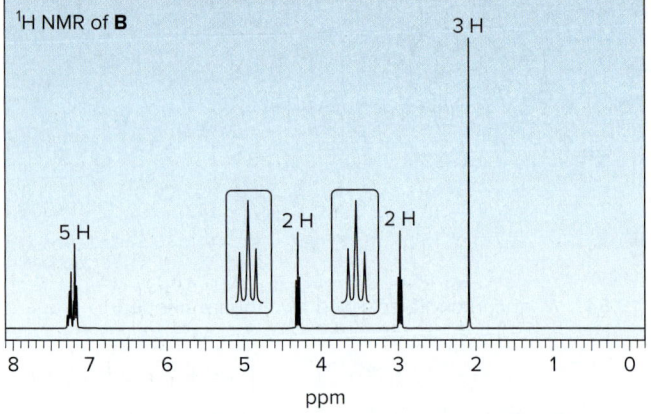

16.61 Phenacetin is an analgesic compound having molecular formula $C_{10}H_{13}NO_2$. Once a common component in over-the-counter pain relievers such as APC (**a**spirin, **p**henacetin, **c**affeine), phenacetin is no longer used because of its liver toxicity. Deduce the structure of phenacetin from its ^{1}H NMR and IR spectra.

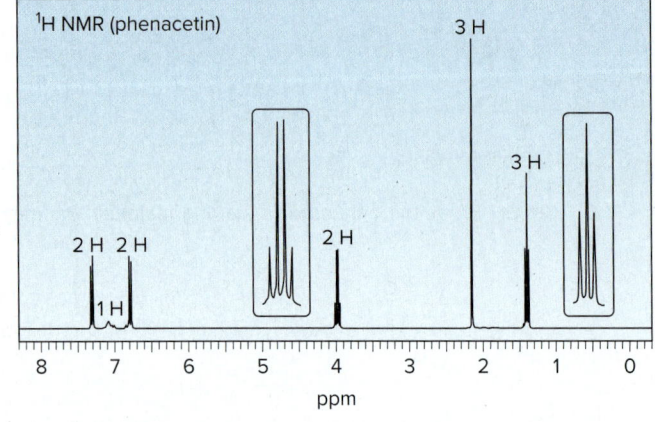

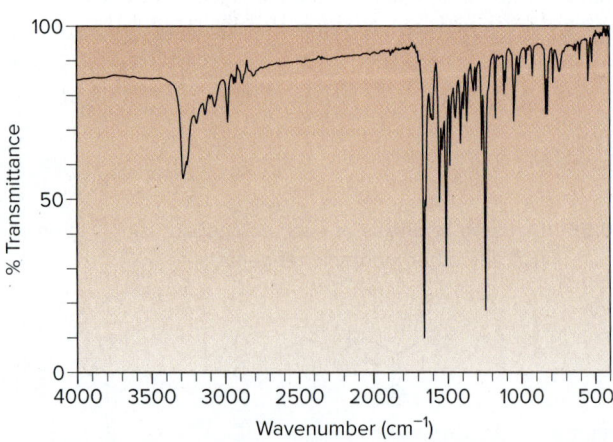

16.62 Identify the structure of compound **C** (molecular formula $C_{11}H_{15}NO_2$), which has an IR absorption at 1699 cm^{-1} and the ^{1}H NMR spectrum shown below.

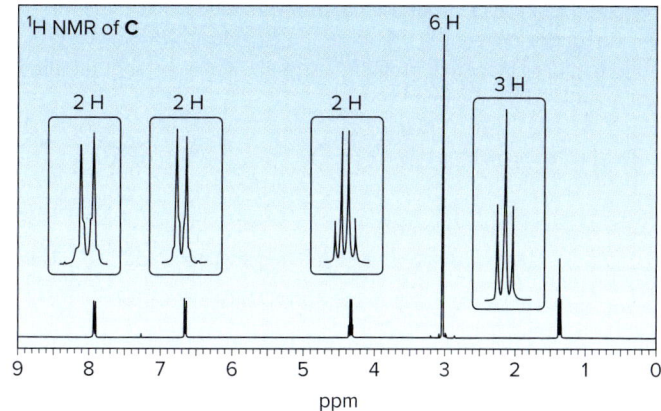

16.63 Identify the structures of **D** and **E**, isomers of molecular formula $C_6H_{12}O_2$, from their IR and ^{1}H NMR data. Signals at 1.35 and 1.60 ppm in the ^{1}H NMR spectrum of **D** and 1.90 ppm in the ^{1}H NMR spectrum of **E** are multiplets.

a. IR absorption for **D** at 1743 cm^{-1}

b. IR absorption for **E** at 1746 cm^{-1}

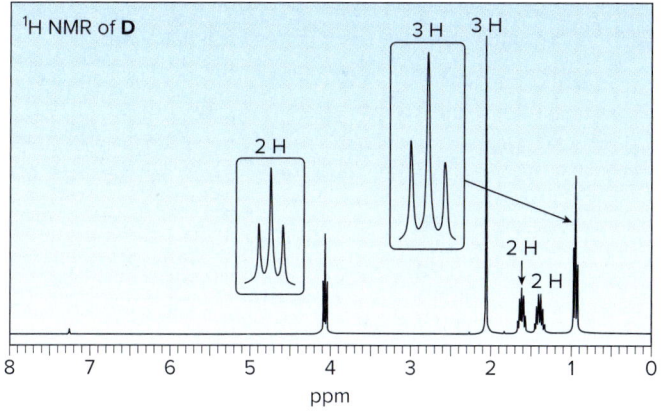
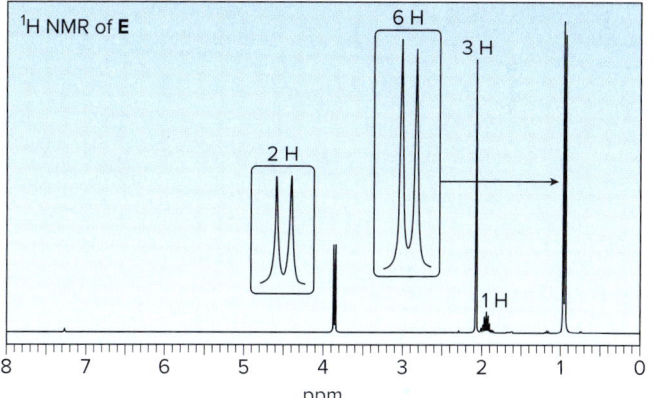

Challenge Problems

16.64 One step in the synthesis of aliskiren, a drug used to treat hypertension (Problems 5.6 and 11.57), involves the conversion of **A** to **B**. Draw a stepwise mechanism for this process that explains the observed stereochemistry.

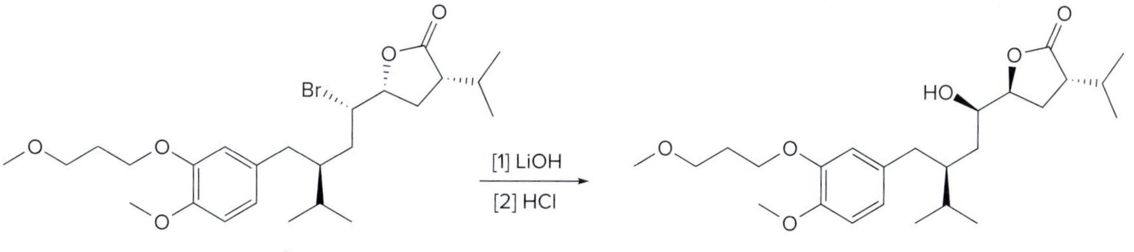

16.65 With reference to amides **A** and **B**, the carbonyl of one amide absorbs at a much higher wavenumber in its IR spectrum than the carbonyl of the other amide. Which absorbs at a higher wavenumber and why?

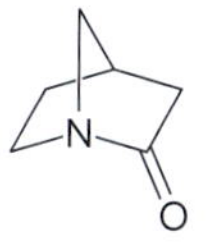

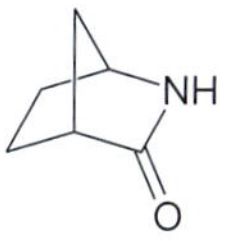

16.66 Compelling evidence for the existence of a tetrahedral intermediate in nucleophilic acyl substitution was obtained in a series of elegant experiments carried out by Myron Bender in 1951. The key experiment was the reaction of aqueous ⁻OH with ethyl benzoate ($C_6H_5COOCH_2CH_3$) labeled at the carbonyl oxygen with ^{18}O. Bender did not allow the hydrolysis to go to completion, and then examined the presence of a label in the *recovered starting material*. He found that some of the recovered ethyl benzoate no longer contained a label at the carbonyl oxygen. With reference to the accepted mechanism of nucleophilic acyl substitution, explain how this provides evidence for a tetrahedral intermediate.

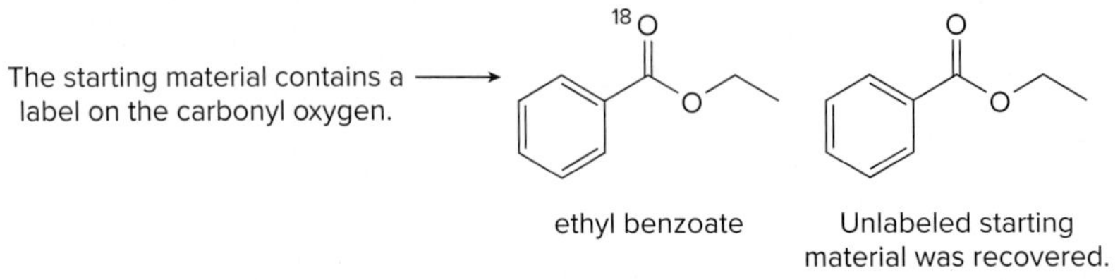

16.67 Draw a stepwise mechanism for the following reaction, the last step in a five-step industrial synthesis of vitamin C that begins with the simple carbohydrate glucose.

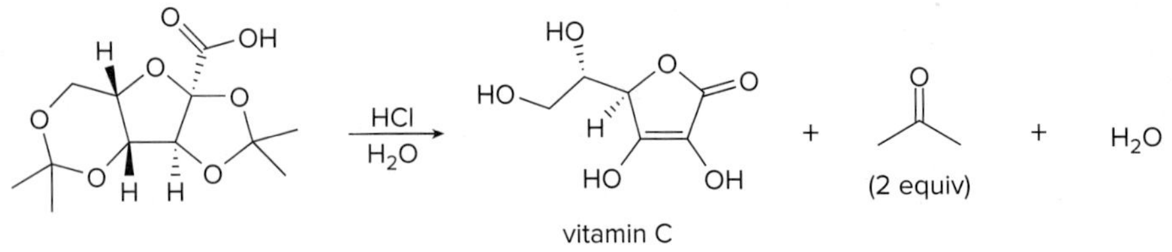

16.68 Draw a stepwise mechanism for the following reaction, a key step in the synthesis of linezolid, an antibacterial agent.

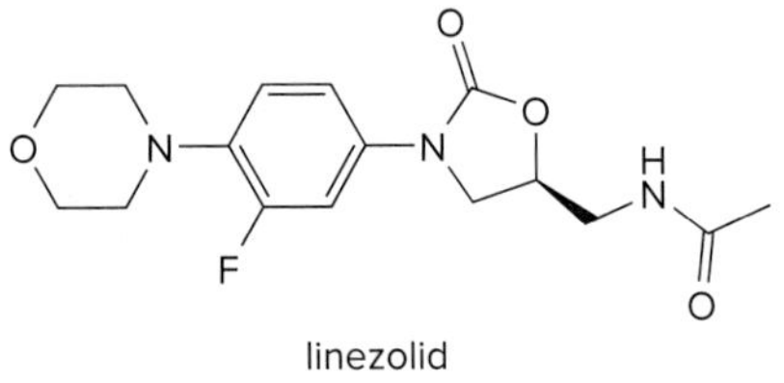

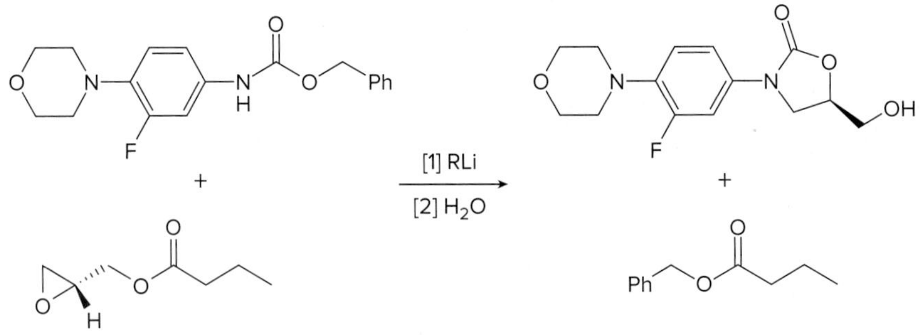

Substitution Reactions of Carbonyl Compounds at the α Carbon

17

Surachetkhamsuk/iStock/Getty Images

- **17.1** Introduction
- **17.2** Enols
- **17.3** Enolates
- **17.4** Enolates of unsymmetrical carbonyl compounds
- **17.5** Racemization at the α carbon
- **17.6** A preview of reactions at the α carbon
- **17.7** Halogenation at the α carbon
- **17.8** Direct enolate alkylation
- **17.9** Malonic ester synthesis
- **17.10** Acetoacetic ester synthesis
- **17.11** Biological decarboxylation

Stemoamide is an amide isolated from the roots of *Stemona tuberosa*, a flowering plant native to China, Southeast Asia, and New Guinea. Extracts from *Stemona tuberosa* have been used in traditional Chinese medicine for treatment of bronchitis and other respiratory illnesses. Stemoamide was isolated in 1992, and its structure was determined by NMR and IR spectroscopy. Stemoamide has been synthesized in the laboratory by several research groups. The last step in a 2011 synthesis involved enolate alkylation, a substitution reaction discussed in Chapter 17.

Why Study...

Reactions at the α Carbon of a Carbonyl Group?

Chapters 17 and 18 focus on reactions that occur at the α carbon to a carbonyl group. These reactions are different from the reactions of Chapters 13, 14, and 16, all of which involved nucleophilic attack at the electrophilic carbonyl carbon. In reactions at the α carbon, the carbonyl compound serves as a *nucleophile* that reacts with a carbon or halogen electrophile to form a new bond to the α carbon.

Chapter 17 concentrates on **substitution reactions at the α carbon,** whereas Chapter 18 concentrates on reactions between two carbonyl compounds, one of which serves as the nucleophile and one of which is the electrophile. Many of the reactions in Chapter 17 form new carbon–carbon bonds, thus adding to your repertoire of reactions that can be used to synthesize more-complex organic molecules from simple precursors. As you will see, the reactions introduced in Chapter 17 have been used to prepare a wide variety of interesting and useful compounds.

17.1 Introduction

Up to now, the discussion of carbonyl compounds has centered on their reactions with nucleophiles at the electrophilic carbonyl carbon. Reactions can also occur at the α carbon to the carbonyl group. These reactions proceed by way of **enols** or **enolates,** two electron-rich intermediates that react with electrophiles, forming a new bond on the α carbon. This reaction results in the **substitution of the electrophile E for hydrogen.**

Hydrogen atoms on the α carbon are called **α hydrogens.**

17.2 Enols

Recall from Chapter 10 that **enol and keto forms are tautomers of the carbonyl group that differ in the position of a double bond and a proton.** These constitutional isomers are in equilibrium with each other.

- A keto tautomer has a C=O and an additional C–H bond.
- An enol tautomer has an O–H group bonded to a C=C.

Equilibrium favors the keto form for most carbonyl compounds largely because a C=O is much stronger than a C=C. For simple carbonyl compounds, < 1% of the enol is present at equilibrium. With unsymmetrical ketones, moreover, two different enols are possible, yet they still total < 1%.

17.2 Enols

With compounds containing two carbonyl groups separated by a single carbon (called β-dicarbonyl compounds or 1,3-dicarbonyl compounds), however, the concentration of the enol form sometimes exceeds the concentration of the keto form.

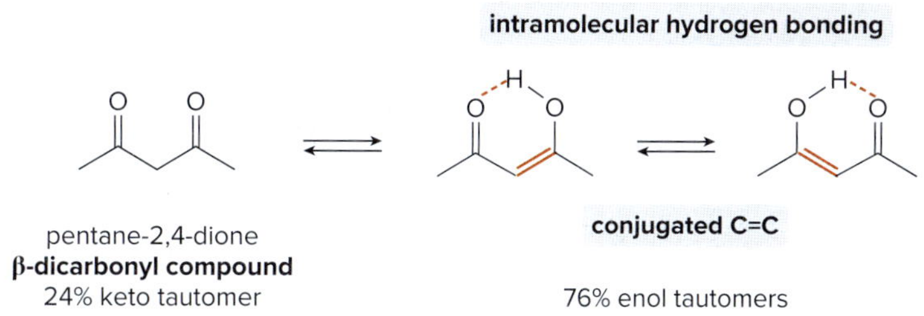

pentane-2,4-dione
β-dicarbonyl compound
24% keto tautomer

76% enol tautomers

Two factors stabilize the enol of β-dicarbonyl compounds: **conjugation** and **intramolecular hydrogen bonding.** The C=C of the enol is conjugated with the carbonyl group, allowing delocalization of the electron density in the π bonds. Moreover, the OH of the enol can hydrogen bond to the oxygen of the nearby carbonyl group. Such intramolecular hydrogen bonds are especially stabilizing when they form a six-membered ring, as in this case.

Sample Problem 17.1 — Interconverting Keto and Enol Tautomers

Convert each compound to its enol or keto tautomer.

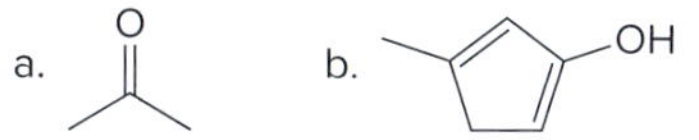

Solution

a. To convert a carbonyl compound to its enol tautomer, **draw a double bond between the carbonyl carbon and the α carbon, and change the C=O to C–OH.** In this case, both α carbons are identical, so only one enol is possible.

b. To convert an enol to its keto tautomer, **change the C–OH to C=O and add a proton to the other end of the C=C.**

Problem 17.1 Draw the enol or keto tautomer(s) of each compound.

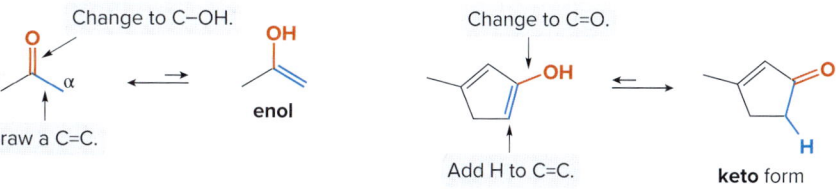

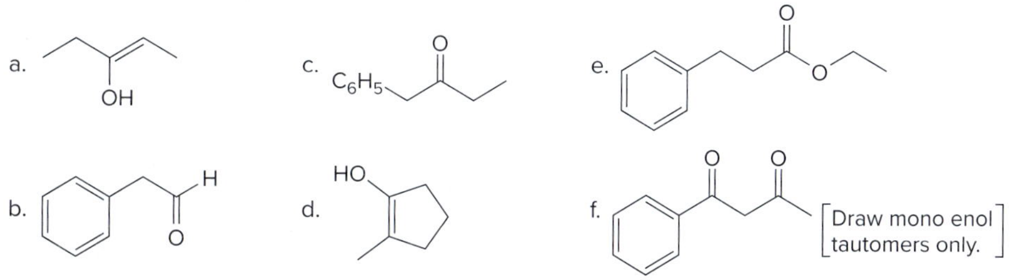

f. [Draw mono enol tautomers only.]

More Practice: Try Problems 17.28, 17.30, 17.31, 17.36.

782 Chapter 17 Substitution Reactions of Carbonyl Compounds at the α Carbon

Problem 17.2 Leptospermone is a herbicide produced by the bottlebrush plant. Draw all possible mono enol tautomers of leptospermone, ignoring stereoisomers. Determine if all the tautomers are similar in stability, or if one tautomer is more or less stable than the others.

leptospermone

Callistimon citrinus, commonly called bottlebrush, is a plant native to Australia and the source of leptospermone (Problem 17.2). *Rafael Santos Rodriguez/Shutterstock*

17.2A The Mechanism of Tautomerization

Tautomerization, the process of converting one tautomer to another, is catalyzed by both acid and base. Tautomerization always requires two steps (**protonation** and **deprotonation**), but the order of these steps depends on whether the reaction takes place in acid or base. In Mechanisms 17.1 and 17.2 for tautomerization, the keto form is converted to the enol form. All of the steps are reversible, though, so they equally apply to the conversion of the enol form to the keto form.

 Mechanism 17.1 Tautomerization in Acid

resonance-stabilized carbocation

① With acid, **protonation** *precedes* **deprotonation**. Protonation of the carbonyl forms a resonance-stabilized carbocation.
② Removal of a proton forms the enol.

 Mechanism 17.2 Tautomerization in Base

resonance-stabilized enolate

① With base, **deprotonation** *precedes* **protonation**. Removal of a proton on the α carbon forms a resonance-stabilized enolate.
② Protonation of the enolate forms the enol.

17.2B Enols in Biological Systems

Key reactions in carbohydrate metabolism involve tautomerizations, and result in the interconversion of α-hydroxy ketones and α-hydroxy aldehydes. In this case, tautomerization generates an **enediol**, because two OH groups are bonded to the C=C.

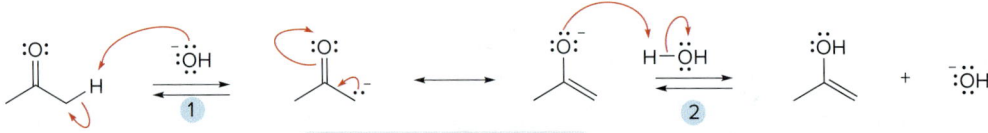

α-hydroxy ketone — enediol — α-hydroxy aldehyde

For example, in the metabolic breakdown of glucose, dihydroxyacetone phosphate is converted to glyceraldehyde 3-phosphate by **two keto–enol tautomerizations.** Although each reaction involves both protonation and deprotonation, both processes are written as a single step in a biological tautomerization.

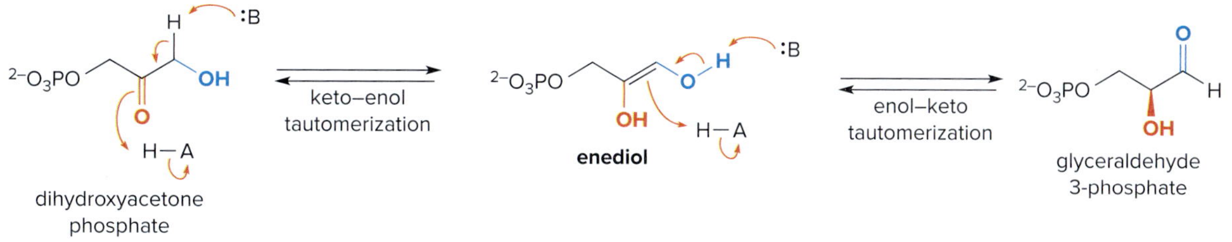

Problem 17.3 One step in the metabolism of glucose involves the isomerization of glucose 6-phosphate to fructose 6-phosphate. (a) Draw a stepwise mechanism for this process if it is carried out in the presence of acid. (b) Use curved arrows to write the reaction as two successive biological tautomerizations using HA as an acid and B: as a base.

glucose 6-phosphate   fructose 6-phosphate

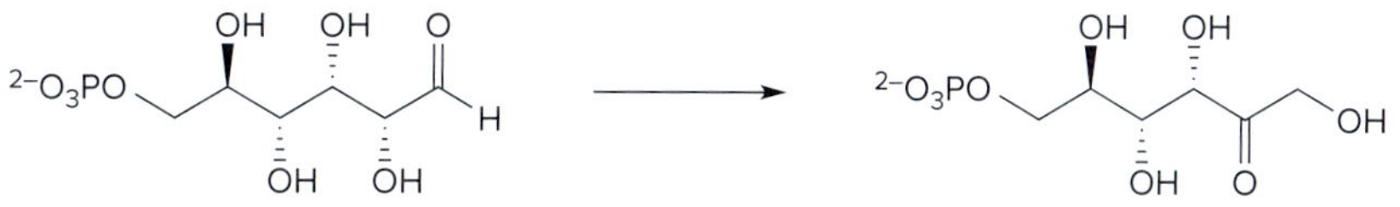

17.2C How Enols React

Like other compounds with carbon–carbon double bonds, **enols are electron rich, so they react as nucleophiles. Enols are even more electron rich than alkenes, though, because the OH group has a powerful electron-donating resonance effect.** A second resonance structure can be drawn for the enol that places a negative charge on one of the carbon atoms. As a result, this carbon atom is especially nucleophilic, and it can react with an electrophile E^+ to form a new bond to carbon. Loss of a proton then forms a neutral product.

two resonance structures for an enol

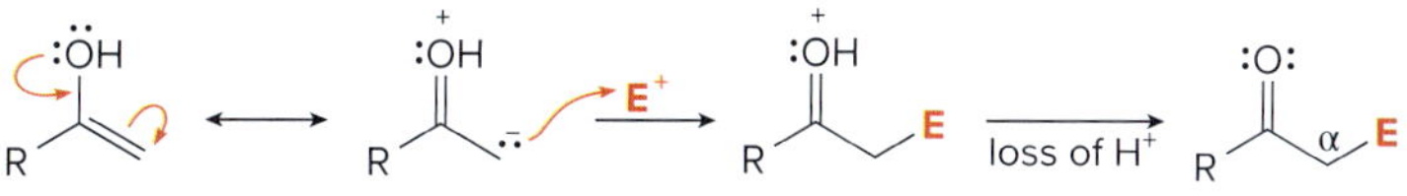

- **Reaction of an enol with an electrophile E^+ forms a new C–E bond on the α carbon. The net result is substitution of H by E on the α carbon.**

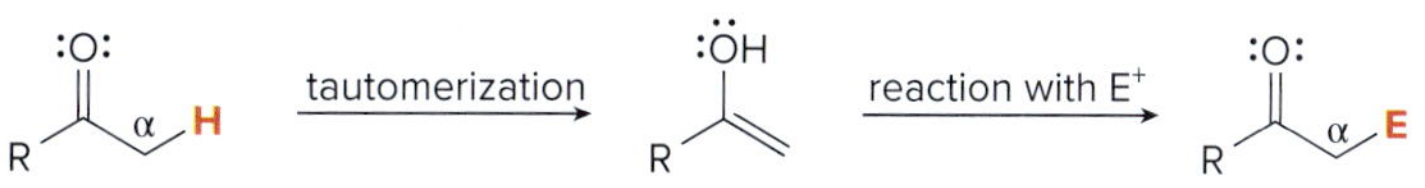

Problem 17.4 When phenylacetaldehyde ($C_6H_5CH_2CHO$) is dissolved in D_2O with added DCl, the hydrogen atoms α to the carbonyl are gradually replaced by deuterium atoms. Write a mechanism for this process that involves enols as intermediates.

17.3 Enolates

Enolates are formed when a base removes a proton on the α carbon to a carbonyl group. A C–H bond on the α carbon is more acidic than many other sp^3 hybridized C–H bonds, because **the resulting enolate is resonance stabilized.** Moreover, one of the resonance structures is especially stable because it places a negative charge on an electronegative oxygen atom.

Forming enolates from carbonyl compounds was first discussed in Section 14.6.

Enolates are always formed by removal of a proton on the **α carbon.**

The pK_a of the α hydrogen in an aldehyde or ketone is ~20. As shown in Table 17.1, this makes it considerably more acidic than the C—H bonds in CH_3CH_3 and $CH_3CH=CH_2$. Although C—H bonds α to a carbonyl are *more acidic* than many other C—H bonds, they are still *less acidic* than O—H bonds that always place the negative charge of the conjugate base on an electronegative oxygen atom (c.f. CH_3CH_2OH and CH_3COOH in Table 17.1).

Table 17.1 A Comparison of pK_a Values

Compound	pK_a	Conjugate base	Structural features of the conjugate base
CH_3CH_3	50	$CH_3\overset{..}{C}H_2$	• The conjugate base has a (−) charge on C, but is not resonance stabilized.
(propene)	43	(allyl anion resonance)	• The conjugate base has a (−) charge on C, and is resonance stabilized.
(acetone)	19.2	(enolate resonance)	• **The conjugate base has two resonance structures, one of which has a (−) charge on O.**
(ethanol)	16	(ethoxide)	• The conjugate base has a (−) charge on O, but is not resonance stabilized.
(acetic acid)	4.8	(carboxylate resonance)	• The conjugate base has two resonance structures, both of which have a (−) charge on O.

Increasing acidity → Increasing stability of the conjugate base

17.3A Examples of Enolates and Related Anions

In addition to enolates from aldehydes and ketones, **enolates from esters and 3° amides can be formed,** although the α hydrogen is somewhat less acidic. **Nitriles** also have acidic protons on the carbon atom adjacent to the cyano group, because the negative charge of the conjugate base is stabilized by delocalization onto an electronegative nitrogen atom.

ester
pK_a ≈ 25

resonance-stabilized enolate

negative charge on O

nitrile
pK_a ≈ 25

resonance-stabilized carbanion

negative charge on N

17.3 Enolates

The protons on the carbon between the two carbonyl groups of a β-dicarbonyl compound are especially acidic because resonance delocalizes the negative charge on two different oxygen atoms. Table 17.2 lists pK_a values for β-dicarbonyl compounds as well as other carbonyl compounds and nitriles.

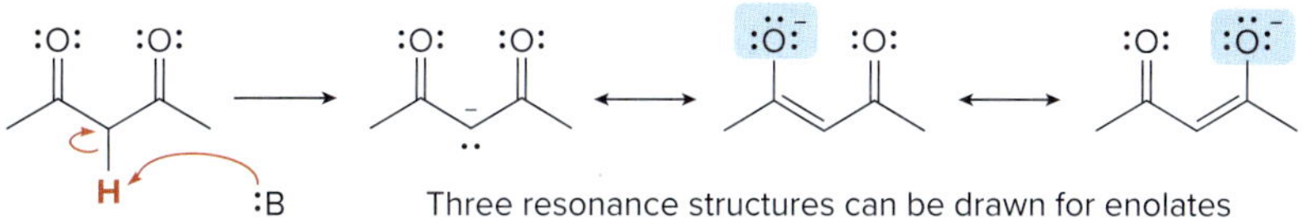

pentane-2,4-dione
pK_a = 9
β-dicarbonyl compound

Three resonance structures can be drawn for enolates derived from β-dicarbonyl compounds.

Table 17.2 pK_a Values for Some Carbonyl Compounds and Nitriles

Compound type	Example	pK_a	Compound type	Example	pK_a
[1] Amide	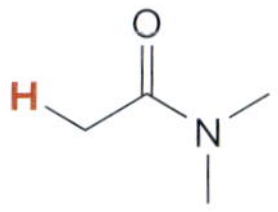 H–CH$_2$–C(O)–N	30	[6] 1,3-Diester	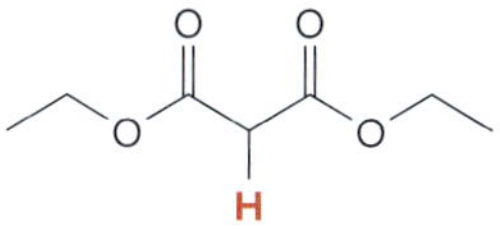	13.3
[2] Nitrile	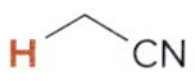 H–CH$_2$–CN	25	[7] 1,3-Dinitrile	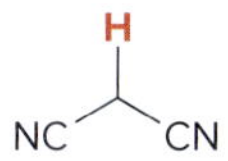 NC–CH$_2$–CN	11
[3] Ester	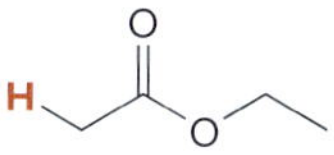 H–CH$_2$–C(O)–O–	25	[8] β-Keto ester	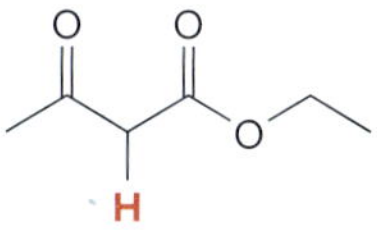	10.7
[4] Ketone	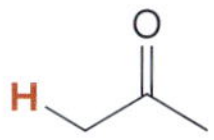 H–CH$_2$–C(O)–	19.2	[9] β-Diketone	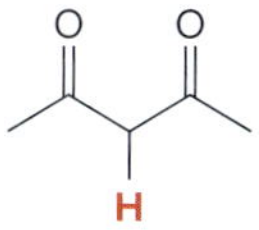	9
[5] Aldehyde	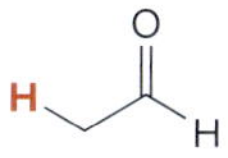 H–CH$_2$–CHO	17			

Problem 17.5 Draw additional resonance structures for each anion.

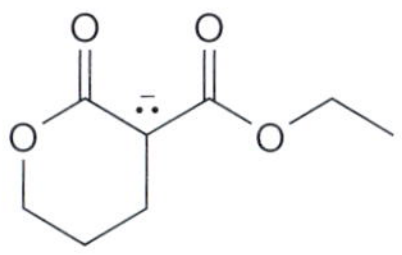

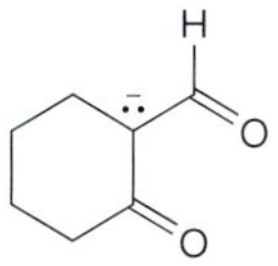

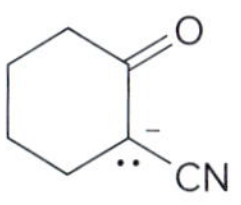

Problem 17.6 Which C–H bonds in the following molecules are acidic because the resulting conjugate base is resonance stabilized?

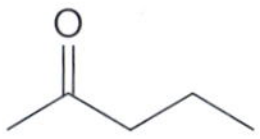

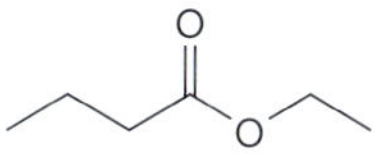

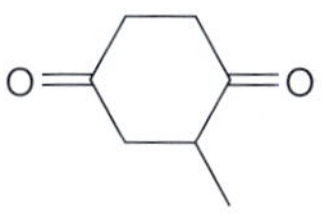

Problem 17.7 Rank the protons in the labeled CH$_2$ groups in order of increasing acidity, and explain why you chose this order.

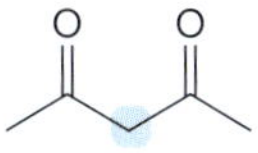

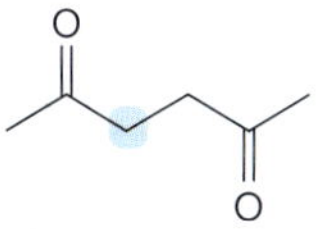

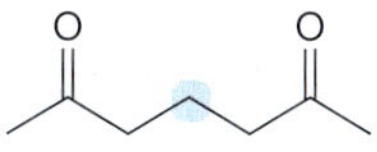

17.3B The Base

The formation of an enolate is an acid–base equilibrium, so the ***stronger*** **the base, the *more* enolate that forms.**

$$R-CO-CH_2-H + :B \rightleftharpoons R-CO-CH_2^- + HB^+$$

acid
$pK_a \approx 20$

conjugate acid

Stronger bases drive the equilibrium to the *right*.

We can predict the extent of an acid–base reaction by comparing the pK_a of the starting acid (the carbonyl compound in this case) with the pK_a of the conjugate acid formed. **The equilibrium favors the side with the *weaker* acid (the acid with the *higher* pK_a value).** The pK_a of many carbonyl compounds is ~20, so a significant amount of enolate will form only if the pK_a of the conjugate acid is > 20.

The common bases used to form enolates are hydroxide ($^-$OH), various alkoxides ($^-$OR), hydride (H^-), and dialkylamides ($^-NR_2$). How much enolate is formed using each of these bases is indicated in Table 17.3.

> We have now used the term *amide* in two different ways—first as a functional group ($RCONH_2$) and now as a base (e.g., $^-NH_2$, which can be purchased as a sodium or lithium salt, $NaNH_2$ or $LiNH_2$, respectively). In Chapter 17 we will use dialkylamides, $^-NR_2$, in which the two H atoms of $^-NH_2$ have been replaced by R groups.

Table 17.3 Enolate Formation with Various Bases:
$RCOCH_3$ ($pK_a \approx 20$) + B: → $RCOCH_2^-$ + HB^+

	Base (B:)	Conjugate acid (HB^+)	pK_a of HB^+	% Enolate
[1]	Na^+ ^-OH	H_2O	15.7	< 1%
[2]	Na^+ $^-OCH_2CH_3$	CH_3CH_2OH	16	< 1%
[3]	K^+ $^-OC(CH_3)_3$	$(CH_3)_3COH$	18	1–10%
[4]	Na^+H^-	H_2	35	100%
[5]	Li^+ $^-N[CH(CH_3)_2]_2$	$HN[CH(CH_3)_2]_2$	40	100%

When the pK_a of the conjugate acid is < 20, as it is for $^-$OH and all $^-$OR (entries 1–3), only a small amount of enolate is formed at equilibrium. These bases are more useful in forming enolates when more acidic 1,3-dicarbonyl compounds are used as starting materials. They are also used when both the enolate and the carbonyl starting material are involved in the reaction, as is the case for reactions described in Chapter 18.

To form an enolate in essentially 100% yield, a much stronger base such as lithium diisopropylamide, Li^+ $^-N[CH(CH_3)_2]_2$, abbreviated as **LDA**, is used (entry 5). **LDA is a strong nonnucleophilic base.** Like the other nonnucleophilic bases (Sections 7.8B and 8.1), its bulky isopropyl groups make the nitrogen atom too hindered to serve as a nucleophile. It is still able, though, to remove a proton in an acid–base reaction.

Enolate formation with LDA is typically carried out at −78 °C, a convenient temperature to maintain in the laboratory because it is the temperature at which dry ice (solid CO_2) sublimes. Immersing a reaction flask in a cooling bath containing dry ice and acetone keeps its contents at a constant low temperature. *Joe Franek/McGraw-Hill Education*

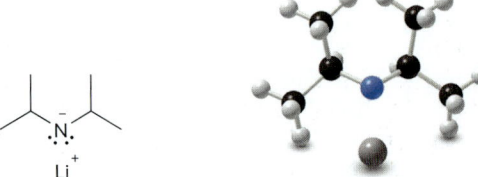

lithium diisopropylamide
LDA

The N atom is too crowded to be a nucleophile.

17.3 Enolates

THF
tetrahydrofuran
a polar aprotic
solvent

LDA quickly deprotonates essentially all of the carbonyl starting material, even at –78 °C, to form the enolate product. THF is the typical solvent for these reactions.

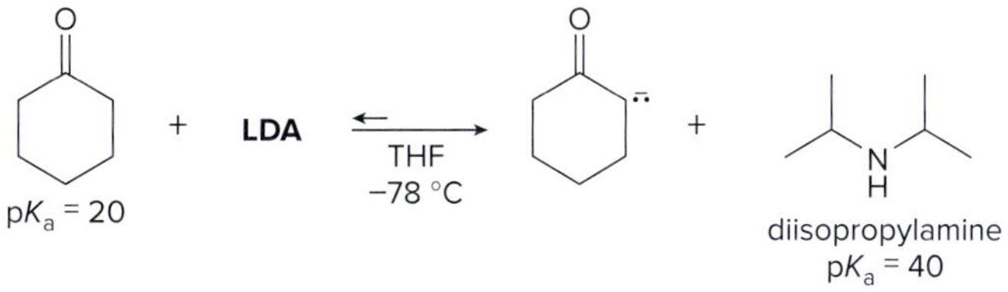

cyclohexanone ($pK_a = 20$) + **LDA** $\underset{-78\ °C}{\overset{THF}{\rightleftharpoons}}$ enolate + diisopropylamine ($pK_a = 40$)

> Equilibrium greatly favors the products.
> **Essentially all of the ketone is converted to enolate.**

LDA can be prepared by deprotonating diisopropylamine with an organolithium reagent such as butyllithium, and then used immediately in a reaction.

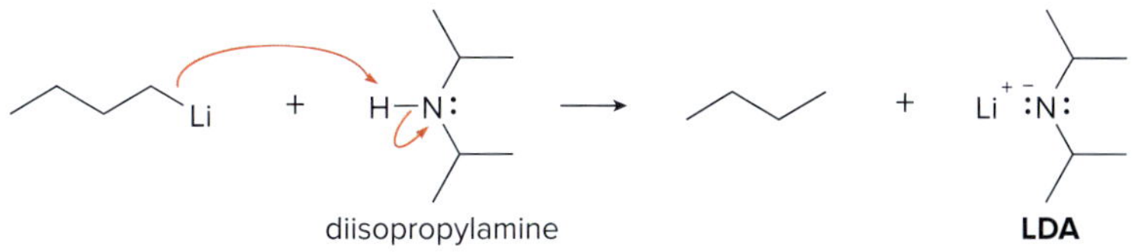

BuLi + H–N(iPr)$_2$ (diisopropylamine) → butane + Li$^+$:N(iPr)$_2^-$ (**LDA**)

Problem 17.8 Draw the product formed when each starting material is treated with LDA in THF solution at –78 °C.

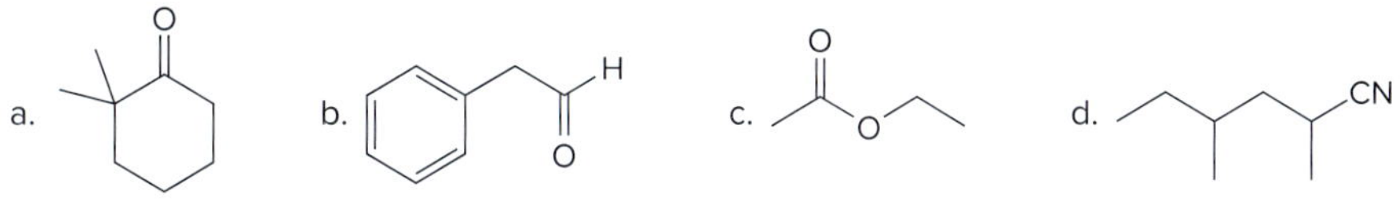

a. 2,2-dimethylcyclohexanone   b. phenylacetaldehyde   c. ethyl acetate   d. 2,4-dimethylhexanenitrile

---

17.3C General Reactions of Enolates

Enolates are nucleophiles, and as such they react with many electrophiles. Because an enolate is resonance stabilized, however, it has two reactive sites—the carbon and oxygen atoms that bear the negative charge. **A nucleophile with two reactive sites is called an *ambident nucleophile.*** In theory, each of these atoms could react with an electrophile to form two different products, one with a new bond to carbon and one with a new bond to oxygen.

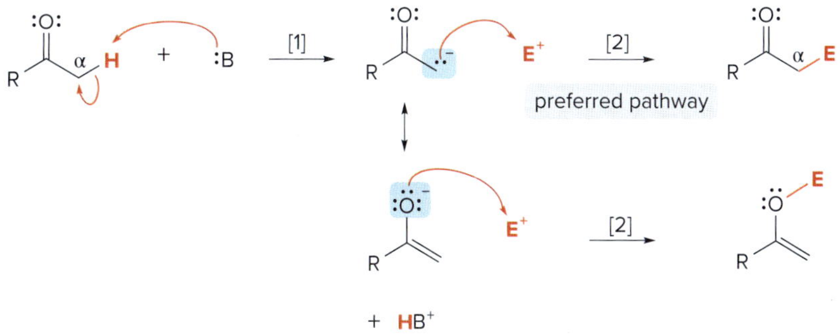

R–C(=O)–CαH + :B $\xrightarrow{[1]}$ enolate $\xrightarrow{E^+}$ $\xrightarrow{[2]}$ R–C(=O)–Cα–E   *preferred pathway*

↕

enolate O-form $\xrightarrow{E^+}$ $\xrightarrow{[2]}$ R–C(OE)=C

+ HB$^+$

> Because enolates usually react at carbon instead of oxygen, the resonance structure that places the negative charge on oxygen will often be omitted in multistep mechanisms.

An enolate usually reacts at the carbon end, however, because this site is more nucleophilic. Thus, **enolates generally react with electrophiles on the α carbon,** so that many reactions in Chapter 17 follow a two-step path:

[1] Reaction of a carbonyl compound with base forms an enolate.
[2] Reaction of the enolate with an electrophile forms a new bond on the α carbon.

17.4 Enolates of Unsymmetrical Carbonyl Compounds

What happens when an unsymmetrical carbonyl compound like 2-methylcyclohexanone is treated with base? **Two enolates are possible,** one formed by removal of a 2° hydrogen, and one formed by removal of a 3° hydrogen.

Path [1] loss of a 2° H → less substituted enolate — This enolate is formed **faster**. (**kinetic enolate**)

Path [2] loss of a 3° H → more substituted enolate (more C's bonded to the C=C) — This enolate is **more stable**. (**thermodynamic enolate**)

Path [1] occurs *faster* than Path [2] because it results in removal of the less hindered 2° hydrogen, forming an enolate on the less substituted α carbon. Path [2] results in removal of a 3° hydrogen, forming the *more stable* enolate with the more substituted double bond. This enolate predominates at equilibrium.

- The kinetic enolate is formed faster because it is the *less substituted* enolate.
- The thermodynamic enolate is lower in energy because it is the *more substituted* enolate.

It is possible to regioselectively form one or the other enolate by the choice of the base, solvent, and reaction temperature.

Kinetic Enolates

The kinetic enolate forms faster, so mild reaction conditions favor it over slower processes with higher energies of activation. It is the less stable enolate, so it must not be allowed to equilibrate to the more stable thermodynamic enolate. **The kinetic enolate is favored by**

[1] **A strong nonnucleophilic base. A bulky base like LDA removes the more accessible proton on the less substituted carbon** much faster than a more hindered proton.

[2] **Polar aprotic solvent.** The solvent must be aprotic so that it does not protonate any enolate that is formed. **THF** is both polar and aprotic.

[3] **Low temperature.** The temperature must be low (**−78 °C**) to prevent the kinetic enolate from equilibrating to the thermodynamic enolate.

2-methylcyclohexanone + LDA, THF, −78 °C → **major product** (kinetic enolate)

- A kinetic enolate is formed with a strong, nonnucleophilic base (LDA) in a polar aprotic solvent (THF) at low temperature (−78 °C).

17.4 Enolates of Unsymmetrical Carbonyl Compounds

Thermodynamic Enolates

A thermodynamic enolate is favored by equilibrating conditions. This is often achieved using a **strong base in a protic solvent.** A strong base yields both enolates, but in a protic solvent, enolates can also be protonated to re-form the carbonyl starting material. At equilibrium, the lower-energy intermediate always wins out, so that **the more stable, more substituted enolate is present in higher concentration.** Thus, the **thermodynamic enolate is favored by**

[1] **A strong base.** $Na^+\ ^-OCH_2CH_3$, $K^+\ ^-OC(CH_3)_3$, or other alkoxides are common.

[2] **Protic solvent.** CH_3CH_2OH or other alcohols.

[3] **Room temperature (25 °C).**

To simplify structures, we use abbreviations:
Me = CH_3, so $NaOCH_3$ = NaOMe
Et = CH_2CH_3, so $NaOCH_2CH_3$ = NaOEt
tBu = $C(CH_3)_3$, so $KOC(CH_3)_3$ = KOtBu

major product
thermodynamic enolate

- A thermodynamic enolate is formed with a strong base (RO⁻) in a polar protic solvent (ROH) at room temperature.

Sample Problem 17.2 Determining the Enolate Formed from an Unsymmetrical Ketone

What is the major enolate formed in each reaction?

Solution

a. **LDA is a strong, nonnucleophilic base** that removes a proton on the less substituted α carbon to form the **kinetic enolate.**

b. **NaOCH₂CH₃ (a strong base) and CH₃CH₂OH (a protic solvent)** favor removal of a proton from the more substituted α carbon to form the **thermodynamic enolate.**

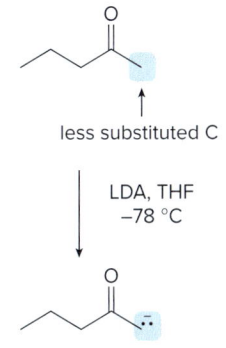

kinetic enolate

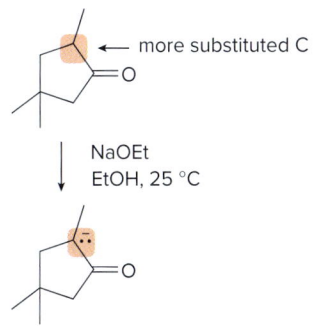

thermodynamic enolate

Problem 17.9 What enolate is formed when each ketone is treated with LDA in THF solution? What enolate is formed when these same ketones are treated with NaOCH₃ in CH₃OH solution?

More Practice: Try Problem 17.33.

17.5 Racemization at the α Carbon

Recall from Section 12.5 that an enolate can be stabilized by the delocalization of electron density only if it possesses the proper geometry and hybridization.

- The electron pair on the carbon adjacent to the C=O must occupy a *p* orbital that overlaps with the two other *p* orbitals of the C=O, making an enolate conjugated.

- Thus, all three atoms of the enolate are sp^2 hybridized and trigonal planar.

These bonding features are shown in the acetone enolate in Figure 17.1.

Figure 17.1
The hybridization and geometry of the acetone enolate $(CH_3COCH_2)^-$

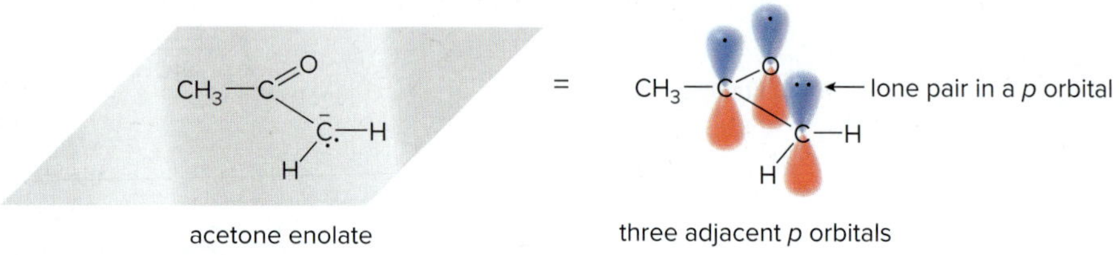

- The O atom and both C's of the enolate are **sp^2 hybridized** and lie in a plane.
- Each atom has a *p* orbital extending above and below the plane; these orbitals overlap to delocalize electron density.

When the α carbon to the carbonyl is a stereogenic center, treatment with aqueous base leads to **racemization** by a two-step process: **deprotonation to form an enolate and protonation to re-form the carbonyl compound.** For example, chiral ketone **A** reacts with aqueous ⁻OH to form an achiral enolate having an sp^2 hybridized α carbon. Because the enolate is planar, it can be protonated with H_2O with equal probability from both directions, yielding a racemic mixture of two ketones.

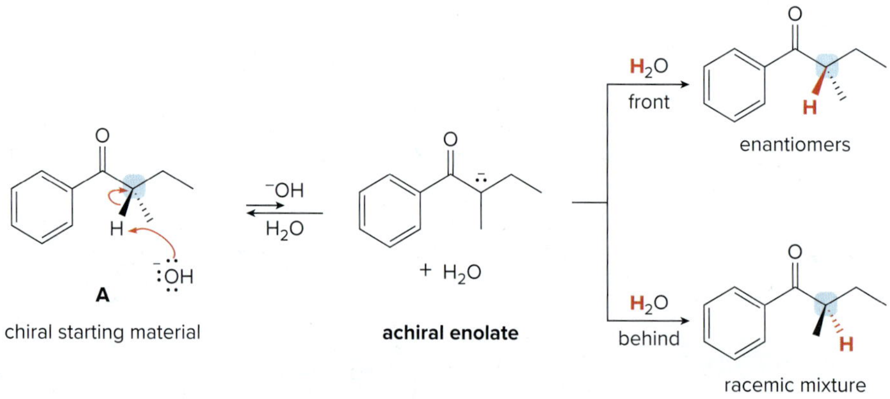

Problem 17.10   Explain each observation: (a) When (*R*)-2-methylcyclohexanone is treated with NaOH in H_2O, the optically active solution gradually loses optical activity. (b) When (*R*)-3-methylcyclohexanone is treated with NaOH in H_2O, the solution remains optically active.

17.6 A Preview of Reactions at the α Carbon

Having learned about the synthesis and properties of enolates, we can now turn our attention to their reactions. Like enols, **enolates are nucleophiles**, but because they are negatively charged, **enolates are much more nucleophilic than neutral enols.** Consequently, they undergo a wider variety of reactions.

Two general types of reactions of enolates—**substitutions** and **reactions with other carbonyl compounds**—will be discussed in the remainder of Chapter 17 and in Chapter 18. Both reactions form new bonds to the carbon α to the carbonyl.

- Enolates react with electrophiles to afford substitution products.

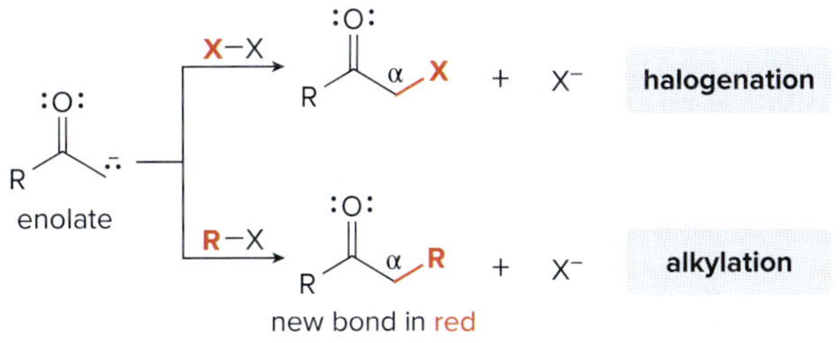

Two different kinds of substitution reactions can occur: **halogenation** with X_2 and **alkylation** with alkyl halides RX. These reactions are detailed in Sections 17.7–17.10.

- Enolates react with other carbonyl groups at the electrophilic carbonyl carbon.

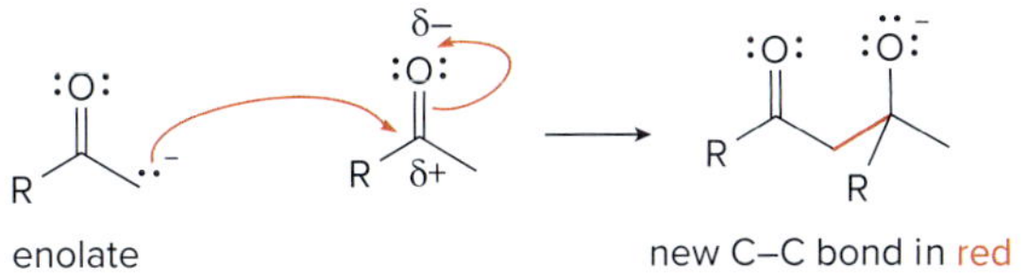

These reactions are more complicated because the initial addition adduct goes on to form different products depending on the structure of the carbonyl group. These reactions form the subject of Chapter 18.

17.7 Halogenation at the α Carbon

The first substitution reaction we examine is **halogenation**. Treatment of a ketone or aldehyde with halogen results in **substitution of X for H on the α carbon,** forming an **α-halo aldehyde or ketone.** Halogenation readily occurs with Cl_2, Br_2, and I_2. Although halogenation can occur in the presence of either acid or base, only halogenation in acid is discussed because it is synthetically much more useful.

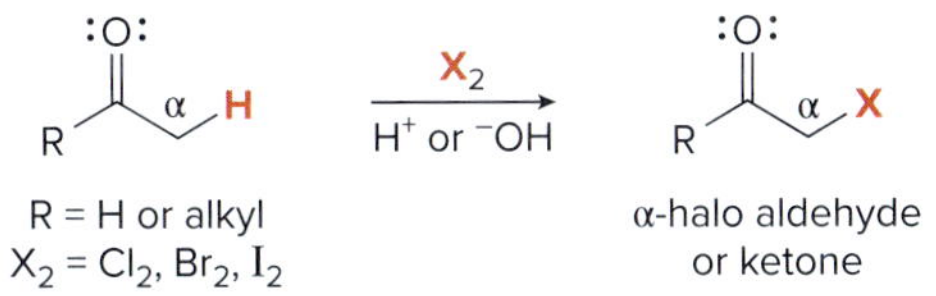

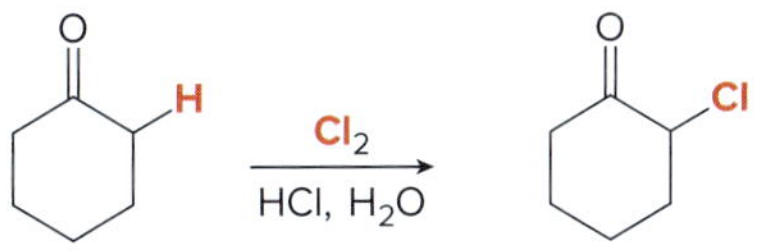

17.7A Halogenation in Acid

Halogenation is often carried out by treating a carbonyl compound with a halogen in acetic acid. In this way, acetic acid is both the solvent and the acid catalyst for the reaction.

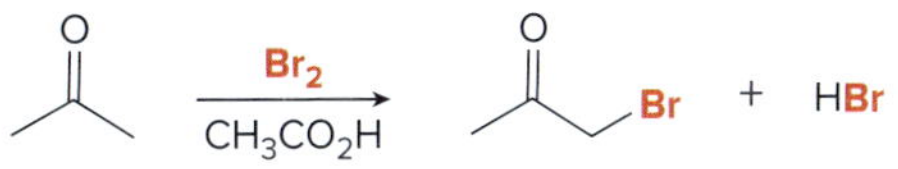

The mechanism of acid-catalyzed halogenation consists of two parts: **tautomerization** of the carbonyl compound to the enol form, and **reaction of the enol with halogen.** Mechanism 17.3 illustrates the reaction of $(CH_3)_2C=O$ with Br_2 in CH_3CO_2H.

792 Chapter 17 Substitution Reactions of Carbonyl Compounds at the α Carbon

Mechanism 17.3 Acid-Catalyzed Halogenation at the α Carbon

1 – 2 The ketone is converted to its **enol tautomer** by the two-step process of protonation followed by deprotonation.
3 – 4 Addition of the halogen to the enol forms a new bond to Br on the α carbon, and deprotonation yields the substitution product.

Problem 17.11   Draw the products of each reaction.

a. cyclopentanone + Cl$_2$, H$_2$O, HCl
b. CH$_3$CH$_2$CHO + Br$_2$, CH$_3$CO$_2$H (propanal)
c. α-tetralone + Br$_2$, CH$_3$CO$_2$H

17.7B Reactions of α-Halo Carbonyl Compounds

α-Halo carbonyl compounds undergo two useful reactions—**elimination** with base and **substitution** with nucleophiles.

For example, treatment of 2-bromocyclohexanone with the base Li$_2$CO$_3$ in the presence of LiBr in the polar aprotic solvent DMF [HCON(CH$_3$)$_2$] affords cyclohex-2-enone by **elimination of the elements of Br and H from the α and β carbons,** respectively. Thus, a two-step method can convert a carbonyl compound such as cyclohexanone to an **α,β-unsaturated carbonyl compound** such as cyclohex-2-enone.

cyclohexanone →[Br$_2$ / CH$_3$CO$_2$H] 2-bromocyclohexanone →[Li$_2$CO$_3$ / LiBr / DMF] cyclohex-2-enone

halogenation   **elimination**

A new π bond is formed in two steps.

α,β-Unsaturated carbonyl compounds undergo a variety of 1,2- and 1,4-addition reactions as discussed in Section 13.15.

[1] Bromination at the α carbon is accomplished with Br$_2$ in CH$_3$CO$_2$H.
[2] Elimination of Br and H occurs with Li$_2$CO$_3$ and LiBr in DMF.

α-Halo carbonyl compounds also react with nucleophiles by S$_N$2 reactions. For example, reaction of 2-bromocyclohexanone with CH$_3$CH$_2$SH affords the substitution product **A**.

2-bromocyclohexanone + CH$_3$CH$_2$SH →[S$_N$2] **A**

Sample Problem 17.3 — Using α-Halo Carbonyl Compounds in Synthesis

What steps are needed to convert ketone **A** to **B** and **C**?

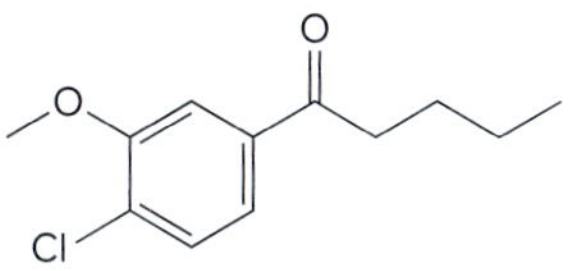

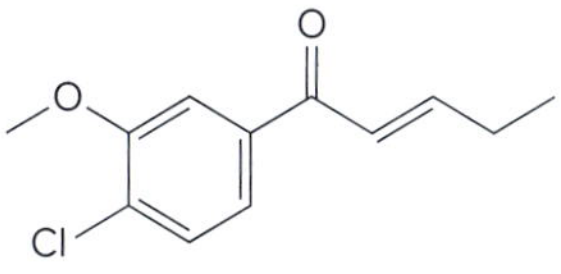

Solution

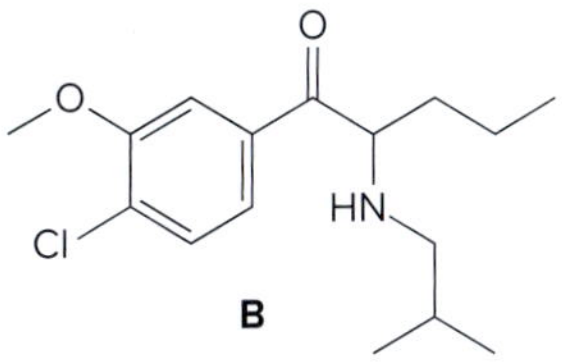

To introduce the N atom on the α carbon to form **B** or the double bond between the α and β carbons to form **C**, we must first convert **A** to an α-halo ketone, **D**. The halogen can then act as a leaving group in a **substitution** reaction to form **B**, or an **elimination** reaction to form **C**.

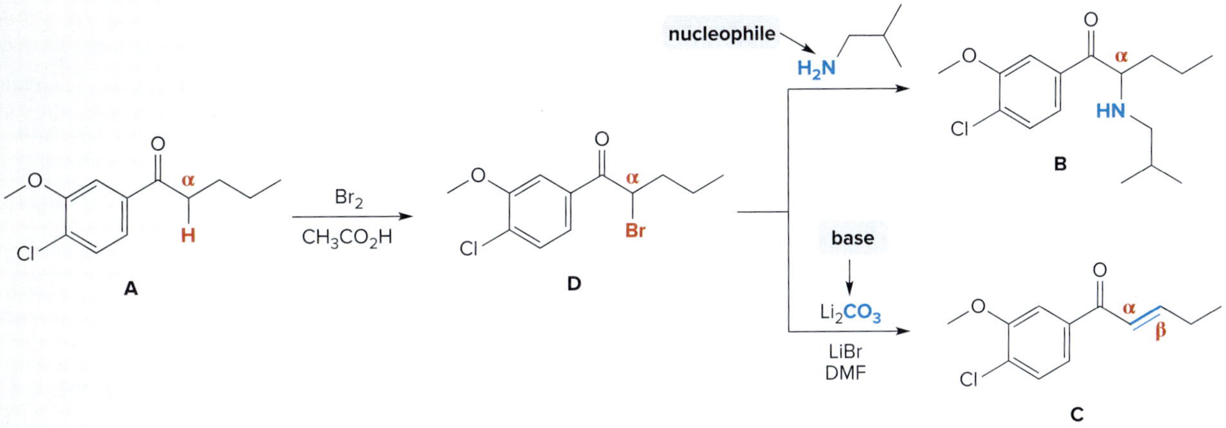

- Reaction of **D** with an amine nucleophile forms the substitution product **B** by an S_N2 mechanism.
- Reaction of **D** with a base forms the elimination product **C**. The elements of Br and H are removed from the α and β carbons to form a π bond.

Problem 17.12 Draw the organic products formed when 2-bromopentan-3-one ($CH_3CH_2COCHBrCH_3$) is treated with each reagent: (a) Li_2CO_3, LiBr, DMF; (b) $CH_3CH_2NH_2$; (c) CH_3SH.

More Practice: Try Problems 17.43c, e; 17.45; 17.46.

Problem 17.13 A key step in a synthesis of the antimalarial drug quinine involves an intramolecular nucleophilic substitution that converts **A** to **B**. Draw the structure of **B** and give the reagents needed to convert **B** to quinine.

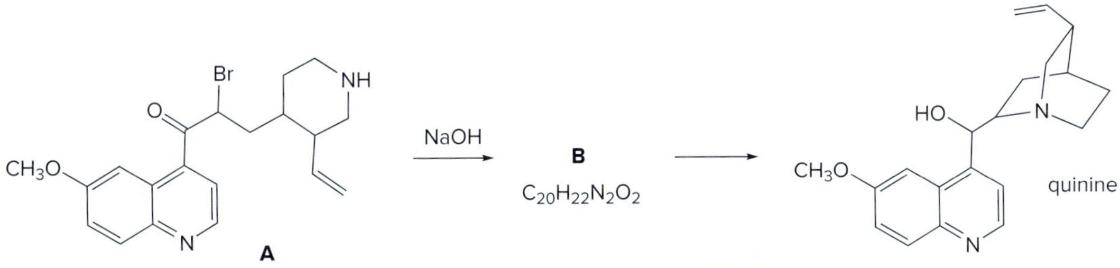

17.8 Direct Enolate Alkylation

Treatment of an aldehyde or ketone with base and an alkyl halide (RX) results in *alkylation*—**the substitution of R for H on the α carbon atom.** Alkylation forms a new carbon–carbon bond on the α carbon.

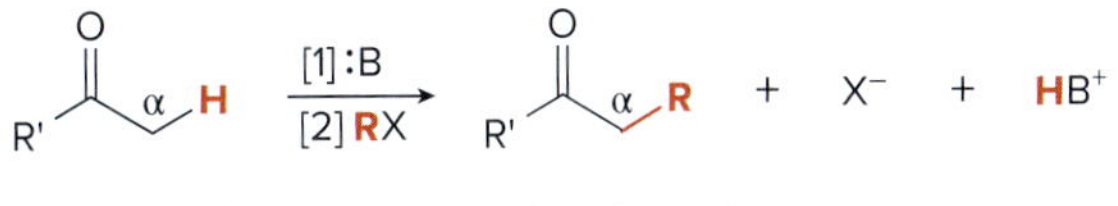

new bond in red

17.8A General Features

We will begin with the most direct method of alkylation, and then (in Sections 17.9 and 17.10) examine two older, multistep methods that are still used today. Direct alkylation is carried out by a two-step process:

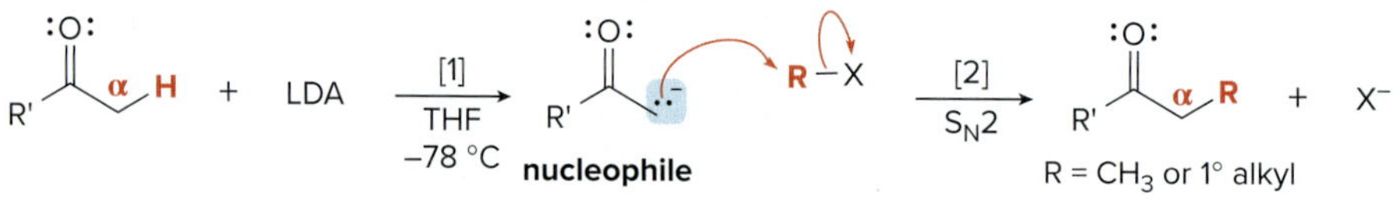

[1] **Deprotonation:** Base removes a proton from the α carbon to generate an enolate. The reaction works best with a strong nonnucleophilic base like LDA in THF solution at low temperature (−78 °C).

[2] **Nucleophilic attack:** The nucleophilic enolate attacks the alkyl halide, displacing the halide (a good leaving group) and forming the alkylation product by an S_N2 reaction.

Because Step [2] is an S_N2 **reaction**, it works best with **unhindered methyl and 1° alkyl halides.** Hindered alkyl halides and those with halogens bonded to sp^2 hybridized carbons do *not* undergo substitution.

R_3CX, $CH_2=CHX$, and C_6H_5X do *not* undergo alkylation reactions with enolates, because they are unreactive in S_N2 reactions.

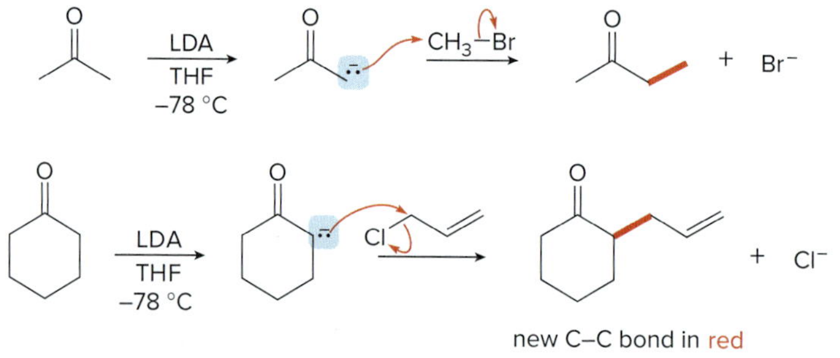

Ester enolates and carbanions derived from nitriles are also alkylated under these conditions.

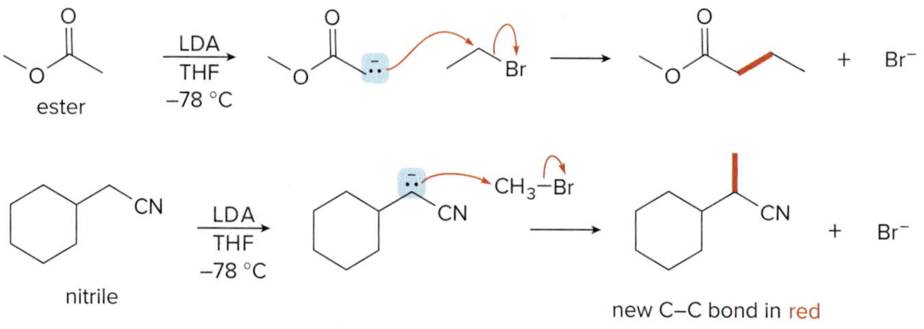

Problem 17.14 What product is formed when each compound is treated first with LDA in THF solution at low temperature, followed by CH_3CH_2I?

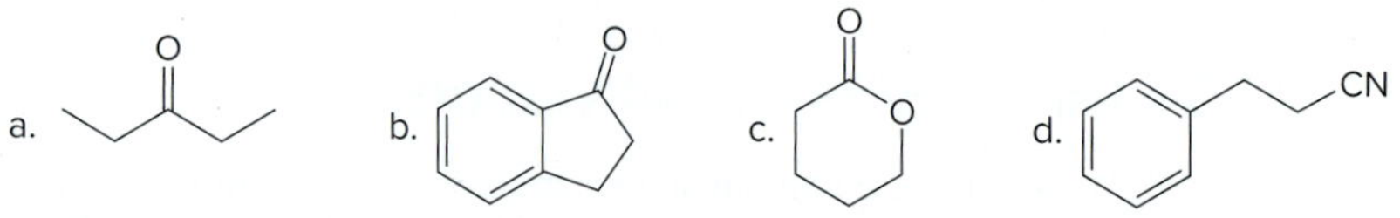

The stereochemistry of enolate alkylation follows the general rule governing the stereochemistry of reactions: **an achiral starting material yields an achiral or racemic product.** For example, when cyclohexanone (an achiral starting material) is converted to 2-ethylcyclohexanone by treatment with base and CH_3CH_2I, a new stereogenic center (labeled

17.8 Direct Enolate Alkylation

in blue) is introduced, and both enantiomers of the product are formed in equal amounts—that is, a **racemic mixture**.

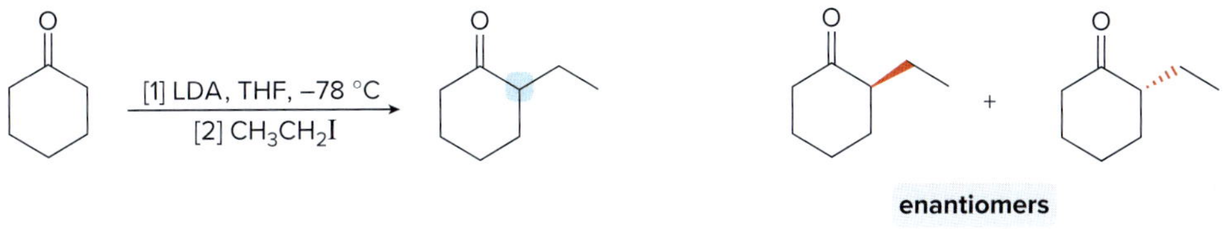

cyclohexanone → [1] LDA, THF, −78 °C; [2] CH₃CH₂I → 2-ethylcyclohexanone (enantiomers)

Problem 17.15 Draw the products obtained (including stereochemistry) when each compound is treated with LDA, followed by CH₃I.

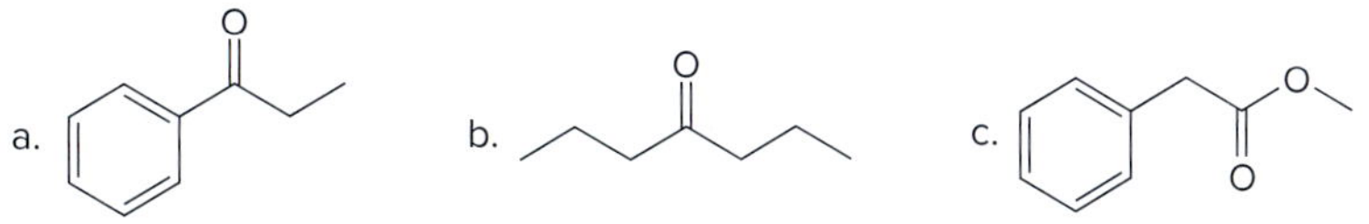

a. propiophenone (PhCOCH₂CH₃)
b. 3-hexanone
c. methyl phenylacetate (PhCH₂CO₂CH₃)

Problem 17.16 The analgesic naproxen can be prepared by a stepwise reaction sequence from ester **A**. Using enolate alkylation in one step, what reagents are needed to convert **A** to naproxen? Draw the structure of each intermediate. Explain why a racemic product is formed.

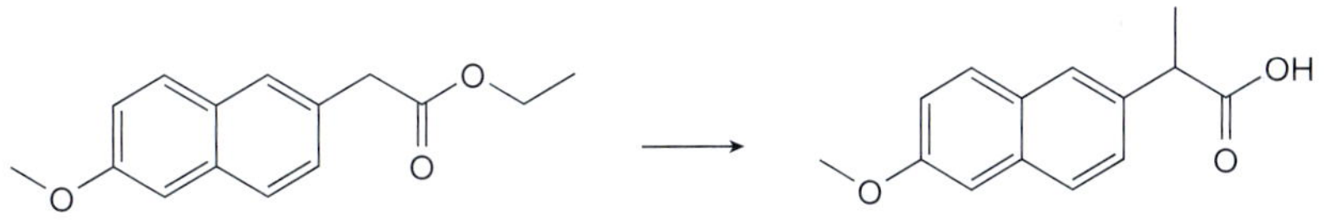

A (methoxynaphthalene-CH₂-CO₂Et) → naproxen

17.8B Alkylation of Unsymmetrical Ketones

An unsymmetrical ketone can be regioselectively alkylated to yield one major product. The strategy depends on the use of the appropriate base, solvent, and temperature to form the kinetic or thermodynamic enolate (Section 17.4), which is then treated with an alkyl halide to form the alkylation product.

An example of this strategy is seen in the intramolecular alkylation of bromo ketone **A** to form either **B** or **C**, depending on the reaction conditions.

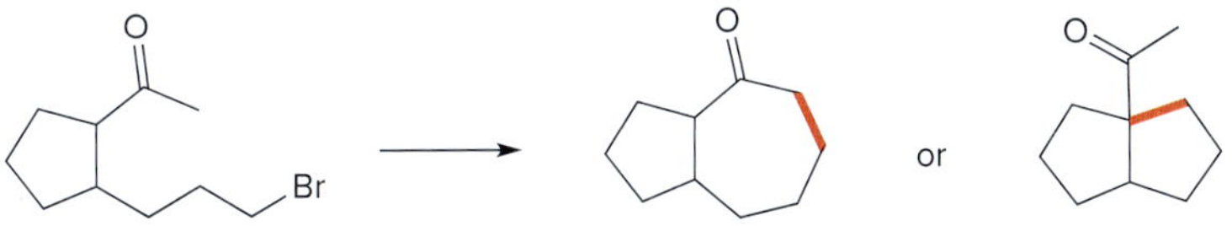

A → **B** or **C**

new C–C bond in red

- Treatment of **A** with LDA in THF at −78° gives the less substituted enolate, which undergoes an intramolecular S_N2 reaction to form the seven-membered ring in **B**.

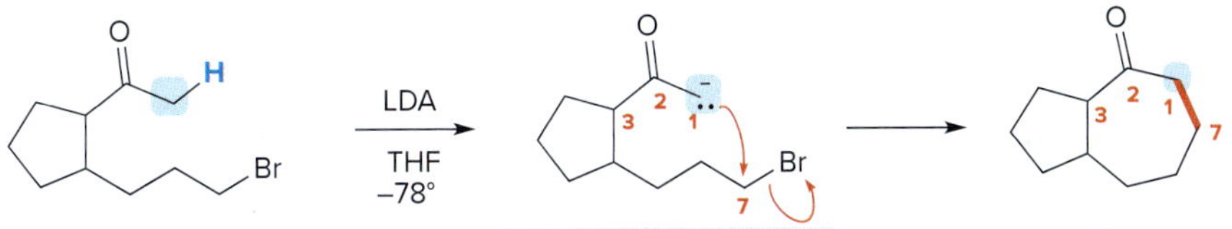

A → LDA, THF, −78° → less substituted enolate (kinetic product) → **B**

- Treatment of **A** with KOC(CH$_3$)$_3$ in (CH$_3$)$_3$COH at room temperature gives the *more* substituted enolate, which undergoes an intramolecular S$_N$2 reaction to form the five-membered ring in **C**.

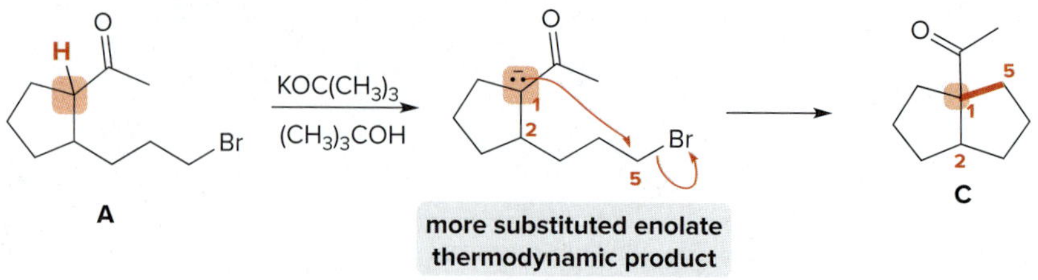

Finally, while enolate alkylation at the less substituted α carbon using LDA is a reliable regioselective reaction, enolate alkylation at the more substituted α carbon with KOC(CH$_3$)$_3$ may lead to mixtures of products. Regioselectivity depends on the identity of the substrate and the experimental parameters, which sometimes must be carefully monitored to maximize the yield of the desired alkylation product.

Problem 17.17 How can pentan-2-one be converted to each compound?

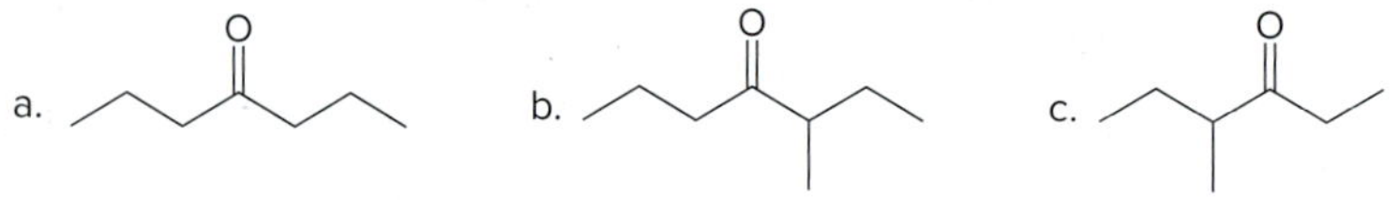

a., b., c.

Problem 17.18 Identify **A**, **B**, and **C**, intermediates in the synthesis of the five-membered ring called an α-methylene-γ-butyrolactone. This heterocyclic ring system is present in some antitumor agents.

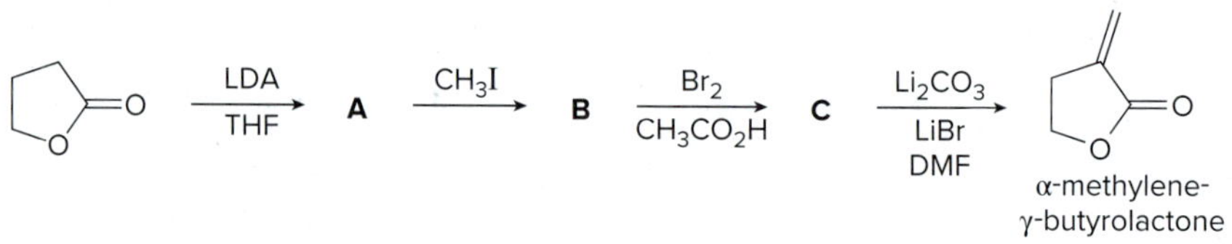

$$\text{lactone} \xrightarrow[\text{THF}]{\text{LDA}} \textbf{A} \xrightarrow{\text{CH}_3\text{I}} \textbf{B} \xrightarrow[\text{CH}_3\text{CO}_2\text{H}]{\text{Br}_2} \textbf{C} \xrightarrow[\substack{\text{LiBr}\\\text{DMF}}]{\text{Li}_2\text{CO}_3} \text{α-methylene-γ-butyrolactone}$$

17.9 Malonic Ester Synthesis

Besides the direct method of enolate alkylation discussed in Section 17.8, a new alkyl group can also be introduced on the α carbon using the malonic ester synthesis and the acetoacetic ester synthesis.

- **The malonic ester synthesis prepares carboxylic acids** having two general structures:

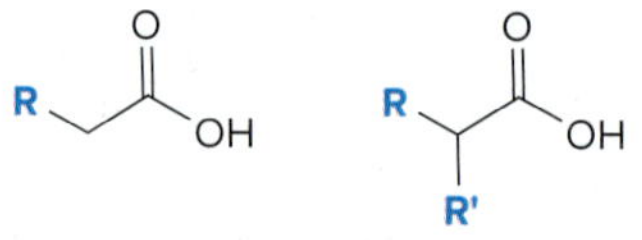

- **The acetoacetic ester synthesis prepares methyl ketones** having two general structures:

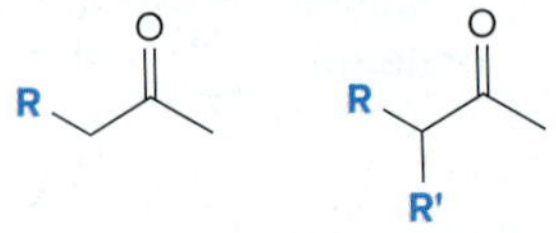

17.9A Background for the Malonic Ester Synthesis

- The malonic ester synthesis is a stepwise method for converting diethyl malonate to a carboxylic acid having one or two alkyl groups on the α carbon.

17.9 Malonic Ester Synthesis

To simplify the structures, the CH_3CH_2 groups of the esters are abbreviated as Et.

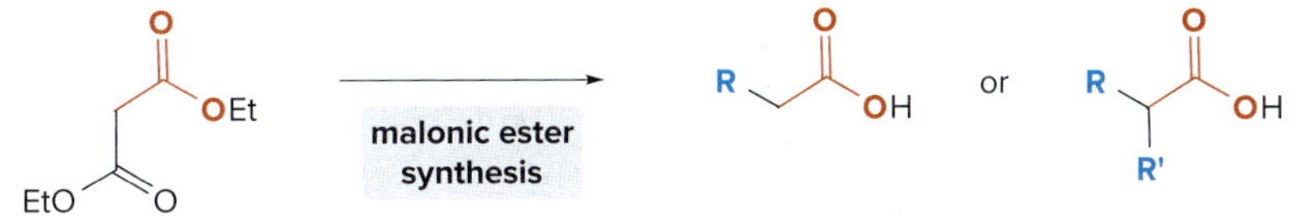

diethyl malonate
Et = CH_3CH_2

Before writing out the steps in the malonic ester synthesis, recall from Section 16.10 that esters are hydrolyzed by aqueous acid. Thus, heating diethyl malonate with acid and water hydrolyzes both esters to carboxy groups, forming a **β-diacid** (1,3-diacid).

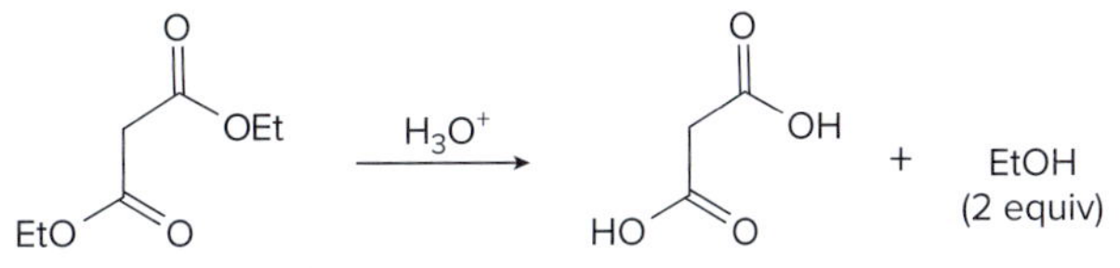

diethyl malonate → malonic acid (β-diacid) + EtOH (2 equiv)

The resulting β-diacids are unstable to heat. They **decarboxylate (lose CO_2)**, resulting in cleavage of a carbon–carbon bond and formation of a carboxylic acid. Decarboxylation is *not* a general reaction of all carboxylic acids. It occurs with β-diacids, however, because CO_2 can be eliminated through a cyclic, six-atom transition state. This forms an enol of a carboxylic acid, which in turn tautomerizes to the more stable keto form.

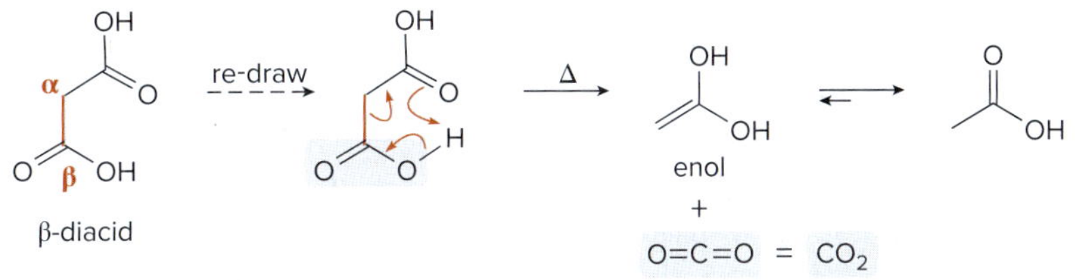

β-diacid → enol + O=C=O = CO_2

The net result of decarboxylation is cleavage of a carbon–carbon bond on the α carbon, with loss of CO_2.

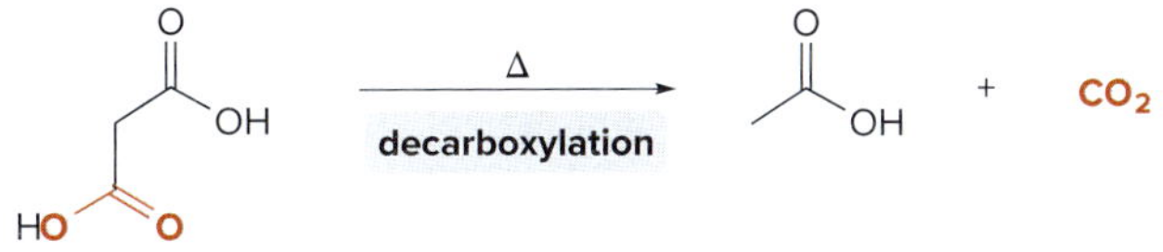

β-diacid → (decarboxylation) + CO_2

Decarboxylation occurs readily whenever a carboxy group (COOH) is bonded to the α carbon of another carbonyl group. For example, β-keto acids also readily lose CO_2 on heating to form ketones.

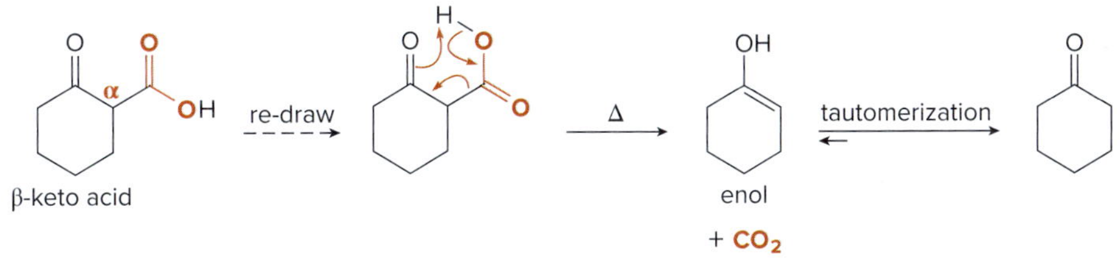

β-keto acid → enol + CO_2 → (tautomerization) ketone

Problem 17.19 Which of the following compounds will readily lose CO_2 when heated?

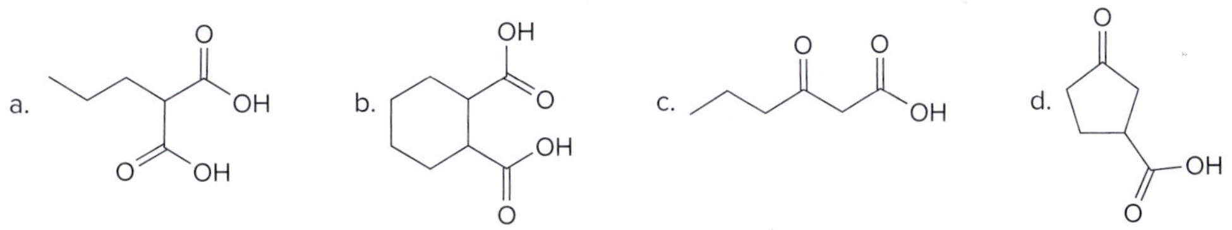

a. b. c. d.

17.9B Steps in the Malonic Ester Synthesis

The malonic ester synthesis converts diethyl malonate to a carboxylic acid in three steps.

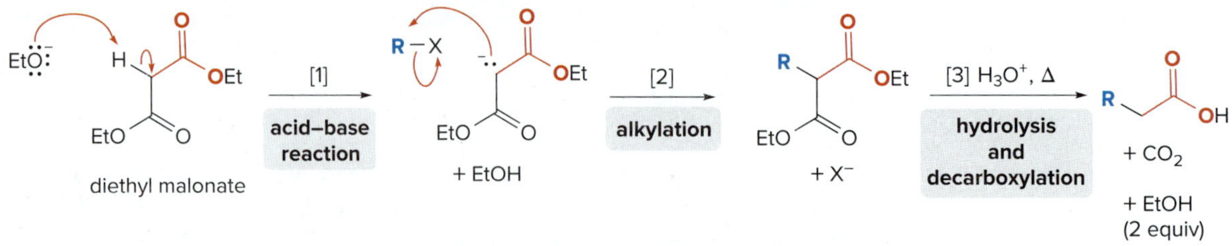

[1] **Deprotonation.** Treatment of diethyl malonate with ⁻OEt removes the acidic α proton between the two carbonyl groups. Recall from Section 17.3A that these protons are more acidic than other α protons because **three resonance structures can be drawn for the enolate,** instead of the usual two. Thus, ⁻OEt, rather than the stronger base LDA, can be used for this reaction.

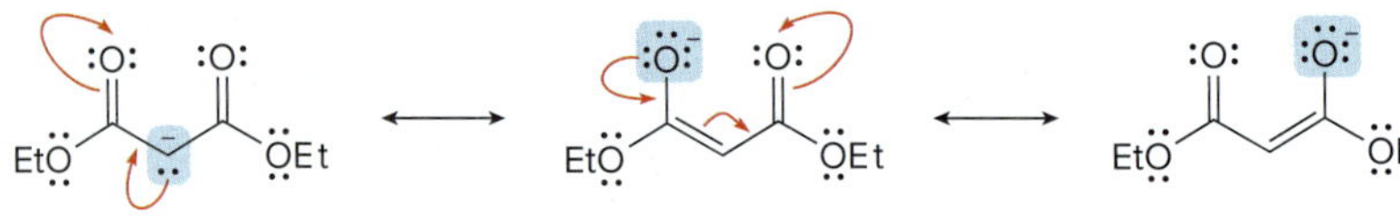

three resonance structures for the conjugate base

[2] **Alkylation.** The nucleophilic enolate reacts with an alkyl halide in an S_N2 reaction to form a substitution product. Because the mechanism is S_N2, the yields are higher when R is CH_3 or a 1° alkyl group.

[3] **Hydrolysis and decarboxylation.** Heating the diester with aqueous acid hydrolyzes the diester to a β-diacid, which loses CO_2 to form a carboxylic acid.

The synthesis of butanoic acid ($CH_3CH_2CH_2COOH$) from diethyl malonate illustrates the basic process:

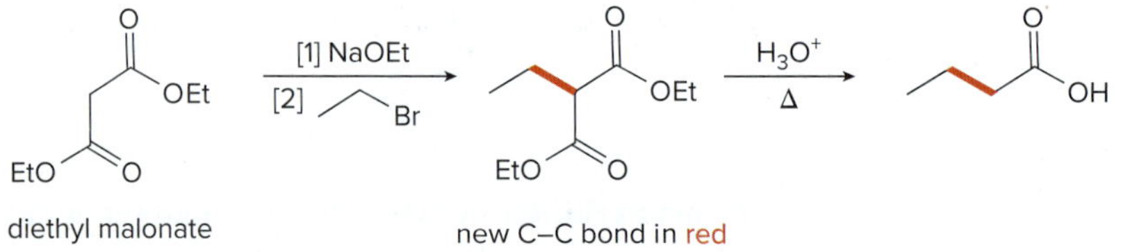

If the first two steps of the reaction sequence are repeated *prior* to hydrolysis and decarboxylation, then a carboxylic acid having *two new alkyl groups* on the α carbon can be synthesized. This is illustrated in the synthesis of 2-benzylbutanoic acid [$CH_3CH_2CH(CH_2C_6H_5)COOH$] from diethyl malonate:

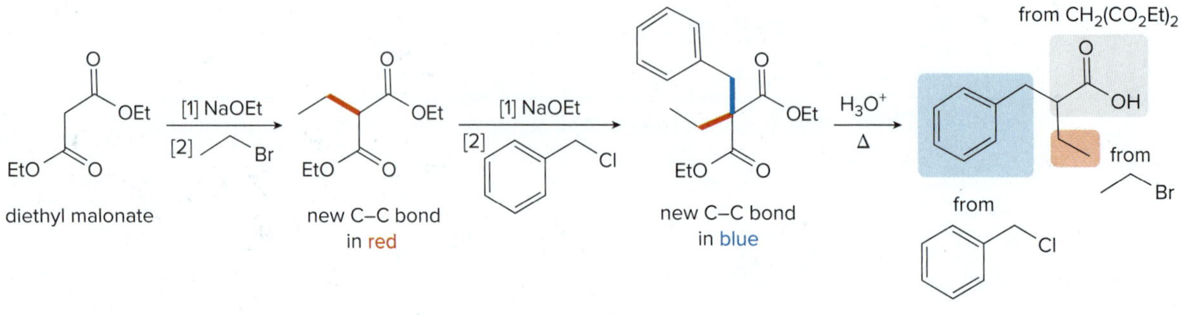

17.9 Malonic Ester Synthesis

An intramolecular malonic ester synthesis can be used to form rings having three to six atoms, provided the appropriate dihalide is used as starting material. For example, cyclopentanecarboxylic acid can be prepared from diethyl malonate and 1,4-dibromobutane ($BrCH_2CH_2CH_2CH_2Br$) by this sequence of reactions:

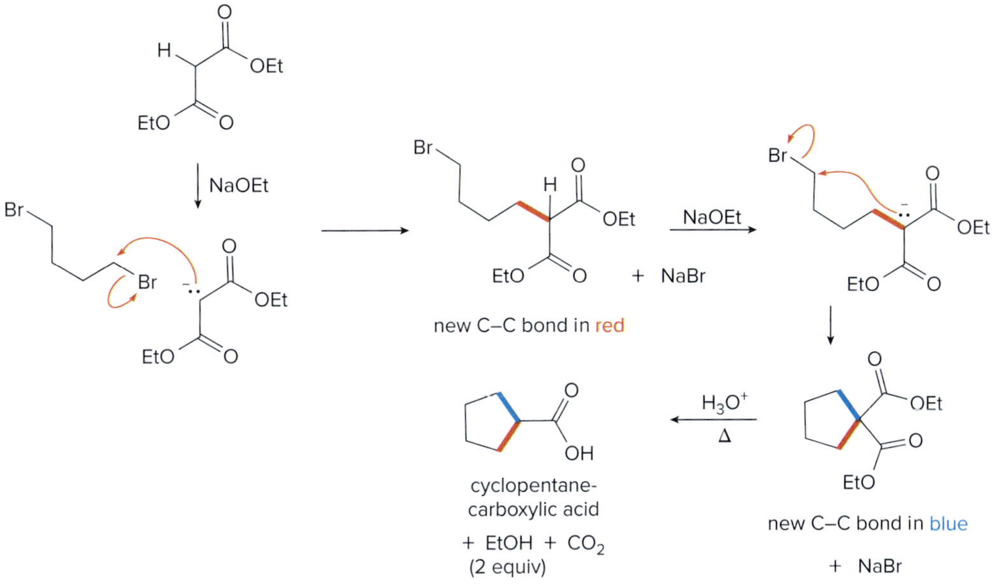

Problem 17.20 Draw the products of each reaction.

a. $CH_2(CO_2Et)_2$ $\xrightarrow{[1]\ NaOEt}_{[2]\ \text{(cyclohexylmethyl bromide)}}$ $\xrightarrow{H_3O^+}_{\Delta}$

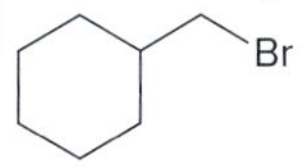

b. $CH_2(CO_2Et)_2$ $\xrightarrow{[1]\ NaOEt}_{[2]\ CH_3Br}$ $\xrightarrow{[1]\ NaOEt}_{[2]\ CH_3Br}$ $\xrightarrow{H_3O^+}_{\Delta}$

Problem 17.21 What cyclic product is formed from each dihalide using the malonic ester synthesis?

a.  Cl~~~Cl   b. Br~~~O~~~Br

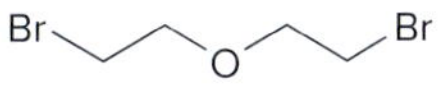

17.9C Retrosynthetic Analysis

To use the malonic ester synthesis, you must be able to determine what starting materials are needed to prepare a given compound—that is, you must **work backwards in the retrosynthetic direction.** This involves a two-step process:

[1] Locate the α carbon to the COOH group, and identify all alkyl groups bonded to the α carbon.
[2] Break the molecule into two (or three) components: Each alkyl group bonded to the α carbon comes from an alkyl halide. The remainder of the molecule comes from $CH_2(COOEt)_2$.

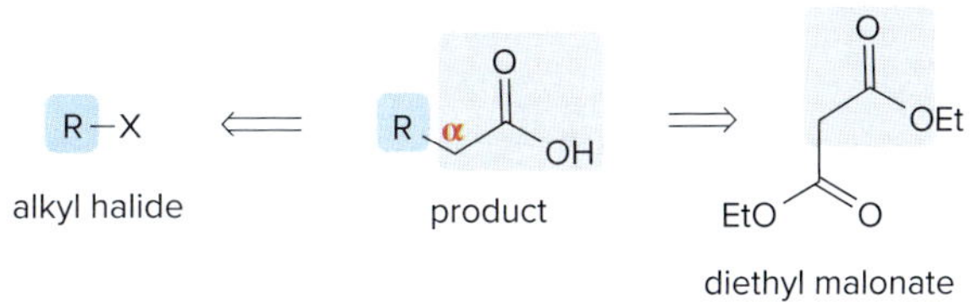

R—X ⟸ R—α—COOH ⟹ diethyl malonate

alkyl halide   product   diethyl malonate

Sample Problem 17.4 — Determining the Starting Materials in a Malonic Ester Synthesis

What starting materials are needed to prepare 2-methylhexanoic acid [CH$_3$CH$_2$CH$_2$CH$_2$CH(CH$_3$)COOH] using a malonic ester synthesis?

Solution

The target molecule has two different alkyl groups bonded to the α carbon, so three components are needed for the synthesis:

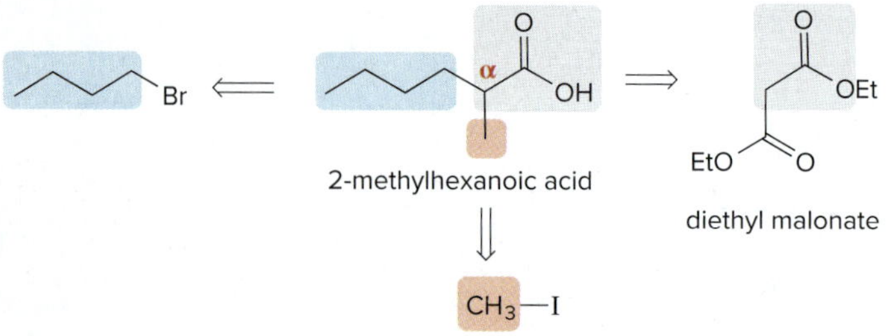

[Retrosynthesis diagram: 2-methylhexanoic acid ⇒ diethyl malonate + butyl bromide + CH$_3$—I]

Writing the synthesis in the synthetic direction:

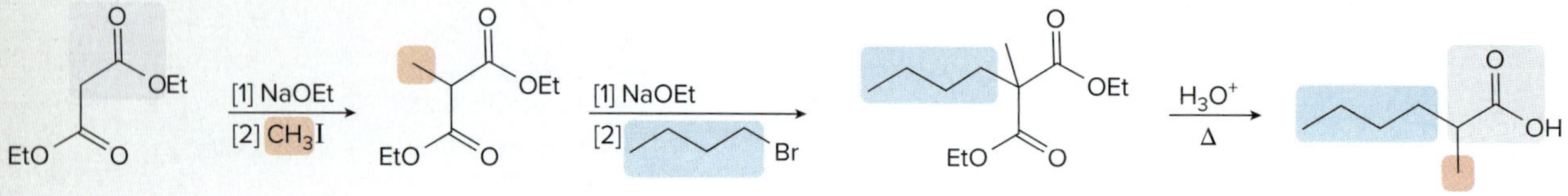

[Synthesis scheme: diethyl malonate →[1] NaOEt, [2] CH$_3$I→ methylated malonate →[1] NaOEt, [2] butyl bromide→ dialkylated malonate →H$_3$O$^+$, Δ→ 2-methylhexanoic acid]

Problem 17.22 What alkyl halides are needed to prepare each carboxylic acid by the malonic ester synthesis?

a. 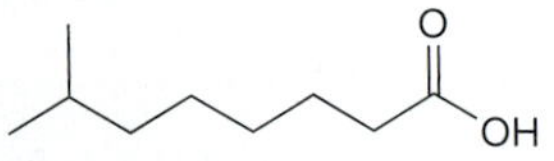 [structure] b. 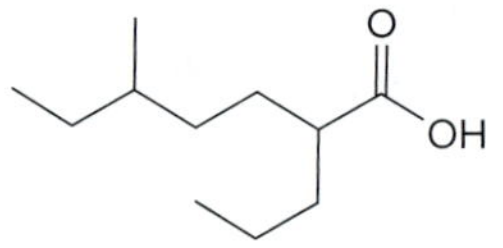 [structure] c. 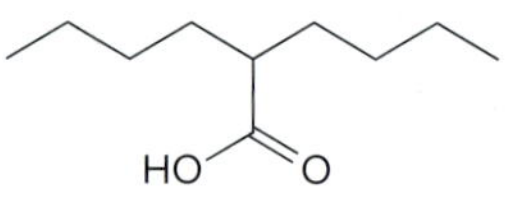 [structure]

More Practice: Try Problems 17.37, 17.38.

Problem 17.23 Explain why each of the following carboxylic acids cannot be prepared by a malonic ester synthesis.

a. 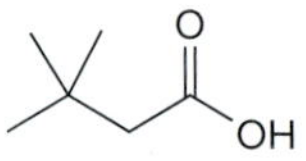 [structure] b. 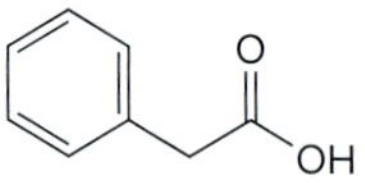 [structure] c. 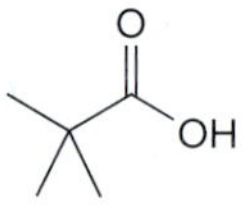 [structure]

17.10 Acetoacetic Ester Synthesis

- **The acetoacetic ester synthesis is a stepwise method for converting ethyl acetoacetate to a ketone having one or two alkyl groups on the α carbon.**

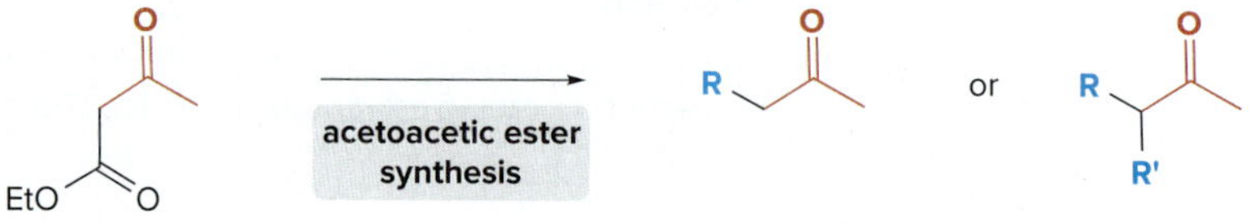

[ethyl acetoacetate] —acetoacetic ester synthesis→ R—C(O)—CH$_3$ or R—C(O)—CH(R')—CH$_3$

17.10A Steps in the Acetoacetic Ester Synthesis

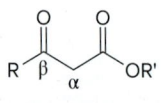

β-keto ester

The steps in the acetoacetic ester synthesis are exactly the same as those in the malonic ester synthesis. Because the starting material, CH$_3$COCH$_2$COOEt, is a β-keto ester, the final product is a **ketone,** not a carboxylic acid.

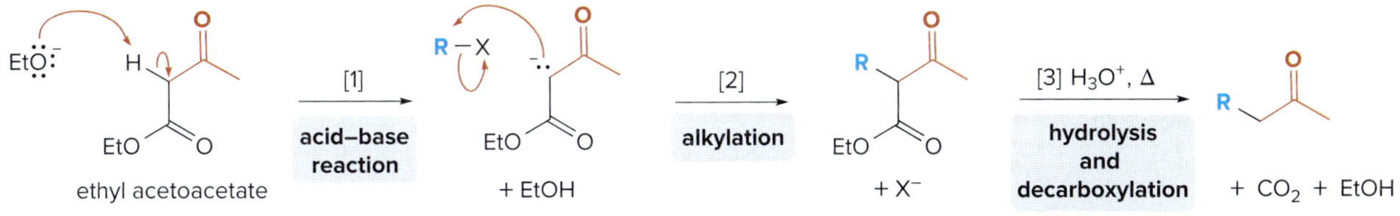

17.10 Acetoacetic Ester Synthesis

[1] **Deprotonation.** Treatment of ethyl acetoacetate with ⁻OEt removes the acidic proton between the two carbonyl groups.

[2] **Alkylation.** The nucleophilic enolate reacts with an alkyl halide (RX) in an S_N2 reaction to form a substitution product. Because the mechanism is S_N2, the yields are higher when R is CH_3 or a 1° alkyl group.

[3] **Hydrolysis and decarboxylation.** Heating the β-keto ester with aqueous acid hydrolyzes the ester to a β-keto acid, which loses CO_2 to form a ketone.

If the first two steps of the reaction sequence are repeated *prior* to hydrolysis and decarboxylation, then a ketone having *two new alkyl groups* on the α carbon can be synthesized.

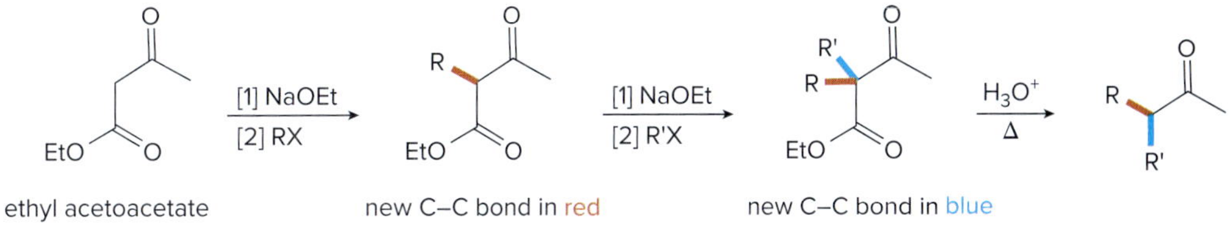

Problem 17.24 What ketones are prepared by the following reactions?

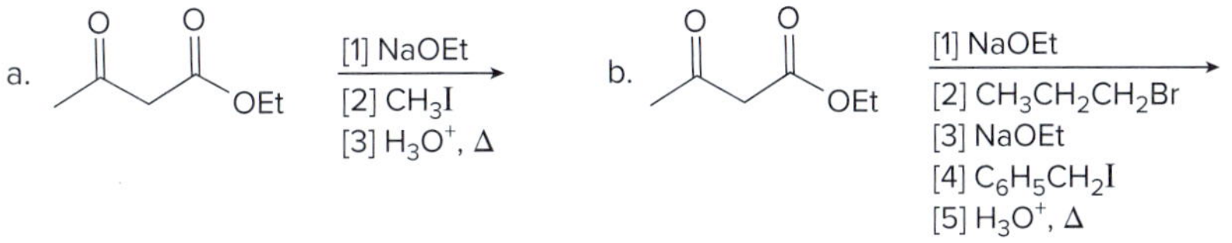

a. [1] NaOEt [2] CH_3I [3] H_3O^+, Δ

b. [1] NaOEt [2] $CH_3CH_2CH_2Br$ [3] NaOEt [4] $C_6H_5CH_2I$ [5] H_3O^+, Δ

17.10B Retrosynthetic Analysis

To determine what starting materials are needed to prepare a given ketone using the acetoacetic ester synthesis, you must again work in the **retrosynthetic** direction. This involves a two-step process:

[1] Identify the alkyl groups bonded to the α carbon to the carbonyl group.
[2] Break the molecule into two (or three) components: Each alkyl group bonded to the α carbon comes from an alkyl halide. The remainder of the molecule comes from CH_3COCH_2COOEt.

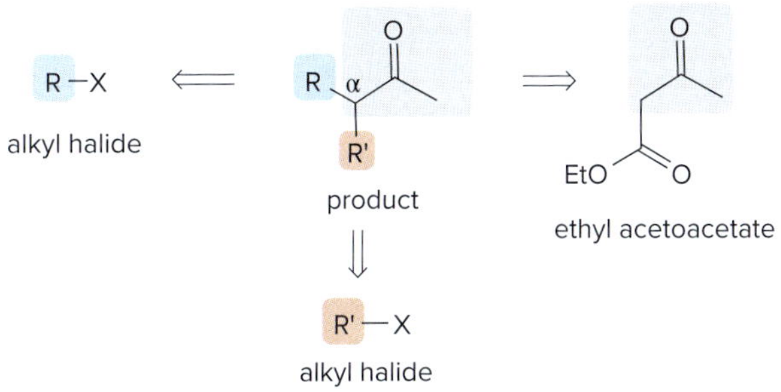

For a ketone with two R groups on the α carbon, three components are needed.

Sample Problem 17.5 Determining the Starting Materials in an Acetoacetic Ester Synthesis

What starting materials are needed to synthesize heptan-2-one using the acetoacetic ester synthesis?

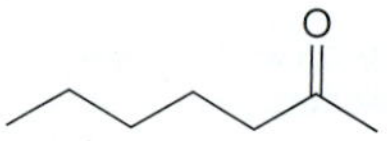

heptan-2-one

Solution

Heptan-2-one has only one alkyl group bonded to the α carbon, so only one alkyl halide is needed in the acetoacetic ester synthesis.

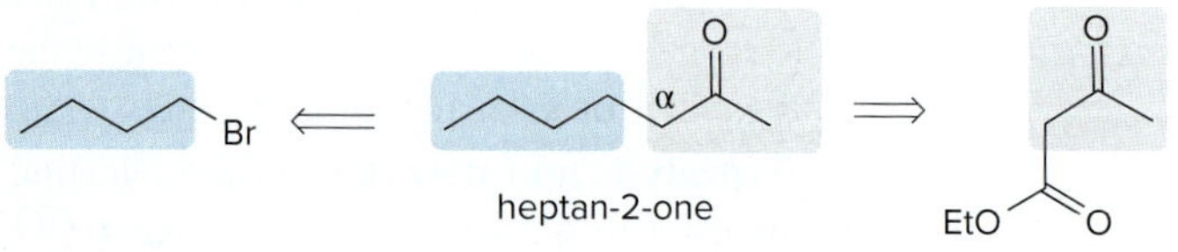

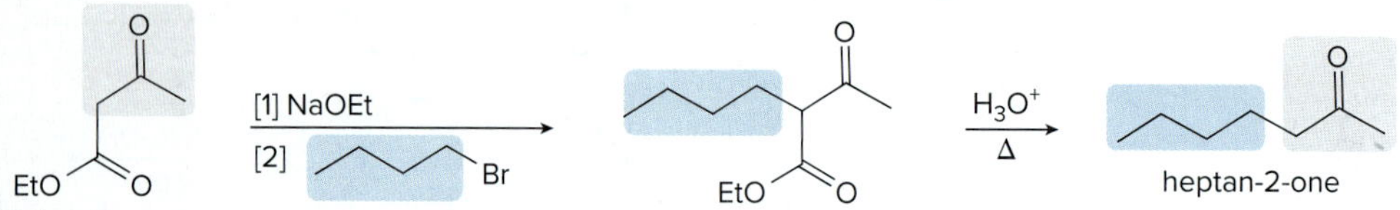

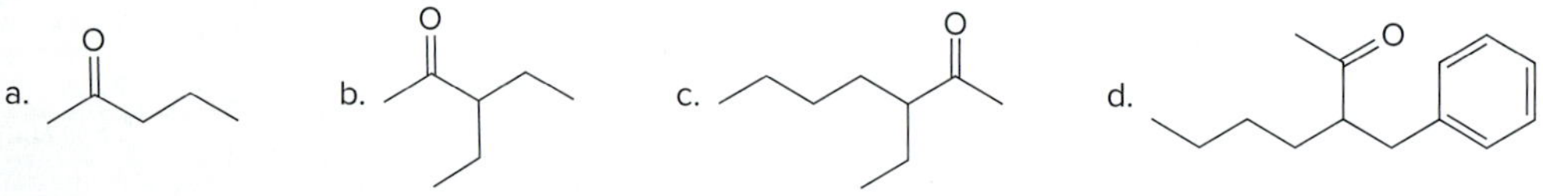

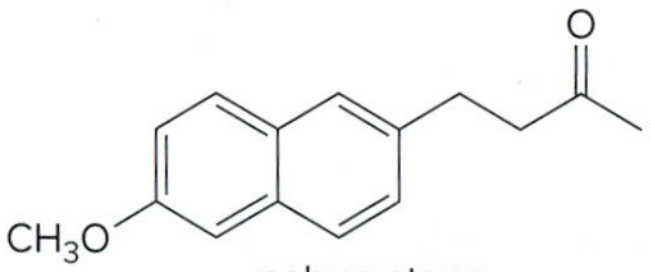

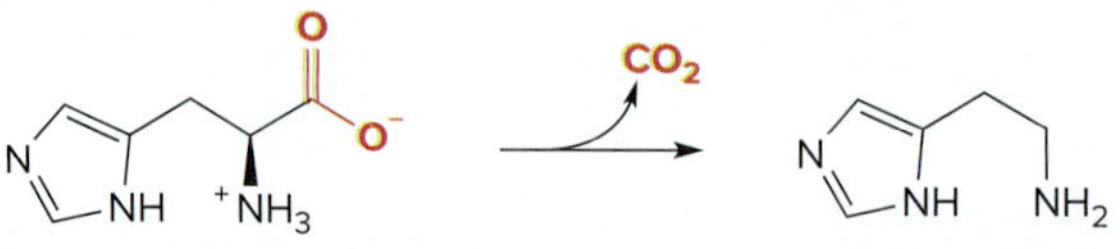

~~~Br ⟸ heptan-2-one ⟹ ethyl acetoacetate

Writing the acetoacetic ester synthesis in the synthetic direction:

ethyl acetoacetate [1] NaOEt / [2] ~~~Br → (alkylated intermediate) H₃O⁺, Δ → heptan-2-one

---

Problem 17.25 What alkyl halides are needed to prepare each ketone using the acetoacetic ester synthesis?

a. b. c. d.

More Practice: Try Problems 17.41, 17.42.

Problem 17.26 Nabumetone is a pain reliever and anti-inflammatory agent sold under the brand name of Relafen.

nabumetone

a. Write out a synthesis of nabumetone from ethyl acetoacetate.
b. What ketone and alkyl halide are needed to synthesize nabumetone by direct enolate alkylation?

17.11 Biological Decarboxylation

Just as decarboxylation is a key step in both the acetoacetic ester synthesis and the malonic ester synthesis, so, too, decarboxylation occurs in many metabolic pathways. Many amines that are central to physiological processes are formed by the **decarboxylation of amino acids.** For example, decarboxylation of the amino acid histidine forms histamine, a triamine involved in the inflammatory response, among other physiological effects (Section 22.5).

histidine $\xrightarrow{-CO_2}$ histamine

Enzymatic decarboxylation occurs with the coenzyme pyridoxal phosphate (PLP, Section 14.13B). As shown previously in Mechanism 14.8, PLP reacts with an amino acid to form an intermediate imine. In the presence of a decarboxylase enzyme, this imine loses CO_2 by cleavage of a carbon–carbon bond (Step [2]). Protonation and hydrolysis form the amine, which has one carbon fewer than the original amino acid.

Problem 17.27 Biological oxidation of the amino acid tyrosine forms L-dopa, which undergoes decarboxylation with PLP and a decarboxylase enzyme to form the neurotransmitter dopamine. Write the steps for the conversion of L-dopa to dopamine.

Chapter 17 REVIEW

KEY REACTIONS

[1] Preparation and reactions of α-halo carbonyl compounds

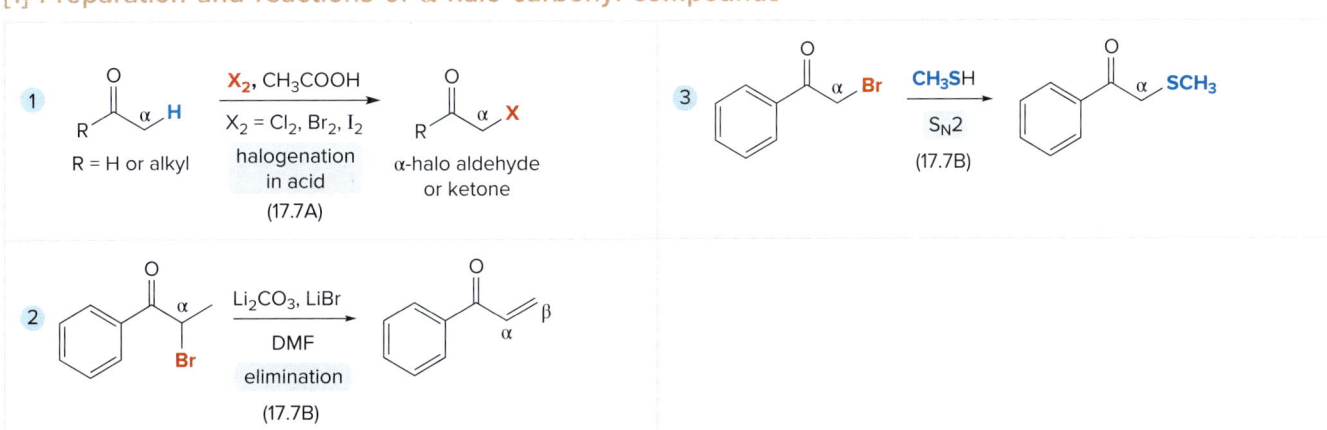

See Sample Problem 17.3. Try Problems 17.43c, e; 17.45; 17.46.

[2] Alkylation reactions at the α carbon

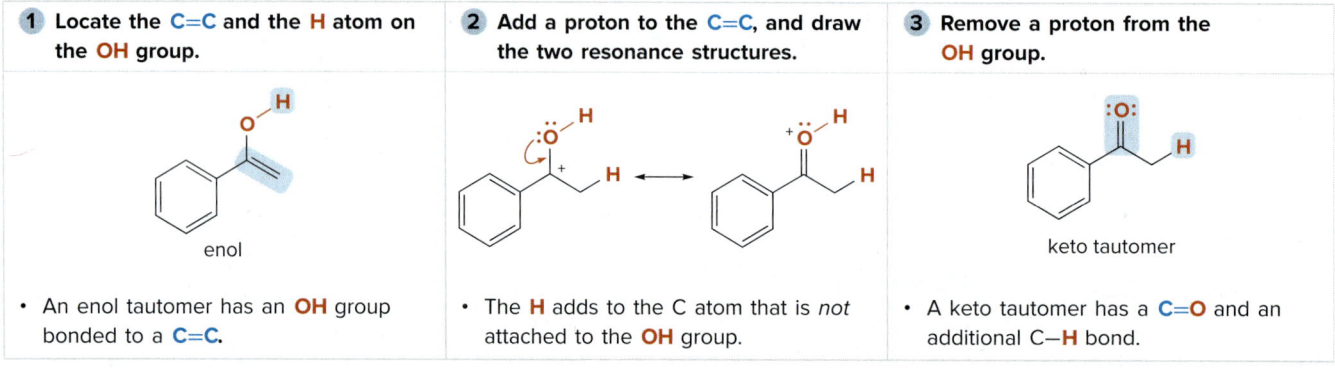

Try Problems 17.43b, d, f; 17.44; 17.49.

KEY SKILLS

[1] Converting an enol to a keto tautomer in acid (17.2)

1 Locate the C=C and the H atom on the OH group.	**2** Add a proton to the C=C, and draw the two resonance structures.	**3** Remove a proton from the OH group.
enol		keto tautomer
• An enol tautomer has an OH group bonded to a C=C.	• The H adds to the C atom that is *not* attached to the OH group.	• A keto tautomer has a C=O and an additional C–H bond.

See Sample Problem 17.1. Try Problems 17.28, 17.30, 17.36.

[2] Determining the major enolate formed in a reaction (17.4)

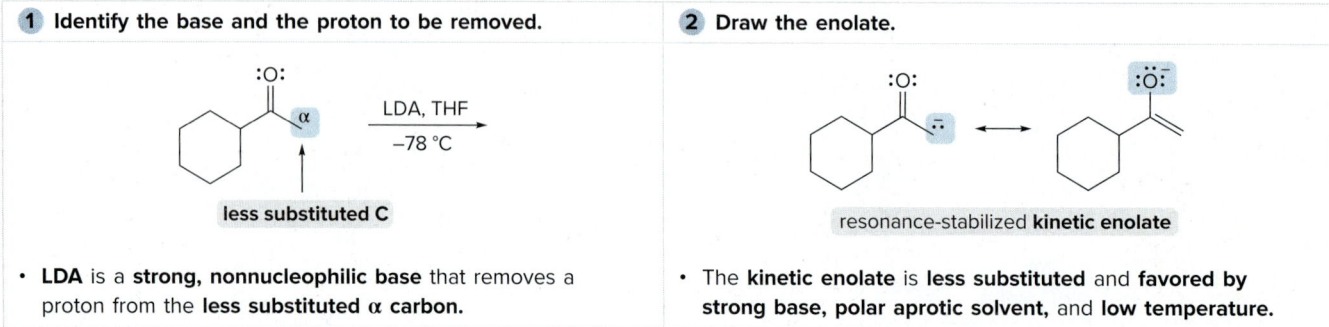

- **LDA** is a **strong, nonnucleophilic base** that removes a proton from the **less substituted** α carbon.
- The **kinetic enolate** is **less substituted** and favored by **strong base, polar aprotic solvent,** and **low temperature**.

See Sample Problem 17.2. Try Problem 17.33.

[3] Determining the major enolate formed in a reaction (17.4)

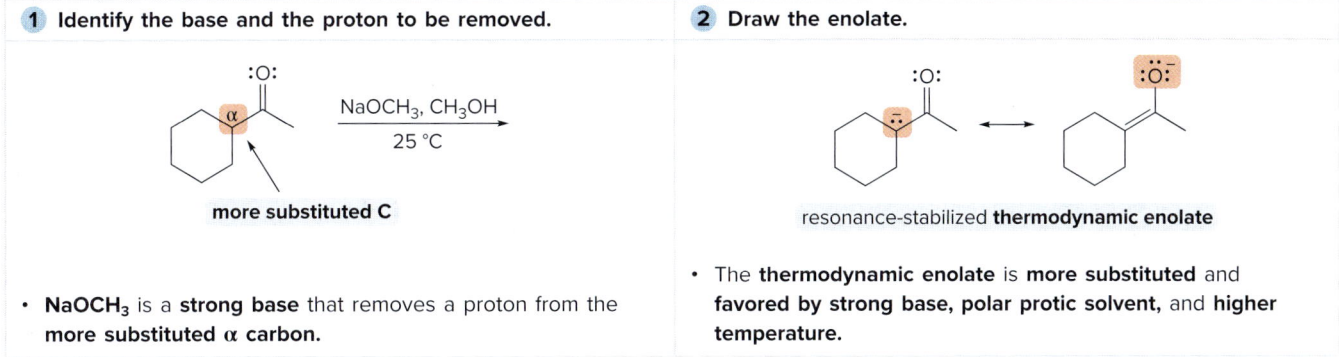

- NaOCH$_3$ is a **strong base** that removes a proton from the more substituted α carbon.

- The **thermodynamic enolate** is **more substituted** and favored by **strong base, polar protic solvent**, and **higher temperature**.

See Sample Problem 17.2.

[4] Preparing a carboxylic acid using a malonic ester synthesis (17.9B); example: 2,4-dimethylpentanoic acid

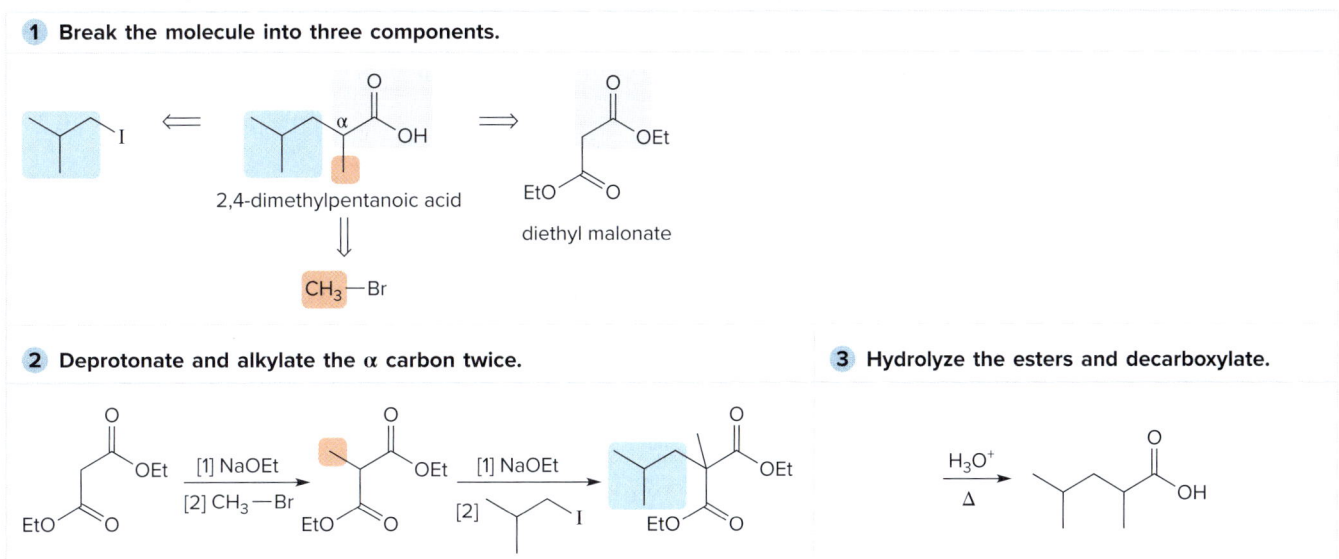

See Sample Problem 17.4. Try Problems 17.37, 17.38.

[5] Preparing a ketone using the acetoacetic ester synthesis (17.10B); example: hexan-2-one

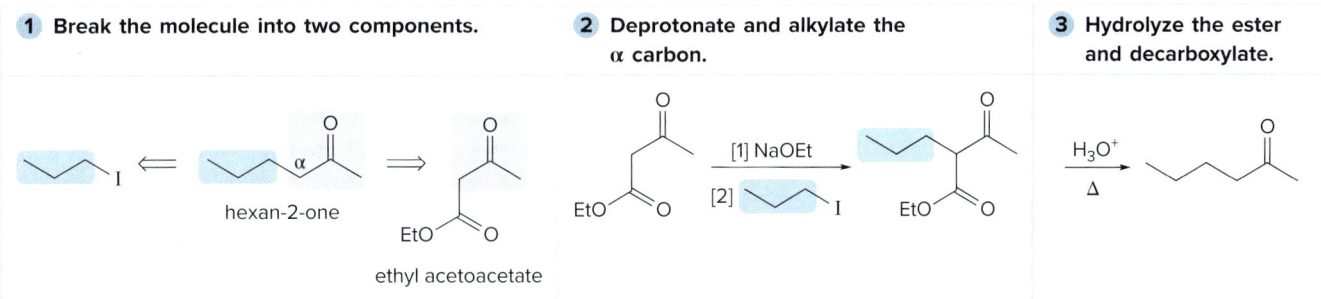

See Sample Problem 17.5. Try Problems 17.41, 17.42.

PROBLEMS

Problems Using Three-Dimensional Models

17.28 Draw enol tautomer(s) for each compound. Ignore stereoisomers.

a. b.

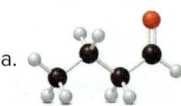

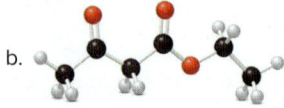

17.29 The cis ketone **A** is isomerized to a trans ketone **B** with aqueous NaOH. A similar isomerization does not occur with ketone **C.** (a) Draw the structure of **B** using a chair cyclohexane. (b) Label the substituents in **C** as cis or trans, and explain the difference in reactivity.

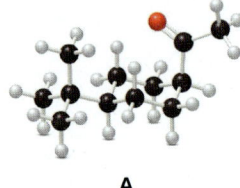

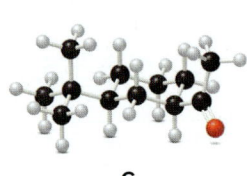

A

C

Enols, Enolates, and Acidic Protons

17.30 Draw enol tautomer(s) for each compound.

a. b. c. (mono enol form)

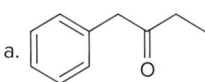

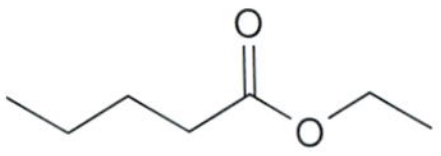

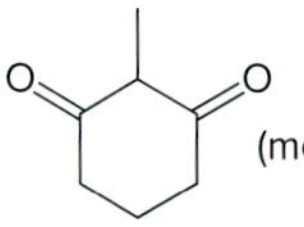

17.31 Which carbonyl compound in each pair exhibits the higher percentage of the enol tautomer?

a. or b. or

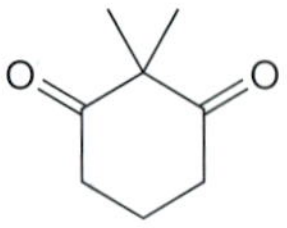

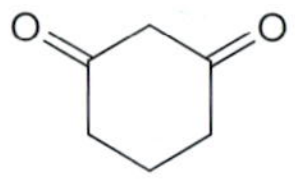

17.32 Rank the labeled protons in each compound in order of increasing acidity.

a. b.

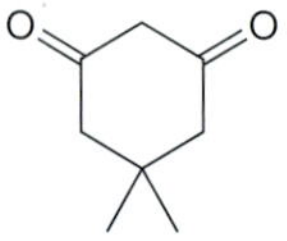

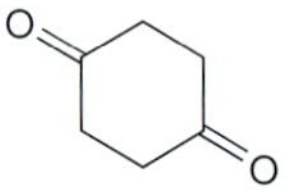

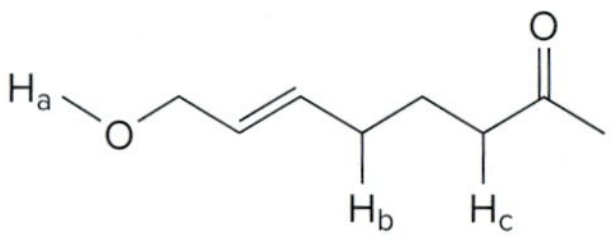

17.33 What is the major enolate (or carbanion) formed when each compound is treated with LDA?

a. b. c. d.

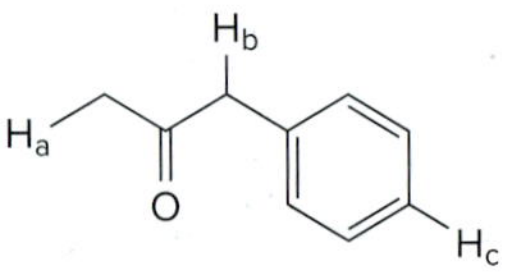

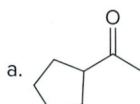

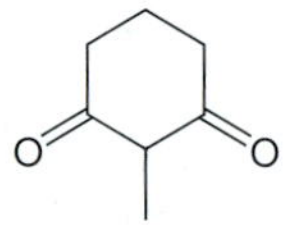

17.34 Why is the pK_a of the H_a protons in 1-acetylcyclohexene higher than the pK_a of the H_b protons?

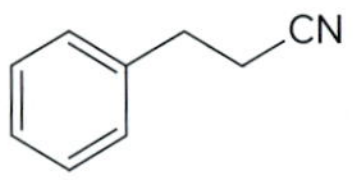

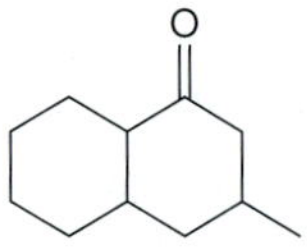

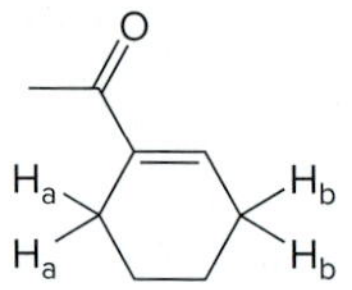

1-acetylcyclohexene

Problems

17.35 Explain why pentane-2,4-dione forms two different alkylation products (**A** or **B**) when the number of equivalents of base is increased from one to two.

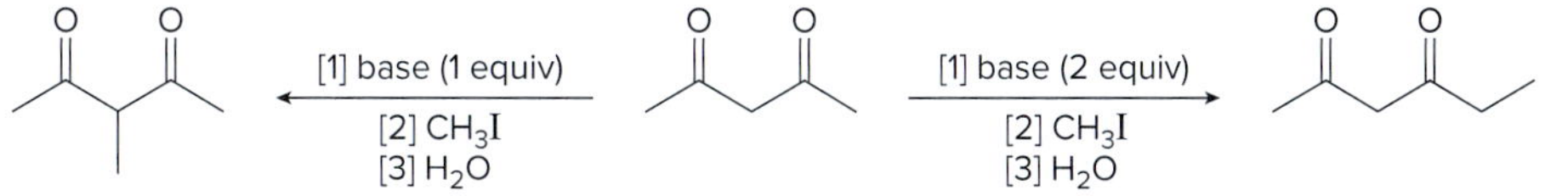

<!-- Reaction scheme: A ← [1] base (1 equiv) [2] CH₃I [3] H₂O — pentane-2,4-dione — [1] base (2 equiv) [2] CH₃I [3] H₂O → B -->

17.36 Vitamin C is a stable enediol. Draw the structure of the two keto tautomers in equilibrium with the enediol, and explain why the enediol is more stable than the other tautomers.

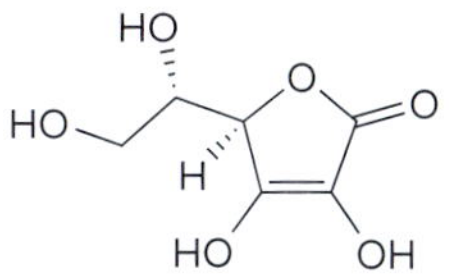

vitamin C
an enediol

Malonic Ester Synthesis

17.37 What alkyl halides are needed to prepare each carboxylic acid using the malonic ester synthesis?

a. 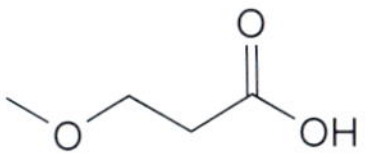 b. 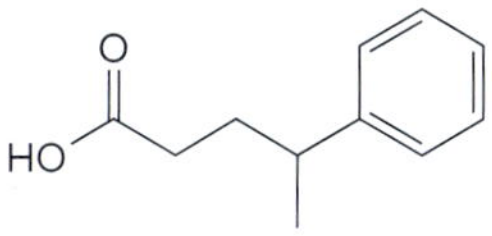 c.

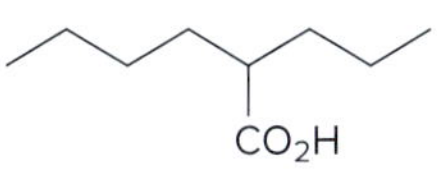

17.38 Use the malonic ester synthesis to prepare each carboxylic acid.

a. 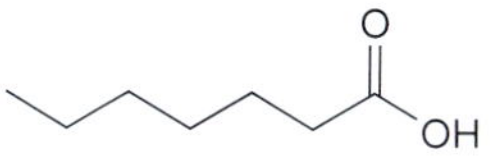 b. 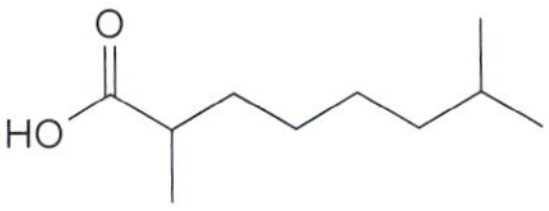 c.

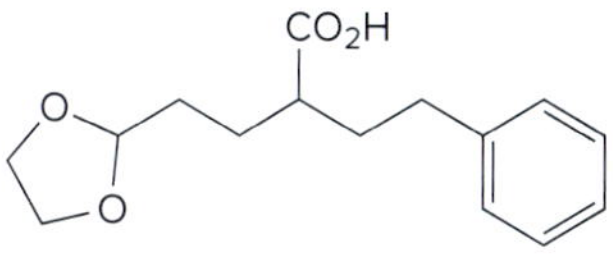

17.39 Synthesize **A** from diethyl malonate and any needed organic compounds and inorganic reagents.

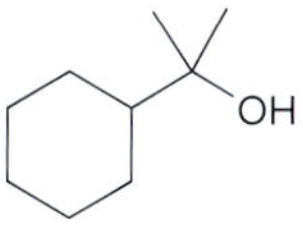

A

17.40 The enolate derived from diethyl malonate reacts with a variety of electrophiles (not just alkyl halides) to form new carbon–carbon bonds. With this in mind, draw the products formed when Na⁺ ⁻CH(CO₂Et)₂ reacts with each electrophile, followed by treatment with H₂O.

a. 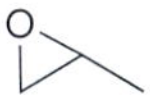 b. 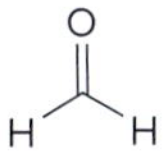 c. 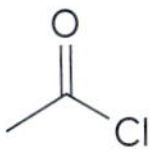 d.

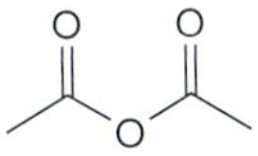

Acetoacetic Ester Synthesis

17.41 What alkyl halides are needed to prepare each ketone using the acetoacetic ester synthesis?

a. 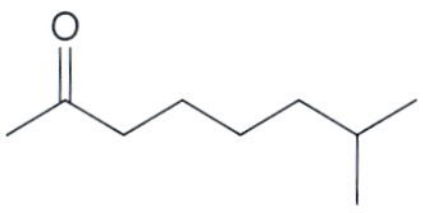 b. 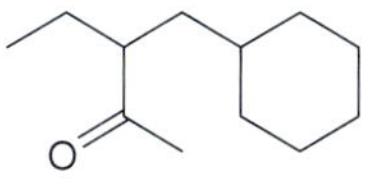 c.

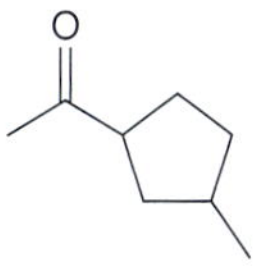

17.42 Synthesize each compound from ethyl acetoacetate. You may use any other organic compounds or inorganic reagents.

a. 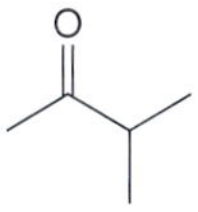 b. 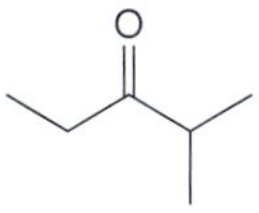 c.

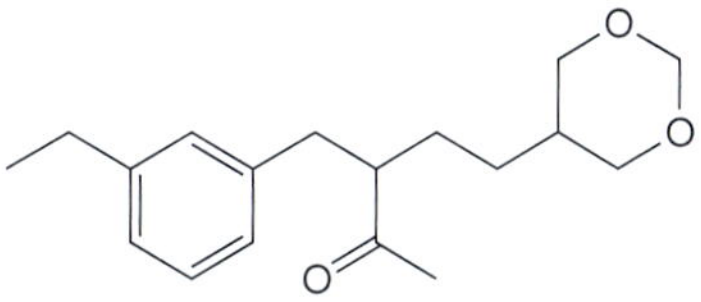

Reactions

17.43 Draw the organic products formed in each reaction.

a. 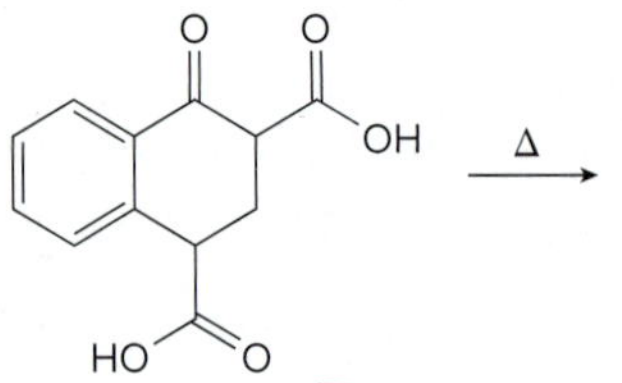 [tetralone-dione with CO₂H groups] →Δ

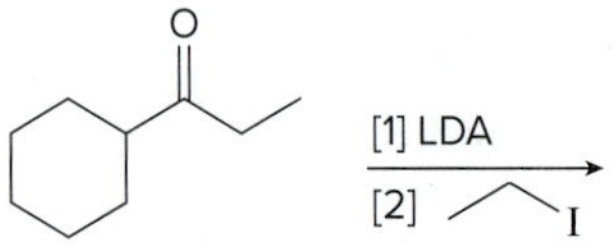

b. 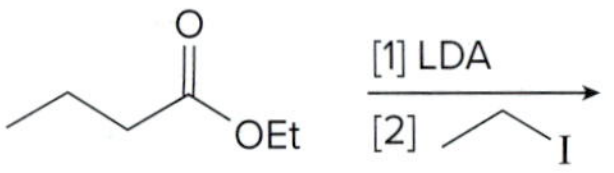 [ethyl butanoate] [1] LDA / [2] allyl iodide

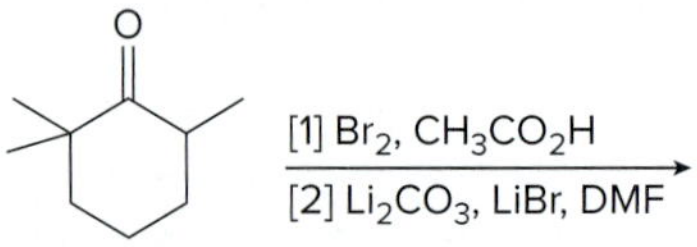

c. 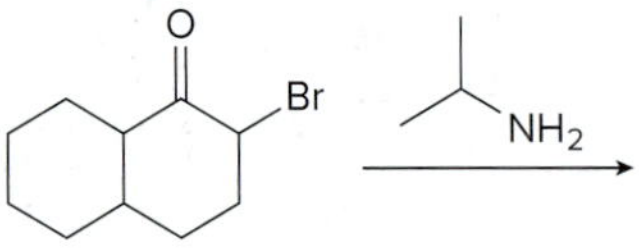 [α-bromo decalone] + isopropylamine (NH₂)

d. [cyclohexyl ethyl ketone] [1] LDA / [2] allyl iodide

e. [2,6-dimethylcyclohexanone] [1] Br₂, CH₃CO₂H / [2] Li₂CO₃, LiBr, DMF

f.  Cl-(CH₂)₄-CN, NaH → C_6H_9N

17.44 Draw the products formed (including stereoisomers) in each reaction.

a. 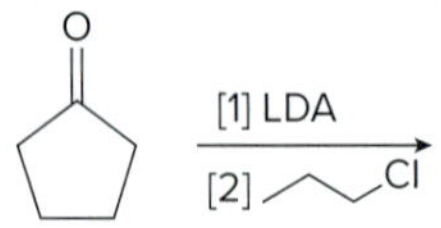 cyclopentanone [1] LDA / [2] allyl chloride

b. 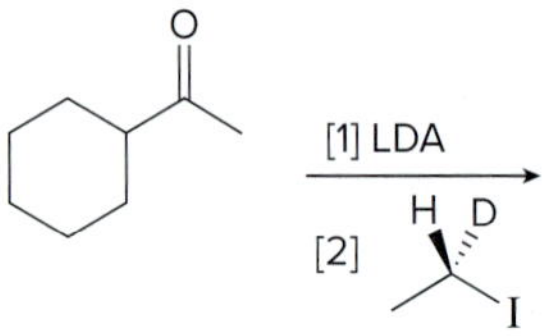 cyclohexyl methyl ketone [1] LDA / [2] CH(H)(D)(I)

c. 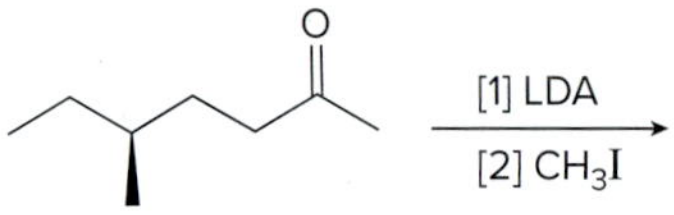 (R)-4-methylheptan-2-one [1] LDA / [2] CH₃I

17.45 Identify **A** in the following reaction, one step in the synthesis of bosentan, a drug used to treat a chronic connective tissue disorder that can cause pulmonary hypertension and open wounds on the fingertips (digital ulcers). Identify the atoms in bosentan that originate in **A**.

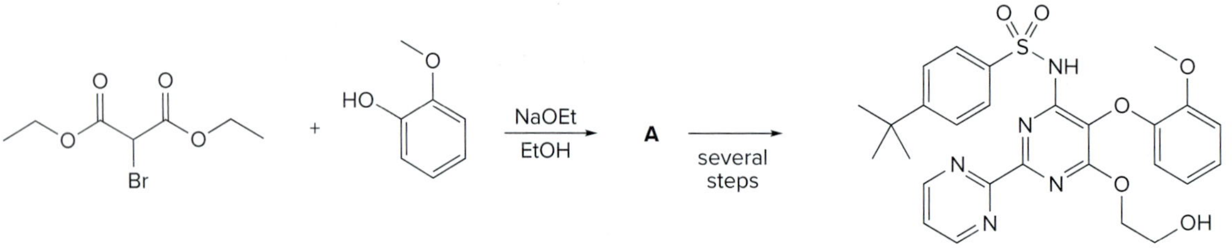

diethyl bromomalonate + guaiacol (2-methoxyphenol) —NaOEt, EtOH→ **A** —several steps→ bosentan

17.46 Identify **C** and **D** in the following reaction scheme, two steps in the synthesis of the cholesterol-lowering drug atorvastatin (trade name Lipitor, Section 25.8).

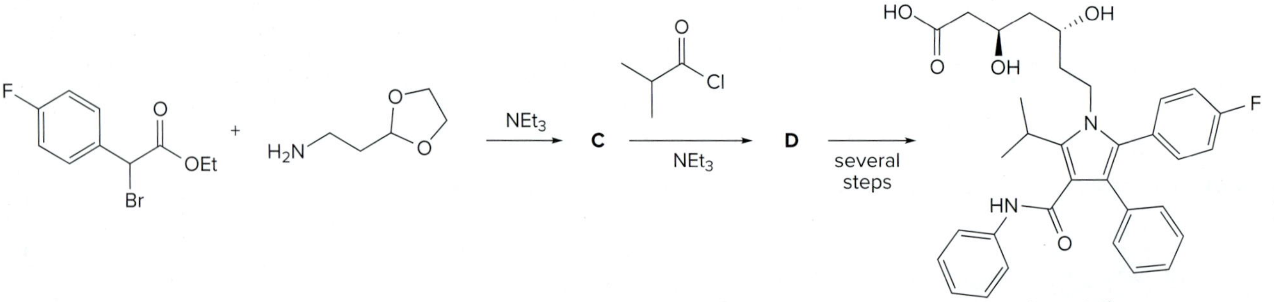

ethyl 2-bromo-2-(4-fluorophenyl)acetate + H₂N-CH₂CH₂-(1,3-dioxolane) —NEt₃→ **C** —isobutyryl chloride, NEt₃→ **D** —several steps→ atorvastatin

17.47 Identify the product in each reaction, and explain why starting materials with identical functional groups give different products.

a. 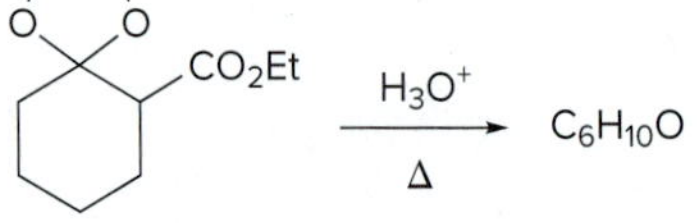 [cyclohexanone ethylene ketal with α-CO₂Et] —H₃O⁺, Δ→ $C_6H_{10}O$

b. 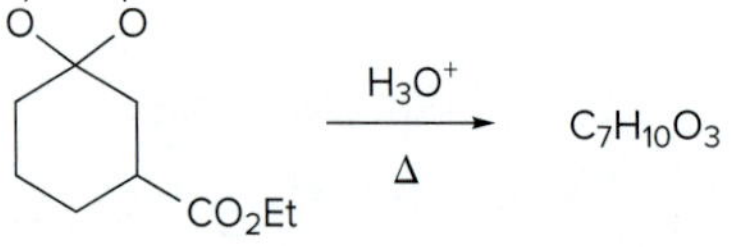 [cyclohexanone ethylene ketal with β-CO₂Et] —H₃O⁺, Δ→ $C_7H_{10}O_3$

17.48 Identify compounds **G** and **H** in the following reaction scheme. **H** represents the structure of stemoamide, the chapter-opening molecule.

17.49 A key step in the synthesis of the narcotic analgesic meperidine (trade name Demerol) is the conversion of phenylacetonitrile to **X**. (a) What is the structure of **X**? (b) What reactions convert **X** to meperidine?

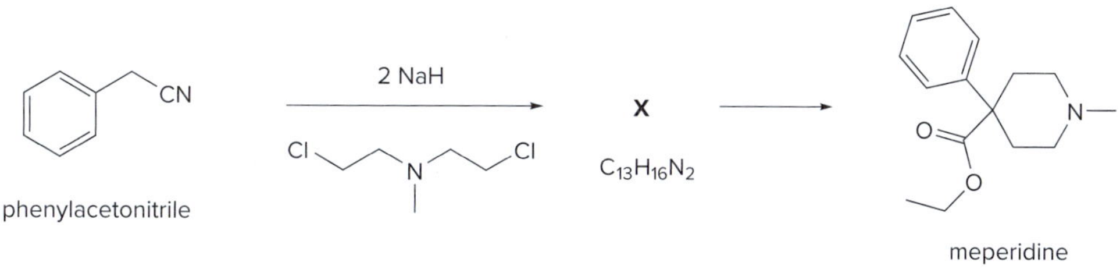

Mechanism

17.50 Although ibuprofen is sold as a racemic mixture, only the *S* enantiomer acts as an analgesic. In the body, however, some of the *R* enantiomer is converted to the *S* isomer by tautomerization to an enol and then protonation to regenerate the carbonyl compound. Write a stepwise mechanism for this isomerization.

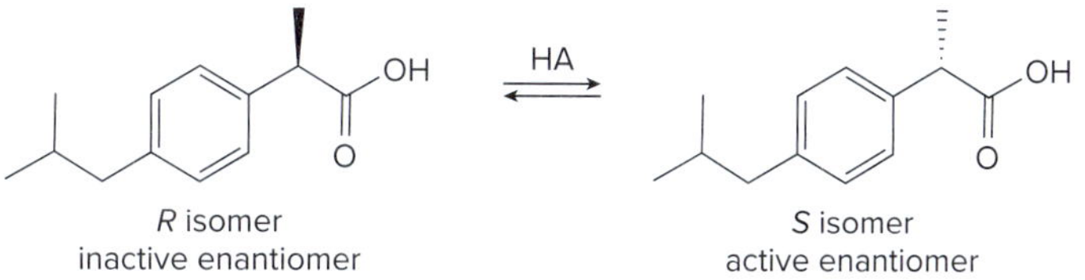

17.51 Use curved arrows to illustrate how the following decarboxylation occurs in the presence of an acid HA. This reaction constitutes one step in the biosynthesis of the amino acid tyrosine.

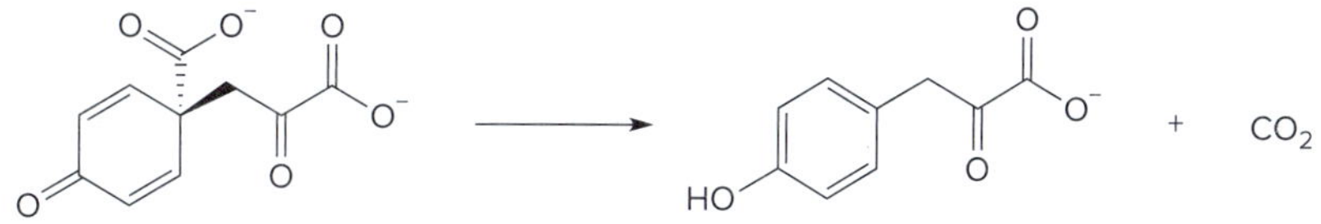

17.52 Write a possible mechanism for the following reaction, one step in the metabolism of glucose by the pentose phosphate pathway. The reaction proceeds by way of an intermediate enediol.

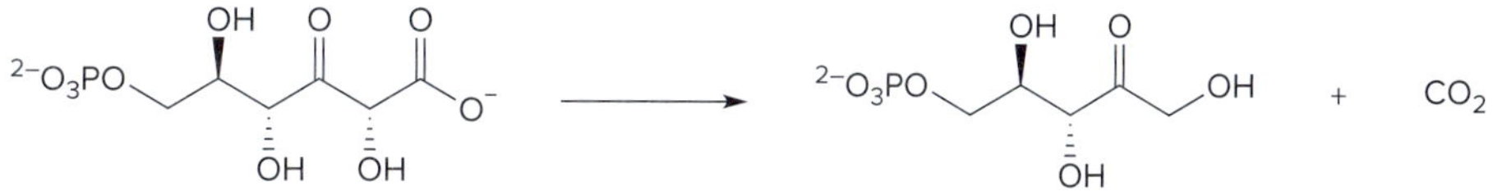

17.53 Draw a stepwise mechanism showing how two alkylation products are formed in the following reaction.

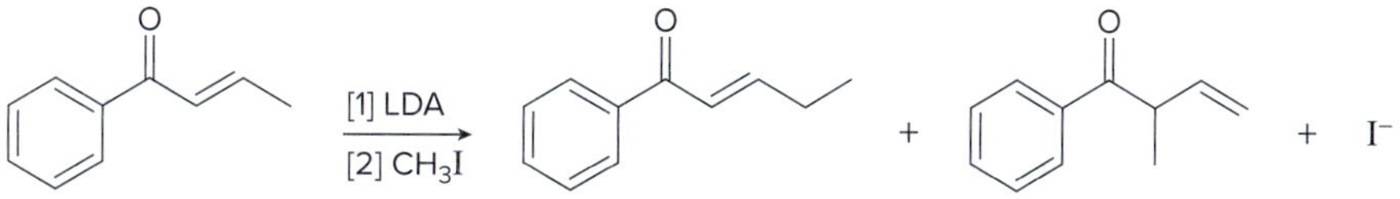

17.54 Treatment of α,β-unsaturated carbonyl compound **X** with base forms the diastereomer **Y**. Write a stepwise mechanism for this reaction. Explain why one stereogenic center changes configuration but the other does not.

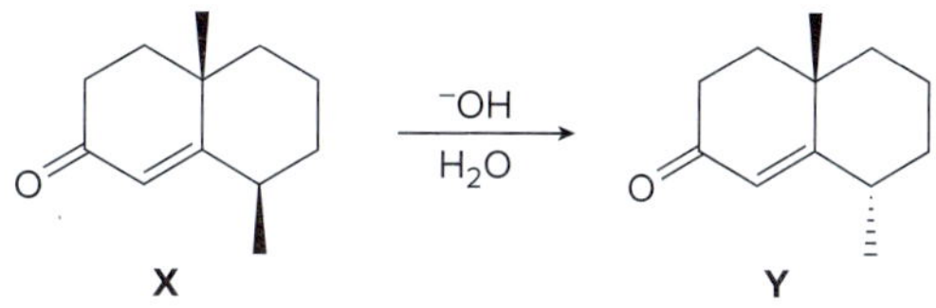

17.55 Draw stepwise mechanisms illustrating how each product is formed.

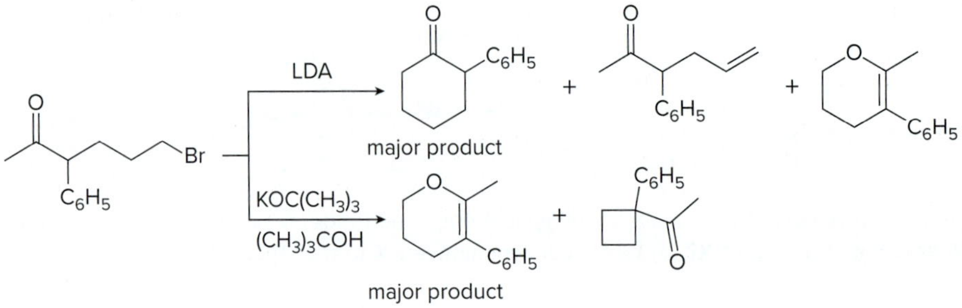

Synthesis

17.56 (a) Draw two different halo ketones that can form **A** by an intramolecular alkylation reaction. (b) How can **A** be synthesized by an acetoacetic ester synthesis?

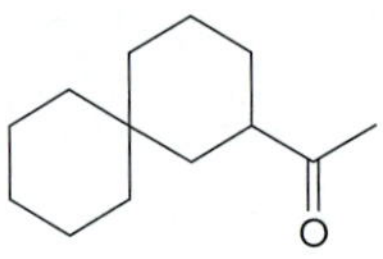

17.57 Synthesize each compound from cyclohexanone and organic halides having ≤ 4 C's. You may use any other inorganic reagents.

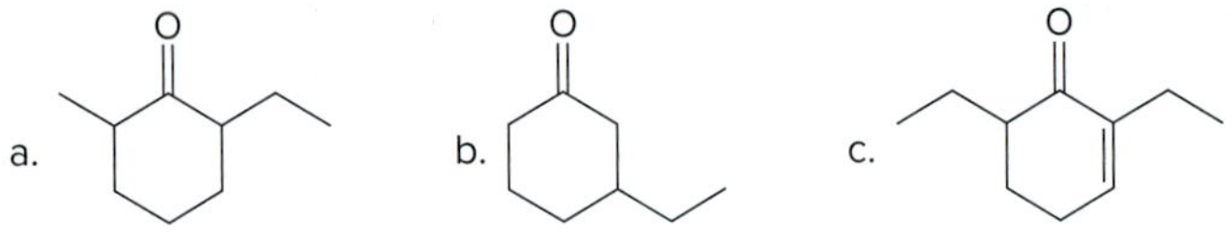

17.58 Synthesize (Z)-hept-5-en-2-one from ethyl acetoacetate (CH₃COCH₂CO₂Et) and the given starting material. You may also use any other organic compounds or required inorganic reagents.

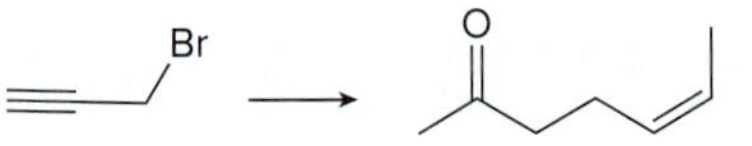

Spectroscopy

17.59 Treatment of **W** with CH₃Li, followed by CH₃I, affords compound **Y** (C₇H₁₄O) as the major product. **Y** shows a strong absorption in its IR spectrum at 1713 cm⁻¹, and its ¹H NMR spectrum is given below. (a) Propose a structure for **Y**. (b) Draw a stepwise mechanism for the conversion of **W** to **Y**.

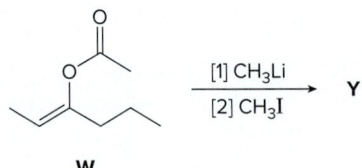

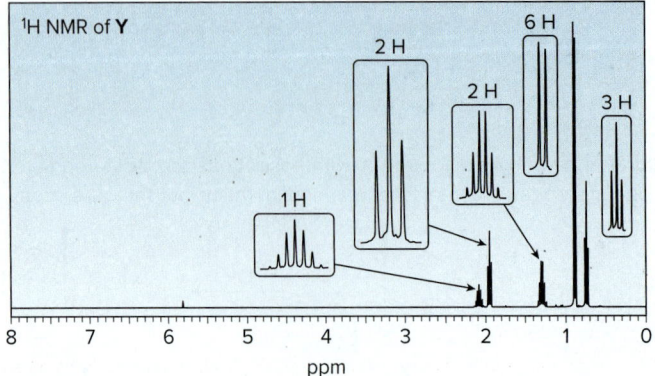

Challenge Problems

17.60 Explain why H_a is much less acidic than H_b. Then draw a mechanism for the following reaction.

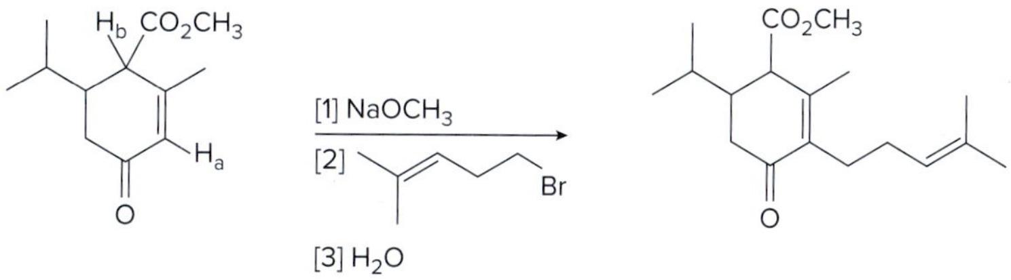

17.61 Devise a stepwise mechanism for the following reaction.

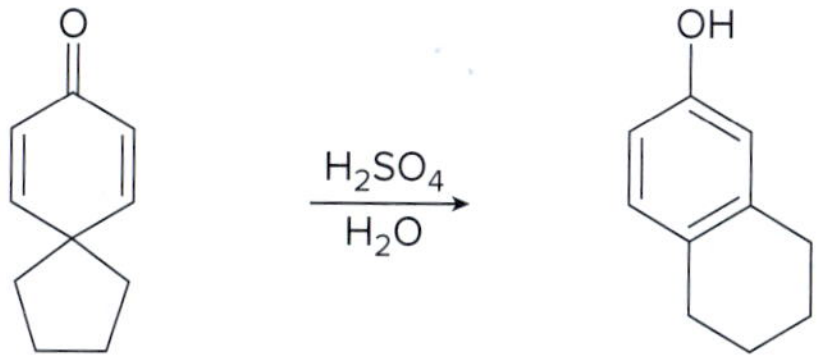

17.62 The last step in the synthesis of β-vetivone, a major constituent of vetiver, a perennial grass found in tropical and subtropical regions of the world, involves treatment of **C** with CH_3Li to form an intermediate **X,** which forms β-vetivone with aqueous acid. Identify the structure of **X** and draw a mechanism for converting **X** to β-vetivone.

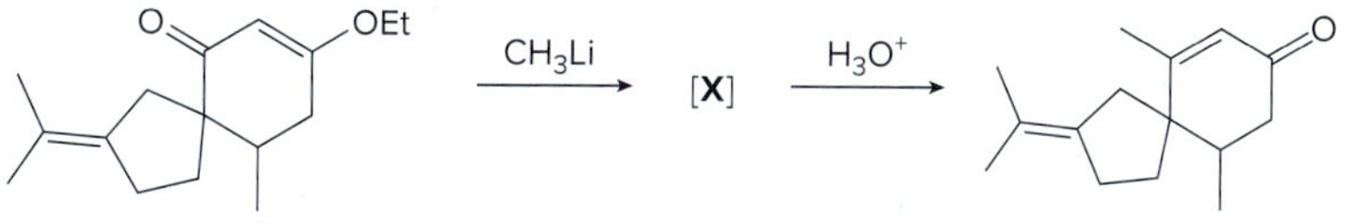

C β-vetivone

17.63 Keeping in mind the mechanism for the dissolving metal reduction of alkynes to trans alkenes in Chapter 11, write a stepwise mechanism for the following reaction, which involves the conversion of an α,β-unsaturated carbonyl compound to a carbonyl compound with a new alkyl group on the α carbon.

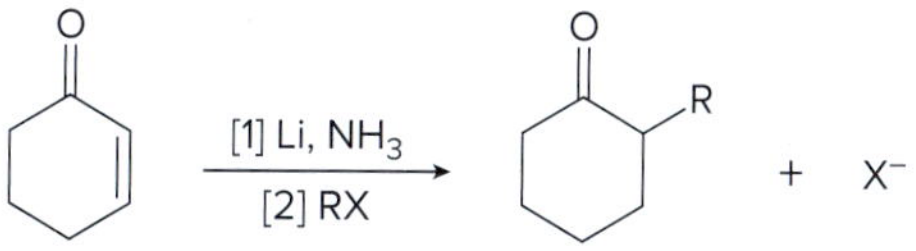

17.64 (−)-Hyoscyamine, an optically active drug used to treat gastrointestinal disorders, is isolated from *Atropa belladonna,* the deadly nightshade plant, by a basic aqueous extraction procedure. If too much base is used during isolation, optically inactive material is isolated. (a) Explain this result by drawing a stepwise mechanism. (b) Explain why littorine, an isomer isolated from the tailflower plant in Australia, can be obtained optically pure regardless of the amount of base used during isolation.

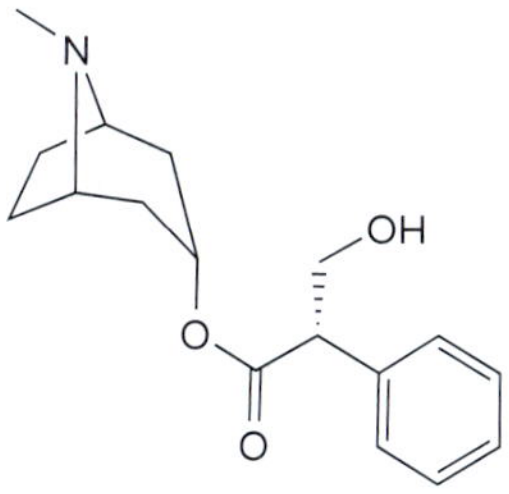

(−)-hyoscyamine

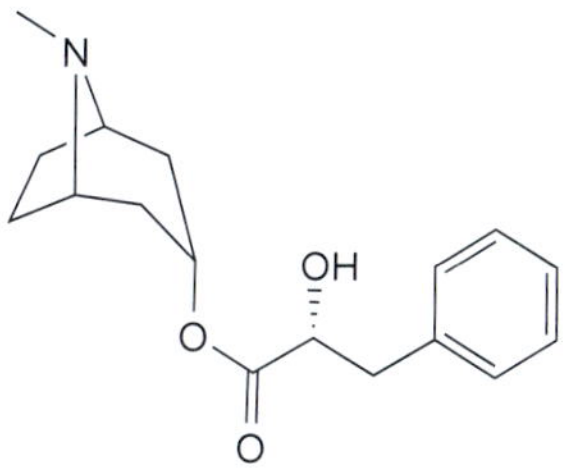

(−)-littorine

18 Carbonyl Condensation Reactions

- 18.1 The aldol reaction
- 18.2 Crossed aldol reactions
- 18.3 Directed aldol reactions
- 18.4 Intramolecular aldol reactions
- 18.5 The Claisen reaction
- 18.6 The crossed Claisen and related reactions
- 18.7 The Dieckmann reaction
- 18.8 Biological carbonyl condensation reactions
- 18.9 The Michael reaction
- 18.10 The Robinson annulation

ar-Turmerone is isolated from turmeric, a tropical flowering perennial in the ginger family grown primarily in Southeast Asia and India. The dried and ground root of the turmeric plant is an essential ingredient in curry. *ar*-Turmerone is an α,β-unsaturated carbonyl compound that can be prepared by a directed aldol reaction between two different carbonyl compounds. In Chapter 18, we learn about carbon–carbon bond-forming reactions between the α carbon of one carbonyl compound and the carbonyl group of another.

Why Study...

Carbonyl Condensation Reactions?

In Chapter 18, we examine **carbonyl condensations**—that is, reactions between two carbonyl compounds—a second type of reaction that occurs at the α carbon of a carbonyl group. Much of what is presented in Chapter 18 applies principles you have already learned. Many of the reactions may look more complicated than those in previous chapters, but they are fundamentally the same. Moreover, a key step in several metabolic pathways involves carbonyl condensations.

Every reaction in Chapter 18 forms a new carbon–carbon bond at the α carbon to a carbonyl group, so these reactions are extremely useful in the synthesis of complex natural products.

18.1 The Aldol Reaction

Chapter 18 concentrates on the second general reaction of enolates—**reaction with other carbonyl compounds.** In these reactions, one carbonyl component serves as the nucleophile and one serves as the electrophile, and a new carbon–carbon bond is formed.

The presence or absence of a leaving group on the electrophilic carbonyl carbon determines the structure of the product. Even though they appear somewhat more complicated, these reactions are often reminiscent of the nucleophilic addition and nucleophilic acyl substitution reactions of Chapters 14 and 16. Four types of reactions are examined:

- **Aldol reaction** (Sections 18.1–18.4)
- **Claisen reaction** (Sections 18.5–18.7)
- **Michael reaction** (Section 18.9)
- **Robinson annulation** (Section 18.10)

18.1A General Features of the Aldol Reaction

In the **aldol reaction,** two molecules of an aldehyde or ketone react with each other in the presence of base to form a **β-hydroxy carbonyl compound.** For example, treatment of acetaldehyde with aqueous ⁻OH forms 3-hydroxybutanal, a **β-hydroxy aldehyde.**

Many aldol products contain an *ald*ehyde and an alcoh*ol*—hence the name *aldol*.

The mechanism of the aldol reaction has **three steps,** as shown in Mechanism 18.1. Carbon–carbon bond formation occurs in Step [2], when the nucleophilic enolate reacts with the electrophilic carbonyl carbon.

The aldol reaction is a reversible equilibrium, so the position of the equilibrium depends on the base and the carbonyl compound. **⁻OH is the base** typically used in an aldol reaction. Recall from Section 17.3B that only a small amount of enolate forms with ⁻OH. In this case, that's appropriate because the starting aldehyde is needed to react with the enolate in the second step of the mechanism.

814 Chapter 18 Carbonyl Condensation Reactions

Mechanism 18.1 The Aldol Reaction

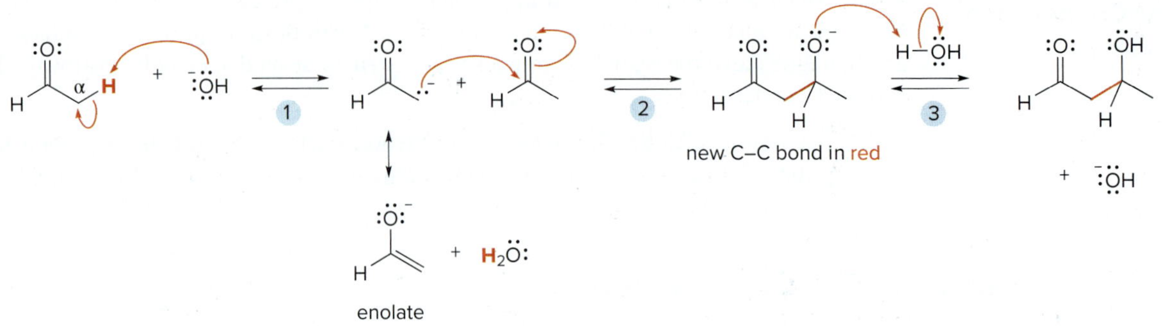

1 The base removes a proton on the α carbon to form a **resonance-stabilized enolate**.
2 **Nucleophilic attack** of the enolate on an electrophilic carbonyl in another molecule of aldehyde forms a new C–C bond.
3 Protonation of the alkoxide forms the **β-hydroxy aldehyde**.

Aldol reactions can be carried out with either aldehydes or ketones. With aldehydes, the equilibrium usually favors the products, but with ketones the equilibrium favors the starting materials. There are ways of driving this equilibrium to the right, however, so we will write aldol products whether the substrate is an aldehyde or a ketone.

- The characteristic reaction of aldehydes and ketones is *nucleophilic addition* (Section 14.7). An aldol reaction is a nucleophilic addition in which an enolate is the nucleophile. See the comparison in Figure 18.1.

Figure 18.1
The aldol reaction—An example of nucleophilic addition

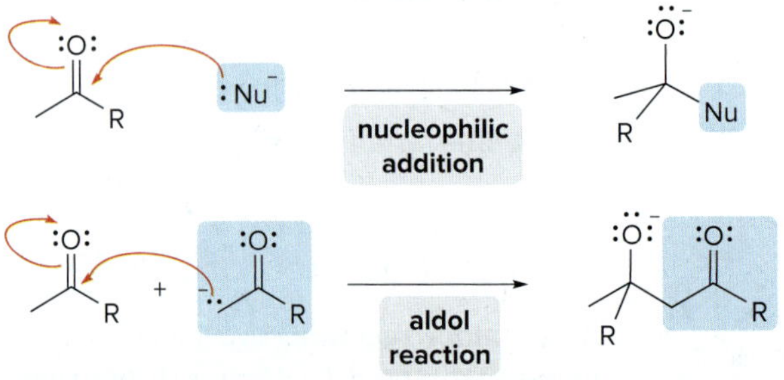

- Aldehydes and ketones react by nucleophilic addition. In an aldol reaction, **an enolate is the nucleophile** that adds to the carbonyl group.

A **second example of an aldol** reaction is shown with propanal as starting material. The two molecules of the aldehyde that participate in the aldol reaction react in opposite ways:

- One molecule of propanal becomes an enolate—an electron-rich *nucleophile*.
- One molecule of propanal serves as the *electrophile* because its carbonyl carbon is electron deficient.

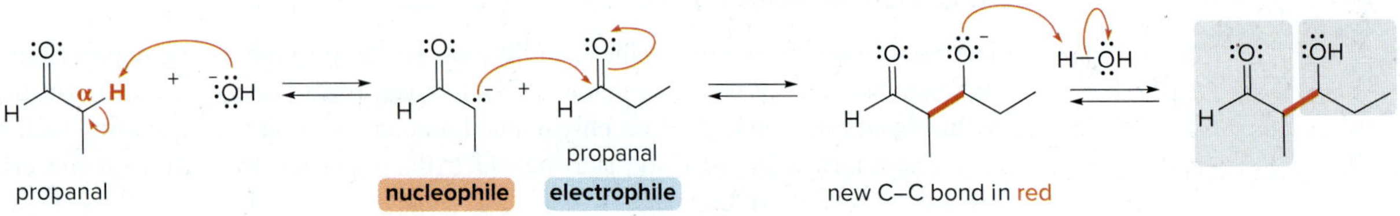

18.1 The Aldol Reaction

These two examples illustrate the general features of the aldol reaction. **The α carbon of one carbonyl component becomes bonded to the carbonyl carbon of the other component.**

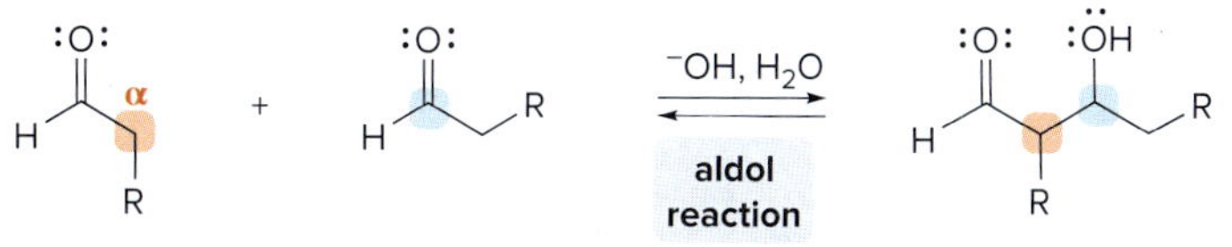

Problem 18.1 Draw the aldol product formed from each compound.

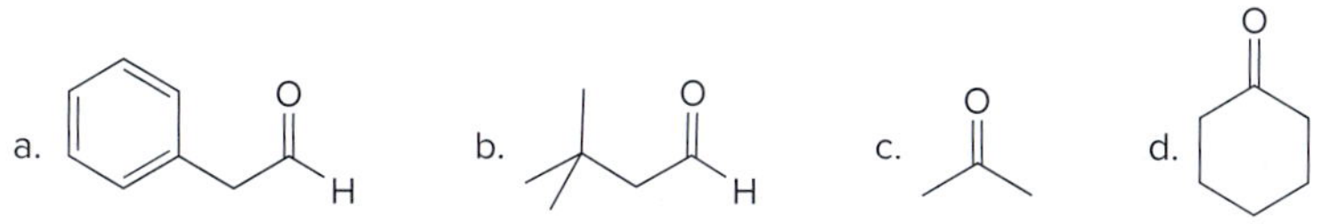

Problem 18.2 Which carbonyl compounds do *not* undergo an aldol reaction when treated with $^-$OH in H_2O?

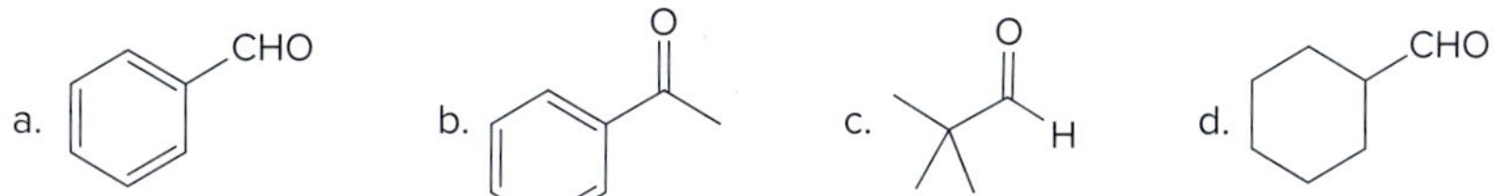

18.1B Retro-Aldol Reaction

Because an aldol reaction is a reversible equilibrium, the β-hydroxy carbonyl products can be re-converted to carbonyl starting materials with heat in the presence of base by a **retro-aldol reaction.** The conversion of 3-hydroxybutanal to acetaldehyde is a retro-aldol reaction, which results in cleavage of the carbon–carbon bond between the α and β carbons.

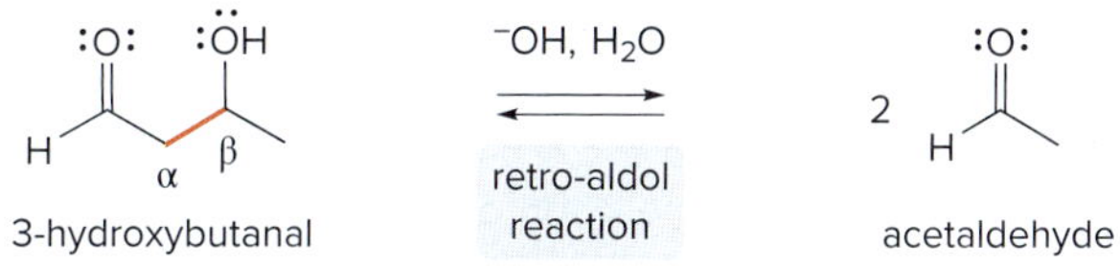

The three-step mechanism of a retro-aldol reaction is just the reverse of an aldol reaction, as shown in Mechanism 18.2. A retro-aldol reaction is a key step in the metabolism of glucose, as we will see in Section 18.8.

Mechanism 18.2 The Retro-Aldol Reaction

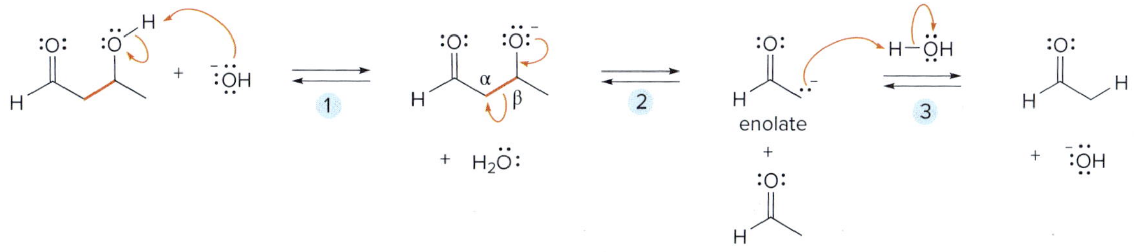

1. The base removes the OH proton to form an alkoxide.
2. An electron pair of the alkoxide is used to form a C=O and the **carbon–carbon bond between the α and β carbons is cleaved.** This process forms an **enolate** and a **molecule of aldehyde.**
3. Protonation of the enolate forms another molecule of aldehyde.

Problem 18.3
What ketone or aldehyde is obtained when each compound is heated in the presence of aqueous base?

18.1C Dehydration of the Aldol Product

All alcohols—including β-hydroxy carbonyl compounds—dehydrate in the presence of *acid*. Only β-hydroxy carbonyl compounds dehydrate in the presence of base.

The β-hydroxy carbonyl compounds formed in the aldol reaction dehydrate more readily than other alcohols. In fact, under the basic reaction conditions, the initial aldol product is often not isolated. Instead, **it loses the elements of H_2O from the α and β carbons to form an α,β-unsaturated carbonyl compound, a conjugated product.**

An aldol reaction is often called an **aldol condensation,** because the β-hydroxy carbonyl compound that is initially formed loses H_2O by dehydration. **A condensation reaction is one in which a small molecule, in this case H_2O, is eliminated during a reaction.**

It may or may not be possible to isolate the β-hydroxy carbonyl compound under the conditions of the aldol reaction.

- When the α,β-unsaturated carbonyl compound is *also conjugated* with a carbon–carbon double bond or a benzene ring, as in the case of Reaction [2], elimination of H_2O is spontaneous and the β-hydroxy carbonyl compound cannot be isolated.

The mechanism of dehydration consists of two steps: **deprotonation followed by loss of ^-OH,** as shown in Mechanism 18.3.

Mechanism 18.3 Dehydration of β-Hydroxy Carbonyl Compounds with Base

resonance-stabilized enolate

1. The base removes a proton on the α carbon to form a **resonance-stabilized enolate.**
2. ^-OH is eliminated as the electron pair of the enolate forms the **new π bond.**

18.1 The Aldol Reaction

Like E1 elimination, E1cB requires **two steps.** Unlike E1, though, the intermediate in E1cB is a *carbanion,* not a carbocation. E1cB stands for **Elimination, unimolecular, conjugate base.**

This elimination mechanism, called the **E1cB mechanism,** differs from the two more general mechanisms of elimination, E1 and E2, which were discussed in Chapter 8. The E1cB mechanism involves two steps and proceeds by way of an **anionic** intermediate.

Regular alcohols dehydrate only in the presence of acid but not base, because hydroxide is a poor leaving group. When the hydroxy group is β to a carbonyl group, however, loss of H and OH from the α and β carbons forms a **conjugated double bond,** and the stability of the conjugated system makes up for having such a poor leaving group.

Dehydration of the initial β-hydroxy carbonyl compound drives the equilibrium of an aldol reaction to the right, thus favoring product formation. Once the conjugated α,β-unsaturated carbonyl compound forms, it is *not* re-converted to the β-hydroxy carbonyl compound.

The E1cB mechanism is especially common in biological pathways. For example, the dehydration of β-hydroxy thioesters to α,β-unsaturated thioesters, a process that occurs during the biosynthesis of fatty acids, follows an E1cB mechanism.

β-hydroxy thioester → α,β-unsaturated thioester (E1cB mechanism)

Problem 18.4 What unsaturated carbonyl compound is formed by dehydration of each β-hydroxy carbonyl compound?

a., b., c.

18.1D Retrosynthetic Analysis

To utilize the aldol reaction in synthesis, you must be able to determine which aldehyde or ketone is needed to prepare a particular β-hydroxy carbonyl compound or α,β-unsaturated carbonyl compound—that is, you must be able to **work backwards, in the retrosynthetic direction.**

How To Synthesize a Compound Using the Aldol Reaction

Example What starting material is needed to prepare each compound by an aldol reaction?

a., b.

Step [1] Locate the α and β carbons of the carbonyl group.

- When a carbonyl group has two different α carbons, **choose the side that contains the OH group** (in a β-hydroxy carbonyl compound) **or is part of the C=C** (in an α,β-unsaturated carbonyl compound).

—Continued

818 Chapter 18 Carbonyl Condensation Reactions

How To, continued...

Step [2] Break the molecule into two components between the α and β carbons.

- The α carbon and all remaining atoms bonded to it belong to one carbonyl component. The β carbon and all remaining atoms bonded to it belong to the other carbonyl component. Both components are identical in all aldols we have thus far examined.

a. Break the molecule into two halves at the labeled bond.

b. Break the molecule into two halves at the labeled bond.

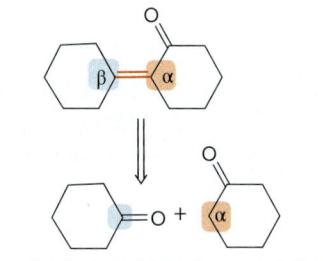

two molecules of cyclohexanone

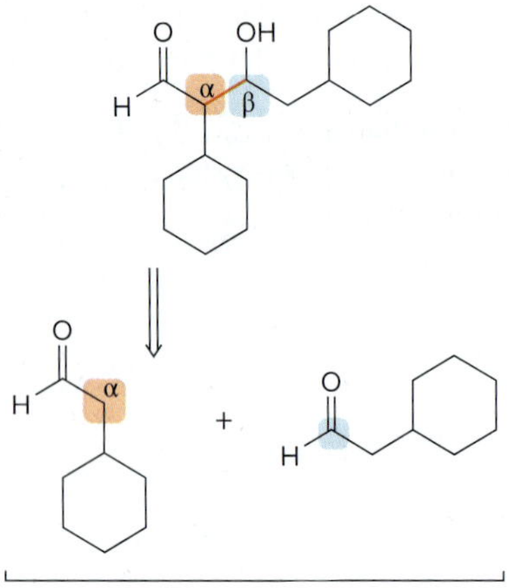

two molecules of the same aldehyde

Problem 18.5 What aldehyde or ketone is needed to prepare each compound by an aldol reaction?

a. 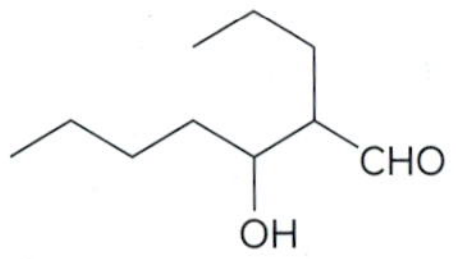 [structure with CHO and OH]   b. C_6H_5–[structure with OH and CHO]–C_6H_5 c. [structure with phenyl ketone]

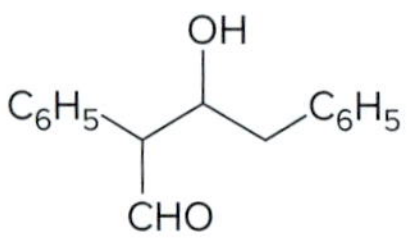

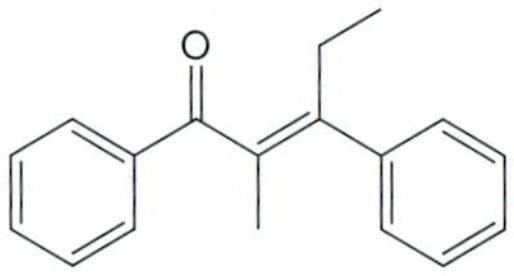

18.2 Crossed Aldol Reactions

In all of the aldol reactions discussed so far, the electrophilic carbonyl and the nucleophilic enolate have originated from the *same* aldehyde or ketone. Sometimes, though, it is possible to carry out an aldol reaction between two *different* carbonyl compounds.

- An aldol reaction between two different carbonyl compounds is called a *crossed aldol* or *mixed aldol reaction*.

18.2A A Crossed Aldol Reaction with Two Different Aldehydes, Both Having α H Atoms

When two different aldehydes, both having α H atoms, are combined in an aldol reaction, *four* different β-hydroxy carbonyl compounds are formed. Four products form, not one, because *both* aldehydes can lose an acidic α hydrogen atom and form an enolate in the presence of base. *Both* enolates can then react with *both* carbonyl compounds, as shown for acetaldehyde and propanal in the following reaction scheme.

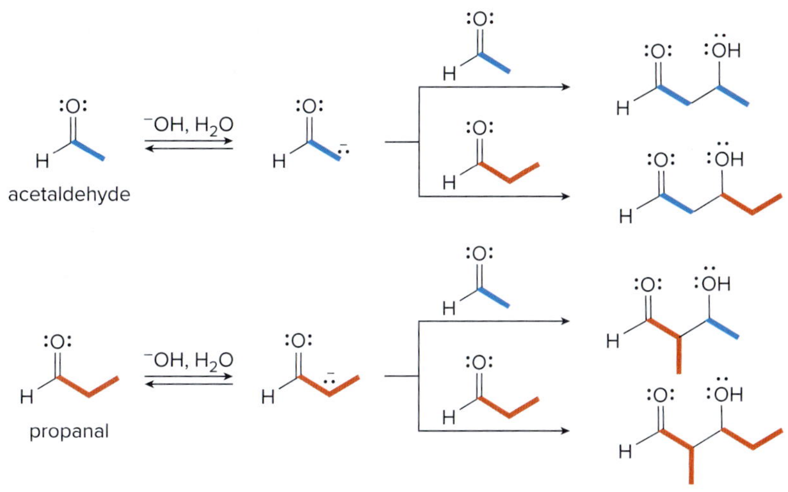

18.2 Crossed Aldol Reactions

[Scheme showing acetaldehyde and propanal each forming enolates with ⁻OH, H₂O, then combining with both acetaldehyde and propanal as electrophiles to give four different products]

- **Conclusion:** When two different aldehydes have α hydrogens, a crossed aldol reaction is *not* synthetically useful.

18.2B Synthetically Useful Crossed Aldol Reactions

Crossed aldols are synthetically useful in two different situations.

- A crossed aldol occurs when only *one* carbonyl component has α H atoms.

When one carbonyl compound has no α hydrogens, a crossed aldol reaction often leads to one product. Two common carbonyl compounds with no α hydrogens used for this purpose are **formaldehyde (CH₂=O)** and **benzaldehyde (C₆H₅CHO)**.

For example, reaction of C₆H₅CHO (as the electrophile) with acetone [(CH₃)₂C=O] in the presence of base forms a single α,β-unsaturated carbonyl compound after dehydration.

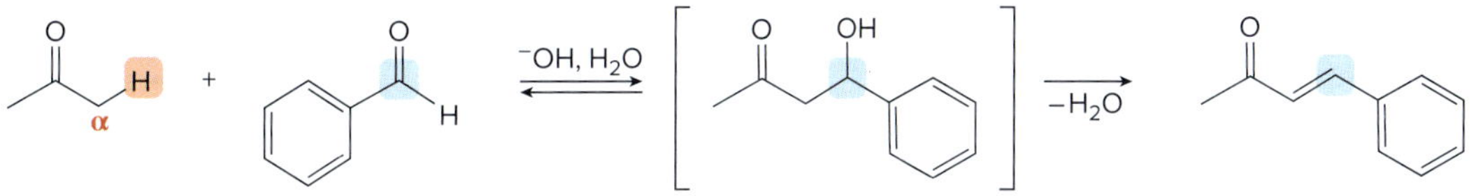

[Scheme: acetone + benzaldehyde →(⁻OH, H₂O) β-hydroxy ketone intermediate →(−H₂O) α,β-unsaturated ketone (+ Z isomer)]

The yield of a single crossed aldol product is increased further if the electrophilic carbonyl component is relatively unhindered (as is the case with most aldehydes) and if it is used in excess.

Problem 18.6 2-Pentylcinnamaldehyde, commonly called flosal, is a perfume ingredient with a jasmine-like odor. Flosal is an α,β-unsaturated aldehyde made by a crossed aldol reaction between benzaldehyde (C₆H₅CHO) and heptanal (CH₃CH₂CH₂CH₂CH₂CH₂CHO), followed by dehydration. Draw a stepwise mechanism for the following reaction that prepares flosal.

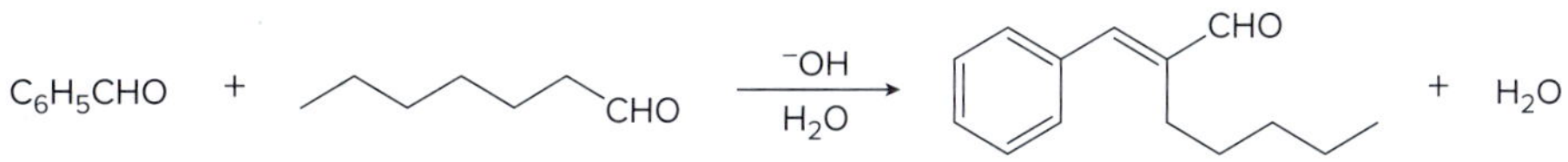

[Reaction: C₆H₅CHO + heptanal →(⁻OH, H₂O) flosal (perfume component) + H₂O]

820 Chapter 18 Carbonyl Condensation Reactions

Problem 18.7 Draw the products formed in each crossed aldol reaction.

a. [butanal] and [formaldehyde (HCHO)]   b. [benzaldehyde] and [cyclohexanone]

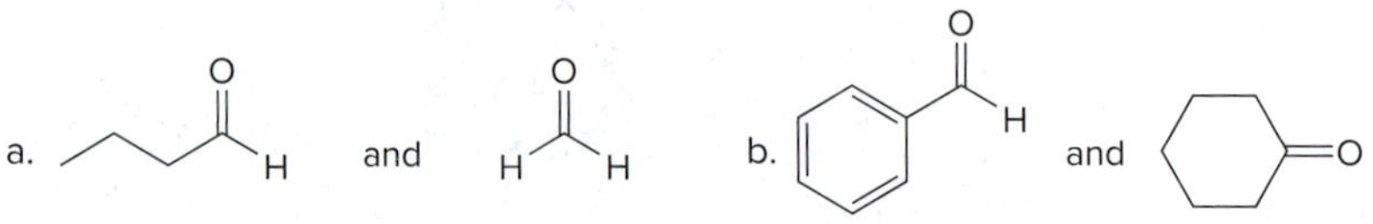

- **A crossed aldol occurs when one carbonyl component has especially acidic α H atoms.**

A useful crossed aldol reaction takes place between an aldehyde or ketone and a β-dicarbonyl (or similar) compound.

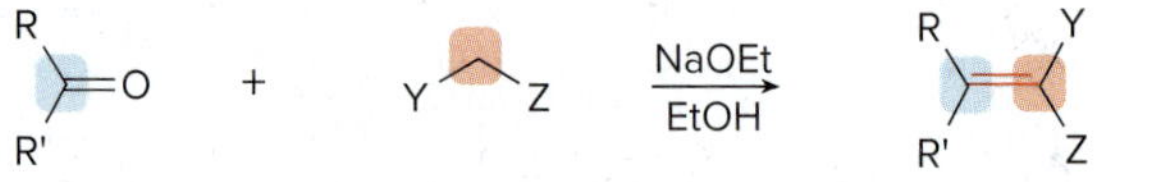

$$\underset{R'=H\text{ or alkyl}}{\overset{R}{\underset{R'}{}}\!\!\!\!\!\!\!\!\!\!C\!=\!O} \;+\; \underset{\substack{Y,Z=COOEt,\,CHO,\\ COR,\,CN\\ \beta\text{-dicarbonyl compound}\\ \text{(and related compounds)}}}{Y\!-\!CH_2\!-\!Z} \;\xrightarrow[\text{EtOH}]{\text{NaOEt}}\; \underset{\text{new C–C σ and π bonds in red}}{\overset{R}{\underset{R'}{}}\!\!\!\!\!\!\!\!\!C\!=\!C\!\!\!\!\overset{Y}{\underset{Z}{}}}$$

[benzaldehyde] + [diethyl malonate CH₂(COOEt)₂] —NaOEt/EtOH→ [PhCH=C(COOEt)₂]

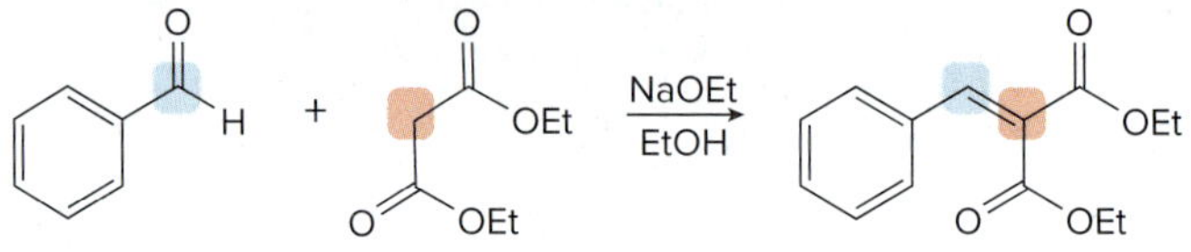

benzaldehyde   diethyl malonate

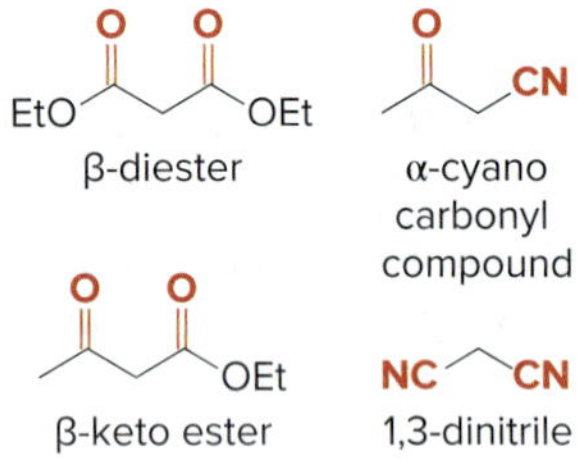

As we learned in Section 17.3, the α hydrogens between two carbonyl groups are especially acidic, so they are more readily removed than other α H atoms. As a result, **the β-dicarbonyl compound always becomes the enolate component of the aldol reaction.** Figure 18.2 shows the steps for the crossed aldol reaction between diethyl malonate and benzaldehyde. In this type of crossed aldol reaction, the initial β-hydroxy carbonyl compound *always* loses water to form the highly conjugated product.

β-Dicarbonyl compounds are sometimes called **active methylene compounds** because they are more reactive toward base than other carbonyl compounds. **1,3-Dinitriles** and **α-cyano carbonyl compounds** are also active methylene compounds.

[EtOOC–CH₂–COOEt β-diester]
[CH₃COCH₂CN α-cyano carbonyl compound]
[CH₃COCH₂COOEt β-keto ester]
[NC–CH₂–CN 1,3-dinitrile]

Figure 18.2
Crossed aldol reaction between benzaldehyde and CH₂(COOEt)₂

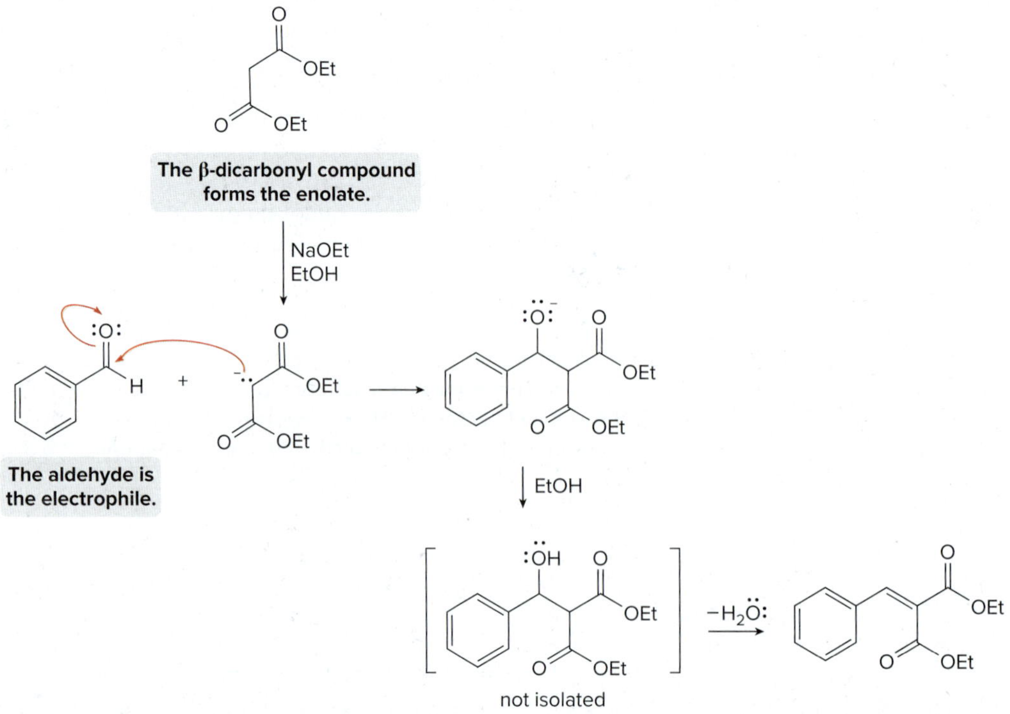

The β-dicarbonyl compound forms the enolate.

↓ NaOEt, EtOH

[PhCHO + ⁻CH(COOEt)₂] → [PhCH(O⁻)–CH(COOEt)₂]

The aldehyde is the electrophile.

↓ EtOH

[PhCH(OH)–CH(COOEt)₂] (not isolated) —H₂O→ [PhCH=C(COOEt)₂]

Problem 18.8 Draw the products formed in the crossed aldol reaction of phenylacetaldehyde ($C_6H_5CH_2CHO$) with each compound: (a) $CH_2(COOEt)_2$; (b) $CH_2(COCH_3)_2$; (c) CH_3COCH_2CN.

Problem 18.9 The first steps in the synthesis of azelnidipine, a calcium channel blocker (Problem 9.59), involves the reaction of β-keto ester **A** with aldehyde **B** in the presence of base. What crossed aldol product is formed in this reaction?

18.3 Directed Aldol Reactions

A **directed aldol reaction** is a variation of the crossed aldol reaction that clearly defines which carbonyl compound becomes the nucleophilic enolate and which reacts at the electrophilic carbonyl carbon. The strategy of a directed aldol reaction is as follows:

[1] Prepare the enolate of one carbonyl component with LDA.
[2] Add the second carbonyl compound (the electrophile) to this enolate.

Because the steps are done sequentially and a strong nonnucleophilic base is used to form the enolate of only one carbonyl component, a variety of carbonyl substrates can be used in the reaction. Both carbonyl components can have α hydrogens because only one enolate is prepared with LDA. Also, when an unsymmetrical ketone is used, LDA selectively forms the **less substituted, kinetic enolate**.

Sample Problem 18.1 illustrates the steps of a directed aldol reaction between a ketone and an aldehyde, both of which have α hydrogens.

Sample Problem 18.1 Determining the Product of a Directed Aldol Reaction

Draw the product of the following directed aldol reaction.

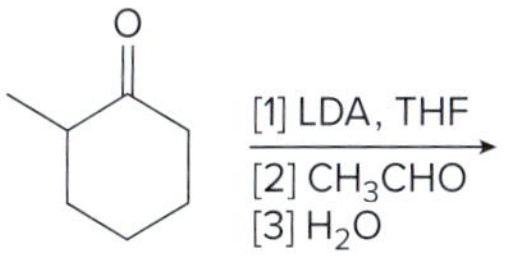

2-methylcyclohexanone [1] LDA, THF [2] CH_3CHO [3] H_2O

Solution

2-Methylcyclohexanone forms an enolate on the less substituted carbon, which then reacts with the electrophile, CH_3CHO.

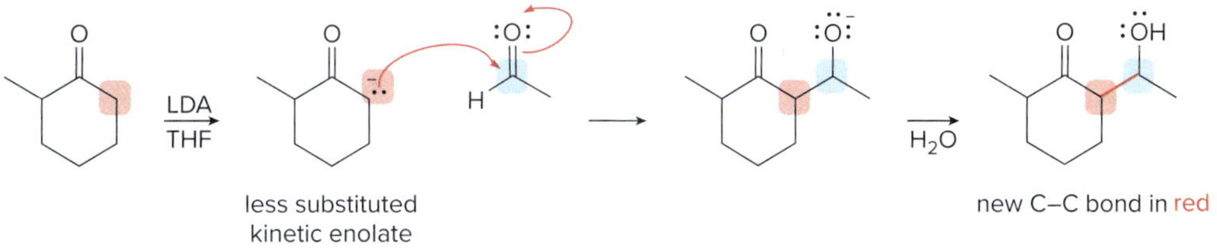

less substituted kinetic enolate → new C–C bond in red

Problem 18.10 Draw the product of each directed aldol reaction.

a. [1] LDA [2] cyclopentanone [3] H_2O

b. [1] LDA [2] CH_3CH_2CHO [3] H_2O

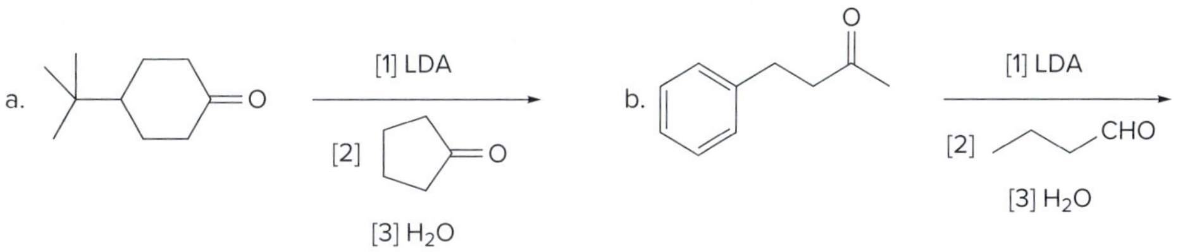

More Practice: Try Problems 18.33, 18.51c.

To determine the needed carbonyl components for a directed aldol, follow the same strategy used for a regular aldol reaction in Section 18.1D, as shown in Sample Problem 18.2.

Sample Problem 18.2 Determining the Starting Materials of a Directed Aldol Reaction

What starting materials are needed to prepare *ar*-turmerone, the chapter-opening molecule, using a directed aldol reaction?

ar-turmerone

Solution

When the desired product is an α,β-unsaturated carbonyl compound, identify the α and β carbons that are part of the C=C, and break the molecule into two components between these carbons.

Break the molecule into two halves.

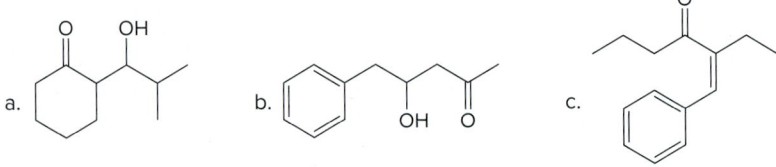

ar-turmerone   Make the enolate here.

Problem 18.11 What carbonyl starting materials are needed to prepare each compound using a directed aldol reaction?

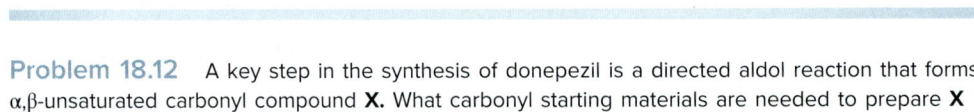

a.   b.   c.

More Practice: Try Problem 18.35.

Problem 18.12 A key step in the synthesis of donepezil is a directed aldol reaction that forms α,β-unsaturated carbonyl compound **X**. What carbonyl starting materials are needed to prepare **X** using a directed aldol reaction? What reagents are needed to convert **X** to donepezil?

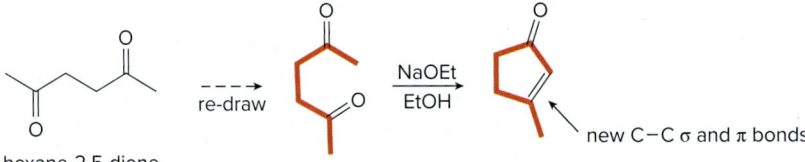

X   donepezil

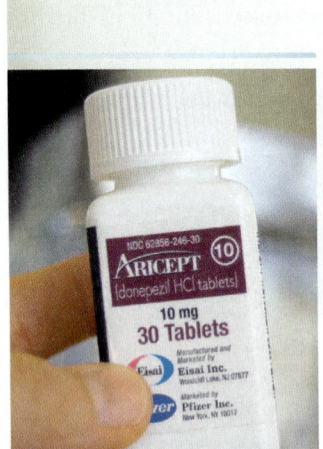

Donepezil (trade name Aricept, Problem 18.12) is a drug used to improve cognitive function in patients suffering from Alzheimer's disease and other types of dementia. *Jill Braaten*

18.4 Intramolecular Aldol Reactions

Aldol reactions with dicarbonyl compounds can be used to make five- and six-membered rings. The enolate formed from one carbonyl group is the nucleophile, and the carbonyl carbon of the other carbonyl group is the electrophile. For example, treatment of hexane-2,5-dione with base forms a five-membered ring.

Hexane-2,5-dione is called a **1,4-dicarbonyl compound** to emphasize the relative position of its carbonyl groups. 1,4-Dicarbonyl compounds are starting materials for synthesizing **five-membered rings.**

hexane-2,5-dione → re-draw → NaOEt, EtOH → new C–C σ and π bonds

The steps in this process, shown in Mechanism 18.4, are no different from the general mechanisms of the aldol reaction and dehydration described in Section 18.1.

Mechanism 18.4 The Intramolecular Aldol Reaction

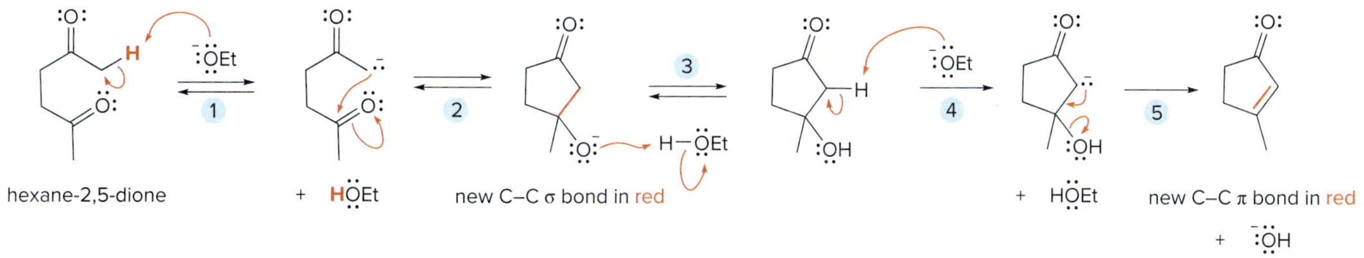

1. The base removes a proton on the α carbon to form a **resonance-stabilized enolate**.
2. **Nucleophilic attack** of the enolate on the electrophilic carbonyl *in the same molecule* forms a new C–C σ bond, generating the five-membered ring.
3. Protonation of the alkoxide forms the **β-hydroxy carbonyl compound**.
4 – 5. **Dehydration occurs by the two-step E1cB mechanism**—loss of a proton to form an enolate and elimination of ⁻OH to form a π bond.

When hexane-2,5-dione is treated with base in Step [1], two different enolates are possible—enolates **A** and **B**, formed by removal of H_a and H_b, respectively. Although enolate **A** goes on to form the five-membered ring, intramolecular cyclization using enolate **B** would lead to a strained three-membered ring.

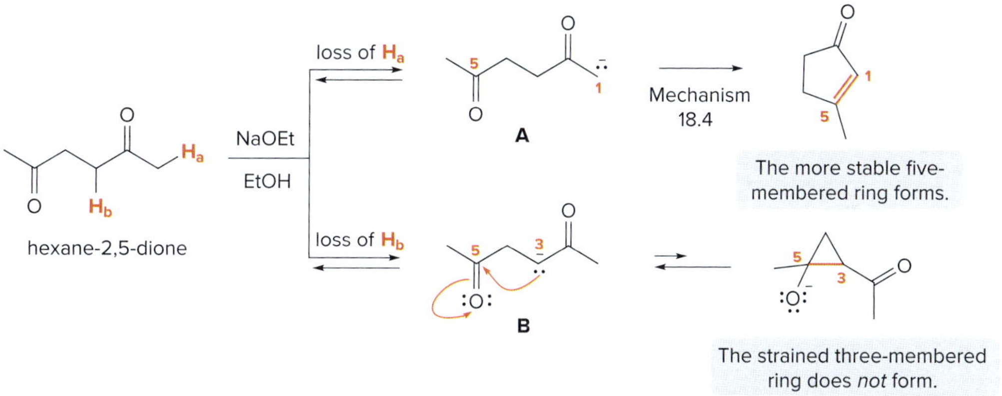

Because the three-membered ring is much higher in energy than the enolate starting material, equilibrium greatly favors the starting materials and the **three-membered ring does not form.** Under the reaction conditions, enolate **B** is re-protonated to form hexane-2,5-dione, because all steps except dehydration are equilibria. **Thus, equilibrium favors formation of the more stable five-membered ring over the much less stable three-membered ring.**

In a similar fashion, six-membered rings can be formed from the intramolecular aldol reaction of **1,5-dicarbonyl compounds.**

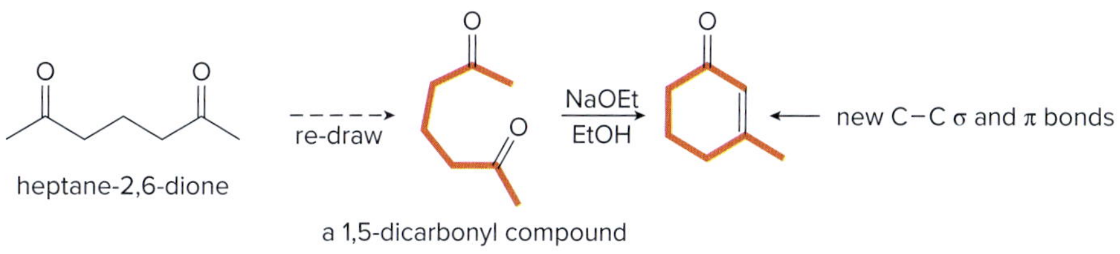

The synthesis of the female sex hormone **progesterone** by W. S. Johnson and co-workers at Stanford University is considered one of the classics in total synthesis. The last six-membered

824 Chapter 18 Carbonyl Condensation Reactions

Figure 18.3
The synthesis of progesterone using an intramolecular aldol reaction

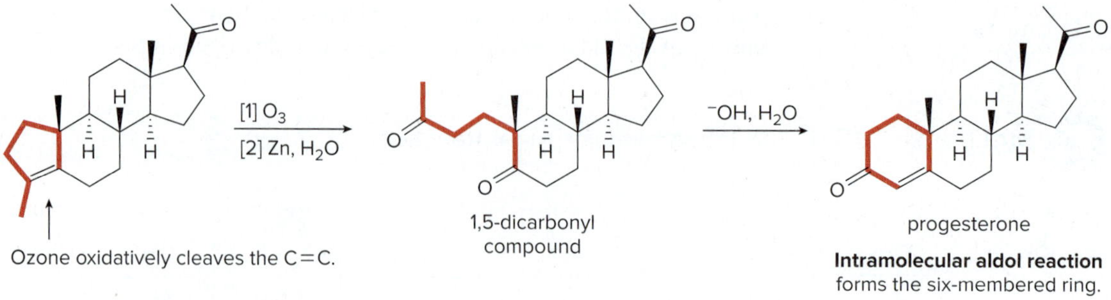

- Oxidative cleavage of the alkene with O_3, followed by Zn, H_2O (Section 11.10), gives the 1,5-dicarbonyl compound.
- Intramolecular aldol reaction of the 1,5-dicarbonyl compound with dilute ^-OH in H_2O solution forms progesterone.
- **This two-step reaction sequence converts a five-membered ring to a six-membered ring.** Reactions that synthesize larger rings from smaller ones are called **ring expansion reactions.**

ring needed in the steroid skeleton was prepared by a two-step sequence using an intramolecular aldol reaction, as shown in Figure 18.3.

Problem 18.13 Draw a stepwise mechanism for the conversion of heptane-2,6-dione to 3-methylcyclohex-2-enone with NaOEt, EtOH.

Sample Problem 18.3 Drawing the Major Product of an Intramolecular Aldol Reaction

What is the major intramolecular aldol product formed when dicarbonyl compound **A** is treated with base?

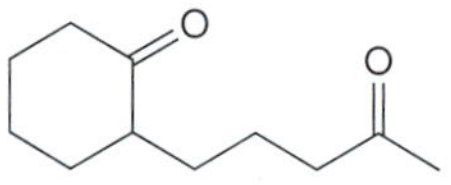

Solution

To draw the products of an intramolecular aldol reaction:
- **Identify all the α carbons** that are bonded to H's.
- Determine how far each α carbon is from the other carbonyl carbon. Reactions that yield **five- or six-membered rings** are favored.
- **α,β-Unsaturated carbonyl systems are favored** over β-hydroxy carbonyl compounds that cannot dehydrate to a conjugated product.

A contains four α carbons that can form enolates. First, consider enolates from the cyclohexanone (at C5 and C7 below) reacting with the acyclic carbonyl (C1). Only the enolate at C5 forms a five-membered ring by intramolecular aldol, but the β-hydroxy ketone **B** can*not* dehydrate to a conjugated system because the α carbon has no H for dehydration. Thus, this path is *not* favored.

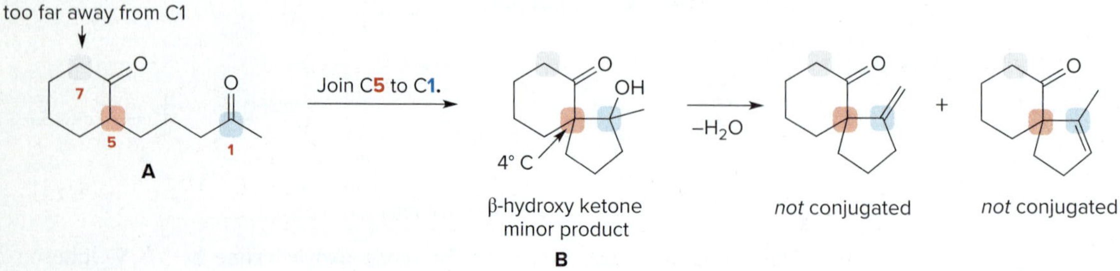

Then, consider enolates from the acyclic ketone (at C5 and C7 below) reacting with the cyclohexanone carbonyl (C1). Only the enolate at C5 forms a five-membered ring by intramolecular aldol, and dehydration forms an α,β-unsaturated carbonyl compound **C**, so **C** is the major product.

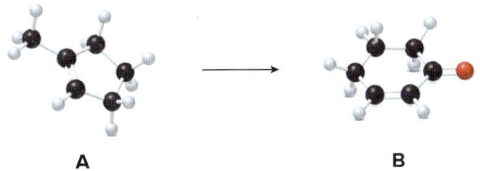

Problem 18.14 What cyclic product is formed when each 1,5-dicarbonyl compound is treated with aqueous ⁻OH?

a. [structure with cyclohexane, acetyl, and CHO groups]   b. [acyclic diketone structure]

More Practice: Try Problem 18.34.

Problem 18.15 Following the two-step reaction sequence depicted in Figure 18.3, write out the steps needed to convert **A** to **B**.

[ball-and-stick models of A and B]

18.5 The Claisen Reaction

The **Claisen reaction** is the second general reaction of enolates with other carbonyl compounds. In the Claisen reaction, two molecules of an ester react with each other in the presence of an alkoxide base to form a **β-keto ester.** For example, treatment of ethyl acetate with NaOEt forms ethyl acetoacetate after protonation with aqueous acid.

Unlike the aldol reaction, which is base-catalyzed, a full equivalent of base is needed to deprotonate the β-keto ester formed in Step [3] of the Claisen reaction.

2 EtO–C(O)–CH₃ $\xrightarrow{\text{[1] NaOEt} \atop \text{[2] H}_3\text{O}^+}$ EtO–C(O)–CH₂–C(O)–CH₃   **β-keto ester**

ethyl acetate   Claisen reaction   new C–C bond in red
                                             ethyl acetoacetate

The mechanism for the Claisen reaction (Mechanism 18.5) resembles the mechanism of an aldol reaction in that it involves nucleophilic addition of an enolate to an electrophilic carbonyl group. Because esters have a leaving group on the carbonyl carbon, however, **loss of a leaving group occurs to form the product of substitution,** *not* **addition.**

826 Chapter 18 Carbonyl Condensation Reactions

 Mechanism 18.5 The Claisen Reaction

Part [1] Formation of the β-keto ester

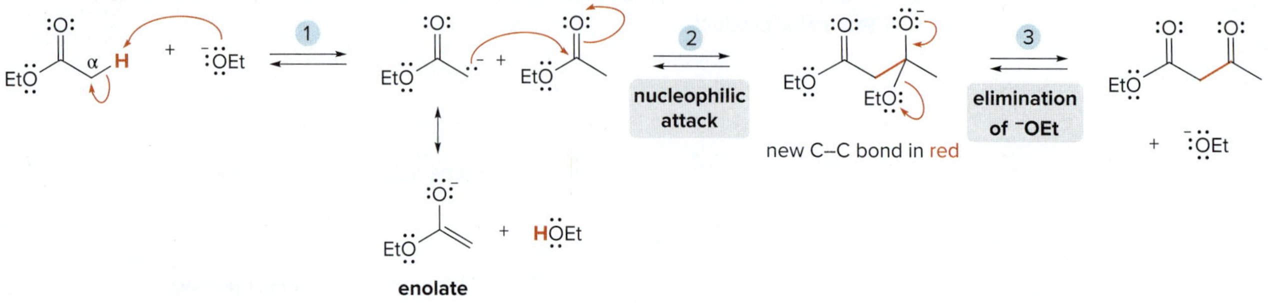

[Mechanism diagram showing steps 1-3: formation of enolate, nucleophilic attack forming new C–C bond in red, and elimination of ⁻OEt]

1. The base removes a proton on the α carbon to form a **resonance-stabilized enolate**.
2. **Nucleophilic attack** of the enolate on an electrophilic carbonyl in another molecule of ester forms a new C–C bond.
3. Loss of the leaving group (⁻OEt) forms a **β-keto ester**.

Part [2] Deprotonation and protonation

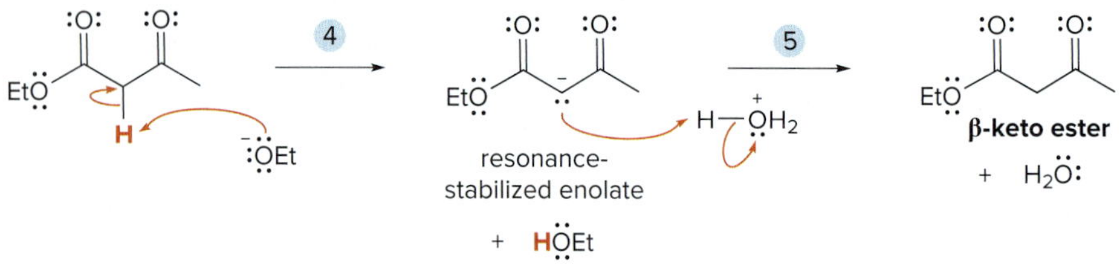

[Mechanism diagram showing steps 4-5: deprotonation to form resonance-stabilized enolate, then protonation with H–OH₂⁺ to give β-keto ester + H₂O]

4. Because the β-keto ester formed in Step [3] has especially acidic protons between its two carbonyl groups, the base removes a proton to form a **resonance-stabilized enolate**.
5. Protonation of the enolate with strong acid re-forms the **β-keto ester**.

Because the generation of a resonance-stabilized enolate from the product β-keto ester drives the Claisen reaction (Step [4] of Mechanism 18.5), **only esters with two or three hydrogens on the α carbon undergo this reaction;** that is, esters must have the general structure CH_3CO_2R' or RCH_2CO_2R'.

- Keep in mind: The characteristic reaction of esters is nucleophilic substitution. A Claisen reaction is a nucleophilic substitution in which an enolate is the nucleophile.

Figure 18.4 compares the general reaction for nucleophilic substitution of an ester with the Claisen reaction. Sample Problem 18.4 reinforces the basic features of the Claisen reaction.

Figure 18.4
The Claisen reaction—
An example of nucleophilic substitution

[Figure showing general nucleophilic substitution of ester (top) compared to Claisen reaction (bottom)]

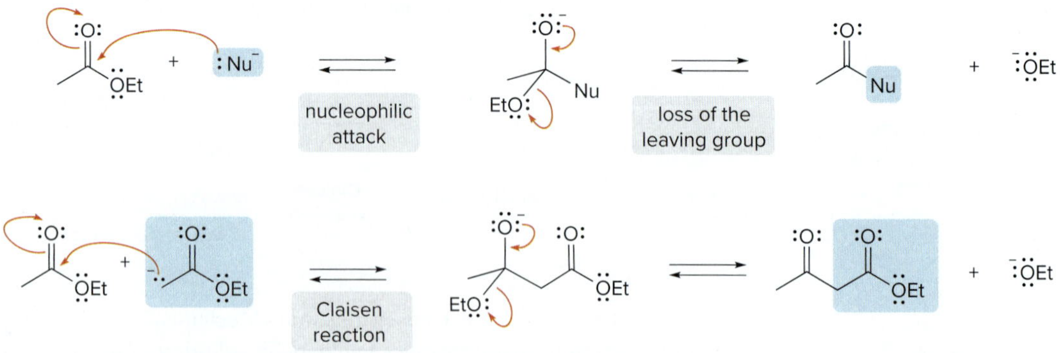

- Esters react by **nucleophilic substitution**. In a Claisen reaction, an **enolate is the nucleophile** that adds to the carbonyl group.

Sample Problem 18.4 — Determining the Product of a Claisen Reaction

Draw the product of the following Claisen reaction.

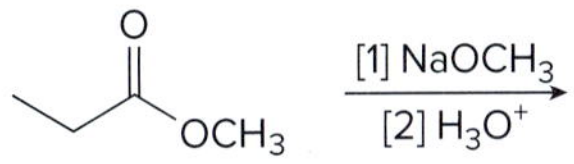

[structure: methyl propanoate] → [1] NaOCH₃; [2] H₃O⁺

Solution

To draw the product of any Claisen reaction, form a new carbon–carbon bond between the α carbon of one ester and the carbonyl carbon of another ester, with elimination of the leaving group ($^-$OCH₃ in this case).

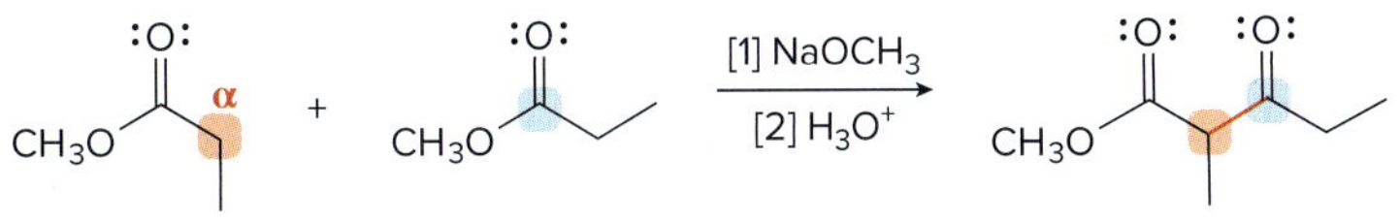

[mechanism scheme showing two molecules of methyl propanoate combining to form the β-keto ester product; new C–C bond in red]

Next, write out the steps of the reaction to verify this product.

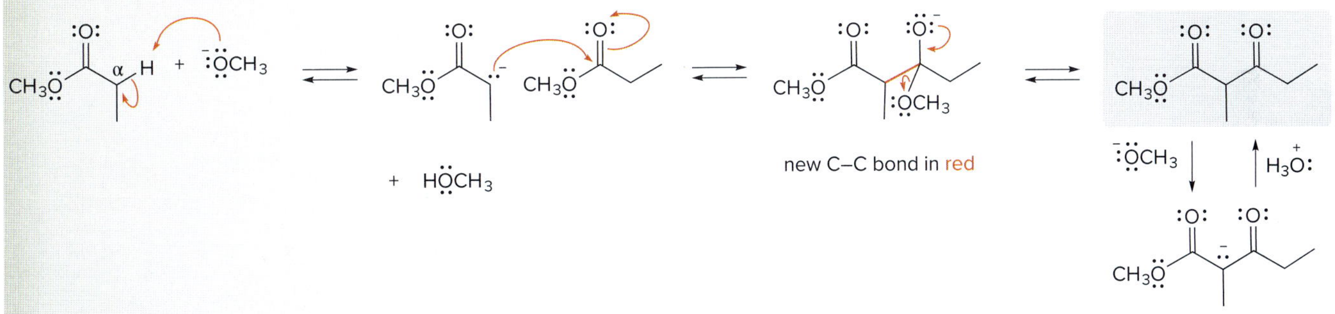

[full mechanism scheme with arrow-pushing showing enolate formation, nucleophilic addition, loss of $^-$OCH₃, and H₃O⁺ workup; new C–C bond in red]

Problem 18.16 What β-keto ester is formed when each ester is used in a Claisen reaction?

a. 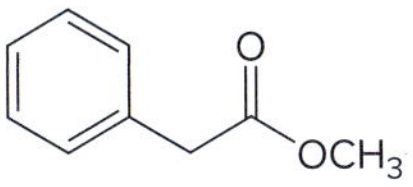 [methyl phenylacetate: PhCH₂C(=O)OCH₃]   b. 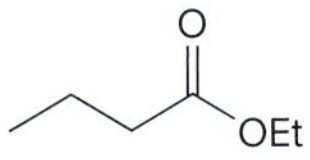 [ethyl butanoate: CH₃CH₂CH₂C(=O)OEt]

18.6 The Crossed Claisen and Related Reactions

Like the aldol reaction, it is sometimes possible to carry out a Claisen reaction with two different carbonyl components as starting materials.

- A Claisen reaction between two different carbonyl compounds is called a *crossed Claisen reaction*.

18.6A Two Useful Crossed Claisen Reactions

A crossed Claisen reaction is synthetically useful in two different instances.

- A crossed Claisen occurs between two different esters when only one has α hydrogens.

828 Chapter 18 Carbonyl Condensation Reactions

When one ester has no α hydrogens, a crossed Claisen reaction often leads to one product. Common esters with no α H atoms include ethyl formate (HCO_2Et) and ethyl benzoate ($C_6H_5CO_2Et$). For example, the reaction of ethyl benzoate (as the electrophile) with ethyl acetate (which forms the enolate) in the presence of base forms predominately one β-keto ester.

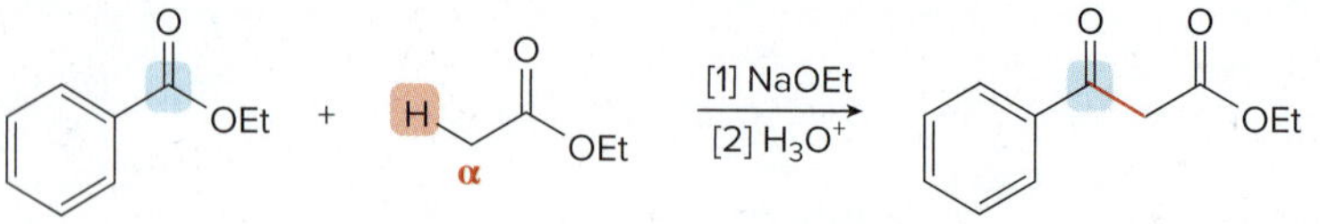

ethyl benzoate + ethyl acetate (Only this ester can form an enolate.) → [1] NaOEt, [2] H_3O^+ → β-keto ester (new C–C bond in red)

- **A crossed Claisen occurs between a ketone and an ester.**

The reaction of a ketone and an ester in the presence of base also forms the product of a crossed Claisen reaction. The enolate is generally formed from the ketone component, and the reaction works best when the ester has no α hydrogens. The product of this crossed Claisen reaction is a **β-dicarbonyl compound,** but *not* a β-keto ester.

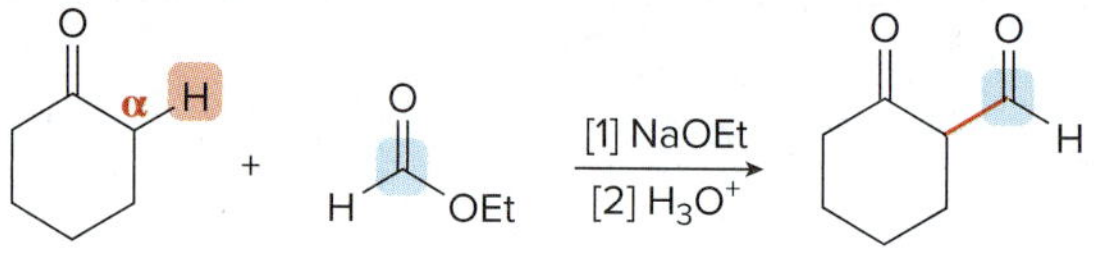

cyclohexanone + HCO_2Et → [1] NaOEt, [2] H_3O^+ → β-dicarbonyl compound (new C–C bond in red)

Problem 18.17 What crossed Claisen product is formed from each pair of compounds?

a. 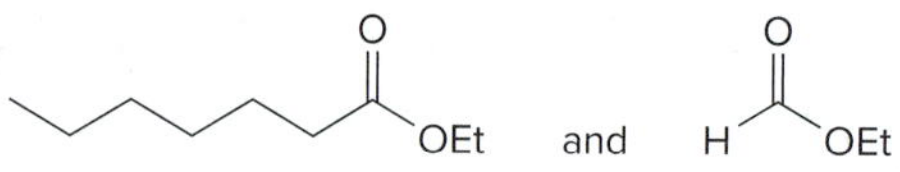 ethyl hexanoate and ethyl formate b. 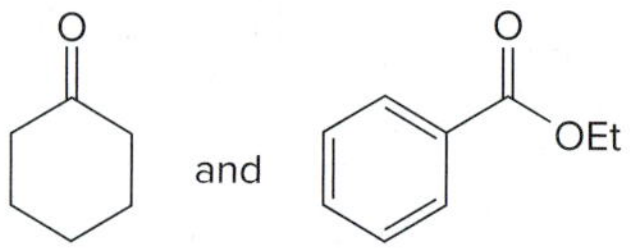 cyclohexanone and ethyl benzoate

Problem 18.18 Avobenzone is a conjugated compound that absorbs ultraviolet light, so it is a common ingredient in commercial sunscreens. Write out two different crossed Claisen reactions that form avobenzone.

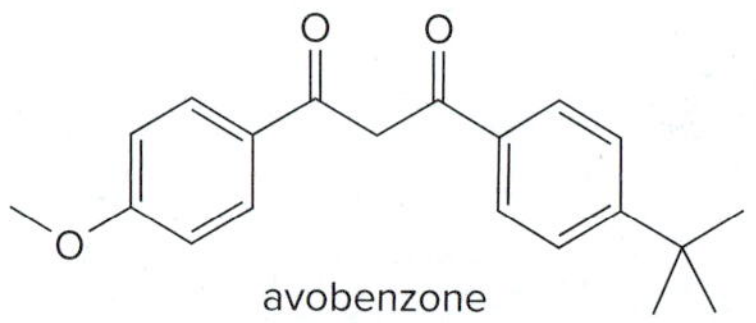

avobenzone

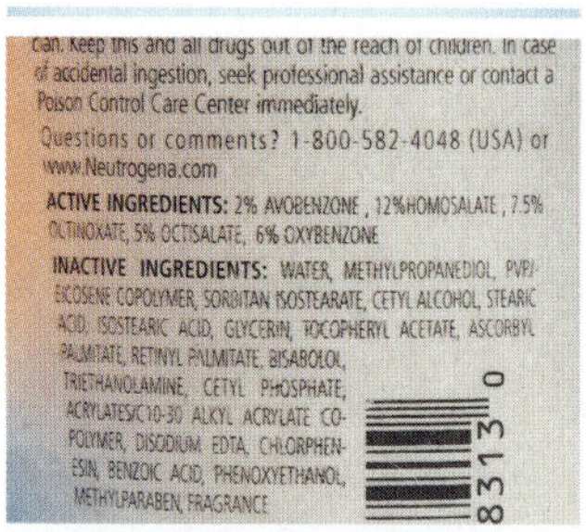

Sunscreen ingredients (Problem 18.18) *Jill Braaten/McGraw-Hill Education*

18.6B Other Useful Variations of the Crossed Claisen Reaction

β-Dicarbonyl compounds are also prepared by reacting an enolate with **ethyl chloroformate** and **diethyl carbonate**.

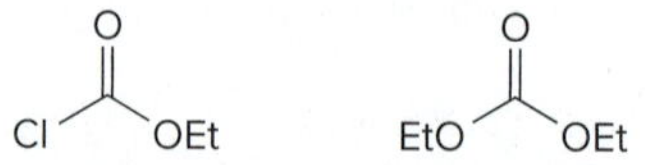

ethyl chloroformate   diethyl carbonate

These reactions resemble a Claisen reaction because they involve the same three steps:

[1] **Formation of an enolate**
[2] **Nucleophilic addition to a carbonyl group**
[3] **Elimination of a leaving group**

18.7 The Dieckmann Reaction

For example, reaction of an ester enolate with diethyl carbonate yields a **β-diester** (Reaction [1]), whereas reaction of a ketone enolate with ethyl chloroformate forms a **β-keto ester** (Reaction [2]). New carbon–carbon bonds are shown in red.

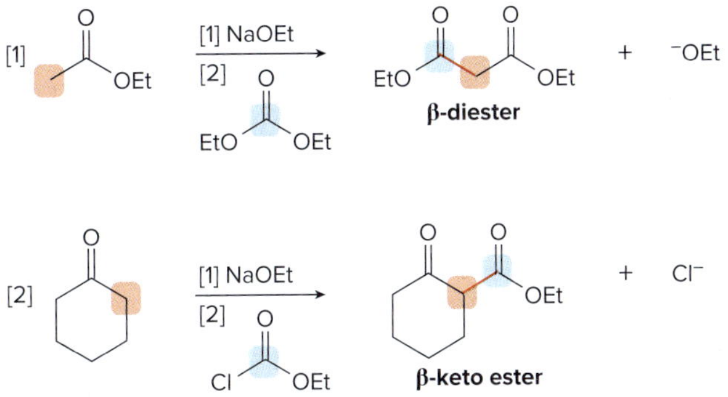

Reaction [2] is noteworthy because it provides easy access to **β-keto esters,** which are useful starting materials in the acetoacetic ester synthesis (Section 17.10). In this reaction, Cl⁻ is eliminated rather than ⁻OEt in Step [3], because Cl⁻ is a better leaving group, as shown in the following steps.

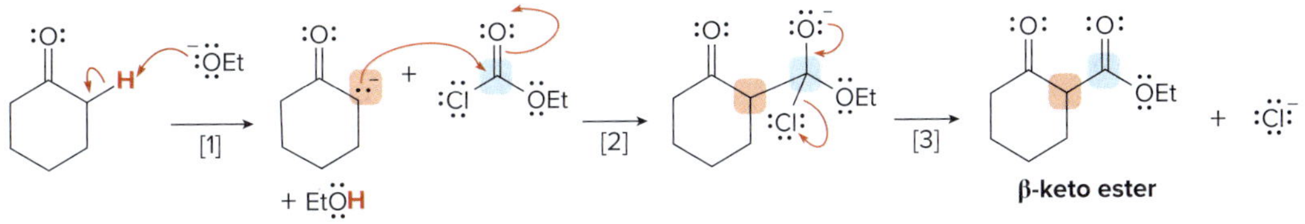

Problem 18.19 Draw the products of each reaction.

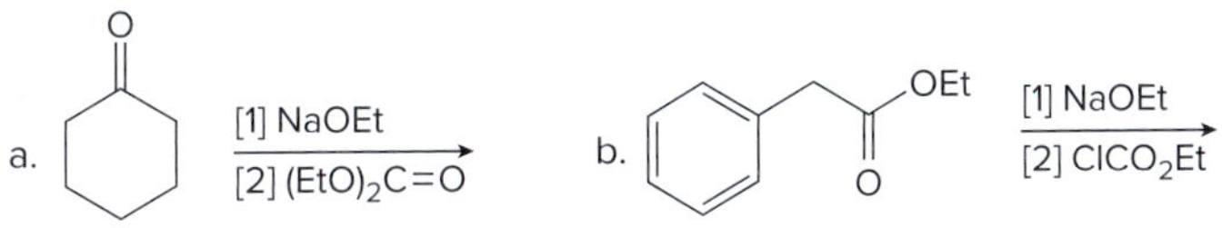

a. cyclohexanone [1] NaOEt [2] (EtO)₂C=O

b. PhCH₂CO₂Et [1] NaOEt [2] ClCO₂Et

Problem 18.20 Two steps in a synthesis of the analgesic ibuprofen include a carbonyl condensation reaction, followed by an alkylation reaction. Identify intermediates **A** and **B** in the synthesis of ibuprofen.

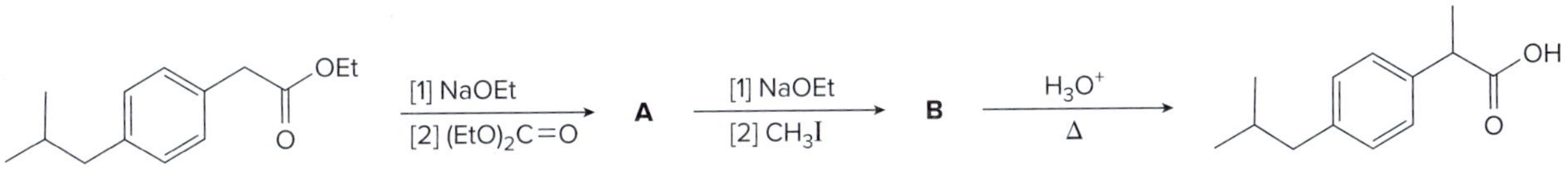

(isobutylphenyl)CH₂CO₂Et → [1] NaOEt, [2] (EtO)₂C=O → **A** → [1] NaOEt, [2] CH₃I → **B** → H₃O⁺, Δ → ibuprofen

---

18.7 The Dieckmann Reaction

Intramolecular Claisen reactions of diesters form five- and six-membered rings. The enolate of one ester is the nucleophile, and the carbonyl carbon of the other is the electrophile. An intramolecular Claisen reaction is called a **Dieckmann reaction.** Two types of diesters give good yields of cyclic products.

- **1,6-Diesters yield five-membered rings by the Dieckmann reaction.**

1,6-diester --re-draw--> [1] NaOEt [2] H₃O⁺ → cyclopentanone β-keto ester (with OEt)

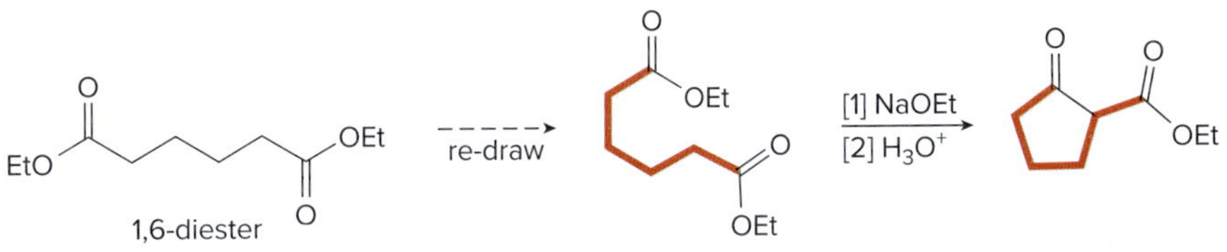

- **1,7-Diesters yield six-membered rings by the Dieckmann reaction.**

$$\text{1,7-diester} \xrightarrow{\text{re-draw}} \xrightarrow[\text{[2] H}_3\text{O}^+]{\text{[1] NaOEt}} \text{β-keto ester (cyclohexanone with CO}_2\text{Et)}$$

The mechanism of the Dieckmann reaction is exactly the same as the mechanism of an intermolecular Claisen reaction (Mechanism 18.5).

Problem 18.21 What two β-keto esters are formed in the Dieckmann reaction of the following diester?

18.8 Biological Carbonyl Condensation Reactions

Aldol, retro-aldol, and Claisen reactions are key steps in several metabolic pathways. In contrast to the stepwise processes illustrated in Mechanisms 18.1, 18.2, and 18.5, the biological reactions, which occur in the presence of enzymes, are often shown with two or more bonds broken or formed at the same time, as shown previously in reactions presented in Sections 11.13B and 17.2B. The acid or base that may be required in a particular reaction, which will be shown with the generic notation **HA** or **:B**, respectively, often comes from a functional group located at or near the active site.

18.8A Biological Aldol Reactions

A key step in the **biosynthesis of glucose** involves the reaction of glyceraldehyde 3-phosphate with dihydroxyacetone phosphate to form fructose 1,6-bisphosphate in the presence of an aldolase enzyme. A new carbon–carbon bond is formed between the aldehyde carbonyl of glyceraldehyde 3-phosphate and the α carbon of dihydroxyacetone phosphate. Fructose 1,6-bisphosphate is drawn as an acyclic ketone in equilibrium with its cyclic hemiacetal (Sections 14.17–14.18).

glyceraldehyde 3-phosphate + dihydroxyacetone phosphate →(aldolase) fructose 1,6-bisphosphate (acyclic form) ⇌ cyclic hemiacetal

The reaction occurs by way of the tautomerization of dihydroxyacetone phosphate to an **enediol**, which adds to the carbonyl group of glyceraldehyde 3-phosphate. This enzyme-catalyzed crossed aldol reaction forms a single stereoisomer in the product, even though two new stereogenic centers are generated.

dihydroxyacetone phosphate → enediol + glyceraldehyde 3-phosphate →(biological aldol) fructose 1,6-bisphosphate
new C–C bond shown in red

18.8 Biological Carbonyl Condensation Reactions

The steps of the citric acid cycle are discussed in detail in Section 27.6.

Another biological aldol reaction occurs during the first step of the **citric acid cycle,** in the reaction of acetyl CoA (Section 16.16) with oxaloacetate to form citrate in the presence of the enzyme citrate synthase. A new carbon–carbon bond is formed between the ketone carbonyl of oxaloacetate and the α carbon of acetyl CoA.

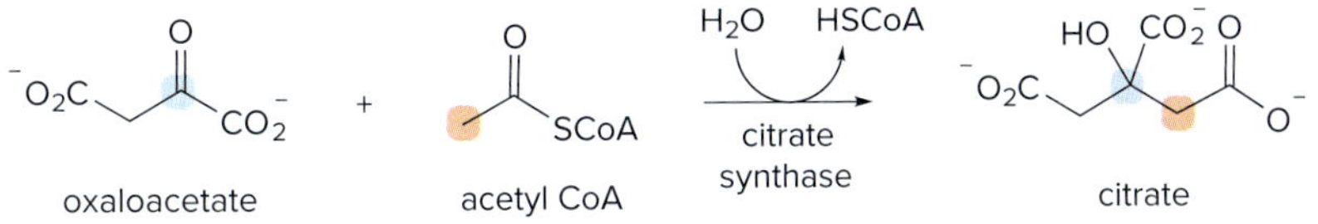

This reaction occurs by way of the tautomerization of the thioester acetyl CoA to an enol-type intermediate, which adds to the carbonyl group of oxaloacetate, forming a new carbon–carbon bond. Aldol condensation first forms the thioester citrate CoA, which is hydrolyzed to citrate.

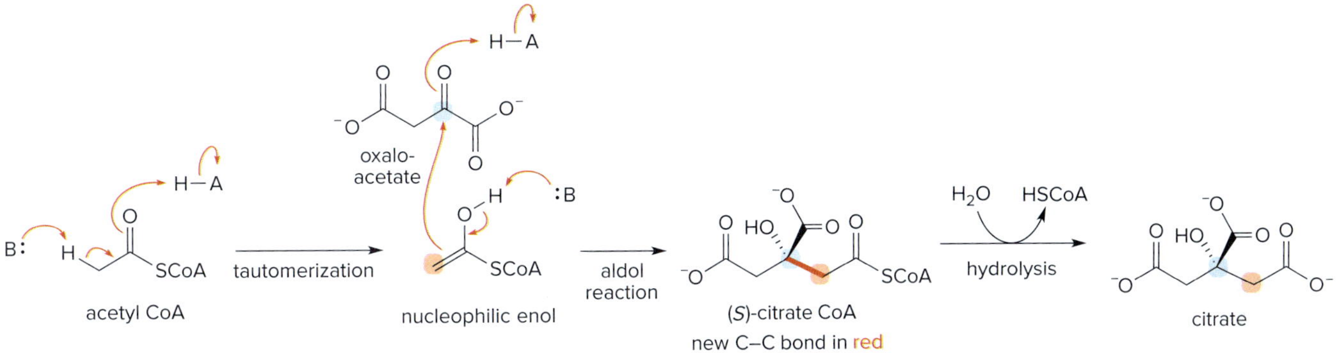

Although these reactions look more complex than the aldol reactions in Section 18.1, they are fundamentally the same.

- Whether an aldol reaction occurs in the laboratory or in a biological system, the α carbon of one carbonyl compound adds to the carbonyl group of another carbonyl compound by nucleophilic addition.

Problem 18.22 (a) Identify the carbon atoms in pyruvate and acetyl CoA that are joined together in the given biological aldol reaction to form citramalate. (b) Use curved arrows to show how this reaction occurs.

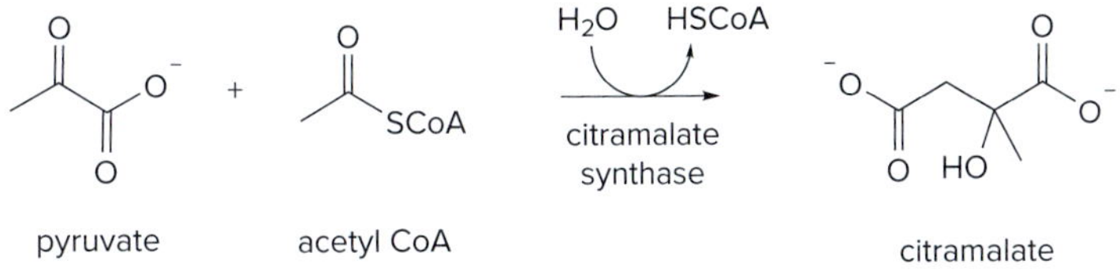

18.8B A Biological Retro-Aldol Reaction

One step in glycolysis, a key 10-step pathway in the metabolism of glucose to CO_2 and H_2O discussed in detail in Section 27.4, is the **retro-aldol** conversion of fructose 1,6-bisphosphate to glyceraldehyde 3-phosphate and dihydroxyacetone phosphate.

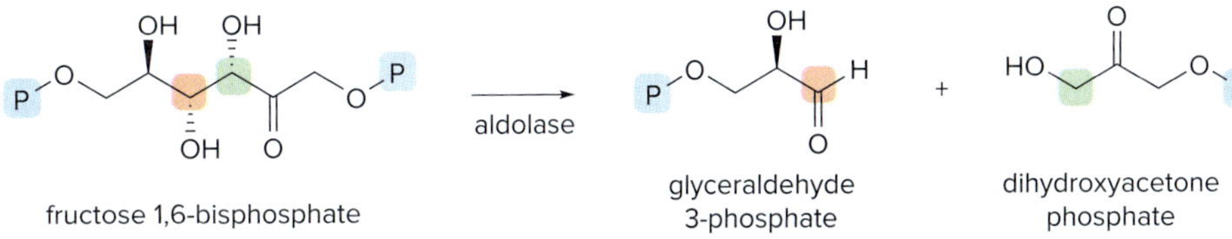

This reaction, catalyzed by an aldolase enzyme, is the reverse of the aldol reaction that prepares fructose 1,6-bisphosphate described in Section 18.8A.

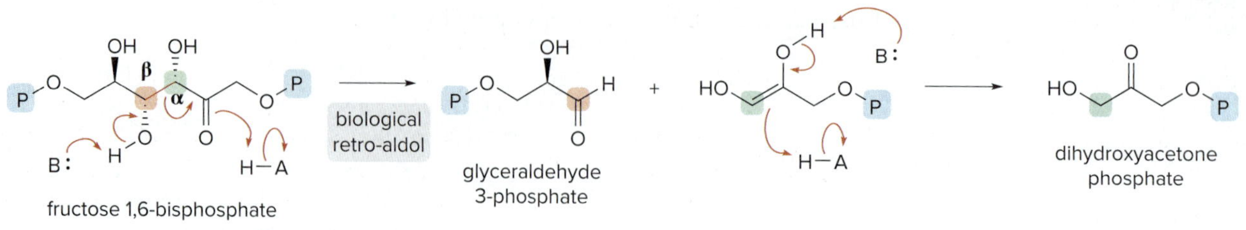

fructose 1,6-bisphosphate
The C–C bond between the α and β C's (shown in red) is cleaved.

biological retro-aldol → glyceraldehyde 3-phosphate + dihydroxyacetone phosphate

- To draw the products of a retro-aldol reaction, always cleave the bond between the α and β carbons to the carbonyl group.

18.8C Biological Claisen Reactions

Biological Claisen reactions constitute the predominant carbon–carbon bond-forming reactions in the biosynthesis of fatty acids. The starting materials in biological systems are *thioesters* (RCOSR'), rather than esters.

The biosynthesis of fatty acids begins with the Claisen condensation of two thioesters (**A** and **B**) to form a four-carbon β-keto thioester with loss of CO_2. A new carbon–carbon bond is formed between the carbonyl carbon of **A** and the α carbon of **B**.

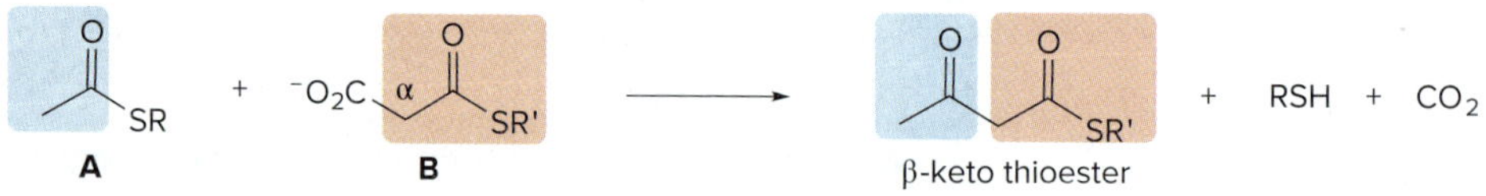

A + **B** → β-keto thioester + RSH + CO_2

It is thought that the mechanism of this reaction begins with decarboxylation of **B** to form an enolate (Step [1]), which adds to the carbonyl group of **A** to generate a tetrahedral intermediate (Step [2]). Loss of the leaving group ($^-$SR) then generates the β-keto thioester (Step [3]). Two carbons of **A** and two carbons of **B** are joined to form a four-carbon β-keto thioester.

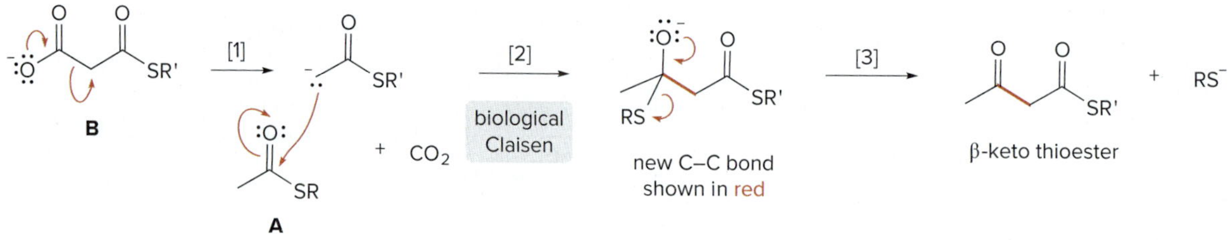

B [1] biological Claisen + CO_2 → [2] new C–C bond shown in red → [3] β-keto thioester + RS$^-$

In the biosynthesis of fatty acids, the β-keto thioester is converted to a thioester by a three-step process.

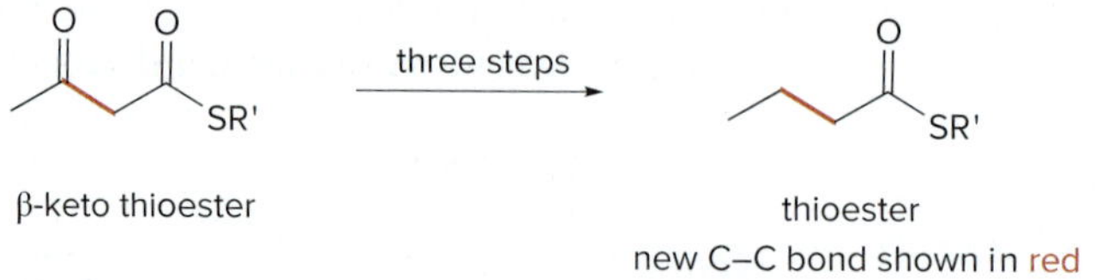

β-keto thioester — three steps → thioester
new C–C bond shown in red

The resulting thioester can then undergo another Claisen reaction with **B** to form a β-keto thioester with two more carbons, which can be converted to a six-carbon thioester, **C**.

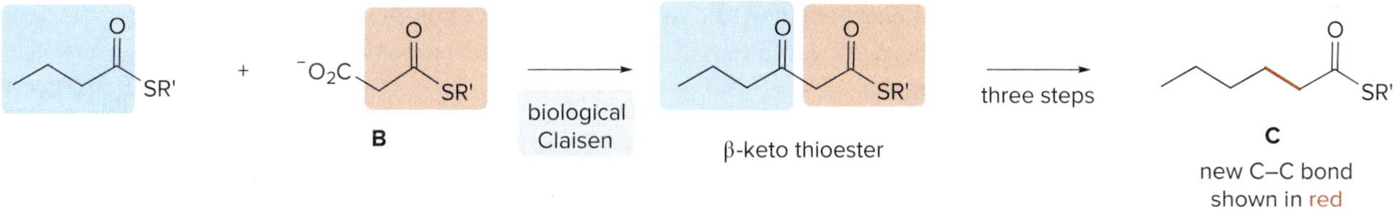

The degradation of fatty acids by β-oxidation, the reverse of fatty acid biosynthesis, is discussed in Section 27.3.

Each time this sequence is carried out, **two more carbons are added to the thioester.** This process illustrates why most fatty acids have an even number of carbon atoms. In stearic acid, eight carbon–carbon bonds are formed by sequential Claisen reactions that add two carbon units at a time.

stearic acid
new C–C bonds formed by the
Claisen reaction shown in red

Problem 18.23 Draw the β-keto thioester formed by the biological Claisen reaction of **B** and **X**.

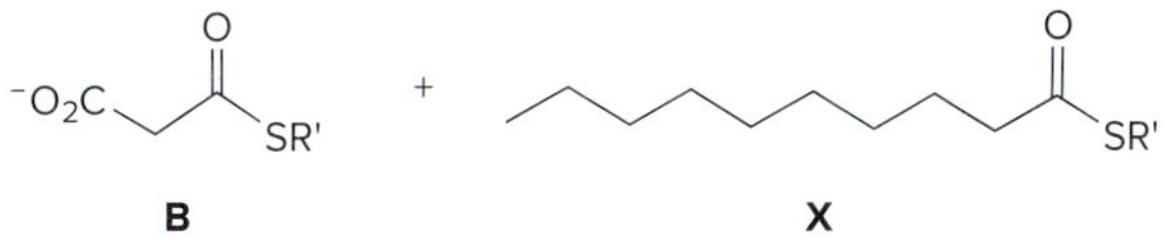

18.9 The Michael Reaction

Like the aldol and Claisen reactions, the **Michael reaction involves two carbonyl components—the enolate of one carbonyl compound and an α,β-unsaturated carbonyl compound.**

Two components of a Michael reaction

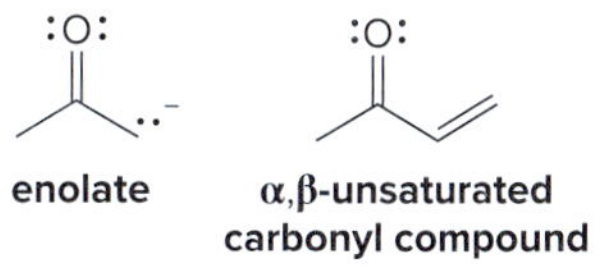

enolate   α,β-unsaturated carbonyl compound

Recall from Section 13.15 that α,β-unsaturated carbonyl compounds are resonance stabilized and have **two electrophilic sites—the carbonyl carbon and the β carbon.**

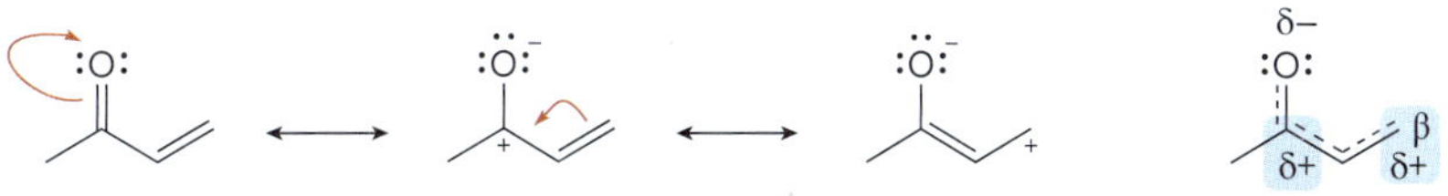

three resonance structures for an
α,β-unsaturated carbonyl compound

hybrid

two electrophilic sites

- **The Michael reaction involves the conjugate addition (1,4-addition) of a resonance-stabilized enolate to the β carbon of an α,β-unsaturated carbonyl system.**

All conjugate additions add the **elements of H and Nu across the α and β carbons.**

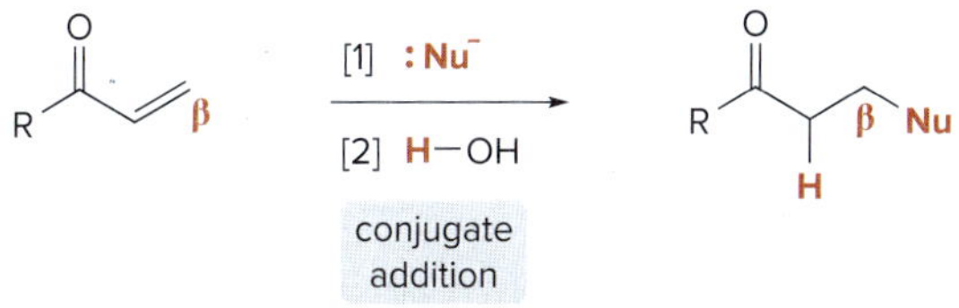

conjugate addition

Chapter 18 Carbonyl Condensation Reactions

In the Michael reaction, the **nucleophile is an enolate.** Enolates of active methylene compounds are particularly common. The α,β-unsaturated carbonyl component is often called a **Michael acceptor.**

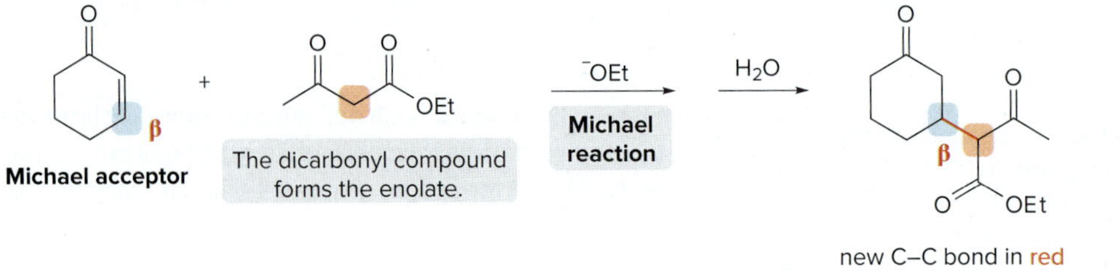

Problem 18.24
Which of the following compounds can serve as Michael acceptors?

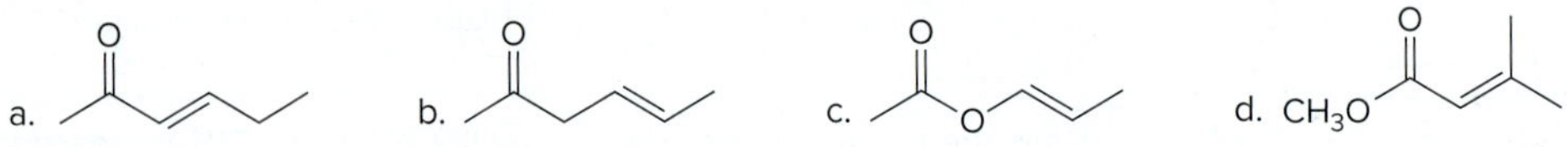

The Michael reaction always forms a new carbon–carbon bond on the β carbon of the Michael acceptor. The key step is nucleophilic addition of the enolate to the β carbon of the Michael acceptor in Step [2], as shown in Mechanism 18.6.

Mechanism 18.6 The Michael Reaction

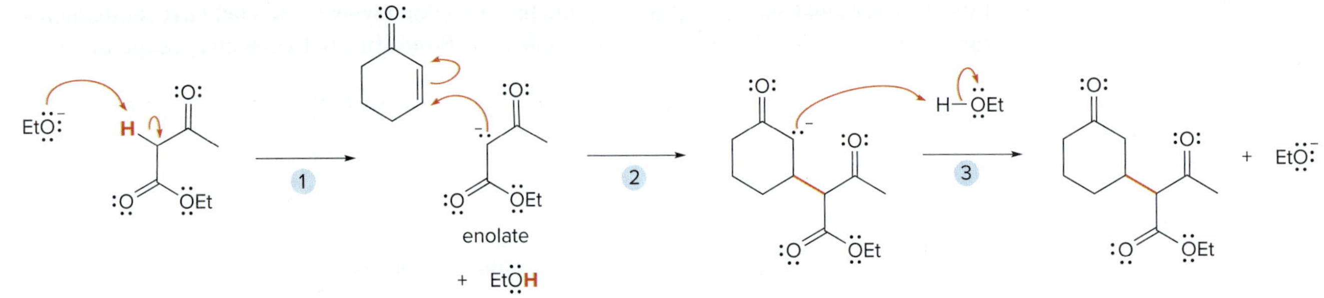

1. The base removes a proton on the carbon between the two carbonyl groups to form an **enolate.**
2. **Nucleophilic addition of the enolate to the β carbon** of the α,β-unsaturated carbonyl compound forms a new carbon–carbon bond and another enolate.
3. Protonation of the enolate forms the **1,4-addition product.**

When the product of a Michael reaction is also a β-keto ester, it can be hydrolyzed and decarboxylated by heating in aqueous acid, as discussed in Section 17.9. This forms a **1,5-dicarbonyl compound.** Figure 18.5 shows a Michael reaction that was a key step in the synthesis of **estrone,** a female sex hormone.

1,5-Dicarbonyl compounds are starting materials for intramolecular aldol reactions, as described in Section 18.4.

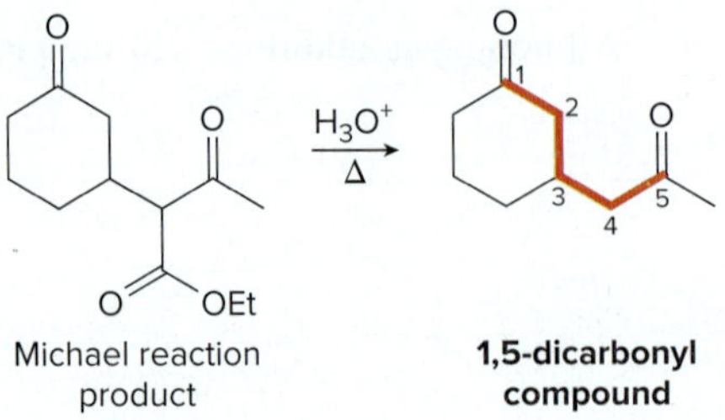

Figure 18.5 Using a Michael reaction in the synthesis of the steroid estrone

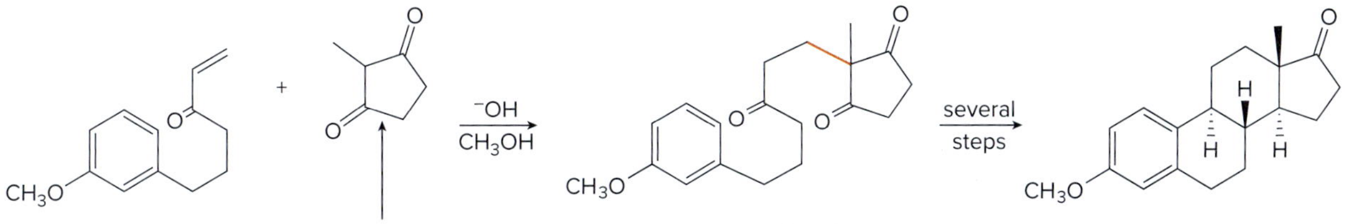

α,β-unsaturated carbonyl compound	carbonyl compound that forms the enolate	new carbon–carbon bond in red	estrone

---

Problem 18.25 What product is formed when each pair of compounds is treated with NaOEt in ethanol?

a. 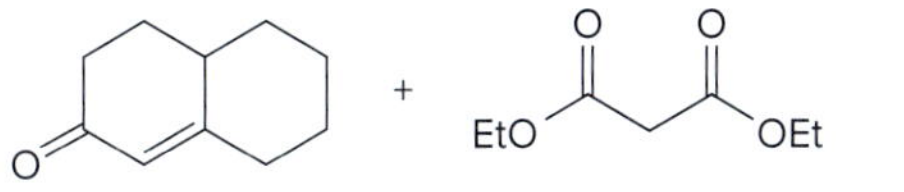 [structure] + EtO₂C-CH₂-CO₂Et   b.  [structure] + NC-CH₂-C(=O)-

---

Problem 18.26 What starting materials are needed to prepare each compound by the Michael reaction?

a. 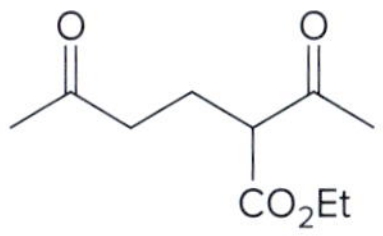 [diketone with CO₂Et]   b. 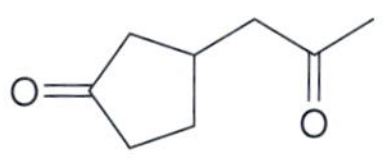 [cyclopentanone derivative]

18.10 The Robinson Annulation

> The word **annulation** comes from the Greek word *annulus* for "ring." The Robinson annulation is named for English chemist Sir Robert Robinson, who was awarded the 1947 Nobel Prize in Chemistry.

The Robinson annulation is a ring-forming reaction that combines a Michael reaction with an intramolecular aldol reaction. Like the other reactions in Chapter 18, it involves enolates and it forms carbon–carbon bonds. The two starting materials for a Robinson annulation are an α,β-unsaturated carbonyl compound and an enolate.

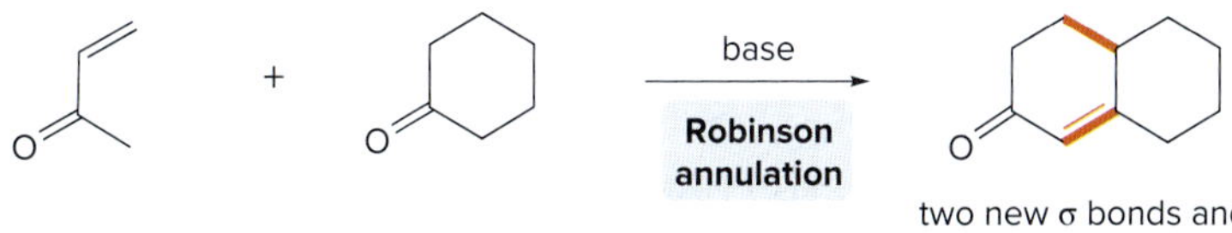

[reaction scheme: α,β-unsaturated carbonyl compound + carbonyl compound that forms an enolate → (base, Robinson annulation) → bicyclic product with two new σ bonds and one new π bond in red]

The Robinson annulation forms a six-membered ring and three new carbon–carbon bonds—two σ bonds and one π bond. The product contains an α,β-unsaturated ketone in a cyclohexane ring—that is, a **cyclohex-2-enone** ring. To generate the enolate component of the Robinson annulation, ⁻OH in H₂O and ⁻OEt in EtOH are typically used.

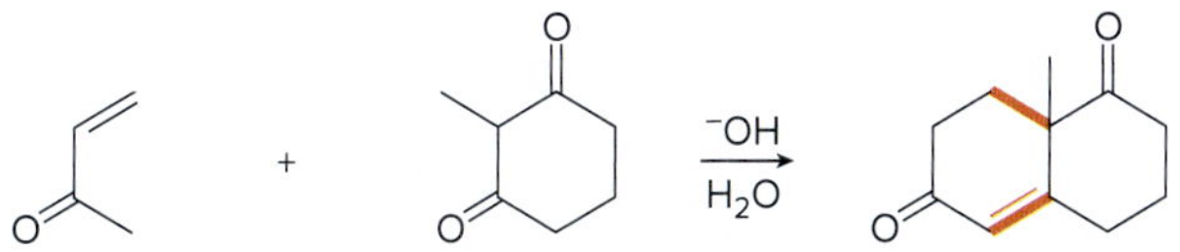

[reaction scheme: methyl vinyl ketone + 2-methylcyclohexane-1,3-dione → (⁻OH, H₂O) → product with new C–C bonds shown in red]

The mechanism of the Robinson annulation consists of a **Michael addition** to the α,β-unsaturated carbonyl compound to form a 1,5-dicarbonyl compound, followed by an

836 Chapter 18 Carbonyl Condensation Reactions

intramolecular aldol reaction to form the six-membered ring. The mechanism is written out in three parts in Mechanism 18.7 for the reaction between methyl vinyl ketone and 2-methylcyclohexane-1,3-dione.

Mechanism 18.7 The Robinson Annulation

Part [1] Michael addition

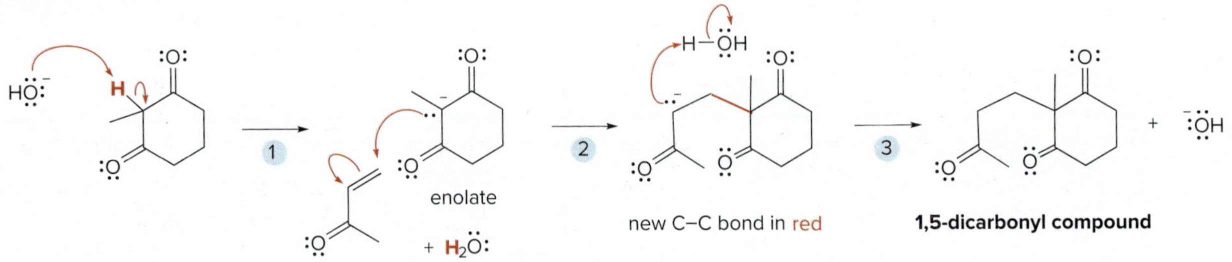

1–2 Base removes the most acidic proton—the proton between the two carbonyl groups—to form an **enolate. Conjugate addition** of the enolate to the α,β-unsaturated carbonyl compound forms a new carbon–carbon bond, generating an enolate.

3 Protonation of the enolate forms a **1,5-dicarbonyl compound**.

Part [2] Intramolecular aldol reaction

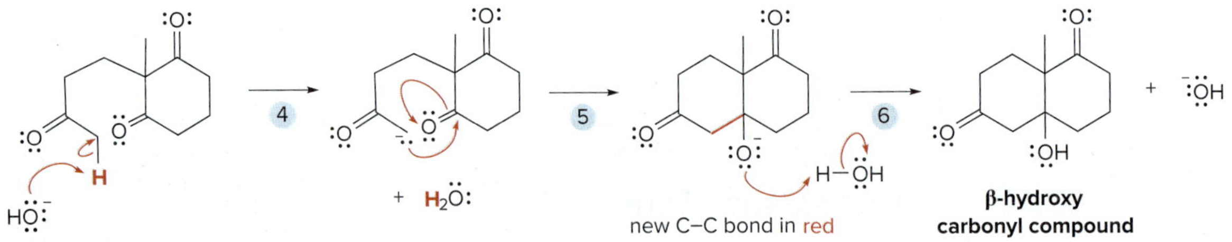

4–5 The base removes a proton to form an **enolate**, which attacks a carbonyl group to form a new C–C σ bond, generating the six-membered ring.

6 Protonation of the alkoxide forms the **β-hydroxy carbonyl compound**.

Part [3] Dehydration of the β-hydroxy carbonyl compound

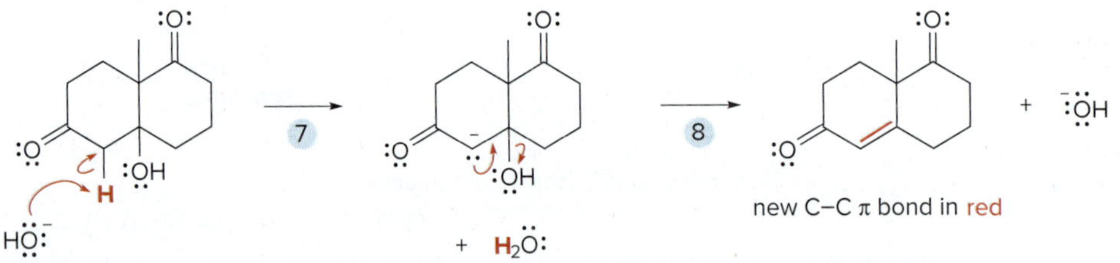

7–8 **Dehydration occurs by the two-step E1cB mechanism**—loss of a proton to form an enolate and elimination of ⁻OH to form a π bond.

The mechanism begins with the three-step **Michael addition** that forms the first carbon–carbon σ bond, generating the 1,5-dicarbonyl compound (Part [1]). An **intramolecular aldol reaction** (Part [2]) forms the second carbon–carbon σ bond, and **dehydration** of the β-hydroxy ketone (Part [3]) forms the π bond.

To draw the product of Robinson annulation without writing out the mechanism each time, **place the α carbon of the compound that becomes the enolate next to the β carbon of the α,β-unsaturated carbonyl compound.** Then, join the appropriate carbons together as shown.

18.10 The Robinson Annulation

If you follow this method of drawing the starting materials, the double bond in the product always ends up in the same position in the six-membered ring.

Sample Problem 18.5   Drawing the Product of a Robinson Annulation

Draw the Robinson annulation product formed from the following starting materials.

Solution

Arrange the starting materials to put the reactive atoms next to each other. For example:

- Place the α,β-unsaturated carbonyl compound *to the left* of the carbonyl compound.
- Determine which α carbon will become the enolate. **The most acidic H is always removed with base first,** which in this case is the H (in red) on the α carbon between the two carbonyl groups. **This α carbon is drawn adjacent to the β carbon of the α,β-unsaturated carbonyl compound.**

Then draw the bonds to form the new six-membered ring.

Problem 18.27

Draw the products when each pair of compounds is treated with $CH_3CH_2O^-$, CH_3CH_2OH in a Robinson annulation reaction.

More Practice: Try Problems 18.49, 18.65d.

838 Chapter 18 Carbonyl Condensation Reactions

To use the Robinson annulation in synthesis, you must be able to determine what starting materials are needed to prepare a given compound, by working in the retrosynthetic direction.

How To Synthesize a Compound Using the Robinson Annulation

Example What starting materials are needed to synthesize the following compound using a Robinson annulation?

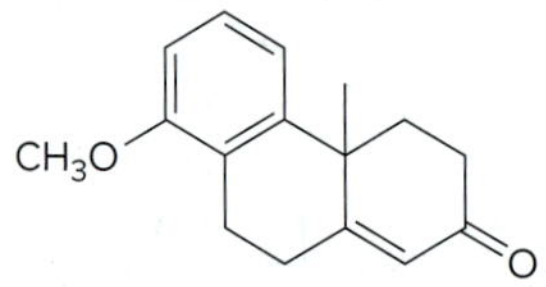

Step [1] Locate the cyclohex-2-enone ring and re-draw the target molecule if necessary.
- To most easily determine the starting materials, always arrange the α,β-unsaturated carbonyl system in the same location. The target compound may have to be flipped or rotated, and you must be careful not to move any bonds to the wrong location during this process.

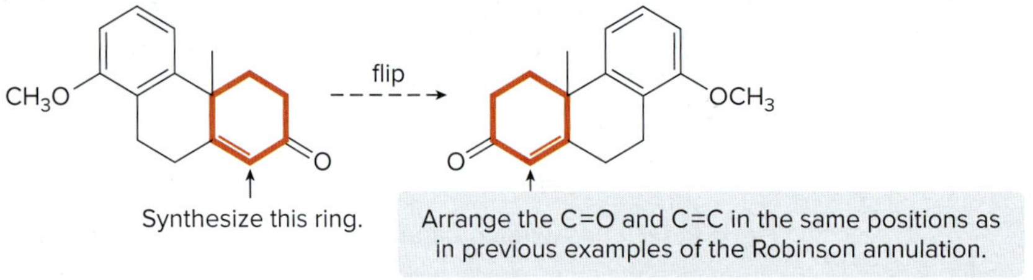

Synthesize this ring.   Arrange the C=O and C=C in the same positions as in previous examples of the Robinson annulation.

Step [2] Break the cyclohex-2-enone ring into two components.
- Break the C=C. One half becomes the carbonyl group of the enolate component.
- Break the bond between the β carbon and the carbon to which it is bonded.

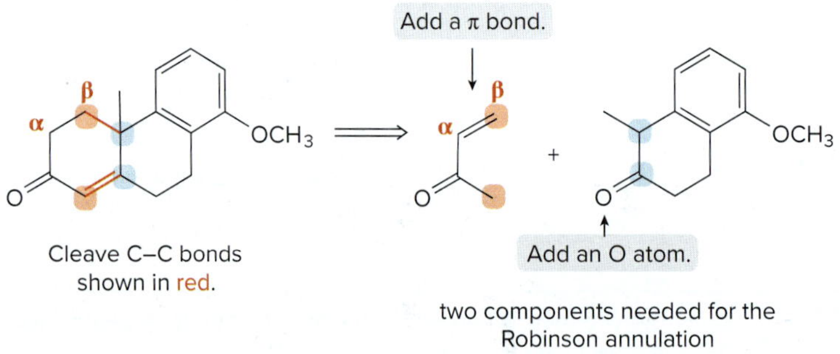

Cleave C–C bonds shown in red.   Add an O atom.

two components needed for the Robinson annulation

Problem 18.28 Which of the following bicyclic ring systems can be prepared by an intermolecular Robinson annulation?

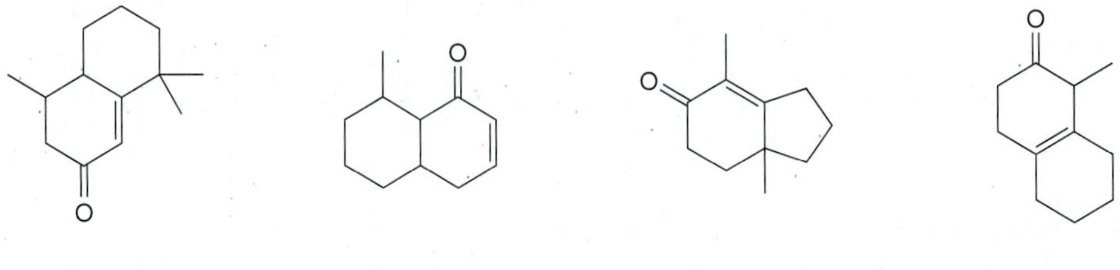

A   B   C   D

Problem 18.29 What starting materials are needed to synthesize each compound by a Robinson annulation?

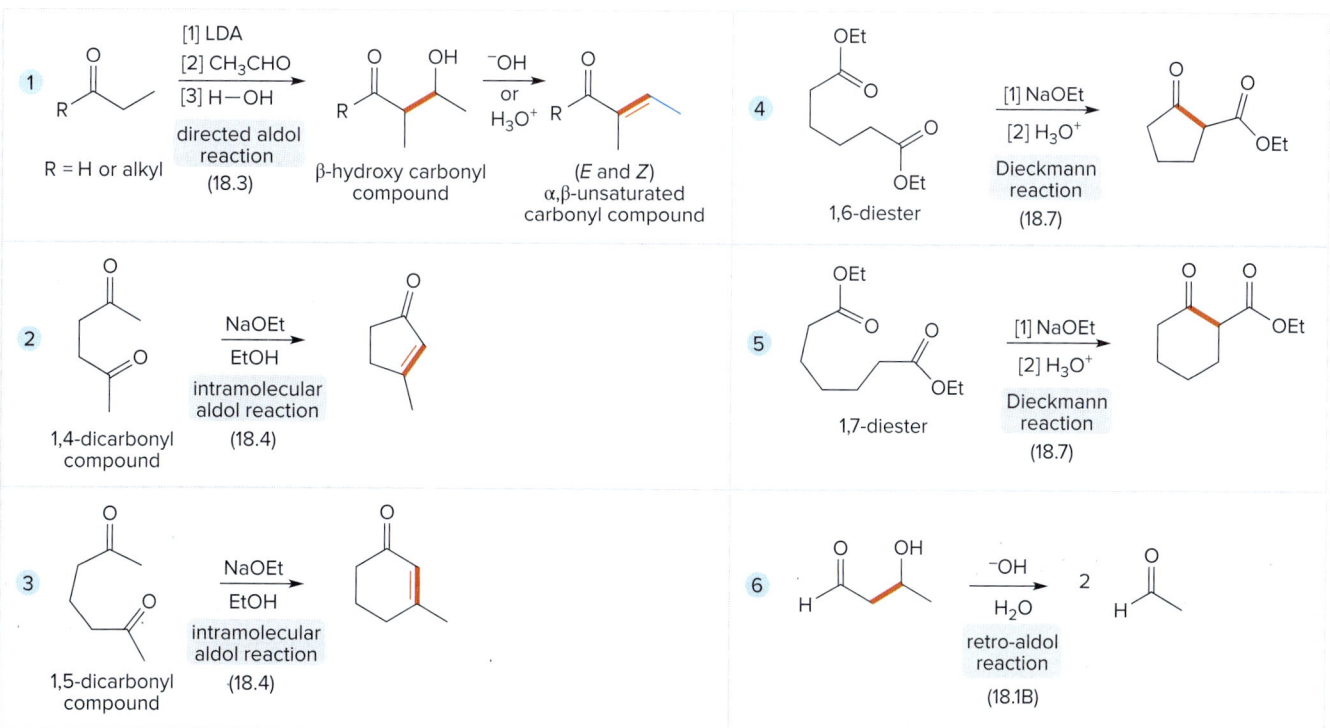

Chapter 18 REVIEW

KEY REACTIONS

[1] The four major carbonyl condensation reactions

1. aldehyde (or ketone) → β-hydroxy carbonyl compound → (E and Z) α,β-unsaturated carbonyl compound — aldol reaction (18.1)

2. ester → β-keto ester — Claisen reaction (18.5)

3. α,β-unsaturated carbonyl compound + carbonyl compound → 1,5-dicarbonyl compound — Michael reaction (18.9)

4. α,β-unsaturated carbonyl compound + cyclohex-2-enone → Robinson annulation (18.10)

Try Problems 18.30; 18.32; 18.46; 18.49; 18.51a, b, d, e.

[2] Useful variations

1. directed aldol reaction (18.3): [1] LDA; [2] CH₃CHO; [3] H–OH → β-hydroxy carbonyl compound → (E and Z) α,β-unsaturated carbonyl compound

2. 1,4-dicarbonyl compound → intramolecular aldol reaction (18.4), NaOEt/EtOH

3. 1,5-dicarbonyl compound → intramolecular aldol reaction (18.4), NaOEt/EtOH

4. 1,6-diester → Dieckmann reaction (18.7), [1] NaOEt; [2] H₃O⁺

5. 1,7-diester → Dieckmann reaction (18.7), [1] NaOEt; [2] H₃O⁺

6. retro-aldol reaction (18.1B), ⁻OH/H₂O

Try Problems 18.33; 18.34; 18.37; 18.51c, f.

KEY SKILLS

[1] Drawing the product of a directed aldol reaction (18.3)

| **1** Prepare the enolate of one carbonyl component with LDA. | **2** Add the second carbonyl compound to this enolate. | **3** Add H₂O, and draw the product. |

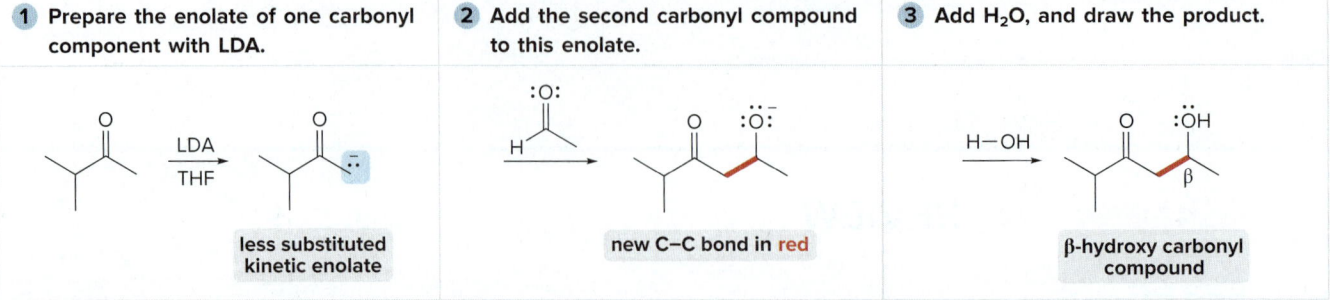

See Sample Problem 18.1. Try Problem 18.33.

[2] Identifying the starting materials to synthesize an α,β-unsaturated carbonyl compound using a directed aldol reaction (18.3)

| **1** Identify the α and β carbons that are part of the C=C. | **2** Break the molecule into two components between these carbons, and add a double bond to oxygen at the β carbon. |

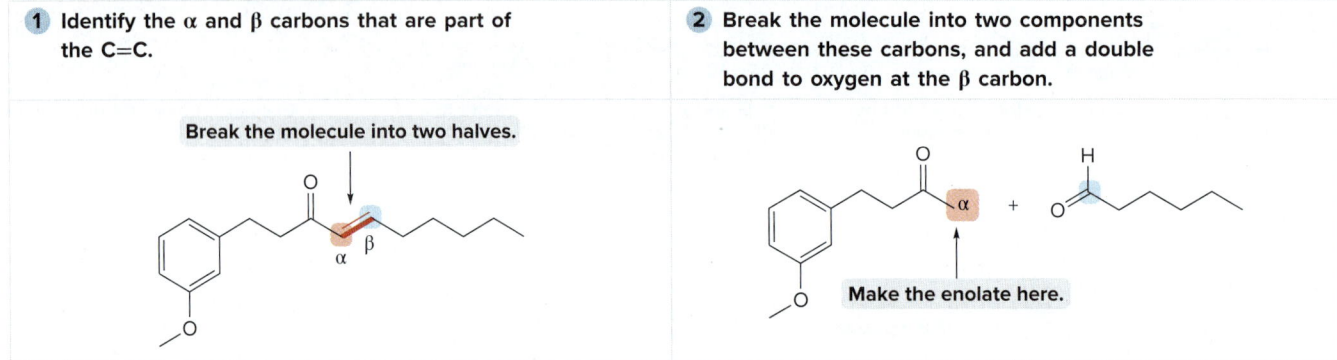

See Sample Problem 18.2. Try Problem 18.35.

[3] Converting a six-membered ring to a five-membered ring using an intramolecular aldol reaction (18.4)

| **1** Treat the alkene with O₃, followed by Zn and H₂O. | **2** React the 1,6-dicarbonyl compound with base. | **3** Draw the product. |

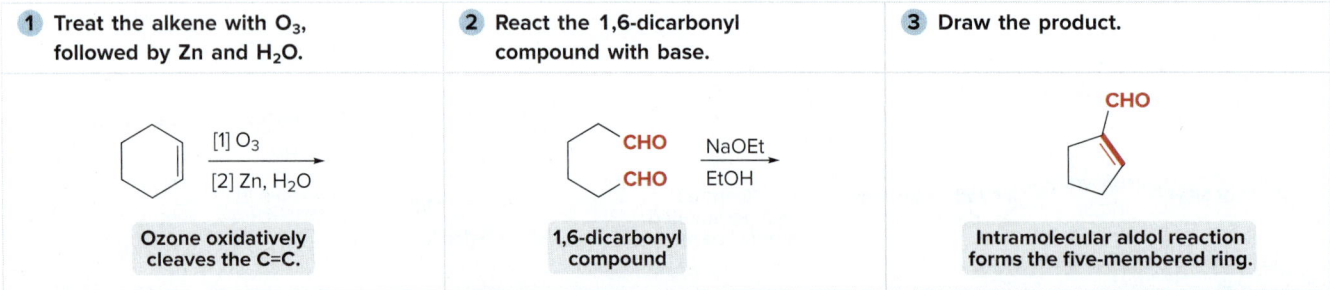

See Figure 18.3. Try Problems 18.31, 18.37.

[4] Drawing the product of a Claisen reaction (18.5)

| **1** Identify the α carbon of one ester and the carbonyl carbon of the other. | **2** Draw the product. |

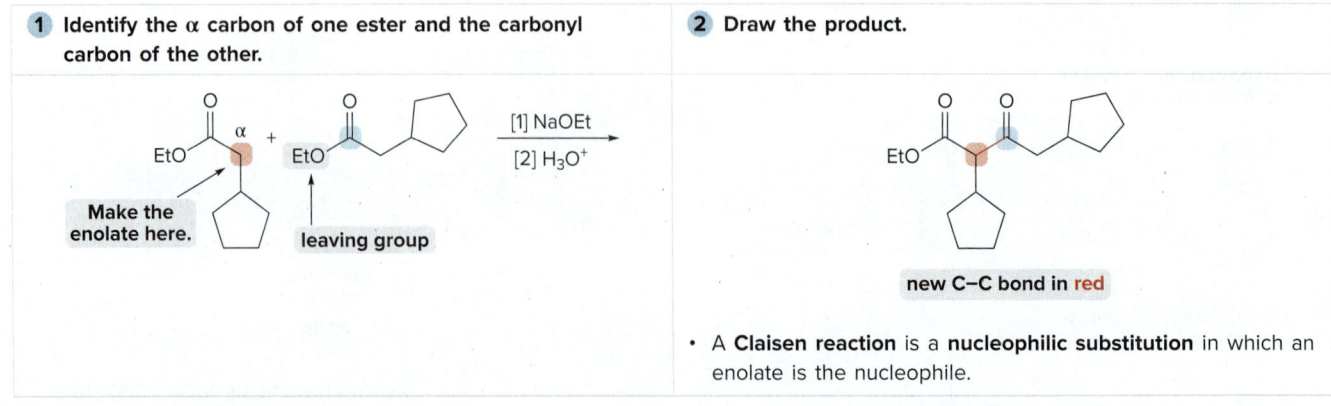

- A **Claisen reaction** is a **nucleophilic substitution** in which an enolate is the nucleophile.

See Sample Problem 18.4. Try Problem 18.38.

[5] Drawing the product of a Robinson annulation (18.10)

1 Arrange the starting materials to put the reactive atoms next to each other.

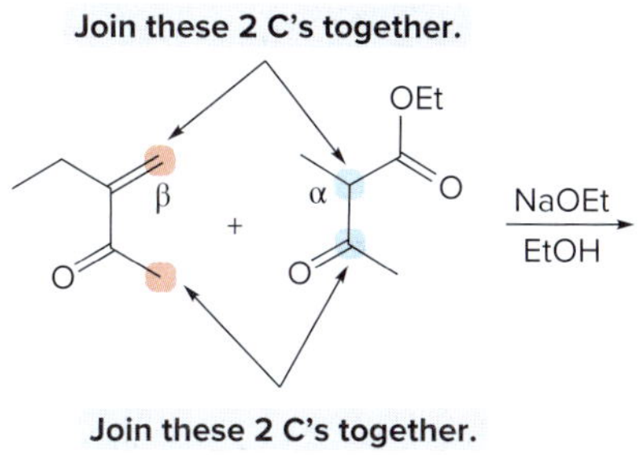

2 Draw the product.

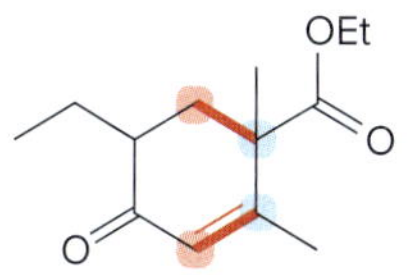

new C–C bonds in red

See Sample Problem 18.5. Try Problems 18.49, 18.65d.

[6] Identifying the starting materials to synthesize a compound using the Robinson annulation (18.10)

1 Locate the cyclohex-2-enone ring, and re-draw the target molecule, if necessary.

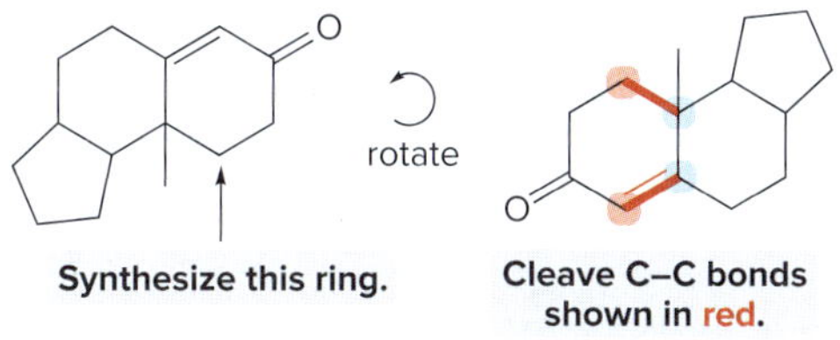

2 Break the cyclohex-2-enone ring into two components.

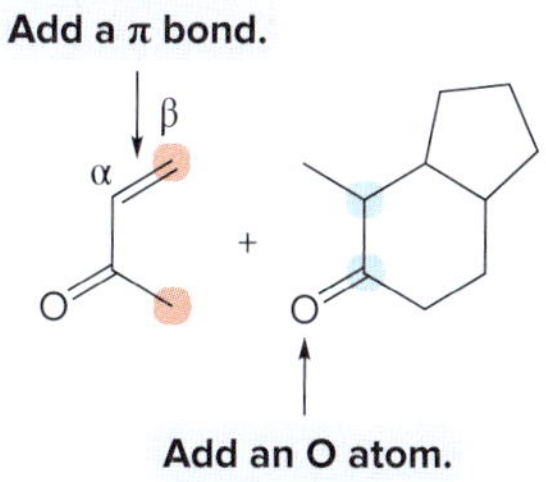

- Arrange the C=O and C=C in the same positions as in previous examples of the Robinson annulation.

See *How To*, p. 838. Try Problem 18.50.

PROBLEMS

Problems Using Three-Dimensional Models

18.30 Draw the aldol product formed from each pair of starting materials using ⁻OH, H₂O.

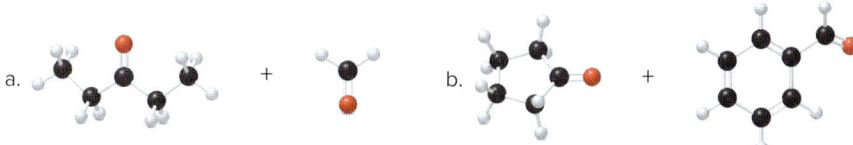

18.31 What steps are needed to convert **A** to **B**?

A   B

Chapter 18 Carbonyl Condensation Reactions

The Aldol Reaction

18.32 Draw the product formed from an aldol reaction with the given starting material(s) using ⁻OH, H₂O.

a. 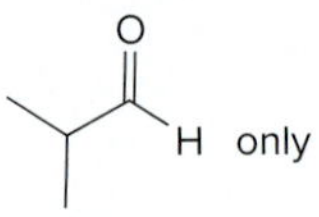 (CH₃)₂CHCHO, H only

b. 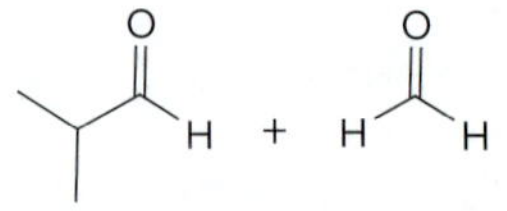 (CH₃)₂CHCHO + HCHO

c. 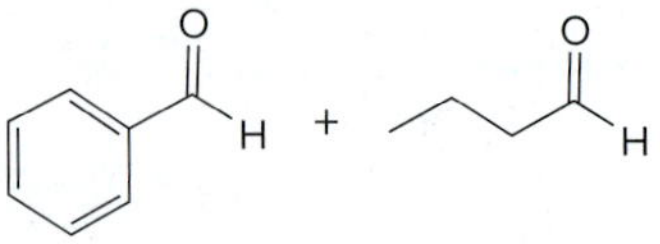 C₆H₅CHO + CH₃CH₂CHO

18.33 Draw the product formed in each directed aldol reaction.

a. 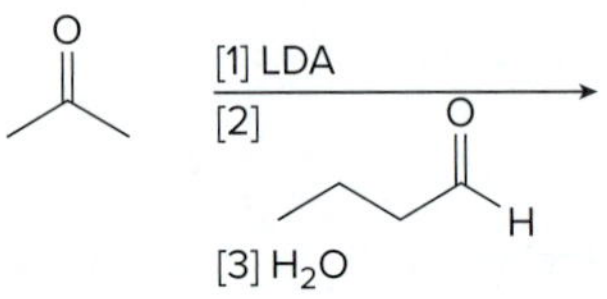 acetone, [1] LDA [2] 3-oxopropanal derivative [3] H₂O

b. 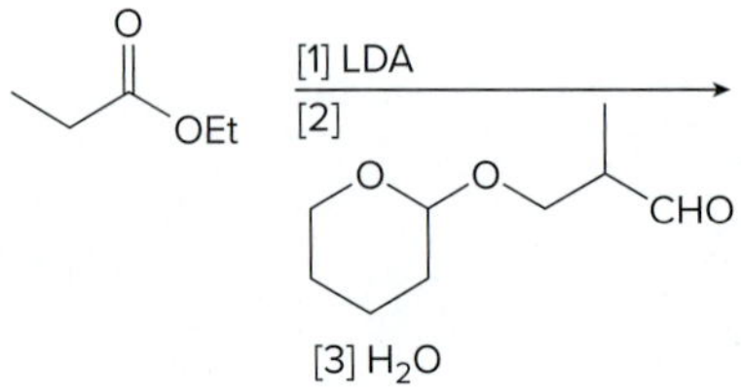 ethyl propanoate, [1] LDA [2] THP-protected hydroxyaldehyde [3] H₂O

18.34 Draw the product formed when each dicarbonyl compound undergoes an intramolecular aldol reaction followed by dehydration, when possible.

a. 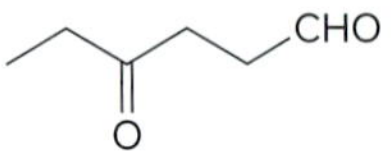 CH₃COCH₂CH₂CH₂CHO

b.  OHC(CH₂)₄CHO

c. 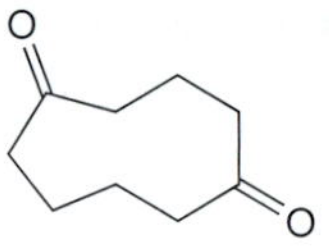 cyclodecane-1,6-dione

d. 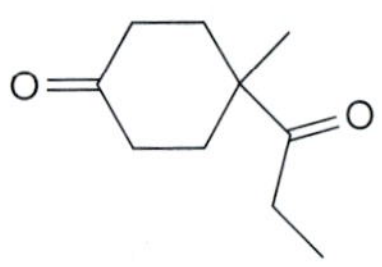 4-acetyl-4-methylcyclohexanone

18.35 What starting materials are needed to synthesize each compound using an aldol or similar reaction?

a. 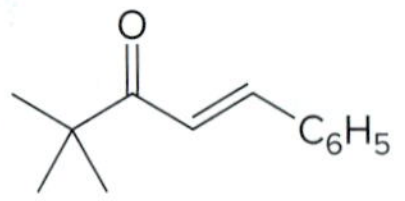 (CH₃)₃C–CO–CH=CH–C₆H₅

b. 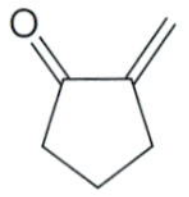 2-methylenecyclopentanone

c. 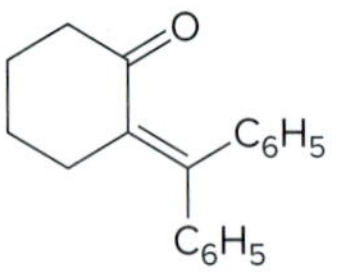 2-(diphenylmethylene)cyclohexanone with C₆H₅ groups

d. 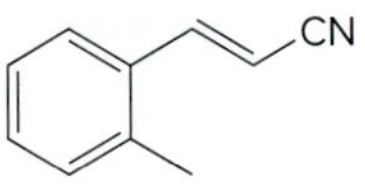 2-methylcinnamonitrile (ortho-methylphenyl CH=CH–CN)

18.36 What dicarbonyl compound is needed to prepare each compound by an intramolecular aldol reaction?

a. 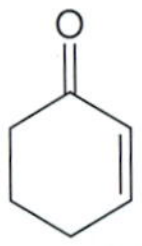 cyclohex-2-enone

b. 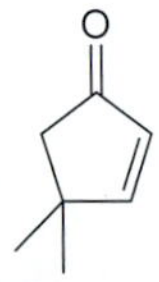 4,4-dimethylcyclopent-2-enone

c. 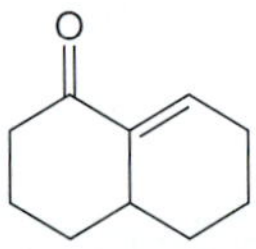 octahydronaphthalenone

d. 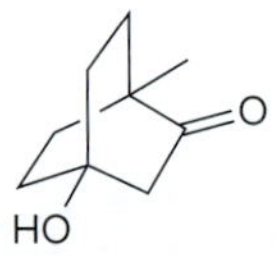 bicyclic hydroxy ketone

18.37 Identify the structures of **C** and **D** in the following reaction sequence.

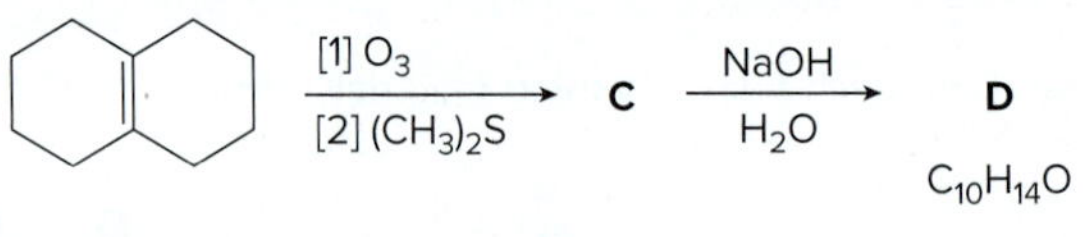

octahydronaphthalene → [1] O₃ [2] (CH₃)₂S → **C** → NaOH, H₂O → **D** (C₁₀H₁₄O)

The Claisen and Dieckmann Reactions

18.38 Draw the product formed from a Claisen reaction with the given starting materials using ⁻OEt, EtOH.

a. 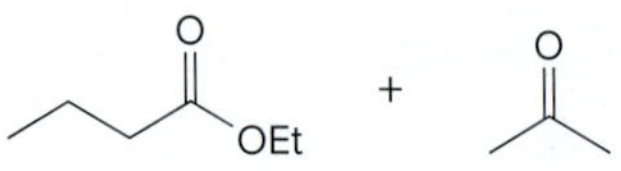 CH₃CH₂CH₂CO₂Et + CH₃CH₂CO₂Et-like (ester)

c. 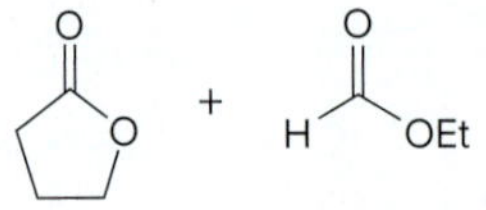 γ-butyrolactone + HCO₂Et

b. 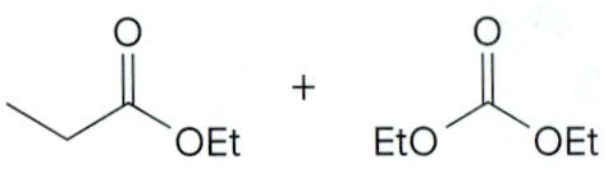 (CH₃)₂CHCO₂Et + EtO₂C–CH₂–CO₂Et (diethyl malonate)

d. 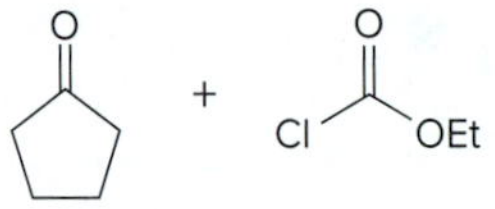 cyclopentanone + ClCO₂Et

18.39 What starting materials are needed to synthesize each compound by a crossed Claisen reaction?

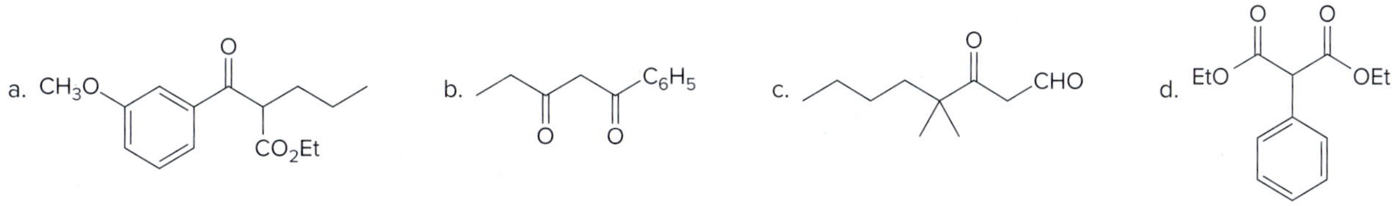

a. [structure: 3-methoxyphenyl ketone with CO₂Et substituent and propyl chain] b. [structure: 1,3-diketone with C₆H₅] c. [structure: ketone with CHO and gem-dimethyl] d. [structure: diethyl phenylmalonate]

18.40 Even though **B** contains three ester groups, a single Dieckmann product results when **B** is treated with NaOCH₃ in CH₃OH, followed by H₃O⁺. Draw the structure and explain why it is the only product formed.

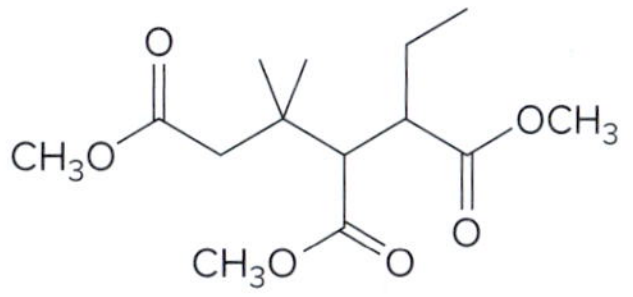

B

Biological Carbonyl Condensation Reactions

18.41 What product is formed when 2-oxoisovalerate reacts with acetyl CoA in a biological aldol reaction?

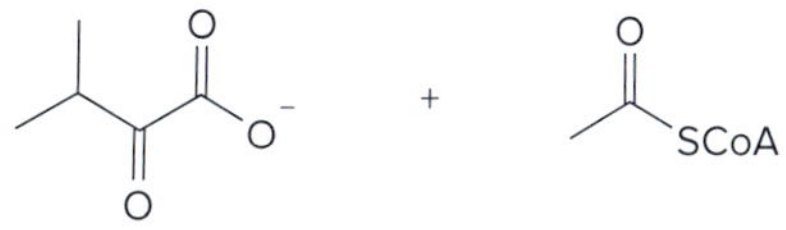

2-oxoisovalerate   acetyl CoA

18.42 What products would be formed by a retro-aldol reaction of **X**?

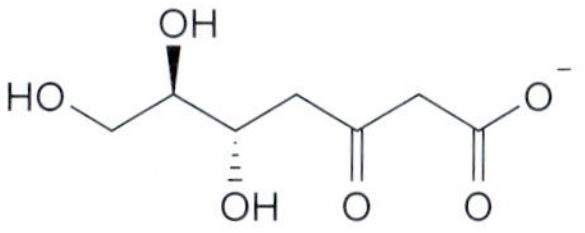

X

18.43 Propose a mechanism for the following reaction, one step in the metabolism of the amino acid leucine.

18.44 Draw a stepwise mechanism for the conversion of two molecules of acetyl CoA to acetoacetyl CoA.

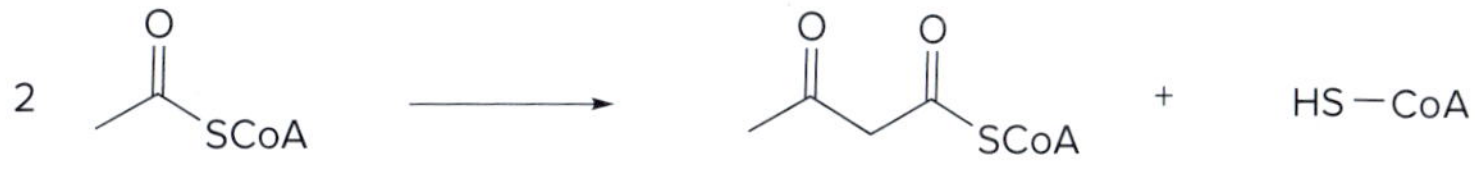

acetyl CoA   acetoacetyl CoA

18.45 One step in the metabolism of fatty acids involves the following retro-Claisen reaction. Draw a stepwise mechanism for this process.

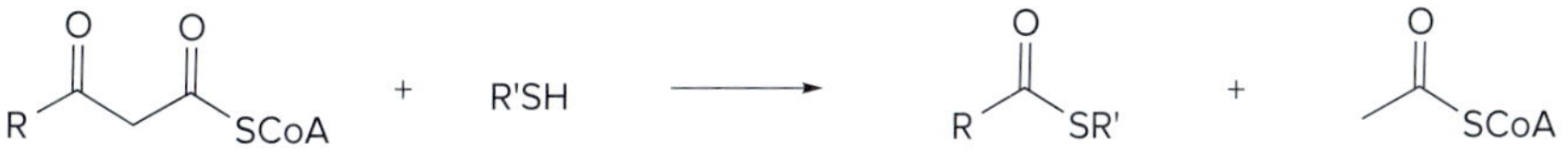

acetyl CoA

Michael Reaction

18.46 Draw the product formed from a Michael reaction with the given starting materials using ⁻OEt, EtOH.

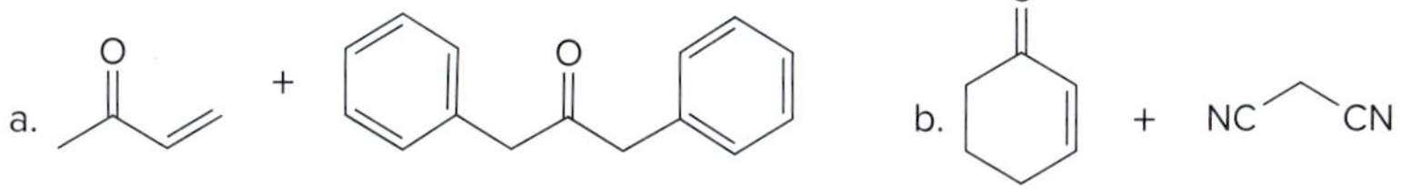

a. [methyl vinyl ketone + 1,3-diphenylacetone]   b. [cyclohex-2-enone + malononitrile NC-CH₂-CN]

844   Chapter 18   Carbonyl Condensation Reactions

18.47 What starting materials are needed to prepare each compound using a Michael reaction?

a., 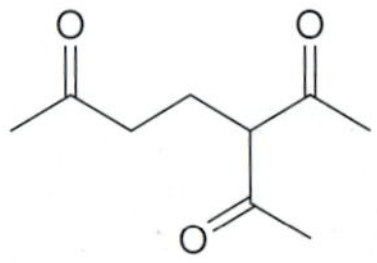 b., 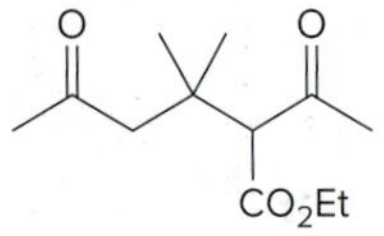 c. 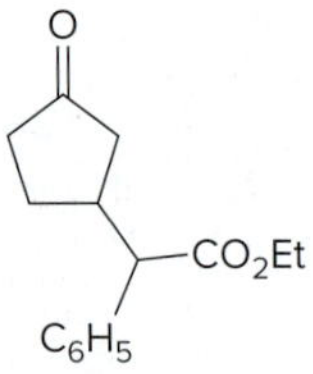 [structures]

18.48 β-Vetivone is isolated from vetiver, a perennial grass that yields a variety of compounds used in traditional eastern medicine, pest control, and fragrance. In one synthesis, ketone **A** is converted to β-vetivone by a two-step process: Michael reaction, followed by intramolecular aldol reaction. (a) What Michael acceptor is needed for the conjugate addition? (b) Draw a stepwise mechanism for the aldol reaction, which forms the six-membered ring.

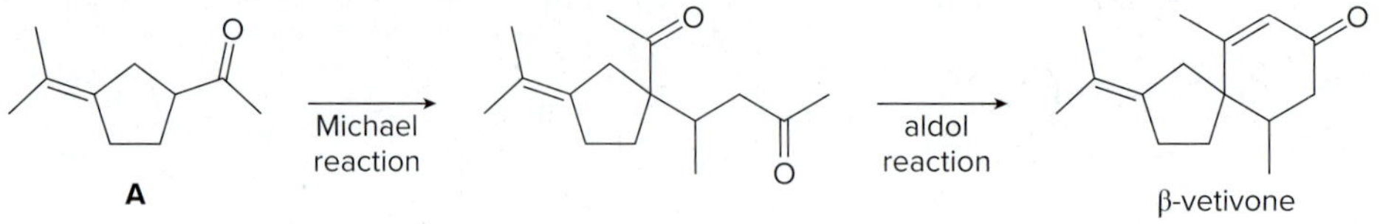

A →(Michael reaction)→ [intermediate] →(aldol reaction)→ β-vetivone

Robinson Annulation

18.49 Draw the product of each Robinson annulation from the given starting materials using ⁻OH in H_2O solution.

a., 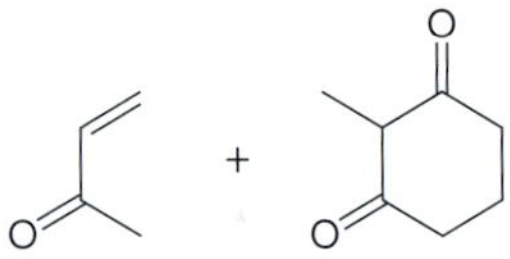 b., c., d. 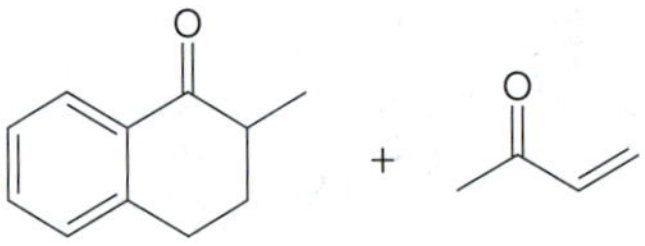 [structures]

18.50 What starting materials are needed to synthesize each compound using a Robinson annulation?

a., 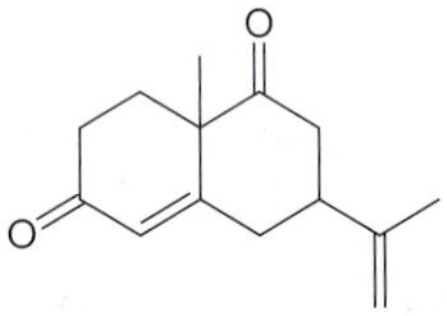 b., 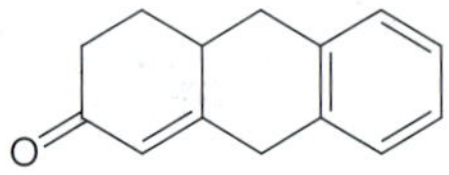 c. 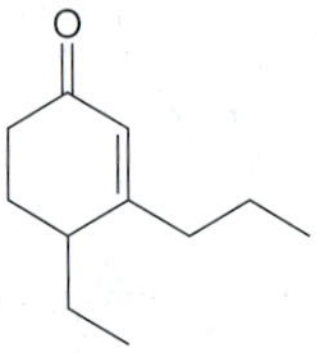 [structures]

Reactions

18.51 Draw the organic products formed in each reaction.

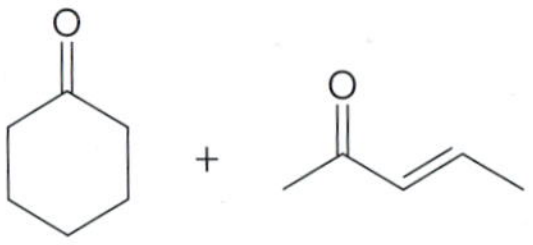

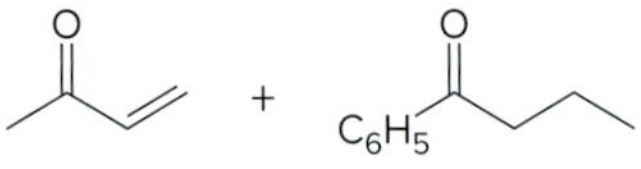

a. 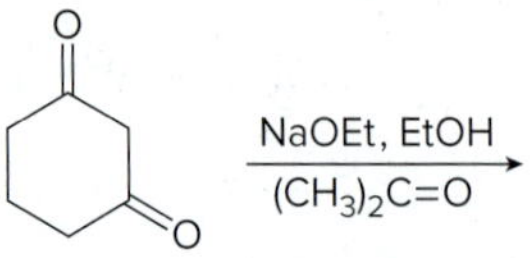 [cyclohexane-1,3-dione] NaOEt, EtOH / $(CH_3)_2C=O$

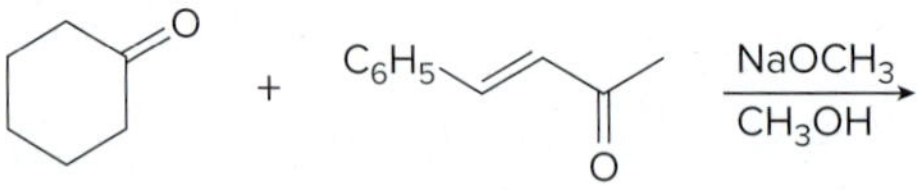

b. 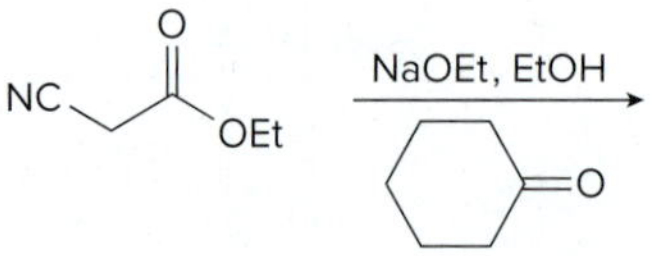 NC–CH₂–CO–OEt + [cyclohexanone] NaOEt, EtOH

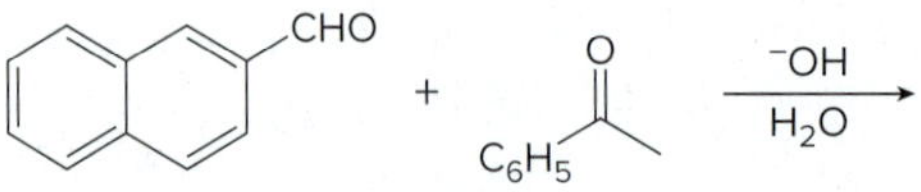

c. 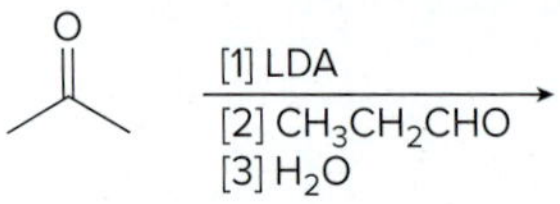 [acetone] [1] LDA [2] CH_3CH_2CHO [3] H_2O

d. [cyclohexanone] + C_6H_5–CH=CH–CO– NaOCH₃ / CH_3OH

e. [naphthalene-CHO] + [ketone-C_6H_5] ⁻OH / H_2O

f. 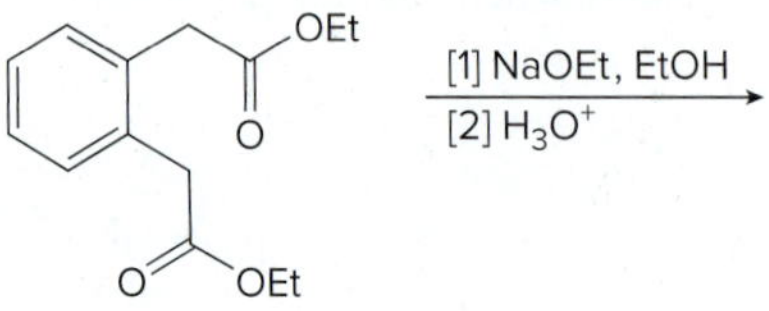 [aryl diester] [1] NaOEt, EtOH [2] H_3O^+

18.52 Identify compounds **A** and **B,** two synthetic intermediates in the 1979 synthesis of the plant growth hormone gibberellic acid by Corey and Smith. Gibberellic acid induces cell division and elongation, thus making plants tall and leaves large.

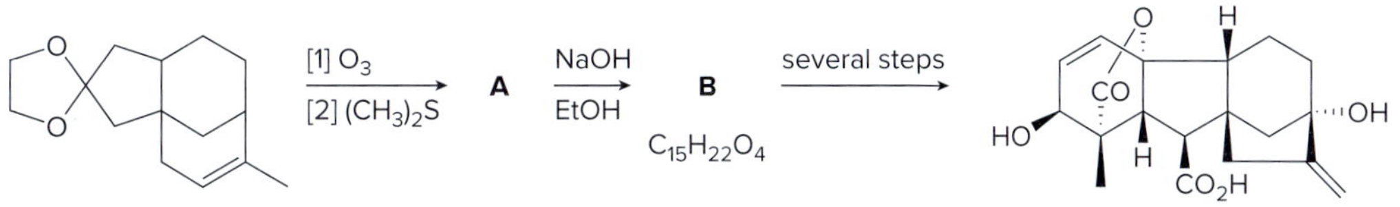

Mechanisms

18.53 In theory, the intramolecular aldol reaction of 6-oxoheptanal could yield the three compounds shown. It turns out, though, that 1-acetylcyclopentene is by far the major product. Why are the other two compounds formed in only minor amounts? Draw a stepwise mechanism to show how all three products are formed.

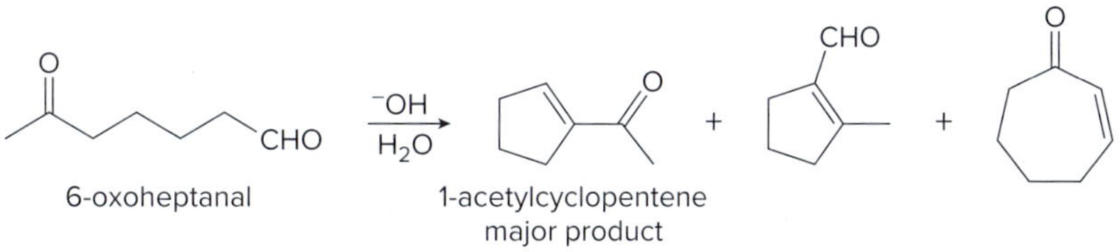

18.54 Draw a stepwise mechanism that illustrates how both products are formed in the following reaction.

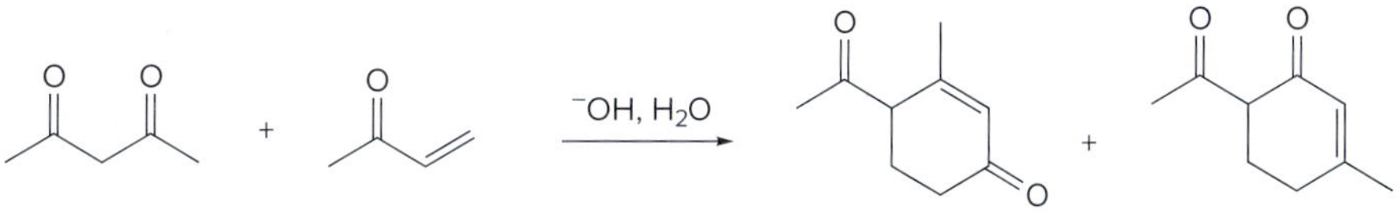

18.55 Biyouyanagin A is an anti-HIV agent isolated from the leaves of a plant of the genus *Hypericum* that is used in traditional Japanese medicine. The six-membered ring in biyouyanagin A was formed in the given reaction. Draw a stepwise mechanism for this process.

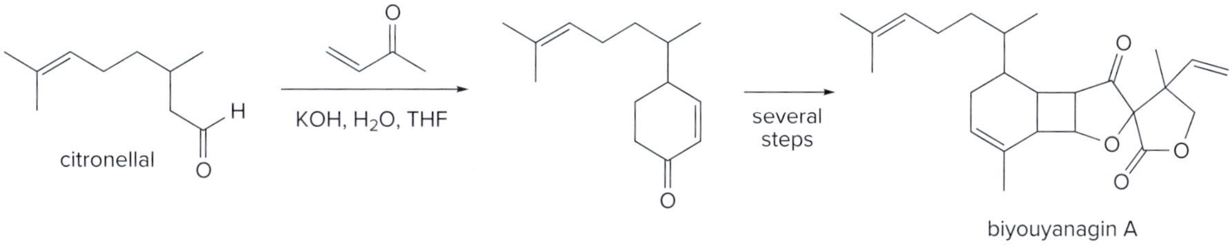

18.56 Jiadifenin is a natural product isolated from the fruit of the Chinese plant *Illicium jiadifengpi,* which has potential for use in treating neurodegenerative disease. The lactone in jiadifenin is formed in the following two-step reaction. Write a stepwise mechanism for the conversion of **M** to **N.**

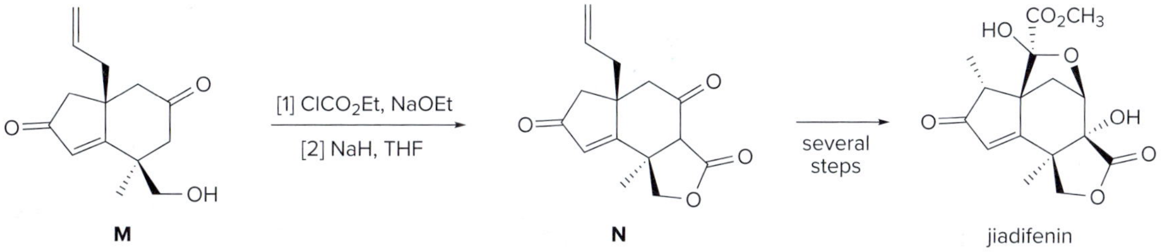

Chapter 18 Carbonyl Condensation Reactions

18.57 Reaction of **X** and phenylacetic acid forms an intermediate **Y**, which undergoes an intramolecular reaction to yield rofecoxib. Rofecoxib is a nonsteroidal anti-inflammatory agent once marketed under the trade name Vioxx, now withdrawn from the market because of increased risk of heart attacks from long-term use in some patients. Identify **Y** and draw a stepwise mechanism for its conversion to rofecoxib.

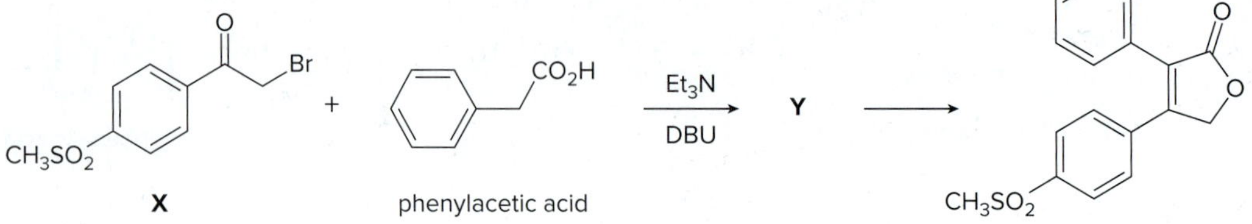

18.58 Coumarin, a naturally occurring compound isolated from lavender, sweet clover, and tonka bean, is made in the laboratory from o-hydroxybenzaldehyde by the reaction depicted below. Draw a stepwise mechanism for this reaction. Coumarin derivatives are useful synthetic anticoagulants.

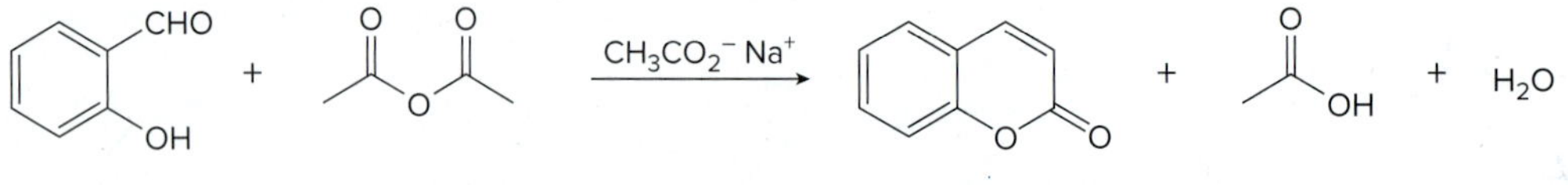

Synthesis

18.59 Devise a synthesis of each compound from cyclopentanone and organic alcohols having ≤ 3 C's. You may also use any required organic or inorganic reagents.

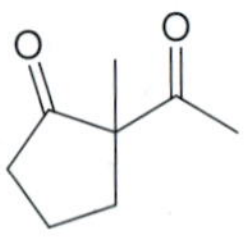

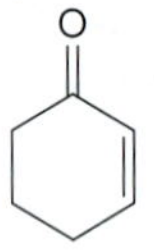

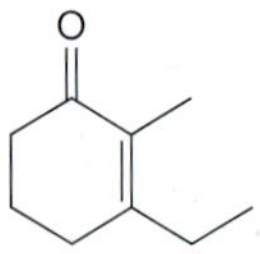

18.60 How would you convert cyclohexanone to each of the following compounds?

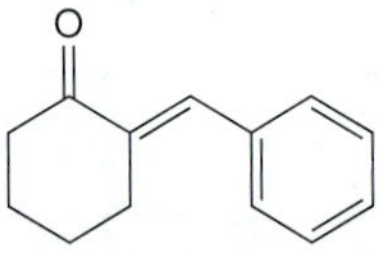

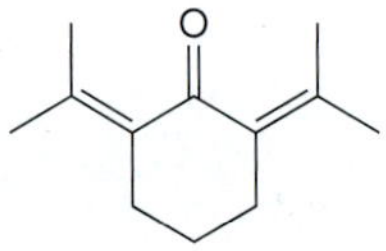

18.61 Devise a synthesis of each compound from $CH_3CH_2CH_2CO_2Et$, $C_6H_5CH_2OH$, CH_3CH_2OH, and CH_3OH. You may also use any required organic or inorganic reagents.

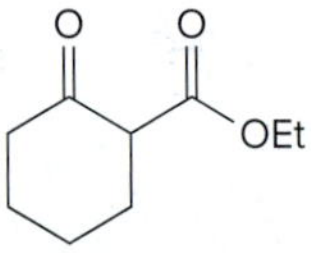

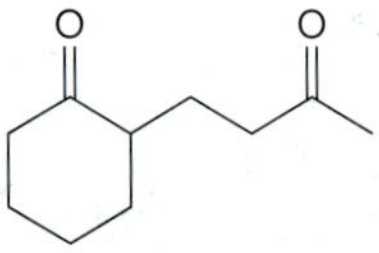

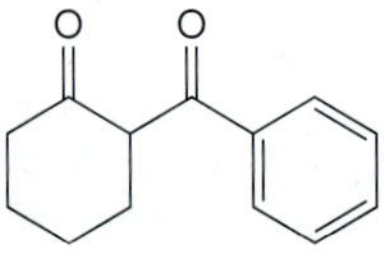

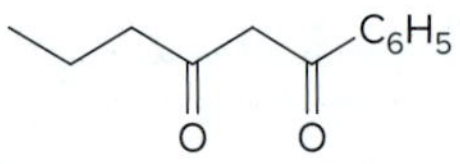

18.62 Devise a synthesis of 2-methylcyclopentanone from cyclohexene. You may also use any required reagents.

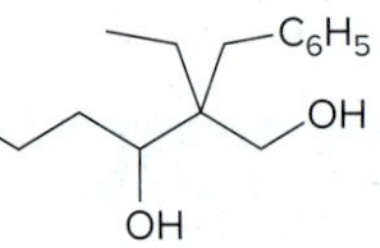

18.63 Devise a synthesis of each compound using acetone [$(CH_3)_2C=O$] as the only source of carbon atoms. You may use any needed organic or inorganic reagents.

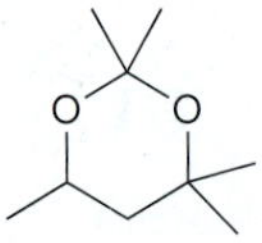

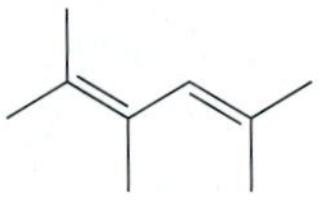

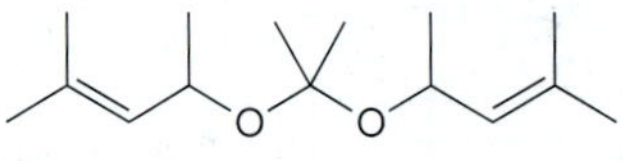

18.64 Octinoxate is an unsaturated ester used as an active ingredient in sunscreens. (a) What carbonyl compounds are needed to synthesize this compound using a condensation reaction? (b) Devise a synthesis of octinoxate from the given organic starting materials and any other needed reagents.

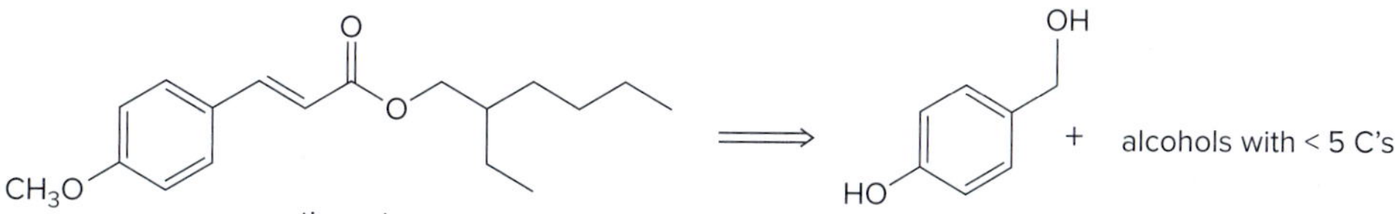

octinoxate

General Problem

18.65 Answer the following questions about 2-acetylcyclopentanone.

a. What starting materials are needed to form 2-acetylcyclopentanone by a Claisen reaction that forms bond (a)?

b. What starting materials are needed to form 2-acetylcyclopentanone by a Claisen reaction that forms bond (b)?

c. What product is formed when 2-acetylcyclopentanone is treated with $NaOCH_2CH_3$, followed by CH_3I?

d. Draw the Robinson annulation product(s) formed by reaction of 2-acetylcyclopentanone with methyl vinyl ketone ($CH_2{=}CHCOCH_3$).

e. Draw the structure of the most stable enol tautomer(s).

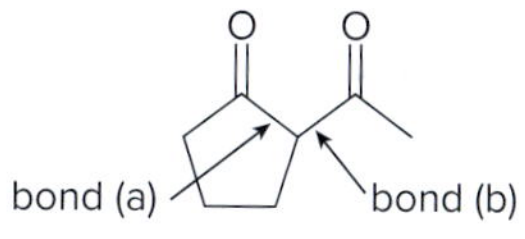

2-acetylcyclopentanone

Challenge Problems

18.66 Draw a stepwise mechanism for the following reaction, which was used in the synthesis of ezetimibe, a drug used to treat patients with high cholesterol.

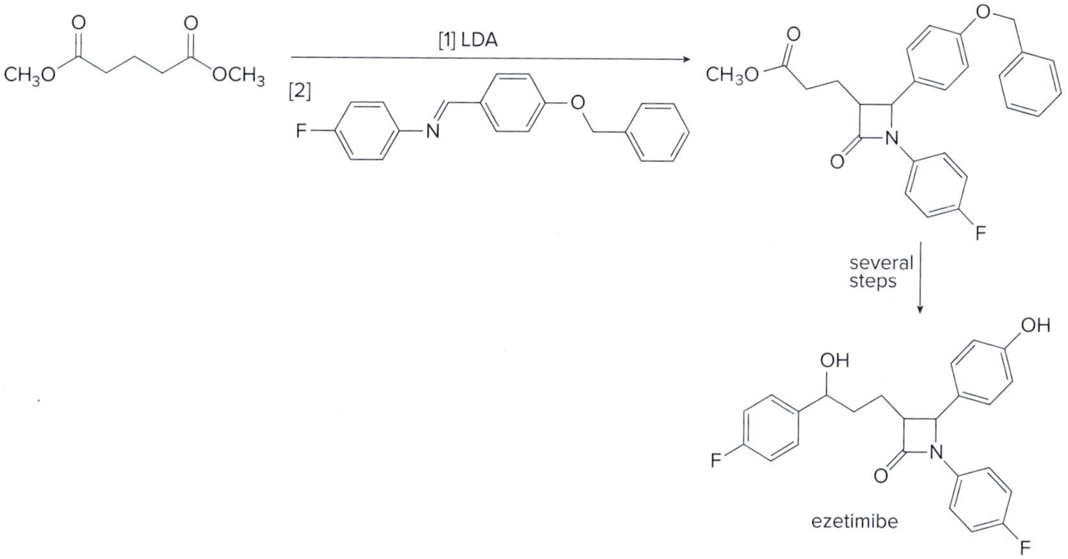

18.67 A key step in a reported synthesis of morphine (Section 2.1), the addictive opiate used to treat severe pain, involves the conversion of **A** to **B.** Draw a stepwise mechanism for this process, which involves both an intramolecular alkylation and an intramolecular aldol reaction.

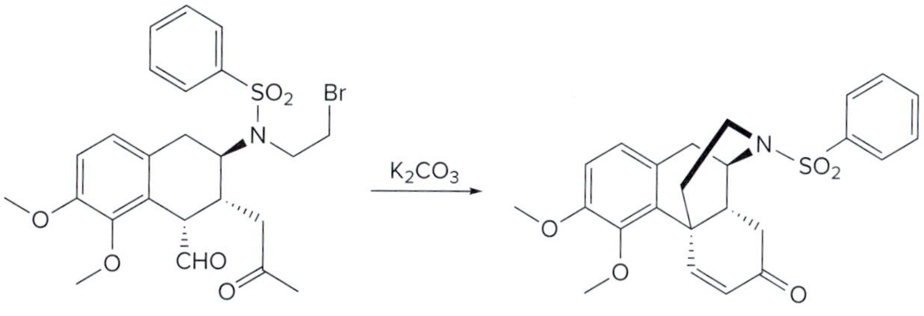

A **B**

18.68 Isophorone is formed from three molecules of acetone [(CH₃)₂C=O] in the presence of base. Draw a mechanism for this process.

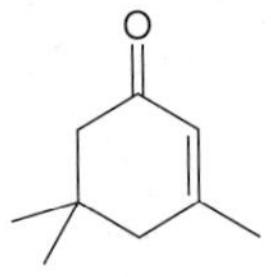

isophorone

18.69 Devise a stepwise mechanism for the following reaction. (Hint: The mechanism begins with the conjugate addition of ⁻OH.)

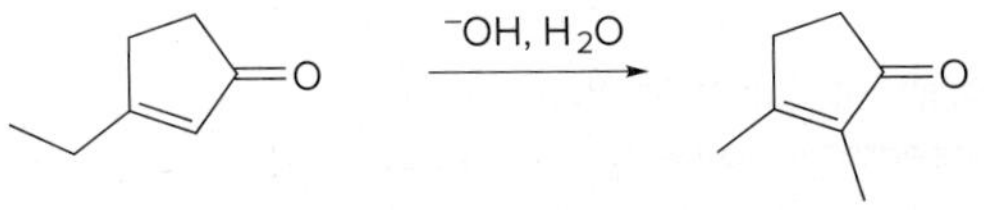

18.70 Draw a stepwise mechanism for the following reaction. (Hint: Two Michael reactions are needed.)

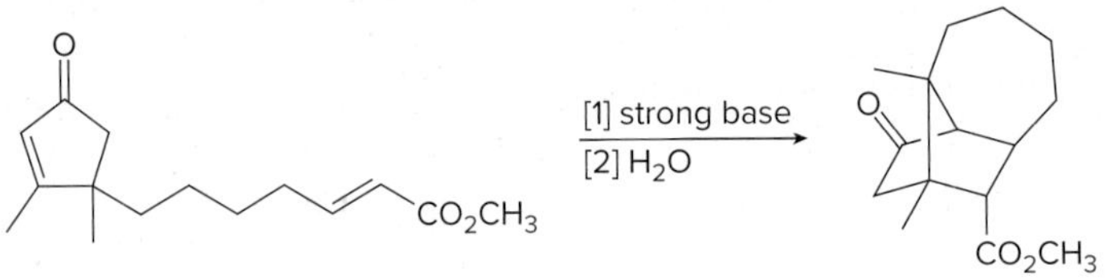

[1] strong base
[2] H₂O

18.71 Draw a stepwise mechanism for the following reaction, one step in the synthesis of the cholesterol-lowering drug pitavastatin, marketed in Japan as a calcium salt under the name Livalo.

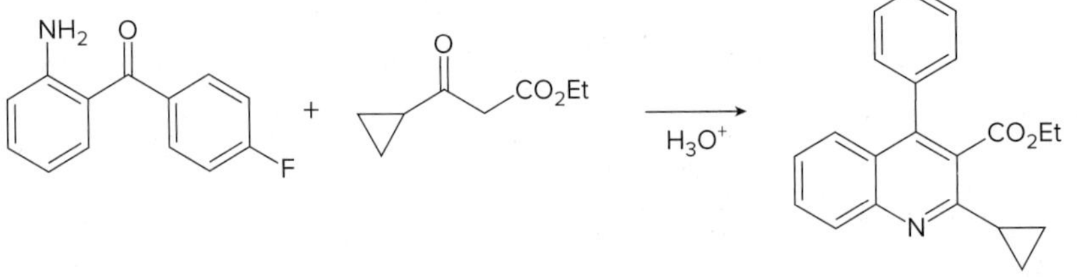

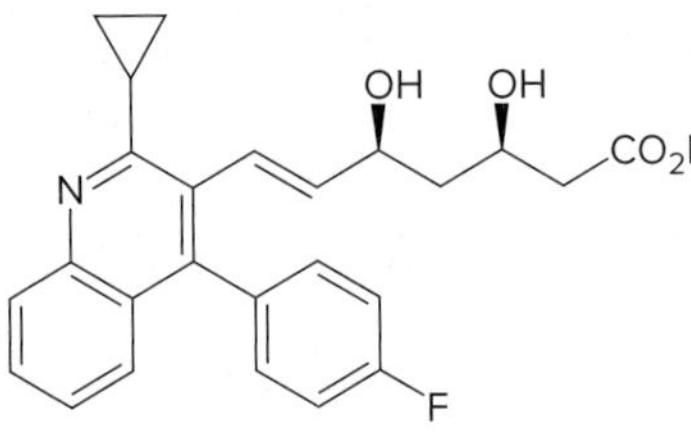

pitavastatin

18.72 Devise a stepwise mechanism for the following reaction, a key step in the synthesis of the antibiotic abyssomicin C. Abyssomicin C was isolated from sediment collected from almost 1000 ft below the surface in the Sea of Japan. (Hint: The mechanism begins with a Dieckmann reaction.)

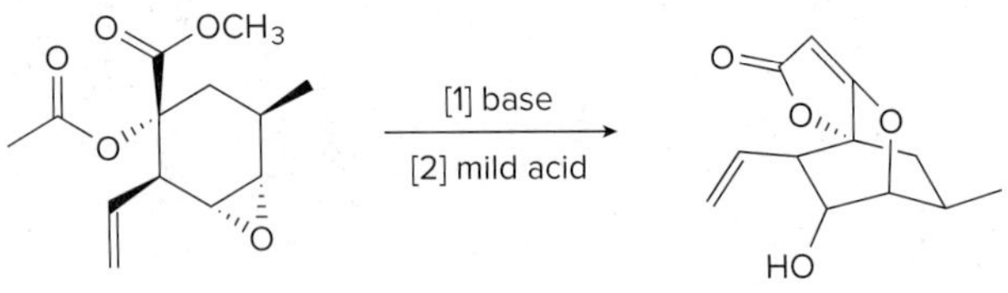

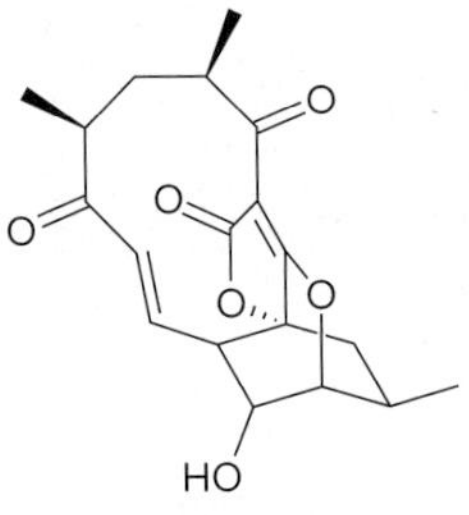

[1] base
[2] mild acid

abyssomicin C

Benzene and Aromatic Compounds

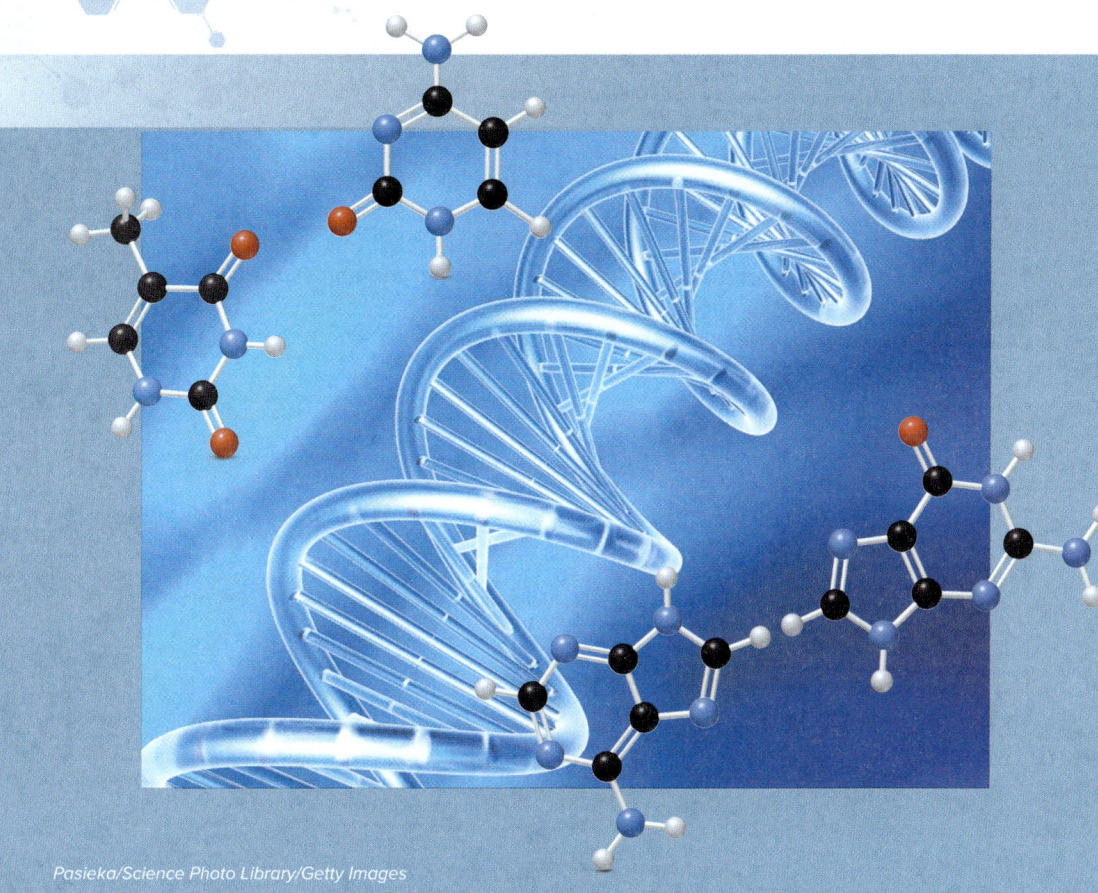

Pasieka/Science Photo Library/Getty Images

- **19.1** Background
- **19.2** The structure of benzene
- **19.3** Nomenclature of benzene derivatives
- **19.4** Spectroscopic properties
- **19.5** Interesting aromatic compounds
- **19.6** Benzene's unusual stability
- **19.7** The criteria for aromaticity—Hückel's rule
- **19.8** Examples of aromatic compounds
- **19.9** Aromatic heterocycles
- **19.10** What is the basis of Hückel's rule?
- **19.11** The inscribed polygon method for predicting aromaticity
- **19.12** Aromatase inhibitors for estrogen-dependent cancer treatment

Each cell in our body contains our genome, the complete set of DNA made up of about three billion aromatic base pairs attached to helical sugar–phosphate backbones. The planar aromatic bases—**cytosine, thymine, adenine,** and **guanine**—that are stacked together hold encrypted, hereditary, genetic instructions directing our development and cellular processes, and the sequence of these bases carries the key to who we are. In Chapter 19, we learn about the unique stability of DNA bases and other aromatic compounds.

Why Study . . . Aromatic Compounds?

The hydrocarbons we have examined thus far—including the alkanes, alkenes, and alkynes, as well as the conjugated dienes and polyenes of Chapter 12—have been aliphatic hydrocarbons. In Chapter 19, we continue our study of conjugated systems with **aromatic hydrocarbons.**

We begin with **benzene** and then examine other cyclic, planar, and conjugated ring systems to learn the modern definition of what it means to be aromatic. Then, in Chapter 20, we will learn about the reactions of aromatic compounds, highly unsaturated hydrocarbons that do not undergo addition reactions like other unsaturated compounds. An explanation of this behavior relies on an understanding of the structure of aromatic compounds presented in Chapter 19. Many naturally occurring compounds contain aromatic rings, and many useful drugs are aromatic.

19.1 Background

> For 6 C's, the maximum number of H's = $2n + 2 = 2(6) + 2 = 14$. Because benzene contains only 6 H's, it has $14 - 6 = 8$ H's fewer than the maximum number. This corresponds to 8 H's/2 H's for each degree of unsaturation = **four degrees of unsaturation in benzene.**

Benzene (C_6H_6) is the simplest aromatic hydrocarbon (or arene). Since its isolation by Michael Faraday from the oily residue remaining in the illuminating gas lines in London in 1825, it has been recognized as an unusual compound. Based on the calculation introduced in Section 10.2, **benzene has four degrees of unsaturation, making it a highly unsaturated hydrocarbon.** But, whereas unsaturated hydrocarbons such as alkenes, alkynes, and dienes readily undergo addition reactions, *benzene does not.* For example, bromine adds to ethylene to form a dibromide, but benzene is inert under similar conditions.

ethylene $\xrightarrow{Br_2}$ Br–CH$_2$–CH$_2$–Br   addition product

C_6H_6 (benzene) $\xrightarrow{Br_2}$ **No reaction**

Benzene *does* react with bromine, but only in the presence of $FeBr_3$ (a Lewis acid), and the reaction is a **substitution,** *not* an addition.

$C_6H_6 \xrightarrow[FeBr_3]{Br_2} C_6H_5Br$ substitution — Br replaces H.

Thus, any structure proposed for benzene must account for its high degree of unsaturation and its lack of reactivity toward electrophilic addition.

In the last half of the nineteenth century August Kekulé proposed structures that were close to the modern description of benzene. In the Kekulé model, benzene was thought to be a rapidly equilibrating mixture of two compounds, each containing a six-membered ring with three alternating π bonds. These structures are now called **Kekulé structures.** In the Kekulé description, the bond between any two carbon atoms is sometimes a single bond and sometimes a double bond.

Kekulé description: An equilibrium

Although benzene is still drawn as a six-membered ring with three alternating π bonds, in reality **there is no equilibrium between two different kinds of benzene molecules.** Instead, current descriptions of benzene are based on resonance and electron delocalization due to orbital overlap, as detailed in Section 19.2.

In the nineteenth century, many other compounds having properties similar to those of benzene were isolated from natural sources. Because these compounds possessed strong and characteristic odors, they were called *aromatic* compounds. It is their chemical properties, though, not their odor that make these compounds special.

- Aromatic compounds resemble benzene—they are unsaturated compounds that do not undergo the addition reactions characteristic of alkenes.

19.2 The Structure of Benzene

Any structure for benzene must account for the following:

- Benzene contains a six-membered ring and three additional degrees of unsaturation.
- Benzene is planar.
- All C—C bond lengths are equal.

Although the Kekulé structures satisfy the first two criteria, they break down with the third, because having three alternating π bonds would mean that benzene should have three short double bonds alternating with three longer single bonds.

This structure implies that the C—C bonds should have **two different lengths**.

- three longer single bonds in red
- three shorter double bonds in black

Resonance

Benzene is conjugated, so we must use resonance and orbitals to describe its structure. The resonance description of benzene consists of two equivalent Lewis structures, each with three double bonds that alternate with three single bonds.

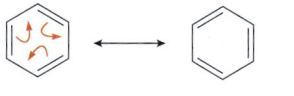

hybrid

The electrons in the π bonds are **delocalized** around the ring.

Some texts draw benzene as a hexagon with an inner circle:

The circle represents the **six π electrons,** distributed over the six atoms of the ring.

The resonance description of benzene matches the Kekulé description with one important exception: **The two Kekulé representations are *not* in equilibrium with each other.** Instead, the true structure of benzene is a resonance **hybrid** of the two Lewis structures, with the dashed lines of the hybrid indicating the position of the π bonds.

We will use one of the two Lewis structures and not the hybrid in drawing benzene, because it is easier to keep track of the electron pairs in the π bonds (the π electrons).

- Because each π bond has two electrons, benzene has six π electrons.

The resonance hybrid of benzene explains why all C—C bond lengths are the same. Each C—C bond is single in one resonance structure and double in the other, so the actual bond length (139 pm) is *intermediate* between a carbon–carbon single bond (153 pm) and a carbon–carbon double bond (134 pm).

The C—C bonds in benzene are equal and intermediate in length.

Hybridization and Orbitals

Each carbon atom in a benzene ring is surrounded by three atoms and no lone pairs of electrons, making it *sp*² **hybridized and trigonal planar with all bond angles 120°.** Each

852   Chapter 19   Benzene and Aromatic Compounds

carbon also has a *p* orbital with one electron that extends above and below the plane of the molecule.

The six adjacent *p* orbitals overlap, delocalizing the six electrons over the six atoms of the ring and making benzene a conjugated molecule. Because each *p* orbital has two lobes, one above and one below the plane of the benzene ring, the overlap of the *p* orbitals creates two "doughnuts" of electron density, as shown in Figure 19.1a. The electrostatic potential plot in Figure 19.1b also shows that the electron-rich region is concentrated above and below the plane of the molecule, where the six π electrons are located.

- Benzene's six π electrons make it electron rich, so it reacts with electrophiles.

Figure 19.1
Two views of the electron density in a benzene ring

a. View of the *p* orbital overlap

b. Electrostatic potential plot

- Overlap of six adjacent *p* orbitals creates two rings of electron density, one above and one below the plane of the benzene ring.

- The electron-rich region (in red) is concentrated above and below the ring carbons, where the six π electrons are located. (The electron-rich region below the plane is hidden from view.)

Problem 19.1   Draw all possible resonance structures for the antihistamine diphenhydramine, the active ingredient in Benadryl.

diphenhydramine

Problem 19.2   What orbitals are used to form the labeled bonds in the following molecule? Of the labeled C–C bonds, which is the shortest?

19.3 Nomenclature of Benzene Derivatives

Many organic molecules contain a benzene ring with one or more substituents, so we must learn how to name them. Many common names are recognized by the IUPAC system, however, so this complicates the nomenclature of benzene derivatives somewhat.

19.3A Monosubstituted Benzenes

To name a benzene ring with one substituent, **name the substituent and add the word *benzene*.** Carbon substituents are named as alkyl groups.

ethylbenzene   ***tert*-butyl**benzene   **chloro**benzene

Many monosubstituted benzenes, such as those with methyl (CH_3-), hydroxy ($-OH$), and amino ($-NH_2$) groups, have common names that you must learn, too.

toluene
(methylbenzene)

phenol
(hydroxybenzene)

aniline
(aminobenzene)

19.3B Disubstituted Benzenes

There are three different ways that two groups can be attached to a benzene ring, so a prefix—**ortho, meta,** or **para**—can be used to designate the relative position of the two substituents. Ortho, meta, and para are also abbreviated as *o, m,* and *p,* respectively.

1,2-Disubstituted benzene
ortho isomer

***o*-dibromobenzene**
or
1,2-dibromobenzene

1,3-Disubstituted benzene
meta isomer

***m*-dibromobenzene**
or
1,3-dibromobenzene

1,4-Disubstituted benzene
para isomer

***p*-dibromobenzene**
or
1,4-dibromobenzene

If the two groups on the benzene ring are different, **alphabetize the names of the substituents** preceding the word *benzene*. If one of the substituents is part of a **common root,** name the **molecule as a derivative of that monosubstituted benzene.**

Alphabetize two different substituent names: **Use a common root name:**

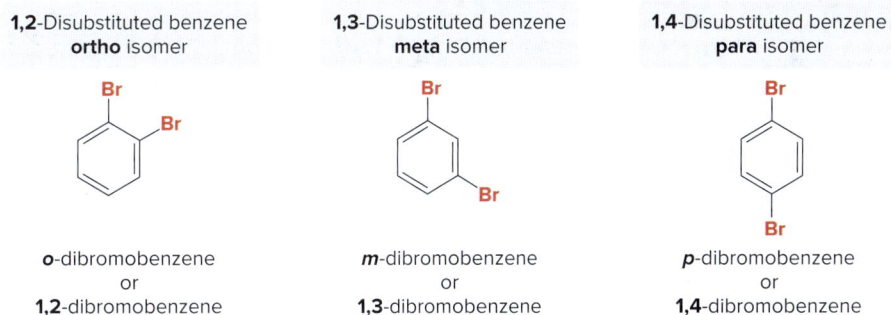

o-bromochloro-
benzene

m-fluoronitro-
benzene

p-bromo**toluene**

o-nitro**phenol**

19.3C Polysubstituted Benzenes

For three or more substituents on a benzene ring:

[1] Number to give the lowest possible set of numbers around the ring.
[2] Alphabetize the substituent names.
[3] When substituents are part of common roots, name the molecule as a derivative of that monosubstituted benzene. The substituent that comprises the common root is located at C1.

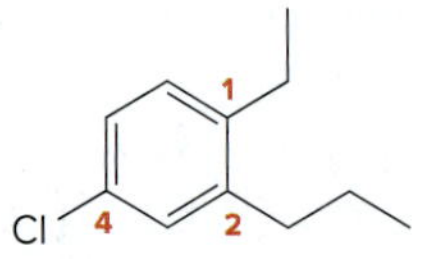

- Assign the lowest possible set of numbers.
- Alphabetize the names of all the substituents.

4-chloro-1-ethyl-2-propylbenzene

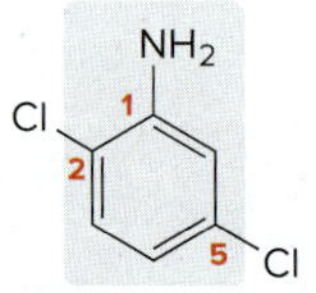

- Name the molecule as a derivative of the common root **aniline**.
- Designate the position of the NH$_2$ group as "1," and then assign the lowest possible set of numbers to the other substituents.

2,5-dichloroaniline

19.3D Naming Aromatic Rings as Substituents

A benzene substituent (C$_6$H$_5$–) is called a **phenyl group**, and it can be abbreviated in a structure as **Ph–**.

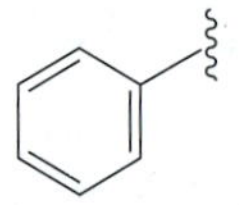

phenyl group
C$_6$H$_5$–   abbreviated as   Ph–

- A phenyl group (C$_6$H$_5$–) is formed by removing one hydrogen from benzene (C$_6$H$_6$).

Benzene, therefore, can be represented as PhH, and phenol (C$_6$H$_5$OH) would be PhOH.

The **benzyl** group contains a benzene ring bonded to a CH$_2$ group. Thus, a benzyl group and a phenyl group differ by the presence of a CH$_2$ group.

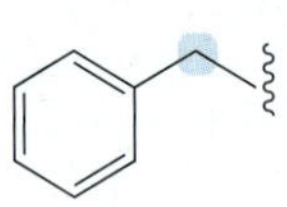

benzyl group
C$_6$H$_5$CH$_2$–

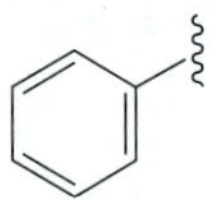

phenyl group
C$_6$H$_5$–

Finally, substituents derived from benzene, as well as all other substituted aromatic rings, are collectively called **aryl groups**, abbreviated as Ar–.

Problem 19.3 Give the IUPAC name for each compound.

a. 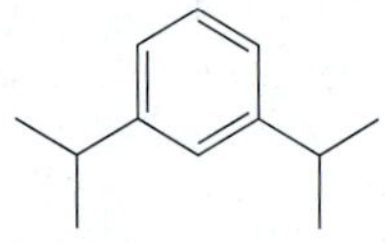   b. 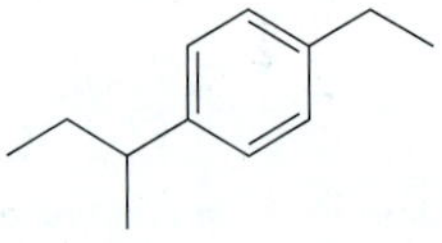   c. 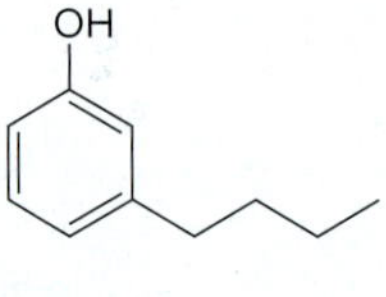   d.

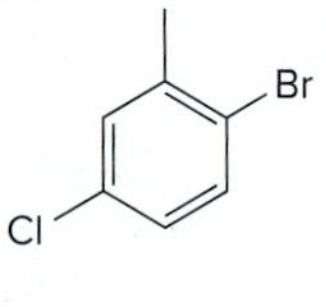

Problem 19.4 Draw the structure corresponding to each name:
a. isobutylbenzene
b. *o*-dichlorobenzene
c. *cis*-1,2-diphenylcyclohexane
d. *m*-bromoaniline
e. 4-chloro-1,2-diethylbenzene
f. 3-*tert*-butyl-2-ethyltoluene

Problem 19.5 What is the structure of propofol, which has the IUPAC name 2,6-diisopropylphenol? Propofol is an intravenous medication used to induce and maintain anesthesia.

19.4 Spectroscopic Properties

The IR spectroscopy of aromatic compounds was discussed in Section B.4A; the NMR spectroscopy of aromatics was presented in Sections C.4 and C.9C. The important IR and NMR absorptions of aromatic compounds are summarized in Table 19.1.

Table 19.1 Characteristic Spectroscopic Absorptions of Benzene Derivatives

Type of spectroscopy	Type of C, H	Absorption
IR absorptions	C_{sp^2}–H C=C (arene)	3150–3000 cm^{-1} 1600, 1500 cm^{-1}
^{1}H NMR absorptions	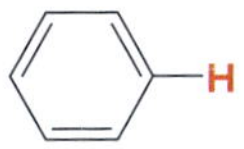 (aryl H)	6.5–8 ppm (highly deshielded protons)
	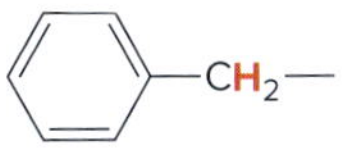 (benzylic H)	1.5–2.5 ppm (somewhat deshielded C_{sp^3}–H)
^{13}C NMR absorption	C_{sp^2} of arenes	120–150 ppm

^{13}C NMR spectroscopy is used to determine the substitution patterns in disubstituted benzenes, because each line in a spectrum corresponds to a different kind of carbon atom. For example, *o*-, *m*-, and *p*-dibromobenzene each exhibit a different number of lines in its ^{13}C NMR spectrum, as shown in Figure 19.2.

Figure 19.2 ^{13}C NMR absorptions of the three isomeric dibromobenzenes

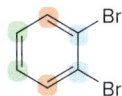

o-dibromobenzene

three types of C's
three ^{13}C NMR signals

m-dibromobenzene

four types of C's
four ^{13}C NMR signals

p-dibromobenzene

two types of C's
two ^{13}C NMR signals

- The number of signals (lines) in the ^{13}C NMR spectrum of a disubstituted benzene with two identical groups indicates whether they are ortho, meta, or para to each other.

Problem 19.6 How many ^{13}C NMR signals does each compound exhibit?

a.

b.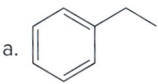

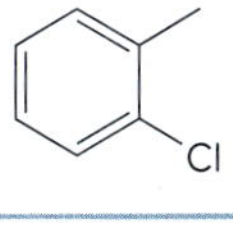

c.

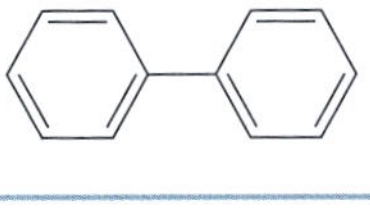

19.5 Interesting Aromatic Compounds

19.5A Polycyclic Aromatic Hydrocarbons

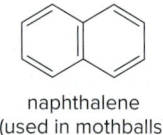

naphthalene
(used in mothballs)

Several compounds containing two or more benzene rings that share carbon–carbon bonds, called **polycyclic aromatic hydrocarbons (PAHs)**, are known. Naphthalene, the simplest PAH, is present in mothballs.

Benzo[*a*]pyrene, a more complicated PAH shown in Figure 19.3, is formed by the incomplete combustion of organic materials. It is found in cigarette smoke, automobile exhaust, and the fumes from charcoal grills. When ingested or inhaled, benzo[*a*]pyrene and other similar PAHs are oxidized to more water-soluble carcinogenic products, which can react with biological nucleophiles that often disrupt normal cell function, leading to cancer or cell death.

Figure 19.3 Benzo[*a*]pyrene, a common PAH

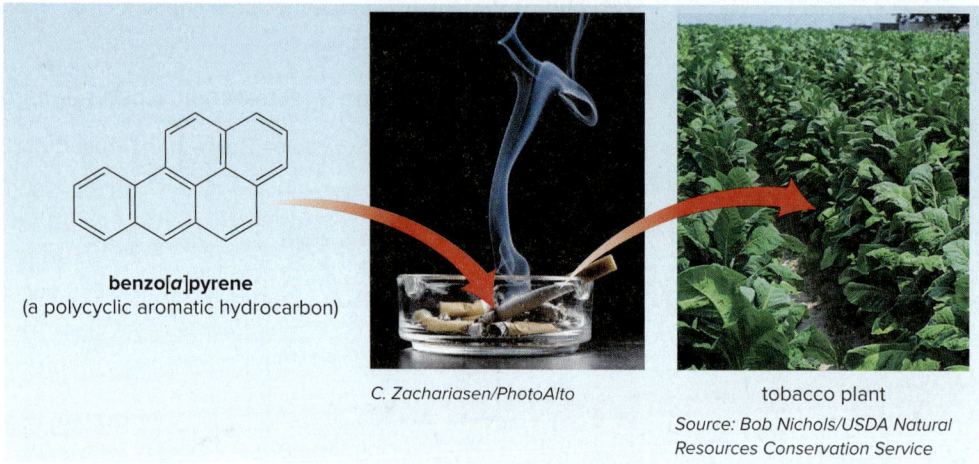

- Benzo[*a*]pyrene, produced by the incomplete oxidation of organic compounds in tobacco, is found in cigarette smoke.

Helicene and **twistoflex** are two synthetic PAHs whose unusual shapes are shown in Figure 19.4. Both helicene and twistoflex are chiral molecules—that is, they are not superimposable on their mirror images, even though neither of them contains a stereogenic center. It's their shape that makes them chiral, not the presence of carbon atoms bonded to four different groups. Each ring system is twisted into a shape that lacks a mirror plane, and each structure is rigid, thus creating the chirality.

Figure 19.4 Helicene and twistoflex—Two synthetic polycyclic aromatic hydrocarbons

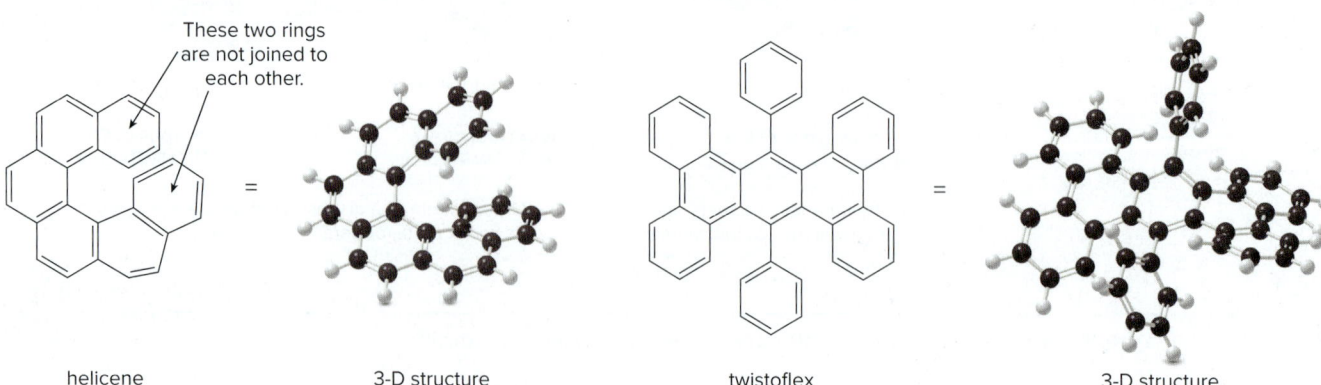

- Helicene consists of six benzene rings. Because the rings at both ends are not bonded to each other, all of the rings twist slightly, creating a rigid helical shape that prevents the hydrogen atoms on both ends from crashing into each other. Similarly, to reduce steric hindrance between the hydrogen atoms on nearby benzene rings, twistoflex is also nonplanar.

19.5B Sunscreens

Commercial sunscreens are given an **SPF** rating (sun protection factor), according to the amount of sunscreen present. The higher the number, the greater the protection.
Jill Braaten/McGraw-Hill Education

Ultraviolet (UV) radiation from the sun is high enough in energy to cleave bonds, forming reactive intermediates that can prematurely age skin and cause skin cancers. Much of the ultraviolet light that filters through the atmosphere is absorbed by **melanin,** the highly conjugated colored pigment in the skin that serves as the body's natural protection against the harmful effects of UV radiation.

Prolonged exposure to the sun can allow more ultraviolet radiation to reach your skin than melanin can absorb. A commercial sunscreen can offer added protection, however, because it contains **conjugated compounds that absorb UV light,** thus shielding your skin (for a time) from the harmful effects of UV radiation. Two sunscreens that have been used for this purpose are *para*-aminobenzoic acid (PABA) and padimate O. Many sunscreens contain more than one component to filter out much of the harmful ultraviolet radiation.

para-aminobenzoic acid (PABA)   padimate O

Problem 19.7 Which of the following compounds might be an ingredient in a commercial sunscreen? Explain why or why not.

a.   b.

19.6 Benzene's Unusual Stability

Considering benzene as the hybrid of two resonance structures adequately explains its equal C—C bond lengths, but does not account for its unusual stability and lack of reactivity toward addition.

Heats of hydrogenation, which were used in Section 12.9 to show that conjugated dienes are more stable than isolated dienes, can also be used to estimate the stability of benzene. Equations [1]–[3] compare the heats of hydrogenation of cyclohexene, cyclohexa-1,3-diene, and benzene, all of which give cyclohexane when treated with excess hydrogen in the presence of a metal catalyst.

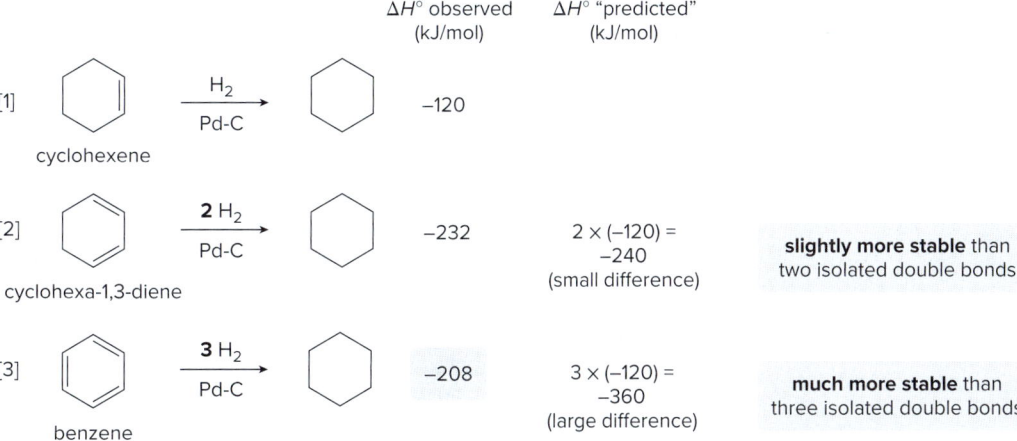

858 Chapter 19 Benzene and Aromatic Compounds

> The relative stability of conjugated dienes versus isolated dienes was first discussed in Section 12.9.

The addition of one mole of H_2 to cyclohexene releases -120 kJ/mol of energy (Equation [1]). If each double bond is worth -120 kJ/mol of energy, then the addition of two moles of H_2 to cyclohexa-1,3-diene (Equation [2]) should release $2 \times (-120 \text{ kJ/mol}) = -240$ kJ/mol of energy. The observed value, however, is -232 kJ/mol. This is *slightly smaller* than expected because cyclohexa-1,3-diene is a conjugated diene, and **conjugated dienes are more stable than two isolated carbon–carbon double bonds.**

The hydrogenations of cyclohexene and cyclohexa-1,3-diene occur readily at room temperature, but benzene can be hydrogenated only under forcing conditions, and even then the reaction is extremely slow. If each double bond is worth -120 kJ/mol of energy, then the addition of three moles of H_2 to benzene should release $3 \times (-120 \text{ kJ/mol}) = -360$ kJ/mol of energy. In fact, the observed heat of hydrogenation is only -208 kJ/mol, which is 152 kJ/mol less than predicted and even *lower* than the observed value for cyclohexa-1,3-diene.

Figure 19.5 compares the hypothetical and observed heats of hydrogenation for benzene.

Figure 19.5
A comparison between the observed and hypothetical heats of hydrogenation for benzene

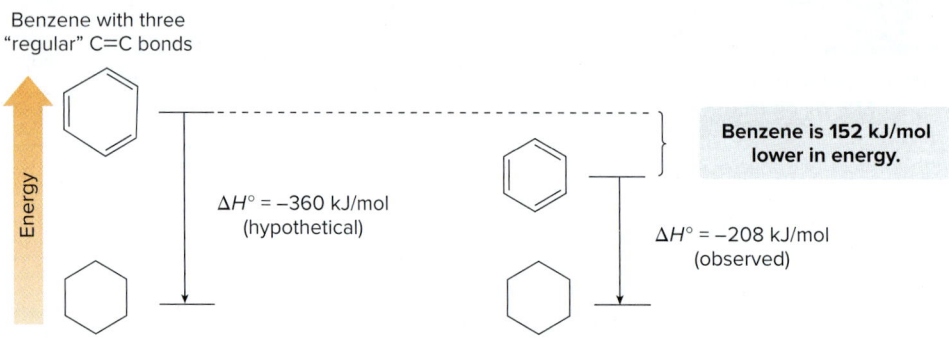

The huge difference between the hypothetical and observed heats of hydrogenation for benzene cannot be explained solely on the basis of resonance and conjugation.

- **The low heat of hydrogenation of benzene means that benzene is *especially stable*, even more so than the conjugated compounds introduced in Chapter 12. This unusual stability is characteristic of aromatic compounds.**

Benzene's unusual behavior in chemical reactions is not limited to hydrogenation. As mentioned in Section 19.1, **benzene does *not* undergo addition reactions typical of other highly unsaturated compounds, including conjugated dienes.** Benzene does not react with Br_2 to yield an addition product. Instead, in the presence of a Lewis acid, bromine *substitutes* for a hydrogen atom, thus yielding a product that retains the benzene ring.

This behavior is characteristic of aromatic compounds. The structural features that distinguish aromatic compounds from the rest are discussed in Section 19.7.

Problem 19.8 Compounds **A** and **B** are both hydrogenated to methylcyclohexane. Which compound has the larger heat of hydrogenation? Which compound is more stable?

A   B

19.7 The Criteria for Aromaticity—Hückel's Rule

Four structural criteria must be satisfied for a compound to be aromatic:

- A molecule must be cyclic, planar, completely conjugated, and contain a particular number of π electrons.

[1] A molecule must be cyclic.

- To be aromatic, each *p* orbital must overlap with *p* orbitals on two adjacent atoms.

The *p* orbitals on all six carbons of benzene continuously overlap, so benzene is aromatic. Hexa-1,3,5-triene has six *p* orbitals, too, but the two on the terminal carbons cannot overlap with each other, so **hexa-1,3,5-triene is *not* aromatic.**

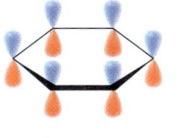   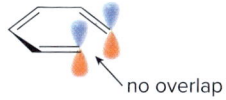

benzene   hexa-1,3,5-triene

Every *p* orbital overlaps with two neighboring *p* orbitals.   There can be no overlap between the *p* orbitals on the two terminal C's.

aromatic   **not aromatic**

[2] A molecule must be planar.

- All adjacent *p* orbitals must be aligned so that the π electron density can be delocalized.

      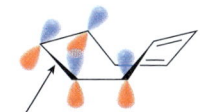

cyclooctatetraene
not aromatic   a tub-shaped, eight-membered ring   Adjacent *p* orbitals cannot overlap. Electrons cannot delocalize.

Cyclooctatetraene resembles benzene in that it is a cyclic molecule with alternating double and single bonds. Cyclooctatetraene is tub shaped, however, **not planar,** so overlap between adjacent π bonds is impossible. **Cyclooctatetraene, therefore, is *not* aromatic,** so it undergoes addition reactions like those of other alkenes.

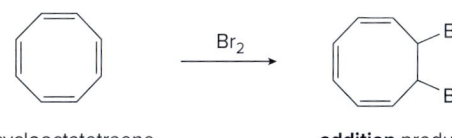

cyclooctatetraene   **addition** product

[3] A molecule must be completely conjugated.

- Aromatic compounds must have a *p* orbital on every atom in the ring.

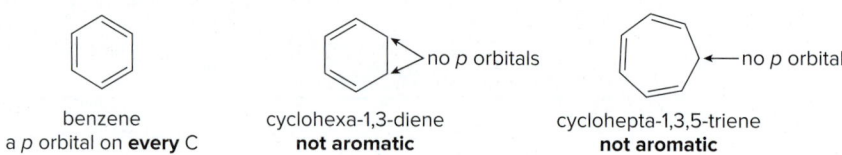

benzene
a *p* orbital on **every** C
aromatic

cyclohexa-1,3-diene
not aromatic

cyclohepta-1,3,5-triene
not aromatic

Both cyclohexa-1,3-diene and cyclohepta-1,3,5-triene contain at least one carbon atom that does not have a *p* orbital, so they are not completely conjugated and therefore **not aromatic.**

[4] A molecule must satisfy Hückel's rule, and contain a particular number of π electrons.

Some compounds satisfy the first three criteria for aromaticity, but still they show none of the stability typical of aromatic compounds. For example, **cyclobutadiene** is so highly reactive that it can be prepared only at extremely low temperatures.

cyclobutadiene

a planar, cyclic, completely conjugated molecule that is *not* aromatic

Hückel's rule refers to the number of π electrons, *not* the number of atoms in a particular ring.

It turns out that in addition to being cyclic, planar, and completely conjugated, a compound needs a particular number of π electrons to be aromatic. Erich Hückel first recognized in 1931 that the following criterion, expressed in two parts and now known as **Hückel's rule,** had to be satisfied, as well:

- An aromatic compound must contain $4n + 2$ π electrons ($n = 0, 1, 2$, and so forth).
- Cyclic, planar, and completely conjugated compounds that contain $4n$ π electrons are especially unstable, and are said to be *antiaromatic*.

Thus, compounds that contain 2, 6, 10, 14, 18, and so forth π electrons are aromatic, as shown in Table 19.2. **Benzene is aromatic and especially stable because it contains 6 π electrons. Cyclobutadiene is antiaromatic and especially unstable because it contains 4 π electrons.**

Table 19.2
The Number of π Electrons That Satisfy Hückel's Rule

n	$4n + 2$
0	2
1	6
2	10
3	14
4, etc.	18

Benzene
An aromatic compound

$4n + 2 = 4(1) + 2 =$
6 π electrons
aromatic

Cyclobutadiene
An antiaromatic compound

$4n = 4(1) =$
4 π electrons
antiaromatic

Considering aromaticity, all compounds can be classified in one of three ways:

[1]	Aromatic	• A cyclic, planar, completely conjugated compound with $4n + 2$ π electrons
[2]	Antiaromatic	• A cyclic, planar, completely conjugated compound with $4n$ π electrons
[3]	Not aromatic or nonaromatic	• A compound that lacks one (or more) of the four requirements to be aromatic or antiaromatic

19.8 Examples of Aromatic Compounds

Many compounds in addition to benzene are aromatic. Several examples are presented in Sections 19.8 and 19.9.

In Section 19.8, we look at three different types of aromatic compounds. Then, in Section 19.9, we examine aromatic heterocycles.

19.8A Aromatic Compounds with a Single Ring

Benzene is the most common aromatic compound having a single ring. **Completely conjugated rings larger than benzene are also aromatic if they are planar and have $4n + 2$ π electrons.**

- Hydrocarbons containing a single ring with alternating double and single bonds are called *annulenes*.

To name an annulene, indicate the number of atoms in the ring in brackets and add the word *annulene*. Thus, benzene is [6]-annulene. Both **[14]-annulene** and **[18]-annulene** are cyclic, planar, completely conjugated molecules that follow Hückel's rule, so they are aromatic.

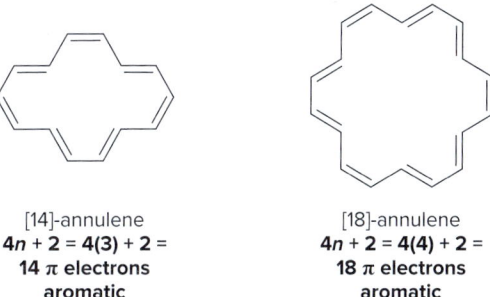

[14]-annulene
$4n + 2 = 4(3) + 2 =$
14 π electrons
aromatic

[18]-annulene
$4n + 2 = 4(4) + 2 =$
18 π electrons
aromatic

[10]-Annulene has 10 π electrons, which satisfies Hückel's rule, but a planar molecule would place the two H atoms inside the ring too close to each other, so the ring puckers to relieve this strain. Because **[10]-annulene is not planar,** the 10 π electrons can't delocalize over the entire ring and it is **not aromatic.**

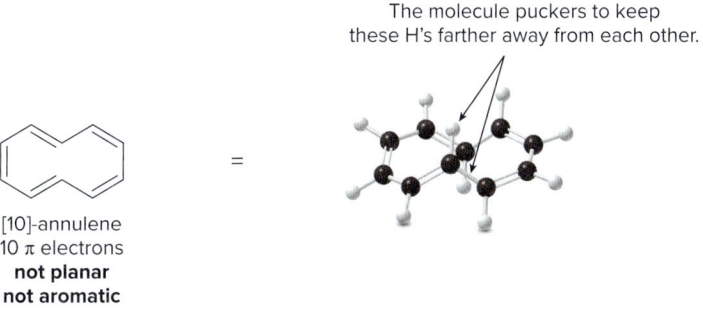

The molecule puckers to keep these H's farther away from each other.

[10]-annulene
10 π electrons
not planar
not aromatic

Problem 19.9 Would [16]-, [20]-, or [22]-annulene be aromatic if each ring is planar?

19.8B Aromatic Compounds with More Than One Ring

Hückel's rule for determining aromaticity can be applied only to monocyclic systems, but many aromatic compounds containing several benzene rings joined together are also known. Two or more six-membered rings with alternating double and single bonds can be fused together to form **polycyclic aromatic hydrocarbons (PAHs).** Joining two benzene rings together forms **naphthalene.** There are two different ways to join three rings together, forming **anthracene** and **phenanthrene,** and many more complex hydrocarbons are known.

naphthalene
10 π electrons

anthracene
14 π electrons

phenanthrene
14 π electrons

As the number of fused benzene rings increases, the number of resonance structures increases as well. Although two resonance structures can be drawn for benzene, naphthalene is a hybrid of three resonance structures.

Problem 19.10 Draw the four resonance structures for anthracene.

19.8C Charged Aromatic Compounds

Both negatively and positively charged ions can also be aromatic if they satisfy all the necessary criteria.

Cyclopentadienyl Anion

The **cyclopentadienyl anion** is a cyclic and planar anion with two double bonds and a nonbonded electron pair. The two π bonds contribute four electrons and the lone pair contributes two more, for a total of six. By Hückel's rule, having **six π electrons confers aromaticity.** The **negatively charged carbon atom must be sp^2 hybridized**, and the **nonbonded electron pair must occupy a p orbital** for the ring to be completely conjugated.

cyclopentadienyl anion
all **sp^2 hybridized C's**
6 π electrons

The lone pair resides in a p orbital.

- The cyclopentadienyl anion is aromatic because it is cyclic, planar, completely conjugated, and has six π electrons.

We can draw **five equivalent resonance structures for the cyclopentadienyl anion**, delocalizing the negative charge over every carbon atom of the ring.

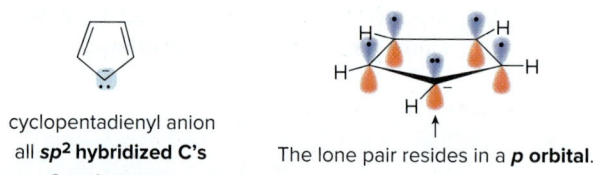

Although five resonance structures can also be drawn for both the **cyclopentadienyl cation** and **radical,** only the cyclopentadienyl anion has six π electrons, a number that satisfies Hückel's rule. The cyclopentadienyl cation has four π electrons, making it antiaromatic and especially unstable. The cyclopentadienyl radical has five π electrons, so it is neither aromatic nor antiaromatic. **Having the "right" number of electrons is necessary for a species to be unusually stable by virtue of aromaticity.**

19.8 Examples of Aromatic Compounds

cyclopentadienyl anion
- **6 π electrons**
- contains **4n + 2 π electrons**

aromatic

cyclopentadienyl cation
- **4 π electrons**
- contains **4n π electrons**

antiaromatic

cyclopentadienyl radical
- **5 π electrons**
- does *not* contain either 4n or 4n + 2 π electrons

nonaromatic

The cyclopentadienyl anion is readily formed from cyclopentadiene by a Brønsted–Lowry acid–base reaction.

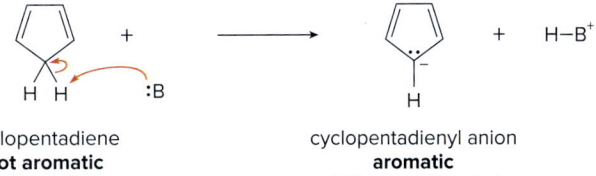

cyclopentadiene
not aromatic
$pK_a = 15$

cyclopentadienyl anion
aromatic
a stabilized conjugate base

Cyclopentadiene itself is not aromatic because it is not fully conjugated. **The cyclopentadienyl anion, however, is aromatic, so it is a very stable base.** As such, it makes cyclopentadiene more acidic than other hydrocarbons. In fact, the pK_a of cyclopentadiene is 15, much *lower* (more acidic) than the pK_a of any C–H bond discussed thus far.

- Cyclopentadiene is more acidic than many hydrocarbons because its conjugate base is aromatic.

Problem 19.11 Draw the product formed when cyclohepta-1,3,5-triene ($pK_a = 39$) is treated with a strong base. Why is its pK_a so much higher than the pK_a of cyclopentadiene?

cyclohepta-1,3,5-triene
$pK_a = 39$

Problem 19.12 Rank the following compounds in order of increasing acidity.

Tropylium Cation

The cyclopentadienyl anion and the tropylium cation both illustrate an important principle: The **number of π electrons determines aromaticity,** not the number of atoms in a ring or the number of *p* orbitals that overlap. The cyclopentadienyl anion and tropylium cation are aromatic because they each have six π electrons.

The **tropylium cation** is a planar carbocation with three double bonds and a positive charge contained in a seven-membered ring. This carbocation is completely conjugated, because the positively charged carbon is sp^2 hybridized and has a vacant *p* orbital that overlaps with the six *p* orbitals from the carbons of the three double bonds. **Because the tropylium cation has three π bonds and no other nonbonded electron pairs, it contains six π electrons,** thereby satisfying Hückel's rule.

tropylium cation
all sp^2 hybridized C's
6 π electrons

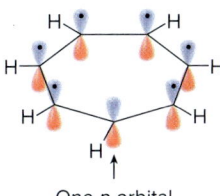
One *p* orbital is vacant.

- The tropylium cation is aromatic because it is cyclic, planar, completely conjugated, and has six π electrons delocalized over the seven atoms of the ring.

Problem 19.13 Draw the seven resonance structures for the tropylium cation.

Problem 19.14 Assuming the rings are planar, label each ion as aromatic, antiaromatic, or not aromatic.

Problem 19.15 Compound **A** exhibits a peak in its ¹H NMR spectrum at 7.6 ppm, indicating that it is aromatic. (a) How are the carbon atoms of the triple bonds hybridized? (b) In what type of orbitals are the π electrons of the triple bonds contained? (c) How many π electrons are delocalized around the ring in **A**?

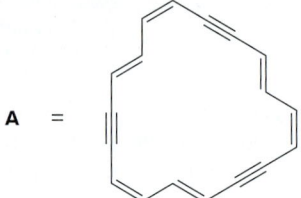

19.9 Aromatic Heterocycles

Recall from Section 9.3 that a **heterocycle** is a ring that contains at least one heteroatom.

Heterocycles containing oxygen, nitrogen, or sulfur—atoms that also have at least one lone pair of electrons—can also be aromatic. With heteroatoms, we must always **determine whether the lone pair is localized on the heteroatom or part of the delocalized π system.** Two examples, **pyridine** and **pyrrole,** illustrate these different possibilities.

19.9A Biological Building Blocks

Pyridine

Pyridine is a heterocycle containing a six-membered ring with three π bonds and one nitrogen atom. Like benzene, two resonance structures (with all neutral atoms) can be drawn.

two resonance structures for pyridine
6 π electrons

Pyridine is cyclic, planar, and completely conjugated, because the three single and three double bonds alternate around the ring. **Pyridine has six π electrons, two from each π bond, thus satisfying Hückel's rule and making pyridine aromatic.** The nitrogen atom of pyridine also has a nonbonded electron pair, which is *localized* on the N atom, so it is *not* part of the delocalized π electron system of the aromatic ring.

How is the nitrogen atom of the pyridine ring hybridized? The N atom is surrounded by three groups (two atoms and a lone electron pair), making it sp^2 **hybridized,** and leaving one unhybridized *p* orbital with one electron that overlaps with adjacent *p* orbitals. The lone pair on N resides in an sp^2 hybrid orbital that is perpendicular to the delocalized π electrons.

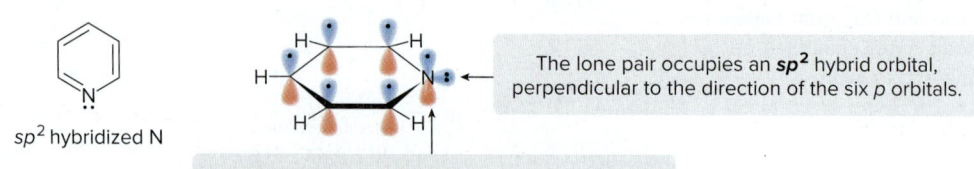

sp^2 hybridized N

The lone pair occupies an sp^2 hybrid orbital, perpendicular to the direction of the six *p* orbitals.

A *p* orbital on N overlaps with adjacent *p* orbitals, making the ring **completely conjugated**.

Pyrrole

Pyrrole contains a five-membered ring with two π bonds and one nitrogen atom. The N atom also has a lone pair of electrons.

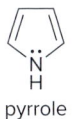
pyrrole

Pyrrole is cyclic and planar, with a total of four π electrons from the two π bonds. Is the nonbonded electron pair localized on N or part of a delocalized π electron system? The lone pair on N is *adjacent* to a double bond. Recall the following general rule from Section 12.5:

- In a system X=Y—Z:, Z is generally sp^2 hybridized and the lone pair occupies a p orbital to make the system conjugated.

If the lone pair on the N atom occupies a p orbital:

- Pyrrole has a p orbital on every adjacent atom, so it is completely conjugated.
- Pyrrole has six π electrons—four from the π bonds and two from the lone pair.

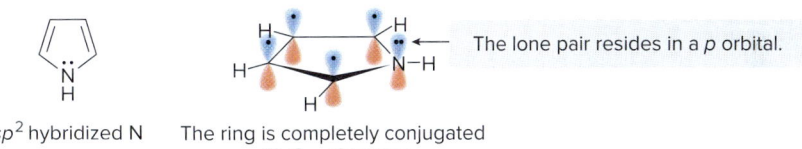

Because pyrrole is cyclic, planar, completely conjugated, and has $4n + 2$ π electrons, **pyrrole is aromatic. The number of electrons—not the size of the ring—determines whether a compound is aromatic.**

Electrostatic potential maps, shown in Figure 19.6 for pyridine and pyrrole, illustrate that the **lone pair in pyridine is localized on N,** whereas the **lone pair in pyrrole is part of the delocalized π system.** Thus, a fundamental difference exists between the N atoms in pyridine and pyrrole.

Figure 19.6
Electrostatic potential maps of pyridine and pyrrole

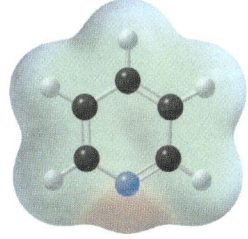
pyridine

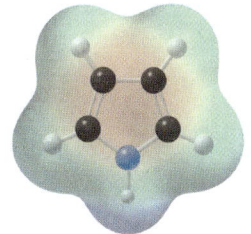
pyrrole

- In pyridine, the nonbonded electron pair is localized on the N atom in an sp^2 hybridized orbital, as shown by the region of high electron density (in red) on N.

- In pyrrole, the nonbonded electron pair is in a p orbital and is delocalized over the ring, so the entire ring is electron rich (red).

- When a heteroatom is already part of a double bond (as in the N of pyridine), its lone pair *cannot* occupy a p orbital, so it *cannot* be delocalized over the ring.
- When a heteroatom is *not* part of a double bond (as in the N of pyrrole), its lone pair can be located in a p orbital and *delocalized* over a ring to make it aromatic.

Histamine

Scombroid fish poisoning, associated with facial flushing, hives, and general itching, is caused by the ingestion of inadequately refrigerated fish, typically mahimahi (pictured) and tuna. Bacteria convert the amino acid histidine (Chapter 23) to histamine, which, when consumed in large amounts, results in this clinical syndrome. *Daniel C. Smith*

Histamine, a biologically active amine formed in many tissues, has an aromatic heterocycle with two N atoms, one of which is similar to the N atom of pyridine and one of which is similar to the N atom of pyrrole.

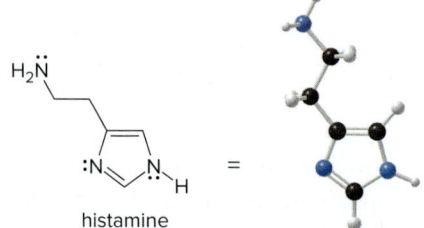

histamine

Histamine has a five-membered ring with two π bonds and two nitrogen atoms, each of which contains a lone pair of electrons. The heterocycle has four π electrons from the two double bonds. **The lone pair on the N in red also occupies a *p* orbital,** making the heterocycle completely conjugated and giving it a total of six π electrons. The lone pair on this N atom is thus delocalized over the five-membered ring and the heterocycle is aromatic. **The lone pair on the N in blue occupies an *sp²* hybrid orbital** perpendicular to the delocalized π electrons.

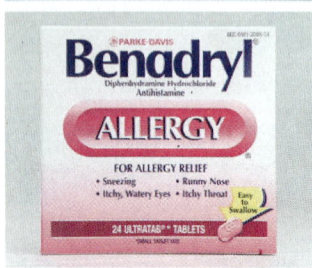

Antihistamines that block the action of histamine on the H1 histamine receptor are used to treat the runny nose and watery eyes of an allergic reaction. *Bob London/Alamy Stock Photo*

- N (in red) resembles the N atom of pyrrole.
- N (in blue) resembles the N atom of pyridine.

N: The lone pair resides in a *p* orbital.

N: The lone pair resides in an *sp²* hybrid orbital.

Histamine produces a wide range of physiological effects in the body. Excess histamine is responsible for the runny nose and watery eyes symptomatic of hay fever. It also stimulates the overproduction of stomach acid and contributes to the formation of hives. These effects result from the interaction of histamine with two different cellular receptors. We will learn more about antihistamines and antiulcer drugs, compounds that block the effects of histamine, in Section 22.5.

Sample Problem 19.1 — Determining the Hybridization of a Heteroatom in an Aromatic Heterocycle

How is each N atom in pilocarpine hybridized, and in what type of orbital does the lone pair on each N reside?

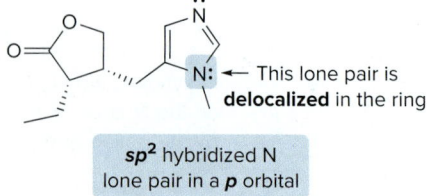

pilocarpine

Pilocarpine (Sample Problem 19.1) is a naturally occurring glaucoma medication isolated from the plant *Pilocarpus microphyllus,* commonly called jaborandi, grown in Brazil. *Schafer & Hill/Photolibrary/Getty Images*

Solution

sp² hybridized N
lone pair in an *sp²* hybrid orbital

← This lone pair is **localized** on the N.

The N atom labeled in red is already part of a double bond, so it is ***sp²*** **hybridized** and its unhybridized *p* orbital is used to form the π bond. As a result, the **lone pair on N occupies one of the *sp²* hybrid orbitals.**

← This lone pair is **delocalized** in the ring.

sp² hybridized N
lone pair in a *p* orbital

The N atom labeled in blue is ***sp²*** **hybridized,** so its **lone pair can occupy a *p* orbital** and delocalize in the five-membered ring. Delocalization gives the ring six π electrons and makes the ring aromatic.

19.9 Aromatic Heterocycles

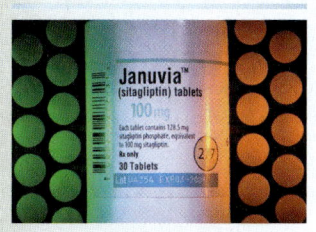

Januvia (Problem 19.16) increases the body's ability to lower blood sugar levels, so it is used alone or in combination with other drugs to treat type 2 diabetes.

Jb Reed/Bloomberg/Getty Images

Problem 19.16 Januvia, the trade name for sitagliptin, was introduced in 2006 for the treatment of type 2 diabetes. (a) Explain why the five-membered ring in sitagliptin is aromatic. (b) Determine the hybridization of each N atom. (c) In what type of orbital does the lone pair on each N atom reside?

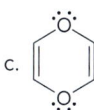

sitagliptin

More Practice: Try Problems 19.34; 19.35; 19.52b, c; 19.53a; 19.54b.

Problem 19.17 Which heterocycles are aromatic?

a. 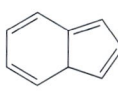 b.  c. 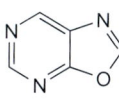 d.

Sample Problem 19.2 — Characterizing a Compound as Aromatic, Antiaromatic, or Not Aromatic

Label each compound as aromatic, antiaromatic, or not aromatic. Assume all completely conjugated rings are planar.

A   **B**   **C**

Solution

Each compound is cyclic, and from the problem statement, we assume that completely conjugated rings are planar, so we must answer just two questions to decide on aromaticity:

- **Is the compound completely conjugated?** If a compound is not completely conjugated, it is *not* aromatic.
- **How many π electrons does the compound contain?** A compound with $4n + 2$ π electrons is *aromatic*; a compound with $4n$ π electrons is *antiaromatic*. A compound with an odd number of π electrons satisfies neither equation and is *not* aromatic.

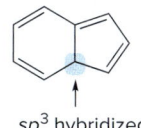

sp^3 hybridized

A

- The C labeled in blue is sp^3 hybridized, so there is no p orbital for overlap and the system is **not completely conjugated. A is not aromatic.**

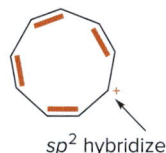

sp^2 hybridized

B

- The ring is completely conjugated because each C of the C=C's and the positively charged C are sp^2 hybridized, so each C has an unhybridized p orbital. The ring has **8 π electrons** from the four C=C's, so it has $4n$ π electrons. **B is antiaromatic.**

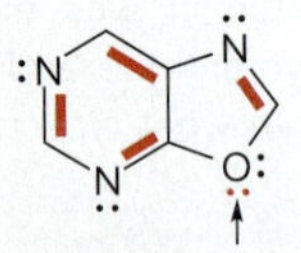

one electron pair
in a *p* orbital

C

- For the ring to be completely conjugated, the O atom must be sp^2 hybridized and one electron pair on O must occupy a *p* orbital. The ring has **10 π electrons** from the four C=C's and the O atom, so it has **4*n* + 2 π** electrons. **C is aromatic.**

Problem 19.18 Label each compound as aromatic, antiaromatic, or not aromatic. Assume all completely conjugated rings are planar.

a. 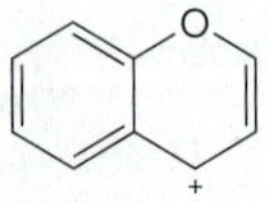 b. 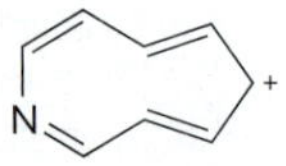 c. 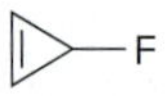 d. 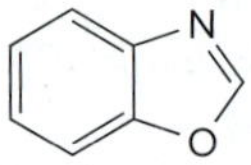 e. 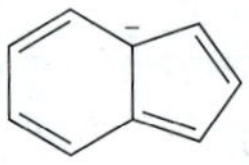

More Practice: Try Problems 19.24, 19.28–19.30.

19.9B Deoxyribonucleic Acid

As we learned in Section 3.9C, deoxyribonucleic acid (**DNA**) has a double helical structure made up of nucleotides, which contain an aromatic nitrogen heterocycle attached to a sugar–phosphate backbone. DNA contains four different bases, derived from the heterocycles **pyrimidine** and **purine.** The π electrons highlighted in red on pyrimidine and purine are delocalized and part of the aromatic system.

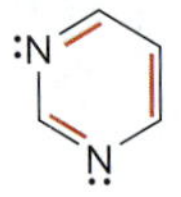

pyrimidine

6 π electrons

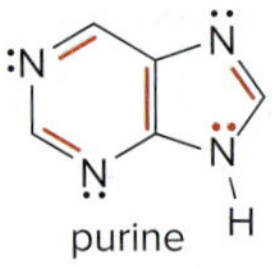

purine

10 π electrons

- **The lone pair electrons on both N atoms in pyrimidine, like pyridine, reside in sp^2 hybridized orbitals perpendicular to the aromatic π system.**
- **Purine, a bicyclic compound that contains four sp^2 hybridized N atoms, has a lone pair (highlighted in red) that resides in a *p* orbital and is part of the aromatic system.**

Two DNA bases—**cytosine** and **thymine**—are derived from pyrimidine and contain six π electrons, and two bases—**adenine** and **guanine**—are derived from purine and contain 10 π electrons.

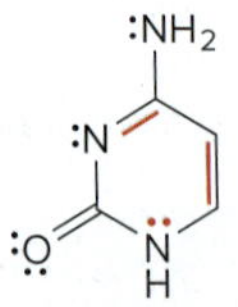

cytosine

6 π electrons

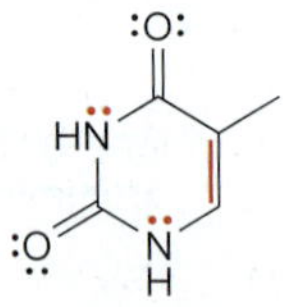

thymine

6 π electrons

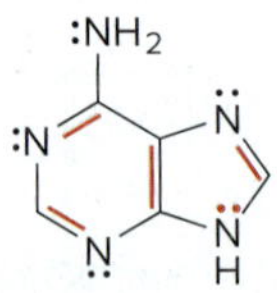

adenine

10 π electrons

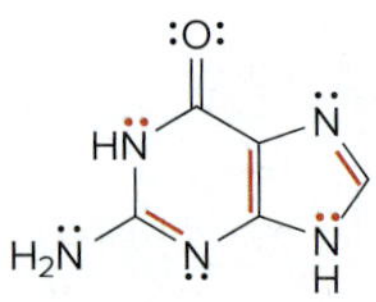

guanine

10 π electrons

pyrimidines **purines**

Three of the bases are typically drawn in a form that contains a C=O, so it may not be obvious that each is aromatic. Because these bases contain a lone pair on N that can be delocalized on the C=O, however, a resonance structure can be drawn that clearly shows that six π electrons are delocalized in the ring. Two resonance structures are shown for cytosine.

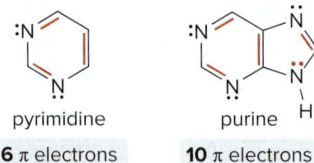

19.10 What Is the Basis of Hückel's Rule?

[Structure showing cytosine resonance: cytosine ↔ 6 π electrons]

Because DNA holds essential, hereditary information that is critical to our growth and development, it is important that the high-molecular-weight nucleic acids are stable. Moreover, they must be packaged compactly because, if uncoiled and tied together, all of the DNA in our body would stretch ~10 billion miles. We will learn much more about the structure of DNA in Sections 26.2 and 26.3.

Problem 19.19 Answer each question for the four bases that comprise DNA. (a) How is each N atom hybridized? (b) In what type of orbital does each lone pair on a N reside?

Problem 19.20 Draw resonance structures for the six-membered thymine and guanine rings that clearly show the π electrons delocalized within the aromatic rings.

19.10 What Is the Basis of Hückel's Rule?

Why does the number of π electrons determine whether a compound is aromatic? Cyclobutadiene is cyclic, planar, and completely conjugated, just like benzene, but why is benzene aromatic and cyclobutadiene antiaromatic?

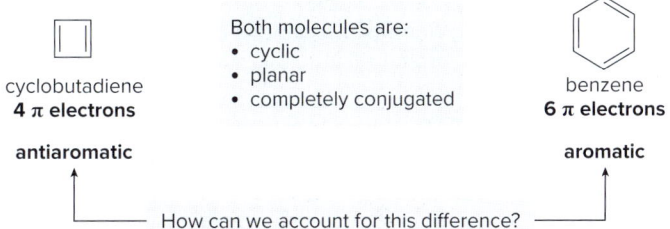

cyclobutadiene
4 π electrons
antiaromatic

Both molecules are:
- cyclic
- planar
- completely conjugated

benzene
6 π electrons
aromatic

How can we account for this difference?

A complete explanation is beyond the scope of an introductory organic chemistry text, but nevertheless, you can better understand the basis of aromaticity by learning more about orbitals and bonding.

19.10A Bonding and Antibonding Orbitals

So far we have used these basic concepts to describe how bonds are formed:

- Hydrogen uses its 1s orbital to form σ bonds with other elements.
- Second-row elements use hybrid orbitals (sp, sp^2, or sp^3) to form σ bonds.
- Second-row elements use p orbitals to form π bonds.

This description of bonding is called **valence bond theory.** In valence bond theory, a covalent bond is formed by the overlap of two atomic orbitals, and the electron pair in the resulting bond is shared by both atoms. Thus, a carbon–carbon double bond consists of a σ bond, formed by overlap of two sp^2 hybrid orbitals, each containing one electron, and a π bond, formed by overlap of two p orbitals, each containing one electron.

This description of bonding works well for most of the organic molecules we have encountered thus far. Unfortunately, it is inadequate for describing systems with many adjacent p orbitals that overlap, as there are in aromatic compounds. To more fully explain the bonding in these systems, we must utilize **molecular orbital (MO) theory.**

MO theory describes bonds as the mathematical combination of atomic orbitals that form a new set of orbitals called **molecular orbitals (MOs).** A molecular orbital occupies a region of space *in a molecule* where electrons are likely to be found. When forming molecular orbitals from atomic orbitals, keep in mind:

> • A set of *n* atomic orbitals forms *n* molecular orbitals.

If *two* atomic orbitals combine, *two* molecular orbitals are formed. This is fundamentally different than valence bond theory. Because aromaticity is based on *p* orbital overlap, what does MO theory predict will happen when two *p* (atomic) orbitals combine?

The two lobes of each *p* orbital are opposite in phase, with a node of electron density at the nucleus. When two *p* orbitals combine, two molecular orbitals should form. The two *p* orbitals can add together constructively—that is, with like phases interacting—or destructively—that is, with opposite phases interacting.

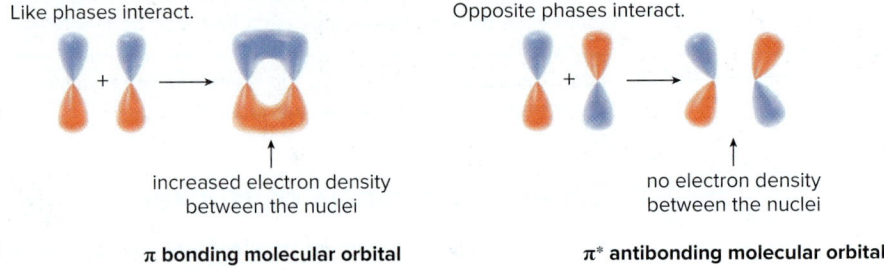

- When two *p* orbitals of *similar* phase overlap side-by-side, a π *bonding* molecular orbital results.
- When two *p* orbitals of *opposite* phase overlap side-by-side, a π* *antibonding* molecular orbital results.

A π bonding MO is lower in energy than the two atomic *p* orbitals from which it is formed because a stable bonding interaction results when orbitals of similar phase combine. A bonding interaction holds nuclei together. Similarly, a π* antibonding MO is higher in energy because a destabilizing node results when orbitals of opposite phase combine. A destabilizing interaction pushes nuclei apart.

If two atomic *p* orbitals each have one electron and then combine to form MOs, the two electrons will occupy the lower-energy π bonding MO, as shown in Figure 19.7.

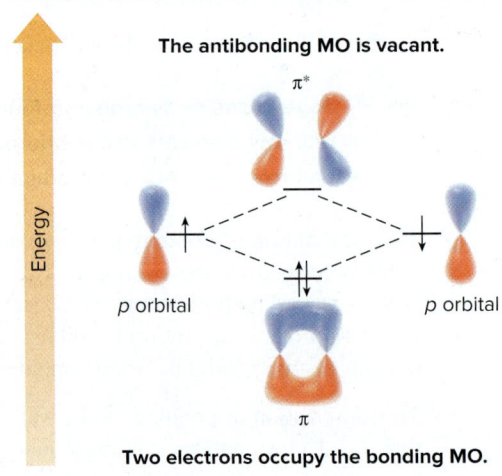

Figure 19.7
Combination of two *p* orbitals to form π and π* molecular orbitals

- Two atomic *p* orbitals combine to form two molecular orbitals. The bonding π MO is *lower* in energy than the two *p* orbitals from which it was formed, and the antibonding π* MO is *higher* in energy than the two *p* orbitals from which it was formed.
- Two electrons fill the lower-energy bonding MO first.

19.10B Molecular Orbitals Formed When More Than Two p Orbitals Combine

The molecular orbital description of benzene is much more complex than the two MOs formed in Figure 19.7. Because each of the six carbon atoms of benzene has a p orbital, **six atomic p orbitals combine to form six π molecular orbitals,** as shown in Figure 19.8. A description of the exact appearance and energies of these six MOs requires more sophisticated mathematics and understanding of MO theory than is presented in this text. Nevertheless, note that the six MOs are labeled ψ_1–ψ_6, with ψ_1 being the lowest in energy and ψ_6 the highest.

Figure 19.8
How the six p orbitals of benzene overlap to form six molecular orbitals

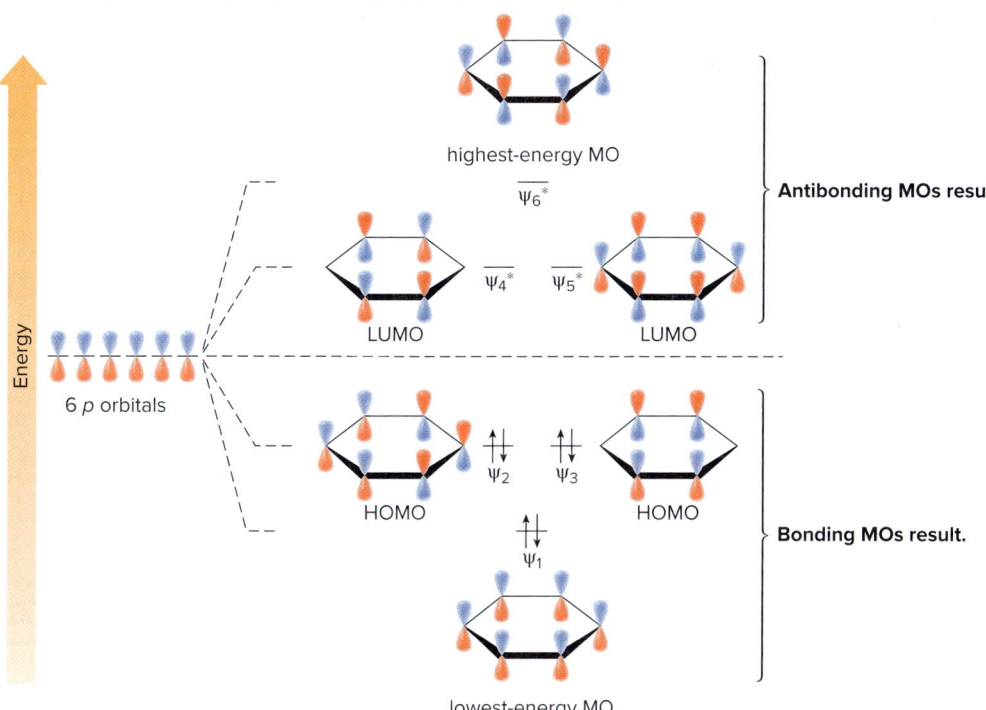

- Depicted in this diagram are the interactions of the six atomic p orbitals of benzene, which form six molecular orbitals. When orbitals of like phase combine, a bonding interaction results. When orbitals of opposite phase combine, a destabilizing node results.

The most important features of the six benzene MOs are as follows:

- **The larger the number of bonding interactions, the lower in energy the MO.** The lowest-energy molecular orbital (ψ_1) has all bonding interactions between the p orbitals.
- **The larger the number of nodes, the higher in energy the MO.** The highest-energy MO (ψ_6^*) has all nodes between the p orbitals.
- Three MOs are lower in energy than the starting p orbitals, making them bonding MOs (ψ_1, ψ_2, and ψ_3), whereas three MOs are higher in energy than the starting p orbitals, making them antibonding MOs (ψ_4^*, ψ_5^*, and ψ_6^*).
- The two pairs of MOs (ψ_2 and ψ_3; ψ_4^* and ψ_5^*) with the same energy are called **degenerate orbitals.**
- **The highest-energy orbital that contains electrons is called the *highest occupied molecular orbital* (HOMO).** For benzene, the degenerate orbitals ψ_2 and ψ_3 are the HOMOs.
- **The lowest-energy orbital that does *not* contain electrons is called the *lowest unoccupied molecular orbital* (LUMO).** For benzene, the degenerate orbitals ψ_4^* and ψ_5^* are the LUMOs.

To fill the MOs, the six electrons are added, two to an orbital, beginning with the lowest-energy orbital. As a result, **the six electrons completely fill the bonding MOs, leaving the antibonding MOs empty.** This is what gives benzene and other aromatic compounds their special stability, and this is why six π electrons satisfies Hückel's $4n + 2$ rule.

- All bonding MOs (and HOMOs) are completely filled in aromatic compounds. No π electrons occupy antibonding MOs.

19.11 The Inscribed Polygon Method for Predicting Aromaticity

An inscribed polygon is also called a **Frost circle.**

To predict whether a compound has π electrons completely filling bonding MOs, we must know how many bonding molecular orbitals and how many π electrons it has. It is possible to predict the relative energies of cyclic, completely conjugated compounds, without sophisticated math (or knowing what the resulting MOs look like) by using the **inscribed polygon method.**

How To Use the Inscribed Polygon Method to Determine the Relative Energies of MOs for Cyclic, Completely Conjugated Compounds

Example Plot the relative energies of the MOs of benzene.

Step [1] Draw the polygon in question inside a circle with its vertices touching the circle and one of the vertices pointing *down*. Mark the points at which the polygon intersects the circle.

- Inscribe a hexagon inside a circle for benzene. The six vertices of the hexagon form six points of intersection, corresponding to the six MOs of benzene. The pattern—a single MO having the lowest energy, two degenerate pairs of MOs, and a single highest-energy MO—matches that found in Figure 19.8.

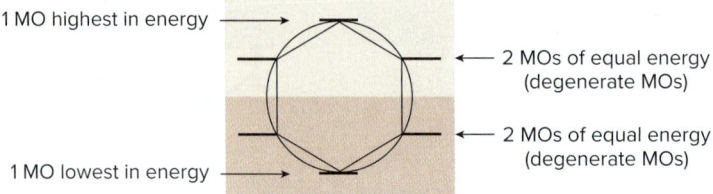

Step [2] Draw a line horizontally through the center of the circle and label MOs as bonding, nonbonding, or antibonding.

- **MOs below this line are bonding,** and lower in energy than the *p* orbitals from which they were formed. Benzene has three bonding MOs.
- **MOs at this line are nonbonding,** and equal in energy to the *p* orbitals from which they were formed. Benzene has no nonbonding MOs.
- **MOs above this line are antibonding,** and higher in energy than the *p* orbitals from which they were formed. Benzene has three antibonding MOs.

Step [3] Add the electrons, beginning with the lowest-energy MO.

- All the bonding MOs (and the HOMOs) are completely filled in aromatic compounds. No π electrons occupy antibonding MOs.
- Benzene is aromatic because it has six π electrons that completely fill the bonding MOs.

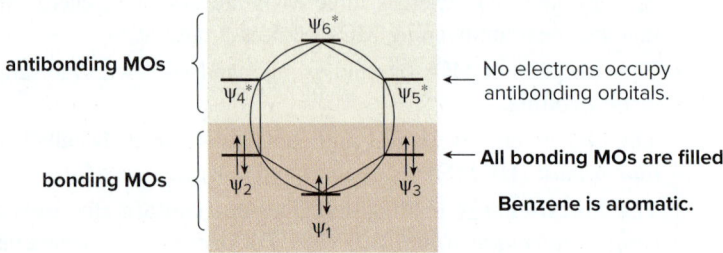

19.11 The Inscribed Polygon Method for Predicting Aromaticity

Figure 19.9 Using the inscribed polygon method for five- and seven-membered rings

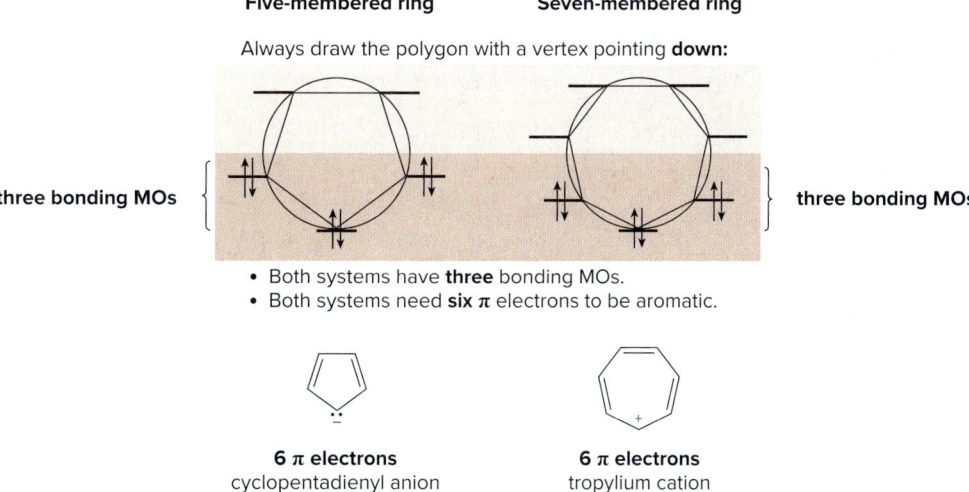

- Both systems have **three** bonding MOs.
- Both systems need **six** π electrons to be aromatic.

6 π electrons
cyclopentadienyl anion

6 π electrons
tropylium cation

This method works for all monocyclic, completely conjugated hydrocarbons regardless of ring size. Figure 19.9 illustrates MOs for completely conjugated five- and seven-membered rings using this method. The total number of MOs always equals the number of vertices of the polygon. **Because both systems have three bonding MOs, each needs six π electrons to fully occupy them,** making the cyclopentadienyl anion and the tropylium cation aromatic, as we learned in Section 19.8C.

The inscribed polygon method is consistent with **Hückel's $4n + 2$ rule**; that is, **there is always one lowest-energy bonding MO** that can hold two π electrons and the **other bonding MOs come in degenerate pairs** that can hold a total of four π electrons. For the compound to be aromatic, these MOs must be completely filled with electrons, so the "magic numbers" for aromaticity fit Hückel's $4n + 2$ rule (Figure 19.10).

Figure 19.10 MO patterns for cyclic, completely conjugated systems

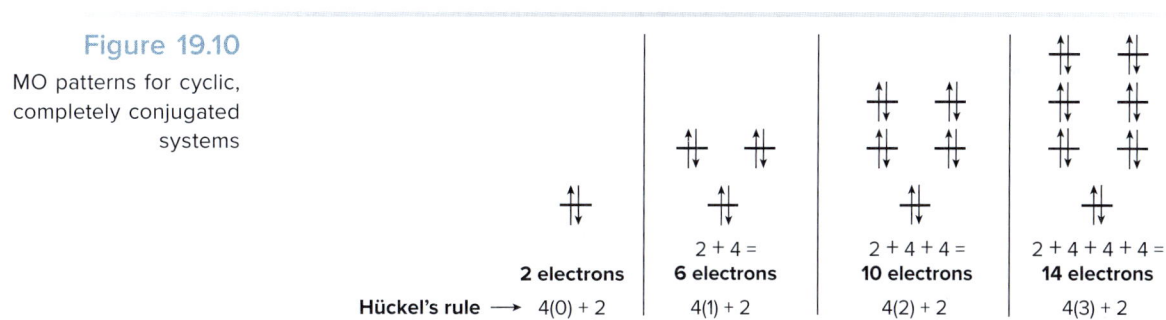

| 2 electrons | 2 + 4 = 6 electrons | 2 + 4 + 4 = 10 electrons | 2 + 4 + 4 + 4 = 14 electrons |

Hückel's rule ⟶ 4(0) + 2   4(1) + 2   4(2) + 2   4(3) + 2

Sample Problem 19.3   Using the Inscribed Polygon Method in Determining Aromaticity

Use the inscribed polygon method to show why cyclobutadiene is not aromatic.

cyclobutadiene
4 π electrons

Solution

Cyclobutadiene has four MOs (formed from its four *p* orbitals), to which its four π electrons must be added.

Step [1] Inscribe a square with a vertex down and mark its four points of intersection with the circle.

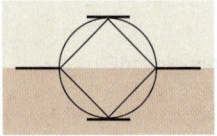

- The four points of intersection correspond to the four MOs of cyclobutadiene.

Steps [2] and [3] Draw a line through the center of the circle, label the MOs, and add the electrons.

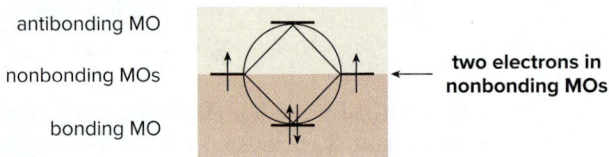

- Cyclobutadiene has four MOs—one bonding, two nonbonding, and one antibonding.
- Adding cyclobutadiene's **four π electrons** to these orbitals places two in the lowest-energy bonding MO and one each in the two nonbonding MOs.
- Separating electrons in two degenerate MOs keeps **like charges farther away from each other.**

Conclusion: Cyclobutadiene is not aromatic because its HOMOs, **two degenerate nonbonding MOs, are not completely filled.**

Problem 19.21 Use the inscribed polygon method to show why the following cation is aromatic.

More Practice: Try Problems 19.45, 19.46.

Problem 19.22 Use the inscribed polygon method to show why the cyclopentadienyl cation and radical are not aromatic.

The procedure followed in Sample Problem 19.3 also illustrates why cyclobutadiene is antiaromatic. Having the two unpaired electrons in nonbonding MOs suggests that cyclobutadiene should be a highly unstable diradical. In fact, antiaromatic compounds resemble cyclobutadiene because their HOMOs contain two unpaired electrons, making them especially unstable.

19.12 Aromatase Inhibitors for Estrogen-Dependent Cancer Treatment

Compounds that contain aromatic rings with loosely held π electrons exhibit a noncovalent attractive force called **π stacking. Parallel π stacking** occurs when two aromatic rings stack on top of each other, and **perpendicular π stacking** occurs when the p orbitals of the two aromatic rings are oriented at a 90° angle. When an electron-rich aromatic ring is attracted to a positively charged ion, **cation–π stacking** occurs.

19.12 Aromatase Inhibitors for Estrogen-Dependent Cancer Treatment

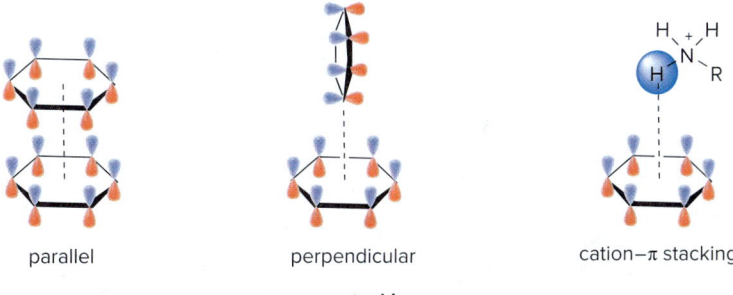

parallel   perpendicular   cation–π stacking

π stacking

π Stacking can be an important attractive force that holds an aromatic substrate or product at the active site of an enzyme. Such interactions are seen in the synthesis of the aromatic ring of the female sex hormones estradiol and estrone from the male sex hormones testosterone and androstenedione, respectively, using the aromatase enzyme.

testosterone → (aromatase enzyme, Fe^{2+}) → estradiol

androstenedione → (aromatase enzyme, Fe^{2+}) → estrone

The active site of the aromatase enzyme contains a **heme** unit, a complex organic compound containing the Fe^{2+} cation complexed to a nitrogen heterocycle called a porphyrin. Heme units are found in several proteins. As we will learn in Chapter 23, the Fe^{2+} ion in the heme of the proteins hemoglobin and myoglobin binds oxygen in the blood.

heme

To synthesize the aromatic ring of estrone, both the heme unit and androstenedione are complexed at the active site of aromatase (Figure 19.11a). The cyclohexanone ring of the substrate is converted to the aromatic ring of estrone by a multistep process, and the newly synthesized aromatic ring forms a cation–π stack with heme (Figure 19.11b) until it is released from the active site.

The aromatase enzyme is the only known enzyme to catalyze the formation of estrogens from male sex hormones. Compounds that inhibit the production of estrogens, so-called aromatase inhibitors, have been and continue to be developed for estrogen-dependent breast cancers (Section 15.4). Many are aromatic compounds that bind to the active site of the aromatase enzyme, resulting in the decreased production of estrogen, which slows the growth of the cancer.

Figure 19.11 Protein ribbon diagram for aromatase with enlargement of the active site

a. Aromatase

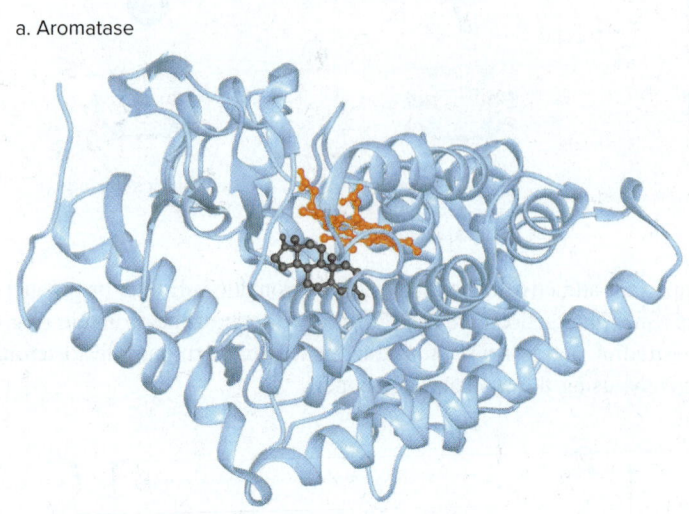

- In the active site of aromatase, androstenedione (gray) is located close to heme (orange) to catalyze the aromatization in the formation of estrone. H atoms are omitted in models.

b.

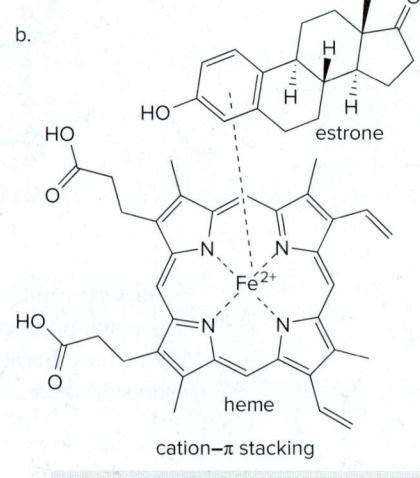

heme–aromatase estrone complex

- The estrone synthesized from androstenedione forms a cation–π stack with heme.

Chapter 19 REVIEW

KEY CONCEPTS

[1] Nomenclature of disubstituted and trisubstituted benzenes (19.3)

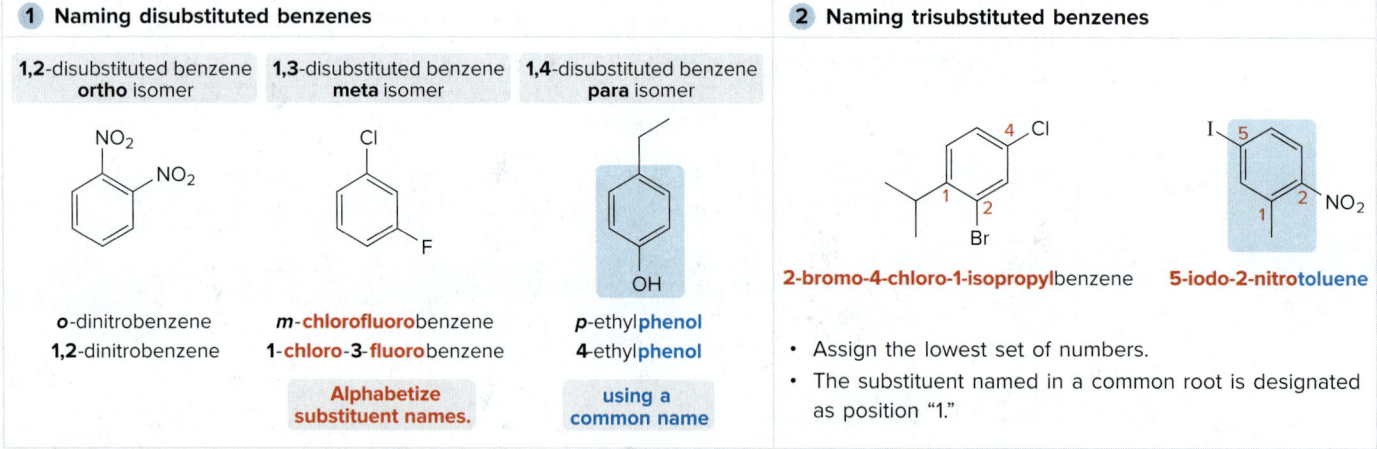

Try Problems 19.23, 19.25, 19.27b.

[2] Examples of aromatic, nonaromatic, and antiaromatic compounds (19.7–19.9)

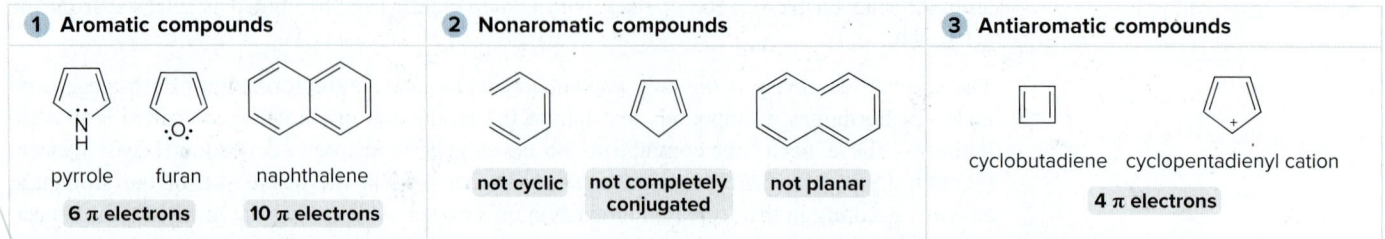

KEY SKILLS

[1] Determining if a cyclic, planar compound is aromatic, antiaromatic, or not aromatic (19.7); example: cyclopentadienyl cation

1 Determine if the molecule is completely conjugated.	**2** Check if the molecule satisfies Hückel's rule.
 ← Empty *p* orbital. **cyclopentadienyl cation** The electrons in bonds shown in red reside in *p* orbitals. The cyclopentadienyl cation is **completely conjugated**. • **Aromatic** and **antiaromatic** compounds must have a ***p* orbital on every atom in the ring**.	2 π electrons →  ← 2 π electrons $4n$ π electrons, where $n = 1$. **4 π electrons** • **Antiaromatic** compounds must contain **$4n$ π electrons** ($n = 0, 1, 2,$ and so forth). **Answer: antiaromatic**

See Sample Problem 19.2. Try Problems 19.24, 19.28–19.30.

---

[2] Determining if a planar compound is aromatic, antiaromatic, or not aromatic (19.8); example: [14]-annulene

1 Determine if the molecule is completely conjugated.	**2** Check if the molecule satisfies Hückel's rule.
 **[14]-annulene** The electrons in bonds shown in red reside in *p* orbitals. [14]-Annulene is **completely conjugated**. • **Aromatic** and **antiaromatic** compounds must have a ***p* orbital on every atom in the ring**.	 $4n + 2$ π electrons, where $n = 3$. **14 π electrons** • **Aromatic** compounds must contain **$4n + 2$ π electrons** ($n = 0, 1, 2,$ and so forth). **Answer: aromatic**

See Sample Problem 19.2, Table 19.2. Try Problems 19.24, 19.28–19.30.

---

[3] Determining if a planar heterocyclic compound is aromatic, antiaromatic, or not aromatic (19.9); example: furan

1 Determine if the molecule is completely conjugated.	**2** Check if the molecule satisfies Hückel's rule.
 One **lone pair** resides in an ***sp²* orbital.** **furan** One **lone pair** resides in a ***p* orbital.** The electrons in bonds shown in red reside in *p* orbitals. Furan is **completely conjugated**. • Count a nonbonded electron pair if it makes the ring aromatic in calculating $4n + 2$.	2 π electrons →  ← 2 π electrons ← 2 π electrons $4n + 2$ π electrons, where $n = 1$. **6 π electrons** • **Aromatic** compounds must contain **$4n + 2$ π electrons** ($n = 0, 1, 2,$ and so forth). **Answer: aromatic**

See Sample Problems 19.1, 19.2. Try Problems 19.24b, c; 19.28; 19.29e; 19.52a; 19.53b; 19.54a.

Chapter 19 Benzene and Aromatic Compounds

[4] Using the inscribed polygon method to determine if a compound is aromatic (19.11); example: the tropylium radical

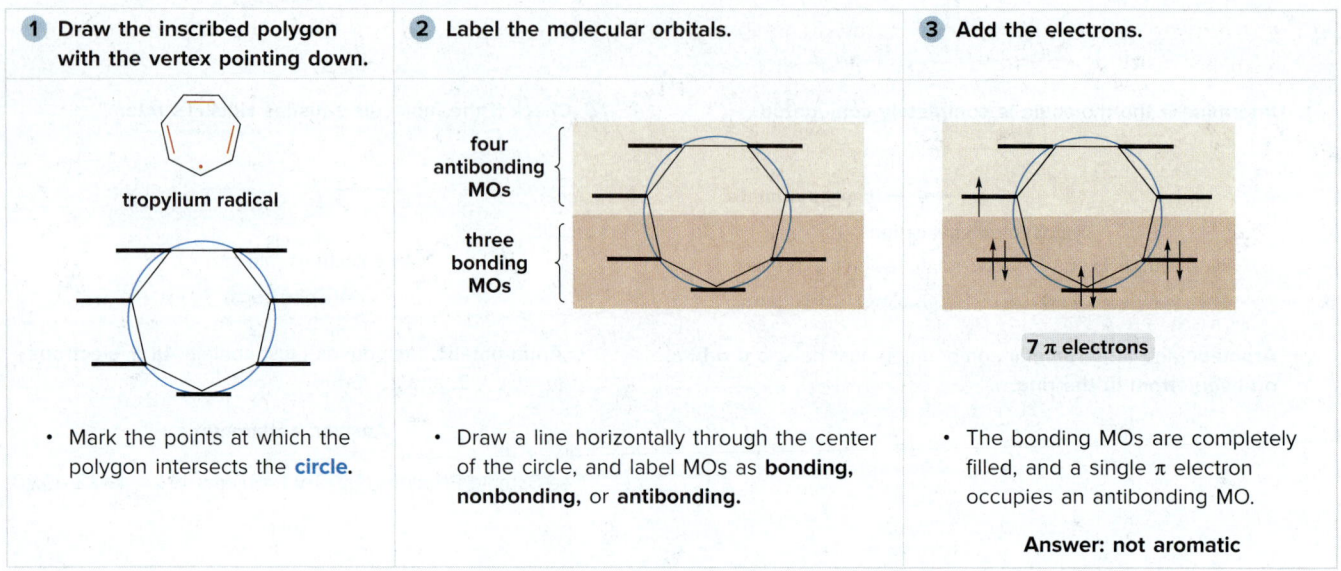

1 Draw the inscribed polygon with the vertex pointing down.	**2** Label the molecular orbitals.	**3** Add the electrons.
• Mark the points at which the polygon intersects the **circle**.	• Draw a line horizontally through the center of the circle, and label MOs as **bonding, nonbonding,** or **antibonding**.	• The bonding MOs are completely filled, and a single π electron occupies an antibonding MO. **Answer: not aromatic**

See *How To* p. 872; Figures 19.9, 19.10; Sample Problem 19.3. Try Problems 19.45, 19.46.

PROBLEMS

Problems Using Three-Dimensional Models

19.23 Name each compound and state how many lines are observed in its ^{13}C NMR spectrum.

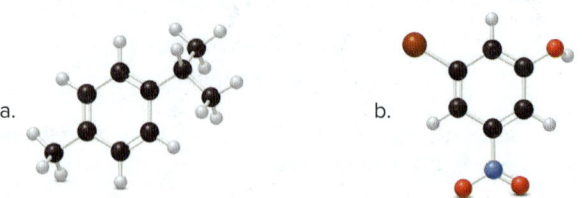

19.24 Classify each compound as aromatic, antiaromatic, or not aromatic.

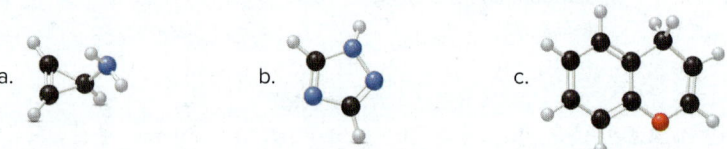

Benzene Structure and Nomenclature

19.25 Give the IUPAC name for each compound.

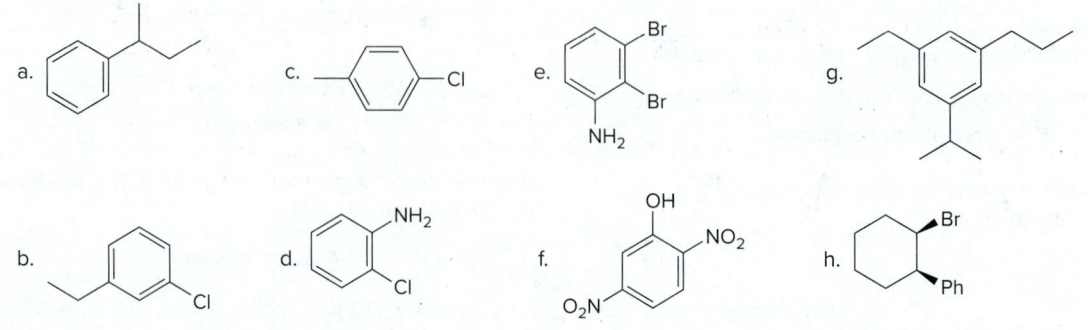

19.26 Draw a structure corresponding to each name.
  a. *p*-dichlorobenzene
  b. *p*-iodoaniline
  c. *o*-bromonitrobenzene
  d. 2,6-dimethoxytoluene
  e. 2-phenylprop-2-en-1-ol
  f. *trans*-1-benzyl-3-phenylcyclopentane

19.27 a. Draw the 14 constitutional isomers of molecular formula C_8H_9Cl that contain a benzene ring.
  b. Name all compounds that contain a trisubstituted benzene ring.
  c. For which compound(s) are stereoisomers possible? Draw all possible stereoisomers.

Aromaticity

19.28 Label each heterocycle as aromatic, antiaromatic, or not aromatic.

a.   b. 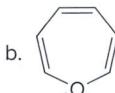 c.   d.

19.29 Label each compound as aromatic, antiaromatic, or not aromatic. Assume all completely conjugated rings are planar.

a. 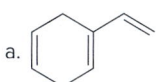 b.   c. 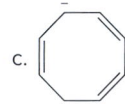 d. 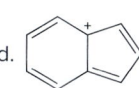 e. 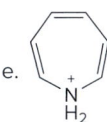 f.

19.30 Label the aromatic rings in each compound, and indicate which electrons are involved in the aromaticity. Thiamine diphosphate is a derivative of vitamin B_1 that is involved in the catalysis of many biological processes. Tetrahydrofolate, a folic acid derivative, has a key role in the metabolism of amino acids and nucleic acids.

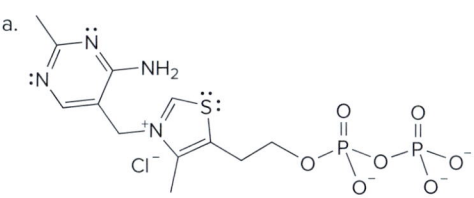

a. thiamine diphosphate

b. tetrahydrofolate

19.31 Three DNA bases exist primarily as the keto tautomer (Section 10.18) because more hydrogen-bonding interactions are possible in this form. However, it was originally thought that the bases existed mainly as the enol tautomers, because they appear to be more traditionally aromatic in this form. Draw the aromatic enol tautomers of the three DNA bases that contain a C=O group.

19.32 Hydrocarbon **A** possesses a significant dipole, even though it is composed of only C—C and C—H bonds. Explain why the dipole arises and use resonance structures to illustrate the direction of the dipole. Which ring is more electron rich?

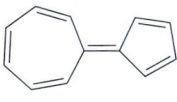

A

19.33 Pentalene, azulene, and heptalene are conjugated hydrocarbons that do not contain a benzene ring. Which hydrocarbons are especially stable or unstable based on the number of π electrons they contain? Explain your choices.

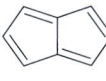 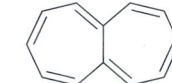

pentalene   azulene   heptalene

19.34 (a) Determine the hybridization of each N atom in tofacitinib, a drug used to treat rheumatoid arthritis. (b) In what type of orbital does the lone pair on each N atom reside?

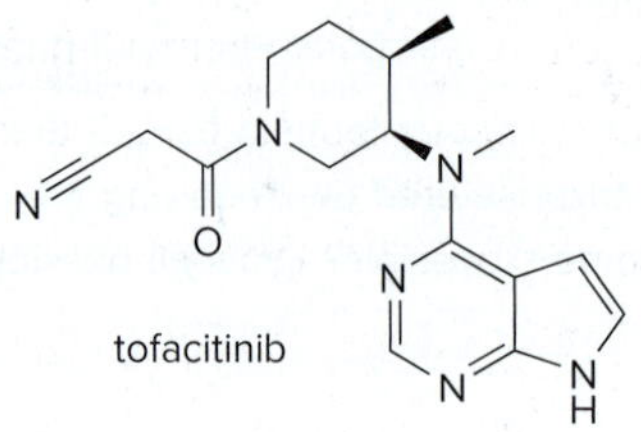

tofacitinib

19.35 (a) Determine the hybridization of each N atom in ibrutinib, a drug used to treat mantle cell lymphoma and chronic lymphocytic leukemia. (b) In what type of orbital does the lone pair on each N atom reside? (c) How many sp^2 hybridized atoms does ibrutinib contain?

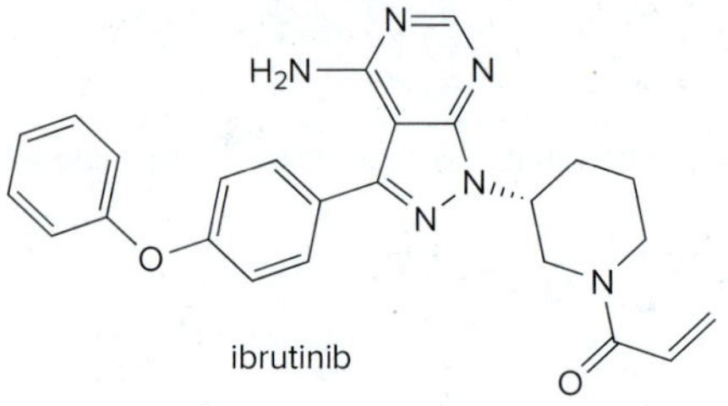

ibrutinib

19.36 AZT was the first drug approved to treat HIV, the virus that causes AIDS. Explain why the six-membered ring of AZT is aromatic.

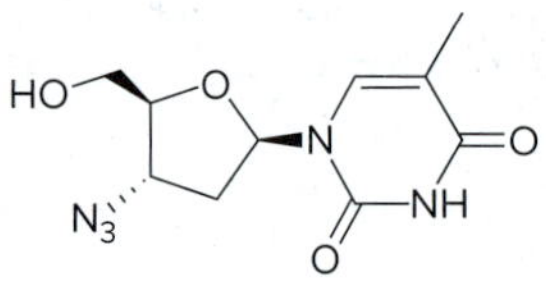

AZT

19.37 Explain the observed rate of reactivity of the following 2° alkyl halides in an S_N1 reaction.

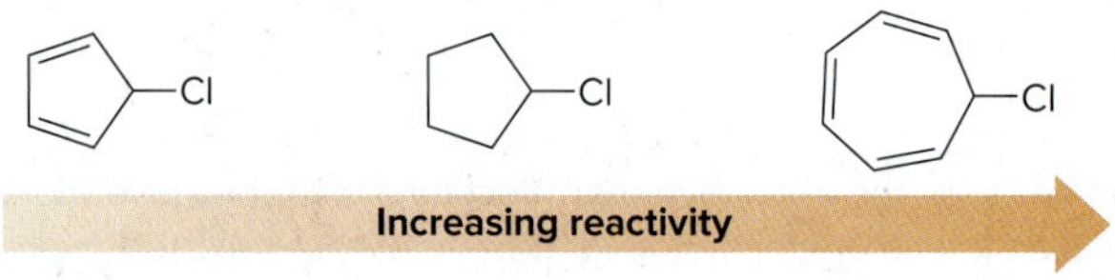

Increasing reactivity

19.38 Draw a stepwise mechanism for the following reaction.

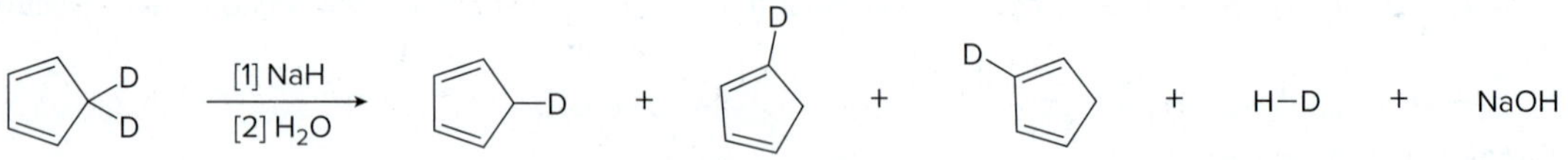

Resonance

19.39 The carbon–carbon bond lengths in naphthalene are not equal. Use a resonance argument to explain why bond (a) is shorter than bond (b).

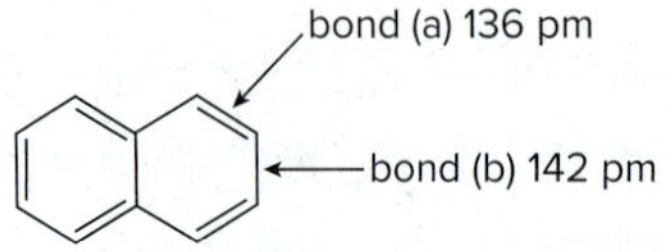

bond (a) 136 pm
bond (b) 142 pm

19.40  a. Draw all reasonable resonance structures for pyrrole, and explain why pyrrole is less resonance stabilized than benzene.

b. Draw all reasonable resonance structures for furan, and explain why furan is less resonance stabilized than pyrrole.

Acidity

19.41 Rank the following compounds in order of increasing acidity.

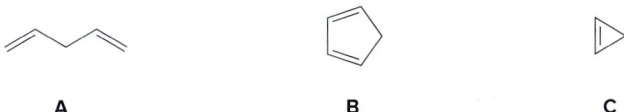

19.42 Treatment of indene with NaNH₂ forms its conjugate base in a Brønsted–Lowry acid–base reaction. Draw all reasonable resonance structures for indene's conjugate base, and explain why the pK_a of indene is lower than the pK_a of most hydrocarbons.

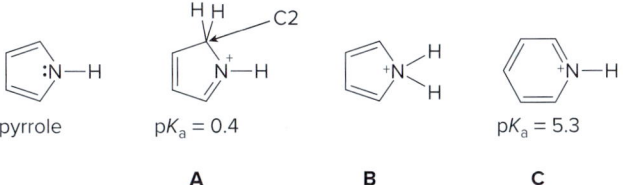

19.43 Draw the conjugate bases of pyrrole and cyclopentadiene. Explain why the sp^3 hybridized C–H bond of cyclopentadiene is more acidic than the N–H bond of pyrrole.

19.44 a. Explain why protonation of pyrrole occurs at C2 to form **A**, rather than on the N atom to form **B**.
b. Explain why **A** is more acidic than **C**, the conjugate acid of pyridine.

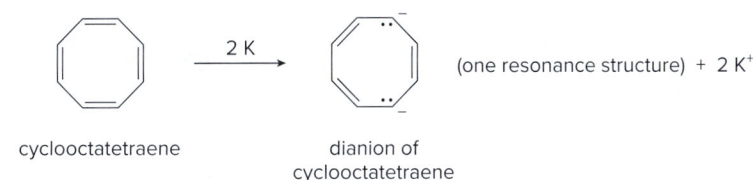

Inscribed Polygon Method

19.45 Use the inscribed polygon method to show the pattern of molecular orbitals in cyclooctatetraene.

a. Label the MOs as bonding, antibonding, or nonbonding.
b. Indicate the arrangement of electrons in these orbitals for cyclooctatetraene, and explain why cyclooctatetraene is not aromatic.
c. Treatment of cyclooctatetraene with potassium forms a dianion. How many π electrons does this dianion contain?
d. How are the π electrons in this dianion arranged in the molecular orbitals?
e. Classify the dianion of cyclooctatetraene as aromatic, antiaromatic, or not aromatic, and explain why this is so.

19.46 Use the inscribed polygon method to show the pattern of molecular orbitals in cyclonona-1,3,5,7-tetraene, and use it to label its cation, radical, and anion as aromatic, antiaromatic, or not aromatic.

Spectroscopy

19.47 How many ^{13}C NMR signals does each compound exhibit?

a. 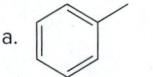   b. 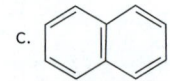   c., d.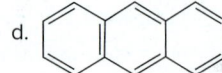

19.48 Which of the diethylbenzene isomers (ortho, meta, or para) corresponds to each set of ^{13}C NMR spectral data?
[A] ^{13}C NMR signals: 16, 29, 125, 127.5, 128.4, and 144 ppm
[B] ^{13}C NMR signals: 15, 26, 126, 128, and 142 ppm
[C] ^{13}C NMR signals: 16, 29, 128, and 141 ppm

19.49 Propose a structure consistent with each set of data.

a. $C_{10}H_{14}$: IR absorptions at 3150–2850, 1600, and 1500 cm^{-1}

b. C_9H_{12}: ^{13}C NMR signals at 21, 127, and 138 ppm

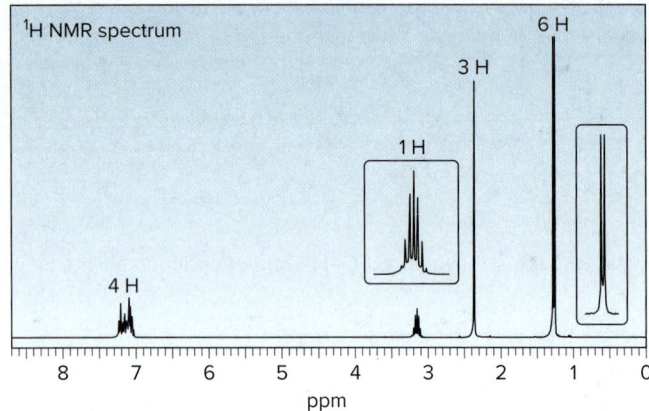
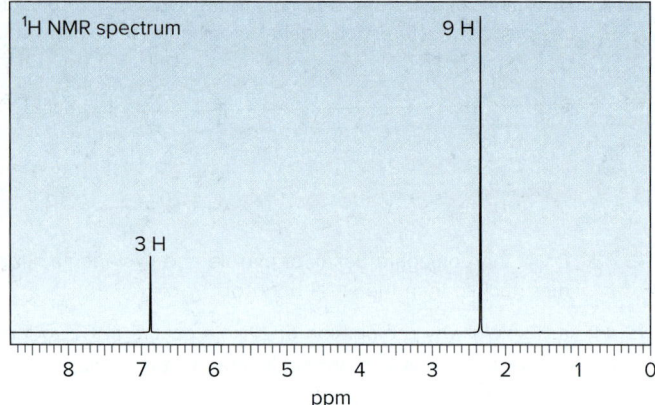

c. C_8H_{10}: IR absorptions at 3108–2875, 1606, and 1496 cm^{-1}

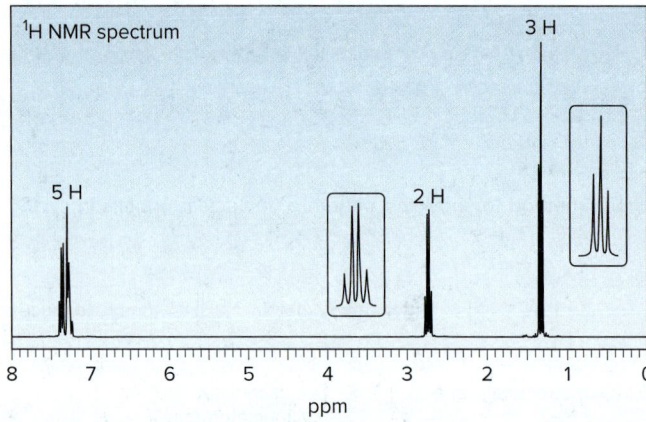

19.50 Thymol (molecular formula $C_{10}H_{14}O$) is the major component of the oil of thyme. Thymol shows IR absorptions at 3500–3200, 3150–2850, 1621, and 1585 cm^{-1}. The ^1H NMR spectrum of thymol is given below. Propose a possible structure for thymol.

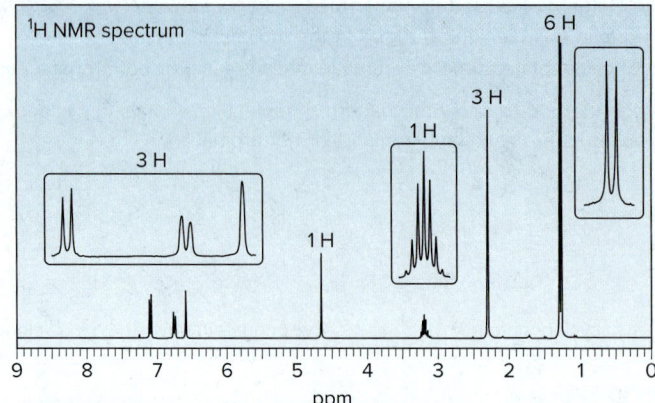

General Problems

19.51 Explain why tetrahydrofuran has a higher boiling point and is much more water soluble than furan, even though both compounds are cyclic ethers containing four carbons.

tetrahydrofuran   furan

19.52 Rizatriptan (trade name Maxalt) is a prescription drug used for the treatment of migraines. (a) How many aromatic rings does rizatriptan contain? (b) Determine the hybridization of each N atom. (c) In what type of orbital does the lone pair on each N reside? (d) Draw all the resonance structures for rizatriptan that contain only neutral atoms. (e) Draw all reasonable resonance structures for the five-membered ring that contains three N atoms.

rizatriptan

19.53 Zolpidem (trade name Ambien) promotes the rapid onset of sleep, making it a widely prescribed drug for treating insomnia.

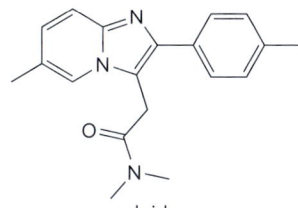

zolpidem

a. In what type of orbital does the lone pair on each N atom in the heterocycle reside?
b. Explain why the bicyclic ring system that contains both N atoms is aromatic.
c. Draw all reasonable resonance structures for the bicyclic ring system.

19.54 Stanozolol is an anabolic steroid that promotes muscle growth. Although stanozolol has been used by athletes and body builders, many physical and psychological problems result from prolonged use and it is banned in competitive sports.

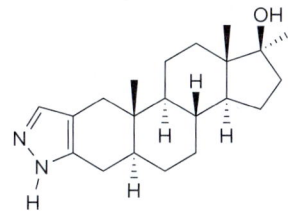

stanozolol

a. Explain why the nitrogen heterocycle—a pyrazole ring—is aromatic.
b. In what type of orbital is the lone pair on each N atom contained?
c. Draw all reasonable resonance structures for stanozolol.
d. Explain why the pK_a of the N–H bond in the pyrazole ring is comparable to the pK_a of the O–H bond, making it considerably more acidic than amines such as CH_3NH_2 (pK_a = 40).

Challenge Problems

19.55 Explain why **A** is aromatic but **B** is not aromatic.

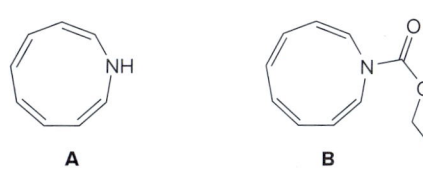

19.56 (a) Which proton in **A** is most acidic? (b) Decide whether **A** is more or less acidic than **B**, and explain your choice.

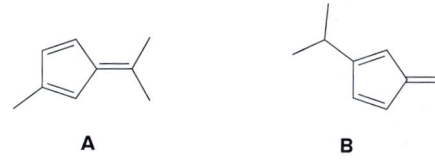

19.57 Use the observed ^1H NMR data to decide whether **C** and its dianion are aromatic, antiaromatic, or not aromatic. **C** shows NMR signals at –4.25 (6 H) and 8.14–8.67 (10 H) ppm. The dianion of **C** shows NMR signals at –3 (10 H) and 21 (6 H) ppm. Why are the signals shifted upfield (or downfield) to such a large extent?

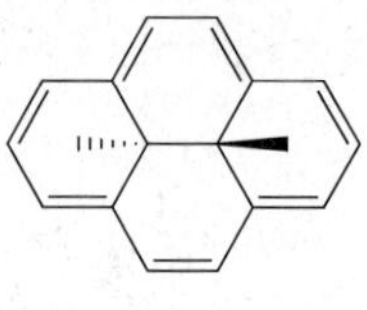

C

19.58 Explain why compound **A** is much more stable than compound **B.**

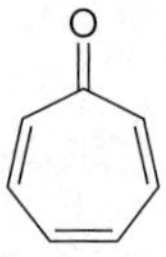
O

A

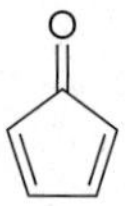
O

B

19.59 (*R*)-Carvone, the major component of the oil of spearmint, undergoes acid-catalyzed isomerization to carvacrol, a major component of the oil of thyme. Draw a stepwise mechanism and explain why this isomerization occurs.

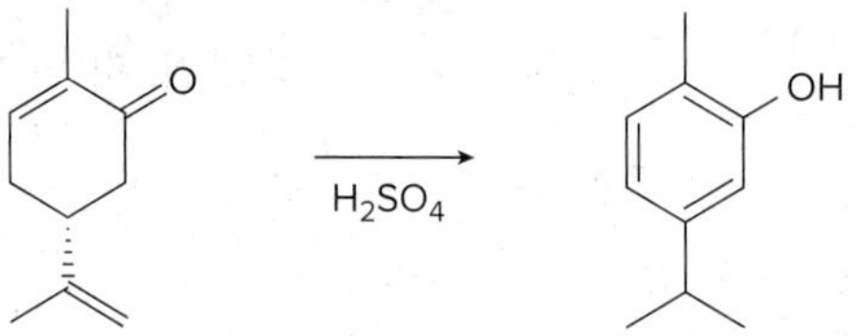
O
H_2SO_4
OH

(*R*)-carvone carvacrol

19.60 Explain why triphenylene resembles benzene in that it does not undergo addition reactions with Br_2, but phenanthrene reacts with Br_2 to yield the addition product drawn. (Hint: Draw resonance structures for both triphenylene and phenanthrene, and use them to determine how delocalized each π bond is.)

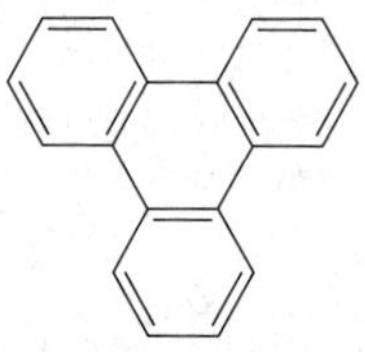

triphenylene

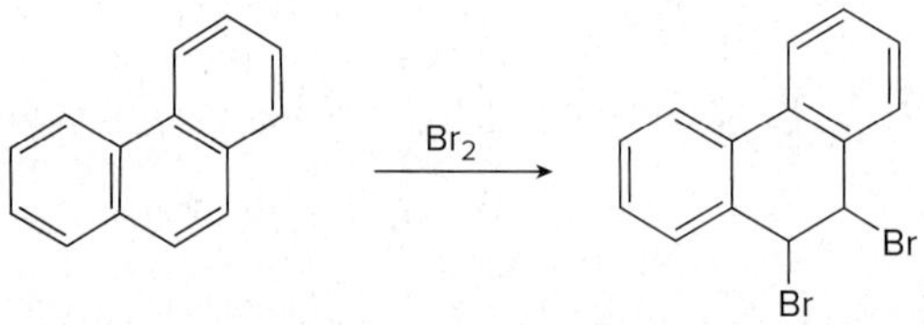
Br_2
Br
Br

phenanthrene

19.61 Although benzene itself absorbs at 128 ppm in its ^{13}C NMR spectrum, the carbons of substituted benzenes absorb either upfield or downfield from this value depending on the substituent. Explain the observed values for the carbon ortho to the given substituent in the monosubstituted benzene derivatives **X** and **Y.**

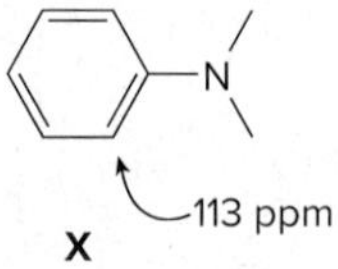
N
113 ppm

X

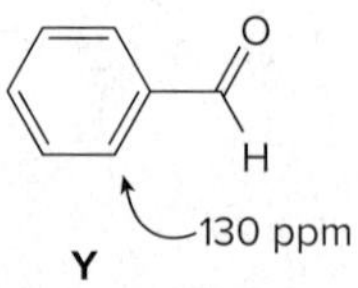
O
H
130 ppm

Y

Reactions of Aromatic Compounds

20

- **20.1** Electrophilic aromatic substitution
- **20.2** The general mechanism
- **20.3** Halogenation
- **20.4** Nitration and sulfonation
- **20.5** Friedel–Crafts alkylation and Friedel–Crafts acylation
- **20.6** Substituted benzenes
- **20.7** Electrophilic aromatic substitution of substituted benzenes
- **20.8** Why substituents activate or deactivate a benzene ring
- **20.9** Orientation effects in substituted benzenes
- **20.10** Limitations on electrophilic substitution reactions with substituted benzenes
- **20.11** Disubstituted benzenes
- **20.12** Synthesis of benzene derivatives
- **20.13** Nucleophilic aromatic substitution
- **20.14** Reactions of substituted benzenes
- **20.15** Multistep synthesis

Jill Braaten

Vitamin K$_1$, phylloquinone, is a fat-soluble vitamin that regulates the synthesis of proteins needed for blood to clot. Dietary sources of vitamin K$_1$ include cauliflower, broccoli, soybeans, leafy greens, and green tea. A severe deficiency of vitamin K$_1$ leads to excessive and sometimes fatal bleeding because of inadequate blood clotting. Vitamin K$_1$ is synthesized by a biological Friedel–Crafts reaction, one of the many examples of electrophilic aromatic substitution, a key reaction of aromatic compounds presented in Chapter 20.

Why Study...
Reactions of Aromatic Compounds?

Chapter 20 discusses the chemical reactions of benzene and other aromatic compounds. Although aromatic rings are unusually stable, making benzene unreactive in most of the reactions discussed so far, benzene acts as a nucleophile with certain electrophiles, yielding substitution products with an intact aromatic ring.

We begin with the basic features and mechanism of electrophilic aromatic substitution (Sections 20.1–20.5), the most prevalent reaction of benzene. Next, we discuss the electrophilic aromatic substitution of substituted benzenes (Sections 20.6–20.12), and conclude with nucleophilic aromatic substitution and other useful reactions of benzene derivatives (Sections 20.13 and 20.14). These reactions have been used to prepare antidepressants, antipsychotics, and drugs to treat diabetes.

20.1 Electrophilic Aromatic Substitution

Based on its structure and properties, what kinds of reactions should benzene undergo? Are any of its bonds particularly weak? Does it have electron-rich or electron-deficient atoms?

- Benzene has six π electrons delocalized in six *p* orbitals that overlap above and below the plane of the ring. These loosely held π electrons make the benzene ring electron rich, so it reacts with *electrophiles*.
- Because benzene's six π electrons satisfy Hückel's rule, benzene is especially stable. Reactions that keep the aromatic ring *intact* are therefore favored.

As a result, **the characteristic reaction of benzene is** *electrophilic aromatic substitution*—**a hydrogen atom is replaced by an electrophile.**

Electrophilic aromatic substitution

C$_6$H$_5$–H + E$^+$ ⟶ C$_6$H$_5$–E + H$^+$

aromatic product

As we learned in Section 19.6, benzene does *not* undergo addition reactions like other unsaturated hydrocarbons, because addition would yield a product that is not aromatic. Substitution of a hydrogen, on the other hand, keeps the aromatic ring intact.

Five specific examples of electrophilic aromatic substitution are shown in Figure 20.1. The basic mechanism, discussed in Section 20.2, is the same in all five cases. The reactions differ only in the identity of the electrophile, E$^+$.

Problem 20.1 Why is benzene less reactive toward electrophiles than an alkene, even though it has more π electrons than an alkene (six versus two)?

Figure 20.1
Five examples of electrophilic aromatic substitution

Reaction		Electrophile
[1] Halogenation—Replacement of H by X (Cl or Br)	PhH + X$_2$ →(FeX$_3$) aryl halide; X = Cl, X = Br	$E^+ = Cl^+$ or Br^+
[2] Nitration—Replacement of H by NO$_2$	PhH + HNO$_3$/H$_2$SO$_4$ → nitrobenzene	$E^+ = \overset{+}{N}O_2$
[3] Sulfonation—Replacement of H by SO$_3$H	PhH + SO$_3$/H$_2$SO$_4$ → benzenesulfonic acid	$E^+ = \overset{+}{S}O_3H$
[4] Friedel–Crafts alkylation—Replacement of H by R	PhH + RCl/AlCl$_3$ → alkyl benzene (arene)	$E^+ = R^+$
[5] Friedel–Crafts acylation—Replacement of H by RCO	PhH + RCOCl/AlCl$_3$ → ketone	$E^+ = R-\overset{+}{C}=\overset{..}{\underset{..}{O}}:$

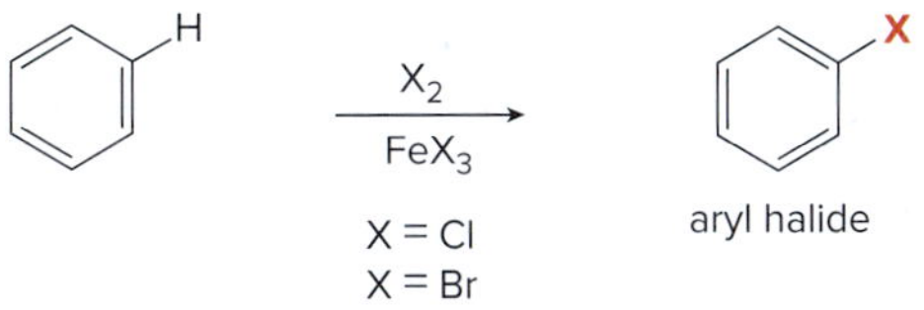

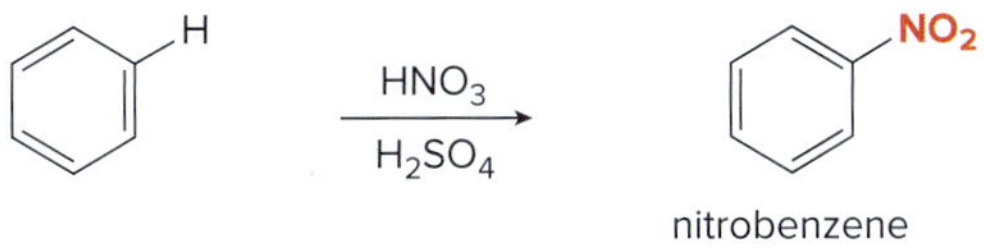

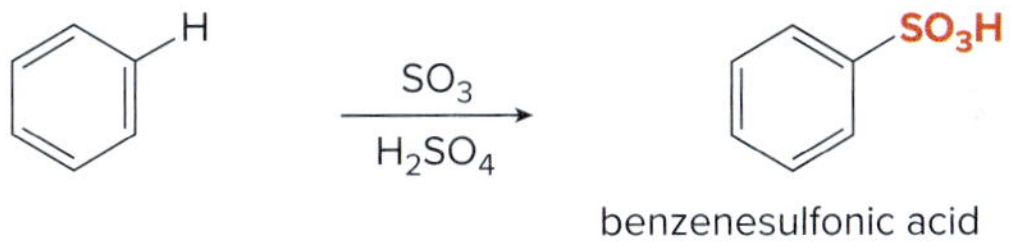

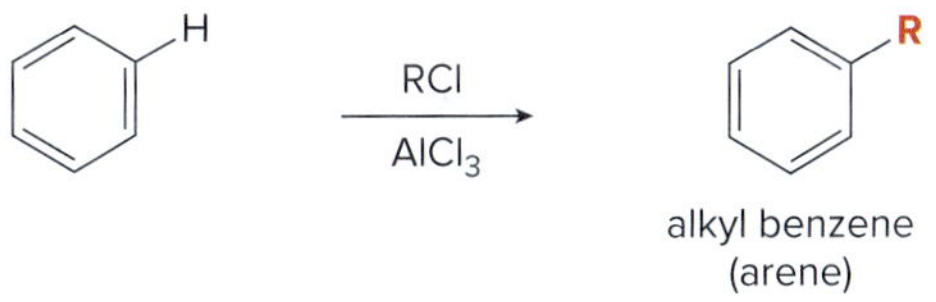

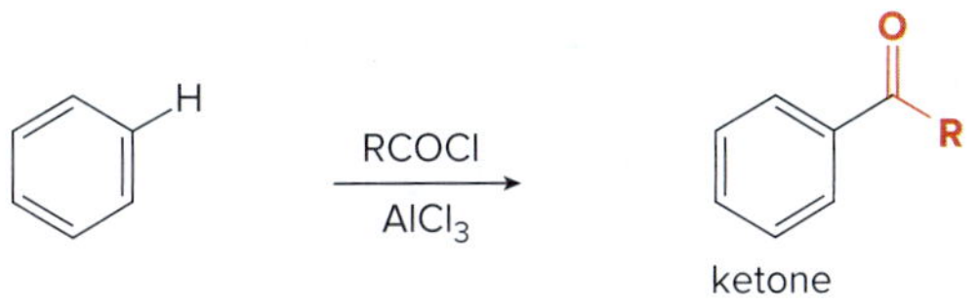

Friedel–Crafts alkylation and acylation, named for Charles Friedel and James Crafts, who discovered the reactions in the nineteenth century, form new carbon–carbon bonds.

20.2 The General Mechanism

No matter what electrophile is used, all electrophilic aromatic substitution reactions occur via a **two-step mechanism**: addition of the electrophile E^+ to form a resonance-stabilized carbocation, followed by deprotonation with base, as shown in Mechanism 20.1.

Mechanism 20.1 General Mechanism—Electrophilic Aromatic Substitution

resonance-stabilized carbocation

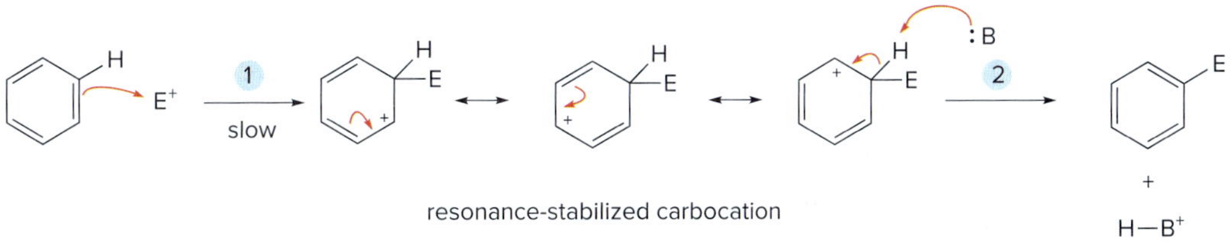

1. Addition of the electrophile E^+ forms a new C—E bond and a **resonance-stabilized carbocation.** This step is rate-determining because the aromaticity of the benzene ring is lost.
2. A base removes the proton **on the carbon bonded to the electrophile,** re-forming the aromatic ring. Any resonance structure can be used to draw the product.

The first step in electrophilic aromatic substitution forms a carbocation, for which three resonance structures can be drawn. To help keep track of the location of the positive charge:

- Always draw in the H atom on the carbon bonded to E. This serves as a reminder that it is the only sp^3 hybridized carbon in the carbocation intermediate.
- Notice that the positive charge in a given resonance structure is always located ortho or para to the new C–E bond. In the hybrid, therefore, the charge is delocalized over three atoms of the ring.

(+) ortho to E   (+) para to E   (+) ortho to E   hybrid

This two-step mechanism for electrophilic aromatic substitution applies to all of the electrophiles in Figure 20.1. **The net result of addition of an electrophile (E^+) followed by elimination of a proton (H^+) is substitution of E for H.**

Problem 20.2 In Step [2] of Mechanism 20.1, loss of a proton to form the substitution product was drawn using only one resonance structure. Use curved arrows to show how the other two resonance structures can be converted to the substitution product (PhE) by removal of a proton with :B.

20.3 Halogenation

The general mechanism outlined in Mechanism 20.1 can now be applied to each of the five specific examples of electrophilic aromatic substitution shown in Figure 20.1. For each mechanism we must learn how to generate a specific electrophile. This step is *different* with each electrophile. Then, the electrophile reacts with benzene by the two-step process of Mechanism 20.1. These two steps are the *same* for all five reactions.

In **halogenation,** benzene reacts with Cl_2 or Br_2 in the presence of a Lewis acid catalyst, such as $FeCl_3$ or $FeBr_3$, to give the **aryl halides** chlorobenzene or bromobenzene, respectively. Analogous reactions with I_2 and F_2 are not synthetically useful because I_2 is too unreactive and F_2 reacts too violently.

benzene $\xrightarrow{Cl_2, FeCl_3}$ chlorobenzene

benzene $\xrightarrow{Br_2, FeBr_3}$ bromobenzene

In bromination (Mechanism 20.2), the Lewis acid $FeBr_3$ reacts with Br_2 to form a **Lewis acid–base complex** that weakens and polarizes the Br–Br bond, making it more electrophilic. This reaction is Step [1] of the mechanism for the bromination of benzene. The remaining two steps follow directly from the general mechanism for electrophilic aromatic substitution: addition of the electrophile (Br^+ in this case) forms a resonance-stabilized carbocation, and loss of a proton regenerates the aromatic ring.

Mechanism 20.2 Bromination of Benzene

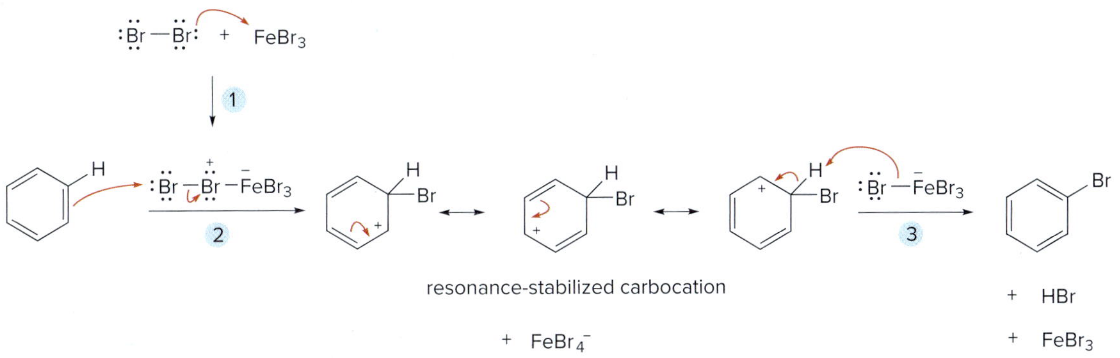

1. Lewis acid–base reaction of Br_2 with $FeBr_3$ forms a species with a weakened Br—Br bond that serves as source of **Br^+**.
2. Addition of the electrophile forms a new C—Br bond and a **resonance-stabilized carbocation.**
3. $FeBr_4^-$ removes the proton **on the carbon bonded to the electrophile,** re-forming the aromatic ring. The Lewis acid catalyst $FeBr_3$ is regenerated for another reaction cycle.

Chlorination proceeds by a similar mechanism. Reactions that introduce a halogen substituent on a benzene ring are widely used, and many halogenated aromatic compounds with a range of biological activity have been synthesized, as shown in Figure 20.2.

Figure 20.2
Examples of biologically active aryl chlorides

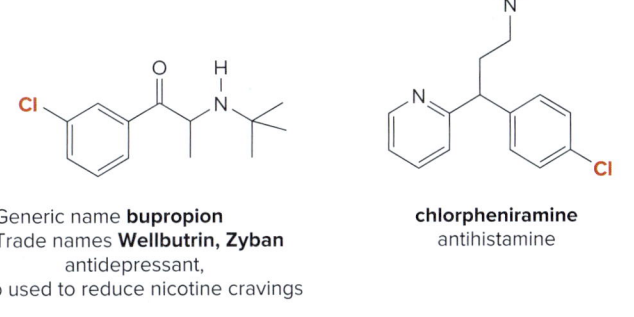

Generic name **bupropion**
Trade names **Wellbutrin, Zyban**
antidepressant,
also used to reduce nicotine cravings

chlorpheniramine
antihistamine

Herbicides were used extensively during the Vietnam War to defoliate dense jungle areas. The concentration of certain herbicide by-products in the soil remains high today.

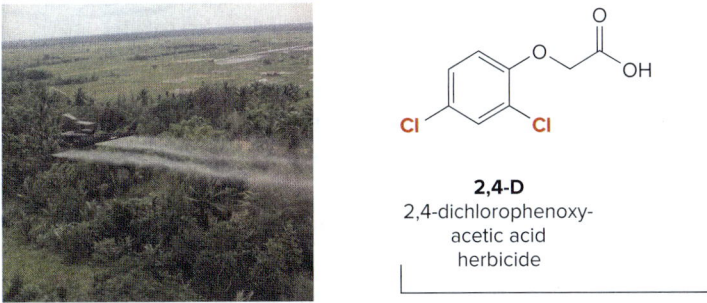

2,4-D
2,4-dichlorophenoxy-
acetic acid
herbicide

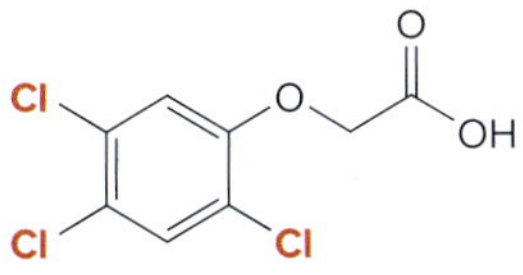

2,4,5-T
2,4,5-trichlorophenoxy-
acetic acid
herbicide

Source: National Archives and Records Administration [NWDNS-111-C-CC59950]

the active components in **Agent Orange,**
a defoliant used in the Vietnam War

Problem 20.3 Draw a detailed mechanism for the chlorination of benzene using Cl_2 and $FeCl_3$.

20.4 Nitration and Sulfonation

Nitration and **sulfonation** of benzene introduce two different functional groups on an aromatic ring. Nitration is an especially useful reaction because a nitro group can then be reduced to an NH_2 group, a common benzene substituent, in a reaction discussed in Section 20.14.

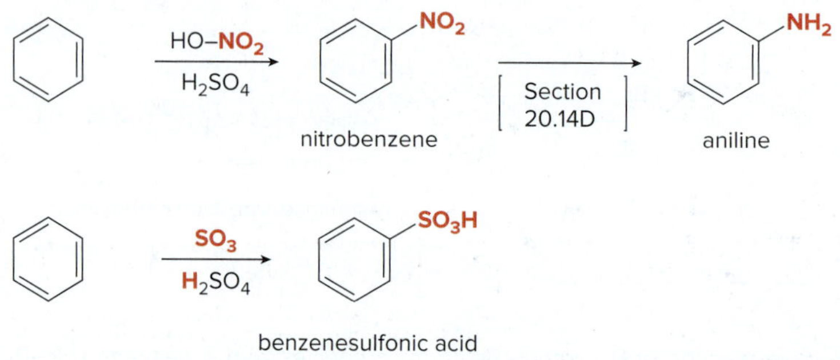

benzene →(HO–NO_2, H_2SO_4)→ nitrobenzene →(Section 20.14D)→ aniline

benzene →(SO_3, H_2SO_4)→ benzenesulfonic acid

Generation of the electrophile in both nitration and sulfonation requires strong acid. In **nitration,** the electrophile is $^+NO_2$ (the **nitronium ion**), formed by protonation of HNO_3 followed by loss of water (Mechanism 20.3).

Mechanism 20.3 Formation of the Nitronium Ion ($^+NO_2$) for Nitration

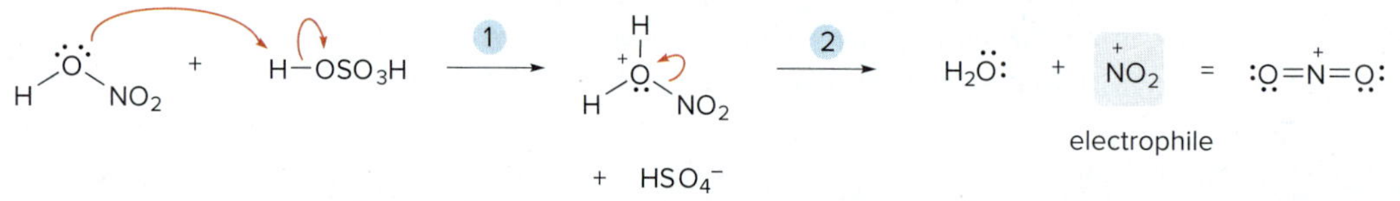

$H\text{-}O\text{-}NO_2$ + $H\text{-}OSO_3H$ →[1]→ $H_2O^+\text{-}NO_2$ + HSO_4^- →[2]→ $H_2\ddot{O}:$ + $^+NO_2$ = $:\ddot{O}=\overset{+}{N}=\ddot{O}:$ (electrophile)

In **sulfonation,** protonation of sulfur trioxide, SO_3, forms a positively charged sulfur species ($^+SO_3H$) that acts as an electrophile (Mechanism 20.4).

Mechanism 20.4 Formation of the Electrophile $^+SO_3H$ for Sulfonation

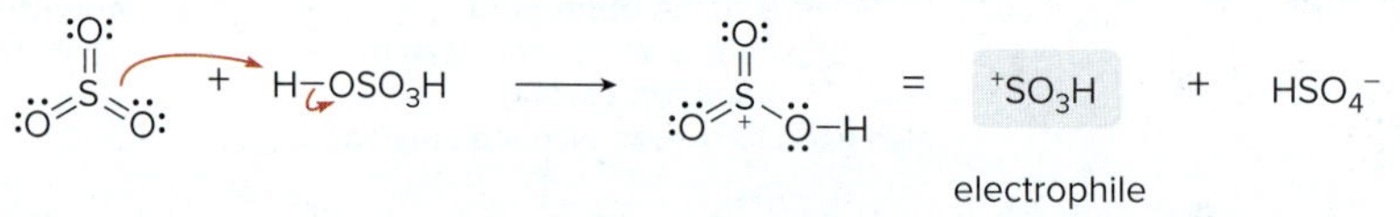

SO_3 + $H\text{-}OSO_3H$ → $^+SO_3\text{-}OH$ = $^+SO_3H$ + HSO_4^- (electrophile)

These steps illustrate how to generate the electrophile E^+ for nitration and sulfonation, the process that begins any mechanism for electrophilic aromatic substitution. To complete either of these mechanisms, you must replace the electrophile E^+ by either $^+NO_2$ or $^+SO_3H$ in the general mechanism (Mechanism 20.1). Thus, **the two-step sequence that replaces H by E is the same regardless of E^+.** This is shown in Sample Problem 20.1 using the reaction of benzene with the nitronium ion.

Sample Problem 20.1 Drawing the Mechanism for Nitration of Benzene

Draw a stepwise mechanism for the nitration of a benzene ring.

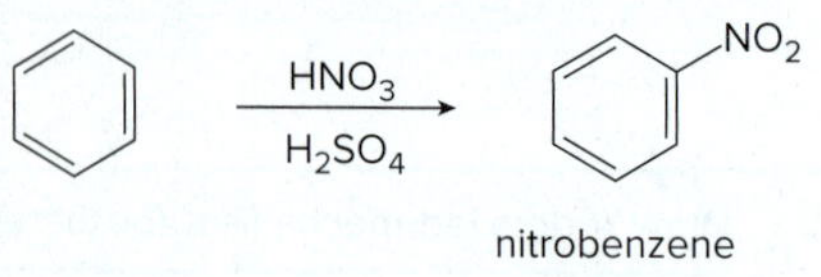

benzene →(HNO_3, H_2SO_4)→ nitrobenzene

Solution

We must first generate the electrophile and then write the two-step mechanism for electrophilic aromatic substitution using it.

Part [1] Generation of the electrophile $^+NO_2$

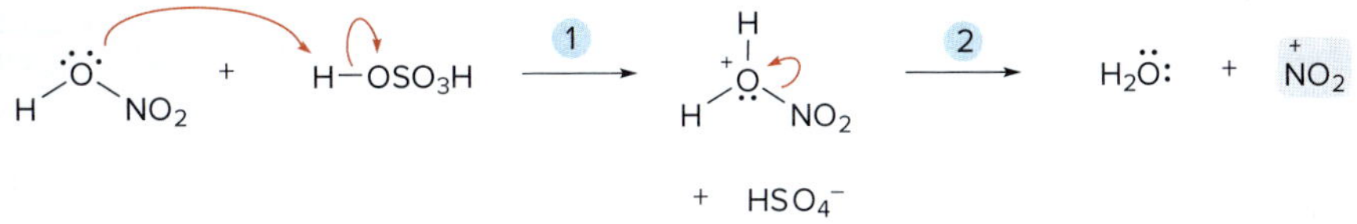

Part [2] Two-step mechanism for electrophilic aromatic substitution

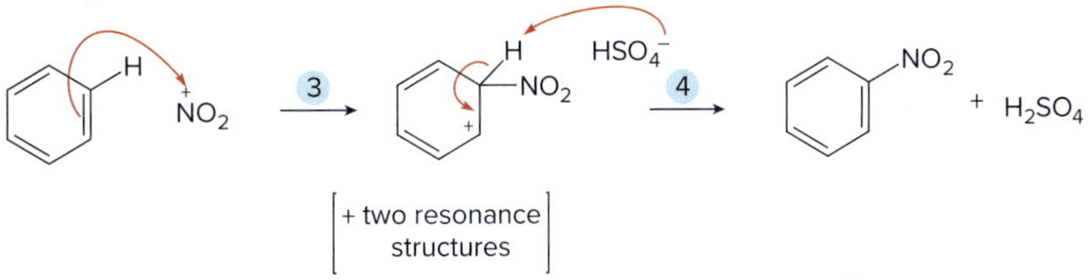

Any species with a lone pair of electrons can be used to remove the proton in the last step. In this case, the mechanism is drawn with HSO_4^-, formed when $^+NO_2$ is generated as the electrophile.

Problem 20.4 Draw a stepwise mechanism for the sulfonation of an alkyl benzene such as **A** to form a substituted benzenesulfonic acid **B.** Treatment of **B** with base forms a sodium salt **C** that can be used as a synthetic detergent to clean away dirt (see Problem 3.23).

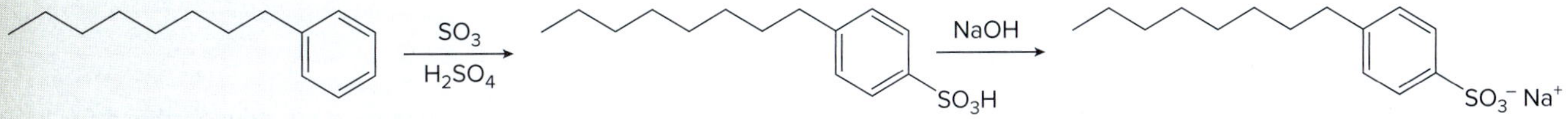

A **B** synthetic detergent **C**

20.5 Friedel–Crafts Alkylation and Friedel–Crafts Acylation

Friedel–Crafts alkylation and **Friedel–Crafts acylation** form new carbon–carbon bonds.

20.5A General Features

In **Friedel–Crafts alkylation,** treatment of benzene with an alkyl halide and a Lewis acid ($AlCl_3$) forms an alkyl benzene. This reaction is an **alkylation** because it results in transfer of an alkyl group from one atom to another (from Cl to benzene).

Friedel–Crafts alkylation

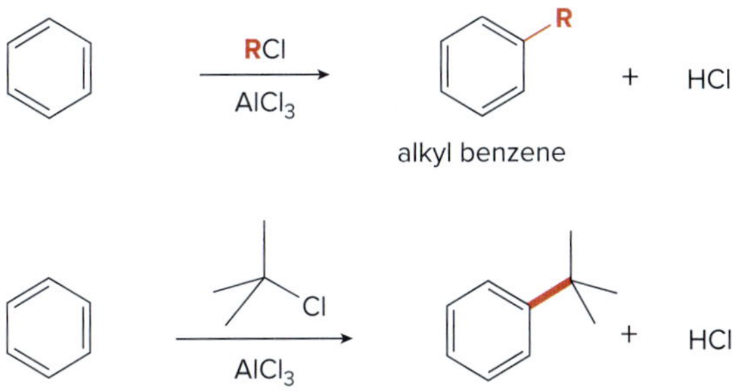

In **Friedel–Crafts acylation,** a benzene ring is treated with an **acid chloride** (RCOCl) and AlCl₃ to form a ketone. This reaction is an **acylation** because it results in the transfer of an acyl group from one atom to another.

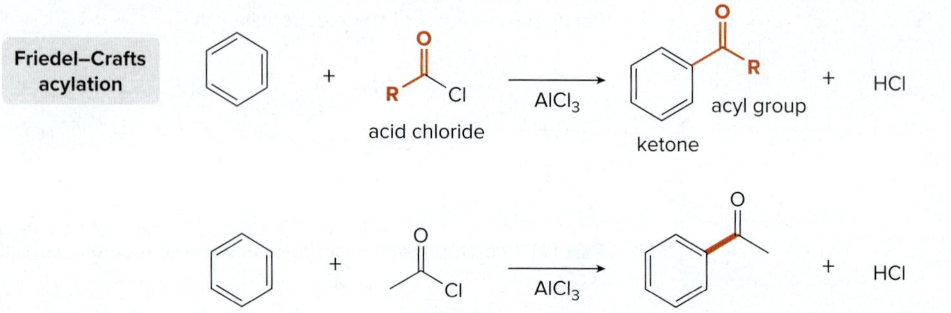

Problem 20.5
What product is formed when benzene is treated with each organic halide in the presence of AlCl₃?

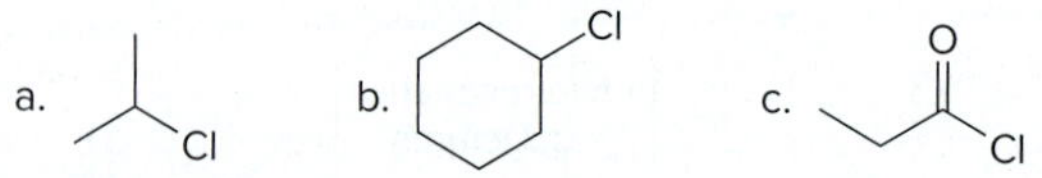

Problem 20.6
What acid chloride would be needed to prepare each of the following ketones from benzene using a Friedel–Crafts acylation?

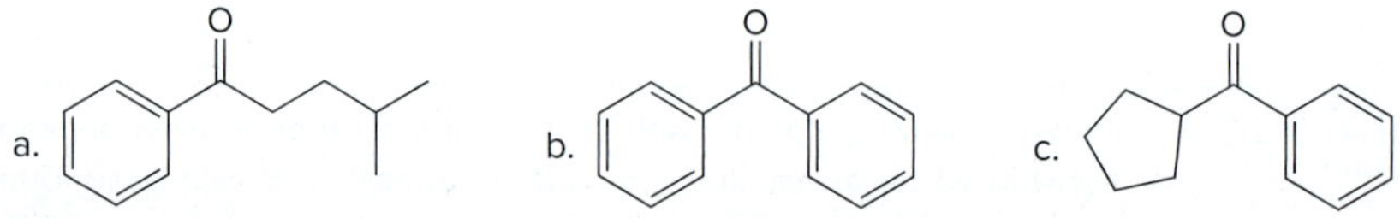

20.5B Mechanism

The mechanisms of alkylation and acylation proceed in a manner analogous to those for halogenation, nitration, and sulfonation. The unique feature in each reaction is how the electrophile is generated.

In **Friedel–Crafts alkylation,** the Lewis acid AlCl₃ reacts with the alkyl chloride to form a **Lewis acid–base complex,** illustrated with CH₃CH₂Cl and (CH₃)₃CCl as alkyl chlorides. The identity of the alkyl chloride determines the exact course of the reaction as shown in Mechanism 20.5.

Mechanism 20.5 Two Possibilities for the Formation of the Electrophile in Friedel–Crafts Alkylation

Possibility [1] For CH₃Cl and 1° RCl

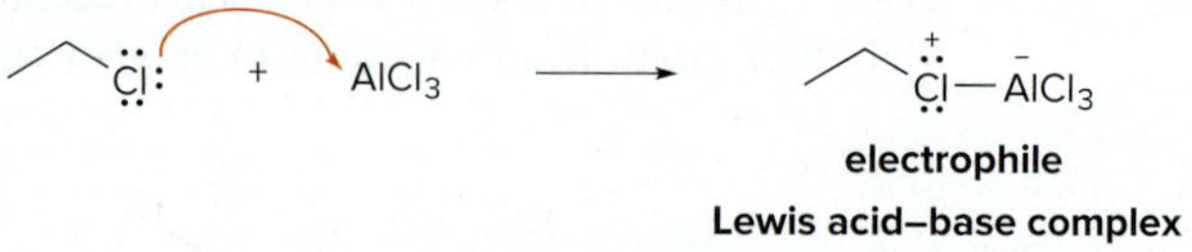

Possibility [2] For 2° and 3° RCl

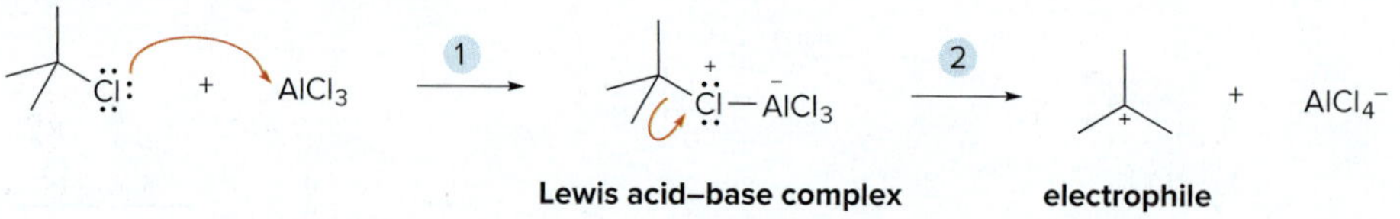

- For CH₃Cl and 1° RCl, the Lewis acid–base complex itself serves as the electrophile for electrophilic aromatic substitution.
- With 2° and 3° RCl, the Lewis acid–base complex reacts further to give a 2° or 3° carbocation, which serves as the electrophile. Carbocation formation occurs only with 2° and 3° alkyl chlorides, because they afford more stable carbocations.

In either case, the electrophile goes on to react with benzene in the two-step mechanism characteristic of electrophilic aromatic substitution, illustrated in Mechanism 20.6 using the 3° carbocation, $(CH_3)_3C^+$.

Mechanism 20.6 Friedel–Crafts Alkylation Using a 3° Carbocation

[mechanism diagram showing benzene + tert-butyl carbocation → arenium intermediate (+ two resonance structures) + :Cl—AlCl₃ → tert-butylbenzene + HCl + AlCl₃]

1 Addition of the carbocation electrophile forms a **new carbon–carbon bond**.

2 AlCl₄⁻ removes a proton on the carbon bearing the new substituent to re-form the aromatic ring.

In **Friedel–Crafts acylation,** the Lewis acid AlCl₃ ionizes the carbon–halogen bond of the acid chloride, thus forming a positively charged carbon electrophile called an **acylium ion,** which is resonance stabilized (Mechanism 20.7). The positively charged carbon atom of the acylium ion then goes on to react with benzene in the two-step mechanism of electrophilic aromatic substitution.

Mechanism 20.7 Formation of the Electrophile in Friedel–Crafts Acylation

[mechanism: R–C(=O)–Cl: + AlCl₃ → Lewis acid–base complex (R–C(=O)–Cl⁺—AlCl₃⁻) → resonance-stabilized acylium ion: R–C⁺=O: ↔ R–C≡O⁺: + AlCl₄⁻]

To complete the mechanism for acylation, insert the electrophile into the general mechanism and draw the last two steps, as illustrated in Sample Problem 20.2.

Sample Problem 20.2 Drawing a Mechanism for a Friedel–Crafts Reaction

Draw a stepwise mechanism for the following Friedel–Crafts acylation.

[reaction: benzene + CH₃C(=O)Cl, AlCl₃ → acetophenone + HCl]

Solution
First generate the **acylium ion,** and then write the two-step mechanism for electrophilic aromatic substitution using it for the electrophile.

Part [1] Generation of the electrophile $(CH_3CO)^+$

[mechanism: CH₃C(=O)Cl: + AlCl₃ → Lewis acid–base complex → resonance-stabilized acylium ion CH₃–C⁺=O: ↔ CH₃–C≡O⁺: + AlCl₄⁻]

Part [2] Two-step mechanism for electrophilic aromatic substitution

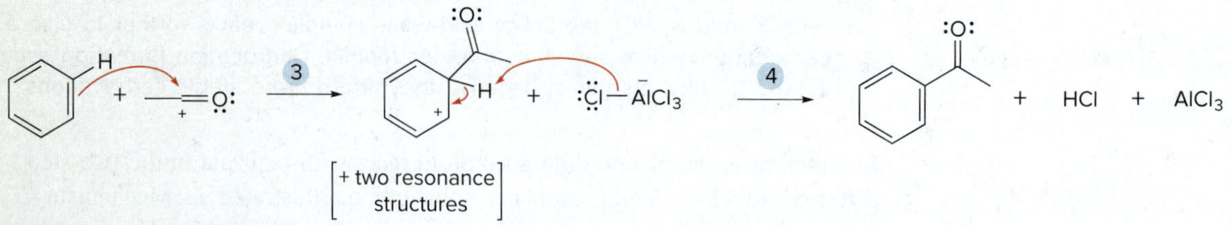

[+ two resonance structures]

Problem 20.7 What acylium ion is formed from each acid chloride?

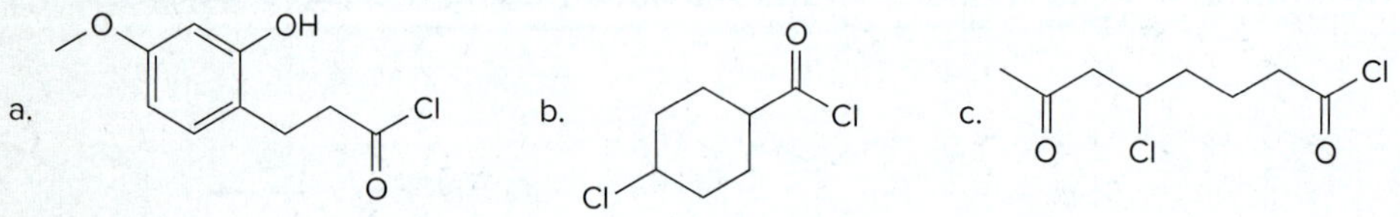

a., b., c.

More Practice: Try Problem 20.11.

20.5C Other Facts About Friedel–Crafts Alkylation

Three additional facts about Friedel–Crafts alkylations must be kept in mind.

[1] **Vinyl halides and aryl halides do *not* react in Friedel–Crafts alkylation.**

Most Friedel–Crafts reactions involve carbocation electrophiles. Because the carbocations derived from vinyl halides and aryl halides are highly unstable and do not readily form, these organic halides do *not* undergo Friedel–Crafts alkylation.

vinyl halide unreactive   aryl halide unreactive

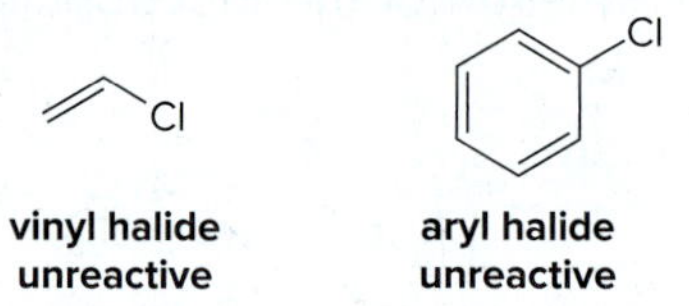

Problem 20.8 Which halides are unreactive in a Friedel–Crafts alkylation reaction?

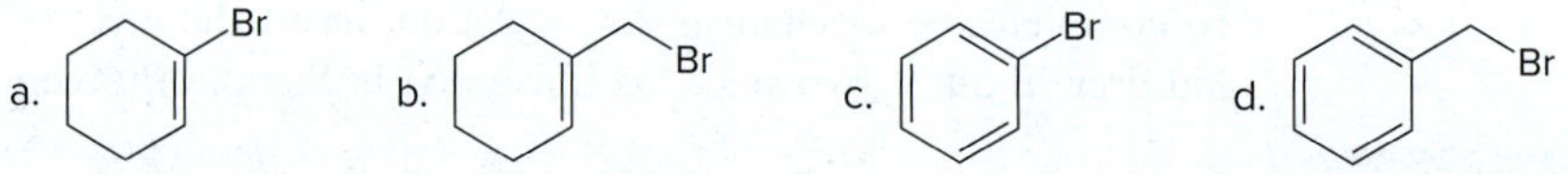

a., b., c., d.

---

[2] **Rearrangements can occur.**

The Friedel–Crafts reaction can yield products having rearranged carbon skeletons when 1° and 2° alkyl halides are used as starting materials, as shown in Equations [1] and [2]. In both reactions, the carbon atom bonded to the halogen in the starting material (labeled in blue) is not bonded to the benzene ring in the product, thus indicating that a rearrangement has occurred.

Recall from Section 9.9 that a 1,2-shift converts a less stable carbocation to a more stable carbocation by shift of a hydrogen atom or an alkyl group.

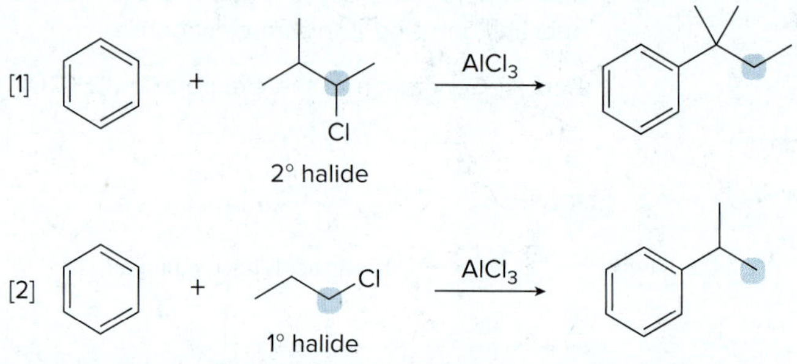

[1] benzene + 2° halide →(AlCl₃)→ product

[2] benzene + 1° halide →(AlCl₃)→ product

20.5 Friedel–Crafts Alkylation and Friedel–Crafts Acylation

The result in Equation [1] is explained by a carbocation rearrangement involving a 1,2-hydride shift: **the less stable 2° carbocation (formed from the 2° halide) rearranges to a more stable 3° carbocation,** as illustrated in Mechanism 20.8.

Mechanism 20.8 Friedel–Crafts Alkylation Involving Carbocation Rearrangement

Part [1] Formation of a 2° carbocation and rearrangement

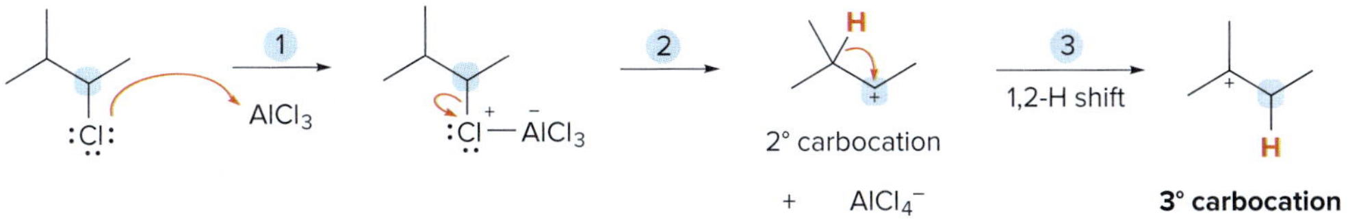

1–2 Lewis acid–base reaction of the alkyl chloride with AlCl₃ and cleavage of the C–Cl bond form a 2° carbocation.

3   **1,2-Hydride shift** converts a 2° carbocation to a **more stable 3° carbocation.**

Part [2] Two-step mechanism for electrophilic aromatic substitution

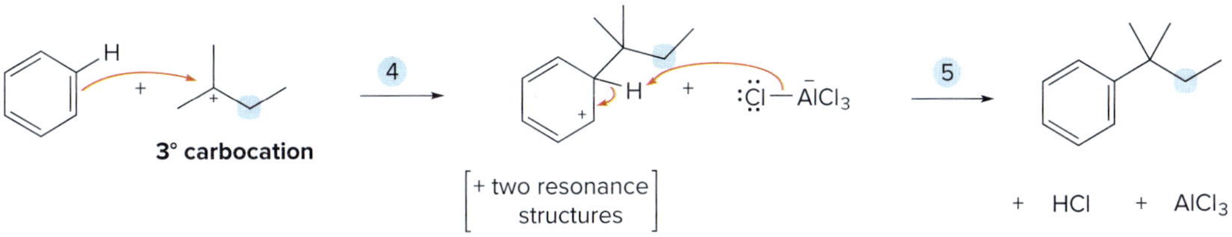

4   Addition of the 3° carbocation forms a new carbon–carbon bond and a **resonance-stabilized carbocation.**

5   AlCl₄⁻ removes a proton on the carbon bearing the new substituent to re-form the aromatic ring.

Rearrangements can occur even when no free carbocation is formed initially. For example, the 1° alkyl chloride in Equation [2] forms a complex with AlCl₃, which does *not* decompose to an unstable 1° carbocation, as shown in Mechanism 20.9. Instead, a **1,2-hydride shift** forms a 2° carbocation, which then serves as the electrophile in the two-step mechanism for electrophilic aromatic substitution.

Mechanism 20.9 A Rearrangement Reaction Beginning with a 1° Alkyl Chloride

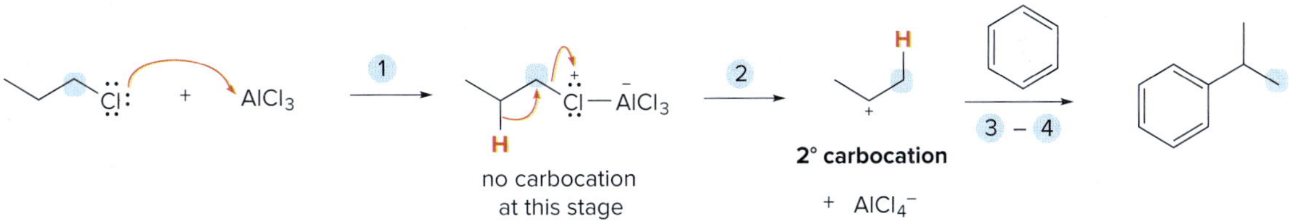

Problem 20.9   Draw a stepwise mechanism for the following reaction.

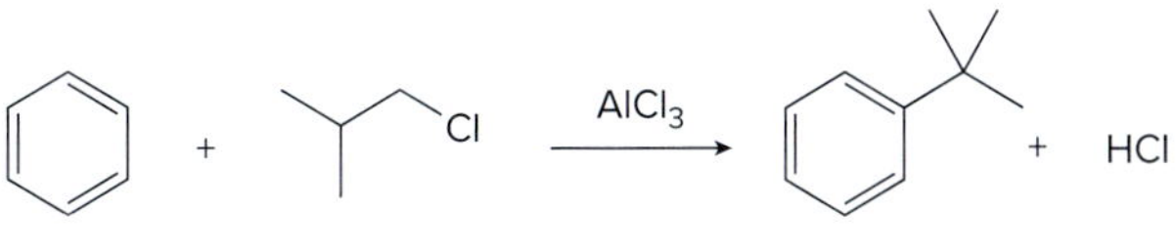

Chapter 20 Reactions of Aromatic Compounds

[3] **Other functional groups that form carbocations can also be used as starting materials.**

Although Friedel–Crafts alkylation works well with alkyl halides, any compound that readily forms a carbocation can be used instead. The two most common alternatives are alkenes and alcohols, both of which afford carbocations in the presence of strong acid.

- Protonation of an alkene forms a carbocation, which can then serve as an electrophile in a Friedel–Crafts alkylation.
- Protonation of an alcohol, followed by loss of water, likewise forms a carbocation.

Each carbocation can then go on to react with benzene to form a product of electrophilic aromatic substitution. For example:

Problem 20.10 Draw the product of each reaction.

a. benzene + cyclohexene $\xrightarrow{H_2SO_4}$

b. benzene + (CH$_3$)$_2$C=CH$_2$ $\xrightarrow{H_2SO_4}$

c. benzene + 2-methylbutan-2-ol $\xrightarrow{H_2SO_4}$

d. benzene + cyclohexanol $\xrightarrow{H_2SO_4}$

20.5D Intramolecular Friedel–Crafts Reactions

All of the Friedel–Crafts reactions discussed thus far have resulted from intermolecular reaction of a benzene ring with an electrophile. Starting materials that contain *both* units are capable of **intramolecular reaction,** and this forms a new ring. Such an intramolecular Friedel–Crafts acylation was a key step in the synthesis of the hallucinogen LSD, as shown in Figure 20.3.

Problem 20.11 Draw a stepwise mechanism for the intramolecular Friedel–Crafts acylation of compound **A** to form **B**. **B** can be converted in one step to the antidepressant sertraline.

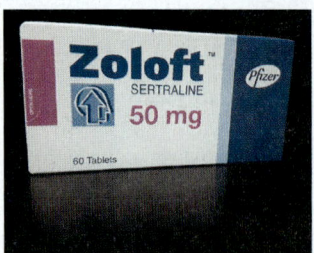

Sertraline (trade name Zoloft, Problem 20.11) is an effective antidepressant because it increases the concentration of the neurotransmitter serotonin in the brain.
Omeletzz/Shutterstock

20.5 Friedel–Crafts Alkylation and Friedel–Crafts Acylation

Figure 20.3 Intramolecular Friedel–Crafts acylation in the synthesis of LSD

Ergot-infected grain, the source of lysergic acid. *Rene Dulhoste/Science Source*

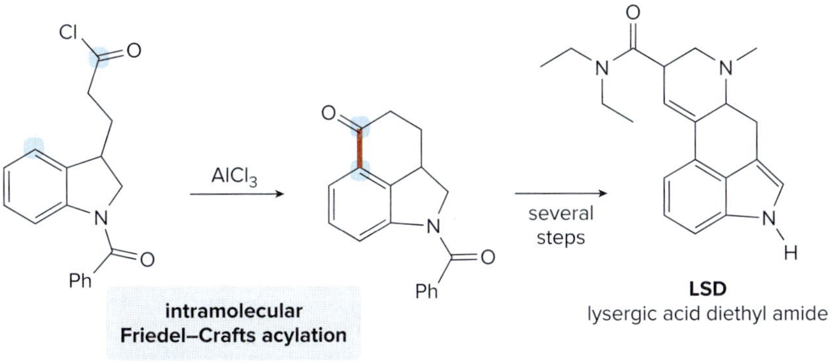

- **Intramolecular Friedel–Crafts acylation** at the labeled carbons formed a product containing a new six-membered ring, which was converted to LSD in several steps.
- LSD was first prepared by Swiss chemist Albert Hofmann in 1938 from a related organic compound isolated from the ergot fungus that attacks rye and other grains. Ergot has a long history as a dreaded poison, affecting individuals who become ill from eating ergot-contaminated bread. The hallucinogenic effects of LSD were first discovered when Hofmann accidentally ingested a small amount of the drug.

Problem 20.12 Intramolecular reactions are also observed in Friedel–Crafts alkylation. Draw the intramolecular alkylation product formed from each of the following reactants. (Watch out for rearrangements!)

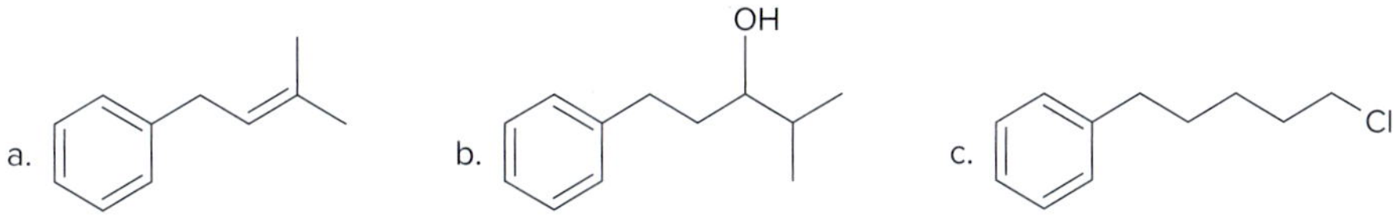

20.5E Biological Friedel–Crafts Reactions

Biological Friedel–Crafts reactions occur as well. As we learned in Section 12.2, allylic diphosphates contain a good leaving group, so they can serve as a source of allylic carbocations. A key step in the biological synthesis of vitamin K_1, the chapter-opening molecule, involves Friedel–Crafts reaction of 1,4-dihydroxynaphthoic acid with phytyl diphosphate to form **X**, which is converted to vitamin K_1 in several steps, as shown in Figure 20.4.

Figure 20.4 Friedel–Crafts reaction in the synthesis of vitamin K_1

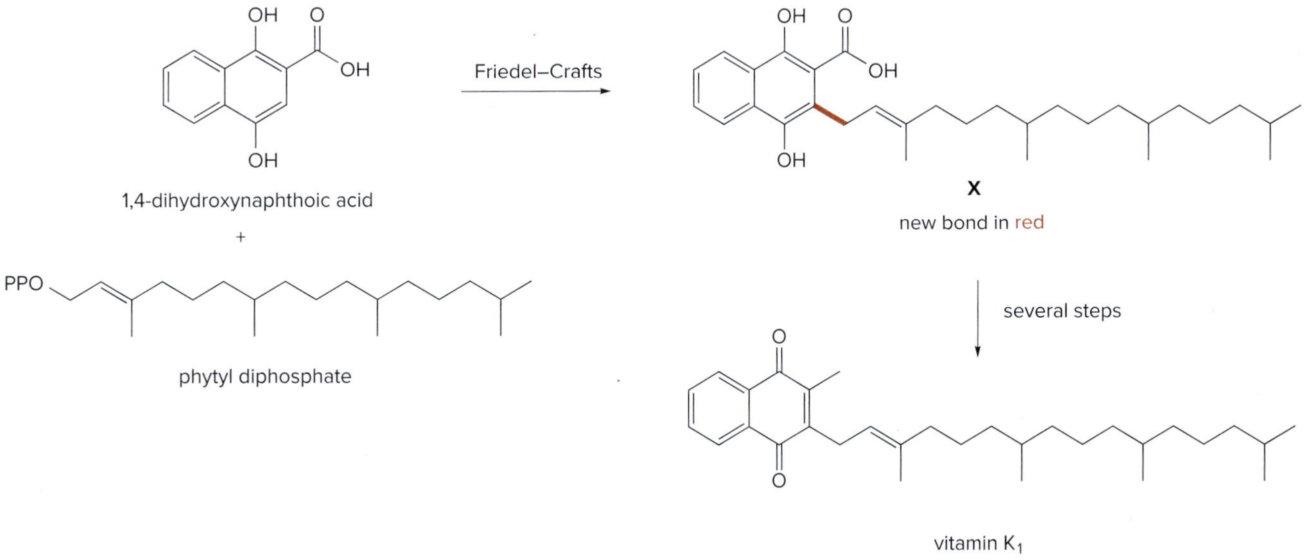

Problem 20.13 (a) Draw resonance structures for the carbocation formed after loss of a leaving group from phytyl diphosphate. (b) Draw the two-step mechanism for Friedel–Crafts alkylation of 1,2-dihydroxynaphthoic acid with this carbocation to form **X**.

20.6 Substituted Benzenes

Many substituted benzene rings undergo electrophilic aromatic substitution. Common substituents include halogens, OH, NH$_2$, alkyl, and many functional groups that contain a carbonyl. Each substituent either increases or decreases the electron density in the benzene ring, and this affects the course of electrophilic aromatic substitution, as we will learn in Section 20.7.

What makes a substituent on a benzene ring electron donating or electron withdrawing? The answer is **inductive effects** and **resonance effects,** both of which can add or remove electron density.

Inductive Effects

Inductive effects stem from the **electronegativity** of the atoms in the substituent and the **polarizability** of the substituent group.

Inductive and resonance effects were first discussed in Sections 2.5B and 2.5C, respectively.

- Atoms more electronegative than carbon—including N, O, and X—pull electron density away from carbon and thus exhibit an electron-*withdrawing* inductive effect.
- Polarizable alkyl groups donate electron density, and thus exhibit an electron-*donating* inductive effect.

Considering inductive effects *only,* an NH$_2$ group withdraws electron density and CH$_3$ donates electron density.

Electron-withdrawing inductive effect

C$_6$H$_5$—NH$_2$

- N is *more electronegative* than C.
- N inductively **withdraws** electron density.

Electron-donating inductive effect

C$_6$H$_5$—CH$_3$

- Alkyl groups are **polarizable,** making them electron-**donating** groups.

Problem 20.14 Which substituents have an electron-withdrawing and which have an electron-donating inductive effect: (a) CH$_3$CH$_2$CH$_2$CH$_2$–; (b) Br–; (c) CH$_3$CH$_2$O–?

Resonance Effects

Resonance effects can either donate or withdraw electron density, depending on whether they place a positive or negative charge on the benzene ring.

- A resonance effect is electron *donating* when resonance structures place a *negative* charge on carbons of the benzene ring.
- A resonance effect is electron *withdrawing* when resonance structures place a *positive* charge on carbons of the benzene ring.

An electron-donating resonance effect is observed whenever an atom Z having a lone pair of electrons is bonded directly to a benzene ring (general structure—C$_6$H$_5$–Z:). Common examples of Z include N, O, and halogen. For example, five resonance structures can be drawn for aniline (C$_6$H$_5$NH$_2$). Because three of them place a *negative* charge on a carbon atom of the benzene ring, an **NH$_2$ group** *donates* **electron density to a benzene ring by a resonance effect.**

[aniline resonance structures]

Three resonance structures place a **(−) charge** on atoms in the ring.

In contrast, **an electron-withdrawing resonance effect is observed in substituted benzenes having the general structure C₆H₅—Y=Z,** where Z is more electronegative than Y. For example, seven resonance structures can be drawn for benzaldehyde (C₆H₅CHO). Because three of them place a *positive* charge on a carbon atom of the benzene ring, a CHO group *withdraws* electron density from a benzene ring by a resonance effect.

[benzaldehyde resonance structures]

Three resonance structures place a **(+) charge** on atoms in the ring.

Problem 20.15 Draw all resonance structures for each compound, and use the resonance structures to determine if the substituent has an electron-donating or electron-withdrawing resonance effect.

a. C₆H₅—OCH₃   b. C₆H₅—C(=O)CH₃

Considering Both Inductive and Resonance Effects

To predict whether a substituted benzene is more or less electron rich than benzene itself, we must consider the **net balance of *both* the inductive and the resonance effects.** Alkyl groups, for instance, donate electrons by an inductive effect, but they have no resonance effect because they lack nonbonded electron pairs or π bonds. As a result,

- An alkyl group is an electron-*donating* group and an alkyl benzene is more electron rich than benzene.

When electronegative atoms, such as N, O, or halogen, are bonded to the benzene ring, they inductively *withdraw* electron density from the ring. All of these groups also have a nonbonded pair of electrons, so they *donate* electron density to the ring by resonance. **The identity of the element determines the net balance of these opposing effects.**

Z = N, O, X

Induction and resonance have **opposite** effects.
- Z inductively *withdraws* electron density.
- Z *donates* electron density by resonance.

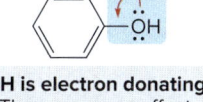

OH is electron donating.
The resonance effect predominates.

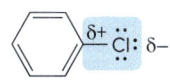

Cl is electron withdrawing.
The inductive effect predominates.

- When a neutral O or N atom is bonded directly to a benzene ring, the resonance effect dominates and the net effect is *electron donation.*
- When a halogen X is bonded to a benzene ring, the inductive effect dominates and the net effect is *electron withdrawal.*

C=O is electron withdrawing.

Thus, **NH₂ and OH are electron-*donating* groups** because the resonance effect predominates, whereas **Cl and Br are electron-*withdrawing* groups** because the inductive effect predominates.

Finally, the inductive and resonance effects in compounds having the general structure $C_6H_5-Y=Z$ (with Z more electronegative than Y) are **both electron withdrawing**; in other words, the two effects *reinforce* each other. This is true for benzaldehyde (C_6H_5CHO) and all other compounds that contain a carbonyl group bonded directly to the benzene ring.

As a result, there are two general structures for a benzene ring with an electron-donating group and two general structures for a benzene ring with an electron-withdrawing group:

Electron-donating groups	Electron-withdrawing groups
R = alkyl Z = N or O	X = halogen Y (δ+ or +)

- Common electron-donating groups are alkyl groups or groups with an N or O atom (with a lone pair) bonded to the benzene ring.
- Common electron-withdrawing groups are halogens or groups with an atom Y bearing a full or partial positive charge (+ or δ+) bonded to the benzene ring.

The net effect of electron donation and withdrawal on the reactions of substituted aromatics is discussed in Sections 20.7–20.9.

Sample Problem 20.3   Classifying a Substituent as Electron Donating or Electron Withdrawing

Classify each substituent as electron donating or electron withdrawing.

a. [phenyl acetate structure]   b. [benzonitrile structure, Ph–CN]

Solution

If necessary, draw out the atoms and bonds of the substituent to clearly see lone pairs and multiple bonds. **Always look at the atom bonded directly to the benzene ring** to determine electron-donating or electron-withdrawing effects.

- An O or N atom with a lone pair of electrons makes a substituent electron *donating*.
- A halogen or an atom with a partial positive charge makes a substituent electron *withdrawing*.

a.
- An O atom with a lone pair is bonded directly to the benzene ring.

an electron-donating group

b.
- An atom with a partial (+) charge is bonded directly to the benzene ring.

an electron-withdrawing group

Problem 20.16 Classify each substituent as electron donating or electron withdrawing.

a. 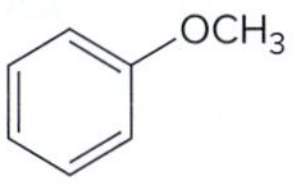 C₆H₅–OCH₃   b. 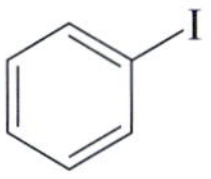 C₆H₅–I   c. 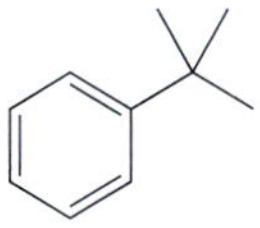 C₆H₅–C(CH₃)₃

More Practice: Try Problem 20.49.

20.7 Electrophilic Aromatic Substitution of Substituted Benzenes

Electrophilic aromatic substitution is a general reaction of *all* aromatic compounds, including polycyclic aromatic hydrocarbons, heterocycles, and substituted benzene derivatives. A substituent affects two aspects of electrophilic aromatic substitution:

- **The rate of reaction:** A substituted benzene reacts faster or slower than benzene itself.
- **The orientation:** The new group is located either ortho, meta, or para to the existing substituent. The identity of the first substituent determines the position of the second substituent.

Toluene ($C_6H_5CH_3$) and nitrobenzene ($C_6H_5NO_2$) illustrate two possible outcomes.

[1] Toluene

Toluene reacts **faster** than benzene in all substitution reactions. Thus, its **electron-donating CH_3 group *activates* the benzene ring** to electrophilic attack. Although three products are possible, compounds with the new group ortho or para to the CH_3 group predominate. The CH_3 group is therefore called an **ortho, para director.**

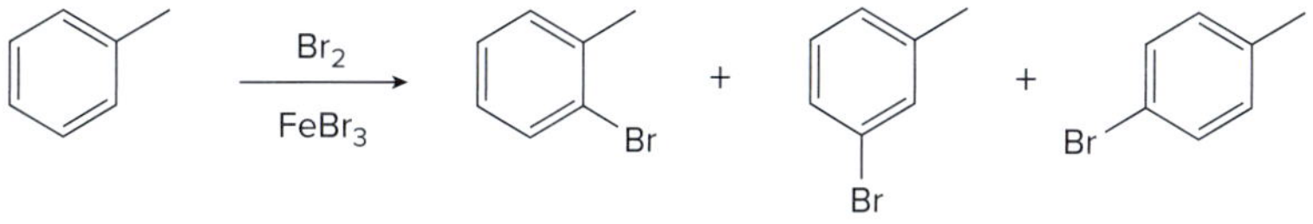

Toluene + Br_2 / $FeBr_3$ → ortho (40%) + meta (trace) + para (60%)

[2] Nitrobenzene

Nitrobenzene reacts **more slowly** than benzene in all substitution reactions. Thus, its **electron-withdrawing NO_2 group *deactivates* the benzene ring** to electrophilic attack. Although three products are possible, the compound with the new group meta to the NO_2 group predominates. The NO_2 group is called a **meta director.**

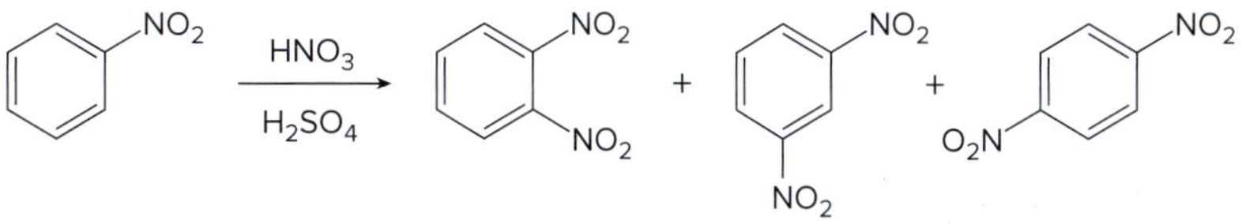

Nitrobenzene + HNO_3 / H_2SO_4 → ortho (7%) + **meta (93%)** + para (trace)

Substituents either activate or deactivate a benzene ring toward electrophiles, and direct selective substitution at specific sites on the ring. **All substituents can be divided into three general types.**

902 Chapter 20 Reactions of Aromatic Compounds

---

[1] Ortho, para directors and activators

- **Substituents that *activate* a benzene ring and direct substitution ortho and para.**

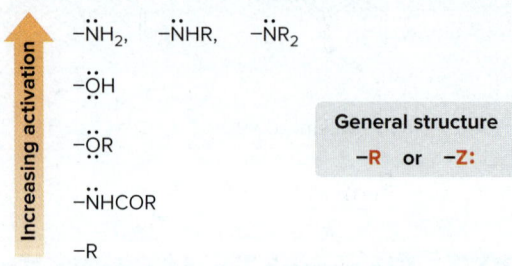

---

[2] Ortho, para deactivators

- **Substituents that *deactivate* a benzene ring and direct substitution ortho and para.**

$$-\ddot{\underset{..}{F}}: \quad -\ddot{\underset{..}{Cl}}: \quad -\ddot{\underset{..}{Br}}: \quad -\ddot{\underset{..}{I}}:$$

---

[3] Meta directors

- **Substituents that direct substitution meta.**
- **All meta directors *deactivate* the ring.**

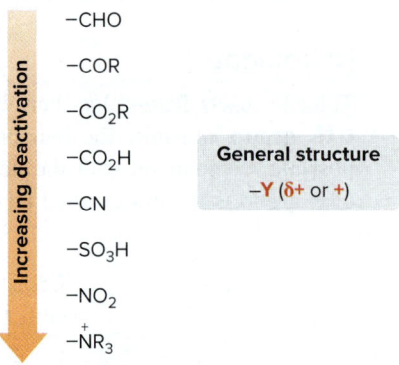

To learn these lists: **Keep in mind that the halogens are in a class by themselves.** Then learn the general structures for each type of substituent.

- All ortho, para directors are R groups or have a nonbonded electron pair on the atom bonded to the benzene ring.

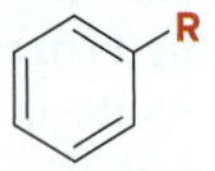

    Ph–R   Ph–Z: Z = N or O --→ The ring is **activated**.
    ortho, para   ortho, para   Z = halogen --→ The ring is **deactivated**.
    director   director

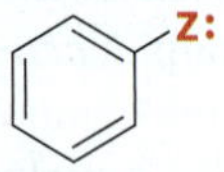

- All meta directors have a full or partial positive charge on the atom bonded to the benzene ring.

    Ph–Y (δ+ or +)

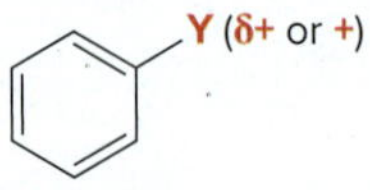

    meta director

Sample Problem 20.4 shows how this information can be used to predict the products of electrophilic aromatic substitution reactions.

20.7 Electrophilic Aromatic Substitution of Substituted Benzenes

Sample Problem 20.4 Drawing the Electrophilic Substitution Products of a Substituted Benzene

Draw the products of each reaction, and state whether the reaction is faster or slower than a similar reaction with benzene.

a. 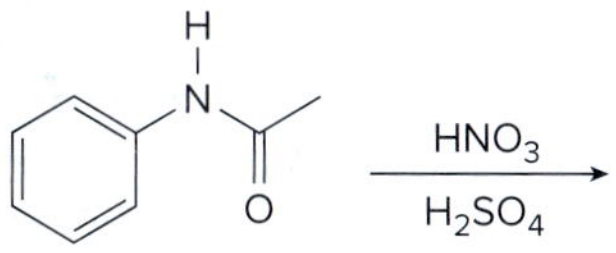 PhNHC(O)CH$_3$ + HNO$_3$/H$_2$SO$_4$

b. 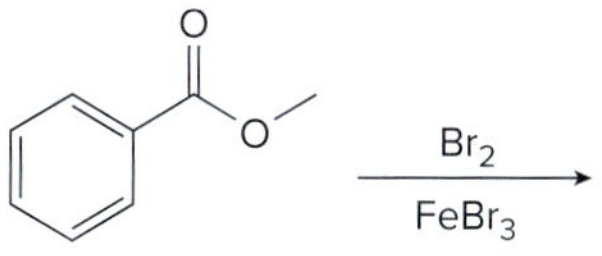 PhC(O)OCH$_3$ + Br$_2$/FeBr$_3$

Solution

To draw the products:

- Draw the Lewis structure for the substituent to see if it has a **lone pair** or **partial positive charge** on the atom bonded to the benzene ring.
- **Classify the substituent**—ortho, para activating; ortho, para deactivating; or meta deactivating—and draw the products.

a. 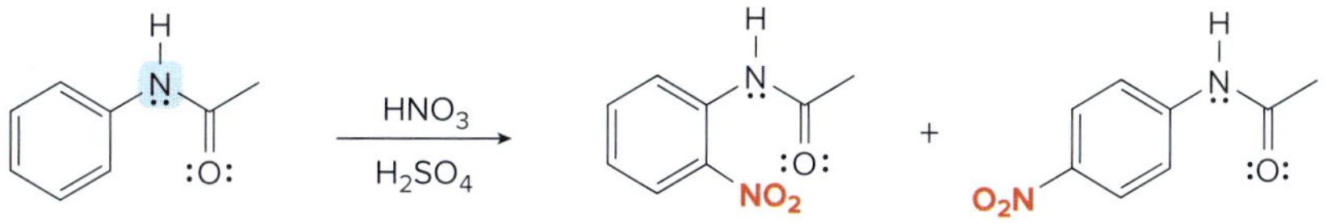 PhNHC(O)CH$_3$ + HNO$_3$/H$_2$SO$_4$ → ortho (NO$_2$) + para (O$_2$N)

The lone pair on N makes this group an **ortho, para activator**. This compound reacts *faster* than benzene.

b. 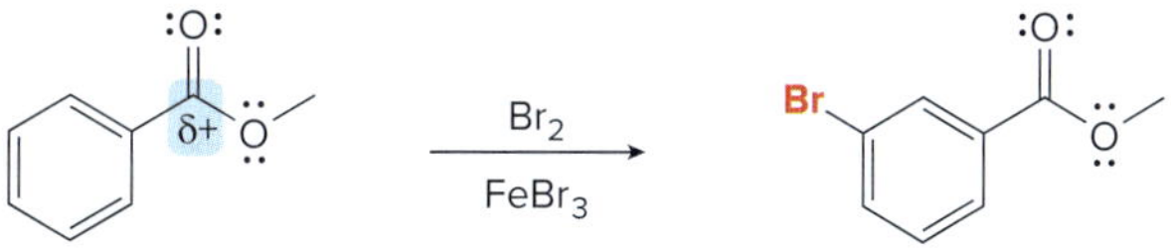 PhC(O)OCH$_3$ (δ+ on C) + Br$_2$/FeBr$_3$ → meta (Br)

The δ+ on the C bonded to the benzene ring makes the group a **meta deactivator**. This compound reacts *more slowly* than benzene.

Problem 20.17 Draw the products formed when each compound is treated with HNO$_3$ and H$_2$SO$_4$. State whether the reaction occurs faster or slower than a similar reaction with benzene.

a. 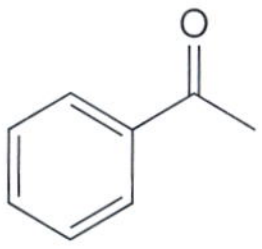 PhC(O)CH$_3$ b. 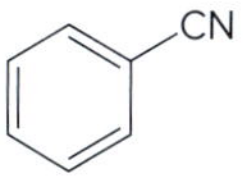 PhCN c. 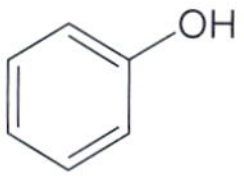 PhOH d. 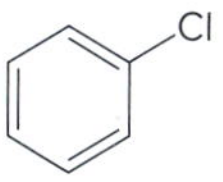 PhCl e. 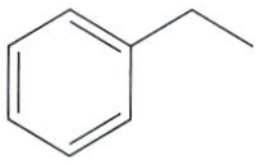 PhCH$_2$CH$_3$

More Practice: Try Problems 20.38, 20.49d.

Sample Problem 20.5 Determining Which Ring in a Polycyclic Compound Is More Reactive in Electrophilic Aromatic Substitution

Which ring in each compound is more reactive toward electrophiles?

a. 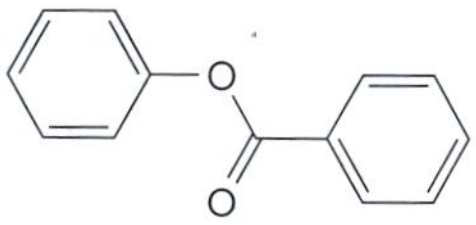 PhOC(O)Ph

b. 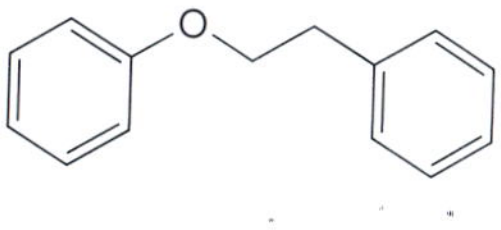 PhOCH$_2$CH$_2$Ph

Solution

Look at the atom bonded *directly* to the aromatic ring to decide on reactivity in electrophilic aromatic substitution.

- An N or O atom with a lone pair makes a ring *more* reactive.
- An alkyl group makes a ring *somewhat more* reactive.
- An atom with a full or partial positive charge makes a ring *less* reactive.

a. Ring **A** is bonded to an O atom with a lone pair, whereas ring **B** is bonded to a C that bears a δ+. Ring **A** is more electron rich and more reactive.

b. Ring **C** is bonded to an O atom with a lone pair, whereas ring **D** is bonded to an alkyl carbon. Ring **C** is more electron rich and more reactive.

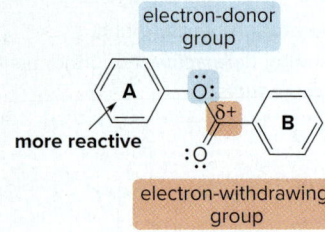

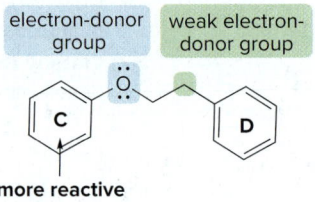

Problem 20.18 Determine which ring in each compound is more reactive toward electrophiles, and explain your choice.

More Practice: Try Problems 20.50, 20.51.

Problem 20.19 Consider the tetracyclic compound with rings labeled **A–D**. (a) Which ring is the *most* reactive in electrophilic aromatic substitution? (b) Which ring is the *least* reactive in electrophilic aromatic substitution?

20.8 Why Substituents Activate or Deactivate a Benzene Ring

- **Why do substituents activate or deactivate a benzene ring?**
- **Why are particular orientation effects observed?** Why are some groups ortho, para directors and some groups meta directors?

To understand why some substituents make a benzene ring react *faster* than benzene itself (activators), whereas others make it react *slower* (deactivators), we must evaluate the rate-determining step (the first step) of the mechanism. Recall from Section 20.2 that the first step in electrophilic aromatic substitution is the addition of an electrophile (E^+) to form a resonance-stabilized carbocation. The Hammond postulate (Section 7.14) makes it possible to predict the relative rate of the reaction by looking at the stability of the carbocation intermediate.

- The more stable the carbocation, the lower in energy the transition state that forms it, and the faster the reaction.

[+ two resonance structures]

Stabilizing the carbocation makes the reaction faster.

The principles of inductive effects and resonance effects, first introduced in Section 20.6, can now be used to predict carbocation stability.

Electron-donor groups **D** stabilize the carbocation.

Electron-withdrawing groups **W** destabilize the carbocation.

- Electron-donating groups *stabilize* the carbocation and *activate* a benzene ring toward electrophilic attack. All activators are R groups, or they have an N or O atom with a lone pair bonded directly to the benzene ring.
- Electron-withdrawing groups *destabilize* the carbocation and *deactivate* a benzene ring toward electrophilic attack. All deactivators are halogens, or they have an atom with a full or partial positive charge bonded directly to the benzene ring.

Problem 20.20 Label each compound as more or less reactive than benzene in electrophilic aromatic substitution.

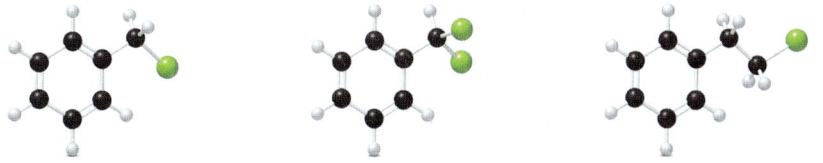

Problem 20.21 Rank the following compounds in order of increasing reactivity in electrophilic aromatic substitution.

20.9 Orientation Effects in Substituted Benzenes

To understand why particular orientation effects arise, you must keep in mind the general structures for ortho, para directors and for meta directors already given in Section 20.7. There are two general types of ortho, para directors and one general type of meta director:

- All ortho, para directors are R groups or have a nonbonded electron pair on the atom bonded to the benzene ring.
- All meta directors have a full or partial positive charge on the atom bonded to the benzene ring.

To evaluate the directing effects of a given substituent, we can follow a stepwise procedure.

How To Determine the Directing Effects of a Particular Substituent

Step [1] Draw all resonance structures for the carbocation formed from attack of an electrophile E⁺ at the ortho, meta, and para positions of a substituted benzene (C₆H₅–A).

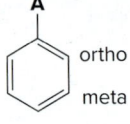

- There are at least three resonance structures for each site of reaction.
- Each resonance structure places a positive charge **ortho** or **para** to the new C–E bond.

Step [2] Evaluate the stability of the intermediate resonance structures. The electrophile attacks at those positions that give the *most stable* carbocation.

Sections 20.9A–C show how this two-step procedure can be used to evaluate the directing effects of the CH₃ group in toluene, the NH₂ group in aniline, and the NO₂ group in nitrobenzene, respectively.

20.9A The CH₃ Group—An ortho, para Director

To understand why a **CH₃ group directs electrophilic aromatic substitution to the ortho and para positions,** first draw all resonance structures that result from electrophilic attack at the ortho, meta, and para positions to the CH₃ group.

Always draw in the H atom at the site of electrophilic attack. This will help you keep track of where the charges go.

The positive charge in all resonance structures is always **ortho or para to the new C–E bond.** It is *not* necessarily ortho or para to the CH₃ group.

In this example, **attack ortho or para to CH₃ generates a resonance structure that places a positive charge on a carbon atom with the CH₃ group.** The electron-donating CH₃ group *stabilizes* the adjacent positive charge. In contrast, attack meta to the CH₃ group does *not* generate any resonance structure stabilized by electron donation. Other alkyl groups are ortho, para directors for the same reason.

- The CH₃ group directs electrophilic attack ortho and para to itself because an electron-donating inductive effect stabilizes the carbocation intermediate.

20.9B The NH₂ Group—An ortho, para Director

To understand why an **amino group (NH₂)** directs electrophilic aromatic substitution to the **ortho and para positions,** follow the same procedure.

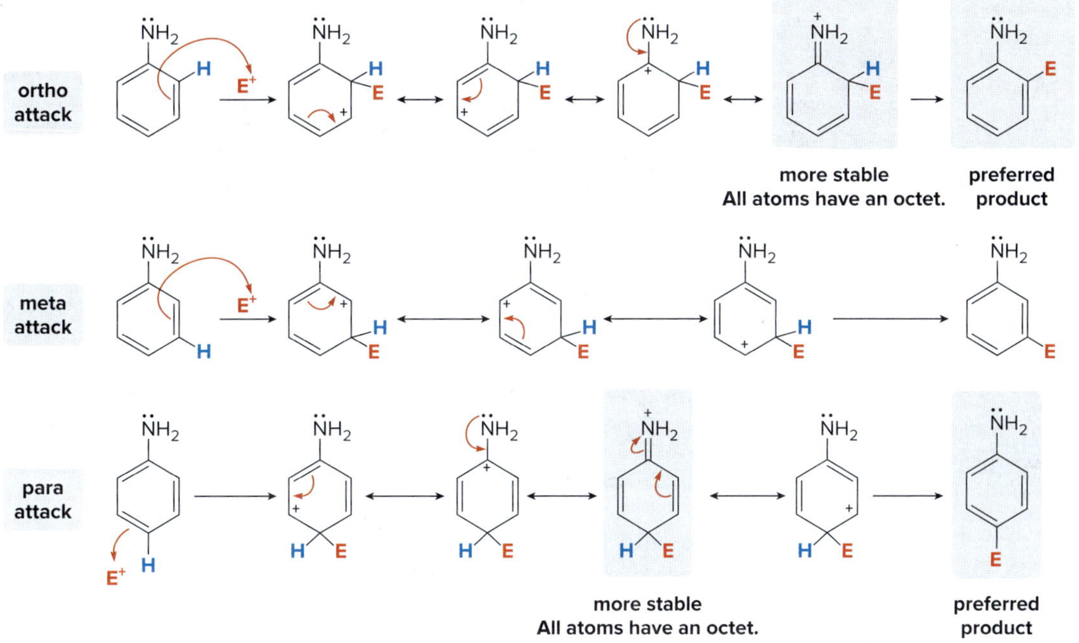

20.9 Orientation Effects in Substituted Benzenes

[ortho attack, meta attack, and para attack resonance structure diagrams for aniline (NH_2 substituent) showing that ortho and para attack generate a fourth resonance structure where all atoms have an octet — labeled "more stable, All atoms have an octet" and "preferred product"]

Attack at the meta position generates the usual three resonance structures. Because of the lone pair on the N atom, attack at the ortho and para positions generates a fourth resonance structure, which is stabilized because **every atom has an octet of electrons. This additional resonance structure can be drawn for all substituents that have an N, O, or halogen atom bonded directly to the benzene ring.**

- The NH_2 group directs electrophilic attack ortho and para to itself because the carbocation intermediate has additional resonance stabilization.

20.9C The NO_2 Group—A meta Director

To understand why a **nitro group (NO_2) directs electrophilic aromatic substitution to the meta position,** follow the same procedure.

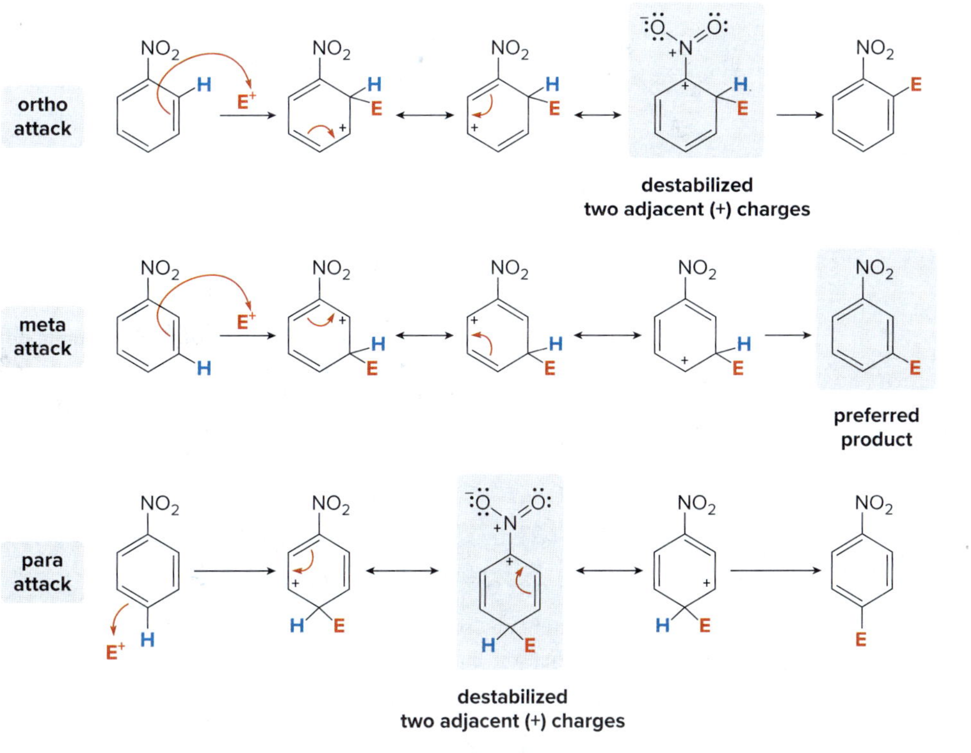

[ortho attack, meta attack, and para attack resonance structure diagrams for nitrobenzene (NO_2 substituent). Ortho and para attack each generate a resonance structure labeled "destabilized, two adjacent (+) charges." Meta attack gives the "preferred product."]

Attack at each position generates three resonance structures. One resonance structure resulting from attack at the ortho and para positions is especially *destabilized*, because it contains a positive charge on two adjacent atoms. Attack at the meta position does not generate any particularly unstable resonance structures.

- With the NO₂ group (and all meta directors), meta attack occurs because attack at the ortho or para position gives a destabilized carbocation intermediate.

Problem 20.22 Draw all resonance structures for the carbocation formed by ortho attack of the electrophile ⁺NO₂ on each starting material. Label any resonance structures that are especially stable or unstable.

a. [tert-butylbenzene] b. [phenol] c. [benzaldehyde]

Figure 20.5 summarizes the reactivity and directing effects of the common substituents on benzene rings.

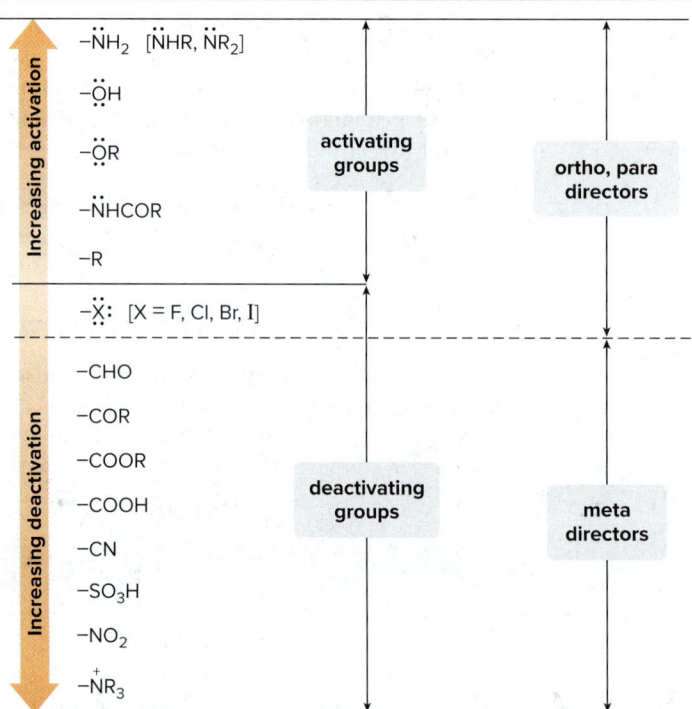

Figure 20.5
The reactivity and directing effects of common substituted benzenes

In summary:
[1] All ortho, para directors except the halogens activate the benzene ring.
[2] All meta directors deactivate the benzene ring.
[3] The halogens deactivate the benzene ring.

20.10 Limitations on Electrophilic Substitution Reactions with Substituted Benzenes

Although electrophilic aromatic substitution works well with most substituted benzenes, halogenation and the Friedel–Crafts reactions have some additional limitations that must be kept in mind.

20.10A Halogenation of Activated Benzenes

Considering all electrophilic aromatic substitution reactions, halogenation occurs the most readily. As a result, benzene rings activated by strong electron-donating groups—OH, NH$_2$, and their alkyl derivatives (OR, NHR, and NR$_2$)—undergo **polyhalogenation** when treated with X$_2$ and FeX$_3$. Aniline (C$_6$H$_5$NH$_2$) and phenol (C$_6$H$_5$OH) both give a tribromo derivative when treated with Br$_2$ and FeBr$_3$. **Substitution occurs at all hydrogen atoms ortho and para to the NH$_2$ and OH groups.**

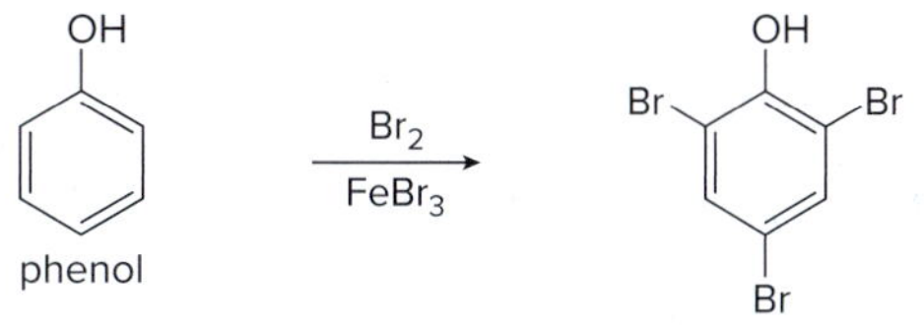

Monosubstitution of H by Br occurs with Br$_2$ *alone* without added catalyst to form a mixture of ortho and para products.

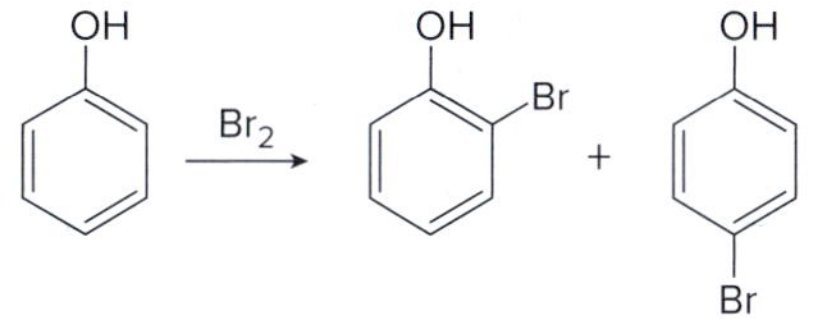

Problem 20.23 Draw the products of each reaction.

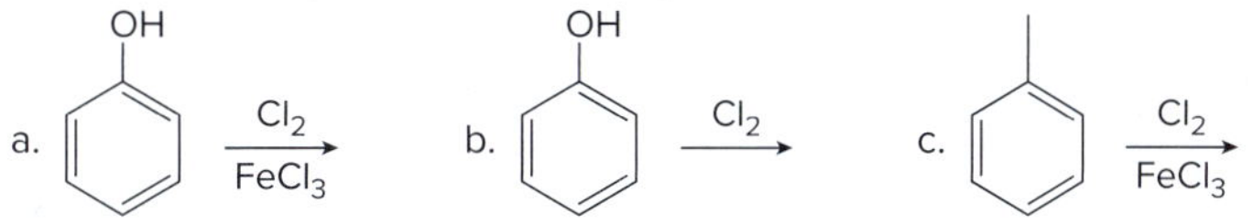

a. phenol + Cl$_2$ / FeCl$_3$ b. phenol + Cl$_2$ c. toluene + Cl$_2$ / FeCl$_3$

20.10B Limitations in Friedel–Crafts Reactions

Friedel–Crafts reactions are the most difficult electrophilic aromatic substitution reactions to carry out in the laboratory. **They do not occur when the benzene ring is substituted with NO$_2$ (or any meta deactivator) or with NH$_2$, NHR, or NR$_2$ (strong activators).**

A benzene ring deactivated by a strong electron-withdrawing group—that is, any of the **meta directors**—is not electron rich enough to undergo Friedel–Crafts reactions.

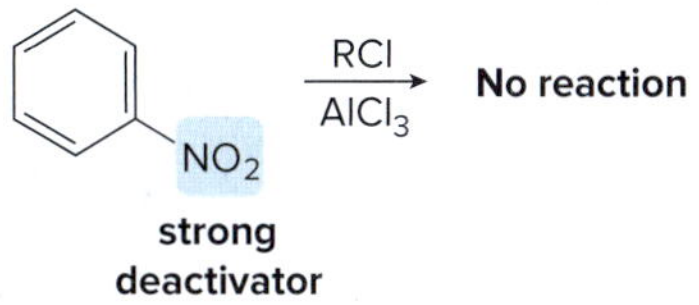

C$_6$H$_5$NO$_2$ (strong deactivator) + RCl / AlCl$_3$ → No reaction

Friedel–Crafts reactions also do not occur with NH$_2$ groups, which are strong activating groups. **NH$_2$ groups are strong Lewis bases** (due to the nonbonded electron pair on N), so they react with AlCl$_3$, the Lewis acid needed for alkylation or acylation. The resulting product contains a positive charge adjacent to the benzene ring, so the **ring is now strongly deactivated** and therefore unreactive in Friedel–Crafts reactions.

C$_6$H$_5$NH$_2$ (Lewis base) + AlCl$_3$ (Lewis acid) → C$_6$H$_5$N$^+$H$_2$AlCl$_3^-$ (strong deactivator) + RCl / AlCl$_3$ → No reaction

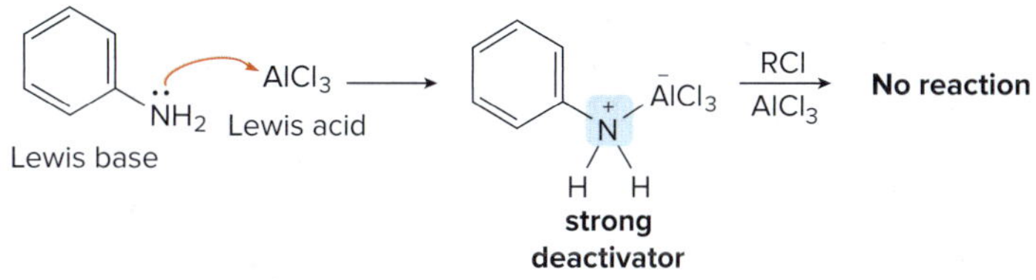

Problem 20.24 Which of the following compounds undergo Friedel–Crafts alkylation with CH₃Cl and AlCl₃? Draw the products formed when a reaction occurs.

a. 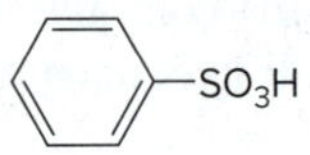 C₆H₅–SO₃H   b. 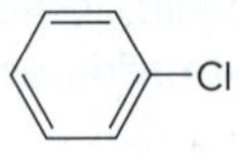 C₆H₅–Cl   c. 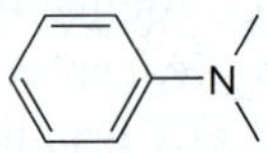 C₆H₅–N(CH₃)₂   d. 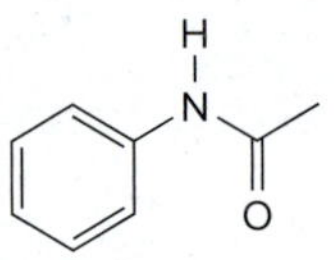 C₆H₅–NH–C(=O)CH₃

20.11 Disubstituted Benzenes

What happens in electrophilic aromatic substitution when a disubstituted benzene ring is used as starting material? **To predict the products, look at the directing effects of *both* substituents and then determine the net result,** using three guidelines.

Rule [1] When the directing effects of two groups *reinforce*, the new substituent is located on the position directed by both groups.

The CH₃ group in *p*-nitrotoluene is an ortho, para director and the NO₂ group is a meta director. These two effects reinforce each other so that one product is formed on treatment with Br₂ and FeBr₃. The position para to the CH₃ group is "blocked" by a nitro group, so no substitution can occur on that carbon.

p-nitrotoluene (ortho, para director CH₃; meta director NO₂) + Br₂ / FeBr₃ → product with Br **ortho to CH₃, meta to NO₂**

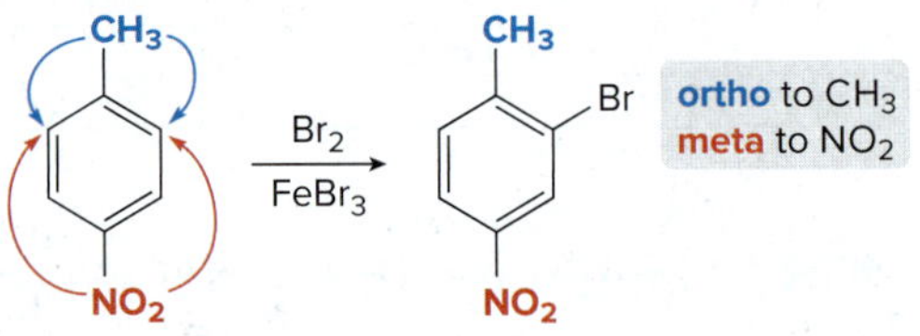

Rule [2] If the directing effects of two groups *oppose* each other, the more powerful activator "wins out."

In compound **A**, the NHCOCH₃ group activates its two ortho positions, and the CH₃ group activates its two ortho positions to reaction with electrophiles. Because the NHCOCH₃ is a stronger activator, substitution occurs ortho to it.

stronger ortho, para director (NHCOCH₃); weaker ortho, para director (CH₃) + Br₂ / FeBr₃ → Br **ortho to the stronger activator**

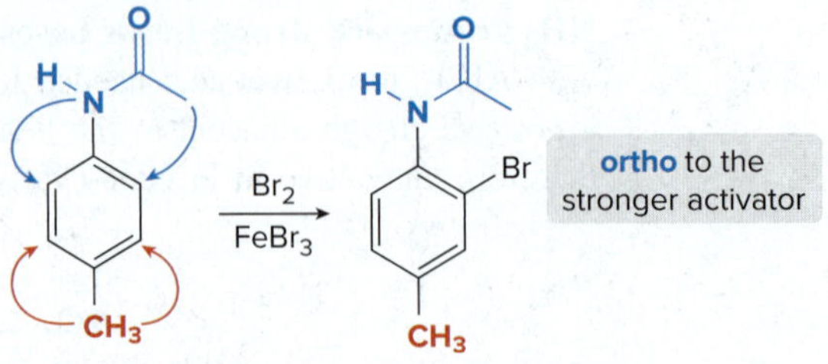

A

20.11 Disubstituted Benzenes

Rule [3] No substitution occurs between two meta substituents because of crowding.

For example, no substitution occurs at the carbon atom between the two CH₃ groups in *m*-xylene, even though two CH₃ groups activate that position.

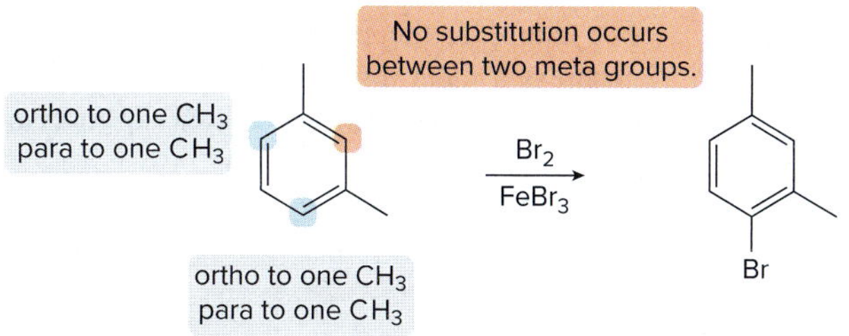

m-xylene (1,3-dimethylbenzene) reacts with Br₂/FeBr₃ to give a bromide product. No substitution occurs between two meta groups.

Sample Problem 20.6 — Drawing the Substitution Products from a Disubstituted Benzene

Draw the products formed from nitration of each compound.

a. 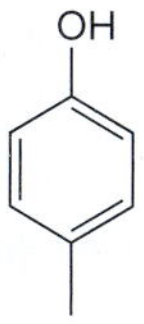 4-methylphenol (para-cresol)
b. 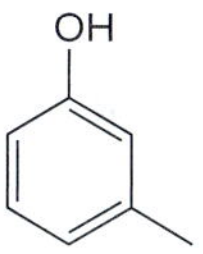 3-methylphenol (meta-cresol)

Solution

a. Both the OH and CH₃ groups are ortho, para directors. Because the **OH group is a stronger activator**, substitution occurs ortho to it.

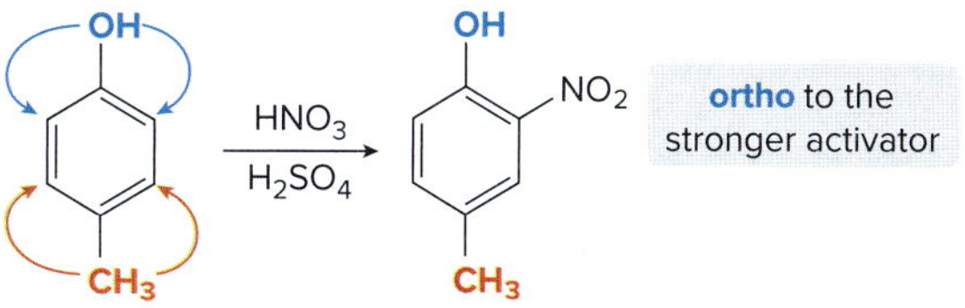

Reaction with HNO₃/H₂SO₄ places NO₂ ortho to the stronger activator (OH).

b. Both the OH and CH₃ groups are ortho, para directors whose directing effects reinforce each other in this case. **No substitution occurs between the two meta substituents,** however, so two products result.

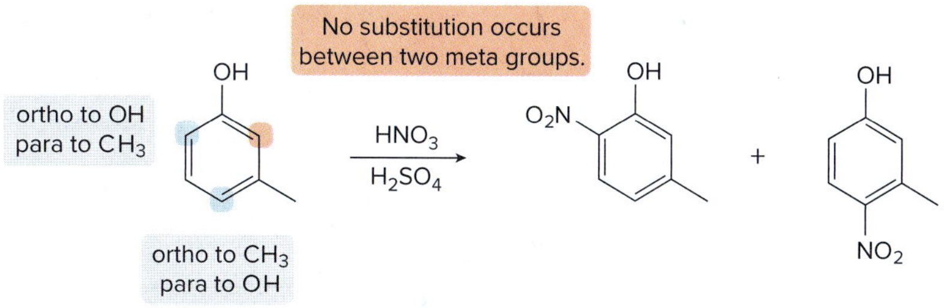

Reaction with HNO₃/H₂SO₄ gives two products: one with NO₂ ortho to OH and para to CH₃, and one with NO₂ ortho to CH₃ and para to OH. No substitution occurs between two meta groups.

Problem 20.25

Draw the products formed when each compound is treated with HNO₃ and H₂SO₄.

a. 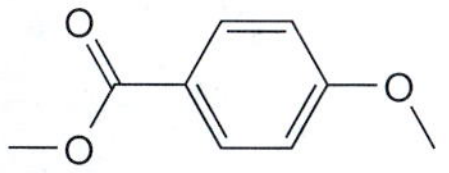 methyl 4-methoxybenzoate
b.  2-bromoanisole
c. 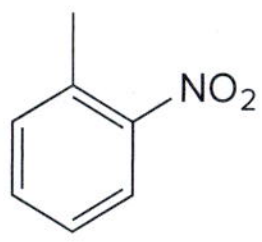 2-nitrotoluene
d. 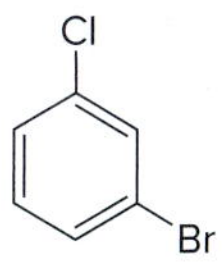 1-chloro-3-bromobenzene

More Practice: Try Problems 20.37, 20.40a–e, 20.42a–c.

20.12 Synthesis of Benzene Derivatives

To synthesize benzene derivatives with more than one substituent, we must always take into account the directing effects of each substituent. In a disubstituted benzene, **the directing effects indicate which substituent must be added to the ring first.**

For example, the Br group in *p*-bromonitrobenzene is an ortho, para director and the NO_2 group is a meta director. Because the two substituents are para to each other, the ortho, para director must be introduced *first* when synthesizing this compound from benzene.

Br   ortho, para director

NO_2 meta director

p-bromonitrobenzene

With two **para** substituents, add the **ortho, para** director *first*.

Thus, Pathway [1], in which bromination precedes nitration, yields the **desired para product,** whereas Pathway [2], in which nitration precedes bromination, yields the **undesired meta isomer.**

Pathway [1]   Bromination before nitration: The **desired para product** is formed.

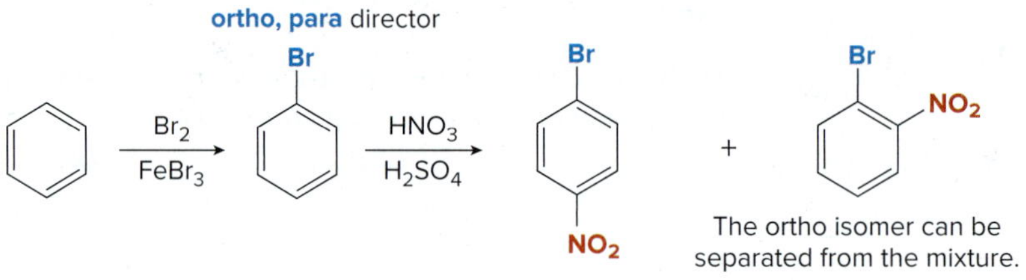

benzene → (Br$_2$/FeBr$_3$) → bromobenzene (ortho, para director, Br) → (HNO$_3$/H$_2$SO$_4$) → para product (Br, NO_2) **desired product** + ortho isomer (Br, NO_2)

The ortho isomer can be separated from the mixture.

Pathway [2]   Nitration before bromination: The **undesired meta isomer** is formed.

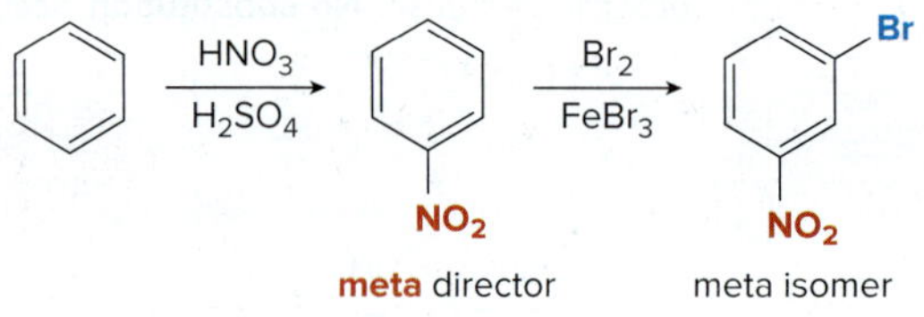

benzene → (HNO$_3$/H$_2$SO$_4$) → nitrobenzene (NO_2, **meta director**) → (Br$_2$/FeBr$_3$) → meta isomer (Br, NO_2)

Pathway [1] yields both the desired para product and the undesired ortho isomer. Because these compounds are constitutional isomers, they are separable. Obtaining such a mixture of ortho and para isomers is often unavoidable.

Sample Problem 20.7 Synthesizing a Disubstituted Benzene

Devise a synthesis of *o*-nitrotoluene from benzene.

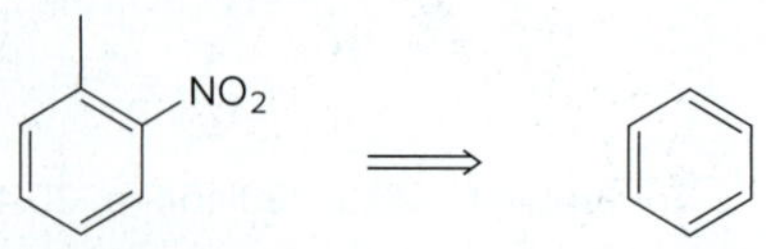

o-nitrotoluene (CH$_3$, NO_2) ⟹ benzene

Solution

The CH₃ group in *o*-nitrotoluene is an ortho, para director and the NO₂ group is a meta director. Because the two substituents are ortho to each other, the **ortho, para director must be introduced first.** The synthesis thus involves two steps: Friedel–Crafts alkylation followed by nitration.

benzene $\xrightarrow{\text{CH}_3\text{Cl}, \text{AlCl}_3}$ toluene $\xrightarrow{\text{HNO}_3, \text{H}_2\text{SO}_4}$ *o*-nitrotoluene + para isomer

---

Problem 20.26 Devise a synthesis of each compound from the indicated starting material.

a. 3-chlorobenzenesulfonic acid ⟹ benzene   b. 3′-nitroacetophenone ⟹ benzene   c. 2-bromo-4-methylphenol ⟹ phenol

---

20.13 Nucleophilic Aromatic Substitution

Although most reactions of aromatic compounds occur by way of electrophilic aromatic substitution, **aryl halides undergo a limited number of substitution reactions with strong nucleophiles.**

Ar–X + :Nu⁻ ⟶ Ar–Nu + :X⁻

X = F, Cl, Br, I
A = H or electron-withdrawing group

- Nucleophilic aromatic substitution results in the substitution of a halogen X on a benzene ring by a nucleophile (:Nu⁻).

As we learned in Section 7.17, these reactions *cannot* occur by an S_N1 or S_N2 mechanism, which take place only at sp^3 hybridized carbons. Instead, two different mechanisms are proposed to explain the results: **addition–elimination** (Section 20.13A) and **elimination–addition** (Section 20.13B).

20.13A Nucleophilic Aromatic Substitution by Addition–Elimination

Aryl halides with strong electron-withdrawing groups (such as NO₂) on the ortho or para positions react with nucleophiles to afford substitution products. Treatment of *p*-chloronitrobenzene with hydroxide (⁻OH) affords *p*-nitrophenol by replacement of Cl by OH.

O₂N–C₆H₄–Cl $\xrightarrow{\text{⁻OH}}$ O₂N–C₆H₄–OH + Cl⁻

p-chloronitrobenzene → *p*-nitrophenol

Nucleophilic aromatic substitution occurs with a variety of strong nucleophiles, including ⁻OH, ⁻OR, ⁻NH₂, ⁻SR, and in some cases, neutral nucleophiles such as NH₃ and RNH₂. The mechanism of these reactions has two steps: **addition of the nucleophile** to form a resonance-stabilized carbanion, followed by **elimination of the halogen leaving group.** Mechanism 20.10 is drawn with an aryl chloride containing a general electron-withdrawing group W.

Mechanism 20.10 Nucleophilic Aromatic Substitution by Addition–Elimination

[Mechanism scheme showing nucleophilic addition to aryl chloride bearing W group, forming resonance-stabilized carbanion, followed by loss of Cl⁻]

resonance-stabilized carbanion

1. Addition of the nucleophile forms a **resonance-stabilized carbanion** and a new C–Nu bond in the rate-determining step.
2. Loss of the leaving group re-forms the aromatic ring.

In nucleophilic aromatic substitution, the following trends in reactivity are observed.

- Increasing the number of electron-withdrawing groups *increases* the reactivity of the aryl halide. Electron-withdrawing groups stabilize the intermediate carbanion and, by the Hammond postulate, lower the energy of the transition state that forms it.
- Increasing the electronegativity of the halogen *increases* the reactivity of the aryl halide. A more electronegative halogen stabilizes the intermediate carbanion by an inductive effect, making aryl fluorides (ArF) much *more* reactive than other aryl halides, which contain less electronegative halogens.

Thus, aryl chloride **B** is more reactive than *o*-chloronitrobenzene (**A**) because it contains *two* electron-withdrawing NO_2 groups. Aryl fluoride **C** is more reactive than **B** because **C** contains the *more electronegative* halogen, fluorine.

[Structures A, B, C shown with increasing reactivity arrow]

A: *o*-chloronitrobenzene
B: 2,4-dinitrochlorobenzene
C: 2,4-dinitrofluorobenzene

Increasing reactivity →

The location of the electron-withdrawing group greatly affects the rate of nucleophilic aromatic substitution. When a nitro group is located ortho or para to the halogen, the negative charge of the intermediate carbanion can be delocalized onto the NO_2 group, thus stabilizing it. With a meta NO_2 group, no such additional delocalization onto the NO_2 group occurs.

[Resonance structures for para NO_2 group case]

para NO_2 group

additional resonance stabilization
The negative charge is *delocalized* on the O atom of the NO_2 group.

[Resonance structures for meta NO_2 group case]

meta NO_2 group — The negative charge is *never* delocalized on the NO_2 group.

20.13 Nucleophilic Aromatic Substitution

Thus, **nucleophilic aromatic substitution by an addition–elimination mechanism occurs only with aryl halides that contain electron-withdrawing substituents at the ortho or para position.**

Problem 20.27 Draw the products of each reaction.

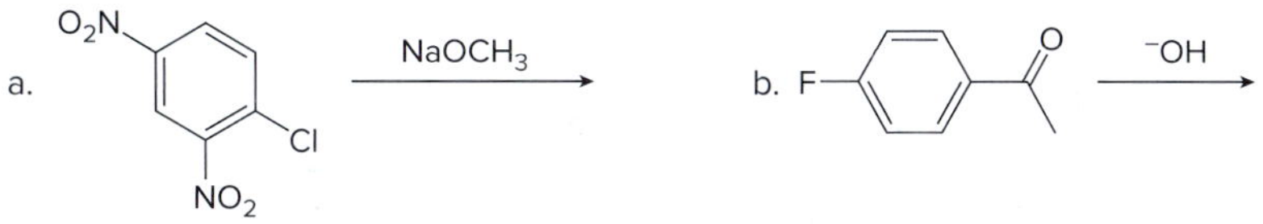

a. [2-chloro-1,4-dinitro... with O_2N, Cl, NO_2] $\xrightarrow{NaOCH_3}$

b. [4-fluoroacetophenone, F–C$_6$H$_4$–C(=O)CH$_3$] $\xrightarrow{^-OH}$

Problem 20.28 Draw a stepwise mechanism for the following reaction that forms ether **D**. **D** can be converted to the antidepressant fluoxetine (trade name Prozac) in a single step.

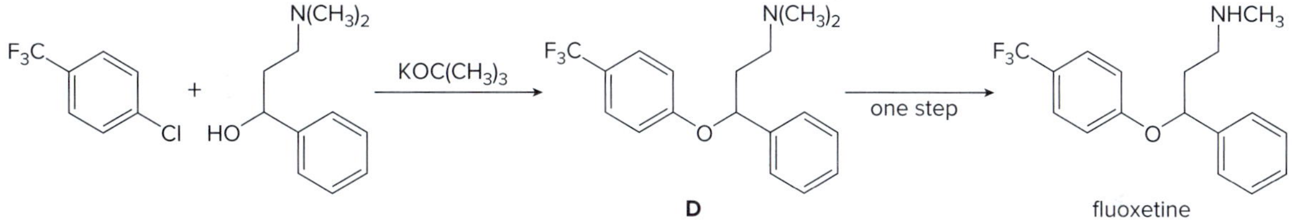

[F$_3$C–C$_6$H$_4$–Cl] + [HO–CH(C$_6$H$_5$)–CH$_2$–CH$_2$–N(CH$_3$)$_2$] $\xrightarrow{KOC(CH_3)_3}$ [F$_3$C–C$_6$H$_4$–O–CH(C$_6$H$_5$)–CH$_2$–CH$_2$–N(CH$_3$)$_2$] $\xrightarrow{\text{one step}}$ [F$_3$C–C$_6$H$_4$–O–CH(C$_6$H$_5$)–CH$_2$–CH$_2$–NHCH$_3$]

D   fluoxetine

20.13B Nucleophilic Aromatic Substitution by Elimination–Addition: Benzyne

Aryl halides that do not contain an electron-withdrawing group generally do *not* react with nucleophiles. **Under extreme reaction conditions, however, nucleophilic aromatic substitution can occur with aryl halides.** For example, heating chlorobenzene with NaOH above 300 °C and 170 atmospheres of pressure affords phenol.

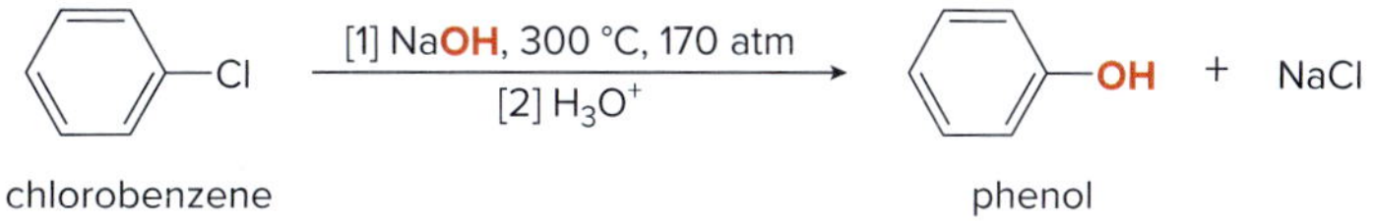

[chlorobenzene, C$_6$H$_5$–Cl] $\xrightarrow[\text{[2] } H_3O^+]{\text{[1] NaOH, 300 °C, 170 atm}}$ [phenol, C$_6$H$_5$–OH] + NaCl

The mechanism proposed to explain this result involves formation of a **benzyne** intermediate (C$_6$H$_4$) by **elimination–addition.** As shown in Mechanism 20.11, **benzyne is a highly reactive, unstable intermediate formed by elimination of HX from an aryl halide.**

Mechanism 20.11 Nucleophilic Aromatic Substitution by Elimination–Addition: Benzyne

[Step 1: chlorobenzene + $^-$OH] $\xrightarrow{1}$ [Step 2: aryl anion with Cl, + H$_2$O] $\xrightarrow{2}$ [**benzyne** + :Cl$^-$] with :Nu$^-$ $\xrightarrow{3}$ [aryl anion with Nu] + H–OH $\xrightarrow{4}$ [Ar–Nu with H] + $^-$:OH

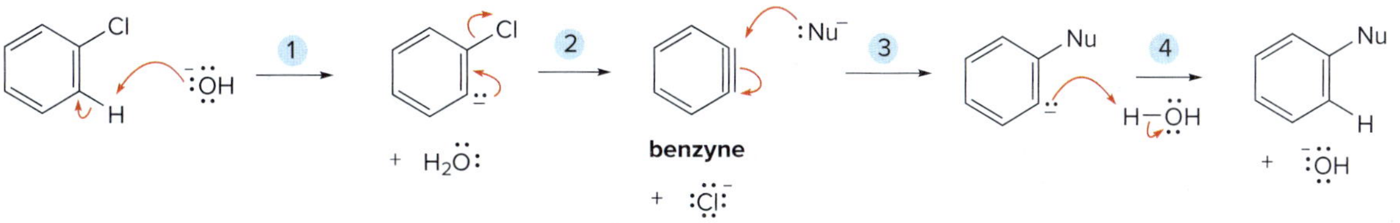

1 – 2 Elimination of H and X from two adjacent atoms forms a reactive **benzyne** intermediate.
3 – 4 Nucleophilic attack and protonation form the substitution product.

Formation of a benzyne intermediate explains why substituted aryl halides form **mixtures** of products. **Nucleophilic aromatic substitution by an elimination–addition mechanism affords substitution on the carbon bonded directly to the leaving group and the carbon

adjacent to it. As an example, treatment of *p*-chlorotoluene with NaNH₂ forms para- and meta-substitution products.

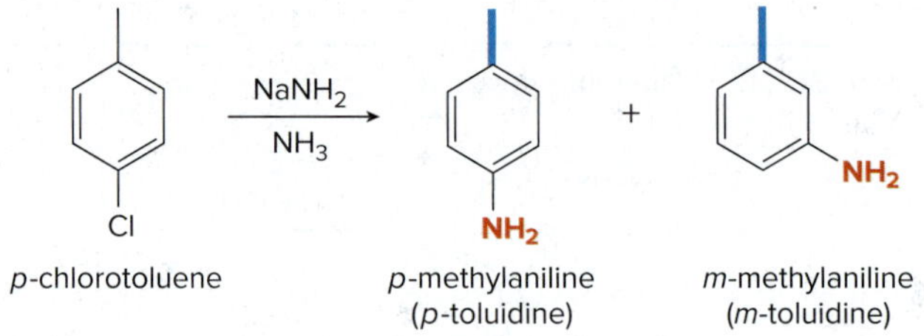

p-chlorotoluene → (NaNH₂, NH₃) → *p*-methylaniline (*p*-toluidine) + *m*-methylaniline (*m*-toluidine)

This result is explained by the fact that nucleophilic attack on the benzyne intermediate may occur at either C3 to form *m*-methylaniline, or C4 to form *p*-methylaniline.

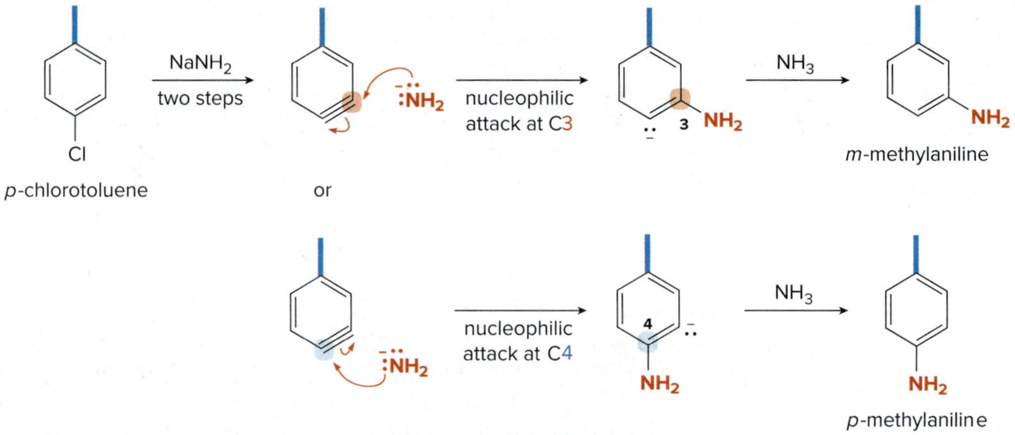

As you might expect, the triple bond in benzyne is unusual. Each carbon of the six-membered ring is sp^2 hybridized, and as a result, the σ bond and two π bonds of the triple bond are formed with the following orbitals:

- The σ bond is formed by overlap of two sp^2 hybrid orbitals.
- One π bond is formed by overlap of two *p* orbitals perpendicular to the plane of the molecule.
- The second π bond is formed by overlap of two sp^2 hybrid orbitals.

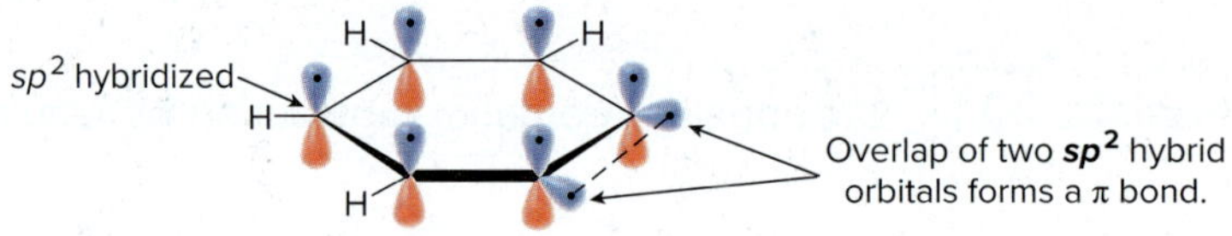

Thus, the second π bond of benzyne differs from all other π bonds seen thus far, because **it is formed by the side-by-side overlap of sp^2 hybrid orbitals, not *p* orbitals.** This π bond, located in the plane of the molecule, is extremely weak.

Problem 20.29 Draw the products of each reaction.

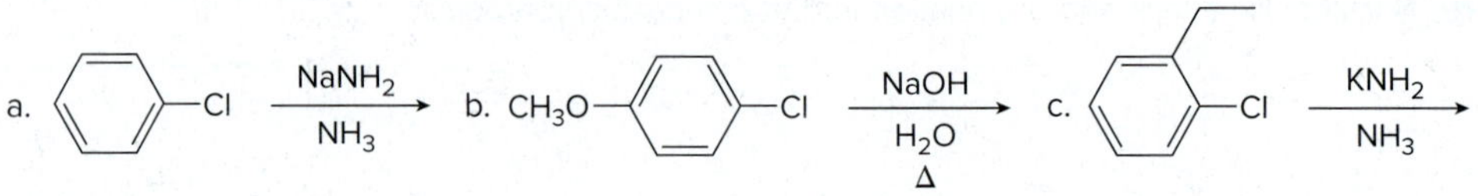

a. PhCl + NaNH₂ / NH₃
b. CH₃O–C₆H₄–Cl + NaOH / H₂O, Δ
c. (2-ethylphenyl)Cl + KNH₂ / NH₃

Problem 20.30 Draw all products formed when *m*-chlorotoluene is treated with KNH₂ in NH₃.

20.14 Reactions of Substituted Benzenes

We finish Chapter 20 by learning some additional reactions of substituted benzenes that greatly expand the ability to synthesize benzene derivatives. In Section 20.14A we learn about halogenation of alkyl benzenes, and in Sections 20.14B–20.14D we examine useful oxidation and reduction reactions. Only reagents and reactions are presented, without reference to the detailed mechanisms.

20.14A Halogenation of Alkyl Benzenes

With proper choice of reaction conditions, **alkyl benzenes undergo selective bromination at the benzylic C—H bond** to form a **benzylic halide.** For example, bromination of ethylbenzene using Br$_2$ in the presence of light or heat forms a benzylic bromide as the sole product. Reaction occurs exclusively at the sp^3 hybridized benzylic carbon.

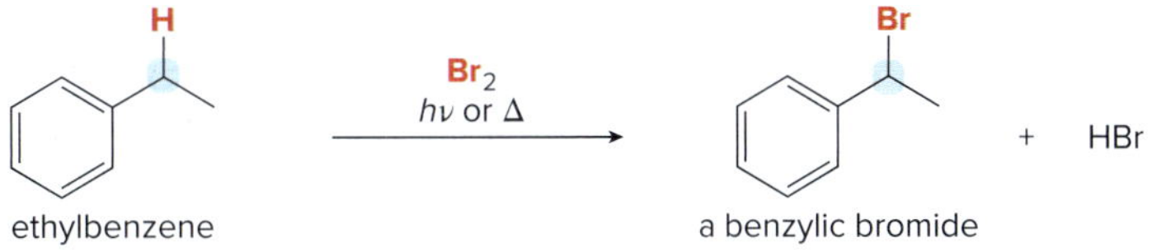

The mechanism of this reaction is different from other mechanisms we have seen thus far, which involve ionic reactive intermediates. Halogenation at the benzylic carbon involves **radical intermediates,** which we will learn about in Chapter 21. In this chapter we concentrate on the use of halogenation in synthesis.

Thus, an alkyl benzene undergoes two useful reactions with Br$_2$, depending on the reaction conditions.

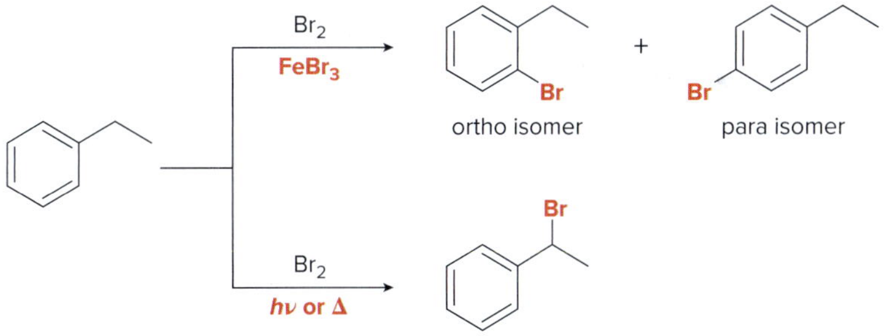

- With Br$_2$ and FeBr$_3$, electrophilic aromatic substitution occurs, resulting in replacement of H by Br on the aromatic ring to form ortho and para isomers.
- With Br$_2$ and light or heat, substitution of H by Br occurs at the *benzylic* carbon of the alkyl group.

The benzylic bromination of alkyl benzenes is a useful reaction because the resulting benzylic halide can serve as starting material for a variety of substitution and elimination reactions, thus making it possible to form many new substituted benzenes. Sample Problem 20.8 illustrates one possibility.

Sample Problem 20.8 Using Benzylic Bromination to Introduce a Double Bond

Design a synthesis of styrene from ethylbenzene.

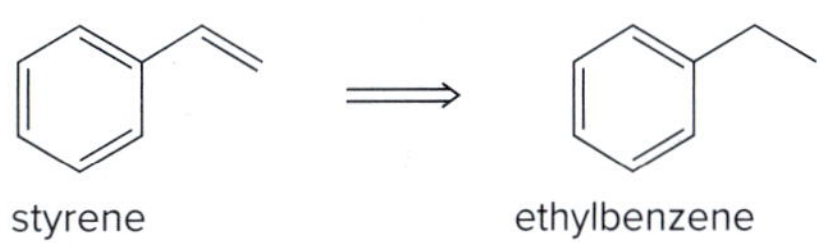

Solution

The double bond can be introduced by a two-step reaction sequence: **bromination** at the benzylic position, followed by **elimination of HBr** with strong base to form the π bond.

ethylbenzene →[Br$_2$, hν or Δ]→ (1-bromoethyl)benzene →[K$^+$ $^-$OC(CH$_3$)$_3$]→ styrene

Problem 20.31 How could you use ethylbenzene to prepare each compound? More than one step is required.

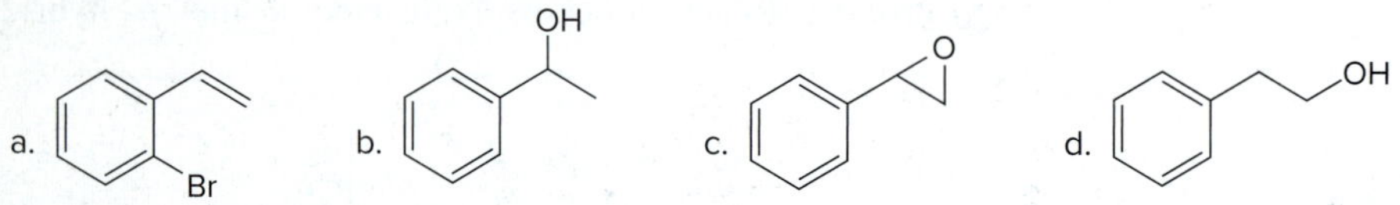

a. 2-bromostyrene
b. 1-phenylethanol
c. styrene oxide
d. 2-phenylethanol

More Practice: Try Problems 20.61a, e, f; 20.62a, c; 20.63.

20.14B Oxidation of Alkyl Benzenes

Arenes containing at least one benzylic C–H bond are oxidized with KMnO$_4$ to benzoic acid, a carboxylic acid with the carboxy group (COOH) bonded directly to the benzene ring. With some alkyl benzenes, this also results in the cleavage of carbon–carbon bonds, so the product has fewer carbon atoms than the starting material.

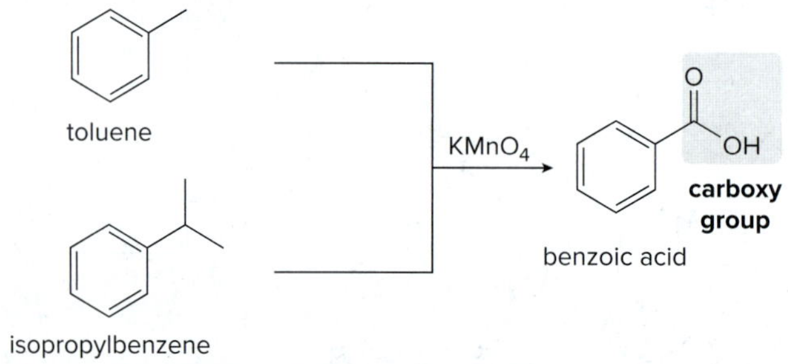

toluene   ⎫
             ⎬ →[KMnO$_4$]→ benzoic acid (carboxy group)
isopropylbenzene ⎭

Substrates with more than one alkyl group are oxidized to dicarboxylic acids. **Compounds without a benzylic C–H bond are inert to oxidation.**

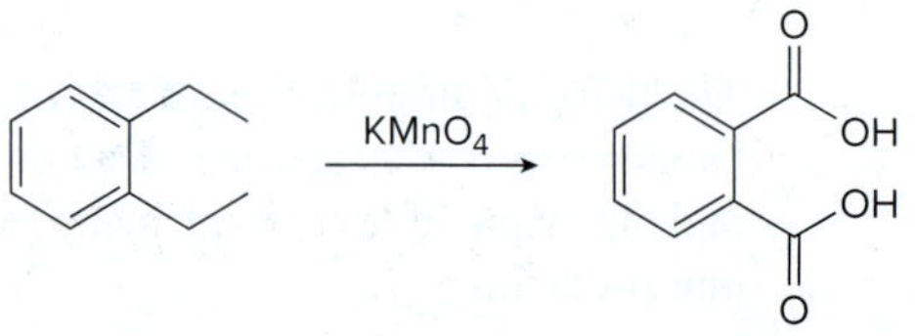

o-diethylbenzene →[KMnO$_4$]→ phthalic acid

tert-butylbenzene →[KMnO$_4$]→ **No reaction**

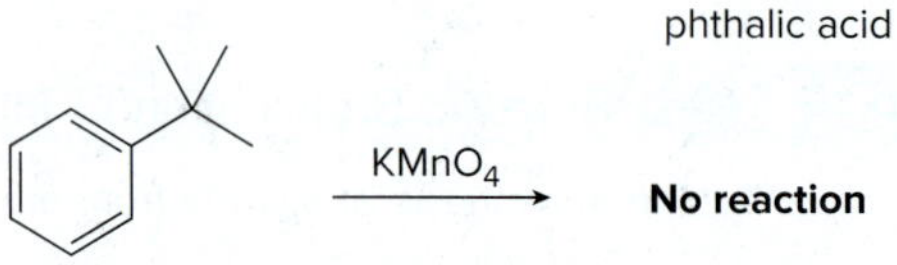

20.14C Reduction of Aryl Ketones to Alkyl Benzenes

Ketones formed as products in Friedel–Crafts acylation can be reduced to alkyl benzenes by two different methods.

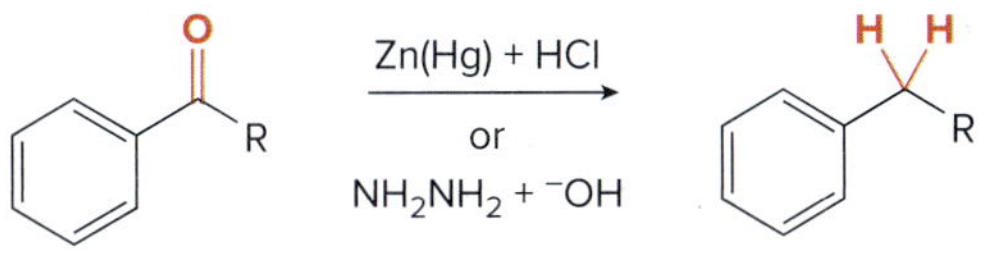

20.14 Reactions of Substituted Benzenes

- The **Clemmensen reduction** uses zinc and mercury in the presence of strong acid.
- The **Wolff–Kishner reduction** uses hydrazine (NH_2NH_2) and strong base (KOH).

Because both C—O bonds in the starting material are converted to C—H bonds in the product, the reduction is difficult and the reaction conditions must be harsh.

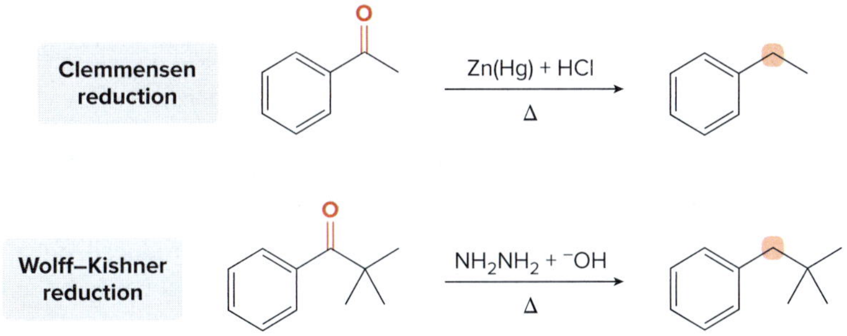

We now know two different ways to introduce an alkyl group on a benzene ring (Figure 20.6):

- **A one-step method using Friedel–Crafts alkylation**
- **A two-step method using Friedel–Crafts acylation to form a ketone, followed by reduction**

Figure 20.6

Two methods to prepare an alkyl benzene

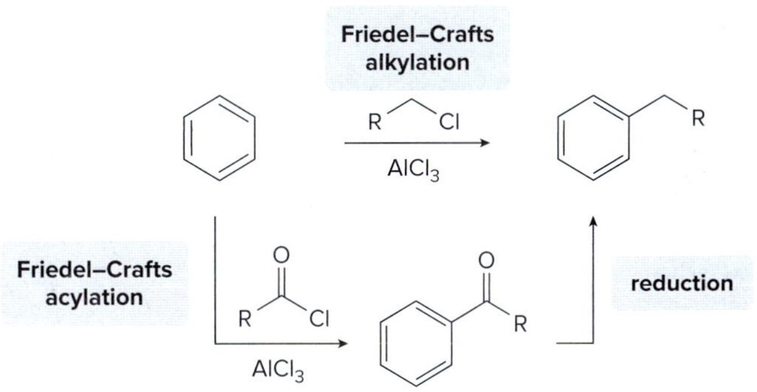

Although the two-step method seems more roundabout, it must be used to synthesize certain alkyl benzenes that cannot be prepared by the one-step Friedel–Crafts alkylation because of rearrangements.

Recall from Section 20.5C that propylbenzene cannot be prepared by a Friedel–Crafts alkylation. Instead, when benzene is treated with 1-chloropropane and $AlCl_3$, isopropylbenzene is formed by a rearrangement reaction. Propylbenzene can be made, however, by a two-step procedure using Friedel–Crafts acylation followed by reduction.

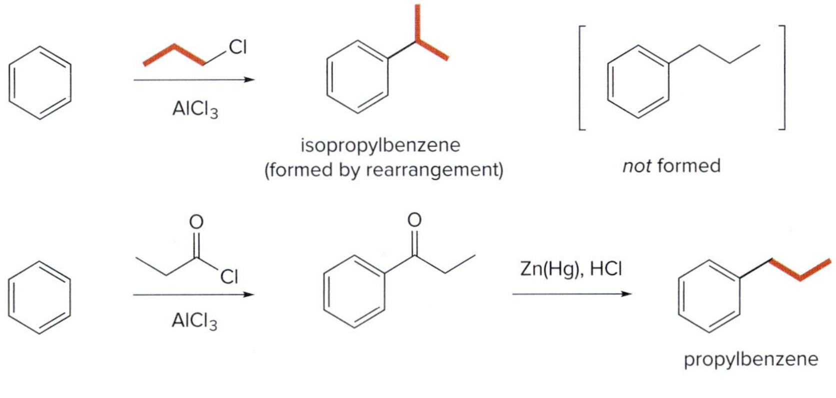

isopropylbenzene
(formed by rearrangement)

not formed

propylbenzene

Chapter 20 Reactions of Aromatic Compounds

Problem 20.32 Write out the two-step sequence that converts benzene to each compound.

a. 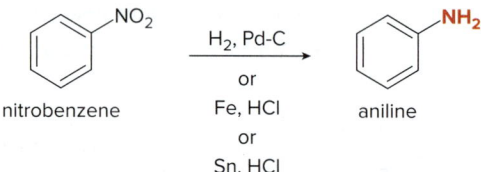   b.

Problem 20.33 What steps are needed to convert benzene to *p*-isobutylacetophenone, a synthetic intermediate used in the synthesis of the anti-inflammatory agent ibuprofen.

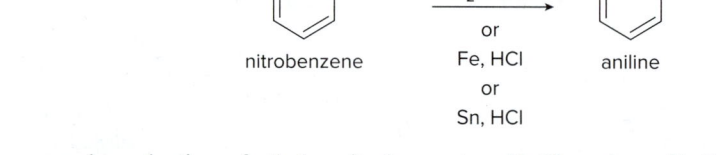

p-isobutylacetophenone → several steps → ibuprofen

20.14D Reduction of Nitro Groups

A nitro group (NO_2) is easily introduced on a benzene ring by nitration with strong acid (Section 20.4). This process is useful because the **nitro group is readily reduced to an amino group (NH_2)** under a variety of conditions. The most common methods use H_2 and a catalyst, or a metal (such as Fe or Sn) and a strong acid like HCl.

nitrobenzene → H_2, Pd-C or Fe, HCl or Sn, HCl → aniline

For example, reduction of ethyl *p*-nitrobenzoate with H_2 and a palladium catalyst forms ethyl *p*-aminobenzoate, a local anesthetic commonly called benzocaine.

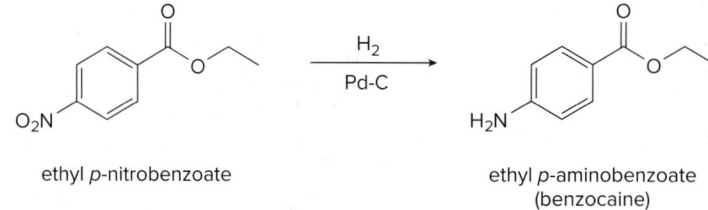

ethyl *p*-nitrobenzoate → H_2, Pd-C → ethyl *p*-aminobenzoate (benzocaine)

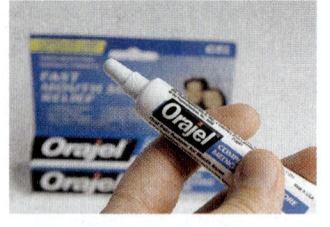

Benzocaine is the active ingredient in the over-the-counter topical anesthetic Orajel. *Jill Braaten/McGraw-Hill Education*

Sample Problem 20.9 illustrates the utility of this process in a short synthesis.

Sample Problem 20.9   Introducing an NH_2 Group in a Synthesis

Design a synthesis of *m*-bromoaniline from benzene.

m-bromoaniline ⟹ benzene

Solution

To devise a retrosynthetic plan, keep in mind:

- The NH_2 group cannot be introduced directly on the ring by electrophilic aromatic substitution. It must be added by a two-step process: **nitration followed by reduction.**
- Both the Br and NH_2 groups are ortho, para directors, but they are located meta to each other on the ring. However, an **NO_2 group (from which an NH_2 group is made) *is* a meta director,** and we can use this fact to our advantage.

20.15 Multistep Synthesis

Retrosynthetic Analysis
Working backwards gives the following **three-step retrosynthetic analysis**:

[diagram: m-bromoaniline ⇐[1] reduction⇐ m-bromonitrobenzene ⇐[2] bromination⇐ nitrobenzene ⇐[3] nitration⇐ benzene]

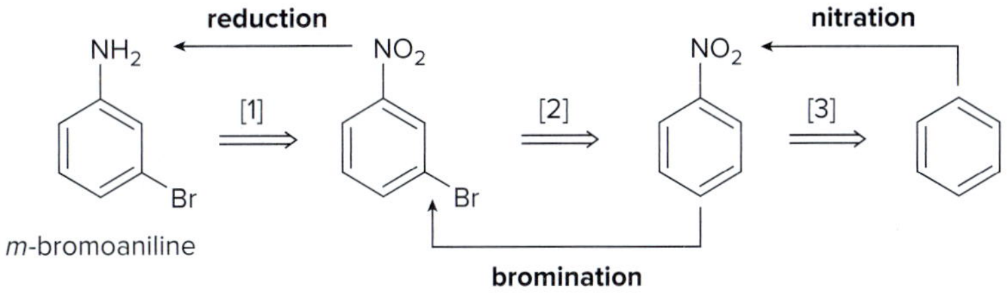

- [1] Form the NH_2 group by reduction of NO_2.
- [2] Introduce the Br group meta to the NO_2 group by halogenation.
- [3] Add the NO_2 group by nitration.

Synthesis
The synthesis involves three steps, and the order is crucial for success. Halogenation (Step [2] of the synthesis) must occur *before* reduction (Step [3]) in order to form the meta-substitution product.

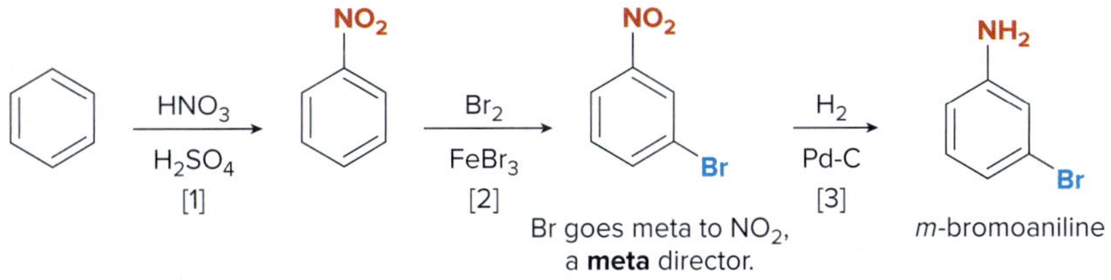

benzene →[1] HNO_3, H_2SO_4 → nitrobenzene →[2] Br_2, $FeBr_3$ → m-bromonitrobenzene (Br goes meta to NO_2, a **meta** director.) →[3] H_2, Pd-C → *m*-bromoaniline

---

Problem 20.34 Synthesize each compound from benzene.

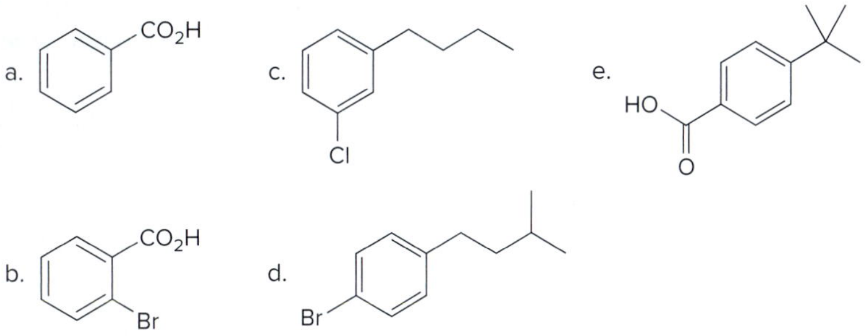

a. benzoic acid (C_6H_5–CO_2H)
b. 2-bromobenzoic acid
c. 1-butyl-3-chlorobenzene
d. 1-bromo-4-(3-methylbutyl)benzene
e. 4-tert-butylbenzoic acid

More Practice: Try Problems 20.61b–d, 20.62b.

20.15 Multistep Synthesis

The reactions learned in Chapter 20 make it possible to synthesize a wide variety of substituted benzenes, as shown in Sample Problems 20.10 and 20.11.

Sample Problem 20.10 — Designing a Multistep Synthesis

Synthesize *p*-nitrobenzoic acid from benzene.

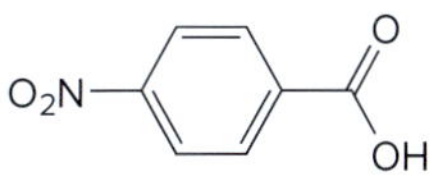

p-nitrobenzoic acid

Solution
Both groups on the ring (NO_2 and COOH) are meta directors. To place these two groups para to each other, remember that the **COOH group is prepared by oxidizing an alkyl group, which is an ortho, para director.**

Retrosynthetic Analysis

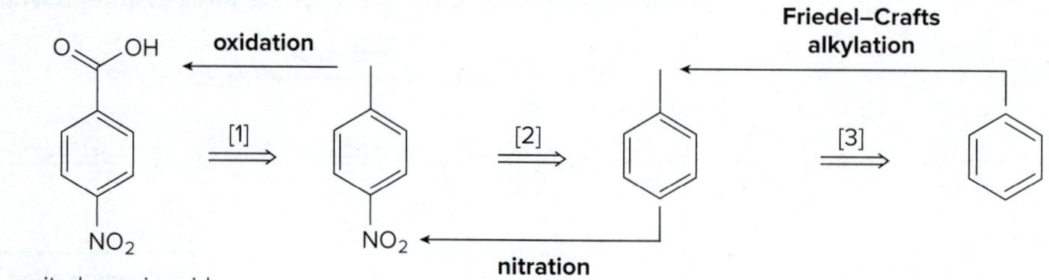

p-nitrobenzoic acid ⟸[1] (oxidation) toluene(NO₂) ⟸[2] (nitration) toluene ⟸[3] (Friedel–Crafts alkylation) benzene

Working backwards:

- [1] Form the COOH group by oxidation of an alkyl group.
- [2] Introduce the NO₂ group para to the CH₃ group (an ortho, para director) by nitration.
- [3] Add the CH₃ group by Friedel–Crafts alkylation.

Synthesis

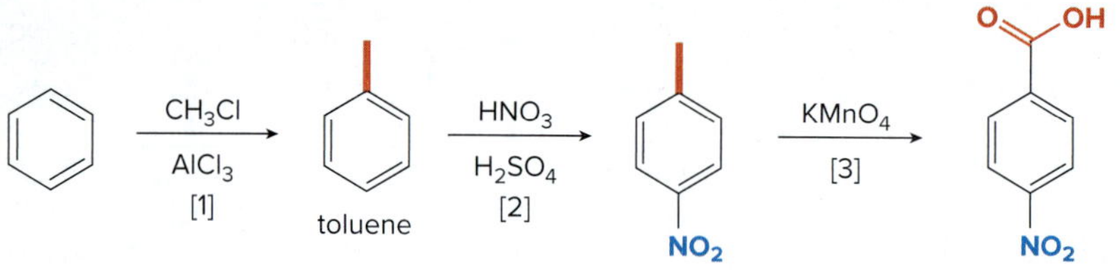

benzene →[CH₃Cl, AlCl₃, [1]] toluene →[HNO₃, H₂SO₄, [2]] *p*-nitrotoluene [+ ortho isomer] →[KMnO₄, [3]] *p*-nitrobenzoic acid

- **Friedel–Crafts alkylation** with CH_3Cl and $AlCl_3$ forms toluene in Step [1]. Because CH_3 is an ortho, para director, nitration yields the desired para product, which can be separated from its ortho isomer (Step [2]).
- **Oxidation with KMnO₄** converts the CH_3 group to a COOH group, giving the desired product in Step [3].

Problem 20.35 Synthesize each compound from benzene.

a.  Cl—C₆H₄—CH=CH₂

b. 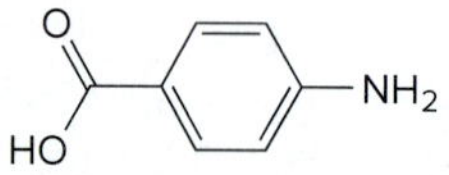 HOOC—C₆H₄—NH₂
   PABA
   sunscreen component

More Practice: Try Problems 20.64–20.66.

Sample Problem 20.11 | Synthesizing a Trisubstituted Benzene

Synthesize the trisubstituted benzene **A** from benzene.

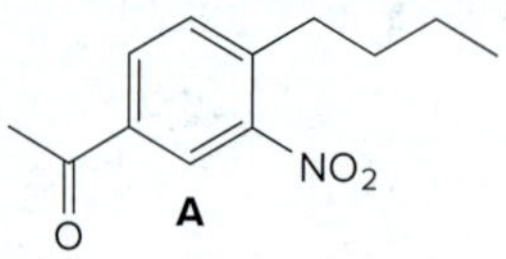

A: benzene with COCH₃, NO₂, and butyl substituents

Solution

Two groups (CH_3CO and NO_2) in **A** are meta directors located meta to each other, and the third substituent, an alkyl group, is an ortho, para director.

Retrosynthetic Analysis

With three groups on the benzene ring, **begin by determining the possible disubstituted benzenes that are immediate precursors of the target compound,** and then eliminate any that cannot be converted to the desired product. For example, three different disubstituted benzenes (**B–D**) can theoretically be precursors to **A**. However, conversion of compounds **B** or **D** to **A** would require a Friedel–Crafts reaction on a deactivated benzene ring, a reaction that does not occur. Thus, only **C** is a feasible precursor of **A**.

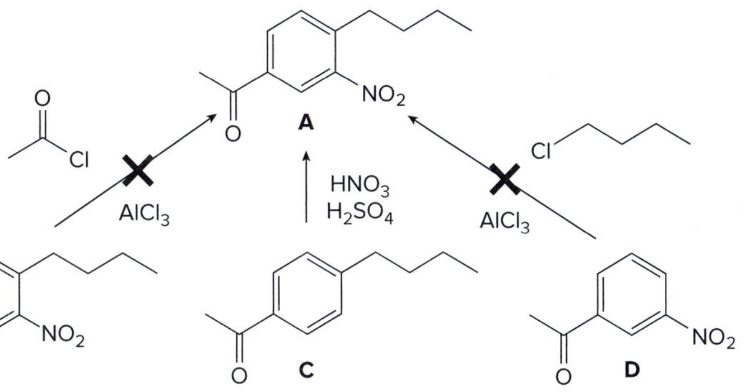

no Friedel–Crafts on a strongly deactivated benzene ring

Only this pathway works.

no Friedel–Crafts on a strongly deactivated benzene ring

To complete the retrosynthetic analysis, prepare **C** from benzene:

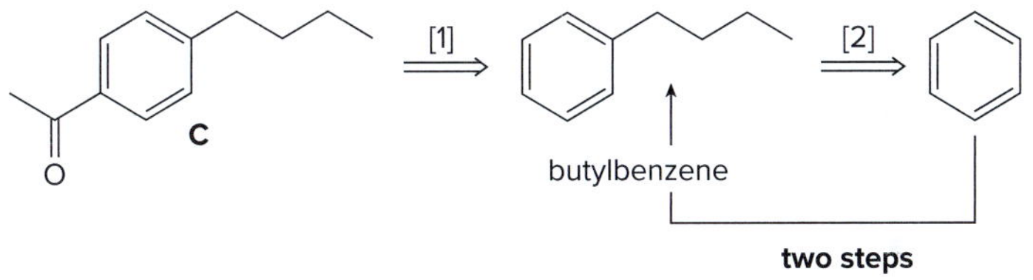

- [1] Add the ketone by Friedel–Crafts acylation.
- [2] Add the alkyl group by the two-step process—Friedel–Crafts acylation followed by reduction. It is not possible to prepare butylbenzene by a one-step Friedel–Crafts alkylation because of a rearrangement reaction (Section 20.14C).

Synthesis

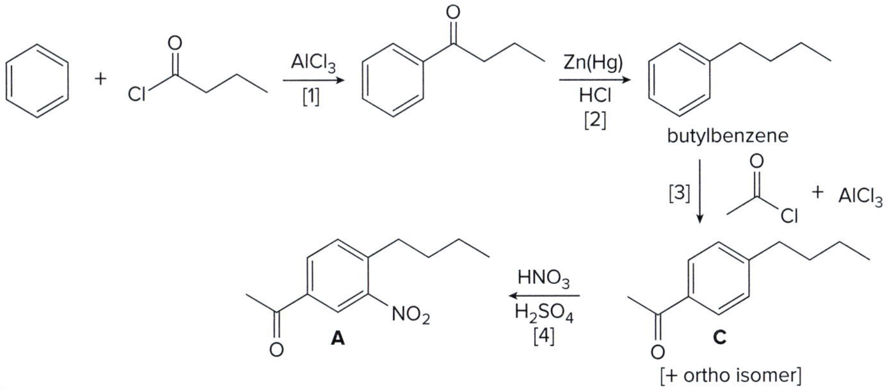

- Friedel–Crafts acylation followed by reduction with Zn(Hg), HCl yields butylbenzene (Steps [1]–[2]).
- Friedel–Crafts acylation gives the para product **C**, which can be separated from its ortho isomer (Step [3]).
- Nitration in Step [4] introduces the NO_2 group ortho to the alkyl group (an ortho, para director) and meta to the CH_3CO group (a meta director).

Problem 20.36 Synthesize each compound from benzene.

a. 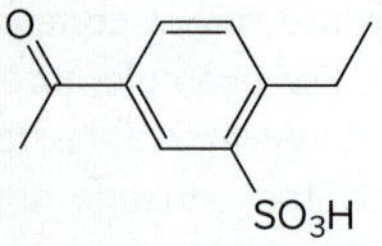 [structure: 1-acetyl-4-ethyl-3-sulfonic acid benzene] b. 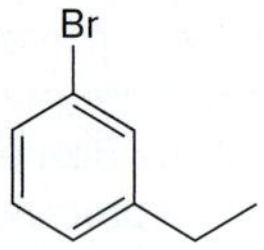 [structure: 1-bromo-3-ethylbenzene] c. 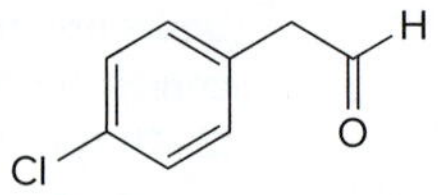 [structure: 4-chlorophenylacetaldehyde]

More Practice: Try Problems 20.61–20.66.

Chapter 20 REVIEW

KEY CONCEPTS

[1] Three rules describing the reactivity and directing effects of common substituents (20.7–20.9)

① Ortho, para directors
- All **ortho, para directors** except the halogens **activate** the benzene ring.

  general structure
  –R or –Z:

  –R –NHCOR –ÖR –ÖH –NR₂

  Increasing activation →

② Meta directors
- All **meta directors deactivate** the benzene ring.

  general structure
  –Y (δ+ or +)

  –CHO –COR –CO₂R –CO₂H –CN –SO₃H –NO₂ –NR₃⁺

  Increasing deactivation →

③ Halogens
- The **halogens deactivate** the benzene ring and **direct ortho, para.**

  general structure
  –X:

  –F: –Cl: –Br: –I:

[2] Summary of substituent effects in electrophilic aromatic substitution (20.6–20.9)

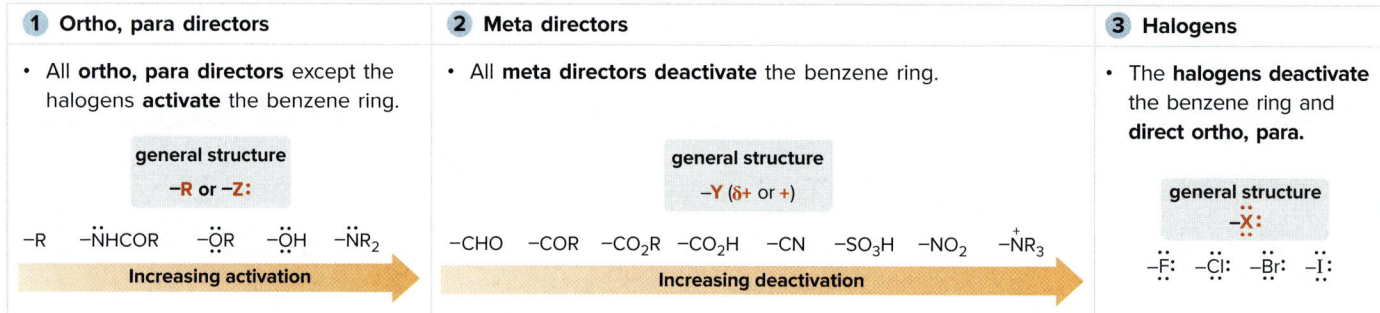

① Substituent	② Inductive effect	③ Resonance effect	④ Reactivity	⑤ Directing effect
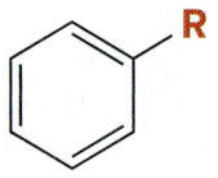 R = alkyl	donating	none	activating	ortho, para
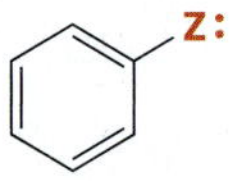 Z: (Z = N or O)	withdrawing	donating	activating	ortho, para
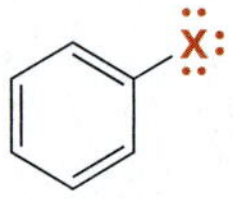 X: (X = halogen)	withdrawing	donating	deactivating	ortho, para
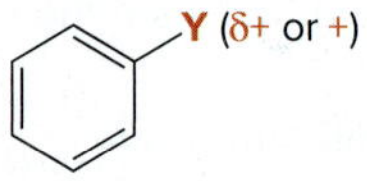 Y (δ+ or +)	withdrawing	withdrawing	deactivating	meta

Try Problems 20.48, 20.49.

KEY REACTIONS

[1] Electrophilic aromatic substitution

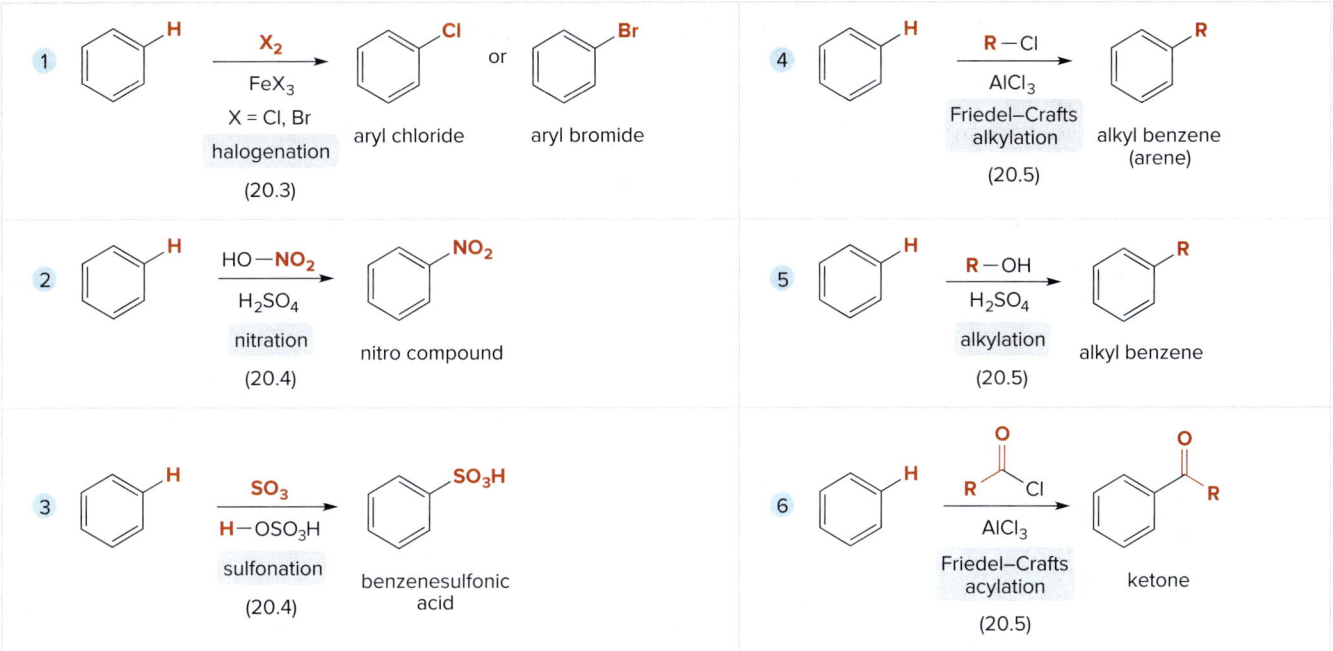

Try Problems 20.37, 20.40a–e.

[2] Nucleophilic aromatic substitution

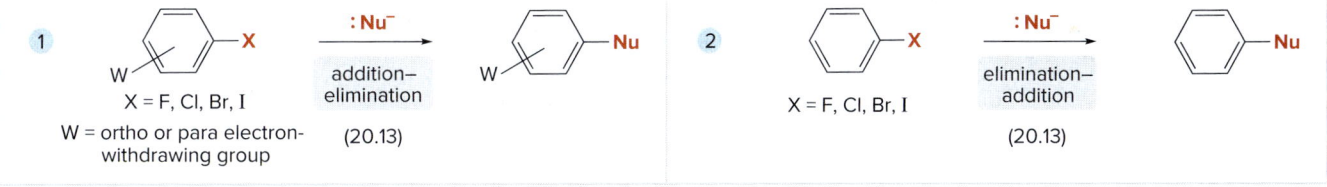

Try Problems 20.40f, 20.42d, 20.44.

[3] Other reactions of benzene derivatives

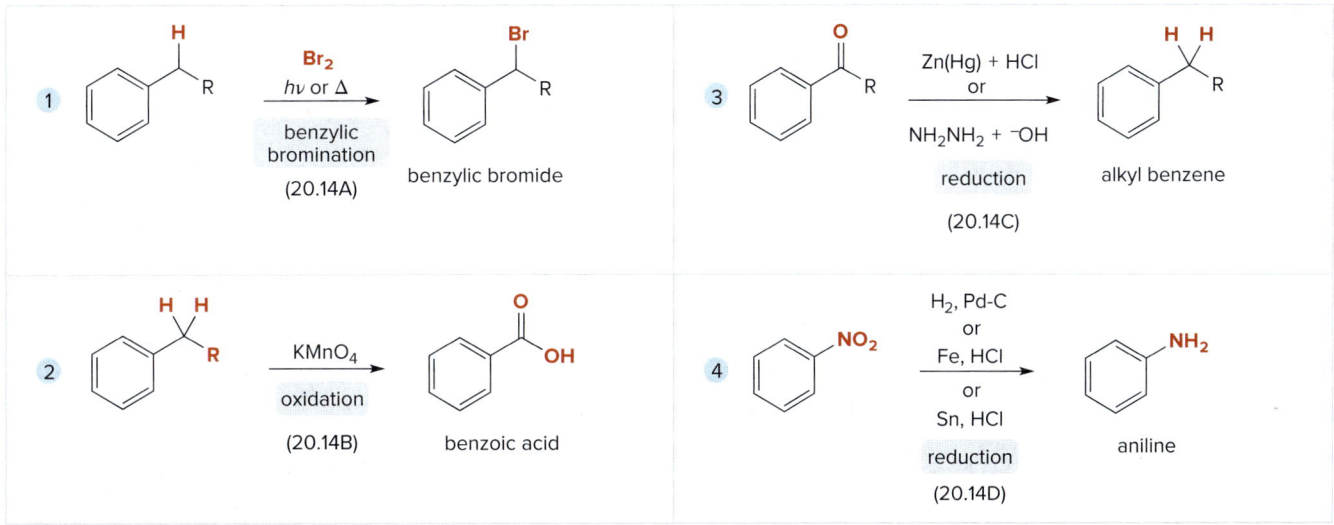

Try Problems 20.38, 20.42a–c.

KEY SKILLS

[1] Classifying substituents as electron donating or electron withdrawing (20.6); two considerations

Draw out the atoms, bonds, and electrons of the substituent, and look at the atom bonded directly to the benzene ring.

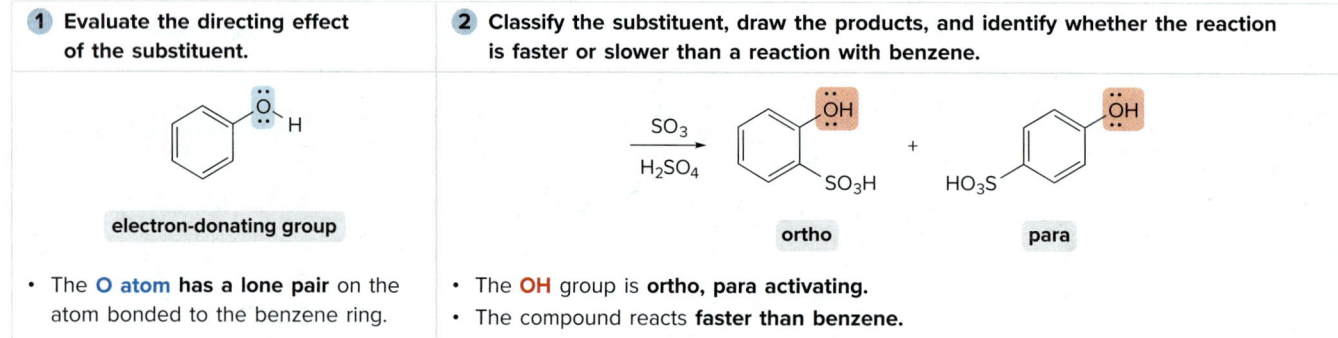

electron-donating group

- An **O** or **N atom** with a **lone pair of electrons** makes a substituent **electron donating**.

electron-withdrawing group

- A **halogen** or an **atom with a partial positive charge** makes a substituent **electron withdrawing**.

See Sample Problem 20.3. Try Problems 20.49–20.51.

[2] Drawing the product(s) from reaction of a monosubstituted benzene with an electrophile (20.7)

① Evaluate the directing effect of the substituent.

② Classify the substituent, draw the products, and identify whether the reaction is faster or slower than a reaction with benzene.

electron-donating group

- The **O atom has a lone pair** on the atom bonded to the benzene ring.

- The **OH** group is **ortho, para activating**.
- The compound reacts **faster than benzene**.

See Sample Problem 20.4. Try Problems 20.38, 20.39, 20.42a.

[3] Drawing the product(s) from reaction of a disubstituted benzene with an electrophile (20.11)

① Evaluate the directing effects, and classify the substituents.

② Determine the net result.

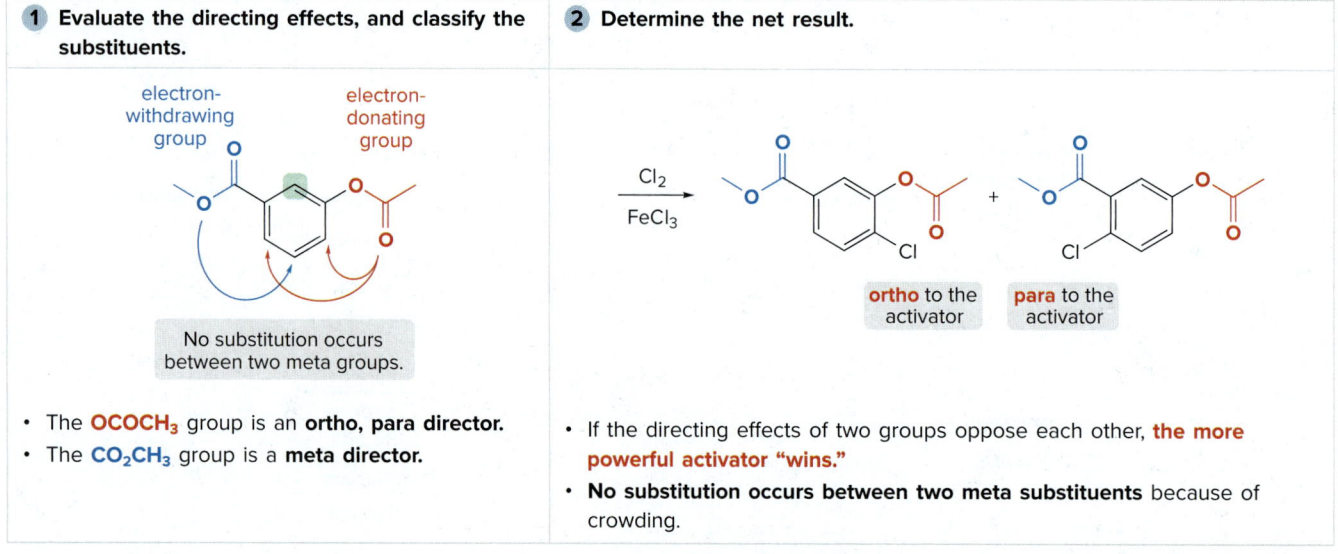

No substitution occurs between two meta groups.

- The **OCOCH₃** group is an **ortho, para director**.
- The **CO₂CH₃** group is a **meta director**.

ortho to the activator   para to the activator

- If the directing effects of two groups oppose each other, **the more powerful activator "wins."**
- **No substitution occurs between two meta substituents** because of crowding.

See Sample Problem 20.6. Try Problems 20.37, 20.40a–e, 20.42c.

Key Mechanism Concepts

[4] Devising a synthesis of a trisubstituted benzene (20.15); example: 4-nitro-2-propylphenol from 1-chloro-4-nitrobenzene

1. Compare the carbon skeletons and functional groups.

- The **Cl** group is converted to an **OH** group.
- The **H** atom is converted to a **CH₂CH₂CH₃** group.

2. Complete the synthesis.

- [1] Convert the **Cl** group to the **OH** group using **addition–elimination**.
- [2] Form the **propyl group** by two steps: **Friedel–Crafts acylation** followed by **reduction**.

- The **Friedel–Crafts acylation** will occur despite the strong **NO₂** deactivator because the **OH** group is a **strong activator**.

See Sample Problem 20.11. Try Problems 20.61–20.66.

KEY MECHANISM CONCEPTS

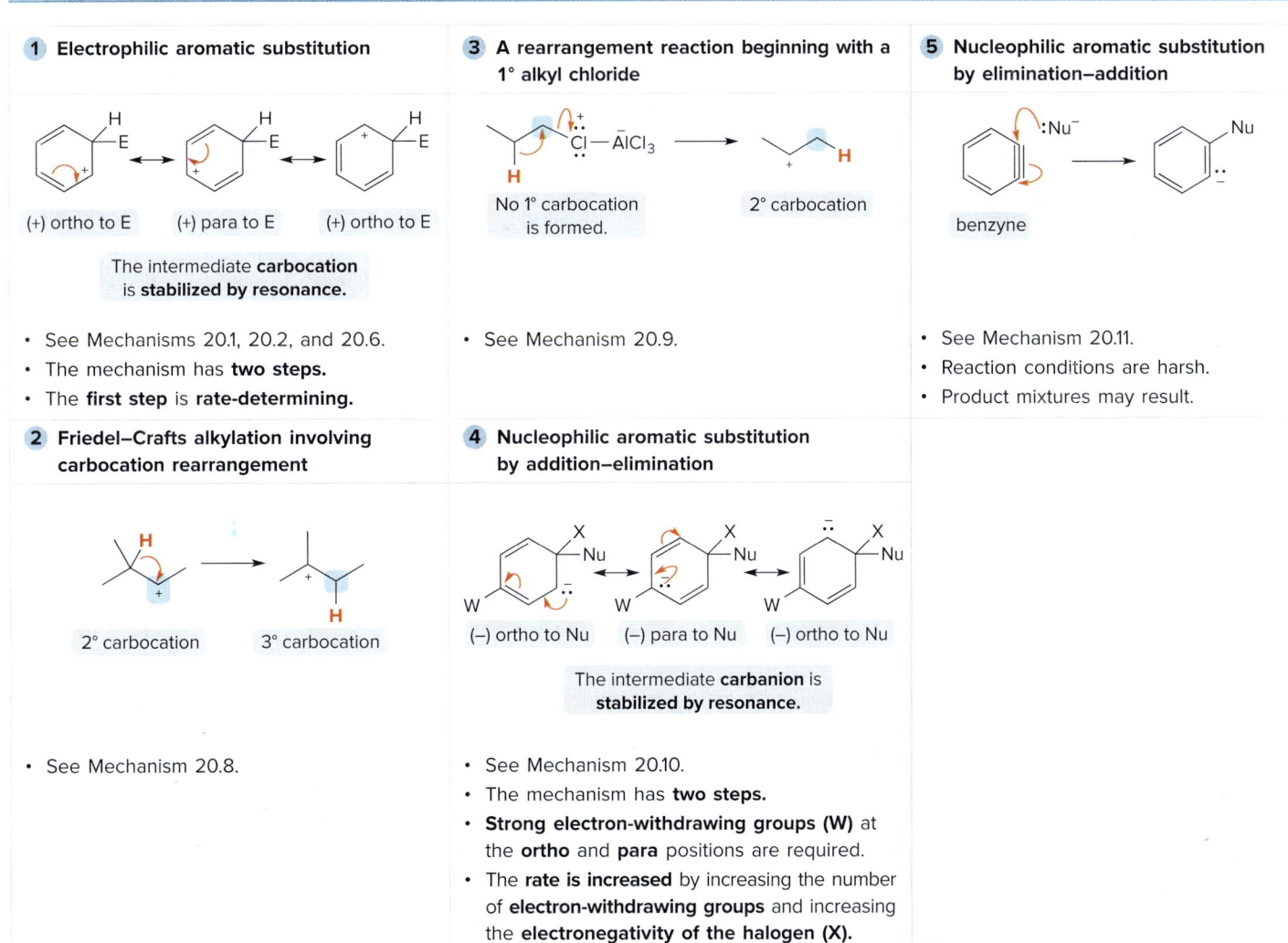

1. Electrophilic aromatic substitution

(+) ortho to E   (+) para to E   (+) ortho to E

The intermediate **carbocation** is **stabilized by resonance**.

- See Mechanisms 20.1, 20.2, and 20.6.
- The mechanism has **two steps**.
- The **first step** is **rate-determining**.

2. Friedel–Crafts alkylation involving carbocation rearrangement

2° carbocation → 3° carbocation

- See Mechanism 20.8.

3. A rearrangement reaction beginning with a 1° alkyl chloride

No 1° carbocation is formed.   2° carbocation

- See Mechanism 20.9.

4. Nucleophilic aromatic substitution by addition–elimination

(−) ortho to Nu   (−) para to Nu   (−) ortho to Nu

The intermediate **carbanion** is **stabilized by resonance**.

- See Mechanism 20.10.
- The mechanism has **two steps**.
- Strong electron-withdrawing groups (**W**) at the **ortho** and **para** positions are required.
- The **rate is increased** by increasing the number of **electron-withdrawing groups** and increasing the **electronegativity of the halogen (X)**.

5. Nucleophilic aromatic substitution by elimination–addition

benzyne

- See Mechanism 20.11.
- Reaction conditions are harsh.
- Product mixtures may result.

Try Problems 20.54–20.60.

PROBLEMS

Problem Using Three-Dimensional Models

20.37 Draw the products formed when **A** and **B** are treated with each of the following reagents: (a) Br$_2$, FeBr$_3$; (b) HNO$_3$, H$_2$SO$_4$; (c) CH$_3$CH$_2$COCl, AlCl$_3$.

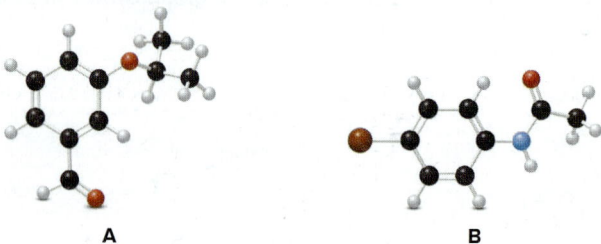

Reactions

20.38 Draw the products formed when phenol (C$_6$H$_5$OH) is treated with each set of reagents.
  a. [1] HNO$_3$, H$_2$SO$_4$; [2] Sn, HCl
  b. [1] (CH$_3$CH$_2$)$_2$CHCOCl, AlCl$_3$; [2] Zn(Hg), HCl
  c. [1] CH$_3$CH$_2$Cl, AlCl$_3$; [2] Br$_2$, $h\nu$
  d. [1] (CH$_3$)$_2$CHCl, AlCl$_3$; [2] KMnO$_4$

20.39 Draw the products formed when each compound is treated with CH$_3$CH$_2$COCl, AlCl$_3$.

20.40 Draw the products of each reaction.

20.41 What products are formed when benzene is treated with each alkyl chloride and AlCl$_3$?

20.42 Draw the products of each reaction.

20.43 You have learned two ways to make an alkyl benzene: Friedel–Crafts alkylation, and Friedel–Crafts acylation followed by reduction. Although some alkyl benzenes can be prepared by both methods, it is often true that only one method can be used to prepare a given alkyl benzene. Which method(s) can be used to prepare each of the following compounds from benzene? Show the steps that would be used.

a. 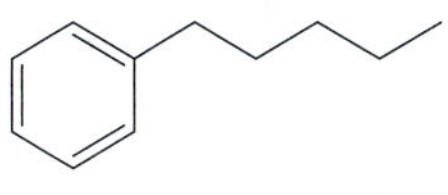   b. 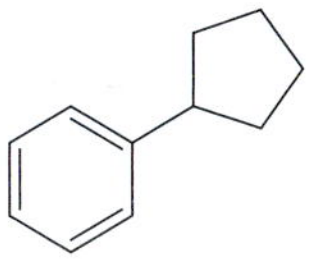   c.

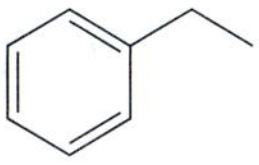

20.44 Identify **X** and **Y,** the products of key steps in two syntheses of pioglitazone, a drug used to treat diabetes.

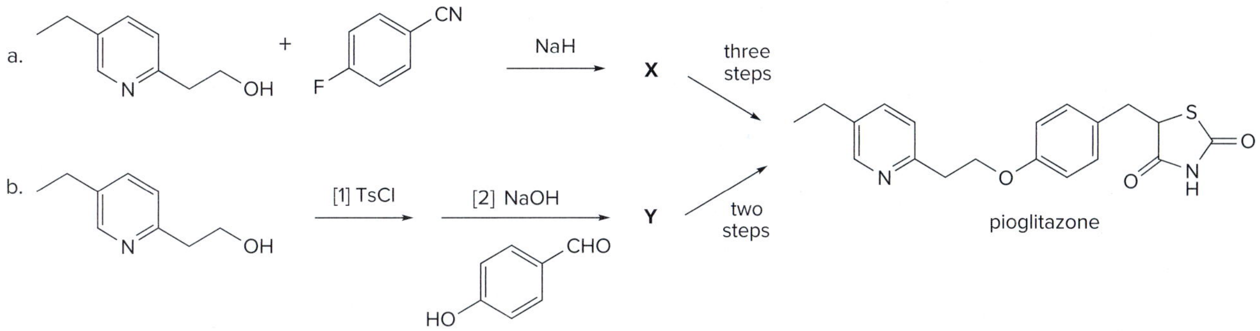

20.45 Identify **M** and **N** in the following reaction sequence, two steps in the original synthesis of the non-sedating antihistamine fexofenadine (Section 22.5B).

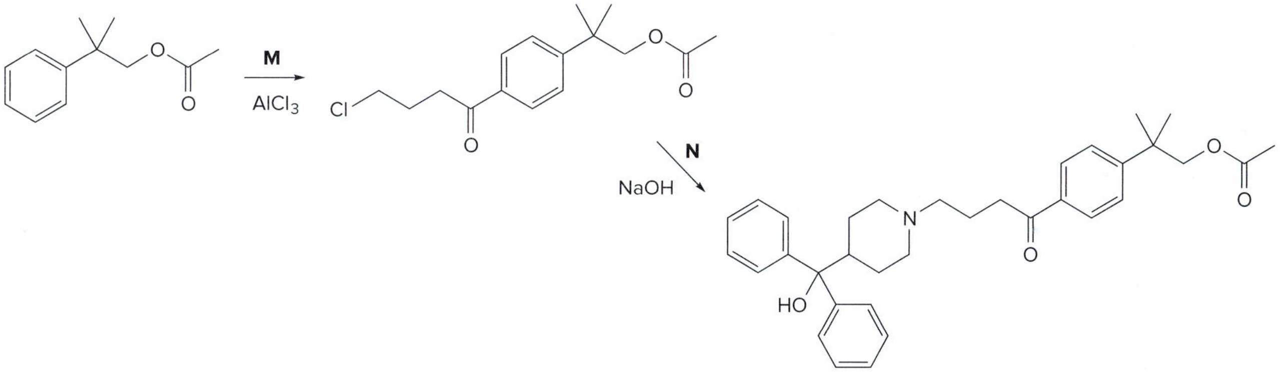

20.46 Draw the structure of **A,** an intermediate in the synthesis of the antipsychotic drug risperidone. Explain why three rings in risperidone are considered aromatic.

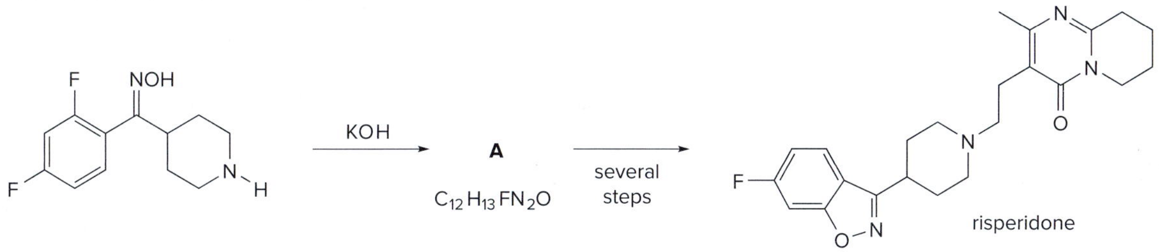

$C_{12}H_{13}FN_2O$

20.47 **D** is an intermediate in the synthesis of rosiglitazone (trade name Avandia), a drug used to treat type 2 diabetes. Suggest two different methods to prepare the ether in **D** by substitution reactions.

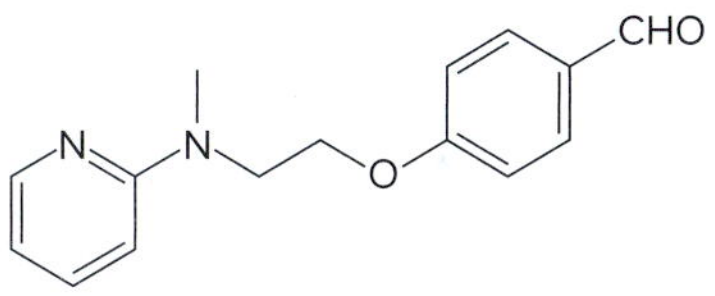

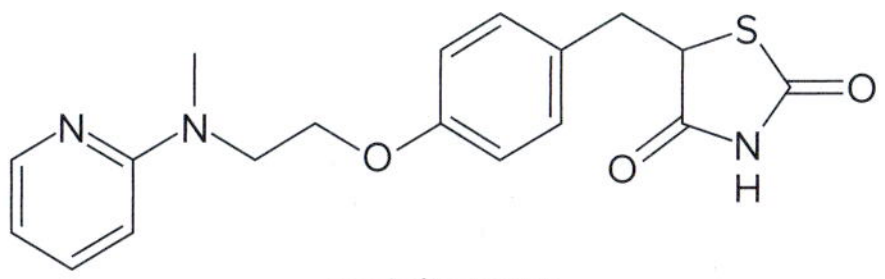

Substituent Effects

20.48 Rank the compounds in each group in order of increasing reactivity in electrophilic aromatic substitution: (a) C_6H_6, C_6H_5Cl, C_6H_5CHO, $C_6H_5OCH_3$; (b) $C_6H_5CH_3$, $C_6H_5NH_2$, $C_6H_5CH_2NH_2$, $C_6H_5CONH_2$.

Chapter 20 Reactions of Aromatic Compounds

20.49 For each of the following substituted benzenes: [1] C_6H_5Br; [2] C_6H_5CN; [3] $C_6H_5OCOCH_3$:
   a. Does the substituent donate or withdraw electron density by an inductive effect?
   b. Does the substituent donate or withdraw electron density by a resonance effect?
   c. On balance, does the substituent make a benzene ring more or less electron rich than benzene itself?
   d. Does the substituent activate or deactivate the benzene ring in electrophilic aromatic substitution?

20.50 Determine which ring in each compound is more reactive in electrophilic aromatic substitution, and draw the product(s) formed when each compound is treated with the general electrophile E^+.

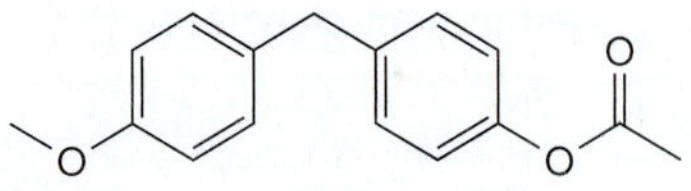

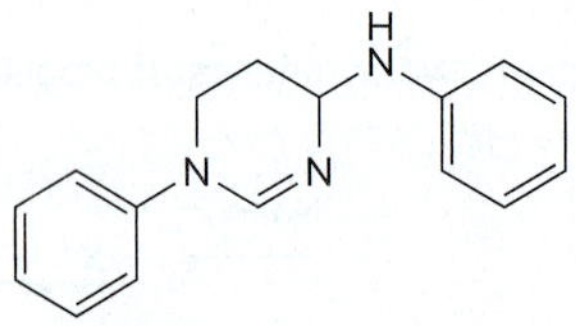

20.51 Consider the tetracyclic aromatic compound drawn below, with rings labeled as **A, B, C,** and **D**. (a) Which of the four rings is *most* reactive in electrophilic aromatic substitution? (b) Which of the four rings is *least* reactive in electrophilic aromatic substitution? (c) What are the major product(s) formed when this compound is treated with one equivalent of Br_2?

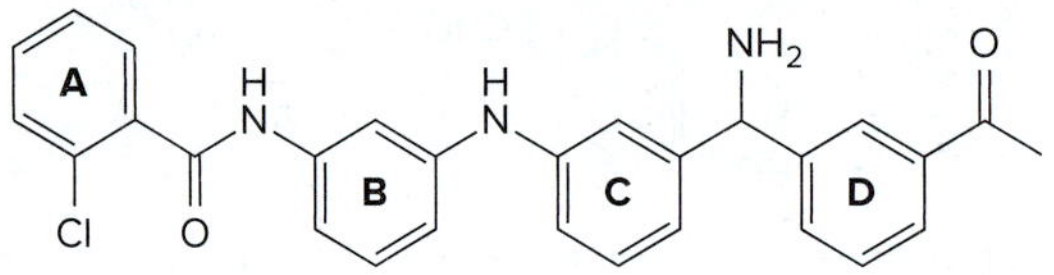

20.52 Explain this observation: Ethyl 3-phenylpropanoate ($C_6H_5CH_2CH_2CO_2CH_2CH_3$) reacts with electrophiles to afford ortho- and para-disubstituted arenes, but ethyl 3-phenylprop-2-enoate ($C_6H_5CH=CHCO_2CH_2CH_3$) reacts with electrophiles to afford meta-disubstituted arenes.

20.53 Rank the aryl halides in each group in order of increasing reactivity in nucleophilic aromatic substitution by an addition–elimination mechanism.
   a. chlorobenzene, *p*-fluoronitrobenzene, *m*-fluoronitrobenzene
   b. 1-fluoro-2,4-dinitrobenzene, 1-fluoro-3,5-dinitrobenzene, 1-fluoro-3,4-dinitrobenzene
   c. 1-fluoro-2,4-dinitrobenzene, 4-chloro-3-nitrotoluene, 4-fluoro-3-nitrotoluene

Mechanisms

20.54 Draw a stepwise, detailed mechanism for the following intramolecular reaction.

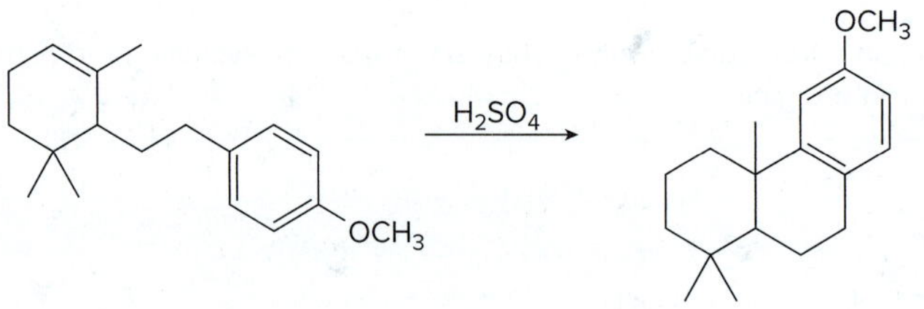

20.55 Draw a stepwise, detailed mechanism for the following reaction.

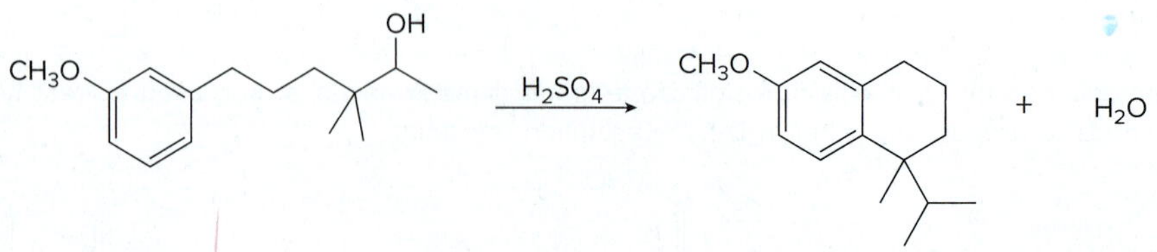

20.56 Draw a stepwise mechanism for the following substitution. Explain why 2-chloropyridine reacts faster than chlorobenzene in this type of reaction.

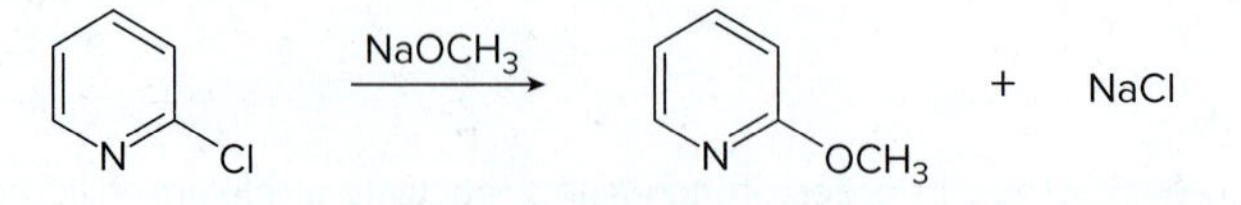

20.57 Although two products (**A** and **B**) are possible when naphthalene undergoes electrophilic aromatic substitution, only **A** is formed. Draw resonance structures for the intermediate carbocation to explain why this is observed.

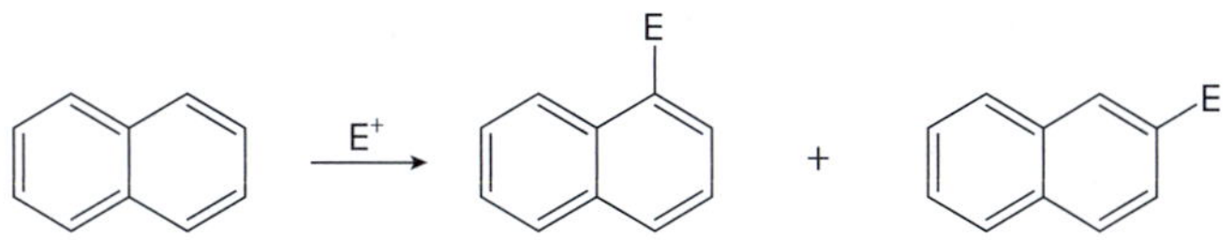

naphthalene → **A** (This product is formed.) + **B** (This product is *not* formed.)

20.58 Draw a stepwise mechanism for the following reaction, which results in the synthesis of bisphenol F (R = H), an additive used in a variety of packaging materials. Bisphenol F is related to BPA (bisphenol A, R = CH$_3$), a reagent used to harden some plastics, now removed from certain baby products because of its estrogen-like activity that can disrupt endocrine pathways.

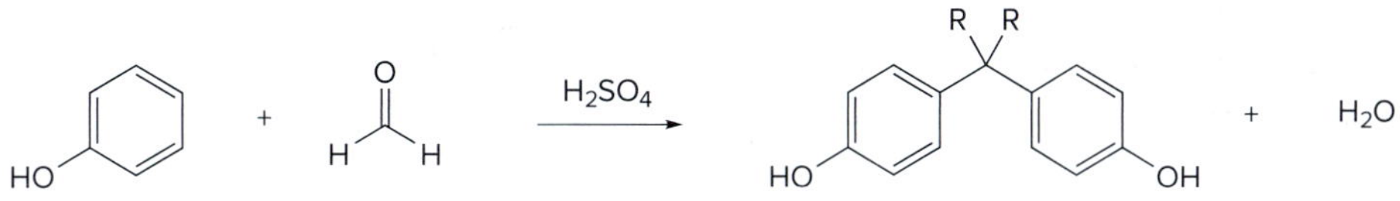

phenol + formaldehyde →(H_2SO_4)→ bisphenol F (R = H) + H_2O

20.59 Draw a stepwise mechanism for the following reaction, one step in the biosynthesis of the amino acid tryptophan. The reaction involves both electrophilic aromatic substitution and decarboxylation in the presence of an acid HA.

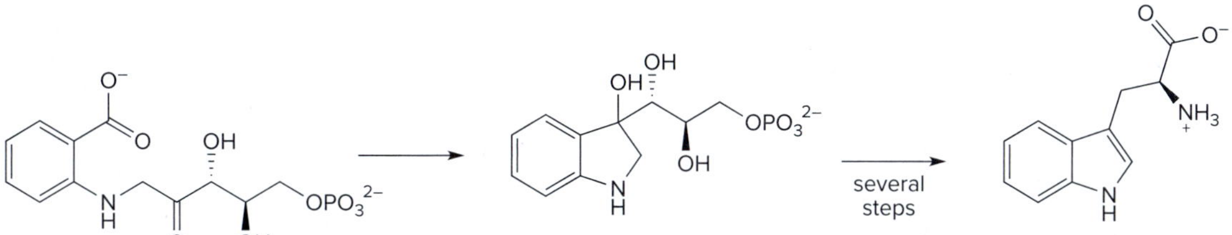

(intermediate) + CO_2 → (several steps) → tryptophan

20.60 One step in the biosynthesis of the ergot alkaloids (Figure 20.3) involves the Friedel–Crafts alkylation of tryptophan with dimethylallyl diphosphate in the presence of a synthase enzyme. Draw a stepwise mechanism for this reaction, including all resonance structures for resonance-stabilized intermediates.

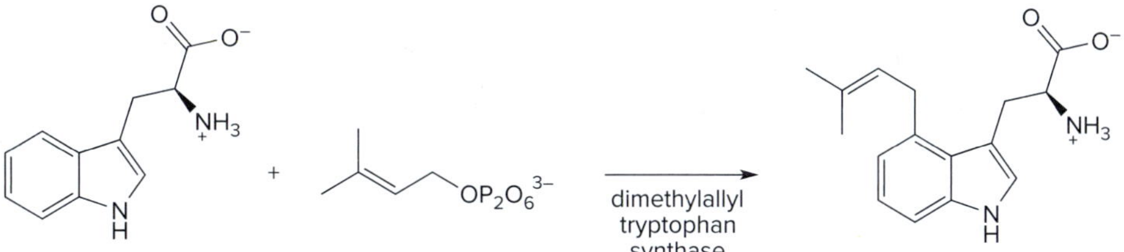

tryptophan + dimethylallyl diphosphate →(dimethylallyl tryptophan synthase)→ product

Synthesis

20.61 Synthesize each compound from benzene, organic halides with < 5 C's, and any other organic or inorganic reagents.

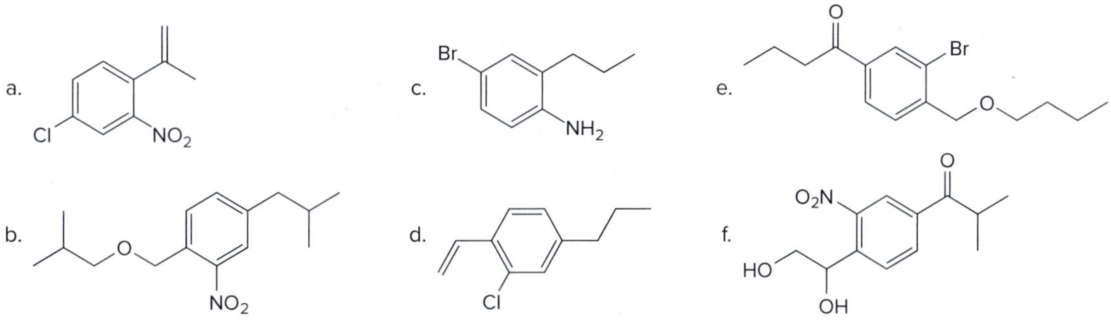

a., b., c., d., e., f.

20.62 Synthesize each compound from toluene ($C_6H_5CH_3$) and any other organic or inorganic reagents.

a. 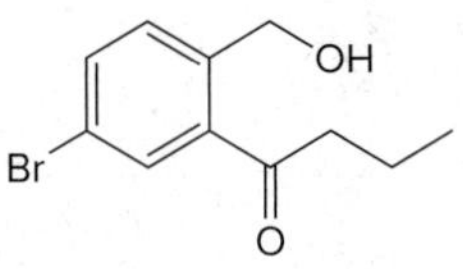 [structure: 4-bromo-2-(hydroxymethyl)phenyl butyl ketone — benzene ring with Br, CH2OH, and C(=O)CH2CH2CH3 substituents]

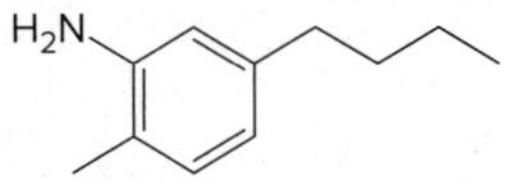

b. [structure: 2-methyl-5-butylaniline — benzene ring with H2N, CH3, and butyl group]

c. [structure: benzene ring with propanoyl group, CHO, and Cl substituents]

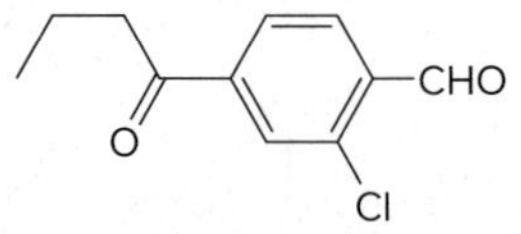

20.63 Use the reactions in this chapter along with those learned in Chapters 10 and 11 to synthesize each compound. You may use benzene, acetylene (HC≡CH), ethanol, ethylene oxide, and any inorganic reagents.

a. 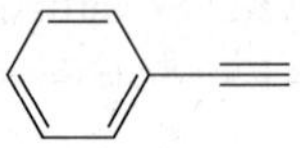 phenylacetylene

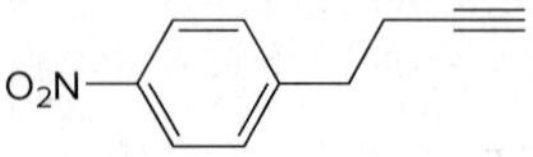

b. 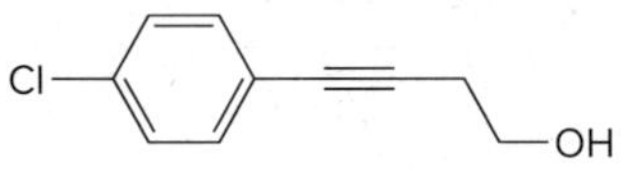 Cl—C6H4—C≡C—CH2CH2—OH

c. O_2N—C6H4—CH2CH2—C≡CH

d. 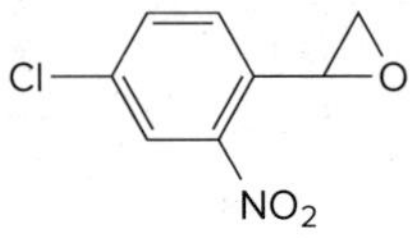 Cl—C6H3(NO2)—epoxide

20.64 Carboxylic acid **X** is an intermediate in the multistep synthesis of proparacaine, a local anesthetic. Devise a synthesis of **X** from phenol and any needed organic or inorganic reagents.

[structure X: 3-nitro-4-propoxybenzoic acid] → several steps → proparacaine [structure: 2-(diethylamino)ethyl 3-amino-4-propoxybenzoate]

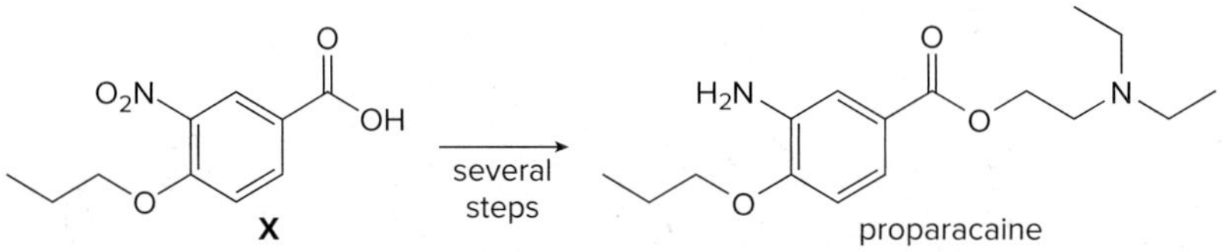

20.65 Devise a synthesis of each compound from the given starting materials. You may also use organic alcohols having four or fewer carbons, and any organic or inorganic reagents.

a. 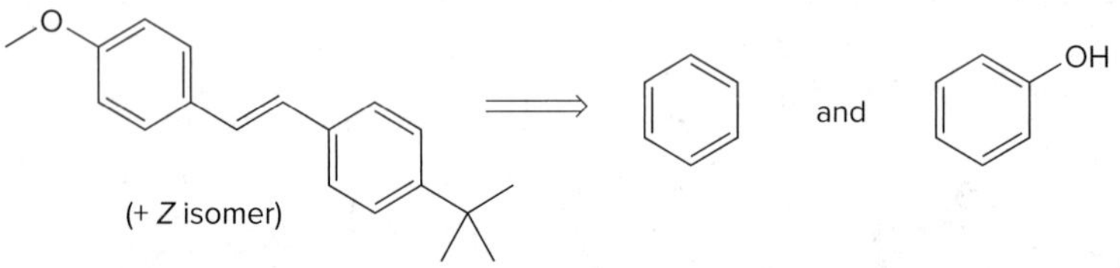 [stilbene derivative: 4-methoxy-4'-tert-butyl stilbene (+ Z isomer)] ⟹ benzene and phenol

b. 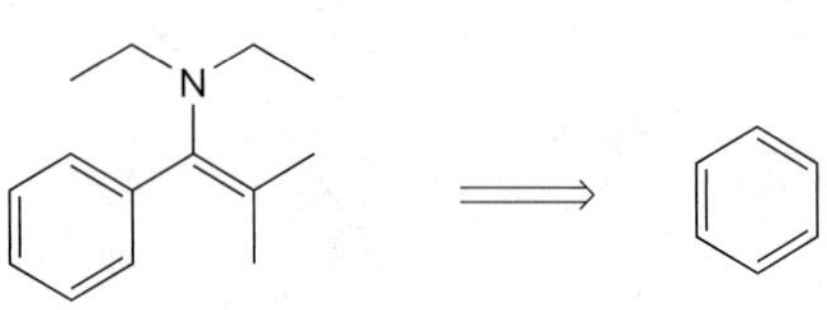 [enamine with phenyl and N(Et)2 groups on alkene] ⟹ benzene

c. 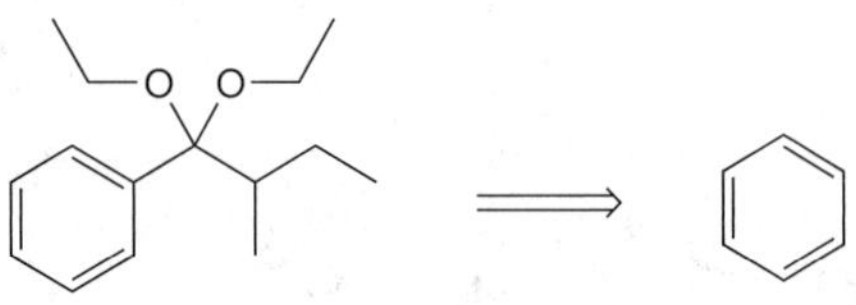 [acetal on benzene] ⟹ benzene

d. 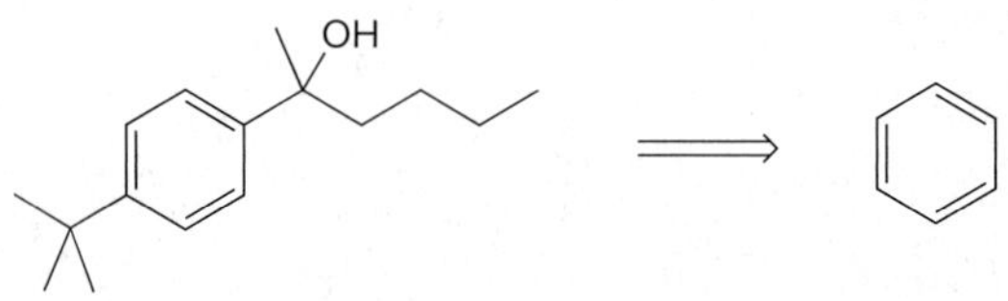 [tert-butyl phenyl carbinol derivative with OH] ⟹ benzene

20.66 Devise a synthesis of each compound from benzene and organic alcohols containing four or fewer carbons. You may also use any required organic or inorganic reagents.

a. 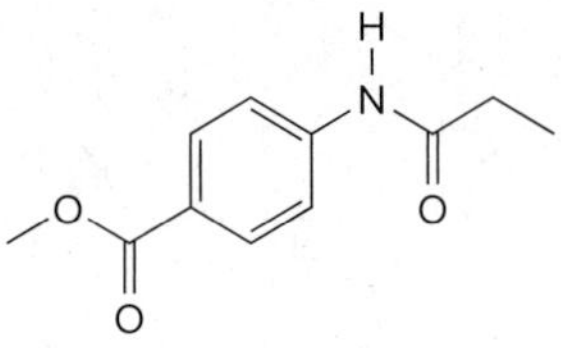 [methyl 4-(propanoylamino)benzoate]

b. 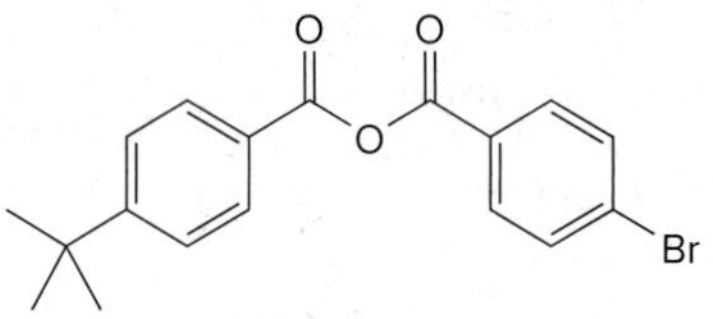 [4-tert-butylbenzoic 4-bromobenzoic anhydride]

c. 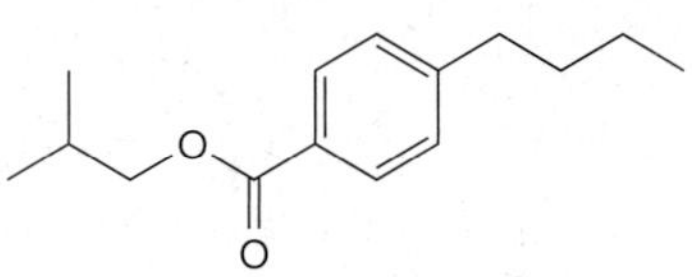 [isobutyl 4-butylbenzoate]

Spectroscopy

20.67 Identify the structures of isomers **A** and **B** (molecular formula C₈H₉Br).

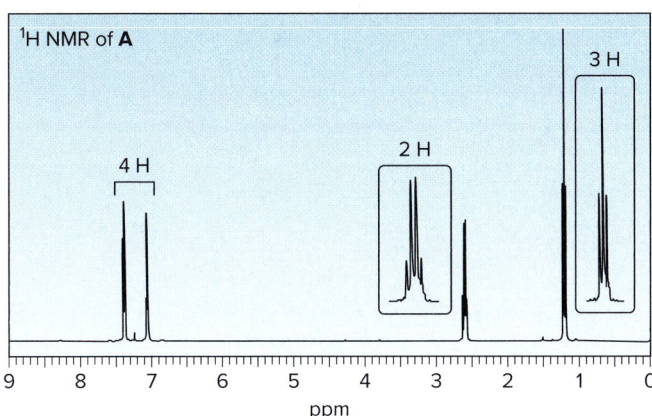
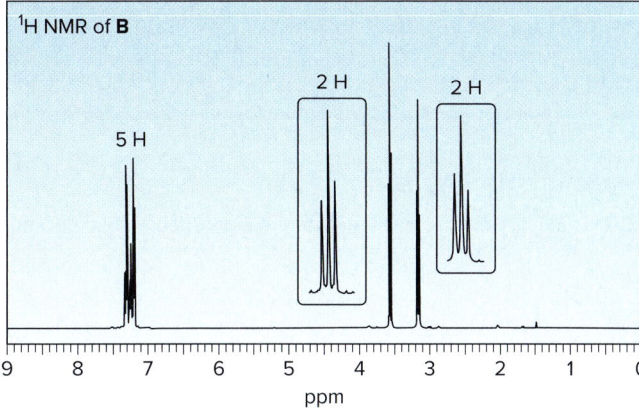

20.68 Propose a structure of compound **C** (molecular formula C₁₀H₁₂O) consistent with the following data. **C** is partly responsible for the odor and flavor of raspberries.

Compound **C**: IR absorption at 1717 cm⁻¹

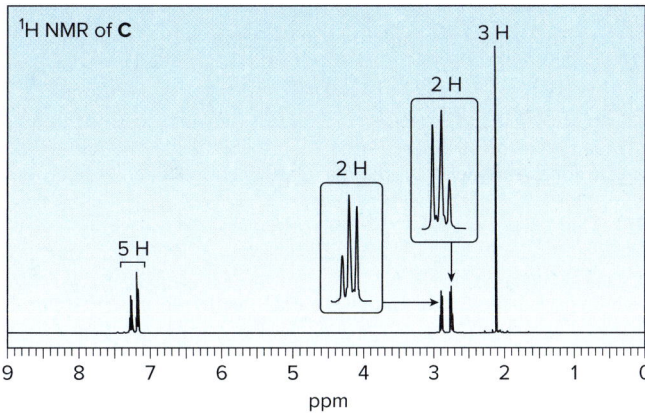

20.69 Compound **X** (molecular formula C₁₀H₁₂O) was treated with NH₂NH₂, ⁻OH to yield compound **Y** (molecular formula C₁₀H₁₄). Based on the ¹H NMR spectra of **X** and **Y** given below, what are the structures of **X** and **Y**?

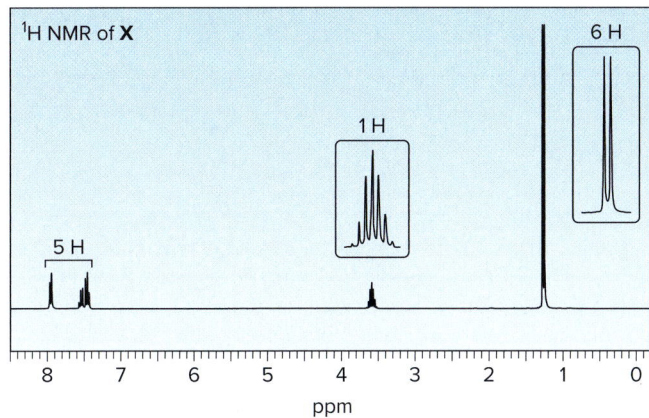
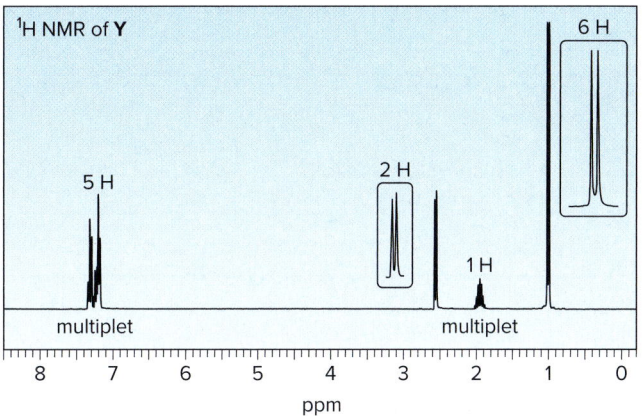

20.70 Reaction of p-cresol with two equivalents of 2-methylprop-1-ene affords BHT, a preservative with molecular formula $C_{15}H_{24}O$. BHT gives the following ^1H NMR spectral data: 1.4 (singlet, 18 H), 2.27 (singlet, 3 H), 5.0 (singlet, 1 H), and 7.0 (singlet, 2 H) ppm. What is the structure of BHT? Draw a stepwise mechanism illustrating how it is formed.

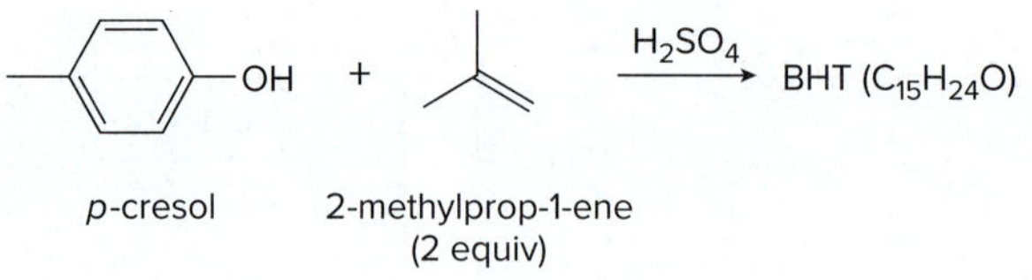

p-cresol + 2-methylprop-1-ene (2 equiv) $\xrightarrow{H_2SO_4}$ BHT ($C_{15}H_{24}O$)

Challenge Problems

20.71 Explain the reactivity and orientation effects observed in each heterocycle.

pyridine $\xrightarrow{E^+}$ (E at C3)

pyrrole $\xrightarrow{E^+}$ (E at C2)

a. Pyridine is less reactive than benzene in electrophilic aromatic substitution and yields 3-substituted products.
b. Pyrrole is more reactive than benzene in electrophilic aromatic substitution and yields 2-substituted products.

20.72 Draw a stepwise mechanism for the dienone–phenol rearrangement, a reaction that forms alkyl-substituted phenols from cyclohexadienones.

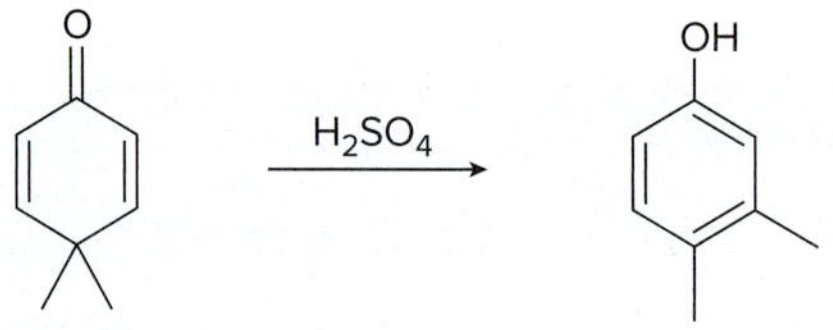

20.73 Draw a stepwise mechanism for the following intramolecular reaction, which is used in the synthesis of the female sex hormone estrone.

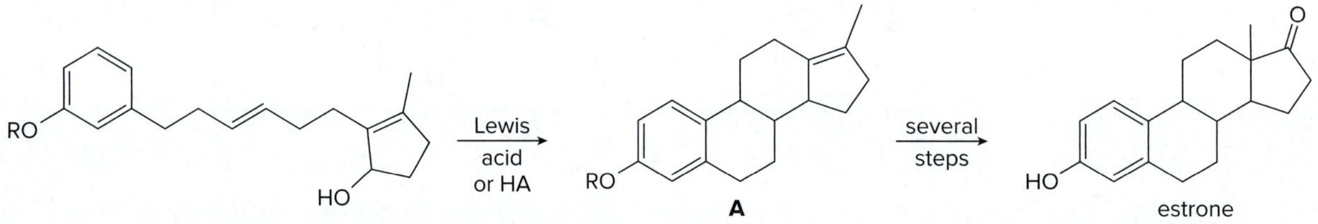

$\xrightarrow{\text{Lewis acid or HA}}$ A $\xrightarrow{\text{several steps}}$ estrone

20.74 Devise a stepwise mechanism for the following reaction. The reaction does not take place by direct electrophilic aromatic substitution at C2. (Hint: The mechanism begins with addition of an electrophile at C3.)

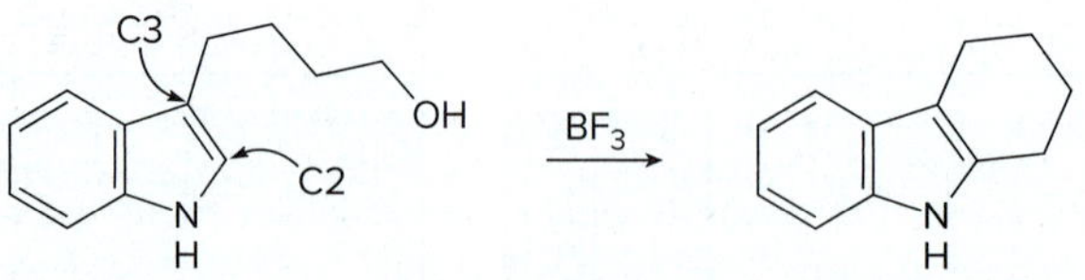

$\xrightarrow{BF_3}$

Radical Reactions

21

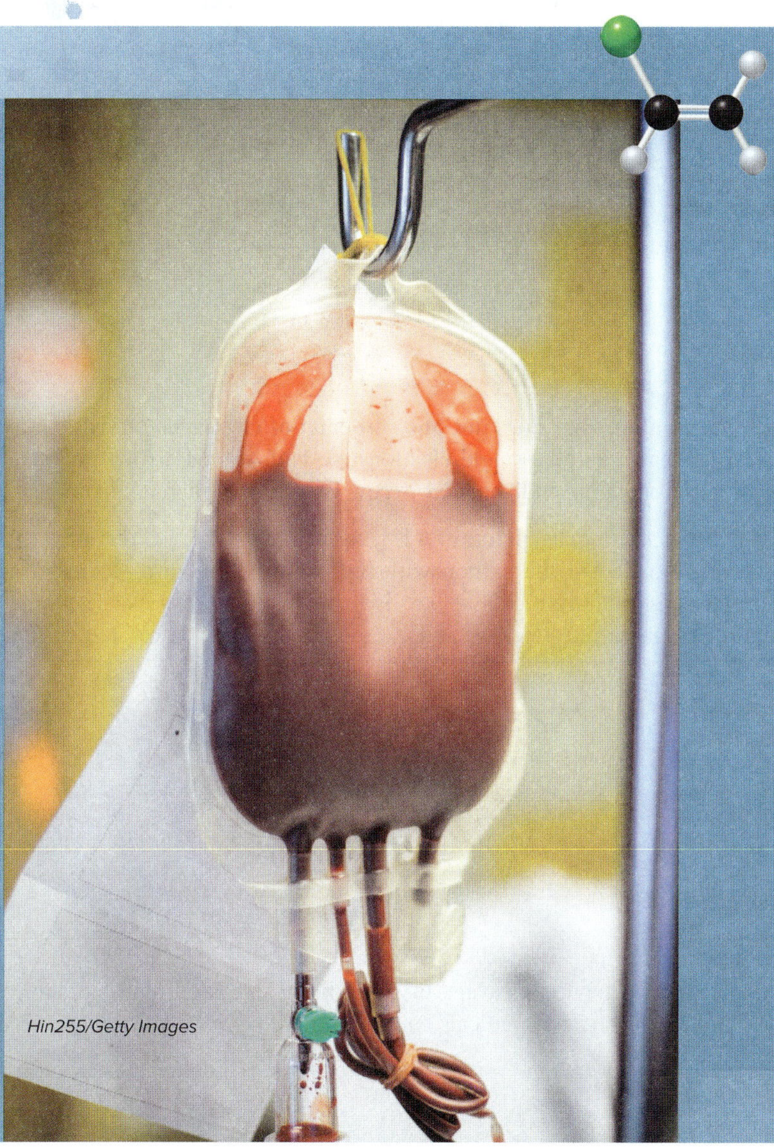

- 21.1 Introduction
- 21.2 General features of radical reactions
- 21.3 Halogenation of alkanes
- 21.4 The mechanism of halogenation
- 21.5 Chlorination of other alkanes
- 21.6 Chlorination versus bromination
- 21.7 The stereochemistry of halogenation reactions
- 21.8 Application: The ozone layer and CFCs
- 21.9 Radical halogenation at an allylic carbon
- 21.10 Application: Oxidation of unsaturated lipids
- 21.11 Application: Antioxidants
- 21.12 Radical addition reactions to double bonds
- 21.13 Polymers and polymerization

Poly(vinyl chloride) (PVC), a synthetic polymer prepared from the monomer **vinyl chloride,** is used in a wide variety of medical, industrial, and home products. Rigid PVC is found in pipes and bottles, whereas flexible PVC is used in blood bags, tubing, and materials needed for hemodialysis and heart bypass. Because PVC is water insoluble, garden hoses, drainpipes, and rain gear are made of PVC. PVC is lightweight, tear resistant, and easily sterilized, and it can be recycled many times before it is no longer usable. In Chapter 21, we learn how polymers like poly(vinyl chloride) are prepared.

Chapter 21 Radical Reactions

Why Study... Radical Reactions?

A **small but significant group** of reactions involves the homolysis of nonpolar bonds to form highly reactive **radical intermediates.** Although they are unlike other organic reactions, radical transformations are important in many biological and industrial processes. The gases O_2 and NO (nitric oxide) are both radicals. Many oxidation reactions with O_2 involve radical intermediates, and biological processes mediated by NO such as blood clotting and neurotransmission may involve radicals. Many useful industrial products such as Styrofoam and polyethylene are prepared by radical processes.

In Chapter 21 we examine the cleavage of nonpolar bonds by radical reactions.

21.1 Introduction

Radicals were first discussed in Section 6.3.

- A *radical* is a reactive intermediate with a single unpaired electron, formed by homolysis of a covalent bond.

$$A\text{—}B \longrightarrow A\cdot \; + \; \cdot B$$
$$\text{radical} \quad \text{radical}$$

Use half-headed curved arrows in radical reactions.

A radical contains an atom that does not have an octet of electrons, making it reactive and unstable. Radical processes involve single electrons, so half-headed arrows are used to show the movement of electrons. **One half-headed arrow is used for each electron.**

Carbon radicals are classified as **primary (1°), secondary (2°),** or **tertiary (3°)** by the number of R groups bonded to the carbon with the unpaired electron. A carbon radical is sp^2 hybridized and **trigonal planar,** like sp^2 hybridized carbocations. The unhybridized p orbital contains the unpaired electron and extends above and below the trigonal planar carbon.

The p orbital contains a single electron.

sp^2 hybridized trigonal planar

Bond dissociation energies for the cleavage of C—H bonds are used as a measure of radical stability. For example, two different radicals can be formed by cleavage of the C—H bonds in $CH_3CH_2CH_3$.

1° H → 1° radical + ·H $\Delta H° = 410$ kJ/mol

2° H → 2° radical + ·H $\Delta H° = 397$ kJ/mol

Cleavage of the **stronger 1° C—H** bond to form the 1° radical ($CH_3CH_2CH_2\cdot$) requires *more* energy than cleavage of the **weaker 2° C—H** bond to form the 2° radical [$(CH_3)_2CH\cdot$]—410 versus 397 kJ/mol. This makes the 2° radical more stable, because less energy is required for its formation, as illustrated in Figure 21.1. Thus, **cleavage of the weaker bond forms the more stable radical,** a specific example of a general trend.

Figure 21.1
The relative stability of 1° and 2° carbon radicals

 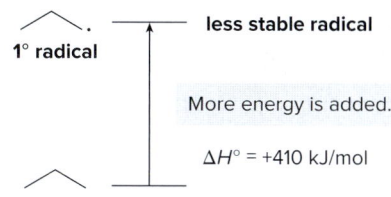 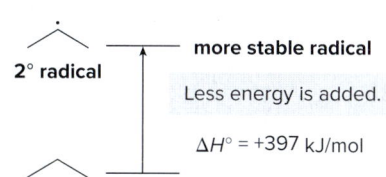

- The stability of a radical increases as the number of alkyl groups bonded to the radical carbon increases.

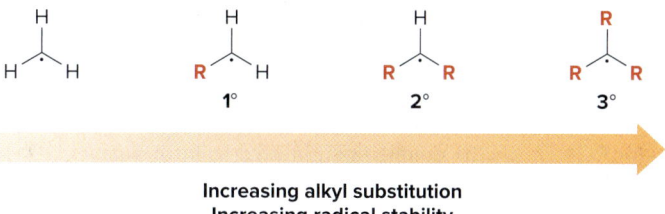

Increasing alkyl substitution
Increasing radical stability

The **lower** the bond dissociation energy for a C–H bond, the **more stable** the resulting carbon radical.

Increasing alkyl substitution increases radical stability in the same way it increases carbocation stability. **Alkyl groups are more polarizable than hydrogen atoms,** so they can more easily donate electron density to the electron-deficient carbon radical, thus increasing stability.

Unlike carbocations, however, **less stable radicals generally do *not* rearrange to more stable radicals.** This difference can be used to distinguish between reactions involving radical intermediates and those involving carbocations.

Problem 21.1 Classify each radical as 1°, 2°, or 3°.

a. b. c. d.

Problem 21.2 Draw the most stable radical that can result from cleavage of a C–H bond in each molecule.

a. b. c. d.

21.2 General Features of Radical Reactions

Radicals are formed from covalent bonds by adding energy in the form of **heat (Δ)** or **light (hν)**. Some radical reactions are carried out in the presence of a **radical initiator,** a compound that contains an especially weak bond that serves as a source of radicals. **Peroxides,** compounds with the general structure **RO–OR,** are the most commonly used radical initiators. Heating a peroxide readily causes homolysis of the weak O–O bond, forming two RO· radicals.

21.2A Two Common Reactions of Radicals

Radicals undergo two main types of reactions: **they react with σ bonds,** and **they add to π bonds,** in both cases achieving an octet of electrons.

[1] Reaction of a Radical X· with a C–H Bond

A radical X· abstracts a hydrogen atom from a C–H σ bond to form H–X and a carbon radical. One electron from the C–H bond is used to form the new H–X bond, and the other electron in the C–H bond remains on carbon.

- One electron in H–X comes from the radical.
- One electron in H–X comes from the C–H bond.

This radical reaction is typically seen with the nonpolar C–H bonds of **alkanes,** which cannot react with polar or ionic electrophiles and nucleophiles.

[2] Reaction of a Radical X· with a C=C

A radical X· also adds to the π bond of a carbon–carbon double bond. One electron from the double bond is used to form a new C–X bond, and the other electron remains on the other carbon originally part of the double bond.

- One electron in C–X comes from the radical.
- One electron in C–X comes from the π bond.

Whenever a radical reacts with a stable single or double bond, **a new radical is formed** in the products.

The electron-rich double bond of an **alkene** reacts with radicals because these reactive intermediates are electron deficient.

21.2B Two Radicals Reacting with Each Other

A radical, once formed, rapidly reacts with whatever is available. Usually that means a stable σ or π bond. Occasionally, however, two radicals come into contact with each other, and they react to form a σ bond.

- One electron in X–X comes from each radical.

The reaction of a radical with oxygen, a diradical in its ground state electronic configuration, is another example of two radicals reacting with each other. In this case, the reaction of O_2 with X· forms a new radical, thus preventing X· from reacting with an organic substrate.

a diradical

Compounds that prevent radical reactions from occurring are called *radical inhibitors* or *radical scavengers*. Besides O_2, vitamin E and related compounds, discussed in Section 21.11, are radical scavengers, too. The fact that these compounds inhibit a reaction often suggests that the reaction occurs via radical intermediates.

Problem 21.3 Draw the products formed when a chlorine atom (Cl·) reacts with each species.

a.   b. $CH_2=CH_2$ c. :C̈l·   d. O_2

21.3 Halogenation of Alkanes

In the presence of light or heat, alkanes react with halogens to form alkyl halides. Halogenation is a **radical substitution reaction,** because a halogen atom X replaces a hydrogen via a mechanism that involves radical intermediates.

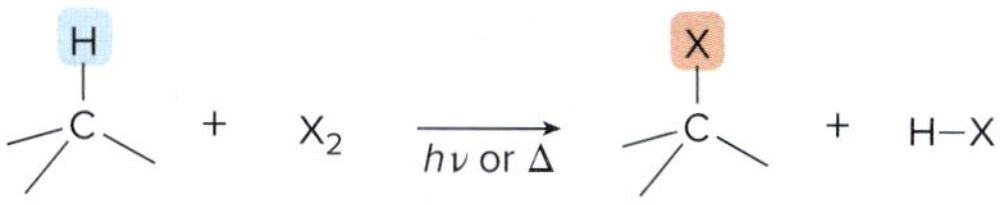

$$\text{R-H} + X_2 \xrightarrow{h\nu \text{ or } \Delta} \text{R-X} + \text{H-X}$$

X = Cl or Br   alkyl halide

Halogenation of alkanes is useful only with Cl_2 and Br_2. Reaction with F_2 is too violent and reaction with I_2 is too slow to be useful. With an alkane that has more than one type of hydrogen atom, a mixture of alkyl halides may result (Reaction [2]).

> When asked to draw the products of halogenation of an alkane, **draw the products of monohalogenation only,** unless specifically directed to do otherwise.

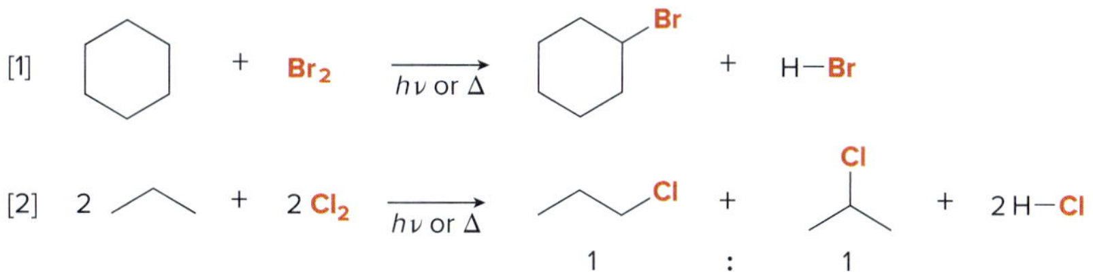

[1] cyclohexane + Br_2 $\xrightarrow{h\nu \text{ or } \Delta}$ bromocyclohexane + H–Br

[2] 2 propane + 2 Cl_2 $\xrightarrow{h\nu \text{ or } \Delta}$ 1-chloropropane + 2-chloropropane + 2 H–Cl
                                                         1   :   1

In these examples of halogenation, a halogen has replaced a single hydrogen atom on the alkane. Can the other hydrogen atoms be replaced, too? Figure 21.2 shows that when CH_4 is treated with *excess* Cl_2, all four hydrogen atoms can be successively replaced by Cl to form CCl_4. **Monohalogenation**—the substitution of a single H by X—can be achieved experimentally by adding halogen X_2 to an excess of alkane.

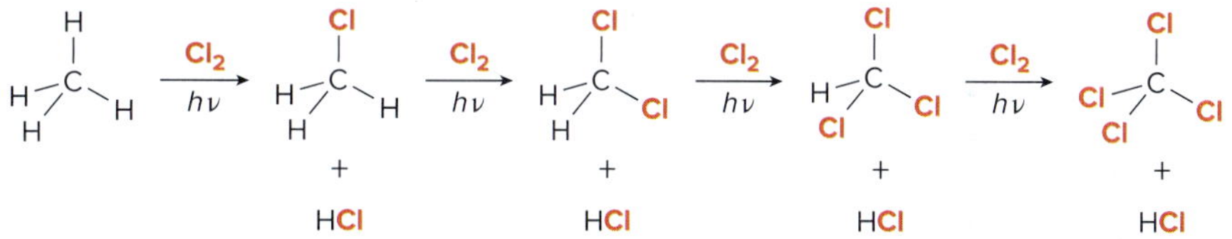

Figure 21.2 Complete halogenation of CH_4 using excess Cl_2

$CH_4 \xrightarrow[h\nu]{Cl_2} CH_3Cl \xrightarrow[h\nu]{Cl_2} CH_2Cl_2 \xrightarrow[h\nu]{Cl_2} CHCl_3 \xrightarrow[h\nu]{Cl_2} CCl_4$
        + HCl   + HCl   + HCl   + HCl

Sample Problem 21.1 Drawing the Products of the Chlorination of an Alkane

Draw all the constitutional isomers formed by monohalogenation of $(CH_3)_2CHCH_2CH_3$ with Cl_2 and $h\nu$.

Solution
Substitute Cl for H on every carbon, and then check to see if any products are identical. The starting material has five C's, but replacement of one H atom on two C's gives the same product. Thus, **$(CH_3)_2CHCH_2CH_3$ affords four monochloro substitution products.**

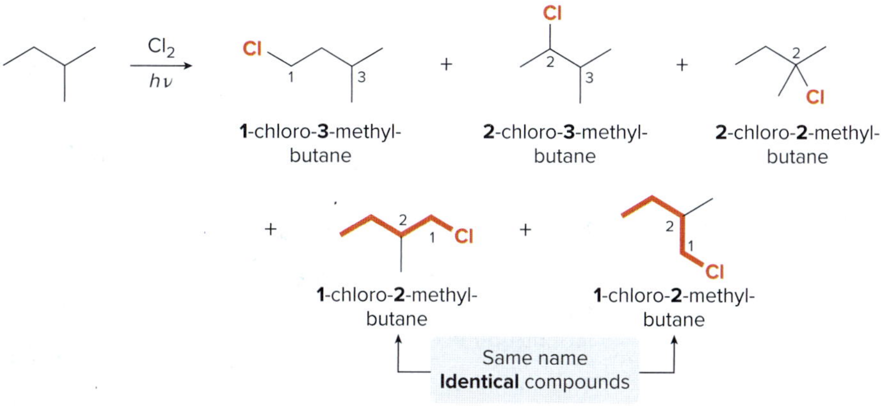

2-methylbutane $\xrightarrow[h\nu]{Cl_2}$

1-chloro-3-methyl-butane + 2-chloro-3-methyl-butane + 2-chloro-2-methyl-butane

+ 1-chloro-2-methyl-butane + 1-chloro-2-methyl-butane

Same name
Identical compounds

Problem 21.4 Draw all constitutional isomers formed by monochlorination of each alkane.

a. ☐   b. ∧∧∧   c. ⊥

More Practice: Try Problems 21.23a, 21.28, 21.38a.

21.4 The Mechanism of Halogenation

Unlike nucleophilic substitution, which proceeds by two different mechanisms depending on the starting material and reagent, all halogenation reactions of alkanes—regardless of the halogen and alkane used—proceed by the *same* mechanism. Three facts about halogenation suggest that the mechanism involves **radical,** not ionic, intermediates.

Fact	Explanation
[1] Light, heat, or added peroxide is necessary for the reaction.	• Light or heat provides the energy needed for homolytic bond cleavage to form radicals. Breaking the weak O–O bond of peroxides initiates radical reactions as well.
[2] O_2 inhibits the reaction.	• The diradical O_2 removes radicals from a reaction mixture, thus preventing reaction.
[3] No rearrangements are observed.	• **Radicals do *not* rearrange.**

21.4A The Steps of Radical Halogenation

The chlorination of cyclopentane illustrates the **three distinct parts of radical halogenation** (Mechanism 21.1):

cyclopentane + Cl_2 $\xrightarrow{h\nu \text{ or } \Delta}$ chlorocyclopentane–Cl + HCl

- *Initiation:* Two radicals are formed by homolysis of a σ bond and this begins the reaction.
- *Propagation:* A radical reacts with another reactant to form a new σ bond and another radical.
- *Termination:* Two radicals combine to form a stable bond. Removing radicals from the reaction mixture without generating any new radicals stops the reaction.

Although initiation generates the Cl· radicals needed to begin the reaction, the **propagation steps ([2] and [3]) form the two reaction products**—chlorocyclopentane and HCl. Once the process has begun, propagation occurs over and over without the need for Step [1] to occur. **A mechanism such as radical halogenation that involves two or more repeating steps is called a *chain mechanism*.** Each propagation step involves a reactive radical abstracting an atom from a stable bond to form a new bond and **another radical that continues the chain.**

Usually a radical reacts with a stable bond to propagate the chain, but occasionally two radicals combine, and this reaction terminates the chain. Depending on the reaction and the reaction conditions, some radical chain mechanisms can repeat thousands of times before termination occurs.

Termination Step [4a] forms Cl_2, a reactant, whereas Step [4c] forms chlorocyclopentane, one of the reaction products. Termination Step [4b] forms **A,** which is neither a reactant nor a

21.4 The Mechanism of Halogenation

Mechanism 21.1 Radical Halogenation of Alkanes

Part [1] Initiation

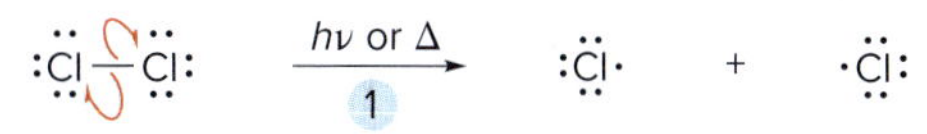

$$:\!\ddot{C}l\!-\!\ddot{C}l\!: \xrightarrow[\;\;1\;\;]{h\nu \text{ or } \Delta} :\!\ddot{C}l\!\cdot\; +\; \cdot\!\ddot{C}l\!:$$

1. **Bond cleavage forms two radicals.** Homolysis of the weakest bond (Cl–Cl) requires light or heat and forms two chlorine radicals.

Part [2] Propagation

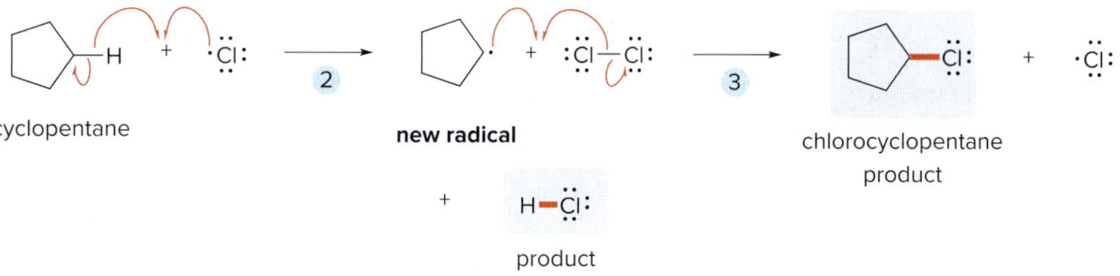

cyclopentane → new radical → chlorocyclopentane product + H–Cl product

2. The **Cl· radical abstracts a hydrogen** from cyclopentane to form HCl (a reaction product) and a new carbon radical.
3. The **carbon radical abstracts a chlorine atom** from Cl$_2$ to form chlorocyclopentane (a reaction product) and Cl·. Because Cl· is a reactant in Step [2], **Steps [2] and [3] can occur repeatedly** without additional initiation (Step [1]).

Part [3] Termination

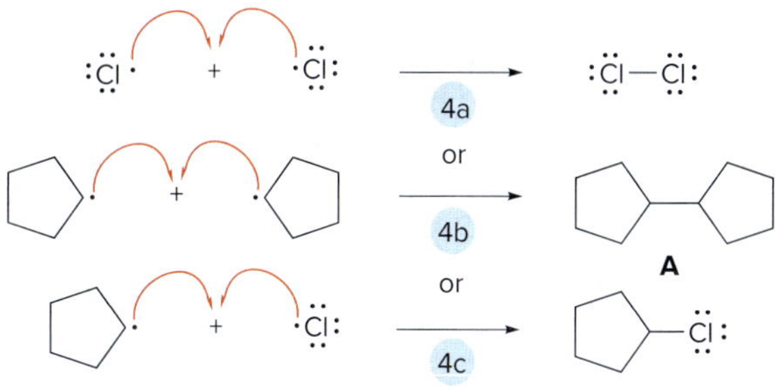

$$:\!\ddot{C}l\!\cdot\; +\; \cdot\!\ddot{C}l\!: \xrightarrow{4a} :\!\ddot{C}l\!-\!\ddot{C}l\!:$$

or (4b) two cyclopentyl radicals combine → bicyclopentyl **A**

or (4c) cyclopentyl radical + Cl· → chlorocyclopentane

4. **Termination** of the chain occurs when any two radicals combine to form a bond.

---

desired product. The formation of a small quantity of **A**, however, is evidence that radicals are formed in the reaction.

The most important steps of radical halogenation are those that lead to product formation—the propagation steps—so subsequent discussion of this reaction concentrates on these steps only.

Problem 21.5 Using Mechanism 21.1 as a guide, write the mechanism for the reaction of CH$_4$ with Br$_2$ to form CH$_3$Br and HBr. Classify each step as initiation, propagation, or termination.

21.4B Energy Changes During the Chlorination of Ethane

The chlorination of ethane illustrates how bond dissociation energies (Section 6.4) can be used to calculate $\Delta H°$ in chain propagation.

$$CH_3CH_3\; +\; Cl_2 \xrightarrow{h\nu \text{ or } \Delta} CH_3CH_2Cl\; +\; HCl$$

Figure 21.3 Energy changes in the propagation steps during the chlorination of ethane

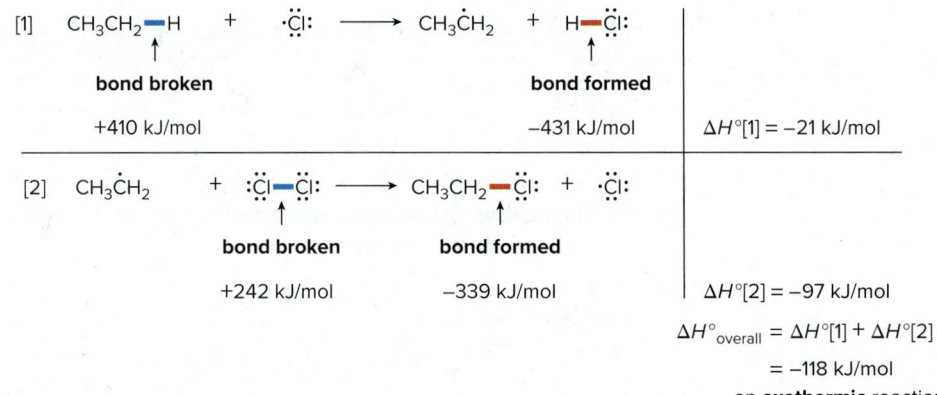

As shown in Figure 21.3, chain propagation consists of the same two steps drawn in Mechanism 21.1: abstraction of a hydrogen atom to form $CH_3CH_2\cdot$ and HCl, followed by abstraction of a chlorine atom by $CH_3CH_2\cdot$ to form CH_3CH_2Cl and a chlorine radical ($Cl\cdot$). The $\Delta H°$ for each step is negative, making the overall $\Delta H°$ negative and the reaction exothermic. Because the transition state for the first propagation step is higher in energy than the transition state for the second propagation step, the **first step is rate-determining.** Both of these facts are illustrated in the energy diagram in Figure 21.4.

Figure 21.4 Energy diagram for the propagation steps in the chlorination of ethane

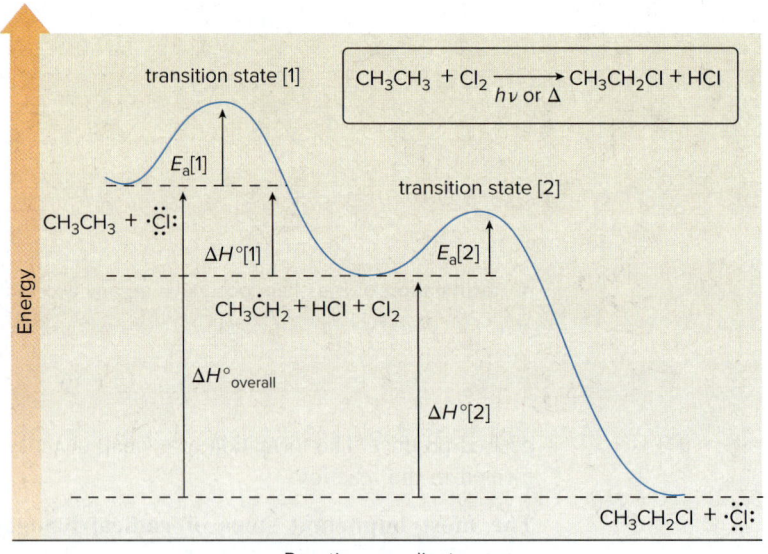

- Because radical halogenation consists of two propagation steps, the energy diagram has two energy barriers.
- The **first step is rate-determining** because its transition state is at higher energy.
- The **reaction is exothermic** because $\Delta H°_{overall}$ is negative.

Problem 21.6 Calculate $\Delta H°$ for the rate-determining step of the reaction of CH_4 with I_2. Explain why this result illustrates that this reaction is extremely slow.

21.5 Chlorination of Other Alkanes

Recall from Section 21.3 that the chlorination of $CH_3CH_2CH_3$ affords a 1:1 mixture of $CH_3CH_2CH_2Cl$ (formed by removal of a 1° hydrogen) and $(CH_3)_2CHCl$ (formed by removal of a 2° hydrogen).

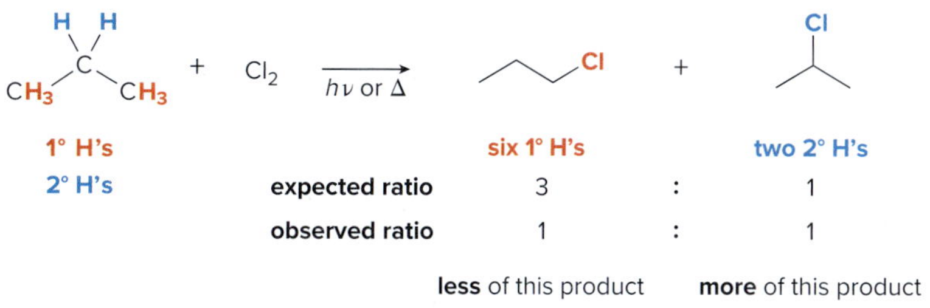

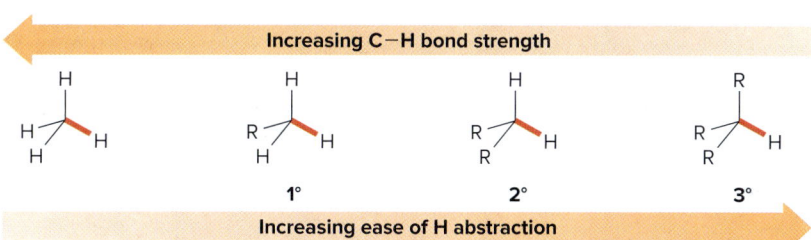

$CH_3CH_2CH_3$ has six 1° hydrogen atoms and only two 2° hydrogens, so the expected product ratio of $CH_3CH_2CH_2Cl$ to $(CH_3)_2CHCl$ (assuming all hydrogens are *equally* reactive) is 3:1. Because the observed ratio is 1:1, however, the 2° C−H bonds must be *more* reactive; that is, **it must be easier to homolytically cleave a 2° C−H bond than a 1° C−H bond.** Recall from Section 21.2 that 2° C−H bonds are *weaker* than 1° C−H bonds. Thus,

- The *weaker* the C−H bond, the *more readily* the hydrogen atom is removed in radical halogenation.

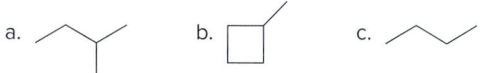

When alkanes react with Cl_2, a mixture of products results, with more product formed by cleavage of the weaker C−H bond than you would expect on statistical grounds.

Problem 21.7 Which C−H bond in each compound is most readily broken during radical halogenation?

a. b. c.

21.6 Chlorination Versus Bromination

Although alkanes undergo radical substitution reactions with both Cl_2 and Br_2, chlorination and bromination exhibit two important differences:

- Chlorination is *faster* than bromination.
- Although chlorination is *unselective*, yielding a mixture of products, bromination is often *selective*, yielding one major product.

For example, propane reacts rapidly with Cl$_2$ to form a 1:1 mixture of 1° and 2° alkyl chlorides. On the other hand, propane reacts with Br$_2$ much more slowly and forms 99% (CH$_3$)$_2$CHBr.

[Reaction scheme: propane + Cl$_2$ (hν or Δ) → 1° alkyl halide (1-chloropropane) + 2° alkyl halide (2-chloropropane), ratio 1 : 1. **Chlorination is fast and unselective.**]

[Reaction scheme: propane + Br$_2$ (hν or Δ) → 1-bromopropane (1%) + 2-bromopropane (99%). **Bromination is slow and selective.**]

- In bromination, the major (and sometimes exclusive) product results from cleavage of the *weakest* C–H bond.

Sample Problem 21.2 — Drawing the Product of Bromination of an Alkane

Draw the major product formed when 3-ethylpentane is heated with Br$_2$.

Solution

Keep in mind: **the *more substituted* the carbon atom, the *weaker* the C–H bond.** The major bromination product in 3-ethylpentane is formed by cleavage of the sole **3° C–H bond**, its weakest C–H bond.

[Reaction: 3-ethylpentane (with 3° H shown) + Br$_2$, Δ → major product with Br on the 3° carbon.]

Problem 21.8

Draw the major product formed when each cycloalkane is heated with Br$_2$.

a. methylcyclohexane b. isopropylcyclohexane c. 1-ethyl-3-methylcyclopentane d. 1,1-dimethylcyclopropane (approx.)

More Practice: Try Problems 21.23b, 21.29, 21.38b.

To explain the difference between chlorination and bromination, we return to the Hammond postulate (Section 7.14). The **rate-determining step in halogenation is the abstraction of a hydrogen atom by the halogen radical,** so we must compare these steps for bromination and chlorination. Keep in mind:

- Transition states in endothermic reactions resemble the *products*. The more stable product is formed faster.
- Transition states in exothermic reactions resemble the *starting materials*. The relative stability of the products does not greatly affect the relative energy of the transition states, so a mixture of products often results.

Bromination: CH$_3$CH$_2$CH$_3$ + Br$_2$

A bromine radical can abstract either a 1° or a 2° hydrogen from propane, generating either a 1° radical or a 2° radical. Calculating $\Delta H°$ using bond dissociation energies reveals that both reactions are **endothermic**, but **it takes *less* energy to form the *more stable* 2° radical.**

21.6 Chlorination Versus Bromination

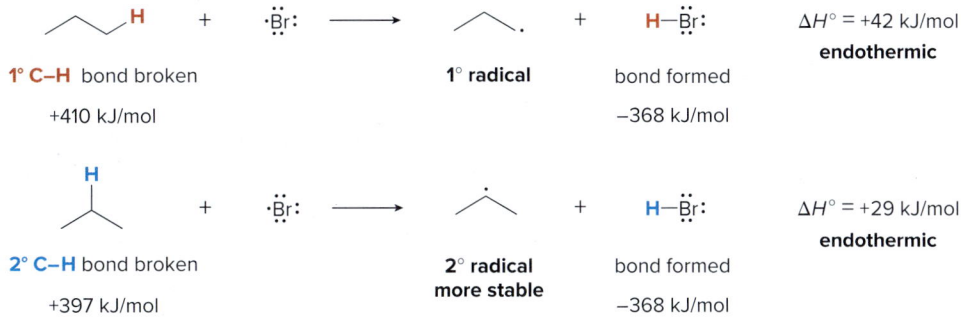

According to the Hammond postulate, the transition state of an endothermic reaction resembles the *products*, so the energy of activation to form the more stable 2° radical is lower and it is formed faster, as shown in the energy diagram in Figure 21.5. Because the 2° radical [$(CH_3)_2CH\cdot$] is converted to 2-bromopropane [$(CH_3)_2CHBr$] in the second propagation step, this **2° alkyl halide is the major product of bromination.**

- Conclusion: Because the rate-determining step in bromination is *endothermic*, the *more stable* radical is formed faster, and often a single radical halogenation product predominates.

Figure 21.5
Energy diagram for a selective endothermic reaction

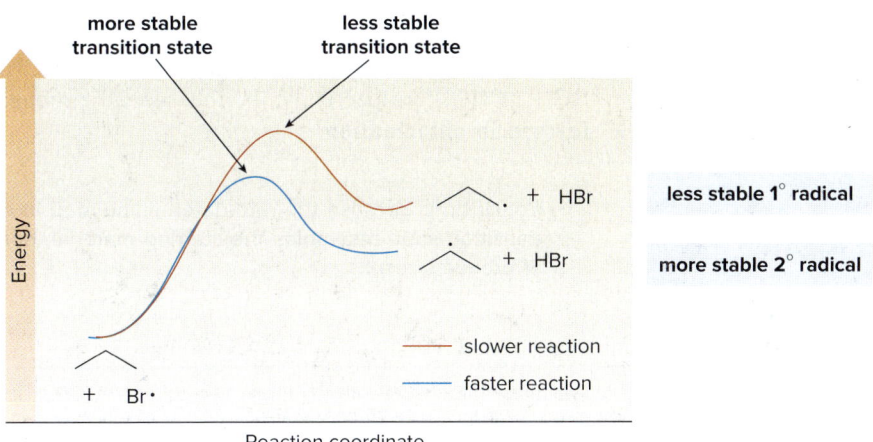

- The transition state to form the less stable 1° radical ($CH_3CH_2CH_2\cdot$) is higher in energy than the transition state to form the more stable 2° radical [$(CH_3)_2CH\cdot$]. Thus, **the 2° radical is formed faster.**

Chlorination: $CH_3CH_2CH_3 + Cl_2$

A chlorine radical can also abstract either a 1° or a 2° hydrogen from propane, generating either a 1° radical or a 2° radical. Calculating $\Delta H°$ using bond dissociation energies reveals that both reactions are *exothermic*.

∧∧H + ·Cl: → ∧∧· + H—Cl:						$\Delta H° = -21$ kJ/mol exothermic
1° C–H bond broken +410 kJ/mol		1° radical		bond formed −431 kJ/mol		
⋎H + ·Cl: → ⋎· + H—Cl:						$\Delta H° = -34$ kJ/mol exothermic
2° C–H bond broken +397 kJ/mol		2° radical		bond formed −431 kJ/mol		

Figure 21.6 Energy diagram for a nonselective exothermic reaction

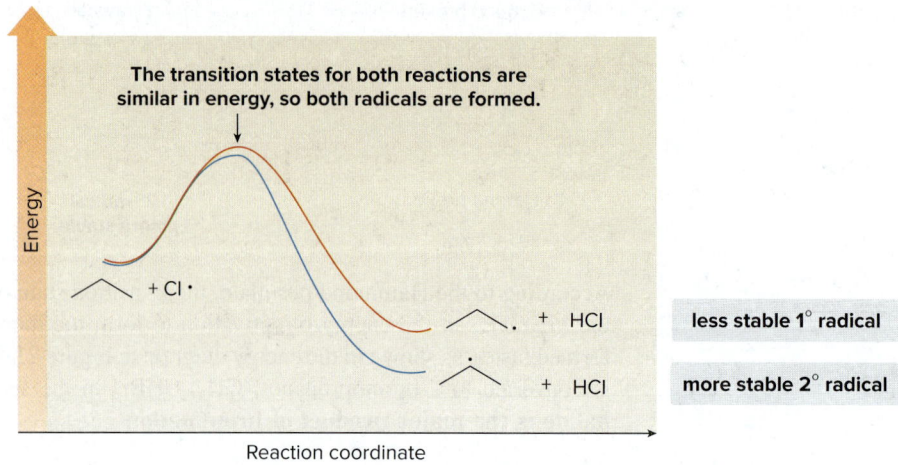

Because chlorination has an *exothermic* rate-determining step, the transition state to form both radicals **resembles the same starting material,** $CH_3CH_2CH_3$. As a result, the relative stability of the two radicals is much less important and **both radicals are formed.** An energy diagram for these processes is drawn in Figure 21.6. Because the 1° and 2° radicals are converted to 1-chloropropane ($CH_3CH_2CH_2Cl$) and 2-chloropropane [$(CH_3)_2CHCl$], respectively, in the second propagation step, **both alkyl halides are formed in chlorination.**

- Conclusion: Because the rate-determining step in chlorination is *exothermic,* the transition state resembles the starting material, both radicals are formed, and a *mixture* of products results.

Problem 21.9 Reaction of $(CH_3)_3CH$ with Cl_2 forms two products: $(CH_3)_2CHCH_2Cl$ (63%) and $(CH_3)_3CCl$ (37%). Why is the major product formed by cleavage of the stronger 1° C–H bond?

21.7 The Stereochemistry of Halogenation Reactions

The stereochemistry of a reaction product depends on whether the reaction occurs at a stereogenic center or at another atom, and whether a new stereogenic center is formed. The rules predicting the stereochemistry of reaction products are summarized in Table 21.1.

Table 21.1 Rules for Predicting the Stereochemistry of Reaction Products

Starting material	Result
Achiral	• An achiral starting material always gives either an achiral or a racemic product.
Chiral	• If a reaction does not occur at a stereogenic center, the configuration at a stereogenic center is *retained* in the product.
	• If a reaction occurs at a stereogenic center, we must know the *mechanism* to predict the stereochemistry of the product.

21.7A Halogenation of an Achiral Starting Material

Halogenation of the **achiral starting material** $CH_3CH_2CH_2CH_3$ forms two constitutional isomers by replacement of either a 1° or 2° hydrogen.

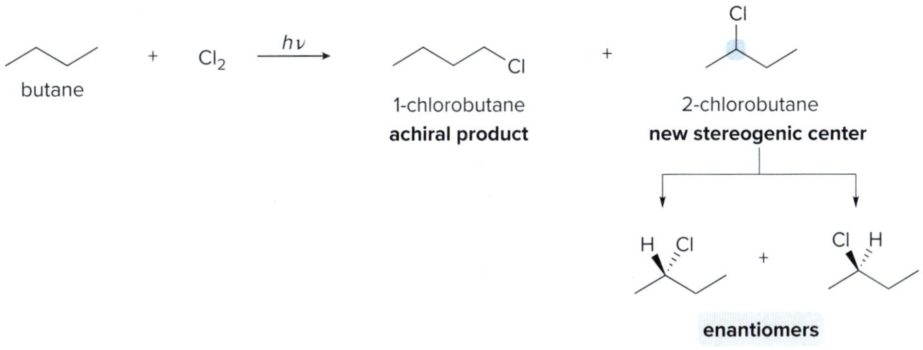

- 1-Chlorobutane ($CH_3CH_2CH_2CH_2Cl$) has no stereogenic center, so it is an **achiral** compound.
- 2-Chlorobutane [$CH_3CH(Cl)CH_2CH_3$] has a new stereogenic center, so an **equal amount of two enantiomers** must form—a racemic mixture.

A racemic mixture results when a new stereogenic center is formed because the first propagation step generates a **planar, sp^2 hybridized radical**. Cl_2 then reacts with the planar radical from either the front or back side to form an equal amount of two enantiomers.

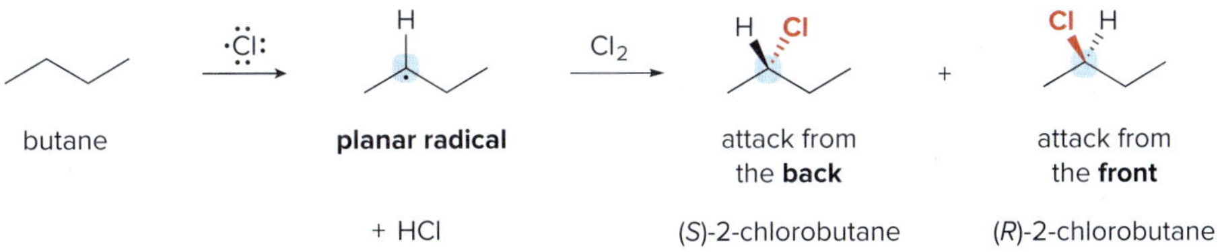

Thus, the achiral starting material butane forms an achiral product (1-chlorobutane) and a racemic mixture of two enantiomers [(R)- and (S)-2-chlorobutane].

21.7B Halogenation of a Chiral Starting Material

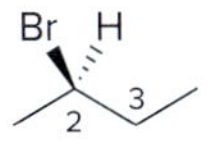

Let's now examine chlorination of the chiral starting material (R)-2-bromobutane at C2 and C3.

Chlorination at C2 occurs at the stereogenic center. Abstraction of a hydrogen atom at C2 forms a trigonal planar sp^2 hybridized radical that is now achiral. This achiral radical then reacts with Cl_2 from either side to form a new stereogenic center, resulting in an **equal amount of two enantiomers—a racemic mixture**.

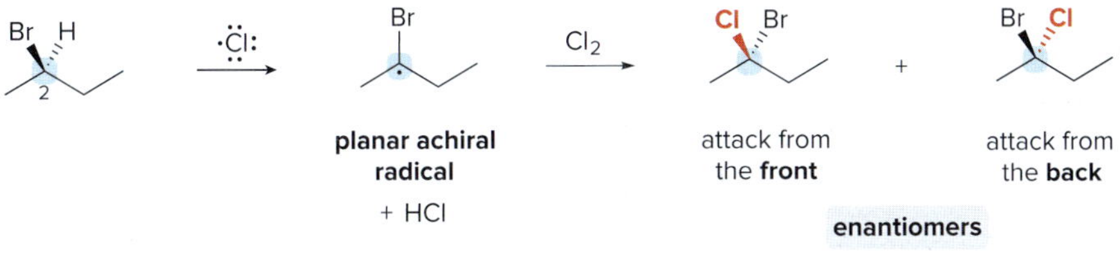

- Radical halogenation reactions occur with *racemization* at a stereogenic center.

Chlorination at C3 does *not* occur at the stereogenic center, but it forms a new stereogenic center. Because no bond is broken to the stereogenic center at C2, **its configuration is *retained*** during the reaction. Abstraction of a hydrogen atom at C3 forms a **trigonal planar** sp^2 hybridized radical that still contains this stereogenic center. Reaction of the radical with Cl_2 from either side forms a new stereogenic center, so the products have two stereogenic centers: the configuration at C2 is the *same* in both compounds, but the configuration at C3 is *different*, making them **diastereomers**.

Thus, four isomers are formed by chlorination of (R)-2-bromobutane at C2 and C3. Attack at the stereogenic center (C2) gives a product with one stereogenic center, resulting in a mixture of enantiomers. Attack at C3 forms a new stereogenic center, giving a mixture of diastereomers.

Sample Problem 21.3 — Drawing All Stereoisomers Formed by Monochlorination

Draw all stereoisomers formed by monochlorination of **A**.

Solution

Look at each C bonded to H's *separately*, and consider whether the reaction occurs *at a stereogenic center* or if it *forms* a stereogenic center. The reactant **A** contains one stereogenic center, making it chiral.

The CH_3 in blue is *not* a stereogenic center, and substitution of a H atom by Cl does *not* form a new stereogenic center. One stereoisomer **B** is formed, and the configuration of the stereogenic center is retained.

The CH_2 in red is *not* a stereogenic center, but substitution of a H atom by Cl forms a *new* stereogenic center, and the new bond to Cl can form from either above or below the planar radical intermediate. Two products, **C** and **D**, are diastereomers.

C — Cl and Br are **trans**.
D — Cl and Br are **cis**.

diastereomers

Thus, chlorination of **A** forms three products, **B, C,** and **D**.

Problem 21.10 What products are formed from monochlorination of (R)-2-bromobutane at C1 and C4? Assign *R* and *S* designations to each stereogenic center.

More Practice: Try Problems 21.41–21.44.

Problem 21.11 Draw the monochlorination products formed when each compound is heated with Cl₂. Include the stereochemistry at any stereogenic center.

a. [CH₃CH₂CH₂CH₂CH₃]   b. [cyclopropyl-CH₃]   c. [CH₃CH₂CH(CH₂CH₃)CH₂CH₃]   d. [H and Cl on wedge/dash, CH₃-CHCl-CH₂CH₃]
(Consider attack at C2 and C3 only.)

21.8 Application: The Ozone Layer and CFCs

Ozone is formed in the upper atmosphere by reaction of oxygen molecules with oxygen atoms. Ozone is also decomposed with sunlight back to these same two species. The overall result of these reactions is to convert high-energy ultraviolet light into heat.

Ozone synthesis: $O_2 + \cdot \ddot{O} \cdot \longrightarrow O_3 \text{ (ozone)} + \text{heat}$

Ozone decomposition: $O_3 \text{ (ozone)} \xrightarrow{h\nu} O_2 + \cdot \ddot{O} \cdot$

The 1995 Nobel Prize in Chemistry was awarded to Mario Molina, Paul Crutzen, and F. Sherwood Rowland for their work in elucidating the interaction of ozone with CFCs.

Propane and butane are now used as propellants in spray cans in place of CFCs.
Jill Braaten/McGraw-Hill Education

Ozone is vital to life; it acts like a shield, protecting the earth's surface from destructive ultraviolet radiation. A decrease in ozone concentration in this protective layer would have some immediate consequences, including an increase in the incidence of skin cancer and eye cataracts. Other long-term effects include a reduced immune response, interference with photosynthesis in plants, and harmful effects on the growth of plankton, the mainstay of the ocean food chain.

Current research suggests that **chlorofluorocarbons** (**CFCs**) are responsible for destroying ozone in the upper atmosphere. **CFCs** are simple halogen-containing organic compounds manufactured under the trade name Freons.

CFCs are inert, odorless, and nontoxic, and they have been used as refrigerants, solvents, and aerosol propellants. Because CFCs are volatile and water insoluble, they readily escape into the upper atmosphere, where they are decomposed by high-energy sunlight to form radicals that destroy ozone by the radical chain mechanism shown in Figure 21.7.

Figure 21.7
CFCs and the destruction of the ozone layer

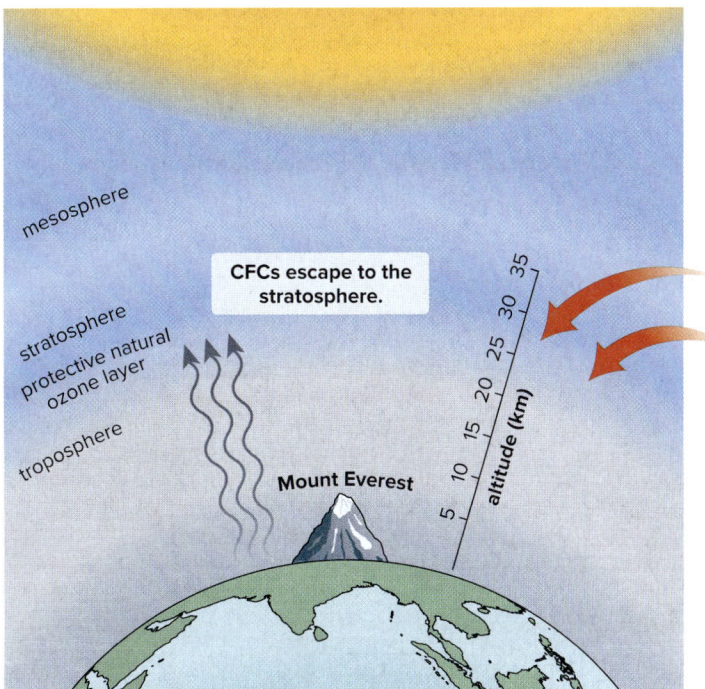

Initiation: CFCs are decomposed by sunlight to form chlorine radicals.

$CFCl_3 \xrightarrow{h\nu} \cdot CFCl_2 + \cdot \ddot{Cl} :$

Propagation: Ozone is destroyed by a chain reaction with radical intermediates.

$\cdot \ddot{Cl} : + O_3 \longrightarrow : \ddot{Cl} - \ddot{O} \cdot + O_2$

$: \ddot{Cl} - \ddot{O} \cdot + \cdot \ddot{O} \cdot \longrightarrow \cdot \ddot{Cl} : + O_2$

- The chain reaction is initiated by homolysis of a C—Cl bond in CFCl₃.
- Propagation consists of two steps. Reaction of Cl· with O₃ forms chlorine monoxide (ClO·), which reacts with oxygen atoms to form O₂ and Cl·.

950 Chapter 21 Radical Reactions

CFCl$_3$
trichlorofluoromethane
CFC 11
Freon 11

CF$_2$Cl$_2$
dichlorodifluoromethane
CFC 12
Freon 12

The overall result is that O$_3$ is consumed as a reactant and O$_2$ molecules are formed. In this way, a small amount of CFC can destroy a large amount of O$_3$. These findings led to a ban on the use of CFCs in aerosol propellants in the United States in 1978 and to the phasing out of their use in refrigeration systems.

Newer alternatives to CFCs are **hydrofluorocarbons (HFCs)** such as CH$_2$FCF$_3$ and **hydrofluoroolefins (HFOs)** such as CH$_2$=CFCF$_3$. These compounds have many properties in common with CFCs, but they are largely decomposed before they reach the stratosphere and therefore have little impact on the ozone layer. HFOs are especially attractive because, unlike CFCs, they also have little global warming potential.

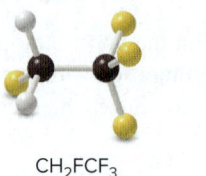

CH$_2$FCF$_3$
HFC-134a

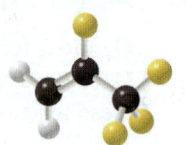

2,3,3,3-tetrafluoropropene
HFO-1234yf

Problem 21.12 CH$_2$FCF$_3$ is decomposed before it reaches the stratosphere by abstraction of a hydrogen atom by the hydroxy radical (·OH). Draw the products of this reaction.

21.9 Radical Halogenation at an Allylic Carbon

Now let's examine radical halogenation at an *allylic carbon*—**the carbon adjacent to a double bond.** Homolysis of the allylic C–H bond of propene generates the **allyl radical,** which has an unpaired electron on the carbon adjacent to the double bond.

propene
allylic C–H (in red) → allyl radical + ·H $\Delta H° = +364$ kJ/mol

The bond dissociation energy for this process (364 kJ/mol) is even less than that for a 3° C–H bond (381 kJ/mol). Because the weaker the C–H bond, the more stable the resulting radical, an **allyl radical is more stable than a 3° radical,** and the following order of radical stability results:

1°   2°   3°   allyl radical

Increasing radical stability →

The allyl radical is more stable than other radicals because two resonance structures can be drawn for it. The "true" structure of the allyl radical is a hybrid of the two resonance structures. In the hybrid, the π bond and the unpaired electron are delocalized.

two resonance structures for the allyl radical   **hybrid**

Problem 21.13 Draw a second resonance structure for each radical. Then draw the hybrid.

a. [CH₃–CH=CH• radical]   b. [cyclohexenyl radical]   c. [cyclohexene with CH₂–CH=CH₂• substituent]   d. [cyclohexene with CH₂–CH=CH₂ and radical]

21.9A Selective Bromination at Allylic C–H Bonds

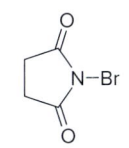

N-bromosuccinimide
NBS

Because allylic C–H bonds are *weaker* than other sp^3 hybridized C–H bonds, the **allylic carbon can be selectively halogenated** by using *N*-bromosuccinimide (**NBS**, Section 10.15) in the presence of light or peroxides. Under these conditions only the allylic C–H bond in cyclohexene reacts to form an allylic halide.

$$\text{cyclohexene} \xrightarrow[h\nu \text{ or ROOR}]{\text{NBS}} \text{3-bromocyclohexene}$$

allylic C → Br (allylic halide)

NBS contains a weak N–Br bond that is homolytically cleaved with light to generate a bromine radical, initiating an allylic halogenation reaction. Propagation then consists of the usual two steps of radical halogenation as shown in Mechanism 21.2.

Mechanism 21.2 Allylic Bromination with NBS

Part [1] Initiation

[NBS] —N–Br: $\xrightarrow{h\nu}$ [succinimidyl radical] •N• + •Br:

1. Homolysis of the weak N–Br bond with light energy forms a **Br• radical** that initiates radical halogenation.

Part [2] Propagation

[cyclohexene]–H + •Br: →(2) [allylic radical] + :Br–Br: (from NBS) →(3) [cyclohexenyl–Br] + •Br:

+ H–Br:

2. The Br• radical abstracts an allylic H to afford an **allylic radical**. (Only one resonance structure is drawn.)
3. The allylic radical reacts with Br₂ to form the **allylic halide**. The radical Br• formed in Step [3] can now react in Step [2], so Steps [2] and [3] can repeatedly occur without additional initiation.

A **low concentration of Br₂** (from NBS) **favors allylic substitution** (over addition) in part because bromine is needed for only *one* step of the mechanism. When Br₂ adds to a double bond, a low Br₂ concentration would first form a low concentration of bridged bromonium ion (Section 10.13), which must then react with more bromine (in the form of Br⁻) in a second step to form a dibromide. **If concentrations of both intermediates— bromonium ion and Br⁻— are low, the overall rate of addition is very slow.**

Besides acting as a source of Br• to initiate the reaction, **NBS generates a low concentration of Br₂** needed in the second chain propagation step (Step [3] of the mechanism). The HBr formed in Step [2] reacts with NBS to form Br₂, which is then used for halogenation in Step [3] of the mechanism.

[NBS] N–Br + HBr → [succinimide] N–H + **Br₂**

NBS   succinimide   used in Step [3] of allylic bromination

Recall from Section 20.14A that alkyl benzenes also undergo two different reactions—electrophilic aromatic substitution or benzylic bromination—depending on the reaction conditions.

Thus, an alkene with allylic C–H bonds undergoes two different reactions depending on the reaction conditions.

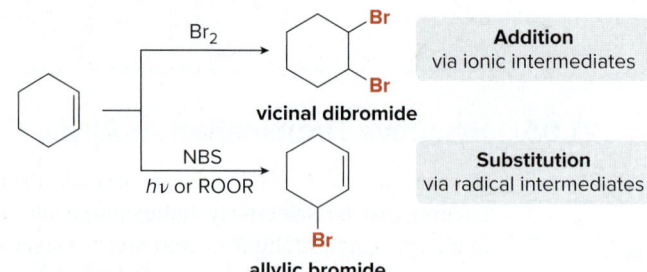

- Treatment of cyclohexene with Br_2 (in an organic solvent like CCl_4) leads to **addition** via **ionic intermediates** (Section 10.13).
- Treatment of cyclohexene with NBS (+ $h\nu$ or ROOR) leads to **allylic substitution,** via **radical intermediates.**

Problem 21.14 Draw the products of each reaction.

a. [cyclopentene] $\xrightarrow{\text{NBS}, h\nu}$ b. [propene] $\xrightarrow{\text{NBS}, h\nu}$ c. [cycloheptene] $\xrightarrow{Br_2}$

21.9B Product Mixtures in Allylic Halogenation

Halogenation at an allylic carbon often results in a mixture of products. For example, bromination of 3-methylbut-1-ene under radical conditions forms a mixture of 3-bromo-3-methylbut-1-ene and 1-bromo-3-methylbut-2-ene.

3-methylbut-1-ene $\xrightarrow{\text{NBS}, h\nu \text{ or ROOR}}$ 3-bromo-3-methylbut-1-ene + 1-bromo-3-methylbut-2-ene

A mixture is obtained because the reaction proceeds by way of a **resonance-stabilized radical.** Abstraction of an allylic hydrogen from the alkene with a Br• radical (from NBS) forms an allylic radical for which **two different Lewis structures** can be drawn.

two nonidentical resonance structures

[resonance structures] + H–Br:

↓ Br_2 ↓ Br_2

3-bromo-3-methylbut-1-ene + 1-bromo-3-methylbut-2-ene + •Br:

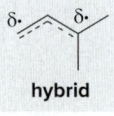

hybrid

As a result, two different C atoms have partial radical character (indicated by δ•), so that Br_2 reacts at two different sites and two allylic halides are formed.

- Whenever two different resonance structures can be drawn for an allylic radical, two *different* allylic halides are formed by radical substitution.

Problem 21.15 Draw all constitutional isomers formed when each alkene is treated with NBS + hv.

a. [structure] b. [structure] c. [structure]

21.10 Application: Oxidation of Unsaturated Lipids

Oils—triacylglycerols having one or more sites of unsaturation in their long carbon chains—are susceptible to oxidation at their allylic carbon atoms. Oxidation occurs by way of a radical chain mechanism, as shown in Figure 21.8.

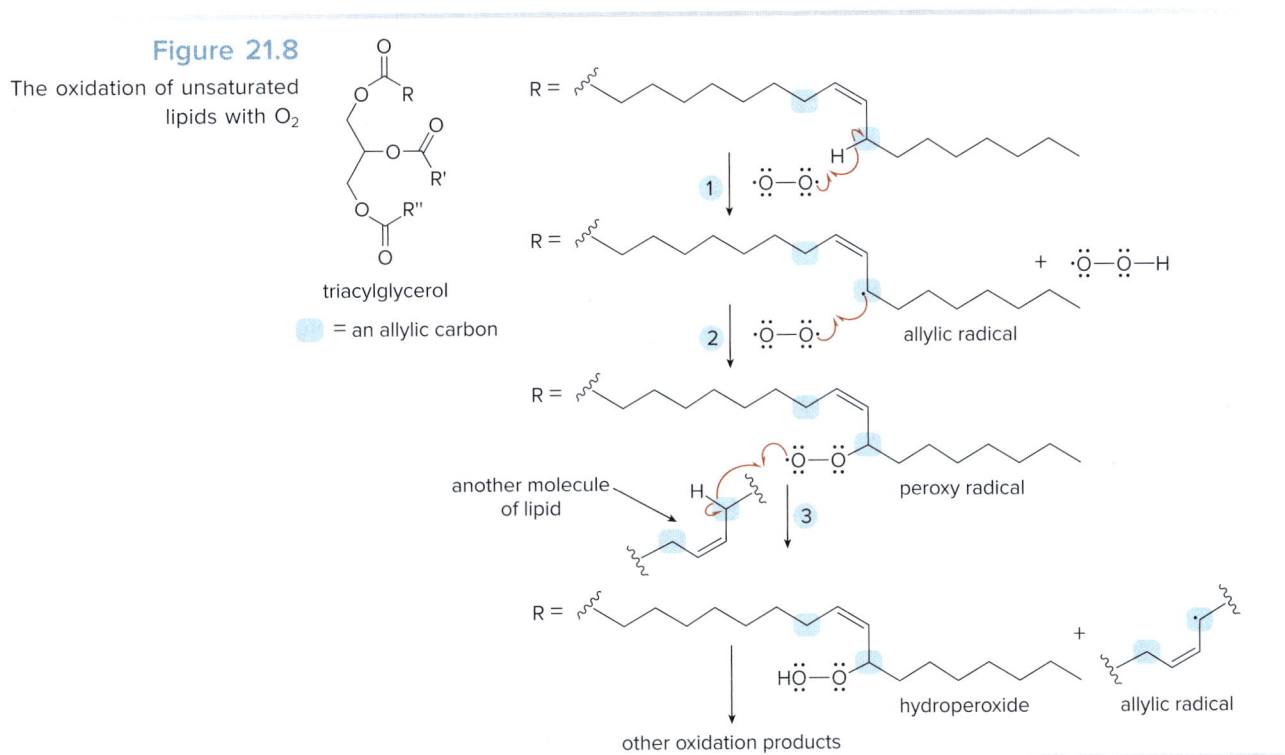

Figure 21.8
The oxidation of unsaturated lipids with O_2

Oxidation is shown at one allylic carbon only. Reaction at the other labeled allylic carbon is also possible.
- **Step 1** Oxygen in the air abstracts an allylic hydrogen atom to form an allylic radical because the allylic C—H bond is weaker than the other C—H bonds.
- **Step 2** The allylic radical reacts with another molecule of O_2 to form a peroxy radical.
- **Step 3** The peroxy radical abstracts an allylic hydrogen from another lipid molecule to form a hydroperoxide and another allylic radical that continues the chain. Steps [2] and [3] can repeat again and again until some other radical terminates the chain.

The hydroperoxides formed by this process are unstable and decompose to other oxidation products, many of which have a disagreeable odor and taste. **This process turns an oil rancid. Unsaturated lipids are more easily oxidized than saturated ones** because they contain **weak allylic C—H bonds** that are readily cleaved in Step [1] of this reaction, forming resonance-stabilized allylic radicals. Because saturated fats have no double bonds and thus no weak allylic C—H bonds, they are much less susceptible to air oxidation, resulting in increased shelf life of products containing them.

954 Chapter 21 Radical Reactions

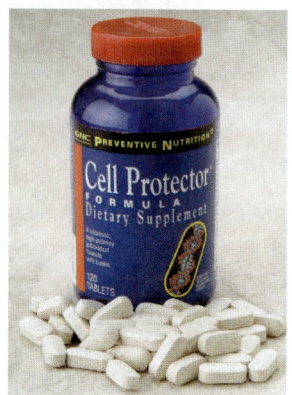

The purported health benefits of antioxidants have made them a popular component in anti-aging formulations.
Elite Images/McGraw-Hill Education

Hazelnuts, almonds, and many other types of nuts are an excellent source of the natural antioxidant vitamin E.
Stockbyte/Corbis

Rosemary extracts contain rosmarinic acid (Problem 21.17), an antioxidant that helps prevent the oxidation of unsaturated vegetable oils.
Pixtal/SuperStock

Problem 21.16 Which C–H bond is most readily cleaved in linoleic acid? Draw all possible resonance structures for the resulting radical. Draw all the hydroperoxides formed by reaction of this resonance-stabilized radical with O_2.

linoleic acid

21.11 Application: Antioxidants

An *antioxidant* is a compound that stops an oxidation reaction from occurring.

- Naturally occurring antioxidants such as **vitamin E** prevent radical reactions that can cause cell damage.
- Synthetic antioxidants such as **BHT**—**b**utylated **h**ydroxy **t**oluene—are added to packaged and prepared foods to prevent oxidation and spoilage.

vitamin E

BHT
(butylated hydroxy toluene)

Vitamin E and BHT are radical inhibitors, so they terminate radical chain mechanisms by reacting with radicals. How do they trap radicals? Both vitamin E and BHT use a hydroxy group bonded to a benzene ring—a general structure called a **phenol.**

Radicals (R·) abstract a hydrogen atom from the OH group of an antioxidant, forming a new resonance-stabilized radical. **This new radical does not participate in chain propagation,** but rather terminates the chain and halts the oxidation process. All phenols (including vitamin E and BHT) inhibit oxidation by this radical process.

phenol

R· abstracts the H atom from the OH group.

R· = a general organic radical

five resonance structures + R—H

The many nonpolar C–C and C–H bonds of vitamin E make it fat soluble, so it dissolves in the nonpolar interior of the cell membrane, where it is thought to inhibit the oxidation of the unsaturated fatty acid residues in the phospholipids. Oxidative damage to lipids in cells via radical mechanisms is thought to play an important role in the aging process. For this reason, many anti-aging formulas with antioxidants like vitamin E are now popular consumer products.

Problem 21.17 Rosmarinic acid is an antioxidant isolated from rosemary. Draw resonance structures for the radical that results from removal of the labeled H atom in rosmarinic acid.

rosmarinic acid

21.12 Radical Addition Reactions to Double Bonds

We now turn our attention to the second common reaction of radicals, addition to double bonds. Because an alkene contains an electron-rich, easily broken π bond, it reacts with an electron-deficient radical.

The π bond is broken.   new radical

Radicals react with alkenes via a radical chain mechanism that consists of initiation, propagation, and termination steps analogous to those discussed previously for radical substitution.

21.12A Addition of HBr

HBr adds to alkenes to form alkyl bromides in the presence of light, heat, or peroxides.

A π bond is broken.   alkyl bromide

The regioselectivity of addition to an unsymmetrical alkene is *different* from the addition of HBr without added light, heat, or peroxides.

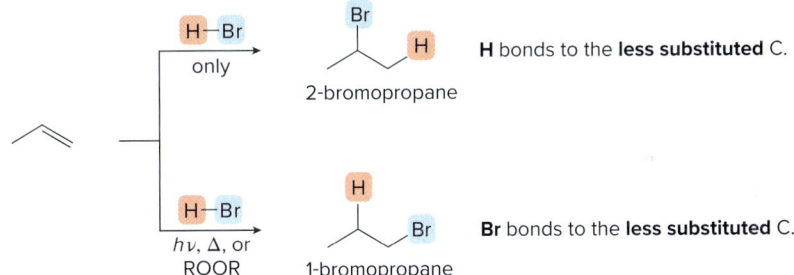

2-bromopropane — H bonds to the **less substituted** C.

1-bromopropane — Br bonds to the **less substituted** C.

- HBr addition to propene *without* added light, heat, or peroxides gives 2-bromopropane: the **H atom is added to the less substituted carbon.** This reaction occurs via **carbocation** intermediates (Section 10.10).
- HBr addition to propene *with* added light, heat, or peroxides gives 1-bromopropane: the **Br atom is added to the less substituted carbon.** This reaction occurs via **radical** intermediates.

Problem 21.18 Draw the product(s) formed when each alkene is treated with either [1] HBr alone; or [2] HBr in the presence of peroxides.

a.   b.   c.

21.12B The Mechanism of the Radical Addition of HBr to an Alkene

In the presence of added light, heat, or peroxides, HBr addition to an alkene forms radical intermediates and, like other radical reactions, proceeds by a mechanism with three distinct parts: initiation, propagation, and termination. Mechanism 21.3 is written for the reaction of $CH_3CH=CH_2$ with HBr and ROOR to form $CH_3CH_2CH_2Br$.

The first propagation step (Step [3] of the mechanism, the addition of Br• to the double bond) is worthy of note. With propene there are two possible paths for this step, depending on which carbon atom of the double bond forms the new bond to bromine. Path [A] forms a less stable

Mechanism 21.3 Radical Addition of HBr to an Alkene

Part [1] Initiation

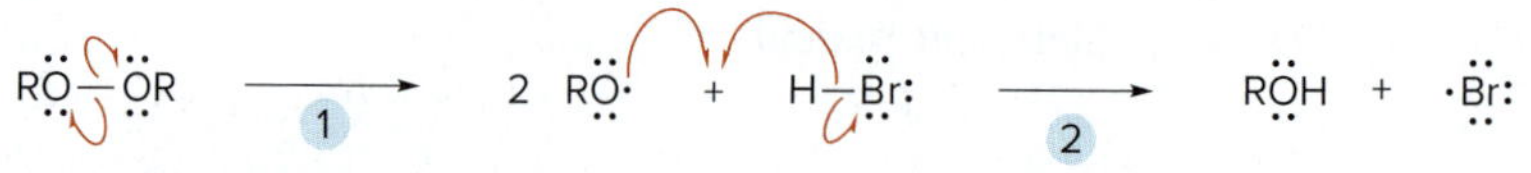

RO—OR
1
2 RO·
+
H—Br:
2
ROH
+
·Br:

1 – 2 Initiation with ROOR occurs in two steps—**homolysis of the weak O–O bond** and abstraction of H to form a bromine radical.

Part [2] Propagation

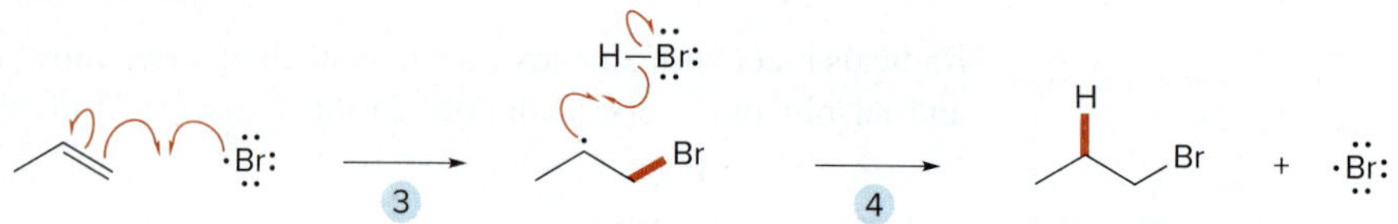

·Br:
3
H—Br:
Br
4
H
Br
+
·Br:

2° radical
new bond shown in red

new bond shown in red

3 Addition of Br· to the terminal carbon forms a **2° radical.**

4 Abstraction of H from HBr forms a new C–H bond and a bromine radical, so Steps [3] and [4] can occur repeatedly.

Part [3] Termination

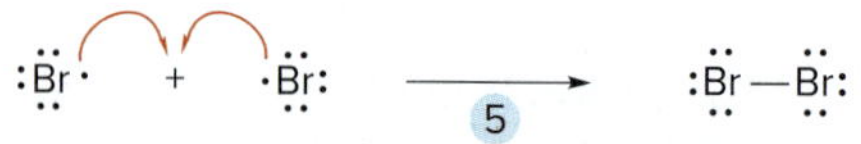

:Br·
+
·Br:
5
:Br—Br:

5 Termination of the chain occurs when any two radicals combine to form a bond.

1° radical, whereas Path [B] forms a more stable 2° radical. **The more stable 2° radical forms faster,** so Path [B] is preferred.

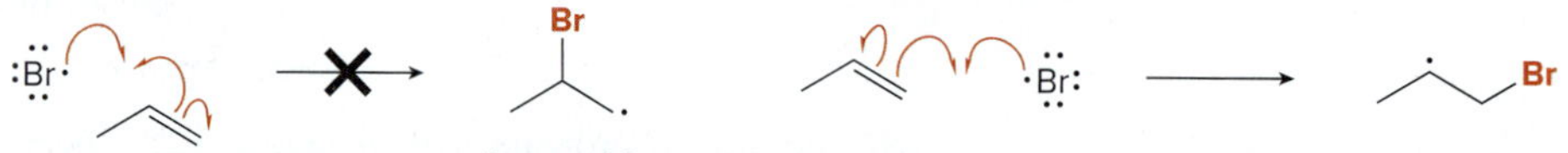

Path [A]:
Does NOT occur

Path [B]:
Preferred path

:Br·
Br
·Br:
Br

less stable
1° radical

more stable
2° radical

The mechanism also illustrates why the regioselectivity of HBr addition is different depending on the reaction conditions. In both reactions, H and Br add to the double bond, but the *order* of addition depends on the mechanism.

Radical addition

·Br:
Br

2° radical
new bond shown in red

Br bonds to the **less substituted** C.

Ionic addition

H—Br
+
H

2° carbocation
new bond shown in red

H bonds to the **less substituted** C.

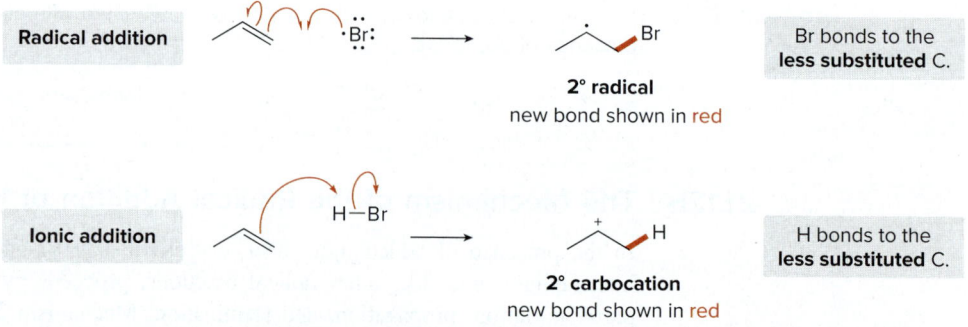

- **In radical addition (HBr with added light, heat, or ROOR),** ***Br· adds first*** **to generate the more stable radical.**
- **In ionic addition (HBr alone),** ***H^+ adds first*** **to generate the more stable carbocation.**

Problem 21.19 When HBr adds to (CH$_3$)$_2$C=CH$_2$ under radical conditions, two radicals are possible products in the first step of chain propagation. Draw the structure of both radicals and indicate which one is formed. Then draw the preferred product from HBr addition under radical conditions.

Problem 21.20 What reagents are needed to convert 1-ethylcyclohexene into (a) 1-bromo-2-ethylcyclohexane; (b) 1-bromo-1-ethylcyclohexane; (c) 1,2-dibromo-1-ethylcyclohexane?

21.13 Polymers and Polymerization

Polymers—**large molecules made up of repeating units of smaller molecules called** *monomers*—include such biologically important compounds as proteins and carbohydrates. They also include such industrially important plastics as polyethylene, poly(vinyl chloride) (PVC, mentioned in the chapter opener), and polystyrene.

21.13A Synthetic Polymers

Many synthetic polymers—that is, those synthesized in the lab—are among the most widely used organic compounds in modern society. Although some synthetic polymers resemble natural substances, many have different and unusual properties that make them more useful than naturally occurring materials. Soft drink bottles, plastic bags, food wrap, compact discs, Teflon, and Styrofoam are all made of synthetic polymers. In this section we examine polymers derived from alkene monomers. Chapter 30 (online) is devoted to a detailed discussion of the synthesis and properties of several different types of synthetic polymers.

- *Polymerization* is the joining together of monomers to make polymers.

For example, joining **ethylene monomers** together forms the polymer **polyethylene**, a plastic used in milk containers and sandwich bags.

HDPE (high-density polyethylene) and **LDPE** (low-density polyethylene) are two common types of polyethylene prepared under different reaction conditions and having different physical properties. HDPE is opaque and rigid, and is used in milk containers and water jugs. LDPE is less opaque and more flexible, and is used in plastic bags and electrical insulation. Products containing HDPE and LDPE (and other plastics) are often labeled with a symbol indicating recycling ease: the lower the number, the easier to recycle.
Jill Braaten/McGraw-Hill Education

Many ethylene derivatives having the general structure **CH$_2$=CHZ** are also used as monomers for polymerization. The identity of Z affects the physical properties of the resulting polymer, making some polymers more suitable for one consumer product (e.g., plastic bags or food wrap) than another (e.g., soft drink bottles or compact discs). Polymerization of CH$_2$=CHZ usually affords polymers with the Z groups on every other carbon atom in the chain.

For example, polymerization of propene forms polypropylene, which is used to make disposable plastic syringes.

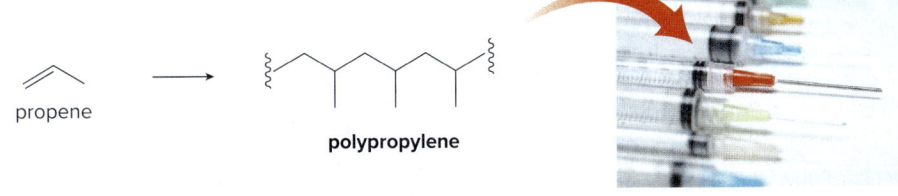

polypropylene syringes
Image Source Trading Ltd/Shutterstock

Sample Problem 21.4 — Drawing the Structure of a Polymer Formed from a Monomer

What polymer is formed when $CH_2=CHCO_2H$ (acrylic acid) is polymerized to form poly(acrylic acid)?

Solution

Draw three or more alkene monomers, **break one bond of each double bond, and join the alkenes together with single bonds.** With unsymmetrical alkenes, substituents are bonded to every other carbon.

Join a C labeled in blue with a C labeled in red.

Poly(acrylic acid) (Sample Problem 21.4) is used in disposable diapers because it absorbs 30 times its weight in water. *Image Source, all rights reserved.*

Problem 21.21 (a) Draw the structure of polystyrene, which is formed by polymerizing the monomer styrene, $C_6H_5CH=CH_2$. (b) What monomer is used to form poly(vinyl acetate), a polymer used in paints and adhesives?

poly(vinyl acetate)

More Practice: Try Problems 21.59, 21.62a.

21.13B Radical Polymerization

The polymers described in Section 21.13A are prepared by polymerization of alkene monomers by **adding a radical to a π bond.** The mechanism resembles the radical addition of HBr to an alkene, except that a **carbon radical rather than a bromine atom is added to the double bond.** Mechanism 21.4 is written with the general monomer $CH_2=CHZ$, and again has three parts: initiation, propagation, and termination.

In radical polymerization, the more substituted radical always adds to the less substituted end of the monomer, a process called **head-to-tail polymerization.**

The **more substituted radical** adds to the **less substituted end** of the double bond.

The new radical is always located on the C bonded to Z.

The polystyrene foam (Problem 21.21a) used in packaging materials and drinking cups for hot beverages is called Styrofoam. Recycled polystyrene can be molded into trays and trash cans. *Jamie Grill/Getty Images*

The alkene monomers used in polymerization are prepared from petroleum.

Problem 21.22 Draw the steps of the mechanism that converts vinyl chloride ($CH_2=CHCl$) to poly(vinyl chloride).

> **Mechanism 21.4** Radical Polymerization of $CH_2=CHZ$

Part [1] Initiation

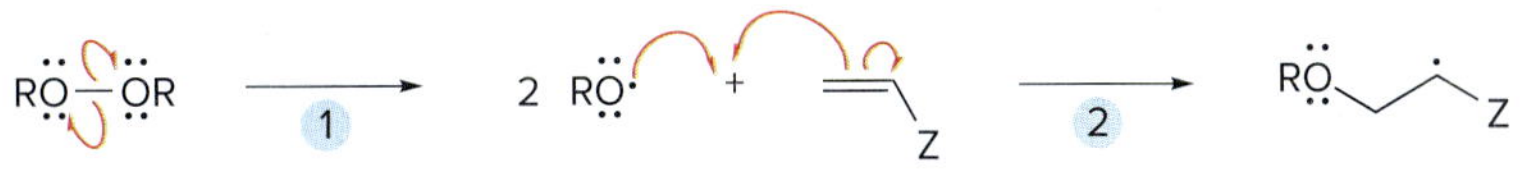

1 – 2 Initiation with ROOR occurs in two steps—homolysis of the **weak O–O bond** and addition of RO· to the alkene to form a carbon radical.

Part [2] Propagation

3 **Chain propagation consists of a single step.** The carbon radical adds to another alkene to form a new C–C bond and another carbon radical. Addition forms the radical with the unpaired electron on the atom with the Z substituent.

Part [3] Termination

4 Termination of the chain occurs when any two radicals combine to form a bond.

Chapter 21 REVIEW

KEY CONCEPTS

Radical stability (21.1)

- A radical is a reactive intermediate with a single unpaired electron.
- The stability of a radical increases as the number of electron-donating groups, such as **alkyl** groups, bonded to the **radical carbon** increases.

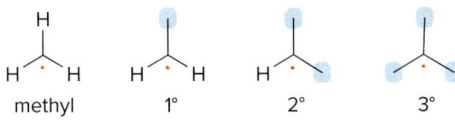

Increasing radical stability

Try Problems 21.25c, 21.26.

KEY REACTIONS

Radical Reactions

1	cyclopentane-H $\xrightarrow[\text{hv or }\Delta]{X_2}$ cyclopentane-X X = Cl, Br **halogenation** (21.4) alkyl halide		3	methylcyclopentene + H—Br $\xrightarrow{\text{hv, }\Delta\text{, or ROOR}}$ 1-bromo-1-methylcyclopentane **addition** (21.12) alkyl bromide	
2	cyclopentene-H $\xrightarrow[\text{hv or ROOR}]{\text{NBS}}$ cyclopentenyl-Br **allylic halogenation** (21.9) allylic halide		4	$CH_2=CHCl$ $\xrightarrow[\text{polymerization}]{\text{ROOR}}$ poly(vinyl chloride) (21.13) polymer	

Try Problems 21.23, 21.28, 21.29, 21.35, 21.36, 21.38, 21.59, 21.62a.

KEY SKILLS

[1] Drawing all the constitutional isomers formed by monochlorination of methylcyclopentane with Cl₂ and heat (21.3)

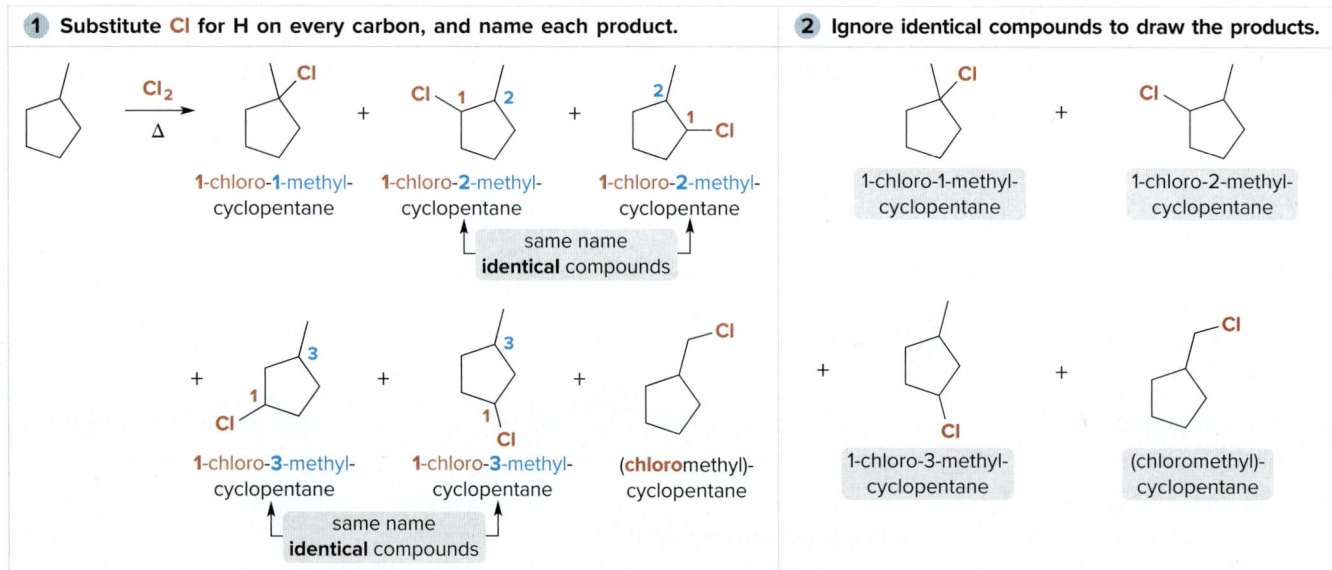

See Sample Problem 21.1. Try Problems 21.23a, 21.28, 21.38a.

[2] Drawing the major product formed by bromination of methylcyclopentane with Br₂ and hv (21.6)

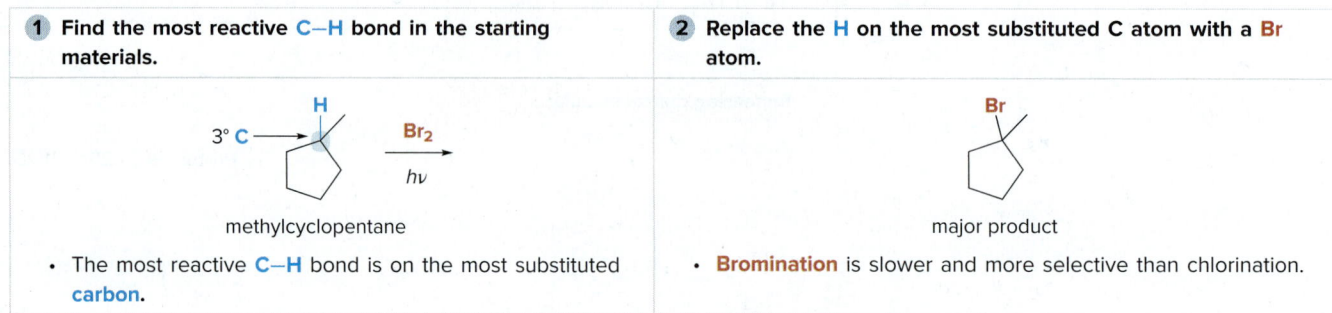

- The most reactive **C—H** bond is on the most substituted **carbon**.
- **Bromination** is slower and more selective than chlorination.

See Sample Problem 21.2. Try Problems 21.23b, 21.29, 21.38b.

Key Skills

[3] Drawing the stereoisomers formed by the monochlorination of a chiral starting material with Cl_2 and heat (21.7)

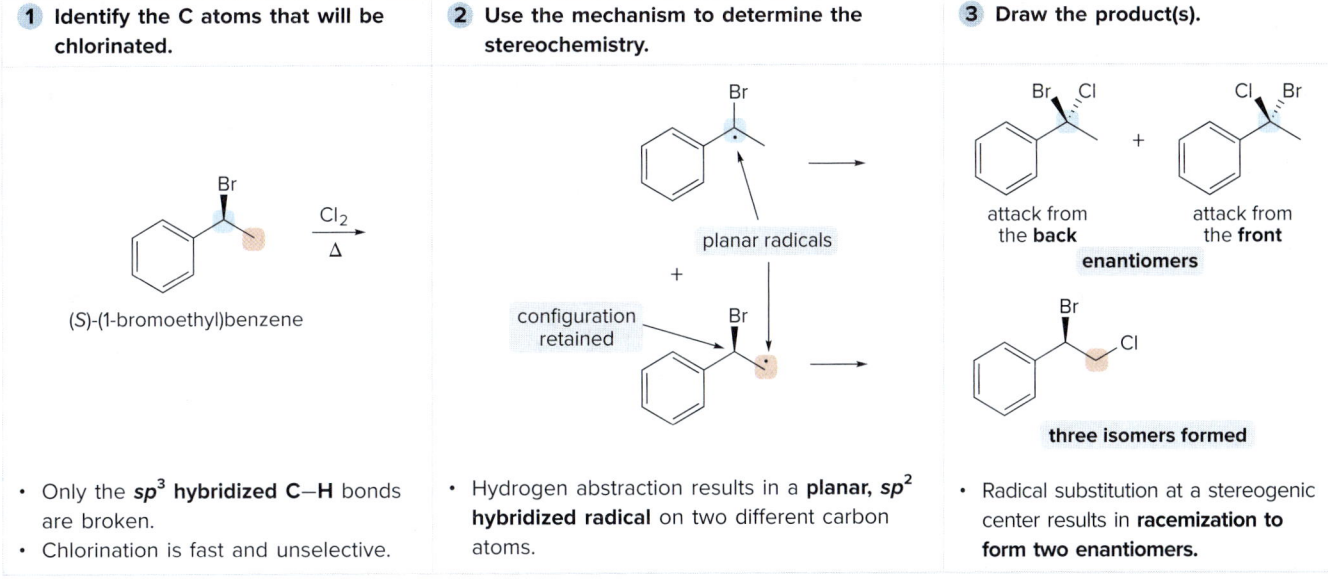

- Only the sp^3 hybridized C—H bonds are broken.
- Chlorination is fast and unselective.

- Hydrogen abstraction results in a **planar**, sp^2 **hybridized radical** on two different carbon atoms.

- Radical substitution at a stereogenic center results in **racemization to form two enantiomers.**

See Sample Problem 21.3. Try Problems 21.41a, d; 21.42; 21.43.

[4] Drawing the products formed when methylenecyclopentane is treated with NBS + ROOR (21.9)

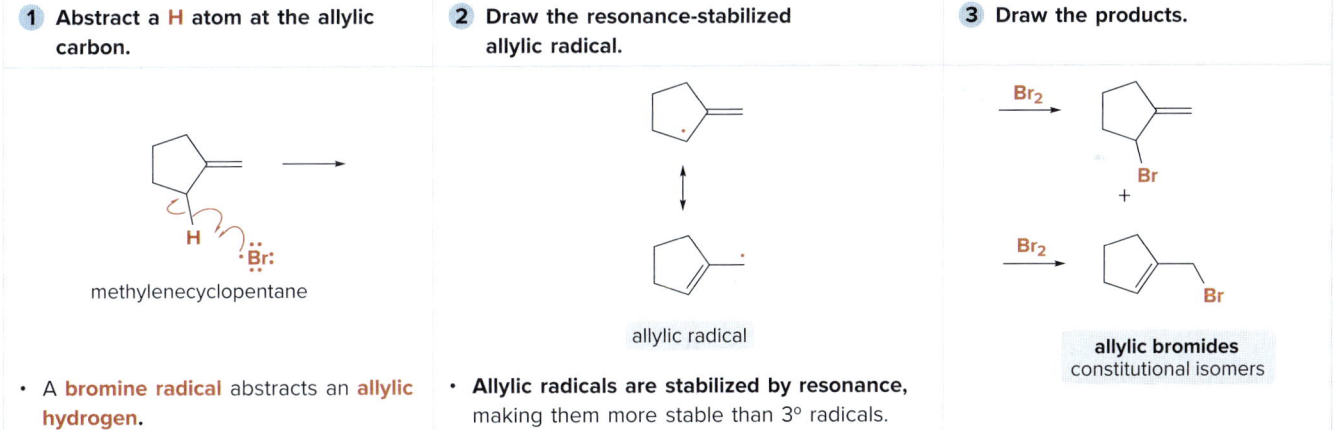

- A **bromine radical** abstracts an **allylic hydrogen**.

- **Allylic radicals** are stabilized by resonance, making them more stable than 3° radicals.

Try Problems 21.35, 21.36, 21.38f.

[5] Drawing the product of a polymerization reaction (21.13); example: polymerization of $CH_2=CHCH_3$

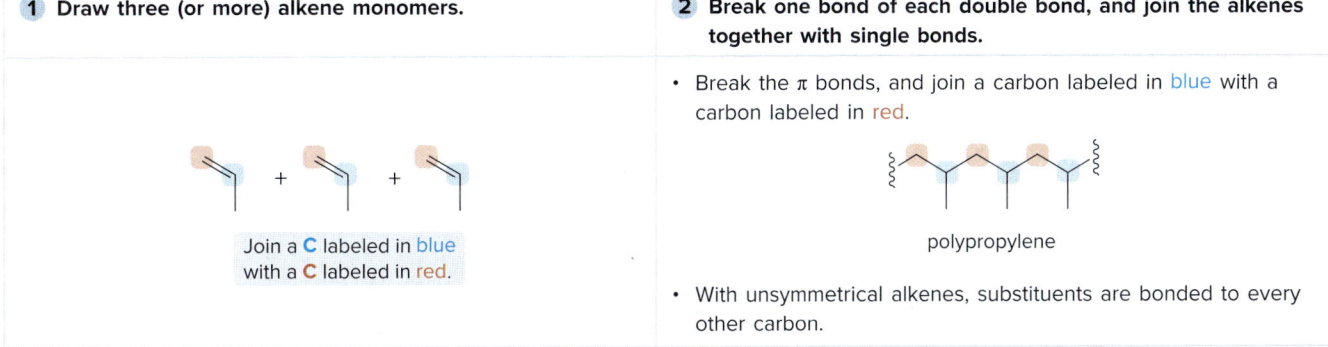

- Break the π bonds, and join a carbon labeled in blue with a carbon labeled in red.

- With unsymmetrical alkenes, substituents are bonded to every other carbon.

See Sample Problem 21.4. Try Problems 21.59, 21.62a.

PROBLEMS

Problems Using Three-Dimensional Models

21.23 (a) Draw all constitutional isomers formed by monochlorination of each alkane with Cl_2 and $h\nu$. (b) Draw the major monobromination product formed by heating each alkane with Br_2.

A   B

21.24 Draw all resonance structures of the radical that results from abstraction of a hydrogen atom from the antioxidant BHA (**b**utylated **h**ydroxy **a**nisole).

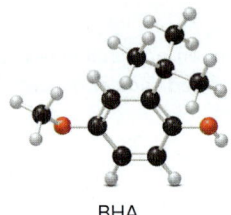

BHA

Radicals and Bond Strength

21.25 With reference to the indicated C–H bonds in the following compound:

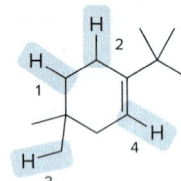

 a. Rank the C–H bonds in order of increasing bond strength.
 b. Draw the radical resulting from cleavage of each C–H bond, and classify it as 1°, 2°, or 3°.
 c. Rank the radicals in order of increasing stability.
 d. Rank the C–H bonds in order of increasing ease of H abstraction in a radical halogenation reaction.

21.26 Rank the following radicals in order of increasing stability.

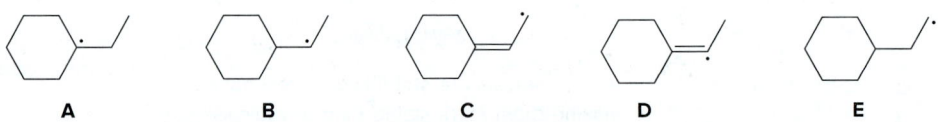

Halogenation of Alkanes

21.27 Rank the indicated hydrogen atoms in order of increasing ease of abstraction in a radical halogenation reaction.

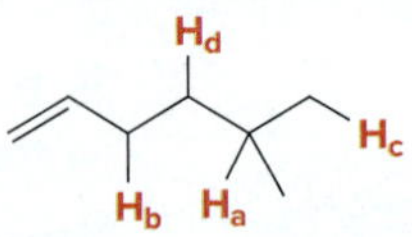

21.28 Draw all constitutional isomers formed by monochlorination of each alkane with Cl_2 and $h\nu$.

 a. 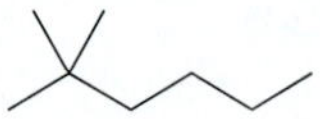   b.    c.

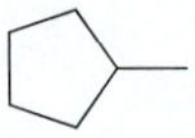

21.29 What is the major monobromination product formed by heating each alkane with Br_2?

 a. 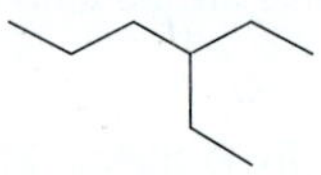   b. 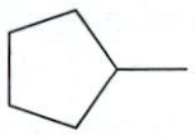   c.

21.30 Five isomeric alkanes (**A–E**) having the molecular formula C_6H_{14} are each treated with $Cl_2 + h\nu$ to give alkyl halides having molecular formula $C_6H_{13}Cl$. **A** yields five constitutional isomers. **B** yields four constitutional isomers. **C** yields two constitutional isomers. **D** yields three constitutional isomers, two of which possess stereogenic centers. **E** yields three constitutional isomers, only one of which possesses a stereogenic center. Identify the structures of **A–E**.

21.31 What alkane is needed to make each alkyl halide by radical halogenation?

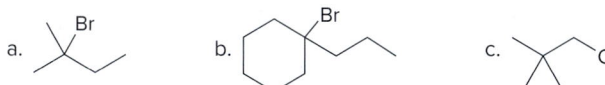

21.32 Which alkyl halides can be prepared in good yield by radical halogenation of an alkane?

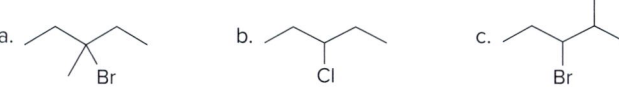

21.33 Explain why radical bromination of *p*-xylene forms **C** rather than **D**.

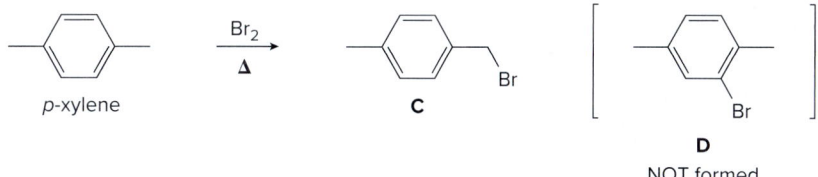

21.34 a. What product(s) (excluding stereoisomers) are formed when **Y** is heated with Cl_2?
b. What product(s) (excluding stereoisomers) are formed when **Y** is heated with Br_2?
c. What steps are needed to convert **Y** to the alkene **Z**?

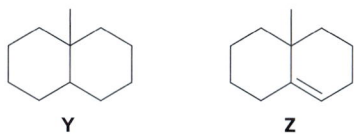

Allylic Halogenation

21.35 Draw the products formed when each alkene is treated with NBS + $h\nu$.

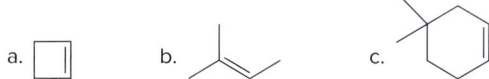

21.36 Draw all constitutional isomers formed when **X** is treated with NBS + $h\nu$.

21.37 Treatment of propylbenzene with NBS + $h\nu$ affords a single constitutional isomer. Suggest a structure for the product and a reason for its formation.

propylbenzene

Reactions

21.38 Draw the organic products formed in each reaction.

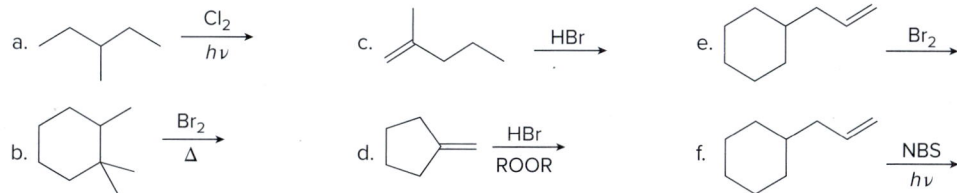

21.39 What reagents are needed to convert cyclopentene to (a) bromocyclopentane; (b) *trans*-1,2-dibromocyclopentane; (c) 3-bromocyclopentene?

21.40 Treatment of a hydrocarbon **A** (molecular formula C_9H_{18}) with Br_2 in the presence of light forms alkyl halides **B** and **C**, both having molecular formula $C_9H_{17}Br$. Reaction of either **B** or **C** with $KOC(CH_3)_3$ forms compound **D** (C_9H_{16}) as the major product. Ozonolysis of **D** forms cyclohexanone and acetone. Identify the structures of **A–D**.

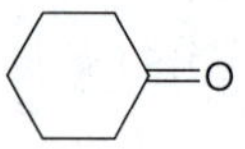

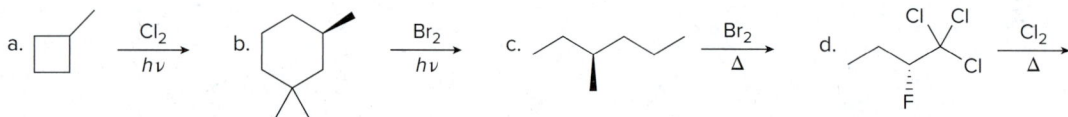

Stereochemistry and Reactions

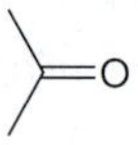

21.41 Draw the products formed in each reaction and include the stereochemistry around any stereogenic centers.

21.42 (a) Draw the products of molecular formula $C_3H_4Cl_2$, including stereoisomers, formed when chlorocyclopropane is heated with Cl_2. (b) Assuming that compounds that have different physical properties are separable, how many fractions would be present if the mixture of products were distilled using an efficient fractional distillation? (c) How many fractions would be optically active?

21.43 (a) Draw all stereoisomers of molecular formula $C_5H_{10}Cl_2$ formed when (*R*)-2-chloropentane is heated with Cl_2. (b) Assuming that products having different physical properties can be separated into fractions by some physical method (such as fractional distillation), how many different fractions would be obtained? (c) Which of these fractions would be optically active?

21.44 (a) Draw all stereoisomers formed by monobromination of the cis and trans isomers of 1,2-dimethylcyclohexane drawn below. (b) How do the products formed from each reactant compare—identical compounds, stereoisomers, or constitutional isomers?

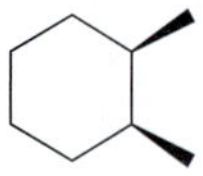

cis-1,2-dimethylcyclohexane 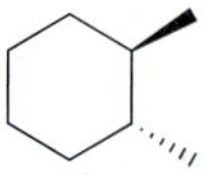   trans-(1R,2S)-dimethylcyclohexane

21.45 Draw the six products (including stereoisomers) formed when **A** is treated with NBS + hv.

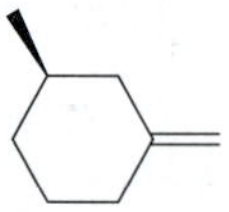

A

21.46 (a) Draw the products (including stereoisomers) formed when 2-methylhex-2-ene is treated with HBr in the presence of peroxides. (b) Draw the products (including stereoisomers) formed when (*S*)-2,4-dimethylhex-2-ene is treated with HBr and peroxides under similar conditions.

Mechanisms

21.47 Consider the following bromination: $(CH_3)_3CH + Br_2 \xrightarrow{\Delta} (CH_3)_3CBr + HBr$.
  a. Calculate $\Delta H°$ for this reaction by using the bond dissociation energies in Table 6.2.
  b. Draw out a stepwise mechanism for the reaction, including the initiation, propagation, and termination steps.
  c. Calculate $\Delta H°$ for each propagation step.
  d. Draw an energy diagram for the propagation steps.
  e. Draw the structure of the transition state of each propagation step.

21.48 Draw a stepwise mechanism for the following reaction.

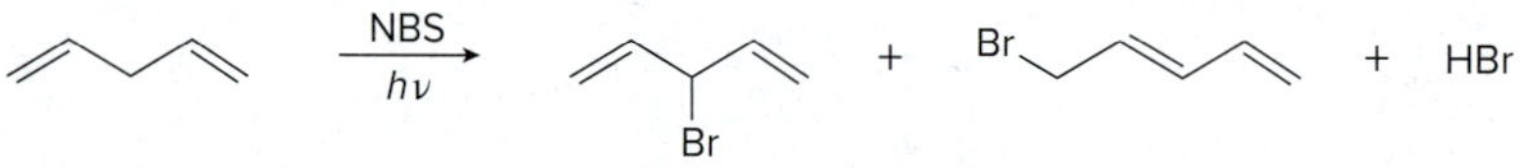

21.49 Like carbocations, radicals formed from compounds that contain another functional group can undergo intramolecular reactions. Draw a stepwise mechanism for the chain-propagating steps of the following intramolecular reaction.

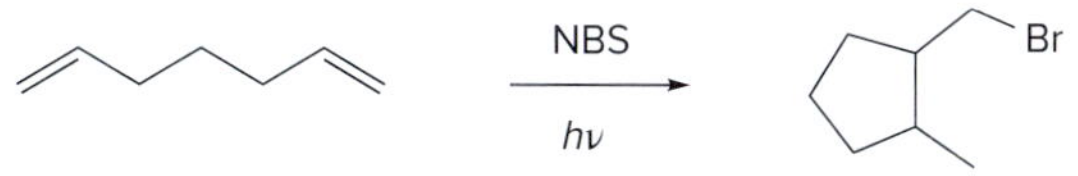

21.50 When 3,3-dimethylbut-1-ene is treated with HBr alone, the major product is 2-bromo-2,3-dimethylbutane. When the same alkene is treated with HBr and peroxide, the sole product is 1-bromo-3,3-dimethylbutane. Explain these results by referring to the mechanisms.

Synthesis

21.51 Devise a synthesis of each compound using CH_3CH_3 as the only source of carbon atoms. You may use any other required organic or inorganic reagents.

a. HC≡CH   b.  (propargyl alcohol)   c. 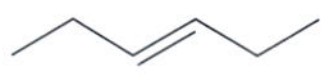 (pent-2-ene)   d. 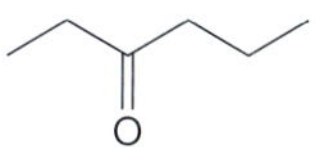 (pentan-2-one)

21.52 Devise a synthesis of $OHC(CH_2)_4CHO$ from cyclohexane using any required organic or inorganic reagents.

21.53 Devise a synthesis of hexane-2,3-diol from propane as the only source of carbon atoms. You may use any other required organic or inorganic reagents.

Radical Oxidation Reactions

21.54 As described in Section 9.17, the leukotrienes, important components in the asthmatic response, are synthesized from arachidonic acid via the hydroperoxide 5-HPETE. Write a stepwise mechanism for the conversion of arachidonic acid to 5-HPETE with O_2.

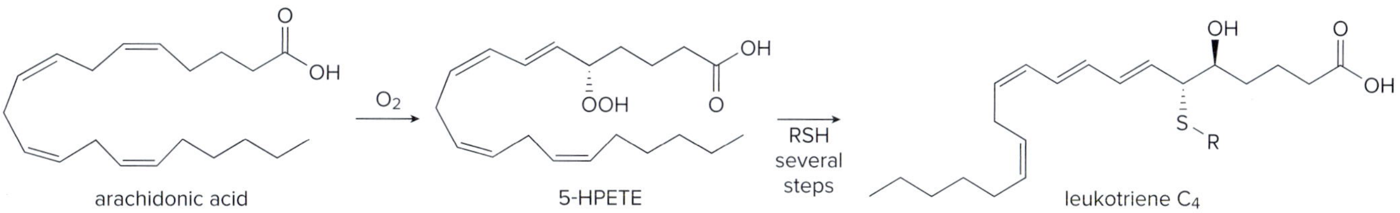

21.55 Ethers are oxidized with O_2 to form hydroperoxides that decompose violently when heated. Draw a stepwise mechanism for this reaction.

Antioxidants

21.56 Resveratrol is an antioxidant found in the skin of red grapes. Its anticancer, anti-inflammatory, and various cardiovascular effects are under active investigation. (a) Draw all resonance structures for the radical that results from homolysis of the OH bond shown in red. (b) Explain why homolysis of this OH bond is preferred to homolysis of either OH bond in the other benzene ring.

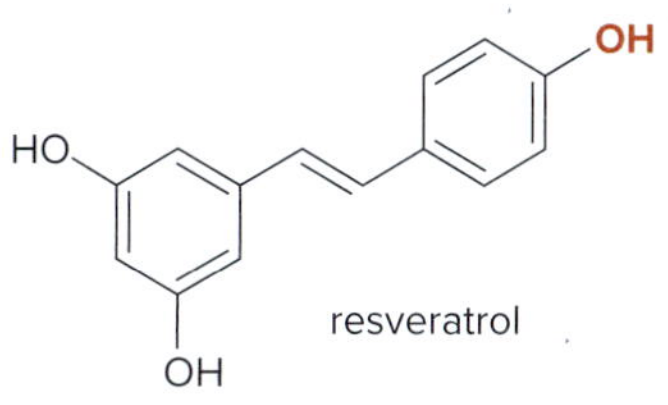

21.57 In cells, vitamin C exists largely as its conjugate base **X**. **X** is an antioxidant because radicals formed in oxidation processes abstract the labeled H atom, forming a new radical that halts oxidation. Draw the structure of the radical formed by H abstraction, and explain why this H atom is most easily removed.

Polymers and Polymerization

21.58 What monomer is needed to form each polymer?

a. polyisobutylene (used to make basketballs)

b. poly(ethyl acrylate) (used in latex paints)

21.59 (a) Hard contact lenses, which first became popular in the 1960s, were made by polymerizing methyl methacrylate [$CH_2=C(CH_3)CO_2CH_3$] to form poly(methyl methacrylate) (PMMA). Draw the structure of PMMA. (b) More-comfortable softer contact lenses introduced in the 1970s were made by polymerizing hydroxyethyl methacrylate [$CH_2=C(CH_3)CO_2CH_2CH_2OH$] to form poly(hydroxyethyl methacrylate) (poly-HEMA). Draw the structure of poly-HEMA. Because neither polymer allows oxygen from the air to pass through to the retina, newer contact lenses that are both comfortable and oxygen-permeable have now been developed.

21.60 Explain why polystyrene is much more readily oxidized by O_2 in the air than polyethylene is. Which H's in polystyrene are most easily abstracted and why?

polystyrene

polyethylene

21.61 Draw a stepwise mechanism for the following polymerization reaction.

21.62 As we will learn in Chapter 30, styrene derivatives such as **A** can be polymerized by way of cationic rather than radical intermediates. Cationic polymerization is an example of electrophilic addition to an alkene involving carbocations.

a. Draw a short segment of the polymer formed by the polymerization of **A**.
b. Why does **A** react faster than styrene ($C_6H_5CH=CH_2$) in a cationic polymerization?

21.63 When two monomers (**X** and **Y**) are polymerized together, a copolymer results. An alternating copolymer is formed when the two monomers **X** and **Y** alternate regularly in the polymer chain. Draw the structure of the alternating copolymer formed when the two monomers, $CH_2=CCl_2$ and $CH_2=CHC_6H_5$, are polymerized together.

Spectroscopy

21.64 Identify the structure of a minor product formed from the radical chlorination of propane, which has molecular formula $C_3H_6Cl_2$ and exhibits the given ^{1}H NMR spectrum.

Challenge Problems

21.65 The triphenylmethyl radical is an unusual persistent radical present in solution in equilibrium with its dimer. For 70 years the dimer was thought to be hexaphenylethane, but in 1970, NMR data showed it to be **A.**

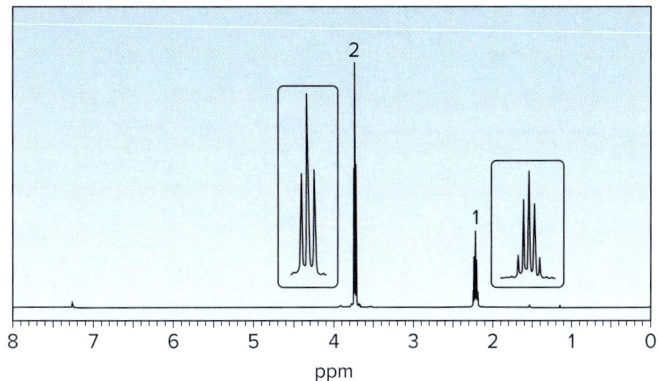

triphenylmethyl radical

hexaphenylethane

A

a. Why is the triphenylmethyl radical more stable than most other radicals?
b. Use curved arrow notation to show how two triphenylmethyl radicals dimerize to form **A.**
c. Propose a reason for the formation of **A** rather than hexaphenylethane.
d. How could ^{1}H and ^{13}C NMR spectroscopy be used to distinguish between hexaphenylethane and **A?**

21.66 Draw a stepwise mechanism for the chain-propagating steps of the following ring-opening reaction.

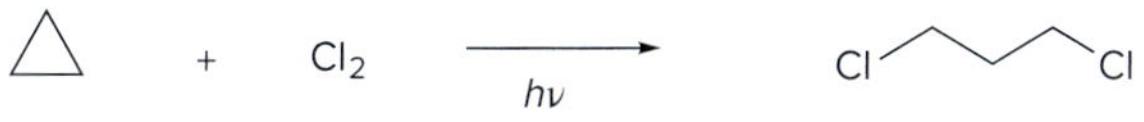

21.67 In the presence of a radical initiator (Z·), tributyltin hydride (R_3SnH, R = $CH_3CH_2CH_2CH_2$) reduces alkyl halides to alkanes: $R'X + R_3SnH \rightarrow R'H + R_3SnX$. The mechanism consists of a radical chain process with an intermediate tin radical:

Initiation: $R_3SnH + Z\cdot \longrightarrow R_3Sn\cdot + HZ$

Propagation:

$R'{-}Br + R_3Sn\cdot \longrightarrow R'\cdot + R_3SnBr$

$R'\cdot + R_3SnH \longrightarrow R'{-}H + R_3Sn\cdot$

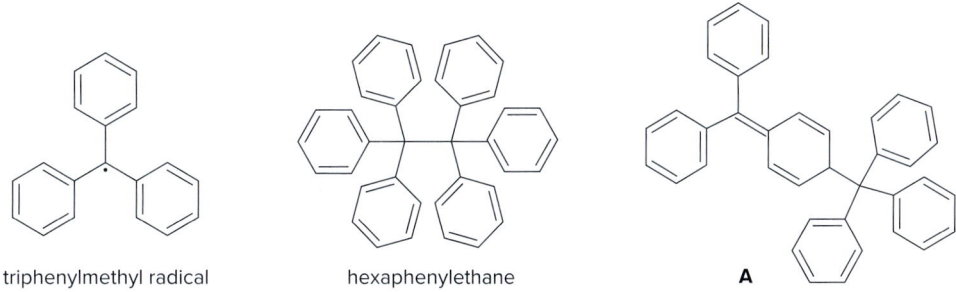

This reaction has been employed in many radical cyclization reactions. Draw a stepwise mechanism for the following reaction.

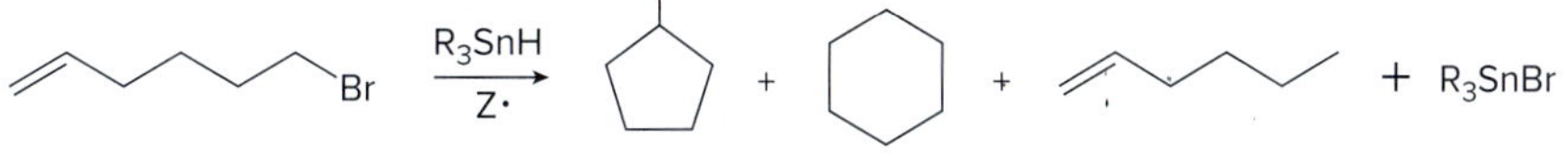

21.68 PGF$_{2\alpha}$ (Section 15.5) is synthesized in cells from arachidonic acid (C$_{20}$H$_{32}$O$_2$) using a cyclooxygenase enzyme that catalyzes a multistep radical pathway. Part of this process involves the conversion of radical **A** to PGG$_2$, an unstable intermediate, which is then transformed to PGF$_{2\alpha}$ and other prostaglandins. Draw a stepwise mechanism for the conversion of **A** to PGG$_2$. (Hint: The mechanism begins with radical addition to a carbon–carbon double bond to form a resonance-stabilized radical.)

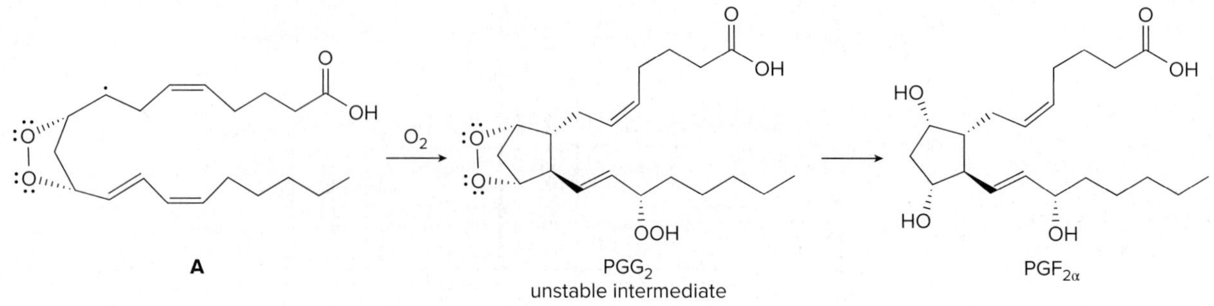

Amines 22

Werner Arnold

22.1 Introduction	**22.7** Reactions of amines—General features	**22.12** Reaction of amines with nitrous acid
22.2 Structure and bonding	**22.8** Amines as bases	**22.13** Substitution reactions of aryl diazonium salts
22.3 Nomenclature	**22.9** Relative basicity of amines and other compounds	**22.14** Coupling reactions of aryl diazonium salts
22.4 Physical and spectroscopic properties	**22.10** Amines as nucleophiles	**22.15** Application: Synthetic dyes and sulfa drugs
22.5 Interesting and useful amines	**22.11** Hofmann elimination	
22.6 Preparation of amines		

Atropine is an alkaloid isolated from *Atropa belladonna,* the deadly nightshade plant. Atropine causes an increase in heart rate, relaxes smooth muscles, and interferes with nerve impulses transmitted by acetylcholine. In higher doses atropine is poisonous, leading to convulsions, coma, and death. Atropine is one of the many naturally occurring amines isolated from a plant source. In Chapter 22, we learn about the properties and reactions of amines.

Why Study... Amines?

We now turn our attention to amines, **organic** derivatives of ammonia (NH₃), formed by replacing one or more hydrogen atoms by alkyl or aryl groups. **Amines are stronger bases and better nucleophiles than other neutral organic compounds,** so much of Chapter 22 focuses on these properties.

Like that of alcohols, the chemistry of amines does not fit neatly into one reaction class, and this can make learning the reactions of amines challenging. Many interesting natural products and widely used drugs are amines, so you also need to know how to introduce this functional group into organic molecules.

22.1 Introduction

Amines **are organic nitrogen compounds,** formed by replacing one or more hydrogen atoms of ammonia (NH₃) with alkyl groups. As discussed in Section 3.2, amines are classified as 1°, 2°, or 3° by the number of alkyl groups bonded to the *nitrogen* atom.

- 1° amine (1 R group on N)
- 2° amine (2 R groups on N)
- 3° amine (3 R groups on N)

Like ammonia, **the amine nitrogen atom has a nonbonded electron pair,** making it both a base and a nucleophile. As a result, amines react with electrophiles to form **ammonium salts**—compounds with a positively charged ammonium ion and an anionic counterion.

base + acid ⟶ ammonium salt

nucleophile + electrophile ⟶ ammonium salt

- The chemistry of amines is dominated by the nonbonded electron pair on the nitrogen atom.

22.2 Structure and Bonding

An amine nitrogen atom is surrounded by three atoms and one nonbonded electron pair, making the N atom sp^3 **hybridized** and **trigonal pyramidal,** with bond angles of approximately 109.5°. Because nitrogen is much more electronegative than carbon or hydrogen, **the C–N and N–H bonds are all polar,** with the N atom electron rich and the C and H atoms electron poor.

CH₃–N(H)(H) 147 pm

CH₃–N(CH₃)(CH₃) 108°

An amine nitrogen atom bonded to an electron pair and three different alkyl groups is technically a stereogenic center, so two nonsuperimposable trigonal pyramids can be drawn.

nonsuperimposable mirror images

This does not mean, however, that such an amine exists as two different enantiomers, because one is rapidly converted to the other at room temperature. The amine flips inside out, passing through a trigonal planar (achiral) transition state. **Because the two enantiomers interconvert, we can ignore the chirality of the amine nitrogen.**

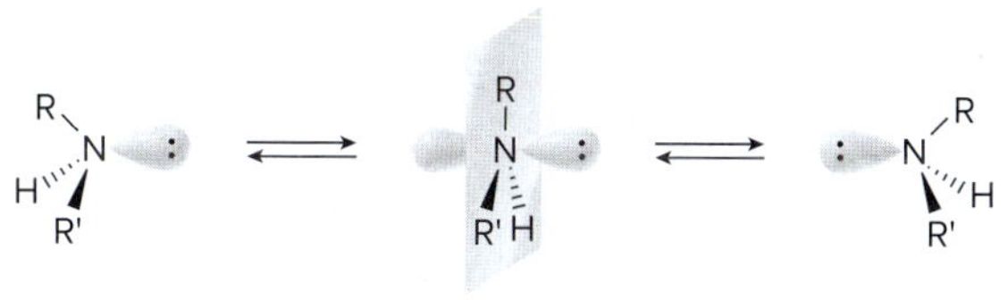

planar transition state

In contrast, **the chirality of an ammonium ion with four different groups on N *cannot* be ignored.** Because there is no nonbonded electron pair on the nitrogen atom, **interconversion cannot occur,** and the N atom is just like a carbon atom with four different groups around it.

chiral ammonium ion

- The N atom of an ammonium ion is a stereogenic center when N is surrounded by four different groups.

Problem 22.1 Label the stereogenic centers in each compound.

a.

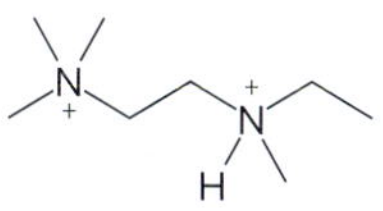

b. 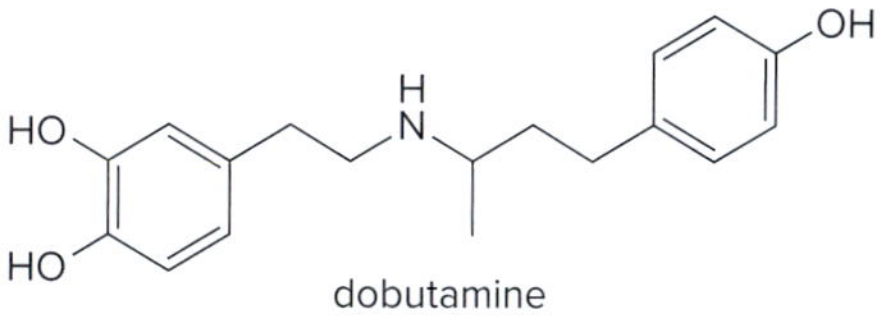 dobutamine
(heart stimulant used in stress tests to measure cardiac fitness)

22.3 Nomenclature

22.3A Primary Amines

Primary amines are named using either systematic or common names.

- To assign the systematic name, find the longest continuous carbon chain bonded to the amine nitrogen, and change the -e ending of the parent alkane to the suffix *-amine.* Then use the usual rules of nomenclature to number the chain and name the substituents.
- To assign a common name, name the alkyl group bonded to the nitrogen atom and add the suffix *-amine,* forming a single word.

CH_3NH_2

Systematic name: **methanamine**
Common name: **methylamine**

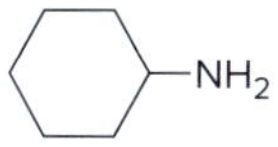

Systematic name: **cyclohexanamine**
Common name: **cyclohexylamine**

22.3B Secondary and Tertiary Amines

Secondary and tertiary amines having identical alkyl groups are named by using the prefix *di-* or *tri-* with the name of the primary amine.

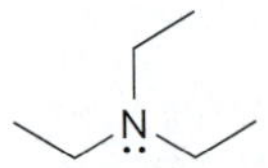

triethylamine

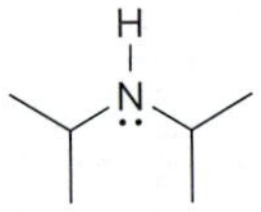

diisopropylamine

Secondary and tertiary amines having more than one kind of alkyl group are named as *N*-substituted primary amines, using the following procedure.

How To Name 2° and 3° Amines with Different Alkyl Groups

Example Name the following 2° amine: $(CH_3)_2CHNHCH_3$.

Step [1] Designate the longest alkyl chain (or largest ring) bonded to the N atom as the parent amine and assign a systematic name.

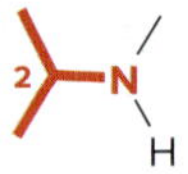

3 C's in the longest chain ---→ propan-2-amine

Step [2] Name the other groups on the N atom as alkyl groups, alphabetize the names, and put the prefix *N-* before the name.

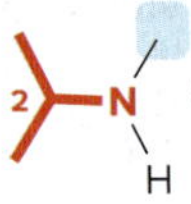

methyl substituent **Answer: N-methylpropan-2-amine**

Sample Problem 22.1 Naming an Amine

Name each amine.

a. 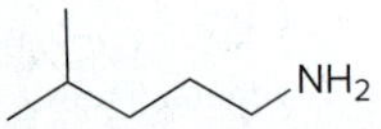 b.

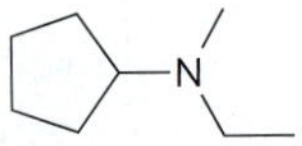

Solution

a. [1] A 1° amine: Find and name the longest chain containing the amine nitrogen.

pentane ---→ pentan*amine*
(5 C's)

[2] Number and name the substituents.

You must use a number to show the location of the NH_2 group.

Answer: 4-methylpentan-1-amine

b. For a 3° amine, one alkyl group on N is the principal R group and the others are substituents.

[1] Name the ring bonded to the N.

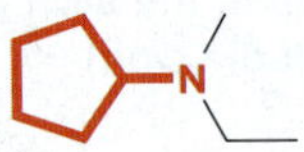

cyclopentanamine

[2] Name the substituents.

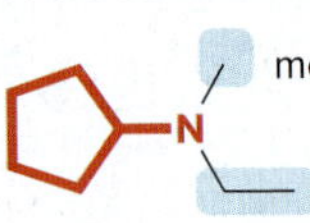

methyl
ethyl

Two N's are needed, one for each alkyl group.

Answer: N-ethyl-N-methylcyclopentanamine

Problem 22.2 Name each amine.

a. [CH3CH(NH2)CH3 structure]
b. [CH3CH2CH2-NH-CH2CH2CH2CH3 structure]
c. [cyclohexyl-N(CH3)2 structure]
d. [CH3CH2CH2CH(NH2)CH2CH(CH3)2 structure]
e. [CH3CH2CH2CH(NHCH2CH3)CH2CH3 structure]
f. [2-methylcyclopentyl-NH-CH2CH2CH3 structure]

More Practice: Try Problems 22.35, 22.37.

22.3C Aromatic Amines

Aromatic amines are named as derivatives of aniline.

[aniline, N-ethylaniline, o-bromoaniline structures]

aniline   N-ethylaniline   o-bromoaniline

22.3D Miscellaneous Nomenclature Facts

An **NH₂** group named as a substituent is called an **amino group.**

There are many different **nitrogen heterocycles,** and each ring type is named differently depending on the number of N atoms in the ring, the ring size, and whether it is aromatic or not. The structures and names of common nitrogen heterocycles are shown in Figure 22.1.

Figure 22.1
Common nitrogen heterocycles

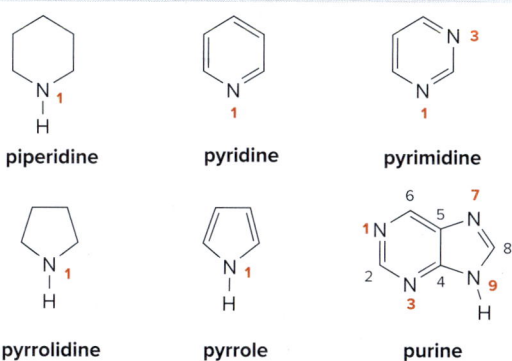

piperidine   pyridine   pyrimidine

pyrrolidine   pyrrole   purine

- Heterocycles with one N atom are numbered to place the N atom at the "1" position.
- Heterocycles with two N atoms are numbered to place one N atom at the "1" position and give the second N atom the lower number.

Problem 22.3 Draw a structure corresponding to each name.

a. 2,4-dimethylhexan-3-amine
b. N-methylpentan-1-amine
c. N-isopropyl-p-nitroaniline
d. N-methylpiperidine
e. N,N-dimethylethanamine
f. 2-aminocyclohexanone
g. N-methylaniline
h. m-ethylaniline

22.4 Physical and Spectroscopic Properties

22.4A Physical Properties

Amines exhibit dipole–dipole interactions because of the polar C—N and N—H bonds. **Primary and secondary amines are also capable of intermolecular hydrogen bonding,** because they contain N—H bonds. Because nitrogen is less electronegative than oxygen, however, intermolecular hydrogen bonds between N and H are *weaker* than those between O and H. How these factors affect the physical properties of amines is summarized in Table 22.1.

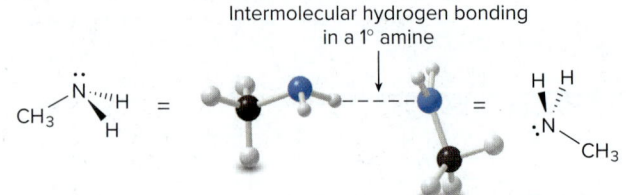

Intermolecular hydrogen bonding in a 1° amine

Table 22.1 Physical Properties of Amines

Property	Observation
Boiling point and melting point	• Primary (1°) and 2° amines have higher bp's than similar compounds (like ethers) incapable of hydrogen bonding, but lower bp's than alcohols that have stronger intermolecular hydrogen bonds. • Tertiary (3°) amines have lower boiling points than 1° and 2° amines of comparable molecular weight, because they have no N—H bonds and are incapable of hydrogen bonding. MW = 73, bp 38 °C, no N—H bond • MW = 73, bp 78 °C, N—H bond • MW = 74, bp 118 °C, O—H bond Increasing intermolecular forces → Increasing boiling point
Solubility	• Amines are soluble in organic solvents regardless of size. • All amines having ≤ 5 C's are H_2O soluble because they can hydrogen bond with H_2O (Section 3.4C). • Amines having > 5 C's are H_2O insoluble because the nonpolar alkyl portion is too large to dissolve in the polar H_2O solvent.

Key: MW = molecular weight

Problem 22.4 Arrange the compounds in order of increasing boiling point.

22.4B Spectroscopic Properties

The spectroscopic properties of amines have been detailed in Spectroscopy Parts A, B, and C.

- Mass spectra: The odd molecular ion in Section A.1B and fragmentation patterns in Section A.4C
- Infrared absorptions: Section B.4C and Table B.2
- ^1H and ^{13}C NMR absorptions: Section C.9A and Tables C.1 and C.5

The general molecular formula for an amine with one N atom is $C_nH_{2n+3}N$.

Key NMR and IR absorptions for amines are summarized in Table 22.2. Figure 22.2 illustrates that the number of N—H peaks in an IR spectrum can be used to distinguish 1°, 2°, and 3° amines.

- 1° Amines show *two* N—H absorptions at 3300–3500 cm^{-1}.
- 2° Amines show *one* N—H absorption at 3300–3500 cm^{-1}.
- 3° Amines do *not* absorb at 3300–3500 cm^{-1} because 3° amines have no N—H bonds.

Table 22.2 Characteristic Spectroscopic Absorptions of Amines

Type of spectroscopy	Type of C, H	Absorption
IR absorption	N—H	3300–3500 cm^{-1} (one or two peaks)
^{1}H NMR absorptions	N—H	0.5–5.0 ppm
	N—C—H	2.3–3.0 ppm
^{13}C NMR absorption	C—N	30–50 ppm

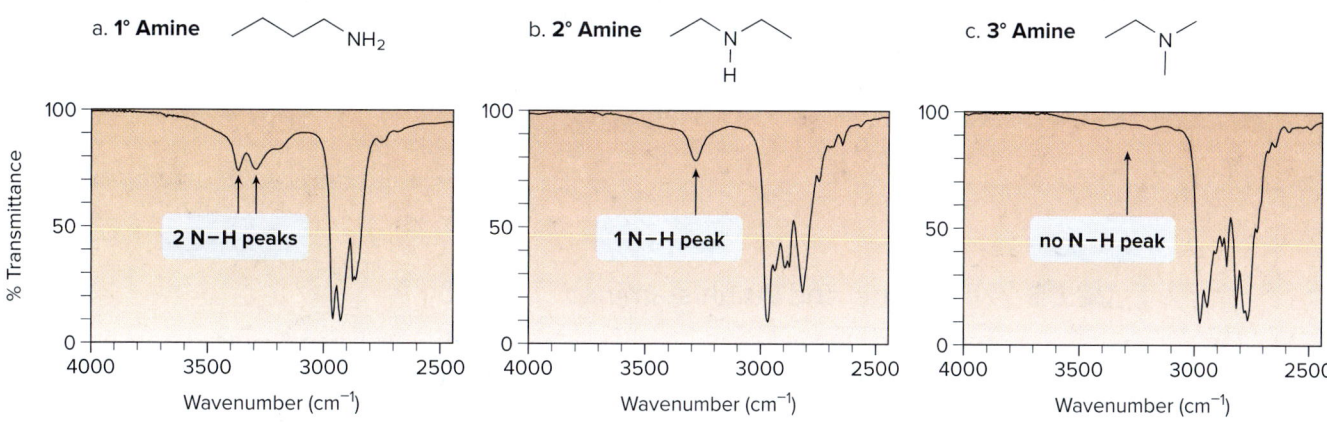

Figure 22.2 The single bond region of the IR spectra for a 1°, 2°, and 3° amine

22.5 Interesting and Useful Amines

A great many simple and complex amines occur in nature, and others with biological activity have been synthesized in the lab.

22.5A Simple Amines and Alkaloids

Many low-molecular-weight amines have *very* foul odors. **Trimethylamine** [$(CH_3)_3N$], formed when enzymes break down certain fish proteins, has the characteristic odor of rotting fish. **Putrescine** ($NH_2CH_2CH_2CH_2CH_2NH_2$) and **cadaverine** ($NH_2CH_2CH_2CH_2CH_2CH_2NH_2$) are both poisonous diamines with putrid odors. They, too, are present in rotting fish and are partly responsible for the odors of semen, urine, and bad breath.

The word *alkaloid* is derived from the word *alkali*, because aqueous solutions of alkaloids are slightly basic.

Naturally occurring amines derived from plant sources are called **alkaloids.** Alkaloids previously encountered in the text include **quinine** (Chapter 8 opener and Problem 17.13), **morphine**

(Section 16.8), and **cocaine** (Chapter 16 opener). Three other common alkaloids are **sparteine, nicotine,** and **scopolamine,** illustrated in Figure 22.3.

Figure 22.3

Three common alkaloids—Sparteine, nicotine, and scopolamine

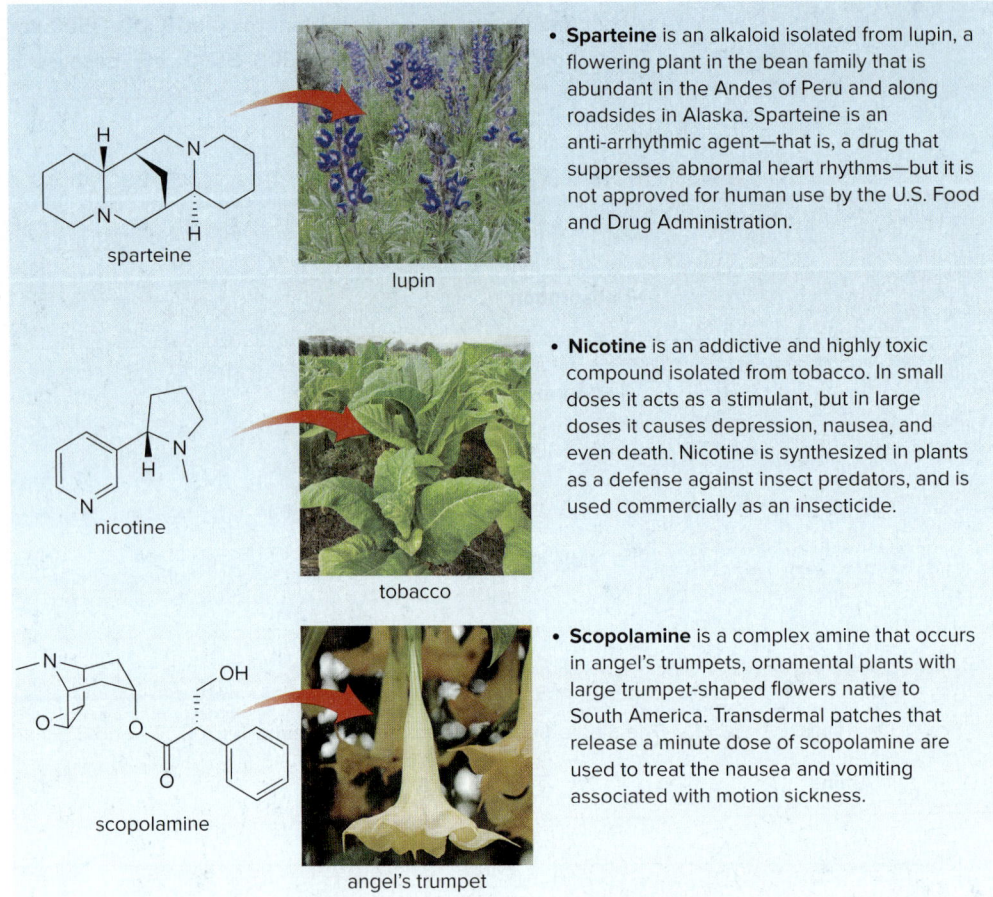

- **Sparteine** is an alkaloid isolated from lupin, a flowering plant in the bean family that is abundant in the Andes of Peru and along roadsides in Alaska. Sparteine is an anti-arrhythmic agent—that is, a drug that suppresses abnormal heart rhythms—but it is not approved for human use by the U.S. Food and Drug Administration.

- **Nicotine** is an addictive and highly toxic compound isolated from tobacco. In small doses it acts as a stimulant, but in large doses it causes depression, nausea, and even death. Nicotine is synthesized in plants as a defense against insect predators, and is used commercially as an insecticide.

- **Scopolamine** is a complex amine that occurs in angel's trumpets, ornamental plants with large trumpet-shaped flowers native to South America. Transdermal patches that release a minute dose of scopolamine are used to treat the nausea and vomiting associated with motion sickness.

Daniel C. Smith; kai4107/Shutterstock; James Forte/Getty Images

22.5B Histamine and Antihistamines

Histamine, a simple triamine first discussed in Section 19.9, is responsible for a wide variety of physiological effects. Histamine is a vasodilator (it dilates capillaries), so it is released at the site of an injury or infection to increase blood flow. It is also responsible for the symptoms of allergies, including a runny nose and watery eyes. In the stomach, histamine stimulates the secretion of acid.

Understanding the central role of histamine in these biochemical processes has helped chemists design drugs to counteract some of its undesirable effects.

Antihistamines bind to the same active site of the enzyme that binds histamine in the cell, but they evoke a different response. An antihistamine like **fexofenadine** (trade name

Allegra), for example, inhibits vasodilation, so it is used to treat the symptoms of the common cold and allergies. Unlike many antihistamines, fexofenadine does not cause drowsiness because it binds to histamine receptors but does not cross the blood–brain barrier, so it does not affect the central nervous system. **Cimetidine** (trade name Tagamet) is a histamine mimic that blocks the secretion of hydrochloric acid in the stomach, so it is used to treat individuals with ulcers.

22.5C Derivatives of 2-Phenylethanamine

A large number of physiologically active compounds are derived from **2-phenylethanamine, $C_6H_5CH_2CH_2NH_2$**. Some of these compounds are synthesized in cells and needed to maintain healthy mental function. Others are isolated from plant sources or are synthesized in the laboratory and have a profound effect on the brain because they interfere with normal neurochemistry. These compounds include **adrenaline, noradrenaline, methamphetamine,** and **mescaline.** Each contains a benzene ring bonded to a two-carbon unit with a nitrogen atom (shown in red).

adrenaline
(epinephrine)
a hormone secreted in response to stress
(Figure 7.17)

noradrenaline
(norepinephrine)
a neurotransmitter that increases heart rate
and dilates air passages

methamphetamine
an addictive stimulant sold as
speed, meth, or crystal meth

mescaline
a hallucinogen isolated from peyote, a cactus native
to the southwestern United States and Mexico

Cocaine, amphetamines, and several other addictive drugs increase the level of dopamine in the brain, which results in a pleasurable "high." With time, the brain adapts to increased dopamine levels, so more drug is required for the same sensation.

Another example, **dopamine,** is a neurotransmitter, a chemical messenger released by one nerve cell (neuron), which then binds to a receptor in a neighboring target cell (Figure 22.4). Dopamine affects brain processes that control movement and emotions, so proper dopamine levels are necessary to maintain an individual's mental and physical health. For example, when dopamine-producing neurons die, the level of dopamine drops, resulting in the loss of motor control symptomatic of Parkinson's disease.

Serotonin is a neurotransmitter that plays an important role in mood, sleep, perception, and temperature regulation. A deficiency of serotonin causes depression. Understanding the central role of serotonin in determining one's mood has led to the development of a variety of drugs for the treatment of depression. The most widely used antidepressants today are selective serotonin reuptake inhibitors (SSRIs). These drugs act by inhibiting the reuptake of serotonin by the neurons that produce it, thus effectively increasing its concentration. Fluoxetine (trade name Prozac) is a common antidepressant that acts in this way.

serotonin

fluoxetine

Figure 22.4 Dopamine—A neurotransmitter

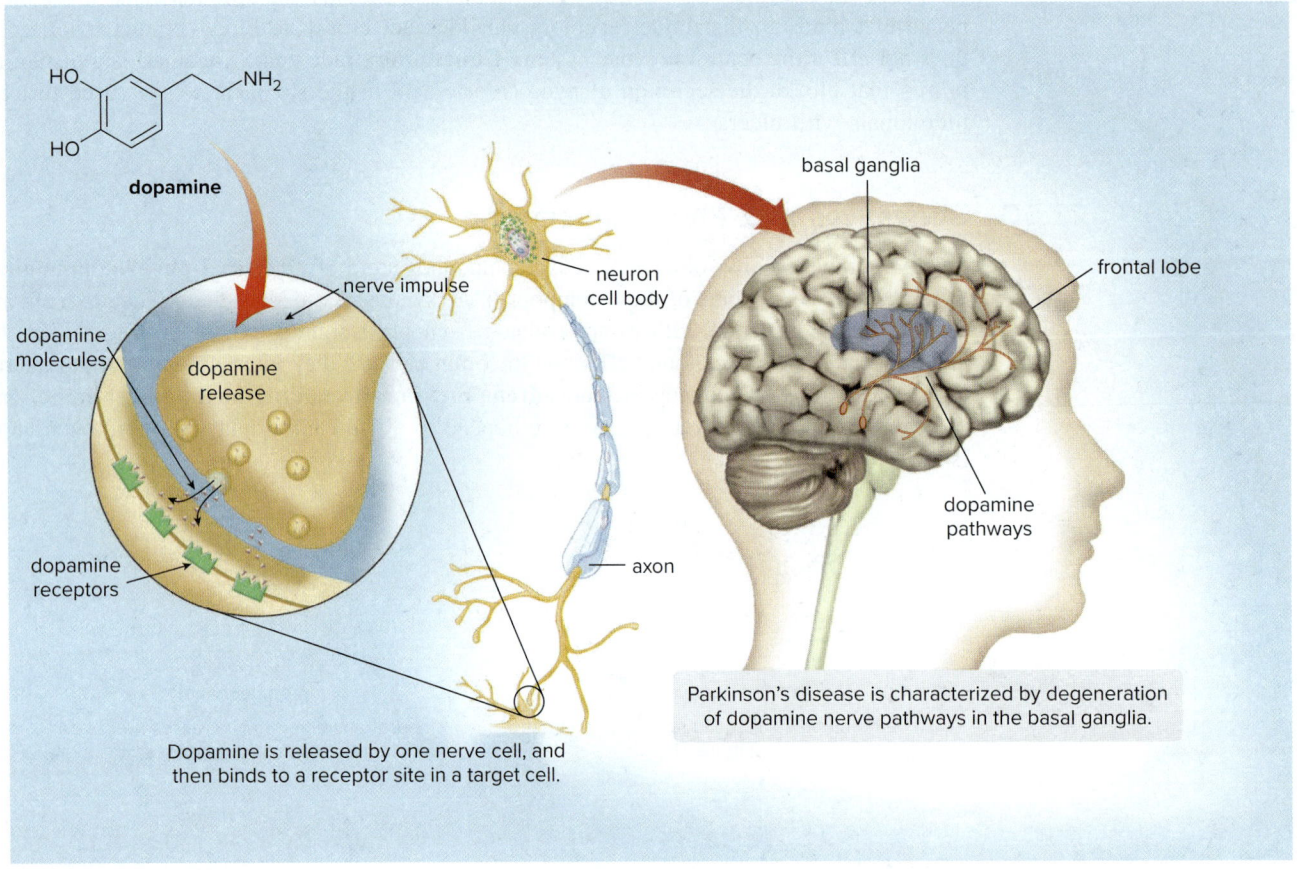

Bufo toads from the Amazon jungle are the source of the hallucinogen bufotenin.
Daniel C. Smith

Drugs that interfere with the metabolism of serotonin have a profound effect on mental state. For example, bufotenin, isolated from *Bufo* toads from the Amazon jungle, and psilocin, isolated from *Psilocybe* mushrooms, are very similar in structure to serotonin and both cause intense hallucinations.

Problem 22.5 LSD (a hallucinogen) and codeine (a narcotic) are structurally more complex derivatives of 2-phenylethanamine. Identify the atoms of 2-phenylethanamine in each of the following compounds.

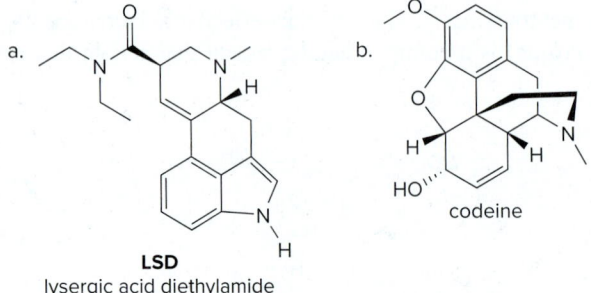

22.6 Preparation of Amines

> In the preparations of a given functional group, many different starting materials form a common product (amines, in this case).

Three types of reactions are used to prepare an amine:

[1] **Nucleophilic substitution** using nitrogen nucleophiles
[2] **Reduction** of other nitrogen-containing functional groups
[3] **Reductive amination** of aldehydes and ketones

22.6A Nucleophilic Substitution Routes to Amines

Nucleophilic substitution is the key step in two different methods for synthesizing amines: direct nucleophilic substitution and the Gabriel synthesis of 1° amines.

Direct Nucleophilic Substitution

Conceptually, the simplest method to synthesize an amine is by **S_N2 reaction of an alkyl halide with NH_3 or an amine.** The method requires two steps:

[1] **Nucleophilic attack** of the nitrogen nucleophile forms an ammonium salt.
[2] **Removal of a proton** on N forms the amine.

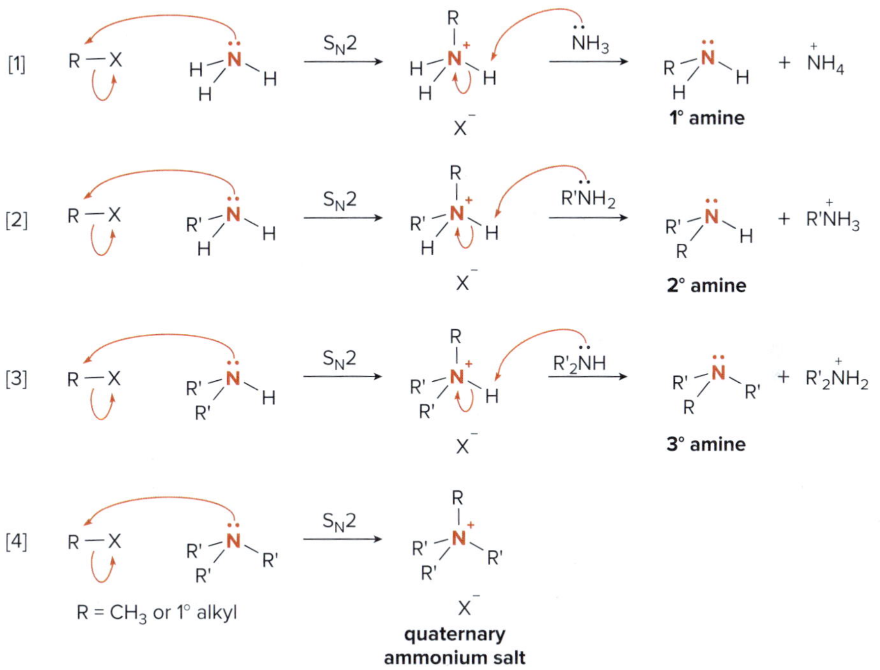

The identity of the nitrogen nucleophile determines the type of amine or ammonium salt formed as product. **One new carbon–nitrogen bond is formed in each reaction.** Because the reaction follows an S_N2 mechanism, the alkyl halide must be unhindered—that is, CH_3X or RCH_2X.

Although this process seems straightforward, polyalkylation of the nitrogen nucleophile limits its usefulness. **Any amine formed by nucleophilic substitution still has a nonbonded electron pair, making it a nucleophile as well.** It will react with remaining alkyl halide to form a more substituted amine. Because of this, a mixture of 1°, 2°, and 3° amines often results. Only the final product—called a **quaternary ammonium salt** because it has four alkyl groups on N—cannot react further, and so the reaction stops.

As a result, this reaction is most useful for preparing 1° amines by using a very large excess of NH_3 (a relatively inexpensive starting material) and for preparing quaternary ammonium salts by alkylating any nitrogen nucleophile with one or more equivalents of alkyl halide.

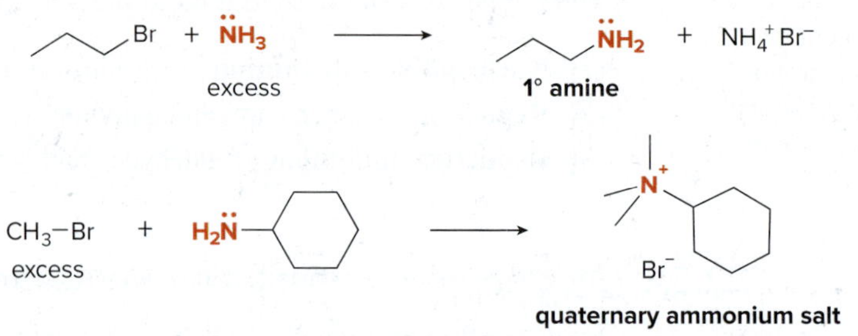

CH₃CH₂CH₂Br + :NH₃ (excess) ⟶ CH₃CH₂CH₂NH₂ (1° amine) + NH₄⁺ Br⁻

CH₃—Br (excess) + H₂N—cyclohexyl ⟶ (CH₃)₃N⁺—cyclohexyl Br⁻ (quaternary ammonium salt)

Problem 22.6 Draw the product of each reaction.

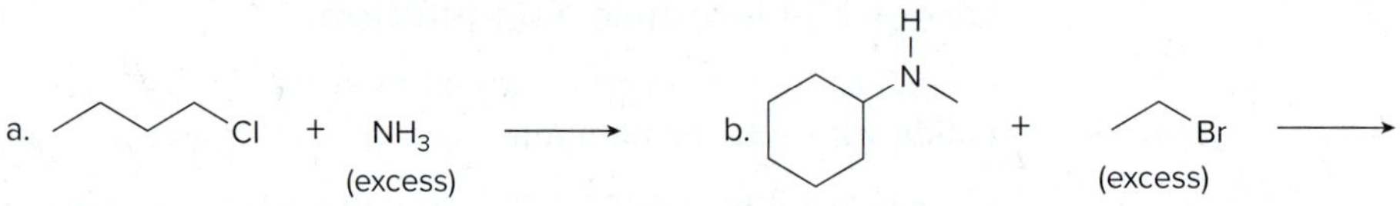

a. CH₃CH₂CH₂CH₂Cl + NH₃ (excess) ⟶

b. N-methylcyclohexylamine + CH₃CH₂Br (excess) ⟶

The Gabriel Synthesis of 1° Amines

To avoid polyalkylation, a nitrogen nucleophile can be used that reacts in a single nucleophilic substitution reaction—that is, the reaction forms a product that does *not* contain a nucleophilic nitrogen atom capable of reacting further.

The **Gabriel synthesis** consists of two steps and uses a resonance-stabilized nitrogen nucleophile to synthesize 1° amines via nucleophilic substitution. The Gabriel synthesis begins with **phthalimide,** one of a group of compounds called **imides.** The N–H bond of an imide is **especially acidic** because the resulting anion is resonance stabilized by the two flanking carbonyl groups.

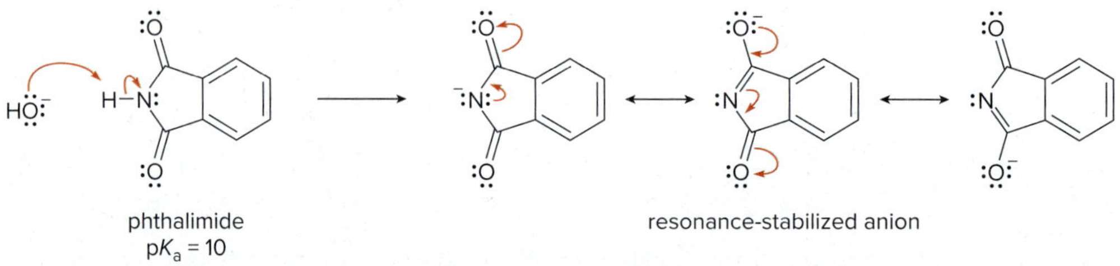

phthalimide
pK_a = 10

resonance-stabilized anion

In the Gabriel synthesis, treatment of phthalimide with ⁻OH forms a nucleophilic anion that can react with an unhindered alkyl halide—that is, CH₃X or RCH₂X—in an **S$_N$2 reaction** to form a substitution product. This alkylated imide is then hydrolyzed with aqueous base to give a 1° amine and a dicarboxylate. This reaction is similar to the hydrolysis of amides to afford carboxylate anions and amines, as discussed in Section 16.12. The overall result of this two-step sequence is **nucleophilic substitution of X by NH₂,** so the Gabriel synthesis can be used to prepare 1° amines only.

Steps in the Gabriel synthesis

R—X + phthalimide anion (nucleophile) —S$_N$2, [1] nucleophilic substitution→ R-N-phthalimide (alkylated imide) + X⁻ —⁻OH, H₂O, [2] hydrolysis→ R—NH₂ (1° amine) + phthalate dicarboxylate by-product

R = CH₃ or 1° alkyl

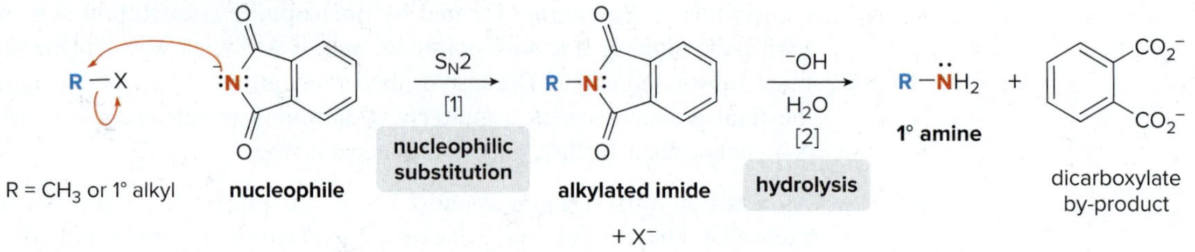

22.6 Preparation of Amines

- The Gabriel synthesis converts an alkyl halide to a 1° amine by a two-step process: nucleophilic substitution followed by hydrolysis.

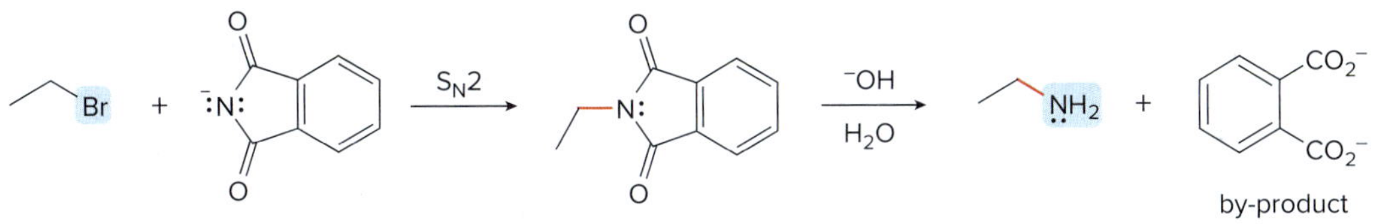

new C–N bond in red

Problem 22.7 What alkyl halide is needed to prepare each 1° amine by the Gabriel synthesis?

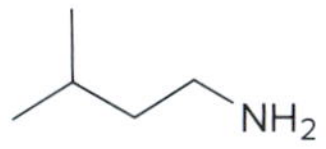

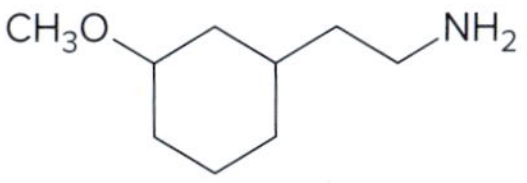

Problem 22.8 Which amines cannot be prepared by the Gabriel synthesis? Explain your choices.

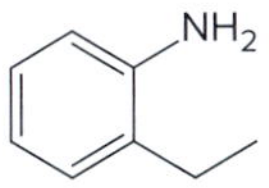

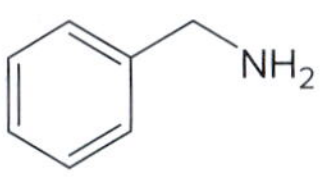

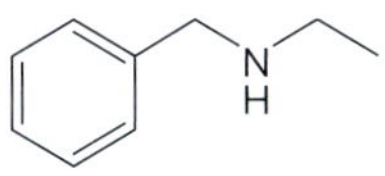

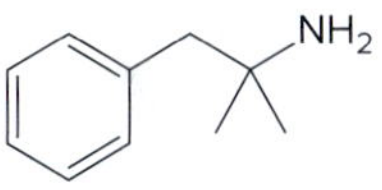

22.6B Reduction of Other Functional Groups That Contain Nitrogen

Amines can be prepared by reduction of nitro compounds, nitriles, and amides. Because the details of these reactions have been discussed previously, they are presented here in summary form only.

[1] From nitro compounds (Section 20.14D)

Nitro groups are reduced to 1° amines using a variety of reducing agents.

$$R-NO_2 \xrightarrow[\text{or Sn, HCl}]{\text{H}_2\text{, Pd-C or Fe, HCl}} R-NH_2 \quad \text{1° amine}$$

[2] From nitriles (Section 15.13B)

Nitriles are reduced to 1° amines with LiAlH$_4$.

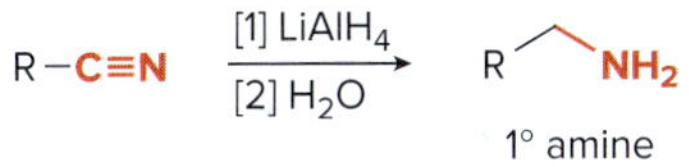

$$R-C\equiv N \xrightarrow[\text{[2] H}_2\text{O}]{\text{[1] LiAlH}_4} R-CH_2-NH_2 \quad \text{1° amine}$$

Because a cyano group is readily introduced by S$_N$2 substitution of alkyl halides with $^-$CN, this provides **a two-step method to convert an alkyl halide to a 1° amine with one more carbon atom.** The conversion of $(CH_3)_2CHCH_2CH_2Br$ to $(CH_3)_2CHCH_2CH_2CH_2NH_2$ illustrates this two-step sequence.

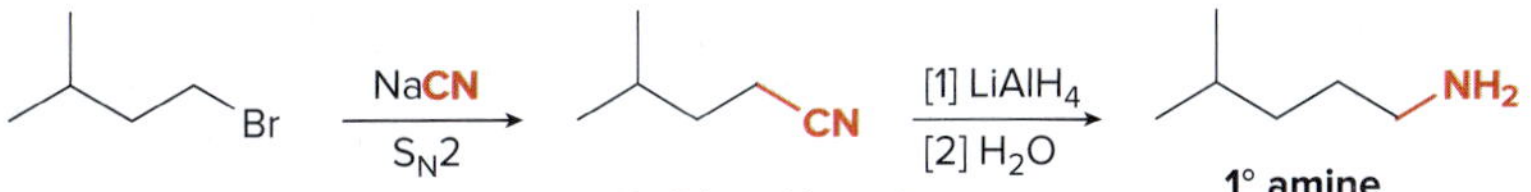

new C–C bond in red

[3] From amides (Section 13.7B)

Primary (1°), 2°, and 3° amides are reduced to 1°, 2°, and 3° amines, respectively, by using LiAlH$_4$.

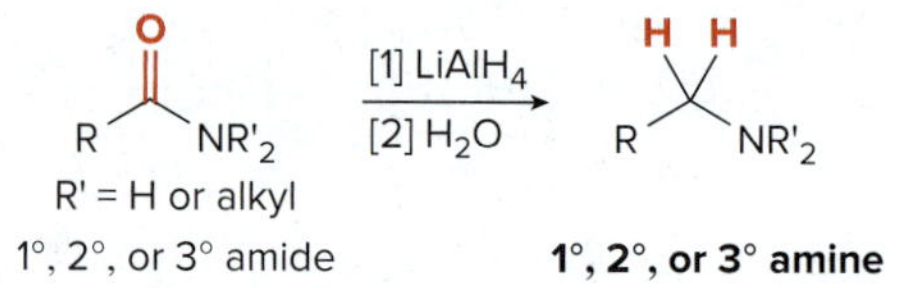

$$R-C(=O)-NR'_2 \xrightarrow[\text{[2] H}_2\text{O}]{\text{[1] LiAlH}_4} R-CH_2-NR'_2$$

R' = H or alkyl
1°, 2°, or 3° amide → 1°, 2°, or 3° amine

Problem 22.9 What nitro compound, nitrile, and amide are reduced to each compound?

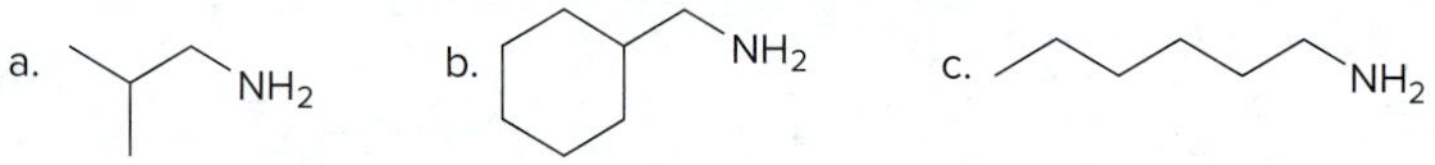

a. (CH$_3$)$_2$CHCH$_2$NH$_2$ b. cyclohexyl-CH$_2$NH$_2$ c. CH$_3$(CH$_2$)$_4$CH$_2$NH$_2$

Problem 22.10 What amine is formed by reduction of each amide?

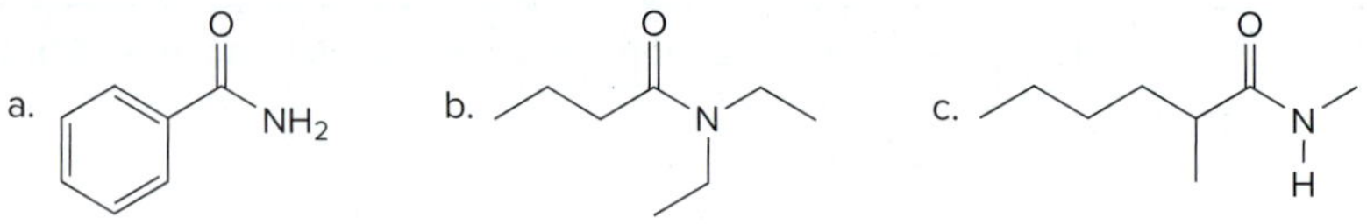

a. PhC(=O)NH$_2$ b. CH$_3$CH$_2$CH$_2$C(=O)N(CH$_2$CH$_3$)$_2$ c. CH$_3$CH$_2$CH$_2$CH$_2$CH(CH$_3$)C(=O)NHCH$_3$

Problem 22.11 Which amines cannot be prepared by reduction of an amide?

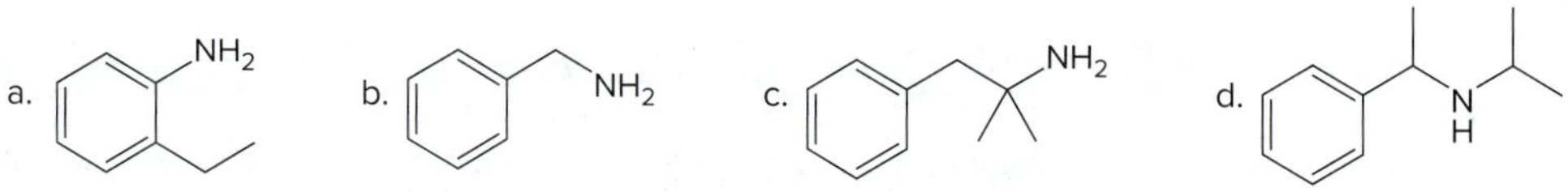

a. 2-ethylaniline   b. benzylamine (PhCH$_2$NH$_2$)   c. PhCH$_2$C(CH$_3$)$_2$NH$_2$ d. PhCH(CH$_3$)NHCH(CH$_3$)$_2$

22.6C Reductive Amination of Aldehydes and Ketones

Reductive amination is a two-step method that converts aldehydes and ketones to 1°, 2°, and 3° amines. Let's first examine this method using NH$_3$ to prepare 1° amines. There are two distinct parts in reductive amination:

[1] **Nucleophilic attack of NH$_3$ on the carbonyl group forms an imine** (Section 14.10), which is not isolated; then,

[2] **Reduction of the imine forms an amine** (Section 13.7B).

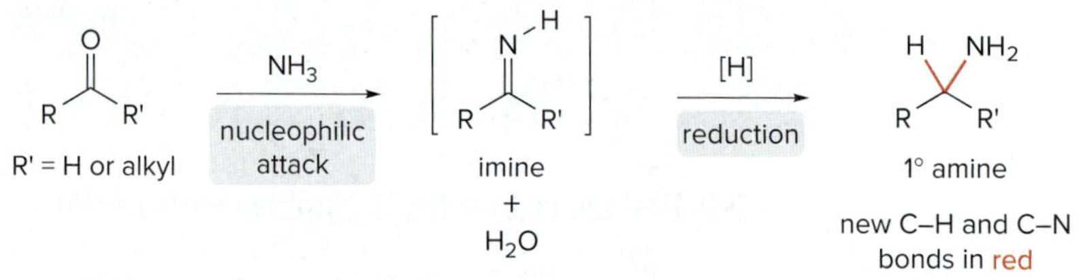

$$R-C(=O)-R' \xrightarrow[\text{nucleophilic attack}]{\text{NH}_3} [R-C(=NH)-R'] \xrightarrow{\text{[H]}} R-CH(NH_2)-R'$$

R' = H or alkyl   imine + H$_2$O   1° amine
                                          new C–H and C–N bonds in red

- **Reductive amination replaces a C=O by a C–H and C–N bond.**

The most effective reducing agent for this reaction is sodium cyanoborohydride (NaBH$_3$CN). This hydride reagent is a derivative of sodium borohydride (NaBH$_4$), formed by replacing one H atom by CN.

NaBH$_3$CN
sodium cyanoborohydride

Reductive amination combines two reactions we have already learned in a different way. Two examples are shown. The second reaction is noteworthy because the product is **amphetamine,** a potent central nervous system stimulant.

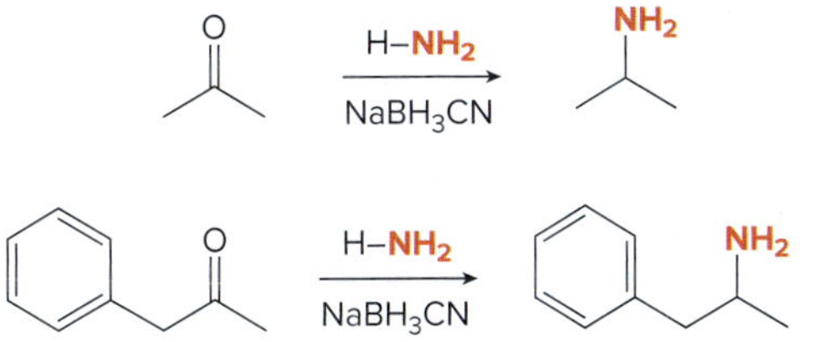

22.6 Preparation of Amines

With a 1° or 2° amine as starting material, reductive amination is used to prepare 2° and 3° amines, respectively. Note the result: **Reductive amination uses an aldehyde or ketone to replace one H atom on a nitrogen atom by an alkyl group,** making a more substituted amine.

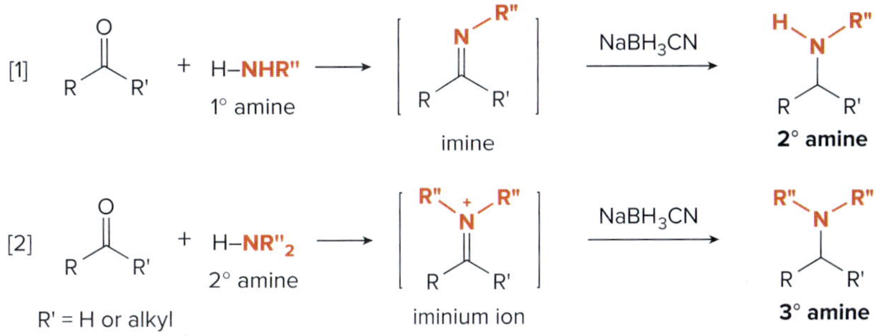

The synthesis of methamphetamine (Section 22.5C) by reductive amination is illustrated in Figure 22.5.

Figure 22.5
Synthesis of methamphetamine by reductive amination

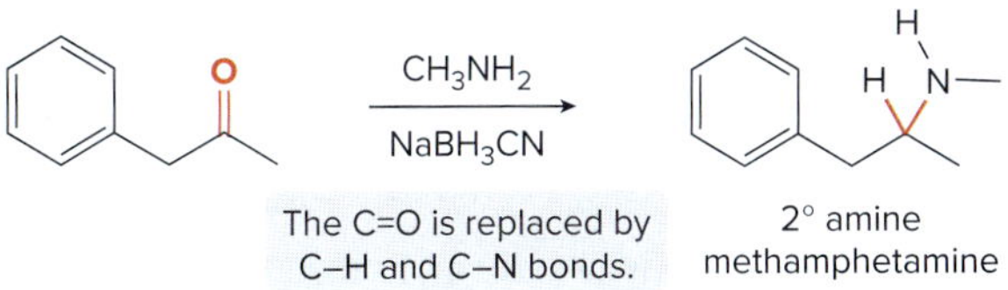

- In reductive amination, one of the H atoms bonded to N is replaced by an alkyl group. As a result, a 1° amine is converted to a 2° amine and a 2° amine is converted to a 3° amine. In this reaction, CH_3NH_2 (a 1° amine) is converted to methamphetamine (a 2° amine).

Problem 22.12 Draw the product of each reaction.

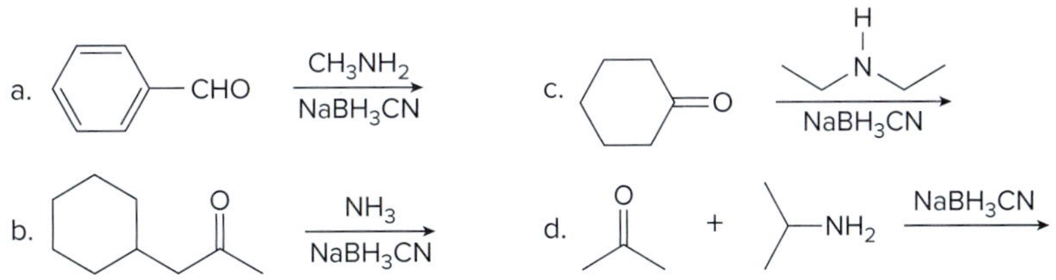

a. PhCH$_2$—CHO + CH_3NH_2 / NaBH$_3$CN

b. cyclohexyl-CH$_2$-C(=O)-CH$_3$ + NH_3 / NaBH$_3$CN

c. cyclohexanone + diethylamine / NaBH$_3$CN

d. acetone + tert-butyl-NH$_2$ / NaBH$_3$CN

Problem 22.13 Maraviroc, a drug used to treat HIV, is prepared by reductive amination of aldehyde **A** with amine **B**. What is the structure of maraviroc, if the most basic N atom of amine **B** is used in reductive amination?

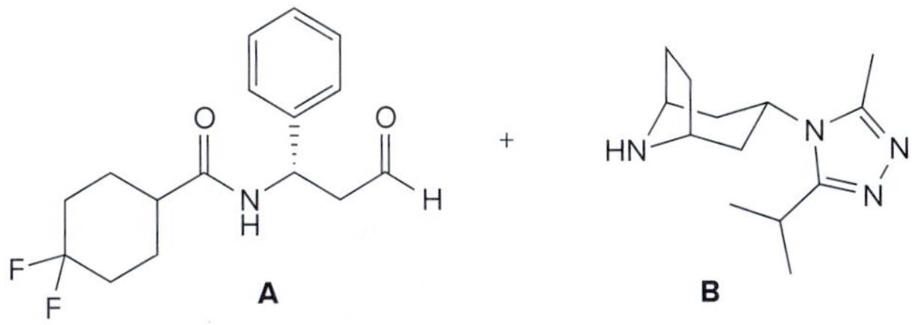

984   Chapter 22   Amines

To use reductive amination in synthesis, you must be able to determine what aldehyde or ketone and nitrogen compound are needed to prepare a given amine—that is, you must work backwards in the retrosynthetic direction. Keep in mind these two points:

- One alkyl group on N comes from the carbonyl compound.
- The remainder of the molecule comes from NH₃ or an amine.

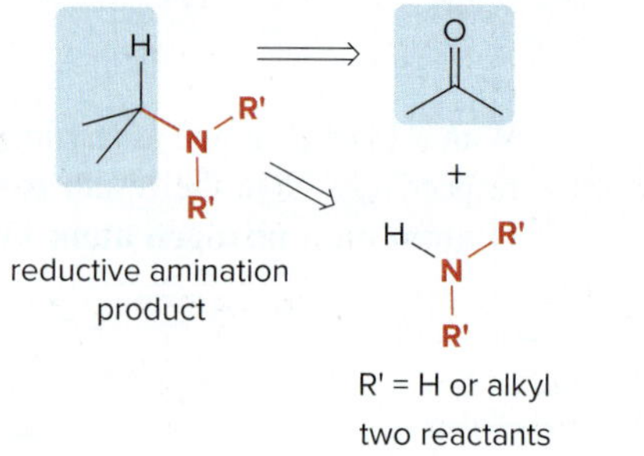

For example, 2-phenylethanamine is a 1° amine, so it has only one alkyl group bonded to N. This alkyl group must come from the carbonyl compound, and the rest of the molecule then comes from the nitrogen component. **For a 1° amine, the nitrogen component must be NH₃.**

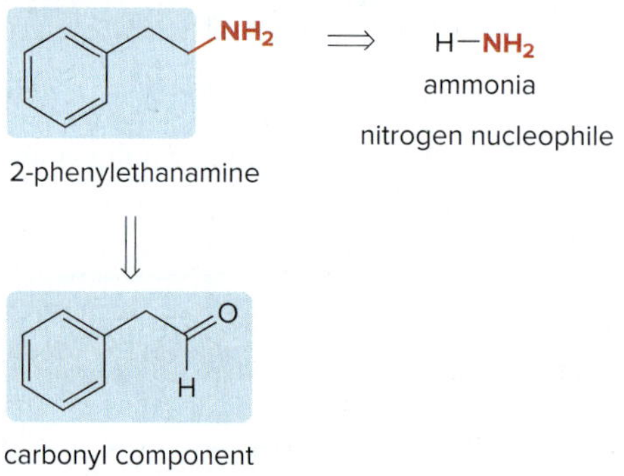

There is usually more than one way to use reductive amination to synthesize 2° and 3° amines, as shown in Sample Problem 22.2 for a 2° amine.

Sample Problem 22.2   Determining the Starting Materials in a Reductive Amination

What aldehyde or ketone and nitrogen component are needed to synthesize *N*-ethylcyclohexanamine by a reductive amination reaction?

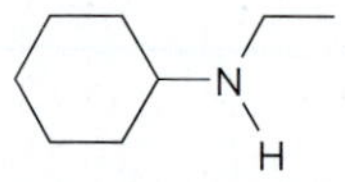

N-ethylcyclohexanamine

Solution

Because *N*-ethylcyclohexanamine has two different alkyl groups bonded to the N atom, either R group can come from the carbonyl component and there are two different ways to form a C–N bond by reductive amination.

Possibility [1]   Use CH₃CH₂NH₂ and cyclohexanone.

Possibility [2]   Use cyclohexanamine and an aldehyde.

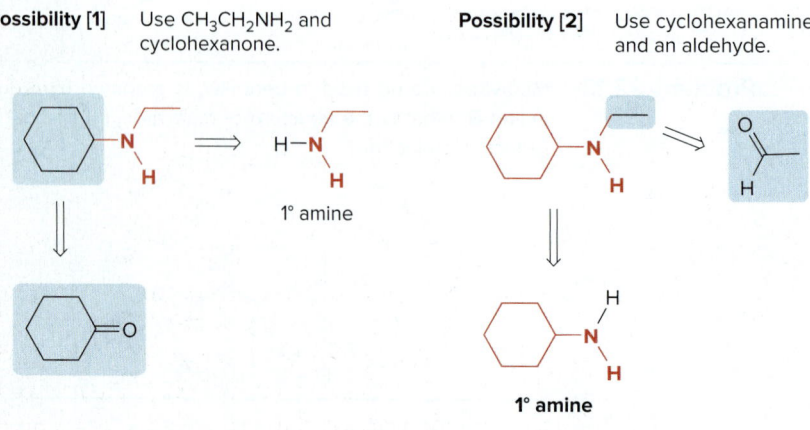

Because **reductive amination adds one R group to a nitrogen atom,** both routes to form the 2° amine begin with a 1° amine.

Problem 22.14 What starting materials are needed to prepare each drug using reductive amination? Give all possible pairs of compounds when more than one route is possible.

a. rimantadine
antiviral used to treat influenza

b. pseudoephedrine
nasal decongestant

More Practice: Try Problems 22.45, 22.51b, 22.52b.

Problem 22.15 (a) Explain why phentermine [PhCH$_2$C(CH$_3$)$_2$NH$_2$] can't be made by a reductive amination reaction.
(b) Give a systematic name for phentermine, one of the components of the banned diet drug fen–phen.

22.7 Reactions of Amines—General Features

- **The chemistry of amines is dominated by the lone pair of electrons on nitrogen.**

Only three elements in the second row of the periodic table have nonbonded electron pairs in neutral organic compounds: nitrogen, oxygen, and fluorine. Because basicity and nucleophilicity decrease across the row, **nitrogen is the most basic and most nucleophilic** of these elements.

Increasing basicity and nucleophilicity

- **Amines are stronger bases and nucleophiles than other neutral organic compounds.**

- **Amines react as *bases* with compounds that contain acidic protons.**
- **Amines react as *nucleophiles* with compounds that contain electrophilic carbons.**

22.8 Amines as Bases

Amines react as bases with a variety of organic and inorganic acids.

$$R\text{-}NH_2 + H\text{-}A \rightleftharpoons R\text{-}\overset{+}{N}H_3 + :A^-$$
base   acid   conjugate acid
                                  p$K_a \approx$ 10–11

To favor the products, the
pK_a of HA must be < 10.

What acids can be used to protonate an amine? Equilibrium favors the products of an acid–base reaction when the weaker acid and base are formed. Because the pK_a of many protonated amines is 10–11, the **pK_a of the starting acid must be less than 10** for equilibrium to favor the products. Amines are thus readily protonated by strong inorganic acids like HCl and H$_2$SO$_4$, and by carboxylic acids as well.

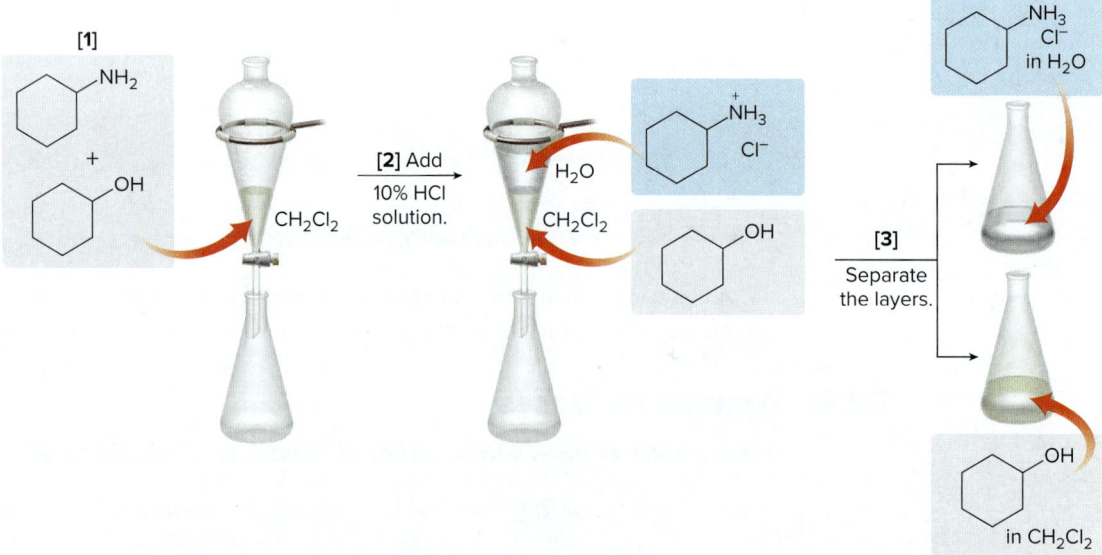

Because amines are protonated by aqueous acid, they can be separated from other organic compounds by extraction using a separatory funnel. **Extraction separates compounds based on solubility differences.** When an amine is protonated by aqueous acid, its solubility properties change.

The principles used in an extraction procedure were detailed in Section 15.10.

For example, when cyclohexanamine is treated with aqueous HCl, it is protonated, forming an ammonium salt. **Because the ammonium salt is ionic, it is soluble in water,** but insoluble in organic solvents. A similar acid–base reaction does not occur with other organic compounds like alcohols, which are much less basic.

cyclohexanamine
- insoluble in H_2O
- soluble in CH_2Cl_2

cyclohexanammonium chloride
- soluble in H_2O
- insoluble in CH_2Cl_2

This difference in acid–base chemistry can be used to separate cyclohexanamine and cyclohexanol by the stepwise extraction procedure illustrated in Figure 22.6.

Figure 22.6 Separation of cyclohexanamine and cyclohexanol by an extraction procedure

Step [1] Dissolve cyclohexanamine and cyclohexanol in CH_2Cl_2.

Step [2] Add 10% HCl solution to form two layers.

Step [3] Separate the layers.

- Both compounds dissolve in the organic solvent CH_2Cl_2.
- Adding 10% aqueous HCl solution forms two layers. When the two layers are mixed, the **HCl protonates the amine (RNH_2) to form $RNH_3^+Cl^-$,** which dissolves in the aqueous layer.
- The cyclohexanol remains in the CH_2Cl_2 layer.
- Draining the lower layer out the bottom stopcock separates the two layers.
- Cyclohexanol (dissolved in CH_2Cl_2) is in one flask. The ammonium salt, $RNH_3^+Cl^-$ (dissolved in water), is in another flask.

- An amine can be separated from other organic compounds by converting it to a water-soluble ammonium salt by an acid–base reaction.

Thus, the water-soluble salt $C_6H_{11}NH_3^+Cl^-$ (obtained by protonation of $C_6H_{11}NH_2$) can be separated from water-insoluble cyclohexanol by an aqueous extraction procedure.

Problem 22.16 Draw the products of each acid–base reaction. Indicate whether equilibrium favors the reactants or products.

a. CH₃CH₂CH₂CH₂—NH₂ + HCl ⇌

b. PhC(O)OH + (CH₃)₂NH ⇌

c. piperidine + H₂O ⇌

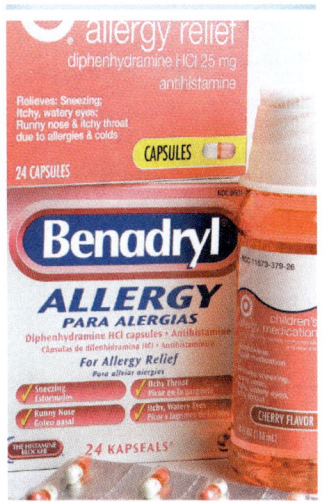

Many antihistamines and decongestants are sold as their ammonium salts.

Jill Braaten/McGraw-Hill Education

Many water-insoluble amines with useful medicinal properties are sold as their water-soluble ammonium salts, which are more easily transported through the body in the aqueous medium of the blood. Benadryl, formed by treating diphenhydramine with HCl, is an over-the-counter antihistamine that is used to relieve the itch and irritation of skin rashes and hives.

diphenhydramine + H—Cl: ⟶ ammonium salt
water insoluble   Benadryl
                         (diphenhydramine hydrochloride)
                         water soluble

Problem 22.17 Write out steps to show how each of the following pairs of compounds can be separated by an extraction procedure.

a. cyclohexyl—NH₂ and methylcyclohexane

b. (CH₃CH₂CH₂CH₂)₂N—CH₂CH₂CH₃ and CH₃CH₂CH₂—O—CH₂CH₂CH₂CH₃

22.9 Relative Basicity of Amines and Other Compounds

The relative acidity of different compounds can be compared using their pK_a values. **The relative *basicity* of different compounds (such as amines) can be compared using the pK_a values of their *conjugate acids*.**

- The *weaker* the conjugate acid, the *higher* its pK_a and the *stronger* the base.

988 Chapter 22 Amines

To compare the basicity of two compounds, keep in mind the following:

- Any factor that *increases* the electron density on the N atom *increases* an amine's basicity.
- Any factor that *decreases* the electron density on N *decreases* an amine's basicity.

22.9A Comparing an Amine and NH₃

Because **alkyl groups are electron donating,** they increase the electron density on nitrogen, which makes an amine like $CH_3CH_2NH_2$ more basic than NH_3. In fact, the pK_a of $CH_3CH_2NH_3^+$ is *higher* than the pK_a of NH_4^+, so **$CH_3CH_2NH_2$ is a *stronger* base than NH_3.**

$pK_a = 9.3$ $H-\overset{+}{N}H_3$ $\overset{..}{N}H_3$

lower pK_a weaker base
stronger acid

$pK_a = 10.8$ $CH_3CH_2\overset{+}{N}H_3$ $CH_3CH_2\overset{..}{N}H_2$

higher pK_a stronger base
weaker acid

> One electron-donor group makes the amine more basic.

The relative basicity of 1°, 2°, and 3° amines depends on additional factors, and will not be considered in this text.

- Primary (1°), 2°, and 3° alkylamines are *more basic* than NH_3 because of the electron-donating inductive effect of the R groups.

Problem 22.18 Which compound in each pair is more basic: (a) $(CH_3)_2NH$ or NH_3; (b) $CH_3CH_2NH_2$ or $ClCH_2CH_2NH_2$?

22.9B Comparing an Alkylamine and an Arylamine

To compare an alkylamine ($CH_3CH_2NH_2$) and an arylamine ($C_6H_5NH_2$, aniline), we must look at the **availability of the nonbonded electron pair on N.** With $CH_3CH_2NH_2$, the electron pair is localized on the N atom. With an arylamine, however, the electron pair is now delocalized on the benzene ring. This *decreases* the electron density on N and makes $C_6H_5NH_2$ less basic than $CH_3CH_2NH_2$.

$CH_3CH_2\overset{..}{N}H_2$

The electron pair is **localized** on N.

stronger base

The electron pair is **delocalized** on the benzene ring.

weaker base

The pK_a values support this reasoning. Because the pK_a of $CH_3CH_2NH_3^+$ is *higher* than the pK_a of $C_6H_5NH_3^+$ (10.8 vs. 4.6), **$CH_3CH_2NH_2$ is a *stronger* base than $C_6H_5NH_2$.**

- Arylamines are *less basic* than alkylamines because the electron pair on N is delocalized.

Substituted anilines are more or less basic than aniline depending on the nature of the substituent.

- Electron-donor groups *add* electron density to the benzene ring, making the arylamine *more basic* than aniline.

D = electron-donor group

D makes the amine *more basic* than aniline.

D
—NH₂
—OH
—OR
—NHCOR
—R

22.9 Relative Basicity of Amines and Other Compounds

- Electron-withdrawing groups *remove* electron density from the benzene ring, making the arylamine *less basic* than aniline.

W = electron-withdrawing group

W makes the amine **less basic** than aniline.

W	
—X	—CN
—CHO	—SO$_3$H
—COR	—NO$_2$
—COOR	—NR$_3^+$
—COOH	

Whether a substituent donates or withdraws electron density depends on the balance of its inductive and resonance effects (Section 20.6 and Figure 20.5).

Sample Problem 22.3 — Determining the Relative Basicity of Anilines

Rank the following compounds in order of increasing basicity.

aniline   p-nitroaniline   p-methylaniline (p-toluidine)

Solution

p-Nitroaniline: NO$_2$ is an electron-withdrawing group, making the amine *less basic* than aniline.

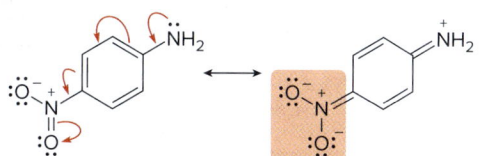

The lone pair on N is **delocalized** on the O atom, *decreasing* the basicity of the amine.

p-Methylaniline: CH$_3$ has an electron-donating inductive effect, making the amine *more basic* than aniline.

CH$_3$ inductively **donates** electron density, *increasing* the basicity of the amine.

p-nitroaniline   aniline   p-methylaniline (p-toluidine)

Increasing basicity →

Problem 22.19
Rank the compounds in each group in order of increasing basicity.

a. aniline, p-methoxyaniline (CH$_3$O—), methyl p-aminobenzoate (CH$_3$O—C(=O)—)

b. cyclohexylamine, aniline, p-nitroaniline

More Practice: Try Problems 22.39b, 22.43.

22.9C Comparing an Alkylamine and an Amide

To compare the basicity of an alkylamine (RNH_2) and an amide ($RCONH_2$), we must once again compare the availability of the nonbonded electron pair on nitrogen. With RNH_2, the electron pair is localized on the N atom. With an amide, however, the electron pair is *delocalized* on the carbonyl oxygen by resonance. This *decreases* the electron density on N, making **an amide much *less basic* than an alkylamine.**

The electron pair on N is **delocalized** on O by resonance.

- **Amides are much less basic than amines because the electron pair on N is delocalized.**

Amides are not much more basic than any carbonyl compound. When an amide is treated with acid, **protonation occurs at the carbonyl oxygen,** *not* **the nitrogen,** because the resulting cation is resonance stabilized.

three resonance structures for the conjugate acid

Problem 22.20 Rank the following compounds in order of increasing basicity.

22.9D Heterocyclic Aromatic Amines

To determine the relative basicity of nitrogen heterocycles that are also aromatic, you must know **whether the nitrogen lone pair is part of the aromatic π system.**

For example, pyridine and pyrrole are both aromatic, but the nonbonded electron pair on the N atom in these compounds is located in different orbitals. Recall from Section 19.9 that the **lone pair of electrons in pyridine occupies an sp^2 hybridized orbital,** perpendicular to the π bonds of the molecule, so it is *not* part of the aromatic system, whereas that of pyrrole resides in a *p* orbital, making it part of the aromatic system. **The lone pair on pyrrole, therefore, is delocalized on all of the atoms of the five-membered ring,** making pyrrole a much *weaker* base than pyridine.

pyridine

pyrrole

The lone pair resides in an sp^2 hybrid orbital.

The lone pair resides in a *p* orbital and is *delocalized* in the ring.

22.9 Relative Basicity of Amines and Other Compounds

> Protonation of pyrrole occurs at a ring *carbon,* not the N atom, as noted in Problem 19.44.

As a result, the pK_a of the conjugate acid of pyrrole is much less than that of the conjugate acid of pyridine.

[pyridinium ion]   pyridine   [pyrrolium ion]   pyrrole

higher pK_a weaker acid   **stronger base**   lower pK_a stronger acid   **weaker base**

pK_a = 5.3   pK_a = 0.4

- Pyrrole is much *less basic* than pyridine because its lone pair of electrons is part of the aromatic π system.

22.9E Hybridization Effects

> The effect of hybridization on the acidity of an H–A bond was first discussed in Section 2.5D.

The hybridization of the orbital that contains an amine's lone pair also affects its basicity. This is illustrated by comparing the basicity of **piperidine** and **pyridine,** two nitrogen heterocycles. The lone pair in piperidine resides in an sp^3 hybrid orbital that has 25% *s*-character. The lone pair in pyridine resides in an sp^2 hybrid orbital that has 33% *s*-character.

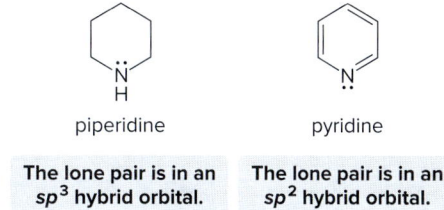

piperidine   pyridine

The lone pair is in an sp^3 hybrid orbital.   The lone pair is in an sp^2 hybrid orbital.

- The *higher* the percent *s*-character of the orbital containing the lone pair, the more tightly the lone pair is held and the *weaker* the base.

Pyridine is a weaker base than piperidine because its nonbonded pair of electrons resides in an sp^2 hybrid orbital. Although pyridine is an aromatic amine, its lone pair is *not* part of the delocalized π system, so its **basicity is determined by the hybridization of its N atom.** As a result, the pK_a value of the conjugate acid of pyridine is much *lower* than that of the conjugate acid of piperidine, making pyridine the *weaker* base.

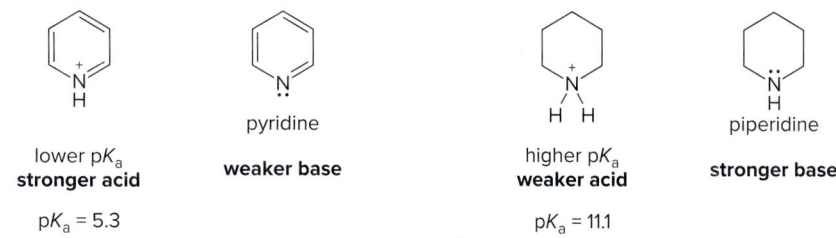

lower pK_a stronger acid   pyridine   **weaker base**   higher pK_a weaker acid   piperidine   **stronger base**

pK_a = 5.3   pK_a = 11.1

Problem 22.21   Rank the following ammonium ions in order of increasing pK_a.

A: 4-methylphenyl-NH_3^+ B: 4-nitrophenyl-NH_3^+ C: 4-methylbenzyl-NH_3^+ D: 4-chlorophenyl-NH_3^+

22.9F Summary of the Factors That Determine Amine Basicity

Acid–base chemistry is central to many processes in organic chemistry, so it has been a constant theme throughout this text. Tables 22.3 and 22.4 organize and summarize the acid–base principles discussed in Section 22.9. The principles in these tables can be used to determine the most basic site in a molecule that has more than one nitrogen atom, as shown in Sample Problem 22.4.

992 Chapter 22 Amines

Table 22.3 Factors That Determine Amine Basicity

Factor		Example
[1]	**Inductive effects:** Electron-donating groups bonded to N *increase* basicity.	• RNH_2, R_2NH, and R_3N are more basic than NH_3.
[2]	**Resonance effects:** Delocalizing the lone pair on N *decreases* basicity.	• Arylamines ($C_6H_5NH_2$) are less basic than alkylamines (RNH_2). • Amides ($RCONH_2$) are much less basic than amines (RNH_2).
[3]	**Aromaticity:** Having the lone pair on N as part of the aromatic π system *decreases* basicity.	• Pyrrole is less basic than pyridine. pyrrole (less basic) pyridine (more basic)
[4]	**Hybridization effects:** Increasing the percent *s*-character in the orbital with the lone pair *decreases* basicity.	• Pyridine is less basic than piperidine. pyridine (less basic) piperidine (more basic)

Table 22.4 Table of pK_a Values of Some Representative Organic Nitrogen Compounds

	Compound	pK_a of the conjugate acid
Ammonia	NH_3	9.3
Alkylamines[a]	piperidine (NH)	11.1
	$(CH_3CH_2)_2NH$	11.1
	$(CH_3CH_2)_3N$	11.0
	$CH_3CH_2NH_2$	10.8
Arylamines[b]	$p\text{-}CH_3OC_6H_4NH_2$	5.3
	$p\text{-}CH_3C_6H_4NH_2$	5.1
	$C_6H_5NH_2$	4.6
	$p\text{-}NO_2C_6H_4NH_2$	1.0
Heterocyclic aromatic amines[c]	pyridine	5.3
	pyrrole (NH)	0.4
Amides	$RCONH_2$	−1

[a] Alkylamines have pK_a values of ~10–11.
[b] The pK_a *decreases* as the electron density of the benzene ring *decreases*.
[c] The pK_a depends on whether the lone pair of N is *localized* or *delocalized*.

Sample Problem 22.4 Determining Which Nitrogen Atom Is the Strongest Base

Which N atom in chloroquine is the strongest base?

chloroquine

Since 1945 chloroquine has been used to treat malaria, an infectious disease caused by a protozoan parasite that is spread by the *Anopheles* mosquito. *Source: James Gathany/CDC*

Solution

Examine the nitrogen atoms in chloroquine, labeled in red, blue, and green, and recall that decreasing the electron density on N decreases basicity.

strongest base → green N

- **N** (red) is bonded to an aromatic ring, so its lone pair is delocalized in the ring like aniline, *decreasing* basicity.
- The lone pair is localized on **N** (blue), but **N** is sp^2 hybridized. Increasing percent *s*-character *decreases* basicity.
- **N** (green) has a localized lone pair and is sp^3 hybridized, making it the *most basic* site in the molecule.

Problem 22.22 Which N atom in each compound is more basic? What product is formed when each compound is treated with HCl? Like sparteine (Figure 22.3), matrine is an alkaloid isolated from lupin. Quinine, the Chapter 8 opening molecule, is an antimalarial drug obtained from the bark of the cinchona tree.

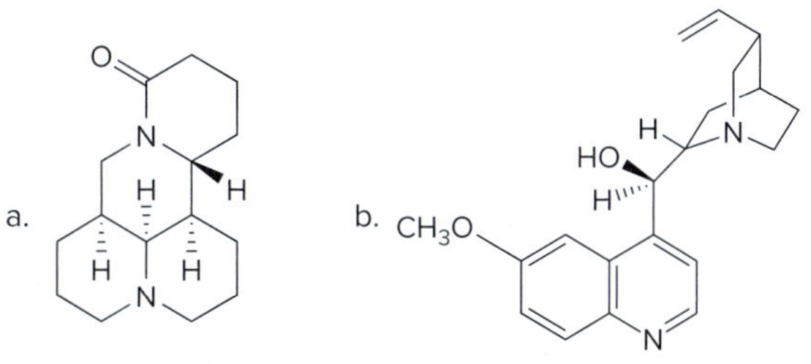

a. matrine
alkaloid in lupin

b. quinine
antimalarial drug

More Practice: Try Problems 22.36a, 22.40–22.42.

22.10 Amines as Nucleophiles

Amines react as nucleophiles with electrophilic carbon atoms. The details of these reactions have been described in Chapters 14 and 16, so they are summarized here only to emphasize the similar role that the amine nitrogen plays.

- Amines attack carbonyl groups to form products of nucleophilic addition or substitution.

The nature of the product depends on the carbonyl electrophile. These reactions are limited to 1° and 2° amines, because only these compounds yield neutral organic products.

[1] Reaction of 1° and 2° amines with aldehydes and ketones (Sections 14.10–14.11)

Aldehydes and ketones react with 1° amines to form imines and with 2° amines to form enamines. Both reactions involve **nucleophilic addition** of the amine to the carbonyl group to form a carbinolamine, which then loses water to form the final product.

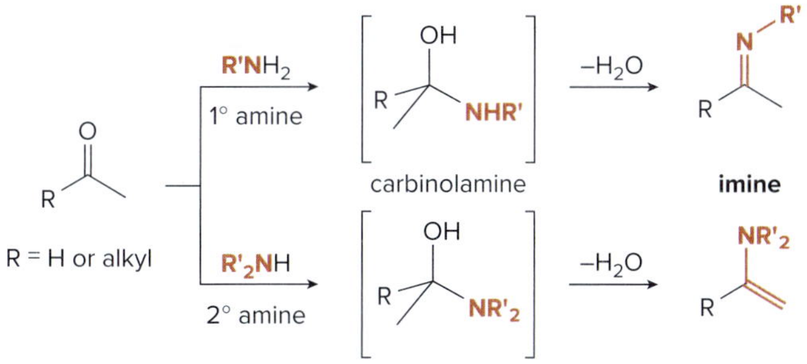

[2] Reaction of NH_3 and 1° and 2° amines with acid chlorides and anhydrides (Sections 16.7–16.8)

Acid chlorides and anhydrides react with NH_3, 1° amines, and 2° amines to form 1°, 2°, and 3° amides, respectively. These reactions involve attack of the nitrogen nucleophile on the carbonyl group followed by elimination of a leaving group (Cl^- or RCO_2^-). The overall result of this reaction is **substitution** of the leaving group by the nitrogen nucleophile.

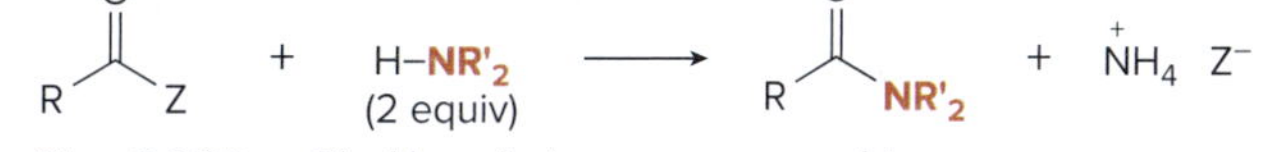

Problem 22.23 Draw the products formed when each carbonyl compound reacts with the following amines:
[1] $CH_3CH_2CH_2NH_2$; [2] $(CH_3CH_2)_2NH$.

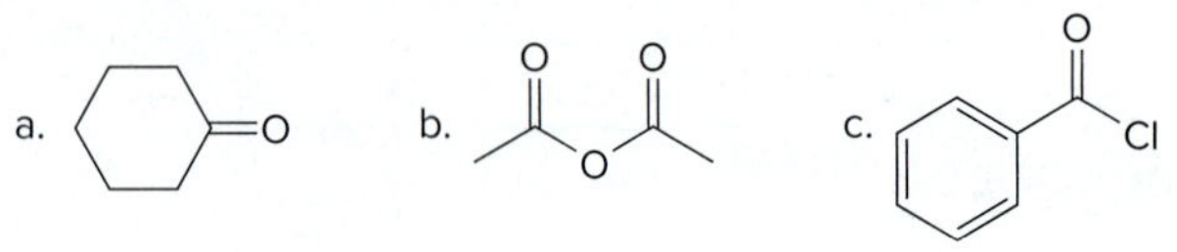

a. cyclohexanone b. acetic anhydride c. benzoyl chloride

The conversion of amines to amides is useful in the synthesis of substituted anilines. For example, aniline itself does not undergo Friedel–Crafts reactions (Section 20.10B). Instead, its basic lone pair on N reacts with the Lewis acid ($AlCl_3$) to form a deactivated complex that does not undergo further reaction.

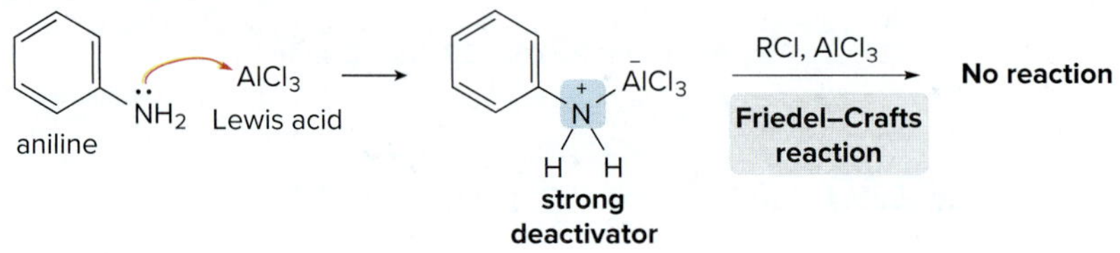

aniline + $AlCl_3$ (Lewis acid) → strong deactivator → RCl, $AlCl_3$, Friedel–Crafts reaction → No reaction

The N atom of an amide, however, is much less basic than the N atom of an amine, so it does not undergo a similar Lewis acid–base reaction with $AlCl_3$. A three-step reaction sequence involving an intermediate amide can thus be used to form the products of the Friedel–Crafts reaction.

[1] **Convert the amine (aniline) into an amide (acetanilide).**
[2] **Carry out the Friedel–Crafts reaction.**
[3] **Hydrolyze the amide** to generate the free amino group.

This three-step procedure is illustrated in Figure 22.7. In this way, **the amide serves as a protecting group for the NH_2 group,** in much the same way that *tert*-butyldimethylsilyl ethers and acetals are used to protect alcohols and carbonyls, respectively (Sections 13.12 and 14.16).

Figure 22.7 An amide as a protecting group for an amine

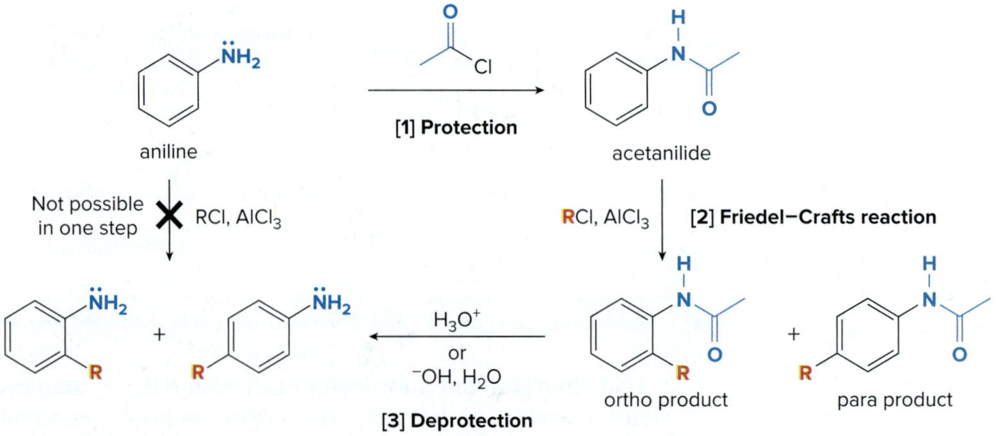

aniline → [1] Protection (CH₃COCl) → acetanilide → [2] Friedel–Crafts reaction (RCl, $AlCl_3$) → ortho product + para product → [3] Deprotection (H_3O^+ or $^-OH, H_2O$) → ortho/para substituted anilines

Not possible in one step: RCl, $AlCl_3$ ✗

A three-step sequence uses an amide as a protecting group.
[1] Treatment of aniline with acetyl chloride (CH_3COCl) forms an **amide** (acetanilide).
[2] Acetanilide, having a much less basic N atom compared to aniline, undergoes **electrophilic aromatic substitution** under Friedel–Crafts conditions, forming a mixture of ortho and para products.
[3] **Hydrolysis of the amide** forms the Friedel–Crafts substitution products.

Problem 22.24 Devise a synthesis of each compound from aniline ($C_6H_5NH_2$).

a. 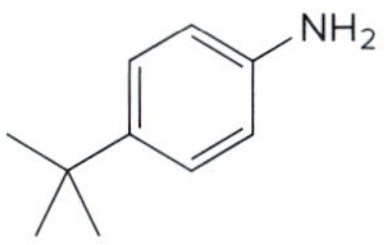 4-tert-butylaniline   b. 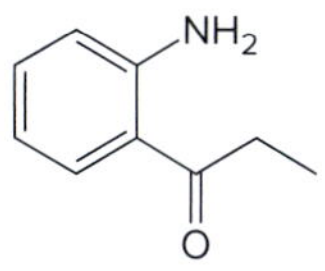 2-aminopropiophenone

22.11 Hofmann Elimination

Amines, like alcohols, contain a poor leaving group. To undergo a β elimination reaction, for example, a 1° amine would need to lose the elements of NH_3 across two adjacent atoms. The leaving group, $^-NH_2$, is such a strong base, however, that this reaction does *not* occur.

The only way around this obstacle is to **convert $^-NH_2$ to a better leaving group.** The most common method to accomplish this is called a **Hofmann elimination,** which converts an amine to a quaternary ammonium salt prior to β elimination.

22.11A Details of the Hofmann Elimination

The **Hofmann elimination** converts an amine to an alkene.

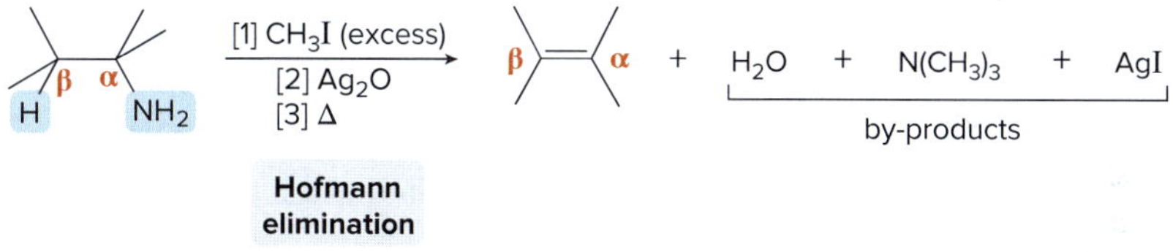

[1] CH_3I (excess)
[2] Ag_2O
[3] Δ

alkene + H_2O + $N(CH_3)_3$ + AgI (by-products)

Hofmann elimination

The Hofmann elimination consists of three steps, as shown for the conversion of propan-1-amine to propene.

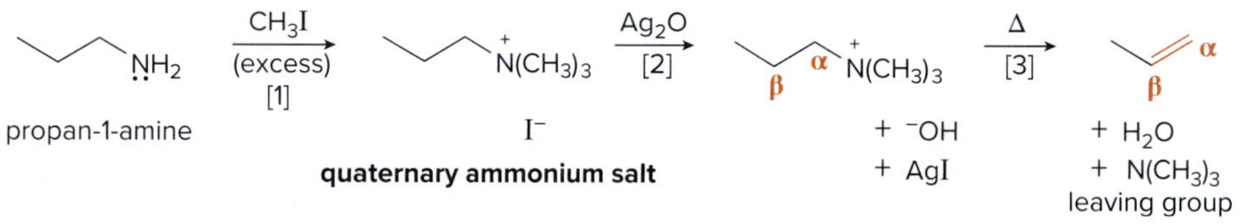

propan-1-amine →[CH_3I (excess) [1]] quaternary ammonium salt (I^-) →[Ag_2O [2]] (^-OH + AgI) →[Δ [3]] propene + H_2O + $N(CH_3)_3$ leaving group

- In Step [1], the amine reacts as a nucleophile in an S_N2 reaction with excess CH_3I to form a quaternary ammonium salt. **The $N(CH_3)_3$ group thus formed is a much better leaving group than $^-NH_2$.**
- Step [2] converts one ammonium salt to another one with a different anion. The silver(I) oxide, Ag_2O, replaces the I^- anion with ^-OH, a strong base.
- When the ammonium salt is heated in Step [3], ^-OH **removes a proton from the β carbon atom,** forming the new π bond of the alkene. The mechanism of elimination is **E2,** so

  - All bonds are broken and formed in a single step.
  - Elimination occurs through an anti periplanar geometry—that is, H and $N(CH_3)_3$ are oriented on opposite sides of the molecule.

The general E2 mechanism for the Hofmann elimination is shown in Mechanism 22.1.

Mechanism 22.1 The E2 Mechanism for the Hofmann Elimination

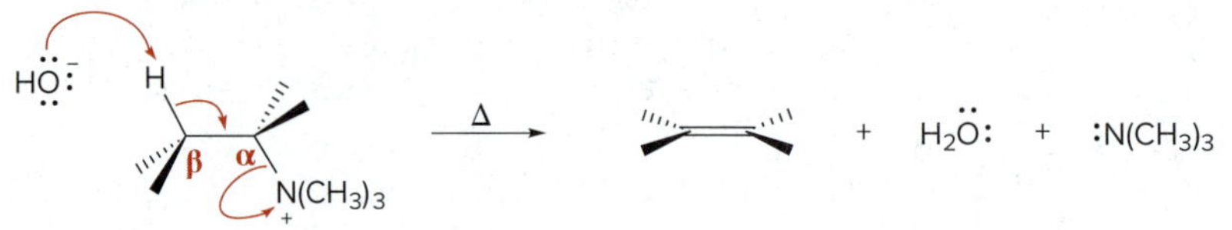

Elimination occurs with an anti periplanar arrangement of H and N(CH$_3$)$_3$. **Base removes a proton on the β carbon,** the electron pair in the C–H bond forms the π bond, and **N(CH$_3$)$_3$ comes off as the leaving group.**

All Hofmann elimination reactions result in the formation of a new π bond between the α and β carbon atoms, as shown for cyclohexanamine and 2-phenylethanamine.

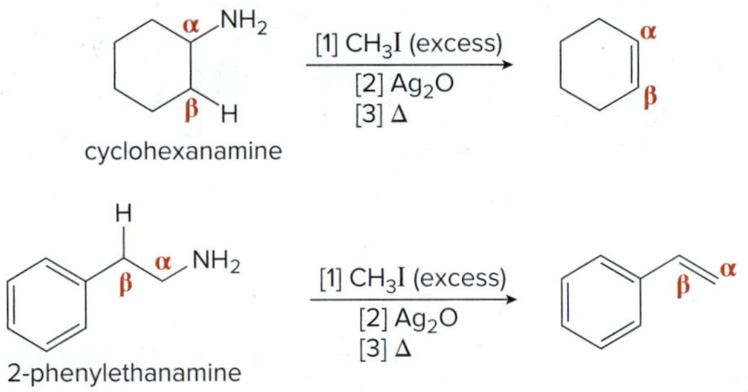

To help remember the reagents needed for the steps of the Hofmann elimination, keep in mind what happens in each step.

- **Step [1]** makes a **good leaving group** by forming a quaternary ammonium salt.
- **Step [2]** provides the **strong base,** ⁻OH, needed for elimination.
- **Step [3]** is the **E2 elimination** that forms the new π bond.

Problem 22.25 Draw the product formed by treating each compound with excess CH$_3$I, followed by Ag$_2$O, and then heat.

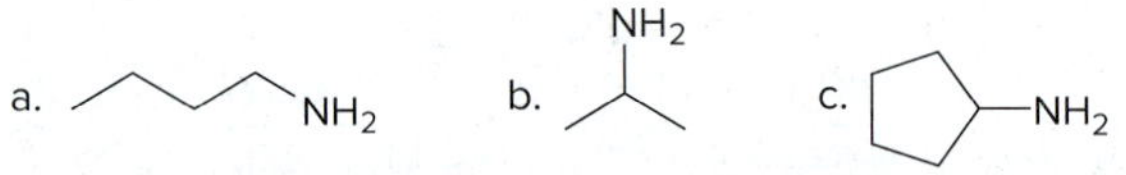

a. (propyl-1,3-diamine structure)   b. (2-amino-2-methylpropane type)   c. (cyclopentylmethanamine)

22.11B Regioselectivity of the Hofmann Elimination

There is one major difference between a Hofmann elimination and other E2 eliminations.

- When constitutional isomers are possible, the major alkene has the *less* substituted double bond in a Hofmann elimination.

For example, Hofmann elimination of the elements of H and N(CH$_3$)$_3$ from 2-methylcyclopentanamine, which has two different β carbons (labeled β$_1$ and β$_2$), yields two constitutional isomers: the disubstituted alkene **A** (the major product) and the trisubstituted alkene **B** (the minor product).

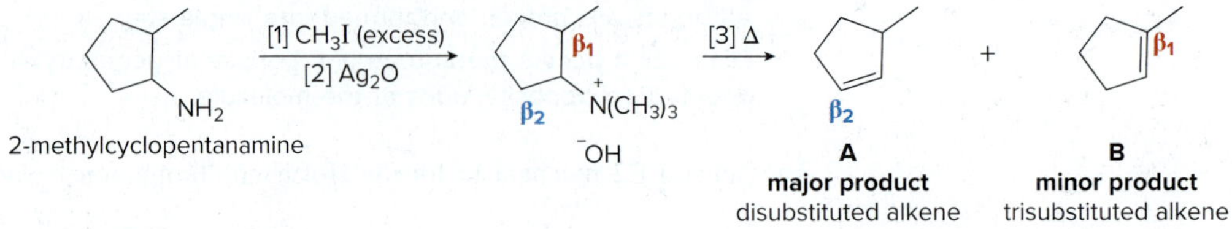

2-methylcyclopentanamine

A major product — disubstituted alkene

B minor product — trisubstituted alkene

This regioselectivity distinguishes a Hofmann elimination from other E2 eliminations, which form the *more* substituted double bond by the Zaitsev rule (Section 8.5). This result is sometimes explained by the size of the leaving group, N(CH$_3$)$_3$. **In a Hofmann elimination, the base removes a proton from the *less* substituted, more accessible β carbon atom, because of the bulky leaving group on the nearby α carbon.**

Sample Problem 22.5 — Drawing the Major Product of a Hofmann Elimination

Draw the major product formed from Hofmann elimination of the following amine.

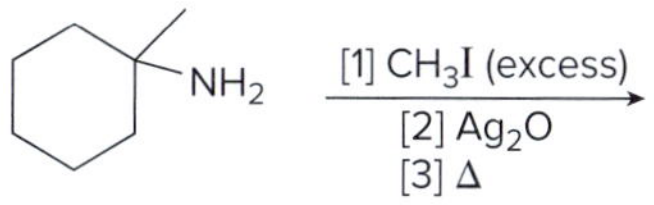

Solution

The amine has three β carbons but two of them are identical, so two alkenes are possible. **Draw elimination products by forming alkenes having a C=C between the α and β carbons.** The major product has the **less substituted double bond**—that is, the alkene with the C=C between the α and β$_1$ carbons in this example.

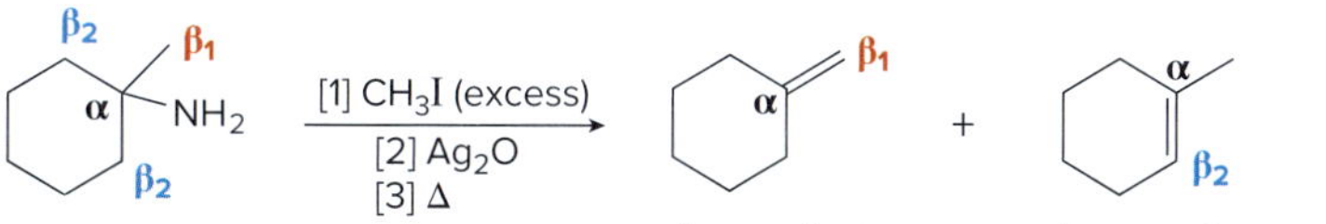

major product — disubstituted alkene
minor product — trisubstituted alkene

Problem 22.26 Draw the major product formed by treating each amine with excess CH$_3$I, followed by Ag$_2$O, and then heat.

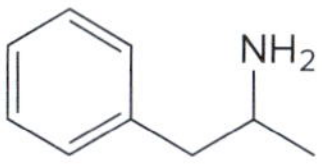

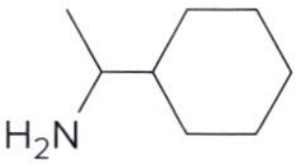

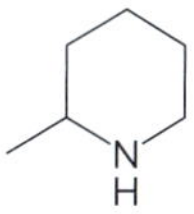

More Practice: Try Problems 22.50, 22.51c, 22.52c, 22.53j, 22.54.

Problem 22.27 Draw the major product formed in each reaction.

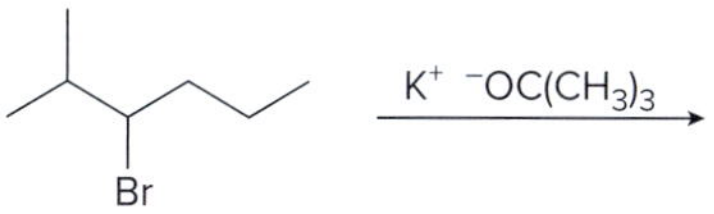

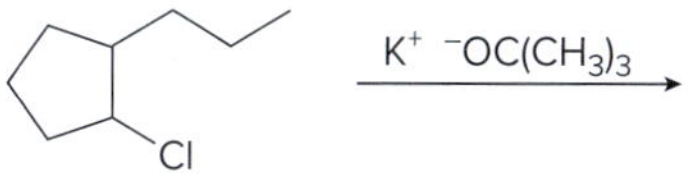

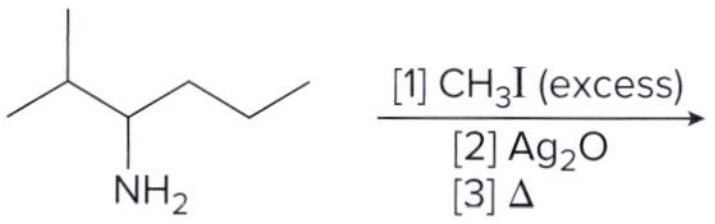

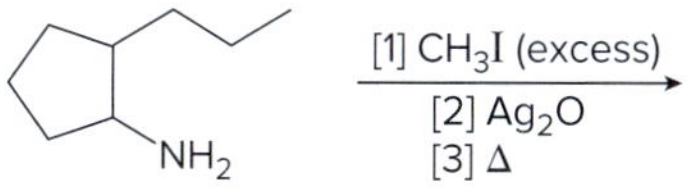

22.12 Reaction of Amines with Nitrous Acid

Nitrous acid, HNO$_2$, is a weak, unstable acid formed from NaNO$_2$ and a strong acid like HCl.

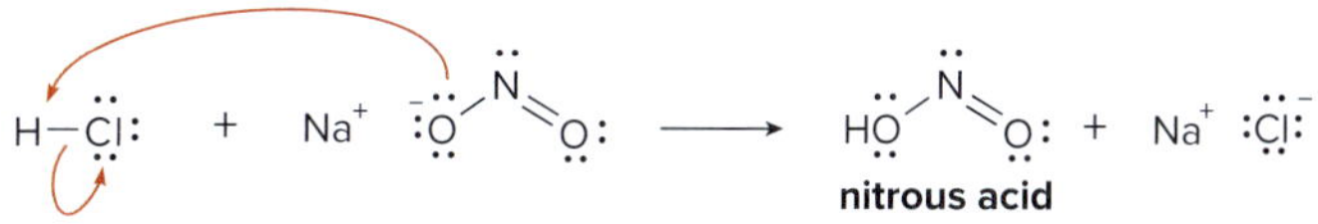

nitrous acid

In the presence of acid, nitrous acid decomposes to $^+$NO, the **nitrosonium ion.** This electrophile then goes on to react with the nucleophilic nitrogen atom of amines to form **diazonium salts (RN$_2^+$Cl$^-$)** from 1° amines and ***N*-nitrosamines (R$_2$NN=O)** from 2° amines.

22.12A Reaction of $^+$NO with 1° Amines

Nitrous acid reacts with 1° alkylamines and arylamines to form diazonium salts. This reaction is called **diazotization.**

The mechanism for this reaction begins with nucleophilic attack of the amine on the nitrosonium ion, and it can conceptually be divided into two parts: formation of an ***N*-nitrosamine,** followed by loss of H$_2$O, as shown in Mechanism 22.2.

Mechanism 22.2 Formation of a Diazonium Salt from a 1° Amine

Part [1] Formation of an *N*-nitrosamine

1 – 2 Nucleophilic attack of the amine on the nitrosonium ion ($^+$NO), followed by loss of a proton, forms an ***N*-nitrosamine.**

Part [2] Formation of a diazonium salt

3 – 5 Three proton transfers form an intermediate with a **good leaving group (H$_2$O).**

6 Loss of water forms a **diazonium ion (RN$_2^+$).** The diazonium salt formed in this reaction consists of the diazonium ion (RN$_2^+$) and a chloride anion (Cl$^-$).

Alkyl diazonium salts are generally not useful compounds. They readily decompose below room temperature to form carbocations with loss of N$_2$, a very good leaving group. These carbocations usually form a complex mixture of substitution, elimination, and rearrangement products.

22.13 Substitution Reactions of Aryl Diazonium Salts

Care must be exercised in handling diazonium salts, because they can explode if allowed to dry.

On the other hand, **aryl diazonium salts are very useful synthetic intermediates.** Although they are rarely isolated and are generally unstable above 0 °C, they are useful starting materials in two general kinds of reactions described in Section 22.13.

22.12B Reaction of $^+$NO with 2° Amines

Secondary alkylamines and arylamines react with nitrous acid to form N-nitrosamines.

$$R_2NH \xrightarrow{\text{NaNO}_2, \text{HCl}} R_2N-N=O$$

2° amine → *N*-nitrosamine

N-nitrosamine in tobacco smoke

Many *N*-nitrosamines are potent carcinogens found in some food and tobacco smoke. Nitrosamines in food can be formed in the same way they are formed in the laboratory: **reaction of a 2° amine with the nitrosonium ion,** formed from nitrous acid (HNO$_2$). The mechanism for this reaction follows the two steps of Part [1] of Mechanism 22.2.

Problem 22.28 Draw the product formed when each compound is treated with NaNO$_2$ and HCl.

a. 2-methylaniline   b. *N*-ethylmethylamine   c. piperidine   d. isopropylamine

22.13 Substitution Reactions of Aryl Diazonium Salts

Aryl diazonium salts undergo two general reactions.

- **Substitution of N$_2$ by an atom or a group of atoms Z forms a variety of substituted benzenes.**

$$C_6H_5-N_2^+ Cl^- \xrightarrow{Z} C_6H_5-Z + N_2 + Cl^-$$

- **Coupling of a diazonium salt with another benzene derivative forms an azo compound,** a compound containing a nitrogen–nitrogen double bond.

$$C_6H_5-N_2^+ Cl^- + C_6H_5-Y \longrightarrow C_6H_5-N=N-C_6H_4-Y + HCl$$

azo compound

Y = NH$_2$, NHR, NR$_2$, OH (a strong electron-donor group)

22.13A Specific Substitution Reactions

Aryl diazonium salts react with a variety of reagents to form products in which **Z (an atom or group of atoms) replaces N$_2$, a very good leaving group.** The mechanism of these reactions varies with the identity of Z, so we will concentrate on the products of the reactions, not the mechanisms.

$$C_6H_5-N_2^+ Cl^- \xrightarrow{Z} C_6H_5-Z + N_2 + Cl^-$$

good leaving group

[1] **Substitution by OH—Synthesis of phenols**

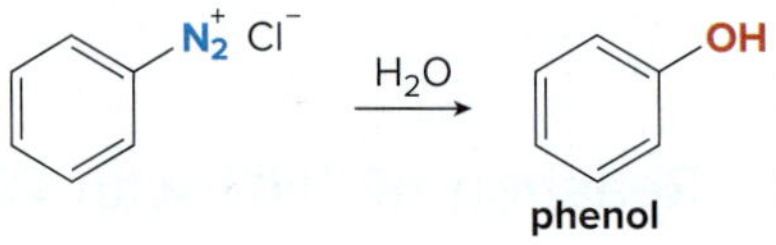

PhN₂⁺Cl⁻ + H₂O → phenol (PhOH)

A diazonium salt reacts with H₂O to form a **phenol.**

[2] **Substitution by Cl or Br—Synthesis of aryl chlorides and bromides**

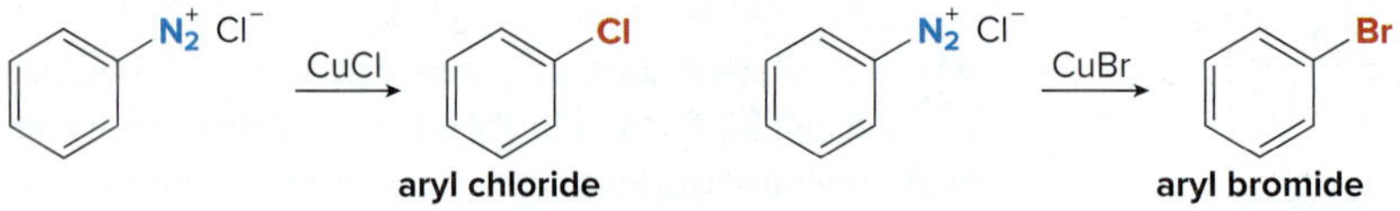

PhN₂⁺Cl⁻ + CuCl → aryl chloride (PhCl)

PhN₂⁺Cl⁻ + CuBr → aryl bromide (PhBr)

A diazonium salt reacts with copper(I) chloride or copper(I) bromide to form an **aryl chloride** or **aryl bromide,** respectively. This is called the **Sandmeyer reaction.** It provides an alternative to direct chlorination and bromination of an aromatic ring using Cl_2 or Br_2 and a Lewis acid catalyst.

[3] **Substitution by F—Synthesis of aryl fluorides**

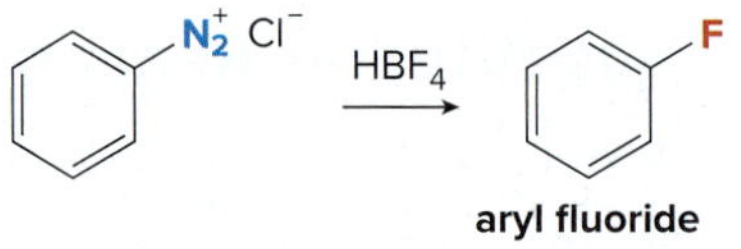

PhN₂⁺Cl⁻ + HBF₄ → aryl fluoride (PhF)

A diazonium salt reacts with fluoroboric acid (HBF_4) to form an **aryl fluoride.** This is a useful reaction because aryl fluorides cannot be produced by direct fluorination with F_2 and a Lewis acid catalyst, because F_2 reacts too violently (Section 20.3).

[4] **Substitution by I—Synthesis of aryl iodides**

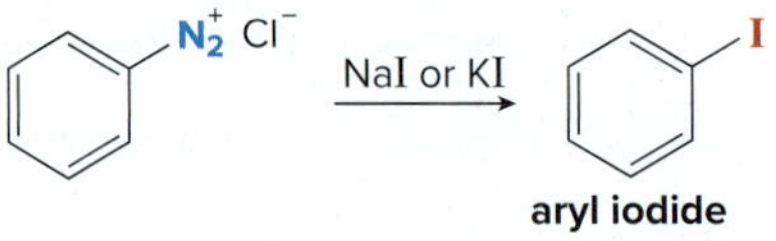

PhN₂⁺Cl⁻ + NaI or KI → aryl iodide (PhI)

A diazonium salt reacts with sodium or potassium iodide to form an **aryl iodide.** This, too, is a useful reaction because aryl iodides cannot be produced by direct iodination with I_2 and a Lewis acid catalyst, because I_2 reacts too slowly (Section 20.3).

[5] **Substitution by CN—Synthesis of benzonitriles**

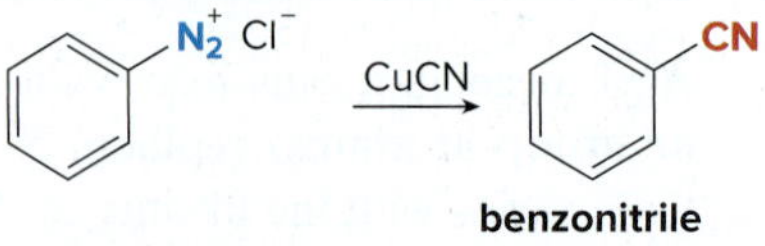

PhN₂⁺Cl⁻ + CuCN → benzonitrile (PhCN)

A diazonium salt reacts with copper(I) cyanide to form a **benzonitrile.** Because a cyano group can be hydrolyzed to a carboxylic acid, reduced to an amine or aldehyde, or converted to a ketone with organometallic reagents, this reaction provides easy access to a wide variety of benzene derivatives using chemistry described in Section 15.13.

22.13 Substitution Reactions of Aryl Diazonium Salts

[6] Substitution by H—Synthesis of benzene

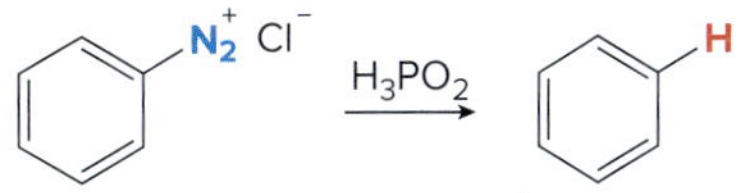

$$C_6H_5N_2^+Cl^- \xrightarrow{H_3PO_2} \text{benzene}$$

A diazonium salt reacts with hypophosphorus acid (H$_3$PO$_2$) to form **benzene.** This reaction has limited utility because it reduces the functionality of the benzene ring by replacing N$_2$ with a hydrogen atom. Nonetheless, this reaction *is* useful in synthesizing compounds that have substitution patterns that are not available by other means.

For example, it is not possible to synthesize 1,3,5-tribromobenzene from benzene by direct bromination. Because Br is an ortho, para director, bromination with Br$_2$ and FeBr$_3$ will not add Br substituents meta to each other on the ring.

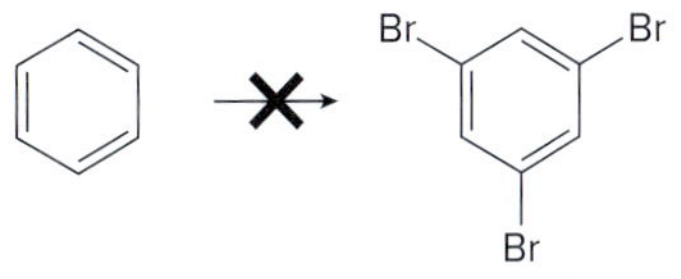

benzene ✗→ 1,3,5-tribromobenzene

It is possible, however, to add three Br atoms meta to each other when aniline is the starting material. Because an NH$_2$ group is a very powerful ortho, para director, three Br atoms are introduced in a single step on halogenation (Section 20.10A). Then, the NH$_2$ group can be removed by diazotization and reaction with H$_3$PO$_2$.

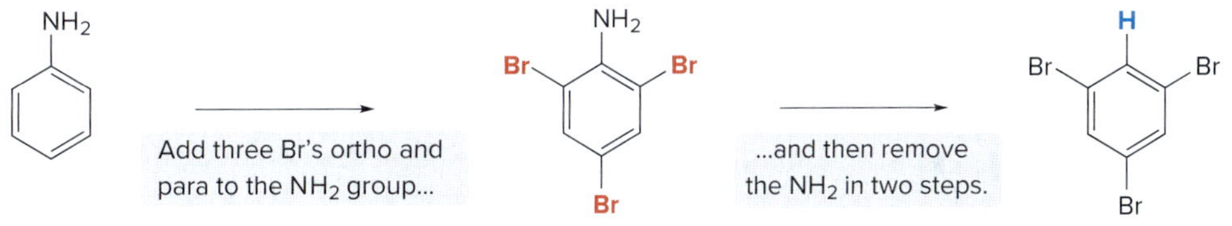

Aniline → (Add three Br's ortho and para to the NH$_2$ group...) 2,4,6-tribromoaniline → (...and then remove the NH$_2$ in two steps.) 1,3,5-tribromobenzene

The complete synthesis of 1,3,5-tribromobenzene from benzene is outlined in Figure 22.8.

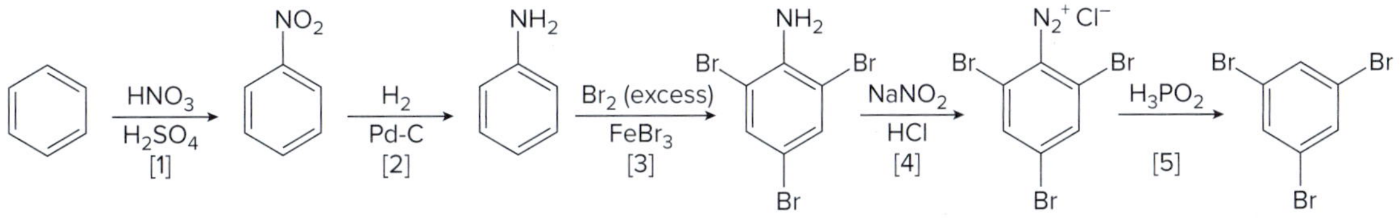

Figure 22.8
The synthesis of 1,3,5-tribromo-benzene from benzene

benzene $\xrightarrow[\text{H}_2\text{SO}_4]{\text{HNO}_3}$ [1] nitrobenzene $\xrightarrow[\text{Pd-C}]{\text{H}_2}$ [2] aniline $\xrightarrow[\text{FeBr}_3]{\text{Br}_2 \text{ (excess)}}$ [3] 2,4,6-tribromoaniline $\xrightarrow[\text{HCl}]{\text{NaNO}_2}$ [4] diazonium salt $\xrightarrow{\text{H}_3\text{PO}_2}$ [5] 1,3,5-tribromobenzene

- Nitration followed by reduction forms aniline (C$_6$H$_5$NH$_2$) from benzene (Steps [1] and [2]).
- Bromination of aniline yields the tribromo derivative in Step [3].
- The NH$_2$ group is removed by a two-step process: diazotization with NaNO$_2$ and HCl (Step [4]), followed by substitution of the diazonium ion by H with H$_3$PO$_2$.

Problem 22.29 Draw the product formed in each reaction.

a. 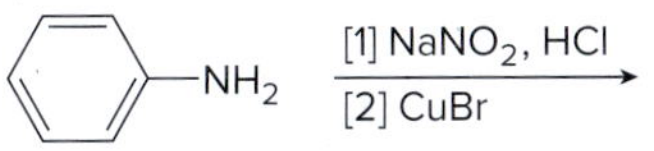 C$_6$H$_5$—NH$_2$ $\xrightarrow{\text{[1] NaNO}_2\text{, HCl}}_{\text{[2] CuBr}}$

b. 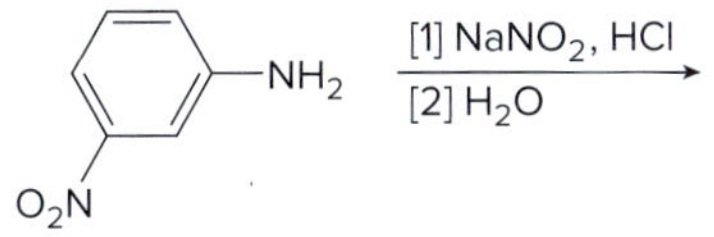 3-O$_2$N-C$_6$H$_4$—NH$_2$ $\xrightarrow{\text{[1] NaNO}_2\text{, HCl}}_{\text{[2] H}_2\text{O}}$

c. CH$_3$O—C$_6$H$_4$—NH$_2$ $\xrightarrow{\text{[1] NaNO}_2\text{, HCl}}_{\text{[2] HBF}_4}$

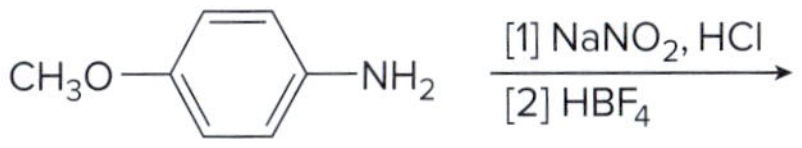

d. 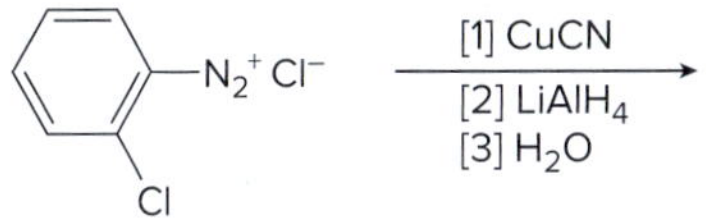 2-Cl-C$_6$H$_4$—N$_2^+$Cl$^-$ $\xrightarrow[\text{[2] LiAlH}_4]{\text{[1] CuCN}}_{\text{[3] H}_2\text{O}}$

22.13B Using Diazonium Salts in Synthesis

Diazonium salts provide easy access to many different benzene derivatives. Keep in mind the following four-step sequence, because it will be used to synthesize many substituted benzenes.

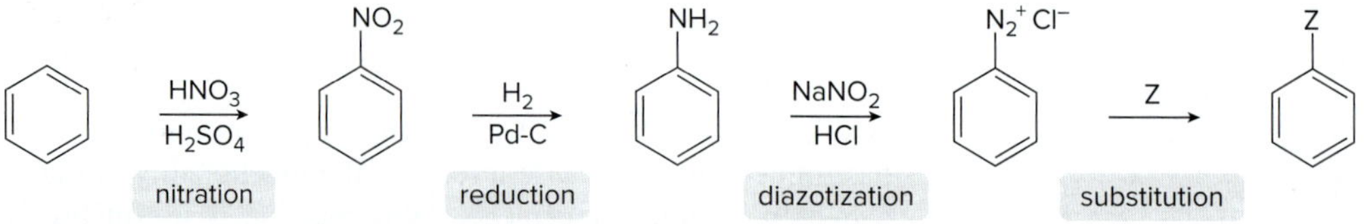

benzene →[HNO$_3$ / H$_2$SO$_4$]→ nitrobenzene (NO$_2$) →[H$_2$ / Pd-C]→ aniline (NH$_2$) →[NaNO$_2$ / HCl]→ diazonium salt (N$_2^+$ Cl$^-$) →[Z]→ Ph–Z

nitration — reduction — diazotization — substitution

Sample Problems 22.6 and 22.7 apply these principles to two different multistep syntheses.

Sample Problem 22.6 Using Diazonium Salts in Synthesis

Synthesize *m*-chlorophenol from benzene.

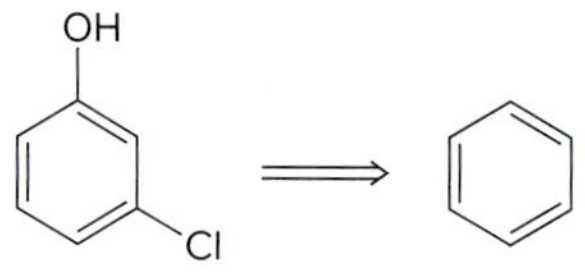

m-chlorophenol ⟹ benzene

Solution
Both OH and Cl are ortho, para directors, but they are located *meta* to each other. The OH group must be formed from a diazonium salt, which can be made from an NO$_2$ group (a meta director) by a stepwise method.

Retrosynthetic Analysis

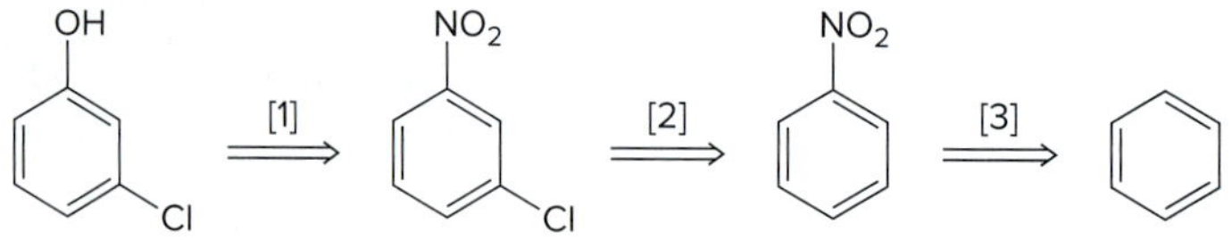

m-chlorophenol ⟹[1] *m*-chloronitrobenzene ⟹[2] nitrobenzene ⟹[3] benzene

Working backwards:
- [1] Form the OH group from NO$_2$ by a three-step procedure using a diazonium salt.
- [2] Introduce Cl meta to NO$_2$ by halogenation.
- [3] Add the NO$_2$ group by nitration.

Synthesis

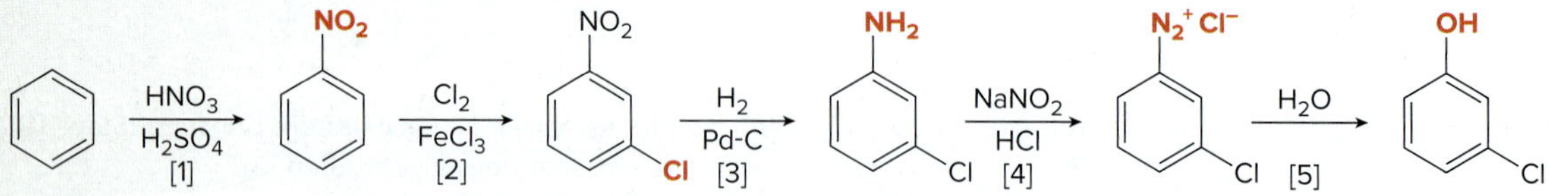

benzene →[HNO$_3$ / H$_2$SO$_4$] [1]→ nitrobenzene →[Cl$_2$ / FeCl$_3$] [2]→ *m*-chloronitrobenzene →[H$_2$ / Pd-C] [3]→ *m*-chloroaniline →[NaNO$_2$ / HCl] [4]→ diazonium salt →[H$_2$O] [5]→ *m*-chlorophenol

- **Nitration followed by chlorination** meta to the NO$_2$ group forms the **meta disubstituted benzene** (Steps [1]–[2]).
- **Reduction of the nitro group** (Step [3]) followed by **diazotization** forms the **diazonium salt** in Step [4], which is then converted to the desired phenol by treatment with **H$_2$O** (Step [5]).

Problem 22.30 Devise a synthesis of each compound from benzene.

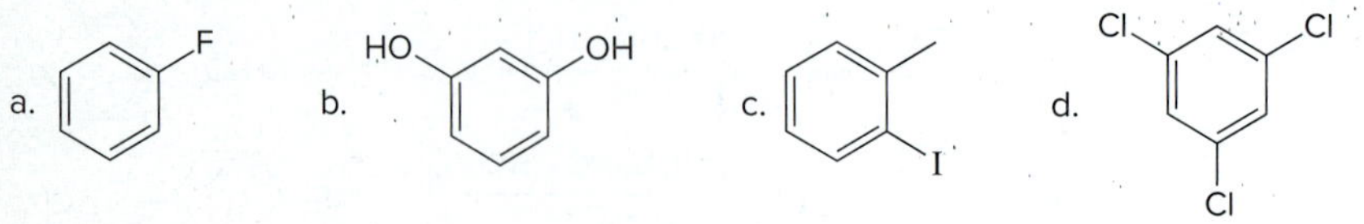

a. fluorobenzene b. resorcinol (1,3-dihydroxybenzene) c. *m*-iodotoluene d. 1,3,5-trichlorobenzene

More Practice: Try Problems 22.61a, b; 22.62a, b; 22.63.

22.13 Substitution Reactions of Aryl Diazonium Salts

Sample Problem 22.7 — Devising a Synthesis with Diazonium Salts

Synthesize *p*-bromobenzaldehyde from benzene.

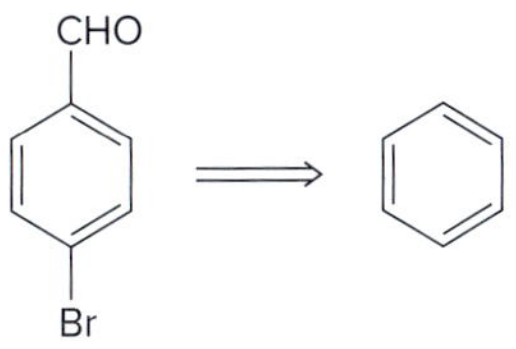

Solution
Because the two groups are located para to each other and Br is an ortho, para director, Br should be added to the ring *first*. **To add the CHO group, recall that it can be formed from CN by reduction.**

Retrosynthetic Analysis

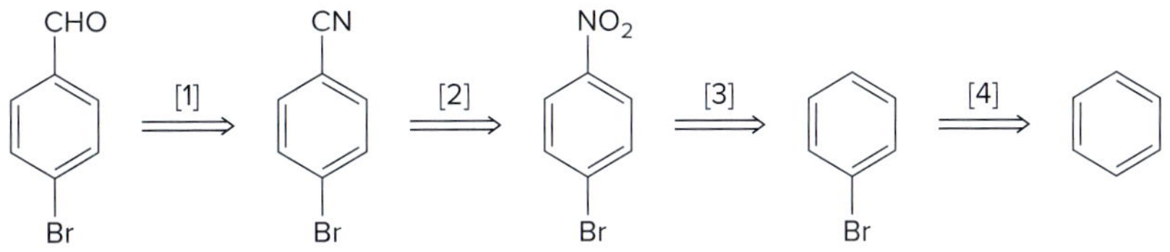

CHO–C₆H₄–Br ⟹[1] CN–C₆H₄–Br ⟹[2] NO₂–C₆H₄–Br ⟹[3] C₆H₅–Br ⟹[4] C₆H₆

Working backwards:
- [1] Form the CHO group by reduction of CN.
- [2] Prepare the CN group from an NO₂ group by a three-step sequence using a diazonium salt.
- [3] Introduce the NO₂ group by nitration, para to the Br atom.
- [4] Introduce Br by bromination with Br₂ and FeBr₃.

Synthesis

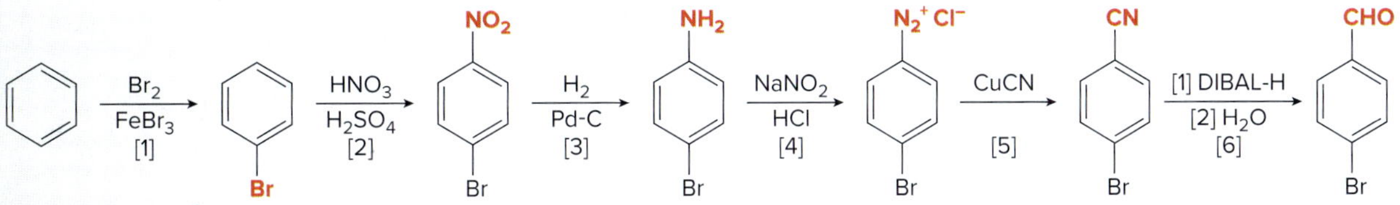

benzene →[Br₂, FeBr₃][1] C₆H₅Br →[HNO₃, H₂SO₄][2] p-O₂N-C₆H₄-Br →[H₂, Pd-C][3] p-H₂N-C₆H₄-Br →[NaNO₂, HCl][4] p-N_2^+ Cl^--C₆H₄-Br →[CuCN][5] p-NC-C₆H₄-Br →[[1] DIBAL-H, [2] H₂O][6] p-OHC-C₆H₄-Br

- **Bromination followed by nitration forms a disubstituted benzene** with two para substituents (Steps [1]–[2]), which can be separated from its undesired ortho isomer.
- **Reduction of the NO₂ group** (Step [3]) followed by **diazotization** forms the diazonium salt in Step [4], which is converted to a nitrile by reaction with CuCN (Step [5]).
- **Reduction of the CN group with DIBAL-H** (a mild reducing agent) **forms the CHO group** and completes the synthesis.

Problem 22.31 Devise a synthesis of each compound from benzene. You may use any other organic or inorganic reagents.

a. 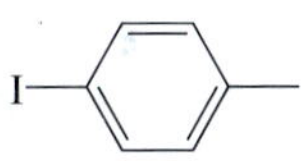 *p*-iodotoluene (I–C₆H₄–CH₃)

b. 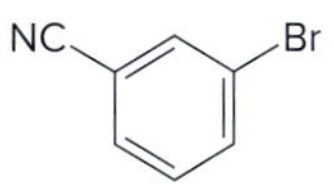 3-bromobenzonitrile (NC–C₆H₄–Br, meta)

c. 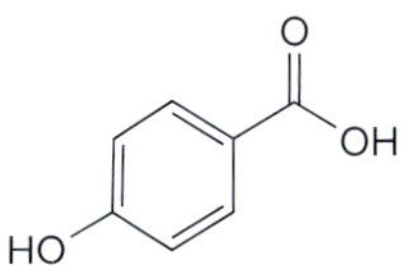 4-hydroxybenzoic acid (HO–C₆H₄–COOH)

More Practice: Try Problems 22.64a, b; 22.65; 22.66.

22.14 Coupling Reactions of Aryl Diazonium Salts

The second general reaction of diazonium salts is **coupling.** When a diazonium salt is treated with an aromatic compound that contains a strong electron-donor group, the two rings join together to form an **azo compound,** a compound with a nitrogen–nitrogen double bond.

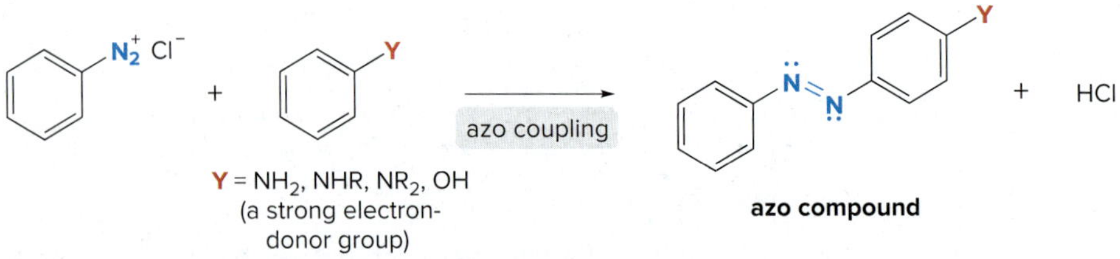

C₆H₅N₂⁺Cl⁻ + C₆H₅–Y →(azo coupling) C₆H₅–N=N–C₆H₄–Y + HCl

Y = NH₂, NHR, NR₂, OH
(a strong electron-donor group)

azo compound

Synthetic dyes are described in more detail in Section 22.15A.

Azo compounds are highly conjugated and colored. Many of these compounds, such as the azo compound "butter yellow," are synthetic dyes. Butter yellow was once used to color margarine.

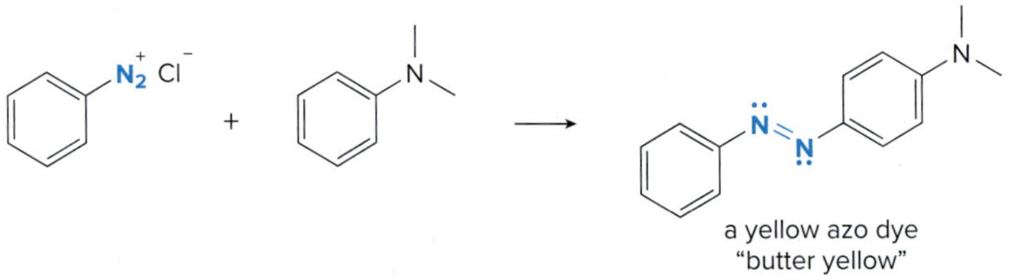

a yellow azo dye
"butter yellow"

This reaction is another example of **electrophilic aromatic substitution,** with the **diazonium salt acting as the electrophile.** Like all electrophilic substitutions (Section 20.2), the mechanism has two steps: **addition of the electrophile** (the diazonium ion) to form a **resonance-stabilized carbocation,** followed by deprotonation, as shown in Mechanism 22.3.

Mechanism 22.3 Azo Coupling

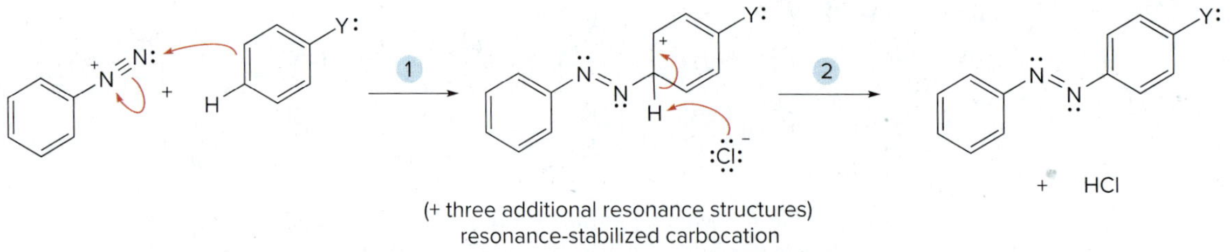

(+ three additional resonance structures)
resonance-stabilized carbocation

1. The diazonium ion reacts with the benzene ring to form a **resonance-stabilized carbocation.**
2. Loss of a proton regenerates the aromatic ring.

Because a diazonium salt is weakly electrophilic, the reaction occurs only when the benzene ring has **a strong electron-donor group Y, where Y = NH₂, NHR, NR₂, or OH.** Although these groups activate both the ortho and para positions, para substitution occurs unless the para position already has another substituent present.

Problem 22.32 Draw the product formed when $C_6H_5N_2^+Cl^-$ reacts with each compound.

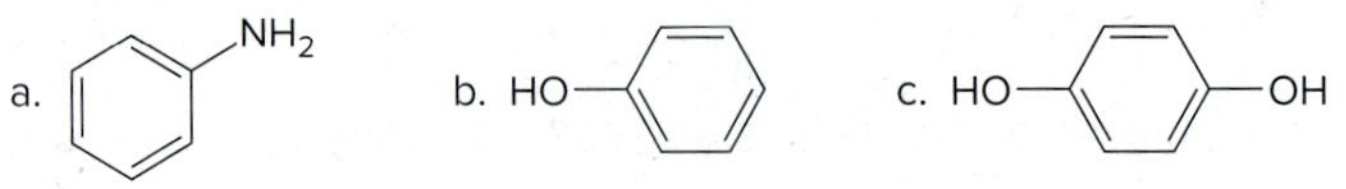

a. C₆H₅–NH₂   b. HO–C₆H₅   c. HO–C₆H₄–OH

To determine what starting materials are needed to synthesize a particular azo compound, always divide the molecule into two components: **one has a benzene ring with a diazonium ion, and one has a benzene ring with a very strong electron-donor group.**

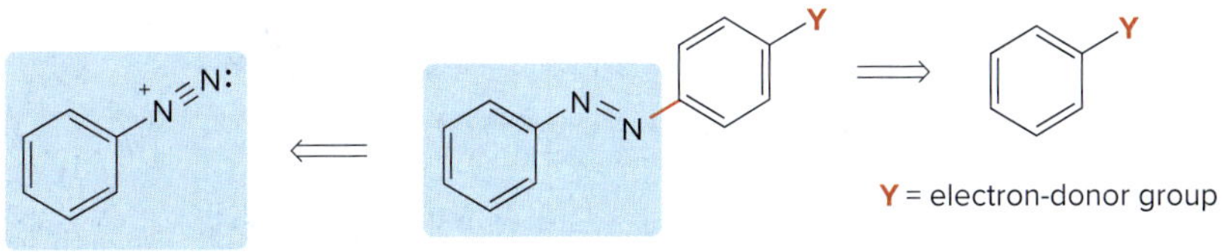

Y = electron-donor group

Sample Problem 22.8 — Synthesizing an Azo Compound

What starting materials are needed to synthesize the following azo compound?

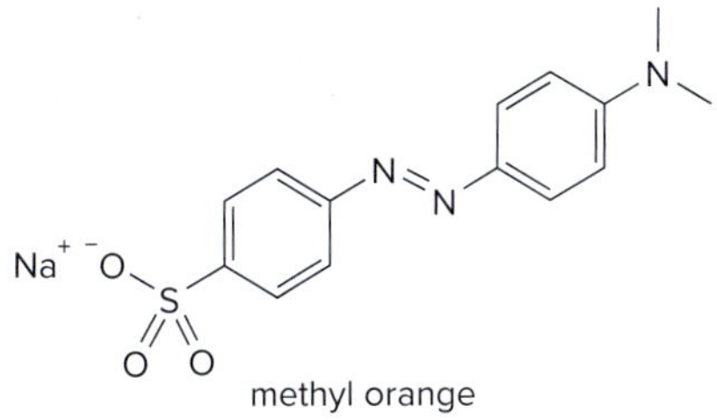

methyl orange
an orange dye

Solution

Both benzene rings in methyl orange have a substituent, but only one group, $N(CH_3)_2$, is a strong electron donor. In determining the two starting materials, the **diazonium ion must be bonded to the ring that is *not* bonded to $N(CH_3)_2$.**

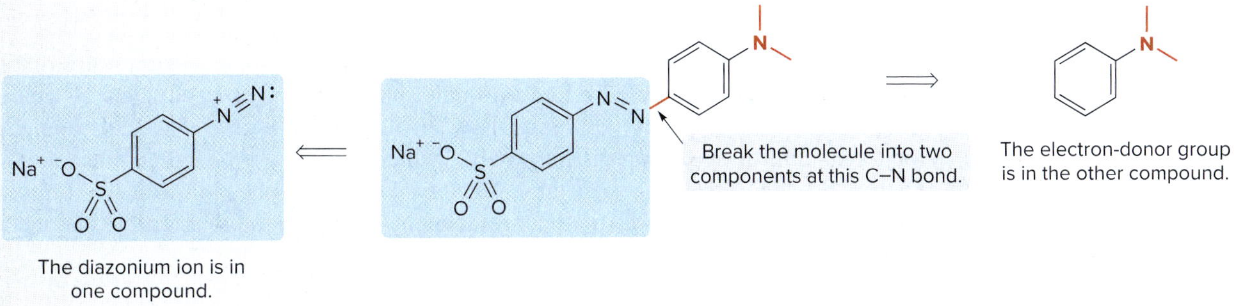

The diazonium ion is in one compound.

Break the molecule into two components at this C–N bond.

The electron-donor group is in the other compound.

Problem 22.33

What starting materials are needed to synthesize each azo compound?

a. b.

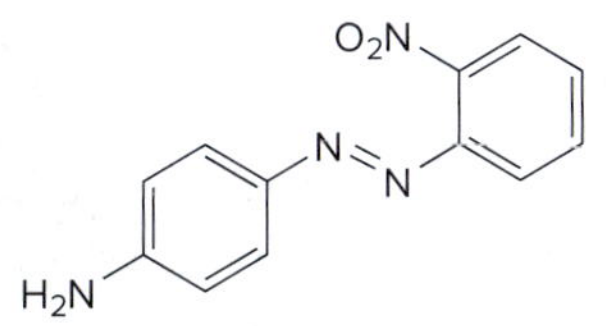

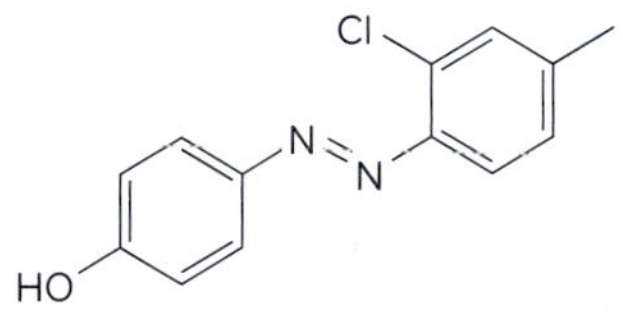

More Practice: Try Problems 22.61c, 22.62c, 22.64c.

22.15 Application: Synthetic Dyes and Sulfa Drugs

Azo compounds have two important applications: as dyes and as sulfa drugs, the first synthetic antibiotics.

22.15A Natural and Synthetic Dyes

Until 1856, all dyes were natural in origin, obtained from plants, animals, or minerals. Three natural dyes known for centuries are **indigo, tyrian purple,** and **alizarin.**

The blue dye **indigo,** derived from the plant *Indigofera tinctoria,* has been used in India for thousands of years. Traders introduced it to the Mediterranean area and then to Europe. **Tyrian purple,** a natural dark purple dye obtained from the mucous gland of a Mediterranean snail of the genus *Murex,* was a symbol of royalty before the collapse of the Roman Empire. **Alizarin,** a bright red dye obtained from madder root (*Rubia tinctorum*), a plant native to India and northeastern Asia, has been found in cloth entombed with Egyptian mummies.

Because all three of these dyes were derived from natural sources, they were difficult to obtain, making them expensive and available only to the privileged. This all changed in 1856 when William Henry Perkin, an 18-year-old student with a makeshift home laboratory, serendipitously prepared a purple dye, which would later be called mauveine, during his failed attempt to synthesize the antimalarial drug quinine. Mauveine is a mixture of two compounds that differ in the presence of only one methyl group on one of the aromatic rings.

A purple shawl dyed with Perkin's mauveine *Science & Society Picture Library/Getty Images*

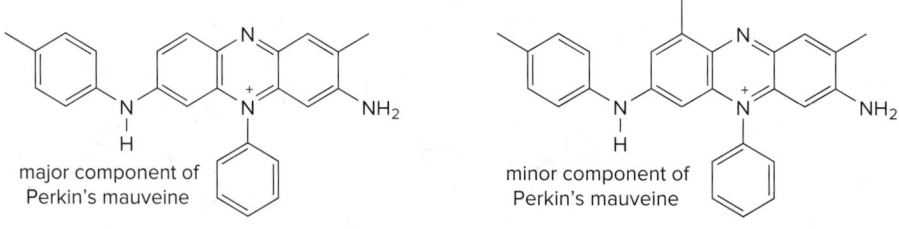

Perkin's discovery marked the beginning of the chemical industry. He patented the dye and went on to build a factory to commercially produce it on a large scale. This event began the surge of research in organic chemistry, not just in the synthesis of dyes, but in the production of perfumes, anesthetics, inks, and drugs as well. Perkin was a wealthy man when he retired at the age of 36 to devote the rest of his life to basic chemical research. The most prestigious award given by the American Chemical Society is named the Perkin Medal in his honor.

Many common synthetic dyes, such as para red and Congo red, are **azo compounds,** prepared by the diazonium coupling reaction described in Section 22.14.

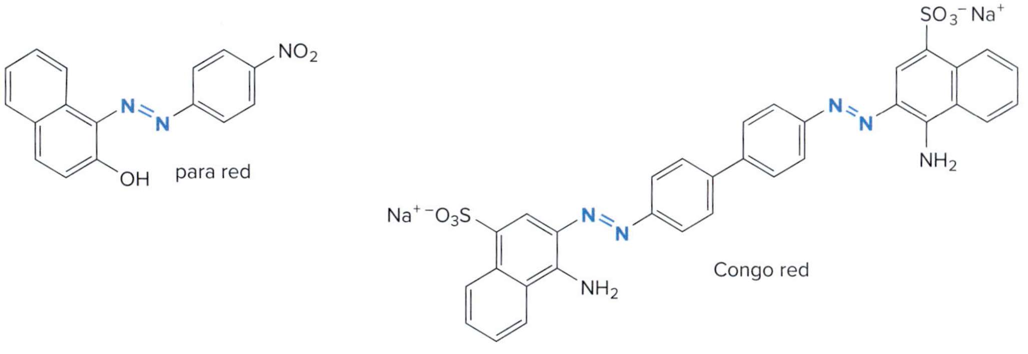

Although natural and synthetic dyes are quite varied in structure, **all of them are colored because they are highly conjugated.** A molecule with many π bonds in conjugation absorbs visible light, taking on the color from the visible spectrum that it does *not* absorb.

Problem 22.34   What two components are needed to prepare para red by azo coupling?

22.15B Sulfa Drugs

Although they may seem quite unrelated, the synthesis of colored dyes led to the development of the first synthetic antibiotics. Much of the early effort in this field was done by the German chemist Paul Ehrlich, who worked with synthetic dyes and used them to stain tissues. This led him on a search for dyes that were lethal to bacteria without affecting other tissue cells, hoping that these dyes could treat bacterial infections. For many years this effort was unsuccessful.

Then, in 1935, Gerhard Domagk, a German physician working for a dye manufacturer, first used a synthetic dye as a drug to kill bacteria. His daughter had contracted a streptococcal infection, and as she neared death, he gave her **prontosil,** an azo dye that inhibited the growth of certain bacteria in mice. His daughter recovered, and the modern era of synthetic antibiotics was initiated. For his pioneering work, Domagk was awarded the Nobel Prize in Physiology or Medicine in 1939.

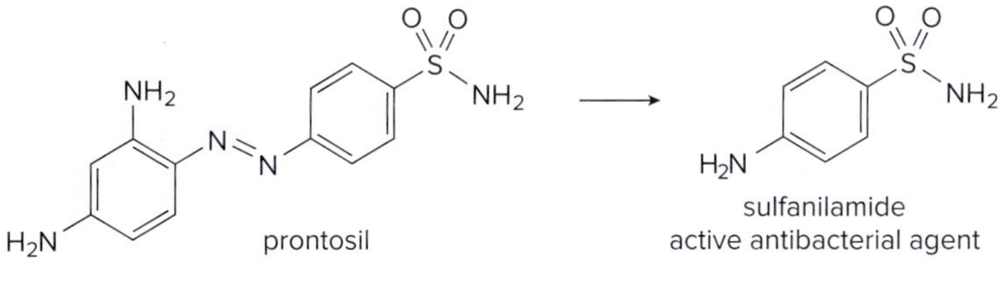

Prontosil and other sulfur-containing antibiotics are collectively called **sulfa drugs.** Prontosil is not the active agent itself. In cells, it is metabolized to **sulfanilamide,** the active drug. To understand how sulfanilamide functions as an antibacterial agent we must examine **folic acid,** which microorganisms synthesize from *p*-aminobenzoic acid.

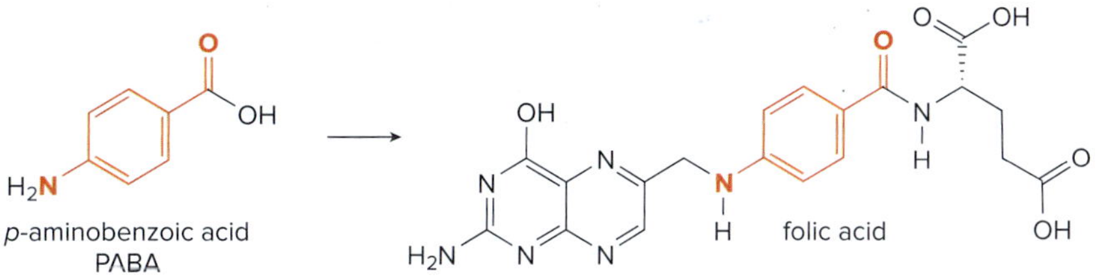

1008   Chapter 22   Amines

Sulfanilamide and *p*-aminobenzoic acid are similar in size and shape and have related functional groups. Thus, when sulfanilamide is administered, bacteria attempt to use it in place of *p*-aminobenzoic acid to prepare folic acid, and this derails folic acid synthesis, so bacteria cannot grow and reproduce. Sulfanilamide affects only bacterial cells, though, because humans do not synthesize folic acid and must obtain it from their diets.

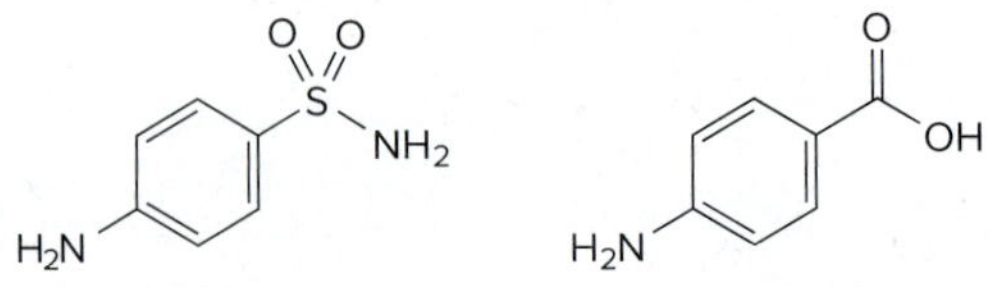

sulfanilamide   *p*-aminobenzoic acid

Many other compounds of similar structure have been prepared and are still widely used as antibiotics. The structures of two other sulfa drugs are shown in Figure 22.9.

Figure 22.9
Two common sulfa drugs

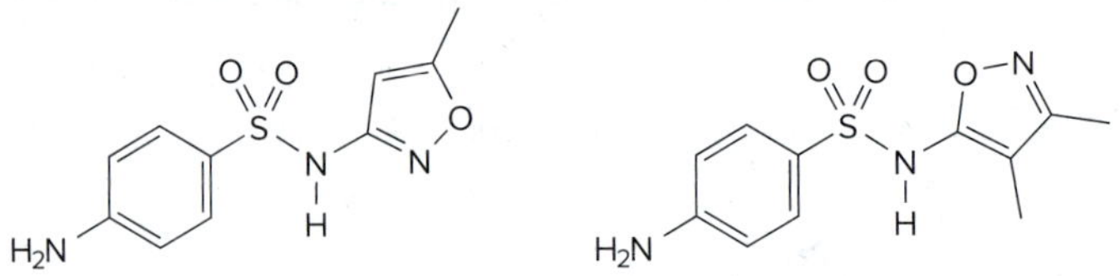

sulfamethoxazole   sulfisoxazole

- Sulfamethoxazole is the sulfa drug in Bactrim, and sulfisoxazole is sold as Gantrisin. Both drugs are commonly used in the treatment of ear and urinary tract infections.

Chapter 22 REVIEW

KEY CONCEPTS

The basicity of amines (22.10)

NH_3 < **CH_3NH_2**

Increasing basicity

- Alkylamines are more basic than NH_3 because of the electron-donating R groups.

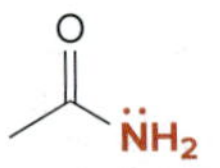

PhNH$_2$ < CH_3NH_2

Increasing basicity

- Alkylamines are more basic than arylamines, which have a delocalized lone pair from the N atom.

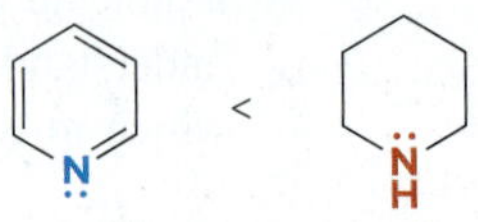

(acetamide) < CH_3NH_2

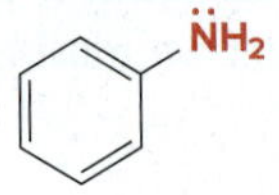

Increasing basicity

- Alkylamines are more basic than amides, which have a delocalized lone pair from the N atom.

pyridine < piperidine

Increasing basicity

- Alkylamines with a lone pair in an sp^3 hybrid orbital are more basic than those with a lone pair in an sp^2 hybrid orbital.

Try Problems 22.36a, 22.39–22.42.

KEY REACTIONS

Preparation of Amines

[1] Direct nucleophilic substitution with NH₃ and amines

1. CH_3-X + $\ddot{N}H_3$ (excess) $\xrightarrow[(22.6A)]{S_N2}$ CH_3-NH_2 + $NH_4^+X^-$
   X = Cl, Br, I ; 1° amine

2. CH_3CH_2-X + $\ddot{N}H_3$ (excess) $\xrightarrow[(22.6A)]{S_N2}$ $(CH_3CH_2)_4N^+\ X^-$
   X = Cl, Br, I ; quaternary ammonium salt

Try Problems 22.49d, 22.53a.

[2] Gabriel synthesis

Phthalimide (H–N:) $\xrightarrow[\text{[2] } CH_3X]{\text{[1] KOH}}$ $\xrightarrow[H_2O]{^-OH}$ CH_3-NH_2 (1° amine) + phthalate dianion

Gabriel synthesis (22.6A)

Try Problem 22.53b.

[3] Reduction methods

1. $R-NO_2$ $\xrightarrow[\text{reduction (22.6B)}]{H_2,\ Pd\text{-}C\ \text{or}\ Fe,\ HCl\ \text{or}\ Sn,\ HCl}$ $R-NH_2$

2. $R-CN$ $\xrightarrow[\text{reduction (22.6B)}]{\text{[1] LiAlH}_4\ \text{[2] H}_2O}$ $R-CH_2-NH_2$ (1° amine)

3. Amide $R'C(=O)NR_2$ $\xrightarrow[\text{reduction (22.6B)}]{\text{[1] LiAlH}_4\ \text{[2] H}_2O}$ amine (R = H or alkyl) → 1°, 2°, and 3° amines

Try Problem 22.53c–e.

[4] Reductive amination

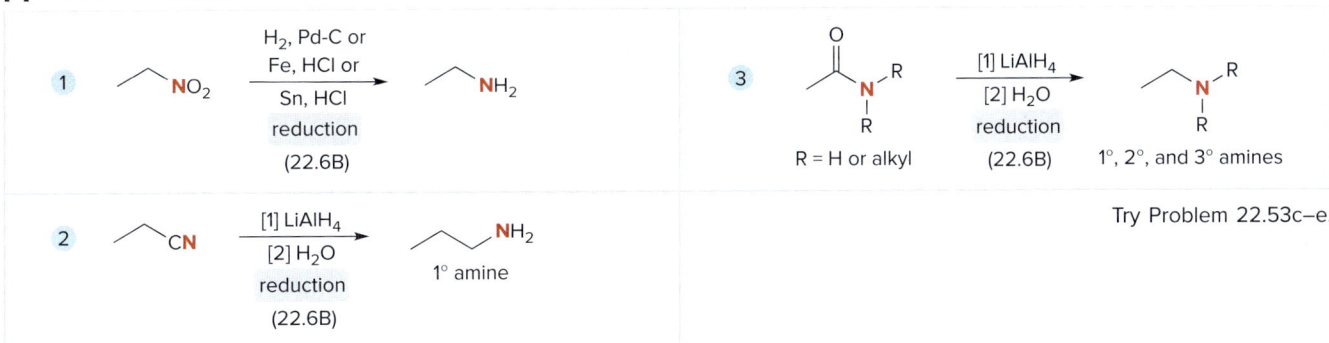

Ketone/aldehyde + R'_2NH $\xrightarrow[\text{reductive amination (22.6C)}]{NaBH_3CN}$ amine (NR'_2 on carbon bearing R)
R, R' = H or alkyl ; 1°, 2°, and 3° amines

Try Problems 22.46, 22.47, 22.49j, 22.53h.

Reactions of Amines

[1] Reaction as a base

$R-\ddot{N}H_2$ + $H-A$ $\underset{(22.8)}{\rightleftarrows}$ $R-\overset{+}{N}H_3$ + $:A^-$
base ; acid ; conjugate acid

Try Problems 22.36b; 22.49a, g.

[2] Nucleophilic addition to aldehydes and ketones

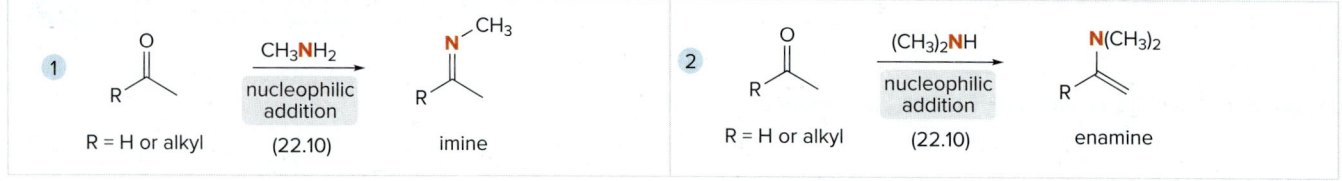

Try Problems 22.49e, 22.53i.

[3] Nucleophilic substitution with acid chlorides and anhydrides

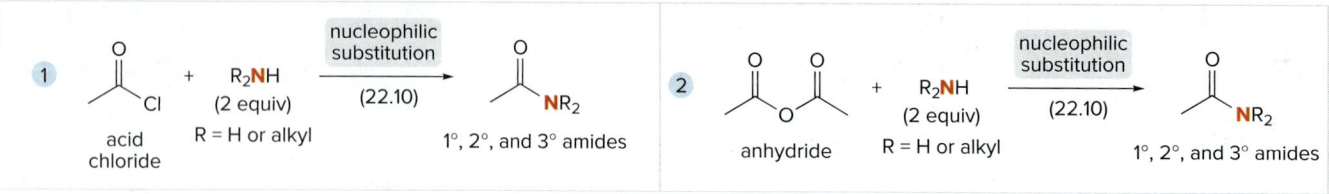

Try Problems 22.49b, c; 22.53f.

[4] Hofmann elimination

[structure with NH₂] → [1] CH₃I (excess), [2] Ag₂O, [3] Δ → alkene + N(CH₃)₃ + H₂O + AgI

Hofmann elimination (22.11) — major product

Try Problems 22.50, 22.51c, 22.52c, 22.53j, 22.54.

[5] Reaction with nitrous acid

1. Ph–NH₂ (1° amine) + NaNO₂ / HCl (22.12) → Ph–N≡N:⁺ Cl⁻ (diazonium salt)

2. pyrrolidine (2° amine) + NaNO₂ / HCl (22.12) → N-nitrosamine

Try Problems 22.49h, 22.53g.

Reactions of Diazonium Salts

[1] Substitution reactions

1. ArN₂⁺ Cl⁻ + H–OH (22.13) → phenol (ArOH)

2. ArN₂⁺ Cl⁻ + CuX, X = Cl, Br (Sandmeyer reaction) (22.13) → aryl halide (ArX)

3. ArN₂⁺ Cl⁻ + HBF₄ (22.13) → aryl fluoride (ArF)

4. ArN₂⁺ Cl⁻ + NaI or KI (22.13) → aryl iodide (ArI)

5. ArN₂⁺ Cl⁻ + CuCN (22.13) → benzonitrile (ArCN)

6. ArN₂⁺ Cl⁻ + H₃PO₂ (22.13) → benzene (ArH)

Try Problem 22.57a, b.

[2] Coupling to form azo compounds

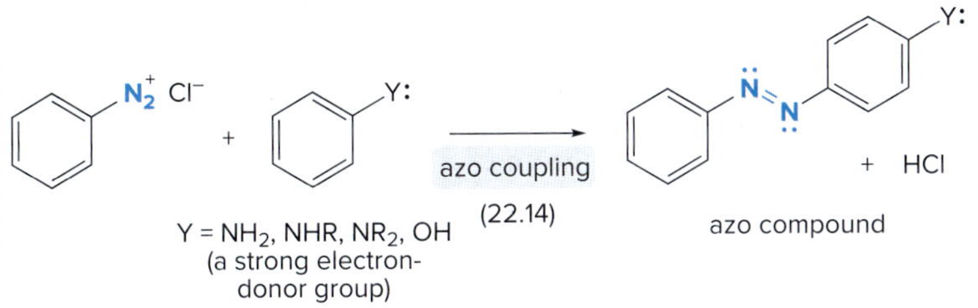

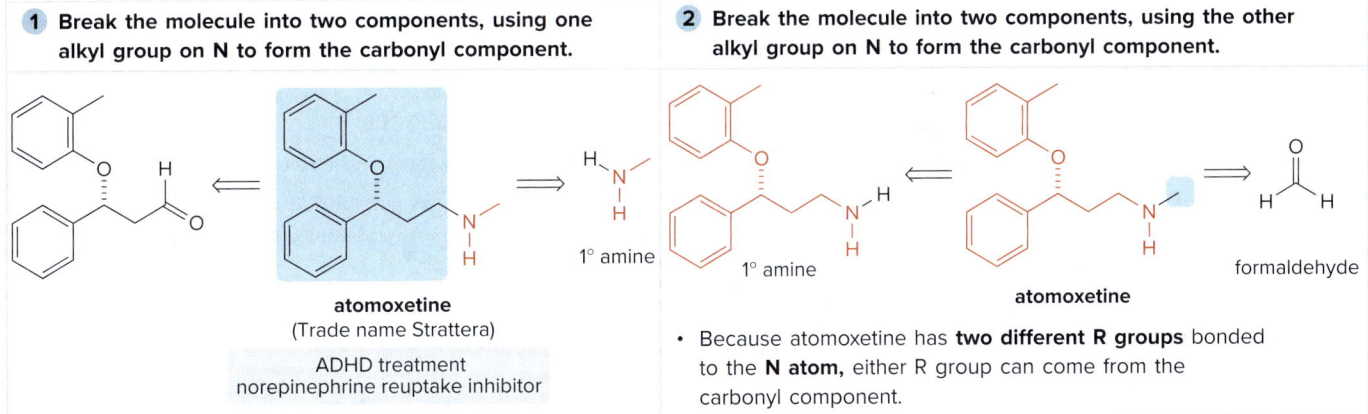

Try Problem 22.57c.

KEY SKILLS

[1] Using retrosynthetic analysis in a reductive amination (22.6C); two possibilities

1 Break the molecule into two components, using one alkyl group on N to form the carbonyl component.

2 Break the molecule into two components, using the other alkyl group on N to form the carbonyl component.

atomoxetine
(Trade name Strattera)
ADHD treatment
norepinephrine reuptake inhibitor

- Because atomoxetine has **two different R groups** bonded to the **N atom,** either R group can come from the carbonyl component.

See Sample Problem 22.2. Try Problems 22.45, 22.51b, 22.52b.

[2] Ranking arylamines in order of increasing basicity (22.9)

1 Identify whether the substituents are **electron donating** or **electron withdrawing**.

2 Rank the compounds.

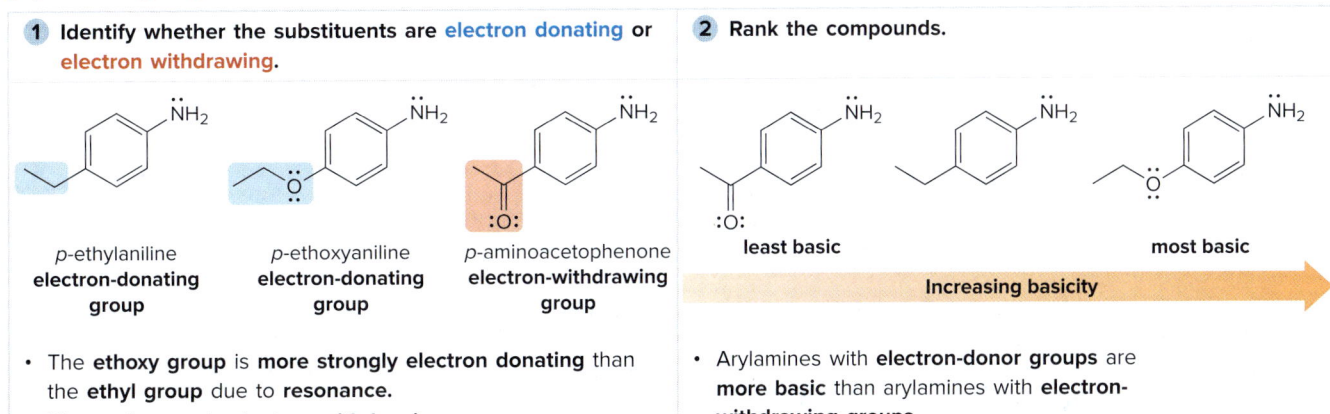

- The **ethoxy group** is **more strongly electron donating** than the **ethyl group** due to **resonance**.
- The **acyl group** is **electron withdrawing**.

- Arylamines with **electron-donor groups** are **more basic** than arylamines with **electron-withdrawing groups**.

See Sample Problem 22.3. Try Problem 22.39b.

[3] Ranking N atoms in order of increasing basicity; example: manzamine C (22.9)

① Identify the different types of nitrogen atoms.	② Evaluate the basicity of the nitrogen atoms.
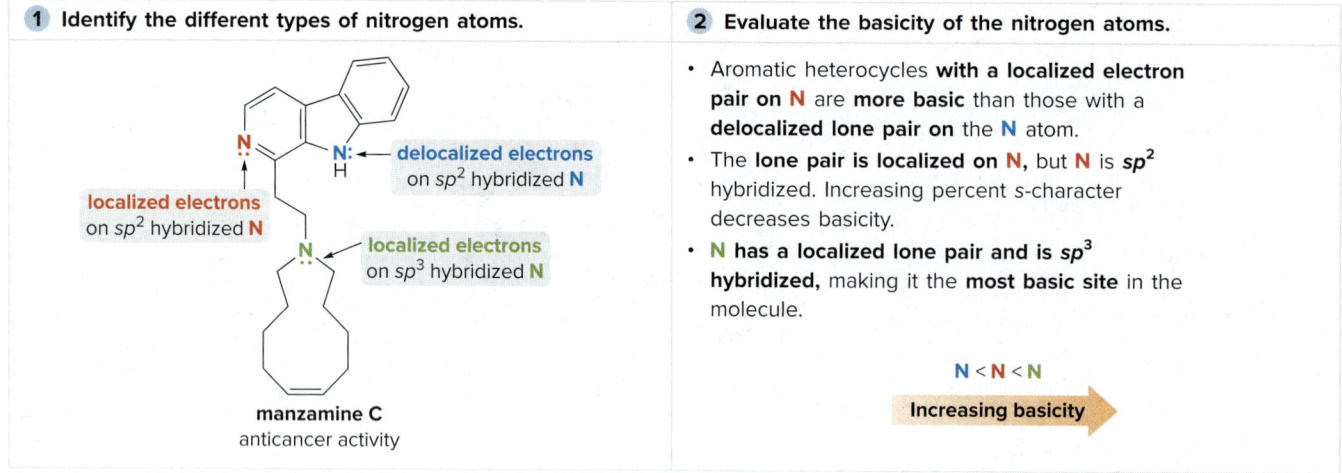	• Aromatic heterocycles **with a localized electron pair on N** are **more basic** than those with a **delocalized lone pair on** the **N** atom. • The **lone pair is localized on N**, but **N** is sp^2 hybridized. Increasing percent s-character decreases basicity. • **N** has a localized lone pair and is sp^3 hybridized, making it the **most basic site** in the molecule. N < N < N **Increasing basicity** →

See Sample Problem 22.4. Try Problems 22.36a, 22.40–22.42.

[4] Drawing the major and minor product formed from Hofmann elimination (22.11)

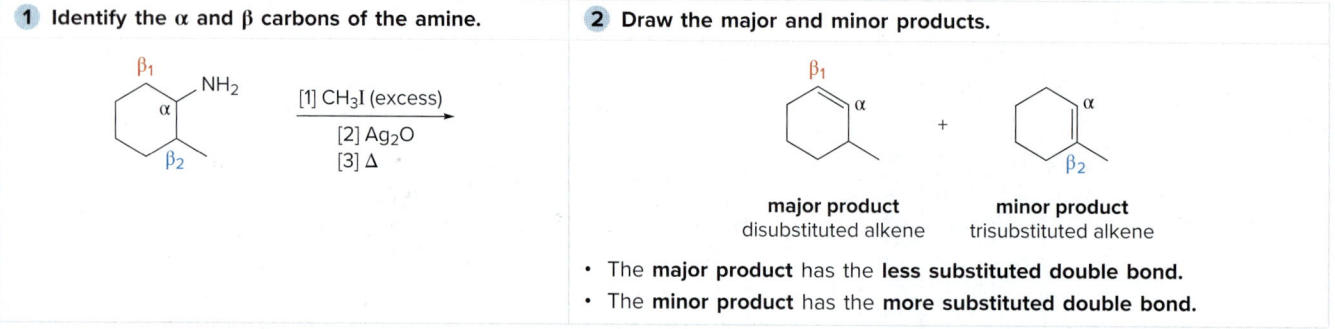

• The **major product** has the **less substituted double bond**.
• The **minor product** has the **more substituted double bond**.

See Sample Problem 22.5. Try Problems 22.50, 22.51c, 22.52c, 22.53j, 22.54.

[5] Devising a synthesis using diazonium salts (22.13)

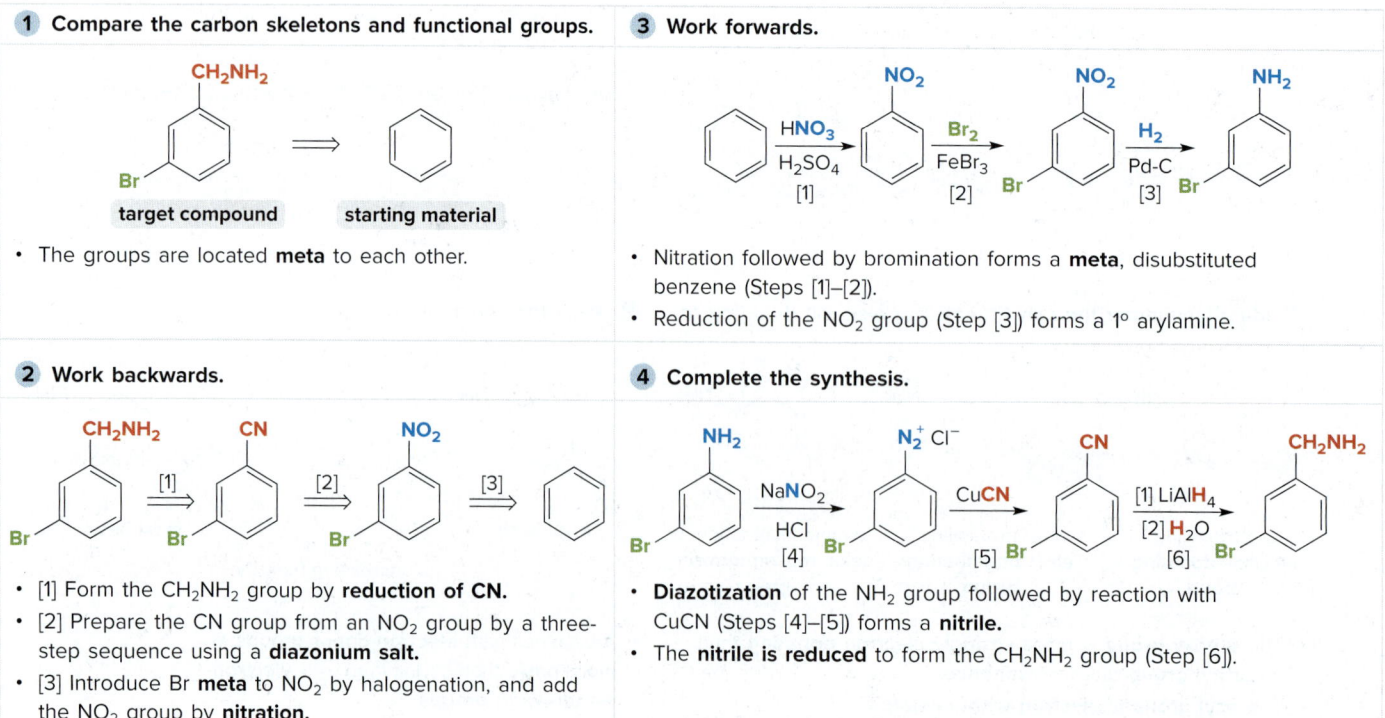

• [1] Form the CH₂NH₂ group by **reduction of CN**.
• [2] Prepare the CN group from an NO₂ group by a three-step sequence using a **diazonium salt**.
• [3] Introduce Br **meta** to NO₂ by halogenation, and add the NO₂ group by **nitration**.

• Nitration followed by bromination forms a **meta**, disubstituted benzene (Steps [1]–[2]).
• Reduction of the NO₂ group (Step [3]) forms a 1° arylamine.

• **Diazotization** of the NH₂ group followed by reaction with CuCN (Steps [4]–[5]) forms a **nitrile**.
• The **nitrile is reduced** to form the CH₂NH₂ group (Step [6]).

See Sample Problems 22.6 and 22.7. Try Problems 22.61–22.66.

[6] Drawing the starting materials needed to synthesize an azo compound (22.14)

1 Break the molecule into two components at the specified C—N bond.

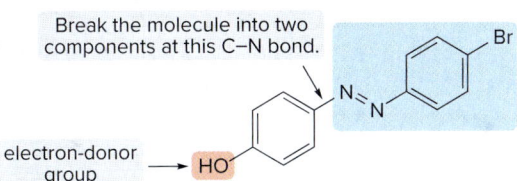

- The **diazonium ion** must be bonded to the ring that is *not* bonded to the electron-donor group.

2 Draw the diazonium ion and the aromatic starting material with a strong electron-donor group.

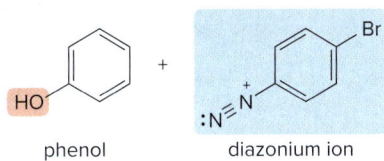

See Sample Problem 22.8. Try Problems 22.61c, 22.62c, 22.64c.

PROBLEMS

Problems Using Three-Dimensional Models

22.35 Give a systematic or common name for each compound.

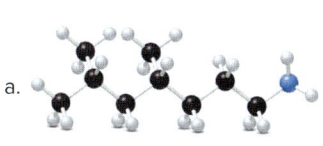

22.36 Varenicline (trade name Chantix) is a drug used to help smokers quit their habit. (a) Which N atom in varenicline is most basic? Explain your choice. (b) What product is formed when varenicline is treated with HCl?

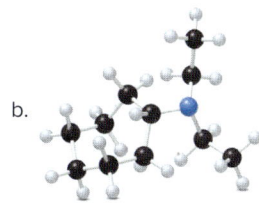

varenicline

Nomenclature

22.37 Give a systematic or common name for each compound.

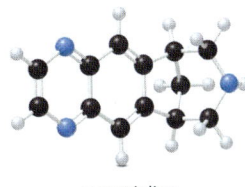

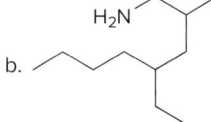

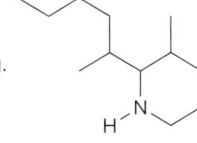

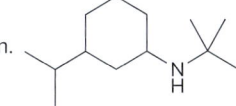

22.38 Draw the structure that corresponds to each name.
a. *N*-isobutylcyclopentanamine
b. tri-*tert*-butylamine
c. *N,N*-diisopropylaniline
d. *N*-methylpyrrole
e. *N*-methylcyclopentanamine
f. 3-methylhexan-2-amine
g. 2-*sec*-butylpiperidine
h. (*S*)-heptan-2-amine

Basicity

22.39 Rank the compounds in each group in order of increasing basicity.

a. 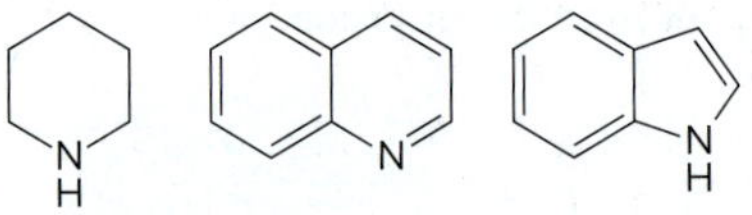 [piperidine, quinoline, indole structures]

b. 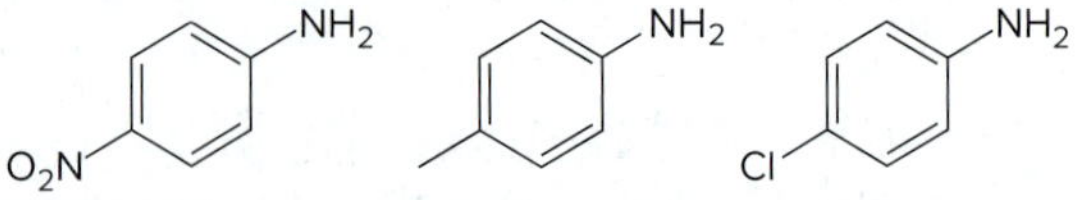 [4-nitroaniline, 4-methylaniline, 4-chloroaniline structures]

22.40 Decide which N atom in each molecule is most basic, and draw the product formed when each compound is treated with CH_3CO_2H. Zolpidem (trade name Ambien) is used to treat insomnia, whereas aripiprazole (trade name Abilify) is used to treat depression, schizophrenia, and bipolar disorders.

a. 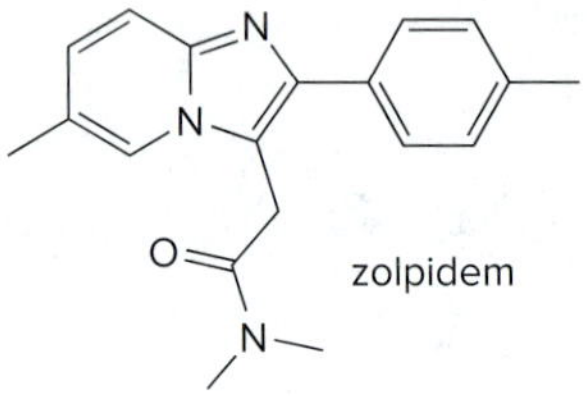 zolpidem

b. 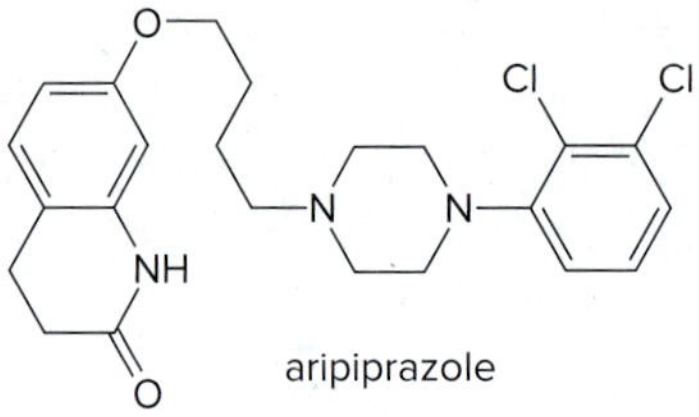 aripiprazole

22.41 Rank the labeled N atoms in the anticancer drug imatinib (trade name Gleevec) in order of increasing basicity. Imatinib, sold as a salt with methanesulfonic acid (CH_3SO_3H), is used for the treatment of chronic myeloid leukemia as well as certain gastrointestinal tumors.

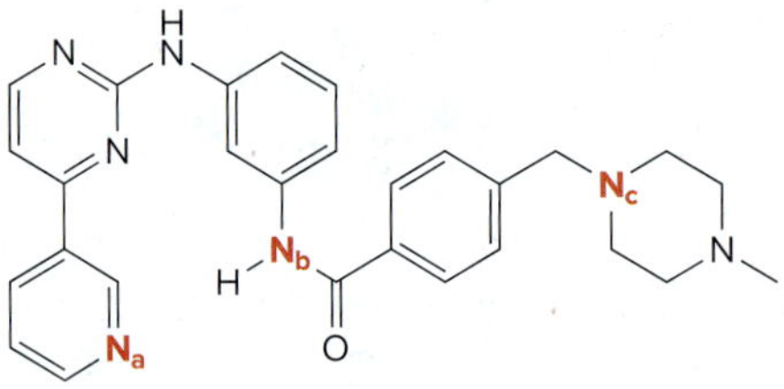

imatinib

22.42 Rank the labeled nitrogen atoms in each compound in order of increasing basicity. Histamine (Section 22.5B) causes the runny nose and watery eyes associated with allergies, and trazodone is a drug used as a sedative and antidepressant.

a. histamine

b. trazodone

22.43 Explain why *m*-nitroaniline is a stronger base than *p*-nitroaniline.

Preparation of Amines

22.44 What amide(s) can be used to prepare each amine by reduction?

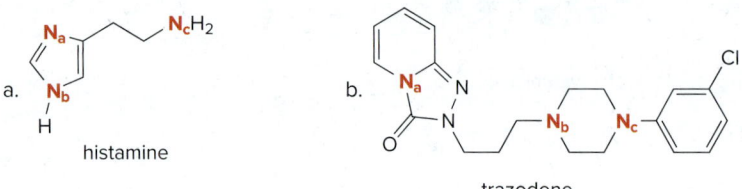

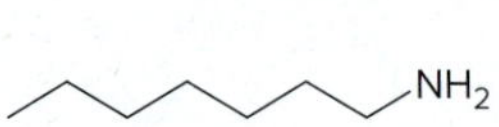

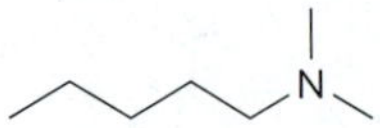

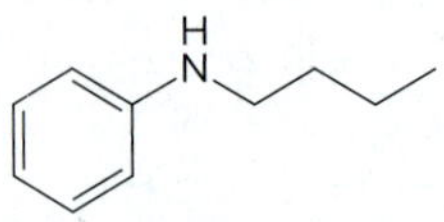

22.45 What carbonyl and nitrogen compounds are needed to make each compound by reductive amination? When more than one set of starting materials is possible, give all possible methods.

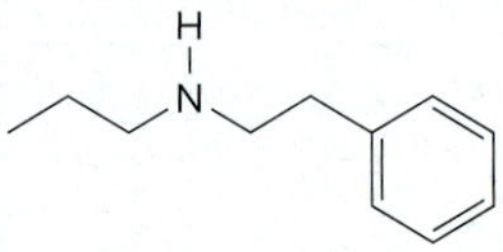

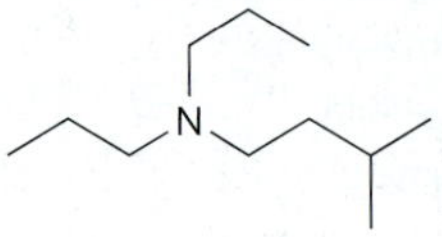

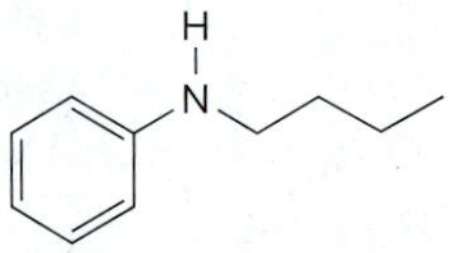

22.46 Draw the product of each reductive amination reaction.

a. C₆H₅—C(=O)—CH₃ + CH₃CH₂CH₂CH₂NH₂, NaBH₃CN

b. cycloheptanone + (CH₃)₂NH, NaBH₃CN

c. C₆H₅CH₂CH₂CHO + NH₃, NaBH₃CN

d. CH₃—C(=O)—CH₂CH₃ + cyclohexyl-NH₂, NaBH₃CN

22.47 One step in the synthesis of lisinopril (Section 5.6, Problem 5.16), a drug used to treat high blood pressure, involves the reaction of **A** with **B** in the presence of a reducing agent to form **C**. What is the structure of **C**?

A (PhCH₂CH₂C(=O)C(=O)OEt) + B (trifluoroacetamido-lysine-proline derivative) → (NaBH₃CN) → **C**

Extraction

22.48 How would you separate toluene (C₆H₅CH₃), benzoic acid (C₆H₅CO₂H), and aniline (C₆H₅NH₂) by an extraction procedure?

Reactions

22.49 Draw the products formed when *p*-methylaniline (*p*-CH₃C₆H₄NH₂) is treated with each reagent.
a. HCl
b. CH₃COCl
c. (CH₃CO)₂O
d. excess CH₃I
e. (CH₃)₂C=O
f. CH₃COCl, AlCl₃
g. CH₃CO₂H
h. NaNO₂, HCl
i. Part (b), then CH₃COCl, AlCl₃
j. CH₃CHO, NaBH₃CN

22.50 Draw the products formed when each amine is treated with [1] CH₃I (excess); [2] Ag₂O; [3] Δ. Indicate the major product when a mixture results.

a. 2-aminooctane
b. N-ethyl-N-isopropylamine
c. 1-amino-1,2-dimethylcyclopentane
d. 2-propylpiperidine

22.51 Answer the following questions about amine **X**, an intermediate in the synthesis of galantamine, a drug used to treat mild to moderate dementia.

a. What amides can be reduced to form **X**?
b. What starting materials can be used to form **X** by reductive amination? Draw all possible methods.
c. What products are formed by Hofmann elimination from **X**?

22.52 Answer the following questions about atomoxetine, a drug used to treat attention deficit hyperactivity disorder (ADHD).

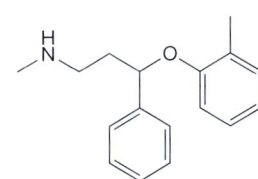

atomoxetine

a. What amides can be reduced to form atomoxetine?
b. What starting materials can be used to form atomoxetine by reductive amination? Draw all possible methods.
c. What products are formed by Hofmann elimination of atomoxetine?

22.53 Draw the organic products formed in each reaction.

a. 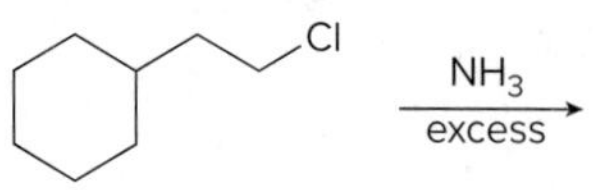 cyclohexyl-CH₂CH₂-Cl + NH₃ (excess) →

b. 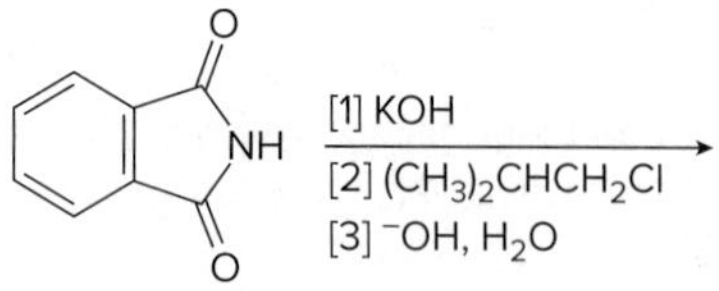 phthalimide [1] KOH [2] (CH₃)₂CHCH₂Cl [3] ⁻OH, H₂O →

c. 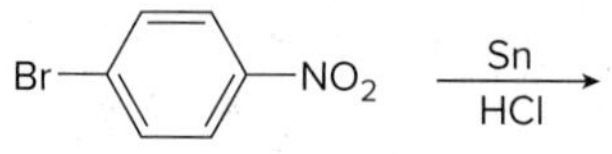 Br—C₆H₄—NO₂ + Sn/HCl →

d. 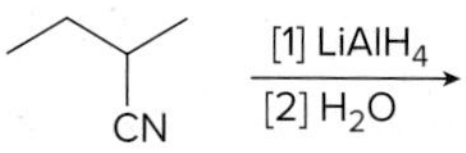 (CH₃)₂CH-CN [1] LiAlH₄ [2] H₂O →

e. 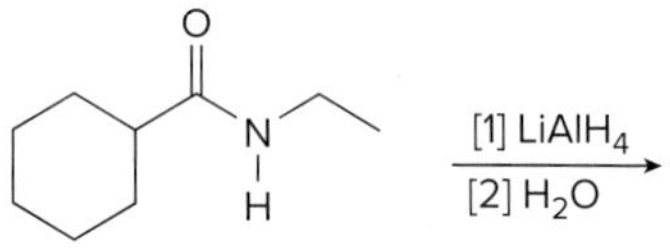 cyclohexyl-C(=O)-NH-CH₂CH₃ [1] LiAlH₄ [2] H₂O →

f. 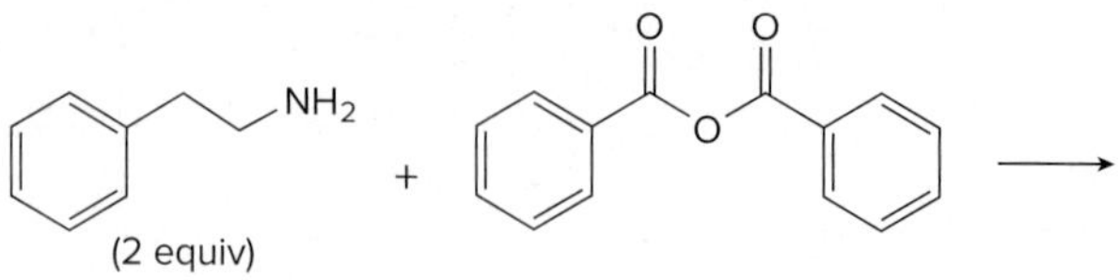 C₆H₅-CH₂CH₂-NH₂ (2 equiv) + (PhCO)₂O →

g. 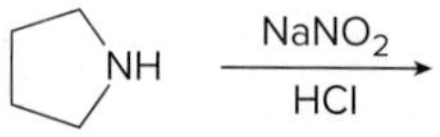 pyrrolidine (NH) + NaNO₂/HCl →

h. 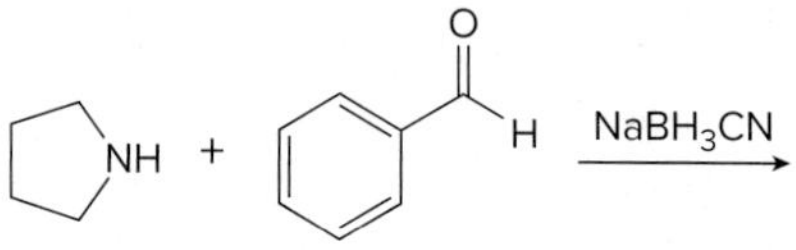 pyrrolidine (NH) + C₆H₅CHO + NaBH₃CN →

i. 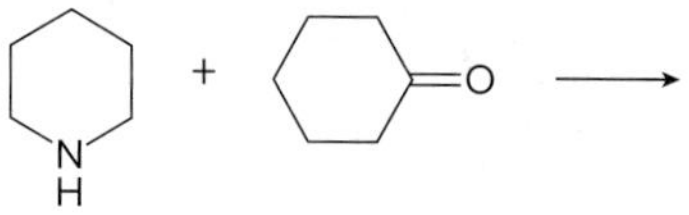 piperidine (NH) + cyclohexanone →

j. 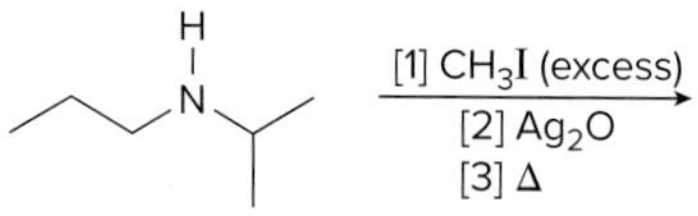 CH₃CH₂CH₂-N(H)-CH(CH₃)₂ [1] CH₃I (excess) [2] Ag₂O [3] Δ →

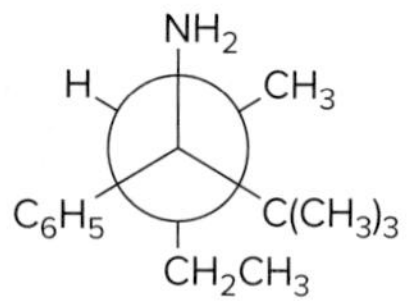

22.54 What is the major Hofmann elimination product formed from each amine?

a. 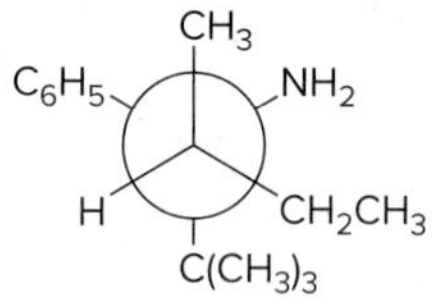 Newman projection: front — C₆H₅, CH₃, NH₂; back — H, CH₂CH₃, C(CH₃)₃

b. Newman projection: front — H, NH₂, CH₃; back — C₆H₅, C(CH₃)₃, CH₂CH₃

c. Newman projection: front — C₆H₅, CH₃, H; back — (CH₃)₃C, NH₂, CH₂CH₃

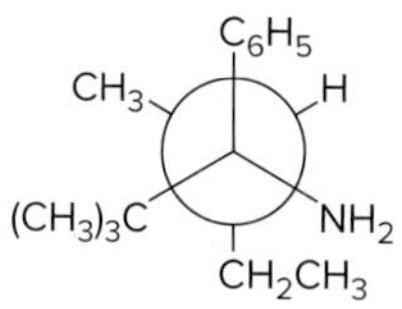

22.55 Identify A, B, and C, three intermediates in the synthesis of the pain reliever and anesthetic fentanyl.

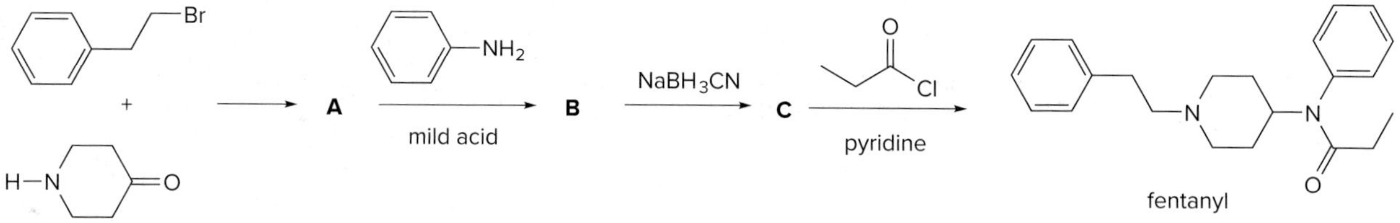

C₆H₅CH₂CH₂Br + H-N(piperidone, C=O) → **A** → (with C₆H₅NH₂, mild acid) → **B** → (NaBH₃CN) → **C** → (CH₃CH₂C(=O)Cl, pyridine) → fentanyl

22.56 Aprepitant (trade name Emend) is used to prevent the acute nausea and vomiting caused by chemotherapy. Identify E and F, intermediates in the synthesis of aprepitant.

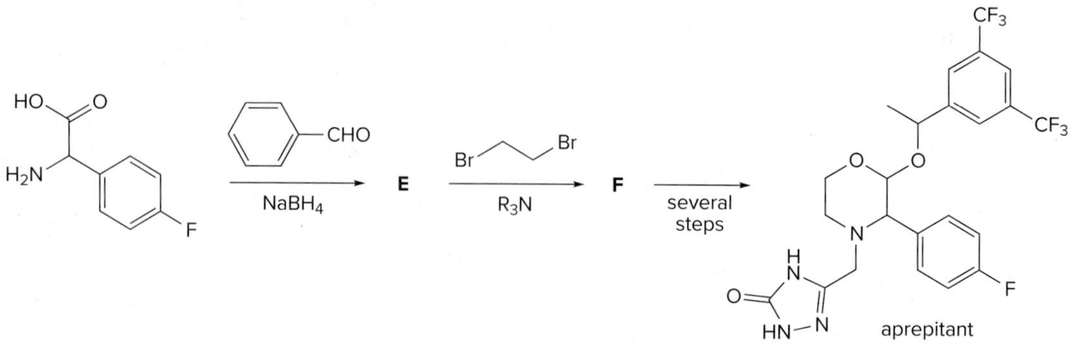

H₂N-CH(4-F-C₆H₄)-COOH + C₆H₅CHO, NaBH₄ → **E** → (Br-CH₂CH₂-Br, R₃N) → **F** → several steps → aprepitant

Problems

22.57 Draw the product formed when **A** is treated with each series of reagents.

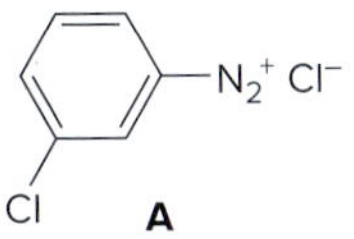

Compound **A**: 3-chlorobenzenediazonium chloride (Ar–N₂⁺ Cl⁻ with Cl at meta position)

a. [1] H₂O; [2] NaH; [3] CH₃Br
b. [1] CuCN; [2] DIBAL-H; [3] H₂O
c. [1] C₆H₅NH₂; [2] CH₃COCl

Mechanism

22.58 Draw a stepwise mechanism for each reaction.

a. 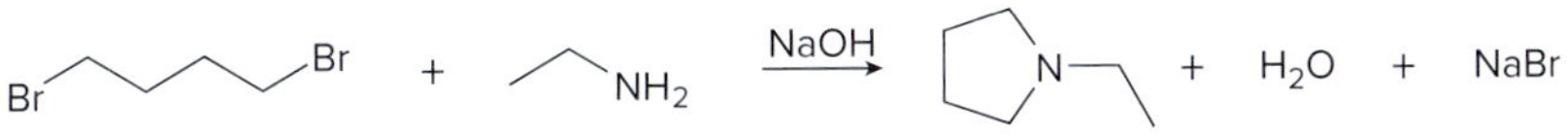 Br(CH₂)₄Br + CH₃CH₂NH₂ →(NaOH) N-ethylpyrrolidine + H₂O + NaBr

b. 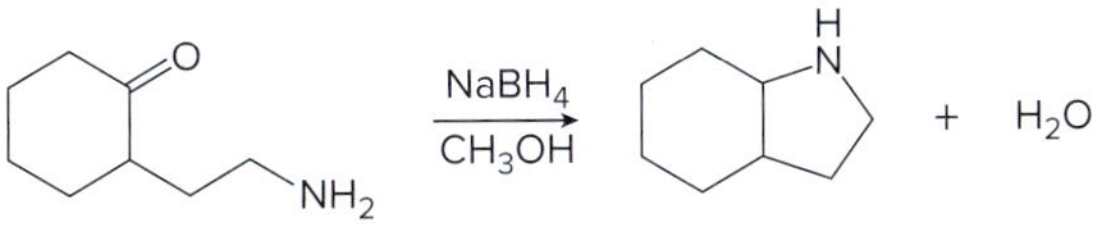 2-(2-aminoethyl)cyclohexanone →(NaBH₄, CH₃OH) octahydroindole + H₂O

22.59 Draw a stepwise mechanism for the following reaction.

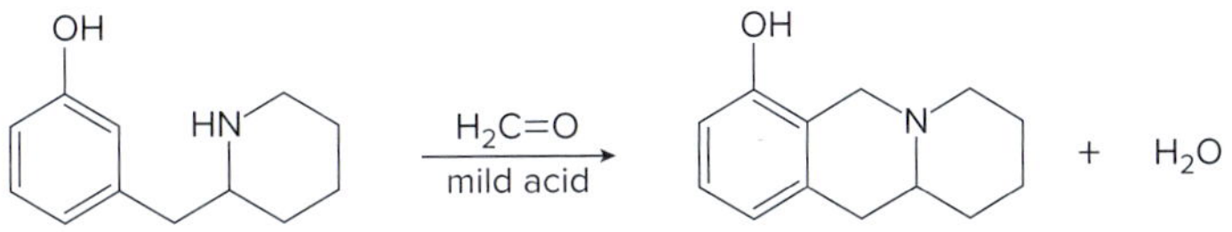

[3-hydroxyphenyl bearing a (piperidin-2-yl)methyl group] + H₂C=O (mild acid) → fused tricyclic hydroxy-quinolizidine + H₂O

22.60 One synthesis of the cholesterol-lowering drug atorvastatin (trade name Lipitor, Section 25.8, Problem 17.46) involves the construction of the pyrrole by reaction of diketone **X** with amine **Y**. Draw a stepwise mechanism for this reaction.

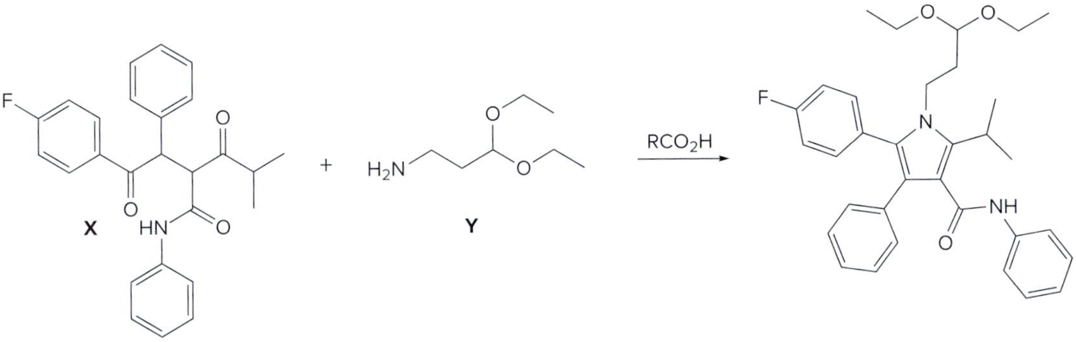

X (diketone with 4-fluorophenyl, phenyl, isopropyl groups, and an anilide) + **Y** (H₂N–CH₂CH₂–CH(OEt)₂) →(RCO₂H) atorvastatin pyrrole core

Synthesis

22.61 Devise a synthesis of each compound from benzene. You may use alcohols with one or two carbons and any inorganic reagents.

a. 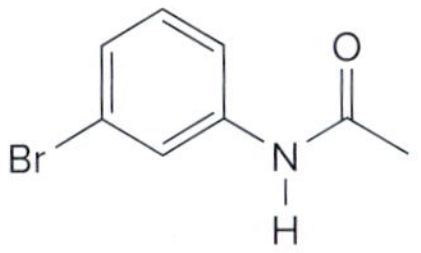 3-bromo-N-acetylaniline (Br on ring, NHC(O)CH₃)
b.  1-bromo-2-ethoxybenzene
c. 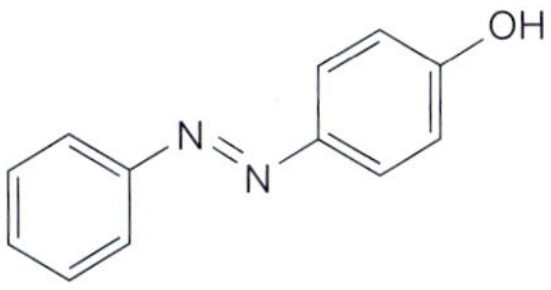 4-hydroxy-azobenzene (C₆H₅–N=N–C₆H₄–OH)

22.62 Devise a synthesis of each compound from aniline (C₆H₅NH₂) as starting material.

a. 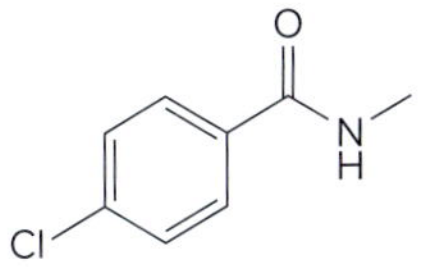 4-chloro-N-methylbenzamide
b. 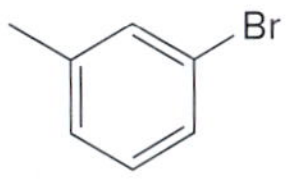 3-bromotoluene
c.  2-hydroxy-5-methyl-azobenzene (C₆H₅–N=N–Ar with OH and CH₃)

22.63 Safrole, which is isolated from sassafras, can be converted to the illegal stimulant MDMA (3,4-methylenedioxymethamphetamine, "Ecstasy") by a variety of methods. (a) Devise a synthesis that begins with safrole and uses a nucleophilic substitution reaction to introduce the amine. (b) Devise a synthesis that begins with safrole and uses reductive amination to introduce the amine.

22.64 Synthesize each compound from benzene. Use a diazonium salt as one of the synthetic intermediates.

22.65 Devise a synthesis of each biologically active compound from benzene.

22.66 Devise a synthesis of each compound from benzene, any organic alcohols having four or fewer carbons, and any required reagents.

Spectroscopy

22.67 Identify the parent and propose a structure for the base peak in the mass spectrum of butan-1-amine.

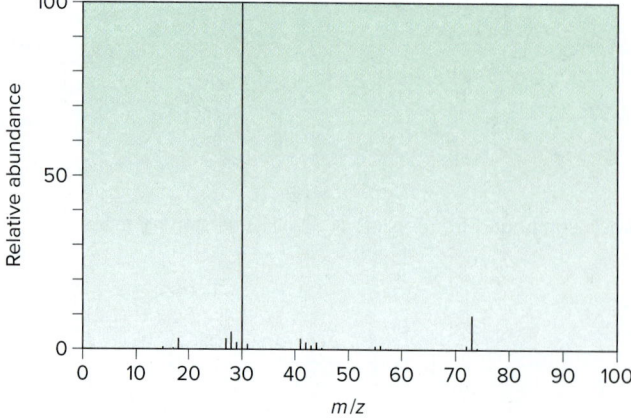

22.68 Three isomeric compounds, **A, B,** and **C,** all have molecular formula $C_8H_{11}N$. The ^1H NMR and IR spectral data of **A, B,** and **C** are given below. What are their structures?

Compound **A:** IR peak at 3400 cm^{-1}

Compound **B:** IR peak at 3310 cm^{-1}

Compound **C:** IR peaks at 3430 and 3350 cm^{-1}

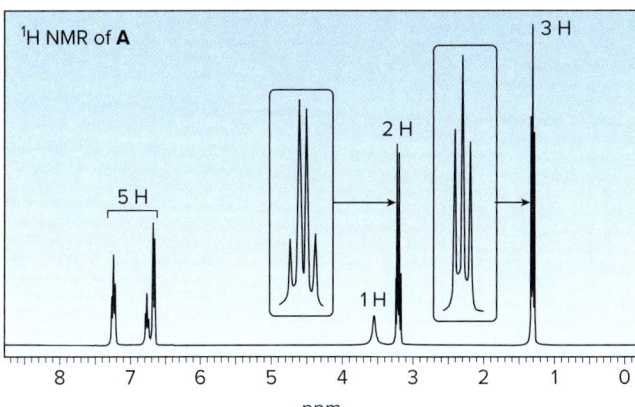

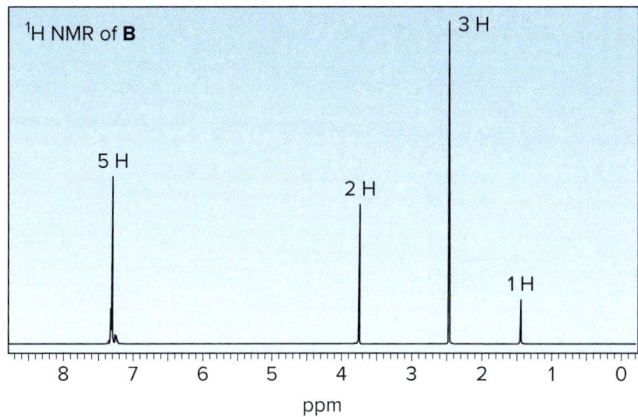

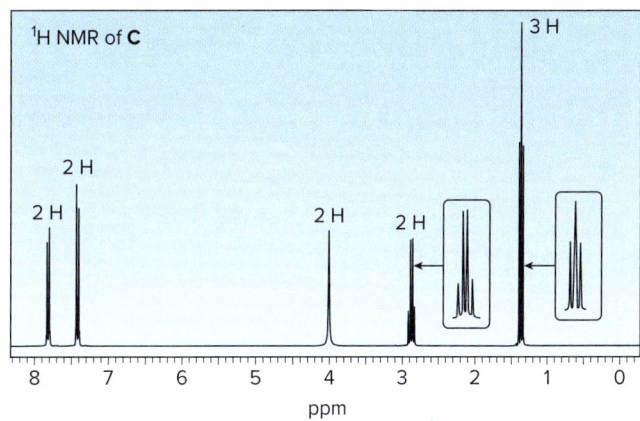

Challenge Problems

22.69 The pK_a of the conjugate acid of guanidine is 13.6, making it one of the strongest neutral organic bases. Offer an explanation.

guanidine pK_a = 13.6

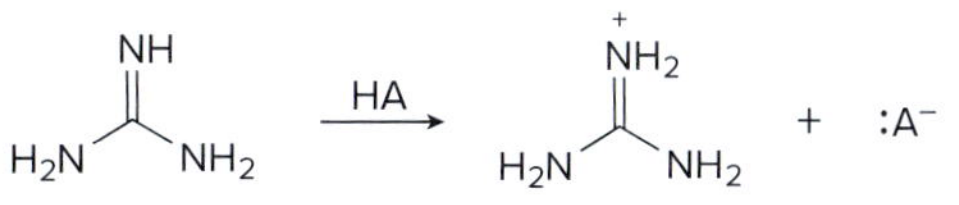

22.70 Rank the following compounds in order of increasing basicity and explain the order you chose.

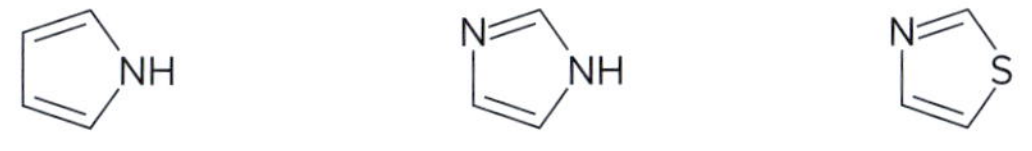

pyrrole imidazole thiazole

22.71 Draw the product **Y** of the following reaction sequence. **Y** was an intermediate in the remarkable synthesis of cyclooctatetraene by Richard Willstatter in 1911.

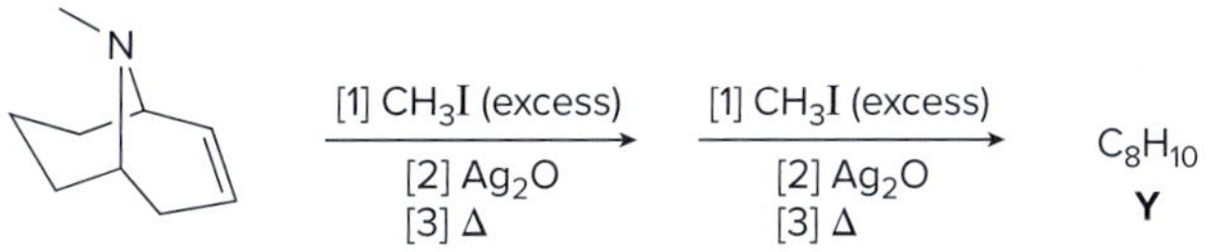

22.72 Devise a synthesis of each compound from the given starting material(s). Albuterol is a bronchodilator and proparacaine is a local anesthetic.

a. 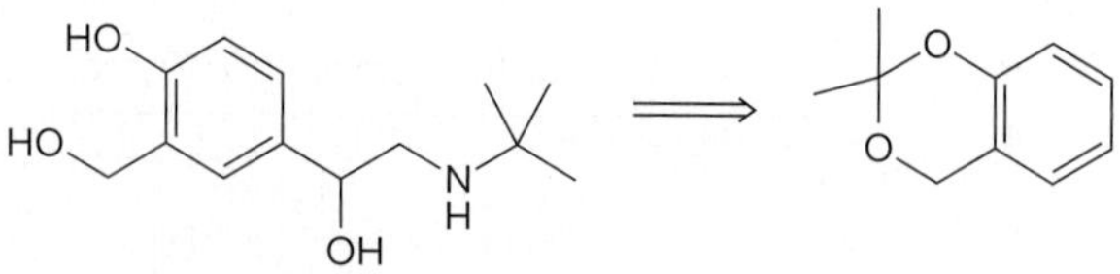 albuterol

b. 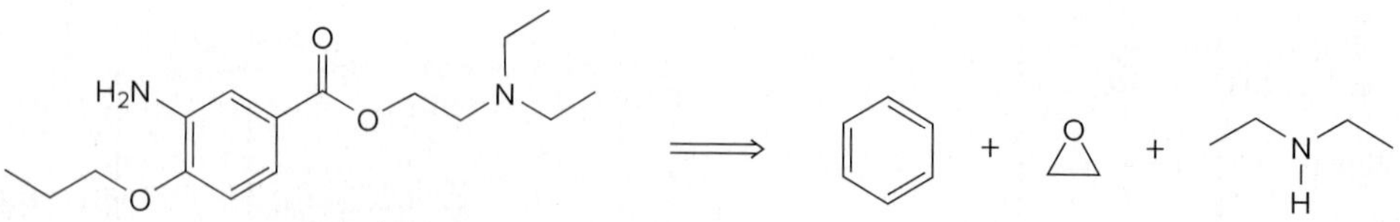 proparacaine

22.73 Heating compound **X** with aqueous formaldehyde forms **Y** ($C_{17}H_{23}NO$), which has been converted to a mixture of lupinine and epilupinine, alkaloids isolated from lupin, a perennial ornamental plant commonly seen on the roadside in parts of Alaska (Figure 22.3). Identify **Y** and explain how it is formed.

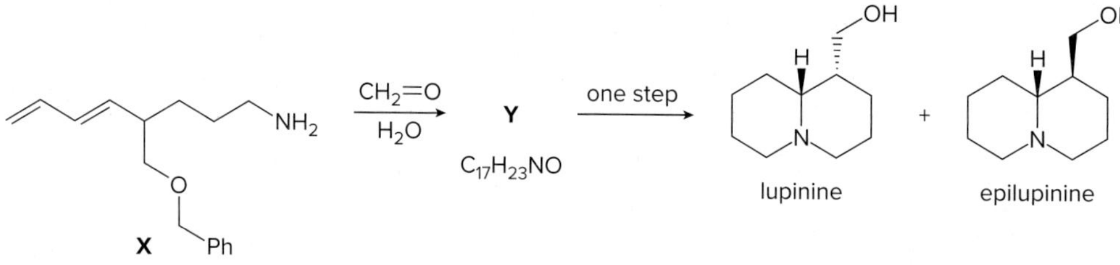

Amino Acids and Proteins

23

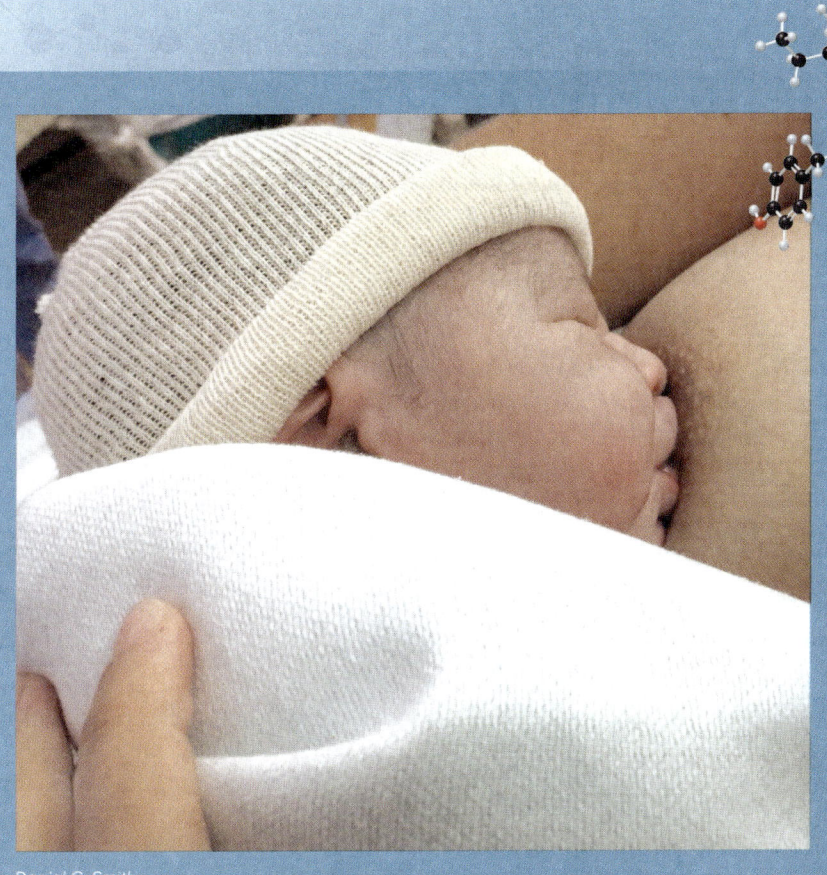

Daniel C. Smith

- **23.1** Amino acids
- **23.2** Separation of amino acids
- **23.3** Enantioselective synthesis of amino acids
- **23.4** Peptides
- **23.5** Peptide sequencing
- **23.6** Peptide synthesis
- **23.7** Automated peptide synthesis
- **23.8** Protein structure
- **23.9** Important proteins
- **23.10** Enzymes

Oxytocin, a peptide consisting of nine amino acids, is a hormone that causes cervical dilation in preparation for childbirth and uterine contractions during labor, and it also stimulates the flow of milk in nursing mothers. Oxytocin was the first peptide hormone synthesized, a feat for which Vincent du Vigneaud was awarded the 1955 Nobel Prize in Chemistry. Oxytocin, sold under the trade name Pitocin, is used to induce labor and to stop bleeding after a delivery. In Chapter 23, we learn about peptides like oxytocin and the amino acids that compose them.

Why Study... Amino Acids and Proteins?

Of the four major groups of biomolecules—lipids, carbohydrates, nucleic acids, and proteins—proteins have the widest array of functions. **Keratin** and **collagen,** for example, are part of a large group of structural proteins that form long insoluble fibers, giving strength and support to tissues. Hair, horns, hooves, and fingernails are all made up of keratin. **Collagen** is found in bone, connective tissue, tendons, and cartilage. **Enzymes** are proteins that catalyze and regulate all aspects of cellular function. **Membrane proteins** transport small organic molecules and ions across cell membranes. **Insulin,** the hormone that regulates blood glucose levels, **fibrinogen** and **thrombin,** which form blood clots, and **hemoglobin,** which transports oxygen from the lungs to tissues, are all proteins.

In Chapter 23 we discuss proteins and their primary components, the amino acids.

23.1 Amino Acids

Amino acids were previously discussed in Sections 3.9A and 15.12.

Naturally occurring amino acids have an amino group (NH_2) bonded to the α carbon of a carboxy group (COOH), so they are called **α-amino acids.**

- All proteins are polyamides formed by joining amino acids together.

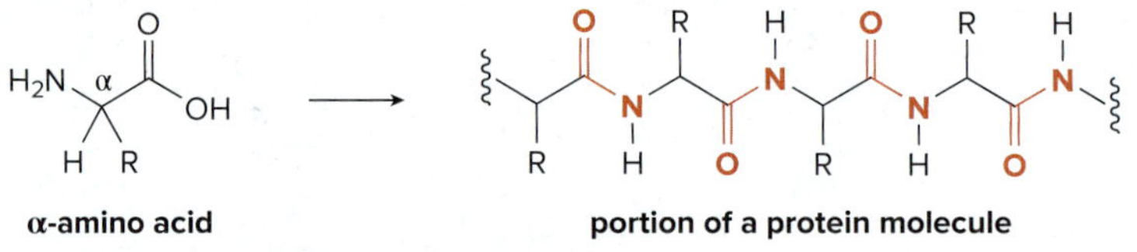

α-amino acid → portion of a protein molecule

23.1A General Features of α-Amino Acids

The 20 amino acids that occur naturally in proteins differ in the identity of the R group bonded to the α carbon. The R group is called the **side chain** of the amino acid.

The simplest amino acid, called glycine, has R = H. **All other amino acids (R ≠ H) have a stereogenic center on the α carbon.** As is true for monosaccharides, the prefixes D and L are used to designate the configuration at the stereogenic center of amino acids. Common, naturally occurring amino acids are called **L-amino acids.** Their enantiomers, D-amino acids, are rarely found in nature. These general structures are shown in Figure 23.1. According to R,S designations, all L-amino acids except cysteine have the ***S* configuration.**

Figure 23.1
The general features of an α-amino acid

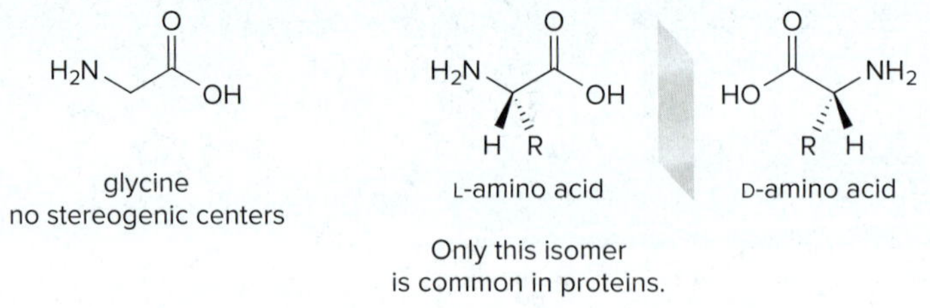

glycine
no stereogenic centers

L-amino acid
Only this isomer is common in proteins.

D-amino acid

All amino acids have common names. These names can be represented by either a one-letter or a three-letter abbreviation. Figure 23.2 is a listing of the 20 naturally occurring amino acids, with their abbreviations. Note the variability in the R groups. A side chain can be a simple alkyl group, or it can have additional functional groups such as OH, SH, COOH, or NH_2.

- Amino acids with an additional COOH group in the side chain are called *acidic* amino acids.
- Those with an additional basic N atom in the side chain are called *basic* amino acids.
- All others are neutral amino acids.

Figure 23.2 The 20 naturally occurring amino acids

Neutral amino acids

Name	Structure	Abbreviations	Name	Structure	Abbreviations
Alanine		Ala A	Phenylalanine*		Phe F
Asparagine		Asn N	Proline	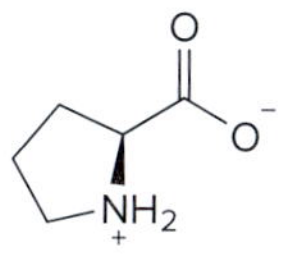	Pro P
Cysteine		Cys C	Serine	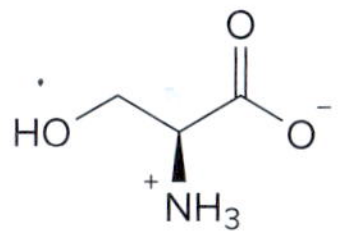	Ser S
Glutamine		Gln Q	Threonine*	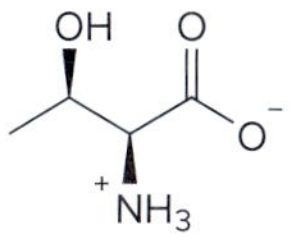	Thr T
Glycine		Gly G	Tryptophan*	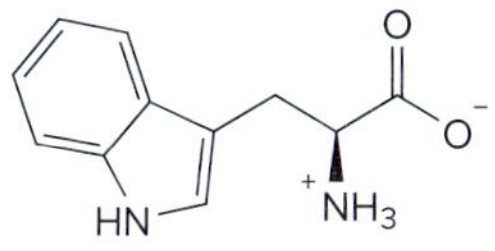	Trp W
Isoleucine*		Ile I	Tyrosine	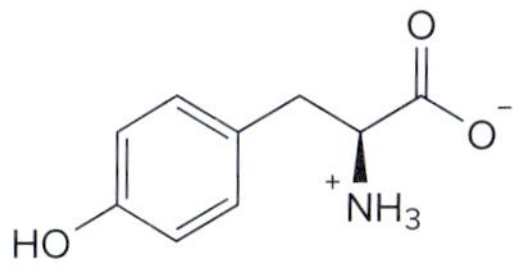	Tyr Y
Leucine*		Leu L	Valine*	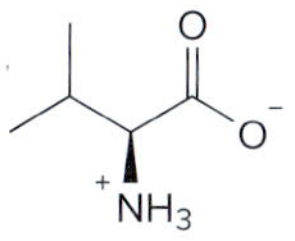	Val V
Methionine*		Met M			

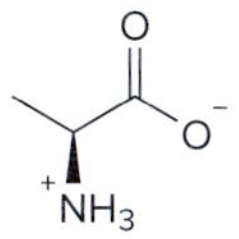

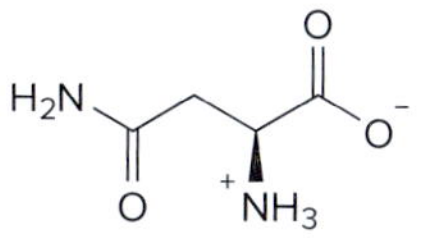

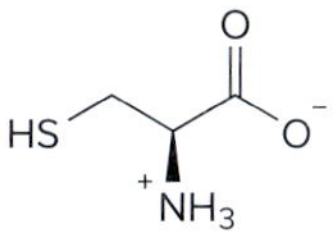

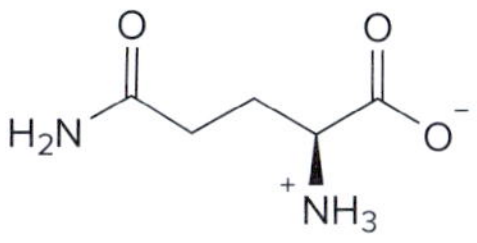

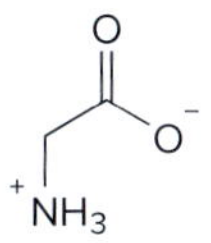

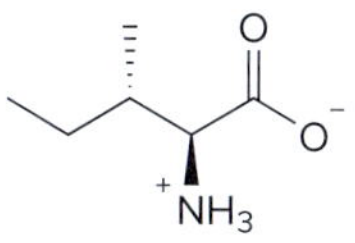

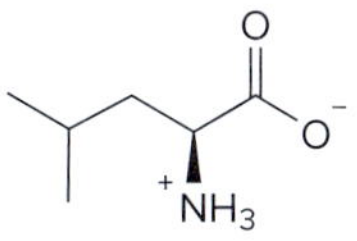

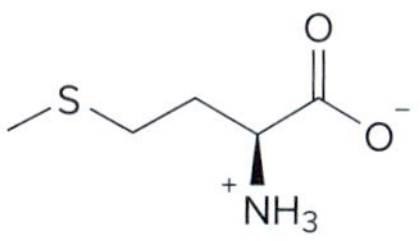

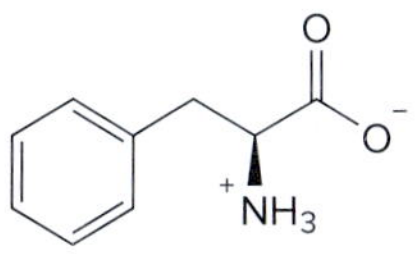

Acidic amino acids

Name	Structure	Abbreviations
Aspartic acid	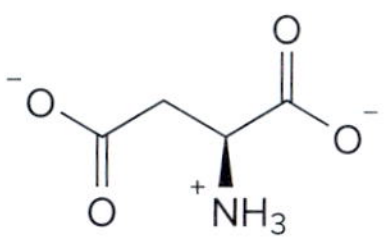	Asp D
Glutamic acid	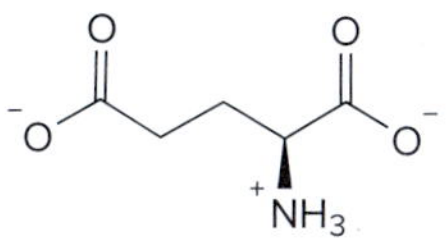	Glu E

Basic amino acids

Name	Structure	Abbreviations
Arginine*	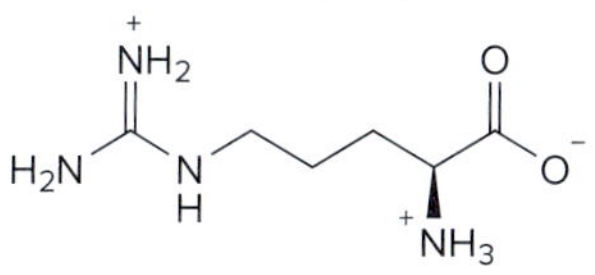	Arg R
Histidine*	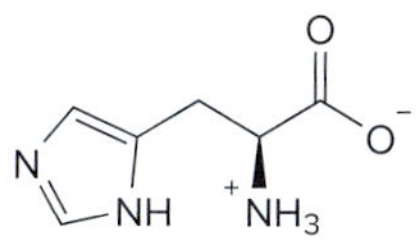	His H
Lysine*	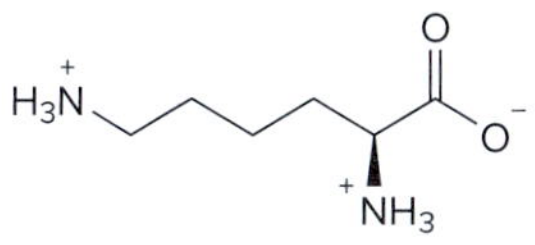	Lys K

Essential amino acids are labeled with an asterisk (*).

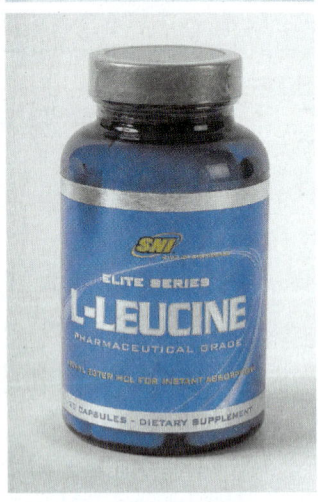

The essential amino acid leucine is sold as a dietary supplement that is used by body builders to help prevent muscle loss and heal muscle tissue after injury. *Jill Braaten*

Look closely at the structures of proline, isoleucine, and threonine.

- **All amino acids are 1° amines except for proline,** which has its N atom in a five-membered ring, making it a **2° amine.**
- **Isoleucine** and **threonine** contain an additional stereogenic center at the β carbon, so there are four possible stereoisomers, only one of which is naturally occurring.

L-proline
2° amine

L-isoleucine

L-threonine

Humans can synthesize only 10 of these 20 amino acids. The remaining 10 are called **essential amino acids** because they must be obtained from the diet. These are labeled with an asterisk in Figure 23.2.

Problem 23.1 Draw the other three stereoisomers of L-isoleucine, and label the stereogenic centers as *R* or *S*.

23.1B Acid–Base Behavior

Recall from Section 3.9A that an amino acid has both an acidic and a basic functional group, so proton transfer forms a salt called a **zwitterion.**

This neutral form of an amino acid does **not** really exist.

The zwitterion is neutral.

This salt is the neutral form of an amino acid.

This form exists at **pH ≈ 6.**

The structures in Figure 23.2 show the charged form of the amino acids at the physiological pH of the blood.

- **Amino acids do not exist to any appreciable extent as uncharged neutral compounds. They exist as salts, giving them high melting points and making them water soluble.**

Amino acids exist in different charged forms, as shown in Figure 23.3, depending on the pH of the aqueous solution in which they are dissolved. For neutral amino acids, the overall charge is +1, 0, or −1. Only at pH ~6 does the zwitterionic form exist.

The —COOH and —NH$_3^+$ groups of an amino acid are ionizable, because they can lose a proton in aqueous solution. As a result, they have different pK_a values. The pK_a of the —COOH group is typically ~2, whereas that of the —NH$_3^+$ group is ~9, as shown in Table 23.1.

Figure 23.3
How the charge of a neutral amino acid depends on the pH

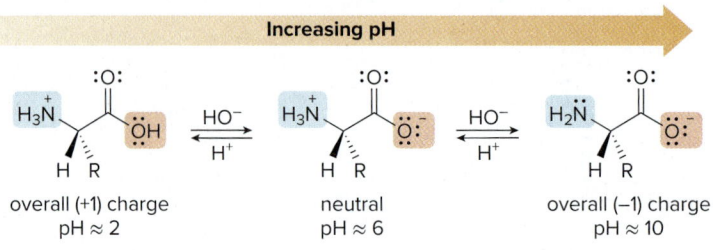

overall (+1) charge
pH ≈ 2

neutral
pH ≈ 6

overall (−1) charge
pH ≈ 10

Table 23.1 pK_a Values for the Ionizable Functional Groups of an α-Amino Acid

Amino acid	α-COOH	α-NH$_3^+$	Side chain	pI
Alanine	2.35	9.87	—	6.11
Arginine	2.01	9.04	12.48	10.76
Asparagine	2.02	8.80	—	5.41
Aspartic acid	2.10	9.82	3.86	2.98
Cysteine	2.05	10.25	8.00	5.02
Glutamic acid	2.10	9.47	4.07	3.08
Glutamine	2.17	9.13	—	5.65
Glycine	2.35	9.78	—	6.06
Histidine	1.77	9.18	6.10	7.64
Isoleucine	2.32	9.76	—	6.04
Leucine	2.33	9.74	—	6.04
Lysine	2.18	8.95	10.53	9.74
Methionine	2.28	9.21	—	5.74
Phenylalanine	2.58	9.24	—	5.91
Proline	2.00	10.60	—	6.30
Serine	2.21	9.15	—	5.68
Threonine	2.09	9.10	—	5.60
Tryptophan	2.38	9.39	—	5.88
Tyrosine	2.20	9.11	10.07	5.63
Valine	2.29	9.72	—	6.00

Some amino acids, such as aspartic acid and lysine, have acidic or basic side chains. These additional ionizable groups complicate somewhat the acid–base behavior of these amino acids. Table 23.1 lists the pK_a values for these acidic and basic side chains as well.

Table 23.1 also lists the isoelectric points (pI) for all of the amino acids. Recall from Section 15.12C that the **isoelectric point is the pH at which an amino acid exists primarily in its neutral form,** and that it can be calculated from the average of the pK_a values of the α-COOH and α-NH$_3^+$ groups (for neutral amino acids only).

Problem 23.2 What form exists at the isoelectric point of each of the following amino acids: (a) valine; (b) leucine; (c) proline; (d) glutamic acid?

Problem 23.3 Explain why the pK_a of the —NH$_3^+$ group of an α-amino acid is lower than the pK_a of the ammonium ion derived from a 1° amine (RNH$_3^+$). For example, the pK_a of the —NH$_3^+$ group of alanine is 9.87 but the pK_a of CH$_3$NH$_3^+$ is 10.63.

Problem 23.4 L-Thyroxine, a thyroid hormone and oral medication used to treat thyroid hormone deficiency, is an amino acid that does not exist in proteins. Draw the zwitterionic form of L-thyroxine.

L-thyroxine

23.2 Separation of Amino Acids

Common methods used to synthesize amino acids in the laboratory yield a racemic mixture. Naturally occurring amino acids exist as a single enantiomer, however, so the two enantiomers obtained must be separated if they are to be used in biological applications. This is not an easy task. Two enantiomers have the same physical properties, so they cannot be separated by common physical methods, such as distillation or chromatography. Moreover, they react in the same way with achiral reagents, so they cannot be separated by chemical reactions either.

Nonetheless, strategies have been devised to separate two enantiomers using physical separation techniques and chemical reactions. We examine two different strategies in Section 23.2. Then, in Section 23.3, we will discuss a method that affords optically active amino acids without the need for separation.

- The separation of a racemic mixture into its component enantiomers is called *resolution*. Thus, a racemic mixture is *resolved* into its component enantiomers.

23.2A Resolution of Amino Acids

The oldest and perhaps still the most widely used method to separate enantiomers exploits the following fact: **enantiomers have the *same* physical properties, but diastereomers have *different* physical properties.** Thus, a racemic mixture can be resolved using the following general strategy.

[1] **Convert a pair of enantiomers to a pair of diastereomers,** which are now separable because they have different melting points and boiling points.

[2] **Separate the diastereomers.**

[3] **Re-convert each diastereomer to the original enantiomer,** now separated from the other.

This general three-step process is illustrated in Figure 23.4.

To resolve a racemic mixture of amino acids such as (R)- and (S)-alanine, the racemate is first treated with acetic anhydride to form **N-acetyl amino acids.** Each of these amides contains one stereogenic center and they are still enantiomers, so they are *still inseparable*.

Both enantiomers of N-acetyl alanine have a free carboxy group that can react with an amine in an acid–base reaction. **If a chiral amine is used, such as (R)-α-methylbenzylamine, the two salts formed are diastereomers, *not* enantiomers.** Diastereomers can be physically separated from each other, so the compound that converts enantiomers to diastereomers is called a **resolving agent.** Either enantiomer of the resolving agent can be used.

23.2 Separation of Amino Acids

Figure 23.4 Resolution of a racemic mixture by converting it to a mixture of diastereomers

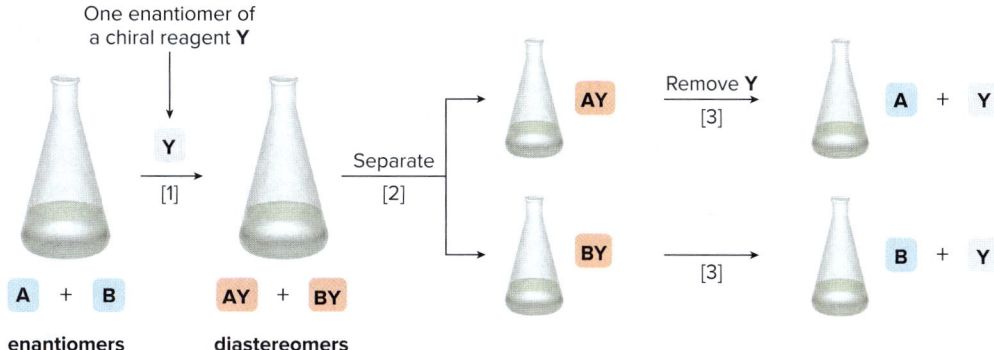

Enantiomers A and B can be separated by reaction with a single enantiomer of a chiral reagent, Y. The process of resolution requires three steps:

[1] Reaction of enantiomers **A** and **B** with **Y** forms two diastereomers, **AY** and **BY**.
[2] Diastereomers **AY** and **BY** have different physical properties, so they can be separated by physical methods such as fractional distillation or crystallization.
[3] **AY** and **BY** are then re-converted to **A** and **B** by a chemical reaction. The two enantiomers **A** and **B** are now separated from each other, and resolution is complete.

How To Use (R)-α-Methylbenzylamine to Resolve a Racemic Mixture of Amino Acids

Step [1] React both enantiomers of an *N*-acetyl amino acid with the *R* isomer of the chiral amine.

These salts have the *same* configuration around one stereogenic center, but the *opposite* configuration about the other stereogenic center.

Step [2] Separate the diastereomers.

—*Continued*

How To, continued...

Step [3] Regenerate the amino acid by hydrolysis of the amide.

[Reaction scheme: Two N-acetyl amide diastereomers each treated with H₂O, ⁻OH to give (S)-alanine (boxed, labeled "The amino acids are now separated.") and (R)-alanine plus the regenerated chiral amine (R)-α-methylbenzylamine ("The chiral amine is also regenerated.")]

Step [1] is just an acid–base reaction in which the racemic mixture of N-acetyl alanines reacts with the same enantiomer of the resolving agent, in this case (R)-α-methylbenzylamine. The salts that form are **diastereomers, *not* enantiomers,** because they have the same configuration about one stereogenic center, but the opposite configuration about the other stereogenic center.

In **Step [2]**, the diastereomers are separated by some physical technique, such as crystallization or distillation.

In **Step [3]**, the amides can be hydrolyzed with aqueous base to regenerate the amino acids. The amino acids are now separated from each other. The optical activity of the amino acids can be measured and compared to their known rotations to determine the purity of each enantiomer.

Problem 23.5 Which of the following amines can be used to resolve a racemic mixture of amino acids?

a. [phenethylamine: PhCH₂CH₂NH₂]
b. [quinuclidine]
c. [(S)-2-aminopropane with wedge bond to NH₂]
d. strychnine (a powerful poison)

Problem 23.6 Write out a stepwise sequence that shows how a racemic mixture of leucine enantiomers can be resolved into optically active amino acids using (R)-α-methylbenzylamine.

23.2B Kinetic Resolution of Amino Acids Using Enzymes

A second strategy used to separate amino acids is based on the fact that two enantiomers react differently with chiral reagents. An **enzyme** is typically used as the chiral reagent.

To illustrate this strategy, we begin again with the two enantiomers of N-acetyl alanine, which were prepared by treating a racemic mixture of (R)- and (S)-alanine with acetic anhydride (Section 23.2A). **Enzymes called acylases hydrolyze amide bonds, such as those found in N-acetyl alanine, but only for amides of L-amino acids.** Thus, when a racemic mixture of N-acetyl alanines is treated with an acylase, only the amide of L-alanine (the S stereoisomer) is hydrolyzed to generate L-alanine, whereas the amide of D-alanine (the R stereoisomer) is untouched. The reaction mixture now consists of one amino acid and one N-acetyl amino acid.

23.3 Enantioselective Synthesis of Amino Acids

Because they have different functional groups with different physical properties, they can be physically separated.

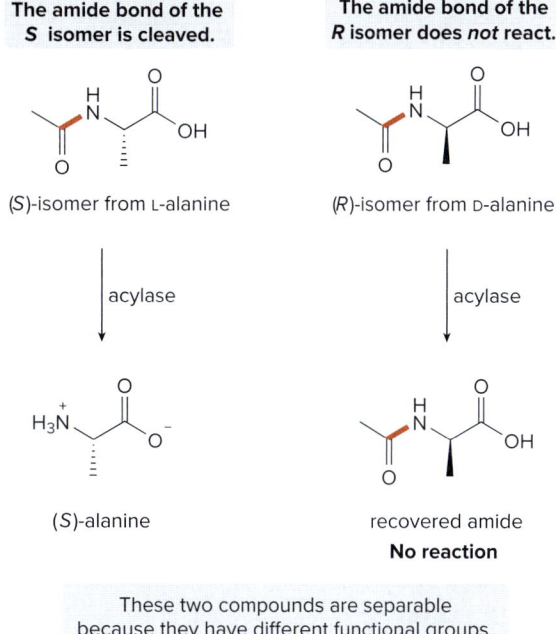

(S)-isomer from L-alanine — The amide bond of the S isomer is cleaved.

(R)-isomer from D-alanine — The amide bond of the R isomer does not react.

↓ acylase

(S)-alanine

recovered amide
No reaction

These two compounds are separable because they have different functional groups.

- Separation of two enantiomers by a chemical reaction that selectively occurs for only one of the enantiomers is called *kinetic resolution*.

Problem 23.7 Draw the organic products formed in the following reaction.

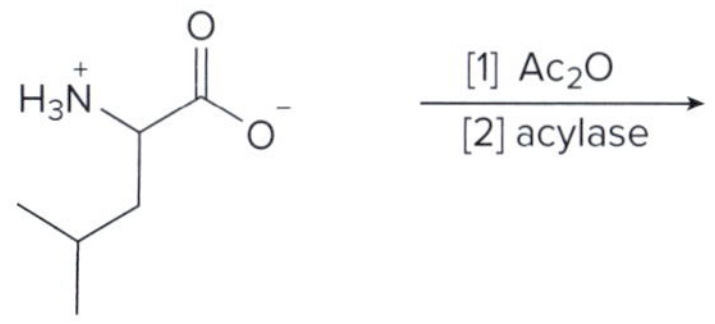

[structure of leucine zwitterion] → [1] Ac$_2$O [2] acylase

(mixture of enantiomers)

23.3 Enantioselective Synthesis of Amino Acids

Although the two methods introduced in Section 23.2 for resolving racemic mixtures of amino acids make enantiomerically pure amino acids available for further research, half of the reaction product is useless because it has the undesired configuration. Moreover, each of these procedures is costly and time-consuming.

If we use a chiral reagent to synthesize an amino acid, however, it is possible to favor the formation of the desired enantiomer over the other, without having to resort to a resolution. For example, single enantiomers of amino acids have been prepared by using **enantioselective (or asymmetric) hydrogenation reactions.** The success of this approach depends on finding a chiral catalyst, in much the same way that a chiral catalyst is used for the Sharpless asymmetric epoxidation (Section 11.14).

The necessary starting material is an alkene. Addition of H$_2$ to the double bond forms an *N*-acetyl amino acid with a new stereogenic center on the α carbon to the carboxy group.

With proper choice of a chiral catalyst, the naturally occurring S configuration can be obtained as product.

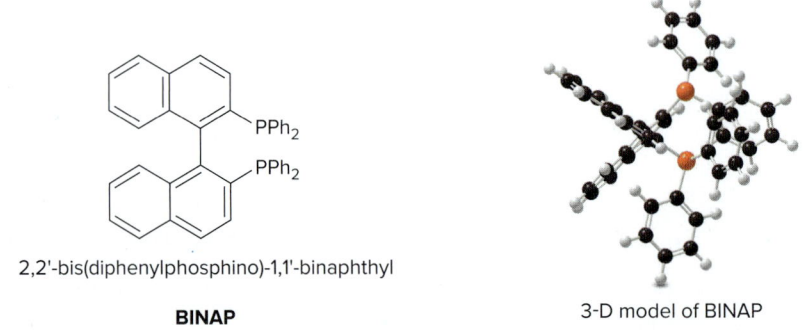

Several chiral catalysts with complex structures have now been developed for this purpose. Many contain **rhodium** as the metal, complexed to a chiral molecule containing one or more phosphorus atoms. One example, abbreviated simply as **Rh***, is drawn below.

This catalyst is synthesized from a rhodium salt and a phosphorus compound, 2,2'-bis(diphenylphosphino)-1,1'-binaphthyl (**BINAP**). It is the BINAP moiety (Figure 23.5) that makes the catalyst chiral.

Figure 23.5
The structure of BINAP

- The two naphthalene rings are oriented at right angles to each other, creating a rigid shape that makes the molecule chiral.

Ryoji Noyori shared the 2001 Nobel Prize in Chemistry for developing methods for asymmetric hydrogenation reactions using the chiral BINAP catalyst.

Twistoflex and helicene (Section 19.5A) are two more aromatic compounds whose shape makes them chiral.

BINAP is one of a small number of molecules that is chiral even though it has no tetrahedral stereogenic centers. Its shape makes it a chiral molecule. The two naphthalene rings of the BINAP molecule are oriented at almost 90° to each other to minimize steric interactions between the hydrogen atoms on adjacent rings. This rigid three-dimensional shape makes BINAP nonsuperimposable on its mirror image, and thus it is a chiral compound.

Enantioselective hydrogenation can be used to synthesize a single stereoisomer of phenylalanine. Treating achiral alkene **A** with H_2 and the chiral rhodium catalyst Rh* forms the S isomer of N-acetyl phenylalanine in 100% ee. Hydrolysis of the acetyl group on nitrogen then yields a single enantiomer of phenylalanine.

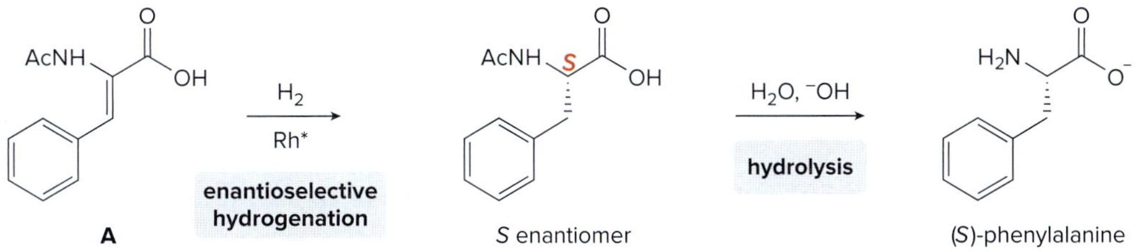

[Reaction scheme: A (AcNH-substituted cinnamic acid derivative) → H₂, Rh* (enantioselective hydrogenation) → S enantiomer, 100% ee → H₂O, ⁻OH (hydrolysis) → (S)-phenylalanine]

Problem 23.8 What alkene is needed to synthesize each amino acid by an enantioselective hydrogenation reaction using H₂ and Rh*: (a) alanine; (b) leucine; (c) glutamine?

23.4 Peptides

When amino acids are joined by amide bonds, they form larger molecules called **peptides** and **proteins**.

- A *dipeptide* has two amino acids joined together by *one* amide bond.
- A *tripeptide* has three amino acids joined together by *two* amide bonds.

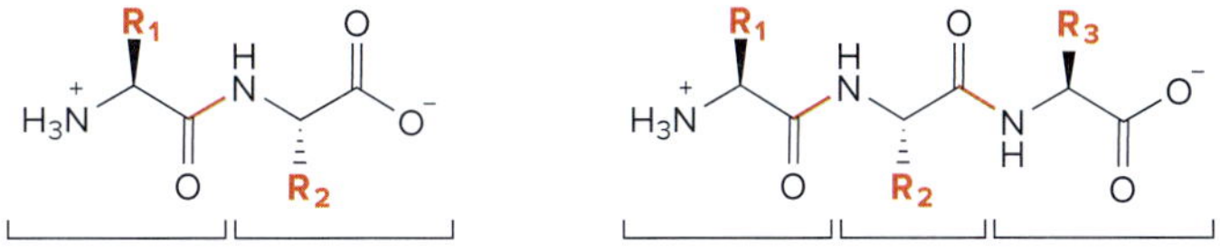

[Structures of a dipeptide and tripeptide with amide (peptide) bonds shown in red]

dipeptide — Two amino acids joined together.   **tripeptide** — Three amino acids joined together.

[Amide bonds are drawn in red.]

Polypeptides and **proteins** both have many amino acids joined in long linear chains, but the term **protein** is usually reserved for polymers of more than 40 amino acids.

- The amide bonds in peptides and proteins are called *peptide bonds*.
- The individual amino acids are called *amino acid residues*.

23.4A Simple Peptides

To form a dipeptide, the amino group of one amino acid forms an amide bond with the carboxy group of another amino acid. Because each amino acid has both an amino group and a carboxy group, **two different dipeptides can be formed.** This is illustrated with alanine and cysteine.

[1] The COO⁻ group of alanine can combine with the NH₃⁺ group of cysteine.

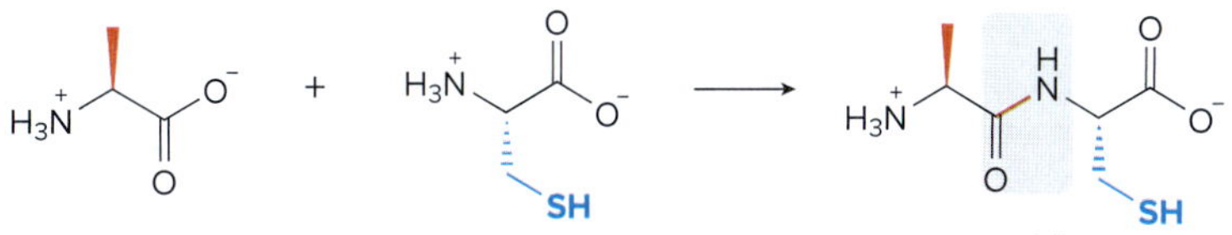

[Reaction: alanine + cysteine → Ala–Cys dipeptide with peptide bond highlighted]

[2] The COO⁻ group of cysteine can combine with the NH₃⁺ group of alanine.

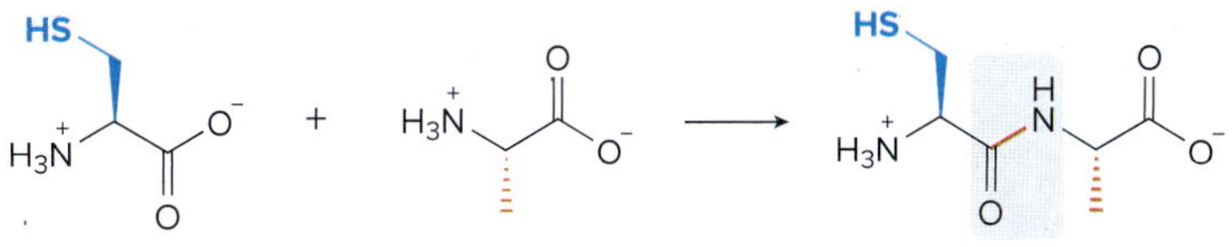

[Reaction: cysteine + alanine → Cys–Ala dipeptide with peptide bond highlighted]

These compounds are **constitutional isomers** of each other. Both have a free amino group (protonated as NH_3^+) at one end of their chains and a free carboxy group (deprotonated as a carboxylate anion, COO^-) at the other.

- The amino acid with the free amino group is called the *N-terminal amino acid*.
- The amino acid with the free carboxy group is called the *C-terminal amino acid*.

By convention, **the N-terminal amino acid is always written at the *left* end of the chain and the C-terminal amino acid at the *right*.** The peptide can be abbreviated by writing the one- or three-letter symbols for the amino acids in the chain from the N-terminal to the C-terminal end. Thus, Ala–Cys has alanine at the N-terminal end and cysteine at the C-terminal end, whereas Cys–Ala has cysteine at the N-terminal end and alanine at the C-terminal end. Sample Problem 23.1 shows how this convention applies to a tripeptide.

Sample Problem 23.1 Drawing the Structure of a Peptide from Three-Letter Symbols

Draw the structure of the following tripeptide, and label its N-terminal and C-terminal amino acids: Ala–Gly–Ser.

Solution

Draw the structures of the amino acids in order from **left to right, placing the COO^- of one amino acid *next* to the NH_3^+ group of the adjacent amino acid. Always draw the NH_3^+ group on the *left* and the COO^- group on the *right*.** Then, join adjacent COO^- and NH_3^+ groups together in amide bonds to form the tripeptide.

tripeptide **Ala–Gly–Ser**
[The new peptide bonds are drawn in red.]

The N-terminal amino acid is **alanine**, and the C-terminal amino acid is **serine**.

Problem 23.9 Draw the structure of each peptide. Label the N-terminal and C-terminal amino acids and all amide bonds.
a. Val–Glu   b. Gly–His–Leu   c. M–A–T–T

More Practice: Try Problems 23.25a; 23.40; 23.41b, d.

The tripeptide in Sample Problem 23.1 has one N-terminal amino acid, one C-terminal amino acid, and two peptide bonds.

- No matter how many amino acid residues are present, there is only **one** N-terminal amino acid and **one** C-terminal amino acid.
- For *n* amino acids in the chain, the number of amide bonds is *n* − 1.

Problem 23.10 Name each peptide using both the one-letter and the three-letter abbreviations for the names of the component amino acids.

23.4B The Peptide Bond

Recall from Section 12.6 that buta-1,3-diene can also exist as s-cis and s-trans conformations. In buta-1,3-diene, the **s-cis conformation** has the two double bonds on the same side of the single bond (dihedral angle = 0°), whereas the **s-trans conformation** has them on opposite sides (dihedral angle = 180°).

The carbonyl carbon of an amide is sp^2 **hybridized** and has **trigonal planar** geometry. A second resonance structure can be drawn that delocalizes the nonbonded electron pair on the N atom. Amides are more resonance stabilized than other acyl compounds, so the **resonance structure having the C=N makes a significant contribution to the hybrid.**

two resonance structures for the peptide bond

Resonance stabilization has important consequences. **Rotation about the C–N bond is restricted** because it has partial double-bond character. As a result, there are two possible conformations.

s-trans   s-cis

- The *s-trans* conformation has the two R groups oriented on *opposite* sides of the C–N bond.
- The *s-cis* conformation has the two R groups oriented on the *same* side of the C–N bond.
- The *s-trans* conformation of a peptide bond is typically more stable than the *s-cis*, because the *s-trans* has the two bulky R groups located farther from each other.

The planar geometry of the peptide bond is analogous to the planar geometry of ethylene (or any other alkene), where the double bond between sp^2 hybridized carbon atoms makes all of the bond angles ~120° and puts all six atoms in the same plane.

A second consequence of resonance stabilization is that **all six atoms involved in the peptide bond lie in the same plane.** All bond angles are ~120°, and the C=O and N–H bonds are oriented 180° from each other.

These six atoms lie in a plane.

The structure of a tetrapeptide illustrates the results of these effects in a long peptide chain.

- The *s-trans* arrangement makes a long chain with a zigzag arrangement.
- In each peptide bond, the N–H and C=O bonds lie parallel and at 180° with respect to each other.

tetrapeptide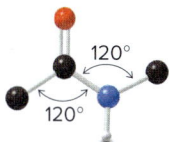

23.4C Interesting Peptides

Even relatively simple peptides can have important biological functions. **Bradykinin**, for example, is a peptide hormone composed of nine amino acids. It stimulates smooth muscle contraction, dilates blood vessels, and causes pain. Bradykinin is a component of bee venom.

Arg–Pro–Pro–Gly–Phe–Ser–Pro–Phe–Arg
bradykinin

Oxytocin and **vasopressin** are nonapeptide hormones, too. Their sequences are identical except for two amino acids, yet this is enough to give them very different biological activities. As mentioned in the chapter opener, oxytocin induces labor by stimulating the contraction of uterine muscles, and it stimulates the flow of milk in nursing mothers. Vasopressin, on the other hand, controls blood pressure by regulating smooth muscle contraction. The N-terminal amino acid in both hormones is a cysteine residue, and the C-terminal residue is glycine. Instead of a free carboxy group, both peptides have an NH_2 group in place of OH, so this is indicated with the additional NH_2 group drawn at the end of the chain.

```
Cys—Tyr—Ile                         Cys—Tyr—Phe
 |                                    |
 S        Gln                         S        Gln
 |         |                          |         |
 S      Asn                           S      Asn
  \   /                                \   /
   Cys                                  Cys
    |                                    |
  Pro—Leu—GlyNH2                       Pro—Arg—GlyNH2
    oxytocin                            vasopressin
```

- The N-terminal amino acid is labeled in red.
- The amino acids that differ are labeled in blue.

> The oxidation of thiols to disulfides was discussed in Section 9.15.

The structure of both peptides includes a **disulfide bond,** a form of covalent bonding in which the —SH groups from two cysteine residues are oxidized to form a sulfur–sulfur bond. In oxytocin and vasopressin, the disulfide bonds make the peptides cyclic.

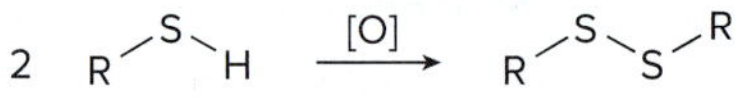

$$2\ R\text{-S-H} \xrightarrow{[O]} R\text{-S-S-}R$$
     thiol   disulfide

The artificial sweetener **aspartame** (Figure 24.10) is the methyl ester of the dipeptide Asp–Phe. This synthetic peptide is 180 times sweeter (on a gram-for-gram basis) than sucrose (common table sugar). Both of the amino acids in aspartame have the naturally occurring L-configuration. If the D-amino acid is substituted for either Asp or Phe, the resulting compound tastes bitter.

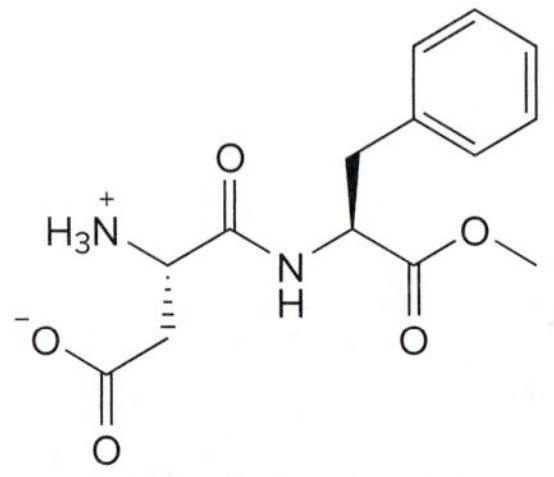

aspartame
the methyl ester of Asp–Phe
a synthetic artificial sweetener

Problem 23.11 Draw the structure of leu-enkephalin, a pentapeptide that acts as an analgesic and opiate, and has the following sequence: Tyr–Gly–Gly–Phe–Leu. (The structure of a related peptide, met-enkephalin, appeared in Section 16.5B.)

Problem 23.12 Glutathione, a powerful antioxidant that destroys harmful oxidizing agents in cells, is composed of glutamic acid, cysteine, and glycine, and has the following structure:

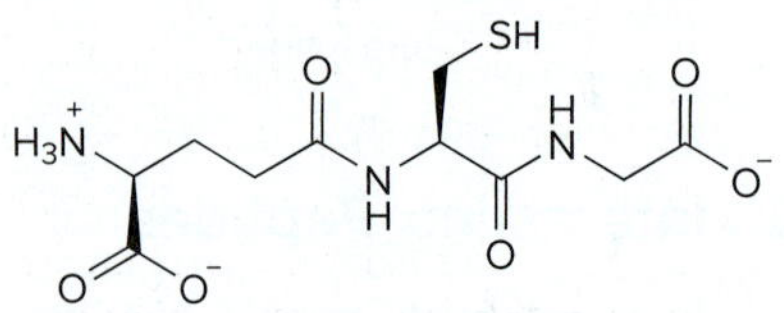

glutathione

a. What product is formed when glutathione reacts with an oxidizing agent?
b. What is unusual about the peptide bond between glutamic acid and cysteine?

23.5 Peptide Sequencing

To determine the structure of a peptide, we must know not only what amino acids compose it, but also the sequence of the amino acids in the peptide chain. Although mass spectrometry has become an increasingly powerful method for the analysis of high-molecular-weight proteins (Section A.5C), chemical methods to determine peptide structure are still widely used and presented in this section.

23.5A Amino Acid Analysis

The structure determination of a peptide begins by analyzing the **total amino acid composition.** The amide bonds are first hydrolyzed by heating with hydrochloric acid for 24 h to form the individual amino acids. The resulting mixture is then separated using high-performance liquid chromatography (HPLC), a technique in which a solution of amino acids is placed on a column and individual amino acids move through the column at characteristic rates, often dependent on polarity.

This process determines both the identity of the individual amino acids and the amount of each present, but it tells nothing about the order of the amino acids in the peptide. For example, complete hydrolysis and HPLC analysis of the tetrapeptide Gly–Gly–Phe–Tyr would indicate the presence of three amino acids—glycine, phenylalanine, and tyrosine—and show that there are twice as many glycine residues as phenylalanine or tyrosine residues. The exact order of the amino acids in the peptide chain must then be determined by additional methods.

23.5B Identifying the N-Terminal Amino Acid—The Edman Degradation

To determine the sequence of amino acids in a peptide chain, a variety of procedures are often combined. One especially useful technique is to **identify the N-terminal amino acid using the Edman degradation.** In the Edman degradation, amino acids are cleaved one at a time from the N-terminal end, the identity of the amino acid determined, and the process repeated until the entire sequence is known. Automated sequencers using this methodology are now available to sequence peptides containing up to about 50 amino acids.

The Edman degradation is based on the reaction of the nucleophilic NH_2 group of the N-terminal amino acid with the electrophilic carbon of phenyl isothiocyanate, $C_6H_5N=C=S$. When the N-terminal amino acid is removed from the peptide chain, two products are formed: **an N-phenylthiohydantoin (PTH) and a new peptide with one *fewer* amino acid.**

The *N*-phenylthiohydantoin derivative contains the atoms of the N-terminal amino acid. **This product identifies the N-terminal amino acid in the peptide** because the PTH derivatives of all 20 naturally occurring amino acids are known and characterized. The new peptide formed in the Edman degradation has one fewer amino acid than the original peptide. Moreover, it contains a new N-terminal amino acid, so the process can be repeated.

Mechanism 23.1 illustrates some of the key steps of the Edman degradation. The nucleophilic N-terminal NH_2 group adds to the electrophilic carbon of phenyl isothiocyanate to form an *N*-phenylthiourea, the product of nucleophilic addition (Part [1]). Intramolecular cyclization followed by elimination results in cleavage of the terminal amide bond in Part [2] to form **a new peptide with one fewer amino acid.** A sulfur heterocycle, called a thiazolinone, is also formed, which rearranges by a multistep pathway to form an *N*-phenylthiohydantoin. **The R group in this product identifies the amino acid located at the N-terminal end.**

Mechanism 23.1 Edman Degradation

Part [1] Formation of an *N*-phenylthiourea

1 – 2 Addition of the amino group of the N-terminal amino acid to phenyl isothiocyanate followed by proton transfer forms an ***N*-phenylthiourea.**

Part [2] Formation of the N-terminal amino acid and *N*-phenylthiohydantoin (PTH)

3 Nucleophilic addition of the S atom to the amide carbonyl forms a five-membered ring.

4 Loss of the amino group forms two products—a thiazolinone ring and **a peptide chain that contains one fewer amino acid than the original peptide.**

5 The thiazolinone rearranges by a multistep pathway to form an ***N*-phenylthiohydantoin (PTH) that contains the original amino acid.**

In theory a protein of any length can be sequenced using the Edman degradation, but in practice, the accumulation of small quantities of unwanted by-products limits sequencing to proteins having fewer than approximately 50 amino acids.

Problem 23.13 Draw the structure of the *N*-phenylthiohydantoin formed by initial Edman degradation of each peptide: (a) Ala–Gly–Phe–Phe; (b) Val–Ile–Tyr.

23.5C Partial Hydrolysis of a Peptide

Additional structural information can be obtained by cleaving some, but not all, of the amide bonds in a peptide. Partial hydrolysis of a peptide with acid forms smaller fragments in a random fashion. Sequencing these peptides and **identifying sites of overlap** can be used to determine the sequence of the complete peptide, as shown in Sample Problem 23.2.

Sample Problem 23.2 Determining the Amino Acid Sequence of a Peptide Using Partial Hydrolysis

Give the amino acid sequence of a hexapeptide that contains the amino acids Ala, Val, Ser, Ile, Gly, Tyr, and forms the following fragments when partially hydrolyzed with HCl: Gly–Ile–Val, Ala–Ser–Gly, and Tyr–Ala.

Solution

Looking for points of overlap in the sequences of the smaller fragments shows how the fragments should be pieced together. In this example, the fragment Ala–Ser–Gly contains amino acids common to the two other fragments, thus showing how the three fragments can be joined together.

common amino acids

Tyr–Ala   Gly–Ile–Val

Ala–Ser–Gly

Answer: ⟶ Tyr–Ala–Ser–Gly–Ile–Val

hexapeptide

Problem 23.14 Give the amino acid sequence of an octapeptide that contains the amino acids Tyr, Ala, Leu (2 equiv), Cys, Gly, Glu, and Val, and forms the following fragments when partially hydrolyzed with HCl: Val–Cys–Gly–Glu, Ala–Leu–Tyr, and Tyr–Leu–Val–Cys.

More Practice: Try Problem 23.46.

Peptides can also be hydrolyzed at specific sites using enzymes. The enzyme carboxypeptidase catalyzes the hydrolysis of the amide bond nearest the C-terminal end, forming the C-terminal amino acid and a peptide with one fewer amino acid. In this way, **carboxypeptidase is used to identify the C-terminal amino acid.**

Other enzymes catalyze the hydrolysis of amide bonds formed with specific amino acids. For example:

- **Trypsin** catalyzes the hydrolysis of amides with a carbonyl group that is part of the basic amino acids arginine and lysine.
- **Chymotrypsin** hydrolyzes amides with carbonyl groups that are part of the aromatic amino acids phenylalanine, tyrosine, and tryptophan.

Chymotrypsin cleaves here.   **Carboxypeptidase** cleaves here.

Ala–Phe–Gly–Leu–Trp–Val–Arg–His–Pro–Pro–Gly

Trypsin cleaves here.

Table 23.2 summarizes these enzyme specificities used in peptide sequencing.

Table 23.2 Cleavage Sites of Specific Enzymes in Peptide Sequencing

Enzyme	Site of cleavage
Carboxypeptidase	Amide bond nearest to the C-terminal amino acid
Chymotrypsin	Amide bond with a carbonyl group from Phe, Tyr, or Trp
Trypsin	Amide bond with a carbonyl group from Arg or Lys

Problem 23.15 (a) What products are formed when each peptide is treated with trypsin? (b) What products are formed when each peptide is treated with chymotrypsin?

[1] Gly–Ala–Phe–Leu–Lys–Ala
[2] Phe–Tyr–Gly–Cys–Arg–Ser
[3] Thr–Pro–Lys–Glu–His–Gly–Phe–Cys–Trp–Val–Val–Phe

Sample Problem 23.3 — Deducing the Sequence of a Peptide

Deduce the sequence of a pentapeptide that contains the amino acids Ala, Glu, Gly, Ser, and Tyr, from the following experimental data. Edman degradation cleaves Gly from the pentapeptide, and carboxypeptidase forms Ala and a tetrapeptide. Treatment of the pentapeptide with chymotrypsin forms a dipeptide and a tripeptide. Partial hydrolysis forms Gly, Ser, and the tripeptide Tyr–Glu–Ala.

Solution
Use each result to determine the location of an amino acid in the pentapeptide.

Experiment	Result
• Edman degradation identifies the N-terminal amino acid—in this case, Gly.	→ Gly– _ _ _ _ _ _
• Carboxypeptidase identifies the C-terminal amino acid (Ala) when it is cleaved from the end of the chain.	→ Gly– _ _ _ _ –Ala
• Chymotrypsin cleaves amide bonds that contain a carbonyl group from an aromatic amino acid—Tyr in this case. Because a dipeptide and a tripeptide are obtained after treatment with chymotrypsin, Tyr must be the C-terminal amino acid of either the di- or tripeptide. As a result, Tyr must be either the second or third amino acid in the pentapeptide chain.	→ Gly–**Tyr**– _ _ _ –Ala or Gly– _ –**Tyr**– _ –Ala
• Partial hydrolysis forms the tripeptide Tyr–Glu–Ala. Because Ala is the C-terminal amino acid, this result identifies the last three amino acids in the chain.	→ Gly– _ –**Tyr**–**Glu**–Ala
• The last amino acid, Ser, must be located at the only remaining position, the second amino acid in the pentapeptide, and the complete sequence is determined.	→ Gly–**Ser**–Tyr–Glu–Ala

Problem 23.16 Deduce the sequence of a heptapeptide that contains the amino acids Ala, Arg, Glu, Gly, Leu, Phe, and Ser, from the following experimental data. Edman degradation cleaves Leu from the heptapeptide, and carboxypeptidase forms Glu and a hexapeptide. Treatment of the heptapeptide with chymotrypsin forms a hexapeptide and a single amino acid. Treatment of the heptapeptide with trypsin forms a pentapeptide and a dipeptide. Partial hydrolysis forms Glu, Leu, Phe, and the tripeptides Gly–Ala–Ser and Ala–Ser–Arg.

More Practice: Try Problems 23.45–23.49.

23.6 Peptide Synthesis

The synthesis of a specific dipeptide, such as Ala–Gly from alanine and glycine, is complicated because both amino acids have two functional groups. As a result, four products—namely, Ala–Ala, Ala–Gly, Gly–Gly, and Gly–Ala—are possible.

23.6 Peptide Synthesis

How do we selectively join the COOH group of alanine with the NH₂ group of glycine?

- Protect the functional groups that we don't want to react, and then form the amide bond.

How To Synthesize a Dipeptide from Two Amino Acids

Example

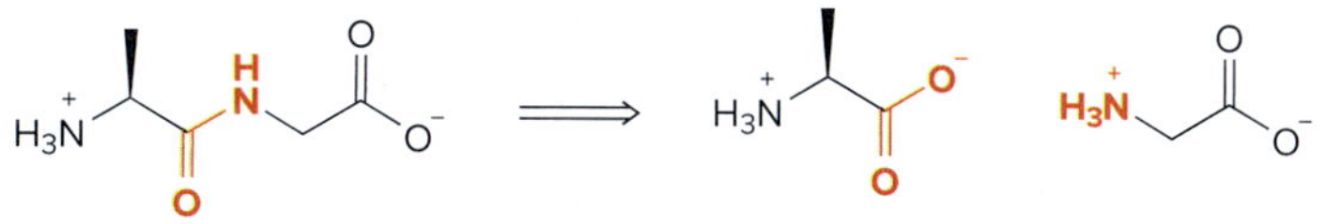

Ala–Gly ⟹ Ala + Gly

Join the functional groups in red.

Step [1] Protect the NH₂ group of alanine.

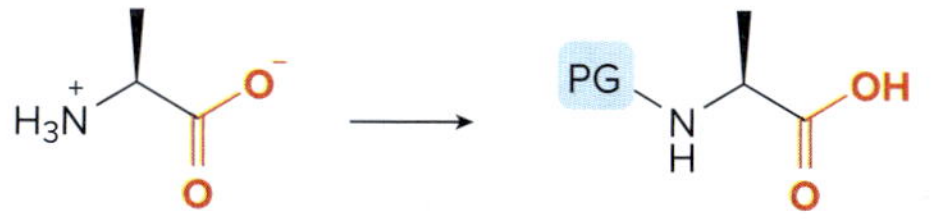

Ala → PG-protected Ala   [PG = protecting group]

- In the neutral amino acid, the NH₂ group exists largely as an ammonium ion, $-NH_3^+$.

Step [2] Protect the COOH group of glycine.

Gly → PG-protected Gly

- In the neutral amino acid, the COOH group exists largely as a carboxylate anion, $-COO^-$.

Step [3] Form the amide bond with DCC.

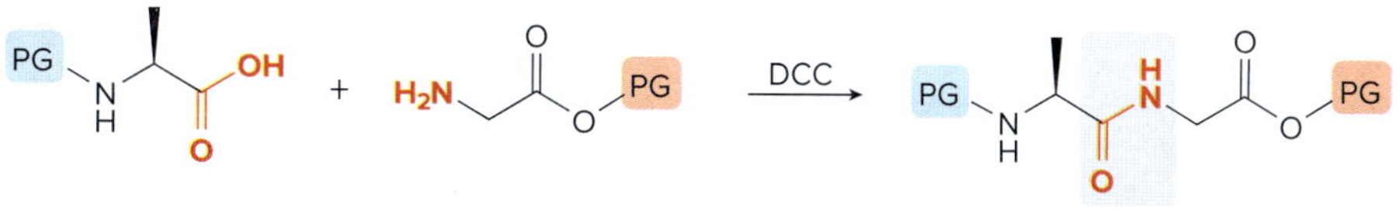

PG-Ala-OH + H₂N-Gly-PG —DCC→ PG-Ala-Gly-PG

Dicyclohexylcarbodiimide (**DCC**) is a reagent commonly used to form amide bonds (see Section 16.9D). DCC makes the OH group of the carboxylic acid a better leaving group, thus **activating the carboxy group toward nucleophilic attack.**

DCC = dicyclohexylcarbodiimide

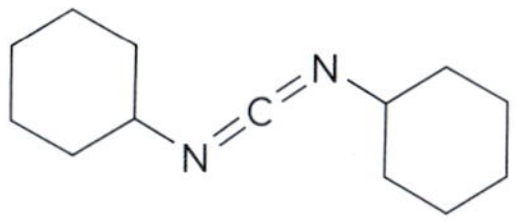

Step [4] Remove one or both protecting groups.

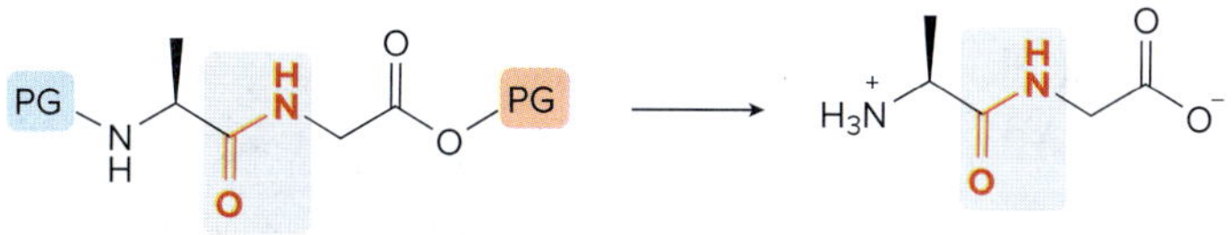

PG-Ala-Gly-PG → Ala–Gly

Two widely used amino protecting groups convert an amine to a **carbamate,** a functional group having a carbonyl bonded to both an oxygen and a nitrogen atom. Because the N atom of the carbamate is bonded to a carbonyl group, the protected amino group is no longer nucleophilic.

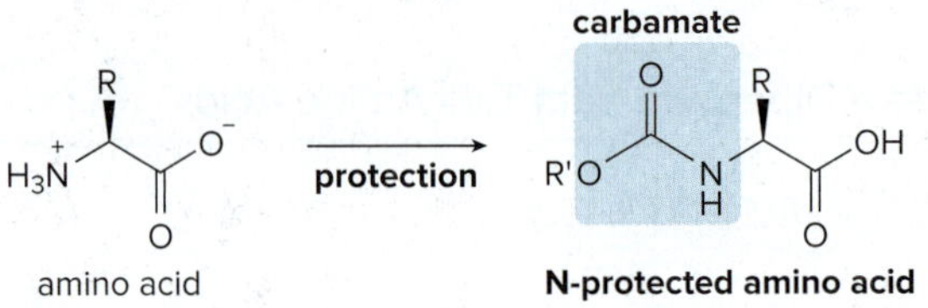

For example, the ***tert*-butoxycarbonyl protecting group,** abbreviated as **Boc,** is formed by reacting the amino acid with di-*tert*-butyl dicarbonate in a nucleophilic acyl substitution reaction.

di-*tert*-butyl dicarbonate

tert-butoxycarbonyl
Boc

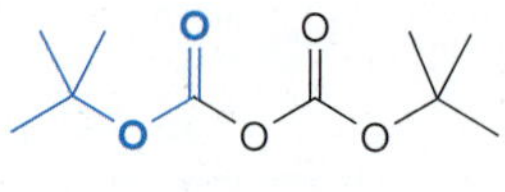

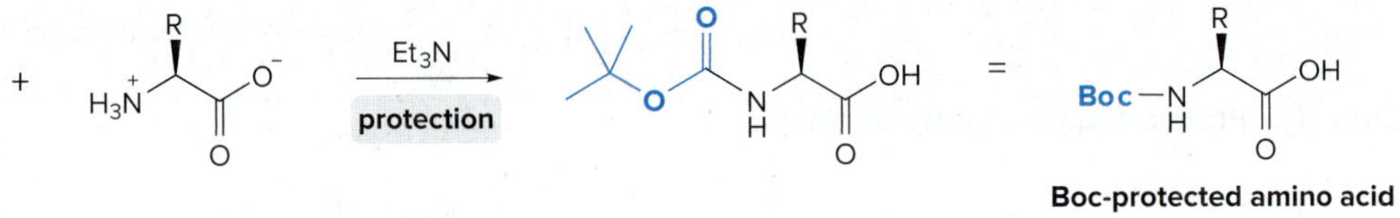

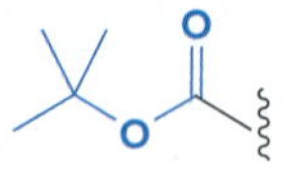

To be a useful protecting group, the Boc group must be removed under reaction conditions that do not affect other functional groups in the molecule. It can be removed with an acid such as **trifluoroacetic acid, HCl,** or **HBr.**

(**Boc**)$_2$O

9-fluorenylmethoxycarbonyl
Fmoc

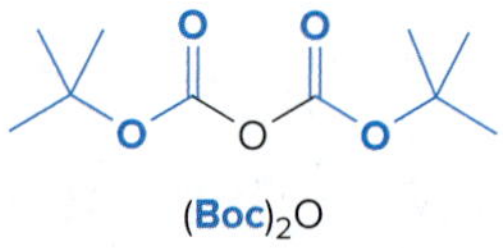

A second amino protecting group, the **9-fluorenylmethoxycarbonyl protecting group,** abbreviated as **Fmoc,** is formed by reacting the amino acid with 9-fluorenylmethyl chloroformate in a nucleophilic acyl substitution reaction.

9-fluorenylmethyl
chloroformate
Fmoc—Cl

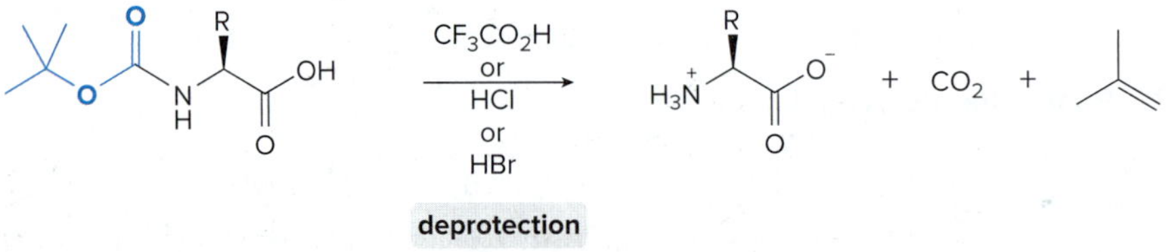

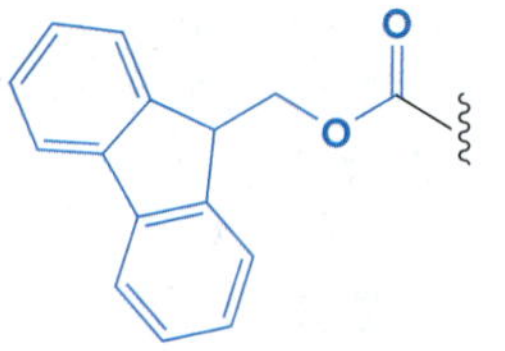

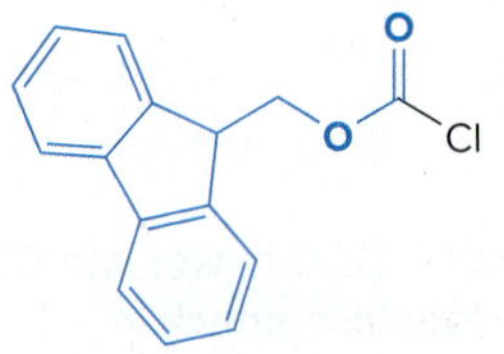

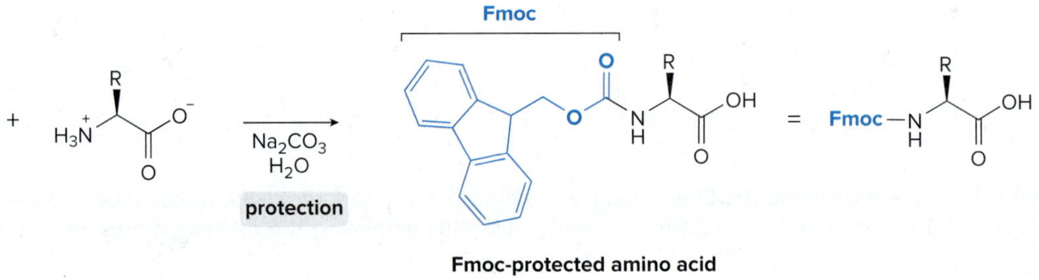

Although the Fmoc protecting group is stable to most acids, it can be removed by treatment with base (NH_3 or an amine).

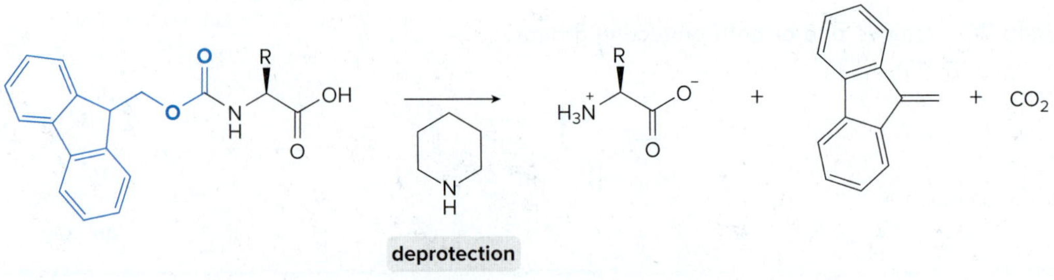

23.6 Peptide Synthesis

The carboxy group is usually protected as a **methyl** or **benzyl ester** by reaction with an alcohol and an acid.

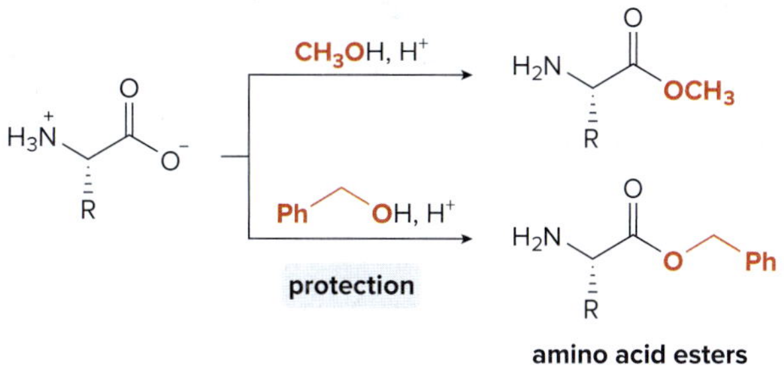

protection → amino acid esters

These esters are usually removed by hydrolysis with aqueous base.

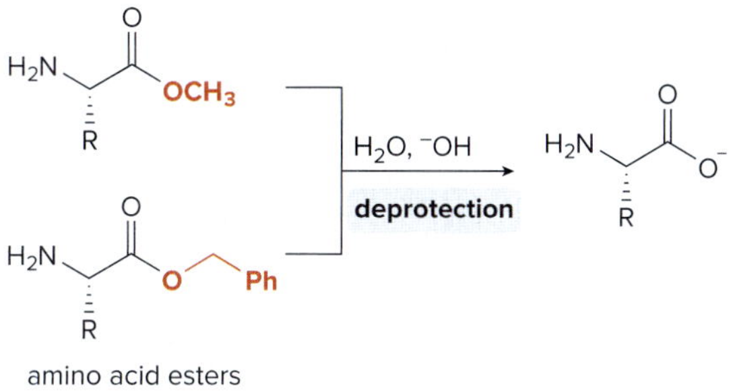

amino acid esters → deprotection

One advantage of using a benzyl ester for protection is that it can also be removed with H_2 in the presence of a Pd catalyst. This process is called **hydrogenolysis**. These conditions are especially mild, because they avoid the use of either acid or base. Benzyl esters can also be removed with HBr in acetic acid.

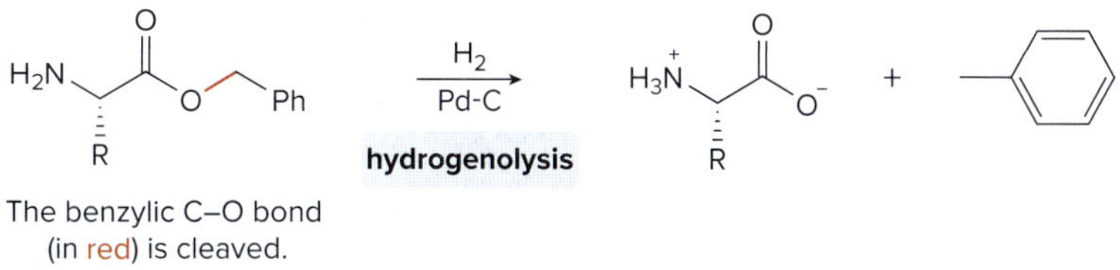

hydrogenolysis

The benzylic C–O bond (in red) is cleaved.

The specific reactions needed to synthesize the dipeptide Ala–Gly are illustrated in Sample Problem 23.4.

Sample Problem 23.4 Devising the Synthesis of a Dipeptide

Draw out the steps in the synthesis of the dipeptide Ala–Gly.

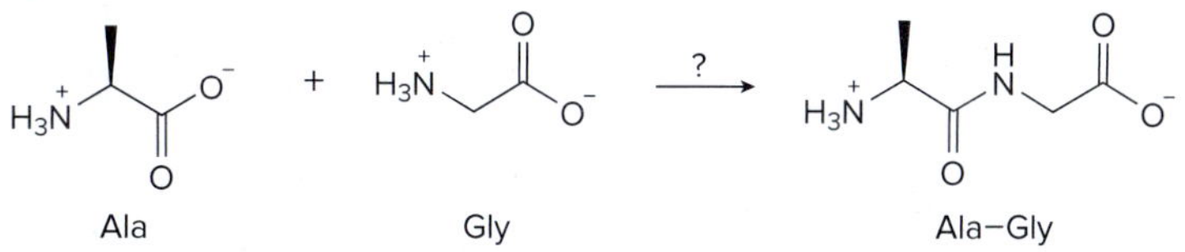

Ala + Gly → Ala–Gly

Solution

Step [1] Protect the NH₂ group of alanine using a Boc group.

$$H_3N^+-Ala-O^- \xrightarrow{(Boc)_2O,\ Et_3N} Boc-NH-Ala-OH$$

Step [2] Protect the COOH group of glycine as a benzyl ester.

$$H_3N^+-Gly-O^- \xrightarrow{PhCH_2OH,\ H^+} H_2N-Gly-OCH_2Ph$$

Step [3] Form the amide bond with DCC.

$$Boc-Ala-OH\ +\ H_2N-Gly-OCH_2Ph \xrightarrow{DCC} Boc-Ala-Gly-OCH_2Ph$$

Step [4] Remove one or both protecting groups.

The protecting groups can be removed in a stepwise fashion or in a single reaction.

Boc–Ala–Gly–OCH₂Ph $\xrightarrow{H_2,\ Pd-C}$ Boc–Ala–Gly   *Remove the benzyl group.*

$\xrightarrow{CF_3COOH}$ Ala–Gly   *Remove the Boc group.*

Boc–Ala–Gly–OCH₂Ph $\xrightarrow{HBr,\ CH_3COOH}$ Ala–Gly   *Remove both protecting groups.*

Problem 23.17 Devise a synthesis of the following dipeptide from amino acid starting materials.

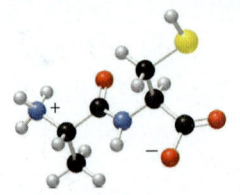

More Practice: Try Problems 23.26, 23.53.

This method can be applied to the synthesis of tripeptides and even larger polypeptides. After the protected dipeptide is prepared in Step [3], only one of the protecting groups is removed, and this dipeptide is coupled to a third amino acid with one of its functional groups protected, as illustrated in the following equations.

Boc–Ala–Gly + Gly–OCH$_2$Ph

N-protected dipeptide   carboxy-protected amino acid

DCC | **Form the amide bond.**

Boc–Ala–Gly–Gly–OCH$_2$Ph

$\xrightarrow[\text{CH}_3\text{COOH}]{\text{HBr}}$ **Remove both protecting groups.**

Ala–Gly–Gly tripeptide

Problem 23.18   Devise a synthesis of each peptide from amino acid starting materials: (a) Leu–Val; (b) Ala–Ile–Gly.

23.7 Automated Peptide Synthesis

Development of the solid phase technique earned Merrifield the 1984 Nobel Prize in Chemistry and has made possible the synthesis of many polypeptides and proteins.

The method described in Section 23.6 works well for the synthesis of small peptides. It is extremely time-consuming to synthesize larger proteins by this strategy, however, because each step requires isolation and purification of the product. The synthesis of larger polypeptides is usually accomplished by using the **solid phase technique** originally developed by R. Bruce Merrifield of Rockefeller University.

In the Merrifield method, an amino acid is attached to an insoluble polymer. Amino acids are sequentially added, one at a time, thereby forming successive peptide bonds. Because impurities and by-products are not attached to the polymer chain, they are removed simply by washing them away with a solvent at each stage of the synthesis.

A commonly used polymer is a **polystyrene derivative** that contains –CH$_2$Cl groups bonded to some of the benzene rings in the polymer chain. The Cl atoms serve as handles that allow attachment of amino acids to the chain.

= Cl–CH$_2$–POLYMER

polystyrene derivative with Cl leaving groups

An Fmoc-protected amino acid is attached to the polymer at its carboxy group by an S$_N$2 reaction.

The amino acid is now bound to the insoluble polymer.

Once the first amino acid is bound to the polymer, additional amino acids can be added sequentially. The steps of the solid phase peptide synthesis technique are illustrated in the accompanying scheme. In the last step, HF cleaves the polypeptide chain from the polymer.

How To Synthesize a Peptide Using the Merrifield Solid Phase Technique

Step [1]
Attach an Fmoc-protected amino acid to the polymer.

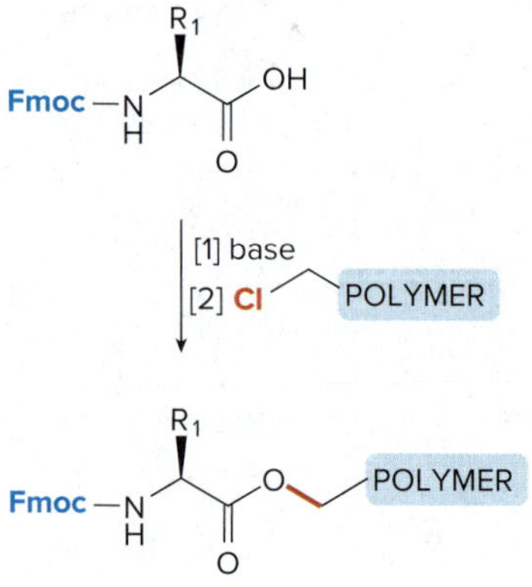

new bond to the polymer in red

Step [2]
Remove the protecting group.

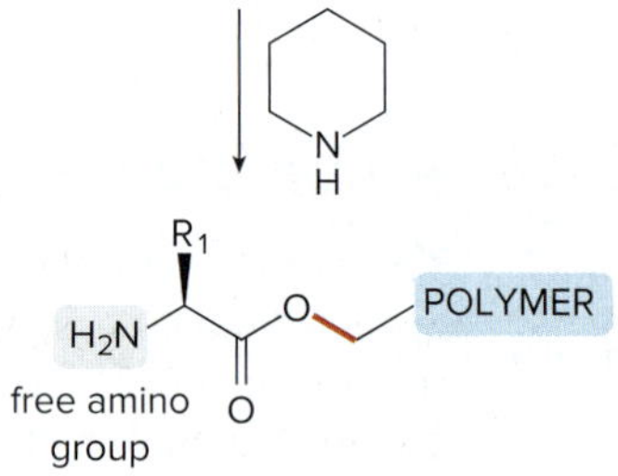

Step [3]
Form the amide bond with DCC.

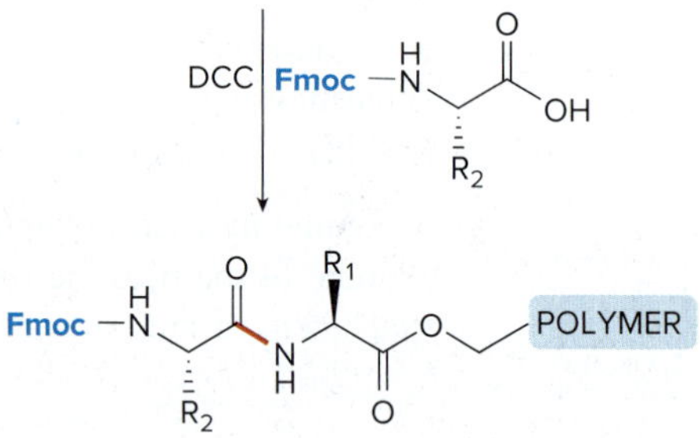

new amide bond in red

Step [4]
Repeat Steps [2] and [3].

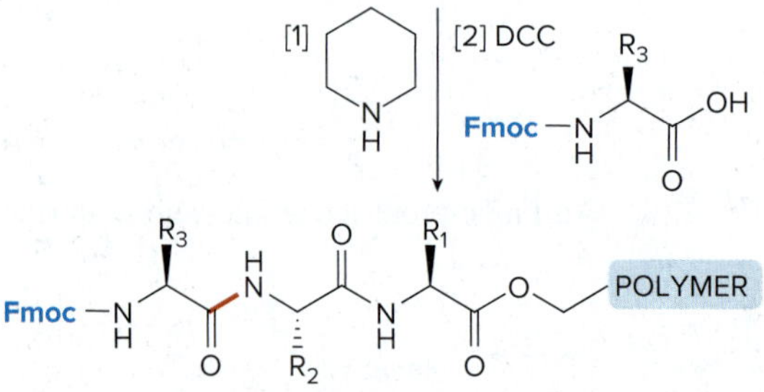

new amide bond in red

—*Continued*

Step [5]
Remove the protecting group and detach the peptide from the polymer.

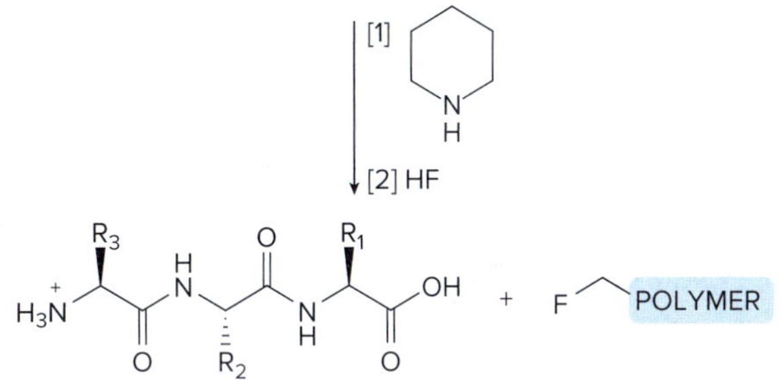

[1] piperidine
[2] HF

tripeptide

The Merrifield method has now been completely automated, so it is possible to purchase peptide synthesizers that automatically carry out all of the above operations and form polypeptides in high yield in a matter of hours, days, or weeks, depending on the length of the chain of the desired product. For example, the protein ribonuclease, which contains 128 amino acids, has been prepared by this technique in an overall yield of 17%. This remarkable synthesis involved 369 separate reactions, and thus the yield of each individual reaction was > 99%.

Problem 23.19 Outline the steps needed to synthesize the tetrapeptide Ala–Leu–Ile–Gly using the Merrifield technique.

23.8 Protein Structure

Now that you have learned some of the chemistry of amino acids, it's time to study proteins, the large polymers of amino acids that are responsible for so much of the structure and function of all living cells. We begin with a discussion of the **primary, secondary, tertiary,** and **quaternary structure** of proteins.

23.8A Primary Structure

The *primary structure* of proteins is the particular sequence of amino acids that is joined by peptide bonds. The most important element of this primary structure is the **amide bond.**

- Rotation around the amide C–N bond is *restricted* because of electron delocalization, and the s-trans conformation is the more stable arrangement.
- In each peptide bond, the N–H and C=O bonds are directed 180° from each other.

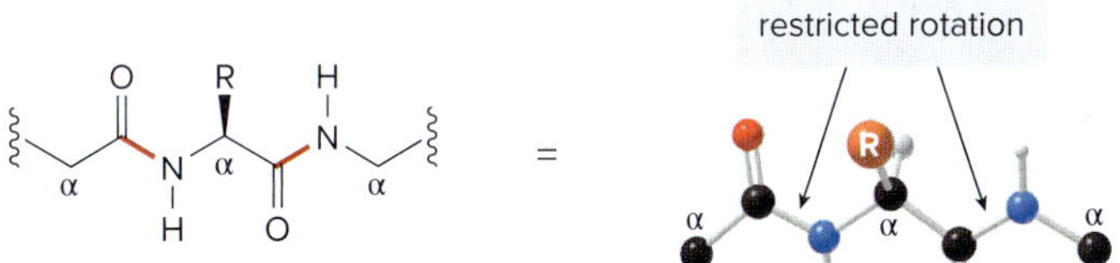

two amide bonds in a peptide chain

restricted rotation

Although rotation about the amide bonds is restricted, **rotation about the other σ bonds in the protein backbone is not.** As a result, the peptide chain can twist and bend into a variety of different arrangements that constitute the secondary structure of the protein.

23.8B Secondary Structure

The three-dimensional conformations of localized regions of a protein are called its *secondary structure*. These regions arise due to hydrogen bonding between the N—H proton of one amide and the C=O oxygen of another. Two arrangements that are particularly stable are called the **α-helix** and the **β-pleated sheet.**

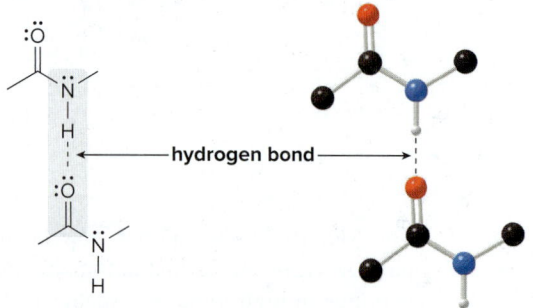

α-Helix

The **α-helix** forms when a peptide chain twists into a right-handed or clockwise spiral, as shown in Figure 23.6. Four important features of the α-helix are as follows:

[1] **Each turn of the helix has 3.6 amino acids.**

[2] **The N—H and C=O bonds point along the axis of the helix.** All C=O bonds point in one direction, and all N—H bonds point in the opposite direction.

[3] **The C=O group of one amino acid is hydrogen bonded to an N—H group four amino acid residues farther along the chain.** Thus, hydrogen bonding occurs between two amino acids *in the same chain*. Note, too, that the hydrogen bonds are parallel to the axis of the helix.

[4] The **R groups of the amino acids extend outward** from the core of the helix.

Figure 23.6
Two different illustrations of the α-helix

a. The right-handed α-helix

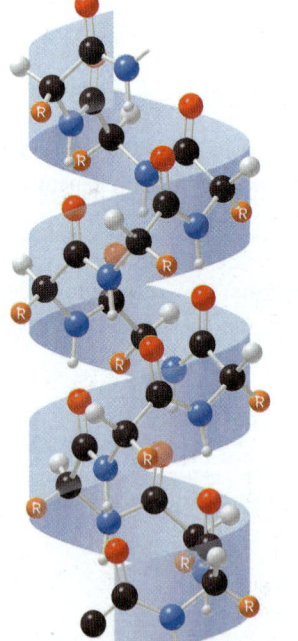

- All atoms of the α-helix are drawn in this representation. All C=O bonds are pointing up and all N—H bonds are pointing down.

b. The backbone of the α-helix

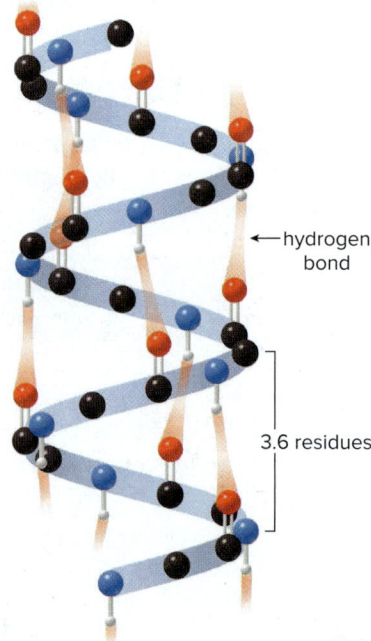

- Only the peptide backbone is drawn in this representation. The hydrogen bonds between the C=O and N—H of amino acids four residues away from each other are shown.

An α-helix can form only if there is rotation about the bonds at the α carbon of the amide carbonyl group, and not all amino acids can do this. For example, proline, the amino acid whose nitrogen atom forms part of a five-membered ring, is more rigid than other amino acids, and its C_α–N bond cannot rotate the necessary amount. Additionally, it has no N–H proton with which to form an intramolecular hydrogen bond to stabilize the helix. Thus, **proline cannot be part of an α-helix.**

Both the myosin in muscle and α-keratin in hair are proteins composed almost entirely of α-helices.

β-Pleated Sheet

The **β-pleated sheet** secondary structure forms when two or more peptide chains, called **strands,** line up side-by-side, as shown in Figure 23.7. All β-pleated sheets have the following characteristics:

[1] **The C=O and N–H bonds lie in the plane of the sheet.**
[2] **Hydrogen bonding often occurs between the N–H and C=O groups of nearby amino acid residues.**
[3] **The R groups are oriented above and below the plane** of the sheet, and alternate from one side to the other along a given strand.

Figure 23.7
Three-dimensional structure of the β-pleated sheet

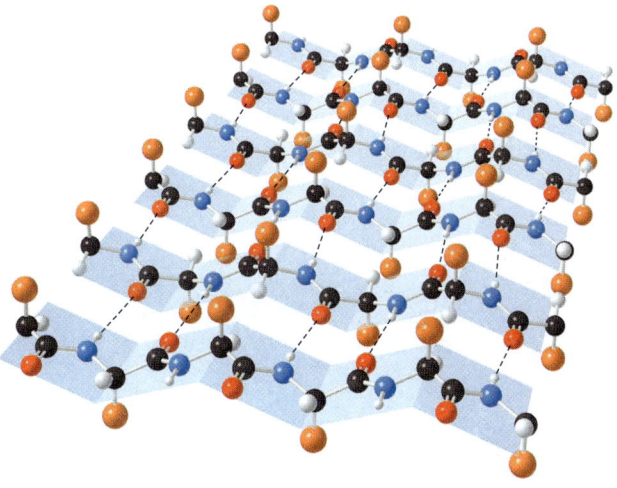

- The β-pleated sheet consists of extended strands of the peptide chains held together by hydrogen bonding. The C=O and N–H bonds lie in the plane of the sheet, and the R groups (shown as orange balls) alternate above and below the plane.

The β-pleated sheet arrangement most commonly occurs with amino acids with small R groups, like alanine and glycine. With larger R groups, steric interactions prevent the chains from getting close together, so the sheet cannot be stabilized by hydrogen bonding.

Most proteins have regions of α-helix and β-pleated sheet, in addition to other regions that cannot be characterized by either of these arrangements. Shorthand symbols are often used to indicate regions of a protein that have α-helix or β-pleated sheet. A **flat helical ribbon** is used for the α-helix, and a **flat wide arrow** is used for the β-pleated sheet. These representations are often used in **ribbon diagrams** to illustrate protein structure.

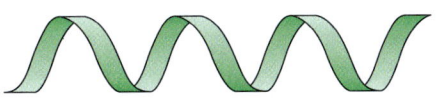
α-helix shorthand

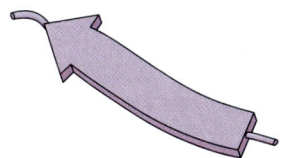
β-pleated sheet shorthand

Proteins are drawn in a variety of ways to illustrate different aspects of their structure. Figure 23.8 illustrates three different representations of the protein lysozyme, an enzyme found in both plants and animals. Lysozyme catalyzes the hydrolysis of bonds in bacterial cell walls, weakening them, often causing the bacteria to burst.

Figure 23.8 Lysozyme

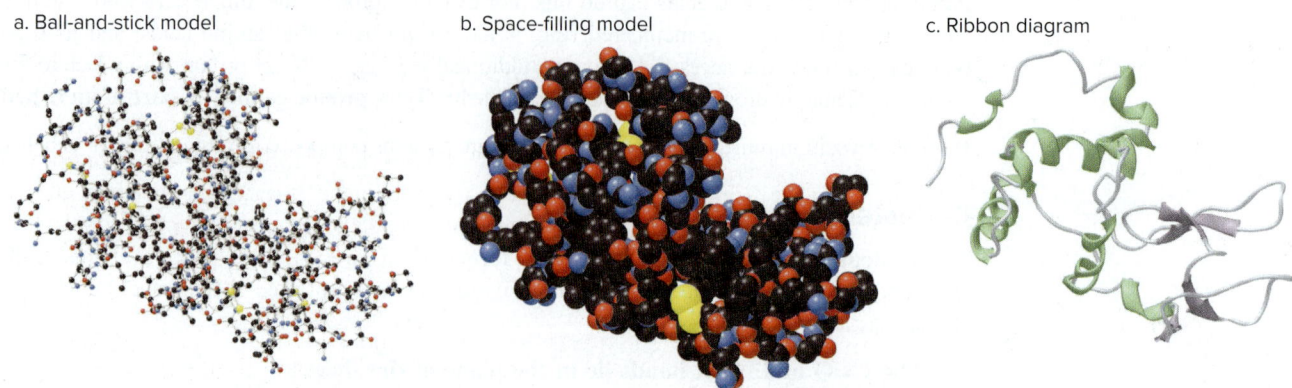

a. Ball-and-stick model b. Space-filling model c. Ribbon diagram

(a) The ball-and-stick model of lysozyme shows the protein backbone with color-coded C, N, O, and S atoms. Individual amino acids are most clearly located using this representation. (b) The space-filling model uses color-coded balls for each atom in the backbone of the enzyme and illustrates how the atoms fill the space they occupy. (c) The ribbon diagram shows regions of α-helix and β-pleated sheet that are not clearly in evidence in the other two representations.

Spider dragline silk is a strong yet elastic protein because it has regions of β-pleated sheet and regions of α-helix (Figure 23.9). α-Helical regions impart elasticity to the silk because the peptide chain is twisted (not fully extended), so it can stretch. β-Pleated sheet regions are almost fully extended, so they can't be stretched further, but their highly ordered three-dimensional structure imparts strength to the silk. Thus, spider silk suits the spider by comprising both types of secondary structure with beneficial properties.

Figure 23.9
Different regions of secondary structure in spider silk

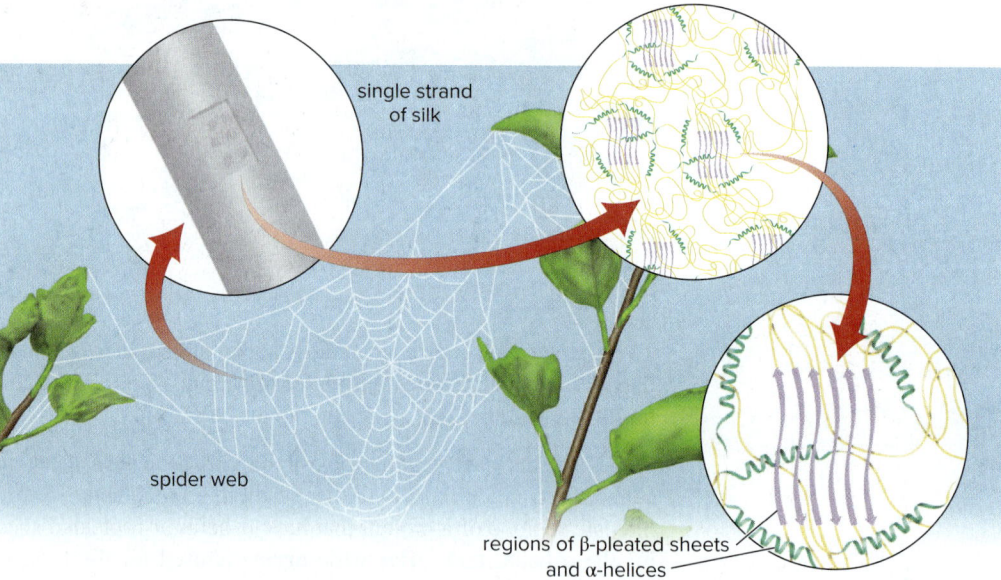

- Spider silk has regions of α-helix and β-pleated sheet that make it both strong and elastic. The green coils represent the α-helical regions, and the purple arrows represent the β-pleated sheet regions. The yellow lines represent other areas of the protein that are neither α-helix nor β-pleated sheet.

23.8C Tertiary and Quaternary Structure

The three-dimensional shape adopted by the entire peptide chain is called its *tertiary structure*. A peptide generally folds into a conformation that maximizes its stability. In the aqueous environment of the cell, proteins often fold in such a way as to place a large number of polar and charged groups on their outer surface, to maximize the dipole–dipole and hydrogen bonding interactions with water. This generally places most of the nonpolar side chains in the interior of the protein, where van der Waals interactions between these hydrophobic groups help stabilize the molecule, too.

In addition, polar functional groups hydrogen bond with each other (not just water), and amino acids with charged side chains like $-COO^-$ and $-NH_3^+$ can stabilize tertiary structure by electrostatic interactions.

Finally, **disulfide bonds are the only covalent bonds that stabilize tertiary structure.** As previously mentioned, these strong bonds form by oxidation of two cysteine residues on either the same polypeptide chain or another polypeptide chain of the same protein.

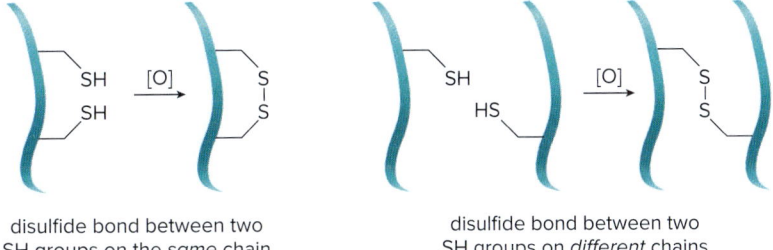

disulfide bond between two SH groups on the *same* chain

disulfide bond between two SH groups on *different* chains

The nonapeptides **oxytocin** and **vasopressin** (Section 23.4C) contain intramolecular disulfide bonds. **Insulin**, on the other hand, consists of two separate polypeptide chains (**A** and **B**) that are covalently linked by two intermolecular disulfide bonds, as shown in Figure 23.10. The **A** chain, which also has an intramolecular disulfide bond, has 21 amino acid residues, whereas the **B** chain has 30.

Figure 23.10
Insulin

A chain: Gly-Ile-Val-Glu-Gln-Cys-Cys-Thr-Ser-Ile-Cys-Ser-Leu-Tyr-Gln-Leu-Glu-Asn-Tyr-Cys-Asn (residue 1 ... 8 9 10 ... 21)

B chain: Phe-Val-Asn-Gln-His-Leu-Cys-Gly-Ser-His-Leu-Val-Glu-Ala-Leu-Tyr-Leu-Val-Cys-Gly-Thr-Lys-Pro-Thr-Tyr-Phe-Phe-Gly-Arg-Glu (residue 1 ... 30)

Insulin is a small protein consisting of two polypeptide chains (designated as the **A** and **B** chains) held together by two disulfide bonds. An additional disulfide bond joins two cysteine residues within the **A** chain.

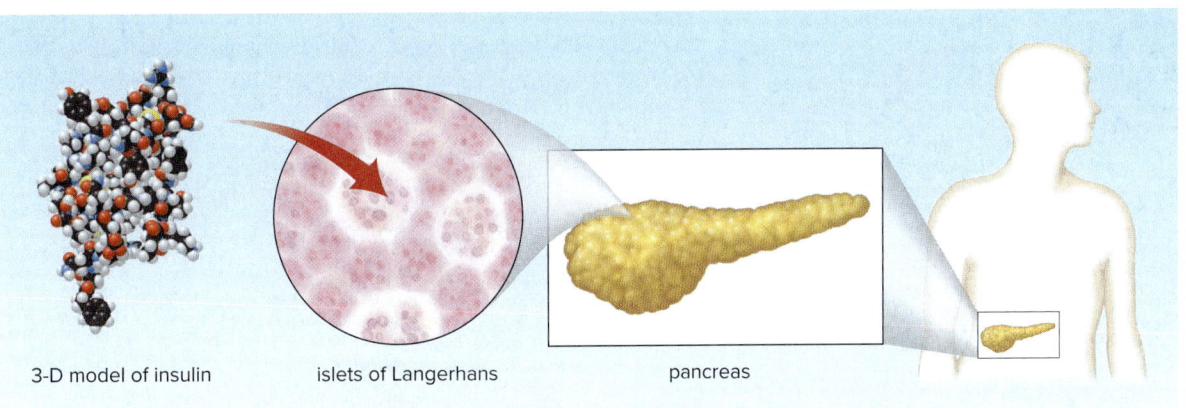

3-D model of insulin   islets of Langerhans   pancreas

Synthesized by groups of cells in the pancreas called the islets of Langerhans, insulin is the protein that regulates the levels of glucose in the blood. Insufficiency of insulin results in diabetes. Many of the abnormalities associated with this disease can be controlled by the injection of insulin. Until the availability of human insulin through genetic engineering techniques, all insulin used by diabetics was obtained from pigs and cattle. The amino acid sequences of these insulin proteins is slightly different from that of human insulin. Pig insulin differs in one amino acid only, whereas bovine insulin has three different amino acids. This is shown in the accompanying table.

Position of residue →	Chain A			Chain B
	8	9	10	30
Human insulin	Thr	Ser	Ile	Thr
Pig insulin	Thr	Ser	Ile	Ala
Bovine insulin	Ala	Ser	Val	Ala

Figure 23.11

The stabilizing interactions in secondary and tertiary protein structure

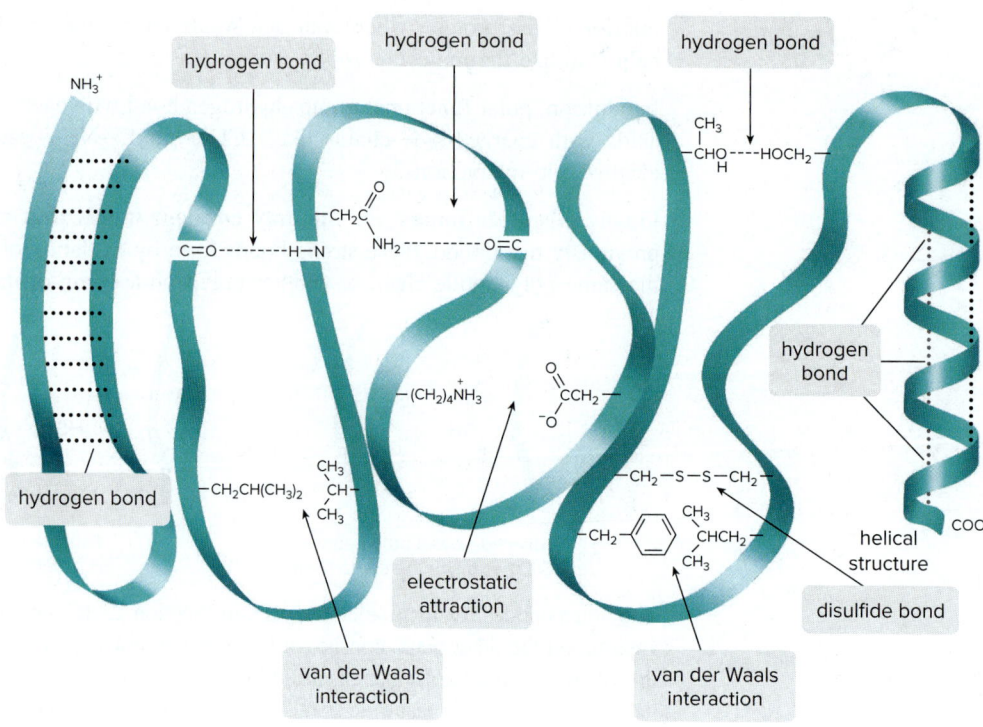

Figure 23.11 schematically illustrates the many different kinds of intramolecular forces that stabilize the secondary and tertiary structures of polypeptide chains.

The shape adopted when two or more folded polypeptide chains aggregate into one protein complex is called the *quaternary structure* of the protein. Each individual polypeptide chain is called a **subunit** of the overall protein. **Hemoglobin,** for example, consists of two α and two β subunits held together by intermolecular forces in a compact three-dimensional shape. The unique function of hemoglobin is possible only when all four subunits are together.

The four levels of protein structure are summarized in Figure 23.12.

Figure 23.12   The primary, secondary, tertiary, and quaternary structure of proteins

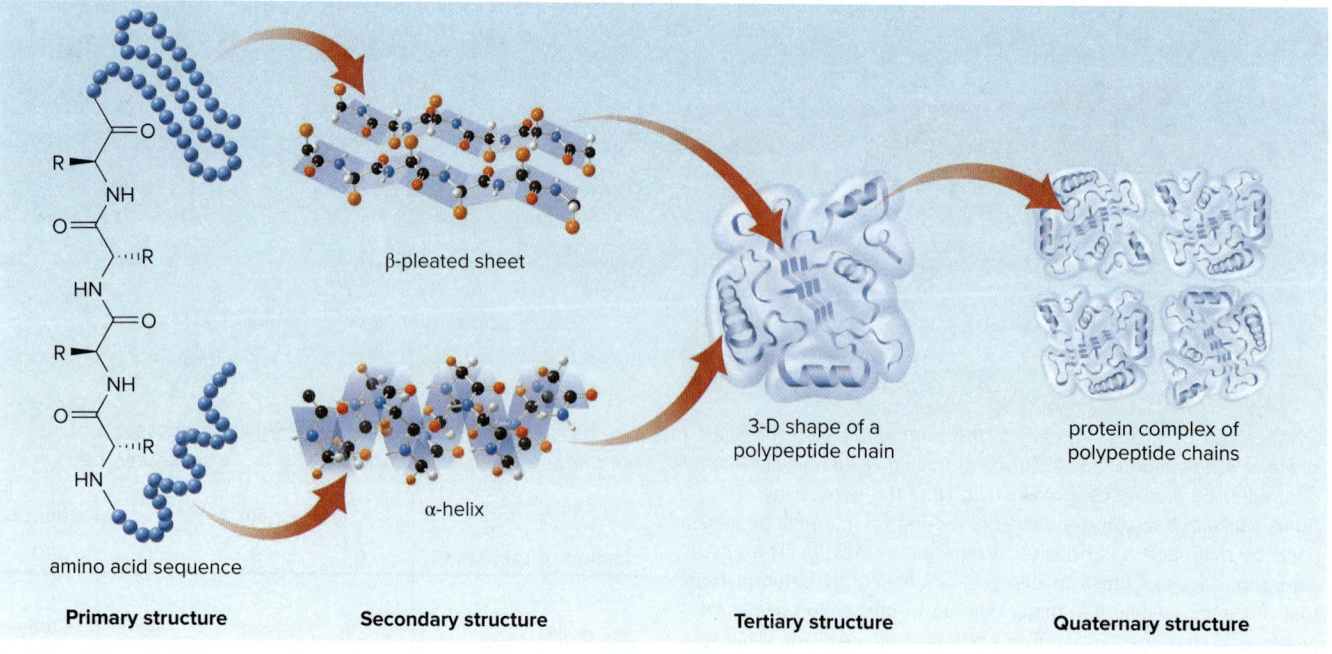

Primary structure   **Secondary structure**   **Tertiary structure**   **Quaternary structure**

Problem 23.20   Which peptide in each pair has side chains that exhibit predominantly van der Waals forces?
a. Met–Gly–Leu–Phe–Gln–Ala or Lys–Gly–Arg–Tyr–Trp–Glu
b. Tyr–Asp–Leu–Lys–His or Phe–Asn–Leu–Leu–Met

Problem 23.21   The fibroin proteins found in silk fibers consist of large regions of β-pleated sheets stacked one on top of another. (a) Explain why having a glycine at every other residue allows the β-pleated sheets to stack on top of each other. (b) Why are silk fibers insoluble in water?

23.8D Protein Denaturation

When the secondary, tertiary, or quaternary structure of a protein is disturbed, the properties of a protein are altered and the biological activity is often lost.

> • **Denaturation** is the process of altering the shape of a protein without breaking the amide bonds that form the primary structure.

High temperature, acid, base, and even agitation can disrupt the noncovalent interactions that hold a protein in a specific shape. Heat breaks up weak van der Waals forces between the nonpolar amino acids. Heat, acid, and base disrupt hydrogen-bonding interactions between polar amino acids, which account for much of the secondary and tertiary structure. A water-soluble coiled protein uncoils into an undefined randomly looped structure that exposes hydrophobic regions and makes the protein less water soluble.

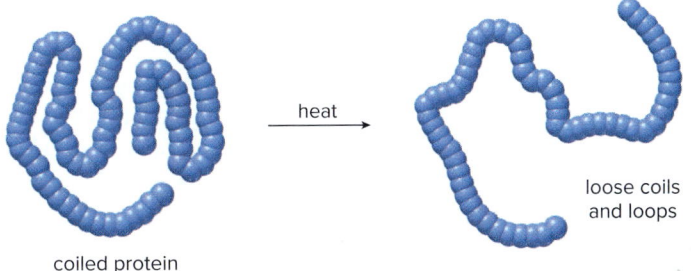

Cooking or whipping egg whites denatures the globular proteins they contain, forming insoluble protein. *Jill Braaten/ McGraw-Hill Education*

We witness many examples of protein denaturation in the kitchen. As milk ages it becomes sour from enzymes that produce lactic acid, which denatures milk proteins that precipitate as an insoluble curd. Ovalbumin, the major protein in egg white, is denatured when an egg is boiled or fried, forming a solid. Even vigorously whipping an egg white denatures its protein, forming the stiff meringue used to top a lemon meringue pie.

23.9 Important Proteins

Proteins are generally classified according to their three-dimensional shapes.

- **Fibrous proteins** are composed of long linear polypeptide chains that are bundled together to form rods or sheets. These proteins are insoluble in water and serve structural roles, giving strength and protection to tissues and cells.
- **Globular proteins** are coiled into compact shapes with hydrophilic outer surfaces that make them water soluble. Enzymes and transport proteins are globular to make them soluble in the blood and other aqueous environments in cells.

23.9A α-Keratins

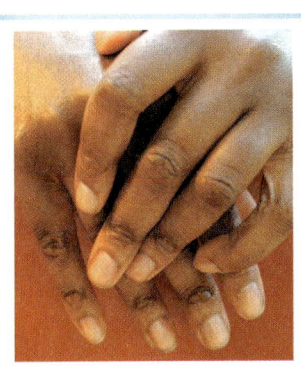

The many disulfide bonds in the proteins that compose fingernails make nails strong and hard. *Diffused Productions/ Alamy Stock Photo*

α-Keratins are the proteins found in hair, hooves, nails, skin, and wool. They are composed almost exclusively of long sections of α-helix units, having large numbers of alanine and leucine residues. Because these nonpolar amino acids extend outward from the α-helix, these proteins are very water insoluble. Two α-keratin helices coil around each other, forming a structure called a **supercoil** or **superhelix**. These, in turn, form larger and larger bundles of fibers, ultimately forming a strand of hair, as shown schematically in Figure 23.13.

Figure 23.13
Anatomy of a hair—
It begins with α-keratin.

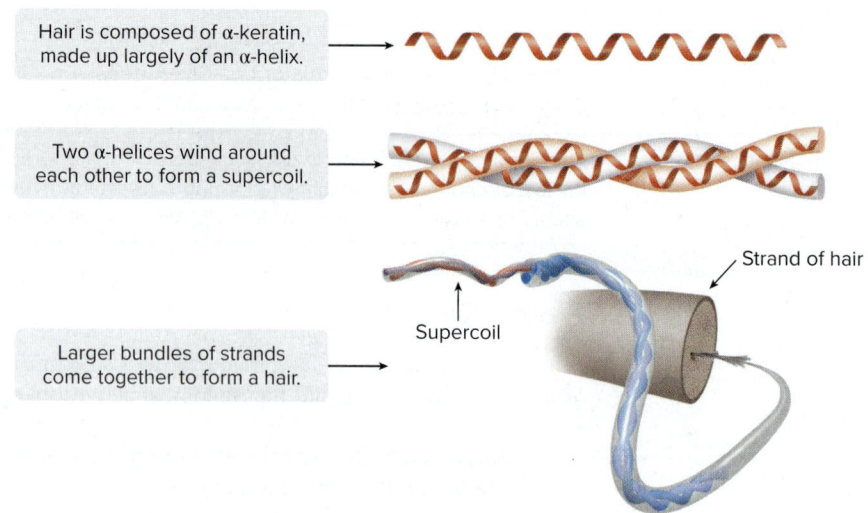

Hair is composed of α-keratin, made up largely of an α-helix.

Two α-helices wind around each other to form a supercoil.

Larger bundles of strands come together to form a hair.

Supercoil

Strand of hair

α-Keratins also have a number of cysteine residues, and because of this, disulfide bonds are formed between adjacent helices. The number of disulfide bridges determines the strength of the material. Claws, horns, and fingernails have extensive networks of disulfide bonds, making them extremely hard.

Straight hair can be made curly by cleaving the disulfide bonds in α-keratin, and then rearranging and re-forming them, as shown schematically in Figure 23.14. First, the disulfide bonds in the straight hair are reduced to thiol groups, so the bundles of α-keratin chains are no longer held in their specific "straight" orientation. Then, the hair is wrapped around curlers and treated with an oxidizing agent that converts the thiol groups back to disulfide bonds, now with twists and turns in the keratin backbone. This makes the hair look curly and is the chemical basis for a "permanent."

Figure 23.14
The chemistry of a "permanent"—Making straight hair curly

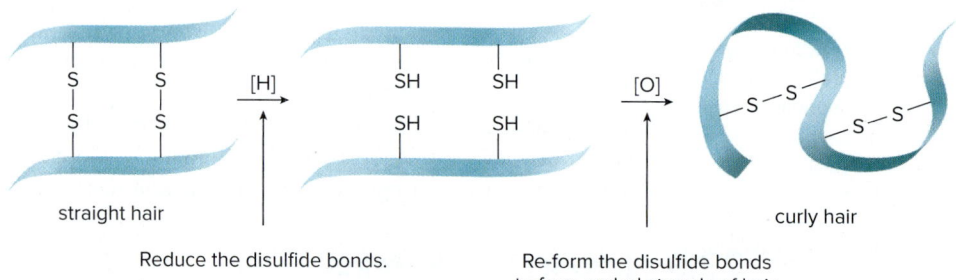

straight hair

Reduce the disulfide bonds.

Re-form the disulfide bonds to form curled strands of hair.

curly hair

- To make straight hair curly, the disulfide bonds holding the α-helical chains together are cleaved by reduction. This forms free thiol groups (–SH). The hair is turned around curlers and then an oxidizing agent is applied. This re-forms the disulfide bonds in the hair, but between different thiol groups, now giving it a curly appearance.

23.9B Collagen

Collagen, the most abundant protein in vertebrates, is found in connective tissues such as bone, cartilage, tendons, teeth, and blood vessels. Glycine and proline account for a large fraction of its amino acid residues, whereas cysteine accounts for very little. Because of the high proline content, it cannot form a right-handed α-helix. Instead, it forms an elongated left-handed helix, and then three of these helices wind around each other to form a right-handed **superhelix** or **triple helix.** The side chain of glycine is only a hydrogen atom, so the high glycine content allows the collagen superhelices to lie compactly next to each other, thus stabilizing the superhelices via hydrogen bonding. Two views of the collagen superhelix are shown in Figure 23.15.

Figure 23.15
Two different representations for the triple helix of collagen

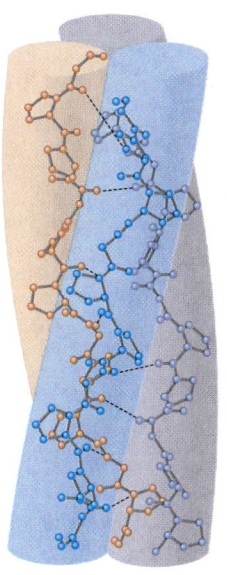
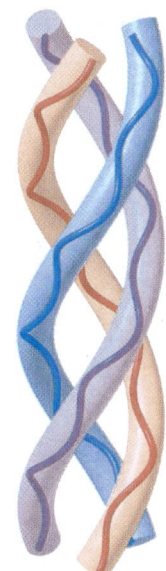

- In collagen, three polypeptide chains having an unusual left-handed helix wind around each other in a right-handed triple helix. The high content of small glycine residues allows the chains to lie close to each other, permitting hydrogen bonding between the chains.

23.9C Hemoglobin and Myoglobin

Hemoglobin and **myoglobin,** two globular proteins, are called **conjugated proteins** because they are composed of a protein unit and a nonprotein molecule called a **prosthetic group.** The prosthetic group in hemoglobin and myoglobin is **heme,** a complex organic compound containing the Fe^{2+} ion first discussed in Section 19.12. The Fe^{2+} ion of hemoglobin and myoglobin binds oxygen in the blood. Hemoglobin, which is present in red blood cells, transports oxygen to wherever it is needed in the body, whereas myoglobin stores oxygen in tissues. Ribbon diagrams for myoglobin and hemoglobin are shown in Figure 23.16.

heme

The high concentration of myoglobin in a whale's muscles allows it to remain underwater for long periods of time. *Daniel C. Smith*

Myoglobin has 153 amino acid residues in a single polypeptide chain. It has eight separate α-helical sections that fold back on one another, with the prosthetic heme group held in a cavity inside the polypeptide. Most of the polar residues are found on the outside of the protein so that they can interact with the water solvent. Spaces in the interior of the protein are filled with nonpolar amino acids. Myoglobin binds oxygen in the blood and stores it in the tissues.

Hemoglobin consists of four polypeptide chains (two α subunits and two β subunits), each of which carries a heme unit. Hemoglobin has more nonpolar amino acids than myoglobin. When each subunit is folded, some of these remain on the surface. The van der Waals attraction between these hydrophobic groups is what stabilizes the quaternary structure of the four subunits.

Figure 23.16
Protein ribbon diagrams for myoglobin and hemoglobin

a. Myoglobin

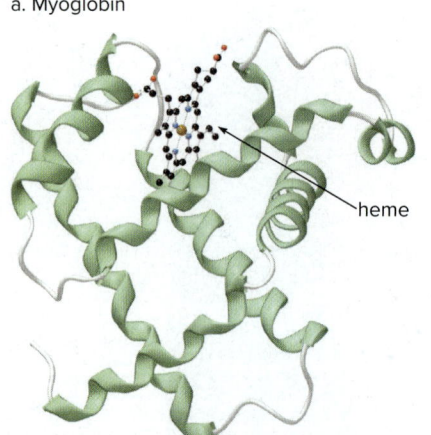

b. Hemoglobin

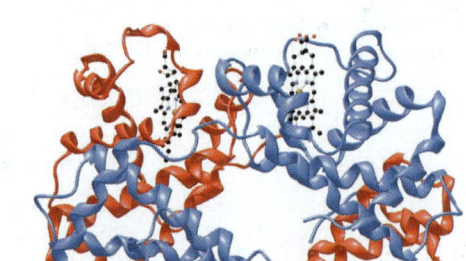

- Myoglobin consists of a single polypeptide chain with a heme unit shown in a ball-and-stick model.

- Hemoglobin consists of two α and two β chains shown in red and blue, respectively, and four heme units shown in ball-and-stick models.

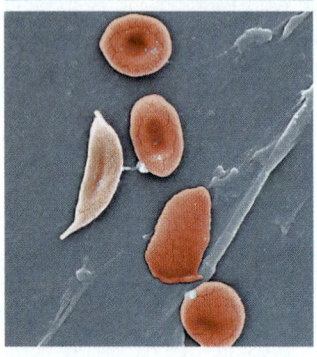

When red blood cells take on a "sickled" shape in persons with sickle cell disease, they occlude capillaries (causing organ injury) and they break easily (leading to profound anemia). This devastating illness results from the change of a single amino acid in hemoglobin. Note the single sickled cell surrounded by red cells with normal morphology.
Source: CDC/Sickle Cell Foundation of Georgia: Jackie George, Beverly Sinclair/ photo by Janice Haney Carr

Carbon monoxide is poisonous because it binds to the Fe^{2+} of hemoglobin more strongly than does oxygen. Hemoglobin complexed with CO cannot carry O_2 from the lungs to the tissues. Without O_2 in the tissues for metabolism, cells cannot function, so they die.

The properties of all proteins depend on their three-dimensional shape, and their shape depends on their primary structure—that is, their amino acid sequence. This is particularly well exemplified by comparing normal hemoglobin with **sickle cell hemoglobin,** a mutant variation in which a single amino acid of both β subunits is changed from glutamic acid to valine. The replacement of one acidic amino acid (Glu) with one nonpolar amino acid (Val) changes the shape of hemoglobin, which has profound effects on its function. Deoxygenated red blood cells with sickle cell hemoglobin become elongated and crescent shaped, and they are unusually fragile. As a result, they do not flow easily through capillaries, causing pain and inflammation, and they break open easily, leading to severe anemia and organ damage. The end result is often a painful and premature death.

This disease, called **sickle cell anemia,** is found almost exclusively among people originating from central and western Africa, where malaria is an enormous health problem. Sickle cell hemoglobin results from a genetic mutation in the DNA sequence that is responsible for the synthesis of hemoglobin. Individuals who inherit this mutation from both parents develop sickle cell anemia, whereas those who inherit it from only one parent are said to have the sickle cell trait. They do not develop sickle cell anemia, and they are more resistant to malaria than individuals without the mutation. This apparently accounts for this detrimental gene being passed on from generation to generation.

23.10 Enzymes

Enzymes are water-soluble globular proteins that serve as biological catalysts for reactions in all living organisms. As we learned in Section 6.11, an enzyme contains an active site that binds a substrate, often in a small cavity that contains amino acids that are attracted to the substrate with various types of intermolecular forces. An enzyme-catalyzed reaction can be 10^6 to 10^{12} times faster than a similar uncatalyzed reaction.

Enzymes are specific. Some enzymes catalyze a single reaction of a single compound. Other enzymes, like trypsin and chymotrypsin (Section 23.5), catalyze a specific type of reaction, the cleavage of peptide bonds involving only certain amino acids.

Enzymes are crucial to the biological reactions that occur in the body, which would otherwise proceed too slowly. In humans, enzymes must catalyze reactions under specific physiological conditions, usually a pH around 7.4 and a temperature of 37 °C.

23.10A Classification of Enzymes

Enzymes are classified into six categories by the type of reaction they catalyze.

- *Oxidoreductases* catalyze oxidation–reduction reactions.
- *Transferases* catalyze the transfer of a group from one molecule to another.
- *Hydrolases* catalyze hydrolysis of esters, amides, and other functional groups that are cleaved when they react with water.
- *Isomerases* catalyze the conversion of one isomer to another.
- *Lyases* catalyze the addition of a molecule to a double bond or the elimination of a molecule to give a double bond.
- *Ligases* catalyze bond formation accompanied by energy release from a hydrolysis reaction.

Table 23.3 summarizes the types of enzymes. Some enzyme classes are further subclassified by the functional group in the substrate or the type of molecule added or removed. For example, a transaminase is a transferase that catalyzes the transfer of an NH_2 group, whereas a kinase is a transferase that catalyzes transfer of a phosphate group.

Table 23.3 Classification of Enzymes

Enzyme Class or Subclass	Reaction Catalyzed
Oxidoreductases	**Oxidation–reduction**
• Oxidases	• Oxidation
• Reductases	• Reduction
• Dehydrogenases	• Addition or removal of 2 H's
Transferases	**Transfer of a group**
• Transaminases	• Transfer of an NH_2 group
• Kinases	• Transfer of a phosphate
Hydrolases	**Hydrolysis**
• Lipases	• Hydrolysis of lipid esters
• Proteases	• Hydrolysis of amide bonds in proteins
• Nucleases	• Hydrolysis of nucleic acids
Isomerases	**Isomerization**
Lyases	**Addition to a double bond or elimination to give a double bond**
• Dehydrases	• Removal of H_2O
• Decarboxylases	• Removal of CO_2
• Synthases	• Addition of a small molecule to a double bond
Ligases	**Bond formation accompanied by ATP hydrolysis**
• Carboxylases	• Bond formation between a substrate and CO_2

Problem 23.22 Classify the enzyme used in each of the following reactions.

a. pyruvate → acetaldehyde + CO_2

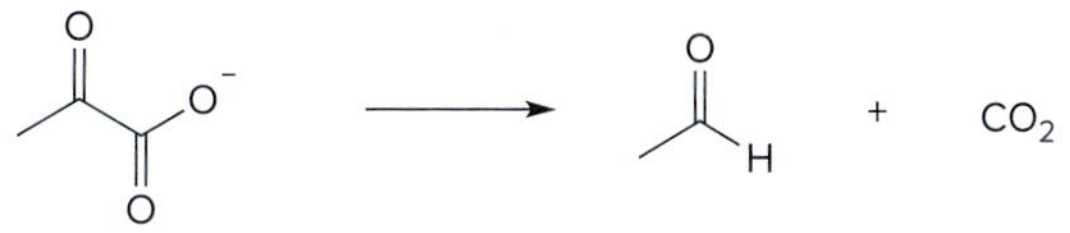

b. maleate → fumarate

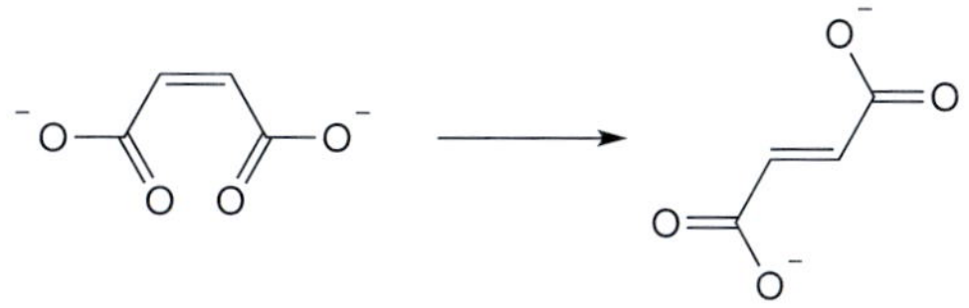

Problem 23.23 To what class do each of the following enzymes belong: (a) chymotrypsin; (b) alcohol dehydrogenase (Section 11.13); (c) phosphofructokinase (Section 27.4A)?

23.10B How Enzymes Work—Serine Proteases

The enormous rate increases that occur in enzyme-catalyzed reactions result from many factors. An enzyme holds the substrate and any reactants in close proximity in the proper orientation. The acidic and basic side chains of the amino acid residues of the enzyme are secured in the precise positions needed to activate functional groups for reaction. Moreover, the binding of the enzyme to the substrate also lowers the energy of the transition state of the reaction to substantially increase the rate of the reaction.

A well-studied example of what happens at an active site is seen with the enzymes trypsin and chymotrypsin, which catalyze the hydrolysis of peptide bonds as discussed in Section 23.5. Trypsin and chymotrypsin are two members a group of enzymes called **serine proteases,** so named because a serine residue in each enzyme plays a key role in the catalysis.

Three amino acid residues, called a **catalytic triad,** are key to the reaction. **The catalytic triad consists of amino acids that contain an acid, a base, and a nucleophile.** The acid and base activate the nucleophile, a polar side chain of an amino acid residue, which then attacks the substrate, a peptide bond, forming a covalent intermediate that is then hydrolyzed to regenerate the enzyme and form the product.

In trypsin and chymotrypsin, the triad is composed of the amino acids aspartate (Asp), histidine (His), and serine (Ser). The enzyme is folded in such a way that these three residues, although located far from each other in the protein, are positioned in close proximity at the active site. The active site of each enzyme has the same three amino acid residues, but the shape of the cavity is somewhat different in trypsin and chymotrypsin, so peptide bonds formed from different amino acids are hydrolyzed by each enzyme.

Mechanism 23.2 illustrates the steps of hydrolysis of a peptide at the active site of a serine protease. The mechanism has two parts, and each part involves the usual two steps of nucleophilic acyl substitution: **nucleophilic attack** followed by **loss of a leaving group.** The mechanism also shows the roles of aspartate and histidine in hydrolysis: aspartate hydrogen bonds to histidine, which acts as a base to activate the nucleophile in each part (the OH group of serine or H_2O).

Mechanism 23.2 Peptide Hydrolysis with a Serine Protease

Part [1] Cleavage of the peptide C–N bond

[Mechanism diagram showing steps 1 and 2 of peptide C–N bond cleavage with Asp, His, and Ser residues, peptide substrate, formation of tetrahedral intermediate, and acyl-enzyme intermediate plus R–NH₂]

The acyl group is covalently bonded to the enzyme.

1. Hydrogen bonding of aspartate to the histidine N–H increases the basicity of histidine, which removes a proton from serine. This activates the serine toward nucleophilic addition to the peptide C=O to form a tetrahedral intermediate.

2. The C=O is re-formed and the C–N bond is cleaved. RNH₂ is formed when the protonated histidine donates H⁺ to generate a good leaving group. The acyl group (RCO–) is now covalently bonded to the enzyme.

Part [2] Formation of the carboxylic acid (RCO₂H)

[Mechanism diagram showing steps 3 and 4 of carboxylic acid formation with Asp, His, Ser residues, water, tetrahedral intermediate, and regeneration of serine plus HO–C(=O)–R]

3. Hydrogen bonding of aspartate to the histidine N–H once again increases the basicity of histidine, which removes a proton from H₂O. Nucleophilic addition of OH to the acyl C=O forms a tetrahedral intermediate.

4. The tetrahedral intermediate collapses to re-form the serine of the enzyme and a carboxylic acid derived from the peptide substrate.

Problem 23.24 Explain why chymotrypsin loses its catalytic activity when the aspartic acid residue of the catalytic triad is replaced by asparagine.

23.10C Using Enzymes to Diagnose and Treat Diseases

Measuring enzyme levels in the blood has aided greatly in diagnosing diseases. The concentration of some enzymes is higher within a cell than in the aqueous fluid outside the cell. When cells are damaged, the cells rupture and die, releasing the enzymes into the bloodstream. Measuring the activity of enzymes in the blood then becomes a powerful tool to diagnose the presence of disease or injury in some organs. For example, a higher-than-normal concentration of creatine phosphokinase (CPK) indicates whether a patient that has chest pain has had a heart attack.

Molecules that inhibit an enzyme can be useful drugs. An effective treatment of the human immunodeficiency virus (HIV), the virus that causes AIDS, uses protease inhibitors. These drugs inhibit the action of the HIV protease enzyme, an essential enzyme needed by HIV to make copies of itself that go on to infect other cells. Deactivating the HIV protease enzyme decreases the virus population, bringing the disease under control. Several protease inhibitors are currently available, and often an individual takes a combination of several drugs to keep the disease in check.

1058 Chapter 23 Amino Acids and Proteins

Another strategy for treating HIV is described in Section 26.9.

Fosamprenavir (trade name Lexiva) is a drug used to treat HIV infections. The body metabolizes fosamprenavir to amprenavir, the active drug that inhibits the HIV protease enzyme, so the virus cannot replicate. A ribbon diagram of the HIV-1 protease enzyme with amprenavir at the active site is show in Figure 23.17.

fosamprenavir → amprenavir

Figure 23.17
Amprenavir at the active site of an HIV protease enzyme

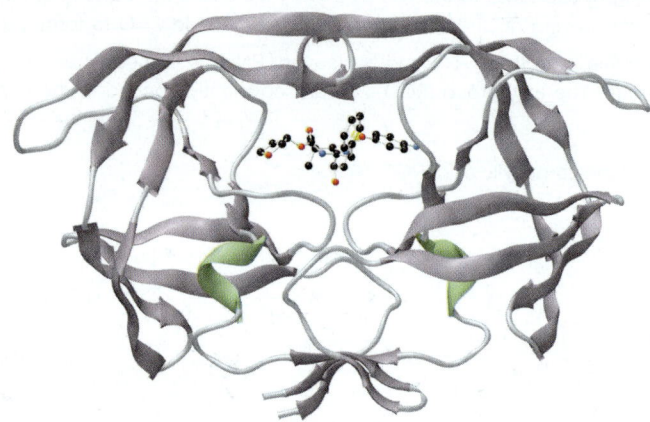

- Amprenavir, the active form of the drug fosamprenavir, inhibits the action of an HIV protease enzyme by binding to the active site.

Chapter 23 REVIEW

KEY REACTIONS

[1] Enantioselective hydrogenation (23.3)

$$\text{AcNH-}\underset{H_2N}{\overset{}{=}}\text{-OH} \xrightarrow{H_2, \text{Rh*}} \text{AcNH-}\underset{H_2N}{\overset{S}{-}}\text{-OH} \xrightarrow{\text{-OH, } H_2O} \overset{+}{H_3N}\underset{H_2N}{\overset{}{-}}\text{-O}^-$$

S enantiomer
Rh* = chiral Rh hydrogenation catalyst

S amino acid

Try Problem 23.38b.

Key Reactions

[2] Adding and removing protecting groups for amino acids (23.6)

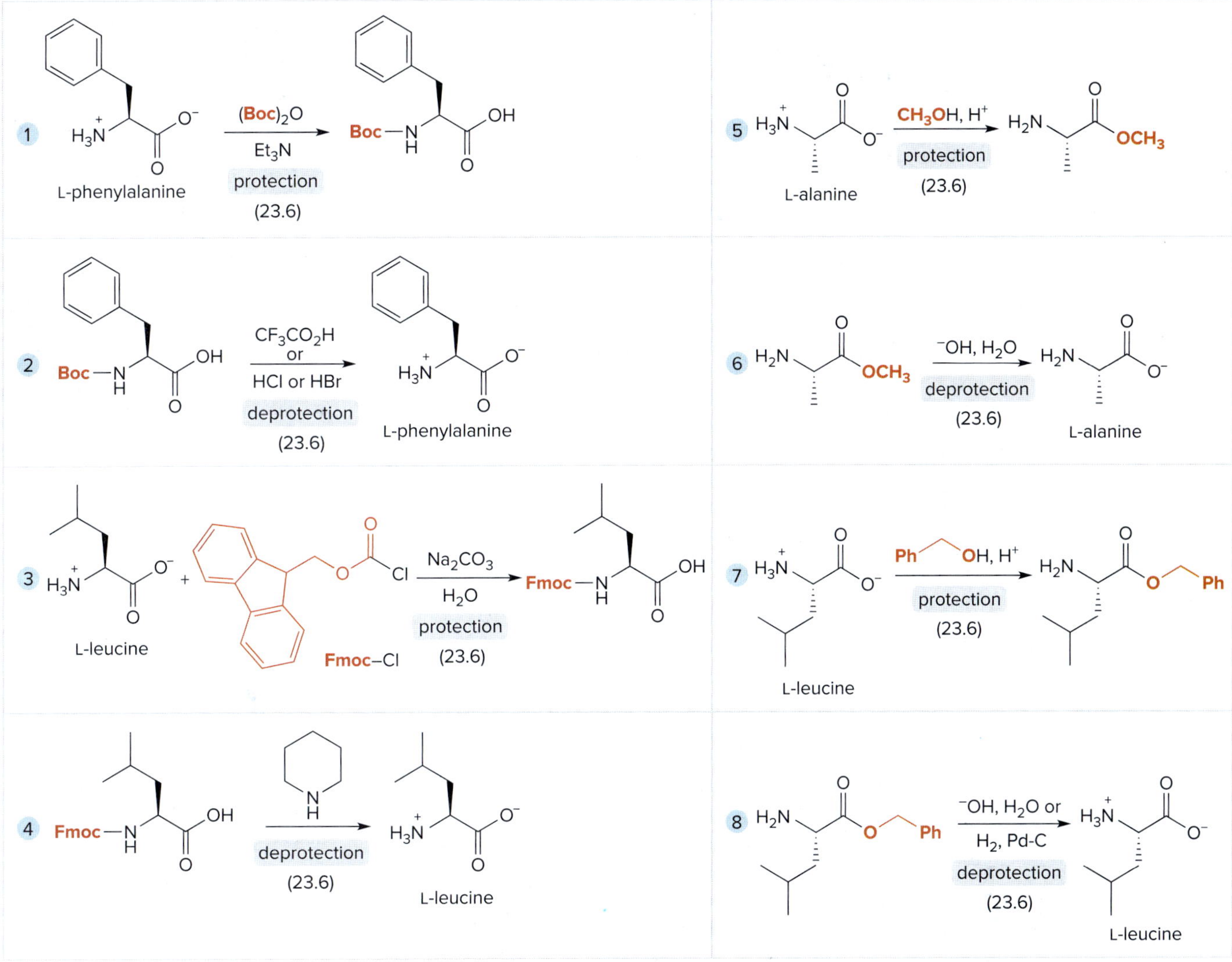

1. L-phenylalanine + (Boc)₂O, Et₃N → Boc-NH-CH(CH₂Ph)-COOH (protection, 23.6)

2. Boc-NH-CH(CH₂Ph)-COOH + CF₃CO₂H or HCl or HBr → L-phenylalanine (deprotection, 23.6)

3. L-leucine + Fmoc–Cl, Na₂CO₃, H₂O → Fmoc-NH-CH(CH₂CH(CH₃)₂)-COOH (protection, 23.6)

4. Fmoc-NH-CH(CH₂CH(CH₃)₂)-COOH + piperidine → L-leucine (deprotection, 23.6)

5. L-alanine + CH₃OH, H⁺ → H₂N-CH(CH₃)-COOCH₃ (protection, 23.6)

6. H₂N-CH(CH₃)-COOCH₃ + ⁻OH, H₂O → L-alanine (deprotection, 23.6)

7. L-leucine + PhCH₂OH, H⁺ → H₂N-CH(CH₂CH(CH₃)₂)-COOCH₂Ph (protection, 23.6)

8. H₂N-CH(CH₂CH(CH₃)₂)-COOCH₂Ph + ⁻OH, H₂O or H₂, Pd-C → L-leucine (deprotection, 23.6)

Try Problems 23.50; 23.52a, d.

[3] Amide formation with DCC

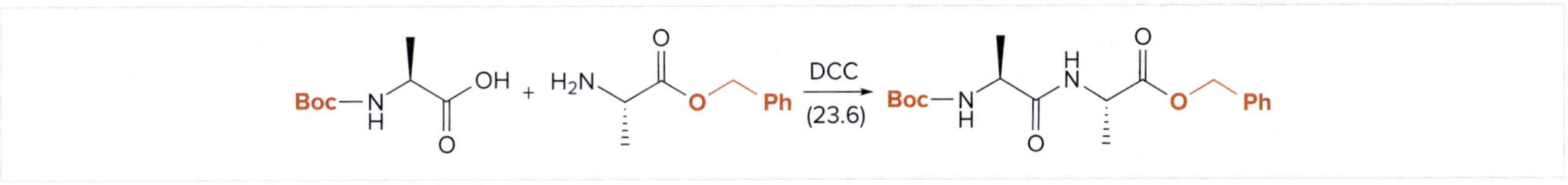

Boc-NH-CH(CH₃)-COOH + H₂N-CH(CH₃)-COOCH₂Ph —DCC (23.6)→ Boc-NH-CH(CH₃)-CO-NH-CH(CH₃)-COOCH₂Ph

KEY SKILLS

[1] Using (R)-α-methylbenzylamine to resolve a racemic mixture of amino acids (23.2A); example: separation of L- and D-leucine

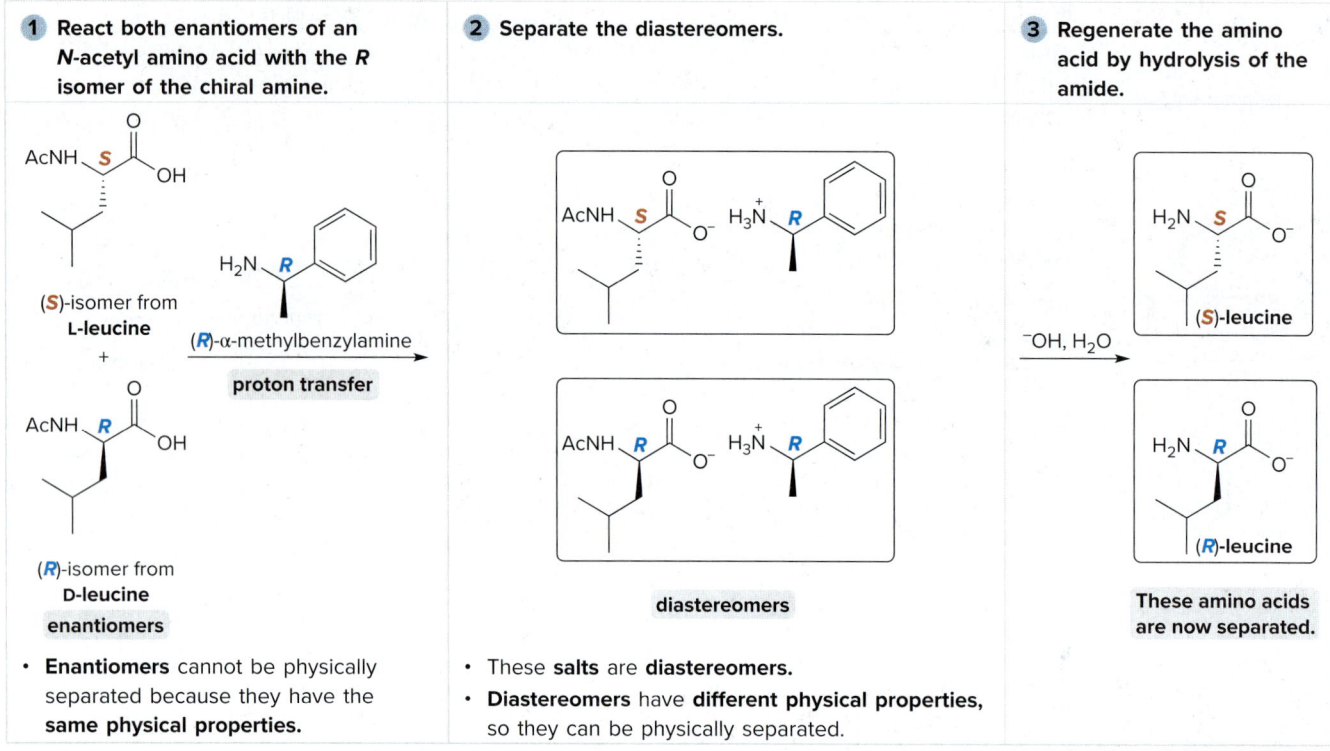

See *How To*, p. 1027. Try Problems 23.35–23.37.

[2] Using enzymes to kinetically resolve a racemic mixture of amino acids (23.2B); example: separation of L- and D-phenylalanine

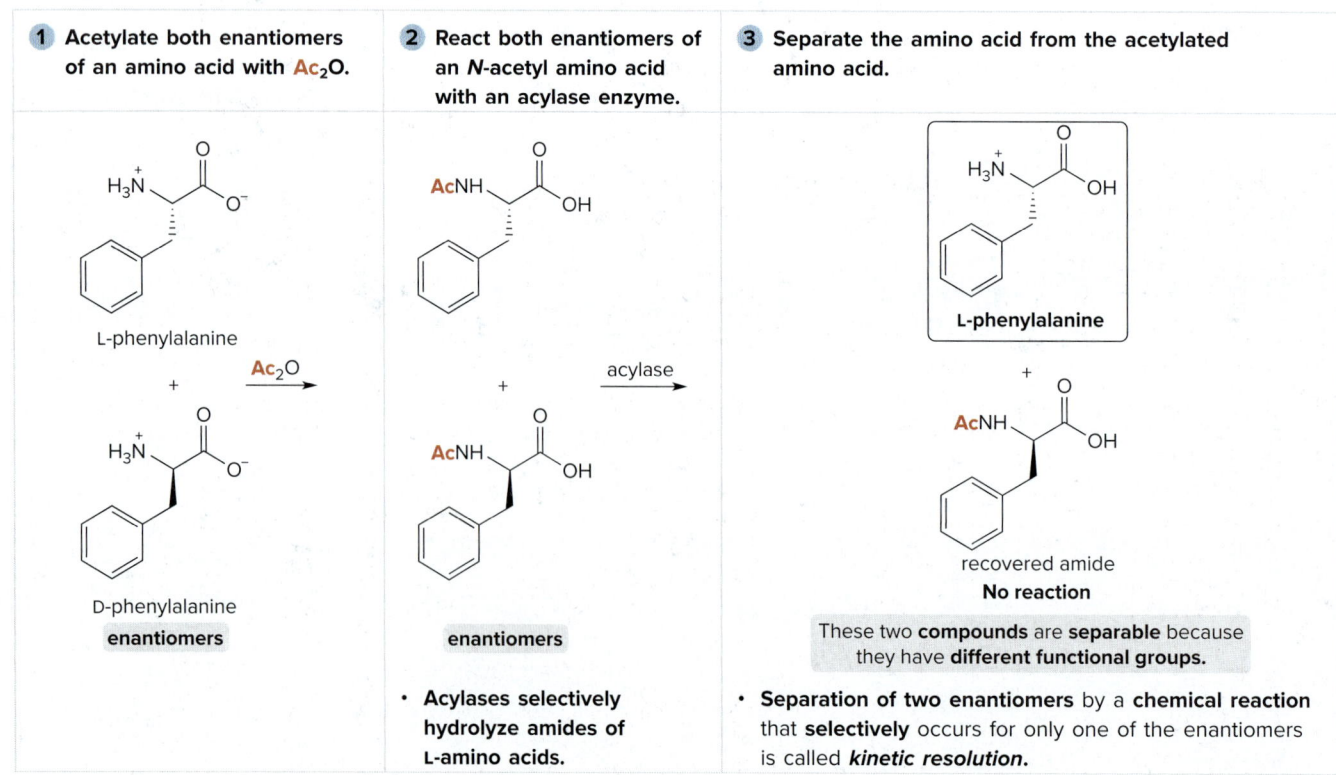

Try Problem 23.38a.

[3] Drawing the structure of a tripeptide, and labeling its N-terminal and C-terminal amino acids (23.4); example: Met–Ala–Arg

1 Draw the structures of the amino acids in order from left to right, placing the COO^- of one amino acid *next* to the NH_3^+ group of the adjacent amino acid.

2 Join adjacent COO^- and NH_3^+ groups together in amide bonds to form the tripeptide.

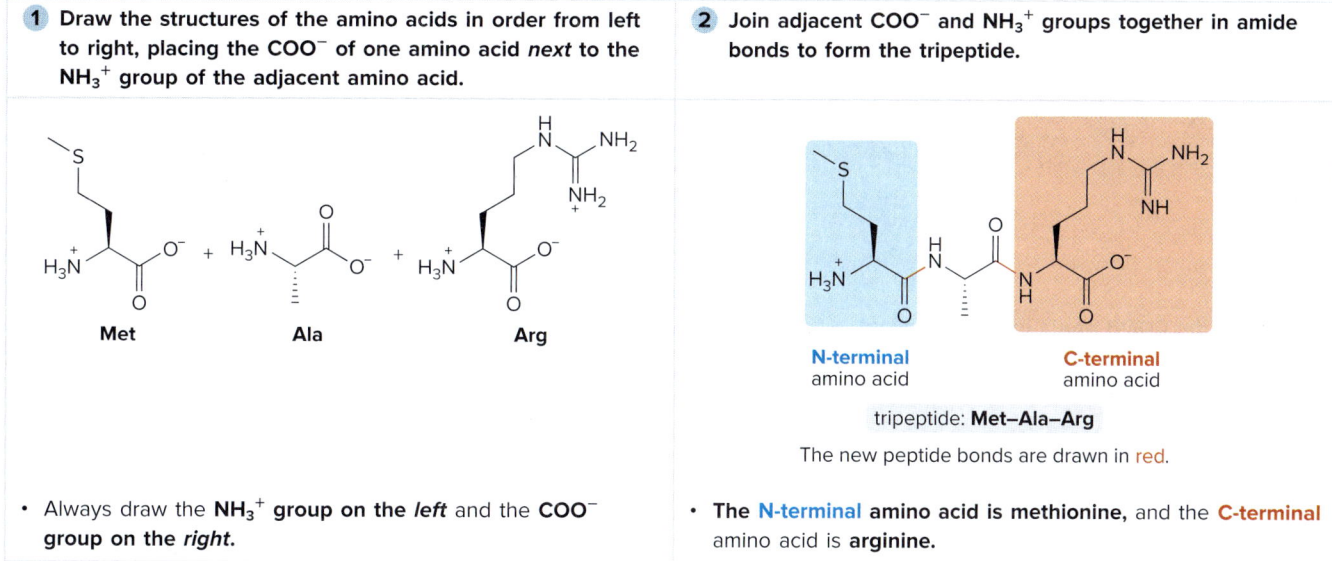

- Always draw the NH_3^+ group on the *left* and the COO^- group on the *right*.

- The **N-terminal** amino acid is methionine, and the **C-terminal** amino acid is arginine.

See Sample Problem 23.1. Try Problems 23.40; 23.41b, d.

[4] Giving the amino acid sequence of a hexapeptide that contains the amino acids Gly, Pro, Val, Ser, Leu, His, and forms the following fragments when partially hydrolyzed with HCl: Ser–Val, Pro–His–Gly, Val–Leu–Pro (23.5)

1 Look for points of overlap.

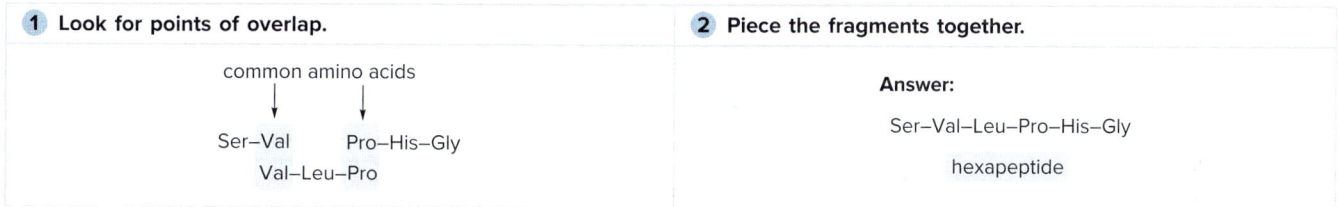

2 Piece the fragments together.

Answer:

Ser–Val–Leu–Pro–His–Gly

hexapeptide

See Sample Problem 23.2. Try Problem 23.46.

[5] Deducing the sequence of a pentapeptide that contains the amino acids Phe, Ile, Ala, Lys, Gly (23.5C)

1 Identify the **N-terminal** amino acid by Edman degradation.	**2** Identify the **C-terminal** amino acid by carboxypeptidase cleavage.	**3** Identify the possible location of Lys or Arg, if applicable, from trypsin cleavage.	**4** Complete the sequence, given the products from partial hydrolysis.
Ala–__–__–__–__	Ala–__–__–__–Ile	• If a tripeptide and a dipeptide are obtained: Ala–Lys–__–__–Ile or Ala–__–Lys–__–Ile	• If Ile, Lys, Ala, and Phe–Gly are obtained: Ala–Lys–Phe–Gly–Ile

See Sample Problem 23.3, Table 23.2. Try Problems 23.45–23.49.

[6] Synthesizing a dipeptide from two amino acids (23.6): example: Leu–Ala

1 Protect the NH_2 group of the N-terminal amino acid.

3 Form the amide bond with DCC.

- **DCC** makes the OH group of the carboxylic acid a better leaving group, thus **activating the carboxy group toward nucleophilic attack.**

2 Protect the COOH group of the C-terminal amino acid.

4 Remove one or both protecting groups.

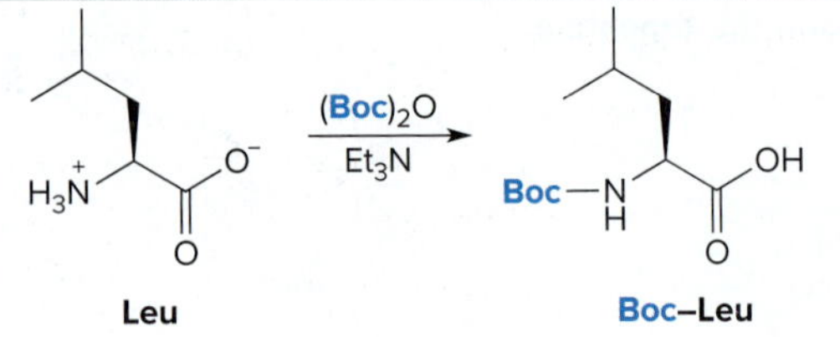

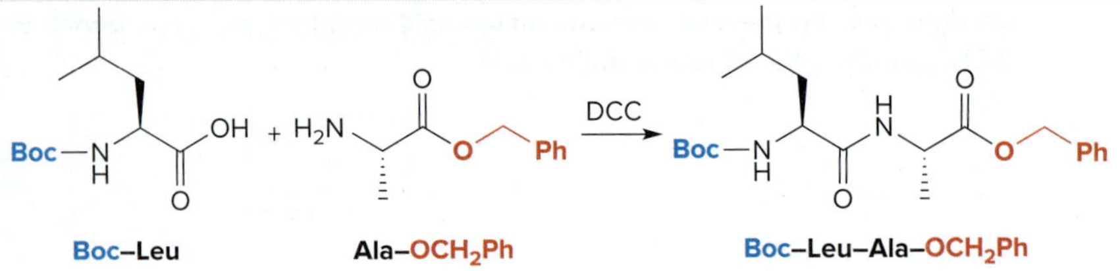

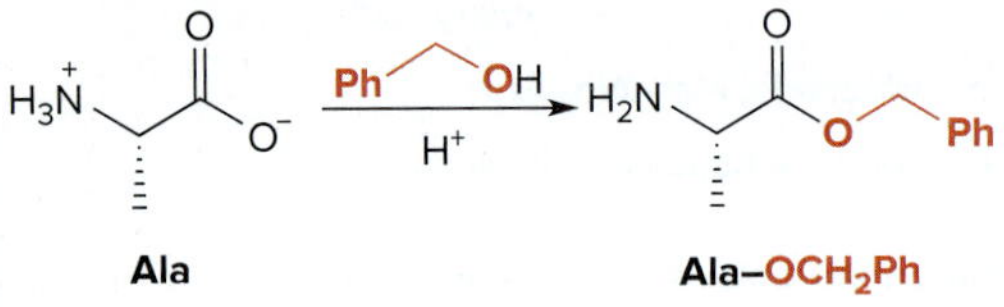

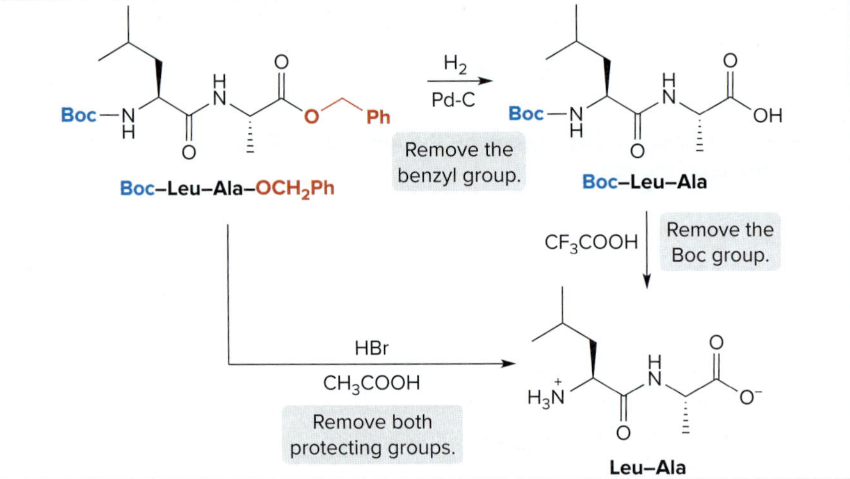

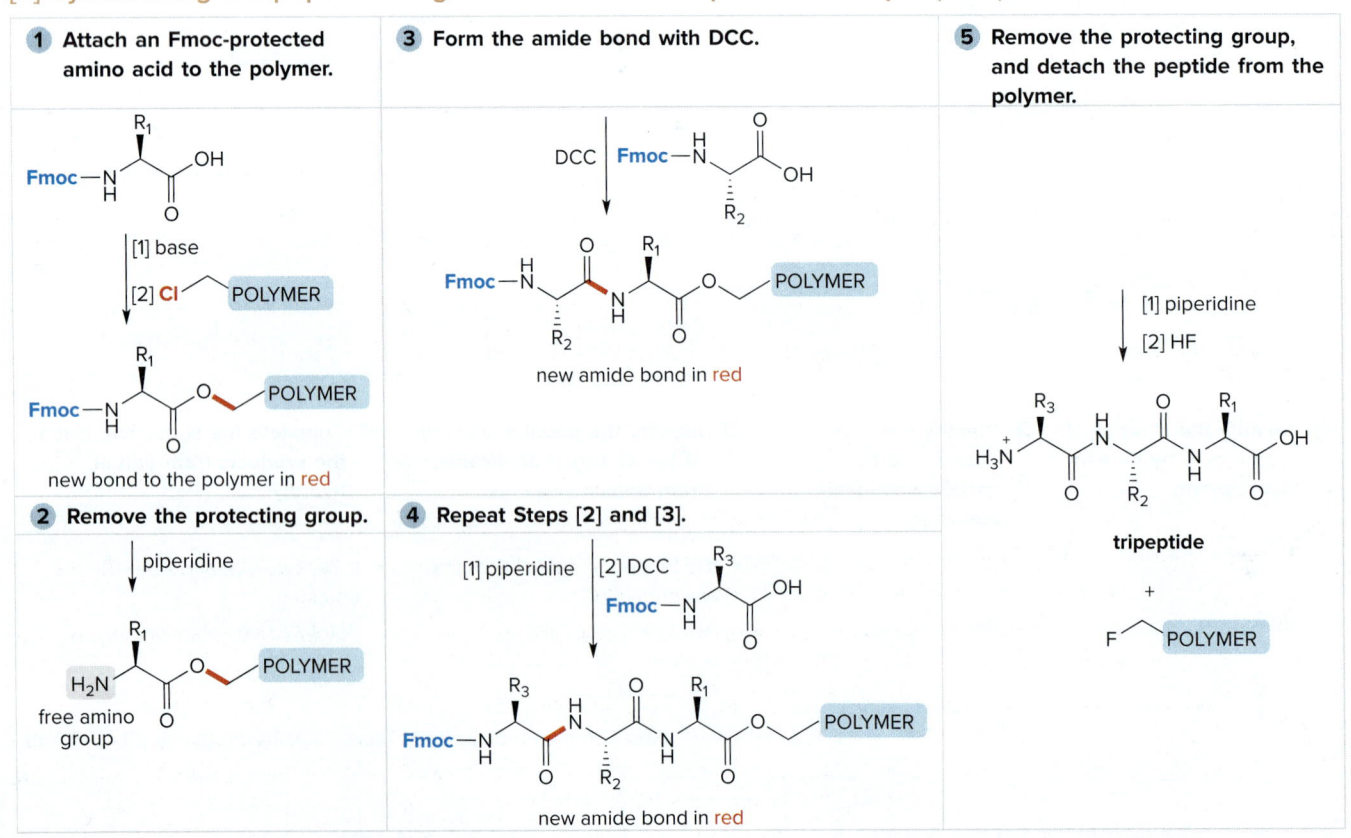

See *How To,* p. 1039, and Sample Problem 23.4. Try Problems 23.26, 23.53.

[7] Synthesizing a tripeptide using the Merrifield solid phase technique (23.7)

1 Attach an Fmoc-protected amino acid to the polymer.

new bond to the polymer in red

2 Remove the protecting group.

3 Form the amide bond with DCC.

new amide bond in red

4 Repeat Steps [2] and [3].

new amide bond in red

5 Remove the protecting group, and detach the peptide from the polymer.

See *How To,* p. 1044. Try Problems 23.27, 23.54.

PROBLEMS

Problems Using Three-Dimensional Models

23.25 With reference to the following peptide: (a) Identify the N-terminal and C-terminal amino acids. (b) Name the peptide using one-letter abbreviations. (c) Label all the amide bonds in the peptide backbone.

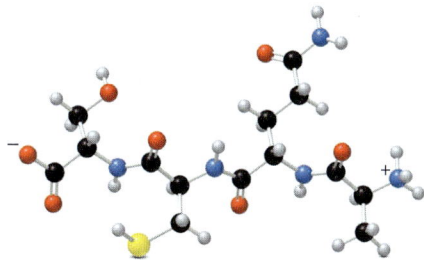

23.26 Devise a synthesis of the following dipeptide from amino acid starting materials.

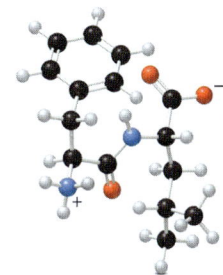

23.27 Write out the steps needed to synthesize the following peptide using the Merrifield method.

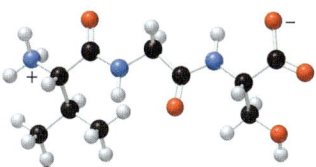

Amino Acids

23.28

![penicillamine structure: HS-C(CH3)2-CH(NH3+)-COO-]
penicillamine

a. (S)-Penicillamine, an amino acid that does not occur in proteins, is used as a copper chelating agent to treat Wilson's disease, an inherited defect in copper metabolism. (R)-Penicillamine is toxic, sometimes causing blindness. Draw the structures of (R)- and (S)-penicillamine.

b. What disulfide is formed from oxidation of (S)-penicillamine?

23.29 Histidine is classified as a basic amino acid because one of the N atoms in its five-membered ring is readily protonated by acid. Which N atom in histidine is protonated and why?

23.30 Tryptophan is not classified as a basic amino acid even though it has a heterocycle containing a nitrogen atom. Why is the N atom in the five-membered ring of tryptophan not readily protonated by acid?

23.31 What is the structure of each amino acid at its isoelectric point: (a) alanine; (b) methionine; (c) aspartic acid; (d) lysine?

23.32 What is the predominant form of each of the following amino acids at pH = 1? What is the overall charge on the amino acid at this pH? (a) threonine; (b) methionine; (c) aspartic acid; (d) arginine

23.33 What is the predominant form of each of the following amino acids at pH = 11? What is the overall charge on the amino acid? (a) valine; (b) proline; (c) glutamic acid; (d) lysine

23.34 a. Draw the structure of the tripeptide A–A–A, and label the two ionizable functional groups.
b. What is the predominant form of A–A–A at pH = 1?
c. The pK_a values for the two ionizable functional groups (3.39 and 8.03) differ considerably from the pK_a values of alanine (2.35 and 9.87; see Table 23.1). Account for the observed pK_a differences.

Resolution; The Synthesis of Chiral Amino Acids

23.35 Write out a scheme for the resolution of the two enantiomers of the antiplatelet drug clopidogrel with 10-camphorsulfonic acid.

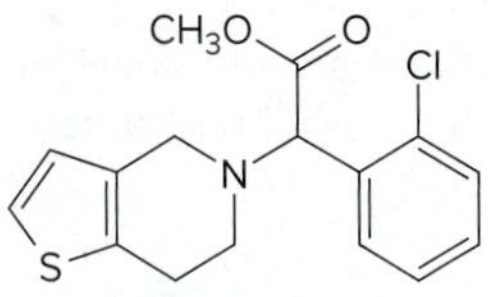

clopidogrel

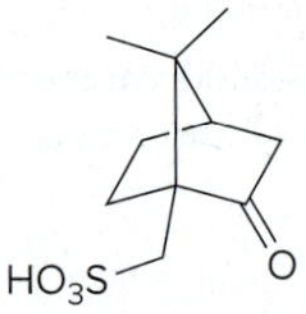

10-camphorsulfonic acid

23.36 Another strategy used to resolve amino acids involves converting the carboxy group to an ester and then using a *chiral carboxylic acid* to carry out an acid–base reaction at the free amino group. Using a racemic mixture of alanine enantiomers and (R)-mandelic acid as resolving agent, write out the steps showing how a resolution process would occur.

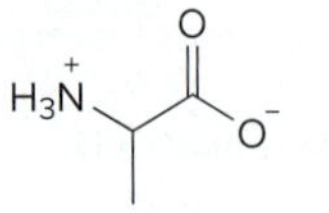

alanine
(two enantiomers)

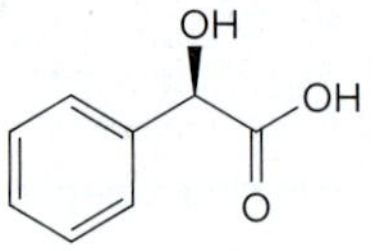

(R)-mandelic acid

23.37 Brucine is a poisonous alkaloid obtained from *Strychnos nux vomica,* a tree that grows in India, Sri Lanka, and northern Australia. Write out a resolution scheme similar to the one given in Section 23.2A, which shows how a racemic mixture of phenylalanine can be resolved using brucine.

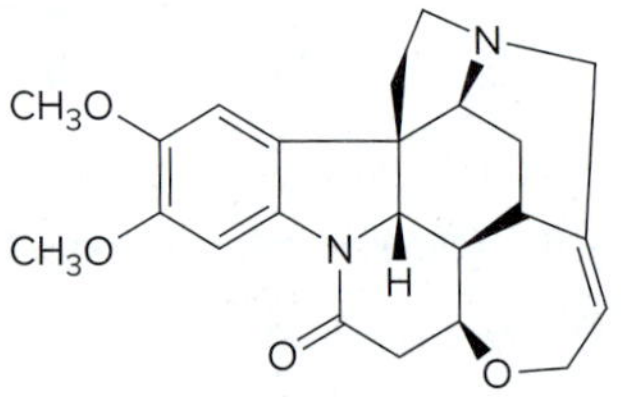

brucine

23.38 Draw the organic products formed in each reaction.

a. 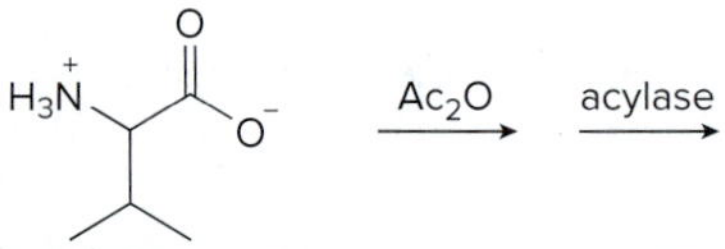 racemic mixture $\xrightarrow{Ac_2O}$ $\xrightarrow{acylase}$

b. 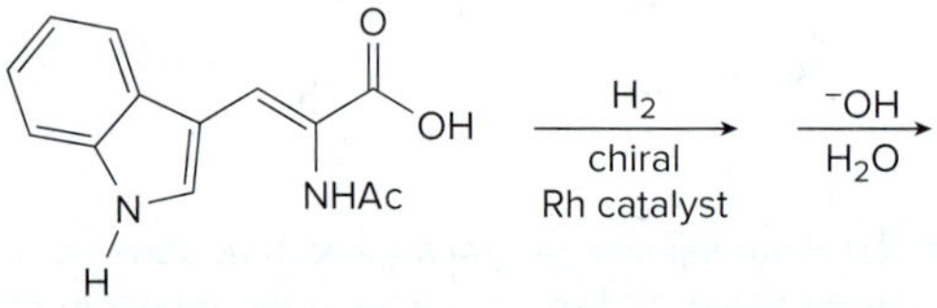 (structure with NHAc) $\xrightarrow[\text{chiral Rh catalyst}]{H_2}$ $\xrightarrow[H_2O]{^-OH}$

23.39 What steps are needed to convert **A** to L-dopa, an uncommon amino acid that is effective in treating Parkinson's disease? These steps are the key reactions in the first commercial asymmetric synthesis using a chiral transition metal catalyst. This process was developed at Monsanto in 1974.

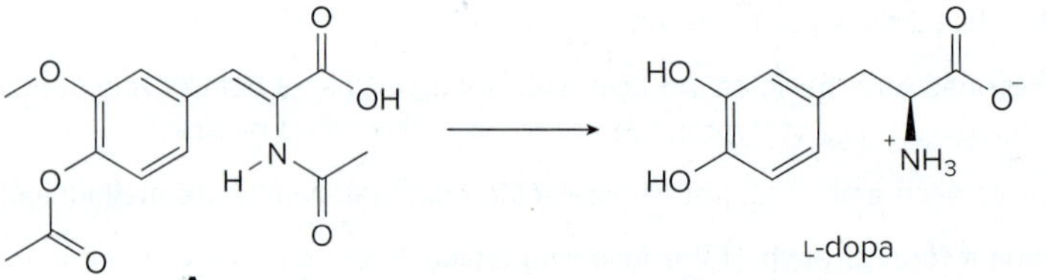

A

L-dopa

Peptide Structure and Sequencing

23.40 Draw the structure for each peptide: (a) Phe–Ala; (b) Gly–Gln; (c) Lys–Gly; (d) R–H.

23.41 For the tetrapeptide Asp–Arg–Val–Tyr:
    a. Name the peptide using one-letter abbreviations.
    b. Draw the structure.
    c. Label all amide bonds.
    d. Label the N-terminal and C-terminal amino acids.

23.42 Name each peptide using both the three-letter and one-letter abbreviations of the component amino acids.

a. 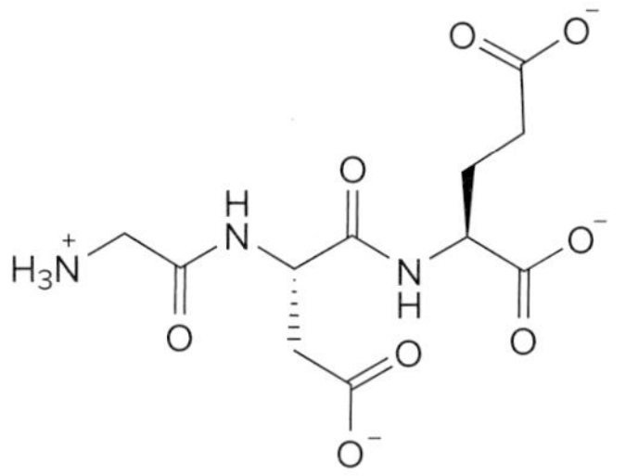 [structure] b. 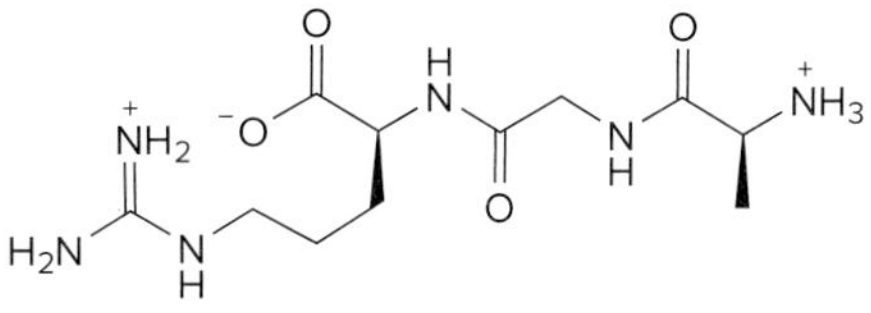 [structure]

23.43 Gramicidin S, a topical antibiotic produced by the bacterium *Bacillus brevis,* is a cyclic decapeptide formed from five amino acids. Draw the structures of the amino acids that form gramicidin S, and explain why this compound possesses two unusual structural features.

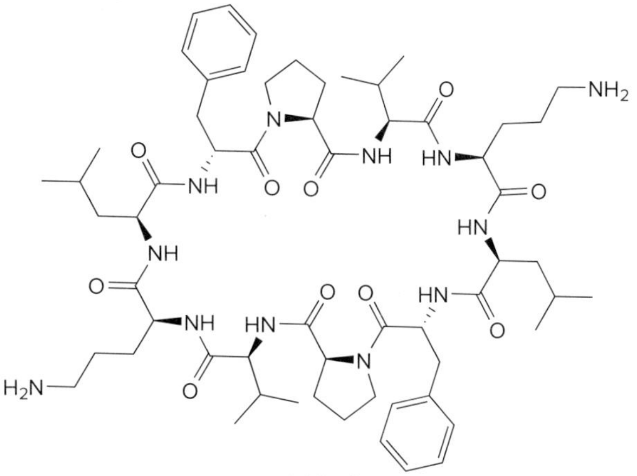

gramicidin S

23.44 The dynorphins are a group of opioid peptides that play an important role in changes in the brain associated with cocaine addiction. One of these peptides, dynorphin A, contains the following amino acid sequence: Tyr–Gly–Gly–Phe–Leu–Arg–Arg–Ile–Arg–Pro–Lys–Leu–Lys. Draw the amino acids and peptide fragments formed when dynorphin A is treated with each reagent or enzyme: (a) chymotrypsin; (b) trypsin; (c) carboxypeptidase; (d) $C_6H_5N=C=S$.

23.45 Consider the decapeptide angiotensin I.

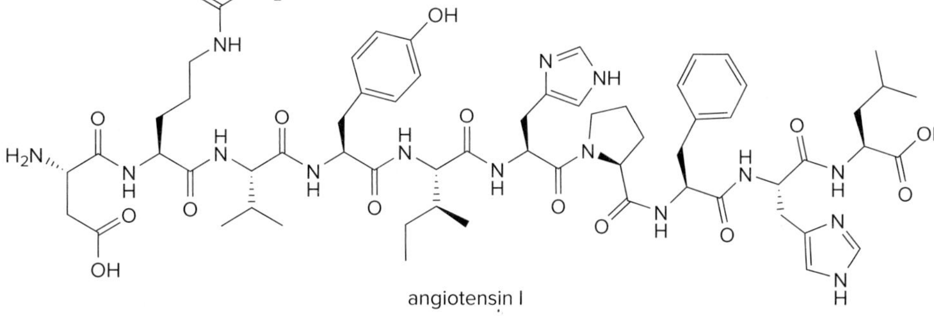

angiotensin I

a. What products are formed when angiotensin I is treated with trypsin?
b. What products are formed when angiotensin I is treated with chymotrypsin?
c. Treatment of angiotensin I with ACE (the angiotensin-converting enzyme) cleaves only the amide bond with the carbonyl group derived from phenylalanine to afford two products. The larger polypeptide is angiotensin II, a hormone that narrows blood vessels and increases blood pressure. Give the amino acid sequence of angiotensin II using three-letter abbreviations. ACE inhibitors are drugs that lower blood pressure by inhibiting the ACE enzyme (Problem 5.16).

23.46 Give the amino acid sequence of each peptide using the fragments obtained by partial hydrolysis of the peptide with acid.
  a. a tetrapeptide that contains Ala, Gly, His, and Tyr, which is hydrolyzed to the dipeptides His–Tyr, Gly–Ala, and Ala–His
  b. a pentapeptide that contains Glu, Gly, His, Lys, and Phe, which is hydrolyzed to His–Gly–Glu, Gly–Glu–Phe, and Lys–His

23.47 Glucagon, a hormone with 29 amino acids, is secreted by the pancreas. When the concentration of glucose in the bloodstream is too low, glucagon stimulates the liver to convert glycogen to glucose, thus increasing the blood glucose concentration. Deduce the amino acid sequence of glucagon from the following data. Treatment of glucagon with chymotrypsin forms: Thr–Ser–Asp–Tyr, Leu–Met–Asn–Thr, His–Ser–Gln–Gly–Thr–Phe, Ser–Lys–Tyr, Val–Gln–Trp, Leu–Asp–Ser–Arg–Arg–Ala–Gln–Asp–Phe. Treatment of glucagon with trypsin forms: Arg, Tyr–Leu–Asp–Ser–Arg, Ala–Gln–Asp–Phe–Val–Gln–Trp–Leu–Met–Asn–Thr, His–Ser–Gln–Gly–Thr–Phe–Thr–Ser–Asp–Tyr–Ser–Lys.

23.48 Use the given experimental data to deduce the sequence of an octapeptide that contains the following amino acids: Ala, Gly (2 equiv), His (2 equiv), Ile, Leu, and Phe. Edman degradation cleaves Gly from the octapeptide, and carboxypeptidase forms Leu and a heptapeptide. Partial hydrolysis forms the following fragments: Ile–His–Leu, Gly, Gly–Ala–Phe–His, and Phe–His–Ile.

23.49 An octapeptide contains the following amino acids: Arg, Glu, His, Ile, Leu, Phe, Tyr, and Val. Carboxypeptidase treatment of the octapeptide forms Phe and a heptapeptide. Treatment of the octapeptide with chymotrypsin forms two tetrapeptides, **A** and **B**. Treatment of **A** with trypsin yields two dipeptides, **C** and **D**. Edman degradation cleaves the following amino acids from each peptide: Glu (octapeptide), Glu (**A**), Ile (**B**), Glu (**C**), and Val (**D**). Partial hydrolysis of tetrapeptide **B** forms Ile–Leu in addition to other products. Deduce the structure of the octapeptide and fragments **A–D**.

Peptide Synthesis

23.50 Draw the organic products formed in each reaction.

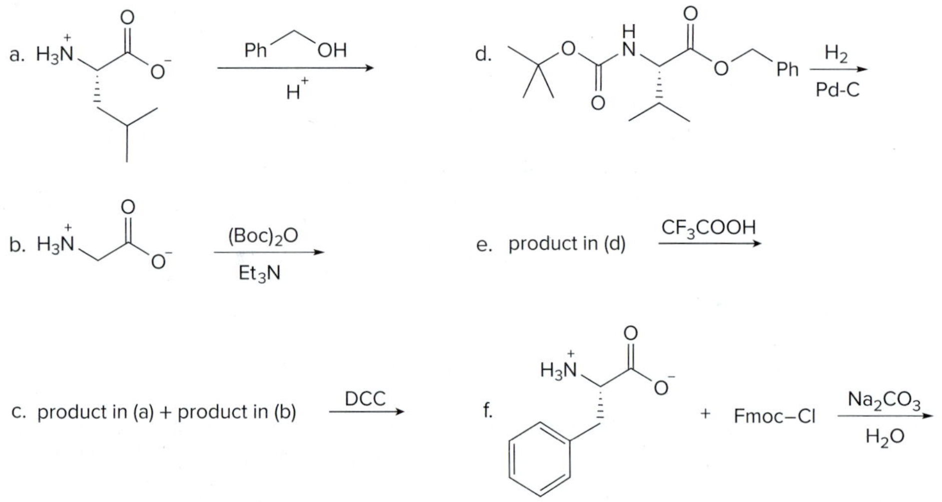

a. H$_3$N$^+$–CH(CH$_2$CH(CH$_3$)$_2$)–COO$^-$ + Ph–CH$_2$–OH, H$^+$

b. H$_3$N$^+$–CH$_2$–COO$^-$ + (Boc)$_2$O, Et$_3$N

c. product in (a) + product in (b), DCC

d. Boc–NH–CH(iPr)–CO–O–CH$_2$–Ph, H$_2$, Pd-C

e. product in (d), CF$_3$COOH

f. H$_3$N$^+$–CH(CH$_2$Ph)–COO$^-$ + Fmoc–Cl, Na$_2$CO$_3$, H$_2$O

23.51 Identify **A–E** in the following reaction sequence.

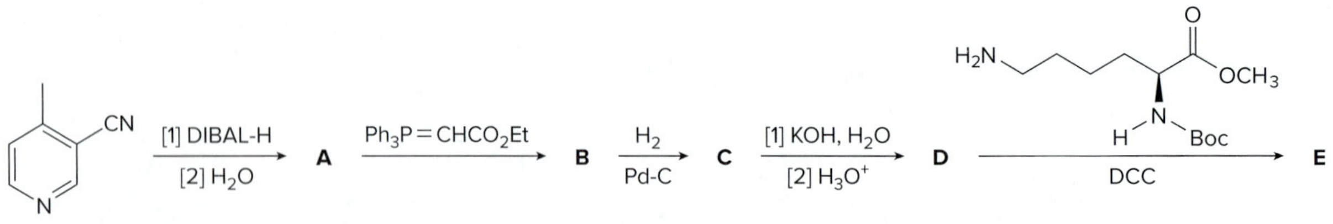

3-cyano-4-methylpyridine →[1] DIBAL-H [2] H$_2$O→ **A** →Ph$_3$P=CHCO$_2$Et→ **B** →H$_2$, Pd-C→ **C** →[1] KOH, H$_2$O [2] H$_3$O$^+$→ **D** + H$_2$N–(CH$_2$)$_4$–CH(NHBoc)–CO–OCH$_3$ →DCC→ **E**

23.52 Draw the product when the following tripeptide is treated with each reagent.

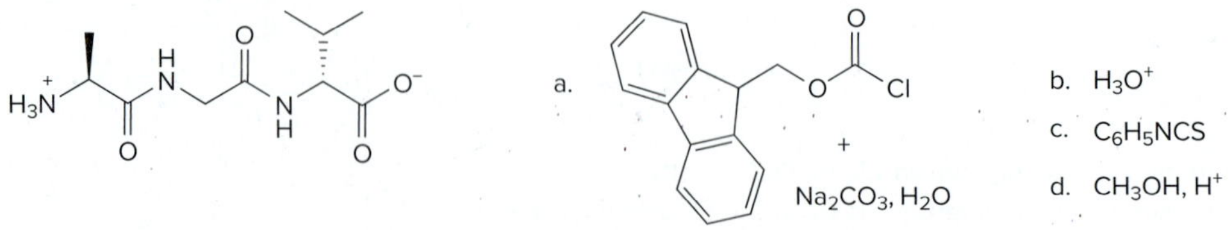

H$_3$N$^+$–Ala–Gly–Val–COO$^-$

a. Fmoc–Cl (fluorenylmethyl chloroformate) + Na$_2$CO$_3$, H$_2$O
b. H$_3$O$^+$
c. C$_6$H$_5$NCS
d. CH$_3$OH, H$^+$

23.53 Draw all the steps in the synthesis of each peptide from individual amino acids:
  (a) Gly–Ala;   (b) Ile–Ala–Phe.

23.54 Write out the steps for the synthesis of each peptide using the Merrifield method:
  (a) Ala–Leu–Phe–Phe;   (b) Phe–Gly–Ala–Ile.

23.55 Another method to form a peptide bond involves a two-step process:

[1] Conversion of a Boc-protected amino acid to a *p*-nitrophenyl ester.
[2] Reaction of the *p*-nitrophenyl ester with an amino acid ester.

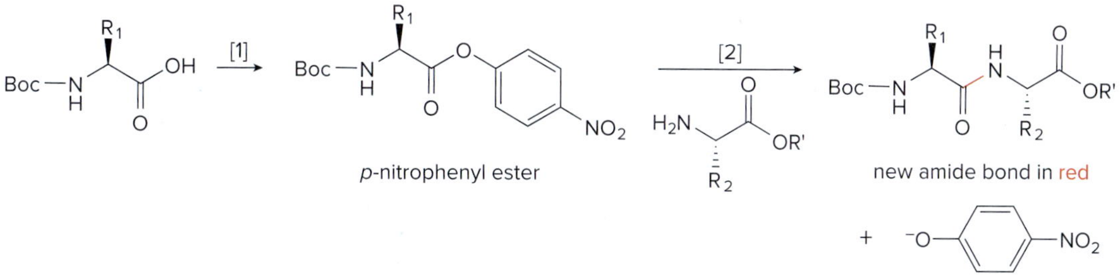

a. Why does a *p*-nitrophenyl ester "activate" the carboxy group of the first amino acid to amide formation?
b. Would a *p*-methoxyphenyl ester perform the same function? Why or why not?

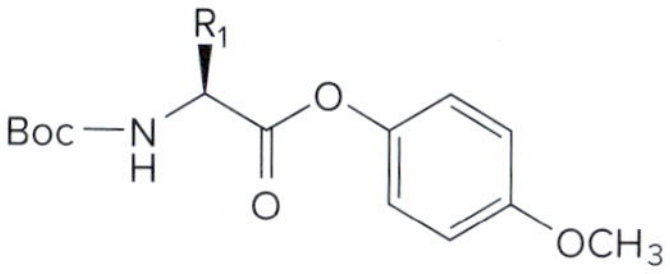

p-methoxyphenyl ester

23.56 Draw the mechanism for the reaction that removes an Fmoc group from an amino acid under these conditions:

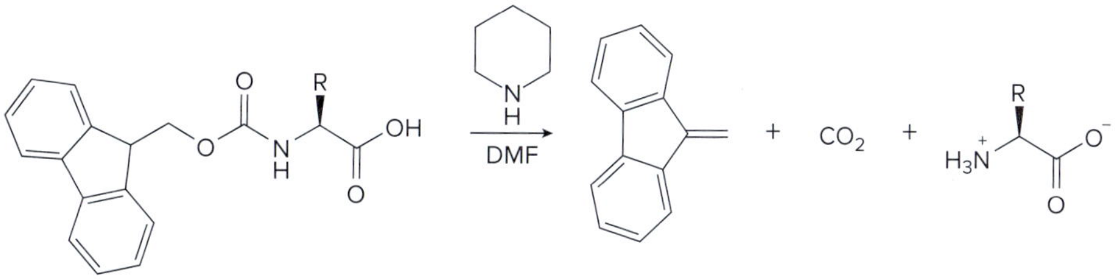

Enzymes and Protein Structure

23.57 Which of the following amino acids are typically found in the interior of a globular protein, and which are typically found on the surface: (a) phenylalanine; (b) aspartic acid; (c) lysine; (d) isoleucine; (e) arginine; (f) glutamic acid?

23.58 Decide if the side chains of the following peptides are nonpolar or polar, and label the hydrophobic and hydrophilic end of each peptide.
 a. VLLFGEDEK b. RKYSFLGAA

23.59 After the peptide chain of collagen has been formed, many of the proline residues are hydroxylated on one of the ring carbon atoms. Why is this process important for the triple helix of collagen?

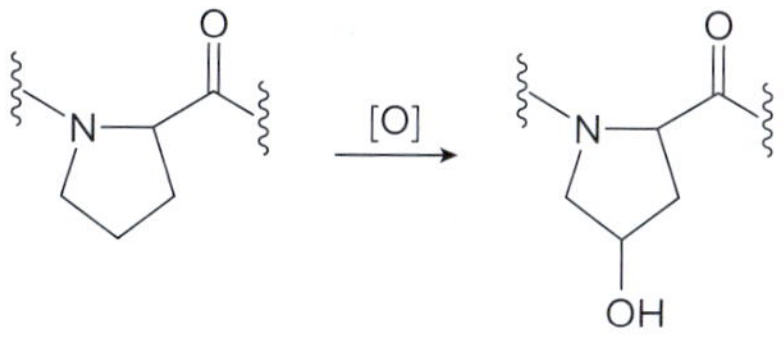

23.60 What class of enzyme catalyzes each reaction?

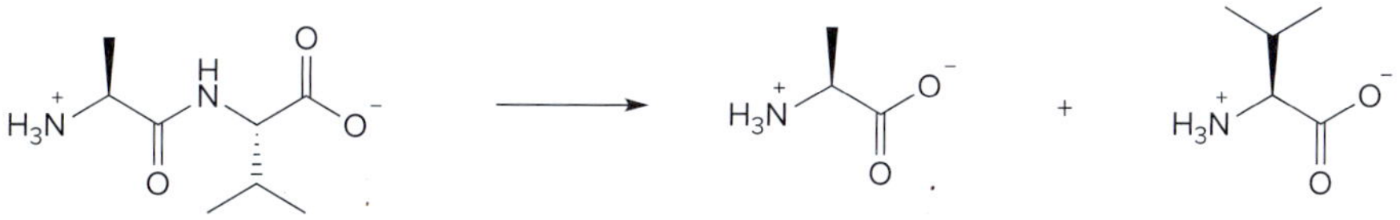

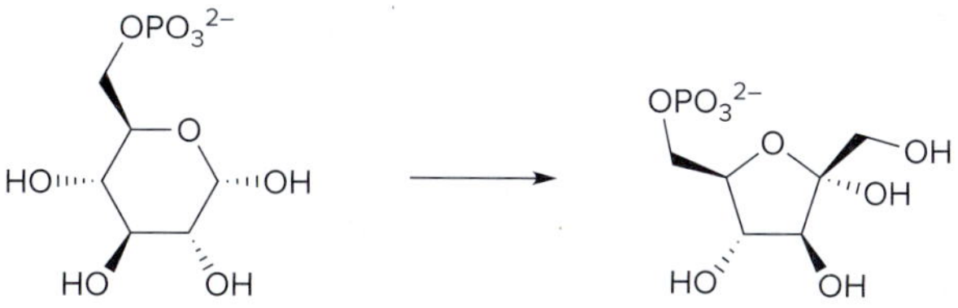

Chapter 23 Amino Acids and Proteins

23.61 Acetylcholinesterase catalyzes the hydrolysis of the neurotransmitter acetylcholine to acetate and choline. The enzyme contains a catalytic triad composed of the amino acids serine, histidine, and glutamate, which catalyzes the hydrolysis in much the same way as the serine proteases discussed in Section 23.5. Draw a stepwise mechanism for this process that illustrates the role of the catalytic triad in the hydrolysis.

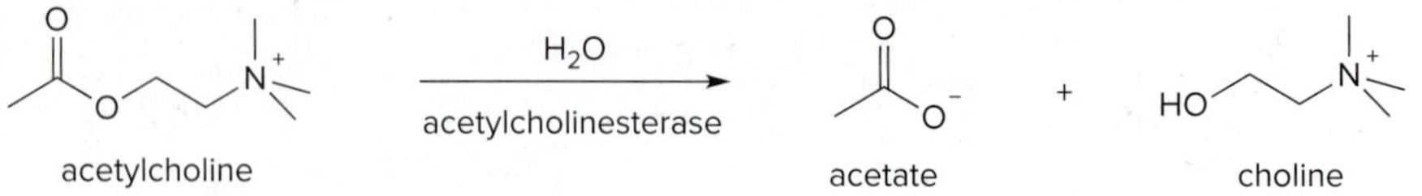

acetylcholine → (H₂O, acetylcholinesterase) → acetate + choline

Challenge Problems

23.62 The anti-obesity drug orlistat works by irreversibly inhibiting pancreatic lipase, an enzyme responsible for the hydrolysis of triacylglycerols in the intestines, so that they are excreted without metabolism. Inhibition occurs by reaction of orlistat with a serine residue of the enzyme, forming a covalently bound, inactive enzyme product. Draw the structure of the product formed during inhibition.

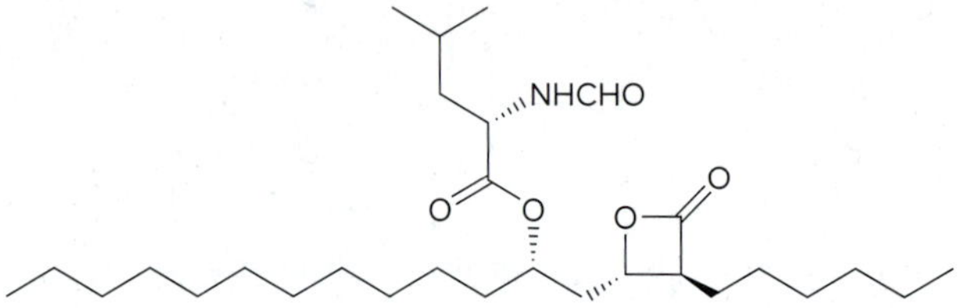

23.63 As shown in Mechanism 23.1, the final steps in the Edman degradation result in rearrangement of a thiazolinone to an N-phenylthiohydantoin. Draw a stepwise mechanism for this acid-catalyzed reaction.

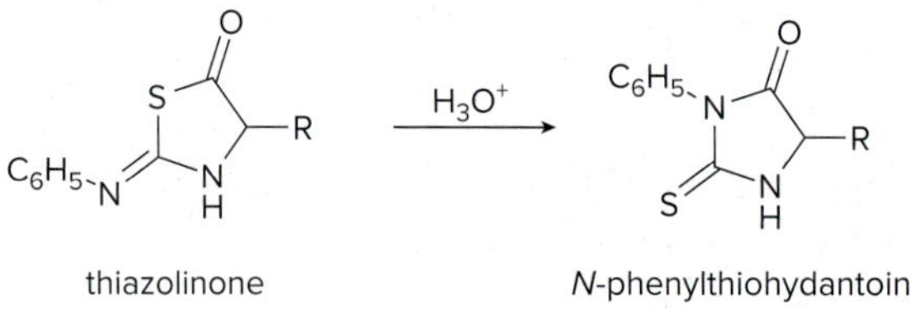

thiazolinone → (H₃O⁺) → N-phenylthiohydantoin

Carbohydrates

MaraZe/Shutterstock

- **24.1** Introduction
- **24.2** Monosaccharides
- **24.3** The family of D-aldoses
- **24.4** The family of D-ketoses
- **24.5** Physical properties of monosaccharides
- **24.6** The cyclic forms of monosaccharides
- **24.7** Glycosides
- **24.8** Reactions of monosaccharides at the OH groups
- **24.9** Reactions at the carbonyl group—Oxidation and reduction
- **24.10** Reactions at the carbonyl group—Adding or removing one carbon atom
- **24.11** Disaccharides
- **24.12** Polysaccharides
- **24.13** Other important sugars and their derivatives

Sucrose, the carbohydrate commonly called table sugar, is composed of two simple sugars, glucose and fructose. Many mammals, birds, and insects use the sucrose in plants as a key food source. Although sugar has been produced for almost 2000 years, sucrose was an expensive luxury until the 1700s when worldwide demand led to the cultivation of large plantations of sugarcane around the globe. In Chapter 24, we learn about the properties and reactions of carbohydrates like sucrose.

Why Study...
Carbohydrates?

Carbohydrates were given their name because molecular formulas of simple carbohydrates could be written as $C_n(H_2O)_n$, making them **hydrates of carbon.**

Carbohydrates such as glucose and cellulose were discussed in Sections 3.9B, 5.1, 6.4, and 14.18.

Although the metabolism of lipids provides more energy per gram than the metabolism of carbohydrates, glucose is the preferred source when a burst of energy is needed during exercise. Glucose is water soluble, so it can be quickly and easily transported through the bloodstream to the tissues.

In Chapter 24, we turn our attention to **carbohydrates,** the largest group of organic molecules in nature, comprising approximately 50% of earth's biomass. Carbohydrates can be simple or complex, having as few as three or as many as thousands of carbon atoms. The glucose metabolized for energy in cells, the sucrose of table sugar, and the cellulose of plant stems and tree trunks are all examples of carbohydrates. Carbohydrates on cell surfaces determine blood type, and carbohydrates form the backbone of DNA, the carrier of all genetic information in the cell. Carbohydrates have many polar functional groups, whose structure and properties can be understood by applying the basic principles of organic chemistry.

24.1 Introduction

Carbohydrates, commonly referred to as sugars and starches, are polyhydroxy aldehydes and ketones, or compounds that can be hydrolyzed to them. The cellulose in plant stems and tree trunks and the chitin in the exoskeletons of arthropods and mollusks are both complex carbohydrates. Four examples are shown in Figure 24.1. They include not only glucose and cellulose, but also doxorubicin (an anticancer drug) and 2′-deoxyadenosine 5′-monophosphate (a nucleotide base from DNA), both of which have a carbohydrate moiety as part of a larger molecule.

Carbohydrates are storehouses of chemical energy. They are synthesized in green plants and algae by **photosynthesis,** a process that uses the energy from the sun to convert carbon dioxide and water to glucose and oxygen. This energy is released when glucose is metabolized. The oxidation of glucose is a multistep process that forms carbon dioxide, water, and a great deal of energy (Section 6.4).

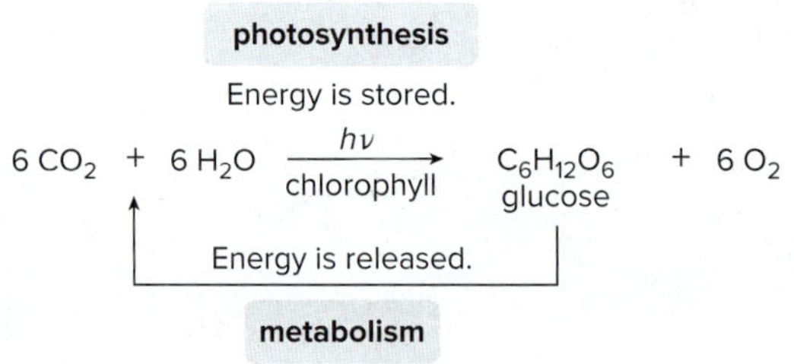

$$6\ CO_2 + 6\ H_2O \xrightarrow[\text{chlorophyll}]{h\nu} C_6H_{12}O_6\ (\text{glucose}) + 6\ O_2$$

photosynthesis: Energy is stored.
metabolism: Energy is released.

Figure 24.1 Some examples of carbohydrates

β-D-glucose
most common simple carbohydrate

cellulose
main component of wood

doxorubicin
an anticancer drug
carbohydrate portion

2′-deoxyadenosine 5′-monophosphate
a nucleotide component of DNA
carbohydrate portion

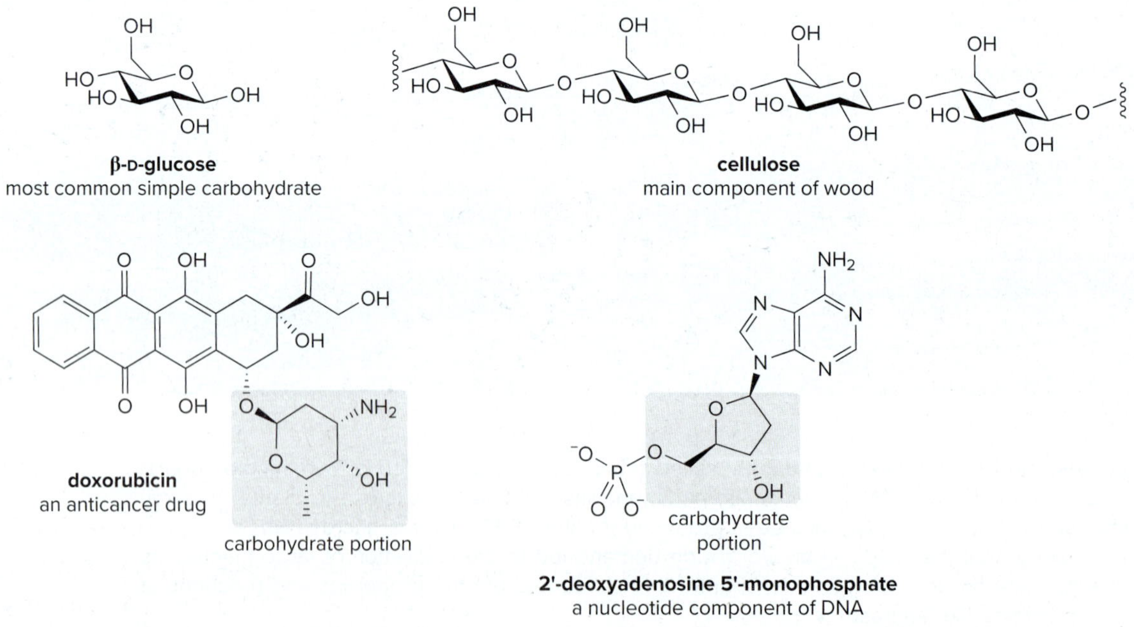

- These compounds illustrate the structural diversity of carbohydrates and their derivatives. **Glucose** is the most common simple sugar, whereas **cellulose,** which comprises wood, plant stems, and grass, is the most common carbohydrate in the plant world. **Doxorubicin,** an anticancer drug that has a carbohydrate ring as part of its structure, has been used in the treatment of leukemia, Hodgkin's disease, and cancers of the breast, bladder, and ovaries. **2′-Deoxyadenosine 5′-monophosphate** is one of the four nucleotides that form DNA.

24.2 Monosaccharides

The simplest carbohydrates are called **monosaccharides** or **simple sugars**. **Monosaccharides have three to seven carbon atoms** in a chain, with a **carbonyl group** at either the terminal carbon (C1) or the carbon adjacent to it (C2). In most carbohydrates, each of the remaining carbon atoms has a **hydroxy group.** Monosaccharides are often drawn vertically, with the carbonyl group at the top. When this convention is used, monosaccharides look different from molecules encountered in prior chapters.

Carbonyl at **C1** - - -→ aldehyde - - -→ aldose

Carbonyl at **C2** - - -→ ketone - - -→ ketose

monosaccharide

D-Fructose is almost twice as sweet as normal table sugar (sucrose) with about the same number of calories per gram. "Lite" food products use only half as much fructose as sucrose for the same level of sweetness, so they have fewer calories. *Jill Braaten*

- Monosaccharides with an aldehyde carbonyl group at C1 are called *aldoses*.
- Monosaccharides with a ketone carbonyl group at C2 are called *ketoses*.

Several examples of simple carbohydrates are shown. D-Glyceraldehyde and dihydroxyacetone have the same molecular formula, so they are **constitutional isomers,** as are D-glucose and D-fructose.

D-glyceraldehyde
an aldehyde
aldose

dihydroxyacetone
a ketone
ketose

D-glucose
an aldehyde
aldose

D-fructose
a ketone
ketose

All carbohydrates have common names. The simplest aldehyde, glyceraldehyde, and the simplest ketone, dihydroxyacetone, are the only monosaccharides whose names do not end in the suffix *-ose.* (The prefix "D-" is explained in Section 24.2C.)

A monosaccharide is called

Dihydroxyacetone is the active ingredient in many artificial tanning agents. *Elite Images/ McGraw-Hill Education*

- a triose if it has 3 C's;
- a tetrose if it has 4 C's;
- a pentose if it has 5 C's;
- a hexose if it has 6 C's, and so forth.

These terms are then combined with the words *aldose* and *ketose* to indicate both the number of carbon atoms in the monosaccharide and whether it contains an aldehyde or ketone. Thus, glyceraldehyde is an **aldotriose** (three C atoms and an aldehyde), glucose is an **aldohexose** (six C atoms and an aldehyde), and fructose is a **ketohexose** (six C atoms and a ketone).

Problem 24.1 Draw the structure of (a) a ketotetrose; (b) an aldopentose; (c) an aldotetrose.

24.2A Fischer Projection Formulas

A striking feature of carbohydrate structure is the presence of stereogenic centers. **All carbohydrates except for dihydroxyacetone contain one or more stereogenic centers.**

The simplest aldehyde, glyceraldehyde, has one stereogenic center, so there are two possible **enantiomers.** Only the enantiomer with the *R* configuration occurs naturally.

(*R*)-glyceraldehyde
naturally occurring enantiomer

(*S*)-glyceraldehyde

The stereogenic centers in sugars are often depicted following a different convention than is usually seen for other stereogenic centers. Instead of drawing a tetrahedron with two bonds in the plane, one in front of the plane and one behind it, the **tetrahedron is tipped so that horizontal bonds come forward (drawn on wedges) and vertical bonds go behind (on dashed wedges).** This structure is then abbreviated by a **cross formula,** also called a **Fischer projection formula.** In a Fischer projection formula:

- A carbon atom is located at the intersection of the two lines of the cross.
- The horizontal bonds come forward, on wedges.
- The vertical bonds go back, on dashed wedges.
- In a carbohydrate, the aldehyde or ketone carbonyl is put at or near the top.

Carbon atoms that are not stereogenic centers are generally drawn in. Using a Fischer projection formula, (*R*)-glyceraldehyde becomes:

Tip red bonds in the plane forward.
(*R*)-glyceraldehyde

- Horizontal bonds come *forward*.
- Vertical bonds go *back*.

Fischer projection formula
(*R*)-glyceraldehyde

Do *not* rotate a Fischer projection formula in the plane of the page, because you might inadvertently convert a compound to its enantiomer. When using Fischer projections, it is usually best to convert them to structures with wedges and dashed wedges, and then manipulate them. Although a Fischer projection formula can be used for the stereogenic center in any compound, it is most commonly used for monosaccharides.

Sample Problem 24.1 Drawing a Fischer Projection Formula

Convert each compound to a Fischer projection formula.

a.

b.

24.2 Monosaccharides

Solution

Rotate and re-draw each molecule to place the horizontal bonds in front of the plane and the vertical bonds behind the plane. Then use a cross to represent the stereogenic center.

a. [Fischer projection redrawing showing H—C—OH with CHO on top and CH₂OH on bottom]

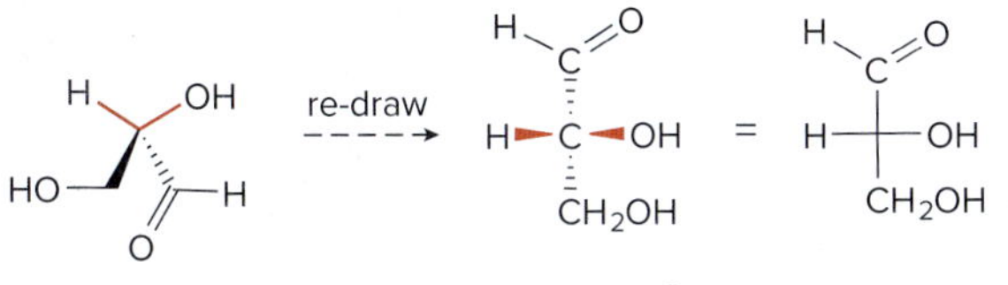

b. [Fischer projection redrawing showing HO—C—H with CHO on top and CH₂OH on bottom]

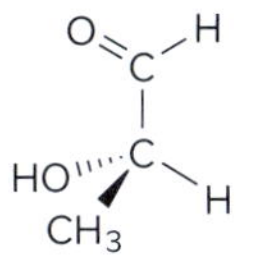

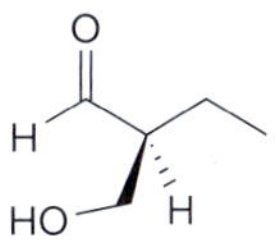

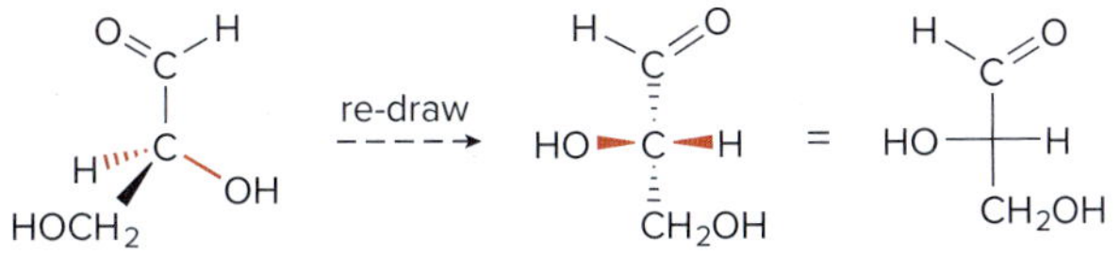

Problem 24.2 Draw each stereogenic center using a Fischer projection formula.

a. CH₃—C(OH)—CH₂CH₂OH with HO and =O
b. HO—C—H with CHO and CH₃
c. Structure with C=O, H, HO, and ethyl group
d. H—C—C with OH, H, CH₃, and CHO

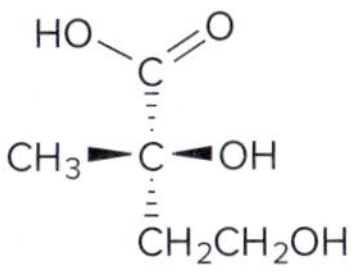

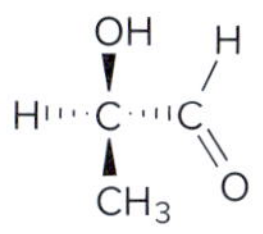

R,S designations can be assigned to any stereogenic center drawn as a Fischer projection formula in the following manner:

[1] Assign priorities (1 → 4) to the four groups bonded to the stereogenic center using the rules detailed in Section 5.6.

[2] When the lowest-priority group occupies a **vertical bond**—that is, it projects *behind* the plane on a dashed wedge—tracing a circle in the **clockwise direction** (from priority group 1 → 2 → 3) gives the ***R* configuration**. Tracing a circle in the **counterclockwise direction** gives the ***S* configuration.**

[3] When the lowest-priority group occupies a **horizontal bond**—that is, it projects *in front of* the plane on a wedge—**reverse the answer** obtained in Step [2] to designate the configuration.

Sample Problem 24.2 — Labeling a Fischer Projection as *R* or *S*

Re-draw each Fischer projection formula using wedges and dashed wedges for the stereogenic center, and label the center as *R* or *S*.

a. Br—C(CH₂OH)(CH₃)—H
b. Cl—C(CHO)(CH₃)—H

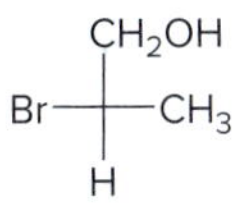

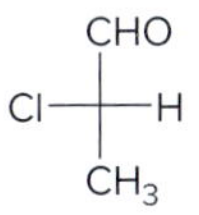

Solution

For each molecule:
[1] Convert the Fischer projection formula to a representation with wedges and dashed wedges.
[2] Assign priorities (Section 5.6).
[3] Determine *R* or *S* in the usual manner. Reverse the answer if priority group [4] is oriented forward (on a wedge).

a. [Stepwise conversion of Br—C(CH₂OH)(CH₃)—H Fischer projection to wedge representation with priorities 1 (Br), 2 (CH₂OH), 3 (CH₃), 4 (H)]

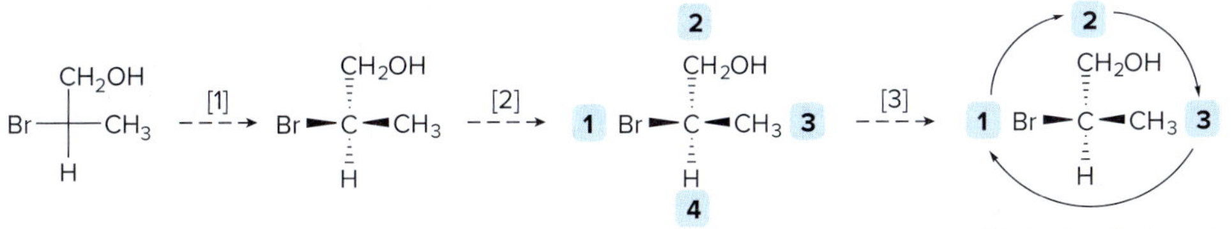

Clockwise circle and group [4] is oriented *behind*: ***R* configuration**

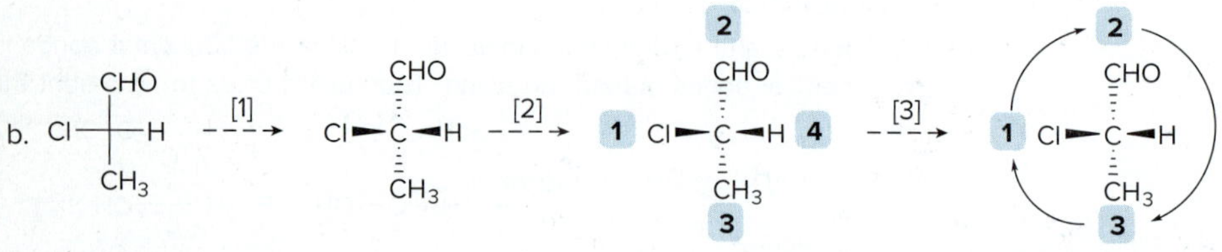

1074   Chapter 24   Carbohydrates

Clockwise circle and group [4] is oriented *forward*: **S configuration**

Problem 24.3 Label each stereogenic center as *R* or *S*.

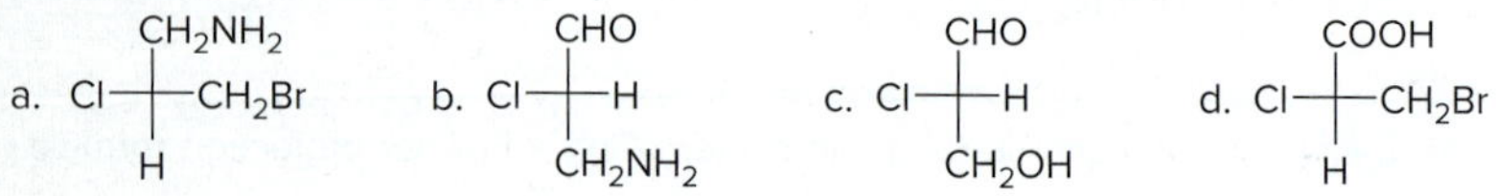

a. Cl—CH₂NH₂ / CH₂Br / H

b. Cl—CHO / H / CH₂NH₂

c. Cl—CHO / H / CH₂OH

d. Cl—COOH / CH₂Br / H

More Practice: Try Problem 24.36.

24.2B Monosaccharides with More Than One Stereogenic Center

The number of possible stereoisomers of a monosaccharide increases exponentially with the number of stereogenic centers present. **An aldohexose has four stereogenic centers, so it has $2^4 = 16$ possible stereoisomers,** or eight pairs of enantiomers.

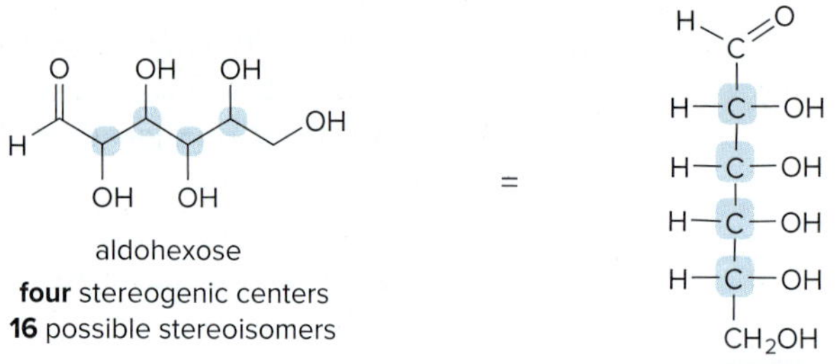

aldohexose
four stereogenic centers
16 possible stereoisomers

= vertical representation

Fischer projection formulas are also used for compounds like aldohexoses that contain several stereogenic centers. In this case, the molecule is drawn with a vertical carbon skeleton and the stereogenic centers are stacked one above another. Using this convention, **all horizontal bonds project *forward* (on wedges).**

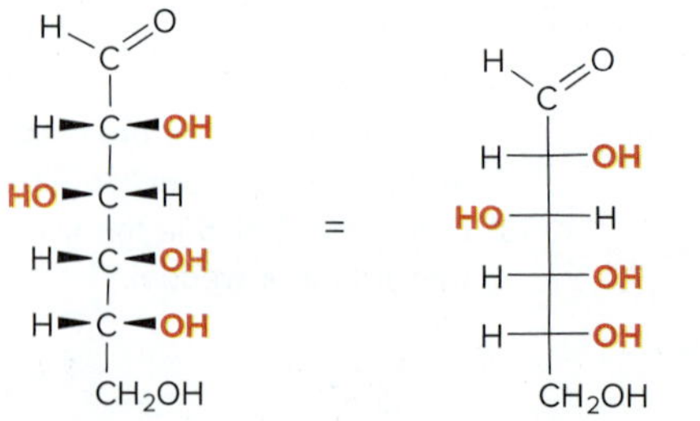

D-glucose
All horizontal bonds are drawn as **wedges**.

= Fischer projection

Although Fischer projections are commonly used to depict monosaccharides with many stereogenic centers, care must be exercised in using them, because they do not give a true picture of the three-dimensional structures they represent. **Each stereogenic center is drawn in the**

Figure 24.2
A Fischer projection and the 3-D structure of glucose

All bonds are eclipsed in a Fischer projection.

D-glucose

Because all bonds are drawn **eclipsed,** the carbon backbone in a Fischer projection would curl around a cylinder.

less stable eclipsed conformation, so the Fischer projection of glucose really represents the molecule in a cylindrical conformation, as shown in Figure 24.2.

Sample Problem 24.3 Converting a Ball-and-Stick Model to a Fischer Projection

Convert the ball-and-stick model to a Fischer projection.

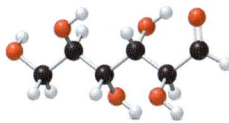

Solution

The ball-and-stick model is shown in the more stable staggered conformation, so it must be converted to the less stable eclipsed conformation used in a Fischer projection.

1. Re-draw the model as a skeletal structure (**A**), and rotate it to place the carbonyl group at the top (**B**).
2. To convert the all-staggered form to the all-eclipsed form, rotate around the bonds in **B** to swing two carbons (labeled in red) 180°, forming **C**.
3. Re-draw **C** so that all bonds to H and OH on the four stereogenic centers are drawn on wedges, forming **D**. Groups that are on wedges in **C** (in red) are on the left side of the carbon skeleton in **D**, and groups on dashed wedges in **C** (in blue) are on the right side of the carbon skeleton in **D**.
4. Replace the wedges with crosses to form the Fischer projection.

A skeletal structure of the ball-and-stick model

B

C

D

Fischer projection

Problem 24.4 Convert the ball-and-stick model to a Fischer projection.

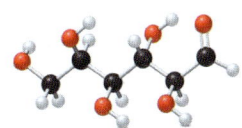

More Practice: Try Problem 24.34.

Problem 24.5 Assign R,S designations to each stereogenic center in glucose.

24.2C D and L Monosaccharides

Although the prefixes *R* and *S* can be used to designate the configuration of stereogenic centers in monosaccharides, an older system of nomenclature uses the prefixes D- and L- instead. **Naturally occurring glyceraldehyde with the *R* configuration is called the D-isomer. Its enantiomer, (*S*)-glyceraldehyde, is called the L-isomer.**

(*R*)-glyceraldehyde   (*S*)-glyceraldehyde
D-glyceraldehyde   L-glyceraldehyde

The letters D and L are used to label all monosaccharides, even those with multiple stereogenic centers. **The configuration of the stereogenic center *farthest* from the carbonyl group determines whether a monosaccharide is D- or L-.**

The two designations, D and *d*, refer to very different phenomena. The "D" designates the configuration around a stereogenic center in a monosaccharide. The "*d*," on the other hand, is an abbreviation for "dextrorotatory"; that is, a *d*-compound rotates the plane of polarized light in the clockwise direction. A D-sugar may be dextrorotatory or it may be levorotatory. **There is no direct correlation between D and *d* or L and *l*.**

- A D-sugar has the OH group on the stereogenic center farthest from the carbonyl on the *right* in a Fischer projection (like D-glyceraldehyde).
- An L-sugar has the OH group on the stereogenic center farthest from the carbonyl on the *left* in a Fischer projection (like L-glyceraldehyde).

OH on the **right**   OH on the **left**
D-sugar   L-sugar

Glucose and all other naturally occurring sugars are D-sugars. L-Glucose, a compound that does not occur in nature, is the enantiomer of D-glucose. **L-Glucose has the opposite configuration at *every* stereogenic center.**

D-glucose
naturally occurring enantiomer

L-glucose

Problem 24.6 (a) Label compounds **A**, **B**, and **C** as D- or L-sugars. (b) How are compounds **A** and **B** related? **A** and **C**? **B** and **C**? Choose from enantiomers, diastereomers, or constitutional isomers.

A   B   C

24.3 The Family of D-Aldoses

The common name of each monosaccharide indicates both the number of atoms it contains and the configuration at each of the stereogenic centers. Because the common names are firmly entrenched in the chemical literature, no systematic method has ever been established to name these compounds.

Beginning with D-glyceraldehyde, one may formulate other D-aldoses having four, five, or six carbon atoms by adding carbon atoms (each bonded to H and OH), one at a time, between C1 and C2. **Two D-aldotetroses can be formed from D-glyceraldehyde,** one with the new OH group on the right and one with the new OH group on the left. Their names are D-erythrose and D-threose. They are two **diastereomers,** each with two stereogenic centers, labeled in blue.

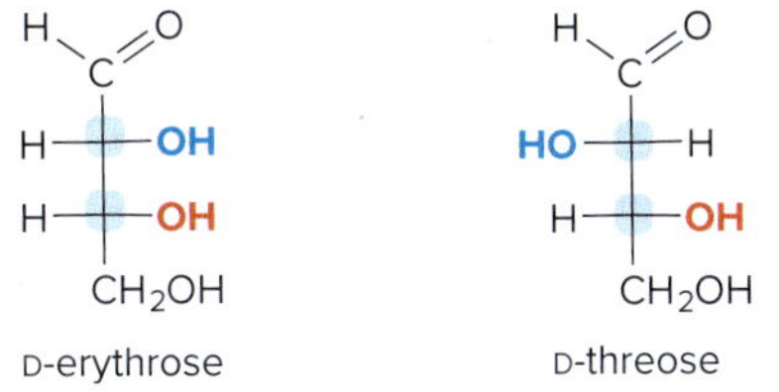

Because each aldotetrose has two stereogenic centers, there are 2^2 or four possible stereoisomers. D-Erythrose and D-threose are two of them. The other two are their enantiomers, called L-erythrose and L-threose, respectively. The configuration around each stereogenic center is exactly the opposite in its enantiomer. All four stereoisomers of the aldotetroses are shown in Figure 24.3.

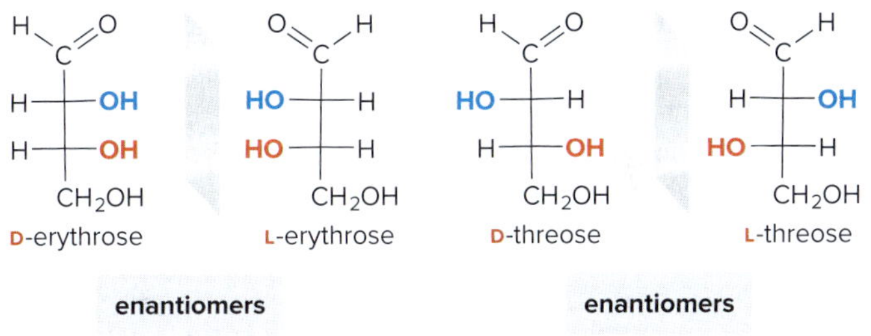

Figure 24.3
The four stereoisomeric aldotetroses

D-Ribose, D-arabinose, and D-xylose are all common aldopentoses in nature. D-Ribose is the carbohydrate component of RNA, the polymer that translates the genetic information of DNA for protein synthesis.

To continue forming the family of D-aldoses, we must add another carbon atom (bonded to H and OH) just below the carbonyl of either tetrose. Because there are *two* D-aldotetroses to begin with, and there are *two* ways to place the new OH (right or left), there are now *four* D-aldopentoses: D-ribose, D-arabinose, D-xylose, and D-lyxose. Each aldopentose now has *three* stereogenic centers, so there are $2^3 = 8$ possible stereoisomers, or four pairs of enantiomers. The D-enantiomer of each pair is shown in Figure 24.4.

Finally, to form the D-aldohexoses, we must add another carbon atom (bonded to H and OH) just below the carbonyl of all the aldopentoses. Because there are *four* D-aldopentoses to begin with, and there are *two* ways to place the new OH (right or left), there are now *eight* D-aldohexoses. Each aldohexose now has *four* stereogenic centers, so there are $2^4 = 16$ possible stereoisomers, or eight pairs of enantiomers. Only the D-enantiomer of each pair is shown in Figure 24.4.

The tree of D-aldoses (Figure 24.4) is arranged in pairs of compounds that are bracketed together. Each pair of compounds, such as D-glucose and D-mannose, has the same configuration around all of its stereogenic centers except for one.

Of the D-aldohexoses, only D-glucose and D-galactose are common in nature. **D-Glucose is by far the most abundant of all D-aldoses.** D-Glucose comes from the hydrolysis of starch and cellulose, and D-galactose comes from the hydrolysis of fruit pectins.

- Two diastereomers that differ in the configuration around only one stereogenic center are called *epimers.*

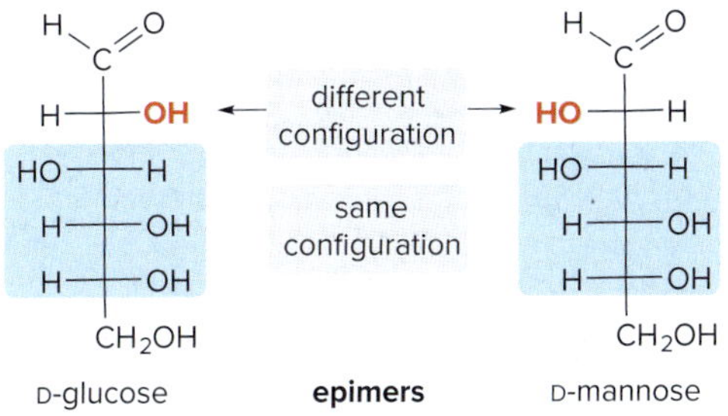

Figure 24.4
The family of D-aldoses having three to six carbon atoms

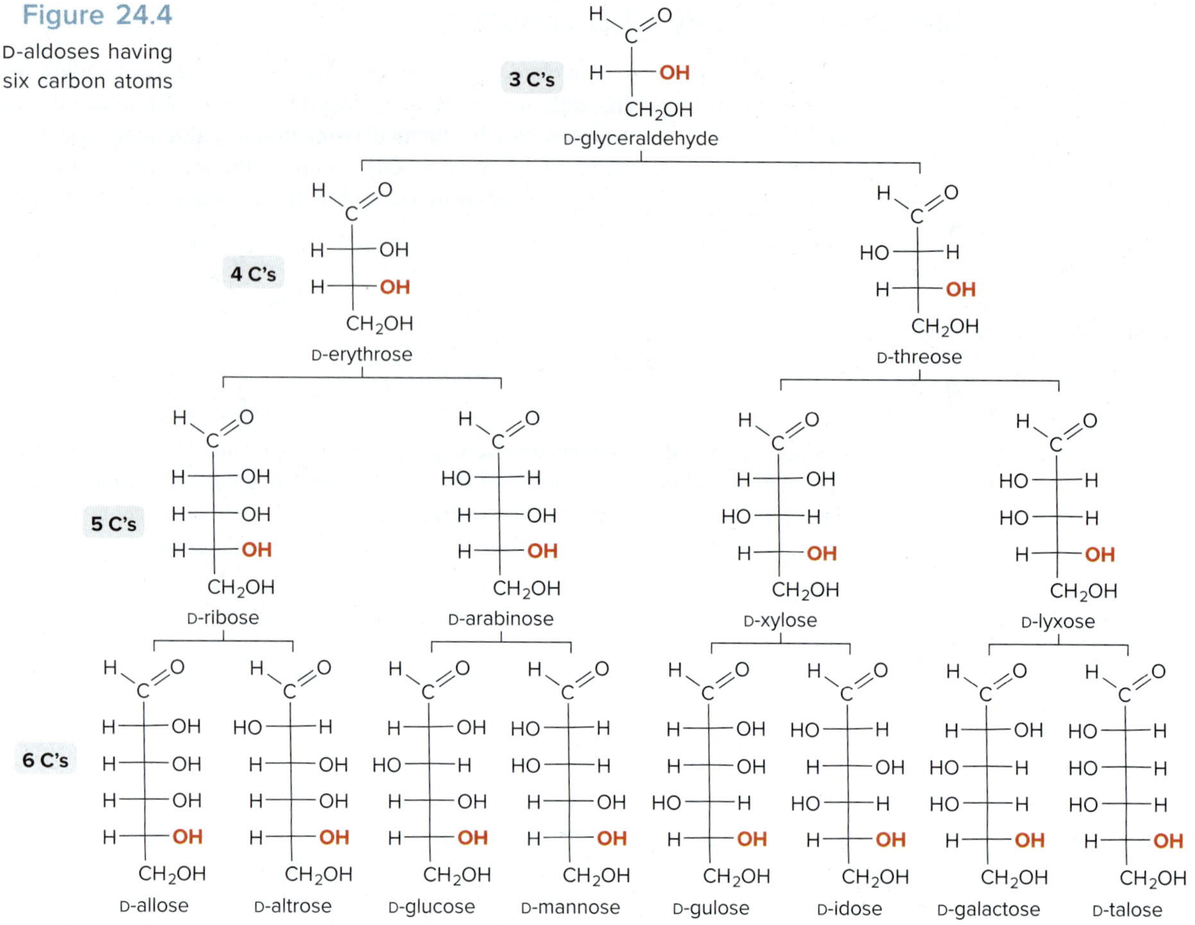

[Figure showing the family tree of D-aldoses from 3 to 6 carbons: D-glyceraldehyde (3 C's); D-erythrose, D-threose (4 C's); D-ribose, D-arabinose, D-xylose, D-lyxose (5 C's); D-allose, D-altrose, D-glucose, D-mannose, D-gulose, D-idose, D-galactose, D-talose (6 C's).]

- All D-aldoses have the OH group on the stereogenic center farthest from the C=O (shown in **red**) on the **right**.

Problem 24.7 How many different aldoheptoses are there? How many are D-sugars? Draw all D-aldoheptoses having the *R* configuration at C2 and C3.

Problem 24.8 Draw two possible epimers of D-erythrose. Name each of these compounds using Figure 24.4.

24.4 The Family of D-Ketoses

The family of D-ketoses, shown in Figure 24.5, is formed from dihydroxyacetone by adding a new carbon (bonded to H and OH) between C2 and C3. Having a carbonyl group at C2 decreases the number of stereogenic centers in these monosaccharides, so that there are only four D-ketohexoses. The most common naturally occurring ketose is D-fructose.

Problem 24.9 Referring to the structures in Figures 24.4 and 24.5, classify each pair of compounds as enantiomers, epimers, diastereomers but not epimers, or constitutional isomers of each other.

    a. D-allose and L-allose
    b. D-altrose and D-gulose
    c. D-galactose and D-talose
    d. D-mannose and D-fructose
    e. D-fructose and D-sorbose
    f. L-sorbose and L-tagatose

Figure 24.5
The family of D-ketoses having three to six carbon atoms

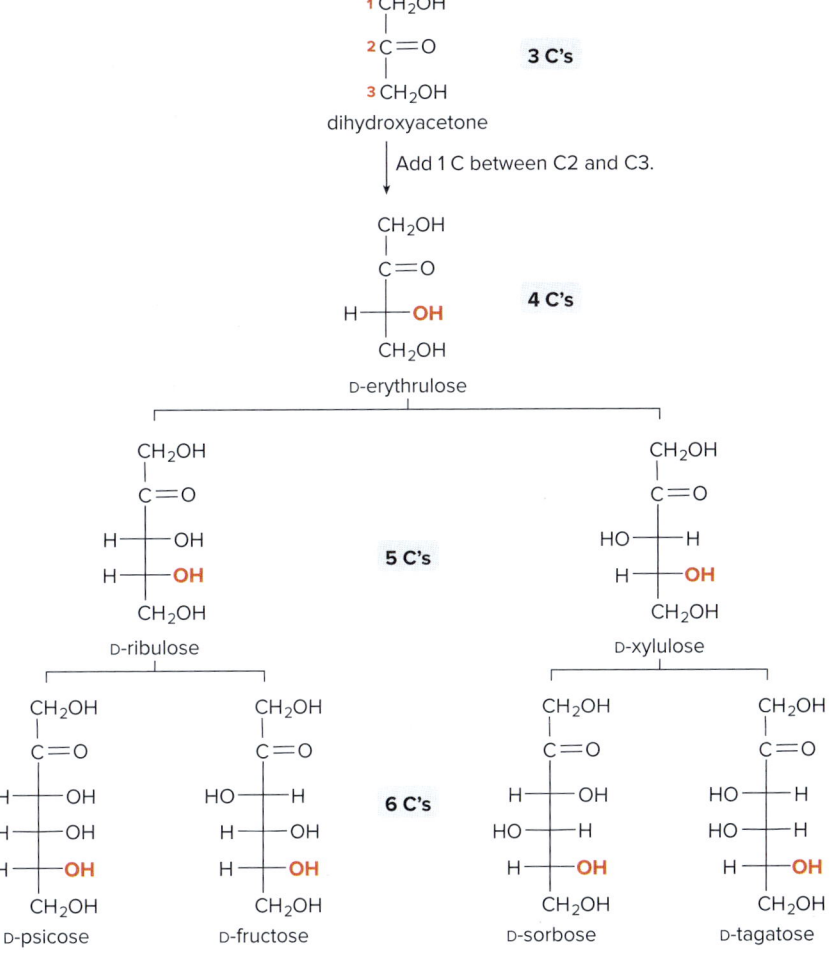

- All D-ketoses have the OH group on the stereogenic center farthest from the C=O (shown in red) on the **right**.

Problem 24.10   a. Draw the enantiomer of D-fructose.
b. Draw an epimer of D-fructose at C4. What is the name of this compound?
c. Draw an epimer of D-fructose at C5. What is the name of this compound?

24.5 Physical Properties of Monosaccharides

Monosaccharides have these physical properties:

- They are all **sweet tasting,** but their relative sweetness varies a great deal.
- They are polar compounds with **high melting points.**
- The presence of so many polar functional groups capable of hydrogen bonding makes them **water soluble.**
- Unlike most other organic compounds, monosaccharides are so polar that they are **insoluble in organic solvents like diethyl ether.**

24.6 The Cyclic Forms of Monosaccharides

Although the monosaccharides in Figures 24.4 and 24.5 are drawn as acyclic carbonyl compounds containing several hydroxy groups, the hydroxy and carbonyl groups of monosaccharides can undergo intramolecular cyclization reactions to form **hemiacetals** having either five or six atoms in the ring. This process was first discussed in Section 14.17.

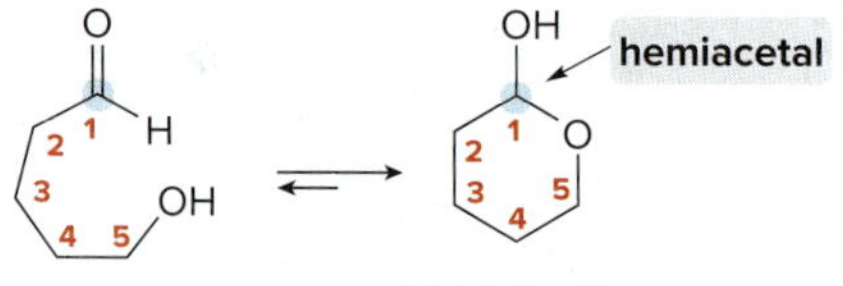

pyranose ring
(a six-membered ring)

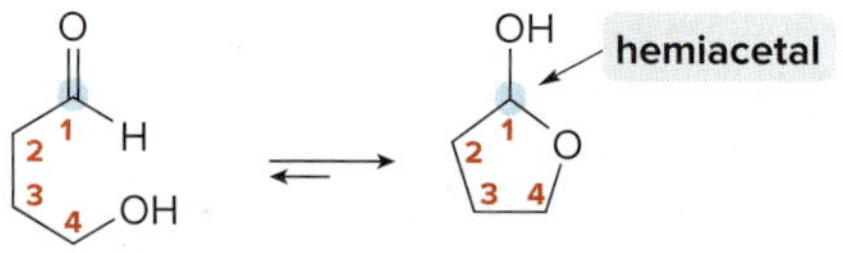

furanose ring
(a five-membered ring)

- A six-membered ring containing an O atom is called a *pyranose* ring.
- A five-membered ring containing an O atom is called a *furanose* ring.

Cyclization of a hydroxy carbonyl compound always forms a stereogenic center at the hemiacetal carbon, called the **anomeric carbon**. The two hemiacetals are called **anomers.**

- Anomers are stereoisomers of a cyclic monosaccharide that differ in the position of the OH group at the hemiacetal carbon.

new stereogenic center at the anomeric carbon

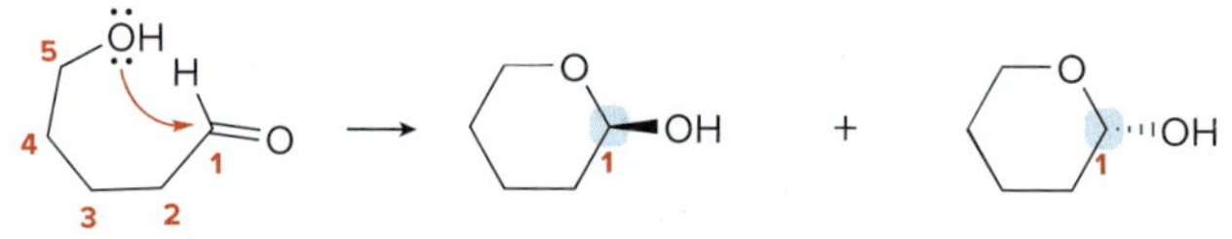

Cyclization forms the more stable ring size in a given molecule. **The most common monosaccharides, the aldohexoses like glucose, typically form a pyranose ring,** so our discussion begins with forming a cyclic hemiacetal from D-glucose.

24.6A Drawing Glucose as a Cyclic Hemiacetal

Which of the five OH groups in glucose is at the right distance from the carbonyl group to form a six-membered ring? The **O atom on the stereogenic center farthest from the carbonyl (C5) is six atoms from the carbonyl carbon, placing it in the proper position for cyclization to form a pyranose ring.**

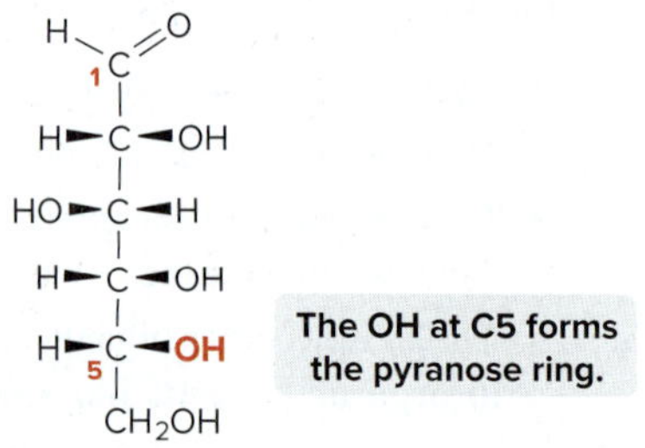

D-glucose

The OH at C5 forms the pyranose ring.

To translate the acyclic form of glucose into a cyclic hemiacetal, we must draw the hydroxy aldehyde in a way that suggests the position of the atoms in the new ring, and then draw the ring. **By convention the O atom in the new pyranose ring is drawn in the upper right corner of the six-membered ring.**

24.6 The Cyclic Forms of Monosaccharides

Rotating the groups on the bottom stereogenic center in **A** places all six atoms needed for the ring (including the OH) in a vertical line (**B**). Re-drawing this representation as a Fischer projection makes the structure appear less cluttered (**C**). Twisting this structure and rotating it 90° forms **D**. Structures **A–D** are four different ways of drawing the same acyclic structure of D-glucose.

We are now set to draw the cyclic hemiacetal formed by nucleophilic attack of the OH group on C5 on the aldehyde carbonyl. Because cyclization creates a new stereogenic center, there are **two cyclic forms of D-glucose, an α anomer and a β anomer**. All the original stereogenic centers retain their configuration in both of the products formed.

- The α anomer of a D monosaccharide has the OH group drawn *down*, trans to the CH$_2$OH group at C5. The α anomer of D-glucose is called α-D-glucose, or α-D-glucopyranose (to emphasize the six-membered ring).
- The β anomer of a D monosaccharide has the OH group drawn *up*, cis to the CH$_2$OH group at C5. The β anomer is called β-D-glucose, or β-D-glucopyranose (to emphasize the six-membered ring).

new stereogenic center at the anomeric carbon (C1)

The **α anomer** in any monosaccharide has the anomeric OH group and the CH$_2$OH group **trans**. The **β anomer** has the anomeric OH group and the CH$_2$OH group **cis**.

These flat, six-membered rings used to represent the cyclic hemiacetals of glucose and other sugars are called **Haworth projections**. The cyclic forms of glucose now have **five stereogenic centers, the four from the starting hydroxy aldehyde and the new anomeric carbon.** α-D-Glucose and β-D-glucose are **diastereomers**, because only the anomeric carbon has a different configuration.

The mechanism for this transformation is exactly the same as the mechanism that converts a hydroxy aldehyde to a cyclic hemiacetal (Mechanism 14.13). **The acyclic aldehyde and two cyclic hemiacetals are all in equilibrium.** Each cyclic hemiacetal can be isolated and crystallized separately, but when any one compound is placed in solution, an equilibrium mixture of all three forms results. This process is called **mutarotation**. At equilibrium, the mixture has 37% of the α anomer, 63% of the β anomer, and only trace amounts of the acyclic hydroxy aldehyde (Figure 24.6). Also shown are representations of the three forms of glucose using wedges and dashed wedges.

Chapter 24 Carbohydrates

Figure 24.6
The three forms of glucose

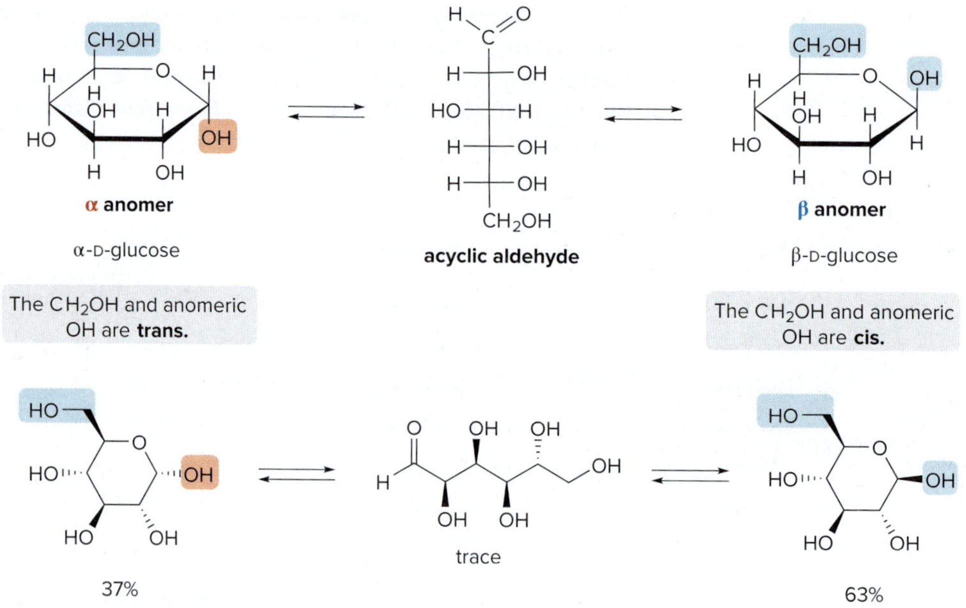

α-D-glucose   acyclic aldehyde   β-D-glucose

α anomer — The CH₂OH and anomeric OH are **trans.**

β anomer — The CH₂OH and anomeric OH are **cis.**

37%   trace   63%

- Bonds above the ring in a Haworth projection are drawn as wedges.
- Bonds below the ring in a Haworth projection are drawn as dashed wedges.

Problem 24.11   Label each Haworth projection as an α or β anomer, and convert the Haworth projection to a six-membered ring with wedges and dashed wedges.

a. 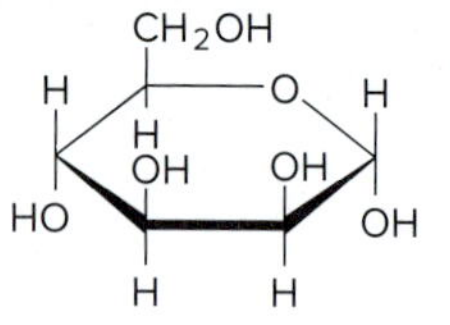   b.

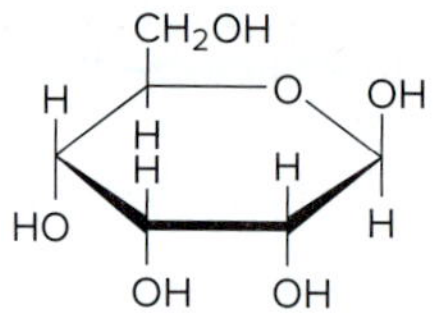

24.6B Haworth Projections

To convert an acyclic monosaccharide to a Haworth projection, follow a stepwise procedure.

How To Draw a Haworth Projection from an Acyclic Aldohexose

Example Convert D-mannose to a Haworth projection.

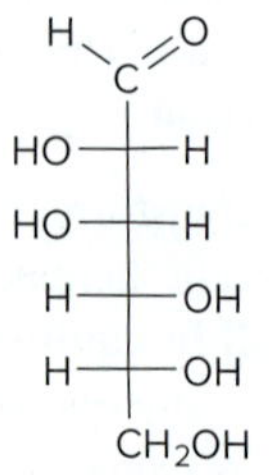

D-mannose

—Continued

Step [1] Place the O atom in the upper right corner of a hexagon, and add the CH₂OH group on the first carbon counterclockwise from the O atom.
- For **D-sugars**, the CH₂OH group is drawn **up**. For **L-sugars**, the CH₂OH group is drawn **down**.

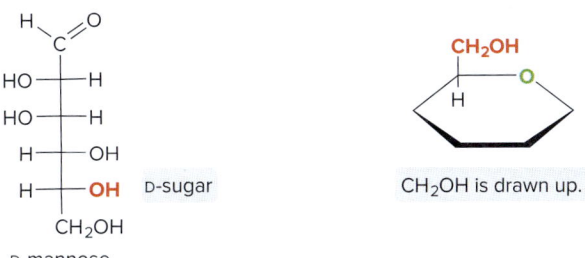

Step [2] Place the anomeric carbon on the first carbon clockwise from the O atom.
- For an α anomer, the **OH** is drawn **down** in a D-sugar.
- For a β anomer, the **OH** is drawn **up** in a D-sugar.

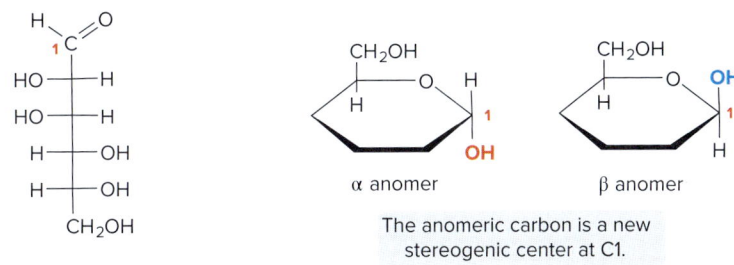

- Remember: **The carbonyl carbon becomes the anomeric carbon** (a new stereogenic center).

Step [3] Add the substituents at the three remaining stereogenic centers clockwise around the ring.
- The substituents on the **right side** of the Fischer projection are drawn **down**.
- The substituents on the **left** are drawn **up**.

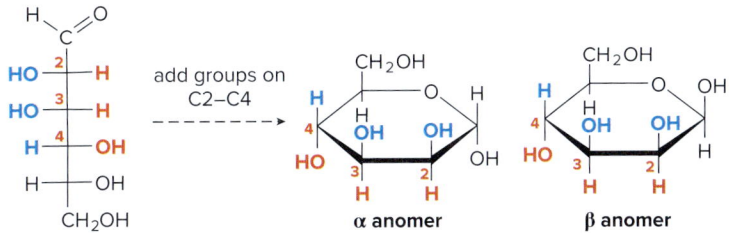

Problem 24.12 Convert each aldohexose to the indicated anomer using a Haworth projection.

a. Draw the α anomer of:

```
    H   O
     \\//
      C
H  ─┼─ OH
H  ─┼─ OH
H  ─┼─ OH
H  ─┼─ OH
    CH₂OH
```

b. Draw the α anomer of:

```
    H   O
     \\//
      C
HO ─┼─ H
HO ─┼─ H
H  ─┼─ OH
HO ─┼─ H
    CH₂OH
```

c. Draw the β anomer of:

```
    H   O
     \\//
      C
HO ─┼─ H
H  ─┼─ OH
H  ─┼─ OH
H  ─┼─ OH
    CH₂OH
```

Chapter 24 Carbohydrates

Sample Problem 24.4 shows how to convert a Haworth projection back to the acyclic form of a monosaccharide. It doesn't matter whether the hemiacetal is the α or β anomer, because **both anomers give the *same* hydroxy aldehyde.**

Sample Problem 24.4 Converting a Haworth Projection to a Fischer Projection

Convert the following Haworth projection to the acyclic form of the aldohexose.

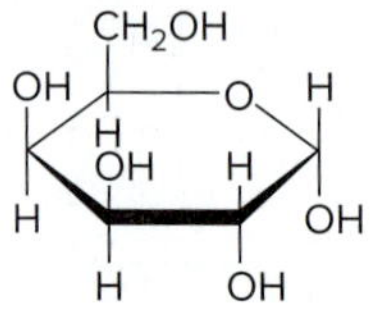

Solution

To convert the substituents to the acyclic form, **start at the pyranose O atom**, and work in a **counterclockwise** fashion around the ring, and from **bottom-to-top** along the chain.

[1] Draw the carbon skeleton, **placing the CHO on the top and the CH$_2$OH on the bottom.**

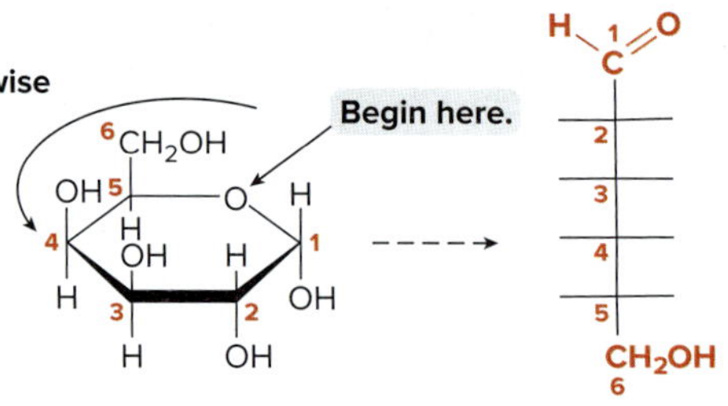

[2] **Classify the sugar as D- or L-.**
- The CH$_2$OH is drawn **up**, so it is a **D-sugar.**
- A D-sugar has the OH group on the bottom stereogenic center on the **right**.

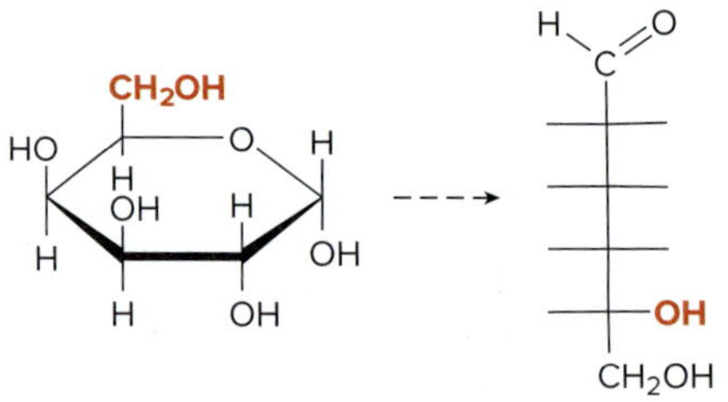

[3] Add the three other stereogenic centers.
- **Up** substituents go on the **left**.
- **Down** substituents go on the **right**.

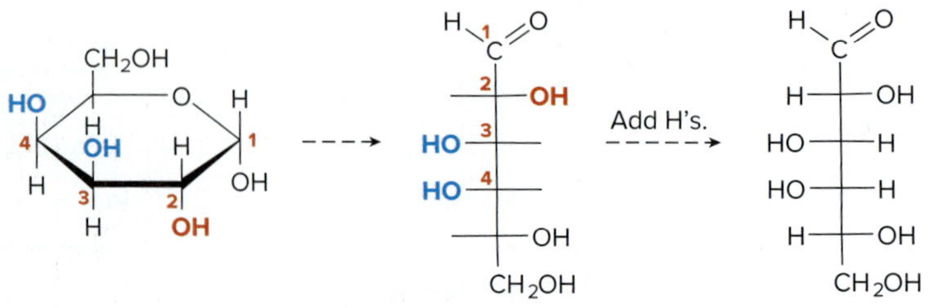

- The anomeric carbon becomes the C=O at C1.

Problem 24.13 Convert each Haworth projection to its acyclic form.

a. [Haworth projection with CH₂OH, HO, OH, H, H, H, OH, OH]

b. [Haworth projection with H, CH₂OH, HO, OH, OH, H, H, OH]

More Practice: Try Problem 24.42.

24.6C Three-Dimensional Representations for D-Glucose

Because the chair form of a six-membered ring gives the truest picture of its three-dimensional shape, we must learn to convert Haworth projections to chair forms.

To convert a Haworth projection to a chair form:

- Draw the pyranose ring with the O atom as an "up" atom.
- The "up" substituents in a Haworth projection become the "up" bonds (either axial or equatorial) on a given carbon atom on a puckered six-membered ring.
- The "down" substituents in a Haworth projection become the "down" bonds (either axial or equatorial) on a given carbon atom on a puckered six-membered ring.

As a result, the three-dimensional chair form of β-D-glucose is drawn in this manner:

[Diagram showing conversion: pyranose ring → Make the O atom an "up" atom → chair → Haworth projection of β-D-glucose → chair form for β-D-glucose]

- "Up" substituents are labeled in red.
- "Down" substituents are labeled in blue.

Glucose has all substituents larger than a hydrogen atom in the more roomy equatorial positions, making it the most stable and thus most prevalent monosaccharide. The β anomer is the major isomer at equilibrium, moreover, because the hemiacetal OH group is in the equatorial position, too. Figure 24.7 shows both anomers of D-glucose drawn as chair conformations.

Problem 24.14 Convert each Haworth projection in Problem 24.13 to a three-dimensional representation using a chair pyranose ring.

Figure 24.7 Three-dimensional representations for both anomers of D-glucose

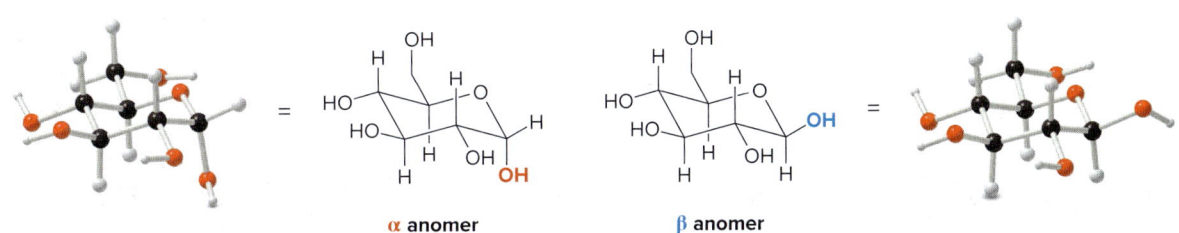

α anomer   β anomer

24.6D Furanoses

Certain monosaccharides—notably aldopentoses and ketohexoses—predominantly form furanose rings, rather than pyranose rings, in solution. The same principles apply to drawing these structures as for drawing pyranose rings, except the ring size is one atom smaller.

- Cyclization always forms a new stereogenic center at the anomeric carbon, so two different anomers are possible. For a D-sugar, the OH group is drawn *down* in the α anomer and *up* in the β anomer.
- Use the same drawing conventions for adding substituents to the five-membered ring. With D-sugars, the CH_2OH group is drawn *up*.

With D-ribose, the OH group used to form the five-membered furanose ring is located on C4. Cyclization yields two anomers at the new stereogenic center, which are called **α-D-ribofuranose** and **β-D-ribofuranose**.

Honey was the first and most popular sweetening agent until it was replaced by sugar (from sugarcane) in modern times. Honey is a mixture consisting largely of D-fructose and D-glucose. *Anastasy Yarmolovich/iStockphoto/Getty Images*

The same procedure can be used to draw the furanose form of D-fructose, the most common ketohexose. Because the carbonyl group is at C2 (instead of C1, as in the aldoses), the OH group at C5 reacts to form the hemiacetal in the five-membered ring. Two anomers are formed.

Problem 24.15 Aldotetroses exist in the furanose form. Draw both anomers of D-erythrose.

24.7 Glycosides

Keep in mind the difference between a hemiacetal and an acetal:

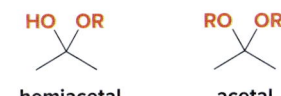

hemiacetal
- one OH group
- one OR group

acetal
- two OR groups

Because monosaccharides exist in solution in an equilibrium between acyclic and cyclic forms, they undergo three types of reactions:

- Reaction of the hemiacetal
- Reaction of the hydroxy groups
- Reaction of the carbonyl group

Even though the acyclic form of a monosaccharide may be present in only trace amounts, the equilibrium can be tipped in its favor by Le Châtelier's principle (Section 9.8). Suppose, for example, that the carbonyl group of the acyclic form reacts with a reagent, thus depleting its equilibrium concentration. The equilibrium will then shift to compensate for the loss, thus producing more of the acyclic form, which can react further.

Note, too, that **monosaccharides have two different types of OH groups.** Most are "regular" alcohols and, as such, undergo reactions characteristic of alcohols. **The anomeric OH group, on the other hand, is part of a hemiacetal, giving it added reactivity.**

24.7A Glycoside Formation

Treatment of a monosaccharide with an alcohol and HCl converts the hemiacetal to an acetal called a glycoside. For example, treatment of α-D-glucose with CH_3OH and HCl forms two glycosides that are diastereomers at the acetal carbon. The α and β labels are assigned in the same way as anomers: with a D-sugar, **an α glycoside has the new OR group (OCH_3 group in this example) *down*, and a β glycoside has the new OR group *up*.**

α-D-glucose → α glycoside + β glycoside

Only the hemiacetal OH reacts.

Mechanism 24.1 explains why **a single anomer forms two glycosides.** The reaction proceeds by way of a **planar carbocation,** which undergoes nucleophilic attack from two different directions to give a mixture of diastereomers. Because both α- and β-D-glucose form the same planar carbocation, each yields the same mixture of two glycosides.

The mechanism also explains why **only the hemiacetal OH group reacts.** Protonation of the hemiacetal OH, followed by loss of H_2O, forms a **resonance-stabilized carbocation** in Step [2]. A resonance-stabilized carbocation is *not* formed by loss of H_2O from any other OH group.

Unlike cyclic hemiacetals, **glycosides are acetals, so they do *not* undergo mutarotation.** When a single glycoside is dissolved in H_2O, it is *not* converted to an equilibrium mixture of α and β glycosides.

- Glycosides are acetals with an alkoxy group (OR) bonded to the anomeric carbon.

Problem 24.16 What glycosides are formed when each monosaccharide is treated with CH_3CH_2OH, HCl: (a) β-D-mannose; (b) α-D-gulose; (c) β-D-fructose?

Chapter 24 Carbohydrates

 Mechanism 24.1 Glycoside Formation

Part [1] Loss of H₂O from the hemiacetal

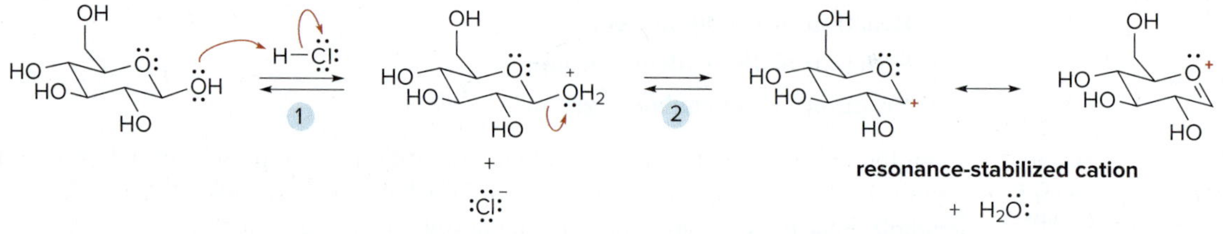

[Reaction scheme showing protonation of hemiacetal OH by HCl, then loss of H₂O to form a resonance-stabilized cation]

1 – 2 Protonation of the hemiacetal OH followed by loss of H₂O forms a **resonance-stabilized carbocation**.

Part [2] Formation of the glycosides

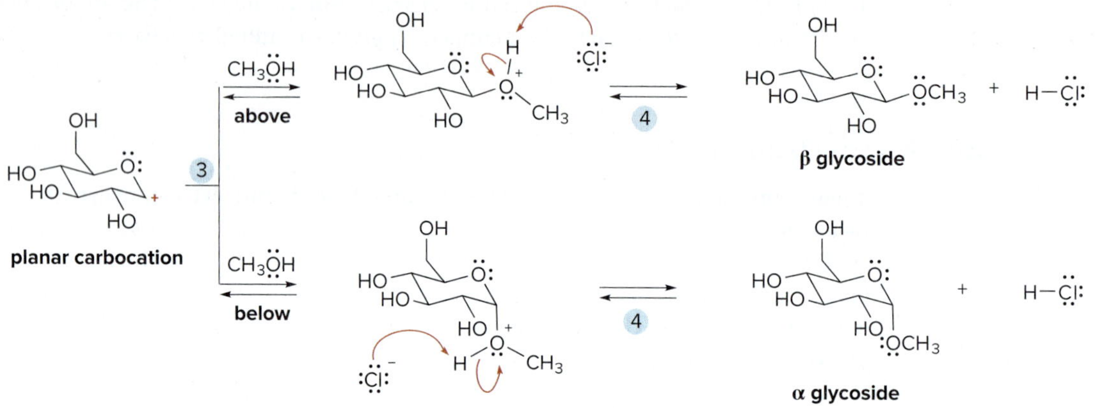

[Reaction scheme showing the planar carbocation being attacked by CH₃OH from above and below, yielding β glycoside and α glycoside]

3 – 4 Nucleophilic attack by **CH₃OH** occurs from both sides of the planar carbocation to yield α and β glycosides after loss of a proton.

24.7B Glycoside Hydrolysis

Because glycosides are acetals, **they are hydrolyzed with acid and water to cyclic hemiacetals and a molecule of alcohol.** A mixture of two anomers is formed from a single glycoside. For example, treatment of methyl α-D-glucopyranoside with aqueous acid forms a mixture of α- and β-D-glucose and methanol.

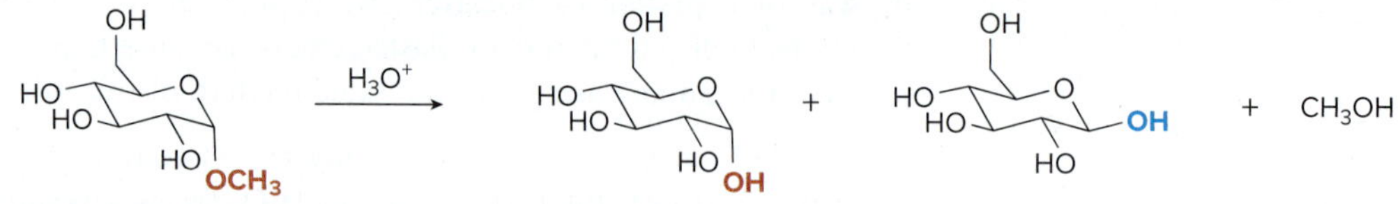

[Reaction scheme: methyl α-D-glucopyranoside + H₃O⁺ → α-D-glucose + β-D-glucose + CH₃OH]

The mechanism for glycoside hydrolysis is just the reverse of glycoside formation. It involves two parts: **formation of a planar carbocation,** followed by **nucleophilic attack of H₂O** to form anomeric hemiacetals, as shown in Mechanism 24.2.

Problem 24.17 Draw a stepwise mechanism for the following reaction.

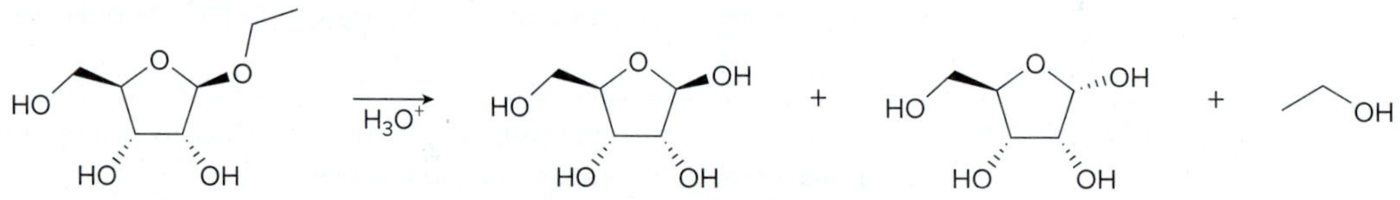

[Reaction scheme showing a furanoside with H₃O⁺ giving two anomeric hemiacetals plus ethanol]

Mechanism 24.2 Glycoside Hydrolysis

Part [1] Loss of CH₃OH from the glycoside

1 – 2 Protonation of the acetal OCH₃ followed by loss of CH₃OH forms a **resonance-stabilized carbocation**.

Part [2] Formation of the hemiacetals

3 – 4 Nucleophilic attack by H₂O occurs from both sides of the planar carbocation to yield α and β anomers after loss of a proton.

24.7C Naturally Occurring Glycosides

Salicin and **solanine** are two naturally occurring compounds that contain glycoside bonds as part of their structure. Salicin is an analgesic isolated from willow bark, and solanine is a poisonous compound produced in the leaves, stem, and green spots on the skin of potatoes. Solanine is also isolated from the berries of the deadly nightshade plant. It is believed that the role of the sugar rings in both salicin and solanine is to increase their water solubility.

The berries of the black nightshade plant (*Solanum nigrum*) are a source of the poisonous alkaloid solanine.
Westend61/Shutterstock

[The O atoms that are part of the glycoside bonds are drawn in red.]

1090 Chapter 24 Carbohydrates

Glycosides are common in nature. All disaccharides and polysaccharides are formed by joining monosaccharides together with glycosidic linkages. These compounds are discussed in detail beginning in Section 24.11.

Problem 24.18 (a) Label all the O atoms that are part of a glycoside in rebaudioside A. Rebaudioside A, marketed under the trade name Truvia, is a sweet glycoside obtained from the stevia plant, which has been used for centuries in Paraguay to sweeten foods. (b) The alcohol or phenol formed from the hydrolysis of a glycoside is called an **aglycon**. What aglycon and monosaccharides are formed by the hydrolysis of rebaudioside A?

Rebaudioside A, a naturally occurring glycoside about 400 times sweeter than table sugar, is obtained from the leaves of the stevia plant, a shrub native to Central and South America.
Linda Hall/Shutterstock

rebaudioside A
Trade name Truvia

24.8 Reactions of Monosaccharides at the OH Groups

Because monosaccharides contain OH groups, they undergo reactions typical of alcohols—that is, they are converted to **ethers** and **esters**. Because the cyclic hemiacetal form of a monosaccharide contains an OH group, this form of a monosaccharide must be drawn as the starting material for any reaction that occurs at an OH group.

All OH groups of a cyclic monosaccharide are converted to ethers by treatment with base and an alkyl halide. For example, α-D-glucose reacts with silver(I) oxide (Ag_2O, a base) and excess CH_3I to form a pentamethyl ether.

hemiacetal OH in red → pentamethyl ether

- The acetal OCH_3 group is in red.
- Ether OCH_3 groups are in blue.

Ag_2O removes a proton from each alcohol, forming an alkoxide (RO⁻), which then reacts with CH_3I in an S_N2 reaction. Because no C—O bonds are broken, the configuration of all substituents in the starting material is **retained**, forming a single product.

The product contains two different types of ether bonds. There are four "regular" ethers formed from the "regular" hydroxyls. The new ether from the hemiacetal is now part of an **acetal**—that is, a **glycoside**.

24.9 Reactions at the Carbonyl Group—Oxidation and Reduction

The four ether bonds that are *not* part of the acetal do not react with any reagents except strong acids like HBr and HI (Section 9.14). **The acetal ether, on the other hand, is hydrolyzed with aqueous acid** (Section 24.7B). Aqueous hydrolysis of a single glycoside (like the pentamethyl ether of α-D-glucose) yields both anomers of the product monosaccharide.

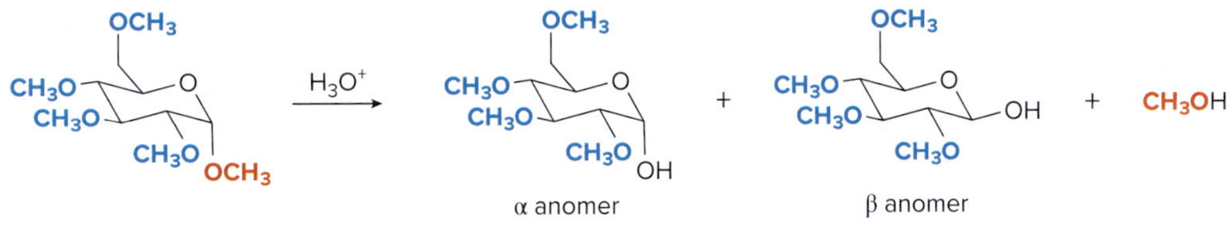

The OH groups of monosaccharides can also be converted to esters. For example, treatment of β-D-glucose with either acetic anhydride or acetyl chloride in the presence of pyridine (a base) converts all OH groups to acetate esters.

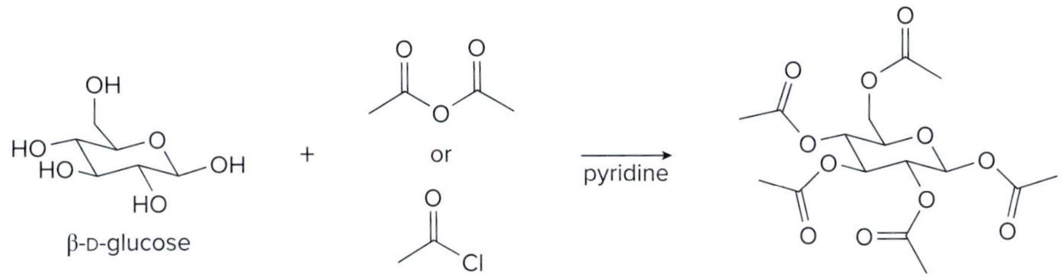

Because it is cumbersome and tedious to draw in all the atoms of the esters, the abbreviation **Ac** is used for the acetyl group, **CH₃C=O**. The esterification of β-D-glucose can then be written as follows:

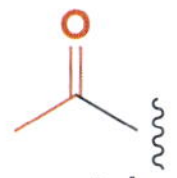

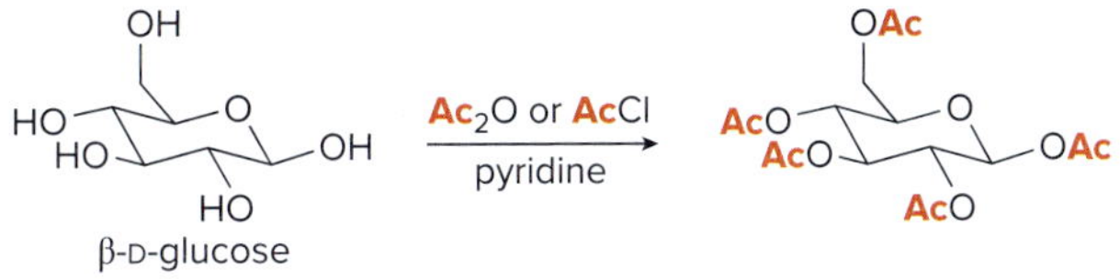

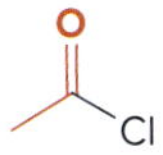

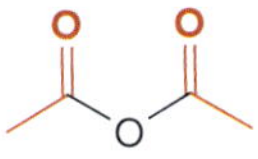

Monosaccharides are so polar that they are insoluble in common organic solvents, making them difficult to isolate and use in organic reactions. **Monosaccharide derivatives** that have five ether or ester groups in place of the OH groups, however, **are readily soluble in organic solvents.**

Problem 24.19 Draw the products formed when β-D-galactose is treated with each reagent.

a. Ag₂O + CH₃I
b. NaH + C₆H₅CH₂Cl
c. The product in (b), then H₃O⁺
d. Ac₂O + pyridine
e. C₆H₅COCl + pyridine
f. The product in (c), then C₆H₅COCl + pyridine

24.9 Reactions at the Carbonyl Group— Oxidation and Reduction

Oxidation and reduction reactions occur at the carbonyl group of monosaccharides, so they all begin with the monosaccharide drawn in the **acyclic form.** We will confine our discussion to aldoses as starting materials.

24.9A Reduction of the Carbonyl Group

Like other aldehydes, the **carbonyl group of an aldose is reduced to a 1° alcohol using NaBH$_4$.** This alcohol is called an **alditol.** For example, reduction of D-glucose with NaBH$_4$ in CH$_3$OH yields glucitol (also called sorbitol).

> Glucitol occurs naturally in some fruits and berries. It is sometimes used as a substitute for sucrose (table sugar). With six polar OH groups capable of hydrogen bonding, glucitol is readily hydrated. It is used as an additive to prevent certain foods from drying out.

D-glucose →(NaBH$_4$, CH$_3$OH)→ glucitol (sorbitol)

Problem 24.20 A 2-ketohexose is reduced with NaBH$_4$ in CH$_3$OH to form a mixture of D-galactitol and D-talitol. What is the structure of the 2-ketohexose?

24.9B Oxidation of Aldoses

Aldoses contain 1° and 2° alcohols and an aldehyde, all of which are oxidizable functional groups. Two different types of oxidation reactions are particularly useful—**oxidation of the aldehyde to a carboxylic acid (an aldonic acid)** and **oxidation of both the aldehyde and the 1° alcohol to a diacid (an aldaric acid).**

aldose →[O]→ aldonic acid or aldaric acid

[1] Oxidation of the aldehyde to a carboxylic acid

The aldehyde carbonyl is the most easily oxidized functional group in an aldose, so a variety of reagents oxidize it to a carboxy group, forming an **aldonic acid.**

Three reagents used for this process produce a characteristic color change because the oxidizing agent is reduced to a colored product that is easily visible. As described in Section 13.8, **Tollens reagent** oxidizes aldehydes to carboxylic acids using Ag$_2$O in NH$_4$OH, and forms a mirror of Ag as a by-product. **Benedict's** and **Fehling's reagents** use a blue Cu^{2+} salt as an oxidizing agent, which is reduced to Cu$_2$O, a brick-red solid. Unfortunately, none of these reagents gives a high yield of aldonic acid. When the aldonic acid is needed to carry on to other reactions, **Br$_2$ + H$_2$O** is used as the oxidizing agent.

D-glucose →(Ag$_2$O, NH$_4$OH or Cu^{2+} or Br$_2$, H$_2$O)→ D-gluconic acid + Ag or Cu$_2$O or Br$^-$

24.9 Reactions at the Carbonyl Group—Oxidation and Reduction

- Any carbohydrate that exists as a *hemiacetal* is in equilibrium with a small amount of acyclic aldehyde, so it is oxidized to an aldonic acid.
- Glycosides are acetals, not hemiacetals, so they are *not* oxidized to aldonic acids.

Carbohydrates that can be oxidized with Tollens, Benedict's, or Fehling's reagent are called **reducing sugars**. Those that do not react with these reagents are called **nonreducing sugars**. Figure 24.8 shows examples of reducing and nonreducing sugars.

Figure 24.8 Examples of reducing and nonreducing sugars

α-D-glucopyranose — hemiacetal — **reducing sugar**

tetramethyl α-D-glucopyranose — hemiacetal — **reducing sugar**

methyl β-D-glucopyranoside — acetal — **nonreducing sugar**

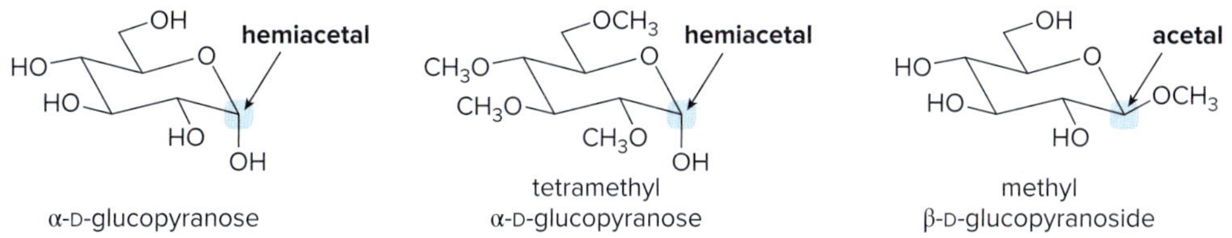

- Carbohydrates containing a hemiacetal are in equilibrium with an acyclic aldehyde, making them **reducing sugars**.
- Glycosides are acetals, so they are *not* in equilibrium with any acyclic aldehyde, making them **nonreducing sugars**.

Problem 24.21 Classify each compound as a reducing or nonreducing sugar.

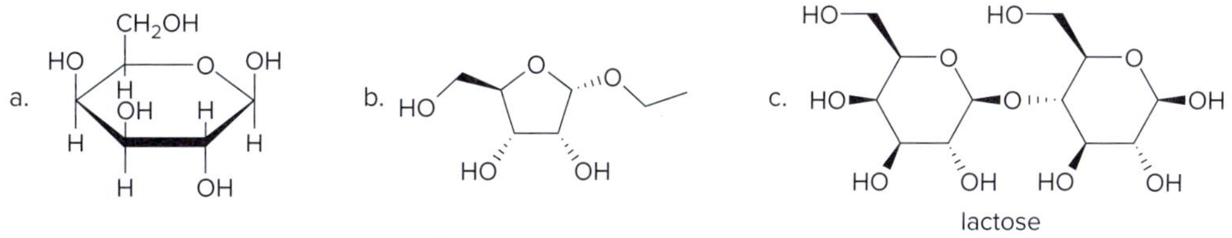

a. [structure]   b. [structure]   c. lactose [structure]

---

[2] Oxidation of both the aldehyde and 1° alcohol to a diacid

Both the aldehyde and 1° alcohol of an aldose are oxidized to carboxy groups by treatment with warm nitric acid, forming an aldaric acid. Under these conditions, D-glucose is converted to D-glucaric acid.

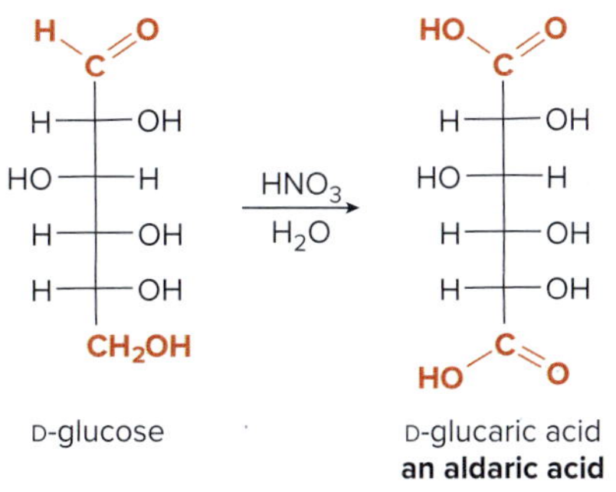

D-glucose $\xrightarrow{\text{HNO}_3 / \text{H}_2\text{O}}$ D-glucaric acid, an aldaric acid

Because aldaric acids have identical functional groups on both terminal carbons, some aldaric acids contain a plane of symmetry, making them achiral molecules. For example, oxidation of D-allose forms an achiral, optically inactive aldaric acid. This contrasts with

1094   Chapter 24   Carbohydrates

D-glucaric acid formed from glucose, which has no plane of symmetry and is thus still optically active.

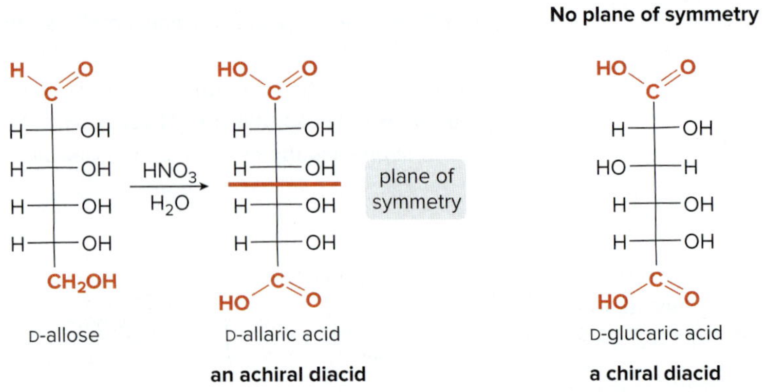

[Structures shown: D-allose → (HNO₃, H₂O) → D-allaric acid (an achiral diacid, with plane of symmetry); D-glucaric acid (a chiral diacid, no plane of symmetry)]

Problem 24.22   Draw the products formed when D-arabinose is treated with each reagent: (a) Ag₂O, NH₄OH; (b) Br₂, H₂O; (c) HNO₃, H₂O.

Problem 24.23   Which aldoses are oxidized to optically inactive aldaric acids: (a) D-erythrose; (b) D-lyxose; (c) D-galactose?

24.10 Reactions at the Carbonyl Group—Adding or Removing One Carbon Atom

Two common procedures in carbohydrate chemistry result in adding or removing one carbon atom from the skeleton of an aldose. The **Wohl degradation** shortens an aldose chain by one carbon, whereas the **Kiliani–Fischer synthesis** lengthens it by one. Both reactions involve cyanohydrins as intermediates. Recall from Section 14.8 that cyanohydrins are formed from aldehydes by addition of the elements of HCN. Cyanohydrins can also be re-converted to carbonyl compounds by treatment with base.

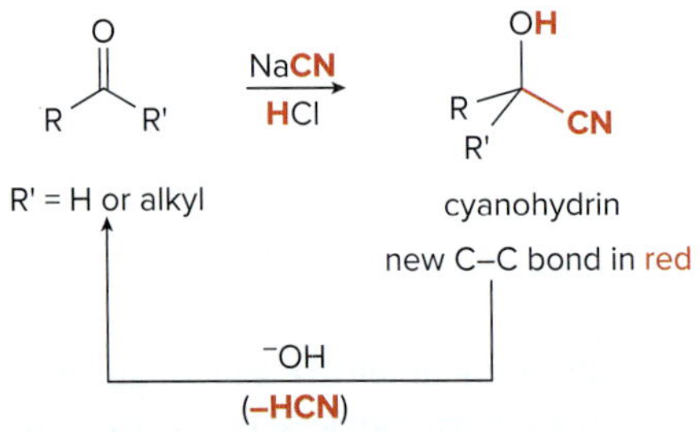

[Reaction scheme: R-C(=O)-R' (R' = H or alkyl) + NaCN/HCl → cyanohydrin (R-C(OH)(CN)-R'), new C–C bond in red; reverse with ⁻OH (–HCN)]

- Forming a cyanohydrin adds one carbon to a carbonyl group.
- Re-converting a cyanohydrin to a carbonyl compound removes one carbon.

24.10A   The Wohl Degradation

The **Wohl degradation** is a stepwise procedure that shortens the length of an aldose chain by cleavage of the C1–C2 bond. As a result, an aldohexose is converted to an aldopentose having the same configuration at its bottom three stereogenic centers (C3–C5). For example, the Wohl degradation converts D-glucose to D-arabinose.

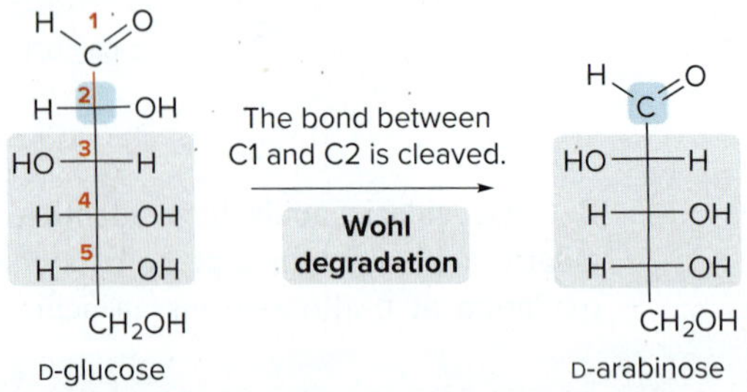

[Structures: D-glucose → Wohl degradation (the bond between C1 and C2 is cleaved) → D-arabinose]

24.10 Reactions at the Carbonyl Group—Adding or Removing One Carbon Atom

The Wohl degradation consists of three steps, illustrated here beginning with D-glucose.

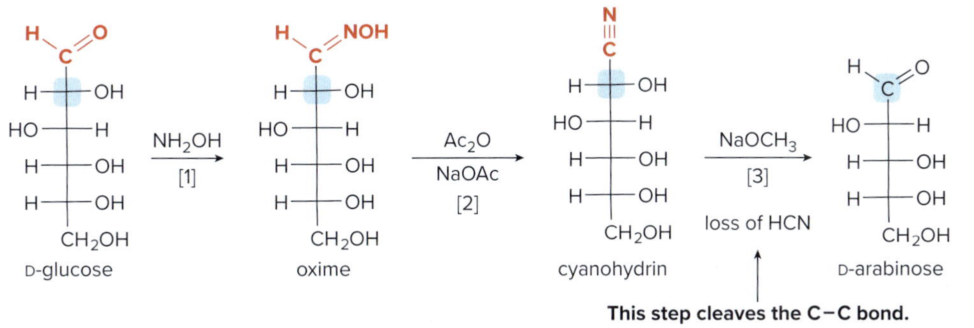

D-glucose →[NH$_2$OH, [1]] oxime →[Ac$_2$O, NaOAc, [2]] cyanohydrin →[NaOCH$_3$, [3], loss of HCN] D-arabinose

↑ This step cleaves the C–C bond.

[1] Treatment of D-glucose with hydroxylamine (NH$_2$OH) forms an **oxime** by nucleophilic addition. This reaction is analogous to the formation of imines discussed in Section 14.10.

[2] Dehydration of the oxime to a nitrile occurs with acetic anhydride (Ac$_2$O) and sodium acetate (NaOAc). The nitrile product is a cyanohydrin.

[3] **Treatment of the cyanohydrin with base results in loss of the elements of HCN to form an aldehyde having one fewer carbon.**

The Wohl degradation converts a stereogenic center at C2 in the original aldose to an sp^2 hybridized C=O. As a result, a pair of aldoses that are epimeric at C2, such as D-galactose and D-talose, yield the *same* aldose (D-lyxose, in this case) upon Wohl degradation.

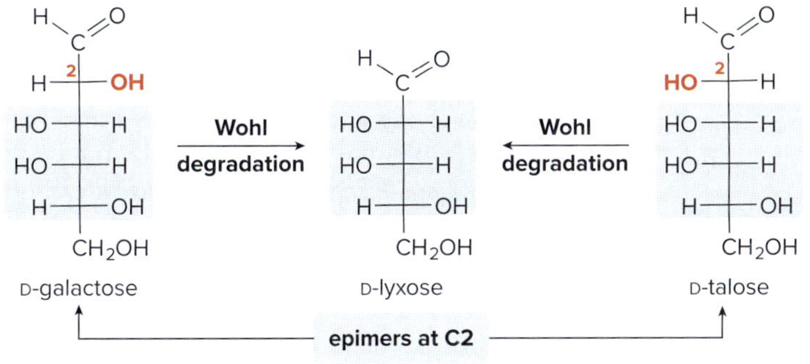

D-galactose →[Wohl degradation] D-lyxose ←[Wohl degradation] D-talose

— epimers at C2 —

Problem 24.24 What two aldoses yield D-xylose on Wohl degradation?

24.10B The Kiliani–Fischer Synthesis

The Kiliani–Fischer synthesis lengthens a carbohydrate chain by adding one carbon to the aldehyde end of an aldose, thus forming a new stereogenic center at C2 of the product. The product consists of epimers that differ only in their configuration about the one new stereogenic center. For example, the Kiliani–Fischer synthesis converts D-arabinose to a mixture of D-glucose and D-mannose.

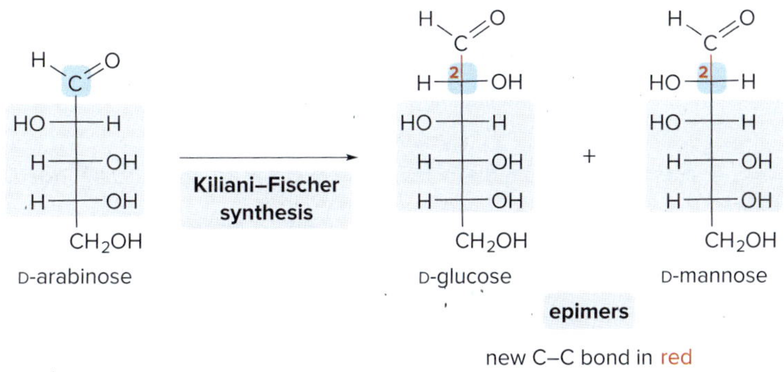

D-arabinose →[Kiliani–Fischer synthesis] D-glucose + D-mannose

epimers

new C–C bond in red

The Kiliani–Fischer synthesis, shown here beginning with D-arabinose, consists of three steps. "Squiggly" lines are meant to indicate that two different stereoisomers are formed at the new stereogenic center. As with the Wohl degradation, **the key intermediate is a cyanohydrin.**

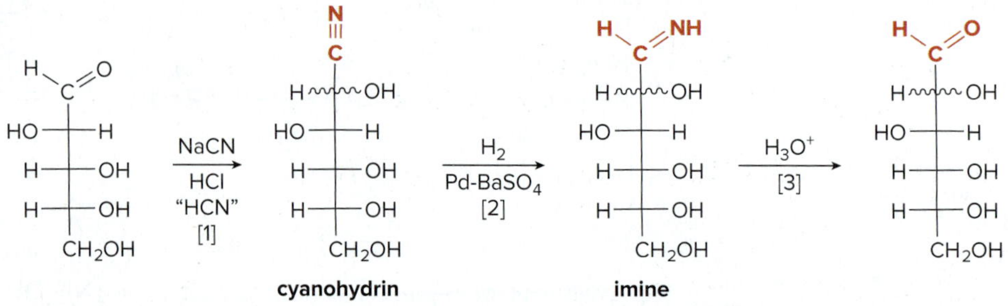

[1] Treating an aldose with NaCN and HCl adds the elements of HCN to the carbonyl group, forming a **cyanohydrin** and a new carbon–carbon bond. Because the sp^2 hybridized carbonyl carbon is converted to an sp^3 hybridized carbon with four different groups, **a new stereogenic center is formed in this step.**

[2] Reduction of the nitrile with H_2 and Pd-BaSO$_4$, a poisoned Pd catalyst, forms an **imine**.

[3] **Hydrolysis of the imine with aqueous acid forms an aldehyde that has one more carbon than the aldose** that began the sequence.

Note that the **Wohl degradation and the Kiliani–Fischer synthesis are conceptually opposite transformations.**

- The Wohl degradation *removes* a carbon atom from the aldehyde end of an aldose. Two aldoses that are epimers at C2 form the *same* product.
- The Kiliani–Fischer synthesis *adds* a carbon to the aldehyde end of an aldose, forming *two epimers* at C2.

Problem 24.25 What aldoses are formed when the following aldoses are subjected to the Kiliani–Fischer synthesis: (a) D-threose; (b) D-ribose; (c) D-galactose?

24.10C Determining the Structure of an Unknown Monosaccharide

The reactions in Sections 24.9–24.10 can be used to determine the structure of an unknown monosaccharide, as shown in Sample Problem 24.5.

Sample Problem 24.5 Determining the Structure of an Unknown Aldose

A D-aldopentose **A** is oxidized to an optically inactive aldaric acid with HNO$_3$. **A** is formed by the Kiliani–Fischer synthesis of a D-aldotetrose **B**, which is also oxidized to an optically inactive aldaric acid with HNO$_3$. What are the structures of **A** and **B**?

Solution

Use each fact to determine the relative orientation of the OH groups in the D-aldopentose.

Fact [1] A D-aldopentose **A** is oxidized to an optically *inactive* aldaric acid with HNO$_3$.

An optically inactive aldaric acid must contain a **plane of symmetry**. Because the **OH group on C4 must be on the right for the D-sugar**, there are only two ways to arrange the OH groups in a five-carbon D-aldaric acid. Thus, only two structures are possible for **A**, labeled **A'** and **A"**.

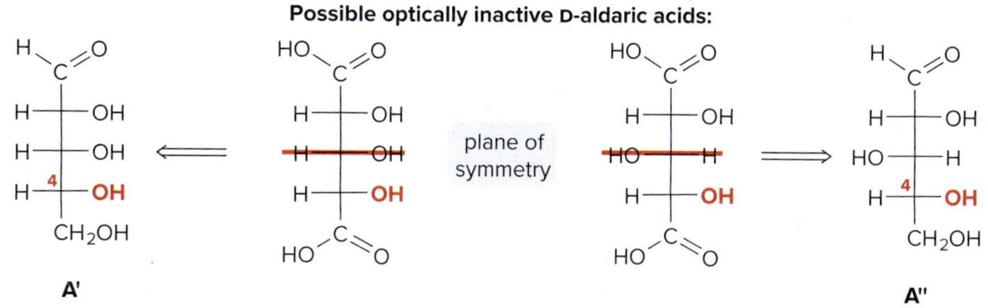

24.10 Reactions at the Carbonyl Group—Adding or Removing One Carbon Atom

Possible optically inactive D-aldaric acids:

[Fischer projections showing four D-aldaric acid structures labeled A′ and A″ with plane of symmetry indicated on the two middle meso aldaric acids]

Fact [2] A is formed by the Kiliani–Fischer synthesis from a D-aldotetrose B.

A′ and A″ are each prepared from a D-aldotetrose (B′ and B″) that has the same configuration at the bottom two stereogenic centers.

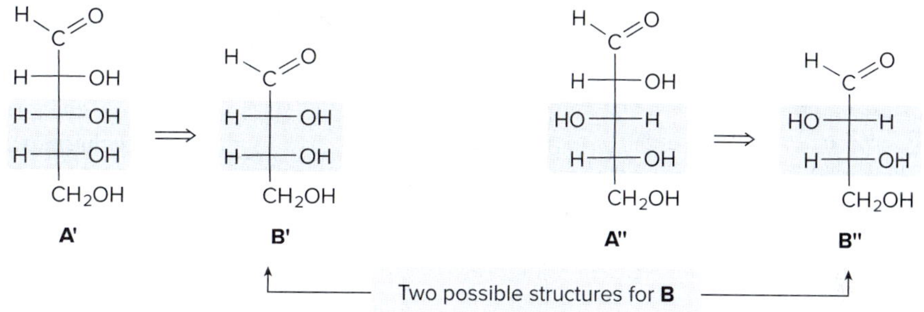

[Fischer projections of A′, B′, A″, B″ — Two possible structures for B]

Fact [3] The D-aldotetrose is oxidized to an optically *inactive* aldaric acid upon treatment with HNO₃.

Only the aldaric acid from B′ has a plane of symmetry, making it optically inactive. Thus, B′ is the correct structure for the D-aldotetrose B, and therefore A′ is the structure of the D-aldopentose A.

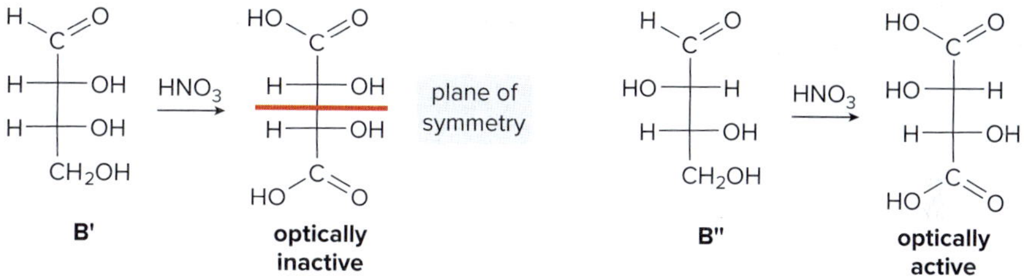

[Fischer projections: B′ → (HNO₃) optically inactive aldaric acid with plane of symmetry; B″ → (HNO₃) optically active aldaric acid]

Answer:

[Fischer projection of aldotetrose = B, and aldopentose = A]

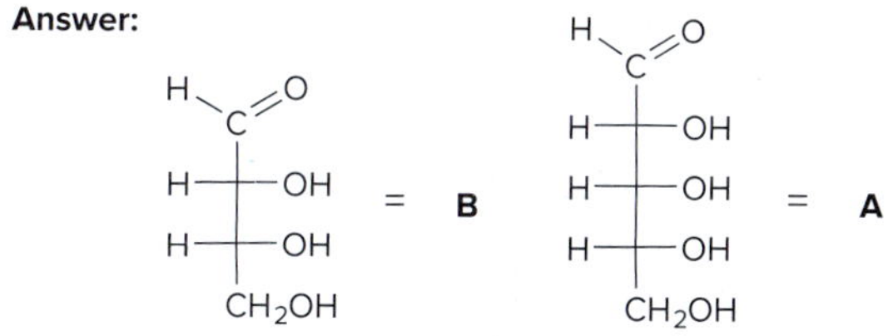

Problem 24.26 A D-aldopentose **A** is oxidized to an optically inactive aldaric acid. On Wohl degradation, **A** forms an aldotetrose **B** that is oxidized to an optically active aldaric acid. What are the structures of **A** and **B**?

More Practice: Try Problems 24.57–24.59.

Problem 24.27 A D-aldohexose **A** is formed from an aldopentose **B** by the Kiliani–Fischer synthesis. Reduction of **A** with NaBH₄ forms an optically inactive alditol. Oxidation of **B** forms an optically active aldaric acid. What are the structures of **A** and **B**?

24.11 Disaccharides

Disaccharides contain two monosaccharides joined by a glycosidic linkage. The general features of a disaccharide include the following:

disaccharide
glycosidic linkage in red
acetal carbon labeled in blue

1→4-β-glycosidic linkage

[1] Two monosaccharide rings may be five- or six-membered, but six-membered rings are much more common. **The two rings are connected by an O atom that is part of an acetal, called a glycosidic linkage,** which may be oriented α or β.

[2] The **glycoside is formed from the anomeric carbon of one monosaccharide and any OH group on the other monosaccharide.** All disaccharides have **one acetal,** plus either a hemiacetal or another acetal.

[3] With pyranose rings, the carbon atoms in each ring are numbered beginning with the anomeric carbon. The most common disaccharides contain two monosaccharides in which the hemiacetal carbon of one ring (Cl) is joined to C4 of the other ring.

The three most abundant disaccharides are **maltose, lactose,** and **sucrose.**

24.11A Maltose

Maltose gets its name from malt, the liquid obtained from barley and other cereal grains.
Mir141/Shutterstock

Maltose, a disaccharide formed by the hydrolysis of starch, is found in germinated grains such as barley. Maltose contains two glucose units joined by a 1→4-α-glycoside bond. Maltose contains one acetal carbon (in red) and one hemiacetal carbon (in blue).

maltose
1→4-α-glycosidic linkage

Because one glucose ring of maltose still contains a hemiacetal, it exists as a mixture of α and β anomers. Only the β anomer is shown. Maltose exhibits two properties of all carbohydrates that contain a hemiacetal: it undergoes **mutarotation,** and it reacts with oxidizing agents, making it a **reducing sugar.**

Hydrolysis of maltose forms two molecules of glucose. The **C1—O** bond is cleaved in this process, and a mixture of glucose anomers forms. The mechanism for this hydrolysis is exactly the same as the mechanism for glycoside hydrolysis in Section 24.7B.

α-D-glucose + β-D-glucose

Problem 24.28 Draw the α anomer of maltose. What products are formed on hydrolysis of this form of maltose?

24.11B Lactose

Lactose is the principal disaccharide found in milk from both humans and cows. Unlike many mono- and disaccharides, lactose is not appreciably sweet. Lactose consists of **one galactose** and **one glucose unit,** joined by a **1→4-β-glycoside bond** from the anomeric carbon of galactose to C4 of glucose.

Milk contains the disaccharide lactose. *Mitch Hrdlicka/Getty Images*

Like maltose, lactose also contains a hemiacetal, so it exists as a mixture of α and β anomers. The β anomer is drawn. Lactose undergoes **mutarotation,** and it reacts with oxidizing agents, making it a **reducing sugar.**

Lactose is digested in the body by first cleaving the 1→4-β-glycoside bond using the enzyme *lactase.* Many individuals, mainly of Asian and African descent, lack adequate amounts of lactase, so they are unable to digest and absorb lactose. This condition, lactose intolerance, is associated with abdominal cramping and recurrent diarrhea when milk and dairy products are ingested.

Problem 24.29 Cellobiose, a disaccharide obtained by the hydrolysis of cellulose, is composed of two glucose units joined by a 1→4-β-glycoside bond. What is the structure of cellobiose?

24.11C Sucrose

Sucrose, the disaccharide mentioned in the chapter opener that is found in sugarcane and used as table sugar (Figure 24.9), is the most common disaccharide in nature. It contains **one glucose unit** and **one fructose unit.**

Figure 24.9 Sucrose

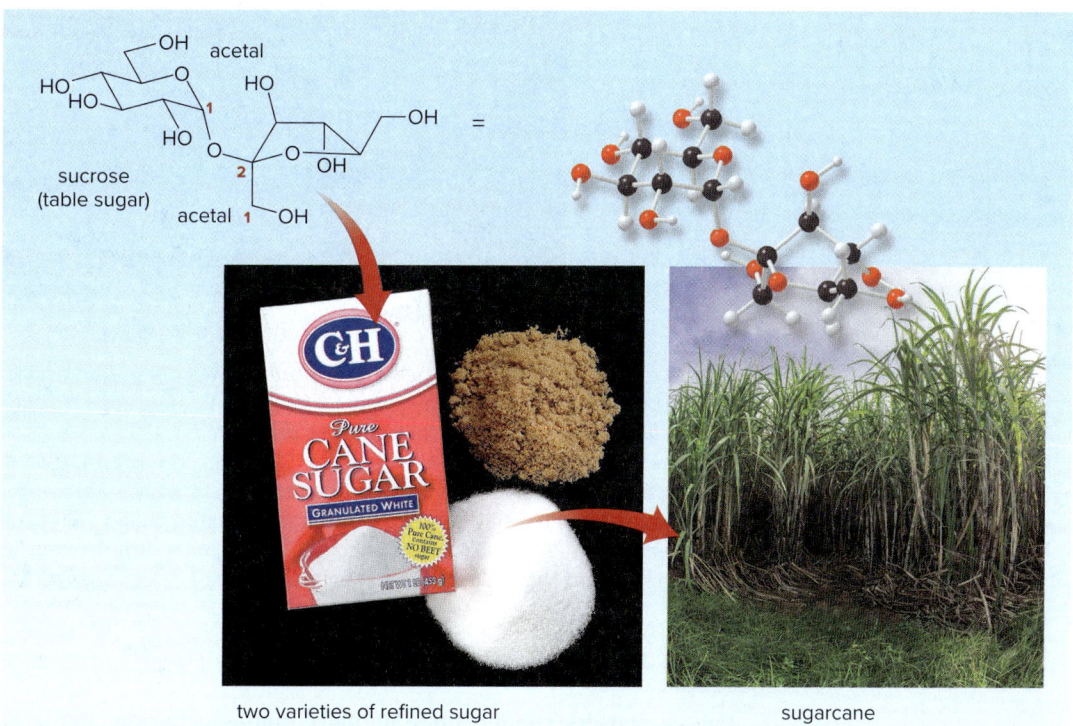

two varieties of refined sugar   sugarcane

Elite Images/McGraw-Hill Education   *SAK_PD/Shutterstock*

The structure of sucrose has several features that make it different from maltose and lactose. Sucrose contains one six-membered ring (glucose) and one five-membered ring (fructose), whereas both maltose and lactose contain two six-membered rings. In sucrose the six-membered glucose ring is joined by an α-glycosidic bond to C2 of a fructofuranose ring. The numbering in a fructofuranose is different from the numbering in a pyranose ring. The anomeric carbon is now designated as C2, so the anomeric carbons of the glucose and fructose rings are both used to form the glycosidic linkage.

As a result, **sucrose contains two acetals but no hemiacetal.** Sucrose, therefore, is a **nonreducing sugar** and **it does *not* undergo mutarotation.**

Sucrose's pleasant sweetness has made it a widely used ingredient in baked goods, cereals, bread, and many other products. It is estimated that the average American ingests 100 lb of sucrose annually. Like other carbohydrates, however, sucrose contains many calories. To reduce caloric intake while maintaining sweetness, a variety of artificial sweeteners have been developed. These include sucralose, aspartame, and saccharin (Figure 24.10). These compounds are much sweeter than sucrose, so only a small amount of each compound is needed to achieve the same level of perceived sweetness.

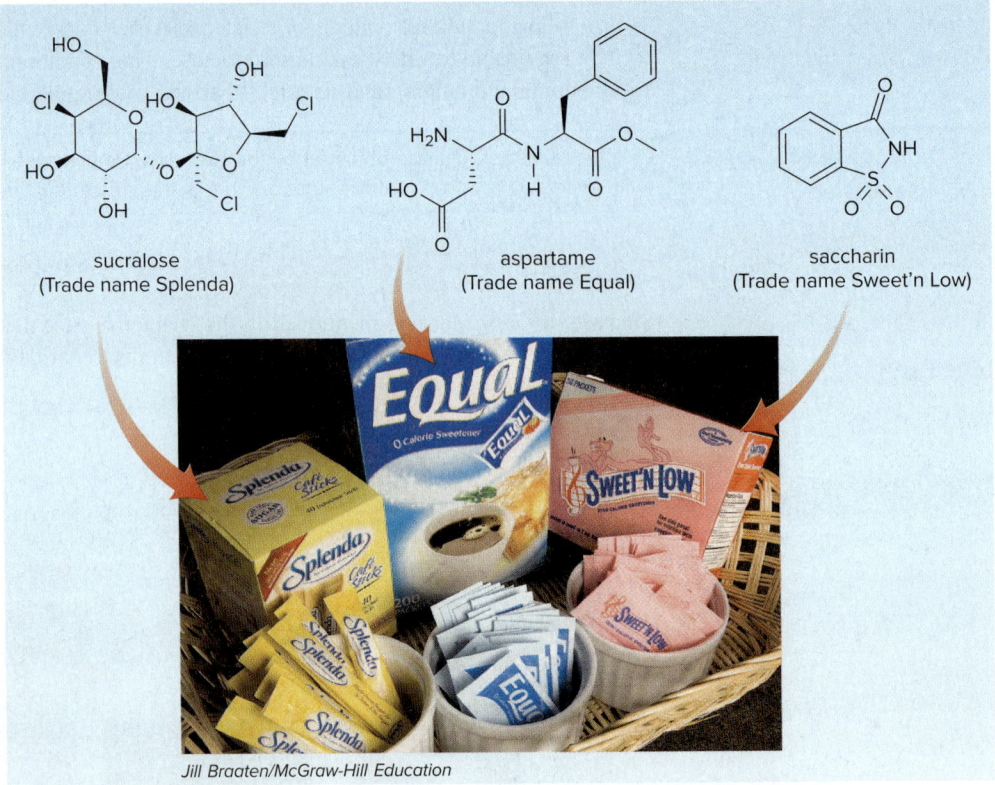

Figure 24.10 Artificial sweeteners

- The sweetness of these three artificial sweeteners was discovered accidentally. The sweetness of sucralose was discovered in 1976 when a chemist misunderstood his superior, and he *tasted* rather than *tested* his compound. Aspartame was discovered in 1965 when a chemist licked his dirty fingers in the lab and tasted its sweetness. Saccharin, the oldest-known artificial sweetener, was discovered in 1879 by a chemist who failed to wash his hands after working in the lab. Saccharin was not used extensively until sugar shortages occurred during World War I. Although there were concerns in the 1970s that saccharin causes cancer, there is no proven link between cancer occurrence and saccharin intake at normal levels.

24.12 Polysaccharides

Polysaccharides contain three or more monosaccharides joined together. Three prevalent polysaccharides in nature are **cellulose, starch,** and **glycogen,** each of which consists of repeating glucose units joined by different glycosidic bonds.

24.12A Cellulose

The structure of cellulose was discussed in Section 5.1.

Cellulose is found in the cell walls of nearly all plants, where it gives support and rigidity to wood and plant stems. Cotton is essentially pure cellulose.

cellulose
1→4-β-glycosidic linkages in red

Cellulose is an unbranched polymer composed of repeating glucose units joined in a 1→4-β-glycosidic linkage. The β-glycosidic linkage forms long linear chains of cellulose molecules that stack in sheets, creating an extensive three-dimensional array. A network of intermolecular hydrogen bonds between the chains and sheets means that only the few OH groups on the surface are available to hydrogen bond to water, making this very polar compound water insoluble.

Ball-and-stick models showing the three-dimensional structures of cellulose and starch were given in Figure 5.2.

Cellulose can be hydrolyzed to glucose by cleaving all the β-glycosidic bonds, yielding both anomers of glucose.

α-D-glucose   β-D-glucose

A **β-glycosidase** is the general name of an enzyme that hydrolyzes a β-glycoside linkage.

In cells, the hydrolysis of cellulose is accomplished by an enzyme called a **β-glucosidase**, which cleaves all the β-glycoside bonds formed from glucose. Humans do not possess this enzyme and therefore cannot digest cellulose. Ruminant animals, on the other hand, such as cattle, deer, and camels, have bacteria containing a β-glucosidase in their digestive systems, so they can derive nutritional benefit from eating grass and leaves.

24.12B Starch

Starch is the main carbohydrate found in the seeds and roots of plants. Corn, rice, wheat, and potatoes are common foods that contain a great deal of starch.

Starch is a polymer composed of repeating glucose units joined in α-glycosidic linkages. Both starch and cellulose are polymers of glucose, but starch contains α glycoside bonds, whereas cellulose contains β glycoside bonds. The two common forms of starch are **amylose** and **amylopectin**.

amylose
(the **linear** form of starch)

1→4-α-glycosidic linkages in red

The 1→6-α-glycosidic linkage forms a branch in the chain.

amylopectin
(the **branched** form of starch)

1→4-α-glycosidic linkages in red
1→6-α-glycosidic linkage in blue

Amylose, which comprises about 20% of starch molecules, has an unbranched skeleton of glucose molecules with **1→4-α-glycoside bonds.** Because of this linkage, an amylose chain adopts a helical arrangement, giving it a very different three-dimensional shape from the linear chains of cellulose. Amylose was first described in Section 5.1.

Amylopectin, which comprises about 80% of starch molecules, likewise consists of a backbone of glucose units joined in **α-glycosidic bonds,** but it also contains considerable branching along the chain. The linear linkages of amylopectin are formed by **1→4-α-glycoside bonds,** similar to amylose. The branches are linked to the chain with **1→6-α-glycosidic linkages.**

Both forms of starch are water soluble. Because the OH groups in these starch molecules are not buried in a three-dimensional network, they are more available for hydrogen bonding with water molecules, leading to greater water solubility than cellulose has.

The ability of amylopectin to form branched polymers is a unique feature of carbohydrates. Other types of polymers in the cell, such as the proteins discussed in Chapter 23, occur in nature only as linear molecules.

Both amylose and amylopectin are hydrolyzed to glucose with cleavage of the glycosidic bonds. The human digestive system has the necessary **α-glucosidase** enzymes needed to catalyze this process. Bread and pasta made from wheat flour, rice, and corn tortillas are all sources of starch that are readily digested.

> **α-Glycosidase** is the general name of an enzyme that hydrolyzes an α-glycoside linkage.

24.12C Glycogen

Glycogen is the major form in which polysaccharides are stored in animals. Glycogen, a polymer of glucose containing **α-glycosidic bonds,** has a branched structure similar to amylopectin, but the branching is much more extensive.

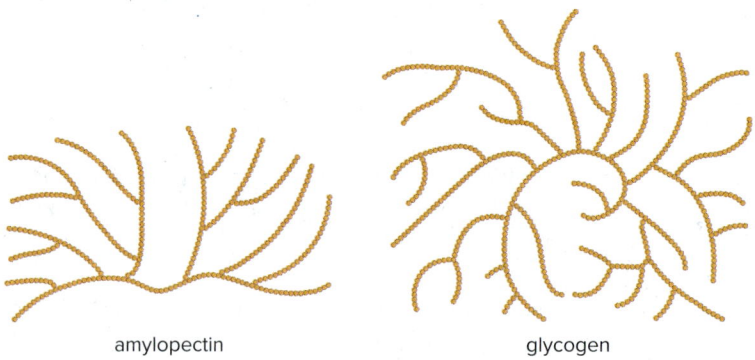

amylopectin   glycogen

Glycogen is stored principally in the liver and muscle. When glucose is needed for energy in the cell, glucose units are hydrolyzed from the ends of the glycogen polymer, and then further metabolized with the release of energy. Because glycogen has a highly branched structure, there are many glucose units at the ends of the branches that can be cleaved whenever the body needs them.

Problem 24.30 Draw the structure of: (a) a polysaccharide formed by joining D-mannose units in 1→4-β-glycosidic linkages; (b) a polysaccharide formed by joining D-glucose units in 1→6-α-glycosidic linkages. The polysaccharide in (b) is dextran, a component of dental plaque.

24.12D Human Milk Oligosaccharides

Human milk oligosaccharides (HMOs), a group of carbohydrates found in breast milk, contain three or four monosaccharides joined together. 2′-Fucosyllactose is the most prevalent component, comprising about 30% of all HMOs. The two glycosidic linkages that join the three monosaccharides together are shown in red.

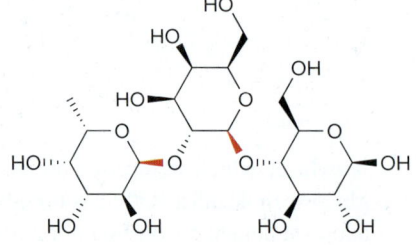

2′-fucosyllactose

The World Health Organization recommends that children are exclusively breast fed until six months of age, and then nursed along with other forms of nutrition until a child is two years old. *Daniel C. Smith*

> An **oligosaccharide** is a carbohydrate with a small number of monosaccharides—generally three to ten—joined together.

HMOs, often called the fiber of breast milk, are not hydrolyzed by the gastric juices of the stomach, nor are they absorbed in the intestines. They nonetheless play a key role in the health of a newborn, by helping to establish the presence of beneficial bacteria in the infant's colon. Moreover, harmful pathogens attach to the surface of HMOs and are eliminated in the feces of the nursing infant.

Ongoing research continues to study the hundreds of unique components of human breast milk in an effort to understand its benefits to both the mother and the child, even after infancy.

Problem 24.31 3-Fucosyllactose is another HMO found in breast milk. (a) Locate any acetal and hemiacetal. (b) What products are formed when 3-fucosyllactose is hydrolyzed in aqueous acid?

3-fucosyllactose

24.13 Other Important Sugars and Their Derivatives

Many other examples of simple and complex carbohydrates with useful properties exist in the biological world. In Section 24.13, we examine some carbohydrates that contain nitrogen atoms.

24.13A Amino Sugars and Related Compounds

Amino sugars contain an NH_2 group instead of an OH group at a non-anomeric carbon. The most common amino sugar in nature, D-glucosamine, is formally derived from D-glucose by replacing the OH at C2 with NH_2. Although it is not classified as a drug, and therefore not regulated by the U.S. Food and Drug Administration, glucosamine is available in many over-the-counter treatments for osteoarthritis.

D-glucosamine → N-acetyl-D-glucosamine **NAG**

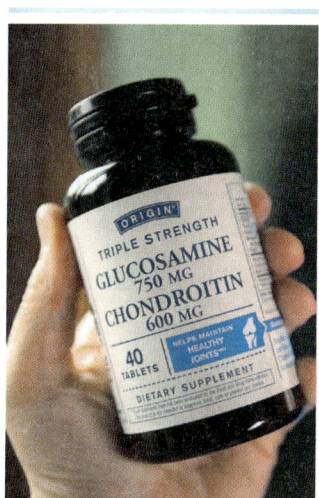

Dietary supplements containing glucosamine are used by individuals suffering from osteoarthritis. *Jill Braaten/McGraw-Hill Education*

Acetylation of glucosamine with acetyl CoA (Section 16.16) forms **N-acetyl-D-glucosamine**, abbreviated as **NAG. Chitin,** the second most abundant carbohydrate polymer, is a polysaccharide formed from NAG units joined together in **1→4-β-glycosidic linkages.** Chitin is identical in structure to cellulose, except that each OH group at C2 is now replaced by $NHCOCH_3$. The exoskeletons of lobsters, crabs, and shrimp are composed of chitin. Like those of cellulose, chitin chains are held together by an extensive network of hydrogen bonds, forming water-insoluble sheets.

1104   Chapter 24   Carbohydrates

The rigidity of a crab shell is due to chitin, a high-molecular-weight carbohydrate molecule. Chitin-based coatings have found several commercial applications, such as extending the shelf life of fruits. Processing plants now convert the shells of crabs, lobsters, and shrimp to chitin and various derivatives for use in many consumer products. *Comstock Images/Getty Images*

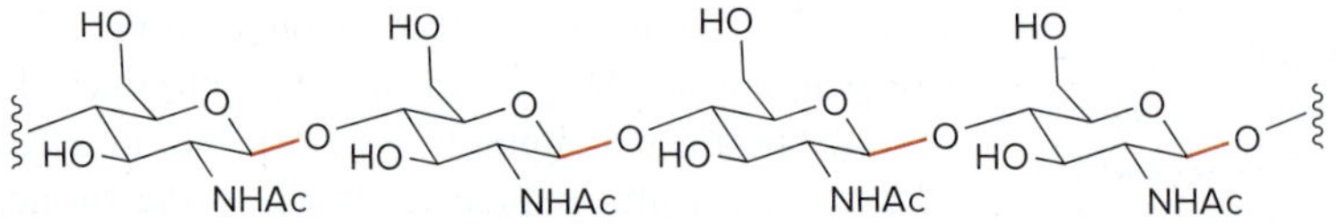

1 → 4-β-glycosidic linkages in red
chitin

Several trisaccharides containing amino sugars are potent antibiotics used in the treatment of certain severe and recurrent bacterial infections. These compounds, such as tobramycin and amikacin, are called **aminoglycoside antibiotics.**

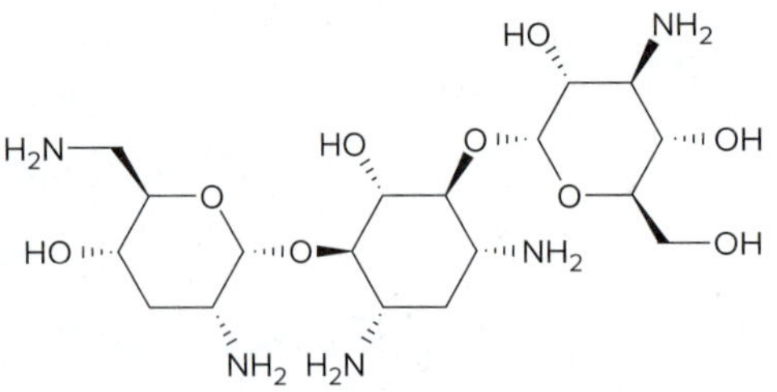

tobramycin

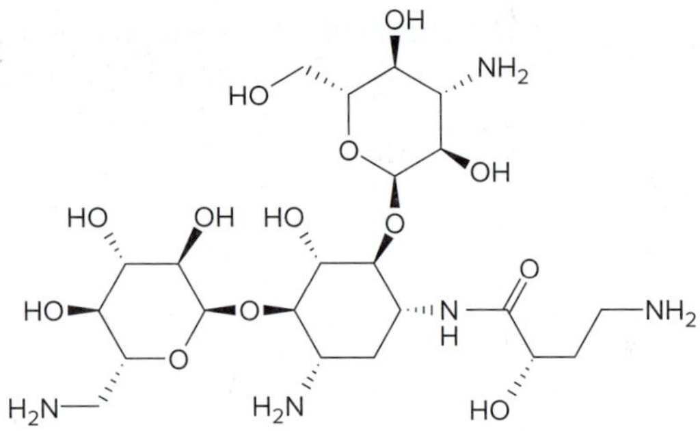

amikacin

Problem 24.32   Treating chitin with H_2O, ^-OH hydrolyzes its amide linkages, forming a compound called chitosan. What is the structure of chitosan? Chitosan has been used in shampoos, fibers for sutures, and wound dressings.

24.13B *N*-Glycosides

***N*-Glycosides are formed when a monosaccharide is reacted with an amine** in the presence of mild acid (Reactions [1] and [2]).

[1] β-D-glucopyranose → α-*N*-glycoside + β-*N*-glycoside

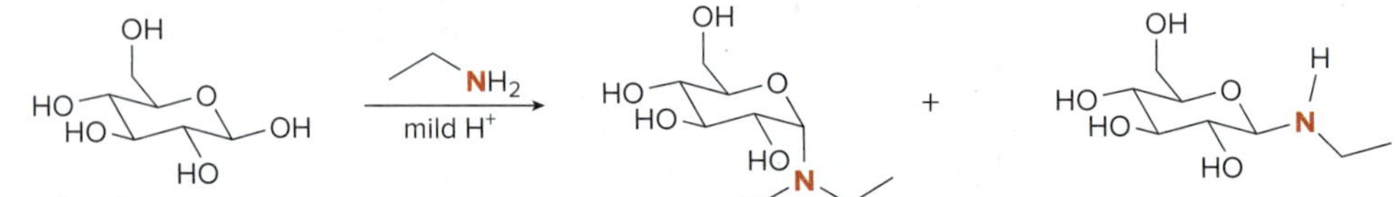

[2] α-D-ribofuranose → (products)

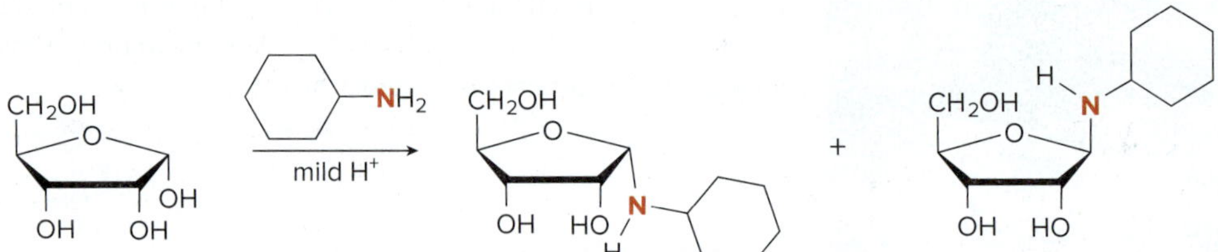

The mechanism of *N*-glycoside formation is analogous to the mechanism for glycoside formation, and both anomers of the *N*-glycoside are formed as products. *N*-Glycosides are key elements in the structures of DNA and RNA, as discussed in Chapter 26.

Problem 24.33   Draw the products of each reaction.

a. (structure) + CH_3NH_2, mild H^+

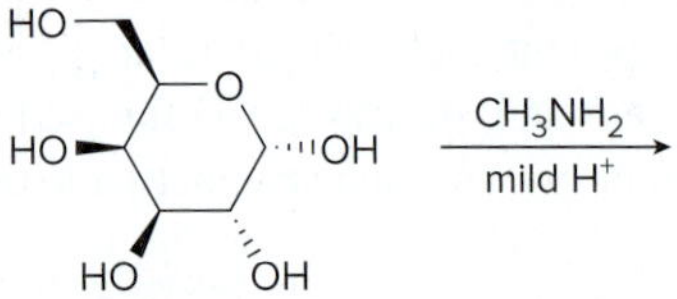

b. (structure) + $C_6H_5NH_2$, mild H^+

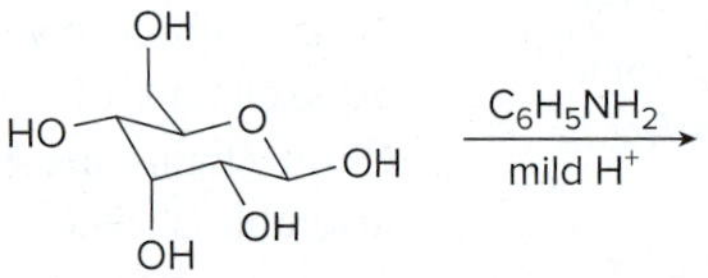

Chapter 24 REVIEW

KEY REACTIONS

[1] Reactions of monosaccharides involving the hemiacetal

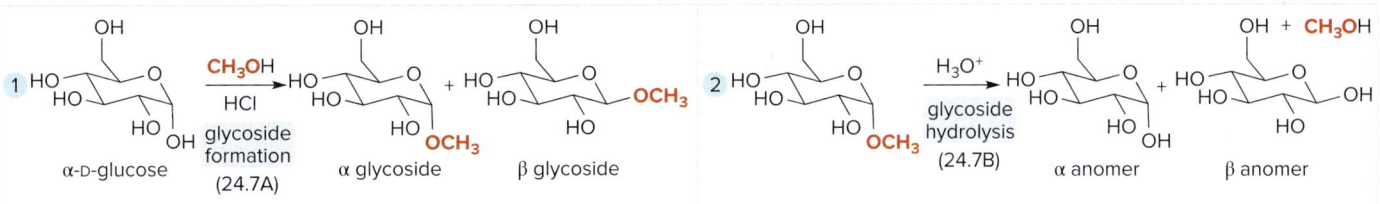

Try Problems 24.45b, d; 24.46a; 24.47.

[2] Reactions of monosaccharides at the OH groups

Try Problems 24.45a, c; 24.46g.

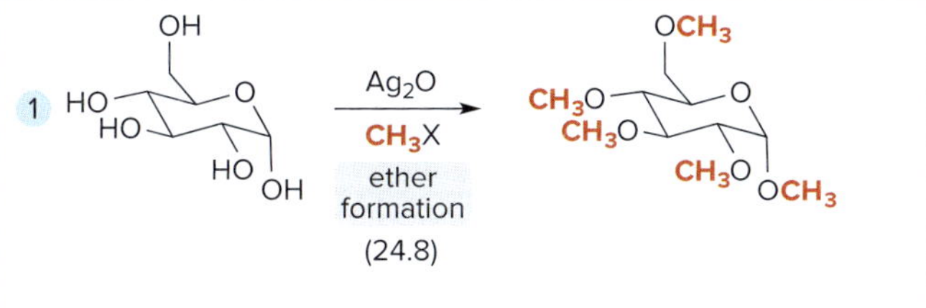

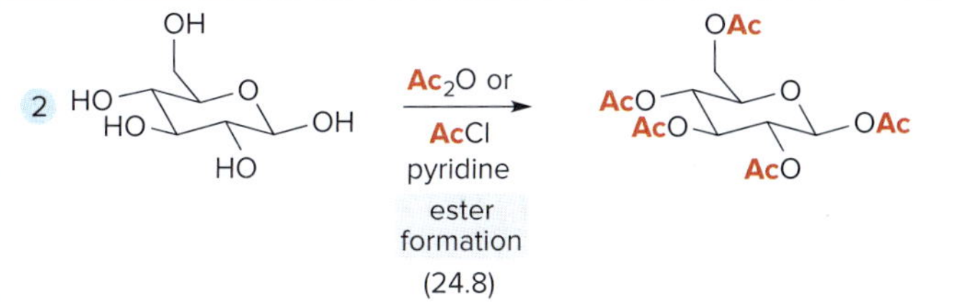

[3] Reactions of monosaccharides at the carbonyl group

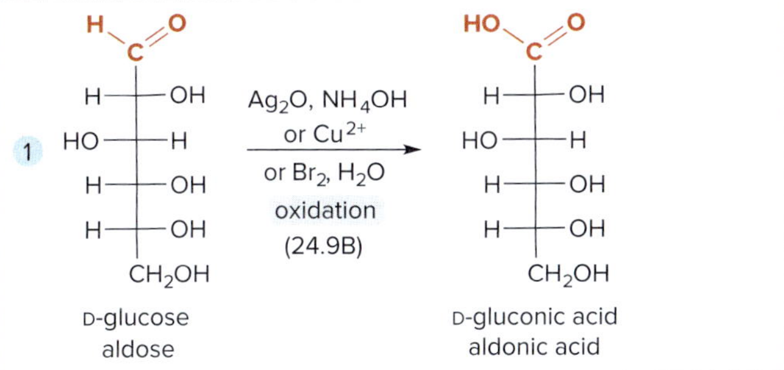

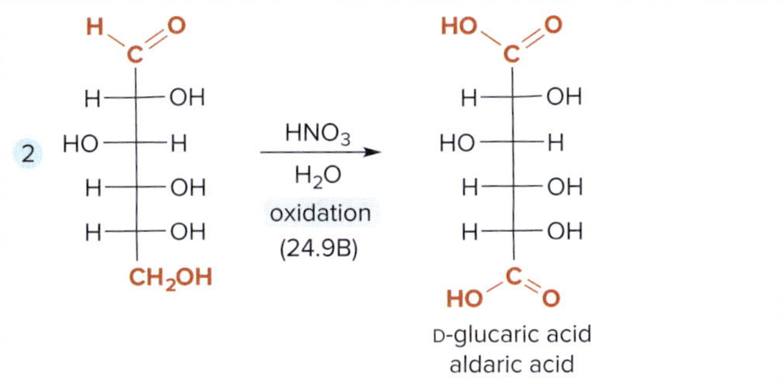

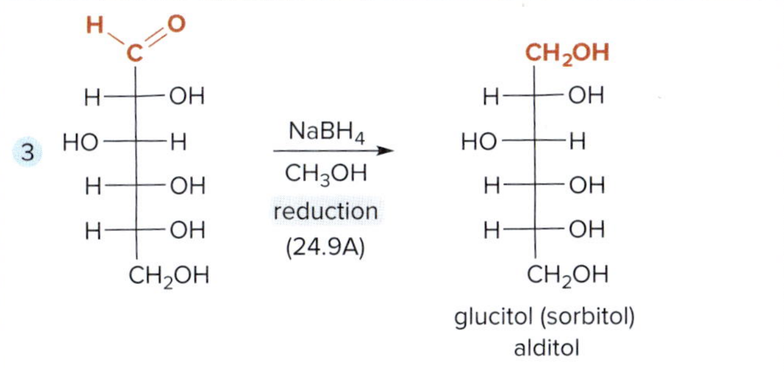

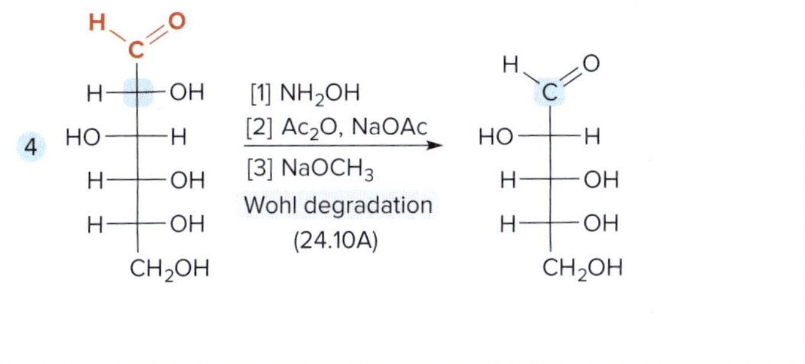

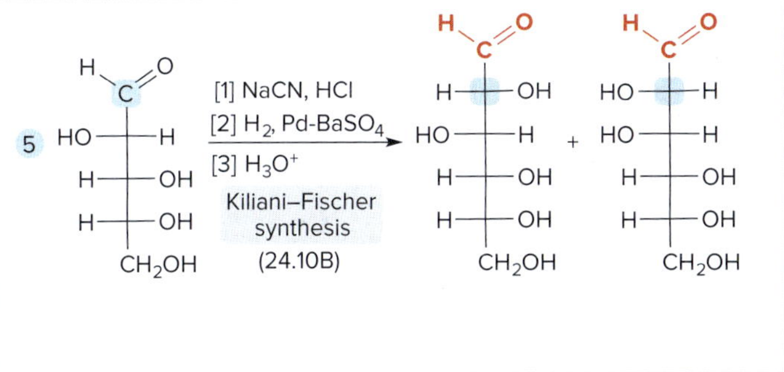

Try Problems 24.46b–f; 24.49c–e; 24.50c, d; 24.51; 24.52.

Chapter 24 Carbohydrates

[4] Other reactions

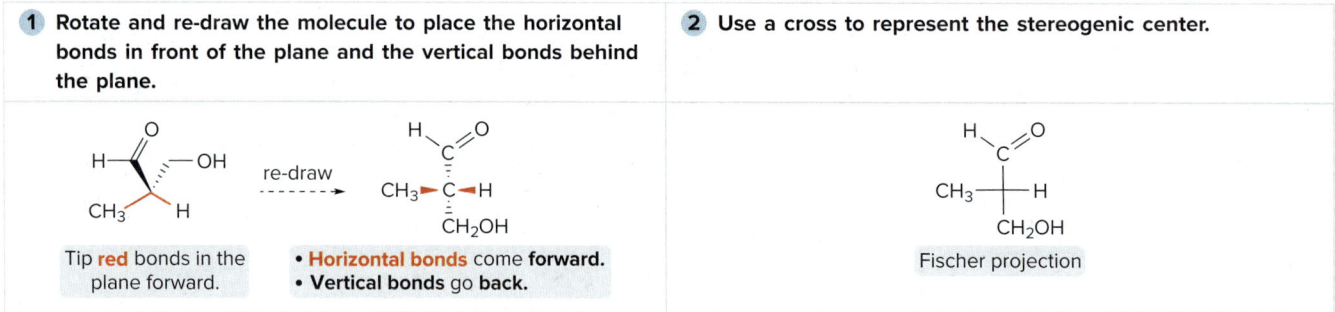

Try Problems 24.46h, 24.47, 24.53, 24.62c, 24.64c.

KEY SKILLS

[1] Converting a compound to a Fischer projection formula (24.2A); example: (S)-3-hydroxy-2-methylpropanal

1 Rotate and re-draw the molecule to place the horizontal bonds in front of the plane and the vertical bonds behind the plane.	2 Use a cross to represent the stereogenic center.
Tip **red** bonds in the plane forward. • **Horizontal bonds** come **forward**. • **Vertical bonds** go **back**.	Fischer projection

See Sample Problem 24.1.

[2] Re-drawing a Fischer projection, and labeling the stereogenic center as R or S (24.2A)

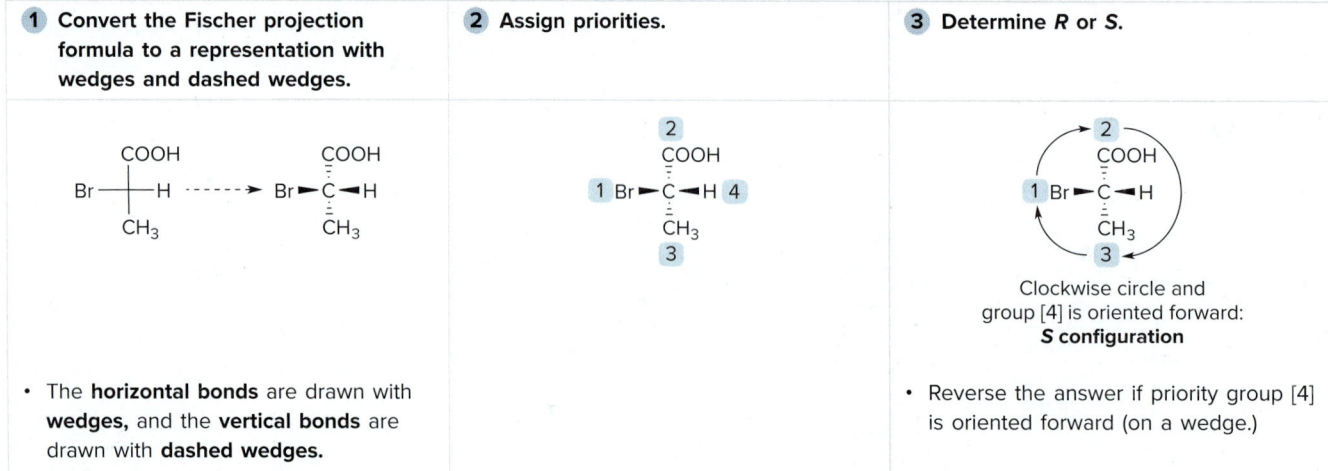

1 Convert the Fischer projection formula to a representation with wedges and dashed wedges.	2 Assign priorities.	3 Determine R or S.
• The **horizontal bonds** are drawn with **wedges**, and the **vertical bonds** are drawn with **dashed wedges**.		Clockwise circle and group [4] is oriented forward: **S configuration** • Reverse the answer if priority group [4] is oriented forward (on a wedge.)

See Sample Problem 24.2. Try Problem 24.36.

[3] Drawing a Haworth projection from an acyclic aldohexose (24.6); example: D-glucose

1 Place the **O atom** in the upper right corner of a hexagon, and add the **CH₂OH** on the first carbon counterclockwise from the **O atom**.

2 Place the anomeric carbon on the first carbon clockwise from the **O atom**.

3 Add the substituents at the three remaining stereogenic centers clockwise around the ring.

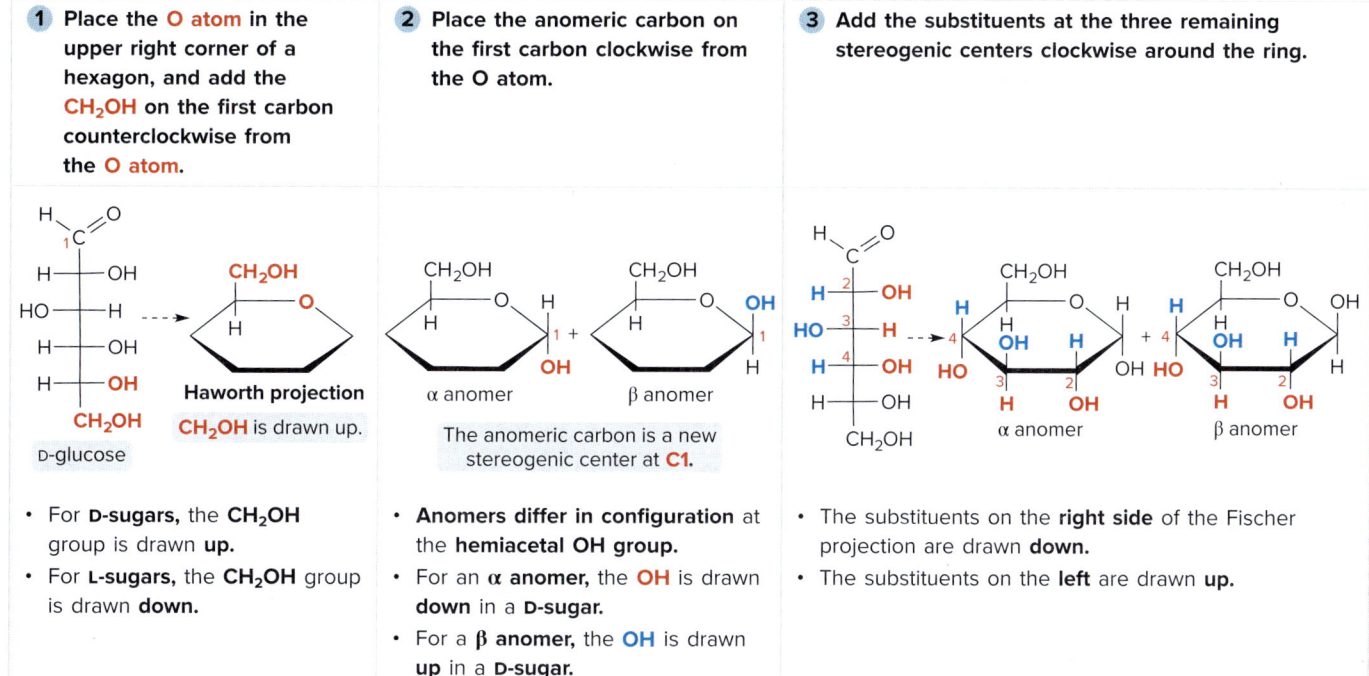

- For **D-sugars**, the **CH₂OH** group is drawn **up**.
- For **L-sugars**, the **CH₂OH** group is drawn **down**.

- **Anomers differ in configuration** at the **hemiacetal OH group**.
- For an **α anomer**, the **OH** is drawn **down** in a D-sugar.
- For a **β anomer**, the **OH** is drawn **up** in a D-sugar.

- The substituents on the **right side** of the Fischer projection are drawn **down**.
- The substituents on the **left** are drawn **up**.

See *How To*, p. 1082. Try Problems 24.39, 24.40a, 24.49a, 24.50a.

[4] Converting a Haworth projection to its acyclic form (24.6B); example: D-glucose

1 Draw the carbon skeleton, placing the **CHO** on the top and the **CH₂OH** on the bottom.

2 Classify the sugar as D- or L-.

3 Add the three other stereogenic centers.

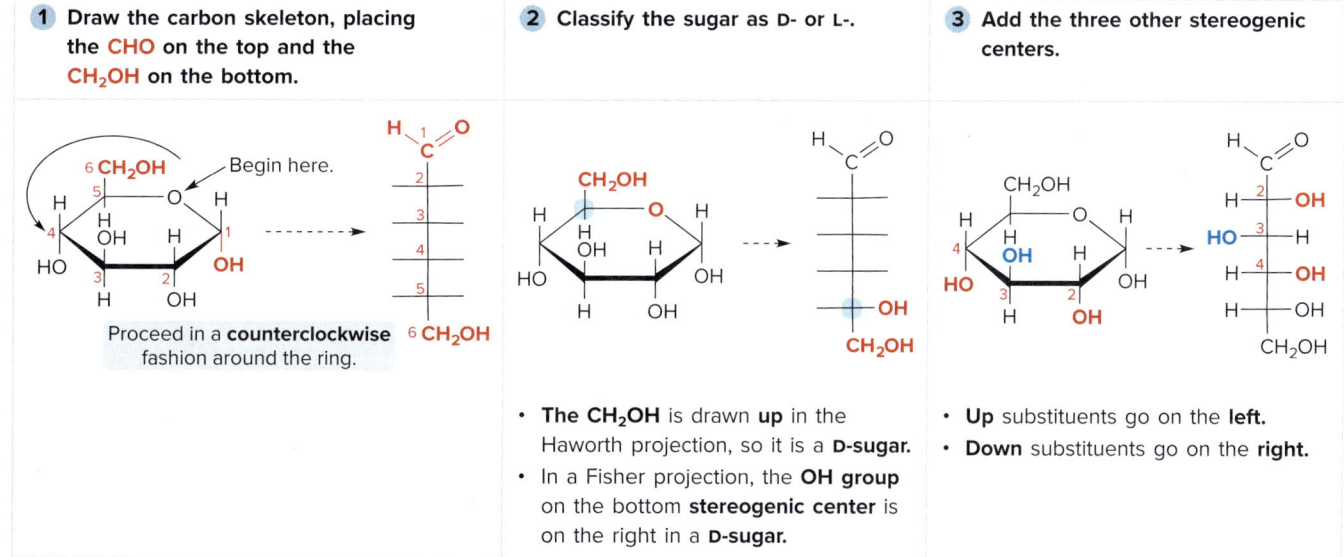

- The **CH₂OH** is drawn **up** in the Haworth projection, so it is a **D-sugar**.
- In a Fisher projection, the **OH group** on the bottom **stereogenic center** is on the right in a **D-sugar**.

- **Up** substituents go on the **left**.
- **Down** substituents go on the **right**.

See Sample Problem 24.4. Try Problems 24.35a, 24.42.

[5] Converting a Haworth projection to a chair form (24.6C); example: D-glucose

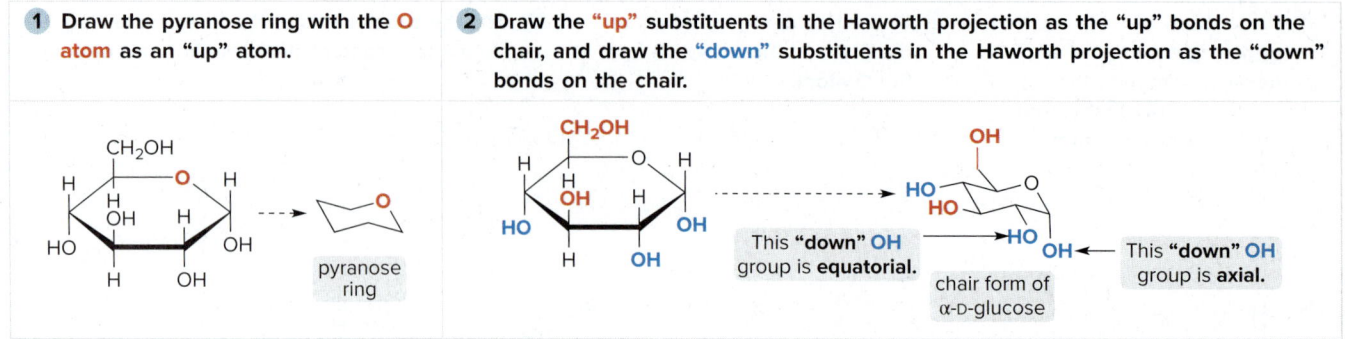

Try Problems 24.41, 24.49b, 24.50b.

[6] Determining the structure of an unknown D-aldopentose given a set of facts (24.9–24.10)

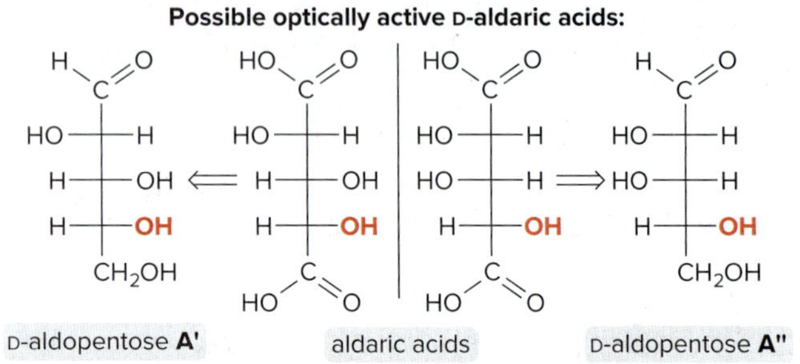

See Sample Problem 24.5. Try Problems 24.57–24.59.

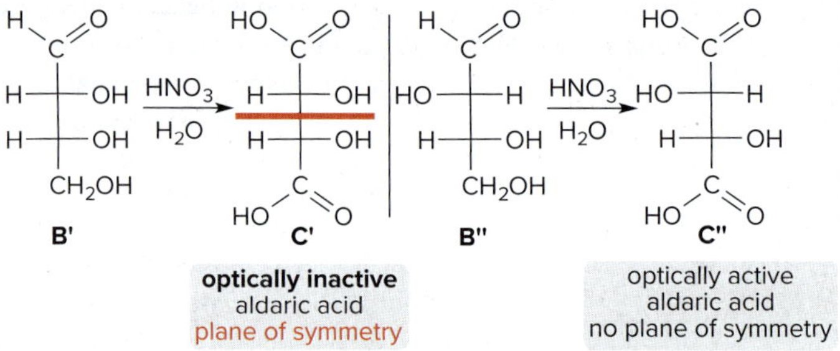

PROBLEMS

Problems Using Three-Dimensional Models

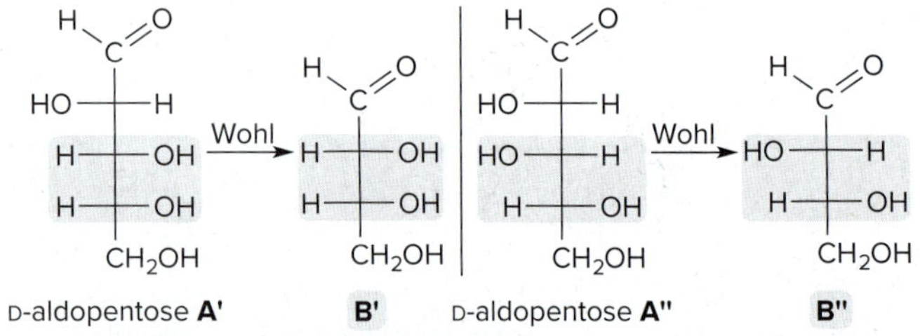

24.34 Convert each ball-and-stick model to a Fischer projection.

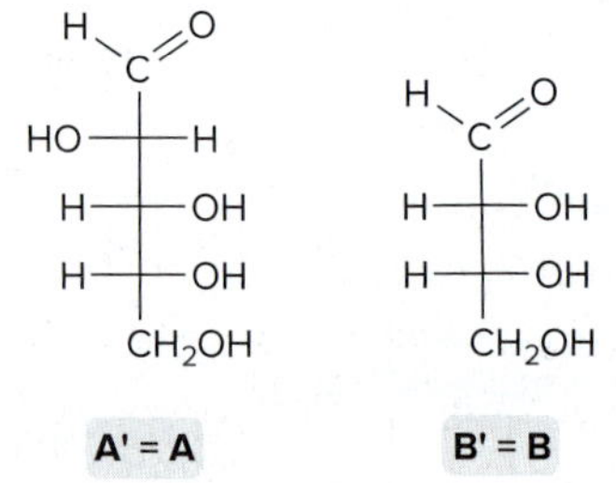

a. 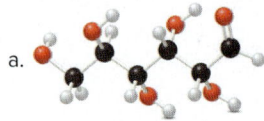   b.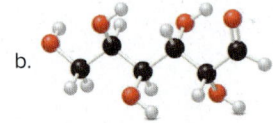

24.35 (a) Convert each cyclic monosaccharide to a Fischer projection of its acyclic form. (b) Name each monosaccharide. (c) Label the anomer as α or β.

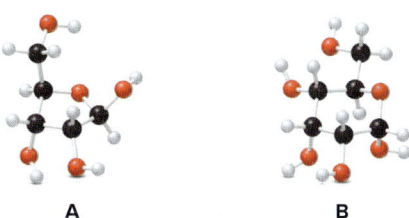

A   B

Fischer Projections

24.36 Convert each compound to a Fischer projection, and label each stereogenic center as R or S.

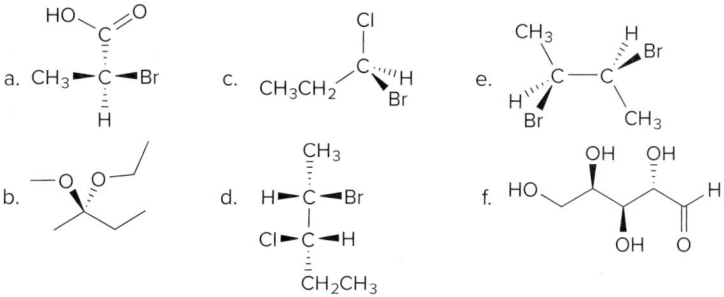

Monosaccharide Structure and Stereochemistry

24.37 For D-arabinose:
a. Draw its enantiomer.
b. Draw an epimer at C3.
c. Draw a diastereomer that is not an epimer.
d. Draw a constitutional isomer that still contains a carbonyl group.

24.38 Consider the following six compounds (**A–F**).

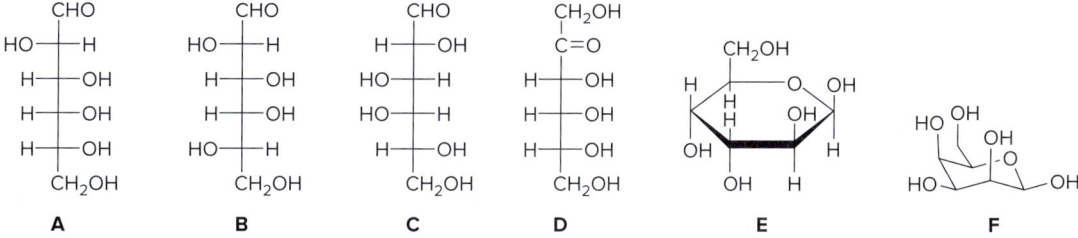

How are the two compounds in each pair related? Choose from enantiomers, epimers, diastereomers but not epimers, constitutional isomers, and identical compounds.
a. **A** and **B**   b. **A** and **C**   c. **B** and **C**   d. **A** and **D**   e. **E** and **F**

24.39 Draw a Haworth projection for each compound using the structures in Figures 24.4 and 24.5.
a. β-D-talopyranose   b. α-D-galactopyranose   c. α-D-tagatofuranose

24.40 Draw the structure of each compound and name it using the information in Figure 24.4.
a. the α anomer of a monosaccharide that is epimeric with D-glucose at C4 using a Haworth projection
b. the β anomer of a monosaccharide that is epimeric with D-gulose at C2 using a chair pyranose

24.41 Draw both pyranose anomers of each aldohexose using a three-dimensional representation with a chair pyranose. Label each anomer as α or β.

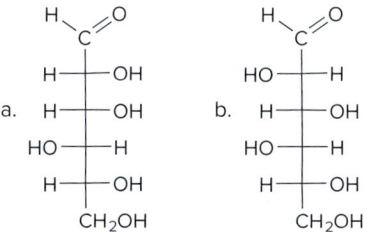

24.42 Convert each cyclic monosaccharide to its acyclic form.

24.43 The most stable conformation of the pyranose ring of most D-aldohexoses places the largest group, CH₂OH, in the equatorial position. An exception to this is the aldohexose D-idose. Draw the two possible chair conformations of either the α or β anomer of D-idose. Explain why the more stable conformation has the CH₂OH group in the axial position.

Monosaccharide Reactions

24.44 Draw the structure (including stereochemistry) of the cyclic hemiacetal(s) formed when each hydroxy carbonyl compound is treated with aqueous acid.

24.45 Draw the products formed when α-D-gulose is treated with each reagent.
 a. CH₃I, Ag₂O
 b. CH₃OH, HCl
 c. Ac₂O, pyridine
 d. The product in (a), then H₃O⁺
 e. The product in (b), then Ac₂O, pyridine
 f. The product in (d), then C₆H₅CH₂Cl, Ag₂O

24.46 Draw the products formed when D-altrose is treated with each reagent.
 a. (CH₃)₂CHOH, HCl
 b. NaBH₄, CH₃OH
 c. Br₂, H₂O
 d. HNO₃, H₂O
 e. [1] NH₂OH; [2] (CH₃CO)₂O, NaOCOCH₃; [3] NaOCH₃
 f. [1] NaCN, HCl; [2] H₂, Pd-BaSO₄; [3] H₃O⁺
 g. CH₃I, Ag₂O
 h. C₆H₅CH₂NH₂, mild H⁺

24.47 What aglycon and monosaccharides are formed when salicin and solanine (Section 24.7C) are each hydrolyzed with aqueous acid?

24.48 Draw a Fischer projection of the monosaccharide from which each of the following glycosides was prepared.

24.49 Answer the following questions about monosaccharide **A**.
 a. Draw the α anomer of **A** in a Haworth projection.
 b. Draw the β anomer of **A** in a three-dimensional representation using a chair conformation.
 c. What two aldoses yield **A** in a Wohl degradation?
 d. What product is formed when **A** undergoes a Wohl degradation?
 e. What product is formed when **A** reacts with Ag₂O in NH₄OH?

24.50 Answer the following questions about monosaccharide **B**.
 a. Draw the β anomer of **B** in a Haworth projection.
 b. Draw the α anomer of **B** in a three-dimensional representation using a chair conformation.
 c. What products are formed when **B** undergoes the Kiliani–Fischer synthesis?
 d. What product is formed when **B** is treated with NaBH₄ in CH₃OH?
 e. Draw the disaccharide formed when two molecules of **B** are joined by a 1→4-β-glycosidic linkage.

24.51 Draw the structure of two different aldohexoses that yield the following aldaric acid when oxidized with HNO₃. Use Figure 24.4 to name each aldohexose.

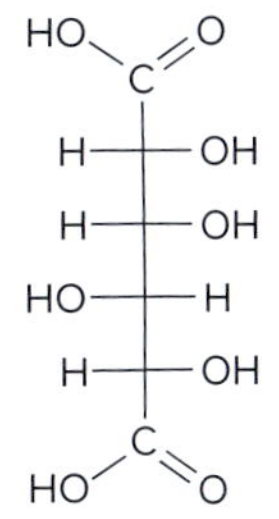

```
       HO    O
         \\ //
          C
    H ── OH
    H ── OH
    HO ── H
    H ── OH
          C
         // \\
       HO    O
```

24.52 What products are formed when each compound undergoes a Kiliani–Fischer synthesis?

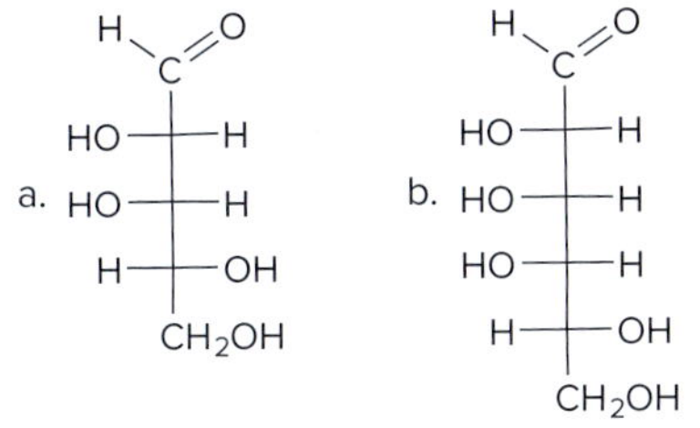

a.
```
       H   O
        \\//
         C
   HO ── H
   HO ── H
    H ── OH
       CH₂OH
```

b.
```
       H   O
        \\//
         C
   HO ── H
   HO ── H
   HO ── H
    H ── OH
       CH₂OH
```

24.53 What products are formed when each compound is treated with aqueous acid?

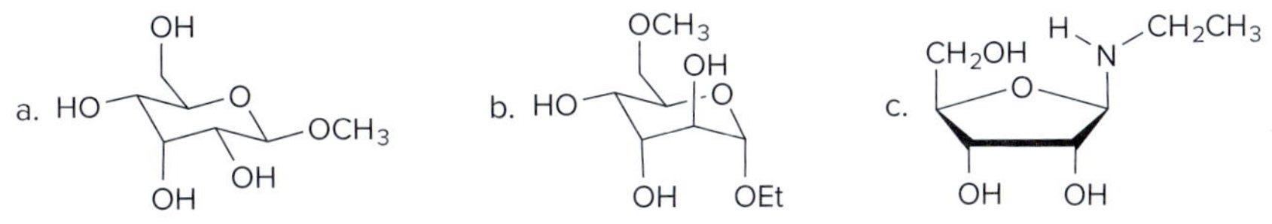

a. (pyranose with OH, HO, OH, OH, OCH₃ substituents)

b. (furanose with OCH₃, OH, HO, OH, OEt substituents)

c. (furanose with CH₂OH, N(H)(CH₂CH₃), OH, OH substituents)

Mechanisms

24.54 Draw a stepwise mechanism for the following reaction.

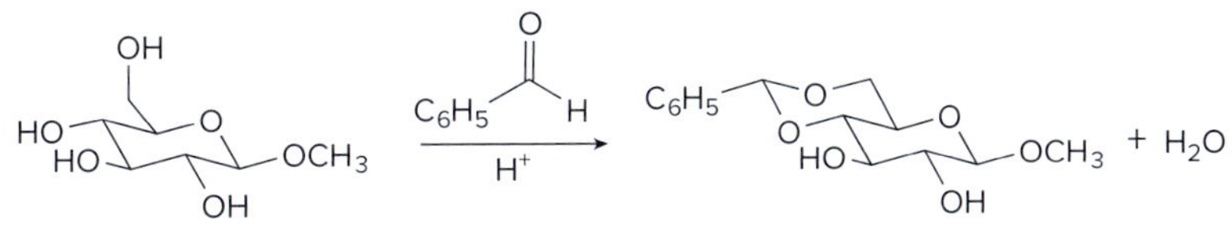

Sugar with OH, HO, HO, OH, OCH₃ + C₆H₅CHO →[H⁺] benzylidene acetal product + H₂O

24.55 Draw a stepwise mechanism for the following hydrolysis.

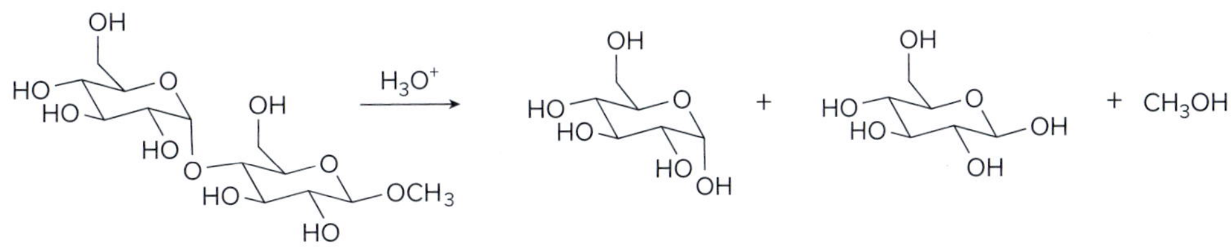

Disaccharide with OCH₃ glycoside →[H₃O⁺] two monosaccharides + CH₃OH

24.56 The following isomerization reaction, drawn using D-glucose as starting material, occurs with all aldohexoses in the presence of base. Draw a stepwise mechanism that illustrates how each compound is formed.

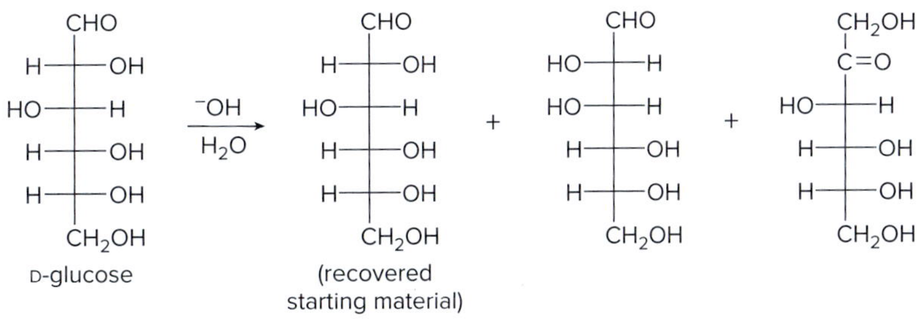

```
    CHO                    CHO                  CHO                 CH₂OH
H ── OH                H ── OH              HO ── H                 C=O
HO ── H     ⁻OH        HO ── H              HO ── H            HO ── H
H ── OH   ─────→       H ── OH       +      H ── OH       +    H ── OH
H ── OH     H₂O        H ── OH              H ── OH            H ── OH
   CH₂OH                  CH₂OH                CH₂OH              CH₂OH
 D-glucose             (recovered
                     starting material)
```

Identifying Monosaccharides

24.57 Which D-aldopentose is oxidized to an optically active aldaric acid and undergoes the Wohl degradation to yield a D-aldotetrose that is oxidized to an optically active aldaric acid?

24.58 Identify compounds **A–D**. A D-aldopentose **A** is oxidized with HNO_3 to an optically inactive aldaric acid **B**. **A** undergoes the Kiliani–Fischer synthesis to yield **C** and **D**. **C** is oxidized to an optically active aldaric acid. **D** is oxidized to an optically inactive aldaric acid.

24.59 A D-aldopentose **A** is reduced to an optically active alditol. Upon Kiliani–Fischer synthesis, **A** is converted to two D-aldohexoses, **B** and **C**. **B** is oxidized to an optically inactive aldaric acid. **C** is oxidized to an optically active aldaric acid. What are the structures of **A–C**?

Disaccharides and Polysaccharides

24.60 Draw the structure of a disaccharide formed from two mannose units joined by a 1→4-α-glycosidic linkage.

24.61
a. Identify the glycosidic linkage in disaccharide **C**, classify the glycosidic bond as α or β, and use numbers to designate its location.
b. Identify the lettered compounds in the following reaction.

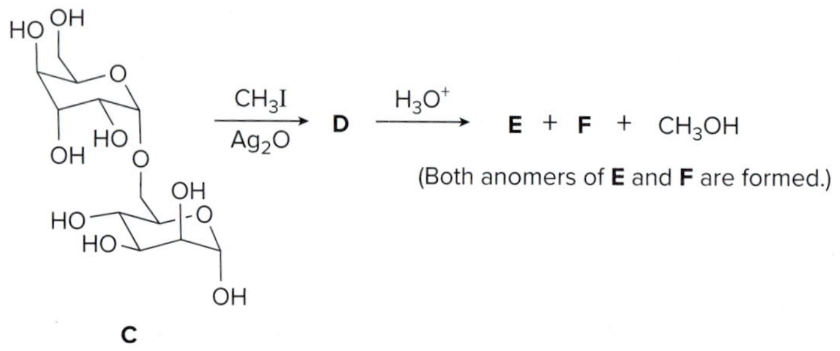

$$C \xrightarrow{CH_3I, Ag_2O} D \xrightarrow{H_3O^+} E + F + CH_3OH$$

(Both anomers of **E** and **F** are formed.)

24.62 Consider the tetrasaccharide stachyose drawn below. Stachyose is found in white jasmine, soybeans, and lentils. Because humans cannot digest it, its consumption causes flatulence.

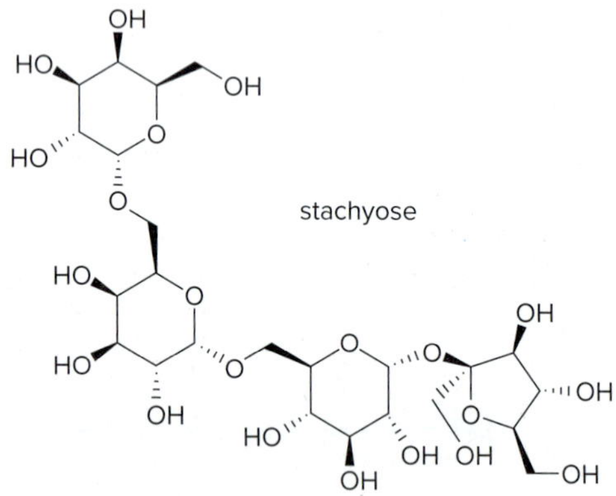

a. Label all glycoside bonds.
b. Classify each glycosidic linkage as α or β and use numbers to designate its location between two rings (e.g., 1→4-β).
c. What products are formed when stachyose is hydrolyzed with H_3O^+?
d. Is stachyose a reducing sugar?
e. What product is formed when stachyose is treated with excess CH_3I, Ag_2O?
f. What products are formed when the product in (e) is treated with H_3O^+?

24.63 Deduce the structure of the disaccharide isomaltose from the following data.

[1] Hydrolysis yields D-glucose exclusively.
[2] Isomaltose is cleaved with α-glycosidase enzymes.
[3] Isomaltose is a reducing sugar.
[4] Methylation with excess CH_3I, Ag_2O and then hydrolysis with H_3O^+ forms two products:

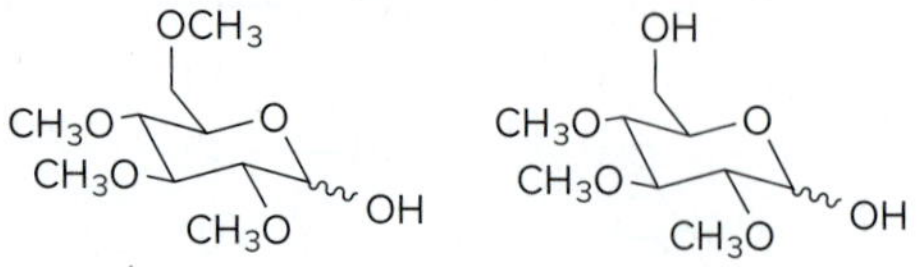

(Both anomers are present.)

24.64 Draw the structure of each of the following compounds.

a. a polysaccharide formed by joining D-glucosamine in 1→6-α-glycosidic linkages

b. a disaccharide formed by joining D-mannose and D-glucose in a 1→4-β-glycosidic linkage using mannose's anomeric carbon

c. an α-*N*-glycoside formed from D-arabinose and $C_6H_5CH_2NH_2$

Challenge Problems

24.65 (a) Draw the more stable chair form of fucose, an essential monosaccharide needed in the diet and a component of carbohydrates on mammalian and plant cell surfaces. (b) Classify fucose as a D- or L-monosaccharide. (c) What two structural features are unusual in fucose?

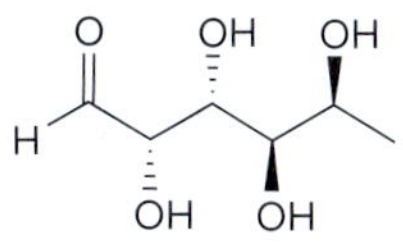

fucose

24.66 Draw a stepwise mechanism for the following reaction.

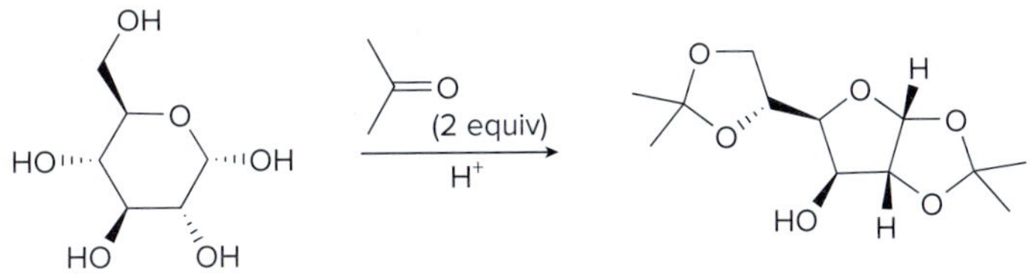

25 Lipids

Iconotec/Glowimages

- 25.1 Introduction
- 25.2 Waxes
- 25.3 Triacylglycerols
- 25.4 Phospholipids
- 25.5 Fat-soluble vitamins
- 25.6 Eicosanoids
- 25.7 Terpenes
- 25.8 Steroids

Stearic acid, a saturated fatty acid first discussed in Section 3.9D, is a key component of cocoa butter, a mixture of triacylglycerols obtained from the cocoa bean that is used to make chocolate. Because cocoa butter is high in saturated fatty acid content, cocoa butter is a solid at room temperature, so it is classified as a fat. Triacylglycerols are the most abundant lipids, water-insoluble biomolecules that contain many carbon–carbon and carbon–hydrogen bonds and few functional groups. In Chapter 25, we learn about the many different types of lipids.

Why Study... Lipids?

In Chapter 25, we turn our attention to **lipids,** biomolecules that are soluble in organic solvents. Unlike the carbohydrates in Chapter 24 and the amino acids and proteins in Chapter 23, lipids contain many carbon–carbon and carbon–hydrogen bonds and few functional groups.

Lipids are the biomolecules that most closely resemble the hydrocarbons we studied in Chapters 4 and 10, so we have already learned many facts that directly explain their properties. Because there is no one functional group that is present in all lipids, however, the chemistry of lipids draws upon knowledge learned in many prior chapters.

25.1 Introduction

The word *lipid* comes from the Greek word *lipos* for "fat."

- *Lipids* are biomolecules that are soluble in organic solvents.

Lipids are unique among organic molecules because their identity is defined on the basis of a *physical property* and not by the presence of a particular functional group. Because of this, lipids come in a wide variety of structures and they have many different functions in the cell. Three examples are given in Figure 25.1.

The large number of **carbon–carbon and carbon–hydrogen σ bonds in lipids makes them very soluble in organic solvents and insoluble in water.** Monosaccharides (from which carbohydrates are formed) and amino acids (from which proteins are formed), on the other hand, are very polar, so they tend to be water soluble. Because lipids share many properties with hydrocarbons, several features of lipid structure and properties have already been discussed. Table 25.1 summarizes sections of the text where aspects of lipid chemistry were covered previously.

Table 25.1 Summary of Lipid Chemistry Discussed Prior to Chapter 25

Topic	Section	Topic	Section
• Vitamin A	3.5	• Steroid synthesis	12.14
• Soap	3.6	• Prostaglandins	15.5
• Phospholipids, the cell membrane	3.7	• Lipid hydrolysis	16.11A
• Fatty acids and triacylglycerols	3.9D	• Soap	16.11B
• Leukotrienes	9.17	• Cholesteryl esters	16.16
• Oral contraceptives	10.5	• Steroid synthesis	18.9
• Fats and oils	10.6	• Lipid oxidation	21.10
• Hydrogenation of oils	11.4	• Vitamin E	21.11

Figure 25.1 Three examples of lipids

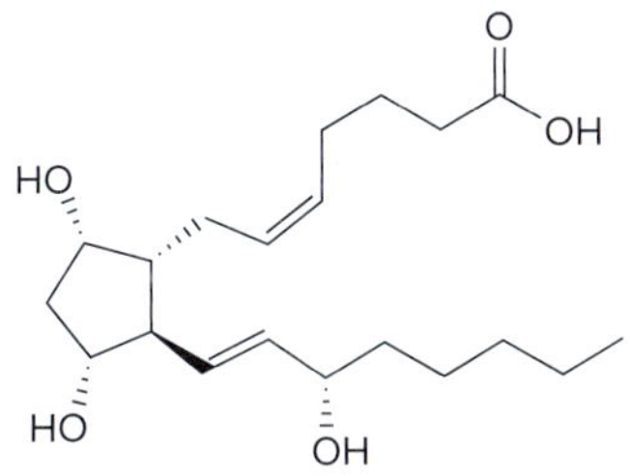

PGF$_{2\alpha}$
a prostaglandin

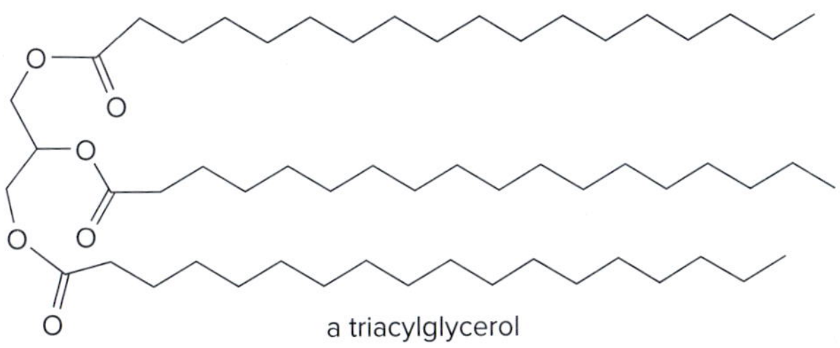

a triacylglycerol

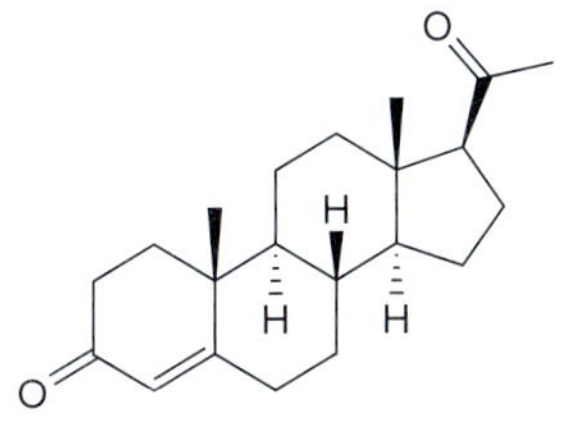

progesterone
a steroid

- All lipids have many C—C and C—H bonds, but there is no one functional group common to all lipids.

Lipids can be categorized as hydrolyzable or nonhydrolyzable.

[1] **Hydrolyzable lipids** **can be cleaved into smaller molecules by hydrolysis with water.** Most hydrolyzable lipids contain an ester unit. We will examine three subgroups: waxes, triacylglycerols, and phospholipids.

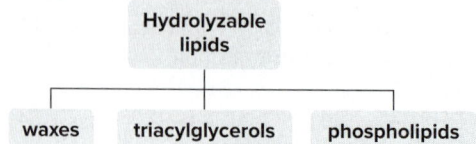

[2] **Nonhydrolyzable lipids** **cannot be cleaved into smaller units by aqueous hydrolysis.** Nonhydrolyzable lipids tend to be more varied in structure. We will examine four different types: fat-soluble vitamins, eicosanoids, terpenes, and steroids.

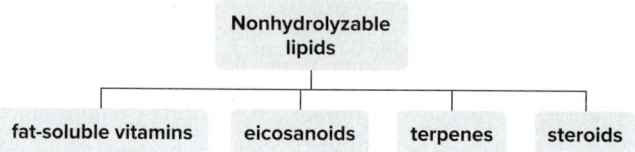

Water beads up on the surface of a leaf because of the leaf's waxy coating. *Daniel C. Smith*

25.2 Waxes

Waxes are the simplest hydrolyzable lipids. **Waxes are esters (RCOOR') formed from a high-molecular-weight alcohol (R'OH) and a fatty acid (RCOOH).**

Because of their long hydrocarbon chains, **waxes are very hydrophobic.** They form a protective coating on the feathers of birds to make them water repellent, and on leaves to prevent water evaporation. **Lanolin,** a wax composed of a complex mixture of high-molecular-weight esters, coats the wool fibers of sheep. **Spermaceti wax,** isolated from the heads of sperm whales, is largely $CH_3(CH_2)_{14}COO(CH_2)_{15}CH_3$. The three-dimensional structure of this compound shows how small the ester group is compared to the long hydrocarbon chains.

The seeds of the jojoba plant grown in the southwestern United States are rich in waxes used in cosmetics and personal care products (Problem 25.1). *Source: USDA, ARS, National Genetic Resources Program*

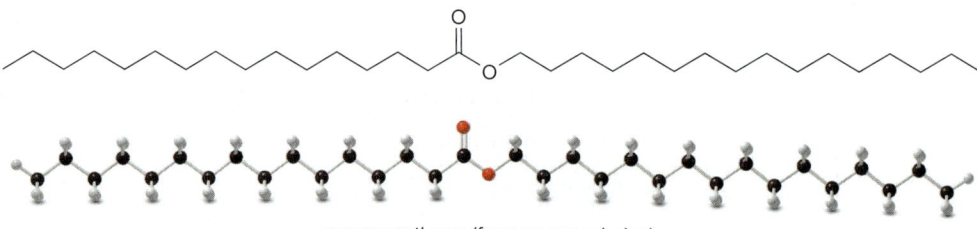

spermaceti wax (from sperm whales)

Problem 25.1 One component of jojoba oil is a wax formed from eicosenoic acid [$CH_3(CH_2)_7CH=CH(CH_2)_9CO_2H$] and $CH_3(CH_2)_7CH=CH(CH_2)_8OH$. Draw the structure of the wax, including the cis geometry of both carbon–carbon double bonds.

25.3 Triacylglycerols

Triacylglycerols, or triglycerides, are the most abundant lipids, and for this reason we have already discussed many of their properties in earlier sections of this text.

- Triacylglycerols are triesters that produce glycerol and three molecules of fatty acid upon hydrolysis.

25.3 Triacylglycerols

[Reaction scheme: triacylglycerol + H₂O (H⁺ or ⁻OH) or enzymes → glycerol + three fatty acids containing 12–20 C's are formed as products.]

triacylglycerol
the most common type of lipid

Simple triacylglycerols are composed of three identical fatty acid side chains, whereas **mixed triacylglycerols** have two or three different fatty acids. Table 25.2 lists the most common fatty acids used to form triacylglycerols.

Table 25.2 The Most Common Fatty Acids in Triacylglycerols

Number of C atoms	Number of C=C bonds	Structure	Name	Mp (°C)
Saturated fatty acids				
12	0	$CH_3(CH_2)_{10}CO_2H$	lauric acid	44
14	0	$CH_3(CH_2)_{12}CO_2H$	myristic acid	58
16	0	$CH_3(CH_2)_{14}CO_2H$	palmitic acid	63
18	0	$CH_3(CH_2)_{16}CO_2H$	stearic acid	69
20	0	$CH_3(CH_2)_{18}CO_2H$	arachidic acid	77
Unsaturated fatty acids				
16	1	$CH_3(CH_2)_5CH=CH(CH_2)_7CO_2H$	palmitoleic acid	1
18	1	$CH_3(CH_2)_7CH=CH(CH_2)_7CO_2H$	oleic acid	4
18	2	$CH_3(CH_2)_4(CH=CHCH_2)_2(CH_2)_6CO_2H$	linoleic acid	−5
18	3	$CH_3CH_2(CH=CHCH_2)_3(CH_2)_6CO_2H$	linolenic acid	−11
20	4	$CH_3(CH_2)_4(CH=CHCH_2)_4(CH_2)_2CO_2H$	arachidonic acid	−49

Line structures of stearic, oleic, linoleic, and linolenic acids can be found in Table 10.1. Ball-and-stick models of these fatty acids are shown in Figure 10.4.

The most common saturated fatty acids are palmitic and stearic acids. The most common unsaturated fatty acid is oleic acid.

Linoleic and linolenic acids are called **essential fatty acids** because we cannot synthesize them and must acquire them in our diets.

Unlike other vegetable oils, oils from palm and coconut trees are very high in saturated fats. Considerable evidence currently suggests that diets high in saturated fats lead to a greater risk of heart disease. For this reason, the demand for coconut and palm oils has decreased in recent years, and some coconut plantations previously farmed in the South Pacific are no longer in commercial operation. *Phiseksit/Shutterstock*

What are the characteristics of these fatty acids?

- All fatty acid chains are unbranched, but they may be saturated or unsaturated.
- Naturally occurring fatty acids have an even number of carbon atoms.
- Double bonds in naturally occurring fatty acids generally have the *Z* configuration.
- The melting point of a fatty acid depends on the degree of unsaturation.

Fats and oils are triacylglycerols; that is, they are triesters of glycerol and these fatty acids.

- Fats have *higher* melting points, making them solids at room temperature.
- Oils have *lower* melting points, making them liquids at room temperature.

This melting point difference correlates with the number of degrees of unsaturation present in the fatty acid side chains. **As the number of double bonds *increases*, the melting point *decreases*, as it does for the constituent fatty acids as well.**

Three-dimensional structures of a saturated and unsaturated triacylglycerol are shown in Figure 25.2. With no double bonds, the three side chains of the saturated lipid lie parallel to each other, making it possible for this compound to pack relatively efficiently in a crystalline lattice, thus leading to a high melting point. In the unsaturated lipid, however, a single *Z* double bond places a kink in the side chain, making it more difficult to pack efficiently in the solid state, thus leading to a lower melting point.

Figure 25.2 Three-dimensional structures of a saturated and unsaturated triacylglycerol

a. A saturated triacylglycerol

- Three saturated side chains lie parallel to each other, making a compact lipid.

b. An unsaturated triacylglycerol

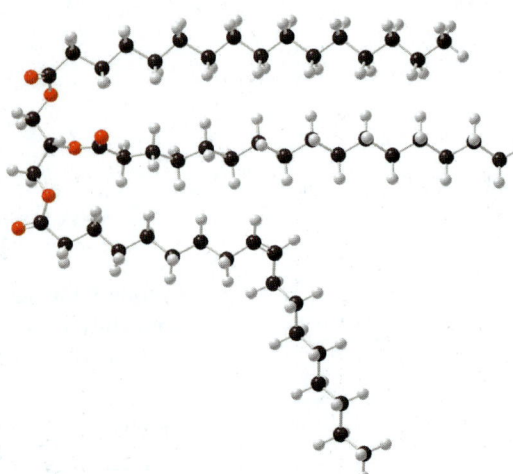

- One Z double bond in a fatty acid side chain produces a twist so that the lipid is less compact.

Fish oils, such as cod liver and herring oils, are very rich in polyunsaturated triacylglycerols. These triacylglycerols pack so poorly that they have very low melting points; thus, they remain liquids even in the cold water inhabited by these fish. *Pixtal/AGE Fotostock*

Solid fats have a relatively high percentage of saturated fatty acids and are generally of animal origin. **Liquid oils have a higher percentage of unsaturated fatty acids** and are generally of vegetable origin. Table 25.3 lists the fatty acid composition of some common fats and oils.

Table 25.3 Fatty Acid Composition of Some Fats and Oils

Source	% Saturated fatty acids	% Oleic acid	% Linoleic acid
beef	49–62	37–43	2–3
milk	37	33	3
coconut	86	7	—
corn	11–16	19–49	34–62
olive	11	84	4
palm	43	40	8
safflower	9	13	78
soybean	15	20	52

Sources: Data from *Merck Index*, 10th ed. Rahway, NJ: Merck and Co.; and Wilson et al., 1967, *Principles of Nutrition*, 2nd ed. New York: Wiley.

The hydrolysis, hydrogenation, and oxidation of triacylglycerols—reactions originally discussed in Chapters 11, 16, and 21—are summarized here for your reference.

[1] Hydrolysis of triacylglycerols (Section 16.11A)

Three ester units are cleaved.

Hydrolysis of a triacylglycerol with water in the presence of either acid, base, or an enzyme yields glycerol and three fatty acids. This cleavage reaction follows the same mechanism as any other ester hydrolysis (Section 16.10). This reaction is the first step in triacylglycerol metabolism.

[2] Hydrogenation of unsaturated fatty acids (Section 11.4)

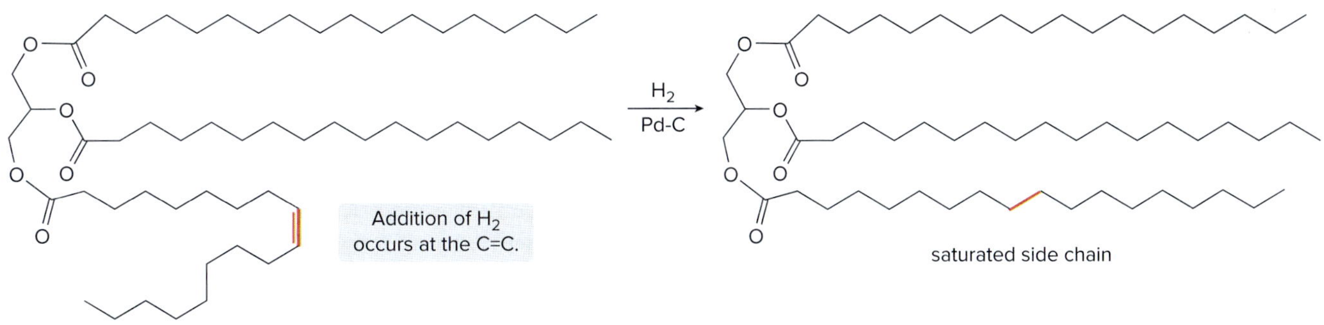

Addition of H_2 occurs at the C=C.

saturated side chain

The double bonds of an unsaturated fatty acid can be hydrogenated by using H_2 in the presence of a transition metal catalyst. Hydrogenation converts a liquid oil to a solid fat. This process, sometimes called **hardening**, is used to prepare margarine from vegetable oils.

[3] Oxidation of unsaturated fatty acids (Section 21.10)

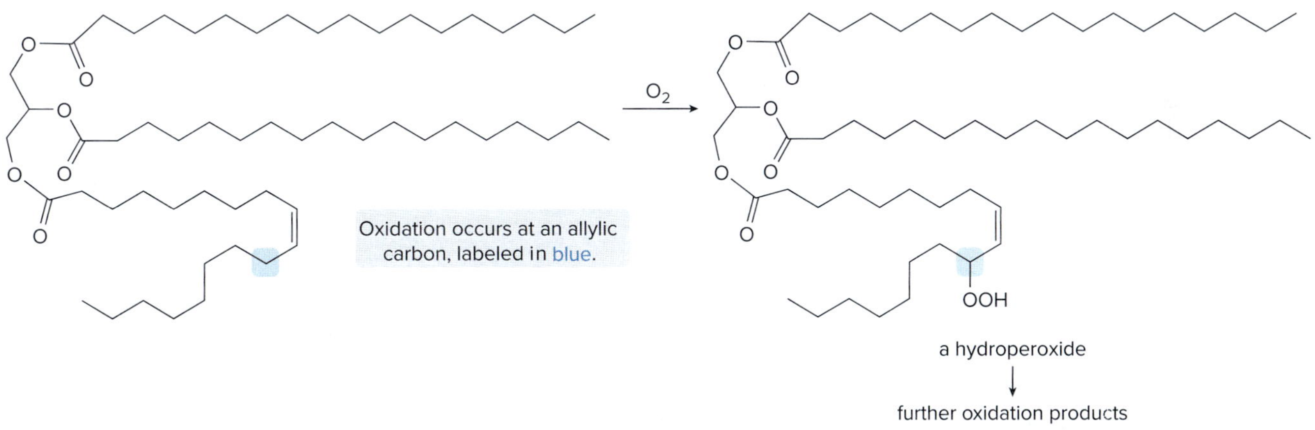

Oxidation occurs at an allylic carbon, labeled in blue.

a hydroperoxide
↓
further oxidation products

Allylic C—H bonds are weaker than other C—H bonds and are thus susceptible to oxidation with molecular oxygen by a radical process. The hydroperoxide formed by this process is unstable, and it undergoes further oxidation to products that often have a disagreeable odor. This oxidation process turns an oil rancid.

In the cell, the principal function of triacylglycerols is energy storage. Complete metabolism of a triacylglycerol yields CO_2 and H_2O, and a great deal of energy. This overall reaction is reminiscent of the combustion of alkanes in fossil fuels, a process that also yields CO_2 and H_2O and provides energy to heat homes and power automobiles (Section 4.14B). Fundamentally both processes convert C—C and C—H bonds to C—O bonds, a highly exothermic reaction.

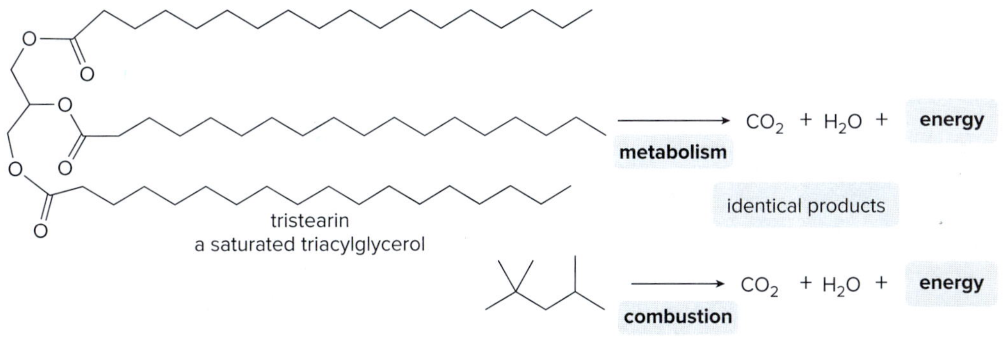

The average body fat content of men and women is ~20% and ~25%, respectively. (For elite athletes, however, the averages are more like < 10% for men and < 15% for women.) This stored fat can fill the body's energy needs for two or three months.

tristearin
a saturated triacylglycerol

$\xrightarrow{\text{metabolism}}$ CO_2 + H_2O + **energy**

identical products

2,2,4-trimethylpentane
(isooctane)
a component of gasoline

$\xrightarrow{\text{combustion}}$ CO_2 + H_2O + **energy**

Ryan McVay/Photodisc/Getty Images

Carbohydrates provide an energy boost, but only for the short term, such as during strenuous exercise. Our long-term energy needs are met by triacylglycerols, because they store ~38 kJ/g, whereas carbohydrates and proteins store only ~16 kJ/g.

Because triacylglycerols release heat on combustion, they can in principle be used as fuels for vehicles. In fact, coconut oil was used as a fuel during both World War I and World War II, when gasoline and diesel supplies ran short. Coconut oil is more viscous than petroleum products and freezes at 24 °C, so engines must be modified to use it and it can't be used in cold climates. Nonetheless, a limited number of trucks and boats can now use vegetable oils, sometimes blended with diesel, as a fuel source. When the price of crude oil is high, the use of these **biofuels** becomes economically attractive.

Problem 25.2 How would you expect the melting point of eicosapentaenoic acid [$CH_3CH_2(CH=CHCH_2)_5(CH_2)_2COOH$] to compare with the melting points of the fatty acids listed in Table 25.2?

Problem 25.3 Draw the products formed when triacylglycerol **A** is treated with each reagent. Rank compounds **A**, **B**, and **C** in order of increasing melting point.

a. H_2O, H^+
b. H_2 (excess), Pd-C → **B**
c. H_2 (1 equiv), Pd-C → **C**

A

Problem 25.4 The main fatty acid component of the triacylglycerols in coconut oil is lauric acid, $CH_3(CH_2)_{10}COOH$. Explain why coconut oil is a liquid at room temperature even though it contains a large fraction of this saturated fatty acid.

Problem 25.5 Unlike many fats and oils, the cocoa butter used to make chocolate is remarkably uniform in composition. All triacylglycerols contain oleic acid esterified to the 2° OH group of glycerol, and either palmitic acid or stearic acid esterified to the 1° OH groups. Draw the structures of two possible triacylglycerols that compose cocoa butter.

25.4 Phospholipids

Phospholipids are hydrolyzable lipids that contain a phosphorus atom. There are two common types of phospholipids: **phosphoacylglycerols** and **sphingomyelins**. Both classes are found almost exclusively in the cell membranes of plants and animals, as discussed in Section 3.7.

25.4A Phosphoacylglycerols

Phosphoacylglycerols (or phosphoglycerides) are the second most abundant type of lipid. They form the principal lipid component of most cell membranes. Their structure resembles that of the triacylglycerols of Section 25.3 with one important difference. In phosphoacylglycerols, only *two* of the hydroxy groups of glycerol are esterified with fatty acids. **The third OH group is part of a phosphodiester,** which is also bonded to another low-molecular-weight alcohol.

There are two prominent types of phosphoacylglycerols. They differ in the identity of the R" group in the phosphodiester.

- When R" = $CH_2CH_2NH_3^+$, the phosphoacylglycerol is called a phosphatidylethanolamine or cephalin.
- When R" = $CH_2CH_2N(CH_3)_3^+$, the phosphoacylglycerol is called a phosphatidylcholine or lecithin.

25.4 Phospholipids

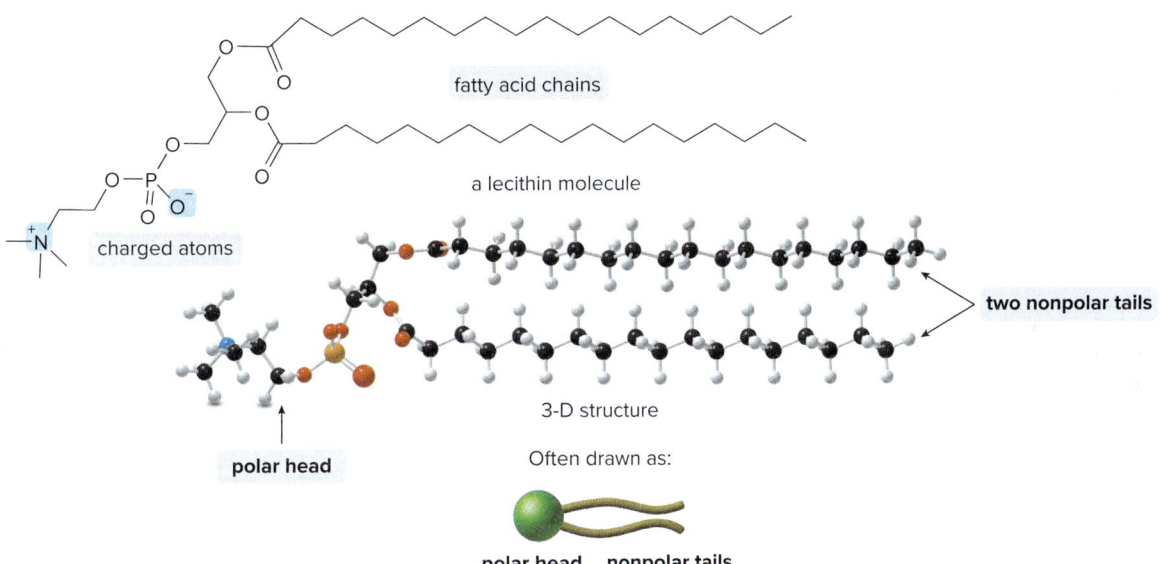

phosphatidylethanolamine or **cephalin**

phosphatidylcholine or **lecithin**

The middle carbon of the glycerol backbone of all of these compounds is a **stereogenic center** (labeled in blue), usually with the R configuration.

The phosphorus side chain of a phosphoacylglycerol makes it different from a triacylglycerol. **The two fatty acid side chains form two nonpolar "tails" that lie parallel to each other, whereas the phosphodiester end of the molecule is a charged or polar "head."** A three-dimensional structure of a phosphoacylglycerol is shown in Figure 25.3.

Figure 25.3 Three-dimensional structure of a phosphoacylglycerol

- A phosphoacylglycerol has two distinct regions: **two nonpolar tails** due to the long-chain fatty acids, and a **very polar head** from the charged phosphodiester.

As discussed in Section 3.7, when these phospholipids are mixed with water, they assemble in an arrangement called a **lipid bilayer. The ionic heads of the phospholipid are oriented on the outside and the nonpolar tails on the inside.** The identity of the fatty acids in the phospholipid determines the rigidity of this bilayer. When the fatty acids are saturated, they pack well in the interior of the lipid bilayer and the membrane is quite rigid. When there are many unsaturated fatty acids, the nonpolar tails cannot pack as well and the bilayer is more fluid. Thus, important characteristics of this lipid bilayer are determined by the three-dimensional structure of the molecules that compose it.

Cell membranes are composed of these lipid bilayers (see Figure 3.5). Proteins and cholesterol are embedded in the membranes as well, but the phospholipid bilayer forms the main fabric of the insoluble barrier that protects the cell.

Problem 25.6 Draw the structure of a lecithin containing oleic acid and palmitic acid as the fatty acid side chains.

Problem 25.7 Phosphoacylglycerols should remind you of soaps (Section 3.6). In what ways are these compounds similar?

25.4B Sphingomyelins

Sphingomyelins, the second major class of phospholipids, are derivatives of the amino alcohol **sphingosine,** in much the same way that triacylglycerols and phosphoacylglycerols are derivatives of glycerol. Other notable features of a sphingomyelin include:

- **A phosphodiester at C1.**
- **An amide formed with a fatty acid at C2.**

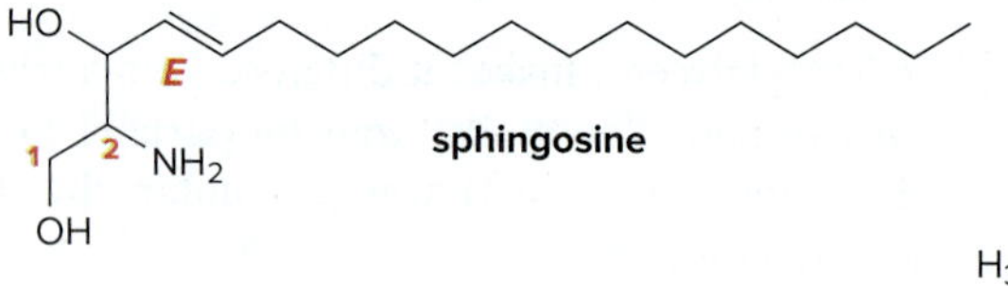

sphingosine

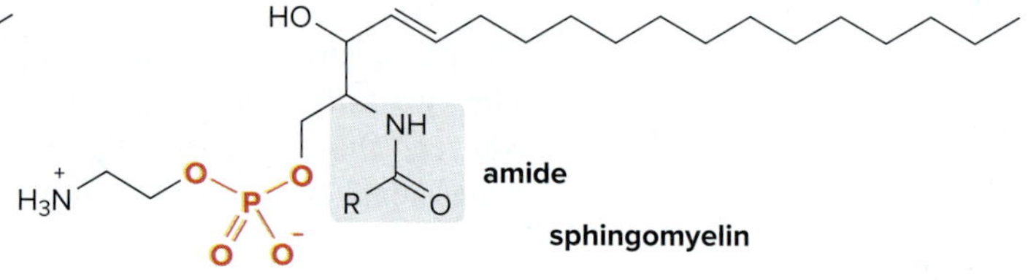

sphingomyelin

Like phosphoacylglycerols, **sphingomyelins are also a component of the lipid bilayer of cell membranes.** The coating that surrounds and insulates nerve cells, the **myelin sheath,** is particularly rich in sphingomyelins and is vital for proper nerve function. Deterioration of the myelin sheath as seen in multiple sclerosis leads to disabling neurological problems.

Figure 25.4 compares the structural features of the most common hydrolyzable lipids: a triacylglycerol, a phosphoacylglycerol, and a sphingomyelin.

Problem 25.8 Why are phospholipids, but not triacylglycerols, found in cell membranes?

Figure 25.4 A comparison of a triacylglycerol, a phosphoacylglycerol, and a sphingomyelin

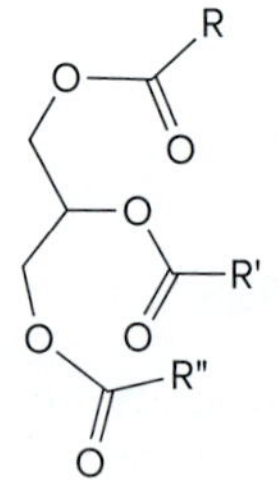

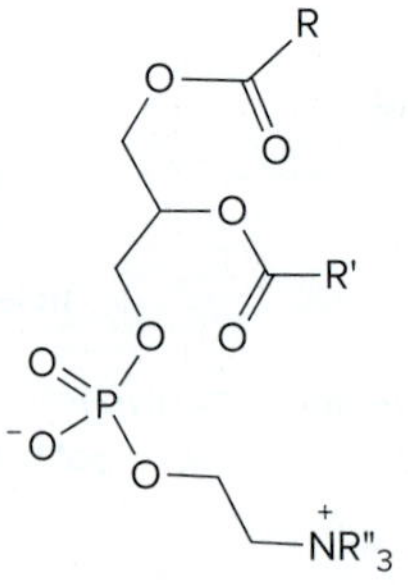

R" = H, CH_3

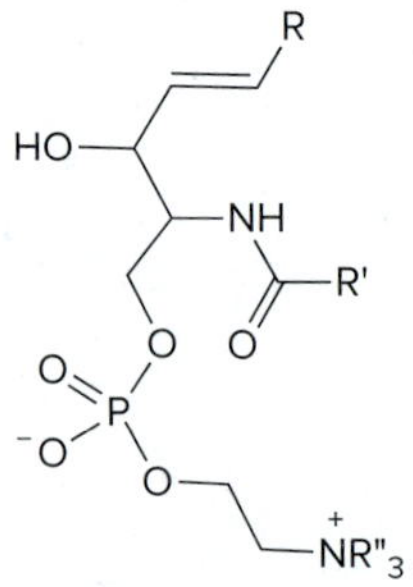

R = $CH_3(CH_2)_{12}$

R" = H, CH_3

- A **triacylglycerol** has three nonpolar side chains.
- The three OH groups of **glycerol** are esterified with three fatty acids.

- A **phosphoacylglycerol** has two nonpolar side chain tails and one ionic head.
- Two OH groups of **glycerol** are esterified with fatty acids.
- A **phosphodiester** is located on a terminal carbon.

- A **sphingomyelin** has two nonpolar side chain tails and one ionic head.
- A sphingomyelin is formed from **sphingosine,** not glycerol. One of the nonpolar tails is an **amide.**
- A **phosphodiester** is located on a terminal carbon.

25.5 Fat-Soluble Vitamins

Vitamins are organic compounds required in small quantities for normal metabolism (Section 3.5). Because our cells cannot synthesize these compounds, they must be obtained in the diet. Vitamins can be categorized as fat soluble or water soluble. **The fat-soluble vitamins are lipids.**

The four fat-soluble vitamins—**A, D, E,** and **K**—are found in fruits and vegetables, fish, liver, and dairy products. Although fat-soluble vitamins must be obtained from the diet, they do not have to be ingested every day. Excess vitamins are stored in fat cells, and then used when needed. Figure 25.5 shows the structure of these vitamins and summarizes their functions.

Figure 25.5 The fat-soluble vitamins

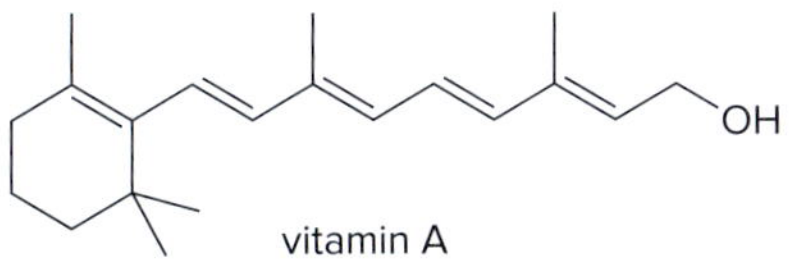

- **Vitamin A** (retinol, Section 3.5) is obtained from fish liver oils and dairy products, and is synthesized from β-carotene, the orange pigment in carrots.
- In the body, vitamin A is converted to 11-*cis*-retinal, the light-sensitive compound responsible for vision in all vertebrates (Section 14.13A). It is also needed for healthy mucous membranes.
- A deficiency of vitamin A causes night blindness, as well as dry eyes and skin.

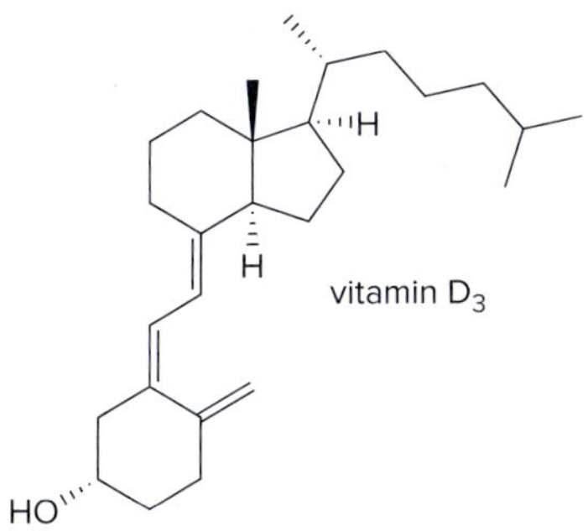

- **Vitamin D_3** is the most abundant of the D vitamins. Strictly speaking, it is not a vitamin because it can be synthesized in the body from cholesterol. Nevertheless, it is classified as such, and many foods (particularly milk) are fortified with vitamin D_3 so that we get enough of this vital nutrient.
- Vitamin D helps regulate both calcium and phosphorus metabolism.
- A deficiency of vitamin D causes rickets, a bone disease characterized by knock-knees, spinal curvature, and other deformities.

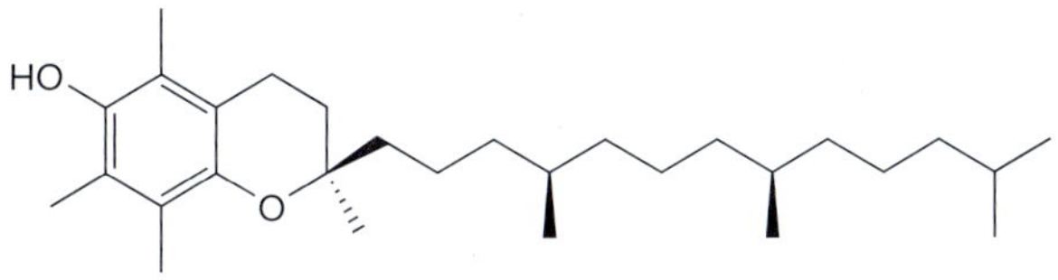

- The term **vitamin E** refers to a group of structurally similar compounds, the most potent being α-tocopherol (Section 21.11).
- Vitamin E is an antioxidant, so it protects unsaturated side chains in fatty acids from oxidation.
- A deficiency of vitamin E causes numerous neurologic problems.

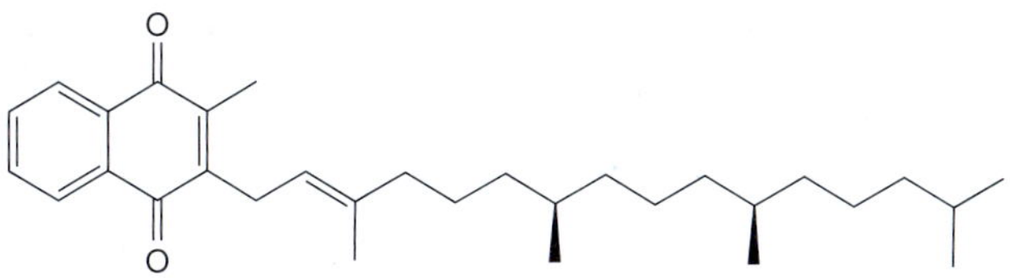

- **Vitamin K** (phylloquinone) regulates the synthesis of prothrombin and other proteins needed for blood to clot.
- A deficiency of vitamin K leads to excessive and sometimes fatal bleeding because of inadequate blood clotting.

Problem 25.9 Explain why regularly ingesting a large excess of a fat-soluble vitamin can lead to severe health problems, whereas ingesting a large excess of a water-soluble vitamin often causes no major health problems.

25.6 Eicosanoids

> The word *eicosanoid* is derived from the Greek word *eikosi*, meaning "20."

The **eicosanoids** are a group of biologically active compounds containing 20 carbon atoms derived from arachidonic acid. The **prostaglandins** (Section 15.5) and the **leukotrienes** (Section 9.17) are two types of eicosanoids. Two others are the **thromboxanes** and **prostacyclins**.

PGF$_{2\alpha}$
a prostaglandin

LTC$_4$
a leukotriene

TXA$_2$
a thromboxane

PGI$_2$
a prostacyclin

All eicosanoids are very potent compounds present in low concentration in cells. They are **local mediators**, meaning that they perform their function in the environment in which they are synthesized. This distinguishes them from **hormones**, which are first synthesized and then transported in the bloodstream to their site of action. Eicosanoids are not stored in cells; rather, they are synthesized from arachidonic acid in response to an external stimulus.

> Other details of the biosynthesis of leukotrienes and prostaglandins were given in Sections 9.17 and 15.5, respectively.

The synthesis of prostaglandins, thromboxanes, and prostacyclins begins with the oxidation of arachidonic acid with O$_2$ by a **cyclooxygenase** enzyme, which forms an unstable cyclic intermediate, PGG$_2$. PGG$_2$ is then converted via different pathways to these three classes of compounds. Leukotrienes are formed by a different pathway, using an enzyme called a **lipoxygenase**. These four paths for arachidonic acid are summarized in Figure 25.6.

Each eicosanoid is associated with specific types of biological activity (Table 25.4). In some cases, the effects oppose one another. For example, thromboxanes are vasoconstrictors that trigger blood platelet aggregation, whereas prostacyclins are vasodilators that inhibit platelet aggregation. The levels of these two eicosanoids must be in the right balance for cells to function properly.

Because of their wide range of biological functions, prostaglandins and their analogues have found several clinical uses. For example, **dinoprostone**, the generic name for **PGE$_2$**, is administered to relax the smooth muscles of the uterus when labor is induced and to terminate pregnancies in the early stages.

PGE$_2$
(dinoprostone)

Figure 25.6
The conversion of arachidonic acid to prostaglandins, thromboxanes, prostacyclins, and leukotrienes

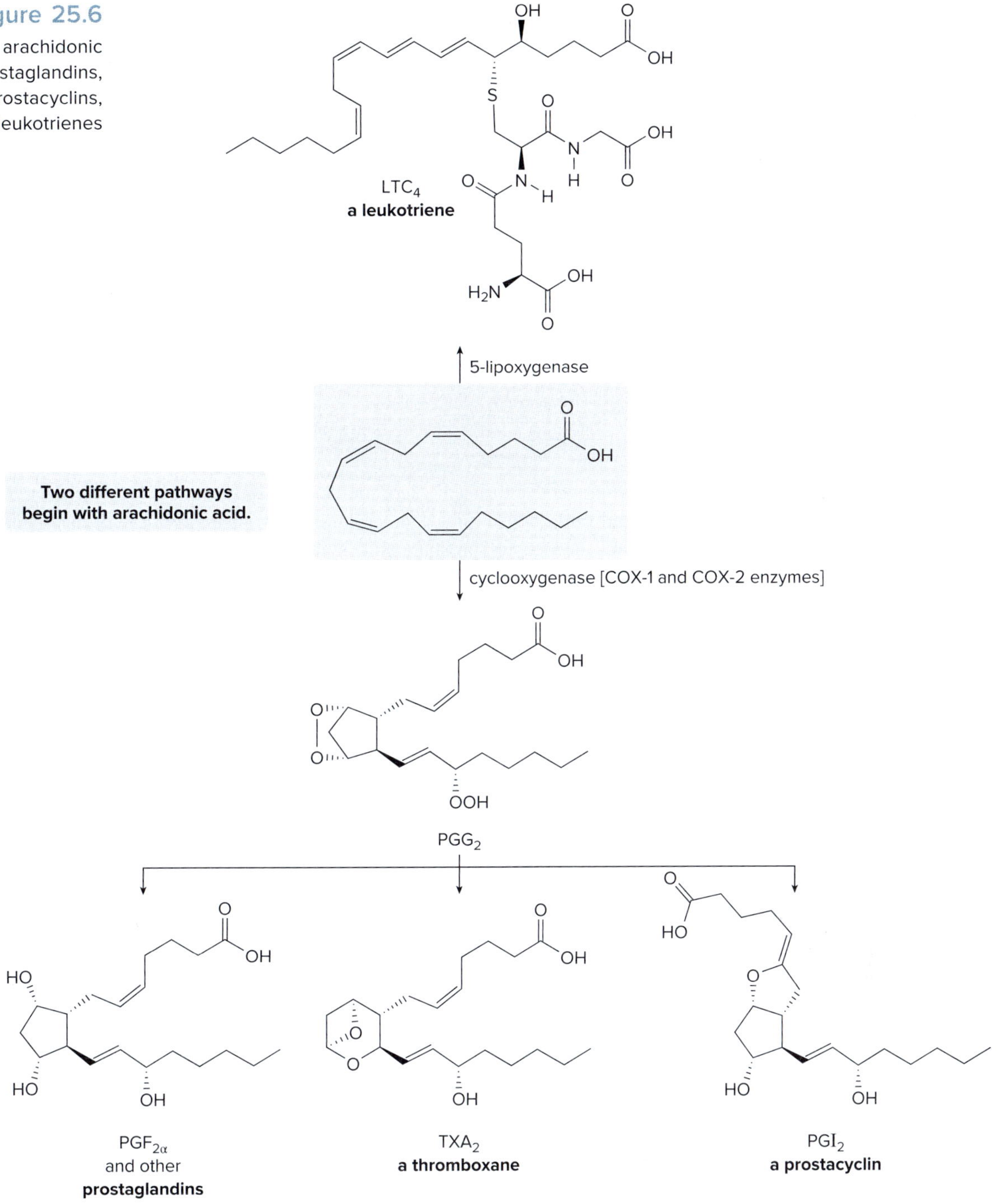

LTC₄ — a leukotriene

Two different pathways begin with arachidonic acid.

↑ 5-lipoxygenase

↓ cyclooxygenase [COX-1 and COX-2 enzymes]

PGG₂

PGF$_{2\alpha}$ and other **prostaglandins**

TXA₂ a thromboxane

PGI₂ a prostacyclin

Table 25.4 Biological Activity of the Eicosanoids

Eicosanoid	Effect	Eicosanoid	Effect
Prostaglandins	• Lower blood pressure	**Thromboxanes**	• Constrict blood vessels
	• Inhibit blood platelet aggregation		• Trigger blood platelet aggregation
	• Control inflammation	**Prostacyclins**	• Dilate blood vessels
	• Lower gastric secretions		• Inhibit blood platelet aggregation
	• Stimulate uterine contractions	**Leukotrienes**	• Constrict smooth muscle, especially in the lungs
	• Relax smooth muscles of the uterus		

Problem 25.10 Which carbons of arachidonic acid become the carbons of the five-membered ring in PGE$_2$?

Because prostaglandins themselves are unstable in the body, often having half-lives of only minutes, more stable analogues have been developed that retain their important biological activity longer. Misoprostol is a prostaglandin analogue administered to prevent gastric ulcers in patients who are at high risk of developing them, and unoprostone isopropyl is used to decrease eye pressure in glaucoma patients.

misoprostol (+ epimer at C16)

unoprostone isopropyl

Studying the biosynthesis of eicosanoids has led to other discoveries as well. For example, aspirin and other nonsteroidal anti-inflammatory drugs (**NSAIDs**) inactivate the cyclooxygenase enzyme needed for prostaglandin synthesis. In this way, NSAIDs block the synthesis of the prostaglandins that cause inflammation (Section 15.5).

More recently, it has been discovered that two *different* cyclooxygenase enzymes, called **COX-1** and **COX-2,** are responsible for prostaglandin synthesis. COX-1 is involved with the usual production of prostaglandins, but COX-2 is responsible for the synthesis of additional prostaglandins in inflammatory diseases like arthritis. **NSAIDs like aspirin and ibuprofen inactivate both the COX-1 and COX-2 enzymes.** This activity also results in an increase in gastric secretions, making an individual more susceptible to ulcer formation.

A group of anti-inflammatory drugs that block only the COX-2 enzyme was developed in the 1990s. These drugs—**rofecoxib, valdecoxib,** and **celecoxib**—do not cause an increase in gastric secretions, and thus were touted as especially effective NSAIDs for patients with arthritis, who need daily doses of these medications. Unfortunately, both rofecoxib and valdecoxib have now been removed from the market, because their use has been associated with an increased risk of heart attack and stroke.

Generic name rofecoxib
Trade name Vioxx

Generic name valdecoxib
Trade name Bextra

Generic name celecoxib
Trade name Celebrex

The discovery of drugs that block prostaglandin synthesis illustrates how basic research in organic chemistry can lead to important practical applications. Elucidating the structure and biosynthesis of prostaglandins began as a project in basic research. It has now resulted in a number of applications that benefit many individuals with various illnesses.

25.7 Terpenes

Terpenes are lipids composed of repeating five-carbon units called isoprene units. An **isoprene unit** has five carbons: four in a row, with a one-carbon branch on a middle carbon.

1 C branch
4 C's in a row
isoprene unit

Terpenes are hydrocarbons that may be acyclic or have one or more rings. The term *terpenoid* is used for compounds that contain isoprene units as well as an oxygen heteroatom. Many **essential oils,** a group of compounds isolated from plant sources by distillation, are terpenes and terpenoids. Examples include myrcene from bayberry and menthol from peppermint.

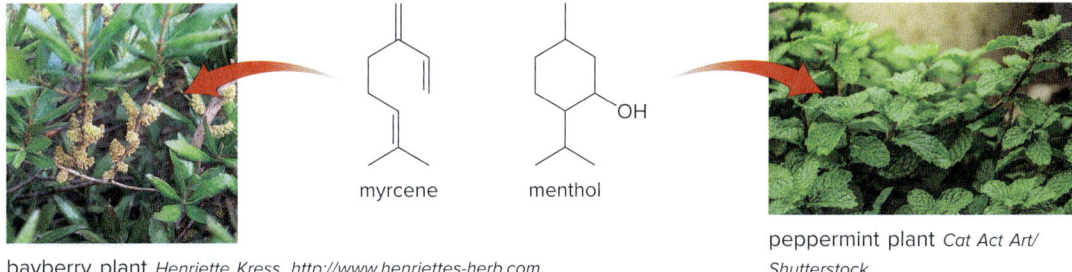

bayberry plant *Henriette Kress, http://www.henriettes-herb.com*

peppermint plant *Cat Act Art/ Shutterstock*

25.7A Locating Isoprene Units

How do we identify the isoprene units in these molecules? Start at one end of the molecule near a branch point. Then **look for a four-carbon chain with a one-carbon branch.** This forms one isoprene unit. Continue along the chain or around the ring until all the carbons are part of an isoprene unit. Keep in mind the following:

- An isoprene unit may be composed of C–C σ bonds only, or there may be π bonds at any position.
- Isoprene units are always connected by one or more carbon–carbon bonds.
- Each carbon atom is part of one isoprene unit only.
- Every isoprene unit has five carbon atoms. Heteroatoms may be present, but their presence is ignored in locating isoprene units.

Myrcene and menthol, for example, each have 10 carbon atoms, so they are composed of two isoprene units.

Several examples, with the isoprene units labeled in red, are given in Figure 25.7.

Figure 25.7

Examples of some common terpenes and terpenoids

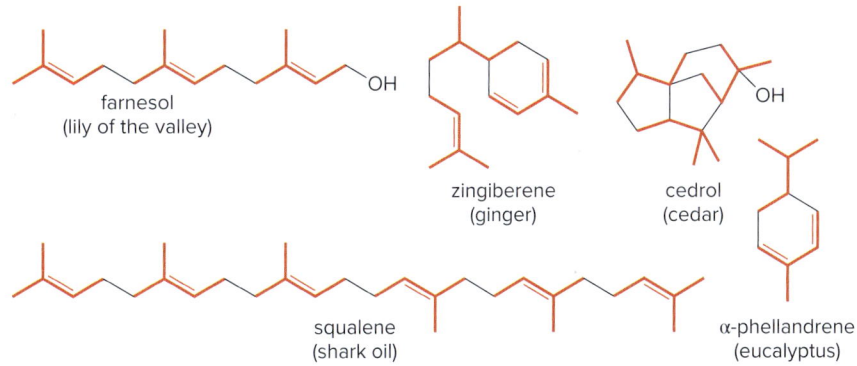

- Isoprene units are labeled in red, with C–C bonds (in black) joining two units.
- The source of each terpene or terpenoid is given in parentheses.

Sample Problem 25.1 Locating Isoprene Units

Locate the isoprene units in each compound.

a. 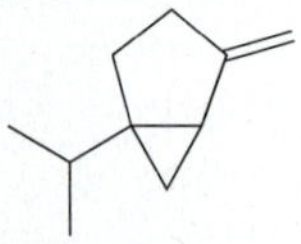 sabinene

b. 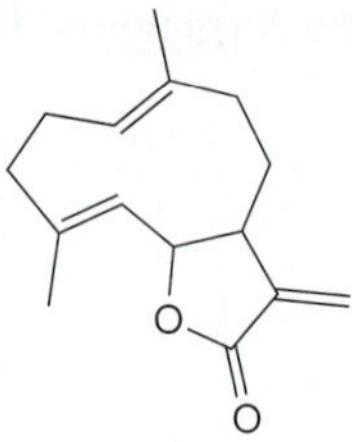 costunolide

Solution

- **Start looking for isoprene units at a *branch* point.** An isoprene unit has four carbons in a row with a one-carbon branch.
- Continue along a chain or around a ring until all carbons are part of *one* isoprene unit.
- **Ignore all heteroatoms.**

a. Sabinene has 10 C's and two isoprene units. **Start at the isopropyl group** to locate isoprene units. Two possibilities are shown.

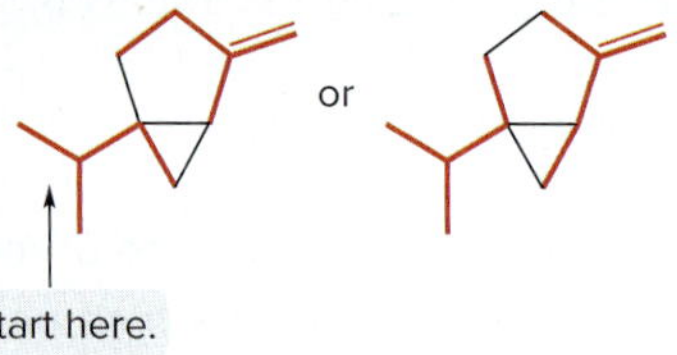

b. Costunolide has 15 C's and three isoprene units. **Start at any branch point and ignore the two O's.** Two possibilities are shown.

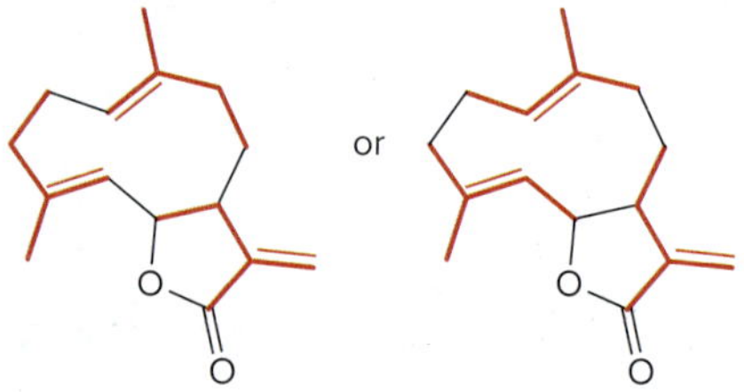

Problem 25.11 Locate the isoprene units in each compound.

a. 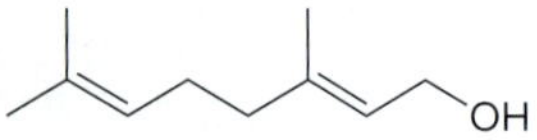 geraniol (roses and geraniums)

b. 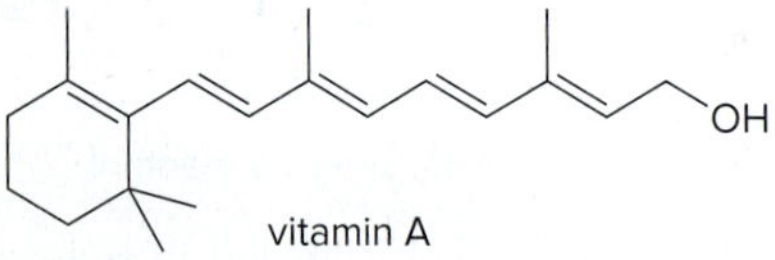 vitamin A

c. 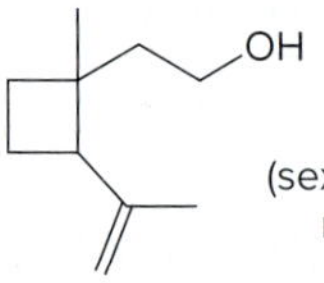 grandisol (sex pheromone of the male boll weevil)

d. 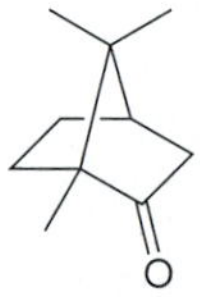 camphor

More Practice: Try Problems 25.18, 25.25, 25.27b, 25.28a.

Terpenes and terpenoids are classified by the number of isoprene units they contain. A *monoterpene* (or *monoterpenoid*) **contains 10 carbons** and has two isoprene units, a *sesquiterpene* (or *sesquiterpenoid*) **contains 15 carbons** and has three isoprene units, and so forth. The different terpene classes are summarized in Table 25.5.

An isoprene unit can be thought of as having a head and a tail. The "head" of the isoprene unit is located at the end of the chain nearest the branch point, and the "tail" is located at the end of the carbon chain farthest from the branch point. Most isoprene units are connected in a "head-to-tail" fashion, as shown for citral, which occurs in lemongrass.

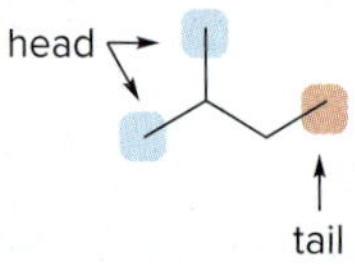

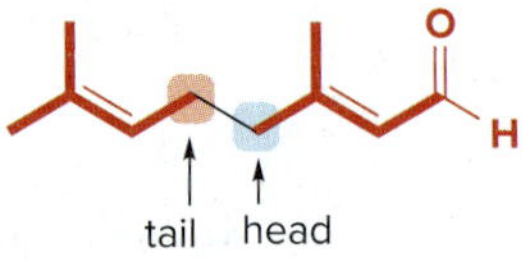

citral

25.7 Terpenes 1129

Table 25.5 Classes of Terpenes and Terpenoids

Name	Number of C atoms	Number of isoprene units
Monoterpene (Monoterpenoid)	10	2
Sesquiterpene (Sesquiterpenoid)	15	3
Diterpene (Diterpenoid)	20	4
Sesterterpene (Sesterterpenoid)	25	5
Triterpene (Triterpenoid)	30	6
Tetraterpene (Tetraterpenoid)	40	8

Amber, fossilized resin that oozed from trees long ago, contains biformene, as well as many other terpenoids called labdanoids (Problem 25.12).
Hjochen/Shutterstock

Problem 25.12 Locate the isoprene units in biformene, a component of amber, and classify biformene as a monoterpene, sesquiterpene, etc.

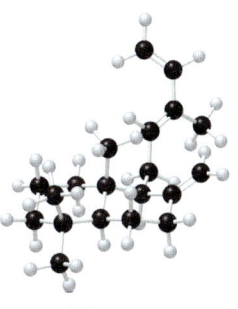

biformene

25.7B The Biosynthesis of Terpenes and Terpenoids

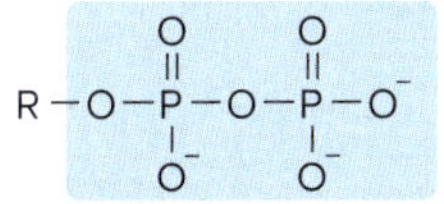

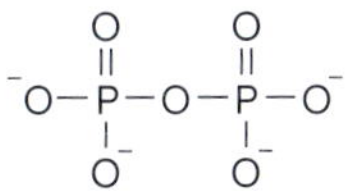

Terpene and terpenoid biosynthesis is an excellent example of how syntheses in nature occur with high efficiency. There are two ways this is accomplished.

[1] **The same reaction is used over and over again to prepare progressively more complex compounds.**

[2] **Key intermediates along the way serve as the starting materials for a wide variety of other compounds.**

All terpenes and terpenoids are synthesized from **dimethylallyl diphosphate** and **isopentenyl diphosphate**. Both of these five-carbon compounds are organic diphosphates (Section 12.2B) with a good leaving group (diphosphate, $P_2O_7^{4-}$, PP_i).

dimethylallyl diphosphate   isopentenyl diphosphate

The overall strategy of biosynthesis from dimethylallyl diphosphate and isopentenyl diphosphate is summarized in Figure 25.8.

There are three basic parts:

[1] The two C_5 diphosphates are converted to **geranyl diphosphate, a C_{10} monoterpenoid.** Geranyl diphosphate is the starting material for all other monoterpenes and monoterpenoids.

[2] Geranyl diphosphate is converted to **farnesyl diphosphate, a C_{15} sesquiterpenoid,** by addition of a five-carbon unit. Farnesyl diphosphate is the starting material for all sesquiterpenes, diterpenes, and related terpenoids.

[3] Two molecules of farnesyl diphosphate are converted to **squalene, a C_{30} triterpene.** Squalene is the starting material for all triterpenes and steroids.

Figure 25.8
An outline of terpene and terpenoid biosynthesis

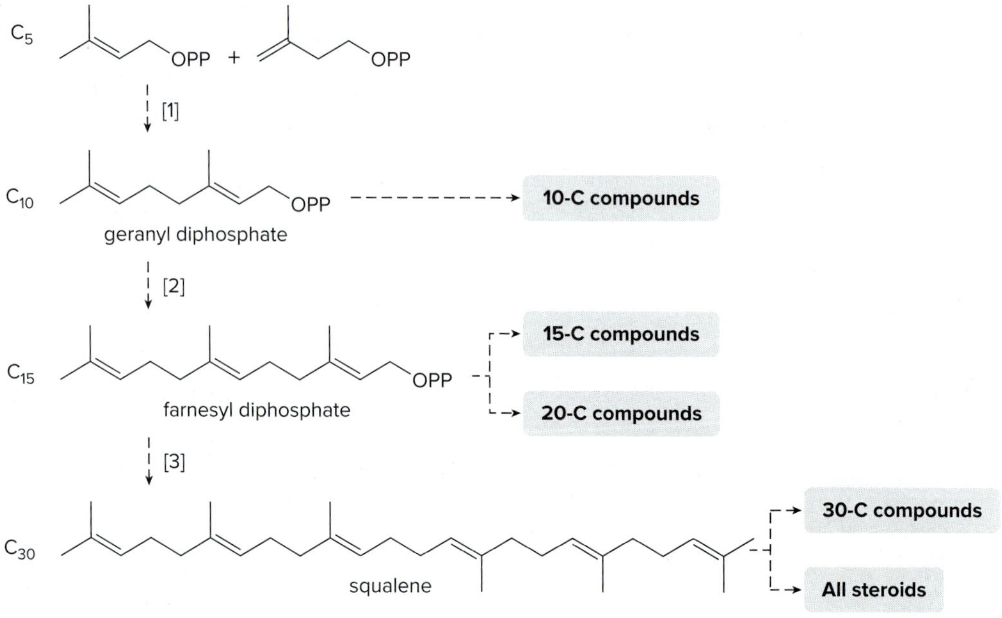

The biological formation of geranyl diphosphate from the two five-carbon diphosphates was shown in Mechanism 12.1. The biological conversion of geranyl diphosphate to farnesyl diphosphate involves a similar pathway, as shown in Mechanism 25.1.

Mechanism 25.1 Biological Formation of Farnesyl Diphosphate

Part [1] Formation of a new carbon–carbon σ bond

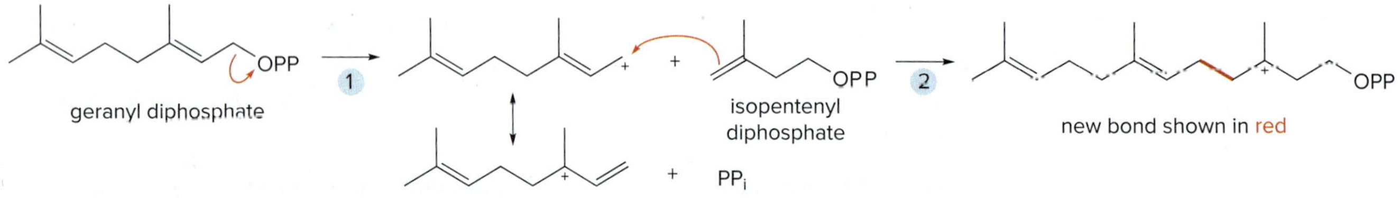

① Loss of the diphosphate leaving group forms a **resonance-stabilized carbocation.**
② Nucleophilic attack of isopentenyl diphosphate on the allylic carbocation forms the new C–C σ bond.

Part [2] Formation of a π bond by loss of a proton

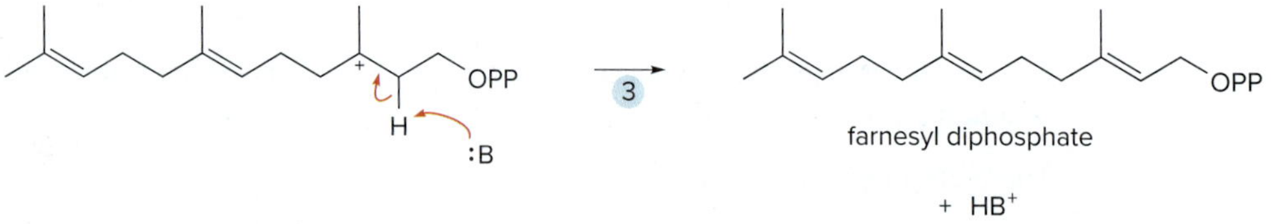

③ Loss of a proton (shown with the general base B:) forms a **new π bond** and farnesyl diphosphate.

Two molecules of farnesyl diphosphate react to form squalene, from which all other triterpenes and steroids are synthesized.

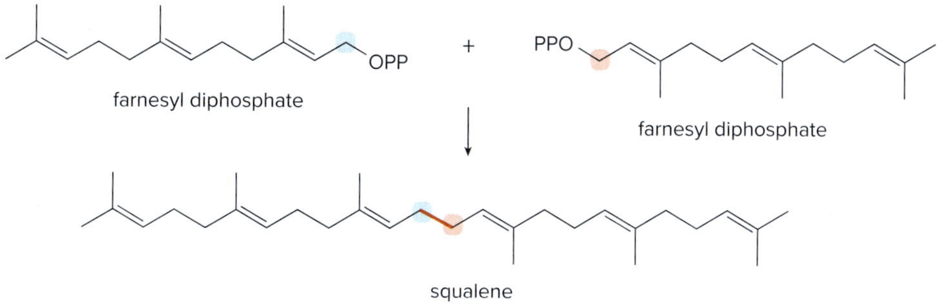

25.7 Terpenes

[Structures: farnesyl diphosphate + farnesyl diphosphate → squalene]

Aqueous hydrolysis of geranyl and farnesyl diphosphates forms the monoterpenoid geraniol and the sesquiterpenoid farnesol, respectively.

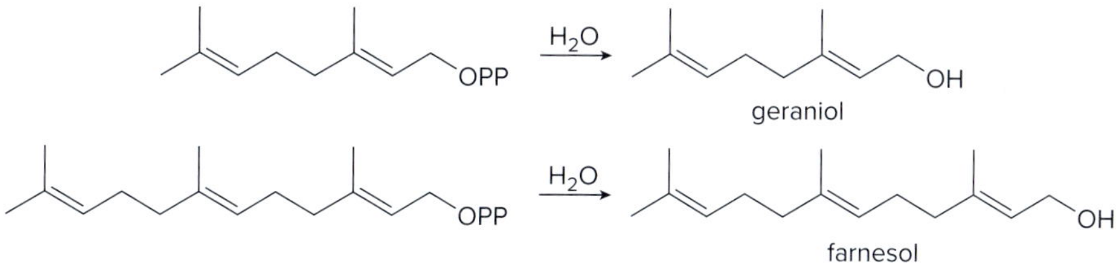

[Structures: geranyl diphosphate + H₂O → geraniol; farnesyl diphosphate + H₂O → farnesol]

All other terpenes and terpenoids are biologically derived from geranyl and farnesyl diphosphates by a series of reactions. Cyclic compounds are formed by intramolecular reactions involving nucleophilic attack of π bonds on intermediate carbocations. To form some cyclic compounds, the *E* double bond in geranyl diphosphate must first isomerize to an isomeric diphosphate with a *Z* double bond, neryl diphosphate, by the process illustrated in Mechanism 25.2. Isomerization forms a substrate with a leaving group and nucleophilic double bond in close proximity so that an intramolecular reaction can occur.

Mechanism 25.2 Isomerization of Geranyl Diphosphate to Neryl Diphosphate

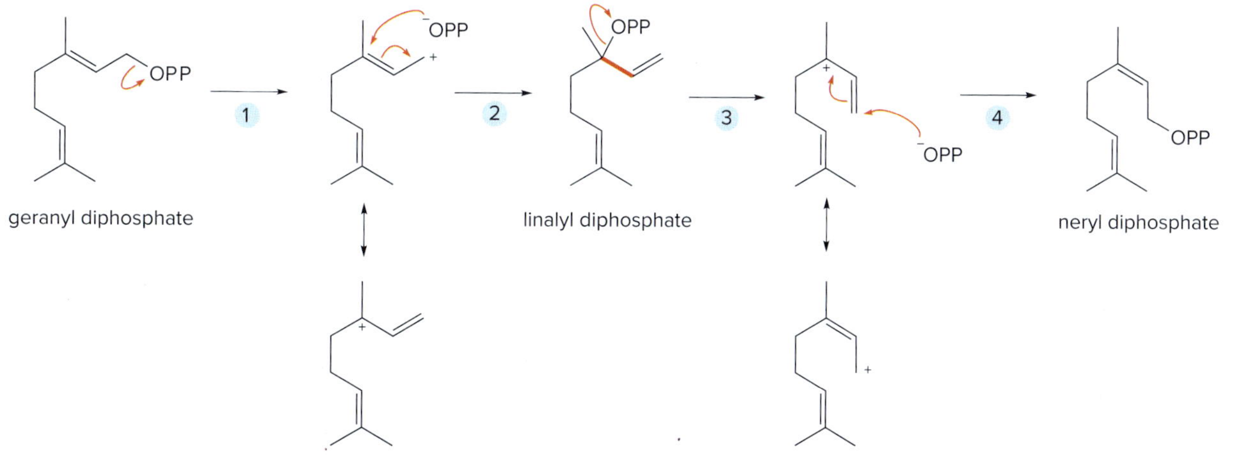

[Mechanism scheme: geranyl diphosphate →(1) carbocation →(2) linalyl diphosphate →(3) carbocation →(4) neryl diphosphate, with resonance structures shown below]

1–2	**Loss of diphosphate** forms a **resonance-stabilized carbocation**, which reacts with the diphosphate anion to form linalyl diphosphate.
3	**Bond rotation** of the single bond shown in red and loss of diphosphate forms a **resonance-stabilized carbocation**.
4	**Nucleophilic attack with diphosphate** forms neryl diphosphate, which has the leaving group and the double bond at the other end of the chain in close proximity for intramolecular cyclization.

In the synthesis of α-terpineol or limonene, for example, geranyl diphosphate isomerizes to form neryl diphosphate (Step [1] in the following reaction sequence). Neryl diphosphate then cyclizes to a 3° carbocation by intramolecular attack (Steps [2]–[3]). Nucleophilic attack of water on this carbocation yields the monoterpenoid α-terpineol (Step [4]), or loss of a proton yields the monoterpene limonene (Step [5]).

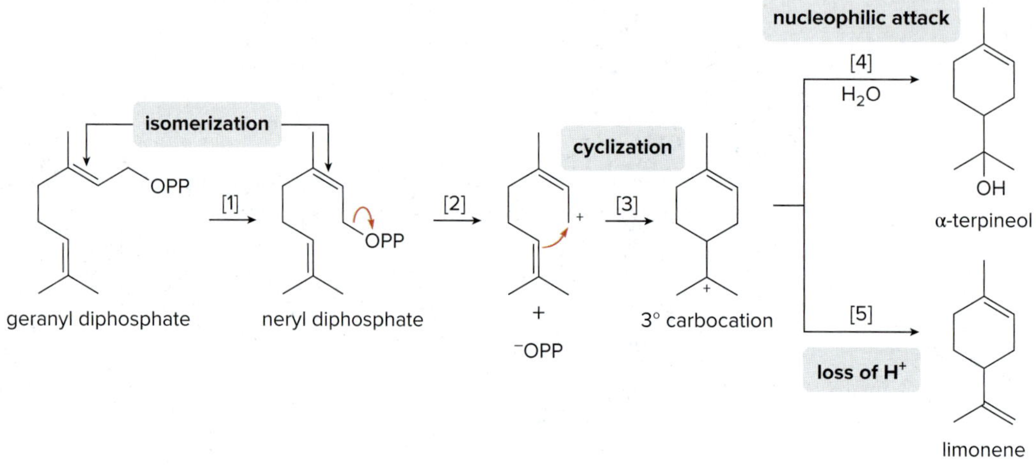

Problem 25.13 Write a stepwise mechanism for the following reaction.

farnesyl diphosphate
+
isopentenyl diphosphate

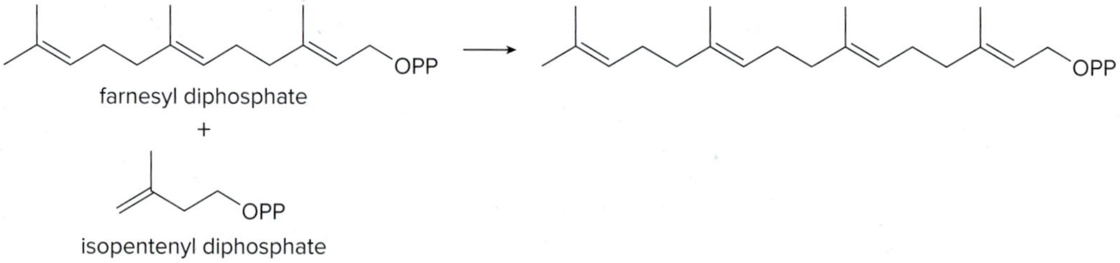

Problem 25.14 Draw a stepwise mechanism for the conversion of geranyl diphosphate to α-terpinene.

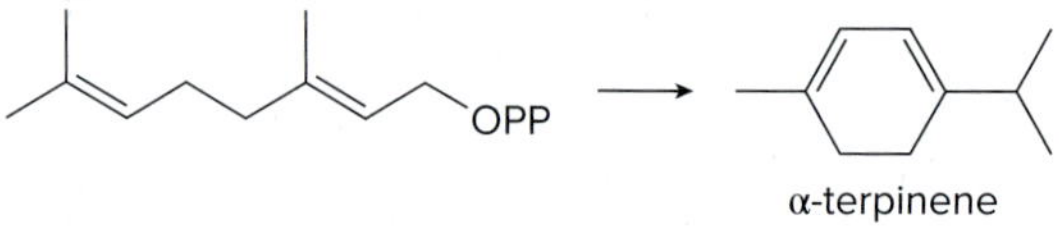

α-terpinene

25.8 Steroids

The steroids are a group of tetracyclic lipids, many of which are biologically active.

25.8A Steroid Structure

Steroids are composed of three six-membered rings and one five-membered ring, joined together as drawn. Many steroids also contain two methyl groups, called **angular methyl groups,** at the two ring junctions indicated. The steroid rings are lettered **A, B, C,** and **D,** and the 17 ring carbons are numbered as shown. The two angular methyl groups are numbered C18 and C19.

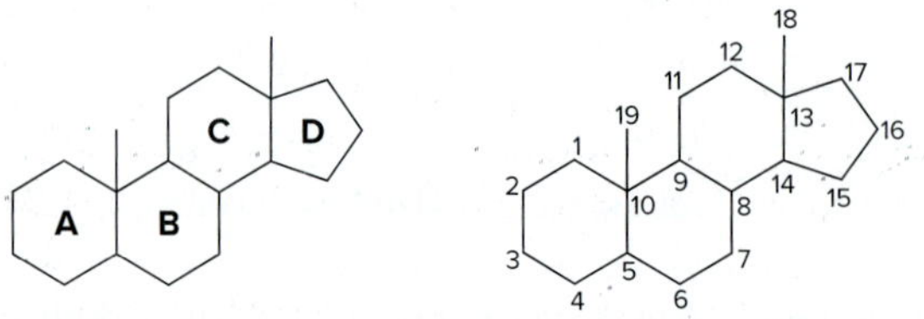

Whenever two rings are fused together, the substituents at the ring fusion can be arranged cis or trans. To see more easily why this is true, consider **decalin,** which consists of two six-membered rings fused together. *trans*-**Decalin** has the two hydrogen atoms at the ring fusion on opposite sides, whereas *cis*-**decalin** has them on the same side.

25.8 Steroids

decalin
Two six-membered rings share a C–C bond.

trans-decalin
2 H's on opposite sides

cis-decalin
2 H's on the same side

Three-dimensional structures of these molecules show how different these two possible arrangements actually are. The two rings of *trans*-decalin lie roughly in the same plane, whereas the two rings of *cis*-decalin are almost perpendicular to each other. **The trans arrangement is *lower* in energy and therefore more stable.**

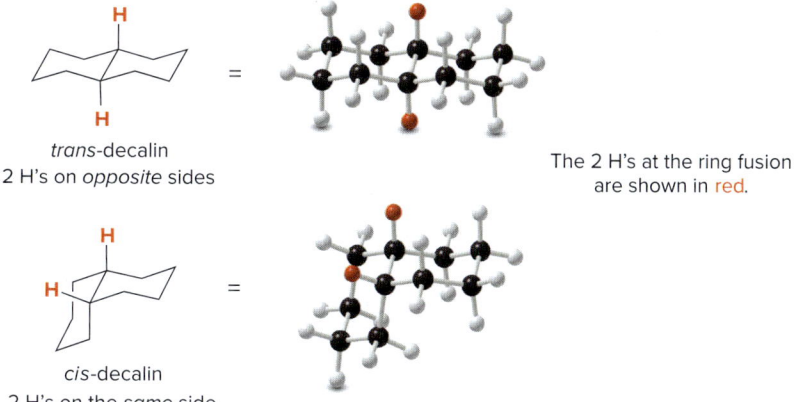

trans-decalin
2 H's on *opposite* sides

The 2 H's at the ring fusion are shown in red.

cis-decalin
2 H's on the *same* side

In steroids, each ring fusion could theoretically have the cis or trans configuration, but by far the most common arrangement is all trans. Because of this, **all four rings of the steroid skeleton lie in the same plane,** and the ring system is fairly rigid. The two angular methyl groups are oriented perpendicular to the plane of the molecule. These methyl groups make one side of the steroid skeleton significantly more hindered than the other, as shown in Figure 25.9.

Figure 25.9
The three-dimensional structure of the steroid nucleus

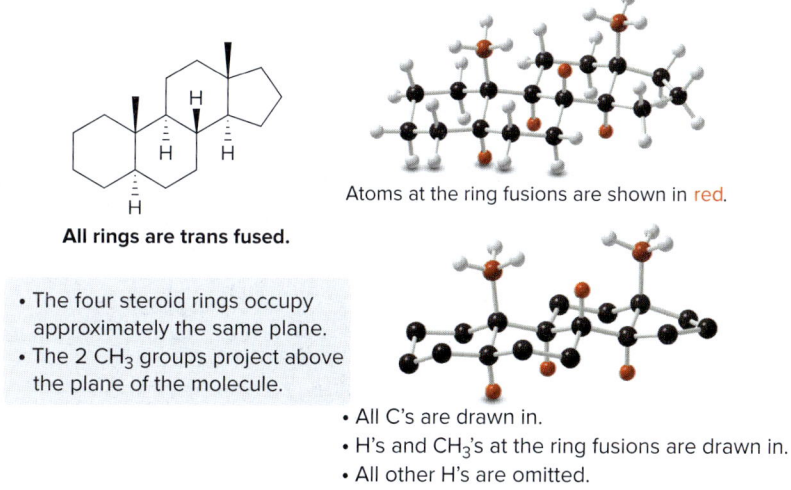

All rings are trans fused.

- The four steroid rings occupy approximately the same plane.
- The 2 CH₃ groups project above the plane of the molecule.

Atoms at the ring fusions are shown in red.

- All C's are drawn in.
- H's and CH₃'s at the ring fusions are drawn in.
- All other H's are omitted.

Although steroids have the same fused-ring arrangement of carbon atoms, they differ in the identity and location of the substituents attached to that skeleton.

Problem 25.15 (a) Draw a skeletal structure of the anabolic steroid 4-androstene-3,17-dione, also called "andro," from the following description. Andro contains the tetracyclic steroid skeleton with carbonyl groups at C3 and C17, a double bond between C4 and C5, and methyl groups bonded to C10 and C13. (b) Add wedges and dashed wedges for all stereogenic centers with the following information: the configuration at C10 is *R*, the configuration at C13 is *S*, and all substituents at ring fusions are trans to each other.

25.8B Cholesterol

The role of cholesterol in plaque formation and atherosclerosis was discussed in Section 16.16.

Cholesterol has the tetracyclic carbon skeleton characteristic of steroids. It also has eight stereogenic carbons (seven on rings and one on a side chain), so there are $2^8 = 256$ possible stereoisomers. In nature, however, only the following stereoisomer exists:

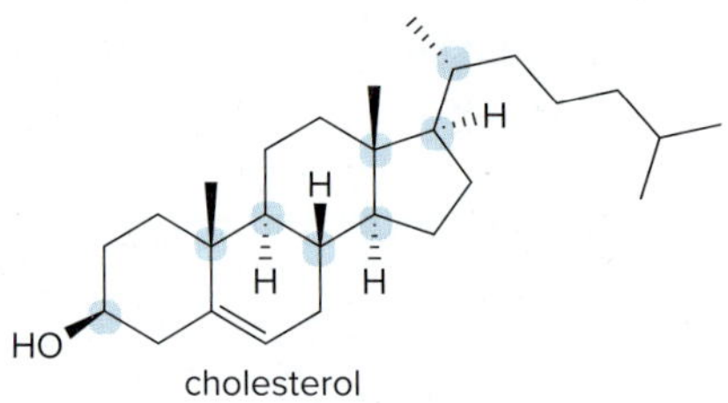

cholesterol

Cholesterol is essential to life because it forms an important component of cell membranes and is the starting material for the synthesis of all other steroids. Humans do not have to ingest cholesterol, because it is synthesized in the liver and then transported to other tissues through the bloodstream. Because cholesterol has only one polar OH group and many nonpolar C–C and C–H bonds, it is **insoluble in water** (and, thus, in the aqueous medium of the blood).

Konrad Bloch and Feodor Lynen shared the 1964 Nobel Prize in Physiology or Medicine for unraveling the complex transformation of squalene to cholesterol.

Cholesterol is synthesized in the body from squalene, a C_{30} triterpene that is itself prepared from smaller terpenes, as discussed in Section 25.7B. Because the biosynthesis of all terpenes begins with acetyl CoA, every one of the 27 carbon atoms of cholesterol comes from the same two-carbon precursor. The major steps in the conversion of squalene to cholesterol are given in Figure 25.10.

Figure 25.10 The biosynthesis of cholesterol

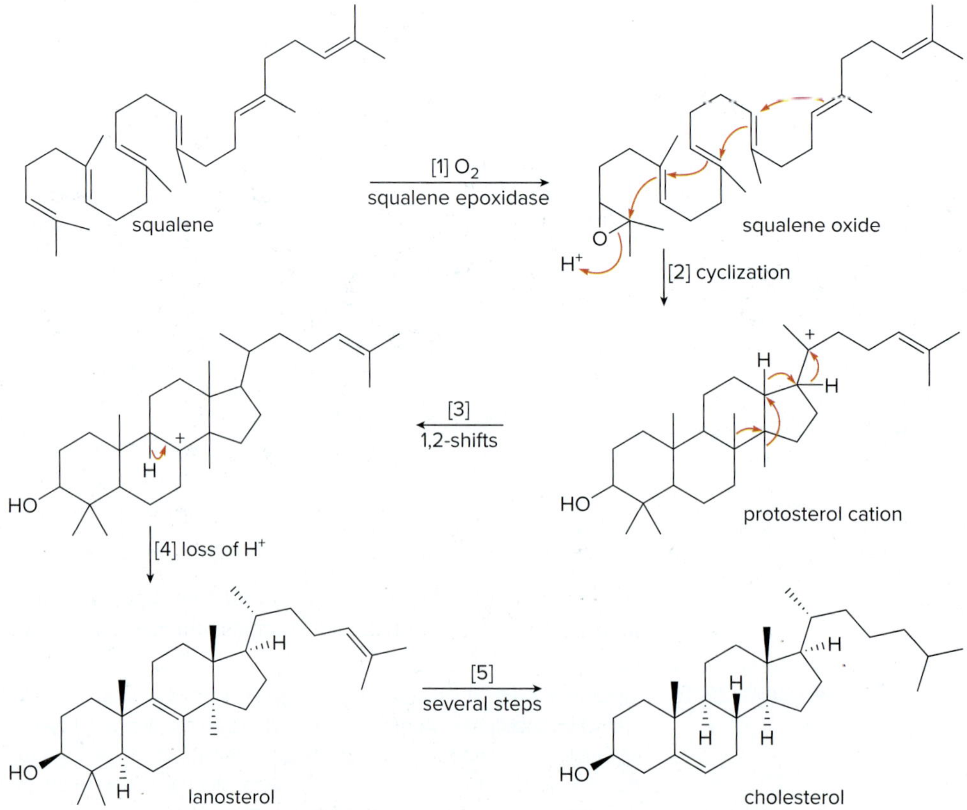

The conversion of squalene to cholesterol consists of five different parts:

[1] **Epoxidation** of squalene with an enzyme, squalene epoxidase, gives squalene oxide, which contains a single epoxide on one of the six double bonds.
[2] **Cyclization** of squalene oxide yields a carbocation, called the protosterol cation. This reaction results in the formation of four new C–C bonds and the tetracyclic ring system.
[3] **The protosterol carbocation rearranges** by a series of 1,2-shifts of either a hydrogen or methyl group to form another 3° carbocation.
[4] **Loss of a proton** gives an alkene called **lanosterol.** Although lanosterol has seven stereogenic centers, a single stereoisomer is formed.
[5] Lanosterol is then converted to cholesterol by a multistep process that results in removal of three methyl groups.

Several drugs called statins are now available to reduce the level of cholesterol in the bloodstream. These compounds act by blocking the biosynthesis of cholesterol at its very early stages. Two examples include atorvastatin (Lipitor) and simvastatin (Zocor), whose structures appear in Figure 25.11.

Figure 25.11
Two cholesterol-lowering drugs

Generic name atorvastatin
Trade name Lipitor

Generic name simvastatin
Trade name Zocor

Problem 25.16 Draw the enantiomer and any two diastereomers of cholesterol. Does the OH group of cholesterol occupy an axial or equatorial position?

Problem 25.17 Lovastatin, which occurs naturally in oyster mushrooms, was the first statin to be marketed. What hydrolysis products are formed when lovastatin is treated with aqueous acid?

Although lovastatin (Problem 25.17) is naturally occurring in oyster mushrooms, the commercial drug is obtained by a fermentation process.
Coxy58/Shutterstock

lovastatin

25.8C Other Steroids

Many other important steroids are hormones secreted by the endocrine glands. Two classes are the **sex hormones** and the **adrenal cortical steroids.**

There are two types of female sex hormones, **estrogens** and **progestins**. The male sex hormones are called **androgens**. The most important members of each hormone type are given in Table 25.6.

Table 25.6 The Female and Male Sex Hormones

Structure	Properties
estradiol, estrone	• **Estradiol** and **estrone** are estrogens synthesized in the ovaries. They control the development of secondary sex characteristics in females and regulate the menstrual cycle.
progesterone	• **Progesterone** is often called the "pregnancy hormone." It is responsible for the preparation of the uterus for implantation of a fertilized egg.
testosterone, androsterone	• **Testosterone** and **androsterone** are androgens synthesized in the testes. They control the development of secondary sex characteristics in males.

Synthetic analogues of these steroids have found important uses, such as ethynylestradiol and norethindrone in oral contraceptives, first mentioned in Section 10.5.

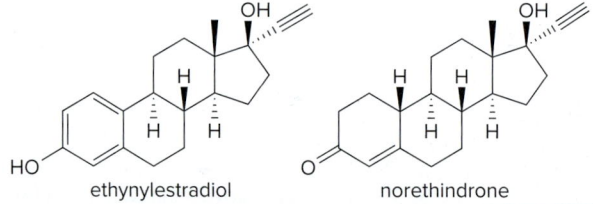

ethynylestradiol norethindrone

Synthetic androgen analogues, called **anabolic steroids,** promote muscle growth. They were first developed to help individuals whose muscles had atrophied from lack of use following surgery. They have since come to be used by athletes and body builders, although their use is not permitted in competitive sports. Many physical and psychological problems result from their prolonged use.

Anabolic steroids, such as stanozolol, nandrolone, and tetrahydrogestrinone have the same effect on the body as testosterone, but they are more stable, so they are not metabolized as quickly. Tetrahydrogestrinone (also called THG or The Clear), the performance-enhancing drug used by track star Marion Jones during the 2000 Sydney Olympics, was considered a "designer steroid" because it was initially undetected in urine tests for doping. After its chemical structure and properties were determined, it was added to the list of banned anabolic steroids in 2004.

Some body builders use anabolic steroids to increase muscle mass. Long-term or excessive use can cause many health problems, including high blood pressure, liver damage, and cardiovascular disease.
Comstock/JupiterImages

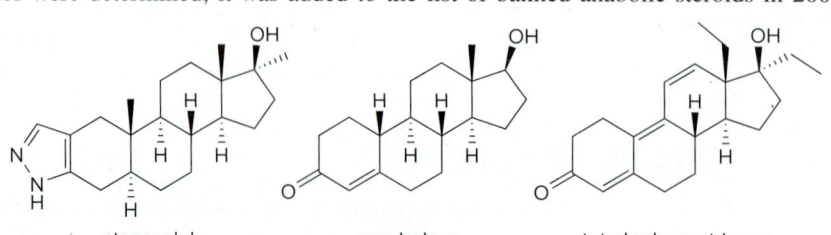

stanozolol nandrolone tetrahydrogestrinone

A second group of steroid hormones includes the **adrenal cortical steroids.** Three examples of these hormones are **cortisone, cortisol,** and **aldosterone.** All of these compounds are synthesized in the outer layer of the adrenal gland. Cortisone and cortisol serve as anti-inflammatory agents, and they also regulate carbohydrate metabolism. Aldosterone regulates blood pressure and volume by controlling the concentration of Na⁺ and K⁺ in body fluids.

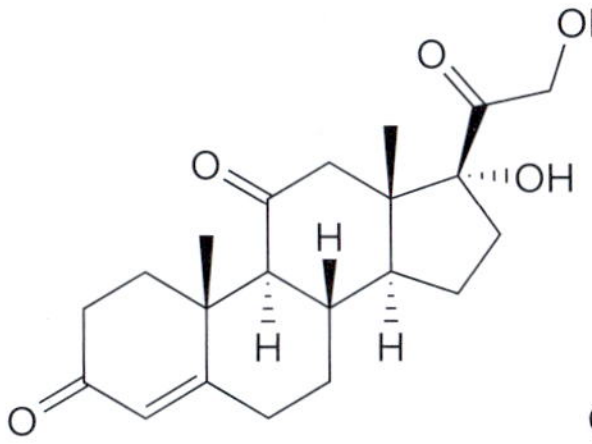

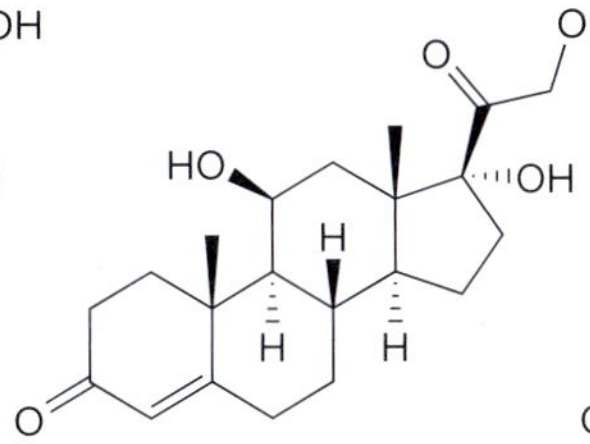

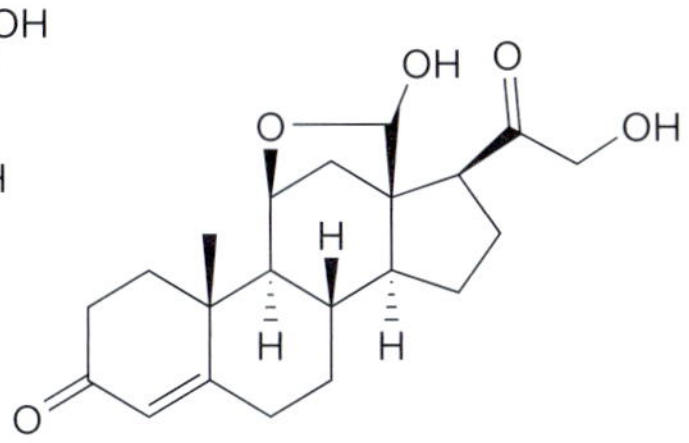

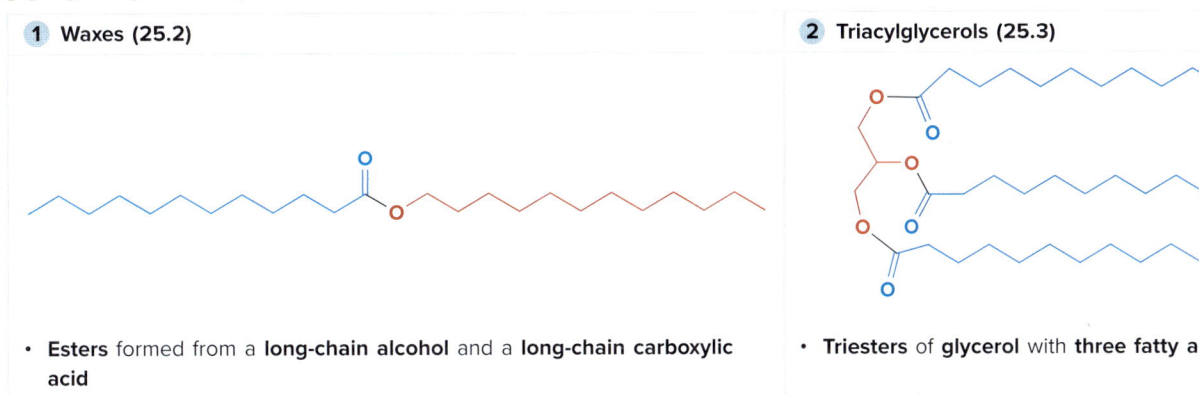

Chapter 25 REVIEW

KEY CONCEPTS

[1] Hydrolyzable lipids

1 Waxes (25.2)

- Esters formed from a **long-chain alcohol** and a **long-chain carboxylic acid**

2 Triacylglycerols (25.3)

- Triesters of **glycerol** with **three fatty acids**

Try Problems 25.21, 25.22.

[2] Hydrolyzable phospholipids (25.4)

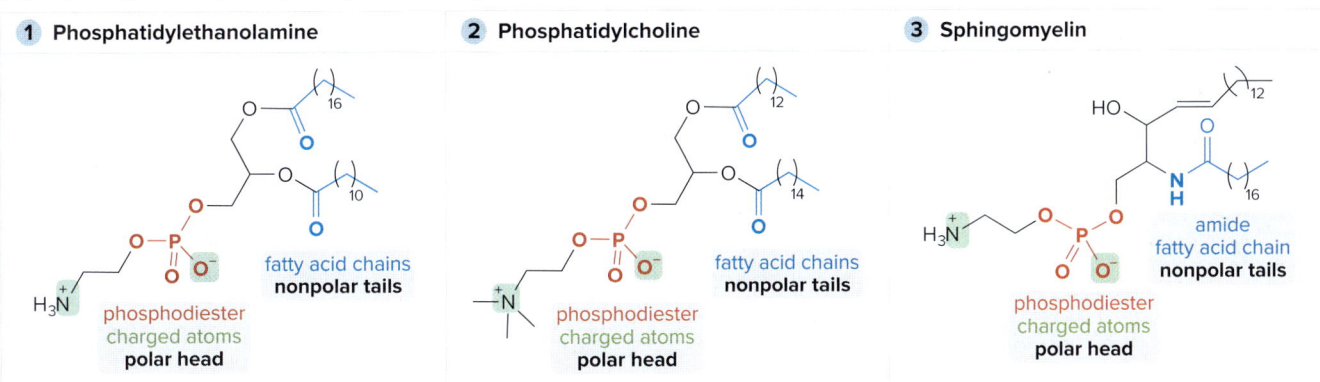

Try Problem 25.24.

[3] Nonhydrolyzable lipids

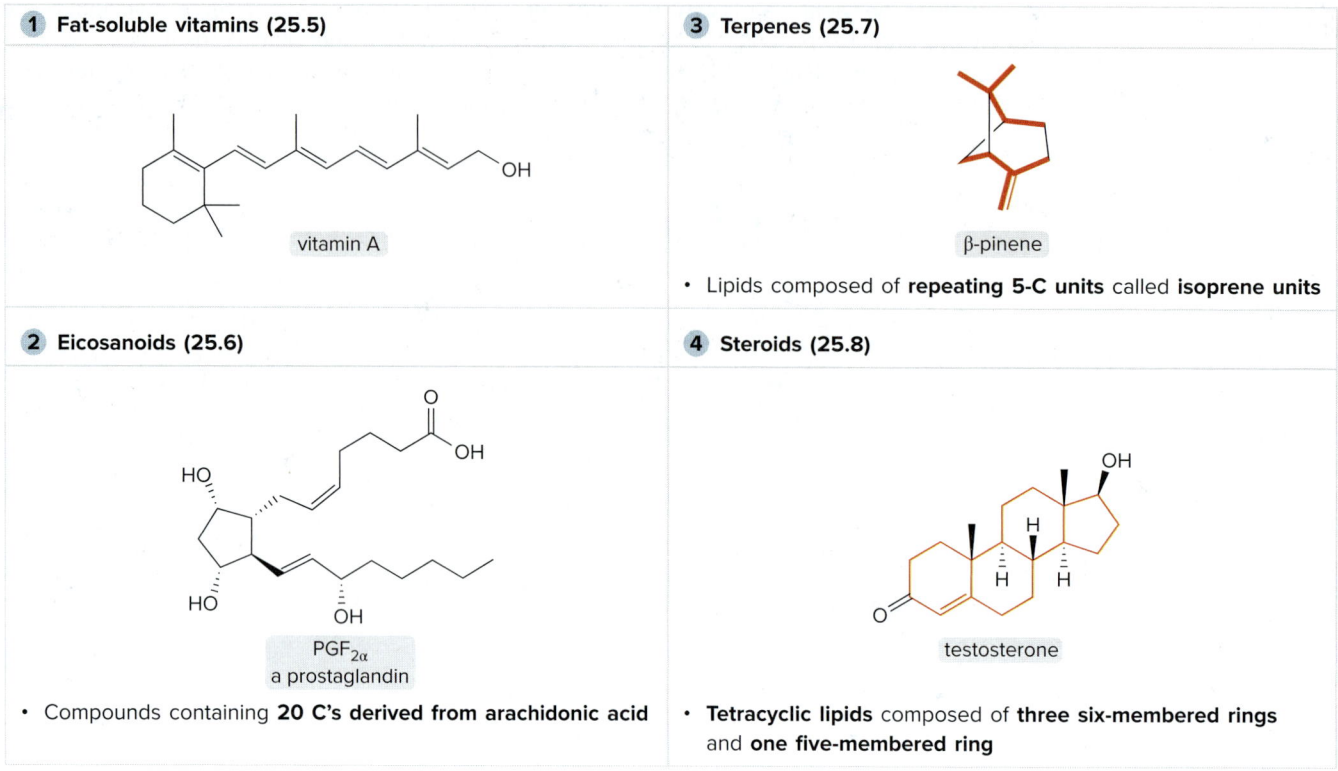

- **1 Fat-soluble vitamins (25.5)**: vitamin A
- **2 Eicosanoids (25.6)**: PGF$_{2\alpha}$, a prostaglandin
  - Compounds containing **20 C's** derived from **arachidonic acid**
- **3 Terpenes (25.7)**: β-pinene
  - Lipids composed of **repeating 5-C units** called **isoprene units**
- **4 Steroids (25.8)**: testosterone
  - **Tetracyclic lipids** composed of **three six-membered rings** and **one five-membered ring**

Try Problems 25.20, 25.37, 25.38, 25.39a.

KEY SKILLS

Locating isoprene units (25.7); example: steviol

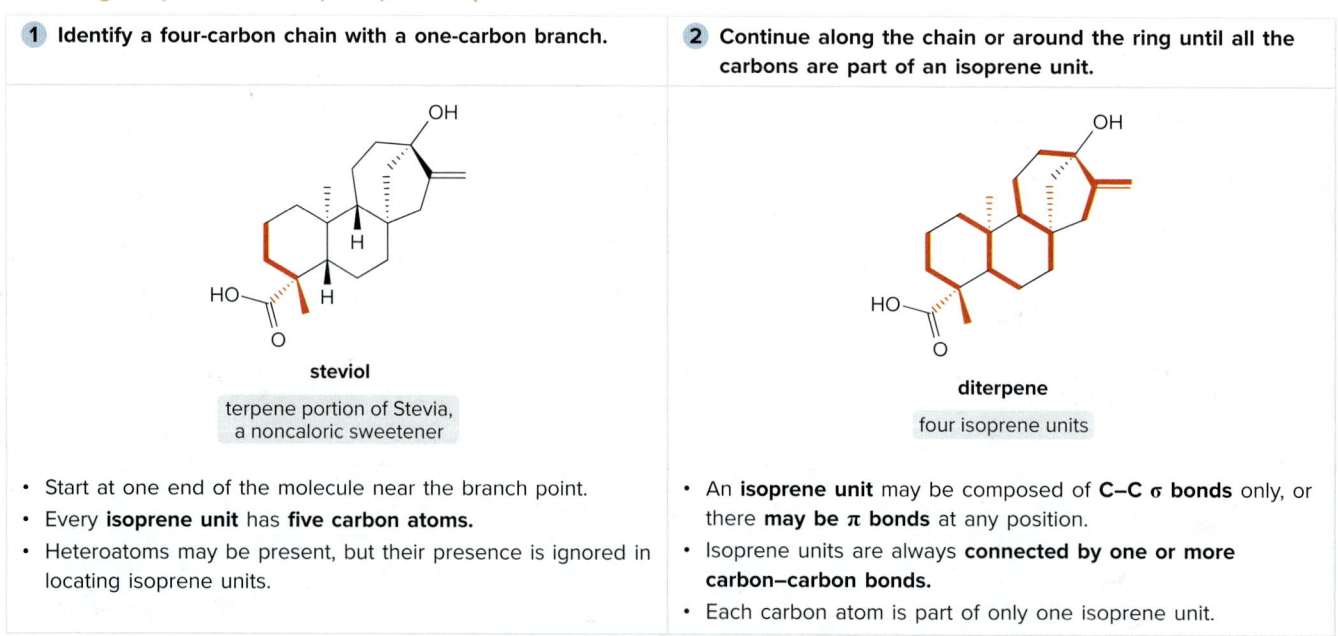

1 Identify a four-carbon chain with a one-carbon branch.

steviol
terpene portion of Stevia, a noncaloric sweetener

- Start at one end of the molecule near the branch point.
- Every **isoprene unit** has **five carbon atoms**.
- Heteroatoms may be present, but their presence is ignored in locating isoprene units.

2 Continue along the chain or around the ring until all the carbons are part of an isoprene unit.

diterpene
four isoprene units

- An **isoprene unit** may be composed of **C–C σ bonds** only, or there **may be π bonds** at any position.
- Isoprene units are always **connected by one or more carbon–carbon bonds**.
- Each carbon atom is part of only one isoprene unit.

See Sample Problem 25.1. Try Problems 25.18, 25.25, 25.27b, 25.28a.

PROBLEMS

Problems Using Three-Dimensional Models

25.18 Locate the isoprene units in each compound.

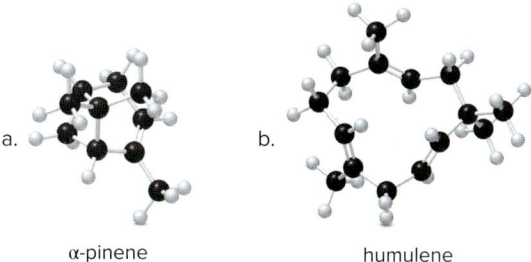

a. α-pinene   b. humulene

25.19 Convert each ball-and-stick model to a skeletal structure that clearly shows the stereochemistry at the ring fusion of these decalin derivatives.

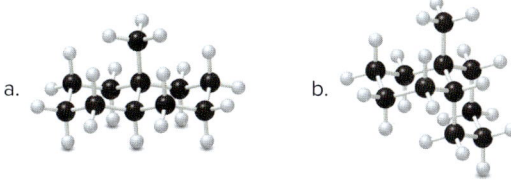

a.   b.

25.20 Convert the ball-and-stick model of androsterone to (a) a skeletal structure using wedges and dashed wedges around all stereogenic centers; and (b) a three-dimensional representation using chair cyclohexane rings.

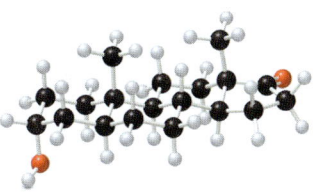

androsterone

Waxes, Triacylglycerols, and Phospholipids

25.21 One component of lanolin, the wax that coats sheep's wool, is derived from cholesterol and stearic acid. Draw its structure, including the correct stereochemistry at all stereogenic centers.

25.22 What is the structure of an optically inactive triacylglycerol that yields two moles of oleic acid and one mole of palmitic acid when hydrolyzed in aqueous acid?

25.23 Triacylglycerol **L** yields compound **M** when treated with excess H_2, Pd-C. Ozonolysis of **L** ([1] O_3; [2] $(CH_3)_2S$) affords compounds **N–P**. What is the structure of **L**?

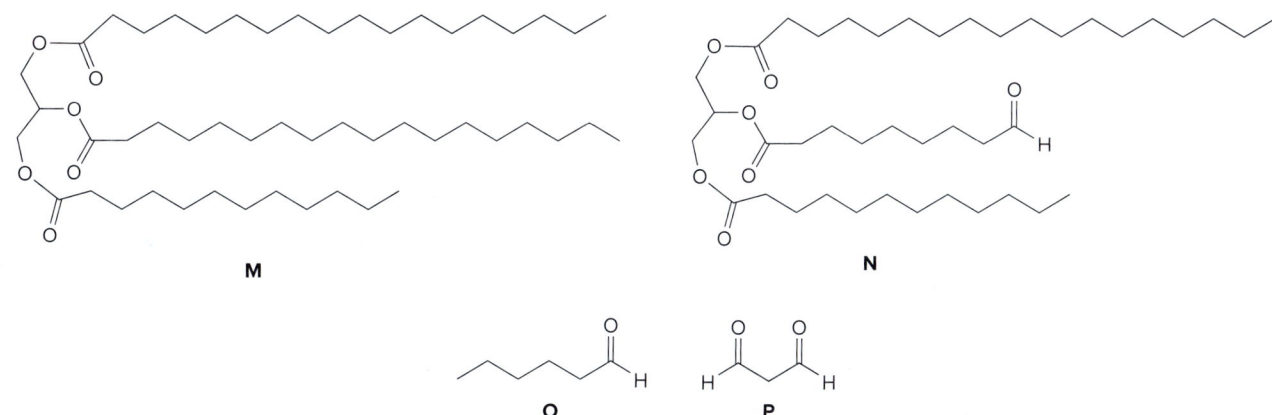

25.24 Draw the structure of these phospholipids:
a. a cephalin formed from two molecules of stearic acid
b. a sphingomyelin formed from palmitic acid

Terpenes and Terpenoids

25.25 Locate the isoprene units in each compound.

a. 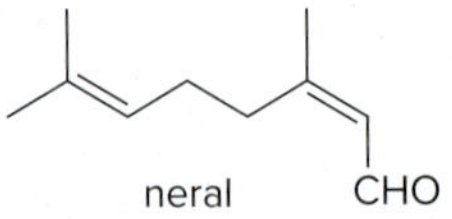 neral

b. 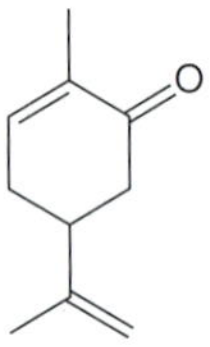 carvone

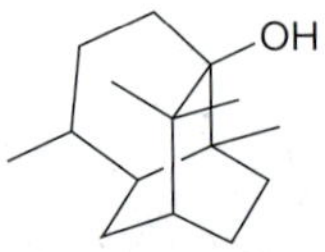

c. 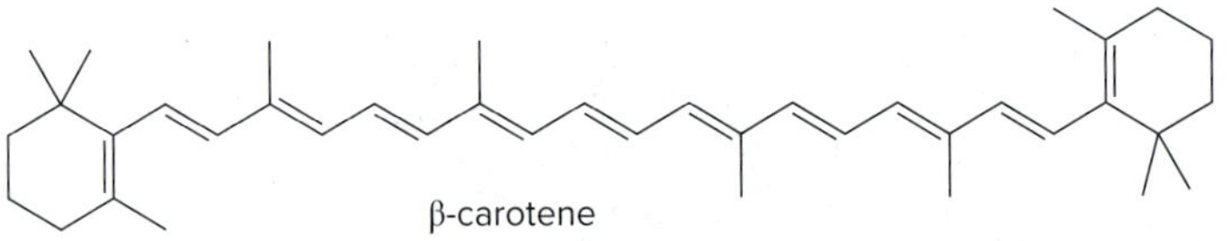 β-carotene

d. patchouli alcohol

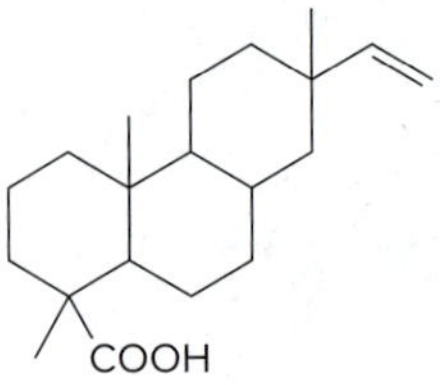

e. periplanone B

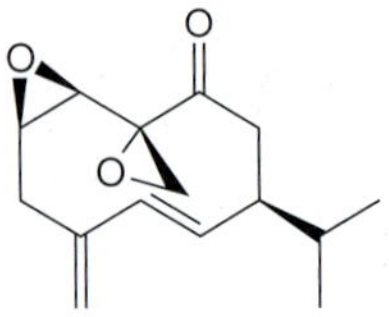

f. dextropimaric acid

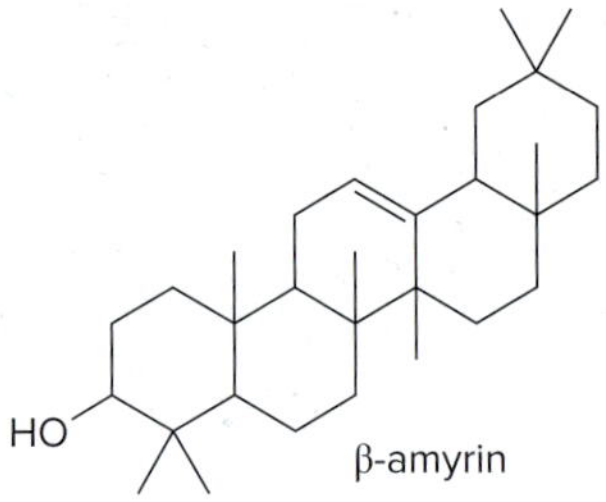

g. β-amyrin

25.26 Classify each terpene and terpenoid in Problem 25.25 (e.g., as a monoterpene, sesquiterpene, etc.).

25.27 Crocin, which occurs naturally in crocus and gardenia flowers, is primarily responsible for the color of saffron. (a) What lipid and monosaccharides are formed by the hydrolysis of crocin? (b) Classify the lipid as a monoterpenoid, diterpenoid, etc., and locate the isoprene units.

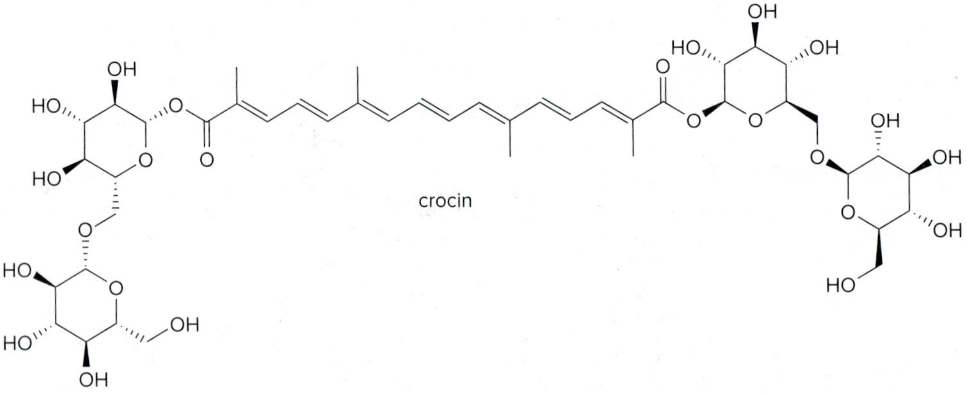

crocin

25.28 (a) Locate the isoprene units in lycopene, the red pigment in tomatoes (Section 12.7). (b) Which isoprene units are connected in a head-to-tail fashion? (c) Label any other isoprene unit as connected in a head-to-head fashion or a tail-to-tail fashion. (d) Classify lycopene as a monoterpene, sesquiterpene, and so on.

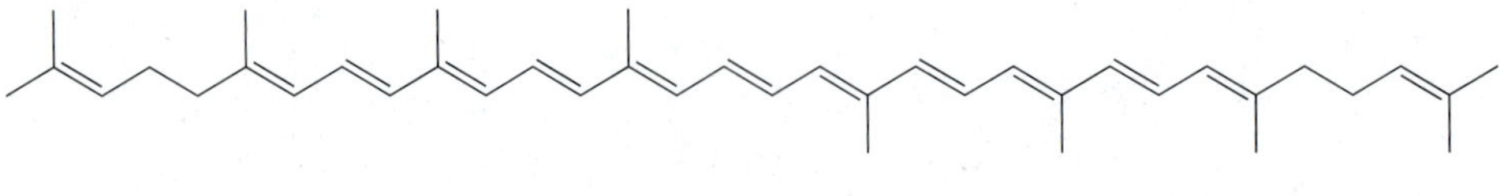

lycopene

25.29 Which of the following compounds are not composed of isoprene units? Locate the isoprene units in each compound that contains them.

a.

b.

c.

d.

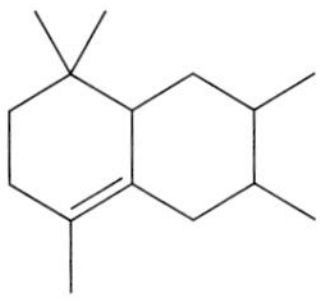

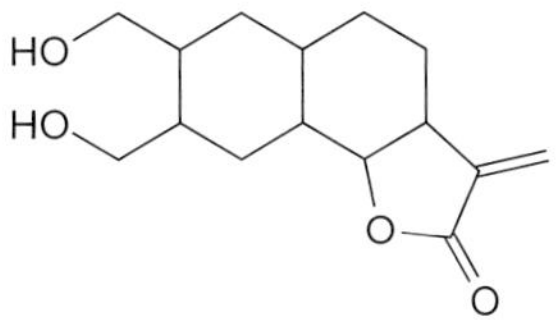

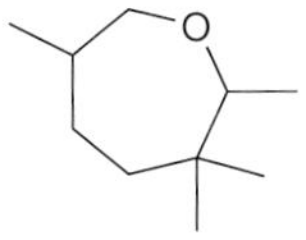

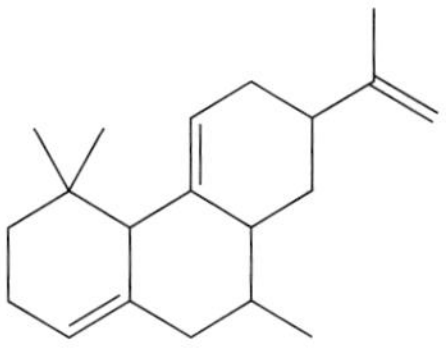

25.30 Guaiol is a sesquiterpene alcohol found in cannabis and other plants. The structure of guaiol is drawn except for a missing CH_3 group. If the isoprene units are joined in a head-to-tail fashion, draw a possible structure for guaiol that contains the additional CH_3 group.

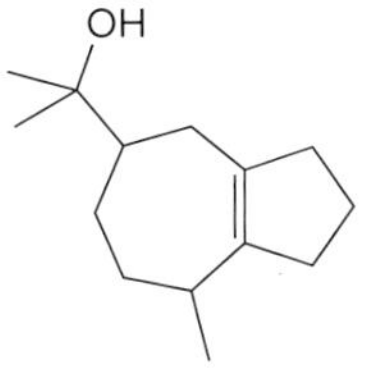

25.31 Draw a stepwise mechanism for the conversion of neryl diphosphate to α-pinene. α-Pinene is a component of pine oil and rosemary oil.

α-pinene

neryl diphosphate

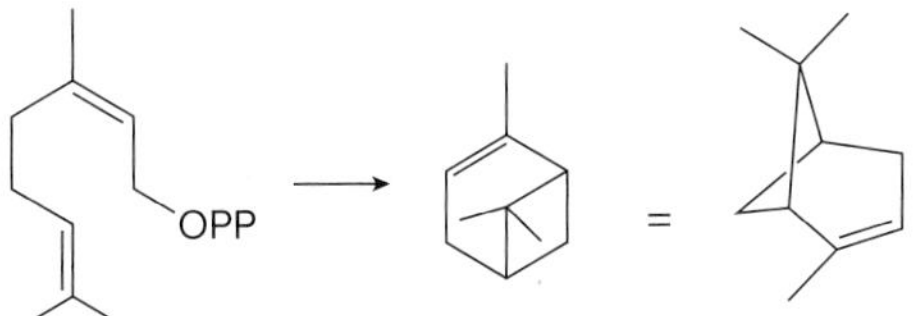

25.32 Flexibilene is a terpene isolated from *Sinularia flexibilis,* a soft coral found in the Indian Ocean. Draw a stepwise mechanism for the formation of flexibilene from farnesyl diphosphate and isopentenyl diphosphate. What is unusual about the cyclization that forms the 15-membered ring of flexibilene?

flexibilene

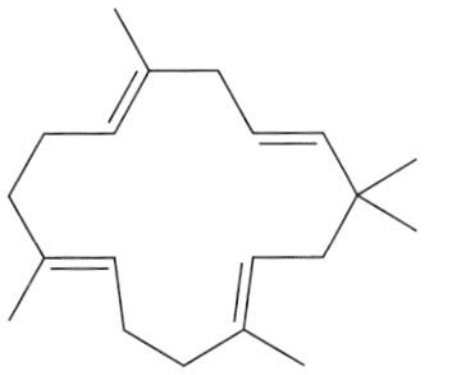

25.33 Draw the structure of a monoterpene that undergoes ozonolysis to yield the following two products. Show all possibilities.

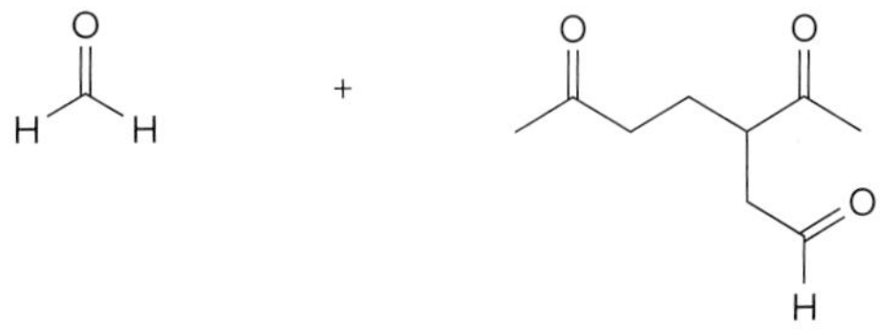

25.34 The biosynthesis of lanosterol from squalene has intrigued chemists since its discovery. It is now possible, for example, to synthesize polycyclic compounds from acyclic or monocyclic precursors by reactions that form several C—C bonds in a single reaction mixture.

  a. Draw a stepwise mechanism for the following reaction.

  b. Show how **X** can be converted to 16,17-dehydroprogesterone. (Hint: See Figure 18.3 for a related conversion.)

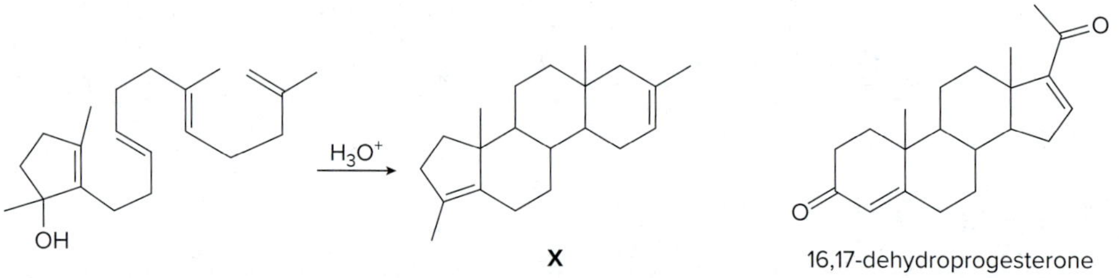

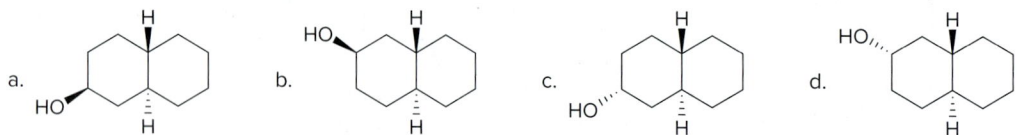

Steroids

25.35 Draw three-dimensional structures for each alcohol. Label the OH groups as occupying axial or equatorial positions.

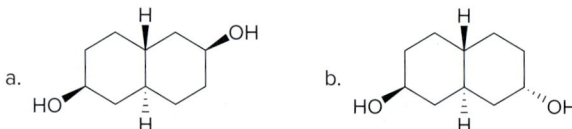

25.36 Axial alcohols are oxidized faster than equatorial alcohols by PCC and other Cr^{6+} oxidants. Which OH group in each compound is oxidized faster?

25.37 (a) Draw a skeletal structure of the anabolic steroid methenolone from the following description. Methenolone contains the tetracyclic steroid skeleton with a carbonyl group at C3, a hydroxyl at C17, a double bond between C1 and C2, and methyl groups bonded to C1, C10, and C13. (b) Add wedges and dashed wedges for all stereogenic centers with the following information: the configuration at C10 is R, the configuration at C13 is S, the configuration at C17 is S, and all substituents at ring fusions are trans to each other. (c) Draw the structure of Primobolan, the product formed when methenolone is treated with $CH_3(CH_2)_5COCl$ and pyridine. Primobolan is an anabolic steroid that can be taken orally or by injection and has been used illegally by well-known Major League Baseball players.

25.38 Betamethasone is a synthetic anti-inflammatory steroid used as a topical cream for itching. Betamethasone is derived from cortisol, with the following structural additions: a C=C between C1 and C2, a fluorine at C9, and a methyl group at C16. The configuration at C9 is R, and the configuration at C16 is S. Draw the structure of betamethasone.

25.39 a. Draw a three-dimensional structure for the following steroid.

  b. What is the structure of the single stereoisomer formed by reduction of this ketone with H_2, Pd-C? Explain why only one stereoisomer is formed.

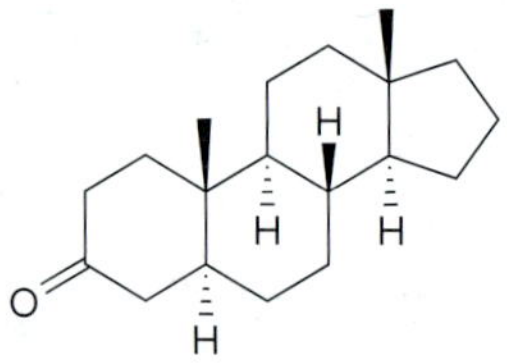

Challenge Problems

25.40 Draw a stepwise mechanism for the following conversion, which forms camphene. Camphene is a component of camphor and citronella oils.

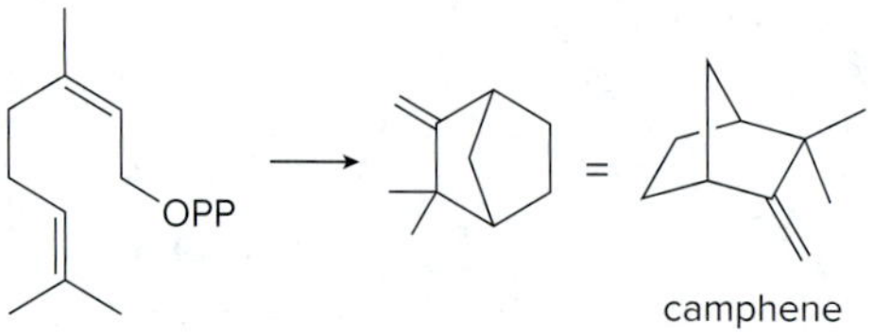

25.41 Draw a stepwise mechanism for the following reaction.

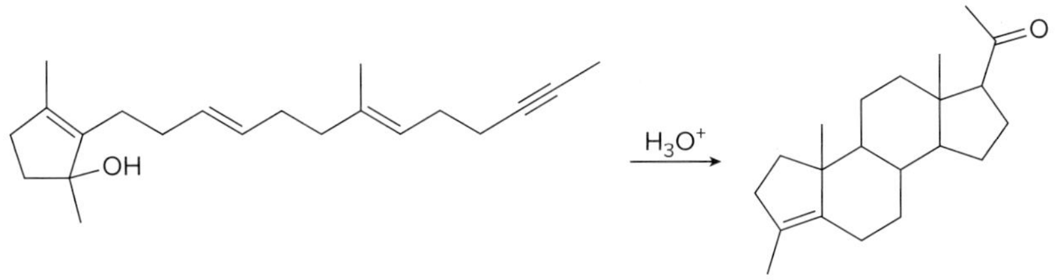

25.42 Farnesyl diphosphate is cyclized to sesquiterpene **A**, which is then converted to the bicyclic product epi-aristolochene. Write a stepwise mechanism for both reactions.

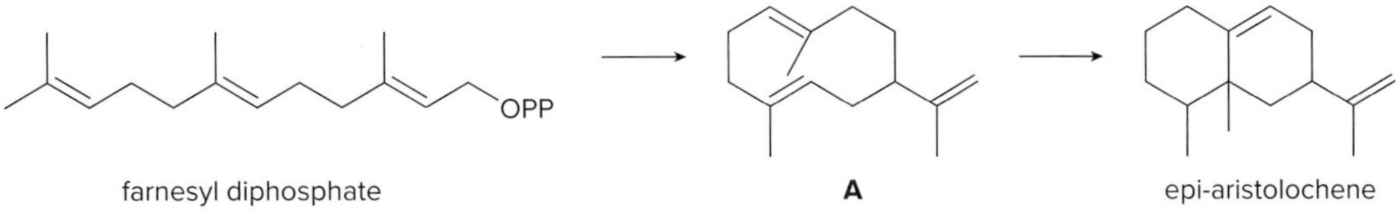

farnesyl diphosphate   **A**   epi-aristolochene

26 Nucleic Acids and Protein Synthesis

Daniel C. Smith

- **26.1** Nucleosides and nucleotides
- **26.2** Nucleic acids
- **26.3** The DNA double helix
- **26.4** Replication
- **26.5** Ribonucleic acids and transcription
- **26.6** The genetic code, translation, and protein synthesis
- **26.7** DNA sequencing
- **26.8** The polymerase chain reaction
- **26.9** Viruses

Marine sponges are a rich source of natural products with promising pharmaceutical potential. Novel biologically active agents isolated from the shallow-water Caribbean sponge *Tectitethya crypta* by Bergmann in 1950 led to the development of **cytarabine**, a drug used to treat various forms of leukemia and non-Hodgkin's lymphoma. Related synthetic nucleosides interfere with the ability of a virus to synthesize nucleic acids, so they are used to treat viral infections. In Chapter 26, we learn about nucleosides and nucleotides, as well as the nucleic acids DNA and RNA, the polymers derived from them, which store and transmit the genetic information of an organism.

Why Study...
Nucleic Acids?

Whether you are tall or short, fair-skinned or dark-complexioned, blue-eyed or brown-eyed, the nucleic acid polymers that reside in your cells determine your unique characteristics. The nucleic acid **DNA** stores the genetic information of a particular organism, whereas the nucleic acid **RNA** translates this genetic information into the synthesis of the proteins needed by cells for proper function and development. Minor alterations in the nucleic acid sequence can have significant effects on an organism, sometimes resulting in devastating diseases like sickle cell anemia and cystic fibrosis.

In Chapter 26, we learn about nucleic acids and the nucleotides from which they are formed.

26.1 Nucleosides and Nucleotides

Nucleic acids **are unbranched polymers composed of repeating monomers called nucleotides.** DNA and RNA are two types of nucleic acids.

- DNA, deoxyribonucleic acid, stores the genetic information of an organism and transmits that information from one generation to another.
- RNA, ribonucleic acid, translates the genetic information contained in DNA into proteins needed for all cellular functions.

26.1A Identifying and Naming Bases, Nucleosides, and Nucleotides

The prefix *deoxy* means *without oxygen*.

The nucleic acids are composed of three components: a monosaccharide, a heterocyclic aromatic base, and a phosphate group. The monosaccharide component of RNA is **D-ribose,** whereas that of DNA is **2'-deoxy-D-ribose,** a monosaccharide that lacks a hydroxy group at C2'. Primes (') are used in numbering the carbons of the monosaccharide components.

D-ribose
(present in RNA)

2'-deoxy-D-ribose
(present in DNA)

The heterocyclic bases in DNA were first discussed in Section 19.9B.

Five common heterocyclic bases are present in nucleic acids. Three bases with one ring (**cytosine, uracil,** and **thymine**) are derived from the parent compound **pyrimidine.** Two bicyclic bases (**adenine** and **guanine**) are derived from the parent compound **purine.** Each base is designated by a one-letter abbreviation.

pyrimidine
parent heterocycle

cytosine
C

uracil
U
(RNA only)

thymine
T
(DNA only)

purine
parent heterocycle

adenine
A

guanine
G

The N atom in **red** bonds to the monosaccharide in a nucleic acid.

Uracil (U) occurs only in RNA, whereas thymine (T) occurs only in DNA. As a result,

- DNA contains the bases A, G, C, and T.
- RNA contains the bases A, G, C, and U.

A *nucleoside* is an *N*-glycoside, formed by joining the anomeric carbon (C1') of the monosaccharide with N1 of a pyrimidine base or N9 of a purine base in a β-glycosidic linkage.

- Joining D-ribose with a base forms a *ribonucleoside*.
- Joining 2'-deoxy-D-ribose with a base forms a *deoxyribonucleoside*.

For example, the ribonucleoside cytidine is formed from D-ribose and cytosine. The deoxyribonucleoside 2'-deoxyadenosine is formed from 2'-deoxy-D-ribose and adenine.

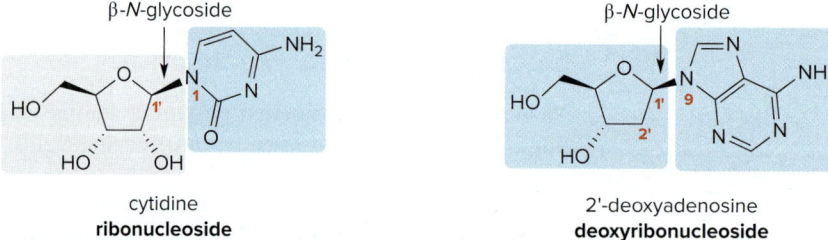

cytidine
ribonucleoside

2'-deoxyadenosine
deoxyribonucleoside

Nucleosides are named as derivatives of the bases from which they are formed.

- To name a nucleoside derived from a pyrimidine base, use the suffix –*idine* (cytosine → cyt*idine*).
- To name a nucleoside derived from a purine base, use the suffix –*osine* (adenine → aden*osine*).
- Add the prefix *deoxy-* for deoxyribonucleosides, as in *deoxy*adenosine.

Problem 26.1 Identify the base and monosaccharide in each nucleoside and then assign a name.

Problem 26.2 Novel antiviral agents isolated from Caribbean sponges led to the development of vidarabine, the first nucleoside drug used to treat herpes infections. From what you learned about monosaccharides in Chapter 24, determine what base and monosaccharide are present in vidarabine.

vidarabine

A *nucleotide* is formed by adding a phosphate group to the 5'-OH group of a nucleoside. Nucleotides are named by adding the term *5'-monophosphate* to the name of the nucleoside from which they are derived. At pH = 7 the phosphate is ionized, so the nucleotide bears a −2 charge.

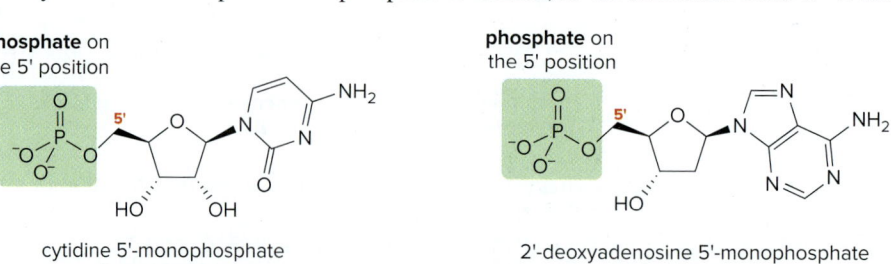

cytidine 5'-monophosphate
ribonucleotide

CMP

2'-deoxyadenosine 5'-monophosphate
deoxyribonucleotide

dAMP

26.1 Nucleosides and Nucleotides

Because of the lengthy names of nucleotides, three- or four-letter abbreviations are commonly used instead. Cytidine 5'-monophosphate is **CMP** and 2'-deoxyadenosine 5'-monophosphate is **dAMP**. Table 26.1 summarizes the names and abbreviations used for the bases, nucleosides, and nucleotides in nucleic acid chemistry.

Table 26.1 Names of Bases, Nucleosides, and Nucleotides in Nucleic Acids

Base	Abbreviation	Nucleoside	Nucleotide	Abbreviation
DNA				
Adenine	A	2'-deoxyadenosine	2'-deoxyadenosine 5'-monophosphate	dAMP
Guanine	G	2'-deoxyguanosine	2'-deoxyguanosine 5'-monophosphate	dGMP
Cytosine	C	2'-deoxycytidine	2'-deoxycytidine 5'-monophosphate	dCMP
Thymine	T	2'-deoxythymidine	2'-deoxythymidine 5'-monophosphate	dTMP
RNA				
Adenine	A	adenosine	adenosine 5'-monophosphate	AMP
Guanine	G	guanosine	guanosine 5'-monophosphate	GMP
Cytosine	C	cytidine	cytidine 5'-monophosphate	CMP
Uracil	U	uridine	uridine 5'-monophosphate	UMP

Sample Problem 26.1 Drawing the Structure of a Nucleotide

Draw the structure of each compound: (a) 2'-deoxyguanosine 5'-monophosphate; (b) UMP. Classify the nucleotide as a ribonucleotide or a deoxyribonucleotide.

Solution

Convert an abbreviation to the name of the nucleotide and then use these steps:

1. Draw the monosaccharide. If the name does not contain the prefix *deoxy*, the monosaccharide is D-ribose, making the compound a ribonucleotide. If the name contains the prefix *deoxy*, the monosaccharide is 2'-deoxy-D-ribose, making the compound a deoxyribonucleotide.
2. Add the base bonded to C1' of the monosaccharide ring, forming an *N*-glycoside with the β configuration.
3. Add the phosphate to the 5'-OH group.

a. For 2'-deoxyguanosine 5'-monophosphate:

2 Form an *N*-glycoside using N9 of guanine.

3 Add phosphate at the 5'-OH.

1 No OH group at C2': 2'-deoxy-D-ribose

2'-deoxyguanosine 5'-monophosphate
deoxyribonucleotide

b. UMP is the abbreviation for uridine 5'-monophosphate.

② Form an *N*-glycoside using N1 of uracil.

③ Add phosphate at the 5'-OH.

① No *deoxy* prefix: the monosaccharide is D-ribose.

uridine 5'-monophosphate
ribonucleotide

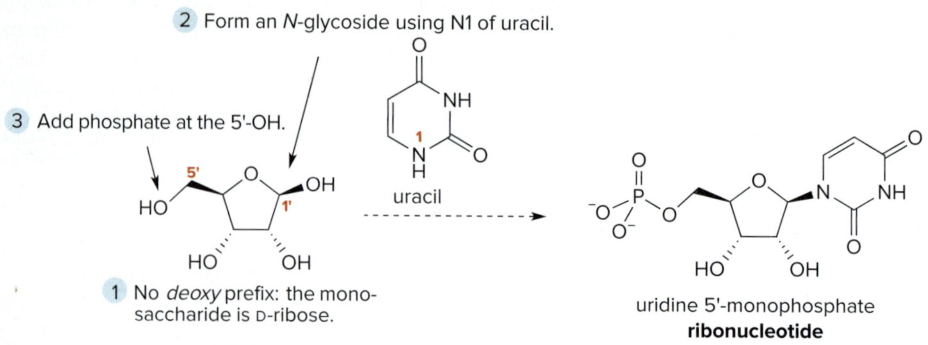

Problem 26.3 Draw the structure of each nucleotide: (a) dTMP; (b) AMP.

More Practice: Try Problems 26.21, 26.22.

DNA molecules contain several million deoxyribonucleotides, whereas RNA molecules are smaller, containing perhaps a few thousand ribonucleotides. DNA is contained in the chromosomes of the nucleus, and the number of chromosomes differs from species to species. Humans have 46 chromosomes (23 pairs). An individual chromosome is composed of many genes. A **gene** is a portion of the DNA molecule responsible for the synthesis of a specific protein.

Problem 26.4 Which nucleic acid (DNA or RNA) contains each of the following components: (a) ribose; (b) 2'-deoxyribose; (c) the base T; (d) the base U; (e) the nucleotide GMP; (f) the nucleotide dCMP?

26.1B Nucleotide Diphosphates and Nucleotide Triphosphates

Reactions that involve ADP and ATP were shown in Section 6.4.

Di- and triphosphates can also be prepared from nucleosides by adding two and three phosphate groups, respectively, to the 5'-OH. For example adenosine can be converted to adenosine 5'-diphosphate (abbreviated as **ADP**) and adenosine 5'-triphosphate (abbreviated as **ATP**). The central role of these phosphates in energy production is discussed in Chapter 27.

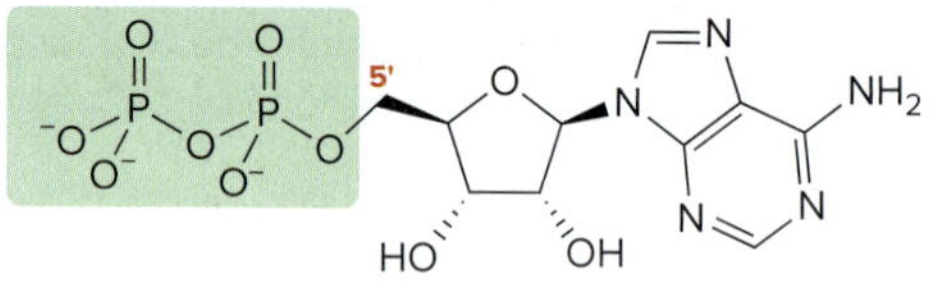

adenosine 5'-diphosphate
ADP

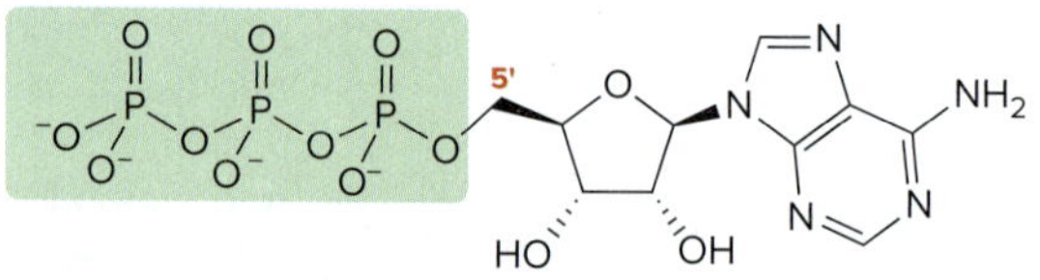

adenosine 5'-triphosphate
ATP

Problem 26.5 Give the name that corresponds to each abbreviation: (a) GTP; (b) dCDP; (c) dTTP; (d) UDP.

26.2 Nucleic Acids

Nucleic acids—both DNA and RNA—are polymers of nucleotides, formed by joining the 3'-OH group of one nucleotide with the 5'-phosphate of a second nucleotide in a **phosphodiester** linkage (Section 3.9).

26.2 Nucleic Acids

For example, joining the 3'-OH group of dCMP and the 5'-phosphate of dAMP forms a **dinucleotide** that contains a 5'-phosphate on one end—the **5' end**—and a 3'-OH group on the other end—the **3' end**.

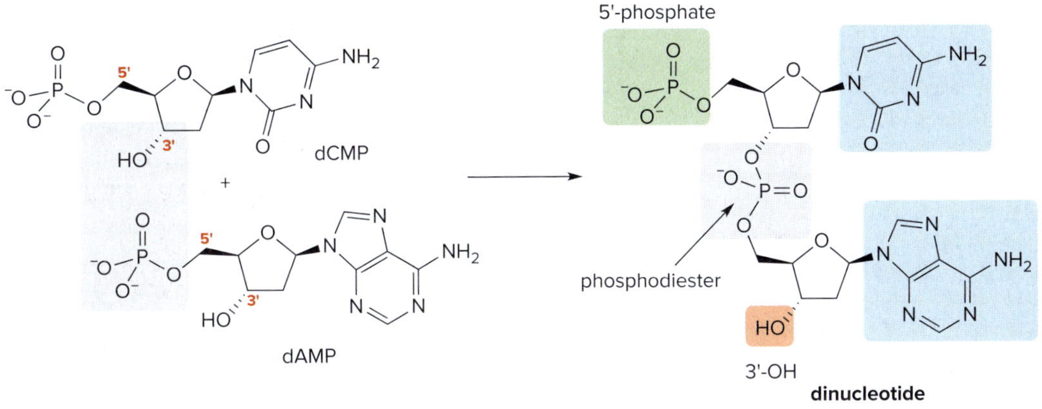

As additional nucleotides are added, the nucleic acid grows, each time forming a new phosphodiester linkage that holds the nucleotides together. **The primary structure of a polynucleotide is the sequence of nucleotides that it contains.** All polynucleotides contain a backbone of alternating sugar and phosphate groups. The identity and order of the bases distinguish one polynucleotide from another.

- A polynucleotide has one free phosphate at the 5' end and a free OH group at the 3' end.
- A polynucleotide is named by the sequence of bases it contains, beginning with the 5' end and using the one-letter abbreviations for the bases.

Figure 26.1 illustrates a polynucleotide that contains three phosphodiesters joining four different nucleotides.

Figure 26.1 Primary structure of a polynucleotide

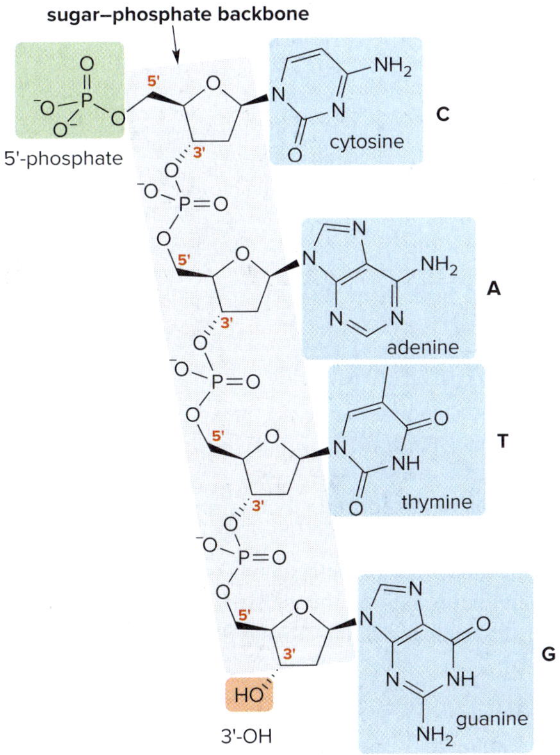

- A polynucleotide contains a backbone consisting of alternating sugar and phosphate groups, highlighted in gray.
- Phosphodiester bonds join the 3'-carbon of one nucleotide to the 5'-carbon of another.
- **A polynucleotide is named from the 5' end (in green) to the 3' end (in red),** using the one-letter abbreviations for the bases it contains.
- The structure of the polynucleotide CATG is drawn.

Sample Problem 26.2 — Drawing the Structure of a Trinucleotide

Draw the structure of the trinucleotide AUG.

Solution

- Because the base U occurs only in **RNA**, draw the structure of the individual nucleotides using **D-ribose** (with an OH group at C2') as the monosaccharide.
- Abbreviations identify the bases of the trinucleotide in order from the 5' end to the 3' end: the nucleotide with the 5'-phosphate contains adenine (A) and the nucleotide with the 3'-OH group contains guanine (G).
- Join the nucleotides together with phosphodiester bonds between the 3'-OH groups and 5'-phosphates to form two phosphodiesters.

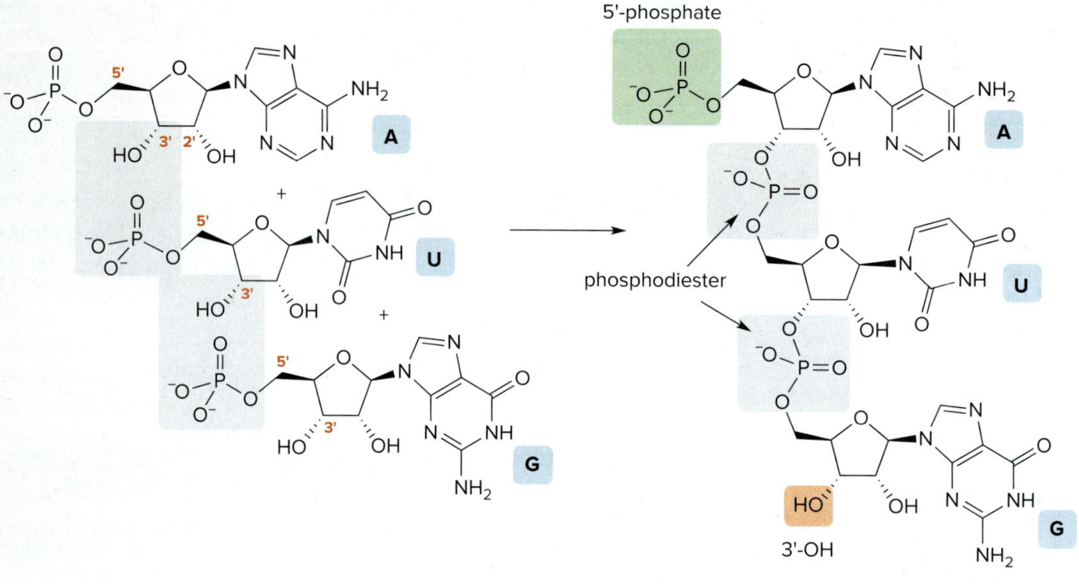

Problem 26.6 Draw the structure of each polynucleotide: (a) CU; (b) TAG.

More Practice: Try Problems 26.29, 26.30.

Problem 26.7 Consider the polynucleotide ATGGCG. (a) How many phosphodiester linkages does the polynucleotide contain? (b) Does the nucleotide at the 5' end contain a purine or pyrimidine base? (c) Could the polynucleotide be part of a DNA or an RNA molecule?

26.3 The DNA Double Helix

Our current understanding of the secondary structure of DNA is based on the model initially proposed by James Watson and Francis Crick in 1953 (Figure 26.2).

- **DNA consists of two polynucleotide strands that wind into a right-handed double helix.**

The sugar–phosphate backbone runs on the outside of the helix and the bases lie on the inside, perpendicular to the axis of the helix. **The two strands of DNA are antiparallel;** that is, one strand runs from the 5' end to the 3' end, while the other strand runs from the 3' end to the 5' end.

The double helix is stabilized by hydrogen bonding between the bases of the two DNA strands as shown in Figure 26.3. A purine base on one strand always hydrogen bonds with a pyrimidine

Figure 26.2 The double helix of DNA

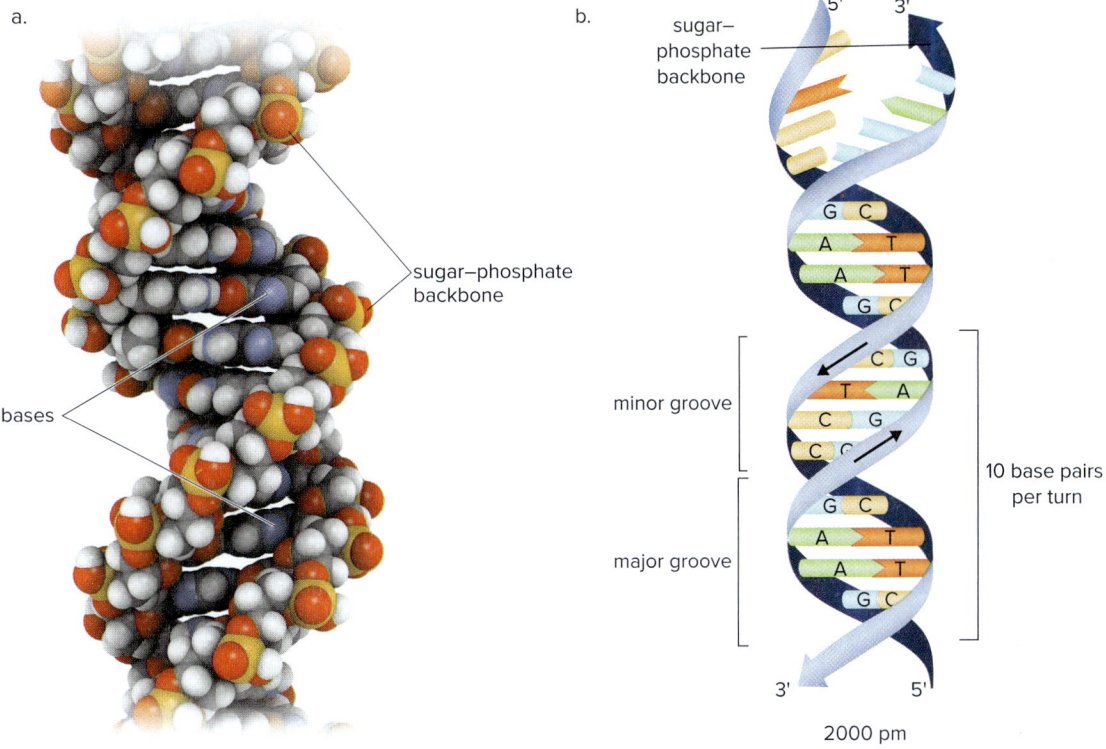

DNA consists of a double helix of polynucleotide chains. In view (a), the three-dimensional model shows the sugar–phosphate backbone visible on the outside of the helix. In the ribbon diagram in view (b), the bases in the interior are labeled, as are the major and minor grooves of the double helix.

Figure 26.3
Hydrogen bonding in the DNA double helix

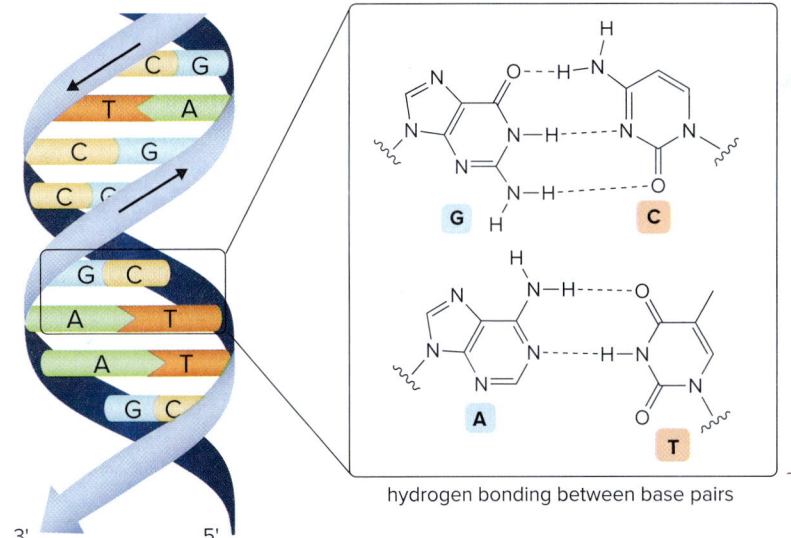

- Hydrogen bonding of base pairs (**A–T** and **G–C**) holds the two strands of DNA together.

base on the other strand. Two bases hydrogen bond in a predictable manner, forming **complementary base pairs.**

- Adenine pairs with thymine using two hydrogen bonds, forming an A–T base pair.
- Cytosine pairs with guanine using three hydrogen bonds, forming a C–G base pair.

The base pairs are stacked one on top of the other, with one complete turn of the helix containing 10 base pairs. The DNA double helix contains two grooves of different sizes, called the **major groove** and the **minor groove,** which run along the length of its cylindrical column (Figure 26.2b). Certain polycyclic aromatic compounds (Section 19.5) bind to the grooves in the DNA double helix.

Problem 26.8 Suggest reasons why the DNA double helix is arranged with the sugar–phosphate backbone on the outside of the double helix, and the base pairs on the inside.

Because of the consistent pairing of bases, we can write the sequence of the complementary strand of DNA when the sequence of one strand is known, as shown in Sample Problem 26.3.

Sample Problem 26.3 Predicting the Sequence of a Complementary Strand of a DNA Molecule

Write the sequence for the complementary strand of the following portion of a DNA molecule: 5'–TAGGCTA–3'.

Solution

The complementary strand runs in the opposite direction, from the 3' end to the 5' end. Use base pairing to determine the corresponding sequence on the complementary strand: A pairs with T and C pairs with G.

Original strand: 5'–T A G G C T A–3'

Complementary strand: 3'–A T C C G A T–5'

Problem 26.9 Write the complementary strand for each of the following strands of DNA.

a. 5'–AAACGTCC–3'   c. 5'–ATTGCACCCGC–3'
b. 5'–TATACGCC–3'   d. 5'–CACTTGATCGG–3'

More Practice: Try Problem 26.31.

Identical twins have the same genetic makeup, so that characteristics determined by DNA—such as hair color, eye color, or complexion—are also identical. Pictured are Matthew and Zachary Smith, identical twin sons of the author. *Daniel C. Smith*

The enormously large DNA molecules that compose the **human genome**—the total DNA content of an individual—pack tightly into the nucleus of the cell. **The genetic information of an organism is stored in the sequence of nucleotides in these DNA molecules.** How is this information transmitted from one generation to another? How, too, is the information in DNA molecules used to direct the synthesis of proteins?

To answer these questions, we must understand three key processes.

- *Replication*—the process by which DNA makes a copy of itself when a cell divides.
- *Transcription*—the ordered synthesis of RNA from DNA. In this process, the genetic information stored in DNA is passed onto RNA.
- *Translation*—the synthesis of proteins from RNA. In this process, the genetic message contained in RNA determines the specific amino acid sequence of a protein.

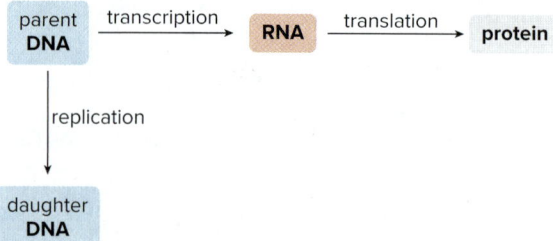

Each chromosome contains many **genes,** those portions of the DNA molecules that result in the synthesis of specific proteins. We say that the genetic message of the DNA molecule is *expressed* in the protein. Only a small fraction (1–2%) of the DNA in a chromosome contains genes that result in protein synthesis.

26.4 Replication

The genetic information in the DNA of a parent cell is passed onto daughter cells by the process of **semiconservative replication.** The strands of DNA separate and each serves as a template for a new strand. The original DNA molecule forms two DNA molecules, each of which contains one strand from the parent DNA and one new strand. The sequence of both strands of the daughter DNA molecules exactly matches the sequence of the parent DNA.

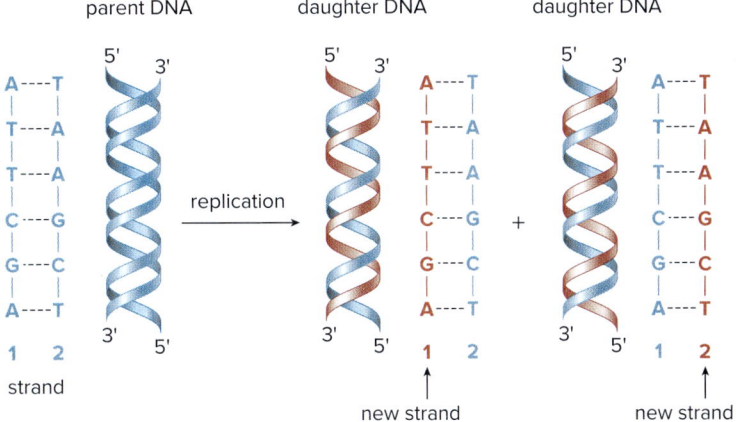

The first step in replication is the unwinding of the DNA helix to expose bases on each strand. Unwinding occurs at many places simultaneously along the helix, creating "bubbles" where replication can occur. Unwinding breaks the hydrogen bonds that hold the two strands of the double helix together (Figure 26.4).

Figure 26.4
DNA replication

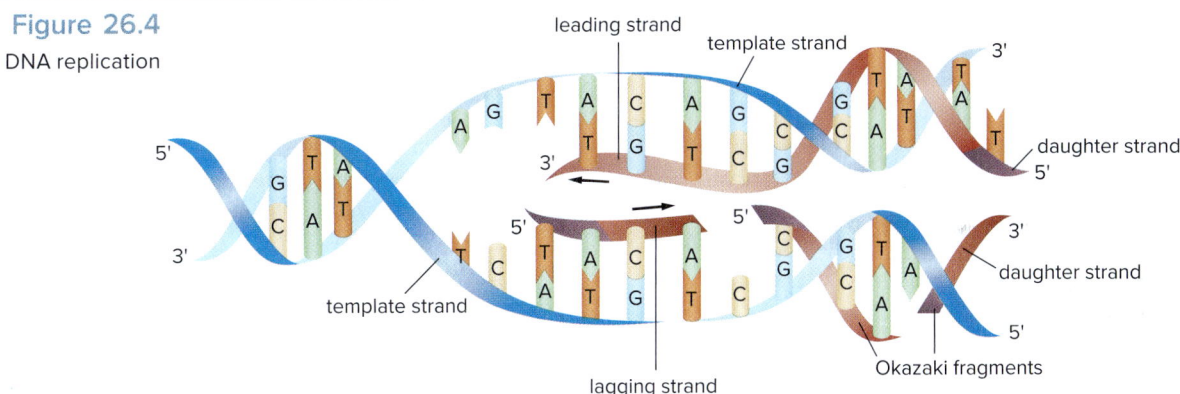

- Replication proceeds along both strands of the unwound DNA. Replication always occurs in the same direction, from the 3' end to the 5' end of the template. The leading strand grows continuously, whereas the lagging strand is synthesized in pieces that are then joined.

Once bases have been exposed in the unwound strands of DNA, the enzyme **DNA polymerase** catalyzes the replication process using the four nucleotide triphosphates (derived from the bases A, T, G, and C). A new phosphodiester bond is formed between the 5'-phosphate of the nucleoside triphosphate and the 3'-OH group of the new DNA strand. Two new strands of DNA grow from the ends of bubbles, called the replication forks.

- The identity of the bases on the template strand determines the order of bases in the new strand: A must pair with T, and G must pair with C.
- Replication occurs in only *one* direction on the template strand, from the 3' end to the 5' end, so the newly synthesized DNA grows from its 5' end to its 3' end.

Because replication proceeds in only one direction, the two new strands of DNA must be synthesized by different techniques. One strand, the **leading strand,** grows continuously from the 5' end to the 3' end, adding bases that are complementary to the template strand. The other strand, the **lagging strand,** is synthesized in small pieces called **Okazaki fragments,** which are joined together by a **DNA ligase** enzyme. The end result is two new strands of DNA, one in each of the daughter DNA molecules, both with complementary base pairs joining the two DNA strands together.

Problem 26.10 What is the sequence of a newly synthesized DNA segment if the template strand has each of the following sequences?

a. 3'–AGAGTCTC–5'
b. 3'–ATTGCTC–5'
c. 3'–ATCCTGTAC–5'
d. 3'–GGCCATACTC–5'

26.5 Ribonucleic Acids and Transcription

26.5A RNA

Ribonucleic acids (RNAs) are composed of nucleotides, but there are significant differences between DNA and RNA. In RNA,

- The sugar is D-ribose.
- Uracil (U) replaces thymine (T) as one of the bases.
- RNA is single stranded.

Although RNA molecules are much smaller than DNA molecules, a single strand of RNA can fold back on itself, forming loops and helical regions that are stabilized by intramolecular hydrogen bonding.

Three different types of RNA are involved in protein synthesis: **ribosomal RNA (rRNA), messenger RNA (mRNA),** and **transfer RNA (tRNA).**

Ribosomal RNA, the most abundant type of RNA, is found in the ribosomes in the cytoplasm of the cell. rRNA provides the site where polypeptides are assembled during protein synthesis.

Messenger RNA is the carrier of information from DNA in the nucleus to the ribosomes in the cytoplasm. Each gene of a DNA molecule corresponds to a specific mRNA molecule. **The sequence of nucleotides in the mRNA molecule determines the amino acid sequence in a particular protein.**

Transfer RNA interprets the genetic information in mRNA and brings specific amino acids to the site of protein synthesis in the ribosome. Each tRNA contains a sequence of three nucleotides called an **anticodon,** which is complementary to three bases in an mRNA molecule, and identifies what amino acid must be added to a growing polypeptide chain. tRNA molecules are often drawn in a cloverleaf fashion (Figure 26.5a). A model that depicts the three-dimensional structure of a tRNA is shown in Figure 26.5b. A particular amino acid may be recognized by one or more tRNA molecules.

Each tRNA also has an acceptor stem at the 3' end that always contains the nucleotides ACC (also shown in Figure 26.5a). The free 3'-OH group at this end is esterified with the α-carboxy group of a specific amino acid.

tRNA bonded to an amino acid

26.5 Ribonucleic Acids and Transcription

Figure 26.5 Transfer RNA

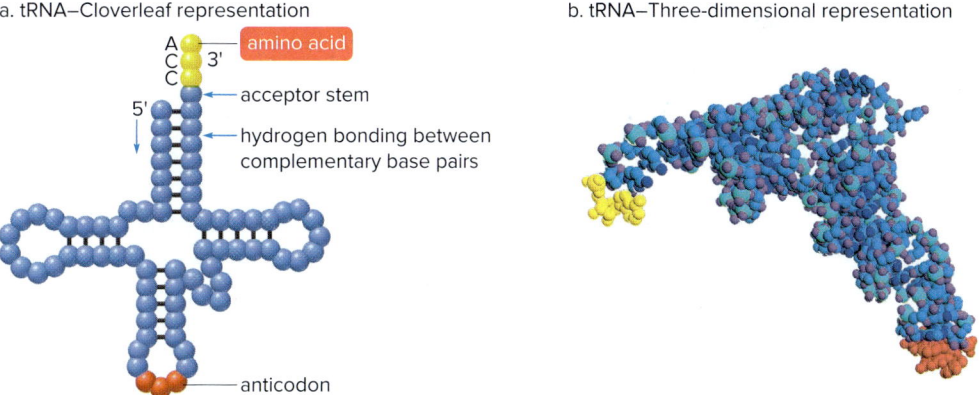

a. tRNA–Cloverleaf representation

b. tRNA–Three-dimensional representation

- tRNAs contain 70–90 nucleotides. Folding of the tRNA molecule creates regions in which complementary base pairs hydrogen bond to each other. Each tRNA binds a specific amino acid to its 3' end and contains an anticodon that identifies that amino acid for protein synthesis.

- In the three-dimensional model of a tRNA, the binding site for the amino acid is shown in yellow and the anticodon is shown in red. *Kenneth Edward/Science Source*

26.5B Transcription

The conversion of the information in DNA to the synthesis of proteins begins with *transcription—the synthesis of mRNA from DNA.*

RNA synthesis begins when the double helix of DNA unwinds, and a complementary strand of mRNA is synthesized from one strand of DNA, called the **template strand.** The strand of DNA not used for mRNA synthesis is called the **coding strand.**

Transcription proceeds from the 3' end to the 5' end of the template strand using an RNA polymerase enzyme (Figure 26.6). Complementary base pairing determines the order of RNA

Figure 26.6 Transcription of DNA to RNA

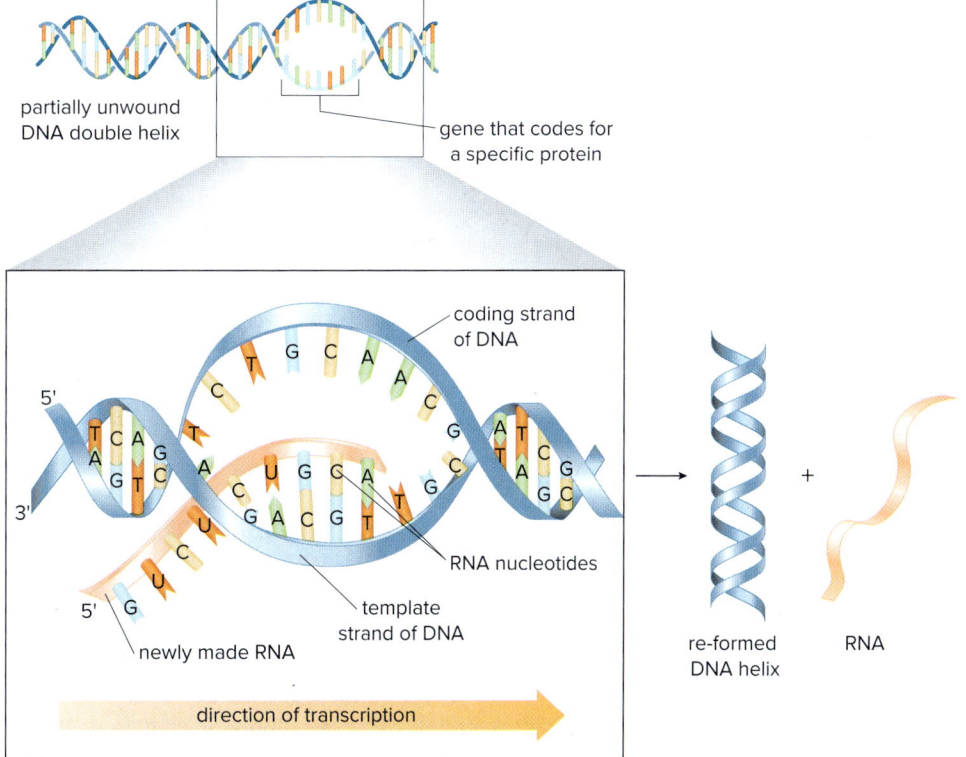

- Transcription proceeds from the 3' end to the 5' end of the template strand, so the mRNA bases are complementary to those in the DNA template.

nucleotides added to the growing RNA chain: **C pairs with G, T pairs with A, and A pairs with U.** Transcription is completed when a particular sequence of bases on the DNA template is reached. The new mRNA molecule is released and the double helix of the DNA molecule is re-formed.

In bacteria, the new mRNA molecule is ready for protein synthesis after it is prepared. In humans, the mRNA molecule first formed is modified before it is ready for protein synthesis by removing and splicing together pieces of mRNA by mechanisms that are not presented here.

- mRNA has a sequence *complementary* to the DNA template strand from which it is prepared.
- mRNA is an *exact copy* of the coding strand of DNA, except that the base U replaces T on the mRNA strand.

Sample Problem 26.4 Determining the Sequence of mRNA from DNA

Write the sequence of mRNA formed from the following template strand of DNA: 3'–CTAGGATAC–5'. Write the sequence of the coding strand of this segment of DNA.

Solution

mRNA has a base sequence that is *complementary* to the template from which it is prepared. On the other hand, mRNA has a base sequence that is *identical* to the coding strand of DNA, except that it contains the base U instead of T.

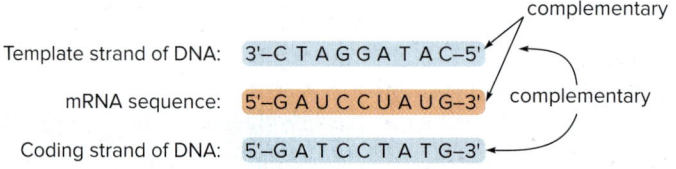

Template strand of DNA: 3'–C T A G G A T A C–5'
mRNA sequence: 5'–G A U C C U A U G–3'
Coding strand of DNA: 5'–G A T C C T A T G–3'

Problem 26.11 For each DNA segment: [1] What is the sequence of the mRNA molecule synthesized from each DNA template? [2] What is the sequence of the coding strand of the DNA molecule?

a. 3'–TGCCTAACG–5'
b. 3'–GACTCC–5'
c. 3'–TTAACGCGA–5'
d. 3'–CAGTGACCGTAC–5'

More Practice: Try Problems 26.36.

Problem 26.12 What is the sequence of the DNA template strand from which each of the following mRNA strands was synthesized?

a. 5'–UGGGGCAUU–3'
b. 5'–GUACCU–3'
c. 5'–CCGACGAUG–3'
d. 5'–GUAGUCACG–3'

26.6 The Genetic Code, Translation, and Protein Synthesis

26.6A The Genetic Code

How can the four different nucleotides in mRNA direct the synthesis of proteins that are formed from 20 amino acids? The answer lies in the **genetic code.**

- The genetic code is the set of three-nucleotide units in mRNA called *codons* that correspond to particular amino acids. As a result, a series of codons in mRNA determines the amino acid sequence in a protein.

26.6 The Genetic Code, Translation, and Protein Synthesis

For example, the codon UCA in mRNA codes for the amino acid serine, whereas the codon UGC codes for cysteine. The same genetic code occurs in almost all organisms, from bacteria to whales to humans.

Given four different nucleotides (A, C, G, and U), there are 64 different ways to combine them into groups of three, so there are 64 different codons. Sixty-one codons code for specific amino acids, so many amino acids correspond to more than one codon, as shown in Table 26.2. For example, GGU, GGC, GGA, and GGG all code for the amino acid glycine. Three codons—UAA, UAG, and UGA—do not correspond to any amino acids; they are called **stop codons** because they signal the stop of protein synthesis.

Table 26.2 The Genetic Code—Triplets in Messenger RNA

First Base (5' end)	Second Base								Third Base (3' end)
	U		C		A		G		
U	UUU	Phe	UCU	Ser	UAU	Tyr	UGU	Cys	U
	UUC	Phe	UCC	Ser	UAC	Tyr	UGC	Cys	C
	UUA	Leu	UCA	Ser	UAA	Stop	UGA	Stop	A
	UUG	Leu	UCG	Ser	UAG	Stop	UGG	Trp	G
C	CUU	Leu	CCU	Pro	CAU	His	CGU	Arg	U
	CUC	Leu	CCC	Pro	CAC	His	CGC	Arg	C
	CUA	Leu	CCA	Pro	CAA	Gln	CGA	Arg	A
	CUG	Leu	CCG	Pro	CAG	Gln	CGG	Arg	G
A	AUU	Ile	ACU	Thr	AAU	Asn	AGU	Ser	U
	AUC	Ile	ACC	Thr	AAC	Asn	AGC	Ser	C
	AUA	Ile	ACA	Thr	AAA	Lys	AGA	Arg	A
	AUG	Met	ACG	Thr	AAG	Lys	AGG	Arg	G
G	GUU	Val	GCU	Ala	GAU	Asp	GGU	Gly	U
	GUC	Val	GCC	Ala	GAC	Asp	GGC	Gly	C
	GUA	Val	GCA	Ala	GAA	Glu	GGA	Gly	A
	GUG	Val	GCG	Ala	GAG	Glu	GGG	Gly	G

A codon is written with the 5' to 3' end of mRNA. **The 5' end of the mRNA molecule codes for the N-terminal amino acid** in a protein, whereas **the 3' end of the mRNA codes for the C-terminal amino acid.**

Problem 26.13 Consider the following mRNA sequence: 5'–CAUAAAACGGAG–3'. (a) What is the N-terminal amino acid coded for by this sequence? (b) What is the C-terminal amino acid?

Problem 26.14 Sometimes codons for amino acids with similar types of side chains (i.e., acidic, basic, hydrophobic, or aromatic) have similarities. Compare the codons for the amino acids aspartic acid, glutamic acid, leucine, phenylalanine, and valine. Comment on the relationship between amino acid structure and codon identity.

26.6B Translation

The translation of the information in mRNA to protein synthesis occurs in the ribosomes at binding sites on rRNA.

mRNA contains the sequence of codons that determines the order of amino acids in the protein. Individual tRNAs bring specific amino acids to add to the peptide chain. Each tRNA contains an **anticodon** of three nucleotides that is complementary to the codon in mRNA and identifies individual amino acids. For example, a codon of UCA in mRNA corresponds to an anticodon of AGU in a tRNA molecule, which identifies serine as the amino acid.

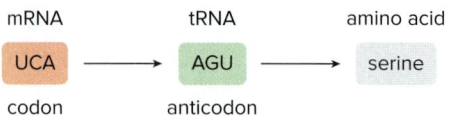

Translation begins when the first codon of an mRNA molecule binds to a ribosome, and a tRNA molecule, which contains the anticodon of the codon, carries the first amino acid of the peptide chain to the binding site. As mentioned in Section 26.5A, each tRNA is esterified to an individual amino acid. The new peptide bond is formed by **nucleophilic acyl substitution** of the amino group of one tRNA-bonded amino acid with the ester carbonyl of another, as shown in Figure 26.7.

Figure 26.7
Translation and the synthesis of the peptide bond

- mRNA contains the codons that determine the sequence of amino acids of a peptide. Each tRNA is bonded to a specific amino acid and contains an anticodon that binds to the mRNA. A peptide bond forms between two amino acids by nucleophilic acyl substitution, and the peptide chain grows until a stop codon is reached. Depicted is the synthesis of a methionine–serine dipeptide.

As each successive codon on the mRNA is read, new tRNAs deliver the next amino acids, peptide bonds are formed, and the protein chain grows until a stop codon signals that synthesis is complete.

Figure 26.8 shows a representative segment of DNA, and the mRNA, tRNA, and amino acid sequences that correspond to it.

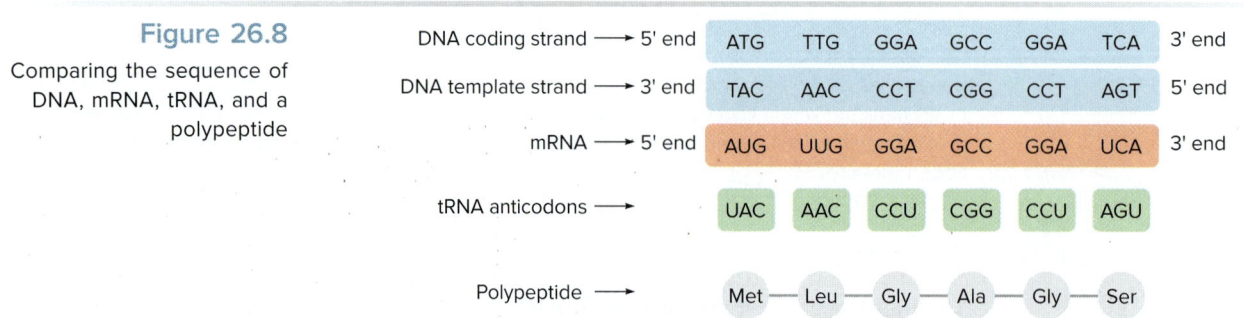

Figure 26.8
Comparing the sequence of DNA, mRNA, tRNA, and a polypeptide

Sample Problem 26.5 Deriving an Amino Acid Sequence from DNA

What polypeptide would be synthesized from the following template strand of DNA:

3'–CGGTGTCTTTTA–5'?

Solution

To determine what polypeptide is synthesized from a DNA template, two steps are needed.
- Use the DNA sequence to determine the transcribed mRNA sequence: C pairs with G, T pairs with A, and A (on DNA) pairs with U (on mRNA).
- Use the codons in Table 26.2 to determine what amino acids are coded for by a given codon in mRNA.

DNA template strand: ⟶ 3' CGG TGT CTT TTA 5'

mRNA: ⟶ 5' GCC ACA GAA AAU 3'

Polypeptide: ⟶ Ala — Thr — Glu — Asn

Problem 26.15 What polypeptide would be synthesized from each of the following template strands of DNA?

a. 3'–TCTCATCGTAATGATTCG–5'   b. 3'–GCTCCTAAATAACACTTA–5'

More Practice: Try Problems 26.39–26.43.

Problem 26.16 What sequence of amino acids would be formed from each mRNA sequence? List the anticodons contained in each of the needed tRNA molecules.

a. 5'–CCACCGGCAAACGAAGCA–3'
b. 5'–GCACCACUAAGAGAC–3'

Problem 26.17 Consider a template strand of DNA with the following sequence: 3'–ATGAAAGCCTTCTGT–5'. (a) What is the coding strand of DNA that corresponds to this template? (b) What mRNA is prepared from this template? (c) What polypeptide is prepared from the mRNA?

Problem 26.18 Fill in the base, codon, anticodon, or amino acid needed to complete the following table that relates the sequences of DNA, mRNA, tRNA, and the resulting polypeptide.

DNA coding strand:	5' end	AAC					3' end
DNA template strand:	3' end		CAT				5' end
mRNA codons:	5' end			UCA		AUG	3' end
tRNA anticodons:					GUG		
Polypeptide:					Thr		

26.7 DNA Sequencing

DNA sequencing has proven to be valuable methodology for determining the sequence of specific genes, individual chromosomes, and even the full genome of an organism. Determining the structure of genes that are associated with specific diseases has allowed scientists to understand how to prevent or cure them.

Because of the large size of DNA molecules, DNA is first cleaved into smaller units and the smaller fragments of DNA are then sequenced individually. Cleavage is carried out with **restriction endonucleases,** enzymes that cleave DNA at specific sequences of bases. Each restriction endonuclease recognizes a particular sequence of bases and cuts *both* strands of DNA in an identical manner.

For example, the enzyme EcoRI recognizes the sequence GAATTC and cuts the DNA molecule between G and A on both strands. The enzyme SmaI, on the other hand, recognizes the sequence CCCGGG and cuts the molecule between C and G.

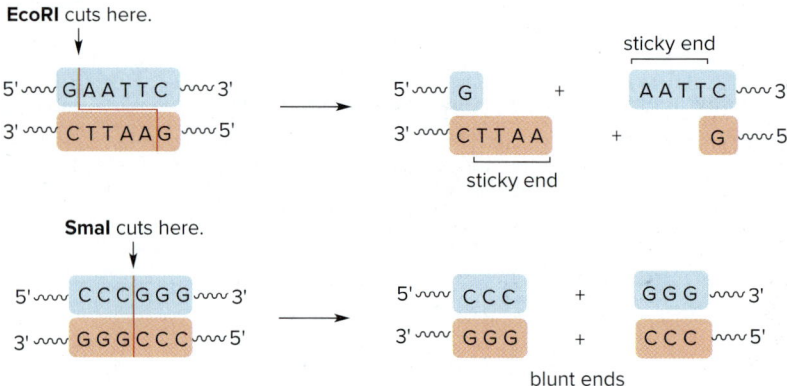

Cleavage with EcoRI affords strands of DNA of different length, with **sticky ends** that have unpaired bases, whereas cleavage with SmaI affords DNA fragments with **blunt ends.** Thousands of restriction enzymes are known and hundreds are commercially available. By cleaving DNA with a variety of restriction endonucleases, sequencing the fragments, and determining overlapping sequences, the DNA sequence of long strands of DNA has been determined.

Problem 26.19 Two other restriction endonucleases are HindIII, which cuts DNA between A and A in the sequence AAGCTT, and HaeIII, which cuts DNA between G and C in the sequence GGCC. Label the cleavage sites in the following segment of DNA. Only one strand of double-stranded DNA is provided.

5'–CGCGAATTGGCCGTAAGCTTACGTCCTAGGGCTACTCCTCGGCCCAATAAAGCTT–3'

In 1980 Frederick Sanger and Walter Gilbert shared the Nobel Prize in Chemistry for the development of methods to sequence DNA.

Early methods of DNA sequencing were developed in the 1970s by Frederick Sanger in Cambridge, England, and Walter Gilbert of Harvard University. Sanger sequencing was the most common method of DNA sequencing for 20 years, and early automated DNA sequencers were based on this technology. DNA sequencing methods have been used to sequence the entire human genome, which consists of 3.1 billion base pairs. It was first reported in preliminary form in 2001 and completed in 2003.

Next-generation DNA synthesizers have been developed since 2000, which have increased the speed and decreased the cost of DNA sequencing. As a comparison, the U.S. government spent $2.7 billion on the Human Genome Project to sequence the human genome from 1990 to 2003. This figure includes the total cost of all activities related to the Human Genome Project, including technology development, ethics research, and program management, as well as determining the framework for organizing the data obtained from sequencing individual segments of DNA. It is estimated that sequencing itself cost somewhere between $500 million and $1 billion.

Using the sophisticated technology available today, as well as the competitive pricing offered by several commercial enterprises, the National Human Genome Research Institute estimated that in 2016, the DNA sequence of an organism could be obtained for under $1000.

26.8 The Polymerase Chain Reaction

In order to study a specific gene, millions of copies of pure gene are needed. In fact, virtually an unlimited number of copies of any gene can be synthesized in just a few hours using a technique called the **polymerase chain reaction (PCR)**. PCR *clones* a segment of DNA; that is, PCR produces exact copies of a fragment of DNA.

- PCR *amplifies* a specific portion of a DNA molecule, producing millions of copies of a single molecule.

Four elements are needed to amplify DNA by PCR:

- The **segment of DNA** that must be copied
- **Two primers**—short polynucleotides that are complementary to the two ends of the segment to be amplified
- A **DNA polymerase enzyme** that will catalyze the synthesis of a complementary strand of DNA from a template strand
- **Nucleoside triphosphates** that serve as the source of the nucleotides A, T, C, and G needed in the synthesis of the new strands of DNA

Each cycle of the polymerase chain reaction involves three steps, illustrated in Figure 26.9.

> PCR was developed by biochemist Kary Mullis of Cetus Corporation, who shared the 1993 Nobel Prize in Chemistry for its discovery.

Figure 26.9 Using the polymerase chain reaction to amplify a sample of DNA

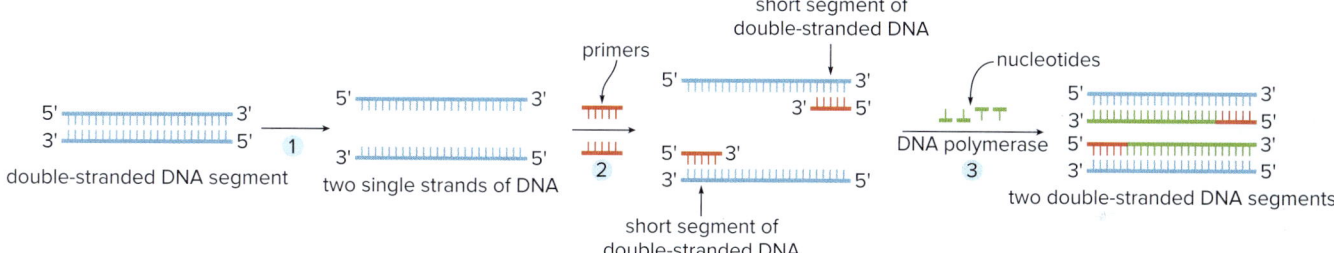

1. The DNA sample is heated to unwind the double helix into two single strands.
2. Primers are added to form a short segment of double-stranded DNA on each strand, to which a DNA polymerase enzyme can add new nucleotides to the 3' end.
3. DNA polymerase is used to add nucleotides to lengthen the DNA segment.
   - DNA polymerase catalyzes the synthesis of a new strand of DNA complementary to the existing strand using nucleoside triphosphates (dATP, dCTP, dGTP, and dTTP) available in the reaction mixture.
   - After one three-step cycle, *one* molecule of double-stranded DNA forms *two* molecules of double-stranded DNA.

Each double-stranded DNA molecule synthesized by this method contains one original strand and one newly synthesized strand. After each cycle, the amount of DNA doubles. After 20 cycles, about one million copies have been made.

Each step of a PCR cycle is carried out at a different temperature. PCR is now a completely automated process using a thermal cycler, an apparatus that controls the heating and cooling needed for each step. A heat-tolerant DNA polymerase called **Taq polymerase** is also typically used, so that new enzyme need not be added as each new cycle begins.

PCR is an indispensable method in clinical chemistry and forensic analysis. PCR is used in diagnosing genetic diseases and in determining paternity. Forensic scientists compare DNA collected from a crime scene with those of a suspect by cleaving DNA from both sources with restriction endonucleases, which are then amplified using the polymerase chain reaction.

26.9 Viruses

A *virus* is an infectious agent consisting of a DNA or RNA molecule that is contained within a protein coating. Because a virus has no enzymes or free nucleotides of its own, it is incapable of replicating until it invades a host organism and takes over the biochemical machinery of the host.

A virus that contains DNA uses the materials in the host organism to replicate DNA, transcribe DNA to RNA, and synthesize a protein coating, thus forming new virus particles that can infect new host cells. The common cold, influenza, and herpes are viral in origin.

A virus that contains RNA is called a **retrovirus.** When a retrovirus invades a host organism, it must first make DNA by the process of **reverse transcription.** Once viral DNA is synthesized, the DNA can transcribe RNA, synthesize protein, and form new retrovirus particles that can infect new host cells.

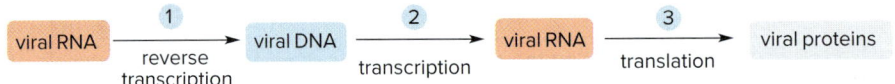

The use of protease inhibitors to treat HIV was discussed in Section 23.10C.

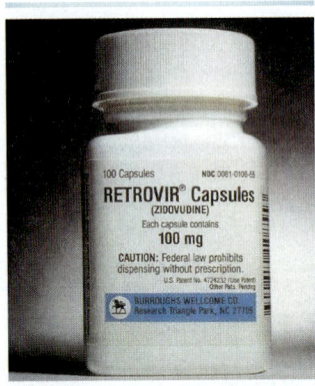

AZT (also known as zidovudine and sold under the trade name Retrovir) has been available since the 1990s for the treatment of HIV. *James Keyser/The LIFE Images Collection/Getty Images*

AIDS (acquired immune deficiency syndrome) is caused by HIV (human immunodeficiency virus), a retrovirus that attacks lymphocytes central to the body's immune response. An individual infected with HIV becomes susceptible to life-threatening bacterial infections.

HIV is currently best treated with a "cocktail" of drugs designed to destroy the virus at different stages of its reproductive cycle. One group of drugs, which includes AZT (azidodeoxythymidine) and ddI (dideoxyinosine), consists of nucleoside analogues designed to interfere with reverse transcription. These drugs are incorporated in viral DNA during reverse transcription, but because each drug lacks a 3'-OH group, no additional nucleotide can be added to the growing DNA chain, halting viral DNA synthesis.

azidodeoxythymidine
AZT

dideoxyinosine
ddI

Problem 26.20 Lamivudine is an antiviral drug formed from heterocycle **A** and cytosine. Draw the structure of lamivudine and explain why it is an effective antiviral agent.

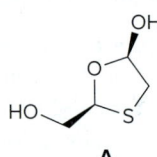

A

Chapter 26 REVIEW

KEY CONCEPTS

A comparison of DNA and RNA (26.1, 26.3, 26.5)

Nucleic acid	DNA	RNA
1 Structure	DNA is composed of a right-handed **double helix with two strands** of deoxyribonucleotides winding in an antiparallel fashion. 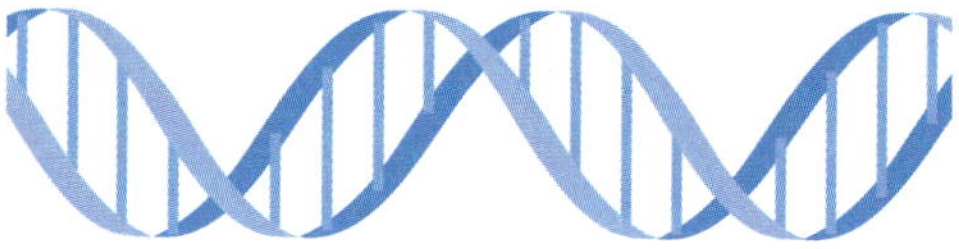	RNA contains a **single strand** of ribonucleotides. 
2 Monosaccharide	The monosaccharide component of DNA is **2'-deoxy-D-ribose**. 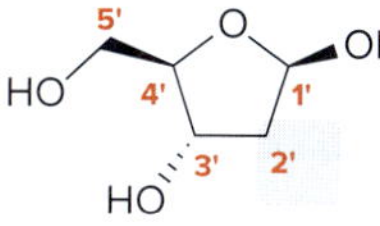	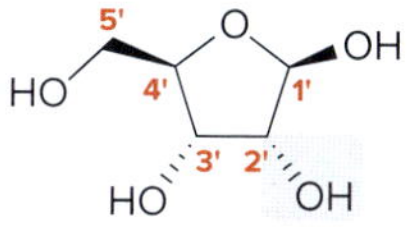 The monosaccharide component of RNA is **D-ribose**.
3 Bases	DNA contains the bases A, G, C, and **T**. 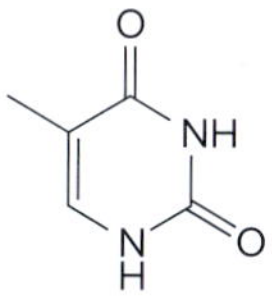 (thymine T) cytosine, C; adenine, A; guanine, G	RNA contains the bases A, G, C, and **U**. 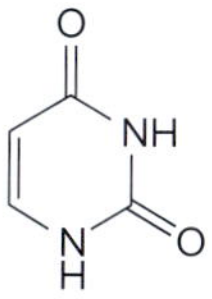 (uracil U) cytosine, C; adenine, A; guanine, G

KEY SKILLS

[1] Drawing the structure of a dinucleotide (26.1, 26.2); example: TC

1 Identify the monosaccharide.

- Bases A, C, and G are present in both DNA and RNA.
- T is present only in DNA and U is present only in RNA.
- Because the dinucleotide is TC, the base is 2'-deoxy-D-ribose.

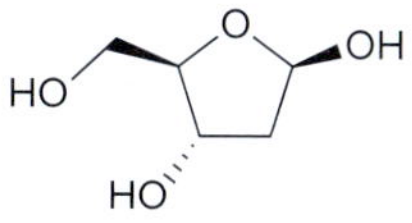

2 Use the abbreviations to draw the nucleotides.

- Abbreviations identify the bases in order from the 5' end (T) to the 3' end (C).
- The nucleotides in TC are dTMP and dCMP.

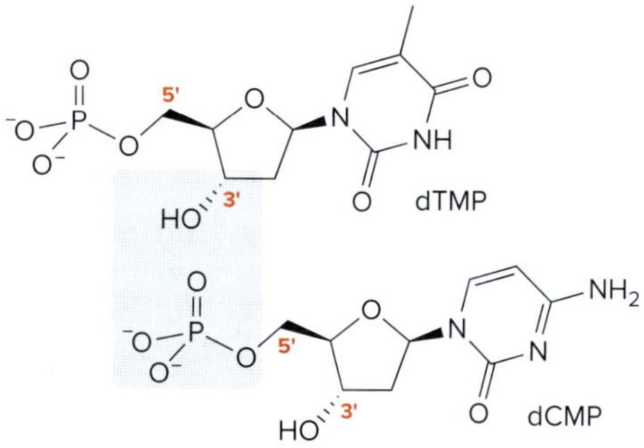

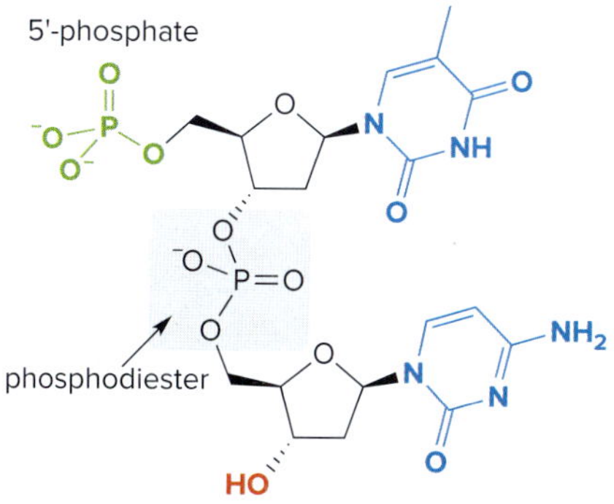

3 Join the 3'-OH group of one nucleotide to the 5'-phosphate of the other in a phosphodiester bond.

dinucleotide TC

See Sample Problem 26.2. Try Problems 26.29, 26.30.

[2] Predicting the sequence of a complementary strand of a DNA segment (26.3); example: 5'–ATCCGTGTA–3'

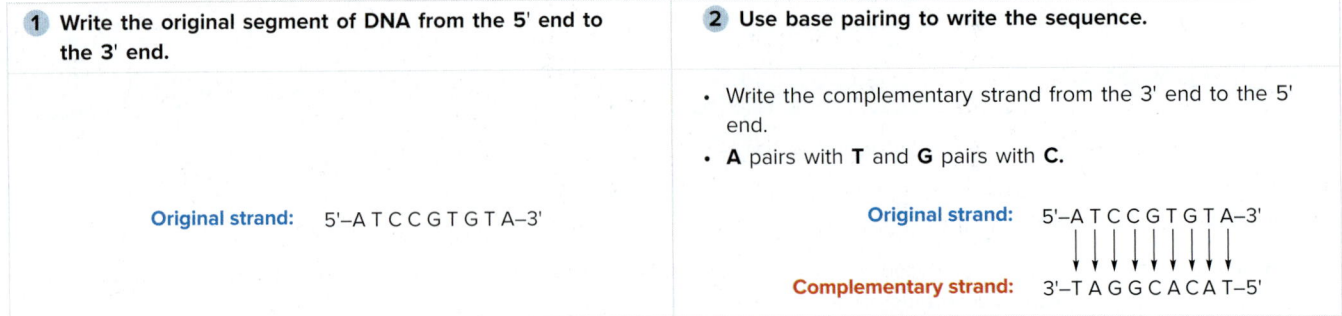

See Sample Problem 26.3. Try Problem 26.31.

[3] Drawing the newly synthesized strand of DNA formed during replication (26.4); example: 5'–GCGATTCCGT–3'

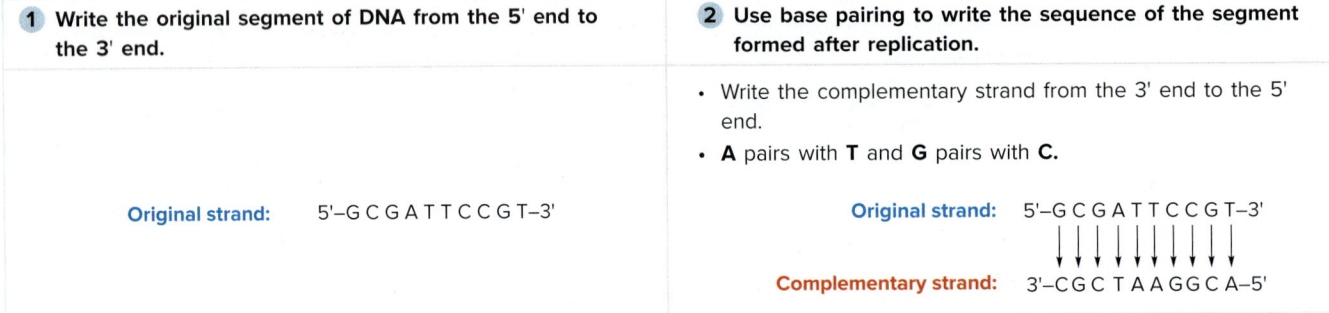

Try Problem 26.35.

[4] Using a DNA template strand to determine an mRNA sequence (after transcription) and the sequence of the coding DNA strand (26.5); example: a DNA template strand with the sequence 3'–AGTATGACG–5'

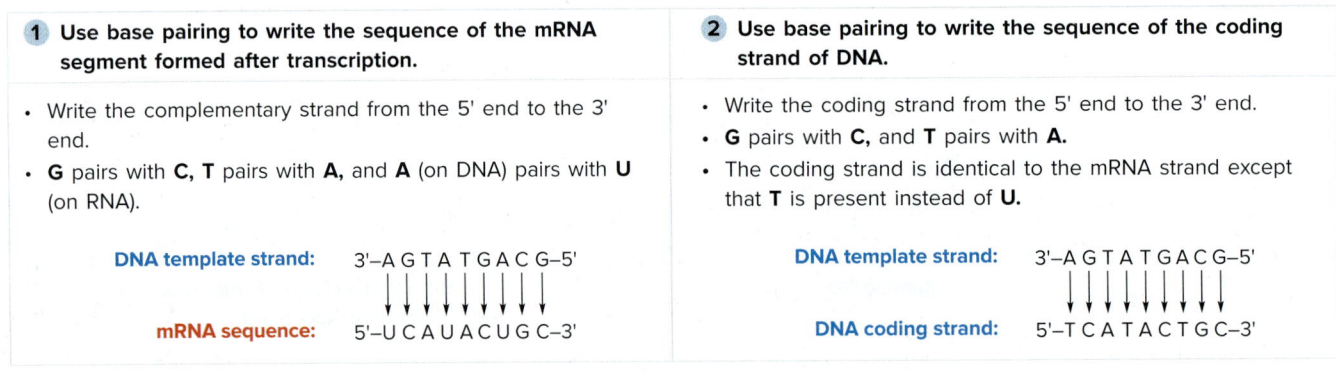

See Sample Problem 26.4. Try Problem 26.36.

[5] Deriving an amino acid sequence from DNA (26.6); example: 3'–CCGTATCTT–5'

1 Use the DNA sequence to determine the transcribed mRNA sequence.

- G pairs with **C**, **T** pairs with **A**, and **A** (on DNA) pairs with **U** (on RNA).

  DNA template strand: 3'– CCG TAT CTT –5'

  mRNA: 5'– GGC AUA GAA –3'

  Triplets correspond to RNA codons.

2 Use the codons in Table 26.2 to determine what amino acids are coded for by a given codon in mRNA.

  mRNA: 5'– GGC AUA GAA –3'

  Polypeptide: Gly – Ile – Glu

See Sample Problem 26.5. Try Problems 26.39–26.43.

PROBLEMS

Problems Using Three-Dimensional Models

26.21 (a) Give the name of each compound shown as a ball-and-stick model. (b) Would the compound be a component of DNA, RNA, or both?

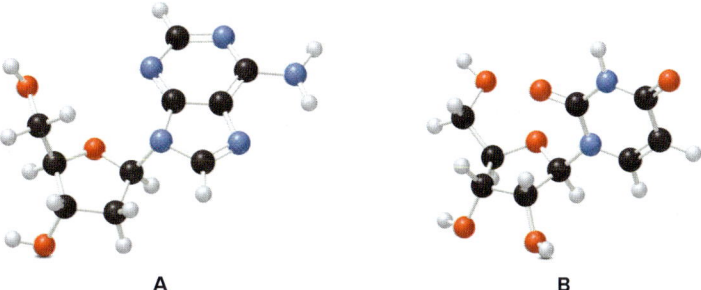

A   B

26.22 Give the name and the three- or four-letter abbreviation for each nucleotide.

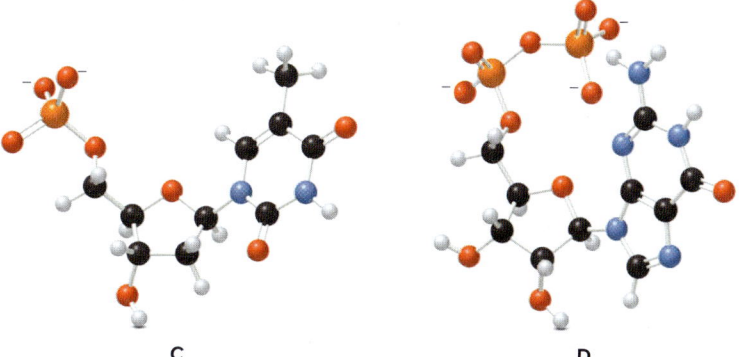

C   D

Bases, Nucleosides, Nucleotides, and Nucleic Acid Structure

26.23 Although the pyrimidine bases could exist as enol tautomers, making them hydroxy pyrimidines that contain a six-membered ring with 6 π electrons, these compounds are more stable as their amide tautomers. (a) Draw three different mono enol tautomers for thymine. (b) Draw a dienol tautomer for uracil. (c) How many enol tautomers can be drawn for caffeine, a natural product that contains a purine ring system? (d) Is caffeine an aromatic compound?

thymine   uracil   caffeine

1166   Chapter 26   Nucleic Acids and Protein Synthesis

26.24 (a) Identify the most acidic proton in thymine and explain your choice. (b) If thymine is treated with two equivalents of very strong base, what dianion is formed?

26.25 Suppose 2,6-diaminopurine replaced adenine as one of the four bases in a nucleic acid. Draw the structure of 2,6-diaminopurine and the hydrogen-bonding interactions that would occur between 2,6-diaminopurine and thymine.

26.26 Idoxuridine is a nucleoside analogue used in ophthalmic solutions or topical ointments to treat herpes infections. (a) Why is the heterocyclic ring system in idoxuridine aromatic? (b) Draw two different enol tautomers of idoxuridine.

idoxuridine

26.27 DNA can be damaged by reaction of its bases with alkylating agents. For example, reaction of guanine with dimethyl sulfate [$(CH_3)_2SO_4$] forms O^6-methylguanine. Explain why alkylation occurs on O. What effect does this alkylation have on the ability to hydrogen bond with cytosine?

guanine → O^6-methylguanine

26.28 Spongothymidine is an *N*-glycoside isolated from *Tectitethya crypta*, a shallow-water Caribbean sponge. Identify the base and monosaccharide that compose spongothymidine, and draw the structure of the monosaccharide using a Fischer projection formula.

spongothymidine

26.29 Draw the structure of the two possible dinucleotides formed from each pair of nucleotides: (a) dTMP and dAMP; (b) uridine 5′-monophosphate and guanosine 5′-monophosphate. Name each dinucleotide.

26.30 Draw the structure of each polynucleotide: (a) GTA; (b) CGU.

26.31 Write the sequence of the complementary strand of each segment of a DNA molecule.
   a. 5′–AAATAAC–3′
   b. 5′–ACTGGACT–3′
   c. 5′–CGATATCCCG–3′
   d. 5′–TTCCCGGGATA–3′

26.32 If 27% of the nucleotides in a sample of DNA contain the base adenine (A), what are the percentages of bases T, G, and C?

26.33 DNA becomes denatured and unwinds when it is heated. Explain why the temperature required for unwinding increases as the G–C content of the double helix increases.

Replication, Transcription, Translation, and Protein Synthesis

26.34 Draw a complete structure of the ribonucleotide codon GCU.

26.35 What is the sequence of a newly synthesized DNA segment if the template strand has the sequence 3′–ATGGCCTATGCGAT–5′?

26.36 For each DNA segment: [1] What is the sequence of the mRNA molecule synthesized from each DNA template? [2] What is the sequence of the coding strand of the DNA molecule?
   a. 3′–ATGGCTTA–5′
   b. 3′–CGGCGCTTA–5′
   c. 3′–GGTATACCG–5′
   d. 3′–TAGGCCGTA–5′

26.37 How is the identity of the second base—whether it is a purine or pyrimidine—in a codon related to the polarity of the side chain of the amino acid it codes for?

26.38 If each of the 61 codons for amino acids occurs with equal frequency in mRNA, which amino acids are least commonly found in proteins?

26.39 Derive the amino acid sequence that is coded for by each mRNA sequence.
   a. 5'–CCAACCUGGGUAGAA–3'
   b. 5'–AUGUUUUUAUGGUGG–3'
   c. 5'–GUCGACGAACCGCAA–3'

26.40 Write a possible mRNA sequence that codes for each peptide.
   a. Ile–Met–Lys–Ser–Tyr
   b. Pro–Gln–Glu–Asp–Phe
   c. Thr–Ser–Asn–Arg

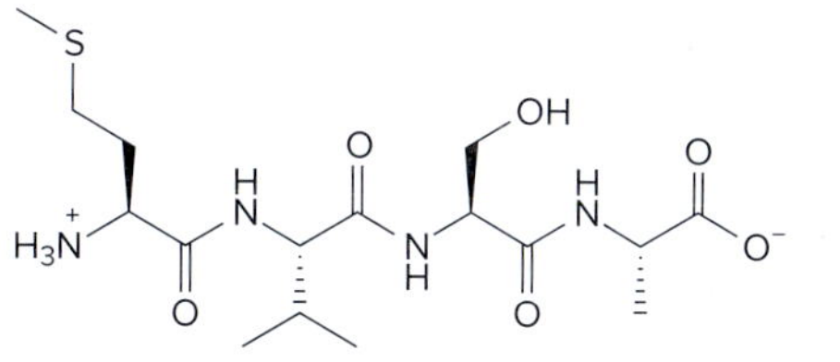

26.41 Considering each nucleotide sequence in an mRNA molecule: [1] write the sequence of the DNA template strand from which the mRNA was synthesized; [2] give the peptide synthesized by the mRNA.
   a. 5'–UAUUCAAUAAAAAAC–3'
   b. 5'–GAUGUAAACAAGCCG–3'

26.42 Using the given DNA template strand, determine the transcribed mRNA sequence and the polypeptide that would be synthesized from the template: 3'–AACGTCCTCACGATT–5'.

26.43 Met-enkephalin (Tyr–Gly–Gly–Phe–Met) is a painkiller and sedative (Section 16.5B). What is a possible nucleotide sequence in the template strand of the gene that codes for met-enkephalin, assuming that every base of the gene is transcribed and then translated?

26.44 Give a possible nucleotide sequence in the template strand of the gene that codes for each peptide.

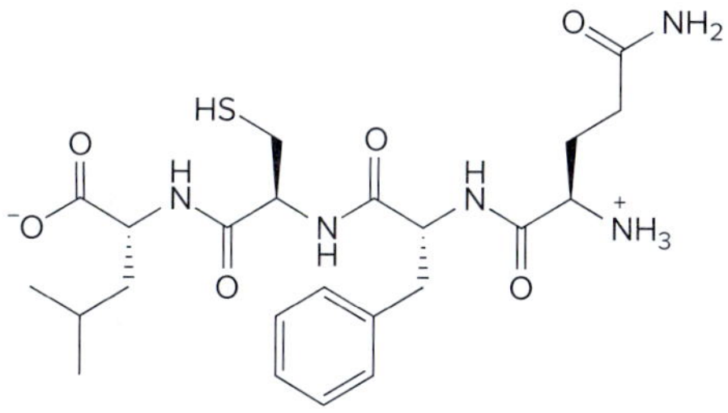

26.45 Draw a complete structure of the template strand of DNA responsible for the synthesis of the dipeptide Met–Trp.

26.46 Give a possible nucleotide sequence in the template strand of the gene that codes for the peptide angiotensin II. As we learned in Problem 23.45, ACE inhibitors are drugs that prevent the formation of angiotensin II, thus decreasing blood pressure.

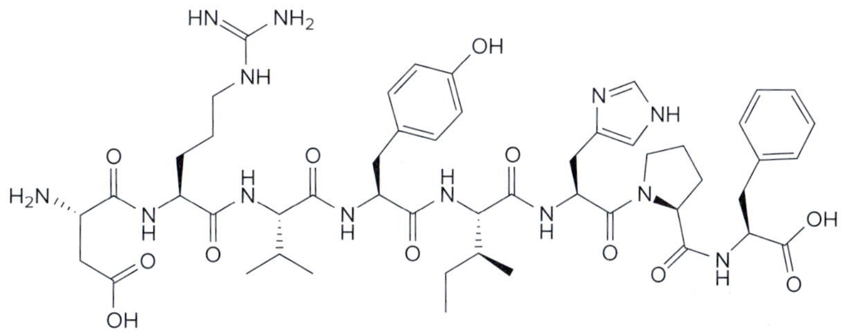

angiotensin II

Mechanism and Synthesis

26.47 Draw a stepwise mechanism for the acid-catalyzed hydrolysis of thymidine to 2'-deoxy-D-ribose and thymine.

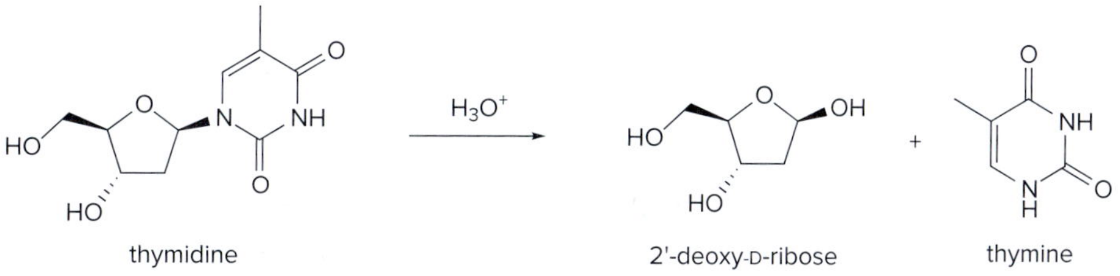

thymidine → 2'-deoxy-D-ribose + α anomer + thymine

Chapter 26 Nucleic Acids and Protein Synthesis

26.48 One way to synthesize uridine involves reaction of **A** with **B** to form **C**, followed by treatment with base. Draw a stepwise mechanism for the formation of **C**.

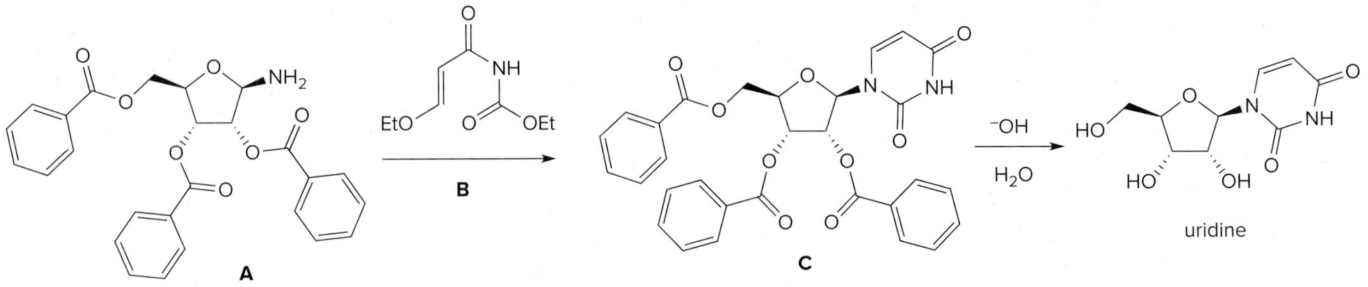

26.49 Acyclovir is an antiviral drug prepared by the following reaction sequence. Identify the structures of **A–C** and acyclovir.

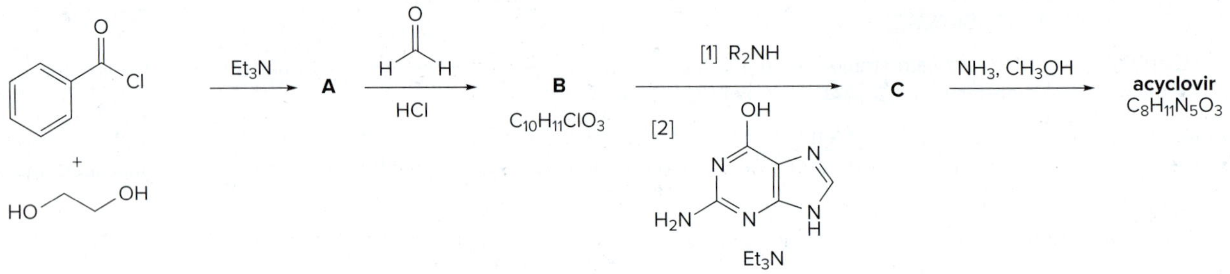

Challenge Problems

26.50 Nucleoside **C** can be synthesized by heating a nucleophilic base (**A**) and an electrophilic monosaccharide derivative (**B**). Suggest a mechanism that explains the stereochemistry of the observed nucleoside. (Hint: The acetate ester at C2' plays a role in the mechanism.)

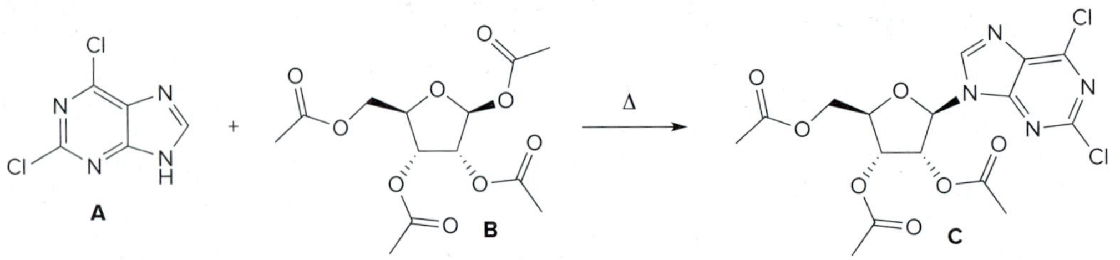

26.51 Identify the nucleoside **X** (including stereochemistry) that is prepared by the following two-step reaction sequence.

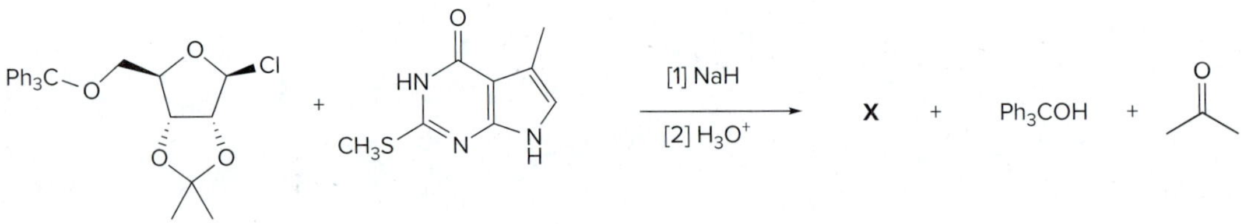

Metabolism

27

Samuel Borges Photography/Shutterstock

27.1 Overview of metabolism
27.2 Key oxidizing and reducing agents in metabolism
27.3 The catabolism of triacylglycerols by β-oxidation
27.4 The catabolism of carbohydrates—Glycolysis
27.5 The fate of pyruvate
27.6 The citric acid cycle and ATP production

Adenosine 5'-triphosphate (ATP) is the nucleoside triphosphate primarily involved in energy production during metabolism. Any process, such as walking, running, swallowing, or breathing, is fueled by the energy release that accompanies the hydrolysis of ATP to adenosine 5'-diphosphate (ADP). Because ATP contains four negatively charged oxygen atoms in close proximity, the electronic repulsion of the like charges drives its hydrolysis to form a product with less electronic repulsion. In Chapter 27, we learn about the interconversion of ATP and ADP, and some of the key pathways that occur during metabolism.

Why Study... Metabolism?

Despite the wide diversity among life forms, virtually all organisms contain the same types of biomolecules—proteins, carbohydrates, lipids, and nucleic acids—and use the same biochemical reactions. Each moment, thousands of reactions occur in a cell: complex biomolecules are broken down into simple components, simple molecules are converted to complex biomolecules, and energy changes occur.

Metabolism is an enormously complex subject encompassing a wide range of biochemical reactions. The details of many metabolic pathways have been determined by the painstaking research efforts of teams of scientists, and new insights into key processes are constantly revealed, as we continue to discover how the biochemical machinery in cells operates. Here in Chapter 27, we concentrate on three key catabolic pathways: the **catabolism of fats** (Section 27.3), the breakdown of glucose to pyruvate by **glycolysis** (Section 27.4), and the **citric acid cycle** (Section 27.6), which converts acetyl CoA to carbon dioxide. Although these reactions constitute only a fraction of those involved in metabolic pathways, they provide an understanding of the complex processes that are constantly taking place within cells.

27.1 Overview of Metabolism

Metabolism **is the sum of all the chemical reactions that take place in an organism.** **Catabolism** is the *breakdown* of large molecules into smaller ones, often releasing energy. **Anabolism** is the *synthesis* of large molecules from smaller ones, often absorbing energy.

Just as gasoline is the fuel that powers most automobiles, food is the fuel that is metabolized by the body to provide energy. Catabolism breaks down the carbohydrates, proteins, and lipids in food into smaller molecules, releasing energy to supply the body's needs. The body can't use the calories of a meal all at once. Energy must be stored in molecules that are readily accessible for use anywhere in the body at any time the energy is needed.

This process involves numerous catabolic pathways that can be organized in four stages, as illustrated in Figure 27.1.

Stage 1

Catabolism begins with **digestion.** The esters in triacylglycerols are hydrolyzed to glycerol and three fatty acids (Section 16.11), the glycosides of carbohydrates are hydrolyzed to monosaccharides (Figure 5.1 and Section 24.12), and the amides of proteins are cleaved to amino acids (Section 23.5A). These small molecules are then absorbed through the intestinal cell wall into the bloodstream and transported to other parts of the body.

Stage 2

Fatty acids, monosaccharides, and amino acids are degraded to acetyl groups (CH_3CO-) that are bonded to coenzyme A (**HSCoA,** Section 3.8), forming the thioester **acetyl CoA** (Section 16.16).

The product of Stage [2] is the *same* for all three types of biomolecules. Two structural features of acetyl CoA make it a key intermediate in a variety of biochemical transformations.

27.1 Overview of Metabolism

Figure 27.1 The four stages of catabolism

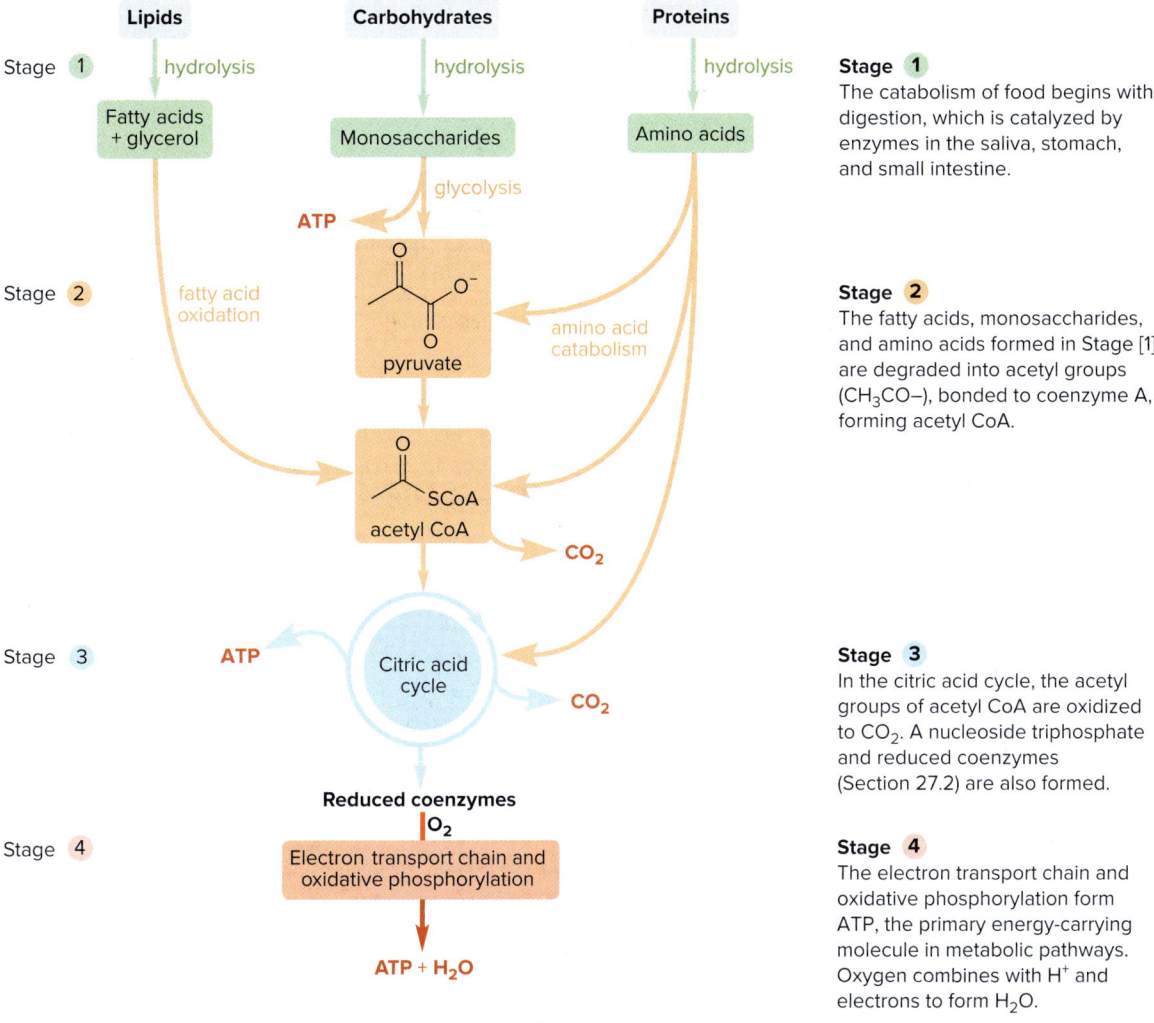

Stage 1
The catabolism of food begins with digestion, which is catalyzed by enzymes in the saliva, stomach, and small intestine.

Stage 2
The fatty acids, monosaccharides, and amino acids formed in Stage [1] are degraded into acetyl groups (CH_3CO-), bonded to coenzyme A, forming acetyl CoA.

Stage 3
In the citric acid cycle, the acetyl groups of acetyl CoA are oxidized to CO_2. A nucleoside triphosphate and reduced coenzymes (Section 27.2) are also formed.

Stage 4
The electron transport chain and oxidative phosphorylation form ATP, the primary energy-carrying molecule in metabolic pathways. Oxygen combines with H^+ and electrons to form H_2O.

- The end result of catabolism is that biomolecules are converted to CO_2 and H_2O and energy is produced and stored in ATP molecules.

- The H's on the α carbon to the carbonyl group are more acidic than the α H's of esters and other acyl derivatives, so the α carbon can more readily act as a *nucleophile* and participate in aldol or Claisen reactions.

- As mentioned in Section 16.16, acetyl CoA has a better leaving group than an ester, so it more readily undergoes *nucleophilic acyl substitution* reactions.

Stage 3

The acetyl group of acetyl CoA is oxidized in the citric acid cycle to CO_2 (Section 27.6). This stage results in the formation of a nucleoside triphosphate and reduced coenzymes (Section 27.2), which enter the last stage of catabolism.

Stage 4

The electron transport chain and oxidative phosphorylation produce **adenosine 5'-triphosphate (ATP)**, the chapter-opening molecule. Oxygen combines with H^+ and electrons (from reduced coenzymes) to form water. Most of the energy obtained from fats, carbohydrates, and proteins is packaged in ATP molecules formed in this stage.

What is the role of ATP in the catabolic pathways and why is ATP synthesis in Stage [4] noteworthy?

As we learned in Sections 6.4 and 6.5B, **ATP is the most prominent member of a group of "high-energy" compounds** that undergo highly exothermic reactions. Recall that this energy can be used in **coupled reactions** to drive a reaction that has an unfavorable energy change. **The interconversion of ATP and ADP is the central method of energy transfer in cells.**

adenosine 5'-triphosphate
ATP

exothermic hydrolysis
H_2O
$HPO_4^{2-} + H^+$
endothermic phosphorylation

adenosine 5'-diphosphate
ADP

- ATP is converted to ADP by a hydrolysis reaction that releases energy. The energy released in this process can be used in coupled reactions that require energy input.
- ATP is synthesized from ADP by a phosphorylation reaction that absorbs energy. ATP synthesis must be coupled with an energy-producing process.

Electrostatic repulsion is one factor used to explain why ATP hydrolysis is energetically favorable. Because ATP contains four negatively charged oxygen atoms in close proximity, the electronic repulsion of the like charges drives its hydrolysis to form ADP, a product with only three negatively charged oxygens, and therefore less electronic repulsion.

Problem 27.1 Explain why the α H's of a thioester like $CH_3COSCH_2CH_3$ have a lower pK_a than those of an ester like $CH_3CO_2CH_2CH_3$.

Problem 27.2 Consider the hydrolysis of ATP to form ADP + HPO_4^{2-}. How does the extent of electron delocalization in the reactant and both products contribute to making this process energetically favorable?

As first discussed in Section 6.5B, coupled reactions are often written with a combination of horizontal and curved arrows. For example, the conversion of creatine phosphate to creatine is a hydrolysis that cleaves a high-energy P–N bond and releases more energy than is needed for the phosphorylation of ADP. Coupling these reactions together forms ATP from ADP. Using curved reaction arrow symbolism, the organic reactant and product are separated by a horizontal arrow, and ADP and ATP are drawn using a curved arrow.

Creatine phosphate is stored in muscle. During strenuous exercise creatine phosphate reacts with ADP to form a new supply of ATP for more energy. Some athletes use creatine supplements to increase the amount of creatine phosphate in their muscle and give themselves a greater energy reserve. *Jill Braaten*

This bond is broken.

creatine phosphate → creatine
ADP → ATP

27.2 Key Oxidizing and Reducing Agents in Metabolism

In Section 27.1, two key compounds in metabolism—acetyl CoA and ATP—were discussed. Coenzymes that serve as oxidizing and reducing agents are also important.

27.2A Nicotinamide Adenine Dinucleotide

Two common coenzymes have been presented in previous chapters.

- NAD^+ (nicotinamide adenine dinucleotide) is a biological *oxidizing* agent. The pyridinium ring of NAD^+ accepts hydride ($H:^-$) from an organic substrate to form NADH (Section 11.13).
- NADH (the reduced form of nicotinamide adenine dinucleotide) is a biological *reducing* agent. NADH transfers $H:^-$ to an organic substrate, forming NAD^+ (Section 13.6).

NAD^+
biological oxidiizng agent

NADH
biological reducing agent

Catabolism generally involves oxidation reactions that produce energy, whereas anabolism generally involves reduction reactions that require energy. Enzymes hold the key reactants in place and contain necessary acidic or basic amino acid side chains, but the coenzymes are the reagents that carry out the oxidation and reduction.

Curved reaction arrow symbolism is used to draw equations for these reactions, as shown with the oxidation of malate to oxaloacetate using NAD^+ in the presence of the enzyme malate dehydrogenase.

malate → oxaloacetate (NAD^+ → NADH + H^+, malate dehydrogenase)

As shown in Table 23.3, an enzyme that adds or removes two hydrogen atoms from a substrate is classified as a **dehydrogenase.** The name of an enzyme often has two words. The first identifies the substrate on which the enzyme acts, and the second identifies the class of reaction catalyzed. Thus, **malate *dehydrogenase*** catalyzes the removal of two H atoms from malate to form oxaloacetate. We return to this reaction when we examine the citric acid cycle in Section 27.6.

Problem 27.3 (a) What products are formed when isocitrate is treated with NAD^+? (b) Write the equation using horizontal and curved arrows. (c) Give a possible name for the enzyme that catalyzes this reaction.

isocitrate

27.2B Flavin Adenine Dinucleotide

Flavin adenine dinucleotide (FAD) is another common biological oxidizing agent. Although its structure is complex, just four atoms of the tricyclic ring system (shown in red) participate in redox reactions. When it acts as an oxidizing agent, FAD is reduced by adding two hydrogen atoms, forming **$FADH_2$, the reduced form of flavin adenine dinucleotide. $FADH_2$, like NADH, is a biological reducing agent.**

FAD
biological oxidizing agent

$FADH_2$
biological reducing agent

Leafy green vegetables, soybeans, and almonds are good sources of riboflavin, vitamin B_2. Since this vitamin is light sensitive, riboflavin-fortified milk contained in glass or clear plastic bottles should be stored in the dark. *Jill Braaten/McGraw-Hill Education*

Flavin is synthesized in cells from vitamin B_2, a yellow, water-soluble vitamin obtained in the diet from leafy greens, soybeans, almonds, and liver. When large quantities of riboflavin are ingested, excess is excreted in the urine, giving it a bright yellow appearance.

riboflavin
vitamin B_2

It is likely that the FAD oxidations follow a variety of mechanisms depending on the substrate and enzyme. One step in the metabolism of fatty acids (Section 27.3) is the removal of two hydrogens from a fatty acyl CoA to form an α,β-unsaturated fatty acyl CoA using FAD.

fatty acyl CoA → α,β-unsaturated fatty acyl CoA

A suggested mechanism for this process involves two steps: loss of a proton from the α carbon to form an enolate, followed by nucleophilic addition of hydride to FAD to form the α,β-unsaturated product and FADH₂, as shown in Mechanism 27.1.

Mechanism 27.1 An Oxidation Reaction with FAD

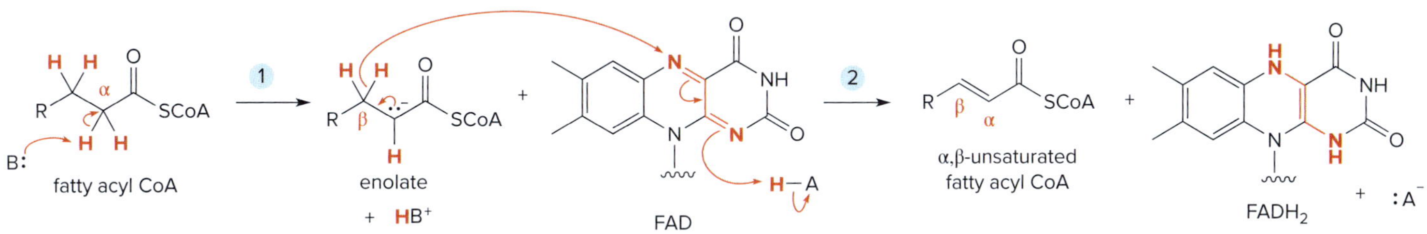

1. Abstraction of a H atom from the α carbon of the fatty acyl CoA forms a resonance-stabilized enolate.
2. H:⁻ is eliminated from the β carbon as the electron pair of the enolate forms the new π bond of the α,β-unsaturated acyl CoA. Nucleophilic attack of H:⁻ on the tricyclic ring system of FAD forms the reduced tricyclic ring system of FADH₂.

We have now seen two coenzymes that can be used for oxidation—**NAD⁺** and **FAD**—and two coenzymes that can be used for reduction—**NADH** and **FADH₂**. How do we know which reactions use NAD⁺/NADH and which use FAD/FADH₂ to carry out oxidation or reduction? Use the following guide:

- Redox reactions involving carbonyl groups generally use NAD⁺/NADH.
- Redox reactions of other functional groups use FAD/FADH₂.

The reduced coenzymes NADH and FADH₂ formed in catabolic oxidations are vital to Stage [4] of catabolism. The electrons from these electron-rich reduced coenzymes are transferred from one molecule to another in the **electron transport chain** and provide the energy to synthesize ATP from ADP by the process of **oxidative phosphorylation.**

- Each NADH that enters the electron transport chain in Stage [4] provides the energy to synthesize 2.5 equivalents of ATP.
- Each FADH₂ provides the energy to synthesize 1.5 equivalents of ATP.

Problem 27.4 (a) By analogy to the oxidation of a fatty acyl CoA with FAD, draw the products formed when succinate reacts with FAD. (b) Write the equation using horizontal and curved arrows. (c) Give a possible name for the enzyme that catalyzes this reaction.

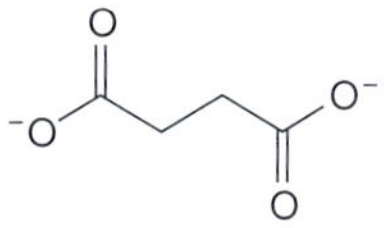

succinate

Problem 27.5 How does the elimination sequence in Mechanism 27.1 compare with mechanisms of elimination—E2, E1, and E1cB—that you have learned in previous chapters?

Problem 27.6 Classify the following reaction as an oxidation or reduction and give the likely coenzyme used and formed. Explain your reasoning.

27.3 The Catabolism of Triacylglycerols by β-Oxidation

The first step in the catabolism of triacylglycerols, the most common lipids, is the hydrolysis of the ester bonds in the presence of a lipase enzyme to form glycerol and fatty acids, which are metabolized in separate pathways.

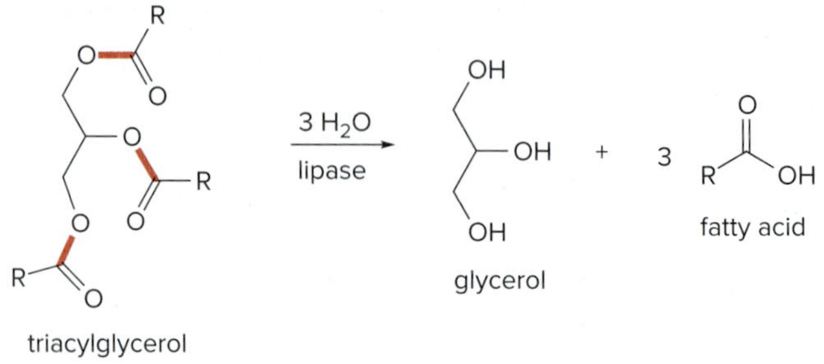

27.3A Glycerol Catabolism

The glycerol formed from triacylglycerol hydrolysis is converted in two steps to dihydroxyacetone phosphate. Phosphorylation of glycerol forms glycerol 3-phosphate, which is then oxidized with NAD^+.

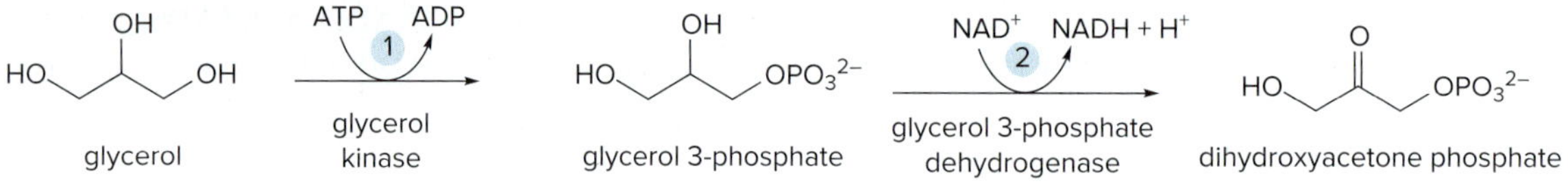

Each step is catalyzed by an enzyme. A **kinase** (Table 23.3) is an enzyme that catalyzes the transfer of a phosphate from one compound to another—in this case, from ATP to glycerol, so the enzyme for Step [1] is named *glycerol kinase*. Because the final product, dihydroxyacetone phosphate, is an intermediate in glycolysis, it is then metabolized in several steps to pyruvate ($CH_3COCO_2^-$), as described in Section 27.4.

27.3B Fatty Acid Catabolism by β-Oxidation

Fatty acids are catabolized by β-oxidation, a process in which two-carbon acetyl CoA units are sequentially cleaved from the fatty acid. Key to this process is the oxidation of the β carbon to the carbonyl group, which then undergoes cleavage between the α and β carbons.

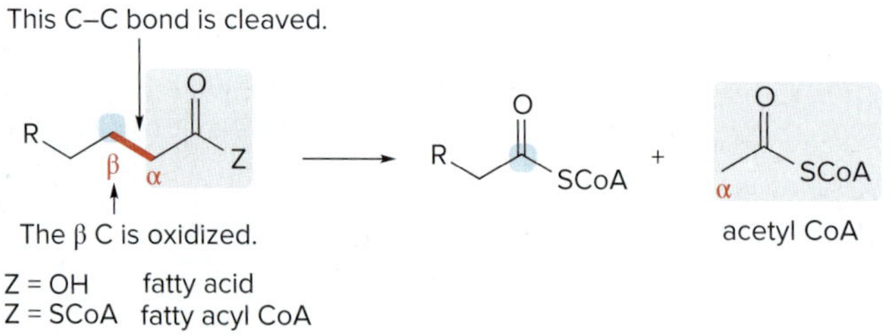

Fatty acid oxidation begins with conversion of the fatty acid to a thioester with coenzyme A, forming a **fatty acyl CoA**. This process requires energy, which comes from the hydrolysis of *two* P–O bonds in ATP to form AMP (adenosine 5′-*mono*phosphate).

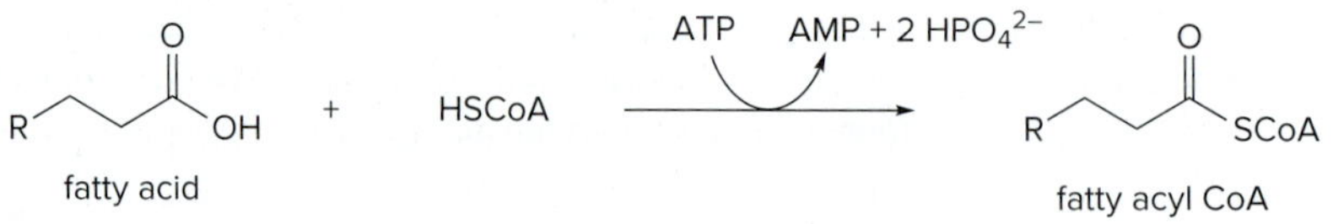

β-Oxidation of the fatty acyl CoA requires a repetitive four-step sequence, as shown in Figure 27.2. Each group of four reactions removes a two-carbon unit from the fatty acyl CoA, and repeats on successively smaller substrates until the carbon chain is completely catabolized to acetyl CoA.

Figure 27.2 The four steps of β-oxidation of a fatty acyl CoA

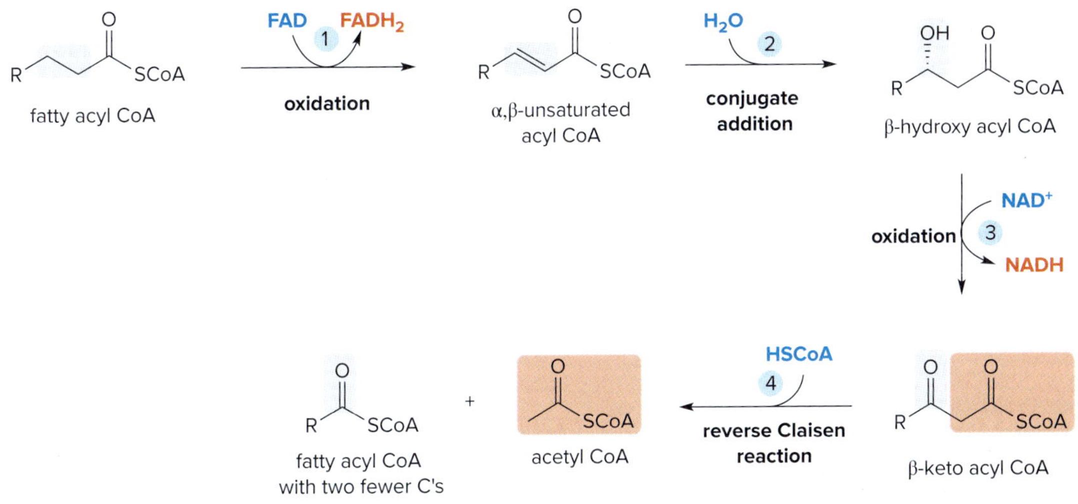

- Oxidation of the fatty acyl CoA with FAD (Step [1]) followed by conjugate addition of H₂O (Step [2]) forms a β-hydroxy acyl CoA. Oxidation of the OH group to a carbonyl (Step [3]) followed by the reverse of a Claisen reaction (Step [4]) forms two products: acetyl CoA and a fatty acyl CoA that has two carbons fewer than the initial substrate. This process repeats until all carbons of the fatty acyl CoA are degraded to acetyl CoA. These reactions are discussed in more detail in the body of the text.

Steps 1 and 2 of β-Oxidation

The first step of fatty acid catabolism involves the FAD-mediated oxidation of a fatty acyl CoA to form an α,β-unsaturated acyl CoA and FADH₂ using an acyl CoA dehydrogenase enzyme. This reaction removes H atoms from the α and β carbons to the carbonyl group by the mechanism shown in Mechanism 27.1.

Conjugate addition of water to the α,β-unsaturated acyl CoA using an enoyl CoA hydratase enzyme yields a β-hydroxy acyl CoA in Step [2]. Water adds to the electrophilic β carbon of the double bond, and the intermediate enolate is protonated on the α carbon.

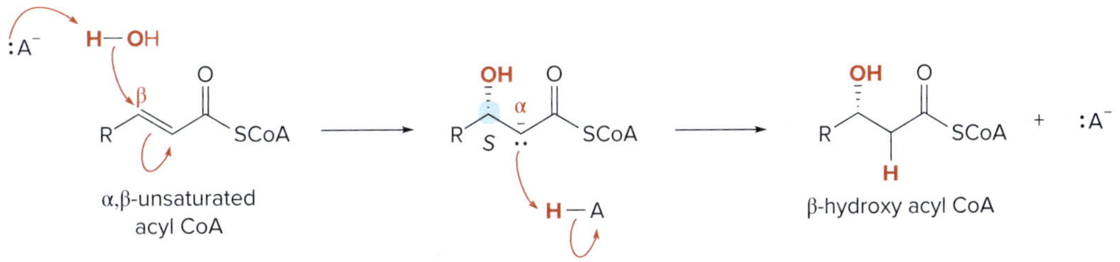

As is the case in many biological reactions, a single enantiomer with the *S* configuration at the newly formed stereogenic center (highlighted in blue) is formed.

Steps 3 and 4 of β-Oxidation

In Step [3], the OH group of the β-hydroxy acyl CoA is oxidized by NAD⁺ to form a β-keto acyl CoA and NADH using a dehydrogenase enzyme. The mechanism of NAD⁺ oxidations was presented in Section 11.13.

The final step in the catabolism of a fatty acid cleaves a carbon–carbon bond in a reaction that is the reverse of the Claisen condensation that you learned about in Chapter 18 (Mechanism 18.5). Addition of the nucleophilic thiol group of coenzyme A is a key feature, as shown in Mechanism 27.2.

Mechanism 27.2 The Reverse Claisen Reaction

[Mechanism diagram showing three steps of the reverse Claisen reaction, producing fatty acyl CoA with two fewer C's and acetyl CoA]

1. Nucleophilic addition of the thiol of coenzyme A forms a tetrahedral intermediate.
2. Loss of the enolate of acetyl CoA as leaving group cleaves the C–C bond, and forms a fatty acyl CoA with two fewer C's than the original fatty acid.
3. Protonation of the enolate forms acetyl CoA.

As a result, a new fatty acyl CoA with two fewer carbons than the original fatty acid is formed in each cycle of β-oxidation. Each four-step sequence also forms one molecule each of acetyl CoA, NADH, and FADH$_2$.

[Overall process of β-oxidation: fatty acyl CoA + FAD + NAD$^+$ + H$_2$O + HSCoA → fatty acyl CoA with two fewer C's + acetyl CoA + FADH$_2$ + NADH + H$^+$]

The fatty acyl CoA formed from one cycle of β-oxidation can then serve as the starting material for a new cycle, and two more carbons are removed as acetyl CoA. For example, the acyl CoA derived from myristic acid, a 14-carbon fatty acid, undergoes β-oxidation to form a 12-carbon acyl CoA, which becomes the substrate for another β-oxidation sequence. The process continues until a four-carbon acyl CoA is cleaved to generate two acetyl CoA molecules.

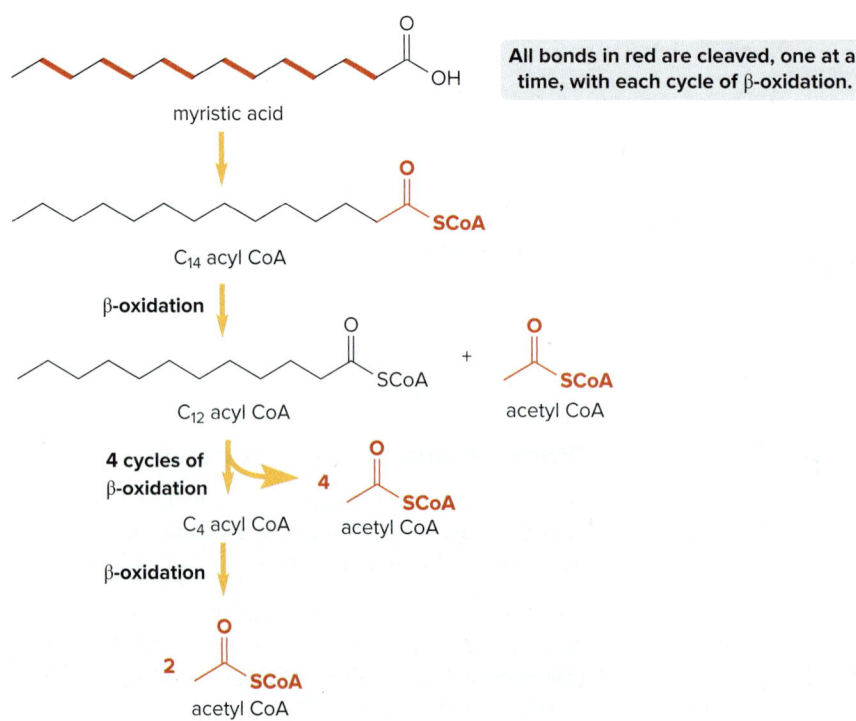

All bonds in red are cleaved, one at a time, with each cycle of β-oxidation.

Myristic acid is the main component of the triacylglycerols derived from nutmeg, a spice obtained from the seed of the nutmeg tree *Myristica fragrans*. Myristic acid is also found in palm kernel oil and coconut oil. *National Geographic Image Collection/Alamy Stock Photo*

As a result:

- A 14-carbon fatty acyl CoA is cleaved to seven two-carbon acetyl CoA molecules.
- A total of *six* cycles of β-oxidation are needed to cleave the six carbon–carbon bonds of myristic acid.

β-Oxidation of unsaturated fatty acids proceeds in a similar fashion, although additional enzyme-catalyzed steps are required. Ultimately, every carbon in the original fatty acid ends up as a carbon atom of acetyl CoA.

After we learn about the citric acid cycle in Section 27.6, we can determine the amount of ATP formed from the products of fatty acid oxidation.

Sample Problem 27.1 Determining the Outcome of β-Oxidation of a Fatty Acid

For lauric acid [$CH_3(CH_2)_{10}CO_2H$]: (a) How many molecules of acetyl CoA are formed from complete oxidation? (b) How many cycles of β-oxidation are needed? (c) How many molecules of NADH and FADH$_2$ are formed?

Solution

- The number of molecules of acetyl CoA equals *one-half* the number of carbons in the original fatty acid.
- Because the final β-oxidation cycle forms *two* molecules of acetyl CoA, the number of cycles is *one fewer* than the number of acetyl CoA molecules formed.
- Each cycle produces one molecule of NADH and one molecule of FADH$_2$, so the number of cycles *equals* the number of molecules of NADH and FADH$_2$ formed.

The 12 carbons of lauric acid form six molecules of acetyl CoA by five cycles of β-oxidation. Five molecules each of NADH and FADH$_2$ are also formed.

lauric acid

Bonds in red are cleaved.

Problem 27.7

For each fatty acid: (a) How many molecules of acetyl CoA are formed from complete oxidation? (b) How many cycles of β-oxidation are needed?

arachidic acid
A

palmitoleic acid
B

More Practice: Try Problems 27.27, 27.28.

Problem 27.8

How many molecules of NADH and FADH$_2$ are formed from the complete catabolism of octacosanoic acid [$CH_3(CH_2)_{26}CO_2H$]?

27.4 The Catabolism of Carbohydrates—Glycolysis

The metabolism of monosaccharides centers around glucose. Whether it is obtained by the hydrolysis of ingested polysaccharides or stored glycogen (Section 24.12C), glucose is the principal monosaccharide used for energy in the human body.

1180 Chapter 27 Metabolism

- Glycolysis is an anaerobic, 10-step pathway that converts glucose to two molecules of pyruvate ($CH_3COCO_2^-$).

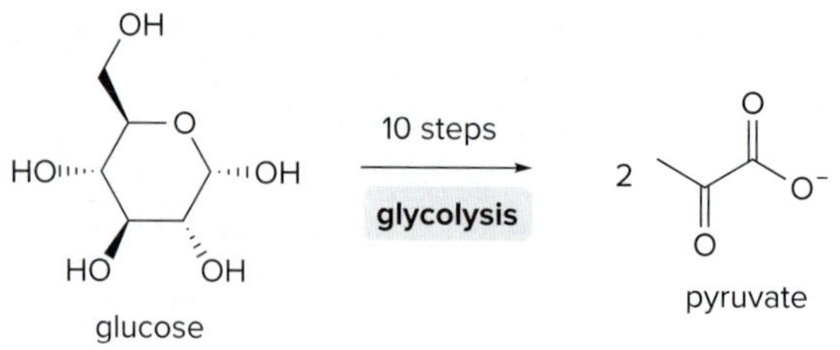

27.4A Glycolysis—Steps [1]–[5]

The first five steps of glycolysis are illustrated in Figure 27.3.

Figure 27.3 Glycolysis Steps [1]–[5]

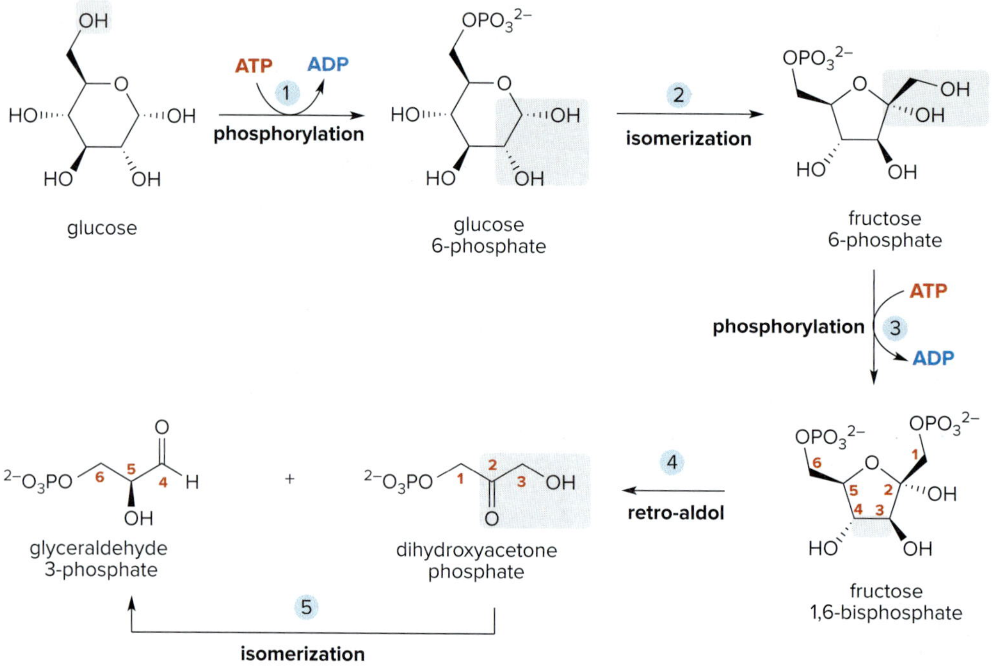

- Steps [1]–[5] of glycolysis convert glucose to **two molecules of glyceraldehyde 3-phosphate**. Steps [1] and [3] involve phosphorylation using ATP, whereas Steps [2] and [5] are isomerizations. The key step is the retro-aldol reaction in Step [4] that cleaves the six-carbon substrate to two three-carbon products. The atoms of a substrate that are changed in a reaction are highlighted in gray.

Steps [1] and [2] of Glycolysis: Phosphorylation and Isomerization

Glycolysis begins with the phosphorylation of glucose to form glucose 6-phosphate, catalyzed by hexokinase. This energetically unfavorable reaction is coupled with the conversion of ATP to ADP to make the reaction energetically favorable.

In Step [2], glucose 6-phosphate is isomerized to fructose 6-phosphate, catalyzed by glucose 6-phosphate isomerase. Isomerization occurs using the ring-opened acyclic form of the glucose hemiacetal **A** (Section 24.6), which undergoes two successive tautomerizations via an **enediol** intermediate **B**. Each tautomerization consists of two operations: protonation (by an acid HA) and deprotonation (by a base B:), as we learned in Sections 10.18 and 17.2B. Tautomerization moves the carbonyl group of the monosaccharide from C1 to C2, generating the acyclic form of fructose 6-phosphate **C**, which is in equilibrium with the hemiacetal form **D**.

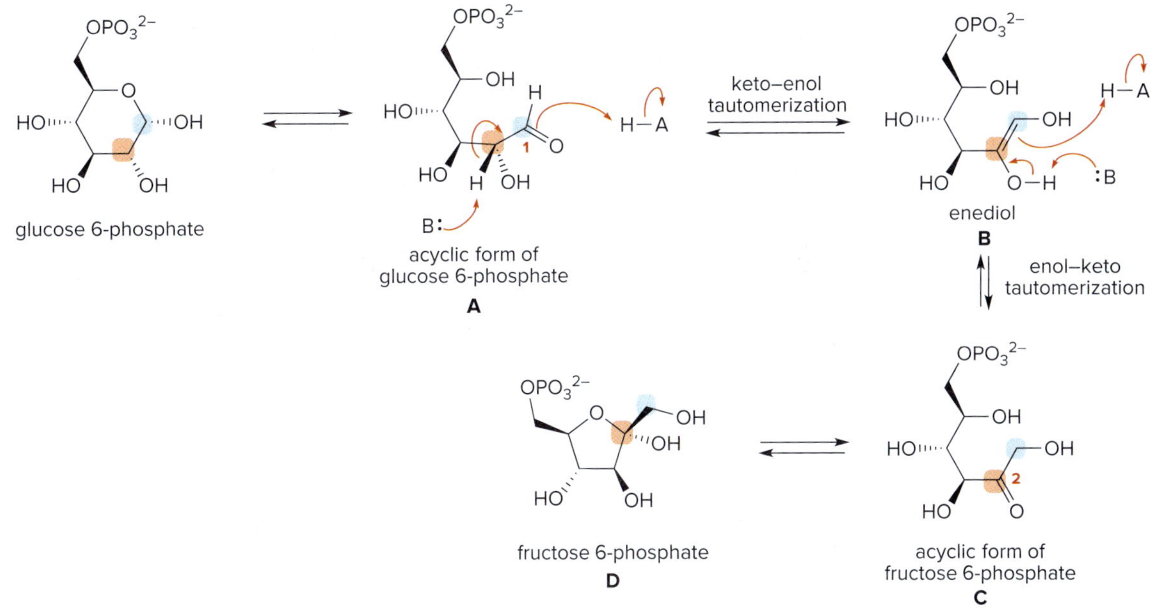

27.4 The Catabolism of Carbohydrates—Glycolysis

> **Problem 27.9** Draw the structure of ATP and use curved arrows to show how glucose reacts with ATP to yield glucose 6-phosphate and ADP in Step [1] of glycolysis.

Steps ③ and ④ of Glycolysis: Phosphorylation and Retro-Aldol Reaction

In Step [3], fructose 6-phosphate is phosphorylated to fructose 1,6-bisphosphate. This reaction is catalyzed by phosphofructokinase. Step [3], like Step [1], is energetically unfavorable, so it is coupled with the conversion of ATP to ADP to make the reaction energetically favorable.

In Step [4], fructose 1,6-bisphosphate is cleaved between C3 and C4 to two three-carbon products using an aldolase enzyme, in a reaction that conceptually resembles the **retro-aldol** reaction discussed in Section 18.8B; that is, **a β-hydroxy carbonyl compound is *cleaved* to two carbonyl compounds.**

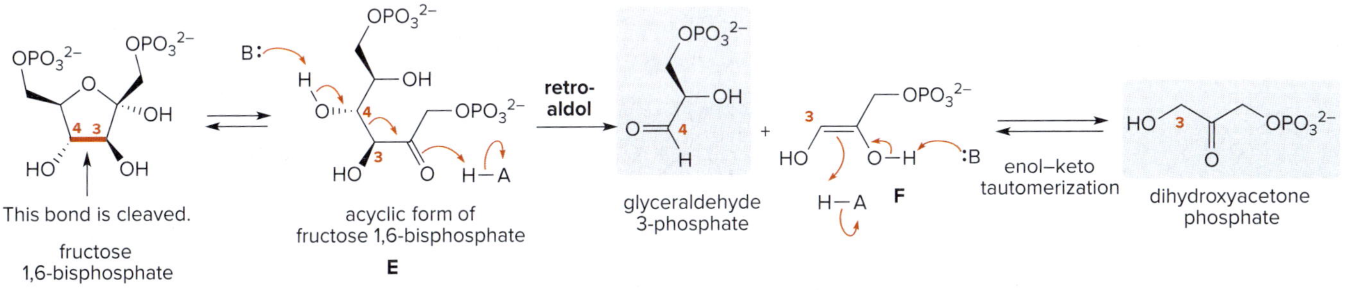

To understand the details of this reaction, we once again consider the ring-opened acyclic form of the fructose 1,6-bisphosphate hemiacetal **E**. Loss of a proton from the OH group at C4 (located β to the carbonyl group) results in cleavage of the C3–C4 bond to form glyceraldehyde 3-phosphate and enediol **F**. Tautomerization of the enediol forms dihydroxyacetone phosphate.

This retro-aldol reaction occurs either directly on the acyclic form of fructose 1,6-bisphosphate, or on an imine derivative formed from **E** to ultimately form the same two products.

Step ⑤ of Glycolysis: Isomerization

Although both glyceraldehyde 3-phosphate and dihydroxyacetone phosphate are formed in Step [4] of glycolysis, only glyceraldehyde 3-phosphate continues on to form pyruvate.

As a result, dihydroxyacetone phosphate is isomerized by triose phosphate isomerase to form a second molecule of glyceraldehyde 3-phosphate. This reaction occurs by way of an enediol intermediate and involves two tautomerizations.

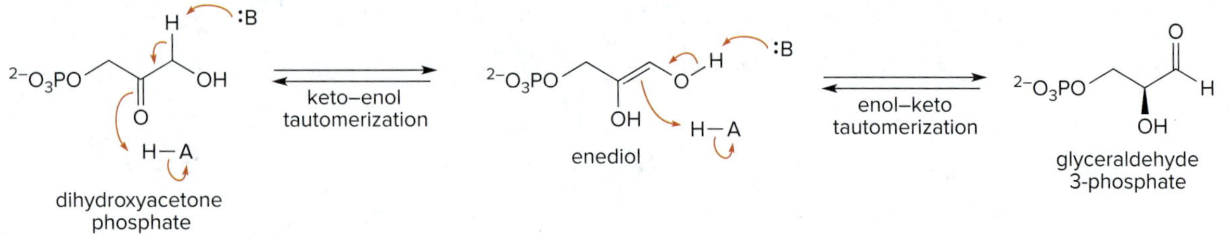

- Steps [1]–[5] of glycolysis convert glucose into *two* molecules of glyceraldehyde 3-phosphate.

Problem 27.10 How are each pair of compounds related: (a) glucose 6-phosphate and fructose 6-phosphate; (b) glyceraldehyde 3-phosphate and dihydroxyacetone phosphate; (c) glyceraldehyde 3-phosphate and enediol **F**. Choose from enantiomers, diastereomers, constitutional isomers, or not isomers of each other.

27.4B Glycolysis—Steps [6]–[10]

Each three-carbon molecule of glyceraldehyde 3-phosphate formed in Step [5] of glycolysis is carried through a series of five reactions that ultimately form pyruvate ($CH_3COCO_2^-$), as shown in Figure 27.4.

Figure 27.4 Glycolysis Steps [6]–[10]

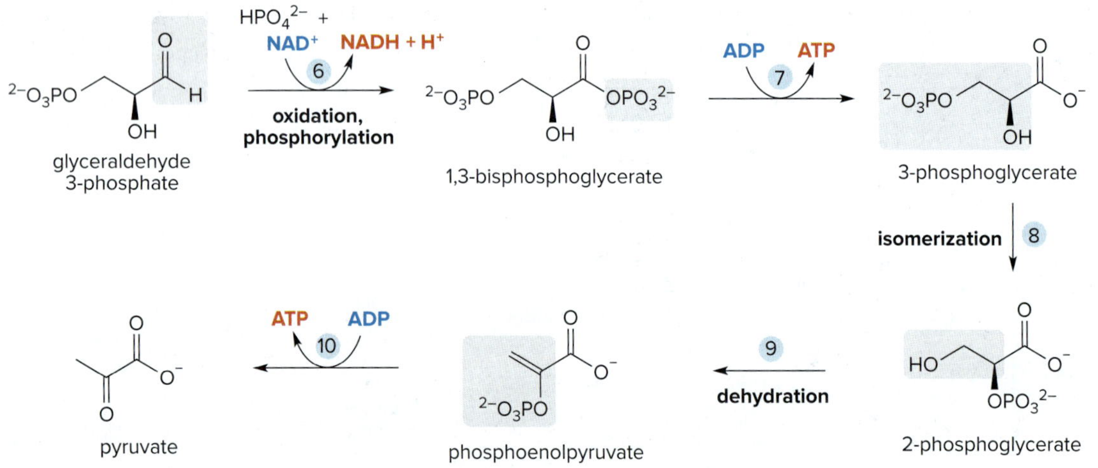

- Steps [6]–[10] of glycolysis convert *two* molecules of glyceraldehyde 3-phosphate to *two* molecules of pyruvate.

Steps [6] and [7] of Glycolysis: Oxidation, Phosphorylation, and Phosphate Transfer

The reaction of acyl phosphates with nucleophiles follows the **two-step mechanism** presented in Mechanism 16.11—**addition** of a nucleophile, followed by **elimination** of a leaving group.

In Step [6], which is catalyzed by glyceraldehyde 3-phosphate dehydrogenase, glyceraldehyde 3-phosphate is oxidized by NAD^+ and phosphorylated with HPO_4^{2-} to form 1,3-bisphosphoglycerate, an **acyl phosphate**.

The acyl phosphate is a reactive substrate, similar to an anhydride, in nucleophilic substitutions. When 1,3-bisphosphoglycerate reacts with ADP in Step [7], the negatively charged phosphate of ADP acts as a nucleophile to attack the electrophilic acyl phosphate. A phosphoryl group (PO_3^{2-}) is transferred to ADP, forming ATP and 3-phosphoglycerate.

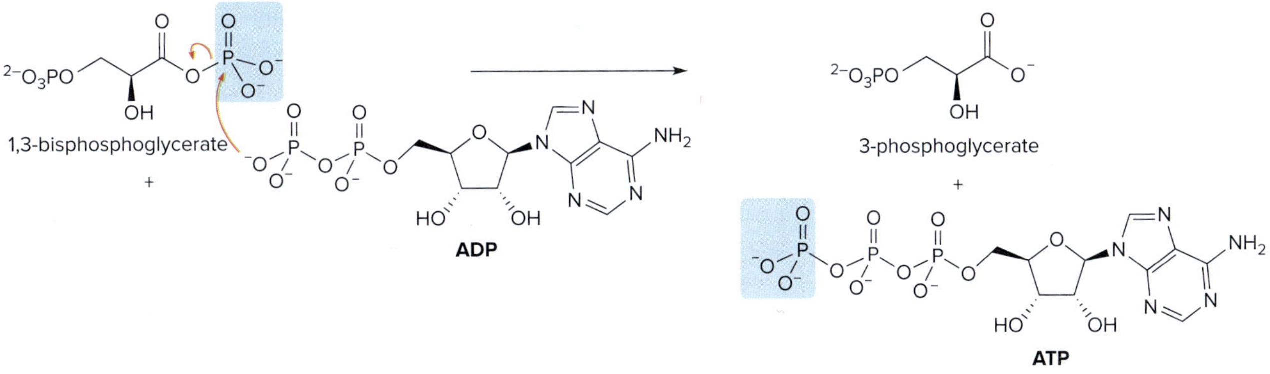

Steps 8 and 9 of Glycolysis: Isomerization and Dehydration

In Step [8], 3-phosphoglycerate is isomerized to 2-phosphoglycerate, catalyzed by phosphoglycerate mutase. This reaction is not simply the transfer of a phosphoryl group from C3 to C2. Instead, 3-phosphoglycerate is converted to 2,3-bisphosphoglycerate, which then loses PO_3^{2-} from C3 to form 2-phosphoglycerate.

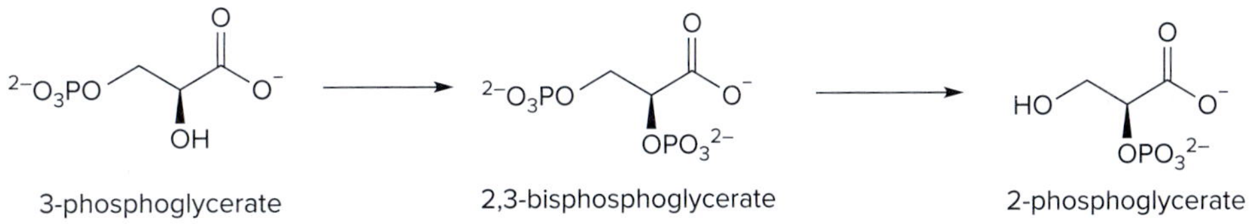

2-Phosphoglycerate is dehydrated to phosphoenolpyruvate using an enolase enzyme. Dehydration follows a two-step E1cB mechanism (Mechanism 18.3), with formation of an intermediate enolate. The overall result is the **loss of water**, forming an α,β-unsaturated carbonyl compound that is also an **enol phosphate**.

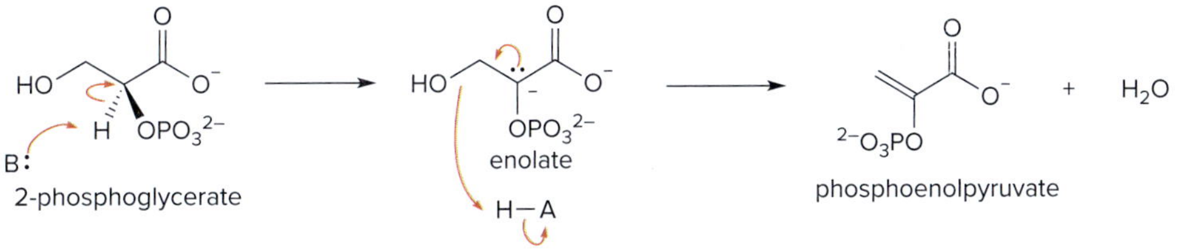

Step 10 of Glycolysis: Phosphate Transfer

Glycolysis is completed when phosphoenolpyruvate transfers PO_3^{2-} to ADP, forming ATP and pyruvate. The reaction is catalyzed by pyruvate kinase.

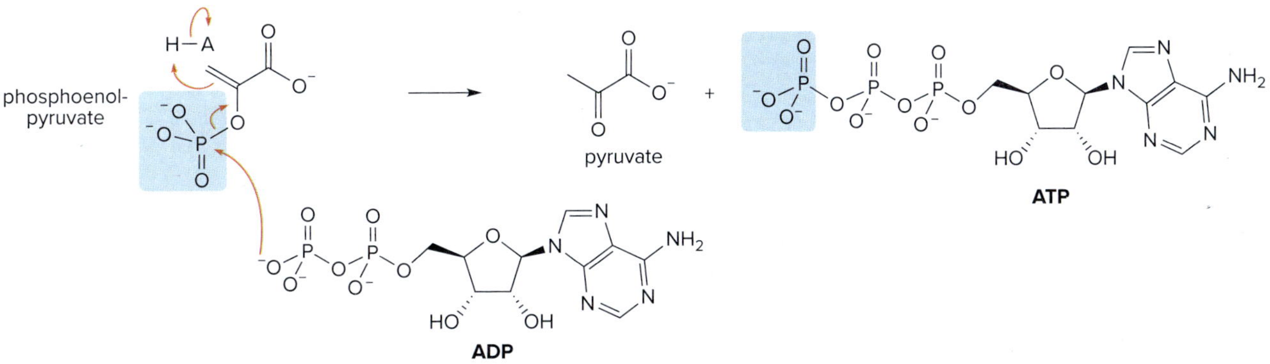

Thus, one NADH molecule is produced in Step [6] and two ATPs are formed in Steps [7] and [10] for *each* glyceraldehyde 3-phosphate.

- Because each glucose molecule forms *two* glyceraldehyde 3-phosphate molecules in Step [5], glycolysis forms *two* pyruvate molecules, *two* NADH molecules, and *four* ATPs in Steps [6]–[10].

What is the net result of all 10 steps of glycolysis? Three major products are formed—ATP, NADH, and pyruvate.

Two ATP molecules are used in Steps [1] and [3], and four ATPs are formed in Steps [7] and [10], so the net result is the synthesis of **two molecules of ATP** from glycolysis. **Two NADH** molecules are formed from two glyceraldehyde molecules during the oxidation in Step [6]. **Two pyruvate molecules** are formed from glucose. The fate of pyruvate depends on oxygen availability, as discussed in Section 27.5.

The overall process of glycolysis can be summarized in the following equation. How the carbon atoms of glucose correlate with the carbons of pyruvate is also shown.

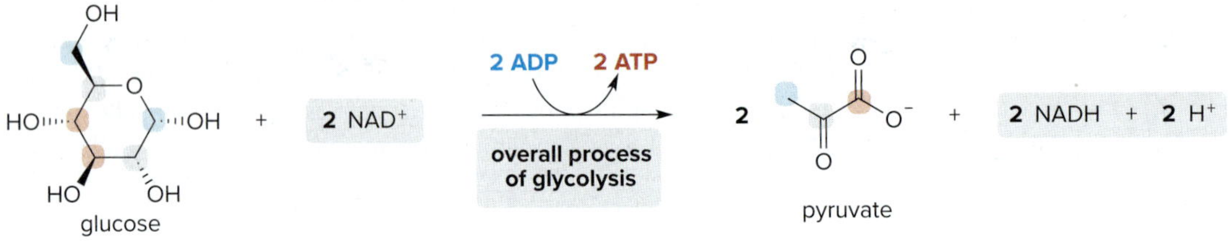

glucose + 2 NAD⁺ $\xrightarrow[\text{of glycolysis}]{\text{overall process}}$ 2 pyruvate + 2 NADH + 2 H⁺

(2 ADP → 2 ATP)

Although glycolysis is an ongoing pathway in cells, the rate of glycolysis depends on the body's need for the products it forms. When ATP levels are high, glycolysis is inhibited at various stages. When ATP levels are depleted during periods of strenuous exercise, glycolysis is activated so that more ATP is synthesized.

Problem 27.11 Despite the fact that 1,3-bisphosphoglycerate has two phosphate groups, ADP reacts at only the indicated position in Step [7] of glycolysis. Offer an explanation for this specificity.

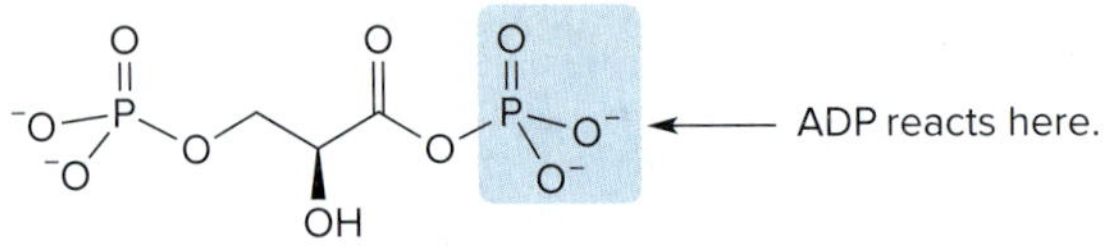

1,3-bisphosphoglycerate — ADP reacts here.

Problem 27.12 For each of the following intermediates in glycolysis, indicate which C atom correlates to C6 of glucose: (a) fructose 6-phosphate; (b) glyceraldehyde 3-phosphate; (c) 2-phosphoglycerate.

27.5 The Fate of Pyruvate

Although pyruvate is the end product of glycolysis, it is not the final product of glucose metabolism. What happens to pyruvate depends on the existing conditions and the organism. Three products are possible:

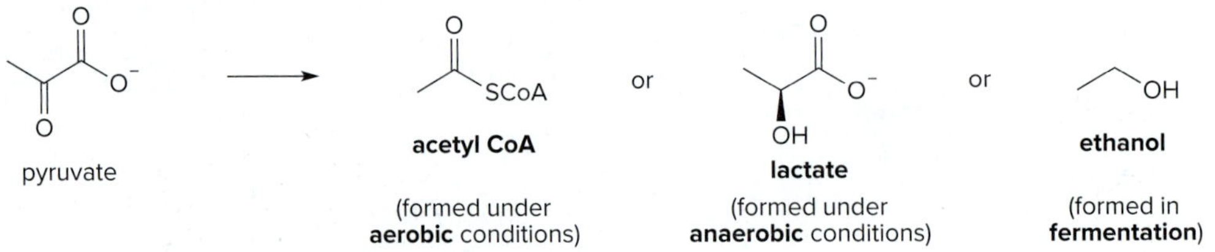

pyruvate → acetyl CoA (formed under **aerobic** conditions) or lactate (formed under **anaerobic** conditions) or ethanol (formed in **fermentation**)

27.5A Conversion of Pyruvate to Acetyl CoA

Under aerobic conditions, oxidation with NAD⁺ in the presence of coenzyme A converts pyruvate to **acetyl CoA,** which then enters the citric acid cycle (Section 27.6). This multistep process is catalyzed by a complex of enzymes and coenzymes, called the **pyruvate dehydrogenase complex.**

pyruvate + HSCoA $\xrightarrow[\text{complex}]{\text{pyruvate dehydrogenase}}$ acetyl CoA + CO_2
(NAD⁺ → NADH + H⁺)

This pathway requires an adequate supply of NAD⁺, which is obtained by the oxidation of NADH (formed in Step [6] of glycolysis) by oxygen. Thus, although oxygen is not needed for this specific reaction, the supply of NAD⁺ is oxygen dependent, and this reaction can occur only when oxygen is plentiful.

Problem 27.13 Which carbons of glucose end up as carbons of CO_2 when pyruvate is converted to acetyl CoA?

27.5B Conversion of Pyruvate to Lactate

When oxygen levels are low and there is insufficient oxygen to re-oxidize NADH back to NAD⁺, cells obtain NAD⁺ by converting **pyruvate to lactate.**

pyruvate $\xrightarrow[\text{lactate dehydrogenase}]{\text{NADH + H}^+ \;\; \text{NAD}^+}$ lactate

NAD⁺ can re-enter glycolysis at Step [6].

Reduction of pyruvate with NADH forms lactate and NAD⁺, which can now re-enter glycolysis and oxidize glyceraldehyde 3-phosphate at Step [6]. Thus, when there are inadequate levels of oxygen, **pyruvate is reduced to lactate for the sole purpose of re-oxidizing NADH to NAD⁺ to maintain glycolysis.**

Anaerobic metabolism leads to an increase in lactate in muscles, which in turn is associated with soreness and cramping. During these periods an "oxygen debt" is created. When vigorous activity ceases, an individual inhales deep breaths of air to repay the oxygen debt, lactate is gradually re-oxidized to pyruvate, and muscle soreness resolves.

Measuring lactate levels in the blood is a common diagnostic tool used by physicians to assess the health of an individual. A higher-than-normal lactate concentration in a resting individual generally indicates inadequate oxygen delivery to some tissues, and can be a sign of lung disease, infection, or another serious condition.

27.5C Conversion of Pyruvate to Ethanol

In yeast and other microorganisms, **pyruvate is converted to ethanol and carbon dioxide** by a two-step process: **decarboxylation** to acetaldehyde (CH_3CHO) followed by **reduction** to ethanol.

pyruvate $\xrightarrow[\text{decarboxylase}]{\text{pyruvate}}$ acetaldehyde + CO_2 $\xrightarrow[\text{dehydrogenase}]{\text{NADH + H}^+ \;\; \text{NAD}^+}$ ethanol

Because the NAD⁺ generated during reduction can enter glycolysis as an oxidizing agent in Step [6], glucose can be metabolized by yeast under *anaerobic* conditions: glycolysis forms pyruvate and two molecules of ATP, and pyruvate is further metabolized to ethanol and CO_2. **The anaerobic conversion of glucose to ethanol and CO_2 is called** *fermentation.*

glucose →[fermentation, 2 ADP → 2 ATP] 2 ethanol + 2 CO_2

The ethanol in beer and wine is obtained by fermenting the carbohydrates in barley malt and grapes, respectively. Fermentation is also key to the production of cheese and yogurt. When yeast is mixed with flour, water, and sugar, the enzymes in the yeast carry out fermentation to produce CO_2, which causes the bread to rise.

Fermentation plays a key role in the production of bread, beer, and cheese. *Tony Robins/Getty Images*

Problem 27.14 Compare the role of the coenzymes NADH and NAD^+ in both processes: (a) pyruvate → acetyl CoA; (b) pyruvate → lactate.

Problem 27.15 How might pyruvate be metabolized in the cornea, which has limited blood supply?

Problem 27.16 Write the equation for the overall conversion of glucose to lactate, including any coenzymes and other key reactants/products.

27.6 The Citric Acid Cycle and ATP Production

The citric acid cycle makes up the third stage of catabolism. In this stage, the acetyl CoA formed from the metabolism of lipids, carbohydrates, and amino acids is converted to carbon dioxide. The citric acid cycle is a *cyclic* metabolic pathway that begins with the addition of acetyl CoA to oxaloacetate and ends when oxaloacetate is formed as a product eight steps later.

acetyl CoA →[citric acid cycle, 8 steps] 2 CO_2 + HSCoA

The citric acid cycle is also called the tricarboxylic acid cycle or the Krebs cycle, named for German chemist and Nobel Laureate Hans Krebs, who worked out the details of these reactions in 1937.

Figure 27.5 illustrates the eight steps of the citric acid cycle. All intermediates are carboxylate anions derived from di- and tricarboxylic acids. The key features of the citric acid cycle include the following:

- The citric acid cycle begins when acetyl CoA reacts with oxaloacetate to form a six-carbon product in Step [1].
- Two carbons are lost as CO_2 in Steps [3] and [4].
- Four molecules of reduced coenzymes (NADH and $FADH_2$) are formed in Steps [3], [4], [6], and [8]. Reduced coenzymes enter the electron transport chain, ultimately forming a great deal of ATP.
- One molecule of GTP, a nucleoside triphosphate analogous to ATP, is synthesized in Step [5].

27.6 The Citric Acid Cycle and ATP Production

Figure 27.5 The steps in the citric acid cycle

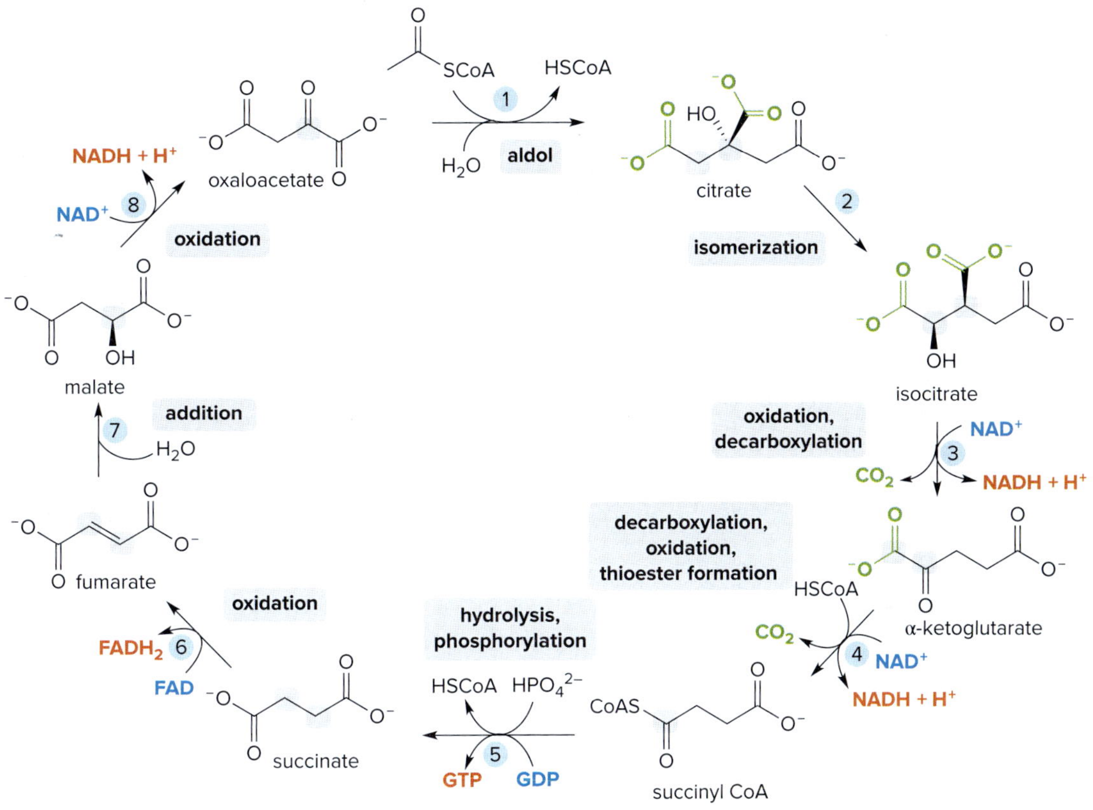

- The citric acid cycle begins with the addition of acetyl CoA to oxaloacetate, and ends eight steps later when oxaloacetate is regenerated.
- Each turn of the cycle forms two molecules of CO_2, four molecules of reduced coenzymes (3 NADH + 1 $FADH_2$), and one GTP.
- The carbons that react in each step are highlighted in gray.
- Each step is enzyme catalyzed: [1] citrate synthase; [2] aconitase; [3] isocitrate dehydrogenase; [4] α-ketoglutarate dehydrogenase; [5] succinyl CoA synthetase; [6] succinate dehydrogenase; [7] fumarase; and [8] malate dehydrogenase.

27.6A The Specific Reactions of the Citric Acid Cycle

Step 1 of the Citric Acid Cycle: Aldol Reaction

The citric acid cycle begins with the nucleophilic addition of acetyl CoA to oxaloacetate to form (S)-citryl CoA. Mechanistically, this reaction occurs by way of tautomerization of acetyl CoA to form a nucleophilic **enol** that adds to the ketone carbonyl of oxaloacetate in an **aldol-type reaction**, yielding (S)-citryl CoA. Hydrolysis of the thioester then forms citrate, the product of Step [1], by a nucleophilic acyl substitution.

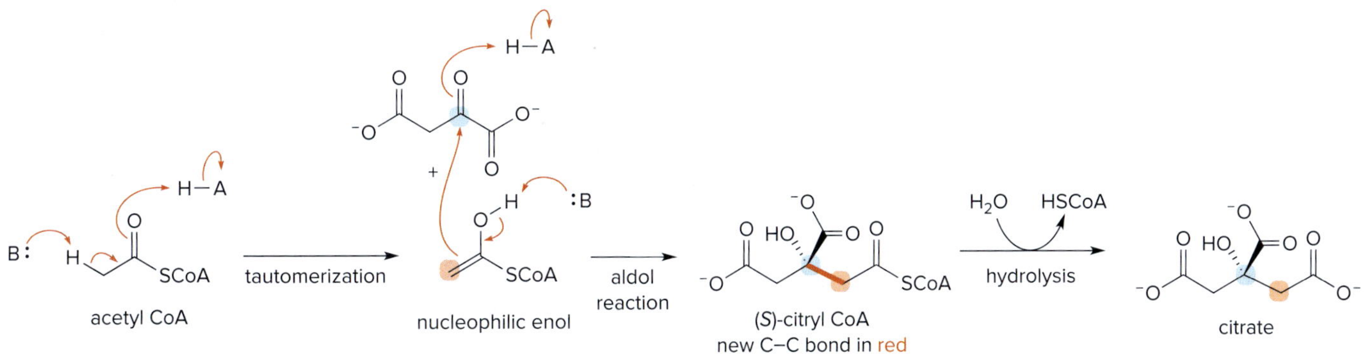

Step 2 of the Citric Acid Cycle: Isomerization

Step [2] of the citric acid cycle involves isomerization of the 3° alcohol in citrate to the 2° alcohol in isocitrate. This process requires two steps: loss of H_2O to form *cis*-aconitate, followed by nucleophilic addition of H_2O to the carbon–carbon double bond.

citrate → *cis*-aconitate → (2R,3S)-isocitrate

This reaction is enantioselective, forming a single stereoisomer of isocitrate.

Step 3 of the Citric Acid Cycle: Oxidation and Decarboxylation

In Step [3], NAD^+ oxidizes isocitrate to a β-keto carboxylate, which undergoes decarboxylation like the β-keto acids discussed in the acetoacetic ester synthesis (Sections 17.10 and 17.11). Decarboxylation forms a resonance-stabilized enolate, which is protonated to α-ketoglutarate.

isocitrate → β-keto carboxylate → enolate → α-ketoglutarate

This step cleaves a carbon–carbon bond to form a five-carbon intermediate and yields a molecule of CO_2.

Step 4 of the Citric Acid Cycle: Decarboxylation, Oxidation, and Thioester Formation

In Step [4], oxidation with NAD^+ in the presence of coenzyme A converts α-ketoglutarate to succinyl CoA and forms the *second* molecule of CO_2. This multistep process is catalyzed by a complex of enzymes and coenzymes, similar to the pyruvate dehydrogenase complex described in Section 27.5A.

α-ketoglutarate → succinyl CoA

Thus, Step [4] cleaves another carbon–carbon bond to form a four-carbon intermediate. No other carbon–carbon σ bonds are broken or formed in the remaining steps of the citric acid cycle.

Step 5 of the Citric Acid Cycle: Hydrolysis and Phosphorylation

In Step [5], succinyl CoA is converted to succinate by a two-part process. First, the thioester undergoes nucleophilic acyl substitution with HPO_4^{2-} to form an acyl phosphate, which then transfers its phosphate to GDP to form a molecule of GTP and succinate.

succinyl CoA → acyl phosphate → succinate

GTP is a high-energy nucleotide triphosphate that serves the same role as ATP. The difference is the base portion of GTP—guanine instead of adenine.

Steps 6–8 of the Citric Acid Cycle: Two Oxidations and Hydration

Steps [6]–[8] are similar to the first three steps of fatty acid catabolism shown in Figure 27.2.

Step [6] involves the FAD-mediated oxidation of succinate to form fumarate and $FADH_2$. This reaction removes H atoms from the α and β carbons to the carbonyl group, generating a product with a **trans double bond**.

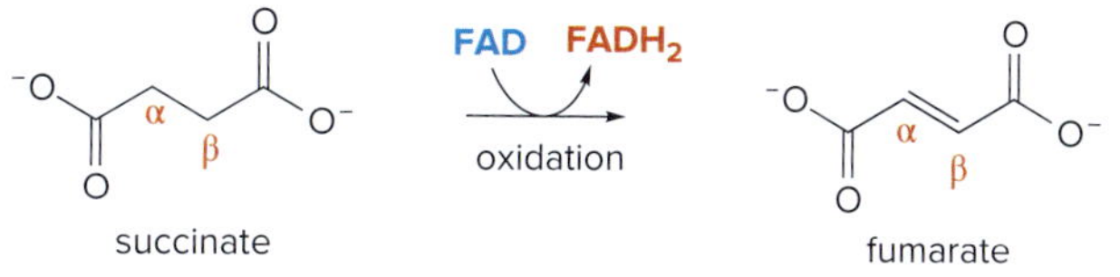

Conjugate addition of water to the α,β-unsaturated dicarboxylic acid yields a single enantiomer of the newly formed 2° alcohol, (*S*)-malate, in Step [7]. Finally, the OH group of the β-hydroxy acyl CoA is **oxidized** by NAD^+ to form oxaloacetate and NADH.

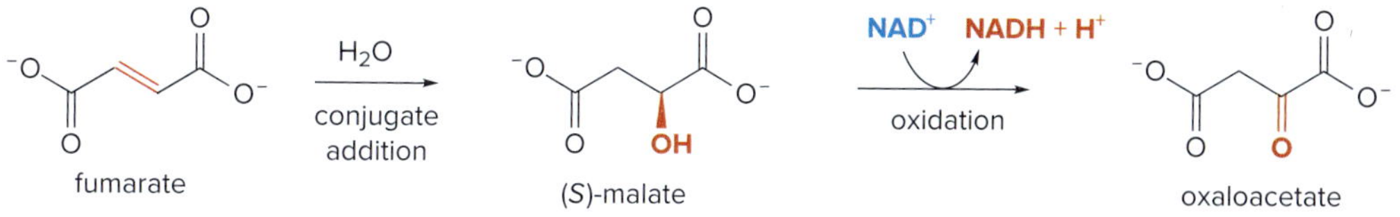

Steps [6]–[8] result in the formation of two more molecules of reduced coenzymes, $FADH_2$ and NADH. Because the product of Step [8] is the starting material for Step [1], the cycle can continue as long as additional acetyl CoA is available for Step [1].

Problem 27.17 Write the steps for the E1cB mechanism that converts citrate to *cis*-aconitate in Step [2] of the citric acid cycle.

Problem 27.18 The rate of the citric acid cycle depends on the body's energy needs. Is this cycle activated or inhibited in each circumstance: (a) high energy demand; (b) high NADH concentration; (c) low supply of ATP?

27.6B The Net Result of the Citric Acid Cycle

Overall the citric acid cycle forms two molecules of CO_2, four molecules of reduced coenzymes (NADH and $FADH_2$), and one molecule of GTP. The net equation for the citric acid cycle is as follows.

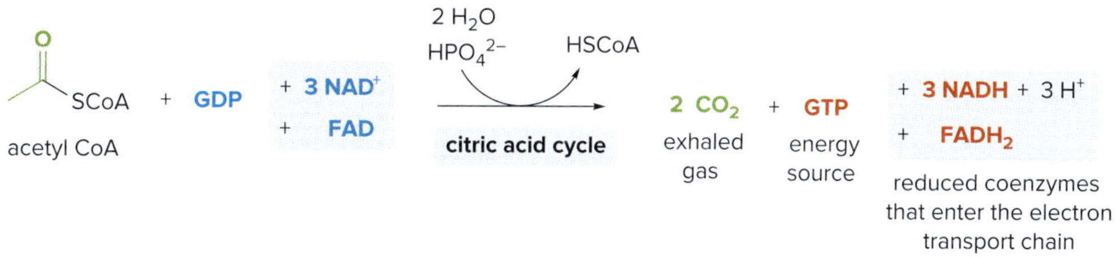

- The main function of the citric acid cycle is to produce reduced coenzymes that enter the electron transport chain and ultimately produce ATP.

In Section 27.2, we learned that electrons from these electron-rich reduced coenzymes provide the energy to synthesize ATP from ADP.

- Each NADH that enters the electron transport chain in Stage [4] provides the energy to synthesize 2.5 equivalents of ATP.
- Each $FADH_2$ provides the energy to synthesize 1.5 equivalents of ATP.

3 NADH × 2.5 ATP/NADH = 7.5 ATP
1 FADH$_2$ × 1.5 ATP/FADH$_2$ = 1.5 ATP
1 GTP = 1 ATP
10 ATP

- Complete catabolism of each acetyl CoA that enters the citric acid cycle results in the synthesis of 10 ATP molecules.

Problem 27.19 What reactions in the citric acid cycle (a) remove CO$_2$; (b) form a carbon–carbon bond; (c) break a carbon–carbon σ bond; (d) oxidize an organic substrate?

27.6C The ATP Yield from the Aerobic Metabolism of Glucose to CO$_2$

How much ATP is generated from the complete catabolism of glucose to CO$_2$? To carry out this calculation, we must consider both the ATP formed directly in reactions, as well as ATP produced from reduced coenzymes from Stage [4] of catabolism.

glucose →[1 glycolysis] 2 pyruvate + 2 NADH → 5 ATP; 2 ATP →[2] 2 acetyl CoA + 2 CO$_2$ + 2 NADH → 5 ATP →[3 citric acid cycle] 4 CO$_2$ + 6 NADH, 2 FADH$_2$, 2 GTP → 15 ATP, 3 ATP, 2 ATP

Total ATP yield = **32 ATP**

① Glycolysis converts glucose to pyruvate and forms **2 ATPs** directly. The two molecules of NADH yield an additional **5 ATPs** after Stage [4] of catabolism.

② When two molecules of pyruvate are oxidized and decarboxylated to two molecules of acetyl CoA, two molecules of NADH are formed, which yield an additional **5 ATPs** after Stage [4].

③ The citric acid cycle converts two molecules of acetyl CoA to two GTPs, the energy equivalent of **2 ATPs**. Six NADH molecules and two FADH$_2$ molecules yield an additional **18 ATPs** from Stage [4]. Thus, **20 ATPs** are formed from two acetyl CoA molecules.

Adding up the ATP formed in each pathway gives a **total of 32 ATP molecules for the complete catabolism of each glucose molecule.** Most of the ATP formed from glucose metabolism comes from the citric acid cycle and Stage [4] of catabolism.

Problem 27.20 How much ATP results from each transformation: (a) glucose → 2 acetyl CoA; (b) 2 pyruvate → 6 CO$_2$?

Problem 27.21 What is the difference in the ATP generation between the aerobic oxidation of glucose to CO$_2$ and the anaerobic conversion of glucose to lactate?

27.6D The ATP Yield from Fatty Acid Oxidation

Now that we have learned about the citric acid cycle, we can determine how much ATP is formed from the complete catabolism of a fatty acid. To determine this quantity, we must take into account the ATP cost for the conversion of a fatty acid to a fatty acyl CoA, as well as the ATP production from coenzymes (NADH and FADH$_2$) and acetyl CoA formed during β-oxidation (Section 27.3B).

For example, in Sample Problem 27.1, we determined that the complete catabolism of lauric acid ($C_{12}H_{24}O_2$) yields 6 acetyl CoA, 5 NADH, and 5 FADH$_2$. Sample Problem 27.2 uses these data to determine the amount of ATP formed from lauric acid.

lauric acid → 6 acetyl-SCoA + **5** NADH + **5** FADH$_2$

Bonds in red are cleaved.

Sample Problem 27.2 — Determining the Amount of ATP Formed during Fatty Acid Catabolism

How much ATP is formed from the complete catabolism of lauric acid ($C_{12}H_{24}O_2$)?

Solution

Add up the ATP used or formed from each operation:
- The conversion of lauric acid to its acyl CoA ($C_{11}H_{23}COSCoA$) *requires* the energy equivalent of 2 ATPs, so the net result is −2 ATPs.
- Determine the ATP formed from the reduced coenzymes after Stage [4] of catabolism.

$$\begin{aligned}
5\ \text{NADH} \times 2.5\ \text{ATP/NADH} &= 12.5\ \text{ATPs} \\
5\ \text{FADH}_2 \times 1.5\ \text{ATP/FADH}_2 &= 7.5\ \text{ATPs} \\
\text{From reduced coenzymes} &= 20\ \text{ATPs}
\end{aligned}$$

- Determine the amount of ATP from each acetyl CoA.

6 acetyl CoA × 10 ATP/acetyl CoA = **60** ATPs

Total the values from each operation: (−2) + 20 + 60 = **78 ATPs** for the complete catabolism of lauric acid.

Problem 27.22 Calculate the number of molecules of ATP formed from the complete catabolism of stearic acid ($C_{18}H_{36}O_2$).

More Practice: Try Problems: 27.23a, 27.35.

Chapter 27 REVIEW

KEY CONCEPTS

[1] Key organic components in metabolism (27.1, 27.2)

① Coenzyme A and acetyl CoA (27.1)
- Coenzyme A (**HSCoA**) is a nucleophilic thiol (3.9) that combines with an acetyl group to form acetyl CoA (**CH$_3$COSCoA**).
- Acetyl CoA is a **thioester** that is the product of Stage [2] of catabolism for lipids, carbohydrates, and amino acids.
- The α carbon of acetyl CoA becomes a nucleophile when an α H is removed.
- Acetyl CoA has a good leaving group, so it undergoes nucleophilic acyl substitutions.
- The acetyl group of acetyl CoA is oxidized in the citric acid cycle to CO_2 (27.6).

② ATP and ADP (27.1)
- The interconversion of ATP and ADP is the central method of energy transfer in cells.
- Hydrolysis of ATP to ADP *releases* energy that can be used to drive reactions that require energy input.
- ATP is synthesized from ADP by a phosphorylation reaction that *requires* energy input.
- Most of the energy obtained from lipids, carbohydrates, and proteins is packaged in ATP molecules formed in Stage [4] of catabolism.

③ NAD$^+$/NADH and FAD/FADH$_2$ (27.2)
- NAD$^+$ and FAD are biological **oxidizing** agents.
- NADH and FADH$_2$ are biological **reducing** agents.
- Redox reactions involving carbonyl groups generally use NAD$^+$/NADH.
- Redox reactions of other functional groups use FAD and FADH$_2$.
- Each NADH that enters Stage [4] of catabolism provides the energy to synthesize **2.5 ATPs.**
- Each FADH$_2$ that enters Stage [4] of catabolism provides the energy to synthesize **1.5 ATPs.**

Try Problems 27.25, 27.26.

[2] Overall reactions for key metabolic pathways

1. A four-step cycle of β-oxidation of a fatty acyl CoA (27.3)

R–CH₂–C(O)–SCoA (fatty acyl CoA) + FAD + NAD⁺ + H₂O + HSCoA → (one cycle of β-oxidation) → R–C(O)–SCoA (fatty acyl CoA with two fewer C's) + CH₃–C(O)–SCoA (acetyl CoA) + FADH₂ + NADH + H⁺

- Each four-step sequence of β-oxidation forms one molecule each of acetyl CoA, FADH₂, and NADH.

2. Glycolysis (27.4)

glucose + 2 NAD⁺ + 2 ADP → (overall process of glycolysis) → 2 pyruvate + 2 ATP + 2 NADH + 2 H⁺

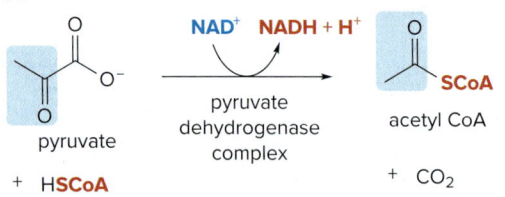

- Glycolysis converts glucose to two molecules of pyruvate by a 10-step pathway.

3. Three possible outcomes for pyruvate (27.5)

[1] Under aerobic conditions:

pyruvate + HSCoA → (pyruvate dehydrogenase complex, NAD⁺ → NADH + H⁺) → acetyl CoA + CO₂

[2] Under anaerobic conditions:

pyruvate → (lactate dehydrogenase, NADH + H⁺ → NAD⁺) → lactate

[3] In fermentation:

pyruvate → (pyruvate decarboxylase) → acetaldehyde → (alcohol dehydrogenase, NADH + H⁺ → NAD⁺) → ethanol + CO₂

4. Citric acid cycle (27.6)

acetyl CoA + GDP + 3 NAD⁺ + FAD + 2 H₂O + HPO₄²⁻ → (citric acid cycle) → HSCoA + 2 CO₂ (exhaled gas) + GTP (energy source) + 3 NADH + 3 H⁺ + FADH₂

- The citric acid cycle is an eight-step cyclic pathway that forms two molecules of CO₂, one molecule of GTP, and four molecules of reduced coenzymes that enter Stage [4] of catabolism and ultimately produce a great deal of ATP.

KEY SKILLS

[1] Determining the products of β-oxidation of a fatty acid (27.3); example: decanoic acid ($C_9H_{19}CO_2H$)

1 Determine the number of molecules of acetyl CoA formed.	**2** Determine the number of molecules of NADH and FADH$_2$ formed.
• The number of molecules of acetyl CoA equals **one-half** the number of C's in the fatty acid. $C_{10}H_{20}O_2 \longrightarrow 5$ acetyl CoA	• The number of cycles of β-oxidation is **one fewer** than the number of acetyl CoA molecules. • The number of cycles **equals** the number of molecules of NADH and FADH$_2$ formed. 4 NADH + 4 FADH$_2$

See Sample Problem 27.1. Try Problems 27.27, 27.28.

[2] Determining the amount of ATP formed during fatty acid catabolism (27.6D); example: decanoic acid ($C_9H_{19}CO_2H$)

1 Determine the ATP used to form the fatty acyl CoA.	**2** Determine the ATP formed from reduced coenzymes.	**3** Determine the amount of ATP from each acetyl CoA and add up the results.
• Forming an acyl CoA requires the energy equivalent of 2 ATPs. $C_{10}H_{20}O_2 \longrightarrow C_9H_{19}COSCoA$ -2 ATP	• β-Oxidation forms 4 NADH and 4 FADH$_2$. 4 NADH × 2.5 ATP/NADH = 10 ATPs 4 FADH$_2$ × 1.5 ATP/FADH$_2$ = 6 ATPs From reduced coenzymes: 16 ATPs	• Each cycle of the citric acid cycle forms 10 ATPs. • Five acetyl CoA molecules are formed from a 10-C fatty acid. 5 acetyl CoA × 10 ATP/acetyl CoA = 50 ATPs • Total ATP production: (−2) + 16 + 50 = 64 ATPs.

See Sample Problem 27.2. Try Problems 27.23a, 27.35.

PROBLEMS

Problems Using Three-Dimensional Models

27.23 (a) Calculate the number of molecules of ATP formed by the complete catabolism of palmitic acid, shown in the ball-and-stick model. (b) How many molecules of CO_2 are formed when palmitic acid is completely catabolized?

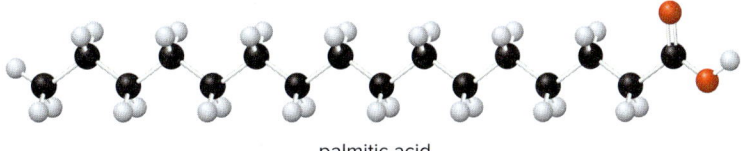

palmitic acid

27.24 (a) What compound forms **A** during glycolysis? (b) What product is formed from **A** during glycolysis? (c) Which carbon of **A** corresponds to C3 of glucose?

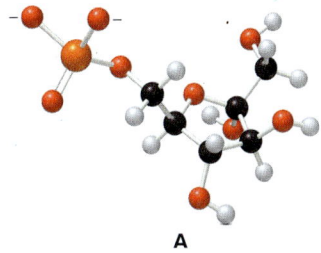

A

Oxidation and Reduction Reactions

27.25 Classify each reaction as an oxidation or reduction and give a possible coenzyme used.

a. [structure of amino acid with $^+NH_3$ converting to imino acid with NH]

b. [phosphate-containing aldehyde with OH converting to bis-phosphate compound]

27.26 Draw a product that could form when 6-hydroxyheptan-2-one is treated with each of the following coenzymes: (a) NAD^+; (b) FAD; (c) NADH.

[structure of 6-hydroxyheptan-2-one]

6-hydroxyheptan-2-one

Metabolic Pathways

27.27 For each of the fatty acids **A** and **B**: (a) How many molecules of acetyl CoA are formed from complete oxidation? (b) How many cycles of β-oxidation are needed? (c) How many molecules of NADH and $FADH_2$ are formed?

$$CH_3(CH_2)_{20}CO_2H \qquad CH_3(CH_2)_{28}CO_2H$$
$$\textbf{A} \qquad\qquad\qquad \textbf{B}$$

27.28 (a) How many molecules of acetyl CoA are formed from complete catabolism of the fatty acids derived from the given triacylglycerol? (b) How many cycles of β-oxidation are needed?

[triacylglycerol structure]

27.29 Where do the labeled atoms of glucose end up when glucose is catabolized to each compound?

[glucose structure with labeled carbon]

a. [3-phosphoglycerate structure with OPO_3^{2-}]
b. [lactate/pyruvate structure with OH]
c. [acetyl-SCoA structure]

glucose

27.30 Glucose is completely catabolized to six molecules of CO_2. What specific reactions generate each molecule of CO_2?

27.31 What is the difference between the amount of ATP generated when glucose is converted to CO_2 compared to when glucose is converted to ethanol in fermentation?

27.32 In fermentation, where do the six carbons of glucose end up?

27.33 (a) Which reactions directly generate ATP (or GTP) when glucose is catabolized to CO_2? (b) How many molecules of ATP are formed directly per glucose molecule? (c) Compare this value to the amount of ATP generated when the reduced coenzymes formed in this catabolic process generate ATP from the electron transport chain and oxidative phosphorylation. Where does most of the ATP come from when glucose is metabolized?

27.34 In Step [3] of the citric acid cycle, oxalosuccinate is decarboxylated to α-ketoglutarate. Why is the particular carboxy group lost as CO_2, when oxalosuccinate has three different carboxy groups that could be removed?

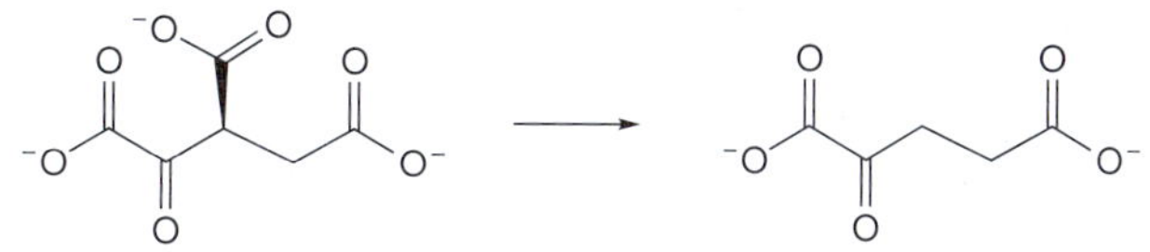

oxalosuccinate α-ketoglutarate

27.35 How many molecules of ATP are formed by complete catabolism of the fatty acids **A** and **B?**

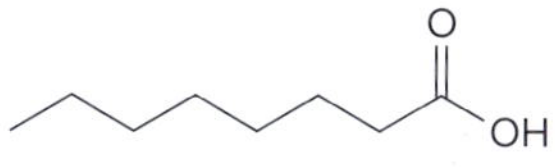

A

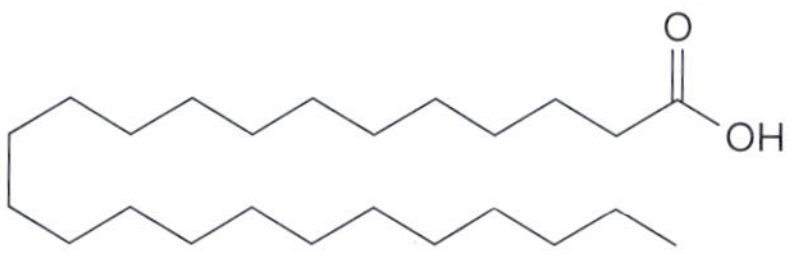

B

27.36 How much ATP is formed by the complete catabolism of glycerol [$HOCH(CH_2OH)_2$]?

27.37 Compare the energy content of glucose and stearic acid by determining the number of ATP molecules formed per carbon for each compound, when it is completely catabolized. Based on these data, are lipids more effective energy-storing molecules than carbohydrates?

Mechanisms

27.38 Draw a stepwise mechanism for the following reaction.

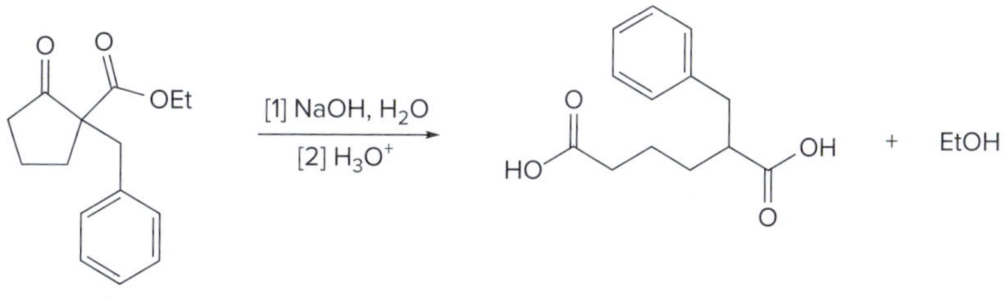

27.39 In Section 27.6 we learned that the reaction of acetyl CoA with oxaloacetate forms citrate in the first step of the citric acid cycle. Another possible reaction of these two starting materials (which does not occur) could form **X.** Draw a stepwise mechanism for this reaction.

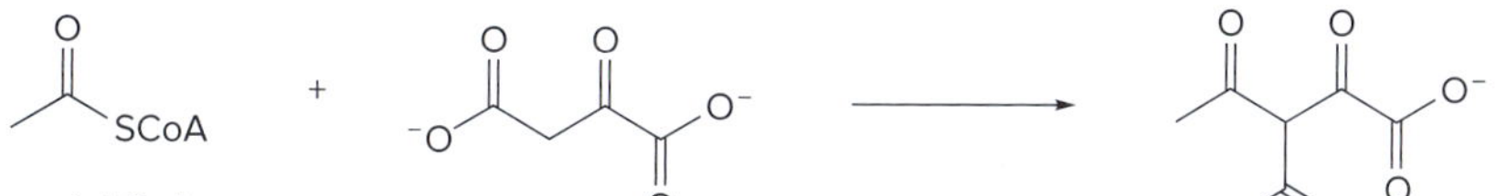

acetyl CoA oxaloacetate **X**

27.40 One step in the biosynthesis of fatty acids involves the reaction of acetyl CoA with malonyl CoA to form acetoacetyl CoA. Draw a stepwise mechanism for this process.

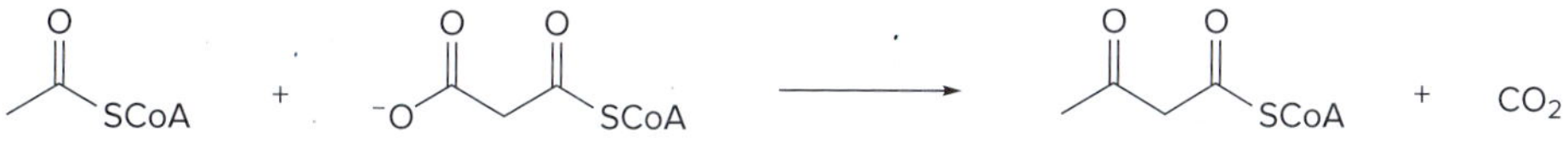

acetyl CoA malonyl CoA acetoacetyl CoA

27.41 Draw a stepwise mechanism for the conjugate addition reaction that converts *cis*-aconitate to isocitrate in Step [2] of the citric acid cycle.

Challenge Problems

27.42 How much ATP is formed from the complete catabolism of one molecule of the given triacylglycerol?

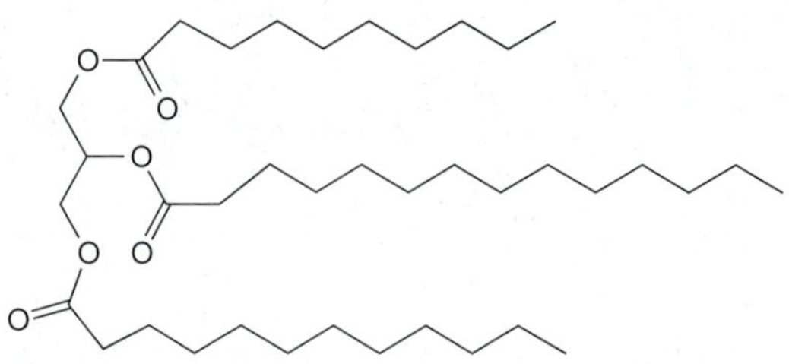

27.43 Mannose is a monosaccharide obtained in the diet from fruits such as cranberries and currants. Catabolism of mannose occurs by the conversion of mannose 6-phosphate to fructose 6-phosphate, an intermediate in glycolysis. Draw a stepwise mechanism for this process.

mannose | mannose 6-phosphate | fructose 6-phosphate

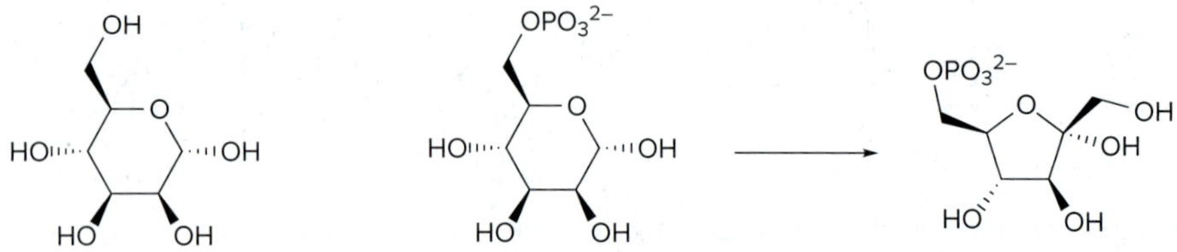

27.44 In the presence of aqueous base, fructose can form monosaccharides **A** and **B,** both of which have a carbonyl group on C3 of the six-carbon skeleton. Draw a stepwise mechanism for this process.

fructose | **A** | **B**

Design elements: Medical molecules concept science vector: © watchara/Shutterstock; Mechanism box (gears): © McGraw-Hill

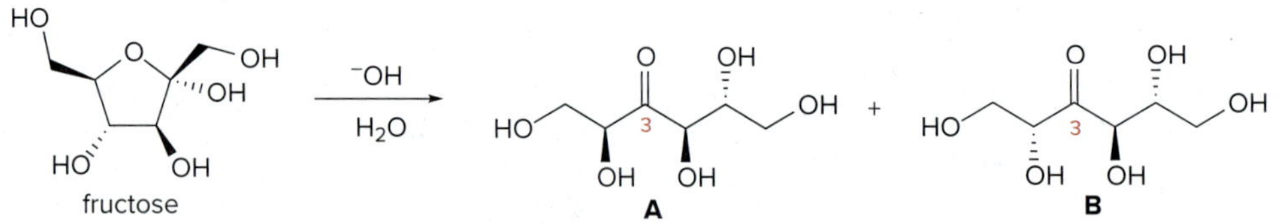

Periodic Table of the Elements

APPENDIX A

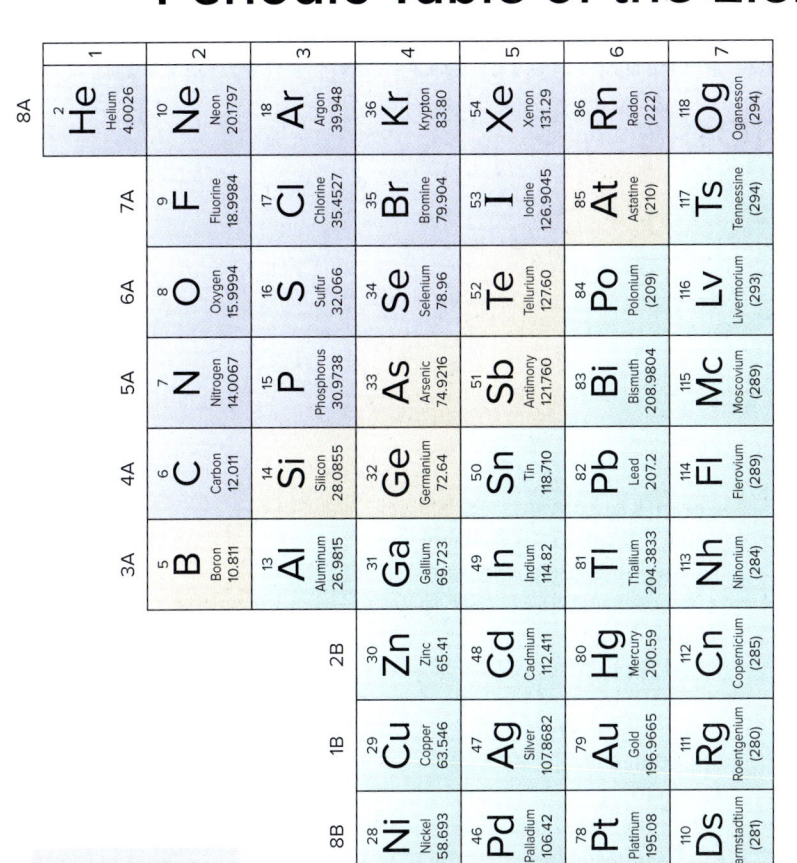
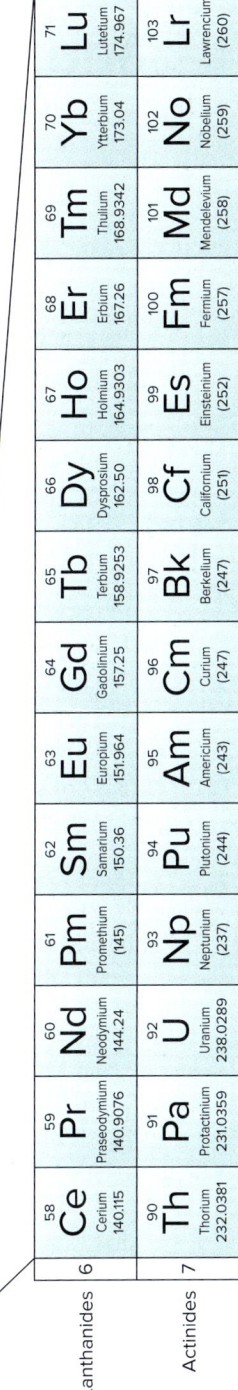

APPENDIX B

Common Abbreviations, Arrows, and Symbols

Abbreviations

Ac	acetyl, CH_3CO-
BBN	9-borabicyclo[3.3.1]nonane
BINAP	2,2'-bis(diphenylphosphino)-1,1'-binaphthyl
Boc	*tert*-butoxycarbonyl, $(CH_3)_3COCO-$
bp	boiling point
Bu	butyl, $CH_3CH_2CH_2CH_2-$
DBN	1,5-diazabicyclo[4.3.0]non-5-ene
DBU	1,8-diazabicyclo[5.4.0]undec-7-ene
DCC	dicyclohexylcarbodiimide
DET	diethyl tartrate
DIBAL-H	diisobutylaluminum hydride, $[(CH_3)_2CHCH_2]_2AlH$
DMF	dimethylformamide, $HCON(CH_3)_2$
DMSO	dimethyl sulfoxide, $(CH_3)_2S=O$
ee	enantiomeric excess
Et	ethyl, CH_3CH_2-
Fmoc	9-fluorenylmethoxycarbonyl
HMPA	hexamethylphosphoramide, $[(CH_3)_2N]_3P=O$
HOMO	highest occupied molecular orbital
IR	infrared
LDA	lithium diisopropylamide, $LiN[CH(CH_3)_2]_2$
LUMO	lowest unoccupied molecular orbital
m-	meta
mCPBA	*m*-chloroperoxybenzoic acid
Me	methyl, CH_3-
MO	molecular orbital
mp	melting point
MS	mass spectrometry
MW	molecular weight
NBS	*N*-bromosuccinimide
NMO	*N*-methylmorpholine *N*-oxide
NMR	nuclear magnetic resonance
o-	ortho
p-	para
PCC	pyridinium chlorochromate
Ph	phenyl, C_6H_5-
ppm	parts per million
Pr	propyl, $CH_3CH_2CH_2-$
RCM	ring-closing metathesis
ROMP	ring-opening metathesis polymerization
TBDMS	*tert*-butyldimethylsilyl

THF	tetrahydrofuran
TMS	tetramethylsilane, $(CH_3)_4Si$
Ts	tosyl, *p*-toluenesulfonyl, $CH_3C_6H_4SO_2-$
TsOH	*p*-toluenesulfonic acid, $CH_3C_6H_4SO_3H$
UV	ultraviolet

Arrows

⟶	reaction arrow
⇌	equilibrium arrows
⟷	double-headed arrow, used between resonance structures
↶	biological reaction arrows
⤴	full-headed curved arrow, showing the movement of an electron pair
⤴	half-headed curved arrow (fishhook), showing the movement of an electron
⟹	retrosynthetic arrow
⤫	no reaction

Symbols

⟷	dipole
$h\nu$	light
Δ	heat
$\delta+$	partial positive charge
$\delta-$	partial negative charge
λ	wavelength
ν	frequency
$\tilde{\nu}$	wavenumber
HA	Brønsted–Lowry acid
B:	Brønsted–Lowry base
:Nu$^-$	nucleophile
E$^+$	electrophile
X	halogen
◂	bond oriented forward
⋯	bond oriented behind
---	partial bond
[ ]‡	transition state
[O]	oxidation
[H]	reduction

Common Element Colors Used in Molecular Art

C H O N F Cl Br I S P

APPENDIX C

pK_a Values for Selected Compounds

Compound	pK_a
HI	−10
HBr	−9
H_2SO_4	−9
(CH$_3$)$_2$C=O$^+$H (protonated acetone)	−7.3
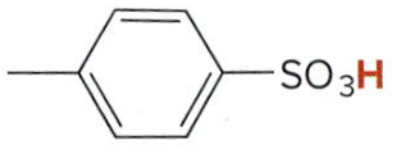 4-methylbenzenesulfonic acid (CH$_3$-C$_6$H$_4$-SO$_3$H)	−7
HCl	−7
[(CH$_3$)$_2$OH]$^+$	−3.8
(CH$_3$OH$_2$)$^+$	−2.5
H$_3$O$^+$	−1.7
CH$_3$SO$_3$H	−1.2
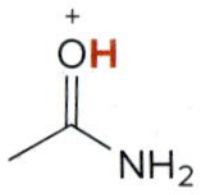 CH$_3$C(=O$^+$H)NH$_2$ (protonated acetamide)	0.0
CF$_3$CO$_2$H	0.2
CCl$_3$CO$_2$H	0.6
 O$_2$N-C$_6$H$_4$-NH$_3^+$	1.0
Cl$_2$CHCO$_2$H	1.3
H$_3$PO$_4$	2.1
FCH$_2$CO$_2$H	2.7
ClCH$_2$CO$_2$H	2.8
BrCH$_2$CO$_2$H	2.9
ICH$_2$CO$_2$H	3.2
HF	3.2
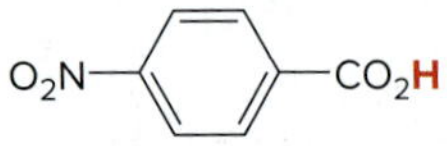 O$_2$N-C$_6$H$_4$-CO$_2$H	3.4
HCO$_2$H	3.8

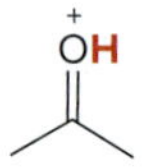

Compound	pK_a
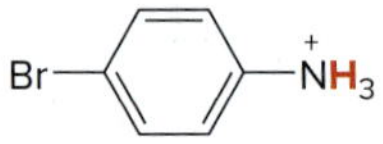 Br-C$_6$H$_4$-NH$_3^+$	3.9
Br-C$_6$H$_4$-CO$_2$H	4.0
C$_6$H$_5$-CO$_2$H	4.2
CH$_3$-C$_6$H$_4$-CO$_2$H	4.3
CH$_3$O-C$_6$H$_4$-CO$_2$H	4.5
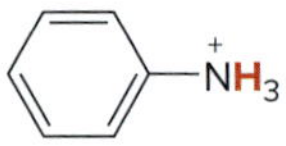 C$_6$H$_5$-NH$_3^+$	4.6
CH$_3$CO$_2$H	4.8
(CH$_3$)$_3$CCO$_2$H	5.0
CH$_3$-C$_6$H$_4$-NH$_3^+$	5.1
pyridinium (C$_5$H$_5$NH$^+$)	5.3
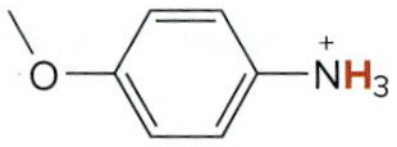 CH$_3$O-C$_6$H$_4$-NH$_3^+$	5.3
H$_2$CO$_3$	6.4
H$_2$PO$_4^-$	6.9
H$_2$S	7.0
O$_2$N-C$_6$H$_4$-OH	7.1
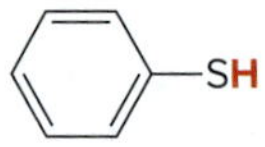 C$_6$H$_5$-SH	7.8

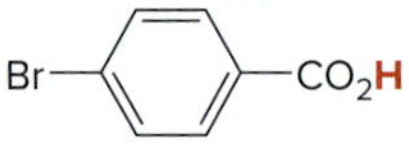

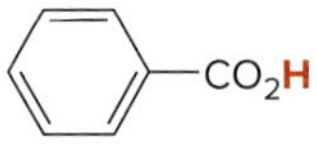

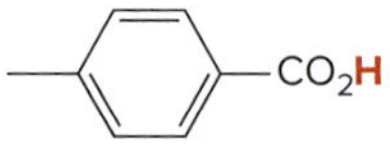

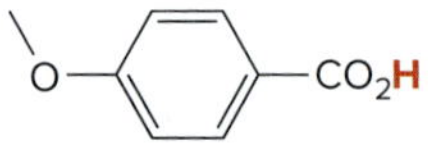

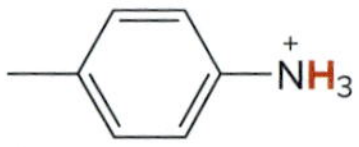

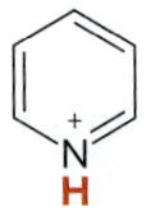

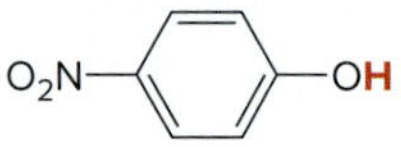

Appendix C pK_a Values for Selected Compounds

Compound	pK_a
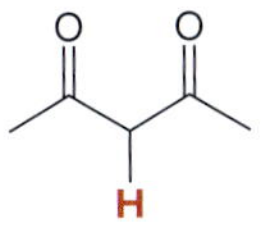 CH$_3$COCH(H)COCH$_3$ (pentane-2,4-dione)	8.9
HC≡N	9.1
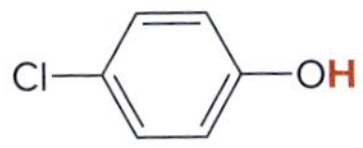 4-Cl-C$_6$H$_4$-OH	9.4
NH$_4^+$	9.4
H$_3$N$^+$CH$_2$CO$_2^-$	9.8
C$_6$H$_5$-OH	10.0
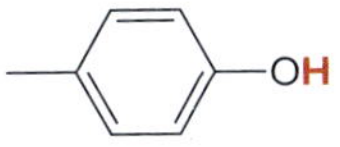 4-CH$_3$-C$_6$H$_4$-OH	10.2
HCO$_3^-$	10.2
CH$_3$NO$_2$	10.2
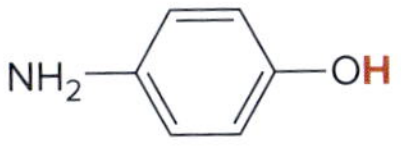 4-NH$_2$-C$_6$H$_4$-OH	10.3
CH$_3$CH$_2$SH	10.5
[(CH$_3$)$_3$NH]$^+$	10.6
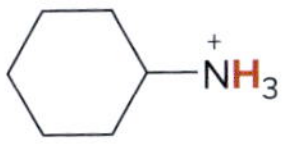 CH$_3$COCH(H)CO$_2$Et	10.7
(CH$_3$NH$_3$)$^+$	10.7
cyclohexyl-NH$_3^+$	10.7
[(CH$_3$)$_2$NH$_2$]$^+$	10.7
CF$_3$CH$_2$OH	12.4
HPO$_4^{2-}$	12.4
EtO$_2$C-CH(H)-CO$_2$Et	13.3

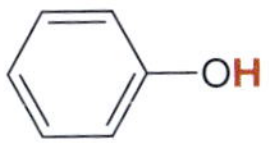

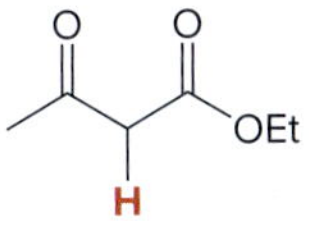

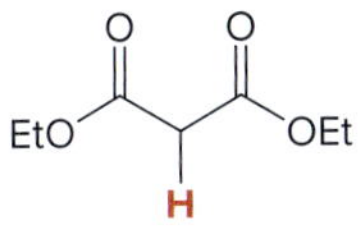

Compound	pK_a
cyclopentadiene-H	15
CH$_3$OH	15.5
H$_2$O	15.7
CH$_3$CH$_2$OH	16
CH$_3$CONH$_2$	16
CH$_3$CHO	17
(CH$_3$)$_3$COH	18
(CH$_3$)$_2$C=O	19.2
CH$_3$CO$_2$CH$_2$CH$_3$	24.5
HC≡CH	25
CH$_3$C≡N	25
CHCl$_3$	25
CH$_3$CON(CH$_3$)$_2$	30
H$_2$	35
NH$_3$	38
CH$_3$NH$_2$	40
C$_6$H$_5$CH$_2$-H	41
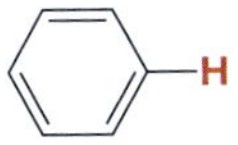 C$_6$H$_5$-H	43
CH$_2$=CHCH$_3$	43
CH$_2$=CH$_2$	44
 cyclopropyl-H	46
CH$_4$	50
CH$_3$CH$_3$	50

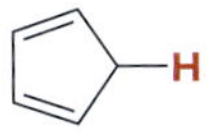

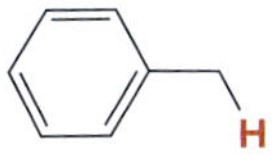

APPENDIX D

Nomenclature

Although the basic principles of nomenclature are presented in the body of this text, additional information is often needed to name many complex organic compounds. Appendix D concentrates on three topics:

- Naming alkyl substituents that contain branching
- Naming polyfunctional compounds
- Naming bicyclic compounds

Naming Alkyl Substituents That Contain Branching

Alkyl groups that contain any number of carbons and no branches are named as described in Section 4.4A: change the *-ane* ending of the parent alkane to the suffix *-yl*. Thus the seven-carbon alkyl group $CH_3CH_2CH_2CH_2CH_2CH_2CH_2-$ is called *heptyl*.

When an alkyl substituent also contains branching, follow a stepwise procedure:

[1] Identify the longest carbon chain of the alkyl group that begins at the point of attachment to the parent. Begin numbering at the point of attachment and use the suffix *-yl* to indicate an alkyl group.

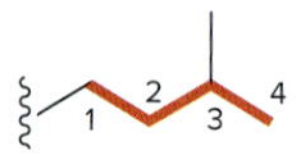

4 C's in the chain ---→ butyl group 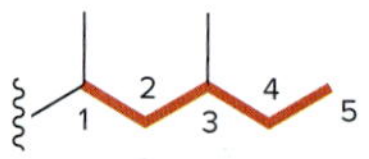   5 C's in the chain ---→ pentyl group

[2] Name all branches off the main alkyl chain and use the numbers from Step [1] to designate their location.

methyl at C3 → 3-methylbutyl   methyls at C1 and C3 → 1,3-dimethylpentyl

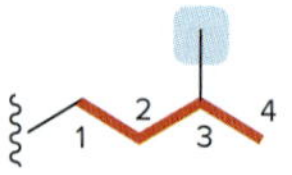

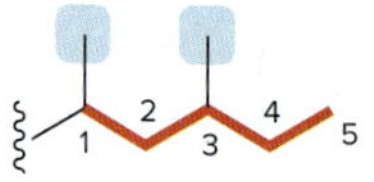

[3] Set the entire name of the substituent in parentheses, and alphabetize this substituent name by the first letter of the complete name.

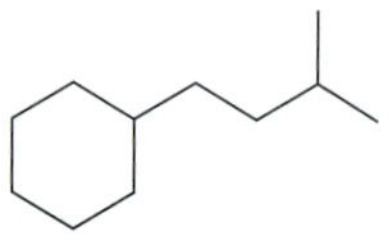

(3-methylbutyl)cyclohexane 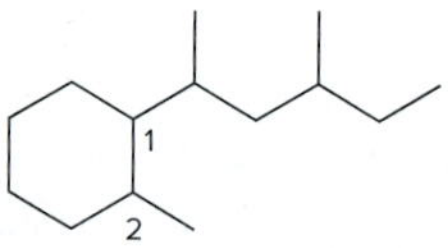   1-(1,3-dimethylpentyl)-2-methylcyclohexane

- Alphabetize the **d** of **d**imethylpentyl before the **m** of **m**ethyl.
- Number the ring to give the lower number to the first substituent alphabetically: place the dimethylpentyl group at C1.

Naming Polyfunctional Compounds

Many organic compounds contain more than one functional group. When one of those functional groups is halo (X—) or alkoxy (RO—), these groups are named as substituents as described in Sections 7.2 and 9.3B. To name other polyfunctional compounds, we must learn which functional group is assigned a higher priority in the rules of nomenclature. Two steps are usually needed:

[1] **Name a compound using the suffix of the highest-priority group,** and name other functional groups as *substituents*. Table D.1 lists the common functional groups in order of *decreasing* priority, as well as the prefixes needed when a functional group must be named as a substituent.

[2] Number the carbon chain to give the lower number to the highest-priority functional group that can be named as a suffix, and then follow all other rules of nomenclature. Examples are shown in Figure D.1.

Table D.1 Summary of Functional Group Nomenclature

Functional group	Suffix	Substituent name (prefix)
Carboxylic acid	-oic acid	carboxy
Ester	-oate	alkoxycarbonyl
Amide	-amide	amido
Nitrile	-nitrile	cyano
Aldehyde	-al	oxo (=O) or formyl (—CHO)
Ketone	-one	oxo
Alcohol	-ol	hydroxy
Amine	-amine	amino
Alkene	-ene	alkenyl
Alkyne	-yne	alkynyl
Alkane	-ane	alkyl
Ether	—	alkoxy
Halide	—	halo

Increasing priority (from bottom to top)

Figure D.1 Examples of nomenclature of polyfunctional compounds

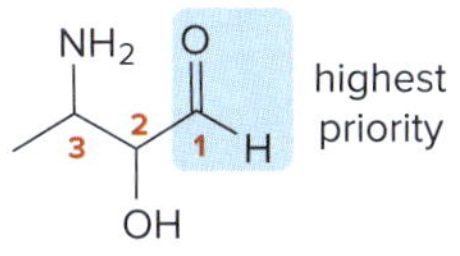

3-amino-2-hydroxybutanal
[Name as a derivative of an **aldehyde**, because CHO is the highest-priority functional group.]

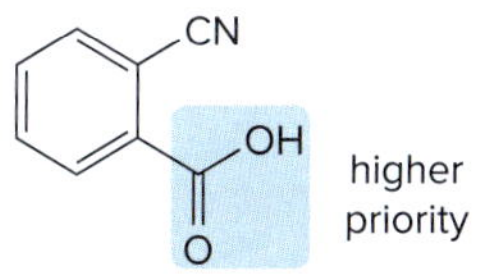

o-cyanobenzoic acid
[Name as a derivative of **benzoic acid**, because COOH is the higher-priority functional group.]

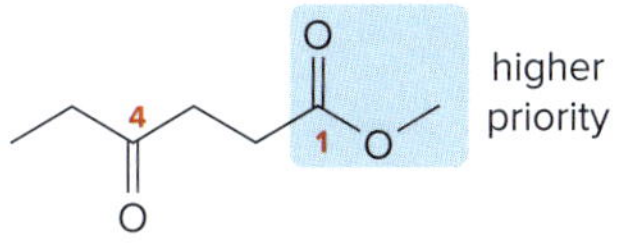

methyl 4-oxohexanoate
[Name as a derivative of an **ester**, because COOR is the higher-priority functional group.]

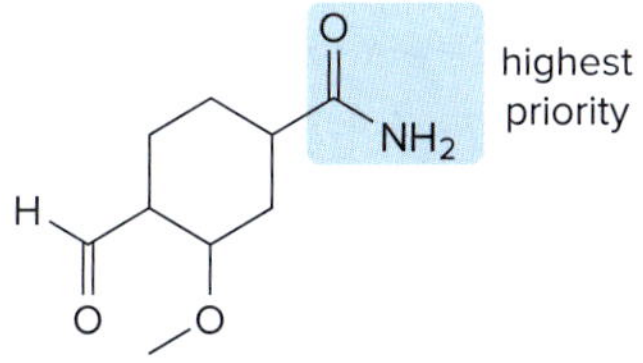

4-formyl-3-methoxycyclohexanecarboxamide
[Name as a derivative of an **amide**, because CONH₂ is the highest-priority functional group.]

Polyfunctional compounds that contain C–C double and triple bonds have characteristic suffixes to identify them, as shown in Table D.2. The higher-priority functional group is assigned the lower number.

Table D.2 Naming Polyfunctional Compounds with C–C Double and Triple Bonds

Functional groups	Suffix	Example
C=C and OH	enol	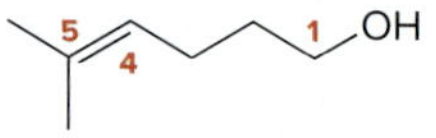 5-methylhex-4-en-1-ol
C=C + C=O (ketone)	enone	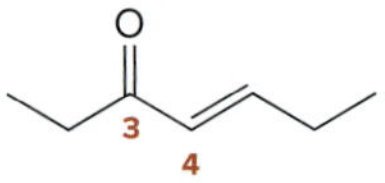 (E)-hept-4-en-3-one
C=C + C≡C	enyne	 hex-1-en-5-yne

Naming Bicyclic Compounds

Bicyclic ring systems—compounds that contain two rings that share one or two carbon atoms—can be bridged, fused, or spiro.

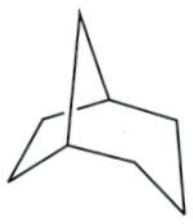

bridged ring 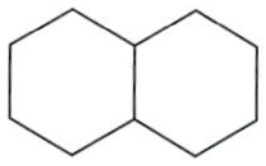   **fused ring** 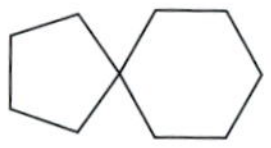   **spiro ring**

- A bridged ring system contains two rings that share two *non-adjacent* carbons.
- A fused ring system contains two rings that share a *common* carbon–carbon bond.
- A spiro ring system contains two rings that share *one carbon atom*.

Fused and bridged ring systems are named as bicyclo[*x.y.z*]alkanes, where the parent alkane corresponds to the total number of carbons in both rings. The numbers *x, y,* and *z* refer to the number of carbons that join the shared carbons together, written in order of *decreasing* size. For a fused ring system, *z* always equals zero, because the two shared carbons are directly joined together. The shared carbons in a bridged ring system are called the **bridgehead carbons.**

8 C's in the ring system → bicyclooctane

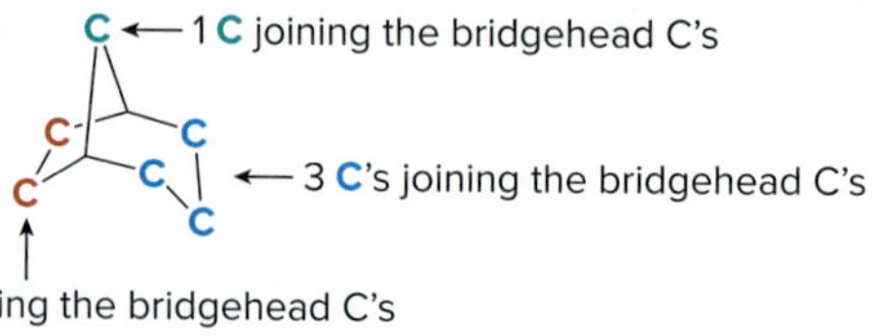

1 C joining the bridgehead C's
3 C's joining the bridgehead C's
2 C's joining the bridgehead C's
Name: bicyclo[3.2.1]octane

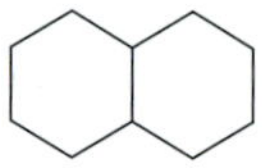

10 C's in the ring system → bicyclodecane

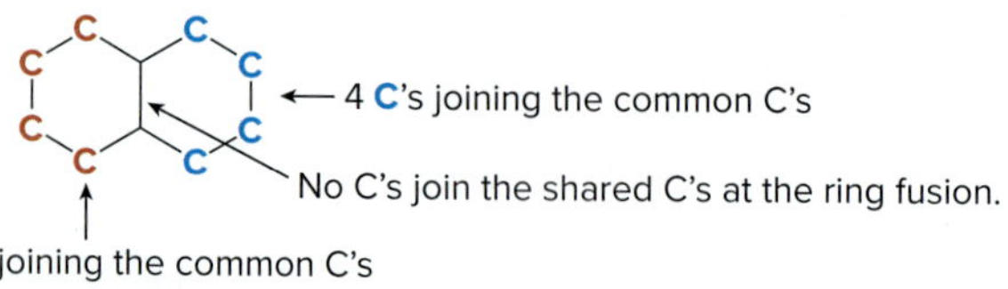

4 C's joining the common C's
No C's join the shared C's at the ring fusion.
4 C's joining the common C's
Name: bicyclo[4.4.0]decane

Rings are numbered beginning at a *shared* carbon, and continuing around the *longest* bridge first, then the next longest, and so forth.

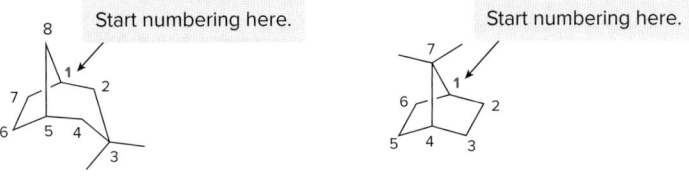

3,3-dimethylbicyclo[3.2.1]octane   7,7-dimethylbicyclo[2.2.1]heptane

Spiro ring systems are named as spiro[*x.y*]alkanes where the parent alkane corresponds to the total number of carbons in both rings, and *x* and *y* refer to the number of carbons that join the shared carbon (the spiro carbon), written in order of *increasing* size. When substituents are present, the rings are numbered beginning with a carbon *adjacent* to the spiro carbon in the *smaller* ring.

10 C's in the ring system   8 C's in the ring system

Name: spiro[4.5]decane   **Name: 2-methylspiro[3.4]octane**

APPENDIX E

Bond Dissociation Energies for Some Common Bonds
[A–B → A• + •B]

Bond	ΔH° kJ/mol	(kcal/mol)
H–Z bonds		
H–F	569	(136)
H–Cl	431	(103)
H–Br	368	(88)
H–I	297	(71)
H–OH	498	(119)
Z–Z bonds		
H–H	435	(104)
F–F	159	(38)
Cl–Cl	242	(58)
Br–Br	192	(46)
I–I	151	(36)
HO–OH	213	(51)
R–H bonds		
CH_3–H	435	(104)
CH_3CH_2–H	410	(98)
$CH_3CH_2CH_2$–H	410	(98)
$(CH_3)_2CH$–H	397	(95)
$(CH_3)_3C$–H	381	(91)
CH_2=CH–H	435	(104)
HC≡C–H	523	(125)
CH_2=$CHCH_2$–H	364	(87)
C_6H_5–H	460	(110)
$C_6H_5CH_2$–H	356	(85)
R–R bonds		
CH_3–CH_3	368	(88)
CH_3–CH_2CH_3	356	(85)
CH_3–CH=CH_2	385	(92)
CH_3–C≡CH	489	(117)

Bond	ΔH° kJ/mol	(kcal/mol)
R–X bonds		
CH_3–F	456	(109)
CH_3–Cl	351	(84)
CH_3–Br	293	(70)
CH_3–I	234	(56)
CH_3CH_2–F	448	(107)
CH_3CH_2–Cl	339	(81)
CH_3CH_2–Br	285	(68)
CH_3CH_2–I	222	(53)
$(CH_3)_2CH$–F	444	(106)
$(CH_3)_2CH$–Cl	335	(80)
$(CH_3)_2CH$–Br	285	(68)
$(CH_3)_2CH$–I	222	(53)
$(CH_3)_3C$–F	444	(106)
$(CH_3)_3C$–Cl	331	(79)
$(CH_3)_3C$–Br	272	(65)
$(CH_3)_3C$–I	209	(50)
R–Z bonds		
CH_3–OH	389	(93)
CH_3CH_2–OH	393	(94)
$CH_3CH_2CH_2$–OH	385	(92)
$(CH_3)_2CH$–OH	401	(96)
$(CH_3)_3C$–OH	401	(96)
CH_3–NH_2	331	(79)
CH_3–SH	305	(73)
Other bonds		
CH_2=CH_2	635	(152)
HC≡CH	837	(200)
O=C=O	535	(128)
O_2	497	(119)

Reactions That Form Carbon–Carbon Bonds

APPENDIX F

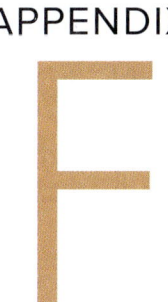

Section	Reaction
10.20A	S_N2 reaction of an alkyl halide with an acetylide anion, $^-C\equiv CR$
10.20B	Opening of an epoxide ring with an acetylide anion, $^-C\equiv CR$
12.12	Diels–Alder reaction
13.10	Reaction of an aldehyde or ketone with a Grignard or organolithium reagent
13.13A	Reaction of an acid chloride with a Grignard or organolithium reagent
13.13A	Reaction of an ester with a Grignard or organolithium reagent
13.13B	Reaction of an acid chloride with an organocuprate reagent
13.14A	Reaction of a Grignard reagent with CO_2
13.14B	Reaction of an epoxide with an organometallic reagent
13.15	Reaction of an α,β-unsaturated carbonyl compound with an organocuprate reagent
14.8	Cyanohydrin formation
14.9	Wittig reaction to form an alkene
15.13	S_N2 reaction of an alkyl halide with NaCN
15.13C	Reaction of a nitrile with a Grignard or organolithium reagent
17.8	Direct enolate alkylation using LDA and an alkyl halide
17.9	Malonic ester synthesis to form a carboxylic acid
17.10	Acetoacetic ester synthesis to form a ketone
18.1	Aldol reaction to form a β-hydroxy carbonyl compound or an α,β-unsaturated carbonyl compound
18.2	Crossed aldol reaction
18.3	Directed aldol reaction
18.5	Claisen reaction to form a β-keto ester
18.6	Crossed Claisen reaction to form a β-dicarbonyl compound
18.7	Dieckmann reaction to form a five- or six-membered ring
18.9	Michael reaction to form a 1,5-dicarbonyl compound
18.10	Robinson annulation to form a cyclohex-2-enone
20.5	Friedel–Crafts alkylation
20.5	Friedel–Crafts acylation
21.13	Radical polymerization of an alkene
22.13	Reaction of a diazonium salt with CuCN
24.10B	Kiliani–Fischer synthesis of an aldose
28.1	Coupling of an organocuprate reagent (R_2CuLi) with an organic halide (R'X)
28.2	The palladium-catalyzed Suzuki reaction of an organic halide with an organoborane
28.3	The palladium-catalyzed Heck reaction of a vinyl or aryl halide with an alkene
28.4	Addition of a dihalocarbene to an alkene to form a cyclopropane
28.5	Simmons–Smith reaction of an alkene with CH_2I_2 and Zn(Cu) to form a cyclopropane
28.6	Olefin metathesis
29.3	Electrocyclic reactions
29.4	Cycloaddition reactions
29.5	Sigmatropic rearrangements
30.2	Chain-growth polymerization
30.4	Polymerization using Ziegler–Natta catalysts

A-11

APPENDIX G

Characteristic IR Absorption Frequencies

Bond	Functional group	Wavenumber (cm^{-1})	Comment
O–H			
	• ROH	3600–3200	broad, strong
	• RCO$_2$H	3500–2500	very broad, strong
N–H			
	• RNH$_2$	3500–3300	two peaks
	• R$_2$NH	3500–3300	one peak
	• RCONH$_2$, RCONHR	3400–3200	one or two peaks; N–H bending also observed at 1640 cm^{-1}
C–H			
	• C$_{sp}$–H	3300	sharp, often strong
	• C$_{sp^2}$–H	3150–3000	medium
	• C$_{sp^3}$–H	3000–2850	strong
	• C$_{sp^2}$–H of RCHO	2830–2700	one or two peaks
C≡C		2250	medium
C≡N		2250	medium
C=O			strong
	• RCOCl	1800	
	• (RCO)$_2$O	1800, 1760	two peaks
	• RCO$_2$R	1745–1735	increasing $\tilde{\nu}$ with decreasing ring size
	• RCHO	1730	
	• RCO$_2$PO$_3^{2-}$	1730–1700	
	• RCOSR'	1720–1690	
	• R$_2$CO	1715	increasing $\tilde{\nu}$ with decreasing ring size
	• RCO$_2$H	1710	
	• R$_2$CO, conjugated	1680	
	• RCONH$_2$, RCONHR, RCONR$_2$	1680–1630	increasing $\tilde{\nu}$ with decreasing ring size
C=C			
	• Alkene	1650	medium
	• Arene	1600, 1500	medium
C=N		1650	medium

APPENDIX H

Characteristic NMR Absorptions

¹H NMR Absorptions

Compound type	Chemical shift (ppm)
Alcohol	
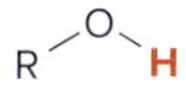 R–O–H	1–5
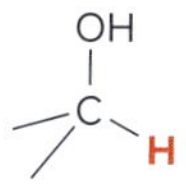 C(OH)–H	3.4–4.0
Aldehyde	
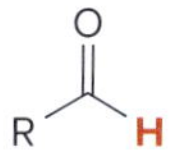 R–C(=O)–H	9–10
Alkane	0.9–2.0
RCH₃	~0.9
R₂CH₂	~1.3
R₃CH	~1.7
Alkene	
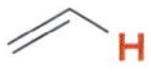 =C–H sp² C–H	4.5–6.0
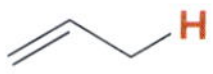 allylic sp³ C–H	1.5–2.5
Alkyl halide	
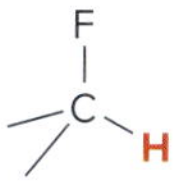 F–C–H	4.0–4.5
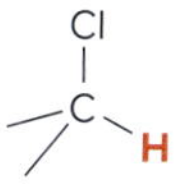 Cl–C–H	3.0–4.0
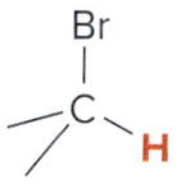 Br–C–H	2.7–4.0
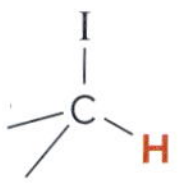 I–C–H	2.2–4.0
Alkyne	
 –C≡C–H	~2.5

Appendix H Characteristic NMR Absorptions

Compound type	Chemical shift (ppm)
Amide — R–C(=O)–N(H)–	7.5–8.5
Amine — R–N(H)–	0.5–5.0
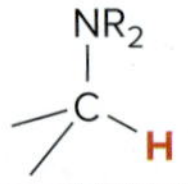 Amine — C–H with NR₂ on C	2.3–3.0
Aromatic compound — sp² C–H	6.5–8
Aromatic — benzylic sp³ C–H	1.5–2.5
Carbonyl compound — sp³ C–H on the α carbon	2.0–2.5
Carboxylic acid — R–C(=O)–OH	10–12
Ether — C–H with OR on C	3.4–4.0

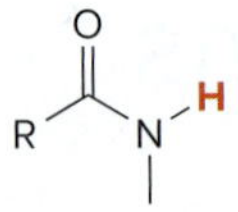

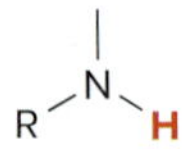

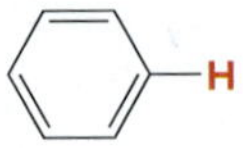

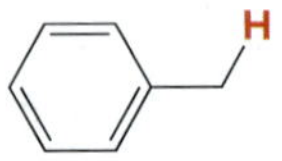

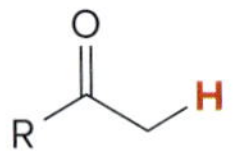

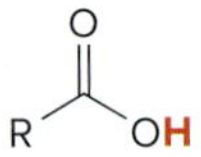

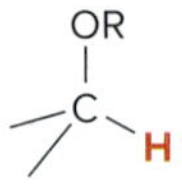

^{13}C NMR Absorptions

Carbon type	Structure	Chemical shift (ppm)
Alkyl, sp³ hybridized C	 C (alkyl)	5–45
Alkyl, sp³ hybridized C bonded to N, O, or X	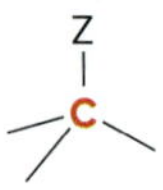 C–Z, Z = N, O, X	30–80
Alkynyl, sp hybridized C	—C≡C—	65–100
Alkenyl, sp² hybridized C	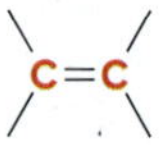 C=C	100–140
Aryl, sp² hybridized C	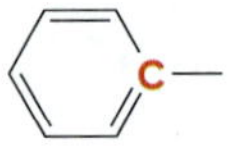 aromatic C	120–150
Carbonyl C	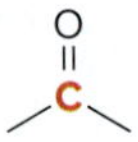 C=O	160–210

APPENDIX

General Types of Organic Reactions

Substitution Reactions

[1] Nucleophilic substitution at an sp^3 hybridized carbon atom

a. Alkyl halides (Chapter 7)

R—X + :Nu⁻ ⟶ R—Nu + X:⁻
          nucleophile

b. Alcohols (Section 9.11)

R—OH + HX ⟶ R—X + H₂O

c. Ethers (Section 9.14)

R—OR' + HX ⟶ R—X + R'—X + H₂O
      X = Br or I

d. Epoxides (Section 9.16)

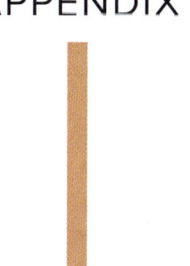

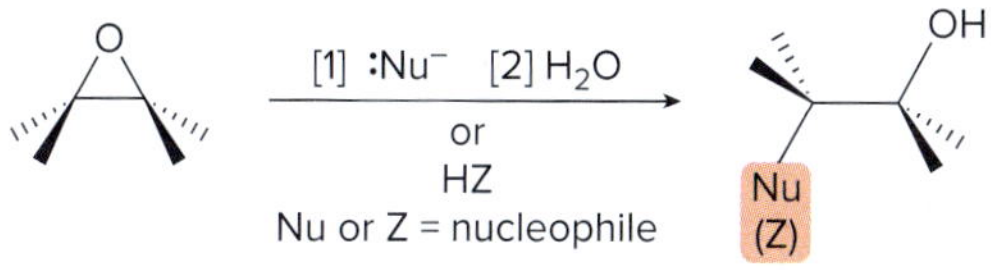

[1] :Nu⁻ [2] H₂O
or
HZ
Nu or Z = nucleophile

[2] Nucleophilic acyl substitution at an sp^2 hybridized carbon atom

Carboxylic acids and their derivatives (Chapter 16)

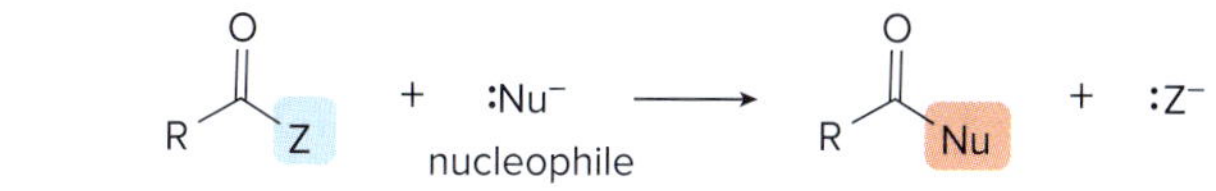

R—C(=O)—Z + :Nu⁻ ⟶ R—C(=O)—Nu + :Z⁻
            nucleophile

Z = OH, Cl, OCOR, OPO₃²⁻,
OR', SR', NR'₂

[3] Radical substitution at an sp^3 hybridized C—H bond

Alkanes (Section 21.3)

R—H + X₂ $\xrightarrow{h\nu \text{ or } \Delta}$ R—X + HX

[4] Electrophilic aromatic substitution

Aromatic compounds (Chapter 20)

C₆H₅—H + E⁺ ⟶ C₆H₅—E + H⁺
      electrophile

[5] Nucleophilic aromatic substitution

Aromatic compounds (Chapter 20)

A—C₆H₄—X + :Nu⁻ ⟶ A—C₆H₄—Nu + X⁻
           nucleophile

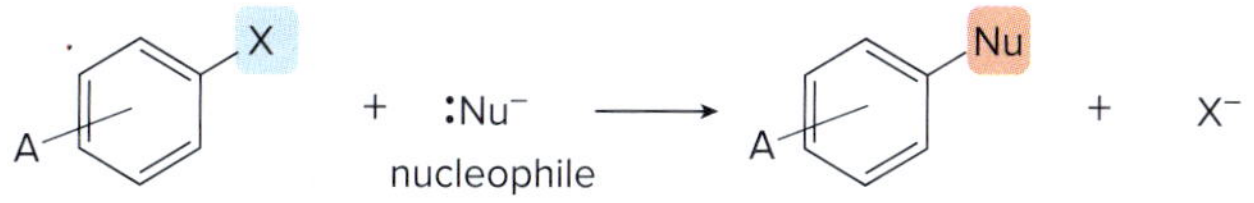

X = F, Cl, Br, I
A = H or electron-withdrawing group

Elimination Reactions

β Elimination at an sp³ hybridized carbon atom

a. Alkyl halides
   (Chapter 8)

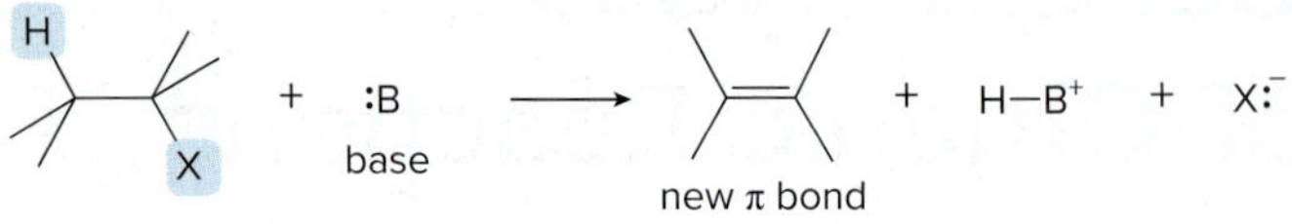

   [structure with H and X] + :B (base) ⟶ [alkene with new π bond] + H–B⁺ + X:⁻

b. Alcohols
   (Section 9.8)

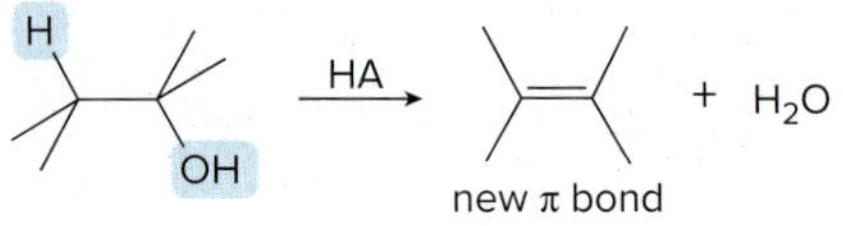

   [structure with H and OH] —HA→ [alkene with new π bond] + H₂O

Addition Reactions

[1] Electrophilic addition to carbon–carbon multiple bonds

a. Alkenes
   (Section 10.8A)

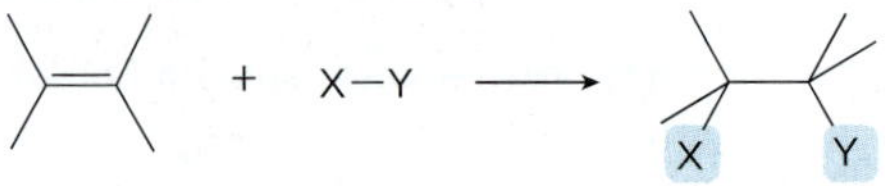

   [alkene] + X–Y ⟶ [product with X and Y added]

b. Alkynes
   (Section 10.8A)

   [alkyne] + X–Y ⟶ [product with X and Y added twice]

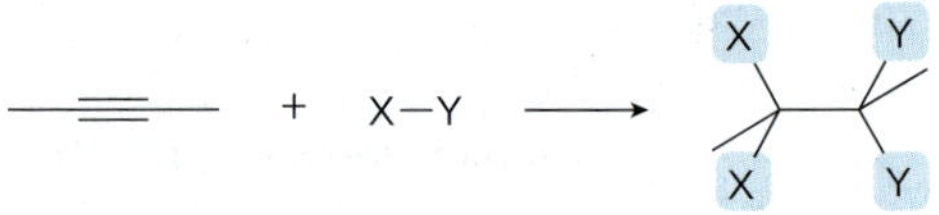

[2] Nucleophilic addition to carbon–oxygen multiple bonds

Aldehydes and ketones
(Chapter 14)

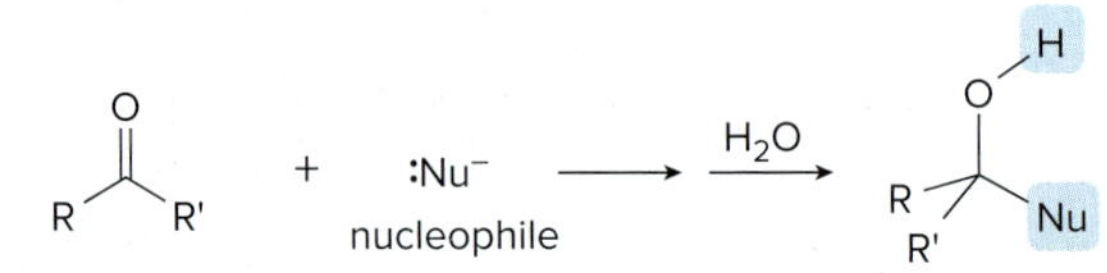

[R(C=O)R'] + :Nu⁻ (nucleophile) —H₂O→ [R–C(OH)(Nu)–R']

R' = H or alkyl

How to Synthesize Particular Functional Groups

APPENDIX J

Acetals
- Reaction of an aldehyde or ketone with two equivalents of an alcohol (14.15)

Acid chlorides
- Reaction of a carboxylic acid with thionyl chloride (16.9)

Acyl phosphates
- Enzyme-catalyzed reaction of a carboxylate with ATP (16.15A)

Alcohols
- Nucleophilic substitution of an alkyl halide with $^-$OH or H_2O (9.6)
- Hydration of an alkene (10.12)
- Hydroboration–oxidation of an alkene (10.16)
- Reduction of an epoxide with $LiAlH_4$ (11.6)
- Reduction of an aldehyde or ketone (13.4)
- Enantioselective biological reduction of an aldehyde or ketone (13.6)
- Reduction of an acid chloride with $LiAlH_4$ (13.7)
- Reduction of an ester with $LiAlH_4$ (13.7)
- Reduction of a carboxylic acid with $LiAlH_4$ (13.7)
- Reaction of an aldehyde or ketone with a Grignard or organolithium reagent (13.10)
- Reaction of an acid chloride with a Grignard or organolithium reagent (13.13)
- Reaction of an ester with a Grignard or organolithium reagent (13.13)
- Reaction of an organometallic reagent with an epoxide (13.14B)

Aldehydes
- Hydroboration–oxidation of a terminal alkyne (10.19)
- Oxidative cleavage of an alkene with O_3 followed by Zn or $(CH_3)_2S$ (11.10)
- Oxidation of a 1° alcohol with PCC (11.12)
- Biological oxidation with NAD^+ (11.13)
- Reduction of an acid chloride with $LiAlH[OC(CH_3)_3]_3$ (13.7)
- Reduction of an ester with DIBAL-H (13.7)
- Hydrolysis of an imine or enamine (14.12)
- Hydrolysis of an acetal (14.15B)
- Reduction of a nitrile (15.13B)

Alkanes
- Catalytic hydrogenation of an alkene with H_2 + Pd-C (11.3)
- Catalytic hydrogenation of an alkyne with two equivalents of H_2 + Pd-C (11.5A)

A-17

- Reduction of an alkyl halide with LiAlH$_4$ (11.6)
- Protonation of an organometallic reagent with H$_2$O, ROH, or acid (13.9)
- Reduction of a ketone to a methylene group (CH$_2$)—the Wolff–Kishner or Clemmensen reaction (20.14C)
- Coupling of an organocuprate reagent (R$_2$CuLi) with an alkyl halide, R'X (28.1)
- Simmons–Smith reaction of an alkene with CH$_2$I$_2$ and Zn(Cu) to form a cyclopropane (28.5)

Alkenes
- Dehydrohalogenation of an alkyl halide with base (8.1)
- Dehydration of an alcohol with acid (9.8)
- Dehydration of an alcohol using POCl$_3$ and pyridine (9.10)
- β Elimination of an alkyl tosylate with base (9.13)
- Catalytic hydrogenation of an alkyne with H$_2$ + Lindlar catalyst to form a cis alkene (11.5B)
- Dissolving metal reduction of an alkyne with Na, NH$_3$ to form a trans alkene (11.5C)
- Wittig reaction (14.9)
- β Elimination of an α-halo carbonyl compound with Li$_2$CO$_3$, LiBr, and DMF (17.7B)
- Hofmann elimination of an amine (22.11)
- Coupling of an organocuprate reagent (R$_2$CuLi) with an organic halide, R'X (28.1)
- The palladium-catalyzed Suzuki reaction of a vinyl or aryl halide with a vinyl- or arylborane (28.2)
- The palladium-catalyzed Heck reaction of a vinyl or aryl halide with an alkene (28.3)
- Olefin metathesis (28.6)

Alkyl halides
- Reaction of an alcohol with HX (9.11)
- Reaction of an alcohol with SOCl$_2$ or PBr$_3$ (9.12)
- Cleavage of an ether with HBr or HI (9.14)
- Hydrohalogenation of an alkene with HX (10.9)
- Halogenation of an alkene with X$_2$ (10.13)
- Hydrohalogenation of an alkyne with two equivalents of HX (10.17A)
- Halogenation of an alkyne with two equivalents of X$_2$ (10.17B)
- Electrophilic addition of HX to a 1,3-diene (12.10)
- Halogenation α to a carbonyl group (17.7)
- Halogenation of an alkyl benzene (20.14A)
- Radical halogenation of an alkane (21.3)
- Radical halogenation at an allylic carbon (21.9)
- Radical addition of HBr to an alkene (21.12)
- Addition of a dihalocarbene to an alkene to form a dihalocyclopropane (28.4)

Alkynes
- Dehydrohalogenation of an alkyl dihalide with base (8.10)
- S$_N$2 reaction of an alkyl halide with an acetylide anion, $^-$C≡CR (10.20)

Amides
- Reaction of an acid chloride with NH$_3$ or an amine (16.7)
- Reaction of an anhydride with NH$_3$ or an amine (16.8)
- Reaction of a carboxylic acid with NH$_3$ or an amine and DCC (16.9)
- Reaction of an ester with NH$_3$ or an amine (16.10)
- Enzyme-catalyzed reaction of an acyl phosphate with an amine (16.15B)

Amines
- Reduction of an amide with LiAlH$_4$ (13.7B)
- Reduction of a nitrile (15.13B)
- Nucleophilic aromatic substitution (20.13)
- Reduction of a nitro group (20.14D)
- S$_N$2 reaction using NH$_3$ or an amine (22.6A)
- Gabriel synthesis (22.6A)
- Reductive amination of an aldehyde or ketone (22.6C)

Amino acids
- Enantioselective hydrogenation using a chiral catalyst (23.3)

Anhydrides
- Reaction of an acid chloride with a carboxylate anion (16.7)
- Dehydration of a dicarboxylic acid (16.9)

Aryl halides
- Halogenation of benzene with X$_2$ + FeX$_3$ (20.3)
- Reaction of a diazonium salt with CuCl, CuBr, HBF$_4$, NaI, or KI (22.13A)

Carboxylic acids and carboxylates
- Oxidative cleavage of an alkyne with ozone (11.11)
- Oxidation of a 1° alcohol with CrO$_3$ (or a similar Cr^{6+} reagent), H$_2$O, H$_2$SO$_4$ (11.12B)
- Oxidation of an aldehyde (13.8)
- Reaction of a Grignard reagent with CO$_2$ (13.14A)
- Hydrolysis of a cyanohydrin (14.8)
- Hydrolysis of a nitrile (15.13A)
- Hydrolysis of an acid chloride (16.7)
- Hydrolysis of an anhydride (16.8)
- Hydrolysis of an ester (16.10)
- Hydrolysis of an amide (16.12)
- Enzyme-catalyzed hydrolysis of a thioester (16.16)
- Malonic ester synthesis (17.9)
- Oxidation of an alkyl benzene with KMnO$_4$ (20.14B)

Cyanohydrins
- Addition of HCN to an aldehyde or ketone (14.8)

1,2-Diols
- Anti dihydroxylation of an alkene with a peroxyacid, followed by ring opening with $^-$OH or H$_2$O (11.9A)
- Syn dihydroxylation of an alkene with KMnO$_4$ or OsO$_4$ (11.9B)

Enamines
- Reaction of an aldehyde or ketone with a 2° amine (14.11)

Epoxides
- Intramolecular S$_N$2 reaction of a halohydrin using base (9.6)
- Epoxidation of an alkene with mCPBA (11.8)
- Enantioselective epoxidation of an allylic alcohol with the Sharpless reagent (11.14)

Esters
- S_N2 reaction of an alkyl halide with a carboxylate anion, RCO_2^- (7.18)
- Reaction of an acid chloride with an alcohol (16.7)
- Reaction of an anhydride with an alcohol (16.8)
- Fischer esterification of a carboxylic acid with an alcohol (16.9)
- Enzyme-catalyzed reaction of a thioester with an alcohol (16.16)

Ethers
- Williamson ether synthesis—S_N2 reaction of an alkyl halide with an alkoxide, ^-OR (9.6)
- Reaction of an alkyl tosylate with an alkoxide, ^-OR (9.13)
- Addition of an alcohol to an alkene in the presence of acid (10.12)
- Anionic polymerization of epoxides to form polyethers (30.3)

Halohydrins
- Reaction of an epoxide with HX (9.16)
- Addition of X and OH to an alkene (10.15)

Imine
- Reaction of an aldehyde or ketone with a 1° amine (14.10)

Ketones
- Hydration of an alkyne with H_2O, H_2SO_4, and $HgSO_4$ (10.18)
- Oxidative cleavage of an alkene with O_3 followed by Zn or $(CH_3)_2S$ (11.10)
- Oxidation of a 2° alcohol with any Cr^{6+} reagent (11.12)
- Biological oxidation of a 2° alcohol (11.13)
- Reaction of an acid chloride with an organocuprate reagent (13.13)
- Hydrolysis of an imine or enamine (14.12)
- Hydrolysis of an acetal (14.15B)
- Reaction of a nitrile with a Grignard or organolithium reagent (15.13C)
- Acetoacetic ester synthesis (17.10)
- Friedel–Crafts acylation (20.5)

Nitriles
- S_N2 reaction of an alkyl halide with NaCN (7.18, 15.13)
- Reaction of an aryl diazonium salt with CuCN (22.13A)

Phenols
- Nucleophilic aromatic substitution (20.13)
- Reaction of an aryl diazonium salt with H_2O (22.13A)

Sulfides
- Reaction of an alkyl halide with ^-SR (9.15)

Thioesters
- Enzyme-catalyzed reaction of an acyl phosphate with a thiol (16.15B)

Thiols
- Reaction of an alkyl halide with ^-SH (9.15)

Glossary

A

Acetal (Section 14.15): A compound having the general structure R$_2$C(OR')$_2$, where R = H, alkyl, or aryl. Acetals are used as protecting groups for aldehydes and ketones.

Acetoacetic ester synthesis (Section 17.10): A stepwise method that converts ethyl acetoacetate to a ketone having one or two carbons bonded to the α carbon.

Acetylation (Section 16.8): A reaction that transfers an acetyl group (CH$_3$CO—) from one atom to another.

Acetyl coenzyme A (Section 16.16): A biochemical thioester that acts as an acetylating reagent. Acetyl coenzyme A is often referred to as acetyl CoA.

Acetyl group (Section 14.2E): A substituent having the structure —COCH$_3$.

Acetylide anion (Sections 10.8B, 13.9B): An anion formed by treating a terminal alkyne with a strong base. Acetylide anions have the general structure R—C≡C$^-$.

Achiral molecule (Section 5.3): A molecule that is superimposable upon its mirror image. An achiral molecule is not chiral.

Acid chloride (Sections 13.1, 16.1): A compound having the general structure RCOCl.

Acidity constant (Section 2.3): A value symbolized by K_a that represents the strength of an acid (HA). The larger the K_a, the stronger the acid.

$$K_a = \frac{[H_3O^+][A{:}^-]}{[H-A]}$$

Active site (Section 6.11): The region of an enzyme that binds the substrate.

Acyclic alkane (Section 4.1): A compound with the general formula C$_n$H$_{2n+2}$. Acyclic alkanes are also called saturated hydrocarbons because they contain the maximum number of hydrogen atoms per carbon.

Acylation (Sections 16.16, 20.5A): A reaction that transfers an acyl group from one atom to another.

Acyl group (Section 14.2E): A substituent having the general structure RCO—.

Acylium ion (Section 20.5B): A positively charged electrophile having the general structure (R—C≡O)$^+$, formed when the Lewis acid AlCl$_3$ ionizes the carbon–halogen bond of an acid chloride.

Acyl phosphate (Section 16.1): A compound having the general structure RCO$_2$PO$_3^{2-}$.

Acyl transfer reaction (Section 16.16): A reaction that transfers an acyl group from one atom to another.

1,2-Addition (Sections 12.10, 13.15): An addition reaction to a conjugated system that adds groups across two adjacent atoms.

1,4-Addition (Sections 12.10, 13.15): An addition reaction that adds groups to the atoms in the 1 and 4 positions of a conjugated system. 1,4-Addition is also called conjugate addition.

Addition polymer (Section 30.1): A polymer prepared by a chain reaction that adds a monomer to the growing end of a polymer chain. Addition polymers are also called chain-growth polymers.

Addition reaction (Sections 6.2C, 10.8): A reaction in which elements are added to a starting material. In an addition reaction, a π bond is broken and two σ bonds are formed.

Aglycon (Section 24.7C): The alcohol formed from hydrolysis of a glycoside.

Alcohol (Section 9.1): A compound having the general structure ROH. An alcohol contains a hydroxy group (OH group) bonded to an sp^3 hybridized carbon atom.

Aldaric acid (Section 24.9B): The dicarboxylic acid formed by the oxidation of the aldehyde and the primary alcohol of an aldose.

Aldehyde (Section 10.19): A compound having the general structure RCHO, where R = H, alkyl, or aryl.

Alditol (Section 24.9A): A compound formed by the reduction of the aldehyde of an aldose to a primary alcohol.

Aldol condensation (Section 18.1C): An aldol reaction in which the initially formed β-hydroxy carbonyl compound loses water by dehydration.

Aldol reaction (Section 18.1A): A reaction in which two molecules of an aldehyde or ketone react with each other in the presence of base to form a β-hydroxy carbonyl compound.

Aldonic acid (Section 24.9B): A compound formed by the oxidation of the aldehyde of an aldose to a carboxylic acid.

Aldose (Section 24.2): A monosaccharide composed of a polyhydroxy aldehyde.

Aliphatic (Section 3.2A): A compound or portion of a compound made up of C—C σ and π bonds but not aromatic bonds.

Alkaloid (Section 22.5A): A basic, nitrogen-containing compound isolated from a plant source.

Alkane (Section 4.1): An aliphatic hydrocarbon having only C—C and C—H σ bonds.

Alkene (Section 8.2A): An aliphatic hydrocarbon that contains a carbon–carbon double bond.

Alkoxide (Sections 8.1, 9.6): An anion having the general structure RO$^-$, formed by deprotonating an alcohol with a base.

Alkoxy group (Section 9.3B): A substituent containing an alkyl group bonded to an oxygen (RO group).

Alkylation (Section 17.8): A reaction that transfers an alkyl group from one atom to another.

Alkyl group (Section 4.4A): A group formed by removing one hydrogen from an alkane. Alkyl groups are named by replacing the suffix *-ane* of the parent alkane with *-yl*.

Alkyl halide (Section 7.1): A compound containing a halogen atom bonded to an sp^3 hybridized carbon atom. Alkyl halides have the general molecular formula C$_n$H$_{2n+1}$X.

1,2-Alkyl shift (Section 9.9): The rearrangement of a less stable carbocation to a more stable carbocation by the shift of an alkyl group from one carbon atom to an adjacent carbon atom.

Alkyl tosylate (Section 9.13): A compound having the general structure ROSO$_2$C$_6$H$_4$CH$_3$. Alkyl tosylates are also called tosylates and are abbreviated as ROTs.

Alkyne (Section 8.10): An aliphatic hydrocarbon that contains a carbon–carbon triple bond.

G-1

Allyl carbocation (Section 12.1B): A carbocation that has a positive charge on the atom adjacent to a carbon–carbon double bond. An allyl carbocation is resonance stabilized.

Allyl group (Section 10.3B): A substituent having the structure —CH$_2$—CH=CH$_2$.

Allylic bromination (Section 21.9): A radical substitution reaction in which bromine replaces a hydrogen atom on the carbon adjacent to a carbon–carbon double bond.

Allylic carbon (Section 21.9): A carbon atom bonded to a carbon–carbon double bond.

Allylic halide (Section 7.1): A molecule containing a halogen atom bonded to the carbon atom adjacent to a carbon–carbon double bond.

Allyl radical (Section 21.9): A radical that has an unpaired electron on the carbon adjacent to a carbon–carbon double bond. An allyl radical is resonance stabilized.

Alpha (α) carbon (Sections 8.1, 14.2B): In an elimination reaction, the carbon that is bonded to the leaving group. In a carbonyl compound, the carbon that is bonded to the carbonyl carbon.

Ambident nucleophile (Section 17.3C): A nucleophile that has two reactive sites.

Amide (Sections 13.1, 16.1): A compound having the general structure RCONR'$_2$, where R' = H or alkyl.

Amide base (Sections 8.10, 17.3B): A nitrogen-containing base formed by deprotonating an amine or ammonia.

Amine (Sections 14.10, 22.1): A basic organic nitrogen compound having the general structure RNH$_2$, R$_2$NH, or R$_3$N. An amine has a nonbonded pair of electrons on the nitrogen atom.

α-Amino acid (Section 3.9A): A compound having the general structure RCH(NH$_2$)COOH. α-Amino acids are the building blocks of proteins.

Amino acid residue (Section 23.5): The individual amino acids in peptides and proteins.

Amino group (Section 22.3D): A substituent having the structure —NH$_2$.

Amino sugar (Section 24.13A): A carbohydrate that contains an NH$_2$ group instead of a hydroxy group at a non-anomeric carbon.

Ammonium salt (Section 22.1): A compound containing a positively charged nitrogen with four σ bonds; for example, R$_4$N$^+$X$^-$.

Anabolism (Section 27.1): In metabolism, the synthesis of large molecules from smaller ones, often absorbing energy.

Angle strain (Section 4.11): An increase in the energy of a molecule resulting when the bond angles of the sp^3 hybridized atoms deviate from the optimum tetrahedral angle of 109.5°.

Angular methyl group (Section 25.8A): A methyl group located at the ring junction of two fused rings of the steroid skeleton.

Anhydride (Section 16.1): A compound having the general structure (RCO)$_2$O.

Aniline (Section 19.3A): A compound having the structure C$_6$H$_5$NH$_2$.

Anion (Section 1.2): A negatively charged ion that results from a neutral atom gaining one or more electrons.

Anionic polymerization (Section 30.2C): Chain-growth polymerization of alkenes substituted by electron-withdrawing groups that stabilize intermediate anions.

Annulation (Section 18.10): A reaction that forms a new ring.

Annulene (Section 19.8A): A hydrocarbon containing a single ring with alternating double and single bonds.

α Anomer (Section 24.6): The stereoisomer of a cyclic monosaccharide in which the anomeric OH and the CH$_2$OH groups are trans. In a D monosaccharide, the hydroxy group on the anomeric carbon is drawn down.

β Anomer (Section 24.6): The stereoisomer of a cyclic monosaccharide in which the anomeric OH and the CH$_2$OH groups are cis. In a D monosaccharide, the hydroxy group on the anomeric carbon is drawn up.

Anomeric carbon (Section 24.6): The stereogenic center at the hemiacetal carbon of a cyclic monosaccharide.

Antarafacial reaction (Section 29.4): A pericyclic reaction that occurs on opposite sides of the two ends of the π electron system.

Anti addition (Section 10.8): An addition reaction in which the two parts of a reagent are added from opposite sides of a double bond.

Antiaromatic compound (Section 19.7): An organic compound that is cyclic, planar, completely conjugated, and has $4n$ π electrons.

Antibonding molecular orbital (Section 19.9A): A high-energy molecular orbital formed when two atomic orbitals of opposite phase overlap.

Anticodon (Section 26.5): A sequence of three nucleotides in a tRNA molecule, which is complementary to three bases in an mRNA molecule and identifies what amino acid must be added to a growing polypeptide chain.

Anti conformation (Section 4.10): A staggered conformation in which the two larger groups on adjacent carbon atoms have a dihedral angle of 180°.

Anti dihydroxylation (Section 11.9A): The addition of two hydroxy groups to opposite faces of a double bond.

Antioxidant (Section 21.11): A compound that stops an oxidation from occurring.

Anti periplanar (Section 8.8A): In an elimination reaction, a geometry where the β hydrogen and the leaving group are on opposite sides of the molecule.

Aromatic compound (Section 19.1): A planar, cyclic organic compound that has p orbitals on all ring atoms and a total of $4n + 2$ π electrons in the orbitals.

Aryl group (Section 19.3D): A substituent formed by removing one hydrogen atom from an aromatic ring.

Aryl halide (Sections 7.1, 20.3): A molecule such as C$_6$H$_5$X, containing a halogen atom X bonded to an aromatic ring.

Asymmetric carbon (Section 5.3): A carbon atom that is bonded to four different groups. An asymmetric carbon is also called a stereogenic center, a chiral center, or a chirality center.

Asymmetric reaction (Sections 11.14, 23.3): A reaction that converts an achiral starting material to predominantly one enantiomer.

Atactic polymer (Section 30.4): A polymer having the substituents randomly oriented along the carbon backbone of an elongated polymer chain.

Atomic number (Section 1.1): The number of protons in the nucleus of an element.

Atomic weight (Section 1.1): The weighted average of the mass of all isotopes of a particular element. The atomic weight is reported in atomic mass units (amu).

Axial bonds (Section 4.12A): Bonds located above or below and perpendicular to the plane of the chair conformation of cyclohexane. Three axial bonds point upwards (on the up carbons) and three axial bonds point downwards (on the down carbons).

Azo compound (Section 22.14): A compound having the general structure RN=NR'.

B

Backside attack (Section 7.11C): Approach of a nucleophile from the side opposite the leaving group.

Barrier to rotation (Section 4.10): The energy difference between the lowest- and highest-energy conformations of a molecule.

Base peak (Section A.1): The peak in the mass spectrum having the greatest abundance value.

Basicity (Section 7.8): A measure of how readily an atom donates its electron pair to a proton.

Benedict's reagent (Section 24.9B): A reagent for oxidizing aldehydes to carboxylic acids using a Cu^{2+} salt, forming brick-red Cu_2O as a side product.

Benzoyl group (Section 14.2E): A substituent having the structure $-COC_6H_5$.

Benzyl group (Section 19.3D): A substituent having the structure $C_6H_5CH_2-$.

Benzylic halide (Sections 7.1, 20.14A): A compound such as $C_6H_5CH_2X$, containing a halogen atom X bonded to a carbon that is bonded to a benzene ring.

Benzyne (Section 20.13B): A reactive intermediate formed by elimination of HX from an aryl halide.

Beta (β) carbon (Sections 8.1, 14.2B): In an elimination reaction, the carbon adjacent to the carbon with the leaving group. In a carbonyl compound, the carbon located two carbons from the carbonyl carbon.

Bimolecular reaction (Sections 6.9B, 7.10, 7.11A): A reaction in which the concentration of both reactants affects the reaction rate and both terms appear in the rate equation. In a bimolecular reaction, two reactants are involved in the only step or the rate-determining step.

Biodegradable polymer (Section 30.9B): A polymer that can be degraded by microorganisms naturally present in the environment.

Biomolecule (Section 3.9): An organic compound found in a biological system.

Boat conformation of cyclohexane (Section 4.12B): An unstable conformation adopted by cyclohexane that resembles a boat. The instability of the boat conformation results from torsional strain and steric strain. The boat conformation of cyclohexane is 30 kJ/mol less stable than the chair conformation.

Boiling point (Section 3.4A): The temperature at which molecules in the liquid phase are converted to the gas phase. Molecules with stronger intermolecular forces have higher boiling points. Boiling point is abbreviated as bp.

Bond dissociation energy (Section 6.4): The amount of energy needed to homolytically cleave a covalent bond.

Bonding (Section 1.2): The joining of two atoms in a stable arrangement. Bonding is a favorable process that leads to lowered energy and increased stability.

Bonding molecular orbital (Section 19.10A): A low-energy molecular orbital formed when two atomic orbitals of similar phase overlap.

Bond length (Section 1.7A): The average distance between the centers of two bonded nuclei. Bond lengths are reported in picometers (pm).

Branched-chain alkane (Section 4.1A): An acyclic alkane that has alkyl substituents bonded to the parent carbon chain.

Bridged ring system (Section 12.13D): A bicyclic ring system in which the two rings share non-adjacent carbon atoms.

Bromination (Sections 10.13, 20.3, 21.6): The reaction of a compound with bromine.

Bromohydrin (Section 10.15): A compound having a bromine and a hydroxy group on adjacent carbon atoms.

Brønsted–Lowry acid (Section 2.1): A proton donor, symbolized by HA. A Brønsted–Lowry acid must contain a hydrogen atom.

Brønsted–Lowry base (Section 2.1): A proton acceptor, symbolized by :B. A Brønsted–Lowry base must be able to form a bond to a proton by donating an available electron pair.

C

^{13}C NMR spectroscopy (Section C.1): A form of nuclear magnetic resonance spectroscopy used to determine the type of carbon atoms in a molecule.

Cahn–Ingold–Prelog system of nomenclature (Section 5.6): The system of designating a stereogenic center as either R or S according to the arrangement of the four groups attached to the center.

Carbamate (Sections 23.6, 30.6): A functional group containing a carbonyl group bonded to both an oxygen and a nitrogen atom. A carbamate is also called a urethane.

Carbanion (Section 2.5D): An ion with a negative charge on a carbon atom.

Carbene (Section 28.4): A neutral reactive intermediate having the general structure $:CR_2$. A carbene contains a divalent carbon surrounded by six electrons, making it a highly reactive electrophile that adds to C—C double bonds.

Carbinolamine (Section 14.6B): An unstable intermediate having a hydroxy group and an amine group on the same carbon. A carbinolamine is formed during the addition of an amine to a carbonyl group.

Carbocation (Section 7.13C): A positively charged carbon atom. A carbocation is sp^2 hybridized and trigonal planar, and contains a vacant p orbital.

Carbohydrate (Sections 3.9B, 14.18, 24.1): A polyhydroxy aldehyde or ketone or a compound that can be hydrolyzed to a polyhydroxy aldehyde or ketone.

Carbonate (Section 30.6D): A compound having the general structure $(RO)_2C=O$.

Carbon backbone (Section 3.1): The C—C and C—H σ bond framework that makes up the skeleton of an organic molecule.

Carbon NMR spectroscopy (Section C.1): A form of nuclear magnetic resonance spectroscopy used to determine the type of carbon atoms in a molecule.

Carbonyl group (Sections 3.2C, 10.18, 13.1): A functional group that contains a carbon–oxygen double bond (C=O). The polar carbon–oxygen bond makes the carbonyl carbon electrophilic.

Carboxy group (Section 15.1): A functional group having the structure COOH.

Carboxylate anion (Section 15.2B): An anion having the general structure RCO_2^-, formed by deprotonating a carboxylic acid with a Brønsted–Lowry base.

Carboxylation (Section 13.14): The reaction of an organometallic reagent with CO_2 to form a carboxylic acid after protonation.

Carboxylic acid (Section 15.1): A compound having the general structure RCO_2H.

Carboxylic acid derivatives (Section 13.1): Compounds having the general structure RCOZ, which can be synthesized from carboxylic acids. Common carboxylic acid derivatives include acid chlorides, anhydrides, esters, and amides.

Catabolism (Section 27.1): In metabolism, the breakdown of large molecules into smaller ones, often releasing energy.

Catalyst (Section 6.10): A substance that speeds up the rate of a reaction, but is recovered unchanged at the end of the reaction and does not appear in the product.

Catalytic hydrogenation (Section 11.3): A reduction reaction involving the addition of H_2 to a π bond in the presence of a metal catalyst.

Catalytic triad (Section 23.10B): A group of three amino acid residues that contains an acid, a base, and a nucleophile, which are needed for an enzyme-catalyzed reaction to occur.

Cation (Section 1.2): A positively charged ion that results from a neutral atom losing one or more electrons.

Cationic polymerization (Section 30.2C): Chain-growth polymerization of alkene monomers involving carbocation intermediates.

Cephalin (Section 25.4A): A phosphoacylglycerol in which the phosphodiester alkyl group is $-CH_2CH_2NH_3^+$. Cephalins are also called phosphatidylethanolamines.

Chain-growth polymer (Section 30.1): A polymer prepared by a chain reaction that adds a monomer to the growing end of a polymer chain. Chain-growth polymers are also called addition polymers.

Chain mechanism (Section 21.4A): A reaction mechanism that involves repeating steps.

Chair conformation of cyclohexane (Section 4.12A): A stable conformation adopted by cyclohexane that resembles a chair. The stability of the chair conformation results from the elimination of angle strain (all C–C–C bond angles are 109.5°) and torsional strain (all groups on adjacent carbon atoms are staggered).

Chemical shift (Section C.1B): The position of an absorption signal on the x axis in an NMR spectrum relative to the reference signal of tetramethylsilane.

Chirality center (Section 5.3): A carbon atom bonded to four different groups. A chirality center is also called a chiral center, a stereogenic center, and an asymmetric center.

Chiral molecule (Section 5.3): A molecule that is not superimposable upon its mirror image.

Chlorination (Sections 10.14, 20.3, 21.5): The reaction of a compound with chlorine.

Chlorofluorocarbons (Sections 7.4, 21.8): Synthetic alkyl halides having the general molecular formula CF_xCl_{4-x}. Chlorofluorocarbons, abbreviated as CFCs, were used as refrigerants and aerosol propellants and contribute to the destruction of the ozone layer.

Chlorohydrin (Section 10.15): A compound having a chlorine and a hydroxy group on adjacent carbon atoms.

Chromate ester (Section 11.12A): An intermediate in the chromium-mediated oxidation of an alcohol having the general structure $R-O-CrO_3H$.

s-Cis (Sections 12.6, 23.4B): The conformation of a 1,3-diene that has the two double bonds on the same side of the single bond that joins them.

Cis isomer (Sections 4.13B, 8.2B): An isomer of a ring or double bond that has two groups on the same side of the ring or double bond.

Citric acid cycle (Section 27.6): A cyclic, eight-step metabolic pathway that begins with the addition of acetyl CoA to oxaloacetate. Overall the citric acid cycle forms two molecules of CO_2, four molecules of reduced coenzymes (NADH and $FADH_2$), and one molecule of GTP.

Claisen reaction (Section 18.5): A reaction between two molecules of an ester in the presence of base to form a β-keto ester.

Claisen rearrangement (Section 29.5): A [3,3] sigmatropic rearrangement of an unsaturated ether to a γ,δ-unsaturated carbonyl compound.

α Cleavage (Section A.4): A fragmentation in mass spectrometry that results in cleavage of a carbon–carbon bond. With aldehydes and ketones, α cleavage results in breaking the bond between the carbonyl carbon and the carbon adjacent to it. With alcohols, α cleavage occurs by breaking a bond between an alkyl group and the carbon that bears the OH group.

Clemmensen reduction (Section 20.14C): A method to reduce aryl ketones to alkyl benzenes using Zn(Hg) in the presence of a strong acid.

Codon (Section 26.6): A set of three nucleotides in mRNA that corresponds to a particular amino acid. The order of codons in an mRNA molecule determines the amino acid sequence of a protein.

Coenzyme (Section 11.13): A compound that acts with an enzyme to carry out a biochemical process.

Combustion (Section 4.14): An oxidation–reduction reaction, in which an alkane or other organic compound reacts with oxygen to form CO_2 and H_2O, releasing energy.

Common name (Section 4.6): The name of a molecule that was adopted prior to and therefore does not follow the IUPAC system of nomenclature.

Compound (Section 1.2): The structure that results when two or more elements are joined together in a stable arrangement.

Concerted reaction (Sections 6.3, 7.11B): A reaction in which all bond forming and bond breaking occurs in one step.

Condensation polymer (Section 30.1): A polymer formed when monomers containing two functional groups come together with loss of a small molecule such as water or HCl. Condensation polymers are also called step-growth polymers.

Condensation reaction (Section 18.1B): A reaction in which a small molecule, often water, is eliminated during the reaction process.

Condensed structure (Section 1.8A): A shorthand representation of the structure of a compound in which all atoms are drawn in but bonds and lone pairs are usually omitted. Parentheses are used to denote similar groups bonded to the same atom.

Configuration (Section 5.2): A particular three-dimensional arrangement of atoms.

Conformations (Section 4.9): The different arrangements of atoms that are interconverted by rotation about single bonds.

Conjugate acid (Section 2.2): The compound that results when a base gains a proton in a proton transfer reaction.

Conjugate addition (Sections 12.10, 13.15): An addition reaction that adds groups to the atoms in the 1 and 4 positions of a conjugated system. Conjugate addition is also called 1,4-addition.

Conjugate base (Section 2.2): The compound that results when an acid loses a proton in a proton transfer reaction.

Conjugated diene (Section 12.1A): A compound that contains two carbon–carbon double bonds joined by a single σ bond. Pi (π) electrons are delocalized over both double bonds. Conjugated dienes are also called 1,3-dienes.

Conjugated protein (Section 23.9C): A structure composed of a protein unit and a non-protein molecule.

Conjugation (Section 12.1): The overlap of p orbitals on three or more adjacent atoms.

Conrotatory rotation (Section 29.3): Rotation of p orbitals in the same direction during electrocyclic ring closure or ring opening.

Constitutional isomers (Sections 1.4, 4.1A, 5.2): Two compounds that have the same molecular formula, but differ in the way the atoms are connected to each other. Constitutional isomers are also called structural isomers.

Coordination polymerization (Section 30.4): A polymerization reaction that uses a homogeneous catalyst that is soluble in the reaction solvents typically used.

Cope rearrangement (Section 29.5): A [3,3] sigmatropic rearrangement of a 1,5-diene to an isomeric 1,5-diene.

Copolymer (Section 30.2D): A polymer prepared by joining two or more different monomers together.

Counterion (Section 2.1): An ion that does not take part in a reaction and is opposite in charge to the ion that does take part in the reaction. A counterion is also called a spectator ion.

Coupled reactions (Section 6.5B): Two reactions paired together to drive an unfavorable process. The energy released by one reaction is used to drive the other reaction.

Coupling constant (Section C.6A): The frequency difference, measured in Hz, between the peaks in a split NMR signal.

Coupling reaction (Section 22.14): A reaction that forms a bond between two discrete molecules.

Covalent bond (Section 1.2): A bond that results from the sharing of electrons between two nuclei. A covalent bond is a two-electron bond.

Crossed aldol reaction (Section 18.2): An aldol reaction in which the two reacting carbonyl compounds are different. A crossed aldol reaction is also called a mixed aldol reaction.

Crossed Claisen reaction (Section 18.6): A Claisen reaction in which the two reacting esters are different.

Crown ether (Section 3.7B): A cyclic ether containing multiple oxygen atoms. Crown ethers bind specific cations depending on the size of their central cavity.

Curved arrow notation (Section 1.6B): A convention that shows the movement of an electron pair. The tail of the arrow begins at the electron pair and the head points to where the electron pair moves.

Cyanide anion (Section 14.8A): An anion having the structure $^-C\equiv N$.

Cyano group (Section 15.1): A functional group consisting of a carbon–nitrogen triple bond ($C\equiv N$).

Cyanohydrin (Section 14.8): A compound having the general structure $RCH(OH)C\equiv N$. A cyanohydrin results from the addition of HCN across the carbonyl of an aldehyde or a ketone.

Cycloaddition (Section 29.1): A pericyclic reaction between two compounds with π bonds to form a cyclic product with two new σ bonds.

Cycloalkane (Sections 4.1, 4.2): A compound that contains carbons joined in one or more rings. Cycloalkanes with one ring have the general formula C_nH_{2n}.

Cyclopropanation (Section 28.4): An addition reaction to a carbon–carbon double bond that forms a cyclopropane.

D

D-Sugar (Section 24.2C): A sugar with the hydroxy group on the stereogenic center farthest from the carbonyl on the right side in the Fischer projection formula.

Decalin (Section 25.8A): Two fused six-membered rings. *cis*-Decalin has the hydrogen atoms at the ring fusion on the same side of the rings, whereas *trans*-decalin has the hydrogen atoms at the ring fusion on opposite sides of the rings.

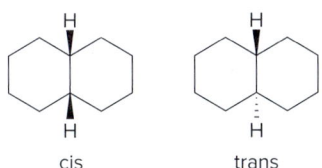

Decarboxylation (Sections 17.9A, 17.11): Loss of CO_2 through cleavage of a carbon–carbon bond.

Degenerate orbitals (Section 19.10B): Orbitals (either atomic or molecular) having the same energy.

Degree of unsaturation (Section 10.2): A ring or a π bond in a molecule. The number of degrees of unsaturation compares the number of hydrogens in a compound to that of a saturated hydrocarbon containing the same number of carbons.

Dehydration (Sections 9.8, 16.9B, 18.1C): A reaction that results in the loss of the elements of water from the reaction components.

Dehydrogenase (Section 27.2A): An enzyme that catalyzes the addition or removal of two hydrogen atoms from a substrate.

Dehydrohalogenation (Section 8.1): An elimination reaction in which the elements of hydrogen and halogen are lost from a starting material.

Delta (δ) scale (Section C.1B): A common scale of chemical shifts used in NMR spectroscopy in which the absorption due to tetramethylsilane (TMS) occurs at zero parts per million.

Denaturation (Section 23.8D): The process of altering the shape of a protein without breaking the amide bonds that form the primary structure.

Deoxy (Section 26.1A): A prefix that means without oxygen.

Deoxyribonucleic acid (DNA, Section 26.1): The nucleic acid that stores the genetic information of an organism and transmits that information from one generation to another.

Deoxyribonucleoside (Section 26.1A): An *N*-glycoside formed by the reaction of 2′-deoxy-D-ribose with certain amine heterocycles.

Deoxyribonucleotide (Section 26.1A): A DNA building block having a deoxyribose and either a purine or pyrimidine base joined together by an *N*-glycosidic linkage, and a phosphate bonded to a hydroxy group of the sugar nucleus.

Deprotection (Section 13.12): A reaction that removes a protecting group, regenerating a functional group.

Deshielding effects (Section C.3A): An effect in NMR caused by a decrease in electron density, thus increasing the strength of the magnetic field felt by the nucleus. Deshielding shifts an absorption downfield.

Dextrorotatory (Section 5.12A): Rotating plane-polarized light in the clockwise direction. The rotation is labeled *d* or (+).

1,3-Diacid (Section 17.9A): A compound containing two carboxylic acids separated by a single carbon atom. 1,3-Diacids are also called β-diacids.

Dialkylamide (Section 17.3B): An amide base having the general structure R_2N^-.

Diastereomers (Section 5.7): Stereoisomers that are not mirror images of each other. Diastereomers have the same *R*,*S* designation for at least one stereogenic center and the opposite *R*,*S* designation for at least one of the other stereogenic centers.

Diastereotopic protons (Section C.2C): Two hydrogen atoms on the same carbon such that substitution of either hydrogen with a group Z forms diastereomers. The two hydrogen atoms are not equivalent and give two NMR signals.

1,3-Diaxial interaction (Section 4.13A): A steric interaction between two axial substituents of the chair form of cyclohexane. Larger axial substituents create unfavorable 1,3-diaxial interactions, destabilizing a cyclohexane conformation.

Diazonium salt (Section 22.12A): An ionic salt having the general structure $(R-N\equiv N)^+Cl^-$.

Diazotization reaction (Section 22.12A): A reaction that converts 1° alkylamines and arylamines to diazonium salts.

1,3-Dicarbonyl compound (Section 17.2): A compound containing two carbonyl groups separated by a single carbon atom.

1,4-Dicarbonyl compound (Section 18.4): A dicarbonyl compound in which the carbonyl groups are separated by three single bonds. 1,4-Dicarbonyl compounds can undergo intramolecular reactions to form five-membered rings.

1,5-Dicarbonyl compound (Section 18.4): A dicarbonyl compound in which the carbonyl groups are separated by four single bonds. 1,5-Dicarbonyl compounds can undergo intramolecular reactions to form six-membered rings.

Dieckmann reaction (Section 18.7): An intramolecular Claisen reaction of a diester to form a ring, typically a five- or six-membered ring.

Diels–Alder reaction (Section 12.12): An addition reaction between a 1,3-diene and a dienophile to form a cyclohexene ring.

1,3-Diene (Section 12.1A): A compound containing two carbon–carbon double bonds joined by a single σ bond. Pi (π) electrons are delocalized over both double bonds. 1,3-Dienes are also called conjugated dienes.

Dienophile (Section 12.12): The alkene component in a Diels–Alder reaction that reacts with a 1,3-diene.

Dihedral angle (Section 4.9): The angle that separates a bond on one atom from a bond on an adjacent atom.

Dihydroxylation (Section 11.9): Addition of two hydroxy groups to a double bond to form a 1,2-diol.

Diol (Section 9.3A): A compound possessing two hydroxy groups. Diols are also called glycols.

Dipeptide (Section 23.4): Two amino acids joined by one amide bond.

Diphosphate (Section 7.16): A good leaving group that is often used in biological systems. Diphosphate ($P_2O_7^{4-}$) is abbreviated as PP_i. The term "diphosphate" is also used to described an organic diphosphate having the general structure $ROP_2O_6^{3-}$.

Dipole (Section 1.12): A partial separation of electronic charge.

Dipole–dipole interaction (Section 3.3B): An attractive intermolecular interaction between the permanent dipoles of polar molecules. The dipoles of adjacent molecules align so that the partial positive and partial negative charges are in close proximity.

Directed aldol reaction (Section 18.3): A crossed aldol reaction in which the enolate of one carbonyl compound is formed, followed by addition of the second carbonyl compound.

Disaccharide (Section 24.11): A carbohydrate containing two monosaccharide units joined by a glycosidic linkage.

Disproportionation (Section 30.2): A method of chain termination in radical polymerization involving the transfer of a hydrogen atom from one polymer radical to another, forming a new C—H bond on one polymer chain and a new double bond on the other.

Disrotatory rotation (Section 29.3): Rotation of *p* orbitals in opposite directions during electrocyclic ring closure or ring opening.

Dissolving metal reduction (Section 11.2): A reduction reaction using alkali metals as a source of electrons and liquid ammonia as a source of protons.

Disubstituted alkene (Section 8.2A): An alkene that has two alkyl groups and two hydrogens bonded to the carbons of the double bond ($R_2C=CH_2$ or $RCH=CHR$).

Disulfide (Sections 9.15A, 23.4C): A compound having the general structure RSSR', often formed between the side chain of two cysteine residues.

Diterpene (Section 25.7A): A terpene that contains 20 carbons and four isoprene units. A diterpenoid contains at least one oxygen atom as well.

Doublet (Section C.6): An NMR signal that is split into two peaks of equal area, caused by one nearby nonequivalent proton.

Doublet of doublets (Section C.8): A splitting pattern of four peaks observed when a signal is split by two different nonequivalent protons.

Downfield shift (Section C.1B): In an NMR spectrum, a term used to describe the relative location of an absorption signal. A downfield shift means the signal is shifted to the left in the spectrum to higher chemical shift on the δ scale.

E

E1 mechanism (Sections 8.3, 8.6): An elimination mechanism that goes by a two-step process involving a carbocation intermediate. E1 is an abbreviation for "Elimination Unimolecular."

E1cB mechanism (Section 18.1C): A two-step elimination mechanism that goes by a carbanion intermediate. E1cB stands for "Elimination Unimolecular, Conjugate Base."

E2 mechanism (Sections 8.3, 8.4): An elimination mechanism that goes by a one-step concerted process, in which both reactants are involved in the transition state. E2 is an abbreviation for "Elimination Bimolecular."

Eclipsed conformation (Section 4.9): A conformation of a molecule where the bonds on one carbon are directly aligned with the bonds on the adjacent carbon.

Edman degradation (Section 23.5B): A procedure used in peptide sequencing in which amino acids are cleaved one at a time from the N-terminal end, the identity of the amino acid determined, and the process repeated until the entire sequence is known.

Eicosanoids (Section 25.6): A group of biologically active compounds containing 20 carbon atoms derived from arachidonic acid.

Elastomer (Section 30.5): A polymer that stretches when stressed but then returns to its original shape.

Electrocyclic ring closure (Section 29.1): An intramolecular pericyclic reaction that forms a cyclic product containing one more σ bond and one fewer π bond than the reactant.

Electrocyclic ring-opening reaction (Section 29.1): A pericyclic reaction in which a σ bond of a cyclic reactant is cleaved to form a conjugated product with one more π bond.

Electromagnetic radiation (Section B.1): Radiant energy having dual properties of both waves and particles. The electromagnetic spectrum contains the complete range of electromagnetic radiation, arbitrarily divided into different regions.

Electron-donating inductive effect (Section 7.13A): An inductive effect in which an electropositive atom or polarizable group donates electron density through σ bonds to another atom.

Electronegativity (Section 1.12): A measure of an atom's attraction for electrons in a bond. Electronegativity indicates how much a particular atom "wants" electrons.

Electron-withdrawing inductive effect (Sections 2.5, 7.13A): An inductive effect in which a nearby electronegative atom pulls electron density toward itself through σ bonds.

Electrophile (Section 2.8): An electron-deficient compound, often symbolized by E^+, which can accept a pair of electrons from an electron-rich compound, forming a covalent bond. Lewis acids are electrophiles.

Electrophilic addition reaction (Section 10.9): An addition reaction in which the first step of the mechanism involves addition of the electrophilic end of a reagent to a π bond.

Electrophilic aromatic substitution (Section 20.1): A characteristic reaction of benzene in which a hydrogen atom on the ring is replaced by an electrophile.

Electrospray ionization (Section A.5C): A method for ionizing large biomolecules in a mass spectrometer. Electrospray ionization is abbreviated as ESI.

Electrostatic potential map (Section 1.12): A color-coded map that illustrates the distribution of electron density in a molecule. Electron-rich regions are indicated in red, and electron-deficient regions are indicated in blue. Regions of intermediate electron density are shown in orange, yellow, and green.

α Elimination (Section 28.4): An elimination reaction involving the loss of two elements from the same atom.

β Elimination (Section 8.1): An elimination reaction involving the loss of elements from two adjacent atoms.

Elimination reaction (Sections 6.2B, 8.1): A chemical reaction in which elements of the starting material are "lost" and a π bond is formed.

Enamine (Section 14.11): A compound having an amine nitrogen atom bonded to a carbon–carbon double bond [$R_2C=CH(NR'_2)$].

Enantiomeric excess (Section 5.12D): A measurement of how much one enantiomer is present in excess of the racemic mixture. Enantiomeric excess (*ee*) is also called optical purity; *ee* = % of one enantiomer − % of the other enantiomer.

Enantiomers (Section 5.3): Stereoisomers that are mirror images but are not superimposable upon each other. Enantiomers have the exact opposite *R,S* designation at every stereogenic center.

Enantioselective reaction (Sections 11.14, 23.3): A reaction that affords predominantly or exclusively one enantiomer. Enantioselective reactions are also called asymmetric reactions.

Enantiotopic protons (Section C.2C): Two hydrogen atoms on the same carbon such that substitution of either hydrogen with a group Z forms enantiomers. The two hydrogen atoms are equivalent and give a single NMR signal.

Endo position (Section 12.13D): A position of a substituent on a bridged bicyclic compound in which the substituent is closer to the longer bridge that joins the two carbons common to both rings.

Endothermic reaction (Section 6.4): A reaction in which the energy of the products is higher than the energy of the reactants. In an endothermic reaction, energy is absorbed and the $\Delta H°$ is a positive value.

Energy of activation (Section 6.7): The energy difference between the transition state and the starting material. The energy of activation, symbolized by E_a, is the minimum amount of energy needed to break bonds in the reactants.

Energy diagram (Section 6.7): A schematic representation of the energy changes that take place as reactants are converted to products. An energy diagram indicates how readily a reaction proceeds, how many steps are involved, and how the energies of the reactants, products, and intermediates compare.

Enolate (Sections 13.15, 17.3): A resonance-stabilized anion formed when a base removes an α hydrogen from the α carbon to a carbonyl group.

Enol tautomer (Sections 9.1, 10.18, 13.15): A compound having a hydroxy group bonded to a carbon–carbon double bond. An enol tautomer [such as $CH_2=C(OH)CH_3$] is in equilibrium with its keto tautomer [$(CH_3)_2C=O$].

Enthalpy change (Section 6.4): The energy absorbed or released in a reaction. Enthalpy change is symbolized by $\Delta H°$ and is also called the heat of reaction.

Entropy (Section 6.6): A measure of the randomness in a system. The more freedom of motion or the more disorder present, the higher the entropy. Entropy is denoted by the symbol $S°$.

Entropy change (Section 6.6): The change in the amount of disorder between reactants and products in a reaction. The entropy change is denoted by the symbol $\Delta S°$. $\Delta S° = S°_{products} − S°_{reactants}$.

Enzyme (Section 6.11): A biochemical catalyst composed of at least one chain of amino acids held together in a very specific three-dimensional shape.

Enzyme–substrate complex (Section 6.11): A structure having a substrate bonded to the active site of an enzyme.

Epoxidation (Section 11.8): Addition of a single oxygen atom to an alkene to form an epoxide.

Epoxide (Section 9.1): A cyclic ether having the oxygen atom as part of a three-membered ring. Epoxides are also called oxiranes.

Epoxy resin (Section 30.6E): A step-growth polymer formed from a fluid prepolymer and a hardener that cross-links polymer chains together.

Equatorial bonds (Section 4.12A): Bonds located in the plane of the chair conformation of cyclohexane (around the equator). Three equatorial bonds point slightly upwards (on the down carbons) and three equatorial bonds point slightly downwards (on the up carbons).

Equilibrium constant (Section 6.5A): A mathematical expression, denoted by the symbol K_{eq}, which relates the amount of starting material and product at equilibrium. K_{eq} = [products]/[starting materials].

Essential oil (Section 25.7): A class of terpenes isolated from plant sources by distillation.

Ester (Sections 13.1, 16.1): A compound having the general structure RCOOR'.

Esterification (Section 16.9C): A reaction that converts a carboxylic acid or a derivative of a carboxylic acid to an ester.

Ether (Section 9.1): A functional group having the general structure ROR'.

Ethynyl group (Section 10.3): An alkynyl substituent having the structure −C≡C−H.

Excited state (Section 25.2): A high-energy electronic state in which one or more electrons have been promoted to a higher-energy orbital by absorption of energy.

Exo position (Section 12.13D): A position of a substituent on a bridged bicyclic compound in which the substituent is closer to the shorter bridge that joins the two carbons common to both rings.

Exothermic reaction (Section 6.4): A reaction in which the energy of the products is lower than the energy of the reactants. In an exothermic reaction, energy is released and the $\Delta H°$ is a negative value.

Extraction (Section 15.10): A laboratory method to separate and purify a mixture of compounds using solubility differences and acid–base principles.

E,Z System of nomenclature (Section 10.3C): A system for unambiguously naming alkene stereoisomers by assigning priorities to the two groups on each carbon of the double bond. The *E* isomer has the two higher-priority groups on opposite sides of the double bond, and the *Z* isomer has them on the same side.

F

Fat (Sections 10.6B, 25.3): A triacylglycerol that is solid at room temperature and composed of fatty acid side chains with a high degree of saturation.

Fatty acid (Section 10.6A): A long-chain carboxylic acid having between 12 and 20 carbon atoms.

Fehling's reagent (Section 24.9B): A reagent for oxidizing aldehydes to carboxylic acids using a Cu^{2+} salt as an oxidizing agent, forming brick-red Cu_2O as a by-product.

Fermentation (Section 27.5C): The anaerobic conversion of glucose to ethanol and CO_2 that occurs in yeast and other microorganisms.

Fibrous proteins (Section 23.9): Long linear polypeptide chains that are bundled together to form rods or sheets.

Fingerprint region (Section B.2B): The region in an IR spectrum at < 1500 cm^{-1}. The region often contains a complex set of peaks and is unique for every compound.

First-order rate equation (Sections 6.9B, 7.10): A rate equation in which the reaction rate depends on the concentration of only one reactant.

Fischer esterification (Section 16.9C): An acid-catalyzed esterification reaction between a carboxylic acid and an alcohol to form an ester.

Fischer projection formula (Section 24.2A): A method for representing stereogenic centers with the stereogenic carbon at the intersection of vertical and horizontal lines. Fischer projections are also called cross formulas.

$$Z\text{—}\overset{W}{\underset{Y}{C}}\text{—}X \quad = \quad Z\text{—}\overset{W}{\underset{Y}{\vert}}\text{—}X$$

Fishhook (Section 6.3B): A half-headed curved arrow used in a reaction mechanism to denote the movement of a single electron.

Flagpole hydrogens (Section 4.12B): Hydrogens in the boat conformation of cyclohexane that are on either end of the "boat" and are forced into close proximity to each other.

Formal charge (Section 1.3C): The electronic charge assigned to individual atoms in a Lewis structure. The formal charge is calculated by subtracting an atom's unshared electrons and half of its shared electrons from the number of valence electrons that a neutral atom would possess.

Formyl group (Section 14.2E): A substituent having the structure —CHO.

Four-centered transition state (Section 10.16): A transition state that involves four atoms.

Fragment (Section A.1): Radicals and cations formed by the decomposition of the molecular ion in a mass spectrometer.

Freons (Sections 7.4, 21.8): Chlorofluorocarbons consisting of simple halogen-containing organic compounds that were once commonly used as refrigerants.

Frequency (Section B.1): The number of waves passing a point per unit time. Frequency is reported in cycles per second (s^{-1}), which is also called hertz (Hz). Frequency is abbreviated with the Greek letter nu (ν).

Friedel–Crafts acylation (Section 20.5A): An electrophilic aromatic substitution reaction in which benzene reacts with an acid chloride in the presence of a Lewis acid to give a ketone.

Friedel–Crafts alkylation (Section 20.5A): An electrophilic aromatic substitution reaction in which benzene reacts with an alkyl halide in the presence of a Lewis acid to give an alkyl benzene.

Frontside attack (Section 7.11C): Approach of a nucleophile from the same side as the leaving group.

Full-headed curved arrow (Section 6.3B): An arrow used in a reaction mechanism to denote the movement of a pair of electrons.

Functional group (Section 3.1): An atom or group of atoms with characteristic chemical and physical properties. The functional group is the reactive part of the molecule.

Functional group interconversion (Section 10.21): A reaction that converts one functional group to another.

Functional group region (Section B.2): The region in an IR spectrum at ≥ 1500 cm^{-1}. Common functional groups show one or two peaks in this region, at a characteristic frequency.

Furanose (Section 24.6): A cyclic five-membered ring of a monosaccharide containing an oxygen atom.

Fused ring system (Section 12.13C): A bicyclic ring system in which the two rings share one bond and two adjacent atoms.

G

Gabriel synthesis (Section 22.6A): A two-step method that converts an alkyl halide to a primary amine using a nucleophile derived from phthalimide.

Gas chromatography (Section A.5B): An analytical technique that separates the components of a mixture based on their boiling points and the rate at which their vapors travel through a column.

Gauche conformation (Section 4.10): A staggered conformation in which the two larger groups on adjacent carbon atoms have a dihedral angle of 60°.

GC–MS (Section A.5B): An analytical instrument that combines a gas chromatograph (GC) and a mass spectrometer (MS) in sequence.

***gem*-Diol** (Section 14.14): A compound having the general structure $R_2C(OH)_2$. *gem*-Diols are also called hydrates.

Geminal dihalide (Section 8.10): A compound that has two halogen atoms on the same carbon atom.

Gene (Section 26.1A): A portion of a DNA molecule responsible for the synthesis of a specific protein.

Genetic code (Section 26.6): The set of three-nucleotide units in mRNA called codons that correspond to particular amino acids. The order of codons in mRNA determines the amino acid sequence in a protein.

Gibbs free energy (Section 6.5A): The free energy of a molecule. Gibbs free energy is denoted by the symbol $G°$.

Gibbs free energy change (Section 6.5A): The overall energy difference between reactants and products. The Gibbs free energy change is denoted by the symbol $\Delta G°$. $\Delta G° = G°_{products} - G°_{reactants}$.

Globular proteins (Section 23.9): Polypeptide chains that are coiled into compact shapes with hydrophilic outer surfaces that make them water soluble.

Glycol (Section 9.3A): A compound possessing two hydroxy groups. Glycols are also called diols.

Glycolysis (Section 27.4): An anaerobic 10-step metabolic pathway that converts glucose to two molecules of pyruvate.

Glycosidase (Section 24.12B): An enzyme that hydrolyzes glycosidic linkages. An α-glycosidase hydrolyzes only α-glycosidic linkages.

Glycoside (Section 24.7A): A monosaccharide with an alkoxy group bonded to the anomeric carbon.

***N*-Glycoside** (Section 24.13B): A monosaccharide containing a nitrogen bonded to the anomeric carbon.

Glycosidic linkage (Section 24.11): An acetal linkage formed between an OH group on one monosaccharide and the anomeric carbon on a second monosaccharide.

Green chemistry (Section 30.8): The use of environmentally benign methods to synthesize compounds.

Grignard reagent (Section 13.9): An organometallic reagent having the general structure RMgX.

Ground state (Section 1.9B): The lowest-energy arrangement of electrons for an atom.

Group number (Section 1.1): The number above a particular column in the periodic table. Group numbers are represented by either an Arabic (1 to 8) or Roman (I to VIII) numeral followed by the letter A or B. The group number of a second-row element is equal to the number of valence electrons in that element.

Grubbs catalyst (Section 28.6): A widely used ruthenium catalyst for olefin metathesis that has the structure $Cl_2(Cy_3P)_2Ru=CHPh$.

Guest molecule (Section 9.5B): A small molecule that can bind to a larger host molecule.

H

^{1}H NMR spectroscopy (Section C.1): A form of nuclear magnetic resonance spectroscopy used to determine the number and type of hydrogen atoms in a molecule. ^{1}H NMR is also called proton NMR spectroscopy.

Half-headed curved arrow (Section 6.3B): An arrow used in a reaction mechanism to denote the movement of a single electron. A half-headed curved arrow is also called a fishhook.

α-Halo aldehyde or ketone (Section 17.7): An aldehyde or ketone with a halogen atom bonded to the α carbon.

Halogenation (Sections 10.13, 20.3, 21.3): The reaction of a compound with a halogen.

Halohydrin (Sections 9.6, 10.15): A compound that has a hydroxy group and a halogen atom on adjacent carbon atoms.

Halonium ion (Section 10.13): A positively charged halogen atom. A bridged halonium ion contains a three-membered ring and is formed in the addition of a halogen (X_2) to an alkene.

Hammond postulate (Section 7.14): A postulate that states that the transition state of a reaction resembles the structure of the species (reactant or product) to which it is closer in energy.

Haworth projection (Section 24.6A): A representation of the cyclic form of a monosaccharide in which the ring is drawn flat.

Head-to-tail polymerization (Section 21.13B): A mechanism of radical polymerization in which the more substituted radical of the growing polymer chain always adds to the less substituted end of the new monomer.

Heat of hydrogenation (Section 11.3A): The $\Delta H°$ of a catalytic hydrogenation reaction equal to the amount of energy released by hydrogenating a π bond.

Heat of reaction (Section 6.4): The energy absorbed or released in a reaction. Heat of reaction is symbolized by $\Delta H°$ and is also called the change in enthalpy.

Heck reaction (Section 28.3): The palladium-catalyzed coupling of a vinyl or aryl halide with an alkene to form a more highly substituted alkene with a new carbon–carbon bond.

α-Helix (Section 23.8B): A secondary structure of a protein formed when a peptide chain twists into a right-handed or clockwise spiral.

Heme (Section 19.12): A complex organic compound containing an Fe^{2+} ion coordinated with a porphyrin.

Hemiacetal (Section 14.15A): A compound that contains an alkoxy group and a hydroxy group bonded to the same carbon atom.

Henderson–Hasselbalch equation (Section 15.8): An expression derived from the equations for K_a and pK_a, which tells us whether a compound will exist in its acidic form (HA) or as its conjugate base ($A:^-$) at a particular pH.

Hertz (Section B.1): A unit of frequency measuring the number of waves passing a point per second.

Heteroatom (Sections 1.6, 3.1): An atom other than carbon or hydrogen. Common heteroatoms in organic chemistry are nitrogen, oxygen, sulfur, phosphorus, and the halogens.

Heterocycle (Section 9.3B): A cyclic compound containing a heteroatom as part of the ring.

Heterolysis (Section 6.3A): The breaking of a covalent bond by unequally dividing the electrons between the two atoms in the bond. Heterolysis generates charged intermediates. Heterolysis is also called heterolytic cleavage.

Hexose (Section 24.2): A monosaccharide containing six carbons.

Highest occupied molecular orbital (Section 19.10B): The molecular orbital with the highest energy that also contains electrons. The highest occupied molecular orbital is abbreviated as HOMO.

High-resolution mass spectrometer (Section A.5A): A mass spectrometer that can measure mass-to-charge ratios to four or more decimal places. High-resolution mass spectra are used to determine the molecular formula of a compound.

Hofmann elimination (Section 22.11): An E2 elimination reaction that converts an amine to a quaternary ammonium salt as the leaving group. The Hofmann elimination gives the less substituted alkene as the major product.

Homologous series (Section 4.1B): A group of compounds that differ by only a CH_2 group in the chain.

Homolysis (Section 6.3A): The breaking of a covalent bond by equally dividing the electrons between the two atoms in the bond. Homolysis generates uncharged radical intermediates. Homolysis is also called homolytic cleavage.

Homopolymer (Section 30.2D): A polymer prepared from a single monomer.

Homotopic protons (Section C.2C): Two equivalent hydrogen atoms such that substitution of either hydrogen with a group Z forms the same product. The two hydrogen atoms give a single NMR signal.

Hooke's law (Section B.3): A physical law that can be used to calculate the frequency of a bond vibration from the strength of the bond and the masses of the atoms attached to it.

Host–guest complex (Section 9.5B): The complex that is formed when a small guest molecule binds to a larger host molecule.

Host molecule (Section 9.5B): A large molecule that can bind a smaller guest molecule.

Hückel's rule (Section 19.7): A principle that states for a compound to be aromatic, it must be cyclic, planar, completely conjugated, and have $4n + 2$ π electrons.

Human genome (Section 26.3): The total DNA content of an individual.

Hybridization (Section 1.9B): The mathematical combination of two or more atomic orbitals (having different shapes) to form the same number of hybrid orbitals (all having the same shape).

Hybrid orbital (Section 1.9B): A new orbital that results from the mathematical combination of two or more atomic orbitals. The hybrid orbital is intermediate in energy compared to the atomic orbitals that were combined to form it.

Hydrate (Sections 11.12B, 14.14): A compound having the general structure $R_2C(OH)_2$. Hydrates are also called *gem*-diols.

Hydration (Sections 10.12, 14.8A): Addition of the elements of water to a molecule.

Hydride (Section 11.2): A negatively charged hydrogen ion ($H:^-$).

1,2-Hydride shift (Section 9.9): Rearrangement of a less stable carbocation to a more stable carbocation by the shift of a hydrogen atom from one carbon atom to an adjacent carbon atom.

Hydroboration (Section 10.16): The addition of the elements of borane (BH_3) to an alkene or alkyne.

Hydrocarbon (Sections 3.2A, 4.1): A compound made up of only the elements of carbon and hydrogen.

Hydrogen bonding (Section 3.3B): An attractive intermolecular interaction that occurs when a hydrogen atom bonded to an O, N, or F atom is electrostatically attracted to a lone pair of electrons on an O, N, or F atom in another molecule.

Hydrogenolysis (Section 23.6): A reaction that cleaves a σ bond using H_2 in the presence of a metal catalyst.

α Hydrogens (Section 17.1): The hydrogen atoms on the carbon bonded to the carbonyl carbon atom (the α carbon).

Hydrohalogenation (Section 10.9): An electrophilic addition of hydrogen halide (HX) to an alkene or alkyne.

Hydrolase (Section 23.10A): An enzyme that catalyzes the hydrolysis of an ester, an amide, or another functional group.

Hydrolysis (Section 14.8A): A cleavage reaction with water.

Hydroperoxide (Section 21.10): An organic compound having the general structure ROOH.

Hydrophilic (Section 3.4C): Attracted to water. The polar portion of a molecule that interacts with polar water molecules is hydrophilic.

Hydrophobic (Section 3.4C): Not attracted to water. The nonpolar portion of a molecule that is not attracted to polar water molecules is hydrophobic.

β-Hydroxy carbonyl compound (Section 18.1A): An organic compound having a hydroxy group on the carbon β to the carbonyl group.

Hydroxy group (Section 9.1): The OH functional group.

Hyperconjugation (Section 7.13B): The overlap of an empty p orbital with an adjacent σ bond.

I

Imide (Section 22.6A): A compound having a nitrogen atom between two carbonyl groups.

Imine (Sections 14.6B, 14.10): A compound with the general structure $R_2C=NR'$. Imines are also called Schiff bases.

Iminium ion (Section 14.10): A resonance-stabilized cation having the general structure $(R_2C=NR'_2)^+$, where R' = H or alkyl.

Inductive effect (Sections 2.5B, 7.13A): The pull of electron density through σ bonds caused by electronegativity differences of atoms.

Infrared (IR) spectroscopy (Section B.2): An analytical technique used to identify the functional groups in a molecule based on their absorption of electromagnetic radiation in the infrared region.

Initiation (Section 21.4A): The initial step in a chain mechanism that forms a reactive intermediate by cleavage of a bond.

Inscribed polygon method (Section 19.11): A method to predict the relative energies of cyclic, completely conjugated compounds to determine which molecular orbitals are filled or empty. The inscribed polygon is also called a Frost circle.

Integration (Section C.5): The area under an NMR signal that is proportional to the number of absorbing nuclei that give rise to the signal.

Intermolecular forces (Section 3.3): The types of interactions that exist between molecules. Functional groups determine the type and strength of these forces. Intermolecular forces are also called noncovalent interactions or nonbonded interactions.

Internal alkyne (Section 10.1): An alkyne that has one carbon atom bonded to each end of the triple bond.

Inversion of configuration (Section 7.11C): The opposite relative stereochemistry of a stereogenic center in the starting material and product of a chemical reaction. In a nucleophilic substitution reaction, inversion results when the nucleophile and leaving group are in the opposite position relative to the three other groups on carbon.

Ionic bond (Section 1.2): A bond that results from the transfer of electrons from one element to another. Ionic bonds result from strong electrostatic interactions between ions with opposite charges. The transfer of electrons forms stable salts composed of cations and anions.

Ionophore (Section 3.7B): An organic molecule that can form a complex with cations so they may be transported across a cell membrane. Ionophores have a hydrophobic exterior and a hydrophilic central cavity that complexes the cation.

Isocyanate (Section 30.6C): A compound having the general structure RN=C=O.

Isoelectric point (Sections 15.12C, 23.1B): The pH at which an amino acid exists primarily in its neutral zwitterionic form. Isoelectric point is abbreviated as pI.

Isolated diene (Section 12.1A): A compound containing two carbon–carbon double bonds joined by more than one σ bond.

Isomerase (Section 23.10A): An enzyme that catalyzes the conversion of one isomer to another.

Isomers (Sections 1.4A, 4.1A, 5.1): Two different compounds that have the same molecular formula.

Isoprene unit (Section 25.7): A five-carbon unit with four carbons in a row and a one-carbon branch on one of the middle carbons.

Isotactic polymer (Section 30.4): A polymer having all the substituents on the same side of the carbon backbone of an elongated polymer chain.

Isotope (Section 1.1): Two or more atoms of the same element having the same number of protons in the nucleus but a different number of neutrons. Isotopes have the same atomic number but different mass numbers.

IUPAC system of nomenclature (Section 4.3): A systematic method for naming compounds developed by the International Union of Pure and Applied Chemistry.

K

K_a (Section 2.3): The symbol that represents the acidity constant of an acid HA. The larger the K_a, the stronger the acid.

$$K_a = \frac{[H_3O^+][A:^-]}{[H-A]}$$

K_{eq} (Section 2.3): The equilibrium constant. K_{eq} = [products]/[starting materials].

Kekulé structures (Section 19.1): Two equilibrating structures for benzene. Each structure contains a six-membered ring and three π bonds alternating with σ bonds around the ring.

Ketal (Section 14.15): A compound having the general structure $R_2C(OR')_2$, where R = alkyl or aryl. Ketals are derived from ketones and constitute a subclass of acetals.

β-Keto ester (Section 17.10): A compound containing a ketone carbonyl on the carbon β to the ester carbonyl group.

Ketone (Section 10.18): A compound with two alkyl groups bonded to the C=O carbon atom, having the general structures $R_2C=O$ or RCOR'.

Ketose (Section 24.2): A monosaccharide composed of a polyhydroxy ketone.

Keto tautomer (Section 10.18): A tautomer of a ketone that has a C=O and a hydrogen bonded to the α carbon. The keto tautomer is in equilibrium with the enol tautomer.

Kiliani–Fischer synthesis (Section 24.10B): A reaction that lengthens the carbon chain of an aldose by adding one carbon to the carbonyl end.

Kinase (Sections 23.10A, 27.3A): An enzyme that catalyzes the transfer of a phosphate from one compound to another.

Kinetic enolate (Section 17.4): The enolate that is formed the fastest—generally the less substituted enolate.

Kinetic product (Section 12.11): In a reaction that can give more than one product, the product that is formed the fastest.

Kinetic resolution (Section 23.2B): The separation of two enantiomers by a chemical reaction that selectively occurs for only one of the enantiomers.

Kinetics (Section 6.5): The study of chemical reaction rates.

L

L-Sugar (Section 24.2C): A sugar with the hydroxy group on the stereogenic center farthest from the carbonyl on the left side in the Fischer projection formula.

Lactam (Section 16.1): A cyclic amide in which the carbonyl carbon–nitrogen σ bond is part of a ring. A β-lactam contains the carbon–nitrogen σ bond in a four-membered ring.

Lactol (Section 14.17): A cyclic hemiacetal.

Lactone (Section 16.1): A cyclic ester in which the carbonyl carbon–oxygen σ bond is part of a ring.

Leaving group (Section 7.6): An atom or group of atoms (Z) that is able to accept the electron density of the C–Z bond during a substitution or elimination reaction.

Leaving group ability (Section 7.7): A measure of how readily a leaving group (Z) can accept the electron density of the C–Z bond during a substitution or elimination reaction.

Le Châtelier's principle (Section 9.8D): The principle that a system at equilibrium will react to counteract any disturbance to the equilibrium.

Lecithin (Section 25.4A): A phosphoacylglycerol in which the phosphodiester alkyl group is $-CH_2CH_2N(CH_3)_3^+$. Lecithins are also called phosphatidylcholines.

Leukotriene (Section 9.17): An unstable and potent biomolecule synthesized in cells by the oxidation of arachidonic acid. Leukotrienes are responsible for biological conditions such as asthma.

Levorotatory (Section 5.12A): Rotating plane-polarized light in the counterclockwise direction. The rotation is labeled l or $(-)$.

Lewis acid (Section 2.8): An electron pair acceptor.

Lewis acid–base reaction (Section 2.8): A reaction that results when a Lewis base donates an electron pair to a Lewis acid.

Lewis base (Section 2.8): An electron pair donor.

Lewis structure (Section 1.3): A representation of a molecule that shows the position of covalent bonds and nonbonding electrons. In Lewis structures, unshared electrons are represented by dots and a two-electron covalent bond is represented by a solid line. Lewis structures are also called electron dot structures.

Ligand (Section 28.2A): A group coordinated to a metal, which donates electron density to or sometimes withdraws electron density from the metal.

Ligase (Section 23.10A): An enzyme that catalyzes bond formation accompanied by energy release from a hydrolysis reaction.

"Like dissolves like" (Section 3.4C): The principle that compounds dissolve in solvents having similar kinds of intermolecular forces; that is, polar compounds dissolve in polar solvents and nonpolar compounds dissolve in nonpolar solvents.

Lindlar catalyst (Section 11.5B): A catalyst for the hydrogenation of an alkyne to a cis alkene. The Lindlar catalyst is Pd adsorbed onto $CaCO_3$ with lead(II) acetate and quinoline.

Lipid (Sections 3.9D, 25.1): A biomolecule with a large number of C–C and C–H σ bonds that is soluble in organic solvents and insoluble in water.

Lone pair of electrons (Section 1.2): A pair of valence electrons that is not shared with another atom in a covalent bond. Lone pairs are also called unshared or nonbonded pairs of electrons.

Lowest unoccupied molecular orbital (Section 19.10B): The molecular orbital with the lowest energy that does not contain electrons. The lowest unoccupied molecular orbital is abbreviated as the LUMO.

Lyase (Section 23.10A): An enzyme that catalyzes the addition of a molecule to a double bond or the elimination of a molecule to give a double bond.

M

M peak (Section A.1): The peak in the mass spectrum that corresponds to the mass of the molecular ion. The M peak is also called the molecular ion peak or the parent peak.

M + 1 peak (Section A.1): The peak in the mass spectrum that corresponds to the mass of the molecular ion plus one. The M + 1 peak is caused by the presence of isotopes that increase the mass of the molecular ion.

M + 2 peak (Section A.2): The peak in the mass spectrum that corresponds to the mass of the molecular ion plus two. The M + 2 peak is caused by the presence of isotopes, typically of a chlorine or a bromine atom.

Magnetic resonance imaging (MRI) (Section C.12): A form of NMR spectroscopy used in medicine.

Malonic ester synthesis (Section 17.9A): A stepwise method that converts diethyl malonate to a carboxylic acid having one or two carbons bonded to the α carbon.

Markovnikov's rule (Section 10.10): The rule that states in the addition of HX to an unsymmetrical alkene, the H atom bonds to the less substituted carbon atom.

Mass number (Section 1.1): The total number of protons and neutrons in the nucleus of a particular atom.

Mass spectrometry (Section A.1): An analytical technique used for measuring the molecular weight and determining the molecular formula of an organic molecule.

Mass-to-charge ratio (Section A.1): A ratio of the mass to the charge of a molecular ion or fragment. Mass-to-charge ratio is abbreviated as m/z.

Megahertz (Section C.1A): A unit used for the frequency of the RF radiation in NMR spectroscopy. Megahertz is abbreviated as MHz; 1 MHz = 10^6 Hz.

Melting point (Section 3.4B): The temperature at which molecules in the solid phase are converted to the liquid phase. Molecules with stronger intermolecular forces and higher symmetry have higher melting points. Melting point is abbreviated as mp.

Merrifield method (Section 23.7): A method for synthesizing polypeptides using insoluble polymer supports.

Meso compound (Section 5.8): An achiral compound that contains two or more tetrahedral stereogenic centers.

Metabolism (Section 27.1): The sum of all the chemical reactions that take place in an organism.

Meta director (Section 20.7): A substituent on a benzene ring that directs a new group to the meta position during electrophilic aromatic substitution.

Meta isomer (Section 19.3B): A 1,3-disubstituted benzene ring. Meta substitution is abbreviated as m-.

Metal hydride reagent (Section 11.2): A reagent containing a polar metal–hydrogen bond that places a partial negative charge on the hydrogen and acts as a source of hydride ions ($H:^-$).

Metathesis (Section 28.6): A reaction between two alkene molecules that results in the interchange of the carbons of their double bonds.

Methylation (Section 7.16): A reaction in which a CH_3 group is transferred from one compound to another.

Methylene group (Sections 4.1B, 10.3B): A CH_2 group bonded to a carbon chain ($-CH_2-$) or part of a double bond ($CH_2=$).

1,2-Methyl shift (Section 9.9): Rearrangement of a less stable carbocation to a more stable carbocation by the shift of a methyl group from one carbon atom to an adjacent carbon atom.

Micelles (Section 3.6): Spherical droplets formed by soap molecules having the ionic heads on the surface and the nonpolar tails packed together in the interior. Grease and oil dissolve in the interior nonpolar region.

Michael acceptor (Section 18.9): The α,β-unsaturated carbonyl compound in a Michael reaction.

Michael reaction (Section 18.9): A reaction in which a resonance-stabilized carbanion (usually an enolate) adds to the β carbon of an α,β-unsaturated carbonyl compound.

Mixed aldol reaction (Section 18.2): An aldol reaction between two different carbonyl compounds. A mixed aldol reaction is also called a crossed aldol reaction.

Mixed anhydride (Section 16.1): An anhydride with two different alkyl groups bonded to the carbonyl carbon atoms.

Molecular ion (Section A.1): The radical cation having the general structure $M^{+\cdot}$, formed by the removal of an electron from an organic molecule. The molecular ion is also called the parent ion.

Molecular orbital theory (Section 19.10A): A theory that describes bonds as the mathematical combination of atomic orbitals to form a new set of orbitals called molecular orbitals. Molecular orbital theory is also called MO theory.

Molecular recognition (Section 9.5B): The ability of a host molecule to recognize and bind specific guest molecules.

Molecule (Section 1.2): A compound containing two or more atoms bonded together with covalent bonds.

Monomers (Sections 5.1, 21.13): Small organic compounds that can be covalently bonded to each other (polymerized) in a repeating pattern.

Monophosphate (Section 7.16): A compound having the general structure $ROPO_3^{2-}$.

Monosaccharide (Section 24.2): A simple sugar having three to seven carbon atoms.

Monosubstituted alkene (Section 8.2A): An alkene that has one alkyl group and three hydrogens bonded to the carbons of the double bond ($RCH=CH_2$).

Monoterpene (Section 25.7A): A terpene that contains 10 carbons and two isoprene units. A monoterpenoid also contains at least one oxygen atom.

Multiplet (Section C.6C): An NMR signal that is split into more than seven peaks.

Mutarotation (Section 24.6A): The process by which a pure anomer of a monosaccharide equilibrates to a mixture of both anomers when placed in solution.

N

n + 1 rule (Section C.6C): The rule that an NMR signal for a proton with *n* nearby nonequivalent protons will be split into *n* + 1 peaks.

Natural product (Section 7.18): A compound isolated from a natural source.

Newman projection (Section 4.9): An end-on representation of the conformation of a molecule. The Newman projection shows the three groups bonded to each carbon atom in a particular C–C bond, as well as the dihedral angle that separates the groups on each carbon.

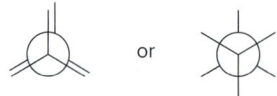 or

Nitration (Section 20.4): An electrophilic aromatic substitution reaction in which benzene reacts with $^+NO_2$ to give nitrobenzene, $C_6H_5NO_2$.

Nitrile (Sections 15.1, 15.13): A compound having the general structure $RC\equiv N$.

Nitronium ion (Section 20.4): An electrophile having the structure $^+NO_2$.

***N*-Nitrosamine** (Section, 22.12B): A compound having the general structure $R_2N-N=O$. Nitrosamines are formed by the reaction of a secondary amine with ^+NO.

Nitrosonium ion (Section 22.12): An electrophile having the structure ^+NO.

NMR peak (Section C.6A): The individual absorptions in a split NMR signal due to nonequivalent nearby protons.

NMR signal (Section C.6A): The entire absorption due to a particular kind of proton in an NMR spectrum.

NMR spectrometer (Section C.1A): An analytical instrument that measures the absorption of RF radiation by certain atomic nuclei when placed in a strong magnetic field.

Nonbonded pair of electrons (Section 1.2): A pair of valence electrons that is not shared with another atom in a covalent bond. Nonbonded electrons are also called unshared or lone pairs of electrons.

Nonbonding molecular orbital (Section 19.11): A molecular orbital having the same energy as the atomic orbitals that formed it.

Nonnucleophilic base (Section 7.8B): A base that is a poor nucleophile due to steric hindrance resulting from the presence of bulky groups.

Nonpolar bond (Section 1.12): A covalent bond in which the electrons are equally shared between the two atoms.

Nonpolar molecule (Section 1.13): A molecule that has no net dipole. A nonpolar molecule has either no polar bonds or multiple polar bonds whose dipoles cancel.

Nonreducing sugar (Section 24.9B): A carbohydrate that cannot be oxidized by Tollens, Benedict's, or Fehling's reagent.

Normal alkane (Section 4.1A): An acyclic alkane that has all of its carbons in a row. A normal alkane is an "*n*-alkane" or a straight-chain alkane.

Nuclear magnetic resonance spectroscopy (Section C.1): A powerful analytical tool that can help identify the carbon and hydrogen framework of an organic molecule.

Nucleic acid (Section 26.2): A polymer of nucleotides, formed by joining the 3'-OH group of one nucleotide with the 5'-phosphate of a second nucleotide in a phosphodiester linkage.

Nucleophile (Sections 2.8, 7.6): An electron-rich compound, symbolized by :Nu⁻, which donates a pair of electrons to an electron-deficient compound, forming a covalent bond. Lewis bases are nucleophiles.

Nucleophilic acyl substitution (Sections 13.2B, 16.1): Substitution of a leaving group by a nucleophile at a carbonyl carbon.

Nucleophilic addition (Section 13.2A): Addition of a nucleophile to the electrophilic carbon of a carbonyl group followed by protonation of the oxygen.

Nucleophilic aromatic substitution (Section 20.13): A substitution reaction of an aryl halide with a strong nucleophile.

Nucleophilicity (Section 7.8A): A measure of how readily an atom donates an electron pair to other atoms.

Nucleophilic substitution (Section 7.6): A reaction in which a nucleophile replaces the leaving group in a molecule.

Nucleoside (Section 26.1A): A biomolecule having a sugar and either a purine or pyrimidine base joined by an *N*-glycosidic linkage.

Nucleotide (Sections 3.9C, 26.1A): A biomolecule having a sugar and either a purine or pyrimidine base joined by an *N*-glycosidic linkage, and a phosphate bonded to a hydroxy group of the sugar nucleus.

O

Observed rotation (Section 5.12A): The angle that a sample of an optically active compound rotates plane-polarized light. The angle is denoted by the symbol α and is measured in degrees (°).

Octet rule (Section 1.2): The general rule governing the bonding process for second-row elements. Through bonding, second-row elements attain a complete outer shell of eight valence electrons.

Oil (Sections 10.6B, 25.3): A triacylglycerol that is liquid at room temperature and composed of fatty acid side chains with a high degree of unsaturation.

Olefin (Section 10.1): An alkene; a compound possessing a carbon–carbon double bond.

Oligosaccharide (Section 24.12D): A carbohydrate with a small number of monosaccharides—generally three to ten—joined together.

Optically active (Section 5.12A): Able to rotate the plane of plane-polarized light as it passes through a solution of a compound.

Optically inactive (Section 5.12A): Not able to rotate the plane of plane-polarized light as it passes through a solution of a compound.

Optical purity (Section 5.12D): A measurement of how much one enantiomer is present in excess of the racemic mixture. Optical purity is also called enantiomeric excess (*ee*); *ee* = % of one enantiomer − % of the other enantiomer.

Orbital (Section 1.1): A region of space around the nucleus of an atom that is high in electron density. There are four different kinds of orbitals, called *s, p, d,* and *f*.

Order of a rate equation (Section 6.9B): The sum of the exponents of the concentration terms in the rate equation of a reaction.

Organoborane (Section 10.16): A compound that contains a carbon–boron bond. Organoboranes have the general structure RBH_2, R_2BH, or R_3B.

Organocopper reagent (Section 13.9): An organometallic reagent having the general structure R_2CuLi. Organocopper reagents are also called organocuprates.

Organolithium reagent (Section 13.9): An organometallic reagent having the general structure RLi.

Organomagnesium reagent (Section 13.9): An organometallic reagent having the general structure RMgX. Organomagnesium reagents are also called Grignard reagents.

Organometallic reagent (Section 13.9): A reagent that contains a carbon atom bonded to a metal.

Organopalladium compound (Section 28.2): An organometallic compound that contains a carbon–palladium bond.

Organophosphorus reagent (Section 14.9A): A reagent that contains a carbon–phosphorus bond.

Ortho isomer (Section 19.3B): A 1,2-disubstituted benzene ring. Ortho substitution is abbreviated as *o*-.

Ortho, para director (Section 20.7): A substituent on a benzene ring that directs a new group to the ortho and para positions during electrophilic aromatic substitution.

Oxaphosphetane (Section 14.9B): An intermediate in the Wittig reaction consisting of a four-membered ring containing a phosphorus–oxygen bond.

Oxidation (Sections 4.14A, 11.1): A process that results in a loss of electrons. For organic compounds, oxidation results in an increase in the number of C—Z bonds or a decrease in the number of C—H bonds; Z = an element more electronegative than carbon.

β-Oxidation (Section 27.3): A catabolic process in which two-carbon units are sequentially cleaved from a fatty acid until all carbons of the fatty acid are degraded to acetyl CoA.

Oxidative addition (Section 28.2A): The addition of a reagent to a metal, often increasing the number of groups around the metal by two.

Oxidative cleavage (Section 11.10): An oxidation reaction that breaks both the σ and π bonds of a multiple bond to form two oxidized products.

Oxidoreductase (Section 23.10A): An enzyme that catalyzes an oxidation–reduction reaction.

Oxime (Section 24.10A): A compound having the general structure $R_2C=NOH$.

Oxirane (Section 9.1): A cyclic ether having the oxygen atom as part of a three-membered ring. Oxiranes are also called epoxides.

Oxy-Cope rearrangement (Section 29.5): A [3,3] sigmatropic rearrangement of a 1,5-dien-3-ol to a δ,ε-unsaturated carbonyl compound.

Ozonolysis (Section 11.10): An oxidative cleavage reaction in which a multiple bond reacts with ozone (O_3) as the oxidant.

Para isomer (Section 19.3B): A 1,4-disubstituted benzene ring. Para substitution is abbreviated as *p*-.

Parent ion (Section A.1): The radical cation having the general structure $M^{+\cdot}$, formed by the removal of an electron from an organic molecule. The parent ion is also called the molecular ion.

Parent name (Section 4.4): The portion of the IUPAC name of an organic compound that indicates the number of carbons in the longest continuous chain in the molecule.

Pentose (Section 24.2): A monosaccharide containing five carbons.

Peptide bond (Section 23.4): The amide bond in peptides and proteins.

Peptides (Sections 16.5B, 23.4): Low-molecular-weight polymers of less than 40 amino acids joined together by amide linkages.

Percent *s*-character (Section 1.11B): The fraction of a hybrid orbital due to the *s* orbital used to form it. As the percent *s*-character increases, a bond becomes shorter and stronger.

Percent transmittance (Section B.2): A measure of how much electromagnetic radiation passes through a sample of a compound and how much is absorbed.

Pericyclic reaction (Section 29.1): A concerted reaction that proceeds through a cyclic transition state.

Peroxide (Section 21.2): A reactive organic compound with the general structure ROOR. Peroxides are used as radical initiators by homolysis of the weak O–O bond.

Peroxyacid (Section 11.7): An oxidizing agent having the general structure RCO_3H.

Peroxy radical (Section 21.10): A radical having the general structure ROO·.

Petroleum (Section 4.7): A fossil fuel containing a complex mixture of compounds, primarily hydrocarbons with 1 to 40 carbon atoms.

Phenol (Sections 9.1, 21.11): A compound such as C_6H_5OH, which contains a hydroxy group bonded to a benzene ring.

Phenyl group (Section 3.2A): A group formed by removal of one hydrogen from benzene, abbreviated as C_6H_5— or Ph—.

Pheromone (Section 4.1): A chemical substance used for communication in an animal or insect species.

Phosphate (Section 7.16): A PO_4^{3-} anion.

Phosphatidylcholine (Section 25.4A): A phosphoacylglycerol in which the phosphodiester alkyl group is $-CH_2CH_2N(CH_3)_3^+$. Phosphatidylcholines are also called lecithins.

Phosphatidylethanolamine (Section 25.4A): A phosphoacylglycerol in which the phosphodiester alkyl group is $-CH_2CH_2NH_3^+$. Phosphatidylethanolamines are also called cephalins.

Phosphoacylglycerols (Section 25.4A): A lipid having a glycerol backbone with two of the hydroxy groups esterified with fatty acids and the third hydroxy group as part of a phosphodiester.

Phosphodiester (Sections 3.2D, 25.4): A functional group having the general formula $ROPO_2OR'$ formed by replacing two of the H atoms in phosphoric acid (H_3PO_4) with alkyl groups.

Phospholipid (Sections 3.7A, 25.4): A hydrolyzable lipid that contains a phosphorus atom.

Phosphonium salt (Section 14.9A): An organophosphorus reagent with a positively charged phosphorus and a suitable counterion; for example, $R_4P^+X^-$. Phosphonium salts are converted to ylides upon treatment with a strong base.

Phosphorane (Section 14.9A): A phosphorus ylide; for example, $Ph_3P=CR_2$.

Photon (Section B.1): A particle of electromagnetic radiation.

Pi (π) bond (Section 1.10B): A bond formed by side-by-side overlap of two *p* orbitals where electron density is not concentrated on the axis joining the two nuclei. Pi (π) bonds are generally weaker than σ bonds.

Pi (π) stacking (Section 19.12): A noncovalent attractive force between aromatic rings with loosely held π electrons.

pK_a (Section 2.3): A logarithmic scale of acid strength. $pK_a = -\log K_a$. The smaller the pK_a, the stronger the acid.

Plane-polarized light (Section 5.12A): Light that has an electric vector that oscillates in a single plane. Plane-polarized light, also called polarized light, arises from passing ordinary light through a polarizer.

Plane of symmetry (Section 5.3): A mirror plane that cuts a molecule in half, so that one half of the molecule is the mirror reflection of the other half.

Plasticizer (Section 30.7): A low-molecular-weight compound added to a polymer to give it flexibility.

β-Pleated sheet (Section 23.8B): A secondary structure of a protein formed when two or more peptide chains line up side by side.

Poisoned catalyst (Section 11.5B): A hydrogenation catalyst with reduced activity that allows selective reactions to occur. The Lindlar catalyst is a poisoned Pd catalyst that converts alkynes to cis alkenes.

Polar aprotic solvent (Section 7.8C): A polar solvent that is incapable of intermolecular hydrogen bonding because it does not contain an O—H or N—H bond.

Polar bond (Section 1.12): A covalent bond in which the electrons are unequally shared between the two atoms. Unequal sharing of electrons results from bonding between atoms of different electronegativity values, usually with a difference of ≥ 0.5 units.

Polarimeter (Section 5.12A): An instrument that measures the degree that a compound rotates plane-polarized light.

Polarity (Section 1.12): A characteristic that results from a dipole. The polarity of a bond is indicated by an arrow with the head of the arrow pointing toward the negative end of the dipole and the tail with a perpendicular line through it at the positive end of the dipole. The polarity of a bond can also be indicated by the symbols δ+ and δ−.

Polarizability (Section 3.3B): A measure of how the electron cloud around an atom responds to changes in its electronic environment.

Polar molecule (Section 1.13): A molecule that has a net dipole. A polar molecule has either one polar bond or multiple polar bonds whose dipoles reinforce.

Polar protic solvent (Section 7.8C): A polar solvent that is capable of intermolecular hydrogen bonding because it contains an O—H or N—H bond.

Polyamide (Section 30.6A): A step-growth polymer that contains many amide bonds. Nylon 6,6 and nylon 6 are polyamides.

Polycarbonate (Section 30.6D): A step-growth polymer that contains many —OC(=O)O— bonds in its backbone, often formed by reaction of $Cl_2C=O$ with a diol.

Polycyclic aromatic hydrocarbon (Section 19.5): An aromatic hydrocarbon containing two or more benzene rings that share carbon–carbon bonds. Polycyclic aromatic hydrocarbons are abbreviated as PAHs.

Polyene (Section 12.7): A compound that contains three or more double bonds.

Polyester (Section 30.6B): A step-growth polymer consisting of many ester bonds between diols and dicarboxylic acids.

Polyether (Sections 9.5B, 30.3): A compound that contains two or more ether linkages.

Polymer (Sections 5.1, 21.13): A large molecule composed of smaller monomer units covalently bonded to each other in a repeating pattern.

Polymerase chain reaction (PCR, Section 26.8): A technique that amplifies a specific portion of DNA, producing millions of copies of a single molecule.

Polymerization (Section 21.13A): The chemical process that joins together monomers to make polymers.

Polysaccharide (Section 24.12): A carbohydrate containing three or more monosaccharide units joined together by glycosidic linkages.

Polyurethane (Section 30.6C): A step-growth polymer that contains many —NHC(=O)O— bonds in its backbone, formed by reaction of a diisocyanate and a diol.

Porphyrin (Section 19.12): A nitrogen-containing heterocycle that can complex metal ions.

Primary (1°) alcohol (Section 3.2): An alcohol having the general structure RCH_2OH.

Primary (1°) alkyl halide (Section 3.2): An alkyl halide having the general structure RCH_2X.

Primary (1°) amide (Section 3.2): An amide having the general structure $RCONH_2$.

Primary (1°) amine (Section 3.2): An amine having the general structure RNH_2.

Primary (1°) carbocation (Section 7.13): A carbocation having the general structure RCH_2^+.

Primary (1°) carbon (Section 3.2): A carbon atom that is bonded to one other carbon atom.

Primary (1°) hydrogen (Section 3.2): A hydrogen that is bonded to a 1° carbon.

Primary protein structure (Section 23.8A): The particular sequence of amino acids joined together by peptide bonds.

Primary (1°) radical (Section 21.1): A radical having the general structure $RCH_2\cdot$.

Prochiral chiral (Section 11.13A): An sp^3 hybridized carbon bonded to two identical groups that can be converted to a stereogenic center by replacement of one of those groups.

Propagation (Section 21.4A): The middle part of a chain mechanism in which one reactive particle is consumed and another is generated. Propagation repeats until a termination step occurs.

Prostaglandin (Section 15.5): A class of lipids containing 20 carbons, a five-membered ring, and a CO_2H group. Prostaglandins possess a wide range of biological activities.

Prosthetic group (Section 23.9C): The non-protein unit of a conjugated protein.

Protecting group (Section 13.12): A blocking group that renders a reactive functional group unreactive, so that it does not interfere with another reaction.

Protection (Section 13.12): The reaction that blocks a reactive functional group with a protecting group.

Proteins (Sections 3.9A, 16.5B, 23.4): High-molecular-weight polymers of 40 or more amino acids joined together by amide linkages.

Proton (Section 2.1): A positively charged hydrogen ion (H^+).

Proton NMR spectroscopy (Section C.1): A form of nuclear magnetic resonance spectroscopy used to determine the number and type of hydrogen atoms in a molecule.

Proton transfer reaction (Section 2.2): A Brønsted–Lowry acid–base reaction; a reaction that results in the transfer of a proton from an acid to a base.

Purine (Sections 22.3, 26.1A): A bicyclic aromatic heterocycle having two nitrogens in each of the rings.

Pyranose (Section 24.6): A cyclic six-membered ring of a monosaccharide containing an oxygen atom.

Pyrimidine (Sections 22.3, 26.1A): A six-membered aromatic heterocycle having two nitrogens in the ring.

Q

Quantum (Section B.1): The discrete amount of energy associated with a particle of electromagnetic radiation (i.e., a photon).

Quartet (Section C.6C): An NMR signal that is split into four peaks having a relative area of 1:3:3:1, caused by three nearby nonequivalent protons.

Quaternary (4°) carbon (Section 3.2): A carbon atom that is bonded to four other carbon atoms.

Quaternary protein structure (Section 23.8C): The shape adopted when two or more folded polypeptide chains aggregate into one protein complex.

Quintet (Section C.6C): An NMR signal that is split into five peaks caused by four nearby nonequivalent protons.

R

Racemic mixture (Section 5.12B): An equal mixture of two enantiomers. A racemic mixture, also called a racemate, is optically inactive.

Racemization (Section 7.12C): The formation of equal amounts of two enantiomers from an enantiomerically pure starting material.

Radical (Sections 6.3B, 21.1): A reactive intermediate with a single unpaired electron, formed by homolysis of a covalent bond.

Radical anion (Section 11.5C): A reactive intermediate containing both a negative charge and an unpaired electron.

Radical cation (Section A.1): A species with an unpaired electron and a positive charge, formed in a mass spectrometer by the bombardment of a molecule with an electron beam.

Radical inhibitor (Section 21.2): A compound that prevents radical reactions from occurring. Radical inhibitors are also called radical scavengers.

Radical initiator (Section 21.2): A compound that contains an especially weak bond that serves as a source of radicals.

Radical polymerization (Section 21.13B): A radical chain reaction involving the polymerization of alkene monomers by adding a radical to a π bond.

Radical scavenger (Section 21.2): A compound that prevents radical reactions from occurring. Radical scavengers are also called radical inhibitors.

Rate constant (Section 6.9B): A constant that is a fundamental characteristic of a reaction. The rate constant, symbolized by k, is a complex mathematical term that takes into account the dependence of a reaction rate on temperature and the energy of activation.

Rate-determining step (Section 6.8): In a multistep reaction mechanism, the step with the highest-energy transition state.

Rate equation (Section 6.9B): An equation that shows the relationship between the rate of a reaction and the concentration of the reactants. The rate equation depends on the mechanism of the reaction and is also called the rate law.

Reaction coordinate (Section 6.7): The x axis in an energy diagram that represents the progress of a reaction as it proceeds from reactant to product.

Reaction mechanism (Section 6.3): A detailed description of how bonds are broken and formed as a starting material is converted to a product.

Reactive intermediate (Section 6.3): A high-energy unstable intermediate formed during the conversion of a stable starting material to a stable product.

Reciprocal centimeter (Section B.2): The unit for wavenumber, which is used to report frequency in IR spectroscopy.

Reducing sugar (Section 24.9B): A carbohydrate that can be oxidized by Tollens, Benedict's, or Fehling's reagent.

Reduction (Sections 4.14A, 11.1): A process that results in the gain of electrons. For organic compounds, reduction results in a decrease in the number of C–Z bonds or an increase in the number of C–H bonds; Z = an element more electronegative than carbon.

Reductive amination (Section 22.6C): A two-step method that converts aldehydes and ketones into amines.

Reductive elimination (Section 28.2A): The elimination of two groups that surround a metal, often forming new carbon–hydrogen or carbon–carbon bonds.

Regioselective reaction (Section 8.5): A reaction that yields predominantly or exclusively one constitutional isomer when more than one constitutional isomer is possible.

Replication (Section 26.3): The process by which DNA makes a copy of itself when a cell divides. The original DNA molecule forms two DNA molecules, each of which contains one strand of DNA from the parent DNA and one new strand.

Resolution (Section 23.2): The separation of a racemic mixture into its component enantiomers.

Resonance (Section C.1A): In NMR spectroscopy, when an atomic nucleus absorbs RF radiation and spin flips to a higher-energy state.

Resonance hybrid (Sections 1.6C, 12.4): A structure that is a weighted composite of all possible resonance structures. The resonance hybrid shows the delocalization of electron density due to the different locations of electrons in individual resonance structures.

Resonance structures (Sections 1.6, 12.2): Two or more structures of a molecule that differ in the placement of π bonds and nonbonded electrons. The placement of atoms and σ bonds stays the same.

Restriction endonuclease (Section 26.7): An enzyme that cleaves DNA at a specific sequence of bases.

Retention of configuration (Section 7.11C): The same relative stereochemistry of a stereogenic center in the reactant and the product of a chemical reaction.

Retention time (Section A.5B): The length of time required for a component of a mixture to travel through a chromatography column.

Retro-aldol reaction (Section 18.1B): The reverse of an aldol reaction in which a β-hydroxy aldehyde or ketone is converted to two carbonyl compounds by cleavage of the carbon–carbon bond between the α and β carbons.

Retro Diels–Alder reaction (Section 12.14B): The reverse of a Diels–Alder reaction in which a cyclohexene is cleaved to give a 1,3-diene and an alkene.

Retrosynthetic analysis (Section 10.21): Working backwards from a product to determine the starting material from which it is made.

RF radiation (Section C.1A): Radiation in the radiofrequency region of the electromagnetic spectrum, characterized by long wavelength and low frequency and energy.

Ribonucleic acid (RNA, Sections 26.1, 26.5): The nucleic acid that translates the genetic information contained in DNA into proteins needed for all cellular functions. Three types of RNA are involved in protein synthesis: ribosomal RNA (rRNA), messenger RNA (mRNA), and transfer RNA (tRNA).

Ribonucleoside (Section 26.1A): An N-glycoside formed by the reaction of D-ribose with certain amine heterocycles.

Ribonucleotide (Section 26.1A): An RNA building block having a ribose and either a purine or pyrimidine base joined by an N-glycosidic linkage, and a phosphate bonded to a hydroxy group of the sugar nucleus.

Ring-closing metathesis (Section 28.6): An intramolecular olefin metathesis reaction using a diene starting material, which results in ring closure.

Ring current (Section C.4): A circulation of π electrons in an aromatic ring caused by the presence of an external magnetic field.

Ring-flipping (Section 4.12B): A stepwise process in which one chair conformation of cyclohexane interconverts with a second chair conformation.

Ring-opening metathesis polymerization (Problem 28.38): An olefin metathesis reaction that forms a high-molecular-weight polymer from certain cyclic alkenes.

Robinson annulation (Section 18.10): A ring-forming reaction that combines a Michael reaction with an intramolecular aldol reaction to form a cyclohex-2-enone.

R,S System of nomenclature (Section 5.6): A system of nomenclature that distinguishes the stereochemistry at a tetrahedral stereogenic center by assigning a priority to each group connected to the stereogenic center. R indicates a clockwise orientation of the three highest-priority groups and S indicates a counterclockwise orientation of the three highest groups. The system is also called the Cahn–Ingold–Prelog system.

Rule of endo addition (Section 12.13D): The rule that the endo product is preferred in a Diels–Alder reaction.

Sandmeyer reaction (Section 22.13A): A reaction between an aryl diazonium salt and a copper(I) halide to form an aryl halide (C_6H_5Cl or C_6H_5Br).

Saponification (Section 16.10B): Basic hydrolysis of an ester to form an alcohol and a carboxylate anion.

Saturated fatty acid (Section 10.6A): A fatty acid having no carbon–carbon double bonds in its long hydrocarbon chain.

Saturated hydrocarbon (Section 4.1): A compound that contains only C–C and C–H σ bonds and no rings, thus having the maximum number of hydrogen atoms per carbon.

Schiff base (Section 14.10): A compound having the general structure $R_2C=NR'$. A Schiff base is also called an imine.

Secondary (2°) alcohol (Section 3.2): An alcohol having the general structure R_2CHOH.

Secondary (2°) alkyl halide (Section 3.2): An alkyl halide having the general structure R_2CHX.

Secondary (2°) amide (Section 3.2): An amide having the general structure $RCONHR'$.

Secondary (2°) amine (Section 3.2): An amine having the general structure R_2NH.

Secondary (2°) carbocation (Section 7.13): A carbocation having the general structure R_2CH^+.

Secondary (2°) carbon (Section 3.2): A carbon atom that is bonded to two other carbon atoms.

Secondary (2°) hydrogen (Section 3.2): A hydrogen that is attached to a 2° carbon.

Secondary protein structure (Section 23.8B): The three-dimensional conformations of localized regions of a protein.

Secondary (2°) radical (Section 21.1): A radical having the general structure $R_2CH\cdot$.

Second-order rate equation (Sections 6.9B, 7.10): A rate equation in which the reaction rate depends on the concentration of two reactants.

Separatory funnel (Section 15.10): An item of laboratory glassware used for extractions.

Septet (Section C.6C): An NMR signal that is split into seven peaks caused by six nearby nonequivalent protons.

Sesquiterpene (Section 25.7A): A terpene that contains 15 carbons and three isoprene units. A sesquiterpenoid also contains at least one oxygen atom.

Sesterterpene (Section 25.7A): A terpene that contains 25 carbons and five isoprene units. A sesterterpenoid also contains at least one oxygen atom.

Sextet (Section C.6C): An NMR signal that is split into six peaks caused by five nearby nonequivalent protons.

Sharpless asymmetric epoxidation (Section 11.14): An enantioselective oxidation reaction that converts the double bond of an allylic alcohol to a predictable enantiomerically enriched epoxide.

Sharpless reagent (Section 11.14): The reagent used in the Sharpless asymmetric epoxidation. The Sharpless reagent consists of *tert*-butyl hydroperoxide, a titanium catalyst, and one enantiomer of diethyl tartrate.

Shielding effects (Section C.3A): An effect in NMR caused by small induced magnetic fields of electrons in the opposite direction to the applied magnetic field. Shielding decreases the strength of the magnetic field felt by the nucleus and shifts an absorption upfield.

1,2-Shift (Section 9.9): Rearrangement of a less stable carbocation to a more stable carbocation by the shift of a hydrogen atom or an alkyl group from one carbon atom to an adjacent carbon atom.

Sigma (σ) bond (Section 1.9A): A cylindrically symmetrical bond that concentrates the electron density on the axis that joins two nuclei. All single bonds are σ bonds.

Sigmatropic rearrangement (Section 29.1): A pericyclic reaction in which a σ bond is broken in the reactant, the π bonds rearrange, and a σ bond is formed in the product.

Silyl ether (Section 13.12): A common protecting group for an alcohol in which the O–H bond is replaced by an O–Si bond.

Simmons–Smith reaction (Section 28.5): Reaction of an alkene with CH_2I_2 and Zn(Cu) to form a cyclopropane.

Singlet (Section C.6A): An NMR signal that occurs as a single peak.

Skeletal structure (Section 1.8B): A shorthand representation of the structure of an organic compound in which carbon atoms and the hydrogen atoms bonded to them are omitted. All heteroatoms and the hydrogens bonded to them are drawn in. Carbon atoms are assumed to be at the junction of any two lines or at the end of a line.

S_N1 mechanism (Sections 7.10, 7.12): A nucleophilic substitution mechanism that goes by a two-step process involving a carbocation intermediate. S_N1 is an abbreviation for "Substitution Nucleophilic Unimolecular."

S_N2 mechanism (Sections 7.10, 7.11): A nucleophilic substitution mechanism that goes by a one-step concerted process, where both reactants are involved in the transition state. S_N2 is an abbreviation for "Substitution Nucleophilic Bimolecular."

Soap (Sections 3.6, 16.11B): The carboxylate salts of long-chain fatty acids prepared by the basic hydrolysis or saponification of a triacylglycerol.

Solubility (Section 3.4C): A measure of the extent to which a compound dissolves in a liquid.

Solute (Section 3.4C): The compound that is dissolved in a liquid solvent.

Solvent (Section 3.4C): The liquid component into which the solute is dissolved.

Specific rotation (Section 5.12C): A standardized physical constant for the amount that a chiral compound rotates plane-polarized light. Specific rotation is denoted by the symbol [α] and defined using a specific sample tube length (*l* in dm), concentration (*c* in g/mL), temperature (25 °C), and wavelength (589 nm). $[\alpha] = \alpha/(l \times c)$

Spectator ion (Section 2.1): An ion that does not take part in a reaction and is opposite in charge to the ion that does take part in a reaction. A spectator ion is also called a counterion.

Spectroscopy (Section A.1): An analytical method using the interaction of electromagnetic radiation with molecules to determine molecular structure.

Sphingomyelin (Section 25.4B): A hydrolyzable phospholipid derived from sphingosine.

Spin flip (Section C.1A): In NMR spectroscopy, when an atomic nucleus absorbs RF radiation and its magnetic field flips relative to the external magnetic field.

Spin–spin splitting (Section C.6): Splitting of an NMR signal into peaks caused by nonequivalent protons on the same carbon or adjacent carbons.

Spiro ring system (Appendix D): A compound having two rings that share a single carbon atom.

Staggered conformation (Section 4.9): A conformation of a molecule in which the bonds on one carbon bisect the R—C—R bond angle on the adjacent carbon.

Step-growth polymer (Section 30.1): A polymer formed when monomers containing two functional groups come together with loss of a small molecule such as water or HCl. Step-growth polymers are also called condensation polymers.

Stereochemistry (Sections 4.9, 5.1): The three-dimensional structure of molecules.

Stereogenic center (Section 5.3): A site in a molecule at which the interchange of two groups forms a stereoisomer. A carbon bonded to four different groups is a tetrahedral stereogenic center. A tetrahedral stereogenic center is also called a chirality center, a chiral center, and an asymmetric center.

Stereoisomers (Sections 4.13B, 5.1): Two isomers that differ only in the way the atoms are oriented in space.

Stereoselective reaction (Section 8.5): A reaction that yields predominantly or exclusively one stereoisomer when two or more stereoisomers are possible.

Stereospecific reaction (Section 10.14): A reaction in which each of two stereoisomers of a starting material yields a particular stereoisomer of a product.

Steric hindrance (Section 7.8B): A decrease in reactivity resulting from the presence of bulky groups at the site of a reaction.

Steric strain (Section 4.10): An increase in energy resulting when atoms in a molecule are forced too close to one another.

Steroid (Sections 12.14C, 25.8): A tetracyclic lipid composed of three six-membered rings and one five-membered ring.

Straight-chain alkane (Section 4.1A): An acyclic alkane that has all of its carbons in a row. Straight-chain alkanes are also called normal alkanes.

Structural isomers (Sections 4.1A, 5.2): Two compounds that have the same molecular formula but differ in the way the atoms are connected to each other. Structural isomers are also called constitutional isomers.

Substituent (Section 4.4): A group or branch attached to the longest continuous chain of carbons in an organic molecule.

Substitution reaction (Section 6.2A): A reaction in which an atom or a group of atoms is replaced by another atom or group of atoms. Substitution reactions involve σ bonds: one σ bond breaks and another is formed at the same atom.

Substrate (Section 6.11): An organic molecule that is transformed by the action of an enzyme.

Sulfide (Section 9.15): A compound having the general structure RSR'.

Sulfonation (Section 20.4): An electrophilic aromatic substitution reaction in which benzene reacts with $^+SO_3H$ to give a benzenesulfonic acid, $C_6H_5SO_3H$.

Suprafacial reaction (Section 29.4): A pericyclic reaction that occurs on the same side of the two ends of the π electron system.

Suzuki reaction (Section 28.2): The palladium-catalyzed coupling of an organic halide (R'X) with an organoborane (RBY$_2$) to form a product R—R'.

Symmetrical anhydride (Section 16.1): An anhydride that has two identical alkyl groups bonded to the carbonyl carbon atoms.

Symmetrical ether (Section 9.1): An ether with two identical alkyl groups bonded to the oxygen.

Syn addition (Section 10.8): An addition reaction in which two parts of a reagent are added from the same side of a double bond.

Syn dihydroxylation (Section 11.9B): The addition of two hydroxy groups to the same face of a double bond.

Syndiotactic polymer (Section 30.4): A polymer having the substituents alternating from one side of the backbone of an elongated polymer chain to the other.

Syn periplanar (Section 8.8): In an elimination reaction, a geometry in which the β hydrogen and the leaving group are on the same side of the molecule.

Systematic name (Section 4.3): The name of a molecule indicating the compound's chemical structure. The systematic name is also called the IUPAC name.

T

Target compound (Section 10.21): The final product of a synthetic scheme.

Tautomerization (Sections 10.18, 17.2A): The process of converting one tautomer to another.

Tautomers (Section 10.18): Constitutional isomers that are in equilibrium and differ in the location of a double bond and a hydrogen atom.

Terminal alkyne (Section 10.1): An alkyne that has the triple bond at the end of the carbon chain.

C-Terminal amino acid (Section 23.4A): The amino acid at the end of a peptide chain with a free carboxy group.

N-Terminal amino acid (Section 23.4A): The amino acid at the end of a peptide chain with a free amino group.

Termination (Section 21.4A): The final step of a chain reaction. In a radical chain mechanism, two radicals combine to form a stable bond.

Terpene (Section 25.7): A hydrocarbon composed of repeating five-carbon isoprene units.

Terpenoid (Section 25.7): A lipid that contains isoprene units as well as at least one oxygen heteroatom.

Tertiary (3°) alcohol (Section 3.2): An alcohol having the general structure R$_3$COH.

Tertiary (3°) alkyl halide (Section 3.2): An alkyl halide having the general structure R$_3$CX.

Tertiary (3°) amide (Section 3.2): An amide having the general structure RCONR'$_2$.

Tertiary (3°) amine (Section 3.2): An amine having the general structure R$_3$N.

Tertiary (3°) carbocation (Section 7.13): A carbocation having the general structure R$_3$C$^+$.

Tertiary (3°) carbon (Section 3.2): A carbon atom that is bonded to three other carbon atoms.

Tertiary (3°) hydrogen (Section 3.2): A hydrogen that is attached to a 3° carbon.

Tertiary protein structure (Section 23.8C): The three-dimensional shape adopted by an entire peptide chain.

Tertiary (3°) radical (Section 21.1): A radical having the general structure $R_3C\cdot$.

Tesla (Section C.1A): A unit used to measure the strength of a magnetic field. Tesla is denoted with the symbol "T."

Tetramethylsilane (Section C.1B): An internal standard used as a reference in NMR spectroscopy. The tetramethylsilane (TMS) reference peak occurs at 0 ppm on the δ scale.

Tetrasubstituted alkene (Section 8.2A): An alkene that has four alkyl groups and no hydrogens bonded to the carbons of the double bond ($R_2C=CR_2$).

Tetraterpene (Section 25.7A): A terpene that contains 40 carbons and eight isoprene units. A tetraterpenoid contains at least one oxygen atom as well.

Tetrose (Section 24.2): A monosaccharide containing four carbons.

Thermodynamic enolate (Section 17.4): The enolate that is lower in energy—generally the more substituted enolate.

Thermodynamic product (Section 12.11): In a reaction that can give more than one product, the product that predominates at equilibrium.

Thermodynamics (Section 6.5): A study of the energy and equilibrium of a chemical reaction.

Thermoplastics (Section 30.7): Polymers that can be melted and then molded into shapes that are retained when the polymer is cooled.

Thermosetting polymer (Section 30.7): A complex network of cross-linked polymer chains that cannot be re-melted to form a liquid phase.

Thioester (Section 16.16): A compound with the general structure RCOSR'.

Thiol (Section 9.15): A compound having the general structure RSH.

Tollens reagent (Sections 13.8, 24.9B): A reagent that oxidizes aldehydes, and consists of silver(I) oxide in aqueous ammonium hydroxide. A Tollens test is used to detect the presence of an aldehyde.

***p*-Toluenesulfonate** (Section 9.13): A very good leaving group having the general structure $CH_3C_6H_4SO_3^-$ and abbreviated as TsO^-. Compounds containing a *p*-toluenesulfonate leaving group are called alkyl tosylates and are abbreviated ROTs.

Torsional energy (Section 4.9): The energy difference between the staggered and eclipsed conformations of a molecule.

Torsional strain (Section 4.9): An increase in the energy of a molecule caused by eclipsing interactions between groups attached to adjacent carbon atoms.

Tosylate (Section 9.13): A very good leaving group having the general structure $CH_3C_6H_4SO_3^-$, and abbreviated as TsO^-.

***s*-Trans** (Sections 12.6, 23.4B): The conformation of a 1,3-diene that has the two double bonds on opposite sides of the single bond that joins them.

Transcription (Section 26.3): The ordered synthesis of RNA from DNA in which the genetic information stored in DNA is passed onto RNA.

Trans diaxial (Section 8.8B): In an elimination reaction of a cyclohexane, a geometry in which the β hydrogen and the leaving group are trans with both in the axial position.

Transferase (Section 23.10A): An enzyme that catalyzes the transfer of a group from one molecule to another.

Trans isomer (Sections 4.13B, 8.3B): An isomer of a ring or double bond that has two groups on opposite sides of the ring or double bond.

Transition state (Section 6.7): An unstable energy maximum as a chemical reaction proceeds from reactants to products. The transition state is at the top of an energy "hill" and can never be isolated.

Translation (Section 26.3): The synthesis of proteins from RNA in which the genetic message contained in RNA determines the specific amino acid sequence of the protein.

Triacylglycerol (Sections 10.6, 16.11A, 25.3): A lipid consisting of the triester of glycerol with three long-chain fatty acids. Triacylglycerols are the lipids that comprise animal fats and vegetable oils. Triacylglycerols are also called triglycerides.

Triose (Section 24.2): A monosaccharide containing three carbons.

Triphosphate (Section 7.16): A good leaving group used in biological systems. Triphosphate ($P_3O_{10}^{5-}$) is abbreviated as PPP. The term "triphosphate" is also used for an organic triphosphate having the general structure $ROP_3O_9^{4-}$.

Triplet (Section C.6): An NMR signal that is split into three peaks having a relative area of 1:2:1, caused by two nearby nonequivalent protons.

Trisubstituted alkene (Section 8.2A): An alkene that has three alkyl groups and one hydrogen bonded to the carbons of the double bond ($R_2C=CHR$).

Triterpene (Section 25.7A): A terpene that contains 30 carbons and six isoprene units. A triterpenoid contains at least one oxygen atom as well.

U

Unimolecular reaction (Sections 6.9B, 7.10, 7.12A): A reaction that has only one reactant involved in the rate-determining step, so the concentration of only one reactant appears in the rate equation.

α,β-Unsaturated carbonyl compound (Section 13.15): A conjugated compound containing a carbonyl group and a carbon–carbon double bond separated by a single σ bond.

Unsaturated fatty acid (Section 10.6A): A fatty acid having one or more carbon–carbon double bonds in its hydrocarbon chain. In natural fatty acids, the double bonds generally have the Z configuration.

Unsaturated hydrocarbon (Section 10.2): A hydrocarbon that has fewer than the maximum number of hydrogen atoms per carbon atom. Hydrocarbons with π bonds or rings are unsaturated.

Unsymmetrical ether (Section 9.1): An ether in which the two alkyl groups bonded to the oxygen are different.

Upfield shift (Section C.1B): In an NMR spectrum, a term used to describe the relative location of an absorption signal. An upfield shift means a signal is shifted to the right in the spectrum to lower chemical shift.

Urethane (Section 30.6C): A compound that contains a carbonyl group bonded to both an OR group and an NHR (or NR_2) group. A urethane is also called a carbamate.

V

Valence bond theory (Section 19.10A): A theory that describes covalent bonding as the overlap of two atomic orbitals with the electron pair in the resulting bond being shared by both atoms.

Valence electrons (Section 1.1): The electrons in the outermost shell of orbitals. Valence electrons determine the properties of a given element. Valence electrons are loosely held and participate in chemical reactions.

Van der Waals forces (Section 3.3B): Very weak intermolecular interactions caused by momentary changes in electron density in molecules. The changes in electron density cause temporary dipoles, which are attracted to temporary dipoles in adjacent molecules. Van der Waals forces are also called London forces.

Vicinal dihalide (Section 8.10): A compound that has two halogen atoms on adjacent carbon atoms.

Vinyl group (Section 10.3B): An alkene substituent having the structure —CH=CH$_2$.

Vinyl halide (Section 7.1): A molecule containing a halogen atom bonded to the sp^2 hybridized carbon of a carbon–carbon double bond.

Virus (Section 26.9): An infectious agent consisting of a DNA or RNA molecule that is contained within a protein coating. A virus replicates when it invades a host organism and takes over the biochemical machinery of the host.

Vitamins (Sections 3.5, 25.5): Organic compounds needed in small amounts by biological systems for normal cell function.

VSEPR theory (Section 1.7B): Valence shell electron pair repulsion theory. A theory that determines the three-dimensional shape of a molecule by the number of groups surrounding a central atom. The most stable arrangement keeps the groups as far away from each other as possible.

W

Walden inversion (Section 7.11C): The inversion of a stereogenic center involved in an S$_N$2 reaction.

Wavelength (Section B.1): The distance from one point of a wave to the same point on the adjacent wave. Wavelength is abbreviated with the Greek letter lambda (λ).

Wavenumber (Section B.2): A unit for the frequency of electromagnetic radiation that is inversely proportional to wavelength. Wavenumber, reported in reciprocal centimeters (cm^{-1}), is used for frequency in IR spectroscopy.

Wax (Section 25.2): A hydrolyzable lipid consisting of an ester formed from a high-molecular-weight alcohol and a fatty acid.

Williamson ether synthesis (Section 9.6): A method for preparing ethers by reacting an alkoxide (RO$^-$) with a methyl or primary alkyl halide.

Wittig reaction (Section 14.9): A reaction of a carbonyl group and an organophosphorus reagent that forms an alkene.

Wittig reagent (Section 14.9A): An organophosphorus reagent having the general structure Ph$_3$P=CR$_2$.

Wohl degradation (Section 24.10A): A reaction that shortens the carbon chain of an aldose by removing one carbon from the aldehyde end.

Wolff–Kishner reduction (Section 20.14C): A method to reduce aryl ketones to alkyl benzenes using hydrazine (NH$_2$NH$_2$) and strong base (KOH).

Woodward–Hoffmann rules (Section 29.3): A set of rules based on orbital symmetry used to explain the stereochemical course of pericyclic reactions.

Y

Ylide (Section 14.9A): A chemical species that contains two oppositely charged atoms bonded to each other, and both atoms have octets of electrons.

Z

Zaitsev rule (Section 8.5): In a β elimination reaction, a rule that states that the major product is the alkene with the most substituted double bond.

Ziegler–Natta catalysts (Section 30.4): Polymerization catalysts prepared from an organoaluminum compound and a Lewis acid such as TiCl$_4$, which afford polymer chains without significant branching and with controlled stereochemistry.

Zwitterion (Sections 3.9A, 15.12A): A neutral compound that contains both a positive and negative charge.

Index

Page numbers followed by "f" indicate figures; those followed by "t" indicate tables. An "A-" before page numbers indicates appendix pages. Online chapters (28–30) are also included within this index and begin on page 1197.

A

Abbreviations
  alkyl groups, 138
  amino acids, 1022, 1023f
  aromatic rings, 854
  list, A-2-3
  nucleic acid bases, 1145
  nucleotides, 1147
ABS (copolymer), 1263
Absorption spectra
  infrared [*See* Infrared (IR) absorptions/spectra]
  NMR (*See* Carbon NMR; Proton NMR)
Acetaldehyde
  formation of, 459–460
  hydrate formation, 673
  3-hydroxybutanal reactions, 813
  nomenclature, 649, 649t
Acetals
  cyclic hemiacetal conversion to, 681–682
  formation of, 675–676
  glycosides, 1087
  hydrolysis of, 677–678
  as protecting groups, 679
Acetamide
  nomenclature, 739
  structure of, 734
Acetaminophen, 761
Acetate
  nucleophilic attack of, 274
  resonance structures, 72, 72f, 478
Acetic acid
  acidity, 72
  benzyl ester removal, 1041
  as Brønsted–Lowry acid, 59f
  as common acid, 76
  deprotonation of, 706
  description of, 702
  dimers, 700f
  in halogenation, 791
  inductive effects in, 710
  melting point, 106
  nomenclature, 698, 737
  pK_a, 707
  as polar protic solvent, 265f
  structure of, 734
Acetic anhydride, 734, 738, 1091
Acetic benzoic anhydride, 738
Acetoacetic ester synthesis, 800–802
Acetone
  boiling point, 104, 104f
  description of, 654
  dipole–dipole interactions, 102
  hydrate formation, 673
  nomenclature, 651
  as polar aprotic solvent, 285t

  solubility, 107–108
  as solvent, 266
Acetone enolate, 790, 790f
Acetonitrile, 699
Acetophenone, 651
N-Acetyl amino acids, 1026
Acetyl AMP, 766
Acetylation, 751
Acetyl chloride, 734, 994f, 1091
Acetylcholine, 107, 767, 1246
Acetyl CoA
  acetylcholine synthesis, 766–767
  in catabolism, 1170–1171, 1171f
  citramalate from, 831
  citrate synthesis, 831
  in citric acid cycle, 1186–1191, 1187f
  glucosamine acetylation, 1103
  pyruvate conversion to, 1184–1185
  synthesis from myristic acid, 1178–1179
Acetylene
  acidity, 73
  as alkyne, 93
  bond length and strength, 42t
  bonds in, 40, 40f, 41t
  internal alkyne formation, 419
  molecular shape, 25
  nomenclature, 389
  pK_a values, 66
N-Acetyl-D-glucosamine (NAG), 1103
Acetyl groups, 651, 1091, 1170–1171
Acetylide anions
  alkyl halides, reactions with, 418–419
  epoxide reactions, 420
  formation of, 398, 417
  as organometallic compounds, 615
  reactions, 417–420
  retrosynthetic analysis, 422–423
  S_N2 substitution reactions, 417–420
  as strong nucleophiles, 398
N-Acetyl phenylalanine, 1030
Acetyl phosphate, 734
Acetylsalicylic acid. *See* Aspirin
Achiral carbocations, 275, 402
Achiral enolates, 790, 794–795
Achiral epoxides, 451
Achiral molecules
  in carbonyl reduction, 606
  definition, 180
  description, 179–182, 182f
  in enolate alkylation, 794
  in epoxidation, 448–449, 460
  in halogenation reactions, 946t, 947
  in hydrohalogenation, 401–402
  meso compounds, 196–197, 199
  as optically inactive, 201
Achiral reactants, 369–370, 402
Acid–base reactions
  in acetoacetic ester synthesis, 801
  amide hydrolysis and, 760–761
  amines in, 985–987
  aspirin, 78
  Brønsted–Lowry, 60–67
  carboxylic acids, 705–708, 755–756
  deprotonation, 66–67, 705–708
  description of, 60–63

  enolate formation, 786–787
  equilibrium position, 66–67
  in extraction technique, 712–713, 986–987
  Henderson–Hasselbalch equation, 708–710, 709f
  identification of acid and base in, 62–63, 63f
  Lewis acid–base reactions, 79–81
Acid-catalyzed reactions
  cyclic hemiacetal formation, 680–681
  dehydration, 351
  Fischer esterification, 753–755
  hydration, 657, 673–675
Acid chlorides
  amide synthesis from, 749–750, 993–994
  description of, 599
  in Friedel–Crafts acylation, 892
  functional group, 97, 98t
  nomenclature, 737, 740t
  nucleophilic substitution, 601–603
  organocuprate reactions with, 626–627
  organometallic reagent reactions with, 616, 625–627
  reduction of, 608–610
  structure of, 734
  as weak base, 737
Acidic amino acids, 1022, 1023f
Acidity
  basicity, inverse relationship with, 64
  element effects, 68–70
  factors determining, 67–76
  hybridization effects, 73
  inductive effects, 70–71
  resonance effects, 72–73
Acidity constant (K_a), 63–64
Acid rain, 59, 76
Acids
  in acetal hydrolysis, 677
  amino (*See* Amino acids)
  Brønsted–Lowry (*See* Brønsted–Lowry acids)
  as catalysts, 243
  common, 76
  conjugate, 60–62, 64
  conversion of alcohols to alkyl halides, 356–359
  dehydration of alcohols to alkenes, 350–355
  in epoxide ring opening reactions, 370–372
  ester hydrolysis to carboxylic acids, 756–757
  origin of word, 59
  reaction of ethers with strong, 364–365
  strength, 63–65
  tautomerization in, 782
Aconitase, 1187f
cis-Aconitate, 1188
Acquired immune deficiency syndrome (AIDS), 1162

Acquired immune deficiency syndrome (AIDS) treatments. *See* HIV treatments
Acrolein, 493
Acropora nasuta, 390
Acrylonitrile, 1261f
Activation energy. *See* Energy of activation (E_a)
Active methylene compounds, 820
Active site, of enzyme, 244
Acyclic acid chlorides, 737
Acyclic acyl phosphates, 737
Acyclic alkanes
  branched-chain, 134
  conformations of, 147–153
  constitutional isomers, 134, 135t
  having more than five carbon atoms, 135
  having one to five carbon atoms, 133–134
  homologous series, 135, 135t
  molecular formula, 133
  nomenclature, 138–142
  normal (*n*) (straight-chain), 134, 135t
  straight-chain, 134
Acyclic alkenes, 386
Acyclic ketones, 650–651
Acyl adenosyl phosphate, 764
Acylase, 1028
Acylation, Friedel–Crafts. *See* Friedel–Crafts acylation
Acyl derivatives. *See* Carboxylic acid derivatives; *specific classes*
Acyl groups
  basicity of, 736
  in carboxylic acids and derivatives, 734
  conjugate acids for, 736t
  in esters, 738
  in Friedel–Crafts acylation, 892
  naming, 651–652
  resonance hybrids of, 736
  in thioesters, 738
Acylium ion, 893–894
Acyl phosphates
  carboxylate anion conversion to, 763–764
  description of, 735
  functional group, 100, 100t
  in glycolysis, 1182
  nomenclature, 737, 740t
  reduction with NADH, 609–610
  structure of, 734
  thioester from, 765–766, 1170–1171
Acyl transfer reactions, 766
Addition polymers, 1256, 1257–1263
  synthesis of (*See* Chain-growth polymerization)
Addition reactions
  acetal formation, 675–678
  in alkenes, A-16
  but-1-yne, 398f, 412–417
  cyclohexene, 397f, 399–412
  cyclopropanation, 1209
  description of, 222

Index

Addition reactions—(Cont.)
  Diels–Alder reactions, 490–498
    (See also Diels–Alder reactions)
  dihydroxylation, 451–453
  electrophilic, 399–404, 403t
  endo addition, 494–495
  epoxidation, 447–450, 460–462
  halogenation, 397f, 398f, 404–407, 406f, 413
  halohydrin formation, 397f, 407–408
  hydration, 397f, 403–404, 412, 673–675
  hydroboration–oxidation, 397f, 408–412
  hydrohalogenation, 397f, 398f, 403t, 413
  ionic, 956
  nucleophilic (See Nucleophilic addition reactions)
  in nucleophilic aromatic substitution, 913–916
  oxidative, 1201–1207
  radical, to double bonds, 955–957
  stereochemistry of, 397
  via ionic intermediates, 951–952
Adenine
  A–T base pair, 1151, 1151f
  as DNA base, 849
  nomenclature, 1146, 1147t
  pi (π) elections in, 868
  in polynucleotide structure, 1149f
  purine-derived, 1145
Adenosine, 1147t
Adenosine 5'-diphosphate (ADP), 715, 1148, 1172, 1175, 1182–1184
Adenosine 5'-monophosphate (AMP), 766, 1147t
Adenosine 5'-triphosphate (ATP)
  aerobic glucose metabolism and, 1190
  carboxylate anion conversion to acyl phosphate, 763–764
  in catabolism, 1171f, 1172
  citric acid cycle and, 1186–1191, 1187f
  fatty acid oxidation and, 1190–1191
  glucose phosphorylation, 233, 234f, 1172
  in glycolysis, 1182–1184
  hydrolysis of, 230, 1172
  oxidative phosphorylation, 1172, 1175
  reactivity, 117
  synthesis from nucleosides, 1148
  triphosphate group in, 100, 368
S-Adenosylmethionine (SAM), 288, 289f
  reactivity, 117
  synthesis, 368
-adiene (suffix), 388
Adipic acid, 1267, 1274–1275, 1275f
ADP. See Adenosine 5'-diphosphate
Adrenal cortical steroids, 1137
Adrenaline
  as amine, 97, 977
  biological synthesis, 288, 289f
Advil, 137, 206
Agent Orange, 889f
Aglycon, 1090
AIDS (acquired immune deficiency syndrome), 1162
AIDS treatments. See HIV treatments
Alanine
  abbreviation for, 716t
  N-acetyl, 1026–1029
  acid–base properties, 717–718
  enantiomers of, 185–186, 1026–1029
  isoelectric point, 719, 1025t
  peptide formation with cysteine, 1031–1032
  physical properties of, 203
  pK_a values, 716t, 1025t
  structure of, 1023f
Albuterol, 183–184, 372, 372f, 373
Alcohols. See also specific alcohols
  acetal formation, 675–677
  alkene synthesis, 350–355
  alkyl halides, conversion to, 356–361
  alkyl tosylates, conversion to, 361–363
  alpha (α) cleavage, 519
  anhydride reactions, 751
  biological oxidation, 458–460
  boiling point, 344t
  carbocation rearrangement during dehydration, 352–355
  in carbonate synthesis, 1270
  classification, 96
  dehydration reactions of, 350–355, 519–520 (See also Dehydration reactions)
  description of, 338
  elimination reactions, 349, 350–355, 362, 364f
  β elimination reactions, 349, 350–355
  E1 mechanism for dehydration, 351–352
  E2 mechanism for dehydration, 352, 355
  epoxide reduction to, 445–446
  ether formation, 404
  in Friedel–Crafts alkylation, 896
  functional group, 95, 96t
  glycosides from, 1087–1088
  hydration of alkenes, 403–404
  hydroboration–oxidation of alkenes, 408–412, 409f, 411t, 412f
  infrared (IR) absorptions/spectra of, 540, 543t
  leaving groups of, 349, 351, 359–360
  Le Châtelier's principle and, 352
  mass spectra of, 519–520
  melting point, 344t
  nomenclature, 339–341, 340f
  nucleophilic substitution, 349, 356–362, 364f
  organometallic reagent reactions with epoxides, 628–629
  oxidation of, 447f, 456–460
  phosphorus tribromide, conversion to alkyl halides with, 360, 360t
  phosphorus oxychloride and pyridine dehydration of, 355
  physical properties of, 343–344, 344t
  primary [See Primary (1°) alcohols]
  reactions of, 346–361
  secondary [See Secondary (2°) alcohols]
  solubility, 344t
  structure of, 339
  synthesis of, 290, 290t, 346–347, 403–404
  tertiary [See Tertiary (3°) alcohols]
  thionyl chloride, conversion to alkyl halides with, 359–360, 360t
Aldaric acids, 1092, 1093, 1096–1097
Aldehydes
  acetal formation from, 675–677
  acetal hydrolysis to, 677–678
  acid chloride reduction to, 608–610, 655t
  alkylation of, 793–795
  amine synthesis from, 982–985
  boiling point, 652t
  catalytic hydrogenation, 606
  crossed aldol reactions, 819–820
  cyanohydrin formation, 660–662
  description of, 599
  in directed aldol reactions, 821–822
  as electron-withdrawing group, 899–900
  enamine formation, 668, 669f, 993
  enamine hydrolysis to, 669–670
  ester reduction to, 608–610, 655t
  formation of, 453, 655, 655t
  functional group, 97, 98t
  general reactions of, 656–658
  α-halo aldehyde, 791
  halogenation at the α carbon, 791
  hydride reduction, 604–608, 659
  hydroboration–oxidation of alkyne to, 417, 655
  α-hydroxy, 782
  β-hydroxy carbonyl compound from, 813–818
  imine formation, 666–667, 669f, 982, 993
  imine hydrolysis to, 669–670
  infrared (IR) absorptions/spectra of, 541, 543t
  interesting compounds, 654
  mass spectrum of, 518–519
  melting point, 652t
  metal hydride reagents, 605
  nitrile reduction synthesis, 721–722
  nomenclature, 648–650
  nucleophilic addition, 600, 601, 617–620, 656–658, 658f, 659–660
  organometallic reagent reactions, 617–620, 659–660
  oxidation of, 613, 705t
  oxidation to carboxylic acid, 1092–1093
  physical properties, 652, 652t
  pK_a values, 785t
  primary alcohol oxidation to, 456–458, 655t
  protecting groups, 679
  reduction to primary alcohol, 440, 604–608, 659
  reductive amination of, 982–985
  solubility, 652t
  spectroscopic properties, 653, 653t
  structure of, 647
  Wittig reaction, 662–666
-aldehyde (suffix), 648
Alder, Kurt, 490
Alditol, 1092
Aldohexoses
  D-aldohexoses, 1077, 1078f
  description of, 1071
  Haworth projections of, 1082–1085
  stereochemistry of, 1075, 1077
Aldol reactions
  biological, 830–831
  in citric acid cycle, 1187
  crossed, 818–821, 820f
  dehydration of aldol product, 816–817
  directed, 821–822
  general features of, 813–815
  intramolecular, 822–825, 835–839
  nucleophilic addition in, 814, 814f
  retro-aldol reactions, 815–816, 831–832, 1180f, 1181
  retrosynthetic analysis, 817–818
  in Robinson annulation, 835–839
Aldonic acids, 1092, 1093
D-Aldopentoses, 1077, 1078f, 1096
Aldoses
  D-aldose family, 1077, 1078f
  in Kiliani–Fischer synthesis, 1094, 1095–1097
  oxidation of, 1092–1094
  reduction of, 1092
  structure of, 1071
  in Wohl degradation, 1094–1096
Aldosterone, 1137
Aldotetroses, 1077, 1078f, 1096–1097
Aldotrioses, 1071
Aleve, 207
Aliphatic carboxylic acids, 710–711
Aliphatic hydrocarbons, 93. See also Alkanes; Alkenes; Alkynes
Aliskiren, 184
Alizarin, 1006
Alkaloids, 975–976, 976f
Alkanes. See also specific alkanes
  acyclic (See Acyclic alkanes)
  alkyl halide reduction to, 445–446, 616
  alkyl halides from, 939–940
  alkyne reduction to, 442
  boiling point, 146, 146t
  bonds in, 94, 133
  bromination of, 943–945
  chlorination of, 939–940, 943–946
  combustion of, 163, 165–166
  cycloalkanes (See Cycloalkanes)
  definition, 93, 133
  description of, 94t
  halogenation of, 939–940
  infrared (IR) absorptions/spectra of, 539, 543t
  melting point, 146, 146t
  natural occurrence of, 145
  nitrile named as derivatives, 699, 699f
  nomenclature, 137–142, 140f
  in olives and olive leaves, 134
  oxidation of, 163–166
  physical properties, 145–146, 146t
  properties of, 145–146
  radical reactions of, 938
  solubility of, 146, 146t
  substituents, 137–138
Alkene metathesis, 1211–1216
Alkene oxides, 342–343
Alkenes
  addition reactions, 396–397, 397f, 399–412, 955–957
  amino acid synthesis from, 1029–1031
  bromination reactions, 952
  classification, 305, 305f
  common names, 389
  constitutional isomers of, 350
  cyclopropanation, 1207–1211
  definition, 93
  dehydration of alcohols to, 350–355, 395
  description of, 385
  dienophiles, 490–498
  dihydroxylation, 447f, 451–453
  disubstituted (See Disubstituted alkenes)
  double bonds in, 304–307
  elimination reactions and, 303, 304–308, 395
  epoxide formation, 447–450, 460–462
  ether formation, 404
  in Friedel–Crafts alkylation, 896

functional group, 94t
halogenation of, 404–407, 406f
halohydrin formation, 407–408
in Heck reaction, 1205–1207
Hofmann elimination synthesis, 995–997
hybrid orbitals in, 304–305, 307
hydration of, 403–404, 412
hydroboration–oxidation of, 408–412, 409f, 411t, 412f
hydrohalogenation of, 399–403, 403t
infrared (IR) absorptions/spectra of, 539, 543t
metathesis, 1211–1216
monosubstituted, 305, 305f, 310–311, 1211
nomenclature, 387–390
oxidation of, 447–455, 447f, 460–462
oxidative cleavage, 447f, 453–455, 655
physical properties, 391
polymerization of, 1257–1267
production by β elimination, 349, 350–355
radical reactions of, 938, 955–957
reduction to alkanes, 435, 436–442
restricted rotation in, 305–306, 306f
solubility, 391
spin–spin splitting, NMR spectra, 575–577
stability of, 307–308
stereoisomers of, 305–307, 389–390
synthesis of, 303, 395, 995–997, 1205–1207
terminal, 1211
tetrasubstituted, 305
trisubstituted, 305, 305f
Wittig reaction, 662–666
Alkenols, 388
Alkoxides
in enolate formation, 786
in epoxy resin synthesis, 1271
Alkoxy groups
in ethers, 341
nomenclature, A-7
structure of, 96t
Alkylamines
pK_a values, 992t
primary, 997–998
relative basicity, 988–990, 992t
secondary, 999
Alkylated imides, 980
Alkylation
in acetoacetic ester synthesis, 801
direct enolate alkylation, 793–796
Friedel–Crafts (See Friedel–Crafts alkylation)
in malonic ester synthesis, 798
Alkyl benzenes
aryl ketone reduction to, 918–920
bromination of, 917–918
electron density of, 899–900
Friedel–Crafts alkylation synthesis, 891, 919
halogenation of, 917–918
oxidation of, 918
sulfonation of, 891
Alkylborane, 408–412
Alkyl bromides
mass spectra of, 515, 516f
synthesis of, 955
Alkyl chlorides
carbocation rearrangement reaction, 895

in Friedel–Crafts alkylation, 892–893, 895
mass spectra of, 514, 515f
Alkyl diazonium salts, 997–998
Alkyl groups
abbreviations for, 138
in alkyl tosylates, 361
in amines, 970
branching molecules, nomenclature, A-6
in bromination, 917
carbocation stability and, 277–278
description, 137
as electron-donating group, 898–900
in esters, 738
in Friedel–Crafts alkylation, 891
in ketones, 647
on nitriles, 696
nomenclature, 137–138, 389, A-6
as ortho, para director, 922
polarizability, 898
radical stability and, 937
in Suzuki reactions, 1203
in thioesters, 738
Alkyl halides. See also specific alkyl halides
acetylide anions, reactions with, 418–419
alcohol conversion to, 356–361
alkanes to, 939–940
allylic halides, 255, 255f
amine synthesis, 979–981
anti periplanar geometry, 318
benzylic halides, 255, 255f
boiling point, 257t
carbon–halogen bonds in, 259
classification, 96, 255
conversion of alcohols to, with HX, 356–359
conversion of alcohols to, with phosphorus tribromide, 359, 360, 360t
conversion of alcohols to, with thionyl chloride, 359–360, 360t
elimination reactions, 255, 259, 303–304, 308–327, 395
in enolate alkylation, 793–794
in ether synthesis from monosaccharides, 1090
examples, 255f
in Friedel–Crafts alkylation, 891
functional group, 95, 96t
hydrohalogenation and, 399–403, 403t
mass spectra, 514–516
mass spectrometry of, 516f
melting point, 257t
molecular formula, 255
nomenclature, 256–257
nucleophilic substitution reactions (See Nucleophilic substitution reactions; S_N1 substitution reactions; S_N2 substitution reactions)
organocuprate coupling in hydrocarbon synthesis, 1199–1200
physical properties of, 257, 257t
primary [See Primary (1°) alkyl halides]
production by nucleophilic substitution, 349, 356–361
reaction mechanism, determining, 323–327, 324t
reduction of, 445–446, 616
retrosynthetic analysis of malonic ester synthesis, 800

secondary [See Secondary (2°) alkyl halides]
simple, 258, 258f
solubility, 257t
as solvents, 258, 258f
tertiary [See Tertiary (3°) alkyl halides]
unreactive, 255, 255f
1,2-Alkyl shift, 353
Alkylthio groups
structure of, 96t
sulfide nomenclature and, 367
Alkyl titanium compounds, 1265
Alkyl tosylates, 361–363, 364f
Alkynes
addition reactions, 396–397, 398f, 412–417
common names, 389
definition, 93
deprotonation, 398, 398t
description of, 385
elimination reactions, 396
functional group, 94t
halogenation, 398f, 413
hydration, 414–416
hydroboration–oxidation, 416–417
hydrohalogenation, 398f, 413
infrared (IR) absorptions/spectra of, 539, 543t
molecular formula, 386
nomenclature, 387–390
oxidation of, 447f, 455–456
oxidative cleavage, 447f, 455–456, 705t
physical properties, 391
reduction to alkanes, 442
reduction to alkenes, 435, 442–445, 444f
reduction to cis alkene, 443
reduction to trans alkene, 443–444
solubility, 391
synthesis of, 303, 322–323, 323f, 396
terminal, 398, 398t, 417–420
Alkynols, 388
D-Allaric acid, 1094
Allegra, 976–977
Allicin, 136
D-Allose, 1078f, 1094
2-Allylcycloheptanol, 389
Allyl groups, 389
Allylic alcohol, 460, 629
Allylic bromination, 951–952
Allylic carbocations
allyl-type resonance, 477–478
in biological reactions, 476–477
conjugation and, 475
in geranyl diphosphate synthesis, 476–477
resonance-stabilized, 487
resonance structures, 476
stability of, 476
Allylic carbon
oxidation of, 440, 953–954, 953f, 1119
radical halogenation at, 950–953
Allylic halides, 255, 255f, 951
Allylic halogenation
description of, 950–951
product mixtures, 952
selective bromination, 951–952
Allylic oxidation, 953–954, 953f
Allylic radicals, 950, 951, 953, 953f
Allylic sigma bonds, 1240–1241
Allylic substitution, 951–952
Allyl-type resonance, 477–478
Alpha (α) elimination reaction, 1208

Alpha (α) carbon
acetoacetic ester synthesis, 800–802
in amino acids, 118, 1022
in dehydrohalogenation, 303–304
nomenclature, 650
reactions at, 656
substitution reactions of carbonyl compounds, 779–803 (See also Enolates; Enols)
Alpha (α) cleavage, 518–520
Alpha (α) helix, 1046–1048, 1046f
Alpha (α) hydrogen, 780, 784, 818–821, 827–828
-al (suffix), 648
Alternating copolymers, 1263
Altocid, 620
Amber, 1129
Ambident nucleophiles, 787
Ambien, 387
Amide bonds, 1028, 1031–1033, 1036–1038, 1037t, 1045, 1267
Amides
acid chlorides to, 749–750
in alkyne synthesis, 322
anhydrides to, 750–751
as bases, 76, 786
carbonyl absorption in, 536
carboxylic acids to, 755–756
classification, 98, 735
cyclic, 735
description of, 599
esters to, 756
functional group, 97, 98t
hydrogen bonds in, 741, 741f
infrared (IR) absorptions/spectra of, 542, 543t
interesting compounds, 744–745
nomenclature, 739, 740t
nucleophilic acyl substitution, 760–761
nucleophilic substitution, 601–603
pK_a values, 785t, 992t
polymers of, 118, 1257, 1267–1268
primary [See Primary (1°) amides]
as protecting group, 994
reduction of, 610–612, 982
relative basicity, 990, 992t
secondary [See Secondary (2°) amides]
in sphingomyelins, 1122, 1122f
as strong base, 737
structure of, 734
tertiary [See Tertiary (3°) amides]
Amide tautomer, 720
Amikacin, 1104
Amine oxides, 452
Amines
alkaloids, 975–976
alkene synthesis, 995–997
amide reduction to, 610–612, 982
amide synthesis, 993–994
anhydride reactions, 751
anhydride reaction with, 750
aromatic (See Aromatic amines)
aryl diazonium salts reactions, 999–1005
as bases, 985–987
basicity of, 987–993, 992t
boiling point, 974t
carboxylic acid reaction with, 755
classification, 97
decarboxylation of amino acids to, 802–803
description of, 970
dyes, 1004–1007
extraction of, 986–987

I-4 Index

Amines—(Cont.)
  formation of, 260
  functional group, 95, 96t
  Gabriel synthesis, 980–981
  N-glycoside synthesis, 1104
  histamine, 802, 866, 976
  Hofmann elimination, 995–997
  imine formation from, 658,
    666–667, 669f
  infrared (IR) absorptions/spectra of,
    543, 543t, 974–975, 975f, 975t
  interesting compounds, 975–978
  mass spectra of, 520
  melting point, 974t
  nitrous acid reactions, 997–999
  as nucleophiles, 993–995
  nucleophilic substitution synthesis,
    979–980
  2-phenylethanamine derivatives,
    977–978
  physical properties, 974t
  pK_a values, 992t
  primary [See Primary (1°) amines]
  reactions of, 985–1008
  reduction reaction synthesis, 981–982
  reductive amination synthesis,
    982–985
  relative basicity, 992t
  secondary [See Secondary (2°) amines]
  simple, 975
  solubility, 974t
  spectroscopic properties, 974–975,
    975f, 975t
  structure of, 970–971
  sulfa drugs, 1007–1008, 1008f
  synthesis of, 979–985
  synthetic dyes, 1004–1007
  tertiary [See Tertiary (3°) amines]
  -amine (suffix), 971
Amino acid residues, 1031
Amino acids. See also Peptides;
    Proteins; specific amino acids
  abbreviations for, 716t
  N-acetyl, 1026–1029
  acid–base properties, 717–718,
    1024–1025
  acidic, 1022, 1023f
  α-amino acids, 118, 715–716,
    1022–1024
  D-amino acids, 1022, 1022f
  L-amino acids, 1022, 1022f, 1028
  basic, 1022, 1023f
  carbon NMR absorptions,
    974, 975t
  in catabolism, 1170–1171, 1171f
  catalytic triad, 1056
  codons, 1156–1157
  C-terminal, 1032
  deamination, 672–673
  decarboxylation to amines, 802–803
  deriving sequence from DNA, 1159
  description of, 118, 715–716, 1022
  enantioselective synthesis of,
    1029–1031
  enzymes, 244
  essential, 695
  human synthesis, 716
  isoelectric point, 718–719, 1025t
  neutral, 1022, 1023f
  nomenclature, 1022, 1023f
  N-terminal, 1032
  pK_a values, 1024–1025, 1025t
  proton NMR absorptions, 974, 975t
  representative examples, 716t
  stereochemistry of, 1022, 1022f
  structure of, 1022–1024, 1023f

  synthesis of, 1026, 1029–1031
  zwitterion form of, 118, 715,
    717–718
Aminobenzene, 853
p-Aminobenzoic acid (PABA), 857,
    1007–1008
Aminoglycoside antibiotics, 1104
Amino groups
  on amino acids, 118
  as electron-donating group,
    898–900, 909
  induction effect of, 898
  nitro reduction to, 920–921
  nomenclature, 973
  as ortho, para director, 906–907
  resonance effect of, 898
  as strong Lewis base, 909
  structure of, 96t
6-Aminohexanoic acid, 1268
3-Amino-2-hydroxybutanal, A-7f
Amino sugars, 1103–1104
Ammonia
  amide synthesis, 749
  in amine synthesis, 979–980
  anhydride reaction with, 750
  carboxylic acid reaction with, 755
  hybrid orbitals in, 36
  molecular shape, 26
  pK_a values, 992t
  in reductive amination, 982, 984
Ammonium ion, 971
Ammonium salts
  formation of, 749, 755, 970, 986
  medicinal preparations, 987
Amoxicillin, 3, 95, 745
AMP (adenosine 5'-monophosphate),
    766, 1147t
Amphetamine, 78, 982–983
Amprenavir, 1058, 1058f
Amygdalin, 661–662
Amylopectin, 1101–1102
Amylose, 120, 1101–1102
Anabolic steroids, 1136
Anabolism, 1170
Analgesics
  aspirin (See Aspirin)
  ibuprofen (See Ibuprofen)
  morphine (See Morphine)
  naproxen, 795
  resiniferatoxin, 598
  salicin, 77f, 703, 1089
  salicylic acid, 703
  sodium salicylate, 703
Anastrozole, 703
Andro, 1133
Androgens, 1136
Androstenedione, 875, 876f
4-Androstene-3,17-dione, 1133
Androsterone, 1136t
Anesthetics
  benzocaine, 920
  diethyl ether, 95, 345
  halothane, 258
  sevoflurane, 114
  xenon, 115
-ane (suffix), 135, 144
Angiotensin, 194
Angle strain
  bridged halonium ions, 405
  in cycloalkanes, 153–154
  in epoxides, 339
Angular methyl groups, 1132–1133
Anhydrides
  acid chloride conversion to, 749–750
  description of, 735
  nomenclature, 738, 740t

  nucleophilic acyl substitution in,
    750–751
  structure of, 734
Aniline
  electrophilic aromatic substitution
    reaction, 890, 906–907, 909
  nitrobenzene reduction to, 920–921
  nomenclature, 853, 973
  polyhalogenation of, 909
  resonance structures, 898–899
  substituted, 994
  in 1,3,5-tribromobenzene
    synthesis, 1001f
Anionic oxy-Cope rearrangements, 1244
Anionic polymerization,
    1261–1264, 1261f
Anions
  acetylide, 398, 418–420, 615
  carboxylate (See Carboxylate anions)
  cyclopentadienyl, 862–863
  definition of, 8
  salt formation, 11
  size, relationship to acidity, 69
  solvation by ion–dipole interactions,
    265–266
Annulation, 835
Annulenes, 861
Anomeric carbon, 1080–1082,
    1086, 1146
Anomers, 1080, 1087–1088
Anopheles mosquito, 992
Antabuse, 460
Antarafacial bond formation,
    1238, 1240t
Antarafacial rearrangement,
    1241–1242, 1242t
Anthracene, 862
Anti addition
  description of, 397
  in dihydroxylation, 451
  in halogenation, 404, 406
  in hydrohalogenation, 402–403, 403t
  reduction of alkyne to trans
    alkene, 443
Antiaromatic compounds, 860,
    867–868
Antibiotics
  amikacin, 1104
  aminoglycosides, 1104
  amoxicillin, 3, 95, 745
  description of, 2f
  β lactams, 528, 544, 744–745, 762
  monensin, 682
  penicillin, 528
  prontosil, 1007
  sulfa drugs, 1007–1008, 1008f
  sulfamethoxazole, 1008f
  sulfanilamide, 1007–1008
  sulfisoxazole, 1008f
  synthetic dyes as, 1007
  tobramycin, 1104
  transport across cell membrane,
    115–116
Antibonding molecular orbital,
    869–872, 870f, 871f
Anticancer drugs
  anastrozole, 703
  aromatase inhibitors, 874–875, 876f
  cytarabine, 1144
  description of, 4–5
  doxorubicin, 1070, 1070f
  epothilone A, 1214
  exemestane, 189
  halomon, 255
  ingenol mebutate, 1197
  paclitaxel, 176, 187–188, 603

  palau'amine, 553
  raloxifene, 272–273
Anticodon, 1154, 1155f
Anti conformation, 150–151
Antidepressant drugs
  bupropion, 889f
  fluoxetine, 3, 206, 915, 977
  paroxetine, 347
  selective serotonin reuptake
    inhibitors, 977
  sertraline, 896
  venlafaxine, 622
Anti dihydroxylation, 451
Antihistamines
  chlorpheniramine, 889f
  cimetidine, 976–977
  description of, 866, 976–977
  diphenhydramine hydrochloride,
    852, 987
  fexofenadine, 976–977
Anti-inflammatory drugs
  aspirin (See Aspirin)
  celecoxib, 1126
  cortisol, 1137
  cortisone, 1137
  dexamethasone, 97
  fluticasone, 258
  ibuprofen (See Ibuprofen)
  nabumetone, 802
  naproxen, 207, 795
  NSAIDS, 1126
  rofecoxib, 1126, 1203
  valdecoxib, 1126
Antimalarial drugs
  chloroquine, 992
  quinine, 311, 793, 975, 993
Antioxidants, 954
Anti periplanar geometry,
    318–321, 318f
Antiulcer drugs
  cimetidine, 976–977
  misoprostol, 1126
Antiviral drugs. See also HIV
    treatments
  lamivudine, 1162
  oseltamivir, 99
  rimantadine, 985
  vidarabine, 1146
Ants, formic acid in, 136
Applied magnetic field, 555, 568–570
D-Arabinose, 1077, 1078f, 1094–1095
Arachidic acid, 1179
Arachidonic acid
  eicosanoid synthesis from,
    1124, 1125f
  hydrogen bonding, 108
  leukotriene synthesis from, 373,
    1124, 1125f
  prostaglandin synthesis from, 704,
    1124, 1125f
Arenes, 850. See also Aromatic
    hydrocarbons; specific aromatic
    hydrocarbons
Arginine
  isoelectric point, 1025t
  pK_a values, 1025t
  structure of, 1023f
Arimidex, 703
Aromacin, 189
Aromatase, 876f
Aromatase inhibitors, 703,
    874–875, 876f
Aromatic amines
  heterocyclic, 990–991, 992t
  nomenclature, 973
Aromatic heterocycles, 864–869

Aromatic hydrocarbons. *See also* Benzene; Benzene derivatives; *specific aromatic hydrocarbons*
annulenes, 861
aromaticity, 859–861, 860t, 872–874, 873f
characterization of, 867–868
charged compounds, 862–864
description of, 93, 94t, 850
electrophilic aromatic substitution reactions (*See* Electrophilic aromatic substitution reactions)
examples of, 861
heterocycles, 864–869
infrared (IR) absorptions/spectra of, 540, 543t, 855t
interesting compounds, 856–857
molecular orbitals, 869–872, 870f, 871f
nucleophilic aromatic substitution reactions, 913–916
polycyclic (*See* Polycyclic aromatic hydrocarbons)
predicting aromaticity, 872–874, 873f
reactions of, 886–924
spectroscopic properties, 855, 855f, 855t
stability of, 858
Aromaticity
basicity, effects on, 992t
criteria, 859–861, 860t
prediction of, 872–874, 873f
Arrow notation
curved arrow, 19–21, 60, 224, 225–227, 226t, 936
double-headed arrow, 18
double reaction arrow, 60
equilibrium arrows, 60, 61, 66
explanation of, A-3
half-headed, 936
open arrow, 421
in organic reactions, 225–227, 226t
polarity, 44–45
Artemisinin, 411, 412f
Artificial/alternative sweeteners, 198, 1034, 1090, 1092, 1100, 1100f
Arylamines
pK_a values, 992t
primary, 998–999
relative basicity, 988–989, 992t
Arylboranes, 1202
Aryl bromides, 1000
Aryl chlorides
biologically active, 889f
reactivity of, 914
synthesis of, 1000
Aryl diazonium salts
azo coupling, 1004–1005
in benzene derivative synthesis, 1002–1005
reactions of, 999–1005
synthesis of, 998–999
Aryl fluorides, 914, 1000
Aryl groups, 854, 970
Aryl halides
alkene synthesis from, 1205–1207
biologically active, 889f
description, 255, 255f
formation of, 888–889
in Heck reactions, 1205
in nucleophilic aromatic substitution, 913–916
stability, 289, 894
Aryl hydrides, 913–916
Aryl iodides, 1000
Ascorbic acid. *See* Vitamin C

Asparagine
isoelectric point, 1025t
pK_a values, 1025t
structure of, 1023f
Aspartame, 1034, 1100, 1100f
Aspartic acid
isoelectric point, 1025t
pK_a values, 1025t
in polyaspartate synthesis, 1279
structure of, 1023f
Aspirin
analgesic action, 703–704
crossing cell membrane, 116
description of, 77–78
hydrogen bonding, 108
reaction rates and, 240
stability, 761
synthesis, 290–291, 291f
Asthma
description of, 373
drugs, 97, 374, 654 (*See also* Bronchodilators)
leukotriene C_4 and, 99
Asymmetric carbon. *See* Stereogenic centers
Asymmetric reactions, 460, 1029–1031
Atactic polymers, 1264–1265
A–T base pair, 1151
Atenolol, 99
-ate (suffix), 698, 738
Atherosclerosis, 768
Atmosphere
carbon dioxide concentration, 165, 166f
CFCs in, 259, 949–950, 949f
nitrous oxide in, 1275
ozone in, 453, 949–950, 949f
Atomic number, 8, 189
Atomic weight, 8
Atoms, 8
Atorvastatin, 1135, 1135f
ATP. *See* Adenosine 5'-triphosphate
ATP hydrolysis, 1055, 1055t
Atropa belladonna, 969
Atropine, 969
Automated peptide synthesis, 1043–1045
Axial bonds, 158
Axial hydrogens, 155, 155f, 156, 156f
Azelnidipine, 821
Azo compounds
formation of, 999, 1004–1005
sulfa drugs, 1007–1008, 1008f
synthetic dyes, 1006–1007
AZT (azidodeoxythymidine), 3, 1162

B

Backside attack
acetylide anion reaction with epoxides, 420
in dihydroxylation, 451
epoxide ring opening, 369, 370–371, 629
in halogenation, 406
in S_N1 substitution reactions, 275
in S_N2 substitution reactions, 269–272, 283, 363
Bactrim, 1008f
Baekeland, Leo, 1273f
Baeyer strain theory, 153
Bakelite, 1273, 1273f
Ball-and-stick model
colors of atoms in, A-3
conversion to Fischer projection, 1075
molecular shape, 25–27

Bananas, 305
Barrier to rotation, 152
Base-catalyzed reactions
aldol reactions, 825
hydration, 673
Base peak, 510
Bases
in acetylide anion formation, 398, 417
in alkyl halide reactions, 323–327, 324t
in alkyne synthesis, 322–323
amide hydrolysis in, 760–761
amines as, 970, 985–987
amino acids as, 717–718
Brønsted–Lowry (*See* Brønsted–Lowry bases)
carbonyl group, 97
carboxylic acid derivatives, 737
common, 76–77, 76f
conjugate, 60–62, 64
in dehydrohalogenation, 303, 303t
in E1 elimination reactions, 316, 316t, 317, 321, 321t
in E2 elimination reactions, 309, 310, 311t, 316, 321, 321t
in enolate formation, 786–787, 786f
ester hydrolysis to carboxylate anion, 756, 757–758
leaving group ability, 262–263
nonnucleophilic, 264
nucleophilic, 323
organometallic reagents as, 614, 615–616
tautomerization in, 782
Basic amino acids, 1022, 1023f
Basicity
acidity, inverse relationship with, 64
of amines, 987–993
aromaticity effects on, 992t
of carbonyl groups, 736
hybridization effects on, 991, 992t
inductive effects on, 992t
leaving group ability and, 262–263, 746–747
nucleophilicity vs., 264–266, 985
relative, 987–993, 992t
resonance effects on, 990, 992t
Basketane, 144f
Bayberry plant, 1127
9-BBN (9-Borabicyclo[3.3.1]nonane), 409
Bee venom, 1033
Benadryl, 852, 987
Benedict's reagent, 1092–1093
Bent molecular structure, 27
Benzaldehyde
in crossed aldol reactions, 819, 820f
electron density, 900
in Grignard reaction, 621
nomenclature, 649, 649t
nucleophilic addition to, 618
resonance structures, 899
Benzamide, 739
Benzene
aromaticity, 859–860
aromatic ring nomenclature, 854
bromination, 888–889
m-bromoaniline synthesis from, 920–921
p-bromobenzaldehyde synthesis from, 1003
p-bromonitrobenzene from, 912
characteristics of, 850
m-chlorophenol synthesis from, 1002
electron density, 852, 852f

electrophilic aromatic substitution reactions, 886–898
Friedel–Crafts acylation, 887f, 891–897 (*See also* Friedel–Crafts acylation)
Friedel–Crafts alkylation, 887f, 891–897 (*See also* Friedel–Crafts alkylation)
as gasoline additive, 522
halogenation, 887f, 888–889
heat of hydrogenation, 857–858, 858f
hybridization, 851–852
molecular orbitals, 871–872, 871f
nitration, 821, 887f, 890–891, 912–913
p-nitrobenzoic acid synthesis from, 921–922
o-nitrotoluene from, 912–913
NMR spectrum of, 565, 578, 578f, 855, 855f, 855t
nomenclature, 853–855
in nylon 6,6 synthesis, 1274, 1275f
orbitals, 851–852
phenyl group in, 93
pi (π) electrons in, 850–851, 860, 869, 886
resonance, 478, 851
spectroscopic properties, 855
stability of, 857–859
structure of, 850–852
sulfonation, 887f, 890–891
synthesis of, 1001
trisubstituted benzene synthesis from, 922–923
Benzenecarboxylic acid, 698
Benzene derivatives
disubstituted (*See* Disubstituted benzene derivatives)
monosubstituted (*See* Monosubstituted benzene derivatives)
NMR spectrum of, 565, 578, 578f, 855, 855f, 855t
nomenclature, 853–855
nucleophilic aromatic substitution reactions, 913–916
polysubstituted, 854
reactions of, 913–921
substituted (*See* Substituted benzene derivatives)
synthesis of, 912–913
Benzenesulfonic acid, 890–891
Benzo[*a*]pyrene, 856, 856f
Benzocaine, 920
Benzoic acid
deprotonation of, 706
extraction from cyclohexanol mixture, 712–713
nomenclature, 698, 737
synthesis of, 921
Benzoic anhydride, 751
Benzonitriles, 1000
Benzophenone, 518, 651
Benzoquinone, 493
Benzoyl chloride, 740t, 749, 752
Benzoyl groups, 651
Benzyl esters, 1041
Benzyl groups, 651, 854
Benzylic bromides, 917
Benzylic bromination, 917–918
Benzylic halides, 255, 255f, 917
Benzylic radical, 1258
Benzyne, 915–916
Bergstrom, Sune, 703
Beryllium, 17

Beta (β) carbon
  in 1,4-addition, 630
  in alcohol dehydration, 350
  in elimination reactions, 303–304
  in E2 mechanism, 312, 319–320
  in fatty acid catabolism, 1176–1177
  in Hofmann elimination, 995–997
  nomenclature, 650
Beta (β) elimination reactions. *See also*
    Elimination reactions
  alcohols, 349, 350–355
  dehydration reactions (*See*
    Dehydration reactions)
  dehydrohalogenation as, 303–304
  Zaitsev rule, 312–313
Beta (β) lactam family of antibiotics,
    528, 544, 744–745, 762
Bextra, 1126
Bicyclic compounds, naming,
    A-8–9
Bicyclic ring systems, 494–495, 494f
Bicyclo[4.4.0]decane, A-8
Bicyclo[3.2.1]octane, A-8
Biformene, 1129
Bimatoprost, 704
Bimolecular reactions
  description of, 241
  elimination reactions (*See* E2
    elimination reactions)
  nucleophilic substitution (*See* S$_N$2
    substitution reactions)
BINAP [2,2'-Bis(diphenylphosphino)-
    1,1'-binaphthyl], 1030
Biocatalysts, 1275
Biodegradable polymers, 1275,
    1278–1279
Biofuels, 1120
Biological reactions
  acyl phosphates in, 763–766
  acyl transfer reactions, 766
  allylic carbocations in, 476–477
  antioxidants, 954
  carboxylate anion conversion to acyl
    phosphate, 763–764
  cholesterol synthesis,
    1134–1135, 1134f
  citric acid cycle, 1186–1190
  Claisen reactions, 832–833
  decarboxylation, 802–803
  dehydration of alcohols, 352
  E1cB elimination, 817
  enantioselective reduction, 607–608
  enols in, 782–783
  Friedel–Crafts reactions, 897–898
  glucose metabolism, 1190
  glycolysis, 1179–1184
  imines in, 671–673
  β lactam antibiotic mechanism, 762
  nucleophilic substitution, 287–288
  oxidation, 458–460
  β-oxidation of fatty acids,
    1176–1178, 1190–1191
  PLP-dependent deamination,
    672–673
  of pyruvate, 1185–1186
  retro-aldol reactions, 831–832,
    1180f, 1181
  tautomerizations, 782–783
  terpene/terpenoid synthesis,
    1129–1132, 1130f
  thioesters in, 766–768
  vision and, 671, 671f
Biomolecules. *See also* Amino acids;
    Carbohydrates; DNA; Lipids;
    Nucleotides; RNA
  chirality, 182

definition, 117
families of, 117
2,2'-Bis(diphenylphosphino)-1,1'-
    binaphthyl (BINAP), 1030
Bisphenol A (BPA), 1270, 1271, 1272f
1,3-Bisphosphoglycerate, 1182–1184
Blattella germanica, 750
Blattellaquinone, 750
Bleaching of corals, 7
Bloch, Konrad, 1134
Blood–brain barrier, 114
Blue Ridge Mountains, 484
Blunt ends of DNA fragments, 1160
Boat conformation, 156–157,
    156f, 157f
Boc (*tert*-butoxycarbonyl) protecting
    group, 1040
Body fat, 1119
Boiling point (bp)
  of alcohols, 344t
  of aldehydes, 652t
  of alkanes, 146, 146t
  of alkenes, 391
  of alkyl halides, 257t
  of alkynes, 391
  of amines, 974t
  of carboxylic acid derivatives, 742t
  of carboxylic acids, 700t
  of diastereomers, 205–206, 205f
  of enantiomers, 201, 203t
  of epoxides, 344t
  of ethers, 344t
  intermolecular forces and, 104–105
  of ketones, 652t
  relative, 105
  of thiols, 366
Bombykol, 1204
Bombyx mori, 1204
Bond angle
  in alcohols, 339
  description of, 24–27, 27t, 41t
  in epoxides, 339
  in ethers, 339
Bond cleavage
  alpha (α) cleavage, 518–520
  in chemical recycling of PET,
    1277–1278
  in citric acid cycle, 1188
  in cyanohydrin formation, 661
  description of, 223–224
  in Diels–Alder reactions, 490–491
  in Edman degradation, 1036
  in E1 elimination reactions, 314–315
  in E2 elimination reactions, 308
  of ethers, 364–365
  in Friedel–Crafts alkylation, 895
  homolysis, 223–224
  kinetic energy and, 240
  in kinetic resolution, 1028–1029
  in malonic ester synthesis, 797
  in nucleophilic substitution, 267,
    268, 274, 357, 364–365
  in peptide hydrolysis, 1057
  in peptide sequencing, 1037–1038,
    1037t
  in peptide synthesis, 1044–1045
  in radical reactions, 936–937, 941,
    943–944
  in retro-aldol reactions, 815
  in sigmatropic rearrangement,
    1240–1242
  in Wohl degradation, 1095
Bond dissociation energy
  allylic halogenation, 950
  bond length and, 229
  bond strength and, 228–229

calculation of enthalpy change,
    229–230, 941–942, 942f
  of common bonds, 42t, 228t, A-10
  definition, 227
  radical stability, 936–937, 937f
Bonding energy, 235
Bonding molecular orbital, 869–872,
    870f, 871f
Bonding (shared) electrons, 12
Bond length
  average for common bonds, 24t
  in benzene, 851
  bond dissociation energy and, 229
  bond strength and, 42–43, 42t
  in buta-1,3-diene, 484–485
  hybridization and, 484
  periodic trends, 24
Bonds and bonding. *See also* specific
    types of bonds
  in alcohols, 339
  bond angle, 24–27, 27t, 41t
  bond dissociation energy, 227–231
  bond length, 24, 24t, 42–43, 42t
  carbonyl group, 97
  chemical shift values, NMR spectra,
    562–563, 564t
  cleavage, 223–225, 224f
  covalent, 11, 41t
  definition of, 10
  degree of atom, 94
  dihedral angle, 147
  electronegativity, 43–45, 44f
  in epoxides, 339
  in ethers, 339
  formation, 225
  α-glycosidic bonds, 1102
  glycosidic linkage, 1098–1104
  hybridization, 34–41 (*See also*
    Hybridization)
  hydrogen, 102, 103t
  infrared absorptions, 533, 533t
  ionic, 11
  Lewis acid–base reactions, 79–81
  Lewis structures, 12–23 (*See also*
    Lewis structures)
  molecular orbitals, 869–874
  molecular orbital theory, 869–870
  molecular shape and, 24–27
  nonpolar, 44
  in nucleophilic substitution, 267,
    268, 274
  organic molecular structures, 28–34
  in organic reactions, 223–227, 224f
  in organic synthesis, 421
  polar, 44–46
  rotation, 38, 39, 147
  valence bond theory, 869
Bond strength
  bond dissociation energy and,
    228–229
  bond length and, 42–43, 42t
  carbon–carbon bonds, 42, 42t
  hydrogen bonding, 102, 103t
  infrared absorptions and, 532–534,
    532f, 533f, 535
9-Borabicyclo[3.3.1]nonane
    (9-BBN), 409
Borane, 409
Boron, 17
bp. *See* Boiling point
BPA (bisphenol A), 1270, 1271, 1272f
Bradykinin, 1033
Brevenal, 339
Bridged halonium ions, 405–408
Bridged ring systems, 494–495, 494f
Bridgehead carbons, A-8

Bromination
  of alkanes, 943–945
  of alkyl benzenes, 917–918
  allylic, 951–952
  of aniline, 1001f
  of benzene, 888–889, 912
  benzene derivative synthesis, 912
  m-bromoaniline synthesis, 920–921
  at α carbon, 793
  product mixtures, 952
  of propane, 944
  selective, 951–952
  of substituted benzenes, 912
Bromine, 515, 516f
Bromine radicals, 951
2-Bromoacetophenone, 651f
α-Bromoacetophenone, 651f
m-Bromoaniline, 920–921
o-Bromoaniline, 973
p-Bromobenzaldehyde, 1003
Bromobenzene, 888
1-Bromobutane, 310–311
2-Bromobutane, 304, 947–948
o-Bromochlorobenzene, 853
Bromocyclohexane, 402
2-Bromocyclohexanone, 792
1-Bromocyclohexene, 1200
Bromoethane, 347
3-Bromohexane
  enantiomers, 185, 185f
  stereogenic centers, 183
(*E*)-1-Bromohex-1-ene, 1199
Bromohydrins, 407
2-Bromo-2-methylbutane, 312
1-Bromo-3-methylbut-2-ene, 952
3-Bromo-3-methylbut-1-ene, 952
cis-1-Bromo-4-methylcyclohexane, 283
1-Bromo-1-methylcyclopentane, 316
2-Bromo-2-methylpropane, 304, 418
p-Bromonitrobenzene, 912
1-Bromopropane
  proton NMR spectrum, 572–573,
    573f
  as radical addition product, 955
2-Bromopropane
  as bromination product, 945
  in ethyl preparation, 347
  mass spectrum, 516f
  proton NMR spectrum, 572, 572f
  as radical addition product, 955
(*Z*)-2-Bromostyrene, 1204
N-Bromosuccinimide (NBS), 407, 951,
    952
p-Bromotoluene, 853
Bronchodilators
  albuterol, 183–184, 372, 372f, 373
  ephedrine, 183
  fluticasone, 258
  tiotropium bromide, 346
Brønsted–Lowry acids
  carboxylic acids, 705–708
  description of, 59–60
  Lewis acids, relation to, 78–79
  reactions, 60–67
Brønsted–Lowry bases
  alkyl halides and, 259
  description of, 59–63
  in elimination reactions, 303
  Lewis acids, relation to, 78
  nucleophiles vs., 263
  reactions, 60–67
  terminology, 263
BTX, 522
Bufotenin, 978
Bufo toads, 978
Bupropion, 889f

Buta-1,3-diene
  carbon–carbon σ bond length, 484–485
  as conjugated diene, 474–475, 475f, 482
  copolymerization of, 1263
  electrocyclic ring opening, 1227, 1231
  electrophilic addition products, 488–490
  molecular orbitals in, 1228, 1229, 1229f
  resonance structures, 478
  Ziegler–Natta polymerization of, 1266
Butanal
  intermolecular forces in, 103
  melting point, 106
Butane
  chlorination of, 946
  conformations, 150–153, 305–306
  isomers, 138
  as normal alkane, 135t
  as propellant, 949
  solubility, 107–108
  as straight-chain alkane, 134
Butanedioic acid, 698
Butane-2,3-diol, 370
Butanoic acid, 542
Butan-1-ol
  intermolecular forces in, 103
  IUPAC nomenclature, 340
  melting point, 106
Butan-2-ol
  (S)-butan-2-ol conversion to its tosylate, 362
  enantiomers of, 184–185, 606
  infrared spectrum, 540
  labeling stereogenic centers, 189–191
tert-Butanol, 265f
Butan-2-one
  boiling point, 652
  infrared spectrum, 541
  nomenclature, 651
  reduction of, 606
But-1-ene, 307, 388
But-2-ene
  alkyne reduction to trans isomer, 443
  epoxidation of isomers, 448–449
  halogenation of, 406
  hydrohalogenation of, 399
  restricted rotation in, 305–306, 306f
  stability, 437
tert-Butoxide, 264
tert-Butoxycarbonyl (Boc) protecting group, 1040
tert-Butoxy groups, 341
Butylated hydroxy toluene (BHT), 954
Butylbenzene, 923
tert-Butylbenzene, 853
tert-Butyldimethylsilyl ether (TBDMS), 623–624
sec-Butyl ethyl sulfide, 367
Butyl groups, 138
sec-Butyl groups, 138
tert-Butyl groups, 138, 159
tert-Butyl hydroperoxide, 446, 461
tert-Butyl iodide, 256
Butyllithium, 77, 664, 787
1-sec-Butyl-3-methylcyclohexane, 143
sec-Butyl methyl ether, 341
tert-Butyl methyl ether (MTBE)
  proton NMR spectrum, 556–558
  solubility of, 109
  synthesis of, 404

5-tert-Butyl-3-methylnonane, 141
But-1-yne
  addition reactions, 398f, 412–417
  halogenation, 398f, 413
  hydration, 398f, 414–417
  hydroboration–oxidation, 398f, 416–417
  hydrohalogenation, 398f, 413
But-2-yne, 443
Butyraldehyde, 649t

C

Cadaverine, 975
Caffeine, 114
Cahn–Ingold–Prelog system, 189
Calan, 703
Calcitriol, 484, 484f
Callistimon citrinus, 782
Calories (cal), 42
Camphor, 1128
Cancer. *See* Anticancer drugs
Candlenuts, 393
Cannabis, 508
Caproaldehyde, 649t
ε-Caprolactam, 1268, 1278
Capsaicin, 4, 223, 598
Caraway, odor of, 208
-carbaldehyde (suffix), 648
Carbamates, 1040, 1269
Carbanions, 73, 77, 225
Carbenes
  addition reaction, 1209
  in cyclopropane synthesis, 1207–1211
  description of, 1208
  dihalocarbene synthesis, 1208
  intermediates in olefin metathesis, 1212–1213
  in Simmons–Smith reaction, 1210–1211
Carbidopa, 48
Carbinolamines, 658, 666–668, 993
Carbocation rearrangement
  in cholesterol biosynthesis, 1135
  conversions of alcohols to alkyl halides, 358–359
  in dehydration of alcohols, 352–355
  in Friedel–Crafts alkylation, 894–895
  in hydration reactions, 404
  in hydrohalogenation, 401, 403t
  1,2-shifts, 353–354, 894–895
  in S_N1 substitution reactions, 358–359
Carbocations
  alkyl diazonium salt decomposition to, 998
  allylic (*See* Allylic carbocations)
  classification of, 277
  definition of, 20
  in E1 elimination reactions, 314
  in electrophilic addition reactions, 399
  in electrophilic aromatic substitution, 887–888, 904–905
  electrostatic potential plots, 278, 278f
  fragmentation and, 516
  in Friedel–Crafts alkylation, 893, 894–895
  heterolysis and, 225
  in hydration reactions, 403–404
  Markovnikov's rule, 400–401
  methyl, 277–279
  planar, 275
  primary [*See* Primary (1°) carbocations]
  protosterol, 1134f, 1135

resonance-stabilized (*See* Resonance-stabilized carbocations)
  secondary [*See* Secondary (2°) carbocations]
  in S_N1 substitution reactions, 274, 275–276, 317
  in tautomerization, 415
  tertiary (3°), 277–279, 895
  tropylium, 863–864
  vinyl, 289
Carbocation stability
  alkyl groups and, 277
  in electrophilic aromatic substitution, 887–888, 904–905
  Hammond postulate, 279, 281–282, 281f
  hyperconjugation and, 278–279
  inductive effects, 277–278, 278f
  reaction rates and, 279, 281–282, 315
  1,2-shifts, 353–354
  S_N1 substitution reactions and, 279, 281–282
Carbohydrates. *See also specific carbohydrates*
  complex, 119
  cyclic acetals and hemiacetals in, 682–683
  description of, 119–120, 1070
  disaccharides (*See* Disaccharides)
  monosaccharides (*See* Monosaccharides)
  nonreducing sugars, 1093, 1093f, 1100
  as polymer feedstock, 1275, 1276t
  polysaccharides (*See* Polysaccharides)
  reducing sugars, 1093, 1093f, 1098, 1099
  simple (*See* Monosaccharides)
Carbon
  allylic, 440, 1119
  anomeric, 1080
  bonding of, 12f, 15, 16t
  bridgehead, A-8
  α carbon [*See* Alpha (α) carbon]
  β carbon [*See* Beta (β) carbon]
  δ-carbon, 650
  γ-carbon, 650
  Ω-carbon, 650
  charge, 16t, 33
  chemical shift values, 583, 584f
  chiral (*See* Stereogenic centers)
  common nucleophile, 266t
  degree of atom, 94–95
  electron-deficient, 71, 92, 116–117, 599
  nomenclature/numbering, 650
  periodic table entry, 9
  prochiral, 458–459
  substitution reactions at α carbon, 656, 779–803
Carbonates, 1270
Carbon backbone. *See* Carbon skeleton
Carbon–carbon bonds
  acetylide anion reaction with alkyl halide, 418–419
  acetylide anion reaction with epoxide, 420
  in aldol reactions, 813–814
  in alkanes, 145
  in alkenes, 304–308, 385
  in alkylation, 793
  in alkynes, 385
  in benzene, 851
  in biological Claisen reactions, 832–833

in buta-1,3-diene, 484–485
  in carbonyl condensation reactions, 813
  in chain-growth polymerizations, 1258
  chemical shift, NMR spectra, 565–566
  in citric acid cycle, 1188
  in Claisen reactions, 825–826
  in cyanohydrin formation, 660–662
  in cyclopropane synthesis, 1207–1211
  degree of atom, 94–95
  in 1,3-dienes, 474
  in dienophiles, 492–493
  double, 14, 80
  in farnesyl diphosphate biosynthesis, 1130
  in fatty acid catabolism, 1176
  fragmentation and, 516–517
  in Friedel–Crafts acylation, 891
  in Friedel–Crafts alkylation, 891, 893
  in geranyl diphosphate synthesis, 477
  in Heck reactions, 1205–1207
  in intramolecular aldol reactions, 822–824
  in isoprene units, 1127
  length and strength, 42, 42t
  in lipids, 1115, 1115f
  in metabolism of triacylglycerols, 1119
  in Michael reaction, 834
  in Newman projections, 148
  in nitrile reduction, 981
  nonpolarity, 44
  in olefin metathesis, 1211–1216, 1212f
  in organic synthesis, 421, 632–634
  in organometallic reagent reactions, 616–618, 625–627, 1198–1207
  in oxidative cleavage, 453–454
  polyfunctional compounds with, nomenclature, A-8t
  in radical polymerization, 958–959
  radical reactions with, 938
  reactions forming, 1198–1216, A-11
  in Robinson annulations, 835
  rotation around, 38, 39 (*See also* Conformations)
  splitting patterns, 575–577
  in Suzuki reactions, 1200–1205
  in triacylglycerols, 393–395
  triple, 40
  in vitamin E, 954
  in Wohl degradation, 1095
  in Ziegler–Natta polymerization, 1265
Carbon dioxide
  atmospheric concentration, 165, 166f
  carboxylation, 628, 705t
  citric acid cycle product, 1189
  combustion product, 165
  electrostatic potential plot, 46f
  glucose metabolism to, ATP yield, 165
  molecular polarity, 45–46
Carbon–hydrogen bonds
  acidity, 68, 69
  in alkanes, 145
  in allylic bromination, 951–952
  in amide reduction, 610–612
  benzylic, 917–918
  in carboxylic acid reduction, 610–612
  in chlorination reactions, 943–944
  degree of atom, 94–95
  in Hofmann elimination, 996
  infrared absorptions, 534
  length and strength, 43
  in lipids, 1115, 1115f

Carbon–hydrogen bonds—(Cont.)
 in metabolism of
  triacylglycerols, 1119
 nonpolarity, 44
 oxidation and, 164, 435, 604
 in oxidative cleavage, 455, 953
 radical reactions with, 938
 in radicals, 936–937
 reduction and, 164, 435, 604
 in reduction of alkyne to alkane, 442
 in reductive amination, 982–983
 in vitamin E, 954
Carbon–metal bonds, 614, 1211–1212
Carbon–nitrogen bonds
 amide classification, 98
 in amide reduction, 611
 amine classification, 97
 in amine synthesis, 979
 in Gabriel synthesis, 981
 in hydrolysis, 661
 in nitriles, 696
 in peptide hydrolysis, 1057
 in primary structure of proteins, 1045
 in reductive amination, 982–983
Carbon NMR
 of aldehydes, 653t
 amines, 974, 975t
 of aromatic compounds, 855, 855f, 855t
 of carboxylic acid derivatives, 743t
 of carboxylic acids, 701t
 characteristic absorptions, A-14
 description of, 581–582
 of ketones, 653t
 of methyl acetate, 584f
 of nitriles, 701t
 number of signals, 582–583
 position of signals, 583–584, 583t, 584f
 of propanal, 653f
 of propanoic acid, 702f
 of propan-1-ol, 584f
Carbon–oxygen bonds
 in alcohols, 338
 in amide reduction, 610–612
 in anionic polymerization of epoxides, 1263–1264
 in carboxylic acid reduction, 610–612
 in epoxides, 338
 in ethers, 338
 functional groups, 97–99
 in α-helix, 1046, 1046f
 in hydration of alkynes, 414–415
 in hydrogenolysis, 1041
 in hydrolysis, 661
 in metabolism of
  triacylglycerols, 1119
 in oxidation, 164
 oxidation and, 435, 435f, 604
 in peptide bond, 1033
 in peptide hydrolysis, 1057
 in β-pleated sheet, 1047, 1047f
 in primary structure of proteins, 1045
 reduction and, 435, 435f, 604
 in reductive amination, 982–983
 in secondary structure of proteins, 1046–1047
Carbon–palladium bonds, 1201
Carbon radicals, 936, 958
Carbon skeleton
 carbon–carbon bond-forming reactions, 1198–1216
 of chair cyclohexane, 154, 155
 description of, 92
 drawing enantiomers, 185

hydrophobic nature of, 109
 of steroids, 497, 1132–1133, 1133f
Carbon tetrachloride, 258
-carbonyl chloride (suffix), 737
Carbonyl compounds. *See also specific classes*
 acetoacetic ester synthesis, 800–802
 acid–base reactions, 622
 alkylation of, 793–796
 biological decarboxylation, 802–803
 classes of, 599
 condensation reactions, 813–839
  (*See also* Carbonyl condensation reactions)
 Cope rearrangement synthesis, 1244
 α-cyano carbonyl compounds, 818–820
 electrophilic substitution, 780
 enolate intermediates in reactions
  (*See* Enolates)
 enol intermediates in reactions
  (*See* Enols)
 α-halo carbonyl compounds, 792–793
 halogenation of, 791–793
 β-hydroxy, 836
 malonic ester synthesis, 796–800
 nucleophilic addition, 600, 601
 nucleophilic substitution, 601–603
 oxidation, 603–604
 ozonide conversion to, 453
 pK_a values, 785t
 reactions of, 600–607, 622, 779–803
 reduction, 603–604, 606–607
 substitution reactions at the α carbon, 656, 780–803
 unsymmetrical, 788–789
Carbonyl condensation reactions
 aldol reactions, 813–818, 831–832
  (*See also* Aldol reactions)
 biological reactions, 830–833
 Claisen reactions, 825–827, 832–833
 crossed aldol reactions, 818–821
 crossed Claisen reactions, 827–829
 description of, 813, 816
 Dieckmann reactions, 829–830
 directed aldol reactions, 821–822
 intramolecular aldol reactions, 822–825
 Michael reactions, 833–835
 Robinson annulations, 835–839
Carbonyl groups
 in aldehydes, 647
 carbon NMR spectra, 583
 description of, 97–98, 599–600
 in dienophiles, 493
 in hydration of alkynes, 414–415
 in hydroboration–oxidation reactions, 416–417
 hydrogenation of, 440
 infrared absorptions, 534–536, 535f
 in ketones, 647
 in monosaccharides, 1071
 oxidation of alcohols, 456
 in oxidative cleavage, 453–454
 reactions of, 1091–1097
 tautomers, 780–783
-carbonyl phosphate (suffix), 737
Carbon–Z σ bonds
 in alkyl halides, 255, 267
 functional groups, 95–97
 homolysis and heterolysis of, 224–225, 224f
 in oxidation and reduction reactions, 164
Carboxy groups
 in amino acids, 118

of benzene derivatives, 918
 in carboxylic acids, 696
Carboxylases, 1055t
Carboxylate anions
 amides to, 760–761
 biological conversion to acyl phosphate, 763–764
 ester hydrolysis to, 756, 757–758
 formation of, 705–708
 nitrile hydrolysis to, 719–721
 nomenclature, 698, 698f
 nucleophilic acyl substitution, 763–766
Carboxylic acid derivatives. *See also* Acid chlorides; Acyl phosphates; Anhydrides; Amides; Esters; Thioesters
 boiling point, 742t
 carbon NMR absorptions, 743t
 description of, 599, 734–735
 infrared absorptions, 743t
 melting point, 742t
 nomenclature, 737–741, 740t
 nucleophilic acyl substitution, 745–768, 763t (*See also* Nucleophilic acyl substitution reactions)
 order of stability, 737
 organometallic reagent reactions, 624–627
 physical properties, 741, 742t
 proton NMR absorptions, 743t
 reactivity, 746–748
 reduction of, 609–610
 solubility, 742t
 spectroscopic properties, 742, 743t
Carboxylic acids
 acid chlorides from, 752–753
 acid chlorides to, 749
 acidity of, 705–708
 aldehyde oxidation to, 613, 1092–1093
 aliphatic, 710–711
 amides from, 755–756
 amides to, 760–761
 anhydrides to, 750
 aspirin (*See* Aspirin)
 benzene derivatives, 918
 boiling point, 700t
 conjugate bases of, 705–708
 deprotonation of, 706–707, 706t
 description of, 599
 diacids, 700
 diethyl malonate to, 796–800
 as dimers, 700, 700f
 ester hydrolysis to, 756–757
 esters from, 753–755
 formation of, 455–458, 628
 functional group, 97, 98t
 infrared (IR) absorptions/spectra of, 542, 543t
 interesting compounds, 702
 melting point, 700t
 nitrile hydrolysis to, 719–721
 nomenclature, 697–698
 nucleophilic acyl substitution, 751–756, 752f, 763t
 nucleophilic substitution, 601–603
 in peptide hydrolysis, 1057
 physical properties of, 700–701, 700t
 pK_a values, 705–706
 reactivity, 746–748
 reduction of, 608–612
 salts of, 702
 solubility, 700t
 spectroscopic properties, 701t, 702f

structure of, 696, 734
 synthesis of, 704, 705t
Carboxypeptidase, 1037–1038, 1037t
Carcinogens, 856, 999
Cardinol, 1276
β-Carotene, 111, 392, 665, 665f, 1123f
Carson, Rachel, 259
Carvone, 208
Cassava, 662
Catabolism
 aerobic glucose metabolism, ATP yield, 1190
 description of, 1170–1172, 1171f
 fatty acid oxidation, ATP yield, 1190–1191
 of glycerol, 1176
 glycolysis, 1179–1184
 of monosaccharides, 1179–1184
 by β-oxidation, 1176–1178
 of triacylglycerols, 1176–1178
Catalysts
 biocatalysts, 1275
 biological, 244, 244f
 chiral, 1029–1030
 description of, 243, 243f
 DNA polymerase, 1161
 enzymes (*See* Enzymes)
 Grubbs catalyst, 1211
 homogeneous, 1265
 Lindlar catalysts, 443
 metal catalysts, 243, 436–437
 palladium, 443, 1200–1205
 pyruvate kinase, 1183
 in reduction reactions, 436, 437
 TsOH, 675
 Ziegler–Natta catalysts, 1265, 1266
Catalytic hydrogenation
 of aldehydes, 606
 of alkenes, 437
 of ketones, 606
Catalytic triad, 1056
Cationic polymerization, 1260, 1261f
Cation-pi (π) stacking, 874–875, 876f
Cations
 counterions, 59
 cyclopentadienyl, 862–863
 definition of, 8
 radical, 510
 salt formation, 11
 solvation by ion–dipole interactions, 265–266
 spectator ions, 59
 tropylium, 863–864
Cecropia moth, 619
Cedar, 1127f
Cedrol, 514, 1127f
Celebrex, 1126
Celecoxib, 1126
Celery ketone, 208
Cell division, replication of DNA, 1152, 1153–1154, 1153f
Cell membrane
 functions, 113
 phosphoacylglycerols in, 1121
 sphingomyelins in, 1122
 structure of, 113–114, 114f
 transport across, 115–116, 115f
Cellobiose, 1099
Cells
 description of, 113
 membrane (*See* Cell membrane)
Cellulose
 configuration of, 178f
 hydrolysis of, 1101
 stereochemistry of, 177–178
 structure of, 177f, 1070f, 1101

Cellulose acetate, 1101
Cephalexin, 745
Cephalin, 1121
Cephalosporins, 745
Cetirizine, 186
CFC 11, 950
CFC 12, 950
CFCs. *See* Chlorofluorocarbons
C–G base pair, 1151
Chain branching, 1259–1260
Chain-growth polymerization
    anionic polymerization, 1261–1264, 1261f
    cationic polymerization, 1260–1261, 1261f
    chain branching in, 1259–1260
    copolymers, 1263
    description of, 1257–1258
    of epoxides, 1263–1264
    ionic polymerization, 1260–1262, 1261f
    radical polymerization, 1258–1260, 1258f
Chain-growth polymers, 1256, 1257–1263
Chain mechanism, 940
Chain termination, 1258–1260, 1262
Chair conformation
    of cyclohexane, 154–156, 155f, 160–162
    of D-glucose, 1085, 1085f
Charge
    Brønsted–Lowry acids and bases, 59, 59f
    on carbon, 33
    dipole, 44–45
    formal, 15–17, 16t
    net charge, 60
    partial, 44, 62, 237
    in resonance hybrid, 22–23
Charged ions. *See* Ions
Chauvin, Yves, 1211
Chemical recycling of polymers, 1277–1278
Chemical shift
    benzene rings, protons on, 565
    carbon–carbon double bonds, protons on, 565
    carbon–carbon triple bonds, protons on, 565–566
    in carbon NMR, 583, 584f
    characteristic values, 563–564, 564t
    equation, 557
    in proton NMR, 556–557, 562–567
    shielding and deshielding effects, 562–563, 562f, 563f, 565–567, 566f
Chiral catalysts, 1029–1030
Chiral centers. *See* Stereogenic centers
Chiral drugs, 206–207
Chirality
    of amines, 971
    of ammonium ion, 971
    description, 179–182
    of molecules with two or more stereogenic centers, 194
    of naturally occurring objects, 182
    principles of, 182f
    prochirality, 458–459
    of synthetic PAHs, 856
    test for, 180
Chirality centers. *See* Stereogenic centers
Chiral molecules
    chemical properties of, 206–208
    definition, 180
    description, 179–182, 182f

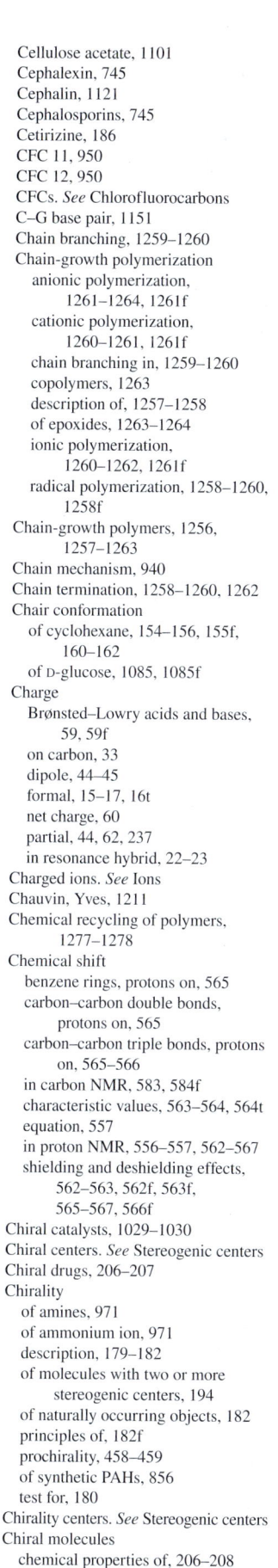

    in halogenation reactions, 946t, 947–948
    optical activity, 202
Chiral reactants, 460
Chitin, 1070, 1103–1104
Chloral, hydration of, 673
Chloral hydrate, 673–674
Chloramphenicol, 183–184
Chlorination
    of alkanes, 939–940, 943–946
    of (*R*)-2-bromobutane, 947–948
    of butane, 946
    of cyclopentane, 940–941
    of ethane, 941–942, 942f
    of propane, 944
Chlorination of benzene, 888–889
Chlorine
    mass spectra and, 514–515, 515f
    as radical, 938, 945, 949f
Chlorine monoxide, 949f
2-Chloroacetic acid, 710
Chlorobenzene
    formation of, 888
    nomenclature, 853
    phenol from, 915
2-Chlorobuta-1,3-diene, 1266
1-Chlorobutane, 946
2-Chlorobutane, 946
2-Chlorobutanoic acid, 711
3-Chlorobutanoic acid, 711
4-Chlorobutanoic acid, 711
Chlorocyclohexane, 319, 319f
Chlorocyclopentane, 940–941
Chlorodesmis fastigiata, 390
4-Chloro-1-ethyl-2-propylbenzene, 854
Chlorofluorocarbons (CFCs)
    description of, 3, 258–259
    ozone layer and, 949–950, 949f
Chloroform, 258, 1208
Chloromethane, 254
1-Chloro-2-methylbutane, 939
1-Chloro-3-methylbutane, 939
2-Chloro-2-methylbutane, 939
2-Chloro-3-methylbutane, 939
1-Chloro-2-methylcyclohexane, 319, 321
2-Chloro-5-methylheptane, 256
1-Chloro-2-methylpropane, 256f
p-Chloronitrobenzene, 913
meta-Chloroperoxybenzoic acid (mCPBA), 446, 447
m-Chlorophenol, 1002
2-Chloropropanal, 650f
1-Chloropropane, 400, 919
2-Chloropropane, 400, 515f
2-Chloropropene, 413
3-Chloropropenoic acid, 575, 575f
α-Chloropropionaldehyde, 650f
Chloroquine, 992
p-Chlorotoluene, 916
1-Chloro-1,3,3-trimethylcyclohexane, 402
Chlorpheniramine, 889f
Cholesterol
    biosynthesis of, 1134–1135, 1134f
    cholesteryl ester formation, 767–768, 767f
    solubility, 109
    specific rotation of, 204
    as steroid, 498, 1134
Cholesterol-lowering drugs
    atorvastatin, 1135, 1135f
    fenofibrate, 758
    lovastatin, 1135
    simvastatin, 189, 484, 484f, 1135
Cholesteryl esters, 767–768, 767f

Chondrocole A, 259
Chromate esters, 457
Chromium(VI) oxide, 446
Chromosomes, 120f
Chronic obstructive pulmonary disease (COPD), 258
Chrysanthemic acid, 583
Chrysanthemum cinerariifolium, 583
Churchane, 144f
Churchill, Winston, 4, 259
Chymotrypsin, 1037–1038, 1037t, 1054, 1056
Cicutoxin, 28
Cimetidine, 976–977
Cinchona tree, 302
Cinnamaldehyde, 110, 654
Cipralex, 703
Cis isomers. *See also* specific isomers
    of alkenes, 307, 391, 437, 443
    of cycloalkanes, 159–163
    of cyclohexane, 160–163
    definition, 159
    of disubstituted cycloalkanes, 198–199
    in S_N1 nucleophilic substitution, 284
    in S_N2 nucleophilic substitution, 271
cis- (prefix), 159, 389
Cis protons, 575
Citral, 1128
Citramalate, 831
Citrate, 222, 767, 1187f
Citrate synthase, 1187f
Citric acid, 110, 710
Citric acid cycle
    aldol reaction in, 831, 1187f
    in catabolism, 1171f, 1172
    citrate formation, 767
    citric acid as intermediate, 710
    decarboxylation in, 1187f, 1188
    hydration in, 1187f, 1189
    hydrolysis in, 1187f, 1188
    isomerization, 1187f, 1188
    net result, 1189
    oxidation in, 1187f, 1188, 1189
    phosphorylation in, 1187f, 1188
    reactions in, 222, 223
    thioester formation in, 1187f, 1188
(*S*)-Citryl CoA, 1187
C_{18} juvenile hormone, 619–620, 1199
Claisen reactions
    biological, 832–833
    crossed, 827–829
    description of, 825–827
    intermolecular, 829–830
    reverse, 1177–1178
Claisen rearrangements, 1243, 1245–1246
The Clear, 1136
Clemmensen reduction, 919
Climate change, 145, 165, 166f
Clockwise *R* isomer, 191–193
Cloning DNA, 1161
CMP (cytidine 5′-monophosphate), 1146–1147
^{13}C NMR. *See* Carbon NMR
Cocaine, 733, 976
Cocoa butter, 1114, 1120
Coconut oil, 395, 760, 1117, 1118t, 1120, 1178
Codeine, 727, 978
Coding strand, DNA, 1155, 1155f
Codons, 1156–1159, 1158f
Coenzymes
    acetyl CoA (*See* Acetyl CoA)
    (*S*)-citryl CoA, 1187
    coenzyme A, 117

fatty acyl CoA (*See* Fatty acyl CoA)
    as oxidation agents, 458
    succinyl CoA, 1188
Collagen, 1022, 1052, 1053f
Colors of elements in molecular art, 25, A-3
Column, in periodic table, 9
Combustion
    of alkanes, 163, 165–166
    of ethanol, 345f
    of gasoline, 1119
Common names
    acid chlorides, 737
    acyl phosphates, 737
    alcohols, 340–341
    aldehydes, 649–650, 649t, 650f
    alkene oxides, 342–343
    alkenes, 389
    alkyl halides, 256, 256f
    alkynes, 389
    amino acids, 1022, 1023f
    anhydrides, 738
    benzene derivatives, 853
    carboxylic acid derivatives, 737
    carboxylic acids, 698
    description of, 144, 144f
    epoxides, 342–343
    epoxyalkanes, 342
    ethers, 341
    ketones, 650–651
    monosaccharides, 1071, 1077
    nitriles, 699, 699f
    oxiranes, 342
    primary amines, 971
Complementary base pairs, DNA, 1151, 1155–1156, 1155f
Compounds, 11–12
Concentration, and reaction rates, 240, 241–242
Concerted reactions
    description of, 223
    Diels–Alder reactions, 491
    E2 elimination reactions, 308–309
    S_N2 substitution reactions, 268–269
Condensation polymers. *See* Step-growth polymers
Condensation reactions, 813–839. *See also* Carbonyl condensation reactions
Condensed structures, 28–30, 29f
Configurations
    amino acids, 1022
    definition, 179
    inversion of (*See* Inversion of configuration)
    prochirality and, 458–459
    retention of (*See* Retention of configuration)
    R or *S* designations, 189–194, 198
Conformations
    of acyclic alkanes, 147–150
    anti, 150–151, 151f, 152f
    of butane, 150–153, 305–306
    of cycloalkanes, 153–163
    of cyclohexane, 154–157, 578
    definition, 147
    description of, 147–150
    of 1,3-dienes, 482–483, 492
    eclipsed (*See* Eclipsed conformations)
    of ethane, 147–150
    gauche, 150–151, 151f, 152f
    of methylcyclohexane, 157–159, 158f
    NMR spectra and, 578
    peptides, 1033

Conformations—(Cont.)
ring-flipping, 156–158, 156f, 160–163
rotation in, 147, 150
Congo red, 1007
Conjugate acids
carboxylic acids as, 706
description of, 60–62, 64
relative basicity comparison, 987–988
Conjugate addition
in citric acid cycle, 1189
description of, 486
in fatty acid catabolism, 1177
in Michael reaction, 833–834
in Robinson annulation, 836
α,β-unsaturated carbonyl compounds, 630–631
Conjugate bases
of carboxylic acids, 705–708
description of, 60–62, 64
Conjugated dienes
addition reactions, 486–498
carbon–carbon bond length, 484–485
comparison with isolated dienes, 484, 485
conformations of, 482–483
conjugation in, 474–475, 475f
Diels–Alder reactions, 490–498 (See also Diels–Alder reactions)
electrophilic addition, 486–488
electrostatic potential, 475, 475f
examples of, 483, 484
hydrogenation of, 485–486
kinetic versus thermodynamic products, 488–490, 489f, 490f
1,2- versus 1,4-addition, 486–490
polymerization of, 1266
reactivity, 492
resonance structures, 478
retrosynthetic analysis, 496–497
stability of, 485–486
stereoisomers of, 482–483
Conjugated double bonds, 478, 817
Conjugated polyenes, 1231, 1236
Conjugated proteins, 1053
Conjugated systems
allyl carbocations, 475–476
aromatic hydrocarbons, 850 (See also Aromatic hydrocarbons; specific aromatic hydrocarbons)
description of, 474
1,3-dienes, 474–475, 482–498
hybridization in, 481
IR absorption shifts, 535
molecular orbital patterns, 873f
Conjugation
aromaticity criteria, 859
carbonyl absorption and, 534–536, 535f
description of, 474
β-dicarbonyl compounds, 781
hybridization and, 481–482
hyperconjugation, 277, 278–279
in synthetic dyes, 1007
Conrotatory, 1232–1236, 1236t
Constitutional isomers
acyclic alkanes, 134, 135t
of alkenes, 350
chlorination of alkane, 939
definition of, 17
description of, 179, 179f, 200, 200f
dipeptides, 1031–1032
of monosaccharides, 1071
in olefin metathesis, 1212f
tautomers (See Tautomers)

from unsymmetrical alkenes, 400–401
Consumer products containing polymers, 1256f
Coordination polymerizations, 1265
COPD (chronic obstructive pulmonary disease), 258
Cope rearrangements, 1243–1245
Copolymers
description of, 1263
in polyester synthesis, 1269
Copper(I) bromide, 1000
Copper(I) chloride, 1000
Copper reagents. See Organocuprate reagents
Corals, 7
Corey, E. J., 5, 373
Corn
ethanol production from, 344, 345f
as polymer feedstock, 1275, 1276t
Cortisol, 1137
Cortisone, 654, 1137
Costunolide, 1128
Cotton, 1101
Counterclockwise S isomer, 191–193
Counterions, 59
Coupled protons, 570
Coupled reactions, 233, 234f, 1172
Coupling constant, 569, 575
Coupling reactions
of aryl diazonium salts, 1004–1005
with organometallic reagents, 1198–1207 (See also Organometallic coupling reactions)
Covalent bonds
in carbon compounds, 41t
definition of, 11
homolysis of, 936
Lewis acid–base reactions, 80
polar, 44–45
vibrational modes, 531, 532–533
Covalent molecules
boiling point, 104
intermolecular forces, 101–103, 103t
melting points of, 105–106
solubility of, 110t
COX-1, 1126
COX-2, 1126
CPK (creatine phosphokinase), 1057
Crafts, James, 887
Creatine phosphate, 1172
Creatine phosphokinase (CPK), 1057
Crick, Francis, 1150
Crossed aldol reactions
description of, 818–821, 820f
directed aldol reactions, 821–822
Crossed Claisen reactions, 827–829
Cross formula, 1072
Cross-linked polymers, 1273
Crown ethers, 115, 345–346, 346f
Crutzen, Paul, 949
Crystallites, 1272–1273
Crystal meth. See Methamphetamine
C-Terminal amino acids, 1032, 1157
Cubane, 144f
Cucumber aldehyde, 652
Curved arrow notation, 19–21, 60, 224, 225–226, 226t
Cyanide, 660–662
o-Cyanobenzoic acid, A-7f
α-Cyano carbonyl compounds, 818–820
2-Cyanocyclohexanecarboxylic acid, 699
Cyano groups
in benzonitrile synthesis, 1000
named as substituent, 699, 699f

in nitrile reduction to amine, 981
on nitriles, 696
Cyanohydrins
formation of, 660–661
in Kiliani–Fischer synthesis, 1094, 1096
naturally-occurring derivatives, 661–662
in Wohl degradation, 1094–1095
Cyclic acetals, 675
Cyclic acid chlorides, 737
Cyclic acyl phosphates, 737
Cyclic amides, 735
Cyclic anhydrides, 735
Cyclic compounds
aromaticity, 859–860
molecular orbital patterns, 873f
stereogenic centers in, 186–189
Cyclic dienophiles, 494
Cyclic esters, 735
Cyclic hemiacetals
acetals from, 681–682
formation of, 680–681, 683
Cyclic ketones, 650–651
Cyclization
in cholesterol biosynthesis, 1134f, 1135
in electrocyclic reactions, 1231–1236
of hydroxy aldehydes, 680–681, 683
in terpene/terpenoid biosynthesis, 1132
Cycloaddition reactions
[2 + 2] cycloadditions, 1237, 1239–1240
[4 + 2] cycloadditions, 1237, 1238–1239
description of, 1227, 1236–1237
Diels–Alder reactions (See Diels–Alder reactions)
orbital symmetry, 1237–1238
Woodward–Hoffmann rules, 1240, 1240t
Cycloalkanes
common names, 144, 144f
conformations of, 153–163
description of, 135–136
disubstituted, 159–160, 198–199
molecular formula, 133, 386
monosubstituted, 157–159, 158f
nomenclature, 142–144, 143f
substituted, 157–163
Cycloalkenes, 388, 388f
Cyclobutadiene, 860, 869, 873–874
Cyclobutane, 136, 153, 154f
Cyclobutene, 1227, 1231
Cyclodecane, 154f
Cycloheptane, 154f
Cyclohepta-1,3,5-triene, 860, 863
Cyclohexa-1,3-diene, 857–858, 860, 1227, 1231
Cyclohexanamine, 971, 986, 996
Cyclohexane
as achiral reactant, 402
boat conformation, 156–157, 156f, 157f
chair conformation, 154–156, 155f, 160–162, 319–320, 319f
conformations of, 154–163, 578
disubstituted, 160–163
E2 elimination in, 319–321
formation with catalyst, 243
in mango, 133
monosubstituted, 157–159, 158f
proton NMR spectrum, 578
ring-flipping, 156–158, 156f, 160–163
skeletal structures, 30, 136

Cyclohexanecarbonyl chloride, 737
1,4-Cyclohexanedimethanol, 1271
cis-Cyclohexane-1,2-diol, 452
Cyclohexanol
dehydration of, 355
extraction from mixture, 712–713, 986
hydration synthesis, 403
mass spectrum, 519f
Cyclohexanone
hybridization in, 41
nucleophilic addition reactions, 658f
synthesis, 792
tertiary alcohol preparation from, 618
Cyclohexene
addition reactions, 397f
cyclohexanol dehydration product, 355
dihydroxylation, 451
in Friedel–Crafts alkylation, 896
halogenation, 397f, 404–407, 406f
halohydrin formation, 397f, 407–408
heat of hydrogenation, 857–858
hydration, 397f, 403–404, 412
hydroboration–oxidation, 397f, 408–412, 409f, 411t, 412f
hydrohalogenation, 397f, 399–403, 403t
as Lewis base, 80
reaction with hydrogen, 243
reduction, 164
Cyclohex-2-enone, 792, 835
Cyclohexylamine, 971
Cyclooctane, 207, 207f
Cyclooctatetraene, 859
Cyclooxygenase, 1124, 1125f, 1126
Cyclopentadiene, 495, 497, 863
Cyclopentadienyl anion, 862–863, 873
Cyclopentadienyl cation, 862–863
Cyclopentadienyl radical, 862–863
Cyclopentane, 136, 154f, 940–941
Cyclopentanecarboxylic acid, 799
Cyclopentane-1,2-diol, 341, 452
Cyclo- (prefix), 135, 142
Cyclopropanation, 1207–1211
Cyclopropane, 136, 153, 1207–1211
Cysteine
C–Z σ bonds in, 97
isoelectric point, 1025t
peptide formation with alanine, 1031–1032
pK_a values, 1025t
structure of, 1023f
Cytarabine, 1144
Cytidine, 1146
Cytidine 5′-monophosphate (CMP), 1146–1147, 1147t
Cytosine, 849, 868–869, 1145–1146, 1147t, 1149f, 1151, 1151f, 1152

D

2,4-D, 889f
Dacron, 1255, 1268
dAMP (2′-deoxyadenosine 5′-monophosphate), 1146–1147
Darvon (propoxyphene), 183
Dashed wedge, 26, 159
Daughter DNA, 1152, 1153, 1153f
DBN, 309
DBU, 309
DCC (dicyclohexylcarbodiimide), 755–756
DDE (dichlorodiphenyldichloroethylene), 304
ddI (dideoxyinosine), 1162

DDT (dichlorodiphenyltrichloroethane)
   degradation of, 304
   description of, 4
   environmental pollutant, 258, 259
   reaction rates and, 240
Deamination, 672–673
Decalin, 1132–1133
Decamethrin, 1207
Decane
   as normal alkane, 135t
   in petroleum, 132
Decarboxylases, 1055t
Decarboxylation
   in acetoacetic ester synthesis, 801
   of amino acids, 802–803
   in biological Claisen reactions, 832
   in citric acid cycle, 1187f, 1188
   in malonic ester synthesis, 797, 798
DEET, 749
Degenerate orbitals, 871
Degrees of unsaturation
   benzene, 850
   calculation of, 386–387
   hydrogenation and, 440, 441f
   hydrogenation data and, 439–440
Dehydrases, 1055t
Dehydration reactions
   of alcohols, 350–355
   aldol products and, 816–817
   alkene synthesis, 350–355
   carbocation rearrangement in, 352–355
   carboxylic acid conversion to amide, 755–756
   defined, 350
   dicarboxylic acid conversion to anhydrides, 753
   E1 mechanism in, 351–352
   E2 mechanism in, 352, 355
   in glycolysis, 1183
   in intramolecular aldol reactions, 823
   Le Châtelier's principle in, 352
   in phosphorus oxychloride and pyridine, 355
   in Robinson annulation, 836
Dehydrogenases, 1055t, 1173
Dehydrohalogenation
   alkyne synthesis, 322–323, 323f
   bases used in, 303, 303t
   description of, 303–304
   E1 mechanism of, 314
   E2 mechanism of, 308, 312
   as stereoselective, 313
   of substituted cyclohexanes, 319–321
Delocalization
   conjugation and, 474
   resonance and, 18, 72, 481
δ (Delta) scale, 556
Demerol, 108
Denaturation of proteins, 1051
Dendrobates histrionicus, 384
2'-Deoxyadenosine, 1146, 1147t
2'-Deoxyadenosine 5'-monophosphate (dAMP), 1070, 1070f, 1146–1147, 1147t
2'-Deoxycytidine, 1147t
2'-Deoxycytidine 5'-monophosphate (dCMP), 1147t
2'-Deoxyguanosine, 1147t
2'-Deoxyguanosine 5'-monophosphate (dGMP), 1147, 1147t
Deoxy- (prefix), 1146
Deoxyribonucleic acid. *See* DNA
Deoxyribonucleosides, 1146
2'-Deoxy-D-ribose, 1145, 1146
2'-Deoxythymidine, 1147t

2'-Deoxythymidine 5'-monophosphate (dTMP), 1147t
Depression. *See* Antidepressant drugs
Deprotection, 622–624, 623f, 994f
Deprotonation
   in acetoacetic ester synthesis, 801
   of alanine, 717–718
   alkynes, 398, 398t
   in cholesterol biosynthesis, 1134f, 1135
   cyanohydrins, 661
   description of, 66–67
   in electrophilic aromatic substitution, 887–888
   in enolate alkylation, 794
   Henderson–Hasselbalch equation, 708–710, 709f
   in malonic ester synthesis, 798
   in racemization at the α carbon, 790
   in tautomerization, 414, 782
Dermabond, 1262
DET (diethyl tartrate), 461
Deuterium, 8
Dexamethasone, 97
Dextrorotatory compounds, 202, 1076
Diacids
   β-diacids, 797
   monosaccharide oxidation synthesis, 1092, 1093–1094
   nomenclature, 698
   in polyester synthesis, 1269
Diacon, 620
Dialkylamides, 786
Dialkylborane, 409, 417
1,4-Diaminobenzene, 1268
1,6-Diaminohexane, 1267, 1274, 1275f
Diastereomers
   of alkenes, 306
   of but-2-ene, 306
   definition, 196, 200, 200f
   of disubstituted cycloalkanes, 199
   vs. enantiomers, 194–196
   epimers, 1077
   of glucose, 1081
   of monosaccharides, 1077
   physical properties of, 205–206
   proton NMR signals, 561
   resolution process, 1026–1028, 1027f
Diastereotopic protons, 561
1,3-Diaxial interactions, 158, 158f
Diazomethane, 1210
Diazonium ion, 998
Diazonium salts
   azo coupling, 1004–1005
   in benzene derivative synthesis, 1000–1003
   reactions of, 999–1005
   synthesis of, 998–999
Diazotization, 998–999, 1001–1003
DIBAL-H (diisobutylaluminum hydride), 608–609, 612t, 721–722
Diborane, 409
Dibromobenzene
   carbon NMR spectra, 855, 855f
   isomers of, 853
   nomenclature, 853
1,4-Dibromobutane, 799
2,3-Dibromobutane, 196–197
1,3-Dibromocyclopentane, 198–199
2,3-Dibromopentane, 194–196
Di-*tert*-butyl dicarbonate, 1040
Dibutyl phthalate, 1274
Dicarbonyl compounds
   in crossed aldol reactions, 820

β-dicarbonyl compounds, 781, 820, 828–829
1,4-dicarbonyl compounds, 822
1,5-dicarbonyl compounds, 823–824, 834, 835–836
   formation of, 828–829
   pK_a values, 781
   tautomers, 781
Dicarboxylic acids
   anhydrides from, 753
   nomenclature, 698
2,5-Dichloroaniline, 854
4,4'-Dichlorobiphenyl, 109
Dichlorocarbenes, 1208
trans-1,2-Dichlorocyclopentane, 405
Dichlorodifluoromethane, 950
Dichlorodiphenyldichloroethylene (DDE), 304
Dichlorodiphenyltrichloroethane. *See* DDT
Dichloromethane, 258f, 457, 711
2,4-Dichlorophenoxyacetic acid, 889f
trans-1,3-Dichloropropene, 577
Dicyclohexylcarbodiimide (DCC), 755–756
Dicyclohexylurea, 756
Dicyclopentadiene, 497
Dideoxyinosine (ddI), 1162
Dieckmann reactions, 829–830
Diels, Otto, 490
Diels–Alder reactions
   description of, 490–491, 1227
   diene reactivity, 492
   dienophile reactivity, 492–493
   LSD synthesis, 473
   retro reaction, 497
   retrosynthetic analysis of product, 496–497
   rule of endo addition, 494–496
   rules governing, 492–496
   stereospecificity in, 493–494, 1238
   steroid synthesis by, 497–498
   as thermal [4 + 2] cycloaddition, 1237
Dienes
   conjugated (*See* Conjugated dienes)
   Cope rearrangements, 1243
   1,3-dienes (*See* Conjugated dienes)
   isolated, 474–475, 475f
   ozonolysis of, 454
Dienophiles, 490–498, 493f
Diepoxides, 1226
Diesters
   in Dieckmann reactions, 829–830
   1,3-diesters, 785t
   pK_a values, 785t
Diethyl carbonate, 828–829
cis-1,2-Diethylcyclopropane, 1211
Diethyl ether
   as anesthetic, 95–96, 345
   as extraction solvent, 711
   Grignard reagent preparation, 614
   infrared spectrum, 540
   nomenclature, 341
   as solvent, 364
Diethyl malonate, 796–800
Diethyl sulfide, 367
Diethyl tartrate (DET), 461
N,N-Diethyl-*m*-toluamide (DEET), 749
Digestion, 1170
Digoxin, 682
Dihalides, 322, 323f
Dihalocarbenes, 1208–1209
Dihedral angle, 147, 150–151, 150f, 152f, 318
Dihydroxyacetone, 1071, 1078

Dihydroxyacetone phosphate, 783, 830–832, 1176, 1180–1182, 1180f
1,4-Dihydroxybenzene, 1271
Dihydroxylation, 447f, 451–453
1,4-Dihydroxynaphthoic acid, 897, 897f
Diisobutylaluminum hydride (DIBAL-H), 608–609, 612t, 721–722
Diisocyanates, 1269
Diisopropylamine, 787, 972
β-Diketone, 785t
Dimers, 700, 700f
1,2-Dimethoxyethane, 557
N,N-Dimethylacetamide, 67, 536
Dimethylallyl diphosphate, 477, 1129
N,N-Dimethylbenzamide, 739
1,3-Dimethylbenzene, 911
2,3-Dimethylbutanal, 648
3,3-Dimethylbutan-2-ol, 352–354
2,2-Dimethylbut-3-enal, 652
cis-3,4-Dimethylcyclobutene, 1231, 1234, 1235
cis-5,6-Dimethylcyclohexa-1,3-diene, 1231
trans-5,6-Dimethylcyclohexa-1,3-diene, 1234, 1235
1,2-Dimethylcyclohexane, 143
1,4-Dimethylcyclohexane, 160–163, 161f
1,6-Dimethylcyclohexene, 388f
1,2-Dimethylcyclopentane, 159
1,3-Dimethylcyclopentane, 206
Dimethyl ether, 17
4,5-Dimethylhexanoic acid, 697
2,4-Dimethylhexan-3-one, 633
2,4-Dimethylhexanoyl phosphate, 740
6,6-Dimethyloct-3-yne, 388
2,2-Dimethyloxirane, 342, 371, 371f
2,3-Dimethyloxirane, 370
2,3-Dimethylpentane, 140, 517
1-(1,3-Dimethylpentyl)-2-methylcyclo-hexane, A-6
2,2-Dimethylpropane, 106, 134
2,2-Dimethylpropanoic acid, 710
2,2-Dimethylpropanoyl phosphate, 737
Dimethyl sulfide, 453
Dimethyl sulfoxide (DMSO)
   in alkyne synthesis, 322
   in halohydrin formation, 407
   octet rule exception, 18
   as polar aprotic solvent, 265, 285t, 286
Dimethyl terephthalate, 1271, 1277–1278
1,3-Dinitriles, 785t, 820
Dinoprostone, 1124
Dinucleotides
   flavin adenine dinucleotide, 194, 1174–1175
   formation of, 1149
   nicotinamide adenine dinucleotide (*See* Nicotinamide adenine dinucleotide)
-dioic acid (suffix), 698
Diols
   description of, 340
   dihydroxylation product, 451–453
   1,2-diols, 451–453
   gem-diols, 673–675
   in polycarbonate synthesis, 1270
   in polyurethane synthesis, 1269
-diol (suffix), 341
Dipeptides
   description of, 1031
   formation of, 1031–1032
   synthesis of, 1038–1043

Diphenhydramine, 852, 987
Diphenhydramine hydrochloride, 852, 987
Diphenyl carbonate, 1270
Diphosphates
  in biological reactions, 476–477
  functional group, 100
  as leaving groups, 287, 477, 1129, 1131
  nucleotides, 1148
Dipole–dipole interactions
  in alcohols, 343
  in aldehydes, 652
  in alkenes, 391
  in alkyl halides, 257
  boiling point and, 104
  in carboxylic acid derivatives, 741
  in carboxylic acids, 700
  description, 102, 103t
  in epoxides, 343
  in ethers, 343
  in ketones, 652
  melting point and, 105
  in nitriles, 700
  polar aprotic solvents, 265, 285
  polar protic solvents, 265, 285
Dipoles
  definition of, 44
  molecular polarity, 45–46
Di- (prefix), 140, 341, 972
Directed aldol reactions, 821–822
Direct enolate alkylation, 793–796
Disaccharides
  description of, 119
  glycosidic linkage in, 1098
  lactose, 1093, 1099
  maltose, 1098
  sucrose, 1099–1100, 1099f, 1100f
Disorder, entropy and, 234, 235
Disparlure, synthesis of, 449–450, 450f
Disproportionation, 1258–1259
Disrotatory, 1232–1236, 1236t
Dissolving metal reduction, 436, 443–444
Disubstituted alkenes
  description of, 305, 305f
  in E2 elimination reactions, 310–311, 395
  in Hofmann elimination, 996–997
  in olefin metathesis, 1211
Disubstituted benzene derivatives
  electrophilic aromatic substitution reactions, 910–911
  nomenclature, 853
  synthesis of, 912–913
  trisubstituted benzene synthesis from, 922–923
Disubstituted cycloalkanes
  conformations, 159–160
  isomers, 198–199, 369, 371
Disulfide bonds
  in α-keratins, 1052, 1052f
  in peptides, 1034
  in rubber, 1266, 1266f
  in tertiary structure of proteins, 1049, 1050f
Disulfides
  functional group, 95, 96t
  thiol oxidation product, 367, 1034
Diterpene/diterpenoid, 1129t
DMSO. See Dimethyl sulfoxide
DNA
  bases, 849, 868–869, 1145, 1151f
  complementary strand, sequence of, 1152
  description of, 120–121, 120f

double helix structure, 1150–1152, 1151f
  human genome in, 1152
  hydrogen bonding in, 102, 102f
  polymerase chain reaction, 1161–1162, 1161f
  replication, 1152, 1153–1154, 1153f
  reverse transcription by viruses, 1162
  sequencing, 1159–1160
DNA ligase, 1154
DNA polymerase, 1153, 1161, 1161f
Dobutamine, 971
Dodecahedrane, 144
Dodecane, 135t
Domagk, Gerhard, 1007
Donepezil, 99, 822
L-Dopa, 186, 803
Dopamine, 97, 803, 977, 978f
Double bonds
  in alkenes, 304–305
  in alkyl halides, 255
  in amide reduction, 611
  in azo compounds, 999, 1004
  in bromination, 917–918
  in buta-1,3-diene, 484–485
  in carbonyl groups, 97–98, 599–600
  in chain-growth polymerizations, 1258
  chemical shift, NMR spectra, 565
  in cholesterol biosynthesis, 1135
  in citric acid cycle, 1188
  composition of, 39
  condensed structures, 29, 29f
  conjugated, 478, 817
  dienes, 388
  in dienophiles, 492–493
  enynes, 388
  enzymes in addition or elimination, 1055, 1055f
  hydrogenation, 438, 440, 1119
  infrared absorptions, 533, 533t
  isomerization, 441
  Lewis acid–base reactions, 80
  in Lewis structures, 14
  melting point of fatty acids and, 394, 394t
  in metathesis catalysts, 1211
  in olefin metathesis, 1211–1216, 1212f
  one atom more electronegative than other, 478
  in oxidative cleavage, 453–454
  in peptide bond, 1045
  in peptide hydrolysis, 1057
  in β-pleated sheet, 1047
  polyfunctional compounds with, nomenclature, A-8t
  radical addition reactions, 955–957
  in radical polymerization, 958
  radical reactions, 938
  radical reactions with, 938
  in reductive amination, 982–983
  resonance structures, 22, 23
  rotation restriction, 39
  in secondary structure of proteins, 1046–1047
  splitting patterns, 575–577
  strength, 42
  trans double bonds, 1189
  in triacylglycerols, 393–395
  Zaitsev rule, 312–313
Double dagger notation, 237
Double-headed arrow notation, 18
Double helix structure of DNA, 1150–1152, 1151f
Double reaction arrow, 60

Doublets, NMR spectra, 568–569, 571, 571t, 572, 575–576, 576f
Doublets of doublets, NMR spectra, 575–576, 576f
Doxorubicin, 1070, 1070f
Drugs. See also names of specific drugs
  nomenclature, 136–137
Drug testing for THC, 522, 522f
du Vigneaud, Vincent, 1021
Dyes, 1004–1007
Dyneema, 1265

E
E. coli, 1275
E1cB mechanism, 817, 823, 1183
Eclipsed conformations
  of butane, 150–153, 151f, 152f
  description of, 147–150
  of 2,3-dibromopentane, 194
  of ethane, 148–149, 148f
  of propane, 148f
EcoRI, 1160
Ecteinascidia turbinata, 5
Ecteinascidin 743, 5
Edman degradation, 1035–1036
E1 elimination reactions
  dehydration of alcohols, 351–352
  description of, 308
  E2 mechanism compared, 315, 321, 321t, 323–327, 324t
  energy diagrams, 314, 315f
  features of, 314, 316t
  kinetics of, 314
  mechanism, 314
  R groups in, 315
  S_N1 substitution reactions compared, 317, 323–327, 324t
  S_N2 substitution reactions compared, 323–327, 324t
  as unimolecular reaction, 314
  Zaitsev rule, 316
E2 elimination reactions
  alkyl halide identity in, 310–311
  alkyl tosylates reactions, 362
  alkyne synthesis, 322–323, 323f
  anti periplanar geometry in, 318–321, 318f
  as concerted reaction, 308–309
  dehydration of alcohols, 352, 355
  E1 mechanism compared, 315, 321, 321t, 323–327, 324t
  energy diagrams, 308–309, 309f
  features of, 308, 311t
  Hofmann elimination, 995–997
  kinetics of, 308
  leaving group and, 310
  mechanism, 308
  as regioselective, 312
  R groups in, 310–311
  S_N1 and S_N2 substitution reactions compared, 323–327, 324t
  stereochemistry of, 318–321
  as stereoselective, 313
  syn periplanar geometry in, 318, 318f
  trans diaxial geometry in, 319–320, 319f
  Zaitsev rule, 312–313
Effexor, 622
E-85 fuel, 344, 345f
Ehrlich, Paul, 1007
Eicosanoids, 1124–1126, 1125f, 1125t
Eicosapentaenoic acid, 110, 1120
Elastomers, 1266
Electrocyclic reactions
  description of, 1227, 1231

orbital symmetry, 1232
  photochemical, 1230, 1234–1236
  stereochemistry of, 1231–1232
  thermal, 1230, 1232–1234
  Woodward–Hoffmann rules for, 1236, 1236t
Electromagnetic radiation, 529–530, 529f
Electromagnetic spectrum, 529, 529f, 554
Electron cloud, 8
Electron-deficient sites/species
  carbon, 71, 116
  carbonyl carbon, 493, 600
  carbonyl group, 599
  in electrostatic potential maps, 45
  functional groups and, 116–117
  heteroatoms and, 92
  Lewis acids, 79
  nucleophile attack on, 263
  reactivity with electron-deficient species, 62
Electron-donating groups, 898–900, 904, 909, 988–989, 992t, 1004–1005, 1200–1201
Electron dot representation. See Lewis structures
Electronegativity
  acidity and, 68–71, 74f
  in aliphatic carboxylic acids, 710–711
  carbon–halogen bonds, 259
  carbon–Z σ bonds (See Carbon–Z σ bonds)
  definition of, 43
  of halogens, 914
  heterolysis and, 224
  organometallic reagents, 614
  periodic trends, 43–44, 44f
  resonance structures and, 478
Electron pair acceptors, 59, 78
Electron pair donors, 59, 62, 78
Electron pairs
  in Brønsted–Lowry acid–base reactions, 60
  in Brønsted–Lowry acids and bases, 59
  curved arrow notation, 224
  in Lewis acid–base reactions, 79–80
  in Lewis acids and bases, 78–79
Electron-rich sites/species
  alkenes, 938
  bases, 259
  benzene ring, 852, 852f
  carbon, 116
  carboxylate anion, 754
  dienes, 475f, 496
  on electrostatic potential maps, 45–46
  functional groups and, 116–117
  Lewis bases, 79
  nucleophiles, 260, 814
  pi (π) bonds, 396, 397f, 955
  reaction with electron-deficient species, 62
  stereochemistry of E2 reaction, 318
Electrons
  bonded (shared), 12
  in bond formation, 225
  curved arrow notation, 224
  delocalized, 18, 22
  density, 9, 45, 45f
  electronegativity, 43–45
  in Lewis structures, 12–17
  lone pairs (See Lone pairs)
  nonbonded (unshared), 12, 12f
  number of, 8, 10, 17
  in resonance structures, 18–23

shielding and deshielding effects in NMR, 562–563, 562f, 563f
valence, 10, 11–12, 16t
Electron transport chain, 1172, 1175
Electron-withdrawing groups, 898–900, 904, 909–910, 913–915, 989
Electrophiles
  carbonyl carbon, 97, 647
  description of, 79
  diazonium salts, 1004
  dihalocarbenes as, 1208
  formation in Friedel–Crafts alkylation, 892–893
  heterolysis and, 225
  reactivity of, 116–117
Electrophilic addition reactions
  1,2- versus 1,4-addition, 486–490
  to alkenes, 399, 403t
  cationic polymerization, 1260
  in conjugated dienes, 486–488
  in dienes, 486–488
  hydration (See Hydration)
  hydrohalogenation (See Hydrohalogenation)
  kinetic versus thermodynamic products, 488–490, 489f, 490f
  list, A-16
  Markovnikov's rule, 400–401
  stereochemistry of, 401–403
  of water, 403–404
Electrophilic aromatic substitution reactions
  azo coupling, 1004–1005
  description of, 886, 887f
  of disubstituted benzenes, 910–911
  Friedel–Crafts acylation, 887f, 891–897, 918–919 (See also Friedel–Crafts acylation)
  Friedel–Crafts alkylation, 887f, 891–897 (See also Friedel–Crafts alkylation)
  general mechanism, 887–888
  halogenation, 887f, 888–889, 909
  limitations on reactions, 908–910
  list, A-15
  nitration, 887f, 890–891, 912, 921
  rate of reactions, 901, 904–905
  of substituted benzenes, 901–913
  sulfonation, 887f, 890–891
Electropositivity, 44
Electrospray ionization (ESI), 523
Electrostatic potential plot
  of acetylene, 396, 397f
  of benzene ring, 852, 852f
  of carbanions, 72f
  of carbocations, 278, 278f
  of carbon dioxide, 46f
  description of, 45, 45f, 46f
  of dienes, 475, 475f
  of enolates, 790, 790f
  of ethylene, 396, 397f
  of pyridine, 865, 865f
  of pyrrole, 865, 865f
  of water, 46f
Element effects on acidity, 68–70, 74f
Elimination reactions
  of alcohols, 349, 350–355, 362, 364f
  alkene synthesis, 303
  alkyl halides, 255, 259, 303–304, 308–327
  description of, 221–222
  dihalocarbene synthesis, 1208
  α elimination, 1208
  β elimination, 303–304
  E1cB mechanism, 817

features of, 303–304
  in glycolysis, 1182
  α-halo carbonyl compounds in, 792–793
  Hofmann elimination, 995–997
  list, A-16
  mechanisms, 308 (See also E1 elimination reactions; E2 elimination reactions)
  in nucleophilic aromatic substitution, 913–916
  in olefin metathesis, 1213
  pi (π) bonds in, 303
  reductive, 1201–1207
  Zaitsev rule on, 312–313
Enals, 652
Enamines
  formation of, 658f, 668, 669f, 993
  hydrolysis of, 669–670
Enantiomeric excess, 203–205, 461
Enantiomers
  amino acid resolution, 1026–1029, 1027f
  of amino acids, 716
  biochemical properties, 206–207
  carbene addition, 1209
  in carbonyl reduction, 606–607
  chemical properties of, 206–208
  chiral drugs, 206–207
  of cyclic compounds, 187–189
  definition, 181, 200, 200f
  vs. diastereomers, 194–196
  in Diels–Alder reactions, 493–494
  of disubstituted cycloalkanes, 199
  drawing, 184–186, 185f
  in enolate alkylation, 794–795
  in enolates, 790
  in epoxidation reactions, 449
  in epoxide ring opening reactions, 369–370
  halogenation products, 405–407, 947
  halohydrin products, 408
  hydrohalogenation products, 402
  meso compounds, 196–197
  of monosaccharides, 1072–1077, 1077f
  nucleophilic addition of organometallic reagent, 618–619
  optical activity, 201–202
  physical properties of, 201–206, 203t
  proton NMR signals, 561
  racemic mixtures, 202–203, 203t
  resolution of, 205, 1026, 1027f
  sense of smell and, 207–208, 207f
  in S_N1 substitution reactions, 275–276
  in S_N2 substitution reactions, 270
Enantioselective reactions, 460–462, 1029–1031
Enantiotopic protons, 561
Endergonic reactions, 235
Endiandric acids, 1254
Endo addition, 494–496
Endo product, 495–496
Endothermic reactions
  bond breakage and, 227–228
  bromination, 944–945
  carbocation formation, 281–282, 281f
  enthalpy change and, 237f
  transition state, 279–280, 280f
Enediols, 782, 1180–1181
Energy
  barrier to rotation, 152
  nonrenewable, 145
  steric strain, 151, 152t
  torsional, 149
Energy of activation (E_a)
  catalysts and, 243

description of, 236–237
  rate constant and, 241
  reaction rate and, 241–242, 280
Energy barrier, 237
Energy diagrams
  chlorination of ethane, 942, 942f
  description of, 236–238, 237f
  E1 elimination reactions, 314, 315f
  E2 elimination reactions, 308–309, 309f
  electrophilic addition reactions, 400
  endothermic reactions, 280f
  kinetic versus thermodynamic products, 488–490, 489f, 490f
  S_N1 substitution reactions, 274, 274f, 281f
  S_N2 substitution reactions, 268, 269f
  of two-step reaction mechanism, 238–240, 239f
-ene (suffix), 387
Enolates
  acetoacetic ester synthesis, 800–802
  achiral, 790
  acid-catalyzed halogenation at the α carbon, 791–792
  aldehydes to, 656
  alkylation, 793–796
  bases in formation of, 786–787, 786t
  biological decarboxylation, 802–803
  description of, 783–784
  electrostatic potential plot, 790, 790f
  esters to, 784
  examples of, 784–785
  formation of, 786–787
  general reactions of, 787, 790–791
  halogenation at the α carbon, 791–793
  ketones to, 656
  kinetic, 788
  malonic ester synthesis, 796–800
  as nucleophiles, 787, 790, 826, 834
  racemization, 790, 790f
  reactions with other carbonyl compounds, 790–791, 813–839 (See also Carbonyl condensation reactions)
  resonance-stabilized, 20, 630, 783–784
  substitution reactions, 790–803
  tertiary amides to, 784
  thermodynamic, 789
  unsymmetrical ketone alkylation, 795–796
Enol phosphates, 1183
Enols
  in acid-catalyzed halogenation, 791–792
  in alkyne hydration, 414–416
  biological reactions, 782–783
  as nucleophiles, 783
  resonance structures, 783
  structure of, 338
  tautomers, 414–416, 630, 780–783, 792
-enol (suffix), A-8
Enones, 652
-enone (suffix), A-8
Enthalpy change ($\Delta H°$)
  bond dissociation energy and, 228–229, 941–942
  bromination of propane, 944–945
  calculation of, 229–230
  chlorination of ethane, 941–942
  chlorination of propane, 945–946
  description of, 227
  free energy change and, 234–236

negative, 227, 229
  positive, 227, 229
Entropy change ($\Delta S°$), 234–236
Environmentally-friendly polymers
  biodegradable, 1275, 1278–1279
  recycling of, 1277–1278, 1277t
  synthesis of, 1274–1275, 1275f, 1276f
-enyne (suffix), A-8
Enzymatic decarboxylation, 802–803
Enzymes. See also Coenzymes
  aconitase, 1187f
  catalysis by, 1056–1057
  as chiral reagent, 1028
  citrate synthase, 1187f
  classification, 1055, 1055t
  description of, 244, 244f, 1054
  diagnosis and treatment of diseases, 1057–1058, 1058f
  in digestion, 1171f
  fumarase, 1187f
  isocitrate dehydrogenase, 1187f
  α-ketoglutarate dehydrogenase, 1187f
  malate dehydrogenase, 1187f
  as oxidation reagents, 458
  peptide hydrolysis, 1037–1038
  succinate dehydrogenase, 1187f
  succinyl CoA synthetase, 1187f
Enzyme–substrate complex, 244
Ephedrine, 183
Epichlorohydrin, 1271, 1272f
Epimers, 1077, 1095–1096
Epinephrine. See Adrenaline
Epothilone A, 1214
Eplerenone (Inspra), 346
Epoxidation
  description of, 447–448, 447f
  disparlure synthesis, 449–450, 450f
  Sharpless epoxidation, 460–462
  of squalene, 1134f, 1135
  stereochemistry of, 448–449
Epoxides. See also specific epoxides
  acetylide anions, reactions with, 420
  acids, reactions with, 370–372
  anionic polymerization, 1263–1264
  boiling point, 344t
  description of, 338
  enantiomers, 369–370
  melting point, 344t
  nomenclature, 342–343
  nucleophilic substitution, 349, 368–372
  organometallic reagent reactions with, 628–629
  physical properties of, 343–344, 344t
  reactions of, 349, 368–372
  reduction of, 445–446
  ring opening reactions, 368–372, 628–629, 1263–1264
  solubility, 344t
  structure of, 339
  synthesis of, 347–348, 447–450
  useful drugs, 346
Epoxyalkanes, 342
1,2-Epoxycyclohexane, 342, 369
1,2-Epoxy-2-methylpropane, 342
cis-2,3-Epoxypentane, 342
Epoxy (prefix), 342
Epoxy resins, 1271, 1272f
E (prefix), 389–390
Equal, 1100f
Equations for organic reactions, writing, 220–221, 220f
Equatorial hydrogens, 155, 155f, 156, 156f
Equatorial position, 157, 161

Index

Equilibrium
  acid–base reactions, 66–67
  Le Châtelier's principle, 352
  in nucleophilic substitution, 263
Equilibrium arrows, 60, 61, 66
Equilibrium constant (K_{eq}), 63, 231–233, 232f, 232t
Ergot fungus, 473
Eribulin mesylate, 4–5
D-Erythrose, 1077, 1078f
Erythroxylon coca, 733
D-Erythrulose, 1079f
Escitalopram, 703
ESI (electrospray ionization), 523
Essential amino acids, 716, 1023f, 1024. *See also specific amino acids*
Essential oils, 1127
Esterification, 753–755, 1269
Esters
  acetoacetic ester synthesis, 800–802
  acid chlorides to, 749
  anhydrides to, 750
  carbonyl absorption in, 536
  carboxylic acids to, 753–755
  of chrysanthemic acid, 583
  cyclic, 735
  description of, 599
  β-dicarbonyl compound from, 828
  in enolate alkylation, 794
  enolates from, 784
  functional group, 97, 98t
  hydrolysis reactions, 756–760
  infrared (IR) absorptions/spectra of, 542, 543t
  interesting compounds, 743–744
  β-keto esters from, 826–829
  malonic ester synthesis, 796–800
  monosaccharide conversion to, 1091
  nomenclature, 738, 740t
  nucleophilic acyl substitution, 756–758
  organometallic reagent reactions with, 616, 625–626
  phosphate esters, 100
  pK_a values, 785t
  reduction of, 608–610
  structure, 393, 734
Estradiol, 392–393, 392f, 1136t
Estrogen-dependent cancer treatment, 874–875, 876f
Estrogens
  aromatase inhibitors, 875
  as oral contraceptive, 392–393, 392f
  as steroids, 1136, 1136t
Estrone
  protecting groups and, 624
  as steroid, 498, 1136t
  synthesis of, 498, 834–835, 875, 876f
Ethambutol, 272
Ethane
  acidity, 73
  as alkane, 93, 133
  bond length and strength, 42t
  bonds in, 38, 41t
  chlorination of, 941–942, 942f
  conformations of, 147–150
  energy, 149, 150f
  molecular formula, 134
  Newman projection, 148f
  nonreactivity of, 92
  as normal alkane, 135t
Ethane-1,2-diol. *See* Ethylene glycol
Ethanethiol, 366
Ethanoic acid. *See* Acetic acid

Ethanol
  acidity, 72
  boiling point, 366
  description of, 3
  isomers of, 17
  oxidation to acetaldehyde, 459–460
  pK_a, 707
  as polar protic solvent, 265f
  prochiral carbon, 458–459
  production by fermentation, 3, 344, 345f
  proton NMR spectrum, 577, 577f
  pyruvate conversion to, 1185–1186
  reactivity of, 92–93
  as renewable fuel source, 344, 345f
  solubility, 109
  structure, 92–93
Ethene, 389. *See also* Ethylene
Ethers. *See also specific ethers*
  boiling point, 344t
  Claisen rearrangement, 1245
  crown, 115, 345–346, 346f
  description of, 338
  functional group, 95, 96t
  infrared (IR) absorptions/spectra of, 540, 543t
  melting point, 344t
  monosaccharide conversion to, 1090–1091
  nomenclature, 341–342
  physical properties of, 343–344, 344t
  products of S_N2 substitution reactions, 290, 290t, 346–347
  reactions of, 349, 364–365
  solubility, 344t
  strong acid, reactions with, 364–365
  structure of, 339
  synthesis, 346–347, 404
  Williamson ether synthesis, 346–347
Ethoxide, 72, 72f, 264, 707
Ethoxy groups, 341
4-Ethoxyoctane, 342
Ethyl, 137
Ethyl acetate
  carbonyl absorption of, 536
  enolate formation, 828
  ethyl acetoacetate from, 826
  nomenclature, 738
  synthesis of, 243
Ethyl acetoacetate, 800–802, 826
Ethyl *p*-aminobenzoate, 920
N-Ethylaniline, 973
Ethylbenzene, 853, 917
Ethyl benzoate, 828
Ethyl butanoate, 743
Ethyl chloride, 256
Ethyl chloroformate, 828–829
Ethyl α-cyanoacrylate, 1261f
Ethylcyclobutane, 143
N-Ethylcyclohexanamine, 984–985
2-Ethylcyclohexanecarbaldehyde, 648
4-Ethyl-3,4-dimethyloctane, 140
5-Ethyl-2,6-dimethyloctane, 138–140
Ethylene
  acidity, 73
  as alkene, 93
  bananas and, 305
  bond length and strength, 42t
  bonds in, 38–39, 39f, 41t
  Lewis structure, 14
  molecular orbitals in, 1228, 1229f
  molecular shape, 25
  nomenclature, 343, 389
  polymerization of, 1258f, 1265
Ethylene derivatives, 957

Ethylene glycol
  in acetal formation, 675
  from chemical recycling of PET, 1277–1278
  nomenclature, 341
  in PET synthesis, 1255, 1257f, 1269, 1277–1278
  in polyethylene furanoate synthesis, 1269
  in polyurethane synthesis, 1269
Ethylene monomers, 957
Ethylene oxide, 343, 1263–1264
1-Ethyl-2-fluorocyclopentane, 256f
N-Ethylformamide, 739
Ethyl formate, 828
Ethyl isopropyl ether, 347
Ethyl isopropyl ketone, 651f
1-Ethyl-3-methylcyclohexane, 142–143
N-Ethyl-*N*-methylcyclopentanamine, 972
Ethyl methyl ketone, 651
4-Ethyl-5-methyloctane, 140
Ethyl *p*-nitrobenzoate, 920
3-Ethylpentane, 944
Ethyl propyl sulfide, 368
3-Ethylthio-5-methyloctane, 367
Ethyne, 389. *See also* Acetylene
Ethynylestradiol, 392–393, 392f, 624, 1136
Ethynyl groups, 389
1-Ethynyl-2-isopropylcyclohexane, 389
Eucalyptus, 1127f
Euphorbia ingens, 1215
Euphorbia peplus, 1197
Euphorbia resinifera, 598
Exemestane, 189
Exergonic reactions, 235
Exo product, 495
Exothermic reactions
  addition reactions, 399
  bond formation and, 227–228
  chlorination of ethane, 942, 942f
  enthalpy change and, 236, 237f
  hydrogenation, 437
  transition state, 280–281, 281f
Extraction
  of amines, 986–987
  technique, 711–714, 713f
E,Z system of nomenclature, 389–390

F

FAD (flavin adenine dinucleotide), 194, 1174–1175
$FADH_2$, 194, 1174–1175, 1189
Faraday, Michael, 850
Farnesol, 1127f, 1131
Farnesyl diphosphate, 477, 1129–1131, 1130f, 1132
Fats
  fatty acid composition of, 1118, 1118t
  properties of, 395
  as triacylglycerols, 1117
Fat-soluble vitamins, 111, 1123, 1123f
Fatty acids. *See also specific fatty acids*
  antioxidants and, 954
  biosynthesis of, 817, 832–833
  catabolism of, 1170–1171, 1171f, 1176–1179, 1177f, 1190–1191
  E1cB biosynthesis, 817
  in lipid bilayer, 1121
  as lipid component, 122
  olestra from, 759
  physical properties of, 394–395
  triacylglycerols and, 122, 122f, 221, 393, 1117t

Fatty acyl AMP, 764, 765
Fatty acyl CoA, 765, 1174, 1176–1179, 1177f
Fehling's reagent, 1092–1093
Fenn, John, 523
Fenofibrate, 758
Fermentation
  alcohol production by, 3, 344, 345f
  in biodegradable polymer synthesis, 1275, 1279
  glucose conversion by, 1185–1186
Fexofenadine, 976–977
Fibrinogen, 1022
Fibrous proteins, 1051
Fingerprint region of IR spectra, 532
First-order rate equations
  description of, 242
  in nucleophilic substitution, 267–268
Fischer esterification, 753–755, 1269
Fischer projection formulas, 185, 1072–1075, 1075t, 1084
Fishhook notation, 224
Five-membered ring
  Dieckmann reaction synthesis, 829
  furanose ring, 1080
  inscribed polygon method for predicting aromaticity, 873, 873f
  intramolecular aldol reaction synthesis, 822–825
  pyranose ring, 1080
  in pyrrole, 865
  in steroids, 1132
Flagpole hydrogens, 156, 157f
Flat helical ribbon, shorthand for α-helix, 1047
Flat wide arrow, shorthand for β-pleated sheet, 1047
Flavin adenine dinucleotide (FAD), 194, 1174
Fleming, Alexander, 528, 744
Flonase, 258
Flosal, 819
9-Fluorenylmethoxycarbonyl (Fmoc) protecting group, 1040, 1043–1044
9-Fluorenylmethyl chloroformate, 1040
Fluoroboric acid, 1000
Fluoromethane, 104, 104f
m-Fluoronitrobenzene, 853
Fluoxetine, 3, 206, 915, 977
Fluticasone, 258
Fmoc (9-Fluorenylmethoxycarbonyl) protecting group, 1040, 1043–1044
Folic acid, 1007–1008
Forensic science, 1162
Formal charge, 15–17, 16t
Formaldehyde
  in Bakelite synthesis, 1273, 1273f
  in crossed aldol reactions, 819
  description of, 654
  hydration of, 673–674
  nomenclature, 649, 649t
  in polymer synthesis, 1276
  primary alcohol preparation, 617–618
Formaldehyde hydrate, 673
Formalin, 654
Formic acid, 136, 698, 737
Formyl groups, 651
4-Formyl-3-methoxycyclohexane-carboxamide, A-7f
Fosamprenavir, 1058, 1058f
Fossil fuels
  alkanes in, 145
  combustion of, 165, 166f

Fragmentation, 510–511, 516–520, 517f, 519f
Free energy change (ΔG°), 231–232, 232f, 232t, 234–236
Freon 11, 950
Freon 12, 950
Frequency
  coupling constant, 569, 575
  definition, 529
  of electromagnetic radiation, 529–530, 529f
  Hooke's law, 532, 532f
  of infrared radiation, 531–532
Friedel, Charles, 887
Friedel–Crafts acylation
  alkyl benzene preparation, 918–919, 919f
  biological reactions, 897–898
  description of, 887f, 891–892
  intramolecular, 896–897
  limitations in, 909–910
  mechanism, 893–894
  substituted aniline synthesis, 994
  trisubstituted benzene synthesis, 922–923
Friedel–Crafts alkylation
  alkyl benzene preparation, 919, 919f
  benzene derivative synthesis, 912–913
  biological reactions, 897–898
  carbocation rearrangements, 894–895
  description of, 887f, 891–892
  limitations in, 909–910
  mechanism, 892–893
  p-nitrobenzoic acid synthesis, 921–922
  starting materials, 896
  substituted aniline synthesis, 994
  trisubstituted benzene synthesis, 922–923
  unreactive halides, 894
Frontalin, 462
Frontside attack
  in halogenation, 947–948
  in S_N1 substitution reactions, 275, 276
  in S_N2 substitution reactions, 269
Frost circle, 872
Fructofuranose, 1086, 1100
Fructokinase, 234
Fructose
  anomers, 1086
  D-fructose, 1071, 1078, 1079f, 1086
  phosphorylation of, 234
  stereogenic centers, 183
  in sucrose, 1099–1100
Fructose 1,6-bisphosphate
  dihydroxyacetone phosphate from, 831–832
  dihydroxyacetone phosphate to, 830
  glyceraldehyde 3-phosphate from, 831–832
  glyceraldehyde 3-phosphate to, 830
  phosphorylation to, 1180f, 1181
Fructose 6-phosphate, 783, 1180–1181, 1180f
2'-Fucosyllactose, 1102
3-Fucosyllactose, 1103
Fuels, renewable, 344, 345f
Fukui, Kenichi, 1228
Full-headed curved arrow notation, 224
Fumarase, 1187f
Fumarate, 1189
Fumaric acid, 493
Functional group region of IR spectra, 532

Functional groups. *See also specific functional groups*
  compounds containing C–O double bonds, 97–99, 98t
  compounds containing C–Z σ bonds, 95–97, 96t
  compounds containing P–O bonds, 100, 100t
  definition, 92
  description, 92–93
  hydrocarbons, 93–95, 94t
  identification of [*See* Infrared (IR) spectroscopy]
  interconversions in organic synthesis, 421
  intermolecular forces and, 101–104
  nomenclature of, A-7t
  organic synthesis and, 290–291
  reactivity and, 116–117
  starting materials in Friedel–Crafts alkylation, 896
  synthesis of, A-17–20
  types of, 93–100
Furanose ring, 1080, 1086
Fused ring system, 494, 494f

G

Gabapentin enacarbil, 188
Gabriel synthesis, 980–981
Galactose
  D-galactose, 1077, 1078f, 1095
  α-D-galactose, 683
  in lactose, 1099
Gantrisin, 1008f
Garcinia subelliptica, 1246
Garlic, allicin in, 136
Garsubellin A, 1246
Gas chromatograph, 522
Gas chromatography–mass spectrometry (GC–MS), 521–522, 521f, 522f
Gasohol, 344, 345f
Gasoline
  combustion, 1119
  ethanol, 344, 345f
  reaction rate, 240
Gauche conformation, 150–151
Geminal dichloride, 396
Geminal dihalides, 322, 396, 413
Geminal protons, 575
Generic names, of drugs, 136–137
Genes, 120f, 1148, 1152
Genetic code, 1156–1157, 1157t
Geranial, 1245
Geraniol, 288, 1128, 1131
Geraniums, 1128
Geranyl acetate, 110
Geranyl diphosphate, 100t, 1129–1132, 1130f
Gibbs free energy, 231
Gilbert, Walter, 1160
Ginger, 392, 492, 1127f
Ginkgo biloba, 702, 744
Ginkgolide B, 744
Glacial acetic acid, 702
Glass transition temperature, 1273
Global warming, 145
Globular proteins, 1051, 1054
D-Glucaric acid, 1093–1094
Glucitol, 1092
D-Gluconic acid, 1092
Glucopyranose
  α-D-glucopyranose, 1081, 1093
  β-D-glucopyranose, 1081, 1104
Glucosamine, 768, 1103

Glucose
  acyclic, 119
  in adipic acid synthesis, 1275
  aerobic metabolism, ATP yield, 1190
  anomers of, 1080–1081, 1082f
  biosynthesis of, 830
  cellulose hydrolysis to, 1101
  cyclic, 119, 683
  energy in, 219, 231
  ester synthesis from, 1091
  ether synthesis from, 1090–1091
  Fischer projection of, 1074–1075, 1075f
  D-glucose, 1071, 1074–1077, 1075f, 1078f, 1085, 1085f, 1092–1095, 1275
  α-D-glucose, 683, 1081, 1082f, 1091, 1101
  β-D-glucose, 683, 1070, 1070f, 1081, 1082f, 1091, 1101
  glycolysis of, 1180–1184
  glycoside formation from, 1087–1088
  hemiacetals in, 682–683
  hydrolysis from starch, 178
  from Kiliani–Fischer synthesis, 1095
  maltose hydrolysis to, 1098
  as monosaccharide, 119
  oxidation of, 1092–1095
  phosphorylation of, 233, 234f
  photosynthesis/metabolism, 1070
  in poly(trimethylene terephthalate) synthesis, 1275, 1276t
  reduction of carbonyl group, 1092
  in sucrose, 1099–1100
  synthesis by glycoside hydrolysis, 1088–1089
  three-dimensional representation of, 1085, 1085f
  in Wohl degradation, 1094–1095
Glucose metabolism. *See also* Glycolysis
  intermediates, 100
  monophosphate dianion intermediates, 714–715
  pyruvate in, 607, 1184–1185
  retro-aldol reaction in, 815, 831–832
  tautomerizations in, 783
Glucose 6-phosphate, 714–715, 783, 1180–1181, 1180f
Glucosidase, 1101, 1102
Glutamate 5-phosphate, 610
Glutamate 5-semialdehyde, 610
Glutamic acid
  isoelectric point, 1025t
  pK_a values, 1025t
  structure of, 1023f
Glutamine
  abbreviation for, 716t
  biological synthesis of, 765
  isoelectric point, 1025t
  pK_a values, 716t, 1025t
  structure of, 1023f
Glutathione, 1034
Glyceraldehyde
  enantiomers of, 1072
  D-glyceraldehyde, 1071, 1076, 1077, 1078f
  L-glyceraldehyde, 1076
  as monosaccharide, 119
  optical rotation of, 202
Glyceraldehyde 3-phosphate, 715
  dihydroxyacetone phosphate conversion to, 783
  fructose 1,6-bisphosphate from, 830
  fructose 1,6-bisphosphate to, 831–832

  in glycolysis, 1182
  octet rule exception, 18
Glycerol
  catabolism of, 1176
  nomenclature, 341
  in phosphoacylglycerols, 1120–1121
  phosphorylation, 1176
  in soap synthesis, 760
  in triacylglycerols, 122, 221, 393, 759, 1116–1117
Glycerol kinase, 1176
Glycerol 3-phosphate, 1176
Glycerol 3-phosphate dehydrogenase, 1176
Glycine
  abbreviation for, 716t
  enantiomers, 716
  isoelectric point, 1025t
  pK_a values, 716t, 1025t
  structure of, 1022f, 1023f
Glycogen, 1102
Glycolic acid, 71
Glycols, 340, 451–453
Glycolysis
  dehydration reactions in, 1183
  isomerization in, 1180–1183, 1180f
  of monosaccharides, 1179–1184
  oxidation, 1182–1183
  phosphate transfer, 1183–1184
  phosphorylation, 1180–1181, 1180f, 1182–1183
  pyruvate, fate of, 1184–1186
  retro-aldol reaction in, 831–832, 1180f, 1181
  steps of, 1180f
Glycopeptide transpeptidase, 762
Glycosidase, 1101, 1102
Glycosides
  formation of, 1087–1088
  hydrolysis of, 1088–1089, 1170
  naturally-occurring compounds, 1089–1090
N-Glycosides
  nucleosides, 1146
  synthesis, 1104
α-Glycosidic bonds, 1102
Glycosidic linkage
  in amino sugars, 1103–1104
  in disaccharides, 1098–1100
  in polysaccharides, 1100–1103
Glycyl phosphate, 735
Goby fish, 390
Goodyear, Charles, 1266
Grandisol, 1128
Grapefruit mercaptan, 367
Green chemistry, 1274
Greenhouse gases
  fossil fuel combustion, 165, 166f
  methane, 145
Green polymer synthesis
  description of, 1274
  feedstock, 1274–1275, 1275f, 1276f
  reagents, 1276
Grignard, Victor, 614
Grignard products, retrosynthetic analysis of, 620–622
Grignard reagents
  in acid chloride reactions, 616, 625
  in aldehyde reactions, 616, 617–620, 659–660
  carboxylation of, 628, 705t
  description of, 614
  in ester reactions, 616, 625–626
  in ketone reactions, 616, 617–620, 659–660

Index

Grignard reagents—(Cont.)
  in nitrile reactions, 722
  preparation of, 614, 616
  protecting groups and, 622–624, 623f
  synthesis and, 632–634
  in α,β-unsaturated carbonyl compound reactions, 631
Ground state, 34
Group number, 9
Groups
  in periodic table, 9
  priority assignment, 189–193, 190f, 192f
  stereogenic center bonding, 183
Grubbs catalyst, 1211–1214
Grubbs, Robert, 1211
Guanine
  C–G base pair, 1151, 1151f
  in citric acid cycle, 1188
  description of, 849
  nomenclature, 1147, 1147t
  pi (π) electrons in, 868
  in polynucleotide structure, 1149f
  purine-derived, 1145
Guanosine, 110, 1147t
Guanosine 5′-monophosphate (GMP), 1147t
Guanosine triphosphate (GTP)
  aerobic glucose metabolism, 1190
  citric acid cycle product, 1187f, 1188, 1189–1190
Gutta-percha, 1266
Gypsy moths, 449–450, 450f

H

HaeIII, 1160
Hair, 1047, 1051, 1052f
Halaven, 5
Half-headed curved arrow notation, 224
Halichondria okadai, 5
Halichondrin B, 5
Halides. *See also* Alkyl halides; Aryl halides; Hydrogen halides; Vinyl halides
  allylic, 255, 255f, 951
  benzylic, 255, 255f, 917
  methyl, 282
  in Suzuki reaction, 1201–1205
Hallucinogens
  bufotenin, 978
  LSD [*See* LSD (lysergic acid diethyl amide)]
  mescaline, 977
  psilocin, 978
Halo alkanes, 256
α-Halo carbonyl compounds, 792–793
Halocyclohexanes, 319–321, 319f
Halogenated ethers, as anesthetics, 345
Halogenation. *See also* Bromination
  acid-catalyzed, 791–792
  of activated benzenes, 909
  of alkanes, 939–949
  of alkenes, 404–406, 406f, 407–408
  of alkyl benzenes, 917–918
  of alkynes, 412–413
  allylic, 950–953
  of benzene, 887f, 888–889
  of but-1-yne, 398f
  of carbonyl compounds at the α carbon, 791–793
  of cyclohexene, 397f, 404–407
  mechanism, 940–942
  racemization at stereogenic center, 947
  radical, 939–949
  stereochemistry of, 405–407, 406f, 946–949, 946t
Halogens
  bonding of, 12f
  common nucleophiles, 266t
  degrees of unsaturation and, 386–387
  electronegativity, 914
  as electron-withdrawing group, 899
  halogenation, 404–407, 406f
  halohydrin formation, 407–408
Halo groups
  nomenclature, A-7
  structure of, 96t
Halohydrin
  epoxide synthesis, 348
  formation, 397f, 407–408
Halomon, 255
Halonium ions, bridged, 405–408
Halothane, 258f
Hammond postulate
  bromination, 945
  E1 elimination reactions, 315
  electrophilic addition reactions, 400–401
  electrophilic aromatic substitution reactions, 904–905
  features of, 279–281
  halogenation, 944
  nucleophilic aromatic substitution reactions, 914
  S_N1 substitution reactions, 279, 281–282, 281f
Handedness, 177, 180, 182. *See also* Stereochemistry
Hansen's disease, 187
Hardeners, in epoxy resins, 1271, 1272f
Hardening of vegetable oil, 440, 1119
Haworth projections, 1081, 1082–1085
Haze, 484
HCN gas, 662
HDL (high-density lipoproteins) particles, 768
HDPE (high-density polyethylene), 957, 1259, 1265
Head-to-tail polymerization, 958
Heat
  in electrocyclic reactions, 1232–1234
  in pericyclic reactions, 1228
  in radical reactions, 937, 940, 955–956
  symbol, 220
Heat of hydrogenation
  benzene, 857–858, 858f
  of dienes, 485–486
  explanation of, 437–438
Heat of reaction. *See* Enthalpy change ($\Delta H°$)
Heck, Richard, 1205
Helicene, 856, 856f, 1030
Helium, 10
Heme, 875, 876f, 1053
Hemiacetals
  in acetal formation, 675
  in acetal hydrolysis, 677–678
  anomers, 1080
  cyclic, 680–682
  monosaccharides, 1079–1086
Hemibrevetoxin B, 96
Hemoglobin, 1022, 1050, 1053–1054, 1054f
Henderson–Hasselbalch equation, 708–710, 709f
Heptanal, 819
Heptane, 135t
Heptan-2-one, 802
(*E*)-Hept-2-ene, 1199
(*E*)-Hept-4-en-3-one, A-8t
Heroin, 114, 115, 751
Heteroatoms
  in condensed structures, 29f
  definition of, 4
  in degree of unsaturation calculation, 386–387
  in functional groups, 116
  in heterocycles, 342
  reactivity of molecule, 92
  resonance structures, 21
  in skeletal structures, 33
  stereogenic N atoms, 184
Heterocycles
  aromatic, 864–869
  definition of, 342
  nitrogen, 973, 973f
  sulfur, 1036
Heterocyclic aromatic amines
  pK_a values, 992t
  relative basicity, 990–991, 992t
Heterogeneous reaction mixtures, 437
Heterolysis
  carbocation formation, 224
  of C–Z bonds, 224f, 225
  in S_N1 substitution reactions, 274, 281
Heterolytic cleavage, 224
Hexachloroethane, 207, 207f
Hexa-2,4-diene, 482, 1231, 1232, 1234, 1235
Hexa-2,4-dienoic acid, 1275
Hexan-3-amine, 520
Hexane
  as extraction solvent, 711
  formation of, 444f
  mass spectrum of, 516, 517f
  as normal alkane, 135t
  in petroleum, 132
  skeletal structure, 30
Hexane-2,5-dione, 822–823
Hexanoic acid
  description of, 702
  nomenclature, 697
Hexan-2-ol, 632, 632f
Hexa-1,3,5-triene
  electrocyclic ring closure, 1227, 1231
  molecular orbitals in, 1230, 1230f
  not aromatic, 859
Hex-1-ene, 539
Hex-3-ene, 389, 444f, 1211
Hex-1-en-5-yne, A-8t
Hexokinase, 234f
Hexoses, 1071
Hex (prefix), 134
Hex-1-yne, 539, 1204
Hex-3-yne, 444f
HFCs (hydrofluorocarbons), 950
HFOs (hydrofluoroolefins), 950
High-density lipoprotein (HDL) particles, 768
High-density polyethylene (HDPE), 957, 1259, 1265, 1277
Highest occupied molecular orbital (HOMO)
  in benzene, 871, 871f
  in buta-1,3-diene, 1229f, 1230
  in cycloaddition reactions, 1238–1239
  description of, 871, 1229
  in electrocyclic reactions, 1232–1233
  in hexa-1,3,5-triene, 1230, 1230f
High-performance liquid chromatography, 1035
High-resolution mass spectrometry, 520–521, 521t
HindIII, 1160
Histamine
  as aromatic heterocycle, 866
  description of, 976
  histidine to, 802
Histidine
  biosynthesis of, 352
  histamine from, 802
  isoelectric point, 1025t
  pK_a values, 1025t
  structure of, 1023f
Histrionicotoxin, 384
HIV (human immunodeficiency virus), 1162
HIV treatments
  amprenavir, 1058, 1058f
  AZT, 3, 1162
  dideoxyinosine, 1162
  fosamprenavir, 1058, 1058f
  protease inhibitors, 1057–1058
HMOs (human milk oligosaccharides), 1102–1103
^{1}H NMR. *See* Proton NMR
Hoffmann, Felix, 703
Hoffmann, Roald, 1228
Hofmann elimination reactions, 995–997
HOMO. *See* Highest occupied molecular orbital
Homocysteine, 717
Homogeneous catalysts, 1265
Homologous series, 135, 135t
Homolysis
  bond cleavage, 223–224
  of C–Z bonds, 224f, 225
  nonpolar bonds in radical reactions (*See* Radical reactions)
Homolytic cleavage, 223–224
Homopolymers, 1263
Homotopic protons, 561
Honey, 1086
Hooke's law, 532, 532f
Hormones
  adrenaline (*See* Adrenaline)
  aldosterone, 1137
  androgens, 1136
  androstenedione, 875
  androsterone, 1136t
  bradykinin, 1033
  calcitriol, 484, 484f
  C_{18} juvenile hormone, 619–620, 1199
  compared to eicosanoids, 1124
  cortisol, 1137
  cortisone, 498, 654, 1137
  endocrine disruptors, 1270
  estradiol, 392–393, 875, 1136t
  estrogens, 1136
  estrone, 498, 624, 834, 875, 1136t
  ethylene, 305
  gibberellic acid, 845
  insulin, 683, 1022
  juvenile, 619–620, 1199
  oxytocin, 735, 1021, 1034, 1049
  prednisone, 654
  progesterone, 392–393, 823–824, 824f, 1136t
  progestins, 1136
  steroids, 1135–1137
  synthetic, 392–393
  testosterone, 875, 1136t
  thyrotropin-releasing hormone, 98
  L-thyroxine, 1025
  vasopressin, 1034
Housane, 144f
5-HPETE (5-hydroperoxyeicosatetraenoic acid), 373
Hückel's rule, 860, 860t, 869–872
Human genome, 1152, 1160

Human Genome Project, 1160
Human immunodeficiency virus (HIV), 1162
Human immunodeficiency virus (HIV) treatments. *See* HIV treatments
Human milk oligosaccharides, 1102–1103
Humulene, 1204
Hybridization, 34–41
  in acetylene, 40
  acidity, effects on, 73, 74f
  basicity, effects on, 991, 992t
  bond length and, 484
  bond strength and, 322
  definition of, 35
  in ethane, 38
  in ethylene, 38–39
  in methane, 34–35
  pi (π) bonds, 39–40
  sigma bonds, 34, 35, 38–40
Hybrid orbitals. *See also specific hybrid orbitals*
  acidity and, 73, 74f
  formation of, 35
  percent *s*-character (*See* Percent *s*-character)
  superscript notation, 37
Hydrates
  of carbon (*See* Carbohydrates)
  formation of, 458, 673–675
Hydration
  but-1-yne, 398f, 414–417
  in citric acid cycle, 1187f, 1189
  cyclohexene, 397f, 403–404, 412
  description of, 673
  kinetics of hydrate formation, 674–675
  Markovnikov's rule, 414
  thermodynamics of hydrate formation, 673–674
Hydrazine, 919
Hydride reduction
  of aldehydes, 604–608, 659
  of ketones, 604–608, 659
  mechanism, 605–606, 659
Hydrides
  as bases, 76
  in biological oxidation of alcohols, 459
  in enolate formation, 786
  as reducing agents, 436
1,2-Hydride shift, 353–354, 359
Hydroboration–oxidation
  but-1-yne, 398f, 416–417
  cyclohexene, 397f, 408–412, 409f, 411t, 412f
  vinylborane synthesis, 1202
Hydrocarbons. *See also specific hydrocarbons*
  alkanes (*See* Alkanes)
  alkenes (*See* Alkenes)
  aromatic (*See* Aromatic hydrocarbons)
  definition, 2, 93
  degrees of unsaturation, 386–387
  description, 93–95, 94t
  fuel type, 145
  organocuprate coupling synthesis, 1199–1200
  oxidation and reduction of, 435, 435f
  polycyclic aromatic hydrocarbons (*See* Polycyclic aromatic hydrocarbons)
  synthesis, 1199–1200
  unsaturated, 386–387
Hydrochloric acid, 1040
  Brønsted–Lowry acids, 80
  as common acid, 76
in epoxide ring opening reactions, 370–371, 371f
  as Lewis acid, 80
Hydrofluorocarbons (HFCs), 950
Hydrofluoroolefins (HFOs), 950
Hydrogen
  alpha (α) hydrogen, 780, 784, 818–821, 827–828
  bonds (*See* Hydrogen bonds)
  degree of atom, 94–95
  electron configuration, 10
  octet rule exception, 17
  prochiral, 458–459
  as reducing agent, 436
Hydrogenation
  in adipic acid synthesis, 1275
  alkene stability, 437–438, 437f
  catalytic (*See* Catalytic hydrogenation)
  degrees of unsaturation and, 439–440
  of dienes, 485–486
  of double bonds, 440, 1119
  enantioselective reactions, 1029–1031
  heat of (*See* Heat of hydrogenation)
  mechanism, 438–439
  of oils, 440–442, 441f, 1119
Hydrogen bonds, 12f
  in alcohols, 343, 344t
  in amides, 741, 741f
  boiling point and, 104
  carboxylic acids, 700, 700f
  description of, 11, 102, 103t
  in β-dicarbonyl compounds, 781
  in DNA, 102f, 1150–1151, 1151f
  infrared absorptions, 533, 533t
  intermolecular, 974
  melting point and, 105
  number of, 12f
  in peptide hydrolysis, 1056–1057
  polar protic solvents, 265
  in secondary structure of proteins, 1046–1047, 1046f, 1050f
  sigma bond, 34
  solubility and, 107–108
  in tertiary structure of proteins, 1048, 1050f
Hydrogen halides
  alcohol conversion to alkyl halides in, 356–358
  electrophilic addition to alkenes, 399–403, 403t
  reactivity of, 356
Hydrogen ions (H^+), in Brønsted–Lowry acids and reactions, 59
Hydrogenolysis, 1041
Hydrogen peroxide, 446
Hydrohalogenation
  of but-1-yne, 398f, 413
  of cyclohexene, 397f, 399–403
  Markovnikov's rule, 400–401
  stereochemistry of, 401–403
Hydrolases, 1055, 1055t
Hydrolysis
  of acetals, 677–678
  in acetoacetic ester synthesis, 801
  of amides, 760–761
  ATP, 230
  in catabolism, 1170, 1171f
  in citric acid cycle, 1187, 1187f, 1188
  of cyanohydrin, 661
  of enamines, 669–670
  in enantioselective synthesis of amino acids, 1030–1031
  in enzymatic decarboxylation, 803
  enzymes in, 1055, 1055t, 1269
  of esters, 756–760
  glucose formation by, 1077, 1098, 1101, 1102
  of glycosides, 1089
  of imines, 669–670, 672–673
  lipids, 758–760
  in malonic ester synthesis, 798
  olestra synthesis, 758–759
  of peptides, 1036–1037
  of rebaudioside A, 1090
  saponification, 757–758, 760
  of starch, 344, 345f
  in substituted aniline synthesis, 994
  in terpene/terpenoid biosynthesis, 1131
  of triacylglycerols, 393, 1118
Hydrolyzable lipids
  definition of, 1116
  phospholipids, 1120–1122, 1121f
  triacylglycerols, 1116–1120, 1117t, 1118f
  waxes, 1116
Hydroperoxides, 953, 953f
5-Hydroperoxyeicosatetraenoic acid (5-HPETE), 373
Hydrophilic, 109, 112
Hydrophobic, 109, 112
Hydroxide (^-OH)
  in enolate formation, 786
  nucleophile in alcohol synthesis, 347
  as strong base, 76
α-Hydroxy acids, 71
Hydroxy aldehydes
  cyclization of, 680–681, 683
  α-hydroxy aldehydes, 782
  β-hydroxy aldehydes, 813
Hydroxybenzene, 853
3-Hydroxybutanal, 813, 815
(*S*)-Hydroxybutanedioic acid, 698
β-Hydroxy carbonyl compounds
  aldehyde conversion to, 813–818
  dehydration of, 816–817
  ketone conversion to, 813–818
  in Robinson annulation, 836
3-Hydroxy carboxylic acids, 1278
Hydroxy groups
  in alcohols, 338
  hydrophilic nature of, 109
  of monosaccharides, 1071, 1090–1091
  in phosphoacylglycerols, 1120
3-Hydroxybutyric acid, 1278
3-Hydroxyvaleric acid, 1278
α-Hydroxy ketones, 782
Hydroxylamine, 1095
5-Hydroxypentanal, 680
5-Hydroxypentan-2-one, 622–624, 623f
β-Hydroxy thioesters, 817
Hymeniacidon agminata, 553
Hyperconjugation, 277, 278–279
Hypophosphorus acid, 1001

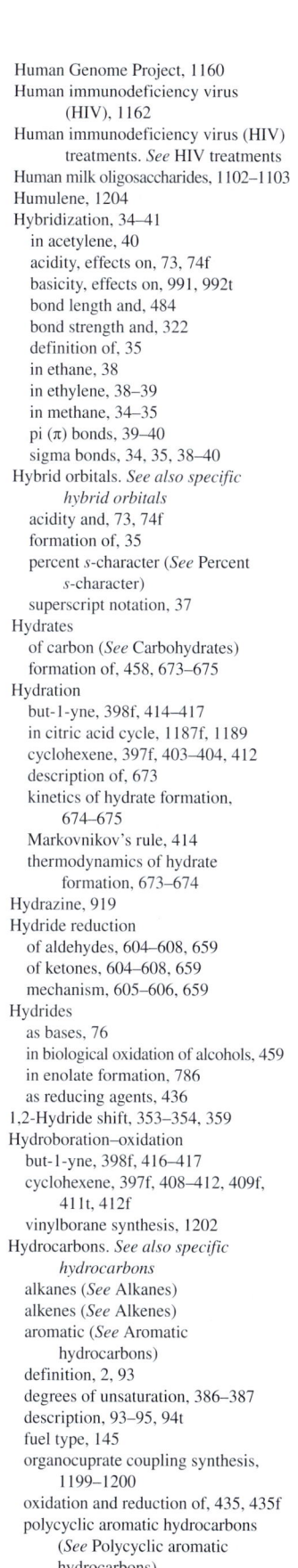

Ibuprofen
  enantiomers, 206
  mode of action, 1126
  nomenclature, 137
  synthesis of, 829, 920
Icosane, 135t
-ic acid (suffix), 698, 699
Identical twins, 1152
Identification of compounds. *See* Infrared (IR) spectroscopy; Mass spectrometry; Nuclear magnetic resonance (NMR) spectroscopy
-ide (suffix), 256
-idine (suffix), 1146
D-Idose, 1078f
Imides, 980
Imidic acid, 720
Imidic acid tautomer, 720
Imines
  in amide reduction, 611
  amino acid degradation, 672–673, 803
  in biological systems, 671–673
  formation of, 658, 666–667, 669f, 993
  hydrolysis of, 669–670
  in Kiliani–Fischer synthesis, 1096
  in nitrile reduction, 721–722
  in organometallic reagent addition to nitriles, 722
  in reductive amination, 982–983
  vision and, 671, 671f
Iminium ion, 667–669, 669f
Indigo, 1006
Indigofera tinctoria, 1006
Inductive effects
  on acidity, 70–71, 74f
  in aliphatic carboxylic acids, 710–711
  on basicity, 992t
  in carbocation stability, 277–278
  definition, 71
  electron-donating, 898–900
  electron-withdrawing, 898–900
  on substituted benzenes, 898–900
-ine (suffix), 256
Infrared (IR) absorptions/spectra
  of alcohols, 540, 543t
  of aldehydes, 541, 543t, 653t
  of alkanes, 539, 543t
  of alkenes, 539, 543t
  of alkynes, 539, 543t
  of amides, 542, 543t
  of amines, 543, 543t, 974–975, 975f, 975t
  analysis of, 536–538
  of aromatic hydrocarbons, 540, 543t, 855, 855t
  bond strength and, 532–534, 532f, 533f
  carboxylic acid derivatives, 743t
  of carboxylic acids, 542, 543t, 701t
  characteristic frequencies, A-12
  characteristics of, 531–532
  of esters, 542, 543t
  of ethers, 540, 543t
  functional group identification, 539–544, 543t
  of ketones, 541, 543t, 653t
  of nitriles, 543, 543t, 701t
  resonance and, 534–536, 535f
  vibration of bonds and, 531–534
Infrared (IR) spectroscopy
  analysis, 536–538
  description of, 509, 529
  electromagnetic radiation and, 529–530, 529f
  functional group identification, 539–544
  IR inactive vibration, 534
  structure determination and, 544–546
  vibration modes, 531
Ingenol, 1197, 1215
Ingenol mebutate, 1197
Ingeo, 1275
Initiator, 1257, 1258, 1260, 1262
Inscribed polygon method for predicting aromaticity, 872–874, 873f

Index

Insecticides
  chrysanthemic acid, 583
  DDT, 4, 259
  decamethrin, 1207
  nicotine, 976
  pyrethrin I, 1207
  telfairine, 255
Inspra, 346
Insulin
  description of, 1022, 1049f
  structure of, 1049
Integral, NMR spectra, 567
Intermolecular forces
  boiling point and, 104–105
  covalent compounds, 101–103
  definition, 101
  dipole–dipole interactions, 102
  hydrogen bonding, 102
  ionic compounds, 101
  melting point and, 105–106
  solubility and, 106–110
  types of, 101–103, 103t
  van der Waals forces, 101
Internal alkynes
  description of, 385
  formation of, 418–419
  hydration of, 414
  hydroboration–oxidation of, 417
  oxidative cleavage of, 455
International Union of Pure and Applied Chemistry, 136. *See also* IUPAC nomenclature
Intramolecular reactions
  aldol reactions, 822–825, 835–839
  Claisen reactions, 829–830
  electrocyclic ring closure, 1227
  Friedel–Crafts reactions, 896–897
  Robinson annulations, 835–839
  sigmatropic rearrangements, 1240–1246
Inversion of configuration
  in alcohol reactions with HX, 357–358
  in alkyl tosylate reactions, 362–363
  in S_N2 substitution reactions, 270–271, 270f, 283
Iodomethane, 104, 104f
1-Iodo-1-methylcyclohexane, 312
(Iodomethyl)zinc iodide, 1210
Ion–dipole interactions, 107, 107f
Ionic addition reactions, 956
Ionic bonding, 101, 103t, 615
Ionic bonds/bonding, 11
Ionic compounds
  boiling point, 104
  intermolecular forces, 101, 103t
  melting point, 105
  solubility of, 107, 107f, 110t
Ionic intermediates, 224f, 952
Ionic polymerization, 1260–1262
Ionophores, 115–116, 115f
Ions. *See also* Anions; Cations; Zwitterions
  bridged halonium ions, 405–408
  definition of, 8
  nitrosonium, 998–999
  transport across cell membranes, 115–116, 115f
IR spectroscopy. *See* Infrared (IR) spectroscopy
Isoamyl acetate, 743
Isobutyl, 138
p-Isobutylacetophenone, 920
Isobutyl chloride, 256f
Isocitrate, 1173, 1187, 1188

Isocitrate dehydrogenase, 1187f
Isocyanate, 1269
Isoelectric point (p*I*), 718–719, 1025, 1025t
Isolated dienes
  comparison with conjugated dienes, 484, 485
  electrophilic addition, 486
  electrostatic potential, 475, 475f
Isoleucine
  isoelectric point, 1025t
  L-isoleucine, 1024
  pK_a values, 1025t
  stereochemistry of, 1024
  structure of, 1023f
Isomerases, 1055, 1055t
Isomerization
  chemistry of vision, 671
  in citric acid cycle, 1188
  enzymes in, 1055, 1055t
  geranyl diphosphate to neryl diphosphate, 1131
  in glycolysis, 783, 1180–1182, 1180f, 1183
Isomers
  constitutional (*See* Constitutional isomers)
  definition of, 17, 134, 179
  DET, 461
  disubstituted benzenes, 853
  of disubstituted cycloalkanes, 198–199
  double bonds, 441
  Lewis structures, 17
  meta, ortho, and para, 912–913
  resonance structures vs., 18
  stereoisomers (*See* Stereoisomers)
  tautomers (*See* Tautomers)
  types of, 200, 200f
Isooctane, 1119
Isopentane. *See* 2-Methylbutane
Isopentenyl diphosphate, 107, 117, 477, 1129, 1130, 1132
Isophthalic acid, 727
Iso- (prefix), 138
Isoprene, 483–484, 1266
Isoprene units, 1126–1129
Isopropyl, 138
Isopropylbenzene, 540, 919
Isopropylcyclopentane, 633–634
S-Isopropyl cyclopentanecarbothioate, 738
3-Isopropyl-4-methylcyclohexanone, 651
Isopropyl 2-methylpentanoate, 741
6-Isopropyl-3,3,7-trimethyldecane, 142
Isotactic polymers, 1264–1265
Isotopes
  of bromine, 515
  of chlorine, 514
  definition of, 8
  exact mass, 521t
  in mass spectrometry, 510
  NMR signals, 556
IUPAC nomenclature
  acid chlorides, 737
  acyl phosphates, 737
  alcohols, 339–340, 340f
  aldehydes, 648–649, 650f
  alkanes, 137–142, 140f
  alkenes, 387–389
  alkyl halides, 256, 256f
  alkynes, 387–389
  amides, 739
  amines, 971–973
  anhydrides, 738
  benzene derivatives, 853–854

carboxylates, 698, 698f
carboxylic acids, 697
chirality centers, 181
common names and, 144
cycloalkanes, 142–144, 143f
deriving structure from, 142
description of, 136–137
dicarboxlic acids, 698
esters, 738
ethers, 341–342
ketones, 650–651
nitriles, 699, 699f
primary amines, 971
secondary amines, 972
substituents, 137–138
sulfides, 367
tertiary amines, 972
thioesters, 738
thiols, 366, 366f

J

Jaborandi, 866
Januvia, 867
cis-Jasmone, 443
Johnson, W. S., 823
Jones, Marion, 1136
Joules (J), 42
Juvenile hormones, 619–620, 619f, 1199

K

K_a (acidity constant), 63–64
Karenia brevis, 339
Keflex, 745
Kekulé, August, 850
Kekulé structures, 850
Kelp, giant, 254
Kelsey, Frances Oldham, 187
Keratins
  description of, 1022
  α-keratins, 1047, 1051–1052, 1052f
Ketals. *See* Acetals
Keto–enol tautomerization, 1180–1181
Keto–enol tautomers, 414–416
β-Keto esters
  esters to, 826–827
  formation of, 829
  ketones from, 800–801
  pK_a values, 785t
α-Ketoglutarate, 1188
α-Ketoglutarate dehydrogenase, 1187f
α-Ketoglutaric acid, 698
Ketohexoses, 1071, 1078, 1079f
Ketones
  acetal formation from, 675–677
  acetal hydrolysis to, 677–678
  acetoacetic ester synthesis to, 800–802
  alkylation of, 793–796
  alkylation of unsymmetrical ketones, 795–796
  amine synthesis from, 982–985
  aryl ketone reduction to alkyl benzenes, 918–919
  benzene to, 892
  boiling point, 652t
  catalytic hydrogenation, 606
  cyanohydrin formation, 660–662
  description of, 599, 647
  β-dicarbonyl compound from, 828
  in directed aldol reactions, 821–822
  enamine formation, 668, 669f, 993
  enamine hydrolysis to, 669–670
  formation of, 453–454

  Friedel–Crafts acylation synthesis, 887f, 891–892, 918–919
  functional group, 97, 98t
  general reactions of, 656–658
  halogenation at the α carbon, 791
  α-halo ketone, 791
  hydration of alkynes to, 414–416, 655t
  hydride reduction, 604–608, 659
  hydroboration–oxidation, 417
  α-hydroxy, 782
  β-hydroxy carbonyl compound from, 813–818
  imine formation, 666–667, 669f, 982, 993
  imine hydrolysis to, 669–670
  infrared (IR) absorptions/spectra of, 541, 543t
  interesting compounds, 654
  mass spectrum of, 518–519
  melting point, 652t
  nomenclature, 650–651
  nucleophilic addition, 600, 601, 617–620, 656–658, 658f, 659–660
  organocuprate reaction with acid chlorides, 626–627, 655t
  organometallic reagent addition to nitriles to, 722
  organometallic reagent reactions and, 617–620, 624–627, 659–660
  physical properties, 652, 652t
  pK_a values, 785t
  preparation of, 655, 655t
  protecting groups, 679
  reduction to secondary alcohol, 440, 604–608, 659
  reductive amination of, 982–985
  secondary alcohol oxidation to, 456, 457, 655t
  solubility, 652t
  spectroscopic properties, 653, 653t
  stereochemistry of carbonyl reduction, 606–607
  Wittig reaction, 662–666
Ketoprofen, 76
Ketoses
  D-ketose family, 1078, 1079f
  structure of, 1071
Keto tautomers, 414
β-Keto thioesters, 832–833
Kevlar, 1268
Kiliani–Fischer synthesis, 1094, 1095–1097
Kinases, 1055t, 1176
Kinetic enolates, 788
Kinetic product of addition reaction, 488–490, 489f
Kinetic resolution, 1028–1029
Kinetics
  definition, 231, 240
  E1 elimination reactions, 314
  E2 elimination reactions, 308
  energy of activation, 236–237, 240–241
  first-order, 242, 268, 273
  hydrate formation, 674–675
  rate equations, 241–242
  second-order, 241, 267, 268, 308
  S_N1 substitution reactions, 273, 277t
  S_N2 substitution reactions, 268, 272t
Krebs, Hans, 1186
Krebs cycle, 1186. *See also* Citric acid cycle
Kukui nuts, 393

L

Labdanoids, 1129
β Lactam family of antibiotics, 528, 544, 744–745, 762
Lactams, 735, 1268
Lactase, 244, 244f, 1099
Lactate, 607–608, 1185
Lactic acid
  description of, 702
  optical rotation of, 202
  specific rotation of, 205
  structure of, 30
Lactols, 680–682
Lactones, 735, 754
Lactose
  breakdown of, 244, 244f
  as disaccharide, 120, 1099
  structure of, 682–683
  type of sugar, 1093
Lactose intolerance, 1099
Laetrile, 661–662
Lagging strand, DNA, 1153f, 1154
Lamivudine, 1162
Lanosterol, 1134f, 1135
Latanoprost, 704
Lauric acid, 1179, 1191
Lavandulol, 622
Lavender oil, 622
LDA. See Lithium diisopropylamide
LDL (low-density lipoproteins) particles, 768
LDPE (low-density polyethylene), 957, 1259
Lead acetate, 443
Leading strand, DNA, 1153–1154, 1153f
Leaving groups
  of alcohols, 349, 351, 356–357, 359–360
  of alkyl tosylates, 362
  of aryl diazonium salts, 999
  of carbonyl compounds, 601–603
  in carboxylic acids and derivatives, 735, 745–747
  in Claisen reactions, 825–826
  in E1 elimination reactions, 314–315, 316t
  in E2 elimination reactions, 310–312, 311t
  in elimination reactions, 303–304
  of epoxides, 349
  of ethers, 349, 365
  good, 261–262, 262t, 284
  in Hofmann elimination, 995–996
  imine formation, 667
  in nucleophilic substitution reactions, 260–263, 269–270, 284, 287–288, 356–357
  in organocuprate reaction with acid chlorides, 626
  in organometallic reagent reaction with carboxylic acid derivatives, 625
  phosphorus-derived, 287–288, 316, 1129
  poor, 262t
  in reduction of carboxylic acids and derivatives, 608–611
Le Châtelier's principle, 352
Lecithin, 1121
Lemongrass, 1128
Length
  bond length (See Bond length)
  units to report wavelength, 529
  wavelength of electromagnetic radiation, 529–530, 529f
Leprosy, 187
Leptospermone, 782
Leucine
  as essential amino acid, 1024
  isoelectric point, 1025t
  pK_a values, 1025t
  structure of, 1023f
Leu-enkephalin, 1034
Leukotriene A_4, 373
Leukotriene C_4, 99, 373–374, 1124, 1125f
Leukotrienes, 373–374, 1124, 1125f, 1125t
Levodopa, 186, 803
Levonorgestrel, 393
Levorotatory compounds, 202, 1076
Lewis acid–base complexes, 892–893
Lewis acid–base reactions
  in bromination, 888–889
  description of, 79–81
  electrophilic addition, 399
  nucleophilic substitution, 260
Lewis acids
  alkyl halides, 260
  in Bakelite synthesis, 1273, 1273f
  borane as, 409
  definition, 78–79
  in Friedel–Crafts reactions, 891–892
Lewis bases
  alkyl halides and, 259
  amino group, 909
  definition, 78
Lewis structures
  condensed structure conversion to, 29–30
  definition of, 12
  drawing procedure, 12–14
  formal charge, 15–17, 16t
  isomers, 17
  molecular shape and, 25–27
  multiple bonds in, 14–15
  octet rule exceptions, 17–18
  resonance structures, 18–23
  skeletal structure conversion to, 32
Lexan, 1256f, 1270, 1276
Lexapro, 703
Lexiva, 1058
Ligands, 1200–1201
Ligases, 1055, 1055t
Light
  in electrocyclic reactions, 1234–1235
  in pericyclic reactions, 1228
  in radical reactions, 937, 940, 955–956
  symbol, 220
"Like dissolves like," 107
Lily of the valley, 1127f
Limonene, 454, 1132
Linalool, 622
Linamarin, 661–662
Lindlar catalyst, 443
Linear molecular structure, 25
Linoleic acid
  bond cleavage, 954
  in candlenuts, 393
  melting point, 394, 394t, 434
  structure of, 394f
Linolenic acid
  in candlenuts, 393
  melting point, 394, 394t, 395
  as omega-3 fatty acid, 395
  structure of, 394f
Lipases, 759, 1055t
Lipid bilayers, 114, 114f, 1121, 1122
Lipids
  cholesterol (See Cholesterol)
  description of, 122, 1115–1116, 1115t
  eicosanoids, 1124–1126, 1125f, 1125t
  examples of, 1115f
  fat-soluble vitamins, 1123, 1123f
  geranyl diphosphate as precursor, 477
  hydrolysis, 758–760
  oxidation at allylic carbon, 953–954, 953f
  phosphoacylglycerols, 1120–1122, 1121f
  phospholipids, 1120–1122
  sphingomyelins, 1122, 1122f
  steroids (See Steroids)
  terpenes/terpenoids, 1126–1132, 1127f
  triacylglycerols (See Triacylglycerols)
  waxes, 1116
Lipitor, 1135, 1135f
Lipoprotein particles, 767–768, 767f
Lipoxygenase, 1124, 1125f
Lisinopril, 194
Lithium acetylide, 615
Lithium aluminum hydride, 436, 604–605, 608–612, 612t, 659, 721; 726, 981–982
Lithium diisopropylamide (LDA)
  as base, 76f, 786t
  in enolate formation, 786–788, 821
  preparation of, 787
Lithium dimethylcuprate, 614, 1210
Lithium tri-tert-butoxyaluminum hydride, 608, 612t
Living polymerization, 1262
Local mediators, 1124
Log, 64
Lone pairs
  allyl-type resonance, 477–478
  in amines, 985
  basicity, effects on, 990–991, 992t
  bond angle and, 24
  Brønsted–Lowry bases, 59, 263
  carbonyl group, 97
  cation with positive charge adjacent to, 478
  C–Z σ bonds, 95
  definition of, 12
  in heteroatoms, 92, 116
  in histamine, 866
  in hybrid orbitals, 36, 36f
  nucleophiles, 263
  number of, 12f
  in pilocarpine, 866
  in pyridine, 864, 865, 865f
  in pyrrole, 865, 865f
  on resonance structures, 19
  on skeletal structures, 33
Lovastatin, 1135
Low-density lipoprotein (LDL) particles, 768
Low-density polyethylene (LDPE), 957, 1259
Lowest unoccupied molecular orbital (LUMO)
  in benzene, 871, 871f
  in buta-1,3-diene, 1229f, 1230
  in cycloaddition reactions, 1238–1239
  description of, 871, 1229
  in hexa-1,3,5-triene, 1230, 1230f
LSD (lysergic acid diethyl amide)
  description of, 473
  as 2-phenylethanamine derivative, 978
  synthesis, 896–897, 897f
Lumigan, 704
LUMO. See Lowest unoccupied molecular orbital
Lyases, 1055, 1055t
Lycopene, 483–484
Lycra, 1270
Lynen, Feodor, 1134
Lysergic acid, 473
Lysergic acid diethyl amide (LSD). See LSD
Lysine, 695, 1023f
Lysozyme, 1047, 1048f
D-Lyxose, 1077, 1078f, 1095

M

Magnetic field, 554–555
Magnetic resonance imaging (MRI), 584–585, 585f
Ma huang, 183
Major contributor, 22–23, 479–480
Major groove in DNA double helix, 1152
Malaria
  artemisinin, 411, 412f
  DDT and, 259
  mefloquine, 387
  quinine treatment, 302, 311
Malate dehydrogenase, 1173, 1187f
Maleic acid, 493
Maleic anhydride, 493
Malic acid, 698
Malonic ester synthesis
  background on, 796–797
  retrosynthetic analysis, 799–800
  steps in, 798–799
Maltose, 1098
D-Mannose, 1077, 1078f, 1082–1083, 1095
Maraviroc, 983
Marine natural products, 4–6
Markovnikov's rule
  in cationic polymerization, 1260
  in electrophilic addition, 400–401, 403t
  in hydrohalogenation, 413
  isolated dienes, 486
Mass number, 8, 190
Mass spectrometers
  description of, 509–510, 509f
  GC–MS instrument, 521–522, 521f
Mass spectrometry (MS)
  of alcohols, 519–520
  of aldehydes, 518–519
  of alkyl halides, 514–516, 515f, 516f
  of amines, 520
  analysis of unknowns using molecular ion, 511–514
  definition of, 509
  description of, 509–511
  electrospray ionization, 523
  fragmentation in, 516–520, 517f, 519f
  gas chromatography and, 521–522, 521f, 522f
  high-resolution, 520–521, 521t
  of ketones, 518–519
  M + 1 peak, 510–511
  M + 2 peak, 514–516
  structure determination and, 545–546
Mass-to-charge (m/z) ratio, 510
Matrine, 993
Mauvine, 1006
mCPBA (meta-Chloroperoxybenzoic acid), 446, 447
Meadowsweet, 703
Medical applications, 2f, 3. See also Antibiotics; Anti-cancer drugs; Bronchodilators

I-20 Index

Mefloquine, 387
Melanin, 857
Melting point (mp)
  of alcohols, 344t
  of aldehydes, 652t
  of alkanes, 146, 146t
  of alkenes, 391
  of alkyl halides, 257t
  of alkynes, 391
  of amines, 974t
  of carboxylic acid derivatives, 742t
  of carboxylic acids, 700t
  of diastereomers, 205–206, 205f
  of enantiomers, 201, 203t
  of epoxides, 344t
  of ethers, 344t
  of fats, 395
  of fatty acids, 394, 394t
  intermolecular forces and, 105–106
  of ketones, 652t
  of monosaccharides, 1079
  of oils, 395, 440, 441f
  of thiols, 366
  of triacylglycerols, 1117, 1117t, 1120
Melt transition temperature, 1273
Membrane proteins, 1022
Menthol, 1127
Meperidine (Demerol), 108
Mercaptans, 366–367
Mercapto groups, 96t, 366–367
Mercury, in Clemmensen reduction, 919
Merrifield, R. Bruce, 1043
Merrifield method, 1043–1045
Mescaline, 977
Meso compounds
  definition, 197
  description of, 196–197
  in Diels–Alder reactions, 493
  of disubstituted cycloalkanes, 199
  in epoxidation, 448
  in halogenation, 406f
Messenger RNA (mRNA)
  description of, 1154
  genetic code, 1156–1157
  genetic code triplets, 1157f
  transcription, 1155–1156, 1155f
  translation to protein synthesis, 1157–1159, 1158f
Metabolism
  aerobic glucose catabolism, ATP yield, 1190
  catabolism in, 1170–1172, 1171f
  citric acid cycle (See Citric acid cycle)
  coupled reactions, 233, 234f
  description of, 1170–1172
  flavin adenine dinucleotide, 1174
  glucose (See Glucose)
  nicotinamide adenine dinucleotide, 1173
  oxidizing and reducing agents in, 1173–1175
  of triacylglycerols, 1119–1120
Meta directors, 901–902, 905, 907–908, 908f, 909–910, 912–913, 920–923
Meta isomers, 853
Metal–carbene intermediates, 1212–1213
Metal catalysts, 243, 436–437, 1211, 1265, 1266
Metal hydride reagents
  aldehyde reduction, 605, 659
  carboxylic acids and derivatives reduction, 608–612, 612t
  description of, 436

  ketone reduction, 605, 659
  in nitrile reduction, 721–722
Metallocyclobutanes, 1213
Metal salts of carboxylate anions, 698
Met-enkephalin, 744, 1034
Methamphetamine, 70, 977, 983, 983f
Methanamine, 971
Methane
  bonding in, 34–35
  complete halogenation, 939f
  description of, 2, 133
  molecular formula, 134
  molecular shape, 25–26
  natural occurrences of, 145
  as normal alkane, 135t
Methanoic acid, 698
Methanol
  hybrid orbitals in, 37
  Lewis structure, 13–14
  MTBE formation, 404
  as polar protic solvent, 265f
Methionine
  abbreviation for, 716t
  as functional group, 97, 117
  isoelectric point, 1025t
  pK_a values, 716t, 1025t
  SAM synthesis from, 368
  structure of, 1023f
Methoprene, 620
Methoxy groups, 341
Methyl acetate
  carbon NMR spectrum, 584f
  structure of, 734
Methyl acrylate, 493f
Methylamine, 971
m-Methylaniline, 916
p-Methylaniline, 916, 989
Methylation of SAM, 288
Methylbenzene, 853
Methyl benzoate, 753
(R)-α-Methylbenzylamine, 1026–1028
2-Methylbuta-1,3-diene, 484
2-Methylbutane
  common name, 144
  melting point, 106
  molecular formula, 134
2-Methylbutanenitrile, 699f
3-Methylbutan-1-thiol, 366
2-Methylbutan-2-ol, 350
2-Methylbutanoyl chloride, 737
3-Methylbut-1-ene, 952
(3-Methylbutyl)cyclohexane, A-6
Methyl carbocation
  hyperconjugation, 278–279
  inductive effects, 277–278
3-Methylcycloheptene, 388f
Methylcyclohexane, 157–159, 158f
2-Methylcyclohexane-1,3-dione, 835–836
2-Methylcyclohexanethiol, 366f
3-Methylcyclohexanol, 340f, 363
trans-4-Methylcyclohexanol, 283
2-Methylcyclohexanone, 788, 790, 821
1-Methylcyclohexene, 1200
3-Methylcyclohexene, 187
2-Methylcyclopentanamine, 996
Methylcyclopentane, 187
2-Methylcyclopentanecarboxamide, 739
(R)-3-Methylcyclopentanone, 659–660
1-Methylcyclopentene, 388f
Methylene, 1210
α-Methylene-γ-butyrolactone, 796
Methylene chloride, 258f
Methylenecyclohexane, 389f
Methylene groups, 135, 389f
Methyl esters, 1034, 1041

3-Methylfentanyl, 512
Methyl α-D-glucopyranoside, 1088
Methyl groups
  angular, 1132–1133
  as electron-donating group, 901, 906
  inductive effect of, 898
  nomenclature, 137
  as ortho, para director, 901, 906
  in steroids, 1132–1133
Methyl halides, 282
6-Methylhept-6-en-2-ol, 388
4-Methylhexane-2-thiol, 366f
2-Methylhexanoic acid, 800
5-Methylhexan-3-ol, 340
5-Methylhex-4-en-1-ol, A-8t
Methyl ketones, 414, 417, 796, 800–802
Methyllithium, 614
Methylmagnesium bromide, 614, 644
Methyl 2-methylbutanoate, 743
N-Methylmorpholine N-oxide (NMO), 452
Methyl 4-oxohexanoate, A-7f
3-Methylpentanal, 650f
4-Methylpentan-1-amine, 972
2-Methylpentane, 149
4-Methylpentane-1,4-diol, 622–624, 623f
2-Methylpentan-3-one, 651f
3-Methylpentan-2-one, 650
(E)-3-Methylpent-2-ene, 390
(Z)-3-Methylpent-2-ene, 390
3-Methylpent-2-ene, 389, 390
4-Methylpent-3-en-2-one, 652
1-Methyl-4-phenyl-4-propionoxy-piperidine, 512
N-Methylpropan-2-amine, 972
2-Methylpropane, 134, 138
Methyl propanoate, 542
2-Methylpropan-1-ol, 291
2-Methylpropan-2-ol, 896
2-Methylpropene, 404, 1261f
2-[4-(2-Methylpropyl)phenyl]propanoic acid, 137
1,2-Methyl shift, 353
2-Methylspiro[3.4]octane, A-9
(S)-3-Methyl-3-sulfanylhexan-1-ol, 366
Methyl thioacetate, 734
Methylthiocyclohexane, 367
β-Methylvaleraldehyde, 650f
Methyl vinyl ketone, 493, 835–836
Micelles, 112, 113f
Michael acceptor, 834
Michael reaction
  description of, 833–835
  estrone synthesis, 835f
  in Robinson annulation, 835–839
Mifepristone, 393
Minor contributor, 22–23, 479–480
Minor groove in DNA double helix, 1152
Mirror images, 177, 179–182
Miscibility, 108
Misoprostol, 1126
Mixed anhydrides, 735, 738, 764
Mixed triacylglycerols, 1117
Mixtures
  allylic halogenation products, 952–953
  extraction of, 711–714, 713f
  racemic (See Racemic mixtures)
  $S_N1/E1$ reactions in, 317
Molecular formula
  of acyclic alkanes, 133
  acyclic alkenes, 386
  of alkanes, 134, 135t
  of alkyl halides, 255
  alkynes, 386

  of chlorofluorocarbons, 258
  of cycloalkanes, 133, 135
  determination of (See Mass spectrometry)
  of ethane, 134
  isomers (See Constitutional isomers; Isomers; Stereoisomers)
  of methane, 134
  of 2-methylbutane, 134
  of propane, 134
  of straight-chain alkanes, 135t
Molecular models, element colors, A-3
Molecular orbital (MO) theory, 869–870, 1228
Molecular orbitals (MOs)
  in benzene, 871–872, 871f
  in buta-1,3-diene, 1229, 1229f
  description of, 869–870
  in ethylene, 1228, 1229f
  features of, 871
  in hexa-1,3,5-triene, 1230, 1230f
  patterns for cyclic, completely conjugated systems, 873f
  in pericyclic reactions, 1228
  prediction/determination of relative energies of, 872–874
Molecular recognition, 345
Molecular shape
  bond angle, 24–27
  bond length, 24, 24t
  four groups around atom, 25
  geometry based on number of groups, 25–27, 27t
  polarity and, 45–46, 46f
  second-row elements, 24, 27
  three groups around atom, 25
  two groups around atom, 25
Molecular weight, determination of. See Mass spectrometry
Molecules
  description of, 11
  free energy of, 231
  isomeric relationships, 200f
  polarity, 45–46
  resonance stabilized, 18
Molina, Mario, 949
Molozonide, 453
Monensin, 682
Monochlorination, 940, 948
Monohalogenation, 939
Monomers
  in chain-growth polymerization, 1257–1258
  description of, 177, 1256
  in radical polymerization, 1258, 1258f
Monophosphates, 100, 100t
5'-monophosphate (suffix), 1146
Monosaccharides. See also specific monosaccharides
  addition or removal of one carbon at carbonyl group, 1094–1097
  in catabolism, 1170–1171, 1171f
  catabolism of, 1179–1184
  cyclic forms of, 1079–1086
  D and L designations, 1076
  description of, 119–120, 1071
  determining structure of unknown, 1096–1097
  in DNA, 1145
  ester synthesis, 1091
  ether synthesis, 1090–1091
  Fischer projection formulas, 1072–1075, 1075f
  glycolysis of, 1179–1184
  glycosides, 1087–1090
  N-glycoside synthesis, 1104

Haworth projections, 1081, 1082–1085
Kiliani–Fischer synthesis, 1094, 1095–1096
melting points, 1079
nomenclature, 1071, 1077
in nucleotides, 120–121
oxidation at carbonyl group, 1092–1094
physical properties, 1079
reactions at carbonyl groups, 1091–1097
reactions at hydroxy groups, 1090–1091
reduction at carbonyl group, 1092
in RNA, 1145
solubility of, 1079, 1091
stereogenic centers in, 1072–1078
taste of, 1079
Wohl degradation, 1094–1096
Monosubstituted alkenes
description of, 305, 305f
in E2 elimination reactions, 310–311
in olefin metathesis, 1211
Monosubstituted benzene derivatives
nomenclature, 853
proton NMR spectra, 578, 578f
Monosubstitution, 909
Monoterpene/monoterpenoid, 1128, 1129t
Morphine, 60, 115, 498, 727, 751, 975
MOs. See Molecular orbitals
Motrin, 137, 206
Moylbdenum, 1211
mp. See Melting point
M peak, 510
M + 1 peak, 510–511
M + 2 peak, 514–516
MPPP, 512
MRI (magnetic resonance imaging), 584–585, 585f
mRNA. See Messenger RNA
MSG, 204
MTBE. See tert-Butyl methyl ether
Mullis, Kary, 1161
Multiple bonds
condensed structures, 29f
priority assignment, 190
resonance structures, 21, 23
Multiplets, NMR spectra, 570, 570t
Murex, 1006
Mutarotation, 1081, 1098, 1099
Myambutol, 272
Myelin sheath, 1122
Mylar, 1269
Myoglobin, 1053, 1054f
Myosin, 1047
Myrcene, 1127
Myristic acid, 1178–1179

N

Nabumetone, 802
NAD⁺ (nicotinamide adenine dinucleotide)
biological oxidation and, 459–460
biological oxidizing agent, 607–608, 1173
in citric acid cycle, 1188
stereogenic centers, 187–188
NADH [nicotinamide adenine dinucleotide (reduced form)]
biological reducing agent, 459–460, 1173
biological reduction and, 607–608
citric acid cycle product, 1189
reduction of acyl phosphates, 609–610
NADPH (nicotinamide adenine dinucleotide phosphate), 608
NAG (*N*-Acetyl-D-glucosamine), 1103
Naked anions, 266
Nandrolone, 1136
Naphthalene, 856, 861
Naprosyn, 207
Naproxen, 207, 795
Narcotics
codeine, 978
meperidine, 108
Natta, Giulio, 1265
Natural dyes, 1006
Natural gas, 145
Natural products. *See also specific compounds and sources*
biologically active compounds, 4–6
rubber, 1265–1266
synthesis of, 290–291
NBS (*N*-bromosuccinimide), 407, 951, 952
Neopentane, 106, 134
Neoprene, 1266
Neral, 652
Neryl diphosphate, 1131–1132
Net reactions, 233
Neuroprotectin D1 (NPD1), 483
Neurotransmitters
acetylcholine, 107, 767, 1246
dopamine, 97, 803, 977, 978f
noradrenaline, 977
serotonin, 896, 977–978
Neutral amino acids, 718–719, 1022, 1023f, 1024–1025
Neutrons, 8
Newman projections
drawing, 147–149
ethane, 148f
propane, 148f
Niacin, 112, 607
Nicotinamide adenine dinucleotide
biological oxidizing agent, 459–460, 1173
biological reducing agent, 607–608, 1173
reduction of acyl phosphates, 609–610
stereogenic centers, 187–188
Nicotinamide adenine dinucleotide phosphate, 608
Nicotine, 114, 976, 976f
Nitration of benzene, 887f, 890–891, 912–913, 921, 922, 1001f, 1002–1003
Nitriles
acidity of, 784
carbon NMR, 701t
description of, 719
in enolate alkylation, 794
hydrolysis of, 719
infrared (IR) absorptions/spectra of, 543, 543t, 701t
interesting compounds, 702–703
nomenclature, 699, 699f
nucleophilic addition in, 719–722
organometallic reagent addition, 722
physical properties of, 700
pK_a values, 785t
products of S_N2 substitution reactions, 290, 290t
reduction of, 721–722, 981
structure of, 696
p-Nitroaniline, 989
Nitrobenzene
electrophilic aromatic substitution reaction, 901, 907–908
formation of, 890–891
reduction to aniline, 890
p-Nitrobenzoic acid, 921–922
Nitrogen
in amines, 970, 985
bonding of, 12f, 16, 16t
common nucleophile, 266t
degrees of unsaturation and, 386–387
formal charge, 16t
mass spectra and, 512
Nitrogen heterocycles, 973, 973f
Nitrogen–hydrogen bonds
acidity, 70
in amines, 974
in α-helix, 1046, 1046f
in peptide bond, 1033
in peptide hydrolysis, 1057
in β-pleated sheet, 1047, 1047f
in primary structure of proteins, 1045
in secondary structure of proteins, 1046–1047
Nitrogen–nitrogen bonds, 999, 1004
Nitro groups
as electron-withdrawing group, 901
as meta director, 901, 907–908, 912, 920–922
reduction of, 920–921, 981
Nitronium ion, 890
o-Nitrophenol, 853
p-Nitrophenol, 913
N-Nitrosamine, 998–999
Nitrosonium ion, 998–999
o-Nitrotoluene, 912–913
p-Nitrotoluene, 910
Nitrous acid, 997–999
Nitrous oxide, 1275
NMO (*N*-methylmorpholine *N*-oxide), 452
NMR spectroscopy. See Nuclear magnetic resonance (NMR) spectroscopy
Node of electron density, 9
Nomenclature. *See also* Common names; IUPAC nomenclature
of alkyl substituents that contain branching, A-6
of bicyclic compounds, A-8–9
of chain-growth polymers, 1261f
D and L designations, 1076
of drugs, 136–137
E,Z system, 389–390
of monosaccharides, 1071, 1076
of nucleosides, 1146
of nucleotides, 1146
of polyfunctional compounds, A-7–8
R,S system, 189–194
Nonactin, 115–116
Nonane, 135t
Nonbonded interactions. *See* Intermolecular forces
Nonbonding (unshared) electrons, 12, 12f
Noncovalent interactions. *See* Intermolecular forces
Nonhydrolyzable lipids, 1116
Nonnucleophilic bases, 264
Nonpolar bonds
cleavage by radical reactions (*See* Radical reactions)
description of, 44
Nonpolar molecules
description of, 45–46
solubility of, 109
Nonpolar tails
phospholipids, 113, 114f, 1121, 1121f
soap, 112, 113f
Nonreducing sugars, 1093, 1093f, 1100
Nonrenewable energy, 145
Nonsteroidal anti-inflammatory drugs (NSAIDs), 1126. *See also* Aspirin; Ibuprofen
Nootkatone, 514
Noradrenaline, 977
Norepinephrine, 977
Norethindrone, 108, 392–393, 392f, 1136
Noyori, Ryoji, 1030
NPD1 (neuroprotectin D1), 483
NSAIDs (nonsteroidal anti-inflammatory drugs), 1126. *See also* Aspirin; Ibuprofen
N-Terminal amino acids, 1032, 1035, 1157
Nuclear magnetic resonance (NMR) spectroscopy
basis of, 554–556
¹³C (carbon) (*See* Carbon NMR)
chemical shift, 556–557
description of, 509
downfield signals, 556
electromagnetic radiation and, 529
general principles, 556–557
¹H (proton) (*See* Proton NMR)
magnetic resonance imaging, 584–585, 585f
nuclear spin in, 554–555
reference signal, 556–557
signal intensity, 556
spectrometers, 555–556, 555f
upfield signals, 556
variables in, 555
Nucleases, 1055t
Nucleic acids
bases, 1145
description of, 120–121, 1145
DNA (*See* DNA)
nomenclature, 1146–1147, 1147t
phosphodiesters, 100
RNA (*See* RNA)
structure of, 1148–1150, 1149f
Nucleophiles
acetylide anions, 398
alkoxide, in ether synthesis, 347
alkyl halides and, 259
ambident, 787
amines as, 970, 979, 993–995
atomic size, 265
carbonyl compounds and, 600
carbonyl group, 97
common nucleophiles, 266t
cyanide anion, 661
description of, 79
enolates, 787
enols, 783
epoxide ring opening with, 369–370
heterolysis and, 225
hydroxide, in alcohol synthesis, 347
as Lewis base, 263
nitrogen as, 985
in nucleophilic addition reactions, 657–658, 658f
nucleophilicity vs. basicity, 264–266
in nucleophilic substitution reactions, 260–261, 269–270, 283–284, 745–746
organometallic reagents as, 613–616
pi (π) bonds and, 92, 116
reactivity of, 116–117

Nucleophilic acyl substitution reactions
  in acid chlorides, 749–750, 993–994
  in amides, 760–761
  in amines, 993–994
  in anhydrides, 750–751
  basicity and, 746–747
  in carbonyl compounds, 601–603
  in carboxylate anions, 763–766
  in carboxylic acids, 751–756, 752f
  chemical recycling of PET, 1277–1278
  description of, 734
  in esters, 756–758
  general mechanism, 745–746
  β lactam antibiotic mechanism, 762
  list of, A-15
  overview of reactions, 748, 762, 763t
  in peptide bond synthesis, 1158, 1158f
Nucleophilic addition reactions
  acid-catalyzed, 657
  1,2- versus 1,4-addition, 486–490, 629–631
  aldehyde reaction with organometallic reagents, 617–620, 659–660
  aldehyde reduction, 605, 659, A-16
  aldol reactions, 814, 814f
  carbonyl compounds, 600, 601
  cyanohydrin formation by, 660–662
  in Edman degradation, 1035–1036
  enamine formation, 658f, 668, 669f, 993
  general mechanism, 656–657
  in glycolysis, 1182
  hydride reduction, 605–606, 609, 659
  imine formation, 658f, 666–667, 669f, 672–673, 993
  ketone reaction with organometallic reagents, 617–620, 659–660
  ketone reduction, 605, 659, A-16
  in Michael reaction, 834
  nitrile reduction, 721–722
  in nitriles, 719–722
  organometallic reagent reactions with carboxylic acid derivatives, 625, 632
  in peptide hydrolysis, 1057
  protecting groups and, 622–624, 623f
  α,β-unsaturated carbonyl compounds, 629–631
  urethane synthesis, 1269
  Wittig reaction, 662–666
Nucleophilic aromatic substitution reactions
  by addition–elimination, 913–915
  description of, 913
  by elimination–addition, 915–916
  list of, A-15
Nucleophilicity
  basicity vs., 264–266, 985
  crown ethers, 345–346, 346f
  solvent effects, 265–266, 265f
  steric hindrance and, 264–265
  substitution reaction factor, 283–284
Nucleophilic substitution reactions. See also Nucleophilic acyl substitution reactions; Nucleophilic aromatic substitution reactions
  in acid chloride reduction, 609
  S-adenosylmethionine, 288, 289f
  alcohol conversion to alkyl halides, 349, 356–361
  of alcohols, 349, 356–363, 364f
  of alkyl halides, 259
  in biological reactions, 287–289

carbocation stability, 277–279
carbonate synthesis, 1270
Claisen reactions, 826, 826f
crown ethers, 345–346, 346f
description, 255
in ester reduction, 609
general features of, 260–261
α-halo carbonyl compounds in, 792–793
Hammond postulate, 279–282
leaving group, 260, 261–263
list of, A-15
mechanisms for, 267–287
nucleophiles in, 263–266
organometallic reagent reactions with carboxylic acid derivatives, 625–626, 632
solvent effects, 265–266
steric effects, 264
Nucleosides, 1146, 1147t
Nucleoside triphosphates, 1161
Nucleotides
  description of, 120, 1145
  formation of, 1146
  nomenclature, 1146–1147, 1147t
Nucleus, 8
Nutmeg, 1178
Nylon
  nylon 6, 1268, 1278
  nylon 6,6, 1256f, 1257, 1267, 1274–1275, 1275f
  nylon 6,10, 1268

Observed rotation (α), 202
Octacosanoic acid, 1179
Octane, 135t
Octanenitrile, 543
(2E,4Z,6E)-Octa-2,4,6-triene, 1231, 1235
(2E,4Z,6Z)-Octa-2,4,6-triene, 1234
Octet rule
  description of, 11
  exceptions to, 17–18
Octylamine, 543
2-Octyl cyanoacrylate, 1262
-oic acid (suffix), 697
Oils
  fatty acid composition of, 1118, 1118t
  hydrogenation of, 434
  insolubility of, 146
  melting point, 395, 440, 441f
  oxidation at allylic carbon, 953–954, 953f
  as triacylglycerols, 1117
  vegetable (See Vegetable oils)
Okazaki fragments, 1154
Olean, 759
Oleander, 678
Oleandrin, 678
Olefin metathesis
  general features of, 1211–1212, 1212f
  mechanism, 1212–1213
  ring-closing metathesis, 1213–1216
Olefins, 385. See also Alkenes
Oleic acid
  melting point, 394, 394t, 434
  structure of, 394f
Oleocanthal, 1214
Olestra, 758–759
Olfactory receptors, 207, 207f
Oligosaccharides, human milk, 1102–1103

Olives and olive leaves, alkanes in, 134
-ol (suffix), 339–340, 366
Omega-3 fatty acids, 395
-one (suffix), 650
-onitrile (suffix), 699
Opsin, 671
Optical activity
  achiral compounds, 201, 370
  chiral compounds, 202
  observed rotation, 202
  racemic mixtures, 202–203, 790
  specific rotation, 203
Optical purity, 203–205
Optical rotation, 201–203, 205–206
Orajel, 920
Oral contraceptives, 392–393, 392f, 1137. See also specific drugs
Orbitals. See also Hybrid orbitals; Molecular orbitals; specific orbitals
  description of, 9–10
  hybridization, 34–41 (See also Hybridization)
Orbital symmetry
  cycloaddition reactions and, 1237–1238
  electrocyclic reactions and, 1232
  sigmatropic rearrangements and, 1241–1243
Organic chemistry
  definition of, 1
  importance of, 1–2, 2f
  synthesis in, 421–423, 632–634
Organic diphosphates, 287, 1129, 1130f
Organic molecules/compounds
  biomolecules, 117–122
  bonding in, 37–40, 38f, 39f, 41t
  cell membrane and, 113–116, 114f, 115f
  common features of, 4
  condensed structures, 28–30, 29f
  drawing structure of, 28–34
  extraction technique, 711–714, 713f
  functional groups, 92–100, 116–117
  intermolecular forces in, 101–103, 103t
  physical properties, 104–110
  reactivity, 116–117
  shape (See Molecular shape)
  skeletal structures, 28, 30–34, 31f
  soap, solubility and, 112–113, 113f
  as solvents, 107–109, 110t
  synthesis, 421–423, 632–634
  vitamins, 111–112 (See also specific vitamins)
Organic monophosphates, 287
Organic reactions. See also specific types of reactions
  acyl transfer reactions, 766
  of alcohols, 349–361
  of aldehydes, 656–682
  of alkenes, 396–420
  of amines, 985–1005
  arrow notation in, 225–227, 226t
  of benzene derivatives, 913–921
  bond dissociation energy, 227–231
  bonds and, 223–227, 224f
  carbon–carbon bond formation, 1198–1216, A-11
  of carbonyl compounds, 600–607, 622, 779–803
  catalysts, 243
  coupled reactions in metabolism, 233, 234f
  endothermic, 227–228
  energy diagrams, 236–240, 237f, 239f

  of enolates, 787, 790–791
  enthalpy change, 227–231, 234–236
  entropy change, 234–236
  enzymes, 244, 244f, 245t
  of epoxides, 349, 368–372
  equations for, 220–221, 220f
  equilibrium constant, 231–233, 232f, 232t
  of ethers, 349, 364–365
  exothermic, 227–228
  free energy change, 231–232, 232f, 232t, 234–236
  general types, A-15–16
  of ketones, 656–682
  kinetics of, 240–242
  laboratory, 245t
  olefin metathesis, 1211–1216
  organocuprate coupling reactions, 1198–1200, 1210
  of organometallic reagents, 631–632
  oxidation and reduction reactions, 164–165, 164f (See also Oxidation; Reduction)
  sequential, 220
  stereospecific, 406
  of substituted benzenes, 898–913, 917–921
  thermodynamics of, 231–236
  two-step reaction mechanism, 238–240
  types of, 221–223, A-15–16 (See also specific types of reactions)
  writing equations, 220–221
Organic solvents, 107–109, 110t
Organic synthesis
  alcohols from acetylide anion and epoxide, 420
  of aspirin, 290–291, 291f
  definition, 290
  development of, 421
  multistep synthesis, 422
  by nucleophilic substitution, 290–291, 290t
  retrosynthetic analysis, 421–423
  terminology, 421
Organic triphosphates, 287
Organoaluminum compounds, 1265
Organoboranes, 409, 417, 1201–1205
Organocuprate coupling reactions
  cyclopropanation, 1210
  hydrocarbon synthesis, 1198, 1199–1200
Organocuprate reagents
  in acid chloride reactions, 626–627, 655
  in coupling reactions, 1198–1200
  description of, 614
  hydrocarbon synthesis, 1198–1200
  preparation of, 614
  in α,β-unsaturated carbonyl compound reactions, 631
Organolithium reagents
  in acid chloride reactions, 616, 625
  in aldehyde reactions, 616, 617–618, 659–660
  in anionic polymerization, 1261
  in arylborane preparation, 1202
  arylborane synthesis from, 1202
  description of, 614
  in ester reactions, 616, 625
  in ketone reactions, 616, 617–618, 659–660
  in LDA preparation, 787
  in nitrile reactions, 722
  preparation of, 614–616

in α,β-unsaturated carbonyl compound reactions, 631
in Wittig reagent synthesis, 664
Organomagnesium reagents, 614. *See also* Grignard reagents
Organometallic coupling reactions
  cyclopropanation, 1210
  general features of, 1198–1199
  Heck reactions, 1205–1207
  hydrocarbon synthesis, 1198, 1199–1200
  Suzuki reactions, 1200–1205
  types of, 1198
Organometallic reagents. *See also* Grignard reagents; Organocuprate reagents; Organolithium reagents
  acetylide anions, 615
  in carboxylic acid derivative reactions, 624–627
  description of, 613–614
  in epoxide reactions, 628–629
  nucleophilic addition and, 617–620, 659–660
  preparation of, 614–615
  reaction as base, 615–616
  reaction as nucleophile, 616
  reactions of, 631–632
  synthesis applications, 619–620
Organopalladium reagents
  in Heck reactions, 1205–1207
  in Suzuki reactions, 1200–1205
Organophosphorus reagent, 663
Organosodium reagents, 615
Orientation of benzene substituents, 901
Orlon fibers, 1261f
Ortho, para activators, 902–903
Ortho, para deactivators, 902
Ortho, para directors, 901–902, 906–907, 908f, 910, 912–913, 921–923, 1001
Ortho isomers
  in bromination reaction, 917
  nomenclature, 853
Oseltamivir, 99
-ose (suffix), 1071
-osine (suffix), 1146
Osmium tetroxide, 446
-o (suffix), 256
Oxalic acid, 702
Oxaloacetate, 831, 1187, 1187f, 1189
Oxaphosphetane, 664
Oxidases, 1055t
Oxidation. *See also* Reduction
  agents, 446, 447f, 456–457, 459–460, 1173, 1174–1175
  of alcohols, 447f, 456–460, 705t
  of aldehydes, 613, 705t
  of aldoses, 1092–1094
  of alkanes, 163–166
  of alkenes, 447–455, 447f
  of alkyl benzenes, 918
  of alkylborane, 410–411
  of alkynes, 447f, 455–456
  allylic, 953–954, 953f
  biological oxidation, 458–460
  carboxylic acid synthesis, 704, 705t
  chiral reactions, 460–462
  in citric acid cycle, 1187f, 1188, 1189
  description of, 435–436, 435f, 603
  dihydroxylation, 447f, 451–453
  enzymes in, 1055, 1055t
  epoxidation, 447–450, 460–462
  FAD as oxidizing agent, 1174–1175
  fatty acid catabolism, 1176–1179, 1177f, 1190–1191

of glucose, 231
in glycolysis, 1182–1183
NAD^+ as oxidizing agent, 459–460, 1173
p-nitrobenzoic acid synthesis, 921–922
of organoborane, 416–417
β-oxidation, 1176–1178
oxidative cleavage (*See* Oxidative cleavage)
reactions, 164–165, 164f, 447–462, 447f
of substituted benzenes, 918
of sulfides, 367
triacylglycerol catabolism, 1176–1178
of vegetable oils, 440, 1119
Oxidative addition reactions
  Heck reactions, 1205–1207
  Suzuki reactions, 1201–1205
Oxidative cleavage
  of alkenes, 447f, 453–455, 655t
  in alkyl benzenes, 918
  of alkynes, 447f, 455–456, 705t
  in progesterone synthesis, 824f
Oxidative phosphorylation, 1172, 1175
Oxidized state, 164
Oxidoreductases, 1055, 1055t
Oxime, 1095
Oxiranes, 338, 342, 343. *See also* Epoxides
2-Oxopentanedioic acid, 698
Oxybenzone
  coral bleaching, 7
  molecular structure, 46–47
Oxy-Cope rearrangements, 1244
Oxygen
  bonding of, 12f, 16t
  common nucleophile, 266t
  degrees of unsaturation and, 386–387
  as diradical, 938, 940
  formal charge, 16t
Oxygen–hydrogen bonds, 70–73
Oxygen–oxygen bonds, 446
-oxy (suffix), 341
Oxytocin, 735, 1021, 1034, 1049
-oyl (suffix), 651
Ozone
  in atmosphere, 259, 453, 949–950, 949f, 1275
  as oxidizing agent, 446
Ozonide, 453
Ozonolysis, 453–454

P

PABA (p-aminobenzoic acid), 857, 1007–1008
Paclitaxel, 176, 187–188, 603
Padimate O, 857
PAHs. *See* Polycyclic aromatic hydrocarbons
Palau'amine, 553, 579
Palladium
  as catalyst, 443
  in Heck reactions, 1205–1207
  in hydrogenolysis, 1041
  in Suzuki reactions, 1200–1205
Palladium(II) acetate, 1205
Palmitic acid, 1117
Palmitoleic acid, 1179
Pantothenic acid, 112
Para isomers
  in bromination reaction, 917
  nomenclature, 853
Parallel pi (π) stacking, 874–875
Para red, 1007

Parent DNA, 1152, 1153
Parent name, 137
Parkinson's disease, 977, 978f
Paroxetine (Paxil), 347
Partial hydrolysis of peptide, 1036–1037
Partially hydrogenated vegetable oils, 440–442, 441f
Patchouli alcohol, 337
Paxil, 347
PCBs (polychlorinated biphenyls), 109
PCC (pyridinium chlorochromate), 446
Peanut butter, 441
PEF (polyethylene furanoate), 1269
PEG [poly(ethylene glycol)], 1263–1264
Penicillin, 528, 544–545, 744–745, 762
Penicillium, 528
Pentadecane, 135t
Penta-1,4-diene, 474–475, 475f, 485
(E)-Penta-1,3-diene, 485
Pentane
  intermolecular forces in, 103
  mass spectrum of, 517
  melting point, 106
  molecular structure, 134
  as normal alkane, 135t
  synthesis of, 485
Pentane-2,4-dione, 781, 785
Pentane-3-thiol, 366f
Pentan-3-ol, 620
Pentan-2-one, 518, 796
Pentan-3-one, 104, 104f
Pent-1-ene, 512
Pentoses, 1071
Pent (prefix), 134
2-Pentylcinnamaldehyde, 819
Pent-1-yne, 512
Peppermint plant, 1127
Peptide bonds
  description of, 1031–1033
  in primary structure of proteins, 1045
  translation and synthesis of, 1158, 1158f
Peptides. *See also* specific peptides
  amino acid analysis, 1035
  automated synthesis of, 1043–1045
  description of, 1031
  dipeptides, 1031–1032, 1038–1043
  Edman degradation, 1035–1036
  interesting compounds, 1033–1034
  N-terminal amino acid identification, 1035–1036
  partial hydrolysis, 1036–1037
  sequencing, 1035–1038
  simple, 1031–1032
  solid phase technique of synthesis, 1043–1045
  structure drawing conventions, 1032
  synthesis of, 1038–1045
  tetrapeptides, 1033
  tripeptides, 1031–1032, 1043–1045
Percent s-character
  acidity and, 73, 74f
  alkene stability and, 307, 391
  basicity and, 991, 992t
  bond length and, 43, 484
  C–H absorptions and, 534
  infrared absorptions and, 534
  stability and, 73
  in vinyl halides, 289
Percent transmittance, 531–532
Pericyclic reactions
  cycloaddition reactions, 1227, 1236–1240 (*See also* Cycloaddition reactions)

Diels–Alder reactions (*See* Diels–Alder reactions)
electrocyclic reactions, 1227, 1231–1236 (*See also* Electrocyclic reactions)
sigmatropic rearrangements, 1227, 1240–1246 (*See also* Sigmatropic rearrangements)
types of, 1227
Period, in periodic table, 9
Periodic table, A-1
  arrangement, 9
  atomic structure and, 8–10
  element effects on acidity, 68–70, 74f
  element location and bonding type, 11
  of elements common in organic chemistry, 9f
  first-row elements, 9f, 10
  second-row elements, 9f, 10, 24, 27
Periodic trends
  acidity, 68–70, 74f
  basicity, 262
  bond dissociation energy, 228–229, 228t
  bond length, 24
  electronegativity, 43–44, 44f
  leaving group ability, 262
  nucleophilicity, 265
Periplanone B, 1226
Perkin, William Henry, 1006
Perlon, 1268, 1278
Peroxides
  in radical addition reactions, 955–956
  as radical indicators, 937
Peroxyacetic acid, 446, 447
Peroxyacids, 446
Peroxy radicals, 953, 953f, 1258
Perpendicular pi (π) stacking, 874–875
Pesticides, 4
PET. *See* Polyethylene terephthalate
Petroleum, 132, 145, 291, 1274, 1275f
Peyote, 977
PGE_2, 1124
$PGF_{2\alpha}$, 703–704, 1115f, 1124, 1125f
PGG_2, 1124, 1125f
Pharmaceuticals. *See also names of specific pharmaceuticals*
  nomenclature, 136–137
PHAs (polyhydroxyalkanoates), 1278–1279
PHB (polyhydroxybutyrate), 1278
PHBV (copolymer), 1278
α-Phellandrene, 1127f
Phenanthrene, 862
Phenols
  antioxidant structure, 954
  in aspirin synthesis, 291
  in Bakelite synthesis, 1273, 1273f
  chlorobenzene to, 915
  Claisen rearrangement synthesis, 1245
  description of, 338
  nomenclature, 853
  pK_a, 707
  polyhalogenation of, 909
  synthesis of, 1000
Phenoxide, 707
Phentermine, 985
Phenylacetaldehyde, 650f, 783
Phenylalanine
  isoelectric point, 1025t
  pK_a values, 1025t
  structure of, 1023f
  synthesis of, 1030–1031
Phenylethanal, 650f
2-Phenylethanamine, 984, 996

2-Phenylethanamine derivatives, 977–978
Phenyl groups, 93, 854
Phenyl isothiocyanate, 1035
Phenylmagnesium bromide, 621
(1Z,3E)-1-Phenylocta-1,3-diene, 1204
N-Phenylthiohydantoin (PTH), 1035–1036
Pheromones
　blatellaquinone, 750
　bombykol, 1204
　disparlure, 449–450, 450f
　(−)-frontalin, 462
　grandisol, 1128
　periplanone B, 1226
　undecane, 133
Phosgene, 1270
Phosphate diesters. See Phosphodiesters
Phosphate esters, 100
Phosphate groups
　as functional group, 100
　as leaving groups, 287, 316
　in nucleotides, 120–121
Phosphatidylcholine, 1121
Phosphatidylethanolamine, 1121
Phosphines, 1201
Phosphoacylglycerols, 1120–1121, 1121f, 1122f
Phosphodiester linkage, 121, 121f, 1148–1150, 1149f
Phosphodiesters, 100, 1122, 1122f
2-Phosphoglycerate, 1183
3-Phosphoglycerate, 1182–1183
Phosphoglycerate mutase, 1183
Phospholipids
　description of, 113–114, 114f, 1120
　phosphoacylglycerols, 1120–1121, 1121f
　sphingomyelins, 1122
Phosphoranes, 663
Phosphoric acid, 714–715
Phosphoric acid esters, 714–715
Phosphorus
　octet rule exception, 17–18
　organic acids containing, 714–715
　in Wittig reactions, 662
Phosphorus oxychloride, 350, 355
Phosphorus–oxygen (P–O) bonds, 100, 100t
Phosphorus tribromide, conversion of alcohols to alkyl halides with, 359, 360, 360t
Phosphorylation
　in citric acid cycle, 1187f, 1188
　of fructose, 234
　of glucose, 233, 234f
　of glycerol, 1176
　in glycolysis, 1182–1183
Photochemical electrocyclic reactions, 1230, 1232, 1234–1236
Photons, 529
Photosynthesis, 344, 345f, 1070
Phthalic acid, 727, 918
Phthalimide, 980
Phylloquinone, 112, 885, 1123f
Physical properties. See Boiling point; Melting point; Optical activity; Solubility
Physiological pH, 709, 714
Phytyl diphosphate, 897, 897f
Pi* (π^*) antibonding molecular orbitals
　in buta-1,3-diene, 1229, 1229f
　description of, 870, 870f
　in ethylene, 1228, 1229f
　in hexa-1,3,5-triene, 1230, 1230f

Pi (π) bonding molecular orbitals
　in buta-1,3-diene, 1229, 1229f
　description of, 870, 870f
　in ethylene, 1228, 1229f
　in hexa-1,3,5-triene, 1230, 1230f
Pi (π) bonds
　in addition reactions, 222, 396
　in aldehyde reduction, 605, 659
　in alkene reduction, 437
　in alkenes, 305, 385
　in alkyne reduction, 442
　in alkynes, 322–323, 323f, 385
　in benzene, 850–852
　of benzyne, 916
　in bromination, 917
　Brønsted–Lowry bases and, 59–60, 59f, 263
　in buta-1,3-diene, 1229, 1229f
　in carbonyl groups, 97–98, 599–600
　chemical shift, NMR spectra, 565–567, 566t
　in cycloaddition reactions, 1227, 1236–1240
　degrees of unsaturation and, 386, 439–440
　in dehydration of aldol product, 816
　in dehydration reactions, 350
　description of, 39–40, 39f, 40f
　in Diels–Alder reactions, 490–491
　in E2 elimination reactions, 309
　in electrocyclic reactions, 1227, 1231–1236
　in elimination reactions, 221–222, 303
　in epoxidation reactions, 447–448
　in ethylene, 1228, 1229f
　in farnesyl diphosphate biosynthesis, 1130
　in functional groups, 116
　in halogenation, 404
　in Heck reactions, 1207
　in hexa-1,3,5-triene, 1230, 1230f
　in Hofmann elimination, 996
　in hydration, 403
　in hydroboration, 409
　in hydrohalogenation, 399, 413
　in intramolecular aldol reactions, 822–823
　in ketone reduction, 605, 659
　in nitriles, 696
　nucleophiles and, 263
　in olefin metathesis, 1211
　in oxidation of alcohols, 457
　in oxidative cleavage, 453, 455
　in pericyclic reactions, 1227
　in polyenes, 482
　in radical polymerization, 958
　in radical reactions, 937, 938, 955
　reactivity of molecule, 92
　relative basicity and, 990
　resonance and, 477–478
　in Robinson annulation, 835
　in sigmatropic rearrangements, 1228, 1240, 1242–1243
　strength, 42
　in synthetic dyes, 1007
　α,β-unsaturated carbonyl compounds, 629
Picato, 1197
Pi (π) electrons
　in adenine, 868
　in annulenes, 861
　in benzene, 850–851, 860, 886
　in cyclopentadienyl ions, 862–863
　in cytosine, 868–869
　in guanine, 868
　Hückel's rule, 860, 860t, 869–872

　in PAHs, 862
　in pyridine, 864
　in pyrrole, 865
　in thymine, 868
　in tropylium cation, 863–864
pI (isoelectric point), 718–719, 1025, 1025t
Pilocarpine, 866
Pilocarpus microphyllus, 866
Piperidine, 973f, 991, 992t
Pi (π) stacking, 874–875
Pitocin, 735, 1021
Pivalic acid, 65
pK_a values
　acid strength and, 64–65, 67–74
　of amines, 987–991, 992t
　of amino acids, 716t, 1024–1025, 1025t
　of carbonyl compounds, 785t
　of carboxylic acids, 705–707
　common compounds, 64t
　conjugate bases and, 784, 784t
　definition, 64
　deprotonation, 66–67
　estimation of, 65
　Henderson–Hasselbalch equation, 708–710, 709f
　list, A-4–5
　of nitriles, 785t
　of nitrogen compounds, 992t
　resonance delocalization and, 72
　terminal alkyne deprotonation, 398, 398t
Planar carbocations, 275–276
Planar molecules
　benzene, 859–860
　Hückel's rule, 869–870
Plan B, 393
Plane of symmetry, 182, 197, 199, 1096–1097
Plane-polarized light, 201–202
PLA [poly(lactic acid)], 1275
Plaque, 768
Plasticizers, 1274
β-Pleated sheet, 1047–1048, 1047f
PLP (pyridoxal phosphate), 646, 672–673
Poison dart frogs, 384
Poisoned catalysts, 443
Polar aprotic solvents, 265–266, 285, 310, 788
Polar bonds
　in alkyl halides, 259–260
　description of, 44–45
　in organometallic reagents, 614
Polar heads
　phospholipids, 113, 114f, 1121, 1121f
　soap, 112, 113f
Polarimeter, 201
Polarizability
　of alkyl groups, 277, 898, 937
　of alkyl halides, 257t
　of alkynes, 385
　boiling point and, 104, 104f
　of halogens, 404
　inductive effects on substituted benzenes, 898
　van der Waals forces and, 101
Polarized light, 201–202
Polar molecules
　description of, 45–46, 46f
　dipole–dipole interactions, 102
　monosaccharides, 1091
　solubility of, 107
Polar protic solvents, 265, 265f, 266, 285, 317, 789

Polar reactions
　description of, 225
　reactive intermediates, 224f
Poly(acrylic acid), 958
Polyacrylonitrile, 1261f
Polyalkylation, 979
Polyamides, 118, 1257, 1267–1268, 1279
Polyaspartate, 1279
Poly(buta-1,3-diene), 1266
Polycarbonates, 1270–1271
Polychlorinated biphenyls (PCBs), 109
Polycyclic aromatic hydrocarbons (PAHs). *See also specific compounds*
　bonding to DNA double helix, 1152
　description of, 856
　examples of, 861–862
　interesting compounds, 856, 856f
Polyesters
　biodegradable, 1278–1279
　olestra, 759
　step-growth polymerization, 1257
　synthesis of, 1268–1269
　Tritan, 1271
Polyethers, 345, 1263–1264
Poly(ethyl α-cyanoacrylate), 1261f
Polyethylene
　chain branching in, 1259
　in consumer products, 1256f
　description, 94
　high-density, 957, 1259, 1265
　low-density, 957, 1259
　as polymer, 957
Polyethylene furanoate (PEF), 1269
Poly(ethylene glycol) (PEG), 1263–1264
Polyethylene terephthalate (PET)
　description of, 1255
　recyclability, 1277–1278, 1277t
　shorthand representation, 1257f
　synthesis of, 1268–1269
　as thermoplastic polymer, 1273
Polyfunctional groups, nomenclature of, A-7–8
Polyhalogenation, 909
Polyhydroxyalkanoates (PHAs), 1278–1279
Polyhydroxybutyrate (PHB), 1278
Polyisobutylene, 1261f
Poly(lactic acid) PLA, 1275
Polymerase chain reaction (PCR), 1161–1162, 1161f
Polymerization
　anionic, 1261–1264, 1261f
　chain branching, 1259–1260
　chain-growth polymers, 1257–1263
　coordination, 1265
　copolymers, 1263
　definition, 957
　description of, 1256
　ionic, 1260–1262, 1261f
　radical, 958–959, 1258–1260, 1258f
　Ziegler–Natta polymerization, 1265–1266
Polymers
　definition of, 177, 957
　in peptide synthesis, 1043–1045
　plasticizers and, 1274
　proteins, 744
　synthetic, 1255–1279 (*See also* Synthetic polymers)
　thermoplastics, 1273
　thermosetting polymers, 1273
Polynucleotides
　description of, 121, 121f, 1149

nomenclature, 1149
structure of, 1149–1150, 1149f
Polypeptides
  description of, 1031
  representative segment of DNA, 1158f
  stabilizing forces, 1050, 1050f
  subunits, 1050
  synthesis of, 1043
Poly (prefix), 1261f
Polypropylene, 957
Polysaccharides
  cellulose, 1101
  description of, 119–120
  glycogen, 1102
  human milk oligosaccharides, 1102–1103
  starch, 1101–1102
Polystyrene
  chain-growth polymerization, 1258
  as polymer, 958
  recycling of, 1277t
  shorthand representation of, 1257f
  as thermoplastic polymer, 1273
Polystyrene derivatives, 1043
Polysubstituted benzene derivatives, 854
Poly(trimethylene terephthalate), 1275, 1276f
Polyurethanes, 1269–1270
Poly(vinyl acetate), 958, 1261f
Poly(vinyl chloride) (PVC), 258, 935, 958, 1256
p orbitals
  allylic carbocations, 475
  in benzene, 851–852
  in benzyne, 916
  in buta-1,3-diene, 1229, 1229f
  conjugation, 474–475
  delocalization of electrons, 481
  description of, 9
  1,3-dienes, 474
  in ethylene, 1228, 1229f
  in hexa-1,3,5-triene, 1230, 1230f
  in histamine, 866
  hybridization of, 34–41
  hyperconjugation, 278–279
  molecular orbitals and, 870–872, 870f, 871f, 1228
  pi (π) stacking, 874–875
  2*p* orbitals, 10, 34–40, 43, 305
  rotation of, 1232
Potassium *tert*-butoxide, 76f, 264, 303, 303t
Potassium dichromate, 446
Potassium hydroxide, 303t
Potassium iodide, 1000
Potassium permanganate, 446, 918
Potassium propanoate, 698f
Potassium sorbate, 702
Precor, 620
Prednisone, 654
Prefixes
  cis-, 159, 389
  cyclo-, 135, 142
  deoxy-, 1146
  di-, 140, 341, 972
  E, 389–390
  epoxy, 342
  hex, 134
  iso-, 138
  in nomenclature, 137
  pent, 134
  poly, 1261f
  R, 189–194, 198
  S, 189–194, 198
  sec-, 138
  tert-, 138

tetra-, 140
trans-, 159, 389
tri-, 140, 972
Z, 389–390
Pregabalin, 186
Prepolymers, 1271, 1272f
Preservatives, 702
Primary (1°) alcohols
  aldehyde reduction to, 440, 604–608, 659
  amide reduction synthesis, 610–612
  boiling point, 344t
  conversion to alkyl halides with HX, 356–358
  conversion to alkyl halides with phosphorus tribromide, 359, 360, 360t
  conversion to alkyl halides with thionyl chloride, 359–360, 360t
  dehydration rate, 350
  description, 96, 338
  E2 mechanism for dehydration, 352
  ester reduction to, 608–610
  hydrogen bonding in, 343
  monosaccharide reduction synthesis, 1092
  oxidation of, 456, 457–458, 705t
  preparation by nucleophilic addition, 617–618, 659–660
Primary (1°) alkylamines, 997–998
Primary (1°) alkyl halides
  acetylide anions, reaction with, 418
  carbocation rearrangement reaction, 895
  description, 96
  in direct enolate alkylation, 794
  E1 elimination reactions, 315
  E2 elimination reactions, 310–311, 324, 324t
  in Friedel–Crafts alkylation, 892–895
  in organocuprate coupling reactions, 1198–1199
  propane to, 944
  reduction of, 445
  S_N2 substitution reactions, 271, 282, 324, 324t
  synthesis of, 356
  Wittig reagent synthesis, 663
Primary (1°) amides
  acid chlorides to, 749, 993–994
  amines to, 993–994
  anhydrides to, 750–751, 993–994
  description, 98
  esters to, 756
  in nitrile hydrolysis, 720
  nomenclature, 739
  proton NMR absorptions, 743t
  structure of, 735
Primary (1°) amines
  amide formation, 993–994
  amide reduction synthesis, 982
  amino acids as, 1024
  boiling point, 974t
  description of, 97, 970
  diazonium salts synthesis, 998–999
  Gabriel synthesis, 980–981
  imine formation from, 666–667, 669f, 993
  infrared (IR) absorptions/spectra of, 975, 975f
  intermolecular hydrogen bonding in, 974
  melting point, 974t
  nitrile reduction synthesis, 721–722, 981
  nitro group reduction synthesis, 981

nomenclature, 971
nucleophilic substitution synthesis, 979–981
reduction reactions in synthesis, 981–982
reductive amination synthesis, 982–985
synthesis of, 979–985
Primary (1°) arylamines, 998–999
Primary (1°) carbocations
  inductive effects, 278
  Markovnikov's rule, 400
  stability, 277, 279, 476
Primary (1°) carbon, 94
Primary (1°) carbon radicals, 936–937, 937f
Primary (1°) hydrogen, 94, 943
Primary structure
  of nucleotides, 1149, 1149f
  of proteins, 1045, 1050f
Priority assignment, stereogenic centers, 189–193, 190f, 192f
Prochirality, 458–459
Progesterone
  as lipid, 1115f
  as oral contraceptive, 392–393, 392f
  synthesis of, 823–824, 824f
Proline
  isoelectric point, 1025t
  pK_a values, 1025t
  L-proline, 1024
  protein structure, 1047
  as secondary amine, 1024
  structure of, 1023f
  zwitterion, 119
Prontosil, 1007
Propanal
  aldol reaction of, 814
  infrared spectrum, 541
  Wittig reaction, 664
Propanamide, 542
Propan-1-amine, 995
Propane
  bromination of, 944–945, 945f
  chlorination of, 944–946, 946f
  molecular formula, 134
  molecular structure, 134
  Newman projection, 148f
  as normal alkane, 135t
  as propellant, 949
  torsional energy of, 149
Propane-1,3-diol, 1275, 1276f
Propane-1-thiol, 366
Propane-1,2,3-triol. See Glycerol
Propanoic acid, 702f
Propan-1-ol
  Carbon NMR spectrum, 584f
  infrared spectrum of, 531–532
Propene, 362, 400, 955, 957
Propionaldehyde, 649t
Propoxyphene, 183
Propyl, 138
Propylbenzene, 919
Propyl tosylate, 362
Prop-2-yn-1-ol, 388
Prostacyclins, 1124, 1125f, 1125t
Prostaglandins
  biological activity of, 1125t
  structure of, 1124
  synthesis blockers, 703–704, 1126
  synthesis of, 704, 1124, 1125f
Prosthetic groups, 1053
Protease inhibitors, 1057–1058, 1058f
Proteases, 1055t
Protecting groups
  acetals as, 679

for alcohols, 623–624
for aldehydes and ketones, 679
amides as, 994
tert-butoxycarbonyl, 1040
description of, 622–623, 623f
9-fluorenylmethoxycarbonyl, 1040, 1043–1044
Protection reactions, 622–624, 623f, 994f
Proteins. *See also* Peptides; *specific proteins*
  amino acids as building blocks, 715, 1031, 1170
  amino acid sequence, 1156–1157
  composition of, 118, 118f
  conjugated, 1053
  denaturation, 1051
  description of, 744, 1022
  enzymes, 1054–1058, 1055t
  expression of, 1152
  fibrous, 1051
  globular, 1051
  α-helix, 1046–1048, 1046f
  important compounds, 1051–1054
  membrane, 1022
  β-pleated sheet, 1047–1048, 1047f
  primary structure, 1045, 1050f
  quaternary structure, 1050, 1050f
  secondary structure, 1046–1048, 1050f
  structures of, 1045–1051, 1050f
  synthesis of, 1155–1159
  tertiary structure, 1048–1050, 1050f
  transcription of DNA to RNA, 1155–1156, 1155f
  translation of RNA in synthesis, 1157–1159, 1158f
  viral DNA synthesis, 1162
Proton acceptors, 59
Protonation. *See also* Nucleophilic addition reactions
  of alcohols, 349, 351, 352, 356–357
  of aldehydes, 601
  in alkyne reduction to trans alkene, 444
  in enzymatic decarboxylation, 803
  of epoxides, 369, 371, 629
  in ester hydrolysis, 757
  of ethers, 364–365
  in glycoside formation, 1088
  in glycoside hydrolysis, 1089
  of ketones, 601
  of pyrrole, 991
  in racemization at the α carbon, 790
  in tautomerization, 414, 782
Proton donors, 59, 63
Proton NMR
  aldehydes, 653t
  alkenes, 575–577, 575f, 576f
  amines, 974, 975t
  benzene derivatives, 565, 578, 855, 855t
  tert-butyl methyl ether spectrum, 556–557
  carboxylic acid derivatives, 743t
  carboxylic acids, 701t
  characteristic absorptions, A-13–14
  characteristic chemical shift values, 562–563, 564t
  characteristics of spectrum, 556–557
  chemical shift, 556–557, 562–567
  cyclohexane conformations, 578
  diastereotopic protons, 561
  doublets, 568–569, 571, 571t, 572, 575–576, 576f
  downfield signals, 556, 562–563
  enantiotopic protons, 561
  equivalent protons in alkenes and cycloalkanes, 559–560

Index

Proton NMR—(Cont.)
  homotopic protons, 561
  identification of unknown
    compound, 579–581
  intensity of signals, 556, 567
  ketones, 653t
  number of peaks, 573–575
  number of signals, 557–561
  OH protons, 577, 577f
  position of signals, 562–567
  propanal, 653f
  propanoic acid, 702f
  quartets, 571, 571t, 577, 580
  septets, 572
  sextets, 573
  shielding and deshielding effects, 562–563, 562f, 563f, 565–567, 566f, 566t
  singlets, 568, 571, 577, 580
  spin–spin splitting, 568–577
  splitting patterns, 570–575, 570t, 571t
  triplets, 568–571, 571t, 577, 580
  upfield signals, 556, 562–563
Protons
  absorbing, 568
  description of, 8
  NMR signals, 557–559, 559f
  number of, 8
  as reducing agents, 436
  relative acidity, 74
  spin, 554
  spin–spin splitting in proton NMR, 568–577
Proton transfer reactions
  deprotonation (See Deprotonation)
  description, 60–67
  zwitterions and, 118, 715, 717–718, 1024
Protosterol carbocation, 1134f, 1135
Proventil, 372f. See also Albuterol
Prozac. See Fluoxetine
Pseudoephedrine, 70, 985
D-Psicose, 1079f
Psilocin, 978
Psilocybe mushrooms, 978
PTH (*N*-phenylthiohydantoin), 1035–1036
PTT [poly(trimethylene terephthalate)], 1275, 1276f
Purification by extraction, 711–714, 713f
Purine, 868, 973f, 1145
Putrescine, 975
PVC. See Poly(vinyl chloride)
Pyrimidines, 868
Pyranose ring, 1080
Pyrethrin I, 1207
Pyridine
  alcohol conversion to alkyl halides, 359–360
  alcohol conversion to alkyl tosylates in, 361, 363
  as aromatic heterocycle, 864
  dehydration of alcohols using, 355
  electrostatic potential map, 865f
  monosaccharide conversion to esters, 1091
  as nitrogen heterocycle, 973f
  relative basicity, 990–991, 992t
  as weak base, 77
Pyridinium chlorochromate (PCC), 446
Pyridoxal phosphate (PLP), 646, 672–673, 802–803
Pyridoxamine phosphate, 672
Pyrimidine, 973f, 1145
Pyrophosphoric acid, 715

Pyrrole
  as aromatic heterocycle, 865
  electrostatic potential map, 865f
  as nitrogen heterocycle, 973f
  relative basicity, 990–991, 992t
Pyrrolidine, 973f
Pyruvate
  acetyl CoA synthesis from, 1184–1185
  in bloodstream, 709
  citramalate from, 831
  ethanol synthesis from, 1185–1186
  glucose conversion to, 1180–1184
  lactate synthesis from, 1185
  reduction to lactate, 607–608
Pyruvate dehydrogenase complex, 1185
Pyruvate kinase, 1183
Pyruvic acid
  deprotonation in bloodstream, 709
  as electrophile, 117

Q

Qing-hao, 411
Qinghaosu, 411
Quantum, 529
Quartets, NMR spectra, 571, 571t, 577, 580
Quaternary ammonium salts, 979–980, 995
Quaternary (4°) carbon, 94
Quaternary structure of proteins, 1050, 1050f
Quinic acid, 58, 75
Quinine, 311, 793, 975
Quinoline, 443

R

Racemic mixtures
  achiral reactants and, 369–370, 946f
  amino acid resolution, 1026–1029, 1027f
  in carbene addition, 1209
  carbonyl reduction and, 606–607
  disparlure synthesis, 450
  electrophilic addition reactions, 402
  in enolate alkylation, 790, 790f, 794–795
  in epoxidation, 449
  halogenation and, 406, 946
  hydroboration–oxidation and, 411
  nucleophilic addition of organometallic reagent, 618–619
  optical properties, 202–203
  resolution, 1026–1029, 1027f
  in S_N1 substitution reactions, 275–276, 275f
Racemization, 275f, 357, 790, 790f
Radiation
  electromagnetic, 529–530, 529f
  infrared, 531
  radiofrequency (RF), 554
Radical cations, 510
Radical indicators, 937
Radical inhibitors, 938, 954
Radical intermediates
  allylic substitution, 952
  description of, 936
  in halogenation, 917, 939, 940
  from homolysis, 224f
  ozone destruction by, 949
  in radical addition, 955
Radical polymerization, 958–959, 1258–1260, 1258f

Radical reactions
  addition reactions to double bonds, 955–957
  allylic bromination, 951–952
  allylic halogenation, 950–953
  allylic oxidation, 953–954, 953f
  antioxidants, 954
  bromination of alkanes, 943–945
  carbon–carbon double bond reactions, 938
  carbon–hydrogen bond reactions, 938
  chlorination of alkanes, 939–940, 943–946
  chlorination of ethane, 941–942, 942f
  description of, 225, 936–937
  features of, 937–938
  halogenation of alkanes, 939–949
  halogenation of alkyl benzenes, 917–918
  mechanism of halogenation, 940–942
  ozone layer and CFCs, 949–950, 949f
  polymerization, 958–959, 1258–1260, 1258f
  stereochemistry of halogenation, 947–949
  two radicals, reaction with each other, 938
Radicals
  allyl, 950
  allylic, 951
  bromine, 951
  chlorine, 938, 945, 949f
  classification of, 936
  cyclopentadienyl, 862–863
  definition of, 225
  oxygen, 938, 940
  reactions of two with each other, 938
  as reactive intermediate, 224f, 225
  resonance-stabilized, 953, 954
  stability of, 936–937, 937f, 950
Radical scavengers, 938
Radical substitution reaction, A-15
Radiofrequency radiation, 554, 584
Raloxifene, 272–273
Rancidity of vegetable oils, 440, 953, 1119
Random copolymers, 1263
Rate constant (k), 241
Rate-determining step
  description of, 239, 239f
  in E1 elimination reactions, 314
  in rate equations, 241–242
  in S_N1 substitution reactions, 274, 281
Rate equations
  first-order, 242, 267–268
  second-order, 241, 267–268, 308
Rate laws, 241
RCM (ring-closing metathesis), 1213–1216
Reaction arrow, 220, 224, 226t
Reaction coordinate, 236, 237f, 238, 239f
Reaction mechanism
  acetal formation, 676–677
  acetal hydrolysis, 677–678
  acetoacetic ester synthesis, 801
  acid-catalyzed halogenation at the α carbon, 791–792
  acid chloride conversion to anhydrides, 750
  1,4-addition to α,β-unsaturated carbonyl compounds, 630
  aldehyde reduction, 605–606
  aldol reaction, 814, 823
  allylic bromination, 951–952
  amide reduction, 611

  amino acid deamination, 672–673
  anhydride conversion to amide, 751
  anionic polymerization, 1262
  azo coupling, 1004–1005
  bromination of benzene, 888–889
  carboxylate anion conversion to acyl phosphate, 763–764
  carboxylation, 628
  carboxylic acid conversion to acid chloride, 752–753
  carboxylic acid conversion to amide, 755–756
  carboxylic acids and derivatives reduction, 608–610
  catalysts, 243
  cationic polymerization, 1260
  Claisen reaction, 826
  cyanohydrin formation, 660–661
  cyclic hemiacetal formation, 680–681
  cyclopropanation, 1209
  dehydration of aldol product, 816–817
  description of, 223
  diazotization, 998
  dichlorocarbene synthesis, 1208
  Edman degradation, 1035–1036
  electrophilic aromatic substitution, 887–888
  elimination reactions, 308 (See also E1 elimination reactions; E2 elimination reactions)
  enamine formation, 668
  enamine hydrolysis, 669
  energy diagrams, 236–240
  farnesyl diphosphate biosynthesis, 1130
  Fischer esterification, 754
  Friedel–Crafts acylation, 893–894
  Friedel–Crafts alkylation, 892–893, 895
  glycoside formation, 1087–1088
  glycoside hydrolysis, 1088–1089
  halogenation, 940–942
  Heck reaction, 1207
  Hofmann elimination, 995–996
  hydrate formation, 674–675
  imine formation, 666–667
  intramolecular aldol reactions, 823
  isomerization of geranyl diphosphate to neryl diphosphate, 1131
  ketone reduction, 605–606
  malonic ester synthesis, 798–799
  multistep, 239
  nitration of benzene, 890–891
  nitrile hydrolysis, 720
  nitrile reduction, 721–722
  nucleophilic acyl substitution, 745–746
  nucleophilic addition, 656–657
  nucleophilic aromatic substitution by addition–elimination, 913–915
  nucleophilic aromatic substitution by elimination–addition, 915–916
  nucleophilic substitution, 267–289, 601–603 (See also S_N2 substitution reactions; S_N1 substitution reactions)
  olefin metathesis, 1213
  organometallic reagent reactions with carboxylic acid derivatives, 625
  peptide hydrolysis with a serine protease, 1056–1057
  radical addition to alkenes, 955–957
  radical halogenation of alkanes, 941

radical polymerization, 958–959, 1259
rate-determining step, 239, 242
rate equation and, 241–242
retro-aldol reaction, 815
reverse Claisen reaction, 1177–1178
Robinson annulation, 835–836
Simmons–Smith reaction, 1210–1211
sulfonation of benzene, 890–891
Suzuki reaction, 1203
tautomerization, 782
two-step, 238–240
Wittig reaction, 664–665
Ziegler–Natta polymerization, 1265
Reaction rates
  carbocation stability and, 279–282, 315
  chlorination vs. bromination, 943–944
  conversion of alcohols to alkyl halides with HX, 356
  E1 elimination reactions, 314–315
  E2 elimination reactions, 308, 310–311
  electrophilic aromatic substitution of substituted benzenes, 901, 904–905
  energy of activation and, 237, 237f, 240–241
  halogenation, 943–944
  Hammond postulate, 279–282, 944
  in radical addition reactions, 956
  rate equations, 241–242, 267–268, 308
  R groups and, 271–273, 276–277
  in S_N1 substitution reactions, 276–277, 279, 281–282
  in S_N2 substitution reactions, 271–273
Reactions. *See specific types of reactions*
  acid–base reactions (*See* Acid–base reactions)
  organic reactions (*See* Organic reactions)
Reactive intermediates
  alkylborane, 409
  bridged halonium ions, 405–406
  carbocation rearrangements, 355
  description of, 223, 421
  heterolysis and, 224
  homolysis and, 224
  in hydration, 404
  in hydrohalogenation, 401
  in S_N1 substitution reactions, 274, 275–276
  in stepwise reaction mechanism, 238
Reactivity
  of aldehydes, 647
  carbonyl compounds, 601
  carboxylic acid derivatives, 602
  of carboxylic acids and derivatives, 746–748
  conjugated dienes, 492
  dienophiles, 492–493
  functional groups and, 116–117
  of ketones, 647
  of organometallic reagents, 614, 632
  proton transfer reactions, 62
  retro Diels–Alder reactions, 497
  in S_N1 substitution reactions, 276–277
  in S_N2 substitution reactions, 271–273
Reagents
  greener, 1276
  metal hydride reagents (*See* Metal hydride reagents)
  organometallic, 613–616
  organosodium reagents, 615

Tollens reagent, 613
  writing organic reactions, 220, 220f
Rebaudioside A, 1090
Receptors, olfactory, 207, 207f
Recycling of polymers, 1255, 1277–1278, 1277t
Red tides, 339
Reduced state, 164
Reducing sugars, 1093, 1093f, 1098, 1099
Reductases, 1055t
Reduction. *See also* Oxidation
  of acid chlorides, 608–610
  agents, 436, 459, 607–608, 1173, 1174
  of aldoses, 1092
  of alkenes, 435, 436–442
  of alkyl halides, 445–446
  of amides, 610–612, 982
  amine synthesis, 981–985
  of aryl ketones to alkyl benzenes, 918–920
  of carboxylic acids, 610–612
  Clemmensen, 919
  of compounds containing polar C–X σ bonds, 445–446
  description of, 164–165, 164f, 435–436, 435f, 603
  dissolving metal reductions, 436
  enzymes in, 1055, 1055t
  of epoxides, 445–446
  of esters, 608–610
  $FADH_2$ as reducing agent, 1174
  hydrogenation, 437–442
  metal hydride reagents, 612, 612t
  NADH as reducing agent, 607–608, 1173
  of nitriles, 721–722, 981
  of nitro groups, 920–921, 981
  reductive amination, 982–985
  of substituted benzenes, 918–921
  Wolff–Kishner, 919
Reductive amination, 982–985
Reductive elimination reactions
  Heck reactions, 1205–1207
  Suzuki reactions, 1201–1205
Refining, of petroleum, 145
Refrigerants, 259
Regioselectivity
  of alcohol dehydration, 350, 395
  of alkylation of unsymmetrical ketones, 795–796
  of E1 elimination reactions, 316
  of E2 elimination reactions, 312, 319
  of electrophilic addition of hydrogen halides, 400, 403t
  of epoxide ring opening reactions, 371–372
  of HBr addition, 955
  of Hofmann elimination, 996–997
  of hydroboration, 410, 411t
  of radical addition to unsymmetrical alkene, 955
  of unsymmetrical ketone alkylation, 795–796
Relafen, 802
Relative basicity, 987–993, 992t
Renewable feedstock for polymer synthesis, 1275, 1276f
Renewable fuels, 344, 345f
Replication of DNA, 1152, 1153–1154, 1153f
Replication forks, 1153
Resiniferatoxin, 598
Resins, epoxy, 1271, 1272f
Resolution
  of amino acids, 1026–1029, 1027f

by conversion to diastereomers, 1026–1028, 1027f
  description of, 205, 1026, 1027f
  kinetic resolution, 1028–1029
Resolving agents, 1026, 1027f
Resonance
  acidity, effects on, 72–74, 706–708
  allyl-type, 477–478
  basicity, effects on, 990, 992t
  benzene, 851
  cations with positive charge adjacent to lone pair, 478
  conjugated double bonds, 478
  double bonds having more electronegative atom than other, 478
  electron-donating effects, 898–900
  electron-withdrawing effects, 898–900
  infrared (IR) absorptions and, 534–536, 535f
  nucleus, 555
  principles of, 18–19
  substituted benzenes, effects on, 898–900
Resonance hybrids
  acetate, 72
  acyl groups, 736
  allylic carbocation, 476
  of allylic radical, 950, 952
  benzene, 851
  description of, 22–23, 479–481
  of naphthalene, 862
  peptide bonds in, 1033
  resonance structures and, 18, 736
Resonance-stabilized acylium ion, 893
Resonance-stabilized anions, 980
Resonance-stabilized carbocations
  allylic, 476, 487
  in azo coupling, 1004
  in bromination, 888–889
  in electrophilic aromatic substitution, 887–888, 904–905, 1004
  in farnesyl diphosphate biosynthesis, 1130
  in glycoside formation, 1087–1088
  in glycoside hydrolysis, 1089
  in halogenation, 888–889
  in isomerization of geranyl diphosphate to neryl diphosphate, 1131
  in nucleophilic aromatic substitution reactions, 914
  tautomerization, 415
Resonance-stabilized cations, 657
Resonance-stabilized enolate anions, 20, 630, 656, 783–784, 826
Resonance-stabilized molecules, 18, 72, 74
Resonance-stabilized radicals, 953, 954
Resonance structures
  acetate, 706
  acyl groups, 736
  allyl-type resonance, 477–478
  antioxidants, 954
  benzylic radical, 1258
  cations with positive charge adjacent to lone pair, 478
  conjugated double bonds, 478
  conjugation and, 476
  cytosine, 868–869
  definition of, 18
  double bonds having more electronegative atom than other, 478
  drawing procedure, 19–22

enolates, 785, 787
enols, 783
relative energy, 480–481
resonance hybrid, 22–23, 479–481
substituted benzenes, 898–899
theory, 18
α,β-unsaturated carbonyl compounds, 629, 833
ylides, 664
Restricted rotation
  in alkenes, 305–306, 306f
  description of, 39
Restriction endonucleases, 1160
Retention of configuration
  in conversion of alcohols to tosylates, 362, 363
  in oxidation, 410–411
  in S_N2 substitution reactions, 269
Retention time, 522
11-*cis*-Retinal, 671, 671f, 1123f
Retinol, 111, 1123f
Retro-aldol reactions
  description of, 815–816
  in glycolysis, 831–832, 1180f, 1181
Retro Diels–Alder reactions, 497
Retronecine, 97
Retrosynthetic analysis
  acetoacetic ester synthesis, 801–802
  with acetylide anions, 422–423
  aldol reactions, 817–818, 822
  of azo compounds, 1005
  of *m*-bromoaniline, 920–921
  of *p*-bromobenzaldehyde, 1003
  of *m*-chlorophenol, 1002
  description of, 421–423
  development of, 421
  diazonium salts in, 1002–1003
  Diels–Alder product, 496–497
  directed aldol reactions, 822
  of disparlure, 449–450, 450f
  of Grignard products, 620–622
  α-halo carbonyl compounds in, 793
  Heck reaction products, 1206
  malonic ester synthesis, 799–800
  of methyl orange, 1005
  of *p*-nitrobenzoic acid, 921–922
  of (1Z,3E)-1-phenylocta-1,3-diene, 1204
  reductive amination, 984–985
  ring-closing metathesis products, 1215
  Robinson annulations, 838–839
  in substituted benzene syntheses, 920–924
  trisubstituted benzene synthesis, 922–923
  Wittig reaction, 665–666
Retrovir, 1162
Retroviruses, 1162
Reverse Claisen reaction, 1177–1178
Reverse transcription of DNA, 1162
RF radiation, 554, 584
R groups
  in alkenes, 307
  in amino acids, 118, 1022
  carbocation stability and, 277–279
  description, 92, 137
  in E1 elimination reactions, 315
  in E2 elimination reactions, 310–311
  in α-helix, 1046, 1046f
  in phosphoacylglycerols, 1120–1121
  in β-pleated sheet, 1047
  in S_N1 substitution reactions, 276–277, 277t
  in S_N2 substitution reactions, 271–273, 272t
  steric hindrance and, 271

Rhodium, 1030
Rhodopsin, 671, 671f
Ribbon diagrams, 1047, 1048f
Riboflavin, 1174
Ribofuranose, 1086
Ribonucleic acid. See RNA
Ribonucleosides, 1146
D-Ribose, 1077, 1078f, 1086, 1145, 1146
Ribosomal RNA (rRNA), 1154, 1157
D-Ribulose, 1079f
Rimantadine, 985
Ring carbon
   degrees of unsaturation and, 386–387, 439–440
   Diels–Alder reactions, 490–491
   stereogenic centers, 186–189
Ring-closing reactions
   electrocyclic reactions, 1227, 1231–1236
   olefin metathesis, 1213–1216
Ring-flipping, 156–158, 156f, 160–163
Ring formation, 835–839
Ring-opening reactions
   electrocyclic reactions, 1227, 1231–1236
   epoxides, 368–372, 628–629, 1263–1264
   in epoxy resin synthesis, 1271
   in nylon 6 synthesis, 1268
Rizatriptan, 273
RNA
   bases of, 1145
   description of, 120–121
   retroviruses, 1162
   transcription, 1155–1156, 1155f
   types of, 1154
Robinson, Sir Robert, 835
Robinson annulation, 835–839
Rocaltrol, 484, 484f
Rofecoxib, 1126, 1203
Roflumilast, 258
Rosemary extracts, 954
Roses, 1128
Rosmarinic acid, 954
Rotation
   around carbon–carbon bonds, 38, 39, 147
   barrier to, 152
   conformations and, 147, 150 (See also Conformations)
   conrotatory, 1232–1236, 1236t
   disrotatory, 1232–1236, 1236t
   optical, 201–203, 205–206
   in primary structure of proteins, 1045
   restricted, 39, 305–306, 306f
Row, in periodic table, 9
Rowland, F. Sherwood, 949
R (prefix), 189–194, 198
rRNA (ribosomal RNA), 1154, 1157
R,S system of nomenclature, 189–194, 458
RU 486, 393
Rubber
   in consumer products, 1256f
   styrene–butadiene rubber (SBR), 1263
   synthesis of, 1265–1267
Rubia tinctorum, 1006
Ruthenium, 1211

S

Sabinene, 1128
Saccharin, 1100, 1100f
Salicin, 77f, 703, 1089
Salicylates, 77
Salicylic acid, 703
Salts
   ammonium (See Ammonium salts)
   Brønsted–Lowry bases, 59
   carboxylic acids, 702
   formation of, 11
   nucleophiles in, 260
SAM-e, 288
SAM (S-adenosylmethionine), 288, 289f
Samuelsson, Bengt, 703
Sandmeyer reaction, 1000
Sanger, Frederick, 1160
Sanger sequencing, 1160
Saponification, 757–758, 760
Saponins, 112
Saran, 1263
Saturated fatty acids, 394, 1118, 1118t
Saturated triacylglycerols, 1117, 1118f
Saytzeff rule. See Zaitsev rule
SBR (styrene–butadiene rubber), 1263
Schiff base, 666. See also Imines
Schrock, Richard, 1211
s-cis conformation
   of conjugated dienes, 482, 492
   of peptide bonds, 1033
Scombroid fish poisoning, 866
Scopolamine, 976, 976t
Seashells, chiral helical, 182
Secondary (2°) alcohols
   boiling point, 344t
   conversion to alkyl halides with HX, 356–357, 359
   conversion to alkyl halides with phosphorus tribromide, 359, 360, 360t
   conversion to alkyl halides with thionyl chloride, 359–360, 360t
   dehydration rate, 350
   description, 96, 338
   E1 mechanism for dehydration, 351–352
   in epoxy resin synthesis, 1271
   hydrogen bonding in, 343
   ketone reduction to, 440, 604–608, 659
   oxidation of, 456, 457
   preparation by nucleophilic addition, 617–618, 659–660
Secondary (2°) alkylamines, 999
Secondary (2°) alkyl halides
   acetylide anions, reaction with, 418
   description, 96, 255
   E1 elimination reactions, 315, 324t, 325
   E2 elimination reactions, 310, 324t, 325
   in Friedel–Crafts alkylation, 892–895
   in nucleophilic substitution reactions, 282
   in organocuprate coupling reactions, 1198–1199
   propane to, 944
   in S_N1 substitution reactions, 276–277, 324t, 325
   in S_N2 substitution reactions, 271, 324t, 325
   synthesis of, 356
Secondary (2°) amides
   acid chlorides to, 749, 993–994
   amines to, 993–994
   amine synthesis, 982
   anhydrides to, 750–751, 993–994
   description, 98
   esters to, 756
   nomenclature, 739
   proton NMR absorptions, 743t
   structure of, 735
Secondary (2°) amines
   amide formation, 993–994
   amide reduction synthesis, 982
   boiling point, 974t
   description of, 97, 970
   enamine formation, 668, 669f, 993
   infrared (IR) absorptions/spectra of, 975, 975f
   intermolecular hydrogen bonding in, 974
   melting point, 974t
   nomenclature, 972
   proline, 1024
   reductive amination synthesis, 982–985
Secondary (2°) arylamines, 999
Secondary (2°) carbocations
   in Friedel–Crafts alkylation, 895
   inductive effects, 277
   in ionic addition, 956
   Markovnikov's rule, 400
   stability, 277, 279, 476
Secondary (2°) carbon, 94
Secondary (2°) carbon radicals, 936–937, 937f
Secondary (2°) hydrogen, 94, 943
Secondary structure of proteins, 1046–1048, 1050f
Second-order rate equations
   description of, 241
   in elimination reactions, 308
   in nucleophilic substitution, 267–268
Sec- (prefix), 138
Semiconservative replication, 1153
Separation of amino acids, 1026–1029, 1027f
Separatory funnel, 712, 713f
Septets, NMR spectra, 572
Serine
   abbreviation for, 716t
   isoelectric point, 1025t
   pK_a values, 716t, 1025t
   structure of, 1023f
Serine proteases, 1056–1057
Serotonin, 977–978
Serotonin reuptake inhibitors (SSRIs), 977
Sertraline, 896
β-Sesquiphellandrene, 492
Sesquiterpene/sesquiterpenoid, 1128, 1129t
Sesterterpene/Sesterterpenoid, 1129t
Seven-membered ring, 873, 873f, 1268
Sevoflurane, 114, 345
Sex hormones, 1135–1136, 1136t. See also specific hormones
Sextets, NMR spectra, 573
Shape of molecule. See Molecular shape
Shark oil, 1127f
Sharpless, K. Barry, 460
Sharpless epoxidation, 460–462
Sharpless reagent, 461
Shielding and deshielding effects in proton NMR, 562–563, 562f, 563f, 565–567, 566f, 566t
Shikimic acid, 99
Sickle cell anemia, 1054
Sickle cell hemoglobin, 1054
Side chain, of amino acid, 118, 1022
Sigma (σ) bonds
   in addition reactions, 222, 396, 399
   in alcohols, 338
   in alkanes, 147
   in alkene reduction, 436
   in alkenes, 305, 385
   in alkynes, 385
   of benzyne, 916
   in buta-1,3-diene, 484–485
   in carbonyl groups, 97–98, 599–600
   compounds containing C–Z σ bonds, 95–97
   compounds containing polar C–X σ bonds, 445–446
   in cycloaddition reactions, 1227, 1236–1240
   description of, 34, 35
   in 1,3-dienes, 474
   in electrocyclic reactions, 1227, 1231–1232
   in elimination reactions, 221–222
   in epoxidation reactions, 447
   in epoxides, 338
   in ethers, 338
   in farnesyl diphosphate biosynthesis, 1130
   in geranyl diphosphate synthesis, 477
   in hydroboration, 409
   hyperconjugation, 278–279
   inductive effects and, 71, 277
   in intramolecular aldol reactions, 822–823
   in lipids, 1115
   in nitriles, 696
   in nucleophilic substitution, 267
   in olefin metathesis, 1211
   in oxidative cleavage, 455
   in primary structure of proteins, 1045
   in radical halogenation, 940
   in radical reactions, 937, 938
   in Robinson annulation, 835
   in sigmatropic rearrangements, 1228, 1240–1245
   strength, 42
   in substitution reactions, 221
Sigmatropic rearrangements
   Claisen rearrangement, 1243, 1245–1246
   Cope rearrangement, 1243–1245
   description of, 1228, 1240–1241
   orbital symmetry, 1241–1243
   [3,3] rearrangement, 1243–1246
   Woodward–Hoffmann rules, 1242, 1242t
Silent Spring (book), 259
Silver(I) oxide
   in aqueous ammonium hydroxide, 613
   deprotonation of glucose, 1090
   as oxidizing agent, 446
Silyl ether, 623–624
Simmons, H. E., 1210
Simmons–Smith reaction, 1210–1211
Simmons–Smith reagent, 1210
Simple sugars. See Monosaccharides
Simvastatin, 189, 484, 484f, 1135
α-Sinensal, 600
Singlets, NMR spectra, 568, 571, 577, 580
Sitagliptin, 867
Six-membered ring
   of benzene, 850–852 (See also Benzene)
   Dieckmann reaction synthesis, 829–830
   formation by cycloaddition, 1238
   intramolecular aldol reaction synthesis, 823–824
   in LSD synthesis, 897f
   in pyridine, 864
   Robinson annulation synthesis, 835–839
   in steroids, 1132

Skeletal structures
  for alkynes, 358
  carbon skeleton (*See* Carbon skeleton)
  charged atoms, 33–34
  description of, 28
  drawing of, 30–34
  interpretation, 31–32, 31f
  sugar–phosphate backbone, in nucleic acids, 121, 121f, 1149f, 1150, 1151f, 1152
SmaI, 1160
Smith, R. D., 1210
S_N1 substitution reactions
  alcohol conversion to alkyl halides with HX, 357
  alkyl halide identity in, 282–283, 282f
  backside attack in, 275
  carbocation rearrangement, 358–359
  carbocation stability, 279, 281–282
  determining factors, 282–287, 285t
  E1 elimination reactions compared, 317, 323–327, 324t
  E2 elimination reactions compared, 323–327, 324t
  energy diagram, 274, 274f
  ethers with strong acids, 364–365
  features of, 273–277, 277t
  frontside attack in, 275
  geraniol biosynthesis, 288
  Hammond postulate, 279, 281–282
  kinetics, 273, 277t
  leaving group in, 284
  mechanism, 267–268, 272t, 274
  nucleophile strength and, 283–284
  polar protic solvents and, 285
  reaction rates, 279–282
  R group in, 276–277, 277t
  S_N2 substitution reactions compared, 282–287, 285t, 323–327, 324t
  stereochemistry of, 275–276, 277t
  as two-step mechanism, 274
S_N2 substitution reactions
  acetylide anions, 417–420
  alcohol conversion to alkyl halides with HX, 357–358
  alcohol conversion to alkyl halides with thionyl chloride, 359–360, 360t
  alcohol synthesis, 290, 290t, 346–347
  alkyl halide identity in, 282–283, 282f
  alkyl halide reactions, 418–419, 979–980
  alkyl tosylates reactions, 362
  amine synthesis, 979–981
  backside attack in, 269–272, 283
  as concerted reaction, 268–269
  determining factors, 282–287, 285t
  E1 elimination reactions compared, 323–327, 324t
  E2 elimination reactions compared, 323–327, 324t
  energy diagram, 268, 269f
  in enolate alkylation, 794
  epoxide ring opening with, 369–370, 420
  epoxy resin synthesis, 1271
  ether production, 346–347
  ethers with strong acids, 364–365
  features of, 268, 272t
  frontside attack in, 269
  α-halo carbonyl compounds in, 792–793
  inversion of configuration, 270–271, 270f
  kinetics, 268, 272t

  leaving group in, 269–270, 284
  mechanism, 267–268, 272t
  molecules synthesized by, 290t
  nucleophile strength and, 283–284
  polar aprotic solvents and, 285
  reduction of polar C–X σ bonds, 445–446
  retention of configuration, 269
  R group in, 271, 272t
  S_N1 substitution reactions compared, 282–285, 285t, 323–327, 324t
  in solid phase technique of peptide synthesis, 1043
  stereochemistry of, 269–271, 270f, 272t
  Wittig reagent synthesis, 663
Soap, 112, 113f, 702, 757, 760
Sodium acetate, 106, 698f
Sodium amide, 76f
Sodium benzoate, 702, 712
Sodium borohydride, 436, 604–605, 612t, 659, 1092
Sodium cyanoborohydride, 982–983
Sodium dichromate, 446
Sodium ethoxide, 76f, 303t, 347
Sodium hydride, 76f
Sodium hydroxide, 76f, 303t, 712
Sodium iodide, 1000
Sodium methoxide, 76f, 303t
Sodium salicylate, 703
Solanine, 1089
Solanum nigrum, 1089
Solid phase technique of peptide synthesis, 1043–1045
Solubility
  of alcohols, 344t
  of aldehydes, 652t
  of alkanes, 146, 146t
  of alkenes, 391
  of alkyl halides, 257t
  of alkynes, 391
  of amines, 974t
  of carboxylic acid derivatives, 742t
  of carboxylic acids, 700t
  definition, 106
  of diastereomers, 205–206, 205f
  of enantiomers, 201, 203t
  of epoxides, 344t
  of ethers, 344t
  extraction technique and, 711–714, 713f
  intermolecular forces and, 106–110
  of ketones, 652t
  "like dissolves like," 107
  of lipids, 1115
  of monosaccharides, 1079, 1091
  soap and, 112, 113f
  of vitamins, 111–112
Solutes, 106
Solvents
  alkyl halides as, 258
  definition, 106
  in E2 elimination reactions, 310
  green polymer synthesis, 1276
  liquid, 220
  nonpolar, 107
  nucleophilicity and, 265–266, 265f, 285
  organic, 107–109, 110t
  polar, 107–110, 110t
  water, 107–110, 110t
s orbitals
  1*s* orbital, 10
  2*s* orbital, 10
  description of, 9
  hybridization of, 34–41

Sorbitol, 1092
D-Sorbose, 1079f
Sorona, 1275, 1276f
Spandex, 1270
Sparteine, 976, 976f
Spearmint, odor of, 208
Specific rotation, 203–205
Spectator ions, 59
Spectrometers
  GC–MS instrument, 521–522, 521f
  mass, 509–510, 509f
  NMR, 555–556, 555f, 567
Spectroscopy, 520
Speed, 977
SPF (sun protection factor) rating, 857
Sphingomyelins, 1122, 1122f
Sphingosine, 1122, 1122f
sp hybrid orbitals
  in acetylene, 40
  in alkynes, 385
  description of, 37t
  formation of, 36–37, 36f
  infrared absorptions and, 534
  in nitriles, 696
  in organometallic reagents, 613
  in oxidative cleavage, 455
  percent *s*-character, 43, 73, 534
sp^2 hybrid orbitals
  in acetone enolate, 790
  in addition of organometallic reagents, 618–619
  in aldehydes, 647
  in alkenes, 307, 385, 391
  in alkyl halides, 255, 255f
  allylic carbocations, 475
  in amides, 1033
  in benzene, 851–852
  in benzyne, 916
  carbon NMR spectra, 583
  in carbon radicals, 936
  in carbonyl groups, 599–600, 736
  in carbonyl reduction, 606
  in carboxylic acids, 696
  delocalized electrons and, 481
  description of, 37t
  enols, 338
  in ethane, 38
  in ethylene, 38–39
  formation of, 36–37, 36f
  in histamine, 866
  infrared absorptions and, 534
  in ketones, 647
  in nucleophilic substitution, 275–276
  in organometallic reagents, 613
  percent *s*-character, 43, 73, 289, 307, 391, 484, 534, 991
  phenols, 338
  in pilocarpine, 866
  in pyridine, 864
  in pyrrole, 865
  relative basicity and, 990–991
  in vinyl halides, 289
sp^3 hybrid orbitals
  in acyclic alkanes, 133
  in addition of organometallic reagents, 618–619
  alcohols, 338, 339, 349
  in alkenes, 307, 391
  in alkyl halides, 255, 255f
  carbon NMR spectra, 583
  in carbonyl reduction, 606
  description of, 37t
  epoxides, 339
  in ethane, 38
  ethers, 339
  infrared absorptions and, 534

  lone pairs in, 36f
  in methane, 35–36, 35f
  in nucleophilic substitution, 260
  percent *s*-character, 43, 73, 289, 307, 391, 484, 534, 991
  prochirality and, 458
Spider silk, 1048, 1048f
Spin–spin splitting in proton NMR
  in alkenes, 575–577, 575f, 576f
  doublets, 568–569
  maximum number of peaks, 573–575
  patterns, 570–575, 571t
  rules, 570–575
  triplets, 569–570
Spiriva, 346
Spiro[4.5]decane, A-9
Splenda, 96, 1100f
Splitting, NMR spectra. *See* Spin–spin splitting in proton NMR
S (prefix), 189–194, 198
Squalene
  biosynthesis of, 1129–1131, 1130f
  cholesterol synthesis from, 1134–1135, 1134f
  as terpene, 1127f
Squalene epoxidase, 1135
SSRIs (serotonin reuptake inhibitors), 977
Staggered conformations
  of butane, 150–153, 151f, 152f
  of chair cyclohexane, 154
  description of, 147–150
  of ethane, 148–149, 148f
  of propane, 148f
Stanozolol, 1136
Starch
  as carbohydrate, 119, 119f, 1070
  configuration of, 178f
  ethanol production from, 344, 345f
  stereochemistry of, 177–178
  structure of, 177f, 1101–1102
Stearic acid
  in cocoa butter, 1114
  as lipid, 122, 1117
  melting point, 394, 394t
  structure of, 394f
Stearidonic acid, 395
Stemoamide, 779
Stemona tuberosa, 779
Step-growth polymers
  description of, 1257
  epoxy resins, 1271, 1272f
  polyamides, 1267–1268
  polycarbonates, 1270–1271
  polyesters, 1268–1269
  polyurethanes, 1269–1270
  types of, 1267
Stepwise reactions, 223, 238–240, 239f
Stereocenters. *See* Stereogenic centers
Stereochemistry
  of addition reactions, 397
  alcohol conversion to alkyl halide with HX, 357–358
  of amino acids, 1022
  of carbonyl reduction, 606–607
  chiral and achiral molecules, 179–182, 182f
  configurations, 179
  conformations (*See* Conformations)
  of dienophiles, 493–494
  of E2 elimination reactions, 318–321
  of electrocyclic reactions, 1231–1236
  of enolate alkylation, 794–795
  of epoxides, 369–370
  explanation of, 147
  of halogenation reactions, 405–407, 406f, 946–949, 946t

Stereochemistry—(Cont.)
 handedness, 177
 of hydroboration–oxidation reactions, 411
 of hydrohalogenation reactions, 401–403
 of nucleophilic addition of organometallic reagent, 618–619
 of pericyclic reactions, 1246, 1246t
 of polymers, 1264–1265
 of S_N1 substitution reactions, 275–276, 277t
 of S_N2 substitution reactions, 269–271, 270f, 272t
 stereogenic centers (See Stereogenic centers)
Stereogenic centers
 in amines, 970–971
 of amino acids, 716, 1022
 on atoms that are not part of ring, 183–184
 in carbonyl reduction, 606–607
 chirality and, 182, 194
 chlorination reactions and, 947–948
 in cyclic compounds, 186–189
 of cyclic hemiacetals, 681–683
 D and L designations, 1076
 definition, 181
 diastereomers, 194–196
 drawing enantiomers, 184–186, 185f
 in enolate alkylation, 794–795
 Fischer projection formulas, 185
 in meso compounds, 196–197
 in monosaccharides, 1072–1078
 number in molecule, 194
 priority assignment, 189–193, 190f, 192f
 prochirality and, 458–459
 racemization at, 947
 R or S assignment, 189–194, 198
 Sharpless epoxidation, 461
 in S_N1 substitution reactions, 275–276
 in S_N2 substitution reactions, 269–270, 357–358
Stereoisomers. See also Stereogenic centers
 of aldohexoses, 1074
 of alkenes, 305–307, 389–390
 of alkynes, 389–390
 amino acids, 1022–1024, 1022f
 anomers, 1080
 chlorination of (R)-2-bromobutane, 947–948
 configuration of, 179
 description of, 159–160, 179, 179f
 diastereomers, 194–196
 disubstituted cycloalkanes, 198–199
 enantiomers (See Enantiomers)
 E,Z system of nomenclature, 389–390
 monochlorination products, 948
 nomenclature, alkenes and alkynes, 389–390
 number possible, 194
 physical properties of, 201–206
 starch and cellulose, 177–178
 stereoselective reactions (See Stereoselective reactions)
 stereospecific reactions (See Stereospecific reactions)
 types of, 200, 200f

Stereoselective reactions
 alkene production, 395
 description of, 313
 reduction of alkyne to cis alkene, 443
Stereospecific reactions
 carbene addition, 1209
 cycloaddition reactions, 1236
 description of, 406
 Diels–Alder reactions, 493–494, 1238
 electrocyclic reactions, 1231
 epoxidation, 448
 Heck reactions, 1205
 organocuprate coupling reactions, 1199
 pericyclic reactions, 1227
 Simmons–Smith reaction, 1211
 Suzuki reaction, 1202
Steric hindrance
 in alcohols, 343
 in alkyl halides, 264, 266, 271, 418
Steric strain
 in acyclic alkanes, 152t
 in alkenes, 307
 in cyclohexane, 156, 157f
 description of, 151
 steric hindrance and, 264
Steroids. See also specific steroids
 adrenal cortical steroids, 1137
 anabolic steroids, 1136
 sex hormones, 1135–1136, 1136t
 structure of, 1132–1133, 1133f
 synthesis by Diels–Alder reaction, 497–498
Stevia, 1090
Sticky ends of DNA fragments, 1160
Stimulants
 amphetamine, 78, 982–983
 cocaine, 733, 976
 dobutamine, 971
 methamphetamine, 977
 nicotine, 976
Stone, Edmund, 77
Stop codons, 1157
Strain
 angle, 153–154
 steric, 151, 152t, 156, 157f
 torsional, 149, 152t, 154, 156, 157f
s-trans conformation
 of conjugated dienes, 482, 492
 of peptide bonds, 1033
Stratospheric ozone layer. See Atmosphere
Streptomyces cinnamonensis, 682
Structural isomers. See Constitutional isomers
Strychnine, 1028
Styrene
 copolymerization of, 1263
 ethylbenzene to, 917
 polymerization of, 958, 1258, 1258f
Styrene–butadiene rubber (SBR), 1263
Substituents, naming of, 137–138, 651–652, 699, 699f
Substituted anilines, synthesis of, 994
Substituted benzene derivatives
 activating and deactivating substituents, 901–905, 908f
 in azo compound formation, 999
 bromination, synthesis by, 917–918
 directing effects of substituents, 901–908, 908f
 electrophilic aromatic substitution, 901–913, 908f
 halogenation of, 909
 inductive effects on, 898–900

 limitations on reactions, 908–910
 multistep synthesis of, 921–924
 orientation effects on, 905–908
 oxidation of, 921
 polysubstituted, 854
 reactions of, 898–913, 917–921
 reduction of, 920–921
 resonance effects on, 898–900
 synthesis of, 921–924, 999–1003
Substituted cycloalkanes, 157–163
N-Substituted primary amines, 972
Substitution reactions
 alkyl halides, 259 (See also Nucleophilic substitution reactions)
 allylic, 951–952
 of aromatic compounds (See Electrophilic aromatic substitution reactions)
 of aryl diazonium salts, 999–1001
 of benzene, 850, 858, 886–898
 at α carbon of carbonyl compounds, 665, 779–803
 description of, 221
 electrophilic aromatic substitution reactions (See Electrophilic aromatic substitution reactions)
 Heck reaction, 1205–1207
 list, A-15
 nucleophilic (See Nucleophilic substitution reactions; S_N1 substitution reactions; S_N2 substitution reactions)
 nucleophilic acyl substitution reactions (See Nucleophilic acyl substitution reactions)
 nucleophilic aromatic (See Nucleophilic aromatic substitution reactions)
 Sandmeyer reaction, 1000
 of substituted benzenes, 898–913
Substrate, 244
Succinate, 1175, 1189
Succinate dehydrogenase, 1187f
Succinic acid, 698, 710
Succinimide, 951
Succinyl CoA, 1188
Succinyl CoA synthetase, 1187f
Sucralose, 96, 1100, 1100f
Sucrose
 as disaccharide, 1069, 1099–1100, 1099f
 olestra synthesis from, 759
 structure of, 95
Sudafed PE, 70
Suffixes
 -adiene, 388
 -al, 648
 -aldehyde, 649
 -amide, 739
 -amine, 971
 -ane, 135, 144
 -ate, 698, 738
 -carbaldehyde, 648
 -carbonyl chloride, 737
 -carbonyl phosphate, 737
 -dioic acid, 698
 -diol, 341
 -ene, 387
 -enol, A-8
 -enone, A-8
 -enyne, A-8
 -ic acid, 698, 699
 -ide, 256
 -idine, 1146

 -ine, 256
 in nomenclature, 137
 -o, 256
 -oic acid, 697
 -ol, 339–340, 366
 -one, 650
 -onitrile, 699
 -ose, 1071
 -osine, 1146
 -oxy, 341
 -oyl, 651
 -thioate, 738
 -thiol, 366
 -yl, 651
 -yl chloride, 737
 -yl phosphate, 737
 -yne, 387
Sugar–phosphate backbone, in nucleic acids, 121, 121f, 1149f, 1150, 1151f, 1152
Sugars. See also Carbohydrates
 amino, 1103–1104
 dietary substitutes (See Artificial/alternative sweeteners)
 simple (See Monosaccharides)
Sulfa drugs, 1007–1008, 1008f
Sulfamethoxazole, 1008f
Sulfanilamide, 1007–1008
Sulfides, 95, 96t, 338, 367–368
Sulfide (suffix), 367
Sulfisoxazole, 1008f
Sulfonation of benzene, 887f, 890–891
Sulfonium ions, 368
Sulfur
 common nucleophile, 266t
 octet rule exception, 17–18
Sulfuric acid
 acid rain formation, 76
 dehydration of alcohols, 350–355, 395
 octet rule exception, 18
Sulfur nucleophile, 368, 373
Sulfur trioxide, 890
Sunscreens, 7, 529, 828, 857
Supercoil of α-keratin, 1051, 1052f
Super glue, 1261f
Superhelix
 collagen, 1052, 1053f
 α-keratin, 1051, 1052f
Suprafacial bond formation, 1238–1240, 1240t
Suprafacial rearrangement, 1241–1243, 1242t
Surface area
 boiling point and, 104, 104f
 van der Waals forces and, 102
Suzuki, Akira, 1205
Suzuki reactions, 1200–1205
Sweeteners, artificial. See Artificial/alternative sweeteners
Sweet'n Low, 1100f
Symbols, list, A-3
Symmetrical anhydrides, 735, 738
Symmetry
 ethers, 338
 melting point and, 105, 106
 plane of, 182, 197, 199
Syn addition
 catalytic hydrogenation, 437
 description of, 397
 in dihydroxylation, 452–453
 in epoxidation, 448
 in hydroboration, 409, 411
 in hydrohalogenation, 402–403, 403t
Syn dihydroxylation, 451
Syndiotactic polymers, 1264–1265

Syn periplanar geometry, 318, 318f
Synthases, 1055t
Synthesis. *See also specific types of compounds and reactions*
  acetylide anions and, 421–423
  carbonyl chemistry in, 632–634, 632f
  retrosynthetic analysis (*See* Retrosynthetic analysis)
Synthetic dyes, 1004–1007
Synthetic intermediates, 421
Synthetic polymers
  amorphous regions, 1272–1273
  atactic, 1264–1265
  biodegradable, 1275, 1278–1279
  chain-growth, 1256, 1257–1263
  crystalline regions, 1272–1273
  description of, 1256
  epoxy resins, 1271, 1272f
  feedstock, 1274–1275, 1275f, 1276f
  green polymer synthesis, 1274–1276, 1275f, 1276f
  isotactic, 1264–1265
  nylon (*See* Nylon)
  plasticizers and, 1274
  polyamides, 1267–1268
  polycarbonates, 1270–1271
  polyesters, 1268–1269
  polyethers, 1263–1264
  polyurethanes, 1269–1270
  poly(vinyl chloride), 258, 935
  properties of, 1273–1274
  recycling of, 1255, 1277–1278, 1277t
  shorthand representation, 1257f
  stereochemistry of, 1264–1265
  structure of, 1272–1274
  syndiotactic, 1264–1265
  thermoplastics, 1273
  thermosetting, 1273
Systematic names, 136–137. *See also* IUPAC nomenclature

T

2,4,5-T, 889f
Tagamet, 976–977
D-Tagatose, 1079f
D-Talose, 1078f, 1095
Tamiflu, 99
Taq polymerase, 1161
Target compound, 421
Tartaric acid
  physical properties of stereoisomers, 205, 205f
  as source of DET, 461
Tautomerization
  in acid-catalyzed halogenation, 791–792
  in citric acid cycle, 1187
  in glycolysis, 1180–1181
  in hydration of alkynes, 414–416
  in hydroboration–oxidation, 416–417
  hydrolysis of nitrile in base, 720
  mechanism, 782–783
  in nitrile hydrolysis, 720
  α,β-unsaturated carbonyl compounds, 630
Tautomers
  amide tautomers, 720–721
  description of, 414
  enol form, 414–416, 630, 780–783, 792
  keto form, 414–416
Taxol, 176, 187–188, 603
TBDMS (*tert*-butyldimethylsilyl ether), 623–624

Tectitethya crypta, 1144
Teflon, 258
Telfairine, 255
Temperature
  boiling point, 104–105
  melting point, 105–106
  polymer behavior and, 1273
  reaction rates and, 240
Template strand, DNA, 1153, 1153f, 1155–1156, 1155f
Teratogens, 187
Terephthalic acid, 1268, 1269
Terminal alkenes, 1211
Terminal alkynes
  as acid, 398
  deprotonation, 398, 398t
  description of, 385
  formation of, 418–419
  hydroboration of, 417
  oxidative cleavage of, 455
Terpenes/Terpenoids
  biosynthesis of, 1129–1132, 1130f
  classification of, 1128, 1129t
  common compounds, 1127f
  description of, 1126–1127
  locating isoprene units, 1127–1129
α-Terpinene, 1132
α-Terpineol, 1132
Tertiary (3°) alcohols
  boiling point, 344t
  in citric acid cycle, 1188
  conversion to alkyl halides with HX, 356–359
  dehydration rate, 350
  description, 96, 338
  E1 mechanism for dehydration, 351–352
  hydrogen bonding in, 343
  organometallic reagent reactions with carboxylic acid derivatives, 625–626
  oxidation of, 456
  preparation by nucleophilic addition, 617–618, 625–626, 659–660
Tertiary (3°) alkyl halides
  acetylide anions, reaction with, 418
  description, 96, 255
  E1 elimination reactions, 315, 317, 324, 324t
  E2 elimination reactions, 310–311, 324, 324t
  in Friedel–Crafts alkylation, 892–893
  S_N1 substitution reactions, 276–277, 282, 317, 324, 324t
  synthesis of, 356
Tertiary (3°) amides
  acid chlorides to, 749, 993–994
  amines to, 993–994
  amine synthesis, 982
  anhydrides to, 750–751, 993–994
  description, 98
  enolates from, 784
  esters to, 756
  nomenclature, 739
  structure of, 735
Tertiary (3°) amines
  amide reduction synthesis, 982
  boiling point, 974t
  description of, 97, 970
  infrared (IR) absorptions/spectra of, 975, 975f
  melting point, 974t
  nomenclature, 972
  reductive amination synthesis, 982–985

Tertiary (3°) carbocations
  in Friedel–Crafts alkylation, 893, 895
  inductive effects, 278
  stability, 277, 279, 476
Tertiary (3°) carbon, 94
Tertiary (3°) carbon radicals, 936
Tertiary (3°) hydrogen, 94
Tertiary structure of proteins, 1048–1050, 1050f
Tert- (prefix), 138
Terylene, 1268
Testosterone, 1136t
Tetrabutylammonium fluoride, 623
Tetrachloromethane, 258
Tetracyclic lipids. *See* Steroids
Tetradecane, 135t
Tetrahalides, 413
Tetrahedral molecular structure
  in acyclic alkanes, 133–134
  in alcohols, 339
  angle strain, 153
  drawing of, 187
  in epoxides, 339
  in ethers, 339
  geometry of, 25–27, 27t
  stereogenic centers, 181, 182–194 (*See also* Stereogenic centers)
Tetrahydrocannabinol. *See* THC
Tetrahydrofuran (THF), 342, 364
Tetrahydrogestrinone (THG), 1136
2,2,4,4-Tetramethylcyclobutane-1,3-diol, 1271
Tetramethyl α-D-glucopyranose, 1093
Tetramethylsilane (TMS), 556–557
Tetrapeptides, 1033
Tetra- (prefix), 140
Tetrasubstituted alkenes, 305
Tetraterpene/tetraterpenoid, 1129t
Tetravalent carbon, 4, 26, 28
Tetroses, 1071
Thalidomide, 187
THC (tetrahydrocannabinol), 76, 508, 522, 522f
Thermal electrocyclic reactions, 1230, 1232–1234
Thermal polyaspartate (TPA), 1279
Thermodynamic enolates, 789
Thermodynamic product of addition reaction, 488–490, 489f
Thermodynamics, 231–236, 673–674
Thermoplastics, 1273
Thermosetting polymers, 1273, 1276
THF (tetrahydrofuran), 342
THG (tetrahydrogestrinone), 1136
Thiazolinone, 1036
-thioate (suffix), 738
Thioesters
  acyl phosphate conversion to, 765–766, 1170–1171
  addition of H_2O to, 222
  biological acylation reactions, 766–768
  in biological Claisen reactions, 832–833
  carboxylate anion conversion to, 763
  in citric acid cycle, 1187f, 1188
  nomenclature, 738, 740t
  structure of, 734
Thiols
  description of, 366–367
  functional group, 95, 96t, 338
  nomenclature, 366f
  as nucleophile, 117, 765
-thiol (suffix), 366

Thionyl chloride
  alcohol conversion to alkyl halide, 359–360, 360t
  carboxylic acid conversion to acid chloride, 752–753
Threonine
  isoelectric point, 1025t
  pK_a values, 1025t
  stereochemistry of, 1024
  structure of, 1023f
  L-threonine, 1024
D-Threose, 1077, 1078f
Thrombin, 1022
Thromboxanes, 1124, 1125f, 1125t
Thymine, 849, 868, 1145, 1147t, 1149f, 1151, 1151f
Thyrotropin-releasing hormone (TRH), 98
L-Thyroxine, 1025
Ticlopidine, 261
Tiotropium bromide (Spiriva), 346
Titanium(IV) isopropoxide, 461
TMS (tetramethylsilane), 556–557
Tobramycin, 1104
α-Tocopherol, 1123f
Tollens reagent, 613, 1092–1093
Toluene
  electrophilic aromatic substitution reaction, 901, 906
  as gasoline additive, 522
  nomenclature, 853
  oxidation to benzoic acid, 918
Toluene 2,6-diisocyanate, 1269
p-Toluenesulfonic acid (TsOH)
  catalyst for acetal formation, 675
  dehydration of alcohols, 350, 395
  as strong acid, 76, 361
p-Toluenesulfonyl chloride (TsCl), 361–363
m-Toluidine, 916
p-Toluidine, 916, 989
Torsional energy, 149
Torsional strain, 149, 152t, 154, 156, 157f
Tosylates, 361–363
Tosyl groups, 361
TPA (thermal polyaspartate), 1279
Trabectedin, 4–5
Trade names, of drugs, 136–137
Trans alkenes, 443–444, 1205
Transaminases, 1055t
Transcription, 1152, 1155–1156, 1155f
Trans dihalides, 413
Trans double bonds, 1189
Trans fats, 441
Trans fatty acids, 441
Transferases, 1055, 1055t
Transfer RNA (tRNA)
  description of, 1154
  structure of, 1155f
  translation to protein synthesis, 1157–1159, 1158f
Trans isomers. *See also specific isomers*
  of alkenes, 307, 391, 437
  of cycloalkanes, 159–163
  of cyclohexane, 160–163
  definition, 159
  of disubstituted cycloalkanes, 198–199
  halogenation products, 405–407
  halohydrin products, 408
  in S_N1 nucleophilic substitution, 284
  in S_N2 nucleophilic substitution, 271

Index

Transition states
 in bromination, 945, 945f
 in chlorination, 946, 946f
 description of, 236–237
 in E1 elimination reactions, 314, 315f
 in E2 elimination reactions, 308–309, 309f
 in electrophilic addition reactions, 400
 endothermic reactions, 279–280, 280f, 944–945, 945f
 energy diagrams, 280f, 281f
 in epoxide ring opening reactions, 371, 371f
 exothermic reactions, 280–281, 281f, 944, 945–946, 946f
 Hammond postulate and, 279–282, 945
 planar, 971
 in S_N1 substitution reactions, 274, 274f
Translation to protein synthesis, 1152, 1157–1159, 1158f
trans- (prefix), 159, 389
Trans protons, 575
TRH (thyrotropin-releasing hormone), 98
Triacylglycerols
 catabolism of, 1176–1178
 description of, 758–759, 1116
 fake fats, 759
 fats, 395
 fatty acids in, 122, 122f, 221, 393–395, 1117t
 formation of, 122, 122f
 hydrolysis of, 394, 1118, 1170
 as lipids, 1115f
 mixed, 1117
 oils (See Oils; Vegetable oils)
 saponification, 760
 structure of, 393, 758f, 1117, 1118f, 1122f
Trialkylborane, 409
Triarylphosphine, 1205
1,3,5-Tribromobenzene, 1001, 1001f
Tricarboxylic acid cycle, 1186. See also Citric acid cycle
Trichlorofluoromethane, 3, 258–259, 950
Trichloromethane, 258
2,4,5-Trichlorophenoxyacetic acid, 889f
Tricyclohexylphosphine, 1201
Tridecane, 135t
Triesters, 122
Triethylamine, 77, 972
1,2,4-Triethylcyclopentane, 143
Trifluoroacetic acid, 1040
Triglycerides. See Triacylglycerols
Trigonal planar molecules
 alkenes, 305, 385
 carbocations, 275–276, 277t, 358
 carbon radical, 936
 carbonyl group, 599–600, 647
 carboxy group, 696
 description of, 24, 25, 27t
 enolates, 790
Trigonal pyramidal molecule, 26
Trimethylamine, 975
Trimethyl borate, 1202
2,5,5-Trimethylcyclohexanecarboxylic acid, 697
2,5,5-Trimethylcyclohexanol, 340f
1,3,3-Trimethylcyclohexene, 402
2,3,5-Trimethylhex-2-ene, 388
2,2,4-Trimethylpentane, 1119
2,3,5-Trimethyl-4-propylheptane, 140
Trinucleotides, 1150
Triols, 340
Trioses, 1071

Tripeptides, 1031–1032, 1043–1045
Triphenylphosphine, 663, 1201
Triphenylphosphine oxide, 664
Triphosphates
 functional group, 100, 100t, 117
 as leaving groups, 287
 nucleotides, 1148
Tri(o-tolyl)phosphine, 1201, 1207
Triple bonds
 of acetylene, 40
 in alkynes, 385, 414, 417
 of benzyne, 915–916
 chemical shift, NMR spectra, 565–566
 diynes, 388
 enynes, 388
 infrared absorptions, 533, 533t
 in nitriles, 696
 polyfunctional compounds with, nomenclature, A-8t
Triple helix of collagen, 1052, 1053f
Triplets, NMR spectra, 568–571, 571t, 577, 580
Tri- (prefix), 972
Tristearin, 122, 1119
Trisubstituted alkenes
 description of, 305, 305f
 in E2 elimination reactions, 395
 in Hofmann elimination, 996–997
Trisubstituted benzene derivative synthesis, 922
Tritan, 1271
Triterpene/triterpenoid, 1129t
tRNA. See Transfer RNA
Tropylium cation, 863–864, 873
Truvia, 1090
Trypsin, 1037, 1037t, 1056
Tryptophan
 isoelectric point, 1025t
 pK_a values, 1025t
 solubility, 110
 structure of, 1023f
TsCl (p-toluenesulfonyl chloride), 361–363
TsOH. See p-Toluenesulfonic acid
Tuberculosis drugs, 272
Tungsten, 1211
ar-Turmerone, 812, 822
Twistoflex, 856, 856f, 1030
Tylenol, 761
Tyrian purple, 1006
Tyrosine, 119, 803, 1023f, 1025t
Tyvek, 1277

U

Ultraviolet radiation
 in electromagnetic spectrum, 529f
 ozone, protection from, 259, 453, 949
 sunscreens, 7, 529, 828, 857
Undecane
 as normal alkane, 135t
 as pheromone, 133
Unhybridized orbitals, 36, 36f, 38, 40
Unimolecular reactions
 E1 elimination reactions, 314
 nucleophilic substitution (See S_N1 substitution reactions)
Unknowns
 mass spectrometry analysis, 511–514
 proton NMR analysis, 579–581
 structure determination, 544–546, 1096–1097
Unoprostone isopropyl, 1126

α,β-Unsaturated carbonyl compounds. See also specific compounds
 1,2-addition to, 629–631
 1,4-addition to, 630–631
 carbonyl compound conversion to, 792
 description of, 629
 in glycolysis, 1183
 β-hydroxy carbonyl compound dehydration to, 816–817
 in Michael reaction, 833–835
 in Robinson annulation, 835–839
Unsaturated fatty acids, 1118, 1118t
 melting point, 394, 395
α,β-Unsaturated fatty acyl CoA, 1174
Unsaturated hydrocarbons
 alkenes (See Alkenes)
 alkynes (See Alkynes)
 benzene, 850 (See also Benzene)
 degrees of unsaturation, 386–387, 439–440
 dienes (See Dienes)
Unsaturated lipids, 953–954, 953f
α,β-Unsaturated thioesters, 817
Unsaturated triacylglycerols, 1117, 1118f
Unsaturation, degrees of. See Degrees of unsaturation
Unsymmetrical alkenes
 halohydrin formation, 407
 hydration of, 404
 hydroboration of, 410
 hydrogen halide addition to, 400–401
Unsymmetrical carbonyl compounds, 788–789
Unsymmetrical epoxides, 369, 371, 445
Unsymmetrical ethers, 338, 346–347
Unsymmetrical ketones, 795–796
Uracil, 1145, 1147t, 1148, 1154
Urethanes, 1040, 1269
Uridine 5'-monophosphate (UMP), 1147t, 1148

V

Valdecoxib, 1126
Valence bond theory, 869
Valence electrons, 10, 11
Valence shell electron pair repulsion (VSEPR) theory, 24
Valeraldehyde, 649t
Valine
 abbreviation for, 716t
 isoelectric point, 1025t
 pK_a values, 716t, 1025t
 structure of, 1023f
 zwitterion, 119
van der Waals forces
 in alkanes, 145
 in alkenes and alkynes, 391
 boiling point and, 104
 denaturation of proteins, 1051
 description of, 101, 103t
 melting point and, 105
 in tertiary structure of proteins, 1049, 1050f
Vane, John, 703
Vanillin, 32, 654
Vanilloids, 4
Vasodilators, 802, 866, 976
Vasopressin, 1034, 1049
Vegetable oils
 hydrogenation of, 440–442, 441f, 1119
 saturated fats in, 1117
 as unsaturated oils, 395

Venlafaxine, 622
Ventolin. See Albuterol
Verapamil, 703
Verelan, 703
Vibration of bonds, infrared spectroscopy and, 531
Vicinal dibromide, 396, 952
Vicinal dihalides, 322, 396, 404
Vidarabine, 1146
Vinyl acetate
 polymerization of, 1258f, 1261f
 proton NMR spectrum, 576, 576f
Vinylboranes, 1202
Vinyl carbanion, 444
Vinyl carbocation, 289, 415
Vinyl chloride
 copolymerization of, 1263
 polymerization of, 1258f
 PVC synthesis from, 935, 958, 1256
1-Vinylcyclopentene, 389
Vinyl groups, 389
Vinyl halides
 alkene synthesis from, 1205–1207
 description, 255, 255f
 1,3-diene synthesis from, 1202
 formation of, 413
 in Heck reactions, 1205
 stability, 289, 894
Vinylidene chloride, 1263
Vioxx, 1126, 1203
Viruses, 1162
Vision, chemistry of, 671, 671f
Vitamin A
 classification, 305
 functions of, 1123f
 isoprene units in, 1128
 solubility, 111
Vitamin B_3, 112, 607
Vitamin B_5, 112
Vitamin B_6, 47–48, 646
Vitamin B_{12}, 1174
Vitamin C, 91, 111–112, 744
Vitamin D_3
 calcitriol and, 484
 classification, 305
 functions of, 1123f
Vitamin E, 938, 954, 1123f
Vitamin K_1, 112, 885, 897, 897f, 1123f
Vitamins, 111, 1123, 1123f
Volatility
 boiling point and, 104
 of plasticizers, 1274
VSEPR (valence shell electron pair repulsion) theory, 24
Vulcanization, 1266
Vulcanized rubber, 1266, 1266f

W

Wald, George, 671
Walden, Paul, 270
Walden inversion, 270
Water
 as achiral molecule, 180
 dehydration (See Dehydration reactions)
 in E1 elimination reactions, 316
 electrophilic addition of, 403–404
 halohydrin formation, 407–408
 hybrid orbitals in, 36
 hydration reactions (See Hydration)
 hydroboration–oxidation (See Hydroboration–oxidation)
 hydrolysis (See Hydrolysis)
 as leaving group, 262

molecular polarity, 45–46, 46f
molecular shape, 26–27
in phenol synthesis, 1000
as polar protic solvent, 265, 265f
in S_N1 nucleophilic substitution, 283–284
as solvent, 107–113, 110t, 113f
Water-soluble vitamins, 111–112
Watson, James, 1150
Wavelength, electromagnetic radiation, 529–530, 529f
Wavenumber scale, 531–532
Wedge symbol, 26, 159, 1072, 1082f
Wellbutrin, 889f
Williamson ether synthesis, 346–347
Willow bark, 77f, 77, 703
Wittig, Georg, 662
Wittig reaction
description of, 662
mechanism, 664–665
reagent, 662–664
retrosynthetic analysis, 665–666
Wittig reagent, 662–664
Wohl degradation, 1094–1096
Wolff–Kishner reduction, 919
Woodward, R. B., 1228
Woodward–Hoffmann rules
for cycloaddition reactions, 1240, 1240t
for electrocyclic reactions, 1236, 1236t
for sigmatropic rearrangement, 1242, 1242t
Woolly foxglove, 682

X

Xalatan, 704
Xenon, 115
m-Xylene, 911
p-Xylene, 522
D-Xylose, 1077, 1078f
D-Xylulose, 1079f

Y

-yl chloride (suffix), 737
Ylides, 663
-yl phosphate (suffix), 737
-yl (suffix), 651
-yne (suffix), 387
Yondelis, 5

Z

Zaitsev rule
dehydration reactions, 350
E1 elimination reactions, 316
E2 elimination reactions, 312–313, 319–320
Z configuration, double bonds, 394
Zestril (lisinopril), 194
Zidovudine, 3, 1162
Ziegler, Karl, 1265
Ziegler–Natta catalysts, 1265, 1266
Ziegler–Natta polymerization, 1265
Zileuton, 374
Zinc, in Clemmensen reduction, 919
Zingiberene, 392, 492, 1127f
Zocor, 189, 484, 484f, 1135
Zoloft, 896
Zolpidem (Ambien), 387
Z (prefix), 389–390
Zwitterions, 118, 715, 717–718, 1024
Zyban, 889f
Zyflo CR, 374